Measure-Theoretic Calculus in Abstract Spaces

Zigang Pan

Measure-Theoretic Calculus in Abstract Spaces

On the Playground of Infinite-Dimensional Spaces

Birkhäuser

Zigang Pan
Mason, OH, USA

ISBN 978-3-031-21911-5 ISBN 978-3-031-21912-2 (eBook)
https://doi.org/10.1007/978-3-031-21912-2

Mathematics Subject Classification: 28-02, 28B05, 28C15, 28C20, 46-02, 46G05, 46G10

This book is published under the imprint Birkhäuser, www.birkhauser-science.com by the registered company Springer Nature Switzerland AG
The registered company address is: Gewerbestrasse 11, 6330 Cham, Switzerland

Preface

This is a book that started out as a reading note of the book Royden (1988) and the MATH 441 & 442 notes by Professor Peter Leob of the University of Illinois at Urbana-Champaign. I am inspired by Luenberger (1969) on the wide applicability of functional analysis tools and its unification of classical results in decision theory, optimization theory, control theory, and numerical approximation. When I took the course MATH 480 with Professor Sean Meyn at the University of Illinois at Urbana-Champaign using the text book Luenberger (1969), I was awestruck by the clean and concise theorems and tight and elegant arguments that leads to the conclusions of the theorems. My background is in control theory, and I find that the tools in Luenberger (1969) require deep knowledge in Lebesgue integral and therefore measure theory. Thus, I started with Royden (1988) to study measure theory and general integration with the goal of studying measure theoretic calculus. Another objective of mine is to set a solid foundation for my control theory research. Thus, I started out on this decade and a half long study that has led to this book.

During this study, I consulted a number of books that are the guiding light that shine upon my way. The foundation of mathematics is set theory, which I find the axiomatic foundation for set theory in my high school award book Mathematics Handbook Editors Group (1979). The book Suppes (1972) clears up most of the paradoxes in my mind about set theory. Professor Leob's notes are excellent in the characterization of equivalence between Axiom of Choice, Zorn's Lemma, Hausdorff's Maximum Principle, and the Well-Ordering Principle. After the completion of the book, I reread the classic (Halmos, 1960) and find it to be very agreeable, which was not the case when I was writing Chap. 2 back then. The concept of nets, as a generalization of sequences, is an essential tool for dealing with various convergence concepts, especially for Riemann and Lebesgue integrals. In Chap. 6, I have included some material from the book Maunder (1996). Chapters 6–10 include a significant amount of material from the book Luenberger (1969). The Alaoglu's Theorem 7.122 in Chap. 7 is, in my view, the key to infinite-dimensional systems. Chapter 9 also references Bartle (1976). The book Bartle (1976) is solid in its mathematical rigor and serves as the starting point of this endeavor. The Global Inverse Function Theorem in Sect. 9.6 is inspired by Ambrosetti and Prodi (1993).

Section 9.9 provides all the essential theorems on analytic functions. Chapter 11 includes significant amount of self-developed material due to lack of reference on this subject. The proof of Radon-Nikodym Theorem 11.169 was adapted from MATH 442 notes by Prof. Peck of the University of Illinois at Urbana-Champaign. The book Royden (1988) offers clues on how to invent the wheel. The calculation of measure (see Sect. 11.7) is a main contribution of this book. As I understand from this study, the definition of a measure, calculating new measures based on already defined measures, and working with the cumulative distribution functions of these measures are essential characteristics in the application of measure theory. In Sect. 11.10, I finally resolve the dual of the space of continuous functions of a compact space to a Banach space. I further correctly defined the space $\mathcal{C}_c(\mathcal{X}, \mathcal{Y})$ and find its dual. This result sets the foundation for the concept of convergence in distribution as the weak* convergence in the dual of $\mathcal{C}_c(\mathcal{X}, \mathcal{Y})$. I could further define $\mathcal{C}_c(\mathcal{X}, \mathcal{Y})$ for $\mathcal{X}$ that are open subset of a finite-dimensional Banach space. I chose to stop there and leave this as an exercise for the reader.

In Chap. 12, differentiation and integration finally comes together. The Tonelli's Theorem is standard. The Fubini's Theorem comes in two flavors: one for general (separable) Banach space valued integrands, and the other for σ-compact conic segment valued integrands. The latter one is as simple as it can get: for absolutely integrable functions of the product measure space, the integral can be calculated as repeated integrals, as long as one sets the intermediate integral to the null vector whenever the intermediate integral is not absolutely integrable. This has direct application to finite-dimensional Banach space valued integrands. The concept of isomeasure is a great discovery. It allows me to equate integrations on different measurable spaces. This has direct application in the Change of Variable Theorem 12.91. There, all I need to do is to show that a certain induced measure is absolutely continuous with respect to the standard Borel measure on $\mathbb{R}^m$. The connection between Riemann-Stieltjes integral and Lebesgue-Stieltjes integral is established in Proposition 12.100 for the real-valued integrand case, and Theorem A.10 for the general Banach space valued integrand case. I attempted to define Sobolev space in a correct fashion, which is different from that of Zeidler (1995) when the dimension of the domain of the function is greater than 1. I feel that my definition is the correct way to go, due to the characterization of absolutely continuous functions on multi-dimensional domains in the Fundamental Theorem of Calculus II (Theorem 12.88). Also, standard results on integral depending on a parameter (Bartle, 1976) are generalized to a more general Banach space setting, in Theorems 12.111, 12.112, 12.115, and 12.116. Here, it is observed that the best practice is to stack the measurable part of the integrand in a function inside the integrand. Then, the assumptions for the integrand are easily formulated and the results are similar to Bartle (1976). The chapter ends with Sect. 12.11 on the topic of manifold. In this section, I studied the topic of manifold where local charts are modeled by some Banach space over the field $\mathbb{K}$. This topic is included due to the necessity to discuss complex logarithm in the proof of Central Limit Theorem 14.63. Here, for a rigorous presentation, one has to make a trip to manifolds. These results are motivated by Isidori (1995), which has a great introduction to differential manifolds in its Appendix A, and by Bishop and Goldberg (1980), which has a proof of the representation of tangent vectors of a finite-dimensional manifold.

In Chap. 13, standard projection theorems are established as well as the dual of a Hilbert space. Examples of complete orthonormal sequences are obtained. The Hermitian adjoint of a linear operator is introduced, which then leads to the definition of positive definite and positive semi-definite operators (including complex versions). Standard results on pseudoinverse of a linear operator between Hilbert spaces is proved. Then, the spectral theory of compact Hermitian linear operators is established. The singular value decomposition is obtained for all compact linear operators between Hilbert spaces. In Chap. 14, standard results on probability theory are obtained. I consulted my MATH 451 & 452 notes of Professor Burkholder of the University of Illinois at Urbana-Champaign, Stark and Woods (1986), Williams (1991), and Billingsley (1995). Professor Burkholder's classes at the University of Illinois at Urbana-Champaign were particularly helpful. The text book used in the classes was Williams (1991), which tells one everything one needs to know about conditional expectations, which I generalized to Banach spaces in Proposition 14.11. Then, Professor Burkholder led us through the definition of the Wiener process and stochastic integration and Markov processes. These topics are beyond my reach if not for MATH 452. After a few iterations, I have finally obtained an elegant version of the Itô's Formula (Theorem 14.122). It applies in infinite dimensional spaces as well, but works like a charm in finite dimensional spaces. Also, I proved the Girsanov's Theorem 14.123 in April 2020 after consulting (Gihman and Skorohod, 1972; Stroock and Varadhan, 1979; Elliott, 1982), and thus completes this decade and a half long study.

After I took my Math 441 & 442 classes on Real Analysis based on the book Royden (1988) at the University of Illinois at Urbana-Champaign, there is something fuzzy about the theory. I am confident that the results are correct, but the arbitrary assignment of $0 \times \infty = 0$ hangs like a dark cloud in the horizon; and the lack of Change of Variable Theorem in the theory for multi-dimensional measure spaces kept me wanting for more results in this field. I did not bother then to study measure theory myself, and pressed on with my PhD study in control theory without solidifying my background. That proved to be a major mistake. After I lost my teaching job for good, I finally settled down to study the basics that underpins my past research. My wife got a secure job and I became an at-home dad. I have all the time in the world. I really don't want to waste my brain away, and pressed on with research to keep my mind sharp and my resumé competitive. God put the thought in me that I could work on foundational mathematics rather than waving my hands at some advanced control problems. I could do both, but the urge to stand on my own feet finally win over my attention. I put down some fruitful control theory research and focused myself on the foundational mathematics that underpins my earlier research. I expected that it might take a few years, but it actually started me on this decade and a half long pursuit. In retrospect, I am happy with my decision of working in mathematics, which is a topic that fascinated me since my youth. The end result is very satisfying for me and I look back at this journey as something that I can be proud of. As a consequence of this study, I find that my past research on linear exponential of quadratic Gaussian control are correct since I finally proved Girsanov's Theorem 14.123. My research in nonlinear stochastic systems can only be regarded as pointers to what could be done in these directions, rather than solid theory. Key integrability assumptions are missing in those research. My research on jump linear systems are correct entirely.

I want to thank my Uncle Desheng Pan and Aunt Deyin Pan for my upbringing. Uncle Sheng gave me a rigorous mathematical mind when I was 10 years old that he drilled me to study Euclidean geometry. I was not happy then, but it gave me a sharp mind ever since. All those classes and lectures in later life are worth much more to me than to any other person, since I constantly think about the "why" behind everything I learned. Aunt Yin and Uncle Sheng shouldered most of the life's burdens when I was in China and gave me and my sister the freedom to study and excel. I thank Aunt Yin especially for all those chores that she did for us, all those clothes that she washed without a washing machine, and all those meals she cooked. One has to remember she did all these in addition to a six day per week job. I want to express my gratitude to the University of Illinois at Urbana-Champaign that gave me a world class education. The friendships I formed during my graduate study will last a lifetime. Most importantly, I am grateful for Professor Tamer Başar of the University of Illinois for his friendship throughout my adult life, first as my advisor and later as my friend. I am also grateful for his time and effort in reading and editing the Preface and the Chap. 1 of this book that resulted in a much more readable book over all. I am grateful for God for showing me the right ways of life at many instances so that I can correct myself in many things and become an independent scholar to finish this foundational book and work on many interesting and important adaptive controls research along this journey of life. God saw the best in me and provided me with a stable and delightful environment, the charming city of Mason, OH, that I can work toward my goal. With God's gentle persuasion, I stayed on the correct path through this book writing endeavor. God, the All Mighty, is my guiding light and my rock. I am forever grateful for God's love. I want to thank Professor Wen-ben Jone of the University of Cincinnati for remaining my friend after I lost my last job. I want to thank Professor Gang Tao of the University of Virginia for helpful comments in reaching the current form of the book. I am grateful to the Birkhäuser (Springer-Nature) staffs for all of their effort and help in bringing this book into publication: Editors Mr. Thomas Hempfling, Ms. Sarah Annette Goob, Mrs. Dana Knowles, and Mr. Christopher Tominich, TEX Support Administrator Suresh Kumar, Project Coordinator Mr. Daniel Ignatius Jagadisan, and Production Contact Ms. Antje Endemann, as well as Production Manager Mrs. Kali Gayathri, and Mr. Periyanayagam Leoselvakumar at Straive. I want to thank my wife for all the help she gave me. I also want to thank my parents for their help over the years and my daughter for her patience (or lack of it) with dad's preachings.

Mason, OH, USA
July 2023

Zigang Pan

Contents

List of Figures

Notations

$\mathbb{N}$, $\mathbb{Z}$, and $\mathbb{Q}$	the sets of natural numbers, integers, and rational numbers, respectively
$\mathbb{R}$ and $\mathbb{C}$	the sets of real numbers and complex numbers, respectively
$\mathbb{K}$	either $\mathbb{R}$ or $\mathbb{C}$
$\mathbb{Z}_+, \mathbb{Z}_-$	$\mathbb{N} \cup \{0\}$, $\mathbb{Z} \setminus \mathbb{N}$, respectively
$\mathbb{R}_+, \mathbb{R}_-, \mathbb{C}_+, \mathbb{C}_-$	$(0, \infty) \subset \mathbb{R}$, $(-\infty, 0) \subset \mathbb{R}$, the open right half of the complex plane, the open left half of the complex plane, respectively
$\overline{\mathbb{R}_+}, \overline{\mathbb{R}_-}, \overline{\mathbb{C}_+}, \overline{\mathbb{C}_-}$	$[0, \infty) \subset \mathbb{R}$, $(-\infty, 0] \subset \mathbb{R}$, the closed right half of the complex plane, the closed left half of the complex plane, respectively
$\mathbb{C}_0$	$\mathbb{C} \setminus \{0\}$
$\in$	belong to
$\notin$	not belong to
$\subseteq$	contained in
$\supseteq$	contains
$\subset$	strict subset of
$\supset$	strict super set of
$\forall$	for all
$\exists$	exists
$\exists!$	exists a unique
$\because$	because
$\therefore$	therefore
$\ni \cdot$	such that
$(x_n)_{n=1}^{\infty}$	the sequence $x_1, x_2, \ldots$
$(x_\alpha)_{\alpha \in \Lambda}$	the ordered collection
id_A	the identity map on a set A

e	the natural number
i	the complex unit
$\|\lambda\|$	the absolute value of a real or complex number λ
$\overline{\lambda}$	the complex conjugate of a complex number λ
$\mathrm{Re}\,(\lambda)$	the real part of a complex number λ
$\mathrm{Im}\,(\lambda)$	the imaginary part of a complex number λ
$\angle\lambda$	the phase angle of a complex number λ
$a \vee b$	the maximum of two real numbers a and b
$a \wedge b$	the minimum of two real numbers a and b
$\mathrm{sgn}(x)$	$\mathrm{sgn}(x) = \begin{cases} -1 & \text{if } x < 0 \\ 0 & \text{if } x = 0 \\ 1 & \text{if } x > 0 \end{cases}, \forall x \in \mathbb{R}$
π	the ratio of the circumference of a circle to its diameter
$\emptyset$	the empty set; See Page 27.
$\{x, y\}$	an unordered pair; See Page 27.
${}^{X}2$	the collection of all subsets of X; See Page 27.
$\cup$	the set union; See Page 27.
(x, y)	an ordered pair; See Page 28.
$X \times Y$	the Cartesian or direct product of sets X and Y; See Page 28.
$A = \{x \in B \mid P(x)\}$	definition of a set A; See Page 28.
$x \sim y$	x and y are related in a relation; See Page 28.
$X/\equiv$	the quotient of the set X with respect to an equivalence relation $\equiv$; See Page 28.
$f : X \to Y$	a function of X to Y; $\{(x, f(x)) \in X \times Y \mid \forall x \in X\}$ is the graph of f; See Page 29.
$\mathrm{graph}\,(f)$	the graph of a function f; See Page 29.
$\mathrm{dom}\,(f)$	the domain of f; See Page 29.
$f(A)$	the image of $A \subseteq X$ under f; See Page 29.
$\mathrm{range}\,(f)$	the range of f, equals to $f(X)$; See Page 29.
$f_{\mathrm{inv}}(A)$	the preimage of $A \subseteq Y$ under f; See Page 29.
onto, surjective	$f(X) = Y$; See Page 29.
1-1, injective	$f(x_1) \neq f(x_2)$ if $x_1, x_2 \in X$ and $x_1 \neq x_2$; See Page 29.
bijective	both surjective and injective; See Page 29.
f_{inv}	the inverse function of f; See Page 29.
$g \circ f$	the composition of $g : Y \to Z$ with $f : X \to Y$; See Page 29.
$f\|_A$	the restriction of f to A; See Page 29.
Y^X	the set of all functions of X to Y; See Page 29.
$\cap$	the set intersection; See Page 30.

$\tilde{A}$	the compliment of a set A, where the whole set is clear from context; See Page 30.
$\setminus$	set minus; See Page 30.
$A \triangle B$	the symmetric difference of A and B, equals to $(A \setminus B) \cup (B \setminus A)$; See Page 30.
$\operatorname{card}(X)$	the number of elements in the finite set X; See Page 31.
$\prod_{\alpha \in \Lambda} X_\alpha$	the Cartesian or direct product of X_α's; See Page 40.
$\pi_\alpha(x)$	the projection of an element in a Cartesian product space to one of the coordinates; See Page 40.
$\overline{A}$	the closure of a set A, where the whole set is clear from context; See Page 41.
A°	the interior of a set A, where the whole set is clear from context; See Page 41.
∂A	the boundary of a set A, where the whole set is clear from context; See Page 41.
$\prod_{\alpha \in \Lambda} (X_\alpha, \mathcal{O}_\alpha)$	the product topological space; See Page 48.
$\mathbf{T}_1$	Tychonoff space; See Page 52.
$\mathbf{T}_2$	Hausdorff space; See Page 52.
$\mathbf{T}_3$	regular space; See Page 53.
$\mathbf{T}_4$	normal space; See Page 53.
$\mathbf{T}_{3\frac{1}{2}}$	completely regular space; See Page 63.
$(x_\alpha)_{\alpha \in \mathcal{A}}$	a net; See Page 64.
$\lim_{\alpha \in \mathcal{A}} x_\alpha$	the limit of a net; See Page 64.
$\lim_{x \to x_0} f(x)$	the limit of $f(x)$ as $x \to x_0$; See Page 68.
$\mathbb{R}_e$	the set of extended real numbers, which equals $\mathbb{R} \cup \{-\infty, +\infty\}$; See Page 71.
$\limsup_{\alpha \in \mathcal{A}} x_\alpha$	the limit superior of a real-valued net; See Page 71.
$\liminf_{\alpha \in \mathcal{A}} x_\alpha$	the limit inferior of a real-valued net; See Page 71.
$\limsup_{x \to x_0} f(x)$	the limit superior of $f(x)$ as $x \to x_0$; See Page 73.
$\liminf_{x \to x_0} f(x)$	the limit inferior of $f(x)$ as $x \to x_0$; See Page 73.
$\mathcal{B}_{\mathcal{X}}(x_0, r)$	the open ball centered at x_0 with radius r; See Page 79.
$\overline{\mathcal{B}}_{\mathcal{X}}(x_0, r)$	the closed ball centered at x_0 with radius r; See Page 79.
$\operatorname{dist}(x_0, S)$	the distance from a point x_0 to a set S in a metric space; See Page 81.
$(X, \rho_X) \times (Y, \rho_Y)$	the finite product metric space; See Page 86.
$\left(\prod_{i=1}^{\infty} X_i, \rho\right)$	the countably infinite product metric space; See Page 89.
$\operatorname{lc}(\mathcal{X})$	The metric space consisting of nonempty closed subsets of $\mathcal{X}$; See Page 101.
$g \circledcirc f$	The lifted composition of the function g and f; See Page 102.
$\operatorname{supp}(f)$	the support of a function; See Page 123.

$\beta(\mathcal{X})$	the Stone-Čech compatification of a completely regular topological space $\mathcal{X}$; See Page 134.
$(\mathcal{M}(A, \mathcal{Y}), \mathcal{F})$	the vector space of $\mathcal{Y}$-valued functions of a set A over the field $\mathcal{F}$; See Page 142.
$\vartheta_{\mathcal{X}}$	the null vector of a vector space $\mathcal{X}$; See Page 144.
$\mathcal{N}(A)$	the null space of a linear operator A; See Page 144.
$\mathcal{R}(A)$	the range space of a linear operator A; See Page 144.
αS	the scalar multiplication by α of a set S in a vector space; See Page 144.
$S + T$	the sum of two sets S and T in a vector space; See Page 144.
$\mathrm{span}\,(A)$	the subspace generated by the set A; See Page 145.
$\mathrm{v}\,(P)$	the linear variety generated by a nonempty set P; See Page 146.
$\mathrm{co}\,(S)$	the convex hull generated by S in a vector space; See Page 147.
$\lVert x \rVert$	the norm of a vector x; See Page 153.
$\lvert x \rvert$	the Euclidean norm of a vector x; See Page 153.
$\mathcal{C}_1([a, b])$	the normed linear space of continuously differentiable real-valued functions on the interval $[a, b]$; See Page 154.
l_p	the normed linear space of real-valued sequences with finite p-norm, $1 \leq p \leq +\infty$; See Page 155.
$\mathrm{l}_p(\mathscr{X})$	the normed linear space of $\mathscr{X}$-valued sequences with finite p-norm, $1 \leq p \leq +\infty$; See Page 159.
$\mathrm{V}\,(P)$	the closed linear variety generated by a nonempty set P; See Page 162.
$\circ P$	the relative interior of a set P; See Page 162.
$\mathscr{X} \times \mathscr{Y}$	the finite Cartesian product normed linear space; See Page 162.
$\mathcal{C}([a, b])$	the Banach space of continuous real-valued functions on the interval $[a, b]$; See Page 166.
$\mathcal{C}(\mathcal{K}, \mathscr{X})$	the normed linear space of continuous $\mathscr{X}$-valued functions on a compact space $\mathcal{K}$; See Page 166.
$\mathscr{X}_{\mathbb{R}}$	the real normed linear space induced by a complex normed linear space $\mathscr{X}$; See Page 170.
$[x]$	the coset of a vector x in a quotient space; See Page 171.
$\mathcal{X}/M$	the quotient space of a vector space $\mathcal{X}$ modulo a subspace M; See Page 171.
$\mathscr{X}/M$	the quotient space of a normed linear space $\mathscr{X}$ modulo a closed subspace M; See Page 171.
$\mathcal{C}_{\mathrm{v}}(\mathcal{X}, \mathscr{Y})$	the vector space of continuous $\mathscr{Y}$-valued functions on $\mathcal{X}$; See Page 173.

$B(\mathcal{X}, \mathcal{Y})$	the set of bounded linear operators of $\mathcal{X}$ to $\mathcal{Y}$; See Page 181.
$\mathcal{X}^*$	the dual of $\mathcal{X}$; See Page 185.
x_*	a vector in the dual; See Page 185.
$\langle\langle x_*, x\rangle\rangle$	the evaluation of a bounded linear functional x_* at the vector x, that is $x_*(x)$; See Page 185.
$c_0(\mathcal{X})$	the subspace of $l_\infty(\mathcal{X})$ consisting of $\mathcal{X}$-valued sequences with limit $\vartheta_{\mathcal{X}}$; See Page 189.
$\mathcal{X}^{**}$	the second dual of $\mathcal{X}$; See Page 197.
$S^\perp$	the orthogonal complement of the set S; See Page 199.
${}^\perp S$	the pre-orthogonal complement of the set S; See Page 200.
A'	the adjoint of a linear operator A; See Page 208.
A''	the adjoint of the adjoint of a linear operator A; See Page 209.
$\mathcal{O}_{\text{weak}}(\mathcal{X})$	the weak topology on a normed linear space $\mathcal{X}$; See Page 211.
$\mathcal{X}_{\text{weak}}$	the weak topological space associated with a normed linear space $\mathcal{X}$; See Page 211.
$\mathcal{O}_{\text{weak}^*}(\mathcal{X}^*)$	the weak* topology on the dual of a normed linear space $\mathcal{X}$; See Page 214.
$\mathcal{X}^*_{\text{weak}^*}$	the weak* topological space associated with the dual of a normed linear space $\mathcal{X}$; See Page 214.
K_{supp}	the support of a convex set K; See Page 226.
$[f, C]$	the epigraph of a convex function $f : C \to \mathbb{R}$; See Page 229.
C_{conj}	the conjugate convex set; See Page 233.
f_{conj}	the conjugate convex functional; See Page 233.
$[f, C]_{\text{conj}}$	the epigraph of the conjugate convex functional; See Page 233.
${}_{\text{conj}}\Gamma$	the pre-conjugate convex set; See Page 236.
${}_{\text{conj}}\varphi$	the pre-conjugate convex functional; See Page 236.
${}_{\text{conj}}[\varphi, \Gamma]$	the epigraph of the pre-conjugate convex functional; See Page 236.
$\geqq$	greater than or equal to (with respect to the positive cone); See Page 247.
$\leqq$	less than or equal to (with respect to the positive cone); See Page 247.
$>$	greater than (with respect to the positive cone); See Page 247.
$<$	less than (with respect to the positive cone); See Page 247.

$S^{\oplus}$	the positive conjugate cone of a set S; See Page 248.
$S^{\ominus}$	the negative conjugate cone of a set S; See Page 248.
$\mathrm{A}_D(x_0)$	the set of admissible deviations in D at x_0; See Page 257.
$f^{(1)}(x_0)$, $\mathrm{D}f(x_0)$	the Fréchet derivative of f at x_0; See Page 258.
$\mathrm{D}f(x_0; u)$	the directional derivative of f at x_0 along u; See Page 258.
$\frac{\partial f}{\partial y}(x_0, y_0)$	partial derivative of f with respect to y at (x_0, y_0); See Page 259.
$\mathrm{ro}(A)(B)$	right operate: $\mathrm{ro}(A)(B) = BA$; See Page 264.
$\mathrm{B}_k(\mathcal{X}, \mathcal{Y})$	the set of bounded multi-linear $\mathcal{Y}$-valued functions on $\mathcal{X}^k$; See Page 271.
$\mathrm{B}_{\mathrm{S}\,k}(\mathcal{X}, \mathcal{Y})$	the set of symmetric bounded multi-linear $\mathcal{Y}$-valued functions on $\mathcal{X}^k$; See Page 271.
$\mathrm{D}^k f(x_0)$, $f^{(k)}(x_0)$	the kth order Fréchet derivative of f at x_0; See Page 272.
$\mathcal{C}_k$, $\mathcal{C}_\infty$	k-times and infinite-times continuously differentiable functions, respectively; See Page 272.
$\frac{\partial^k f}{\partial x_{i_k} \cdots \partial x_{i_1}}$	kth-order partial derivative of f; See Page 277.
$\mathrm{Sm}(D)$	$= \bigcup_{\alpha \in \mathbb{K}} \alpha D$; See Page 277.
$\mathcal{C}_k(\Omega, \mathcal{Y})$	the normed linear space of k-times continuously differentiable $\mathcal{Y}$-valued functions on a compact set $\Omega \subseteq \mathcal{X}$; See Page 313.
$\mathcal{C}_{\mathrm{b}}(\mathscr{X}, \mathcal{Y})$	the normed linear space of bounded continuous $\mathcal{Y}$-valued functions on a topological space $\mathscr{X}$; See Page 314.
$\mathcal{C}_{\mathrm{b}\,k}(\Omega, \mathcal{Y})$	the normed linear space of k-times bounded continuously differentiable $\mathcal{Y}$-valued functions on a set $\Omega \subseteq \mathcal{X}$; See Page 315.
$A^{T_{n_1,\ldots,n_m}}$	the transpose of an mth order tensor A with the permutation $(n_1, \ldots, n_m)$; See Page 316.
$A \otimes B$	the outer product of two tensors; See Page 317.
$\mathbf{0}_{m_1 \times \cdots \times m_n}$	an nth order $\mathbb{K}$-valued tensor in $\mathrm{B}(\mathbb{K}^{m_n}, \ldots, \mathrm{B}(\mathbb{K}^{m_1}, \mathbb{K}) \cdots)$ with all elements equal to 0
$\mathbf{1}_{m_1 \times \cdots \times m_n}$	an nth order $\mathbb{K}$-valued tensor in $\mathrm{B}(\mathbb{K}^{m_n}, \ldots, \mathrm{B}(\mathbb{K}^{m_1}, \mathbb{K}) \cdots)$ with all elements equal to 1
$\exp(A)$	the exponential function of a linear operator A on a Banach space; See Page 329.
$\mathcal{S}_{+\,\mathcal{X}}$, $\mathcal{S}_{\mathrm{psd}\,\mathcal{X}}$	sets of positive definite and positive semi-definite operators over the real normed linear space $\mathcal{X}$, respectively; See Page 335.
$\mathcal{S}_{-\,\mathcal{X}}$, $\mathcal{S}_{\mathrm{nsd}\,\mathcal{X}}$	sets of negative definite and negative semi-definite operators over the real normed linear space $\mathcal{X}$, respectively; See Page 335.
$\mathcal{S}_{\mathcal{X}}$	$\mathrm{B}_{\mathrm{S}\,2}(\mathcal{X}, \mathbb{R})$; See Page 336.

$(\mathbb{R}, \mathcal{B}_L, \mu_L)$	Lebesgue measure space; See Page 370.
μ_{Lo}	Lebesgue outer measure; See Page 370.
$\mathcal{B}_B(\mathcal{X})$	Borel sets; See Page 371.
$\mathcal{A}(\mathcal{X})$	the algebra generated by the topology of a topological space $\mathcal{X}$; See Page 371.
μ_B	the Borel measure on $\mathbb{R}$; See Page 371.
$\mathbb{R}$	$((\mathbb{R}, \mathbb{R}, \lvert\cdot\rvert), \mathcal{B}_B(\mathbb{R}), \mu_B)$; See Page 375.
P a.e. in $\mathcal{X}$	P holds almost everywhere in $\mathcal{X}$; See Page 379.
$P(x)$ a.e. $x \in \mathcal{X}$	P holds almost everywhere in $\mathcal{X}$; See Page 379.
$\chi_{A,X}$	the indicator function of $A \subseteq X$. Function of X to $\{0, 1\}$; See Page 387.
$\mathcal{P} \circ f$	$\mathcal{P} \circ f : X \to [0, \infty) \subset \mathbb{R}$ defined by $\mathcal{P} \circ f(x) = \lVert f(x) \rVert$, $\forall x \in X$; See Page 388.
$\mathfrak{R}(\mathfrak{X})$	the collection of all representation of $\mathfrak{X}$; See Page 391.
$\mathfrak{I}(\mathfrak{X})$	the integration system on $\mathfrak{X}$; See Page 391.
$\int_X f \, d\mu$	the integral of a function f on a set X with respect to measure μ; See Page 392.
$\int_X f(x) \, d\mu(x)$	the integral of a function f on a set X with respect to measure μ; See Page 392.
$\mathcal{P} \circ \mu$	the total variation of a Banach space valued pre-measure μ; See Page 421.
$\mathcal{P} \circ \mu$	the total variation of a Banach space valued measure μ; See Page 426.
$\mu_1 + \mu_2$	the $\mathcal{Y}$-valued measure that equals to the sum of two $\mathcal{Y}$-valued measures on the same measurable space; See Page 455.
$\alpha\mu$	the $\mathcal{Y}$-valued measure that equals to the scalar product of $\alpha \in \mathbb{K}$ and $\mathcal{Y}$-valued measure μ; See Page 456.
μy	the $\mathcal{Y}$-valued measure that equals to scalar product of a $\mathbb{K}$-valued measure μ and $y \in \mathcal{Y}$; See Page 456.
$A\mu$	the $\mathcal{Z}$-valued measure that equals to product of an bounded linear operator A and a $\mathcal{Y}$ valued measure μ; See Page 456.
$\mathcal{M}_\sigma(X, \mathcal{B}, \mathcal{Y})$	the vector space of σ-finite $\mathcal{Y}$-valued measures on the measurable space $(X, \mathcal{B})$; See Page 463.
$\mathcal{M}_f(X, \mathcal{B}, \mathcal{Y})$	the normed linear space of finite $\mathcal{Y}$-valued measures on the measurable space $(X, \mathcal{B})$; See Page 464.
$\lim_{n \in \mathbb{N}} \mu_n = \nu$	a sequence of σ-finite ($\mathcal{Y}$-valued) measures $(\mu_n)_{n=1}^{\infty}$ converges to a σ-finite ($\mathcal{Y}$-valued) measure ν; See Page 466.
$\mu_1 \leq \mu_2$	the measures μ_1 and μ_2 on the measurable space $(X, \mathcal{B})$ can be compared if $\mu_1(E) \leq \mu_2(E)$, $\forall E \in \mathcal{B}$; See Page 467.

$\begin{bmatrix} \mu_{1,1} & \cdots & \mu_{1,m} \\ \vdots & & \vdots \\ \mu_{n,1} & \cdots & \mu_{n,m} \end{bmatrix}$	the vector measure; See Page 469.
$\mu_1 \ll \mu_2$	the measure μ_1 is absolutely continuous with respect to the measure μ_2; See Page 480.
$\mu_1 \perp \mu_2$	the measures μ_1 and μ_2 are mutually singular; See Page 480.
$\frac{\mathrm{d}\nu}{\mathrm{d}\mu}$	the Radon–Nikodym derivative of the σ-finite $\mathcal{Y}$-valued measure ν with respect to the σ-finite $\mathbb{K}$-valued measure μ; See Page 489.
$\mathcal{P}_p \circ f$	the function $\\|f(\cdot)\\|^p$; See Page 499.
ess sup	the essential supremum; See Page 501.
$\lim_{n\in\mathbb{N}} z_n \doteq z$ in $\bar{\mathrm{L}}_p$	the sequence $(z_n)_{n=1}^{\infty} \subseteq \bar{\mathrm{L}}_p$ converges to $z \in \bar{\mathrm{L}}_p$ in $\bar{\mathrm{L}}_p$ pseudo-norm; See Page 503.
$\mathcal{M}_{ft}(\mathcal{X}, \mathcal{Y})$	the normed linear space of finite $\mathcal{Y}$-valued topological measures on $\mathcal{X}$; See Page 519.
$\mathcal{M}_{\sigma t}(\mathcal{X}, \mathcal{Y})$	the vector space of σ-finite $\mathcal{Y}$-valued topological measures on $\mathcal{X}$; See Page 519.
$\mathcal{M}_{\sigma}(X, \mathcal{B})$	the set of σ-finite measures on the measurable space $(X, \mathcal{B})$; See Page 520.
$\mathcal{M}_{f}(X, \mathcal{B})$	the set of finite measures on the measurable space $(X, \mathcal{B})$; See Page 520.
$\mathcal{M}_{\sigma t}(\mathcal{X})$	the set of σ-finite topological measures on the topological space $\mathcal{X}$; See Page 520.
$\mathcal{M}_{ft}(\mathcal{X})$	the set of finite topological measures on the topological space $\mathcal{X}$; See Page 520.
$\mathcal{C}_c(\mathfrak{X}, \mathcal{Y})$	the normed linear space of $\mathcal{Y}$-valued continuous function of a finite-dimensional Banach space, that converges to $\vartheta_{\mathcal{Y}}$ as $\\|x\\|_{\mathfrak{X}} \to \infty$; See Page 538.
$\rightsquigarrow$	convergence in distribution; See Page 542.
$\prod_{j=1}^{m} \mu_j$	the product measure of $\mu_1, \ldots, \mu_m$; See Page 560.
$E_{x_1}, E_{\langle x_2 \rangle}$	for a set $E \subseteq X_1 \times X_2$, E_{x_1} is the section of the set E with the first coordinate fixed at x_1, and $E_{\langle x_2 \rangle}$ is the section of the set E with the second coordinate fixed at x_2; See Page 568.
$f_{x_1}, f_{\langle x_2 \rangle}$	for a function $f : X_1 \times X_2 \to Y$, $f_{x_1} : X_2 \to Y$ is the section of the function f with the first coordinate fixed at x_1, and $f_{\langle x_2 \rangle} : X_1 \to Y$ is the section of the function f with the second coordinate fixed at x_2; See Page 570.
$\prod_{i=1}^{m} \mathbb{X}_i$	product topological measure space of $\mathbb{X}_1, \ldots, \mathbb{X}_m$, where $m \in \mathbb{N}$; See Page 576.

r_{x_1,x_2}	the semi-open rectangle in $\mathbb{R}^m$ with corners x_1 and x_2; See Page 585.
$\bar{\mathrm{r}}_{x_1,x_2}$	the closed rectangle in $\mathbb{R}^m$ with corners x_1 and x_2; See Page 585.
$\mathrm{r}^{\circ}_{x_1,x_2}$	the open rectangle in $\mathbb{R}^m$ with corners x_1 and x_2; See Page 585.
$\mathrm{P}(\Omega)$	the principal of a region Ω; See Page 585.
$\mathrm{VRect}_{i,x_1,x_2}$	the set of vertexes of r_{x_1,x_2} with i coordinates equal to that of x_1; See Page 587.
$\Delta_F(\mathrm{r}_{x_1,x_2})$	the increment of F on r_{x_1,x_2}; See Page 587.
$T_F\left(\mathrm{r}_{x_1,x_2}\right)$	the total variation of a function F on the semi-open rectangle r_{x_1,x_2}; See Page 587.
$(\mathbb{R}^m, \mathcal{B}_{\mathrm{L}m}, \mu_{\mathrm{L}m}), \mu_{\mathrm{L}mo}$	m-dimensional Lebesgue measure space and the m-dimensional Lebesgue outer measure; See Page 606.
$\int_U g(x)\,\mathrm{d}F(x)$	the integral of function g with respect to $\mathcal{Y}$-valued measure space $(\mathrm{P}(\Omega), \mathcal{B}_{\mathrm{B}}(\mathrm{P}(\Omega)), \mu)$ whose distribution function is $F : \Omega \to \mathcal{Y}$ over the set $U \in \mathcal{B}_{\mathrm{B}}(\mathrm{P}(\Omega))$; See Page 607.
$\pi_{\widehat{i_0}} : \mathbb{R}^m \to \mathbb{R}^{m-1}$	$\pi_{\widehat{i_0}}(x) = (\pi_1(x), \ldots, \pi_{i_0-1}(x), \pi_{i_0+1}(x), \ldots, \pi_m(x))$, $\forall x \in \mathbb{R}^m$; See Page 607.
$\int_a^b f(x)\,\mathrm{d}x$	the integral of function f from $a \in \mathbb{R}$ to $b \in \mathbb{R}$ with respect to μ_{B}; See Page 618.
$\mathrm{dia}(S)$	the diameter of a subset of a metric space; See Page 625.
$e_{m,i}$	the ith unit vector in $\mathbb{R}^m$; See Page 628.
$\mathbb{S}_{n-1}$	the finite compact separable metric measure space on the n-dimensional unit sphere, where the measure of a set is equal to the "surface area" of the set; See Page 663.
$W_{p,1,x_0}(\mathbb{X}, \mathcal{Y})$	1st order Sobolev space of $\mathbb{X}$ to $\mathcal{Y}$ under p-norm with respect to $x_0 \in \mathbb{X}$; See Page 664.
$\mathrm{Ln} : \mathbb{C}_0 \to \{a + \mathrm{i}b \in \mathbb{C} \mid -\pi < b \leq \pi\}$	the principal value of logarithm function; See Page 691.
$\overline{\exp}$	the exponential function of a manifold to $\mathbb{C}_0$; See Page 692.
$\overline{\ln}$	the logarithm function of $\mathbb{C}_0$ to a manifold; See Page 692.
$\mathcal{C}_\infty(p, \mathbb{K})$	the collection of all functionals that are smooth in a neighborhood of p on a manifold $\mathcal{X}$; See Page 693.
$\mathsf{T}_p\mathcal{X}$	the space of tangent vectors on a manifold $\mathcal{X}$ at $p \in \mathcal{X}$; See Page 694.
$F_{\star p}$	the differential of a mapping F at p on a manifold $\mathcal{X}$; See Page 694.
$\mathsf{T}^*_p\mathcal{X}$	the space of tangent covectors on a manifold $\mathcal{X}$ at $p \in \mathcal{X}$; See Page 695.

$\mathsf{L}_f\lambda$	the Lie derivative of λ along the vector field f; See Page 695.
$\mathsf{V}(\mathcal{X})$	the set of all smooth vector fields on the manifold $\mathcal{X}$; See Page 695.
$[f, g]$	the Lie bracket of vector fields f and g; See Page 697.
$\langle x, y\rangle$	the inner product of vectors x and y; See Page 699.
$x \perp y$	the vectors x and y are orthogonal in a pre-Hilbert space; See Page 701.
$x \perp S$	the vector x is orthogonal to the set S in a pre-Hilbert space; See Page 701.
x^*	$x^* \in \mathfrak{X}^*$ that satisfies $\langle\langle x^*, y\rangle\rangle = \langle y, x\rangle$, $\forall y \in \mathfrak{X}$; See Page 705.
$S^{\perp_{\mathrm{h}}}$	the orthogonal complement in Hilbert spaces; See Page 707.
$M \oplus N$	the direct sum of M and N; See Page 707.
A^*	the Hermitian adjoint of A; See Page 708.
$\mathrm{Gram}(y_1, \ldots, y_n)$	the Gram matrix of $y_1, \ldots, y_n$; See Page 711.
$\mathrm{gram}(y_1, \ldots, y_n)$	the Gram determinant of $y_1, \ldots, y_n$; See Page 711.
$\bar{\mathcal{S}}_{\mathfrak{X}}$	set of Hermitian operators on a real or complex Hilbert space; See Page 724.
$\bar{\mathcal{S}}_{+\,\mathfrak{X}}$, $\bar{\mathcal{S}}_{\mathrm{psd}\,\mathfrak{X}}$	sets of positive definite and positive semi-definite operators over real or complex Hilbert space $\mathfrak{X}$, respectively; See Page 724.
$\bar{\mathcal{S}}_{-\,\mathfrak{X}}$, $\bar{\mathcal{S}}_{\mathrm{nsd}\,\mathfrak{X}}$	sets of negative definite and negative semi-definite operators over real or complex Hilbert space $\mathfrak{X}$, respectively; See Page 724.
$A^\dagger$	the pseudoinverse of $A \in \mathrm{B}(\mathfrak{X}, \mathfrak{Y})$; See Page 729.
$\mathrm{K}(\mathfrak{X}, \mathfrak{Y})$	the set of compact linear bounded operators of $\mathfrak{X}$ to $\mathfrak{Y}$; See Page 735.
$\boldsymbol{E}(x)$	expectation of x; See Page 751.
$\boldsymbol{E}(x\|\hat{\mathcal{B}})$	conditional expectation of x given $\hat{\mathcal{B}}$; See Page 753.
$\boldsymbol{P}(F\|\hat{\mathcal{B}})$	the conditional probability of event F happens given $\hat{\mathcal{B}}$, which is defined to be $\boldsymbol{E}(\chi_{F,\Omega}\|\hat{\mathcal{B}})$. ; See Page 761.
$\boldsymbol{P}(F\|E)$	the conditional probability of event F happening given the event E happened; See Page 763.
L_x	the law for the random variable x; See Page 771.
$x \sim N(\bar{x}, K)$	Gaussian (normal) random variable (vector) with mean $\bar{x}$ and covariance K; See Page 776.
$\aleph := (\mathbb{Z}_+, {}^{\mathbb{Z}_+}2, \mu)$	the topological measure space on $\mathbb{Z}_+$ with the counting measure and the partial ordering of $\leq$; See Page 779.
Ψ_x	the characteristic function of the random variable x; See Page 791.

$\widehat{\int}_{\bar{\mathrm{r}}_{a,b}} \phi_s \, \mathrm{d}w_s$, $\widehat{\int}_a^b \phi_s \, \mathrm{d}w_s$	Riemann-Stieltjes integral of ϕ with respect to the Wiener process w on the interval $\bar{\mathrm{r}}_{a,b}$; See Page 843.
$\widetilde{\int}_{\bar{\mathrm{r}}_{a,b}} \phi(s) \, \mathrm{d}w_s$, $\widetilde{\int}_a^b \phi(s) \, \mathrm{d}w_s$	Wiener integral of ϕ (with respect to the Wiener process w) on the interval $\bar{\mathrm{r}}_{a,b}$; See Page 849.
$\hat{\mathcal{Q}}(\bar{\mathrm{r}}_{a,b} \times \boldsymbol{\Omega}, \mathcal{Y})$	the set of all $\mathcal{Y}$-valued simple predictable step functions on $\bar{\mathrm{r}}_{a,b} \times \boldsymbol{\Omega}$; See Page 850.
$\overline{\int}_{\bar{\mathrm{r}}_{a,b}} \phi_s \, \mathrm{d}w_s$ $\overline{\int}_a^b \phi_s \, \mathrm{d}w_s$	the Itô integral of ϕ (with respect to Wiener process w) on the interval $\bar{\mathrm{r}}_{a,b}$; See Page 857.
$\mathcal{U}(\bar{\mathrm{r}}_{a,b} \times \boldsymbol{\Omega}, \mathcal{Y})$	This set contains Itô integrable processes, whose Itô integral can be approximated by integrating the sampled and hold process; See Page 860.
$\hat{\mathfrak{R}}\left(\bar{\mathrm{r}}_{a,b}\right)$	the collection of all sampled partitions of a compact interval $\bar{\mathrm{r}}_{a,b} \subseteq \mathbb{R}$; See Page 907.
$\hat{\mathfrak{I}}\left(\bar{\mathrm{r}}_{a,b}\right)$	the Riemann integration system on a compact interval $\bar{\mathrm{r}}_{a,b} \subset \mathbb{R}$; See Page 907.
$\widehat{\int}_a^b f \, \mathrm{d}g$, $\widehat{\int}_{\bar{\mathrm{r}}_{a,b}} f(s) \, \mathrm{d}g(s)$	the Riemann-Stieltjes integral of f with respect to g on $\bar{\mathrm{r}}_{a,b}$; See Page 907.

Chapter 1
Introduction

1.1 The Tour of the Book

When I started on this mathematics endeavor, I knew I had to start with set theory. Various paradoxes hang in the air but I knew this subject is beyond my ability. I have to start with a set of axioms that are solid and extensive. I thought that I studied so much math, and did so many problems, I never have to question my own understanding of mathematics. So, what I know by heart should be correct. I just have to find an axiomatic system that conforms with my understanding that avoids the various paradoxes that signals warnings in certain directions. I find the Zermelo–Fraenkel–Cantor (ZFC) axiomatic set theory from my Chinese award book that I received during my high school years Mathematics Handbook Editors Group (1979). This axiomatic system claims to be consistent. It agrees with my understanding in set theory. Most importantly, I see that it avoids the obvious pitfalls of the paradoxes. I took a leap of faith here and based my Chap. 2 on this axiomatic system. My understanding here is that in my entire life, all the sets I encounter are defined via these axioms except Axiom 4. Axiom 4 is something that I am not familiar with, and maybe the source of the paradoxes when violated. I feel that I am safe that I define sets only using the rest of the axioms. So, there it is. I had the understanding that in set definition using Axiom 5, one must specify clearly the set A always and then the paradoxes are avoided. I hope that I am correct with my decision. I am confident that I am correct. Throughout the book, I never had those weird set definitions that calls for suspicion. So, even if this axiomatic system is not entirely consistent, I feel that my work will not be of void. The major theorem in Chap. 2 is the equivalence of the Axiom of Choice, the Hausdorff's Maximal Principle, the Zorn's Lemma, and the Well-Ordering Principle. The proof of this equivalence (Theorem 2.18) is based on the teaching notes of Professor Peter Loeb on Real Analysis at the University of Illinois at Urbana-Champaign. There is no more leap of faith in the rest of the book. Everything follows from reasoning, and Axiom 4 is never invoked. Then, the book Suppes (1972) clears up the mysteries in

Z. Pan, *Measure-Theoretic Calculus in Abstract Spaces*,
https://doi.org/10.1007/978-3-031-21912-2_1

the paradoxes, and reinforced my confidence in my presentation. The only thing that I am not so confident about is that I may need to take the natural numbers as given objects rather than defining them using set, which might cause some confusion if one think too much about it. But, this approach has its own problems, in the sense that there are objects that are not sets. This might cause a lot of trouble in establishing the consistency of the set theory. Anyway, I will stay with what I have in the book here. After I have finished my book, I reread the classic book Halmos (1960) in set theory and find its treatment of set theory to be very agreeable with my understanding of the subject. This was not the case when I read it during the time of my writing Chap. 2. I think that it is safe to say that Halmos (1960) can be the foundation for set theory as well.

Chapter 3 is a study of topological spaces. A topological space is a set together with a collection of subsets of the set which is called the topology on the set. Any set in the topology is considered an open set. Any set whose complement is an open set is considered a closed set. Then, we can talk about set closure, interior, and boundary. Continuity is the key concept for a function between topological spaces. Fundamental notions of a basis of a topological space, a basis at a particular point in the topological space, the first and second countability of a topological space, the various classes of topological spaces depending on the separability axioms, the category theory of topological spaces, the connectedness concepts are important building blocks for this chapter. There is the proof of Urysohn's Lemma (Theorem 3.55) which establishes the existence of real-valued continuous functions with certain properties on a normal topological space. Furthermore, Tietze's Extension Theorem 3.57 allows us to extend a real-valued function defined on a closed subset of a normal topological space to the entire space. At the end of the chapter, the key notion of "net" is introduced which is a generalization of the concept of a sequence. A sequence has only one direction: forward. A net is like the water flow in a river, it has many vortexes locally, but in the big picture, it has only one direction: forward. With the definition of a net, we can talk about limits and convergence. If we have a net of real (or extended real) numbers, one can talk about limit superior and limit inferior. The concept of continuity is fundamentally linked to the concept of net in topological spaces. It will be seen that the definition of Lebesgue and Riemann integrals are all based on the convergence of the nets.

Metric spaces are studied in Chap. 4. A metric space is a set together with a function that defines the distance (metric) between each pair of points in the set. It induces a topology on the set, and therefore we can talk about open and closed sets, continuity and convergence (in terms of the metric). The concept of completeness is essential in metric spaces. A complete metric space is one in which every Cauchy sequence converges, and therefore the space does not have "holes" in it. In addition to these topological concepts, we can further talk about uniform continuity, uniform convergence, equi-continuity, etc., in general uniformity. The fundamental difference between a complete metric space and an incomplete one is delineated in Baire Category Theorem 4.41. Any given metric space can be embedded as a dense subset of a complete metric space (Theorem 4.49). Because of the additional properties of a metric space instead of a topological space, we are interested in the question of when a topological space is metrizable. This question is resolved through the Urysohn Metrization Theorem 4.53. Because of the ability to talk about uniform convergence, we provide conditions when two limit operations in series can be interchanged without affecting the final answer. This is the topic of Sect. 4.9. The limit operation in series can be interchanged if the joint limit exists. The joint limit exists if certain convergence is uniform.

The concept of compactness of a set is pervasive in analysis. Chapter 5 is devoted to the study of compact spaces and locally compact spaces. The concept of compactness involves actually four concepts: countably compact, Bolzano–Weierstrass property, sequentially compact, and compact for general topological spaces. A countably compact space is equivalent to the space having the Bolzano–Weierstrass property. Otherwise, these compactness properties are distinct in general topological spaces. Here, metric space again demonstrates its superiority over topological spaces, where all these compactness notions are equivalent in metric spaces as demonstrated in the Borel-Lebesgue Theorem 5.37. It is well-known that a continuous real-valued function on a nonempty (countably) compact space achieves its maximum and minimum on the space. Ascoli-Arzelá Theorem 5.44 then says that for metric space valued equi-continuous functions, pointwise convergence implies uniform convergence, which in turn implies that the limit function is continuous. In some sense, this is a sequential compactness theorem for equi-continuous family of functions. In Sect. 5.6, there is the Tychonoff Theorem 5.47, which states that any arbitrary product of compact spaces is compact. The chapter then turns to study locally compact spaces in Sect. 5.7. A main result for locally compact Hausdorff spaces is that a compact set in such a space admits partition of unity with respect to any open covering among the class of functions that satisfy certain standard properties. Another main result for locally compact Hausdorff spaces is that they admit the Alexandroff one-point compactification. We make use of the Alexandroff one-point compactification later in Chap. 11 to define the space $\mathcal{C}_c(\mathcal{X}, \mathcal{Y})$ and find its dual. Then, the concept of proper functions is introduced. The existence of a proper continuous function on a locally compact Hausdorff space is equivalent to σ-compactness of the space. Then, the concept of paracompactness is introduced and it is proved that metric spaces are paracompact. This chapter ends with a discussion of Stone-Čech compactification.

Starting in Chap. 6, we bring binary operations into the study of spaces. A set with a binary operation (addition) is called a group if the binary operation is associative, admits a unit element (0), and every element in the set admits an additive inverse. If the group operation is commutative, then we have an abelian group. On the other hand, a semi-group is something less than a group. It will become a group if it admits inverses for every element in the set. A step further of the concept of a group is a ring. A ring is an abelian group together with another binary operation defined on the set: multiplication. On a ring, the multiplication must be associative, and distributive with respect to addition (left and right). A ring can be commutative if the multiplication operation is commutative. It can have a 1 element if the multiplication operation admits an identity element. A commutative ring with an identity element (for multiplication) is further called a field, if every element in the set except 0 has a multiplicative inverse. The concept of field leads us to the familiar entities: $\mathbb{Q}$, $\mathbb{R}$, $\mathbb{C}$. A vector space over a field is a set together with vector addition, defined on the product of the vector space with itself, and scalar multiplication, defined on the product of the vector space and the field, further satisfying a list of conditions for scalar multiplication and vector addition. A vector space reminds us of $\mathbb{R}^n$ or $\mathbb{R}^{n_1\times n_2\times\cdots\times n_p}$ (but with less technical capabilities). We can define linear operators of a vector space to another (or the same) vector space. The concept of linear combination, subspaces, product space, linear variety, linear independence, and dimension are introduced in vector spaces. For real or complex vector spaces (vector spaces over the field $\mathbb{R}$ or $\mathbb{C}$), the concept of convexity can be introduced.

A vector space is a space with vector operations but without a topology. A normed linear space is a vector space together with a nonnegative real-valued function defined on the set: which is called the norm. The norm is like the concept of the length of an arrow. The arrow has a direction and a length. With the definition of a norm, the space admits a metric space structure, where the distance between two points is simply the length of the difference of the two points (vectors). Thus, we have a topology on the space and can talk about open and closed sets, continuity, and convergence. Since it is a metric space, we can talk about uniform convergence and uniformity in general. A complete normed linear space is called a Banach space. (Remember that completeness is a concept introduced for metric spaces). The Banach space is the focus of study in Chap. 7. In $\mathbb{R}^n$, compactness of a set is equivalent to the set being closed and bounded. For finite-dimensional Banach spaces, compactness of a set is again equivalent to the set being closed and bounded. But, for infinite-dimensional Banach spaces, compactness is hard to come by. But, a lot can be said for a set that is compact in a Banach space. (Then, it is countably compact, has Bolzano–Weierstrass property, sequentially compact. Remember the Borel-Lebesgue Theorem). Since a Banach space has the vector space structure, we can talk about the quotient space with respect to a given subspace. The quotient space is a Banach space if the given subspace is closed and we define norm on the quotient space in a specific way. The given subspace can be the nullspace of a pseudo-norm (which is less than a norm in the sense that the pseudo-norm can be 0 for non-null vectors in the vector space). The vector space with a pseudo-norm can become a normed linear space if we take the quotient of the vector space with respect to the nullspace of the pseudo-norm. Then, we switch gear and prove the Stone–Weierstrass Theorem 7.56 for the Banach space of real-valued functions on a compact space.

Starting in Sect. 7.8, we study linear operators defined on a normed linear space. The set of bounded linear operators of a normed linear space to a Banach space itself forms a Banach space. In particular, bounded linear operators of a Banach space to its associated field $\mathbb{K}$ (stands for $\mathbb{R}$ or $\mathbb{C}$) are called functionals on the original Banach space, which form a Banach space of special interest: it is called the dual of the original Banach space. A Banach space together with its dual allows for many optimization problems to be solved. A fundamental result concerning a normed linear space and its dual is the Hahn–Banach Theorem. In this chapter (Chap. 7), we discuss the extension form of the Hahn–Banach Theorem. In simple words, Hahn–Banach Theorem guarantees the existence of an extension of a bounded linear functional defined on a subspace of the normed linear space to a bounded linear functional on the entire normed linear space without increasing the norm of the functional. This key existence result is proved via a direct application of the Axiom of Choice (or its equivalent). It allows for proving the existence of the optimum in many optimization problem formulated in the dual space. With this theorem, we know that not all formulations of an optimization problem are equal. A crafty formulation allows the solution to be found, while a naive formulation leads to nowhere (literally). The dual of the dual of a Banach space in general contains the original Banach space as a subspace. It should be noted that there are some Banach spaces whose duals are unknown to us (even though it exists in theory). We cannot comprehend such a dual space. The dual of the dual of a Banach space may be the original Banach space, in which case, we say the original Banach space is reflexive. Reflexive Banach spaces are simpler to deal with. We will restrict ourselves to such spaces if certain proofs do not go through for general Banach spaces. In the space of $\mathbb{R}^n$, we have a powerful theorem called the projection theorem. Minimum norm from a vector to a subspace can be found by the projection theorem. In Banach spaces, the concept of orthogonality can be generalized, except that, for a set in the Banach space, its orthogonal complement is a subset in the dual of the Banach space. For a subset in the dual of a Banach space, its orthogonal complement is a subset of the dual of the dual of the Banach space. To simplify things, we define the pre-orthogonal complement of this set (in the dual) to be a set in the original Banach space. With these concepts, minimum norm problems of a point to a subspace can be characterized using the orthogonal complement or the pre-orthogonal complement (see Propositions 7.97 and 7.99). A strong result concerning a bijective bounded linear operator between two Banach spaces is that its inverse is a bounded linear operator. This is a deep result that is a consequence of the Baire Category Theorem. This theorem demonstrates the additional advantage of working in a Banach space, rather than a normed linear space. For a bounded linear operator A of a Banach space $\mathcal{X}$ to another Banach space $\mathcal{Y}$, there is a corresponding a bounded linear operator $A' : \mathcal{Y}^* \to \mathcal{X}^*$, which is called the adjoint of A that maps the dual of $\mathcal{Y}$ to the dual of $\mathcal{X}$. In terms of matrices, the adjoint of a matrix is its transpose, if the definition of linear functional is standard. The range spaces and null spaces of A and A' are closely related as delineated in Propositions 7.112 and 7.114. These are the infinite-dimensional counterparts of the matrix fact $\dim(\mathcal{R}(A)) = n - \dim(\mathcal{N}(A))$, where A is an $m \times n$-dimensional matrix, but are strictly weaker conclusions.

Chapter 7 concludes with a discussion of weak topology on a Banach space and the weak* topology on the dual of a Banach space. On a Banach space, the norm induces a topology on the space, which is called the strong topology. Then, there is the weak topology, which is the weakest topology such that all bounded linear functionals of the Banach space are continuous. A weak open set is a strong open set, and a weak closed set is a strong closed set. But, a strong open set may or may not be a weak open set, and a strong closed set may or may not be a weak closed set. Remember that topology is the collection of subsets of the space that we call open sets. Weak topology is a subset of strong topology. With the definition of the weak topology on the Banach space, we can talk about continuity and convergence in the weak topology. It is true that vector addition and scalar multiplication are continuous in the weak topology. For finite-dimensional Banach spaces, the weak topology is the same as the strong topology. So, the study of the weak topology is for infinite-dimensional Banach spaces only. On the dual of a Banach space, (remember it is also a Banach space), we have the weak topology, and we will define a possibly even weaker topology: the weak* topology: it is the weakest topology on the dual such that all elements of the Banach space viewed as bounded linear functionals on the dual are continuous. When the Banach space is reflexive, then the weak* topology coincides with the weak topology on the dual of the Banach space. Again, it is true that vector addition and scalar multiplication are continuous in the weak* topology. The main reason why we study these topologies is the Alaoglu Theorem 7.122 that says that the closed unit ball of the dual of the Banach space is a weak* compact set. This should not rush the reader to the wrong conclusion that every sequence in the closed unit ball has a weak* convergent subsequence! Remember this is weak* topology, not a metric space, and thus weak* compactness does not imply weak* sequential compactness. The right conclusion that can be made here is that every sequence in the closed unit ball has a weak* cluster point (Bolzano–Weierstrass property). Another piece of useful result in the weak topology is that a bounded linear operator A between Banach spaces is continuous as an operator between the weak topologies of these Banach spaces. I think that Alaoglu Theorem 7.122 holds the key to infinite-dimensional system theory where one can have the tools of compactness available.

Chapter 8 studies convex optimization problems in real Banach spaces. The concept of a hyperplane is defined to be a maximal proper linear variety in a normed linear space. Every closed hyperplane can be identified by a set $\{x \in \mathcal{X} \mid f(x) = c\}$, where f is a non-null bounded linear functional in the dual of $\mathcal{X}$ and $c \in \mathbb{R}$. A direct consequence of the Hahn–Banach Theorem is Mazur's Theorem 8.7 which says that given a linear variety and a convex set with nonempty interior in a real normed linear space, if the linear variety does not intersect the interior of the convex set, then there is a hyperplane containing the linear variety such that the convex set is contained in exactly one of the two closed half spaces associated with the hyperplane. Another direct consequence of Hahn–Banach Theorem is the Eidelheit Separation Theorem 8.9 that says given two nonempty convex sets K_1 and K_2 in a real normed linear space $\mathcal{X}$, if the interior of K_1 is nonempty, and the interior of K_1 does not intersect K_2, then we may find a hyperplane that separates K_1 and K_2. A direct consequence of these theorems is that a closed convex subset of a real normed linear space is weakly closed (remember the weak topology). This, combined with the Alaoglu Theorem, yields that, for a reflexive real Banach space $\mathcal{X}$, a closed bounded convex subset in the dual of $\mathcal{X}$ is weak* (weakly) compact. The infimization of the norm of a point to a nonempty convex set in a real normed linear space is equivalent to the maximization problem involving the support functional of the convex set over the support of the convex set (Proposition 8.15). The same problem, if formulated in the dual of a reflexive real Banach space, admits a minimizer. For more general problems where the optimizing functional is convex but not necessarily a norm expression, we introduce the concept of conjugate convex functional. For convex functionals defined on a convex subset of the dual of a real normed linear space, we introduce the concept of pre-conjugate convex functional. This culminates in Propositions 8.32 and 8.33, where the original function is equal to the supremization over bounded linear functionals in the conjugate convex set of a functional involving the conjugate convex functional. Under stronger assumptions, the supremization becomes maximization in Proposition 8.34. A step further in this line of study is the Fenchel Duality Theorem 8.35. The key game theory result is obtained in Proposition 8.39. For constrained convex optimization problems, we first introduce the concept of positive cone and its dual notion of positive conjugate cone. The chapter ends with presentation of two variants of the Lagrange multiplier theory (Propositions 8.57 and 8.58), as well as sufficient conditions of optimality.

In a Banach space, we can do vector addition and scalar multiplication and have the concept of distance. All the necessary ingredients are present for the study of calculus on the space. In Chap. 9, we embark on the study of differentiation on Banach spaces. This topic is relatively well understood and has significant body of coverage in Luenberger (1969). I follow closely the results on Fréchet derivatives for functions between Banach spaces. I take extra care in the results for functions with general domains (not simply open domains). There, an extra condition is needed for the local shape of the domain at the point of differentiation. For high order derivatives (including partial derivatives), we inevitably deal with tensor operators. The Chain Rule (Theorem 9.18) and Mean Value Theorems are established for first-order differentiations. A function f is $\mathcal{C}_k$ at $x_0 \in \mathcal{X}$ and $\mathrm{dom}\,(f)$ is locally convex implies that the kth order derivative $f^{(k)}(x_0)$ is a k-fold symmetric operator on $\mathcal{X}$ (Proposition 9.28). The calculation of high order derivatives of a function can be carried out according to Proposition 9.27, once it is shown that it is sufficiently many times differentiable. Interchange of order of differentiation can be carried out according to Propositions 9.31 and 9.32. These results are standard (Bartle, 1976) if the domain of the function is open. When this assumption is not valid, then the technical assumptions on the domain that allow the interchange of order of differentiation are delineated in these propositions. The derivatives of useful functions are obtained as well as some basic properties governing the high order derivatives. A key take away from Sect. 9.4 is that a $\mathcal{C}_k$ function composed with another $\mathcal{C}_k$ function again yields a $\mathcal{C}_k$ function. In particular, this result holds at each point in the domain (see Proposition 9.45). For a point in the interior of the domain, the function is $\mathcal{C}_k$ at the point if all partial derivatives up to kth order are continuous at the point (see Proposition 9.47). The study of high order derivative ends with the well-known Taylor's Theorem 9.48.

The usefulness of differentiation in Banach spaces lies in the existence of various mapping theorems in Sect. 9.5 and analytic functions in Sect. 9.9. In Sect. 9.5, we present the Contraction Mapping Theorem, Injective Mapping Theorem, Surjective Mapping Theorem, Open Mapping Theorem, Inverse Function Theorem, and two versions of Implicit Function Theorems (for continuous and differentiable cases). These results have been well motivated in Luenberger (1969). The global version of the Inverse Function Theorem is motivated by Ambrosetti and Prodi (1993). The result is that for a continuous and countably proper function F between Hausdorff topological spaces, if it is locally continuously invertible at each point in the domain, dom (F) is arcwise connected and the range (F) is simply connected, then F is a homeomorphism between dom (F) and range (F). (A continuous function is said to be countably proper if the inverse image of any compact set is countably compact). The technicalities all disappear once we deal with metric spaces. In Sect. 9.7, we present sufficient conditions under which the limit operation and differentiation operation in series can be interchanged. The result becomes standard (Bartle, 1976) if the domain of the function is an open set. In Sect. 9.8, we present the definition and properties of basic tensor operations. Section 9.9 includes all essential ingredients of analytic functions on Banach spaces: the definition of an analytic function, the Taylor series expansion of an analytic function, the composition of analytic functions yields an analytic function, the frequently encountered analytic functions in a Banach space, the inversion of bounded linear operators between Banach spaces as an analytic function, and the analytic versions of Inverse Function Theorem and Implicit Function Theorem. It also includes the discussion of the analytic function: the exponentiation of a bounded linear operator on a Banach space. The chapter ends with a section on Newton's Method on the numerical computation of the root of a function, motivated by Luenberger (1969).

In Chap. 10, we derive local necessary, as well as sufficient conditions for the unconstrained optimization of a functional, the constrained optimization of a functional subject to equality constraints, as well as inequality constraints, on real Banach spaces. We define the concept of positive definite, positive semi-definite, negative definite, and negative semi-definite operators on a real Banach space. We derive necessary and sufficient conditions for Fréchet differentiable functionals to be a convex functional. In constrained optimization problems, the constraints are mappings between Banach spaces, and the inequality constraints are in terms of positive cones. Lagrange multiplier theory is presented for constrained optimization problems. The generalized Kuhn–Tucker Theorem for optimization problems with infinite-dimensional inequality constraints is presented. The results are standard and follow Luenberger (1969).

The first five sections of Chap. 11 are devoted to the study of measure theory, measurable functions on a measure space, general integration theory on a measure space, and the interchange of integration and limit via the Lebesgue Dominated Convergence Theorems. As the study proceeded, I decided that it is unfruitful to deal primarily with complete measure spaces. Rather we focus on general measure space that may not be complete. The focus here is on Borel measure spaces, when dealing with measures on a topological space. The foundation of measure theory is the Carathéodory Extension Theorem. In the definition of the Lebesgue measure on $\mathbb{R}$, we observe its key feature: any Lebesgue measurable set E can be contained by some open set U with arbitrarily small Lebesgue measure $\mu_{\mathrm{L}}(U \setminus E)$; or equivalently, it can contain some closed set F with arbitrarily small Lebesgue measure $\mu_{\mathrm{L}}(E \setminus F)$. We restrict the coverage to only those Borel measurable sets $\mathcal{B}_{\mathrm{B}}(\mathbb{R})$. The restriction of the Lebesgue measure to these Borel measurable sets is defined to be the Borel measure on $\mathbb{R}$, μ_{B}. This measure clearly inherits the key feature from Lebesgue measure. This leads us to define topological measure space $\mathbb{X}$ on a topological space $\mathcal{X} := (X, \mathcal{O})$ to be a the space $((X, \mathcal{O}), \mathcal{B}, \mu)$ where μ is a measure on the measurable space of $(X, \mathcal{B} := \mathcal{B}_{\mathrm{B}}(\mathcal{X}))$, where $\mathcal{B}$ is the σ-algebra generated by the open subsets of X, such that any (Borel) measurable set $E \in \mathcal{B}$ can be contained by some open set $U \in \mathcal{O}$ with arbitrarily small measure $\mu(U \setminus E)$. A metric measure space is a topological measure space on top of a metric space, and Banach measure space is a topological measure space on top of a Banach space, and so on and so forth. One simple relationship due to this definition is that if for any set $E \in \mathcal{B}$, we let $(E, \mathcal{O}_E)$ be the topological subspace of $\mathcal{X}$, and $(E, \mathcal{B}_E, \mu_E)$ be the measure subspace of $(X, \mathcal{B}, \mu)$, then $\mathcal{B}_E$ is the σ-algebra generated by $\mathcal{O}_E$ and μ_E satisfies the preceding condition such that $\mathbb{E} := ((E, \mathcal{O}_E), \mathcal{B}_E, \mu_E)$ forms a topological measure space. Thus, we call $\mathbb{E}$ the topological measure subspace of $\mathbb{X}$. On a topological measure space, all measurable subsets admit a topological measure structure. With the focus on Borel measurable sets, a function is measurable if the inverse image of any open set (and hence any Borel measurable set) is Borel measurable. These concepts are not changed. But, the concept of two functions being almost everywhere equal to each other is significantly changed. We say that a property holds almost everywhere if the set on which the property does not make sense or does not hold *is a Borel measurable set and has measure zero*. The burden here is to guarantee that this negligible set is (Borel) measurable, which can be ensured by the measurability of the functions on both sides of the equality and the range of the functions is a separable metric space. The pointwise limit of measurable functions is again measurable. But for almost everywhere limit of measurable functions, one has to be careful: the limiting function is measurable but not necessarily defined everywhere. To get a measurable function that is defined on the entire domain as an almost everywhere limit of the sequence of measurable functions, one can simply set the function to an arbitrary constant whenever the pointwise limit does not exist. By careful arguments in the proof, the standard machinery of measure theory carries through to Borel measurable spaces. The Littlewood's three principles are cleanly established and appropriately generalized to Borel measurable spaces.

The integration of a measurable function on a measure space is carefully defined. Thus, $0 \times \infty = 0$ is completely banished from the theory. First, we define a directed system, which is called the integration system, on the range of the integrand (which is a normed linear space). Next, integration on finite measure spaces is defined. This directed system results in a net for the corresponding integrand, whose convergence determines the integrability of the integrand. Special care is taken for cases when the integrand is possibly unbounded on the domain of integration, where the integration system has special rules for the sample point that represents a measurable set in the range of the integrand. Based on this definition, we further define the integration of a measurable function on a measure space of ∞ measure to be the limit of the net of integrals on finite measure subspaces of the measure space with the partial ordering being set containment for the finite measure subspaces. Then, we prove that these definitions lead to expected results for the integration of simple functions. Then, the Bounded Convergence Theorem 11.77 is established for general integrands over finite measure spaces. Then, we prove that our definition leads to the same integration formula for nonnegative real-valued measurable functions over any measure space, Proposition 11.78, as Royden (1988). Now, we are fully sure that our definition of integration is correct. We then prove the Monotone Convergence Theorem and establish all well-known properties for Lebesgue integral of nonnegative extended real-valued functions. Finally, we proceed in Sect. 11.5 to establish Lebesgue Dominated Convergence Theorem for Banach space valued measurable functions, which culminates in Theorem 11.91. Thus, under the assumption that the integrands are absolutely integrable over the measure space and the Banach space being separable, all important properties of integration for Banach space valued functions are established in Proposition 11.92. The integral version of Jensen's Inequality is established in Theorem 11.98.

With these developments in integration theory, I couldn't help wondering whether a measure can be Banach space valued. The integrations on measurable subset over a measure space of a Banach space valued measurable function seems to generate a Banach space valued measure on the original measurable space. The question is then how to define such a measure. The definition is relatively straight forward if for every measurable set, the Banach space valued measure generates a vector in the Banach space. This leads to definition of pre-measure on a measurable space. For the pre-measure to be countably additive, one must assume that the measures on countable collections of disjoint measurable sets must be absolutely summable. Taking this idea to the limit when the collection of disjoint sets is small in size but covers the entire measurable space, we inevitably stumble upon the total variation of the Banach space valued pre-measure. This total variation must be a finite measure on the measurable space. Now, the intuition is obtained and the rest is just working out the details. To define a Banach space valued measure μ on a measurable space, there must be first a measure ν on the measurable space, which is known as the total variation of μ, to be denoted by $\mathcal{P} \circ \mu$. Whenever a measurable set E has $\nu(E) = \infty$, $\mu(E)$ is undefined; on the other hand, when $\nu(E) < \infty$, then $\mu(E)$ is a point in the Banach space, and satisfies the countable additivity for its subsets. $\nu(E)$ must then relate to $\mu(E)$ as the set of E is broken down to ever smaller disjoint pieces $(E_i)_{i=1}^{n}$ and the sum of the $\|\mu(E_i)\|$'s must give us back $\nu(E)$. This definition is very different from its obvious counterpart in Royden (1988), the signed measure. But, this is the correct path forward. We confirm the intuition that integrations on measurable subsets over a measure space of a measurable function $f(\cdot)$ indeed generate a Banach space valued measure on the original measurable space, where the total variation is simply the integration on the measurable subsets over the measure space of the measurable function $\|f(\cdot)\|$ (see Proposition 11.116). To generate a σ-finite Banach space valued measure space, we may piece together countably many finite Banach space valued measure spaces as long as they are compatible on the common domain of definition. The resulting σ-finite Banach space valued measure space admits all of the finite Banach space valued measure spaces as measure subspaces. This is the so-called generation process (Proposition 11.118), and the resulting σ-finite Banach space valued measure space is furthermore unique. There will be a few opportunities in the book to apply the generation process to yield desired σ-finite Banach space valued measure spaces. Following that, we define integration over Banach space valued measure spaces along the same line as for measure spaces. The integral of simple functions over a Banach space valued measure space is determined and is as expected. Then, the Lebesgue Dominated Convergence Theorem 11.131 is established. We define a function to be absolutely integrable over a Banach space valued measure space if its norm function is integrable over the total variation of Banach space valued measure. Under the assumption of absolute integrability of integrands, most important properties of integration hold for integration over Banach space valued measure spaces (see Proposition 11.132).

Now, we have Banach space valued measure spaces. For a fixed measurable space $\mathcal{X} := (X, \mathcal{B})$ and a fixed Banach space $\mathcal{Y}$, the collection of all finite $\mathcal{Y}$-valued measures on $\mathcal{X}$ form a Banach space $\mathcal{M}_f(\mathcal{X}, \mathcal{Y})$, where the norm of each finite $\mathcal{Y}$-valued measure is the total variation of the set X. The key here is the definition of vector addition and scalar multiplication for finite $\mathcal{Y}$-valued measures. The vector addition and scalar multiplication are defined for σ-finite $\mathcal{Y}$-valued measures as well via the generation process. Then, the collection of all σ-finite $\mathcal{Y}$-valued measures on $\mathcal{X}$ forms a vector space $\mathcal{M}_\sigma(\mathcal{X}, \mathcal{Y})$. We can further define a topology on $\mathcal{M}_\sigma(\mathcal{X}, \mathcal{Y})$ whose restriction on $\mathcal{M}_f(\mathcal{X}, \mathcal{Y})$ is exactly the one induced by the norm on $\mathcal{M}_f(\mathcal{X}, \mathcal{Y})$. Then, $\mathcal{M}_\sigma(\mathcal{X}, \mathcal{Y})$ is a vector space with a topology, we can then talk about continuity and convergence in $\mathcal{M}_\sigma(\mathcal{X}, \mathcal{Y})$. Then, we prove the Lebesgue Dominated Convergence Theorem 11.151 in the general setting of $\mathcal{M}_\sigma(\mathcal{X}, \mathcal{Y})$, when the underlying measure on the measurable space is also changing but converges to some σ-finite $\mathcal{Y}$-valued measure μ according to the topology on $\mathcal{M}_\sigma(\mathcal{X}, \mathcal{Y})$. The result of Proposition 11.116 is further generalized to σ-finite $\mathcal{Y}$-valued measure spaces. The integrations on measurable subsets over a σ-finite $\mathcal{Y}$-valued measure space of a measurable function $f(\cdot)$ generate a σ-finite Banach space valued measure on the original measurable space, where the total variation of the new measure is upper bounded by the integration on the measurable subsets over the total variation of $\mathcal{Y}$-valued measure space of the measurable function $\|f(\cdot)\|$ (see Proposition 11.153). The upper bound of the total variation is exactly the total variation if we restrict $\mathcal{Y}$ to be $\mathbb{K}$ (see Proposition 11.154).

Hence, up to this point, we have the following: given a measurable function over a σ-finite ($\mathbb{K}$-valued) measure space, it is possible to integrate the function to generate a σ-finite $\mathcal{Z}$-valued measure on the measurable space, where $\mathcal{Z}$ is a Banach space which contains the range of the measurable function. The inverse question of the above is interesting and is resolved by the Radon-Nikodym Theorem. The fundamental condition that allows this inverse operation (Radon-Nikodym derivative) is that the σ-finite $\mathcal{Z}$-valued measure must be absolutely continuous with respect to the σ-finite $\mathbb{K}$-valued measure. Once the Radon-Nikodym derivative is known, integration with respect to the $\mathcal{Z}$-valued measure can be easily carried out with respect to the $\mathbb{K}$-valued measure. This is carried out in Sect. 11.8, which culminates in Theorems 11.169 and 11.171, where the first one is for measures and the second one is for Banach space valued measures. Section 11.9 studies the $L_p(\mathcal{X}, \mathcal{Y})$, where $p \in [1, \infty] \subset \mathbb{R}_e$, $\mathcal{X}$ is a σ-finite measure space, and $\mathcal{Y}$ is a separable Banach space. Minkowski's Inequality and Hölder's Inequality are then presented, which guarantee that $\bar{L}_p(\mathcal{X}, \mathcal{Y})$ is a pseudo-normed linear space. If we take the quotient space of $\bar{L}_p(\mathcal{X}, \mathcal{Y})$ with respect to the nullspace of the pseudo-norm, we have the normed linear space $L_p(\mathcal{X}, \mathcal{Y})$, which a Banach space if $\mathcal{Y}$ is a Banach space. The main objective is to characterize the dual space of $L_p(\mathcal{X}, \mathcal{Y})$. This leads to the Riesz Representation Theorem 11.186, which says that $(L_p(\mathcal{X}, \mathcal{Y}))^* = L_q(\mathcal{X}, \mathcal{Y}^*)$, where $p \in [1, \infty)$ and $q \in (1, \infty]$ with $1/p + 1/q = 1$, $\mathcal{X}$ is a σ-finite measure space, and $\mathcal{Y}$ is a separable reflexive Banach space with $\mathcal{Y}^*$ being separable.

Chapter 11 ends with Sect. 11.10 that contains the study of the Riesz Representation Theorem on the characterization of the duals of $\mathcal{C}(\mathcal{X}, \mathcal{Y})$, where $\mathcal{X}$ is a compact Hausdorff topological space, and $\mathcal{Y}$ is a normed linear space. Under certain conditions, we have that $(\mathcal{C}(\mathcal{X}, \mathcal{Y}))^*$ equals the space of all finite $\mathcal{Y}^*$-valued *topological measures* on $\mathcal{X}$, $\mathcal{M}_{ft}(\mathcal{X}, \mathcal{Y}^*)$ (see Theorem 11.201). The condition in this theorem is easily satisfied if the measure space $\mathcal{X}$ is a compact rectangle in $\mathbb{R}^m$ (see Theorem 11.204) or if $\mathcal{Y}$ is finite-dimensional (see Theorem 11.205). Along the way, we provide sufficient conditions under which $\mathcal{C}(\mathcal{X}, \mathcal{Y})$ is separable (see Proposition 11.206), and prove that for a locally compact separable metric space $\mathfrak{X}$, $\mathcal{M}_f(\mathcal{X}, \mathcal{B}_B(\mathcal{X}), \mathcal{Y}) = \mathcal{M}_{ft}(\mathcal{X}, \mathcal{Y})$, i. e., a finite $\mathcal{Y}$-valued measure on the Borel measurable space is automatically a $\mathcal{Y}$-valued metric (topological) measure (see Theorem 11.198). The notation $\mathcal{C}_c(\mathfrak{X}, \mathcal{Y})$ appears in the literature often, which is defined to be the space of continuous functions of a finite-dimensional Banach space $\mathfrak{X}$ to a finite-dimensional Banach space $\mathcal{Y}$ that has a compact support. This puzzle is resolved in this section and it has been clarified that $\mathcal{C}_c(\mathfrak{X}, \mathcal{Y})$ is the space of all $\mathcal{Y}$-valued continuous functions on $\mathfrak{X}$ that converges to $\vartheta_{\mathcal{Y}}$ as $\|x\|_{\mathfrak{X}} \to \infty$. We characterized the dual $(\mathcal{C}_c(\mathfrak{X}, \mathcal{Y}))^* = \mathcal{M}_{ft}(\mathfrak{X}, \mathcal{Y}^*) = \mathcal{M}_f(\mathfrak{X}, \mathcal{B}_B(\mathfrak{X}), \mathcal{Y}^*)$ and obtained the fact $\mathcal{C}_c(\mathfrak{X}, \mathcal{Y})$ is separable in Theorem 11.209. This characterization allows for a definition of the weak* convergence in $\mathcal{M}_{ft}(\mathfrak{X}, \mathcal{Y}^*)$ as the notion of convergence in distribution on the space of measures. This definition (Definition 11.210) puts the convergence in distribution concept on solid ground, as it is exactly the weak* convergence in $\mathcal{M}_{ft}(\mathfrak{X}, \mathcal{Y}^*)$. Note that the space of probability measures on $\mathfrak{X}$ is a subset of $\mathcal{M}_{ft}(\mathfrak{X}, \mathbb{R})$.

Chapter 12 is the climax of this mathematical journey, linking integration, Radon-Nikodym derivative, and Fréchet derivative together that allow us to go smoothly from one to the other, and solve the puzzle of the connection between multi-dimensional Riemann integral with Lebesgue integral that we have defined, and derive the Change of Variable formula that we are so familiar with in Riemann integration in the Lebesgue integration setting. Since Banach space valued measures were introduced, to obtain product measure spaces, it becomes necessary to derive a generalized Carathéodory Extension Theorem 12.4 to handle these measure spaces. Section 12.2 introduces the concept of isomeasure: there are two measurable spaces $\mathcal{X}$ and $\bar{\mathcal{X}}$, and each endowed with a (Banach space valued) measure μ and $\bar{\mu}$, respectively. The two measures are said to be isomeasuric if there is a bijective mapping $g : \mathcal{X} \to \bar{\mathcal{X}}$ such that the inverse image of any measurable set in $\bar{\mathcal{X}}$

is measurable in $\mathcal{X}$, the image of any measurable set in $\mathcal{X}$ is measurable in $\bar{\mathcal{X}}$, the two total variations of the two measures agree (equal to each other) for any measurable set in $\bar{\mathcal{X}}$ and its inverse image under g in $\mathcal{X}$, and the two measures agree (equal to each other whenever one of them is defined, or both remain undefined when the common total variation is ∞) for any measurable set in $\bar{\mathcal{X}}$ and its inverse image under g in $\mathcal{X}$. Two (Banach space valued) measure spaces being isomeasuric imply that they are equivalent up to relabeling of elements. Isomeasure preserves completeness, finiteness, and σ-finiteness of measure spaces. A step further, we introduce the concept of homeomorphic isomeasure between two (Banach space valued) topological measure spaces, which is basically a bijective mapping that is an isomeasure and a homeomorphism at the same time. Two topological measure spaces are equivalent up to relabeling of elements if they are homeomorphically isomeasuric. Given a topological measure space, a Borel measurable space on another topological space, and a homeomorphism between the two topological spaces, the induced measure on the measurable space based on the bijective mapping and the topological measure space, is such that the topological measure space and the topological measurable space with the induced measure are homeomorphically isomeasuric under the given mapping. It is straightforward to prove that integration on one of the measure spaces can be easily computed via integration on the other measure space after the integrand is composed with the bijective mapping. As a prelude to the ultimate Change of Variable Theorem, we obtain Theorem 12.16: we have two measure spaces and a bijective mapping between them; the mapping induces a measure on the second measure space based on the first measure space. Now, if we know that the induced measure admits a Radon-Nikodym derivative with respect to the measure on the second measure space, then we may easily calculate the integral on the first measure space by calculating the integral on the second measure space with respect to the induced measure, which can be expressed as an integration over the second measure space as long as we multiply the Radon-Nikodym derivative into the integrand. The ultimate theorem on change of variables will delineate the conditions such that the induced measure admits the Radon-Nikodym derivative as desired. This task is postponed until Sect. 12.6.

In Sect. 12.3, we establish the existence of the product ($\mathbb{K}$-valued) measure space of finitely many ($\mathbb{K}$-valued) measure spaces. These results are motivated by the Math 442 notes of Professor Peck of the University of Illinois at Urbana-Champaign. Special care is taken to deal with Borel measure spaces, rather than Lebesgue measure spaces. Here, things are actually simpler. A measurable set in the product measure space has its sections in the multiplier measure spaces to be all measurable. Then a measurable function in the product measure space has all of its sections in the multiplier measure spaces to be measurable. Tonelli's Theorem 12.29 allows the integral in product measure spaces of nonnegative real-valued integrands to be written as repeated integrals, where we have to allow integrals of nonnegative extended real-valued functions, introduced in Definition 11.79. Tonelli's Theorem is usually used to prove a function to be absolutely integrable over the product measure space. Under the assumption that the integrand is absolutely integrable over the product measure space, Fubini's Theorem then gives conditions that allow the joint integral to be computed iteratively. It is observed that general integration can be done if the range of the integrand is a subset of a separable Banach space. But, for product measure spaces, if the integration system is set up on a separable Banach space, there is no guarantee that iterative integrals will make sense, that first integrating with respect to one multiplier measure space may not render a measurable function for the next integration. Here, we can say that the product integral can be written as repeated integrals if we appropriately define the intermediate integrand so that it is measurable and equals the integral with respect to the first multiplier measure space almost everywhere whenever the first iterative integral is absolutely integrable. Fubini's Theorem 12.30 promises that such expression as iterative integrals always exist, and if one can make the integrand for the second iterative integration step to be measurable and equal to the integral with respect to the first multiplier measure space almost everywhere, then one can write the product integral as repeated integrals using such an integrand. The picture for integrands whose range is a subset of a σ-compact conic segment of a Banach space is much simpler. All one needs to do (see Fubini's Theorem 12.31) is to write the product integration as repeated integral, and in the integration with respect to the first multiplier measure space, just set the integration result to be the null vector if the integrand is not absolutely integrable (with respect to the first multiplier measure space), then automatically the second integration has an absolutely integrable integrand. This result readily applies to finite-dimensional Banach space valued integrands. In this section, we further prove that the finite product measure space of second countable topological measure spaces is a second countable topological measure space (with respect to the product topological space). Things are nice here. Same can be said when the multiplier topological measure spaces are $\mathbb{K}$-valued topological measure spaces. Fubini's Theorem 12.35 generalizes the Theorem 12.30 to $\mathbb{K}$-valued product measure spaces.

We see that (Banach space valued) measures are important objects to deal with. The question is how to keep track of them, even when their domain is $\mathbb{R}^m$. It is known in the one-dimensional case, a measure can be represented by its cumulative distribution function, which is a function of bounded variation when the measure is finite. In Sect. 12.4, we embark on a study to characterize so-called functions of locally bounded variation, which are functions on $\mathbb{R}^m$ that will correspond to the cumulative distribution function of (Banach space valued) measures on $\mathbb{R}^m$. The first observation is that for a measure on the closed interval $[1, 2]$, the cumulative distribution function must live on an interval $(1 - \epsilon, 2]$ unless the rule of the cumulative distribution function is modified at end points of the interval. This modification seems to be difficult to characterize for measures on $\mathbb{R}^m$. Also, things will be weird if the domain of the cumulative distribution function is non-standard and too complex. As motivated by the one-dimensional case, we must define these cumulative distribution functions to be continuous on the right. (Continuous on the left works as well, if we insist that cumulative distribution functions are continuous on the left on every point in the domain. We just have to pick one and stick with it.) Then, we proceed first to define the notion of a region Ω, of which the cumulative distribution functions are defined, and its principal $\mathrm{P}(\Omega)$ on which the measure is defined. The region Ω is the countable union of closed rectangles in $\mathbb{R}^m$; while the principal $\mathrm{P}(\Omega)$ is the countable disjoint union of the corresponding semi-open rectangles in $\mathbb{R}^m$ that further satisfies the assumption that for any closed rectangle contained in Ω, its corresponding semi-open rectangle must be contained in $\mathrm{P}(\Omega)$. As defined, there is a unique $\mathrm{P}(\Omega)$ for each region Ω, but not vice versa. It is easy to see that a region and its principal are (Borel) measurable sets. The principal enlarges if the region enlarges. For a rectangle in $\mathbb{R}^m$, it must be a region, and its principal is a rectangle that can be easily characterized (Proposition 12.39). The countable union of semi-open rectangles in $\mathbb{R}^m$ must be a region, whose principal is itself. Product of regions is a region, whose principal is the product of the principals of the regions. An open subset of $\mathbb{R}^m$ is a region whose principal is itself.

Then, we proceed to define the function of locally bounded variation on a region. These are functions that are continuous on the right first of all. We define the increment of such a function on a semi-open rectangle (whose closure is a subset of the region) to be an alternate sum of the values of the function on vertices of the rectangle. We define the total variation of such a function on a semi-open rectangle (whose closure is a subset of the region) to be the supremum of the sum of the norms of the increments of the function on semi-open rectangles whose disjoint union is the given semi-open rectangle. Second of all, for a function to be of locally bounded variation, it must have finite total variation on any semi-open rectangle whose closure is a subset of the region. Lastly, on any closed rectangle that is a subset of the domain (a region), we may define the total variation function of the semi-open rectangle with opposite corners consisting of the lower right corner of the rectangle and any point in the rectangle, this function must be continuous from the right. So, there are three conditions for a function to be of locally bounded variation. The total variation on semi-open rectangles reminds us of the total variation of the Banach space valued measure, and the last condition on the right continuity of the total variation function reminds us of the fact that the total variation must be a measure and therefore must have a right continuous cumulative distribution function. The function is said to be of bounded variation if its total variation on the domain (region) is finite. We hope that we have nailed this definition. To prove this, we define a cumulative distribution function for a σ-finite Banach space valued measure on the principal of the region to be one such that for any semi-open rectangle whose closure is a subset of the region, the measure of the semi-open rectangle equals the increment of the cumulative distribution function on the semi-open rectangle. In Theorem 12.50, we establish the result that for any Banach space ($\mathcal{Y}$) valued function of locally bounded variation defined on a region that is a subset of $\mathbb{R}^m$, there exists a unique σ-finite $\mathcal{Y}$-valued measure on the principal of the region such that the function is a cumulative distribution function of the $\mathcal{Y}$-valued measure, and the total variation of measure on any semi-open rectangle whose closure is a subset of the region is equal to total variation of the function on the semi-open rectangle, and the total variation of the measure on the entire principal of the region is equal to the total variation of the function. This gives one way of the correspondence. For any finite $\mathcal{Y}$-valued Borel measure on the principal of the region in $\mathbb{R}^m$, we can always find a function of locally bounded variation defined on the region that is the cumulative distribution function of the measure (see Proposition 12.51). So, the converse is valid for finite measures. For σ-finite $\mathcal{Y}$-valued measures, a cumulative distribution function may not exist and is not unique. The converse holds (see Proposition 12.52) if the region in $\mathbb{R}^m$ is a rectangle Ω (which must be a region by Proposition 12.39 as we have discussed) and the $\mathcal{Y}$-valued measure is on the principal of Ω, if the total variation of the measure on any semi-open rectangle, whose closure is a subset of the rectangle (region), is finite, then for any point x_0 in the region, we may find the cumulative distribution function, that is defined on the region, of the measure with origin x_0, which is a function of locally bounded variation; for the total variation of the $\mathcal{Y}$-valued measure, we may find the cumulative distribution function of the total variation with origin x_0, and this function is also of locally bounded variation and its value on any semi-open rectangle whose closure is a subset of the region equals the total variation of the cumulative distribution function for the $\mathcal{Y}$-valued measure on the semi-open rectangle. For more general regions, a cumulative distribution function may not be unique or even exist since the increment of a function has many degrees of freedom that may not be captured by just specifying the origin x_0.

With these converses done, the chapter brings up the question of whether a function of locally bounded variation or a cumulative distribution function of a $\mathcal{Y}$-valued measure is measurable on $\mathbb{R}^m$. Propositions 12.54 and 12.55 show that the cumulative distribution function on a region defined for finite $\mathcal{Y}$-valued measures and cumulative distribution function with origin x_0 defined on a rectangle for σ-finite $\mathcal{Y}$-valued measures are measurable. For arbitrary functions of locally bounded variation, it should be clear that one-dimensional case is fine. For higher dimensional domains, there is too much freedom in the function to allow for a positive conclusion. An example on $\mathbb{R}^m$, which is the m-dimensional product of the Borel measure space $\mathbb{R}$, demonstrates that the m-dimensional outer measure induced by the m-dimensional Borel measure of a set E is equal to the infimum over the sum of m-dimensional Borel measures of any countable open covering of E. The measurable sets with respect to this outer measure (m-dimensional Lebesgue outer measure) are the m-dimensional Lebesgue measurable sets. Since the Borel measure μ_B on $\mathbb{R}$ is translational invariant, so is the m-dimensional Borel measure μ_{Bm}, as well as the m-dimensional Lebesgue measure μ_{Lm}. Section 12.4 ends with the definition of Lebesgue-Stieltjes Integral: the integral on a region in $\mathbb{R}^m$ of a measurable function with respect to some function of locally bounded variation is defined to be the integral on the principal of the region of the measurable function with respect to the unique σ-finite Banach space valued measure that corresponds to the function of locally bounded variation, which is prescribed in Theorem 12.50.

Section 12.5 investigates the properties of cumulative distribution functions for Banach space valued measures that are absolutely continuous (admits Radon-Nikodym derivative) with respect to the m-dimensional Borel measure μ_{Bm}. In essence, we seek to define absolutely continuous functions on $\mathbb{R}^m$. In $\mathbb{R}$, the corresponding definition has been widely known. On higher dimensional spaces, the m-dimensional integral of the Radon-Nikodym derivative still allows additional degrees of freedom in the result in terms of integration constant (functions) in the lower dimensional surface of the region. A function can be the cumulative distribution function for an absolutely continuous measure (with respect to μ_{Bm}) without being continuous. This is obviously an undesirable aspect. Thus, we define a function on a rectangle to be absolutely continuous at a point x_0 if freezing any subset of the m-coordinates yields an absolutely continuous function in a lower dimension; and in the m-dimensional space, $\forall \epsilon > 0$, there exists a $\delta > 0$, for any pairwise disjoint semi-open rectangles in a vicinity of x_0, such that their μ_{Bm} measure summing to less than δ, implies that the norms of the increments of the function on the semi-open rectangles sum to less than ϵ. This seems to be a natural way of defining absolutely continuous functions on higher dimensions. The reason why in this definition the domain is restricted to be a rectangle is that the ϵ-δ language does not make sense if the domain is full of holes or has weird boundaries. A possible alternative to this choice is to consider open domain, which should work just as well. But, the former allows for piecing together absolutely continuous functions on closed rectangles, which would result in an absolutely continuous function on the joint rectangle. The rectangle domain allows investigation of absolute continuity on the boundary of a set. In view of this, one just needs to have a big set on which the function is absolutely continuous. A smaller region in the big set might be investigated for physical phenomena. It is proved in the section that a function is absolutely continuous at a point implies that it is continuous at that point. Vectorizing two absolutely continuous functions on the same domain yields an absolutely continuous function. Absolute continuity is preserved under restriction of the domain. Piecing together absolutely continuous functions on two smaller rectangles yields an absolutely continuous function on the joint rectangle if the smaller rectangles are both relatively open or both relatively closed.

Closely related to absolutely continuous functions on intervals is the Lipschitz continuous functions. A Lipschitz continuous function composed with an absolutely continuous function on an interval is again an absolutely continuous function. This fact is the key to the integrability of nonlinear ordinary differential equations. There is no counterpart of this result in higher dimensions, which determines the difficulty in the solvability of general partial differential equations. The product of two absolutely continuous functions on intervals remains absolutely continuous. This is the reason why Integration by Part Theorem exists. Again, there is no counterpart of this result in higher dimensions. The composition of bounded linear operator with an absolutely continuous function on $\mathbb{R}^m$ is again absolutely continuous. The key word here is "linear", as it does not work for nonlinear operators. An absolutely continuous function is automatically of locally bounded variation. Therefore, there exists a unique σ-finite $\mathcal{Y}$-valued measure that admits the function as a cumulative distribution function (see Proposition 12.73). This $\mathcal{Y}$-valued measure is absolutely continuous with respect to μ_{Bm} and therefore admits a Radon-Nikodym derivative f if $\mathcal{Y}$ is reflexive and separable with $\mathcal{Y}^*$ being separable. Then, the $\mathcal{Y}$-valued measure is such a measure with kernel f over $\mathbb{R}^m$. Then, the increment of the function on a semi-open rectangle is equal to the integral of f over the semi-open rectangle with respect to μ_{Bm}. The converse direction of the preceding discussion works like this: start with a rectangle in $\mathbb{R}^m$ and a σ-finite $\mathcal{Y}$-valued measure on its principal (as a topological subspace of $\mathbb{R}^m$), assume that $\mathcal{Y}$-valued measure is absolutely continuous with respect to μ_{Bm} on the principal; then the cumulative distribution function of the $\mathcal{Y}$-valued measure with any origin x_0 in the rectangle is absolutely continuous (Proposition 12.75).

So far, we have been only working with integration and Radon-Nikodym derivative. Section 12.6 studies the line between integration and Fréchet derivative. The key here is the Vitali's Lemma. For higher dimensional spaces (beyond 1), the key is to adapt the Vitali's Lemma to rectangles rather than balls. Then, we may define Lebesgue point of an open set in $\mathbb{R}^m$ and show that the set of Lebesgue points is Lebesgue measurable (not Borel), and its complement in the original open set has (m-dimensional) Lebesgue measure 0 (Proposition 12.80). Given a right continuous function on $\mathbb{R}^m$, it directly leads to that the partial derivatives with respect to coordinate variables are μ_{Bm}-measurable, and the Fréchet derivative of the function is μ_{Bm}-measurable. The question is when do these derivatives exist almost everywhere, and whether it allows the interchange of integration and differentiation. Standard result in the Riemann integral case directly carries over to Lebesgue integral (only in one-dimension, see Theorems 12.82 and 12.83). Here, I also consulted the book Spivak (1965). In Fundamental Theorem of Calculus I, Theorem 12.86, we show that the function defined as a m-dimensional integral of a measurable function that is absolutely integrable over any semi-open rectangles in the open rectangle domain must admit all partial derivatives along all coordinate variables almost everywhere μ_{Bm}. The partial derivatives equal to (almost everywhere μ_{Bm}) the $(m-1)$-dimensional integral subject to measurability concerns. If the function is finite-dimensional Banach space valued, then the partial derivatives along coordinate variables are equal to (almost everywhere μ_{Bm}) the $(m-1)$-dimensional integral whenever the integral is absolutely integrable and set to the null vector whenever it is not.

In Fundamental Theorem of Calculus II, Theorem 12.88, a $\mathcal{Y}$-valued absolutely continuous function defined on an open rectangle of $\mathbb{R}^m$ can be expressed as the sum of $2^m - 1$ integrals of $2^m - 1$ measurable functions and the function value at x_0 in the domain. These measurable functions are unique up to a set of measure 0 on their domain of definition once x_0 is chosen. Therefore, these $2^m - 1$ measurable functions are called the stream functions of the original function with respect to the point x_0. The partial derivatives along coordinate variables are equal to (almost everywhere μ_B) some one-dimensional stream function with respect to some point. If the stream functions are locally bounded, then the Fréchet derivative of the original function exists almost everywhere in the open rectangle. The stream functions of the original function with respect to any point $\bar{x}$ can be obtained from the stream functions with respect to x_0. When $\mathcal{Y}$ is finite-dimensional, this transformation of stream functions can be calculated whenever the integrands in the transformation are absolutely integrable but sets to null vector of $\mathcal{Y}$ if any of the integrands is not absolutely integrable over its domain of integration. Integration by Parts Theorem 12.89 allows for absolutely continuous functions on intervals instead of requiring $\mathcal{C}_1$ functions. The Change of Variable Theorem 12.91 specifies the conditions under which the induced measure is absolutely continuous with respect

μ_{Bm} and proved the usual formula, which allows for absolutely integrable integrand functions and homeomorphisms that may not be $\mathcal{C}_1$.

In Sect. 12.7, we include some incomplete characterization of the dual of $\mathcal{C}_k(\mathcal{X}, \mathcal{Y})$. Then, we present the specialization of the Change of Variable Theorem to two-dimensional and three-dimensional common domain transformations (Examples 12.95–12.97), and one-dimensional general domain transformation (Proposition 12.98). There is also a result that states the equivalence of Riemann-Stieltjes and Lebesgue-Stieltjes integrals for bounded measurable real-valued integrands with respect to monotone integrator (Proposition 12.100). The surface measure on the unit sphere in $\mathbb{R}^m$ is defined in Example 12.104. In Sect. 12.8, we introduce a new definition of Sobolev space that agrees with the classical definition (Zeidler, 1995) in one-dimensional case. The driving reason here is that for functions with generalized derivatives, one must look at absolutely continuous functions. These functions can be expressed as sum of integrals of the stream functions with respect to some point x_0 in the domain exactly. If we define the norm of an absolutely continuous function to include the $\bar{\mathrm{L}}_p$ norms of the function as well as of its stream functions, then one can show that the space is a Banach space. It is a space of (absolutely) continuous functions with all integral norms that is complete. When $p = 2$, this space is a Hilbert space with integral inner products.

In Sect. 12.9, we study continuity and Fréchet differentiability with respect to $x \in D \subseteq \mathcal{X}$ of a function defined as the integral of an integrand function over a finite or σ-finite Banach space valued measure space with x as a parameter. For the finite Banach space valued measure space case, continuity of the function is established in Theorem 12.111. Fréchet differentiability is established in Theorem 12.112, where we allow the domain D to be somewhat general. The result becomes standard if we assume that the domain D is open. This then allows us to prove the Leibniz Formula Theorem 12.113. For the σ-finite Banach space valued measure space case, continuity of the function is established in Theorems 12.114 and 12.115. Fréchet differentiability is established in Theorem 12.116. One take away from these results is that it is best to lump the measurable part of the integrand into a signal that the integrand depends on continuously; then the result follows elegantly with easy generalization to the $\mathcal{C}_k$ case. We then switch gear to show that a Lipschitz continuous function composed with a function of locally bounded variation on an interval again yields a function of locally bounded variation (Proposition 12.117). Vectorizing two functions of locally bounded variation on an interval yields a function of locally bounded variation (Proposition 12.118). We prove that any function in $\bar{\mathrm{L}}_p(\mathbb{X}, \mathcal{Y})$, where $\mathbb{X}$ is a compact interval in $\mathbb{R}$, can be approximated by an absolutely continuous function with any prescribed precision in $\bar{\mathrm{L}}_p(\mathbb{X}, \mathcal{Y})$. The end of the section is Taylor's Theorem 12.122 with the integral form of the remainder. Section 12.10 studies the condition under which an integration on a compact rectangle of $\mathbb{R}^m$ can be calculated as repeated integrals. Here, the goal is to obtain sufficient conditions for the result to hold, that is easily checked and free of technicalities. Theorem 12.124 is an obvious answer to this quest. Then, going deeper, we arrive at the Iterated Integral Theorem 12.127, which is very general and practical for applied mathematics calculations.

The last section of the chapter is connected with the topic of manifolds where local charts are modeled by some Banach space over the field $\mathbb{K}$. This topic is included here due to the necessity to discuss complex logarithms later in Chap. 14 in the proof of the Central Limit Theorem 14.63. Here, for a rigorous presentation, one has to make a trip to the manifolds. These results are motivated by Isidori (1995), which has an excellent introduction to differential manifolds in its Appendix A. The book Bishop and Goldberg (1980) has a proof of the representation of tangent vector at a point of a smooth manifold of finite dimensions. In Sect. 12.11.1, we provide the basic definition of a smooth and analytic manifold of variant $\mathcal{Y}$, where $\mathcal{Y}$ is a Banach space over $\mathbb{K}$. We give a few examples of smooth and analytic manifolds. We then define the logarithm function on the space $\mathbb{C}_0 := \mathbb{C} \setminus \{0\}$, which takes value in a manifold (Example 12.139). Properties of this $\overline{\ln}$ function are obtained in Proposition 12.140. In Sect. 12.11.2, we define tangent vector at a point of a smooth manifold. The collection of all such vectors form the tangent space at the point of the manifold, which is isomorphic to the reflexive Banach space $\mathcal{Y}$. We then define the differential of a smooth mapping between two smooth manifolds at a point and prove that the differential is a linear map (Theorem 12.144). This leads to the Chain Rule (Theorem 12.145) in the context of manifolds. At the end of this subsection, we define the tangent covectors at a point of a smooth manifold. In Sect. 12.11.3, we define vector fields, the Lie derivative, and the Lie bracket and establish their properties in Theorem 12.150 that includes the Jacobi Identity. Due to my determination that I will not deal with the existence and uniqueness of ordinary differential equations in this book, we have to quit here in the pursuit of smooth manifolds.

Chapter 13 contains key results that are specific to Hilbert spaces. Pre-Hilbert space is a vector space together with the binary operation, called the inner product, that induces the norm on the space. A Hilbert space is a pre-Hilbert space that is further complete. So, it is a special kind of a Banach space. All of the results for Banach spaces are valid in Hilbert spaces. The special feature of Hilbert spaces is that the Projection Theorem 13.13 holds. The dual of a Hilbert space is easily characterized in Riesz-Fréchet Theorem 13.15, and a Hilbert space is reflexive. The dual is isometrically isomorphic to itself if the underlying field is real. When the underlying field is complex, then the dual is isometric (but not isomorphic) to the Hilbert space. There exists a norm preserving conjugate linear mapping from a Hilbert space to it dual, which is further invertible. With this mapping, we can define two vectors in the Hilbert space to be orthogonal if one of the vectors is orthogonal (as defined for Banach space and its dual) to the bounded linear functional that is the image of the other vector under the conjugate map. In Sect. 13.4, we define the Hermitian adjoint of a bounded linear operator between Hilbert spaces and obtain its properties. In Sect. 13.5, we present the Gram–Schmidt procedure and other results related to orthonormal sequences in Hilbert spaces. Legendre polynomials are discussed in Example 13.34. The L_2 theory for Fourier series is presented in Example 13.36. The Projection Theorem is revisited in Sect. 13.6, and the minimum norm problem for convex sets is covered. Then, we define positive-definite and positive semi-definite operators among Hermitian operators on a Hilbert space and derive their properties in Sect. 13.7.

Section 13.8 is devoted to the study of projection operators on Hilbert spaces, and the pseudoinverse of bounded linear operators between Hilbert spaces. The last section of this chapter, Sect. 13.9, presents the study of eigenvector and eigenvalues for a bounded linear operator on a Hilbert space. In general, eigenvector or eigenvalue may not exist. But, once we restrict ourselves to compact Hermitian linear operators, then this concept is very useful and generalizes common understandings for Hermitian matrices. In Spectral Theory, Theorem 13.52, the singular value decomposition of a compact linear operator between Hilbert spaces is obtained and the eigenstructure of a compact Hermitian linear operator is obtained. This topic is motivated by Zeidler (1995) and the well-known theory for matrices. It is shown in Example 13.53 that linear integral operators with continuous kernels are compact operators between L_2 spaces. In Proposition 13.55, certain matrix like properties are established for compact linear operators. For compact Hermitian linear operators, Fredholm Alternative Proposition 13.56 establishes certain matrix like properties for these operators. In Example 13.57, we establish that first order Sobolev space $W_{2,1,x_0}(\mathbb{X}, \mathcal{Y})$ is a Hilbert space when $\mathbb{X}$ is a σ-finite metric measure subspace of $\mathbb{R}^m$ and $\mathcal{Y}$ is separable Hilbert space with $\mathcal{Y}^*$ being separable.

Chapter 14 is devoted to probability theory. The material in this chapter is inspired mainly by Williams (1991) and MATH 451 and 452 notes of Professor Burkholder of the University of Illinois at Urbana-Champaign. The fundamental notions in probability theory are introduced in Sect. 14.1: random variables, Banach space valued random variables, the expectation, independence of σ-algebras, independence of random variables, conditional expectation, the law of a random variable, and its probability density function. The conditional expectation of a random variable given a sub-σ-algebra is defined to be the Radon-Nikodym derivative of appropriately defined measures (see Definition 14.7). The main properties are listed in Proposition 14.11, which is motivated by Williams (1991). We then prove the Fundamental Theorem of Modeling, Theorem 14.15, which allows all probability problems to be set up as measure theoretic problems. After a section on Gaussian random variables and vectors (Sect. 14.2), the Weak Law of Large Numbers, Theorem 14.29, is presented in Sect. 14.3. Then, we directly turn to the study of Martingale theory in Sect. 14.4, defining a stochastic process as a measurable function on the product measure space of the probability measure space and the σ-finite measure space of time domain. Proofs of Doob's Optional Stopping Theorem 14.36, Doob's Upcrossing Lemma 14.39, and Doob's Forward Convergence Theorem 14.41 are given. Here, the Law of Large Numbers is revisited, and the Strong Law of Large Numbers (Theorem 14.44) is proved. We then turn in Sect. 14.5 to study Banach space valued Martingales indexed by $\mathbb{Z}_+$. We present the generalized Doob's Optional Stopping Theorem 14.48, Kolmogorov 0 1 Law, Theorem 14.50, and some results on the tail σ-algebra Proposition 14.51. Then, we present results (especially Lévy's Inversion Formula, Theorem 14.55) related to the characteristic function for a real-valued random variable in Sect. 14.6.

In Sect. 14.7, we discuss the mode of convergence and formally apply the Definition 11.210 in the probability measure space context, Definition 14.56. Then, the Modes of Convergence Theorem 14.57 follows readily from the definition. We introduce the concept of uniformly growth boundedness in Definition 14.58, which is also known as the tightness condition in Williams (1991). The convergence in distribution notion we defined is weaker than that of Williams (1991). Proposition 14.60 then gives the conditions to bridge the classical notion of convergence in distribution and our notion of convergence in distribution. The uniformly growth boundedness condition plays a key role in this result, which is dependent on Skorokhod Representation Theorem 14.59. In Sect. 14.8, we proceed to present the Central Limit Theorem 14.63, which is dependent on Helly's Lemma 14.61 and Lévy's Convergence Theorem 14.62. Results on the characteristic function and its inversion formula have direct application to Fourier transform. In Sect. 14.9, we present some results on uniform integrable Martingales. One of my goals in this mathematics study is to establish the existence of Wiener process so as to provide a solid foundation for my past control research. It was sad that I was not able to do this before late July 2018, when I opened my MATH 452 notes of Professor Burkholder of the University of Illinois at Urbana-Champaign and found that the proof for Wiener process is right there, which was not there the last time I checked the notes. So, finally, God opened the door to probability theory for me then. Section 14.10 contains a rigorous proof of the existence of the Wiener process. In Theorem 14.70, we prove that the conditional expectation is the least square estimator of a random variable given any observation σ-algebra. Proposition 14.71 is an effort to generalize Proposition 14.60 to $\mathbb{R}^m$-valued random variables. It is used in the proof of Proposition 14.72, which allows us to form infinite vector sums of independent identically distributed Gaussian random variables and conclude that the sum is an $\mathbb{R}^m$-valued Gaussian random variable. The construction of the Wiener process begins with the Haar functions (Example 14.73). Then, we construct the Wiener process on the unit interval, then on the $[0, \infty) \subset \mathbb{R}$. Then, we define and construct the m-dimensional ($\mathbb{R}^m$-valued) standard Wiener process (Definition 14.78 and Example 14.79). Also included is the proof that the Wiener process is nowhere differentiable with probability 1.

In Sect. 14.11, we study continuous time Martingale processes showing that a particular transformation of the Wiener process yields a class of Martingale processes in the continuous time domain. Lemma 14.85 is a key technical lemma that allows us to analyze the time maximum of the absolute value of the Wiener process. The main result of this section is the Law of Iterated Logarithms (Theorem 14.87). I then study continuous time stopping times and prove a result on the stopped Wiener process Proposition 14.94. The climax of this chapter is the last three sections on stochastic integration. Here, stochastic integrals are defined with respect to standard Wiener process when the integrand is of bounded variation, and therefore the integral exists in the Riemann-Stieltjes integral sense. Then the integrand can be generalized to the $\bar{L}_2$ deterministic signal case, which yields the Wiener integral Proposition 14.98; and also to the $\bar{L}_2$ random signal case, which yields the Itô integral Definition 14.104. Yet, in these more general integrals, the resulting function may not be a stochastic process. It was hinted in Professor Burkholder's notes that modification of resulting function on sets of measure zero may be performed that renders the resulting function a stochastic process, i. e., Itô processes. These results are beyond my grasp before June 2020. After reading Professor Burkholder's notes again, I understood his approach to Itô process. Results on the resulting Itô process are summarized in Sect. 14.13. In these sections, Proposition 14.107 shows that if the integrand process is in the subspace $\mathcal{U}(\bar{\mathrm{r}}_{a,b} \times \Omega, \mathcal{Y})$ then the integral can be approximated by integrals of the sample-and-hold version of the integrand process. As it turns out, the Itô process exists for general $\bar{\mathrm{L}}_2$ random integrands that are Itô integrable, and is unique up to a set of probability zero. But, for the Itô's Formula, Theorem 14.122, to hold, we need to impose some mild integrability assumptions for various technical reasons. Thus, we need the integrand process to be in $\mathcal{U}(\bar{\mathrm{r}}_{a,b} \times \Omega, \mathcal{Y})$. The main difficulty is with the quadratic variation term in the 2-order Taylor expansion of the desired Itô process. For this term to converge to the desired formula, I have to impose growth conditions. The formula applies to general separable Hilbert space valued stochastic processes. The result is elegant and specializes nicely when the involved Hilbert spaces are finite-dimensional. The key here is to make the assumption that the integrand processes satisfy the Riemann Criterion for Integrability, which is necessary and sufficient in the finite-dimensional case but is only sufficient for the general Hilbert space case. In April 2020, I finally proved the Girsanov's Theorem 14.123 after consulting Gihman and Skorohod (1972), Stroock and Varadhan (1979), Elliott (1982). Thus, my earlier research on linear exponential of quadratic Gaussian control are on solid ground if the Girsanov's Theorem holds, in particular, the exponential formula $\xi_0^{t_f}(f)$ is a Martingale, which is always a super Martingale to begin with. I had trouble to derive a sufficient condition for $\xi_0^{t_f}(f)$ to be a Martingale. Then, I spotted Theorem 13.27 and Example 13.32 in Elliott (1982). The proofs for these results require localization results for stochastic integrals that is beyond my grasp. So, I think that I will stop here in my pursuit of probability theory. My earlier research on jump linear systems are entirely correct. My earlier research on nonlinear stochastic systems are just pointers to what one can do in these nonlinear systems but are not in general rigorous in terms of theory. Key integrability assumptions are missing in those results.

Appendix A includes some results in elementary calculus and some results on Riemann-Stieltjes integral that is generalized to Banach space valued integrand and integrator. These results are necessary for the discussion on stochastic integrations and motivated by the excellent results of Bartle (1976). We define the Riemann-Stieltjes integral as a limit of an appropriate net, and then prove Cauchy Criterion for Integrability, the bilinearity of the Riemann-Stieltjes integral, and Integration by Parts (Theorem A.8). The Integration by Parts Theorem is where the Riemann-Stieltjes integral really shines. It holds under the single assumption that the one of the integrals is well-defined. Then, we continue to prove the Riemann Criterion for Integration (Theorem A.9). This then leads to the Integrability Theorem A.10, which connects the Riemann-Stieltjes integral to the Lebesgue-Stieltjes integral that we have been discussing in the main text of this book. It also establishes a sufficient condition when Riemann-Stieltjes integral converges in gauge mode and gives bounds on Riemann-Stieltjes integral. In Theorem A.11, we prove that the Riemann-Stieltjes integral must converge in gauge mode whenever it exists if the integrator is $\mathrm{id}_{\mathbb{R}}$. In Theorem A.12, we give a sufficient condition under which the Riemann-Stieltjes integral is continuous with respect to the terminal time of the integral. In Theorem A.13, we give results on the absolute Riemann-Stieltjes integrability of the integrand function with respect to the integrator when the integrand is Riemann-Stieltjes integrable with respect to the integrator. Then, the Riemann Criterion for Integrability is revisited in Theorem A.14 on the necessity of this criterion when the integrand is finite-dimensional and the integrator is $\mathrm{id}_{\mathbb{R}}$. In Theorem A.15, it is proved that if the integrand is bounded and $\mathcal{B}_{\mathrm{B}}(\mathbb{R})$-measurable, and the integrator is of bounded variation, and furthermore the Riemann-Stieltjes integral satisfies the Riemann Criterion for Integrability, then the Riemann-Stieltjes integral equals the corresponding Lebesgue-Stieltjes integral. Corollaries A.16 and A.17 are sufficiency results of certain Riemann-Stieltjes integrals satisfying the Riemann Criterion for Integrability. The appendix ends with Theorem A.18 which states that if two integrands both satisfy the Riemann Criterion for Integrability with respect to a $\mathbb{K}$-valued integrator of bounded variation, then the product of the integrands also satisfies the Riemann Criterion for Integrability with respect to the $\mathbb{K}$-valued integrator.

1.2 How to Use the Book

The book is written in a linear order mostly. Sections A.1 and A.2 can be read first. Then, one may proceed from Chap. 2 through 11 and 12. At this point, Sect. A.3 is now accessible. This section is a required reading before one can start reading Sect. 14.12. After Chap. 12, one can proceed to Chaps. 13 and 14 and thus complete the book. The book can also be used as a reference book. Every result is stated independently. Thus, one can locate the result that he/she is interested in and then check for required definitions and terminologies. The proof for each result is self-contained. Therefore one may dig deeper into the interested result by reading its proof. These are the main ways of reading this book. This book can serve as a graduate level textbook for mathematics-loving students in the subject of analysis. The prerequisite for the book is the undergraduate calculus course where some exposure to the Riemann integral is presented. Since the book is self-contained and touches on most of the abstract spaces that are prevalent in analysis, a student completing this book will have a good understanding of what is available in analysis and tackle a range of applied mathematics problems.

1.3 What This Book Does Not Include

This book does not include many topics on manifolds, which is an interesting and useful topic. It has direct ramification in our physical world. Interested readers are encouraged to further explore this topic. I personally would like very much to explore this field but I have to go back to my controls research, which is the true calling to me. In probability theory, one may further investigate Markov processes and bring the results on Poisson process and Markov chains to light.

Chapter 2
Set Theory

2.1 Axiomatic Foundations of Set Theory

We will list the nine axioms of Zermelo–Fraenkel–Cantor axiom system. The ninth axiom, which is the axiom of choice, will be introduced in Sect. 2.7. Let A and B be sets and x and y be objects (which is another name for sets).

Axiom 1 (Axiom of Extensionality) *$A = B$ if $\forall x \in A$, we have $x \in B$; and $\forall x \in B$, we have $x \in A$.*

Axiom 2 (Axiom of Empty Set) *There exists an empty set $\emptyset$ that does not contain any element.*

Axiom 3 (Axiom of Pairing) *For any objects x and y, there exists a set $\{x, y\}$ that contains only x and y.*

Axiom 4 (Axiom of Regularity) *Any nonempty set $A \neq \emptyset$, there exists $a \in A$, such that $\forall b \in A$, and we have $b \notin a$.*

Axiom 5 (Axiom of Replacement) *$\forall x \in A$, let there be one and only one y to form an ordered pair (x, y). Then, the collection of all such y's is a set B.*

Axiom 6 (Axiom of Power Set) *The collection of all subsets of A is a set denoted by ${}^A2$.*

Axiom 7 (Axiom of Union) *For any collection of sets $(A_\lambda)_{\lambda \in \Lambda}$, where Λ is a set, then $\bigcup_{\lambda \in \Lambda} A_\lambda$ is a well-defined set.*

Axiom 8 (Axiom of Infinity) *There exists a set A such that $\emptyset \in A$, and $\forall x \in A$, we have $\{\emptyset, x\} \in A$.*

By Axiom 2, there exists the empty set $\emptyset$, which we may call 0. Now, by Axiom 3, there exists the set $\{\emptyset\}$, which is nonempty and we may call 1. Again, by Axiom 3,

Z. Pan, *Measure-Theoretic Calculus in Abstract Spaces*,
https://doi.org/10.1007/978-3-031-21912-2_2

there exists the set $\{\emptyset, \{\emptyset\}\}$, which we will call 2. After we define n, we may define $n+1 := \{0, n\}$, which exists by axiom of pairing. This allows us to define all natural numbers. By Axiom 8, these natural numbers can form the set, $\mathbb{N} := \{1, 2, \ldots\}$, which is the set of natural numbers. Furthermore, by Axiom 6, we may define the set of all real numbers, $\mathbb{R}$.

For any $x \in A$ and $y \in B$, we may apply Axiom 3 to define the ordered pair $(x, y) := \{\{\{\{x\}\}, 1\}, \{\{\{y\}\}, 2\}\}$. Then, the set $A \times B$ is defined by $\bigcup_{x \in A} \bigcup_{y \in B} \{(x, y)\}$, which is a valid set by Axiom 7. By Axiom 5, any portion A_s of a well-defined set A is again a set, which is called a *subset* of A, and we will write $A_s \subseteq A$. Thus, the formula

$$\{x \in A \mid p(x) \text{ is true.}\}$$

defines a set as long as A is a set and $p(x)$ is unambiguous logic expression.

2.2 Relations and Equivalence

Definition 2.1 Let A and B be sets. A *relation* R from A to B is a subset of $A \times B$. $\forall x \in A, \forall y \in B$, we say $x \sim y$ if $(x, y) \in R$. We will say that R is a *relation on* A if it is a relation from A to A. We define

$$\begin{aligned} \text{dom}\,(R) &:= \{x \in A \mid \exists y \in B, \text{ such that } x \sim y\} \\ \text{range}\,(R) &:= \{y \in B \mid \exists x \in A, \text{ such that } x \sim y\} \end{aligned}$$

which are well-defined subsets. ◇

Definition 2.2 Let A be a set and R be a relation on A. $\forall x, y, z \in A$:

1. R is *reflexive* if $x \sim x$.
2. R is *symmetric* if $x \sim y$ implies $y \sim x$.
3. R is *transitive* if $x \sim y$ and $y \sim z$ implies $x \sim z$.
4. R is an *equivalence* relationship if it is reflexive, symmetric, and transitive, and will be denoted by "$\equiv$."
5. R is *antisymmetric* if $x \sim y$ and $y \sim x$ implies $x = y$. ◇

Let $\equiv$ be an equivalence relationship on A; then, it partitions A into disjoint equivalence classes $A_x := \{y \in A \mid x \equiv y\}$, $\forall x \in A$. The collection of all equivalence classes, $A/\equiv := \{A_x \subseteq A \mid x \in A\}$, is called the *quotient* of A with respect to $\equiv$.

2.3 Function

Definition 2.3 Let X and Y be sets and $D \subseteq X$. A *function* f of D to Y, denoted by $f : D \to Y$, is a relation from D to Y such that $\forall x \in D$, there is exactly one $y \in Y$, such that $(x, y) \in f$; we will denote that y as $f(x)$. The *graph* of f is the set $\operatorname{graph}(f) := \{(x, f(x)) \in X \times Y \mid x \in D\}$. The *domain* of f is $\operatorname{dom}(f) = D$. $\forall A \subseteq X$, the *image* under f of A is $f(A) := \{y \in Y \mid \exists x \in A \cap D$ such that $f(x) = y\}$, which is a subset of Y. The *range* of f is $\operatorname{range}(f) = f(X)$. $\forall B \subseteq Y$, the *inverse image* under f of B is $f_{\text{inv}}(B) := \{x \in D \mid f(x) \in B\}$, which is a subset of D. f is said to be *surjective* if $f(X) = Y$; and f is said to be *injective* if $f(x_1) \neq f(x_2)$, $\forall x_1, x_2 \in D$ with $x_1 \neq x_2$; f is said to be *bijective* if it is both surjective and injective, in which case it is *invertible* and the inverse function is denoted by $f_{\text{inv}} : Y \to D$. We will say that f is a function from X to Y. ◇

Let $f : D \to Y$ and $g : Y \to Z$ be functions, we may define a function $h : D \to Z$ by $h(x) = g(f(x))$, and then h is called the *composition* of g with f and is denoted by $g \circ f$. Let $A \subseteq X$. We may define a function $l : A \cap D \to Y$ by $l(x) = f(x)$, $\forall x \in A \cap D$. This function is called the *restriction* of f to A, and denoted by $f|_A$. Let $f : D \to Y$, $g : Y \to Z$, and $h : Z \to W$; we have $(h \circ g) \circ f = h \circ (g \circ f)$. Let $f : D \to D$ and $k \in \mathbb{Z}_+$; we will write $f^k := \underbrace{f \circ \cdots \circ f}_{k}$, where $f^0 := \mathrm{id}_D$.

A function $f : X \to Y$ is a subset of $X \times Y$. Then, $f \in {}^{X \times Y}2$. The collection of all functions of X to Y is then a set given by

$$Y^X := \{f \in {}^{X \times Y}2 \mid \forall x \in X,\ \exists!\, y \in Y \ni \cdot (x, y) \in f\}$$

We have the following result concerning the inverse of a function.

Proposition 2.4 *Let $\phi : X \to Y$, where X and Y are sets. Then, ϕ is bijective if, and only if, $\exists \psi_i : Y \to X$, $i = 1, 2$, such that $\phi \circ \psi_1 = \mathrm{id}_Y$ and $\psi_2 \circ \phi = \mathrm{id}_X$. Furthermore, $\phi_{\text{inv}} = \psi_1 = \psi_2$.*

Proof "Sufficiency" Let $\psi_i : Y \to X$, $i = 1, 2$, exist. $\forall y \in Y$, $\phi \circ \psi_1(y) = \mathrm{id}_Y(y) = y$, which implies that $y \in \operatorname{range}(\phi)$, and hence, ϕ is surjective. Suppose that ϕ is not injective, then $\exists x_1, x_2 \in X$ with $x_1 \neq x_2$ such that $\phi(x_1) = \phi(x_2)$. Then, we have

$$x_1 = \mathrm{id}_X(x_1) = \psi_2(\phi(x_1)) = \psi_2(\phi(x_2)) = \mathrm{id}_X(x_2) = x_2$$

which is a contradiction. Hence, ϕ is injective. This proves that ϕ is bijective.

"Necessity" Let ϕ be bijective. Then, $\phi_{\text{inv}} : Y \to X$ exists. $\forall x \in X$, let $y = \phi(x)$, then $x = \phi_{\text{inv}}(y)$; hence, $\phi_{\text{inv}}(\phi(x)) = x$. Therefore, we have $\phi_{\text{inv}} \circ \phi = \mathrm{id}_X$. $\forall y \in Y$, let $x = \phi_{\text{inv}}(y)$; then $y = \phi(x)$; hence, $\phi(\phi_{\text{inv}}(y)) = y$. Therefore, we have $\phi \circ \phi_{\text{inv}} = \mathrm{id}_Y$. Hence, $\psi_1 = \psi_2 = \phi_{\text{inv}}$.

Let ψ_1 and ψ_2 satisfy the assumption of the proposition, and ϕ_{inv} be the inverse function of ϕ. Then, we have

$$\psi_1 = \text{id}_X \circ \psi_1 = (\phi_{\text{inv}} \circ \phi) \circ \psi_1 = \phi_{\text{inv}} \circ (\phi \circ \psi_1) = \phi_{\text{inv}} \circ \text{id}_Y = \phi_{\text{inv}}$$

$$\psi_2 = \psi_2 \circ \text{id}_Y = \psi_2 \circ (\phi \circ \phi_{\text{inv}}) = (\psi_2 \circ \phi) \circ \phi_{\text{inv}} = \text{id}_X \circ \phi_{\text{inv}} = \phi_{\text{inv}}$$

This completes the proof of the proposition. □

For bijective functions $f : X \to Y$ and $g : Y \to Z$, $g \circ f$ is also bijective and $(g \circ f)_{\text{inv}} = f_{\text{inv}} \circ g_{\text{inv}}$.

2.4 Set Operations

Let X be a set and ${}^X2$ be the set consisting of all subsets of X. $\forall A, B \subseteq X$, we will define

$$A \cup B := \{x \in X \mid x \in A \text{ or } x \in B\}$$
$$A \cap B := \{x \in X \mid x \in A \text{ and } x \in B\}$$
$$\tilde{A} := \{x \in X \mid x \notin A\}$$
$$A \setminus B := \{x \in A \mid x \notin B\} = A \cap \tilde{B}$$
$$A \triangle B := (A \setminus B) \cup (B \setminus A)$$

We have the following results.

Proposition 2.5 *Let $A, B, D, A_\lambda \in {}^X2$, $f : D \to Y$, $C, E, C_\lambda \in {}^Y2$, where X and Y are sets, $\lambda \in \Lambda$, and Λ is an index set. Then, we have:*

1. $A \cup B = B \cup A$ *and* $A \cap B = B \cap A$.
2. $A \subseteq A \cup B$ *and* $A = A \cup B$ *if, and only if,* $B \subseteq A$.
3. $A \cup \emptyset = A$, $A \cap \emptyset = \emptyset$, $A \cup X = X$, *and* $A \cap X = A$.
4. $\tilde{\emptyset} = X$, $\tilde{\tilde{A}} = A$, $A \cup \tilde{A} = X$, $A \cap \tilde{A} = \emptyset$, *and* $A \subseteq B$ *if, and only if,* $\tilde{B} \subseteq \tilde{A}$.
5. *The De Morgan's Laws:*

$$\Big(\bigcup_{\lambda \in \Lambda} A_\lambda\Big)^{\sim} = \bigcap_{\lambda \in \Lambda} \widetilde{A_\lambda}; \qquad \Big(\bigcap_{\lambda \in \Lambda} A_\lambda\Big)^{\sim} = \bigcup_{\lambda \in \Lambda} \widetilde{A_\lambda}.$$

6. $B \cup \big(\bigcap_{\lambda \in \Lambda} A_\lambda\big) = \bigcap_{\lambda \in \Lambda} (B \cup A_\lambda)$ *and* $B \cap \big(\bigcup_{\lambda \in \Lambda} A_\lambda\big) = \bigcup_{\lambda \in \Lambda} (B \cap A_\lambda)$.
7. $f\big(\bigcup_{\lambda \in \Lambda} A_\lambda\big) = \bigcup_{\lambda \in \Lambda} f(A_\lambda)$ *and* $f\big(\bigcap_{\lambda \in \Lambda} A_\lambda\big) \subseteq \bigcap_{\lambda \in \Lambda} f(A_\lambda)$.

8. $f_{\text{inv}}\big(\bigcup_{\lambda\in\Lambda} C_\lambda\big) = \bigcup_{\lambda\in\Lambda} f_{\text{inv}}(C_\lambda)$ *and* $f_{\text{inv}}\big(\bigcap_{\lambda\in\Lambda} C_\lambda\big) = \bigcap_{\lambda\in\Lambda} f_{\text{inv}}(C_\lambda)$.
9. $f_{\text{inv}}(C\setminus E) = f_{\text{inv}}(C)\setminus f_{\text{inv}}(E)$, $f(f_{\text{inv}}(C)) = C\cap \text{range}\,(f)$, *and* $f_{\text{inv}}(f(A)) \supseteq A\cap \text{dom}\,(f) = A\cap D$.

The proof of the above results is standard and is therefore omitted.

2.5 Algebra of Sets

Definition 2.6 A set X is said to be *finite* if it is either empty or the range of a function of $\{1, 2, \ldots, n\}$, with $n \in \mathbb{N}$. In this case, card (X) denotes the number of elements in X. It is said to be *countable* if it is either empty or the range of a function of $\mathbb{N}$. ◇

Definition 2.7 Let X be a set and $\mathcal{A} \subseteq {}^X2$. $\mathcal{A}$ is said to be an *algebra* of sets on X (or a *Boolean algebra* on X) if:

(i) $\emptyset, X \in \mathcal{A}$.
(ii) $\forall A, B \in \mathcal{A}$, $A\cup B \in \mathcal{A}$, and $\tilde{A} \in \mathcal{A}$.

$\mathcal{A}$ is said to be a *σ-algebra on X* if it is an algebra on X and countable union of sets in $\mathcal{A}$ is again in $\mathcal{A}$. ◇

Let $\mathcal{M} \subseteq {}^X2$, where X is a set; then, there exists a smallest algebra on X, $\mathcal{A}_0 \subseteq {}^X2$, containing $\mathcal{M}$, which means that any algebra on X, $\mathcal{A}_1 \subseteq {}^X2$, that contains $\mathcal{M}$, and we have $\mathcal{A}_0 \subseteq \mathcal{A}_1$. This algebra $\mathcal{A}_0$ is said to be the *algebra on X generated by* $\mathcal{M}$. Also, there exists a smallest σ-algebra on X, $\mathcal{A} \subseteq {}^X2$, containing $\mathcal{M}$, which is said to be the *σ-algebra on X generated by* $\mathcal{M}$.

Proposition 2.8 *Let X be a set, $\mathcal{E}$ be a nonempty collection of subsets of X, $\mathcal{A}$ be the algebra on X generated by $\mathcal{E}$, and*

$$\bar{\mathcal{A}} := \Big\{ A \subseteq X \;\Big|\; \exists n, m \in \mathbb{N}, \forall i_1, \ldots, i_{2n} \in \{1, \ldots, m\}, \exists F_{i_1,\ldots,i_{2n}} \subseteq X$$
$$\text{with } F_{i_1,\ldots,i_{2n}} \in \mathcal{E} \text{ or } \big(F_{i_1,\ldots,i_{2n}}\big)^{\sim} \in \mathcal{E}, \text{ such that}$$
$$A = \bigcup_{i_1=1}^{m}\bigcap_{i_2=1}^{m}\cdots\bigcap_{i_{2n}=1}^{m} F_{i_1,\ldots,i_{2n}} \Big\}$$

Then, $\mathcal{A} = \bar{\mathcal{A}}$.

Proof $\forall E \in \mathcal{E}$, let $n = 1$, $m = 1$, and $F_{1,1} = E$. Then, $E = \bigcup_{i_1=1}^{1}\bigcap_{i_2=1}^{1} F_{i_1,i_2} \in \bar{\mathcal{A}}$. Hence, we have $\mathcal{E} \subseteq \bar{\mathcal{A}}$. It is clear that $\bar{\mathcal{A}} \subseteq \mathcal{A}$. All we need to show is that $\bar{\mathcal{A}}$ is an algebra on X. Then, $\mathcal{A} \subseteq \bar{\mathcal{A}}$ and the result follows.

Fix $E \in \mathcal{E} \neq \emptyset$. $E \subseteq X$. Let $n = 1$, $m = 2$, $F_{1,1} = E$, $F_{1,2} = \tilde{E}$, $F_{2,1} = E$, and $F_{2,2} = \tilde{E}$. Then, $\emptyset = \bigcup_{i_1=1}^{2} \bigcap_{i_2=1}^{2} F_{i_1,i_2} \in \bar{\mathcal{A}}$. Let $n = 1$, $m = 2$, $F_{1,1} = E$, $F_{1,2} = E$, $F_{2,1} = \tilde{E}$, and $F_{2,2} = \tilde{E}$. Then, $X = \bigcup_{i_1=1}^{2} \bigcap_{i_2=1}^{2} F_{i_1,i_2} \in \bar{\mathcal{A}}$.

$\forall A, B \in \bar{\mathcal{A}}$, $\exists n_A, m_A \in \mathbb{N}$, $\forall i_1, \ldots, i_{2n_A} \in \{1, \ldots, m_A\}$, $\exists F^{\langle A \rangle}_{i_1,\ldots,i_{2n_A}} \subseteq X$ with $F^{\langle A \rangle}_{i_1,\ldots,i_{2n_A}} \in \mathcal{E}$ or $\left(F^{\langle A \rangle}_{i_1,\ldots,i_{2n_A}}\right)^{\sim} \in \mathcal{E}$ such that $A = \bigcup_{i_1=1}^{m_A} \bigcap_{i_2=1}^{m_A} \cdots \bigcap_{i_{2n_A}=1}^{m_A} F^{\langle A \rangle}_{i_1,\ldots,i_{2n_A}}$, and $\exists n_B, m_B \in \mathbb{N}$, $\forall i_1, \ldots, i_{2n_B} \in \{1, \ldots, m_B\}$, $\exists F^{\langle B \rangle}_{i_1,\ldots,i_{2n_B}} \subseteq X$ with $F^{\langle B \rangle}_{i_1,\ldots,i_{2n_B}} \in \mathcal{E}$ or $\left(F^{\langle B \rangle}_{i_1,\ldots,i_{2n_B}}\right)^{\sim} \in \mathcal{E}$ such that $B = \bigcup_{i_1=1}^{m_B} \bigcap_{i_2=1}^{m_B} \cdots \bigcap_{i_{2n_B}=1}^{m_B} F^{\langle B \rangle}_{i_1,\ldots,i_{2n_B}}$.

Note that $\tilde{A} = \bigcap_{i_1=1}^{m_A} \bigcup_{i_2=1}^{m_A} \cdots \bigcup_{i_{2n_A}=1}^{m_A} \left(F^{\langle A \rangle}_{i_1,\ldots,i_{2n_A}}\right)^{\sim}$. Let $n = n_A + 1$, $m = m_A$, $\forall i_1, \ldots, i_{2n} \in \{1, \ldots, m\}$, $G_{i_1,\ldots,i_{2n}} = \left(F^{\langle A \rangle}_{i_2,\ldots,i_{2n-1}}\right)^{\sim}$. Then, $\tilde{A} = \bigcup_{i_1=1}^{m} \bigcap_{i_2=1}^{m} \cdots \bigcap_{i_{2n}=1}^{m} G_{i_1,\ldots,i_{2n}} \in \bar{\mathcal{A}}$.

Without loss of generality, assume $n_A \geq n_B$. Let $n = n_A$ and $m = m_A + m_B$. Define $\bar{i} = 1 + (i \bmod m_A)$ and $\tilde{i} = 1 + (i \bmod m_B)$, $\forall i \in \mathbb{N}$. $\forall i_1, \ldots, i_{2n} \in \{1, \ldots, m\}$, let $G_{i_1,\ldots,i_{2n}} = \begin{cases} F^{\langle A \rangle}_{i_1,\bar{i}_2,\ldots,\bar{i}_{2n_A}} & \text{if } i_1 \leq m_A \\ F^{\langle B \rangle}_{i_1-m_A,\tilde{i}_2,\ldots,\tilde{i}_{2n_B}} & \text{if } i_1 > m_A \end{cases}$. Then, it is easy to check that $A \cup B = \bigcup_{i_1=1}^{m} \bigcap_{i_2=1}^{m} \cdots \bigcap_{i_{2n}=1}^{m} G_{i_1,\ldots,i_{2n}} \in \bar{\mathcal{A}}$. Hence, $\bar{\mathcal{A}}$ is an algebra on X.

This completes the proof of the proposition. □

2.6 Partial Ordering and Total Ordering

Definition 2.9 Let A be a set and $\preceq$ be a relation on A. $\preceq$ will be called a *partial ordering* if it is reflexive and transitive. It will be called a *total ordering* if it is an antisymmetric partial ordering and satisfies $\forall x, y \in A$ with $x \neq y$, and we have either $x \preceq y$ or $y \preceq x$ (not both). ◇

As an example, the set containment "$\subseteq$" is a partial ordering on any collection of sets, while "$\leq$" is a total ordering on any subset of $\mathbb{R}$.

Definition 2.10 Let A be a set with a partial ordering "$\preceq$":

1. $a \in A$ is said to be *minimal* if, $\forall x \in A$, $x \preceq a$ implies $a \preceq x$.
2. $a \in A$ is said to be the *least element* if, $\forall x \in A$, $a \preceq x$, and $x \preceq a$ implies that $x = a$.
3. $a \in A$ is said to be *maximal* if, $\forall x \in A$, $a \preceq x$ implies $x \preceq a$.
4. $a \in A$ is said to be the *greatest element* if, $\forall x \in A$, $x \preceq a$, and $a \preceq x$ implies that $x = a$. ◇

Definition 2.11 Let A be a set with a partial ordering "$\preceq$" and $E \subseteq A$:

1. $a \in A$ is said to be an *upper bound* of E if $x \preceq a$, $\forall x \in E$. It is the *least upper bound* of E if it is the least element in the set of all upper bounds of E.

2. $a \in A$ is said to be a *lower bound* of E if $a \preceq x$, $\forall x \in E$. It is the *greatest lower bound* of E if it is the greatest element in the set of all lower bounds of E. ◇

We have the following results.

Proposition 2.12 *Let A be a set with a partial ordering "$\preceq$." Then, the following holds:*

(i) *If $a \in A$ is the least element, then it is minimal.*
(ii) *There is at most one least element in A.*
(iii) *Define a relation $\succeq$ by $\forall x, y \in A$, $x \succeq y$ if $y \preceq x$. Then, $\succeq$ is a partial ordering on A. Furthermore, $\succeq$ is antisymmetric if $\preceq$ is antisymmetric:*

1. *$a \in A$ is the least element for $(A, \preceq)$ if, and only if, it is the greatest element for $(A, \succeq)$.*
2. *$a \in A$ is minimal for $(A, \preceq)$ if, and only if, it is maximal for $(A, \succeq)$.*

(iv) *If $a \in A$ is the greatest element, then it is maximal.*
(v) *There is at most one greatest element in A.*
(vi) *If $\preceq$ is antisymmetric, then $a \in A$ is minimal if, and only if, there does not exist $x \in A$ such that $x \preceq a$ and $x \neq a$.*
(vii) *If $\preceq$ is antisymmetric, then $a \in A$ is maximal if, and only if, there does not exist $x \in A$ such that $a \preceq x$ and $x \neq a$.*
(viii) *If $\preceq$ is antisymmetric, then it is a total ordering if, and only if, $\forall x_1, x_2 \in A$, and we have $x_1 \preceq x_2$ or $x_2 \preceq x_1$.*

Proof (i) is straightforward from Definition 2.10.

For (ii), let a_1 and a_2 be least elements of A. By a_1 being the least element, we have $a_1 \preceq a_2$. By a_2 being the least element, we then have $a_1 = a_2$. Hence, the least element is unique if it exists.

For (iii), $\forall x, y, z \in A$, since $x \preceq x$ implies $x \succeq x$, then $\succeq$ is reflexive. If $x \succeq y$ and $y \succeq z$, we have $y \preceq x$ and $z \preceq y$, which implies $z \preceq x$, and hence, $x \succeq z$. This shows that $\succeq$ is transitive. Hence, $\succeq$ is a partial ordering on A.

When $\preceq$ is antisymmetric, $x \succeq y$ and $y \succeq x$ imply that $x \preceq y$ and $y \preceq x$, and therefore $x = y$. Hence, $\succeq$ is also antisymmetric.

For 1, "only if" let $a \in A$ be the least element in $(A, \preceq)$. $\forall x \in A$, we have $a \preceq x$. This implies $x \succeq a$, $\forall x \in A$. $\forall x \in A$ with $a \succeq x$, we have $x \preceq a$, by a being the least element in $(A, \preceq)$, and we have $x = a$. Hence, a is the greatest element in $(A, \succeq)$. The "if" part is similar to the "only if" part.

For 2, "only if" let $a \in A$ be a minimal element for $(A, \preceq)$. Then, $\forall x \in A$ with $a \succeq x$ implies $x \preceq a$, and hence, $a \preceq x$, which yields $x \succeq a$. Hence, a is a maximal element for $(A, \succeq)$. The "if" part is similar to the "only if" part.

(iv) is straightforward from Definition 2.10.

For (v), let a_1 and a_2 be greatest elements of A. By a_1 being the greatest element, we have $a_2 \preceq a_1$. By a_2 being the greatest element, we then have $a_1 = a_2$. Hence, the greatest element is unique if it exists.

For (vi), "if," $\forall x \in A$ with $x \preceq a$, then we have $a = x$, which means that $a \preceq x$; hence, a is minimal. "Only if," suppose that $\exists x \in A$ such that $x \preceq a$ and $x \neq a$.

Note that $a \preceq x$ since a is minimal. Then, $a = x$, since $\preceq$ is antisymmetric, which is a contradiction.

For (vii), "if," $\forall x \in A$ with $a \preceq x$, then we have $a = x$, which means that $x \preceq a$; hence, a is maximal. "Only if," suppose that $\exists x \in A$ such that $a \preceq x$ and $x \neq a$. Note that $x \preceq a$ since a is maximal. Then, $a = x$, since $\preceq$ is antisymmetric, which is a contradiction.

For (viii), "if," $\forall x_1, x_2 \in A$ with $x_1 \neq x_2$, we must have $x_1 \preceq x_2$ or $x_2 \preceq x_1$. They cannot hold at the same time since, otherwise, $x_1 = x_2$, which is a contradiction. "Only if," $\forall x_1, x_2 \in A$, when $x_1 = x_2$, then $x_1 \preceq x_2$; when $x_1 \neq x_2$, then $x_1 \preceq x_2$ or $x_2 \preceq x_1$; hence, in both cases, we have $x_1 \preceq x_2$ or $x_2 \preceq x_1$.

This completes the proof of the proposition. □

2.7 Basic Principles

Now, we introduce the last axiom in Zermelo–Fraenkel–Cantor axiom system.

Axiom 9 (Axiom of Choice) *Let $(A_\lambda)_{\lambda \in \Lambda}$ be a collection of nonempty sets, Λ is a set (this collection is a set by Axiom 5), then there exists a function $f : \Lambda \to \bigcup_{\lambda \in \Lambda} A_\lambda$, such that, $\forall \lambda \in \Lambda$, and we have $f(\lambda) \in A_\lambda$.*

With Axioms 1–8 holding, the axiom of choice is equivalent to the following three results.

Theorem 2.13 (Hausdorff Maximal Principle) *Let $\preceq$ be a partial ordering on a set E. Then, there exists a maximal (with respect to set containment $\subseteq$) subset $F \subseteq E$, such that $\preceq$ is a total ordering on F.*

Theorem 2.14 (Zorn's Lemma) *Let $\preceq$ be an antisymmetric partial ordering on a nonempty set E. If every nonempty totally order subset F of E has an upper bound in E, then there is a maximal element in E.*

Definition 2.15 A *well-ordering* of a set is a total ordering such that every nonempty subset has a least element. ◇

Theorem 2.16 (Well-Ordering Principle) *Every set can be well-ordered.*

To prove the equivalence we described above, we need the following result.

Lemma 2.17 *Let E be a nonempty set and $\preceq$ is an antisymmetric partial ordering on E. Assume that every nonempty subset S of E, on which $\preceq$ is a total ordering, has a least upper bound in E. Let $f : E \to E$ be a mapping such that $x \preceq f(x)$, $\forall x \in E$. Then, f has a fixed point on E, i. e., $\exists w \in E$, $f(w) = w$.*

Proof Fix a point $a \in E$, since $E \neq \emptyset$. We define a collection of "good" sets:

$$\mathcal{B} = \Big\{B \subseteq E \;\Big|\; (i)\, a \in B \quad (ii)\, f(B) \subseteq B \quad (iii)\, \forall F \subseteq B,\, F \neq \emptyset$$
$$F \text{ is totally ordered with } \preceq \text{ implies that the least upper}$$
$$\text{bound of } F \text{ belongs to } B.\Big\}$$

Consider the set

$$B_0 := \{x \in E \mid a \preceq x\}$$

Clearly, B_0 is nonempty since $a \in B_0$ and

$$f(B_0) = \{f(x) \in E \mid a \preceq x \preceq f(x)\} \subseteq B_0$$

since f satisfies $x \preceq f(x)$, $\forall x \in E$.

For any $F \subseteq B_0$, such that F is totally ordered with $\preceq$ and $F \neq \emptyset$. Let e_0 be the least upper bound of F in E. Then, $\exists x_0 \in F$ such that $a \preceq x_0 \preceq e_0$, and therefore $e_0 \in B_0$.

This shows that $B_0 \in \mathcal{B}$ and $\mathcal{B}$ is nonempty.

The following result holds for the collection $\mathcal{B}$.

Claim 2.17.1 Let $\{B_\alpha \mid \alpha \in \Lambda\}$ be any nonempty subcollection of $\mathcal{B}$; then $\bigcap_{\alpha \in \Lambda} B_\alpha \in \mathcal{B}$.

Proof of Claim (i) $a \in B_\alpha$, $\forall \alpha \in \Lambda$. This implies $a \in \bigcap_{\alpha \in \Lambda} B_\alpha$.

(ii) By Proposition 2.5, we have $f(\bigcap_{\alpha \in \Lambda} B_\alpha) \subseteq \bigcap_{\alpha \in \Lambda} f(B_\alpha) \subseteq \bigcap_{\alpha \in \Lambda} B_\alpha$, where the last $\subseteq$ follows from the fact $f(B_\alpha) \subseteq B_\alpha$, $\forall \alpha \in \Lambda$.

(iii) Let $F \subseteq \bigcap_{\alpha \in \Lambda} B_\alpha$, which is totally ordered by $\preceq$ and $F \neq \emptyset$.

For any $\alpha \in \Lambda$, $F \subseteq B_\alpha$ implies that the least upper bound of F is an element of B_α. Therefore, the least upper bound of F is in the intersection $\bigcap_{\alpha \in \Lambda} B_\alpha$.

This establishes $\bigcap_{\alpha \in \Lambda} B_\alpha \in \mathcal{B}$ and completes the proof of the claim. □

The claim shows that the collection $\mathcal{B}$ is closed under arbitrary intersection, as long as the collection is nonempty. Define $A := \bigcap_{B \in \mathcal{B}} B$. By the above claim, we have $A \in \mathcal{B}$, i. e., A is the smallest set in $\mathcal{B}$.

Hence, $A \subseteq B_0$, i. e., the set A satisfies, in addition to (i)–(iii).

(iv) $\forall x \in A, a \preceq x$.

Define the relation $\prec$ on E as $\forall x, y \in E, x \prec y$ if, and only if, $x \preceq y$ and $x \neq y$. Define the set P by

$$P = \{x \in A \mid \forall y \in A,\;\; y \prec x \;\Rightarrow\; f(y) \preceq x\}$$

Clearly, $a \in P$, since there does not exist any $y \in A$ such that $y \prec a$, by $\preceq$ being antisymmetric. Therefore, P is nonempty.

We claim that:

Claim 2.17.2 (v) $\forall x \in P, \forall z \in A$, then $z \preceq x$ or $f(x) \preceq z$.

Proof of Claim Fix $x \in P$, and let

$$B := \{z \in A \mid z \preceq x\} \cup \{z \in A \mid f(x) \preceq z\}$$

We will show that $B \in \mathcal{B}$:
(i) $a \in A$, $x \in P \subseteq A$, by (iv), $a \preceq x$, which further implies that $a \in B$.
(ii) $\forall z \in B \subseteq A$, then $f(z) \in A$ since $A \in \mathcal{B}$. There are three exhaustive scenarios. If $z \prec x$, since $x \in P$ and $z \in B \subseteq A$, then $f(z) \preceq x$. This implies that $f(z) \in B$. If $z = x$, then $f(x) \preceq f(x) = f(z)$. This implies that $f(z) \in B$. If $f(x) \preceq z$, then $f(x) \preceq z \preceq f(z)$. This again implies that $f(z) \in B$. Hence, in all three scenarios, we have $f(z) \in B$. Then, $f(B) \subseteq B$ by the arbitrariness of $z \in B$.
(iii) Let $F \neq \emptyset$ be any totally ordered subset of B and $e_0 \in E$ be the least upper bound of F. Since $F \subseteq B \subseteq A$ and $A \in \mathcal{B}$, then $e_0 \in A$. There are two exhaustive scenarios. If there exists $y \in F$ such that $f(x) \preceq y$, then $f(x) \preceq y \preceq e_0$. This implies $e_0 \in B$. If, for any $y \in F$, $y \preceq x$, then $F \subseteq \{z \in A \mid z \preceq x\}$. This implies that x is an upper bound of F and $e_0 \preceq x$, since e_0 is the least upper bound of F. Therefore, $e_0 \in B$. In both of the cases, we have $e_0 \in B$.

This establishes that $B \in \mathcal{B}$. By A being the smallest set in $\mathcal{B}$, we have $A = B$. Therefore, the claim is proven. □

Now, we show that $P \in \mathcal{B}$:
(i) $a \in P$ and therefore $P \neq \emptyset$.
(ii) Fix an $x \in P \subseteq A$. Then, $f(x) \in A$. $\forall y \in A$ such that $y \prec f(x)$. We need to show that $f(y) \preceq f(x)$, which then implies $f(x) \in P$. By (v), there are two exhaustive scenarios. If $y \preceq x$, then $y \preceq x$. If $f(x) \preceq y$, then $f(x) \preceq y \prec f(x)$ forms a contradiction by $\preceq$ being antisymmetric. Therefore, we must have $y \preceq x$, which results in the following two exhaustive scenarios. If $y \prec x$, then $f(y) \preceq x$ since $x \in P$. This implies that $f(y) \preceq x \preceq f(x)$. If $y = x$, then $f(y) = f(x) \preceq f(x)$. In both cases, we have $f(y) \preceq f(x)$. By the arbitrariness of y, we have $f(x) \in P$, which further implies $f(P) \subseteq P$ by the arbitrariness of $x \in P$.
(iii) Let $F \neq \emptyset$ be a totally ordered subset in P. Let $e_0 \in E$ be the least upper bound of F. We have $F \subseteq A$ implies that $e_0 \in A$ by $A \in \mathcal{B}$. $\forall z \in A$ with $z \prec e_0$, implies that z must not be an upper bound of F. Therefore, $\exists x_0 \in F$ such that $x_0 \npreceq z$. By (v), we have $z \prec x_0$. Hence, by $x_0 \in F \subseteq P$, $z \in A$, and $z \prec x_0$, we have $f(z) \preceq x_0$. Therefore, $f(z) \preceq e_0$ since e_0 is an upper bound of F. This further implies that $e_0 \in P$ by the arbitrariness of z.

This proves that $P \in \mathcal{B}$.

Since $P \subseteq A$ and A is the smallest set in $\mathcal{B}$, then $P = A$.

The set A satisfies properties (i)–(v).

For any $x_1, x_2 \in A$, by (v), there are two exhaustive scenarios. If $x_1 \preceq x_2$, then x_1 and x_2 are related through $\preceq$. If $f(x_2) \preceq x_1$, then $x_2 \preceq f(x_2) \preceq x_1$, which implies that x_1 and x_2 are related through $\preceq$. Therefore, x_1 and x_2 are related through $\preceq$ in both cases. Then, by Proposition 2.12 (viii), A is totally ordered by $\preceq$ and nonempty. Let $w \in E$ be the least upper bound of A. Then, $w \in A$, since $A \in \mathcal{B}$.

Therefore, $f(w) \in A$ by $f(A) \subseteq A$, which implies that $f(w) \preceq w$. This coupled with $w \preceq f(w)$ yields $f(w) = w$, since $\preceq$ is antisymmetric.

This completes the proof of the lemma. $\square$

Theorem 2.18 *Under the Axioms 1–8, the following are equivalent:*

1. *Axiom of choice*
2. *Hausdorff maximum principle*
3. *Zorn's lemma*
4. *Well-ordering principle*

Proof 1. ⇒ 2. Define

$$\mathcal{E} := \{A \subseteq E \mid \preceq \text{ defines a total ordering on } A\}$$

Clearly, $\emptyset \in \mathcal{E}$, then $\mathcal{E} \neq \emptyset$. Define a partial ordering on $\mathcal{E}$ by $\subseteq$, which is set containment. This partial ordering $\subseteq$ is clearly reflexive, transitive, and antisymmetric.

$\forall A \in \mathcal{E}$, define a collection

$$\mathcal{A}_A := \begin{cases} \{B \in \mathcal{E} \mid A \subset B\} & \text{if } \exists B \in \mathcal{E} \text{ such that } A \subset B \\ \{A\} & \text{otherwise} \end{cases}$$

Clearly, $\mathcal{A}_A \neq \emptyset$. By axiom of choice, $\exists T : \mathcal{E} \to \mathcal{E}$ such that $T(A) = B \in \mathcal{A}_A$, $\forall A \in \mathcal{E}$.

We will show that T admits a fixed point by Lemma 2.17. Let $\mathcal{B} \subseteq \mathcal{E}$ be any nonempty subset on which $\subseteq$ is a total ordering. Let $C := \bigcup_{B \in \mathcal{B}} B$. Clearly, $C \subseteq E$. We will show $\preceq$ is a total ordering on C. Since $\preceq$ is a partial ordering on E, then it is a partial ordering on C. $\forall x_1, x_2 \in C$, $\exists B_1, B_2 \in \mathcal{B}$ such that $x_1 \in B_1$ and $x_2 \in B_2$. Since $\subseteq$ is a total ordering on $\mathcal{B}$, then we may without loss of generality assume $B_1 \subseteq B_2$. Then, $x_1, x_2 \in B_2$. Since $B_2 \in \mathcal{B} \subseteq \mathcal{E}$, then $\preceq$ is a total ordering on B_2, which means that we have $x_1 \preceq x_2$ or $x_2 \preceq x_1$. Furthermore, if $x_1 \preceq x_2$ and $x_2 \preceq x_1$, then $x_1 = x_2$ by $\preceq$ being antisymmetric on B_2. Therefore, by Proposition 2.12, $\preceq$ is a total ordering on C. Hence, $C \in \mathcal{E}$. This shows that $\mathcal{B}$ admits least upper bound C in $\mathcal{E}$ with respect to $\subseteq$. By the definition of T, it is clear that $A \subseteq T(A)$, $\forall A \in \mathcal{E}$. By Lemma 2.17, T has a fixed point on $\mathcal{E}$, i. e., $\exists A_0 \in \mathcal{E}$ such that $T(A_0) = A_0$.

By the definitions of T and $\mathcal{A}_{A_0}$, there does not exist $B \in \mathcal{E}$ such that $A_0 \subset B$. Hence, by Proposition 2.12 (vii), A_0 is maximal in $\mathcal{E}$ with respect to $\subseteq$.

2. ⇒ 3. Let E be a nonempty set with an antisymmetric partial ordering $\preceq$. By Hausdorff maximum principle, there exists a maximal (with respect to $\subseteq$) totally ordered (with respect to $\preceq$) subset $F \subseteq E$. We must have $F \neq \emptyset$; otherwise, let $x_0 \in E$ (since $E \neq \emptyset$), $F \subset \{x_0\} \subseteq E$, and $\{x_0\}$ is totally ordered by $\preceq$, which violates the fact that F is maximal (with respect to $\subseteq$). Then, F has an upper bound $e_0 \in E$.

Claim 2.18.1 $e_0 \in F$.

Proof of Claim Suppose $e_0 \notin F$. Define $A := F \cup \{e_0\} \subseteq E$. Clearly, $F \subseteq A$ and $F \neq A$. We will show that $\preceq$ is a total ordering on A. Clearly, $\preceq$ is an antisymmetric partial ordering on A since it is an antisymmetric partial ordering on E. $\forall x_1, x_2 \in A$, we will distinguish 4 exhaustive and mutually exclusive cases: Case 1: $x_1, x_2 \in F$; Case 2: $x_1 \in F$, $x_2 = e_0$; Case 3: $x_1 = e_0$, $x_2 \in F$; Case 4: $x_1 = x_2 = e_0$. In Case 1, we have $x_1 \preceq x_2$ or $x_2 \preceq x_1$ since $\preceq$ is a total ordering on F. In Case 2, we have $x_1 \preceq x_2 = e_0$ since e_0 is an upper bound of F. In Case 3, we have $x_2 \preceq x_1 = e_0$. In Case 4, we have $x_1 = e_0 \preceq e_0 = x_2$. Hence, $\preceq$ is a total ordering on A. Note that $F \subseteq A$ and $F \neq A$. By Proposition 2.12 (vii), this contradicts with the fact that F is maximal with respect to $\subseteq$. Therefore, we must have $e_0 \in F$. This completes the proof of the claim. □

$\forall e_1 \in E$ such that $e_0 \preceq e_1$. $\forall x \in F$, we have $x \preceq e_0 \preceq e_1$. Hence, e_1 is an upper bound of F. By Claim 2.18.1, we must have $e_1 \in F$. Then, $e_1 \preceq e_0$ since e_0 is an upper bound of F. This shows that e_0 is maximal in E with respect to $\preceq$.

3. $\Rightarrow$ 4. Let E be a set. It is clear that $\emptyset \subseteq E$ is well-ordered by the empty relation. Define

$$\mathcal{E} := \{(A_\alpha, \trianglelefteq_\alpha) \mid A_\alpha \subseteq E,\ A_\alpha \text{ is well-ordered by } \trianglelefteq_\alpha\}$$

Then, $\mathcal{E} \neq \emptyset$. Define an ordering $\preceq$ on $\mathcal{E}$ by $\forall (A_1, \trianglelefteq_1), (A_2, \trianglelefteq_2) \in \mathcal{E}$, we say $(A_1, \trianglelefteq_1) \preceq (A_2, \trianglelefteq_2)$ if the following three conditions hold: (i) $A_1 \subseteq A_2$; (ii) $\trianglelefteq_2 = \trianglelefteq_1$ on A_1; (iii) $\forall x_1 \in A_1$, $\forall x_2 \in A_2 \setminus A_1$, and we have $x_1 \trianglelefteq_2 x_2$.

Now, we will show that $\preceq$ defines an antisymmetric partial ordering on $\mathcal{E}$. $\forall (A_1, \trianglelefteq_1), (A_2, \trianglelefteq_2), (A_3, \trianglelefteq_3) \in \mathcal{E}$. Clearly, $(A_1, \trianglelefteq_1) \preceq (A_1, \trianglelefteq_1)$. Hence, $\preceq$ is reflexive. If $(A_1, \trianglelefteq_1) \preceq (A_2, \trianglelefteq_2)$ and $(A_2, \trianglelefteq_2) \preceq (A_3, \trianglelefteq_3)$, we have $A_1 \subseteq A_2 \subseteq A_3$, and (i) holds; (ii) $\trianglelefteq_3 = \trianglelefteq_2$ on A_2 and $\trianglelefteq_2 = \trianglelefteq_1$ on A_1 imply that $\trianglelefteq_3 = \trianglelefteq_1$ on A_1; (iii) $\forall x_1 \in A_1$, $\forall x_2 \in A_3 \setminus A_1$, we have 2 exhaustive senarios: if $x_2 \in A_2$, then $x_2 \in A_2 \setminus A_1$, which implies $x_1 \trianglelefteq_2 x_2$ and hence $x_1 \trianglelefteq_3 x_2$; if $x_2 \in A_3 \setminus A_2$, then we have $x_1 \in A_2$ and $x_1 \trianglelefteq_3 x_2$, and thus, we have $x_1 \trianglelefteq_3 x_2$ in both cases. Therefore, $(A_1, \trianglelefteq_1) \preceq (A_3, \trianglelefteq_3)$, and hence, $\preceq$ is transitive. If $(A_1, \trianglelefteq_1) \preceq (A_2, \trianglelefteq_2)$ and $(A_2, \trianglelefteq_2) \preceq (A_1, \trianglelefteq_1)$, then $A_1 \subseteq A_2 \subseteq A_1 \Rightarrow A_1 = A_2$ and $\trianglelefteq_2 = \trianglelefteq_1$ on A_1. Hence, $(A_1, \trianglelefteq_1) = (A_2, \trianglelefteq_2)$, which shows that $\preceq$ is antisymmetric. Therefore, $\preceq$ defines an antisymmetric partial ordering on $\mathcal{E}$.

Let $\mathcal{A} \subseteq \mathcal{E}$ be any nonempty subset totally ordered by $\preceq$. Take $\mathcal{A} = \{(A_\alpha, \trianglelefteq_\alpha) \mid \alpha \in \Lambda\}$, where $\Lambda \neq \emptyset$ is an index set. Define $A := \bigcup_{\alpha \in \Lambda} A_\alpha$. Define an ordering $\trianglelefteq$ on A by: $\forall x_1, x_2 \in A$, $\exists (A_1, \trianglelefteq_1), (A_2, \trianglelefteq_2) \in \mathcal{A}$ such that $x_1 \in A_1$ and $x_2 \in A_2$, without loss of generality, assume that $(A_1, \trianglelefteq_1) \preceq (A_2, \trianglelefteq_2)$ since $\mathcal{A}$ is totally ordered by $\preceq$, then $x_1, x_2 \in A_2$, and we will say that $x_1 \trianglelefteq x_2$ if $x_1 \trianglelefteq_2 x_2$. We will now show that this ordering is uniquely defined independent of $(A_2, \trianglelefteq_2) \in \mathcal{A}$. Let $(A_3, \trianglelefteq_3) \in \mathcal{A}$ be such that $x_1, x_2 \in A_3$. Since $\mathcal{A}$ is totally ordered by $\preceq$, then there are two exhaustive cases: Case 1: $(A_3, \trianglelefteq_3) \preceq (A_2, \trianglelefteq_2)$; Case 2: $(A_2, \trianglelefteq_2) \preceq (A_3, \trianglelefteq_3)$. In Case 1, we have $A_3 \subseteq A_2$ and $\trianglelefteq_3 = \trianglelefteq_2$ on A_3, which implies that $x_1 \trianglelefteq x_2 \Leftrightarrow x_1 \trianglelefteq_2 x_2 \Leftrightarrow x_1 \trianglelefteq_3 x_2$. In Case 2, we have $A_2 \subseteq A_3$

and $\unlhd_3 = \unlhd_2$ on A_2, which implies that $x_1 \unlhd x_2 \Leftrightarrow x_1 \unlhd_2 x_2 \Leftrightarrow x_1 \unlhd_3 x_2$. Hence, the ordering $\unlhd$ is well-defined on A.

Next, we will show that $\unlhd$ is a total ordering on A. $\forall x_1, x_2, x_3 \in A$. $\exists (A_i, \unlhd_i) \in \mathcal{A}$ such that $x_i \in A_i$, $i = 1, 2, 3$. Since $\mathcal{A}$ is totally ordered by $\preceq$, then, without loss of generality, assume that $(A_1, \unlhd_1) \preceq (A_2, \unlhd_2) \preceq (A_3, \unlhd_3)$. Then, $x_1, x_2, x_3 \in A_3$. Clearly, $x_1 \unlhd x_1$ since $x_1 \unlhd_3 x_1$, which implies that $\unlhd$ is reflective. If $x_1 \unlhd x_2$ and $x_2 \unlhd x_3$, then $x_1 \unlhd_3 x_2 \unlhd_3 x_3$, which implies $x_1 \unlhd_3 x_3$ since $\unlhd_3$ is transitive on A_3, and hence, $x_1 \unlhd x_3$. This shows that $\unlhd$ is transitive. If $x_1 \unlhd x_2$ and $x_2 \unlhd x_1$, then $x_1 \unlhd_3 x_2$ and $x_2 \unlhd_3 x_1$, which implies that $x_1 = x_2$ since $\unlhd_3$ is antisymmetric on A_3. This shows that $\unlhd$ is antisymmetric. Since $\unlhd_3$ is a well-ordering on A_3, then we must have $x_1 \unlhd_3 x_2 \Leftrightarrow x_1 \unlhd x_2$ or $x_2 \unlhd_3 x_1 \Leftrightarrow x_2 \unlhd x_1$. Hence, $\unlhd$ defines a total ordering on A.

Next, we will show that $\unlhd$ is a well-ordering on A. $\forall B \subseteq A$ with $B \neq \emptyset$. Fix $x_0 \in B$. Then, $\exists (A_1, \unlhd_1) \in \mathcal{A}$ such that $x_0 \in A_1$. Note that $\emptyset \neq B \cap A_1 \subseteq A_1$. Since A_1 is well-ordered by $\unlhd_1$, then $\exists e \in B \cap A_1$, which is the least element of $B \cap A_1$. $\forall y \in B \subseteq A$, $\exists (A_2, \unlhd_2) \in \mathcal{A}$ such that $y \in A_2$. We have 2 exhaustive and mutually exclusive cases: Case 1: $y \in A_1$; Case 2: $y \in A_2 \setminus A_1$. In Case 1, $e \unlhd_1 y$ since e is the least element of $B \cap A_1$, which implies that $e \unlhd y$. In Case 2, since $\mathcal{A}$ is totally ordered by $\preceq$, we must have $(A_1, \unlhd_1) \preceq (A_2, \unlhd_2)$, which implies that $e \unlhd_2 y$, by (iii) in the definition of $\preceq$, and hence, $e \unlhd y$. In both cases, we have shown that $e \unlhd y$. Since $\unlhd$ is a total ordering on A, then e is the least element of B. Therefore, $\unlhd$ is a well-ordering on A, which implies $(A, \unlhd) \in \mathcal{E}$.

$\forall (A_1, \unlhd_1) \in \mathcal{A}$. (i) $A_1 \subseteq A$. (ii) $\forall x_1, x_2 \in A_1$, $x_1 \unlhd_1 x_2 \Leftrightarrow x_1 \unlhd x_2$; hence, $\unlhd = \unlhd_1$ on A_1. (iii) $\forall x_1 \in A_1$, $\forall x_2 \in A \setminus A_1$, $\exists (A_2, \unlhd_2) \in \mathcal{A}$ such that $x_2 \in A_2 \setminus A_1$; since $\mathcal{A}$ is totally ordered by $\preceq$, then we must have $(A_1, \unlhd_1) \preceq (A_2, \unlhd_2)$; hence, $x_1 \unlhd_2 x_2$ and $x_1 \unlhd x_2$. Therefore, we have shown $(A_1, \unlhd_1) \preceq (A, \unlhd)$. Hence, $(A, \unlhd) \in \mathcal{E}$ is an upper bound of $\mathcal{A}$.

By Zorn's Lemma, there is a maximal element $(F, \unlhd_F) \in \mathcal{E}$. We claim that $F = E$. We will prove this by an argument of contradiction. Suppose $F \subset E$, then $\exists x_0 \in E \setminus F$. Let $H := F \cup \{x_0\}$. Define an ordering $\unlhd_H$ on H by: $\forall x_1, x_2 \in H$, if $x_1, x_2 \in F$, we say $x_1 \unlhd_H x_2$ if $x_1 \unlhd_F x_2$; if $x_1 \in F$ and $x_2 = x_0$, then we let $x_1 \unlhd_H x_2$; if $x_1 = x_2 = x_0$, we let $x_1 \unlhd_H x_2$. Now, we will show that $\unlhd_H$ is a well-ordering on H. $\forall x_1, x_2, x_3 \in H$. If $x_1 \in F$, then $x_1 \unlhd_F x_1$ and $x_1 \unlhd_H x_1$; if $x_1 = x_0$, then $x_1 \unlhd_H x_1$. Hence, $\unlhd_H$ is reflexive. If $x_1 \unlhd_H x_2$ and $x_2 \unlhd_H x_3$, we have 4 exhaustive and mutually exclusive cases: Case 1: $x_1, x_3 \in F$; Case 2: $x_1 \in F$ and $x_3 = x_0$; Case 3: $x_3 \in F$ and $x_1 = x_0$; Case 4: $x_1 = x_3 = x_0$. In Case 1, we must have $x_2 \in F$ and then $x_1 \unlhd_F x_2$ and $x_2 \unlhd_F x_3$, which implies that $x_1 \unlhd_F x_3$, and hence, $x_1 \unlhd_H x_3$. In Case 2, we have $x_1 \unlhd_H x_3$. In Case 3, we must have $x_2 = x_0$, which leads to a contradiction $x_0 \unlhd_H x_3$; hence, this case is impossible. In Case 4, we have $x_1 \unlhd_H x_3$. In all cases except that is impossible, we have $x_1 \unlhd_H x_3$. Hence, $\unlhd_H$ is transitive. If $x_1 \unlhd_H x_2$ and $x_2 \unlhd_H x_1$, we have 4 exhaustive and mutually exclusive cases: Case 1: $x_1, x_2 \in F$; Case 2: $x_1 \in F$ and $x_2 = x_0$; Case 3: $x_2 \in F$ and $x_1 = x_0$; Case 4: $x_1 = x_2 = x_0$. In Case 1, we have $x_1 \unlhd_F x_2$ and $x_2 \unlhd_F x_1$, which implies that $x_1 = x_2$ since $\unlhd_F$ is antisymmetric on F. In Case 2, we have $x_0 \unlhd_H x_1$, which is a contradiction, and hence, this case is impossible. In Case 3, we have $x_0 \trianglelefteq_H x_2$, which is a contradiction, and hence, this case is impossible. In Case 4, we have $x_1 = x_2$. In all cases except those impossible, we have $x_1 = x_2$. Hence, $\trianglelefteq_H$ is antisymmetric. When $x_1, x_2 \in F$, then, we must have $x_1 \trianglelefteq_F x_2$ or $x_2 \trianglelefteq_F x_1$ since $\trianglelefteq_F$ is a well-ordering on F, and hence, $x_1 \trianglelefteq_H x_2$ or $x_2 \trianglelefteq_H x_1$. When $x_1 \in F$ and $x_2 = x_0$, then $x_1 \trianglelefteq_H x_2$. When $x_2 \in F$ and $x_1 = x_0$, then $x_2 \trianglelefteq_H x_1$. When $x_1 = x_2 = x_0$, then $x_1 \trianglelefteq_H x_2$. This shows that $\trianglelefteq_H$ is a total ordering on H. $\forall B \subseteq H$ with $B \neq \emptyset$. We will distinguish two exhaustive and mutually exclusive cases: Case 1: $B = \{x_0\}$; Case 2: $B \neq \{x_0\}$. In Case 1, x_0 is the least element of B. In Case 2, $B \setminus \{x_0\} \subseteq F$ and is nonempty and hence admits a least element $e_0 \in B \setminus \{x_0\} \subseteq F$ with respect to $\trianglelefteq_F$. $\forall x \in B$, if $x \in B \setminus \{x_0\}$, then $e_0 \trianglelefteq_F x$ and hence $e_0 \trianglelefteq_H x$; if $x = x_0$, then $e_0 \trianglelefteq_H x$. Hence, e_0 is the least element of B since $\trianglelefteq_H$ is a total ordering on H. Therefore, $\trianglelefteq_H$ is a well-ordering on H and $(H, \trianglelefteq_H) \in \mathcal{E}$.

Clearly, $F \subset H$, $\trianglelefteq_F = \trianglelefteq_H$ on F, and $\forall x_1 \in F$ and $\forall x_2 \in H \setminus F$, we have $x_2 = x_0$ and $x_1 \trianglelefteq_H x_2$. This implies that $(F, \trianglelefteq_F) \preceq (H, \trianglelefteq_H)$. Since $(F, \trianglelefteq_F)$ is maximal in $\mathcal{E}$ with respect to $\preceq$, we must have $(H, \trianglelefteq_H) \preceq (F, \trianglelefteq_F)$, and hence, $H \subseteq F$. This is a contradiction. Therefore, $F = E$ and E is well-ordered by $\trianglelefteq_F$.

4. ⇒ 1. Let $(A_\lambda)_{\lambda \in \Lambda}$ be a collection of nonempty sets, and Λ is a set. Let $A := \bigcup_{\lambda \in \Lambda} A_\lambda$. By well-ordering principle, A may be well-ordered by $\trianglelefteq$. $\forall \lambda \in \Lambda$, $A_\lambda \subseteq A$ is nonempty and admits the least element $e_\lambda \in A_\lambda$. This defines a function $f : \Lambda \to A$ by $f(\lambda) = e_\lambda \in A_\lambda$, $\forall \lambda \in \Lambda$.

This completes the proof of the theorem. □

Example 2.19 Let Λ be an index set and $(A_\alpha)_{\alpha \in \Lambda}$ be a collection of sets. We will try to define the *Cartesian (direct) product* $\prod_{\alpha \in \Lambda} A_\alpha$. Let $A = \bigcup_{\alpha \in \Lambda} A_\alpha$, which is a set by the axiom of union. Then, as we discussed in Sect. 2.3, A^Λ is a set, which consists of all functions of Λ to A. Define the *projection* functions $\pi_\alpha : A^\Lambda \to A$, $\forall \alpha \in \Lambda$, by, $\forall f \in A^\Lambda$, $\pi_\alpha(f) = f(\alpha)$. Then, we may define the set

$$\prod_{\alpha \in \Lambda} A_\alpha := \{f \in A^\Lambda | \pi_\alpha(f) \in A_\alpha,\ \forall \alpha \in \Lambda\}$$

When all of A_α's are nonempty, then, by axiom of choice, the product $\prod_{\alpha \in \Lambda} A_\alpha$ is also nonempty. ◇

Chapter 3
Topological Spaces

3.1 Fundamental Notions

Definition 3.1 A *topological space* $(X, \mathcal{O})$ consists of a set X and a collection $\mathcal{O}$ of subsets (namely, *open subsets*) of X such that

(i) $\emptyset, X \in \mathcal{O}$.
(ii) $\forall O_1, O_2 \in \mathcal{O}$, we have $O_1 \cap O_2 \in \mathcal{O}$.
(iii) $\forall (O_\alpha)_{\alpha \in \Lambda} \subseteq \mathcal{O}$, where Λ is an index set, we have $\bigcup_{\alpha \in \Lambda} O_\alpha \in \mathcal{O}$.

The collection $\mathcal{O}$ is called a *topology* for the set X. ◇

Definition 3.2 Let $(X, \mathcal{O})$ be a topological space and $F \subseteq X$. The *complement* of F is $\widetilde{F} := X \setminus F$. F is said to be *closed* if $\widetilde{F} \in \mathcal{O}$. The *closure* of F is given by $\overline{F} := \bigcap_{\substack{F \subseteq B \\ \widetilde{B} \in \mathcal{O}}} B$, which is clearly a closed set. The *interior* of F is given by $F^\circ := \bigcup_{\substack{B \subseteq F \\ B \in \mathcal{O}}} B$, which is clearly an open set. A *point of closure* of F is a point in $\overline{F}$. An *interior point* of F is a point in F°. A *boundary point* of F is a point $x \in X$ such that $\forall O \in \mathcal{O}$ with $x \in O$, we have $O \cap F \neq \emptyset$ and $O \cap \widetilde{F} \neq \emptyset$. The *boundary* of F, denoted by ∂F, is the set of all boundary points of F. An *exterior point* of F is a point in $(\widetilde{F})^\circ$, where $\widetilde{F}^\circ$ is called the *exterior* of F. An *accumulation point* of F is a point $x \in X$ such that $\forall O \in \mathcal{O}$ with $x \in O$, we have $O \cap (F \setminus \{x\}) \neq \emptyset$. ◇

Clearly, $\emptyset$ and X are both closed and open.

Proposition 3.3 *Let $(X, \mathcal{O})$ be a topological space and A, B, and E are subsets of X. Then,*

(i) $E \subseteq \overline{E}$, $\overline{\overline{E}} = \overline{E}$, $E^\circ \subseteq E$, $(E^\circ)^\circ = E^\circ$, *and* $\widetilde{E}^\circ = \widetilde{\overline{E}}$.
(ii) $\forall x \in X$, *x is a point of closure of E if, and only if,* $\forall O \in \mathcal{O}$ *with* $x \in O$, *we have* $O \cap E \neq \emptyset$.

Z. Pan, *Measure-Theoretic Calculus in Abstract Spaces*,
https://doi.org/10.1007/978-3-031-21912-2_3

(iii) *$\forall x \in X$, x is an interior point of E if, and only if, $\exists O \in \mathcal{O}$ with $x \in O$ such that $O \subseteq E$.*
(iv) $\overline{A \cup B} = \overline{A} \cup \overline{B}$, $(A \cap B)^\circ = A^\circ \cap B^\circ$.
(v) *E is closed if, and only if, $E = \overline{E}$.*
(vi) $\overline{E} = E^\circ \cup \partial E$.
(vii) *X equals the disjoint union $E^\circ \cup \partial E \cup (\widetilde{E})^\circ$.*

Proof (i) Clearly, $E \subseteq \overline{E}$. Then, $\overline{E} \subseteq \overline{\overline{E}}$. $\forall C \supseteq E$ with $\widetilde{C} \in \mathcal{O}$, we have $C \supseteq \overline{E}$. Then, $\overline{E} = \bigcap_{\substack{C \supseteq E \\ \widetilde{C} \in \mathcal{O}}} C \supseteq \bigcap_{\substack{C \supseteq \overline{E} \\ \widetilde{C} \in \mathcal{O}}} C = \overline{\overline{E}}$. Hence, we have $\overline{E} = \overline{\overline{E}}$.

Clearly, $E^\circ \subseteq E$. Note that

$$\widetilde{\overline{E}} = \Bigg(\bigcap_{\substack{B \supseteq E \\ \widetilde{B} \in \mathcal{O}}} B\Bigg)^{\sim} = \bigcup_{\substack{B \supseteq E \\ \widetilde{B} \in \mathcal{O}}} \widetilde{B} = \bigcup_{\substack{O \subseteq \widetilde{E} \\ O \in \mathcal{O}}} O = \widetilde{E}^\circ$$

Furthermore,

$$(E^\circ)^\circ = \left(\widetilde{\widetilde{E}}^\circ\right)^\circ = \left(\widetilde{\overline{\widetilde{E}}}\right)^\circ = \widetilde{\overline{\overline{\widetilde{E}}}} = \widetilde{\overline{\widetilde{E}}} = \left(\widetilde{\widetilde{E}}\right)^\circ = E^\circ$$

(ii) "Only if" $\forall x \in \overline{E}$, we have $x \in \bigcap_{\substack{B \supseteq E \\ \widetilde{B} \in \mathcal{O}}} B$. $\forall O \in \mathcal{O}$ with $x \in O$, let $O_1 := O \cup \widetilde{\overline{E}} \in \mathcal{O}$. Note that $x \in O_1$ and $E \cap \widetilde{\overline{E}} = \emptyset$. Suppose $O \cap E = \emptyset$. Then, $E \cap O_1 = \emptyset$, which further implies that $E \subseteq \widetilde{O_1}$. Then, $\overline{E} \subseteq \widetilde{O_1}$ and $x \in \widetilde{O_1}$. This contradicts with $x \in O$. Hence, $O \cap E \neq \emptyset$.
"If" $\forall x \in \widetilde{\overline{E}} = \widetilde{E}^\circ \in \mathcal{O}$. Then, $\exists O := \widetilde{E}^\circ \in \mathcal{O}$ such that $E \cap O = \emptyset$. Hence, the result holds.

(iii) "Only if" $\forall x \in E^\circ \subseteq E$, then $E^\circ \in \mathcal{O}$.
"If" $\forall x \in X$, $\exists O \in \mathcal{O}$ such that $x \in O \subseteq E$. Then, $x \in O \subseteq \bigcup_{\substack{B \subseteq E \\ B \in \mathcal{O}}} B = E^\circ$. Hence, the result holds.

(iv) Let $\mathcal{B} := \{O \in \mathcal{O} \mid O \subseteq A \cap B\}$, $\mathcal{B}_A := \{O \in \mathcal{O} \mid O \subseteq A\}$, and $\mathcal{B}_B := \{O \in \mathcal{O} \mid O \subseteq B\}$. $\forall O_1 \in \mathcal{B}_A$ and $\forall O_2 \in \mathcal{B}_B$, then, $O_1 \cap O_2 \in \mathcal{B}$. On the other hand, $\forall O \in \mathcal{B}$, we have $O = O \cap O$ and $O \in \mathcal{B}_A$ and $O \in \mathcal{B}_B$. Then,

$$(A \cap B)^\circ = \bigcup_{\substack{O \subseteq A \cap B \\ O \in \mathcal{O}}} O = \bigcup_{\substack{O_1 \subseteq A,\, O_2 \subseteq B \\ O_1, O_2 \in \mathcal{O}}} (O_1 \cap O_2)$$

$$* = \Bigg(\bigcup_{\substack{O_1 \subseteq A \\ O_1 \in \mathcal{O}}} O_1\Bigg) \cap \Bigg(\bigcup_{\substack{O_2 \subseteq B \\ O_2 \in \mathcal{O}}} O_2\Bigg) = A^\circ \cap B^\circ$$

We also have

$$\overline{A \cup B} = \left(\left(\widetilde{A \cup B}\right)^{\circ}\right)^{\sim} = ((\widetilde{A} \cap \widetilde{B})^{\circ})^{\sim} = ((\widetilde{A})^{\circ} \cap (\widetilde{B})^{\circ})^{\sim}$$

$$= \widetilde{\widetilde{\overline{A}} \cap \widetilde{\overline{B}}} = \overline{A} \cup \overline{B}$$

(v) "If" E is closed since $E = \overline{E}$ and $\overline{E}$ is closed.

"Only if" Since E is closed, then $\overline{E} \subseteq E$. Then, we have $E = \overline{E}$. Hence, the result holds.

(vi) This result follows directly from (ii), (iii), and Definition 3.2.

(vii) Note that $X = \overline{E} \cup \widetilde{\overline{E}} = E^{\circ} \cup \partial E \cup \widetilde{E}^{\circ}$. By (iii) and Definition 3.2, E° and ∂E are disjoint. It is obvious that $\widetilde{\overline{E}}$ is disjoint with $E^{\circ} \cup \partial E = \overline{E}$. Hence, the result holds. □

To simplify notation in the theory, we will abuse the notation to write $x \in \mathcal{X}$ when $x \in X$ and $A \subseteq \mathcal{X}$ when $A \subseteq X$ for a topological space $\mathcal{X} := (X, \mathcal{O})$. We will later simply discuss a topological space $\mathcal{X}$ without further reference to components of $\mathcal{X}$, where the topology is understood to be $\mathcal{O}_{\mathcal{X}}$. When it is clear from the context, we will neglect the subscript $\mathcal{X}$.

Proposition 3.4 *Let* $(X, \mathcal{O})$ *be a topological space and* $A \subseteq X$. *A admits the* subset topology $\mathcal{O}_A := \{O \cap A \mid O \in \mathcal{O}\}$.

Proof Clearly, $\mathcal{O}_A$ is a collection of subsets of A. $\emptyset = \emptyset \cap A \in \mathcal{O}_A$ and $A = X \cap A \in \mathcal{O}_A$. $\forall O_{A1}, O_{A2} \in \mathcal{O}_A$, $\exists O_1, O_2 \in \mathcal{O}$ such that $O_{A1} = O_1 \cap A$ and $O_{A2} = O_2 \cap A$. Then, $O_1 \cap O_2 \in \mathcal{O}$ since $\mathcal{O}$ is a topology. Then, $O_{A1} \cap O_{A2} = (O_1 \cap O_2) \cap A \in \mathcal{O}_A$. $\forall (O_{A\alpha})_{\alpha \in \Lambda} \subseteq \mathcal{O}_A$, where Λ is an index set, we have, $\forall \alpha \in \Lambda$, $\exists O_\alpha \in \mathcal{O}$ such that $O_{A\alpha} = O_\alpha \cap A$. Then, $\bigcup_{\alpha \in \Lambda} O_\alpha \in \mathcal{O}$ since $\mathcal{O}$ is a topology. Therefore, $\bigcup_{\alpha \in \Lambda} O_{A\alpha} = \left(\bigcup_{\alpha \in \Lambda} O_\alpha\right) \cap A \in \mathcal{O}_A$.

Hence, $\mathcal{O}_A$ is a topology on A. □

Let $(X, \mathcal{O})$ be a topological space and $A \subseteq X$. The property of a set $E \subseteq A$ being open or closed is relative with respect to $(X, \mathcal{O})$, that is, this property may change if we consider the subset topology $(A, \mathcal{O}_A)$.

Proposition 3.5 *Let* $\mathcal{X}$ *be a topological space,* $A \subseteq \mathcal{X}$ *be endowed with the subset topology* $\mathcal{O}_A$, *and* $E \subseteq A$. *Then,*

(1) *E is closed in* $\mathcal{O}_A$ *if, and only if,* $E = A \cap F$, *where* $F \subseteq \mathcal{X}$ *is closed in* $\mathcal{O}_{\mathcal{X}}$.
(2) *The* closure of E relative to $(A, \mathcal{O}_A)$ (the closure of E in $\mathcal{O}_A$) *is equal to* $\overline{E} \cap A$, *where* $\overline{E}$ *is the closure of E relative to* $\mathcal{X}$.

Proof Here, the set complementation and the set closure operation are relative to $\mathcal{X}$.

(1) "If" $A \setminus E = A \setminus (A \cap F) = A \cap \widetilde{A \cap F} = A \cap \widetilde{F}$. Since F is closed in $\mathcal{O}_{\mathcal{X}}$, then $\widetilde{F} \in \mathcal{O}_{\mathcal{X}}$. Then, $A \setminus E \in \mathcal{O}_A$. Hence, E is closed in $\mathcal{O}_A$.

"Only if" $A \setminus E \in \mathcal{O}_A$. Then, $\exists O \in \mathcal{O}_\mathcal{X}$ such that $A \setminus E = A \cap O$. Then, $E = A \setminus (A \setminus E) = A \cap \widetilde{A \cap O} = A \cap \widetilde{O}$. Hence, the result holds.

(2) By (1), $\overline{E} \cap A$ is closed in $\mathcal{O}_A$. Then, the closure of E relative to $(A, \mathcal{O}_A)$ is contained in $\overline{E} \cap A$. On the other hand, by Proposition 3.3, if $x \in \mathcal{X}$ is a point of closure of E relative to $\mathcal{X}$, then it is a point of closure of E in $\mathcal{O}_A$ if $x \in A$. Then, $\overline{E} \cap A$ is contained in the closure of E relative to $(A, \mathcal{O}_A)$. Hence, the result holds.

This completes the proof of the proposition. □

Definition 3.6 For two topologies over the same set X, $\mathcal{O}_1$ and $\mathcal{O}_2$, we will say that $\mathcal{O}_1$ is *stronger (finer)* than $\mathcal{O}_2$ if $\mathcal{O}_1 \supset \mathcal{O}_2$, in which case, $\mathcal{O}_2$ is said to be *weaker (coarser)* than $\mathcal{O}_1$. ◇

Proposition 3.7 *Let X be a set and $\mathcal{A} \subseteq {}^X2$. Then, there exists the weakest topology $\mathcal{O}$ on X such that $\mathcal{A} \subseteq \mathcal{O}$. This topology is called the* topology generated by $\mathcal{A}$.

Proof Let $\mathcal{M} := \{\mathcal{X} \subseteq {}^X2 \mid \mathcal{A} \subseteq \mathcal{X}$ and $\mathcal{X}$ is a topology on $X\}$ and $\mathcal{O} = \bigcap_{\mathcal{X} \in \mathcal{M}} \mathcal{X}$. Clearly, ${}^X2 \in \mathcal{M}$ and hence $\mathcal{O}$ is well-defined. Then,

(i) $\emptyset, X \in \mathcal{X}, \forall \mathcal{X} \in \mathcal{M}$. Hence, $\emptyset, X \in \mathcal{O}$.
(ii) $\forall A_1, A_2 \in \mathcal{O}$, we have $A_1, A_2 \in \mathcal{X}, \forall \mathcal{X} \in \mathcal{M}$. Then, $A_1 \cap A_2 \in \mathcal{X}, \forall \mathcal{X} \in \mathcal{M}$. Hence, $A_1 \cap A_2 \in \mathcal{O}$.
(iii) $\forall (A_\alpha)_{\alpha \in \Lambda} \subseteq \mathcal{O}$, where Λ is an index set, we have, $\forall \alpha \in \Lambda, \forall \mathcal{X} \in \mathcal{M}, A_\alpha \in \mathcal{X}$. Then, $\bigcup_{\alpha \in \Lambda} A_\alpha \in \mathcal{X}, \forall \mathcal{X} \in \mathcal{M}$. Hence, we have $\bigcup_{\alpha \in \Lambda} A_\alpha \in \mathcal{O}$.

Therefore, $\mathcal{O}$ is a topology on X. Clearly, $\mathcal{A} \subseteq \mathcal{O}$ since $\mathcal{A} \subseteq \mathcal{X}, \forall \mathcal{X} \in \mathcal{M}$. Therefore, $\mathcal{O}$ is the weakest topology containing $\mathcal{A}$. □

3.2 Continuity

Definition 3.8 Let $(X, \mathcal{O}_X)$ and $(Y, \mathcal{O}_Y)$ be topological spaces, $D \subseteq X$ with the subset topology $\mathcal{O}_D$, and $f : D \to Y$ (or $f : (D, \mathcal{O}_D) \to (Y, \mathcal{O}_Y)$ to be more specific). Then, f is said to be *continuous* if, $\forall O_Y \in \mathcal{O}_Y$, we have $f_{\mathrm{inv}}(O_Y) \in \mathcal{O}_D$. f is said to be *continuous at* $x_0 \in D$ if, $\forall O_Y \in \mathcal{O}_Y$ with $f(x_0) \in O_Y$, $\exists U \in \mathcal{O}_X$ with $x_0 \in U$ such that $f(U) \subseteq O_Y$. f is said to be *continuous on* $E \subseteq D$ if it is continuous at x, $\forall x \in E$. ◇

Proposition 3.9 *Let $\mathcal{X}$ and $\mathcal{Y}$ be topological spaces, $D \subseteq \mathcal{X}$ with the subset topology $\mathcal{O}_D$, and $f : D \to \mathcal{Y}$. f is continuous if, and only if, $\forall x_0 \in D$, f is continuous at x_0.*

Proof "If" $\forall O_Y \in \mathcal{O}_\mathcal{Y}$, $\forall x \in f_{\mathrm{inv}}(O_Y) \subseteq D$. Since f is continuous at x, then $\exists U_x \in \mathcal{O}_\mathcal{X}$ with $x \in U_x$ such that $f(U_x) \subseteq O_Y$, which implies, by Proposition 2.5, that $U_x \cap D \subseteq f_{\mathrm{inv}}(O_Y)$. Then, $f_{\mathrm{inv}}(O_Y) = \bigcup_{x \in f_{\mathrm{inv}}(O_Y)} (U_x \cap D) = (\bigcup_{x \in f_{\mathrm{inv}}(O_Y)} U_x) \cap D \in \mathcal{O}_D$. Hence, f is continuous.

"Only if" $\forall x_0 \in D$, $\forall O_Y \in \mathcal{O}_\mathcal{Y}$ with $f(x_0) \in O_Y$, let $U = f_{\text{inv}}(O_Y) \in \mathcal{O}_D$. By Proposition 3.4, $\exists \bar{U} \in \mathcal{O}_\mathcal{X}$ such that $U = \bar{U} \cap D$. Then, $x_0 \in U$. By Proposition 2.5, $f(\bar{U}) = f(U) \subseteq O_Y$. Hence, f is continuous at x_0.

This completes the proof of the proposition. □

Proposition 3.10 *Let $\mathcal{X}$ and $\mathcal{Y}$ be topological spaces and $f : \mathcal{X} \to \mathcal{Y}$. f is continuous if, and only if, $\forall B \subseteq \mathcal{Y}$ with $\widetilde{B} \in \mathcal{O}_\mathcal{Y}$, we have $\widetilde{f_{\text{inv}}(B)} \in \mathcal{O}_\mathcal{X}$, that is, the inverse image of any closed set in $\mathcal{Y}$ is closed in $\mathcal{X}$.*

Proof "If" $\forall O \in \mathcal{O}_\mathcal{Y}$, we have, by Proposition 2.5, $f_{\text{inv}}(O) = \widetilde{f_{\text{inv}}(\widetilde{O})} \in \mathcal{O}_\mathcal{X}$. Hence, f is continuous.

"Only if" $\forall B \subseteq \mathcal{Y}$ with $\widetilde{B} \in \mathcal{O}_\mathcal{Y}$. Since f is continuous, then, by Proposition 2.5, $\widetilde{f_{\text{inv}}(B)} = f_{\text{inv}}(\widetilde{B}) \in \mathcal{O}_\mathcal{X}$. Hence, the result holds.

This completes the proof of the proposition. □

Theorem 3.11 *Let $\mathcal{X}$ and $\mathcal{Y}$ be topological spaces, $f : \mathcal{X} \to \mathcal{Y}$, and $\mathcal{X} = X_1 \cup X_2$, where X_1 and X_2 are both open or both closed. Let X_1 and X_2 be endowed with subset topologies $\mathcal{O}_{X_1}$ and $\mathcal{O}_{X_2}$, respectively. Assume that $f|_{X_1} : X_1 \to \mathcal{Y}$ and $f|_{X_2} : X_2 \to \mathcal{Y}$ are continuous. Then, f is continuous.*

Proof Consider the case that X_1 and X_2 are both open. $\forall x_0 \in \mathcal{X}$. Without loss of generality, assume $x_0 \in X_1$. $\forall O \in \mathcal{O}_\mathcal{Y}$ with $f(x_0) \in O$. Since $f|_{X_1}$ is continuous, then, by Proposition 3.9, $\exists U \in \mathcal{O}_{X_1}$ with $x_0 \in U$ such that $f|_{X_1}(U) \subseteq O$. Since $X_1 \in \mathcal{O}_\mathcal{X}$, then $U \in \mathcal{O}_\mathcal{X}$. Note that $f(U) = f|_{X_1}(U) \subseteq O$, since $U \subseteq X_1$. Hence, f is continuous at x_0. By the arbitrariness of x_0 and Proposition 3.9, f is continuous.

Consider the case that X_1 and X_2 are both closed. For all closed subset $B \subseteq \mathcal{Y}$, we have $f_{\text{inv}}(B) \subseteq X$. Then, $f_{\text{inv}}(B) \cap X_1 = (f|_{X_1})_{\text{inv}}(B)$ is closed in $\mathcal{O}_{X_1}$, by Proposition 3.10 and the continuity of $f|_{X_1}$. Similarly, $f_{\text{inv}}(B) \cap X_2 = (f|_{X_2})_{\text{inv}}(B)$ is closed in $\mathcal{O}_{X_2}$. Since X_1 and X_2 are closed sets in $\mathcal{O}_\mathcal{X}$, then, $f_{\text{inv}}(B) \cap X_1$ and $f_{\text{inv}}(B) \cap X_2$ are closed in $\mathcal{O}_\mathcal{X}$, by Proposition 3.5. Then, $f_{\text{inv}}(B) = (f_{\text{inv}}(B) \cap X_1) \cup (f_{\text{inv}}(B) \cap X_2)$ is closed in $\mathcal{O}_\mathcal{X}$. By Proposition 3.10, f is continuous.

This completes the proof of the theorem. □

Proposition 3.12 *Let $\mathcal{X}$, $\mathcal{Y}$, and $\mathcal{Z}$ be topological spaces, $f : \mathcal{X} \to \mathcal{Y}$, $g : \mathcal{Y} \to \mathcal{Z}$, and $x_0 \subset \mathcal{X}$. Assume that f is continuous at x_0 and g is continuous at $y_0 := f(x_0)$. Then, $g \circ f : \mathcal{X} \to \mathcal{Z}$ is continuous at x_0.*

Proof $\forall O_Z \in \mathcal{O}_\mathcal{Z}$ with $g(f(x_0)) \in O_Z$. Since g is continuous at $f(x_0)$, then $\exists O_Y \in \mathcal{O}_\mathcal{Y}$ with $f(x_0) \in O_Y$ such that $g(O_Y) \subseteq O_Z$. Since f is continuous at x_0, then $\exists O_X \in \mathcal{O}_\mathcal{X}$ with $x_0 \in O_X$ such that $f(O_X) \subseteq O_Y$. Then, $g(f(O_X)) \subseteq O_Z$. Hence, $g \circ f$ is continuous at x_0. This completes the proof of the proposition. □

Definition 3.13 Let $\mathcal{X}$ and $\mathcal{Y}$ be topological spaces and $f : \mathcal{X} \to \mathcal{Y}$. f is said to be a *homeomorphism* between $\mathcal{X}$ and $\mathcal{Y}$ if it is bijective and continuous and $f_{\text{inv}} : \mathcal{Y} \to \mathcal{X}$ is also continuous. The spaces $\mathcal{X}$ and $\mathcal{Y}$ are said to be *homeomorphic* if there exists a homeomorphism between them. ◇

Any properties invariant under homeomorphisms are called *topological properties*.

Homeomorphisms preserve topological properties in topological spaces. Isomorphisms preserve algebraic properties in algebraic systems. Isometries preserve metric properties in metric spaces.

Definition 3.14 Let $\mathcal{X}$ be a topological space, $D \subseteq \mathcal{X}$ with the subset topology $\mathcal{O}_D$, and $f : D \to \mathbb{R}$. f is said to be *upper semicontinuous* if $\forall a \in \mathbb{R}$, $f_{\text{inv}}((-\infty, a)) \in \mathcal{O}_D$. f is said to be *upper semicontinuous at* $x_0 \in \mathcal{X}$ if $\forall \epsilon \in (0, \infty) \subset \mathbb{R}, \exists U \in \mathcal{O}_{\mathcal{X}}$ with $x_0 \in U$ such that $f(x) < f(x_0)+\epsilon, \forall x \in U \cap D$. f is said to be *lower semicontinuous* if $-f$ is upper semicontinuous. ◇

Proposition 3.15 *Let $\mathcal{X}$ be a topological space, $D \subseteq \mathcal{X}$ with the subset topology $\mathcal{O}_D$, and $f : D \to \mathbb{R}$. f is upper semicontinuous if, and only if, f is upper semicontinuous at x_0, $\forall x_0 \in D$.*

Proof This is straightforward and is therefore omitted. □

Proposition 3.16 *Let $\mathcal{X}$ and $\mathcal{Y}$ be topological spaces, $f : \mathcal{X} \to \mathbb{R}$ and $g : \mathcal{X} \to \mathbb{R}$ be upper semicontinuous at $x_0 \in \mathcal{X}$, $h : \mathcal{Y} \to \mathcal{X}$ be continuous at $y_0 \in \mathcal{Y}$, and $h(y_0) = x_0$. Then, $f + g$ is upper semicontinuous at x_0 and $f \circ h$ is upper semicontinuous at y_0. Furthermore, if f is also lower semicontinuous at x_0, then f is continuous at x_0.*

Proof $\forall \epsilon \in (0, \infty) \subset \mathbb{R}, \exists U_f \in \mathcal{O}_{\mathcal{X}}$ with $x_0 \in U_f$ such that $f(x) < f(x_0) + \epsilon$, $\forall x \in U_f$, by the upper semicontinuity of f. By the upper semicontinuity of g, $\exists U_g \in \mathcal{O}_{\mathcal{X}}$ with $x_0 \in U_g$ such that $g(x) < g(x_0) + \epsilon, \forall x \in U_g$. Then, $x_0 \in U := U_f \cap U_g \in \mathcal{O}_{\mathcal{X}}$ and $(f + g)(x) = f(x) + g(x) < f(x_0) + \epsilon + g(x_0) + \epsilon = (f + g)(x_0) + 2\epsilon, \forall x \in U$. Hence, $f + g$ is upper semicontinuous at x_0.

$\forall \epsilon \in (0, \infty) \subset \mathbb{R}, \exists U_f \in \mathcal{O}_{\mathcal{X}}$ with $x_0 \in U_f$ such that $f(x) < f(x_0) + \epsilon$, $\forall x \in U_f$, by the upper semicontinuity of f. By the continuity of h, $\exists U_h \in \mathcal{O}_{\mathcal{Y}}$ with $y_0 \in U_h$ such that $h(y) \in U_f, \forall y \in U_h$. Then, $(f \circ h)(y) < f(h(y_0))+\epsilon, \forall y \in U_h$. Hence, $f \circ h$ is upper semicontinuous at y_0.

$\forall \epsilon \in (0, \infty) \subset \mathbb{R}$, by the upper semicontinuity of f, $\exists U_1 \in \mathcal{O}_{\mathcal{X}}$ with $x_0 \in U_1$ such that $f(x) < f(x_0)+\epsilon, \forall x \in U_1$. By the lower semicontinuity of f, $\exists U_2 \in \mathcal{O}_{\mathcal{X}}$ with $x_0 \in U_2$ such that $f(x) > f(x_0)-\epsilon, \forall x \in U_2$. Then, $x_0 \in U := U_1 \cap U_2 \in \mathcal{O}_{\mathcal{X}}$ and $|f(x) - f(x_0)| < \epsilon, \forall x \in U$. Hence, f is continuous at x_0.

This completes the proof of the proposition. □

3.3 Basis and Countability

Definition 3.17 Let $(X, \mathcal{O})$ be a topological space and $\mathcal{B} \subseteq \mathcal{O}$. $\mathcal{B}$ is said to be a *basis* of the topological space if, $\forall O \in \mathcal{O}, \forall x \in O, \exists B \in \mathcal{B}$ such that $x \in B \subseteq O$. $\mathcal{B}_x \subseteq \mathcal{O}$ with $x \in B, \forall B \in \mathcal{B}_x$, is said to be a *basis at* $x \in X$ if $\forall O \in \mathcal{O}$ with $x \in O, \exists B \in \mathcal{B}_x$ such that $x \in B \subseteq O$. ◇

If $\mathcal{B}$ is a basis for the topology $\mathcal{O}$, then the topology generated by $\mathcal{B}$ is $\mathcal{O}$ and, $\forall O \in \mathcal{O}$, $O = \bigcup_{x \in O} B_x$, where $B_x \in \mathcal{B}$ is such that $x \in B_x \subseteq O$.

Proposition 3.18 *Let X be a set and $\mathcal{B} \subseteq {}^X2$. Then, $\mathcal{B}$ is a basis for the topology generated by $\mathcal{B}$, $\mathcal{O}$, if, and only if, the following two conditions hold:*

(i) *$\forall x \in X$, $\exists B \in \mathcal{B}$ such that $x \in B$; (that is, $\bigcup_{B \in \mathcal{B}} B = X$).*
(ii) *$\forall B_1, B_2 \in \mathcal{B}$, $\forall x \in B_1 \cap B_2$, $\exists B_3 \in \mathcal{B}$ such that $x \in B_3 \subseteq B_1 \cap B_2$.*

Proof "Only if" Let $\mathcal{B}$ be a basis for $\mathcal{O}$. Since $X \in \mathcal{O}$, then $\forall x \in X$, $\exists B \in \mathcal{B}$ such that $x \in B \subseteq X$. So (i) is true. $\forall B_1, B_2 \in \mathcal{B} \subseteq \mathcal{O}$, we have $B_1 \cap B_2 \in \mathcal{O}$. $\forall x \in B_1 \cap B_2$, $\exists B_3 \in \mathcal{B}$ such that $x \in B_3 \subseteq B_1 \cap B_2$. Hence, (ii) is also true.

"If" Define $\bar{\mathcal{O}} := \{O \subseteq X \mid \forall x \in O,\ \exists B \in \mathcal{B} \text{ such that } x \in B \subseteq O\}$. Clearly, $\mathcal{B} \subseteq \bar{\mathcal{O}}$ and $\emptyset \in \bar{\mathcal{O}}$. $X \in \bar{\mathcal{O}}$ since (i). $\forall O_1, O_2 \in \bar{\mathcal{O}}$, $\forall x \in O_1 \cap O_2$, we have $x \in O_1$ and $x \in O_2$. Then, by the definition of $\bar{\mathcal{O}}$, $\exists B_1, B_2 \in \mathcal{B}$ such that $x \in B_1 \subseteq O_1$ and $x \in B_2 \subseteq O_2$. Then, $x \in B_1 \cap B_2$. By (ii), $\exists B_3 \in \mathcal{B}$ such that $x \in B_3 \subseteq B_1 \cap B_2 \subseteq O_1 \cap O_2$. Then, $O_1 \cap O_2 \in \bar{\mathcal{O}}$. $\forall (O_\alpha)_{\alpha \in \Lambda} \subseteq \bar{\mathcal{O}}$, where Λ is an index set, let $O = \bigcup_{\alpha \in \Lambda} O_\alpha$. $\forall x \in O$, $\exists \alpha \in \Lambda$ such that $x \in O_\alpha$. By the definition of $\bar{\mathcal{O}}$, $\exists B \in \mathcal{B}$ such that $x \in B \subseteq O_\alpha \subseteq O$. Hence, $O \in \bar{\mathcal{O}}$. Therefore, $\bar{\mathcal{O}}$ is a topology on X. By the definition of $\bar{\mathcal{O}}$, $\mathcal{B}$ is a basis for $\bar{\mathcal{O}}$. Note that $\bar{\mathcal{O}} \supseteq \mathcal{O}$ since $\mathcal{O}$ is the weakest topology containing $\mathcal{B}$. Then, $\mathcal{B}$ is a basis for $\mathcal{O}$.

This completes the proof of the proposition. □

Example 3.19 For the real line $\mathbb{R}$, let $\mathcal{A} := \{\text{interval } (a, b) \mid a, b \in \mathbb{R},\ a < b\}$. Then, the topology generated by $\mathcal{A}$, $\mathcal{O}_{\mathbb{R}}$, is the usual topology on $\mathbb{R}$ as we know before. By Proposition 3.18, $\mathcal{A}$ is a basis for this topology. ◇

Definition 3.20 A topological space $(X, \mathcal{O})$ is said to satisfy the *first axiom of countability* if there exists a countable basis at each $x \in X$, i.e., $\forall x \in X$, $\exists \mathcal{B}_x \subseteq \mathcal{O}$, which is countable, such that, $\forall O \in \mathcal{O}$ with $x \in O$, we have $\exists B \in \mathcal{B}_x$ with $x \in B \subseteq O$, and $\forall B \in \mathcal{B}_x$, we have $x \in B$. In this case, we will say that the topological space is *first countable*. The topological space is said to satisfy the *second axiom of countability* if there exists a countable basis $\mathcal{B}$ for $\mathcal{O}$, in which case, we will say that it is *second countable*. ◇

Clearly, a second countable topological space is also first countable.

Example 3.21 The real line is first countable, where a countable basis at any $x \in \mathbb{R}$ consists of intervals of the form $(x - r, x + r)$ with $r \in \mathbb{Q}$ and $r > 0$. The real line is also second countable, where a countable basis for the topology consists of intervals of the form (r_1, r_2) with $r_1, r_2 \in \mathbb{Q}$ and $r_1 < r_2$. ◇

When basis are available on topological spaces $\mathcal{X}$ and $\mathcal{Y}$, in Definitions 3.8 and 3.14, we may restrict the open sets O_Y and U to be basis open sets without changing the meaning of the definition. In Proposition 3.3, we may restrict O to be a basis open set and the results still hold.

Definition 3.22 A collection $(A_\alpha)_{\alpha\in\Lambda}$ of sets, where Λ is an index set, is said to be a *covering* of a set X if $X \subseteq \bigcup_{\alpha\in\Lambda} A_\alpha$. It is an *open covering* if A_α's are open sets in a specific topological space. ◇

Definition 3.23 A topological space $\mathcal{X}$ is said to be *Lindelöf* if any open covering of $\mathcal{X}$ has a countable subcovering. ◇

Proposition 3.24 *A second countable topological space $\mathcal{X}$ is Lindelöf.*

Proof Let $(B_i)_{i\in N}$ be a countable basis for $\mathcal{X}$, where N is a countable index set. Let $(O_\alpha)_{\alpha\in\Lambda}$ be an open covering of $\mathcal{X}$, where Λ is an index set. $\forall \alpha \in \Lambda$, $O_\alpha = \bigcup_{i\in N_\alpha} B_i$, where $N_\alpha \subseteq N$. Then, $\mathcal{X} = \bigcup_{\alpha\in\Lambda}\bigcup_{i\in N_\alpha} B_i$. We may determine a subcollection $(B_i)_{i\in\bar{N}}$, where $\bar{N} \subseteq N$ such that these B_i's appear at least once in the previous union. Then, $\mathcal{X} = \bigcup_{i\in\bar{N}} B_i$. $\forall i \in \bar{N}$, the collection $\mathcal{A}_i := \{O_\alpha \mid \alpha \in \Lambda, B_i \subseteq O_\alpha\}$ is nonempty. Then, by Axiom of Choice, there exists an assignment $\left(O_{\alpha_i}\right)_{i\in\bar{N}}$ such that $O_{\alpha_i} \in \mathcal{A}_i$, $\forall i \in \bar{N}$. Then, we have $\left(O_{\alpha_i}\right)_{i\in\bar{N}}$ is a countable subcover. This completes the proof of the proposition. □

3.4 Products of Topological Spaces

Proposition 3.25 *Let Λ be an index set and $(X_\alpha, \mathcal{O}_\alpha)$ be a topological space, $\forall \alpha \in \Lambda$. The* product topology *$\mathcal{O}$ on $\prod_{\alpha\in\Lambda} X_\alpha$ is the topology generated by*

$$\mathcal{B} := \left\{ \prod_{\alpha\in\Lambda} O_\alpha \;\middle|\; O_\alpha \in \mathcal{O}_\alpha,\ \forall \alpha \in \Lambda,\ \text{and } O_\alpha = X_\alpha \text{ for all but a finite number of } \alpha\text{'s} \right\}$$

The collection $\mathcal{B}$ forms a basis for the topology $\mathcal{O}$. We will also write $\left(\prod_{\alpha\in\Lambda} X_\alpha, \mathcal{O}\right) = \prod_{\alpha\in\Lambda}(X_\alpha, \mathcal{O}_\alpha)$. When $(X_\alpha, \mathcal{O}_\alpha) = (X, \mathcal{O}) =: \mathcal{X}$, $\forall \alpha \in \Lambda$, we will denote $\prod_{\alpha\in\Lambda}(X_\alpha, \mathcal{O}_\alpha)$ by $\mathcal{X}^\Lambda$.

Proof We will prove this by Proposition 3.18. Since $\prod_{\alpha\in\Lambda} X_\alpha \in \mathcal{B}$, then (i) is true. $\forall B_1, B_2 \in \mathcal{B}$, $B_1 = \prod_{\alpha\in\Lambda} O_{1\alpha}$, $O_{1\alpha} \in \mathcal{O}_\alpha$, $\forall \alpha \in \Lambda$, and $O_{1\alpha} = X_\alpha$, $\forall \alpha \in \Lambda \setminus \Lambda_1$, where $\Lambda_1 \subseteq \Lambda$ is a finite set; $B_2 = \prod_{\alpha\in\Lambda} O_{2\alpha}$, $O_{2\alpha} \in \mathcal{O}_\alpha$, $\forall \alpha \in \Lambda$, and $O_{2\alpha} = X_\alpha$, $\forall \alpha \in \Lambda \setminus \Lambda_2$, where $\Lambda_2 \subseteq \Lambda$ is a finite set. Let $B_3 = B_1 \cap B_2 = \prod_{\alpha\in\Lambda}(O_{1\alpha} \cap O_{2\alpha})$. Clearly, $O_{1\alpha} \cap O_{2\alpha} \in \mathcal{O}_\alpha$, $\forall \alpha \in \Lambda$, and $O_{1\alpha} \cap O_{2\alpha} = X_\alpha$, $\forall \alpha \in \Lambda \setminus (\Lambda_1 \cup \Lambda_2)$, where $\Lambda_1 \cup \Lambda_2 \subseteq \Lambda$ is a finite set. Hence, $B_3 \in \mathcal{B}$. Then, (ii) also holds. By Proposition 3.18, $\mathcal{B}$ is a basis for $\mathcal{O}$. This completes the proof of the proposition.

□

Definition 3.26 Let $\mathcal{X}_\alpha$ be a topological space, $\forall \alpha \in \Lambda$, where Λ is an index set, and $\mathcal{Y} = \prod_{\alpha\in\Lambda} \mathcal{X}_\alpha$ be the product topological space. Define a collection of *projection* functions $\pi_\alpha : \mathcal{Y} \to \mathcal{X}_\alpha$, $\forall \alpha \in \Lambda$, by $\pi_\alpha(y) = y_\alpha$ for $y = (y_\alpha)_{\alpha\in\Lambda} \in \mathcal{Y}$. ◇

Proposition 3.27 *Let $(X_\alpha, \mathcal{O}_\alpha)$ be a topological space, $\forall \alpha \in \Lambda$, where Λ is an index set, and $(Y, \mathcal{O}) = \prod_{\alpha \in \Lambda}(X_\alpha, \mathcal{O}_\alpha)$ be the product topological space. Then, $\mathcal{O}$ is the weakest topology on which the projection functions π_α, $\forall \alpha \in \Lambda$, are continuous.*

Proof We will distinguish two exhaustive and mutually exclusive cases: Case 1: $\Lambda = \emptyset$, and Case 2: $\Lambda \neq \emptyset$. Case 1: $\Lambda = \emptyset$. $(Y, \mathcal{O}) = (\{\emptyset\}, \{\emptyset, \{\emptyset\}\})$. Then, $\mathcal{O}$ is the only topology on Y. Hence, the result is true.

Case 2: $\Lambda \neq \emptyset$. Define

$$\mathcal{A} := \Big\{ O \subseteq Y \;\Big|\; O = \prod_{\alpha \in \Lambda} O_\alpha,\ \exists \alpha_0 \in \Lambda \text{ such that } O_{\alpha_0} \in \mathcal{O}_{\alpha_0} \text{ and}$$
$$O_\alpha = X_\alpha,\ \forall \alpha \in \Lambda \setminus \{\alpha_0\} \Big\}$$

Clearly, $\mathcal{A} \neq \emptyset$ is the collection of inverse images of open sets under π_α, $\forall \alpha \in \Lambda$. It is also clear that $\mathcal{A} \subseteq \mathcal{B} \subseteq \mathcal{O}$, where $\mathcal{B}$ is the basis for $\mathcal{O}$ that we introduced in Proposition 3.25. $\forall B \in \mathcal{B}$, we have $B = \bigcap_{i=1}^{k} A_i$ for some $k \in \mathbb{N}$ and $A_1, \ldots, A_k \in \mathcal{A}$. Hence, the topology generated by $\mathcal{A}$ contains $\mathcal{B}$. Then, the topology generated by $\mathcal{A}$ equals $\mathcal{O}$. Hence, $\mathcal{O}$ is the weakest topology containing $\mathcal{A}$. This completes the proof of the proposition. □

Proposition 3.28 *Let $\mathcal{X}_i := (X_i, \mathcal{O}_i)$ be a second countable topological space, $\forall i \in N \subseteq \mathbb{N}$, and N is a countable index set. Then, the product topological space $\mathcal{X} := (X, \mathcal{O}) := \prod_{i \in N} \mathcal{X}_i$ is second countable.*

Proof $\forall i \in N$, let $\mathcal{B}_i \subseteq \mathcal{O}_i$ be a countable basis for $\mathcal{X}_i$, and without loss of generality, assume $X_i \in \mathcal{B}_i$. Let $\mathcal{B}$ be the basis for the product topological space $\mathcal{X}$ as defined in Proposition 3.25. Define

$$\bar{\mathcal{B}} := \Big\{ \prod_{i \in N} B_i \subseteq X \;\Big|\; B_i \in \mathcal{B}_i, \forall i \in N, \text{ and } B_i = X_i \text{ for all but a finite}$$
$$\text{number of } i\text{'s} \Big\} \subseteq \mathcal{B} \subseteq \mathcal{O}$$

Since $\mathcal{B}_i$'s are countable, then $\bar{\mathcal{B}}$ is countable. We will show that $\bar{\mathcal{B}}$ is a basis for $\mathcal{O}$. $\forall O \in \mathcal{O}$, $\forall x \in O$, $\exists O_B \in \mathcal{B}$ such that $x \in O_B \subseteq O$, where $O_B = \prod_{i \in N} O_{Bi}$, $O_{Bi} \in \mathcal{O}_i$, $\forall i \in N$, $O_{Bi} = X_i$, $\forall i \in N \setminus \bar{N}$, and $\bar{N} \subseteq N$ is a finite set. $\forall i \in \bar{N}$, $\pi_i(x) \in O_{Bi}$. Then, $\exists B_i \in \mathcal{B}_i$ such that $\pi_i(x) \in B_i \subseteq O_{Bi}$. Let $B_i := X_i$, $\forall i \in N \setminus \bar{N}$. Let $B := \prod_{i \in N} B_i \in \bar{\mathcal{B}}$. Clearly, $x \in B \subseteq O_B \subseteq O$. Hence, $\bar{\mathcal{B}}$ is a basis for $\mathcal{X}$. Then, $\mathcal{X}$ is second countable. This completes the proof of the proposition. □

Proposition 3.29 *Let $\mathcal{X}_\alpha := (X_\alpha, \mathcal{O}_\alpha)$ be a topological space, $\forall \alpha \in \Lambda$, where Λ is an index set. Let $\mathcal{X} = (X, \mathcal{O}) := \prod_{\alpha\in\Lambda} \mathcal{X}_\alpha$ be the product space. Assume that $A_\alpha \subseteq \mathcal{X}_\alpha$, $\forall \alpha \in \Lambda$. Then, $\overline{\prod_{\alpha\in\Lambda} A_\alpha} = \prod_{\alpha\in\Lambda} \overline{A_\alpha}$.*

Proof Note that

$$\Big(\prod_{\alpha\in\Lambda} \overline{A_\alpha}\Big)^{\sim} = \bigcup_{\alpha\in\Lambda} \pi_{\alpha\,\mathrm{inv}}(\widetilde{\overline{A_\alpha}})$$

where $\pi_\alpha : \mathcal{X} \to \mathcal{X}_\alpha$ is the projection function, $\forall \alpha \in \Lambda$. $\forall \alpha \in \Lambda$, by Definition 3.2, $\widetilde{\overline{A_\alpha}} \in \mathcal{O}_\alpha$. By Proposition 3.27, π_α is continuous. Then, $\pi_{\alpha\,\mathrm{inv}}(\widetilde{\overline{A_\alpha}}) \in \mathcal{O}$. Then, $\prod_{\alpha\in\Lambda} \overline{A_\alpha}$ is a closed set in $\mathcal{X}$. Clearly, $\prod_{\alpha\in\Lambda} A_\alpha \subseteq \prod_{\alpha\in\Lambda} \overline{A_\alpha}$. Then, we have $\overline{\prod_{\alpha\in\Lambda} A_\alpha} \subseteq \prod_{\alpha\in\Lambda} \overline{A_\alpha}$.

On the other hand, $\forall x \in \prod_{\alpha\in\Lambda} \overline{A_\alpha}$, for all basis open set $O \in \mathcal{O}$ with $x \in O$, by Proposition 3.25, we have $O = \prod_{\alpha\in\Lambda} O_\alpha$, where $O_\alpha \in \mathcal{O}_\alpha$, $\forall \alpha \in \Lambda$, and $O_\alpha = X_\alpha$ for all but finitely many α's. $\forall \alpha \in \Lambda$, $\pi_\alpha(x) \in O_\alpha$ and $\pi_\alpha(x) \in \overline{A_\alpha}$. By Proposition 3.3, $O_\alpha \cap A_\alpha \neq \emptyset$. Then, by Axiom of Choice, we have $O \cap (\prod_{\alpha\in\Lambda} A_\alpha) = \prod_{\alpha\in\Lambda}(O_\alpha \cap A_\alpha) \neq \emptyset$. Thus, by Proposition 3.3, $x \in \overline{\prod_{\alpha\in\Lambda} A_\alpha}$. Hence, we have $\prod_{\alpha\in\Lambda} \overline{A_\alpha} \subseteq \overline{\prod_{\alpha\in\Lambda} A_\alpha}$.

Therefore, $\overline{\prod_{\alpha\in\Lambda} A_\alpha} = \prod_{\alpha\in\Lambda} \overline{A_\alpha}$. This completes the proof of the proposition. □

An immediate consequence of the above proposition is that if $A_\alpha \subseteq \mathcal{X}_\alpha$ is closed, $\forall \alpha \in \Lambda$, where $\mathcal{X}_\alpha$'s are topological spaces, then $\prod_{\alpha\in\Lambda} A_\alpha$ is closed in the product topology.

Proposition 3.30 *Let $\left(\Lambda_\beta\right)_{\beta\in\Gamma}$ be a collection of pairwise disjoint sets, where Γ is an index set. Let $(X_\alpha, \mathcal{O}_\alpha)$ be a topological space, $\forall \alpha \in \bigcup_{\beta\in\Gamma} \Lambda_\beta$. $\forall \beta \in \Gamma$, $(\prod_{\alpha\in\Lambda_\beta} X_\alpha, \mathcal{O}^{\langle\beta\rangle})$ is the product space with the product topology. $(\prod_{\beta\in\Gamma}\prod_{\alpha\in\Lambda_\beta} X_\alpha, \mathcal{O}^{\langle\Gamma\rangle})$ is the product space of product spaces with the product topology. Let $(\prod_{\alpha\in\bigcup_{\beta\in\Gamma}\Lambda_\beta} X_\alpha, \mathcal{O})$ be the product space with the product topology. Then, $(\prod_{\beta\in\Gamma}\prod_{\alpha\in\Lambda_\beta} X_\alpha, \mathcal{O}^{\langle\Gamma\rangle})$ and $(\prod_{\alpha\in\bigcup_{\beta\in\Gamma}\Lambda_\beta} X_\alpha, \mathcal{O})$ are homeomorphic.*

Proof Define a mapping $E : \prod_{\beta\in\Gamma}\prod_{\alpha\in\Lambda_\beta} X_\alpha \to \prod_{\alpha\in\bigcup_{\beta\in\Gamma}\Lambda_\beta} X_\alpha$ by, $\forall x \in \prod_{\beta\in\Gamma}\prod_{\alpha\in\Lambda_\beta} X_\alpha$, $\forall \alpha \in \bigcup_{\beta\in\Gamma}\Lambda_\beta$, $\exists!\ \beta_\alpha \in \Gamma \ni\cdot\ \alpha \in \Lambda_{\beta_\alpha}$, $\pi_\alpha(E(x)) = \pi_\alpha^{\langle\beta_\alpha\rangle}(\pi_{\beta_\alpha}^{\langle\Gamma\rangle}(x))$.

$\forall y \in \prod_{\alpha\in\bigcup_{\beta\in\Gamma}\Lambda_\beta} X_\alpha$. $\forall \beta \in \Gamma$, define $x_\beta \in \prod_{\alpha\in\Lambda_\beta} X_\alpha$ be such that $\pi_\alpha^{\langle\beta\rangle}(x_\beta) = \pi_\alpha(y)$, $\forall \alpha \in \Lambda_\beta$. Define $x \in \prod_{\beta\in\Gamma}\prod_{\alpha\in\Lambda_\beta} X_\alpha$ by $\pi_\beta^{\langle\Gamma\rangle}(x) = x_\beta$, $\forall \beta \in \Gamma$. Then, $\forall \alpha \in \bigcup_{\beta\in\Gamma}\Lambda_\beta$, $\exists!\ \beta_\alpha \ni\cdot\ \alpha \in \Lambda_{\beta_\alpha}$, and $\pi_\alpha(E(x)) = \pi_\alpha^{\langle\beta_\alpha\rangle}(\pi_{\beta_\alpha}^{\langle\Gamma\rangle}(x)) = \pi_\alpha(y)$ and hence $E(x) = y$. Hence, E is surjective.

$\forall x, z \in \prod_{\beta\in\Gamma}\prod_{\alpha\in\Lambda_\beta} X_\alpha$ with $E(x) = E(z)$. $\forall \alpha \in \bigcup_{\beta\in\Gamma}\Lambda_\beta$, $\exists!\ \beta_\alpha \ni\cdot\ \alpha \in \Lambda_{\beta_\alpha}$. Then, $\pi_\alpha^{\langle\beta_\alpha\rangle}(\pi_{\beta_\alpha}^{\langle\Gamma\rangle}(x)) = \pi_\alpha(E(x)) = \pi_\alpha(E(z)) = \pi_\alpha^{\langle\beta_\alpha\rangle}(\pi_{\beta_\alpha}^{\langle\Gamma\rangle}(z))$. Hence,

$\forall \beta \in \Gamma$, $\forall \alpha \in \Lambda_\beta$, $\pi_\alpha^{\langle\beta\rangle}(\pi_\beta^{\langle\Gamma\rangle}(x)) = \pi_\alpha^{\langle\beta\rangle}(\pi_\beta^{\langle\Gamma\rangle}(z))$, which implies that $\pi_\beta^{\langle\Gamma\rangle}(x) = \pi_\beta^{\langle\Gamma\rangle}(z)$. Hence, $x = z$. This shows that E is injective.

Hence, E is bijective and admits an inverse E_{inv}.

Next, we show that E is continuous by showing that E is continuous at x_0, $\forall x_0 \in \prod_{\beta\in\Gamma}\prod_{\alpha\in\Lambda_\beta} X_\alpha$. $\forall x_0 \in \prod_{\beta\in\Gamma}\prod_{\alpha\in\Lambda_\beta} X_\alpha$. Let $y_0 = E(x_0)$. $\forall B \in \mathcal{O}$ which is a basis open set with $y_0 \in B$. Then, $B = \prod_{\alpha\in\bigcup_{\beta\in\Gamma}\Lambda_\beta} B_\alpha$, $B_\alpha \in \mathcal{O}_\alpha$, $\forall \alpha \in \bigcup_{\beta\in\Gamma}\Lambda_\beta$, and $B_\alpha = X_\alpha$ for all α's except finitely many α's, say $\alpha \in \Lambda_N$. $\forall \alpha \in \bigcup_{\beta\in\Gamma}\Lambda_\beta$, $\exists! \beta_\alpha \in \Gamma \ni \cdot \alpha \in \Lambda_{\beta_\alpha}$. Let $\Gamma_N = \{\beta_\alpha \mid \alpha \in \Lambda_N\}$. Then, Γ_N is a finite set. $\forall \beta \in \Gamma_N$, let $\Lambda_{N\beta} := \Lambda_N \cap \Lambda_\beta$, which is a nonempty finite set. Then, Λ_N equals the disjoint union of $\bigcup_{\beta\in\Gamma_N}\Lambda_{N\beta}$. Define $B^{\langle\beta\rangle} := \prod_{\alpha\in\Lambda_\beta} B_\alpha$, $\forall \beta \in \Gamma$. Then, $\forall \beta \in \Gamma$, $B^{\langle\beta\rangle}$ is a basis open set in $\left(\prod_{\alpha\in\Lambda_\beta} X_\alpha, \mathcal{O}^{\langle\beta\rangle}\right)$. Define $B^{\langle\Gamma\rangle} := \prod_{\beta\in\Gamma} B^{\langle\beta\rangle}$. $\forall \beta \in \Gamma \setminus \Gamma_N$, $B^{\langle\beta\rangle} = \prod_{\alpha\in\Lambda_\beta} X_\alpha$. Clearly, $B^{\langle\Gamma\rangle}$ is a basis open set in $\mathcal{O}^{\langle\Gamma\rangle}$. $\forall \beta \in \Gamma_N$, $\forall \alpha \in \Lambda_{N\beta}$, $\pi_\alpha^{\langle\beta\rangle}(\pi_\beta^{\langle\Gamma\rangle}(x_0)) = \pi_\alpha(y_0) \in \pi_\alpha(B) = B_\alpha$. Then, $x_0 \in B^{\langle\Gamma\rangle}$. $\forall x \in B^{\langle\Gamma\rangle}$, $\forall \beta \in \Gamma_N$, $\forall \alpha \in \Lambda_{N\beta}$, we have $\pi_\alpha(E(x)) = \pi_\alpha^{\langle\beta\rangle}(\pi_\beta^{\langle\Gamma\rangle}(x)) \in \pi_\alpha^{\langle\beta\rangle}(\pi_\beta^{\langle\Gamma\rangle}(B^{\langle\Gamma\rangle})) = \pi_\alpha^{\langle\beta\rangle}(B^{\langle\beta\rangle}) = B_\alpha$. Hence, $E(x) \in B$, which implies that $E(B^{\langle\Gamma\rangle}) \subseteq B$. Hence, E is continuous at x_0. Then, by the arbitrariness of x_0 and Proposition 3.9, E is continuous.

Finally, we will show E_{inv} is continuous by showing that, $\forall y_0 \in \prod_{\alpha\in\bigcup_{\beta\in\Gamma}\Lambda_\beta} X_\alpha$, E_{inv} is continuous at y_0. $\forall y_0 \in \prod_{\alpha\in\bigcup_{\beta\in\Gamma}\Lambda_\beta} X_\alpha$. Let $x_0 = E_{\text{inv}}(y_0) \in \prod_{\beta\in\Gamma}\prod_{\alpha\in\Lambda_\beta} X_\alpha$. For any basis open set, $B^{\langle\Gamma\rangle} \in \mathcal{O}^{\langle\Gamma\rangle}$ with $x_0 \in B^{\langle\Gamma\rangle}$. Then, $B^{\langle\Gamma\rangle} = \prod_{\beta\in\Gamma} \bar{B}^{\langle\beta\rangle}$, $\bar{B}^{\langle\beta\rangle} \in \mathcal{O}^{\langle\beta\rangle}$, $\forall \beta \in \Gamma$, and $\bar{B}^{\langle\beta\rangle} = \prod_{\alpha\in\Lambda_\beta} X_\alpha$ for all $\beta \in \Gamma$ except finitely many β's, say $\beta \in \Gamma_N$. $\forall \beta \in \Gamma_N$, $\pi_\beta^{\langle\Gamma\rangle}(x_0) \in \bar{B}^{\langle\beta\rangle}$. Then, there exists a basis open set $B^{\langle\beta\rangle} \in \mathcal{O}^{\langle\beta\rangle}$ such that $\pi_\beta^{\langle\Gamma\rangle}(x_0) \in B^{\langle\beta\rangle} \subseteq \bar{B}^{\langle\beta\rangle}$. Then, $B^{\langle\beta\rangle} = \prod_{\alpha\in\Lambda_\beta} B_\alpha$, $B_\alpha \in \mathcal{O}_\alpha$, $\forall \alpha \in \Lambda_\beta$, and $B_\alpha = X_\alpha$ for all $\alpha \in \Lambda_\beta$ except finitely many α's, say $\alpha \in \Lambda_{N\beta}$. $\forall \alpha \in \Lambda_{N\beta}$, $\pi_\alpha(y_0) = \pi_\alpha^{\langle\beta\rangle}(\pi_\beta^{\langle\Gamma\rangle}(x_0)) \in B_\alpha$. Let $\Lambda_N = \bigcup_{\beta\in\Gamma_N}\Lambda_{N\beta}$, which is a finite set and the union is pairwise disjoint. Let $B_\alpha = X_\alpha$, $\forall \alpha \in \Lambda_\beta$, $\forall \beta \in \Gamma \setminus \Gamma_N$. Define $B := \prod_{\alpha\in\bigcup_{\beta\in\Gamma}\Lambda_\beta} B_\alpha$. Clearly, B is a basis open set in $\mathcal{O}$ and $y_0 \in B$. $\forall y \in B$. Let $x = E_{\text{inv}}(y)$, then, $y = E(x)$. $\forall \beta \in \Gamma$, $\forall \alpha \in \Lambda_\beta$, $\pi_\alpha(y) = \pi_\alpha(E(x)) = \pi_\alpha^{\langle\beta\rangle}(\pi_\beta^{\langle\Gamma\rangle}(x)) \in B_\alpha$. Then, $\forall \beta \in \Gamma_N$, $\pi_\beta^{\langle\Gamma\rangle}(x) \in \prod_{\alpha\in\Lambda_\beta} B_\alpha = B^{\langle\beta\rangle} \subseteq \bar{B}^{\langle\beta\rangle}$. Then, $x \in B^{\langle\Gamma\rangle}$. Hence, $E_{\text{inv}}(B) \subseteq B^{\langle\Gamma\rangle}$. Hence, E_{inv} is continuous at y_0. By the arbitrariness of y_0 and Proposition 3.9, we have E_{inv} is continuous.

This shows that E is a homeomorphism of $(\prod_{\beta\in\Gamma}\prod_{\alpha\in\Lambda_\beta} X_\alpha, \mathcal{O}^{\langle\Gamma\rangle})$ to $(\prod_{\alpha\in\bigcup_{\beta\in\Gamma}\Lambda_\beta} X_\alpha, \mathcal{O})$ and completes the proof of the proposition. □

Proposition 3.31 *Let Λ be an index set, $\mathcal{X}_\alpha := (X_\alpha, \mathcal{O}_{X\alpha})$ and $\mathcal{Y}_\alpha := (Y_\alpha, \mathcal{O}_{Y\alpha})$ be topological spaces, $\alpha \in \Lambda$. Assume that $\mathcal{X}_\alpha$ and $\mathcal{Y}_\alpha$ are homeomorphic, $\forall \alpha \in \Lambda$. Then, the product topological spaces $\mathcal{X} = (\prod_{\alpha\in\Lambda} X_\alpha, \mathcal{O}_X) := \prod_{\alpha\in\Lambda}\mathcal{X}_\alpha$ and $\mathcal{Y} = (\prod_{\alpha\in\Lambda} Y_\alpha, \mathcal{O}_Y) := \prod_{\alpha\in\Lambda}\mathcal{Y}_\alpha$ are homeomorphic.*

Proof $\forall \alpha \in \Lambda$, since $\mathcal{X}_\alpha$ and $\mathcal{Y}_\alpha$ are homeomorphic, then, $\exists H_\alpha : \mathcal{X}_\alpha \to \mathcal{Y}_\alpha$ such that H_α is bijective and both H_α and $H_{\alpha\,\mathrm{inv}}$ are continuous. Define mapping $H : \mathcal{X} \to \mathcal{Y}$ by, $\forall x \in \prod_{\alpha \in \Lambda} X_\alpha$, $\pi_\alpha^{\langle Y \rangle}(H(x)) = H_\alpha(\pi_\alpha^{\langle X \rangle}(x))$, $\forall \alpha \in \Lambda$.

$\forall y \in \mathcal{Y}$, define $x \in \mathcal{X}$ by $\pi_\alpha^{\langle X \rangle}(x) = H_{\alpha\,\mathrm{inv}}(\pi_\alpha^{\langle Y \rangle}(y))$, $\forall \alpha \in \Lambda$. Then, $H(x) = y$. Hence, H is surjective.

$\forall x_1, x_2 \in \mathcal{X}$ with $x_1 \neq x_2$. Then, $\exists \alpha_0 \in \Lambda$ such that $\pi_{\alpha_0}^{\langle X \rangle}(x_1) \neq \pi_{\alpha_0}^{\langle X \rangle}(x_2)$. Then, $\pi_{\alpha_0}^{\langle Y \rangle}(H(x_1)) = H_{\alpha_0}(\pi_{\alpha_0}^{\langle X \rangle}(x_1)) \neq H_{\alpha_0}(\pi_{\alpha_0}^{\langle X \rangle}(x_2)) = \pi_{\alpha_0}^{\langle Y \rangle}(H(x_2))$, since H_{α_0} is injective. Hence, $H(x_1) \neq H(x_2)$. Therefore, H is injective. Therefore, H is invertible with inverse H_{inv}.

Next, we show that H is continuous. For any basis open set, $O_Y \in \mathcal{O}_Y$. Then, $O_Y = \prod_{\alpha \in \Lambda} O_{Y\alpha}$ with $O_{Y\alpha} \in \mathcal{O}_{Y\alpha}$, $\forall \alpha \in \Lambda$, and $O_{Y\alpha} = Y_\alpha$ for all α's except finitely many α's, say $\alpha \in \Lambda_N$. Then, $H_{\mathrm{inv}}(O_Y) = \prod_{\alpha \in \Lambda} H_{\alpha\,\mathrm{inv}}(O_{Y\alpha}) \in \mathcal{O}_X$. Hence, H is continuous.

Finally, we show that H_{inv} is continuous. For any basis open set, $O_X \in \mathcal{O}_X$. Then, $O_X = \prod_{\alpha \in \Lambda} O_{X\alpha}$ with $O_{X\alpha} \in \mathcal{O}_{X\alpha}$, $\forall \alpha \in \Lambda$, and $O_{X\alpha} = X_\alpha$ for all α's except finitely many α's, say $\alpha \in \Lambda_N$. Then, $H(O_X) = \prod_{\alpha \in \Lambda} H_\alpha(O_{X\alpha}) \in \mathcal{O}_Y$. Hence, H_{inv} is continuous.

Hence, H is a homeomorphism. This completes the proof of the proposition. □

Proposition 3.32 *Let Λ be an index set, $\mathcal{X}$ be a topological space, $\mathcal{Y}_\alpha := (Y_\alpha, \mathcal{O}_\alpha)$ be a topological space, and $f_\alpha : \mathcal{X} \to \mathcal{Y}_\alpha$, $\forall \alpha \in \Lambda$. Let $\mathcal{Y} = (Y, \mathcal{O}) := \prod_{\alpha \in \Lambda} \mathcal{Y}_\alpha$ be the product topological space and $f : \mathcal{X} \to \mathcal{Y}$ be given by, $\forall x \in \mathcal{X}$, $\pi_\alpha(f(x)) = f_\alpha(x)$, $\forall \alpha \in \Lambda$. Then, f is continuous at $x_0 \in \mathcal{X}$ if, and only if, f_α is continuous at x_0, $\forall \alpha \in \Lambda$.*

Proof "Sufficiency" $\forall O \in \mathcal{O}$ with $f(x_0) \in O$. By Proposition 3.25 and Definition 3.17, $\exists$ a basis open set $B = \prod_{\alpha \in \Lambda} B_\alpha \in \mathcal{O}$ such that $f(x_0) \in B \subseteq O$, where $B_\alpha \in \mathcal{O}_\alpha$, $\forall \alpha \in \Lambda$, and $B_\alpha = Y_\alpha$ for all α's except finitely many α's, say $\alpha \in \Lambda_N$. $\forall \alpha \in \Lambda_N$, $f_\alpha(x_0) = \pi_\alpha(f(x_0)) \in \pi_\alpha(B) = B_\alpha$. By the continuity of f_α at x_0, $\exists U_\alpha \in \mathcal{O}_\mathcal{X}$ with $x_0 \in U_\alpha$ such that $f_\alpha(U_\alpha) \subseteq B_\alpha$. Let $U := \bigcap_{\alpha \in \Lambda_N} U_\alpha \in \mathcal{O}_\mathcal{X}$. Clearly, $x_0 \in U$. $\forall x \in U$, $\forall \alpha \in \Lambda_N$, $\pi_\alpha(f(x)) = f_\alpha(x) \in f_\alpha(U) \subseteq f_\alpha(U_\alpha) \subseteq B_\alpha$. Hence, $f(x) \in B$ and $f(U) \subseteq B \subseteq O$. Then, f is continuous at x_0.

"Necessity" $\forall \alpha \in \Lambda$, the projection function π_α is continuous, by Proposition 3.27. Note that $f_\alpha = \pi_\alpha \circ f$. Then, f_α is continuous at x_0 by Proposition 3.12.

This completes the proof of the proposition. □

3.5 The Separation Axioms

Definition 3.33 Let $(X, \mathcal{O})$ be a topological space. It is said to be

T$_1$ (Tychonoff): $\forall x, y \in X$ with $x \neq y$, $\exists O \in \mathcal{O}$ such that $x \in O$ and $y \in \widetilde{O}$.

T$_2$ (Hausdorff): $\forall x, y \in X$ with $x \neq y$, $\exists O_1, O_2 \in \mathcal{O}$ such that $x \in O_1$, $y \in O_2$, and $O_1 \cap O_2 = \emptyset$.

T$_3$ (regular): it is Tychonoff and, $\forall x \in X$, $\forall F \subseteq X$ with F being closed and $x \notin F$, $\exists O_1, O_2 \in \mathcal{O}$ such that $x \in O_1$, $F \subseteq O_2$, and $O_1 \cap O_2 = \emptyset$.

T$_4$ (normal): it is Tychonoff and, $\forall F_1, F_2 \subseteq X$ with F_1 and F_2 being closed and $F_1 \cap F_2 = \emptyset$, $\exists O_1, O_2 \in \mathcal{O}$ such that $F_1 \subseteq O_1$, $F_2 \subseteq O_2$, and $O_1 \cap O_2 = \emptyset$. $\diamond$

Note that $(X, \mathcal{O})$ is Tychonoff which implies that, $\forall x \in X$, the singleton set $\{x\}$ is closed since $\widetilde{\{x\}} = \bigcup_{y \in X, y \neq x} O_y$, where $O_y \in \mathcal{O}$ and $y \in O_y$ and $x \notin O_y$. Then, it is clear that $\mathbf{T}_4 \Rightarrow \mathbf{T}_3 \Rightarrow \mathbf{T}_2 \Rightarrow \mathbf{T}_1$.

Proposition 3.34 *A topological space $(X, \mathcal{O})$ is Tychonoff if, and only if, $\forall x \in X$, the singleton set $\{x\}$ is closed.*

Proof "Only if" $\forall x \in X$, $\forall y \in X$ with $y \neq x$, we have $\exists O_y \in \mathcal{O}$ such that $y \in O_y$ and $x \in \widetilde{O_y}$. Then, $\widetilde{\{x\}} = \bigcup_{y \in \widetilde{\{x\}}} O_y \in \mathcal{O}$. Hence, $\{x\}$ is closed.

"If" $\forall x, y \in X$ with $x \neq y$, we have $\widetilde{\{y\}} \in \mathcal{O}$. Then, $x \in \widetilde{\{y\}}$ and $y \notin \widetilde{\{y\}}$. Hence, $(X, \mathcal{O})$ is Tychonoff.

This completes the proof of the proposition. □

Proposition 3.35 *Let $(X, \mathcal{O})$ be a Tychonoff topological space. It is normal if, and only if, for all closed subset $F \subseteq X$ and any open subset $O \in \mathcal{O}$ with $F \subseteq O$, $\exists U \in \mathcal{O}$ such that $F \subseteq U \subseteq \overline{U} \subseteq O$.*

Proof "Necessity" Since $(X, \mathcal{O})$ is normal, F and $\widetilde{O}$ are closed, and $F \cap \widetilde{O} = \emptyset$, then $\exists O_1, O_2 \in \mathcal{O}$ such that $F \subseteq O_1$, $\widetilde{O} \subseteq O_2$, and $O_1 \cap O_2 = \emptyset$. Then, we have $F \subseteq O_1 \subseteq \widetilde{O_2} \subseteq O$. Since $\widetilde{O_2}$ is closed, then, $\overline{O_1} \subseteq \widetilde{O_2}$. Therefore, we have $F \subseteq O_1 \subseteq \overline{O_1} \subseteq \widetilde{O_2} \subseteq O$. So, $U = O_1$.

"Sufficiency" For any closed subsets $F_1, F_2 \subseteq X$ with $F_1 \cap F_2 = \emptyset$, we have $F_1 \subseteq \widetilde{F_2}$. Then, $\exists U \in \mathcal{O}$ such that $F_1 \subseteq U \subseteq \overline{U} \subseteq \widetilde{F_2}$. Let $O_1 = U \in \mathcal{O}$ and $O_2 = \widetilde{\overline{U}} \in \mathcal{O}$. Clearly, $O_1 \cap O_2 = \emptyset$, $F_1 \subseteq O_1$, and $F_2 \subseteq O_2$. Hence, $(X, \mathcal{O})$ is normal.

This completes the proof of the proposition. □

It is easy to show that the product topological space of Hausdorff topological spaces is Hausdorff.

3.6 Category Theory

Definition 3.36 In a topological space $(X, \mathcal{O})$, a subset $D \subseteq X$ is said to be *dense* if $\overline{D} = X$. $(X, \mathcal{O})$ is said to be *separable* if there exists a countable dense subset $D \subseteq X$. A subset $M \subseteq X$ is said to be *nowhere dense* if $\left(\widetilde{M}\right)^\circ = \widetilde{\overline{M}}$ is dense in X. A subset $F \subseteq X$ is said to be of *first category* (or *meager*) if F is the countable union of nowhere dense subsets of X. A subset $S \subseteq X$ is said to be of *second category* (or *nonmeager*) if S is not meager. A subset $H \subseteq X$ is said to be a *residual set* (or *comeager*) if $\widetilde{H}$ is meager. $(X, \mathcal{O})$ is said to be *second category everywhere* if every nonempty open subset of X is of second category. ◇

Proposition 3.37 *Let $\mathcal{X}$ be a topological space and $Y \subseteq \mathcal{X}$ be dense. Then, $\forall O \in \mathcal{O}$, $\overline{O \cap Y} = \overline{O}$.*

Proof Clearly, $\overline{O} \supseteq O \cap Y$ is closed. Then, we have $\overline{O \cap Y} \subseteq \overline{O}$. $\forall x \in \overline{O}$, $\forall U \in \mathcal{O}$ with $x \in U$, by Proposition 3.3, $U \cap O \neq \emptyset$. Then, $(U \cap O) \cap Y \neq \emptyset$ since $\overline{Y} = \mathcal{X}$ and $U \cap O \in \mathcal{O}$. Hence, we have $U \cap (O \cap Y) \neq \emptyset$. This implies that $x \in \overline{O \cap Y}$, by Proposition 3.3. Hence, we have $\overline{O} \subseteq \overline{O \cap Y}$. Therefore, $\overline{O} = \overline{O \cap Y}$. This completes the proof of the proposition. □

Proposition 3.38 *Let $\mathcal{X}$ be a topological space. Then, $\mathcal{X}$ is second category everywhere if, and only if, countable intersection of open dense subsets is dense.*

Proof "Only if" Let $(O_n)_{n=1}^{\infty}$ be a sequence of open dense subsets of $\mathcal{X}$. Let $F_n := \widetilde{O_n}$, $\forall n \in \mathbb{N}$. Clearly, F_n is closed and nowhere dense, $\forall n \in \mathbb{N}$, since $\overline{\widetilde{\overline{F_n}}} = \overline{\widetilde{F_n}} = \overline{O_n} = \mathcal{X}$. Note that $\bigcap_{n=1}^{\infty} O_n = \left(\bigcup_{n=1}^{\infty} F_n\right)^{\sim}$. Now, suppose that $\bigcap_{n=1}^{\infty} O_n$ is not dense. Then, $\exists O \in \mathcal{O}_{\mathcal{X}}$ which is nonempty such that $\left(\bigcap_{n=1}^{\infty} O_n\right) \cap O = \emptyset$. Then, $\left(\left(\bigcup_{n=1}^{\infty} F_n\right)^{\sim}\right) \cap O = \emptyset$. Hence, $O = \bigcup_{n=1}^{\infty}(F_n \cap O)$ and is of first category. This contradicts with the fact that O is of second category. Therefore, $\bigcap_{n=1}^{\infty} O_n$ must be dense.

The case of finite intersection can be converted to the above scenario by padding $\mathcal{X}$ as additional open dense subsets.

"If" $\forall O \in \mathcal{O}$ with $O \neq \emptyset$. Fix any countable collection $(E_\alpha)_{\alpha \in \Lambda}$ of nowhere dense subsets of $\mathcal{X}$. $\forall \alpha \in \Lambda$, $\widetilde{\overline{E_\alpha}}$ is open and dense. Then, $\bigcap_{\alpha \in \Lambda} \widetilde{\overline{E_\alpha}}$ is dense. Then, $O \cap \left(\bigcap_{\alpha \in \Lambda} \widetilde{\overline{E_\alpha}}\right) \neq \emptyset$, which further implies that $O \not\subseteq \bigcup_{\alpha \in \Lambda} \overline{E_\alpha}$. Hence, $O \not\subseteq \bigcup_{\alpha \in \Lambda} E_\alpha$. Hence, O is of second category by the arbitrariness of $(E_\alpha)_{\alpha \in \Lambda}$. Hence, $\mathcal{X}$ is second category everywhere by the arbitrariness of O.

This completes the proof of the proposition. □

Proposition 3.39 *Let $(X, \mathcal{O})$ be a topological space. $F, O \subseteq X$ with $O, \widetilde{F} \in \mathcal{O}$. Then,*

(i) *$\overline{O} \setminus O$ and $F \setminus F^\circ$ are nowhere dense.*
(ii) *If $(X, \mathcal{O})$ is second category everywhere and F is of first category, then F is nowhere dense.*

Proof (i) $\overline{O} \setminus O = \overline{O} \cap \widetilde{O}$ is closed. Note that

$$\overline{\widetilde{\overline{\overline{O} \setminus O}}} = \overline{\widetilde{\overline{O} \cap \widetilde{O}}} = \overline{\widetilde{\overline{O}} \cup O} = \overline{\widetilde{\overline{O}}} \cup \overline{O} = X$$

where we have applied Proposition 3.3 in the above. Hence $\overline{O} \setminus O$ is nowhere dense.

$F \setminus F^\circ = F \cap \widetilde{F^\circ}$ is closed. Note that

$$\overline{\widetilde{\overline{F \setminus F^\circ}}} = \overline{\widetilde{F \cap \widetilde{F^\circ}}} = \overline{\widetilde{F} \cup F^\circ} = \overline{\widetilde{F}} \cup \overline{F^\circ} = \widetilde{F^\circ} \cup \overline{F^\circ} = X$$

where we have applied Proposition 3.3 in the above. Hence $F \setminus F^\circ$ is nowhere dense.

(ii) Let $F = \bigcup_{n=1}^{\infty} F_n$, where F_n's are nowhere dense subsets of X. Then, $\widetilde{\overline{F_n}}$ is open and dense in $(X, \mathcal{O})$. By Proposition 3.38, we have $\bigcap_{n=1}^{\infty} \widetilde{\overline{F_n}}$ is dense.

Since F is closed, then $\overline{F} = F$. $\forall m \in \mathbb{N}$. Note that $F \supseteq \bigcup_{n=1}^{m} F_n$. Then, $\overline{F} \supseteq \overline{\bigcup_{n=1}^{m} F_n} = \bigcup_{n=1}^{m} \overline{F_n}$, by Proposition 3.3. Therefore,

$$F = \overline{F} \supseteq \bigcup_{n=1}^{\infty} \overline{F_n} \supseteq \bigcup_{n=1}^{\infty} F_n = F$$

This implies $\overline{F} = \bigcup_{n=1}^{\infty} \overline{F_n}$. Hence, we have $\widetilde{\overline{F}} = \bigcap_{n=1}^{\infty} \widetilde{\overline{F_n}}$, which is dense. Hence, F is nowhere dense.

This completes the proof of the proposition. □

Proposition 3.40 *Let $(X, \mathcal{O})$ be a topological space. Then,*

(i) *A closed set $F \subseteq X$ is nowhere dense if, and only if, it contains no nonempty open subset.*
(ii) *A subset $E \subseteq X$ is nowhere dense if, and only if, $\forall O \in \mathcal{O}$ with $O \neq \emptyset$, $\exists U \in \mathcal{O}$ with $U \neq \emptyset$ such that $U \subseteq O \setminus E$.*

Proof (i) "Only if" Suppose $\exists U \in \mathcal{O}$ with $U \neq \emptyset$ and $U \subseteq F$. Then, $\widetilde{F} \cap U = \emptyset$. This contradicts with the fact that $\overline{\widetilde{\overline{F}}} = \overline{\widetilde{F}} = X$, by Proposition 3.3. Hence, the result holds.

"If" $\forall x \in X$, $\forall U \in \mathcal{O}$ with $x \in U$. Then, $U \not\subseteq F$ implies $U \cap \widetilde{F} \neq \emptyset$. This implies that x is a point of closure of $\widetilde{F}$, by Proposition 3.3. Hence, we have $\overline{\widetilde{\overline{F}}} = \overline{\widetilde{F}} = X$. Hence, F is nowhere dense.

(ii) "Only if" $\forall O \in \mathcal{O}$ with $O \neq \emptyset$. Note that

$$O \setminus E = O \cap \widetilde{E} \supseteq O \cap \widetilde{\overline{E}}$$

Since E is nowhere dense, then $\widetilde{\overline{E}}$ is open and dense, and hence $O \cap \widetilde{\overline{E}}$ is open and nonempty by the nonemptyness of O and Proposition 3.3. Then, $\exists U = O \cap \widetilde{\overline{E}}$.

"If" $\forall x \in X$, $\forall O \in \mathcal{O}$ with $x \in O$. Then, $\exists U \in \mathcal{O}$ with $U \neq \emptyset$ such that $U \subseteq O \setminus E$. Note that $U = U^\circ \subseteq \left(O \cap \widetilde{E}\right)^\circ = O^\circ \cap \widetilde{E}^\circ = O \cap \widetilde{\overline{E}}$, where we have made use of Proposition 3.3. Hence, $O \cap \widetilde{\overline{E}}$ is nonempty. By the arbitrariness of O and Proposition 3.3, x is a point of closure for $\widetilde{\overline{E}}$. By the arbitrariness of x, E is nowhere dense.

This completes the proof of the proposition. □

Theorem 3.41 (Uniform Boundedness Principle) *Let $(X, \mathcal{O})$ be a topological space that is second category everywhere. Let $\mathcal{F}$ be a family of continuous real-valued functions of X. $\forall x \in X$, $\exists M_x \in [0, \infty) \subset \mathbb{R}$ such that $|f(x)| \leq M_x$, $\forall f \in \mathcal{F}$. Then, $\exists U \in \mathcal{O}$ with $U \neq \emptyset$ and $\exists M \in [0, \infty) \subset \mathbb{R}$ such that $|f(x)| \leq M$, $\forall x \in U$ and $\forall f \in \mathcal{F}$.*

Proof Let $E_{m,f} := f_{\text{inv}}([-m, m]) \subseteq X$, $\forall f \in \mathcal{F}$ and $\forall m \in \mathbb{Z}_+$. Since f in continuous and $[-m, m]$ is a closed interval, then, by Proposition 3.10, $E_{m,f}$'s are closed. $\forall m \in \mathbb{Z}_+$, let $E_m := \bigcap_{f \in \mathcal{F}} E_{m,f}$, which is closed. $\forall x \in X$, $\exists m \in \mathbb{Z}_+$ with $m \geq M_x$ such that $|f(x)| \leq m$, $\forall f \in \mathcal{F}$. Then, $x \in E_{m,f}$, $\forall f \in \mathcal{F}$, and hence $x \in E_m$. Therefore, $X = \bigcup_{m=0}^{\infty} E_m$. Since X is second category everywhere, then E_m's are not all nowhere dense. Then, $\exists n \in \mathbb{Z}_+$ such that E_n is not nowhere dense, that is, $\overline{\widetilde{\widetilde{E_n}}} = \overline{\widetilde{E_n}} \neq X$. Then, by Proposition 3.40, E_n contains a nonempty open subset $U \in \mathcal{O}$. Then, $U \subseteq E_n$ and $U \neq \emptyset$. $\forall x \in U$, $\forall f \in \mathcal{F}$, we have $|f(x)| \leq n$. This completes the proof of the theorem. □

3.7 Connectedness

Definition 3.42 A topological space $\mathcal{X}$ is said to be *connected* if there do not exist nonempty open sets O_1 and O_2 such that $\mathcal{X} = O_1 \cup O_2$ and $O_1 \cap O_2 = \emptyset$. Such a pair of O_1 and O_2 is called a *separation* of $\mathcal{X}$ if it exists. ◇

Proposition 3.43 *Let $\mathcal{X}$ and $\mathcal{Y}$ be topological spaces and $f : \mathcal{X} \to \mathcal{Y}$ be a surjective continuous function. If $\mathcal{X}$ is connected, then $\mathcal{Y}$ is connected.*

Proof Suppose $\mathcal{Y}$ is not connected. Then, $\exists O_1, O_2 \in \mathcal{O}_\mathcal{Y}$ with $O_1 \neq \emptyset$ and $O_2 \neq \emptyset$ such that $O_1 \cup O_2 = \mathcal{Y}$ and $O_1 \cap O_2 = \emptyset$. Since f is surjective, then $f_{\text{inv}}(O_1) \neq \emptyset$ and $f_{\text{inv}}(O_2) \neq \emptyset$. Since f is continuous, then $f_{\text{inv}}(O_1), f_{\text{inv}}(O_2) \in \mathcal{O}_\mathcal{X}$. By Proposition 2.5, $f_{\text{inv}}(O_1) \cup f_{\text{inv}}(O_2) = f_{\text{inv}}(O_1 \cup O_2) = \mathcal{X}$ and $f_{\text{inv}}(O_1) \cap f_{\text{inv}}(O_2) = f_{\text{inv}}(O_1 \cap O_2) = \emptyset$. This contradicts with the assumption that $\mathcal{X}$ is connected. Therefore, $\mathcal{Y}$ is connected. This completes the proof of the proposition. □

Theorem 3.44 (Mean Value Theorem) *Let $\mathcal{X}$ be a topological space and $f : \mathcal{X} \to \mathbb{R}$ be a continuous function. Assume that $\mathcal{X}$ is connected and $\exists x, y \in \mathcal{X}$ such that $f(x) < c < f(y)$ for some $c \in \mathbb{R}$. Then, $\exists z \in \mathcal{X}$ such that $f(z) = c$.*

Proof Suppose the result is false. Then, $f_{\text{inv}}(\{c\}) = \emptyset$. Let $O_1 := f_{\text{inv}}((c, \infty))$ and $O_2 := f_{\text{inv}}((-\infty, c))$. Then, $x \in O_2 \in \mathcal{O}_\mathcal{X}$ and $y \in O_1 \in \mathcal{O}_\mathcal{X}$, by assumptions of the theorem. By Proposition 2.5, $O_1 \cap O_2 = \emptyset$ and $O_1 \cup O_2 = \mathcal{X}$. This contradicts with the assumption that $\mathcal{X}$ is connected. Hence, the result is true. This completes the proof of the theorem. □

Proposition 3.45 *Let $\mathcal{X}$ be a topological space, $U \subseteq V \subseteq \overline{U} \subseteq \mathcal{X}$, and U be connected in the subset topology $\mathcal{O}_U$. Then, V is connected in the subset topology $\mathcal{O}_V$.*

Proof Suppose V is not connected in its subset topology. Then, $\exists O_{V1}, O_{V2} \in \mathcal{O}_V$ with $O_{V1} \neq \emptyset$ and $O_{V2} \neq \emptyset$, such that $O_{V1} \cup O_{V2} = V$ and $O_{V1} \cap O_{V2} = \emptyset$. By Proposition 3.4, $\exists O_1, O_2 \in \mathcal{O}$ such that $O_{V1} = O_1 \cap V$ and $O_{V2} = O_2 \cap V$. Let $x_1 \in O_{V1}$ and $x_2 \in O_{V2}$. Then, $x_i \in \overline{U} \cap O_i$, $i = 1, 2$. By Proposition 3.3, $\exists \bar{x}_i \in U \cap O_i =: O_{Ui} \neq \emptyset$, $i = 1, 2$. By Proposition 3.4, $O_{U1}, O_{U2} \in \mathcal{O}_U$. Note that $O_{U1} \cap O_{U2} = U \cap O_1 \cap O_2 = U \cap V \cap O_1 \cap O_2 = U \cap (O_1 \cap V) \cap (O_2 \cap V) = U \cap O_{V1} \cap O_{V2} = \emptyset$ and $O_{U1} \cup O_{U2} = U \cap (O_1 \cup O_2) = U \cap V \cap (O_1 \cup O_2) = U \cap ((V \cap O_1) \cup (V \cap O_2)) = U \cap (O_{V1} \cup O_{V2}) = U \cap V = U$. Hence, the pair (O_{U1}, O_{U2}) is a separation of U. This implies that U is not connected in its subset topology. This contradicts with the assumption. Hence, V is connected. □

Definition 3.46 Let $\mathcal{X}$ be a topological space, $x_0 \in \mathcal{X}$. Let

$$\mathcal{M} := \{M \subseteq \mathcal{X} \mid x_0 \in M, M \text{ is connected in the subset topology.}\}$$

The *component containing* x_0 is defined by $A := \bigcup_{M \in \mathcal{M}} M$. ◇

Clearly, $\mathcal{X}$ is the union of its components.

Proposition 3.47 *Let $\mathcal{X}$ be a topological space and $x_0 \in \mathcal{X}$. Then, the component A of $\mathcal{X}$ containing x_0 is connected and closed.*

Proof Let $\mathcal{M}$ be as defined in Definition 3.46. Suppose A is not connected. Let $\mathcal{O}_A$ be the subset topology on A. Then, $\exists O_1, O_2 \in \mathcal{O}_A$ with $O_1 \neq \emptyset$ and $O_2 \neq \emptyset$ such that $O_1 \cup O_2 = A$ and $O_1 \cap O_2 = \emptyset$. Without loss of generality, assume that $x_0 \in O_1$. By the definition of A, $\exists A_0 \in \mathcal{M}$ such that $A_0 \cap O_2 \neq \emptyset$. Note that $A_0 = (O_1 \cap A_0) \cup (O_2 \cap A_0)$ and $O_1 \cap A_0 \ni x_0$ and $O_2 \cap A_0$ are nonempty and open in the subset topology on A_0. Furthermore, $(O_1 \cap A_0) \cap (O_2 \cap A_0) = \emptyset$. This shows that A_0 is not connected, which contradicts with the fact that $A_0 \in \mathcal{M}$. Hence, A is connected.

By Proposition 3.45, $\overline{A}$ is connected. Then, we have $\overline{A} \in \mathcal{M}$ and $\overline{A} = A$. By Proposition 3.3, A is closed in $\mathcal{O}$.

This completes the proof of the proposition. □

Proposition 3.48 *Let $\mathcal{X} := (X, \mathcal{O})$ be a topological space and $A_\alpha \subseteq X$ be connected (in subset topology), $\forall \alpha \in \Lambda$, where Λ is an index set. Assume that $A_{\alpha_1} \cap A_{\alpha_2} \neq \emptyset$, $\forall \alpha_1, \alpha_2 \in \Lambda$. Then, $A := \bigcup_{\alpha \in \Lambda} A_\alpha$ is connected (in subset topology).*

Proof Suppose A is not connected. Then, $\exists O_1, O_2 \in \mathcal{O}$ such that $O_1 \cap A \neq \emptyset \neq O_2 \cap A$, $(O_1 \cap A) \cap (O_2 \cap A) = \emptyset$, and $(O_1 \cap A) \cup (O_2 \cap A) = (O_1 \cup O_2) \cap A = A$. Let $x_i \in O_i \cap A$, $i = 1, 2$. Then, $\exists \alpha_i \in \Lambda$ such that $x_i \in A_{\alpha_i}$, $i = 1, 2$. By the assumption, let $x_0 \in A_{\alpha_1} \cap A_{\alpha_2} \neq \emptyset$. Without loss of generality, assume $x_0 \in O_1$. Then, $x_0 \in O_1 \cap A_{\alpha_2} \neq \emptyset$, $x_2 \in O_2 \cap A_{\alpha_2} \neq \emptyset$, $(O_1 \cap A_{\alpha_2}) \cap (O_2 \cap A_{\alpha_2}) \subseteq$

$(O_1 \cap A) \cap (O_2 \cap A) = \emptyset$, and $(O_1 \cap A_{\alpha_2}) \cup (O_2 \cap A_{\alpha_2}) = (O_1 \cup O_2) \cap A_{\alpha_2} = A_{\alpha_2}$. Hence, $O_1 \cap A_{\alpha_2}$ and $O_2 \cap A_{\alpha_2}$ form a separation of A_{α_2}. This implies that A_{α_2} is not connected. This is a contradiction. Therefore, A is connected. This completes the proof of the proposition. □

Definition 3.49 Let $\mathcal{X}$ be a topological space. It is said to be *locally connected* if there exists a basis $\mathcal{B}$ such that B is connected (in the subset topology), $\forall B \in \mathcal{B}$. ◇

Proposition 3.50 *Any component of a locally connected topological space is open.*

Proof Let $\mathcal{X}$ be a locally connected topological space and $\mathcal{B}$ be a basis made up of connected sets. $\forall x_0 \in \mathcal{X}$, let A be the component containing x_0. By Proposition 3.47, A is closed and connected. $\forall x \in A$, $\exists B \in \mathcal{B}$ such that $x \in B$. Since B is connected, then, by Proposition 3.48 and Definition 3.46, $B \subseteq A$. Hence, A is open. This completes the proof of the proposition. □

Proposition 3.51 *Let $(X_\alpha, \mathcal{O}_\alpha)$ be a connected topological space, $\forall \alpha \in \Gamma$, where Γ is an index set. Let $(X, \mathcal{O})$ be the product topological space $\prod_{\alpha \in \Gamma}(X_\alpha, \mathcal{O}_\alpha)$. Then, $(X, \mathcal{O})$ is connected.*

Proof Suppose that X is not connected. Then, $\exists O_1, O_2 \in \mathcal{O}$ with $O_1 \neq \emptyset$ and $O_2 \neq \emptyset$ such that $O_1 \cup O_2 = X$ and $O_1 \cap O_2 = \emptyset$. Let $\mathcal{B}$ be the basis defined in Proposition 3.25. Then, there exist nonempty disjoint index sets Λ_1 and Λ_2 such that $O_1 = \bigcup_{\lambda \in \Lambda_1} B_\lambda$ and $O_2 = \bigcup_{\lambda \in \Lambda_2} C_\lambda$, where $B_\lambda \in \mathcal{B}$, $\forall \lambda \in \Lambda_1$, and $C_\lambda \in \mathcal{B}$, $\forall \lambda \in \Lambda_2$. $\forall \lambda \in \Lambda_1$, $B_\lambda = \prod_{\alpha \in \Gamma} B_{\lambda\alpha}$, where $B_{\lambda\alpha} \in \mathcal{O}_\alpha$, $\forall \alpha \in \Gamma$, and $B_{\lambda\alpha} = X_\alpha$ for all except finitely many α's, say $\alpha \in \Gamma_\lambda$. Note that $\Gamma_\lambda \neq \emptyset$, since $B_\lambda \subseteq O_1 \subset X$. $\forall \lambda \in \Lambda_2$, $C_\lambda = \prod_{\alpha \in \Gamma} C_{\lambda\alpha}$, where $C_{\lambda\alpha} \in \mathcal{O}_\alpha$, $\forall \alpha \in \Gamma$, and $C_{\lambda\alpha} = X_\alpha$ for all except finitely many α's, say $\alpha \in \Gamma_\lambda$. Note that $\Gamma_\lambda \neq \emptyset$, since $C_\lambda \subseteq O_2 \subset X$. Note that

$$\emptyset = O_1 \cap O_2 = \bigcup_{\lambda \in \Lambda_1} \bigcup_{\gamma \in \Lambda_2} (B_\lambda \cap C_\gamma)$$

Therefore, $B_\lambda \cap C_\gamma = \emptyset$, $\forall \lambda \in \Lambda_1$, $\forall \gamma \in \Lambda_2$, which implies that $\exists \alpha_{\lambda\gamma} \in \Gamma \ni \cdot$ $B_{\lambda\alpha_{\lambda\gamma}} \cap C_{\gamma\alpha_{\lambda\gamma}} = \emptyset$.

Fix $x_1 \in O_1$ and $x_2 \in O_2$. Then, $x_1 \neq x_2$. $\exists \lambda_1 \in \Lambda_1$ such that $x_1 \in B_{\lambda_1}$. Let $\Gamma_1 := \Gamma_{\lambda_1}$. $\exists \lambda_2 \in \Lambda_2$ such that $x_2 \in C_{\lambda_2}$. Let $\Gamma_2 := \Gamma_{\lambda_2}$. Note that $\forall x \in X$ with $\pi_\alpha(x) = \pi_\alpha(x_1)$, $\forall \alpha \in \Gamma_1$, we have $x \in B_{\lambda_1} \subseteq O_1$. Similarly, $\forall x \in X$ with $\pi_\alpha(x) = \pi_\alpha(x_2)$, $\forall \alpha \in \Gamma_2$, we have $x \in C_{\lambda_2} \subseteq O_2$. Therefore, starting with $x_1 \in O_1$ and switch, one by one, its coordinate $\pi_\alpha(x_1)$ to $\pi_\alpha(x_2)$, for all $\alpha \in \Gamma_2$, we will end up with a point $x_3 \in O_2$. Therefore, there must exist a step in this process such that switching one coordinate $\pi_{\alpha_0}(x_1)$ to $\pi_{\alpha_0}(x_2)$, for some $\alpha_0 \in \Gamma_2$, leads to the change of set membership from $\bar{x}_1 \in O_1$ before the switch to $\bar{x}_2 \in O_2$ after the switch. In summary, there exist $\bar{x}_1 \in O_1$, $\bar{x}_2 \in O_2$, and $\alpha_0 \in \Gamma$ such that $\pi_\alpha(\bar{x}_1) = \pi_\alpha(\bar{x}_2)$, $\forall \alpha \in \Gamma \setminus \{\alpha_0\}$. Since $O_1 \cap O_2 = \emptyset$, we must have $\pi_{\alpha_0}(\bar{x}_1) \neq \pi_{\alpha_0}(\bar{x}_2)$.

Define

$$\Lambda_{10} := \{\lambda \in \Lambda_1 \mid B_\lambda = \prod_{\alpha \in \Gamma} B_{\lambda\alpha}, \pi_\alpha(\bar{x}_1) \in B_{\lambda\alpha}, \forall \alpha \in \Gamma \setminus \{\alpha_0\}\}$$

and

$$\Lambda_{20} := \{\lambda \in \Lambda_2 \mid C_\lambda = \prod_{\alpha \in \Gamma} C_{\lambda\alpha}, \pi_\alpha(\bar{x}_2) \in C_{\lambda\alpha}, \forall \alpha \in \Gamma \setminus \{\alpha_0\}\}$$

Let $M := \prod_{\alpha \in \Gamma} M_\alpha \subseteq X$, where $M_\alpha = \{\pi_\alpha(\bar{x}_1)\}$, $\forall \alpha \in \Gamma \setminus \{\alpha_0\}$, and $M_{\alpha_0} = X_{\alpha_0}$. Note that $M \subseteq X = O_1 \cup O_2 = \left(\bigcup_{\lambda \in \Lambda_1} B_\lambda\right) \cup \left(\bigcup_{\gamma \in \Lambda_2} C_\gamma\right)$ implies that

$$M \subseteq \Big(\bigcup_{\lambda \in \Lambda_{10}} B_\lambda\Big) \cup \left(\bigcup_{\gamma \in \Lambda_{20}} C_\gamma\right) = \bigcup_{\lambda \in \Lambda_{10}} \bigcup_{\gamma \in \Lambda_{20}} (B_\lambda \cup C_\gamma)$$

Therefore, we have $X_{\alpha_0} = \pi_{\alpha_0}(M) \subseteq \bigcup_{\lambda \in \Lambda_{10}} \bigcup_{\gamma \in \Lambda_{20}} (\pi_{\alpha_0}(B_\lambda) \cup \pi_{\alpha_0}(C_\gamma)) = \bigcup_{\lambda \in \Lambda_{10}} \bigcup_{\gamma \in \Lambda_{20}} (B_{\lambda\alpha_0} \cup C_{\gamma\alpha_0}) \subseteq X_{\alpha_0}$, by Proposition 2.5. Then,

$$\begin{aligned} X_{\alpha_0} &= \bigcup_{\lambda \in \Lambda_{10}} \bigcup_{\gamma \in \Lambda_{20}} (B_{\lambda\alpha_0} \cup C_{\gamma\alpha_0}) = \Big(\bigcup_{\lambda \in \Lambda_{10}} B_{\lambda\alpha_0}\Big) \cup \left(\bigcup_{\gamma \in \Lambda_{20}} C_{\gamma\alpha_0}\right) \\ &* =: D_1 \cup D_2 \end{aligned}$$

By derivations in the first paragraph of this proof, we have $\forall \lambda \in \Lambda_{10}$, $\forall \gamma \in \Lambda_{20}$, $B_\lambda \cap C_\gamma = \prod_{\alpha \in \Gamma} (B_{\lambda\alpha} \cap C_{\gamma\alpha}) = \emptyset$. Note that, $\forall \alpha \in \Gamma \setminus \{\alpha_0\}$, $\pi_\alpha(\bar{x}_1) = \pi_\alpha(\bar{x}_2) \in B_{\lambda\alpha} \cap C_{\gamma\alpha} \neq \emptyset$. Then, we must have $B_{\lambda\alpha_0} \cap C_{\gamma\alpha_0} = \emptyset$. Therefore, we have

$$D_1 \cap D_2 = \bigcup_{\lambda \in \Lambda_{10}} \bigcup_{\gamma \in \Lambda_{20}} (B_{\lambda\alpha_0} \cap C_{\gamma\alpha_0}) = \emptyset$$

Clearly, $D_1, D_2 \in \mathcal{O}_{\alpha_0}$, $\pi_{\alpha_0}(\bar{x}_1) \in D_1 \neq \emptyset$, and $\pi_{\alpha_0}(\bar{x}_2) \in D_2 \neq \emptyset$ since $\bar{x}_1 \in O_1$ and $\bar{x}_2 \in O_2$. This shows that D_1 and D_2 form a separation of X_{α_0}. This contradicts with the assumption that $(X_\alpha, \mathcal{O}_\alpha)$ is connected, $\forall \alpha \in \Gamma$. Therefore, $(X, \mathcal{O})$ is connected.

This completes the proof of the proposition. □

Definition 3.52 Let $\mathcal{X}$ be a topological space and $I := [0, 1] \subset \mathbb{R}$ be endowed with the subset topology of $\mathbb{R}$. A *curve* in $\mathcal{X}$ is a continuous mapping $\gamma : I \to \mathcal{X}$. $\gamma(0)$ is called the *beginning point* and $\gamma(1)$ is called the *end point*. A *closed curve* is such that $\gamma(0) = \gamma(1)$. Two closed curves γ_1 and γ_2 are said to be *homotopic* to each other if there exists a continuous function $\phi : I \times I \to \mathcal{X}$ such that $\phi(t, 0) = \gamma_1(t)$, $\phi(t, 1) = \gamma_2(t)$, and $\phi(0, t) = \phi(1, t)$, $\forall t \in I$. ◇

Definition 3.53 Let $\mathcal{X}$ be a topological space. It is said to be *arcwise connected* if $\forall x_1, x_2 \in \mathcal{X}$, there exists a curve γ in $\mathcal{X}$ such that $\gamma(0) = x_1$ and $\gamma(1) = x_2$. $\mathcal{X}$ is said to be *simply connected* if it is arcwise connected and any closed curve is homotopic to a single point (that is, a degenerate curve γ with $\gamma(t) = x \in \mathcal{X}$, $\forall t \in [0, 1] \subset \mathbb{R}$.) $\diamond$

Proposition 3.54 *Let $(X, \mathcal{O})$ be a topological space. Then, it is connected if it is arcwise connected.*

Proof Suppose $(X, \mathcal{O})$ is not connected. Then, $\exists O_1, O_2 \in \mathcal{O}$ with $O_1 \neq \emptyset$ and $O_2 \neq \emptyset$ such that $O_1 \cup O_2 = X$ and $O_1 \cap O_2 = \emptyset$. Fix $x_1 \in O_1$ and $x_2 \in O_2$. Then, $x_1 \neq x_2$. Since $(X, \mathcal{O})$ is arcwise connected, then there exists a curve γ such that $\gamma(0) = x_1$ and $\gamma(1) = x_2$. Define $t := \sup\{s \in [0, 1] \subset \mathbb{R} \mid \gamma([0, s]) \subseteq O_1\}$. Then, $t \in [0, 1] \subset \mathbb{R}$.

Claim 3.54.1 $\gamma(t) \in O_1$.

Proof of Claim Suppose $\gamma(t) \notin O_1$, then $\gamma(t) \in O_2$. Then, $t > 0$, since $\gamma(0) = x_1 \in O_1$. Since γ is continuous, then $\exists t_1 \in [0, t) \subset \mathbb{R}$ such that $\gamma((t_1, t]) \subseteq O_2$. $\forall s \in (t_1, 1] \subset \mathbb{R}$, $\gamma([0, s]) \cap O_2 \neq \emptyset$. Then, $s \notin \{s \in [0, 1] \subset \mathbb{R} \mid \gamma([0, s]) \subseteq O_1\}$. Hence, $s \geq t$, which implies that $t_1 \geq t$. This contradicts $t_1 < t$. Hence, $\gamma(t) \in O_1$. This completes the proof of the claim. $\square$

Since $\gamma(1) = x_2 \in O_2$, then $t < 1$. By the continuity of γ, $\exists t_2 \in (t, 1] \subset \mathbb{R}$ such that $\gamma([t, t_2)) \subseteq O_1$. $\forall s \in [0, t)$, $\exists s_1 \in (s, t]$ such that $s_1 \in \{s \in [0, 1] \subset \mathbb{R} \mid \gamma([0, s]) \subseteq O_1\}$. Then, $\gamma([0, s_1]) \subseteq O_1$ and hence $\gamma(s) \in O_1$. Therefore, $\gamma([0, t)) \subseteq O_1$. This coupled with $\gamma([t, t_2)) \subseteq O_1$, implies $\gamma([0, t_2)) \subseteq O_1$. Then, $t \geq t_2$. This contradicts with $t < t_2$.

Therefore, $(X, \mathcal{O})$ is connected. This completes the proof of the proposition. $\square$

3.8 Continuous Real-Valued Functions

Theorem 3.55 (Urysohn's Lemma) *Let $(X, \mathcal{O})$ be a normal topological space, $A, B \subseteq X$ be closed subsets, and $A \cap B = \emptyset$. Then, there exists a continuous real-valued function $f : X \to [0, 1] \subset \mathbb{R}$ such that $f(x) = 0$, $\forall x \in A$, and $f(x) = 1$, $\forall x \in B$.*

Proof Since the set $Q := \mathbb{Q} \cap [0, 1]$ is countable, then, by recursively applying Proposition 3.35, we may find $(O_r)_{r \in Q} \subseteq \mathcal{O}$ such that the following two properties are satisfied:

1. $\forall r \in Q$, $A \subseteq O_r \subseteq \overline{O_r} \subseteq \widetilde{B}$.
2. $\forall r, s \in Q$ with $r < s$, $\overline{O_r} \subseteq O_s$.

Define the real-valued function $f : X \to \mathbb{R}$ by

$$f(x) = \inf(\{r \in Q \mid x \in O_r\} \cup \{1\})$$

Clearly, $f : X \to [0,1]$, $f(x) = 0$, $\forall x \in O_0$, and $f(x) = 1$, $\forall x \in \widetilde{O_1}$. By 1, we have $A \subseteq O_0$ and $O_1 \subseteq \widetilde{B}$. Hence, all we need to show is that f is continuous. $\forall x_0 \in X$, we will show that f is continuous at x_0. Let $a_0 = f(x_0) \in [0,1]$. $\forall U \subseteq \mathbb{R}$ with U being open and $a_0 \in U$, $\exists a_1, a_2, a_3, a_4 \in \mathbb{Q}$ such that $a_1 < a_2 < a_0 < a_3 < a_4$ and $(a_1, a_4) \subseteq U$. Let $\bar{a}_2 = \max\{a_2, 0\}$ and $\bar{a}_3 = \min\{a_3, 1\}$. Then, we must have $a_1 < \bar{a}_2 \le a_0 \le \bar{a}_3 < a_4$ and $\bar{a}_2, \bar{a}_3 \in Q$. We will distinguish three exhaustive and mutually exclusive cases: Case 1: $a_0 \in (0,1)$, Case 2: $a_0 = 0$, and Case 3: $a_0 = 1$.

Case 1: $a_0 \in (0,1)$. Then, we must have $a_1 < \bar{a}_2 < a_0 < \bar{a}_3 < a_4$. Let $V = \widetilde{\overline{O_{\bar{a}_2}}} \cap O_{\bar{a}_3} \in \mathcal{O}$. $\forall x \in V$, we have $x \in O_{\bar{a}_3}$ and $f(x) \le \bar{a}_3$. Also, $x \in \widetilde{\overline{O_{\bar{a}_2}}}$ implies that $f(x) \ge \bar{a}_2$. Hence, $f(V) \subseteq [\bar{a}_2, \bar{a}_3] \subset (a_1, a_4) \subseteq U$. $f(x_0) = a_0 < \bar{a}_3$ implies that $x_0 \in O_{\bar{a}_3}$. $f(x_0) = a_0 > \bar{a}_2$ implies that $\exists \tilde{a}_2 \in (\bar{a}_2, a_0) \cap Q$ such that $x_0 \in \widetilde{O_{\tilde{a}_2}} \subseteq \widetilde{\overline{O_{\bar{a}_2}}}$. Therefore $x_0 \in V$. This shows that $\exists V \in \mathcal{O}$ with $x_0 \in V$ such that $f(V) \subseteq U$.

Case 2 $a_0 = 0$. Then, we must have $a_1 < 0 = a_0 < \bar{a}_3 < a_4$. Take $V = O_{\bar{a}_3} \in \mathcal{O}$. We must have $x_0 \in V$. $\forall x \in V$, $0 \le f(x) \le \bar{a}_3$. Hence, $f(V) \subseteq [0, \bar{a}_3] \subset (a_1, a_4) \subseteq U$. Hence, $\exists V \in \mathcal{O}$ with $x_0 \in V$ such that $f(V) \subseteq U$.

Case 3 $a_0 = 1$. Then, we must have $a_1 < \bar{a}_2 < a_0 = 1 < a_4$. Take $V = \widetilde{\overline{O_{\bar{a}_2}}} \in \mathcal{O}$. Since $f(x_0) = a_0 = 1$, then $x_0 \in \widetilde{O_{\frac{1+\bar{a}_2}{2}}} \subseteq \widetilde{\overline{O_{\bar{a}_2}}} = V$. $\forall x \in V$, $f(x) \ge \bar{a}_2$. Hence, $f(V) \subseteq [\bar{a}_2, 1] \subset (a_1, a_4) \subseteq U$. Hence, $\exists V \in \mathcal{O}$ with $x_0 \in V$ such that $f(V) \subseteq U$.

Therefore, in all cases, $\exists V \in \mathcal{O}$ with $x_0 \in V$ such that $f(V) \subseteq U$. Hence, f is continuous at x_0. By the arbitrariness of x_0 and Proposition 3.9, f is continuous. This completes the proof of the theorem. □

Proposition 3.56 *Let $\mathcal{X}$ and $\mathcal{Y}$ be topological spaces and $\mathcal{Y}$ be Hausdorff. $f_1 : \mathcal{X} \to \mathcal{Y}$ and $f_2 : \mathcal{X} \to \mathcal{Y}$ are continuous. Let $D \subseteq \mathcal{X}$ be dense. Assume that $f_1|_D = f_2|_D$. Then, $f_1 = f_2$.*

Proof Suppose $f_1 \neq f_2$. Then, $\exists x \in \mathcal{X}$ such that $f_1(x) \neq f_2(x)$. Since $\mathcal{Y}$ is Hausdorff, then $\exists O_1, O_2 \in \mathcal{O}_\mathcal{Y}$ such that $f_1(x) \in O_1$, $f_2(x) \in O_2$, and $O_1 \cap O_2 = \emptyset$. Since f_1 and f_2 are continuous, we have $U_1 := f_{1\,\mathrm{inv}}(O_1) \in \mathcal{O}_\mathcal{X}$ and $U_2 := f_{2\,\mathrm{inv}}(O_2) \in \mathcal{O}_\mathcal{X}$. Note that $x \in U_1 \cap U_2 \in \mathcal{O}_\mathcal{X}$ and $x \in \overline{D}$, then, by Proposition 3.3, $\exists \bar{x} \in D \cap U_1 \cap U_2$. Then, $f_1(\bar{x}) \in O_1$ and $f_2(\bar{x}) \in O_2$, which implies that $f_1|_D(\bar{x}) \neq f_2|_D(\bar{x})$. This is a contradiction. Hence, we must have $f_1 = f_2$.

This completes the proof of the proposition. □

Theorem 3.57 (Tietze's Extension Theorem) *Let $(X, \mathcal{O})$ be a normal topological space, $A \subseteq X$ be closed, and $h : A \to \mathbb{R}$. Let A be endowed with the subset topology $\mathcal{O}_A$. Assume that h is continuous. Then, there exists a continuous function $k : X \to \mathbb{R}$ such that $k|_A = h$.*

Proof Let $f := \dfrac{h}{1 + |h|}$. Then, $|f(x)| < 1$, $\forall x \in A$, and by Proposition 3.12, f is continuous.

Claim 3.57.1 Let $l : A \to \mathbb{R}$ be a continuous function such that $|l(x)| \leq c_1 \in \mathbb{R}$, $\forall x \in A$, where $c_1 > 0$. Then, there exists a continuous function $g : X \to \mathbb{R}$ such that $|g(x)| \leq c_1/3$, $\forall x \in X$, and $|l(x) - g(x)| \leq 2c_1/3$, $\forall x \in A$.

Proof of Claim Let $B := \{x \in A \mid l(x) \leq -c_1/3\}$ and $C := \{x \in A \mid l(x) \geq c_1/3\}$. Then, B and C are closed sets in $\mathcal{O}_A$, by the continuity of l and Proposition 3.10. Since A is closed, then B and C are closed in $\mathcal{O}$, by Proposition 3.5. Clearly, $B \cap C = \emptyset$. By Urysohn's Lemma, there exists a continuous function $g : X \to \mathbb{R}$ such that $|g(x)| \leq c_1/3$, $\forall x \in X$, $g(x) = -c_1/3$, $\forall x \in B$, and $g(x) = c_1/3$, $\forall x \in C$. Hence, $|l(x) - g(x)| \leq 2c_1/3$, $\forall x \in A$. This completes the proof of the claim. □

By repeated application of Claim 3.57.1, we may define $f_i : X \to \mathbb{R}$, $\forall i \in \mathbb{N}$, such that f_i is continuous, $|f_i(x)| \leq \frac{2^{i-1}}{3^i}$, $\forall x \in X$, and $\left|f(x) - \sum_{k=1}^{i} f_k(x)\right| \leq \frac{2^i}{3^i}$, $\forall x \in A$.

Define $g : X \to \mathbb{R}$ by $g(x) = \lim_{i \in \mathbb{N}} \sum_{k=1}^{i} f_k(x)$, $\forall x \in X$. Clearly, g is well-defined, $g|_A = f$, and $|g(x)| \leq \sum_{i=1}^{\infty} \frac{2^{i-1}}{3^i} = 1$, $\forall x \in X$. $\forall x_0 \in X$. $\forall \epsilon \in (0, \infty) \subset \mathbb{R}$. $\exists N \in \mathbb{N}$ such that $\left|\sum_{i=N+1}^{\infty} f_i(x)\right| < \epsilon/3$, $\forall x \in X$. By the continuity of $f_1, \ldots, f_N$ and Proposition 3.9, $\exists U \in \mathcal{O}$ with $x_0 \in U$ such that $\left|\sum_{i=1}^{N} f_i(x) - \sum_{i=1}^{N} f_i(x_0)\right| < \epsilon/3$, $\forall x \in U$. Then, we have, $\forall x \in U$,

$$\begin{aligned} |g(x) - g(x_0)| &\leq \left|g(x) - \sum_{i=1}^{N} f_i(x)\right| + \left|\sum_{i=1}^{N} f_i(x) - \sum_{i=1}^{N} f_i(x_0)\right| \\ &\quad + \left|\sum_{i=1}^{N} f_i(x_0) - g(x_0)\right| < \epsilon \end{aligned}$$

Therefore, g is continuous at x_0. Then, g is continuous, by the arbitrariness of x_0 and Proposition 3.9.

Let $D := \{x \in X \mid |g(x)| = 1\}$. Clearly, D is a closed set, by Proposition 3.10. Note that $A \cap D = \emptyset$, since $g|_A = f$ and $|f(x)| < 1$, $\forall x \in A$. Then, by Urysohn's Lemma, there exists a continuous function $\bar{g} : X \to [0, 1]$ such that $\bar{g}|_A = 1$ and $\bar{g}|_D = 0$. Define $k : X \to \mathbb{R}$ by $k(x) = \dfrac{\bar{g}(x)g(x)}{1 - \bar{g}(x)\,|g(x)|}$, $\forall x \in X$. By Propositions 3.12 and 3.32 and the fact that $1 - \bar{g}(x)|g(x)| \neq 0$, $\forall x \in X$, we have k is continuous. $\forall x \in A$, $k(x) = \dfrac{g(x)}{1 - |g(x)|} = h(x)$. Hence, $k|_A = h$.

This completes the proof of the theorem. □

Definition 3.58 Let X be a set and $\mathcal{F}$ be a collection of real-valued functions of X. Then, there is the weakest topology on X such that all functions in $\mathcal{F}$ are continuous. This topology is called the *weak topology generated by* $\mathcal{F}$. ◇

Let X be a set, $I := [0, 1] \subset \mathbb{R}$, and $\mathcal{F}$ be a collection of functions of X to I such that $\forall x, y \in X$ with $x \neq y$, $\exists f \in \mathcal{F}$, we have $f(x) \neq f(y)$. Each $f \in \mathcal{F}$

is a point in I^X and $\mathcal{F}$ can be identified with a subset of I^X. The topology that $\mathcal{F}$ inherits as a subspace of I^X is called the *topology of pointwise convergence*. Now, X can be identified with a subset of $I^{\mathcal{F}}$ by, $\forall x \in X$, $\pi_f(x) = f(x)$, $\forall f \in \mathcal{F}$. Then, the topology of X as a subset of $I^{\mathcal{F}}$ is the weak topology generated by $\mathcal{F}$.

Proposition 3.59 *Let $\mathcal{X}$ be a topological space, $I := [0, 1] \subset \mathbb{R}$, and $\mathcal{F}$ be a collection of continuous functions of $\mathcal{X}$ to I such that $\forall x, y \in \mathcal{X}$ with $x \neq y$, $\exists f \in \mathcal{F}$, we have $f(x) \neq f(y)$. Let $E : \mathcal{X} \to I^{\mathcal{F}}$ be the equivalence map given by, $\forall x \in \mathcal{X}$, $\pi_f(E(x)) = f(x)$, $\forall f \in \mathcal{F}$. Then, E is continuous. Furthermore, if $\forall$ closed set $F \subseteq \mathcal{X}$ and $\forall x \in \mathcal{X}$ with $x \notin F$, $\exists f \in \mathcal{F}$ with $f(x) = 1$ and $f|_F = 0$, then $E : \mathcal{X} \to E(\mathcal{X})$ is a homeomorphism.*

Proof $\forall x_0 \in \mathcal{X}$. Fix a basis open set O in $I^{\mathcal{F}}$ with $E(x_0) \in O$. By Proposition 3.25, $O = \prod_{f \in \mathcal{F}} O_f$, where $O_f \in \mathcal{O}_I$ with $\mathcal{O}_I$ being the subset topology on I, $\forall f \in \mathcal{F}$, and $O_f = I$ for all f's except finitely many f's, say $f \in \mathcal{F}_N$. Let $U = \bigcap_{f \in \mathcal{F}_N} f_{\text{inv}}(O_f) \in \mathcal{O}_{\mathcal{X}}$. By $E(x_0) \in O$, we have $x_0 \in U$. $\forall x \in U$, we have $\pi_f(E(x)) \in O_f = I$, $\forall f \in \mathcal{F} \setminus \mathcal{F}_N$, and $\pi_f(E(x)) = f(x) \in O_f$, $\forall f \in \mathcal{F}_N$. Hence, $E(x) \in O$. Then, $E(U) \subseteq O$. Therefore, E is continuous at x_0. By the arbitrariness of x_0 and Proposition 3.9, E is continuous.

Under the additional assumption on $\mathcal{F}$, we need to show that E is a homeomorphism between $\mathcal{X}$ and $E(\mathcal{X})$. $\forall x, y \in \mathcal{X}$ with $x \neq y$, $\exists f \in \mathcal{F}$ such that $\pi_f(E(x)) = f(x) \neq f(y) = \pi_f(E(y))$. Then, $E(x) \neq E(y)$. Hence, $E : \mathcal{X} \to E(\mathcal{X})$ is injective. Clearly, $E : \mathcal{X} \to E(\mathcal{X})$ is surjective. Then, $E : \mathcal{X} \to E(\mathcal{X})$ is bijective and admits inverse $E_{\text{inv}} : E(\mathcal{X}) \to \mathcal{X}$. $\forall x_0 \in \mathcal{X}$, we will show that E_{inv} is continuous at $E(x_0)$. $\forall O \in \mathcal{O}_{\mathcal{X}}$ with $x_0 \in O$. $\widetilde{O}$ is closed and $x_0 \notin \widetilde{O}$. Then, $\exists f_0 \in \mathcal{F}$ such that $f_0(x_0) = 1$ and $f_0|_{\widetilde{O}} = 0$. Define $U = \prod_{f \in \mathcal{F}} U_f \subseteq I^{\mathcal{F}}$ by $U_f = I$, $\forall f \in \mathcal{F} \setminus \{f_0\}$ and $U_{f_0} = (1/2, 1] \in \mathcal{O}_I$. Clearly, U is open in $I^{\mathcal{F}}$. Clearly, $E(x_0) \in U$. $\forall x \in X$ with $E(x) \in U$, we have $\pi_{f_0}(E(x)) = f_0(x) > 1/2$. Then, $x \notin \widetilde{O}$ and $x \in O$. This shows that $E_{\text{inv}}(E(\mathcal{X}) \cap U) \subseteq O$. Hence, E_{inv} is continuous at $E(x_0)$. By the arbitrariness of x_0 and Proposition 3.9, $E_{\text{inv}} : E(\mathcal{X}) \to \mathcal{X}$ is continuous. This implies that $E : \mathcal{X} \to E(\mathcal{X})$ is a homeomorphism.

This completes the proof of the proposition. □

Definition 3.60 A topological space $\mathcal{X}$ is said to be *completely regular* (or $\mathbf{T}_{3\frac{1}{2}}$) if it is Tychonoff and $\forall x_0 \in \mathcal{X}$ and for all closed set $F \subseteq \mathcal{X}$ with $x_0 \notin F$, there exists a continuous real-valued function $f : \mathcal{X} \to [0, 1]$ such that $f(x_0) = 1$ and $f|_F = 0$. ◇

Proposition 3.61 *A normal topological space is completely regular. A completely regular topological space is regular.*

Proof Let $\mathcal{X}$ be a normal topological space. Then, $\mathcal{X}$ is Tychonoff. $\forall x_0 \in \mathcal{X}$ and for all closed set $F \subseteq \mathcal{X}$ with $x_0 \notin F$, we have $\{x_0\}$ is closed, by Proposition 3.34. By Urysohn's Lemma, there exists a continuous real-valued function $f : \mathcal{X} \to [0, 1]$ such that $f(x_0) = 1$ and $f|_F = 0$. Hence, $\mathcal{X}$ is completely regular.

Let $\mathcal{X}$ be a completely regular topological space. Then, $\mathcal{X}$ is Tychonoff. $\forall x_0 \in \mathcal{X}$ and for all closed set $F \subseteq \mathcal{X}$ with $x_0 \notin F$, there exists a continuous real-valued function $f : \mathcal{X} \to [0, 1]$ such that $f(x_0) = 1$ and $f|_F = 0$. Let $O_1 := \{x \in \mathcal{X} \mid f(x) > 1/2\}$ and $O_2 := \{x \in \mathcal{X} \mid f(x) < 1/2\}$. Then, $O_1, O_2 \in \mathcal{O}$ by the continuity of f. Clearly, $x_0 \in O_1$, $F \subseteq O_2$, and $O_1 \cap O_2 = \emptyset$. Hence, $\mathcal{X}$ is regular.

This completes the proof of the proposition. □

Corollary 3.62 *Let $\mathcal{X}$ be a completely regular topological space, $I = [0, 1] \subset \mathbb{R}$, and $\mathcal{F} := \{f : \mathcal{X} \to I \mid f$ is continuous$\}$. Then, the equivalence map: $E : \mathcal{X} \to I^{\mathcal{F}}$ defined by $\pi_f(E(x)) = f(x)$, $\forall x \in \mathcal{X}$, $\forall f \in \mathcal{F}$, is a homeomorphism between $\mathcal{X}$ and $E(\mathcal{X}) \subseteq I^{\mathcal{F}}$.*

Proof Since $\mathcal{X}$ is completely regular, then $\mathcal{X}$ is Tychonoff and all singleton subset of $\mathcal{X}$ is closed. Then, it is easy to check that all assumptions in Proposition 3.59 are satisfied. Then, the result follows. This completes the proof of the corollary. □

3.9 Nets and Convergence

Definition 3.63 A *directed system* is a nonempty set A and a relation on A, $\prec$, such that

(i) $\prec$ is transitive.
(ii) $\forall \alpha, \beta \in A$, $\exists \gamma \in A$ such that $\alpha \prec \gamma$ and $\beta \prec \gamma$.

A *net* is a mapping of a directed system $\mathcal{A} := (A, \prec)$ to a topological space $\mathcal{X}$. $\forall \alpha \in A$, the image is x_α. The net is denoted by $(x_\alpha)_{\alpha \in \mathcal{A}}$, where we have abuse the notation to say $\alpha \in \mathcal{A}$ when $\alpha \in A$. It is understood that the relation for $\mathcal{A}$ is $\prec_{\mathcal{A}}$, where we will ignore the subscript $\mathcal{A}$ if no confusion arises.

A point $x \in \mathcal{X}$ is a *limit* of the net $(x_\alpha)_{\alpha \in \mathcal{A}}$ if $\forall O \in \mathcal{O}$ with $x \in O$, $\exists \alpha_0 \in \mathcal{A} \ni \cdot \forall \alpha \in \mathcal{A}$ with $\alpha_0 \prec \alpha$, we have $x_\alpha \in O$. We also say that $(x_\alpha)_{\alpha \in \mathcal{A}}$ converges to x.

A point $x \in \mathcal{X}$ is a *cluster point* of $(x_\alpha)_{\alpha \in \mathcal{A}}$ if $\forall O \in \mathcal{O}$ with $x \in O$, $\forall \alpha \in \mathcal{A}$, $\exists \beta \in \mathcal{A}$ with $\alpha \prec \beta \ni \cdot\ x_\beta \in O$. ◇

Clearly, a limit point of a net is a cluster point of the net.

In Definition 3.63, we may restrict O to be a basis open set without changing the meaning of the definition.

Example 3.64 $(\mathbb{N}, \le)$ is a directed system. A net over $(\mathbb{N}, \le)$ corresponds to a sequence. ◇

Proposition 3.65 *Let $\mathcal{X}$ be a topological space. Then, the following statements hold:*

(i) *$\mathcal{X}$ is Hausdorff if, and only if, for all net $(x_\alpha)_{\alpha \in \mathcal{A}} \subseteq \mathcal{X}$, there exists at most one limit point for the net. We then write $x = \lim_{\alpha \in \mathcal{A}} x_\alpha$ when the limit exists.*

(ii) *If $\mathcal{X}$ is Hausdorff, any convergent net $(x_\alpha)_{\alpha\in\mathcal{A}} \subseteq \mathcal{X}$ with limit $x \in \mathcal{X}$ has exactly one cluster point, which is x.*

Proof (i) "Only if" Suppose there exists a net $(x_\alpha)_{\alpha\in\mathcal{A}} \subseteq \mathcal{X}$ such that $\exists x_A, x_B \in \mathcal{X}$ with $x_A \neq x_B$ and x_A and x_B are limit points of the net. Since $\mathcal{X}$ is Hausdorff, then $\exists O_1, O_2 \in \mathcal{O}$ such that $x_A \in O_1$, $x_B \in O_2$, and $O_1 \cap O_2 = \emptyset$. Since x_A is the limit of the net, then $\exists \alpha_1 \in \mathcal{A}$, $\forall \alpha \in \mathcal{A}$ with $\alpha_1 \prec \alpha$, we have $x_\alpha \in O_1$. Similarly, since x_B is the limit of the net, then $\exists \alpha_2 \in \mathcal{A}$, $\forall \alpha \in \mathcal{A}$ with $\alpha_2 \prec \alpha$, we have $x_\alpha \in O_2$. Since $\mathcal{A}$ is a directed system, $\exists \alpha_3 \in \mathcal{A}$ such that $\alpha_1 \prec \alpha_3$ and $\alpha_2 \prec \alpha_3$. Then, we have $x_{\alpha_3} \in O_1$ and $x_{\alpha_3} \in O_2$, which implies that $O_1 \cap O_2 \neq \emptyset$, which is a contradiction. Therefore, every net in $\mathcal{X}$ has at most one limit point.

"If" Suppose $\mathcal{X}$ is not Hausdorff. Then, $\exists x_A, x_B \in \mathcal{X}$ with $x_A \neq x_B$ such that $\forall O_A, O_B \in \mathcal{O}$ with $x_A \in O_A$ and $x_B \in O_B$, we have $O_A \cap O_B \neq \emptyset$. Let $\Lambda := \{(O_A, O_B) \,|\, x_A \in O_A \in \mathcal{O}, x_B \in O_B \in \mathcal{O}\}$. Clearly, $(\mathcal{X}, \mathcal{X}) \in \Lambda$, then $\Lambda \neq \emptyset$. Define a relation $\prec$ on Λ by, $\forall (O_{A1}, O_{B1}), (O_{A2}, O_{B2}) \in \Lambda$, we say $(O_{A1}, O_{B1}) \prec (O_{A2}, O_{B2})$ if $O_{A1} \supseteq O_{A2}$ and $O_{B1} \supseteq O_{B2}$. Clearly, $\prec$ is transitive on Λ. $\forall (O_{A1}, O_{B1}), (O_{A2}, O_{B2}) \in \Lambda$, we have $x_A \in O_{A3} := O_{A1} \cap O_{A2} \in \mathcal{O}$ and $x_B \in O_{B3} := O_{B1} \cap O_{B2} \in \mathcal{O}$. Then, we have $(O_{A3}, O_{B3}) \in \Lambda$, $(O_{A1}, O_{B1}) \prec (O_{A3}, O_{B3})$, and $(O_{A2}, O_{B2}) \prec (O_{A3}, O_{B3})$. Hence, $\mathcal{A} := (\Lambda, \prec)$ is a directed system. $\forall (O_A, O_B) \in \Lambda$, $O_A \cap O_B \neq \emptyset$. By Axiom of Choice, we may have a mapping $x_{(O_A, O_B)} \in O_A \cap O_B$, $\forall (O_A, O_B) \in \Lambda$. Then, the net $\left(x_{(O_A,O_B)}\right)_{(O_A,O_B)\in\mathcal{A}} \subseteq \mathcal{X}$. $\forall O_{A1} \in \mathcal{O}$ with $x_A \in O_{A1}$. Fix $O_{B1} := \mathcal{X} \in \mathcal{O}$ with $x_B \in O_{B1}$. Then, $(O_{A1}, O_{B1}) \in \Lambda$. $\forall (O_{A2}, O_{B2}) \in \Lambda$ with $(O_{A1}, O_{B1}) \prec (O_{A2}, O_{B2})$, we have $x_{(O_{A2}, O_{B2})} \in O_{A2} \cap O_{B2} \subseteq O_{A1} \cap O_{B1} = O_{A1}$. Hence, x_A is a limit point of $\left(x_{(O_A,O_B)}\right)_{(O_A,O_B)\in\mathcal{A}}$. $\forall O_{B1} \in \mathcal{O}$ with $x_B \in O_{B1}$. Fix $O_{A1} := \mathcal{X} \in \mathcal{O}$ with $x_A \in O_{A1}$. Then, $(O_{A1}, O_{B1}) \in \Lambda$. $\forall (O_{A2}, O_{B2}) \in \Lambda$ with $(O_{A1}, O_{B1}) \prec (O_{A2}, O_{B2})$, we have $x_{(O_{A2}, O_{B2})} \in O_{A2} \cap O_{B2} \subseteq O_{A1} \cap O_{B1} = O_{B1}$. Hence, x_B is a limit point of $\left(x_{(O_A,O_B)}\right)_{(O_A,O_B)\in\mathcal{A}}$. This contradicts with the assumption that every net has at most one limit point. Therefore, $\mathcal{X}$ is Hausdorff.

(ii) Let $\mathcal{X}$ be Hausdorff and net $(x_\alpha)_{\alpha\in\mathcal{A}} \subseteq \mathcal{X}$ satisfy $\lim_{\alpha\in\mathcal{A}} x_\alpha = x \in \mathcal{X}$. Clearly, x is a cluster point of the net $(x_\alpha)_{\alpha\in\mathcal{A}}$. Let $y \in \mathcal{X}$ with $y \neq x$ be another cluster point of the net $(x_\alpha)_{\alpha\in\mathcal{A}}$. Then, there exist $O_1, O_2 \in \mathcal{O}$ with $x \in O_1$ and $y \in O_2$ and $O_1 \cap O_2 = \emptyset$. Since x is the limit of the net, then $\exists \alpha_1 \in \mathcal{A}$ such that $\forall \alpha \in \mathcal{A}$ with $\alpha_1 \prec \alpha$, we have $x_\alpha \in O_1$. Then, $x_\alpha \notin O_2$. This contradicts the definition that y is cluster point of $(x_\alpha)_{\alpha\in\mathcal{A}}$. Hence, the net has exactly one cluster point, which is x.

This completes the proof of the proposition. □

Proposition 3.66 *Let $\mathcal{X}$ and $\mathcal{Y}$ be topological spaces, $D \subseteq \mathcal{X}$ with subset topology $\mathcal{O}_D$, and $f : D \to \mathcal{Y}$. Then, the following are equivalent:*

(i) *f is continuous at $x_0 \in D$.*
(ii) *$\forall$ net $(x_\alpha)_{\alpha\in\mathcal{A}} \subseteq D$ with x_0 as a limit point, we have that the net $(f(x_\alpha))_{\alpha\in\mathcal{A}}$ has a limit point $f(x_0)$.*
(iii) *$\forall$ net $(x_\alpha)_{\alpha\in\mathcal{A}} \subseteq D$ with x_0 as a cluster point, we have that the net $(f(x_\alpha))_{\alpha\in\mathcal{A}} \subseteq \mathcal{Y}$ has a cluster point $f(x_0)$.*

Proof (i) $\Rightarrow$ (ii). Fix a net $(x_\alpha)_{\alpha\in\mathcal{A}} \subseteq D$ with $x_0 \in D$ as a limit point. $\forall O_Y \in \mathcal{O}_\mathcal{Y}$ with $f(x_0) \in O_Y$. By the continuity of f at x_0, $\exists O_X \in \mathcal{O}_\mathcal{X}$ with $x_0 \in O_X$ such that $f(O_X) \subseteq O_Y$. Since x_0 is a limit point of $(x_\alpha)_{\alpha\in\mathcal{A}}$, then $\exists\alpha_0 \in \mathcal{A}$ such that, $\forall\alpha \in \mathcal{A}$ with $\alpha_0 \prec \alpha$, we have $x_\alpha \in O_X$. Then, $f(x_\alpha) \in O_Y$. Hence, we have $f(x_0)$ is a limit point of $(f(x_\alpha))_{\alpha\in\mathcal{A}}$.

(ii) $\Rightarrow$ (i). Suppose f is not continuous at $x_0 \in D$. Then, $\exists O_{Y0} \in \mathcal{O}_\mathcal{Y}$ with $f(x_0) \in O_{Y0}$ such that, $\forall O_X \in \mathcal{O}_\mathcal{X}$ with $x_0 \in O_X$, we have $f(O_X) \not\subseteq O_{Y0}$. Let $\mathcal{M} := \{O \in \mathcal{O}_\mathcal{X} \mid x_0 \in O\}$. Clearly, $\mathcal{X} \in \mathcal{M}$ and $\mathcal{M} \neq \emptyset$. Define a relation $\prec$ on $\mathcal{M}$ by, $\forall O_1, O_2 \in \mathcal{M}$, we say $O_1 \prec O_2$ if $O_1 \supseteq O_2$. Clearly, $\prec$ is transitive on $\mathcal{M}$. $\forall O_1, O_2 \in \mathcal{M}$, let $O_3 = O_1 \cap O_2 \in \mathcal{O}_\mathcal{X}$ and $x_0 \in O_3$. Then, $O_3 \in \mathcal{M}$, $O_1 \prec O_3$, and $O_2 \prec O_3$. Hence, $\mathcal{A} := (\mathcal{M}, \prec)$ is a directed system. $\forall O \in \mathcal{M}$, $f(O) \setminus O_{Y0} \neq \emptyset$. By Axiom of Choice, we may define a net $(x_O)_{O\in\mathcal{A}}$ by $x_O \in O \cap D$ with $f(x_O) \notin O_{Y0}$. Clearly, x_0 is a limit point of $(x_O)_{O\in\mathcal{A}}$. Yet, $f(x_0) \in O_{Y0} \in \mathcal{O}_\mathcal{Y}$ and $f(x_O) \notin O_{Y0}$, $\forall O \in \mathcal{M}$. Then, $f(x_0)$ is not a limit point of the net $(f(x_O))_{O\in\mathcal{A}}$. This contradicts with the assumption. Therefore, f is continuous at x_0.

(i) $\Rightarrow$ (iii). Fix a net $(x_\alpha)_{\alpha\in\mathcal{A}} \subseteq D$ with x_0 as a cluster point. $\forall O_Y \in \mathcal{O}_\mathcal{Y}$ with $f(x_0) \in O_Y$, by the continuity of f at x_0, $\exists U \in \mathcal{O}_\mathcal{X}$ with $x_0 \in U$ such that $f(U) \subseteq O_Y$. By Definition 3.63, $\forall\alpha \in \mathcal{A}$, $\exists\alpha_0 \in \mathcal{A}$ with $\alpha \prec \alpha_0$, $x_{\alpha_0} \in U$. Then, $f(x_{\alpha_0}) \in O_Y$. Hence, $f(x_0)$ is a cluster point of the net $(f(x_\alpha))_{\alpha\in\mathcal{A}}$.

(iii) $\Rightarrow$ (i). Suppose f is not continuous at x_0. Let $\mathcal{M} := \{O \in \mathcal{O}_\mathcal{X} \mid x_0 \in O\}$. Clearly, $\mathcal{A} := (\mathcal{M}, \supseteq)$ is a directed system. $\exists O_{Y0} \in \mathcal{O}_\mathcal{Y}$ with $f(x_0) \in O_{Y0}$ such that $\forall U \in \mathcal{M}$, we have $f(U) \not\subseteq O_{Y0}$. By Axiom of Choice, we may assign to each $U \in \mathcal{M}$ an $x_U \in U \cap D$ such that $f(x_U) \in \widetilde{O_{Y0}}$. Consider the net $(x_U)_{U\in\mathcal{A}} \subseteq D$. Clearly, x_0 is a limit point of the net and therefore is a cluster point of the net. Consider the net $(f(x_U))_{U\in\mathcal{A}}$. For the open set $O_{Y0} \ni f(x_0)$, $\forall U \in \mathcal{A}$, $f(x_U) \in \widetilde{O_{Y0}}$. Then, $f(x_0)$ is not a cluster point of $(f(x_U))_{U\in\mathcal{A}}$. This contradicts with the assumption. Therefore, f must be continuous at x_0.

This completes the proof of the proposition. □

Proposition 3.67 *Let $(X_\alpha, \mathcal{O}_\alpha)$ be a topological space, $\forall\alpha \in \Lambda$, where Λ is an index set. Let $(X, \mathcal{O})$ be the product space $\prod_{\alpha\in\Lambda}(X_\alpha, \mathcal{O}_\alpha)$. Let $(x_\beta)_{\beta\in\mathcal{A}} \subseteq X$ be a net. Then, $x_0 \in X$ is a limit point of $(x_\beta)_{\beta\in\mathcal{A}}$ if, and only if, $\forall\alpha \in \Lambda$, $\pi_\alpha(x_0) \in X_\alpha$ is a limit point of $(\pi_\alpha(x_\beta))_{\beta\in\mathcal{A}}$.*

Proof "Only if" $\forall\alpha \in \Lambda$, by Proposition 3.27, π_α is continuous. Then, π_α is continuous at $x_0 \in X$, by Proposition 3.9. By Proposition 3.66, $\pi_\alpha(x_0)$ is a limit point of the net $(\pi_\alpha(x_\beta))_{\beta\in\mathcal{A}}$.

"If" Suppose that $x_0 \in \mathcal{X}$ is not a limit point of the net $(x_\beta)_{\beta\in\mathcal{A}}$. Then, $\exists$ a basis open set $B \in \mathcal{O}$ with $x_0 \in B$ such that, $\forall\beta_0 \in \mathcal{A}$, $\exists\beta \in \mathcal{A}$ with $\beta_0 \prec \beta$, we have $x_\beta \notin B$. Then, $B = \prod_{\alpha\in\Lambda} O_\alpha$, $O_\alpha \in \mathcal{O}_\alpha$, $\forall\alpha \in \Lambda$, and $O_\alpha = X_\alpha$ for all α's except finitely many α's, say $\alpha \in \Lambda_N$. Then, $\forall\beta_0 \in \mathcal{A}$, $\exists\beta \in \mathcal{A}$ with $\beta_0 \prec \beta$, we have $x_\beta \notin B$. This implies that $\pi_{\alpha_\beta}(x_\beta) \notin O_{\alpha_\beta}$, for some $\alpha_\beta \in \Lambda_N$. Then, by an argument of contradiction, we may show that $\exists\alpha_0 \in \Lambda_N$ such that, $\forall\beta_0 \in \mathcal{A}$, $\exists\beta \in \mathcal{A}$ with $\beta_0 \prec \beta$, we have $\pi_{\alpha_0}(x_\beta) \notin O_{\alpha_0}$. Hence, $\pi_{\alpha_0}(x_0) \in O_{\alpha_0}$ is not the limit of the net $\left(\pi_{\alpha_0}(x_\beta)\right)_{\beta\in\mathcal{A}}$. This contradicts with the assumption. Hence, we have x_0 is a limit point of $\left(x_\beta\right)_{\beta\in\mathcal{A}}$.

This completes the proof of the proposition. □

Proposition 3.68 *Let $(X, \mathcal{O})$ be a topological space, $E \subseteq X$, and $x \in X$. $x \in \overline{E}$ if, and only if, $\exists$ a net $(x_\alpha)_{\alpha\in\mathcal{A}} \subseteq E$ such that x is a limit point of the net.*

Proof "Only if" Let $\mathcal{M} := \{O \in \mathcal{O} \mid x \in O\}$. Clearly, $X \in \mathcal{M}$, then $\mathcal{M} \neq \emptyset$. Clearly, $\mathcal{A} := (\mathcal{M}, \supseteq)$ is a directed system. Since $x \in \overline{E}$, then, by Proposition 3.3, $\forall O \in \mathcal{A}$, $O \cap E \neq \emptyset$. By Axiom of Choice, $\exists$ a net $(x_O)_{O\in\mathcal{A}} \subseteq E$ such that $x_O \in O \cap E$, $\forall O \in \mathcal{A}$. $\forall O \in \mathcal{O}$ with $x \in O$, then $O \in \mathcal{A}$. $\forall O_1 \in \mathcal{A}$ with $O \supseteq O_1$, we have $x_{O_1} \in O_1 \cap E \subseteq O$. Hence, x is a limit point of $(x_O)_{O\in\mathcal{A}}$.

"If" Let $(x_\alpha)_{\alpha\in\mathcal{A}} \subseteq E$ be the net such that x is a limit point of the net. $\forall O \in \mathcal{O}$ with $x \in O$, $\exists \alpha_0 \in \mathcal{A}$, $\forall \alpha \in \mathcal{A}$ with $\alpha_0 \prec \alpha$, we have $x_\alpha \in E \cap O$. Since $(x_\alpha)_{\alpha\in\mathcal{A}}$ is a net, then $\exists \alpha_1 \in \mathcal{A}$ with $\alpha_0 \prec \alpha_1$. Then, $x_{\alpha_1} \in O \cap E \neq \emptyset$. By Proposition 3.3, $x \in \overline{E}$.

This completes the proof of the proposition. □

Definition 3.69 Let $(X, \mathcal{O})$ be a topological space, $\mathcal{A} := (A, \prec)$ be a directed system, and $(x_\alpha)_{\alpha\in\mathcal{A}} \subseteq X$ be a net. Let $A_s \subseteq A$ be a subset with the same relation $\prec$ as A such that $\forall \alpha \in \mathcal{A}$, $\exists \alpha_s \in \mathcal{A}_s$ such that $\alpha \prec \alpha_s$. Then, $\mathcal{A}_s := (A_s, \prec)$ is a directed system and $(x_\alpha)_{\alpha\in\mathcal{A}_s}$ is a net, which is called a *subnet* of $(x_\alpha)_{\alpha\in\mathcal{A}}$. ◇

Proposition 3.70 *Let $(X, \mathcal{O})$ be a topological space and $(x_\alpha)_{\alpha\in\mathcal{A}} \subseteq X$ be a net. Then, $x_0 \in X$ is a limit point of $(x_\alpha)_{\alpha\in\mathcal{A}}$ if, and only if, any subnet $(x_\alpha)_{\alpha\in\mathcal{A}_s}$ has a limit point x_0.*

Proof "Only if" Since $x_0 \in X$ is a limit point of $(x_\alpha)_{\alpha\in\mathcal{A}}$, then $\forall O \in \mathcal{O}$ with $x_0 \in O$, $\exists \alpha_0 \in \mathcal{A}$ such that, $\forall \alpha \in \mathcal{A}$ with $\alpha_0 \prec \alpha$, we have $x_\alpha \in O$. Let $(x_\alpha)_{\alpha\in\mathcal{A}_s}$ be a subnet. Then, $\exists \alpha_{s0} \in \mathcal{A}_s$ such that $\alpha_0 \prec \alpha_{s0}$. $\forall \alpha_s \in \mathcal{A}_s$ with $\alpha_{s0} \prec \alpha_s$, we have $\alpha_0 \prec \alpha_s$ and $x_{\alpha_s} \in O$. Hence, x_0 is a limit point of the subnet.

"If" Since $(x_\alpha)_{\alpha\in\mathcal{A}}$ is a subnet of itself, then it has limit x_0.

This completes the proof of the proposition. □

A cluster point of a subnet is clearly a cluster point of the net.

Proposition 3.71 *Let $(X, \mathcal{O})$ be a topological space and $(x_\alpha)_{\alpha\in\mathcal{A}} \subseteq X$ be a net. Then, $x_0 \in X$ is a limit point of $(x_\alpha)_{\alpha\in\mathcal{A}}$ if, and only if, for every subnet $(x_\alpha)_{\alpha\in\mathcal{A}_s}$ of $(x_\alpha)_{\alpha\in\mathcal{A}}$, there exists a subsubnet $(x_\alpha)_{\alpha\in\mathcal{A}_{ss}}$ that has a limit point x_0.*

Proof "Sufficiency" We assume that every subnet $(x_\alpha)_{\alpha\in\mathcal{A}_s}$ of $(x_\alpha)_{\alpha\in\mathcal{A}}$, there exists a subsubnet $(x_\alpha)_{\alpha\in\mathcal{A}_{ss}}$ that has a limit point x_0. We will prove the result using an argument of contradiction. Suppose x_0 is not a limit point of $(x_\alpha)_{\alpha\in\mathcal{A}}$. Then, $\exists O_0 \in \mathcal{O}$ with $x_0 \in O_0$, $\forall \alpha_0 \in \mathcal{A}$, $\exists \alpha \in \mathcal{A}$ with $\alpha_0 \prec \alpha$ such that $x_\alpha \in \widetilde{O_0}$. Define $\mathcal{A}_s := \left(\left\{\alpha \in \mathcal{A} \,\middle|\, x_\alpha \in \widetilde{O_0}\right\}, \prec\right)$. Clearly, $(x_\alpha)_{\alpha\in\mathcal{A}_s}$ is a subnet of $(x_\alpha)_{\alpha\in\mathcal{A}}$. Any subsubnet $(x_\alpha)_{\alpha\in\mathcal{A}_{ss}}$ of $(x_\alpha)_{\alpha\in\mathcal{A}_s}$, $\forall \alpha_{ss} \in \mathcal{A}_{ss}$, we have $x_{\alpha_{ss}} \in \widetilde{O_0}$. Then, x_0 is not a limit of $(x_\alpha)_{\alpha\in\mathcal{A}_{ss}}$. This contradicts the assumption. Therefore, x_0 is a limit point of $(x_\alpha)_{\alpha\in\mathcal{A}}$.

"Necessity" Let x_0 be a limit point of $(x_\alpha)_{\alpha\in\mathcal{A}}$ and $(x_\alpha)_{\alpha\in\mathcal{A}_s}$ be a subnet. By Proposition 3.70, x_0 is a limit point of $(x_\alpha)_{\alpha\in\mathcal{A}_s}$, which is a subsubnet of itself. Then, the result holds.

This completes the proof of the proposition. □

Definition 3.72 Let $\mathcal{X} := (X, \mathcal{O}_X)$ and $\mathcal{Y} := (Y, \mathcal{O}_Y)$ be topological spaces, $D \subseteq \mathcal{X}$, $f : D \to \mathcal{Y}$, and $x_0 \in \mathcal{X}$ be an accumulation point of D. $y_0 \in \mathcal{Y}$ is said to be a *limit point of* $f(x)$ *as* $x \to x_0$ if $\forall O_Y \in \mathcal{O}_Y$ with $y_0 \in O_Y$, $\exists U \in \mathcal{O}_X$ with $x_0 \in U$ such that $f(U \setminus \{x_0\}) = f((D \cap U) \setminus \{x_0\}) \subseteq O_Y$. We will also say that $f(x)$ *converges to* y_0 *as* $x \to x_0$. ◇

When basis are available on topological spaces $\mathcal{X}$ and $\mathcal{Y}$, in Definition 3.72, we may restrict the open sets O_Y and U to be basis open sets without changing the meaning of the definition.

Proposition 3.73 *Let $\mathcal{X} := (X, \mathcal{O}_X)$ and $\mathcal{Y} := (Y, \mathcal{O}_Y)$ be topological spaces, $D \subseteq \mathcal{X}$, $f : D \to \mathcal{Y}$, and $x_0 \in \mathcal{X}$ be an accumulation point of D. If $\mathcal{Y}$ is Hausdorff, then there is at most one limit point of $f(x)$ as $x \to x_0$. In this case, we will write* $\lim_{x\to x_0} f(x) = y_0 \in \mathcal{Y}$ *when the limit exists.*

Proof Suppose $f(x)$ admits limit points $y_A, y_B \in \mathcal{Y}$ as $x \to x_0$ with $y_A \neq y_B$. Since $\mathcal{Y}$ is Hausdorff, then $\exists U_A, U_B \in \mathcal{O}_Y$ such that $y_A \in U_A$, $y_B \in U_B$, and $U_A \cap U_B = \emptyset$. Since y_A is a limit point of $f(x)$ as $x \to x_0$, then $\exists V_A \in \mathcal{O}_X$ with $x_0 \in V_A$ such that $f(V_A \setminus \{x_0\}) \subseteq U_A$. Since y_B is a limit point of $f(x)$ as $x \to x_0$, then $\exists V_B \in \mathcal{O}_X$ with $x_0 \in V_B$ such that $f(V_B \setminus \{x_0\}) \subseteq U_B$. Then, $x_0 \in V := V_A \cap V_B \in \mathcal{O}$. Since x_0 is an accumulation point of D, then $\exists x \in (D \cap V) \setminus \{x_0\}$. Then, we have $f(x) \in U_A$ since $x \in (D \cap V_A) \setminus \{x_0\}$ and $f(x) \in U_B$ since $x \in (D \cap V_B) \setminus \{x_0\}$. Then, $f(x) \in U_A \cap U_B \neq \emptyset$. This contradicts with $U_A \cap U_B = \emptyset$. Hence, the result holds. This completes the proof of the proposition. □

Proposition 3.74 *Let $\mathcal{X} := (X, \mathcal{O}_X)$ and $\mathcal{Y} := (Y, \mathcal{O}_Y)$ be topological spaces, $D \subseteq X$ with subset topology $\mathcal{O}_D$, $f : D \to \mathcal{Y}$, and $x_0 \in D$. Then, the following statements are equivalent:*

(i) *f is continuous at x_0.*
(ii) *If x_0 is an accumulation point of D, then $f(x_0)$ is a limit point of $f(x)$ as $x \to x_0$.*

Proof (i) ⇒ (ii). This is straightforward.

(ii) ⇒ (i). We will distinguish two exhaustive and mutually exclusive cases: Case 1: x_0 is not an accumulation point of D, and Case 2: x_0 is an accumulation point of D. Case 1: x_0 is not an accumulation point of D. $\exists V \in \mathcal{O}_X$ with $x_0 \in V$ such that $V \cap D = \{x_0\}$. $\forall U \in \mathcal{O}_Y$ with $f(x_0) \in U$, we have $f(V) = \{f(x_0)\} \subseteq U$. Hence, f is continuous at x_0.

Case 2: x_0 is an accumulation point of D. $\forall U \in \mathcal{O}_Y$ with $f(x_0) \in U$, $\exists V \in \mathcal{O}_X$ with $x_0 \in V$ such that $f(V \setminus \{x_0\}) \subseteq U$. Then, we have $f(V) \subseteq U$. Hence, f is continuous at x_0.

In both cases, f is continuous at x_0.

This completes the proof of the proposition. □

Proposition 3.75 *Let $\mathcal{X} := (X, \mathcal{O}_X)$, $\mathcal{Y} := (Y, \mathcal{O}_Y)$, and $\mathcal{Z} := (Z, \mathcal{O}_Z)$ be topological spaces, $D \subseteq \mathcal{X}$, $f : D \to \mathcal{Y}$, $x_0 \in \mathcal{X}$ be an accumulation point of D, $y_0 \in \mathcal{Y}$ be a limit point of $f(x)$ as $x \to x_0$, and $g : \mathcal{Y} \to \mathcal{Z}$ be continuous at y_0. Then, $g(y_0) \in \mathcal{Z}$ is a limit point of $g(f(x))$ as $x \to x_0$. When $\mathcal{Y}$ and $\mathcal{Z}$ are Hausdorff, then we may write* $\lim_{x \to x_0} g(f(x)) = g(\lim_{x \to x_0} f(x))$.

Proof $\forall O_Z \in \mathcal{O}_Z$ with $g(y_0) \in O_Z$, by the continuity of g at y_0, $\exists O_Y \in \mathcal{O}_Y$ with $y_0 \in O_Y$ such that $g(O_Y) \subseteq O_Z$. Since y_0 is the limit of $f(x)$ as $x \to x_0$, then $\exists O_X \in \mathcal{O}_X$ with $x_0 \in O_X$ such that $f(O_X \setminus \{x_0\}) \subseteq O_Y$. Then, $g(f(O_X \setminus \{x_0\})) \subseteq O_Z$. Hence, $g(f(x))$ converges to $g(y_0)$ as $x \to x_0$. This completes the proof of the proposition. □

Proposition 3.76 *Let $\mathcal{X}$ be a topological space, $\bar{D} \subseteq \mathcal{X}$, $x_0 \in \mathcal{X}$ be an accumulation point of $\bar{D}$, $\mathcal{Y}$ and $\mathcal{Z}$ be Hausdorff topological spaces, $D \subseteq \mathcal{Y}$, $y_0 \in \mathcal{Y}$ be an accumulation point of D, $f : \bar{D} \to D$, and $g : D \to \mathcal{Z}$. Assume that*

(i) *$\exists O_0 \in \mathcal{O}_{\mathcal{X}}$ with $x_0 \in O_0$ such that $f(O_0 \setminus \{x_0\}) \subseteq D \setminus \{y_0\}$.*
(ii) *$\lim_{x \to x_0} f(x) = y_0$ and $\lim_{y \to y_0} g(y) = z_0 \in \mathcal{Z}$.*

Then, $\lim_{x \to x_0} g(f(x)) = z_0$.

Proof $\forall O_Z \in \mathcal{O}_{\mathcal{Z}}$ with $z_0 \in O_Z$, by $\lim_{y \to y_0} g(y) = z_0$, $\exists O_Y \in \mathcal{O}_{\mathcal{Y}}$ with $y_0 \in O_Y$ such that $g(O_Y \setminus \{y_0\}) = g((O_Y \cap D) \setminus \{y_0\}) \subseteq O_Z$. By $\lim_{x \to x_0} f(x) = y_0$, $\exists O_X \in \mathcal{O}_{\mathcal{X}}$ with $x_0 \in O_X$ such that $f(O_X \setminus \{x_0\}) = f((O_X \cap \bar{D}) \setminus \{x_0\}) \subseteq O_Y$. Let $O_1 := O_0 \cap O_X \in \mathcal{O}_{\mathcal{X}}$. Clearly, $x_0 \in O_1$. Then, $\forall x \in (O_1 \cap \bar{D}) \setminus \{x_0\}$, we have $f(x) \in O_Y \cap (D \setminus \{y_0\}) = (O_Y \cap D) \setminus \{y_0\}$ and $g(f(x)) \in O_Z$. Hence, $g(f((O_1 \cap \bar{D}) \setminus \{x_0\})) \subseteq O_Z$. Hence, $\lim_{x \to x_0} g(f(x)) = z_0$. This completes the proof of the proposition. □

Proposition 3.77 *Let $\mathcal{X}$, $\mathcal{Y}$, and $\mathcal{Z}$ be Hausdorff topological spaces, $\bar{D} \subseteq \mathcal{X}$, $x_0 \in \mathcal{X}$ be an accumulation point of $\bar{D}$, $D \subseteq \mathcal{Y}$, $y_0 \in \mathcal{Y}$ be an accumulation point of D, $f : \bar{D} \to D$ be bijective, and $g : D \to \mathcal{Z}$. Assume that* $\lim_{x \to x_0} f(x) = y_0$ *and* $\lim_{y \to y_0} f_{\text{inv}}(y) = x_0$. *Then,* $\lim_{x \to x_0} g(f(x)) = \lim_{y \to y_0} g(y)$ *whenever one of the limits exists in $\mathcal{Z}$.*

Proof We will prove the result by distinguishing two exhaustive cases: Case 1: $\lim_{y \to y_0} g(y) = z_0 \in \mathcal{Z}$, and Case 2: $\lim_{x \to x_0} g(f(x)) = z_0 \in \mathcal{Z}$.

Case 1: $\lim_{y \to y_0} g(y) = z_0 \in \mathcal{Z}$. We will further distinguish three exhaustive and mutually exclusive subcases: Case 1a: $y_0 \notin D$, Case 1b: $y_0 \in D$ and $x_0 = f_{\text{inv}}(y_0)$, and Case 1c: $y_0 \in D$ and $x_0 \neq \bar{x}_0 := f_{\text{inv}}(y_0)$. Case 1a: $y_0 \notin D$. Then, (i) of Proposition 3.76 is satisfied with $O_0 := \mathcal{X}$. By Proposition 3.76, we have $\lim_{x \to x_0} g(f(x)) = z_0 \in \mathcal{Z}$. Case 1b: $y_0 \in D$ and $x_0 = f_{\text{inv}}(y_0)$. Let $O_0 := \mathcal{X}$. $\forall x \in (O_0 \cap \bar{D}) \setminus \{x_0\}$, we have $D \ni f(x) \neq f(x_0) = y_0$ by f being bijective. This implies that $f(O_0 \setminus \{x_0\}) \subseteq D \setminus \{y_0\}$. Then, $\lim_{x \to x_0} g(f(x)) = z_0 \in \mathcal{Z}$ by Proposition 3.76. Case 1c: $y_0 \in D$ and $x_0 \neq \bar{x}_0 := f_{\text{inv}}(y_0)$. By $\mathcal{X}$ being Hausdorff, $\exists O_0 \in \mathcal{O}_{\mathcal{X}}$ such that $x_0 \in O_0$ and $\bar{x}_0 \notin O_0$. This leads to $\forall x \in (O_0 \cap \bar{D}) \setminus \{x_0\}$, we have $D \ni f(x) \neq f(\bar{x}_0) = y_0$ by f being bijective. This implies that $f(O_0 \setminus$

$\{x_0\}) \subseteq D \setminus \{y_0\}$. Then, $\lim_{x \to x_0} g(f(x)) = z_0 \in \mathcal{Z}$ by Proposition 3.76. Hence, $\lim_{x \to x_0} g(f(x)) = \lim_{y \to y_0} g(y) = z_0 \in \mathcal{Z}$ in all three subcases. Hence, the result holds in this case.

Case 2 $\lim_{x \to x_0} g(f(x)) = z_0 \in \mathcal{Z}$. Define $h : \bar{D} \to \mathcal{Z}$ by $h(x) = g(f(x))$, $\forall x \in \bar{D}$. Then, $\lim_{x \to x_0} h(x) = z_0$. By Case 1, we have $\lim_{y \to y_0} h(f_{\text{inv}}(y)) = z_0 \in \mathcal{Z}$. Then, $\lim_{y \to y_0} g(f(f_{\text{inv}}(y))) = \lim_{y \to y_0} g(y) = z_0$. Hence, the result holds in this case.

This completes the proof of the proposition. □

Example 3.78 Let $g : \mathbb{R} \to \mathbb{R}$. It is desired to calculate $\lim_{y \to +\infty} g(y)$. We will apply Proposition 3.77 to this calculation. Take $\mathcal{X} = \mathbb{R}$, $\mathcal{Y} = \mathbb{R}_e$, and $\mathcal{Z} = \mathbb{R}_e$. Let $\bar{D} := (-\infty, -1] \cup (0, +\infty) \subset \mathbb{R} = \mathcal{X}$, $x_0 = 0$, $D := \mathbb{R} \subset \mathbb{R}_e = \mathcal{Y}$, and $y_0 = +\infty$. Then, $g : D \to \mathcal{Z}$. Define $f : \bar{D} \to D$ by $f(x) = \begin{cases} 1/x & x > 0 \\ x+1 & x \leq -1 \end{cases}$, $\forall x \in \bar{D}$. Clearly, f is bijective with $f_{\text{inv}} : D \to \bar{D}$ given by $f_{\text{inv}}(y) = \begin{cases} 1/y & y > 0 \\ y-1 & y \leq 0 \end{cases}$, $\forall y \in D$. Clearly, $\lim_{x \to 0} f(x) = +\infty$ and $\lim_{y \to +\infty} f_{\text{inv}}(y) = 0$. Then, by Proposition 3.77, $\lim_{y \to +\infty} g(y) = \lim_{x \to 0} g(f(x))$ whenever one of the limits exists in $\mathcal{Z}$. ◇

Proposition 3.79 *Let $\mathcal{X} := (X, \mathcal{O}_X)$ and $\mathcal{Y} := (Y, \mathcal{O}_Y)$ be topological spaces, $D \subseteq \mathcal{X}$, $x_0 \in \mathcal{X}$ be an accumulation point of D, $y_0 \in \mathcal{Y}$, and $f : D \to \mathcal{Y}$. Then, the following statements are equivalent:*

(i) *y_0 is a limit point of $f(x)$ as $x \to x_0$.*
(ii) *$\forall$ net $(x_\alpha)_{\alpha \in \mathcal{A}} \subseteq D \setminus \{x_0\}$ with x_0 as a limit, we have y_0 is a limit point of the net $(f(x_\alpha))_{\alpha \in \mathcal{A}}$.*

Proof (i) ⇒ (ii). Fix any net $(x_\alpha)_{\alpha \in \mathcal{A}} \subseteq D \setminus \{x_0\}$ with x_0 as a limit. $\forall U \in \mathcal{O}_Y$ with $y_0 \in U$, by (i), $\exists V \in \mathcal{O}_X$ with $x_0 \in V$ such that $f(V \setminus \{x_0\}) \subseteq U$. $\exists \alpha_0 \in \mathcal{A}$ such that $\forall \alpha \in \mathcal{A}$ with $\alpha_0 \prec \alpha$, we have $x_\alpha \in V$. Then, $x_\alpha \in (D \cap V) \setminus \{x_0\}$ and $f(x_\alpha) \in U$. Hence, $(f(x_\alpha))_{\alpha \in \mathcal{A}}$ has a limit y_0.

(ii) ⇒ (i). Suppose y_0 is not a limit point of $f(x)$ as $x \to x_0$. Then, $\exists U_0 \in \mathcal{O}_Y$ with $y_0 \in U_0$, $\forall V \in \mathcal{O}_X$ with $x_0 \in V$, we have $f(V \setminus \{x_0\}) \nsubseteq U_0$. Then, $\exists x_V \in (D \cap V) \setminus \{x_0\}$ such that $f(x_V) \notin U_0$. Let $M := \{V \in \mathcal{O}_X \mid x_0 \in V\}$ and $\mathcal{A} := (M, \supseteq)$. Clearly, $\mathcal{A}$ is a directed system. By Axiom of Choice, we may construct a net $(x_V)_{V \in \mathcal{A}} \subseteq D \setminus \{x_0\}$. Clearly, x_0 is a limit point of this net. But, $\forall V \in \mathcal{A}$, $f(x_V) \notin U_0$. Hence, y_0 is not a limit point of the net $(f(x_V))_{V \in \mathcal{A}}$. This contradicts with (ii). Therefore, y_0 is a limit point of $f(x)$ as $x \to x_0$.

This completes the proof of the proposition. □

Example 3.80 The extended real line is $\mathbb{R}_e := \mathbb{R} \cup \{\infty\} \cup \{-\infty\}$. We assume that $\forall x \in \mathbb{R}$, $-\infty < x < \infty$. We define the usual operations: $\forall x \in \mathbb{R}$,

$$x + \infty = \infty \qquad x - \infty = -\infty$$

$$x \cdot \infty = \infty; \quad \text{if } x > 0$$

$$x \cdot (-\infty) = -\infty; \quad \text{if } x > 0$$

$$\infty + \infty = \infty \qquad -\infty - \infty = -\infty$$

$$\infty \cdot \infty = \infty \qquad \infty \cdot (-\infty) = -\infty$$

The operations $\infty - \infty$ and $0 \cdot \infty$ are undefined.

On $\mathbb{R}_e$, we introduce the countable collection of subsets of $\mathbb{R}_e$, $\mathcal{B}_{\mathbb{R}_e}$, as follows. $\emptyset, \mathbb{R}, \mathbb{R}_e \in \mathcal{B}_{\mathbb{R}_e}$. $\forall r_1, r_2 \in \mathbb{Q}$ with $r_1 < r_2$, $[-\infty, r_1)$, (r_1, r_2), $(r_2, +\infty] \in \mathcal{B}_{\mathbb{R}_e}$.

By Proposition 3.18, it is easy to show that $\mathcal{B}_{\mathbb{R}_e}$ is a basis for a topology on $\mathbb{R}_e$. This topology is denoted $\mathcal{O}_{\mathbb{R}_e}$, which is the usual topology on $\mathbb{R}_e$. It is easy to show that $(\mathbb{R}_e, \mathcal{O}_{\mathbb{R}_e})$ is an arcwise connected second countable Hausdorff topological space.

An important property of $\mathbb{R}_e$ is that $\forall E \subseteq \mathbb{R}_e$, $\sup_{x\in E} x \in \mathbb{R}_e$ and $\inf_{x\in E} x \in \mathbb{R}_e$.

It is easy to see that $\mathbb{R}$ as a subset of $\mathbb{R}_e$ admits the subset topology $\mathcal{O}$ that equals the usual topology $\mathcal{O}_{\mathbb{R}}$ on $\mathbb{R}$. ◇

Proposition 3.81 *Let X be a set, $f_1 : X \to \mathbb{R}$, $f_2 : X \to \mathbb{R}$, $g_1 : X \to \mathbb{R}_e$, and $g_2 : X \to \mathbb{R}_e$. Assume that $g_1(x) + g_2(x) \in \mathbb{R}_e$ is well-defined, $\forall x \in X$. Then,*

(i) $\sup_{x\in X} g_1(x) \le M \in \mathbb{R}_e$ *if, and only if,* $\forall x \in X$, $g_1(x) \le M$.
(ii) $\inf_{x\in X} g_1(x) \ge m \in \mathbb{R}_e$ *if, and only if,* $\forall x \in X$, $g_1(x) \ge m$.
(iii) $\sup_{x\in X} g_1(x) > M \in \mathbb{R}_e$ *if, and only if,* $\exists x \in X$, $g_1(x) > M$.
(iv) $\inf_{x\in X} g_1(x) < m \in \mathbb{R}_e$ *if, and only if,* $\exists x \in X$, $g_1(x) < m$.
(v) $\sup_{x\in X}(-g_1)(x) = -\inf_{x\in X} g_1(x)$.
(vi) $\sup_{x\in X}(f_1 + f_2)(x) \le \sup_{x\in X} f_1(x) + \sup_{x\in X} f_2(x)$.
(vii) $\forall \alpha \in (0, \infty) \subset \mathbb{R}$, $\sup_{x\in X}(\alpha g_1(x)) = \alpha \sup_{x\in X} g_1(x)$; $\forall \alpha \in [0, \infty) \subset \mathbb{R}$, $\sup_{x\in X}(\alpha f_1(x)) = \alpha \sup_{x\in X} f_1(x)$ *when* $\sup_{x\in X} f_1(x) \in \mathbb{R}$.
(viii) $\sup_{x\in X}(g_1 + g_2)(x) \le \sup_{x\in X} g_1(x) + \sup_{x\in X} g_2(x)$, *when the right-hand side is well-defined.*

Proof This is straightforward and is therefore omitted. □

Definition 3.82 Let $(x_\alpha)_{\alpha\in\mathcal{A}} \subseteq \mathbb{R}_e$ be a net. The *limit superior* and *limit inferior* of the net are defined by

$$\limsup_{\alpha\in\mathcal{A}} x_\alpha = \inf_{\alpha\in\mathcal{A}} \sup_{\beta\in\mathcal{A} \text{ with } \alpha\prec\beta} x_\beta \in \mathbb{R}_e$$

$$\liminf_{\alpha\in\mathcal{A}} x_\alpha = \sup_{\alpha\in\mathcal{A}} \inf_{\beta\in\mathcal{A} \text{ with } \alpha\prec\beta} x_\beta \in \mathbb{R}_e$$

◇

Proposition 3.83 *Let* $(x_\alpha)_{\alpha\in\mathcal{A}} \subseteq \mathbb{R}_e$*,* $(y_\alpha)_{\alpha\in\mathcal{A}} \subseteq \mathbb{R}$*, and* $(z_\alpha)_{\alpha\in\mathcal{A}} \subseteq \mathbb{R}$ *be nets over the same directed system. Then, we have*

(i) $\liminf_{\alpha\in\mathcal{A}} x_\alpha \leq \limsup_{\alpha\in\mathcal{A}} x_\alpha$.
(ii) $-\liminf_{\alpha\in\mathcal{A}} x_\alpha = \limsup_{\alpha\in\mathcal{A}}(-x_\alpha)$.
(iii) $\liminf_{\alpha\in\mathcal{A}} x_\alpha = \limsup_{\alpha\in\mathcal{A}} x_\alpha = L \in \mathbb{R}_e$ *if, and only if,* $\lim_{\alpha\in\mathcal{A}} x_\alpha = L$.
(iv) $\limsup_{\alpha\in\mathcal{A}}(y_\alpha + z_\alpha) \leq \limsup_{\alpha\in\mathcal{A}} y_\alpha + \limsup_{\alpha\in\mathcal{A}} z_\alpha$*, when the right-hand side makes sense.*
(v) *if* $\lim_{\alpha\in\mathcal{A}} y_\alpha = y \in \mathbb{R}$*, then* $\limsup_{\alpha\in\mathcal{A}}(y_\alpha + z_\alpha) = y + \limsup_{\alpha\in\mathcal{A}} z_\alpha$.

Proof Let $V_\alpha := \{\beta \in \mathcal{A} \,|\, \alpha \prec \beta\}$, $\forall\alpha \in \mathcal{A}$. Then, $V_\alpha \neq \emptyset$, $\forall\alpha \in \mathcal{A}$, since $\mathcal{A}$ is a directed system, and $V_\alpha \supseteq V_\beta$, $\forall\alpha, \beta \in \mathcal{A}$ with $\alpha \prec \beta$.

(i) Let $l := \liminf_{\alpha\in\mathcal{A}} x_\alpha \in \mathbb{R}_e$ and $L := \limsup_{\alpha\in\mathcal{A}} x_\alpha \in \mathbb{R}_e$. $\forall m \in \mathbb{R}$ with $m < l$, $\sup_{\alpha\in\mathcal{A}} \inf_{\beta\in\mathcal{A} \text{ with } \alpha\prec\beta} x_\beta > m$ implies that, by Proposition 3.81, $\exists\alpha_0 \in \mathcal{A}$ such that $\inf_{\beta\in V_{\alpha_0}} x_\beta > m$. Then, $\forall\alpha \in \mathcal{A}$, $\exists\alpha_1 \in \mathcal{A}$ such that $\alpha_0 \prec \alpha_1$ and $\alpha \prec \alpha_1$. Then, $m < \inf_{\beta\in V_{\alpha_0}} x_\beta \leq \inf_{\beta\in V_{\alpha_1}} x_\beta \leq \sup_{\beta\in V_{\alpha_1}} x_\beta \leq \sup_{\beta\in V_\alpha} x_\beta$. Hence, $L \geq m$. By the arbitrariness of m, we have $L \geq l$.

(ii) Note that, by Proposition 3.81,

$$\limsup_{\alpha\in\mathcal{A}}(-x_\alpha) = \inf_{\alpha\in\mathcal{A}} \sup_{\beta\in V_\alpha}(-x_\beta) = \inf_{\alpha\in\mathcal{A}}\big(-\inf_{\beta\in V_\alpha} x_\beta\big)$$
$$= -\sup_{\alpha\in\mathcal{A}} \inf_{\beta\in V_\alpha} x_\beta = -\liminf_{\alpha\in\mathcal{A}} x_\alpha$$

(iii) "If" $\forall m \in \mathbb{R}$ with $m > L$, $\exists\alpha_0 \in \mathcal{A}$ such that $x_\alpha \in (-\infty, m)$, $\forall\alpha \in \mathcal{A}$ with $\alpha_0 \prec \alpha$. Then, $\sup_{\beta\in V_{\alpha_0}} x_\beta \leq m$ and $\limsup_{\alpha\in\mathcal{A}} x_\alpha \leq m$. By the arbitrariness of m, we have $\limsup_{\alpha\in\mathcal{A}} x_\alpha \leq L$. By (ii), we have $-L = \lim_{\alpha\in\mathcal{A}}(-x_\alpha) \geq \limsup_{\alpha\in\mathcal{A}}(-x_\alpha) = -\liminf_{\alpha\in\mathcal{A}} x_\alpha$. Then, by (i), we have $L \leq \liminf_{\alpha\in\mathcal{A}} x_\alpha \leq \limsup_{\alpha\in\mathcal{A}} x_\alpha \leq L$.

"Only if" We will distinguish three exhaustive and mutually exclusive cases: Case 1: $L = -\infty$, Case 2: $L \in \mathbb{R}$, and Case 3: $L = +\infty$. Case 1: $L = -\infty$. $\forall m \in \mathbb{R}$, $\limsup_{\alpha\in\mathcal{A}} x_\alpha < m$ implies that $\exists\alpha_0 \in \mathcal{A}$ such that $\sup_{\beta\in V_{\alpha_0}} x_\beta < m$. Then, $x_\beta \in (-\infty, m)$, $\forall\beta \in V_{\alpha_0}$. Hence, we have $\lim_{\alpha\in\mathcal{A}} x_\alpha = -\infty = L$. Case 2: $L \in \mathbb{R}$. $\forall\epsilon \in (0, \infty) \subset \mathbb{R}$, $L = \liminf_{\alpha\in\mathcal{A}} x_\alpha$ implies that $\exists\alpha_1 \in \mathcal{A}$ such that $\inf_{\beta\in V_{\alpha_1}} x_\beta > L - \epsilon$. Then, $x_\beta \in (L-\epsilon, +\infty) \subset \mathbb{R}$, $\forall\beta \in V_{\alpha_1}$. $L = \limsup_{\alpha\in\mathcal{A}} x_\alpha$ implies that $\exists\alpha_2 \in \mathcal{A}$ such that $\sup_{\beta\in V_{\alpha_2}} x_\beta < L+\epsilon$. Then, $x_\beta \in (-\infty, L+\epsilon) \subset \mathbb{R}$, $\forall\beta \in V_{\alpha_2}$. Let $\alpha_0 \in \mathcal{A}$ with $\alpha_1 \prec \alpha_0$ and $\alpha_2 \prec \alpha_0$. Then, $x_\beta \in (L-\epsilon, L+\epsilon)$, $\forall\beta \in \mathcal{A}$ with $\alpha_0 \prec \beta$. Therefore, $\lim_{\alpha\in\mathcal{A}} x_\alpha = L$. Case 3: $L = +\infty$. $\forall M \in \mathbb{R}$, $\liminf_{\alpha\in\mathcal{A}} x_\alpha > M$ implies that $\exists\alpha_0 \in \mathcal{A}$ such that $\inf_{\beta\in V_{\alpha_0}} x_\beta > M$. Then, $x_\beta \in (M, \infty)$, $\forall\beta \in V_{\alpha_0}$. Hence, we have $\lim_{\alpha\in\mathcal{A}} x_\alpha = +\infty = L$.

(iv) Note that, $\forall\alpha \in \mathcal{A}$, by Proposition 3.81,

$$\sup_{\beta\in V_\alpha}(y_\beta + z_\beta) \leq \sup_{\beta\in V_\alpha} y_\beta + \sup_{\beta\in V_\alpha} z_\beta$$

Then, by Proposition 3.81, we have, $\forall \alpha \in \mathcal{A}$,

$$\limsup_{\alpha \in \mathcal{A}} (y_\alpha + z_\alpha) \le \sup_{\beta \in V_\alpha} y_\beta + \sup_{\beta \in V_\alpha} z_\beta$$

$\forall \gamma \in \mathcal{A}$, $\exists \alpha_0 \subset \mathcal{A}$ with $\alpha \prec \alpha_0$ and $\gamma \prec \alpha_0$. Then,

$$\limsup_{\alpha \in \mathcal{A}} (y_\alpha + z_\alpha) \le \sup_{\beta \in V_{\alpha_0}} y_\beta + \sup_{\beta \in V_{\alpha_0}} z_\beta \le \sup_{\beta \in V_\alpha} y_\beta + \sup_{\beta \in V_\gamma} z_\beta$$

Hence, we have $\limsup_{\alpha \in \mathcal{A}} (y_\alpha + z_\alpha) \le \limsup_{\alpha \in \mathcal{A}} y_\alpha + \limsup_{\alpha \in \mathcal{A}} z_\alpha$, when the right-hand side makes sense.

(v) By (iv) and (iii), we have $\limsup_{\alpha \in \mathcal{A}} (y_\alpha + z_\alpha) \le y + \limsup_{\alpha \in \mathcal{A}} z_\alpha$. Note that $\limsup_{\alpha \in \mathcal{A}} z_\alpha = \limsup_{\alpha \in \mathcal{A}} (y_\alpha + z_\alpha - y_\alpha) \le \limsup_{\alpha \in \mathcal{A}} (y_\alpha + z_\alpha) + \limsup_{\alpha \in \mathcal{A}} (-y_\alpha)$. Then, we have $\limsup_{\alpha \in \mathcal{A}} (y_\alpha + z_\alpha) \ge y + \limsup_{\alpha \in \mathcal{A}} z_\alpha$. Hence, we have $\limsup_{\alpha \in \mathcal{A}} (y_\alpha + z_\alpha) = y + \limsup_{\alpha \in \mathcal{A}} z_\alpha$.

This completes the proof of the proposition. □

Definition 3.84 Let $\mathcal{X} := (X, \mathcal{O}_X)$ be topological spaces, $D \subseteq \mathcal{X}$, $f : D \to \mathbb{R}$, and $x_0 \in \mathcal{X}$ be an accumulation point of D. Then the *limit superior* and *limit inferior* of $f(x)$ as $x \to x_0$ are defined by

$$\limsup_{x \to x_0} f(x) = \inf_{O \in \mathcal{O} \text{ with } x_0 \in O} \ \sup_{x \in (D \cap O) \setminus \{x_0\}} f(x) \in \mathbb{R}_e$$

$$\liminf_{x \to x_0} f(x) = \sup_{O \in \mathcal{O} \text{ with } x_0 \in O} \ \inf_{x \in (D \cap O) \setminus \{x_0\}} f(x) \in \mathbb{R}_e$$

◇

Proposition 3.85 *Let $\mathcal{X}$ be a topological space, $D \subseteq \mathcal{X}$, $x_0 \in \mathcal{X}$ be an accumulation point of D, $f : D \to \mathbb{R}$, and $g : D \to \mathbb{R}$. Then, we have*

(i) $\liminf_{x \to x_0} f(x) \le \limsup_{x \to x_0} f(x)$.
(ii) $-\liminf_{x \to x_0} f(x) = \limsup_{x \to x_0} (-f)(x)$.
(iii) $\liminf_{x \to x_0} f(x) = \limsup_{x \to x_0} f(x) = L \in \mathbb{R}_e$ *if, and only if,* $\lim_{x \to x_0} f(x) = L$.
(iv) $\limsup_{x \to x_0} (f + g)(x) \le \limsup_{x \to x_0} f(x) + \limsup_{x \to x_0} g(x)$, *when the right-hand side makes sense.*
(v) *if* $\lim_{x \to x_0} f(x) = y \in \mathbb{R}$, *then* $\limsup_{x \to x_0} (f + g)(x) = \limsup_{x \to x_0} g(x) + y$.

Proof (i) Let $l := \liminf_{x \to x_0} f(x) \in \mathbb{R}_e$ and $L := \limsup_{x \to x_0} f(x) \in \mathbb{R}_e$. $\forall m \in \mathbb{R}$ with $m < l$, $\sup_{O \in \mathcal{O} \text{ with } x_0 \in O} \inf_{x \in (D \cap O) \setminus \{x_0\}} f(x) > m$ implies that $\exists U \in \mathcal{O}$ with $x_0 \in U$ such that $\inf_{x \in (D \cap U) \setminus \{x_0\}} f(x) > m$. $\forall O \in \mathcal{O}$ with $x_0 \in O$, we have $x_0 \in V := O \cap U \in \mathcal{O}$ and $(D \cap V) \setminus \{x_0\} \neq \emptyset$, since x_0 is an accumulation point of D. Then, $m < \inf_{x \in (D \cap U) \setminus \{x_0\}} f(x) \le \inf_{x \in (D \cap V) \setminus \{x_0\}} f(x) \le \sup_{x \in (D \cap V) \setminus \{x_0\}} f(x) \le \sup_{x \in (D \cap O) \setminus \{x_0\}} f(x)$. Hence, $L \ge m$. By the arbitrariness of m, we have $L \ge l$.

(ii) Note that, by Proposition 3.81,

$$\begin{aligned}\limsup_{x\to x_0}(-f)(x) &= \inf_{O\in\mathcal{O}\text{ with }x_0\in O}\ \sup_{x\in(D\cap O)\setminus\{x_0\}}(-f(x))\\ &= \inf_{O\in\mathcal{O}\text{ with }x_0\in O}\Big(-\inf_{x\in(D\cap O)\setminus\{x_0\}}f(x)\Big)\\ &= -\sup_{O\in\mathcal{O}\text{ with }x_0\in O}\ \inf_{x\in(D\cap O)\setminus\{x_0\}}f(x)\\ &= -\liminf_{x\to x_0}f(x)\end{aligned}$$

(iii) "If" $\forall m \in \mathbb{R}$ with $m > L$, $\exists V \in \mathcal{O}$ with $x_0 \in V$ such that $f(V \setminus \{x_0\}) \subseteq (-\infty, m)$. Then, $\sup_{x\in(D\cap V)\setminus\{x_0\}} f(x) \leq m$ and $\limsup_{x\to x_0} f(x) \leq m$. By the arbitrariness of m, we have $\limsup_{x\to x_0} f(x) \leq L$. By (ii), we have $-L = \lim_{x\to x_0}(-f)(x) \geq \limsup_{x\to x_0}(-f)(x) = -\liminf_{x\to x_0} f(x)$. Then, by (i), we have $L \leq \liminf_{x\to x_0} f(x) \leq \limsup_{x\to x_0} f(x) \leq L$.

"Only if" We will distinguish three exhaustive and mutually exclusive cases: Case 1: $L = -\infty$, Case 2: $L \in \mathbb{R}$, and Case 3: $L = +\infty$. Case 1: $L = -\infty$. $\forall m \in \mathbb{R}$, $\limsup_{x\to x_0} f(x) < m$ implies that $\exists V \in \mathcal{O}$ with $x_0 \in V$ such that $\sup_{x\in(D\cap V)\setminus\{x_0\}} f(x) < m$. Then, $f(V \setminus \{x_0\}) \subseteq (-\infty, m)$. Hence, we have $\lim_{x\to x_0} f(x) = -\infty = L$. Case 2: $L \in \mathbb{R}$. $\forall \epsilon \in (0, \infty) \subset \mathbb{R}$, $L = \liminf_{x\to x_0} f(x)$ implies that $\exists V_1 \in \mathcal{O}$ with $x_0 \in V_1$ such that $\inf_{x\in(D\cap V_1)\setminus\{x_0\}} f(x) > L - \epsilon$. Then, $f(V_1 \setminus \{x_0\}) \subseteq (L - \epsilon, +\infty) \subseteq \mathbb{R}$. $L = \limsup_{x\to x_0} f(x)$ implies that $\exists V_2 \in \mathcal{O}$ with $x_0 \in V_2$ such that $\sup_{x\in(D\cap V_2)\setminus\{x_0\}} f(x) < L + \epsilon$. Then, $f(V_2 \setminus \{x_0\}) \subseteq (-\infty, L + \epsilon)$. Let $V := V_1 \cap V_2 \in \mathcal{O}$. Clearly, $x_0 \in V$, and, by Proposition 2.5, $f(V \setminus \{x_0\}) \subseteq f(V_1 \setminus \{x_0\}) \cap f(V_2 \setminus \{x_0\}) \subseteq (L - \epsilon, L + \epsilon)$. Therefore, $\lim_{x\to x_0} f(x) = L$. Case 3: $L = +\infty$. $\forall M \in \mathbb{R}$, $\liminf_{x\to x_0} f(x) > M$ implies that $\exists V \in \mathcal{O}$ with $x_0 \in V$ such that $\inf_{x\in(D\cap V)\setminus\{x_0\}} f(x) > M$. Then, $f(V \setminus \{x_0\}) \subseteq (M, +\infty)$. Hence, we have $\lim_{x\to x_0} f(x) = +\infty = L$.

(iv) Note that, $\forall O \in \mathcal{O}$ with $x_0 \in O$, by Proposition 3.81,

$$\sup_{x\in(D\cap O)\setminus\{x_0\}}(f+g)(x) \leq \Big(\sup_{x\in(D\cap O)\setminus\{x_0\}} f(x)\Big) + \Big(\sup_{x\in(D\cap O)\setminus\{x_0\}} g(x)\Big)$$

Then, by Proposition 3.81, we have, $\forall O \in \mathcal{O}$ with $x_0 \in O$,

$$\limsup_{x\to x_0}(f+g)(x) \leq \Big(\sup_{x\in(D\cap O)\setminus\{x_0\}} f(x)\Big) + \Big(\sup_{x\in(D\cap O)\setminus\{x_0\}} g(x)\Big)$$

$\forall U \in \mathcal{O}$ with $x_0 \in U$, we have $x_0 \in V := U \cap O \in \mathcal{O}$. Then,

$$\limsup_{x\to x_0}(f+g)(x) \leq \Big(\sup_{x\in(D\cap V)\setminus\{x_0\}} f(x)\Big) + \Big(\sup_{x\in(D\cap V)\setminus\{x_0\}} g(x)\Big)$$
$$\leq \Big(\sup_{x\in(D\cap O)\setminus\{x_0\}} f(x)\Big) + \Big(\sup_{x\in(D\cap U)\setminus\{x_0\}} g(x)\Big)$$

Hence, we have $\limsup_{x\to x_0}(f+g)(x) \leq \limsup_{x\to x_0} f(x) + \limsup_{x\to x_0} g(x)$, when the right-hand side makes sense.

(v) By (iv) and (iii), $\limsup_{x\to x_0}(f+g)(x) \leq y + \limsup_{x\to x_0} g(x)$. Note that $\limsup_{x\to x_0} g(x) = \limsup_{x\to x_0}(f+g-f)(x) \leq \limsup_{x\to x_0}(f+g)(x) + \limsup_{x\to x_0}(-f)(x)$. Then, we have $\limsup_{x\to x_0}(f+g)(x) \geq y + \limsup_{x\to x_0} g(x)$. Hence, we have $\limsup_{x\to x_0}(f+g)(x) = \limsup_{x\to x_0} g(x)+y$.

This completes the proof of the proposition. □

Proposition 3.86 *Let $\mathcal{X} := (X, \mathcal{O}_X)$ be a topological space, $D \subseteq X$ with the subset topology $\mathcal{O}_D$, $f : D \to \mathbb{R}$, and $x_0 \in D$. Then, the following statements are equivalent:*

(i) *f is upper semicontinuous at x_0.*
(ii) *If x_0 is an accumulation point of D, then* $\limsup_{x\to x_0} f(x) \leq f(x_0)$.

Proof (i) ⇒ (ii). Let x_0 be an accumulation point of D. By the upper semicontinuity of f at x_0, $\forall \epsilon \in (0,\infty) \subset \mathbb{R}$, $\exists O \in \mathcal{O}_X$ with $x_0 \in O$ such that $f(x) < f(x_0)+\epsilon$, $\forall x \in O \cap D$. Then, $\sup_{x\in(O\cap D)\setminus\{x_0\}} f(x) \leq f(x_0)+\epsilon$, and $\limsup_{x\to x_0} f(x) \leq f(x_0)+\epsilon$. By the arbitrariness of ϵ, we have $\limsup_{x\to x_0} f(x) \leq f(x_0)$.

(ii) ⇒ (i). We will distinguish two exhaustive and mutually exclusive cases: Case 1: x_0 is not an accumulation point of D, and Case 2: x_0 is an accumulation point of D. Case 1: x_0 is not an accumulation point of D. $\exists V \in \mathcal{O}_X$ with $x_0 \in V$ such that $V \cap D = \{x_0\}$. $\forall \epsilon \in (0,\infty) \subset \mathbb{R}$, we have $f(x) = f(x_0) < f(x_0)+\epsilon$, $\forall x \in V \cap D$. Hence, f is upper semicontinuous at x_0.

Case 2: x_0 is an accumulation point of D. $\forall \epsilon \in \mathbb{R}_+$, $\limsup_{x\to x_0} f(x) < f(x_0)+\epsilon$ implies that $\exists O \in \mathcal{O}$ with $x_0 \in O$ such that $\sup_{x\in(O\cap D)\setminus\{x_0\}} f(x) < f(x_0)+\epsilon$. Then, $f(x) < f(x_0)+\epsilon$, $\forall x \in O \cap D$. Hence, f is upper semicontinuous at x_0.

In both cases, f is upper semicontinuous at x_0.

This completes the proof of the proposition. □

Proposition 3.87 *Let $\mathcal{X}$ be a topological space, $D \subseteq \mathcal{X}$, $x_0 \in \mathcal{X}$ be an accumulation point of D, and $f : D \to \mathbb{R}$. Assume there exists a net $(x_\alpha)_{\alpha\in\mathcal{A}} \subseteq D \setminus \{x_0\}$ such that $\lim_{\alpha\in\mathcal{A}} x_\alpha = x_0$ and $\liminf_{\alpha\in\mathcal{A}} f(x_\alpha) = c \in \mathbb{R}_e$. Then,* $\liminf_{x\to x_0} f(x) \leq c$.

Proof By Definitions 3.84 and 3.82, we have

$$\liminf_{x\to x_0} f(x) = \sup_{O\in\mathcal{O} \text{ with } x_0\in O} \inf_{x\in(D\cap O)\setminus\{x_0\}} f(x)$$

$$\liminf_{\alpha\in\mathcal{A}} f(x_\alpha) = \sup_{\alpha\in\mathcal{A}} \inf_{\beta\in\mathcal{A} \text{ with } \alpha\prec\beta} f(x_\beta) = c$$

$\forall O \in \mathcal{O}$ with $x_0 \in O$, since $\lim_{\alpha\in\mathcal{A}} x_\alpha = x_0$, then $\exists \alpha_0 \in \mathcal{A}$ such that $\forall \alpha \in \mathcal{A}$ with $\alpha_0 \prec \alpha$, we have $x_\alpha \in O$. Then, $x_\alpha \in (O \cap D) \setminus \{x_0\}$. This leads to

$$\inf_{x\in(D\cap O)\setminus\{x_0\}} f(x) \le \inf_{\beta\in\mathcal{A} \text{ with } \alpha_0\prec\beta} f(x_\beta) \le c$$

Hence, we have $\liminf_{x\to x_0} f(x) \le c$. This completes the proof of the proposition. □

Definition 3.88 Let $\mathcal{A}_i := (A_i, \prec_i)$ be a directed system, $i = 1, 2$, and $\mathcal{X} := (X, \mathcal{O})$ be a topological space. Define a relation $\prec$ on $A_1 \times A_2$ by, $\forall (\alpha_1, \beta_1), (\alpha_2, \beta_2) \in A_1 \times A_2$, we say $(\alpha_1, \beta_1) \prec (\alpha_2, \beta_2)$ if $\alpha_1 \prec_1 \alpha_2$ and $\beta_1 \prec_2 \beta_2$. It is easy to verify that $\mathcal{A} := (A_1 \times A_2, \prec) =: \mathcal{A}_1 \times \mathcal{A}_2$ is a directed system. A *joint net* is a mapping of the directed system $\mathcal{A}_1 \times \mathcal{A}_2$ to $\mathcal{X}$, denoted by $\left(x_{\alpha,\beta}\right)_{(\alpha,\beta)\in\mathcal{A}_1\times\mathcal{A}_2}$. The joint net is said to admit a *(joint) limit point* $\hat{x} \in \mathcal{X}$ if it admits a limit point $\hat{x}$ when viewed as a net over the directed system $\mathcal{A}$. When $\mathcal{X}$ is Hausdorff, by Proposition 3.65, the joint net admits at most one joint limit point, which will be denoted by $\lim_{(\alpha,\beta)\in\mathcal{A}_1\times\mathcal{A}_2} x_{\alpha,\beta} \in \mathcal{X}$ if it exists. ◇

Proposition 3.89 *Let $X := [0, \infty] \subset \mathbb{R}_e$ with subset topology $\mathcal{O}$, $\mathcal{X} := (X, \mathcal{O})$, and $c \in (0, \infty) \subset \mathbb{R}$. Then, $+ : \mathcal{X} \times \mathcal{X} \to \mathcal{X}$ and $c\cdot : \mathcal{X} \to \mathcal{X}$ are continuous.*

Proof We will prove this using Proposition 3.9. $\forall (x_1, x_2) \in X \times X$. We will distinguish four exhaustive and mutually exclusive cases. Case 1: $x_1 < \infty$ and $x_2 < \infty$. Then, $x_1 + x_2 < \infty$. $\forall$ basis open set $U := (r_1, r_2) \cap X \in \mathcal{O}$ with $x_1 + x_2 \in U$, take $V_1 := (x_1 - (x_1 + x_2 - r_1)/2, x_1 + (r_2 - x_1 - x_2)/2) \cap X \in \mathcal{O}$ and $V_2 := (x_2 - (x_1 + x_2 - r_1)/2, x_2 + (r_2 - x_1 - x_2)/2) \cap X \in \mathcal{O}$. Then, $V_1 \times V_2$ is an open set in $\mathcal{X} \times \mathcal{X}$ with $(x_1, x_2) \in V_1 \times V_2$ and $\forall (\bar{x}_1, \bar{x}_2) \in V_1 \times V_2$, we have $\bar{x}_1 + \bar{x}_2 \in U$. Hence, $+ : \mathcal{X} \times \mathcal{X} \to \mathcal{X}$ is continuous at (x_1, x_2).

Case 2: $x_1 < \infty$ and $x_2 = \infty$. Then, $x_1 + x_2 = \infty$. $\forall$ basis open set $U := (r_1, \infty] \cap X \in \mathcal{O}$ with $x_1 + x_2 \in U$, take $V_1 := (x_1 - 1, x_1 + 1) \cap X \in \mathcal{O}$ and $V_2 := (r_1 - x_1 + 1, \infty] \cap X \in \mathcal{O}$. Then, $V_1 \times V_2$ is an open set in $\mathcal{X} \times \mathcal{X}$ with $(x_1, x_2) \in V_1 \times V_2$ and $\forall (\bar{x}_1, \bar{x}_2) \in V_1 \times V_2$, we have $\bar{x}_1 + \bar{x}_2 \in U$. Hence, $+ : \mathcal{X} \times \mathcal{X} \to \mathcal{X}$ is continuous at (x_1, x_2).

Case 3: $x_1 = \infty$ and $x_2 < \infty$. By Case 2 and symmetry, $+ : \mathcal{X} \times \mathcal{X} \to \mathcal{X}$ is continuous at (x_1, x_2).

Case 4: $x_1 = \infty$ and $x_2 = \infty$. Then, $x_1 + x_2 = \infty$. $\forall$ basis open set $U := (r_1, \infty] \cap X \in \mathcal{O}$ with $x_1 + x_2 \in U$, take $V_1 := (r_1/2, \infty] \cap X \in \mathcal{O}$ and $V_2 := (r_1/2, \infty] \cap X \in \mathcal{O}$. Then, $V_1 \times V_2$ is an open set in $\mathcal{X} \times \mathcal{X}$ with $(x_1, x_2) \in V_1 \times V_2$

and $\forall(\bar{x}_1, \bar{x}_2) \in V_1 \times V_2$, we have $\bar{x}_1 + \bar{x}_2 \in U$. Hence, $+ : \mathcal{X} \times \mathcal{X} \to \mathcal{X}$ is continuous at (x_1, x_2).

In all cases, we have $+ : \mathcal{X} \times \mathcal{X} \to \mathcal{X}$ is continuous at (x_1, x_2). By Proposition 3.9, $+ : \mathcal{X} \times \mathcal{X} \to \mathcal{X}$ is continuous.

$\forall x \in X$. We will distinguish two exhaustive and mutually exclusive cases. Case 1: $x < \infty$. Then, $cx < \infty$. $\forall$ basis open set $U := (r_1, r_2) \cap X \in \mathcal{O}$ with $cx \in U$, take $V := (r_1/c, r_2/c) \cap X \in \mathcal{O}$. Then, $x \in V$ and $\forall \bar{x} \in V$, we have $c\bar{x} \in U$. Hence, $c\cdot : \mathcal{X} \to \mathcal{X}$ is continuous at x.

Case 2: $x = \infty$. Then, $cx = \infty$. $\forall$ basis open set $U := (r_1, \infty] \cap X \in \mathcal{O}$ with $cx \in U$, take $V := (r_1/c, \infty] \cap X \in \mathcal{O}$. Then, $x \in V$ and $\forall \bar{x} \in V$, we have $c\bar{x} \in U$. Hence, $c\cdot : \mathcal{X} \to \mathcal{X}$ is continuous at x.

In both cases, we have $c\cdot : \mathcal{X} \to \mathcal{X}$ is continuous at x. By Proposition 3.9, $c\cdot : \mathcal{X} \to \mathcal{X}$ is continuous. This completes the proof of the proposition. □

Proposition 3.90 *Let $\mathcal{X} := (X, \mathcal{O}_X)$ and $\mathcal{Y} := (Y, \mathcal{O}_Y)$ be topological spaces, $\mathcal{X}$ be separable, and $f : \mathcal{X} \to \mathcal{Y}$ be continuous. Then, $f(X) \subseteq \mathcal{Y}$ is separable (in its subset topology).*

Proof Since $\mathcal{X}$ is separable, then it has a countable subset $D \subseteq X$ such that $\overline{D} = X$. Then, $f(D) \subseteq f(X)$ and $f(D)$ is a countable set. $\forall y_0 \in f(X)$, $\exists x_0 \in X$ such that $y_0 = f(x_0)$. By Proposition 3.68, there exists a net $(x_\alpha)_{\alpha \in \mathcal{A}} \subseteq D$ such that $\lim_{\alpha \in \mathcal{A}} x_\alpha = x_0$. By Proposition 3.66, we have the net $(f(x_\alpha))_{\alpha \in \mathcal{A}} \subseteq f(D)$ and $\lim_{\alpha \in \mathcal{A}} f(x_\alpha) = f(x_0) = y_0$. Thus, by Proposition 3.68, we have $y_0 \subseteq \overline{f(D)}$, where the closure is with respect to $\mathcal{Y}$. By the arbitrariness of y_0, we have $f(X) \subseteq \overline{f(D)}$. By Proposition 3.5, we have $f(D)$ is dense in $f(X)$, which implies that $f(X)$ is separable. This completes the proof of the proposition. □

Chapter 4
Metric Spaces

4.1 Fundamental Notions

Definition 4.1 A *metric space* (X, ρ) is a set X together with a *metric* $\rho : X \times X \to \mathbb{R}$ such that, $\forall x, y, z \in X$,

(i) $\rho(x, y) \geq 0$.
(ii) $\rho(x, y) = 0 \Leftrightarrow x = y$.
(iii) $\rho(x, y) = \rho(y, x)$.
(iv) $\rho(x, y) \leq \rho(x, z) + \rho(z, y)$. ◇

Let $S \subseteq X$. Then, $(S, \rho|_{S \times S})$ is also a metric space.

Example 4.2 $(\mathbb{R}, \rho)$, with $\rho(x, y) = |x - y|$, $\forall x, y \in \mathbb{R}$, is a metric space. $(\mathbb{R}^n, \rho)$, with $\rho(x, y) = (\sum_{i=1}^{n} |x_i - y_i|^2)^{1/2}$, $\forall x = (x_1, \ldots, x_n)$, $y = (y_1, \ldots, y_n) \in \mathbb{R}^n$, is a metric space, where $n \in \mathbb{N}$. $(\mathbb{R}^n, \rho)$, with $\rho(x, y) = \sum_{i=1}^{n} |x_i - y_i|$, $\forall x = (x_1, \ldots, x_n)$, $y = (y_1, \ldots, y_n) \in \mathbb{R}^n$, is a metric space, where $n \in \mathbb{N}$. ◇

Proposition 4.3 *Let* $\mathcal{X} := (X, \rho)$ *be a metric space, and an* open ball centered at $x_0 \in X$ with radius $r \in (0, \infty) \subset \mathbb{R}$ *is defined by* $\mathcal{B}_{\mathcal{X}}(x_0, r) := \{x \in X \mid \rho(x, x_0) < r\}$.

The metric space generates a natural topology $\mathcal{O}$ *on* X *with the basis collection* $\mathcal{B}$ *given by* $\mathcal{B} := \{\mathcal{B}_{\mathcal{X}}(x_0, r) \mid x_0 \in X, r \in (0, \infty) \subset \mathbb{R}\}$.

The closed ball centered at $x_0 \in X$ with radius $r \in [0, \infty) \subset \mathbb{R}$ *is defined by* $\overline{\mathcal{B}}_{\mathcal{X}}(x_0, r) := \{x \in X \mid \rho(x, x_0) \leq r\}$, *which is closed.*

Proof We will show that $\mathcal{B}$ is the basis for its generated topology by Proposition 3.18. (i) $\forall x \in X$, $x \in \mathcal{B}_{\mathcal{X}}(x, 1)$. (ii) $\forall \mathcal{B}_{\mathcal{X}}(x_1, r_1), \mathcal{B}_{\mathcal{X}}(x_2, r_2) \in \mathcal{B}$, let $x \in \mathcal{B}_{\mathcal{X}}(x_1, r_1) \cap \mathcal{B}_{\mathcal{X}}(x_2, r_2)$. Then, we have $\rho(x, x_1) < r_1$ and $\rho(x, x_2) < r_2$. Let $r := \min\{r_1 - \rho(x, x_1), r_2 - \rho(x, x_2)\} \in (0, \infty) \subset \mathbb{R}$. Then, $x \in \mathcal{B}_{\mathcal{X}}(x, r) \in \mathcal{B}$. $\forall x_3 \in \mathcal{B}_{\mathcal{X}}(x, r)$, we have $\rho(x_3, x_1) \leq \rho(x_3, x) + \rho(x, x_1) < r_1$ and $\rho(x_3, x_2) \leq \rho(x_3, x) + \rho(x, x_2) < r_2$. Hence, we have $x_3 \in \mathcal{B}_{\mathcal{X}}(x_1, r_1) \cap \mathcal{B}_{\mathcal{X}}(x_2, r_2)$.

Z. Pan, *Measure-Theoretic Calculus in Abstract Spaces*,
https://doi.org/10.1007/978-3-031-21912-2_4

Therefore, we have $\mathcal{B}_{\mathcal{X}}(x, r) \subseteq \mathcal{B}_{\mathcal{X}}(x_1, r_1) \cap \mathcal{B}_{\mathcal{X}}(x_2, r_2)$. Hence, the assumptions of Proposition 3.18 are satisfied, and then $\mathcal{B}$ is a basis for $\mathcal{O}$.

Next, we show that $\overline{\mathcal{B}}_{\mathcal{X}}(x_0, r)$ is a closed set. $\forall x \in \left(\overline{\mathcal{B}}_{\mathcal{X}}(x_0, r)\right)^{\sim}$, we have $\rho(x, x_0) > r$. Let $r_1 := \rho(x, x_0) - r \in (0, \infty) \subset \mathbb{R}$. $\forall x_1 \in \mathcal{B}_{\mathcal{X}}(x, r_1)$, we have $\rho(x_1, x_0) \geq \rho(x, x_0) - \rho(x, x_1) > r$. Hence, we have $x \in \mathcal{B}_{\mathcal{X}}(x, r_1) \subseteq \left(\overline{\mathcal{B}}_{\mathcal{X}}(x_0, r)\right)^{\sim}$. Therefore, $\left(\overline{\mathcal{B}}_{\mathcal{X}}(x_0, r)\right)^{\sim}$ is open and $\overline{\mathcal{B}}_{\mathcal{X}}(x_0, r)$ is closed.

This completes the proof of the proposition. □

We will sometimes talk about a metric space $\mathcal{X} := (X, \rho)$ without referring the components of $\mathcal{X}$, where the metric is understood to be $\rho_{\mathcal{X}}$ and the natural topology is understood to be $\mathcal{O}_{\mathcal{X}}$. When it is clear from the context, we will also neglect the subscript $\mathcal{X}$. We will abuse the notation and say $x \in \mathcal{X}$ and $A \subseteq \mathcal{X}$ when $x \in X$ and $A \subseteq X$.

On a metric space, we can talk about open and closed sets, and all those concepts defined in Chap. 3, all with respect to the natural topology.

A metric space is clearly first countable, where a countable basis at $x_0 \in \mathcal{X}$ is $\{\mathcal{B}(x_0, r) \mid r \in \mathbb{Q}, r > 0\}$.

Proposition 4.4 *A metric space (X, ρ) is separable if, and only if, it is second countable.*

Proof "Only if" Let $D \subseteq X$ be a countable dense set. Let $\mathcal{M} := \{\mathcal{B}(x, r) \mid x \in D, r \in \mathbb{Q}, \text{ and } r > 0\}$. Clearly, $\mathcal{M}$ is countable and $\mathcal{M} \subseteq \mathcal{O}$. $\forall O \in \mathcal{O}$, $\forall x \in O$. $\exists r \in (0, \infty) \cap \mathbb{Q}$ such that $\mathcal{B}(x, r) \subseteq O$. Since D is dense, then $\exists x_1 \in \mathcal{B}(x, r/2) \cap D$. Let $r_1 = r/2 \in (0, \infty) \cap \mathbb{Q}$. Then, we have $x \in \mathcal{B}(x_1, r_1) \subseteq \mathcal{B}(x, r) \subseteq O$ and $\mathcal{B}(x_1, r_1) \in \mathcal{M}$. Hence, $\mathcal{M}$ is a basis for $\mathcal{O}$. Hence, $(X, \mathcal{O})$ is second countable.

"If" Let $\mathcal{O}$ has a countable basis $\mathcal{B}$. By Axiom of Choice, we may assign a $x_B \in B$, $\forall B \in \mathcal{B}$ with $B \neq \emptyset$. Let $D = \{x_B \in X \mid B \in \mathcal{B}, B \neq \emptyset\}$. Then, D is countable. $\forall x \in X$, $\forall O \in \mathcal{O}$ with $x \in O$, $\exists B \in \mathcal{B}$ such that $x \in B \subseteq O$. Then, $x_B \in D \cap B \subseteq D \cap O \neq \emptyset$. Hence, by Proposition 3.3, we have $x \in \overline{D}$. Therefore, by the arbitrariness of x, we have D is dense. Hence, $(X, \mathcal{O})$ is separable.

This completes the proof of the proposition. □

Proposition 4.5 *Let $\mathcal{X}$ be a topological space and $\mathcal{Y}$ and $\mathcal{Z}$ be metric spaces. Let $f : \mathcal{X} \to \mathcal{Y}$, $g : \mathcal{Y} \to \mathcal{Z}$, $h : \mathcal{Y} \to \mathcal{X}$, $x_0 \in \mathcal{X}$, and $y_0 \in \mathcal{Y}$. Then, the following statements hold:*

1. *f is continuous at x_0 if, and only if, $\forall \epsilon \in (0, \infty) \subset \mathbb{R}$, $\exists U \in \mathcal{O}_{\mathcal{X}}$ with $x_0 \in U$ such that $\rho_{\mathcal{Y}}(f(x), f(x_0)) < \epsilon$, $\forall x \in U$.*
2. *g is continuous at y_0 if, and only if, $\forall \epsilon \in (0, \infty) \subset \mathbb{R}$, $\exists \delta \in (0, \infty) \subset \mathbb{R}$ such that $\rho_{\mathcal{Z}}(g(y), g(y_0)) < \epsilon$, $\forall y \in \mathcal{B}_{\mathcal{Y}}(y_0, \delta)$.*
3. *h is continuous at y_0 if, and only if, $\forall U \in \mathcal{O}_{\mathcal{X}}$ with $h(y_0) \in U$, $\exists \delta \in (0, \infty) \subset \mathbb{R}$ such that $h(y) \in U$, $\forall y \in \mathcal{B}_{\mathcal{Y}}(y_0, \delta)$.*

Proof The proof is straightforward and is therefore omitted. □

Definition 4.6 Let $\mathcal{X}$ and $\mathcal{Y}$ be metric spaces and $f : \mathcal{X} \to \mathcal{Y}$ be a homeomorphism. f is said to be an *isometry* between $\mathcal{X}$ and $\mathcal{Y}$ if $\rho_{\mathcal{Y}}(f(x_1), f(x_2)) = \rho_{\mathcal{X}}(x_1, x_2)$, $\forall x_1, x_2 \in \mathcal{X}$. Then, the two metric spaces are said to be *isometric*. ◇

Definition 4.7 Let X be a set and ρ_1 and ρ_2 be two metrics on X. ρ_1 and ρ_2 are said to be *equivalent* if the identity map from (X, ρ_1) to (X, ρ_2) is a homeomorphism. ◇

When two metrics are equivalent, then the natural topologies generated by them are equal to each other.

Clearly, a metric space is Hausdorff.

4.2 Convergence and Completeness

Proposition 4.8 *Let $\mathcal{X}$ be a metric space and $(x_\alpha)_{\alpha\in\mathcal{A}} \subseteq \mathcal{X}$ be a net. Then, $\lim_{\alpha\in\mathcal{A}} x_\alpha = x \in \mathcal{X}$ if, and only if, $\forall\epsilon \in (0,\infty) \subset \mathbb{R}$, $\exists\alpha_0 \in \mathcal{A}$, $\forall\alpha \in \mathcal{A}$ with $\alpha_0 \prec \alpha$, we have $\rho(x_\alpha, x) < \epsilon$.*

Proof This is straightforward and is omitted. □

Since metric spaces are Hausdorff, then the limit is unique if it exists.

Definition 4.9 Let $\mathcal{X}$ be a metric space, $x_0 \in \mathcal{X}$, and $S \subseteq \mathcal{X}$. The *distance* from x_0 to S is $\text{dist}(x_0, S) := \inf_{s\in S} \rho(x_0, s) \in [0,\infty] \subset \mathbb{R}_e$. ◇

$\text{dist}(x_0, S) = \infty$ if, and only if, $S = \emptyset$.

Proposition 4.10 *Let $\mathcal{X}$ be a metric space, $x_0 \in \mathcal{X}$, $S \subseteq \mathcal{X}$, and S is closed. Then, $x_0 \in S$ if, and only if, $\text{dist}(x_0, S) = 0$.*

Proof "Only if" This is obvious.

"If" By the fact that $\text{dist}(x_0, S) = 0$, $\forall n \in \mathbb{N}$, $\exists x_n \in S$ such that $\rho(x_0, x_n) < 1/n$. Then, $\lim_{n\in\mathbb{N}} x_n = x_0$. By Proposition 3.68, $x_0 \in \overline{S}$. Since S is closed, then, by Proposition 3.3, $S = \overline{S}$. Hence, $x_0 \in S$.

This completes the proof of the proposition. □

Proposition 4.11 *A metric space with its natural topology is normal.*

Proof Let $\mathcal{X}$ be the metric space. Clearly, $\mathcal{X}$ is Hausdorff. For all closed sets $F_1, F_2 \subseteq \mathcal{X}$ with $F_1 \cap F_2 = \emptyset$. We will distinguish two exhaustive and mutually exclusive cases: Case 1: $F_1 = \emptyset$ or $F_2 = \emptyset$, and Case 2: $F_1 \neq \emptyset$ and $F_2 \neq \emptyset$. Case 1: $F_1 = \emptyset$ or $F_2 = \emptyset$. Without loss of generality, assume $F_1 = \emptyset$. Take $O_1 = \emptyset \in \mathcal{O}$ and $O_2 = \mathcal{X} \in \mathcal{O}$, then $F_1 \subseteq O_1$, $F_2 \subseteq O_2$, $O_1 \cap O_2 = \emptyset$. Case 2: $F_1 \neq \emptyset$ and $F_2 \neq \emptyset$. $\forall x \in F_1$, $\text{dist}(x, F_2) \in (0,\infty) \subset \mathbb{R}$, by Proposition 4.10. Define $O_1 \in \mathcal{O}$ by $O_1 := \bigcup_{x\in F_1} \mathcal{B}(x, \text{dist}(x, F_2)/3)$. $\forall x \in F_2$, $\text{dist}(x, F_1) \in (0,\infty) \subset \mathbb{R}$ by Proposition 4.10. Define $O_2 \in \mathcal{O}$ by $O_2 := \bigcup_{x\in F_2} \mathcal{B}(x, \text{dist}(x, F_1)/3)$. Clearly, $F_1 \subseteq O_1$ and $F_2 \subseteq O_2$. Note that $O_1 \cap O_2 = \emptyset$, since otherwise, $\exists x_0 \in O_1 \cap O_2$, $\exists x_1 \in F_1$ such that $x_0 \in \mathcal{B}(x_1, \text{dist}(x_1, F_2)/3)$, $\exists x_2 \in F_2$ such that $x_0 \in \mathcal{B}(x_2, \text{dist}(x_2, F_1)/3)$, without loss of generality, assume $\text{dist}(x_1, F_2) \leq \text{dist}(x_2, F_1)$, then $\text{dist}(x_2, F_1) \leq \rho(x_2, x_1) \leq \rho(x_2, x_0) + \rho(x_0, x_1) < \text{dist}(x_2, F_1)/3 + \text{dist}(x_1, F_2)/3 \leq 2\,\text{dist}(x_2, F_1)/3$, which is a contradiction.

Hence, in both cases, $\exists O_1, O_2 \in \mathcal{O}$ such that $F_1 \subseteq O_1$, $F_2 \subseteq O_2$, $O_1 \cap O_2 = \emptyset$. Hence, $\mathcal{X}$ is normal. This completes the proof of the proposition. □

Definition 4.12 Let $\mathcal{X}$ be a metric space and $(x_n)_{n=1}^{\infty} \subseteq \mathcal{X}$. The sequence is said to be a *Cauchy sequence* if $\forall \epsilon \in (0, \infty) \subset \mathbb{R}$, $\exists N \in \mathbb{N}$, $\forall n, m \geq N$, $\rho(x_n, x_m) < \epsilon$. ◇

Clearly, every convergent sequence in a metric space is a Cauchy sequence.

Proposition 4.13 *Let $\mathcal{X}$ be a metric space, $E \subseteq \mathcal{X}$, and $x_0 \in \mathcal{X}$. Then, $x_0 \in \overline{E}$ if, and only if, $\exists (x_n)_{n=1}^{\infty} \subseteq E$ such that $\lim_{n \in \mathbb{N}} x_n = x_0$.*

Proof "Only if" $\forall n \in \mathbb{N}$, since $x_0 \in \overline{E}$, then, by Proposition 3.3, $\exists x_n \in E \cap \mathcal{B}(x_0, 1/n)$. Clearly, $(x_n)_{n=1}^{\infty} \subseteq E$ and $\lim_{n \in \mathbb{N}} x_n = x_0$.

"If" This is immediate by Proposition 3.68.

This completes the proof of the proposition. □

Definition 4.14 A metric space is said to be *complete* if every Cauchy sequence in the metric space converges to a point in the space. ◇

Proposition 4.15 *Let $\mathcal{X} := (X, \rho)$ be a metric space, $\mathcal{Y} := (Y, \mathcal{O})$ be a topological space, $f : \mathcal{X} \to \mathcal{Y}$, and $x_0 \in \mathcal{X}$. Then, the following statements are equivalent:*

(i) *f is continuous at x_0.*
(ii) *if x_0 is an accumulation point of $\mathcal{X}$, then $f(x)$ converges to $f(x_0)$ as $x \to x_0$.*
(iii) *$\forall (x_n)_{n=1}^{\infty} \subseteq \mathcal{X}$ with $\lim_{n \in \mathbb{N}} x_n = x_0$, we have $(f(x_n))_{n=1}^{\infty} \subseteq \mathcal{Y}$ converges to $f(x_0)$.*
(iv) *$\forall (x_n)_{n=1}^{\infty} \subseteq \mathcal{X}$ with x_0 as a cluster point, we have that $(f(x_n))_{n=1}^{\infty} \subseteq \mathcal{Y}$ admits a cluster point $f(x_0)$.*

Proof (i) ⇔ (ii). This follows from Proposition 3.74.

(i) ⇒ (iii). This follows from Proposition 3.66.

(iii) ⇒ (iv). $\forall (x_n)_{n=1}^{\infty} \subseteq \mathcal{X}$ with x_0 as a cluster point. Since $\mathcal{X}$ is a metric space and therefore first countable, then $\exists$ a subsequence $\left(x_{n_i}\right)_{i=1}^{\infty}$ of $(x_n)_{n=1}^{\infty}$ such that $\lim_{i \in \mathbb{N}} x_{n_i} = x_0$. Then, by (iii), $\left(f(x_{n_i})\right)_{i=1}^{\infty}$ converges to $f(x_0)$. Then, $f(x_0)$ is a cluster point of $\left(f(x_{n_i})\right)_{i=1}^{\infty}$ and therefore a cluster point of $(f(x_n))_{n=1}^{\infty}$.

(iv) ⇒ (i). Suppose f is not continuous at x_0. $\exists O_{Y0} \in \mathcal{O}_{\mathcal{Y}}$ with $f(x_0) \in O_{Y0}$ such that $\forall n \in \mathbb{N}$, we have $f(\mathcal{B}(x_0, 1/n)) \not\subseteq O_{Y0}$. Then, $\exists x_n \in \mathcal{B}(x_0, 1/n)$ such that $f(x_n) \in \widetilde{O_{Y0}}$. Consider the sequence $(x_n)_{n=1}^{\infty}$. Clearly, $x_0 = \lim_{n \in \mathbb{N}} x_n$ and therefore is a cluster point of the sequence. Consider the sequence $(f(x_n))_{n=1}^{\infty}$. For the open set $O_{Y0} \ni f(x_0)$, $\forall n \in \mathbb{N}$, $f(x_n) \in \widetilde{O_{Y0}}$. Then, $f(x_0)$ is not a cluster point of $(f(x_n))_{n=1}^{\infty}$. This contradicts with the assumption. Therefore, f must be continuous at x_0.

This completes the proof of the proposition. □

Proposition 4.16 *Let $\mathcal{X} := (X, \rho)$ be a metric space, $\mathcal{Y} := (Y, \mathcal{O})$ be a topological space, $D \subseteq \mathcal{X}$, $f : D \to \mathcal{Y}$, $x_0 \in \mathcal{X}$ be an accumulation point of D, and $y_0 \in \mathcal{Y}$. Then, the following statements are equivalent:*

(i) *$f(x)$ converges to y_0 as $x \to x_0$.*
(ii) *$\forall (x_n)_{n=1}^{\infty} \subseteq D \setminus \{x_0\}$ with $\lim_{n\in\mathbb{N}} x_n = x_0$, we have that y_0 is a limit point of $(f(x_n))_{n=1}^{\infty}$.*

Proof (i) $\Rightarrow$ (ii). By (i), $\forall O \in \mathcal{O}$ with $y_0 \in O$, $\exists \delta \in (0, \infty) \subset \mathbb{R}$ such that $\forall x \in (D \cap \mathcal{B}_{\mathcal{X}}(x_0, \delta)) \setminus \{x_0\}$, $f(x) \in O$. $\forall (x_n)_{n=1}^{\infty} \subseteq D \setminus \{x_0\}$ with $\lim_{n\in\mathbb{N}} x_n = x_0$, $\exists N \in \mathbb{N}$ such that $\forall n \geq N$, $x_n \in \mathcal{B}_{\mathcal{X}}(x_0, \delta)$. Then, $x_n \in (D \cap \mathcal{B}_{\mathcal{X}}(x_0, \delta)) \setminus \{x_0\}$ and $f(x_n) \in O$. Hence, y_0 is a limit point of $(f(x_n))_{n=1}^{\infty}$.

(ii) $\Rightarrow$ (i). We will show this by an argument of contradiction. Suppose (i) does not hold. Then, $\exists O_0 \in \mathcal{O}$ with $y_0 \in O_0$, $\forall n \in \mathbb{N}$, $\exists x_n \in (D \cap \mathcal{B}_{\mathcal{X}}(x_0, 1/n)) \setminus \{x_0\}$ such that $f(x_n) \in \mathcal{X} \setminus O_0$. Clearly, the sequence $(x_n)_{n=1}^{\infty} \subseteq D \setminus \{x_0\}$ and $\lim_{n\in\mathbb{N}} x_n = x_0$. But, $f(x_n) \notin O_0$, $\forall n \in \mathbb{N}$. Then, y_0 is not a limit point of $(f(x_n))_{n=1}^{\infty}$. This contradicts (ii). Hence, (i) must hold.

This completes the proof of the proposition. □

Proposition 4.17 *Let $\mathcal{X}$ be a metric space, $D \subseteq \mathcal{X}$, $f : D \to \mathbb{R}$, and $x_0 \in \mathcal{X}$ be an accumulation point of D. Then, we have*

$$\limsup_{x\to x_0} f(x) = \inf_{\epsilon\in(0,\infty)\subset\mathbb{R}} \sup_{x\in(D\cap\mathcal{B}(x_0,\epsilon))\setminus\{x_0\}} f(x)$$

$$\liminf_{x\to x_0} f(x) = \sup_{\epsilon\in(0,\infty)\subset\mathbb{R}} \inf_{x\in(D\cap\mathcal{B}(x_0,\epsilon))\setminus\{x_0\}} f(x)$$

Proof Let $L = \limsup_{x\to x_0} f(x) \in \mathbb{R}_e$ and $\bar{L} := \inf_{\epsilon\in(0,\infty)\subset\mathbb{R}} \sup_{x\in(D\cap\mathcal{B}(x_0,\epsilon))\setminus\{x_0\}} f(x) \in \mathbb{R}_e$. $\forall m \in \mathbb{R}$ with $m < L$, we have $m < \sup_{x\in(D\cap V)\setminus\{x_0\}} f(x)$, $\forall V \in \mathcal{O}_{\mathcal{X}}$ with $x_0 \in V$. Then, $\forall \epsilon \in (0, \infty) \subset \mathbb{R}$, $\sup_{x\in(D\cap\mathcal{B}(x_0,\epsilon))\setminus\{x_0\}} f(x) > m$. Hence, $m \leq \bar{L}$. By the arbitrariness of m, we have $L \leq \bar{L}$. $\forall m \in \mathbb{R}$ with $m > L$, $\exists V \in \mathcal{O}_{\mathcal{X}}$ with $x_0 \in V$ such that $\sup_{x\in(D\cap V)\setminus\{x_0\}} f(x) < m$. Then, $\exists \epsilon \in (0, \infty) \subset \mathbb{R}$ such that $\mathcal{B}(x_0, \epsilon) \subseteq V$. Then, $\sup_{x\in(D\cap\mathcal{B}(x_0,\epsilon))\setminus\{x_0\}} f(x) \leq \sup_{x\in(D\cap V)\setminus\{x_0\}} f(x) < m$. Then, we have $\bar{L} < m$. By the arbitrariness of m, we have $\bar{L} \leq L$. Hence, $L = \bar{L}$.

Note that, by Propositions 3.85 and 3.81,

$$\begin{aligned}\liminf_{x\to x_0} f(x) &= -\limsup_{x\to x_0}(-f)(x) = -\inf_{\epsilon\in(0,\infty)\subset\mathbb{R}} \sup_{x\in(D\cap\mathcal{B}(x_0,\epsilon))\setminus\{x_0\}} (-f)(x) \\ &= \sup_{\epsilon\in(0,\infty)\subset\mathbb{R}} \inf_{x\in(D\cap\mathcal{B}(x_0,\epsilon))\setminus\{x_0\}} f(x)\end{aligned}$$

This completes the proof of the proposition. □

4.3 Uniform Continuity and Uniformity

Definition 4.18 Let $\mathcal{X}$ and $\mathcal{Y}$ be metric spaces and $f : \mathcal{X} \to \mathcal{Y}$. f is said to be *uniformly continuous* if $\forall \epsilon \in (0, \infty) \subset \mathbb{R}$, $\exists \delta \in (0, \infty) \subset \mathbb{R}$ such that $\rho_{\mathcal{Y}}(f(x_1), f(x_2)) < \epsilon$, $\forall x_1, x_2 \in \mathcal{X}$ with $\rho_{\mathcal{X}}(x_1, x_2) < \delta$. ◇

Clearly, a function f is uniformly continuous implies that it is continuous.

Definition 4.19 Let $\mathcal{X}$ and $\mathcal{Y}$ be metric spaces and $f : \mathcal{X} \to \mathcal{Y}$. f is said to be a *uniform homeomorphism* if f is bijective and both f and f_{inv} are uniformly continuous. ◇

Properties preserved under uniform homeomorphisms are called uniform properties. These include Cauchy sequences, completeness, uniform continuity, and total boundedness.

Definition 4.20 Let X be a set and ρ_1 and ρ_2 be two metrics defined on X. Then, the two metrics are said to be *uniformly equivalent* if the identity map from (X, ρ_1) to (X, ρ_2) is a uniform homeomorphism. ◇

Proposition 4.21 *Let (X, ρ) be a metric space. Define $\sigma : X \times X \to \mathbb{R}$ by $\sigma(x_1, x_2) = \dfrac{\rho(x_1, x_2)}{1 + \rho(x_1, x_2)}$, $\forall x_1, x_2 \in X$. Then, σ is a metric on X and ρ and σ are uniformly equivalent.*

Proof $\forall x_1, x_2, x_3 \in X$, $\sigma(x_1, x_2) \in [0, 1)$; $\sigma(x_1, x_2) = 0 \Leftrightarrow \rho(x_1, x_2) = 0 \Leftrightarrow x_1 = x_2$; $\sigma(x_1, x_2) = \sigma(x_2, x_1)$; by the monotonicity of $\frac{s}{1+s}$ on $s > -1$, we have

$$\begin{aligned}\sigma(x_1, x_2) &\le \frac{\rho(x_1, x_3) + \rho(x_3, x_2)}{1 + \rho(x_1, x_3) + \rho(x_3, x_2)} \le \frac{\rho(x_1, x_3)}{1 + \rho(x_1, x_3)} + \frac{\rho(x_3, x_2)}{1 + \rho(x_3, x_2)} \\ * &= \sigma(x_1, x_3) + \sigma(x_3, x_2)\end{aligned}$$

Hence, σ defines a metric on X.

$\forall \epsilon \in (0, \infty) \subset \mathbb{R}$, $\forall x_1, x_2 \in X$ with $\rho(x_1, x_2) < \epsilon$, we have $\sigma(x_1, x_2) \le \rho(x_1, x_2) < \epsilon$. Hence, $\text{id}_X : (X, \rho) \to (X, \sigma)$ is uniformly continuous.

On the other hand, $\forall \epsilon \in (0, \infty) \subset \mathbb{R}$, $\forall x_1, x_2 \in X$ with $\sigma(x_1, x_2) < \frac{\epsilon}{1+\epsilon}$, we have $\rho(x_1, x_2) < \epsilon$. Hence, $\text{id}_X : (X, \sigma) \to (X, \rho)$ is uniformly continuous.

Therefore, ρ and σ are uniformly equivalent. □

Definition 4.22 A metric space $\mathcal{X}$ is said to be *totally bounded* if $\forall \epsilon \in (0, \infty) \subset \mathbb{R}$, there exist finitely many open balls with radius ϵ that cover $\mathcal{X}$. ◇

Proposition 4.23 *Let $\mathcal{X}$, $\mathcal{Y}$, $\mathcal{W}$, and $\mathcal{Z}$ be metric spaces, $(x_n)_{n=1}^{\infty} \subseteq \mathcal{X}$ be a Cauchy sequence, $f : \mathcal{X} \to \mathcal{Y}$ and $g : \mathcal{Y} \to \mathcal{Z}$ be uniformly continuous functions, and $h : \mathcal{X} \to \mathcal{W}$ be a uniform homeomorphism. Then, the following statements hold:*

(i) *$(f(x_n))_{n=1}^{\infty}$ is a Cauchy sequence.*
(ii) *$g \circ f$ is uniformly continuous.*

(iii) *If $\mathcal{X}$ is complete, then $\mathcal{W}$ is complete.*
(iv) *If $\mathcal{X}$ is totally bounded and f is surjective, then $\mathcal{Y}$ is totally bounded.*

Proof (i) $\forall \epsilon \in (0, \infty) \subset \mathbb{R}$, by the uniform continuity of f, $\exists \delta \in (0, \infty) \subset \mathbb{R}$ such that $\rho_{\mathcal{Y}}(f(x_a), f(x_b)) < \epsilon$, $\forall x_a, x_b \in \mathcal{X}$ with $\rho_{\mathcal{X}}(x_a, x_b) < \delta$. Since $(x_i)_{i=1}^{\infty}$ is Cauchy, then $\exists N \in \mathbb{N}$ such that $\rho_{\mathcal{X}}(x_n, x_m) < \delta$, $\forall n, m \geq N$. Then, $\rho_{\mathcal{Y}}(f(x_n), f(x_m)) < \epsilon$. Hence, $(f(x_i))_{i=1}^{\infty}$ is a Cauchy sequence.

(ii) $\forall \epsilon \in (0, \infty) \subset \mathbb{R}$, by the uniform continuity of g, $\exists \delta_1 \in (0, \infty) \subset \mathbb{R}$ such that $\rho_{\mathcal{Z}}(g(y_1), g(y_2)) < \epsilon$, $\forall y_1, y_2 \in \mathcal{Y}$ with $\rho_{\mathcal{Y}}(y_1, y_2) < \delta_1$. By the uniform continuity of f, $\exists \delta \in (0, \infty) \subset \mathbb{R}$ such that $\rho_{\mathcal{Y}}(f(x_a), f(x_b)) < \delta_1$, $\forall x_a, x_b \in \mathcal{X}$ with $\rho_{\mathcal{X}}(x_a, x_b) < \delta$. Then, we have $\rho_{\mathcal{Z}}(g(f(x_a)), g(f(x_b))) < \epsilon$. Hence, $g \circ f$ is uniformly continuous.

(iii) $\forall$ Cauchy sequence $(w_i)_{i=1}^{\infty} \subseteq \mathcal{W}$. By (i), $(h_{\text{inv}}(w_i))_{i=1}^{\infty} \subseteq \mathcal{X}$ is a Cauchy sequence. Since $\mathcal{X}$ is complete, then $\lim_{i \in \mathbb{N}} h_{\text{inv}}(w_i) = x_0 \in \mathcal{X}$. By Proposition 3.66, we have $\lim_{i \in \mathbb{N}} w_i = \lim_{i \in \mathbb{N}} h(h_{\text{inv}}(w_i)) = h(x_0) \in \mathcal{W}$. Hence, $\mathcal{W}$ is complete.

(iv) $\forall \epsilon \in (0, \infty) \subset \mathbb{R}$, by the uniform continuity of f, $\exists \delta \in (0, \infty) \subset \mathbb{R}$ such that $\rho_{\mathcal{Y}}(f(x_a), f(x_b)) < \epsilon$, $\forall x_a, x_b \in \mathcal{X}$ with $\rho_{\mathcal{X}}(x_a, x_b) < \delta$. By the total boundedness of $\mathcal{X}$, there exists a finite set $X_N \subseteq \mathcal{X}$, such that $\bigcup_{x \in X_N} \mathcal{B}_{\mathcal{X}}(x, \delta) = \mathcal{X}$. Then, by the surjectiveness of f and Proposition 2.5, we have $\bigcup_{x \in X_N} f(\mathcal{B}_{\mathcal{X}}(x, \delta)) = \mathcal{Y}$. Note that $f(\mathcal{B}_{\mathcal{X}}(x, \delta)) \subseteq \mathcal{B}_{\mathcal{Y}}(f(x), \epsilon)$, $\forall x \in \mathcal{X}$. Then, we have $\bigcup_{x \in X_N} \mathcal{B}_{\mathcal{Y}}(f(x), \epsilon) = \mathcal{Y}$. Hence, $\mathcal{Y}$ is totally bounded.

This completes the proof of the proposition. □

Definition 4.24 Let X be a set and $\mathcal{Y} := (Y, \rho)$ be a metric space. Let $(f_\alpha)_{\alpha \in \mathcal{A}}$ be a net of functions of X to $\mathcal{Y}$. Then, the net is said to *converge uniformly* to a function $f : X \to \mathcal{Y}$ if $\forall \epsilon \in (0, \infty) \subset \mathbb{R}$, $\exists \alpha_0 \in \mathcal{A}$, $\forall \alpha \in \mathcal{A}$ with $\alpha_0 \prec \alpha$, we have $\rho(f_\alpha(x), f(x)) < \epsilon$, $\forall x \in X$. ◇

Definition 4.25 Let X be a set and $\mathcal{Y} := (Y, \rho)$ be a metric space. Let $(f_n)_{n=1}^{\infty}$ be a sequence of functions of X to $\mathcal{Y}$. Then, the sequence is said to be a *uniform Cauchy sequence* if $\forall \epsilon \in (0, \infty) \subset \mathbb{R}$, $\exists N \in \mathbb{N}$, $\forall n, m \in \mathbb{N}$ with $n, m \geq N$, we have $\rho(f_n(x), f_m(x)) < \epsilon$, $\forall x \in X$. ◇

A uniformly convergent sequence $(f_n)_{n=1}^{\infty}$ is a uniform Cauchy sequence. A uniform Cauchy sequence in a complete metric space is uniformly convergent.

Proposition 4.26 *Let $\mathcal{X} := (X, \mathcal{O})$ be a topological space and $\mathcal{Y} := (Y, \rho)$ be a metric space. Let $(f_n)_{n=1}^{\infty}$ be a uniformly convergent sequence of functions of $\mathcal{X}$ to $\mathcal{Y}$ whose limit is $f : \mathcal{X} \to \mathcal{Y}$. Assume that, $\forall n \in \mathbb{N}$, f_n is continuous at $x_0 \in \mathcal{X}$. Then, f is continuous at x_0.*

Proof $\forall \epsilon \in (0, \infty) \subset \mathbb{R}$, $\exists N \in \mathbb{N}$, we have $\rho(f_N(x), f(x)) < \epsilon/3$, $\forall x \in \mathcal{X}$. Since f_N is continuous at x_0, then $\exists U \in \mathcal{O}$ with $x_0 \in U$, $\forall x \in U$, we have $\rho(f_N(x), f_N(x_0)) < \epsilon/3$. Then, $\forall x \in U$, we have

$$\begin{aligned}\rho(f(x), f(x_0)) &\leq \rho(f(x), f_N(x)) + \rho(f_N(x), f_N(x_0)) + \rho(f_N(x_0), f(x_0)) \\ &< \epsilon\end{aligned}$$

Therefore, f is continuous at x_0.

This completes the proof of the proposition. □

Definition 4.27 Let $\mathcal{X} := (X, \mathcal{O})$ be a topological space and $\mathcal{Y} := (Y, \rho)$ be a metric space. Let $\mathcal{F}$ be a family of continuous functions of $\mathcal{X}$ to $\mathcal{Y}$. $\mathcal{F}$ is said to be *equicontinuous at* $x_0 \in \mathcal{X}$ if $\forall \epsilon \in (0, \infty) \subset \mathbb{R}$, $\exists U \in \mathcal{O}$ with $x_0 \in U$, $\forall x \in U$, we have $\rho(f(x), f(x_0)) < \epsilon$, $\forall f \in \mathcal{F}$. $\mathcal{F}$ is said to be *equicontinuous* if it is equicontinuous at x_0, $\forall x_0 \in \mathcal{X}$. ◇

4.4 Product Metric Spaces

Definition 4.28 Let (X, ρ_X) and (Y, ρ_Y) be metric spaces. Then, the *product metric space* $(X \times Y, \rho) := (X, \rho_X) \times (Y, \rho_Y)$ is defined by

$$\rho((x_1, y_1), (x_2, y_2)) := ((\rho_X(x_1, x_2))^2 + (\rho_Y(y_1, y_2))^2)^{1/2}$$

$\forall x_1, x_2 \in X$, $\forall y_1, y_2 \in Y$. ρ is called the *Cartesian metric*. ◇

In the above definition, it is straightforward to show that ρ defines a metric on $X \times Y$.

Proposition 4.29 *Let $\mathcal{X} := (X, \rho_X)$ be a metric space with natural topology $\mathcal{O}_X$. Let $\mathcal{Y} := (Y, \rho_Y)$ be a metric space with natural topology $\mathcal{O}_Y$. Let $\mathcal{X} \times \mathcal{Y} := (X \times Y, \rho)$ be the product metric space with the natural topology $\mathcal{O}_m$. Let $(X \times Y, \mathcal{O})$ be the product topological space $(X, \mathcal{O}_X) \times (Y, \mathcal{O}_Y)$. Then, $\mathcal{O} = \mathcal{O}_m$.*

Proof A basis for $\mathcal{O}$ is $\mathcal{M} := \{O_X \times O_Y \mid O_X \in \mathcal{O}_X, O_Y \in \mathcal{O}_Y\}$. A basis for $\mathcal{O}_m$ is $\mathcal{M}_m := \{\mathcal{B}_{\mathcal{X}\times\mathcal{Y}}((x, y), r) \mid x \in X, y \in Y, r \in (0, \infty) \subset \mathbb{R}\}$.

$\forall \mathcal{B}_{\mathcal{X}\times\mathcal{Y}}((x_0, y_0), r) \in \mathcal{M}_m$. $\forall (x, y) \in \mathcal{B}_{\mathcal{X}\times\mathcal{Y}}((x_0, y_0), r)$. Let $d := r - \rho((x, y), (x_0, y_0)) \in (0, \infty) \subset \mathbb{R}$. $\forall (\bar{x}, \bar{y}) \in \mathcal{B}_{\mathcal{X}}\left(x, d/\sqrt{2}\right) \times \mathcal{B}_{\mathcal{Y}}\left(y, d/\sqrt{2}\right) \in \mathcal{M}$, we have

$$\begin{aligned}\rho((\bar{x}, \bar{y}), (x_0, y_0)) &\leq \rho((\bar{x}, \bar{y}), (x, y)) + \rho((x, y), (x_0, y_0)) \\ &= ((\rho_X(\bar{x}, x))^2 + (\rho_Y(\bar{y}, y))^2)^{1/2} + r - d < r\end{aligned}$$

Then, $(\bar{x}, \bar{y}) \in \mathcal{B}_{\mathcal{X}\times\mathcal{Y}}((x_0, y_0), r)$. This yields $\mathcal{B}_{\mathcal{X}}\left(x, d/\sqrt{2}\right) \times \mathcal{B}_{\mathcal{Y}}\left(y, d/\sqrt{2}\right) \subseteq \mathcal{B}_{\mathcal{X}\times\mathcal{Y}}((x_0, y_0), r)$. This implies that $\mathcal{B}_{\mathcal{X}\times\mathcal{Y}}((x_0, y_0), r) \in \mathcal{O}$. Hence, we have $\mathcal{O}_m \subseteq \mathcal{O}$.

On the other hand, $\forall O_X \times O_Y \in \mathcal{M}$. $\forall (x, y) \in O_X \times O_Y$, we have $x \in O_X$ and $y \in O_Y$. Then, $\exists d_1 \in (0, \infty) \subset \mathbb{R}$ such that $\mathcal{B}_{\mathcal{X}}(x, d_1) \subseteq O_X$ and $\exists d_2 \in (0, \infty) \subset \mathbb{R}$ such that $\mathcal{B}_{\mathcal{Y}}(y, d_2) \subseteq O_Y$. Let $d = \min\{d_1, d_2\} \in (0, \infty) \subset \mathbb{R}$. $\forall (\bar{x}, \bar{y}) \in \mathcal{B}_{\mathcal{X}\times\mathcal{Y}}((x, y), d) \in \mathcal{M}_m$, we have $\rho_X(\bar{x}, x) < d$ and $\rho_Y(\bar{y}, y) < d$, which implies that $\bar{x} \in \mathcal{B}_{\mathcal{X}}(x, d_1)$ and $\bar{y} \in \mathcal{B}_{\mathcal{Y}}(y, d_2)$. Then, $\mathcal{B}_{\mathcal{X}\times\mathcal{Y}}((x, y), d) \subseteq O_X \times O_Y$. Hence, we have $O_X \times O_Y \in \mathcal{O}_m$. Thus, $\mathcal{O} \subseteq \mathcal{O}_m$.

Hence, $\mathcal{O} = \mathcal{O}_m$. This completes the proof of the proposition. □

The above proposition shows that the natural topology induced by the product metric defined in Definition 4.28 is the product topology on $X \times Y$.

Proposition 4.30 *Let $\mathcal{X}$ be a metric space. Then the metric $\rho_{\mathcal{X}}$ is a uniformly continuous function on $\mathcal{X} \times \mathcal{X}$.*

Proof $\forall \epsilon \in (0, \infty) \subset \mathbb{R}$. Let $\delta = \epsilon/\sqrt{2} \in (0, \infty) \subset \mathbb{R}$. $\forall (x_1, x_2), (y_1, y_2) \in \mathcal{X} \times \mathcal{X}$ with $\rho_{\mathcal{X}\times\mathcal{X}}((x_1, x_2), (y_1, y_2)) < \delta$. Then, we have $\delta > ((\rho_{\mathcal{X}}(x_1, y_1))^2 + (\rho_{\mathcal{X}}(x_2, y_2))^2)^{1/2} \geq (\rho_{\mathcal{X}}(x_1, y_1) + \rho_{\mathcal{X}}(x_2, y_2))/\sqrt{2}$. Note that

$$\rho_{\mathcal{X}}(x_1, x_2) \leq \rho_{\mathcal{X}}(x_1, y_1) + \rho_{\mathcal{X}}(y_1, x_2)$$
$$\rho_{\mathcal{X}}(y_1, x_2) \leq \rho_{\mathcal{X}}(x_1, x_2) + \rho_{\mathcal{X}}(x_1, y_1)$$
$$\rho_{\mathcal{X}}(y_1, x_2) \leq \rho_{\mathcal{X}}(y_1, y_2) + \rho_{\mathcal{X}}(x_2, y_2)$$
$$\rho_{\mathcal{X}}(y_1, y_2) \leq \rho_{\mathcal{X}}(x_2, y_1) + \rho_{\mathcal{X}}(y_2, x_2)$$

This implies that

$$\begin{aligned}
-\epsilon = -\sqrt{2}\delta < -\rho_{\mathcal{X}}(x_1, y_1) - \rho_{\mathcal{X}}(x_2, y_2) \leq \rho_{\mathcal{X}}(x_1, x_2) - \rho_{\mathcal{X}}(x_2, y_1) \\
+\rho_{\mathcal{X}}(x_2, y_1) - \rho_{\mathcal{X}}(y_1, y_2) = \rho_{\mathcal{X}}(x_1, x_2) - \rho_{\mathcal{X}}(y_1, y_2) \leq \rho_{\mathcal{X}}(x_1, y_1) \\
+\rho_{\mathcal{X}}(x_2, y_1) + \rho_{\mathcal{X}}(x_2, y_2) - \rho_{\mathcal{X}}(x_2, y_1) = \rho_{\mathcal{X}}(x_1, y_1) + \rho_{\mathcal{X}}(x_2, y_2) < \epsilon
\end{aligned}$$

Hence, we have $|\rho_{\mathcal{X}}(x_1, x_2) - \rho_{\mathcal{X}}(y_1, y_2)| < \epsilon$. Hence, $\rho_{\mathcal{X}}$ is uniformly continuous on $\mathcal{X} \times \mathcal{X}$. This completes the proof of the proposition. □

Proposition 4.31 *Let $\mathcal{X}$ and $\mathcal{Y}$ be complete metric spaces and $\mathcal{Z} = \mathcal{X} \times \mathcal{Y}$ be the product metric space with the Cartesian metric ρ. Then, $\mathcal{Z}$ is complete.*

Proof Fix any Cauchy sequence $((x_n, y_n))_{n=1}^{\infty} \subseteq \mathcal{Z}$. $\forall \epsilon \in (0, \infty) \subset \mathbb{R}$, $\exists N \in \mathbb{N}$ such that $\rho((x_n, y_n), (x_m, y_m)) < \epsilon$, $\forall n, m \geq N$. Then, we have $\rho_{\mathcal{X}}(x_n, x_m) < \epsilon$ and $\rho_{\mathcal{Y}}(y_n, y_m) < \epsilon$. Hence, $(x_n)_{n=1}^{\infty} \subseteq \mathcal{X}$ and $(y_n)_{n=1}^{\infty} \subseteq \mathcal{Y}$ are Cauchy sequences. By the completeness of $\mathcal{X}$ and $\mathcal{Y}$, $\exists x_0 \in \mathcal{X}$ and $\exists y_0 \in \mathcal{Y}$ such that $\lim_{n\in\mathbb{N}} x_n = x_0$ and $\lim_{n\in\mathbb{N}} y_n = y_0$. By Proposition 3.67, we have $\lim_{n\in\mathbb{N}}(x_n, y_n) = (x_0, y_0) \in \mathcal{Z}$. Hence, $\mathcal{Z}$ is complete. This completes the proof of the proposition. □

Clearly, Definition 4.28 and Propositions 4.29 and 4.31 may be easily generalized to the case of $\mathcal{X}_1 \times \cdots \times \mathcal{X}_n$, where $n \in \mathbb{N}$ and $\mathcal{X}_i$ are metric spaces, $i = 1, \ldots, n$. When $n = 0$, it should be noted that $\prod_{\alpha \in \emptyset} \mathcal{X}_\alpha = (\{\emptyset\}, \rho)$, where $\rho(\emptyset, \emptyset) = 0$.

Proposition 4.32 *Let $\mathcal{X}_\alpha$ be a metric space, $\alpha \in \Lambda$, where Λ is a finite set. Let $\Lambda = \bigcup_{\beta \in \Gamma} \Lambda_\beta$, where Λ_β's are pairwise disjoint and finite and Γ is also finite. $\forall \beta \in \Gamma$, let $\mathcal{X}^{\langle \beta \rangle} := \prod_{\alpha \in \Lambda_\beta} \mathcal{X}_\alpha$ be the product metric space. Let $\mathcal{X}^{\langle \Gamma \rangle} := \prod_{\beta \in \Gamma} \prod_{\alpha \in \Lambda_\beta} \mathcal{X}_\alpha$ be the product metric space of product metric spaces, and $\mathcal{X} := \prod_{\alpha \in \Lambda} \mathcal{X}_\alpha$ be the product metric space. Then, $\mathcal{X}$ and $\mathcal{X}^{\langle \Gamma \rangle}$ are isometric.*

Proof Define $E : \prod_{\beta \in \Gamma} \prod_{\alpha \in \Lambda_\beta} \mathcal{X}_\alpha \to \prod_{\alpha \in \Lambda} \mathcal{X}_\alpha$ by, $\forall x \in \mathcal{X}^{\langle \Gamma \rangle}$, $\forall \alpha \in \Lambda$, $\exists!\ \beta_\alpha \in \Gamma \ni \cdot\ \alpha \in \Lambda_{\beta_\alpha}$, $\pi_\alpha(E(x)) = \pi_\alpha^{\langle \beta_\alpha \rangle}(\pi_{\beta_\alpha}^{\langle \Gamma \rangle}(x))$. By Proposition 3.30, E is a homeomorphism. $\forall x, y \in \mathcal{X}^{\langle \Gamma \rangle}$, we have

$$\rho(E(x), E(y)) = \Big(\sum_{\alpha \in \Lambda} (\rho_\alpha(\pi_\alpha(E(x)), \pi_\alpha(E(y))))^2 \Big)^{1/2}$$

$$= \Big(\sum_{\beta \in \Gamma} \sum_{\alpha \in \Lambda_\beta} (\rho_\alpha(\pi_\alpha(E(x)), \pi_\alpha(E(y))))^2 \Big)^{1/2}$$

$$= \Big(\sum_{\beta \in \Gamma} \sum_{\alpha \in \Lambda_\beta} (\rho_\alpha(\pi_\alpha^{\langle \beta \rangle}(\pi_\beta^{\langle \Gamma \rangle}(x)), \pi_\alpha^{\langle \beta \rangle}(\pi_\beta^{\langle \Gamma \rangle}(y))))^2 \Big)^{1/2}$$

$$= \Big(\sum_{\beta \in \Gamma} \Big(\Big(\sum_{\alpha \in \Lambda_\beta} (\rho_\alpha(\pi_\alpha^{\langle \beta \rangle}(\pi_\beta^{\langle \Gamma \rangle}(x)), \pi_\alpha^{\langle \beta \rangle}(\pi_\beta^{\langle \Gamma \rangle}(y))))^2 \Big)^{1/2} \Big)^2 \Big)^{1/2}$$

$$= \Big(\sum_{\beta \in \Gamma} (\rho^{\langle \beta \rangle}(\pi_\beta^{\langle \Gamma \rangle}(x), \pi_\beta^{\langle \Gamma \rangle}(y)))^2 \Big)^{1/2} = \rho^{\langle \Gamma \rangle}(x, y)$$

Hence, E is an isometry. This completes the proof of the proposition. □

Proposition 4.33 *Let $\mathcal{X}_\alpha := (X_\alpha, \rho_{X\alpha})$ and $\mathcal{Y}_\alpha := (Y_\alpha, \rho_{Y\alpha})$ be uniformly homeomorphic metric spaces, $\forall \alpha \in \Lambda$, where Λ is a finite index set. Define the product metric spaces $\mathcal{X} := (X, \rho_X) := \prod_{\alpha \in \Lambda} \mathcal{X}_\alpha$ and $\mathcal{Y} := (Y, \rho_Y) := \prod_{\alpha \in \Lambda} \mathcal{Y}_\alpha$, where ρ_X and ρ_Y are the Cartesian metric. Then, $\mathcal{X}$ and $\mathcal{Y}$ are uniformly homeomorphic.*

Proof Let $F_\alpha : \mathcal{X}_\alpha \to \mathcal{Y}_\alpha$ be a uniform homeomorphism, $\forall \alpha \in \Lambda$. Define $F : \mathcal{X} \to \mathcal{Y}$ by, $\forall x \in \mathcal{X}$, $\pi_\alpha^{\langle Y \rangle}(F(x)) = F_\alpha(\pi_\alpha^{\langle X \rangle}(x))$, $\forall \alpha \in \Lambda$. By Propositions 4.29 and 3.31, F is a homeomorphism between $\mathcal{X}$ and $\mathcal{Y}$. We need only to show that F and F_{inv} are uniformly continuous.

Let $m \in \mathbb{Z}_+$ be the number of elements in Λ. $\forall \epsilon \in (0, \infty) \subset \mathbb{R}$, $\forall \alpha \in \Lambda$, $\exists \delta_\alpha \in (0, \infty) \subset \mathbb{R}$ such that, $\forall x_{\alpha 1}, x_{\alpha 2} \in \mathcal{X}_\alpha$ with $\rho_{X\alpha}(x_{\alpha 1}, x_{\alpha 2}) < \delta_\alpha$, we have $\rho_{Y\alpha}(F_\alpha(x_{\alpha 1}), F_\alpha(x_{\alpha 2})) < \epsilon/\sqrt{1+m}$, by the uniform continuity of F_α. Let $\delta = \min\{\inf_{\alpha\in\Lambda} \delta_\alpha, 1\} \in (0, \infty) \subset \mathbb{R}$. $\forall x_1, x_2 \in \mathcal{X}$ with $\rho_X(x_1, x_2) < \delta$, we have, $\forall \alpha \in \Lambda$,

$$\rho_{X\alpha}(\pi_\alpha^{\langle X \rangle}(x_1), \pi_\alpha^{\langle X \rangle}(x_2)) < \delta \leq \delta_\alpha$$

This implies that $\rho_{Y\alpha}(F_\alpha(\pi_\alpha^{\langle X \rangle}(x_1)), F_\alpha(\pi_\alpha^{\langle X \rangle}(x_2))) < \epsilon/\sqrt{1+m}$. Hence, we have

$$\begin{aligned} \rho_Y(F(x_1), F(x_2)) &= \Big(\sum_{\alpha\in\Lambda}(\rho_{Y\alpha}(\pi_\alpha^{\langle Y \rangle}(F(x_1)), \pi_\alpha^{\langle Y \rangle}(F(x_2))))^2\Big)^{1/2} \\ &= \Big(\sum_{\alpha\in\Lambda}(\rho_{Y\alpha}(F_\alpha(\pi_\alpha^{\langle X \rangle}(x_1)), F_\alpha(\pi_\alpha^{\langle X \rangle}(x_2))))^2\Big)^{1/2} < \epsilon \end{aligned}$$

Hence, F is uniformly continuous.

It is easy to see that $F_{\text{inv}} : \mathcal{Y} \to \mathcal{X}$ is given by, $\forall y \in \mathcal{Y}$, $\pi_\alpha^{\langle X \rangle}(F_{\text{inv}}(y)) = F_{\alpha\,\text{inv}}(\pi_\alpha^{\langle Y \rangle}(y))$, $\forall \alpha \in \Lambda$. Then, it is straightforward to apply a similar argument as above to show that F_{inv} is uniformly continuous.

This completes the proof of the proposition. □

Proposition 4.34 *Let $\mathcal{X}_i := (X_i, \rho_i)$ be a metric space with natural topology $\mathcal{O}_i$, $\forall i \in \mathbb{N}$. On $X := \prod_{i=1}^\infty X_i$, define the metric, $\forall x, y \in X$,*

$$\rho(x, y) = \sum_{i=1}^{\infty} 2^{-i} \frac{\rho_i(\pi_i(x), \pi_i(y))}{1 + \rho_i(\pi_i(x), \pi_i(y))}$$

Then, the natural topology, $\mathcal{O}_m$, on $\mathcal{X} := (X, \rho)$ induced by the metric ρ equals the product topology $\mathcal{O}$, where $(X, \mathcal{O}) = \prod_{i=1}^\infty (X_i, \mathcal{O}_i)$.

Proof We will first show that ρ defines a metric on X. $\forall x, y, z \in X$. Clearly, we have $0 \leq \rho(x, y) < 1$ and $\rho(x, y) = \rho(y, x)$. When $x = y$, we have $\rho(x, y) = 0$. If $\rho(x, y) = 0$, then $\rho_i(\pi_i(x), \pi_i(y)) = 0$, $\forall i \in \mathbb{N}$, which implies that $\pi_i(x) = \pi_i(y)$,

$\forall i \in \mathbb{N}$, and hence $x = y$. Note that the function $\frac{s}{1+s} = 1 - \frac{1}{1+s}$ is strictly increasing on $s > -1$. Then, we have

$$\begin{aligned}\rho(x, y) &\leq \sum_{i=1}^{\infty} 2^{-i} \frac{\rho_i(\pi_i(x), \pi_i(z)) + \rho_i(\pi_i(z), \pi_i(y))}{1 + \rho_i(\pi_i(x), \pi_i(z)) + \rho_i(\pi_i(z), \pi_i(y))} \\ &\leq \sum_{i=1}^{\infty} 2^{-i} \frac{\rho_i(\pi_i(x), \pi_i(z))}{1 + \rho_i(\pi_i(x), \pi_i(z))} + \sum_{i=1}^{\infty} 2^{-i} \frac{\rho_i(\pi_i(z), \pi_i(y))}{1 + \rho_i(\pi_i(z), \pi_i(y))} \\ &= \rho(x, z) + \rho(z, y)\end{aligned}$$

Hence, ρ is a metric on X.

Fix any basis open set $B = \mathcal{B}_{\mathcal{X}}(x, r) \in \mathcal{O}_m$, where $x \in X$ and $r \in (0, \infty) \subset \mathbb{R}$. $\forall y \in B$. Let $\delta := r - \rho(x, y) > 0$ and $N \in \mathbb{N}$ be such that $2^{-N} < \delta/2$. Consider the set $C_y = \prod_{i=1}^{\infty} C_{yi} \subseteq X$ given by $C_{yi} = \mathcal{B}_{\mathcal{X}_i}(\pi_i(y), \delta/2)$, $i = 1, \ldots, N$, and $C_{yi} = X_i$, $i = N+1, N+2, \ldots$. Clearly, $y \in C_y \in \mathcal{O}$. $\forall z \in C_y$, we have $\rho(y, z) < \sum_{i=1}^{N} 2^{-i-1}\delta + \sum_{i=N+1}^{\infty} 2^{-i} < \delta/2 + 2^{-N} < \delta$. Then, $\rho(x, z) \leq \rho(x, y) + \rho(y, z) < r$. Hence, $z \in B$. Then, $C_y \subseteq B$. This shows that $B \in \mathcal{O}$. Hence, $\mathcal{O}_m \subseteq \mathcal{O}$.

Fix any basis open set $C \in \mathcal{O}$. Then, $C = \prod_{i=1}^{\infty} C_i$, where $C_i \in \mathcal{O}_i$, $i = 1, \ldots, N$, and $C_i = X_i$, $i = N+1, N+2, \ldots$, for some $N \in \mathbb{N}$. $\forall x \in C$, we have $\pi_i(x) \in C_i$, $\forall i \in \mathbb{N}$. Then, $\forall i = 1, \ldots, N$, $\exists \delta_i \in (0, \infty) \subset \mathbb{R}$ such that $\mathcal{B}_{\mathcal{X}_i}(\pi_i(x), \delta_i) \subseteq C_i$. Let $\delta_x := \min_{1 \leq i \leq N} 2^{-i} \frac{\delta_i}{1+\delta_i} > 0$. Let $B = \mathcal{B}_{\mathcal{X}}(x, \delta_x) \in \mathcal{O}_m$. $\forall y \in B$, $\rho(x, y) < \delta_x$ implies that $2^{-i} \frac{\rho_i(\pi_i(x), \pi_i(y))}{1+\rho_i(\pi_i(x), \pi_i(y))} < 2^{-i} \frac{\delta_i}{1+\delta_i}$ and $\rho_i(\pi_i(x), \pi_i(y)) < \delta_i$, $i = 1, \ldots, N$. Hence, $y \in C$. Then, we have $x \in B \subseteq C$. This shows that $C \in \mathcal{O}_m$. Hence, $\mathcal{O} \subseteq \mathcal{O}_m$.

Hence, $\mathcal{O} = \mathcal{O}_m$. This completes the proof of the proposition. □

Proposition 4.35 *Let $\mathcal{X}_i := (X_i, \rho_i)$ be a complete metric space, $\forall i \in \mathbb{N}$. On $X := \prod_{i=1}^{\infty} X_i$, define the metric ρ as in Proposition 4.34. Then, the product metric space $\mathcal{X} := (X, \rho)$ is complete.*

Proof For all Cauchy sequence $(x_n)_{n=1}^{\infty} \subseteq \mathcal{X}$, clearly, $(\pi_i(x_n))_{n=1}^{\infty} \subseteq \mathcal{X}_i$ is a Cauchy sequence, $\forall i \in \mathbb{N}$. By the completeness of $\mathcal{X}_i$, $\exists y_i \in \mathcal{X}_i$ such that $\lim_{n \in \mathbb{N}} \pi_i(x_n) = y_i$, $\forall i \in \mathbb{N}$. Define $y \in \mathcal{X}$ by $\pi_i(y) = y_i$, $\forall i \in \mathbb{N}$. By Propositions 3.67 and 4.34, we have y is the limit of $(x_n)_{n=1}^{\infty}$. Hence, $\mathcal{X}$ is complete. This completes the proof of the proposition. □

Proposition 4.36 *Let $\mathcal{X}_i := (X_i, \rho_{Xi})$ and $\mathcal{Y}_i := (Y_i, \rho_{Yi})$ be uniformly homeomorphic metric spaces, $\forall i \in \mathbb{N}$. Define the infinite product metric spaces $\mathcal{X} := (X, \rho_X) := \left(\prod_{i=1}^{\infty} X_i, \rho_X\right)$ and $\mathcal{Y} := (Y, \rho_Y) := \left(\prod_{i=1}^{\infty} Y_i, \rho_Y\right)$, where ρ_X and ρ_Y are defined as in Proposition 4.34. Then, $\mathcal{X}$ and $\mathcal{Y}$ are uniformly homeomorphic.*

Proof Let $F_i : \mathcal{X}_i \to \mathcal{Y}_i$ be a uniform homeomorphism, $\forall i \in \mathbb{N}$. Define $F : \mathcal{X} \to \mathcal{Y}$ by, $\forall x \in \mathcal{X}$, $\pi_i^{\langle Y \rangle}(F(x)) = F_i(\pi_i^{\langle X \rangle}(x))$, $\forall i \in \mathbb{N}$. By Propositions 4.34 and 3.31,

F is a homeomorphism between $\mathcal{X}$ and $\mathcal{Y}$. We need only to show that F and F_{inv} are uniformly continuous.

$\forall \epsilon \in (0,\infty) \subset \mathbb{R}$, $\exists N \in \mathbb{N}$ such that $2^{-N} < \epsilon/2$. $\forall i = 1,\ldots,N$, $\exists \delta_i \in (0,\infty) \subset \mathbb{R}$ such that, $\forall x_{i1}, x_{i2} \in X_i$ with $\rho_{Xi}(x_{i1}, x_{i2}) < \delta_i$, we have $\rho_{Yi}(F_i(x_{i1}), F_i(x_{i2})) < \epsilon/2$, by the uniform continuity of F_i. Let $\delta = \min_{1\le i\le N} 2^{-i}\frac{\delta_i}{1+\delta_i} > 0$. $\forall x_1, x_2 \in \mathcal{X}$ with $\rho_X(x_1, x_2) < \delta$, we have, $\forall i = 1,\ldots,N$,

$$\frac{\rho_{Xi}(\pi_i^{\langle X\rangle}(x_1), \pi_i^{\langle X\rangle}(x_2))}{1+\rho_{Xi}(\pi_i^{\langle X\rangle}(x_1), \pi_i^{\langle X\rangle}(x_2))} < 2^i\delta \le \frac{\delta_i}{1+\delta_i}$$

This implies that $\rho_{Xi}(\pi_i^{\langle X\rangle}(x_1), \pi_i^{\langle X\rangle}(x_2)) < \delta_i$, which further implies that $\rho_{Yi}(F_i(\pi_i^{\langle X\rangle}(x_1)), F_i(\pi_i^{\langle X\rangle}(x_2))) < \epsilon/2$. Hence, we have

$$\begin{aligned}
\rho_Y(F(x_1), F(x_2)) &= \sum_{i=1}^{\infty} 2^{-i}\frac{\rho_{Yi}(\pi_i^{\langle Y\rangle}(F(x_1)), \pi_i^{\langle Y\rangle}(F(x_2)))}{1+\rho_{Yi}(\pi_i^{\langle Y\rangle}(F(x_1)), \pi_i^{\langle Y\rangle}(F(x_2)))} \\
&= \sum_{i=1}^{\infty} 2^{-i}\frac{\rho_{Yi}(F_i(\pi_i^{\langle X\rangle}(x_1)), F_i(\pi_i^{\langle X\rangle}(x_2)))}{1+\rho_{Yi}(F_i(\pi_i^{\langle X\rangle}(x_1)), F_i(\pi_i^{\langle X\rangle}(x_2)))} \\
&< \sum_{i=1}^{N} 2^{-i}\frac{\epsilon}{2} + 2^{-N} < \epsilon
\end{aligned}$$

Hence, F is uniformly continuous.

It is easy to see that $F_{\text{inv}} : \mathcal{Y} \to \mathcal{X}$ is given by, $\forall y \in \mathcal{Y}$, $\pi_i^{\langle X\rangle}(F_{\text{inv}}(y)) = F_{i\,\text{inv}}(\pi_i^{\langle Y\rangle}(y))$, $\forall i \in \mathbb{N}$. Then, it is straightforward to apply a similar argument as above to show that F_{inv} is uniformly continuous.

This completes the proof of the proposition. □

4.5 Subspaces

Proposition 4.37 *Let $\mathcal{X} := (X, \rho)$ be a metric space with natural topology $\mathcal{O}$ and $A \subseteq \mathcal{X}$ be a subset. $\mathcal{A} := (A, \rho)$ is a metric space with natural topology $\mathcal{O}_A$. A may also be endowed with the subset topology $\mathcal{O}_s$ with respect to $(X, \mathcal{O})$. Then, $\mathcal{O}_A = \mathcal{O}_s$.*

Proof A basis for the natural topology $\mathcal{O}_A$ is

$$\begin{aligned}
\mathcal{B} &:= \{\mathcal{B}_{\mathcal{A}}(x, r) \mid x \in A, r \in (0,\infty) \subseteq \mathbb{R}\} \\
&= \{\mathcal{B}_{\mathcal{X}}(x, r) \cap A \mid x \in A, r \in (0,\infty) \subseteq \mathbb{R}\}
\end{aligned}$$

$\mathcal{O}_s = \{O \cap A \mid O \in \mathcal{O}\}$. Clearly, we have $\mathcal{B} \subseteq \mathcal{O}_s$. Then, $\mathcal{O}_A \subseteq \mathcal{O}_s$. On the other hand, $\forall O_s \in \mathcal{O}_s$, $O_s = O \cap A$ with $O \in \mathcal{O}$. $\forall x \in O_s$, $x \in A$ and $\exists r \in (0, \infty) \subset \mathbb{R}$ such that $\mathcal{B}_{\mathcal{X}}(x, r) \subseteq O$. Then, $\mathcal{B}_{\mathcal{A}}(x, r) \subseteq O_s$. Then, $O_s \in \mathcal{O}_A$. Then, $\mathcal{O}_s \subseteq \mathcal{O}_A$. Hence, $\mathcal{O}_A = \mathcal{O}_s$.

This completes the proof of the proposition. □

Proposition 4.38 *Let $\mathcal{X}$ be a metric space and $S \subseteq \mathcal{X}$. If $\mathcal{X}$ is separable, then $\mathcal{S} := (S, \rho)$ is separable. If S is separable, then $\overline{S}$ is separable.*

Proof Since $\mathcal{X}$ is separable, by Proposition 4.4, there exists a countable basis $\mathcal{B}$ for $\mathcal{X}$. Let $\mathcal{B}_S := \{B \cap S \mid B \in \mathcal{B}\}$, which is a countable basis for $\mathcal{S}$. By Proposition 4.4, $\mathcal{S}$ is separable.

Let S be separable. Then, there exists a countable dense subset $D \subseteq S$. The closure of D in $\mathcal{X}$ is $\overline{D}$. By Proposition 3.5, $S \subseteq \overline{D}$. By Proposition 3.3, $\overline{D} \subseteq \overline{S} \subseteq \overline{D}$. Hence, $\overline{D} = \overline{S}$ and D is dense in $\overline{S}$. Hence, $\overline{S}$ is separable. This completes the proof of the proposition. □

Proposition 4.39 *Let $\mathcal{X} := (X, \rho)$ be a metric space and $S \subseteq X$.*

(i) *If (S, ρ) is complete, then S is closed in $\mathcal{X}$.*
(ii) *If $\mathcal{X}$ is complete and S is closed in $\mathcal{X}$, then (S, ρ) is complete.*

Proof (i) Any convergent sequence is Cauchy. Then, by Proposition 4.13, (S, ρ) is complete implies that any point of closure of S is in S. Then, $S = \overline{S}$ and S is closed. (ii) Any Cauchy sequence in S will converge in $\mathcal{X}$ since $\mathcal{X}$ is complete. Then, by Proposition 4.13, the limit lies in S since S is closed. Hence, (S, ρ) is complete. This completes the proof of the proposition. □

4.6 Baire Category

Theorem 4.40 (Baire) *Let $\mathcal{X}$ be a complete metric space and $(O_k)_{k=1}^{\infty}$ be a sequence of open dense sets in $\mathcal{X}$. Then, $\bigcap_{k=1}^{\infty} O_k$ is dense in $\mathcal{X}$.*

Proof $\forall U \in \mathcal{O}$ with $U \neq \emptyset$. Since O_1 is dense, then $\exists x_1 \in O_1 \cap U$. By the openness of U and O_1, $\exists r_1 \in (0, \infty) \subset \mathbb{R}$ such that $\mathcal{B}(x_1, r_1) \subseteq O_1 \cap U$. $\forall n \in \mathbb{N}$ with $n > 1$, since O_n is dense, then $\exists x_n \in O_n \cap \mathcal{B}(x_{n-1}, r_{n-1}/2)$. Since $O_n, \mathcal{B}(x_{n-1}, r_{n-1}) \in \mathcal{O}$, then $\exists r_n \in (0, r_{n-1}/2] \subseteq \mathbb{R}$ such that $\mathcal{B}(x_n, r_n) \subseteq O_n \cap \mathcal{B}(x_{n-1}, r_{n-1})$.

Note that $\forall n \in \mathbb{N}$, $r_n \le 2^{1-n} r_1$ and, $\forall m > n$,

$$\begin{aligned} \rho(x_n, x_m) &\le \rho(x_n, x_{n+1}) + \cdots + \rho(x_{m-1}, x_m) < \frac{1}{2}(r_n + \cdots + r_{m-1}) \\ &\le 2^{-n} r_1 + \cdots + 2^{1-m} r_1 < 2^{1-n} r_1 \end{aligned}$$

Hence, $(x_n)_{n=1}^{\infty}$ is a Cauchy sequence, which converges to $x_0 \in \mathcal{X}$, by the completeness of $\mathcal{X}$. By Propositions 3.66, 3.67, and 4.30, we have, $\forall n \in \mathbb{N}$,

$$\rho(x_n, x_0) = \lim_{m \in \mathbb{N}} \rho(x_n, x_m) < \frac{1}{2}(r_n + 2^{-1}r_n + 2^{-2}r_n + \cdots) = r_n$$

Then, $x_0 \in \mathcal{B}(x_n, r_n) \subseteq O_n$. Hence, $x_0 \in \left(\bigcap_{k=1}^{\infty} O_k\right) \cap U \neq \emptyset$. Hence, $\bigcap_{k=1}^{\infty} O_k$ is dense in $\mathcal{X}$. This completes the proof of the theorem. □

Theorem 4.41 (Baire Category Theorem) *Let $\mathcal{X}$ be a complete metric space, then it is second category everywhere, that is, no nonempty open subset of $\mathcal{X}$ is of first category.*

Proof We will prove this theorem by using Proposition 3.38. Let $(O_\alpha)_{\alpha \in \Lambda}$ be a countable collection of open dense sets in $\mathcal{X}$. When Λ is finite, we may add the open dense set $\mathcal{X}$ into the collection to make it infinite and countable. By Baire's Theorem, we then have $\bigcap_{\alpha \in \Lambda} O_\alpha$ is dense in $\mathcal{X}$. Then, by Proposition 3.38, $\mathcal{X}$ is second category everywhere. This completes the proof of the theorem. □

4.7 Completion of Metric Spaces

Proposition 4.42 *Let $\mathcal{X} := (X, \mathcal{O})$ be a topological space and $\mathcal{Y} := (Y, \rho)$ be a complete metric space. Let $(f_n)_{n=1}^{\infty}$ be a uniform Cauchy sequence of continuous functions of $\mathcal{X}$ to $\mathcal{Y}$. Then, there exists a continuous function $f : \mathcal{X} \to \mathcal{Y}$ such that $(f_n)_{n=1}^{\infty}$ converges uniformly to f.*

Proof $\forall x \in \mathcal{X}$, $(f_n(x))_{n=1}^{\infty}$ is a Cauchy sequence in $\mathcal{Y}$. By the completeness of $\mathcal{Y}$ and Proposition 3.65, $\exists!\, y \in \mathcal{Y}$ such that $\lim_{n \in \mathbb{N}} f_n(x) = y$. This defines a function $f : \mathcal{X} \to \mathcal{Y}$.

Since $(f_n)_{n=1}^{\infty}$ is a uniform Cauchy sequence, then, $\forall \epsilon \in (0, \infty) \subset \mathbb{R}$, $\exists N \in \mathbb{N}$, $\forall n, m \in \mathbb{N}$ with $n, m \geq N$, we have $\rho(f_n(x), f_m(x)) < \epsilon$, $\forall x \in \mathcal{X}$. By Propositions 4.30, 3.66, and 3.67, we have $\rho(f_n(x), f(x)) = \lim_{m \in \mathbb{N}} \rho(f_n(x), f_m(x)) \leq \epsilon$, $\forall x \in \mathcal{X}$. Hence, $(f_n)_{n=1}^{\infty}$ converges uniformly to f. By Proposition 4.26, f is continuous.

This completes the proof of the proposition. □

Definition 4.43 Let $\mathcal{X}$ be a metric space. A net $(x_\alpha)_{\alpha \in \mathcal{A}} \subseteq \mathcal{X}$ is said to be *Cauchy* if $\forall \epsilon \in (0, \infty) \subset \mathbb{R}$, $\exists \alpha_0 \in \mathcal{A}$, $\forall \alpha_1, \alpha_2 \in \mathcal{A}$ with $\alpha_0 \prec \alpha_1$ and $\alpha_0 \prec \alpha_2$, we have $\rho(x_{\alpha_1}, x_{\alpha_2}) < \epsilon$. ◇

Any convergent net in a metric space is Cauchy.

Proposition 4.44 *Let $\mathcal{X}$ be a metric space. Then, $\mathcal{X}$ is complete if, and only if, every Cauchy net has a limit in $\mathcal{X}$.*

Proof "If" A Cauchy sequence is a Cauchy net and admits a limit in $\mathcal{X}$. Hence, $\mathcal{X}$ is complete.

"Only if" Let $(x_\alpha)_{\alpha\in\mathcal{A}} \subseteq \mathcal{X}$ be a Cauchy net. $\exists \alpha_1 \in \mathcal{A} \ni \cdot\ \forall \bar{\alpha}_1, \bar{\alpha}_2 \in \mathcal{A}$ with $\alpha_1 \prec \bar{\alpha}_1$ and $\alpha_1 \prec \bar{\alpha}_2$, we have $\rho(x_{\bar{\alpha}_1}, x_{\bar{\alpha}_2}) < 1$. $\forall n \in \mathbb{N}$ with $n > 1$, $\exists \alpha_n \in \mathcal{A}$ with $\alpha_{n-1} \prec \alpha_n$, $\forall \bar{\alpha}_1, \bar{\alpha}_2 \in \mathcal{A}$ with $\alpha_n \prec \bar{\alpha}_1$ and $\alpha_n \prec \bar{\alpha}_2$, we have $\rho(x_{\bar{\alpha}_1}, x_{\bar{\alpha}_2}) < 1/n$. $\forall n \in \mathbb{N}$, $\forall m_1, m_2 \in \mathbb{N}$ with $m_1, m_2 > n$, we have $\alpha_n \prec \alpha_{n+1} \prec \cdots \prec \alpha_{m_1}$ and $\alpha_n \prec \alpha_{n+1} \prec \cdots \prec \alpha_{m_2}$. Then, $\rho(x_{\alpha_{m_1}}, x_{\alpha_{m_2}}) < 1/n$. Hence, $\left(x_{\alpha_n}\right)_{n=1}^{\infty} \subseteq \mathcal{X}$ is a Cauchy sequence. By the completeness of $\mathcal{X}$, $\exists x_0 \in \mathcal{X}$ such that $\lim_{n\in\mathbb{N}} x_{\alpha_n} = x_0$. $\forall n \in \mathbb{N}$, $\forall \alpha \in \mathcal{A}$ with $\alpha_n \prec \alpha$, $\forall m \in \mathbb{N}$ with $m > n$, we have $\alpha_n \prec \alpha_{n+1} \prec \cdots \prec \alpha_m$ and $\rho(x_\alpha, x_{\alpha_m}) < 1/n$. Then, by Propositions 4.30, 3.66, and 3.67, we have $\rho(x_\alpha, x_0) = \lim_{m\in\mathbb{N}} \rho(x_\alpha, x_{\alpha_m}) \le 1/n$. Hence, we have $\lim_{\alpha\in\mathcal{A}} x_\alpha = x_0$.

This completes the proof of the proposition. □

Proposition 4.45 *Let $\mathcal{X} := (X, \mathcal{O})$ be a topological space, $\mathcal{Y} := (Y, \rho)$ be a complete metric space, $D \subseteq X$, $x_0 \in X$ be an accumulation point of D, and $f : D \to \mathcal{Y}$. Then, $\lim_{x\to x_0} f(x) \in \mathcal{Y}$ if, and only if, $\forall \epsilon \in (0, \infty) \subset \mathbb{R}$, $\exists O \in \mathcal{O}$ with $x_0 \in O$, $\forall \bar{x}, \hat{x} \in (D \cap O) \setminus \{x_0\}$, we have $\rho(f(\bar{x}), f(\hat{x})) < \epsilon$.*

Proof "Sufficiency" Assume that $\forall \epsilon \in (0, \infty) \subset \mathbb{R}$, $\exists O \in \mathcal{O}$ with $x_0 \in O$, $\forall \bar{x}, \hat{x} \in (D \cap O) \setminus \{x_0\}$, we have $\rho(f(\bar{x}), f(\hat{x})) < \epsilon$. Define $\mathcal{M} := \{O \in \mathcal{O} \mid x_0 \in O\}$. Clearly, $X \in \mathcal{M}$ and $\mathcal{M} \neq \emptyset$. It is easy to see that $\mathcal{A} := (\mathcal{M}, \supseteq)$ is a directed system. Since x_0 is an accumulation point of D, then $\forall O \in \mathcal{A}$, $(D \cap O) \setminus \{x_0\} \neq \emptyset$. By Axiom of Choice, $\exists$ a net $(x_O)_{O\in\mathcal{A}} \subseteq X$ such that $x_O \in (D\cap O)\setminus\{x_0\}$, $\forall O \in \mathcal{A}$. This also defines a net $(f(x_O))_{O\in\mathcal{A}} \subseteq Y$ by Axiom of Replacement. $\forall \epsilon \in (0, \infty) \subset \mathbb{R}$, by the assumption, $\exists \hat{O} \in \mathcal{A}$, $\forall \bar{x}, \hat{x} \in (D\cap\hat{O})\setminus\{x_0\}$, we have $\rho(f(\bar{x}), f(\hat{x})) < \epsilon$. $\forall O_1, O_2 \in \mathcal{A}$ with $\hat{O} \supseteq O_1$ and $\hat{O} \supseteq O_2$, $x_{O_i} \in (D \cap O_i) \setminus \{x_0\} \subseteq (D \cap \hat{O}) \setminus \{x_0\}$, $i = 1, 2$. This implies that $\rho(f(x_{O_1}), f(x_{O_2})) < \epsilon$. This shows that the net $(f(x_O))_{O\in\mathcal{A}} \subseteq Y$ is Cauchy. By Proposition 4.44, $\lim_{O\in\mathcal{A}} f(x_O) = y_0 \in Y$. $\forall \epsilon \in (0, \infty) \subset \mathbb{R}$, $\exists O_1 \in \mathcal{A}$, $\forall O \in \mathcal{A}$ with $O_1 \supseteq O$, we have $\rho(y_0, f(x_O)) < \epsilon/2$. By the assumption, $\exists O_2 \in \mathcal{A}$, $\forall \bar{x}, \hat{x} \in (D \cap O_2) \setminus \{x_0\}$, we have $\rho(f(\bar{x}), f(\hat{x})) < \epsilon/2$. Let $O_3 := O_1 \cap O_2 \in \mathcal{A}$. Then, $O_1 \supseteq O_3$ and $x_{O_3} \in (D \cap O_3) \setminus \{x_0\} \subseteq (D \cap O_2) \setminus \{x_0\}$. $\forall x \in (D \cap O_3) \setminus \{x_0\}$, we have $\rho(y_0, f(x)) \le \rho(y_0, f(x_{O_3})) + \rho(f(x_{O_3}), f(x)) < \epsilon$. Hence, we have $\lim_{x\to x_0} f(x) = y_0 \in \mathcal{Y}$.

"Necessity" Let $\lim_{x\to x_0} f(x) = y_0 \in \mathcal{Y}$. Then, $\forall \epsilon \in (0, \infty) \subset \mathbb{R}$, $\exists O \in \mathcal{O}$ with $x_0 \in O$, $\forall x \in (D \cap O) \setminus \{x_0\}$, we have $\rho(f(x), y_0) < \epsilon/2$. Hence, $\forall \bar{x}, \hat{x} \in (D \cap O) \setminus \{x_0\}$, $\rho(f(\bar{x}), f(\hat{x})) \le \rho(f(\bar{x}), y_0) + \rho(y_0, f(\hat{x})) < \epsilon$.

This completes the proof of the proposition. □

Proposition 4.46 *Let (X, ρ_X) and (Y, ρ_Y) be metric spaces, (Y, ρ_Y) be complete, $E \subseteq X$, and $f : E \to Y$ be uniformly continuous. Then, there is a unique continuous extension $g : \overline{E} \to Y$. Furthermore, g is uniformly continuous.*

Proof $\forall (x_i)_{i=1}^{\infty} \subseteq E$ with $\lim_{i\in\mathbb{N}} x_i = x_0 \in X$, by Proposition 4.23 and the uniform continuity of f, $(f(x_i))_{i=1}^{\infty}$ is a Cauchy sequence in Y. Since (Y, ρ_Y) is complete, then $\exists y_0 \in Y$ such that $\lim_{i\in\mathbb{N}} f(x_i) = y_0$. Let $(\bar{x}_i)_{i=1}^{\infty} \subseteq E$ be any other sequence with $\lim_{i\in\mathbb{N}} \bar{x}_i = x_0$. Then, the sequence $(x_1, \bar{x}_1, x_2, \bar{x}_2, \ldots)$ converges to x_0. By

Proposition 4.23 and the uniform continuity of f, $(f(x_1), f(\bar{x}_1), f(x_2), f(\bar{x}_2), \ldots)$ is a Cauchy sequence in Y, which converges since (Y, ρ_Y) is complete. The limit for this Cauchy sequence must be y_0 by Proposition 3.70. Hence, we have $\lim_{i\in\mathbb{N}} f(\bar{x}_i) = y_0$. Hence, y_0 is dependent only on x_0 but not on the sequence $(x_i)_{i=1}^{\infty}$.

By Proposition 4.13, we may define a function $g : \overline{E} \to Y$ by $g(x_0) = y_0$, $\forall x_0 \in \overline{E}$. $\forall x_0 \in E$, choose a sequence $(x_0, x_0, \ldots)$, which converges to x_0, then $y_0 = f(x_0)$. Hence, we have $g|_E = f$.

Next, we show that g is uniformly continuous. $\forall \epsilon \in (0, \infty) \subset \mathbb{R}$, by the uniform continuity of f on E, $\exists \delta \in (0, \infty) \subset \mathbb{R}$ such that $\rho_Y(f(x_1), f(x_2)) < \epsilon$, $\forall x_1, x_2 \in E$ with $\rho_X(x_1, x_2) < \delta$. $\forall x_1, x_2 \in \overline{E}$ with $\rho_X(x_1, x_2) < \delta$. By Proposition 4.13, $\exists \left(x_i^{\langle 1\rangle}\right)_{i=1}^{\infty}, \left(x_i^{\langle 2\rangle}\right)_{i=1}^{\infty} \subseteq E$ such that $\lim_{i\in\mathbb{N}} x_i^{\langle j\rangle} = x_j$, $j = 1, 2$. By the definition of g, we have $\lim_{i\in\mathbb{N}} f(x_i^{\langle j\rangle}) = g(x_j)$, $j = 1, 2$. By Proposition 3.67, we have $\lim_{i\in\mathbb{N}}(f(x_i^{\langle 1\rangle}), f(x_i^{\langle 2\rangle})) = (g(x_1), g(x_2))$. By Propositions 3.66 and 4.30, we have $\lim_{i\in\mathbb{N}} \rho_Y(f(x_i^{\langle 1\rangle}), f(x_i^{\langle 2\rangle})) = \rho_Y(g(x_1), g(x_2))$. Hence, we have $\rho_Y(g(x_1), g(x_2)) \leq \epsilon$. Hence, g is uniformly continuous.

Let $h : \overline{E} \to Y$ be any continuous mapping such that $h|_E = f$. $\forall \bar{x} \in \overline{E}$, by Proposition 4.13, $\exists (x_i)_{i=1}^{\infty} \subseteq E$ such that $\lim_{i\in\mathbb{N}} x_i = \bar{x}$. By continuity of h and g and Proposition 3.66, we have $h(\bar{x}) = \lim_{i\in\mathbb{N}} h(x_i) = \lim_{i\in\mathbb{N}} f(x_i) = \lim_{i\in\mathbb{N}} g(x_i) = g(\bar{x})$. Then, $h = g$. This shows that g is unique in the class of continuous functions that extends f.

This completes the proof of the proposition. □

Definition 4.47 A *pseudo-metric* ρ on a set X satisfies (i), (iii), and (iv) in Definition 4.1 and $\rho(x, x) = 0$, $\forall x \in X$, but not necessarily (ii). ◇

Lemma 4.48 *Let X be a set and ρ be a pseudo-metric on X. Define an equivalence relation $\equiv$ on X by $x_1 \equiv x_2$ if $\rho(x_1, x_2) = 0$, $\forall x_1, x_2 \in X$. Let $\mathcal{Y}$ be the quotient set $X/\equiv$, that is, $\mathcal{Y} := \{F \subseteq X \mid F \neq \emptyset, \forall x_1, x_2 \in F, x_1 \equiv x_2, \forall x_3 \in X \setminus F, x_1 \not\equiv x_3\}$. Define a mapping $\rho_Y : \mathcal{Y} \times \mathcal{Y} \to \mathbb{R}$ by $\rho_Y(Y_1, Y_2) = \rho(x_1, x_2)$, $\forall Y_1, Y_2 \in \mathcal{Y}$ and $x_1 \in Y_1$ and $x_2 \in Y_2$. Then, $(\mathcal{Y}, \rho_Y)$ is a metric space and said to be the* quotient space of (X, ρ) modulo ρ.

Proof We first show that $\equiv$ is an equivalence relation. Clearly, $\equiv$ is reflexive and symmetric.

Claim 4.48.1 $\forall x, y, z \in X$, if $\rho(y, z) = 0$, then $\rho(x, y) = \rho(x, z)$.

Proof of Claim $\rho(x, z) \leq \rho(x, y) + \rho(y, z) = \rho(x, y) \leq \rho(x, z) + \rho(y, z) = \rho(x, z)$. This completes the proof of the claim. □

By Claim 4.48.1, $\equiv$ is transitive. Hence, $\equiv$ is a equivalence relation.

Again by Claim 4.48.1, the mapping ρ_Y is well-defined.

To show that ρ_Y is a metric on $\mathcal{Y}$, we note that, $\forall Y_1, Y_2, Y_3 \in \mathcal{Y}$, (i) $0 \leq \rho_Y(Y_1, Y_2) < \infty$, (ii) $\rho_Y(Y_1, Y_2) = 0 \Leftrightarrow \forall x_1 \in Y_1, \forall x_2 \in Y_2, \rho(x_1, x_2) = 0$ and $x_1 \equiv x_2 \Leftrightarrow Y_1 = Y_2$, (iii) $\rho_Y(Y_1, Y_2) = \rho(x_1, x_2) = \rho(x_2, x_1) = \rho_Y(Y_2, Y_1)$, for some $x_1 \in Y_1$ and $x_2 \in Y_2$, and (iv) $\rho_Y(Y_1, Y_2) = \rho(x_1, x_2) \le \rho(x_1, x_3) + \rho(x_3, x_2) = \rho_Y(Y_1, Y_3) + \rho_Y(Y_3, Y_2)$, for some $x_1 \in Y_1$, $x_2 \in Y_2$, and $x_3 \in Y_3$. Hence, ρ_Y is a metric on $\mathcal{Y}$.

This completes the proof of the lemma. □

Theorem 4.49 *Let (X, ρ) be a metric space. Then,*

(i) *There exists a complete metric space $(\bar{X}, \bar{\rho})$ such that X is isometrically embedded as a dense subset in $\bar{X}$.*
(ii) *Let (Y, ρ_Y) be any complete metric space with $X \subseteq Y$ and $\rho_Y|_{X\times X} = \rho$. Then, $(\bar{X}, \bar{\rho})$ is isometric with $\overline{X}$ in Y.*

Proof (i) Let $\mathcal{X} := \{(x_n)_{n=1}^\infty \subseteq X \mid (x_n)_{n=1}^\infty$ is a Cauchy sequence$\}$. Define $\sigma : \mathcal{X} \times \mathcal{X} \to \mathbb{R}$ by $\sigma\left(\left(x_n^{\langle 1\rangle}\right)_{n=1}^\infty, \left(x_n^{\langle 2\rangle}\right)_{n=1}^\infty\right) = \lim_{n\in\mathbb{N}} \rho(x_n^{\langle 1\rangle}, x_n^{\langle 2\rangle})$, $\forall \left(x_n^{\langle 1\rangle}\right)_{n=1}^\infty, \left(x_n^{\langle 2\rangle}\right)_{n=1}^\infty \in \mathcal{X}$. The mapping σ is well defined by Propositions 4.30 and 4.23 and the fact that $\left((x_n^{\langle 1\rangle}, x_n^{\langle 2\rangle})\right)_{n=1}^\infty$ is a Cauchy sequence.

$\forall \left(x_n^{\langle 1\rangle}\right)_{n=1}^\infty, \left(x_n^{\langle 2\rangle}\right)_{n=1}^\infty, \left(x_n^{\langle 3\rangle}\right)_{n=1}^\infty \in \mathcal{X}$, $\sigma\left(\left(x_n^{\langle 1\rangle}\right)_{n=1}^\infty, \left(x_n^{\langle 2\rangle}\right)_{n=1}^\infty\right) \in \overline{\mathbb{R}_+}$; $\sigma\left(\left(x_n^{\langle 1\rangle}\right)_{n=1}^\infty, \left(x_n^{\langle 2\rangle}\right)_{n=1}^\infty\right) = \lim_{n\in\mathbb{N}} \rho(x_n^{\langle 1\rangle}, x_n^{\langle 2\rangle}) = \lim_{n\in\mathbb{N}} \rho(x_n^{\langle 2\rangle}, x_n^{\langle 1\rangle}) = \sigma\left(\left(x_n^{\langle 2\rangle}\right)_{n=1}^\infty, \left(x_n^{\langle 1\rangle}\right)_{n=1}^\infty\right)$;

$$\begin{aligned}
&\sigma\left(\left(x_n^{\langle 1\rangle}\right)_{n=1}^\infty, \left(x_n^{\langle 2\rangle}\right)_{n=1}^\infty\right)\\
&= \lim_{n\in\mathbb{N}} \rho(x_n^{\langle 1\rangle}, x_n^{\langle 2\rangle}) \le \lim_{n\in\mathbb{N}} (\rho(x_n^{\langle 1\rangle}, x_n^{\langle 3\rangle}) + \rho(x_n^{\langle 3\rangle}, x_n^{\langle 2\rangle}))\\
&= \sigma\left(\left(x_n^{\langle 1\rangle}\right)_{n=1}^\infty, \left(x_n^{\langle 3\rangle}\right)_{n=1}^\infty\right) + \sigma\left(\left(x_n^{\langle 3\rangle}\right)_{n=1}^\infty, \left(x_n^{\langle 2\rangle}\right)_{n=1}^\infty\right)
\end{aligned}$$

and also $0 = \lim_{n\in\mathbb{N}} \rho(x_n^{\langle 1\rangle}, x_n^{\langle 1\rangle}) = \sigma\left(\left(x_n^{\langle 1\rangle}\right)_{n=1}^\infty, \left(x_n^{\langle 1\rangle}\right)_{n=1}^\infty\right)$. Hence, σ defines a pseudo-metric on $\mathcal{X}$.

By Lemma 4.48, we can define the equivalence relation $\equiv$ on $\mathcal{X}$ by $\left(x_n^{\langle 1\rangle}\right)_{n=1}^\infty \equiv \left(x_n^{\langle 2\rangle}\right)_{n=1}^\infty$ if $\lim_{n\in\mathbb{N}} \rho(x_n^{\langle 1\rangle}, x_n^{\langle 2\rangle}) = 0$, $\forall \left(x_n^{\langle 1\rangle}\right)_{n=1}^\infty, \left(x_n^{\langle 2\rangle}\right)_{n=1}^\infty \in \mathcal{X}$. Then, define the set

$$\begin{aligned}
\bar{X} := \mathcal{X}/\equiv := \Bigg\{ F \subseteq \mathcal{X} \;\Bigg|\; & F \ne \emptyset,\\
&\forall \left(x_n^{\langle 1\rangle}\right)_{n=1}^\infty, \left(x_n^{\langle 2\rangle}\right)_{n=1}^\infty \in F, \left(x_n^{\langle 1\rangle}\right)_{n=1}^\infty \equiv \left(x_n^{\langle 2\rangle}\right)_{n=1}^\infty,\\
&\forall \left(x_n^{\langle 3\rangle}\right)_{n=1}^\infty \in \mathcal{X} \setminus F, \left(x_n^{\langle 1\rangle}\right)_{n=1}^\infty \not\equiv \left(x_n^{\langle 3\rangle}\right)_{n=1}^\infty \Bigg\}
\end{aligned}$$

A metric on $\bar{X}$ is $\bar{\rho}$, defined by $\bar{\rho}(F_1, F_2) = \sigma\left(\left(x_n^{\langle 1\rangle}\right)_{n=1}^{\infty}, \left(x_n^{\langle 2\rangle}\right)_{n=1}^{\infty}\right)$, $\forall F_1, F_2 \in \bar{X}$, for some $\left(x_n^{\langle 1\rangle}\right)_{n=1}^{\infty} \in F_1$ and $\left(x_n^{\langle 2\rangle}\right)_{n=1}^{\infty} \in F_2$.

Define mapping $T : X \to \bar{X}$ by, $\forall x \in X$, $T(x) = F \in \bar{X}$ such that the Cauchy sequence $(x, x, \ldots) \in F$. Let X_I be the image of X under T. Then, $T : X \to X_I$ is surjective. $\forall x_1, x_2 \in X$ with $T(x_1) = T(x_2)$, we have $(x_1, x_1, \ldots) \equiv (x_2, x_2, \ldots)$, which implies that $\sigma((x_1, x_1, \ldots), (x_2, x_2, \ldots)) = \rho(x_1, x_2) = 0$, and hence $x_1 = x_2$. Therefore, T is injective. Hence, $T : X \to X_I$ is bijective and admits an inverse $T_{\text{inv}} : X_I \to X$. $\forall x_1, x_2 \in X$, $\bar{\rho}(T(x_1), T(x_2)) = \sigma((x_1, x_1, \ldots), (x_2, x_2, \ldots)) = \rho(x_1, x_2)$. Hence, T is metric preserving. Then, both T and T_{inv} are uniformly continuous. Hence, T is an isometry between X and X_I.

Therefore, (X, ρ) is isometrically embedded in $(\bar{X}, \bar{\rho})$.

Next, we show that $(\bar{X}, \bar{\rho})$ is complete. Fix a Cauchy sequence $(F_n)_{n=1}^{\infty} \subseteq \bar{X}$. $\forall n \in \mathbb{N}$, let $\left(x_i^{\langle n\rangle}\right)_{i=1}^{\infty} \in F_n$. Note that $\left(x_i^{\langle n\rangle}\right)_{i=1}^{\infty} \subseteq X$ is a Cauchy sequence. $\forall n, m \in \mathbb{N}$, $\bar{\rho}(F_n, F_m) = \sigma\left(\left(x_i^{\langle n\rangle}\right)_{i=1}^{\infty}, \left(x_i^{\langle m\rangle}\right)_{i=1}^{\infty}\right) = \lim_{i\in\mathbb{N}} \rho(x_i^{\langle n\rangle}, x_i^{\langle m\rangle})$. Define a sequence $\left(x_i^{\langle 0\rangle}\right)_{i=1}^{\infty} \subseteq X$ as follows. Set $n_0 = 0$. $\forall i \in \mathbb{N}$, choose $n_i \in \mathbb{N}$ with $n_i > n_{i-1}$ such that $\forall m_1, m_2 \geq n_i$, we have $\rho(x_{m_1}^{\langle i\rangle}, x_{m_2}^{\langle i\rangle}) < 1/i$, and set $x_i^{\langle 0\rangle} = x_{n_i}^{\langle i\rangle}$.

Claim 4.49.1 $\left(x_i^{\langle 0\rangle}\right)_{i=1}^{\infty} \in \mathcal{X}$.

Proof of Claim $\forall \epsilon \in (0, \infty) \subset \mathbb{R}$, $\exists N_1 \in \mathbb{N}$ such that $\forall n, m \geq N_1$, we have $\bar{\rho}(F_n, F_m) < \epsilon$. This is valid since $(F_n)_{n=1}^{\infty}$ is Cauchy in $\bar{X}$. Let $N_2 \in \mathbb{N}$ be such that $1/N_2 < \epsilon$. Let $N = \max\{N_1, N_2\} \in \mathbb{N}$. $\forall i, j \geq N$, $\rho(x_i^{\langle 0\rangle}, x_j^{\langle 0\rangle}) = \rho(x_{n_i}^{\langle i\rangle}, x_{n_j}^{\langle j\rangle})$. Since $\bar{\rho}(F_i, F_j) = \lim_{l\in\mathbb{N}} \rho(x_l^{\langle i\rangle}, x_l^{\langle j\rangle}) < \epsilon$, then $\exists i_0 \in \mathbb{N}$ with $i_0 \geq \max\{n_i, n_j\}$ such that $0 \leq \rho(x_{i_0}^{\langle i\rangle}, x_{i_0}^{\langle j\rangle}) < \bar{\rho}(F_i, F_j) + \epsilon < 2\epsilon$. Then, we have $\rho(x_i^{\langle 0\rangle}, x_j^{\langle 0\rangle}) \leq \rho(x_{n_i}^{\langle i\rangle}, x_{i_0}^{\langle i\rangle}) + \rho(x_{i_0}^{\langle i\rangle}, x_{i_0}^{\langle j\rangle}) + \rho(x_{i_0}^{\langle j\rangle}, x_{n_j}^{\langle j\rangle}) < 1/i + 2\epsilon + 1/j < 4\epsilon$. Hence, $\left(x_i^{\langle 0\rangle}\right)_{i=1}^{\infty} \in \mathcal{X}$. This completes the proof of the claim. □

Let $F_0 \in \bar{X}$ be such that $\left(x_i^{\langle 0\rangle}\right)_{i=1}^{\infty} \in F_0$.

Claim 4.49.2 $\lim_{n\in\mathbb{N}} F_n = F_0$.

Proof of Claim $\forall \epsilon \in (0, \infty) \subset \mathbb{R}$, $\exists N_1 \in \mathbb{N}$ such that $\forall n, m \geq N_1$, we have $\bar{\rho}(F_n, F_m) < \epsilon$. Let $N_2 \in \mathbb{N}$ be such that $1/N_2 < \epsilon$. Let $N := \max\{N_1, N_2\} \in \mathbb{N}$. $\forall m \geq N$, fix $i_0 := n_m \geq m \geq N$. $\forall i \geq i_0$, $\exists i_1 > n_i \geq i \geq n_m$ such that $0 \leq \rho(x_{i_1}^{\langle m\rangle}, x_{i_1}^{\langle i\rangle}) < \bar{\rho}(F_m, F_i) + \epsilon < 2\epsilon$. Then,

$$\begin{aligned}
0 &\leq \rho(x_i^{\langle m\rangle}, x_i^{\langle 0\rangle}) = \rho(x_i^{\langle m\rangle}, x_{n_i}^{\langle i\rangle}) \\
&\leq \rho(x_i^{\langle m\rangle}, x_{i_1}^{\langle m\rangle}) + \rho(x_{i_1}^{\langle m\rangle}, x_{i_1}^{\langle i\rangle}) + \rho(x_{i_1}^{\langle i\rangle}, x_{n_i}^{\langle i\rangle}) < 1/m + 2\epsilon + 1/i < 4\epsilon
\end{aligned}$$

Hence, $0 \leq \bar{\rho}(F_m, F_0) = \lim_{i\in\mathbb{N}} \rho(x_i^{\langle m\rangle}, x_i^{\langle 0\rangle}) \leq 4\epsilon$. Therefore, we have $\lim_{n\in\mathbb{N}} \bar{\rho}(F_n, F_0) = 0$. This completes the proof of the claim. □

Hence, we have shown that any Cauchy sequence $(F_n)_{n=1}^{\infty} \subseteq \bar{X}$ admits a limit $F_0 \in \bar{X}$. Hence, $(\bar{X}, \bar{\rho})$ is complete.

To complete the proof of (i), we need only to show that X_I is dense in $\bar{X}$. $\forall F \in \bar{X}$, let $(x_n)_{n=1}^{\infty} \in F$. $\forall \epsilon \in (0, \infty) \subset \mathbb{R}$, $\exists N \in \mathbb{N}$ such that $\forall n \in \mathbb{N}$ with $n \geq N$, we have $\rho(x_N, x_n) < \epsilon/2$. Let $F_N := T(x_N) \in X_I$. Then, by Propositions 4.30, 3.66, and 3.67, $0 \leq \bar{\rho}(F_N, F) = \lim_{n\in\mathbb{N}} \rho(x_N, x_n) \leq \epsilon/2 < \epsilon$. Hence, $F_N \in \mathcal{B}_{(\bar{X},\bar{\rho})}(F, \epsilon) \cap X_I \neq \emptyset$. Hence, X_I is dense in $\bar{X}$, by Proposition 3.3.

(ii) Let X_c be the closure of X in (Y, ρ_Y). Since T is an isometry between X and X_I, then $T : X \to \bar{X}$ is uniformly continuous. By Proposition 4.46, $\exists G : X_c \to \bar{X}$ such that $G|_X = T$ and G is uniformly continuous. Note that $T_{\text{inv}} : X_I \to X \subseteq Y$ is also uniformly continuous, since T_{inv} is metric preserving. Then, by Propositions 4.39 and 4.46, $\exists H : \bar{X} \to X_c$ such that $H|_{X_I} = T_{\text{inv}}$ and H is uniformly continuous. $G \circ H : \bar{X} \to \bar{X}$ is uniformly continuous, by Proposition 4.23. Note that $(G \circ H)|_{X_I} = \text{id}_{\bar{X}}\big|_{X_I}$. By Proposition 3.56, we have $G \circ H = \text{id}_{\bar{X}}$. $H \circ G : X_c \to X_c$ is uniformly continuous, by Proposition 4.23. Note that $(H \circ G)|_X = \text{id}_{X_c}\big|_X$. By Proposition 3.56, we have $H \circ G = \text{id}_{X_c}$. Therefore, $G : X_c \to \bar{X}$ is bijective with inverse $H : \bar{X} \to X_c$, by Proposition 2.4. Hence, X_c and $\bar{X}$ are uniformly homeomorphic.

$\forall y_1, y_2 \in X_c$, $\exists \left(x_i^{\langle 1\rangle}\right)_{i=1}^{\infty}, \left(x_i^{\langle 2\rangle}\right)_{i=1}^{\infty} \in \mathcal{X}$ such that $\lim_{i\in\mathbb{N}} x_i^{\langle j\rangle} = y_j$, $j = 1, 2$, by Proposition 4.13. Then, by Proposition 3.66, we have $G(y_j) = \lim_{i\in\mathbb{N}} G(x_i^{\langle j\rangle}) = \lim_{i\in\mathbb{N}} T(x_i^{\langle j\rangle})$, $j = 1, 2$. By Propositions 3.66, 3.67, and 4.30 and the fact that T is metric preserving on X, we have $\bar{\rho}(G(y_1), G(y_2)) = \lim_{i\in\mathbb{N}} \bar{\rho}(T(x_i^{\langle 1\rangle}), T(x_i^{\langle 2\rangle})) = \lim_{i\in\mathbb{N}} \rho(x_i^{\langle 1\rangle}, x_i^{\langle 2\rangle}) = \lim_{i\in\mathbb{N}} \rho_Y(x_i^{\langle 1\rangle}, x_i^{\langle 2\rangle}) = \rho_Y(y_1, y_2)$. Hence, $G : X_c \to \bar{X}$ is also metric preserving. Hence, G is an isometry between X_c and $\bar{X}$.

This completes the proof of the theorem. □

4.8 Metrization of Topological Spaces

Definition 4.50 A topological space $(X, \mathcal{O})$ is said to be *metrizable* if there is a metric ρ on X whose natural topology is exactly the same as $\mathcal{O}$. ◇

Proposition 4.51 *Let $\mathcal{X}_\alpha := (X_\alpha, \mathcal{O}_\alpha)$ be metrizable topological spaces, $\forall \alpha \in \Lambda$, where Λ is a countable set. Then, the product topological space $\mathcal{X} = (X, \mathcal{O}) := \prod_{\alpha\in\Lambda} \mathcal{X}_\alpha$ is metrizable.*

Proof $\forall \alpha \in \Lambda$, let ρ_α be a metric on X_α whose natural topology is exactly $\mathcal{O}_\alpha$. We will distinguish three exhaustive and mutually exclusive cases: Case 1: $\Lambda = \emptyset$, Case 2: $\Lambda \neq \emptyset$ and is finite, and Case 3: Λ is countably infinite.

Case 1: $\Lambda = \emptyset$. Then, $(X, \mathcal{O}) = (\{\emptyset\}, \{\emptyset, \{\emptyset\}\})$. It is clearly metrizable, where the metric ρ may be defined by $\rho(\emptyset, \emptyset) = 0$.

Case 2: $\Lambda \neq \emptyset$ and is finite. We may define a metric ρ on X to be the Cartesian metric as in Definition 4.28. By Proposition 4.29, the natural topology induced by ρ is exactly the same as the product topology $\mathcal{O}$. Hence, $\mathcal{X}$ is metrizable.

Case 3: Λ is countably infinite. The result follows by Proposition 4.34.

This completes the proof of the proposition. □

Definition 4.52 Let $I := [0, 1] \subset \mathbb{R}$ and A be a set. I^A is called a *cube*. Associate with I the subset topology of $\mathbb{R}$. Denote by $\mathcal{X}^{\mathbb{N}}$ the *countably infinite product space* of $\mathcal{X}$. The cube $I^{\mathbb{N}}$ is metrizable, by Proposition 4.51, and is called a *Hilbert cube*. ◇

Theorem 4.53 (Urysohn Metrization Theorem) *Every normal topological space satisfying the second axiom of countability is metrizable.*

Proof Let $\mathcal{X} := (X, \mathcal{O})$ be a second countable normal topological space. Let $(B_\alpha)_{\alpha\in\Lambda}$ be a countable basis for the topology $\mathcal{O}$. $\forall \alpha_1, \alpha_2 \in \Lambda$, if $\widetilde{B_{\alpha_1}} \cap \overline{B_{\alpha_2}} = \emptyset$, then, by Urysohn's Lemma, $\exists$ a continuous function $f_{\alpha_1\alpha_2} : \mathcal{X} \to [0, 1] =: I \subset \mathbb{R}$ such that $f_{\alpha_1\alpha_2}|_{\widetilde{B_{\alpha_1}}} = 0$ and $f_{\alpha_1\alpha_2}|_{\overline{B_{\alpha_2}}} = 1$.

Define $\mathcal{F} := \left\{ f_{\alpha_1\alpha_2} : \mathcal{X} \to I \;\middle|\; \alpha_1, \alpha_2 \in \Lambda, \widetilde{B_{\alpha_1}} \cap \overline{B_{\alpha_2}} = \emptyset \right\}$. Clearly, $\mathcal{F}$ is a countable set. $\forall x_0 \in \mathcal{X}$, for all closed set $F \subseteq \mathcal{X}$ with $x_0 \notin F$, we have $\exists \alpha_1 \in \Lambda$ such that $x_0 \in B_{\alpha_1} \subseteq \widetilde{F}$. By Definition 3.33 and Propositions 3.34 and 3.35, $\exists U \in \mathcal{O}$ such that $\{x_0\} \subseteq U \subseteq \overline{U} \subseteq B_{\alpha_1}$. Then, $\exists \alpha_2 \in \Lambda$ such that $x_0 \in B_{\alpha_2} \subseteq U$. Then, we have $\{x_0\} \subseteq B_{\alpha_2} \subseteq \overline{B_{\alpha_2}} \subseteq \overline{U} \subseteq B_{\alpha_1}$, and hence $\widetilde{B_{\alpha_1}} \cap \overline{B_{\alpha_2}} = \emptyset$. Then, $\exists f_{\alpha_1\alpha_2} \in \mathcal{F}$ such that $f_{\alpha_1\alpha_2}|_{\widetilde{B_{\alpha_1}}} = 0$ and $f_{\alpha_1\alpha_2}|_{\overline{B_{\alpha_2}}} = 1$, which implies that $f_{\alpha_1\alpha_2}|_F = 0$ and $f_{\alpha_1\alpha_2}(x_0) = 1$. Hence, $\mathcal{F}$ is a countable collection of continuous $[0, 1]$-valued functions and satisfies $\forall x_0 \in \mathcal{X}$, for all closed set $F \subseteq \mathcal{X}$ with $x_0 \notin F$, $\exists f \in \mathcal{F} \ni \cdot\ f(x_0) = 1$ and $f|_F = 0$. Since $\mathcal{X}$ is normal, then it is Tychonoff and any singleton subset is closed. $\forall x, y \in \mathcal{X}$ with $x \neq y$, we have $\{y\}$ is closed and therefore, $\exists f \in \mathcal{F} \ni \cdot\ f(x) = 1 \neq 0 = f(y)$.

By Proposition 3.59, the equivalence map $E : \mathcal{X} \to I^{\mathcal{F}}$ is a homeomorphism between $\mathcal{X}$ and $E(\mathcal{X})$. Since I is metrizable and $\mathcal{F}$ is countable, then, by Proposition 4.51, $I^{\mathcal{F}}$ is metrizable with a metric $\rho_{I^{\mathcal{F}}}$. By Proposition 4.37, the metric $\rho_{I^{\mathcal{F}}}$ generates the subset topology on $E(\mathcal{X})$. Since E is a homeomorphism between $\mathcal{X}$ and $E(\mathcal{X})$, then $\mathcal{X}$ is metrizable with a metric $\rho_{\mathcal{X}} : \mathcal{X} \times \mathcal{X} \to \mathbb{R}$ given by $\rho_{\mathcal{X}}(x, y) := \rho_{I^{\mathcal{F}}}(E(x), E(y)), \forall x, y \in \mathcal{X}$.

This completes the proof of the theorem. □

4.9 Interchange Limits

Theorem 4.54 (Joint Limit Theorem) *Let $\mathcal{Y} := (Y, \rho)$ be a metric space and $(y_{\alpha,\beta})_{(\alpha,\beta)\in\mathcal{A}_1\times\mathcal{A}_2} \subseteq \mathcal{Y}$ be a joint net. Assume that $\lim_{(\alpha,\beta)\in\mathcal{A}_1\times\mathcal{A}_2} y_{\alpha,\beta} = \hat{y} \in \mathcal{Y}$ and, $\forall \alpha \in \mathcal{A}_1$, $\lim_{\beta\in\mathcal{A}_2} y_{\alpha,\beta} = \bar{y}_\alpha \in \mathcal{Y}$. Then, $\lim_{\alpha\in\mathcal{A}_1} \bar{y}_\alpha = \lim_{\alpha\in\mathcal{A}_1} \lim_{\beta\in\mathcal{A}_2} y_{\alpha,\beta} = \lim_{(\alpha,\beta)\in\mathcal{A}_1\times\mathcal{A}_2} y_{\alpha,\beta}$.*

Proof $\forall \epsilon \in (0, \infty) \subset \mathbb{R}$, by Definition 3.88 and Proposition 4.8, $\exists (\alpha_0, \beta_0) \in \mathcal{A}_1 \times \mathcal{A}_2$, $\forall (\alpha, \beta) \in \mathcal{A}_1 \times \mathcal{A}_2$ with $(\alpha_0, \beta_0) \prec (\alpha, \beta)$, we have $\rho(y_{\alpha,\beta}, \hat{y}) < \epsilon/2$. Fix any $\alpha \in \mathcal{A}_1$ with $\alpha_0 \prec_1 \alpha$. By the fact that $\lim_{\beta \in \mathcal{A}_2} y_{\alpha,\beta} = \bar{y}_\alpha \in \mathcal{Y}$, $\exists \beta_\alpha \in \mathcal{A}_2$, $\forall \beta \in \mathcal{A}_2$ with $\beta_\alpha \prec_2 \beta$, we have $\rho(y_{\alpha,\beta}, \bar{y}_\alpha) < \epsilon/2$. Since $\mathcal{A}_2$ is a directed system, then $\exists \beta_1 \in \mathcal{A}_2$ such that $\beta_0 \prec_2 \beta_1$ and $\beta_\alpha \prec_2 \beta_1$. Then, we have $\rho(\bar{y}_\alpha, \hat{y}) \le \rho(\bar{y}_\alpha, y_{\alpha,\beta_1}) + \rho(y_{\alpha,\beta_1}, \hat{y}) < \epsilon$. This shows that $\lim_{\alpha \in \mathcal{A}_1} \bar{y}_\alpha = \hat{y}$. Hence, we have $\lim_{\alpha \in \mathcal{A}_1} \bar{y}_\alpha = \lim_{\alpha \in \mathcal{A}_1} \lim_{\beta \in \mathcal{A}_2} y_{\alpha,\beta} = \lim_{(\alpha,\beta) \in \mathcal{A}_1 \times \mathcal{A}_2} y_{\alpha,\beta}$. This completes the proof of the theorem. □

Corollary 4.55 *Let $\mathcal{Y} := (Y, \rho)$ be a metric space and $(y_{\alpha,\beta})_{(\alpha,\beta) \in \mathcal{A}_1 \times \mathcal{A}_2} \subseteq \mathcal{Y}$ be a joint net. Assume that $\lim_{(\alpha,\beta) \in \mathcal{A}_1 \times \mathcal{A}_2} y_{\alpha,\beta} = \hat{y} \in \mathcal{Y}$, $\forall \alpha \in \mathcal{A}_1$, $\lim_{\beta \in \mathcal{A}_2} y_{\alpha,\beta} = \bar{y}_\alpha \in \mathcal{Y}$, and, $\forall \beta \in \mathcal{A}_2$, $\lim_{\alpha \in \mathcal{A}_1} y_{\alpha,\beta} = \widetilde{y}_\beta \in \mathcal{Y}$. Then, $\lim_{\alpha \in \mathcal{A}_1} \bar{y}_\alpha = \lim_{\alpha \in \mathcal{A}_1} \lim_{\beta \in \mathcal{A}_2} y_{\alpha,\beta} = \lim_{(\alpha,\beta) \in \mathcal{A}_1 \times \mathcal{A}_2} y_{\alpha,\beta} = \lim_{\beta \in \mathcal{A}_2} \widetilde{y}_\beta = \lim_{\beta \in \mathcal{A}_2} \lim_{\alpha \in \mathcal{A}_1} y_{\alpha,\beta}$.*

Proof This is straightforward by Joint Limit Theorem 4.54. □

Theorem 4.56 (Iterated Limit Theorem) *Let $\mathcal{Y} := (Y, \rho)$ be a complete metric space and $(y_{\alpha,\beta})_{(\alpha,\beta) \in \mathcal{A}_1 \times \mathcal{A}_2} \subseteq \mathcal{Y}$ be a joint net. Assume that*

(i) $\forall \alpha \in \mathcal{A}_1$, $\lim_{\beta \in \mathcal{A}_2} y_{\alpha,\beta} = \bar{y}_\alpha \in \mathcal{Y}$.
(ii) *The nets $(y_{\alpha,\beta})_{\alpha \in \mathcal{A}_1}$ converge uniformly to $\widetilde{y}_\beta \in \mathcal{Y}$, $\forall \beta \in \mathcal{A}_2$.*

Then, $\lim_{\alpha \in \mathcal{A}_1} \bar{y}_\alpha = \lim_{\alpha \in \mathcal{A}_1} \lim_{\beta \in \mathcal{A}_2} y_{\alpha,\beta} = \lim_{(\alpha,\beta) \in \mathcal{A}_1 \times \mathcal{A}_2} y_{\alpha,\beta} = \lim_{\beta \in \mathcal{A}_2} \widetilde{y}_\beta = \lim_{\beta \in \mathcal{A}_2} \lim_{\alpha \in \mathcal{A}_1} y_{\alpha,\beta}$.

Proof By (ii), $\forall \epsilon \in (0, \infty) \subset \mathbb{R}$, $\exists \alpha_0 \in \mathcal{A}_1$, $\forall \alpha \in \mathcal{A}_1$ with $\alpha_0 \prec_1 \alpha$, $\forall \beta \in \mathcal{A}_2$, we have $\rho(y_{\alpha,\beta}, \widetilde{y}_\beta) < \epsilon/4$. Fix an $\alpha_1 \in \mathcal{A}_1$ with $\alpha_0 \prec_1 \alpha_1$. Since $\lim_{\beta \in \mathcal{A}_2} y_{\alpha_1,\beta} = \bar{y}_{\alpha_1}$, then $\exists \beta_0 \in \mathcal{A}_2$, $\forall \beta \in \mathcal{A}_2$ with $\beta_0 \prec_2 \beta$, we have $\rho(y_{\alpha_1,\beta}, \bar{y}_{\alpha_1}) < \epsilon/4$. Then, $\forall \beta_1, \beta_2 \in \mathcal{A}_2$ with $\beta_0 \prec_2 \beta_1$ and $\beta_0 \prec_2 \beta_2$, we have $\rho(\widetilde{y}_{\beta_1}, \widetilde{y}_{\beta_2}) \le \rho(\widetilde{y}_{\beta_1}, y_{\alpha_1,\beta_1}) + \rho(y_{\alpha_1,\beta_1}, \bar{y}_{\alpha_1}) + \rho(\bar{y}_{\alpha_1}, y_{\alpha_1,\beta_2}) + \rho(y_{\alpha_1,\beta_2}, \widetilde{y}_{\beta_2}) < \epsilon$. Hence, the net $(\widetilde{y}_\beta)_{\beta \in \mathcal{A}_2}$ is a Cauchy net. By Proposition 4.44, $\lim_{\beta \in \mathcal{A}_2} \widetilde{y}_\beta = \hat{y} \in \mathcal{Y}$.

Then, $\exists \bar{\beta} \in \mathcal{A}_2$, $\forall \beta \in \mathcal{A}_2$ with $\bar{\beta} \prec_2 \beta$, we have $\rho(\widetilde{y}_\beta, \hat{y}) < \epsilon/2$. Note that $\forall (\alpha, \beta) \in \mathcal{A}_1 \times \mathcal{A}_2$ with $(\alpha_0, \bar{\beta}) \prec (\alpha, \beta)$, we have $\rho(y_{\alpha,\beta}, \hat{y}) \le \rho(y_{\alpha,\beta}, \widetilde{y}_\beta) + \rho(\widetilde{y}_\beta, \hat{y}) < \epsilon$. Hence, $\lim_{(\alpha,\beta) \in \mathcal{A}_1 \times \mathcal{A}_2} y_{\alpha,\beta} = \hat{y}$. Finally, by Joint Limit Theorem 4.54, we have $\lim_{\alpha \in \mathcal{A}_1} \lim_{\beta \in \mathcal{A}_2} y_{\alpha,\beta} = \lim_{\alpha \in \mathcal{A}_1} \bar{y}_\alpha = \hat{y}$.

This completes the proof of the theorem. □

Proposition 4.57 *Let $\mathcal{X} := (X, \mathcal{O})$ be a topological space, $\mathcal{Y} := (Y, \rho)$ be a complete metric space, $D \subseteq \mathcal{X}$, $x_0 \in \mathcal{X} \setminus D$ be an accumulation point of D, and $(f_\alpha)_{\alpha \in \mathcal{A}}$ be a net of functions of D to $\mathcal{Y}$. Assume that $(f_\alpha)_{\alpha \in \mathcal{A}}$ converges uniformly to function $f : D \to \mathcal{Y}$ and $\lim_{x \to x_0} f_\alpha(x) = y_\alpha \in \mathcal{Y}$, $\forall \alpha \in \mathcal{A}$. Then, $\lim_{\alpha \in \mathcal{A}} y_\alpha = \lim_{\alpha \in \mathcal{A}} \lim_{x \to x_0} f_\alpha(x) = \lim_{x \to x_0} \lim_{\alpha \in \mathcal{A}} f_\alpha(x) = \lim_{x \to x_0} f(x) \in \mathcal{Y}$.*

Proof Fix any net $(x_\beta)_{\beta \in \hat{\mathcal{A}}} \subseteq D$ with x_0 as a limit. Such net exists by Proposition 3.68. By the assumption that $(f_\alpha)_{\alpha \in \mathcal{A}}$ converges uniformly to f, then, for the joint net $(f_\alpha(x_\beta))_{(\alpha,\beta) \in \mathcal{A} \times \hat{\mathcal{A}}}$, we have the nets $(f_\alpha(x_\beta))_{\alpha \in \mathcal{A}}$ converge uniformly to $f(x_\beta) \in \mathcal{Y}$, $\forall \beta \in \hat{\mathcal{A}}$. By Proposition 3.79, we have

$\lim_{\beta\in\hat{\mathcal{A}}} f_\alpha(x_\beta) = y_\alpha$, $\forall\alpha \in \mathcal{A}$. By Iterated Limit Theorem 4.56, we have $\lim_{\alpha\in\mathcal{A}} y_\alpha = \lim_{\alpha\in\mathcal{A}} \lim_{\beta\in\hat{\mathcal{A}}} f_\alpha(x_\beta) = \lim_{(\alpha,\beta)\in\mathcal{A}\times\hat{\mathcal{A}}} f_\alpha(x_\beta) = \lim_{\beta\in\hat{\mathcal{A}}} \lim_{\alpha\in\mathcal{A}} f_\alpha(x_\beta) = \lim_{\beta\in\hat{\mathcal{A}}} f(x_\beta) = \hat{y} \in \mathcal{Y}$. By the arbitrariness of $(x_\beta)_{\beta\in\hat{\mathcal{A}}}$ and Proposition 3.79, we have $\hat{y} = \lim_{x\to x_0} f(x)$. Hence, we have $\lim_{\alpha\in\mathcal{A}} y_\alpha = \lim_{\alpha\in\mathcal{A}} \lim_{x\to x_0} f_\alpha(x) = \hat{y} = \lim_{x\to x_0} f(x) = \lim_{x\to x_0} \lim_{\alpha\in\mathcal{A}} f_\alpha(x) \in \mathcal{Y}$. This completes the proof of the proposition. □

Definition 4.58 Let X be a set and $\rho : X \times X \to [0, +\infty] \subseteq \mathbb{R}_e$. Then, (X, ρ) is said to be an *extended metric space* if $\forall x, y, z \in X$,

(i) $\rho(x, y) \geq 0$.
(ii) $\rho(x, y) = 0 \Leftrightarrow x = y$.
(iii) $\rho(x, y) = \rho(y, x)$.
(iv) $\rho(x, y) \leq \rho(x, z) + \rho(z, y)$. ◇

The only difference between an extended metric space and a metric space is that $\rho(x, y)$ may equal ∞. It is easy to show that in an extended metric space, the relation $x \sim y$ if $\rho(x, y) < \infty$ defines an equivalence relationship. On each equivalent set $E_{y_0} := \{x \in X \mid \rho(x, y_0) < \infty\}$, where $y_0 \in X$, $(E_{y_0}, \rho|_{E_{y_0}\times E_{y_0}})$ defines a metric space.

Proposition 4.59 *Let $\mathcal{X} := (X, \rho)$ be a metric space, and $Y := \{E \subseteq X \mid \emptyset \neq E$, and E is closed in $\mathcal{X}\} \subseteq {}^{X}2$. Define $\bar{\rho} : Y \times Y \to [0, +\infty] \subseteq \mathbb{R}_e$ by $\bar{\rho}(E_1, E_2) := \sup_{x_1\in E_1} \operatorname{dist}(x_1, E_2) + \sup_{x_2\in E_2} \operatorname{dist}(x_2, E_1)$. Then, $(Y, \bar{\rho}) =: \operatorname{lc}(\mathcal{X})$ is an extended metric space.*

Proof Fix any $E_1, E_2, E_3 \in Y$. (i) $\forall x_1 \in E_1$, $\operatorname{dist}(x_1, E_2) \in [0, \infty) \subset \mathbb{R}$ since $E_2 \neq \emptyset$. Then, $\sup_{x_1\in E_1} \operatorname{dist}(x_1, E_2) \in [0, \infty] \subset \mathbb{R}_e$. By symmetry, we have $\bar{\rho}(E_1, E_2) \in [0, \infty] \subset \mathbb{R}_e$.

(iii) This is immediate by the definition of $\bar{\rho}$.

(ii) If $E_1 = E_2$, clearly, we have $\bar{\rho}(E_1, E_2) = 0$. On the other hand, if $\bar{\rho}(E_1, E_2) = 0$, then $\sup_{x_1\in E_1} \operatorname{dist}(x_1, E_2) = 0$ and $\sup_{x_2\in E_2} \operatorname{dist}(x_2, E_1) = 0$. Fix any $x_1 \in E_1$, we have $\operatorname{dist}(x_1, E_2) = 0$. By Proposition 4.10, we have $x_1 \in E_2$. By the arbitrariness of x_1, we have $E_1 \subseteq E_2$. By symmetry, we have $E_2 \subseteq E_1$. Hence, $E_1 = E_2$.

(iv) Note that

$$\sup_{x_1\in E_1} \operatorname{dist}(x_1, E_2) = \sup_{x_1\in E_1} \inf_{x_2\in E_2} \rho(x_1, x_2)$$

$$= \sup_{x_1\in E_1} \inf_{x_2\in E_2} \inf_{x_3\in E_3} \rho(x_1, x_2) \leq \sup_{x_1\in E_1} \inf_{x_2\in E_2} \inf_{x_3\in E_3} (\rho(x_1, x_3) + \rho(x_3, x_2))$$

where the last inequality follows from the triangular inequality. Fix any $\epsilon \in (0, \infty) \subset \mathbb{R}$, $\forall x_1 \in E_1$, $\exists \bar{x}_3(x_1) \in E_3$ such that $\rho(x_1, \bar{x}_3(x_1)) < \text{dist}(x_1, E_3) + \frac{\epsilon}{2}$. Then,

$$
\begin{aligned}
\sup_{x_1 \in E_1} \text{dist}(x_1, E_2) &\le \sup_{x_1 \in E_1} \inf_{x_2 \in E_2} (\rho(x_1, \bar{x}_3(x_1)) + \rho(x_2, \bar{x}_3(x_1))) \\
&\le \sup_{x_1 \in E_1} \inf_{x_2 \in E_2} (\text{dist}(x_1, E_3) + \frac{\epsilon}{2} + \rho(x_2, \bar{x}_3(x_1))) \\
&= \sup_{x_1 \in E_1} (\text{dist}(x_1, E_3) + \frac{\epsilon}{2} + \inf_{x_2 \in E_2} \rho(x_2, \bar{x}_3(x_1))) \\
&\le \sup_{x_1 \in E_1} \text{dist}(x_1, E_3) + \frac{\epsilon}{2} + \sup_{x_1 \in E_1} \inf_{x_2 \in E_2} \rho(x_2, \bar{x}_3(x_1))) \\
&\le \sup_{x_1 \in E_1} \text{dist}(x_1, E_3) + \frac{\epsilon}{2} + \sup_{x_3 \in E_3} \text{dist}(x_3, E_2)
\end{aligned}
$$

where the first inequality follows from Proposition 3.81, the second inequality follows from the choice of $\bar{x}_3(x_1)$, the equality follows from Proposition 3.81, the third inequality follows from Proposition 3.81, and the last inequality follows from Proposition 3.81.

By symmetry, we have $\sup_{x_2 \in E_2} \text{dist}(x_2, E_1) \le \sup_{x_2 \in E_2} \text{dist}(x_2, E_3) + \frac{\epsilon}{2} + \sup_{x_3 \in E_3} \text{dist}(x_3, E_1)$. Adding these two inequality together to arrive at $\bar{\rho}(E_1, E_2) \le \sup_{x_1 \in E_1} \text{dist}(x_1, E_3) + \frac{\epsilon}{2} + \sup_{x_3 \in E_3} \text{dist}(x_3, E_2) + \sup_{x_2 \in E_2} \text{dist}(x_2, E_3) + \frac{\epsilon}{2} + \sup_{x_3 \in E_3} \text{dist}(x_3, E_1) = \bar{\rho}(E_1, E_3) + \bar{\rho}(E_3, E_2) + \epsilon$. By the arbitrariness of ϵ, we have (iv).

This completes the proof of the proposition. □

Definition 4.60 Let $\mathcal{X}$ and $\mathcal{Y}$ be metric spaces and $f : \mathcal{X} \to \mathcal{Y}$. f is said to be *uniformly continuous on* $E \subseteq \mathcal{X}$ if $\forall \epsilon \in (0, \infty) \subset \mathbb{R}$, $\exists \delta \in (0, \infty) \subset \mathbb{R}$ such that $\forall x_0 \in E$, $\forall x_1 \in \mathcal{B}_{\mathcal{X}}(x_0, \delta)$, we have $\rho_{\mathcal{Y}}(f(x_0), f(x_1)) < \epsilon$. Clearly, f is *uniformly continuous* (according to Definition 4.18) if it is uniformly continuous on $\mathcal{X} = \text{dom}(f)$. ◇

Definition 4.61 Let X and Y be sets and $f : X \to {}^{Y}2$. Then, f is said to be a *Y-valued set-valued map*. Let Z be a set, and $g : Y \to Z$. We will define a Z-valued set-valued map $h : X \to {}^{Z}2$ by $h(x) := g(f(x)) \subseteq Z$, $\forall x \in X$. This h is denoted by $g \circledcirc f$ and said to the *lifted composition of g and f*. ◇

Proposition 4.62 *Let $\mathcal{X}$, $\mathcal{Y}$, and $\mathcal{Z}$ be metric spaces, $f : \mathcal{X} \to \text{lc}(\mathcal{Y})$, $g : \mathcal{Y} \to \mathcal{Z}$, $h := g \circledcirc f$. Assume that $h : \mathcal{X} \to \text{lc}(\mathcal{Z})$, f is continuous at x_0, and g is uniformly continuous on $f(x_0) \subseteq \mathcal{Y}$. Then, h is continuous at x_0.*

Proof Let $\mathcal{X}$, $\mathcal{Y}$, and $\mathcal{Z}$ have metrics $\rho_{\mathcal{X}}$, $\rho_{\mathcal{Y}}$, and $\rho_{\mathcal{Z}}$, respectively, and $\text{lc}(\mathcal{Y})$ and $\text{lc}(\mathcal{Z})$ have metrics $\bar{\rho}_{\mathcal{Y}}$ and $\bar{\rho}_{\mathcal{Z}}$, respectively.

Fix any $\epsilon \in (0, \infty) \subset \mathbb{R}$. By g being uniformly continuous on $f(x_0) \subseteq \mathcal{Y}$, $\exists \delta_1(\epsilon) \in (0, \infty) \subset \mathbb{R}$, $\forall y_0 \in f(x_0) \subseteq \mathcal{Y}$, $\forall y_1 \in \mathcal{B}_{\mathcal{Y}}(y_0, \delta_1(\epsilon))$, we have $\rho_{\mathcal{Z}}(g(y_0), g(y_1)) < \epsilon$. By f being continuous at x_0, $\exists \delta_2(\epsilon) \in (0, \infty) \subset \mathbb{R}$, $\forall x_1 \in \mathcal{B}_{\mathcal{X}}(x_0, \delta_2(\epsilon))$, we have $\bar{\rho}_{\mathcal{Y}}(f(x_0), f(x_1)) < \delta_1(\epsilon)$.

Then, fix any $x_1 \in \mathcal{B}_{\mathcal{X}}(x_0, \delta_2(\epsilon))$. $\forall y_0 \in f(x_0)$, by $\bar{\rho}_{\mathcal{Y}}(f(x_0), f(x_1)) < \delta_1(\epsilon)$, we have $\text{dist}(y_0, f(x_1)) < \delta_1(\epsilon)$. Then, $\exists y_1 \in f(x_1)$ such that $\rho_{\mathcal{Y}}(y_0, y_1) < \delta_1(\epsilon)$. Thus, $\rho_{\mathcal{Z}}(g(y_0), g(y_1)) < \epsilon$. Therefore, $\text{dist}(g(y_0), h(x_1)) < \epsilon$. Then, we have $\sup_{y_0 \in f(x_0)} \text{dist}(g(y_0), h(x_1)) \leq \epsilon$, which is equivalent to

$$\sup_{z_0 \in h(x_0)} \text{dist}(z_0, h(x_1)) \leq \epsilon$$

On the other hand, $\forall y_1 \in f(x_1)$, by $\bar{\rho}_{\mathcal{Y}}(f(x_0), f(x_1)) < \delta_1(\epsilon)$, we have $\text{dist}(y_1, f(x_0)) < \delta_1(\epsilon)$. Then, $\exists y_0 \in f(x_0)$ such that $\rho_{\mathcal{Y}}(y_0, y_1) < \delta_1(\epsilon)$. Thus, $\rho_{\mathcal{Z}}(g(y_0), g(y_1)) < \epsilon$. Therefore, $\text{dist}(g(y_1), h(x_0)) < \epsilon$. Then, we have $\sup_{y_1 \in f(x_1)} \text{dist}(g(y_1), h(x_0)) \leq \epsilon$, which is equivalent to

$$\sup_{z_1 \in h(x_1)} \text{dist}(z_1, h(x_0)) \leq \epsilon$$

Thus, we have

$$\bar{\rho}_{\mathcal{Z}}(h(x_0), h(x_1)) = \sup_{z_0 \in h(x_0)} \text{dist}(z_0, h(x_1)) + \sup_{z_1 \in h(x_1)} \text{dist}(z_1, h(x_0)) \leq 2\epsilon$$

Hence, h is continuous at x_0. This completes the proof of the proposition. □

Chapter 5
Compact and Locally Compact Spaces

5.1 Compact Spaces

Definition 5.1 A topological space $(X, \mathcal{O})$ is called *compact* if every open covering has a finite subcovering, that is, $\forall (O_\alpha)_{\alpha\in\Lambda} \subseteq \mathcal{O}$ with $X \subseteq \bigcup_{\alpha\in\Lambda} O_\alpha$, where Λ is an index set and $\exists$ a finite set $\Lambda_N \subseteq \Lambda$ such that $X \subseteq \bigcup_{\alpha\in\Lambda_N} O_\alpha$. ◇

A subset $K \subseteq X$ is called compact if it is compact in its subset topology. This is equivalent to that any open covering of K in X has a finite subcovering of K.

Definition 5.2 Let X be a set and $\mathcal{F}$ be a collection of subsets in X. $\mathcal{F}$ is said to admit the *finite intersection property* if any finite subcollection of $\mathcal{F}$ has a nonempty intersection. ◇

Proposition 5.3 *A topological space $\mathcal{X}$ is compact if, and only if, any collection of closed sets $\mathcal{F}$ with the finite intersection property has a nonempty intersection.*

Proof "If" Suppose $\mathcal{X}$ is not compact. Then, there exists an open covering $(O_\alpha)_{\alpha\in\Lambda} \subseteq \mathcal{O}$ of $\mathcal{X}$, which does not have any finite subcovering. Then, $(\widetilde{O_\alpha})_{\alpha\in\Lambda}$ is a collection of closed sets with the finite intersection property. Yet, $\bigcap_{\alpha\in\Lambda} \widetilde{O_\alpha} = \left(\bigcup_{\alpha\in\Lambda} O_\alpha\right)^{\sim} = \emptyset$. This contradicts the assumption. Hence, $\mathcal{X}$ is compact.

"Only if" Suppose that the result does not hold. Then, there exists a collection of closed sets $\mathcal{F}$ with finite intersection property such that $\bigcap_{F\in\mathcal{F}} F = \emptyset$. Then, we have $\bigcup_{F\in\mathcal{F}} \widetilde{F} = \left(\bigcap_{F\in\mathcal{F}} F\right)^{\sim} = \mathcal{X}$. By the compactness of $\mathcal{X}$, there exists a finite collection $\mathcal{F}_N \subseteq \mathcal{F}$ such that $\bigcup_{F\in\mathcal{F}_N} \widetilde{F} = \mathcal{X}$. Then, $\bigcap_{F\in\mathcal{F}_N} F = \emptyset$. This contradicts with the assumption that $\mathcal{F}$ has finite intersection property. Hence, the result holds.

This completes the proof of the proposition. □

Proposition 5.4 *A topological space $\mathcal{X}$ is compact if, and only if, $\forall$ net $(x_\alpha)_{\alpha\in\mathcal{A}} \subseteq \mathcal{X}$ has a cluster point in $\mathcal{X}$.*

Z. Pan, *Measure-Theoretic Calculus in Abstract Spaces*,
https://doi.org/10.1007/978-3-031-21912-2_5

Proof "Only if" Fix any net $(x_\alpha)_{\alpha\in\mathcal{A}} \subseteq \mathcal{X}$. $\forall\alpha \in \mathcal{A}$, let $F_\alpha := \{x_\beta | \beta \in \mathcal{A}, \alpha \prec \beta\}$. Let $\overline{F_\alpha}$ be the closure of F_α. For any finite set $A \subseteq \mathcal{A}$, $\exists\alpha_0 \in \mathcal{A}$ such that $\alpha \prec \alpha_0$, $\forall\alpha \in A$, since $\mathcal{A}$ is a directed system. Then, $x_{\alpha_0} \in F_\alpha$, $\forall\alpha \in A$. Hence, we have $x_{\alpha_0} \in \bigcap_{\alpha\in A} \overline{F_\alpha} \neq \emptyset$. Hence, $(\overline{F_\alpha})_{\alpha\in\mathcal{A}}$ is a collection of closed sets in $\mathcal{X}$ with the finite intersection property. By Proposition 5.3 and the compactness of $\mathcal{X}$, we have $\bigcap_{\alpha\in\mathcal{A}} \overline{F_\alpha} \neq \emptyset$. Fix an $x \in \bigcap_{\alpha\in\mathcal{A}} \overline{F_\alpha}$. $\forall O \in \mathcal{O}$ with $x \in O$, $\forall\alpha_0 \in \mathcal{A}$, we have $x \in \overline{F_{\alpha_0}}$ and $O \cap F_{\alpha_0} \neq \emptyset$. Hence, $\exists\alpha \in \mathcal{A}$ with $\alpha_0 \prec \alpha$ such that $x_\alpha \in O \cap F_{\alpha_0} \subseteq O$. Hence, x is a cluster point of the net $(x_\alpha)_{\alpha\in\mathcal{A}}$.

"If" Suppose $\mathcal{X}$ is not compact. Then, there exists an open covering $(O_\alpha)_{\alpha\in\Lambda} \subseteq \mathcal{O}$ of $\mathcal{X}$, where Λ is an index set, such that $\mathcal{X}$ is not covered by any finite subcollection of sets in $(O_\alpha)_{\alpha\in\Lambda}$. Let $\Gamma := \{B \subseteq \Lambda | B \text{ is a finite set}\}$. Define a relation $\prec$ on Γ by $\forall B_1, B_2 \in \Gamma$, $B_1 \prec B_2$ if $B_1 \subseteq B_2$. Clearly, $\mathcal{A} := (\Gamma, \prec)$ is a directed system. $\forall B \in \Gamma$, let $x_B \in \left(\bigcup_{\alpha\in B} O_\alpha\right)^\sim$. This constructs a net $(x_B)_{B\in\mathcal{A}} \subseteq \mathcal{X}$. By the assumption, $\exists x_0 \in \mathcal{X}$ such that x_0 is a cluster point of the net. $\forall\alpha_0 \in \Lambda$, $\{\alpha_0\} \in \Gamma$. $\forall O \in \mathcal{O}$ with $x_0 \in \mathcal{O}$, $\exists B \in \Gamma$ with $\{\alpha_0\} \prec B$ such that $x_B \in O$. Then, $x_B \in O \cap \left(\bigcup_{\alpha\in B} O_\alpha\right)^\sim \subseteq O \cap \widetilde{O_{\alpha_0}} \neq \emptyset$. Hence, by the arbitrariness of O and Proposition 3.3, $x_0 \in \overline{\widetilde{O_{\alpha_0}}} = \widetilde{O_{\alpha_0}}$. Then, we have $x_0 \in \bigcap_{\alpha_0\in\Lambda} \widetilde{O_{\alpha_0}} = \left(\bigcup_{\alpha\in\Lambda} O_\alpha\right)^\sim = \emptyset$. This is a contradiction. Hence, $\mathcal{X}$ must be compact.

This completes the proof of the proposition. □

Proposition 5.5 *A closed subset of a compact space is compact. A compact subset of a Hausdorff space is closed.*

Proof Let $(X, \mathcal{O})$ be a compact topological space and $K \subseteq X$ be closed. $\forall$ open covering $(O_\alpha)_{\alpha\in\Lambda} \subseteq \mathcal{O}$ of K, where Λ is an index set. We have $X = \left(\bigcup_{\alpha\in\Lambda} O_\alpha\right) \cup \widetilde{K}$, which forms an open covering of X. By the compactness of X, there exists a finite set $\Lambda_N \subseteq \Lambda$ such that $X = \left(\bigcup_{\alpha\in\Lambda_N} O_\alpha\right) \cup \widetilde{K}$. Then, we have $K \subseteq \bigcup_{\alpha\in\Lambda_N} O_\alpha$. Therefore, K is compact.

Let $(X, \mathcal{O})$ be a Hausdorff space and $K \subseteq X$ be compact. Suppose that K is not closed. Then, by Proposition 3.3, $\exists x_0 \in \overline{K} \setminus K$. $\forall x \in K$, we have $x \neq x_0$. Since $(X, \mathcal{O})$ is Hausdorff, $\exists O_1^{\langle x\rangle}, O_2^{\langle x\rangle} \in \mathcal{O}$ such that $x_0 \in O_1^{\langle x\rangle}$, $x \in O_2^{\langle x\rangle}$, and $O_1^{\langle x\rangle} \cap O_2^{\langle x\rangle} = \emptyset$. Then, $\left(O_2^{\langle x\rangle}\right)_{x\in K}$ forms an open covering of K. By the compactness of K, $\exists$ a finite set $K_N \subseteq K$ such that $\left(O_2^{\langle x\rangle}\right)_{x\in K_N}$ forms an open covering of K. Then, $x_0 \in O := \bigcap_{x\in K_N} O_1^{\langle x\rangle} \in \mathcal{O}$, and $O \cap K \subseteq O \cap \left(\bigcup_{x\in K_N} O_2^{\langle x\rangle}\right) = \emptyset$. This contradicts with $x_0 \in \overline{K}$, by Proposition 3.3. Hence, K is closed.

This completes the proof of the proposition. □

Theorem 5.6 (Heine–Borel) *Let A be a subset of $\mathbb{R}$. A is compact if, and only if, A is closed and bounded.*

Proof "If" Consider first the special case $A = [a, b]$, where $a, b \in \mathbb{R}$ and $a < b$. Let $(O_\alpha)_{\alpha\in\Lambda} \subseteq \mathcal{O}_{\mathbb{R}}$ be an arbitrary open covering of A, where Λ is an index set. Let

$B := \{x \in A \mid \text{The interval } [a, x] \text{ can be covered byfinitely many sets in } (O_\alpha)_{\alpha \in \Lambda}\}$. Clearly, $a \in B$. Let $c = \sup B$. Then, $c \in A$. $\exists \alpha_0 \in \Lambda$ such that $c \in O_{\alpha_0}$. $\exists \delta \in (0, \infty) \subset \mathbb{R}$ such that $[c - \delta, c + \delta] \subseteq O_{\alpha_0}$. By the definition of c, $\exists d \in B$ such that $d \in [c - \delta, c]$. By $d \in B$, $[a, d]$ can be covered by finitely many sets in $(O_\alpha)_{\alpha \in \Lambda}$. Now, adding O_{α_0} to this finitely many sets, then $[a, c + \delta]$ is covered by finitely many sets in $(O_\alpha)_{\alpha \in \Lambda}$. Hence, $c \in B$. Note that we must have $c = b$ since otherwise $c < b$ and $\min\{c + \delta, b\} \in B$, which contradicts the definition of c. Therefore, there exists a finite subcovering of $[a, b] = A$. This shows that A is compact.

Now, let A be any closed and bounded subset of $\mathbb{R}$. Let $(O_\alpha)_{\alpha \in \Lambda} \subseteq \mathcal{O}_{\mathbb{R}}$ be an arbitrary open covering of A, where Λ is an index set. Since A is bounded, then $A \subseteq [a, b]$, for some $a, b \in \mathbb{R}$ with $a < b$. Note that $\left(\bigcup_{\alpha \in \Lambda} O_\alpha\right) \cup \widetilde{A} \supseteq [a, b]$. By the compactness of $[a, b]$, there exists a finite set $\Lambda_N \subseteq \Lambda$ such that $[a, b] \subseteq \left(\bigcup_{\alpha \in \Lambda_N} O_\alpha\right) \cup \widetilde{A}$. Then, $A \subseteq \bigcup_{\alpha \in \Lambda_N} O_\alpha$. Hence, A is compact.

"Only if" Since $\mathbb{R}$ is Hausdorff, by Proposition 5.5, A is closed. Let $I_n := (-n, n) \subset \mathbb{R}$, $\forall n \in \mathbb{N}$. Then, $\bigcup_{n \in \mathbb{N}} I_n \supseteq A$. By the compactness of A, $\exists N \in \mathbb{N}$ such that $A \subseteq \bigcup_{1 \le n \le N} I_n = I_N$. Hence, A is bounded.

This completes the proof of the theorem. □

Proposition 5.7 *Let $\mathcal{X}$ and $\mathcal{Y}$ be topological spaces, $\mathcal{X}$ be compact, and $f : \mathcal{X} \to \mathcal{Y}$ be continuous. Then, $f(\mathcal{X}) \subseteq \mathcal{Y}$ is compact.*

Proof Let $(O_{Y\alpha})_{\alpha \in \Lambda} \subseteq \mathcal{O}_{\mathcal{Y}}$ be an arbitrary open covering of $f(\mathcal{X})$, where Λ is an index set. Then, $(f_{\text{inv}}(O_{Y\alpha}))_{\alpha \in \Lambda} \subseteq \mathcal{O}_{\mathcal{X}}$ is an open covering of $\mathcal{X}$, by the continuity of f. By the compactness of $\mathcal{X}$, there exists a finite set $\Lambda_N \subseteq \Lambda$ such that $(f_{\text{inv}}(O_{Y\alpha}))_{\alpha \in \Lambda_N}$ is a subcovering of $\mathcal{X}$. Then, by Proposition 2.5, $(O_{Y\alpha})_{\alpha \in \Lambda_N}$ is a finite subcovering of $f(\mathcal{X})$. Hence, $f(\mathcal{X})$ is compact. This completes the proof of the proposition. □

Proposition 5.8 *Let $\mathcal{X}$ be a compact space, $\mathcal{Y}$ be a Hausdorff topological space, and $f : \mathcal{X} \to \mathcal{Y}$ be a bijective continuous function. Then, f is a homeomorphism.*

Proof By Proposition 5.7, $\mathcal{Y}$ is compact. By the assumption of the proposition, f is invertible with inverse $f_{\text{inv}} : \mathcal{Y} \to \mathcal{X}$. $\forall$ closed set $F \subseteq \mathcal{X}$. By Proposition 5.5, F is compact. By Proposition 5.7, $f(F)$ is compact. By Proposition 5.5, $f(F)$ is closed. By Proposition 3.10, f_{inv} is continuous. Hence, f is a homeomorphism. This completes the proof of the proposition. □

Definition 5.9 Let $(X, \mathcal{O})$ be a topological space and $\mathcal{U} \subseteq \mathcal{O}$ be an open covering of X. $\mathcal{V} \subseteq \mathcal{O}$ is said to be an *open refinement* of $\mathcal{U}$ (or to *refine* $\mathcal{U}$) if $\mathcal{V}$ is an open covering of X and $\forall V \in \mathcal{V}$, $\exists U \in \mathcal{U}$ such that $V \subseteq U$. ◇

Proposition 5.10 *Let $\mathcal{X} := (X, \mathcal{O})$ be a topological space. $\mathcal{X}$ is compact if, and only if, any open covering $\mathcal{U} \subseteq \mathcal{O}$ has a finite open refinement $\mathcal{V}$.*

Proof "Only if" Let $\mathcal{U} \subseteq \mathcal{O}$ be an open covering of X. By the compactness of $\mathcal{X}$, there exists a finite set $\mathcal{V} \subseteq \mathcal{U}$ that covers X. Then, $\mathcal{V}$ is a finite open refinement of $\mathcal{U}$.

"If" Let $\mathcal{U} \subseteq \mathcal{O}$ be an arbitrary open covering of X. Then, there exists a finite open refinement $\mathcal{V}$ of $\mathcal{U}$. $\forall V \in \mathcal{V}$, $\exists U_V \in \mathcal{U}$ such that $V \subseteq U_V$. By Axiom of Choice and Axiom of Replacement, we may define a set $\bar{\mathcal{U}} := \{U_V \in \mathcal{U} \mid V \in \mathcal{V}\}$. Clearly, $\bar{\mathcal{U}} \subseteq \mathcal{U}$ is a finite subcovering of X. Hence, $\mathcal{X}$ is compact.

This completes the proof of the proposition. □

Proposition 5.11 *Let $(X, \mathcal{O})$ be a Hausdorff topological space and $(K_n)_{n=1}^{\infty}$ be a sequence of compact subsets of X with $K_{n+1} \subseteq K_n$, $\forall n \in \mathbb{N}$. Let $O \in \mathcal{O}$ and $\bigcap_{n\in\mathbb{N}} K_n \subseteq O$. Then, $\exists n_0 \in \mathbb{N}$ such that $K_{n_0} \subseteq O$.*

Proof By Proposition 5.5, K_n is closed, $\forall n \in \mathbb{N}$. Let $O_n := O \cup \widetilde{K_n}$, $\forall n \in \mathbb{N}$. Then, we have

$$\bigcup_{n=1}^{\infty} O_n = O \cup \left(\bigcup_{n=1}^{\infty} \widetilde{K_n}\right) = O \cup \left(\bigcap_{n=1}^{\infty} K_n\right)^{\sim} = X \supseteq K_1$$

By the compactness of K_1, $\exists N \in \mathbb{N}$ such that $K_1 \subseteq \bigcup_{n=1}^{N} O_n = O \cup \left(\bigcap_{n=1}^{N} K_n\right)^{\sim} = O \cup \widetilde{K_N}$. Then, $K_N \subseteq O$. This completes the proof of the proposition. □

Proposition 5.12 *Let $\mathcal{X}$ be a Hausdorff topological space and $K_\alpha \subseteq \mathcal{X}$ be a compact subset, $\forall \alpha \in \Lambda$, where Λ is an index set. Assume that $\bigcap_{\alpha\in\Lambda} K_\alpha = \emptyset$. Then, there exists a finite set $\Lambda_N \subseteq \Lambda$ such that $\bigcap_{\alpha\in\Lambda_N} K_\alpha = \emptyset$.*

Proof We will distinguish two exhaustive and mutually exclusive cases: Case 1: Λ is finite; Case 2: Λ is infinite. Case 1: Λ is finite. Choose $\Lambda_N := \Lambda$ and the result follows.

Case 2: Λ is infinite. $\forall \alpha \in \Lambda$, K_α is closed in $\mathcal{X}$ by Proposition 5.5. Note that $\bigcup_{\alpha\in\Lambda} \widetilde{K_\alpha} = \left(\bigcap_{\alpha\in\Lambda} K_\alpha\right)^{\sim} = \mathcal{X} \supseteq K_{\alpha_0}$, for some $\alpha_0 \in \Lambda$. By the compactness of K_{α_0}, there exists a finite set $\bar{\Lambda}_N \subseteq \Lambda$ such that $K_{\alpha_0} \subseteq \bigcup_{\alpha\in\bar{\Lambda}_N} \widetilde{K_\alpha}$. Let $\Lambda_N := \bar{\Lambda}_N \cup \{\alpha_0\}$, which is a finite set. Then, $\bigcap_{\alpha\in\Lambda_N} K_\alpha = K_{\alpha_0} \cap \left(\bigcap_{\alpha\in\bar{\Lambda}_N} K_\alpha\right) = \emptyset$. This completes the proof of the proposition. □

Proposition 5.13 *Let $\mathcal{X} := (X, \mathcal{O})$ be a Hausdorff topological space, $K_1, K_2 \subseteq \mathcal{X}$ be compact subsets, and $K_1 \cap K_2 = \emptyset$. Then, $\exists O_1, O_2 \in \mathcal{O}$ such that $K_1 \subseteq O_1$, $K_2 \subseteq O_2$, and $O_1 \cap O_2 = \emptyset$.*

Proof Since $(X, \mathcal{O})$ is Hausdorff, then $\forall x_1 \in K_1$ and $\forall x_2 \in K_2$, and we must have $x_1 \neq x_2$ and $\exists O^{\langle 1\rangle}_{x_1,x_2}, O^{\langle 2\rangle}_{x_1,x_2} \in \mathcal{O}$ such that $x_j \in O^{\langle j\rangle}_{x_1,x_2}$, $j = 1, 2$, and $O^{\langle 1\rangle}_{x_1,x_2} \cap O^{\langle 2\rangle}_{x_1,x_2} = \emptyset$. Then, $\forall x_1 \in K_1$, $K_2 \subseteq \bigcup_{x_2\in K_2} O^{\langle 2\rangle}_{x_1,x_2}$. By the compactness of K_2, there exists a finite set $K_{2,x_1} \subseteq K_2$ such that $K_2 \subseteq \bigcup_{x_2\in K_{2,x_1}} O^{\langle 2\rangle}_{x_1,x_2} =: O^{\langle 2\rangle}_{x_1} \in \mathcal{O}$. It is clear that $x_1 \in \bigcap_{x_2\in K_{2,x_1}} O^{\langle 1\rangle}_{x_1,x_2} =: O^{\langle 1\rangle}_{x_1} \in \mathcal{O}$ and $O^{\langle 1\rangle}_{x_1} \cap O^{\langle 2\rangle}_{x_1} = \emptyset$. Then, $K_1 \subseteq \bigcup_{x_1\in K_1} O^{\langle 1\rangle}_{x_1}$. By the compactness of K_1, there exists a finite set $K_{1N} \subseteq K_1$

such that $K_1 \subseteq \bigcup_{x_1 \in K_{1N}} O_{x_1}^{\langle 1 \rangle} =: O_1 \in \mathcal{O}$. Note that $K_2 \subseteq \bigcap_{x_1 \in K_{1N}} O_{x_1}^{\langle 2 \rangle} =: O_2 \in \mathcal{O}$. Clearly, $O_1 \cap O_2 = \emptyset$. This completes the proof of the proposition. □

Proposition 5.14 *A compact Hausdorff topological space is normal.*

Proof Let $(X, \mathcal{O})$ be a compact Hausdorff topological space. Fix any closed sets $F_1, F_2 \subseteq X$ with $F_1 \cap F_2 = \emptyset$. By Proposition 5.5, F_1 and F_2 are compact. By Proposition 5.13, $\exists O_1, O_2 \in \mathcal{O}$ such that $F_1 \subseteq O_1$, $F_2 \subseteq O_2$, and $O_1 \cap O_2 = \emptyset$. Hence, $(X, \mathcal{O})$ is normal. This completes the proof of the proposition. □

Proposition 5.15 *Let X be a set and $\mathcal{O}$ and $\mathcal{O}_1$ be topologies on X. Assume that $\mathcal{O}_1$ is weaker than $\mathcal{O}$, that is, $\mathcal{O}_1 \subset \mathcal{O}$, and $(X, \mathcal{O})$ is compact. Then, $(X, \mathcal{O}_1)$ is compact.*

Proof Let $\mathcal{U}_1 \subseteq \mathcal{O}_1$ be any open covering of $(X, \mathcal{O}_1)$. Then, it is an open covering of $(X, \mathcal{O})$. By the compactness of $(X, \mathcal{O})$, there exists finite subcovering $\mathcal{U}_N \subseteq \mathcal{U}_1$ of $(X, \mathcal{O})$. Clearly, $\mathcal{U}_N \subseteq \mathcal{O}_1$ is a finite subcovering of $(X, \mathcal{O}_1)$. Therefore, $(X, \mathcal{O}_1)$ is compact. This completes the proof of the proposition. □

Proposition 5.16 *Let X be a set and $\mathcal{O}$ and $\mathcal{O}_1$ be topologies on X. Assume that $\mathcal{O}_1$ is stronger than $\mathcal{O}$, that is, $\mathcal{O} \subset \mathcal{O}_1$ and $(X, \mathcal{O})$ is Hausdorff. Then, $(X, \mathcal{O}_1)$ is Hausdorff.*

Proof $\forall x, y \in X$ with $x \neq y$, $\exists O_1, O_2 \in \mathcal{O}$ such that $x \in O_1$, $y \in O_2$, and $O_1 \cap O_2 = \emptyset$, since $(X, \mathcal{O})$ is Hausdorff. Note that $O_1, O_2 \in \mathcal{O}_1$ since $\mathcal{O} \subseteq \mathcal{O}_1$. Hence, $(X, \mathcal{O}_1)$ is Hausdorff. This completes the proof of the proposition. □

Proposition 5.17 *Let $(X, \mathcal{O})$ be a compact Hausdorff space. Then, any weaker topology $\mathcal{O}_1 \subset \mathcal{O}$ is not Hausdorff, and any stronger topology $\mathcal{O}_2 \supset \mathcal{O}$ is not compact.*

Proof Since $\mathcal{O}_1 \subset \mathcal{O}$, then $\exists O_0 \in \mathcal{O} \setminus \mathcal{O}_1$. Then, $\widetilde{O_0}$ is closed in $(X, \mathcal{O})$. By Proposition 5.5, $\widetilde{O_0}$ is compact in $(X, \mathcal{O})$. By Proposition 5.15, $\widetilde{O_0}$ is compact in $(X, \mathcal{O}_1)$. Suppose $(X, \mathcal{O}_1)$ is Hausdorff. Then, by Proposition 5.5, $\widetilde{O_0}$ is closed in $(X, \mathcal{O}_1)$. This implies that $O_0 \in \mathcal{O}_1$, which contradicts with $O_0 \in \mathcal{O} \setminus \mathcal{O}_1$. Hence, $(X, \mathcal{O}_1)$ is not Hausdorff.

Since $\mathcal{O}_2 \supset \mathcal{O}$, then $\exists O_0 \in \mathcal{O}_2 \setminus \mathcal{O}$. Suppose $(X, \mathcal{O}_2)$ is compact. Then, by Proposition 5.5, $\widetilde{O_0}$ is compact in $(X, \mathcal{O}_2)$. By Proposition 5.15, $\widetilde{O_0}$ is compact in $(X, \mathcal{O})$. By Proposition 5.5, $\widetilde{O_0}$ is closed in $(X, \mathcal{O})$. Then, $O_0 \in \mathcal{O}$, which contradicts with $O_0 \in \mathcal{O}_2 \setminus \mathcal{O}$. Hence, $(X, \mathcal{O}_2)$ is not compact.

This completes the proof of the proposition. □

Proposition 5.18 *Let $\mathcal{X}$ be a compact topological space, $\mathcal{Y}$ be a Hausdorff topological space, $\mathcal{Z}$ be a topological space, $f : \mathcal{X} \to \mathcal{Y}$ be surjective and continuous, and $g : \mathcal{Y} \to \mathcal{Z}$ be such that $T := g \circ f : \mathcal{X} \to \mathcal{Z}$ is continuous. Then, g is continuous.*

Proof By Proposition 5.7 and the compactness of $\mathcal{X}$, $\mathcal{Y} = f(\mathcal{X})$ is compact. Then, $\mathcal{Y}$ is compact and Hausdorff.

Claim 5.18.1 $\forall O_{\mathcal{Z}} \subseteq \mathcal{Z}$, $g_{\text{inv}}(O_{\mathcal{Z}}) = f(T_{\text{inv}}(O_{\mathcal{Z}}))$.

Proof of Claim $\forall y_0 \in g_{\text{inv}}(O_{\mathcal{Z}}) \subseteq \mathcal{Y}$, $\exists x_0 \in \mathcal{X}$ such that $y_0 = f(x_0)$ since f is surjective. Then, $g(y_0) = g(f(x_0)) = T(x_0) \in O_{\mathcal{Z}}$. This implies that $x_0 \in T_{\text{inv}}(O_{\mathcal{Z}})$. Then, $y_0 \in f(T_{\text{inv}}(O_{\mathcal{Z}}))$. Hence, $g_{\text{inv}}(O_{\mathcal{Z}}) \subseteq f(T_{\text{inv}}(O_{\mathcal{Z}}))$.

On the other hand, $\forall y_0 \in f(T_{\text{inv}}(O_{\mathcal{Z}}))$, $\exists x_0 \in T_{\text{inv}}(O_{\mathcal{Z}})$ such that $y_0 = f(x_0)$. Then, $O_{\mathcal{Z}} \ni T(x_0) = g(f(x_0)) = g(y_0)$. This implies that $y_0 \in g_{\text{inv}}(O_{\mathcal{Z}})$. Hence, $f(T_{\text{inv}}(O_{\mathcal{Z}})) \subseteq g_{\text{inv}}(O_{\mathcal{Z}})$.

Therefore, we have $g_{\text{inv}}(O_{\mathcal{Z}}) = f(T_{\text{inv}}(O_{\mathcal{Z}}))$. This completes the proof of the claim. □

$\forall$ closed set $F \subseteq \mathcal{Z}$. $T_{\text{inv}}(F)$ is closed in $\mathcal{X}$ by Proposition 3.10 and the continuity of T. Then, $T_{\text{inv}}(F)$ is compact by Proposition 5.5. This implies that, by Proposition 5.7, $f(T_{\text{inv}}(F))$ is compact. By Proposition 5.5, $g_{\text{inv}}(F) = f(T_{\text{inv}}(F))$ is closed. Then, by Proposition 3.10 and the arbitrariness of F, g is continuous.

This completes the proof of the proposition. □

Definition 5.19 A compact and connected Hausdorff topological space containing more than one point is called a *continuum*. ◇

Proposition 5.20 *Let $\mathcal{X} := (X, \mathcal{O})$ be a Hausdorff topological space, $(K_n)_{n=1}^{\infty}$ be a sequence of compact and connected subsets of $\mathcal{X}$, and $K_{n+1} \subseteq K_n$, $\forall n \in \mathbb{N}$. Then, $\bigcap_{n=1}^{\infty} K_n =: K$ is compact and connected.*

Proof By Proposition 5.5, K_n is closed, $\forall n \in \mathbb{N}$. Then, K is closed. Then, K is closed in the subset topology of K_1 by Proposition 3.5. By Proposition 5.5, K is compact.

Suppose K is not connected. Then, $\exists O_1, O_2 \in \mathcal{O}$ such that $K = O_{K1} \cup O_{K2} := (O_1 \cap K) \cup (O_2 \cap K)$, $O_{K1} \neq \emptyset$, $O_{K2} \neq \emptyset$, and $O_{K1} \cap O_{K2} = \emptyset$. Define $\bar{K}_i := K \setminus O_{Ki}$, $i = 1, 2$. Then, $i = 1, 2$, $\bar{K}_i = K \cap \widetilde{O_i}$, which is closed and therefore compact by Proposition 5.5. Note that $O_{K1} = \bar{K}_2$ and $O_{K2} = \bar{K}_1$. Then, O_{K1} and O_{K2} are disjoint compact sets. By Proposition 5.13, $\exists \bar{O}_1, \bar{O}_2 \in \mathcal{O}$ such that $O_{Ki} \subseteq \bar{O}_i$, $i = 1, 2$, and $\bar{O}_1 \cap \bar{O}_2 = \emptyset$. Then, $\bar{O}_1 \cup \bar{O}_2 \supseteq K$, by Proposition 5.11, $\exists n_0 \in \mathbb{N}$ such that $\bar{O}_1 \cup \bar{O}_2 \supseteq K_{n_0}$. Note that $\bar{O}_i \cap K_{n_0} \supseteq O_{Ki} \neq \emptyset$ and is open in the subset topology of K_{n_0}, $i = 1, 2$, and then, $\bar{O}_1 \cap K_{n_0}$ and $\bar{O}_2 \cap K_{n_0}$ form a separation of K_{n_0}. This contradicts with the assumption that K_{n_0} is connected. Hence, K is connected.

This completes the proof of the proposition. □

We note that the union of finitely many compact subsets of a topological space is compact.

5.2 Countable and Sequential Compactness

Definition 5.21 Let $(X, \mathcal{O})$ be a topological space. A subset $K \subseteq X$ is said to be *countably compact* if any countable open covering of K admits a finite subcovering. ◇

Proposition 5.22 *A topological space is compact if, and only if, it is Lindelöf and countably compact.*

Proof This is straightforward and therefore omitted. □

Proposition 5.23 *A second countable topological space is compact if, and only if, it is countably compact.*

Proof A second countable topological space is Lindelöf, by Proposition 3.24. Then, the result follows from Proposition 5.22. □

Proposition 5.24 *Let $\mathcal{X}$ be a countably compact topological space, $\mathcal{Y}$ be a topological space, and $f : \mathcal{X} \to \mathcal{Y}$ be continuous. Then, $f(\mathcal{X})$ is countably compact.*

Proof Fix any countable open covering $(O_\alpha)_{\alpha\in\Lambda} \subseteq \mathcal{O}_\mathcal{Y}$ of $f(\mathcal{X})$, where Λ is a countable index set. By Proposition 2.5 and the continuity of f, $(f_{\text{inv}}(O_\alpha))_{\alpha\in\Lambda}$ is a countable open covering of $\mathcal{X}$. By the countable compactness of $\mathcal{X}$, there exists a finite set $\Lambda_N \subseteq \Lambda$ such that $\bigcup_{\alpha\in\Lambda_N} f_{\text{inv}}(O_\alpha) = \mathcal{X}$. Then, we have $\bigcup_{\alpha\in\Lambda_N} O_\alpha \supseteq \bigcup_{\alpha\in\Lambda_N} f(f_{\text{inv}}(O_\alpha)) = f\left(\bigcup_{\alpha\in\Lambda_N} f_{\text{inv}}(O_\alpha)\right) = f(\mathcal{X})$, by Proposition 2.5. Hence, $f(\mathcal{X})$ is countably compact. This completes the proof of the proposition. □

Definition 5.25 Let $\mathcal{X}$ be a topological space. It is said to have the *Bolzano–Weierstrass property* if $\forall (x_n)_{n=1}^\infty \subseteq \mathcal{X}$, there exists a cluster point $x \in \mathcal{X}$ of the sequence. ◇

Proposition 5.26 *A topological space $\mathcal{X}$ is countably compact if, and only if, it has the Bolzano–Weierstrass property.*

Proof "If" Suppose $\mathcal{X}$ is not countably compact. Then, there exists an open covering $(O_n)_{n=1}^\infty \subseteq \mathcal{O}$ of $\mathcal{X}$ that does not have any finite subcovering. $\forall n \in \mathbb{N}$, $\exists x_n \in \widetilde{\bigcup_{i=1}^n O_i} = \bigcap_{i=1}^n \widetilde{O_i}$. Then, $(x_n)_{n=1}^\infty \subseteq \mathcal{X}$. By the Bolzano–Weierstrass property, $\exists x_0 \in \mathcal{X}$ such that x_0 is a cluster point of $(x_n)_{n=1}^\infty$. By Proposition 3.3, $x_0 \in \overline{(x_n)_{n=1}^\infty}$. $\forall n \in \mathbb{N}$, since $x_0 \in \overline{(x_i)_{i=n}^\infty}$, $(x_i)_{i=n}^\infty \subseteq \bigcap_{i=1}^n \widetilde{O_i}$, and $\bigcap_{i=1}^n \widetilde{O_i}$ is closed, then $x_0 \in \bigcap_{i=1}^n \widetilde{O_i}$. Then, $x_0 \in \bigcap_{n=1}^\infty \widetilde{O_i}$, which contradicts with the fact that $(O_n)_{n=1}^\infty$ is an open covering of $\mathcal{X}$. Therefore, $\mathcal{X}$ is countably compact.

"Only if" Fix any sequence $(x_n)_{n=1}^\infty \subseteq \mathcal{X}$. Let $B_n := (x_i)_{i=n}^\infty$, $\forall n \in \mathbb{N}$. Suppose $\bigcap_{n=1}^\infty \overline{B_n} = \emptyset$. Then, $\left(\widetilde{\overline{B_n}}\right)_{n\in\mathbb{N}}$ form a countable open covering of $\mathcal{X}$. By the countable compactness of $\mathcal{X}$, $\exists N \in \mathbb{N}$ such that $\bigcup_{n=1}^N \widetilde{\overline{B_n}} = \mathcal{X}$. Hence, we have $\bigcap_{n=1}^N \overline{B_n} = \emptyset$. This contradicts with the fact that $x_{N+1} \in \bigcap_{n=1}^N \overline{B_n}$. Therefore, $\bigcap_{n=1}^\infty \overline{B_n} \neq \emptyset$. Let $x_0 \in \bigcap_{n=1}^\infty \overline{B_n} \subseteq \mathcal{X}$. $\forall O \in \mathcal{O}$ with $x_0 \in O$, $\forall n \in \mathbb{N}$, $x_0 \in \overline{B_n}$ implies that $O \cap B_n \neq \emptyset$ by Proposition 3.3. Then, $\exists m \in \mathbb{N}$ with $m \geq n$ such that $x_m \in O$. This shows that x_0 is a cluster point of $(x_n)_{n=1}^\infty$. Hence, $\mathcal{X}$ has the Bolzano–Weierstrass property.

This completes the proof of the proposition. □

Definition 5.27 A topological space $\mathcal{X}$ is said to be *sequentially compact* if $\forall (x_n)_{n=1}^{\infty} \subseteq \mathcal{X}$, and there exists a subsequence $\left(x_{n_i}\right)_{i=1}^{\infty}$ that converges to some $x_0 \in \mathcal{X}$. ◇

Proposition 5.28 *Let $\mathcal{X}$ be a topological space. Then, the following statements hold:*

(i) *If it is sequentially compact, then $\mathcal{X}$ is countably compact.*
(ii) *If it is first countable and countably compact, then $\mathcal{X}$ is sequentially compact.*

Proof (i) $\mathcal{X}$ is sequentially compact implies that $\mathcal{X}$ has the Bolzano–Weierstrass property since the limit of a convergent subsequence is a cluster point for the subsequence and hence a cluster point for the original sequence. Then, $\mathcal{X}$ is countably compact by Proposition 5.26.

(ii) Let $\mathcal{X}$ be countably compact and first countable. Fix any $(x_n)_{n=1}^{\infty} \subseteq \mathcal{X}$. By Proposition 5.26, $\mathcal{X}$ has the Bolzano–Weierstrass property. Then, $\exists x_0 \in \mathcal{X}$ such that x_0 is a cluster point of $(x_n)_{n=1}^{\infty}$. Let $\mathcal{B}_{x_0} := (B_n)_{n=1}^{\infty} \subseteq \mathcal{O}$ be a countable basis at x_0. (In case $\mathcal{B}_{x_0}$ is finite, we will pad $\mathcal{X}$ as additional basis sets to make it countably infinite.) Let $n_0 = 0$. $\forall i \in \mathbb{N}$, $\exists n_i \in \mathbb{N}$ with $n_i > n_{i-1}$ such that $x_{n_i} \in \bigcap_{j=1}^{i} B_j \in \mathcal{O}$. Clearly, $\left(x_{n_i}\right)_{i=1}^{\infty}$ is a subsequence of $(x_n)_{n=1}^{\infty}$ and admits a limit point x_0. Hence, $\mathcal{X}$ is sequentially compact.

This completes the proof of the proposition. □

The relationships between different compactness concepts are: compactness implies countable compactness, which is equivalent to the Bolzano–Weierstrass property; sequential compactness implies countable compactness.

5.3 Real-Valued Functions and Compactness

Proposition 5.29 *Let $\mathcal{X}$ be a countably compact topological space and $f : \mathcal{X} \to \mathbb{R}$ be continuous. Then, f is bounded, that is, $\exists a, b \in \mathbb{R}$ such that $a \leq f(x) \leq b$, $\forall x \in \mathcal{X}$. Furthermore, if $\mathcal{X} \neq \emptyset$, $\exists x_m, x_M \in \mathcal{X}$ such that $f(x_m) = \min_{\bar{x} \in \mathcal{X}} f(\bar{x}) \leq f(x) \leq \max_{\bar{x} \in \mathcal{X}} f(\bar{x}) = f(x_M)$, $\forall x \in \mathcal{X}$, i. e., f achieves its minimum and maximum on $\mathcal{X}$.*

Proof Note that $\mathbb{R} = \bigcup_{n \in \mathbb{N}} (-n, n)$. Then, $\mathcal{X} = \bigcup_{n \in \mathbb{N}} f_{\text{inv}}((-n, n))$, by Proposition 2.5. By the continuity of f, $f_{\text{inv}}((-n, n))$ is open, $\forall n \in \mathbb{N}$. By the countable compactness of $\mathcal{X}$, $\exists N \in \mathbb{N}$ such that $\mathcal{X} = \bigcup_{n=1}^{N} f_{\text{inv}}((-n, n)) = f_{\text{inv}}((-N, N))$. By Proposition 2.5, $f(\mathcal{X}) \subseteq (-N, N)$. Hence, f is bounded.

Let $\mathcal{X} \neq \emptyset$. Define $M := \sup_{x \in \mathcal{X}} f(x)$ and $m := \inf_{x \in \mathcal{X}} f(x)$. Then, $-N \leq m \leq M \leq N$. $\forall n \in \mathbb{N}$, $\exists x_n \in \mathcal{X}$ such that $M - 1/n < f(x_n) \leq M$. We thus obtain a sequence $(x_n)_{n=1}^{\infty} \subseteq \mathcal{X}$. By Proposition 5.26 and the countable compactness of $\mathcal{X}$, $(x_n)_{n=1}^{\infty}$ admits a cluster point $x_M \in \mathcal{X}$. By Proposition 3.66, the sequence $(f(x_n))_{n=1}^{\infty}$ admits a cluster point $f(x_M)$. By construction, $\lim_{n \in \mathbb{N}} f(x_n) = M$, which is the only cluster point of $(f(x_n))_{n=1}^{\infty}$. Then, we have $f(x_M) = M$. By an argument that is similar to the above, $\exists x_m \in \mathcal{X}$ such that $f(x_m) = m$. This completes the proof of the proposition. □

Proposition 5.30 *Let $\mathcal{X}$ be a countably compact topological space and $f : \mathcal{X} \to \mathbb{R}$ be upper semicontinuous. Then, f is bounded from above, that is, $\exists b \in \mathbb{R}$ such that $f(x) \le b$, $\forall x \in \mathcal{X}$. Furthermore, if $\mathcal{X} \neq \emptyset$, then $\exists x_M \in \mathcal{X}$ such that $f(x) \le \max_{\bar{x} \in \mathcal{X}} f(\bar{x}) = f(x_M)$, $\forall x \in \mathcal{X}$, i. e., f achieves its maximum on $\mathcal{X}$.*

Proof Note that $\mathbb{R} = \bigcup_{n \in \mathbb{N}} (-\infty, n)$. Then, $\mathcal{X} = \bigcup_{n \in \mathbb{N}} f_{\text{inv}}((-\infty, n))$, by Proposition 2.5. By the upper semicontinuity of f, $f_{\text{inv}}((-\infty, n))$ is open, $\forall n \in \mathbb{N}$. By the countable compactness of $\mathcal{X}$, $\exists N \in \mathbb{N}$ such that $\mathcal{X} = \bigcup_{n=1}^{N} f_{\text{inv}}((-\infty, n)) = f_{\text{inv}}((-\infty, N))$. By Proposition 2.5, $f(\mathcal{X}) \subseteq (-\infty, N)$.

Let $\mathcal{X} \neq \emptyset$. Define $M := \sup_{x \in \mathcal{X}} f(x)$. Then, $-\infty < M \le N$. $\forall n \in \mathbb{N}$, $\exists x_n \in \mathcal{X}$ such that $M - 1/n < f(x_n) \le M$. We thus obtain a sequence $(x_n)_{n=1}^{\infty} \subseteq \mathcal{X}$. By Proposition 5.26 and the countable compactness of $\mathcal{X}$, $(x_n)_{n=1}^{\infty}$ admits a cluster point $x_M \in \mathcal{X}$. By construction, $\lim_{n \in \mathbb{N}} f(x_n) = M$. By Definition 3.14, $\forall \epsilon \in \mathbb{R}_+$, $\exists U \in \mathcal{O}$ with $x_M \in U$, such that $\forall x \in U$, we have $f(x) - \epsilon < f(x_M)$. This implies that $\exists n \ge \left\lceil \frac{1}{\epsilon} \right\rceil$ such that $x_n \in U$, since x_M is a cluster point of $(x_n)_{n=1}^{\infty}$. Thus, we have $M - 2\epsilon \le M - \frac{1}{n} - \epsilon < f(x_n) - \epsilon < f(x_M) \le M$. By the arbitrariness of ϵ, we have $f(x_M) = M$. This completes the proof of the proposition. □

Lemma 5.31 (Dini's Lemma) *Let $\mathcal{X} := (X, \mathcal{O})$ be a countably compact topological space and $(f_n)_{n=1}^{\infty}$ be a sequence of upper semicontinuous real-valued functions on $\mathcal{X}$. Assume that $\forall x \in \mathcal{X}$, $(f_n(x))_{n=1}^{\infty} \subseteq \mathbb{R}$ is nonincreasing and converges to 0. Then, $(f_n)_{n=1}^{\infty}$ converges to the zero function uniformly on $\mathcal{X}$.*

Proof $\forall \epsilon \in (0, \infty) \subset \mathbb{R}$, let $O_n := \{x \in \mathcal{X} | f_n(x) < \epsilon\}$, $\forall n \in \mathbb{N}$. By the upper semicontinuity of f_n, $O_n \in \mathcal{O}$, $\forall n \in \mathbb{N}$. Since $\lim_{n \in \mathbb{N}} f_n(x) = 0$, $\forall x \in \mathcal{X}$, then $\mathcal{X} = \bigcup_{n \in \mathbb{N}} O_n$. By the countable compactness of $\mathcal{X}$, $\exists N \in \mathbb{N}$ such that $\mathcal{X} = \bigcup_{n=1}^{N} O_n$. $\forall n \in \mathbb{N}$ with $n \ge N$, $\forall x \in \mathcal{X}$, $\exists i \in \{1, \ldots, N\}$ such that $x \in O_i$, and then $0 \le f_n(x) \le f_i(x) < \epsilon$. Hence, the result is established. This completes the proof of the lemma. □

Proposition 5.32 *Let $\mathcal{X} := (X, \mathcal{O})$ be a topological space and $(f_n)_{n=1}^{\infty}$ be a sequence of upper semicontinuous real-valued functions on $\mathcal{X}$. Assume that $\forall x \in \mathcal{X}$, $(f_n(x))_{n=1}^{\infty} \subseteq \mathbb{R}$ is nonincreasing and converges to $f_0(x)$. Then:*

(i) *f_0 is upper semicontinuous.*
(ii) *If, in addition, $\mathcal{X}$ is countably compact and f_0 is lower semicontinuous, then $(f_n)_{n=1}^{\infty}$ converges to f_0 uniformly on $\mathcal{X}$.*

Proof (i) $\forall a \in \mathbb{R}$, since $\forall x \in \mathcal{X}$, $f_0(x) < a \Leftrightarrow f_n(x) < a$ for all $n \ge m_x$, for some $m_x \in \mathbb{N}$; and $f_{i\,\text{inv}}((-\infty, a)) \subseteq f_{j\,\text{inv}}((-\infty, a))$, $\forall i \le j$ with $i, j \in \mathbb{N}$, and then $f_{0\,\text{inv}}((-\infty, a)) = \bigcup_{n \in \mathbb{N}} f_{n\,\text{inv}}((-\infty, a))$. By the upper semicontinuity of f_n, $f_{n\,\text{inv}}((-\infty, a)) \in \mathcal{O}$, $\forall n \in \mathbb{N}$. Hence, $f_{0\,\text{inv}}((-\infty, a)) \in \mathcal{O}$. Therefore, f_0 is upper semicontinuous.

(ii) f_0 is continuous by Proposition 3.16. Then, by Proposition 3.16, $(f_n - f_0)_{n=1}^{\infty}$ is a sequence of upper semicontinuous functions, which is nonincreasing in n and converges to 0 pointwise. By Dini's Lemma, $(f_n - f_0)_{n=1}^{\infty}$ converges to the zero function uniformly. Then, $(f_n)_{n=1}^{\infty}$ converges to f_0 uniformly.

This completes the proof of the proposition. □

Proposition 5.33 *Let $\mathcal{X}$ be a topological space and $(f_n)_{n=1}^{\infty}$ be a sequence of upper semicontinuous real-valued functions on $\mathcal{X}$ that converges uniformly to $f : \mathcal{X} \to \mathbb{R}$. Then, f is upper semicontinuous.*

Proof $\forall x_0 \in \mathcal{X}$. $\forall \epsilon \in (0, \infty) \subset \mathbb{R}$, $\exists N \in \mathbb{N}$, we have $|f_N(x) - f(x)| < \epsilon/3$, $\forall x \in \mathcal{X}$. Since f_N is upper semicontinuous, then, by Proposition 3.15, $\exists U \in \mathcal{O}$ with $x_0 \in U$, $\forall x \in U$, we have $f_N(x) < f_N(x_0) + \epsilon/3$. Then, $\forall x \in U$, we have

$$f(x) - f(x_0) = f(x) - f_N(x) + f_N(x) - f_N(x_0) + f_N(x_0) - f(x_0) < \epsilon$$

Therefore, f is upper semicontinuous at x_0 by Definition 3.14. By Proposition 3.15, f is upper semicontinuous.

This completes the proof of the proposition. □

5.4 Compactness in Metric Spaces

Lemma 5.34 *A sequentially compact metric space is totally bounded.*

Proof Let $\mathcal{X}$ be a sequentially compact metric space. Suppose $\mathcal{X}$ is not totally bounded. Then, $\exists \epsilon_0 \in (0, \infty) \subset \mathbb{R}$, and $\mathcal{X}$ cannot be covered by finitely many open balls with radius ϵ_0. Clearly, $\mathcal{X} \neq \emptyset$. Fix an $x_1 \in \mathcal{X}$. $\forall n \in \mathbb{N}$ with $n \geq 2$, $\exists x_n \in \mathcal{X} \setminus \left(\bigcup_{i=1}^{n-1} \mathcal{B}(x_i, \epsilon_0) \right)$. The sequence $(x_n)_{n=1}^{\infty}$ is such that $\rho(x_n, x_m) \geq \epsilon_0$, $\forall n, m \in \mathbb{N}$ with $n \neq m$. Clearly, this sequence does not have any convergent subsequence. This contradicts with the assumption that $\mathcal{X}$ is sequentially compact. Therefore, $\mathcal{X}$ must be totally bounded. This completes the proof of the lemma. □

Definition 5.35 Let $\mathcal{X}$ be a sequentially compact metric space and $(O_\alpha)_{\alpha \in \Lambda}$ be an open covering of $\mathcal{X}$, where Λ is an index set. The *Lebesgue number* of $(O_\alpha)_{\alpha \in \Lambda}$ is defined by

$$\begin{aligned} \epsilon := \sup \big\{ \delta \in (0, \infty) \subset \mathbb{R} \bigm| \forall x \in \mathcal{X}, \exists \bar{x} \in \mathcal{X} \ni \cdot\, \rho(x, \bar{x}) \geq \delta \\ \text{and } \mathcal{B}(x, \delta) \subseteq O_\alpha \text{ for some } \alpha \in \Lambda \} \end{aligned}$$

◇

Lemma 5.36 *Let $\mathcal{X}$ be a sequentially compact metric space containing at least two distinct points and $(O_\alpha)_{\alpha \in \Lambda}$ be an open covering of $\mathcal{X}$, where Λ is an index set. Then, the Lebesgue number ϵ of $(O_\alpha)_{\alpha \in \Lambda}$ is positive and belongs to $\mathbb{R}$.*

Proof Define a function $\phi : \mathcal{X} \to \mathbb{R}$ by, $\forall x \in \mathcal{X}$,

$$\phi(x) = \sup\{c \in (0, \infty) \subset \mathbb{R} \mid \mathcal{B}(x, c) \subseteq O_\alpha \text{ for some } \alpha \in \Lambda \text{ and } \exists \bar{x} \in \mathcal{X} \ni \cdot \rho(\bar{x}, x) \geq c\}$$

By Lemma 5.34, $\mathcal{X}$ is totally bounded. Then, $\forall x \in \mathcal{X}$, $\exists c_x \in (0, \infty) \subset \mathbb{R}$, such that $\nexists \bar{x} \in \mathcal{X}$ with $\rho(\bar{x}, x) \geq c_x$. Then, $\phi(x) \leq c_x$. Since $(O_\alpha)_{\alpha\in\Lambda}$ is an open cover of $\mathcal{X}$, $\exists c_{1,x} \in (0, \infty) \subset \mathbb{R}$ such that $\mathcal{B}(x, c_{1,x}) \subseteq O_\alpha$ for some $\alpha \in \Lambda$. Since $\mathcal{X}$ contains at least two distinct points, then $\exists \bar{x} \in \mathcal{X}$ with $\bar{x} \neq x$ such that $\rho(\bar{x}, x) =: c_{2,x} > 0$. Then, $\bar{c}_x := \min\{c_{1,x}, c_{2,x}\} > 0$ and $\phi(x) \geq \bar{c}_x$. Hence, $0 < \phi(x) < +\infty$, $\forall x \in \mathcal{X}$.

Next, we will show that ϕ is continuous. $\forall x, y \in \mathcal{X}$, we will distinguish two exhaustive and mutually exclusive cases: Case 1: $\phi(x) - \rho(x, y) \leq 0$; Case 2: $\phi(x) - \rho(x, y) =: l > 0$. Case 1: $\phi(x) - \rho(x, y) \leq 0$. Then, we have $\phi(y) > 0 \geq \phi(x) - \rho(x, y)$. Case 2: $\phi(x) - \rho(x, y) =: l > 0$. Then, $\forall \delta \in (0, l)$, $\exists \alpha \in \Lambda$, and $\exists \bar{x} \in \mathcal{X}$ such that $\mathcal{B}(x, \phi(x) - \delta) \subseteq O_\alpha$ and $\rho(x, \bar{x}) \geq \phi(x) - \delta$. Then, $\rho(\bar{x}, y) \geq \rho(x, \bar{x}) - \rho(x, y) \geq \phi(x) - \delta - \rho(x, y) = l - \delta$. $\forall \bar{y} \in \mathcal{B}(y, l - \delta)$, we have $\rho(x, \bar{y}) \leq \rho(x, y) + \rho(y, \bar{y}) < l - \delta + \rho(x, y) = \phi(x) - \delta$. Then, $\bar{y} \in \mathcal{B}(x, \phi(x) - \delta) \subseteq O_\alpha$. This implies that $\mathcal{B}(y, l - \delta) \subseteq O_\alpha$. Hence, we have $\phi(y) \geq l - \delta$. By the arbitrariness of δ, we have $\phi(y) \geq l = \phi(x) - \rho(x, y)$. Hence, in both cases, we have arrived at $\phi(y) \geq \phi(x) - \rho(x, y)$. Then, $\phi(x) - \phi(y) \leq \rho(x, y)$. This further implies that $\phi(y) - \phi(x) \leq \rho(y, x)$ by symmetry. Hence, we have $|\phi(x) - \phi(y)| \leq \rho(x, y)$. Hence, ϕ is continuous.

By the sequential compactness of $\mathcal{X}$, $\mathcal{X} \neq \emptyset$, and Propositions 5.28 and 5.29, $\exists x_m \in \mathcal{X}$ such that $\phi(x_m) = \min_{x\in\mathcal{X}} \phi(x) \in (0, \infty) \subset \mathbb{R}$. Then, the Lebesgue number $\epsilon = \phi(x_m)$ is positive and belongs to $\mathbb{R}$. This completes the proof of the lemma. □

Theorem 5.37 (Borel–Lebesgue Theorem) *Let $\mathcal{X}$ be a metric space. Then, the following are equivalent:*

(i) *$\mathcal{X}$ is compact.*
(ii) *$\mathcal{X}$ is countably compact.*
(iii) *$\mathcal{X}$ is sequentially compact.*
(iv) *$\mathcal{X}$ has the Bolzano–Weierstrass property.*

Proof (i) ⇒ (ii). This follows directly from Definitions 5.1 and 5.21.

(ii) ⇒ (iii). Clearly, $\mathcal{X}$ is first countable. By Proposition 5.28, $\mathcal{X}$ is sequentially compact.

(iii) ⇒ (i). We will distinguish three exhaustive and mutually exclusive cases: Case 1: $\mathcal{X} = \emptyset$; Case 2: $\mathcal{X}$ is a singleton set; Case 3: $\mathcal{X}$ contains at least two distinct points. Case 1: $\mathcal{X} = \emptyset$. Clearly, $\mathcal{X}$ is compact. Case 2: $\mathcal{X}$ is a singleton set. Clearly, $\mathcal{X}$ is compact. Case 3: $\mathcal{X}$ contains at least two distinct points. Fix any open covering $(O_\alpha)_{\alpha\in\Lambda} \subseteq \mathcal{O}$ of $\mathcal{X}$. Let ϵ be the Lebesgue number of the covering, which is a positive real number by Lemma 5.36. By Lemma 5.34, $\mathcal{X}$ is totally bounded. Then, there exists a finite set $X_N \subseteq \mathcal{X}$ such that $\mathcal{X} = \bigcup_{x\in X_N} \mathcal{B}(x, \epsilon/2)$. $\forall x \in X_N$,

$\mathcal{B}(x, \epsilon/2) \subseteq O_{\alpha_x}$ for some $\alpha_x \in \Lambda$. Hence, $\mathcal{X} = \bigcup_{x \in X_N} O_{\alpha_x}$, which is a finite subcovering. Therefore, $\mathcal{X}$ is compact.

(ii) $\Leftrightarrow$ (iv). This is proved in Proposition 5.26.

This completes the proof of the theorem. □

Proposition 5.38 *A metric space $\mathcal{X}$ is compact if, and only if, it is complete and totally bounded.*

Proof "Only if" Fix any Cauchy sequence $(x_n)_{n=1}^{\infty} \subseteq \mathcal{X}$. By the compactness of $\mathcal{X}$ and Borel–Lebesgue Theorem, there exists a subsequence $\left(x_{n_i}\right)_{i=1}^{\infty}$ that converges to $x_0 \in \mathcal{X}$. Then, it is straightforward to show that $\lim_{n \in \mathbb{N}} x_n = x_0$. (Here, the notation $\lim_{n \in \mathbb{N}} x_n$ makes sense since metric spaces are Hausdorff.) Hence, $\mathcal{X}$ is complete. By Borel–Lebesgue Theorem and Lemma 5.34, $\mathcal{X}$ is totally bounded.

"If" Fix any sequence $(x_n)_{n=1}^{\infty} \subseteq \mathcal{X}$. We will construct subsequences as following. Take $\left(x_n^{\langle 1 \rangle}\right)_{n=1}^{\infty} = (x_n)_{n=1}^{\infty}$. $\forall m \in \mathbb{N}$ with $m \geq 2$, since $\mathcal{X}$ is covered by finite many balls of radius $1/m$, we may find a subsequence $\left(x_n^{\langle m \rangle}\right)_{n=1}^{\infty}$ of $\left(x_n^{\langle m-1 \rangle}\right)_{n=1}^{\infty}$ such that $\left(x_n^{\langle m \rangle}\right)_{n=1}^{\infty} \subseteq \mathcal{B}\left(x^{\langle m \rangle}, 1/m\right)$ for some $x^{\langle m \rangle} \in \mathcal{X}$. Now, consider the sequence of diagonal elements $\left(x_n^{\langle n \rangle}\right)_{n=1}^{\infty} \subseteq \mathcal{X}$, which is a subsequence of the original sequence $(x_n)_{n=1}^{\infty}$. Clearly, this sequence is Cauchy by construction. By the completeness of $\mathcal{X}$, $\lim_{n \in \mathbb{N}} x_n^{\langle n \rangle} = x_0 \in \mathcal{X}$. Hence, we have shown that $\mathcal{X}$ is sequentially compact. By Borel–Lebesgue theorem, $\mathcal{X}$ is compact.

This completes the proof of the proposition. □

Proposition 5.39 *Let $\mathcal{X} := (X, \rho)$ be a compact metric space, $\mathcal{Y} := (Y, \sigma)$ be a metric space, and $f : \mathcal{X} \to \mathcal{Y}$ be continuous. Then, f is uniformly continuous.*

Proof $\forall \epsilon \in (0, \infty) \subset \mathbb{R}$, $\forall x \in \mathcal{X}$, by the continuity of f, $\exists \delta_x \in (0, \infty) \subset \mathbb{R}$ such that $\sigma(f(x), f(\bar{x})) < \epsilon/2$, $\forall \bar{x} \in \mathcal{B}(x, \delta_x)$. Then, $\mathcal{X} = \bigcup_{x \in \mathcal{X}} \mathcal{B}(x, \delta_x/2)$. By the compactness of $\mathcal{X}$, there exists a finite set $X_N \subseteq \mathcal{X}$ such that $\mathcal{X} = \bigcup_{x \in X_N} \mathcal{B}(x, \delta_x/2)$. Let $\delta = \min\{\inf_{x \in X_N} \delta_x/2, 1\} \in (0, \infty) \subset \mathbb{R}$. $\forall x_1, x_2 \in \mathcal{X}$ with $\rho(x_1, x_2) < \delta$, $\exists x_0 \in X_N$ such that $x_1 \in \mathcal{B}\left(x_0, \delta_{x_0}/2\right)$. Then, $x_2 \in \mathcal{B}\left(x_0, \delta_{x_0}\right)$. This implies that $x_1, x_2 \in \mathcal{B}\left(x_0, \delta_{x_0}\right)$. Then, $\sigma(f(x_1), f(x_2)) \leq \sigma(f(x_1), f(x_0)) + \sigma(f(x_0), f(x_2)) < \epsilon$. Hence, f is uniformly continuous. This completes the proof of the proposition. □

Proposition 5.40 *Let $n \in \mathbb{Z}_+$ and $K \subseteq \mathbb{R}^n$. K is compact if, and only if, K is closed and bounded.*

Proof When $n \in \mathbb{N}$, note that $\mathbb{R}^n$ is a complete metric space. Then, the result follows from Propositions 5.38 and 4.39. When $n = 0$, note that $\mathbb{R}^n$ is a singleton set and is compact. Then, the result follows immediately. □

5.5 The Ascoli–Arzelá Theorem

Lemma 5.41 *Let $(f_n)_{n=1}^{\infty}$ be a sequence of functions of a countable set D to a metric space $\mathcal{Y}$ such that $\forall x \in D$, $\overline{(f_n(x))_{n\in\mathbb{N}}}$ is compact. Then, there exists a subsequence $\left(f_{n_k}\right)_{k=1}^{\infty}$ such that $\forall x \in D$, and the sequence $\left(f_{n_k}(x)\right)_{k=1}^{\infty}$ converges to some point $f_0(x) \in \mathcal{Y}$.*

Proof We will distinguish three exhaustive and mutually exclusive cases: Case 1: $D = \emptyset$; Case 2: $D \neq \emptyset$ and is finite; Case 3: D is countably infinite. Case 1: $D = \emptyset$. Then, $(f_n)_{n=1}^{\infty}$ is the subsequence we seek. Case 2: $D \neq \emptyset$ and is finite. Take $D = \{x_1, \ldots, x_m\}$ for some $m \in \mathbb{N}$. Let $\left(f_n^{\langle 0\rangle}\right)_{n=1}^{\infty} = (f_n)_{n=1}^{\infty}$. $\forall k = 1, \ldots, m$, $\overline{\left(f_n^{\langle k-1\rangle}(x_k)\right)_{n\in\mathbb{N}}} \subseteq \overline{(f_n(x_k))_{n\in\mathbb{N}}}$ is a closed set and, therefore, a compact set by Proposition 5.5. By Borel–Lebesgue theorem, there exists a subsequence $\left(f_n^{\langle k\rangle}\right)_{n=1}^{\infty}$ of $\left(f_n^{\langle k-1\rangle}\right)_{n=1}^{\infty}$ such that $\left(f_n^{\langle k\rangle}(x_k)\right)_{n=1}^{\infty}$ converges. Hence, the sequence $\left(f_n^{\langle m\rangle}\right)_{n=1}^{\infty}$ is the subsequence we seek. Case 3: D is countably infinite. Take $D = (x_m)_{m=1}^{\infty}$. Let $\left(f_n^{\langle 0\rangle}\right)_{n=1}^{\infty} = (f_n)_{n=1}^{\infty}$. $\forall k \in \mathbb{N}$, $\overline{\left(f_n^{\langle k-1\rangle}(x_k)\right)_{n\in\mathbb{N}}} \subseteq \overline{(f_n(x_k))_{n\in\mathbb{N}}}$ is a closed set and, therefore, a compact set by Proposition 5.5. By Borel–Lebesgue theorem, there exists a subsequence $\left(f_n^{\langle k\rangle}\right)_{n=1}^{\infty}$ of $\left(f_n^{\langle k-1\rangle}\right)_{n=1}^{\infty}$ such that $\left(f_n^{\langle k\rangle}(x_k)\right)_{n=1}^{\infty}$ converges. Now, the diagonal sequence $\left(f_k^{\langle k\rangle}\right)_{k=1}^{\infty}$ is the subsequence we seek. This completes the proof of the lemma. □

Lemma 5.42 *Let $\mathcal{X} := (X, \mathcal{O})$ be a compact topological space, $\mathcal{Y} := (Y, \rho)$ be a metric space, $\mathcal{F}$ be an equicontinuous family of functions of $\mathcal{X}$ to $\mathcal{Y}$, and $(f_n)_{n=1}^{\infty} \subseteq \mathcal{F}$ be such that $\lim_{n\in\mathbb{N}} f_n(x) = f_0(x)$, $\forall x \in \mathcal{X}$, for some $f_0 : \mathcal{X} \to \mathcal{Y}$. Then, f_0 is continuous and $(f_n)_{n=1}^{\infty}$ converges to f_0 uniformly.*

Proof $\forall \epsilon \in (0, \infty) \subset \mathbb{R}$, $\forall x_0 \in \mathcal{X}$, by the equicontinuity of $\mathcal{F}$, $\exists U_{x_0,\epsilon} \in \mathcal{O}$ with $x_0 \in U_{x_0,\epsilon}$ such that $\forall x \in U_{x_0,\epsilon}$, $\forall f \in \mathcal{F}$, $\rho(f(x), f(x_0)) < \epsilon$.

By $\lim_{n\in\mathbb{N}} f_n(x_0) = f_0(x_0)$, $\exists N_{x_0,\epsilon} \in \mathbb{N}$, $\forall n \in \mathbb{N}$ with $n \geq N_{x_0,\epsilon}$, we have $\rho(f_n(x_0), f_0(x_0)) < \epsilon$.

$\forall x_0 \in \mathcal{X}$, $\forall \epsilon \in (0, \infty) \subset \mathbb{R}$, $\forall x \in U_{x_0,\epsilon}$, we have, by Propositions 4.30, 3.66, and 3.67, $\rho(f_0(x), f_0(x_0)) = \lim_{n\in\mathbb{N}} \rho(f_n(x), f_n(x_0)) \leq \epsilon$. Hence, f_0 is continuous at x_0. By the arbitrariness of x_0 and Proposition 3.9, f_0 is continuous.

$\forall \epsilon \in (0, \infty) \subset \mathbb{R}$, $\mathcal{X} \subseteq \bigcup_{x\in X} U_{x,\epsilon}$. By the compactness of $\mathcal{X}$, there exists a finite set $X_N \subseteq X$ such that $\mathcal{X} \subseteq \bigcup_{x\in X_N} U_{x,\epsilon}$. Let $N := \max\{\sup_{x\in X_N} N_{x,\epsilon}, 1\} \in \mathbb{N}$, $\forall n \in \mathbb{N}$ with $n \geq N$, $\forall x \in \mathcal{X}$, $\exists x_0 \in X_N$ such that $x \in U_{x_0,\epsilon}$. Then, we have

$$\rho(f_n(x), f_0(x)) \leq \rho(f_n(x), f_n(x_0)) + \rho(f_n(x_0), f_0(x_0)) + \rho(f_0(x_0), f_0(x)) < 3\epsilon$$

Hence, we have $(f_n)_{n=1}^{\infty}$ converges to f_0 uniformly.

This completes the proof of the lemma. □

Lemma 5.43 *Let $\mathcal{X} := (X, \mathcal{O})$ be a topological space, $\mathcal{Y} := (Y, \rho)$ be a metric space, $(f_n)_{n=1}^{\infty}$ be an equicontinuous sequence of functions of $\mathcal{X}$ to $\mathcal{Y}$, and $(f_n(x))_{n=1}^{\infty}$ converge at every $x \in D$, where $D \subseteq \mathcal{X}$ is dense. Assume that $\forall x \in \mathcal{X}$, $\overline{(f_n(x))_{n=1}^{\infty}} \subseteq \mathcal{Y}$ is complete. Then, $\exists$ continuous function $f_0 : \mathcal{X} \to \mathcal{Y}$ such that $\lim_{n \in \mathbb{N}} f_n(x) = f_0(x)$, $\forall x \in \mathcal{X}$.*

Proof $\forall \epsilon \in (0, \infty) \subset \mathbb{R}$, $\forall x \in \mathcal{X}$, since $(f_n)_{n=1}^{\infty}$ is equicontinuous, then $\exists O \in \mathcal{O}$ with $x \in O$ such that, $\forall \bar{x} \in O$, $\rho(f_n(x), f_n(\bar{x})) < \epsilon$, $\forall n \in \mathbb{N}$.

Since D is dense, then $\exists x_0 \in D \cap O$. $(f_n(x_0))_{n=1}^{\infty} \subseteq \mathcal{Y}$ is Cauchy since it is convergent. Then, $\exists N \in \mathbb{N}$, $\forall n, m \geq N$, and we have $\rho(f_n(x_0), f_m(x_0)) < \epsilon$. Note that

$$\begin{aligned}\rho(f_n(x), f_m(x)) \leq \rho(f_n(x), f_n(x_0)) + \rho(f_n(x_0), f_m(x_0)) \\ +\rho(f_m(x_0), f_m(x)) < 3\epsilon\end{aligned}$$

Therefore, $(f_n(x))_{n=1}^{\infty} \subseteq \overline{(f_n(x))_{n=1}^{\infty}} \subseteq \mathcal{Y}$ is a Cauchy sequence, which converges since $\overline{(f_n(x))_{n=1}^{\infty}}$ is complete. Then, we may define $f_0 : \mathcal{X} \to \mathcal{Y}$ by $f_0(x) = \lim_{n \in \mathbb{N}} f_n(x)$, $\forall x \in \mathcal{X}$.

By Proposition 4.30, 3.66, and 3.67, we have, $\forall \bar{x} \in O$, $\rho(f_0(x), f_0(\bar{x})) = \lim_{n \in \mathbb{N}} \rho(f_n(x), f_n(\bar{x})) \leq \epsilon$. Hence, f_0 is continuous at x. By the arbitrariness of x and Proposition 3.9, f_0 is continuous. This completes the proof of the lemma. □

Theorem 5.44 (Ascoli–Arzelá Theorem) *Let $\mathcal{X} := (X, \mathcal{O})$ be a separable topological space, $\mathcal{Y} := (Y, \rho)$ be a metric space, $\mathcal{F}$ be an equicontinuous family of functions of $\mathcal{X}$ to $\mathcal{Y}$, and $(f_n)_{n=1}^{\infty} \subseteq \mathcal{F}$ be such that $\overline{(f_n(x))_{n \in \mathbb{N}}} \subseteq \mathcal{Y}$ is compact, $\forall x \in \mathcal{X}$. Then, there exists a subsequence $\left(f_{n_k}\right)_{k=1}^{\infty}$ of $(f_n)_{n=1}^{\infty}$ that converges pointwise to a continuous function $f_0 : \mathcal{X} \to \mathcal{Y}$, and the convergence is uniform on any compact subset of $\mathcal{X}$.*

Proof Let $D \subseteq \mathcal{X}$ be a countable dense set. By Lemma 5.41, there exists a subsequence $\left(f_{n_k}\right)_{k=1}^{\infty}$ of $(f_n)_{n=1}^{\infty}$ that converges pointwise on D to a function $\bar{f} : D \to \mathcal{Y}$. Note that $\forall x \in \mathcal{X}$, $\overline{(f_n(x))_{n \in \mathbb{N}}} \subseteq \mathcal{Y}$ is compact implies that $\overline{(f_n(x))_{n \in \mathbb{N}}}$ is complete by Proposition 5.38. Then, by Lemma 5.43, there exists a continuous function $f_0 : \mathcal{X} \to \mathcal{Y}$ to which $\left(f_{n_k}\right)_{k=1}^{\infty}$ converges pointwise. By Lemma 5.42, $\left(f_{n_k}\right)_{k=1}^{\infty}$ converges to f_0 uniformly on compact subsets of $\mathcal{X}$. This completes the proof of the theorem. □

5.6 Product Spaces

Lemma 5.45 *Let $\mathcal{A}$ be a collection of subsets of a set X with the finite intersection property. Then, there is a collection $\mathcal{M}$ of subsets in X such that $\mathcal{A} \subseteq \mathcal{M}$, $\mathcal{M}$ has the finite intersection property, and $\mathcal{M}$ is maximal with respect to this property, that is, $\forall \mathcal{C} \subseteq {}^X2$ with the finite intersection property and $\mathcal{M} \subseteq \mathcal{C}$, we have $\mathcal{C} = \mathcal{M}$.*

Proof Clearly, $X \neq \emptyset$ by the fact that $\mathcal{A}$ has the finite intersection property. Define $M := \{\mathcal{C} \subseteq {}^{X}2 \mid \mathcal{A} \subseteq \mathcal{C}$ and $\mathcal{C}$ has the finite intersection property.$\}$ Set containment $\subseteq$ forms an antisymmetric partial ordering on M and $\mathcal{A} \in M \neq \emptyset$. Let $N \subseteq M$ be any nonempty totally ordered subset. Let $\mathcal{B} := \bigcup_{\mathcal{C} \in N} \mathcal{C} \subseteq {}^{X}2$. Fix any finite subcollection $\mathcal{B}_N \subseteq \mathcal{B}$. We will distinguish two exhaustive and mutually exclusive cases: $\mathcal{B}_N = \emptyset$; Case 2: $\mathcal{B}_N \neq \emptyset$. Case 1: $\mathcal{B}_N = \emptyset$. Then, $\bar{B} := \bigcap_{B \in \mathcal{B}_N} B = X \neq \emptyset$. Case 2: $\mathcal{B}_N \neq \emptyset$. Take $\mathcal{B}_N = \{B_1, \ldots, B_n\}$ for some $n \in \mathbb{N}$. $\forall B_i \in \mathcal{B}_N$, $\exists \mathcal{C}_i \in N$ such that $B_i \in \mathcal{C}_i$. Since N is totally ordered by $\subseteq$, without loss of generality, assume that $\mathcal{C}_1 \subseteq \mathcal{C}_2 \subseteq \cdots \subseteq \mathcal{C}_n$. Then, $B_i \in \mathcal{C}_n$, $\forall B_i \in \mathcal{B}_N$. By the definition of the set M, we have $\bigcap_{i=1}^{n} B_i \neq \emptyset$. In both cases, we have arrived at $\bigcap_{B \in \mathcal{B}_N} B \neq \emptyset$. Hence, $\mathcal{B}$ has finite intersection property. Since $N \neq \emptyset$, then $\mathcal{A} \subseteq \mathcal{B}$. Hence, $\mathcal{B} \in M$ and is an upper bound of N. By Zorn's Lemma, there exists a maximal element $\mathcal{M}$ of M. This completes the proof of the lemma. □

Lemma 5.46 *Let $\mathcal{B}$ be a collection of subsets of a set X that is maximal with respect to the finite intersection property. Then, each intersection of a finite number of sets in $\mathcal{B}$ is again in $\mathcal{B}$, and each set that meets every set in $\mathcal{B}$ is itself in $\mathcal{B}$.*

Proof Fix any finite subcollection $\mathcal{B}_N \subseteq \mathcal{B}$. Let $\bar{B} := \bigcap_{B \in \mathcal{B}_N} B$. Then, $\bar{B} \neq \emptyset$. Let $\bar{\mathcal{B}} := \mathcal{B} \cup \{\bar{B}\} \supseteq \mathcal{B}$. It is easy to see that $\bar{\mathcal{B}}$ has the finite intersection property. Then, by maximality of $\mathcal{B}$, we have $\bar{\mathcal{B}} = \mathcal{B}$. Hence, $\bar{B} \in \mathcal{B}$.

Let $C \subseteq X$ be such that $C \cap B \neq \emptyset$, $\forall B \in \mathcal{B}$. Let $\bar{\mathcal{B}} := \mathcal{B} \cup \{C\} \supseteq \mathcal{B}$. Fix any finite subcollection $\mathcal{M} \subseteq \bar{\mathcal{B}}$. We will distinguish two exhaustive and mutually exclusive cases: Case 1: $C \notin \mathcal{M}$; Case 2: $C \in \mathcal{M}$. Case 1: $C \notin \mathcal{M}$. Then, $\mathcal{M} \subseteq \mathcal{B}$. Then, $\bigcap_{B \in \mathcal{M}} B \neq \emptyset$ since $\mathcal{B}$ has the finite intersection property. Case 2: $C \in \mathcal{M}$. Then, $\bar{\mathcal{M}} := \mathcal{M} \setminus \{C\} \subseteq \mathcal{B}$ and is a finite set. Then, $\bigcap_{B \in \bar{\mathcal{M}}} B \in \mathcal{B}$. Then, $\bigcap_{B \in \mathcal{M}} B = C \cap \left(\bigcap_{B \in \bar{\mathcal{M}}} B\right) \neq \emptyset$. Hence, in both cases, we have $\bigcap_{B \in \mathcal{M}} B \neq \emptyset$. This shows that $\bar{\mathcal{B}}$ has the finite intersection property. By the maximality of $\mathcal{B}$, we have $\bar{\mathcal{B}} = \mathcal{B}$. Hence, $C \in \mathcal{B}$.

This completes the proof of the lemma. □

Theorem 5.47 (Tychonoff Theorem) *Let $\mathcal{X}_\alpha := (X_\alpha, \mathcal{O}_\alpha)$ be a compact topological space, $\forall \alpha \in \Lambda$, where Λ is an index set. Then, the product topological space $\mathcal{X} := (X, \mathcal{O}) := \prod_{\alpha \in \Lambda} \mathcal{X}_\alpha$ is compact.*

Proof *(Bourbaki)* We will distinguish two exhaustive and mutually exclusive cases. Case 1: $\Lambda = \emptyset$; Case 2: $\Lambda \neq \emptyset$. Case 1: $\Lambda = \emptyset$. Then, X is a singleton set, which is clearly compact.

Case 2: $\Lambda \neq \emptyset$. We will prove this case by Proposition 5.3. Note that a basis for $\mathcal{X}$ is given by $\mathcal{B} := \left\{\prod_{\alpha \in \Lambda} O_\alpha \subseteq X \;\middle|\; O_\alpha \in \mathcal{O}_\alpha, \forall \alpha \in \Lambda, \text{ and } O_\alpha = X_\alpha \text{ for all } \alpha\text{'s except finitely many } \alpha\text{'s}\right\}$. Fix a collection $\mathcal{A}$ of closed sets in $\mathcal{X}$ with the finite intersection property. By Lemma 5.45, let $\mathcal{C}$ be a collection of subsets of $\mathcal{X}$ such that $\mathcal{A} \subseteq \mathcal{C}$ and $\mathcal{C}$ is maximal with respect to the finite intersection property.

$\forall \alpha \in \Lambda$, let $\mathcal{C}_\alpha := \left\{C_\alpha \subseteq X_\alpha \;\middle|\; \exists C \in \mathcal{C} \ni \cdot\, \pi_\alpha(C) = C_\alpha\right\}$. $\forall$ finite subcollection $\mathcal{C}_{\alpha N} \subseteq \mathcal{C}_\alpha$, $\forall C_\alpha \in \mathcal{C}_{\alpha N}$, $\exists C_{C_\alpha} \in \mathcal{C}$ such that $C_\alpha = \pi_\alpha(C_{C_\alpha})$. Since $\mathcal{C}$ has the finite intersection property, then $\bigcap_{C_\alpha \in \mathcal{C}_{\alpha N}} C_{C_\alpha} \neq \emptyset$. Then, by Proposition 2.5,

$\bigcap_{C_\alpha \in \mathcal{C}_{\alpha N}} C_\alpha = \bigcap_{C_\alpha \in \mathcal{C}_{\alpha N}} \pi_\alpha(C_{C_\alpha}) \supseteq \pi_\alpha\left(\bigcap_{C_\alpha \in \mathcal{C}_{\alpha N}} C_{C_\alpha}\right) \neq \emptyset$. Hence, $\mathcal{C}_\alpha$ has finite intersection property. Let $\bar{\mathcal{C}_\alpha} := \left\{\overline{C_\alpha} \,\middle|\, C_\alpha \in \mathcal{C}_\alpha\right\}$. Then, $\bar{\mathcal{C}_\alpha}$ has finite intersection property. By the compactness of $\mathcal{X}_\alpha$ and Proposition 5.3, $\exists x_\alpha \in \bigcap_{C_\alpha \in \mathcal{C}_\alpha} \overline{C_\alpha}$.

Let $x \in \mathcal{X}$ be given by $\pi_\alpha(x) = x_\alpha$, $\forall \alpha \in \Lambda$. Consider a set S of the form $S = \pi_{\alpha\,\mathrm{inv}}(O_\alpha)$ for some $\alpha \in \Lambda$ and for some $O_\alpha \in \mathcal{O}_\alpha$ with $x_\alpha \in O_\alpha$. Then, $S \cap C \neq \emptyset$, $\forall C \in \mathcal{C}$, since $x_\alpha \in \overline{\pi_\alpha(C)}$. Hence, $S \in \mathcal{C}$ by Lemma 5.46 and the maximality of $\mathcal{C}$. $\forall B \in \mathcal{B}$ with $x \in B$, $\exists n \in \mathbb{N}$, $\exists \alpha_1, \ldots, \alpha_n \in \Lambda$, and $\exists O_{\alpha_i} \in \mathcal{O}_{\alpha_i}$, $i = 1, \ldots, n$, such that $x_{\alpha_i} \in O_{\alpha_i}$ and $B = \bigcap_{i=1}^n \pi_{\alpha_i\,\mathrm{inv}}(O_{\alpha_i})$. Then, $B \in \mathcal{C}$ by Lemma 5.46.

$\forall F \in \mathcal{A}$, F is closed. Then, $F \cap B \neq \emptyset$, $\forall B \in \mathcal{B}$ with $x \in B$, by the finite intersection property of $\mathcal{C}$. By Proposition 3.3, $x \in \overline{F} = F$. Hence, $x \in \bigcap_{F \in \mathcal{A}} F \neq \emptyset$. By Proposition 5.3, $\mathcal{X}$ is compact.

This completes the proof of the theorem. □

Proposition 5.48 *Let $\mathcal{X}_\alpha := (X_\alpha, \mathcal{O}_\alpha)$ be a sequentially compact topological space, $\forall \alpha \in \Lambda$, where Λ is a countable index set. Then, the product topological space $\mathcal{X} := (X, \mathcal{O}) := \prod_{\alpha \in \Lambda} \mathcal{X}_\alpha$ is sequentially compact.*

Proof We will distinguish three exhaustive and mutually exclusive cases: Case 1: $\Lambda = \emptyset$; Case 2: $\Lambda \neq \emptyset$ and is finite; Case 3: Λ is countably infinite. Case 1: $\Lambda = \emptyset$. Then, $X = \{\emptyset\}$ is a singleton set. Clearly, $\mathcal{X}$ is sequentially compact.

Case 2: $\Lambda \neq \emptyset$ and is finite. Without loss of generality, assume that $\Lambda = \{1, \ldots, m\}$ for some $m \in \mathbb{N}$. Fix a sequence $\left(x_n^{\langle 0\rangle}\right)_{n=1}^\infty \subseteq \mathcal{X}$. $\forall k \in \{1, \ldots, m\}$, $\left(\pi_k(x_n^{\langle k-1\rangle})\right)_{n=1}^\infty \subseteq \mathcal{X}_k$. By the sequential compactness of $\mathcal{X}_k$, there exists a subsequence $\left(x_n^{\langle k\rangle}\right)_{n=1}^\infty$ of $\left(x_n^{\langle k-1\rangle}\right)_{n=1}^\infty$ such that $\left(\pi_k(x_n^{\langle k\rangle})\right)_{n=1}^\infty$ converges to $x_k \in \mathcal{X}_k$. Let $x \in \mathcal{X}$ be given by $\pi_k(x) = x_k$, $\forall k \in \{1, \ldots, m\}$. The sequence $\left(x_n^{\langle m\rangle}\right)_{n=1}^\infty \subseteq \mathcal{X}$ is a subsequence of $\left(x_n^{\langle 0\rangle}\right)_{n=1}^\infty$. Clearly, $\forall k \in \{1, \ldots, m\}$, $\left(\pi_k(x_n^{\langle m\rangle})\right)_{n=1}^\infty$ converges to x_k. By Proposition 3.67, $\left(x_n^{\langle m\rangle}\right)_{n=1}^\infty$ converges to x. Hence, $\mathcal{X}$ is sequentially compact.

Case 3: Λ is countably infinite. Without loss of generality, assume that $\Lambda = \mathbb{N}$. Fix a sequence $\left(x_n^{\langle 0\rangle}\right)_{n=1}^\infty \subseteq \mathcal{X}$. $\forall k \in \mathbb{N}$, $\left(\pi_k(x_n^{\langle k-1\rangle})\right)_{n=1}^\infty \subseteq \mathcal{X}_k$. By the sequential compactness of $\mathcal{X}_k$, there exists a subsequence $\left(x_n^{\langle k\rangle}\right)_{n=1}^\infty$ of $\left(x_n^{\langle k-1\rangle}\right)_{n=1}^\infty$ such that $\left(\pi_k(x_n^{\langle k\rangle})\right)_{n=1}^\infty$ converges to $x_k \in \mathcal{X}_k$. Let $x \in \mathcal{X}$ be given by $\pi_k(x) = x_k$, $\forall k \in \mathbb{N}$. Now consider the diagonal sequence $\left(x_n^{\langle n\rangle}\right)_{n=1}^\infty \subseteq \mathcal{X}$, which is a subsequence of $\left(x_n^{\langle 0\rangle}\right)_{n=1}^\infty$. Clearly, $\forall k \in \mathbb{N}$, $\left(\pi_k(x_n^{\langle n\rangle})\right)_{n=1}^\infty$ converges to x_k. By Proposition 3.67, $\left(x_n^{\langle n\rangle}\right)_{n=1}^\infty$ converges to x. Hence, $\mathcal{X}$ is sequentially compact.

This completes the proof of the proposition. □

5.7 Locally Compact Spaces

5.7.1 Fundamental Notion

Definition 5.49 A topological space $\mathcal{X}$ is *locally compact* if $\forall x \in \mathcal{X}, \exists O \in \mathcal{O}$ with $x \in O$ such that $\overline{O}$ is compact. ◇

Clearly, a compact space is locally compact.

Example 5.50 $\mathbb{R}^n$ is a locally compact space but not a compact space, $\forall n \in \mathbb{N}$. ◇

Proposition 5.51 *A topological space $\mathcal{X}$ is locally compact if, and only if, the collection $\mathcal{B}_l := \{O \in \mathcal{O} \mid \overline{O}$ is compact$\}$ forms a basis for $\mathcal{O}$.*

Proof "Only if" Let $\mathcal{X}$ be locally compact. $\forall O \in \mathcal{O}, \forall x \in O, \exists B \in \mathcal{B}_l$ such that $x \in B$ and $\overline{B}$ is compact. Let $B_1 := O \cap B$. Then, $x \in B_1 \in \mathcal{O}$ and $B_1 \subseteq O$. Note that $\overline{B_1} \subseteq \overline{B}$ and is closed. By Proposition 5.5, $\overline{B_1}$ is compact. Then, $B_1 \in \mathcal{B}_l$. Therefore, $\mathcal{B}_l$ is a basis for $\mathcal{O}$.

"If" This is straightforward.

This completes the proof of the proposition. □

Proposition 5.52 *Let $\mathcal{X}$ be a locally compact topological space and $K \subseteq \mathcal{X}$ be compact. Then, $\exists O \in \mathcal{O}$, such that $K \subseteq O \subseteq \overline{O}$ and $\overline{O}$ is compact.*

Proof $\forall x \in K$, by the local compactness of $\mathcal{X}$, $\exists O_x \in \mathcal{O}$ such that $x \in O_x \subseteq \overline{O_x}$ and $\overline{O_x}$ is compact. Then, $K \subseteq \bigcup_{x \in K} O_x$. By the compactness of K, there exists a finite set $K_N \subseteq K$ such that $K \subseteq \bigcup_{x \in K_N} O_x =: O \in \mathcal{O}$. By Proposition 3.3, $\overline{O} = \overline{\bigcup_{x \in K_N} O_x} = \bigcup_{x \in K_N} \overline{O_x}$, which is compact. This completes the proof of the proposition. □

Proposition 5.53 *Let $\mathcal{X}$ be a locally compact Hausdorff topological space and $\mathcal{B}_l$ be the basis of $\mathcal{X}$ defined in Proposition 5.51. $\forall O \in \mathcal{O}, \forall x \in O, \exists B \in \mathcal{B}_l$ such that $x \in B \subseteq \overline{B} \subseteq O$ and $\overline{B}$ is compact.*

Proof $\forall O \in \mathcal{O}, \forall x \in O$, we will distinguish two exhaustive and mutually exclusive cases: Case 1: $O = \mathcal{X}$; Case 2: $O \subset \mathcal{X}$. Case 1: $O = \mathcal{X}$. The result holds by Definition 5.49. Case 2: $O \subset \mathcal{X}$. $\forall y \in \widetilde{O}$, since $\mathcal{X}$ is Hausdorff, then $\exists O_y^{\langle 1 \rangle}, O_y^{\langle 2 \rangle} \in \mathcal{O}$ such that $x \in O_y^{\langle 1 \rangle}$, $y \in O_y^{\langle 2 \rangle}$, and $O_y^{\langle 1 \rangle} \cap O_y^{\langle 2 \rangle} = \emptyset$. By Proposition 5.51 and Definition 3.17, $\exists B_y \in \mathcal{B}_l$ such that $x \in B_y \subseteq O_y^{\langle 1 \rangle}$. Then, $B_y \subseteq \widetilde{O_y^{\langle 2 \rangle}}$ and $\overline{B_y} \subseteq \widetilde{O_y^{\langle 2 \rangle}}$, since $\widetilde{O_y^{\langle 2 \rangle}}$ is closed in $\mathcal{X}$. This implies that $y \notin \overline{B_y}$. Note that $\widetilde{O}$ is closed in $\mathcal{X}$, and then $\widetilde{O} \cap \overline{B_y}$ is closed in $\mathcal{X}$. Since $\overline{B_y}$ is compact, then, by Propositions 3.5 and 5.5, $\widetilde{O} \cap \overline{B_y}$ is compact. Note that $\bigcap_{y \in \widetilde{O}} (\widetilde{O} \cap \overline{B_y}) = \emptyset$. By Proposition 5.12, there exists a finite set $D_N \subseteq \widetilde{O}$ such that $\bigcap_{y \in D_N} (\widetilde{O} \cap \overline{B_y}) = \emptyset$. Clearly, $D_N \neq \emptyset$ since $O \subset \mathcal{X} \neq \emptyset$. Then, we have $x \in \bigcap_{y \in D_N} B_y \subseteq \bigcap_{y \in D_N} \overline{B_y} \subseteq O$. Since $\bigcap_{y \in D_N} B_y \in \mathcal{O}$, then $\exists B \in \mathcal{B}_l$ such that $x \in B \subseteq \bigcap_{y \in D_N} B_y$ and $\overline{B}$ is compact. Note that $\overline{B} \subseteq \bigcap_{y \in D_N} \overline{B_y} \subseteq O$. This completes the proof of the proposition. □

Proposition 5.54 *Let $\mathcal{X}$ be a locally compact Hausdorff topological space, $U \in \mathcal{O}$, and $K \subseteq U$ be compact. Then, $\exists V \in \mathcal{O}$ such that $K \subseteq V \subseteq \overline{V} \subseteq U$ and $\overline{V}$ is compact.*

Proof $\forall x \in K$, by Proposition 5.53, $\exists V_x \in \mathcal{O}$ such that $x \in V_x \subseteq \overline{V_x} \subseteq U$ and $\overline{V_x}$ is compact. Then, $K \subseteq \bigcup_{x \in K} V_x$. By the compactness of K, there exists a finite set $K_N \subseteq K$ such that $K \subseteq \bigcup_{x \in K_N} V_x =: V \in \mathcal{O}$. By Proposition 3.3, $\overline{V} = \bigcup_{x \in K_N} \overline{V_x} \subseteq U$ and is compact. This completes the proof of the proposition. □

Proposition 5.55 *Let $\mathcal{X}$ be a locally compact space and $F \subseteq \mathcal{X}$. Then, F is closed if, and only if, for any closed and compact set $K \subseteq \mathcal{X}$, we have $F \cap K$ is closed.*

Proof "Only if" Let F be closed. For any closed and compact $K \subseteq \mathcal{X}$, $F \cap K$ is closed.

"If" $\forall x \in \overline{F}$. By local compactness of $\mathcal{X}$, $\exists O \in \mathcal{O}$ such that $x \in O \subseteq \overline{O}$ and $\overline{O}$ is compact. By the assumption, $\overline{O} \cap F$ is closed. $\forall U \in \mathcal{O}$ with $x \in U$, we have $x \in O \cap U \in \mathcal{O}$. By Proposition 3.3, $(O \cap U) \cap F \neq \emptyset$, since $x \in \overline{F}$. Then, we have $U \cap (\overline{O} \cap F) \neq \emptyset$. Hence, we have $x \in \overline{\overline{O} \cap F} = \overline{O} \cap F \subseteq F$, by Proposition 3.3. Hence, $\overline{F} \subseteq F$ and F is closed.

This completes the proof of the proposition. □

Proposition 5.56 *Let $\mathcal{X}$ be a locally compact Hausdorff topological space and $(O_n)_{n=1}^{\infty} \subseteq \mathcal{O}$ be a sequence of open dense subsets in $\mathcal{X}$. Then, $\bigcap_{n=1}^{\infty} O_n$ is dense. Therefore, $\mathcal{X}$ is second category everywhere.*

Proof $\forall U \in \mathcal{O}$ with $U \neq \emptyset$, since O_1 is dense, then $\exists x_1 \in U \cap O_1$. By Proposition 5.53, $\exists V_1 \in \mathcal{O}$ such that $x_1 \in V_1 \subseteq \overline{V_1} \subseteq U \cap O_1$ and $\overline{V_1}$ is compact. $\forall n \in \mathbb{N}$ with $n \geq 2$, since O_n is dense, then $\exists x_n \in V_{n-1} \cap O_n$. By Proposition 5.53, $\exists V_n \in \mathcal{O}$ such that $x_n \in V_n \subseteq \overline{V_n} \subseteq V_{n-1} \cap O_n$ and $\overline{V_n}$ is compact. Hence, $\left(\overline{V_n}\right)_{n=1}^{\infty}$ is a sequence of nonempty closed sets that is nonincreasing (that is, $\overline{V_{n+1}} \subseteq \overline{V_n}$, $\forall n \in \mathbb{N}$). Clearly, this sequence has the finite intersection property. By the compactness of $\overline{V_1}$ and Proposition 5.3, $\bigcap_{n=1}^{\infty} \overline{V_n} \neq \emptyset$. Clearly, $\bigcap_{n=1}^{\infty} \overline{V_n} \subseteq U \cap \left(\bigcap_{n=1}^{\infty} O_n\right) \neq \emptyset$. Therefore, $\bigcap_{n=1}^{\infty} O_n$ is dense in $\mathcal{X}$, by the arbitrariness of U.

By Proposition 3.38, $\mathcal{X}$ is second category everywhere. This completes the proof of the proposition. □

Proposition 5.57 *Let $\mathcal{X} := (X, \mathcal{O})$ be a Hausdorff space, $Y \subseteq X$ be a dense subset, and $\mathcal{O}_Y$ be the subset topology on Y. Assume that $\mathcal{Y} := (Y, \mathcal{O}_Y)$ is locally compact. Then, Y is open in $\mathcal{X}$.*

Proof $\forall y \in Y$, by the local compactness of $\mathcal{Y}$, $\exists U_y \in \mathcal{O}_Y$ such that $y \in U_y \subseteq U_{yc} \subseteq Y$, where U_{yc} is the closure of U_y in $\mathcal{Y}$, and U_{yc} is compact. Then, U_{yc} is compact in $\mathcal{X}$ and is closed, by Proposition 5.5. Then, $U_{yc} \supseteq \overline{U_y}$. On the other hand, by Proposition 3.5, $U_{yc} = \overline{U_y} \cap Y \subseteq \overline{U_y}$. Hence, we have $U_{yc} = \overline{U_y}$. Since $U_y \in \mathcal{O}_Y$, $\exists O \in \mathcal{O}$ such that $U_y = O \cap Y$. By Proposition 3.37, we have $U_{yc} =$

$\overline{O \cap Y} = \overline{O}$. Hence, we have $y \in U_y \subseteq O \subseteq \overline{O} = U_{yc} \subseteq Y$. Hence, Y is open in $\mathcal{X}$. This completes the proof of the proposition. □

Lemma 5.58 *Let $\mathcal{X} := (X, \mathcal{O})$ be a locally compact Hausdorff topological space, $Y \in \mathcal{O}$, and $\mathcal{O}_Y$ be the subset topology on Y. Then, $\mathcal{Y} := (Y, \mathcal{O}_Y)$ is locally compact.*

Proof $\forall x \in Y$, by Proposition 5.53, $\exists U \in \mathcal{O}$ such that $x \in U \subseteq \overline{U} \subseteq Y$ and $\overline{U}$ is compact. Then, $U \in \mathcal{O}_Y$, and the closure of U in $\mathcal{Y}$ is $\overline{U}$, by Proposition 3.5. Then, $\overline{U}$ is compact in $\mathcal{Y}$. Hence, $\mathcal{Y}$ is locally compact. This completes the proof of the lemma. □

Lemma 5.59 *Let $\mathcal{X} := (X, \mathcal{O})$ be a locally compact Hausdorff topological space, $Y \subseteq \mathcal{X}$ be closed, and $\mathcal{O}_Y$ be the subset topology on Y. Then, $\mathcal{Y} := (Y, \mathcal{O}_Y)$ is locally compact.*

Proof $\forall x \in Y$, by the local compactness of $\mathcal{X}$, $\exists U \in \mathcal{O}$ such that $x \in U \subseteq \overline{U}$ and $\overline{U}$ is compact. Then, $\bar{U} := U \cap Y \in \mathcal{O}_Y$. Let $\bar{U}_c$ be the closure of $\bar{U}$ in $\mathcal{Y}$. Then, $\bar{U}_c \subseteq \overline{U} \cap Y \subseteq \overline{U}$. Note that $\bar{U}_c$ is a closed set in $\mathcal{Y}$ and therefore a closed set in $\mathcal{X}$ since Y is closed in $\mathcal{X}$ and Proposition 3.5. Then, $\bar{U}_c$ is compact by Proposition 5.5. Thus, we have $x \in \bar{U} \subseteq \bar{U}_c \subseteq Y$ and $\bar{U}_c$ is compact. Hence, $\mathcal{Y}$ is locally compact. This completes the proof of the lemma. □

Proposition 5.60 *Let $\mathcal{X} := (X, \mathcal{O})$ be a locally compact Hausdorff topological space, $Y \subseteq X$, and $\mathcal{O}_Y$ be the subset topology on Y. $\mathcal{Y} := (Y, \mathcal{O}_Y)$ is locally compact if, and only if, Y is relatively open in $\overline{Y}$.*

Proof "Only if" Let $\mathcal{O}_{\overline{Y}}$ be the subset topology on $\overline{Y}$. Then, $(\overline{Y}, \mathcal{O}_{\overline{Y}})$ is Hausdorff. By Proposition 5.57, Y is relative open in $\mathcal{O}_{\overline{Y}}$.

"If" Since $\overline{Y}$ is closed in $\mathcal{X}$, then, by Lemma 5.59, $(\overline{Y}, \mathcal{O}_{\overline{Y}})$ is locally compact. Clearly, $(\overline{Y}, \mathcal{O}_{\overline{Y}})$ is Hausdorff. In $(\overline{Y}, \mathcal{O}_{\overline{Y}})$, by Lemma 5.58, $\mathcal{Y}$ is locally compact.

This completes the proof of the proposition. □

5.7.2 *Partition of Unity*

Definition 5.61 Let $\mathcal{X}$ be a topological space. For a function $f : \mathcal{X} \to \mathbb{R}$, the *support* of f is the set $\text{supp}(f) := \overline{\{x \in \mathcal{X} \mid f(x) \neq 0\}}$. A collection of real-valued functions $(\phi_\alpha)_{\alpha \in \Gamma}$ on $\mathcal{X}$ is said to be *subordinate* to a collection of open sets $(O_\lambda)_{\lambda \in \Lambda}$, if $\forall \alpha \in \Gamma$, $\exists \lambda \in \Lambda$ such that $\text{supp}(\phi_\alpha) \subseteq O_\lambda$. ◇

Proposition 5.62 *Let $\mathcal{X}$ be a locally compact Hausdorff topological space, $U \in \mathcal{O}$, and $K \subseteq U$ be compact. Then, there exists a continuous function $f : \mathcal{X} \to [0, 1] \subset \mathbb{R}$ such that $f|_K = 1$ and $f|_{\tilde{U}} = 0$. Furthermore,* $\text{supp}(f) \subseteq U$ *is compact.*

Proof Since the set $Q := \mathbb{Q} \cap [0, 1]$ is countable, then, by recursively applying Proposition 5.54, we may find $(O_r)_{r \in Q} \subseteq \mathcal{O}$ such that the following two properties are satisfied:

1. $\forall r \in Q$, $K \subseteq O_r \subseteq \overline{O_r} \subseteq U$ and $\overline{O_r}$ is compact.
2. $\forall r, s \in Q$ with $r < s$, $\overline{O_r} \subseteq O_s$.

Define the real-valued function $\bar{f} : X \to \mathbb{R}$ by

$$\bar{f}(x) = \inf(\{r \in Q \mid x \in O_r\} \cup \{1\})$$

Clearly, $\bar{f} : X \to [0, 1]$, $\bar{f}(x) = 0$, $\forall x \in O_0$, and $\bar{f}(x) = 1$, $\forall x \in \widetilde{O_1}$. By 1, we have $K \subseteq O_0$ and $O_1 \subseteq U$. Next, we will show that $\bar{f}$ is continuous. $\forall x_0 \in X$, we will show that $\bar{f}$ is continuous at x_0. Let $a_0 = \bar{f}(x_0) \in [0, 1]$. $\forall B \subseteq \mathbb{R}$ with B being open and $a_0 \in B$, $\exists a_1, a_2, a_3, a_4 \in \mathbb{Q}$ such that $a_1 < a_2 < a_0 < a_3 < a_4$ and $(a_1, a_4) \subseteq B$. Let $\bar{a}_2 = \max\{a_2, 0\}$ and $\bar{a}_3 = \min\{a_3, 1\}$. Then, we must have $a_1 < \bar{a}_2 \le a_0 \le \bar{a}_3 < a_4$ and $\bar{a}_2, \bar{a}_3 \in Q$. We will distinguish three exhaustive and mutually exclusive cases: Case 1: $a_0 \in (0, 1)$; Case 2: $a_0 = 0$; Case 3: $a_0 = 1$.

Case 1: $a_0 \in (0, 1)$. Then, we must have $a_1 < \bar{a}_2 < a_0 < \bar{a}_3 < a_4$. Let $V = \widetilde{\overline{O_{\bar{a}_2}}} \cap O_{\bar{a}_3} \in \mathcal{O}$. $\forall x \in V$, we have $x \in O_{\bar{a}_3}$ and $\bar{f}(x) \le \bar{a}_3$. Also, $x \in \widetilde{\overline{O_{\bar{a}_2}}}$ implies that $\bar{f}(x) \ge \bar{a}_2$. Hence, $\bar{f}(V) \subseteq [\bar{a}_2, \bar{a}_3] \subset (a_1, a_4) \subseteq B$. $\bar{f}(x_0) = a_0 < \bar{a}_3$ implies that $x_0 \in O_{\bar{a}_3}$. $\bar{f}(x_0) = a_0 > \bar{a}_2$ implies that $\exists \widetilde{a}_2 \in (\bar{a}_2, a_0) \cap Q$ such that $x_0 \in \widetilde{O_{\widetilde{a}_2}} \subseteq \widetilde{\overline{O_{\bar{a}_2}}}$. Therefore, $x_0 \in V$. This shows that $\exists V \in \mathcal{O}$ with $x_0 \in V$ such that $\bar{f}(V) \subseteq B$.

Case 2: $a_0 = 0$. Then, we must have $a_1 < 0 = a_0 < \bar{a}_3 < a_4$. Take $V = O_{\bar{a}_3} \in \mathcal{O}$. We must have $x_0 \in V$. $\forall x \in V$, $0 \le \bar{f}(x) \le \bar{a}_3$. Hence, $\bar{f}(V) \subseteq [0, \bar{a}_3] \subset (a_1, a_4) \subseteq B$. Hence, $\exists V \in \mathcal{O}$ with $x_0 \in V$ such that $\bar{f}(V) \subseteq B$.

Case 3: $a_0 = 1$. Then, we must have $a_1 < \bar{a}_2 < a_0 = 1 < a_4$. Take $V = \widetilde{\overline{O_{\bar{a}_2}}} \in \mathcal{O}$. Since $\bar{f}(x_0) = a_0 = 1$, then $x_0 \in \widetilde{O_{\frac{1+\bar{a}_2}{2}}} \subseteq \widetilde{\overline{O_{\bar{a}_2}}} = V$. $\forall x \in V$, $\bar{f}(x) \ge \bar{a}_2$. Hence, $\bar{f}(V) \subseteq [\bar{a}_2, 1] \subset (a_1, a_4) \subseteq B$. Hence, $\exists V \in \mathcal{O}$ with $x_0 \in V$ such that $\bar{f}(V) \subseteq B$.

Therefore, in all cases, $\exists V \in \mathcal{O}$ with $x_0 \in V$ such that $\bar{f}(V) \subseteq B$. Then, $\bar{f}$ is continuous at x_0. By the arbitrariness of x_0 and Proposition 3.9, $\bar{f}$ is continuous. Define $f : \mathcal{X} \to [0, 1] \subset \mathbb{R}$ by $f(x) = 1 - \bar{f}(x)$, $\forall x \in \mathcal{X}$. Clearly, f is continuous by Proposition 3.12, $f|_K = 1$, and $f|_{\widetilde{U}} = 0$. Note that $\operatorname{supp}(f) = \overline{\{x \in \mathcal{X} \mid \bar{f}(x) < 1\}} \subseteq \overline{O_1} \subseteq U$ and is compact by Proposition 5.5. This completes the proof of the proposition. □

Theorem 5.63 (Partition of Unity) *Let $\mathcal{X}$ be a locally compact Hausdorff topological space, $K \subseteq \mathcal{X}$ be compact, and $(O_\lambda)_{\lambda \in \Lambda} \subseteq \mathcal{O}$ be an open covering of K, where Λ is an index set. Let $\mathcal{F}$ be a collection of continuous real-valued functions on $\mathcal{X}$ such that:*

(i) *The constantly zero function is in $\mathcal{F}$.*
(ii) *$\forall f, g \in \mathcal{F}$, $f + g \in \mathcal{F}$.*

(iii) $\forall f, g \in \mathcal{F}$ *with* $\text{supp}(f) \subseteq M := \{x \in \mathcal{X} \mid g(x) \neq 0\}$, *define* $f/g : \mathcal{X} \to \mathbb{R}$ *by*

$$(f/g)(x) = \begin{cases} f(x)/g(x) & \forall x \in \text{supp}(f) \\ 0 & \forall x \in \widetilde{\text{supp}(f)} \end{cases}$$

and then $f/g \in \mathcal{F}$.

(iv) $\forall O \in \mathcal{O}$, $\forall x_0 \in O$, $\exists f \in \mathcal{F}$ *such that* $f(x_0) = 1$, $f|_{\widetilde{O}} = 0$, *and* $f : \mathcal{X} \to [0, 1]$.

Then, there exists a finite collection $\Phi \subseteq \mathcal{F}$ *of continuous nonnegative real-valued functions on* $\mathcal{X}$, *which is subordinate to* $(O_\lambda)_{\lambda \in \Lambda}$, *such that* $\left(\sum_{\phi \in \Phi} \phi\right)\Big|_K = 1$, $\sum_{\phi \in \Phi} \phi(x) \in [0, 1] \subset \mathbb{R}$, $\forall x \in \mathcal{X}$, *and* $\text{supp}(\phi)$ *is compact*, $\forall \phi \in \Phi$.

Remark 5.64 Note that in (iii) above f/g is always continuous since $(f/g)|_{\widetilde{\text{supp}(f)}} = 0$ is continuous, $(f/g)|_M$ is continuous, $\widetilde{\text{supp}(f)}$ and M are open, $\widetilde{\text{supp}(f)} \cup M = \mathcal{X}$, and Theorem 3.11 implies the result. ◇

Proof Let $O := \bigcup_{\lambda \in \Lambda} O_\lambda \supseteq K$. By Proposition 5.54, there exists $U \in \mathcal{O}$ such that $K \subseteq U \subseteq \overline{U} \subseteq O$ and $\overline{U}$ is compact. $\forall \lambda \in \Lambda$, let $U_\lambda := U \cap O_\lambda \in \mathcal{O}$. Then, $K \subseteq \bigcup_{\lambda \in \Lambda} U_\lambda = U$.

$\forall x_0 \in K$, $\exists \lambda_0 \in \Lambda$ such that $x_0 \in U_{\lambda_0}$. By Proposition 5.53, $\exists \bar{U}_{\lambda_0} \in \mathcal{O}$ such that $x_0 \in \bar{U}_{\lambda_0} \subseteq \overline{\bar{U}_{\lambda_0}} \subseteq U_{\lambda_0}$ and $\overline{\bar{U}_{\lambda_0}}$ is compact. By (iv), there exists a continuous function $f_{x_0} : \mathcal{X} \to [0, 1]$ in $\mathcal{F}$ such that $f_{x_0}(x_0) = 1$ and $f_{x_0}|_{\widetilde{\bar{U}_{\lambda_0}}} = 0$. Let $V_{x_0} := \{x \in \mathcal{X} | f_{x_0}(x) > 0\}$. Then, $x_0 \in V_{x_0} \in \mathcal{O}$ since f_{x_0} is continuous. Note that $V_{x_0} \subseteq \bar{U}_{\lambda_0}$, which implies that $\text{supp}(f_{x_0}) = \overline{V_{x_0}} \subseteq \overline{\bar{U}_{\lambda_0}} \subseteq U_{\lambda_0} \subseteq O_{\lambda_0}$, which is compact by Proposition 5.5. Hence, f_{x_0} has compact support.

$\forall x_0 \in \overline{U} \setminus K$, $\widetilde{K} \in \mathcal{O}$, by Proposition 5.5. By (iv), there exists a continuous function $g_{x_0} : \mathcal{X} \to [0, 1]$ in $\mathcal{F}$ such that $g_{x_0}(x_0) = 1$ and $g_{x_0}|_K = 0$. Let $W_{x_0} := \{x \in \mathcal{X} | g_{x_0}(x) > 0\}$. Then, $x_0 \in W_{x_0} \in \mathcal{O}$ since g_{x_0} is continuous. Clearly, $W_{x_0} \subseteq \widetilde{K}$. Note that $\overline{U} \subseteq \left(\bigcup_{x \in K} V_x\right) \cup \left(\bigcup_{x \in \overline{U} \setminus K} W_x\right)$. By the compactness of $\overline{U}$, there exist finite sets $K_N \subseteq K$ and $U_N \subseteq \overline{U} \setminus K$ such that $\overline{U} \subseteq \left(\bigcup_{x \in K_N} V_x\right) \cup \left(\bigcup_{x \in U_N} W_x\right)$. By the construction of W_x's, $K \subseteq \bigcup_{x \in K_N} V_x$. Let $f := \sum_{x \in K_N} f_x \in \mathcal{F}$ and $g := \sum_{x \in U_N} g_x \in \mathcal{F}$ (here $f, g \in \mathcal{F}$ by (i) and (ii)). Then, f is nonnegative, $f(x) > 0$, $\forall x \in K$, g is nonnegative, and $f(x) + g(x) > 0$, $\forall x \in \overline{U}$.

Define $\Phi := \{\phi : \mathcal{X} \to [0, 1] \mid \phi = f_x/(f + g), x \in K_N\}$, which is a finite set. Note that $\forall x \in K_N$, $\text{supp}(f_x) \subseteq U_{\lambda_0} \subseteq U \subseteq \overline{U} \subseteq \{x \in \mathcal{X} \mid (f + g)(x) \neq 0\}$, for some $\lambda_0 \in \Lambda$. Then, $f_x/(f + g) \in \mathcal{F}$. Hence, $\Phi \subseteq \mathcal{F}$. $\forall \phi \in \Phi$, $\exists x \in K_N$ such that $\phi = f_x/(f + g)$. Then, ϕ is continuous by the fact that $\phi \in \mathcal{F}$. $\text{supp}(\phi) = \text{supp}(f_x)$ is compact and $\text{supp}(\phi) \subseteq O_\lambda$, for some $\lambda \in \Lambda$. Hence, Φ is subordinate to $(O_\lambda)_{\lambda\in\Lambda}$. Clearly, ϕ is nonnegative, $\forall\phi \in \Phi$. $\forall x \in K$, we have

$$\sum_{\phi\in\Phi} \phi(x) = \frac{f(x)}{f(x)+g(x)} = \frac{f(x)}{f(x)} = 1$$

and $\forall x \in \mathcal{X}$, we have

$$0 \leq \sum_{\phi\in\Phi} \phi(x) = (f/(f+g))(x) \leq 1$$

This completes the proof of the theorem. □

Corollary 5.65 *Let $\mathcal{X}$ be a locally compact Hausdorff topological space, $K \subseteq \mathcal{X}$ be compact, and $(O_\lambda)_{\lambda\in\Lambda} \subseteq \mathcal{O}$ be an open covering of K, where Λ is an index set. Then, there exists a finite collection Φ of continuous nonnegative real-valued functions on $\mathcal{X}$, which is subordinate to $(O_\lambda)_{\lambda\in\Lambda}$, such that $\left(\sum_{\phi\in\Phi}\phi\right)\Big|_K = 1$, $\sum_{\phi\in\Phi}\phi(x) \in [0,1] \subset \mathbb{R}$, $\forall x \in \mathcal{X}$, and* supp$(\phi)$ *is compact, $\forall\phi \in \Phi$.*

Proof Let $\mathcal{F}$ be the collection of continuous real-valued functions on $\mathcal{X}$. Clearly, $\mathcal{F}$ satisfies (i)–(iii) in Theorem 5.63. Note that $\{x_0\}$ is compact, and by Proposition 5.62, (iv) of Theorem 5.63 is also satisfied by $\mathcal{F}$. Then, the result follows. □

5.7.3 *The Alexandroff One-point Compactification*

Theorem 5.66 (Alexandroff One-Point Compactification) *Let $\mathcal{X} := (X, \mathcal{O})$ be a locally compact Hausdorff topological space. The* Alexandroff one-point compactification *of $\mathcal{X}$ is the set $X_c := X \cup \{\omega\}$ with the topology $\mathcal{O}_c := \{O_c \subseteq X_c \mid O_c \in \mathcal{O}$ or $X_c \setminus O_c$ is compact in $\mathcal{X}\}$. Then, $\mathcal{X}_c := (X_c, \mathcal{O}_c)$ is a compact Hausdorff space, and the identity map* id $: X \to X_c \setminus \{\omega\}$ *is a homeomorphism. The element ω is called the* point at infinity *in $\mathcal{X}_c$.*

Proof We first show that $\mathcal{O}_c$ is a topology on X_c. (i) $\emptyset \in \mathcal{O}$, and then $\emptyset \in \mathcal{O}_c$; $X_c \setminus X_c = \emptyset$ is compact in $\mathcal{X}$, and then $X_c \in \mathcal{O}_c$. (ii) $\forall O_{c1}, O_{c2} \in \mathcal{O}_c$, we will distinguish four exhaustive and mutually exclusive cases: Case 1: $O_{c1}, O_{c2} \in \mathcal{O}$; Case 2: $X_c \setminus O_{c1}$ and $X_c \setminus O_{c2}$ are compact in $\mathcal{X}$; Case 3: $X_c \setminus O_{c1}$ is compact in $\mathcal{X}$ and $O_{c2} \in \mathcal{O}$; Case 4: $O_{c1} \in \mathcal{O}$ and $X_c \setminus O_{c2}$ is compact in $\mathcal{X}$. Case 1: $O_{c1}, O_{c2} \in \mathcal{O}$. Then, $O_{c1} \cap O_{c2} \in \mathcal{O}$, and hence, $O_{c1} \cap O_{c2} \in \mathcal{O}_c$. Case 2: $X_c \setminus O_{c1}$ and $X_c \setminus O_{c2}$ are compact in $\mathcal{X}$. Then, $X_c \setminus (O_{c1} \cap O_{c2}) = (X_c \setminus O_{c1}) \cup (X_c \setminus O_{c2})$, which is compact in $\mathcal{X}$. This implies that $O_{c1} \cap O_{c2} \in \mathcal{O}_c$. Case 3: $X_c \setminus O_{c1}$ is compact in $\mathcal{X}$ and $O_{c2} \in \mathcal{O}$. Let $\bar{O}_{c1} := O_{c1} \setminus \{\omega\}$; then $\bar{O}_{c1} = X \setminus (X_c \setminus O_{c1})$. Since $X_c \setminus O_{c1}$ is compact in $\mathcal{X}$ and $\mathcal{X}$ is Hausdorff, then, by Proposition 5.5, $X_c \setminus O_{c1}$ is closed in $\mathcal{X}$. Then, $\bar{O}_{c1} \in \mathcal{O}$. Note that $O_{c1} \cap O_{c2} = \bar{O}_{c1} \cap O_{c2} \in \mathcal{O}$. Then, $O_{c1} \cap O_{c2} \in \mathcal{O}_c$. Case 4: $O_{c1} \in \mathcal{O}$ and $X_c \setminus O_{c2}$ is compact in $\mathcal{X}$. By an argument that is similar to Case 3, we have $O_{c1} \cap O_{c2} \in \mathcal{O}_c$. Hence, in all four cases, we have $O_{c1} \cap O_{c2} \in \mathcal{O}_c$.

(iii) $\forall (O_{c\lambda})_{\lambda \in \Lambda} \subseteq \mathcal{O}_c$, where Λ is an index set, we will distinguish two exhaustive and mutually exclusive cases: Case A: $(O_{c\lambda})_{\lambda \in \Lambda} \subseteq \mathcal{O}$; Case B: $\exists \lambda_0 \in \Lambda$ such that $\omega \in O_{c\lambda_0}$. Case A: $(O_{c\lambda})_{\lambda \in \Lambda} \subseteq \mathcal{O}$. Then, $\bigcup_{\lambda \in \Lambda} O_{c\lambda} \in \mathcal{O}$, and hence, $\bigcup_{\lambda \in \Lambda} O_{c\lambda} \in \mathcal{O}_c$. Case B: $\exists \lambda_0 \in \Lambda$ such that $\omega \in O_{c\lambda_0}$. Then, $X_c \setminus O_{c\lambda_0}$ is compact in $\mathcal{X}$. We may partition Λ into two disjoint sets Λ_1 and Λ_2 such that $\Lambda = \Lambda_1 \cup \Lambda_2$, $\Lambda_1 \cap \Lambda_2 = \emptyset$, $\forall \lambda \in \Lambda_1$, $O_{c\lambda} \in \mathcal{O}$, $\forall \lambda \in \Lambda_2$, $X_c \setminus O_{c\lambda}$ is compact in $\mathcal{X}$. Note that

$$X_c \setminus \left(\bigcup_{\lambda \in \Lambda} O_{c\lambda} \right) = \left(\bigcap_{\lambda \in \Lambda_1} (X_c \setminus O_{c\lambda}) \right) \cap \left(\bigcap_{\lambda \in \Lambda_2} (X_c \setminus O_{c\lambda}) \right)$$

$$= \left(\bigcap_{\lambda \in \Lambda_1} (X \setminus O_{c\lambda}) \right) \cap \left(\bigcap_{\lambda \in \Lambda_2} (X_c \setminus O_{c\lambda}) \right) \cap (X_c \setminus O_{c\lambda_0})$$

$\forall \lambda \in \Lambda_1$, $X \setminus O_{c\lambda}$ is a closed set in $\mathcal{X}$. $\forall \lambda \in \Lambda_2$, $X_c \setminus O_{c\lambda}$ is compact in $\mathcal{X}$ and therefore closed in $\mathcal{X}$ by Proposition 5.5. Hence, $X_c \setminus \left(\bigcup_{\lambda \in \Lambda} O_{c\lambda} \right)$ is a closed subset of $X_c \setminus O_{c\lambda_0}$ and hence compact in $\mathcal{X}$ by Proposition 5.5. Then, $\bigcup_{\lambda \in \Lambda} O_{c\lambda} \in \mathcal{O}_c$. Hence, in both cases, we have $\bigcup_{\lambda \in \Lambda} O_{c\lambda} \in \mathcal{O}_c$. Summarizing the above, $\mathcal{O}_c$ is a topology on X_c.

Next, we show that $\mathcal{X}_c$ is compact. Fix an open covering $(O_{c\lambda})_{\lambda \in \Lambda} \subseteq \mathcal{O}_c$ of $\mathcal{X}_c$. We may partition Λ into two disjoint sets Λ_1 and Λ_2 such that $\Lambda = \Lambda_1 \cup \Lambda_2$, $\Lambda_1 \cap \Lambda_2 = \emptyset$, $\forall \lambda \in \Lambda_1$, $O_{c\lambda} \in \mathcal{O}$, $\forall \lambda \in \Lambda_2$, $X_c \setminus O_{c\lambda}$ is compact in $\mathcal{X}$. Since $\omega \in X_c$, then $\exists \lambda_0 \in \Lambda_2$ such that $\omega \in O_{c\lambda_0}$. Then, $X_c \setminus O_{c\lambda_0}$ is compact in $\mathcal{X}$. $\forall \lambda \in \Lambda_2$, let $\bar{O}_{c\lambda} := O_{c\lambda} \setminus \{\omega\}$, and then $\bar{O}_{c\lambda} = X \setminus (X_c \setminus O_{c\lambda})$. By Proposition 5.5 and the compactness of $X_c \setminus O_{c\lambda}$, $\bar{O}_{c\lambda} \in \mathcal{O}$. Note that

$$X_c = \left(\bigcup_{\lambda \in \Lambda_1} O_{c\lambda} \right) \cup \left(\bigcup_{\lambda \in \Lambda_2} O_{c\lambda} \right) \cup O_{c\lambda_0} = O_{c\lambda_0} \cup (X_c \setminus O_{c\lambda_0})$$

Then, $X_c \setminus O_{c\lambda_0} \subseteq \left(\bigcup_{\lambda \in \Lambda_1} O_{c\lambda} \right) \cup \left(\bigcup_{\lambda \in \Lambda_2} \bar{O}_{c\lambda} \right)$. By the compactness of $X_c \setminus O_{c\lambda_0}$, there exist finite sets $\Lambda_{N1} \subseteq \Lambda_1$ and $\Lambda_{N2} \subseteq \Lambda_2$ such that $X_c \setminus O_{c\lambda_0} \subseteq \left(\bigcup_{\lambda \in \Lambda_{N1}} O_{c\lambda} \right) \cup \left(\bigcup_{\lambda \in \Lambda_{N2}} \bar{O}_{c\lambda} \right)$. Then, $X_c = \left(\bigcup_{\lambda \in \Lambda_{N1}} O_{c\lambda} \right) \cup \left(\bigcup_{\lambda \in \Lambda_{N2}} O_{c\lambda} \right) \cup O_{c\lambda_0}$. Hence, $\mathcal{X}_c$ is compact.

Next, we show that $\mathcal{X}_c$ is Hausdorff. $\forall x_1, x_2 \in \mathcal{X}_c$ with $x_1 \neq x_2$. We will distinguish two exhaustive and mutually exclusive cases: Case 1: $x_1, x_2 \in \mathcal{X}$; Case 2: $x_1 = \omega$ or $x_2 = \omega$. Case 1: $x_1, x_2 \in \mathcal{X}$. Since $\mathcal{X}$ is Hausdorff, then $\exists O_1, O_2 \in \mathcal{O}$ such that $x_1 \in O_1$, $x_2 \in O_2$, and $O_1 \cap O_2 = \emptyset$. Clearly, $O_1, O_2 \in \mathcal{O}_c$. Case 2: $x_1 = \omega$ or $x_2 = \omega$. Without loss of generality, assume $x_2 = \omega$. Then, $x_1 \in \mathcal{X}$. By local compactness of $\mathcal{X}$, $\exists O_1 \in \mathcal{O}$ such that $x_1 \in O_1 \subseteq \overline{O_1} \subseteq \mathcal{X}$ and $\overline{O_1}$ is compact. Then, $O_1 \in \mathcal{O}_c$ and $O_2 := X_c \setminus \overline{O_1} \in \mathcal{O}_c$. Clearly, $x_2 \in O_2$ and $O_1 \cap O_2 = \emptyset$.

Hence, in both cases, we have obtained $O_1, O_2 \in \mathcal{O}_c$ such that $x_1 \in O_1$, $x_2 \in O_2$, and $O_1 \cap O_2 = \emptyset$. Hence, $\mathcal{X}_c$ is Hausdorff.

Finally, we show that $\mathrm{id} : \mathcal{X} \to \mathcal{X}_c \setminus \{\omega\}$ is a homeomorphism. Clearly, id is bijective. $\forall O_c \in \mathcal{O}_c$, we either have $O_c \in \mathcal{O}$, which implies that $O_c \cap (X_c \setminus \{\omega\}) = O_c \in \mathcal{O}$; or we have $X_c \setminus O_c$ is compact, which implies that $\bar{O}_c := O_c \setminus \{\omega\} = X \setminus (X_c \setminus O_c) \in \mathcal{O}$, by Proposition 5.5, and hence, $O_c \cap (X_c \setminus \{\omega\}) = \bar{O}_c \in \mathcal{O}$. Hence, the subset topology $\mathcal{O}_{\bar{c}}$ on $\mathcal{X}_c \setminus \{\omega\}$ with respect to $\mathcal{X}_c$ is contained in $\mathcal{O}$. It is easy to see that $\mathcal{O} \subseteq \mathcal{O}_{\bar{c}}$. Then, $\mathcal{O} = \mathcal{O}_{\bar{c}}$. Hence, $\mathrm{id} : \mathcal{X} \to \mathcal{X}_c \setminus \{\omega\}$ is a homeomorphism.

This completes the proof of the theorem. □

5.7.4 Proper Functions

Definition 5.67 Let $\mathcal{X}$ and $\mathcal{Y}$ be topological spaces and $f : \mathcal{X} \to \mathcal{Y}$ be continuous. f is said to be *proper* if $\forall$ compact set $K \subseteq \mathcal{Y}$, we have $f_{\mathrm{inv}}(K) \subseteq \mathcal{X}$ is compact. f is said to be *countably proper* if $\forall$ compact set $K \subseteq \mathcal{Y}$, we have $f_{\mathrm{inv}}(K) \subseteq \mathcal{X}$ is countably compact. ◇

Proposition 5.68 *Let $\mathcal{X} := (X, \mathcal{O}_X)$ and $\mathcal{Y} := (Y, \mathcal{O}_Y)$ be locally compact Hausdorff topological spaces and $f : \mathcal{X} \to \mathcal{Y}$ be continuous. Let $\mathcal{X}_c := (X_c, \mathcal{O}_{Xc})$ and $\mathcal{Y}_c := (Y_c, \mathcal{O}_{Yc})$ be the Alexandroff one-point compactifications of $\mathcal{X}$ and $\mathcal{Y}$, respectively, where $X_c = X \cup \{\omega_x\}$ and $Y_c = Y \cup \{\omega_y\}$. Define a function $f_c : \mathcal{X}_c \to \mathcal{Y}_c$ by $f_c(x) = f(x)$, $\forall x \in X$, and $f_c(\omega_x) = \omega_y$. Then, f is proper if, and only if, f_c is continuous.*

Proof "Only if" Let f be proper. $\forall O_{Yc} \in \mathcal{O}_{Yc}$, we will distinguish two exhaustive and mutually exclusive cases: Case 1: $O_{Yc} \in \mathcal{O}_Y$; Case 2: $Y_c \setminus O_{Yc}$ is compact in $\mathcal{Y}$. Case 1: $O_{Yc} \in \mathcal{O}_Y$. Then, $f_{c\,\mathrm{inv}}(O_{Yc}) = f_{\mathrm{inv}}(O_{Yc}) \in \mathcal{O}_X$. This implies that $f_{c\,\mathrm{inv}}(O_{Yc}) \in \mathcal{O}_{Xc}$. Case 2: $Y_c \setminus O_{Yc}$ is compact in $\mathcal{Y}$. Then, $f_{c\,\mathrm{inv}}(Y_c \setminus O_{Yc}) = f_{\mathrm{inv}}(Y_c \setminus O_{Yc})$ is compact in $\mathcal{X}$ by the properness of f. Then, by Proposition 2.5, $f_{c\,\mathrm{inv}}(O_{Yc}) = X_c \setminus f_{c\,\mathrm{inv}}(Y_c \setminus O_{Yc}) \in \mathcal{O}_{Xc}$. Hence, in both cases, we have $f_{c\,\mathrm{inv}}(O_{Yc}) \in \mathcal{O}_{Xc}$. Hence, f_c is continuous.

"If" Let f_c be continuous. Fix a compact set $K \subseteq \mathcal{Y}$. Then, $\omega_y \in Y_c \setminus K \in \mathcal{O}_{Yc}$. By the continuity of f_c, we have $\omega_x \in f_{c\,\mathrm{inv}}(Y_c \setminus K) \in \mathcal{O}_{Xc}$. This implies that $X_c \setminus f_{c\,\mathrm{inv}}(Y_c \setminus K)$ is compact in $\mathcal{X}$. By Proposition 2.5, $f_{\mathrm{inv}}(K) = f_{c\,\mathrm{inv}}(K) = X_c \setminus f_{c\,\mathrm{inv}}(Y_c \setminus K)$. Hence, f is proper.

This completes the proof of the proposition. □

Proposition 5.69 *Let $\mathcal{X}$ be a topological space, $\mathcal{Y}$ be a metric space, $F \subseteq \mathcal{X}$ be closed, and $f : \mathcal{X} \to \mathcal{Y}$ be continuous and proper. Then, $f(F) \subseteq \mathcal{Y}$ is closed.*

Proof $\forall y \in \overline{f(F)}$, by Proposition 4.13, $\exists (y_n)_{n=1}^{\infty} \subseteq f(F)$ such that $\lim_{n\in\mathbb{N}} y_n = y$. Then, $\forall n \in \mathbb{N}$, $\exists x_n \in F$ such that $f(x_n) = y_n$. Let $K = \{y\} \cup (y_n)_{n=1}^{\infty}$. Then, K is compact. By the properness of f, $f_{\mathrm{inv}}(K)$ is compact. By Propositions 3.5 and 5.5, $F \cap f_{\mathrm{inv}}(K)$ is compact. Note that $(x_n)_{n=1}^{\infty} \subseteq F \cap f_{\mathrm{inv}}(K)$. Then, by

Proposition 5.4, $(x_n)_{n=1}^{\infty}$ admits a cluster point $x \in F \cap f_{\text{inv}}(K)$. By the continuity of f and Proposition 3.66, we have the sequence $(f(x_n))_{n=1}^{\infty} = (y_n)_{n=1}^{\infty}$ admits a cluster point $f(x)$. Since $\mathcal{Y}$ is Hausdorff and $\lim_{n\in\mathbb{N}} y_n = y$, then $y = f(x) \in f(F)$. Therefore, $\overline{f(F)} \subseteq f(F)$ and $f(F)$ is closed. This completes the proof of the proposition. □

In the above proposition, the assumption on $\mathcal{Y}$ may be relaxed to first countable Hausdorff space.

Proposition 5.70 *Let $\mathcal{X}$ be a topological space, $\mathcal{Y}$ be a locally compact Hausdorff topological space, $F \subseteq \mathcal{X}$ be closed, and $f : \mathcal{X} \to \mathcal{Y}$ be continuous and proper. Then, $f(F) \subseteq \mathcal{Y}$ is closed.*

Proof $\forall y \in \overline{f(F)}$, by the local compactness of $\mathcal{Y}$, $\exists O \in \mathcal{O}_{\mathcal{Y}}$ such that $y \in O$ and $\overline{O}$ is compact in $\mathcal{Y}$. Then, $y \in \overline{f(F)} \cap O$. $\forall U \in \mathcal{O}_{\mathcal{Y}}$ with $y \in U$, by Proposition 3.3, we have $\emptyset \neq (U \cap O) \cap f(F) = U \cap (O \cap f(F))$. Hence, $y \in \overline{f(F) \cap O}$ by Proposition 3.3. By Proposition 3.68, there exists a net $(y_\alpha)_{\alpha\in\mathcal{A}} \subseteq f(F) \cap O$ such that $\lim_{\alpha\in\mathcal{A}} y_\alpha = y$. $\forall \alpha \in \mathcal{A}$, $\exists x_\alpha \in F$ such that $y_\alpha = f(x_\alpha)$. Then, $x_\alpha \in F \cap f_{\text{inv}}(O)$. Then, the net $(x_\alpha)_{\alpha\in\mathcal{A}} \subseteq F \cap f_{\text{inv}}(\overline{O})$. By the properness of f, we have $f_{\text{inv}}(\overline{O}) \subseteq \mathcal{X}$ is compact. By Proposition 3.5, $F \cap f_{\text{inv}}(\overline{O})$ is closed relative to $f_{\text{inv}}(\overline{O})$, which further implies that $F \cap f_{\text{inv}}(\overline{O})$ is compact by Proposition 5.5. By Proposition 5.4, the net $(x_\alpha)_{\alpha\in\mathcal{A}}$ admits a cluster point $x \in F \cap f_{\text{inv}}(\overline{O})$. By the continuity of f and Proposition 3.66, $f(x)$ is a cluster point of the net $(f(x_\alpha))_{\alpha\in\mathcal{A}} = (y_\alpha)_{\alpha\in\mathcal{A}}$. Since $\mathcal{Y}$ is Hausdorff and $\lim_{\alpha\in\mathcal{A}} y_\alpha = y$, then $y = f(x) \in f(F)$. Hence, $\overline{f(F)} \subseteq f(F)$ and $f(F)$ is closed. This completes the proof of the proposition. □

5.8 σ-Compact Spaces

Definition 5.71 A topological space is said to be *σ-compact* if it is the union of countably infinitely many compact sets. ◇

Proposition 5.72 *Let $\mathcal{X}$ be a locally compact topological space. Then, the following statements are equivalent:*

(i) *$\mathcal{X}$ is Lindelöf.*
(ii) *$\mathcal{X}$ is σ-compact.*
(iii) *$\exists (O_n)_{n=1}^{\infty} \subseteq \mathcal{O}$ such that $\forall n \in \mathbb{N}$, $\overline{O_n} \subseteq O_{n+1}$ is compact, and $\mathcal{X} = \bigcup_{n=1}^{\infty} O_n$. This sequence is called an* exhaustion *of $\mathcal{X}$.*

Furthermore, if $\mathcal{X}$ is Hausdorff, then the above is equivalent to:

(iv) *$\exists \phi : \mathcal{X} \to [0, \infty) \subset \mathbb{R}$ that is proper and continuous.*

Proof (i) ⇒ (ii). $\forall x \in \mathcal{X}$, by local compactness of $\mathcal{X}$, $\exists O_x \in \mathcal{O}$ such that $x \in O_x \subseteq \overline{O_x}$ and $\overline{O_x}$ is compact. $\mathcal{X} = \bigcup_{x\in\mathcal{X}} O_x$. Since $\mathcal{X}$ is Lindelöf, then $\exists$ a countable set $X_C \subseteq \mathcal{X}$ such that $\mathcal{X} = \bigcup_{x \in X_C} O_x$. We will distinguish three exhaustive and mutually exclusive cases: Case 1: $X_C = \emptyset$; Case 2: $X_C \neq \emptyset$ is finite; Case 3: X_C is countably infinite. Case 1: $X_C = \emptyset$. Let $K_n = \emptyset$, $\forall n \in \mathbb{N}$, which are clearly compact. Then, $\mathcal{X} = \emptyset = \bigcup_{n=1}^{\infty} K_n$. Hence, $\mathcal{X}$ is σ-compact. Case 2: $X_C \neq \emptyset$ is finite. Without loss of generality, assume that $X_C = \{x_1, \ldots, x_n\}$ for some $n \in \mathbb{N}$. Let $K_i = \overline{O_{x_i}}$, $i = 1, \ldots, n$, and $K_i = \emptyset$, $i = n+1, n+2, \ldots$. Clearly, K_i's are compact and $\mathcal{X} = \bigcup_{i=1}^{\infty} K_i$. Hence, $\mathcal{X}$ is σ-compact. Case 3: X_C is countably infinite. Without loss of generality, assume that $X_C = \{x_1, x_2, \ldots\}$. Let $K_i = \overline{O_{x_i}}$, $\forall i \in \mathbb{N}$. Clearly, K_i's are compact and $\mathcal{X} = \bigcup_{i=1}^{\infty} K_i$. Hence, $\mathcal{X}$ is σ-compact. In all cases, $\mathcal{X}$ is σ-compact.

(ii) $\Rightarrow$ (iii). Let $\mathcal{X} = \bigcup_{n=1}^{\infty} K_n$, where K_n is compact, $\forall n \in \mathbb{N}$. Without loss of generality, we assume that $K_n \subseteq K_{n+1}$, $\forall n \in \mathbb{N}$, since, otherwise, we may let $\bar{K}_n = \bigcup_{i=1}^{n} K_n$ and consider $\mathcal{X} = \bigcup_{n=1}^{\infty} \bar{K}_n$ instead. Let $O_0 := \emptyset \in \mathcal{O}$. Then, $\overline{O_0} = \emptyset$ is compact. $\forall n \in \mathbb{N}$, $K_n \cup \overline{O_{n-1}}$ is compact, and by Proposition 5.52, $\exists O_n \in \mathcal{O}$ such that $K_n \cup \overline{O_{n-1}} \subseteq O_n \subseteq \overline{O_n}$ and $\overline{O_n}$ is compact. Then, $(O_n)_{n=1}^{\infty}$ is an exhaustion of $\mathcal{X}$ that we seek.

(iii) $\Rightarrow$ (i). Let $(U_n)_{n=1}^{\infty}$ be an exhaustion of $\mathcal{X}$. Fix any open covering $(O_\alpha)_{\alpha \in \Lambda} \subseteq \mathcal{O}$ of $\mathcal{X}$, where Λ is an index set. $\forall n \in \mathbb{N}$, $\overline{U_n} \subseteq \bigcup_{\alpha \in \Lambda} O_\alpha$. By the compactness of $\overline{U_n}$, there exists a finite set $\Lambda_{Nn} \subseteq \Lambda$ such that $\overline{U_n} \subseteq \bigcup_{\alpha \in \Lambda_{Nn}} O_\alpha$. Then, $\mathcal{X} = \bigcup_{n=1}^{\infty} \overline{U_n} \subseteq \bigcup_{n=1}^{\infty} \bigcup_{\alpha \in \Lambda_{Nn}} O_\alpha$, which is a countable subcovering. Hence, $\mathcal{X}$ is Lindelöf.

Now, assume that $\mathcal{X}$ is Hausdorff.

(iii) $\Rightarrow$ (iv). Let $(O_n)_{n=1}^{\infty}$ be an exhaustion of $\mathcal{X}$. $\forall n \in \mathbb{N}$, $\overline{O_n} \subseteq O_{n+1}$ is compact. Then, by Proposition 5.62, there exists a continuous function $\phi_n : \mathcal{X} \to [0, 1]$ such that $\phi_n|_{\overline{O_n}} = 1$ and $\phi_n|_{\widetilde{O_{n+1}}} = 0$. Let $\phi := \sum_{n=1}^{\infty}(1 - \phi_n)$. Clearly, $\phi : \mathcal{X} \to [0, \infty) \subset \mathbb{R}$. $\forall x_0 \in \mathcal{X}$, $\exists n_0 \in \mathbb{N}$ such that $x_0 \in O_{n_0}$. Then, $\phi|_{O_{n_0}} = \sum_{i=1}^{n_0-1}(1 - \phi_i)\Big|_{O_{n_0}}$, which is clearly continuous. $\forall U \in \mathcal{O}_{\mathbb{R}}$ with $\phi(x_0) \in U$, $\exists V \in \mathcal{O}$ with $x_0 \in V \subseteq O_{n_0}$ such that $\phi|_{O_{n_0}}(V) \subseteq U$. Then, $\phi(V) \subseteq U$. Hence, ϕ is continuous at x_0. By the arbitrariness of x_0 and Proposition 3.9, ϕ is continuous. Fix any compact subset $K \subseteq \mathbb{R}$. Then, K is closed and bounded, by Heine–Borel theorem. $\exists N \in \mathbb{N}$ such that $K \subseteq [-N, N] \subset \mathbb{R}$. $\forall x \in \widetilde{O_{N+2}}$, we have $x \in \widetilde{O_n}$, $n = 1, \ldots, N+2$, and hence, $\phi_n(x) = 0$, $n = 1, \ldots, N+1$, and $\phi(x) \geq N + 1 > N$. Then, $\phi_{\text{inv}}(K) \subseteq O_{N+2} \subseteq \overline{O_{N+2}}$. By Proposition 3.10, $\phi_{\text{inv}}(K)$ is closed in $\mathcal{X}$. By the compactness of $\overline{O_{N+2}}$ and Proposition 5.5, $\phi_{\text{inv}}(K)$ is compact. Hence, ϕ is proper.

(iv) $\Rightarrow$ (ii). Let $K_n = [-n, n] \subset \mathbb{R}$, $\forall n \in \mathbb{N}$. Then, K_n's are compact in $\mathbb{R}$ and $\mathbb{R} \subseteq \bigcup_{n=1}^{\infty} K_n$. By the properness of ϕ, $\phi_{\text{inv}}(K_n)$ is compact in $\mathcal{X}$, $\forall n \in \mathbb{N}$. By Proposition 2.5, $\mathcal{X} = \bigcup_{n=1}^{\infty} \phi_{\text{inv}}(K_n)$. Hence, $\mathcal{X}$ is σ-compact.

This completes the proof of the proposition. □

5.9 Paracompact Spaces

Definition 5.73 Let $\mathcal{X}$ be a topological space and $\mathcal{A}$ be a collection of subsets in $\mathcal{X}$. $\mathcal{A}$ is said to be *locally finite* if $\forall x \in \mathcal{X}$, $\exists U \in \mathcal{O}$ with $x \in U$ such that U meets only finitely many members of $\mathcal{A}$, that is, $U \cap A \neq \emptyset$ for finitely many $A \in \mathcal{A}$. ◇

Proposition 5.74 *Let $\mathcal{X}$ be a topological space and $(E_\lambda)_{\lambda \in \Lambda}$ be a locally finite collection of subsets of $\mathcal{X}$, where Λ is an index set:*

(i) *Let $E = \bigcup_{\lambda \in \Lambda} E_\lambda$. Then, $\overline{E} = \bigcup_{\lambda \in \Lambda} \overline{E_\lambda}$.*
(ii) *Let $K \subseteq \mathcal{X}$ be compact. Then, K meets only finitely many members of $(E_\lambda)_{\lambda \in \Lambda}$.*

Proof (i) $\forall x \in \overline{E}$, by local finiteness of $(E_\lambda)_{\lambda \in \Lambda}$, $\exists U \in \mathcal{O}$ with $x \in U$, and then U meets only finitely many members of $(E_\lambda)_{\lambda \in \Lambda}$. Let $\Lambda_N \subseteq \Lambda$ be the finite set such that $U \cap E_\lambda \neq \emptyset$, $\forall \lambda \in \Lambda_N$. Suppose $x \notin \bigcup_{\lambda \in \Lambda} \overline{E_\lambda}$. Then, $\forall \lambda \in \Lambda_N$, $x \notin \overline{E_\lambda}$. $\exists U_\lambda \in \mathcal{O}$ with $x \in U_\lambda$ such that $U_\lambda \cap E_\lambda = \emptyset$, by Proposition 3.3. Let $O := \left(\bigcap_{\lambda \in \Lambda_N} U_\lambda\right) \cap U \in \mathcal{O}$. Then, $x \in O$ and $O \cap E_\lambda = \emptyset$, $\forall \lambda \in \Lambda$. Then, $O \cap E = \emptyset$. This contradicts with $x \in \overline{E}$, by Proposition 3.3. Hence, $x \in \bigcup_{\lambda \in \Lambda} \overline{E_\lambda}$. Then, $\overline{E} \subseteq \bigcup_{\lambda \in \Lambda} \overline{E_\lambda}$.

On the other hand, $\forall \lambda \in \Lambda$, $E_\lambda \subseteq E$. This implies that $\overline{E_\lambda} \subseteq \overline{E}$. Therefore, we have $\bigcup_{\lambda \in \Lambda} \overline{E_\lambda} \subseteq \overline{E}$.

Hence, $\overline{E} = \bigcup_{\lambda \in \Lambda} \overline{E_\lambda}$.

(ii) Let $K \subseteq \mathcal{X}$ be compact. $\forall x \in K$, by the local finiteness of $(E_\lambda)_{\lambda \in \Lambda}$, $\exists U_x \in \mathcal{O}$ with $x \in U_x$ such that U_x meets only finitely many members of $(E_\lambda)_{\lambda \in \Lambda}$. Then, $K \subseteq \bigcup_{x \in K} U_x$. By the compactness of K, there exists a finite set $K_N \subseteq K$ such that $K \subseteq \bigcup_{x \in K_N} U_x$. Clearly, $\bigcup_{x \in K_N} U_x$ meets only finitely many members in $(E_\lambda)_{\lambda \in \Lambda}$. Hence, K meets only finitely many members in $(E_\lambda)_{\lambda \in \Lambda}$.

This completes the proof of the proposition. □

Definition 5.75 A topological space $\mathcal{X}$ is said to be *paracompact* if every open covering of $\mathcal{X}$ has a locally finite open refinement. ◇

Proposition 5.76 *A closed subset of a paracompact space is paracompact in the subset topology.*

Proof Let $\mathcal{X}$ be a paracompact space, $F \subseteq \mathcal{X}$ be closed, and $\mathcal{O}_F$ be the subset topology on F. Let $(O_{F\alpha})_{\alpha \in \Lambda} \subseteq \mathcal{O}_F$ be any open covering of F, where Λ is an index set (open in the subset topology $\mathcal{O}_F$). By Proposition 3.4, $\forall \alpha \in \Lambda$, $\exists O_\alpha \in \mathcal{O}$ such that $O_{F\alpha} = O_\alpha \cap F$. Then, $F \subseteq \bigcup_{\alpha \in \Lambda} O_\alpha$ and $\mathcal{X} \subseteq \left(\bigcup_{\alpha \in \Lambda} O_\alpha\right) \cup \widetilde{F}$. By the paracompactness of $\mathcal{X}$, there exists a locally finite open refinement $\mathcal{V}$ of $(O_\alpha)_{\alpha \in \Lambda} \cup \{\widetilde{F}\}$. Let $\bar{\mathcal{V}} := \{V \in \mathcal{V} \,|\, V \subseteq O_\alpha, \text{ for some } \alpha \in \Lambda\}$. Then, $F \subseteq \bigcup_{V \in \bar{\mathcal{V}}} V$, since other V's in $\mathcal{V}$ are subset of $\widetilde{F}$. Clearly, $\bar{\mathcal{V}}$ is locally finite. Then,

$\{V \cap F \mid V \in \bar{\mathcal{V}}\} \subseteq \mathcal{O}_F$ is a locally finite open refinement of $(O_{F\alpha})_{\alpha\in\Lambda}$ (open in the subset topology $\mathcal{O}_F$). Hence, $(F, \mathcal{O}_F)$ is paracompact. This completes the proof of the proposition. □

Theorem 5.77 *Metric spaces are paracompact.*

Proof Let $\mathcal{X}$ be a metric space. Let $(O_\alpha)_{\alpha\in\Lambda}$ be any open covering of $\mathcal{X}$, where Λ is an index set. By well-ordering principle, Λ may be well-ordered by $\trianglelefteq$. We will construct an open refinement of $(O_\alpha)_{\alpha\in\Lambda}$ in the following steps. Let $X_0 = \mathcal{X}$.

Step n ($n \in \mathbb{N}$): $\forall\alpha \in \Lambda$, let $P_{\alpha n} := \{x \in O_\alpha \cap X_{n-1} \mid \alpha$ is the least element of $\{\beta \in \Lambda \mid x \in O_\beta\}\}$. Let $Q_{\alpha n} := \{x \in P_{\alpha n} \mid \mathcal{B}(x, 3/2^n) \subseteq O_\alpha\}$ and $D_{\alpha n} := \bigcup_{x\in Q_{\alpha n}} \mathcal{B}(x, 1/2^n) \in \mathcal{O}$. Let $X_n := X_{n-1} \setminus \big(\bigcup_{\alpha\in\Lambda} D_{\alpha n}\big)$.

We will now show that $\mathcal{V} := \{D_{\alpha n} \mid \alpha \in \Lambda, n \in \mathbb{N}\}$ is a locally finite open refinement of $(O_\alpha)_{\alpha\in\Lambda}$. $\forall x_0 \in \mathcal{X}$. Then, $\{\beta \in \Lambda \mid x_0 \in O_\beta\} \neq \emptyset$. Let α_0 be the least element of the set, which exists since Λ is well-ordered. Then, $x_0 \in O_{\alpha_0}$. Then, $\exists n_0 \in \mathbb{N}$ such that $\mathcal{B}(x_0, 3/2^{n_0}) \subseteq O_{\alpha_0}$, since O_{α_0} is open. Then, $x_0 \in D_{\alpha_0 n_0}$ or $x_0 \in D_{\beta n}$ for some $\beta \in \Lambda$ and for some $n \in \mathbb{N}$ with $n < n_0$. Hence, $\mathcal{V}$ covers $\mathcal{X}$. Clearly, $D_{\alpha n} \subseteq O_\alpha$ and is open by construction, $\forall\alpha \in \Lambda$ and $\forall n \in \mathbb{N}$. Then, $\mathcal{V}$ is an open refinement of $(O_\alpha)_{\alpha\in\Lambda}$.

Fix any $x_0 \in \mathcal{X}$. Consider the set $\{\beta \in \Lambda \mid \exists n \in \mathbb{N}, x_0 \in D_{\beta n}\} \neq \emptyset$. Let α_1 be the least element of this set. Then, $x_0 \in D_{\alpha_1 n_0}$ for some $n_0 \in \mathbb{N}$. Furthermore, $\exists j_0 \in \mathbb{N}$ such that $\mathcal{B}\big(x_0, 2^{-j_0}\big) \subseteq D_{\alpha_1 n_0}$. Consider the open set $\mathcal{B}\big(x_0, 2^{-j_0-n_0}\big) \ni x_0$.

$\forall n \geq n_0 + j_0$, $\forall\alpha \in \Lambda$, since $\mathcal{B}\big(x_0, 2^{-j_0}\big) \subseteq D_{\alpha_1 n_0} \subseteq O_{\alpha_1}$, then $\forall x \in \mathcal{B}\big(x_0, 2^{-j_0-n_0}\big)$, we have $\mathcal{B}\big(x, 2^{-j_0} - 2^{-j_0-n_0}\big) \subseteq \mathcal{B}\big(x_0, 2^{-j_0}\big) \subseteq D_{\alpha_1 n_0}$. $\forall y \in D_{\alpha n}$, since $n > n_0$, then $\exists\bar{x} \in Q_{\alpha n}$ with $\bar{x} \notin D_{\alpha_1 n_0}$ such that $y \in \mathcal{B}\big(\bar{x}, 2^{-n}\big)$. Then, $\exists\bar{x} \in \mathcal{B}\big(y, 2^{-n}\big)$ such that $\bar{x} \in D_{\alpha n} \setminus D_{\alpha_1 n_0}$. Note that $2^{-n} \leq 2^{-j_0-n_0} \leq 2^{-j_0} - 2^{-j_0-n_0}$. Therefore, $x \notin D_{\alpha n}$. Hence, $\mathcal{B}\big(x_0, 2^{-j_0-n_0}\big) \cap D_{\alpha n} = \emptyset$. Thus, $\mathcal{B}\big(x_0, 2^{-j_0-n_0}\big)$ does not intersect any $D_{\alpha n}$, $\forall\alpha \in \Lambda$, $\forall n \geq n_0 + j_0$.

Claim 5.77.1 $\forall n < n_0 + j_0$, $\mathcal{B}\big(x_0, 2^{-j_0-n_0}\big)$ intersects at most one of the set $D_{\alpha n}$'s, $\alpha \in \Lambda$.

Proof of Claim Suppose the result is not true. Then, $\exists\alpha, \beta \in \Lambda$ such that $\mathcal{B}\big(x_0, 2^{-j_0-n_0}\big) \cap D_{\alpha n} \neq \emptyset$, $\mathcal{B}\big(x_0, 2^{-j_0-n_0}\big) \cap D_{\beta n} \neq \emptyset$, and $\alpha \neq \beta$. Without loss of generality, assume that $\alpha \trianglelefteq \beta$. Let $p \in \mathcal{B}\big(x_0, 2^{-j_0-n_0}\big) \cap D_{\alpha n}$ and $q \in \mathcal{B}\big(x_0, 2^{-j_0-n_0}\big) \cap D_{\beta n}$. By the definition of $D_{\alpha n}$, $\exists\bar{p} \in Q_{\alpha n} \subseteq D_{\alpha n}$ such that $p \in \mathcal{B}\big(\bar{p}, 2^{-n}\big) \subseteq D_{\alpha n}$. Similarly, $\exists\bar{q} \in Q_{\beta n} \subseteq D_{\beta n}$ such that $q \in \mathcal{B}\big(\bar{q}, 2^{-n}\big) \subseteq D_{\beta n}$. Then, by the definition of $Q_{\alpha n}$, we have $\mathcal{B}(\bar{p}, 3/2^n) \subseteq O_\alpha$. Similarly, $\mathcal{B}(\bar{q}, 3/2^n) \subseteq O_\beta$. Note that

$$\begin{aligned}\rho(\bar{p}, \bar{q}) &\leq \rho(\bar{p}, p) + \rho(p, x_0) + \rho(x_0, q) + \rho(q, \bar{q}) \\ &< 2^{-n} + 2^{-j_0-n_0} + 2^{-j_0-n_0} + 2^{-n} \leq 3/2^n\end{aligned}$$

Then, $\bar{q} \in O_\alpha$. But, $\bar{q} \in Q_{\beta n} \subseteq P_{\beta n}$, $\alpha \trianglelefteq \beta$, and $\alpha \neq \beta$ implies that $\bar{q} \notin O_\alpha$. This is a contradiction. Therefore, the result must hold. This completes the proof of the claim. □

Hence, $\mathcal{B}(x_0, 2^{-j_0-n_0})$ can meet at most $n_0 + j_0 - 1$ sets in $\mathcal{V}$. Hence, $\mathcal{V}$ is locally finite.

Therefore, we have obtained a locally finite open refinement of $(O_\alpha)_{\alpha\in\Lambda}$. Hence, $\mathcal{X}$ is paracompact. This completes the proof of the theorem. □

Definition 5.78 Let $\mathcal{X}$ be a topological space and $(E_\alpha)_{\alpha\in\Lambda}$ be a collection of subsets of $\mathcal{X}$, where Λ is an index set. $(E_\alpha)_{\alpha\in\Lambda}$ is said to be *star-finite* if $\forall\alpha_0 \in \Lambda$, E_{α_0} meets only finitely many members in the collection. ◇

It is easy to see that a star-finite open covering of $\mathcal{X}$ is locally finite. But the converse need not hold.

Proposition 5.79 *A σ-compact locally compact space is paracompact.*

Proof Let $\mathcal{X}$ be a σ-compact locally compact space. Fix any open covering $(O_\alpha)_{\alpha\in\Lambda}$ of $\mathcal{X}$, where Λ is an index set. By Proposition 5.72, there exists an exhaustion $(U_n)_{n=1}^{\infty}$ of $\mathcal{X}$. Let $U_{-1} = \emptyset$ and $U_0 = \emptyset$. Let $V_{\alpha n} := O_\alpha \cap (U_{n+1} \setminus \overline{U_{n-2}}) \in \mathcal{O}$, $\forall\alpha \in \Lambda$, $\forall n \in \mathbb{N}$. It is easy to see that $\mathcal{V} := \{V_{\alpha n} \mid \alpha \in \Lambda, n \in \mathbb{N}\}$ is an open refinement of $(O_\alpha)_{\alpha\in\Lambda}$ that covers $\mathcal{X}$. $\forall n \in \mathbb{N}$, by the compactness of $\overline{U_n}$ and Proposition 5.5, $\overline{U_n} \setminus U_{n-1} \subseteq \bigcup_{\alpha\in\Lambda} V_{\alpha n}$ is compact. Then, there exists a finite set $\mathcal{V}_n \subseteq \{V_{\alpha n} \mid \alpha \in \Lambda\}$ such that $\overline{U_n} \setminus U_{n-1} \subseteq \bigcup_{V\in\mathcal{V}_n} V$. Then, $\mathcal{V}_L := \bigcup_{n=1}^{\infty} \mathcal{V}_n$ is an open refinement of $(O_\alpha)_{\alpha\in\Lambda}$ that covers $\mathcal{X}$ since $\bigcup_{n=1}^{\infty}(\overline{U_n} \setminus U_{n-1}) = \mathcal{X}$. $\forall V \in \mathcal{V}_L$, $\exists n_0 \in \mathbb{N}$ such that $V \in \mathcal{V}_{n_0}$. Then, V does not intersect any member of $\mathcal{V}_n$ with $n \in \mathbb{N}$ and $|n - n_0| > 2$. Hence, $\mathcal{V}_L$ is star-finite. Then, $\mathcal{V}_L$ is a locally finite open refinement of $(O_\alpha)_{\alpha\in\Lambda}$. Then, $\mathcal{X}$ is paracompact. This completes the poof of the proposition. □

Proposition 5.80 *Let $\mathcal{X}$ be a locally compact Hausdorff topological space. $\mathcal{X}$ is paracompact if, and only if, any open covering of $\mathcal{X}$ has a star-finite open refinement (that covers $\mathcal{X}$).*

Proof "Only if" Let $\mathcal{U}$ be any open covering of $\mathcal{X}$. $\forall x \in \mathcal{X}$, $\exists U_x \in \mathcal{U}$ such that $x \in U_x \in \mathcal{O}$. By Proposition 5.53, $\exists O_x \in \mathcal{O}$ such that $x \in O_x \subseteq \overline{O_x} \subset U_x$ and $\overline{O_x}$ is compact. Hence, $(O_x)_{x\in\mathcal{X}}$ is an open refinement of $\mathcal{U}$. By the paracompactness of $\mathcal{X}$, there exists a locally finite open refinement $\mathcal{V}$ of $(O_x)_{x\in\mathcal{X}}$ that covers $\mathcal{X}$. $\forall V \in \mathcal{V}$, $\exists x \in \mathcal{X}$ such that $V \subseteq O_x$. Then, $\overline{V} \subseteq \overline{O_x}$. By Propositions 5.5 and 3.5, we have $\overline{V}$ is compact. By Proposition 5.74, $\overline{V}$ meets only finitely many members of $\mathcal{V}$. Hence, $\mathcal{V}$ is star-finite. Therefore, $\mathcal{V}$ is a star-finite open refinement of $\mathcal{U}$.

"If" Let $\mathcal{U}$ be any open covering of $\mathcal{X}$. Then, there is a star-finite open refinement $\mathcal{V}$ of $\mathcal{U}$. Then, $\mathcal{V}$ is a locally finite open refinement of $\mathcal{U}$. Hence, $\mathcal{X}$ is paracompact.

This completes the proof of the proposition. □

Proposition 5.81 *A paracompact Hausdorff topological space is normal.*

Proof Let $\mathcal{X}$ be a paracompact Hausdorff topological space. Clearly, $\mathcal{X}$ is Tychonoff. $\forall x_0 \in \mathcal{X}$ and $\forall$ closed set $F \subseteq \mathcal{X}$ with $x_0 \notin F$, we will show that $\exists O_1, O_2 \in \mathcal{O}$ such that $x_0 \in O_1$, $F \subseteq O_2$, and $O_1 \cap O_2 = \emptyset$. $\forall x \in F$, then $x \neq x_0$. Since $\mathcal{X}$ is Hausdorff, $\exists O_x^{\langle 1\rangle}, O_x^{\langle 2\rangle} \in \mathcal{O}$ such that $x_0 \in O_x^{\langle 1\rangle}$, $x \in O_x^{\langle 2\rangle}$, and $O_x^{\langle 1\rangle} \cap O_x^{\langle 2\rangle} = \emptyset$. Then, $\mathcal{X} \subseteq \widetilde{F} \cup \left(\bigcup_{x\in F} O_x^{\langle 2\rangle}\right)$. By the paracompactness of $\mathcal{X}$, there exists a locally finite open refinement $\mathcal{V} \subseteq \mathcal{O}$ of $\{\widetilde{F}\} \cup \left\{O_x^{\langle 2\rangle} \,\middle|\, x \in F\right\}$. Let $\bar{\mathcal{V}} := \left\{V \in \mathcal{V} \,\middle|\, V \subseteq O_x^{\langle 2\rangle},\text{ for some } x \in F\right\}$. Then, $\bar{\mathcal{V}}$ is an open covering of F, since $\forall V \in \mathcal{V} \setminus \bar{\mathcal{V}}$ we have $V \subseteq \widetilde{F}$. Furthermore, $\bar{\mathcal{V}}$ is locally finite. $\forall V \in \bar{\mathcal{V}}$, $\exists x \in F$ such that $V \subseteq O_x^{\langle 2\rangle}$, $x_0 \in O_x^{\langle 1\rangle}$, and $O_x^{\langle 2\rangle} \cap O_x^{\langle 1\rangle} = \emptyset$. Then, $V \subseteq O_x^{\langle 2\rangle} \subseteq \widetilde{O_x^{\langle 1\rangle}}$. Hence, we have $\overline{V} \subseteq \widetilde{O_x^{\langle 1\rangle}}$ and $x_0 \notin \overline{V}$. By Proposition 5.74, $\overline{\bigcup_{V\in\bar{\mathcal{V}}} V} = \bigcup_{V\in\bar{\mathcal{V}}} \overline{V} =: \widetilde{O_1}$. Then, $O_1 \in \mathcal{O}$, $F \subseteq \bigcup_{V\in\bar{\mathcal{V}}} V =: O_2 \in \mathcal{O}$, $O_2 \subseteq \widetilde{O_1}$, and $x_0 \in O_1$. Hence, $\exists O_1, O_2 \in \mathcal{O}$ such that $x_0 \in O_1$, $F \subseteq O_2$, and $O_1 \cap O_2 = \emptyset$. Then, $\mathcal{X}$ is regular.

Next, we show that $\mathcal{X}$ is normal. Fix any closed sets $F_1, F_2 \subseteq \mathcal{X}$ with $F_1 \cap F_2 = \emptyset$. $\forall x \in F_2$, then $x \notin F_1$. Since $\mathcal{X}$ is regular, $\exists O_x^{\langle 1\rangle}, O_x^{\langle 2\rangle} \in \mathcal{O}$ such that $F_1 \subseteq O_x^{\langle 1\rangle}$, $x \in O_x^{\langle 2\rangle}$, and $O_x^{\langle 1\rangle} \cap O_x^{\langle 2\rangle} = \emptyset$. Then, $\mathcal{X} \subseteq \widetilde{F_2} \cup \left(\bigcup_{x\in F_2} O_x^{\langle 2\rangle}\right)$. By the paracompactness of $\mathcal{X}$, there exists a locally finite open refinement $\mathcal{V} \subseteq \mathcal{O}$ of $\{\widetilde{F_2}\} \cup \left\{O_x^{\langle 2\rangle} \,\middle|\, x \in F_2\right\}$. Let $\bar{\mathcal{V}} := \{V \in \mathcal{V} \mid V \subseteq O_x^{\langle 2\rangle},\text{ for some } x \in F_2\}$. Then, $\bar{\mathcal{V}}$ is an open covering of F_2, since $\forall V \in \mathcal{V}\setminus\bar{\mathcal{V}}$ we have $V \subseteq \widetilde{F_2}$. Furthermore, $\bar{\mathcal{V}}$ is locally finite. $\forall V \in \bar{\mathcal{V}}$, $\exists x \in F_2$ such that $V \subseteq O_x^{\langle 2\rangle}$, $F_1 \subseteq O_x^{\langle 1\rangle}$, and $O_x^{\langle 2\rangle} \cap O_x^{\langle 1\rangle} = \emptyset$. Then, $V \subseteq O_x^{\langle 2\rangle} \subseteq \widetilde{O_x^{\langle 1\rangle}}$. Hence, we have $\overline{V} \subseteq \widetilde{O_x^{\langle 1\rangle}}$ and $F_1 \cap \overline{V} = \emptyset$. By Proposition 5.74, $\overline{\bigcup_{V\in\bar{\mathcal{V}}} V} = \bigcup_{V\in\bar{\mathcal{V}}} \overline{V} =: \widetilde{O_1}$. Then, $O_1 \in \mathcal{O}$, $F_2 \subseteq \bigcup_{V\in\bar{\mathcal{V}}} V =: O_2 \in \mathcal{O}$, $O_2 \subseteq \widetilde{O_1}$, and $F_1 \cap O_1 = \emptyset$. Hence, $\exists O_1, O_2 \in \mathcal{O}$ such that $F_1 \subseteq O_1$, $F_2 \subseteq O_2$, and $O_1 \cap O_2 = \emptyset$. Then, $\mathcal{X}$ is normal.

This completes the proof of the proposition. □

5.10 The Stone–Čech Compactification

Definition 5.82 Let $\mathcal{X} := (X, \mathcal{O})$ be a completely regular topological space, $I = [0, 1] \subseteq \mathbb{R}$, and $\mathcal{F}$ be the family of continuous functions of $\mathcal{X}$ to I. By Corollary 3.62, $\mathcal{X}$ is homeomorphic to $E(\mathcal{X}) \subseteq I^{\mathcal{F}}$, where E is the equivalence map. Let $F = \overline{E(\mathcal{X})}$ and $\mathcal{O}_F$ be the subset topology of F. Then, $\beta(\mathcal{X}) := (F, \mathcal{O}_F)$ is a compact Hausdorff topological space, by Tychonoff Theorem and Proposition 5.5. $\beta(\mathcal{X})$ is said to be the *Stone–Čech compactification* of $\mathcal{X}$. ◇

Proposition 5.83 *Let $\mathcal{X} := (X, \mathcal{O})$ be a completely regular topological space, $I = [0, 1] \subseteq \mathbb{R}$, $\mathcal{F}$ be the family of continuous functions of $\mathcal{X}$ to I, and $E : \mathcal{X} \to$*

$E(\mathcal{X}) \subseteq I^{\mathcal{F}}$ be the equivalence map. Then, there exists a unique compact Hausdorff topological space $\beta(\mathcal{X})$ with the following properties:

(i) *The space $E(\mathcal{X})$ is dense in $\beta(\mathcal{X})$.*
(ii) *Each bounded continuous real-valued function on $E(\mathcal{X})$ extends to a bounded continuous real-valued function on $\beta(\mathcal{X})$.*
(iii) *If $\mathcal{X}$ is a dense subset of a compact Hausdorff topological space $\mathcal{Y}$, then there exists a unique continuous mapping $\phi : \beta(\mathcal{X}) \to \mathcal{Y}$ such that ϕ is surjective and $\phi|_{E(\mathcal{X})} = E_{\text{inv}}$.*

Furthermore, if $\mathcal{X}$ is locally compact, then $E(\mathcal{X})$ is an open subset of $\beta(\mathcal{X})$.

Proof We will first show that $\beta(\mathcal{X})$ defined in Definition 5.82 satisfies properties (i), (ii), and (iii). (i) By Proposition 3.5, $E(\mathcal{X})$ is dense in $F = \overline{E(\mathcal{X})} = \beta(\mathcal{X})$. (ii) Fix any bounded continuous real-valued function $g : E(\mathcal{X}) \to \mathbb{R}$. Then, $\exists N \in \mathbb{N}$ such that $g : E(\mathcal{X}) \to [-N, N] \subset \mathbb{R}$. Then, $\bar{g} := ((g + N)/(2N)) \circ E \in \mathcal{F}$ by Proposition 3.12. Now, consider the function $\bar{h} := \pi_{\bar{g}}|_{\beta(\mathcal{X})} : \beta(\mathcal{X}) \to [0, 1] \subset \mathbb{R}$, which is continuous by Proposition 3.27. Note that $\bar{h}|_{E(\mathcal{X})} = \bar{g} \circ E_{\text{inv}} = (g + N)/(2N)$. Then, by Proposition 3.12, $h := 2N\bar{h} - N$ is a continuous function of $\beta(\mathcal{X})$ to $[-N, N]$. Furthermore, $h|_{E(\mathcal{X})} = g$. Hence, h is the desired extension that we seek. Clearly, $h : \beta(\mathcal{X}) \to [-N, N] \subset \mathbb{R}$.

(iii) Let $\mathcal{Y} := (Y, \mathcal{O}_{\mathcal{Y}})$ be a compact Hausdorff space such that $\mathcal{X}$ is dense in $\mathcal{Y}$. Let $\mathcal{G}$ be the family of continuous real-valued functions of $\mathcal{Y}$ to I. Let $E^{\langle\mathcal{Y}\rangle} : \mathcal{Y} \to I^{\mathcal{G}}$ be the equivalence map and $\pi_g^{\langle\mathcal{G}\rangle} : I^{\mathcal{G}} \to I$ be the projection function, $\forall g \in \mathcal{G}$. By Propositions 5.14 and 3.61, $\mathcal{Y}$ is completely regular. By Corollary 3.62, $E^{\langle\mathcal{Y}\rangle} : \mathcal{Y} \to E^{\langle\mathcal{Y}\rangle}(\mathcal{Y}) \subseteq I^{\mathcal{G}}$ is a homeomorphism. Define a mapping $\psi : \mathcal{G} \to \mathcal{F}$ by $\psi(g) = g|_{\mathcal{X}}$, $\forall g \in \mathcal{G}$. By Proposition 3.56, ψ is injective. By Tychonoff theorem, $I^{\mathcal{G}}$ is a compact Hausdorff space. Define a mapping $\Psi : I^{\mathcal{F}} \to I^{\mathcal{G}}$ by $\pi_g^{\langle\mathcal{G}\rangle}(\Psi(i_f)) = \pi_{\psi(g)}(i_f)$, $\forall i_f \in I^{\mathcal{F}}$, $\forall g \in \mathcal{G}$. $\forall x \in \mathcal{X}$, $E^{\langle\mathcal{Y}\rangle}(x) \in I^{\mathcal{G}}$ satisfies $\pi_g^{\langle\mathcal{G}\rangle}(E^{\langle\mathcal{Y}\rangle}(x)) = g(x) = g|_{\mathcal{X}}(x) = \psi(g)(x) = \pi_{\psi(g)}(E(x))$, $\forall g \in \mathcal{G}$. Then, we have $E^{\langle\mathcal{Y}\rangle}(x) = \Psi(E(x))$, $\forall x \in \mathcal{X}$. Hence, $\Psi \circ E = E^{\langle\mathcal{Y}\rangle}|_{\mathcal{X}}$.

Claim 5.83.1 Ψ is continuous.

Proof of Claim Fix any basis open set $U \subseteq I^{\mathcal{G}}$. Then, $U = \prod_{g\in\mathcal{G}} U_g$, where $U_g \subseteq I$ is open, $\forall g \in \mathcal{G}$, and $U_g = I$ for all g's except finitely many g's, say $g \in \mathcal{G}_N \subseteq \mathcal{G}$. Let $V := \prod_{f\in\mathcal{F}} V_f \subseteq I^{\mathcal{F}}$ be given by $V_f = I$, $\forall f \notin \text{range}(\psi)$, and $V_f = U_g$, $\forall f \in \text{range}(\psi)$ and $f = \psi(g)$. The set V is well-defined since ψ is injective. $\forall i_f \in V$, $\forall g \in \mathcal{G}$, $\pi_g^{\langle\mathcal{G}\rangle}(\Psi(i_f)) = \pi_{\psi(g)}(i_f) \in V_{\psi(g)} = U_g$. Hence, we have $\Psi(i_f) \in U$. Then, $V \subseteq \Psi_{\text{inv}}(U)$. $\forall i_f \in I^{\mathcal{F}} \setminus V$, $\exists f_0 \in \text{range}(\psi)$ such that $\pi_{f_0}(i_f) \notin V_{f_0}$. Then, $f_0 = \psi(g_0)$, where $g_0 \in \mathcal{G}_N$. Note that $\pi_{g_0}^{\langle\mathcal{G}\rangle}(\Psi(i_f)) = \pi_{\psi(g_0)}(i_f) \notin V_{\psi(g_0)} = U_{g_0}$. Then, $\Psi(i_f) \notin U$. This shows that $I^{\mathcal{F}} \setminus V \subseteq \Psi_{\text{inv}}(I^{\mathcal{G}} \setminus U)$. Hence, we have $V = \Psi_{\text{inv}}(U)$, by Proposition 2.5. Clearly, V is a basis open set in $I^{\mathcal{F}}$. Hence, Ψ is continuous. This completes the proof of the claim. □

Since $\mathcal{X}$ is dense in $\mathcal{Y}$ and $E^{\langle\mathcal{Y}\rangle}$ is a homeomorphism between $\mathcal{Y}$ and $E^{\langle\mathcal{Y}\rangle}(\mathcal{Y})$, then $E^{\langle\mathcal{Y}\rangle}(\mathcal{X})$ is dense in $E^{\langle\mathcal{Y}\rangle}(\mathcal{Y})$. Since $\mathcal{Y}$ is compact and $E^{\langle\mathcal{Y}\rangle}$ is continuous, then $E^{\langle\mathcal{Y}\rangle}(\mathcal{Y})$ is compact in $I^{\mathcal{G}}$, by Proposition 5.7. Furthermore, by Proposition 5.5, $E^{\langle\mathcal{Y}\rangle}(\mathcal{Y})$ is closed in $I^{\mathcal{G}}$. Then, $\overline{E^{\langle\mathcal{Y}\rangle}(\mathcal{X})} = E^{\langle\mathcal{Y}\rangle}(\mathcal{Y})$, where $\overline{E^{\langle\mathcal{Y}\rangle}(\mathcal{X})}$ is the closure of $E^{\langle\mathcal{Y}\rangle}(\mathcal{X})$ in $I^{\mathcal{G}}$. By the compactness of $\beta(\mathcal{X})$, the continuity of Ψ, and Proposition 5.7, $\Psi(\beta(\mathcal{X})) \subseteq I^{\mathcal{G}}$ is compact. Furthermore, by Proposition 5.5, $\Psi(\beta(\mathcal{X}))$ is closed in $I^{\mathcal{G}}$. Note that $E^{\langle\mathcal{Y}\rangle}(\mathcal{X}) = \Psi(E(\mathcal{X})) \subseteq \Psi(\beta(\mathcal{X}))$. Then, $\Psi(\beta(\mathcal{X})) \supseteq E^{\langle\mathcal{Y}\rangle}(\mathcal{Y})$.

Claim 5.83.2 $\Psi(\beta(\mathcal{X})) = E^{\langle\mathcal{Y}\rangle}(\mathcal{Y})$.

Proof of Claim Suppose $E^{\langle\mathcal{Y}\rangle}(\mathcal{Y}) \subset \Psi(\beta(\mathcal{X}))$, then $\Psi_{\text{inv}}(E^{\langle\mathcal{Y}\rangle}(\mathcal{Y})) \subset \beta(\mathcal{X})$ and $\beta(\mathcal{X}) \cap (I^{\mathcal{F}} \setminus \Psi_{\text{inv}}(E^{\langle\mathcal{Y}\rangle}(\mathcal{Y}))) \neq \emptyset$. Note that $\Psi_{\text{inv}}(E^{\langle\mathcal{Y}\rangle}(\mathcal{Y}))$ is closed in $I^{\mathcal{F}}$ by the closedness of $E^{\langle\mathcal{Y}\rangle}(\mathcal{Y})$, the continuity of Ψ, and Proposition 3.10, which further implies that $I^{\mathcal{F}} \setminus \Psi_{\text{inv}}(E^{\langle\mathcal{Y}\rangle}(\mathcal{Y})) \in \mathcal{O}_{I^{\mathcal{F}}}$. By the denseness of $E(\mathcal{X})$ in $\beta(\mathcal{X})$, we have $E(\mathcal{X}) \cap (I^{\mathcal{F}} \setminus \Psi_{\text{inv}}(E^{\langle\mathcal{Y}\rangle}(\mathcal{Y}))) \neq \emptyset$, and hence, $\exists x \in \mathcal{X}$ such that $E(x) \in I^{\mathcal{F}} \setminus \Psi_{\text{inv}}(E^{\langle\mathcal{Y}\rangle}(\mathcal{Y}))$. By Proposition 2.5, $I^{\mathcal{F}} \setminus \Psi_{\text{inv}}(E^{\langle\mathcal{Y}\rangle}(\mathcal{Y})) = \Psi_{\text{inv}}(I^{\mathcal{G}} \setminus E^{\langle\mathcal{Y}\rangle}(\mathcal{Y}))$. Then, $\Psi(E(x)) \in I^{\mathcal{G}} \setminus E^{\langle\mathcal{Y}\rangle}(\mathcal{Y})$. This contradicts with the fact that $\Psi(E(x)) = E^{\langle\mathcal{Y}\rangle}(x) \in E^{\langle\mathcal{Y}\rangle}(\mathcal{Y})$. Therefore, we must have $\Psi(\beta(\mathcal{X})) = E^{\langle\mathcal{Y}\rangle}(\mathcal{Y})$. This completes the proof of the claim. □

Define $\phi : \beta(\mathcal{X}) \to \mathcal{Y}$ by $\phi = E^{\langle\mathcal{Y}\rangle}{}_{\text{inv}} \circ \Psi|_{\beta(\mathcal{X})}$. Clearly, ϕ is continuous and surjective by Proposition 3.12. Clearly, $\phi \circ E = E^{\langle\mathcal{Y}\rangle}{}_{\text{inv}} \circ E^{\langle\mathcal{Y}\rangle}|_{\mathcal{X}} = \text{id}|_{\mathcal{X}}$. Hence, $\phi|_{E(\mathcal{X})} = E_{\text{inv}}$.

Let $\bar{\phi} : \beta(\mathcal{X}) \to \mathcal{Y}$ be any continuous and surjective mapping such that $\bar{\phi}|_{E(\mathcal{X})} = E_{\text{inv}} = \phi|_{E(\mathcal{X})}$. By Proposition 3.56 and the denseness of $E(\mathcal{X})$ in $\beta(\mathcal{X})$, we have $\phi = \bar{\phi}$. Hence, ϕ is unique.

Thus, we have shown that the Stone–Čech compactification $\beta(\mathcal{X})$ satisfies (i), (ii), and (iii). Next, we will show that $\beta(\mathcal{X})$ is unique. Let $\mathcal{Y}$ be any compact Hausdorff space that satisfies (i), (ii), and (iii). Since $\mathcal{X}$ and $E(\mathcal{X})$ are homeomorphic, we may identify $\mathcal{X}$ as $E(\mathcal{X})$. Then, by (iii), $\exists!\, \phi : \beta(\mathcal{X}) \to \mathcal{Y}$, which is continuous and surjective, such that $\phi|_{\mathcal{X}} = \text{id}_{\mathcal{X}}$. Since $\mathcal{Y}$ satisfies (iii), then $\exists!\, \lambda : \mathcal{Y} \to \beta(\mathcal{X})$, which is continuous and surjective, such that $\lambda|_{\mathcal{X}} = \text{id}_{\mathcal{X}}$. Then, $\phi \circ \lambda : \mathcal{Y} \to \mathcal{Y}$ is continuous and satisfies $(\phi \circ \lambda)|_{\mathcal{X}} = \text{id}_{\mathcal{X}}$. By Proposition 3.56, we have $\phi \circ \lambda = \text{id}_{\mathcal{Y}}$. On the other hand, $\lambda \circ \phi : \beta(\mathcal{X}) \to \beta(\mathcal{X})$ is continuous and satisfies $(\lambda \circ \phi)|_{\mathcal{X}} = \text{id}_{\mathcal{X}}$. Then, by Proposition 3.56, we have $\lambda \circ \phi = \text{id}_{\beta(\mathcal{X})}$. By Proposition 2.4, we have $\lambda = \phi_{\text{inv}}$. Hence, $\beta(\mathcal{X})$ and $\mathcal{Y}$ are homeomorphic. Hence, $\beta(\mathcal{X})$ is unique.

If, in addition, $\mathcal{X}$ is locally compact, then, by Proposition 5.57, $\mathcal{X}$ is open in $\beta(\mathcal{X})$. This completes the proof of the proposition. □

Example 5.84 $\mathbb{R}_e$ with the topology $\mathcal{O}_{\mathbb{R}_e}$ introduced in Example 3.80 is Hausdorff and second countable. It is easy to show that $\mathbb{R}_e$ is compact. By Proposition 5.14, $\mathbb{R}_e$ is a normal topological space. By Urysohn Metrization Theorem 4.53, $\mathbb{R}_e$ is metrizable. Hence, $\mathbb{R}_e$ is a second countable metrizable compact Hausdorff topological space. ◇

Chapter 6
Vector Spaces

6.1 Group

Definition 6.1 A *group* is the triple $(G, +, e)$ that consists a nonempty set G, an operation $+ : G \times G \to G$, and a unit element $e \in G$, satisfying, $\forall g_1, g_2, g_3 \in G$:

(i) $(g_1 + g_2) + g_3 = g_1 + (g_2 + g_3)$ (associativeness).
(ii) $e + g_1 = g_1 + e = g_1$ (unit element).
(iii) $\exists (-g_1) \in G$ such that $g_1 + (-g_1) = e = (-g_1) + g_1$ (existence of inverse, which is clearly unique). ◇

We have the following result.

Proposition 6.2 *Let $(G, +, e)$ be a group and $H \subseteq G$ be nonempty. Then, $(H, +, e)$ is a group (which will be called a subgroup of $(G, +, e)$) if, and only if, $\forall g_1, g_2 \in H$, we have $g_1 + (-g_2) \in H$.*

Proof "Necessity" This is obvious.

"Sufficiency" Since $H \neq \emptyset$, then $\exists g \in H$. Then, $e = g + (-g) \in H$. $\forall g_1, g_2 \in H$, $e + (-g_2) = (-g_2) \in H$. Then, $g_1 + g_2 = g_1 + (-(-g_2)) \in H$. Hence, H is closed under $+$. Then, it is straightforward to check that $(H, +, e)$ satisfies all the properties of Definition 6.1. Hence, it is a group.

This completes the proof of the proposition. □

Proposition 6.3 *Let $(G, +, e)$ be a group and $g_1, g_2, g_3 \in G$. If $g_1 + g_2 = g_1 + g_3$, then $g_2 = g_3$. On the other hand, if $g_1 + g_3 = g_2 + g_3$, then $g_1 = g_2$.*

Proof If $g_1 + g_2 = g_1 + g_3$, then we have

$$
\begin{aligned}
g_2 &= e + g_2 = ((-g_1) + g_1) + g_2 = (-g_1) + (g_1 + g_2) \\
&= (-g_1) + (g_1 + g_3) = ((-g_1) + g_1) + g_3 = e + g_3 = g_3
\end{aligned}
$$

Z. Pan, *Measure-Theoretic Calculus in Abstract Spaces*,
https://doi.org/10.1007/978-3-031-21912-2_6

If $g_1 + g_3 = g_2 + g_3$, then we have

$$\begin{aligned} g_1 &= g_1 + e = g_1 + (g_3 + (-g_3)) = (g_1 + g_3) + (-g_3) \\ &= (g_2 + g_3) + (-g_3) = g_2 + (g_3 + (-g_3)) = g_2 + e = g_2 \end{aligned}$$

This completes the proof of the proposition. □

Definition 6.4 Let $(G, +, e)$ be a group; the *order* of the group is the number of elements in G, if G is finite. $\forall g \in G$, the *order* of g is the integer $n > 0$ such that $\underbrace{g + \cdots + g}_{n} = e$. ◇

Definition 6.5 Let $(G, +_G, e_G)$ and $(H, +_H, e_H)$ be two groups, and $T : G \to H$. T is said to be a *homomorphism* if, $\forall g_1, g_2 \in G$, we have $T(g_1) +_H T(g_2) = T(g_1 +_G g_2)$. T is said to be an *isomorphism* if it is bijective and a homomorphism; in this case, the two groups are said to be *isomorphic*. ◇

Let $T : G \to H$ be a homomorphism; then $T(e_G) = e_H$.

Definition 6.6 Let $(G, +, e)$ be a group. $g_1, g_2 \in G$ are said to be *conjugate* if $\exists g_3 \in G$ such that $g_2 = (-g_3) + g_1 + g_3$. Let $(H, +, e)$ be a subgroup of $(G, +, e)$. It is said to be *normal* (*self-conjugate*) if, $\forall h \in H$, $\forall g \in G$, we have $(-g) + h + g \in H$. ◇

Definition 6.7 Let $(G, +, \mathbf{0})$ be a group. It is said to be *abelian* if, $\forall g_1, g_2 \in G$, we have $g_1 + g_2 = g_2 + g_1$ (commutativeness). Then, the unit element $\mathbf{0}$ is also called the *zero element*. ◇

Sometimes, we are interested in an algebraic structure that is weaker than a group. For example, the structure of functions $f : X \to X$ with respect to the function composition operation. This leads us to the following definition.

Definition 6.8 A *semigroup* is the triple $(G, \circ, e)$ that consists of a nonempty set G, an operation $\circ : G \times G \to G$, and a unit element $e \in G$, satisfying, $\forall g_1, g_2, g_3 \in G$:

(i) $(g_1 \circ g_2) \circ g_3 = g_1 \circ (g_2 \circ g_3)$ (associativeness).
(ii) $e \circ g_1 = g_1 = g_1 \circ e$ (unit element).

Furthermore, a semigroup that satisfies $g_1 \circ g_2 = g_2 \circ g_1$, $\forall g_1, g_2 \in G$, is called an *abelian semigroup*. ◇

6.2 Ring

Definition 6.9 A *ring* $(R, +, \times, \mathbf{0})$ is an abelian group $(R, +, \mathbf{0})$ with an operation $\times : R \times R \to R$ such that, $\forall r_1, r_2, r_3 \in R$:

(i) $(r_1 \times r_2) \times r_3 = r_1 \times (r_2 \times r_3)$ (associativeness).
(ii) $r_1 \times (r_2 + r_3) = r_1 \times r_2 + r_1 \times r_3$ (right distributiveness).
(iii) $(r_1 + r_2) \times r_3 = r_1 \times r_3 + r_2 \times r_3$ (left distributiveness).

A ring is *commutative* if, $\forall r_1, r_2 \in R$, we have $r_1 \times r_2 = r_2 \times r_1$. A ring is *with identity element* if $\exists \mathbf{1} \in R$ such that, $\forall r \in R$, $\mathbf{1} \times r = r \times \mathbf{1} = r$. (The identity element is unique if it exists.) ◇

Proposition 6.10 *Let $(R, +, \times, \mathbf{0})$ be a ring. $S \subseteq R$. Then, $(S, +, \times, \mathbf{0})$ is a ring (which will be called a* subring*) if $(S, +, \mathbf{0})$ is a subgroup of $(R, +, \mathbf{0})$ and $s_1 \times s_2 \in S$, $\forall s_1, s_2 \in S$.*

Proof Clearly, $(S, +, \mathbf{0})$ is an abelian group. Then, it is straightforward to show that $(S, +, \times, \mathbf{0})$ is a ring. □

Proposition 6.11 *Let $(R, +, \times, \mathbf{0})$ be a ring, and $r_1, r_2 \in R$; then, we have:*

(i) $r_1 \times \mathbf{0} = \mathbf{0} \times r_1 = \mathbf{0}$.
(ii) $r_1 \times (-r_2) = (-r_1) \times r_2 = -(r_1 \times r_2)$.
(iii) $(-r_1) \times (-r_2) = r_1 \times r_2$.

Proof Note that $\mathbf{0} + \mathbf{0} \times r_1 = \mathbf{0} \times r_1 = (\mathbf{0} + \mathbf{0}) \times r_1 = \mathbf{0} \times r_1 + \mathbf{0} \times r_1$. By Proposition 6.3, we have $\mathbf{0} \times r_1 = \mathbf{0}$. Similarly, we can show that $r_1 \times \mathbf{0} = \mathbf{0}$.

Note that $r_1 \times r_2 + r_1 \times (-r_2) = r_1 \times (r_2 + (-r_2)) = r_1 \times \mathbf{0} = \mathbf{0}$. Then, we have $r_1 \times (-r_2) = -(r_1 \times r_2)$. Similarly, we can show that $(-r_1) \times r_2 = -(r_1 \times r_2)$.

Note that $(-r_1) \times (-r_2) + (-(r_1 \times r_2)) = (-r_1) \times (-r_2) + r_1 \times (-r_2) = ((-r_1) + r_1) \times (-r_2) = \mathbf{0} \times (-r_2) = \mathbf{0}$. Then, we have $(-r_1) \times (-r_2) = r_1 \times r_2$.

This completes the proof of the proposition. □

Definition 6.12 Let $(R, +, \times, \mathbf{0})$ be a ring and $S \subseteq R$. Then, S is said to be an *ideal* if $(S, +, \mathbf{0})$ is a subgroup of $(R, +, \mathbf{0})$ and $r \times s, s \times r \in S$, $\forall r \in R$ and $\forall s \in S$. ◇

Definition 6.13 Let $(R, +_R, \times_R, \mathbf{0}_R)$ and $(S, +_S, \times_S, \mathbf{0}_S)$ be two rings and $T : R \to S$. T is said to be a *ring homomorphism* if it is a homomorphism and $T(r_1) \times_S T(r_2) = T(r_1 \times_R r_2)$, $\forall r_1, r_2 \in R$. T is said to be a *ring isomorphism* if it is a bijective ring homomorphism; in this case, the two rings are said to be *isomorphic*. ◇

6.3 Field

Definition 6.14 Let $(F, +, \times, \mathbf{0})$ be a commutative ring with identity element $\mathbf{1}$. Then, the quintuple $(F, +, \times, \mathbf{0}, \mathbf{1})$ is said to be a *field* if $(F \setminus \{\mathbf{0}\}, \times, \mathbf{1})$ forms an abelian group. ◇

Proposition 6.15 *Let $(F, +, \times, \mathbf{0}, \mathbf{1})$ be a field, $A \subseteq F$, and $\mathbf{0}, \mathbf{1} \in A$. Then, $(A, +, \times, \mathbf{0}, \mathbf{1})$ is a field (which will be called a* subfield*) if, and only if, $\forall a_1, a_2 \in A$, $a_1 + (-a_2) \in A$, and $a_1 \times a_2^{-1} \in A$ when $a_2 \neq \mathbf{0}$, where a_2^{-1} denotes the multiplicative inverse of a_2.*

Proof "Necessity" This is straightforward.

"Sufficiency" By Proposition 6.2, $(A, +, \mathbf{0})$ is a group. Since $(F, +, \mathbf{0})$ is abelian, then $(A, +, \mathbf{0})$ is also an abelian group. $\forall a_1 \in A$ with $a_1 \neq \mathbf{0}$. By the assumption of the proposition, we have $\mathbf{1} \in A$, by the property of field, we have $\mathbf{1} \neq \mathbf{0}$. Then, $a_1^{-1} = \mathbf{1} \times a_1^{-1} \in A$. $\forall a_1, a_2 \in A$. If $a_2 = \mathbf{0}$, then, by Proposition 6.11, $a_1 \times a_2 = \mathbf{0} \in A$. On the other hand, if $a_2 \neq \mathbf{0}$, we have $a_2^{-1} \in A$ and $a_1 \times a_2 = a_1 \times (a_2^{-1})^{-1} \in A$. Hence, by Proposition 6.10, $(A, +, \times, \mathbf{0})$ is a ring. Since $(F, +, \times, \mathbf{0})$ is a commutative ring with identity element $\mathbf{1}$ and $\mathbf{1} \in A$, then $(A, +, \times, \mathbf{0})$ is also a commutative ring with identity element $\mathbf{1}$. Note that $\mathbf{1} \in A \setminus \{\mathbf{0}\}$, and then, $A \setminus \{\mathbf{0}\} \neq \emptyset$ and $A \setminus \{\mathbf{0}\} \subseteq F \setminus \{\mathbf{0}\}$. $\forall a_1, a_2 \in A \setminus \{\mathbf{0}\}$, $a_1 \times a_2^{-1} \in A$. We claim that $a_1 \times a_2^{-1} \neq \mathbf{0}$ since, otherwise, $a_1 = a_1 \times \mathbf{1} = a_1 \times (a_2^{-1} \times a_2) = (a_1 \times a_2^{-1}) \times a_2 = \mathbf{0}$ by Proposition 6.11, which is a contradiction. Hence, $a_1 \times a_2^{-1} \in A \setminus \{\mathbf{0}\}$. Since $(F \setminus \{\mathbf{0}\}, \times, \mathbf{1})$ is an abelian group, then, by Proposition 6.2, $(A \setminus \{\mathbf{0}\}, \times, \mathbf{1})$ is also a group and is further abelian. Therefore, $(A, +, \times, \mathbf{0}, \mathbf{1})$ is a field.

This completes the proof of the proposition. □

6.4 Vector Spaces

Associated with every vector space is a set of scalars. This set of scalars can be any algebraic field $\mathcal{F} := (F, +, \cdot, \mathbf{0}, \mathbf{1})$. Examples of fields are the rational numbers $\mathbb{Q}$, the real numbers $\mathbb{R}$, and the complex numbers $\mathbb{C}$. Here, we will abuse the notation to say $x \in \mathcal{F}$ when $x \in F$.

Definition 6.16 A *vector space $\mathcal{X}$ over a field $\mathcal{F} := (F, +, \cdot, \mathbf{0}, \mathbf{1})$* is a set X of elements called vectors together with two operations $\oplus$ and $\otimes$. $\oplus : X \times X \to X$ is called vector addition. It associates any two vectors $x, y \in X$ with a vector $x \oplus y \in X$, the sum of x and y. $\otimes : F \times X \to X$ is called scalar multiplication. It associates a scalar $\alpha \in F$ and a vector $x \in X$ with a vector $\alpha \otimes x \in X$, the scalar multiple of x by α. Furthermore, the following properties hold for $\forall x, y, z \in X$ and $\forall \alpha, \beta \in F$:

(i) $x \oplus y = y \oplus x$ (commutative law).
(ii) $(x \oplus y) \oplus z = x \oplus (y \oplus z)$ (associative law).
(iii) $\exists$ a null vector $\vartheta \in X$ such that $x \oplus \vartheta = x$, $\forall x \in X$.
(iv) $\alpha \otimes (x \oplus y) = (\alpha \otimes x) \oplus (\alpha \otimes y) = \alpha \otimes x \oplus \alpha \otimes y$, where we neglected the parenthesis in the last equality since we assume that $\otimes$ takes precedence over $\oplus$ (distributive law).
(v) $(\alpha + \beta) \otimes x = \alpha \otimes x \oplus \beta \otimes x$ (distributive law).
(vi) $(\alpha\beta) \otimes x = \alpha \otimes (\beta \otimes x)$ (associative law).
(vii) $\mathbf{0} \otimes x = \vartheta$ and $\mathbf{1} \otimes x = x$.

We thus denote the quadruple $(X, \oplus, \otimes, \vartheta)$ by $\mathcal{X}$. The vector space is denoted by $(\mathcal{X}, \mathcal{F})$. $\diamond$

For convenience, $(-\mathbf{1}) \otimes x =: \ominus x$ and called the negative of the vector x. Note that

$$(\ominus x) \oplus x = x \oplus (\ominus x) = \mathbf{1} \otimes x \oplus (-\mathbf{1}) \otimes x = (\mathbf{1} + (-\mathbf{1})) \otimes x = \mathbf{0} \otimes x = \vartheta$$

We will also denote $x \ominus y := x \oplus (\ominus y)$. We will abuse the notation to say $x \in \mathcal{X}$ when $x \in X$. Note that $(\mathcal{X}, \oplus, \vartheta)$ forms an abelian group.

Proposition 6.17 *Let $\mathcal{X} := (X, \oplus, \otimes, \vartheta)$ be a vector space over the field $\mathcal{F} := (F, +, \cdot, \mathbf{0}, \mathbf{1})$. $\forall x, y, z \in \mathcal{X}$, $\forall \alpha, \beta \in \mathcal{F}$, we have:*

1. $x \oplus y = x \oplus z \Rightarrow y = z$ *(cancellation law).*
2. $\alpha \otimes x = \alpha \otimes y$ *and* $\alpha \neq \mathbf{0} \Rightarrow x = y$ *(cancellation law).*
3. $\alpha \otimes x = \beta \otimes x$ *and* $x \neq \vartheta \Rightarrow \alpha = \beta$ *(cancellation law).*
4. $(\alpha - \beta) \otimes x = \alpha \otimes x \ominus \beta \otimes x$ *(distributive law).*
5. $\alpha \otimes (x \ominus y) = \alpha \otimes x \ominus \alpha \otimes y$ *(distributive law).*
6. $\alpha \otimes \vartheta = \vartheta$.

We will call ϑ the origin.

Example 6.18 $X = \{\vartheta\}$ with $\vartheta \oplus \vartheta = \vartheta$ and $\alpha \otimes \vartheta = \vartheta$, $\forall \alpha \in \mathcal{F}$. Then, $(X, \oplus, \otimes, \vartheta)$ is a vector space over $\mathcal{F}$. $\diamond$

Example 6.19 Let $\mathcal{F} := (F, +, \cdot, \mathbf{0}, \mathbf{1})$ be a field. Then, $(F, +, \cdot, \mathbf{0})$ is a vector space over $\mathcal{F}$. We will abuse the notation and say that $\mathcal{F}$ is a vector space over $\mathcal{F}$. $\diamond$

Example 6.20 Let $\mathcal{F} := (F, +, \cdot, \mathbf{0}, \mathbf{1})$ be a field, $\mathcal{Y} := (Y, \oplus_Y, \otimes_Y, \vartheta_Y)$ be a vector space over $\mathcal{F}$, and A be a set. $X = \{f : A \to \mathcal{Y}\}$, that is, X is the set of all $\mathcal{Y}$-valued functions on A. Define vector addition and scalar multiplication by, $\forall x, y \in X$, $\forall \alpha \in \mathcal{F}$, $z_a := x \oplus y \in X$ is given by $z_a(u) = x(u) \oplus_Y y(u)$, $\forall u \in A$, $z_s := \alpha \otimes x \in X$ is given by $z_s(u) = \alpha \otimes_Y x(u)$, $\forall u \in A$. Let $\vartheta \in X$ be given by $\vartheta(u) = \vartheta_Y$, $\forall u \in A$.

Now, we will show that $\mathcal{X} := (X, \oplus, \otimes, \vartheta)$ is a vector space over $\mathcal{F}$. $\forall x, y, z \in X$, $\forall \alpha, \beta \in \mathcal{F}$, $\forall u \in A$, we have:

(i) $(x \oplus y)(u) = x(u) \oplus_Y y(u) = y(u) \oplus_Y x(u) = (y \oplus x)(u) \Rightarrow x \oplus y = y \oplus x$.

(ii) $((x \oplus y) \oplus z)(u) = (x \oplus y)(u) \oplus_Y z(u) = (x(u) \oplus_Y y(u)) \oplus_Y z(u) = x(u) \oplus_Y (y(u) \oplus_Y z(u)) = x(u) \oplus_Y (y \oplus z)(u) = (x \oplus (y \oplus z))(u) \Rightarrow (x \oplus y) \oplus z = x \oplus (y \oplus z)$.

(iii) $(x \oplus \vartheta)(u) = x(u) \oplus_Y \vartheta_Y = x(u) \Rightarrow x \oplus \vartheta = x$.

(iv) $(\alpha \otimes (x \oplus y))(u) = \alpha \otimes_Y (x \oplus y)(u) = \alpha \otimes_Y (x(u) \oplus_Y y(u)) = \alpha \otimes_Y x(u) \oplus_Y \alpha \otimes_Y y(u) = (\alpha \otimes x)(u) \oplus_Y (\alpha \otimes y)(u) = (\alpha \otimes x \oplus \alpha \otimes y)(u) \Rightarrow \alpha \otimes (x \oplus y) = \alpha \otimes x \oplus \alpha \otimes y$.

(v) $((\alpha + \beta) \otimes x)(u) = (\alpha + \beta) \otimes_Y x(u) = \alpha \otimes_Y x(u) \oplus_Y \beta \otimes_Y x(u) = (\alpha \otimes x)(u) \oplus_Y (\beta \otimes x)(u) = (\alpha \otimes x \oplus \beta \otimes x)(u) \Rightarrow (\alpha + \beta) \otimes x = \alpha \otimes x \oplus \beta \otimes x$.

(vi) $((\alpha\beta) \otimes x)(u) = (\alpha\beta) \otimes_Y x(u) = \alpha \otimes_Y (\beta \otimes_Y x(u)) = \alpha \otimes_Y (\beta \otimes x)(u) = (\alpha \otimes (\beta \otimes x))(u) \Rightarrow (\alpha\beta) \otimes x = \alpha \otimes (\beta \otimes x)$.

(vii) $(\mathbf{0}\otimes x)(u) = \mathbf{0}\otimes_Y x(u) = \vartheta_Y = \vartheta(u) \Rightarrow \mathbf{0}\otimes x = \vartheta$; $(\mathbf{1}\otimes x)(u) = \mathbf{1}\otimes_Y x(u) = x(u) \Rightarrow \mathbf{1}\otimes x = x$.

Therefore, $\mathcal{X}$ is a vector space over $\mathcal{F}$. This vector space will be denoted by $(\mathcal{M}(A, \mathcal{Y}), \mathcal{F})$. ◇

Example 6.21 Let $\mathcal{F} := (F, +, \cdot, \mathbf{0}, \mathbf{1})$ be a field. $X = F^n$ with $n \in \mathbb{N}$. Define vector addition and scalar multiplication by $x \oplus y := (\xi_1 + \eta_1, \dots, \xi_n + \eta_n) \in X$, $\alpha \otimes x := (\alpha\xi_1, \dots, \alpha\xi_n) \in X$, $\forall x = (\xi_1, \dots, \xi_n)$, $y = (\eta_1, \dots, \eta_n) \in X$, $\forall \alpha \in \mathcal{F}$. Let $\vartheta := (\mathbf{0}, \dots, \mathbf{0}) \in X$. Then, it is straightforward to check that $\mathcal{F}^n := (X, \oplus, \otimes, \vartheta)$ is a vector space over $\mathcal{F}$. ◇

Example 6.22 Let $\mathcal{F} := (F, +, \cdot, \mathbf{0}, \mathbf{1})$ be a field. $X = F^{m\times n} := \{m \times n$-dimensional F-valued matrices$\}$ with $m, n \in \mathbb{N}$. Define vector addition and scalar multiplication by $x \oplus y := (\xi_{ij} + \eta_{ij})_{m\times n} \in X$, $\alpha \otimes x := (\alpha\xi_{ij})_{m\times n} \in X$, $\forall x = (\xi_{ij})_{m\times n}$, $y = (\eta_{ij})_{m\times n} \in X$, $\forall \alpha \in \mathcal{F}$. Let $\vartheta := (\mathbf{0})_{m\times n} \in X$. Then, it is straightforward to check that $\mathcal{F}^{m\times n} := (X, \oplus, \otimes, \vartheta)$ is a vector space over $\mathcal{F}$. ◇

Example 6.23 Let $\mathcal{F} := (F, +, \cdot, \mathbf{0}, \mathbf{1})$ be a field. $X = \{(\xi_k)_{k=1}^\infty \mid \xi_k \in \mathcal{F}, \forall k \in \mathbb{N}\}$. Define vector addition and scalar multiplication by $x \oplus y := (\xi_k + \eta_k)_{k=1}^\infty \in X$, $\alpha \otimes x := (\alpha\xi_k)_{k=1}^\infty \in X$, $\forall x = (\xi_k)_{k=1}^\infty$, $y = (\eta_k)_{k=1}^\infty \in X$, $\forall \alpha \in \mathcal{F}$. Let $\vartheta := (\mathbf{0}, \mathbf{0}, \dots) \in X$. Then, it is straightforward to check that $(X, \oplus, \otimes, \vartheta)$ is a vector space over $\mathcal{F}$. ◇

6.5 Product Spaces

Proposition 6.24 *Let* $\mathcal{X} := (X, \oplus_X, \otimes_X, \vartheta_X)$ *and* $\mathcal{Y} := (Y, \oplus_Y, \otimes_Y, \vartheta_Y)$ *be vector spaces over the field* $\mathcal{F} := (F, +, \cdot, \mathbf{0}, \mathbf{1})$. *The* Cartesian product *of* $\mathcal{X}$ *and* $\mathcal{Y}$, *denoted by* $\mathcal{X} \times \mathcal{Y}$, *is the quadruple* $(X \times Y, \oplus, \otimes, (\vartheta_X, \vartheta_Y))$, *where the vector addition* $\oplus : (X \times Y) \times (X \times Y) \to X \times Y$ *and the scalar multiplication* $\otimes : \mathcal{F} \times (X \times Y) \to X \times Y$ *are given by,* $\forall (x_1, y_1), (x_2, y_2) \in X \times Y$, $\forall \alpha \in \mathcal{F}$, $(x_1, y_1) \oplus (x_2, y_2) := (x_1 \oplus_X x_2, y_1 \oplus_Y y_2)$ *and* $\alpha \otimes (x_1, y_1) := (\alpha \otimes_X x_1, \alpha \otimes_Y y_1)$. *Then,* $(\mathcal{X} \times \mathcal{Y}, \mathcal{F})$ *is a vector space.*

Proof Let $\vartheta := (\vartheta_X, \vartheta_Y) \in X \times Y$. $\forall (x_1, y_1), (x_2, y_2), (x_3, y_3) \in X \times Y$, $\forall \alpha, \beta \in \mathcal{F}$:

(i) $(x_1, y_1) \oplus (x_2, y_2) = (x_1 \oplus_X x_2, y_1 \oplus_Y y_2) = (x_2 \oplus_X x_1, y_2 \oplus_Y y_1) = (x_2, y_2) \oplus (x_1, y_1)$.

(ii) $((x_1, y_1) \oplus (x_2, y_2)) \oplus (x_3, y_3) = (x_1 \oplus_X x_2, y_1 \oplus_Y y_2) \oplus (x_3, y_3) = ((x_1 \oplus_X x_2) \oplus_X x_3, (y_1 \oplus_Y y_2) \oplus_Y y_3) = (x_1 \oplus_X (x_2 \oplus_X x_3), y_1 \oplus_Y (y_2 \oplus_Y y_3)) = (x_1, y_1) \oplus (x_2 \oplus_X x_3, y_2 \oplus_Y y_3) = (x_1, y_1) \oplus ((x_2, y_2) \oplus (x_3, y_3))$.

(iii) $(x_1, y_1) \oplus \vartheta = (x_1 \oplus_X \vartheta_X, y_1 \oplus_Y \vartheta_Y) = (x_1, y_1)$.

(iv) $\alpha \otimes ((x_1, y_1) \oplus (x_2, y_2)) = \alpha \otimes (x_1 \oplus_X x_2, y_1 \oplus_Y y_2) = (\alpha \otimes_X (x_1 \oplus_X x_2), \alpha \otimes_Y (y_1 \oplus_Y y_2)) = (\alpha \otimes_X x_1 \oplus_X \alpha \otimes_X x_2, \alpha \otimes_Y y_1 \oplus_Y \alpha \otimes_Y y_2) = (\alpha \otimes_X x_1, \alpha \otimes_Y y_1) \oplus (\alpha \otimes_X x_2, \alpha \otimes_Y y_2) = \alpha \otimes (x_1, y_1) \oplus \alpha \otimes (x_2, y_2)$.

(v) $(\alpha + \beta) \otimes (x_1, y_1) = ((\alpha + \beta) \otimes_X x_1, (\alpha + \beta) \otimes_Y y_1) = (\alpha \otimes_X x_1 \oplus_X \beta \otimes_X x_1, \alpha \otimes_Y y_1 \oplus_Y \beta \otimes_Y y_1) = (\alpha \otimes_X x_1, \alpha \otimes_Y y_1) \oplus (\beta \otimes_X x_1, \beta \otimes_Y y_1) = \alpha \otimes (x_1, y_1) \oplus \beta \otimes (x_1, y_1)$.

(vi) $(\alpha\beta) \otimes (x_1, y_1) = ((\alpha\beta) \otimes_X x_1, (\alpha\beta) \otimes_Y y_1) = (\alpha \otimes_X (\beta \otimes_X x_1), \alpha \otimes_Y (\beta \otimes_Y y_1)) = \alpha \otimes (\beta \otimes_X x_1, \beta \otimes_Y y_1) = \alpha \otimes (\beta \otimes (x_1, y_1))$.

(vii) $\mathbf{0} \otimes (x_1, y_1) = (\mathbf{0} \otimes_X x_1, \mathbf{0} \otimes_Y y_1) = (\vartheta_X, \vartheta_Y) = \vartheta$; $\mathbf{1} \otimes (x_1, y_1) = (\mathbf{1} \otimes_X x_1, \mathbf{1} \otimes_Y y_1) = (x_1, y_1)$.

Hence, $\mathcal{X} \times \mathcal{Y}$ is a vector space over $\mathcal{F}$. □

With the above definition, it is easy to generalize to $\mathcal{X}_1 \times \mathcal{X}_2 \times \cdots \times \mathcal{X}_n$, where $n \in \mathbb{N}$. We will also write $\mathcal{X}^n = \underbrace{\mathcal{X} \times \cdots \times \mathcal{X}}_{n}$, where $n \in \mathbb{N}$. When $n = 0$, $\prod_{i=1}^{n} \mathcal{X}_i$ is given by the vector space defined in Example 6.18.

6.6 Subspaces

Proposition 6.25 *Let $\mathcal{X} := (X, \oplus, \otimes, \vartheta)$ be a vector space over the field $\mathcal{F} := (F, +, \cdot, \mathbf{0}, \mathbf{1})$, and $M \subseteq X$ with $M \neq \emptyset$. Then, $\mathcal{M} := (M, \oplus, \otimes, \vartheta)$ is a vector space over $\mathcal{F}$ (which will be called a* subspace *of $(\mathcal{X}, \mathcal{F})$) if, and only if, $\forall x, y \in M$, $\forall \alpha, \beta \in \mathcal{F}$, we have $\alpha \otimes x \oplus \beta \otimes y \in M$. We will also abuse the notation to say M is a subspace of $(\mathcal{X}, \mathcal{F})$. $\mathcal{M}$ is said to be a* proper subspace *of $(\mathcal{X}, \mathcal{F})$ if $M \subset X$.*

Proof "If" Since $M \neq \emptyset$, then $\exists x_0 \in M$. $\vartheta = \vartheta \oplus \vartheta = \mathbf{0} \otimes x_0 \oplus \mathbf{0} \otimes x_0 \in M$. $\forall x, y \in M$, $\forall \alpha, \beta \in \mathcal{F}$. $x \oplus y = \mathbf{1} \otimes x \oplus \mathbf{1} \otimes y \in M$. Hence, M is closed under vector addition. $\alpha \otimes x = (\alpha + \mathbf{0}) \otimes x = \alpha \otimes x \oplus \mathbf{0} \otimes x \in M$. Hence, M is closed under scalar multiplication. Then, it is straightforward to show that $\mathcal{M}$ is a vector space over $\mathcal{F}$. "Only if" This is straightforward. This completes the proof of the proposition. □

Example 6.26 We present the following list of examples of subspaces:

1. Let $(\mathcal{X}, \mathcal{F})$ be a vector space. Then, the singleton set $M = \{\vartheta\}$ is a subspace.
2. Consider the vector space $(\mathbb{R}^3, \mathbb{R})$. Any straight line or plane that passes through the origin is a subspace.
3. Consider the vector space $(\mathbb{R}^n, \mathbb{R})$, $n \in \mathbb{N}$. Let $a \in \mathbb{R}^n$. The set $M := \{x \in \mathbb{R}^n \mid \langle a, x \rangle = 0\}$ is a subspace.
4. Let $X := \{(\xi_k)_{k=1}^{\infty} \mid \xi_k \in \mathbb{R}, k \in \mathbb{N}\}$, $\oplus$ and $\otimes$ be the usual addition and scalar multiplication, and $\vartheta = (0, 0, \ldots)$. By Example 6.23, $\mathcal{X} := (X, \oplus, \otimes, \vartheta)$ is a vector space over $\mathbb{R}$. Let $M := \{(\xi_k)_{k=1}^{\infty} \in \mathcal{X} \mid \lim_{k \in \mathbb{N}} \xi_k \in \mathbb{R}\}$ be a subspace.
5. Let $X := \{f : (0, 1] \to \mathbb{R}^n\}$, $\oplus$ and $\otimes$ be the usual addition and scalar multiplication, and $\vartheta : (0, 1] \to \mathbb{R}^n$ be given by $\vartheta(t) = \vartheta_{\mathbb{R}^n}$, $\forall t \in (0, 1]$. By Example 6.20, $\mathcal{X} := (X, \oplus, \otimes, \vartheta)$ is a vector space over $\mathbb{R}$. Let $M := \{f \in \mathcal{X} \mid f \text{ is continuous}\}$ be a subspace. ◇

To simplify notation in the theory, we will later simply discuss a vector space $(\mathcal{X}, \mathcal{F})$ without further reference to components of $\mathcal{X}$, where the operations are understood to be $\oplus_{\mathcal{X}}$ and $\otimes_{\mathcal{X}}$ and the null vector is understood to be $\vartheta_{\mathcal{X}}$. When it is clear from the context, we will neglect the subscript $\mathcal{X}$. Also, we will write $x_1 + x_2$ for $x_1 \oplus x_2$ and αx_1 for $\alpha \otimes x_1$, $\forall x_1, x_2 \in \mathcal{X}$, $\forall \alpha \in \mathcal{F}$.

Definition 6.27 Let $\mathcal{X}$ be a vector space over the field $\mathcal{F}$. $f : \mathcal{X} \to \mathcal{F}$ is said to be a *functional*. ◇

Definition 6.28 Let $\mathcal{X}$ and $\mathcal{Y}$ be vector spaces over the field $\mathcal{F}$. $A : \mathcal{X} \to \mathcal{Y}$ is said to be *linear* if $\forall x_1, x_2 \in \mathcal{X}$, $\forall \alpha, \beta \in \mathcal{F}$, $A(\alpha x_1 + \beta x_2) = \alpha A(x_1) + \beta A(x_2)$. Then, A is called a *(vector space) homomorphism* or a *linear operator*. Furthermore, if it is bijective, then A is said to be a *(vector space) isomorphism*. The *null space* of A is $\mathcal{N}(A) := \{x \in \mathcal{X} \mid A(x) = \vartheta_{\mathcal{Y}}\}$. The *range space* of A is $\mathcal{R}(A) := \text{range}\,(A)$. $B : \mathcal{X} \to \mathcal{Y}$ is said to be an *affine operator* if $B(x) = A(x) + y_0$, $\forall x \in \mathcal{X}$, where $A : \mathcal{X} \to \mathcal{Y}$ is a linear operator and $y_0 \in \mathcal{Y}$. ◇

Example 6.29 A row vector $v \in \mathbb{R}^{1\times n}$ is a linear functional on $\mathbb{R}^n$. A matrix $A \in \mathbb{R}^{m\times n}$ is a linear function of $\mathbb{R}^n$ to $\mathbb{R}^m$. ◇

For linear operators, we will adopt the following convention. Let $A : \mathcal{X} \to \mathcal{Y}$, $B : \mathcal{Y} \to \mathcal{Z}$, and $x \in \mathcal{X}$, where $\mathcal{X}, \mathcal{Y}, \mathcal{Z}$ are vector spaces and A and B are linear operators. We will write Ax for $A(x)$ and BA for $B \circ A$. Clearly, $\mathcal{N}(A)$ is a subspace of $\mathcal{X}$ and $\mathcal{R}(A)$ is a subspace of $\mathcal{Y}$. When A and B are bijective, we will denote A_{inv} by A^{-1} and B_{inv} by B^{-1}, where A^{-1} and B^{-1} are linear operators, then BA is also bijective, and $(BA)^{-1} = A^{-1}B^{-1}$.

Definition 6.30 Let $(\mathcal{X}, \mathcal{F})$ be a vector space, $\alpha \in \mathcal{F}$, and $S, T \subseteq \mathcal{X}$. The sets αS and $S + T$ are defined by

$$\alpha S := \{\alpha s \in \mathcal{X} \mid s \in S\}; \qquad S + T := \{s + t \in \mathcal{X} \mid s \in S,\ t \in T\}$$ ◇

This concept is illustrated in Fig. 6.1.

We should note that $S + T = T + S$, $\emptyset + S = \emptyset$, $\{\vartheta\} + S = S$, and $\alpha\emptyset = \emptyset$. Thus, $S - T := S + (-T)$.

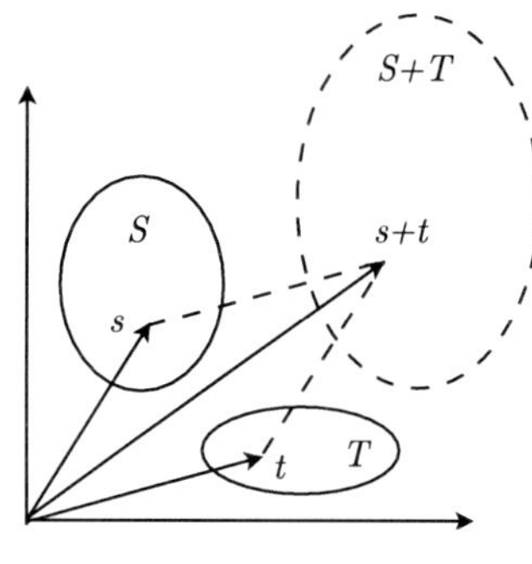

Fig. 6.1 The sum of two sets

Proposition 6.31 *Let M and N be subspaces of a vector space $(\mathcal{X}, \mathcal{F})$ and $\bar{\alpha} \in \mathcal{F}$. Then, $M \cap N$, $M + N$, and $\bar{\alpha}M$ are subspaces of $(\mathcal{X}, \mathcal{F})$.*

Proof Since M and N are subspaces, then $\vartheta \in M$ and $\vartheta \in N$. Hence, $\vartheta \in M \cap N \neq \emptyset$, $\vartheta \in M + N \neq \emptyset$, and $\vartheta = \bar{\alpha}\vartheta \in \bar{\alpha}M \neq \emptyset$.

$\forall x, y \in M \cap N$, $\forall \alpha, \beta \in \mathcal{F}$, $\alpha x + \beta y \in M$ and $\alpha x + \beta y \in N$. Then, $\alpha x + \beta y \in M \cap N$. Hence, $M \cap N$ is a subspace.

$\forall x, y \in M + N$, $\forall \alpha, \beta \in \mathcal{F}$, we have $x = x_1 + x_2$ and $y = y_1 + y_2$, where $x_1, y_1 \in M$ and $x_2, y_2 \in N$. Then, $\alpha x_1 + \beta y_1 \in M$ and $\alpha x_2 + \beta y_2 \in N$. This implies that $\alpha x + \beta y = (\alpha x_1 + \beta y_1) + (\alpha x_2 + \beta y_2) \in M + N$. Hence, $M + N$ is a subspace.

$\forall x, y \in \bar{\alpha}M$, $\forall \alpha, \beta \in \mathcal{F}$, we have $x = \bar{\alpha}\bar{x}$ and $y = \bar{\alpha}\bar{y}$, where $\bar{x}, \bar{y} \in M$. Then, $\alpha x + \beta y = \alpha\bar{\alpha}\bar{x} + \beta\bar{\alpha}\bar{y} = \bar{\alpha}(\alpha\bar{x} + \beta\bar{y}) \in \bar{\alpha}M$. Hence, $\bar{\alpha}M$ is a subspace.

This completes the proof of the proposition. □

We should note that $M \cup N$ is in general not a subspace.

Definition 6.32 A *linear combination* of vectors $x_1, \ldots, x_n$, where $n \in \mathbb{Z}_+$, in a vector space $(\mathcal{X}, \mathcal{F})$ is a sum of the form $\sum_{i=1}^{n} \alpha_i x_i := \alpha_1 x_1 + \cdots + \alpha_n x_n$, where $\alpha_1, \ldots, \alpha_n \in \mathcal{F}$. ◇

Note that $+$ is defined for two vectors. To sum n vectors, one must add two at a time. By the definition of vector space, the simplified notion is not ambiguous. When $n = 0$, we take the sum to be $\vartheta_{\mathcal{X}}$. A linear combination can only involve finitely many vectors.

Definition 6.33 Let $(\mathcal{X}, \mathcal{F})$ be a vector space and $S \subseteq \mathcal{X}$.

$$\begin{aligned} \operatorname{span}(S) &:= \{\textstyle\sum_{i=1}^{n} \alpha_i x_i \mid x_i \in S, \alpha_i \in \mathcal{F}, i = 1, \ldots, n, n \in \mathbb{Z}_+\} \\ &= \{\text{linear combination of vectors in } S\} \end{aligned}$$

is called the *subspace generated by S*. ◇

Proposition 6.34 *Let $(\mathcal{X}, \mathcal{F})$ be a vector space and $S \subseteq \mathcal{X}$. Then,* span (S) *is the smallest subspace containing S.*

Proof Clearly, $\vartheta \in \operatorname{span}(S) \neq \emptyset$. $\forall x, y \in \operatorname{span}(S)$, $\forall \alpha, \beta \in \mathcal{F}$, It is easy to show that $\alpha x + \beta y \in \operatorname{span}(S)$. Hence, span (S) is a subspace. Clearly, $\operatorname{span}(S) \supseteq S$.

$\forall M \subseteq \mathcal{X}$ such that $S \subseteq M$ and M is a subspace. Clearly, $\vartheta \in M$ by the proof of Proposition 6.25. $\forall y$ equals to a linear combination of vectors in S. Then, $y = \sum_{i=1}^{n} \alpha_i x_i$ with $n \in \mathbb{Z}_+$, $x_1, \ldots, x_n \in S$, and $\alpha_1, \ldots, \alpha_n \in \mathcal{F}$. This implies that $x_1, \ldots, x_n \in M$ and $y \in M$. Hence, $\operatorname{span}(S) \subseteq M$.

Hence, span (S) is the smallest subspace containing S. This completes the proof of the proposition. □

Definition 6.35 Let $(\mathcal{X}, \mathcal{F})$ be a vector space, $M \subseteq \mathcal{X}$ be a subspace, and $x_0 \in \mathcal{X}$. Then, $V := \{x_0\} + M$ is called a *linear variety*. ◇

The translation of a subspace is a linear variety. We will abuse the notation to write $x_0 + M$ for $\{x_0\} + M$. $\forall \bar{x} \in V$, $V - \bar{x} := V - \{\bar{x}\}$ is a subspace.

Definition 6.36 Let $(\mathcal{X}, \mathcal{F})$ be a vector space, $S \subseteq \mathcal{X}$, and $S \neq \emptyset$. The *linear variety generated by* S, denoted by $\mathrm{v}(S)$, is defined as the intersection of all linear varieties in $\mathcal{X}$ that contain S. ◇

Proposition 6.37 *Let $(\mathcal{X}, \mathcal{F})$ be a vector space and $\emptyset \neq S \subseteq \mathcal{X}$. Then, $\mathrm{v}(S)$ is a linear variety given by $\mathrm{v}(S) = x_0 + \mathrm{span}(S - x_0)$, where x_0 is any vector in S.*

Proof Note that $S = S - x_0 + x_0 \subseteq x_0 + \mathrm{span}(S - x_0)$ and $x_0 + \mathrm{span}(S - x_0)$ is a linear variety. Hence, $\mathrm{v}(S) \subseteq x_0 + \mathrm{span}(S - x_0)$.

$\forall V \subseteq \mathcal{X}$ such that $S \subseteq V$ and V is a linear variety. Then, $x_0 \in V$ and $V - x_0$ is a subspace. Clearly, $S - x_0 \subseteq V - x_0$, which implies that, by Proposition 6.34, $\mathrm{span}(S - x_0) \subseteq V - x_0$. Therefore, $x_0 + \mathrm{span}(S - x_0) \subseteq V$. Hence, $x_0 + \mathrm{span}(S - x_0) \subseteq \mathrm{v}(S)$. Therefore, $\mathrm{v}(S) = x_0 + \mathrm{span}(S - x_0)$.

This completes the proof of the proposition. □

6.7 Convex Sets

Denote $\mathbb{K}$ to be either $\mathbb{R}$ or $\mathbb{C}$.

Definition 6.38 Let $(\mathcal{X}, \mathbb{K})$ be a vector space and $C \subseteq \mathcal{X}$. C is said to be *convex* if, $\forall x_1, x_2 \in C$, $\forall \alpha \in [0, 1] \subset \mathbb{R}$, we have $\alpha x_1 + (1 - \alpha)x_2 \in C$. ◇

Subspaces and linear varieties are convex, so is $\emptyset$. See Fig. 6.2 for illustration.

Proposition 6.39 *Let $(\mathcal{X}, \mathbb{K})$ be a vector space and $K, G \subseteq \mathcal{X}$ be convex sets. Then:*

1. *λK is convex, $\forall \lambda \in \mathbb{K}$.*
2. *$K + G$ is convex.*

Proof $\forall \lambda \in \mathbb{K}$. $\forall x_1, x_2 \in \lambda K$, $\forall \alpha \in [0, 1] \subset \mathbb{R}$. $\exists k_1, k_2 \in K$ such that $x_1 = \lambda k_1$ and $x_2 = \lambda k_2$. Then, $\alpha x_1 + (1 - \alpha)x_2 = \alpha\lambda k_1 + (1 - \alpha)\lambda k_2 = \lambda(\alpha k_1 + (1 - \alpha)k_2)$. Since K is convex, then $\alpha k_1 + (1 - \alpha)k_2 \in K$. This implies that $\alpha x_1 + (1 - \alpha)x_2 \in \lambda K$. Hence, λK is convex.

$\forall x_1, x_2 \in K + G$, $\forall \alpha \in [0, 1] \subset \mathbb{R}$. $\exists k_1, k_2 \in K$ and $\exists g_1, g_2 \in G$ such that $x_i = k_i + g_i$, $i = 1, 2$. Note that

$$\begin{aligned}
\alpha x_1 + (1 - \alpha)x_2 &= \alpha(k_1 + g_1) + (1 - \alpha)(k_2 + g_2) \\
&= \alpha k_1 + \alpha g_1 + (1 - \alpha)k_2 + (1 - \alpha)g_2 \\
&= (\alpha k_1 + (1 - \alpha)k_2) + (\alpha g_1 + (1 - \alpha)g_2)
\end{aligned}$$

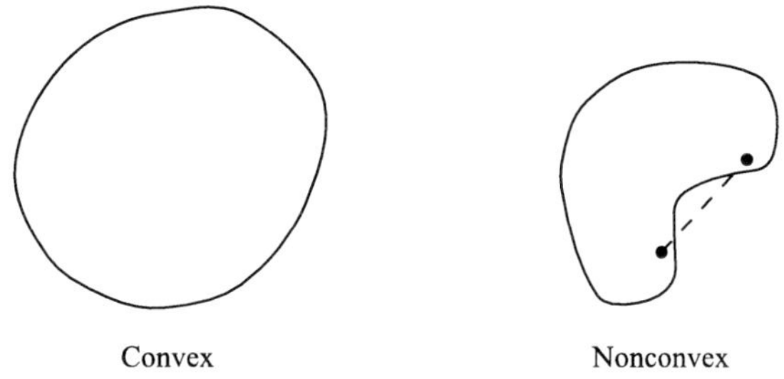

Fig. 6.2 Convex and nonconvex sets

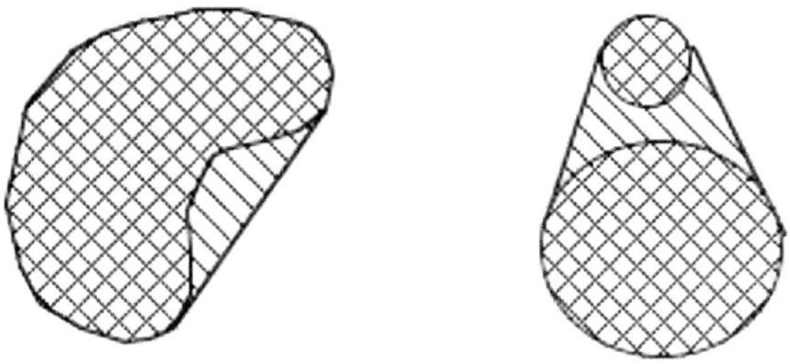

Fig. 6.3 Convex hulls

Since K and G are convex, then $(\alpha k_1 + (1-\alpha)k_2) \in K$ and $(\alpha g_1 + (1-\alpha)g_2) \in G$. This implies that $\alpha x_1 + (1-\alpha)x_2 \in K + G$. Hence, $K + G$ is convex.

This completes the proof of the proposition. □

Proposition 6.40 *Let $(\mathcal{X}, \mathbb{K})$ be a vector space and $\{C_\lambda\}_{\lambda \in \Lambda}$ be a collection of convex subsets of $\mathcal{X}$. Then, $C := \bigcap_{\lambda \in \Lambda} C_\lambda$ is convex.*

Proof $\forall x_1, x_2 \in C$, $\forall \alpha \in [0, 1] \subset \mathbb{R}$. $\forall \lambda \in \Lambda$, $x_1, x_2 \in C_\lambda$. Since C_λ is convex, then $\alpha x_1 + (1-\alpha)x_2 \in C_\lambda$. This implies that $\alpha x_1 + (1-\alpha)x_2 \in C$. Hence, C is convex. This completes the proof of the proposition. □

Definition 6.41 Let $(\mathcal{X}, \mathbb{K})$ be a vector space and $S \subseteq \mathcal{X}$. The *convex hull generated by* S, denoted by $\text{co}(S)$, is the smallest convex set containing S (Fig. 6.3). ◇

Justification of the existence of convex hull rests with Proposition 6.40.

Definition 6.42 Let $(\mathcal{X}, \mathbb{K})$ be a vector space. A *convex combination* of vectors $x_1, \ldots, x_n \in \mathcal{X}$, where $n \in \mathbb{N}$, is a linear combination $\sum_{i=1}^{n} \alpha_i x_i$ with $\alpha_i \in [0, 1] \subset \mathbb{R}$, $\forall i = 1, \ldots, n$, and $\sum_{i=1}^{n} \alpha_i = 1$. ◇

Proposition 6.43 *Let $(\mathcal{X}, \mathbb{K})$ be a vector space and $S \subseteq \mathcal{X}$. Then,*

$$\text{co}(S) = \{\textit{convex combinations of vectors in } S\}$$

Proof We need the follow result.

Claim 6.43.1 Let $G \subseteq \mathcal{X}$ be a convex subset. Then any convex combination of vectors in G belongs to G.

Proof of Claim We need to show: $\forall n \in \mathbb{N}$, $\forall x_1, \ldots, x_n \in G$, $\forall \alpha_1, \ldots, \alpha_n \in \mathbb{R}$, such that $\alpha_i \in [0, 1] \subset \mathbb{R}$, $i = 1, \ldots, n$, $\sum_{i=1}^{n} \alpha_i = 1$ implies $\sum_{i=1}^{n} \alpha_i x_i \in G$. We will prove this by mathematical induction on n:

1° Consider $n = 1$. Then, $\alpha_1 = 1$ and $\sum_{i=1}^{n} \alpha_i x_i = x_1 \in G$. The result holds.

2° Assume that the result holds for $n = k \in \mathbb{N}$.

3° Consider the case $n = k + 1$. Without loss of generality, assume $\alpha_1 > 0$. By the induction hypothesis, we have $\sum_{i=1}^{k} \frac{\alpha_i}{\alpha_1 + \cdots + \alpha_k} x_i \in G$. Then,

$$\sum_{i=1}^{k+1} \alpha_i x_i = (\alpha_1 + \cdots + \alpha_k) \sum_{i=1}^{k} \frac{\alpha_i}{\alpha_1 + \cdots + \alpha_k} x_i + \alpha_{k+1} x_{k+1}$$

$$= \alpha_{k+1} x_{k+1} + (1 - \alpha_{k+1}) \sum_{i=1}^{k} \frac{\alpha_i}{\alpha_1 + \cdots + \alpha_k} x_i$$

By the convexity of G, we have $\sum_{i=1}^{k+1} \alpha_i x_i \in G$. Hence, the result holds for $n = k + 1$.

This completes the induction process and the proof of the claim. □

$\forall x_1, x_2 \in K :=$ {convex combinations of vectors in S}, $\forall \alpha \in [0, 1] \subset \mathbb{R}$. By the definition of K, $\forall i = 1, 2$, $\exists n_i \in \mathbb{N}$, $\exists y_{i,1}, \ldots, y_{i,n_i} \in S$, $\exists \alpha_{i,1}, \ldots, \alpha_{i,n_i} \in [0, 1] \subset \mathbb{R}$, such that $\sum_{j=1}^{n_i} \alpha_{i,j} = 1$ and $x_i = \sum_{j=1}^{n_i} \alpha_{i,j} y_{i,j}$. Then,

$$\alpha x_1 + (1 - \alpha) x_2 = \alpha \sum_{j=1}^{n_1} \alpha_{1,j} y_{1,j} + (1 - \alpha) \sum_{j=1}^{n_2} \alpha_{2,j} y_{2,j}$$

$$= \sum_{j=1}^{n_1} \alpha \alpha_{1,j} y_{1,j} + \sum_{j=1}^{n_2} (1 - \alpha) \alpha_{2,j} y_{2,j}$$

Note that $\alpha \alpha_{1,j} \geq 0$, $j = 1, \ldots, n_1$, and $(1 - \alpha)\alpha_{2,j} \geq 0$, $j = 1, \ldots, n_2$, and $\sum_{j=1}^{n_1} \alpha \alpha_{1,j} + \sum_{j=1}^{n_2} (1 - \alpha) \alpha_{2,j} = \alpha \sum_{j=1}^{n_1} \alpha_{1,j} + (1 - \alpha) \sum_{j=1}^{n_2} \alpha_{2,j} = \alpha + (1 - \alpha) = 1$. Hence, $\alpha x_1 + (1 - \alpha) x_2 \in K$. This shows that K is convex. Clearly, $S \subseteq K$.

On the other hand, fix any convex set G in the vector space, satisfying $S \subseteq G$. $\forall p \in K$, by Claim 6.43.1, $p \in G$ since $S \subseteq G$. Then, $K \subseteq G$.

The above implies that K is the smallest convex set containing S. Hence, $K = \mathrm{co}\,(S)$. This completes the proof of the proposition. □

Definition 6.44 Let $(\mathcal{X}, \mathbb{K})$ be a vector space and $C \subseteq \mathcal{X}$. C is said to be a *cone with vertex at origin* if $\vartheta \in C$ and, $\forall x \in C$, $\forall \alpha \in [0, \infty) \subset \mathbb{R}$, we have $\alpha x \in C$. C is said to be a *cone with vertex* $p \in \mathcal{X}$ if $C = p + D$, where D is a cone with vertex

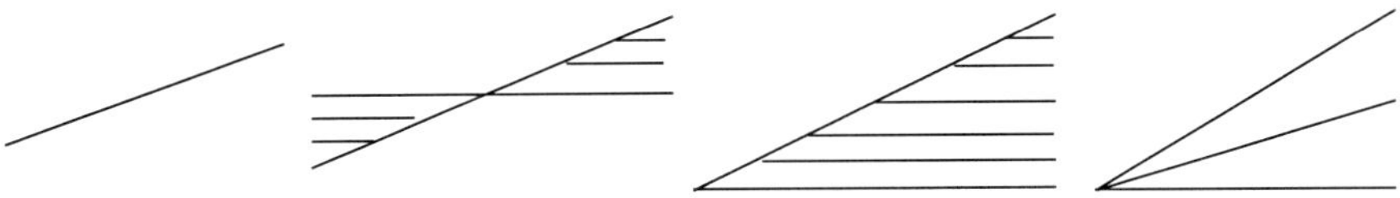

Fig. 6.4 Cones

at origin. C is said to be a *conic segment* if $\vartheta \in C$ and, $\forall x \in C$, $\forall \alpha \in [0, 1] \subset \mathbb{R}$, we have $\alpha x \in C$. ◇

If vertex is not explicitly mentioned, it is assumed to be at origin (Fig. 6.4).

Convex cones arise in connection with positive vectors. In $\mathbb{R}^n$ with $n \in \mathbb{N}$, the positive cone may be defined as

$$P = \{x = (\xi_1, \ldots, \xi_n) \in \mathbb{R}^n \mid \xi_i \geq 0,\ i = 1, \ldots, n\}$$

6.8 Linear Independence and Dimensions

Definition 6.45 Let $(\mathcal{X}, \mathcal{F})$ be a vector space, $x \in \mathcal{X}$, and $S \subseteq \mathcal{X}$. The vector x is said to be *linearly dependent upon* S if $x \in \operatorname{span}(S)$. Otherwise, x is said to be *linearly independent* of S. S is said to be a *linearly independent set* if, $\forall y \in S$, y is linearly independent of $S \setminus \{y\}$. ◇

Note that $\emptyset$ is a linearly independent set; $\{x\}$ is a linearly independent set if, and only if, $x \neq \vartheta$; and $\{x_1, x_2\}$ is a linearly independent set if, and only if, x_1 and x_2 do not lie on a common line through the origin.

Theorem 6.46 *Let $\mathcal{X}$ be a vector space over the field $\mathcal{F} := (F, +, \cdot, \mathbf{0}, \mathbf{1})$ and $S \subseteq \mathcal{X}$. Then, S is a linearly independent set if, and only if, $\forall n \in \mathbb{N}$, $\forall \alpha_1, \ldots, \alpha_n \in \mathcal{F}$, $\forall x_1, \ldots, x_n \in S$, which are distinct, we have $\sum_{i=1}^{n} \alpha_i x_i = \vartheta$, which implies that $\alpha_i = \mathbf{0}$, $i = 1, \ldots, n$.*

Proof "Sufficiency" We will prove it using an argument of contradiction. Suppose S is not a linearly independent set. Then, $\exists y \subset S$ such that y is linearly dependent upon $S \setminus \{y\}$. So, $y \in \operatorname{span}(S \setminus \{y\})$. $\exists n \in \mathbb{N}$, $\exists x_2, \ldots, x_n \in S \setminus \{y\}$, and $\exists \alpha_2, \ldots, \alpha_n \in \mathcal{F}$ such that $y = \sum_{i=2}^{n} \alpha_i x_i$ (when $n = 1$, then $y = \vartheta$). Without loss of generality, we may assume that $x_2, \ldots, x_n$ are distinct. Let $x_1 = y$ and $\alpha_1 = -\mathbf{1} \neq \mathbf{0}$. Then, we have $\sum_{i=1}^{n} \alpha_i x_i = \vartheta$ with $\alpha_1 \neq \mathbf{0}$ and $x_1, \ldots, x_n$ are distinct. This is a contradiction. Hence, the sufficiency result holds.

"Necessity" We again prove this by an argument of contradiction. Suppose the result does not hold. $\exists n \in \mathbb{N}$, $\exists \alpha_1, \ldots, \alpha_n \in \mathcal{F}$, and $\exists x_1, \ldots, x_n \in S$ which are distinct such that $\sum_{i=1}^{n} \alpha_i x_i = \vartheta$ and $\exists i_0 \in \{1, \ldots, n\}$ such that $\alpha_{i_0} \neq \mathbf{0}$. Without loss of generality, we may assume $i_0 = 1$. Then, we have $\alpha_1 x_1 = -\sum_{i=2}^{n} \alpha_i x_i$. Hence, $x_1 = \sum_{i=2}^{n} (-\alpha_1^{-1} \alpha_i) x_i$. Note that $x_i \neq x_1$ implies that $x_i \in S \setminus \{x_1\}$,

$i = 2, \ldots, n$. Hence, x_1 is linearly dependent upon $S \setminus \{x_1\}$. This is a contradiction. Then, the necessity result holds.

This completes the proof of this theorem. □

Corollary 6.47 *Let $\mathcal{X}$ be a vector space over the field $\mathcal{F} := (F, +, \cdot, \mathbf{0}, \mathbf{1})$, $\{x_1, \ldots, x_n\} \subseteq X$ be a linearly independent set, x_i's are distinct, $\alpha_1, \ldots, \alpha_n$, $\beta_1, \ldots, \beta_n \in \mathcal{F}$, and $n \in \mathbb{Z}_+$. Assume that $\sum_{i=1}^{n} \alpha_i x_i = \sum_{i=1}^{n} \beta_i x_i$. Then, $\alpha_i = \beta_i$, $i = 1, \ldots, n$.*

Proof The assumption implies $\sum_{i=1}^{n}(\alpha_i - \beta_i)x_i = \vartheta$. When $n = 0$, clearly the result holds. When $n \in \mathbb{N}$, by Theorem 6.46, we have $\alpha_i - \beta_i = \mathbf{0}$, $i = 1, \ldots, n$. This completes the proof of the corollary. □

Definition 6.48 Let $\mathcal{X}$ be a vector space over the field $\mathcal{F} := (F, +, \cdot, \mathbf{0}, \mathbf{1})$, $n \in \mathbb{Z}_+$, $x_1, \ldots, x_n \in \mathcal{X}$. The vectors $x_1, \ldots, x_n$ are *linearly independent* if, $\forall \alpha_1, \ldots, \alpha_n \in \mathcal{F}$, $\sum_{i=1}^{n} \alpha_i x_i = \vartheta$, which implies $\alpha_1 = \cdots = \alpha_n = \mathbf{0}$. Otherwise, these vectors are said to be *linearly dependent*. ◇

Lemma 6.49 *Let $\mathcal{X}$ be a vector space over the field $\mathcal{F} := (F, +, \cdot, \mathbf{0}, \mathbf{1})$. Then:*

1. *$x_1, \ldots, x_n \in \mathcal{X}$ are linearly independent, where $n \in \mathbb{Z}_+$, if, and only if, they are distinct and the set $\{x_1, \ldots, x_n\}$ is a linearly independent set.*
2. *$S \subseteq \mathcal{X}$ is a linearly independent set if, and only if, $\forall n \in \mathbb{N}$, $\forall x_1, \ldots, x_n \in S$ that are distinct, which implies that $x_1, \ldots, x_n$ are linearly independent.*

Proof 1. "Sufficiency" When $n = 0$, $x_1, \ldots, x_n$ are linearly independent and $\emptyset$ is a linearly independent set. Hence, the result holds. When $n \in \mathbb{N}$, this is straightforward from Theorem 6.46. "Necessity" We will prove this using an argument of contradiction. Suppose the result does not hold. We will distinguish two exhaustive cases: Case 1: $x_1, \ldots, x_n$ are not distinct; Case 2: the set $\{x_1, \ldots, x_n\}$ is not a linearly independent set. Case 1. Without loss of generality, assume $x_1 = x_2$. Set $\alpha_1 = \mathbf{1}$, $\alpha_2 = -\mathbf{1}$ and the rest of α_i's to $\mathbf{0}$. Then, we have $\sum_{i=1}^{n} \alpha_i x_i = \vartheta$, and hence, $x_1, \ldots, x_n$ is linearly dependent. This is a contradiction. Case 2. By Theorem 6.46, $\exists m \in \mathbb{N}$, $\exists \alpha_1, \ldots, \alpha_m \in \mathcal{F}$, which are not all $\mathbf{0}$'s, and $\exists y_1, \ldots, y_m \in \{x_1, \ldots, x_n\}$ that are distinct, such that $\sum_{i=1}^{m} \alpha_i y_i = \vartheta$. Clearly, $m \leq n$; otherwise, y_i's are not distinct. Without loss of generality, assume $y_1 = x_1, \ldots, y_m = x_m$. Set $\alpha_{m+1} = \cdots = \alpha_n = \mathbf{0}$. Then, we have $\sum_{i=1}^{n} \alpha_i x_i = \vartheta$ with $\alpha_1, \ldots, \alpha_n$ not all $\mathbf{0}$'s. This is a contradiction. Thus, we have arrived at a contradiction in every case. Hence, the necessity result holds.

2. This is straightforward from Theorem 6.46.

This completes the proof of the lemma. □

Definition 6.50 Let $(\mathcal{X}, \mathcal{F})$ be a vector space and $S \subseteq \mathcal{X}$ be a linearly independent set with $n \in \mathbb{Z}_+$ elements. S is said to be a *basis* of the vector space if span $(S) = \mathcal{X}$. In this case, the vector space is said to be *finite-dimensional* with *dimension* n. All other vector spaces are said to be *infinite-dimensional*. ◇

Theorem 6.51 *Let $\mathcal{X}$ be a finite-dimensional vector space over the field $\mathcal{F} := (F, +, \cdot, \mathbf{0}, \mathbf{1})$. Then, the dimension $n \in \{0\} \cup \mathbb{N}$ is unique. Furthermore, any linearly independent set of n vectors forms a basis of the vector space.*

Proof Let $n \in \mathbb{Z}_+$ be the minimum of dimensions for $\mathcal{X}$. Then, there exists a set $S_1 \subseteq \mathcal{X}$ with n elements such that S_1 is a basis for the vector space. We will show that $\forall y_1, \ldots, y_m \in \mathcal{X}$ with $m > n$, then $y_1, \ldots, y_m$ are linearly dependent. This implies that any subset with more than n elements cannot be a linearly independent set by Lemma 6.49. Henceforth, the dimension of the vector space is unique.

$\forall y_1, \ldots, y_m \in \mathcal{X}$ with $m > n$. There are two exhaustive and mutually exclusive cases: Case 1: $n = 0$; Case 2: $n > 0$. Case 1: $n = 0$. Then, $S_1 = \emptyset$, and $\mathcal{X}$ contains a single vector ϑ. Hence, $y_1 = \cdots = y_m = \vartheta$. Clearly, $y_1, \ldots, y_m$ are linearly dependent. This case is proven. Case 2: $n > 0$. Take $S_1 = \{x_1, \ldots, x_n\}$. Since S_1 is a basis for the vector space, then, $\forall i \in \{1, \ldots, m\}$, $y_i \in \operatorname{span}(S_1)$, that is, $\exists \alpha_{ij} \in \mathcal{F}$, $j = 1, \ldots, n$, such that $y_i = \sum_{j=1}^{n} \alpha_{ij} x_j$. Consider the $m \times n$-dimensional matrix $A := (\alpha_{ij})_{m \times n}$. Since $m > n$, then $\operatorname{rank}(A) < m$, which implies that the row vectors of A are linearly dependent. $\exists \beta_1, \ldots, \beta_m \in \mathcal{F}$, which are not all $\mathbf{0}$'s, such that $\sum_{i=1}^{m} \beta_i \alpha_{ij} = \mathbf{0}$, $j = 1, \ldots, n$. This implies $\sum_{i=1}^{m} \beta_i y_i = \sum_{i=1}^{m} \sum_{j=1}^{n} \beta_i \alpha_{ij} x_j = \sum_{j=1}^{n} (\sum_{i=1}^{m} \beta_i \alpha_{ij}) x_j = \vartheta$. Hence, $y_1, \ldots, y_m$ are linearly dependent. This case is also proven.

Hence, the dimension of the vector space is unique.

$\forall S_2 \subseteq \mathcal{X}$ with n vectors that is a linearly independent set, we will show that S_2 is a basis for the vector space. We will distinguish two exhaustive and mutually exclusive cases: Case 1: $n = 0$; Case 2: $n > 0$. Case 1: $n = 0$. Then, $S_2 = \emptyset$, and $\mathcal{X}$ is a singleton set consisting of the null vector, which equals to $\operatorname{span}(S_2)$. Hence, S_2 is a basis for the vector space. Case 2: $n > 0$. Take $S_2 = \{z_1, \ldots, z_n\}$. $\forall x \in \mathcal{X}$, $x, z_1, \ldots, z_n$ (for a total of $n + 1$ vectors), by the preceding proof, are linearly dependent. Then, $\exists \alpha, \beta_1, \ldots, \beta_n \in \mathcal{F}$, which are not all $\mathbf{0}$'s, such that $\alpha x + \sum_{i=1}^{n} \beta_i z_i = \vartheta$. Suppose $\alpha = \mathbf{0}$. Then, $\sum_{i=1}^{n} \beta_i z_i = \vartheta$. Since S_2 is a linearly independent set, by Theorem 6.46, then $\beta_1 = \cdots = \beta_n = \mathbf{0}$. This contradicts with the fact that $\alpha, \beta_1, \ldots, \beta_n$ are not all $\mathbf{0}$'s. Hence, $\alpha \neq \mathbf{0}$. Then, $x = \sum_{i=1}^{n} (-\alpha^{-1} \beta_i) z_i$ and $x \in \operatorname{span}(S_2)$. Therefore, we have $\mathcal{X} = \operatorname{span}(S_2)$. Hence, S_2 is a basis of the vector space.

This completes the proof of the theorem. □

Finite-dimensional spaces are simpler to analyze. Many results of finite-dimensional spaces may be generalized to infinite-dimensional spaces. We endeavor to stress the similarity between the finite- and infinite-dimensional spaces.

Chapter 7
Banach Spaces

7.1 Normed Linear Spaces

Vector spaces admit algebraic properties, but they lack topological properties.

Denote $\mathbb{K}$ to be either $\mathbb{R}$ or $\mathbb{C}$.

Definition 7.1 Let $(\mathcal{X}, \mathbb{K})$ be a vector space. A *norm* on the vector space is a real-valued function $\|\cdot\| : \mathcal{X} \to [0, \infty) \subset \mathbb{R}$ that satisfies the following properties: $\forall x, y \in \mathcal{X}$, $\forall \alpha \in \mathbb{K}$,

(i) $0 \leq \|x\| < +\infty$ and $\|x\| = 0 \Leftrightarrow x = \vartheta$.
(ii) $\|x + y\| \leq \|x\| + \|y\|$. (triangle inequality)
(iii) $\|\alpha x\| = |\alpha| \|x\|$.

A *real (complex) normed linear space* is a vector space over the filed $\mathbb{R}$ (or $\mathbb{C}$) together with a norm defined on it. A normed linear space consisting of the triple $(\mathcal{X}, \mathbb{K}, \|\cdot\|)$. ◇

To simplify notation in the theory, we may later simply discuss a normed linear space $\mathfrak{X} := (\mathcal{X}, \mathbb{K}, \|\cdot\|)$ without further reference to components of $\mathfrak{X}$, where the operations are understood to be $\oplus_{\mathfrak{X}}$ and $\otimes_{\mathfrak{X}}$, the null vector is understood to be $\vartheta_{\mathfrak{X}}$, and the norm is understood to be $\|\cdot\|_{\mathfrak{X}}$. When it is clear from the context, we will neglect the subscript $\mathfrak{X}$.

Example 7.2 Let $n \in \mathbb{Z}_+$. The Euclidean space $(\mathbb{R}^n, \mathbb{R}, |\cdot|)$ is a normed linear space, where the norm is defined by $|(\xi_1, \ldots, \xi_n)| = \sqrt{\xi_1^2 + \cdots + \xi_n^2}$, $\forall (\xi_1, \ldots, \xi_n) \in \mathbb{R}^n$. Note that we specifically denote the Euclidean norm as $|\cdot|$, rather than $\|\cdot\|$, to distinguish it from other norms on $\mathbb{R}^n$. ◇

Example 7.3 Let $n \in \mathbb{Z}_+$. The space $(\mathbb{C}^n, \mathbb{C}, |\cdot|)$ is a normed linear space, where the norm is defined by $|(\xi_1, \ldots, \xi_n)| = \sqrt{\sum_{i=1}^n |\xi_i|^2}$, $\forall (\xi_1, \ldots, \xi_n) \in \mathbb{C}^n$. Note that

Z. Pan, *Measure-Theoretic Calculus in Abstract Spaces*,
https://doi.org/10.1007/978-3-031-21912-2_7

we specifically denote the norm as $|\cdot|$, rather than $\|\cdot\|$, to distinguish it from other norms on $\mathbb{C}^n$. ◇

Example 7.4 Let $X := \{f : [a, b] \rightarrow \mathbb{R}\}$, where $a, b \in \mathbb{R}$ with $a \leq b$, $\oplus$ and $\otimes$ be the usual vector addition and scalar multiplication, and $\vartheta : [a, b] \rightarrow \mathbb{R}$ be given by $\vartheta(t) = 0$, $\forall t \in [a, b]$. By Example 6.20, $\mathcal{X} := (X, \oplus, \otimes, \vartheta)$ is a vector space over $\mathbb{R}$. Let $M := \{f \in \mathcal{X} \mid f \text{ is continuous}\}$ is a subspace by Proposition 6.25. Let $\mathcal{M} := (M, \oplus, \otimes, \vartheta)$. Then, $(\mathcal{M}, \mathbb{R})$ is a vector space. Introduce a norm on this space by $\|x\| = \max_{a \leq t \leq b} |x(t)|$, $\forall x \in \mathcal{M}$. Now, we verify the properties of the norm. $\forall x, y \in \mathcal{M}$, $\forall \alpha \in \mathbb{R}$.

(i) Since $[a, b]$ is a nonempty compact set and x is continuous, then, $\|x\| = |x(t)| \in [0, \infty) \subset \mathbb{R}$, for some $t \in [a, b]$. Clearly, $\|x\| = 0 \Leftrightarrow x = \vartheta$.

(ii) $\|x + y\| = \max\limits_{a \leq t \leq b} |x(t) + y(t)| \leq \max\limits_{a \leq t \leq b} (|x(t)| + |y(t)|) \leq \max\limits_{a \leq t \leq b} |x(t)| + \max\limits_{a \leq t \leq b} |y(t)| = \|x\| + \|y\|$.

(iii) $\|\alpha x\| = \max\limits_{a \leq t \leq b} |\alpha x(t)| = \max\limits_{a \leq t \leq b} |\alpha||x(t)| = |\alpha| \max\limits_{a \leq t \leq b} |x(t)| = |\alpha|\|x\|$, where have made use of Proposition 3.81 and the fact that $\|x\| \in \mathbb{R}$ in the third equality.

Hence, $\mathcal{C}([a, b]) := (\mathcal{M}, \mathbb{R}, \|\cdot\|)$ is a normed linear space. ◇

Example 7.5 Let $X := \{f : [a, b] \rightarrow \mathbb{R}\}$, where $a, b \in \mathbb{R}$ with $a < b$, $\oplus$ and $\otimes$ be the usual vector addition and scalar multiplication, and $\vartheta : [a, b] \rightarrow \mathbb{R}$ be given by $\vartheta(t) = 0$, $\forall t \in [a, b]$. By Example 6.20, $\mathcal{X} := (X, \oplus, \otimes, \vartheta)$ is a vector space over $\mathbb{R}$. Let $M := \{f \in \mathcal{X} \mid f \text{ is continuously differentiable}\}$ is a subspace by Proposition 6.25. Let $\mathcal{M} := (M, \oplus, \otimes, \vartheta)$. Then, $(\mathcal{M}, \mathbb{R})$ is a vector space. Introduce a norm on this space by $\|x\| = \max_{a \leq t \leq b}|x(t)| + \max_{a \leq t \leq b}\left|x^{(1)}(t)\right|$, $\forall x \in \mathcal{M}$. Now, we verify the properties of the norm. $\forall x, y \in \mathcal{M}$, $\forall \alpha \in \mathbb{R}$.

(i) Since $[a, b]$ is a nonempty compact set and x is continuously differentiable, then, $\|x\| = |x(t_1)| + \left|x^{(1)}(t_2)\right| \in [0, \infty) \subset \mathbb{R}$, for some $t_1, t_2 \in [a, b]$. Clearly, $\|x\| = 0 \Leftrightarrow x = \vartheta$.

(ii) $\|x + y\| = \max\limits_{a \leq t \leq b} |x(t) + y(t)| + \max\limits_{a \leq t \leq b} |x^{(1)}(t) + y^{(1)}(t)| \leq \max\limits_{a \leq t \leq b} |x(t)| + \max\limits_{a \leq t \leq b} |y(t)| + \max\limits_{a \leq t \leq b} |x^{(1)}(t)| + \max\limits_{a \leq t \leq b} |y^{(1)}(t)| = \|x\| + \|y\|$.

(iii) $\|\alpha x\| = \max_{a \leq t \leq b} |\alpha x(t)| + \max_{a \leq t \leq b} \left|\alpha x^{(1)}(t)\right| = |\alpha|\|x\|$.

Hence, $\mathcal{C}_1([a, b]) := (\mathcal{M}, \mathbb{R}, \|\cdot\|)$ is a normed linear space. ◇

Example 7.6 Let $X := \{(\xi_k)_{k=1}^{\infty} \mid \xi_k \in \mathbb{R}, \ \forall k \in \mathbb{N}\}$, $\oplus$ and $\otimes$ be the usual vector addition and scalar multiplication, and $\vartheta := (0, 0, \ldots) \in X$. By Example 6.23, $\mathcal{X} := (X, \oplus, \otimes, \vartheta)$ is a vector space over $\mathbb{R}$. For $p \in [1, \infty) \subset \mathbb{R}$, let $M_p := \{(\xi_k)_{k=1}^{\infty} \in X \mid \sum_{k=1}^{\infty} |\xi_k|^p < +\infty\}$. Define the norm $\|x\|_p = \left(\sum_{k=1}^{\infty} |\xi_k|^p\right)^{1/p}$, $\forall x = (\xi_k)_{k=1}^{\infty} \in M_p$. Let $M_\infty := \{(\xi_k)_{k=1}^{\infty} \in X \mid \sup_{k \geq 1} |\xi_k| < +\infty\}$. Define the norm $\|x\|_\infty = \sup_{k \geq 1} |\xi_k|$, $\forall x = (\xi_k)_{k=1}^{\infty} \in M_\infty$. $\forall p \in [1, +\infty] \subset \mathbb{R}_e$, we will show that $\mathcal{M}_p := (M_p, \oplus, \otimes, \vartheta)$ is a subspace in $(\mathcal{X}, \mathbb{R})$ and $\mathrm{l}_p := (\mathcal{M}_p, \mathbb{R}, \|\cdot\|_p)$ is a normed linear space.

$\forall p \in [1, +\infty] \subset \mathbb{R}_e$. $\vartheta \in \mathrm{l}_p \neq \emptyset$. $\forall x, y \in \mathcal{M}_p$, $\forall \alpha, \beta \in \mathbb{R}$. It is easy to check that $\|\alpha x\|_p = |\alpha| \|x\|_p$. Then, by Theorem 7.9, $\|\alpha x + \beta y\|_p \leq \|\alpha x\|_p + \|\beta y\|_p = |\alpha| \|x\|_p + |\beta| \|y\|_p < +\infty$. This implies that $\alpha x + \beta y \subset \mathcal{M}_p$. Hence, $\mathcal{M}_p$ is a subspace of $(\mathcal{X}, \mathbb{R})$. It is easy to check that $\|x\|_p \in [0, \infty) \subset \mathbb{R}$, $\forall x \in \mathrm{l}_p$, and $\|x\|_p = 0 \Leftrightarrow x = \vartheta$. Therefore, l_p is a normed linear space. ◇

Lemma 7.7 *$\forall a, b \in [0, \infty) \subset \mathbb{R}$, $\forall \lambda \in (0, 1) \subset \mathbb{R}$, we have*

$$a^\lambda b^{1-\lambda} \leq \lambda a + (1 - \lambda) b$$

where equality holds if, and only if, $a = b$.

Proof Define $f : [0, \infty) \to \mathbb{R}$ by $f(t) = t^\lambda - \lambda t + \lambda - 1$, $\forall t \in [0, \infty) \subset \mathbb{R}$. Then, f is continuous, and is differentiable on $(0, \infty)$. $f^{(1)}(t) = \lambda t^{\lambda - 1} - \lambda$, $\forall t \in (0, \infty)$. Then, $f^{(1)}(t) > 0$, $\forall t \in (0, 1)$ and $f^{(1)}(t) < 0$, $\forall t \in (1, \infty)$. Then, $f(t) \leq f(1) = 0$, $\forall t \in [0, \infty)$, where equality holds if, and only if, $t = 1$.

We will distinguish two exhaustive and mutually exclusive cases: Case 1: $b = 0$; Case 2: $b > 0$. Case 1: $b = 0$. We have $a^\lambda b^{1-\lambda} = 0 \leq \lambda a = \lambda a + (1 - \lambda) b$, where equality holds if, and only if, $a = b = 0$. This case is proved. Case 2: $b > 0$. Since $f(a/b) \leq 0$, we immediately obtain the desired inequality, where equality holds if, and only if, $a/b = 1 \Leftrightarrow a = b$. This case is also proved.

This completes the proof of the lemma. □

Theorem 7.8 (Hölder's Inequality) *Let $p \in [1, +\infty) \subset \mathbb{R}$ and $q \in (1, +\infty] \subset \mathbb{R}_e$ with $1/p + 1/q = 1$. Then, $\forall x = (\xi_k)_{k=1}^\infty \in \mathrm{l}_p$, $\forall y = (\eta_k)_{k=1}^\infty \in \mathrm{l}_q$, we have*

$$\sum_{k=1}^\infty |\xi_k \eta_k| \leq \|x\|_p \|y\|_q$$

When $q < \infty$, equality holds if, and only if, $\exists \alpha, \beta \in \mathbb{R}$, which are not both zeros, such that $\alpha |\xi_k|^p = \beta |\eta_k|^q$, $k = 1, 2, \ldots$. When $q = \infty$, equality holds if, and only if, $|\eta_k| = \|y\|_\infty$ for any $k \in \mathbb{N}$ with $|\xi_k| > 0$.

Proof We will distinguish two exhaustive and mutually exclusive cases: Case 1: $q = \infty$; Case 2: $1 < q < +\infty$. Case 1: $q = \infty$. Then, $p = 1$. We have

$$\sum_{k=1}^\infty |\xi_k \eta_k| = \sum_{k=1}^\infty |\xi_k| |\eta_k| \leq \sum_{k=1}^\infty |\xi_k| \|y\|_\infty = \|x\|_1 \|y\|_\infty$$

where equality holds if, and only if, $|\eta_k| = \|y\|_\infty$ for any $k \in \mathbb{N}$ with $|\xi_k| > 0$. This case is proved.

Case 2: $1 < q < +\infty$. Then, $1 < p < +\infty$. We will further distinguish two exhaustive and mutually exclusive cases: Case 2a: $\|x\|_p \|y\|_q = 0$; Case 2b: $\|x\|_p \|y\|_q > 0$. Case 2a: $\|x\|_p \|y\|_q = 0$. Without loss of generality, assume $\|y\|_q = 0$. Then, $y = \vartheta$ and

$$\sum_{k=1}^{\infty} |\xi_k \eta_k| = 0 = \|x\|_p \|y\|_q$$

Equality holds $\Leftrightarrow \alpha = 0$ and $\beta = 1$ and $\alpha|\xi_k|^p = \beta|\eta_k|^q$, $k = 1, 2, \ldots$. This subcase is proved. Case 2b: $\|x\|_p \|y\|_q > 0$. Then, $\|x\|_p > 0$ and $\|y\|_q > 0$. $\forall k \in \mathbb{N}$, by Lemma 7.7 with $a = \left(\frac{|\xi_k|}{\|x\|_p}\right)^p$, $b = \left(\frac{|\eta_k|}{\|y\|_q}\right)^q$, and $\lambda = 1/p$, we have

$$\frac{|\xi_k \eta_k|}{\|x\|_p \, \|y\|_q} \leq \frac{1}{p} \frac{|\xi_k|^p}{\|x\|_p^p} + \frac{1}{q} \frac{|\eta_k|^q}{\|y\|_q^q}$$

with equality if, and only if, $\frac{|\xi_k|^p}{\|x\|_p^p} = \frac{|\eta_k|^q}{\|y\|_q^q}$. Summing the above inequality for all $k \in \mathbb{N}$, we have

$$\frac{\sum_{k=1}^{\infty} |\xi_k \eta_k|}{\|x\|_p \, \|y\|_q} \leq \frac{1}{p} + \frac{1}{q} = 1$$

with equality if, and only if, $\frac{|\xi_k|^p}{\|x\|_p^p} = \frac{|\eta_k|^q}{\|y\|_q^q}$, $k = 1, 2, \ldots$. Equality $\Rightarrow \alpha = 1/\|x\|_p^p$ and $\beta = 1/\|y\|_q^q$ and $\alpha|\xi_k|^p = \beta|\eta_k|^q$, $k = 1, 2, \ldots$. On the other hand, if $\exists \alpha, \beta \in \mathbb{R}$, which are not both zeros, such that $\alpha|\xi_k|^p = \beta|\eta_k|^q$, $k = 1, 2, \ldots$, then, without loss of generality, assume $\beta \neq 0$. Let $\alpha_1 = \alpha/\beta$. Then, $\alpha_1|\xi_k|^p = |\eta_k|^q$, $k = 1, 2, \ldots$. Hence, $\alpha_1 \|x\|_p^p = \|y\|_q^q$, which further implies that $\alpha_1 = \|y\|_q^q / \|x\|_p^p$. Hence, $\frac{|\xi_k|^p}{\|x\|_p^p} = \frac{|\eta_k|^q}{\|y\|_q^q}$, $k = 1, 2, \ldots$. This implies equality. Therefore, equality if, and only if, $\exists \alpha, \beta \in \mathbb{R}$, which are not both zeros, such that $\alpha|\xi_k|^p = \beta|\eta_k|^q$, $k = 1, 2, \ldots$. This subcase is proved.

This completes the proof of the theorem. □

When $p = 2 = q$, the Hölder's inequality becomes the well-known *Cauchy–Schwarz Inequality*:

$$\sum_{k=1}^{\infty} |\xi_k \eta_k| \leq \left(\sum_{k=1}^{\infty} |\xi_k|^2\right)^{1/2} \left(\sum_{k=1}^{\infty} |\eta_k|^2\right)^{1/2}$$

Theorem 7.9 (Minkowski's Inequality) *Let $p \in [1, \infty] \subset \mathbb{R}_e$. $\forall x, y \in \mathrm{l}_p$, then, $x + y \in \mathrm{l}_p$ and*

$$\|x + y\|_p \leq \|x\|_p + \|y\|_p$$

When $1 < p < \infty$, equality holds if, and only if, $\exists \alpha, \beta \in [0, \infty) \subset \mathbb{R}$, which are not both zeros, such that $\alpha x = \beta y$.

Proof $\forall x = (\xi_k)_{k=1}^{\infty}, y = (\eta_k)_{k=1}^{\infty} \in \mathrm{l}_p$. We will distinguish three exhaustive and mutually exclusive cases: Case 1: $p = 1$; Case 2: $p = \infty$; Case 3: $1 < p < \infty$.

Case 1: $p = 1$. $\|x + y\|_1 = \sum_{k=1}^{\infty} |\xi_k + \eta_k| \leq \sum_{k=1}^{\infty} (|\xi_k| + |\eta_k|) = \|x\|_1 + \|y\|_1 < +\infty$. Equality holds if, and only if, $\xi_k \eta_k \geq 0$, $k = 1, 2, \ldots$. Hence, $x + y \in \mathrm{l}_1$.

Case 2: $p = \infty$. $\|x + y\|_\infty = \sup_{k \in \mathbb{N}} |\xi_k + \eta_k| \leq \sup_{k \in \mathbb{N}} (|\xi_k| + |\eta_k|) \leq \sup_{k \in \mathbb{N}} |\xi_k| + \sup_{k \in \mathbb{N}} |\eta_k| = \|x\|_\infty + \|y\|_\infty < +\infty$. Hence, $x + y \in \mathrm{l}_\infty$.

Case 3: $1 < p < \infty$. We will further distinguish two exhaustive and mutually exclusive cases: Case 3a: $\|x\|_p \|y\|_p = 0$; Case 3b: $\|x\|_p \|y\|_p > 0$. Case 3a: $\|x\|_p \|y\|_p = 0$. Without loss of generality, assume $\|y\|_p = 0$. Then, $y = \vartheta$ and $\eta_k = 0, k = 1, 2, \ldots$. Hence, $x + y = x \in \mathrm{l}_p$ and $\|x + y\|_p = \|x\|_p = \|x\|_p + \|y\|_p$. Let $\alpha = 0$ and $\beta = 1$, we have $\alpha x = \beta y$. Hence, equality holds if, and only if, $\exists \alpha, \beta \in [0, \infty) \subset \mathbb{R}$, which are not both zeros, such that $\alpha x = \beta y$. This subcase is proved.

Case 3b: $\|x\|_p \|y\|_p > 0$. Then, $\|x\|_p > 0$ and $\|y\|_p > 0$. Let $\lambda = \frac{\|x\|_p}{\|x\|_p + \|y\|_p} \in (0, 1) \subset \mathbb{R}$. $\forall k \in \mathbb{N}$.

$$\frac{|\xi_k + \eta_k|^p}{(\|x\|_p + \|y\|_p)^p} \leq \left(\frac{|\xi_k| + |\eta_k|}{\|x\|_p + \|y\|_p} \right)^p = \left(\lambda \frac{|\xi_k|}{\|x\|_p} + (1 - \lambda) \frac{|\eta_k|}{\|y\|_p} \right)^p$$

Since $1 < p < \infty$, then the function t^p is strictly convex on $[0, \infty) \subset \mathbb{R}$. Then, we have

$$\frac{|\xi_k + \eta_k|^p}{(\|x\|_p + \|y\|_p)^p} \leq \lambda \frac{|\xi_k|^p}{\|x\|_p^p} + (1 - \lambda) \frac{|\eta_k|^p}{\|y\|_p^p}$$

where equality holds $\Leftrightarrow \xi_k \eta_k \geq 0$ and $\frac{|\xi_k|}{\|x\|_p} = \frac{|\eta_k|}{\|y\|_p} \Leftrightarrow \frac{\xi_k}{\|x\|_p} = \frac{\eta_k}{\|y\|_p}$.

Summing the above inequalities for all $k \in \mathbb{N}$, we have

$$\frac{\|x+y\|_p^p}{(\|x\|_p+\|y\|_p)^p} = \sum_{k=1}^{\infty} \frac{|\xi_k+\eta_k|^p}{(\|x\|_p+\|y\|_p)^p}$$
$$\leq \lambda \frac{\sum_{k=1}^{\infty}|\xi_k|^p}{\|x\|_p^p} + (1-\lambda)\frac{\sum_{k=1}^{\infty}|\eta_k|^p}{\|y\|_p^p} = 1$$

Therefore,

$$\|x+y\|_p \leq \|x\|_p + \|y\|_p < +\infty$$

where equality holds $\Leftrightarrow \dfrac{\xi_k}{\|x\|_p} = \dfrac{\eta_k}{\|y\|_p}$, $k = 1, 2, \ldots \Leftrightarrow (1/\|x\|_p)x = (1/\|y\|_p)y$. Equality implies that $\alpha = 1/\|x\|_p$ and $\beta = 1/\|y\|_p$ and $\alpha x = \beta y$. On the other hand, if $\exists \alpha, \beta \in [0, \infty) \subset \mathbb{R}$, which are not both zeros, such that $\alpha x = \beta y$, then, without loss of generality, assume $\beta \neq 0$. Then, $y = \alpha_1 x$ with $\alpha_1 = \alpha/\beta$. Easy to show that $\|y\|_p = \alpha_1\|x\|_p$. Then, $\alpha_1 = \|y\|_p/\|x\|_p$. This implies that $(1/\|x\|_p)x = (1/\|y\|_p)y$, which further implies equality. Hence, equality holds if, and only if, $\exists \alpha, \beta \in [0, \infty)$, which are not both zeros, such that $\alpha x = \beta y$. Clearly, $x + y \in \mathbf{l}_p$. This subcase is proved.

This completes the proof of the theorem. □

Example 7.10 Let $\mathcal{X}$ be a real (complex) normed linear space. Let $Y := \{(\xi_k)_{k=1}^{\infty} \mid \xi_k \in \mathcal{X}, \forall k \in \mathbb{N}\}$. By Example 6.20, $(Y, \oplus, \otimes, \vartheta)$ is a vector space over $\mathbb{K}$, where $\oplus$, $\otimes$, and ϑ are defined as in the example. For $p \in [1, \infty) \subset \mathbb{R}$, let $M_p := \{(\xi_k)_{k=1}^{\infty} \in Y \mid \sum_{k=1}^{\infty}\|\xi_k\|_{\mathcal{X}}^p < +\infty\}$. Define the norm $\|x\|_p = \left(\sum_{k=1}^{\infty}\|\xi_k\|_{\mathcal{X}}^p\right)^{1/p}$, $\forall x = (\xi_k)_{k=1}^{\infty} \in M_p$. Let $M_\infty := \{(\xi_k)_{k=1}^{\infty} \in Y \mid \sup_{k\geq 1}\|\xi_k\|_{\mathcal{X}} < +\infty\}$. Define the norm $\|x\|_\infty = \sup_{k\geq 1}\|\xi_k\|_{\mathcal{X}}$, $\forall x = (\xi_k)_{k=1}^{\infty} \in M_\infty$.

$\forall p \in [1, \infty] \subset \mathbb{R}_e$, $\forall x = (\xi_k)_{k=1}^{\infty}$, $y = (\eta_k)_{k=1}^{\infty} \in M_p$.

$$\|x+y\|_p$$
$$= \begin{cases} \left(\sum_{k=1}^{\infty}\|\xi_k+\eta_k\|_{\mathcal{X}}^p\right)^{1/p} \leq \left(\sum_{k=1}^{\infty}(\|\xi_k\|_{\mathcal{X}}+\|\eta_k\|_{\mathcal{X}})^p\right)^{1/p} & p \in [1,\infty) \\ \sup_{k\geq 1}\|\xi_k+\eta_k\|_{\mathcal{X}} \leq \sup_{k\geq 1}(\|\xi_k\|_{\mathcal{X}}+\|\eta_k\|_{\mathcal{X}}) & p = \infty \end{cases}$$
$$\leq \begin{cases} \left(\sum_{k=1}^{\infty}\|\xi_k\|_{\mathcal{X}}^p\right)^{1/p} + \left(\sum_{k=1}^{\infty}\|\eta_k\|_{\mathcal{X}}^p\right)^{1/p} & p \in [1,\infty) \\ \sup_{k\geq 1}\|\xi_k\|_{\mathcal{X}} + \sup_{k\geq 1}\|\eta_k\|_{\mathcal{X}} & p = \infty \end{cases}$$
$$= \|x\|_p + \|y\|_p < +\infty$$

where we have made use of the Minkowski's Inequality. In the preceding inequality, when $1 < p < \infty$, equality holds if, and only if, $\|\xi_k + \eta_k\|_{\mathcal{X}} = \|\xi_k\|_{\mathcal{X}} + \|\eta_k\|_{\mathcal{X}}$, $\forall k \in \mathbb{N}$, and $\exists \alpha, \beta \in [0, \infty) \subset \mathbb{R}$, which are not both zeros, such that $\alpha\|\xi_k\|_{\mathcal{X}} = \beta\|\eta_k\|_{\mathcal{X}}$, $\forall k \in \mathbb{N}$.

$\forall p \in [1, \infty] \subset \mathbb{R}_e$. Note that $\vartheta = (\vartheta_{\mathcal{X}}, \vartheta_{\mathcal{X}}, \ldots) \in M_p \neq \emptyset$. $\forall x = (\xi_k)_{k=1}^{\infty} \in M_p$, $\forall \alpha \in \mathbb{K}$, we have

$$\|\alpha x\|_p = \begin{cases} \left(\sum_{k=1}^{\infty} \|\alpha \xi_k\|_{\mathcal{X}}^p\right)^{1/p} = \left(\sum_{k=1}^{\infty} |\alpha|^p \|\xi_k\|_{\mathcal{X}}^p\right)^{1/p} & p \in [1, \infty) \\ \sup_{k \geq 1} \|\alpha \xi_k\|_{\mathcal{X}} = \sup_{k \geq 1} |\alpha| \|\xi_k\|_{\mathcal{X}} & p = \infty \end{cases}$$

$$= \begin{cases} |\alpha| \left(\sum_{k=1}^{\infty} \|\xi_k\|_{\mathcal{X}}^p\right)^{1/p} & p \in [1, \infty) \\ |\alpha| \sup_{k \geq 1} \|\xi_k\|_{\mathcal{X}} & p = \infty \end{cases} = |\alpha| \|x\|_p < +\infty$$

where we have made use of Proposition 3.81. Then, $\forall x, y \in M_p$, $\forall \alpha, \beta \in \mathbb{K}$, we have $\|\alpha x + \beta y\|_p \leq \|\alpha x\|_p + \|\beta y\|_p = |\alpha|\|x\|_p + |\beta|\|y\|_p < +\infty$. Then, $\alpha x + \beta y \in M_p$. Hence, $\mathcal{M}_p := (M_p, \oplus, \otimes, \vartheta)$ is a subspace in $(Y, \oplus, \otimes, \vartheta)$. It is easy to check that, $\forall x \in \mathcal{M}_p$, $\|x\|_p \in [0, \infty) \subset \mathbb{R}$, and $\|x\|_p = 0 \Leftrightarrow x = (\vartheta_{\mathcal{X}}, \vartheta_{\mathcal{X}}, \ldots) = \vartheta$. Therefore, $\mathbf{l}_p(\mathcal{X}) := (\mathcal{M}_p, \mathbb{K}, \|\cdot\|_p)$ is a real (complex) normed linear space, $\forall p \in [1, \infty] \subset \mathbb{R}_e$. ◇

Example 7.11 Let $X := \{f : [a, b] \to \mathbb{R}\}$, where $a, b \in \mathbb{R}$ with $a < b$, $\oplus$ and $\otimes$ be the usual vector addition and scalar multiplication, and $\vartheta : [a, b] \to \mathbb{R}$ be given by $\vartheta(t) = 0$, $\forall t \in [a, b]$. By Example 6.20, $\mathcal{X} := (X, \oplus, \otimes, \vartheta)$ is a vector space over $\mathbb{R}$. Let $M := \{f \in \mathcal{X} \mid f \text{ is continuous}\}$ is a subspace by Proposition 6.25. Let $\mathcal{M} := (M, \oplus, \otimes, \vartheta)$. Then, $(\mathcal{M}, \mathbb{R})$ is a vector space. Introduce a norm on this space by $\|x\| = \int_a^b |x(t)|\, dt$, $\forall x \in \mathcal{M}$. Now, we verify the properties of the norm. $\forall x, y \in \mathcal{M}$, $\forall \alpha \in \mathbb{R}$.

(i) Since x is continuous, then, $|x(\cdot)|$ is continuous and therefore integrable on $[a, b]$. Hence, $0 \leq \|x\| < +\infty$. $x = \vartheta \Rightarrow \|x\| = 0$. On the other hand, $x \neq \vartheta \Rightarrow \exists t_1 \in [a, b]$ such that $|x(t_1)| > 0$. By continuity of x, $\exists \delta \in (0, \infty) \subset \mathbb{R}$ such that $|x(t)| > |x(t_1)|/2$, $\forall t \in [t_1 - \delta, t_1 + \delta] \cap [a, b] \Rightarrow \|x\| > 0$. Hence, $\|x\| = 0 \Leftrightarrow x = \vartheta$.

(ii) $\|x + y\| = \int_a^b |x(t) + y(t)|\, dt \leq \int_a^b (|x(t)| + |y(t)|)\, dt = \|x\| + \|y\|$.

(iii) $\|\alpha x\| = \int_a^b |\alpha x(t)|\, dt = |\alpha| \|x\|$.

Hence, $(\mathcal{M}, \mathbb{R}, \|\cdot\|)$ is a normed linear space. ◇

Definition 7.12 Let X be a set, $\mathcal{Y}$ be a normed linear space, $f : X \to \mathcal{Y}$. f is said to be *bounded* if $\exists M \in [0, \infty) \subset \mathbb{R}$ such that $\|f(x)\| \leq M$, $\forall x \in X$. $S \subseteq \mathcal{Y}$ is said to be *bounded* if $\exists M \in [0, \infty) \subset \mathbb{R}$ such that $\|s\| \leq M$, $\forall s \in S$. ◇

Proposition 7.13 *Let* $\mathcal{X} := (\mathcal{X}, \mathbb{K}, \|\cdot\|)$ *be a normed linear space and* $M \subseteq \mathcal{X}$ *be a subspace. Then,* $(M, \mathbb{K}, \|\cdot\|)$ *is also a normed linear space.*

Proof This is straightforward and is therefore omitted. □

7.2 The Natural Metric

A normed linear space is actually a metric space.

Proposition 7.14 *A normed linear space* $\mathcal{X} := (\mathcal{X}, \mathbb{K}, \|\cdot\|)$ *admits the natural metric* $\rho : \mathcal{X} \times \mathcal{X} \to [0, \infty) \subset \mathbb{R}$ *given by* $\rho(x, y) = \|x - y\|$, $\forall x, y \in \mathcal{X}$.

Proof $\forall x, y, z \in \mathcal{X}$, we have (i) $\rho(x, y) \in [0, \infty)$; (ii) $\rho(x, y) = \|x - y\| = 0 \Leftrightarrow x - y = \vartheta \Leftrightarrow x = y$; (iii) $\rho(x, y) = \|x - y\| = \|(-1)(x - y)\| = \|y - x\| = \rho(y, x)$; and (iv) Note that

$$\begin{aligned}\rho(x, z) &= \|x - z\| = \|(x - y) + (y - z)\| \le \|x - y\| + \|y - z\| \\ &= \rho(x, y) + \rho(y, z)\end{aligned}$$

Hence, ρ is a metric. This completes the proof of the proposition. □

Now, we can talk about topological properties and metric properties of a normed linear space. When we refer to these properties of a normed linear space, we are referring to the above metric specifically.

Proposition 7.15 *Let* $\mathcal{X}$ *be a normed linear space and* $C \subseteq \mathcal{X}$ *be convex. Then,* $\overline{C}$ *and* C° *are convex.*

Proof $\forall x_1, x_2 \in \overline{C}, \forall \alpha \in [0, 1] \subset \mathbb{R}$, we need to show that $\alpha x_1 + (1-\alpha)x_2 \in \overline{C}$. By Proposition 3.3, $\forall r \in (0, \infty) \subset \mathbb{R}$, $\mathcal{B}(x_i, r) \cap C \neq \emptyset$, $i = 1, 2$. Let $p_i \in \mathcal{B}(x_i, r) \cap C$, $i = 1, 2$. Then, by convexity of C, we have $\alpha p_1 + (1 - \alpha)p_2 \in C$. Note that

$$\begin{aligned}&\|(\alpha x_1 + (1 - \alpha)x_2) - (\alpha p_1 + (1 - \alpha)p_2)\| \\ &= \|\alpha(x_1 - p_1) + (1 - \alpha)(x_2 - p_2)\| \\ &\le \alpha\|x_1 - p_1\| + (1 - \alpha)\|x_2 - p_2\| < r\end{aligned}$$

Then, $\alpha p_1 + (1-\alpha)p_2 \in \mathcal{B}(\alpha x_1 + (1 - \alpha)x_2, r) \cap C \neq \emptyset$. Hence, $\alpha x_1 + (1-\alpha)x_2 \in \overline{C}$ by the arbitrariness of r and Proposition 3.3. Then, $\overline{C}$ is convex.

$\forall x_1, x_2 \in C^\circ$, $\forall \alpha \in [0, 1] \subset \mathbb{R}$, we need to show that $\alpha x_1 + (1 - \alpha)x_2 \in C^\circ$. $\exists r_i \in (0, \infty) \subset \mathbb{R}$ such that $\mathcal{B}(x_i, r_i) \subseteq C$, $i = 1, 2$. Let $r := \min\{r_1, r_2\} > 0$. $\forall p \in \mathcal{B}(\alpha x_1 + (1 - \alpha)x_2, r)$, let $w := p - \alpha x_1 - (1 - \alpha)x_2$. Then, $\|w\| < r$ and $x_i + w \in \mathcal{B}(x_i, r_i) \subseteq C$, $i = 1, 2$. By the convexity of C, we have $C \ni \alpha(x_1 + w) + (1 - \alpha)(x_2 + w) = p$. Hence, we have $\mathcal{B}(\alpha x_1 + (1 - \alpha)x_2, r) \subseteq C$ and $\alpha x_1 + (1 - \alpha)x_2 \in C^\circ$. Hence, C° is convex.

This completes the proof of the proposition. □

Proposition 7.16 *Let $\mathcal{X}$ be a normed linear space, $x_0 \in \mathcal{X}$, $S \subseteq \mathcal{X}$, and $P = x_0 + S$. Then, $\overline{P} = x_0 + \overline{S}$ and $P^\circ = x_0 + S^\circ$.*

Proof $\forall x \in \overline{P}$, by Proposition 3.3, $\forall r \in (0, \infty) \subset \mathbb{R}$, $\exists p_0 \in P \cap \mathcal{B}(x, r)$. Then, $p_0 - x_0 \in S \cap \mathcal{B}(x - x_0, r) \neq \emptyset$. Hence, by Proposition 3.3, $x - x_0 \in \overline{S}$ and $x \in x_0 + \overline{S}$. Hence, $\overline{P} \subseteq x_0 + \overline{S}$.

On the other hand, $\forall x \in x_0 + \overline{S}$, we have $x - x_0 \in \overline{S}$ and, by Proposition 3.3, $\forall r \in (0, \infty) \subset \mathbb{R}$, $\exists s_0 \in S \cap \mathcal{B}(x - x_0, r) \neq \emptyset$. Then, $x_0 + s_0 \in P \cap \mathcal{B}(x, r) \neq \emptyset$. By Proposition 3.3, $x \in \overline{P}$. Hence, $x_0 + \overline{S} \subseteq \overline{P}$.

Therefore, we have $\overline{P} = x_0 + \overline{S}$.

$\forall x \in P^\circ$, $\exists r \in (0, \infty) \subset \mathbb{R}$ such that $\mathcal{B}(x, r) \subseteq P$. Then, $\mathcal{B}(x - x_0, r) = \mathcal{B}(x, r) - x_0 \subseteq P - x_0 = S$. Hence, $x - x_0 \in S^\circ$, $x \in x_0 + S^\circ$ and $P^\circ \subseteq x_0 + S^\circ$.

On the other hand, $\forall x \in x_0 + S^\circ$, we have $x - x_0 \in S^\circ$. Then, $\exists r \in (0, \infty) \subset \mathbb{R}$ such that $\mathcal{B}(x - x_0, r) \subseteq S$. Hence, $\mathcal{B}(x, r) = \mathcal{B}(x - x_0, r) + x_0 \subseteq x_0 + S = P$. Therefore, $x \in P^\circ$ and $x_0 + S^\circ \subseteq P^\circ$.

Hence, we have $P^\circ = x_0 + S^\circ$. This completes the proof of the proposition. □

Proposition 7.17 *Let $\mathcal{X}$ be a normed linear space over $\mathbb{K}$. Then, the following statements hold.*

(i) *If $M \subseteq \mathcal{X}$ is a subspace, then $\overline{M}$ is a subspace.*
(ii) *If $V \subseteq \mathcal{X}$ is a linear variety, then $\overline{V}$ is a linear variety.*
(iii) *If $C \subseteq \mathcal{X}$ is a cone with vertex $p \in \mathcal{X}$, then $\overline{C}$ is a cone with vertex p.*
(iv) *Let $x_0 \in \mathcal{X}$ and $r \in (0, \infty) \subset \mathbb{R}$. Then, $\overline{\mathcal{B}(x_0, r)} = \overline{\mathcal{B}}(x_0, r)$.*

Proof (i) Clearly, we have $\vartheta \in M \subseteq \overline{M} \neq \emptyset$. $\forall x, y \in \overline{M}$, $\forall \alpha \in \mathbb{K}$, we will show that $x + y, \alpha x \in \overline{M}$. Then, $\overline{M}$ is a subspace. $\forall r \in (0, \infty) \subset \mathbb{R}$, by Proposition 3.3, $\exists x_0 \in \mathcal{B}(x, r/2) \cap M \neq \emptyset$ and $\exists y_0 \in \mathcal{B}(y, r/2) \cap M \neq \emptyset$. Since M is a subspace, then $x_0 + y_0 \in M$ and $x_0 + y_0 \in \mathcal{B}(x + y, r)$. Hence, we have $M \cap \mathcal{B}(x + y, r) \neq \emptyset$, which implies that $x + y \in \overline{M}$. $\forall r \in (0, \infty) \subset \mathbb{R}$, by Proposition 3.3, $\exists x_0 \in \mathcal{B}(x, r/(1 + |\alpha|)) \cap M \neq \emptyset$. Since M is a subspace, then $\alpha x_0 \in M$ and $\alpha x_0 \in \mathcal{B}(\alpha x, r)$. Hence, we have $M \cap \mathcal{B}(\alpha x, r) \neq \emptyset$, which implies that $\alpha x \in \overline{M}$. Hence, $\overline{M}$ is a subspace.

(ii) Note that $V = x_0 + M$, where $x_0 \in \mathcal{X}$ and $M \subseteq \mathcal{X}$ is a subspace. By Proposition 7.16, $\overline{V} = x_0 + \overline{M}$. By (i), $\overline{M}$ is a subspace. Then, $\overline{V}$ is a linear variety.

(iii) First, we will prove the special case $p = \vartheta$. Clearly, $\vartheta \in C \subset \overline{C}$. $\forall x \in \overline{C}$, $\forall \alpha \in [0, \infty) \subset \mathbb{R}$, we will show that $\alpha x \in \overline{C}$. $\forall r \in (0, \infty) \subset \mathbb{R}$, by Proposition 3.3, $\exists x_0 \in C \cap \mathcal{B}(x, r/(1 + |\alpha|)) \neq \emptyset$. Then, we have $\alpha x_0 \in C \cap \mathcal{B}(\alpha x, r) \neq \emptyset$. By Proposition 3.3, $\alpha x \in \overline{C}$. Hence, $\overline{C}$ is a cone with vertex at the origin.

For general $p \in \mathcal{X}$, we have $C = p + C_0$, where C_0 is a cone with vertex at the origin. By Proposition 7.16, $\overline{C} = p + \overline{C_0}$. By the special case we have shown, $\overline{C_0}$ is a cone with vertex at the origin. Hence, $\overline{C}$ is a cone with vertex p.

(iv) First, we will prove the special case $x_0 = \vartheta$. By Proposition 4.3, $\overline{\mathcal{B}}(\vartheta, r)$ is a closed set and $\overline{\mathcal{B}}(\vartheta, r) \supseteq \mathcal{B}(\vartheta, r)$. Then, we have $\overline{\mathcal{B}(\vartheta, r)} \subseteq \overline{\mathcal{B}}(\vartheta, r)$. On the other hand, $\forall x \in \overline{\mathcal{B}}(\vartheta, r)$, define $(x_k)_{k=1}^\infty \subseteq \mathcal{X}$ by $x_k := \frac{k}{1+k} x$, $\forall k \in \mathbb{N}$. Then, $\|x_k\| =$

$\frac{k}{1+k}\|x\| \leq \frac{k}{1+k}r < r$, $\forall k \in \mathbb{N}$. Therefore, $(x_k)_{k=1}^{\infty} \subseteq \mathcal{B}(\vartheta, r)$. It is obvious that $\lim_{k\in\mathbb{N}} x_k = x$, then, by Proposition 4.13, $x \in \overline{\mathcal{B}(\vartheta, r)}$. Then, we have $\overline{\mathcal{B}}(\vartheta, r) \subseteq \overline{\mathcal{B}(\vartheta, r)}$. Hence, the result holds for $x_0 = \vartheta$.

For arbitrary $x_0 \in \mathcal{X}$, by Proposition 7.16, $\overline{\mathcal{B}(x_0, r)} = \overline{x_0 + \mathcal{B}(\vartheta, r)} = x_0 + \overline{\mathcal{B}(\vartheta, r)} = x_0 + \overline{\mathcal{B}}(\vartheta, r) = \overline{\mathcal{B}}(x_0, r)$. Hence, the result holds.

This completes the proof of the proposition. □

Proposition 7.18 *Let $\mathcal{X}$ be a normed linear space, $P \subseteq \mathcal{X}$, and $P \neq \emptyset$. The closed linear variety generated by P, denoted by $\mathrm{V}(P)$, is the intersection of all closed linear varieties containing P. Then, $\mathrm{V}(P) = \overline{\mathrm{v}(P)}$.*

Proof By Proposition 7.17, $\overline{\mathrm{v}(P)}$ is a closed linear variety containing P. Then, we have $\mathrm{V}(P) \subseteq \overline{\mathrm{v}(P)}$. On the other hand, $\mathrm{V}(P)$ is a closed linear variety containing P, and then $\mathrm{v}(P) \subseteq \mathrm{V}(P)$. Hence, $\overline{\mathrm{v}(P)} \subseteq \mathrm{V}(P)$. Therefore, the result holds. This completes the proof of the proposition. □

The justification of the definition of $\mathrm{V}(P)$ is that intersection of linear varieties is a linear variety when the intersection is nonempty.

Definition 7.19 Let $\mathcal{X}$ be a normed linear space and $P \subseteq \mathcal{X}$ be nonempty. $x \in P$ is said to be a *relative interior point* of P if it is an interior point of P relative to the subset topology of $\mathrm{V}(P)$. The set of all relative interior points of P is called the relative interior of P, and denoted by $^{\circ}P$. P is said to be *relatively open* if P is open in the subset topology of $\mathrm{V}(P)$. ◇

Proposition 7.20 *Let $\mathcal{X}$ be a normed linear space and $(x_\alpha)_{\alpha\in\mathcal{A}} \subseteq \mathcal{X}$ be a net. Then, $\lim_{\alpha\in\mathcal{A}} x_\alpha = x \in \mathcal{X}$ if, and only if, $\forall \epsilon \in (0, \infty) \subset \mathbb{R}$, $\exists \alpha_0 \in \mathcal{A}$, $\forall \alpha \in \mathcal{A}$ with $\alpha_0 \prec \alpha$, we have $\|x - x_\alpha\| < \epsilon$.*

Proposition 7.21 *Let $\mathcal{X}$ be a normed linear space. Then, $\|\cdot\|$ is a uniformly continuous function on $\mathcal{X}$.*

Proof This follows directly from Propositions 4.30 and 7.14. □

7.3 Product Spaces

Proposition 7.22 *Let $\mathcal{X} := (\mathcal{X}, \mathbb{K}, \|\cdot\|_{\mathcal{X}})$ and $\mathcal{Y} := (\mathcal{Y}, \mathbb{K}, \|\cdot\|_{\mathcal{Y}})$ be normed linear spaces. By Proposition 6.24, $\mathcal{X} \times \mathcal{Y}$ is a vector space over $\mathbb{K}$. Define a function $\|\cdot\| : \mathcal{X} \times \mathcal{Y} \to [0, \infty) \subset \mathbb{R}$ by $\|(x, y)\| := (\|x\|_{\mathcal{X}}^2 + \|y\|_{\mathcal{Y}}^2)^{1/2}$, $\forall (x, y) \in \mathcal{X} \times \mathcal{Y}$. Then, $(\mathcal{X} \times \mathcal{Y}, \mathbb{K}, \|\cdot\|)$ is a normed linear space. This normed linear space will be called the Cartesian product of $\mathcal{X}$ and $\mathcal{Y}$ and be denoted by $\mathcal{X} \times \mathcal{Y}$.*

Proof $\forall (x_1, y_1), (x_2, y_2) \in \mathcal{X} \times \mathcal{Y}$, $\forall \alpha \in \mathbb{K}$, we have $\|(x_1, y_1)\| = (\|x_1\|_{\mathcal{X}}^2 + \|y_1\|_{\mathcal{Y}}^2)^{1/2} \in [0, \infty) \subset \mathbb{R}$ and $\|(x_1, y_1)\| = 0 \Leftrightarrow \|x_1\|_{\mathcal{X}} = 0$ and $\|y_1\|_{\mathcal{Y}} = 0 \Leftrightarrow x_1 = \vartheta_{\mathcal{X}}$ and $y_1 = \vartheta_{\mathcal{Y}} \Leftrightarrow (x_1, y_1) = \vartheta$; $\|(x_1, y_1) + (x_2, y_2)\| = \|(x_1 + x_2, y_1 + y_2)\| = (\|x_1 + x_2\|_{\mathcal{X}}^2 + \|y_1 + y_2\|_{\mathcal{Y}}^2)^{1/2} \leq$

$\left((\|x_1\|_{\mathcal{X}} + \|x_2\|_{\mathcal{X}})^2 + (\|y_1\|_{\mathcal{Y}} + \|y_2\|_{\mathcal{Y}})^2\right)^{1/2} \leq (\|x_1\|^2_{\mathcal{X}} + \|y_1\|^2_{\mathcal{Y}})^{1/2} + (\|x_2\|^2_{\mathcal{X}} + \|y_2\|^2_{\mathcal{Y}})^{1/2} = \|(x_1, y_1)\| + \|(x_2, y_2)\|$, where the first inequality follows from the fact that $\mathcal{X}$ and $\mathcal{Y}$ are normed linear spaces and the second inequality follows from the Minkowski's Inequality; $\|\alpha\,(x_1, y_1)\| = \|(\alpha x_1, \alpha y_1)\| = (\|\alpha x_1\|^2_{\mathcal{X}} + \|\alpha y_1\|^2_{\mathcal{Y}})^{1/2} = (|\alpha|^2\|x_1\|^2_{\mathcal{X}} + |\alpha|^2\|y_1\|^2_{\mathcal{Y}})^{1/2} = |\alpha|(\|x_1\|^2_{\mathcal{X}} + \|y_1\|^2_{\mathcal{Y}})^{1/2} = |\alpha|\|(x_1, y_1)\|$. Hence, $\|\cdot\|$ is a norm on $\mathcal{X} \times \mathcal{Y}$. This completes the proof of the proposition. □

Clearly, the natural metric for the Cartesian product $\mathcal{X} \times \mathcal{Y}$ is the Cartesian metric defined in Definition 4.28.

The above proposition may also be generalized to the case of $\mathcal{X}_1 \times \mathcal{X}_2 \times \cdots \times \mathcal{X}_n$, where $n \in \mathbb{N}$ and $\mathcal{X}_i$'s are normed linear spaces over the same field $\mathbb{K}$. When $n = 0$, it should be noted that $\prod_{\alpha \in \emptyset} \mathcal{X}_\alpha = (\{\emptyset =: \vartheta\}, \mathbb{K}, \|\cdot\|)$, where $\|\vartheta\| = 0$.

Proposition 7.23 *Let $\mathcal{X} := (\mathcal{X}, \mathbb{K}, \|\cdot\|)$ be a normed linear space. Then, the vector addition $\oplus_{\mathcal{X}} : \mathcal{X} \times \mathcal{X} \to \mathcal{X}$ is uniformly continuous; and the scalar multiplication $\otimes_{\mathcal{X}} : \mathbb{K} \times \mathcal{X} \to \mathcal{X}$ is continuous.*

Proof $\forall \epsilon \in (0, \infty) \subset \mathbb{R}$, $\forall (x_1, x_2), (y_1, y_2) \in \mathcal{X} \times \mathcal{X}$ with

$$\|(x_1, x_2) - (y_1, y_2)\|_{\mathcal{X}\times\mathcal{X}} < \epsilon/\sqrt{2}$$

we have

$$\begin{aligned}&\|(x_1 + x_2) - (y_1 + y_2)\| \leq \|x_1 - y_1\| + \|x_2 - y_2\| \\ &\leq \sqrt{2}\,(\|x_1 - y_1\|^2 + \|x_2 - y_2\|^2)^{1/2} = \sqrt{2}\|(x_1, x_2) - (y_1, y_2)\|_{\mathcal{X}\times\mathcal{X}} < \epsilon\end{aligned}$$

Hence, the vector addition $\oplus_{\mathcal{X}}$ is uniformly continuous.

$\forall \alpha_0 \in \mathbb{K}$, $\forall x_0 \in \mathcal{X}$, $\forall \epsilon \in (0, \infty) \subset \mathbb{R}$, let $\delta = \epsilon/(1 + \epsilon + |\alpha_0| + \|x_0\|) \in (0, 1) \subset \mathbb{R}$. $\forall (\alpha, x) \in \mathcal{B}_{\mathbb{K}\times\mathcal{X}}((\alpha_0, x_0), \delta)$, we have

$$\begin{aligned}\|\alpha x - \alpha_0 x_0\| &\leq \|\alpha x - \alpha x_0\| + \|\alpha x_0 - \alpha_0 x_0\| \\ &= |\alpha|\|x - x_0\| + \|x_0\||\alpha - \alpha_0| \\ &\leq \left(|\alpha|^2 + \|x_0\|^2\right)^{1/2} \left(\|x - x_0\|^2 + |\alpha - \alpha_0|^2\right)^{1/2} \\ &\leq (|\alpha| + \|x_0\|)\delta \leq (1 + |\alpha_0| + \|x_0\|)\delta < \epsilon\end{aligned}$$

where we have made use of Cauchy–Schwarz Inequality in the second inequality. Hence, $\otimes_{\mathcal{X}}$ is continuous at (α_0, x_0). By the arbitrariness of (α_0, x_0), we have that $\otimes_{\mathcal{X}}$ is continuous.

This completes the proof of the proposition. □

Definition 7.24 Let $\mathcal{X} := (\mathscr{X}, \mathbb{K}, \|\cdot\|_{\mathcal{X}})$ and $\mathcal{Y} := (\mathscr{Y}, \mathbb{K}, \|\cdot\|_{\mathcal{Y}})$ be two normed linear spaces over the same field $\mathbb{K}$ and $A : \mathcal{X} \to \mathcal{Y}$ be a vector space isomorphism. A is said to be an *isometrical isomorphism* if $\|Ax\|_{\mathcal{Y}} = \|x\|_{\mathcal{X}}$, $\forall x \in \mathcal{X}$. The $\mathcal{X}$ and $\mathcal{Y}$ are said to be *isometrically isomorphic*. ◇

Let $A : \mathcal{X} \to \mathcal{Y}$ be an isometrical isomorphism between $\mathcal{X}$ and $\mathcal{Y}$. Then, A is an isometry between $\mathcal{X}$ and $\mathcal{Y}$. Both A and A_{inv} are uniformly continuous. $\mathcal{X}$ and $\mathcal{Y}$ are equal to each other up to a relabeling of their vectors.

Proposition 7.25 *Let $\mathcal{X}_\alpha$, $\alpha \in \Lambda$, be normed linear spaces over $\mathbb{K}$, where Λ is a finite index set. Let $\Lambda = \bigcup_{\beta \in \Gamma} \Lambda_\beta$, where Λ_β's are pairwise disjoint and finite and Γ is also finite. $\forall \beta \in \Gamma$, let $\mathcal{X}^{\langle\beta\rangle} := \prod_{\alpha \in \Lambda_\beta} \mathcal{X}_\alpha$ be the Cartesian product space. Let $\mathcal{X}^{\langle\Gamma\rangle} := \prod_{\beta \in \Gamma} \prod_{\alpha \in \Lambda_\beta} \mathcal{X}_\alpha$ be the Cartesian product space of product spaces, and $\mathcal{X} := \prod_{\alpha \in \Lambda} \mathcal{X}_\alpha$ be the Cartesian product space. Then, $\mathcal{X}$ and $\mathcal{X}^{\langle\Gamma\rangle}$ are isometrically isomorphic.*

Proof Define $E : \prod_{\beta \in \Gamma} \prod_{\alpha \in \Lambda_\beta} \mathcal{X}_\alpha \to \prod_{\alpha \in \Lambda} \mathcal{X}_\alpha$ by, $\forall x \in \mathcal{X}^{\langle\Gamma\rangle}$, $\forall \alpha \in \Lambda$, $\exists!\, \beta_\alpha \in \Gamma \ni \cdot\ \alpha \in \Lambda_{\beta_\alpha}$, $\pi_\alpha(E(x)) = \pi_\alpha^{\langle\beta_\alpha\rangle}(\pi_{\beta_\alpha}^{\langle\Gamma\rangle}(x))$. By Proposition 4.32, E is a isometry. It is clear that E is a linear operator since the projection functions are linear for vector spaces. Hence, E is a vector space isomorphism. Then, E is an isometrical isomorphism. This completes the proof of the proposition. □

7.4 Banach Spaces

Definition 7.26 If a normed linear space is complete with respect to the natural metric, then it is called a *Banach space*. ◇

Clearly, a Cauchy sequence in a normed linear space is bounded.

Proposition 7.27 *A normed linear space $\mathcal{X}$ is complete if, and only if, every absolutely summable series is summable, that is $\forall (x_n)_{n=1}^\infty \subseteq \mathcal{X}$, $\sum_{n=1}^\infty \|x_n\| \in \mathbb{R}$ $\Rightarrow \sum_{n=1}^\infty x_n \in \mathcal{X}$.*

Proof "Only if" Let $\mathcal{X}$ be complete and $(x_n)_{n=1}^\infty \subseteq \mathcal{X}$ be absolutely summable. Then, $\sum_{n=1}^\infty \|x_n\| \in \mathbb{R}$ and $\forall \epsilon \in (0, \infty) \subset \mathbb{R}$, $\exists N \in \mathbb{N}$ such that, $\forall n, m \in \mathbb{N}$ with $n \geq m$, we have $\sum_{i=m}^n \|x_i\| < \epsilon$. Then, $\left\|\sum_{i=m}^n x_i\right\| < \epsilon$. Let $s_n := \sum_{i=1}^n x_i$, $\forall n \in \mathbb{N}$, be the partial sum. Then, $(s_n)_{n=1}^\infty \subseteq \mathcal{X}$ is a Cauchy sequence. By the completeness of $\mathcal{X}$, $\sum_{n=1}^\infty x_n = \lim_{n \in \mathbb{N}} s_n \in \mathcal{X}$.

"If" Let $(x_n)_{n=1}^\infty \subseteq \mathcal{X}$ be a Cauchy sequence. Let $n_0 = 0$. $\forall k \in \mathbb{N}$, $\exists n_k \in \mathbb{N}$ with $n_k > n_{k-1}$ such that $\|x_n - x_m\| < 2^{-k}$, $\forall n, m \geq n_k$. Then, $\left(x_{n_k}\right)_{k=1}^\infty$ is a subsequence of $(x_n)_{n=1}^\infty$. Let $y_1 := x_{n_1}$ and $y_k := x_{n_k} - x_{n_{k-1}}$, $\forall k \geq 2$. Then, the sequence $(y_k)_{k=1}^\infty \subseteq \mathcal{X}$ and its kth partial sum is x_{n_k}. Note that $\|y_k\| < 2^{-k+1}$, $\forall k \geq 2$. Then, $\sum_{k=1}^\infty \|y_k\| < \|y_1\| + 1$, which implies that $\sum_{k=1}^\infty y_k = x_0 \in \mathcal{X}$. Hence, we have $\lim_{k \in \mathbb{N}} x_{n_k} = x_0 \in \mathcal{X}$.

We will now show that $\lim_{n\in\mathbb{N}} x_n = x_0$. $\forall k \in \mathbb{N}$, $\exists N \in \mathbb{N}$ with $N \geq k$ such that $\|x_{n_i} - x_0\| < 2^{-k}$, $\forall i \geq N$. $\forall n \geq n_N$, we have $\|x_n - x_0\| \leq \|x_n - x_{n_N}\| + \|x_{n_N} - x_0\| < 2^{-N} + 2^{-k} \leq 2^{-k+1}$. Hence, $\lim_{n\in\mathbb{N}} x_n = x_0$. Therefore, $\mathcal{X}$ is complete.

This completes the proof of the proposition. □

We frequently take great care to formulate problems arising in applications as equivalent problems in Banach spaces rather than other incomplete spaces. The principal advantage of Banach spaces in optimization problems is that when seeking an optimal vector, we often construct a sequence (net) of vectors, each member of which is superior to preceding ones. The desired optimal vector is then the limit of the sequence (net). In order for this scheme to be effective, there must be available a test for convergence which can be applied when the limit is unknown. The Cauchy criterion for convergence meets this requirement provided the space is complete.

Example 7.28 Consider Example 7.11 with $a = 0$ and $b = 1$. We will show that the space $(\mathcal{M}, \mathbb{R}, \|\cdot\|)$ is incomplete. Take $(x_n)_{n=1}^{\infty} \subseteq \mathcal{M}$ to be, $\forall n \in \mathbb{N}$,

$$x_n(t) = \begin{cases} 0 & 0 \leq t \leq \frac{1}{2} - \frac{1}{n+1} \\ (n+1)t - \frac{n+1}{2} + 1 & \frac{1}{2} - \frac{1}{n+1} < t \leq \frac{1}{2} \\ 1 & t > \frac{1}{2} \end{cases}$$

Clearly, x_n is continuous (See Fig. 7.1) and $\|x_n\| < 1$, $\forall n \in \mathbb{N}$. $\forall n, m \in \mathbb{N}$, we have

$$\|x_n - x_m\| = \int_0^1 |x_n(t) - x_m(t)|\, dt = \left| \int_0^1 (x_n(t) - x_m(t))\, dt \right|$$
$$= \left| \frac{1}{2(n+1)} - \frac{1}{2(m+1)} \right|$$

Then, $(x_n)_{n=1}^{\infty}$ is Cauchy. Yet, it is obvious that there is no continuous function $x \in \mathcal{M}$ such that $\lim_{n\in\mathbb{N}} x_n = x$, i.e., $\lim_{n\in\mathbb{N}} \int_0^1 |x_n(t) - x(t)|\, dt = 0$. Hence, the space is incomplete. ◇

Example 7.29 $\mathbb{K}^n$ with norm defined as in Example 7.2 or Example 7.3 is a Banach space. ◇

Example 7.30 Consider the real normed linear space $\mathcal{C}([a, b])$ defined in Example 7.4, where $a, b \in \mathbb{R}$ and $a \leq b$. We will show that it is complete. Take a Cauchy sequence $(x_n)_{n=1}^{\infty} \subseteq \mathcal{C}([a, b])$. $\forall \epsilon \in (0, \infty) \subset \mathbb{R}$, $\exists N \in \mathbb{N}$ such that $\forall n, m \geq N$, $0 \leq |x_n(t) - x_m(t)| \leq \|x_n - x_m\| < \epsilon$, $\forall t \in [a, b]$. This shows that, $\forall t \in [a, b]$, $(x_n(t))_{n=1}^{\infty} \subseteq \mathbb{R}$ is a Cauchy sequence, which converges to $x(t) \in \mathbb{R}$ since $\mathbb{R}$ is complete. This defines a function $x : [a, b] \to \mathbb{R}$. It is easy to show that $(x_n)_{n=1}^{\infty}$, viewed as a sequence of functions of $[a, b]$ to $\mathbb{R}$, converges uniformly to

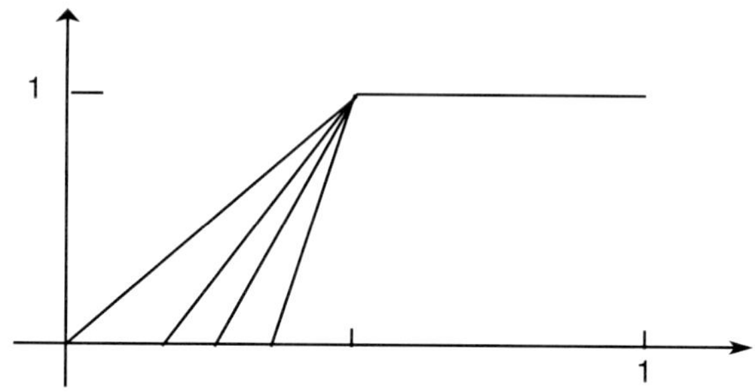

Fig. 7.1 Sequence for Example 7.28

x. By Proposition 4.26, x is continuous. Hence, $x \in \mathcal{C}([a, b])$. It is easy to see that $\lim_{n\in\mathbb{N}} \|x_n - x\| = 0$. Hence, $\lim_{n\in\mathbb{N}} x_n = x$. Hence, $\mathcal{C}([a, b])$ is a Banach space. ◇

Example 7.31 Let $\mathcal{K}$ be a countably compact topological space, $\mathcal{X}$ be a normed linear space over the field $\mathbb{K}$, $Y := \{f : \mathcal{K} \to \mathcal{X}\}$. Define the usual vector addition $\oplus$ and scalar multiplication $\otimes$ and null vector ϑ on Y as in Example 6.20. Then, $\mathcal{Y} := (Y, \oplus, \otimes, \vartheta)$ is a vector space over $\mathbb{K}$. Let $M := \{f \in Y \mid f \text{ is continuous}\}$. Then, by Propositions 3.32, 6.25, and 7.23, $\mathcal{M} := (M, \oplus, \otimes, \vartheta)$ is a subspace of $(\mathcal{Y}, \mathbb{K})$. Define a function $\|\cdot\| : \mathcal{M} \to [0, \infty) \subset \mathbb{R}$ by $\|f\| = \max\left\{\sup_{k\in\mathcal{K}} \|f(k)\|_{\mathcal{X}}, 0\right\}$, $\forall f \in \mathcal{M}$. This function is well-defined by Propositions 7.21, 3.12, and 5.29. We will show that $\|\cdot\|$ defines a norm on $\mathcal{M}$. We will distinguish two exhaustive and mutually exclusive cases: Case 1: $\mathcal{K} = \emptyset$; Case 2: $\mathcal{K} \neq \emptyset$. Case 1: $\mathcal{K} = \emptyset$. Then, $\mathcal{M}$ is a singleton set $\{\emptyset\}$ and $\|\emptyset\| = 0$. Clearly, $(\mathcal{M}, \mathbb{K}, \|\cdot\|)$ is a normed linear space. Case 2: $\mathcal{K} \neq \emptyset$. $\forall f, g \in \mathcal{M}$, $\forall \alpha \in \mathbb{K}$, $\|f\| = \max_{k\in\mathcal{K}} \|f(k)\|_{\mathcal{X}}$ by Propositions 7.21, 3.12, and 5.29. $\|f\| = 0 \Leftrightarrow \|f(k)\|_{\mathcal{X}} = 0$, $\forall k \in \mathcal{K}$ $\Leftrightarrow f(k) = \vartheta_{\mathcal{X}}$, $\forall k \in \mathcal{K} \Leftrightarrow f = \vartheta$. $\|f + g\| = \max_{k\in\mathcal{K}} \|f(k) + g(k)\|_{\mathcal{X}} \leq \max_{k\in\mathcal{K}}(\|f(k)\|_{\mathcal{X}} + \|g(k)\|_{\mathcal{X}}) \leq \max_{k\in\mathcal{K}} \|f(k)\|_{\mathcal{X}} + \max_{k\in\mathcal{K}} \|g(k)\|_{\mathcal{X}} = \|f\| + \|g\|$. $\|\alpha f\| = \max_{k\in\mathcal{K}} \|\alpha f(k)\|_{\mathcal{X}} = \max_{k\in\mathcal{K}} |\alpha| \|f(k)\|_{\mathcal{X}} = |\alpha| \|f\|$. Hence, $(\mathcal{M}, \mathbb{K}, \|\cdot\|)$ is a normed linear space. In both case, we have shown that $\mathcal{C}(\mathcal{K}, \mathcal{X}) := (\mathcal{M}, \mathbb{K}, \|\cdot\|)$ is a normed linear space. ◇

Example 7.32 Let $\mathcal{K}$ be a countably compact topological space, $\mathcal{X}$ be a Banach space over the field $\mathbb{K}$ (with norm $\|\cdot\|_{\mathcal{X}}$). Consider the normed linear space $\mathcal{C}(\mathcal{K}, \mathcal{X})$ (with norm $\|\cdot\|$) defined in Example 7.31. We will show that this space is a Banach space. We will distinguish two exhaustive and mutually exclusive cases: Case 1: $\mathcal{K} = \emptyset$; Case 2: $\mathcal{K} \neq \emptyset$. Case 1: $\mathcal{K} = \emptyset$. Then, $\mathcal{C}(\mathcal{K}, \mathcal{X})$ is a singleton set. Hence, any Cauchy sequence must converge. Thus, $\mathcal{C}(\mathcal{K}, \mathcal{X})$ is a Banach space. Case 2: $\mathcal{K} \neq \emptyset$. Take a Cauchy sequence $(x_n)_{n=1}^{\infty} \subseteq \mathcal{C}(\mathcal{K}, \mathcal{X})$. $\forall \epsilon \in (0, \infty) \subset \mathbb{R}$, $\exists N \in \mathbb{N}$ such that $\forall n, m \geq N$, $0 \leq \|x_n(k) - x_m(k)\|_{\mathcal{X}} \leq \|x_n - x_m\| < \epsilon$, $\forall k \in \mathcal{K}$. This shows that, $\forall k \in \mathcal{K}$, $(x_n(k))_{n=1}^{\infty} \subseteq \mathcal{X}$ is a Cauchy sequence, which converges to $x(k) \in \mathcal{X}$ since $\mathcal{X}$ is complete. This defines a function $x : \mathcal{K} \to \mathcal{X}$. It is easy to show that $(x_n)_{n=1}^{\infty}$, viewed as a sequence of functions of $\mathcal{K}$ to $\mathcal{X}$, converges uniformly to x. By Proposition 4.26, x is continuous. Hence, $x \in \mathcal{C}(\mathcal{K}, \mathcal{X})$. It is easy to see that $\lim_{n\in\mathbb{N}} \|x_n - x\| = 0$. Hence, $\lim_{n\in\mathbb{N}} x_n = x$. Hence, $\mathcal{C}(\mathcal{K}, \mathcal{X})$ is a Banach space.

In both cases, we have shown that $\mathcal{C}(\mathcal{K}, \mathcal{X})$ is a Banach space when $\mathcal{X}$ is a Banach space. ◇

Example 7.33 Let $\mathcal{X}$ be a Banach space over the field $\mathbb{K}$. Consider the normed linear space $\mathrm{l}_p(\mathcal{X})$ (with norm $\|\cdot\|_p$) defined in Example 7.10, where $p \in [1, \infty] \subset \mathbb{R}_e$. We will show that this space is a Banach space. We will distinguish two exhaustive and mutually exclusive cases: Case 1: $p \in [1, \infty)$; Case 2: $p = \infty$.

Case 1: $p \in [1, \infty)$. Take a Cauchy sequence $(x_n)_{n=1}^{\infty} \subseteq \mathrm{l}_p(\mathcal{X})$, where $x_n = (\xi_{n,k})_{k=1}^{\infty} \subseteq \mathcal{X}$, $\forall n \in \mathbb{N}$. $\forall \epsilon \in (0, \infty) \subset \mathbb{R}$, $\exists N_\epsilon \in \mathbb{N}$ such that $\forall n, m \geq N_\epsilon$, we have $\|x_n - x_m\|_p < \epsilon$. $\forall k \in \mathbb{N}$, $\|\xi_{n,k} - \xi_{m,k}\|_{\mathcal{X}} \leq \left(\sum_{i=1}^{\infty} \|\xi_{n,i} - \xi_{m,i}\|_{\mathcal{X}}^p\right)^{1/p} = \|x_n - x_m\|_p < \epsilon$. Hence, $\forall k \in \mathbb{N}$, $(\xi_{n,k})_{n=1}^{\infty} \subseteq \mathcal{X}$ is a Cauchy sequence, which converges to some $\xi_k \in \mathcal{X}$ since $\mathcal{X}$ is a Banach space. Let $x := (\xi_k)_{k=1}^{\infty} \subseteq \mathcal{X}$. We will show that $x \in \mathrm{l}_p(\mathcal{X})$ and $\lim_{n\in\mathbb{N}} x_n = x$. Since $(x_n)_{n=1}^{\infty}$ is a Cauchy sequence, then it is bounded, that is $\exists M \in [0, \infty) \subset \mathbb{R}$ such that $\|x_n\|_p \leq M$, $\forall n \in \mathbb{N}$. Then, we have $\|x_n\|_p^p = \sum_{i=1}^{\infty} \|\xi_{n,i}\|_{\mathcal{X}}^p \leq M^p$, $\forall n \in \mathbb{N}$. Then, $\forall n, k \in \mathbb{N}$, we have $\sum_{i=1}^{k} \|\xi_{n,i}\|_{\mathcal{X}}^p \leq M^p$. By Propositions 7.21 and 3.66, we have $\sum_{i=1}^{k} \|\xi_i\|_{\mathcal{X}}^p = \lim_{n\in\mathbb{N}} \sum_{i=1}^{k} \|\xi_{n,i}\|_{\mathcal{X}}^p \leq M^p$. Hence, we have $\sum_{i=1}^{\infty} \|\xi_i\|_{\mathcal{X}}^p \leq M^p$ and $\|x\|_p = \left(\sum_{i=1}^{\infty} \|\xi_i\|_{\mathcal{X}}^p\right)^{1/p} \leq M$. Hence, we have $x \in \mathrm{l}_p(\mathcal{X})$. $\forall \epsilon \in (0, \infty) \subset \mathbb{R}$, $\forall n, m \in \mathbb{N}$ with $n, m \geq N_\epsilon$, we have $\|x_n - x_m\|_p < \epsilon$. Then, $\forall k \in \mathbb{N}$, $\sum_{i=1}^{k} \|\xi_{n,i} - \xi_{m,i}\|_{\mathcal{X}}^p < \epsilon^p$. Taking limit as $m \to \infty$, by Propositions 7.21, 3.66, and 7.23, we have $\sum_{i=1}^{k} \|\xi_{n,i} - \xi_i\|_{\mathcal{X}}^p = \lim_{m\in\mathbb{N}} \sum_{i=1}^{k} \|\xi_{n,i} - \xi_{m,i}\|_{\mathcal{X}}^p \leq \epsilon^p$, $\forall k \in \mathbb{N}$. Hence, we have $\|x_n - x\|_p = \left(\sum_{i=1}^{\infty} \|\xi_{n,i} - \xi_i\|_{\mathcal{X}}^p\right)^{1/p} \leq \epsilon$. This shows that $\lim_{n\in\mathbb{N}} x_n = x$. Hence, $\mathrm{l}_p(\mathcal{X})$ is complete and therefore a Banach space.

Case 2: $p = \infty$. Take a Cauchy sequence $(x_n)_{n=1}^{\infty} \subseteq \mathrm{l}_\infty(\mathcal{X})$, where $x_n = (\xi_{n,k})_{k=1}^{\infty} \subseteq \mathcal{X}$, $\forall n \in \mathbb{N}$. $\forall \epsilon \in (0, \infty) \subset \mathbb{R}$, $\exists N_\epsilon \in \mathbb{N}$ such that $\forall n, m \geq N_\epsilon$, we have $\|x_n - x_m\|_\infty < \epsilon$. $\forall k \in \mathbb{N}$, $\|\xi_{n,k} - \xi_{m,k}\|_{\mathcal{X}} \leq \sup_{i\in\mathbb{N}} \|\xi_{n,i} - \xi_{m,i}\|_{\mathcal{X}} = \|x_n - x_m\|_\infty < \epsilon$. Hence, $\forall k \in \mathbb{N}$, $(\xi_{n,k})_{n=1}^{\infty} \subseteq \mathcal{X}$ is a Cauchy sequence, which converges to some $\xi_k \in \mathcal{X}$ since $\mathcal{X}$ is a Banach space. Let $x := (\xi_k)_{k=1}^{\infty} \subseteq \mathcal{X}$. We will show that $x \in \mathrm{l}_\infty(\mathcal{X})$ and $\lim_{n\in\mathbb{N}} x_n = x$. Since $(x_n)_{n=1}^{\infty}$ is a Cauchy sequence, then it is bounded, that is $\exists M \in [0, \infty) \subset \mathbb{R}$ such that $\|x_n\|_\infty \leq M$, $\forall n \in \mathbb{N}$. Then, we have $\|\xi_{n,k}\|_{\mathcal{X}} \leq \sup_{i\in\mathbb{N}} \|\xi_{n,i}\|_{\mathcal{X}} = \|x_n\|_\infty \leq M$, $\forall n, k \in \mathbb{N}$. By Propositions 7.21 and 3.66, we have $\|\xi_k\|_{\mathcal{X}} = \lim_{n\in\mathbb{N}} \|\xi_{n,k}\|_{\mathcal{X}} \leq M$. Hence, $\|x\|_\infty = \sup_{i\in\mathbb{N}} \|\xi_i\|_{\mathcal{X}} \leq M$. Therefore, $x \in \mathrm{l}_\infty(\mathcal{X})$. $\forall \epsilon \in (0, \infty) \subset \mathbb{R}$, $\forall n, m \in \mathbb{N}$ with $n, m \geq N_\epsilon$, we have $\|x_n - x_m\|_\infty < \epsilon$. Then, $\forall k \in \mathbb{N}$, $\|\xi_{n,k} - \xi_{m,k}\|_{\mathcal{X}} < \epsilon$. Taking limit as $m \to \infty$, by Propositions 7.21, 3.66, and 7.23, we have $\|\xi_{n,k} - \xi_k\|_{\mathcal{X}} = \lim_{m\in\mathbb{N}} \|\xi_{n,k} - \xi_{m,k}\|_{\mathcal{X}} \leq \epsilon$, $\forall k \in \mathbb{N}$. Hence, we have $\|x_n - x\|_\infty = \sup_{i\in\mathbb{N}} \|\xi_{n,i} - \xi_i\|_{\mathcal{X}} \leq \epsilon$. This shows that $\lim_{n\in\mathbb{N}} x_n = x$. Hence, $\mathrm{l}_\infty(\mathcal{X})$ is complete and therefore a Banach space.

In summary, we have shown that $\mathrm{l}_p(\mathcal{X})$ is a Banach space, $\forall p \in [1, \infty] \subset \mathbb{R}_e$, when $\mathcal{X}$ is a Banach space. ◇

Definition 7.34 Let $\mathcal{X}$ be a normed linear space and $S \subseteq \mathcal{X}$. S is said to be complete if S with the natural metric forms a complete metric space. ◇

By Proposition 4.39, a subset in a Banach space is complete if, and only if, it is closed.

Proposition 7.35 *Let $\mathcal{Y}$ be a normed linear space over $\mathbb{K}$, $S_1, S_2 \subseteq \mathcal{Y}$ be separable subsets, and $\alpha \in \mathbb{K}$. Then,* $\text{span}(S_1)$, $\overline{\text{span}(S_1)}$, $S_1 + S_2$, αS_1, $S_1 \cap S_2$, *and* $S_1 \cup S_2$ *are separable subsets of* $\mathcal{Y}$.

Proof Let $\mathbb{K}_{\mathbb{Q}} := \mathbb{Q}$ if $\mathbb{K} = \mathbb{R}$; and $\mathbb{K}_{\mathbb{Q}} := \{a + ib \in \mathbb{C} \mid a, b \in \mathbb{Q}\}$, if $\mathbb{K} = \mathbb{C}$. Clearly, $\mathbb{K}_{\mathbb{Q}}$ is a countable dense set in $\mathbb{K}$. Let $D \subseteq S_1$ be a countable dense subset. Let $\hat{D} := \{\sum_{i=1}^{n} \alpha_i y_i \in \mathcal{Y} \mid n \in \mathbb{Z}_+,\ \alpha_i \in \mathbb{K}_{\mathbb{Q}},\ y_i \in D,\ i = 1, \ldots, n\}$, which is a countable set. Clearly, $\hat{D} \subseteq \text{span}(S_1)$ is a dense subset. Hence, $\text{span}(S_1) \subseteq \mathcal{Y}$ is separable. By Proposition 4.38, $\overline{\text{span}(S_1)}$ is separable. It is straightforward to show that $S_1 + S_2$, αS_1, $S_1 \cap S_2$, and $S_1 \cup S_2$ are separable. □

Theorem 7.36 *In a normed linear space $\mathcal{X}$, any finite-dimensional subspace $M \subseteq \mathcal{X}$ is complete.*

Proof Let $\mathcal{X}$ be a normed linear space over the field $\mathbb{K}$. Let $\bar{n} \in \mathbb{Z}_+$ be the dimension of M, which is well-defined by Theorem 6.51. We will prove the theorem by mathematical induction on $\bar{n}$.

1° $\bar{n} = 0$. Then, $M = \{\vartheta\}$. Clearly, any Cauchy sequence in M must converge to $\vartheta \in M$. Hence, M is complete. $\bar{n} = 1$. Let $\{e_1\} \subseteq M$ be a basis for M. Clearly, $e_1 \neq \vartheta$ and $\|e_1\| > 0$. Fix any Cauchy sequence $(x_n)_{n=1}^{\infty} \subseteq M$. Then, $x_n = \alpha_n e_1$ for some $\alpha_n \in \mathbb{K}$, $\forall n \in \mathbb{N}$. Since $(x_n)_{n=1}^{\infty}$ is Cauchy, then $\forall \epsilon \in (0, \infty) \subset \mathbb{R}$, $\exists N \in \mathbb{N}$, $\forall n, m \geq N$, we have $\|x_n - x_m\| < \|e_1\|\epsilon$. Then, we have $|\alpha_n - \alpha_m| = \|x_n - x_m\|/\|e_1\| < \epsilon$. Hence, $(\alpha_n)_{n=1}^{\infty}$ is a Cauchy sequence in $\mathbb{K}$. Then, $\lim_{n\in\mathbb{N}} \alpha_n = \alpha \in \mathbb{K}$ since $\mathbb{K}$ is complete. It is easy to show that $\lim_{n\in\mathbb{N}} x_n = \alpha e_1 \in M$. Hence, M is complete.

2° Assume M is complete when $\bar{n} = k - 1 \in \mathbb{N}$.

3° Consider the case $\bar{n} = k \in \{2, 3, \ldots\} \subset \mathbb{N}$. Let $\{e_1, \ldots, e_k\} \subseteq M$ be a basis for M. Define $M_i := \text{span}(\{e_1, \ldots, e_k\} \setminus \{e_i\})$ and $\delta_i := \text{dist}(e_i, M_i)$, $i = 1, \ldots, k$. $\forall i = 1, \ldots, k$, we have $\delta_i \in [0, \infty) \subset \mathbb{R}$. M_i, which is a $k - 1$-dimensional subspace of $\mathcal{X}$. By inductive assumption M_i is complete. By Proposition 4.39, M_i is closed. Clearly, $e_i \notin M_i$. By Proposition 4.10, $\delta_i > 0$.

Let $\delta := \min\{\delta_1, \ldots, \delta_k\} > 0$. Fix any Cauchy sequence $(x_n)_{n=1}^{\infty} \subseteq M$. $\forall n \in \mathbb{N}$, x_n admits a unique representation $x_n = \sum_{i=1}^{k} \lambda_{n,i} e_i$ where $\lambda_{n,i} \in \mathbb{K}$, $i = 1, \ldots, k$, by Definition 6.50 and Corollary 6.47. $\forall \epsilon \in (0, \infty) \subset \mathbb{R}$, $\exists N \in \mathbb{N}$ such that $\forall n, m \geq N$, we have $\|x_n - x_m\| < \delta\epsilon$. $\forall i \in \{1, \ldots, k\}$, we have $\epsilon > \|x_n - x_m\|/\delta = \left\|\sum_{j=1}^{k} (\lambda_{n,j} - \lambda_{m,j}) e_j\right\|/\delta \geq \left|\lambda_{n,i} - \lambda_{m,i}\right| \delta_i/\delta \geq \left|\lambda_{n,i} - \lambda_{m,i}\right|$. Hence, $\left(\lambda_{n,i}\right)_{n=1}^{\infty} \subseteq \mathbb{K}$ is a Cauchy sequence, $\forall i \in \{1, \ldots, k\}$. Then, $\lim_{n\in\mathbb{N}} \lambda_{n,i} = \lambda_i \in \mathbb{K}$. Let $x := \sum_{i=1}^{k} \lambda_i e_i \in M$. Now, it is straightforward to show that $\lim_{n\in\mathbb{N}} \|x_n - x\| = 0$. Then, $\lim_{n\in\mathbb{N}} x_n = x \in M$. Hence, M is complete.

This completes the induction process. This completes the proof of the theorem. □

Definition 7.37 Let $\mathcal{X}$ be a vector space over the field $\mathbb{K}$ and $\|\cdot\|_1$ and $\|\cdot\|_2$ be two norms defined on $\mathcal{X}$. These two norms are said to be *equivalent* if $\exists K \in (0, \infty) \subset \mathbb{R}$ such that $\|x\|_1/K \leq \|x\|_2 \leq K\|x\|_1$, $\forall x \in \mathcal{X}$. ◇

Clearly, two norms are equivalent implies that the natural metrics are uniformly equivalent.

Theorem 7.38 *Let $\mathcal{X}$ be a finite-dimensional vector space over the field $\mathbb{K}$. Any two norms on $\mathcal{X}$ are equivalent.*

Proof Let $n \in \mathbb{Z}_+$ be the dimension of $\mathcal{X}$, which is well-defined by Theorem 6.51. Let $\|\cdot\|_1$ and $\|\cdot\|_2$ be two norms defined on $\mathcal{X}$. We will distinguish three exhaustive and mutually exclusive cases: Case 1: $n = 0$; Case 2: $n = 1$; Case 3: $n \geq 2$. Case 1: $n = 0$. Then, $\mathcal{X} = \{\vartheta\}$ and $\|\vartheta\|_1 = 0 = \|\vartheta\|_2$. Hence, the two norms are equivalent. Case 2: $n = 1$. Let $\{e_1\}$ be a basis for $\mathcal{X}$. Clearly, $e_1 \neq \vartheta$. Let $\delta := \|e_1\|_1/\|e_1\|_2 \in (0, \infty) \subset \mathbb{R}$ and $K = \max\{\delta, \delta^{-1}\} \in (0, \infty) \subset \mathbb{R}$. $\forall x \in \mathcal{X}$, $\exists!\ \alpha \in \mathbb{K}$ such that $x = \alpha e_1$. Then, $\|x\|_1 = |\alpha|\|e_1\|_1 = |\alpha|\|e_1\|_2\delta = \delta\|x\|_2$ and $\|x\|_1/K \leq \|x\|_2 \leq K\|x\|_1$. Hence, the two norms are equivalent.

Case 3: $n \geq 2$. Let $\{e_1, \ldots, e_n\} \subseteq \mathcal{X}$ be a basis for $\mathcal{X}$. $\forall x \in \mathcal{X}$, by Definition 6.50 and Corollary 6.47, $\exists!\ \alpha_1, \ldots, \alpha_n \in \mathbb{K}$ such that $x = \sum_{i=1}^{n} \alpha_i e_i$. We will show that $\exists K_1 \in (0, \infty) \subset \mathbb{R}$ such that $(\sum_{i=1}^{n} |\alpha_i|)/K_1 \leq \|x\|_1 \leq K_1(\sum_{i=1}^{n} |\alpha_i|)$. By a similar argument, $\exists K_2 \in (0, \infty) \subset \mathbb{R}$ such that $(\sum_{i=1}^{n} |\alpha_i|)/K_2 \leq \|x\|_2 \leq K_2(\sum_{i=1}^{n} |\alpha_i|)$. Then, $\|x\|_1/K \leq \|x\|_2 \leq K\|x\|_1$, where $K := K_1K_2$. Hence, the two norms are equivalent.

Clearly, $e_i \neq \vartheta$ and $\|e_i\|_1 > 0$, $i = 1, \ldots, n$. Let $\delta := \max_{1 \leq i \leq n} \|e_i\|_1 \in (0, \infty) \subset \mathbb{R}$. Then, $\|x\|_1 \leq \sum_{i=1}^{n} |\alpha_i|\|e_i\|_1 \leq \delta(\sum_{i=1}^{n} |\alpha_i|)$. Define $M_i = \operatorname{span}(\{e_1, \ldots, e_n\} \setminus \{e_i\})$, $i = 1, \ldots, n$, and $\delta_i = \operatorname{dist}(e_i, M_i)$ (with respect to $\|\cdot\|_1$). Since $\{e_1, \ldots, e_n\}$ is a basis for $\mathcal{X}$, then $e_i \notin M_i$. $\forall i \in \{1, \ldots, k\}$, by Theorem 7.36, M_i is complete (with respect to $\|\cdot\|_1$). By Proposition 4.39, M_i is closed (with respect to $\|\cdot\|_1$). By Proposition 4.10, we have $\delta_i \in (0, \infty) \subset \mathbb{R}$. Let $\bar{\delta} := \min\{\delta_1, \ldots, \delta_n\} \in (0, \infty) \subset \mathbb{R}$. Then, $\|x\|_1 = \left\|\sum_{i=1}^{n} \alpha_i e_i\right\|_1 \geq |\alpha_i|\delta_i \geq |\alpha_i|\bar{\delta}$, $\forall i \in \{1, \ldots, n\}$. Hence, we have $\|x\|_1 \geq (\bar{\delta}/n)(\sum_{i=1}^{n} |\alpha_i|)$. Let $K_1 = \max\{\delta, n/\bar{\delta}\} > 0$. Then, we have $(\sum_{i=1}^{n} |\alpha_i|)/K_1 \leq \|x\|_1 \leq K_1(\sum_{i=1}^{n} |\alpha_i|)$.

This completes the proof of the theorem. □

7.5 Compactness

Definition 7.39 Let $\mathcal{X}$ be a normed linear space and $S \subseteq \mathcal{X}$. S is said to be *compact* if S together with the natural metric forms a compact metric space. ◇

Proposition 5.29 says that a continuous function achieves its minimum and maximum on nonempty countably compact spaces. This has immediate generalization to infinite-dimensional spaces. Yet, the compactness restriction is so severe in infinite-dimensional spaces that it is applicable in minority of problems.

Lemma 7.40 *Let $(\mathcal{X}, \mathbb{C})$ be a vector space. Then $\mathcal{X}$ is also a vector space over the field $\mathbb{R}$. Furthermore, if $\mathfrak{X} := (\mathcal{X}, \mathbb{C}, \|\cdot\|)$ is a normed linear space, then $\mathfrak{X}_{\mathbb{R}} := (\mathcal{X}, \mathbb{R}, \|\cdot\|)$ is also a normed linear space. $\mathfrak{X}$ and $\mathfrak{X}_{\mathbb{R}}$ are isometric and admit the same metric space properties. In particular, $\mathfrak{X}$ is a Banach space if, and only if, $\mathfrak{X}_{\mathbb{R}}$ is a Banach space.*

Proof Let $(\mathcal{X}, \mathbb{C})$ be a vector space. Note that $\mathbb{R} \subset \mathbb{C}$. Then, $\forall x, y, z \in \mathcal{X}$, $\forall \alpha, \beta \in \mathbb{R}$, we have (i) $x + y = y + x$; (ii) $(x + y) + z = x + (y + z)$; (iii) $\vartheta_{\mathcal{X}} + x = x$; (iv) $\alpha(x + y) = \alpha x + \alpha y$; (v) $(\alpha + \beta)x = \alpha x + \beta x$; (vi) $(\alpha\beta)x = \alpha(\beta x)$; (vii) $0 \cdot x = \vartheta_{\mathcal{X}}$ and $1 \cdot x = x$. Hence, $(\mathcal{X}, \mathbb{R})$ is a vector space.

Let $\mathfrak{X} := (\mathcal{X}, \mathbb{C}, \|\cdot\|)$ be a normed linear space. Then, $(\mathcal{X}, \mathbb{R})$ is a vector space. $\forall x, y \in \mathfrak{X}$, $\forall \alpha \in \mathbb{R}$, we have (i) $\|x\| \in [0, \infty) \subset \mathbb{R}$ and $\|x\| = 0 \Leftrightarrow x = \vartheta_{\mathcal{X}}$; (ii) $\|x + y\| \leq \|x\| + \|y\|$; $\|\alpha x\| = |\alpha|\|x\|$. Hence, $\mathfrak{X}_{\mathbb{R}} := (\mathcal{X}, \mathbb{R}, \|\cdot\|)$ is a normed linear space.

Clearly, $\mathrm{id}_{\mathfrak{X}} : \mathfrak{X} \to \mathfrak{X}_{\mathbb{R}}$ is an isometry and the natural metrics induced by $\mathfrak{X}$ and $\mathfrak{X}_{\mathbb{R}}$ on $\mathcal{X}$ are identical. Hence, $\mathfrak{X}$ and $\mathfrak{X}_{\mathbb{R}}$ admits the same metric space properties. Then, $\mathfrak{X}$ is a Banach space if, and only if, $\mathfrak{X}$ is complete if, and only if, $\mathfrak{X}_{\mathbb{R}}$ is complete if, and only if, $\mathfrak{X}_{\mathbb{R}}$ is a Banach space.

This completes the proof of the lemma. □

Proposition 7.41 *Let $K \subseteq \mathbb{C}^n$ with $n \in \mathbb{Z}_+$. Then, K is compact if, and only if, K is closed and bounded.*

Proof "Necessity" By Proposition 5.38, K is complete and totally bounded. Then, K is bounded. By Proposition 4.39, K is closed.

"Sufficiency" Let $|\cdot|$ be the norm on $\mathbb{C}^n$. By Lemma 7.40, $\mathfrak{X} := (\mathbb{C}^n, \mathbb{R}, |\cdot|)$ admits the same metric space property as $\mathbb{C}^n$. Then, $K \subseteq \mathfrak{X}$ is closed and bounded. Note that $\mathfrak{X}$ is isometrically isomorphic to $\mathbb{R}^{2n}$. Hence, $K \subseteq \mathfrak{X}$ is compact. Then, $K \subseteq \mathbb{C}^n$ is compact. This completes the proof of the proposition. □

Proposition 7.42 *Let $\mathfrak{X}$ be a finite-dimensional normed linear space over the field $\mathbb{K}$. $K \subseteq \mathfrak{X}$ is compact if, and only if, K is closed and bounded.*

Proof "Necessity" By Proposition 5.38, K is complete and totally bounded. Then, K is bounded. By Proposition 4.39, K is closed.

"Sufficiency" Let $S \subseteq \mathfrak{X}$ be a basis for $\mathfrak{X}$, that is, S is linearly independent and $\mathrm{span}(S) = \mathfrak{X}$. Since $\mathfrak{X}$ is finite-dimensional, then let $n \in \mathbb{Z}_+$ be the dimension of $\mathfrak{X}$, which is well-defined by Theorem 6.51. We will distinguish two exhaustive and mutually exclusive cases: Case 1: $n = 0$; Case 2: $n \in \mathbb{N}$. Case 1: $n = 0$. Then, $\mathfrak{X}$ is a singleton set. Clearly K is compact. Case 2: $n \in \mathbb{N}$. Let $S = \{e_1, \ldots, e_n\}$. $\forall x \in \mathfrak{X}$, x can be uniquely expressed as $\sum_{i=1}^{n} \alpha_i e_i$ for some $\alpha_1, \ldots, \alpha_n \in \mathbb{K}$. This allows us to define a bijective mapping $\psi : \mathfrak{X} \to \mathbb{K}^n$ by $\psi(x) = (\alpha_1, \ldots, \alpha_n)$, $\forall x = \sum_{i=1}^{n} \alpha_i e_i \in \mathfrak{X}$. Define a alternative norm $\|\cdot\|_1$ on $\mathfrak{X}$ by $\|x\|_1 = \left\|\sum_{i=1}^{n} \alpha_i e_i\right\|_1 = \sqrt{\sum_{i=1}^{n} |\alpha_i|^2}$, $\forall x \in \mathfrak{X}$. It is easy to show that $\|\cdot\|_1$ is a norm. By Theorem 7.38, there exists $\xi \in [1, \infty) \subset \mathbb{R}$ such that $\|x\|_1/\xi \leq \|x\| \leq \xi\|x\|_1$, $\forall x \in \mathfrak{X}$. Hence, ψ is a homeomorphism. By Proposition 3.10, $\psi(K)$ is closed. By the equivalence of the two norms, $\psi(K)$ is bounded. By Propositions 5.40 or 7.41, $\psi(K)$ is compact. By Proposition 5.7, K is compact.

This completes the proof of the proposition. □

7.6 Quotient Spaces

Proposition 7.43 *Let $M \subseteq \mathcal{X}$ be a subspace of a vector space $\mathcal{X}$ over a field $\mathcal{F} := (F, +, \cdot, \mathbf{0}, \mathbf{1})$. $x_1, x_2 \in \mathcal{X}$ are said to be equivalent modulo M if $x_1 - x_2 \in M$. This equivalence relationship (easy to show) partitions the space $\mathcal{X}$ into disjoint subsets, or classes, of equivalent elements: namely, the linear varieties that are distinct translations of the subspace M. These classes are often called the cosets of M. $\forall x \in \mathcal{X}$, there is a unique coset of M, $[x](= x + M)$, such that $x \in [x]$. The quotient of $\mathcal{X}$ modulo M is defined to be the set N of all cosets of M. Define vector addition and scalar multiplication on N by, $\forall [x_1], [x_2] \in N$, $\forall \alpha \in \mathcal{F}$, $[x_1] + [x_2] := [x_1 + x_2]$ and $\alpha[x_1] := [\alpha x_1]$. Let the null vector on N be $[\vartheta_{\mathcal{X}}]$. Then, N together with the vector addition, scalar multiplication, and the null vector form a vector space over $\mathcal{F}$. This vector space will be called the quotient space of $\mathcal{X}$ modulo M and denoted by $\mathcal{X}/M$. Define a function $\phi : \mathcal{X} \to \mathcal{X}/M$ by $\phi(x) = [x]$, $\forall x \in \mathcal{X}$. Then, ϕ is a linear function and will be called the natural homomorphism.*

Proof Fix any $[x_1], [x_2], [x_3] \in \mathcal{X}/M$ and any $\alpha, \beta \in \mathcal{F}$. We will first show that vector addition and scalar multiplication are uniquely defined. $\forall y_1 \in [x_1]$, $\forall y_2 \in [x_2]$, we have $x_1 - y_1, x_2 - y_2 \in M$; and $(x_1 + x_2) - (y_1 + y_2) \in M$, since M is a subspace. Then, $[x_1] + [x_2] = [x_1 + x_2] = [y_1 + y_2] = [y_1] + [y_2]$. Hence, the vector addition is uniquely defined. $\alpha x_1 - \alpha y_1 = \alpha(x_1 - y_1) \in M$. Then, $\alpha[x_1] = [\alpha x_1] = [\alpha y_1] = \alpha[y_1]$. Hence, the scalar multiplication is uniquely defined.

(i) $[x_1] + [x_2] = [x_1 + x_2] = [x_2 + x_1] = [x_2] + [x_1]$; (ii) $([x_1] + [x_2]) + [x_3] = [(x_1 + x_2) + x_3] = [x_1 + (x_2 + x_3)] = [x_1] + ([x_2] + [x_3])$; (iii) $[x_1] + [\vartheta_{\mathcal{X}}] = [x_1 + \vartheta_{\mathcal{X}}] = [x_1]$; (iv) $\alpha([x_1] + [x_2]) = \alpha[x_1 + x_2] = [\alpha(x_1 + x_2)] = [\alpha x_1 + \alpha x_2] = [\alpha x_1] + [\alpha x_2] = \alpha[x_1] + \alpha[x_2]$; (v) $(\alpha + \beta)[x_1] = [(\alpha + \beta)x_1] = [\alpha x_1 + \beta x_1] = [\alpha x_1] + [\beta x_1] = \alpha[x_1] + \beta[x_1]$; (vi) $(\alpha\beta)[x_1] = [(\alpha\beta)x_1] = [\alpha(\beta x_1)] = \alpha[\beta x_1] = \alpha(\beta[x_1])$; (vii) $\mathbf{0}[x_1] = [\mathbf{0}x_1] = [\vartheta_{\mathcal{X}}]$; $\mathbf{1}[x_1] = [x_1]$. Hence, $\mathcal{X}/M$ is a vector space over $\mathcal{F}$.

Clearly, ϕ is a linear function. This completes the proof of the proposition. □

Proposition 7.44 *Let $\mathfrak{X} := (\mathcal{X}, \mathbb{K}, \|\cdot\|_{\mathfrak{X}})$ be a normed linear space and $M \subseteq \mathfrak{X}$ be a closed subspace, and $\mathcal{X}/M$ be the quotient space of $\mathcal{X}$ modulo M. Define a norm $\|\cdot\|$ on $\mathcal{X}/M$ by, $\forall [x] \in \mathcal{X}/M$, $\|[x]\| := \inf_{m \in M} \|x - m\|_{\mathfrak{X}} = \mathrm{dist}(x, M)$. Then, $\mathfrak{X}/M := (\mathcal{X}/M, \mathbb{K}, \|\cdot\|)$ is a normed linear space, which will be called the quotient space of $\mathfrak{X}$ modulo M.*

Proof By Proposition 7.43, $\mathcal{X}/M$ is a vector space over $\mathbb{K}$. Here, we only need to show that $\|\cdot\|$ defines a norm on $\mathcal{X}/M$. First, we show that $\|\cdot\| : \mathcal{X}/M \to [0, \infty) \subset \mathbb{R}$ is uniquely defined. $\forall [x] \in \mathcal{X}/M$, $\forall y \in [x]$, $y - x \in M$. Then, we have $\|[y]\| =$

$\inf_{m\in M} \|y - m\|_{\mathcal{X}} = \inf_{m\in M} \|x - y + y - m\|_{\mathcal{X}} = \inf_{m\in M} \|x - m\|_{\mathcal{X}} = \|[x]\| \leq \|x\|_{\mathcal{X}} < +\infty$. Hence, $\|\cdot\|$ is uniquely defined.

Next, we show that $\|\cdot\|$ defines a norm on $\mathcal{X}/M$. $\forall [x_1], [x_2] \in \mathcal{X}/M$, $\forall \alpha \in \mathbb{K}$, we have (i) $\|[x_1]\| \in [0, \infty) \subset \mathbb{R}$, since $\|[x_1]\| \leq \|x_1\|_{\mathcal{X}} < +\infty$. $[x_1] = [\vartheta_{\mathcal{X}}] \Rightarrow \|[x_1]\| = \inf_{m\in M} \|m\|_{\mathcal{X}} = 0$ and $\|[x_1]\| = 0 \Rightarrow \text{dist}(x_1, M) = 0$, by Proposition 4.10, we have $x_1 \in M \Rightarrow [x_1] = [\vartheta_{\mathcal{X}}]$. Hence $[x_1] = [\vartheta_{\mathcal{X}}] \Leftrightarrow \|[x_1]\| = 0$.

(ii) $\|[x_1] + [x_2]\| = \|[x_1 + x_2]\| = \inf_{m\in M} \|x_1 + x_2 - m\|_{\mathcal{X}}$. Note that $\|[x_i]\| = \inf_{m\in M} \|x_i - m\|_{\mathcal{X}}$, $i = 1, 2$. $\forall \epsilon \in (0, \infty) \subset \mathbb{R}$, $i = 1, 2$, $\exists m_i \in M$ such that $\|x_i - m_i\|_{\mathcal{X}} < \|[x_i]\| + \epsilon$. Then, we have $\|x_1 + x_2 - m_1 - m_2\|_{\mathcal{X}} \leq \|x_1 - m_1\|_{\mathcal{X}} + \|x_2 - m_2\|_{\mathcal{X}} \leq \|[x_1]\| + \|[x_2]\| + 2\epsilon$. Hence, we have $\|[x_1] + [x_2]\| \leq \|[x_1]\| + \|[x_2]\| + 2\epsilon$. By the arbitrariness of ϵ, we have $\|[x_1] + [x_2]\| \leq \|[x_1]\| + \|[x_2]\|$.

(iii) $\|\alpha[x_1]\| = \|[\alpha x_1]\| = \inf_{m\in M} \|\alpha x_1 - m\|_{\mathcal{X}}$. We will distinguish two exhaustive and mutually exclusive cases: Case 1: $\alpha = 0$; Case 2: $\alpha \neq 0$. Case 1: $\alpha = 0$. $\|\alpha[x_1]\| = \inf_{m\in M} \|m\|_{\mathcal{X}} = 0 = |\alpha| \|[x_1]\|$. Case 2: $\alpha \neq 0$. $\|\alpha[x_1]\| = \inf_{m\in M} \|\alpha x_1 - \alpha m\|_{\mathcal{X}} = \inf_{m\in M} |\alpha| \|x_1 - m\|_{\mathcal{X}} = |\alpha| \|[x_1]\|$. Hence, in both cases, we have $\|\alpha[x_1]\| = |\alpha| \|[x_1]\|$.

This shows that $\|\cdot\|$ is a norm on $\mathcal{X}/M$. Hence, $\mathcal{X}/M$ is a normed linear space. This completes the proof of the proposition. □

Proposition 7.45 *Let $\mathcal{X}$ be a Banach space and $M \subseteq \mathcal{X}$ be a closed subspace. Then, the quotient space $\mathcal{X}/M$ is a Banach space.*

Proof By Proposition 7.44, $\mathcal{X}/M$ is a normed linear space. Here, we need only to show that $\mathcal{X}/M$ is complete. Let $\|\cdot\|_{\mathcal{X}}$ be the norm on $\mathcal{X}$ and $\|\cdot\|$ be the norm on $\mathcal{X}/M$. We will prove this using Proposition 7.27. Fix any absolutely summable series $([x_n])_{n=1}^{\infty} \subseteq \mathcal{X}/M$. Then, $\sum_{n=1}^{\infty} \|[x_n]\| \in \mathbb{R}$. $\forall n \in \mathbb{N}$, $\exists y_n \in [x_n]$ such that $\|y_n\|_{\mathcal{X}} < \|[x_n]\| + 2^{-n}$. Then, $\sum_{i=1}^{\infty} \|y_i\|_{\mathcal{X}} < \sum_{i=1}^{\infty} (\|[x_i]\| + 2^{-i}) = \sum_{i=1}^{\infty} \|[x_i]\| + 1$. Then, $(y_n)_{n=1}^{\infty} \subseteq \mathcal{X}$ is absolutely summable. By Proposition 7.27, $\sum_{n=1}^{\infty} y_n = y \in \mathcal{X}$, since $\mathcal{X}$ is complete. Now, it is easy to show that $\sum_{n=1}^{\infty} [x_n] = \sum_{n=1}^{\infty} [y_n] = [y] \in \mathcal{X}/M$. Hence, $\mathcal{X}/M$ is complete by Proposition 7.27. This completes the proof of the proposition. □

Definition 7.46 Let $(\mathcal{X}, \mathbb{K})$ be a vector space and $\|\cdot\| : \mathcal{X} \to [0, \infty) \subset \mathbb{R}$. Assume that $\|\cdot\|$ satisfies (ii) and (iii) of Definition 7.1 and $\|\vartheta\| = 0$, but not necessarily (i). Then, $\|\cdot\|$ is called a *pseudo-norm*. ◇

Proposition 7.47 *Let $(\mathcal{X}, \mathbb{K})$ be a vector space and $\|\cdot\| : \mathcal{X} \to [0, \infty) \subset \mathbb{R}$ be a pseudo-norm on $\mathcal{X}$. Then, the set $M := \{x \in \mathcal{X} \mid \|x\| = 0\}$ is a subspace of $(\mathcal{X}, \mathbb{K})$. On the quotient space $\mathcal{X}/M$, define a norm $\|\cdot\|_1 : \mathcal{X}/M \to [0, \infty) \subset \mathbb{R}$ by $\|[x]\|_1 = \|x\|$, $\forall [x] \in \mathcal{X}/M$. Then, the space $(\mathcal{X}/M, \mathbb{K}, \|\cdot\|_1)$ is a normed linear space, which will be called the quotient space of $(\mathcal{X}, \mathbb{K})$ modulo $\|\cdot\|$.*

Proof We need the following claim.

Claim 7.47.1 $\forall x \in \mathcal{X}$, $\forall m \in M$, we have $\|x\| = \|x + m\|$.

Proof of Claim Note that $\|x\| \leq \|x+m\| + \|-m\| = \|x+m\| \leq \|x\| + \|m\| = \|x\|$. This completes the proof of the claim. □

Clearly, $\vartheta_{\mathcal{X}} \in M \neq \emptyset$. $\forall m_1, m_2 \in M$, $\forall \alpha \in \mathbb{K}$, we have $\alpha m_1 \in M$, by the properties of the pseudo-norm, and $m_1 + m_2 \in M$ by Claim 7.47.1. Hence, M is a subspace of $(\mathcal{X}, \mathbb{K})$. By Proposition 7.43, $\mathcal{X}/M$ is a vector space over $\mathbb{K}$.

By Claim 7.47.1, $\|\cdot\|_1 : \mathcal{X}/M \to [0, \infty) \subset \mathbb{R}$ is uniquely defined.

Next, we show that $\|\cdot\|_1$ is a norm on $\mathcal{X}/M$. $\forall [x_1], [x_2] \in \mathcal{X}/M$, $\forall \alpha \in \mathbb{K}$, $\|[x_1]\|_1 = \|x_1\| \in [0, \infty) \subset \mathbb{R}$. $[x_1] = [\vartheta_{\mathcal{X}}]$ implies that $\|[x_1]\|_1 = \|[\vartheta_{\mathcal{X}}]\|_1 = \|\vartheta_{\mathcal{X}}\| = 0$. $\|[x_1]\|_1 = 0$ implies that $\|x_1\| = 0$ and $x_1 \in M$, which further implies that $x_1 \in [\vartheta_{\mathcal{X}}]$ and $[x_1] = [\vartheta_{\mathcal{X}}]$. $\|[x_1] + [x_2]\|_1 = \|[x_1 + x_2]\|_1 = \|x_1 + x_2\| \leq \|x_1\| + \|x_2\| = \|[x_1]\|_1 + \|[x_2]\|_1$. $\|\alpha[x_1]\|_1 = \|[\alpha x_1]\|_1 = \|\alpha x_1\| = |\alpha|\|x_1\| = |\alpha|\|[x_1]\|_1$. Hence, $\|\cdot\|_1$ is a norm on $\mathcal{X}/M$. Therefore, $(\mathcal{X}/M, \mathbb{K}, \|\cdot\|_1)$ is a normed linear space. This completes the proof of the proposition. □

7.7 The Stone-Weierstrass Theorem

Definition 7.48 Let X be a set, $f : X \to \mathbb{R}$, and $g : X \to \mathbb{R}$. $f \vee g : X \to \mathbb{R}$ and $f \wedge g : X \to \mathbb{R}$ are defined by $(f \vee g)(x) = \max\{f(x), g(x)\}$ and $(f \wedge g)(x) = \min\{f(x), g(x)\}$, $\forall x \in \mathcal{X}$. ◇

Proposition 7.49 *Let $\mathcal{X}$ be a topological space and $f : \mathcal{X} \to \mathbb{R}$ and $g : \mathcal{X} \to \mathbb{R}$ be continuous at $x_0 \in \mathcal{X}$. Then, $f \vee g$ and $f \wedge g$ are continuous at x_0. If furthermore f and g are continuous, then $f \vee g$ and $f \wedge g$ are continuous.*

Proof This is straightforward. □

Example 7.50 Let $\mathcal{X}$ be a topological space and $\mathcal{Y}$ be a normed linear space over the field $\mathbb{K}$. Let $(\mathcal{M}(\mathcal{X}, \mathcal{Y}), \mathbb{K})$ be the vector space defined in Example 6.20. Let $V := \{f \in \mathcal{M}(\mathcal{X}, \mathcal{Y}) \mid f \text{ is continuous}\}$. Then, by Propositions 3.12, 3.32, 6.25, and 7.23, V is a subspace of $(\mathcal{M}(\mathcal{X}, \mathcal{Y}), \mathbb{K})$. This subspace will be denoted by $\mathcal{C}_v(\mathcal{X}, \mathcal{Y})$. ◇

Definition 7.51 Let $\mathcal{X}$ be a topological space, $\mathcal{C}_v(\mathcal{X}, \mathbb{R})$ be the vector space defined in Example 7.50. $L \subseteq \mathcal{C}_v(\mathcal{X}, \mathbb{R})$ is said to be a *lattice* if $L \neq \emptyset$ and $\forall f, g \in L$, $f \vee g \in L$ and $f \wedge g \in L$. A subspace $M \subseteq \mathcal{C}_v(\mathcal{X}, \mathbb{R})$ is said to be an *algebra* if $\forall f, g \in M$, $fg \in M$, where fg is the product of functions f and g. ◇

Proposition 7.52 *Let $\mathcal{X} := (X, \mathcal{O})$ be a compact space, $\mathcal{C}(\mathcal{X}, \mathbb{R})$ be the Banach space as defined in Example 7.32, and $L \subseteq \mathcal{C}(\mathcal{X}, \mathbb{R})$ be a lattice. Assume $h : \mathcal{X} \to \mathbb{R}$ defined by $h(x) = \inf_{f \in L} f(x)$, $\forall x \in \mathcal{X}$ is continuous. Then, $\forall \epsilon \in (0, \infty) \subset \mathbb{R}$, $\exists g \in L$ such that $0 \leq g(x) - h(x) < \epsilon$, $\forall x \in \mathcal{X}$.*

Proof We will distinguish two exhaustive and mutually exclusive cases: Case 1: $\mathcal{X} = \emptyset$; Case 2: $\mathcal{X} \neq \emptyset$. Case 1: $\mathcal{X} = \emptyset$. $\mathcal{C}(\mathcal{X}, \mathbb{R})$ is a singleton set and $L = \mathcal{C}(\mathcal{X}, \mathbb{R})$ since L is a lattice and therefore nonempty. Then, the result holds. Case 2: $\mathcal{X} \neq \emptyset$. $\forall \epsilon \in (0, \infty) \subset \mathbb{R}$, $\forall x \in \mathcal{X}$, since $h(x) \in \mathbb{R}$, then $\exists f_x \in L$ such that

$0 \leq f_x(x) - h(x) < \epsilon/3$. Since f_x and h are continuous, then $\exists O_x \in \mathcal{O}$ with $x \in O_x$ such that $\forall y \in O_x$, we have $|f_x(y) - f_x(x)| < \epsilon/3$ and $|h(y) - h(x)| < \epsilon/3$. Then, $|f_x(y) - h(y)| \leq |f_x(y) - f_x(x)| + |f_x(x) - h(x)| + |h(x) - h(y)| < \epsilon$, $\forall y \in O_x$. Then, $\mathcal{X} \subseteq \bigcup_{x \in \mathcal{X}} O_x$. Since $\mathcal{X}$ is compact, then there exists a finite set $X_N \subseteq \mathcal{X}$ such that $\mathcal{X} \subseteq \bigcup_{x \in X_N} O_x$. Since $X \neq \emptyset$ then X_N must be nonempty. Let $g := \bigwedge_{x \in X_N} f_x \in L$. $\forall x \in \mathcal{X}$, $\exists x_0 \in X_N$ such that $x \in O_{x_0}$ and $0 \leq g(x) - h(x) \leq f_{x_0}(x) - h(x) < \epsilon$. This completes the proof of the proposition. □

Proposition 7.53 *Let $\mathcal{X} := (X, \mathcal{O})$ be a compact space, $\mathcal{C}(\mathcal{X}, \mathbb{R})$ be the Banach space as defined in Example 7.32, and $L \subseteq \mathcal{C}(\mathcal{X}, \mathbb{R})$ be a lattice satisfying the following conditions:*

(i) *L separates points, that is $\forall x, y \in \mathcal{X}$ with $x \neq y$, $\exists f \in L$ such that $f(x) \neq f(y)$.*
(ii) *$\forall f \in L$, $\forall c \in \mathbb{R}$, $c + f, cf \in L$.*

Then, $\forall h \in \mathcal{C}(\mathcal{X}, \mathbb{R})$, $\forall \epsilon \in (0, \infty) \subset \mathbb{R}$, there exists $g \in L$ such that $0 \leq g(x) - h(x) < \epsilon$, $\forall x \in \mathcal{X}$.

Proof We will distinguish two exhaustive and mutually exclusive cases: Case 1: $\mathcal{X} = \emptyset$; Case 2: $\mathcal{X} \neq \emptyset$. Case 1: $\mathcal{X} = \emptyset$. $\mathcal{C}(\mathcal{X}, \mathbb{R})$ is a singleton set and $L = \mathcal{C}(\mathcal{X}, \mathbb{R})$ since L is a lattice and therefore nonempty. Then, the result holds.

Case 2: $\mathcal{X} \neq \emptyset$. We need the following two results.

Claim 7.53.1 $\forall a, b \in \mathbb{R}$, $\forall x_1, x_2 \in \mathcal{X}$ with $x_1 \neq x_2$, $\exists f \in L$ such that $f(x_1) = a$ and $f(x_2) = b$.

Proof of Claim Let $g \in L$ be such that $g(x_1) \neq g(x_2)$, which exists by (i). Let

$$f = \frac{a - b}{g(x_1) - g(x_2)} g + \frac{b g(x_1) - a g(x_2)}{g(x_1) - g(x_2)} \in L$$

Then, f is the desired function. □

Claim 7.53.2 $\forall a, b \in \mathbb{R}$ with $a \leq b$, $\forall$ closed set $F \subseteq \mathcal{X}$, $\forall x_0 \in \mathcal{X}$ with $x_0 \notin F$, $\exists f \in L$ such that $f(x_0) = a$, $f(x) \geq a$, $\forall x \in \mathcal{X}$, and $f(x) > b$, $\forall x \in F$.

Proof of Claim $\forall x \in F$, we have $x \neq x_0$. By Claim 7.53.1, $\exists f_x \in L$ such that $f_x(x_0) = a$ and $f_x(x) = b + 1$. Let $O_x := \{\bar{x} \in \mathcal{X} \mid f_x(\bar{x}) > b\}$. Then, $O_x \in \mathcal{O}$ since f_x is continuous. Clearly, $x \in O_x$. $F \subseteq \bigcup_{x \in F} O_x$. By Proposition 5.5, F is compact. Then, there exists a finite set $F_N \subseteq F$ such that $F \subseteq \bigcup_{x \in F_N} O_x$. Take $g \in L \neq \emptyset$. Then, $a = 0g + a \in L$ by (ii). Let $f := a \vee \left(\bigvee_{x \in F_N} f_x\right) \in L$. Clearly, $f(x_0) = a$ and $f(x) \geq a$, $\forall x \in \mathcal{X}$. $\forall x \in F$, $\exists x_0 \in F_N$ such that $x \in O_{x_0}$. Then, $f(x) \geq f_{x_0}(x) > b$. Hence, f is the desired function. This completes the proof of the claim. □

$\forall h \in \mathcal{C}(\mathcal{X}, \mathbb{R})$, let $\bar{L} := \{f \in L \mid f(x) \geq h(x), \forall x \in \mathcal{X}\}$. By the continuity of h, the compactness of $\mathcal{X}$, and Proposition 5.29, $\exists b \in \mathbb{R}$ such that $h(x) \leq b$, $\forall x \in \mathcal{X}$. Then, the constant function $b \in \bar{L} \neq \emptyset$ ($b \in L$ by (ii)). $\forall f_1, f_2 \in \bar{L}$, it is easy to show that $f_1 \vee f_2 \in \bar{L}$ and $f_1 \wedge f_2 \in \bar{L}$. Hence, $\bar{L}$ is a lattice. We will show that $h(x) = \inf_{f \in \bar{L}} f(x)$, $\forall x \in \mathcal{X}$. Then, the result follows from Proposition 7.52. $\forall x_0 \in \mathcal{X}$, $\forall \eta \in (0, \infty) \subset \mathbb{R}$, let $F_{x_0,\eta} := \{x \in \mathcal{X} \mid h(x) \geq h(x_0) + \eta\}$. By the continuity of h and Proposition 3.10, $F_{x_0,\eta}$ is closed. Clearly, $x_0 \notin F_{x_0,\eta}$. Now, by Claim 7.53.2, $\exists f_{x_0,\eta} \in L$ such that $f_{x_0,\eta}(x_0) = h(x_0) + \eta$, $f_{x_0,\eta}(x) \geq h(x_0) + \eta$, $\forall x \in \mathcal{X}$, and $f_{x_0,\eta}(x) > b$, $\forall x \in F_{x_0,\eta}$. It is clear that $f_{x_0,\eta}(x) > h(x)$, $\forall x \in \mathcal{X}$. Hence, $f_{x_0,\eta} \in \bar{L}$. Then, $\inf_{f \in \bar{L}} f(x_0) \leq f_{x_0,\eta}(x_0) = h(x_0) + \eta$. By the definition of $\bar{L}$ and the arbitrariness of η, we have $h(x_0) \leq \inf_{f \in \bar{L}} f(x_0) \leq h(x_0)$. Hence, we have $h(x_0) = \inf_{f \in \bar{L}} f(x_0)$.

This completes the proof of the proposition. □

Lemma 7.54 $\forall \epsilon \in (0, \infty) \subset \mathbb{R}$, *there exists a polynomial* $P(s)$ *in one variable such that* $|P(s) - |s|| < \epsilon$, $\forall s \in [-1, 1] \subset \mathbb{R}$.

Proof This result is a special case of Bernsteĭn Approximation Theorem (Bartle, 1976, pg. 171). For a generalization of the Bernsteĭn Approximation Theorem to multi-variable case, see Page 532. □

Lemma 7.55 *Let* $\mathcal{X}$ *be a nonempty compact space and* $\mathcal{A} \subseteq \mathcal{C}(\mathcal{X}, \mathbb{R})$ *be an algebra. Then,* $\overline{\mathcal{A}}$ *is an algebra.*

Proof Note that $\mathcal{A}$ is a subspace of $\mathcal{C}(\mathcal{X}, \mathbb{R})$ by Definition 7.51. Then, by Proposition 7.17, $\overline{\mathcal{A}}$ is a subspace of $\mathcal{C}(\mathcal{X}, \mathbb{R})$. $\forall f, g \in \overline{\mathcal{A}}$, $\forall \epsilon \in (0, 1) \subset \mathbb{R}$, $\exists \bar{f}, \bar{g} \in \mathcal{A}$ such that $\|f - \bar{f}\| < \frac{\epsilon}{2(1+\|g\|)}$ and $\|g - \bar{g}\| < \frac{\epsilon}{2(1+\|f\|)}$. Then, $\bar{f}\bar{g} \in \mathcal{A}$ since $\mathcal{A}$ is an algebra, and $\|fg - \bar{f}\bar{g}\| = \max_{x \in \mathcal{X}} |f(x)g(x) - \bar{f}(x)\bar{g}(x)|$ $\leq \max_{x \in \mathcal{X}} \left(|f(x)g(x) - f(x)\bar{g}(x)| + |f(x)\bar{g}(x) - \bar{f}(x)\bar{g}(x)|\right) \leq \max_{x \in \mathcal{X}} |f(x)|$ $|g(x) - \bar{g}(x)| + \max_{x \in \mathcal{X}} |\bar{g}(x)| \cdot |f(x) - \bar{f}(x)| \leq \|f\|\|g - \bar{g}\| + \|g + \bar{g} - g\|$ $\|f - \bar{f}\| < \frac{\epsilon\|f\|}{2(1+\|f\|)} + \frac{\epsilon(\|g\|+\|g-\bar{g}\|)}{2(1+\|g\|)} < \epsilon$. Hence, $fg \in \overline{\mathcal{A}}$ by the arbitrariness of ϵ. This shows that $\overline{\mathcal{A}}$ is an algebra. This completes the proof of the lemma. □

Theorem 7.56 (Stone-Weierstrass Theorem) *Let* $\mathcal{X} := (X, \mathcal{O})$ *be a compact space and* $\mathcal{C}(\mathcal{X}, \mathbb{R})$ *be the Banach space defined in Example 7.32. Assume that* $\mathcal{A} \subseteq \mathcal{C}(\mathcal{X}, \mathbb{R})$ *is an algebra and satisfies*

(i) $\mathcal{A}$ *separates points, that is,* $\forall x_1, x_2 \in \mathcal{X}$ *with* $x_1 \neq x_2$*, then* $\exists f \in \mathcal{A}$ *such that* $f(x_1) \neq f(x_2)$.
(ii) $\mathcal{A}$ *contains all constant functions.*

Then, $\overline{\mathcal{A}} = \mathcal{C}(\mathcal{X}, \mathbb{R})$*, that is* $\mathcal{A}$ *is dense in* $\mathcal{C}(\mathcal{X}, \mathbb{R})$.

Proof We will distinguish two exhaustive and mutually exclusive cases: Case 1: $\mathcal{X} = \emptyset$; Case 2: $\mathcal{X} \neq \emptyset$. Case 1: $\mathcal{X} = \emptyset$. $\mathcal{C}(\mathcal{X}, \mathbb{R})$ is a singleton set and $\mathcal{A} = \mathcal{C}(\mathcal{X}, \mathbb{R})$ since $\mathcal{A}$ is a subspace and therefore nonempty. Then, the result holds.

Case 2: $\mathcal{X} \neq \emptyset$. By Lemma 7.55, $\overline{\mathcal{A}}$ is an algebra. We need the following claim.

Claim 7.56.1 $\forall f \in \overline{\mathcal{A}}$, then $|f| \in \overline{\mathcal{A}}$.

Proof of Claim Fix $f \in \overline{\mathcal{A}}$. $\forall \epsilon \in (0, \infty) \subset \mathbb{R}$, $\exists g \in \mathcal{A}$ such that $\|f - g\| < \epsilon/2$. By Proposition 5.29 and Definitions 5.1 and 5.21, $\exists N \in (0, \infty) \subset \mathbb{R}$ such that $\|g\| < N$. Then, $g/N \in \mathcal{A}$ since $\mathcal{A}$ is an algebra. By Lemma 7.54, $\exists$ a polynomial $P(s)$ such that $|P(s) - |s|| < \frac{\epsilon}{2N}$, $\forall s \in [-1, 1] \subset \mathbb{R}$. Hence, we have $P \circ (g/N) \in \mathcal{A}$ since $\mathcal{A}$ is an algebra. Furthermore, $|P(g(x)/N) - |g(x)|/N| < \frac{\epsilon}{2N}$, $\forall x \in \mathcal{X}$. Let $h = NP \circ (g/N) \in \mathcal{A}$. Then, we have $\|h - |g|\| < \epsilon/2$. Note that $\||f| - h\| \leq \||f| - |g|\| + \||g| - h\| \leq \|f - g\| + \||g| - h\| < \epsilon$. Hence, $|f| \in \overline{\mathcal{A}}$, by the arbitrariness of ϵ. This completes the proof of the claim. □

$\forall f, g \in \overline{\mathcal{A}}$, we have

$$f \vee g = \frac{1}{2}(f + g) + \frac{1}{2}|f - g| \in \overline{\mathcal{A}}$$

$$f \wedge g = \frac{1}{2}(f + g) - \frac{1}{2}|f - g| \in \overline{\mathcal{A}}$$

Hence, $\overline{\mathcal{A}}$ is a lattice. Clearly, $\overline{\mathcal{A}}$ separates points since $\mathcal{A}$ separates points. By Proposition 7.53, $\forall h \in \mathcal{C}(\mathcal{X}, \mathbb{R})$, $\forall \epsilon \in (0, \infty) \subset \mathbb{R}$, there exists $g \in \overline{\mathcal{A}}$ such that $0 \leq g(x) - h(x) < \epsilon$, $\forall x \in \mathcal{X}$. Then, $\|g - h\| < \epsilon$, and $h \in \overline{\overline{\mathcal{A}}} = \overline{\mathcal{A}}$, where the last equality follows from Proposition 3.3. Hence, we have $\overline{\mathcal{A}} = \mathcal{C}(\mathcal{X}, \mathbb{R})$. This completes the proof of the theorem. □

Corollary 7.57 *Let $X \subseteq \mathbb{R}^n$ be a closed and bounded set with the subset topology $\mathcal{O}$, where $n \in \mathbb{Z}_+$, $\mathcal{X} := (X, \mathcal{O})$, and $f : \mathcal{X} \to \mathbb{R}$ be a continuous function. Then, $\forall \epsilon \in (0, \infty) \subset \mathbb{R}$, there exists a polynomial $P : \mathcal{X} \to \mathbb{R}$ in n variables such that $|f(x) - P(x)| < \epsilon$, $\forall x \in \mathcal{X}$.*

Proof By Proposition 5.40, $\mathcal{X}$ is a compact space. Let $\mathcal{C}(\mathcal{X}, \mathbb{R})$ be the Banach space defined in Example 7.32. Then, $f \in \mathcal{C}(\mathcal{X}, \mathbb{R})$. Let $\mathcal{A}$ be all polynomial in n variables on $\mathcal{X}$. Clearly, $\mathcal{A} \subseteq \mathcal{C}(\mathcal{X}, \mathbb{R})$ and is a linear subspace of $\mathcal{C}(\mathcal{X}, \mathbb{R})$. Then, it is easy to show that $\mathcal{A}$ is an algebra. $\forall x_1, x_2 \in \mathcal{X}$ with $x_1 \neq x_2$, there exists a coordinate $i_0 \in \{1, \ldots, n\}$ such that $\pi_{i_0}(x_1) \neq \pi_{i_0}(x_2)$. Then, the polynomial $f \in \mathcal{A}$ given by $f(x) = \pi_{i_0}(x)$, $\forall x \in \mathcal{X}$, separates x_1 and x_2. Hence, $\mathcal{A}$ separates points. Clearly, $\mathcal{A}$ contains all the constant functions. By the Stone-Weierstrass Theorem, $\overline{\mathcal{A}} = \mathcal{C}(\mathcal{X}, \mathbb{R})$. This completes the proof of the corollary. □

Corollary 7.58 *Let $S_1 = [0, 2\pi] \subset \mathbb{R}$ and $f \in \mathcal{C}(S_1, \mathbb{R})$. Assume that $f(0) = f(2\pi)$. Let $\mathcal{M} = \{g \in \mathcal{C}(S_1, \mathbb{R}) \mid g(x) = \cos(nx), \forall x \in S_1$ or $g(x) = \sin(nx), \forall x \in S_1$, where $n \in \mathbb{Z}_+\}$. Let $\mathcal{A} = \operatorname{span}(\mathcal{M}) \subseteq \mathcal{C}(S_1, \mathbb{R})$. Then, $f \in \overline{\mathcal{A}}$.*

Proof Let $S_2 \subset \mathbb{R}^2$ be the unit circle: $S_2 := \{(y_1, y_2) \in \mathbb{R}^2 \mid y_1^2 + y_2^2 = 1\}$. Define a mapping $\Psi : S_1 \to S_2$ by $\Psi(x) = (\cos(x), \sin(x))$, $\forall x \in S_1$. Clearly, Ψ is surjective and continuous. Note that S_1 is compact and S_2 is Hausdorff. It is obvious that we may define a function $\Phi : S_2 \to \mathbb{R}$ such that $\Phi \circ \Psi = f$, which is continuous. By Proposition 5.18, we have Φ is continuous. Note that S_2 is closed and bounded in $\mathbb{R}^2$, then S_2 is compact by Proposition 5.40. Hence, $\Phi \in \mathcal{C}(S_2, \mathbb{R})$. By Corollary 7.57, $\forall \epsilon \in (0, \infty) \subset \mathbb{R}$, there exists a polynomial $P : S_2 \to \mathbb{R}$ such that $|\Phi(y_1, y_2) - P(y_1, y_2)| < \epsilon$, $\forall y_1, y_2 \in S_2$. Then, $|f(x) - P \circ \Psi(x)| = |\Phi \circ \Psi(x) - P \circ \Psi(x)| < \epsilon$, $\forall x \in S_1$. Note that $P \circ \Psi \in \mathcal{A}$, since, $\forall \gamma, \theta \in \mathbb{R}$,

$$\begin{aligned}
(\sin(\theta))^2 &= \frac{1}{2}(1 - \cos(2\theta)); \qquad (\cos(\theta))^2 = \frac{1}{2}(1 + \cos(2\theta));\\
\sin(\gamma)\cos(\theta) &= \frac{1}{2}(\sin(\gamma + \theta) + \sin(\gamma - \theta));\\
\sin(\gamma)\sin(\theta) &= \frac{1}{2}(\cos(\gamma - \theta) - \cos(\gamma + \theta));\\
\cos(\gamma)\cos(\theta) &= \frac{1}{2}(\cos(\gamma - \theta) + \cos(\gamma + \theta))
\end{aligned}$$

Hence, $f \in \overline{\mathcal{A}}$, by the arbitrariness of ϵ. This completes the proof of the corollary. □

For ease of presentation below, we will define two functions: sqr : $\mathbb{R} \to \mathbb{R}$ by $\text{sqr}(x) = x^2$, $\forall x \in \mathbb{R}$, and sqrt : $[0, \infty) \to [0, \infty)$ by $\text{sqrt}(x) = \sqrt{x}$, $\forall x \in [0, \infty)$.

Corollary 7.59 *Let $\mathcal{X} := (X, \mathcal{O})$ be a compact space and $\mathcal{A} \subseteq \mathcal{C}(\mathcal{X}, \mathbb{R})$ be an algebra that satisfies*

(i) *$\mathcal{A}$ separates points.*
(ii) *$\exists f_0 \in \mathcal{A}$ such that $f_0(x) \neq 0$, $\forall x \in \mathcal{X}$.*

Then, the constant function $1 \in \overline{\mathcal{A}} = \mathcal{C}(\mathcal{X}, \mathbb{R})$.

Proof We will distinguish two exhaustive and mutually exclusive cases: Case 1: $\mathcal{X} = \emptyset$; Case 2: $\mathcal{X} \neq \emptyset$. Case 1: $\mathcal{X} = \emptyset$. $\mathcal{C}(\mathcal{X}, \mathbb{R})$ is a singleton set and $\mathcal{A} = \mathcal{C}(\mathcal{X}, \mathbb{R})$ since $\mathcal{A}$ is a subspace and therefore nonempty. Then, the result holds.

Case 2: $\mathcal{X} \neq \emptyset$. By Lemma 7.55, $\overline{\mathcal{A}}$ is an algebra.

Since $\mathcal{A}$ is an algebra, then, $\text{sqr} \circ f_0 \in \mathcal{A}$, which satisfies that $\text{sqr} \circ f_0(x) > 0$, $\forall x \in \mathcal{X}$. Furthermore, $g_0 := \text{sqr} \circ f_0 / \|\text{sqr} \circ f_0\| \in \mathcal{A}$, which satisfies $g_0 : \mathcal{X} \to (0, 1] \subset \mathbb{R}$.

By Corollary 7.57, $\forall \epsilon \in (0, \infty) \subset \mathbb{R}$, there exists a polynomial Q_ϵ in one variable such that $\left|\sqrt{s} - Q_\epsilon(s)\right| < \epsilon$, $\forall s \in [0, 1] \subset \mathbb{R}$, and $Q_\epsilon(0) = 0$.

$\forall f : \mathcal{X} \to [0, 1] \subset \mathbb{R}$ with $f \in \overline{\mathcal{A}}$, $Q_\epsilon \circ f \in \overline{\mathcal{A}}$ since $\overline{\mathcal{A}}$ is an algebra and $Q_\epsilon(0) = 0$. Then, $\text{sqrt} \circ f \in \overline{\overline{\mathcal{A}}} = \overline{\mathcal{A}}$. Recursively, we may conclude that $\text{sqrt}^n \circ g_0 \in \overline{\mathcal{A}}$, $\forall n \in \mathbb{N}$.

By Proposition 5.29, $\exists x_m \in \mathcal{X}$ such that $g_0(x) \geq g_0(x_m) =: \gamma > 0$, $\forall x \in \mathcal{X}$. $\forall \epsilon \in (0, 1) \subset \mathbb{R}$, $\exists n_0 \in \mathbb{N}$ such that $\left|1 - \gamma^{2^{-n_0}}\right| < \epsilon$. Then, $\|1 - \text{sqrt}^{n_0} \circ g_0\| < \epsilon$. Hence, the constant function $1 \in \overline{\overline{\mathcal{A}}} = \overline{\mathcal{A}}$.

Clearly, $\overline{\mathcal{A}}$ separates points since $\mathcal{A}$ does. Hence, by Theorem 7.56, $\overline{\mathcal{A}} = \overline{\overline{\mathcal{A}}} = \mathcal{C}(\mathcal{X}, \mathbb{R})$. This completes the proof of the corollary. □

Proposition 7.60 *Let $\mathcal{X}$ be a compact space and $\mathcal{A} \subseteq \mathcal{C}(\mathcal{X}, \mathbb{R})$ be an algebra that separates points on $\mathcal{X}$. Then, either $\overline{\mathcal{A}} = \mathcal{C}(\mathcal{X}, \mathbb{R})$ or $\exists x_0 \in \mathcal{X}$ such that $\overline{\mathcal{A}} = \{f \in \mathcal{C}(\mathcal{X}, \mathbb{R}) \mid f(x_0) = 0\}$.*

Proof We will distinguish two exhaustive and mutually exclusive cases: Case 1: $\mathcal{X} = \emptyset$; Case 2: $\mathcal{X} \neq \emptyset$. Case 1: $\mathcal{X} = \emptyset$. $\mathcal{C}(\mathcal{X}, \mathbb{R})$ is a singleton set and $\mathcal{A} = \mathcal{C}(\mathcal{X}, \mathbb{R})$ since $\mathcal{A}$ is a subspace and therefore nonempty. Then, the result holds.

Case 2: $\mathcal{X} \neq \emptyset$. By Lemma 7.55, $\overline{\mathcal{A}}$ is an algebra. We will further distinguish two exhaustive and mutually exclusive cases: Case 2a: $1 \in \overline{\mathcal{A}}$; Case 2b: $1 \notin \overline{\mathcal{A}}$. Case 2a: $1 \in \overline{\mathcal{A}}$. Clearly, $\overline{\mathcal{A}}$ is an algebra that separates points. By Stone-Weierstrass Theorem, $\mathcal{C}(\mathcal{X}, \mathbb{R}) = \overline{\overline{\mathcal{A}}} = \overline{\mathcal{A}}$. Then, the result holds.

Case 2b: $1 \notin \overline{\mathcal{A}}$. Then, $\overline{\mathcal{A}} \subset \mathcal{C}(\mathcal{X}, \mathbb{R})$. By Corollary 7.59, $\forall f \in \overline{\mathcal{A}}$, there exists $x \in \mathcal{X}$ such that $f(x) = 0$. Then, we have the following claims.

Claim 7.60.1 $\forall f \in \overline{\mathcal{A}}$, $|f| \in \overline{\mathcal{A}}$. Hence, $\overline{\mathcal{A}}$ is a lattice.

Proof of Claim $\forall f \in \overline{\mathcal{A}}$, by the compactness of $\mathcal{X}$ and Proposition 5.29, $\exists M \in (0, \infty) \subset \mathbb{R}$ such that $\|f\| \leq M$. Let $g := f/M \in \overline{\mathcal{A}}$. Then, $g : \mathcal{X} \to [-1, 1] \subset \mathbb{R}$. $\forall \epsilon \in (0, \infty) \subset \mathbb{R}$, by Lemma 7.54, there exists a polynomial $P : [-1, 1] \to \mathbb{R}$ with $P(0) = 0$ such that $|P(s) - |s|| < \epsilon/M$, $\forall s \in [-1, 1] \subset \mathbb{R}$. Since $\overline{\mathcal{A}}$ is an algebra, then $M \cdot P \circ g \in \overline{\mathcal{A}}$ and $\||f| - M \cdot P \circ g\| = M\||g| - P \circ g\| < \epsilon$. Hence, $|f| \in \overline{\overline{\mathcal{A}}} = \overline{\mathcal{A}}$.

$\forall f, g \in \overline{\mathcal{A}}$, we have

$$f \vee g = \frac{1}{2}(f + g) + \frac{1}{2}|f - g| \in \overline{\mathcal{A}}$$

$$f \wedge g = \frac{1}{2}(f + g) - \frac{1}{2}|f - g| \in \overline{\mathcal{A}}$$

Hence, $\overline{\mathcal{A}}$ is a lattice. This completes the proof of the claim. □

Claim 7.60.2 $\exists x_0 \in \mathcal{X}$ such that $f(x_0) = 0$, $\forall f \in \overline{\mathcal{A}}$.

Proof of Claim We will prove this using an argument of contradiction. Suppose the claim is false. $\forall x \in \mathcal{X}$, $\exists f_x \in \overline{\mathcal{A}}$ such that $f_x(x) \neq 0$. By Claim 7.60.1, $|f_x| \in \overline{\mathcal{A}}$ and $|f_x(x)| > 0$. Let $O_x := \{\bar{x} \in \mathcal{X} \mid |f_x(\bar{x})| > 0\}$. Since $f_x \in \overline{\mathcal{A}} \subseteq \mathcal{C}(\mathcal{X}, \mathbb{R})$, then $x \in O_x \in \mathcal{O}_{\mathcal{X}}$. Hence, we have $\mathcal{X} \subseteq \bigcup_{x \in \mathcal{X}} O_x$. By the compactness of $\mathcal{X}$, there exists a finite set $X_N \subseteq \mathcal{X}$ such that $\mathcal{X} \subseteq \bigcup_{x \in X_N} O_x$. Clearly, $X_N \neq \emptyset$ since $\mathcal{X} \neq \emptyset$. Let $f := \sum_{x \in X_N} |f_x| \in \overline{\mathcal{A}}$. Hence, $f(x) > 0$, $\forall x \in \mathcal{X}$. By Corollary 7.59, we have $1 \in \overline{\overline{\mathcal{A}}} = \overline{\mathcal{A}}$, which is a contradiction. Hence, the claim is true. This completes the proof of the claim. □

Claim 7.60.3 $\forall$ closed set $F \subseteq \mathcal{X}$ with $x_0 \notin F$, we have

(i) Let $\mathcal{A}_F := \{h \in \mathcal{C}(F, \mathbb{R}) \mid \exists f \in \overline{\mathcal{A}} \text{ such that } h = f|_F\}$, then $\overline{\mathcal{A}_F} = \mathcal{C}(F, \mathbb{R})$.
(ii) $\forall \epsilon \in (0, 1) \subset \mathbb{R}$, $\exists g \in \overline{\mathcal{A}}$ such that $g : \mathcal{X} \to [0, 1] \subset \mathbb{R}$ and $g(x) \geq 1 - \epsilon$, $\forall x \in F$.

Proof of Claim Clearly, $\mathcal{A}_F$ is an algebra since $\overline{\mathcal{A}}$ is an algebra. Furthermore, $\mathcal{A}_F$ separate points on F since $\overline{\mathcal{A}}$ separates points on $\mathcal{X}$.

By Proposition 5.5, F with the subset topology is compact.

$\forall x \in F$, we have $x \neq x_0$. $\exists f_x \in \mathcal{A}$ such that $f_x(x) \neq f_x(x_0) = 0$. By Claim 7.60.1, $|f_x| \in \overline{\mathcal{A}}$ and $|f_x(x)| > 0$. Let $O_x := \{\bar{x} \in \mathcal{X} \mid |f_x(\bar{x})| > 0\}$. Then, we have $x \in O_x \in \mathcal{O}_{\mathcal{X}}$. Hence, $F \subseteq \bigcup_{x \in F} O_x$. By the compactness of F, there exists a finite set $F_N \subseteq F$ such that $F \subseteq \bigcup_{x \in F_N} O_x$. Let $f := \sum_{x \in F_N} |f_x|$. Then, $f \in \overline{\mathcal{A}}$ and $f(x) > 0$, $\forall x \in F$.

Then, $h := f|_F \in \mathcal{A}_F$ and $h(x) > 0$, $\forall x \in F$. By Corollary 7.59, we have $\overline{\mathcal{A}_F} = \mathcal{C}(F, \mathbb{R})$. Hence, (i) is true.

$\exists M \in (0, \infty) \subset \mathbb{R}$ such that $\|f\| \leq M$. Then, $g_0 := f/M \in \overline{\mathcal{A}}$ and $g_0 : \mathcal{X} \to [0, 1] \subset \mathbb{R}$.

By Corollary 7.57, $\forall \epsilon \in (0, \infty) \subset \mathbb{R}$, there exists a polynomial Q_ϵ in one variable such that $\left|\sqrt{s} - Q_\epsilon(s)\right| < \epsilon$, $\forall s \in [0, 1] \subset \mathbb{R}$, and $Q_\epsilon(0) = 0$.

$\forall \bar{f} : \mathcal{X} \to [0, 1] \subset \mathbb{R}$ with $\bar{f} \in \overline{\mathcal{A}}$, $Q_\epsilon \circ \bar{f} \in \overline{\mathcal{A}}$ since $\overline{\mathcal{A}}$ is an algebra and $Q_\epsilon(0) = 0$. Then, $\text{sqrt} \circ \bar{f} \in \overline{\overline{\mathcal{A}}} = \overline{\mathcal{A}}$. Recursively, we may conclude that $\text{sqrt}^n \circ g_0 \in \overline{\mathcal{A}}$, $\forall n \in \mathbb{N}$.

By the compactness of F and Proposition 5.29, $\exists \gamma \in (0, 1] \subset \mathbb{R}$ such that $g_0(x) \geq \gamma$, $\forall x \in F$. $\forall \epsilon \in (0, 1) \subset \mathbb{R}$, $\exists n_0 \in \mathbb{N}$ such that $\gamma^{2^{-n_0}} \geq 1 - \epsilon$. Then, $g := \text{sqrt}^{n_0} \circ g_0 \in \overline{\mathcal{A}}$, $g : \mathcal{X} \to [0, 1] \subset \mathbb{R}$ and $g(x) \geq 1 - \epsilon$, $\forall x \in F$. Hence, (ii) is true.

This completes the proof of the claim. □

Claim 7.60.4 $\forall g \in \mathcal{C}(\mathcal{X}, \mathbb{R})$ with $g(x_0) = 0$ and $g(x) \geq 0$, $\forall x \in \mathcal{X}$, we have $g \in \overline{\mathcal{A}}$.

Proof of Claim $\forall \epsilon \in (0, \infty) \subset \mathbb{R}$, let $O := \{x \in \mathcal{X} \mid g(x) < \epsilon/2\}$. Then, $x_0 \in O \in \mathcal{O}_{\mathcal{X}}$ since g is continuous. Let $M := \|g\| \in [0, \infty) \subset \mathbb{R}$.

By Claim 7.60.3, $\exists h_1 \in \overline{\mathcal{A}}$ such that $h_1 : \mathcal{X} \to [0, 1] \subset \mathbb{R}$ and $h_1(x) \geq 2/3$, $\forall x \in \tilde{O}$. Note that $h_1(x_0) = 0$ by Claim 7.60.2. Let $h_2 := 3Mh_1/2 \in \overline{\mathcal{A}}$. Then, we have $0 \leq h_2(x) \leq 3M/2$, $\forall x \in \mathcal{X}$, and $h_2(x) \geq M \geq g(x)$, $\forall x \in \tilde{O}$.

Let $U := \{x \in O \mid h_2(x) < \epsilon/2\}$. Then, $x_0 \in U \in \mathcal{O}_{\mathcal{X}}$ since $O \in \mathcal{O}_{\mathcal{X}}$, h_2 is continuous, and $h_2(x_0) = 0$. Furthermore, $g|_{\tilde{U}} \in \mathcal{C}(\tilde{U}, \mathbb{R})$. By Claim 7.60.3, $\exists h_3 \in \overline{\mathcal{A}}$ such that $\left|h_3|_{\tilde{U}}(x) - g|_{\tilde{U}}(x)\right| < \epsilon$, $\forall x \in \tilde{U}$. Define $h_4 := (h_3 \vee 0) \wedge h_2$. Then, $h_4 \in \overline{\mathcal{A}}$ by Claim 7.60.1. $\forall x \in \mathcal{X}$, we will show that $|g(x) - h_4(x)| < \epsilon$ by distinguishing three exhaustive and mutually exclusive cases: Case A: $x \in U$; Case B: $x \in O \setminus U$; Case C: $x \in \tilde{O}$.

Case A: $x \in U$. Then, we have $0 \leq h_2(x) < \epsilon/2$, $0 \leq g(x) < \epsilon/2$. Then, $0 \leq h_4(x) = (h_3(x) \vee 0) \wedge h_2(x) \leq h_2(x) < \epsilon/2$. Therefore, $|g(x) - h_4(x)| < \epsilon$.

Case B: $x \in O \setminus U$. Then, we have $|h_3(x) - g(x)| < \epsilon$, $h_2(x) \geq 0$, and $0 \leq g(x) < \epsilon/2$. Hence, $0 \leq h_3(x) \vee 0 < g(x) + \epsilon$. Then, $0 \leq h_4(x) < g(x) + \epsilon$ and $-\epsilon/2 < -g(x) \leq h_4(x) - g(x) < \epsilon$. Therefore, we have $|g(x) - h_4(x)| < \epsilon$.

Case C: $x \in \tilde{O}$. Then, we have $|h_3(x) - g(x)| < \epsilon$ and $h_2(x) \geq M \geq g(x) \geq 0$. Hence, $g(x) - \epsilon < h_3(x) < g(x) + \epsilon$, $g(x) - \epsilon < h_3(x) \vee 0 < g(x) + \epsilon$, and $g(x) - \epsilon < h_4(x) < g(x) + \epsilon$. Therefore, $|g(x) - h_4(x)| < \epsilon$.

Hence, in all three cases, we have $|g(x) - h_4(x)| < \epsilon$. Hence $\|g - h_4\| < \epsilon$. By the arbitrariness of ϵ, we have $g \in \overline{\overline{\mathcal{A}}} = \overline{\mathcal{A}}$. □

$\forall f \in \mathcal{C}(\mathcal{X}, \mathbb{R})$ with $f(x_0) = 0$, let $f_+ := f \vee 0$ and $f_- := (-f) \vee 0$. Clearly, $f_+, f_- \in \mathcal{C}(\mathcal{X}, \mathbb{R})$, $f_+(x_0) = 0$, $f_-(x_0) = 0$, and $f = f_+ - f_-$. By Claim 7.60.4, $f_+, f_- \in \overline{\mathcal{A}}$. Then, $f \in \overline{\mathcal{A}}$ since $\overline{\mathcal{A}}$ is an algebra. This coupled with Claim 7.60.2, we have that $\overline{\mathcal{A}} = \{f \in \mathcal{C}(\mathcal{X}, \mathbb{R}) \mid f(x_0) = 0\}$. Hence, the result holds in this case.

This completes the proof of the proposition. □

7.8 Linear Operators

Definition 7.61 Let $\mathfrak{X}$ and $\mathfrak{Y}$ be normed linear spaces over the field $\mathbb{K}$. A linear operator $A : \mathfrak{X} \to \mathfrak{Y}$ is said to be *bounded* if $\exists M \in [0, \infty) \subset \mathbb{R}$ such that $\|Ax\|_{\mathfrak{Y}} \leq M\|x\|_{\mathfrak{X}}$, $\forall x \in \mathfrak{X}$. ◇

Here, the boundedness concept is different from Definition 7.12. We are talking about bounded linear operators here, not bounded operators. Definition 7.61 only applies to linear operators.

Proposition 7.62 *Let $\mathfrak{X}$ and $\mathfrak{Y}$ be normed linear spaces over $\mathbb{K}$ and $A : \mathfrak{X} \to \mathfrak{Y}$ be a linear operator. Then,*

(i) *If A is bounded then A is uniformly continuous.*
(ii) *If A is continuous at some $x_0 \in \mathfrak{X}$ then A is bounded.*

Proof (i) $\exists M \in [0, \infty) \subset \mathbb{R}$ such that $\|Ax\|_{\mathfrak{Y}} \leq M\|x\|_{\mathfrak{X}}$, $\forall x \in \mathfrak{X}$. $\forall \epsilon \in (0, \infty) \subset \mathbb{R}$, let $\delta = \epsilon/(1 + M) \in (0, \infty) \subset \mathbb{R}$. $\forall x_1, x_2 \in \mathfrak{X}$ with $\|x_1 - x_2\|_{\mathfrak{X}} < \delta$, we have $\|Ax_1 - Ax_2\|_{\mathfrak{Y}} = \|A(x_1 - x_2)\|_{\mathfrak{Y}} \leq M\|x_1 - x_2\|_{\mathfrak{X}} < \epsilon$. Hence A is uniformly continuous.

(ii) $\forall \epsilon \in (0, \infty) \subset \mathbb{R}$, $\exists \delta \in (0, \infty) \subset \mathbb{R}$, $\forall x \in \mathcal{B}_{\mathfrak{X}}(x_0, \delta)$, we have $\|Ax - Ax_0\|_{\mathfrak{Y}} < \epsilon$. $\forall \bar{x} \in \mathcal{B}_{\mathfrak{X}}(\vartheta_{\mathfrak{X}}, \delta)$, we have $\|A\bar{x}\|_{\mathfrak{Y}} = \|A(x_0 + \bar{x}) - Ax_0\|_{\mathfrak{Y}} < \epsilon$. Hence, A is continuous at $\vartheta_{\mathfrak{X}}$. Then, $\exists \delta_0 \in (0, \infty) \subset \mathbb{R}$ such that $\|Ax\|_{\mathfrak{Y}} < 1$, $\forall x \in \mathcal{B}_{\mathfrak{X}}(\vartheta_{\mathfrak{X}}, \delta_0)$. Then, $\forall x \in \mathfrak{X}$, we have two possibilities: (a) $x = \vartheta_{\mathfrak{X}}$, then $\|Ax\|_{\mathfrak{Y}} = 0 \leq \frac{2}{\delta_0}\|x\|_{\mathfrak{X}}$; (b) $x \neq \vartheta_{\mathfrak{X}}$, then, $\frac{\delta_0}{2\|x\|_{\mathfrak{X}}}x \in \mathcal{B}_{\mathfrak{X}}(\vartheta_{\mathfrak{X}}, \delta_0)$ and we have $\|Ax\|_{\mathfrak{Y}} = \left\|\frac{2\|x\|_{\mathfrak{X}}}{\delta_0}A\left(\frac{\delta_0}{2\|x\|_{\mathfrak{X}}}x\right)\right\|_{\mathfrak{Y}} = \frac{2\|x\|_{\mathfrak{X}}}{\delta_0}\left\|A\left(\frac{\delta_0}{2\|x\|_{\mathfrak{X}}}x\right)\right\|_{\mathfrak{Y}} < \frac{2}{\delta_0}\|x\|_{\mathfrak{X}}$. Hence, A is bounded.

This completes the proof of the proposition. □

Proposition 7.63 *Let $\mathcal{X}$ and $\mathcal{Y}$ be normed linear spaces over the field $\mathbb{K}$. Let $(\mathcal{M}(\mathcal{X}, \mathcal{Y}), \mathbb{K})$ be the vector space defined in Example* 6.20 *with the null vector ϑ. Let $P := \{A \in \mathcal{M}(\mathcal{X}, \mathcal{Y}) \mid A$ is linear. $\}$. Then, P is a subspace of $\mathcal{M}(\mathcal{X}, \mathcal{Y})$. Define a functional $\|\cdot\|$ on P by $\|A\| := \inf\{M \in [0, \infty) \subset \mathbb{R} \mid \|Ax\|_{\mathcal{Y}} \le M\|x\|_{\mathcal{X}},\ \forall x \in \mathcal{X}\}$, $\forall A \in P$. Then, $\forall A \in P$, $\|A\|$ may be equivalently expressed as*

$$\begin{aligned} \|A\| &= \sup_{x \in \mathcal{X}, \|x\|_{\mathcal{X}} \le 1} \|Ax\|_{\mathcal{Y}} \\ &= \max\left\{ \sup_{\substack{x \in \mathcal{X} \\ x \neq \vartheta_{\mathcal{X}}}} \frac{\|Ax\|_{\mathcal{Y}}}{\|x\|_{\mathcal{X}}}, 0 \right\} = \max\left\{ \sup_{\substack{x \in \mathcal{X} \\ \|x\|_{\mathcal{X}} = 1}} \|Ax\|_{\mathcal{Y}}, 0 \right\} \end{aligned}$$

Let $N := \{A \in P \mid \|A\| < +\infty\}$. Then, $(N, \mathbb{K}, \|\cdot\|) =: \mathrm{B}(\mathcal{X}, \mathcal{Y})$ is a normed linear space.

Proof It is easy to check that P is closed under vector addition and scalar multiplication in $\mathcal{M}(\mathcal{X}, \mathcal{Y})$. Clearly, $\vartheta \in P \neq \emptyset$. Hence, P is a subspace of $\mathcal{M}(\mathcal{X}, \mathcal{Y})$.

Next, we prove the four equivalent definitions of $\|A\|$. $\forall A \in P$, define the set $B_A := \{M \in [0, \infty) \subset \mathbb{R} \mid \|Ax\|_{\mathcal{Y}} \le M\|x\|_{\mathcal{X}},\ \forall x \in \mathcal{X}\}$. Let $\delta_{A1} := \sup_{x \in \mathcal{X}, \|x\|_{\mathcal{X}} \le 1} \|Ax\|_{\mathcal{Y}}$. Since $\|\vartheta_{\mathcal{X}}\|_{\mathcal{X}} = 0$ and $\|A\vartheta_{\mathcal{X}}\|_{\mathcal{Y}} = \|\vartheta_{\mathcal{Y}}\|_{\mathcal{Y}} = 0$, then $\delta_{A1} \ge 0$. $\forall M \in B_A$, we have $\|Ax\|_{\mathcal{Y}} \le M\|x\|_{\mathcal{X}} \le M$, $\forall x \in \mathcal{X}$ with $\|x\|_{\mathcal{X}} \le 1$. Hence, we have $\delta_{A1} \le M$ and $\delta_{A1} \le \|A\|$.

Let $\delta_{A2} := \max\left\{\sup_{x \in \mathcal{X},\ x \neq \vartheta_{\mathcal{X}}} \frac{\|Ax\|_{\mathcal{Y}}}{\|x\|_{\mathcal{X}}}, 0\right\}$. $\forall x \in \mathcal{X}$ with $x \neq \vartheta_{\mathcal{X}}$, we have $\delta_{A1} \ge \left\|A\left(\frac{x}{\|x\|_{\mathcal{X}}}\right)\right\|_{\mathcal{Y}} = \frac{\|Ax\|_{\mathcal{Y}}}{\|x\|_{\mathcal{X}}}$. Hence, we have $\delta_{A2} \le \delta_{A1}$. Let $\delta_{A3} := \max\left\{\sup_{x \in \mathcal{X},\ \|x\|_{\mathcal{X}} = 1} \|Ax\|_{\mathcal{Y}}, 0\right\}$. Clearly, we have $0 \le \delta_{A3} \le \delta_{A2}$.

On the other hand, $\forall x \in \mathcal{X}$, we have either $x = \vartheta_{\mathcal{X}}$, then $\|Ax\|_{\mathcal{Y}} = 0 \le \delta_{A3}\|x\|_{\mathcal{X}}$; or $x \neq \vartheta_{\mathcal{X}}$, then $\|x\|_{\mathcal{X}} > 0$ and $\delta_{A3} \ge \left\|A\left(\frac{x}{\|x\|_{\mathcal{X}}}\right)\right\|_{\mathcal{Y}} = \frac{\|Ax\|_{\mathcal{Y}}}{\|x\|_{\mathcal{X}}}$, and hence, $\|Ax\|_{\mathcal{Y}} \le \delta_{A3}\|x\|_{\mathcal{X}}$. Thus, $\forall x \in \mathcal{X}$, we have $\|Ax\|_{\mathcal{Y}} \le \delta_{A3}\|x\|_{\mathcal{X}}$. If $\delta_{A3} = +\infty$, then $\|A\| \le \delta_{A3}$. If $\delta_{A3} < +\infty$, then $\delta_{A3} \in B_A$ and $\|A\| \le \delta_{A3}$. Therefore, $\|A\| \le \delta_{A3}$.

Hence, $\|A\| = \delta_{A1} = \delta_{A2} = \delta_{A3}$.

It is easy to check that N is closed under vector addition and scalar multiplication in P. Clearly, $\vartheta \in N \neq \emptyset$. Hence, N is a subspace of P.

Finally, we will show that $\|\cdot\|$ defines a norm on N. $\forall A_1, A_2 \in N$, $\forall \alpha \in \mathbb{K}$.

(i) $\|A_1\| \in [0, \infty) \subset \mathbb{R}$. If $A_1 = \vartheta$, then $\|A_1\| = \sup_{x \in \mathcal{X}, \|x\|_{\mathcal{X}} \le 1} \|A_1 x\|_{\mathcal{Y}} = \sup_{x \in \mathcal{X}, \|x\|_{\mathcal{X}} \le 1} 0 = 0$. On the other hand, if $\|A_1\| = 0$, then, $\forall x \in \mathcal{X}$, $0 \le \|A_1 x\|_{\mathcal{Y}} \le \|A_1\|\|x\|_{\mathcal{X}} = 0$, which implies that $A_1 x = \vartheta_{\mathcal{Y}}$ and $A_1 = \vartheta$. Hence, $\|A_1\| = 0 \Leftrightarrow A_1 = \vartheta$.

(ii) $\|A_1 + A_2\| = \sup_{x \in \mathcal{X},\ \|x\|_{\mathcal{X}} \le 1} \|(A_1 + A_2)x\|_{\mathcal{Y}} = \sup_{x \in \mathcal{X},\ \|x\|_{\mathcal{X}} \le 1} \|A_1 x + A_2 x\|_{\mathcal{Y}} \le \sup_{x \in \mathcal{X},\ \|x\|_{\mathcal{X}} \le 1} (\|A_1 x\|_{\mathcal{Y}} + \|A_2 x\|_{\mathcal{Y}}) \le \sup_{x \in \mathcal{X},\ \|x\|_{\mathcal{X}} \le 1} \|A_1 x\|_{\mathcal{Y}} + \sup_{x \in \mathcal{X},\ \|x\|_{\mathcal{X}} \le 1} \|A_2 x\|_{\mathcal{Y}} = \|A_1\| + \|A_2\|$.

(iii) $\|\alpha A_1\| = \sup_{x \in \mathcal{X},\ \|x\|_{\mathcal{X}} \leq 1} \|(\alpha A_1)x\|_{\mathcal{Y}} = \sup_{x \in \mathcal{X},\ \|x\|_{\mathcal{X}} \leq 1} \|\alpha(A_1 x)\|_{\mathcal{Y}} = \sup_{x \in \mathcal{X},\ \|x\|_{\mathcal{X}} \leq 1} |\alpha| \|A_1 x\|_{\mathcal{Y}} = |\alpha| \sup_{x \in \mathcal{X},\ \|x\|_{\mathcal{X}} \leq 1} \|A_1 x\|_{\mathcal{Y}} = |\alpha| \|A_1\|$, where we have made use of Proposition 3.81 in the fourth equality.

Hence, $\mathrm{B}(\mathcal{X}, \mathcal{Y})$ is a normed linear space. □

Proposition 7.64 *Let $\mathcal{X}$, $\mathcal{Y}$, and $\mathcal{Z}$ be normed linear spaces over the field $\mathbb{K}$, $A \in \mathrm{B}(\mathcal{X}, \mathcal{Y})$, and $B \in \mathrm{B}(\mathcal{Y}, \mathcal{Z})$. Then, $\forall x \in \mathcal{X}$, we have $\|Ax\| \leq \|A\| \|x\|$, where the three norms are over three different normed linear spaces. Furthermore, $\|BA\| \leq \|B\| \|A\|$, where the three norms are over three different normed linear spaces.*

Proof This is straightforward and is therefore omitted. □

Proposition 7.65 *Let $\mathcal{X}$ and $\mathcal{Y}$ be normed linear spaces over the field $\mathbb{K}$ and $\mathrm{B}(\mathcal{X}, \mathcal{Y})$ be the normed linear space of bounded linear operators of $\mathcal{X}$ to $\mathcal{Y}$. Define an operator $\psi : \mathrm{B}(\mathcal{X}, \mathcal{Y}) \times \mathcal{X} \to \mathcal{Y}$ by $\psi(A, x) = Ax$, $\forall A \in \mathrm{B}(\mathcal{X}, \mathcal{Y})$, $\forall x \in \mathcal{X}$. Then, ψ is continuous.*

Proof $\forall (A_0, x_0) \in \mathrm{B}(\mathcal{X}, \mathcal{Y}) \times \mathcal{X}$, $\forall \epsilon \in (0, \infty) \subset \mathbb{R}$, let $\delta = \frac{\epsilon}{\epsilon + \|x_0\|} \wedge \frac{\epsilon}{\epsilon + \|A_0\|} \wedge \epsilon \in (0, \infty) \subset \mathbb{R}$. $\forall (A, x) \in \mathcal{B}_{\mathrm{B}(\mathcal{X},\mathcal{Y}) \times \mathcal{X}}((A_0, x_0), \delta)$, we have $\|\psi(A, x) - \psi(A_0, x_0)\| = \|Ax - A_0 x_0\| = \|Ax - Ax_0 + Ax_0 - A_0 x_0\| \leq \|A(x - x_0)\| + \|(A - A_0)x_0\| \leq \|A\| \|x - x_0\| + \|A - A_0\| \|x_0\| \leq (\|A_0\| + \|A - A_0\|)\delta + \|x_0\|\delta \leq (\delta + \|A_0\|)\delta + \|x_0\|\delta \leq \epsilon + \epsilon = 2\epsilon$. Hence, ψ is continuous. This completes the proof of the proposition. □

Proposition 7.66 *Let $\mathcal{X}$ and $\mathcal{Y}$ be normed linear spaces over the field $\mathbb{K}$. Let $\mathrm{B}(\mathcal{X}, \mathcal{Y})$ be the normed linear space defined in Proposition 7.63. If $\mathcal{Y}$ is a Banach space, then $\mathrm{B}(\mathcal{X}, \mathcal{Y})$ is also a Banach space.*

Proof All we need to show is that $\mathrm{B}(\mathcal{X}, \mathcal{Y})$ is complete. Take a Cauchy sequence $(A_n)_{n=1}^{\infty} \subseteq \mathrm{B}(\mathcal{X}, \mathcal{Y})$. $\forall x \in \mathcal{X}$, we have $\|A_n x - A_m x\| \leq \|A_n - A_m\| \|x\|$, $\forall n, m \in \mathbb{N}$, by Proposition 7.64. Hence, $(A_n x)_{n=1}^{\infty} \subseteq \mathcal{Y}$ is a Cauchy sequence. Since $\mathcal{Y}$ is a Banach space, then $\exists !\ y_x \in \mathcal{Y}$ such that $\lim_{n \in \mathbb{N}} A_n x = y_x$. Hence, we may define a function $f : \mathcal{X} \to \mathcal{Y}$ by $f(x) = y_x$, $\forall x \in \mathcal{X}$.

$\forall x_1, x_2 \in \mathcal{X}$, $\forall \alpha, \beta \in \mathbb{K}$, by Propositions 7.23, 4.15, and 3.67, we have $f(\alpha x_1 + \beta x_2) = \lim_{n \in \mathbb{N}} A_n(\alpha x_1 + \beta x_2) = \lim_{n \in \mathbb{N}} (\alpha A_n x_1 + \beta A_n x_2) = \alpha \lim_{n \in \mathbb{N}} A_n x_1 + \beta \lim_{n \in \mathbb{N}} A_n x_2 = \alpha f(x_1) + \beta f(x_2)$. Hence, f is linear.

Since, $(A_n)_{n=1}^{\infty}$ is Cauchy, then, $\forall \epsilon \in (0, \infty) \subset \mathbb{R}$, $\exists N \in \mathbb{N}$, $\forall n, m \geq N$, $\|A_n - A_m\| < \epsilon$. $\forall x \in \mathcal{X}$, $\|A_n x - A_m x\| \leq \|A_n - A_m\| \|x\| \leq \epsilon \|x\|$. Then, by Propositions 3.66, 7.21, and 7.23, $\|f(x) - A_m x\| = \lim_{n \in \mathbb{N}} \|A_n x - A_m x\| \leq \lim_{n \in \mathbb{N}} \|A_n - A_m\| \|x\| \leq \epsilon \|x\|$. Therefore, $\|f(x)\| \leq \|f(x) - A_N x\| + \|A_N x\| \leq \epsilon \|x\| + \|A_N\| \|x\| = (\epsilon + \|A_N\|) \|x\|$. This implies that f is bounded. Hence, $f \in \mathrm{B}(\mathcal{X}, \mathcal{Y})$. Note that $0 \leq \lim_{n \in \mathbb{N}} \|A_n - f\| = \lim_{n \in \mathbb{N}} \sup_{x \in \mathcal{X},\ \|x\| \leq 1} \|(A_n - f)(x)\| = \lim_{n \in \mathbb{N}} \sup_{x \in \mathcal{X},\ \|x\| \leq 1} \|A_n x - f(x)\| \leq \epsilon$. By the arbitrariness of ϵ, we have $\lim_{n \in \mathbb{N}} \|A_n - f\| = 0$ and hence $\lim_{n \in \mathbb{N}} A_n = f \in \mathrm{B}(\mathcal{X}, \mathcal{Y})$. Hence, $\mathrm{B}(\mathcal{X}, \mathcal{Y})$ is complete.

This completes the proof of the proposition. □

Proposition 7.67 *Let $\mathcal{X}$ and $\mathcal{Y}$ be normed linear spaces over the field $\mathbb{K}$, $\mathcal{X}$ be finite-dimensional with dimension $n \in \mathbb{Z}_+$, and $A : \mathcal{X} \to \mathcal{Y}$ be a linear operator. Then, $A \in \mathrm{B}(\mathcal{X}, \mathcal{Y})$.*

Proof Let $X_N \subseteq \mathcal{X}$ be a basis for $\mathcal{X}$, which then contains exactly n vectors. We will distinguish two exhaustive and mutually exclusive cases: Case 1: $n = 0$; Case 2: $n \in \mathbb{N}$. Case 1: $n = 0$. Then, $\mathcal{X} = \{\vartheta_{\mathcal{X}}\}$. Then, by Proposition 7.63, we have $\|A\| = \sup_{x\in\mathcal{X},\ \|x\|\leq 1} \|Ax\| = 0$. Hence, $A \in \mathrm{B}(\mathcal{X}, \mathcal{Y})$.

Case 2: $n \in \mathbb{N}$. Let $X_N = \{x_1, \ldots, x_n\}$. $\forall x \in \mathcal{X}$, by Corollary 6.47 and Definition 6.50, $\exists!\ \alpha_1, \ldots, \alpha_n \in \mathbb{K}$ such that $x = \sum_{i=1}^n \alpha_i x_i$. Then, we may define an alternative norm $\|\cdot\|_1$ on $\mathcal{X}$ by $\|x\|_1 = \sqrt{\sum_{i=1}^n |\alpha_i|^2}$. It is easy to check that $\|\cdot\|_1$ defines a norm on $\mathcal{X}$. By Theorem 7.38, $\exists M \in (0, \infty) \subset \mathbb{R}$ such that $\|x\|/M \leq \|x\|_1 \leq M\|x\|$, $\forall x \in \mathcal{X}$. Define $r_i := \|Ax_i\|$, $i = 1, \ldots, n$. Then, by Proposition 7.63,

$$\begin{aligned}
\|A\| &= \sup_{x\in\mathcal{X},\ \|x\|\leq 1} \|Ax\| = \sup_{\alpha_1,\ldots,\alpha_n\in\mathbb{K},\ \|\sum_{i=1}^n \alpha_i x_i\|\leq 1} \left\| A\left(\sum_{i=1}^n \alpha_i x_i\right)\right\| \\
&\leq \sup_{\alpha_1,\ldots,\alpha_n\in\mathbb{K},\ \|\sum_{i=1}^n \alpha_i x_i\|\leq 1} \sum_{i=1}^n |\alpha_i| r_i \\
&\leq \sup_{\alpha_1,\ldots,\alpha_n\in\mathbb{K},\ \|\sum_{i=1}^n \alpha_i x_i\|\leq 1} \left(\sum_{i=1}^n r_i^2\right)^{1/2} \left(\sum_{i=1}^n |\alpha_i|^2\right)^{1/2} \\
&= \sup_{x\in\mathcal{X},\ \|x\|\leq 1} \left(\sum_{i=1}^n r_i^2\right)^{1/2} \|x\|_1 \leq \sup_{x\in\mathcal{X},\ \|x\|\leq 1} \left(\sum_{i=1}^n r_i^2\right)^{1/2} M\|x\| \\
&\leq M\left(\sum_{i=1}^n r_i^2\right)^{1/2}
\end{aligned}$$

where we have applied the Cauchy–Schwarz Inequality in the second inequality. Hence, $A \in \mathrm{B}(\mathcal{X}, \mathcal{Y})$.

This completes the proof of the proposition. □

Proposition 7.68 *Let $\mathcal{X}$ and $\mathcal{Y}$ be normed linear spaces over the field $\mathbb{K}$ and $A : \mathcal{X} \to \mathcal{Y}$ be a linear operator. If A is bounded, then $\mathcal{N}(A) \subseteq \mathcal{X}$ is closed. On the other hand, if $\mathcal{N}(A)$ is closed and $\mathcal{R}(A) \subseteq \mathcal{Y}$ is finite-dimensional, then A is bounded.*

Proof Let A be bounded. By Proposition 7.62, A is continuous. Note that the set $\{\vartheta_{\mathcal{Y}}\} \subseteq \mathcal{Y}$ is closed by Proposition 3.34. By Proposition 3.10, $\mathcal{N}(A) = A_{\mathrm{inv}}(\{\vartheta_{\mathcal{Y}}\})$ is closed.

Let $\mathcal{N}(A)$ be closed and $\mathcal{R}(A)$ be finite-dimensional. Let $n \in \mathbb{Z}_+$ be the dimension of $\mathcal{R}(A)$. We will distinguish two exhaustive and mutually exclusive cases: Case 1: $n = 0$; Case 2: $n \in \mathbb{N}$. Case 1: $n = 0$. Then, $\mathcal{R}(A) = \{\vartheta_{\mathcal{Y}}\}$. Hence, $A = \vartheta_{\mathrm{B}(\mathcal{X},\mathcal{Y})}$, which is clearly bounded.

Case 2: $n \in \mathbb{N}$. Then, $\exists X_N = \{x_1, \ldots, x_n\} \subseteq \mathcal{X}$ with exactly n elements such that $A(X_N) \subseteq \mathcal{Y}$ is a basis of $\mathcal{R}(A)$. Let $\mathcal{X}/\mathcal{N}(A)$ be the quotient normed linear space as defined in Proposition 7.44. $\forall x \in \mathcal{X}$, $Ax \in \mathcal{R}(A) = \operatorname{span}(A(X_N))$. Then, by Definition 6.50 and Corollary 6.47, $\exists!\ \alpha_1, \ldots, \alpha_n \in \mathbb{K}$ such that $Ax = \sum_{i=1}^{n} \alpha_i Ax_i$. Then, we have $A\big(x - \sum_{i=1}^{n} \alpha_i x_i\big) = \vartheta_{\mathcal{Y}}$ and $x - \sum_{i=1}^{n} \alpha_i x_i \in \mathcal{N}(A)$. Hence, $[x] = \big[\sum_{i=1}^{n} \alpha_i x_i\big] = \sum_{i=1}^{n} \alpha_i [x_i]$. This implies that $\mathcal{X}/\mathcal{N}(A) \subseteq \operatorname{span}(\{[x_1], \ldots, [x_n]\})$. Hence, $\mathcal{X}/\mathcal{N}(A)$ is finite-dimensional.

Define a mapping $\bar{A} : \mathcal{X}/\mathcal{N}(A) \to \mathcal{Y}$ by $\bar{A}[x] = Ax$, $\forall [x] \in \mathcal{X}/\mathcal{N}(A)$. Note that $\forall \bar{x} \in [x]$, we have $x - \bar{x} \in \mathcal{N}(A)$ and $Ax = A\bar{x}$. Hence, $\bar{A}$ is uniquely defined. $\forall [x_1], [x_2] \in \mathcal{X}/\mathcal{N}(A)$, $\forall \alpha, \beta \in \mathbb{K}$, we have $\bar{A}(\alpha[x_1] + \beta[x_2]) = \bar{A}([\alpha x_1 + \beta x_2]) = A(\alpha x_1 + \beta x_2) = \alpha Ax_1 + \beta Ax_2 = \alpha\bar{A}([x_1]) + \beta\bar{A}([x_2])$. Hence, $\bar{A}$ is a linear operator. By Proposition 7.67, $\bar{A}$ is bounded. $\forall x \in \mathcal{X}$, $\|Ax\| = \|\bar{A}[x]\| \leq \|\bar{A}\|\,\|[x]\| \leq \|\bar{A}\|\,\|x\|$. Hence, $A \in \mathrm{B}(\mathcal{X}, \mathcal{Y})$.

This completes the proof of the proposition. □

Proposition 7.69 *Let $\mathcal{X}$ be a normed linear space and $M \subseteq \mathcal{X}$ be a closed subspace. Let $\mathcal{X}/M$ be the quotient normed linear space and $\phi : \mathcal{X} \to \mathcal{X}/M$ be the natural homomorphism. Then, $\|\phi\| \leq 1$.*

Proof By Proposition 7.43, ϕ is linear. $\forall x \in \mathcal{X}$, $\|\phi(x)\| = \inf_{m \in M} \|x - m\| \leq \|x\|$. Then, $\phi \in \mathrm{B}(\mathcal{X}, \mathcal{X}/M)$ and $\|\phi\| \leq 1$. This completes the proof of the proposition. □

Proposition 7.70 *Let $\mathcal{X}$ and $\mathcal{Y}$ be normed linear spaces over the field $\mathbb{K}$, $A \in \mathrm{B}(\mathcal{X}, \mathcal{Y})$, and $\phi : \mathcal{X} \to \mathcal{X}/\mathcal{N}(A)$ be the natural homomorphism. Then, $\exists B \in \mathrm{B}(\mathcal{X}/\mathcal{N}(A), \mathcal{Y})$ such that $A = B \circ \phi$, B is injective, and $\|A\| = \|B\|$.*

Proof By Proposition 7.68, $\mathcal{N}(A)$ is closed. Then, by Proposition 7.44, $\mathcal{X}/\mathcal{N}(A)$ is a normed linear space. Define $B : \mathcal{X}/\mathcal{N}(A) \to \mathcal{Y}$ by $B([x]) = Ax$, $\forall [x] \in \mathcal{X}/\mathcal{N}(A)$. Note that $\forall \bar{x} \in [x]$, we have $x - \bar{x} \in \mathcal{N}(A)$ and $Ax = A\bar{x}$. Hence, B is uniquely defined. $\forall [x_1], [x_2] \in \mathcal{X}/\mathcal{N}(A)$, $\forall \alpha, \beta \in \mathbb{K}$, we have $B(\alpha[x_1] + \beta[x_2]) = B([\alpha x_1 + \beta x_2]) = A(\alpha x_1 + \beta x_2) = \alpha Ax_1 + \beta Ax_2 = \alpha B([x_1]) + \beta B([x_2])$. Hence, B is linear. $\forall [x_1], [x_2] \in \mathcal{X}/\mathcal{N}(A)$, $B[x_1] = B[x_2]$ implies that $Ax_1 = Ax_2$, $x_1 - x_2 \in \mathcal{N}(A)$, and hence, $[x_1] = [x_2]$. This shows that B is injective. $\forall x \in \mathcal{X}$, $B \circ \phi(x) = B[x] = Ax$. Hence, $A = B \circ \phi$.

$\forall [x] \in \mathcal{X}/\mathcal{N}(A)$, $\forall m \in \mathcal{N}(A)$, $\|B[x]\| = \|Ax\| = \|A(x - m)\| \leq \|A\|\,\|x - m\|$. This implies that $\|B[x]\| \leq \|A\|\,\|[x]\|$. Hence, $B \in \mathrm{B}(\mathcal{X}/\mathcal{N}(A), \mathcal{Y})$ and $\|B\| \leq \|A\|$. By Propositions 7.64 and 7.69, $\|A\| = \|B \circ \phi\| \leq \|B\|\,\|\phi\| \leq \|B\|$. Then, $\|A\| = \|B\|$. This completes the proof of the proposition. □

7.9 Dual Spaces

7.9.1 Basic Concepts

Definition 7.71 Let $\mathcal{X}$ be a normed linear space over the field $\mathbb{K}$. Then, the space $\mathrm{B}(\mathcal{X}, \mathbb{K})$, which consists of all bounded linear functionals on $\mathcal{X}$, is called the *dual* of $\mathcal{X}$ and denoted by $\mathcal{X}^*$. We will denote the vectors in $\mathcal{X}^*$ by x_* and denote $x_*(x)$ by $\langle\langle x_*, x\rangle\rangle$. ◇

Proposition 7.72 *Let $\mathcal{X}$ be a normed linear space over the field $\mathbb{K}$ and $\mathcal{X}^*$ be its dual. Then, the following statements hold.*

(i) *If a linear functional $f : \mathcal{X} \to \mathbb{K}$ is continuous at some $x_0 \in \mathcal{X}$, then $f \in \mathcal{X}^*$.*
(ii) *$\forall x_* \in \mathcal{X}^*$, x_* is uniformly continuous.*
(iii) *For any linear functional $f : \mathcal{X} \to \mathbb{K}$, we have $\|f\| = \inf\{M \in [0, \infty) \subset \mathbb{R} \mid |f(x)| \le M\|x\|_{\mathcal{X}}, \forall x \in \mathcal{X}\} = \sup_{x\in\mathcal{X}, \|x\|_{\mathcal{X}}\le 1} |f(x)| = \max\left\{\sup_{x\in\mathcal{X},\, x\neq\vartheta_{\mathcal{X}}} \frac{|f(x)|}{\|x\|_{\mathcal{X}}}, 0\right\} = \max\left\{\sup_{x\in\mathcal{X},\, \|x\|_{\mathcal{X}}=1} |f(x)|, 0\right\}$.*
(iv) *$|\langle\langle x_*, x\rangle\rangle| \le \|x_*\|\|x\|$, $\forall x \in \mathcal{X}$, $\forall x_* \in \mathcal{X}^*$.*
(v) *$\langle\langle \cdot, \cdot\rangle\rangle$ is a continuous function on $\mathcal{X}^* \times \mathcal{X}$.*
(vi) *$\mathcal{X}^*$ is a Banach space.*
(vii) *A linear functional $f : \mathcal{X} \to \mathbb{K}$ is bounded if, and only if, $\mathcal{N}(f)$ is closed.*

Proof (i) and (ii) are direct consequences of Proposition 7.62. (iii) is a direct consequence of Proposition 7.63. (iv) follows from Proposition 7.64. (v) follows from Proposition 7.65. (vi) follows from Proposition 7.66 since $\mathbb{K}$ is a Banach space. Finally, (vii) follows from Proposition 7.68. This completes the proof of the proposition. □

7.9.2 Duals of Some Common Banach Spaces

Example 7.73 Let us consider the dual of $\mathcal{X} := (\{\vartheta_{\mathcal{X}}\}, \mathbb{K}, \|\cdot\|)$. Clearly, there is a single linear functional on $\mathcal{X}$ given by $f : \mathcal{X} \to \mathbb{K}$ and $f(\vartheta_{\mathcal{X}}) = 0$. Clearly, this linear functional is bounded with norm 0. Hence, $\mathcal{X}^* = (\{f\}, \mathbb{K}, \|\cdot\|_1)$, which is isometrically isomorphic to $\mathcal{X}$. ◇

Example 7.74 Let us consider the dual of $\mathcal{X} := \mathbb{K}^n$, $n \in \mathbb{N}$. $\forall x := (\xi_1, \ldots, \xi_n) \in \mathbb{K}^n$, any functional of the form $f(x) = \sum_{i=1}^n \eta_i \xi_i$ with $\eta_1, \ldots, \eta_n \in \mathbb{K}$ is clearly linear. By Cauchy–Schwarz Inequality, we have $|f(x)| = \left|\sum_{i=1}^n \eta_i \xi_i\right| \le \sum_{i=1}^n |\eta_i||\xi_i| \le \sqrt{\sum_{i=1}^n |\eta_i|^2}\sqrt{\sum_{i=1}^n |\xi_i|^2} = \sqrt{\sum_{i=1}^n |\eta_i|^2}|x|$. Hence, f is bounded and $\|f\| \le \sqrt{\sum_{i=1}^n |\eta_i|^2}$. Since the equality is achieved at $x = (\overline{\eta_1}, \ldots, \overline{\eta_n})$ in the above inequality, where $\overline{\eta_i}$ denotes the complex conjugate of η_i, then $\|f\| =$

$\sqrt{\sum_{i=1}^{n}|\eta_i|^2}$. Clearly, for different n-tuple $(\eta_1,\ldots,\eta_n)$, the linear functional f is distinct. Now, let f be a bounded linear functional on $\mathcal{X}$. Let $e_i \in \mathcal{X}$ be the ith unit vector (all components of e_i are zero except a 1 at the ith component), $i = 1,\ldots,n$. Let $\eta_i = f(e_i) \in \mathbb{K}$, $i = 1,\ldots,n$. Then, $\forall x = (\xi_1,\ldots,\xi_n) \in \mathcal{X}$, we have $x = \sum_{i=1}^{n}\xi_i e_i$ and $f(x) = \sum_{i=1}^{n}\eta_i\xi_i$. Define a function $\Psi : \mathcal{X}^* \to \mathbb{K}^n$ by $\Psi(f) = (\eta_1,\ldots,\eta_n)$, $\forall f \in \mathcal{X}^*$. Clearly, Ψ is linear, bijective and norm preserving. Hence, the dual of $\mathbb{K}^n$ is (isometrically isomorphic to) $\mathbb{K}^n$. ◇

Lemma 7.75 *Let $\mathcal{X}$ be a normed linear space over $\mathbb{K}$, $\mathcal{X}^*$ be its dual, and $x_* \in \mathcal{X}^*$. Then, $\forall \epsilon \in (0,1) \subset \mathbb{R}$, $\exists x \in \mathcal{X}$ with $\|x\| \le 1$ such that $\langle\langle x_*, x\rangle\rangle \in \mathbb{R}$ and $\langle\langle x_*, x\rangle\rangle \ge (1-\epsilon)\|x_*\|$.*

Proof $\forall \epsilon \in (0,1) \subset \mathbb{R}$. We will distinguish two exhaustive and mutually exclusive cases: Case 1: $\|x_*\| = 0$; Case 2: $\|x_*\| > 0$. Case 1: $\|x_*\| = 0$. Then, $x_* = \vartheta_*$. Take $x = \vartheta$ and the result holds. Case 2: $\|x_*\| > 0$. Note that $\|x_*\| = \sup_{x\in\mathcal{X},\ \|x\|\le 1}|\langle\langle x_*, x\rangle\rangle|$. Then, $\exists \bar{x} \in \mathcal{X}$ with $\|\bar{x}\| \le 1$ such that $|\langle\langle x_*, \bar{x}\rangle\rangle| \ge (1-\epsilon)\|x_*\| > 0$. Take $x := \frac{|\langle\langle x_*,\bar{x}\rangle\rangle|}{\langle\langle x_*,\bar{x}\rangle\rangle}\bar{x} \in \mathcal{X}$. Then, $\|x\| = \|\bar{x}\| \le 1$ and $\langle\langle x_*, x\rangle\rangle = \left\langle\left\langle x_*, \frac{|\langle\langle x_*,\bar{x}\rangle\rangle|}{\langle\langle x_*,\bar{x}\rangle\rangle}\bar{x}\right\rangle\right\rangle = \frac{|\langle\langle x_*,\bar{x}\rangle\rangle|}{\langle\langle x_*,\bar{x}\rangle\rangle}\langle\langle x_*, \bar{x}\rangle\rangle = |\langle\langle x_*, \bar{x}\rangle\rangle| \in \mathbb{R}$ and $\langle\langle x_*, x\rangle\rangle \ge (1-\epsilon)\|x_*\|$. This completes the proof of the lemma. □

Example 7.76 Let $\mathcal{Y}_i$ be a normed linear space over the field $\mathbb{K}$, $\mathcal{Y}_i^*$ be its dual, $i = 1,\ldots,n$, $n \in \mathbb{N}$. We will study the dual of $\mathcal{X} := \prod_{i=1}^{n}\mathcal{Y}_i$. We will show that $\mathcal{X}^* = \prod_{i=1}^{n}\mathcal{Y}_i^*$ isometrically isomorphically.

Let $f \in \mathcal{X}^*$. $\forall i = 1,\ldots,n$, define $A_i : \mathcal{Y}_i \to \mathcal{X}$ by $x = (y_1,\ldots,y_n) = A_i y$, $\forall y \in \mathcal{Y}_i$, and $y_j = \vartheta_{\mathcal{Y}_j}$, $\forall j \in 1,\ldots,n$ with $j \ne i$, and $y_i = y$. Clearly, A_i is well-defined and linear and bounded with $\|A_i\| \le 1$. Define $f_i : \mathcal{Y}_i \to \mathbb{K}$ by $f_i = f \circ A_i$. Then, f_i is a bounded linear functional on $\mathcal{Y}_i$ with $\|f_i\| \le \|f\|\|A_i\| \le \|f\|$. Hence, $f_i \in \mathcal{Y}_i^*$. Denote $f_i =: y_{*i}$. $\forall x = (y_1,\ldots,y_n) \in \mathcal{X}$, we have $y_i \in \mathcal{Y}_i$, $i = 1,\ldots,n$. By the linearity of f, we have

$$f(x) = f\left(\sum_{i=1}^{n}A_i y_i\right) = \sum_{i=1}^{n}f(A_i y_i) = \sum_{i=1}^{n}\langle\langle y_{*i}, y_i\rangle\rangle \tag{7.1}$$

Let $x_* := (y_{*1},\ldots,y_{*n}) \in \prod_{i=1}^{n}\mathcal{Y}_i^*$. $\forall \epsilon \in (0,1) \subset \mathbb{R}$, $\forall i = 1,\ldots,n$, by Lemma 7.75, $\exists \bar{y}_i \in \mathcal{Y}$ with $\|\bar{y}_i\| \le 1$ such that $\langle\langle y_{*i}, \bar{y}_i\rangle\rangle \in \mathbb{R}$ and $\langle\langle y_{*i}, \bar{y}_i\rangle\rangle \ge (1-\epsilon)\|y_{*i}\|$. Let $\hat{y}_i = \|y_{*i}\|\bar{y}_i$. Then, $\langle\langle y_{*i}, \hat{y}_i\rangle\rangle \ge (1-\epsilon)\|y_{*i}\|^2$. Then, we have $\left|f\left(\sum_{i=1}^{n}A_i\hat{y}_i\right)\right| = \left|\sum_{i=1}^{n}\langle\langle y_{*i}, \hat{y}_i\rangle\rangle\right| = \sum_{i=1}^{n}\langle\langle y_{*i}, \hat{y}_i\rangle\rangle \ge (1-\epsilon)\sum_{i=1}^{n}\|y_{*i}\|^2 = (1-\epsilon)\|x_*\|^2$. Since $f \in \mathcal{X}^*$, then, by Proposition 7.72, we have $(1-\epsilon)\|x_*\|^2 \le \|f\|\left\|\sum_{i=1}^{n}A_i\hat{y}_i\right\| = \|f\|\left(\sum_{i=1}^{n}\|\hat{y}_i\|^2\right)^{1/2} \le \|f\|\left(\sum_{i=1}^{n}\|y_{*i}\|^2\right)^{1/2} = \|f\|\|x_*\|$. By the arbitrariness of ϵ, we have $\|x_*\| \le \|f\|$.

Based on the preceding analysis, we may define a function $\psi : \mathcal{X}^* \to \prod_{i=1}^{n}\mathcal{Y}_i^*$ by $\psi(f) = (f \circ A_1,\ldots,f \circ A_n)$, $\forall f \in \mathcal{X}^*$. Then, $\|\psi(f)\| \le \|f\|$. Clearly, ψ is linear. $\forall f_1, f_2 \in \mathcal{X}^*$ with $\psi(f_1) = \psi(f_2)$. Then, we have $f_1 \circ A_i = f_2 \circ A_i$, $\forall i \in$

$\{1, \ldots, n\}$. $\forall x := (y_1, \ldots, y_n) \in \mathcal{X}$, by (7.1), we have $f_1(x) = \sum_{i=1}^{n} f_1(A_i y_i) = \sum_{i=1}^{n} f_2(A_i y_i) = f_2(x)$. Then, we have $f_1 = f_2$. Hence, ψ is injective. $\forall x_* := (y_{*1}, \ldots, y_{*n}) \in \prod_{i=1}^{n} \mathcal{Y}_i^*$, define $\phi(x_*) : \mathcal{X} \to \mathbb{K}$ by $\phi(x_*)(x) = \sum_{i=1}^{n} \langle\langle y_{*i}, y_i \rangle\rangle$, $\forall x := (y_1, \ldots, y_n) \in \mathcal{X}$. Note that, $\forall x := (y_1, \ldots, y_n) \in \mathcal{X}$, $\sum_{i=1}^{n} |\langle\langle y_{*i}, y_i \rangle\rangle| \leq \sum_{i=1}^{n} \|y_{*i}\| \|y_i\| \leq \|x_*\| \|x\|$, where we have made use of Proposition 7.72 in the first inequality and Cauchy–Schwarz Inequality in the second inequality. Clearly, $\phi(x_*)$ is linear and bounded with

$$\begin{aligned} \|\phi(x_*)\| &= \sup_{x \in \mathcal{X},\ \|x\| \leq 1} |\phi(x_*)(x)| \leq \sup_{x=(y_1,\ldots,y_n) \in \mathcal{X},\ \|x\| \leq 1} \sum_{i=1}^{n} |\langle\langle y_{*i}, y_i \rangle\rangle| \\ &\leq \sup_{x \in \mathcal{X},\ \|x\| \leq 1} \|x_*\| \|x\| \leq \|x_*\| \end{aligned} \tag{7.2}$$

Hence, $\phi(x_*) \in \mathcal{X}^*$ and

$$\psi(\phi(x_*)) = (\phi(x_*) \circ A_1, \ldots, \phi(x_*) \circ A_n) = (y_{*1}, \ldots, y_{*n}) = x_*$$

Then, $\psi \circ \phi = \mathrm{id}_{\prod_{i=1}^{n} \mathcal{Y}_i^*}$. Hence, ψ is surjective. Therefore, ψ is bijective and admits inverse ψ_{inv}. By Proposition 2.4, $\phi = \psi_{\mathrm{inv}}$. $\forall f \in \mathcal{X}^*$, by (7.2), we have $\|f\| = \|\phi(\psi(f))\| \leq \|\psi(f)\| \leq \|f\|$. Then, $\|\psi(f)\| = \|f\|$ and ψ is an isometry. Hence, ψ is an isometrical isomorphism. ◇

Example 7.77 Let us consider the dual of $\mathcal{X} := \mathrm{l}_p(\mathcal{Y})$, $p \in [1, \infty) \subset \mathbb{R}$, $\mathcal{Y}$ is a normed linear space over $\mathbb{K}$. Let $\mathcal{Y}^*$ be the dual of $\mathcal{Y}$ and $q \in (1, \infty] \subset \mathbb{R}_e$ with $1/p + 1/q = 1$. We will show that $\mathcal{X}^*$ is isometrically isomorphic to $\mathrm{l}_q(\mathcal{Y}^*)$.

Let $f \in \mathcal{X}^*$. $\forall i \in \mathbb{N}$, define $A_i : \mathcal{Y} \to \mathcal{X}$ by $x = (y_1, y_2, \ldots) = A_i y$, $\forall y \in \mathcal{Y}$, and $y_j = \vartheta_{\mathcal{Y}}$, $\forall j \in \mathbb{N}$ with $j \neq i$, and $y_i = y$. Clearly, A_i is well-defined, linear, and bounded with $\|A_i\| \leq 1$. Define $f_i : \mathcal{Y} \to \mathbb{K}$ by $f_i = f \circ A_i$. Then, f_i is a bounded linear functional on $\mathcal{Y}$ with $\|f_i\| \leq \|f\| \|A_i\| \leq \|f\|$. Hence, $f_i \in \mathcal{Y}^*$. Denote $f_i =: y_{*i}$. $\forall x = (y_1, y_2, \ldots) \in \mathcal{X}$, we have $y_i \in \mathcal{Y}$, $\forall i \in \mathbb{N}$, and $\sum_{i=1}^{\infty} \|y_i\|^p < +\infty$. Then, $\lim_{n \in \mathbb{N}} \sum_{i=n+1}^{\infty} \|y_i\|^p = 0$. This implies that $\lim_{n \in \mathbb{N}} \left\| x - \sum_{i=1}^{n} A_i y_i \right\| = \lim_{n \in \mathbb{N}} \left(\sum_{i=n+1}^{\infty} \|y_i\|^p \right)^{1/p} = 0$. Hence, $\lim_{n \in \mathbb{N}} \sum_{i=1}^{n} A_i y_i = x$. By the continuity of f and Proposition 3.66, we have

$$f(x) = \lim_{n \in \mathbb{N}} f\left(\sum_{i=1}^{n} A_i y_i \right) = \lim_{n \in \mathbb{N}} \sum_{i=1}^{n} f(A_i y_i) = \lim_{n \in \mathbb{N}} \sum_{i=1}^{n} \langle\langle y_{*i}, y_i \rangle\rangle \tag{7.3}$$

Claim 7.77.1 $x_* := (y_{*1}, y_{*2}, \ldots) \in \mathrm{l}_q(\mathcal{Y}^*)$ and $\|x_*\| \leq \|f\|$.

Proof of Claim We will distinguish two exhaustive and mutually exclusive cases: Case 1: $1 < p < +\infty$; Case 2: $p = 1$.

Case 1: $1 < p < +\infty$. Then, $1 < q < +\infty$. $\forall \epsilon \in (0, 1) \subset \mathbb{R}$, $\forall i \in \mathbb{N}$, by Lemma 7.75, $\exists \bar{y}_i \in \mathcal{Y}$ with $\|\bar{y}_i\| \leq 1$ such that $\langle\langle y_{*i}, \bar{y}_i \rangle\rangle \in \mathbb{R}$ and $\langle\langle y_{*i}, \bar{y}_i \rangle\rangle \geq (1 - \epsilon)\|y_{*i}\|$. Let $\hat{y}_i = \|y_{*i}\|^{q/p} \bar{y}_i$, then $\left\| \hat{y}_i \right\| \leq \|y_{*i}\|^{q/p}$,

$\langle\langle y_{*i}, \hat{y}_i \rangle\rangle \in \mathbb{R}$, and $\langle\langle y_{*i}, \hat{y}_i \rangle\rangle \geq (1-\epsilon)\|y_{*i}\|^{q/p+1} = (1-\epsilon)\|y_{*i}\|^q$. Then, $\forall n \in \mathbb{N}$, $\left|f(\sum_{i=1}^n A_i \hat{y}_i)\right| = \left|\sum_{i=1}^n \langle\langle y_{*i}, \hat{y}_i \rangle\rangle\right| = \sum_{i=1}^n \langle\langle y_{*i}, \hat{y}_i \rangle\rangle \geq (1-\epsilon)\sum_{i=1}^n \|y_{*i}\|^q$. Since $f \in \mathcal{X}^*$, then, by Proposition 7.72, we have $(1-\epsilon)\sum_{i=1}^n \|y_{*i}\|^q \leq \|f\|\left\|\sum_{i=1}^n A_i \hat{y}_i\right\| \leq \|f\|\left(\sum_{i=1}^n \left\|\hat{y}_i\right\|^p\right)^{1/p} \leq \|f\| \cdot \left(\sum_{i=1}^n \|y_{*i}\|^q\right)^{1/p}$. Then, we have $\left(\sum_{i=1}^n \|y_{*i}\|^q\right)^{1/q} \leq \|f\|/(1-\epsilon)$. By the arbitrariness of n, we have $\left(\sum_{i=1}^\infty \|y_{*i}\|^q\right)^{1/q} \leq \|f\|/(1-\epsilon)$. By the arbitrariness of ϵ, we have $\left(\sum_{i=1}^\infty \|y_{*i}\|^q\right)^{1/q} \leq \|f\|$. Hence, the result holds in this case.

Case 2: $p = 1$. Then, $q = +\infty$. $\forall \epsilon \in (0, 1) \subset \mathbb{R}$, $\forall i \in \mathbb{N}$, by Lemma 7.75, $\exists \hat{y}_i \in \mathcal{Y}$ with $\left\|\hat{y}_i\right\| \leq 1$ such that $\langle\langle y_{*i}, \hat{y}_i \rangle\rangle \in \mathbb{R}$ and $\langle\langle y_{*i}, \hat{y}_i \rangle\rangle \geq (1-\epsilon)\|y_{*i}\|$. Then, $(1-\epsilon)\|y_{*i}\| \leq \left|\langle\langle y_{*i}, \hat{y}_i \rangle\rangle\right| = \left|f(A_i \hat{y}_i)\right| \leq \|f\|\|A_i\|\left\|\hat{y}_i\right\| \leq \|f\|$. By the arbitrariness of n, we have $\sup_{i \geq 1} \|y_{*i}\| \leq \|f\|/(1-\epsilon)$. By the arbitrariness of ϵ, we have $\sup_{i \geq 1} \|y_{*i}\| \leq \|f\|$. Hence, the result holds in this case.

This completes the proof of the claim. □

The preceding analysis shows that we may define a function $\psi : \mathcal{X}^* \to \mathrm{l}_q(\mathcal{Y}^*)$ by $\psi(f) = (f \circ A_1, f \circ A_2, \ldots)$, $\forall f \in \mathcal{X}^*$. Clearly, ψ is linear. $\forall f_1, f_2 \in \mathcal{X}^*$ with $\psi(f_1) = \psi(f_2)$. Then, we have $f_1 \circ A_i = f_2 \circ A_i$, $\forall i \in \mathbb{N}$. $\forall x := (y_1, y_2, \ldots) \in \mathcal{X}$, by (7.3), we have $f_1(x) = \lim_{n \in \mathbb{N}} \sum_{i=1}^n f_1(A_i y_i) = \lim_{n \in \mathbb{N}} \sum_{i=1}^n f_2(A_i y_i) = f_2(x)$. Then, we have $f_1 = f_2$. Hence, ψ is injective. $\forall x_* := (y_{*1}, y_{*2}, \ldots) \in \mathrm{l}_q(\mathcal{Y}^*)$, define $\phi(x_*) : \mathcal{X} \to \mathbb{K}$ by $\phi(x_*)(x) = \sum_{i=1}^\infty \langle\langle y_{*i}, y_i \rangle\rangle$, $\forall x := (y_1, y_2, \ldots) \in \mathcal{X}$. Note that, $\forall x := (y_1, y_2, \ldots) \in \mathcal{X}$, $\sum_{i=1}^\infty |\langle\langle y_{*i}, y_i \rangle\rangle| \leq \sum_{i=1}^\infty \|y_{*i}\|\|y_i\| \leq \|x_*\|\|x\|$, where we have made use of Proposition 7.72 in the first inequality and Hölder's Inequality in the second inequality. Hence, $\phi(x_*)(x)$ is well-defined. Hence, $\phi(x_*)$ is well-defined. Clearly, $\phi(x_*)$ is linear and bounded with

$$
\begin{aligned}
\|\phi(x_*)\| = \sup_{x \in \mathcal{X},\ \|x\| \leq 1} |\phi(x_*)(x)| &\leq \sup_{x=(y_1,y_2,\ldots) \in \mathcal{X},\ \|x\| \leq 1} \sum_{i=1}^\infty |\langle\langle y_{*i}, y_i \rangle\rangle| \\
&\leq \sup_{x \in \mathcal{X},\ \|x\| \leq 1} \|x_*\|\|x\| \leq \|x_*\|
\end{aligned} \tag{7.4}
$$

Hence, $\phi(x_*) \in \mathcal{X}^*$ and

$$
\psi(\phi(x_*)) = (\phi(x_*) \circ A_1, \phi(x_*) \circ A_2, \ldots) = (y_{*1}, y_{*2}, \ldots) = x_*
$$

Then, $\psi \circ \phi = \mathrm{id}_{\mathrm{l}_q(\mathcal{Y}^*)}$. Hence, ψ is surjective. Therefore, ψ is bijective and admits inverse ψ_{inv}. By Proposition 2.4, $\phi = \psi_{\mathrm{inv}}$. $\forall f \in \mathcal{X}^*$, by Claim 7.77.1 and (7.4), we have $\|f\| = \|\phi(\psi(f))\| \leq \|\psi(f)\| \leq \|f\|$. Then, $\|\psi(f)\| = \|f\|$ and ψ is an isometry. Hence, ψ is an isometrical isomorphism. ◇

Example 7.78 Let $\mathcal{Y}$ be a normed linear space over the field $\mathbb{K}$, $\mathcal{Y}^*$ be its dual, and $M = \{x = (y_1, y_2, \ldots) \in \mathrm{l}_\infty(\mathcal{Y}) \mid \lim_{n \in \mathbb{N}} y_n = \vartheta_\mathcal{Y}\}$. Clearly, M is a subspace of $\mathrm{l}_\infty(\mathcal{Y})$. By Proposition 7.13, M is a normed linear space over $\mathbb{K}$, which will be denoted by $c_0(\mathcal{Y})$. Next, we will study the dual of $\mathcal{X} := c_0(\mathcal{Y})$. We will show that $\mathcal{X}^* = l_1(\mathcal{Y}^*)$ isometrically isomorphically.

Let $f \in \mathcal{X}^*$. $\forall i \in \mathbb{N}$, define $A_i : \mathcal{Y} \to \mathcal{X}$ by $x = (y_1, y_2, \ldots) = A_i y$, $\forall y \in \mathcal{Y}$, and $y_j = \vartheta_{\mathcal{Y}}$, $\forall j \in \mathbb{N}$ with $j \neq i$, and $y_i = y$. Clearly, A_i is well-defined and linear and bounded with $\|A_i\| \leq 1$. Define $f_i : \mathcal{Y} \to \mathbb{K}$ by $f_i = f \circ A_i$. Then, f_i is a bounded linear functional on $\mathcal{Y}$ with $\|f_i\| \leq \|f\|\|A_i\| \leq \|f\|$. Hence, $f_i \in \mathcal{Y}^*$. Denote $f_i =: y_{*i}$. $\forall x = (y_1, y_2, \ldots) \in \mathcal{X}$, we have $y_i \in \mathcal{Y}$, $\forall i \in \mathbb{N}$, $\lim_{n\in\mathbb{N}} y_i = \vartheta_{\mathcal{Y}}$, and $\lim_{n\in\mathbb{N}} \|y_i\| = 0$ (by Proposition 7.21). Then, $\lim_{n\in\mathbb{N}} \sup_{i\geq n+1} \|y_i\| = 0$. This implies that $\lim_{n\in\mathbb{N}} \left\|x - \sum_{i=1}^n A_i y_i\right\| = \lim_{n\in\mathbb{N}} \sup_{k\geq n+1} \|y_k\| = \lim_{n\in\mathbb{N}} \max_{k\geq n+1} \|y_k\| = 0$. Hence, $\lim_{n\in\mathbb{N}} \sum_{i=1}^n A_i y_i = x$. By the continuity of f and Proposition 3.66, we have

$$f(x) = \lim_{n\in\mathbb{N}} f\left(\sum_{i=1}^n A_i y_i\right) = \lim_{n\in\mathbb{N}} \sum_{i=1}^n f(A_i y_i) = \lim_{n\in\mathbb{N}} \sum_{i=1}^n \langle\langle y_{*i}, y_i\rangle\rangle \tag{7.5}$$

Let $x_* := (y_{*1}, y_{*2}, \ldots)$. $\forall \epsilon \in (0, 1) \subset \mathbb{R}$, $\forall i \in \mathbb{N}$, by Lemma 7.75, $\exists \hat{y}_i \in \mathcal{Y}$ with $\|\hat{y}_i\| \leq 1$ such that $\langle\langle y_{*i}, \hat{y}_i\rangle\rangle \in \mathbb{R}$ and $\langle\langle y_{*i}, \hat{y}_i\rangle\rangle \geq (1-\epsilon)\|y_{*i}\|$. Then, $\forall n \in \mathbb{N}$, $(1-\epsilon)\sum_{i=1}^n \|y_{*i}\| \leq \left|\sum_{i=1}^n \langle\langle y_{*i}, \hat{y}_i\rangle\rangle\right| = \left|\sum_{i=1}^n f(A_i\hat{y}_i)\right| = \left|f\left(\sum_{i=1}^n A_i\hat{y}_i\right)\right| \leq \|f\|\left\|\sum_{i=1}^n A_i\hat{y}_i\right\| \leq \|f\|$. By the arbitrariness of n, we have $\sum_{i=1}^\infty \|y_{*i}\| \leq \|f\|/(1-\epsilon)$. By the arbitrariness of ϵ, we have

$$\sum_{i=1}^\infty \|y_{*i}\| \leq \|f\| \tag{7.6}$$

Hence, $x_* \in l_1(\mathcal{Y}^*)$.

Based on the preceding analysis, we may define a function $\psi : \mathcal{X}^* \to l_1(\mathcal{Y}^*)$ by $\psi(f) = (f \circ A_1, f \circ A_2, \ldots)$, $\forall f \in \mathcal{X}^*$. Clearly, ψ is linear. $\forall f_1, f_2 \in \mathcal{X}^*$ with $\psi(f_1) = \psi(f_2)$. Then, we have $f_1 \circ A_i = f_2 \circ A_i$, $\forall i \in \mathbb{N}$. $\forall x := (y_1, y_2, \ldots) \in \mathcal{X}$, by (7.5), we have $f_1(x) = \lim_{n\in\mathbb{N}} \sum_{i=1}^n f_1(A_i y_i) = \lim_{n\in\mathbb{N}} \sum_{i=1}^n f_2(A_i y_i) = f_2(x)$. Then, we have $f_1 = f_2$. Hence, ψ is injective. $\forall x_* := (y_{*1}, y_{*2}, \ldots) \in l_1(\mathcal{Y}^*)$, define $\phi(x_*) : \mathcal{X} \to \mathbb{K}$ by $\phi(x_*)(x) = \sum_{i=1}^\infty \langle\langle y_{*i}, y_i\rangle\rangle$, $\forall x := (y_1, y_2, \ldots) \in \mathcal{X}$. Note that, $\forall x := (y_1, y_2, \ldots) \in \mathcal{X}$, $\sum_{i=1}^\infty |\langle\langle y_{*i}, y_i\rangle\rangle| \leq \sum_{i=1}^\infty \|y_{*i}\|\|y_i\| \leq \|x_*\|\|x\|$, where we have made use of Proposition 7.72 in the first inequality and Hölder's Inequality in the second inequality. Hence, $\phi(x_*)(x)$ is well-defined. Hence, $\phi(x_*)$ is well-defined. Clearly, $\phi(x_*)$ is linear and bounded with

$$\begin{aligned}\|\phi(x_*)\| &= \sup_{x\in\mathcal{X},\ \|x\|\leq 1} |\phi(x_*)(x)| \leq \sup_{x=(y_1,y_2,\ldots)\in\mathcal{X},\ \|x\|\leq 1} \sum_{i=1}^\infty |\langle\langle y_{*i}, y_i\rangle\rangle| \\ &\leq \sup_{x\in\mathcal{X},\ \|x\|\leq 1} \|x_*\|\|x\| \leq \|x_*\|\end{aligned} \tag{7.7}$$

Hence, $\phi(x_*) \in \mathfrak{X}^*$ and

$$\psi(\phi(x_*)) = (\phi(x_*) \circ A_1, \phi(x_*) \circ A_2, \ldots) = (y_{*1}, y_{*2}, \ldots) = x_*$$

Then, $\psi \circ \phi = \mathrm{id}_{l_1(\mathcal{Y}^*)}$. Hence, ψ is surjective. Therefore, ψ is bijective and admits inverse ψ_{inv}. By Proposition 2.4, $\phi = \psi_{\mathrm{inv}}$. $\forall f \in \mathfrak{X}^*$, by (7.6) and (7.7), we have $\|f\| = \|\phi(\psi(f))\| \leq \|\psi(f)\| \leq \|f\|$. Then, $\|\psi(f)\| = \|f\|$ and ψ is an isometry. Hence, ψ is an isometrical isomorphism. ◇

7.9.3 Extension Form of Hahn–Banach Theorem

Definition 7.79 Let $\mathcal{X}$ be a vector space over $\mathbb{K}$. A *sublinear functional* is $p : \mathcal{X} \to \mathbb{R}$ satisfying, $\forall x_1, x_2 \in \mathcal{X}$, $\forall \alpha \in \mathbb{R}$ with $\alpha \geq 0$, (i) $p(x_1 + x_2) \leq p(x_1) + p(x_2)$; and (ii) $p(\alpha x_1) = \alpha p(x_1)$. ◇

Note that any $\|\cdot\|$ on $\mathcal{X}$ is a sublinear functional. We introduce the above definition to illustrate the full generality of the Hahn–Banach Theorem.

Theorem 7.80 (Extension Form of Hahn–Banach Theorem) *Let $\mathcal{X}$ be a vector space over the field $\mathbb{R}$, $p : \mathcal{X} \to \mathbb{R}$ be a sublinear functional, $M \subseteq \mathcal{X}$ be a subspace, and $f : M \to \mathbb{R}$ be a linear functional on M satisfying $f(m) \leq p(m)$, $\forall m \in M$. Then, $\exists$ a linear functional $F : \mathcal{X} \to \mathbb{R}$ such that $F|_M = f$ and $F(x) \leq p(x)$, $\forall x \in \mathcal{X}$. Furthermore, if $\mathcal{X}$ is normed with $\|\cdot\|$ and p is continuous at $\vartheta_{\mathcal{X}}$, then F is continuous.*

Proof We will prove the theorem using Zorn's Lemma. Define a collection of extensions of f, $\mathcal{E}$, by

$$\mathcal{E} := \{(g, N) \mid N \subseteq \mathcal{X} \text{ is a subspace, } M \subseteq N,\ g : N \to \mathbb{R} \text{ is a linear functional, such that } G|_M = f \text{ and } g(n) \leq p(n), \forall n \in N\}$$

Clearly, $(f, M) \in \mathcal{E} \neq \emptyset$. Define a relation $\preceq$ on $\mathcal{E}$ by $\forall (g_1, N_1), (g_2, N_2) \in \mathcal{E}$, we say $(g_1, N_1) \preceq (g_2, N_2)$ if $N_1 \subseteq N_2$ and $g_2|_{N_1} = g_1$. Clearly, $\preceq$ is reflexive, antisymmetric, and transitive. Hence, $\preceq$ is an antisymmetric partial ordering on $\mathcal{E}$.

For any nonempty subcollection $\mathcal{C} \subseteq \mathcal{E}$ such that $\preceq$ is a total ordering on $\mathcal{C}$, let $N_c = \bigcup_{(g,N)\in\mathcal{C}} N$. Then, $\vartheta_{\mathcal{X}} \in M \subseteq N_c \subseteq \mathcal{X}$. $\forall x_1, x_2 \in N_c$, $\forall \alpha, \beta \in \mathbb{R}$, $\exists (g_1, N_1), (g_2, N_2) \in \mathcal{C}$ such that $x_1 \in N_1$ and $x_2 \in N_2$. Since $\mathcal{C}$ is totally ordered, by Proposition 2.12, then, without loss of generality, we may assume that $(g_1, N_1) \preceq (g_2, N_2)$. Then, $N_1 \subseteq N_2$ and $x_1, x_2 \in N_2$. This implies that $\alpha x_1 + \beta x_2 \in N_2 \subseteq N_c$, since N_2 is a subspace. The above shows that N_c is a subspace of $\mathcal{X}$.

Define a functional $g_c : N_c \to \mathbb{R}$ by $\forall x \in N_c$, $\exists (g, N) \in \mathcal{C}$ such that $x \in N$, we assign $g_c(x) := g(x)$. Such a functional is uniquely defined because of the following reasoning. $\forall x \in N_c$, $\forall (g_1, N_1), (g_2, N_2) \in \mathcal{C}$ such that $x \in N_1 \cap N_2$. By the total ordering of $\mathcal{C}$ and Proposition 2.12, we may assume that, without loss of generality, $(g_1, N_1) \preceq (g_2, N_2)$. Then, $x \in N_1 \subseteq N_2$ and $g_2|_{N_1} = g_1$. Hence, $g_1(x) = g_2(x)$. Therefore, g_c is well-defined.

$\forall x_1, x_2 \in N_c$, $\forall \alpha, \beta \in \mathbb{R}$, $\exists (g_1, N_1), (g_2, N_2) \in \mathcal{C}$ such that $x_1 \in N_1$ and $x_2 \in N_2$. By the total ordering of $\mathcal{C}$ and Proposition 2.12, we may assume that, without loss of generality, $(g_1, N_1) \preceq (g_2, N_2)$. Then, $x_1, x_2 \in N_2$ and $g_c(\alpha x_1 + \beta x_2) = g_2(\alpha x_1 + \beta x_2) = \alpha g_2(x_1) + \beta g_2(x_2) = \alpha g_c(x_1) + \beta g_c(x_2)$. Hence, g_c is a linear functional. $g_c(x_1) = g_2(x_1) \leq p(x_1)$. $\forall m \in M$, $g_c(m) = g_2(m) = f(m)$. Therefore, $(g_c, N_c) \in \mathcal{E}$.

$\forall (g, N) \in \mathcal{C}$, we have $N \subseteq N_c$ and $\forall n \in N$, $g_c(n) = g(n)$. Then, $(g, N) \preceq (g_c, N_c)$. Hence, (g_c, N_c) is an upper bound of $\mathcal{C}$.

By Zorn's Lemma, $\exists (g_M, N_M) \in \mathcal{E}$, which is maximal with respect to $\preceq$. Now, we are going to show that $N_M = \mathcal{X}$. Suppose $N_M \subset \mathcal{X}$. Then, $\exists x_0 \in \mathcal{X} \setminus N_M$. Let $N_e = \{x \in \mathcal{X} \mid \exists \alpha \in \mathbb{R}, \exists n \in N_M \ni \cdot\, x = \alpha x_0 + n\}$. Clearly, $N_M \subset N_e$, $x_0 \in N_e$, and N_e is a subspace of $\mathcal{X}$. $\forall x \in N_e$, $\exists!\, \alpha \in \mathbb{R}$ and $\exists!\, n \in N_M$ such that $x = \alpha x_0 + n$ (otherwise, we may deduce that $x_0 \in N_M$). Define $g_e : N_e \to \mathbb{R}$ by $g_e(x) = g_e(\alpha x_0 + n) = g_M(n) + \alpha g_e(x_0)$, $\forall x = \alpha x_0 + n \in N_e$, $\alpha \in \mathbb{R}$, and $n \in N_M$, where $g_e(x_0) \in \mathbb{R}$ is a constant to be determined. Clearly, g_e is well-defined and is a linear functional on N_e. $\forall n \in N_M$, we have $g_e(n) = g_M(n)$, and hence, $g_e|_{N_M} = g_M$. Then, $g_e|_M = f$.

We will show that $(g_e, N_e) \in \mathcal{E}$ by finding an admissible constant $g_e(x_0)$ such that $g_e(x) \leq p(x)$, $\forall x \in N_e$. $\forall n_1, n_2 \in N_M$, we have $g_M(n_1) + g_M(n_2) = g_M(n_1 + n_2) \leq p(n_1 + n_2) \leq p(n_1 + x_0) + p(n_2 - x_0)$. Then, $g_M(n_2) - p(n_2 - x_0) \leq p(n_1 + x_0) - g_M(n_1)$. By the arbitrariness of n_1 and n_2, we have $\sup_{n_2 \in N_M} (g_M(n_2) - p(n_2 - x_0)) \leq \inf_{n_1 \in N_M} (p(n_1 + x_0) - g_M(n_1))$. Since $\vartheta_{\mathcal{X}} \in N_M$, then $\exists c \in \mathbb{R}$ such that

$$
\begin{aligned}
-\infty < \sup_{n_2 \in N_M} (g_M(n_2) - p(n_2 - x_0)) &\leq c \\
&\leq \inf_{n_1 \in N_M} (p(n_1 + x_0) - g_M(n_1)) < +\infty
\end{aligned}
$$

Let $g_e(x_0) := c$.

$\forall x \in N_e$, $x = \alpha x_0 + n$, where $\alpha \in \mathbb{R}$ and $n \in N_M$. We will distinguish three exhaustive and mutually exclusive cases: Case 1: $\alpha > 0$; Case 2: $\alpha = 0$; Case 3: $\alpha < 0$. Case 1: $\alpha > 0$. $g_e(x) = \alpha c + g_M(n) = \alpha(c + g_M(n/\alpha)) \leq \alpha(p(n/\alpha + x_0) - g_M(n/\alpha) + g_M(n/\alpha)) = \alpha p(x/\alpha) = p(x)$. Case 2: $\alpha = 0$. Then, $x \in N_M$ and $g_e(x) = g_M(n) \leq p(n) = p(x)$. Case 3: $\alpha < 0$. $g_e(x) = \alpha c + g_M(n) = \alpha(c + g_M(n/\alpha)) \leq \alpha(g_M(-n/\alpha) - p(-n/\alpha - x_0) + g_M(n/\alpha)) = -\alpha p(-x/\alpha) = p(x)$. Hence, we have $g_e(x) \leq p(x)$ in all three cases. Therefore, $(g_e, N_e) \in \mathcal{E}$. We have shown that $(g_M, N_M) \preceq (g_e, N_e)$ and $(g_M, N_M) \neq (g_e, N_e)$. This contradicts the fact that (g_M, N_M) is maximal in $\mathcal{E}$ by Proposition 2.12.

Hence, $N_M = \mathcal{X}$ and $F = g_M$.

If, in addition, $\mathcal{X}$ is normed with $\|\cdot\|$ and p is continuous at $\vartheta_{\mathcal{X}}$, then, $\forall \epsilon \in (0, \infty) \subset \mathbb{R}$, $\exists \delta \in (0, \infty) \subset \mathbb{R}$ such that $|p(x)| < \epsilon$, $\forall x \in \mathcal{B}(\vartheta_{\mathcal{X}}, \delta)$. $\forall x \in$

$\mathcal{B}(\vartheta_{\mathcal{X}}, \delta)$, we will distinguish two exhaustive and mutually exclusive cases: Case 1: $F(x) \geq 0$; Case 2: $F(x) < 0$. Case 1: $F(x) \geq 0$. Then, $0 \leq F(x) \leq p(x) < \epsilon$ and $|F(x)| < \epsilon$. Case 2: $F(x) < 0$. Then, $|F(x)| = -F(x) = F(-x) \leq p(-x) < \epsilon$. Hence, in both cases, we have $|F(x)| < \epsilon$. Then, F is continuous at $\vartheta_{\mathcal{X}}$. Thus, by Proposition 7.72, F is a bounded linear functional, and therefore continuous.

This completes the proof of the theorem. □

To obtain the counterpart result for complex vector spaces, we need the following result.

Lemma 7.81 *Let $(\mathcal{X}, \mathbb{C})$ be a vector space, $f : \mathcal{X} \to \mathbb{C}$ be a functional on $(\mathcal{X}, \mathbb{C})$, and $g : \mathcal{X} \to \mathbb{R}$ be given by $g(x) = \operatorname{Re}(f(x))$, $\forall x \in \mathcal{X}$. Then, f is a linear functional on $(\mathcal{X}, \mathbb{C})$ if, and only if, g is a linear functional on $(\mathcal{X}, \mathbb{R})$ and $f(x) = g(x) - \mathrm{i}g(\mathrm{i}x)$, $\forall x \in \mathcal{X}$.*

Proof By Lemma 7.40, $(\mathcal{X}, \mathbb{R})$ is a vector space. "Necessity" $\forall x_1, x_2 \in \mathcal{X}$, $\forall \alpha, \beta \in \mathbb{R}$, we have $g(\alpha x_1 + \beta x_2) = \operatorname{Re}(f(\alpha x_1 + \beta x_2)) = \operatorname{Re}(\alpha f(x_1) + \beta f(x_2)) = \alpha \operatorname{Re}(f(x_1)) + \beta \operatorname{Re}(f(x_2)) = \alpha g(x_1) + \beta g(x_2)$. Hence, g is a linear functional on $(\mathcal{X}, \mathbb{R})$. $\forall x \in \mathcal{X}$, let $f(x) = g(x) + \mathrm{i} \operatorname{Im}(f(x))$. Since f is a linear functional on $(\mathcal{X}, \mathbb{C})$, then $f(\mathrm{i}x) = g(\mathrm{i}x) + \mathrm{i} \operatorname{Im}(f(\mathrm{i}x)) = \mathrm{i}f(x) = -\operatorname{Im}(f(x)) + \mathrm{i}g(x)$. Then, $\operatorname{Im}(f(x)) = -g(\mathrm{i}x)$ and $f(x) = g(x) - \mathrm{i}g(\mathrm{i}x)$.

"Sufficiency" $\forall x_1, x_2 \in \mathcal{X}$, $\forall \alpha := \alpha_r + \mathrm{i}\alpha_i, \beta := \beta_r + \mathrm{i}\beta_i \in \mathbb{C}$, where $\alpha_r, \alpha_i, \beta_r, \beta_i \in \mathbb{R}$, we have

$$\begin{aligned}
&f(\alpha x_1 + \beta x_2) = g(\alpha x_1 + \beta x_2) - \mathrm{i}g(\mathrm{i}\alpha x_1 + \mathrm{i}\beta x_2)\\
&= g(\alpha_r x_1 + \mathrm{i}\alpha_i x_1 + \beta_r x_2 + \mathrm{i}\beta_i x_2) - \mathrm{i}g(\alpha_r \mathrm{i}x_1 - \alpha_i x_1 + \beta_r \mathrm{i}x_2 - \beta_i x_2)\\
&= \alpha_r g(x_1) + \alpha_i g(\mathrm{i}x_1) + \beta_r g(x_2) + \beta_i g(\mathrm{i}x_2) - \mathrm{i}\alpha_r g(\mathrm{i}x_1) + \mathrm{i}\alpha_i g(x_1)\\
&\quad - \mathrm{i}\beta_r g(\mathrm{i}x_2) + \mathrm{i}\beta_i g(x_2)\\
&= \alpha_r f(x_1) + \mathrm{i}\alpha_i f(x_1) + \beta_r f(x_2) + \mathrm{i}\beta_i f(x_2) = \alpha f(x_1) + \beta f(x_2)
\end{aligned}$$

Hence, f is a linear functional on $(\mathcal{X}, \mathbb{C})$.

This completes the proof of the lemma. □

Theorem 7.82 (Extension Form of Hahn–Banach Theorem) *Let $\mathcal{X}$ be a vector space over the field $\mathbb{C}$, $p : \mathcal{X} \to \mathbb{R}$ be a sublinear functional, $M \subseteq \mathcal{X}$ be a subspace (of $(\mathcal{X}, \mathbb{C})$), and $f : M \to \mathbb{C}$ be a linear functional satisfying* $\operatorname{Re}(f(m)) \leq p(m)$, *$\forall m \in M$. Then, there exists a linear functional $F : \mathcal{X} \to \mathbb{C}$, such that $F|_M = f$ and* $\operatorname{Re}(F(x)) \leq p(x)$, *$\forall x \in \mathcal{X}$. Furthermore, if $\mathcal{X}$ is normed with $\|\cdot\|$ and p is continuous at $\vartheta_{\mathcal{X}}$, then F is continuous.*

Proof By Lemma 7.40, $(\mathcal{X}, \mathbb{R})$ is a vector space and $(M, \mathbb{R})$ is also a vector space. It is easy to show that $(M, \mathbb{R})$ is a subspace of $(\mathcal{X}, \mathbb{R})$. Define $g : M \to \mathbb{R}$ by $g(m) = \operatorname{Re}(f(m))$, $\forall m \in M$. By Lemma 7.81, g is a linear functional on $(M, \mathbb{R})$ and $f(m) = g(m) - \mathrm{i}g(\mathrm{i}m)$, $\forall m \in M$. Note that p is a sublinear functional on $(\mathcal{X}, \mathbb{C})$, then it is a sublinear functional on $(\mathcal{X}, \mathbb{R})$. Furthermore, $g(m) =$

$\text{Re}\,(f(m)) \leq p(m)$, $\forall m \in M$. Then, by Hahn–Banach Theorem, Theorem 7.80, $\exists$ a linear functional G on $(\mathcal{X}, \mathbb{R})$ such that $G|_M = g$ and $G(x) \leq p(x)$, $\forall x \in \mathcal{X}$.

Define a functional F on $(\mathcal{X}, \mathbb{C})$ by $F(x) = G(x) - \mathrm{i}G(\mathrm{i}x)$, $\forall x \in \mathcal{X}$. By Lemma 7.81, F is a linear functional on $(\mathcal{X}, \mathbb{C})$. $\forall m \in M$, $\mathrm{i}m \in M$, since M is a subspace of $(\mathcal{X}, \mathbb{C})$. Then, $F(m) = G(m) - \mathrm{i}G(\mathrm{i}m) = g(m) - \mathrm{i}g(\mathrm{i}m) = f(m)$. Hence, we have $F|_M = f$. $\forall x \in \mathcal{X}$, $\text{Re}\,(F(x)) = G(x) \leq p(x)$. Hence, F is the functional we seek.

Furthermore, if $(\mathcal{X}, \mathbb{C})$ is normed with $\|\cdot\|$ and p is continuous at $\vartheta_{\mathcal{X}}$, then, by Lemma 7.40, $\mathfrak{X}_{\mathbb{R}} := (\mathcal{X}, \mathbb{R}, \|\cdot\|)$ is a normed linear space. By Theorem 7.80, G is continuous. Then, $G : \mathfrak{X}_{\mathbb{R}} \to \mathbb{R}$ is continuous. By Propositions 3.12, 3.32, and 7.23, $F : \mathfrak{X}_{\mathbb{R}} \to \mathbb{C}$ is continuous. By Lemma 7.40, F is continuous on $\mathfrak{X} := (\mathcal{X}, \mathbb{C}, \|\cdot\|)$.

This completes the proof of the theorem. □

Theorem 7.83 (Simple Version of Hahn–Banach Theorem) *Let $\mathfrak{X}$ be a normed linear space over $\mathbb{K}$, $M \subseteq \mathfrak{X}$ be a subspace, $f : M \to \mathbb{K}$ be a linear functional which is bounded (on M), that is $\|f\|_M := \sup_{m \in M,\, \|m\| \leq 1} |f(m)| \in [0, \infty) \subset \mathbb{R}$. Then, there exists a $F \in \mathfrak{X}^*$ such that $F|_M = f$ and $\|F\| = \|f\|_M$.*

Proof Define a functional $p : \mathfrak{X} \to \mathbb{R}$ by $p(x) = \|f\|_M \|x\|$, $\forall x \in \mathfrak{X}$. It is easy to check that p is a sublinear functional. We are going to distinguish two exhaustive and mutually exclusive cases: Case 1: $\mathbb{K} = \mathbb{R}$; Case 2: $\mathbb{K} = \mathbb{C}$.

Case 1: $\mathbb{K} = \mathbb{R}$. Clearly, by Proposition 7.72, we have $f(m) \leq |f(m)| \leq p(m)$, $\forall m \in M$, and p is continuous. By Hahn–Banach Theorem, Theorem 7.80, $\exists$ a linear functional $F : \mathfrak{X} \to \mathbb{R}$ such that $F|_M = f$ and $F(x) \leq p(x)$, $\forall x \in \mathfrak{X}$. Furthermore, F is continuous. Hence, $F \in \mathfrak{X}^*$, by Proposition 7.72. Note that $\|f\|_M = \sup_{m \in M,\, \|m\| \leq 1} |F(m)| \leq \sup_{x \in \mathfrak{X},\, \|x\| \leq 1} |F(x)| = \|F\|$. On the other hand, we have, $\forall x \in \mathfrak{X}$, $-F(x) = F(-x) \leq p(-x) = \|f\|_M \|x\|$ and $F(x) \leq p(x) = \|f\|_M \|x\|$. Hence, $|F(x)| \leq \|f\|_M \|x\|$. Then, $\|F\| \leq \|f\|_M$. Hence, $\|F\| = \|f\|_M$.

Case 2: $\mathbb{K} = \mathbb{C}$. Note that, $\forall m \in M$, $\text{Re}\,(f(m)) \leq |\text{Re}\,(f(m))| \leq |f(m)| \leq p(m)$, by Proposition 7.72. By Hahn–Banach Theorem, Theorem 7.82, there exists a linear functional $F : \mathfrak{X} \to \mathbb{C}$ such that $F|_M = f$ and $\text{Re}\,(F(x)) \leq p(x)$, $\forall x \in \mathfrak{X}$. Furthermore, since p is continuous, then F is continuous. By Proposition 7.72, $F \in \mathfrak{X}^*$. Note that $\|f\|_M = \sup_{m \in M,\, \|m\| \leq 1} |F(m)| \leq \sup_{x \in \mathfrak{X},\, \|x\| \leq 1} |F(x)| = \|F\|$. On the other hand, $\forall x \in \mathfrak{X}$, we have either $F(x) = 0$, then $|F(x)| \leq \|f\|_M \|x\|$; or $F(x) \neq 0$, then $|F(x)| = \frac{|F(x)|}{F(x)} F(x) = F\left(\frac{|F(x)|}{F(x)} x\right) = \text{Re}\left(F\left(\frac{|F(x)|}{F(x)} x\right)\right) \leq p\left(\frac{|F(x)|}{F(x)} x\right) = \|f\|_M \left\|\frac{|F(x)|}{F(x)} x\right\| = \|f\|_M \left|\frac{|F(x)|}{F(x)}\right| \cdot \|x\| = \|f\|_M \|x\|$. Thus, we have $|F(x)| \leq \|f\|_M \|x\|$, $\forall x \in \mathfrak{X}$. Therefore, $\|F\| \leq \|f\|_M$. Hence, $\|F\| = \|f\|_M$.

This completes the proof of the theorem. □

Corollary 7.84 *Let $\mathcal{X}$ be a vector space over the field $\mathbb{K}$, $M \subseteq \mathcal{X}$ be a subspace, $p : \mathcal{X} \to \mathbb{R}$ be a sublinear functional, $f : M \to \mathbb{K}$ be a linear functional on M, and $(\mathcal{G}, \circ, E)$ be an abelian semigroup of linear operators of $\mathcal{X}$ to $\mathcal{X}$. Assume that the following conditions hold.*

(i) $Am \in M$, $\forall A \in \mathcal{G}$ *and* $\forall m \in M$.
(ii) $p(Ax) \leq p(x)$, $\forall x \in \mathcal{X}$ *and* $\forall A \in \mathcal{G}$.
(iii) $f(Am) = f(m)$, $\forall A \in \mathcal{G}$ *and* $\forall m \in M$.
(iv) $\text{Re}\,(f(m)) \leq p(m)$, $\forall m \in M$.

Then, there exists a linear functional $F : \mathcal{X} \to \mathbb{K}$ such that $F|_M = f$, $F(Ax) = F(x)$, $\forall A \in \mathcal{G}$ and $\forall x \in \mathcal{X}$, and $\text{Re}\,(F(x)) \leq p(x)$, $\forall x \in \mathcal{X}$. Furthermore, if, in addition, $\mathcal{X}$ is normed with $\|\cdot\|$ and p is continuous at $\vartheta_{\mathcal{X}}$, then F is continuous.

Proof Define a functional $q : \mathcal{X} \to \mathbb{R}_e$ by

$$q(x) = \inf_{\substack{n \in \mathbb{N} \\ A_1,\ldots,A_n \in \mathcal{G}}} \frac{1}{n} p\Big(\sum_{i=1}^{n} A_i x\Big); \qquad \forall x \in \mathcal{X}$$

Since $p(\vartheta_{\mathcal{X}}) = 0$, then $q(\vartheta_{\mathcal{X}}) = 0$. Since $E \in \mathcal{G}$, then, $\forall x \in \mathcal{X}$, $q(x) \leq p(Ex) \leq p(x) < +\infty$. $\forall x_1, x_2 \in \mathcal{X}$, $\forall n, k \in \mathbb{N}$, $\forall A_1, \ldots, A_n \in \mathcal{G}$, $\forall B_1, \ldots, B_k \in \mathcal{G}$, we have

$$q(x_1 + x_2) \leq \frac{1}{nk} p\Big(\sum_{i=1}^{n}\sum_{j=1}^{k} A_i \circ B_j(x_1 + x_2)\Big)$$

$$= \frac{1}{nk} p\Big(\sum_{i=1}^{n}\sum_{j=1}^{k} A_i \circ B_j(x_1) + \sum_{i=1}^{n}\sum_{j=1}^{k} A_i \circ B_j(x_2)\Big)$$

$$= \frac{1}{nk} p\Big(\sum_{i=1}^{n} A_i\Big(\sum_{j=1}^{k} B_j(x_1)\Big) + \sum_{j=1}^{k} B_j\Big(\sum_{i=1}^{n} A_i(x_2)\Big)\Big)$$

$$\leq \frac{1}{nk} p\Big(\sum_{i=1}^{n} A_i\Big(\sum_{j=1}^{k} B_j(x_1)\Big)\Big) + \frac{1}{nk} p\Big(\sum_{j=1}^{k} B_j\Big(\sum_{i=1}^{n} A_i(x_2)\Big)\Big)$$

$$\leq \frac{1}{nk} \sum_{i=1}^{n} p\Big(A_i\Big(\sum_{j=1}^{k} B_j(x_1)\Big)\Big) + \frac{1}{nk} \sum_{j=1}^{k} p\Big(B_j\Big(\sum_{i=1}^{n} A_i(x_2)\Big)\Big)$$

$$\leq \frac{1}{nk}\sum_{i=1}^{n} p\Big(\sum_{j=1}^{k} B_j(x_1)\Big) + \frac{1}{nk}\sum_{j=1}^{k} p\Big(\sum_{i=1}^{n} A_i(x_2)\Big)$$

$$= \frac{1}{k} p\Big(\sum_{j=1}^{k} B_j(x_1)\Big) + \frac{1}{n} p\Big(\sum_{i=1}^{n} A_i(x_2)\Big)$$

where the first two equalities follows from the fact that $\mathcal{G}$ is an abelian semigroup of linear operators on $\mathcal{X}$, the second and third inequalities follows from the fact that p is sublinear, the fourth inequality follows from (ii) in the assumption. Then, by the definition of q, we have

$$q(x_1 + x_2) \leq q(x_1) + q(x_2)$$

and the right-hand side makes sense since $q(x_1) < +\infty$ and $q(x_2) < +\infty$. Note that $0 = q(\vartheta_{\mathcal{X}}) \leq q(x_1) + q(-x_1)$. Then, $q(x_1) > -\infty$. Hence, $q : \mathcal{X} \to \mathbb{R}$.

$\forall \alpha \in [0, \infty) \subset \mathbb{R}$, $\forall x \in \mathcal{X}$, we have

$$q(\alpha x) = \inf_{\substack{n \in \mathbb{N} \\ A_1,\ldots,A_n \in \mathcal{G}}} \frac{1}{n} p\Big(\sum_{i=1}^{n} A_i(\alpha x)\Big) = \inf_{\substack{n \in \mathbb{N} \\ A_1,\ldots,A_n \in \mathcal{G}}} \frac{1}{n} p\Big(\alpha \sum_{i=1}^{n} A_i x\Big)$$

$$= \inf_{\substack{n \in \mathbb{N} \\ A_1,\ldots,A_n \in \mathcal{G}}} \Big(\alpha \frac{1}{n} p\Big(\sum_{i=1}^{n} A_i x\Big)\Big) = \alpha q(x)$$

where the fourth equality follows from Proposition 3.81 and the fact that $q(x) \in \mathbb{R}$. Hence, we have shown that q is a sublinear functional.

$\forall m \in M$, $\forall n \in \mathbb{N}$, $\forall A_1, \ldots, A_n \in \mathcal{G}$, we have

$$\operatorname{Re}\Big(f(m)\Big) = \frac{1}{n}\sum_{i=1}^{n} \operatorname{Re}\Big(f(A_i m)\Big) = \frac{1}{n}\operatorname{Re}\Big(f\Big(\sum_{i=1}^{n} A_i m\Big)\Big)$$

$$\leq \frac{1}{n} p\Big(\sum_{i=1}^{n} A_i m\Big)$$

where the first equality follows from (i) and (iii) in the assumptions, the second equality follows from the linearity of f, and the first inequality follows from (iv) in the assumptions. Hence, we have $\operatorname{Re}(f(m)) \leq q(m)$, $\forall m \in M$.

By the extension forms of Hahn–Banach Theorem (Theorem 7.80 for $\mathbb{K} = \mathbb{R}$ and Theorem 7.82 for $\mathbb{K} = \mathbb{C}$), there exists a linear functional $F : \mathcal{X} \to \mathbb{K}$ such that $F|_M = f$ and $\operatorname{Re}(F(x)) \leq q(x)$, $\forall x \in \mathcal{X}$.

$\forall x \in \mathcal{X}, \forall A \in \mathcal{G}, \forall n \in \mathbb{N}$, we have

$$q(x - Ax) \le \frac{1}{n} p\Big(E(x - Ax) + A(x - Ax) + \cdots + A^{n-1}(x - Ax)\Big)$$
$$= \frac{1}{n} p(Ex - A^n x) \le \frac{1}{n}\big(p(Ex) + p(A^n(-x))\big) \le \frac{1}{n}(p(x) + p(-x))$$

where the first inequality follows from the definition of q, the second inequality follows from the fact that p is sublinear, and the third inequality follows from (ii) in the assumption. Hence, by the arbitrariness of n, we have $q(x - Ax) \le 0$. We will show that $F(Ax) = F(x)$ by distinguishing two exhaustive and mutually exclusive cases: Case 1: $\mathbb{K} = \mathbb{R}$; Case 2: $\mathbb{K} = \mathbb{C}$.

Case 1: $\mathbb{K} = \mathbb{R}$. We have $F(x) - F(Ax) = F(x - Ax) \le q(x - Ax) \le 0$. On the other hand, we have $F(Ax) - F(x) = F(-x) - F(A(-x)) = F((-x) - A(-x)) \le q((-x) - A(-x)) \le 0$. Hence, $F(x) = F(Ax)$.

Case 2: $\mathbb{K} = \mathbb{C}$. Define $H : \mathcal{X} \to \mathbb{R}$ by $H(x) = \operatorname{Re}(F(x))$, $\forall x \in \mathcal{X}$. By Lemma 7.40, $(\mathcal{X}, \mathbb{R})$ is a vector space. By Lemma 7.81, H is a linear functional on $(\mathcal{X}, \mathbb{R})$ and $F(x) = H(x) - \mathrm{i}H(\mathrm{i}x)$, $\forall x \in \mathcal{X}$. Note that $H(x) - H(Ax) = H(x - Ax) = \operatorname{Re}(F(x - Ax)) \le q(x - Ax) \le 0$. On the other hand, we have $H(Ax) - H(x) = H(-x) - H(A(-x)) = H((-x) - A(-x)) = \operatorname{Re}(F((-x) - A(-x))) \le q((-x) - A(-x)) \le 0$. Hence, $H(x) = H(Ax)$. Then, $F(Ax) = H(Ax) - \mathrm{i}H(\mathrm{i}Ax) = H(Ax) - \mathrm{i}H(A(\mathrm{i}x)) = H(x) - \mathrm{i}H(\mathrm{i}x) = F(x)$.

Furthermore, if, in addition, $\mathcal{X}$ is normed with $\|\cdot\|$ and p is continuous at $\vartheta_{\mathcal{X}}$, then $\forall \epsilon \in (0, \infty) \subset \mathbb{R}$, $\exists \delta \in (0, \infty) \subset \mathbb{R}$, $\forall x \in \mathcal{B}(\vartheta_{\mathcal{X}}, \delta)$, we have $|p(x) - p(\vartheta_{\mathcal{X}})| = |p(x)| < \epsilon$. Note that $q(x) \le p(x) < \epsilon$. Then, $q(-x) \le p(-x) < \epsilon$. Since q is a sublinear functional, we have $0 = q(\vartheta_{\mathcal{X}}) \le q(x) + q(-x)$. Hence, we have $q(x) \ge -q(-x) > -\epsilon$. Therefore, $|q(x) - q(\vartheta_{\mathcal{X}})| = |q(x)| < \epsilon$. Hence, q is continuous at $\vartheta_{\mathcal{X}}$. Therefore, by the extension forms of Hahn–Banach Theorem, F is continuous.

This completes the proof of the corollary. □

Proposition 7.85 *Let $\mathfrak{X}$ be a normed linear space over the field $\mathbb{K}$ and $x_0 \in \mathfrak{X}$ with $x_0 \neq \vartheta_{\mathfrak{X}}$. Then, $\exists x_* \in \mathfrak{X}^*$ with $\|x_*\| = 1$ such that $\langle\langle x_*, x_0 \rangle\rangle = \|x_*\|\|x_0\|$.*

Proof Consider the subspace $M := \operatorname{span}(\{x_0\})$. Define $f : M \to \mathbb{K}$ by $f(\alpha x_0) = \alpha \|x_0\|$, $\forall \alpha \in \mathbb{K}$. Clearly, f is uniquely defined by Proposition 6.17. Clearly f is a linear functional on M and $\|f\|_M := \sup_{m \in M, \|m\| \le 1} |f(m)| = \sup_{\alpha \in \mathbb{K}, \|\alpha x_0\| \le 1} |\alpha \|x_0\|| = 1$. By the simple version of Hahn–Banach Theorem, there exists $x_* \in \mathfrak{X}^*$ such that $x_*|_M = f$ and $\|x_*\| = \|f\|_M = 1$. Then, $\langle\langle x_*, x_0 \rangle\rangle = f(x_0) = \|x_0\| = \|x_*\|\|x_0\|$. This completes the proof of the proposition. □

Example 7.86 Let $\mathcal{Y}$ be a normed linear space over the field $\mathbb{K}$, $\mathfrak{X} := \mathrm{l}_\infty(\mathcal{Y})$ be the normed linear space as defined in Example 7.10, and $y_0 \in \mathcal{Y}$ with $y_0 \neq \vartheta_{\mathcal{Y}}$. We will show that $\mathfrak{X}^* \neq \mathrm{l}_1(\mathcal{Y}^*)$ by Hahn–Banach Theorem.

Let $M := \{x := (\xi_1, \xi_2, \ldots) \in \mathcal{X} \mid \lim_{n\in\mathbb{N}} \xi_n \in \mathcal{Y}\}$. Clearly, M is a subspace of $\mathcal{X}$. By Proposition 7.85, there exists $y_* \in \mathcal{Y}^*$ such that $\|y_*\| = 1$ and $\langle\langle y_*, y_0\rangle\rangle = \|y_0\| > 0$. Define $f : M \to \mathbb{K}$ by $f(m) = \langle\langle y_*, \lim_{n\in\mathbb{N}} \xi_n\rangle\rangle$, $\forall m := (\xi_1, \xi_2, \ldots) \in M$. It is easy to show that f is a linear functional on M. $\forall m = (\xi_1, \xi_2, \ldots) \in M$ with $\|m\| \leq 1$, we have

$$\begin{aligned}|f(m)| &= \left|\left\langle\left\langle y_*, \lim_{n\in\mathbb{N}} \xi_n\right\rangle\right\rangle\right| \leq \left\| \lim_{n\in\mathbb{N}} \xi_n \right\| = \lim_{n\in\mathbb{N}} \|\xi_n\| \leq \sup_{n\geq 1} \|\xi_n\| \\ &= \|m\| \leq 1\end{aligned}$$

where we have applied Propositions 7.72, 7.21, and 3.66. Hence, we have $\|f\|_M := \sup_{m\in M,\ \|m\|\leq 1} |f(m)| \leq 1$. Consider $m_0 = (y_0, y_0, \ldots) \in M$, we have $|f(m_0)| = |\langle\langle y_*, y_0\rangle\rangle| = \|y_0\| = \|m_0\| > 0$. Hence, by Proposition 7.72, $\|f\|_M = 1$. By the simple version of Hahn–Banach Theorem, there exists $x_* \in \mathcal{X}^*$ such that $x_*|_M = f$ and $\|x_*\| = \|f\|_M = 1$. Clearly, there does not exist $(\eta_{*1}, \eta_{*2}, \ldots) \in \mathrm{l}_1(\mathcal{Y}^*)$ such that x_* is given by $\langle\langle x_*, x\rangle\rangle = \sum_{n=1}^{\infty} \langle\langle \eta_{*n}, \xi_n\rangle\rangle$, $\forall x = (\xi_1, \xi_2, \ldots) \in \mathcal{X}$. Hence, $\mathcal{X}^* \neq \mathrm{l}_1(\mathcal{Y}^*)$. ◇

7.9.4 Second Dual Space

Definition 7.87 Let $\mathcal{X}$ be a normed linear space over the field $\mathbb{K}$ and $\mathcal{X}^*$ be its dual. The dual space of $\mathcal{X}^*$ is called the *second dual* of $\mathcal{X}$ and is denoted by $\mathcal{X}^{**}$. ◇

Remark 7.88 Let $\mathcal{X}$ be a normed linear space over the field $\mathbb{K}$, $\mathcal{X}^*$ be its dual, and $\mathcal{X}^{**}$ be its second dual. Then, $\mathcal{X}$ is isometrically isomorphic to a dense subset of a Banach space, which can be taken as a subspace of $\mathcal{X}^{**}$. This can be proved as follows.

By Proposition 7.72, $\mathcal{X}^{**}$ is a Banach space over $\mathbb{K}$. $\forall x \in \mathcal{X}$, define a functional $f : \mathcal{X}^* \to \mathbb{K}$ by $f(x_*) = \langle\langle x_*, x\rangle\rangle$, $\forall x_* \in \mathcal{X}^*$. Clearly, f is a linear functional on $\mathcal{X}^*$. Note that $|f(x_*)| = |\langle\langle x_*, x\rangle\rangle| \leq \|x\|\|x_*\|$, $\forall x_* \in \mathcal{X}^*$, by Proposition 7.72. Then, $\|f\| \leq \|x\|$. When $x = \vartheta_{\mathcal{X}}$, then $\|f\| = 0 = \|x\|$. When $x \neq \vartheta_{\mathcal{X}}$, then, by Proposition 7.85, $\exists x_{*0} \in \mathcal{X}^*$ with $\|x_{*0}\| = 1$ such that $\langle\langle x_{*0}, x\rangle\rangle = \|x\|$. Then, $\|f\| = \sup_{x_*\in\mathcal{X}^*,\ \|x_*\|\leq 1} |f(x_*)| \geq |f(x_{*0})| = \|x\|$. Hence, we have $\|f\| = \|x\|$. Hence, $f \in \mathcal{X}^{**}$. Thus, we may define a *natural mapping* $\phi : \mathcal{X} \to \mathcal{X}^{**}$ by $\phi(x) = f$, $\forall x \in \mathcal{X}$.

$\forall x_1, x_2 \in \mathcal{X}$, $\forall \alpha, \beta \in \mathbb{K}$, $\forall x_* \in \mathcal{X}^*$, we have $\phi(\alpha x_1 + \beta x_2)(x_*) = \langle\langle x_*, \alpha x_1 + \beta x_2\rangle\rangle = \alpha\langle\langle x_*, x_1\rangle\rangle + \beta\langle\langle x_*, x_2\rangle\rangle = \alpha\phi(x_1)(x_*) + \beta\phi(x_2)(x_*)$. Hence, ϕ is a linear function. $\forall x \in \mathcal{X}$, we have $\|\phi(x)\| = \|x\|$. Hence, ϕ is an isometry and ϕ is injective. Therefore, ϕ is an isometrical isomorphism between $\mathcal{X}$ and $\phi(\mathcal{X})$. Clearly, $\phi(\mathcal{X}) \subseteq \mathcal{X}^{**}$ is a subspace. Then, by Proposition 7.17, $\overline{\phi(\mathcal{X})} \subseteq \mathcal{X}^{**}$ is a closed subspace. By Proposition 4.39, $\overline{\phi(\mathcal{X})}$ is complete. Hence, $\overline{\phi(\mathcal{X})}$ is a Banach space. By Proposition 3.5, $\phi(\mathcal{X})$ is dense in $\overline{\phi(\mathcal{X})}$. This completes the proof of the remark. ◇

Definition 7.89 Let $\mathcal{X}$ be a normed linear space over the field $\mathbb{K}$, $\mathcal{X}^{**}$ be its second dual, and $\phi : \mathcal{X} \to \mathcal{X}^{**}$ be the natural mapping as defined in Remark 7.88. $\mathcal{X}$ is said to be reflexive if $\phi(\mathcal{X}) = \mathcal{X}^{**}$, that is $\mathcal{X}$ and $\mathcal{X}^{**}$ are isometrically isomorphic. Then, we may label $\mathcal{X}^{**}$ such that $\mathcal{X}^{**} = \mathcal{X}$ and $\phi = \mathrm{id}_{\mathcal{X}}$. ◇

A reflexive normed linear space is clearly a Banach space.

Proposition 7.90 *Let $\mathcal{X}$ be a Banach space over the field $\mathbb{K}$ and $\mathcal{X}^*$ be its dual. Then, $\mathcal{X}$ is reflexive if, and only if, $\mathcal{X}^*$ is reflexive.*

Proof Let $\mathcal{X}^{**}$ be the second dual of $\mathcal{X}$ and $\mathcal{X}^{***}$ be the dual of $\mathcal{X}^{**}$.

"Necessity" Let $\mathcal{X}$ be reflexive. Then, $\mathcal{X}^{**} = \mathcal{X}$ isometrically isomorphically. Then, $\mathcal{X}^* = (\mathcal{X}^{**})^* = \mathcal{X}^{***}$ isometrically isomorphically. Hence, $\mathcal{X}^*$ is reflexive.

"Sufficiency" Let $\mathcal{X}^*$ be reflexive. Then, $\mathcal{X}^{***} = \phi_*(\mathcal{X}^*)$, where $\phi_* : \mathcal{X}^* \to \mathcal{X}^{***}$ is the natural mapping on $\mathcal{X}^*$. We will show that $\mathcal{X}$ is reflexive by an argument of contradiction. Suppose $\mathcal{X}$ is not reflexive. Let $\phi : \mathcal{X} \to \mathcal{X}^{**}$ be the natural mapping. ϕ is an isometrical isomorphism between $\mathcal{X}$ and $\phi(\mathcal{X})$. Since $\mathcal{X}$ is complete, then $\phi(\mathcal{X})$ is also complete. Then, by Proposition 4.39, $\phi(\mathcal{X})$ is a closed subspace of $\mathcal{X}^{**}$. Since $\mathcal{X}$ is not reflexive, then there exists $y_{**} \in \mathcal{X}^{**} \setminus \phi(\mathcal{X})$. Then, $y_{**} \neq \vartheta_{**} = \phi(\vartheta)$. Let $\delta := \inf_{m_{**} \in \phi(\mathcal{X})} \|y_{**} - m_{**}\|$. By Proposition 4.10, $\delta \in (0, \infty) \subset \mathbb{R}$. Consider the subspace $N := \{m_{**} + \alpha y_{**} \in \mathcal{X}^{**} \mid \alpha \in \mathbb{K}, m_{**} \in \phi(\mathcal{X})\}$. Since $y_{**} \notin \phi(\mathcal{X})$, then $\forall n_{**} \in N$, $\exists!\, \alpha \in \mathbb{K}$ and $\exists!\, m_{**} \in \phi(\mathcal{X})$ such that $n_{**} = \alpha y_{**} + m_{**}$. Define a linear functional $f : N \to \mathbb{K}$ by $f(n_{**}) = \alpha$, $\forall n_{**} = \alpha y_{**} + m_{**} \in N$, where $m_{**} \in \phi(\mathcal{X})$ and $\alpha \in \mathbb{K}$. $\forall n_{**} \in N$, let $n_{**} = \alpha y_{**} + m_{**}$, where $m_{**} \in \phi(\mathcal{X})$ and $\alpha \in \mathbb{K}$. We have $|f(n_{**})| = |\alpha| = |\alpha\delta|/\delta \leq \frac{1}{\delta}\|\alpha y_{**} + m_{**}\| = \frac{1}{\delta}\|n_{**}\|$. Hence, $\|f\|_N \leq 1/\delta$. By the simple version of Hahn–Banach Theorem, there exists $F \in \mathcal{X}^{***}$ such that $F|_N = f$ and $\|F\| = \|f\|_N$. Since $\mathcal{X}^*$ is reflexive, then $\phi_*(\mathcal{X}^*) = \mathcal{X}^{***}$. Then, $\exists y_* \in \mathcal{X}^*$ such that $F = \phi_*(y_*)$. Then, $\forall x_{**} \in \mathcal{X}^{**}$, we have $F(x_{**}) = \phi_*(y_*)(x_{**}) = \langle\langle x_{**}, y_* \rangle\rangle$. $\forall m_{**} \in \phi(\mathcal{X})$, we have $\langle\langle m_{**}, y_* \rangle\rangle = F(m_{**}) = f(m_{**}) = 0$. Hence, $\forall x \in \mathcal{X}$, $\langle\langle y_*, x \rangle\rangle = \phi(x)(y_*) = \langle\langle \phi(x), y_* \rangle\rangle = 0$. Hence, $y_* = \vartheta_*$. Then, $F = \phi_*(\vartheta_*) = \vartheta_{***}$. This contradicts with the fact that $F(y_{**}) = f(y_{**}) = 1 \neq 0$. Therefore, $\mathcal{X}$ must be reflexive.

This completes the proof of the proposition. □

Example 7.91 Clearly, the normed linear space $(\{\vartheta_{\mathcal{X}}\}, \mathbb{K}, \|\cdot\|)$ as defined in Example 7.73 is reflexive. Clearly $\mathbb{K}^n$, $n \in \mathbb{N}$, are reflective. Let $\mathcal{Y}$ be a reflexive Banach space. Then, $\mathrm{l}_p(\mathcal{Y})$, $p \in (1, \infty) \subset \mathbb{R}$, are reflexive. $\mathrm{l}_1(\mathcal{Y})$ is not reflexive by Example 7.86 (when $\mathcal{Y}$ is nontrivial). Then, by Proposition 7.90, $\mathrm{l}_\infty(\mathcal{Y}^*)$ is not reflexive either. ◇

Proposition 7.92 *Let $\mathcal{X}$ be a finite-dimensional normed linear space over the field $\mathbb{K}$. Then, $\mathcal{X}$ is reflexive.*

Proof Let $n \in \mathbb{Z}_+$ be the dimension of $\mathcal{X}$ and $\phi : \mathcal{X} \to \mathcal{X}^{**}$ be the natural mapping. We will distinguish two exhaustive and mutually exclusive cases: Case 1: $n = 0$; Case 2: $n \in \mathbb{N}$. Case 1: $n = 0$. Then, $\mathcal{X} = (\{\vartheta\}, \mathbb{K}, \|\cdot\|)$. By Example 7.73, $\mathcal{X}^* = (\{\vartheta_*\}, \mathbb{K}, \|\cdot\|_*)$. Then $\mathcal{X}^{**} = (\{\vartheta_{**}\}, \mathbb{K}, \|\cdot\|_{**})$. Clearly, $\phi(\vartheta) = \vartheta_{**}$. Hence, $\phi(\mathcal{X}) = \mathcal{X}^{**}$ and $\mathcal{X}$ is reflexive.

Case 2: $n \in \mathbb{N}$. Let $\{e_1, \ldots, e_n\} \subseteq \mathcal{X}$ be a basis of $\mathcal{X}$. $\forall x \in \mathcal{X}$, by Corollary 6.47, x can be uniquely expressed as $\sum_{j=1}^{n} \alpha_j e_j$ for some $\alpha_1, \ldots, \alpha_n \in \mathbb{K}$. $\forall i = 1, \ldots, n$, define $f_i : \mathcal{X} \to \mathbb{K}$ by $f_i(x) = \alpha_i$, $\forall x = \sum_{j=1}^{n} \alpha_j e_j \in \mathcal{X}$. Clearly, f_i is well-defined and a linear functional. By Proposition 7.67, f_i is continuous and $f_i \in \mathcal{X}^*$. Denote f_i by $e_{*i} \in \mathcal{X}^*$. $\forall x_* \in \mathcal{X}^*$, $\forall i = 1, \ldots, n$, let $\beta_i = \langle\langle x_*, e_i \rangle\rangle \in \mathbb{K}$. $\forall x = \sum_{j=1}^{n} \alpha_j e_j \in \mathcal{X}$, we have

$$\begin{aligned} \langle\langle x_*, x \rangle\rangle &= \sum_{j=1}^{n} \alpha_j \langle\langle x_*, e_j \rangle\rangle = \sum_{j=1}^{n} \beta_j \alpha_j = \sum_{k=1}^{n} \beta_j \langle\langle e_{*j}, x \rangle\rangle \\ &= \left\langle\!\!\left\langle \sum_{k=1}^{n} \beta_j e_{*j}, x \right\rangle\!\!\right\rangle \end{aligned}$$

Hence, $x_* = \sum_{j=1}^{n} \beta_j e_{*j}$. Therefore, $\mathcal{X}^* = \operatorname{span}(\{e_{*1}, \ldots, e_{*n}\})$. $\forall x_{**} \in \mathcal{X}^{**}$, $\forall i = 1, \ldots, n$, let $\gamma_i = \langle\langle x_{**}, e_{*i} \rangle\rangle \in \mathbb{K}$. $\forall x_* = \sum_{j=1}^{n} \beta_j e_{*j} \in \mathcal{X}^*$, we have

$$\begin{aligned} \langle\langle x_{**}, x_* \rangle\rangle &= \sum_{j=1}^{n} \beta_j \langle\langle x_{**}, e_{*j} \rangle\rangle = \sum_{j=1}^{n} \beta_j \gamma_j = \sum_{j=1}^{n} \beta_j \left\langle\!\!\left\langle e_{*j}, \sum_{k=1}^{n} \gamma_k e_k \right\rangle\!\!\right\rangle \\ &= \left\langle\!\!\left\langle x_*, \sum_{k=1}^{n} \gamma_k e_k \right\rangle\!\!\right\rangle = \left\langle\!\!\left\langle \phi\left(\sum_{k=1}^{n} \gamma_k e_k\right), x_* \right\rangle\!\!\right\rangle \end{aligned}$$

Hence, $x_{**} = \phi\left(\sum_{k=1}^{n} \gamma_k e_k\right)$ and $\mathcal{X}^{**} \subseteq \phi(\mathcal{X})$. This shows that $\mathcal{X}$ is reflexive.

This completes the proof of the proposition. □

7.9.5 Alignment and Orthogonal Complements

Definition 7.93 Let $\mathcal{X}$ be a normed linear space, $x \in \mathcal{X}$, $\mathcal{X}^*$ be the dual, and $x_* \in \mathcal{X}^*$. We say that x_* is *aligned* with x if $\langle\langle x_*, x \rangle\rangle = \|x_*\| \|x\|$. ◇

Clearly, ϑ aligns with any vector in the dual and ϑ_* is aligned with any vector in $\mathcal{X}$. By Proposition 7.85, $\forall x \in \mathcal{X}$ with $x \neq \vartheta$, there exists a $x_* \in \mathcal{X}^*$ with $x_* \neq \vartheta_*$ that is aligned with it.

Definition 7.94 Let $\mathcal{X}$ be a normed linear space, $x \in \mathcal{X}$, $\mathcal{X}^*$ be the dual, and $x_* \in \mathcal{X}^*$. We say that x and x_* are *orthogonal* if $\langle\langle x_*, x \rangle\rangle = 0$. ◇

Definition 7.95 Let $\mathcal{X}$ be a normed linear space, $S \subseteq \mathcal{X}$, and $\mathcal{X}^*$ be the dual. The *orthogonal complement* of S, denoted by $S^\perp$, consists of all $x_* \in \mathcal{X}^*$ that is orthogonal to every vector in S, that is $S^\perp := \{x_* \in \mathcal{X}^* \mid \langle\langle x_*, x \rangle\rangle = 0, \forall x \in S\}$. ◇

Definition 7.96 Let $\mathcal{X}$ be a normed linear space, $\mathcal{X}^*$ be the dual, and $S \subseteq \mathcal{X}^*$. The *pre-orthogonal complement* of S, denoted by ${}^{\perp}S$, consists of all $x \in \mathcal{X}$ that is orthogonal to every vector in S, that is ${}^{\perp}S := \{x \in \mathcal{X} \mid \langle\langle x_*, x \rangle\rangle = 0, \forall x_* \in S\}$. $\diamond$

Proposition 7.97 *Let $\mathcal{X}$ be a normed linear space over the field $\mathbb{K}$, $M \subseteq \mathcal{X}$ be a subspace, and $y \in \mathcal{X}$. Then, the following statements hold.*

(i) $\delta := \inf_{m \in M} \|y - m\| = \max_{x_* \in M^\perp,\ \|x_*\| \leq 1} \operatorname{Re}(\langle\langle x_*, y \rangle\rangle)$, *where the maximum is achieved at some $x_{*0} \in M^\perp$ with $\|x_{*0}\| \leq 1$. If the infimum is achieved at $m_0 \in M$ then $y - m_0$ is aligned with x_{*0}.*

(ii) *If $\exists m_0 \in M$ and $\exists x_{*0} \in M^\perp$ with $\|x_{*0}\| = 1$ such that $y - m_0$ is aligned with x_{*0}, then the infimum is achieved at m_0 and the maximum is achieved at x_{*0}, that is $\delta = \|y - m_0\| = \langle\langle x_{*0}, y \rangle\rangle = \operatorname{Re}(\langle\langle x_{*0}, y \rangle\rangle)$.*

Proof (i) $\forall x_* \in M^\perp$ with $\|x_*\| \leq 1$, $\forall m \in M$, we have

$$\operatorname{Re}(\langle\langle x_*, y \rangle\rangle) = \operatorname{Re}(\langle\langle x_*, y - m \rangle\rangle) \leq |\operatorname{Re}(\langle\langle x_*, y - m \rangle\rangle)|$$
$$\leq |\langle\langle x_*, y - m \rangle\rangle| \leq \|x_*\| \|y - m\| \leq \|y - m\|$$

where we have applied Proposition 7.72 in the third inequality. Then, we have

$$\sup_{x_* \in M^\perp,\ \|x_*\| \leq 1} \operatorname{Re}(\langle\langle x_*, y \rangle\rangle) \leq \inf_{m \in M} \|y - m\| = \delta$$

We will distinguish two exhaustive and mutually exclusive cases: Case 1: $\delta = 0$; Case 2: $\delta > 0$. Case 1: $\delta = 0$. Take $x_{*0} := \vartheta_*$. Clearly, $x_{*0} \in M^\perp$ and $\|x_{*0}\| = 0 \leq 1$. Then, $\delta = \operatorname{Re}(\langle\langle x_{*0}, y \rangle\rangle) \leq \sup_{x_* \in M^\perp,\ \|x_*\| \leq 1} \operatorname{Re}(\langle\langle x_*, y \rangle\rangle)$. Therefore, we have $\delta = \inf_{m \in M} \|y - m\| = \max_{x_* \in M^\perp,\ \|x_*\| \leq 1} \operatorname{Re}(\langle\langle x_*, y \rangle\rangle)$ and the maximum is achieved at some $x_{*0} \in M^\perp$ with $\|x_{*0}\| \leq 1$. If the infimum is achieved at $m_0 \in M$, then, $0 = \delta = \|y - m_0\|$ and $y = m_0$. Then, clearly $y - m_0$ is aligned with x_{*0}. This case is proved.

Case 2: $\delta > 0$. Then, $y \in \mathcal{X} \setminus M$. Consider the subspace $\operatorname{span}(M \cup \{y\}) =: \bar{M}$. $\forall x \in \bar{M}$, there exists a unique $\alpha \in \mathbb{K}$ and a unique $m \in M$ such that $x = \alpha y + m$. Then, we may define a functional $f : \bar{M} \to \mathbb{K}$ by $f(x) = f(\alpha y + m) = \alpha\delta$. Clearly, f is a linear functional on $\bar{M}$. $\|f\|_{\bar{M}} := \sup_{x \in \bar{M},\ \|x\| \leq 1} |f(x)| \leq \sup_{\alpha \in \mathbb{K}, m \in M, \|\alpha y + m\| \leq 1} |\alpha|\delta \leq \sup_{\alpha \in \mathbb{K}, m \in M, \|\alpha y + m\| \leq 1} \|\alpha y + m\| \leq 1$. By the simple version of Hahn–Banach Theorem, $\exists x_{*0} \in \mathcal{X}^*$ such that $x_{*0}|_{\bar{M}} = f$ and $\|x_{*0}\| = \|f\|_{\bar{M}} \leq 1$. $\forall m \in M$, we have $\langle\langle x_{*0}, m \rangle\rangle = f(m) = 0$. Hence, $x_{*0} \in M^\perp$. Note that $\delta = f(y) = \langle\langle x_{*0}, y \rangle\rangle$. Hence, $\delta = \inf_{m \in M} \|y - m\| = \max_{x_* \in M^\perp,\ \|x_*\| \leq 1} \operatorname{Re}(\langle\langle x_*, y \rangle\rangle)$ and the maximum is achieved at some $x_{*0} \in M^\perp$ with $\|x_{*0}\| \leq 1$. If the infimum is achieved at $m_0 \in M$, then $\delta = \|y - m_0\| = \langle\langle x_{*0}, y \rangle\rangle = \langle\langle x_{*0}, y - m_0 \rangle\rangle \leq \|x_{*0}\| \|y - m_0\| \leq \|y - m_0\|$, where the first inequality follows from Proposition 7.72. Hence, $\langle\langle x_{*0}, y - m_0 \rangle\rangle = \|x_{*0}\| \|y - m_0\|$ and x_{*0} is aligned with $y - m_0$. This case is proved.

(ii) Note that $\delta \leq \|y - m_0\| = \|x_{*0}\|\|y - m_0\| = \langle\langle x_{*0}, y - m_0\rangle\rangle = \langle\langle x_{*0}, y\rangle\rangle = \operatorname{Re}(\langle\langle x_{*0}, y\rangle\rangle) \leq \delta$. Then, the result follows.

This completes the proof of the proposition. □

Proposition 7.98 *Let $\mathcal{X}$ be a normed linear space over $\mathbb{K}$, $A, C \subseteq \mathcal{X}$, and $B, D \subseteq \mathcal{X}^*$. Then, the following statements holds.*

(i) *If $A \subseteq C$, then $C^\perp \subseteq A^\perp$. Similarly, if $B \subseteq D$, then ${}^\perp D \subseteq {}^\perp B$.*
(ii) *$A^\perp \subseteq \mathcal{X}^*$ is a closed subspace.*
(iii) *${}^\perp B \subseteq \mathcal{X}$ is a closed subspace.*
(iv) *${}^\perp(A^\perp) = \overline{\operatorname{span}(A)}$.*

Proof (i) This follows directly from Definition 7.95.

(ii) $\vartheta_* \in A^\perp \neq \emptyset$. $\forall x_{*1}, x_{*2} \in A^\perp$, $\forall \alpha, \beta \in \mathbb{K}$, $\forall x \in A$, we have $\langle\langle \alpha x_{*1} + \beta x_{*2}, x\rangle\rangle = \alpha\langle\langle x_{*1}, x\rangle\rangle + \beta\langle\langle x_{*2}, x\rangle\rangle = 0$. Hence, $\alpha x_{*1} + \beta x_{*2} \in A^\perp$ and $A^\perp$ is a subspace. $\forall x_* \in \overline{A^\perp}$, by Proposition 4.13, $\exists (x_{*n})_{n=1}^\infty \subseteq A^\perp$ such that $\lim_{n\in\mathbb{N}} x_{*n} = x_*$. Then, $\forall x \in A$, by Propositions 7.72 and 3.66, we have $\langle\langle x_*, x\rangle\rangle = \lim_{n\in\mathbb{N}} \langle\langle x_{*n}, x\rangle\rangle = 0$. Then, $x_* \in A^\perp$ and $\overline{A^\perp} \subseteq A^\perp$. By Proposition 3.3, $A^\perp = \overline{A^\perp}$ and $A^\perp$ is closed. Hence, (ii) follows.

(iii) This can be shown by a similar argument as (ii).

(iv) Clearly, $A \subseteq {}^\perp(A^\perp)$. Then, by (iii), we have $M := \overline{\operatorname{span}(A)} \subseteq {}^\perp(A^\perp)$. On the other hand, $A \subseteq M$, then, by (i), we have $A^\perp \supseteq M^\perp$ and ${}^\perp(A^\perp) \subseteq {}^\perp(M^\perp)$. $\forall x_0 \in {}^\perp(M^\perp)$, by Proposition 7.97, we have

$$\inf_{m\in M} \|x_0 - m\| = \max_{x_*\in M^\perp,\ \|x_*\|\leq 1} \operatorname{Re}(\langle\langle x_*, x_0\rangle\rangle) = 0$$

Since M is closed, by Proposition 4.10, $x_0 \in M$. Hence, we have ${}^\perp(A^\perp) \subseteq {}^\perp(M^\perp) \subseteq M$. Hence, $M = {}^\perp(A^\perp)$. Hence, (iv) follows.

This completes the proof of the proposition. □

Proposition 7.99 *Let $\mathcal{X}$ be a normed linear space over the field $\mathbb{K}$, $M \subseteq \mathcal{X}$ be a subspace, and $y_* \in \mathcal{X}^*$. Then, the following statements holds.*

(i) $\delta := \min_{x_*\in M^\perp} \|y_* - x_*\| = \sup_{m\in M,\ \|m\|\leq 1} \operatorname{Re}(\langle\langle y_*, m\rangle\rangle) =: \bar{\delta}$, *where the minimum is achieved at some $x_{*0} \in M^\perp$. If the supremum is achieved at $m_0 \in M$ with $\|m_0\| \leq 1$, then m_0 is aligned with $y_* - x_{*0}$.*
(ii) *If $\exists m_0 \in M$ with $\|m_0\| = 1$ and $\exists x_{*0} \in M^\perp$ such that $y_* - x_{*0}$ is aligned with m_0, then the minimum is achieved at x_{*0} and the supremum is achieved at m_0, that is $\delta = \|y_* - x_{*0}\| = \langle\langle y_*, m_0\rangle\rangle = \operatorname{Re}(\langle\langle y_*, m_0\rangle\rangle)$.*

Proof (i) $\forall x_* \in M^\perp$, $\forall m \in M$ with $\|m\| \leq 1$, we have

$$\begin{aligned}\operatorname{Re}(\langle\langle y_*, m\rangle\rangle) &= \operatorname{Re}(\langle\langle y_* - x_*, m\rangle\rangle) \leq |\operatorname{Re}(\langle\langle y_* - x_*, m\rangle\rangle)| \\ &\leq |\langle\langle y_* - x_*, m\rangle\rangle| \leq \|y_* - x_*\|\|m\| \leq \|y_* - x_*\|\end{aligned}$$

where we have applied Proposition 7.72 in the third inequality. Then, we have

$$\delta = \inf_{x_* \in M^\perp} \|y_* - x_*\| \geq \sup_{m \in M,\ \|m\| \leq 1} \operatorname{Re}(\langle\langle y_*, m\rangle\rangle) = \bar{\delta}$$

We will distinguish two exhaustive and mutually exclusive cases: Case 1: $\delta = 0$; Case 2: $\delta > 0$. Case 1: $\delta = 0$. By Proposition 7.98, $M^\perp$ is a closed subspace. By Proposition 4.10, we have $y_* \in M^\perp$ the minimum is achieved at a unique vector $x_{*0} := y_*$. Take $m_0 = \vartheta$ with $\|m_0\| = 0 \leq 1$. Then, $\delta = \operatorname{Re}(\langle\langle y_*, m_0\rangle\rangle) \leq \sup_{m \in M,\ \|m\| \leq 1} \operatorname{Re}(\langle\langle y_*, m\rangle\rangle) = \bar{\delta} \leq \delta$. Hence, $\delta = \min_{x_* \in M^\perp} \|y_* - x_*\| = \sup_{m \in M,\ \|m\| \leq 1} \operatorname{Re}(\langle\langle y_*, m\rangle\rangle) = \bar{\delta}$ and the minimum is achieved at $x_{*0} = y_* \in M^\perp$. If the supremum is achieved at $m_0 \in M$ with $\|m_0\| \leq 1$. Then, clearly m_0 is aligned with $y_* - x_{*0}$. This case is proved.

Case 2: $\delta > 0$. Then, $y_* \notin M^\perp$. Consider the subspace M. Then, we may define a functional $f : M \to \mathbb{K}$ by $f(m) = \langle\langle y_*, m\rangle\rangle$. Clearly, f is a linear functional on M. $\forall m \in M$ with $\|m\| \leq 1$, we have either $f(m) = 0$, then $|f(m)| = 0 \leq \sup_{m \in M,\ \|m\| \leq 1} \operatorname{Re}(\langle\langle y_*, m\rangle\rangle) = \bar{\delta}$; or $f(m) \neq 0$, then $|f(m)| = \frac{|f(m)|}{f(m)} f(m) = \operatorname{Re}\left(\frac{|f(m)|}{f(m)} f(m)\right) = \operatorname{Re}\left(f\left(\frac{|f(m)|}{f(m)} m\right)\right) = \operatorname{Re}\left(\left\langle\left\langle y_*, \frac{|f(m)|}{f(m)} m\right\rangle\right\rangle\right) \leq \bar{\delta}$. Hence, $\|f\|_M := \sup_{m \in M,\ \|m\| \leq 1} |f(m)| \leq \bar{\delta} \leq \delta < +\infty$. By the simple version of Hahn–Banach Theorem, $\exists y_{*0} \in \mathcal{X}^*$ such that $y_{*0}|_M = f$ and $\|y_{*0}\| = \|f\|_M \leq \bar{\delta}$. Let $x_{*0} := y_* - y_{*0}$. $\forall m \in M$, we have $\langle\langle x_{*0}, m\rangle\rangle = \langle\langle y_*, m\rangle\rangle - \langle\langle y_{*0}, m\rangle\rangle = 0$. Hence, $x_{*0} \in M^\perp$. Note that $\delta \geq \bar{\delta} \geq \|f\|_M = \|y_{*0}\| = \|y_* - x_{*0}\| \geq \inf_{x_* \in M^\perp} \|y_* - x_*\| = \delta$. Hence, $\delta = \min_{x_* \in M^\perp} \|y_* - x_*\| = \sup_{m \in M,\ \|m\| \leq 1} \operatorname{Re}(\langle\langle y_*, m\rangle\rangle) = \bar{\delta}$ and the minimum is achieved at some $x_{*0} \in M^\perp$. If the supremum is achieved at $m_0 \in M$ with $\|m_0\| \leq 1$, then $\delta = \|y_* - x_{*0}\| \geq \|y_* - x_{*0}\|\|m_0\| \geq |\langle\langle y_* - x_{*0}, m_0\rangle\rangle| = |\langle\langle y_*, m_0\rangle\rangle| \geq \operatorname{Re}(\langle\langle y_*, m_0\rangle\rangle) = \delta$, where the second inequality follows from Proposition 7.72. Hence, $\langle\langle y_* - x_{*0}, m_0\rangle\rangle = \|y_* - x_{*0}\|\|m_0\|$ and $y_* - x_{*0}$ is aligned with m_0. This case is proved.

(ii) Note that $\delta \leq \|y_* - x_{*0}\| = \|y_* - x_{*0}\|\|m_0\| = \langle\langle y_* - x_{*0}, m_0\rangle\rangle = \langle\langle y_*, m_0\rangle\rangle = \operatorname{Re}(\langle\langle y_*, m_0\rangle\rangle) \leq \delta$. Hence, the result follows.

This completes the proof of the proposition. □

Proposition 7.100 *Let $\mathcal{X}$ be a normed linear space over the field $\mathbb{K}$ and $S \subseteq \mathcal{X}$ be a subspace. By Proposition 7.13, S is a normed linear space over $\mathbb{K}$. Then, the following statement holds.*

(i) *S^* is isometrically isomorphic to $\mathcal{X}^*/S^\perp$.*
(ii) *If $\mathcal{X}$ is reflexive and S is closed, then S is reflexive.*

Proof (i) Define a mapping $A : \mathcal{X}^* \to S^*$ by $\langle\langle A(x_*), s\rangle\rangle = \langle\langle x_*, s\rangle\rangle$, $\forall x_* \in \mathcal{X}^*$ and $\forall s \in S$. $\forall x_* \in \mathcal{X}^*$, we will show that $A(x_*) \in S^*$. Clearly, $A(x_*)$ is a linear functional on S. $\|A(x_*)\| := \sup_{s \in S, \|s\| \leq 1} |\langle\langle A(x_*), s\rangle\rangle| = \sup_{s \in S, \|s\| \leq 1} |\langle\langle x_*, s\rangle\rangle| \leq \sup_{s \in S, \|s\| \leq 1} \|x_*\|\|s\| \leq \|x_*\| < +\infty$. Hence, $A(x_*) \in S^*$ and A is well-defined.

$\forall x_{*1}, x_{*2} \in \mathcal{X}^*$, $\forall \alpha, \beta \in \mathbb{K}$, $\forall s \in S$, we have

$$\begin{aligned}\langle\langle A(\alpha x_{*1} + \beta x_{*2}), s\rangle\rangle &= \langle\langle \alpha x_{*1} + \beta x_{*2}, s\rangle\rangle = \alpha\langle\langle x_{*1}, s\rangle\rangle + \beta\langle\langle x_{*2}, s\rangle\rangle \\ &= \alpha\langle\langle A(x_{*1}), s\rangle\rangle + \beta\langle\langle A(x_{*2}), s\rangle\rangle = \langle\langle \alpha A(x_{*1}) + \beta A(x_{*2}), s\rangle\rangle\end{aligned}$$

Hence, A is a linear function. Since $\|Ax_*\| \leq \|x_*\|$, $\forall x_* \in \mathcal{X}^*$, then $A \in \mathrm{B}(\mathcal{X}^*, S^*)$ with $\|A\| \leq 1$.

By Proposition 7.68, $\mathcal{N}(A) \subseteq \mathcal{X}^*$ is a closed subspace. $\forall x_* \in \mathcal{N}(A)$, we have $Ax_* = \vartheta_{S^*}$. $\forall s \in S$, $0 = \langle\langle Ax_*, s\rangle\rangle = \langle\langle x_*, s\rangle\rangle$. Hence, $x_* \in S^\perp$. Therefore, $\mathcal{N}(A) \subseteq S^\perp$. On the other hand, $\forall x_* \in S^\perp$, $\forall s \in S$, we have $0 = \langle\langle x_*, s\rangle\rangle = \langle\langle Ax_*, s\rangle\rangle$. Then, $Ax_* = \vartheta_{S^*}$. Hence, $x_* \in \mathcal{N}(A)$. Thus, $S^\perp \subseteq \mathcal{N}(A)$. In conclusion, $S^\perp = \mathcal{N}(A)$.

By Proposition 7.45, $\mathcal{X}^*/S^\perp$ is a Banach space. Let $\phi : \mathcal{X}^* \to \mathcal{X}^*/S^\perp$ be the natural homomorphism. By Proposition 7.70, there exists $A_D \in \mathrm{B}\big(\mathcal{X}^*/S^\perp, S^*\big)$ such that $A = A_D \circ \phi$, A_D is injective, and $\|A_D\| = \|A\| \leq 1$.

$\forall s_* \in S^*$, by the simple version of Hahn–Banach Theorem, there exists $x_* \in \mathcal{X}^*$ such that $x_*|_S = s_*$ and $\|x_*\| = \|s_*\|$. $\forall s \in S$, we have $\langle\langle s_*, s\rangle\rangle = \langle\langle x_*, s\rangle\rangle = \langle\langle Ax_*, s\rangle\rangle = \langle\langle A_D(\phi(x_*)), s\rangle\rangle = \langle\langle A_D([x_*]), s\rangle\rangle$. Hence, $s_* = A_D([x_*])$. Then, A_D is surjective. Thus, A_D is bijective.

$\forall [x_*] \in \mathcal{X}^*/S^\perp$, $\|[x_*]\| = \inf_{y_* \in S^\perp} \|x_* - y_*\|$, by Proposition 7.44. By Proposition 7.99, we have

$$\begin{aligned}\|[x_*]\| &= \min_{y_* \in S^\perp} \|x_* - y_*\| = \sup_{s \in S,\ \|s\| \leq 1} \mathrm{Re}\,(\langle\langle x_*, s\rangle\rangle) \\ &= \sup_{s \in S,\ \|s\| \leq 1} \mathrm{Re}\,(\langle\langle Ax_*, s\rangle\rangle) = \sup_{s \in S,\ \|s\| \leq 1} \mathrm{Re}\,(\langle\langle A_D(\phi(x_*)), s\rangle\rangle) \\ &= \sup_{s \in S,\ \|s\| \leq 1} \mathrm{Re}\,(\langle\langle A_D[x_*], s\rangle\rangle) \\ &\leq \sup_{s \in S,\ \|s\| \leq 1} |\mathrm{Re}\,(\langle\langle A_D[x_*], s\rangle\rangle)| \leq \sup_{s \in S,\ \|s\| \leq 1} |\langle\langle A_D[x_*], s\rangle\rangle| \\ &\leq \sup_{s \in S,\ \|s\| \leq 1} \|A_D[x_*]\|\|s\| \leq \|A_D[x_*]\| = \|Ax_*\| \\ &= \inf_{y_* \in \mathcal{N}(A)} \|A(x_* - y_*)\| = \inf_{y_* \in S^\perp} \|A(x_* - y_*)\| \\ &\leq \inf_{y_* \in S^\perp} \|A\|\|x_* - y_*\| \leq \|[x_*]\|\end{aligned}$$

where we have applied Proposition 7.72 in the third inequality and Proposition 7.64 in the fifth inequality. Therefore, we have $\|[x_*]\| = \|A_D[x_*]\|$ and A_D is an isometry. Thus, A_D is an isometrical isomorphism between $\mathcal{X}^*/S^\perp$ and S^*. Hence, (i) is established.

(ii) Let $\mathcal{X}$ be reflexive and S be a closed subspace. Let $\psi : S \to S^{**}$ be the natural mapping. All we need to show is that $\psi(S) = S^{**}$. Fix a $s_{**} \in S^{**}$. Define a functional $\tau : \mathcal{X}^* \to \mathbb{K}$ by $\tau(x_*) = \langle\langle s_{**}, Ax_*\rangle\rangle$, $\forall x_* \in \mathcal{X}^*$. It is easy to show that τ is a linear functional on $\mathcal{X}^*$. $\forall x_* \in \mathcal{X}^*$ with $\|x_*\| \leq 1$, we have $|\tau(x_*)| = |\langle s_{**}, Ax_*\rangle| \leq \|s_{**}\|\|Ax_*\| \leq \|s_{**}\|\|A\|\|x_*\| \leq \|s_{**}\| < +\infty$, where we have applied Propositions 7.72 and 7.64. Hence, $\tau \in \mathcal{X}^{**}$.

Since $\mathcal{X}$ is reflexive, then, by Remark 7.88 and Definition 7.89, $\exists x_0 \in \mathcal{X}$ such that $\tau(x_*) = \langle\langle x_*, x_0\rangle\rangle$, $\forall x_* \in \mathcal{X}^*$. $\forall y_* \in S^\perp$, we have $\langle\langle y_*, x_0\rangle\rangle = \tau(y_*) = \langle\langle s_{**}, Ay_*\rangle\rangle = \langle\langle s_{**}, \vartheta_{S^*}\rangle\rangle = 0$, where the third equality follows from the fact that $S^\perp = \mathcal{N}(A)$. Hence, $x_0 \in {}^\perp(S^\perp) = S$, by Proposition 7.98. Hence, $\forall s_* \in S^*$, $\exists x_* \in \mathcal{X}^*$ such that $Ax_* = A_D[x_*] = s_*$ since A_D is an isometrical isomorphism. Then, we have $\langle\langle s_{**}, s_*\rangle\rangle = \langle\langle s_{**}, Ax_*\rangle\rangle = \tau(x_*) = \langle\langle x_*, x_0\rangle\rangle = \langle\langle Ax_*, x_0\rangle\rangle = \langle\langle s_*, x_0\rangle\rangle$. This implies that $s_{**} = \psi(x_0)$ and $\psi(S) = S^{**}$. Hence, (ii) is established.

This completes the proof of the proposition. □

7.10 The Open Mapping Theorem

Definition 7.101 Let $\mathcal{X} := (X, \mathcal{O}_X)$ and $\mathcal{Y} := (Y, \mathcal{O}_Y)$ be topological spaces and $A : \mathcal{X} \to \mathcal{Y}$. A is called an *open mapping* if $\forall O_X \in \mathcal{O}_X$, we have $A(O_X) \in \mathcal{O}_Y$, that is the image of each open set is open. ◇

Proposition 7.102 *Let $\mathcal{X}$ be a normed linear space over the field $\mathbb{K}$, $S, T \subseteq \mathcal{X}$, and $\alpha \in \mathbb{K}$. Then, the following statements hold.*

(i) $\overline{\alpha S} = \alpha \overline{S}$.
(ii) *If $\alpha \neq 0$, then $\widetilde{\alpha S} = \alpha \widetilde{S}$.*
(iii) *If $\alpha \neq 0$, then $(\alpha S)^\circ = \alpha S^\circ$.*
(iv) $\overline{S} + \overline{T} \subseteq \overline{S + T}$.
(v) $S^\circ + T^\circ \subseteq (S + T)^\circ$.

Proof (i) We will distinguish three exhaustive and mutually exclusive cases: Case 1: $\alpha = 0$ and $S = \emptyset$; Case 2: $\alpha = 0$ and $S \neq \emptyset$; Case 3: $\alpha \neq 0$. Case 1: $\alpha = 0$ and $S = \emptyset$. Then, $\overline{S} = \emptyset$ and $\overline{\alpha S} = \emptyset = \alpha \overline{S}$. Case 2: $\alpha = 0$ and $S \neq \emptyset$. Then, $\overline{S} \neq \emptyset$ and $\overline{\alpha S} = \{\vartheta\} = \alpha \overline{S}$.

Case 3: $\alpha \neq 0$. $\forall x \in \overline{\alpha S}$, by Proposition 4.13, $\exists (x_n)_{n=1}^\infty \subseteq \alpha S$ such that $\lim_{n\in\mathbb{N}} x_n = x$. Note that $\left(\alpha^{-1}x_n\right)_{n=1}^\infty \subseteq S$ and $\lim_{n\in\mathbb{N}} \alpha^{-1}x_n = \alpha^{-1}x$ by Propositions 3.66 and 7.23. By Proposition 4.13, $\alpha^{-1}x \in \overline{S}$. Then, $x \in \alpha\overline{S}$. This shows that $\overline{\alpha S} \subseteq \alpha \overline{S}$.

On the other hand, $\forall x \in \alpha\overline{S}$, then $\alpha^{-1}x \in \overline{S}$. By Proposition 4.13, $\exists (x_n)_{n=1}^\infty \subseteq S$ such that $\lim_{n\in\mathbb{N}} x_n = \alpha^{-1}x$. Note that $(\alpha x_n)_{n=1}^\infty \subseteq \alpha S$ and $\lim_{n\in\mathbb{N}} \alpha x_n = x$ by Propositions 3.66 and 7.23. By Proposition 4.13, $x \in \overline{\alpha S}$. Then, $\alpha\overline{S} \subseteq \overline{\alpha S}$. Hence, $\overline{\alpha S} = \alpha \overline{S}$.

(ii) $\forall x \in \widetilde{\alpha S}$ if, and only if, $\alpha^{-1}x \in \widetilde{S}$ if, and only if, $x \in \alpha\widetilde{S}$. Hence, $\widetilde{\alpha S} = \alpha\widetilde{S}$.

(iii) $\forall x \in (\alpha S)^\circ$, $\exists \delta \in (0, \infty) \subset \mathbb{R}$ such that $\mathcal{B}(x, \delta) \subseteq \alpha S$. Then, $\mathcal{B}\left(\alpha^{-1}x, \delta/|\alpha|\right) = \alpha^{-1}\mathcal{B}(x, \delta) \subseteq S$. Hence, $\alpha^{-1}x \in S^\circ$. Then, $x \in \alpha S^\circ$. This shows that $(\alpha S)^\circ \subseteq \alpha S^\circ$.

On the other hand, $\forall x \in \alpha S^\circ$, we have $\alpha^{-1}x \in S^\circ$. $\exists \delta \in (0, \infty) \subset \mathbb{R}$ such that $\mathcal{B}(\alpha^{-1}x, \delta) \subseteq S$. Then, $\mathcal{B}(x, |\alpha|\delta) = \alpha\mathcal{B}(\alpha^{-1}x, \delta) \subseteq \alpha S$. Hence, $x \in (\alpha S)^\circ$. This shows that $\alpha S^\circ \subseteq (\alpha S)^\circ$. Hence, we have $(\alpha S)^\circ = \alpha S^\circ$.

(iv) $\forall \bar{x} \in \overline{S} + \overline{T}$, $\exists \bar{s} \in \overline{S}$ and $\exists \bar{t} \in \overline{T}$ such that $\bar{x} = \bar{s} + \bar{t}$. By Proposition 4.13, $\exists (s_n)_{n=1}^\infty \subseteq S$ and $\exists (t_n)_{n=1}^\infty \subseteq T$ such that $\lim_{n\in\mathbb{N}} s_n = \bar{s}$ and $\lim_{n\in\mathbb{N}} t_n = \bar{t}$. Then, $(s_n + t_n)_{n=1}^\infty \subseteq S + T$ and $\lim_{n\in\mathbb{N}}(s_n + t_n) = \bar{s} + \bar{t} = \bar{x}$, by Propositions 7.23, 3.66, and 3.67. By Proposition 4.13, $\bar{x} \in \overline{S+T}$. Hence, $\overline{S} + \overline{T} \subseteq \overline{S+T}$.

(v) $\forall x \in S^\circ + T^\circ$, $\exists s_0 \in S^\circ$ and $\exists t_0 \in T^\circ$ such that $x = s_0 + t_0$. Then, $\exists r_s, r_t \in (0, \infty) \subset \mathbb{R}$ such that $\mathcal{B}(s_0, r_s) \subseteq S$ and $\mathcal{B}(t_0, r_t) \subseteq T$. Thus, we have $\mathcal{B}(x, r_s + r_t) = \mathcal{B}(s_0, r_s) + \mathcal{B}(t_0, r_t) \subseteq S + T$ and $x \in (S+T)^\circ$. Hence, $S^\circ + T^\circ \subseteq (S+T)^\circ$.

This completes the proof of the proposition. □

Theorem 7.103 (Open Mapping Theorem) *Let $\mathcal{X}$ and $\mathcal{Y}$ be Banach spaces over the field $\mathbb{K}$ and $A \in \mathrm{B}(\mathcal{X}, \mathcal{Y})$ be surjective. Then, A is an open mapping. Furthermore, if A is injective, then $A_{\mathrm{inv}} \in \mathrm{B}(\mathcal{Y}, \mathcal{X})$.*

Proof We need the following claim.

Claim 7.103.1 The image of the unit ball in $\mathcal{X}$ under A contains an open ball centered at the origin in $\mathcal{Y}$, that is $\exists \delta \in (0, \infty) \subset \mathbb{R}$ such that $\mathcal{B}_\mathcal{Y}(\vartheta_\mathcal{Y}, \delta) \subseteq A(\mathcal{B}_\mathcal{X}(\vartheta_\mathcal{X}, 1))$.

Proof of Claim Let $S_n := \mathcal{B}_\mathcal{X}(\vartheta_\mathcal{X}, 2^{-n})$, $\forall n \in \mathbb{Z}_+$. Since A is linear and surjective and $\bigcup_{k=1}^\infty kS_1 = \mathcal{X}$, then, by Proposition 2.5, we have $\mathcal{Y} = \bigcup_{k=1}^\infty A(kS_1) = \bigcup_{k=1}^\infty kA(S_1)$. Since $\mathcal{Y}$ is a complete metric space, then, by Baire Category Theorem, $\mathcal{Y}$ is second category everywhere. Then, $\mathcal{Y}$ is not of first category, that is $\mathcal{Y}$ is not countable union of nowhere dense sets. Then, by Proposition 7.102, $A(S_1) \subseteq \mathcal{Y}$ is not nowhere dense. Then, $\overline{A(S_1)} \subseteq \mathcal{Y}$ is not nowhere dense. By Proposition 3.40, $\exists \bar{y} \in \mathcal{Y}$ and $\exists \bar{\delta} \in (0, \infty) \subset \mathbb{R}$ such that $\mathcal{B}_\mathcal{Y}(\bar{y}, \bar{\delta}) \subseteq \overline{A(S_1)}$. Then, we have $\mathcal{B}_\mathcal{Y}(\vartheta_\mathcal{Y}, \bar{\delta}) = \mathcal{B}_\mathcal{Y}(\bar{y}, \bar{\delta}) - \bar{y} \subseteq \overline{A(S_1)} - \overline{A(S_1)}$. Note that, by Proposition 7.102 and the linearity of A, $-\overline{A(S_1)} = \overline{-A(S_1)} = \overline{A(-S_1)} = \overline{A(S_1)}$. Then, again by Proposition 7.102 and the linearity of A, $\mathcal{B}_\mathcal{Y}(\vartheta_\mathcal{Y}, \bar{\delta}) \subseteq \overline{A(S_1)} + \overline{A(S_1)} \subseteq \overline{A(S_1) + A(S_1)} = \overline{A(S_1 + S_1)} = \overline{A(S_0)}$. Thus, $\overline{A(S_0)} \subseteq \mathcal{Y}$ contains a ball center at the origin with radius $\bar{\delta}$. By Proposition 7.102 and linearity of A, $\mathcal{B}_\mathcal{Y}(\vartheta_\mathcal{Y}, 2^{-n}\bar{\delta}) = 2^{-n}\mathcal{B}_\mathcal{Y}(\vartheta_\mathcal{Y}, \bar{\delta}) \subseteq 2^{-n}\overline{A(S_0)} = \overline{A(2^{-n}S_0)} = \overline{A(S_n)} \subseteq \mathcal{Y}$, $\forall n \in \mathbb{Z}_+$.

Now, we proceed to show that $\mathcal{B}_\mathcal{Y}(\vartheta_\mathcal{Y}, \bar{\delta}/2) \subseteq A(S_0)$. Fix an arbitrary vector $y \in \mathcal{B}_\mathcal{Y}(\vartheta_\mathcal{Y}, \bar{\delta}/2)$. Then, $y \in \overline{A(S_1)}$ and $\exists x_1 \in S_1$, such that $\|y - Ax_1\| < 2^{-2}\bar{\delta}$. This implies that $y - Ax_1 \in \mathcal{B}_\mathcal{Y}(\vartheta_\mathcal{Y}, 2^{-2}\bar{\delta}) \subseteq \overline{A(S_2)}$. Recursively, $\forall n \in \mathbb{N}$ with $n \geq 2$, $y - \sum_{k=1}^{n-1} Ax_k \in \overline{A(S_n)}$. Then, $\exists x_n \in S_n$ such that $\left\|y - \sum_{k=1}^n Ax_k\right\| < 2^{-n-1}\bar{\delta}$. This implies that $y - \sum_{k=1}^n Ax_k \in \mathcal{B}_\mathcal{Y}(\vartheta_\mathcal{Y}, 2^{-n-1}\bar{\delta}) \subseteq \overline{A(S_{n+1})}$. Since $x_n \in S_n$ and $\|x_n\| < 2^{-n}$, $\forall n \in \mathbb{N}$, then $\sum_{k=1}^\infty \|x_n\| < 1$. By Proposition 7.27, we have $\sum_{k=1}^\infty x_n = x \in \mathcal{Y}$. It is easy to show that $x \in S_0$. Then, $y = \lim_{n\in\mathbb{N}} \sum_{k=1}^n Ax_k = \lim_{n\in\mathbb{N}} A\left(\sum_{k=1}^n x_k\right) = Ax$, where we have applied Proposition 3.66. Thus, $y \in A(S_0)$ and $\mathcal{B}_\mathcal{Y}(\vartheta_\mathcal{Y}, \bar{\delta}/2) \subseteq A(S_0)$.

This completes the proof of the claim. □

Fix any open set $O \subseteq \mathcal{X}$ and any $y \in A(O)$. Let $x \in O$ be such that $Ax = y$. Then, there exists $r \in (0, \infty) \subset \mathbb{R}$ such that $\mathcal{B}_{\mathcal{X}}(x, r) \subseteq O$. By Claim 7.103.1, $\exists \delta \in (0, \infty) \subset \mathbb{R}$ such that $\mathcal{B}_{\mathcal{Y}}(\vartheta_{\mathcal{Y}}, \delta) \subseteq A(\mathcal{B}_{\mathcal{X}}(\vartheta_{\mathcal{X}}, r))$. By the linearity of A, we have $\mathcal{B}_{\mathcal{Y}}(y, \delta) = y + \mathcal{B}_{\mathcal{Y}}(\vartheta_{\mathcal{Y}}, \delta) \subseteq Ax + A(\mathcal{B}_{\mathcal{X}}(\vartheta_{\mathcal{X}}, r)) = A(\mathcal{B}_{\mathcal{X}}(x, r)) \subseteq A(O)$. By the arbitrariness of y, $A(O) \subseteq \mathcal{Y}$ is open.

If, in addition, A is injective, then A is bijective and A_{inv} exists. Since A is open mapping, then A_{inv} is continuous. It is obvious that $A_{\text{inv}} : \mathcal{Y} \to \mathcal{X}$ is linear since A is linear. Therefore, $A_{\text{inv}} \in \mathrm{B}(\mathcal{Y}, \mathcal{X})$.

This completes the proof of the theorem. □

Proposition 7.104 *Let $\mathcal{X}$ be a vector space over $\mathbb{K}$ and $\|\cdot\|_1$ and $\|\cdot\|_2$ be two norms on $\mathcal{X}$ such that $\mathcal{X}_1 := (\mathcal{X}, \mathbb{K}, \|\cdot\|_1)$ and $\mathcal{X}_2 := (\mathcal{X}, \mathbb{K}, \|\cdot\|_2)$ are Banach spaces. If $\exists M \in [0, \infty) \subset \mathbb{R}$ such that $\|x\|_2 \leq M\|x\|_1$, $\forall x \in \mathcal{X}$. Then, the two norms are equivalent.*

Proof Consider the mapping $A := \mathrm{id}_{\mathcal{X}} : \mathcal{X}_1 \to \mathcal{X}_2$. Clearly A is a linear bijective function. By the assumption of the proposition, $A \in \mathrm{B}(\mathcal{X}_1, \mathcal{X}_2)$. By Open Mapping Theorem, $A_{\text{inv}} \in \mathrm{B}(\mathcal{X}_2, \mathcal{X}_1)$. Then, $\exists \bar{M} \in [0, \infty) \subset \mathbb{R}$ such that $\|x\|_1 = \|A_{\text{inv}}x\|_1 \leq \bar{M}\|x\|_2$, $\forall x \in \mathcal{X}$. Now, take $K = \max\{M, \bar{M}\} + 1 \in (0, \infty) \subset \mathbb{R}$. Then, we have $\|x\|_1/K \leq \|x\|_2 \leq K\|x\|_1$, $\forall x \in \mathcal{X}$. Hence, the two norms are equivalent. This completes the proof of the proposition. □

Theorem 7.105 (Closed Graph Theorem) *Let $\mathcal{X} := (\mathcal{X}, \mathbb{K}, \|\cdot\|)$ and $\mathcal{Y}$ be Banach spaces over the field $\mathbb{K}$ and $A : \mathcal{X} \to \mathcal{Y}$ be a linear operator. Assume that A satisfies that $\forall (x_n)_{n=1}^{\infty} \subseteq \mathcal{X}$ with $\lim_{n\in\mathbb{N}} x_n = x_0 \in \mathcal{X}$ and $\lim_{n\in\mathbb{N}} Ax_n = y_0 \in \mathcal{Y}$, we have $y_0 = Ax_0$. Then, $A \in \mathrm{B}(\mathcal{X}, \mathcal{Y})$.*

Proof Define a functional $\|\cdot\|_1 : \mathcal{X} \to \mathbb{R}$ by $\|x\|_1 := \|x\| + \|Ax\|$, $\forall x \in \mathcal{X}$. It is easy to show that $\|\cdot\|_1$ defines a norm on $\mathcal{X}$. Let $\mathcal{X}_1 := (\mathcal{X}, \mathbb{K}, \|\cdot\|_1)$ be the normed linear space. We will show that $\mathcal{X}_1$ is complete. Fix any Cauchy sequence $(x_n)_{n=1}^{\infty} \subseteq \mathcal{X}_1$. It is easy to see that $(x_n)_{n=1}^{\infty} \subseteq \mathcal{X}$ is a Cauchy sequence; and that $(Ax_n)_{n=1}^{\infty} \subseteq \mathcal{Y}$ is a Cauchy sequence. By the completeness of $\mathcal{X}$ and $\mathcal{Y}$, there exists $x_0 \in \mathcal{X}$ and $y_0 \in \mathcal{Y}$ such that $\lim_{n\in\mathbb{N}} \|x_n - x_0\| = 0$ and $\lim_{n\in\mathbb{N}} \|Ax_n - y_0\| = 0$. By the assumption of the theorem, we have $y_0 = Ax_0$. Hence, we have $\lim_{n\in\mathbb{N}} \|x_n - x_0\|_1 = 0$. Then, $\lim_{n\in\mathbb{N}} x_n = x_0$ in $\mathcal{X}_1$. Thus, $\mathcal{X}_1$ is a Banach space. Clearly, $\forall x \in \mathcal{X}$, we have $\|x\| \leq \|x\|_1$. By Proposition 7.104, $\|\cdot\|$ and $\|\cdot\|_1$ are equivalent. Then, $\exists M \in [0, \infty) \subset \mathbb{R}$ such that $\|x\|_1 \leq M\|x\|$, $\forall x \in \mathcal{X}$. Then, we have $\|Ax\| \leq M\|x\|$, $\forall x \in \mathcal{X}$. Hence, $A \in \mathrm{B}(\mathcal{X}, \mathcal{Y})$. This completes the proof of the theorem. □

Recall that the graph of a function $f : X \to Y$ is the set $\{(x, y) \in X \times Y \mid x \in X,\ y = f(x)\}$. Then, we have the following alternative statement of the Closed Graph Theorem.

Theorem 7.106 (Closed Graph Theorem) *Let $\mathcal{X}$ and $\mathcal{Y}$ be Banach spaces over the field $\mathbb{K}$ and $A : \mathcal{X} \to \mathcal{Y}$ be a linear operator. Then, $A \in \mathrm{B}(\mathcal{X}, \mathcal{Y})$ if, and only if, the graph of A is a closed set in $\mathcal{X} \times \mathcal{Y}$.*

Proof "Sufficiency" Let the graph of A be $\mathrm{graph}(A)$. Fix any sequence $(x_n)_{n=1}^{\infty} \subseteq \mathcal{X}$ with $\lim_{n\in\mathbb{N}} x_n = x_0 \in \mathcal{X}$ and $\lim_{n\in\mathbb{N}} Ax_n = y_0 \in \mathcal{Y}$. Note that $((x_n, Ax_n))_{n=1}^{\infty} \subseteq$

graph $(A) \subseteq \mathcal{X} \times \mathcal{Y}$ and $\lim_{n\in\mathbb{N}}(x_n, Ax_n) = (x_0, y_0)$, by Proposition 3.67. Then, by Proposition 4.13, we have $(x_0, y_0) \in \overline{\text{graph}(A)} = \text{graph}(A)$, where the equality follows from Proposition 3.3. Hence, we have $y_0 = Ax_0$. By the Closed Graph Theorem, Theorem 7.105, we have $A \in B(\mathcal{X}, \mathcal{Y})$.

"Necessity" $\forall (x_0, y_0) \in \overline{\text{graph}(A)}$, by Proposition 4.13, there exists $((x_n, Ax_n))_{n=1}^{\infty} \subseteq \text{graph}(A)$ such that $\lim_{n\in\mathbb{N}}(x_n, Ax_n) = (x_0, y_0)$. By Proposition 3.67, we have $\lim_{n\in\mathbb{N}} x_n = x_0$ and $\lim_{n\in\mathbb{N}} Ax_n = y_0$. By Proposition 3.66, we have $y_0 = \lim_{n\in\mathbb{N}} Ax_n = Ax_0$. Hence, $(x_0, y_0) \in \text{graph}(A)$. By Proposition 3.3, graph (A) is closed in $\mathcal{X} \times \mathcal{Y}$.

This completes the proof of the theorem. □

Proposition 7.107 *Let $\mathcal{X}$ be a Banach space over the field $\mathbb{K}$, $\mathcal{Y}$ be a normed linear space over the same field, and $\mathcal{F} \subseteq B(\mathcal{X}, \mathcal{Y})$. Assume that $\forall x \in \mathcal{X}$, $\exists M_x \in [0, \infty) \subset \mathbb{R}$ such that $\|Tx\| \leq M_x$, $\forall T \in \mathcal{F}$. Then, $\exists M \in [0, \infty) \subset \mathbb{R}$, such that $\|T\| \leq M$, $\forall T \in \mathcal{F}$.*

Proof By Baire Category Theorem, $\mathcal{X}$ is second category everywhere. $\forall T \in \mathcal{F}$, let $f : \mathcal{X} \to \mathbb{R}$ be given by $f(x) = \|Tx\|$, $\forall x \in \mathcal{X}$. By Propositions 7.21 and 3.12, f is a continuous real-valued function. By Uniform Boundedness Principle, Theorem 3.41, there exists an open set $O \subseteq \mathcal{X}$ with $O \neq \emptyset$ and $\bar{M} \in [0, \infty) \subset \mathbb{R}$ such that $\|Tx\| \leq \bar{M}$, $\forall T \in \mathcal{F}$, $\forall x \in O$. Since O is nonempty and open, then $\exists \mathcal{B}_{\mathcal{X}}(x_0, r) \subseteq O$ for some $x_0 \in \mathcal{X}$ and some $r \in (0, \infty) \subset \mathbb{R}$. $\forall x \in \mathcal{X}$ with $\|x\| < r$, $x + x_0 \in \mathcal{B}_{\mathcal{X}}(x_0, r) \subseteq O$ and $\forall T \in \mathcal{F}$, we have

$$\|Tx\| = \|T(x + x_0) - Tx_0\| \leq \|T(x + x_0)\| + \|Tx_0\| \leq \bar{M} + M_{x_0}$$

Hence, $\forall T \in \mathcal{F}$, we have, $\forall \epsilon \in (0, r) \subset \mathbb{R}$,

$$\|T\| = \sup_{x\in\mathcal{X},\ \|x\|\leq 1} \|Tx\| = \sup_{x\in\mathcal{X},\ \|x\|\leq 1} (r - \epsilon)^{-1}\|T((r - \epsilon)x)\| \leq \frac{\bar{M} + M_{x_0}}{r - \epsilon}$$

By the arbitrariness of ϵ, we have $\|T\| \leq (\bar{M} + M_{x_0})/r =: M < +\infty$. This completes the proof of the proposition. □

Proposition 7.108 *Let $\mathcal{X}$ be a Banach space over the field $\mathbb{K}$, $\mathcal{Y}$ be a normed linear space over the same field, and $(T_n)_{n=1}^{\infty} \subseteq B(\mathcal{X}, \mathcal{Y})$. Assume that $\forall x \in \mathcal{X}$, $\lim_{n\in\mathbb{N}} T_n x = T(x) \in \mathcal{Y}$. Then, $T \in B(\mathcal{X}, \mathcal{Y})$.*

Proof $\forall x_1, x_2 \in \mathcal{X}$, $\forall \alpha, \beta \in \mathbb{K}$, we have

$$T(\alpha x_1 + \beta x_2) = \lim_{n\in\mathbb{N}} T_n(\alpha x_1 + \beta x_2) = \lim_{n\in\mathbb{N}} (\alpha T_n x_1 + \beta T_n x_2) = \alpha T x_1 + \beta T x_2$$

by Propositions 7.23 and 3.66. Hence, T is linear. $\forall x \in \mathcal{X}$, by Propositions 7.21 and 3.66, $\lim_{n\in\mathbb{N}} \|T_n x\| = \|Tx\| < +\infty$. Then, $\exists M_x \in [0, \infty) \subset \mathbb{R}$ such that $\|T_n x\| \leq M_x$, $\forall n \in \mathbb{N}$. By Proposition 7.107, $\exists M \in [0, \infty) \subset \mathbb{R}$ such that $\|T_n\| \leq M$, $\forall n \in \mathbb{N}$. $\forall x \in \mathcal{X}$ with $\|x\| \leq 1$, by Proposition 7.64, $\|Tx\| = \lim_{n\in\mathbb{N}} \|T_n x\| \leq$

$\limsup_{n\in\mathbb{N}} \|T_n\| \|x\| \leq M < +\infty$. Hence, $\|T\| \leq M < +\infty$. Therefore, $T \in \mathrm{B}(\mathcal{X}, \mathcal{Y})$. This completes the proof of the proposition. □

7.11 The Adjoints of Linear Operators

Proposition 7.109 *Let $\mathcal{X}$ and $\mathcal{Y}$ be normed linear spaces over the field $\mathbb{K}$ and $A \in \mathrm{B}(\mathcal{X}, \mathcal{Y})$. The adjoint operator of A is $A' : \mathcal{Y}^* \to \mathcal{X}^*$ defined by*

$$\langle\langle A'(y_*), x\rangle\rangle = \langle\langle y_*, Ax\rangle\rangle; \qquad \forall x \in \mathcal{X},\ \forall y_* \in \mathcal{Y}^*$$

Then, $A' \in \mathrm{B}(\mathcal{Y}^*, \mathcal{X}^*)$ with $\|A'\| = \|A\|$.

Proof First, we will show that A' is well-defined. $\forall y_* \in \mathcal{Y}^*$, $\forall x \in \mathcal{X}$. $\langle\langle y_*, Ax\rangle\rangle \in \mathbb{K}$. Hence, $f : \mathcal{X} \to \mathbb{K}$ defined by $f(x) = \langle\langle y_*, Ax\rangle\rangle$, $\forall x \in \mathcal{X}$, is a functional on $\mathcal{X}$. By the linearity of A and y_*, f is a linear functional. $\forall x \in \mathcal{X}$, we have, by Proposition 7.64

$$|f(x)| \leq \|y_*\| \|Ax\| \leq \|y_*\| \|A\| \|x\|$$

Hence, f is a bounded linear functional with $\|f\| \leq \|A\| \|y_*\|$. The above shows that $A'(y_*) = f \in \mathcal{X}^*$. Hence, A' is well-defined.

It is straightforward to show that A' is a linear operator. By the fact that $\|A' y_*\| = \|f\| \leq \|A\| \|y_*\|$, $\forall y_* \in \mathcal{Y}^*$, we have $A' \in \mathrm{B}(\mathcal{Y}^*, \mathcal{X}^*)$ and $\|A'\| \leq \|A\|$.

On the other hand, $\forall x \in \mathcal{X}$, we have either $Ax = \vartheta_{\mathcal{Y}}$, then $\|Ax\| = 0 \leq \|A'\| \|x\|$; or $Ax \neq \vartheta_{\mathcal{Y}}$, then, by Proposition 7.85, $\exists y_* \in \mathcal{Y}^*$ with $\|y_*\| = 1$ such that $\|Ax\| = \langle\langle y_*, Ax\rangle\rangle$, which implies that $\|Ax\| = \langle\langle A' y_*, x\rangle\rangle \leq \|A' y_*\| \|x\| \leq \|A'\| \|y_*\| \|x\| = \|A'\| \|x\|$. Hence, we must have $\|Ax\| \leq \|A'\| \|x\|$. This implies that $\|A\| \leq \|A'\|$.

Then, $\|A\| = \|A'\|$. This completes the proof of the proposition. □

Proposition 7.110 *Let $\mathcal{X}$, $\mathcal{Y}$, and $\mathcal{Z}$ be normed linear spaces over the field $\mathbb{K}$. Then, the following statements hold.*

(i) $\mathrm{id}'_{\mathcal{X}} = \mathrm{id}_{\mathcal{X}^*}$.
(ii) *If $A_1, A_2 \in \mathrm{B}(\mathcal{X}, \mathcal{Y})$, then $(A_1 + A_2)' = A'_1 + A'_2$.*
(iii) *If $A \in \mathrm{B}(\mathcal{X}, \mathcal{Y})$ and $\alpha \in \mathbb{K}$, then $(\alpha A)' = \alpha A'$.*
(iv) *If $A_1 \in \mathrm{B}(\mathcal{X}, \mathcal{Y})$ and $A_2 \in \mathrm{B}(\mathcal{Y}, \mathcal{Z})$, then $(A_2 A_1)' = A'_1 A'_2$.*
(v) *If $A \in \mathrm{B}(\mathcal{X}, \mathcal{Y})$ and A has a bounded inverse, then $(A^{-1})' = (A')^{-1}$.*

Proof (i) $\forall x_* \in \mathcal{X}^*$, $\forall x \in \mathcal{X}$, $\langle\langle \mathrm{id}'_{\mathcal{X}}(x_*), x\rangle\rangle = \langle\langle x_*, \mathrm{id}_{\mathcal{X}}(x)\rangle\rangle = \langle\langle x_*, x\rangle\rangle = \langle\langle \mathrm{id}_{\mathcal{X}^*}(x_*), x\rangle\rangle$. Hence, the result follows.

(ii) $\forall y_* \in \mathcal{Y}^*$, $\forall x \in \mathcal{X}$, we have $\langle\langle (A_1 + A_2)' y_*, x\rangle\rangle = \langle\langle y_*, (A_1 + A_2)x\rangle\rangle = \langle\langle y_*, A_1 x\rangle\rangle + \langle\langle y_*, A_2 x\rangle\rangle = \langle\langle A'_1 y_*, x\rangle\rangle + \langle\langle A'_2 y_*, x\rangle\rangle = \langle\langle A'_1 y_* + A'_2 y_*, x\rangle\rangle = \langle\langle (A'_1 + A'_2) y_*, x\rangle\rangle$. Hence, the result follows.

(iii) $\forall y_* \in \mathcal{Y}^*$, $\forall x \in \mathcal{X}$, we have $\langle\langle (\alpha A)' y_*, x \rangle\rangle = \langle\langle y_*, (\alpha A)x \rangle\rangle = \alpha \langle\langle y_*, Ax \rangle\rangle = \alpha \langle\langle A' y_*, x \rangle\rangle = \langle\langle \alpha(A' y_*), x \rangle\rangle = \langle\langle (\alpha A') y_*, x \rangle\rangle$. Hence, the result follows.

(iv) $\forall z_* \in \mathcal{Z}^*$, $\forall x \in \mathcal{X}$, we have $\langle\langle (A_2 A_1)' z_*, x \rangle\rangle = \langle\langle z_*, (A_2 A_1)x \rangle\rangle = \langle\langle z_*, A_2(A_1 x) \rangle\rangle = \langle\langle A_2' z_*, A_1 x \rangle\rangle = \langle\langle A_1'(A_2' z_*), x \rangle\rangle = \langle\langle (A_1' A_2') z_*, x \rangle\rangle$. Hence, the result follows.

(v) By (i) and (iv), we have $(A^{-1})' A' = (A A^{-1})' = \mathrm{id}_{\mathcal{Y}}' = \mathrm{id}_{\mathcal{Y}^*}$ and $A'(A^{-1})' = (A^{-1} A)' = \mathrm{id}_{\mathcal{X}}' = \mathrm{id}_{\mathcal{X}^*}$. By Proposition 2.4, we have $(A^{-1})' = (A')^{-1}$.

This completes the proof of the proposition. □

Proposition 7.111 *Let $\mathcal{X}$ and $\mathcal{Y}$ be normed linear spaces over the field $\mathbb{K}$, $A \in B(\mathcal{X}, \mathcal{Y})$, $\phi_{\mathcal{X}} : \mathcal{X} \to \mathcal{X}^{**}$ and $\phi_{\mathcal{Y}} : \mathcal{Y} \to \mathcal{Y}^{**}$ be the natural mappings on $\mathcal{X}$ and $\mathcal{Y}$, respectively, and $A'' : \mathcal{X}^{**} \to \mathcal{Y}^{**}$ be the adjoint of the adjoint of A. Then, we have $A'' \circ \phi_{\mathcal{X}} = \phi_{\mathcal{Y}} \circ A$.*

Proof $\forall x \in \mathcal{X}$, $\forall y_* \in \mathcal{Y}^*$, we have

$$\begin{aligned} \langle\langle A''(\phi_{\mathcal{X}}(x)), y_* \rangle\rangle &= \langle\langle \phi_{\mathcal{X}}(x), A' y_* \rangle\rangle = \langle\langle A' y_*, x \rangle\rangle = \langle\langle y_*, Ax \rangle\rangle \\ &= \langle\langle \phi_{\mathcal{Y}}(Ax), y_* \rangle\rangle \end{aligned}$$

Hence, the desired result follows. This completes the proof of the proposition. □

By Proposition 7.111 and Remark 7.88, $\forall A, B \in B(\mathcal{X}, \mathcal{Y})$, we have $A' = B'$ if, and only if, $A = B$.

Proposition 7.112 *Let $\mathcal{X}$ and $\mathcal{Y}$ be normed linear spaces over the field $\mathbb{K}$ and $A \in B(\mathcal{X}, \mathcal{Y})$. Then, the following statements hold.*

(i) $(\mathcal{R}(A))^\perp = \mathcal{N}(A')$.
(ii) ${}^\perp(\mathcal{R}(A')) = \mathcal{N}(A)$.
(iii) *If $\mathcal{R}(A)$ is closed, then $\mathcal{R}(A) = {}^\perp(\mathcal{N}(A'))$.*

Proof (i) Fix a vector $y_* \in \mathcal{N}(A')$. $\forall y \in \mathcal{R}(A)$, $\exists x \in \mathcal{X}$ such that $y = Ax$. Then, $\langle\langle y_*, y \rangle\rangle = \langle\langle y_*, Ax \rangle\rangle = \langle\langle A' y_*, x \rangle\rangle = \langle\langle \vartheta_{\mathcal{X}^*}, x \rangle\rangle = 0$. Hence, $y_* \in (\mathcal{R}(A))^\perp$. This shows that $\mathcal{N}(A') \subseteq (\mathcal{R}(A))^\perp$.

On the other hand, fix a vector $y_* \in (\mathcal{R}(A))^\perp$. $\forall x \in \mathcal{X}$, we have $\langle\langle A' y_*, x \rangle\rangle = \langle\langle y_*, Ax \rangle\rangle = 0$. Hence, we have $A' y_* = \vartheta_{\mathcal{X}^*}$. This shows that $(\mathcal{R}(A))^\perp \subseteq \mathcal{N}(A')$.

Hence, we have $(\mathcal{R}(A))^\perp = \mathcal{N}(A')$.

(ii) Fix a vector $x \in \mathcal{N}(A)$. $\forall x_* \in \mathcal{R}(A')$, $\exists y_* \in \mathcal{Y}^*$ such that $x_* = A' y_*$. Then, $\langle\langle x_*, x \rangle\rangle = \langle\langle y_*, Ax \rangle\rangle = \langle\langle y_*, \vartheta_{\mathcal{Y}} \rangle\rangle = 0$. Hence, $x \in {}^\perp(\mathcal{R}(A'))$. This shows that $\mathcal{N}(A) \subseteq {}^\perp(\mathcal{R}(A'))$.

On the other hand, fix a vector $x \in {}^\perp(\mathcal{R}(A'))$. $\forall y_* \in \mathcal{Y}^*$, we have $0 = \langle\langle A' y_*, x \rangle\rangle = \langle\langle y_*, Ax \rangle\rangle$. Hence, by Proposition 7.85, we have $Ax = \vartheta_{\mathcal{Y}}$ and $x \in \mathcal{N}(A)$. This shows that ${}^\perp(\mathcal{R}(A')) \subseteq \mathcal{N}(A)$.

Hence, we have ${}^\perp(\mathcal{R}(A')) = \mathcal{N}(A)$.

(iii) Fix $y \in \mathcal{R}(A)$. Then $\exists x \in \mathcal{X}$ such that $y = Ax$. $\forall y_* \in \mathcal{N}(A')$, we have $\langle\langle y_*, y\rangle\rangle = \langle\langle A'y_*, x\rangle\rangle = \langle\langle \vartheta_{\mathcal{X}^*}, x\rangle\rangle = 0$. Hence, we have $y \in {}^{\perp}(\mathcal{N}(A'))$ and $\mathcal{R}(A) \subseteq {}^{\perp}(\mathcal{N}(A'))$.

On the other hand, fix $y \in {}^{\perp}(\mathcal{N}(A'))$. By Proposition 7.97, $\delta := \inf_{k\in\mathcal{R}(A)} \|y - k\| = \max_{y_*\in(\mathcal{R}(A))^{\perp},\, \|y_*\|\leq 1} \operatorname{Re}(\langle\langle y_*, y\rangle\rangle)$, where the maximum is achieved at some $y_{*0} \in (\mathcal{R}(A))^{\perp}$ with $\|y_{*0}\| \leq 1$. By (i), we have $(\mathcal{R}(A))^{\perp} = \mathcal{N}(A') \ni y_{*0}$. Then, $\delta = \operatorname{Re}(\langle\langle y_{*0}, y\rangle\rangle) = 0$. By Proposition 4.10, $y \in \mathcal{R}(A)$ since $\mathcal{R}(A) \subseteq \mathcal{Y}$ is closed. This shows that ${}^{\perp}(\mathcal{N}(A')) \subseteq \mathcal{R}(A)$.

Hence, we have $\mathcal{R}(A) = {}^{\perp}(\mathcal{N}(A'))$. This completes the proof of the proposition.

□

The dual version of (ii) in the above proposition is deeper, which requires both Open Mapping Theorem and Hahn–Banach Theorem. Toward this end, we need the following result.

Proposition 7.113 *Let $\mathcal{X}$ and $\mathcal{Y}$ be Banach spaces over the field $\mathbb{K}$ and $A \in \mathrm{B}(\mathcal{X}, \mathcal{Y})$. Assume that $\mathcal{R}(A) \subseteq \mathcal{Y}$ is closed. Then, there exists $K \in [0, \infty) \subset \mathbb{R}$ such that, $\forall y \in \mathcal{R}(A)$, there exists $x \in \mathcal{X}$ such that $y = Ax$ and $\|x\| \leq K\|y\|$.*

Proof Since A is continuous, then $\mathcal{N}(A) \subseteq \mathcal{X}$ is closed. By Proposition 7.45, $\mathcal{X}/\mathcal{N}(A)$ is a Banach space. Let $\phi : \mathcal{X} \to \mathcal{X}/\mathcal{N}(A)$ be the natural homomorphism, which is a bounded linear function by Proposition 7.69. By Proposition 7.70, there exists a bounded linear function $A_D : \mathcal{X}/\mathcal{N}(A) \to \mathcal{R}(A)$ such that $A = A_D \circ \phi$, $\|A\| = \|A_D\|$ and A_D is injective. By Proposition 4.39, $\mathcal{R}(A)$ is complete. Then, by Proposition 7.13, $\mathcal{R}(A)$ is a Banach space. The mapping A_D is surjective to $\mathcal{R}(A)$. Hence, $A_D : \mathcal{X}/\mathcal{N}(A) \to \mathcal{R}(A)$ is a bijective bounded linear operator. By Open Mapping Theorem, $A_D^{-1} \in \mathrm{B}(\mathcal{R}(A), \mathcal{X}/\mathcal{N}(A))$. $\forall y \in \mathcal{R}(A)$, Let $[x] := A_D^{-1} y \in \mathcal{X}/\mathcal{N}(A)$. Then, by Proposition 7.64, $\|[x]\| \leq \|A_D^{-1}\|\|y\|$. We will distinguish two exhaustive and mutually exclusive cases: Case 1: $\|[x]\| = 0$; Case 2: $\|[x]\| > 0$. Case 1: $\|[x]\| = 0$. Then, $[x] = [\vartheta_{\mathcal{X}}]$. Take $x = \vartheta_{\mathcal{X}}$, we have $y = A_D[x] = Ax = \vartheta_{\mathcal{Y}}$ and $\|x\| = 0 = 2\|A_D^{-1}\|\|y\|$. Case 2: $\|[x]\| > 0$. Note that $\|[x]\| = \inf_{x\in[x]} \|x\|$. Then, $\exists x \in [x]$ such that $\|x\| \leq 2\|[x]\|$. Then, $y = A_D[x] = Ax$ and $\|x\| \leq 2\|[x]\| \leq 2\|A_D^{-1}\|\|y\|$. Hence, the desired result holds in both cases with $K = 2\|A_D^{-1}\|$. This completes the proof of the proposition.

□

Proposition 7.114 *Let $\mathcal{X}$ and $\mathcal{Y}$ be Banach spaces over the field $\mathbb{K}$ and $A \in \mathrm{B}(\mathcal{X}, \mathcal{Y})$. Assume that $\mathcal{R}(A)$ is closed in $\mathcal{Y}$. Then,*

$$\mathcal{R}(A') = (\mathcal{N}(A))^{\perp}$$

Proof Fix an $x_* \in \mathcal{R}(A')$, then $\exists y_* \in \mathcal{Y}^*$ such that $x_* = A'y_*$. $\forall x \in \mathcal{N}(A)$, we have $\langle\langle x_*, x\rangle\rangle = \langle\langle A'y_*, x\rangle\rangle = \langle\langle y_*, Ax\rangle\rangle = \langle\langle y_*, \vartheta_{\mathcal{Y}}\rangle\rangle = 0$. Then, $x_* \in (\mathcal{N}(A))^{\perp}$. This shows that $\mathcal{R}(A') \subseteq (\mathcal{N}(A))^{\perp}$.

On the other hand, fix an $x_* \in (\mathcal{N}(A))^\perp$. Let $K \in [0, \infty) \subset \mathbb{R}$ be the constant described in Proposition 7.113. $\forall y \in \mathcal{R}(A)$, $\forall x \in \mathcal{X}$ with $Ax = y$, $\langle\langle x_*, x\rangle\rangle$ assumes a constant value that is dependent on y only. So, we may define a functional $f : \mathcal{R}(A) \to \mathbb{K}$ by $f(y) = \langle\langle x_*, x\rangle\rangle$, $\forall y \in \mathcal{R}(A)$ and $\forall x \in \mathcal{X}$ with $Ax = y$. Clearly, f is a linear functional. By Proposition 7.113, $\exists x_0 \in \mathcal{X}$ with $Ax_0 = y$, such that $\|x_0\| \leq K\|y\|$. Then, by Proposition 7.64, $|f(y)| = |\langle\langle x_*, x_0\rangle\rangle| \leq \|x_*\|\|x_0\| \leq \|x_*\|K\|y\|$. Hence, $\|f\|_{\mathcal{R}(A)} := \sup_{y\in\mathcal{R}(A),\ \|y\|\leq 1} |f(y)| \leq K\|x_*\| < +\infty$. By the simple version of Hahn–Banach Theorem 7.83, $\exists y_* \in \mathcal{Y}^*$ such that $y_*|_{\mathcal{R}(A)} = f$ and $\|y_*\| = \|f\|_{\mathcal{R}(A)}$. $\forall x \in \mathcal{X}$, we have

$$\langle\langle A' y_*, x\rangle\rangle = \langle\langle y_*, Ax\rangle\rangle = f(Ax) = \langle\langle x_*, x\rangle\rangle$$

This implies that $A'y_* = x_* \in \mathcal{R}(A')$. Hence, we have $(\mathcal{N}(A))^\perp \subseteq \mathcal{R}(A')$.

Therefore, we have $\mathcal{R}(A') = (\mathcal{N}(A))^\perp$. This completes the proof of the proposition. □

7.12 Weak Topology

Definition 7.115 Let $\mathcal{X}$ be a normed linear space over the field $\mathbb{K}$. The *weak topology* on $\mathcal{X}$, denoted by $\mathcal{O}_{\text{weak}}(\mathcal{X}) \subseteq {}^{\mathcal{X}}2$, is the weak topology generated by $\mathcal{X}^*$, that is the weakest topology on $\mathcal{X}$ such that $x_* : \mathcal{X} \to \mathbb{K}$, $\forall x_* \in \mathcal{X}^*$, are continuous. ◇

For a normed linear space $\mathcal{X}$, let $\mathcal{O}_{\mathcal{X}}$ be the natural topology induced by the norm on $\mathcal{X}$. Then, $\mathcal{O}_{\text{weak}}(\mathcal{X}) \subseteq \mathcal{O}_{\mathcal{X}}$. We usually call the topology $\mathcal{O}_{\mathcal{X}}$ the strong topology. A set $S \subseteq \mathcal{X}$ is weakly open, that is $S \in \mathcal{O}_{\text{weak}}(\mathcal{X})$, then S is strongly open, that is $S \in \mathcal{O}_{\mathcal{X}}$, and if S is weakly closed, then it is strongly closed, but not conversely. We will denote the topological space $(\mathcal{X}, \mathcal{O}_{\text{weak}}(\mathcal{X}))$ by $\mathcal{X}_{\text{weak}}$.

Proposition 7.116 *Let $\mathcal{X}$ be a normed linear space over the field $\mathbb{K}$. Then, the following statements hold.*

(i) *A basis for $\mathcal{O}_{\text{weak}}(\mathcal{X})$ consists of all sets of the form*

$$\{x \in \mathcal{X} \mid \langle\langle x_{*i}, x\rangle\rangle \in O_i, i = 1, \ldots, n\} \tag{7.8}$$

where $n \in \mathbb{N}$, $O_i \in \mathcal{O}_{\mathbb{K}}$ and $x_{*i} \in \mathcal{X}^*$, $i = 1, \ldots, n$, and $\mathcal{O}_{\mathbb{K}}$ is the natural topology on $\mathbb{K}$. A basis at $x_0 \in \mathcal{X}$ for $\mathcal{O}_{\text{weak}}(\mathcal{X})$ consists of all sets of the form

$$\{x \in \mathcal{X} \mid |\langle\langle x_{*i}, x - x_0\rangle\rangle| < \epsilon, i = 1, \ldots, n\} \tag{7.9}$$

where $\epsilon \in (0, \infty) \subset \mathbb{R}$, $n \in \mathbb{N}$, and $x_{*i} \in \mathcal{X}^*$, $i = 1, \ldots, n$.

(ii) $\mathcal{X}_{\text{weak}}$ *is completely regular* ($\mathbf{T}_{3\frac{1}{2}}$).

(iii) *For a sequence* $(x_k)_{k=1}^{\infty} \subseteq \mathcal{X}_{\text{weak}}$, $x_0 \in \mathcal{X}$ *is the limit point of the sequence in the weak topology if, and only if,* $\lim_{k\in\mathbb{N}} \langle\langle x_*, x_k\rangle\rangle = \langle\langle x_*, x_0\rangle\rangle$, $\forall x_* \in \mathcal{X}^*$. *In this case, we will write* $\lim_{k\in\mathbb{N}} x_k = x_0$ weakly *and say that* $(x_k)_{k=1}^{\infty}$ *converges weakly to* x_0.

Proof (i) By Definition 7.115, $\mathcal{O}_{\text{weak}}(\mathcal{X})$ is the topology generated by sets of the form (7.8). We will show that these sets form a basis for the topology by Proposition 3.18. Take $B = \{x \in \mathcal{X} \mid \langle\langle \vartheta_*, x\rangle\rangle \in \mathbb{K}\}$, which is of the form (7.8) and $B = \mathcal{X}$. $\forall x \in \mathcal{X}$, we have $x \in B$. For any $B_1, B_2 \subseteq \mathcal{X}$, which are of the form (7.8), clearly, $B_1 \cap B_2$ is again of the form (7.8). Hence, Proposition 3.18 applies and the sets of the form (7.8) form a basis for $\mathcal{O}_{\text{weak}}(\mathcal{X})$.

Let $B \subseteq \mathcal{X}$ be any set of the form (7.9). Clearly, $x_0 \in B \in \mathcal{O}_{\text{weak}}(\mathcal{X})$. $\forall O \in \mathcal{O}_{\text{weak}}(\mathcal{X})$ with $x_0 \in O$, by Definition 3.17, there exists a basis open set $B_1 := \{x \in \mathcal{X} \mid \langle\langle x_{*i}, x\rangle\rangle \in O_i, i = 1, \ldots, n\}$, for some $n \in \mathbb{N}$ and some $x_{*i} \in \mathcal{X}^*$ and $O_i \in \mathcal{O}_{\mathbb{K}}$, $i = 1, \ldots, n$, such that $x_0 \in B_1 \subseteq O$. For each $i = 1, \ldots, n$, $c_i := \langle\langle x_{*i}, x_0\rangle\rangle \in O_i \in \mathcal{O}_{\mathbb{K}}$. Then, $\exists \epsilon_i \in (0, \infty) \subset \mathbb{R}$ such that $\mathcal{B}_{\mathbb{K}}(c_i, \epsilon_i) \subseteq O_i$. Take $\epsilon = \min_{1\le i\le n} \epsilon_i \in (0, \infty) \subset \mathbb{R}$ and $B_2 := \{x \in \mathcal{X} \;\Big|\; |\langle\langle x_{*i}, x - x_0\rangle\rangle| < \epsilon, i = 1, \ldots, n\}$. Clearly, B_2 is of the form (7.9) and $x_0 \in B_2 \subseteq B_1 \subseteq O$. Hence, sets of the form (7.9) form a basis at x_0 for $\mathcal{O}_{\text{weak}}(\mathcal{X})$. Thus, (i) holds.

(ii) $\forall x_1, x_2 \in \mathcal{X}$ with $x_1 \neq x_2$, we have $x_1 - x_2 \neq \vartheta$. By Proposition 7.85, $\exists x_* \in \mathcal{X}^*$ with $\|x_*\| = 1$ such that $\langle\langle x_*, x_1 - x_2\rangle\rangle = \|x_1 - x_2\| > 0$. Let $O_1 := \{a \in \mathbb{K} \mid \operatorname{Re}(a) > \operatorname{Re}(\langle\langle x_*, x_2\rangle\rangle) + \|x_1 - x_2\|/2\}$ and $O_2 := \{a \in \mathbb{K} \mid \operatorname{Re}(a) < \operatorname{Re}(\langle\langle x_*, x_2\rangle\rangle) + \|x_1 - x_2\|/2\}$. Then, $O_1, O_2 \in \mathcal{O}_{\mathbb{K}}$ and $O_1 \cap O_2 = \emptyset$. Let $B_1 := \{x \in \mathcal{X} \mid \langle\langle x_*, x\rangle\rangle \in O_1\}$ and $B_2 := \{x \in \mathcal{X} \mid \langle\langle x_*, x\rangle\rangle \in O_2\}$. Then, $B_1, B_2 \in \mathcal{O}_{\text{weak}}(\mathcal{X})$, $x_1 \in B_1$, $x_2 \in B_2$, and $B_1 \cap B_2 = \emptyset$. This shows that $\mathcal{X}_{\text{weak}}$ is Hausdorff.

Next, we show that $\mathcal{X}_{\text{weak}}$ is completely regular. Fix any weakly closed set $F \subseteq \mathcal{X}_{\text{weak}}$ and $x_0 \in \widetilde{F} \in \mathcal{O}_{\text{weak}}(\mathcal{X})$. Then, there exists a basis open set $B = \{x \in \mathcal{X} \mid \langle\langle x_{*i}, x\rangle\rangle \in O_i, i = 1, \ldots, n\}$, for some $n \in \mathbb{N}$ and some $x_{*i} \in \mathcal{X}^*$ and $O_i \in \mathcal{O}_{\mathbb{K}}$, $i = 1, \ldots, n$, such that $x_0 \in B \subseteq \widetilde{F}$. Let $O := \prod_{i=1}^{n} O_i \subseteq \mathbb{K}^n$ which is open. Let $p_0 := (\langle\langle x_{*1}, x_0\rangle\rangle, \ldots, \langle\langle x_{*n}, x_0\rangle\rangle) \in \mathbb{K}^n$. Then, $p_0 \in O$. By Propositions 4.11 and 3.61, $\mathbb{K}^n$ is normal and therefore completely regular. Then, there exists a continuous function $f : \mathbb{K}^n \to [0, 1] \subset \mathbb{R}$ such that $f|_{\tilde{O}} = 0$ and $f(p_0) = 1$. Define $g : \mathcal{X}_{\text{weak}} \to [0, 1] \subset \mathbb{R}$ by $g(x) = f(\langle\langle x_{*1}, x\rangle\rangle, \ldots, \langle\langle x_{*n}, x\rangle\rangle)$, $\forall x \in \mathcal{X}_{\text{weak}}$. By Propositions 3.12 and 3.32, g is a continuous real-valued function on $\mathcal{X}_{\text{weak}}$, $g(x_0) = f(p_0) = 1$, and $g|_F = 0$. Hence, $\mathcal{X}_{\text{weak}}$ is completely regular. Thus, (ii) holds.

(iii) "Only if" $\forall x_* \in \mathcal{X}^*$, $x_* : \mathcal{X} \to \mathbb{K}$ is weakly continuous. By Proposition 3.66, we have $\lim_{k\in\mathbb{N}} \langle\langle x_*, x_k\rangle\rangle = \langle\langle x_*, x_0\rangle\rangle$.

"If" Let $(x_k)_{k=1}^{\infty}$ satisfy that $\lim_{k\in\mathbb{N}} \langle\langle x_*, x_k\rangle\rangle = \langle\langle x_*, x_0\rangle\rangle$, $\forall x_* \in \mathcal{X}^*$. For any basis open set $B = \{x \in \mathcal{X} \mid \langle\langle x_{*i}, x\rangle\rangle \in O_i, i = 1, \ldots, n\}$, for some $n \in \mathbb{N}$ and some $x_{*i} \in \mathcal{X}^*$ and $O_i \in \mathcal{O}_{\mathbb{K}}$, $i = 1, \ldots, n$, with $x_0 \in B$, we have that $\forall i = 1, \ldots, n$, $\exists N_i \in \mathbb{N}$ such that $\langle\langle x_{*i}, x_k\rangle\rangle \in O_i$, $\forall k \ge N_i$. Take $N = \max_{1\le i\le n} N_i \in$

$\mathbb{N}$. $\forall k \geq N$, $x_k \in B$. This shows that $(x_k)_{k=1}^{\infty}$ converges weakly to x_0. Hence, (iii) holds.

This completes the proof of the proposition. □

Proposition 7.117 *Let $\mathcal{X}$ be a normed linear space over the field $\mathbb{K}$ and $\mathcal{X}_{\text{weak}}$ be the topological space of $\mathcal{X}$ endowed with the weak topology. Then, vector addition $\oplus : \mathcal{X}_{\text{weak}} \times \mathcal{X}_{\text{weak}} \to \mathcal{X}_{\text{weak}}$ is continuous; and scalar multiplication $\otimes : \mathbb{K} \times \mathcal{X}_{\text{weak}} \to \mathcal{X}_{\text{weak}}$ is continuous.*

Proof Fix $(x_0, y_0) \in \mathcal{X}_{\text{weak}} \times \mathcal{X}_{\text{weak}}$. We will show that $\oplus$ is continuous at (x_0, y_0). Fix a basis open set $B \in \mathcal{O}_{\text{weak}}(\mathcal{X})$ with $x_0 + y_0 \in B$. Then, by Proposition 7.116, $B = \{z \in \mathcal{X} \mid |\langle\langle x_{*i}, z - x_0 - y_0 \rangle\rangle| < \epsilon, i = 1, \ldots, n\}$, for some $n \in \mathbb{N}$, for some $\epsilon \in (0, \infty) \subset \mathbb{R}$, and for some $x_{*i} \in \mathcal{X}^*$, $i = 1, \ldots, n$. Let $B_1 := \{x \in \mathcal{X} \mid |\langle\langle x_{*i}, x - x_0 \rangle\rangle| < \epsilon/2, i = 1, \ldots, n\}$ and $B_2 := \{y \in \mathcal{X} \mid |\langle\langle x_{*i}, y - y_0 \rangle\rangle| < \epsilon/2, i = 1, \ldots, n\}$. Clearly, $B_1, B_2 \in \mathcal{O}_{\text{weak}}(\mathcal{X})$, $B_1 \times B_2 \in \mathcal{O}_{\mathcal{X}_{\text{weak}} \times \mathcal{X}_{\text{weak}}}$, and $(x_0, y_0) \in B_1 \times B_2$. $\forall (x, y) \in B_1 \times B_2$, $\forall i = 1, \ldots, n$, $|\langle\langle x_{*i}, x + y - x_0 - y_0 \rangle\rangle| \leq |\langle\langle x_{*i}, x - x_0 \rangle\rangle| + |\langle\langle x_{*i}, y - y_0 \rangle\rangle| < \epsilon$. Hence, $x + y \in B$. This shows that $\oplus$ is continuous at (x_0, y_0). By the arbitrariness of (x_0, y_0) and Proposition 3.9, $\oplus$ is continuous.

Fix $(\alpha_0, x_0) \in \mathbb{K} \times \mathcal{X}_{\text{weak}}$. We will show that $\otimes$ is continuous at (α_0, x_0). Fix a basis open set $B \in \mathcal{O}_{\text{weak}}(\mathcal{X})$ with $\alpha_0 x_0 \in B$. Then, by Proposition 7.116, $B = \{z \in \mathcal{X} \mid |\langle\langle x_{*i}, z - \alpha_0 x_0 \rangle\rangle| < \epsilon, i = 1, \ldots, n\}$, for some $n \in \mathbb{N}$, for some $\epsilon \in (0, \infty) \subset \mathbb{R}$, and for some $x_{*i} \in \mathcal{X}^*$, $i = 1, \ldots, n$. Let $M := \max_{1 \leq i \leq n} \|x_{*i}\| \in [0, \infty) \subset \mathbb{R}$. Let $B_1 := \{\alpha \in \mathbb{K} \mid |\alpha - \alpha_0| < \frac{\epsilon}{\epsilon + 2M\|x_0\|}\}$ and $B_2 := \{x \in \mathcal{X} \mid |\langle\langle x_{*i}, x - x_0 \rangle\rangle| < \frac{\epsilon}{2(1+|\alpha_0|)}, i = 1, \ldots, n\}$. Clearly, $B_1 \in \mathcal{O}_{\mathbb{K}}$ and $B_2 \in \mathcal{O}_{\text{weak}}(\mathcal{X})$, $B_1 \times B_2 \in \mathcal{O}_{\mathbb{K} \times \mathcal{X}_{\text{weak}}}$, and $(\alpha_0, x_0) \in B_1 \times B_2$. $\forall (\alpha, x) \in B_1 \times B_2$, $\forall i = 1, \ldots, n$, $|\langle\langle x_{*i}, \alpha x - \alpha_0 x_0 \rangle\rangle| \leq |\langle\langle x_{*i}, \alpha x - \alpha x_0 \rangle\rangle| + |\langle\langle x_{*i}, \alpha x_0 - \alpha_0 x_0 \rangle\rangle| = |\alpha||\langle\langle x_{*i}, x - x_0 \rangle\rangle| + |\alpha - \alpha_0||\langle\langle x_{*i}, x_0 \rangle\rangle| \leq (|\alpha_0| + |\alpha - \alpha_0|)|\langle\langle x_{*i}, x - x_0 \rangle\rangle| + |\alpha - \alpha_0|\|x_{*i}\|\|x_0\| \leq (|\alpha_0| + 1)|\langle\langle x_{*i}, x - x_0 \rangle\rangle| + |\alpha - \alpha_0| M \|x_0\| < \epsilon$. Hence, $\alpha x \in B$. This shows that $\otimes$ is continuous at (α_0, x_0). By the arbitrariness of (α_0, x_0) and Proposition 3.9, $\otimes$ is continuous.

This completes the proof of the proposition. □

Proposition 7.118 *Let $\mathcal{X}$ be a finite-dimensional normed linear space over the field $\mathbb{K}$, $\mathcal{O}_{\mathcal{X}}$ be the strong topology on $\mathcal{X}$, and $\mathcal{O}_{\text{weak}}(\mathcal{X})$ be the weak topology on $\mathcal{X}$. Then, $\mathcal{O}_{\mathcal{X}} = \mathcal{O}_{\text{weak}}(\mathcal{X})$.*

Proof Clearly, $\mathcal{O}_{\text{weak}}(\mathcal{X}) \subseteq \mathcal{O}_{\mathcal{X}}$. Fix any basis open set $O = \mathcal{B}_{\mathcal{X}}(x_0, r) \in \mathcal{O}_{\mathcal{X}}$ with $x_0 \in \mathcal{X}$ and $r \in (0, \infty) \subset \mathbb{R}$. We will show that $O \in \mathcal{O}_{\text{weak}}(\mathcal{X})$. Let $n \in \mathbb{Z}_+$ be the dimension of $\mathcal{X}$. We will distinguish two exhaustive and mutually exclusive cases: Case 1: $n = 0$; Case 2: $n \in \mathbb{N}$. Case 1: $n = 0$. Then, $\mathcal{X}$ is a singleton set. $O = \mathcal{X} \in \mathcal{O}_{\text{weak}}(\mathcal{X})$. This case is proved.

Case 2: $n \in \mathbb{N}$. Let $\{e_1, \ldots, e_n\} \subseteq \mathcal{X}$ be a basis of $\mathcal{X}$ with $\|e_i\| = 1$, $i = 1, \ldots, n$. $\forall i = 1, \ldots, n$, let $f_i : \mathcal{X} \to \mathbb{K}$ be defined by $f_i(x) = \alpha_i$, $\forall x = \sum_{j=1}^{n} \alpha_j e_j \in \mathcal{X}$. Clearly, f_i is well-defined and is a linear functional. By Proposition 7.67, f_i is continuous. Denote f_i by $e_{*i} \in \mathcal{X}^*$. $\forall x_1 \in O$, let $\delta = (r - \|x_1 - x_0\|)/n \in (0, \infty) \subset$

$\mathbb{R}$. Let $B := \{x \in \mathcal{X} \mid |\langle\langle e_{*i}, x - x_1\rangle\rangle| < \delta, i = 1, \ldots, n\}$. By Proposition 7.116, $B \in \mathcal{O}_{\text{weak}}(\mathcal{X})$. $\forall x \in B$, let $x - x_1 = \sum_{j=1}^{n} \alpha_j e_j$. Then, $\forall i = 1, \ldots, n$, $|\alpha_i| = |\langle\langle e_{*i}, x - x_1\rangle\rangle| < \delta$. Thus, $\|x - x_1\| \leq \sum_{j=1}^{n} |\alpha_j| \|e_j\| < n\delta = r - \|x_1 - x_0\|$. Then, we have $\|x - x_0\| \leq \|x - x_1\| + \|x_1 - x_0\| < r$ and $x \in O$. Hence, $x_1 \in B \subseteq O$. Therefore, $O \in \mathcal{O}_{\text{weak}}(\mathcal{X})$.

In both cases, we have shown that $O \in \mathcal{O}_{\text{weak}}(\mathcal{X})$. Then, $\mathcal{O}_{\mathcal{X}} \subseteq \mathcal{O}_{\text{weak}}(\mathcal{X})$. Hence, $\mathcal{O}_{\mathcal{X}} = \mathcal{O}_{\text{weak}}(\mathcal{X})$. This completes the proof of the proposition. □

Given a normed linear space $\mathcal{X}$, its dual $\mathcal{X}^*$ is a Banach space. On $\mathcal{X}^*$, we can also talk about the notion of weak topology. The weak topology of $\mathcal{X}^*$ is the weakest topology on $\mathcal{X}^*$ such that all of the functional in $\mathcal{X}^{**}$ are continuous. This topology turns out to be less useful than the weak topology for $\mathcal{X}^*$ generated by $\mathcal{X}$ (or more precisely by $\phi(\mathcal{X})$, where $\phi : \mathcal{X} \to \mathcal{X}^{**}$ is the natural mapping). This leads us to the following definition.

Definition 7.119 Let $\mathcal{X}$ be a normed linear space over the field $\mathbb{K}$, $\mathcal{X}^*$ be its dual, and $\phi : \mathcal{X} \to \mathcal{X}^{**}$ be the natural mapping. The *weak* topology* on $\mathcal{X}^*$, denoted by $\mathcal{O}_{\text{weak}^*}(\mathcal{X}^*) \subseteq 2^{\mathcal{X}^*}$, is the weak topology generated by $\phi(\mathcal{X})$, that is the weakest topology on $\mathcal{X}^*$ such that $\phi(x) : \mathcal{X}^* \to \mathbb{K}$, $\forall x \in \mathcal{X}$, are continuous. ◇

For a normed linear space $\mathcal{X}$, let $\mathcal{O}_{\mathcal{X}^*}$ be the natural topology induced by the norm on $\mathcal{X}^*$. Then, $\mathcal{O}_{\text{weak}^*}(\mathcal{X}^*) \subseteq \mathcal{O}_{\text{weak}}(\mathcal{X}^*) \subseteq \mathcal{O}_{\mathcal{X}^*}$. Therefore, the weak* topology is weaker than the weak topology on $\mathcal{X}^*$, which is further weaker than the strong topology on $\mathcal{X}^*$, $\mathcal{O}_{\mathcal{X}^*}$. Clearly, if $\mathcal{X}$ is reflexive, then the weak topology and the weak* topology coincide. We will denote the topological space $(\mathcal{X}^*, \mathcal{O}_{\text{weak}^*}(\mathcal{X}^*))$ by $\mathcal{X}^*_{\text{weak}^*}$. We have the following two basic results for the weak* topology, which are counterpart results for Propositions 7.116 and 7.117.

Proposition 7.120 *Let $\mathcal{X}$ be a normed linear space over the field $\mathbb{K}$ and $\phi : \mathcal{X} \to \mathcal{X}^{**}$ be the natural mapping. Then, the following statements hold.*

(i) *A basis for $\mathcal{O}_{\text{weak}^*}(\mathcal{X}^*)$ consists of all sets of the form*

$$\{x_* \in \mathcal{X}^* \mid \langle\langle x_*, x_i\rangle\rangle = \langle\langle \phi(x_i), x_*\rangle\rangle \in O_i, i = 1, \ldots, n\} \tag{7.10}$$

where $n \in \mathbb{N}$, $O_i \in \mathcal{O}_{\mathbb{K}}$ and $x_i \in \mathcal{X}$, $i = 1, \ldots, n$, and $\mathcal{O}_{\mathbb{K}}$ is the natural topology on $\mathbb{K}$. A basis at $x_{*0} \in \mathcal{X}^*$ for $\mathcal{O}_{\text{weak}^*}(\mathcal{X}^*)$ consists of all sets of the form

$$\{x_* \in \mathcal{X}^* \mid |\langle\langle x_* - x_{*0}, x_i\rangle\rangle| = |\langle\langle \phi(x_i), x_* - x_{*0}\rangle\rangle| < \epsilon, i = 1, \ldots, n\} \tag{7.11}$$

where $\epsilon \in (0, \infty) \subset \mathbb{R}$, $n \in \mathbb{N}$, and $x_i \in \mathcal{X}$, $i = 1, \ldots, n$.

(ii) *$\mathcal{X}^*_{\text{weak}^*}$ is completely regular ($\mathbf{T}_{3\frac{1}{2}}$).*

(iii) *For a sequence $(x_{*k})_{k=1}^{\infty} \subseteq \mathcal{X}^*_{\text{weak}^*}$, $x_{*0} \in \mathcal{X}^*$ is the limit point of the sequence in the weak* topology if, and only if, $\lim_{k \in \mathbb{N}} \langle\langle x_{*k}, x\rangle\rangle = \langle\langle x_{*0}, x\rangle\rangle$, $\forall x \in \mathcal{X}$.*

In this case, we will write $\lim_{k\in\mathbb{N}} x_{*k} = x_{*0}$ weak* *and say that* $(x_{*k})_{k=1}^{\infty}$ *converges weak** *to* x_{*0}.

Proof (i) By Definition 7.119, $\mathcal{O}_{\text{weak}^*}(\mathcal{X}^*)$ is the topology generated by sets of the form (7.10). We will show that these sets form a basis for the topology by Proposition 3.18. Take $B = \{x_* \in \mathcal{X}^* \mid \langle\langle x_*, \vartheta\rangle\rangle \in \mathbb{K}\}$, which is of the form (7.10) and $B = \mathcal{X}^*$. $\forall x_* \in \mathcal{X}^*$, we have $x_* \in B$. For any $B_1, B_2 \subseteq \mathcal{X}^*$, which are of the form (7.10), clearly, $B_1 \cap B_2$ is again of the form (7.10). Hence, Proposition 3.18 applies and the sets of the form (7.10) form a basis for $\mathcal{O}_{\text{weak}^*}(\mathcal{X}^*)$.

Let $B \subseteq \mathcal{X}^*$ be any set of the form (7.11). Clearly, $x_{*0} \in B \in \mathcal{O}_{\text{weak}^*}(\mathcal{X}^*)$. $\forall O \in \mathcal{O}_{\text{weak}^*}(\mathcal{X}^*)$ with $x_{*0} \in O$, by Definition 3.17, there exists a basis open set $B_1 := \{x_* \in \mathcal{X}^* \mid \langle\langle x_*, x_i\rangle\rangle \in O_i, i = 1, \ldots, n\}$, for some $n \in \mathbb{N}$ and some $x_i \in \mathcal{X}$ and $O_i \in \mathcal{O}_{\mathbb{K}}$, $i = 1, \ldots, n$, such that $x_{*0} \in B_1 \subseteq O$. For each $i = 1, \ldots, n$, $c_i := \langle\langle x_{*0}, x_i\rangle\rangle \in O_i \in \mathcal{O}_{\mathbb{K}}$. Then, $\exists \epsilon_i \in (0, \infty) \subset \mathbb{R}$ such that $\mathcal{B}_{\mathbb{K}}(c_i, \epsilon_i) \subseteq O_i$. Take $\epsilon = \min_{1\le i\le n} \epsilon_i \in (0, \infty) \subset \mathbb{R}$ and $B_2 := \{x_* \in \mathcal{X}^* \mid |\langle\langle x_* - x_{*0}, x_i\rangle\rangle| < \epsilon, i = 1, \ldots, n\}$. Clearly, B_2 is of the form (7.11) and $x_{*0} \in B_2 \subseteq B_1 \subseteq O$. Hence, sets of the form (7.11) form a basis at x_{*0} for $\mathcal{O}_{\text{weak}^*}(\mathcal{X}^*)$. Thus, (i) holds.

(ii) $\forall x_{*1}, x_{*2} \in \mathcal{X}^*$ with $x_{*1} \neq x_{*2}$, we have $x_{*1} - x_{*2} \neq \vartheta_*$. By Lemma 7.75, $\exists x \in \mathcal{X}$ with $\|x\| \le 1$ such that $\langle\langle x_{*1} - x_{*2}, x\rangle\rangle \in \mathbb{R}$ and $\langle\langle x_{*1} - x_{*2}, x\rangle\rangle \ge \|x_{*1} - x_{*2}\|/2 > 0$. Let $O_1 := \{a \in \mathbb{K} \mid \operatorname{Re}(a) > \operatorname{Re}(\langle\langle x_{*2}, x\rangle\rangle) + \|x_{*1} - x_{*2}\|/4\}$ and $O_2 := \{a \in \mathbb{K} \mid \operatorname{Re}(a) < \operatorname{Re}(\langle\langle x_{*2}, x\rangle\rangle) + \|x_{*1} - x_{*2}\|/4\}$. Then, $O_1, O_2 \in \mathcal{O}_{\mathbb{K}}$ and $O_1 \cap O_2 = \emptyset$. Let $B_1 := \{x_* \in \mathcal{X}^* \mid \langle\langle x_*, x\rangle\rangle \in O_1\}$ and $B_2 := \{x_* \in \mathcal{X}^* \mid \langle\langle x_*, x\rangle\rangle \in O_2\}$. Then, $B_1, B_2 \in \mathcal{O}_{\text{weak}^*}(\mathcal{X}^*)$, $x_{*1} \in B_1$, $x_{*2} \in B_2$, and $B_1 \cap B_2 = \emptyset$. This shows that $\mathcal{X}^*_{\text{weak}^*}$ is Hausdorff.

Next, we show that $\mathcal{X}^*_{\text{weak}^*}$ is completely regular. Fix any weak* closed set $F \subseteq \mathcal{X}^*_{\text{weak}^*}$ and $x_{*0} \in \widetilde{F} \in \mathcal{O}_{\text{weak}^*}(\mathcal{X}^*)$. Then, there exists a basis open set $B = \{x_* \in \mathcal{X}^* \mid \langle\langle x_*, x_i\rangle\rangle \in O_i, i = 1, \ldots, n\}$, for some $n \in \mathbb{N}$ and some $x_i \in \mathcal{X}$ and $O_i \in \mathcal{O}_{\mathbb{K}}$, $i = 1, \ldots, n$, such that $x_{*0} \in B \subseteq \widetilde{F}$. Let $O := \prod_{i=1}^{n} O_i \subseteq \mathbb{K}^n$ which is open. Let $p_0 := (\langle\langle x_{*0}, x_1\rangle\rangle, \ldots, \langle\langle x_{*0}, x_n\rangle\rangle) \in \mathbb{K}^n$. Then, $p_0 \in O$. By Propositions 4.11 and 3.61, $\mathbb{K}^n$ is normal and therefore completely regular. Then, there exists a continuous function $f : \mathbb{K}^n \to [0, 1] \subset \mathbb{R}$ such that $f|_{\widetilde{O}} = 0$ and $f(p_0) = 1$. Define $g : \mathcal{X}^*_{\text{weak}^*} \to [0, 1] \subset \mathbb{R}$ by $g(x_*) = f(\langle\langle x_*, x_1\rangle\rangle, \ldots, \langle\langle x_*, x_n\rangle\rangle)$, $\forall x_* \in \mathcal{X}^*_{\text{weak}^*}$. By Propositions 3.12 and 3.32, g is a continuous real-valued function on $\mathcal{X}^*_{\text{weak}^*}$, $g(x_{*0}) = f(p_0) = 1$, and $g|_F = 0$. Hence, $\mathcal{X}^*_{\text{weak}^*}$ is completely regular. Thus, (ii) holds.

(iii) "Only if" $\forall x \in \mathcal{X}$, $\phi(x) : \mathcal{X}^* \to \mathbb{K}$ is weak* continuous. By Proposition 3.66, we have $\lim_{k\in\mathbb{N}} \langle\langle x_{*k}, x\rangle\rangle = \lim_{k\in\mathbb{N}} \langle\langle \phi(x), x_{*k}\rangle\rangle = \langle\langle \phi(x), x_{*0}\rangle\rangle = \langle\langle x_{*0}, x\rangle\rangle$.

"If" Let $(x_{*k})_{k=1}^{\infty}$ satisfy that $\lim_{k\in\mathbb{N}} \langle\langle x_{*k}, x\rangle\rangle = \langle\langle x_{*0}, x\rangle\rangle$, $\forall x \in \mathcal{X}$. For any basis open set $B = \{x_* \in \mathcal{X}^* \mid \langle\langle x_*, x_i\rangle\rangle \in O_i, i = 1, \ldots, n\}$, for some $n \in \mathbb{N}$ and some $x_i \in \mathcal{X}$ and $O_i \in \mathcal{O}_{\mathbb{K}}$, $i = 1, \ldots, n$, with $x_{*0} \in B$, we have that $\forall i = 1, \ldots, n$, $\exists N_i \in \mathbb{N}$ such that $\langle\langle x_{*k}, x_i\rangle\rangle \in O_i$, $\forall k \ge N_i$. Take $N = \max_{1\le i\le n} N_i \in \mathbb{N}$. $\forall k \ge N$, $x_{*k} \in B$. This shows that $(x_{*k})_{k=1}^{\infty}$ converges weak* to x_{*0}. Hence, (iii) holds.

This completes the proof of the proposition. □

Proposition 7.121 *Let $\mathcal{X}$ be a normed linear space over the field $\mathbb{K}$ and $\mathcal{X}^*_{\text{weak}^*}$ be the topological space of $\mathcal{X}^*$ endowed with the weak* topology. Then, vector addition $\oplus : \mathcal{X}^*_{\text{weak}^*} \times \mathcal{X}^*_{\text{weak}^*} \to \mathcal{X}^*_{\text{weak}^*}$ is continuous; and scalar multiplication $\otimes : \mathbb{K} \times \mathcal{X}^*_{\text{weak}^*} \to \mathcal{X}^*_{\text{weak}^*}$ is continuous.*

Proof Fix $(x_{*0}, y_{*0}) \in \mathcal{X}^*_{\text{weak}^*} \times \mathcal{X}^*_{\text{weak}^*}$. We will show that $\oplus$ is continuous at (x_{*0}, y_{*0}). Fix a basis open set $B \in \mathcal{O}_{\text{weak}^*}(\mathcal{X}^*)$ with $x_{*0} + y_{*0} \in B$. Then, by Proposition 7.120, $B = \{z_* \in \mathcal{X}^* \mid |\langle\langle z_* - x_{*0} - y_{*0}, x_i \rangle\rangle| < \epsilon, i = 1, \ldots, n\}$, for some $n \in \mathbb{N}$, for some $\epsilon \in (0, \infty) \subset \mathbb{R}$, and for some $x_i \in \mathcal{X}$, $i = 1, \ldots, n$. Let $B_1 := \{x_* \in \mathcal{X}^* \mid |\langle\langle x_* - x_{*0}, x_i \rangle\rangle| < \epsilon/2, i = 1, \ldots, n\}$ and $B_2 := \{y_* \in \mathcal{X}^* \mid |\langle\langle y_* - y_{*0}, x_i \rangle\rangle| < \epsilon/2, i = 1, \ldots, n\}$. Clearly, $B_1, B_2 \in \mathcal{O}_{\text{weak}^*}(\mathcal{X}^*)$, $B_1 \times B_2 \in \mathcal{O}_{\mathcal{X}^*_{\text{weak}^*} \times \mathcal{X}^*_{\text{weak}^*}}$, and $(x_{*0}, y_{*0}) \in B_1 \times B_2$. $\forall (x_*, y_*) \in B_1 \times B_2, \forall i = 1, \ldots, n$, $|\langle\langle x_* + y_* - x_{*0} - y_{*0}, x_i \rangle\rangle| \leq |\langle\langle x_* - x_{*0}, x_i \rangle\rangle| + |\langle\langle y_* - y_{*0}, x_i \rangle\rangle| < \epsilon$. Hence, $x_* + y_* \in B$. This shows that $\oplus$ is continuous at (x_{*0}, y_{*0}). By the arbitrariness of (x_{*0}, y_{*0}) and Proposition 3.9, $\oplus$ is continuous.

Fix $(\alpha_0, x_{*0}) \in \mathbb{K} \times \mathcal{X}^*_{\text{weak}^*}$. We will show that $\otimes$ is continuous at (α_0, x_{*0}). Fix a basis open set $B \in \mathcal{O}_{\text{weak}^*}(\mathcal{X}^*)$ with $\alpha_0 x_{*0} \in B$. Then, by Proposition 7.120, $B = \{z_* \in \mathcal{X}^* \mid |\langle\langle z_* - \alpha_0 x_{*0}, x_i \rangle\rangle| < \epsilon, i = 1, \ldots, n\}$, for some $n \in \mathbb{N}$, for some $\epsilon \in (0, \infty) \subset \mathbb{R}$, and for some $x_i \in \mathcal{X}$, $i = 1, \ldots, n$. Let $M := \max_{1 \leq i \leq n} \|x_i\| \in [0, \infty) \subset \mathbb{R}$. Let $B_1 := \{\alpha \in \mathbb{K} \mid |\alpha - \alpha_0| < \frac{\epsilon}{\epsilon + 2M\|x_{*0}\|}\}$ and $B_2 := \{x_* \in \mathcal{X}^* \mid |\langle\langle x_* - x_{*0}, x_i \rangle\rangle| < \frac{\epsilon}{2(1+|\alpha_0|)}, i = 1, \ldots, n\}$. Clearly, $B_1 \in \mathcal{O}_{\mathbb{K}}$ and $B_2 \in \mathcal{O}_{\text{weak}^*}(\mathcal{X}^*)$, $B_1 \times B_2 \subset \mathcal{O}_{\mathbb{K} \times \mathcal{X}^*_{\text{weak}^*}}$, and $(\alpha_0, x_{*0}) \in B_1 \times B_2$. $\forall (\alpha, x_*) \in B_1 \times B_2, \forall i = 1, \ldots, n$, $|\langle\langle \alpha x_* - \alpha_0 x_{*0}, x_i \rangle\rangle| \leq |\langle\langle \alpha x_* - \alpha x_{*0}, x_i \rangle\rangle| + |\langle\langle \alpha x_{*0} - \alpha_0 x_{*0}, x_i \rangle\rangle| = |\alpha||\langle\langle x_* - x_{*0}, x_i \rangle\rangle| + |\alpha - \alpha_0||\langle\langle x_{*0}, x_i \rangle\rangle| \leq (|\alpha_0| + |\alpha - \alpha_0|)|\langle\langle x_* - x_{*0}, x_i \rangle\rangle| + |\alpha - \alpha_0|\|x_{*0}\|\|x_i\| \leq (|\alpha_0| + 1)|\langle\langle x_* - x_{*0}, x_i \rangle\rangle| + |\alpha - \alpha_0| M \|x_{*0}\| < \epsilon$. Hence, $\alpha x_* \in B$. This shows that $\otimes$ is continuous at (α_0, x_{*0}). By the arbitrariness of (α_0, x_{*0}) and Proposition 3.9, $\otimes$ is continuous.

This completes the proof of the proposition. □

The importance of the weak* topology stems from the following Theorem.

Theorem 7.122 (Alaoglu Theorem) *Let $\mathcal{X}$ be a normed linear space over the field $\mathbb{K}$ and $S = \overline{\mathcal{B}}_{\mathcal{X}^*}(\vartheta_*, r)$, for some $r \in \overline{\mathbb{R}_+}$. Then, S is weak* compact.*

Proof Let the weak* topology on S be $\mathcal{O}_{\text{weak}^*}(S)$. Denote the topological space $(S, \mathcal{O}_{\text{weak}^*}(S))$ by $\mathcal{S}$. $\forall x \in \mathcal{X}$, $\forall x_* \in S$, by Proposition 7.72, we have $|\langle\langle x_*, x \rangle\rangle| \leq \|x_*\|\|x\| \leq r\|x\|$. Let $I_x := \overline{\mathcal{B}}_{\mathbb{K}}(0, r\|x\|)$. Then, $\langle\langle x_*, x \rangle\rangle \in I_x$. Let $I_x \subseteq \mathbb{K}$ be endowed with the subset topology $\mathcal{O}_x$. Denote the topological space $(I_x, \mathcal{O}_x)$ by $\mathcal{I}_x$. By Proposition 5.40 or 7.41, $\mathcal{I}_x$ is compact. Let $\mathcal{P} := \prod_{x \in \mathcal{X}} \mathcal{I}_x$, whose topology is denoted by $\mathcal{O}_{\mathcal{P}}$. By Tychonoff Theorem 5.47, $\mathcal{P}$ is compact. Define the equivalence map $E : \mathcal{S} \to \mathcal{P}$ by $\pi_x(E(x_*)) = \langle\langle x_*, x \rangle\rangle$, $\forall x \in \mathcal{X}$, $\forall x_* \in S$.

Claim 7.122.1 $E : \mathcal{S} \to E(\mathcal{S}) \subseteq \mathcal{P}$ is a homeomorphism.

Proof of Claim Clearly, $E : \mathcal{S} \to E(\mathcal{S})$ is surjective. $\forall x_{*1}, x_{*2} \in S$ with $x_{*1} \neq x_{*2}$, then $\exists x \in \mathcal{X}$ such that $\pi_x(E(x_{*1})) = \langle\langle x_{*1}, x \rangle\rangle \neq \langle\langle x_{*2}, x \rangle\rangle = \pi_x(E(x_{*2}))$.

Then, $E(x_{*1}) \neq E(x_{*2})$. Hence, E is injective. Therefore, $E : \mathcal{S} \to E(\mathcal{S})$ is bijective and admits inverse $E_{\text{inv}} : E(\mathcal{S}) \to \mathcal{S}$.

Fix any $x_{*0} \in \mathcal{S}$. Fix any basis open set $O \in \mathcal{O}_{\mathcal{P}}$ with $E(x_{*0}) \in O$. By Proposition 3.25, $O = \prod_{x \in \mathcal{X}} O_x$ where $O_x \in \mathcal{O}_x$, $\forall x \in \mathcal{X}$, and $O_x = I_x$ for all x's except finitely many x's, say $x \in X_N$. We will distinguish two exhaustive and mutually exclusive cases: Case 1: $X_N = \emptyset$; Case 2: $X_N \neq \emptyset$. Case 1: $X_N = \emptyset$. Take $x_1 = \vartheta$ and $O_1 = \mathbb{K}$. Let $B := \{x_* \in \mathcal{S} \mid \langle\langle x_*, x_1 \rangle\rangle \in O_1\} = S$. Clearly, $x_{*0} \in B \in \mathcal{O}_{\text{weak}^*}(S)$ and $\forall x_* \in B$, $\forall x \in \mathcal{X}$, $\pi_x(E(x_*)) = \langle\langle x_*, x \rangle\rangle \in I_x = O_x$. Then, $E(x_*) \in O$ and $E(B) \subseteq O$. Case 2: $X_N \neq \emptyset$. $\forall x \in X_N$, by Proposition 3.4, let $\bar{O}_x \in \mathcal{O}_{\mathbb{K}}$ be such that $\bar{O}_x \cap I_x = O_x$. Let $B := \{x_* \in \mathcal{S} \mid \langle\langle x_*, x \rangle\rangle \in \bar{O}_x, \forall x \in X_N\}$. Clearly, $x_{*0} \in B \in \mathcal{O}_{\text{weak}^*}(S)$. $\forall x_* \in B$, $\forall x \in X_N$, $\pi_x(E(x_*)) = \langle\langle x_*, x \rangle\rangle \in \bar{O}_x \cap I_x = O_x$. Hence, $E(x_*) \in O$ and $E(B) \subseteq O$. In both cases, we have shown that $\exists B \in \mathcal{O}_{\text{weak}^*}(S)$ with $x_{*0} \in B$ such that $E(B) \subseteq O$. Hence, E is continuous at x_{*0}. By the arbitrariness of x_{*0} and Proposition 3.9, E is continuous.

Fix any $p_0 \in E(\mathcal{S})$. Let $x_{*0} = E_{\text{inv}}(p_0) \in \mathcal{S}$. Fix any basis open set $B \in \mathcal{O}_{\text{weak}^*}(S)$ with $x_{*0} \in B$. By Proposition 7.120, $B = \{x_* \in \mathcal{S} \mid \langle\langle x_*, x_i \rangle\rangle \in O_i, i = 1, \ldots, n\}$, for some $n \in \mathbb{N}$ and some $x_i \in \mathcal{X}$ and $O_i \in \mathcal{O}_{\mathbb{K}}$, $i = 1, \ldots, n$. Without loss of generality, we may assume that $x_1, \ldots, x_n$ are distinct (otherwise, say $x_n = x_{n-1}$, then set $\bar{O}_{n-1} = O_n \cap O_{n-1}$ and $\bar{O}_i = O_i$, $i = 1, \ldots, n-2$, and consider $x_1, \ldots, x_{n-1}$ with $\bar{O}_1, \ldots, \bar{O}_{n-1}$). $\forall x \in \mathcal{X}$, if $x = x_i$ for some $i \in \{1, \ldots, n\}$, then let $O_x = O_i \cap I_x \in \mathcal{O}_x$; otherwise, let $O_x = I_x \in \mathcal{O}_x$. Let $O := \prod_{x \in \mathcal{X}} O_x$. Clearly, $p_0 \in O$. $\forall p \in O \cap E(\mathcal{S})$, let $x_* = E_{\text{inv}}(p) \in \mathcal{S}$. $\forall i = 1, \ldots, n$, $\langle\langle x_*, x_i \rangle\rangle = \pi_{x_i}(E(x_*)) \in O_{x_i} = O_i \cap I_{x_i}$. Then, $x_* \in B$ and $E_{\text{inv}}(O \cap E(\mathcal{S})) \subseteq B$. This shows that E_{inv} is continuous at p_0. By the arbitrariness of p_0 and Proposition 3.9, E_{inv} is continuous.

Hence, $E : \mathcal{S} \to E(\mathcal{S})$ is a homeomorphism. This completes the proof of the claim. □

Next, we will show that $E(\mathcal{S})$ is closed. $\forall p_0 \in \overline{E(\mathcal{S})} \subseteq \mathcal{P}$, define $f_0 : \mathcal{X} \to \mathbb{K}$ by $f_0(x) = \pi_x(p_0)$, $\forall x \in \mathcal{X}$. $\forall x_1, x_2 \in \mathcal{X}$, $\forall \alpha, \beta \in \mathbb{K}$, let $z = \alpha x_1 + \beta x_2$. $\forall \epsilon \in (0, \infty) \subset \mathbb{R}$, $\forall x \in \mathcal{X}$, if $x \in \{x_1, x_2, z\}$, let $O_x = \mathcal{B}_{\mathbb{K}}(\pi_x(p_0), \epsilon) \cap I_x \in \mathcal{O}_x$, otherwise, let $O_x = I_x \in \mathcal{O}_x$. Let $O := \prod_{x \in \mathcal{X}} O_x \in \mathcal{O}_{\mathcal{P}}$. Clearly, $p_0 \in O$. By Proposition 3.3, $\exists p_1 \in E(\mathcal{S}) \cap O$. Let $x_* = E_{\text{inv}}(p_1) \in \mathcal{S}$. $\forall x \in \{x_1, x_2, z\}$, $\langle\langle x_*, x \rangle\rangle = \pi_x(p_1) \in O_x = \mathcal{B}_{\mathbb{K}}(\pi_x(p_0), \epsilon) \cap I_x \subseteq \mathcal{B}_{\mathbb{K}}(f_0(x), \epsilon)$. Hence, we have $|\langle\langle x_*, x_1 \rangle\rangle - f_0(x_1)| < \epsilon$, $|\langle\langle x_*, x_2 \rangle\rangle - f_0(x_2)| < \epsilon$, and $|\langle\langle x_*, z \rangle\rangle - f_0(z)| < \epsilon$. This implies that

$$
\begin{aligned}
|f_0(z) - \alpha f_0(x_1) - \beta f_0(x_2)| &= |f_0(z) - \langle\langle x_*, z \rangle\rangle - \alpha f_0(x_1) \\
&\quad + \alpha \langle\langle x_*, x_1 \rangle\rangle - \beta f_0(x_2) + \beta \langle\langle x_*, x_2 \rangle\rangle | < \epsilon + |\alpha|\epsilon + |\beta|\epsilon
\end{aligned}
$$

By the arbitrariness of ϵ, we have $f_0(z) = \alpha f_0(x_1) + \beta f_0(x_2)$. Hence, f_0 is linear. $\forall x \in \mathcal{X}$, $f_0(x) = \pi_x(p_0) \in I_x$, then $|f_0(x)| \leq r\|x\|$. Therefore, $f_0 \in \mathcal{X}^*$ and

$\|f_0\| \le r$. Hence, $f_0 \in S$. $\forall x \in \mathcal{X}$, $\pi_x(E(f_0)) = f_0(x) = \pi_x(p_0)$. Hence, $E(f_0) = p_0$ and $p_0 \in E(\mathcal{S})$. This shows that $\overline{E(\mathcal{S})} \subseteq E(\mathcal{S})$. By Proposition 3.3, $E(\mathcal{S})$ is closed.

By Proposition 5.5 and the compactness of $\mathcal{P}$, we have $E(\mathcal{S})$ is compact. By Proposition 5.7, $\mathcal{S}$ is compact. This completes the proof of the theorem. □

Proposition 7.123 *Let $\mathcal{X}$ and $\mathcal{Y}$ be normed linear spaces over $\mathbb{K}$, $A \in \mathrm{B}(\mathcal{X}, \mathcal{Y})$, and $\mathcal{X}_{\text{weak}}$ and $\mathcal{Y}_{\text{weak}}$ be the topological spaces of $\mathcal{X}$ and $\mathcal{Y}$ endowed with the weak topology. Then, $A : \mathcal{X}_{\text{weak}} \to \mathcal{Y}_{\text{weak}}$ is continuous.*

Proof Fix any basis open set $O_Y \in \mathcal{O}_{\text{weak}}(\mathcal{Y})$, we will show that $A_{\text{inv}}(O_Y) \in \mathcal{O}_{\text{weak}}(\mathcal{X})$. By Proposition 7.116, $O_Y = \{y \in \mathcal{Y} \mid \langle\langle y_{*i}, y\rangle\rangle \in O_i,\ i = 1, \dots, n\}$, where $n \in \mathbb{N}$, $O_i \in \mathcal{O}_{\mathbb{K}}$, $y_{*i} \in \mathcal{Y}^*$, $i = 1, \dots, n$. Then, $A_{\text{inv}}(O_Y) = \{x \in \mathcal{X} \mid \langle\langle y_{*i}, Ax\rangle\rangle \in O_i,\ i = 1, \dots, n\} = \{x \in \mathcal{X} \mid \langle\langle A' y_{*i}, x\rangle\rangle \in O_i,\ i = 1, \dots, n\} \in \mathcal{O}_{\text{weak}}(\mathcal{X})$. Hence, $A : \mathcal{X}_{\text{weak}} \to \mathcal{Y}_{\text{weak}}$ is continuous. This completes the proof of the proposition. □

Definition 7.124 Let $\mathcal{X}$ be a normed linear space and $D \subseteq \mathcal{X}$. D is said to be *locally convex at* $x_0 \in D$ if $\exists \delta_0 \in (0, \infty) \subset \mathbb{R}$, such that $D_0 := D \cap \mathcal{B}_{\mathcal{X}}(x_0, \delta_0)$ is convex. D is said to be *locally convex* if it is locally convex at any $x \in D$. ◇

Then, $D \subseteq \mathcal{X}$ is locally convex at $x_0 \in D$ if, and only if, there exists a basis $\mathcal{B} \subseteq \mathcal{O}$ at x_0 for the subset topological space $(D, \mathcal{O})$ such that each $B \in \mathcal{B}$ is convex. Furthermore, $D \subseteq \mathcal{X}$ is locally convex if, and only if, there exists a basis $\mathcal{B} \subseteq \mathcal{O}$ for the subset topological space $(D, \mathcal{O})$ such that each $B \in \mathcal{B}$ is convex.

Proposition 7.125 *Let $\mathcal{X}$ be a Banach space over $\mathbb{K}$. Then, $\mathcal{X}$ is σ-compact if, and only if, $\mathcal{X}$ is finite-dimensional.*

Proof "Sufficiency" Assume that $\mathcal{X}$ is finite-dimensional. Let $K_n := \overline{\mathcal{B}}_{\mathcal{X}}(\vartheta_{\mathcal{X}}, n)$, $\forall n \in \mathbb{N}$. By Proposition 7.42, K_n is compact, $\forall n \in \mathbb{N}$. Clearly, $\bigcup_{n=1}^{\infty} K_n = \mathcal{X}$. Hence, $\mathcal{X}$ is σ-compact.

"Necessity" Assume that $\mathcal{X}$ is σ-compact. Then, $\mathcal{X} = \bigcup_{n=1}^{\infty} K_n$, where $K_n \subseteq \mathcal{X}$ is compact, $\forall n \in \mathbb{N}$. By Proposition 5.5, K_n is closed, $\forall n \in \mathbb{N}$. By Baire Category Theorem 4.40, $\mathcal{X}$ is second category everywhere. Then, $\mathcal{X} \neq \emptyset$ is not of first category. This implies that K_n are not all nowhere dense. By Proposition 3.40, $\exists n_0 \in \mathbb{N}$, $\exists x_0 \in K_{n_0}$, $\exists \delta \in (0, \infty) \subset \mathbb{R}$, such that $\mathcal{B}_{\mathcal{X}}(x_0, \delta) \subseteq K_{n_0}$. Introduce the transformation $F : \mathcal{X} \to \mathcal{X}$ by $F(x) = \frac{2}{\delta}(x - x_0)$, $\forall x \in \mathcal{X}$. Clearly, F is a homeomorphism. Then, $\overline{\mathcal{B}}_{\mathcal{X}}(\vartheta_{\mathcal{X}}, 1) \subseteq F(K_{n_0})$. By K_{n_0} being compact, we have $F(K_{n_0})$ is compact since F is a homeomorphism. Then, $\overline{\mathcal{B}}_{\mathcal{X}}(\vartheta_{\mathcal{X}}, 1)$ is a closed subset of $F(K_{n_0})$ and therefore compact by Proposition 5.5.

We conclude from the previous paragraph that $K := \overline{\mathcal{B}}_{\mathcal{X}}(\vartheta_{\mathcal{X}}, 1)$ is compact. Suppose that $\mathcal{X}$ is not finite-dimensional. Then, there exists $(y_n)_{n=1}^{\infty} \subseteq \mathcal{X}$ such that y_n's are distinct and the set $E := \{y_n \mid n \in \mathbb{N}\} \subseteq \mathcal{X}$ is a linearly independent set. Without loss of generality, we may assume that $\|y_n\| = 1$, $\forall n \in \mathbb{N}$. Let $x_1 := y_1 \in K$, $z_n \in \arg\min_{z \in \operatorname{span}(\{x_1, \dots, x_{n-1}\})} \|y_n - z\|$, and $x_n := \frac{1}{\|y_n - z_n\|}(y_n - z_n) \in K$, $n = 2, 3, \dots$. Clearly, z_n exists since the minimization is over finite-dimensional subspaces of $\mathcal{X}$. By E being a linearly independent set, we have $\|y_n - z_n\| > 0$, $\forall n \in \mathbb{N}$. Hence, $x_n \in \mathcal{X}$ is well-defined, $\forall n \in \mathbb{N}$. Clearly, $\|x_n\| = 1$ and $x_n \in K$, $\forall n \in \mathbb{N}$. By K being compact and Borel-Lebesgue Theorem 5.37, $\exists x_0 \in K$ and a subsequence $(x_{n_i})_{i=1}^{\infty}$ of $(x_n)_{n=1}^{\infty}$ such that $\lim_{i\in\mathbb{N}} x_{n_i} = x_0$. Since $x_{n_i} \in \operatorname{span}(E)$, $\forall i \in \mathbb{N}$, then $x_0 \in \overline{\operatorname{span}(E)}$. This implies that $x_0 = \sum_{j=1}^{\infty} a_j x_j$, where $a_j \in \mathbb{K}$, $\forall j \in \mathbb{N}$. Note that $x_0 = \lim_{i\in\mathbb{N}} x_{n_i} = \lim_{i\in\mathbb{N}} \sum_{j=1}^{n_i-1} a_j x_j$ and thus $\lim_{i\in\mathbb{N}}(x_{n_i} - \sum_{j=1}^{n_i-1} a_j x_j) = \vartheta_{\mathcal{X}}$. But, $\forall i \in \mathbb{N}$, $\left\|x_{n_i} - \sum_{j=1}^{n_i-1} a_j x_j\right\| = \frac{1}{\|y_{n_i}-z_{n_i}\|}\left\|y_{n_i} - \sum_{j=1}^{n_i-1}\left(\|y_{n_i} - z_{n_i}\|\right)a_j x_j - z_{n_i}\right\|$. The vector $\sum_{j=1}^{n_i-1}\left(\|y_{n_i} - z_{n_i}\|\right)a_j x_j - z_{n_i} \in \operatorname{span}\left(\{x_1, \ldots, x_{n_i-1}\}\right)$. By the definition of z_{n_i}, we have $\left\|x_{n_i} - \sum_{j=1}^{n_i-1} a_j x_j\right\| = \frac{1}{\|y_{n_i}-z_{n_i}\|}\left\|y_{n_i} - \sum_{j=1}^{n_i-1}\left(\|y_{n_i} - z_{n_i}\|\right)a_j x_j - z_{n_i}\right\| \geq \frac{1}{\|y_{n_i}-z_{n_i}\|}\|y_{n_i} - z_{n_i}\| = 1$. This contradicts with $\lim_{i\in\mathbb{N}}(x_{n_i} - \sum_{j=1}^{n_i-1} a_j x_j) = \vartheta_{\mathcal{X}}$. Hence, the hypothesis does not hold, and $\mathcal{X}$ must be finite-dimensional.

This completes the proof of the proposition. □

Proposition 7.126 *Let $\mathcal{X} := (X, \mathcal{O})$ be a separable topological space, $\mathcal{Y}$ be a Banach space, and $f : X \to \mathcal{Y}$ be continuous. Then, $\mathcal{W} := \operatorname{span}(f(\mathcal{X})) \subseteq \mathcal{Y}$ is a separable normed linear subspace of $\mathcal{Y}$, and $\overline{\mathcal{W}} \subseteq \mathcal{Y}$ is a separable Banach subspace of $\mathcal{Y}$.*

Proof $\mathcal{W}$ is a subspace of $\mathcal{Y}$ and hence a normed linear subspace of $\mathcal{Y}$. $\overline{\mathcal{W}}$ is a closed subspace of $\mathcal{Y}$ and hence a complete normed linear subspace of $\mathcal{Y}$, by Proposition 4.39. Then, $\overline{\mathcal{W}}$ is a Banach subspace of $\mathcal{Y}$. By Proposition 3.90, $f(\mathcal{X})$ is separable. Then, by Proposition 7.35, $\mathcal{W}$ and $\overline{\mathcal{W}}$ are separable subsets of $\mathcal{Y}$. This completes the proof of the proposition. □

Proposition 7.127 *Let $\mathcal{Y} := (Y, \rho)$ be a metric space, and $S_i \subseteq \mathcal{Y}$ be a separable subset, $\forall i \in N \subseteq \mathbb{N}$. Then, $U := \bigcap_{i\in N} S_i$ and $V := \bigcup_{i\in N} S_i$ are separable subsets of $\mathcal{Y}$.*

Proof Since $U \subseteq S_1$ and S_1 is separable. By Proposition 4.38, U is separable.

Let $D_i \subseteq S_i$ be a countable dense subset of S_i, $\forall i \in N$. Then, $D := \bigcup_{i\in N} D_i \subseteq V$ and D is countable by Cantor's diagonal argument. $\forall y_0 \in V$, then $y_0 \in S_{i_0}$ for some $i_0 \in N$. By Proposition 4.13, $\exists (y_j)_{j=1}^{\infty} \subseteq D_{i_0} \subseteq D$ such that $y_0 = \lim_{j\in\mathbb{N}} y_j$. Then, again by Proposition 4.13, $y_0 \in \overline{D}$. By the arbitrariness of y_0, we have $V \subseteq \overline{D}$. Then, D is dense in V, which implies that V is separable.

This completes the proof of the proposition. □

Proposition 7.128 *Let $\mathcal{X} := (X, \mathcal{O})$ be a separable topological space, $\mathcal{Y}$ be a Banach space, and $f_n : X \to \mathcal{Y}$ be continuous, $\forall n \in \mathbb{N}$. Then, $\mathcal{W}_n := \operatorname{span}(f_n(\mathcal{X})) \subseteq \mathcal{Y}$ is a separable normed linear subspace of $\mathcal{Y}$, and $\mathcal{W} := \overline{\operatorname{span}\left(\bigcup_{n\in\mathbb{N}} \mathcal{W}_n\right)} \subseteq \mathcal{Y}$ is a separable Banach subspace of $\mathcal{Y}$.*

Proof This follows directly from Propositions 7.126, 7.127, 7.35, and 4.39. □

Fig. 7.2 Modes of convergence in $\mathcal{Y} = \mathcal{X}^*$

$$\begin{array}{c} \lim_{n\in\mathbb{N}} y_n = y_0 \text{ in } \mathcal{Y} \\ \Downarrow \\ \lim_{n\in\mathbb{N}} y_n = y_0 \text{ weakly} \\ \Downarrow \quad \Uparrow \; \mathcal{X} \text{ is reflexive} \\ \lim_{n\in\mathbb{N}} y_n = y_0 \text{ weak}^* \end{array}$$

Here, we will summarize different modes of convergence in a Banach space in the following figure. Let $\mathcal{X}$ be a Banach space over $\mathbb{K}$, $\mathcal{Y} := \mathcal{X}^*$, and $\mathcal{Z} := \mathcal{Y}^* = \mathcal{X}^{**}$. A sequence $(y_n)_{n=1}^{\infty} \subseteq \mathcal{Y}$ can converge to some $y_0 \in \mathcal{Y}$ in norm, or weakly, or weak*. Then, we have the diagram in Fig. 7.2.

Chapter 8
Global Theory of Optimization

In this chapter, we are going to develop a number of tools for optimization in real normed linear spaces, including the geometric form of Hahn–Banach Theorem, minimum norm duality for convex sets, Fenchel Duality Theorem, and Lagrange multiplier theory for convex programming. In this chapter, we will restrict our attention to real spaces, rather than complex ones.

8.1 Hyperplanes and Convex Sets

Definition 8.1 Let $\mathcal{X}$ be a real vector space. A *hyperplane* H is a maximal proper linear variety in $\mathcal{X}$, that is, a linear variety $H \subset \mathcal{X}$, and if $V \supseteq H$ is a linear variety, then either $V = H$ or $V = \mathcal{X}$. ◇

Proposition 8.2 *Let $\mathcal{X}$ be a real vector space. $H \subseteq \mathcal{X}$ is a hyperplane if, and only if, there exists a linear functional $f : \mathcal{X} \to \mathbb{R}$ and $c \in \mathbb{R}$ with f being not identically equal to zero, such that $H = \{x \in \mathcal{X} \mid f(x) = c\}$.*

Proof "Necessity" Let H be a hyperplane. Then, H is a linear variety. There exists a subspace M and $x_0 \in \mathcal{X}$ such that $H = x_0 + M$. We will distinguish two exhaustive and mutually exclusive cases: Case 1. $x_0 \notin M$; Case 2: $x_0 \in M$. Case 1: $x_0 \notin M$. Let $\bar{M} := \text{span}(M \cup \{x_0\})$. Clearly, $\bar{M} \supset H$ and is a linear variety. Then, $\bar{M} = \mathcal{X}$ by Definition 8.1. $\forall x \in \mathcal{X}$, x can be uniquely written as $\alpha x_0 + m$, where $\alpha \in \mathbb{R}$ and $m \in M$. Define $f : \mathcal{X} \to \mathbb{R}$ by $f(x) = f(\alpha x_0 + m) = \alpha$, $\forall x \in \mathcal{X}$. Clearly, f is a linear functional and is not identically equal to zero. It is straightforward to verify that $H = \{x \in \mathcal{X} \mid f(x) = 1\}$. Case 2: $x_0 \in M$. Then, $H = M$. By Definition 8.1, $\exists x_1 \in \mathcal{X} \setminus H$. Let $\bar{M} := \text{span}(M \cup \{x_1\})$. Clearly, $\bar{M} \supset H$ and is a linear variety. Then, $\bar{M} = \mathcal{X}$ by Definition 8.1. $\forall x \in \mathcal{X}$, x can be uniquely written as $\alpha x_1 + m$, where $\alpha \in \mathbb{R}$ and $m \in M$. Define $f : \mathcal{X} \to \mathbb{R}$ by $f(x) = f(\alpha x_1 + m) = \alpha$,

Z. Pan, *Measure-Theoretic Calculus in Abstract Spaces*,
https://doi.org/10.1007/978-3-031-21912-2_8

$\forall x \in \mathcal{X}$. Clearly, f is a linear functional and is not identically equal to zero. It is straightforward to verify that $H = \{x \in \mathcal{X} \mid f(x) = 0\}$.

"Sufficiency" Let $H = \{x \in \mathcal{X} \mid f(x) = c\}$, where $f : \mathcal{X} \to \mathbb{R}$ is a linear functional, $c \in \mathbb{R}$, and f is not identically equal to zero. Let $M := \mathcal{N}(f)$. Clearly, M is a proper subspace of $\mathcal{X}$. Since f is not identically equal to zero, then $\exists x_0 \in \mathcal{X} \setminus M$ such that $f(x_0) = 1$. $\forall x \in \mathcal{X}$, $f(x - f(x)x_0) = 0$. Then, $x - f(x)x_0 \in M$ and $x \in \text{span}(M \cup \{x_0\})$. Hence, $\mathcal{X} = \text{span}(M \cup \{x_0\})$ and M is a maximal proper subspace. Then, $H = cx_0 + M$ is a hyperplane.

This completes the proof of the proposition. □

Consider a hyperplane H in a real normed linear space $\mathcal{X}$. By Proposition 7.17, $\overline{H}$ is a linear variety. By Definition 8.1, $\overline{H} = \mathcal{X}$ or $\overline{H} = H$. Thus, a hyperplane in a real normed linear space must either be dense or closed.

Proposition 8.3 *Let $\mathcal{X}$ be a real normed linear space and $H \subseteq \mathcal{X}$. H is a closed hyperplane if, and only if, there exist $x_* \in \mathcal{X}^*$ and $c \in \mathbb{R}$ with $x_* \neq \vartheta_*$ such that $H = \{x \in \mathcal{X} \mid \langle\langle x_*, x\rangle\rangle = c\}$.*

Proof "Necessity" Let H be a closed hyperplane. By Proposition 8.2, there exists a linear functional $f : \mathcal{X} \to \mathbb{R}$ and $c \in \mathbb{R}$ with f being not identically equal to zero such that $H = \{x \in \mathcal{X} \mid f(x) = c\}$. All we need to show is that $f \in \mathcal{X}^*$. Since H is a linear variety, then $H \neq \emptyset$. Fix $x_0 \in H$. It is easy to show that $M := H - x_0 = \mathcal{N}(f)$. By Proposition 7.16, M is closed. By Proposition 7.72, $f \in \mathcal{X}^*$.

"Sufficiency" By Proposition 8.2, H is a hyperplane. By the continuity of x_* and Proposition 3.10, H is closed. Hence, H is a closed hyperplane.

This completes the proof of the proposition. □

For a real normed linear space $\mathcal{X}$ and a closed hyperplane $H \subset \mathcal{X}$. Then, $H = \{x \in \mathcal{X} \mid \langle\langle x_*, x\rangle\rangle = c\}$, where $\vartheta_* \neq x_* \in \mathcal{X}^*$ and $c \in \mathbb{R}$. We associate four sets with H: (a) $\{x \in \mathcal{X} \mid \langle\langle x_*, x\rangle\rangle \leq c\}$; (b) $\{x \in \mathcal{X} \mid \langle\langle x_*, x\rangle\rangle < c\}$; (c) $\{x \in \mathcal{X} \mid \langle\langle x_*, x\rangle\rangle \geq c\}$; (d) $\{x \in \mathcal{X} \mid \langle\langle x_*, x\rangle\rangle > c\}$, which are called half-spaces. The first two are negative half-spaces. The last two are positive half-spaces. The first and third are closed. The second and fourth are open.

Definition 8.4 Let $\mathcal{X}$ be a real normed linear space and $K \subseteq \mathcal{X}$ be convex with $\vartheta \in K^\circ$. The *Minkowski functional* $p : \mathcal{X} \to \mathbb{R}$ of K is defined by $p(x) = \inf\{r \in \mathbb{R} \mid r^{-1}x \in K, r > 0\}$, $\forall x \in \mathcal{X}$. ◇

Proposition 8.5 *Let $\mathcal{X}$ be a real normed linear space and $K \subseteq \mathcal{X}$ be convex with $\vartheta \in K^\circ$. Then, the Minkowski functional $p : \mathcal{X} \to \mathbb{R}$ of K satisfies*

(i) $0 \leq p(x) < +\infty$, $\forall x \in \mathcal{X}$.
(ii) $p(\alpha x) = \alpha p(x)$, $\forall x \in \mathcal{X}$, $\forall \alpha \in [0, \infty) \subset \mathbb{R}$.
(iii) $p(x_1 + x_2) \leq p(x_1) + p(x_2)$, $\forall x_1, x_2 \in \mathcal{X}$.
(iv) *p is uniformly continuous.*
(v) $\overline{K} = \{x \in \mathcal{X} \mid p(x) \leq 1\}$; $K^\circ = \{x \in \mathcal{X} \mid p(x) < 1\}$.

Furthermore, (ii) *and* (iii) *imply that p is a sublinear functional.*

Proof (i) Since $\vartheta \in K^\circ$, then $\exists \epsilon_0 \in (0, \infty) \subset \mathbb{R}$ such that $\mathcal{B}(\vartheta, \epsilon_0) \subseteq K$. $\forall x \in \mathcal{X}$, we have either $x = \vartheta$, then $p(\vartheta) = 0$; or $x \neq \vartheta$, then $0 \leq p(x) \leq \|x\|/\epsilon_0 < +\infty$.

(ii) $\forall x \in \mathcal{X}$, $\forall \alpha \in [0, \infty) \subset \mathbb{R}$, we will distinguish three exhaustive and mutually exclusive cases: Case 1: $x = \vartheta$; Case 2: $x \neq \vartheta$ and $\alpha = 0$; Case 3: $x \neq \vartheta$ and $\alpha > 0$. Case 1: $x = \vartheta$. We have $p(\alpha x) = p(\vartheta) = 0 = \alpha p(x)$. Case 2: $x \neq \vartheta$ and $\alpha = 0$. Then, we have $p(\alpha x) = p(\vartheta) = 0 = \alpha p(x)$. Case 3: $x \neq \vartheta$ and $\alpha > 0$. $\forall r \in \{r \in \mathbb{R} \mid r^{-1} x \in K, r > 0\}$, we have $\alpha r > 0$ and $\alpha r \in \{r \in \mathbb{R} \mid r^{-1}(\alpha x) \in K, r > 0\}$. Hence, $\alpha p(x) \geq p(\alpha x)$. On the other hand, $\forall r \in \{r \in \mathbb{R} \mid r^{-1}(\alpha x) \in K, r > 0\}$, we have $r/\alpha \in \{r \in \mathbb{R} \mid r^{-1} x \in K, r > 0\}$. Hence, $\alpha^{-1} p(\alpha x) \geq p(x)$. Therefore, we have $\alpha p(x) = p(\alpha x)$.

(iii) $\forall x_1, x_2 \in \mathcal{X}$, $\forall r_1 \in \{r \in \mathbb{R} \mid r^{-1} x_1 \in K, r > 0\}$, $\forall r_2 \in \{r \in \mathbb{R} \mid r^{-1} x_2 \in K, r > 0\}$, we have $r_1^{-1} x_1, r_2^{-1} x_2 \in K$ By the convexity of K, we have

$$(r_1 + r_2)^{-1}(x_1 + x_2) = \frac{r_1}{r_1 + r_2} r_1^{-1} x_1 + \frac{r_2}{r_1 + r_2} r_2^{-1} x_2 \in K$$

Then, $r_1 + r_2 \in \{r \in \mathbb{R} \mid r^{-1}(x_1 + x_2) \in K, r > 0\}$. Hence, we have $r_1 + r_2 \geq p(x_1 + x_2)$ and $p(x_1) + p(x_2) \geq p(x_1 + x_2)$.

(iv) $\forall x \in \mathcal{X}$, $0 \leq p(x) \leq \|x\|/\epsilon_0$. $\forall \epsilon \in (0, \infty) \subset \mathbb{R}$, let $\delta = \epsilon_0 \epsilon \in (0, \infty) \subset \mathbb{R}$. $\forall x_1, x_2 \in \mathcal{X}$ with $\|x_1 - x_2\| < \delta$, we have $p(x_1) \leq p(x_2) + p(x_1 - x_2) \leq p(x_2) + \|x_1 - x_2\|/\epsilon_0 < p(x_2) + \epsilon$ and $p(x_2) \leq p(x_1) + p(x_2 - x_1) \leq p(x_1) + \|x_2 - x_1\|/\epsilon_0 < p(x_1) + \epsilon$. Hence, $|p(x_1) - p(x_2)| < \epsilon$. This shows that p is uniformly continuous.

(v) $\forall x \in K^\circ$, we have either $x = \vartheta$, then $p(x) = 0 < 1$; or $x \neq \vartheta$, then $\exists \epsilon \in (0, \infty) \subset \mathbb{R}$ such that $\mathcal{B}(x, \epsilon\|x\|) \subseteq K$, which implies that $p(x) \leq 1/(1 + \epsilon) < 1$. Therefore, $K^\circ \subseteq \{x \in \mathcal{X} \mid p(x) < 1\}$.

$\forall x \in \mathcal{X}$ with $p(x) < 1$, by the continuity of p, $\exists \delta \in (0, \infty) \subset \mathbb{R}$ such that $p(y) < 1$, $\forall y \in \mathcal{B}(x, \delta)$. $\forall y \in \mathcal{B}(x, \delta)$, $y \in K$ by the convexity of K, $p(y) < 1$, and the fact that $\vartheta \in K$. Then, $\mathcal{B}(x, \delta) \subseteq K$ and $x \in K^\circ$. Therefore, we have $\{x \in \mathcal{X} \mid p(x) < 1\} \subseteq K^\circ$. Thus, $K^\circ = \{x \in \mathcal{X} \mid p(x) < 1\}$.

$\forall x \in K$, we have $p(x) \leq 1$. Then, $K \subseteq \{x \in \mathcal{X} \mid p(x) \leq 1\}$. By the continuity of p and Proposition 3.10, $\{x \in \mathcal{X} \mid p(x) \leq 1\}$ is closed. Then, $\overline{K} \subseteq \{x \in \mathcal{X} \mid p(x) \leq 1\}$.

$\forall x \in \mathcal{X}$ with $p(x) \leq 1$, $\forall \rho \in (0, 1) \subset \mathbb{R}$, $p(\rho x) = \rho p(x) < 1$. Then, $\rho x \in K^\circ$. Then, by Proposition 4.13, $x \in \overline{K^\circ} \subseteq \overline{K}$. Hence, $\{x \in \mathcal{X} \mid p(x) \leq 1\} \subseteq \overline{K}$. Therefore, $\overline{K} = \{x \in \mathcal{X} \mid p(x) \leq 1\}$.

This completes the proof of the proposition. □

Proposition 8.6 *Let $\mathcal{X}$ be a real normed linear space and $K \subseteq \mathcal{X}$ be convex with $K^\circ \neq \emptyset$. Then, $\overline{K} = \overline{K^\circ}$.*

Proof Clearly, $K^\circ \subseteq K$. Then, $\overline{K^\circ} \subseteq \overline{K}$.

To show that $\overline{K} \subseteq \overline{K^\circ}$, we will distinguish two exhaustive and mutually exclusive cases: Case 1: $\vartheta \in K^\circ$; Case 2: $\vartheta \notin K^\circ$. Case 1: $\vartheta \in K^\circ$. Let $p : \mathcal{X} \to [0, \infty) \subset \mathbb{R}$ be the Minkowski functional of K. By Proposition 8.5, p is a continuous sublinear functional, $K^\circ = \{x \in \mathcal{X} \mid p(x) < 1\}$, and $\overline{K} = \{x \in \mathcal{X} \mid p(x) \leq 1\}$. $\forall x \in \overline{K}$,

$\forall \epsilon \in (0, \infty) \subset \mathbb{R}$, $p(x) \le 1$. By the convexity of K, $y := \frac{\|x\|}{\epsilon+\|x\|} x \in K$. Then, $p(y) = \frac{\|x\|}{\epsilon+\|x\|} p(x) < 1$. Then, $y \in K^\circ$. Note that $\|x - y\| \le \frac{\epsilon}{\epsilon+\|x\|}\|x\| < \epsilon$. Hence, $x \in \overline{K^\circ}$. Therefore, $\overline{K} \subseteq \overline{K^\circ}$. This case is proved.

Case 2: $\vartheta \notin K^\circ$. Let $x_0 \in K^\circ \neq \emptyset$. Let $K_1 = K - x_0$. By Propositions 7.16 and 6.39, $K_1^\circ = K^\circ - x_0 \neq \emptyset$ and K_1 is convex. By Case 1, $\overline{K_1} \subseteq \overline{K_1^\circ}$. By Proposition 7.16, $\overline{K} = \overline{x_0 + K_1} = x_0 + \overline{K_1} \subseteq x_0 + \overline{K_1^\circ} = \overline{x_0 + K_1^\circ} = \overline{(x_0 + K_1)^\circ} = \overline{K^\circ}$. This case is proved.

Hence, $\overline{K} = \overline{K^\circ}$. This completes the proof of the proposition. □

8.2 Geometric Form of Hahn–Banach Theorem

The next theorem is the geometric form of Hahn–Banach Theorem.

Theorem 8.7 (Mazur's Theorem) *Let $\mathcal{X}$ be a real normed linear space, $K \subseteq \mathcal{X}$ be a convex set with nonempty interior, and $V \subseteq \mathcal{X}$ is a linear variety with $V \cap K^\circ = \emptyset$. Then, there exists a closed hyperplane H containing V but no interior point of K and K is contained in one of the closed half-spaces associated with H; that is, $\exists c \in \mathbb{R}$ and $\exists x_* \in \mathcal{X}^*$ with $x_* \neq \vartheta_*$ such that $\langle\langle x_*, v\rangle\rangle = c$, $\forall v \in V$, $\langle\langle x_*, k\rangle\rangle < c$, $\forall k \in K^\circ$, and $\langle\langle x_*, k\rangle\rangle \le c$, $\forall k \in \overline{K}$.*

Proof We will distinguish two exhaustive and mutually exclusive cases: Case 1: $\vartheta \in K^\circ$; Case 2: $\vartheta \notin K^\circ$. Case 1: $\vartheta \in K^\circ$. Let $p : \mathcal{X} \to [0, \infty) \subset \mathbb{R}$ be the Minkowski functional of K. By Proposition 8.5, p is a continuous sublinear functional. Let $M := \operatorname{span}(V)$. Since V is a linear variety and $\vartheta \notin V$, then $\exists x_1 \in V$ such that $V - x_1 =: \bar{M}$ is a subspace in M and $x_1 \notin \bar{M}$. $\forall m \in M$, $\exists!\, \alpha \in \mathbb{R}$ and $\exists!\, \bar{m} \in \bar{M}$ such that $m = \alpha x_1 + \bar{m}$. Then, we may define a functional $f : M \to \mathbb{R}$ by $f(m) = f(\alpha x_1 + \bar{m}) = \alpha$, $\forall m \in M$. Clearly, f is a linear functional on M and $f(v) = 1$, $\forall v \in V$. $\forall v \in V$, we have $v \notin K^\circ$ and $p(v) \ge 1 = f(v)$. $\forall m = \alpha x_1 + \bar{m} \in M$, we have either $\alpha > 0$, then $f(m) = \alpha f(x_1 + \alpha^{-1}\bar{m}) = \alpha \le \alpha p(x_1 + \alpha^{-1}\bar{m}) = p(m)$, where the inequality follows from the fact that $x_1 + \alpha^{-1}\bar{m} \in V$; or $\alpha \le 0$, then $f(m) = \alpha \le 0 \le p(m)$. Thus, $\forall m \in M$, we have $f(m) \le p(m)$. By the extension form of Hahn–Banach Theorem, there exists $x_* \in \mathcal{X}^*$ such that $x_*|_M = f$ and $\langle\langle x_*, x\rangle\rangle \le p(x)$, $\forall x \in \mathcal{X}$. Clearly, $x_* \neq \vartheta_*$. $\forall v \in V$, $\langle\langle x_*, v\rangle\rangle = f(v) = 1 =: c \in \mathbb{R}$. $\forall k \in \overline{K}$, we have $\langle\langle x_*, k\rangle\rangle \le p(k) \le 1$. $\forall k \in K^\circ$, we have $\langle\langle x_*, k\rangle\rangle \le p(k) < 1$. Hence, the closed hyperplane $H := \{x \in \mathcal{X} \mid \langle\langle x_*, x\rangle\rangle = c\}$ is the one that we seek.

Case 2: $\vartheta \notin K^\circ$. Let $x_0 \in K^\circ$. Then, by Proposition 6.39, $K_1 := K - x_0$ is a convex set. $V_1 := V - x_0$ is a linear variety. By Proposition 7.16, $K_1^\circ = K^\circ - x_0$. Then, $\vartheta \in K_1^\circ$ and $V_1 \cap K_1^\circ = \emptyset$. By Case 1, there exist $c_1 \in \mathbb{R}$ and $x_* \in \mathcal{X}^*$ with $x_* \neq \vartheta_*$ such that $\langle\langle x_*, v_1\rangle\rangle = c_1$, $\forall v_1 \in V_1$, $\langle\langle x_*, k_1\rangle\rangle < c_1$, $\forall k_1 \in K_1^\circ$, and $\langle\langle x_*, k_1\rangle\rangle \le c_1$, $\forall k_1 \in \overline{K_1}$. Let $c := c_1 + \langle\langle x_*, x_0\rangle\rangle \in \mathbb{R}$. Then, $\forall v \in V$, $\langle\langle x_*, v\rangle\rangle = c$. $\forall k \in K^\circ$, $\langle\langle x_*, k\rangle\rangle < c$. By Proposition 7.16, $\overline{K_1} = \overline{K} - x_0$. Then,

$\forall k \in \overline{K}$, $\langle\langle x_*, k\rangle\rangle \le c$. Thus, the closed hyperplane $H := \{x \in \mathcal{X} \mid \langle\langle x_*, x\rangle\rangle = c\}$ is the one that we seek.

This completes the proof of the theorem. □

Definition 8.8 Let $\mathcal{X}$ be a real normed linear space. A closed hyperplane $H := \{x \in \mathcal{X} \mid \langle\langle x_*, x\rangle\rangle = c\}$ with $x_* \in \mathcal{X}^*$, $x_* \neq \vartheta_*$, and $c \in \mathbb{R}$ is called a *supporting hyperplane* of a convex set $K \subseteq \mathcal{X}$ if either $\inf_{k\in K} \langle\langle x_*, k\rangle\rangle = c$ or $\sup_{k\in K} \langle\langle x_*, k\rangle\rangle = c$. ◇

Clearly, for a convex set K in a real normed linear space, if K admits interior points, then, by Mazur's Theorem, there exists a supporting hyperplane of K passing through each boundary point of K.

Theorem 8.9 (Eidelheit Separation Theorem) *Let $\mathcal{X}$ be a real normed linear space and $K_1, K_2 \subseteq \mathcal{X}$ be nonempty convex sets with $K_1^\circ \neq \emptyset$ and $K_1^\circ \cap K_2 = \emptyset$. Then, there exists a closed hyperplane H that separates K_1 and K_2, that is $\exists x_* \in \mathcal{X}^*$ with $x_* \neq \vartheta_*$ and $\exists c \in \mathbb{R}$ such that*

$$\sup_{k_1\in K_1} \langle\langle x_*, k_1\rangle\rangle \le c \le \inf_{k_2\in K_2} \langle\langle x_*, k_2\rangle\rangle$$

Proof Let $K := K_1^\circ - K_2$. By Propositions 7.15 and 6.39, K is convex. Clearly, $K^\circ \neq \emptyset$ since $K_1^\circ \neq \emptyset$ and $K_2 \neq \emptyset$. Since $K_1^\circ \cap K_2 = \emptyset$, then $\vartheta \notin K$. Let $V = \{\vartheta\}$. Then, V is a linear variety and $V \cap K^\circ = \emptyset$. By Mazur's Theorem, $\exists x_* \in \mathcal{X}^*$ with $x_* \neq \vartheta_*$ such that $\langle\langle x_*, k\rangle\rangle \le \langle\langle x_*, \vartheta\rangle\rangle = 0$, $\forall k \in K$. $\forall k_1 \in K_1^\circ$, $\forall k_2 \in K_2$, $k_1 - k_2 \in K$ and $\langle\langle x_*, k_1 - k_2\rangle\rangle \le 0$. Then, $\langle\langle x_*, k_1\rangle\rangle \le \langle\langle x_*, k_2\rangle\rangle$. Hence, $-\infty < \sup_{k_1\in K_1^\circ} \langle\langle x_*, k_1\rangle\rangle \le c \le \inf_{k_2\in K_2} \langle\langle x_*, k_2\rangle\rangle < +\infty$ for some $c \in \mathbb{R}$. By Proposition 8.6, $\overline{K_1} = \overline{K_1^\circ}$. By Proposition 4.13, $\forall k_1 \in \overline{K_1}$, $\langle\langle x_*, k_1\rangle\rangle \le \sup_{\bar{k}_1\in K_1^\circ} \langle\langle x_*, \bar{k}_1\rangle\rangle \le c$. Hence, we have

$$\sup_{k_1\in K_1} \langle\langle x_*, k_1\rangle\rangle = \sup_{k_1\in K_1^\circ} \langle\langle x_*, k_1\rangle\rangle \le c \le \inf_{k_2\in K_2} \langle\langle x_*, k_2\rangle\rangle$$

Thus, the closed hyperplane $H := \{x \in \mathcal{X} \mid \langle\langle x_*, x\rangle\rangle = c\}$ is the one we seek. This completes the proof of the theorem. □

Proposition 8.10 *Let $\mathcal{X}$ be a real normed linear space and $K \subseteq \mathcal{X}$ be a closed convex set. Then, $\forall x_0 \in \mathcal{X} \setminus K$. Then, there exists a $x_* \in \mathcal{X}^*$ such that $\langle\langle x_*, x_0\rangle\rangle < \inf_{k\in K} \langle\langle x_*, k\rangle\rangle$. Hence, K is weakly closed.*

Proof We will distinguish two exhaustive and mutually exclusive cases: Case 1: $K = \emptyset$; Case 2: $K \neq \emptyset$. Case 1: $K = \emptyset$. Let $x_* = \vartheta_*$. Then, $\langle\langle x_*, x_0\rangle\rangle = 0 < +\infty = \inf_{k\in K} \langle\langle x_*, k\rangle\rangle$. Clearly, K is weakly closed. This case is proved.

Case 2: $K \neq \emptyset$. $\forall x_0 \in \mathcal{X} \setminus K$, by Proposition 4.10, $\operatorname{dist}(x_0, K) = \inf_{k\in K} \|x_0 - k\| =: d \in (0, \infty) \subset \mathbb{R}$. Then, $K_1 := \mathcal{B}_{\mathcal{X}}(x_0, d)$ is a convex set with nonempty interior and $K_1 \cap K = \emptyset$. By Eidelheit Separation Theorem, there exists $x_* \in \mathcal{X}^*$ with $x_* \neq \vartheta_*$ and $\exists c \in \mathbb{R}$ such that $\sup_{k_1\in K_1} \langle\langle x_*, k_1\rangle\rangle \le c \le$

$\inf_{k\in K}\langle\langle x_*, k\rangle\rangle$. By Lemma 7.75, $\langle\langle x_*, x_0\rangle\rangle < \sup_{k_1\in K_1}\langle\langle x_*, k_1\rangle\rangle \le c$. Then, $\langle\langle x_*, x_0\rangle\rangle < \inf_{k\in K}\langle\langle x_*, k\rangle\rangle$.

Let $\bar{K}$ be the closure of K in the weak topology $\mathcal{O}_{\text{weak}}(\mathcal{X})$. $x_0 \in \{x \in \mathcal{X} \mid \langle\langle x_*, x\rangle\rangle < c\} =: O$. Clearly, O is a weakly open set and $O \cap K = \emptyset$. Then, by Proposition 3.3, $x_0 \notin \bar{K}$. Thus, we have shown that $\forall x_0 \in \widetilde{K}$, $x_0 \in \widetilde{\bar{K}}$. Hence, $\bar{K} \subseteq K$. Hence, K is weakly closed by Proposition 3.3. This case is proved.

This completes the proof of the proposition. □

Proposition 8.11 *Let $\mathcal{X}$ be a reflexive real normed linear space and $K \subseteq \mathcal{X}^*$ be a bounded closed convex set. Then, K is weak* compact.*

Proof Since K is bounded, then there exists $n \in \mathbb{Z}_+$ such that $K \subseteq \overline{\mathcal{B}}_{\mathcal{X}^*}(\vartheta_*, n) =: S$. By Alaoglu Theorem, S is weak* compact. Since K is closed and convex, then, by Proposition 8.10, K is weakly closed. Since $\mathcal{X}$ is reflexive, then the weak topology and the weak* topology on $\mathcal{X}^*$ coincide. Then, K is weak* closed. By Proposition 5.5, K is weak* compact. This completes the proof of the proposition. □

8.3 Duality in Minimum Norm Problems

Definition 8.12 Let $\mathcal{X}$ be a real normed linear space and $K \subseteq \mathcal{X}$ be a nonempty convex set. The *support* of K is the set $K_{\text{supp}} := \{x_* \in \mathcal{X}^* \mid \sup_{k\in K}\langle\langle x_*, k\rangle\rangle < +\infty\}$. ◇

Proposition 8.13 *Let $\mathcal{X}$ be a real normed linear space and $K, G \subseteq \mathcal{X}$ be nonempty convex sets. Then, the following statements hold.*

(i) *$K_{\text{supp}} \subseteq \mathcal{X}^*$ is a convex cone.*

(ii) *If K is closed, then $K = \bigcap_{x_*\in K_{\text{supp}}} \{x \in \mathcal{X} \mid \langle\langle x_*, x\rangle\rangle \le \sup_{k\in K}\langle\langle x_*, k\rangle\rangle\}$ $= \bigcap_{x_*\in K_{\text{supp}},\ x_*\neq\vartheta_*} \{x \in \mathcal{X} \mid \langle\langle x_*, x\rangle\rangle \le \sup_{k\in K}\langle\langle x_*, k\rangle\rangle\}$, that is, K equals to the intersection of all closed half-spaces containing K.*

(iii) *$(K + G)_{\text{supp}} = K_{\text{supp}} \cap G_{\text{supp}}$.*

(iv) *$(K \cap G)_{\text{supp}} \supseteq K_{\text{supp}} + G_{\text{supp}}$, if $K \cap G \neq \emptyset$.*

Proof (i) Clearly, $\vartheta_* \in K_{\text{supp}}$. $\forall x_{*1}, x_{*2} \in K_{\text{supp}}$, $\forall \alpha \in [0, \infty) \subset \mathbb{R}$, we have either $\alpha = 0$, then $\alpha x_{*1} = \vartheta_* \in K_{\text{supp}}$; or $\alpha > 0$, then $\sup_{k\in K}\langle\langle \alpha x_{*1}, k\rangle\rangle = \sup_{k\in K}\alpha\langle\langle x_{*1}, k\rangle\rangle = \alpha\sup_{k\in K}\langle\langle x_{*1}, k\rangle\rangle < +\infty$, where we have applied Proposition 3.81, which further implies that $\alpha x_{*1} \in K_{\text{supp}}$. This shows that K_{supp} is a cone. Note that $\sup_{k\in K}\langle\langle x_{*1} + x_{*2}, k\rangle\rangle = \sup_{k\in K}(\langle\langle x_{*1}, k\rangle\rangle + \langle\langle x_{*2}, k\rangle\rangle) \le \sup_{k\in K}\langle\langle x_{*1}, k\rangle\rangle + \sup_{k\in K}\langle\langle x_{*2}, k\rangle\rangle < +\infty$, where we have applied Proposition 3.81. Then, $x_{*1} + x_{*2} \in K_{\text{supp}}$. This coupled with the fact that K_{supp} is a cone implies that K_{supp} is a convex cone.

(ii) Clearly $K \subseteq \bigcap_{x_* \in K_{\text{supp}}} \{x \in \mathcal{X} \mid \langle\langle x_*, x\rangle\rangle \leq \sup_{k\in K} \langle\langle x_*, k\rangle\rangle\} =: \bar{K}$. $\forall x \in \mathcal{X}\setminus K$, by Proposition 8.10, $\exists x_{*0} \in \mathcal{X}^*$ such that $\langle\langle x_{*0}, x\rangle\rangle < \inf_{k\in K} \langle\langle x_{*0}, k\rangle\rangle$. Then, $\sup_{k\in K} \langle\langle -x_{*0}, k\rangle\rangle < \langle\langle -x_{*0}, x\rangle\rangle < +\infty$. Hence, $-x_{*0} \in K_{\text{supp}}$ and $x \in \mathcal{X} \setminus \bar{K}$. This shows that $\mathcal{X} \setminus K \subseteq \mathcal{X} \setminus \bar{K}$ and $\bar{K} \subseteq K$. Hence, $K = \bar{K}$. Note that, for $\vartheta_* \in K_{\text{supp}}$, $\{x \in \mathcal{X} \mid \langle\langle \vartheta_*, x\rangle\rangle \leq \sup_{k\in K} \langle\langle \vartheta_*, k\rangle\rangle\} = \mathcal{X}$. Then, $K = \bigcap_{x_* \in K_{\text{supp}},\, x_* \neq \vartheta_*} \{x \in \mathcal{X} \mid \langle\langle x_*, x\rangle\rangle \leq \sup_{k\in K} \langle\langle x_*, k\rangle\rangle\}$. Let $\acute{K}$ be the intersection of all closed half-spaces containing K. Clearly, $K \subseteq \acute{K}$. Any closed half-space containing K can be expressed as $\{x \in \mathcal{X} \mid \langle\langle x_*, x\rangle\rangle \leq c\}$ for some $x_* \in \mathcal{X}^*$ with $x_* \neq \vartheta_*$ and for some $c \in \mathbb{R}$. Then, $c \geq \sup_{k\in K} \langle\langle x_*, k\rangle\rangle \in \mathbb{R}$. Hence, $x_* \in K_{\text{supp}}$ and $\acute{K} \subseteq \bigcap_{x_* \in K_{\text{supp}},\, x_* \neq \vartheta_*} \bigcap_{c \geq \sup_{k\in K} \langle\langle x_*, k\rangle\rangle} \{x \in \mathcal{X} \mid \langle\langle x_*, x\rangle\rangle \leq c\} = \bigcap_{x_* \in K_{\text{supp}},\, x_* \neq \vartheta_*} \{x \in \mathcal{X} \mid \langle\langle x_*, x\rangle\rangle \leq \sup_{k\in K} \langle\langle x_*, k\rangle\rangle\} = K$. Hence, $K = \acute{K}$.

(iii) $\forall x_* \in K_{\text{supp}} \cap G_{\text{supp}}$, then $\sup_{k\in K} \langle\langle x_*, k\rangle\rangle =: c_1 < +\infty$ and $\sup_{g\in G} \langle\langle x_*, g\rangle\rangle =: c_2 < +\infty$. $\forall x \in K + G$, $x = k + g$ for some $k \in K$ and some $g \in G$. Then, $\langle\langle x_*, x\rangle\rangle = \langle\langle x_*, k\rangle\rangle + \langle\langle x_*, g\rangle\rangle \leq c_1 + c_2 < +\infty$. This shows that $x_* \in (K + G)_{\text{supp}}$. Hence, $K_{\text{supp}} \cap G_{\text{supp}} \subseteq (K + G)_{\text{supp}}$.

On the other hand, $\forall x_* \in (K + G)_{\text{supp}}$, $\sup_{x\in K+G} \langle\langle x_*, x\rangle\rangle =: c < +\infty$. Fix $k_0 \in K$ and $g_0 \in G$, since K and G are nonempty. Then, $\sup_{k\in K} \langle\langle x_*, k\rangle\rangle = \sup_{x\in g_0+K} \langle\langle x_*, x\rangle\rangle - \langle\langle x_*, g_0\rangle\rangle \leq c - \langle\langle x_*, g_0\rangle\rangle < +\infty$ and $\sup_{g\in G} \langle\langle x_*, g\rangle\rangle = \sup_{x\in k_0+G} \langle\langle x_*, x\rangle\rangle - \langle\langle x_*, k_0\rangle\rangle \leq c - \langle\langle x_*, k_0\rangle\rangle < +\infty$. Hence, $x_* \in K_{\text{supp}} \cap G_{\text{supp}}$. This shows that $(K + G)_{\text{supp}} \subseteq K_{\text{supp}} \cap G_{\text{supp}}$. Therefore, we have $(K + G)_{\text{supp}} = K_{\text{supp}} \cap G_{\text{supp}}$.

(iv) $\forall x_* \in K_{\text{supp}} + G_{\text{supp}}$, let $x_* = x_{*1} + x_{*2}$ with $x_{*1} \in K_{\text{supp}}$ and $x_{*2} \in G_{\text{supp}}$. $\forall x \in K \cap G$, we have $\langle\langle x_*, x\rangle\rangle = \langle\langle x_{*1}, x\rangle\rangle + \langle\langle x_{*2}, x\rangle\rangle \leq \sup_{k\in K} \langle\langle x_{*1}, k\rangle\rangle + \sup_{g\in G} \langle\langle x_{*2}, g\rangle\rangle < +\infty$. Hence, $x_* \in (K \cap G)_{\text{supp}}$. Hence, we have $K_{\text{supp}} + G_{\text{supp}} \subseteq (K \cap G)_{\text{supp}}$.

This completes the proof of the proposition. □

Definition 8.14 Let $\mathcal{X}$ be a real normed linear space and $K \subseteq \mathcal{X}$ be a nonempty convex set. The *support functional* of K is $h : K_{\text{supp}} \to \mathbb{R}$ given by $h(x_*) = \sup_{k\in K} \langle\langle x_*, k\rangle\rangle$, $\forall x_* \in K_{\text{supp}}$. ◇

In the above definition, h takes value in $\mathbb{R}$ since $K \neq \emptyset$ and $x_* \in K_{\text{supp}}$.

Proposition 8.15 *Let $\mathcal{X}$ be a real normed linear space, $K \subseteq \mathcal{X}$ be a nonempty convex set, $h : K_{\text{supp}} \to \mathbb{R}$ be the support functional of K, and $y \in \mathcal{X}$. Then, the following statements holds.*

(i) $\delta := \inf_{k\in K} \|y - k\| = \max_{x_* \in K_{\text{supp}},\, \|x_*\| \leq 1} (\langle\langle x_*, y\rangle\rangle - h(x_*))$, *where the maximum is achieved at some $x_{*0} \in K_{\text{supp}}$ with $\|x_{*0}\| \leq 1$. If the infimum is achieved at $k_0 \in K$ then $y - k_0$ is aligned with x_{*0} and $\langle\langle x_{*0}, k_0\rangle\rangle = h(x_{*0})$.*

(ii) *If $\exists k_0 \in K$ and $\exists x_{*0} \in K_{\text{supp}}$ with $\|x_{*0}\| = 1$ such that $y - k_0$ is aligned with x_{*0} and $\langle\langle x_{*0}, k_0\rangle\rangle = h(x_{*0})$, then the infimum is achieved at k_0 and the maximum is achieved at x_{*0}, that is $\delta = \|y - k_0\| = \langle\langle x_{*0}, y\rangle\rangle - h(x_{*0})$.*

Proof (i) $\forall x_* \in K_{\text{supp}}$ with $\|x_*\| \leq 1$, $\forall k \in K$, we have $\|y - k\| \geq \|x_*\| \|y - k\| \geq \langle\langle x_*, y - k\rangle\rangle = \langle\langle x_*, y\rangle\rangle - \langle\langle x_*, k\rangle\rangle \geq \langle\langle x_*, y\rangle\rangle - h(x_*)$. Hence, $\delta = \inf_{k\in K} \|y - k\| \geq \sup_{x_*\in K_{\text{supp}},\ \|x_*\|\leq 1}(\langle\langle x_*, y\rangle\rangle - h(x_*))$.

We will distinguish two exhaustive and mutually exclusive cases: Case 1: $\delta = 0$; Case 2: $\delta > 0$. Case 1: $\delta = 0$. Take $x_{*0} = \vartheta_* \in K_{\text{supp}}$. $\|x_{*0}\| = 0 \leq 1$. Then, $\delta = 0 = \langle\langle x_{*0}, y\rangle\rangle - h(x_{*0})$. Then, $\delta := \inf_{k\in K} \|y - k\| = \max_{x_*\in K_{\text{supp}},\ \|x_*\|\leq 1}(\langle\langle x_*, y\rangle\rangle - h(x_*))$. If the infimum is achieved at $k_0 \in K$, then $y - k_0$ is aligned with x_{*0} and $\langle\langle x_{*0}, k_0\rangle\rangle = 0 = h(x_{*0})$. This case is proved.

Case 2: $\delta > 0$. Since $K \neq \emptyset$, then $\delta < +\infty$. Take $K_1 = \mathcal{B}_{\mathcal{X}}(y, \delta)$. Then, K_1 is a convex set with nonempty interior. $K \cap K_1^\circ = \emptyset$. By Eidelheit Separation Theorem, there exists $x_* \in \mathcal{X}^*$ with $x_* \neq \vartheta_*$ and $\exists c \in \mathbb{R}$ such that $\sup_{k_1\in K_1} \langle\langle x_*, k_1\rangle\rangle \leq c \leq \inf_{k\in K} \langle\langle x_*, k\rangle\rangle$. Take $x_{*0} = -\|x_*\|^{-1} x_*$. Then, by Proposition 3.81, we have $\inf_{k_1\in K_1} \langle\langle x_{*0}, k_1\rangle\rangle \geq -c/\|x_*\| \geq \sup_{k\in K} \langle\langle x_{*0}, k\rangle\rangle$. This shows, $x_{*0} \in K_{\text{supp}}$ and $\|x_{*0}\| = 1$. The above inequality is equivalent to $\langle\langle x_{*0}, y\rangle\rangle + \inf_{x\in\mathcal{B}_{\mathcal{X}}(\vartheta,\delta)} \langle\langle x_{*0}, x\rangle\rangle \geq h(x_{*0})$. By Lemma 7.75, we have $\langle\langle x_{*0}, y\rangle\rangle - h(x_{*0}) \geq \delta$. Hence, we have $\delta = \langle\langle x_{*0}, y\rangle\rangle - h(x_{*0}) = \max_{x_*\in K_{\text{supp}},\ \|x_*\|\leq 1}(\langle\langle x_*, y\rangle\rangle - h(x_*))$. If the infimum is achieved at some $k_0 \in K$, then $\delta = \|y - k_0\| = \langle\langle x_{*0}, y\rangle\rangle - h(x_{*0}) \leq \langle\langle x_{*0}, y\rangle\rangle - \langle\langle x_{*0}, k_0\rangle\rangle = \langle\langle x_{*0}, y - k_0\rangle\rangle \leq \|x_{*0}\| \|y - k_0\| \leq \|y - k_0\|$, where the second inequality follows from Proposition 7.72. Hence, $\langle\langle x_{*0}, y - k_0\rangle\rangle = \|x_{*0}\| \|y - k_0\|$, x_{*0} is aligned with $y - k_0$, and $\langle\langle x_{*0}, k_0\rangle\rangle = h(x_{*0})$. This case is proved.

(ii) Note that $\delta \leq \|y - k_0\| = \|x_{*0}\| \|y - k_0\| = \langle\langle x_{*0}, y - k_0\rangle\rangle = \langle\langle x_{*0}, y\rangle\rangle - h(x_{*0}) \leq \delta$. Then, the result follows.

This completes the proof of the proposition. □

Now, we will state a proposition that guarantees the existence of a minimizing solution to a minimum norm problem. This proposition is based on the following result.

Proposition 8.16 *Let $\mathcal{X}$ be a real normed linear space. Then, $\|\cdot\| : \mathcal{X} \to [0, \infty) \subset \mathbb{R}$ is weakly lower semicontinuous.*

Proof By Definition 3.14, all we need to show is that, $\forall a \in \mathbb{R}$, the set $S_a := \{x \in \mathcal{X} \mid \|x\| > a\}$ is weakly open. Note that $\widetilde{S_a}$ is strongly closed and convex. Then, by Proposition 8.10, $\widetilde{S_a}$ is weakly closed. Then, S_a is weakly open. Hence, the norm is a weakly lower semicontinuous functional on $\mathcal{X}$. This completes the proof of the proposition. □

Proposition 8.17 *Let $\mathcal{X}$ be a reflexive real normed linear space, $x_* \in \mathcal{X}^*$, and $K \subseteq \mathcal{X}^*$ be a nonempty closed convex set. Then, $\delta = \min_{k_*\in K} \|x_* - k_*\|$ and the minimum is achieved at some $k_{*0} \in K$.*

Proof Fix a $k_{*1} \in K \neq \emptyset$. Let $\mu := \|x_* - k_{*1}\| \in [0, \infty) \subset \mathbb{R}$ and $d = \mu + \|x_*\| + 1 \in (0, \infty) \subset \mathbb{R}$. Then, $\forall k_* \in K$ with $\|k_*\| > d$, we have $\|x_* - k_*\| \geq \|k_*\| - \|x_*\| > \mu + 1$. Thus, $\delta = \inf_{k_*\in K} \|x_* - k_*\| = \inf_{k_*\in K\cap\overline{\mathcal{B}}_{\mathcal{X}^*}(\vartheta_*,d)} \|x_* - k_*\|$ and any k_{*0} achieving the infimum for the original problem must be in the set $K \cap \overline{\mathcal{B}}_{\mathcal{X}^*}(\vartheta_*, d) =: K_1$. By Proposition 6.40, K_1 is bounded closed and convex. Note that $k_{*1} \in K_1 \neq \emptyset$. Then, by Proposition 8.11, K_1 is weak* compact. By Propositions 7.121, 8.16, and 3.16, we have $f : \mathcal{X}^* \to \mathbb{R}$ given by $f(k_*) = \|x_* - k_*\|$, $\forall k_* \in \mathcal{X}^*$, is weakly lower semicontinuous. By Proposition 7.90, K_1 is weakly compact since $\mathcal{X}$ is reflexive. By Proposition 5.30 and Definition 3.14, there exists $k_{*0} \in K_1$ that achieves the infimum on K_1. Such k_{*0} achieves the infimum on K. This completes the proof of the proposition. □

8.4 Convex and Concave Functionals

Definition 8.18 Let $\mathcal{X}$ be a real vector space, $C \subseteq \mathcal{X}$ be a convex set, $f : C \to \mathbb{R}$ and $g : C \to \mathbb{R}$. f is said to be *convex*, if $\forall x_1, x_2 \in C$, $\forall \alpha \in [0, 1] \subset \mathbb{R}$, we have

$$f(\alpha x_1 + (1 - \alpha)x_2) \leq \alpha f(x_1) + (1 - \alpha) f(x_2)$$

f is said to be *strictly convex* if $\forall x_1, x_2 \in C$ with $x_1 \neq x_2$ and $\forall \alpha \in (0, 1) \subset \mathbb{R}$, we have that the strict inequality holds in the above. g is said to be *(strictly) concave* if $-g$ is (strictly) convex. ◇

Definition 8.19 Let $\mathcal{X}$ be a real vector space, $C \subseteq \mathcal{X}$, and $f : C \to \mathbb{R}$. The *epigraph* of f over C is the set

$$[f, C] := \{(r, x) \in \mathbb{R} \times \mathcal{X} \mid x \in C, f(x) \leq r\}$$

◇

Proposition 8.20 *Let $\mathcal{X}$ be a real vector space, $C \subseteq \mathcal{X}$ be convex, and $f : C \to \mathbb{R}$. Then, f is convex if, and only if, the epigraph $[f, C]$ is convex.*

Proof "Sufficiency" $\forall x_1, x_2 \in C$, $\forall \alpha \in [0, 1] \subset \mathbb{R}$, we have $(f(x_1), x_1)$, $(f(x_2), x_2) \in [f, C]$. By the convexity of the epigraph, then $\alpha\,(f(x_1), x_1) + (1 - \alpha)\,(f(x_2), x_2) = (\alpha f(x_1) + (1 - \alpha) f(x_2), \alpha x_1 + (1 - \alpha)x_2) \in [f, C]$. Then, $\alpha f(x_1) + (1 - \alpha) f(x_2) \geq f(\alpha x_1 + (1 - \alpha)x_2)$. Hence, f is convex.

"Necessity" Let f be convex. $\forall (r_1, x_1), (r_2, x_2) \in [f, C]$, $\forall \alpha \in [0, 1] \subset \mathbb{R}$, we have $r_1 \geq f(x_1)$ and $r_2 \geq f(x_2)$. By the convexity of f and C, we have $\alpha x_1 + (1 - \alpha)x_2 \in C$ and $\alpha r_1 + (1 - \alpha)r_2 \geq \alpha f(x_1) + (1 - \alpha) f(x_2) \geq f(\alpha x_1 + (1 - \alpha)x_2)$. Then, we have $\alpha\,(r_1, x_1) + (1 - \alpha)\,(r_2, x_2) = (\alpha r_1 + (1 - \alpha)r_2, \alpha x_1 + (1 - \alpha)x_2) \in [f, C]$. Hence, the epigraph is convex.

This completes the proof of the proposition. □

Proposition 8.21 *Let $\mathcal{X}$ be a real normed linear space, $C \subseteq \mathcal{X}$ be nonempty, and $f : C \to \mathbb{R}$. Then,* $\mathrm{V}([f, C]) = \mathbb{R} \times \mathrm{V}(C)$.

Proof We will first show that $\mathrm{v}([f, C]) = \mathbb{R} \times \mathrm{v}(C)$. Fix $x_0 \in C$. Then, $(f(x_0), x_0) \in [f, C]$. $\forall (r, x) \in \mathrm{v}([f, C]) = (f(x_0), x_0) + \mathrm{span}([f, C] - (f(x_0), x_0))$, where the equality follows from Proposition 6.37, then $\exists n \in \mathbb{Z}_+$, $\exists (r_1, x_1), \ldots, (r_n, x_n) \in [f, C]$, and $\exists \alpha_1, \ldots, \alpha_n \in \mathbb{R}$ such that $(r - f(x_0), x - x_0) = \sum_{i=1}^{n} \alpha_i (r_i - f(x_0), x_i - x_0)$. Then, $x \in \mathrm{v}(C) = x_0 + \mathrm{span}(C - x_0)$ and $(r, x) \in \mathbb{R} \times \mathrm{v}(C)$. Hence, $\mathrm{v}([f, C]) \subseteq \mathbb{R} \times \mathrm{v}(C)$.

On the other hand, $\forall (r, x) \in \mathbb{R} \times \mathrm{v}(C)$, then, by Proposition 6.37, $x \in \mathrm{v}(C) = x_0 + \mathrm{span}(C - x_0)$. Then, $\exists n \in \mathbb{Z}_+$, $\exists x_1, \ldots, x_n \in C$, and $\exists \alpha_1, \ldots, \alpha_n \in \mathbb{R}$ such that $x - x_0 = \sum_{i=1}^{n} \alpha_i (x_i - x_0)$. $\exists r_0 \in (0, \infty) \subset \mathbb{R}$ such that $r_0 > f(x_0)$. Then, $(r_0, x_0) \in [f, C]$. Let $\alpha_0 = (r - f(x_0) - \sum_{i=1}^{n} \alpha_i (f(x_i) - f(x_0)))/(r_0 - f(x_0))$. Now it is easy to check that $(r - f(x_0), x - x_0) = \sum_{i=1}^{n} \alpha_i (f(x_i) - f(x_0), x_i - x_0) + \alpha_0 (r_0 - f(x_0), x_0 - x_0)$. Note that $(f(x_i), x_i) \in [f, C]$, $\forall i = 1, \ldots, n$. Then, we have $(r, x) \in (f(x_0), x_0) + \mathrm{span}([f, C] - (f(x_0), x_0)) = \mathrm{v}([f, C])$. Hence, $\mathbb{R} \times \mathrm{v}(C) \subseteq \mathrm{v}([f, C])$. Therefore, we have $\mathrm{v}([f, C]) = \mathbb{R} \times \mathrm{v}(C)$.

By Proposition 7.18, $\mathrm{V}([f, C]) = \overline{\mathrm{v}([f, C])} = \overline{\mathbb{R} \times \mathrm{v}(C)}$. By Proposition 3.29, we have $\overline{\mathbb{R} \times \mathrm{v}(C)} = \overline{\mathbb{R}} \times \overline{\mathrm{v}(C)} = \mathbb{R} \times \mathrm{V}(C)$. Therefore, we have $\mathrm{V}([f, C]) = \mathbb{R} \times \mathrm{V}(C)$. This completes the proof of the proposition. □

Proposition 8.22 *Let $\mathcal{X}$ be a real normed linear space, $C \subseteq \mathcal{X}$ be convex, $f : C \to \mathbb{R}$ be convex, and $^\circ C \neq \emptyset$. Then, $[f, C]$ has a relative interior point (r_0, x_0) if, and only if, f is continuous at x_0, $x_0 \in {}^\circ C$, and $r_0 \in (f(x_0), +\infty) \subset \mathbb{R}$.*

Proof "Sufficiency" Fix $x_0 \in {}^\circ C$ and $r_0 \in (f(x_0), +\infty) \subset \mathbb{R}$. Let f be continuous at x_0. Then, for $\epsilon = (r_0 - f(x_0))/2 \in (0, \infty) \subset \mathbb{R}$, $\exists \delta \in (0, r_0 - f(x_0) - \epsilon] \subset \mathbb{R}$ such that $\forall x \in \mathcal{B}_{\mathcal{X}}(x_0, \delta) \cap \mathrm{V}(C)$, we have $x \in C$ and $|f(x) - f(x_0)| < \epsilon$. Clearly, $(r_0, x_0) \in [f, C]$. $\forall (r, x) \in \mathcal{B}_{\mathbb{R} \times \mathcal{X}}((r_0, x_0), \delta) \cap \mathrm{V}([f, C]) = \mathcal{B}_{\mathbb{R} \times \mathcal{X}}((r_0, x_0), \delta) \cap (\mathbb{R} \times \mathrm{V}(C))$, we have $x \in \mathcal{B}_{\mathcal{X}}(x_0, \delta) \cap \mathrm{V}(C)$ and $r \in \mathcal{B}(r_0, \delta)$. Then, $x \in C$ and $r > r_0 - \delta \geq f(x_0) + \epsilon > f(x)$. Hence, $(r, x) \in [f, C]$. This shows that $(r_0, x_0) \in {}^\circ[f, C]$.

"Necessity" Let $(r_0, x_0) \in {}^\circ[f, C]$. Then, $\exists \epsilon_0 \in (0, \infty) \subset \mathbb{R}$ such that $(\mathcal{B}_{\mathbb{R}}(r_0, \epsilon_0) \times \mathcal{B}_{\mathcal{X}}(x_0, \epsilon_0)) \cap \mathrm{V}([f, C]) \subseteq [f, C]$. By Proposition 8.21, we have $\mathcal{B}_{\mathbb{R}}(r_0, \epsilon_0) \times (\mathcal{B}_{\mathcal{X}}(x_0, \epsilon_0) \cap \mathrm{V}(C)) \subseteq [f, C]$. Therefore, $\mathcal{B}_{\mathcal{X}}(x_0, \epsilon_0) \cap \mathrm{V}(C) \subseteq C$, $x_0 \in {}^\circ C$, $f(x) \leq r_0 - \epsilon_0$, $\forall x \in \mathcal{B}_{\mathcal{X}}(x_0, \epsilon_0) \cap \mathrm{V}(C)$, and $f(x_0) \leq r_0 - \epsilon_0 < r_0$.

$\forall \epsilon \in (0, \infty) \subset \mathbb{R}$, Let $\delta = \min\{\epsilon/(r_0 - f(x_0)), 1\} \in (0, 1] \subset \mathbb{R}$. $\forall x \in \mathcal{B}_{\mathcal{X}}(x_0, \delta\epsilon_0) \cap C$, we have $x, x_0, x_0 + \delta^{-1}(x - x_0), x_0 - \delta^{-1}(x - x_0) \in \mathcal{B}_{\mathcal{X}}(x_0, \epsilon_0) \cap \mathrm{V}(C) \subseteq C$. By the convexity of f, we have

$$\begin{aligned}
f(x) &= f((1 - \delta)x_0 + \delta(x_0 + \delta^{-1}(x - x_0))) \\
&\leq (1 - \delta)f(x_0) + \delta f(x_0 + \delta^{-1}(x - x_0)) \\
&\leq f(x_0) + (r_0 - \epsilon_0 - f(x_0))\delta < f(x_0) + \epsilon \\
f(x_0) &= f(\frac{1}{1 + \delta}x + \frac{\delta}{1 + \delta}(x_0 - \delta^{-1}(x - x_0))) \\
&\leq \frac{1}{1 + \delta}f(x) + \frac{\delta}{1 + \delta}f(x_0 - \delta^{-1}(x - x_0)) \\
&\leq \frac{1}{1 + \delta}f(x) + \frac{\delta}{1 + \delta}(r_0 - \epsilon_0)
\end{aligned}$$

$$\Rightarrow (1+\delta)f(x_0) \le f(x) + (r_0 - \epsilon_0)\delta$$
$$\Rightarrow f(x) \ge f(x_0) - (r_0 - f(x_0) - \epsilon_0)\delta > f(x_0) - \epsilon$$

Hence, $|f(x) - f(x_0)| < \epsilon$. This shows that f is continuous at x_0.

This completes the proof of the proposition. □

Proposition 8.23 *Let $\mathcal{X}$ be a real normed linear space, $C \subseteq \mathcal{X}$ be convex, $x_0 \in {}^\circ C \neq \emptyset$, and $f : C \to \mathbb{R}$ be convex. If f is continuous at x_0, then f is continuous at x, $\forall x \in {}^\circ C$.*

Proof Fix any $x \in {}^\circ C$. We will distinguish two exhaustive and mutually exclusive cases: Case 1: $x = x_0$; Case 2: $x \neq x_0$. Case 1: $x = x_0$. Then f is continuous at x. Case 2: $x \neq x_0$. $\forall \epsilon \in (0, \infty) \subset \mathbb{R}$, by the continuity of f at x_0 and the fact that $x_0 \in {}^\circ C$, $\exists \delta \in (0, \infty) \subset \mathbb{R}$ such that $|f(y) - f(x_0)| < \epsilon$, $\forall y \in \mathcal{B}_{\mathcal{X}}(x_0, \delta) \cap \mathrm{V}(C) \subseteq C$. Since $x \in {}^\circ C$, then $\exists \delta_1 \in (0, \delta] \subset \mathbb{R}$ such that $\mathcal{B}_{\mathcal{X}}(x, \delta_1) \cap \mathrm{V}(C) \subseteq C$. Take $\beta = \frac{\|x - x_0\| + \delta_1/2}{\|x - x_0\|} \in (1, \infty) \subset \mathbb{R}$. Then, $\beta(x - x_0) + x_0 \in \mathcal{B}_{\mathcal{X}}(x, \delta_1) \cap \mathrm{V}(C) \subseteq C$. $\forall y \in \mathcal{B}_{\mathcal{X}}(x, (\beta - 1)\delta_1/\beta) \cap C$, $y = (1 - 1/\beta)(\beta/(\beta - 1)(y - x) + x_0) + (1/\beta)(\beta(x - x_0) + x_0)$. Note that $\beta/(\beta - 1)(y - x) + x_0 \in \mathcal{B}_{\mathcal{X}}(x_0, \delta) \cap \mathrm{V}(C) \subseteq C$. By the convexity of f, we have

$$\begin{aligned} f(y) &\le (1 - 1/\beta) f(\beta/(\beta - 1)(y - x) + x_0) + (1/\beta) f(\beta(x - x_0) + x_0) \\ &< (1 - 1/\beta)(f(x_0) + \epsilon) + (1/\beta) f(\beta(x - x_0) + x_0) =: r_1 \end{aligned}$$

Define $r := r_1 + (\beta - 1)\delta_1/\beta$. Then, by Proposition 8.21, $\mathcal{B}_{\mathbb{R}\times\mathcal{X}}((r, x), (\beta - 1)\delta_1/\beta) \cap \mathrm{V}([f, C]) = \mathcal{B}_{\mathbb{R}\times\mathcal{X}}((r, x), (\beta - 1)\delta_1/\beta) \cap (\mathbb{R} \times \mathrm{V}(C)) \subseteq [f, C]$. Hence, $(r, x) \in {}^\circ[f, C]$. By Proposition 8.22, f is continuous at x. This completes the proof of the proposition. □

Proposition 8.24 *Let $\mathcal{X}$ be a finite-dimensional real normed linear space, $C \subseteq \mathcal{X}$ be a convex set, and $f : C \to \mathbb{R}$ be convex. Then, $\forall x_0 \in {}^\circ C$, f is continuous at x_0.*

Proof We will distinguish two exhaustive and mutually exclusive cases: Case 1: ${}^\circ C = \emptyset$; Case 2: ${}^\circ C \neq \emptyset$. Case 1: ${}^\circ C = \emptyset$. This is trivial.

Case 2: ${}^\circ C \neq \emptyset$. Fix any $x_0 \in {}^\circ C$. Then, $\exists \delta \in (0, \infty) \subset \mathbb{R}$ such that $\bar{K} := \overline{\mathcal{B}}_{\mathcal{X}}(x_0, \delta) \cap \mathrm{V}(C) \subseteq C$. Let $M := \mathrm{V}(C) - x_0$, which is a closed subspace. Let $n \in \mathbb{Z}_+$ be the dimension of M, which is well-defined by Theorem 6.51. We will further distinguish two exhaustive and mutually exclusive cases: Case 2a: $n = 0$; Case 2b: $n \in \mathbb{N}$. Case 2a: $n = 0$. Then, $C = \{x_0\}$. Clearly, f is continuous at x_0.

Case 2b: $n \in \mathbb{N}$. Let $\{e_1, \ldots, e_n\} \subseteq M$ be a basis in M such that $\|e_i\| = 1$, $\forall i = 1, \ldots, n$. Then, $K := \mathrm{co}(\{x_0 \pm \delta e_i \mid i = 1, \ldots, n\}) \subseteq \mathrm{co}(\bar{K}) \subseteq C$ since C is convex. $\forall x \in K$, by Proposition 6.43, $x = \sum_{i=1}^n \alpha_i (x_0 + \delta e_i) + \sum_{i=1}^n \beta_i (x_0 - \delta e_i)$ for some $\alpha_i, \beta_i \in [0, 1] \subset \mathbb{R}$, $i = 1, \ldots, n$, with $\sum_{i=1}^n (\alpha_i + \beta_i) = 1$. By the convexity of f, we have

$$f(x) \le \sum_{i=1}^{n} \alpha_i f(x_0 + \delta e_i) + \sum_{i=1}^{n} \beta_i f(x_0 - \delta e_i)$$

$$\le \sum_{i=1}^{n} |f(x_0 + \delta e_i)| + \sum_{i=1}^{n} |f(x_0 - \delta e_i)| =: r_1$$

It is easy to see that $p(m) := \sum_{i=1}^{n} |\alpha_i|$, $\forall m = \sum_{i=1}^{n} \alpha_i e_i \in M$ defines a norm on M. By Theorem 7.38, $\exists \xi \in [1, \infty) \subset \mathbb{R}$ such that $p(m)/\xi \le \|m\| \le \xi p(m)$, $\forall m \in M$. $\forall x \in \mathcal{B}_{\mathcal{X}}(x_0, \delta/\xi) \cap \mathrm{V}(C)$, we have $x = x_0 + \sum_{i=1}^{n} \alpha_i \delta e_i$ for some $\alpha_1, \ldots, \alpha_n \in \mathbb{R}$. Then, $\delta/\xi > \|x - x_0\| \ge p(\sum_{i=1}^{n} \alpha_i \delta e_i)/\xi = \delta/\xi \sum_{i=1}^{n} |\alpha_i|$. Then, we have $\sum_{i=1}^{n} |\alpha_i| < 1$. Then, x can be expressed as a convex combination of vectors in $\{x_0 \pm \delta e_i \mid i = 1, \ldots, n\}$ and $x \in K$. Hence, $\mathcal{B}_{\mathcal{X}}(x_0, \delta/\xi) \cap \mathrm{V}(C) \subseteq K \subseteq C$. Take $r = \delta/\xi + r_1$. It is easy to show that $\mathcal{B}_{\mathbb{R}\times\mathcal{X}}((r, x_0), \delta/\xi) \cap \mathrm{V}([f, C]) \subseteq [f, C]$, by Proposition 8.21. Then, $(r, x_0) \in {}^{\circ}[f, C]$. By Proposition 8.22, f is continuous at x_0.

This completes the proof of the proposition. □

Proposition 8.25 *Let $\mathcal{X}$ be a real normed linear space, $C \subseteq \mathcal{X}$, and $f : C \to \mathbb{R}$. Then, the following statements hold.*

(i) *If $[f, C]$ is closed, then, f is lower semicontinuous.*
(ii) *If C is closed and f is lower semicontinuous, then $[f, C]$ is closed.*

Proof (i) $\forall a \in \mathbb{R}$, $V_a := \{(a, x) \in \mathbb{R} \times \mathcal{X} \mid x \in \mathcal{X}\}$ is closed. Then, $[f, C] \cap V_a = \{(a, x) \in \mathbb{R} \times \mathcal{X} \mid x \in C, f(x) \le a\}$ is closed. Hence, by Proposition 4.13, $T_a := \{x \in C \mid f(x) \le a\}$ is closed. Note that

$$\{x \in C \mid -f(x) < a\} = C \setminus \{x \in C \mid f(x) \le -a\} = C \setminus T_{-a}$$

Clearly, $C \setminus T_{-a}$ is open in the subset topology of C. Then, $-f$ is upper semicontinuous and f is lower semicontinuous.

(ii) $\forall (r_0, x_0) \in \overline{[f, C]}$, by Proposition 4.13, $\exists ((r_n, x_n))_{n=1}^{\infty} \subseteq [f, C]$ such that $\lim_{n\in\mathbb{N}} (r_n, x_n) = (r_0, x_0)$. By Proposition 3.67, we have $\lim_{n\in\mathbb{N}} r_n = r_0$ and $\lim_{n\in\mathbb{N}} x_n = x_0$. By Definition 8.19, we have $(x_n)_{n=1}^{\infty} \subseteq C$ and $r_n \ge f(x_n)$, $\forall n \in \mathbb{N}$. By Proposition 4.13, $x_0 \in \overline{C} = C$.

We will distinguish two exhaustive and mutually exclusive cases: Case 1: there exists infinitely many $n \in \mathbb{N}$ such that $x_n = x_0$; Case 2: there exists only finitely many $n \in \mathbb{N}$ such that $x_n = x_0$. Case 1: there exists infinitely many $n \in \mathbb{N}$ such that $x_n = x_0$. Then, without loss of generality, assume $x_n = x_0$, $\forall n \in \mathbb{N}$. Then, $r_n \ge f(x_0)$, $\forall n \in \mathbb{N}$. Hence, $f(x_0) \le \lim_{n\in\mathbb{N}} r_n = r_0$. Then, $(r_0, x_0) \in [f, C]$.

Case 2: there exists only finitely many $n \in \mathbb{N}$ such that $x_n = x_0$. Then, without loss of generality, assume $x_n \ne x_0$, $\forall n \in \mathbb{N}$. Then, x_0 is an accumulation point of C. By Propositions 3.86 and 3.85 and Definition 3.14, $-f(x_0) \ge$

$\limsup_{x \to x_0}(-f(x)) = -\liminf_{x \to x_0} f(x)$. By Proposition 3.87, we have $f(x_0) \leq \liminf_{x \to x_0} f(x) \leq \liminf_{n \in \mathbb{N}} f(x_n) \leq \liminf_{n \in \mathbb{N}} r_n = r_0$. Then, $(r_0, x_0) \in [f, C]$.

In both cases, we have $(r_0, x_0) \in [f, C]$. Then, $\overline{[f, C]} \subseteq [f, C]$. By Proposition 3.3, $[f, C]$ is closed.

This completes the proof of the proposition. □

Proposition 8.26 *Let $\mathcal{X}$ be a real normed linear space, $C \subseteq \mathcal{X}$ be convex, and $f : C \to \mathbb{R}$ be convex. Assume that $[f, C]$ is closed. Then, f is weakly lower semicontinuous.*

Proof $\forall a \in \mathbb{R}$, $V_a := \{(a, x) \in \mathbb{R} \times \mathcal{X} \mid x \in \mathcal{X}\}$ is closed. Then, $[f, C] \cap V_a = \{(a, x) \in \mathbb{R} \times \mathcal{X} \mid x \in C, f(x) \leq a\}$ is closed. Hence, by Proposition 4.13, $T_a := \{x \in C \mid f(x) \leq a\}$ is closed. Since f is convex, then T_a is convex. By Proposition 8.10, T_a is weakly closed. Note that

$$\{x \in C \mid -f(x) < a\} = C \setminus \{x \in C \mid f(x) \leq -a\} = C \setminus T_{-a}$$

Clearly, $C \setminus T_{-a}$ is weakly open in the subset topology of C. Then, $-f$ is weakly upper semicontinuous and f is weakly lower semicontinuous. This completes the proof of the proposition. □

8.5 Conjugate Convex Functionals

Definition 8.27 Let $\mathcal{X}$ be a real normed linear space, $C \subseteq \mathcal{X}$ be a nonempty convex set, and $f : C \to \mathbb{R}$ be convex. The *conjugate set* C_{conj} is defined as

$$C_{\text{conj}} := \{x_* \in \mathcal{X}^* \mid \sup_{x \in C} (\langle\langle x_*, x \rangle\rangle - f(x)) < +\infty\}$$

and the *functional* $f_{\text{conj}} : C_{\text{conj}} \to \mathbb{R}$ *conjugate to* f is defined by

$$f_{\text{conj}}(x_*) = \sup_{x \in C} (\langle\langle x_*, x \rangle\rangle - f(x)), \qquad \forall x_* \in C_{\text{conj}}$$

We will use a compact notation $[f, C]_{\text{conj}}$ for $\left[f_{\text{conj}}, C_{\text{conj}}\right]$. ◇

In the above definition, f_{conj} takes value in $\mathbb{R}$ since $x_* \in C_{\text{conj}}$ and $C \neq \emptyset$.

Proposition 8.28 *Let $\mathcal{X}$ be a real normed linear space, $C \subseteq \mathcal{X}$ be a nonempty convex set, $f : C \to \mathbb{R}$ be convex. Then, $C_{\text{conj}} \subseteq \mathcal{X}^*$ is convex, $f_{\text{conj}} : C_{\text{conj}} \to \mathbb{R}$ is convex, and $\left[f_{\text{conj}}, C_{\text{conj}}\right] \subseteq \mathbb{R} \times \mathcal{X}^*$ is a closed convex set.*

Proof $\forall x_{*1}, x_{*2} \in C_{\text{conj}}$, $\forall \alpha \in [0, 1] \subset \mathbb{R}$, let $M_i := f_{\text{conj}}(x_{*i}) = \sup_{x \in C} (\langle\langle x_{*i}, x \rangle\rangle - f(x)) \in \mathbb{R}$, $i = 1, 2$. $\forall x \in C$, we have $\langle\langle \alpha x_{*1} + (1 - \alpha) x_{*2}, x \rangle\rangle - f(x) = \alpha (\langle\langle x_{*1}, x \rangle\rangle - f(x)) + (1 - \alpha) (\langle\langle x_{*2}, x \rangle\rangle - f(x)) \leq \alpha M_1 + (1 - \alpha) M_2 <$

$+\infty$. Then, $\sup_{x\in C}(\langle\langle \alpha x_{*1}+(1-\alpha)x_{*2}, x\rangle\rangle - f(x)) \le \alpha M_1+(1-\alpha)M_2$. Hence, $\alpha x_{*1}+(1-\alpha)x_{*2}\in C_{\text{conj}}$ and C_{conj} is convex.

The above implies that $f_{\text{conj}}(\alpha x_{*1}+(1-\alpha)x_{*2}) \le \alpha f_{\text{conj}}(x_{*1}) + (1-\alpha)f_{\text{conj}}(x_{*2})$. Hence, f_{conj} is convex.

By Proposition 8.20, $[f_{\text{conj}}, C_{\text{conj}}]$ is convex. $\forall (s_0, x_{*0}) \in \overline{[f_{\text{conj}}, C_{\text{conj}}]}$, by Proposition 4.13, there exists $((s_k, x_{*k}))_{k=1}^{\infty} \subseteq [f_{\text{conj}}, C_{\text{conj}}]$ such that $\lim_{k\in\mathbb{N}}(s_k, x_{*k}) = (s_0, x_{*0})$. By Proposition 3.67, we have $\lim_{k\in\mathbb{N}} s_k = s_0$ and $\lim_{k\in\mathbb{N}} x_{*k} = x_{*0}$. $\forall x\in C$, $\forall k\in\mathbb{N}$, since $(s_k, x_{*k}) \in [f_{\text{conj}}, C_{\text{conj}}]$, then, $\langle\langle x_{*k}, x\rangle\rangle - f(x) \le f_{\text{conj}}(x_{*k}) \le s_k$. By Propositions 7.72 and 3.66, we have $\langle\langle x_{*0}, x\rangle\rangle - f(x) = \lim_{k\in\mathbb{N}}(\langle\langle x_{*k}, x\rangle\rangle - f(x)) \le \lim_{k\in\mathbb{N}} s_k = s_0 < +\infty$. Hence, $x_{*0}\in C_{\text{conj}}$ and $f_{\text{conj}}(x_{*0}) \le s_0$. Then, $(s_0, x_{*0}) \in [f_{\text{conj}}, C_{\text{conj}}]$. This shows that $\overline{[f_{\text{conj}}, C_{\text{conj}}]} \subseteq [f_{\text{conj}}, C_{\text{conj}}]$. By Proposition 3.3, $[f_{\text{conj}}, C_{\text{conj}}]$ is closed.

This completes the proof of the proposition. □

Geometric interpretation of $f_{\text{conj}} : C_{\text{conj}} \to \mathbb{R}$. Fix $x_*\in C_{\text{conj}}$. $\forall (r, x)\in [f, C]$, we have $f(x)\le r$. Then, $\langle\langle(-1, x_*), (r, x)\rangle\rangle = \langle\langle x_*, x\rangle\rangle - r \le f_{\text{conj}}(x_*)$. It is easy to recognize that $\sup_{(r,x)\in[f,C]}\langle\langle(-1, x_*), (r, x)\rangle\rangle = f_{\text{conj}}(x_*)$. Hence, the closed hyperplane $H := \{(r, x)\in \mathbb{R}\times\mathcal{X} \mid \langle\langle(-1, x_*), (r, x)\rangle\rangle = f_{\text{conj}}(x_*)\}$ is a supporting hyperplane of $[f, C]$.

Proposition 8.29 *Let $\mathcal{X}$ be a real normed linear space, $C\subseteq\mathcal{X}$ be a nonempty convex set, and $f : C\to\mathbb{R}$ be a convex functional. Assume that $[f, C] =: K \subseteq \mathbb{R}\times\mathcal{X}$ is closed. Then, $\exists x_{*0}\in\mathcal{X}^*$ such that*

$$\sup_{(r,x)\in K}\langle\langle(-1, x_{*0}), (r, x)\rangle\rangle < +\infty$$

Then, $x_{*0}\in C_{\text{conj}} \neq \emptyset$.

Proof Fix $x_0\in C\neq\emptyset$. By Proposition 8.20, K is convex. Clearly, $K\neq\emptyset$. By the assumption of the proposition, K is closed. Let $r_0 := f(x_0) - 1 \in\mathbb{R}$. Then, $(r_0, x_0)\notin K$. By Proposition 8.10 and Example 7.76, $\exists(\bar{s}_0, \bar{x}_{*0})\in\mathbb{R}\times\mathcal{X}^*$ such that

$$\langle\langle(\bar{s}_0, \bar{x}_{*0}), (r_0, x_0)\rangle\rangle < \inf_{(r,x)\in K}\langle\langle(\bar{s}_0, \bar{x}_{*0}), (r, x)\rangle\rangle \tag{8.1}$$

Since $(f(x_0), x_0)\in K$, then $\bar{s}_0 r_0 + \langle\langle\bar{x}_{*0}, x_0\rangle\rangle < \bar{s}_0 f(x_0) + \langle\langle\bar{x}_{*0}, x_0\rangle\rangle$, which implies that $\bar{s}_0 > 0$. Let $x_{*0} = -\bar{s}_0^{-1}\bar{x}_{*0}\in\mathcal{X}^*$. Then, (8.1) is equivalent to

$$\langle\langle(-1, x_{*0}), (r_0, x_0)\rangle\rangle > \sup_{(r,x)\in K}\langle\langle(-1, x_{*0}), (r, x)\rangle\rangle$$

The above implies that $\langle\langle(-1, x_{*0}), (r_0, x_0)\rangle\rangle > \sup_{x\in C}(\langle\langle x_{*0}, x\rangle\rangle - f(x))$. Hence, $x_{*0}\in C_{\text{conj}}\neq\emptyset$.

This completes the proof of the proposition. □

Proposition 8.30 *Let $\mathcal{X}$ be a real normed linear space and $K\subseteq\mathbb{R}\times\mathcal{X} =: \mathcal{W}$ be a closed convex set.*

Assume that there exists a nonvertical hyperplane such that K is contained in one of the half-spaces associated with the hyperplane, that is $\exists (s_1, x_{*1}) \in \mathcal{W}^* = \mathbb{R} \times \mathcal{X}^*$ with $s_1 \neq 0$ such that $\sup_{(r,x) \in K} \langle\langle (s_1, x_{*1}), (r, x) \rangle\rangle < +\infty$.

Then, $\forall (r_0, x_0) \in \mathcal{W} \setminus K$, there exists a nonvertical hyperplane separating (r_0, x_0) and K, that is, $\exists x_{*0} \in \mathcal{X}^*$ such that either

$$\sup_{(r,x) \in K} \langle\langle (-1, x_{*0}), (r, x) \rangle\rangle < \langle\langle (-1, x_{*0}), (r_0, x_0) \rangle\rangle \tag{8.2a}$$

or

$$\inf_{(r,x) \in K} \langle\langle (-1, x_{*0}), (r, x) \rangle\rangle > \langle\langle (-1, x_{*0}), (r_0, x_0) \rangle\rangle \tag{8.2b}$$

Proof Fix $(r_0, x_0) \in \mathcal{W} \setminus K$. By Proposition 8.10, $\exists (\bar{s}_0, \bar{x}_{*0}) \in \mathbb{R} \times \mathcal{X}^*$ such that

$$\langle\langle (\bar{s}_0, \bar{x}_{*0}), (r_0, x_0) \rangle\rangle < \inf_{(r,x) \in K} \langle\langle (\bar{s}_0, \bar{x}_{*0}), (r, x) \rangle\rangle \tag{8.3}$$

We will distinguish three exhaustive and mutually exclusive cases: Case 1: $\bar{s}_0 > 0$; Case 2: $\bar{s}_0 < 0$; Case 3: $\bar{s}_0 = 0$. Case 1: $\bar{s}_0 > 0$. Let $x_{*0} = -\bar{s}_0^{-1} \bar{x}_{*0} \in \mathcal{X}^*$. Then, (8.3) is equivalent to

$$-r_0 + \langle\langle x_{*0}, x_0 \rangle\rangle > \sup_{(r,x) \in K} (-r + \langle\langle x_{*0}, x \rangle\rangle)$$

Hence, (8.2a) holds.

Case 2: $\bar{s}_0 < 0$. Let $x_{*0} = -\bar{s}_0^{-1} \bar{x}_{*0} \in \mathcal{X}^*$. Then, (8.3) is equivalent to

$$-r_0 + \langle\langle x_{*0}, x_0 \rangle\rangle < \inf_{(r,x) \in K} (-r + \langle\langle x_{*0}, x \rangle\rangle)$$

Hence, (8.2b) holds.

Case 3: $\bar{s}_0 = 0$. We will further distinguish two exhaustive and mutually exclusive cases: Case 3a: $K = \emptyset$; Case 3b: $K \neq \emptyset$. Case 3a: $K = \emptyset$. Take $x_{*0} = \vartheta_{\mathcal{X}^*}$. Then,

$$-r_0 > -\infty = \sup_{(r,x) \in K} \langle\langle (-1, x_{*0}), (r, x) \rangle\rangle$$

Hence, (8.2a) holds.

Case 3b: $K \neq \emptyset$. Define

$$M_1 := \sup_{(r,x)\in K} \langle\langle(s_1, x_{*1}), (r, x)\rangle\rangle - \langle\langle(s_1, x_{*1}), (r_0, x_0)\rangle\rangle \in \mathbb{R}$$

$$M_2 := \inf_{(r,x)\in K} \langle\langle(\bar{s}_0, \bar{x}_{*0}), (r, x)\rangle\rangle - \langle\langle(\bar{s}_0, \bar{x}_{*0}), (r_0, x_0)\rangle\rangle \in (0, \infty) \subset \mathbb{R}$$

Let $\delta := M_2/(1 + |M_1|) \in (0, \infty) \subset \mathbb{R}$ and $(\tilde{s}_0, \tilde{x}_{*0}) = (\bar{s}_0, \bar{x}_{*0}) - \delta(s_1, x_{*1}) \in \mathbb{R} \times \mathcal{X}^*$. Then, we have

$$\begin{aligned}
&\inf_{(r,x)\in K} \langle\langle(\tilde{s}_0, \tilde{x}_{*0}), (r, x)\rangle\rangle - \langle\langle(\tilde{s}_0, \tilde{x}_{*0}), (r_0, x_0)\rangle\rangle \\
&= \inf_{(r,x)\in K} (\langle\langle(\bar{s}_0, \bar{x}_{*0}), (r, x)\rangle\rangle - \delta\langle\langle(s_1, x_{*1}), (r, x)\rangle\rangle) \\
&\quad -\langle\langle(\tilde{s}_0, \tilde{x}_{*0}), (r_0, x_0)\rangle\rangle \\
&\geq \inf_{(r,x)\in K} \langle\langle(\bar{s}_0, \bar{x}_{*0}), (r, x)\rangle\rangle + \inf_{(r,x)\in K} (-\delta\langle\langle(s_1, x_{*1}), (r, x)\rangle\rangle) \\
&\quad -\langle\langle(\tilde{s}_0, \tilde{x}_{*0}), (r_0, x_0)\rangle\rangle \\
&= \inf_{(r,x)\in K} \langle\langle(\bar{s}_0, \bar{x}_{*0}), (r, x)\rangle\rangle - \delta \sup_{(r,x)\in K} \langle\langle(s_1, x_{*1}), (r, x)\rangle\rangle \\
&\quad -\langle\langle(\tilde{s}_0, \tilde{x}_{*0}), (r_0, x_0)\rangle\rangle \\
&= M_2 - \delta M_1 > 0
\end{aligned}$$

where we have applied Proposition 3.81 in the above. Hence, we have obtained an alternative $(\bar{s}_0, \bar{x}_{*0}) := (\tilde{s}_0, \tilde{x}_{*0})$ such that (8.3) holds. For this alternative pair, $\tilde{s}_0 = -\delta s_1 \neq 0$. Hence, this case can be solved by Case 1 or Case 2 with the alternative pair.

This completes the proof of the proposition. □

Definition 8.31 Let $\mathcal{X}$ be a real normed linear space, $\Gamma \subseteq \mathcal{X}^*$ be a nonempty convex set, and $\varphi : \Gamma \to \mathbb{R}$ be convex. The *pre-conjugate set* ${}_{\text{conj}}\Gamma$ is defined as

$${}_{\text{conj}}\Gamma := \{x \in \mathcal{X} \mid \sup_{x_*\in\Gamma} (\langle\langle x_*, x\rangle\rangle - \varphi(x_*)) < +\infty\}$$

and the *functional* ${}_{\text{conj}}\varphi : {}_{\text{conj}}\Gamma \to \mathbb{R}$ *pre-conjugate to* φ is defined by

$${}_{\text{conj}}\varphi(x) = \sup_{x_*\in\Gamma} (\langle\langle x_*, x\rangle\rangle - \varphi(x_*)), \qquad \forall x \in {}_{\text{conj}}\Gamma$$

We will use a compact notation ${}_{\text{conj}}[\varphi, \Gamma]$ for $\left[{}_{\text{conj}}\varphi, {}_{\text{conj}}\Gamma\right]$. ◇

In the above definition, ${}_{\text{conj}}\varphi$ takes value in $\mathbb{R}$ since $x \in {}_{\text{conj}}\Gamma$ and $\Gamma \neq \emptyset$.

Next, we state three duality results regarding conjugate convex functionals.

Proposition 8.32 *Let $\mathcal{X}$ be a real normed linear space, $C \subseteq \mathcal{X}$ be a nonempty convex set, and $f : C \to \mathbb{R}$ be a convex functional. Assume that $[f, C]$ is closed. Then, $[f, C] = {}_{\text{conj}}\big[[f, C]_{\text{conj}}\big]$. Therefore, $\forall x_0 \in C$, $f(x_0) = \sup_{x_* \in C_{\text{conj}}}(\langle\langle x_*, x_0\rangle\rangle - f_{\text{conj}}(x_*))$.*

Proof Since $C \neq \emptyset$, then C_{conj} and f_{conj} are well-defined. By Proposition 8.29, $C_{\text{conj}} \neq \emptyset$. By Proposition 8.28, C_{conj} is convex and f_{conj} is convex. Then, ${}_{\text{conj}}(C_{\text{conj}})$ and ${}_{\text{conj}}(f_{\text{conj}})$ are well-defined. Thus, ${}_{\text{conj}}\big[[f, C]_{\text{conj}}\big]$ is a well-defined set.

We will first show that $[f, C] \subseteq {}_{\text{conj}}\big[f_{\text{conj}}, C_{\text{conj}}\big] = {}_{\text{conj}}\big[[f, C]_{\text{conj}}\big]$. Fix any $(r, x) \in [f, C]$. $\forall x_* \in C_{\text{conj}}$, we have $f_{\text{conj}}(x_*) \geq \langle\langle x_*, x\rangle\rangle - f(x)$. Hence, we have $r \geq f(x) \geq \langle\langle x_*, x\rangle\rangle - f_{\text{conj}}(x_*)$. Thus, we have

$$r \geq \sup_{x_* \in C_{\text{conj}}} \langle\langle x_*, x\rangle\rangle - f_{\text{conj}}(x_*) = {}_{\text{conj}}\big(f_{\text{conj}}\big)(x)$$

Hence, $(r, x) \in {}_{\text{conj}}\big[f_{\text{conj}}, C_{\text{conj}}\big]$ and $[f, C] \subseteq {}_{\text{conj}}\big[f_{\text{conj}}, C_{\text{conj}}\big]$.

On the other hand, fix any $(r_0, x_0) \in (\mathbb{R} \times \mathcal{X}) \setminus [f, C]$. By Proposition 8.29, $\exists \bar{x}_{*0} \in \mathcal{X}^*$ such that

$$\sup_{(r,x) \in [f,C]} \langle\langle(-1, \bar{x}_{*0}), (r, x)\rangle\rangle < +\infty$$

Note that $[f, C]$ is convex, by Proposition 8.20. By Proposition 8.30, there exists a nonvertical hyperplane separating (r_0, x_0) and $K := [f, C]$, that is, $\exists x_{*0} \in \mathcal{X}^*$ such that either

$$\sup_{(r,x) \in K} \langle\langle(-1, x_{*0}), (r, x)\rangle\rangle < \langle\langle(-1, x_{*0}), (r_0, x_0)\rangle\rangle \tag{8.4a}$$

or

$$\inf_{(r,x) \in K} \langle\langle(-1, x_{*0}), (r, x)\rangle\rangle > \langle\langle(-1, x_{*0}), (r_0, x_0)\rangle\rangle \tag{8.4b}$$

Since K is the epigraph of f, then (8.4b) is impossible since the left-hand side equals to $-\infty$. Therefore, (8.4a) must hold. Then, $c := \sup_{(r,x) \in K} \langle\langle(-1, x_{*0}), (r, x)\rangle\rangle = \sup_{x \in C}(\langle x_{*0}, x\rangle - f(x)) = f_{\text{conj}}(x_{*0}) \in \mathbb{R}$ and $x_{*0} \in C_{\text{conj}}$. Then, $(c, x_{*0}) \in \big[f_{\text{conj}}, C_{\text{conj}}\big]$. But, $c < -r_0 + \langle\langle x_{*0}, x_0\rangle\rangle$ implies that $r_0 < \langle\langle x_{*0}, x_0\rangle\rangle - f_{\text{conj}}(x_{*0}) \leq \sup_{x_* \in C_{\text{conj}}} \langle\langle x_*, x_0\rangle\rangle - f_{\text{conj}}(x_*)$. Therefore, $(r_0, x_0) \notin {}_{\text{conj}}\big[f_{\text{conj}}, C_{\text{conj}}\big]$. Hence, ${}_{\text{conj}}\big[f_{\text{conj}}, C_{\text{conj}}\big] \subseteq [f, C]$.

This yields $[f, C] = {}_{\text{conj}}\big[[f, C]_{\text{conj}}\big]$, which implies that f and ${}_{\text{conj}}\big(f_{\text{conj}}\big)$ admit the same domain of definition and equal to each other on C. Therefore, $\forall x_0 \in C$, $f(x_0) = {}_{\text{conj}}\big(f_{\text{conj}}\big)(x_0) = \sup_{x_* \in C_{\text{conj}}}(\langle\langle x_*, x_0\rangle\rangle - f_{\text{conj}}(x_*))$. □

Proposition 8.33 *Let $\mathcal{X}$ be a real normed linear space, $C \subseteq \mathcal{X}$ be a nonempty convex set, $f : C \to \mathbb{R}$ be a convex functional, f be lower semicontinuous at $x_0 \in C$, $C_{\text{conj}} \subseteq \mathcal{X}^*$ be the conjugate set, and $f_{\text{conj}} : C_{\text{conj}} \to \mathbb{R}$ be the conjugate functional. Then, $f(x_0) = \sup_{x_* \in C_{\text{conj}}}(\langle\langle x_*, x_0\rangle\rangle - f_{\text{conj}}(x_*))$.*

Proof $\forall x_* \in C_{\text{conj}}$, we have $\langle\langle x_*, x_0\rangle\rangle - f_{\text{conj}}(x_*) = \langle\langle x_*, x_0\rangle\rangle + \inf_{x\in C}(f(x) - \langle\langle x_*, x\rangle\rangle) \leq f(x_0)$. Then,

$$f(x_0) \geq \sup_{x_* \in C_{\text{conj}}} (\langle\langle x_*, x_0\rangle\rangle - f_{\text{conj}}(x_*))$$

$\forall \epsilon \in (0, \infty) \subset \mathbb{R}$, by the lower semicontinuity of f at x_0, $\exists \delta \in (0, \epsilon/2] \subset \mathbb{R}$ such that $f(x) > f(x_0) - \epsilon/2$, $\forall x \in \mathcal{B}_{\mathcal{X}}(x_0, \delta) \cap C$. Consider the real normed linear space $\mathcal{W} := \mathbb{R} \times \mathcal{X}$. Define $K_2 := [f, C]$, which is clearly nonempty. By Proposition 8.20, K_2 is convex. Define $K_1 := \mathcal{B}_{\mathcal{W}}((f(x_0) - \epsilon, x_0), \delta)$. Then, K_1 is convex and $K_1^\circ = K_1 \neq \emptyset$.

Claim 8.33.1 $K_1^\circ \cap K_2 = \emptyset$.

Proof of Claim Suppose $\exists (r, x) \in K_1^\circ \cap K_2$. Then, $x \in \mathcal{B}_{\mathcal{X}}(x_0, \delta) \cap C$, $r < f(x_0) - \epsilon + \delta \leq f(x_0) - \epsilon/2$ and $r \geq f(x)$. This leads to $f(x) < f(x_0) - \epsilon/2$, which is a contradiction. Hence, $K_1^\circ \cap K_2 = \emptyset$. This completes the proof of the claim. □

By Eidelheit Separation Theorem and Example 7.76, $\exists c \in \mathbb{R}$ and $\exists (\bar{s}_0, \bar{x}_{*0}) \in \mathcal{W}^* = \mathbb{R} \times \mathcal{X}^*$ with $(\bar{s}_0, \bar{x}_{*0}) \neq (0, \vartheta_{\mathcal{X}^*})$ such that

$$\sup_{(r,x)\in K_1} \langle\langle (\bar{s}_0, \bar{x}_{*0}), (r, x)\rangle\rangle \leq c \leq \inf_{(r,x)\in K_2} \langle\langle (\bar{s}_0, \bar{x}_{*0}), (r, x)\rangle\rangle \tag{8.5}$$

The second inequality in (8.5) implies that $-\infty < c \leq \inf_{(r,x)\in K_2}(\bar{s}_0 r + \langle\langle \bar{x}_{*0}, x\rangle\rangle$. Since $K_2 = [f, C]$, then $\bar{s}_0 \geq 0$.

Claim 8.33.2 $\bar{s}_0 > 0$.

Proof of Claim Suppose $\bar{s}_0 = 0$. By the fact that $(\bar{s}_0, \bar{x}_{*0}) \neq (0, \vartheta_{\mathcal{X}^*})$, we have $\bar{x}_{*0} \neq \vartheta_{\mathcal{X}^*}$. Then, (8.5) implies that $\langle\langle \bar{x}_{*0}, x_0\rangle\rangle \geq c \geq \sup_{(r,x)\in K_1} \langle\langle \bar{x}_{*0}, x\rangle\rangle = \sup_{x\in\mathcal{B}_{\mathcal{X}}(x_0,\delta)} \langle\langle \bar{x}_{*0}, x\rangle\rangle$. This is impossible since $\bar{x}_{*0} \neq \vartheta_{\mathcal{X}^*}$. Therefore, $\bar{s}_0 > 0$. This completes the proof of the claim. □

Let $x_{*0} := -\bar{s}_0^{-1}\bar{x}_{*0} \in \mathcal{X}^*$. Then, (8.5) is equivalent to

$$\begin{aligned}\inf_{(r,x)\in K_1} (\langle\langle x_{*0}, x\rangle\rangle - r) \geq -\bar{s}_0^{-1} c &\geq \sup_{(r,x)\in K_2} (\langle\langle x_{*0}, x\rangle\rangle - r) \\ &= \sup_{x\in C}(\langle\langle x_{*0}, x\rangle\rangle - f(x))\end{aligned}$$

Hence, $x_{*0} \in C_{\text{conj}}$ and $f_{\text{conj}}(x_{*0}) \leq \langle\langle x_{*0}, x_0 \rangle\rangle - f(x_0) + \epsilon$. Therefore, we have

$$\langle\langle x_{*0}, x_0 \rangle\rangle - f_{\text{conj}}(x_{*0}) \geq f(x_0) - \epsilon$$

By the arbitrariness of ϵ, $f(x_0) = \sup_{x_* \in C_{\text{conj}}}(\langle\langle x_*, x_0 \rangle\rangle - f_{\text{conj}}(x_*))$.

This completes the proof of the proposition. □

Proposition 8.34 *Let $\mathcal{X}$ be a real normed linear space, $C \subseteq \mathcal{X}$ be a nonempty convex set, $C^\circ \neq \emptyset$, $f : C \to \mathbb{R}$ be a convex functional, f be continuous at $\bar{x} \in C^\circ$, $C_{\text{conj}} \subseteq \mathcal{X}^*$ be the conjugate set, and $f_{\text{conj}} : C_{\text{conj}} \to \mathbb{R}$ be the conjugate functional. Then, $\forall x_0 \in C^\circ$, $f(x_0) = \max_{x_* \in C_{\text{conj}}}(\langle\langle x_*, x_0 \rangle\rangle - f_{\text{conj}}(x_*))$.*

Proof $\forall x_* \in C_{\text{conj}}$, we have $\langle\langle x_*, x_0 \rangle\rangle - f_{\text{conj}}(x_*) = \langle\langle x_*, x_0 \rangle\rangle + \inf_{x \in C}(f(x) - \langle\langle x_*, x \rangle\rangle) \leq f(x_0)$. Then,

$$f(x_0) \geq \sup_{x_* \in C_{\text{conj}}} (\langle\langle x_*, x_0 \rangle\rangle - f_{\text{conj}}(x_*))$$

Consider the real normed linear space $\mathcal{W} := \mathbb{R} \times \mathcal{X}$. Let $V = \{(f(x_0), x_0)\} \subseteq \mathcal{W}$. Clearly, V is a linear variety. Define $K := [f, C]$. By Proposition 8.20, K is convex. By Proposition 8.22, K admits relative interior point $(\bar{r}, \bar{x})$ for some $\bar{r} > f(\bar{x})$. By Proposition 8.21, $\text{V}([f, C]) = \mathbb{R} \times \text{V}(C) = \mathbb{R} \times \mathcal{X} = \mathcal{W}$ since $C^\circ \neq \emptyset$. Hence, $(\bar{r}, \bar{x}) \in K^\circ \neq \emptyset$. Note that $(f(x_0) - \delta, x_0) \notin [f, C]$, $\forall \delta \in (0, \infty) \subset \mathbb{R}$. Then, $(f(x_0), x_0) \notin K^\circ$. Hence, $V \cap K^\circ = \emptyset$. By Mazur's Theorem and Example 7.76, $\exists c \in \mathbb{R}$ and $\exists (\bar{s}_0, \bar{x}_{*0}) \in \mathcal{W}^* = \mathbb{R} \times \mathcal{X}^*$ with $(\bar{s}_0, \bar{x}_{*0}) \neq (0, \vartheta_{\mathcal{X}^*})$ such that

$$\langle\langle (\bar{s}_0, \bar{x}_{*0}), (f(x_0), x_0) \rangle\rangle = c \geq \sup_{(r,x) \in K} \langle\langle (\bar{s}_0, \bar{x}_{*0}), (r, x) \rangle\rangle$$

which is equivalent to

$$\bar{s}_0 f(x_0) + \langle\langle \bar{x}_{*0}, x_0 \rangle\rangle \geq \sup_{(r,x) \in K} (\bar{s}_0 r + \langle\langle \bar{x}_{*0}, x \rangle\rangle) \tag{8.6}$$

Since $K = [f, C]$, then $\bar{s}_0 \leq 0$, otherwise the right-hand side of (8.6) equals to $+\infty$.

Claim 8.34.1 $\bar{s}_0 < 0$.

Proof of Claim Suppose $\bar{s}_0 = 0$. By the fact that $(\bar{s}_0, \bar{x}_{*0}) \neq (0, \vartheta_{\mathcal{X}^*})$, we have $\bar{x}_{*0} \neq \vartheta_{\mathcal{X}^*}$. Then, (8.6) implies that $\langle\langle \bar{x}_{*0}, x_0 \rangle\rangle \geq \sup_{x \in C} \langle\langle \bar{x}_{*0}, x \rangle\rangle$. Note that $x_0 \in C^\circ$ and $\bar{x}_{*0} \neq \vartheta_{\mathcal{X}^*}$ implies that $\sup_{x \in C} \langle\langle \bar{x}_{*0}, x \rangle\rangle > \langle\langle \bar{x}_{*0}, x_0 \rangle\rangle$. This is a contradiction. Therefore, $\bar{s}_0 < 0$. This completes the proof of the claim. □

Let $x_{*0} := |\bar{s}_0|^{-1} \bar{x}_{*0} \in \mathcal{X}^*$. Then, (8.6) is equivalent to

$$- f(x_0) + \langle\langle x_{*0}, x_0 \rangle\rangle \geq \sup_{(r,x) \in K} (-r + \langle\langle x_{*0}, x \rangle\rangle) = \sup_{x \in C}(\langle\langle x_{*0}, x \rangle\rangle - f(x))$$

Hence, $x_{*0} \in C_{\text{conj}}$ and $-f(x_0) + \langle\langle x_{*0}, x_0 \rangle\rangle \geq f_{\text{conj}}(x_{*0})$.

Therefore, we have

$$f(x_0) \le \langle\langle x_{*0}, x_0\rangle\rangle - f_{\text{conj}}(x_{*0}) \le \sup_{x_* \in C_{\text{conj}}} (\langle\langle x_*, x_0\rangle\rangle - f_{\text{conj}}(x_*)) \le f(x_0)$$

Hence, $f(x_0) = \max_{x_* \in C_{\text{conj}}}(\langle\langle x_*, x_0\rangle\rangle - f_{\text{conj}}(x_*))$, where the maximum is achieved at x_{*0}. □

8.6 Fenchel Duality Theorem

Let $\mathcal{X}$ be a real normed linear space, $C_1, C_2 \subseteq \mathcal{X}$ be nonempty convex sets, and $f_1 : C_1 \to \mathbb{R}$ and $f_2 : C_2 \to \mathbb{R}$ be convex functionals. We consider the problem of

$$\mu := \inf_{x \in C_1 \cap C_2} (f_1(x) + f_2(x))$$

We assume that $f_{1\text{conj}}$, $C_{1\text{conj}}$, $f_{2\text{conj}}$, $C_{2\text{conj}}$ are easily characterized yet $(f_1 + f_2)_{\text{conj}}$ and $(C_1 \cap C_2)_{\text{conj}}$ are difficult to determine. Then, the above infimum can be equivalently calculated by

$$\mu = \inf_{(r_1,x)\in[f_1,C_1],(r_2,x)\in[f_2,C_2]} (r_1 + r_2)$$

The idea of Fenchel Duality Theorem can be illustrated in Fig. 8.1.

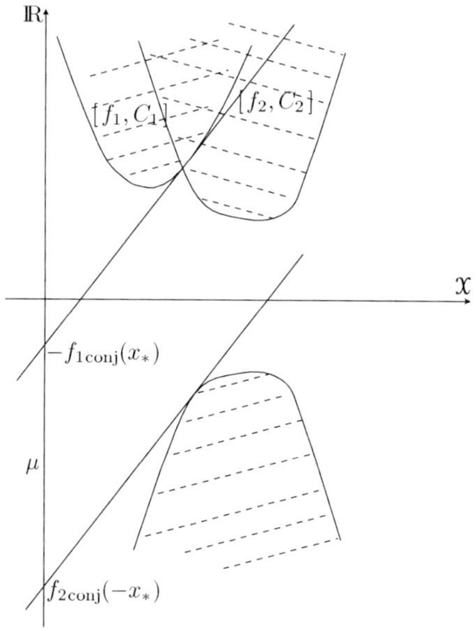

Fig. 8.1 Fenchel duality

Theorem 8.35 (Fenchel Duality Theorem) *Let $\mathcal{X}$ be a real normed linear space, $C_1, C_2 \subseteq \mathcal{X}$ be nonempty convex sets, $f_1 : C_1 \to \mathbb{R}$ and $f_2 : C_2 \to \mathbb{R}$ be convex functionals, and $C_2 \cap C_1^\circ \neq \emptyset$. Assume that f_1 is continuous at $\bar{x} \in C_1^\circ$ and $\mu := \inf_{x \in C_1 \cap C_2}(f_1(x) + f_2(x)) \in \mathbb{R}$. Let $f_{1\text{conj}} : C_{1\text{conj}} \to \mathbb{R}$ and $f_{2\text{conj}} : C_{2\text{conj}} \to \mathbb{R}$ be conjugate functionals of f_1 and f_2, respectively. Then, the following statements hold.*

(i) $\mu = \inf_{x \in C_1 \cap C_2}(f_1(x) + f_2(x)) = \max_{x_* \in C_{1\text{conj}} \cap (-C_{2\text{conj}})}(-f_{1\text{conj}}(x_*) - f_{2\text{conj}}(-x_*))$, *where the maximum is achieved at some $x_{*0} \in C_{1\text{conj}} \cap (-C_{2\text{conj}})$. If the infimum is achieved by some $x_0 \in C_1 \cap C_2$, then, we have*

$$f_{1\text{conj}}(x_{*0}) = \langle\langle x_{*0}, x_0\rangle\rangle - f_1(x_0) \tag{8.7a}$$

$$f_{2\text{conj}}(-x_{*0}) = \langle\langle -x_{*0}, x_0\rangle\rangle - f_2(x_0) \tag{8.7b}$$

(ii) *If there exists $x_{*0} \in C_{1\text{conj}} \cap (-C_{2\text{conj}})$ and $x_0 \in C_1 \cap C_2$ such that* (8.7) *holds, then the infimum is achieved at x_0 and the maximum is achieved at x_{*0}.*

Proof $\forall x_* \in C_{1\text{conj}} \cap (-C_{2\text{conj}})$, we have

$$\begin{aligned} -f_{1\text{conj}}(x_*) &= -\sup_{x \in C_1}(\langle\langle x_*, x\rangle\rangle - f_1(x)) = \inf_{x \in C_1}(-\langle\langle x_*, x\rangle\rangle + f_1(x)) \\ &\leq \inf_{x \in C_1 \cap C_2}(-\langle\langle x_*, x\rangle\rangle + f_1(x)) \\ -f_{2\text{conj}}(-x_*) &= -\sup_{x \in C_2}(\langle\langle -x_*, x\rangle\rangle - f_2(x)) \\ &= \inf_{x \in C_2}(\langle\langle x_*, x\rangle\rangle + f_2(x)) \leq \inf_{x \in C_1 \cap C_2}(\langle\langle x_*, x\rangle\rangle + f_2(x)) \end{aligned}$$

Then, we have

$$\begin{aligned} -f_{1\text{conj}}(x_*) - f_{2\text{conj}}(-x_*) &\leq \inf_{x \in C_1 \cap C_2}(-\langle\langle x_*, x\rangle\rangle + f_1(x)) \\ &+ \inf_{x \in C_1 \cap C_2}(\langle\langle x_*, x\rangle\rangle + f_2(x)) \leq \inf_{x \in C_1 \cap C_2}(f_1(x) + f_2(x)) = \mu \end{aligned} \tag{8.8}$$

(i) Consider the sets $K_1 := [f_1 - \mu, C_1]$ and $K_2 := \{(r, x) \in \mathbb{R} \times \mathcal{X} \mid x \in C_2, r \leq -f_2(x)\} = \{(r, x) \in \mathbb{R} \times \mathcal{X} \mid (-r, x) \in [f_2, C_2]\}$. By Proposition 8.20, K_1 and K_2 are nonempty convex sets. By Proposition 8.22, K_1 admits relative interior point $(\bar{r}, \bar{x})$ for some $\bar{r} \in \mathbb{R}$. By Proposition 8.21, $\mathrm{V}([f_1 - \mu, C_1]) = \mathbb{R} \times \mathrm{V}(C_1) = \mathbb{R} \times \mathcal{X}$. Then, $(\bar{r}, \bar{x}) \in K_1^\circ \neq \emptyset$.

We will show that $K_1^\circ \cap K_2 = \emptyset$ by an argument of contradiction. Suppose $(r, x) \in K_1^\circ \cap K_2 \neq \emptyset$. Then, we have $(r, x) \in K_1^\circ$, which implies that $x \in C_1$ and $r > f_1(x) - \mu$; and $(r, x) \in K_2$, which implies that $x \in C_2$ and $r \leq -f_2(x)$. Then, $x \in C_1 \cap C_2$ and $f_1(x) + f_2(x) < \mu$. This contradicts with the definition of μ. Hence, $K_1^\circ \cap K_2 = \emptyset$.

By Eidelheit Separation Theorem and Example 7.76, $\exists(\bar{s}_0, \bar{x}_{*0}) \in \mathbb{R} \times \mathcal{X}^*$ with $(\bar{s}_0, \bar{x}_{*0}) \neq \vartheta_{\mathbb{R}\times\mathcal{X}^*}$ such that

$$-\infty < \sup_{(r,x)\in K_1} (\langle\langle \bar{x}_{*0}, x\rangle\rangle + r\bar{s}_0) \leq \inf_{(r,x)\in K_2} (\langle\langle \bar{x}_{*0}, x\rangle\rangle + r\bar{s}_0) < +\infty \tag{8.9}$$

We will show that $\bar{s}_0 < 0$ by an argument of contradiction. Suppose $\bar{s}_0 \geq 0$. We will distinguish two exhaustive and mutually exclusive cases: Case 1: $\bar{s}_0 > 0$; Case 2: $\bar{s}_0 = 0$. Case 1: $\bar{s}_0 > 0$. Then, $\sup_{(r,x)\in K_1} (\langle\langle \bar{x}_{*0}, x\rangle\rangle + r\bar{s}_0) = +\infty$, which contradicts (8.9). Case 2: $\bar{s}_0 = 0$. Then, $\bar{x}_{*0} \neq \vartheta_{\mathcal{X}^*}$. By (8.9), we have

$$-\infty < \sup_{x\in C_1} \langle\langle \bar{x}_{*0}, x\rangle\rangle \leq \inf_{x\in C_2} \langle\langle \bar{x}_{*0}, x\rangle\rangle < +\infty$$

Let $\acute{x} \in C_1^\circ \cap C_2 \neq \emptyset$. Then, $\sup_{x\in C_1} \langle\langle \bar{x}_{*0}, x\rangle\rangle \leq \langle\langle \bar{x}_{*0}, \acute{x}\rangle\rangle$. This is not possible since $\acute{x} \in C_1^\circ$ and $\bar{x}_{*0} \neq \vartheta_{\mathcal{X}^*}$. Hence, we have a contradiction in both cases. Therefore, $\bar{s}_0 < 0$.

Let $x_{*0} = |\bar{s}_0|^{-1} \bar{x}_{*0}$. Then, (8.9) is equivalent to

$$-\infty < \sup_{(r,x)\in K_1} (\langle\langle x_{*0}, x\rangle\rangle - r) \leq \inf_{(r,x)\in K_2} (\langle\langle x_{*0}, x\rangle\rangle - r) < +\infty \tag{8.10}$$

Note that

$$\begin{aligned}
+\infty > \sup_{(r,x)\in K_1} (\langle\langle x_{*0}, x\rangle\rangle - r) &= \sup_{x\in C_1} (\langle\langle x_{*0}, x\rangle\rangle - f_1(x) + \mu) \\
&\geq \sup_{x\in C_1\cap C_2} (\langle\langle x_{*0}, x\rangle\rangle - f_1(x)) + \mu =: d_1 > -\infty
\end{aligned}$$

Hence, $x_{*0} \in C_{1\text{conj}}$. Note also that

$$\begin{aligned}
-\infty &< \inf_{(r,x)\in K_2} (\langle\langle x_{*0}, x\rangle\rangle - r) = \inf_{x\in C_2} (\langle\langle x_{*0}, x\rangle\rangle + f_2(x)) \\
&= -\sup_{x\in C_2} (\langle\langle -x_{*0}, x\rangle\rangle - f_2(x)) \leq \inf_{x\in C_1\cap C_2} (\langle\langle x_{*0}, x\rangle\rangle + f_2(x)) \\
&=: d_2 < +\infty
\end{aligned}$$

Hence, $x_{*0} \in -C_{2\text{conj}}$. By (8.10), $d_1 \leq d_2$. On the other hand, we have

$$\begin{aligned}
&d_2 - d_1 \\
&= \inf_{x\in C_1\cap C_2} (\langle\langle x_{*0}, x\rangle\rangle + f_2(x)) + \inf_{x\in C_1\cap C_2} (-\langle\langle x_{*0}, x\rangle\rangle + f_1(x)) - \mu \\
&\leq \inf_{x\in C_1\cap C_2} (f_1(x) + f_2(x)) - \mu = 0
\end{aligned}$$

Hence, we have

$$d_1 = \sup_{x \in C_1} (\langle\langle x_{*0}, x \rangle\rangle - f_1(x) + \mu) = - \sup_{x \in C_2} (\langle\langle -x_{*0}, x \rangle\rangle - f_2(x)) = d_2$$

This implies that $d_1 = f_{1\text{conj}}(x_{*0}) + \mu = -f_{2\text{conj}}(-x_{*0}) = d_2$. Therefore, we have $\mu = -f_{1\text{conj}}(x_{*0}) - f_{2\text{conj}}(-x_{*0})$ and $x_{*0} \in C_{1\text{conj}} \cap (-C_{2\text{conj}})$.

This coupled with (8.8) implies

$$\mu = \inf_{x \in C_1 \cap C_2} (f_1(x) + f_2(x)) = \max_{x_* \in C_{1\text{conj}} \cap (-C_{2\text{conj}})} (-f_{1\text{conj}}(x_*) - f_{2\text{conj}}(-x_*))$$

where the maximum is achieved at some $x_{*0} \in C_{1\text{conj}} \cap (-C_{2\text{conj}})$.

If the infimum is achieved at $x_0 \in C_1 \cap C_2$, then we have

$$\begin{aligned}
-f_{1\text{conj}}(x_{*0}) &\leq \inf_{x \in C_1 \cap C_2} (-\langle\langle x_{*0}, x \rangle\rangle + f_1(x)) \leq -\langle\langle x_{*0}, x_0 \rangle\rangle + f_1(x_0) \\
-f_{2\text{conj}}(-x_{*0}) &\leq \inf_{x \in C_1 \cap C_2} (\langle\langle x_{*0}, x \rangle\rangle + f_2(x)) \leq \langle\langle x_{*0}, x_0 \rangle\rangle + f_2(x_0) \\
\mu = -f_{1\text{conj}}(x_{*0}) &- f_{2\text{conj}}(-x_{*0}) \leq f_1(x_0) + f_2(x_0) = \mu
\end{aligned}$$

Hence, (8.7) holds.

(ii) If there exists $x_{*0} \in C_{1\text{conj}} \cap (-C_{2\text{conj}})$ and $x_0 \in C_1 \cap C_2$ such that (8.7) holds, then, by (8.8), we have

$$\begin{aligned}
\mu &\geq -f_{1\text{conj}}(x_{*0}) - f_{2\text{conj}}(-x_{*0}) = -\langle\langle x_{*0}, x_0 \rangle\rangle + f_1(x_0) \\
&\quad + \langle\langle x_{*0}, x_0 \rangle\rangle + f_2(x_0) = f_1(x_0) + f_2(x_0) \geq \mu
\end{aligned}$$

Hence, the infimum is achieved at x_0 and the maximum is achieved at x_{*0}.

This completes the proof of the theorem. □

Proposition 8.36 *Let Ω be a set, $J : \Omega \to \mathbb{R}_e$, Λ be an index set. $A : \Lambda \to {}^{\Omega}2$. Assume $\bigcup_{\lambda \in \Lambda} A(\lambda) = \Omega$. Then,*

$$\inf_{\omega \in \Omega} J(\omega) = \inf_{\lambda \in \Lambda} \inf_{\omega \in A(\lambda)} J(\omega); \qquad \sup_{\omega \in \Omega} J(\omega) = \sup_{\lambda \in \Lambda} \sup_{\omega \in A(\lambda)} J(\omega)$$

Proof $\forall \omega_o \in \Omega$, there exists $\lambda_o \in \Lambda$, such that $\omega_o \in A(\lambda_o)$. Then,

$$J(\omega_o) \geq \inf_{\omega \in A(\lambda_o)} J(\omega) \geq \inf_{\lambda \in \Lambda} \inf_{\omega \in A(\lambda)} J(\omega)$$

Hence, by Proposition 3.81, we have

$$\inf_{\omega \in \Omega} J(\omega) \geq \inf_{\lambda \in \Lambda} \inf_{\omega \in A(\lambda)} J(\omega)$$

On the other hand, $\forall \lambda \in \Lambda$, $A(\lambda) \subseteq \Omega$ implies that

$$\inf_{\omega \in \Omega} J(\omega) \leq \inf_{\omega \in A(\lambda)} J(\omega)$$

Then,

$$\inf_{\omega \in \Omega} J(\omega) \leq \inf_{\lambda \in \Lambda} \inf_{\omega \in A(\lambda)} J(\omega)$$

Combining the above arguments, we have

$$\inf_{\omega \in \Omega} J(\omega) = \inf_{\lambda \in \Lambda} \inf_{\omega \in A(\lambda)} J(\omega)$$

Substituting $-J$ into this inequality, by Proposition 3.81, yields the other equality.

This completes the proof of the proposition. □

Proposition 8.37 *Let X and Y be sets, $g_1 : X \to \mathbb{R}_e$, $g_2 : Y \to \mathbb{R}_e$. Assume that, $\forall x \in X$, $\forall y \in Y$, $g_1(x) + g_2(y) \in \mathbb{R}_e$ is well-defined. Then,*

$$\inf_{(x,y) \in X \times Y} (g_1(x) + g_2(y)) = \inf_{x \in X} g_1(x) + \inf_{y \in Y} g_2(y)$$

when the right-hand side makes sense.

Proof Define $\mu_l := \inf_{(x,y) \in X \times Y} (g_1(x)+g_2(y))$ and $\mu_r := \inf_{x \in X} g_1(x)+ \inf_{y \in Y} g_2(y)$. We will distinguish two exhaustive and mutually exclusive cases: Case 1: $X \times Y = \emptyset$; Case 2: $X \times Y \neq \emptyset$. Case 1: $X \times Y = \emptyset$. Without loss of generality, assume $X = \emptyset$. Then, $\mu_l = +\infty = \inf_{x \in X} g_1(x)$. Since $\mu_r \in \mathbb{R}_e$ is well-defined, then $\inf_{y \in Y} g_2(y) > -\infty$. Hence, the desired equality holds.

Case 2: $X \times Y \neq \emptyset$. Then, $X \neq \emptyset$ and $Y \neq \emptyset$. This implies that $\inf_{x \in X} g_1(x) = \inf_{(x,y) \in X \times Y} g_1(x)$ and $\inf_{y \in Y} g_2(y) = \inf_{(x,y) \in X \times Y} g_2(y)$. By Proposition 3.81, we have

$$\mu_l \geq \inf_{(x,y) \in X \times Y} g_1(x) + \inf_{(x,y) \in X \times Y} g_2(y) = \mu_r$$

We will further distinguish two exhaustive and mutually exclusive cases: Case 2a: $\mu_r = +\infty$; Case 2b: $\mu_r < +\infty$. Case 2a: $\mu_r = +\infty$. Then, $+\infty \geq \mu_l \geq \mu_r = +\infty$. Hence, the desired equality holds.

Case 2b: $\mu_r < +\infty$. $\forall m \in \mathbb{R}$ with $m > \mu_r$, by Proposition 3.81, $\exists x_0 \in X$ and $\exists y_0 \in Y$ such that $g_1(x_0) + g_2(y_0) < m$. Then, again by Proposition 3.81, we have $\mu_l < m$. This implies that $\mu_l \leq \mu_r$. Hence, the desired equality holds.

This completes the proof of the proposition. □

Proposition 8.38 *Let $\mathcal{X}$ be a topological space, Y be a set, and $V : \mathcal{X} \times Y \to \mathbb{R}$. Assume that V satisfies the following two conditions.*

(i) $\forall x_1 \in \mathcal{X}$, $W(x_1) := \inf_{y \in Y} V(x_1, y) \in \mathbb{R}$. *This defines the function $W : \mathcal{X} \to \mathbb{R}$.*
(ii) $\forall y \in Y$, *define the function $f_y : \mathcal{X} \to \mathbb{R}$ by $f_y(x) = V(x, y)$, $\forall x \in \mathcal{X}$. The collection of functions $\{f_y \mid y \in Y\}$ is equicontinuous.*

Then, W is continuous.

Proof $\forall x_0 \in \mathcal{X}$, $\forall \epsilon \in (0, \infty) \subset \mathbb{R}$, by (ii), $\exists U \in \mathcal{O}_{\mathcal{X}}$ with $x_0 \in U$ such that, $\forall \bar{x} \in U$, $\forall y \in Y$, we have $\left|f_y(\bar{x}) - f_y(x_0)\right| = |V(\bar{x}, y) - V(x_0, y)| < \epsilon$. By (i), $\exists \bar{y}, y_0 \in Y$ such that $|W(\bar{x}) - V(\bar{x}, \bar{y})| < \epsilon$ and $|W(x_0) - V(x_0, y_0)| < \epsilon$. Note that

$$\begin{aligned}
-2\epsilon &< -|W(\bar{x}) - V(\bar{x}, \bar{y})| - |V(\bar{x}, \bar{y}) - V(x_0, \bar{y})| \leq W(\bar{x}) - V(x_0, \bar{y}) \\
&\leq W(\bar{x}) - V(x_0, \bar{y}) + V(x_0, \bar{y}) - W(x_0) = W(\bar{x}) - W(x_0) \\
&= W(\bar{x}) - V(\bar{x}, y_0) + V(\bar{x}, y_0) - W(x_0) \leq V(\bar{x}, y_0) - W(x_0) \\
&\leq |V(\bar{x}, y_0) - V(x_0, y_0)| + |V(x_0, y_0) - W(x_0)| < 2\epsilon
\end{aligned}$$

Hence, we have $|W(\bar{x}) - W(x_0)| < 2\epsilon$ and W is continuous at x_0. By Proposition 3.9, W is continuous.

This completes the proof of the proposition. □

Next, we present a result on game theory.

Proposition 8.39 *Let $\mathfrak{X}$ be a reflexive real normed linear space and $A \subseteq \mathfrak{X}$ and $B \subseteq \mathfrak{X}^*$ be nonempty bounded closed convex sets. Then,*

$$\min_{x \in A} \max_{x_* \in B} \langle\langle x_*, x \rangle\rangle = \max_{x_* \in B} \min_{x \in A} \langle\langle x_*, x \rangle\rangle$$

Proof Let $\mu := \inf_{x \in A} \sup_{x_* \in B} \langle\langle x_*, x \rangle\rangle$. Let $h : \mathfrak{X} \times B \to \mathbb{R}$ be given by $h(x, x_*) = \langle\langle x_*, x \rangle\rangle$, $\forall x \in \mathfrak{X}$, $\forall x_* \in B$. By Proposition 7.72, h is continuous. Define $f_1 : \mathfrak{X} \to \mathbb{R}_e$ by $f_1(x) = \sup_{x_* \in B} h(x, x_*)$, $\forall x \in \mathfrak{X}$. $\forall x \in \mathfrak{X}$, define $h_x : \mathfrak{X}^* \to \mathbb{R}$ by $h_x(x_*) = h(x, x_*) = \langle\langle x_*, x \rangle\rangle$, $\forall x_* \in \mathfrak{X}^*$. Then, h_x is weak* continuous. By Proposition 8.11, B is weak* compact. Then, by Proposition 5.29, $\exists \bar{x}_* \in B$ such that $f_1(x) = h_x(\bar{x}_*) = \max_{x_* \in B} h_x(x_*) = \max_{x_* \in B} h(x, x_*) \in \mathbb{R}$. Hence, $f_1 : \mathfrak{X} \to \mathbb{R}$ takes value in $\mathbb{R}$.

Since B is bounded, then $\exists M_B \in [0, \infty) \subset \mathbb{R}$ such that $\|x_*\| \leq M_B$, $\forall x_* \in B$. $\forall x_* \in B$, define $h_{x_*} : \mathfrak{X} \to \mathbb{R}$ by $h_{x_*}(x) = h(x, x_*)$, $\forall x \in \mathfrak{X}$. $\forall x_0 \in \mathfrak{X}$, $\forall \epsilon \in (0, \infty) \subset \mathbb{R}$, let $\delta = \epsilon/(1 + M_B) \in (0, \infty) \subset \mathbb{R}$. $\forall x \in \mathcal{B}_{\mathfrak{X}}(x_0, \delta)$, $\forall x_* \in B$, we have $\left|h_{x_*}(x) - h_{x_*}(x_0)\right| = |h(x, x_*) - h(x_0, x_*)| = |\langle\langle x_*, x - x_0 \rangle\rangle| \leq \|x_*\| \|x - x_0\| \leq M_B \delta < \epsilon$, where we have applied Proposition 7.72. Hence, $\{h_{x_*} \mid x_* \in B\}$ is equicontinuous. By Proposition 8.38, $f_1 : \mathfrak{X} \to \mathbb{R}$ is continuous.

$\forall x_1, x_2 \in \mathcal{X}$, $\forall \alpha \in [0, 1] \subset \mathbb{R}$, we have

$$f_1(\alpha x_1 + (1-\alpha)x_2) = \max_{x_* \in B} \langle\langle x_*, \alpha x_1 + (1-\alpha)x_2 \rangle\rangle$$
$$\leq \sup_{x_* \in B} \alpha\langle\langle x_*, x_1\rangle\rangle + \sup_{x_* \in B} (1-\alpha)\langle\langle x_*, x_2\rangle\rangle$$
$$= \alpha \max_{x_* \in B} \langle\langle x_*, x_1\rangle\rangle + (1-\alpha) \max_{x_* \in B} \langle\langle x_*, x_2\rangle\rangle$$
$$= \alpha f_1(x_1) + (1-\alpha) f_1(x_2)$$

where we have applied Proposition 3.81 in the second equality. Hence, f_1 is convex.

Since A is bounded, then, $\exists M_A \in [0, \infty) \subset \mathbb{R}$ such that $\|x\| \leq M_A$, $\forall x \in A$. $\forall x \in A$, $\forall x_* \in B$, by Proposition 7.72, we have $\langle\langle x_*, x\rangle\rangle \geq -\|x_*\|\|x\| \geq -M_A M_B$. Then, $f_1(x) \geq -M_A M_B$ since $B \neq \emptyset$. Then, $\mu \geq -M_A M_B$. Since $A \neq \emptyset$, then $\mu < +\infty$. Hence, μ is finite.

Define $f_2 : A \to \mathbb{R}$ by $f_2(x) = 0$, $\forall x \in A$. Clearly, f_2 is convex. Note $\mu = \inf_{x \in \mathcal{X} \cap A}(f_1(x) + f_2(x))$. Now, it is easy to check that all assumptions for the Fenchel Duality Theorem are satisfied. Then,

$$\mu = \max_{x_* \in \mathcal{X}_{\text{conj}} \cap (-A_{\text{conj}})} (-f_{1\text{conj}}(x_*) - f_{2\text{conj}}(-x_*))$$

Claim 8.39.1 $\mathcal{X}_{\text{conj}} = B$ and $f_{1\text{conj}}(x_*) = 0$, $\forall x_* \in B$.

Proof of Claim $\mathcal{X}_{\text{conj}} = \{x_* \in \mathcal{X}^* \mid \sup_{x \in \mathcal{X}}(\langle\langle x_*, x\rangle\rangle - f_1(x)) < +\infty\}$.

$\forall x_{*0} \in B$, $\forall x \in \mathcal{X}$, we have $f_1(x) = \max_{x_* \in B} \langle\langle x_*, x\rangle\rangle \geq \langle\langle x_{*0}, x\rangle\rangle$. Then, $\langle\langle x_{*0}, x\rangle\rangle - f_1(x) \leq 0$. Hence, $x_{*0} \in \mathcal{X}_{\text{conj}}$. Hence, $B \subseteq \mathcal{X}_{\text{conj}}$. The above also implies that $f_{1\text{conj}}(x_{*0}) = \sup_{x \in \mathcal{X}}(\langle\langle x_{*0}, x\rangle\rangle - f_1(x)) = 0$.

On the other hand, $\forall x_{*0} \in \mathcal{X}^* \setminus B$, by Proposition 8.10, $\exists x_{**0} \in \mathcal{X}^{**}$ such that $\langle\langle x_{**0}, x_{*0}\rangle\rangle < \inf_{x_* \in B} \langle\langle x_{**0}, x_*\rangle\rangle$. Since $\mathcal{X}$ is reflexive, then $x_{**0} = \phi(x_0)$ for some $x_0 \in \mathcal{X}$, where $\phi : \mathcal{X} \to \mathcal{X}^{**}$ is the natural mapping. Then, $\langle\langle x_{*0}, x_0\rangle\rangle < \inf_{x_* \in B} \langle\langle x_*, x_0\rangle\rangle$ and $\langle\langle x_{*0}, -x_0\rangle\rangle > \sup_{x_* \in B} \langle\langle x_*, -x_0\rangle\rangle = f_1(-x_0)$. Then, $\langle\langle x_{*0}, -\alpha x_0\rangle\rangle - f_1(-\alpha x_0) = \alpha\,(\langle\langle x_{*0}, -x_0\rangle\rangle - f_1(-x_0)) > 0$, $\forall \alpha \in (0, \infty) \subset \mathbb{R}$. Hence, $\sup_{x \in \mathcal{X}}(\langle\langle x_{*0}, x\rangle\rangle - f_1(x)) = +\infty$ and $x_{*0} \in \mathcal{X}^* \setminus \mathcal{X}_{\text{conj}}$. This implies that $\mathcal{X}_{\text{conj}} \subseteq B$. Therefore, we have $\mathcal{X}_{\text{conj}} = B$.

This completes the proof of the claim. □

Claim 8.39.2 $A_{\text{conj}} = \mathcal{X}^*$ and $f_{2\text{conj}}(x_*) = \max_{x \in A} \langle\langle x_*, x\rangle\rangle$, $\forall x_* \in \mathcal{X}^*$.

Proof of Claim $A_{\text{conj}} = \{x_* \in \mathcal{X}^* \mid \sup_{x \in A} \langle\langle x_*, x\rangle\rangle < +\infty\}$.

Since $\mathcal{X}$ is reflexive, then, by Definition 7.89, $\phi : \mathcal{X} \to \mathcal{X}^{**}$ is an isometrical isomorphism. Then, $\phi(A) \subseteq \mathcal{X}^{**}$ is a nonempty bounded closed convex set. By Proposition 7.90, $\mathcal{Y} := \mathcal{X}^*$ is a reflexive real normed linear space. By Proposition 8.11, $\phi(A)$ is weak* compact. $\forall x_* \in \mathcal{X}^*$, $h_{x_*} \circ \phi_{\text{inv}} : \mathcal{X}^{**} \to \mathbb{R}$ is weak* continuous. By Proposition 5.29, we have

$$\sup_{x\in A} \langle\langle x_*, x\rangle\rangle = \sup_{x_{**}\in\phi(A)} \langle\langle x_{**}, x_*\rangle\rangle = \langle\langle x_{**0}, x_*\rangle\rangle = \langle\langle x_*, \phi_{\text{inv}}(x_{**0})\rangle\rangle$$
$$= \langle\langle x_*, x_0\rangle\rangle \in \mathbb{R}$$

for some $x_{**0} \in \phi(A)$ and $x_0 = \phi_{\text{inv}}(x_{**0}) \in A$. Then, $x_* \in A_{\text{conj}}$ and $A_{\text{conj}} = \mathcal{X}^*$. Clearly, $f_{2\text{conj}}(x_*) = \langle\langle x_*, x_0\rangle\rangle = \max_{x\in A} \langle\langle x_*, x\rangle\rangle$. This completes the proof of the claim. □

Then,

$$\mu = \inf_{x\in A} \max_{x_*\in B} \langle\langle x_*, x\rangle\rangle = \max_{x_*\in B\cap\mathcal{X}^*} \left(-\max_{x\in A} \langle\langle -x_*, x\rangle\rangle\right) = \max_{x_*\in B} \min_{x\in A} \langle\langle x_*, x\rangle\rangle$$

To show that the infimum in the above equation is actually achieved, we note that

$$-\mu = \min_{x_*\in B} \max_{x\in(-A)} \langle\langle x_*, x\rangle\rangle = \min_{x_*\in B} \max_{x_{**}\in\phi(-A)} \langle\langle x_{**}, x_*\rangle\rangle$$

By Proposition 7.90, $\mathcal{Y} := \mathcal{X}^*$ is a reflexive real normed linear space and $\phi(\mathcal{X}) = \mathcal{X}^{**}$. Since ϕ is an isometrical isomorphism, then $\phi(-A)$ is a nonempty bounded closed convex set. Applying the result that we have obtained in this proof to the above, we have

$$-\mu = \max_{x_{**}\in\phi(-A)} \min_{x_*\in B} \langle\langle x_{**}, x_*\rangle\rangle = \max_{x\in(-A)} \min_{x_*\in B} \langle\langle x_*, x\rangle\rangle$$

which is equivalent to $\mu = \min_{x\in A} \max_{x_*\in B} \langle\langle x_*, x\rangle\rangle$.

This completes the proof of the proposition. □

8.7 Positive Cones and Convex Mappings

Definition 8.40 Let $\mathcal{X}$ be a normed linear space and $P \subseteq \mathcal{X}$ be a closed convex cone. For $x, y \in \mathcal{X}$, we will write $x \geqq y$ (with respect to P) if $x - y \in P$. The cone P defining this relation is called the *positive cone* in $\mathcal{X}$. The cone $N = -P$ is called the *negative cone* in $\mathcal{X}$ and we write $y \leqq x$ if $y - x \in N$. We will write $x > y$ ($y < x$) if $x - y \in P^\circ$ ($y - x \in N^\circ = -P^\circ$). ◇

It is easy to check that relations $\leqq$ and $\geqq$ are reflexive and transitive.

Proposition 8.41 *Let $\mathcal{X}$ be a normed linear space with positive cone P. $\forall x_1, x_2, x_3, x_4 \in \mathcal{X}$, $\forall \alpha \in [0, \infty) \subset \mathbb{R}$, we have*

(i) *$x_1, x_2 \in P$ implies $x_1 + x_2 \in P$, $\alpha x_1 \in P$, and $-\alpha x_1 \leqq \vartheta$.*
(ii) *$x_1 \leqq x_1$.*

(iii) $x_1 \leqqdot x_2$ *and* $x_2 \leqqdot x_3$ *implies* $x_1 \leqqdot x_3$.
(iv) $x_1 \leqqdot x_2$ *and* $x_3 \leqqdot x_4$ *implies* $x_1 + x_3 \leqqdot x_2 + x_4$.
(v) $x_1 \leqqdot x_2$ *implies* $\alpha x_1 \leqqdot \alpha x_2$.
(vi) $x_1 \gtrdot \vartheta$ *and* $x_2 \geqqdot \vartheta$ *implies* $x_1 + x_2 \gtrdot \vartheta$.
(vii) $x_1 \lessdot x_2$ *and* $\alpha > 0$ *implies* $\alpha x_1 \lessdot \alpha x_2$.

Proof This is straightforward. □

Definition 8.42 Let $\mathcal{X}$ be a real normed linear space and $S \subseteq \mathcal{X}$. The set $S^\oplus := \{x_* \in \mathcal{X}^* \mid \langle\langle x_*, x\rangle\rangle \ge 0,\ \forall x \in S\}$ is called the *positive conjugate cone* of S. The set $S^\ominus := \{x_* \in \mathcal{X}^* \mid \langle\langle x_*, x\rangle\rangle \le 0,\ \forall x \in S\}$ is called the *negative conjugate cone* of S. Clearly, $S^\ominus = -S^\oplus$. ◇

Proposition 8.43 *Let $\mathcal{X}$ be a real normed linear space and $S, T \subseteq \mathcal{X}$. Then,*

(i) *$S^\oplus \subseteq \mathcal{X}^*$ is a closed convex cone.*
(ii) *If $S \subseteq T$, then $T^\oplus \subseteq S^\oplus$.*

Proof (i) Clearly, $\vartheta_* \in S^\oplus$. $\forall x_* \in S^\oplus$, $\forall \alpha \in [0, \infty) \subset \mathbb{R}$, $\forall x \in S$, we have $\langle\langle \alpha x_*, x\rangle\rangle = \alpha\langle\langle x_*, x\rangle\rangle \ge 0$. Hence, $\alpha x_* \in S^\oplus$. Therefore, $S^\oplus$ is a cone with vertex at origin. $\forall x_{*1}, x_{*2} \in S^\oplus$, $\forall x \in S$, we have $\langle\langle x_{*1} + x_{*2}, x\rangle\rangle = \langle\langle x_{*1}, x\rangle\rangle + \langle\langle x_{*2}, x\rangle\rangle \ge 0$. Hence, $x_{*1} + x_{*2} \in S^\oplus$. Therefore, $S^\oplus$ is a convex cone.

$\forall x_* \in \overline{S^\oplus}$, by Proposition 4.13, $\exists (x_{*n})_{n=1}^\infty \subseteq S^\oplus$ such that $\lim_{n\in\mathbb{N}} x_{*n} = x_*$. $\forall x \in S$, by Propositions 7.72 and 3.66, $\langle\langle x_*, x\rangle\rangle = \lim_{n\in\mathbb{N}} \langle\langle x_{*n}, x\rangle\rangle \ge 0$. Therefore, $x_* \in S^\oplus$ and $\overline{S^\oplus} \subseteq S^\oplus$. By Proposition 3.3, $S^\oplus$ is closed.

(ii) This is straightforward.

This completes the proof of the proposition. □

Proposition 8.44 *Let $\mathcal{X}$ and $\mathcal{Y}$ be real normed linear spaces, $A \in B(\mathcal{X}, \mathcal{Y})$, and $S \subseteq \mathcal{X}$. Then, $(A(S))^\oplus = A'_{\text{inv}}(S^\oplus)$.*

Proof $\forall y_* \in (A(S))^\oplus \subseteq \mathcal{Y}^*$, $\forall s \in S$, we have $\langle\langle A'y_*, s\rangle\rangle = \langle\langle y_*, As\rangle\rangle \ge 0$. Then, $A'y_* \in S^\oplus$ and $y_* \in A'_{\text{inv}}(S^\oplus)$. Hence, $(A(S))^\oplus \subseteq A'_{\text{inv}}(S^\oplus)$.

On the other hand, $\forall y_* \in A'_{\text{inv}}(S^\oplus)$, $A'y_* \in S^\oplus$. $\forall s \in S$, we have $\langle\langle y_*, As\rangle\rangle = \langle\langle A'y_*, s\rangle\rangle \ge 0$. Therefore, we have $y_* \in (A(S))^\oplus$ and $A'_{\text{inv}}(S^\oplus) \subseteq (A(S))^\oplus$.

Hence, $(A(S))^\oplus = A'_{\text{inv}}(S^\oplus)$. This completes the proof of the proposition. □

Definition 8.45 Let $\mathcal{X}$ be a real normed linear space and $P \subseteq \mathcal{X}$ be the positive cone. We will define $P^\oplus \subseteq \mathcal{X}^*$ to be the *positive cone* in the dual. ◇

Proposition 8.46 *Let $\mathcal{X}$ be a real normed linear space with the positive cone P. If $x_0 \in \mathcal{X}$ satisfies that $\langle\langle x_*, x_0\rangle\rangle \ge 0$, $\forall x_* \in P^\oplus$ (or $x_* \geqqdot \vartheta_*$), then $x_0 \in P$.*

Proof Suppose $x_0 \notin P$, by Proposition 8.10, $\exists x_* \in \mathcal{X}^*$, such that $-\infty < \langle\langle x_*, x_0\rangle\rangle < \inf_{x\in P} \langle\langle x_*, x\rangle\rangle$. Since P is a cone, then $\inf_{x\in P} \langle\langle x_*, x\rangle\rangle = 0$ (it must be greater than or equal to 0 since otherwise the infimum must be $-\infty$; and it must be less than or equal to 0 since $\vartheta_{\mathcal{X}} \in P$). Hence, $x_* \in P^\oplus$ and $\langle\langle x_*, x_0\rangle\rangle < 0$. This contradicts with the assumption. Therefore, we must have $x_0 \in P$. This completes the proof of the proposition. □

Proposition 8.47 *Let $\mathcal{X}$ be a real normed linear space with positive cone P. $\forall x_* \in \mathcal{X}^*$, $\forall x \in \mathcal{X}$, we have*

(i) $x \geqq \vartheta$ *and* $x_* \geqq \vartheta_*$ *implies* $\langle\langle x_*, x\rangle\rangle \geq 0$.
(ii) $x \leqq \vartheta$ *and* $x_* \geqq \vartheta_*$ *implies* $\langle\langle x_*, x\rangle\rangle \leq 0$.
(iii) $x > \vartheta$, $x_* \geqq \vartheta_*$, *and* $x_* \neq \vartheta_*$ *implies* $\langle\langle x_*, x\rangle\rangle > 0$.
(iv) $x \geqq \vartheta$, $x \neq \vartheta$, *and* $x_* > \vartheta_*$ *implies* $\langle\langle x_*, x\rangle\rangle > 0$.

Proof This is straightforward. □

Definition 8.48 Let $\mathcal{X}$ be a real vector space, $\Omega \subseteq \mathcal{X}$, and $\mathcal{Z}$ be a real normed linear space with the positive cone $P \subseteq \mathcal{Z}$. A mapping $G : \Omega \to \mathcal{Z}$ is said to be *convex* if Ω is convex and $\forall x_1, x_2 \in \Omega$, $\forall \alpha \in [0, 1] \subset \mathbb{R}$, we have $G(\alpha x_1 + (1-\alpha)x_2) \leqq \alpha G(x_1) + (1-\alpha)G(x_2)$. ◇

We note that the convexity of a mapping depends on the definition of the positive cone P.

Proposition 8.49 *Let $\mathcal{X}$ be a real vector space, $\Omega \subseteq \mathcal{X}$ be convex, $\mathcal{Z}$ be a real normed linear space with the positive cone $P \subseteq \mathcal{Z}$, and $G : \Omega \to \mathcal{Z}$ be a convex mapping. Then, $\forall z \in \mathcal{Z}$, the set $\Omega_z := \left\{x \in \Omega \,\middle|\, G(x) \leqq z\right\}$ is convex.*

Proof Fix any $z \in \mathcal{Z}$. $\forall x_1, x_2 \in \Omega_z$, $\forall \alpha \in [0, 1] \subset \mathbb{R}$, we have $G(x_1) \leqq z$ and $G(x_2) \leqq z$. Since P is a convex cone, then, $\alpha G(x_1) + (1-\alpha)G(x_2) \leqq \alpha z + (1-\alpha)z = z$. By the convexity of G, we have $G(\alpha x_1 + (1-\alpha)x_2) \leqq \alpha G(x_1) + (1-\alpha)G(x_2)$. By Proposition 8.41, we have $G(\alpha x_1 + (1-\alpha)x_2) \leqq z$. Hence, $\alpha x_1 + (1-\alpha)x_2 \in \Omega_z$. Hence, Ω_z is convex. This completes the proof of the proposition. □

8.8 Lagrange Multipliers

The basic problem to be considered in this section is

$$\mu_0 := \inf_{x \in \Omega,\ G(x) \leqq \vartheta_{\mathcal{Z}}} f(x) \tag{8.11}$$

where $\mathcal{X}$ is a real vector space, $\Omega \subseteq \mathcal{X}$ is a nonempty convex set, $f : \Omega \to \mathbb{R}$ is a convex functional, $\mathcal{Z}$ is a real normed linear space with positive cone $P \subseteq \mathcal{Z}$, and $G : \Omega \to \mathcal{Z}$ is a convex mapping.

Toward a solution to the above problem, we consider the following class of problems:

$$\Gamma := \left\{ z \in \mathcal{Z} \;\middle|\; \exists x \in \Omega \ni \cdot\, G(x) \stackrel{\leq}{=} z \right\} \tag{8.12a}$$

$$\omega(z) := \inf_{x \in \Omega,\ G(x) \stackrel{\leq}{=} z} f(x); \qquad \forall z \in \Gamma \tag{8.12b}$$

where $\omega : \Gamma \to \mathbb{R}_e$ is the *primal functional*. To guarantee that ω is real-valued, we make the following assumption.

Assumption 8.50 *$\vartheta_{\mathcal{Z}} \in \Gamma$ and* (i) *$\exists \bar{z} \in {}^{\circ}\Gamma$ such that $\omega(\bar{z}) \in \mathbb{R}$ or* (ii) *$\mu := \inf_{x \in \Omega} f(x) > -\infty$ holds.*

Fact 8.51 *$\Gamma \subseteq \mathcal{Z}$ is convex.*

Proof $\forall z_1, z_2 \in \Gamma$, $\forall \alpha \in [0, 1] \subset \mathbb{R}$, there exist $x_i \in \Omega$ such that $G(x_i) \stackrel{\leq}{=} z_i$, $i = 1, 2$. By the convexity of Ω, we have $\alpha x_1 + (1 - \alpha)x_2 \in \Omega$. Then, by the convexity of G and Proposition 8.41, $G(\alpha x_1 + (1 - \alpha)x_2) \stackrel{\leq}{=} \alpha G(x_1) + (1 - \alpha)G(x_2) \stackrel{\leq}{=} \alpha z_1 + (1 - \alpha)z_2$. Hence, $\alpha z_1 + (1 - \alpha)z_2 \in \Gamma$. This completes the proof of the fact. □

Fact 8.52 *Under Assumption 8.50, $\omega : \Gamma \to \mathbb{R}$ is real-valued, convex, and nonincreasing, that is, $\forall z_1, z_2 \in \Gamma$ with $z_1 \stackrel{\leq}{=} z_2$, we have $\omega(z_1) \geq \omega(z_2)$.*

Proof We will first show that $\omega(z) \in \mathbb{R}$, $\forall z \in \Gamma$. Let (i) in Assumption 8.50 hold. Fix any $z \in \Gamma$. We will distinguish two exhaustive and mutually exclusive cases: Case 1: $z = \bar{z}$; Case 2: $z \neq \bar{z}$. Case 1: $z = \bar{z}$. Then, $\omega(z) = \omega(\bar{z}) \in \mathbb{R}$.

Case 2: $z \neq \bar{z}$. By the definition of Γ and the fact that f is real-valued, we have $\omega(z) < +\infty$. By (i) of Assumption 8.50, $\exists \delta \in (0, \infty) \subset \mathbb{R}$ such that $\mathcal{B}_{\mathcal{Z}}(\bar{z}, \delta) \cap \mathrm{V}(\Gamma) \subseteq \Gamma$. Let $\bar{z}_1 := \bar{z} + \frac{\delta}{2\|\bar{z} - z\|}(\bar{z} - z) \in \mathcal{B}_{\mathcal{Z}}(\bar{z}, \delta) \cap \mathrm{V}(\Gamma) \subseteq \Gamma$ and $\bar{\alpha} = \frac{\|\bar{z} - z\|}{\|\bar{z} - z\| + \delta/2} \in (0, 1) \subset \mathbb{R}$. It is easy to verify that $\bar{z} = \bar{\alpha}\bar{z}_1 + (1 - \bar{\alpha})z$. Then, we have

$$\begin{aligned}
-\infty < \omega(\bar{z}) &= \inf_{\bar{x} \in \Omega,\ G(\bar{x}) \stackrel{\leq}{=} \bar{\alpha}\bar{z}_1 + (1 - \bar{\alpha})z} f(\bar{x}) \\
&\leq \inf_{\bar{x} = \bar{\alpha}\bar{x}_1 + (1 - \bar{\alpha})x,\ \bar{x}_1 \in \Omega,\ G(\bar{x}_1) \stackrel{\leq}{=} \bar{z}_1,\ x \in \Omega,\ G(x) \stackrel{\leq}{=} z} f(\bar{x}) \\
&\leq \inf_{\bar{x}_1 \in \Omega,\ G(\bar{x}_1) \stackrel{\leq}{=} \bar{z}_1,\ x \in \Omega,\ G(x) \stackrel{\leq}{=} z} (\bar{\alpha} f(\bar{x}_1) + (1 - \bar{\alpha}) f(x)) \\
&= \inf_{\bar{x}_1 \in \Omega,\ G(\bar{x}_1) \stackrel{\leq}{=} \bar{z}_1} \bar{\alpha} f(\bar{x}_1) + \inf_{x \in \Omega,\ G(x) \stackrel{\leq}{=} z} (1 - \bar{\alpha}) f(x) \\
&= \bar{\alpha} \inf_{\bar{x}_1 \in \Omega,\ G(\bar{x}_1) \stackrel{\leq}{=} \bar{z}_1} f(\bar{x}_1) + (1 - \bar{\alpha}) \inf_{x \in \Omega,\ G(x) \stackrel{\leq}{=} z} f(x) \\
&= \bar{\alpha}\omega(\bar{z}_1) + (1 - \bar{\alpha})\omega(z)
\end{aligned}$$

where the second equality follows from Proposition 8.37, and the third equality follows from Proposition 3.81. Then, $\omega(z) > -\infty$. Hence, $\omega(z) \in \mathbb{R}$.

In both cases, we have $\omega(z) \in \mathbb{R}$. Therefore, $\omega : \Gamma \to \mathbb{R}$ is real-valued.

Let (ii) in Assumption 8.50 hold. Fix any $z \in \Gamma$. Then, $\omega(z) \geq \mu > -\infty$. By the definition of Γ, we have $\omega(z) < +\infty$. Hence, $\omega(z) \in \mathbb{R}$.

Thus, under Assumption 8.50, $\omega : \Gamma \to \mathbb{R}$ is real-valued.

$\forall z_1, z_2 \in \Gamma$, $\forall \alpha \in [0, 1] \subset \mathbb{R}$, we have

$$\begin{aligned}
\omega(\alpha z_1 + (1-\alpha)z_2) &= \inf_{x\in\Omega,\ G(x)\stackrel{\leq}{=}\alpha z_1+(1-\alpha)z_2} f(x) \\
&\leq \inf_{x=\alpha x_1+(1-\alpha)x_2,\ x_1\in\Omega,\ G(x_1)\stackrel{\leq}{=}z_1,\ x_2\in\Omega,\ G(x_2)\stackrel{\leq}{=}z_2} f(x) \\
&\leq \inf_{x_1\in\Omega,\ G(x_1)\stackrel{\leq}{=}z_1,\ x_2\in\Omega,\ G(x_2)\stackrel{\leq}{=}z_2} (\alpha f(x_1) + (1-\alpha) f(x_2)) \\
&= \inf_{x_1\in\Omega,\ G(x_1)\stackrel{\leq}{=}z_1} \alpha f(x_1) + \inf_{x_2\in\Omega,\ G(x_2)\stackrel{\leq}{=}z_2} (1-\alpha) f(x_2) \\
&= \alpha \inf_{x_1\in\Omega,\ G(x_1)\stackrel{\leq}{=}z_1} f(x_1) + (1-\alpha) \inf_{x_2\in\Omega,\ G(x_2)\stackrel{\leq}{=}z_2} f(x_2) \\
&= \alpha\omega(z_1) + (1-\alpha)\omega(z_2)
\end{aligned}$$

where the second equality follows from Proposition 8.37, and the third equality follows from Proposition 3.81. Hence, ω is convex.

It is obvious that ω is nonincreasing. This completes the proof of the fact. □

Now, let the functional $\omega_{\text{conj}} : \Gamma_{\text{conj}} \to \mathbb{R}$ be conjugate to $\omega : \Gamma \to \mathbb{R}$.

Fact 8.53 *Under Assumption 8.50, we have $\Gamma_{\text{conj}} \subseteq P^{\ominus}$ and $-\Gamma_{\text{conj}} \subseteq P^{\oplus}$.*

Proof Since $\vartheta_{\mathcal{Z}} \in \Gamma$ implies that $\exists x_1 \in \Omega$ such that $G(x_1) \stackrel{\leq}{=} \vartheta_{\mathcal{Z}}$, then $\Gamma \supseteq G(x_1) + P$. $\forall z_* \in \Gamma_{\text{conj}}$, by Definition 8.27, we have

$$\begin{aligned}
+\infty > \sup_{z\in\Gamma}(\langle\langle z_*, z\rangle\rangle - \omega(z)) &\geq \sup_{z\in G(x_1)+P} (\langle\langle z_*, z\rangle\rangle - \omega(z)) \\
&\geq \sup_{z\in G(x_1)+P} (\langle\langle z_*, z\rangle\rangle - \omega(G(x_1))) \\
&= \sup_{\bar{z}\in P} \langle\langle z_*, \bar{z}\rangle\rangle + \langle\langle z_*, G(x_1)\rangle\rangle - \omega(G(x_1))
\end{aligned}$$

where the third inequality follows from Fact 8.52, and the equality follows from Proposition 8.37. Hence, $\sup_{z\stackrel{\geq}{=}\vartheta_{\mathcal{Z}}} \langle\langle z_*, z\rangle\rangle < +\infty$. Since P is a cone, then $\sup_{z\stackrel{\geq}{=}\vartheta_{\mathcal{Z}}} \langle\langle z_*, z\rangle\rangle = 0$. This implies that $z_* \in P^{\ominus}$, $\Gamma_{\text{conj}} \subseteq P^{\ominus}$, and $-\Gamma_{\text{conj}} \subseteq P^{\oplus}$. This completes the proof of the fact. □

Fact 8.54 *Let Assumption 8.50 hold. Define $\bar{\omega} : P^{\ominus} \to \mathbb{R}_e$ by*

$$\bar{\omega}(z_*) = \sup_{z\in\Gamma}(\langle\langle z_*, z\rangle\rangle - \omega(z)); \qquad \forall z_* \in P^{\ominus}$$

Then, $\sup_{z_* \in \Gamma_{\text{conj}}} -\omega_{\text{conj}}(z_*) = \sup_{z_* \in P^{\ominus}} -\bar{\omega}(z_*)$. $\forall z_* \in P^{\oplus}$*, we have that*

$$-\bar{\omega}(-z_*) = \inf_{x \in \Omega} (f(x) + \langle\langle z_*, G(x)\rangle\rangle) =: \varphi(z_*)$$

where $\varphi : P^{\oplus} \to \mathbb{R}_e$ *is called the* dual functional.

Proof Clearly, $\forall z_* \in \Gamma_{\text{conj}}$, $\bar{\omega}(z_*) = \omega_{\text{conj}}(z_*)$ and, by the definition of Γ_{conj}, $\forall z_* \in P^{\ominus} \setminus \Gamma_{\text{conj}}$, $\bar{\omega}(z_*) = +\infty$. Hence, $\sup_{z_* \in \Gamma_{\text{conj}}} -\omega_{\text{conj}}(z_*) = \sup_{z_* \in P^{\ominus}} -\bar{\omega}(z_*)$.

$\forall z_* \in P^{\oplus}$, $-z_* \leqq \vartheta_{\mathcal{Z}^*}$,

$$\begin{aligned}
-\bar{\omega}(-z_*) &= -\sup_{z \in \Gamma}(\langle\langle -z_*, z\rangle\rangle - \omega(z)) = \inf_{z \in \Gamma}(\langle\langle z_*, z\rangle\rangle + \omega(z)) \\
&= \inf_{z \in \Gamma}(\langle\langle z_*, z\rangle\rangle + \inf_{x \in \Omega,\ G(x) \leqq z} f(x)) = \inf_{z \in \Gamma} \inf_{x \in \Omega,\ G(x) \leqq z} (\langle\langle z_*, z\rangle\rangle + f(x)) \\
&= \inf_{x \in \Omega,\ z \in \Gamma,\ G(x) \leqq z} (\langle\langle z_*, z\rangle\rangle + f(x)) = \inf_{x \in \Omega} \inf_{z \in \Gamma,\ z \geqq G(x)} (\langle\langle z_*, z\rangle\rangle + f(x)) \\
&= \inf_{x \in \Omega} \inf_{z \geqq G(x)} (\langle\langle z_*, z\rangle\rangle + f(x)) = \inf_{x \in \Omega}(f(x) + \inf_{z \geqq G(x)} \langle\langle z_*, z\rangle\rangle) \\
&= \inf_{x \in \Omega}(f(x) + \langle\langle z_*, G(x)\rangle\rangle)
\end{aligned}$$

where the second equality follows from Proposition 3.81; the fourth equality follows from Proposition 8.37; the fifth and sixth equality follows from Proposition 8.36; and the eighth equality follows from Proposition 8.37. This completes the proof of the fact. □

The desired theory follows by applying either duality results for convex functionals, which are Propositions 8.33 and 8.34. This leads to two different regularity conditions. To apply Proposition 8.33, we assume that

Assumption 8.55 *ω is lower semicontinuous at $\vartheta_{\mathcal{Z}}$.*

To apply Proposition 8.34, we assume that

Assumption 8.56 *$\exists x_1 \in \Omega$ such that $G(x_1) \lessdot \vartheta_{\mathcal{Z}}$.*

Now, we state the two Lagrange duality results.

Proposition 8.57 *Let $\mathcal{X}$ be a real vector space, $\Omega \subseteq \mathcal{X}$ be nonempty and convex, $f : \Omega \to \mathbb{R}$ be a convex functional, $\mathcal{Z}$ be a real normed linear space with the positive cone $P \subseteq \mathcal{Z}$, $G : \Omega \to \mathcal{Z}$ be a convex mapping, and μ_0 be defined as in* (8.11). *Let Assumptions* 8.50 *and* 8.55 *hold. Then,*

$$\mu_0 = \sup_{z_* \geqq \vartheta_{\mathcal{Z}^*}} \inf_{x \in \Omega} (f(x) + \langle\langle z_*, G(x)\rangle\rangle) = \sup_{z_* \geqq \vartheta_{\mathcal{Z}^*}} \varphi(z_*) \tag{8.13}$$

Furthermore, if the supremum in (8.13) *is achieved at* $z_{*0} \in P^{\oplus}$, *that is,*

$$\mu_0 = \inf_{x \in \Omega} (f(x) + \langle\langle z_{*0}, G(x)\rangle\rangle) \tag{8.14}$$

then the following statement holds: the infimum in (8.11) *is achieved at* $x_0 \in \Omega$ *with* $G(x_0) \stackrel{\lessdot}{=} \vartheta_{\mathcal{Z}}$ *if, and only if, the infimum in* (8.14) *is achieved at* $x_0 \in \Omega$ *with* $G(x_0) \stackrel{\lessdot}{=} \vartheta_{\mathcal{Z}}$ *and* $\langle\langle z_{*0}, G(x_0)\rangle\rangle = 0$.

Proof By Facts 8.51 and 8.52, Assumption 8.55, and Proposition 8.33, we have

$$\mu_0 = \omega(\vartheta_{\mathcal{Z}}) = \sup_{z_* \in \Gamma_{\text{conj}}} -\omega_{\text{conj}}(z_*)$$

Then, by Facts 8.53 and 8.54, we have

$$\mu_0 = \sup_{z_* \in P^{\ominus}} -\bar{\omega}(z_*) = \sup_{z_* \stackrel{\gtrdot}{=} \vartheta_{\mathcal{Z}^*}} -\bar{\omega}(-z_*) = \sup_{z_* \stackrel{\gtrdot}{=} \vartheta_{\mathcal{Z}^*}} \varphi(z_*)$$

Therefore, (8.13) holds.

Let the supremum in (8.13) be achieved at $z_{*0} \in P^{\oplus}$, that is, (8.14) holds.

If the infimum in (8.11) is achieved at $x_0 \in \Omega$ with $G(x_0) \stackrel{\lessdot}{=} \vartheta_{\mathcal{Z}}$, then, we have $\mu_0 = f(x_0) \geq f(x_0) + \langle\langle z_{*0}, G(x_0)\rangle\rangle \geq \inf_{x \in \Omega}(f(x) + \langle\langle z_{*0}, G(x)\rangle\rangle) = \mu_0$. Hence, the infimum in (8.14) is achieved at x_0 and $\langle\langle z_{*0}, G(x_0)\rangle\rangle = 0$.

On the other hand, if the infimum in (8.14) is achieved at $x_0 \in \Omega$ with $G(x_0) \stackrel{\lessdot}{=} \vartheta_{\mathcal{Z}}$ and $\langle\langle z_{*0}, G(x_0)\rangle\rangle = 0$, then $\mu_0 = f(x_0) + \langle\langle z_{*0}, G(x_0)\rangle\rangle = f(x_0)$. Hence, the infimum in (8.11) is achieved at x_0.

This completes the proof of the proposition. □

Proposition 8.58 *Let* $\mathcal{X}$ *be a real vector space,* $\Omega \subseteq \mathcal{X}$ *be nonempty and convex,* $f : \Omega \to \mathbb{R}$ *be a convex functional,* $\mathcal{Z}$ *be a real normed linear space with the positive cone* $P \subseteq \mathcal{Z}$, $P^{\circ} \neq \emptyset$, $G : \Omega \to \mathcal{Z}$ *be a convex mapping. Let Assumptions* 8.50 *and* 8.56 *hold. Then,*

$$\mu_0 = \max_{z_* \stackrel{\gtrdot}{=} \vartheta_{\mathcal{Z}^*}} \inf_{x \in \Omega} (f(x) + \langle\langle z_*, G(x)\rangle\rangle) = \max_{z_* \stackrel{\gtrdot}{=} \vartheta_{\mathcal{Z}^*}} \varphi(z_*) \tag{8.15}$$

where the maximum is achieved at $z_{*0} \in P^{\oplus}$, *that is,*

$$\mu_0 = \inf_{x \in \Omega} (f(x) + \langle\langle z_{*0}, G(x)\rangle\rangle) \tag{8.16}$$

Furthermore, the infimum in (8.11) *is achieved at* $x_0 \in \Omega$ *with* $G(x_0) \stackrel{\lessdot}{=} \vartheta_{\mathcal{Z}}$ *if, and only if, the infimum in* (8.16) *is achieved at* $x_0 \in \Omega$ *with* $G(x_0) \stackrel{\lessdot}{=} \vartheta_{\mathcal{Z}}$ *and* $\langle\langle z_{*0}, G(x_0)\rangle\rangle = 0$.

Proof By Assumption 8.56, $G(x_1) \lessdot \vartheta_{\mathcal{Z}}$. Then, $-G(x_1) \in P^{\circ}$ and $G(x_1) + P \subseteq \Gamma$. By Proposition 7.16, $\Gamma^{\circ} \supseteq G(x_1) + P^{\circ}$ and $\vartheta_{\mathcal{Z}} \in \Gamma^{\circ}$. $\forall z \in G(x_1) + P$, we have $z \stackrel{\geq}{=} G(x_1)$ and, by Fact 8.52, $\omega(z) \leq \omega(G(x_1))$. Since $G(x_1) \lessdot \vartheta_{\mathcal{Z}}$, then $\exists \delta \in (0, \infty) \subset \mathbb{R}$ such that $\mathcal{B}_{\mathcal{Z}}(\vartheta_{\mathcal{Z}}, \delta) \subseteq G(x_1) + P$. Take $r_0 := \omega(G(x_1)) + \delta \in \mathbb{R}$. It is easy to check that $\mathcal{B}_{\mathbb{R}\times\mathcal{Z}}((r_0, \vartheta_{\mathcal{Z}}), \delta) \subseteq [\omega, \Gamma]$. Hence, $(r_0, \vartheta_{\mathcal{Z}}) \in [\omega, \Gamma]^\circ$. By Proposition 8.22, ω is continuous at $\vartheta_{\mathcal{Z}}$. By Facts 8.51 and 8.52 and Proposition 8.34, we have

$$\mu_0 = \omega(\vartheta_{\mathcal{Z}}) = \max_{z_* \in \Gamma_{\text{conj}}} -\omega_{\text{conj}}(z_*)$$

Then, by Facts 8.53 and 8.54, we have

$$\mu_0 = \max_{z_* \in P^{\ominus}} -\bar{\omega}(z_*) = \max_{z_* \stackrel{\geq}{=} \vartheta_{\mathcal{Z}^*}} -\bar{\omega}(-z_*) = \max_{z_* \stackrel{\geq}{=} \vartheta_{\mathcal{Z}^*}} \varphi(z_*)$$

Therefore, (8.15) holds and the maximum is achieved at $z_{*0} \in P^{\oplus}$.

If the infimum in (8.11) is achieved at $x_0 \in \Omega$ with $G(x_0) \stackrel{\leq}{=} \vartheta_{\mathcal{Z}}$, then, we have $\mu_0 = f(x_0) \geq f(x_0) + \langle\langle z_{*0}, G(x_0)\rangle\rangle \geq \inf_{x\in\Omega}(f(x) + \langle\langle z_{*0}, G(x)\rangle\rangle) = \mu_0$. Hence, the infimum in (8.16) is achieved at x_0 and $\langle\langle z_{*0}, G(x_0)\rangle\rangle = 0$.

On the other hand, if the infimum in (8.16) is achieved at $x_0 \in \Omega$ with $G(x_0) \stackrel{\leq}{=} \vartheta_{\mathcal{Z}}$ and $\langle\langle z_{*0}, G(x_0)\rangle\rangle = 0$, then $\mu_0 = f(x_0) + \langle\langle z_{*0}, G(x_0)\rangle\rangle = f(x_0)$. Hence, the infimum in (8.11) is achieved at x_0.

This completes the proof of the proposition. □

In the Propositions 8.57 and 8.58, z_{*0} is called the *Lagrange multiplier*. Assumption 8.50 guarantees that the primal functional is real-valued and convex. Assumption 8.56 guarantees the existence of a Lagrange multiplier. This assumption is restrictive. On the other hand, Assumption 8.55 guarantees the duality but not the existence of a Lagrange multiplier. This condition is more relaxed.

Corollary 8.59 *Let $\mathcal{X}$ be a real vector space, $\Omega \subseteq \mathcal{X}$ be nonempty and convex, $f : \Omega \to \mathbb{R}$ be a convex functional, $\mathcal{Z}$ be a real normed linear space with the positive cone $P \subseteq \mathcal{Z}$, $P^\circ \neq \emptyset$, $G : \Omega \to \mathcal{Z}$ be a convex mapping. Let Assumptions 8.50 and 8.56 hold and $x_0 \in \Omega$ with $G(x_0) \stackrel{\leq}{=} \vartheta_{\mathcal{Z}}$ achieves the infimum in (8.11). Then, there exists $z_{*0} \in \mathcal{Z}^*$ with $z_{*0} \stackrel{\geq}{=} \vartheta_{\mathcal{Z}^*}$ such that the Lagrangian $L : \Omega \times P^{\oplus} \to \mathbb{R}$ defined by*

$$L(x, z_*) := f(x) + \langle\langle z_*, G(x)\rangle\rangle; \qquad \forall x \in \Omega,\ \forall z_* \in P^{\oplus}$$

admits a saddle point at (x_0, z_{*0}), i.e.,

$$L(x_0, z_*) \leq L(x_0, z_{*0}) \leq L(x, z_{*0}); \qquad \forall x \in \Omega,\ \forall z_* \in P^{\oplus}$$

Proof By Proposition 8.58, there exists $z_{*0} \in \mathcal{Z}^*$ with $z_{*0} \stackrel{\geq}{=} \vartheta_{\mathcal{Z}^*}$ such that $L(x_0, z_{*0}) \leq L(x, z_{*0})$, $\forall x \in \Omega$ and $\langle\langle z_{*0}, G(x_0)\rangle\rangle = 0$. Then, $\forall z_* \in P^{\oplus}$, we have $L(x_0, z_*) - L(x_0, z_{*0}) = \langle\langle z_*, G(x_0)\rangle\rangle - \langle\langle z_{*0}, G(x_0)\rangle\rangle = \langle\langle z_*, G(x_0)\rangle\rangle \leq$

0. Hence, the saddle-point condition holds. This completes the proof of the corollary. □

Next, we present two sufficiency results on Lagrange multipliers.

Proposition 8.60 *Let $\mathcal{X}$ be a real vector space, $\Omega \subseteq \mathcal{X}$ be nonempty, $f : \Omega \to \mathbb{R}$, $\mathcal{Z}$ be a real normed linear space with the positive cone $P \subseteq \mathcal{Z}$, $G : \Omega \to \mathcal{Z}$.*

Assume that there exist $x_0 \in \Omega$ and $z_{*0} \in \mathcal{Z}^*$ with $z_{*0} \geqq \vartheta_{\mathcal{Z}^*}$ such that

$$f(x_0) + \langle\langle z_{*0}, G(x_0)\rangle\rangle \leq f(x) + \langle\langle z_{*0}, G(x)\rangle\rangle; \qquad \forall x \in \Omega$$

Then,

$$f(x_0) = \inf_{x \in \Omega,\ G(x) \leqq G(x_0)} f(x)$$

Proof $\forall x \in \Omega$ with $G(x) \leqq G(x_0)$, we have $\langle\langle z_{*0}, G(x)\rangle\rangle \leq \langle\langle z_{*0}, G(x_0)\rangle\rangle$, since $z_{*0} \in P^{\oplus}$. By the assumption of the proposition, $f(x_0) + \langle\langle z_{*0}, G(x_0)\rangle\rangle \leq f(x) + \langle\langle z_{*0}, G(x)\rangle\rangle$. Then, $f(x_0) \leq f(x)$. This completes the proof of the proposition.

□

Proposition 8.61 *Let $\mathcal{X}$ be a real vector space, $\Omega \subseteq \mathcal{X}$ be nonempty, $f : \Omega \to \mathbb{R}$, $\mathcal{Z}$ be a real normed linear space with the positive cone $P \subseteq \mathcal{Z}$, $G : \Omega \to \mathcal{Z}$.*

Assume that there exist $x_0 \in \Omega$ and $z_{*0} \in \mathcal{Z}^*$ with $z_{*0} \geqq \vartheta_{\mathcal{Z}^*}$ such that the Lagrangian $L : \Omega \times P^{\oplus} \to \mathbb{R}$ given by

$$L(x, z_*) = f(x) + \langle\langle z_*, G(x)\rangle\rangle; \qquad \forall x \in \Omega,\ \forall z_* \in P^{\oplus}$$

Admits a saddle point at (x_0, z_{*0}), i.e.,

$$L(x_0, z_*) \leq L(x_0, z_{*0}) \leq L(x, z_{*0}); \qquad \forall x \in \Omega,\ \forall z_* \in P^{\oplus}$$

Then, $G(x_0) \leqq \vartheta_{\mathcal{Z}}$ and

$$f(x_0) = L(x_0, z_{*0}) = \inf_{x \in \Omega,\ G(x) \leqq \vartheta_{\mathcal{Z}}} f(x)$$

Proof By the first inequality in the saddle-point condition, we have

$$\langle\langle z_*, G(x_0)\rangle\rangle \leq \langle\langle z_{*0}, G(x_0)\rangle\rangle; \qquad \forall z_* \in P^{\oplus}$$

$\forall z_* \in P^{\oplus}$, we have $z_* \geqq \vartheta_{\mathcal{Z}^*}$ and $z_* + z_{*0} \geq \vartheta_{\mathcal{Z}^*}$. Then, $\langle\langle z_* + z_{*0}, G(x_0)\rangle\rangle \leq \langle\langle z_{*0}, G(x_0)\rangle\rangle$ and $\langle\langle z_*, G(x_0)\rangle\rangle \leq 0$. By Proposition 8.46, $G(x_0) \in (-P)$ and $G(x_0) \leqq \vartheta_{\mathcal{Z}}$. Furthermore, $0 = \langle\langle \vartheta_{\mathcal{Z}^*}, G(x_0)\rangle\rangle \leq \langle\langle z_{*0}, G(x_0)\rangle\rangle \leq 0$ implies that $\langle\langle z_{*0}, G(x_0)\rangle\rangle = 0$.

$\forall x \in \Omega$ with $G(x) \stackrel{\leq}{=} \vartheta_{\mathcal{Z}}$, we have $f(x) \geq f(x) + \langle\langle z_{*0}, G(x)\rangle\rangle \geq f(x_0) + \langle\langle z_{*0}, G(x_0)\rangle\rangle = f(x_0)$. This completes the proof of the proposition. □

Next, we present a result on the sensitivity of the infimization problem.

Proposition 8.62 *Let $\mathcal{X}$ be a real vector space, $\Omega \subseteq \mathcal{X}$ be nonempty, $f : \Omega \to \mathbb{R}$, $\mathcal{Z}$ be a real normed linear space with the positive cone $P \subseteq \mathcal{Z}$, $G : \Omega \to \mathcal{Z}$.*

Let $z_i \in \mathcal{Z}$, $\mu_i = \inf_{x\in\Omega,\ G(x)\stackrel{\leq}{=}z_i} f(x)$, $i = 0, 1$. Let $z_{*0} \in P^{\oplus} \subseteq \mathcal{Z}^*$ be the Lagrange multiplier associated with μ_0, that is

$$\mu_0 = \inf_{x\in\Omega} (f(x) + \langle\langle z_{*0}, G(x) - z_0\rangle\rangle)$$

Assume that $\mu_0 \in \mathbb{R}$. Then, we have

$$\mu_1 - \mu_0 \geq \langle\langle -z_{*0}, z_1 - z_0\rangle\rangle$$

Proof $\forall x \in \Omega$ with $G(x) \stackrel{\leq}{=} z_1$, we have $\mu_0 \leq f(x) + \langle\langle z_{*0}, G(x) - z_0\rangle\rangle$, which implies that

$$f(x) \geq \mu_0 - \langle\langle z_{*0}, G(x) - z_0\rangle\rangle \geq \mu_0 - \langle\langle z_{*0}, z_1 - z_0\rangle\rangle$$

Hence, $\mu_1 \geq \mu_0 + \langle\langle -z_{*0}, z_1 - z_0\rangle\rangle$. This completes the proof of the proposition. □

Chapter 9
Differentiation in Banach Spaces

In this chapter, we are going to develop the concept of derivative in normed linear spaces.

9.1 Fundamental Notion

Definition 9.1 Let $\mathcal{X}$ be a normed linear space over the field $\mathbb{K}$, $D \subseteq \mathcal{X}$, and $x_0 \in \overline{D}$. $u \in \mathcal{X}$ is said to be an *admissible deviation in* D *at* x_0 if $\forall \epsilon \in (0, \infty) \subset \mathbb{R}$, we have $\{x_0 + r\bar{u} \mid r \in (0, \epsilon) \subset \mathbb{R},\ \bar{u} \in \mathcal{B}(u, \epsilon)\} \cap D \neq \emptyset$. Let $\mathrm{A}_D(x_0)$ be the *set of admissible deviations in* D *at* x_0. ◇

Clearly, if $x_0 \in D^\circ$, then $\mathrm{A}_D(x_0) = \mathcal{X}$. Another fact is that when $D_1 \subseteq D_2 \subseteq \mathcal{X}$ and $x_0 \in \overline{D_1}$, then $\mathrm{A}_{D_1}(x_0) \subseteq \mathrm{A}_{D_2}(x_0)$. Yet another fact is that when $x_0 \in \overline{D_1}$, $D_1, D_2 \subseteq \mathcal{X}$, and $\exists \delta \in (0, \infty) \subset \mathbb{R}$ such that $D_1 \cap \mathcal{B}(x_0, \delta) = D_2 \cap \mathcal{B}(x_0, \delta)$, then $\mathrm{A}_{D_1}(x_0) = \mathrm{A}_{D_2}(x_0)$.

Proposition 9.2 *Let* $\mathcal{X}$ *be a normed linear space over the field* $\mathbb{K}$, $D \subseteq \mathcal{X}$, *and* $x_0 \in \overline{D}$. *Then,* $\mathrm{A}_D(x_0) \subseteq \mathcal{X}$ *is a closed cone.*

Proof Clearly, if $x_0 \in D$, then $\vartheta \in \mathrm{A}_D(x_0)$. On the other hand, if $x_0 \in \overline{D} \setminus D$, then, by Proposition 4.13, x_0 is an accumulation point of D. $\forall \epsilon \in (0, \infty) \subset \mathbb{R}$, $\exists \bar{x} \in (\mathcal{B}(x_0, \epsilon^2) \cap D) \setminus \{x_0\}$. Let $v := \bar{x} - x_0$. Then, $\delta := \sqrt{\|v\|} \in (0, \epsilon) \subset \mathbb{R}$. Let $\bar{v} := \delta^{-1} v$. Then, $\bar{v} \in \mathcal{B}(\vartheta, \epsilon)$ and $\bar{x} = x_0 + \delta \bar{v} \in \{x_0 + r\bar{u} \mid r \in (0, \epsilon) \subset \mathbb{R},\ \bar{u} \in \mathcal{B}(\vartheta, \epsilon)\} \cap D \neq \emptyset$. Hence, $\vartheta \in \mathrm{A}_D(x_0)$. Therefore, $\vartheta \in \mathrm{A}_D(x_0)$ if $x_0 \in \overline{D}$.

$\forall u \in \mathrm{A}_D(x_0)$, $\forall \alpha \in [0, \infty) \subset \mathbb{R}$, we will show that $\alpha u \in \mathrm{A}_D(x_0)$ by distinguishing two exhaustive and mutually exclusive cases: Case 1: $\alpha = 0$; Case 2: $\alpha > 0$. Case 1: $\alpha = 0$. Then, $\alpha u = \vartheta \in \mathrm{A}_D(x_0)$. Case 2: $\alpha > 0$. $\forall \epsilon \in (0, \infty) \subset \mathbb{R}$, let $\bar{\epsilon} = \min\{\alpha\epsilon, \epsilon/\alpha\} \in (0, \infty) \subset \mathbb{R}$. By $u \in \mathrm{A}_D(x_0)$, $\exists \bar{x} \in \{x_0 + r\bar{u} \mid r \in (0, \bar{\epsilon}) \subset \mathbb{R},\ \bar{u} \in \mathcal{B}(u, \bar{\epsilon})\} \cap D$. Hence, $\bar{x} \in D$ and $\bar{x} = x_0 + r\bar{u} = x_0 + (r/\alpha)(\alpha\bar{u})$ with $r \in (0, \bar{\epsilon}) \subset \mathbb{R}$ and $\bar{u} \in \mathcal{B}(u, \bar{\epsilon})$. Then, $r/\alpha \in (0, \epsilon) \subset \mathbb{R}$ and $\alpha\bar{u} \in \mathcal{B}(\alpha u, \epsilon)$.

Z. Pan, *Measure-Theoretic Calculus in Abstract Spaces*,
https://doi.org/10.1007/978-3-031-21912-2_9

Hence, $\bar{x} \in \{x_0 + \tilde{r}\tilde{u} \mid \tilde{r} \in (0, \epsilon) \subset \mathbb{R},\ \tilde{u} \in \mathcal{B}(\alpha u, \epsilon)\} \cap D \neq \emptyset$. This implies that $\alpha u \in \mathrm{A}_D(x_0)$. Hence, $\mathrm{A}_D(x_0)$ is a cone.

$\forall u \in \overline{\mathrm{A}_D(x_0)}$, $\forall \epsilon \in (0, \infty) \subset \mathbb{R}$, by Proposition 3.3, $\exists u_1 \in \mathrm{A}_D(x_0) \cap \mathcal{B}(u, \epsilon/2)$. By $u_1 \in \mathrm{A}_D(x_0)$, $\exists \bar{x} \in \{x_0 + r\bar{u} \mid r \in (0, \epsilon/2) \subset \mathbb{R},\ \bar{u} \in \mathcal{B}(u_1, \epsilon/2)\} \cap D$. Then, $\bar{x} \in \{x_0 + r\bar{u} \mid r \in (0, \epsilon) \subset \mathbb{R},\ \bar{u} \in \mathcal{B}(u, \epsilon)\} \cap D \neq \emptyset$. Hence, $u \in \mathrm{A}_D(x_0)$. Then, $\overline{\mathrm{A}_D(x_0)} \subseteq \mathrm{A}_D(x_0)$ and $\mathrm{A}_D(x_0)$ is closed. This completes the proof of the proposition. □

Definition 9.3 Let $\mathcal{X}$ and $\mathcal{Y}$ be normed linear spaces over $\mathbb{K}$, $D \subseteq \mathcal{X}$, $f : D \to \mathcal{Y}$, and $x_0 \in D$. Assume that $\overline{\mathrm{span}(\mathrm{A}_D(x_0))} = \mathcal{X}$. Let $L \in \mathrm{B}(\mathcal{X}, \mathcal{Y})$ be such that $\forall \epsilon \in (0, \infty) \subset \mathbb{R}$, $\exists \delta \in (0, \infty) \subset \mathbb{R}$, $\forall x \in D \cap \mathcal{B}_{\mathcal{X}}(x_0, \delta)$, we have

$$\| f(x) - f(x_0) - L(x - x_0)\| \le \epsilon \|x - x_0\|$$

Then, L is called the *(Fréchet) derivative of* f *at* x_0 and denoted by $f^{(1)}(x_0)$ or $\mathrm{D}f(x_0)$. When L exists, we will say that f is *(Fréchet) differentiable at* x_0. $\mathrm{D}f$ or $f^{(1)}$ will denote the $\mathrm{B}(\mathcal{X}, \mathcal{Y})$-valued function whose domain of definition is $\mathrm{dom}\left(f^{(1)}\right) := \{x \in D \mid \mathrm{D}f(x) \in \mathrm{B}(\mathcal{X}, \mathcal{Y}) \text{ exists}\}$. If f is differentiable at x_0, $\forall x_0 \in D$, we say f is *(Fréchet) differentiable*. In this case, $\mathrm{D}f : D \to \mathrm{B}(\mathcal{X}, \mathcal{Y})$ or $f^{(1)} : D \to \mathrm{B}(\mathcal{X}, \mathcal{Y})$. ◇

Clearly, when $\mathcal{X} = \mathcal{Y} = \mathbb{R}$ and $D = [a, b] \subseteq \mathcal{X}$ with $a < b$, then $Df(t)$ is simply the derivative of f at $t \in [a, b]$, as we know before.

Definition 9.4 Let $\mathcal{X}$ and $\mathcal{Y}$ be normed linear spaces over $\mathbb{K}$, $D \subseteq \mathcal{X}$, $f : D \to \mathcal{Y}$, $x_0 \in D$, and $u \in \mathrm{A}_D(x_0)$. Let $M_D(x_0) := \{(r, \bar{u}) \in \mathbb{R} \times \mathcal{X} \mid r \in (0, +\infty) \subset \mathbb{R},\ \bar{u} \in \mathcal{X},\ x_0 + r\bar{u} \in D\}$, and $\bar{g} : M_D(x_0) \to \mathcal{Y}$ be given by $\bar{g}(r, \bar{u}) = r^{-1}(f(x_0 + r\bar{u}) - f(x_0))$, $\forall (r, \bar{u}) \in M_D(x_0)$. Clearly, $(0, u)$ is an accumulation point of $M_D(x_0)$ since $u \in \mathrm{A}_D(x_0)$. The *directional derivative of* f *at* x_0 *along* u, denoted by $\mathrm{D}f(x_0; u)$, is the limit $\lim_{(r,\bar{u}) \to (0,u)} \bar{g}(r, \bar{u})$, when it exists. ◇

Clearly, the directional derivative is unique when it exists, since $\mathcal{Y}$ is Hausdorff.

Proposition 9.5 *Let* $\mathcal{X}$ *and* $\mathcal{Y}$ *be normed linear spaces over* $\mathbb{K}$, $D \subseteq \mathcal{X}$, $f : D \to \mathcal{Y}$, $x_0 \in D$, $\overline{\mathrm{span}(\mathrm{A}_D(x_0))} = \mathcal{X}$, $L \in \mathrm{B}(\mathcal{X}, \mathcal{Y})$ *be the Fréchet derivative of* f *at* x_0*, and* $u \in \mathrm{A}_D(x_0)$. *Then,* $\mathrm{D}f(x_0; u) = Lu$.

Proof $\forall \epsilon \in (0, \infty) \subset \mathbb{R}$, let $\bar{\epsilon} = \epsilon/(2 + 2\|u\|) \in (0, \infty) \subset \mathbb{R}$. By $\mathrm{D}f(x_0) = L$, $\exists \delta \in (0, \infty) \subset \mathbb{R}$ such that $\|f(x) - f(x_0) - L(x - x_0)\| \le \bar{\epsilon}\|x - x_0\|$, $\forall x \in \mathcal{B}_{\mathcal{X}}(x_0, \delta) \cap D$. Let $M_D(x_0)$ and $\bar{g} : M_D(x_0) \to \mathcal{Y}$ be as defined in Definition 9.4. Let $\bar{\delta} = \min\{\epsilon/(2 + 2\|L\|), \delta/(1 + \|u\|), 1\} \in (0, \infty) \subset \mathbb{R}$. $\forall (r, \bar{u}) \in \left(M_D(x_0) \cap \mathcal{B}_{\mathbb{R}\times\mathcal{X}}\left((0, u), \bar{\delta}\right)\right) \setminus \{(0, u)\}$, we have $\bar{x} := x_0 + r\bar{u} \in D \cap \mathcal{B}_{\mathcal{X}}(x_0, \delta)$. This implies that

$$\begin{aligned}
\|\bar{g}(r, \bar{u}) - Lu\| &= \|f(x_0 + r\bar{u}) - f(x_0) - rLu\|/r \\
&= \|f(\bar{x}) - f(x_0) - L(\bar{x} - x_0) + rL(\bar{u} - u)\|/r \\
&\le \|f(\bar{x}) - f(x_0) - L(\bar{x} - x_0)\|/r + \|L(\bar{u} - u)\|
\end{aligned}$$

$$\leq \bar{\epsilon}\|\bar{x} - x_0\|/r + \|L\|\|(\bar{u} - u)\| = \bar{\epsilon}\|\bar{u}\| + \|L\|\|\bar{u} - u\|$$
$$\leq \bar{\epsilon}(\|u\| + \|\bar{u} - u\|) + \|L\|\|\bar{u} - u\| \leq \bar{\epsilon}(\|u\| + \bar{\delta}) + \|L\|\bar{\delta}$$
$$< \epsilon$$

where the second inequality follows from Proposition 7.64. Hence, we have $\mathrm{D}f(x_0; u) = \lim_{(r,\bar{u})\to(0,u)} \bar{g}(r, \bar{u}) = Lu$. This completes the proof of the proposition. □

Proposition 9.6 *Let $\mathcal{X}$ and $\mathcal{Y}$ be normed linear spaces over $\mathbb{K}$, $D \subseteq \mathcal{X}$, $f : D \to \mathcal{Y}$, $x_0 \in D$, and $\overline{\mathrm{span}\,(\mathrm{A}_D\,(x_0))} = \mathcal{X}$. Then, f has at most one derivative at x_0.*

Proof We will prove the result by an argument of contradiction. Suppose, $\exists L_1, L_2 \in \mathrm{B}(\mathcal{X}, \mathcal{Y})$ with $L_1 \neq L_2$ such that L_1 and L_2 are derivatives of f at x_0. By $\overline{\mathrm{span}\,(\mathrm{A}_D\,(x_0))} = \mathcal{X}$ and Proposition 3.56, $\exists \bar{u} \in \mathrm{span}\,(\mathrm{A}_D\,(x_0))$ such that $L_1\bar{u} \neq L_2\bar{u}$. Since L_1 and L_2 are linear, then $\exists u_0 \in \mathrm{A}_D\,(x_0)$ such that $L_1u_0 \neq L_2u_0$. By Proposition 9.5, $\mathrm{D}f(x_0; u_0) = L_1u_0 = L_2u_0$. This is a contradiction. Hence, the result holds. This completes the proof of the proposition. □

Proposition 9.7 *Let $\mathcal{X}$ and $\mathcal{Y}$ be normed linear spaces over $\mathbb{K}$, $D \subseteq \mathcal{X}$, $f : D \to \mathcal{Y}$, $x_0 \in D$, and $\overline{\mathrm{span}\,(\mathrm{A}_D\,(x_0))} = \mathcal{X}$. Assume that $\mathrm{D}f(x_0) \in \mathrm{B}(\mathcal{X}, \mathcal{Y})$ exists; then f is continuous at x_0.*

Proof This is straightforward and is therefore omitted. □

Definition 9.8 Let X be a set, $\mathcal{Y}$ and $\mathcal{Z}$ be normed linear spaces over $\mathbb{K}$, $D \subseteq X \times \mathcal{Y}$, $f : D \to \mathcal{Z}$, and $(x_0, y_0) \in D$. f is said to be *partial (Fréchet) differentiable with respect to y at (x_0, y_0)* if $g : D_1 \to \mathcal{Z}$ given by $g(y) = f(x_0, y)$, $\forall y \in D_1$, where $D_1 := \{y \in \mathcal{Y} \,|\, (x_0, y) \in D\}$, is differentiable at y_0. $\mathrm{D}g(y_0) \in \mathrm{B}(\mathcal{Y}, \mathcal{Z})$ is called the *partial derivative of f with respect to y at (x_0, y_0)* and is denoted by $\frac{\partial f}{\partial y}(x_0, y_0)$. $\frac{\partial f}{\partial y}$ will denote the $\mathrm{B}(\mathcal{Y}, \mathcal{Z})$-valued function whose domain of definition is $\mathrm{dom}\left(\frac{\partial f}{\partial y}\right) := \left\{(x, y) \in D \,\middle|\, \frac{\partial f}{\partial y}(x, y) \in \mathrm{B}(\mathcal{Y}, \mathcal{Z}) \text{ exists}\right\}$. When $\mathrm{dom}\left(\frac{\partial f}{\partial y}\right) = D$, we say that f is *partial differentiable with respect to y*. ◇

For notational simplicity, we will adopt the "matrix" notation for our later developments. Let $\mathcal{X}_1, \ldots, \mathcal{X}_n$ be normed linear spaces over $\mathbb{K}$, $n \in \mathbb{N}$. For a vector $(x_1, \ldots, x_n) \in \mathcal{X}_1 \times \cdots \times \mathcal{X}_n$, we will use the notation of a column vector

$$\begin{bmatrix} x_1 \\ \vdots \\ x_n \end{bmatrix}$$

to alternatively denote it. Let $\mathcal{Y}_1, \ldots, \mathcal{Y}_m$ be normed linear spaces over $\mathbb{K}$, $m \in \mathbb{N}$. $\forall L \in \mathrm{B}(\mathcal{X}_1 \times \cdots \times \mathcal{X}_n, \mathcal{Y}_1 \times \cdots \times \mathcal{Y}_m)$ can be equivalently written as

$$\begin{bmatrix} L_{11} & \cdots & L_{1n} \\ \vdots & & \vdots \\ L_{m1} & \cdots & L_{mn} \end{bmatrix}$$

where $L_{ij} \in \mathrm{B}(\mathcal{X}_j, \mathcal{Y}_i)$. Then, $(y_1, \ldots, y_m) = L(x_1, \ldots, x_n)$ can be equivalently expressed as

$$\begin{bmatrix} y_1 \\ \vdots \\ y_m \end{bmatrix} = \begin{bmatrix} L_{11} & \cdots & L_{1n} \\ \vdots & & \vdots \\ L_{m1} & \cdots & L_{mn} \end{bmatrix} \begin{bmatrix} x_1 \\ \vdots \\ x_n \end{bmatrix} = \begin{bmatrix} \sum_{j=1}^{n} L_{1j} x_j \\ \vdots \\ \sum_{j=1}^{n} L_{mj} x_j \end{bmatrix}$$

Let $\mathcal{Z}_1, \ldots, \mathcal{Z}_p$ be normed linear spaces over $\mathbb{K}$, $p \in \mathbb{N}$, and $\bar{L} \in \mathrm{B}(\mathcal{Y}_1 \times \cdots \times \mathcal{Y}_m, \mathcal{Z}_1 \times \cdots \times \mathcal{Z}_p)$ be given by

$$\begin{bmatrix} \bar{L}_{11} & \cdots & \bar{L}_{1m} \\ \vdots & & \vdots \\ \bar{L}_{p1} & \cdots & \bar{L}_{pm} \end{bmatrix}$$

Then, $\bar{L}L \in \mathrm{B}(\mathcal{X}_1 \times \cdots \times \mathcal{X}_n, \mathcal{Z}_1 \times \cdots \times \mathcal{Z}_p)$ and

$$\begin{aligned} \bar{L}L &= \begin{bmatrix} \bar{L}_{11} & \cdots & \bar{L}_{1m} \\ \vdots & & \vdots \\ \bar{L}_{p1} & \cdots & \bar{L}_{pm} \end{bmatrix} \begin{bmatrix} L_{11} & \cdots & L_{1n} \\ \vdots & & \vdots \\ L_{m1} & \cdots & L_{mn} \end{bmatrix} \\ &= \begin{bmatrix} \sum_{j=1}^{m} \bar{L}_{1j} L_{j1} & \cdots & \sum_{j=1}^{m} \bar{L}_{1j} L_{jn} \\ \vdots & & \vdots \\ \sum_{j=1}^{m} \bar{L}_{pj} L_{j1} & \cdots & \sum_{j=1}^{m} \bar{L}_{pj} L_{jn} \end{bmatrix} \end{aligned}$$

It is easy to show that the adjoint of L is

$$L' = \begin{bmatrix} L'_{11} & \cdots & L'_{m1} \\ \vdots & & \vdots \\ L'_{1n} & \cdots & L'_{mn} \end{bmatrix}$$

Proposition 9.9 *Let $\mathcal{X}$, $\mathcal{Y}$, and $\mathcal{Z}$ be normed linear spaces over $\mathbb{K}$, $D \subseteq \mathcal{X} \times \mathcal{Y}$, $f : D \to \mathcal{Z}$, and $(x_0, y_0) \in D$. Assume $\mathrm{D}f(x_0, y_0) \in \mathrm{B}(\mathcal{X} \times \mathcal{Y}, \mathcal{Z})$ exists. Let*

$D_{x_0} := \{y \in \mathcal{Y} \mid (x_0, y) \in D\}$. *Assume that* $\overline{\text{span}\left(\text{A}_{D_{x_0}}(y_0)\right)} = \mathcal{Y}$. *Then, f is partial differentiable with respect to y at (x_0, y_0), and* $\frac{\partial f}{\partial y}(x_0, y_0) \in \text{B}(\mathcal{Y}, \mathcal{Z})$ *is given by* $\frac{\partial f}{\partial y}(x_0, y_0)(k) = \text{D}f(x_0, y_0)(\vartheta_{\mathcal{X}}, k)$, $\forall k \in \mathcal{Y}$.

Let $D_{y_0} := \{x \in \mathcal{X} \mid (x, y_0) \in D\}$. *Assume that* $\overline{\text{span}\left(\text{A}_{D_{y_0}}(x_0)\right)} = \mathcal{X}$. *By symmetry, f is partial differentiable with respect to x at (x_0, y_0), and* $\frac{\partial f}{\partial x}(x_0, y_0) \in \text{B}(\mathcal{X}, \mathcal{Z})$ *is given by* $\frac{\partial f}{\partial x}(x_0, y_0)(h) = \text{D}f(x_0, y_0)(h, \vartheta_{\mathcal{Y}})$, $\forall h \in \mathcal{X}$.

Hence, we have $\text{D}f(x_0, y_0) = \left[\frac{\partial f}{\partial x}(x_0, y_0) \; \frac{\partial f}{\partial y}(x_0, y_0)\right]$ *in "matrix" notation.*

Proof Let $\bar{g} : D_{x_0} \to \mathcal{Z}$ be given by $\bar{g}(y) = f(x_0, y)$, $\forall y \in D_{x_0}$. Clearly, $y_0 \in D_{x_0}$. By the fact that f is differentiable at (x_0, y_0), then $\forall \epsilon \in (0, \infty) \subset \mathbb{R}$, $\exists \delta \in (0, \infty) \subset \mathbb{R}$, $\forall (x, y) \in D \cap \mathcal{B}_{\mathcal{X}\times\mathcal{Y}}((x_0, y_0), \delta)$, we have $\|f(x, y) - f(x_0, y_0) - \text{D}f(x_0, y_0)(x - x_0, y - y_0)\| \le \epsilon\|(x - x_0, y - y_0)\|$. Then, $\forall y \in D_{x_0} \cap \mathcal{B}_{\mathcal{Y}}(y_0, \delta)$, we have $(x_0, y) \in D \cap \mathcal{B}_{\mathcal{X}\times\mathcal{Y}}((x_0, y_0), \delta)$ and $\epsilon\|y - y_0\| \ge \|f(x_0, y) - f(x_0, y_0) - \text{D}f(x_0, y_0)(\vartheta_{\mathcal{X}}, y - y_0)\| = \|\bar{g}(y) - \bar{g}(y_0) - \text{D}f(x_0, y_0)(\vartheta_{\mathcal{X}}, y - y_0)\|$. Let $L : \mathcal{Y} \to \mathcal{Z}$ be given by $L(k) = \text{D}f(x_0, y_0)(\vartheta_{\mathcal{X}}, k)$, $\forall k \in \mathcal{Y}$. Clearly, L is a linear operator. Note that

$$\|L\| = \sup_{k\in\mathcal{Y},\ \|k\|\le 1} \|Lk\| = \sup_{k\in\mathcal{Y},\ \|k\|\le 1} \|\text{D}f(x_0, y_0)(\vartheta_{\mathcal{X}}, k)\|$$
$$\le \sup_{k\in\mathcal{Y},\ \|k\|\le 1} \|\text{D}f(x_0, y_0)\|\|(\vartheta_{\mathcal{X}}, k)\| \le \|\text{D}f(x_0, y_0)\| < +\infty$$

where the first inequality follows from Proposition 7.64 and the last inequality follows from the fact that $\text{D}f(x_0, y_0) \in \text{B}(\mathcal{X} \times \mathcal{Y}, \mathcal{Z})$. Hence, $L \in \text{B}(\mathcal{Y}, \mathcal{Z})$. Then, $\|\bar{g}(y) - \bar{g}(y_0) - L(y - y_0)\| \le \epsilon\|y - y_0\|$, $\forall y \in D_{x_0} \cap \mathcal{B}_{\mathcal{Y}}(y_0, \delta)$. This implies that $\text{D}\bar{g}(y_0) = L$. By Definition 9.8, $\frac{\partial f}{\partial y}(x_0, y_0) = L$.

By symmetry, f is partial differentiable with respect to x at (x_0, y_0), and $\frac{\partial f}{\partial x}(x_0, y_0) \in \text{B}(\mathcal{X}, \mathcal{Z})$ is given by $\frac{\partial f}{\partial x}(x_0, y_0)(h) = \text{D}f(x_0, y_0)(h, \vartheta_{\mathcal{Y}})$, $\forall h \in \mathcal{X}$.

Note that, $\forall h \in \mathcal{X}$ and $\forall k \in \mathcal{Y}$,

$$\text{D}f(x_0, y_0)(h, k) = \frac{\partial f}{\partial x}(x_0, y_0)(h) + \frac{\partial f}{\partial y}(x_0, y_0)(k)$$

Then,

$$\text{D}f(x_0, y_0)\begin{bmatrix} h \\ k \end{bmatrix} = \left[\frac{\partial f}{\partial x}(x_0, y_0) \; \frac{\partial f}{\partial y}(x_0, y_0)\right]\begin{bmatrix} h \\ k \end{bmatrix}$$

Hence, the desired "matrix" notation follows. This completes the proof of the proposition. □

9.2 The Derivatives of Some Common Functions

Proposition 9.10 *Let $\mathcal{X}$ and $\mathcal{Y}$ be normed linear spaces over $\mathbb{K}$, $D \subseteq \mathcal{X}$, $f : D \to \mathcal{Y}$, $x_0 \in D$, and $\overline{\operatorname{span}(A_D(x_0))} = \mathcal{X}$. Assume that $\exists \delta_0 \in (0, \infty) \subset \mathbb{R}$ and $\exists y_0 \in \mathcal{Y}$ such that $f(x) = y_0$, $\forall x \in D \cap \mathcal{B}_{\mathcal{X}}(x_0, \delta_0)$. Then, f is Fréchet differentiable at x_0 and $\mathrm{D}f(x_0) = \vartheta_{\mathrm{B}(\mathcal{X},\mathcal{Y})}$.*

Proof This is straightforward and is therefore omitted. □

Proposition 9.11 *Let $\mathcal{X}$ and $\mathcal{Y}$ be normed linear spaces over $\mathbb{K}$, $x_0 \in D_2 \subseteq D_1 \subseteq \mathcal{X}$, and $f : D_1 \to \mathcal{Y}$. If f is Fréchet differentiable at x_0 and $\overline{\operatorname{span}\left(A_{D_2}(x_0)\right)} = \mathcal{X}$, then $g := f|_{D_2}$ is Fréchet differentiable at x_0 and $\mathrm{D}g(x_0) = \mathrm{D}f(x_0)$. On the other hand, if g is Fréchet differentiable at x_0 and $\exists \delta_0 \in (0, \infty) \subset \mathbb{R}$ such that $D_1 \cap \mathcal{B}_{\mathcal{X}}(x_0, \delta_0) = D_2 \cap \mathcal{B}_{\mathcal{X}}(x_0, \delta_0)$, then f is Fréchet differentiable at x_0 and $\mathrm{D}f(x_0) = \mathrm{D}g(x_0)$.*

Proof This is straightforward and is therefore omitted. □

Proposition 9.12 *Let $\mathcal{X}$ be a normed linear space over $\mathbb{K}$, and $f : \mathcal{X} \to \mathcal{X}$ be given by $f = \mathrm{id}_{\mathcal{X}}$, that is, $f(x) = x$, $\forall x \in \mathcal{X}$. Then, f is Fréchet differentiable and $\mathrm{D}f(x)(h) = h$, $\forall x \in \mathcal{X}$ and $\forall h \in \mathcal{X}$. Then, $\mathrm{D}f(x) = \mathrm{id}_{\mathcal{X}}$, $\forall x \in \mathcal{X}$.*

Proof This is straightforward and is therefore omitted. □

Proposition 9.13 *Let $\mathcal{X}$ and $\mathcal{Y}$ be normed linear spaces over $\mathbb{K}$, and $f : \mathcal{X} \times \mathcal{Y} \to \mathcal{X}$ be given by $f = \pi_{\mathcal{X}}$, that is, $f(x, y) = x$, $\forall (x, y) \in \mathcal{X} \times \mathcal{Y}$. Then, f is Fréchet differentiable and $\mathrm{D}f(x, y)(h, k) = h$, $\forall (x, y) \in \mathcal{X} \times \mathcal{Y}$ and $\forall (h, k) \in \mathcal{X} \times \mathcal{Y}$. In "matrix" notation, we have $\mathrm{D}f(x, y) = \left[\, \mathrm{id}_{\mathcal{X}} \;\; \vartheta_{\mathrm{B}(\mathcal{Y},\mathcal{X})} \,\right]$, $\forall (x, y) \in \mathcal{X} \times \mathcal{Y}$.*

Proof This is straightforward and is therefore omitted. □

Proposition 9.14 *Let $\mathcal{X}$ be a normed linear space over $\mathbb{K}$, and $f : \mathcal{X} \times \mathcal{X} \to \mathcal{X}$ be given by $f(x_1, x_2) = x_1 + x_2$, $\forall (x_1, x_2) \in \mathcal{X} \times \mathcal{X}$. Then, f is Fréchet differentiable, and $\mathrm{D}f : \mathcal{X} \times \mathcal{X} \to \mathrm{B}(\mathcal{X} \times \mathcal{X}, \mathcal{X})$ is given by $\forall (x_1, x_2) \in \mathcal{X} \times \mathcal{X}$, $\forall (h_1, h_2) \in \mathcal{X} \times \mathcal{X}$, $\mathrm{D}f(x_1, x_2)(h_1, h_2) = h_1 + h_2$. In "matrix" notation, we have $\mathrm{D}f(x_1, x_2) = \left[\, \mathrm{id}_{\mathcal{X}} \;\; \mathrm{id}_{\mathcal{X}} \,\right]$.*

Proof $\forall (x_1, x_2) \in \mathcal{X} \times \mathcal{X}$, let $L : \mathcal{X} \times \mathcal{X} \to \mathcal{X}$ be given by $L(h_1, h_2) = h_1 + h_2$, $\forall (h_1, h_2) \in \mathcal{X} \times \mathcal{X}$. Clearly, L is a linear operator. Note that

$$\begin{aligned}
\|L\| &= \sup_{(h_1,h_2)\in\mathcal{X}\times\mathcal{X},\ \|(h_1,h_2)\|\le 1} \|L(h_1, h_2)\| \\
&\le \sup_{(h_1,h_2)\in\mathcal{X}\times\mathcal{X},\ \|(h_1,h_2)\|\le 1} \|h_1\| + \|h_2\| \\
&\le \sup_{(h_1,h_2)\in\mathcal{X}\times\mathcal{X},\ \|(h_1,h_2)\|\le 1} \sqrt{2}\left(\|h_1\|^2 + \|h_2\|^2\right)^{1/2} \le \sqrt{2}
\end{aligned}$$

where the second inequality follows from Cauchy–Schwarz inequality. Hence, $L \in \mathrm{B}(\mathcal{X} \times \mathcal{X}, \mathcal{X})$.

Clearly, $\mathrm{A}_{\mathcal{X}\times\mathcal{X}}(x_1, x_2) = \mathcal{X} \times \mathcal{X}$ since $(x_1, x_2) \in (\mathcal{X} \times \mathcal{X})^\circ$. $\forall \epsilon \in (0, \infty) \subset \mathbb{R}$, set $\delta = 1 \in \mathbb{R}$, $\forall (\bar{x}_1, \bar{x}_2) \in (\mathcal{X} \times \mathcal{X}) \cap \mathcal{B}_{\mathcal{X}\times\mathcal{X}}((x_1, x_2), \delta)$, and we have $\| f(\bar{x}_1, \bar{x}_2) - f(x_1, x_2) - L(\bar{x}_1 - x_1, \bar{x}_2 - x_2)\| = 0 \leq \epsilon \|(\bar{x}_1 - x_1, \bar{x}_2 - x_2)\|$. By Definition 9.3, $\mathrm{D}f(x_1, x_2) = L$. This completes the proof of the proposition. □

Proposition 9.15 *Let $\mathcal{X}$ and $\mathcal{Y}$ be normed linear spaces over $\mathbb{K}$, $D \subseteq \mathcal{X}$, $f_1 : D \to \mathcal{Y}$, $f_2 : D \to \mathcal{Y}$, $x_0 \in D$, $\alpha_1, \alpha_2 \in \mathbb{K}$, and $g : D \to \mathcal{Y}$ be given by $g(x) = \alpha_1 f_1(x) + \alpha_2 f_2(x)$, $\forall x \in D$. Assume that $f_1^{(1)}(x_0)$ and $f_2^{(1)}(x_0)$ exist. Then, g is Fréchet differentiable at x_0 and $g^{(1)}(x_0) = \alpha_1 f_1^{(1)}(x_0) + \alpha_2 f_2^{(1)}(x_0)$.*

Proof Define $L := \alpha_1 f_1^{(1)}(x_0) + \alpha_2 f_2^{(1)}(x_0) \in \mathrm{B}(\mathcal{X}, \mathcal{Y})$. By assumption, $\overline{\mathrm{span}\,(\mathrm{A}_D\,(x_0))} = \mathcal{X}$. $\forall \epsilon \in (0, \infty) \subset \mathbb{R}$, by the differentiability of f_1 at x_0, $\exists \delta_1 \in (0, \infty) \subset \mathbb{R}$ such that $\forall x \in D \cap \mathcal{B}_{\mathcal{X}}(x_0, \delta_1)$, we have $\left\| f_1(x) - f_1(x_0) - f_1^{(1)}(x_0)(x - x_0)\right\| \leq \epsilon \|x - x_0\|$. By the differentiability of f_2 at x_0, $\exists \delta_2 \in (0, \infty) \subset \mathbb{R}$ such that $\forall x \in D \cap \mathcal{B}_{\mathcal{X}}(x_0, \delta_2)$, we have $\left\| f_2(x) - f_2(x_0) - f_2^{(1)}(x_0)(x - x_0)\right\| \leq \epsilon \|x - x_0\|$. Let $\delta := \min\{\delta_1, \delta_2\} > 0$. $\forall x \in D \cap \mathcal{B}_{\mathcal{X}}(x_0, \delta)$, we have

$$
\begin{aligned}
&\|g(x) - g(x_0) - L(x - x_0)\| \\
&= \|\alpha_1 \left(f_1(x) - f_1(x_0) - f_1^{(1)}(x_0)(x - x_0)\right) + \alpha_2 \left(f_2(x) - f_2(x_0)\right. \\
&\quad \left. - f_2^{(1)}(x_0)(x - x_0)\right)\| \leq (|\alpha_1| + |\alpha_2|)\epsilon \|x - x_0\|
\end{aligned}
$$

Hence, $g^{(1)}(x_0) = L$. This completes the proof of the proposition. □

Proposition 9.16 *Let $\mathcal{X}$ be a normed linear space over $\mathbb{K}$ and $f : \mathbb{K} \times \mathcal{X} \to \mathcal{X}$ be given by $f(\alpha, x) = \alpha x$, $\forall (\alpha, x) \in \mathbb{K} \times \mathcal{X}$. Then, f is Fréchet differentiable and $\mathrm{D}f : \mathbb{K} \times \mathcal{X} \to \mathrm{B}(\mathbb{K} \times \mathcal{X}, \mathcal{X})$ is given by $\forall (\alpha, x) \in \mathbb{K} \times \mathcal{X}$, $\forall (d, h) \in \mathbb{K} \times \mathcal{X}$, $\mathrm{D}f(\alpha, x)(d, h) = \alpha h + dx$. Thus, $\mathrm{D}f(\alpha, x) = \left[\, x \;\; \alpha\, \mathrm{id}_{\mathcal{X}} \,\right]$ in "matrix" notation.*

Proof $\forall (\alpha, x) \in \mathbb{K} \times \mathcal{X}$, let $L : \mathbb{K} \times \mathcal{X} \to \mathcal{X}$ be given by $L(d, h) = \alpha h + dx$, $\forall (d, h) \in \mathbb{K} \times \mathcal{X}$. Clearly, L is a linear operator. Note that

$$
\begin{aligned}
\|L\| &= \sup_{(d,h)\in\mathbb{K}\times\mathcal{X},\ \|(d,h)\|\leq 1} \|L(d, h)\| \\
&\leq \sup_{(d,h)\in\mathbb{K}\times\mathcal{X},\ \|(d,h)\|\leq 1} |\alpha|\|h\| + |d|\|x\| \\
&\leq \sup_{(d,h)\in\mathbb{K}\times\mathcal{X},\ \|(d,h)\|\leq 1} \left(|\alpha|^2 + \|x\|^2\right)^{1/2} \left(|d|^2 + \|h\|^2\right)^{1/2} \\
&\leq \|(\alpha, x)\| < +\infty
\end{aligned}
$$

where the second inequality follows from Cauchy–Schwarz inequality. Hence, $L \in \mathrm{B}(\mathbb{K} \times \mathcal{X}, \mathcal{X})$.

Clearly, $\mathrm{A}_{\mathbb{K}\times\mathcal{X}}(\alpha, x) = \mathbb{K} \times \mathcal{X}$ since $(\alpha, x) \in (\mathbb{K} \times \mathcal{X})^\circ$. $\forall \epsilon \in (0, \infty) \subset \mathbb{R}$, set $\delta = 2\epsilon \in (0, \infty) \subset \mathbb{R}$, and $\forall (\bar{\alpha}, \bar{x}) \in (\mathbb{K} \times \mathcal{X}) \cap \mathcal{B}_{\mathbb{K}\times\mathcal{X}}((\alpha, x), \delta)$, we have

$$\begin{aligned}
&\| f(\bar{\alpha}, \bar{x}) - f(\alpha, x) - L(\bar{\alpha} - \alpha, \bar{x} - x)\| \\
&= \|\bar{\alpha}\bar{x} - \alpha x - \alpha(\bar{x} - x) - (\bar{\alpha} - \alpha)x\| = \|(\bar{\alpha} - \alpha)(\bar{x} - x)\| \\
&\leq \frac{1}{2}(|\bar{\alpha} - \alpha|^2 + \|\bar{x} - x\|^2) = \frac{1}{2}\|(\bar{\alpha} - \alpha, \bar{x} - x)\|^2 \\
&\leq \epsilon \|(\bar{\alpha} - \alpha, \bar{x} - x)\|
\end{aligned}$$

By Definition 9.3, $\mathrm{D}f(\alpha, x) = L$. This completes the proof of the proposition. □

In the previous proposition, we have abused the notation using the "matrix" notation that we identify xd as dx.

To state the next proposition, we will introduce a new notation. Let $A \in \mathrm{B}(\mathcal{X}, \mathcal{Y})$ and $B \in \mathrm{B}(\mathcal{Y}, \mathcal{Z})$, where $\mathcal{X}$, $\mathcal{Y}$, and $\mathcal{Z}$ are normed linear spaces over $\mathbb{K}$. Clearly, $f(A, B) := BA \in \mathrm{B}(\mathcal{X}, \mathcal{Z})$. Let $g(A) : \mathrm{B}(\mathcal{Y}, \mathcal{Z}) \to \mathrm{B}(\mathcal{X}, \mathcal{Z})$ be given by $g(A)(B) = BA$, $\forall B \in \mathrm{B}(\mathcal{Y}, \mathcal{Z})$. Clearly, $g(A)$ is a bounded linear operator. It is easy to see that $g : \mathrm{B}(\mathcal{X}, \mathcal{Y}) \to \mathrm{B}(\mathrm{B}(\mathcal{Y}, \mathcal{Z}), \mathrm{B}(\mathcal{X}, \mathcal{Z}))$ is a bounded linear operator with $\|g\| \leq 1$. This operator g is needed in compact "matrix" notation for many linear operators. We will use the ro to denote g for any normed linear spaces $\mathcal{X}$, $\mathcal{Y}$, and $\mathcal{Z}$. The meaning of ro is "right operate." For $x \in \mathcal{X}$, we will identify $\mathcal{X}$ with $\mathrm{B}(\mathbb{R}, \mathcal{X})$ and write $\mathrm{ro}(x)(A) = Ax$.

This brings us to the next proposition.

Proposition 9.17 *Let $\mathcal{X}$ and $\mathcal{Y}$ be normed linear spaces over $\mathbb{K}$ and $f : \mathrm{B}(\mathcal{X}, \mathcal{Y}) \times \mathcal{X} \to \mathcal{Y}$ be given by $f(A, x) = Ax$, $\forall (A, x) \in \mathrm{B}(\mathcal{X}, \mathcal{Y}) \times \mathcal{X}$. Then, f is Fréchet differentiable, and $\mathrm{D}f : \mathrm{B}(\mathcal{X}, \mathcal{Y}) \times \mathcal{X} \to \mathrm{B}(\mathrm{B}(\mathcal{X}, \mathcal{Y}) \times \mathcal{X}, \mathcal{Y})$ is given by $\forall (A, x) \in \mathrm{B}(\mathcal{X}, \mathcal{Y}) \times \mathcal{X}$, $\forall (\Delta, h) \in \mathrm{B}(\mathcal{X}, \mathcal{Y}) \times \mathcal{X}$, $\mathrm{D}f(A, x)(\Delta, h) = Ah + \Delta x$. In "matrix" notation, $\mathrm{D}f(A, x) = \begin{bmatrix} \mathrm{ro}(x) & A \end{bmatrix}$.*

Proof $\forall (A, x) \in \mathrm{B}(\mathcal{X}, \mathcal{Y}) \times \mathcal{X}$, let $L : \mathrm{B}(\mathcal{X}, \mathcal{Y}) \times \mathcal{X} \to \mathcal{Y}$ be given by $L(\Delta, h) = Ah + \Delta x$, $\forall (\Delta, h) \in \mathrm{B}(\mathcal{X}, \mathcal{Y}) \times \mathcal{X}$. Clearly, L is a linear operator. Note that

$$\begin{aligned}
\|L\| &= \sup_{(\Delta, h) \in \mathrm{B}(\mathcal{X}, \mathcal{Y}) \times \mathcal{X},\ \|(\Delta, h)\| \leq 1} \|L(\Delta, h)\| \\
&\leq \sup_{(\Delta, h) \in \mathrm{B}(\mathcal{X}, \mathcal{Y}) \times \mathcal{X},\ \|(\Delta, h)\| \leq 1} \|A\|\|h\| + \|\Delta\|\|x\| \\
&\leq \sup_{(\Delta, h) \in \mathrm{B}(\mathcal{X}, \mathcal{Y}) \times \mathcal{X},\ \|(\Delta, h)\| \leq 1} \left(\|A\|^2 + \|x\|^2\right)^{1/2} \left(\|\Delta\|^2 + \|h\|^2\right)^{1/2} \\
&\leq \|(A, x)\| < +\infty
\end{aligned}$$

where the first inequality follows from Proposition 7.64, and the second inequality follows from Cauchy–Schwarz inequality. Hence, $L \in \mathrm{B}(\mathrm{B}(\mathcal{X}, \mathcal{Y}) \times \mathcal{X}, \mathcal{Y})$.

Clearly, $\mathrm{A}_{\mathrm{B}(\mathcal{X},\mathcal{Y})\times\mathcal{X}}(A, x) = \mathrm{B}(\mathcal{X}, \mathcal{Y}) \times \mathcal{X}$ since $(A, x) \in (\mathrm{B}(\mathcal{X}, \mathcal{Y}) \times \mathcal{X})^\circ = \mathrm{B}(\mathcal{X}, \mathcal{Y}) \times \mathcal{X}$. $\forall \epsilon \in (0, \infty) \subset \mathbb{R}$, set $\delta = 2\epsilon \in (0, \infty) \subset \mathbb{R}$, and $\forall (\bar{A}, \bar{x}) \in (\mathrm{B}(\mathcal{X}, \mathcal{Y}) \times \mathcal{X}) \cap \mathcal{B}_{\mathrm{B}(\mathcal{X},\mathcal{Y})\times\mathcal{X}}((A, x), \delta)$, we have

$$
\begin{aligned}
&\left\| f(\bar{A}, \bar{x}) - f(A, x) - L(\bar{A} - A, \bar{x} - x) \right\| \\
&= \left\| \bar{A}\bar{x} - Ax - A(\bar{x} - x) - (\bar{A} - A)x \right\| = \left\| (\bar{A} - A)(\bar{x} - x) \right\| \\
&\leq \frac{1}{2}(\left\| \bar{A} - A \right\|^2 + \|\bar{x} - x\|^2) = \frac{1}{2}\left\| (\bar{A} - A, \bar{x} - x) \right\|^2 \\
&\leq \epsilon \left\| (\bar{A} - A, \bar{x} - x) \right\|
\end{aligned}
$$

where the first inequality follows from Proposition 7.64. By Definition 9.3, $\mathrm{D}f(A, x) = L$. This completes the proof of the proposition. □

9.3 Chain Rule and Mean Value Theorem

Theorem 9.18 (Chain Rule) *Let $\mathcal{X}$, $\mathcal{Y}$, and $\mathcal{Z}$ be normed linear spaces over $\mathbb{K}$, $D_x \subseteq \mathcal{X}$, $D_y \subseteq \mathcal{Y}$, $f : D_x \to D_y$, $g : D_y \to \mathcal{Z}$, $x_0 \in D_x$, $y_0 := f(x_0) \in D_y$, and $h := g \circ f : D_x \to \mathcal{Z}$. Assume that f is Fréchet differentiable at x_0 with $\mathrm{D}f(x_0) \in \mathrm{B}(\mathcal{X}, \mathcal{Y})$ and g is Fréchet differentiable at y_0 with $\mathrm{D}g(y_0) \in \mathrm{B}(\mathcal{Y}, \mathcal{Z})$. Then, h is differentiable at x_0 and $\mathrm{D}h(x_0) \in \mathrm{B}(\mathcal{X}, \mathcal{Z})$ is given by*

$$
\mathrm{D}h(x_0) = \mathrm{D}g(f(x_0)) \circ \mathrm{D}f(x_0) = \mathrm{D}g(y_0)\mathrm{D}f(x_0)
$$

Proof Define $L : \mathcal{X} \to \mathcal{Z}$ by, $\forall \bar{x} \in \mathcal{X}$, $L(\bar{x}) = \mathrm{D}g(y_0)\mathrm{D}f(x_0)\bar{x} = \mathrm{D}g(y_0) \cdot (\mathrm{D}f(x_0)(\bar{x}))$. Clearly, L is a linear operator and, by Proposition 7.64, $L \in \mathrm{B}(\mathcal{X}, \mathcal{Z})$.

$\forall \epsilon \in (0, \infty) \subset \mathbb{R}$, let $\bar{\epsilon} = (\sqrt{4\epsilon + (\|\mathrm{D}f(x_0)\| + \|\mathrm{D}g(y_0)\|)^2} - \|\mathrm{D}f(x_0)\| - \|\mathrm{D}g(y_0)\|)/2 \in (0, \infty) \subset \mathbb{R}$. By the fact that $\mathrm{D}g(y_0) \in \mathrm{B}(\mathcal{Y}, \mathcal{Z})$, $\exists \delta_1 \in (0, \infty) \subset \mathbb{R}$ such that $\forall y \in D_y \cap \mathcal{B}_{\mathcal{Y}}(y_0, \delta_1)$, we have

$$
\|g(y) - g(y_0) - \mathrm{D}g(y_0)(y - y_0)\| \leq \bar{\epsilon}\|y - y_0\|
$$

By the fact that $\mathrm{D}f(x_0) \in \mathrm{B}(\mathcal{X}, \mathcal{Y})$, we have $\overline{\mathrm{span}\left(\mathrm{A}_{D_x}(x_0)\right)} = \mathcal{X}$ and $\exists \delta_2 \in (0, \delta_1/(\bar{\epsilon} + \|\mathrm{D}f(x_0)\|)] \subset \mathbb{R}$ such that $\forall x \in D_x \cap \mathcal{B}_{\mathcal{X}}(x_0, \delta_2)$, we have

$$
\|f(x) - f(x_0) - \mathrm{D}f(x_0)(x - x_0)\| \leq \bar{\epsilon}\|x - x_0\|
$$

Then, by Proposition 7.64, $\|f(x) - y_0\| = \|f(x) - f(x_0)\| \leq \bar{\epsilon}\|x - x_0\| + \|\mathrm{D}f(x_0)\|\|x - x_0\| \leq (\bar{\epsilon} + \|\mathrm{D}f(x_0)\|)\|x - x_0\| < \delta_1$. This implies that $f(x) \in$

$D_y \cap \mathcal{B}_{\mathcal{Y}}(y_0, \delta_1)$. Then, we have

$$\begin{aligned} &\|h(x) - h(x_0) - L(x - x_0)\| = \|g(f(x)) - g(y_0) - L(x - x_0)\| \\ &\leq \|g(f(x)) - g(y_0) - \mathrm{D}g(y_0)(f(x) - y_0)\| \\ &\quad + \|\mathrm{D}g(y_0)(f(x) - f(x_0) - \mathrm{D}f(x_0)(x - x_0))\| \\ &\leq \bar{\epsilon}\|f(x) - y_0\| + \|\mathrm{D}g(y_0)\|\|f(x) - f(x_0) - \mathrm{D}f(x_0)(x - x_0)\| \\ &\leq \bar{\epsilon}\,(\bar{\epsilon} + \|\mathrm{D}f(x_0)\|)\|x - x_0\| + \bar{\epsilon}\|\mathrm{D}g(y_0)\|\,\|x - x_0\| \\ &= \bar{\epsilon}\,(\bar{\epsilon} + \|\mathrm{D}f(x_0)\| + \|\mathrm{D}g(y_0)\|)\|x - x_0\| = \epsilon\|x - x_0\| \end{aligned}$$

where the second inequality follows from Proposition 7.64. By Definition 9.3, $\mathrm{D}h(x_0) = L$. This completes the proof of the proposition. □

Proposition 9.19 *Let $\mathcal{X}$, $\mathcal{Y}$, and $\mathcal{Z}$ be normed linear spaces over $\mathbb{K}$, $D \subseteq \mathcal{X}$, $f_1 : D \to \mathcal{Y}$, $f_2 : D \to \mathcal{Z}$, $x_0 \in D$, and $g : D \to \mathcal{Y} \times \mathcal{Z}$ be given by $g(x) = (f_1(x), f_2(x))$, $\forall x \in D$. Then, the following statement holds. g is Fréchet differentiable at x_0 if, and only if, f_1 and f_2 are Fréchet differentiable at x_0. In this case, $\mathrm{D}g(x_0)(h) = (\mathrm{D}f_1(x_0)(h), \mathrm{D}f_2(x_0)(h))$, $\forall h \in \mathcal{X}$. In "matrix" notation, $\mathrm{D}g(x_0) = \begin{bmatrix} \mathrm{D}f_1(x_0) \\ \mathrm{D}f_2(x_0) \end{bmatrix}$.*

Proof "Sufficiency" By the differentiability of f_1 and f_2 at x_0, we have $\mathrm{span}\,(\mathrm{A}_D(x_0)) = \mathcal{X}$. Define $L : \mathcal{X} \to \mathcal{Y} \times \mathcal{Z}$ by, $\forall h \in \mathcal{X}$, $L(h) = (\mathrm{D}f_1(x_0)(h), \mathrm{D}f_2(x_0)(h))$. Clearly, L is a linear operator. Note that

$$\begin{aligned} \|L\| &= \sup_{h \in \mathcal{X},\ \|h\| \leq 1} \|L(h)\| \\ &= \sup_{h \in \mathcal{X},\ \|h\| \leq 1} \sqrt{\|\mathrm{D}f_1(x_0)(h)\|^2 + \|\mathrm{D}f_2(x_0)(h)\|^2} \\ &\leq \sup_{h \in \mathcal{X},\ \|h\| \leq 1} \sqrt{\|\mathrm{D}f_1(x_0)\|^2\|h\|^2 + \|\mathrm{D}f_2(x_0)\|^2\|h\|^2} \\ &\leq \sqrt{\|\mathrm{D}f_1(x_0)\|^2 + \|\mathrm{D}f_2(x_0)\|^2} < +\infty \end{aligned}$$

where the first inequality follows from Proposition 7.64 and the last inequality follows from the fact $\mathrm{D}f_1(x_0) \in \mathrm{B}(\mathcal{X}, \mathcal{Y})$ and $\mathrm{D}f_2(x_0) \in \mathrm{B}(\mathcal{X}, \mathcal{Z})$. Hence, $L \in \mathrm{B}(\mathcal{X}, \mathcal{Y} \times \mathcal{Z})$.

Since f_1 and f_2 are differentiable at x_0, then $\forall \epsilon \in (0, \infty) \subset \mathbb{R}$, $\exists \delta_1 \in (0, \infty) \subset \mathbb{R}$ such that $\forall x \in D \cap \mathcal{B}_{\mathcal{X}}(x_0, \delta_1)$, we have $\|f_1(x) - f_1(x_0) - \mathrm{D}f_1(x_0)(x - x_0)\| \leq \epsilon/\sqrt{2}\|x - x_0\|$; $\exists \delta_2 \in (0, \infty) \subset \mathbb{R}$ such that $\forall x \in D \cap \mathcal{B}_{\mathcal{X}}(x_0, \delta_2)$, we have $\|f_2(x) - f_2(x_0) - \mathrm{D}f_2(x_0)(x - x_0)\| \leq \epsilon/\sqrt{2}\|x - x_0\|$. Let $\delta = \min\{\delta_1, \delta_2\} \in (0, \infty) \subset \mathbb{R}$. $\forall x \in D \cap \mathcal{B}_{\mathcal{X}}(x_0, \delta)$, we have

$$
\begin{aligned}
&\|g(x) - g(x_0) - L(x - x_0)\| \\
&= \|(f_1(x), f_2(x)) - (f_1(x_0), f_2(x_0)) \\
&\quad -(\mathrm{D}f_1(x_0)(x - x_0), \mathrm{D}f_2(x_0)(x - x_0))\| \\
&= \|(f_1(x) - f_1(x_0) - \mathrm{D}f_1(x_0)(x - x_0), f_2(x) \\
&\quad - f_2(x_0) - \mathrm{D}f_2(x_0)(x - x_0))\| \\
&= \left(\|f_1(x) - f_1(x_0) - \mathrm{D}f_1(x_0)(x - x_0)\|^2\right. \\
&\quad \left.+\|f_2(x) - f_2(x_0) - \mathrm{D}f_2(x_0)(x - x_0)\|^2\right)^{1/2} \\
&\leq \epsilon\|x - x_0\|
\end{aligned}
$$

Hence, g is differentiable at x_0 and $\mathrm{D}g(x_0) = L$.

"Necessity" By the differentiability of g at x_0, we have $\overline{\mathrm{span}\,(\mathrm{A}_D\,(x_0))} = \mathcal{X}$. Note that $f_1 = \pi_{\mathcal{Y}} \circ g$. By chain rule and Proposition 9.13, f_1 is Fréchet differentiable at x_0 and

$$\mathrm{D}f_1(x_0) = \left[\,\mathrm{id}_{\mathcal{Y}}\ \ \vartheta_{\mathrm{B}(\mathcal{Z},\mathcal{Y})}\,\right]\mathrm{D}g(x_0)$$

By symmetry, f_2 is Fréchet differentiable at x_0 and

$$\mathrm{D}f_2(x_0) = \left[\,\vartheta_{\mathrm{B}(\mathcal{Y},\mathcal{Z})}\ \ \mathrm{id}_{\mathcal{Z}}\,\right]\mathrm{D}g(x_0)$$

Then, $\mathrm{D}g(x_0) = \begin{bmatrix} \mathrm{D}f_1(x_0) \\ \mathrm{D}f_2(x_0) \end{bmatrix}$.

This completes the proof of the proposition. □

Theorem 9.20 (Mean Value Theorem) *Let $\mathcal{X}$ be a real normed linear space, $D \subseteq \mathcal{X}$, $f : D \to \mathbb{R}$, $x_1, x_2 \in D$, and $\varphi : [0, 1] \to D$ be given by $\varphi(t) = tx_1 + (1-t)x_2$, $\forall t \in I := [0, 1] \subset \mathbb{R}$. Assume that f is continuous at $\varphi(t)$, $\forall t \in I$, and f is Fréchet differentiable at $\varphi(t)$, $\forall t \in I^\circ$. Then, there exists $t_0 \in I^\circ$ such that*

$$f(x_1) - f(x_2) = \mathrm{D}f(\varphi(t_0))(x_1 - x_2)$$

Proof By Propositions 9.19, 9.16, and 9.15 and chain rule, φ is Fréchet differentiable and $\mathrm{D}\varphi(t)(h) = h(x_1 - x_2)$, $\forall h \in \mathbb{R}$, $\forall t \in I$. Define $g : I \to \mathbb{R}$ by $g(t) = f(\varphi(t))$, $\forall t \in I$. By chain rule, g is Fréchet differentiable at t, $\forall t \in I^\circ$, and $\mathrm{D}g(t)(h) = \mathrm{D}f(\varphi(t))(\mathrm{D}\varphi(t)(h)) = h\mathrm{D}f(\varphi(t))(x_1 - x_2)$, $\forall h \in \mathbb{R}$, $\forall t \in I^\circ$. Then, $\mathrm{D}g(t) = \mathrm{D}f(\varphi(t))(x_1 - x_2)$, $\forall t \in I^\circ$. By Proposition 3.12, g is continuous. By mean value theorem (Bartle, 1976, see Theorem 27.6), we have

$$f(x_1) - f(x_2) = g(1) - g(0) = \mathrm{D}g(t_0)(1-0)$$
$$= \mathrm{D}f(t_0x_1 + (1-t_0)x_2)(x_1 - x_2)$$

for some $t_0 \in I^\circ$. This completes the proof of the theorem. □

Toward a general mean value theorem for vector-valued functions on possibly complex normed linear spaces, we present the following two lemmas.

Lemma 9.21 *Let* $D := \{a + \mathrm{i}0 \mid a \in I := [0,1] \subset \mathbb{R}\} \subset \mathbb{C}$, $x_0 := x_r + \mathrm{i}x_i \in D$, *where* $x_r \in I$ *and* $x_i = 0$, $f : D \to \mathbb{C}$ *be Fréchet differentiable at* x_0, $\mathrm{D}f(x_0) = d_r + \mathrm{i}d_i \in \mathbb{C}$, *where* $d_r, d_i \in \mathbb{R}$, *and* $g : I \to \mathbb{R}$ *be given by* $g(t) = \mathrm{Re}\,(f(t + \mathrm{i}0))$, $\forall t \in I$. *Then,* g *is Fréchet differentiable at* x_r *and* $\mathrm{D}g(x_r) = d_r = \mathrm{Re}\,(\mathrm{D}f(x_0)) \in \mathbb{R}$.

Proof Note that $\mathrm{span}\,(\mathrm{A}_I\,(x_r)) = \mathbb{R}$. $\forall \epsilon \in (0,\infty) \subset \mathbb{R}$, by the differentiability of f at x_0, $\exists \delta \in (0,\infty) \subset \mathbb{R}$ such that $\forall \bar{x} = \bar{x}_r + \mathrm{i}\bar{x}_i \in D \cap \mathcal{B}_{\mathbb{C}}(x_0, \delta)$, we have $|f(\bar{x}) - f(x_0) - \mathrm{D}f(x_0)(\bar{x} - x_0)| \le \epsilon|\bar{x} - x_0|$. Note that $\bar{x}_r \in I$ and $\bar{x}_i = 0$. Then, the above implies that $|(\mathrm{Re}\,(f(\bar{x}_r)) - \mathrm{Re}\,(f(x_r)) - (\bar{x}_r - x_r)d_r)$ $+\mathrm{i}(\mathrm{Im}\,(f(\bar{x}_r)) - \mathrm{Im}\,(f(x_r)) - (\bar{x}_r - x_r)d_i)| \le \epsilon|\bar{x}_r - x_r|$. This further implies that $|\mathrm{Re}\,(f(\bar{x}_r)) - \mathrm{Re}\,(f(x_r)) - (\bar{x}_r - x_r)d_r| \le \epsilon|\bar{x}_r - x_r|$. Then, $\forall t \in I \cap \mathcal{B}_{\mathbb{R}}(x_r, \delta)$, $\bar{x} := t + \mathrm{i}0 \in D \cap \mathcal{B}_{\mathbb{C}}(x_0, \delta)$ and $|g(t) - g(x_r) - (t - x_r)d_r| \le \epsilon|t - x_r|$. Hence, $\mathrm{D}g(x_r) = d_r$. This completes the proof of the lemma. □

Lemma 9.22 *Let* $D := \{a + \mathrm{i}0 \mid a \in I := [0,1] \subset \mathbb{R}\} \subset \mathbb{C}$, $f : D \to \mathbb{C}$ *be continuous, and* f *be Fréchet differentiable at* $a + \mathrm{i}0$, $\forall a \in I^\circ$. *Then,* $\exists t_0 \in I^\circ$ *such that* $\mathrm{Re}\,(f(1) - f(0)) = \mathrm{Re}\,(\mathrm{D}f(t_0))$.

Proof Let $g : I \to \mathbb{R}$ be given by $g(t) = \mathrm{Re}\,(f(t + \mathrm{i}0))$, $\forall t \in I$. Clearly, g is continuous since f is continuous. By Lemma 9.21, g is Fréchet differentiable at t, $\forall t \in I^\circ$, and $\mathrm{D}g(t) = \mathrm{Re}\,(\mathrm{D}f(t + \mathrm{i}0))$, $\forall t \in I^\circ$. By mean value theorem (Bartle, 1976, see Theorem 27.6), $\exists t_0 \in I^\circ$ such that $g(1) - g(0) = \mathrm{D}g(t_0) = \mathrm{Re}\,(\mathrm{D}f(t_0 + \mathrm{i}0))$. Then, $\mathrm{Re}\,(f(1) - f(0)) = g(1) - g(0) = \mathrm{Re}\,(\mathrm{D}f(t_0 + \mathrm{i}0)) = \mathrm{Re}\,(\mathrm{D}f(t_0))$. This completes the proof of the lemma. □

Theorem 9.23 (Mean Value Theorem) *Let* $\mathcal{X}$ *and* $\mathcal{Y}$ *be normed linear spaces over* $\mathbb{K}$, $D \subseteq \mathcal{X}$, $f : D \to \mathcal{Y}$, $x_1, x_2 \in D$, *and the line segment connecting* x_1 *and* x_2 *be contained in* D. *Assume that* f *is continuous at* $x = tx_1 + (1-t)x_2$, $\forall t \in I := [0,1] \subset \mathbb{R}$, *and* $\mathrm{D}f(x) \in \mathrm{B}(\mathcal{X}, \mathcal{Y})$ *exists at* $x = tx_1 + (1-t)x_2$, $\forall t \in I^\circ$. *Then,* $\exists t_0 \in I^\circ$ *such that* $\|f(x_1) - f(x_2)\| \le \|\mathrm{D}f(t_0x_1 + (1-t_0)x_2)(x_1 - x_2)\|$.

Proof We will distinguish three exhaustive and mutually exclusive cases: Case 1: $f(x_1) = f(x_2)$; Case 2: $f(x_1) \neq f(x_2)$ and $\mathbb{K} = \mathbb{R}$; Case 3: $f(x_1) \neq f(x_2)$ and $\mathbb{K} = \mathbb{C}$. Case 1: $f(x_1) = f(x_2)$. Take t_0 to be any point in I°. The desired result follows.

Case 2: $f(x_1) \neq f(x_2)$ and $\mathbb{K} = \mathbb{R}$. By Proposition 7.85, $\exists y_* \in \mathcal{Y}^*$ with $\|y_*\| = 1$ such that $\langle\langle y_*, f(x_1) - f(x_2)\rangle\rangle = \|f(x_1) - f(x_2)\|$. Define $\varphi : I \to D$ by

$\varphi(t) = tx_1 + (1-t)x_2$, $\forall t \in I$. By Propositions 9.19, 9.16, and 9.15 and chain rule, φ is Fréchet differentiable. Define $g : I \to \mathbb{R}$ by $g(t) = \langle\langle y_*, f(\varphi(t))\rangle\rangle$, $\forall t \in I$. By Proposition 3.12, g is continuous. By chain rule and Propositions 9.17 and 9.19, g is Fréchet differentiable at t, $\forall t \in I^\circ$, and $\mathrm{D}g(t)(d) = \langle\langle y_*, \mathrm{D}f(\varphi(t))(\mathrm{D}\varphi(t)(d))\rangle\rangle = \langle\langle y_*, \mathrm{D}f(\varphi(t))(d\,(x_1 - x_2))\rangle\rangle = \langle\langle y_*, \mathrm{D}f(\varphi(t))(x_1 - x_2)\rangle\rangle d$, $\forall d \in \mathbb{R}$ and $\forall t \in I^\circ$. Hence, $\mathrm{D}g(t) = \langle\langle y_*, \mathrm{D}f(\varphi(t))(x_1 - x_2)\rangle\rangle$, $\forall t \in I^\circ$. By mean value theorem (Bartle, 1976, see Theorem 27.6), there exists $t_0 \in I^\circ$ such that $g(1) - g(0) = \mathrm{D}g(t_0)$. Then, we have $\|f(x_1) - f(x_2)\| = \langle\langle y_*, f(x_1) - f(x_2)\rangle\rangle = g(1) - g(0) = \mathrm{D}g(t_0) \leq |\mathrm{D}g(t_0)| = |\langle\langle y_*, \mathrm{D}f(\varphi(t_0))(x_1 - x_2)\rangle\rangle| \leq \|\mathrm{D}f(\varphi(t_0))(x_1 - x_2)\|$, where the last inequality follows from Proposition 7.72. The desired result follows.

Case 3: $f(x_1) \neq f(x_2)$ and $\mathbb{K} = \mathbb{C}$. By Proposition 7.85, $\exists y_* \in \mathcal{Y}^*$ with $\|y_*\| = 1$ such that $\langle\langle y_*, f(x_1) - f(x_2)\rangle\rangle = \|f(x_1) - f(x_2)\|$. Let $\bar{D} := \{a + \mathrm{i}0 \mid a \in I\} \subset \mathbb{C}$. Define $\varphi : \bar{D} \to D$ by $\varphi(t) = tx_1 + (1-t)x_2$, $\forall t \in \bar{D}$. By Propositions 9.19, 9.16, and 9.15 and chain rule, φ is Fréchet differentiable and $\mathrm{D}\varphi(t)(d) = d\,(x_1 - x_2)$, $\forall d \in \mathbb{C}$ and $\forall t \in \bar{D}$. Define $g : \bar{D} \to \mathbb{C}$ by $g(t) = \langle\langle y_*, f(\varphi(t))\rangle\rangle$, $\forall t \in \bar{D}$. By Proposition 3.12, g is continuous. By chain rule and Propositions 9.17 and 9.19, g is Fréchet differentiable at $a + \mathrm{i}0$, $\forall a \in I^\circ$, and $\mathrm{D}g(a + \mathrm{i}0)(d) = \langle\langle y_*, \mathrm{D}f(\varphi(a + \mathrm{i}0))(\mathrm{D}\varphi(a + \mathrm{i}0)(d))\rangle\rangle = \langle\langle y_*, \mathrm{D}f(\varphi(a))(d\,(x_1 - x_2))\rangle\rangle = \langle\langle y_*, \mathrm{D}f(\varphi(a))(x_1 - x_2)\rangle\rangle d$, $\forall d \in \mathbb{C}$ and $\forall a \in I^\circ$. Hence, $\mathrm{D}g(a + \mathrm{i}0) = \langle\langle y_*, \mathrm{D}f(\varphi(a + \mathrm{i}0))(x_1 - x_2)\rangle\rangle$, $\forall a \in I^\circ$. Note that $g(1) - g(0) = \langle\langle y_*, f(x_1)\rangle\rangle - \langle\langle y_*, f(x_2)\rangle\rangle = \|f(x_1) - f(x_2)\| \in \mathbb{R}$. By Lemma 9.22, there exists $t_0 \in I^\circ$ such that $g(1) - g(0) = \mathrm{Re}\,(\mathrm{D}g(t_0))$. Then, we have $\|f(x_1) - f(x_2)\| = g(1) - g(0) \leq |\mathrm{D}g(t_0)| = |\langle\langle y_*, \mathrm{D}f(\varphi(t_0))(x_1 - x_2)\rangle\rangle| \leq \|\mathrm{D}f(\varphi(t_0))(x_1 - x_2)\|$, where the last inequality follows from Proposition 7.72. The desired result follows.

This completes the proof of the theorem. □

Proposition 9.24 *Let $\mathcal{X}$, $\mathcal{Y}$, and $\mathcal{Z}$ be normed linear spaces over $\mathbb{K}$, $D \subseteq \mathcal{X} \times \mathcal{Y}$, $f : D \to \mathcal{Z}$, $(x_0, y_0) \in D$. Assume that the following conditions hold:*

(i) $\exists \delta_0 \in (0, \infty) \subset \mathbb{R}$, $\forall (x, y) \in D \cap \mathcal{B}_{\mathcal{X}\times\mathcal{Y}}((x_0, y_0), \delta_0)$, *we have* $(x, ty + (1-t)y_0) \in D$, $\forall t \in I := [0, 1] \subset \mathbb{R}$.

(ii) *f is partial differentiable with respect to x at (x_0, y_0) and* $\frac{\partial f}{\partial x}(x_0, y_0) \in \mathrm{B}(\mathcal{X}, \mathcal{Z})$.

(iii) $\forall (x, y) \in D \cap \mathcal{B}_{\mathcal{X}\times\mathcal{Y}}((x_0, y_0), \delta_0)$, *$f$ is partial differentiable with respect to y at (x, y) and $\frac{\partial f}{\partial y}$ is continuous at (x_0, y_0).*

Then, f is Fréchet differentiable at (x_0, y_0), and $\mathrm{D}f(x_0, y_0) \in \mathrm{B}(\mathcal{X} \times \mathcal{Y}, \mathcal{Z})$ is given by $\mathrm{D}f(x_0, y_0)(h, k) = \frac{\partial f}{\partial x}(x_0, y_0)(h) + \frac{\partial f}{\partial y}(x_0, y_0)(k)$, $\forall (h, k) \in \mathcal{X} \times \mathcal{Y}$. In "matrix" notation, $\mathrm{D}f(x_0, y_0) = \left[\, \frac{\partial f}{\partial x}(x_0, y_0) \;\; \frac{\partial f}{\partial y}(x_0, y_0) \,\right]$.

Proof We will first show that $\overline{\mathrm{span}\,(\mathrm{A}_D\,(x_0, y_0))} = \mathcal{X} \times \mathcal{Y}$. Define $D_{x_0} := \{y \in \mathcal{Y} \mid (x_0, y) \in D\}$ and $D_{y_0} := \{x \in \mathcal{X} \mid (x, y_0) \in D\}$. By the partial differentiability of f with respect to x at (x_0, y_0), we have $\overline{\mathrm{span}\left(\mathrm{A}_{D_{y_0}}(x_0)\right)} = \mathcal{X}$. By the partial differentiability of f with respect to y at (x_0, y_0), we have $\overline{\mathrm{span}\left(\mathrm{A}_{D_{x_0}}(y_0)\right)} = \mathcal{Y}$. $\forall u \in \mathrm{A}_{D_{y_0}}(x_0)$, $\forall \epsilon \in (0, \infty) \subset \mathbb{R}$, $\exists \bar{x} := x_0 + r\bar{u} \in D_{y_0}$ with $0 < r < \epsilon$ and $\bar{u} \in \mathcal{B}_{\mathcal{X}}(u, \epsilon)$. Then, $(\bar{x}, y_0) = (x_0, y_0) + r(\bar{u}, \vartheta_{\mathcal{Y}}) \in D$ and $(\bar{u}, \vartheta_{\mathcal{Y}}) \in \mathcal{B}_{\mathcal{X}\times\mathcal{Y}}((u, \vartheta_{\mathcal{Y}}), \epsilon)$. Hence, $(u, \vartheta_{\mathcal{Y}}) \in \mathrm{A}_D\,(x_0, y_0)$. Then, $\mathrm{A}_{D_{y_0}}(x_0) \times \{\vartheta_{\mathcal{Y}}\} \subseteq \mathrm{A}_D\,(x_0, y_0)$, which implies that $\mathcal{X} \times \{\vartheta_{\mathcal{Y}}\} \subseteq \overline{\mathrm{span}\,(\mathrm{A}_D\,(x_0, y_0))}$. By symmetry, we have $\{\vartheta_{\mathcal{X}}\} \times \mathcal{Y} \subseteq \overline{\mathrm{span}\,(\mathrm{A}_D\,(x_0, y_0))}$. By Proposition 7.17, $\overline{\mathrm{span}\,(\mathrm{A}_D\,(x_0, y_0))} = \mathcal{X} \times \mathcal{Y}$.

Define $L : \mathcal{X} \times \mathcal{Y} \to \mathcal{Z}$ by $L(h, k) = \frac{\partial f}{\partial x}(x_0, y_0)(h) + \frac{\partial f}{\partial y}(x_0, y_0)(k)$, $\forall (h, k) \in \mathcal{X} \times \mathcal{Y}$. Clearly, L is a linear operator. Note that

$$\begin{aligned}
\|L\| &= \sup_{(h,k)\in\mathcal{X}\times\mathcal{Y},\ \|(h,k)\|\le 1} \|L(h, k)\| \\
&\le \sup_{(h,k)\in\mathcal{X}\times\mathcal{Y},\ \|(h,k)\|\le 1} \left(\left\|\frac{\partial f}{\partial x}(x_0, y_0)\right\| \|h\| + \left\|\frac{\partial f}{\partial y}(x_0, y_0)\right\| \|k\|\right) \\
&\le \sup_{(h,k)\in\mathcal{X}\times\mathcal{Y},\ \|(h,k)\|\le 1} \sqrt{\left\|\frac{\partial f}{\partial x}(x_0, y_0)\right\|^2 + \left\|\frac{\partial f}{\partial y}(x_0, y_0)\right\|^2} \\
&\quad \cdot \sqrt{\|h\|^2 + \|k\|^2} \le \sqrt{\left\|\frac{\partial f}{\partial x}(x_0, y_0)\right\|^2 + \left\|\frac{\partial f}{\partial y}(x_0, y_0)\right\|^2} < +\infty
\end{aligned}$$

where the first inequality follows from Proposition 7.64 and the second inequality follows from Cauchy–Schwarz inequality. Hence, $L \in \mathrm{B}(\mathcal{X} \times \mathcal{Y}, \mathcal{Z})$.

$\forall \epsilon \in (0, \infty) \subset \mathbb{R}$, by the partial differentiability of f with respect to x at (x_0, y_0), $\exists \delta_1 \in (0, \delta_0] \subset \mathbb{R}$ such that $\forall (x, y) \in D \cap \mathcal{B}_{\mathcal{X}\times\mathcal{Y}}((x_0, y_0), \delta_1)$, we have $\left\|f(x, y_0) - f(x_0, y_0) - \frac{\partial f}{\partial x}(x_0, y_0)(x - x_0)\right\| \le \epsilon/\sqrt{2}\|x - x_0\|$. By the continuity of $\frac{\partial f}{\partial y}$ at (x_0, y_0), $\exists \delta_2 \in (0, \delta_1] \subset \mathbb{R}$, such that $\forall (\bar{x}, \bar{y}) \in D \cap \mathcal{B}_{\mathcal{X}\times\mathcal{Y}}((x_0, y_0), \delta_2)$, we have $\left\|\frac{\partial f}{\partial y}(\bar{x}, \bar{y}) - \frac{\partial f}{\partial y}(x_0, y_0)\right\| < \epsilon/\sqrt{2}$.

Claim 9.24.1 $\left\|f(x, y) - f(x, y_0) - \frac{\partial f}{\partial y}(x_0, y_0)(y - y_0)\right\| \le \epsilon/\sqrt{2}\|y - y_0\|$, $\forall (x, y) \in D \cap \mathcal{B}_{\mathcal{X}\times\mathcal{Y}}((x_0, y_0), \delta_2)$.

Proof of Claim Fix any $(x, y) \in D \cap \mathcal{B}_{\mathcal{X}\times\mathcal{Y}}((x_0, y_0), \delta_2)$. Let $\bar{D} = I$ if $\mathbb{K} = \mathbb{R}$; or $\bar{D} = \{a + \mathrm{i}0 \mid a \in I\} \subset \mathbb{C}$ if $\mathbb{K} = \mathbb{C}$. Define $\psi : \bar{D} \to \mathcal{Z}$ by $\psi(t) = f(x, ty + (1 - t)y_0) - \frac{\partial f}{\partial y}(x_0, y_0)(t\,(y - y_0))$, $\forall t \in \bar{D}$. By chain rule, each term in the definition of ψ is Fréchet differentiable. By Proposition 9.15, ψ is Fréchet differentiable. By mean value theorem, $\exists t_0 \in I^\circ$,

$$\left\| f(x,y) - f(x,y_0) - \frac{\partial f}{\partial y}(x_0,y_0)(y-y_0) \right\| = \|\psi(1) - \psi(0)\| \leq \|\mathrm{D}\psi(t_0)\|$$

$$= \left\| \frac{\partial f}{\partial y}(x, t_0 y + (1-t_0)y_0)(y-y_0) - \frac{\partial f}{\partial y}(x_0,y_0)(y-y_0) \right\|$$

$$\leq \left\| \frac{\partial f}{\partial y}(x, t_0 y + (1-t_0)y_0) - \frac{\partial f}{\partial y}(x_0,y_0) \right\| \|y-y_0\| \leq \epsilon/\sqrt{2}\|y-y_0\|$$

where the second inequality follows from Proposition 7.64. This completes the proof of the claim. □

Therefore, $\forall (x,y) \in D \cap \mathcal{B}_{\mathcal{X}\times\mathcal{Y}}((x_0,y_0),\delta_2)$, we have

$$\|f(x,y) - f(x_0,y_0) - L(x-x_0, y-y_0)\|$$
$$\leq \left\| f(x,y) - f(x,y_0) - \frac{\partial f}{\partial y}(x_0,y_0)(y-y_0) \right\|$$
$$+ \left\| f(x,y_0) - f(x_0,y_0) - \frac{\partial f}{\partial x}(x_0,y_0)(x-x_0) \right\|$$
$$\leq \epsilon/\sqrt{2}\|y-y_0\| + \epsilon/\sqrt{2}\|x-x_0\| \leq \epsilon\|(x-x_0, y-y_0)\|$$

where the last inequality follows from Cauchy–Schwarz inequality. Hence, $\mathrm{D}f(x_0,y_0) = L$. This completes the proof of the proposition. □

We observe that Condition (i) of Proposition 9.24 is easily satisfied when $(x_0,y_0) \in D^\circ$.

9.4 Higher Order Derivatives

9.4.1 *Basic Concept*

We introduce the following notation. Let $\mathcal{X}$ and $\mathcal{Y}$ be normed linear spaces over $\mathbb{K}$. Denote $\mathrm{B}(\mathcal{X},\mathcal{Y})$ by $\mathrm{B}_1(\mathcal{X},\mathcal{Y})$. Recursively, denote $\mathrm{B}(\mathcal{X}, \mathrm{B}_k(\mathcal{X},\mathcal{Y}))$ by $\mathrm{B}_{k+1}(\mathcal{X},\mathcal{Y})$, $\forall k \in \mathbb{N}$. Note that $\mathrm{B}_k(\mathcal{X},\mathcal{Y})$ is the set of bounded multi-linear $\mathcal{Y}$-valued functions on $\mathcal{X}^k$, $\forall k \in \mathbb{N}$. Define the subset of symmetric functions by $\mathrm{B_S}_k(\mathcal{X},\mathcal{Y}) := \{L \in \mathrm{B}_k(\mathcal{X},\mathcal{Y}) \mid L(h_k)\cdots(h_1) = L(v_k)\cdots(v_1), \forall (h_1,\ldots,h_k) \in \mathcal{X}^k,\ \forall (v_1,\ldots,v_k) = \text{a permutation of } (h_1,\ldots,h_k)\}$. Note that $\mathrm{B_S}_k(\mathcal{X},\mathcal{Y})$ is a closed subspace of $\mathrm{B}_k(\mathcal{X},\mathcal{Y})$. Then, by Proposition 7.13, $\mathrm{B_S}_k(\mathcal{X},\mathcal{Y})$ is a normed linear space over $\mathbb{K}$. If $\mathcal{Y}$ is a Banach space, then, by Proposition 7.66, $\mathrm{B}_k(\mathcal{X},\mathcal{Y})$ is a Banach space. Then, by Proposition 4.39, $\mathrm{B_S}_k(\mathcal{X},\mathcal{Y})$ is a Banach space. For notational consistency, we will denote $\mathrm{B_S}_0(\mathcal{X},\mathcal{Y}) := \mathrm{B}_0(\mathcal{X},\mathcal{Y}) := \mathcal{Y}$.

Definition 9.25 Let $\mathcal{X}$ and $\mathcal{Y}$ be normed linear spaces over $\mathbb{K}$, $D \subseteq \mathcal{X}$, $f : D \to \mathcal{Y}$, and $x_0 \in D$. Let $f^{(1)}$ be defined with domain of definition $\text{dom}\left(f^{(1)}\right)$. We may consider the derivative of $f^{(1)}$. If $f^{(1)}$ is differentiable at $x_0 \in \text{dom}\left(f^{(1)}\right)$, then f is said to be *twice Fréchet differentiable at* x_0. The *second order derivative of* f *at* x_0 is $\text{D}(\text{D}f)(x_0) =: \text{D}^2 f(x_0) =: f^{(2)}(x_0) \in \text{B}(\mathcal{X}, \text{B}(\mathcal{X}, \mathcal{Y})) = \text{B}_2(\mathcal{X}, \mathcal{Y})$. $\text{D}^2 f$ or $f^{(2)}$ will denote the $\text{B}_2(\mathcal{X}, \mathcal{Y})$-valued function whose domain of definition is $\text{dom}\left(f^{(2)}\right) := \left\{x \in \text{dom}\left(f^{(1)}\right) \mid f^{(2)}(x) \in \text{B}_2(\mathcal{X}, \mathcal{Y}) \text{ exists}\right\}$. Recursively, if $f^{(k)}$ is Fréchet differentiable at $x_0 \in \text{dom}\left(f^{(k)}\right)$, then f is said to be $(k+1)$-*times Fréchet differentiable at* x_0 and the $(k+1)$*th order derivative of* f *at* x_0 is $\text{D}f^{(k)}(x_0) =: \text{D}^{k+1} f(x_0) =: f^{(k+1)}(x_0) \in \text{B}_{k+1}(\mathcal{X}, \mathcal{Y})$, where $k \in \mathbb{N}$. $\text{D}^{k+1} f$ or $f^{(k+1)}$ will denote the $\text{B}_{k+1}(\mathcal{X}, \mathcal{Y})$-valued function whose domain of definition is $\text{dom}\left(f^{(k+1)}\right) := \left\{x \in \text{dom}\left(f^{(k)}\right) \mid f^{(k+1)}(x) \in \text{B}_{k+1}(\mathcal{X}, \mathcal{Y}) \text{ exists}\right\}$. For notational consistency, we will let $f^{(0)} = f$. ◇

Note that $\text{dom}\left(f^{(k+1)}\right) \subseteq \text{dom}\left(f^{(k)}\right) \subseteq D$, $\forall k \in \mathbb{N}$.

Definition 9.26 Let $\mathcal{X}$ and $\mathcal{Y}$ be normed linear spaces over $\mathbb{K}$, $D \subseteq \mathcal{X}$, $f : D \to \mathcal{Y}$, and $x_0 \in D$. Assume that $\exists \delta_0 \in (0, \infty) \subset \mathbb{R}$ such that f is k-times Fréchet differentiable at x, $\forall x \in D \cap \mathcal{B}_{\mathcal{X}}(x_0, \delta_0)$, where $k \in \mathbb{N}$, that is, $D \cap \mathcal{B}_{\mathcal{X}}(x_0, \delta_0) \subseteq \text{dom}\left(f^{(k)}\right)$, and $f^{(k)}$ is continuous at x_0. Then, we say that f is $\mathcal{C}_k$ *at* x_0. If f is $\mathcal{C}_k$ at x, $\forall x \in D$, then we say f is $\mathcal{C}_k$. If f is $\mathcal{C}_k$ at $\forall x \in D \cap \mathcal{B}_{\mathcal{X}}(x_0, \delta_0)$, $\forall k \in \mathbb{N}$, then we say that f is $\mathcal{C}_\infty$ *at* x_0. If f is $\mathcal{C}_\infty$ at x, $\forall x \in D$, then we say that f is $\mathcal{C}_\infty$. ◇

Note that f being $(k+1)$-times differentiable at $x_0 \in D$ does not imply that f is $\mathcal{C}_k$ at x_0 since $\text{dom}\left(f^{(k)}\right)$ may not contain $D \cap \mathcal{B}_{\mathcal{X}}(x_0, \delta)$, $\forall \delta \in (0, \infty) \subset \mathbb{R}$. When $\text{dom}\left(f^{(k)}\right) \supseteq D \cap \mathcal{B}_{\mathcal{X}}(x_0, \delta_0)$, for some $\delta_0 \in (0, \infty) \subset \mathbb{R}$, and f is $(k+1)$-times differentiable at x_0, then f is $\mathcal{C}_k$ at x_0. In particular, if f is $\mathcal{C}_{k+1}$ at x_0, then f is $\mathcal{C}_k$ at x, $\forall x \in D \cap \mathcal{B}_{\mathcal{X}}(x_0, \delta_0)$, for some $\delta_0 \in (0, \infty) \subset \mathbb{R}$. If f is infinitely many times differentiable at x, $\forall x \in D$, then f is $\mathcal{C}_\infty$.

Proposition 9.27 *Let $\mathcal{X}$ and $\mathcal{Y}$ be normed linear spaces over $\mathbb{K}$, $D \subseteq \mathcal{X}$, and $f : D \to \mathcal{Y}$ be $\mathcal{C}_{n+m-1}$ at x_0 and $(n+m)$-times Fréchet differentiable at $x_0 \in D$, where $n, m \in \mathbb{N}$. Fix $(h_1, \ldots, h_n) \in \mathcal{X}^n$. Define the function $g : \text{dom}\left(f^{(n)}\right) \to \mathcal{Y}$ by $g(x) = f^{(n)}(x)(h_n) \cdots (h_1)$, $\forall x \in D_n := \text{dom}\left(f^{(n)}\right)$. Then, the following statements hold:*

(i) *g is m-times Fréchet differentiable at x_0, and $g^{(m)}(x_0) \in \text{B}_m(\mathcal{X}, \mathcal{Y})$ is given by $g^{(m)}(x_0)(h_{n+m}) \cdots (h_{n+1}) = f^{(n+m)}(x_0)(h_{n+m}) \cdots (h_1)$, $\forall (h_{n+1}, \ldots, h_{n+m}) \in \mathcal{X}^m$.*
(ii) *If f is $\mathcal{C}_{n+m}$ at x_0, then g is $\mathcal{C}_m$ at x_0.*

Proof We will first prove (i) using mathematical induction on m.
1° $m = 1$. Since f is $\mathcal{C}_n$ at x_0, then $\exists \delta_0 \in (0, \infty) \subset \mathbb{R}$ such that $D \cap \mathcal{B}_{\mathcal{X}}(x_0, \delta_0) \subseteq D_n$. By the $(n+1)$-times differentiability of f at x_0, we have $\overline{\text{span}\left(\text{A}_{D_n}(x_0)\right)} = \mathcal{X}$. Define $L : \mathcal{X} \to \mathcal{Y}$ by, $\forall h \in \mathcal{X}$, $L(h) = f^{(n+1)}(x_0)(h)(h_n) \cdots (h_1)$. Clearly, L is a

linear operator. Note that

$$\|L\| = \sup_{h\in\mathcal{X},\ \|h\|\le 1} \|L(h)\| = \sup_{h\in\mathcal{X},\ \|h\|\le 1} \left\| f^{(n+1)}(x_0)(h)(h_n)\cdots(h_1)\right\|$$
$$\le \sup_{h\in\mathcal{X},\ \|h\|\le 1} \left\| f^{(n+1)}(x_0)\right\| \|h\| \|h_n\|\cdots\|h_1\|$$
$$\le \left\| f^{(n+1)}(x_0)\right\| \|h_n\|\cdots\|h_1\| < +\infty$$

where the first inequality follows from Proposition 7.64. Hence, $L \in \mathrm{B}(\mathcal{X}, \mathcal{Y})$.

$\forall \epsilon \in (0, \infty) \subset \mathbb{R}$, by the differentiability of $f^{(n)}$ at x_0, $\exists \delta \in (0, \delta_0] \subset \mathbb{R}$ such that $\forall x \in D_n \cap \mathcal{B}_{\mathcal{X}}(x_0, \delta)$, we have $\left\| f^{(n)}(x) - f^{(n)}(x_0) - f^{(n+1)}(x_0)(x - x_0)\right\| \le \epsilon/(1 + \|h_n\|\cdots\|h_1\|)\|x - x_0\|$. Then, we have

$$\begin{aligned}
&\|g(x) - g(x_0) - L(x - x_0)\|\\
&= \| f^{(n)}(x)(h_n)\cdots(h_1) - f^{(n)}(x_0)(h_n)\cdots(h_1)\\
&\quad - f^{(n+1)}(x_0)(x - x_0)(h_n)\cdots(h_1)\|\\
&= \left\|(f^{(n)}(x) - f^{(n)}(x_0) - f^{(n+1)}(x_0)(x - x_0))(h_n)\cdots(h_1)\right\|\\
&\le \left\| f^{(n)}(x) - f^{(n)}(x_0) - f^{(n+1)}(x_0)(x - x_0)\right\| \|h_n\|\cdots\|h_1\|\\
&\le \epsilon\|x - x_0\|
\end{aligned}$$

where the first inequality follows from Proposition 7.64. Hence, $\mathrm{D}g(x_0) = L$ and g is Fréchet differentiable at x_0.

2° Assume that (i) holds for $m \le k$, $\forall k \in \mathbb{N}$.

3° Consider the case $m = k + 1$. Since f is $\mathcal{C}_{n+k}$ at x_0, then $\exists \delta_0 \in (0, \infty) \subset \mathbb{R}$ such that f is $(n + k)$-times Fréchet differentiable at $\bar{x}$ and is $\mathcal{C}_{n+k-1}$ at $\bar{x}$, $\forall \bar{x} \in D \cap \mathcal{B}_{\mathcal{X}}(x_0, \delta_0)$. By inductive assumption, g is k-times Fréchet differentiable at $\bar{x}$ and $g^{(k)}(\bar{x})(h_{n+k})\cdots(h_{n+1}) = f^{(n+k)}(\bar{x})(h_{n+k})\cdots(h_1)$, $\forall (h_{n+1}, \ldots, h_{n+k}) \in \mathcal{X}^k$. Hence, $\bar{x} \in \bar{D}_{n+k} := \mathrm{dom}\left(g^{(k)}\right)$. Then, we have $D \cap \mathcal{B}_{\mathcal{X}}(x_0, \delta_0) \subseteq \bar{D}_{n+k} \subseteq D_n \subseteq D$. Note that $\bar{D}_{n+k} \cap \mathcal{B}_{\mathcal{X}}(x_0, \delta_0) = D \cap \mathcal{B}_{\mathcal{X}}(x_0, \delta_0)$. Then, $\overline{\mathrm{span}\left(\mathrm{A}_{\bar{D}_{n+k}}(x_0)\right)} = \overline{\mathrm{span}\left(\mathrm{A}_D(x_0)\right)} = \mathcal{X}$, since f is differentiable at x_0.

Define $L : \mathcal{X} \to \mathrm{B}_k(\mathcal{X}, \mathcal{Y})$ by, $\forall h \in \mathcal{X}$, $\forall (h_{n+1}, \ldots, h_{n+k}) \in \mathcal{X}^k$, $L(h)(h_{n+k})\cdots(h_{n+1}) = f^{(n+k+1)}(x_0)(h)(h_{n+k})\cdots(h_1)$. Clearly, L is a linear operator. Note that

$$\|L\| = \sup_{h\in\mathcal{X},\ \|h\|\le 1} \|L(h)\|$$
$$= \sup_{h\in\mathcal{X},\ \|h\|\le 1,\ h_{n+i}\in\mathcal{X},\ \|h_{n+i}\|\le 1,\ i=1,\ldots,k} \|L(h)(h_{n+k})\cdots(h_{n+1})\|$$

$$= \sup_{h\in\mathcal{X},\ \|h\|\le 1,\ h_{n+i}\in\mathcal{X},\ \|h_{n+i}\|\le 1,\ i=1,\dots,k} \left\| f^{(n+k+1)}(x_0)(h)(h_{n+k})\cdots(h_1)\right\|$$
$$\le \left\| f^{(n+k+1)}(x_0)\right\| \|h_n\|\cdots\|h_1\| < +\infty$$

where the first inequality follows from Proposition 7.64. Then, $L \in \mathrm{B}_{k+1}(\mathcal{X},\mathcal{Y})$.

$\forall \epsilon \in (0,\infty) \subset \mathbb{R}$, by the differentiability of $f^{(n+k)}$ at x_0, $\exists \delta \in (0,\delta_0] \subset \mathbb{R}$ such that $\forall x \in \operatorname{dom}\left(f^{(n+k)}\right) \cap \mathcal{B}_{\mathcal{X}}(x_0,\delta) = D \cap \mathcal{B}_{\mathcal{X}}(x_0,\delta)$, we have $\left\| f^{(n+k)}(x) - f^{(n+k)}(x_0) - f^{(n+k+1)}(x_0)(x-x_0)\right\| \le \epsilon/(1+\|h_n\|\cdots\|h_1\|)$ $\|x-x_0\|$. Then, $\forall x \in \bar{D}_{n+k} \cap \mathcal{B}_{\mathcal{X}}(x_0,\delta) = D \cap \mathcal{B}_{\mathcal{X}}(x_0,\delta)$,

$$\left\| g^{(k)}(x) - g^{(k)}(x_0) - L(x-x_0)\right\|$$
$$= \sup_{h_{n+i}\in\mathcal{X},\ \|h_{n+i}\|\le 1,\ i=1,\dots,k} \left\| (g^{(k)}(x) - g^{(k)}(x_0) - L(x-x_0))(h_{n+k})\cdots(h_{n+1})\right\|$$
$$= \sup_{h_{n+i}\in\mathcal{X},\ \|h_{n+i}\|\le 1,\ i=1,\dots,k} \left\| f^{(n+k)}(x)(h_{n+k})\cdots(h_1) - f^{(n+k)}(x_0)(h_{n+k})\cdots(h_1) - f^{(n+k+1)}(x_0)(x-x_0)(h_{n+k})\cdots(h_1)\right\|$$
$$\le \left\| f^{(n+k)}(x) - f^{(n+k)}(x_0) - f^{(n+k+1)}(x_0)(x-x_0)\right\| \|h_n\|\cdots\|h_1\|$$
$$\le \epsilon\|x-x_0\|$$

where the first inequality follows from Proposition 7.64. Hence, $g^{(k+1)}(x_0) = \mathrm{D}g^{(k)}(x_0) = L$. Therefore, g is $(k+1)$-times differentiable at x_0.

This completes the induction process and the proof of (i).

For (ii), let f be $\mathcal{C}_{n+m}$ at x_0. By Definition 9.26, $\exists \delta_0 \in (0,\infty) \subset \mathbb{R}$ such that $D \cap \mathcal{B}_{\mathcal{X}}(x_0,\delta_0) \subseteq D_{n+m} := \operatorname{dom}\left(f^{(n+m)}\right) \subseteq D$ and $f^{(n+m)}$ is continuous at x_0. $\forall x \in D_{n+m} \cap \mathcal{B}_{\mathcal{X}}(x_0,\delta_0) = D \cap \mathcal{B}_{\mathcal{X}}(x_0,\delta_0)$, f is $\mathcal{C}_{n+m-1}$ at x and $(n+m)$-times differentiable at x. By (i), g is m-times Fréchet differentiable at x, and $\forall (h_{n+1},\dots,h_{n+m}) \in \mathcal{X}^m$, we have $g^{(m)}(x)(h_{n+m})\cdots(h_{n+1}) = f^{(n+m)}(x)(h_{n+m})\cdots(h_1)$. Then, $D \cap \mathcal{B}_{\mathcal{X}}(x_0,\delta_0) \subseteq \operatorname{dom}\left(g^{(m)}\right) \subseteq D_n \subseteq D$. $\forall \epsilon \in (0,\infty) \subset \mathbb{R}$, by the continuity of $f^{(n+m)}$ at x_0, $\exists \delta \in (0,\delta_0] \subset \mathbb{R}$ such that $\left\| f^{(n+m)}(x) - f^{(n+m)}(x_0)\right\| < \epsilon/(1+\|h_n\|\cdots\|h_1\|)$, $\forall x \in D_{n+m} \cap \mathcal{B}_{\mathcal{X}}(x_0,\delta) = D \cap \mathcal{B}_{\mathcal{X}}(x_0,\delta)$. $\forall \bar{x} \in \operatorname{dom}\left(g^{(m)}\right) \cap \mathcal{B}_{\mathcal{X}}(x_0,\delta) = D \cap \mathcal{B}_{\mathcal{X}}(x_0,\delta)$, we have

$$\left\| g^{(m)}(x) - g^{(m)}(x_0)\right\|$$
$$= \sup_{h_{n+i}\in\mathcal{X},\ \|h_{n+i}\|\le 1,\ i=1,\dots,m} \left\| (g^{(m)}(x) - g^{(m)}(x_0))(h_{n+m})\cdots(h_{n+1})\right\|$$

$$= \sup_{h_{n+i} \in \mathcal{X},\ \|h_{n+i}\| \leq 1,\ i=1,\ldots,m} \Big\| f^{(n+m)}(x)(h_{n+m}) \cdots (h_1)$$
$$- f^{(n+m)}(x_0)(h_{n+m}) \cdots (h_1) \Big\|$$
$$\leq \Big\| f^{(n+m)}(x) - f^{(n+m)}(x_0) \Big\| \|h_n\| \cdots \|h_1\| < \epsilon$$

where the first inequality follows from Proposition 7.64. Hence, $g^{(m)}$ is continuous at x_0. Therefore, g is $\mathcal{C}_m$ at x_0.

This completes the proof of the proposition. □

Proposition 9.28 *Let $\mathcal{X}$ and $\mathcal{Y}$ be normed linear spaces over $\mathbb{K}$, $D \subseteq \mathcal{X}$, $x_0 \in D$, and $f : D \to \mathcal{Y}$ be $\mathcal{C}_n$ at x_0, where $n \in \mathbb{N}$. Assume that D is locally convex at x_0. Then, $f^{(n)}(x_0) \in \mathrm{B_S}_n(\mathcal{X}, \mathcal{Y})$.*

Proof By the assumption, $\exists \delta_0 \in (0, \infty) \subset \mathbb{R}$ such that $D \cap \mathcal{B}_{\mathcal{X}}(x_0, \delta_0)$ is convex. Without loss of generality, assume f is n-times differentiable at x, $\forall x \in D \cap \mathcal{B}_{\mathcal{X}}(x_0, \delta_0)$. We will prove the proposition by mathematical induction on n.

1° $n = 1$. Clearly, $f^{(1)}(x_0) \in \mathrm{B}(\mathcal{X}, \mathcal{Y}) = \mathrm{B}_1(\mathcal{X}, \mathcal{Y}) = \mathrm{B_S}_1(\mathcal{X}, \mathcal{Y})$.

Next, we consider $n = 2$. We will prove this case using an argument of contradiction. Suppose $f^{(2)}(x_0)$ is not symmetric. Then, $\exists \bar{h}_0, \bar{l}_0 \in \mathcal{X}$ such that $f^{(2)}(x_0)(\bar{h}_0)(\bar{l}_0) \neq f^{(2)}(x_0)(\bar{l}_0)(\bar{h}_0)$. By the differentiability of f at x_0, we have $\overline{\mathrm{span}\,(\mathrm{A}_D\,(x_0))} = \mathcal{X}$. Then, by $f^{(2)}(x_0) \in \mathrm{B}_2(\mathcal{X}, \mathcal{Y})$ and Proposition 3.56, $\exists \tilde{h}_0, \tilde{l}_0 \in \mathrm{span}\,(\mathrm{A}_D\,(x_0))$ such that $f^{(2)}(x_0)(\tilde{h}_0)(\tilde{l}_0) \neq f^{(2)}(x_0)(\tilde{l}_0)(\tilde{h}_0)$. Since $f^{(2)}(x_0)$ is multi-linear, then $\exists \hat{h}_0, \hat{l}_0 \in \mathrm{A}_D\,(x_0)$ such that $f^{(2)}(x_0)(\hat{h}_0)(\hat{l}_0) \neq f^{(2)}(x_0)(\hat{l}_0)(\hat{h}_0)$. By continuity of $f^{(2)}(x_0)$, $\exists \epsilon_1 \in (0, \infty) \subset \mathbb{R}$ such that $\forall \check{h} \in \mathcal{B}_{\mathcal{X}}\big(\hat{h}_0, \epsilon_1\big)$, $\forall \check{l} \in \mathcal{B}_{\mathcal{X}}\big(\hat{l}_0, \epsilon_1\big)$, we have $f^{(2)}(x_0)(\check{h})(\check{l}) \neq f^{(2)}(x_0)(\check{l})(\check{h})$. By $\hat{h}_0, \hat{l}_0 \in \mathrm{A}_D\,(x_0)$, $\exists r_h, r_l \in (0, \epsilon_1) \subset \mathbb{R}$, $\exists \check{h}_0 \in \mathcal{B}_{\mathcal{X}}\big(\hat{h}_0, \epsilon_1\big)$, $\exists \check{l}_0 \in \mathcal{B}_{\mathcal{X}}\big(\hat{l}_0, \epsilon_1\big)$, such that $x_0 + r_h \check{h}_0, x_0 + r_l \check{l}_0 \in D \cap \mathcal{B}_{\mathcal{X}}(x_0, \delta_0)$. Clearly, we have $f^{(2)}(x_0)(r_h \check{h}_0)(r_l \check{l}_0) = r_h r_l f^{(2)}(x_0)(\check{h}_0)(\check{l}_0) \neq r_h r_l f^{(2)}(x_0)(\check{l}_0)(\check{h}_0) = f^{(2)}(x_0)(r_l \check{l}_0)(r_h \check{h}_0)$. Let $h_0 := r_h/2\check{h}_0$ and $l_0 := r_l/2\check{l}_0$. Then, by the convexity of the set $D \cap \mathcal{B}_{\mathcal{X}}(x_0, \delta_0)$, we have $x_0, x_0 + h_0, x_0 + l_0, x_0 + h_0 + l_0 \in D \cap \mathcal{B}_{\mathcal{X}}(x_0, \delta_0)$, and $f^{(2)}(x_0)(h_0)(l_0) \neq f^{(2)}(x_0)(l_0)(h_0)$. Clearly, $h_0 \neq \vartheta_{\mathcal{X}}$ and $l_0 \neq \vartheta_{\mathcal{X}}$. Let $\epsilon_0 := \big\| f^{(2)}(x_0)(h_0)(l_0) - f^{(2)}(x_0)(l_0)(h_0) \big\| / (\|h_0\| \|l_0\|) \in (0, \infty) \subset \mathbb{R}$. Since f is $\mathcal{C}_2$ at x_0, then $\exists \delta_1 \in (0, \delta_0] \subset \mathbb{R}$ such that $\forall x \in D \cap \mathcal{B}_{\mathcal{X}}(x_0, \delta_1)$, we have $\big\| f^{(2)}(x) - f^{(2)}(x_0) \big\| < \epsilon_0/2$. By proper scaling of h_0 and l_0, we may assume that $\forall t_1, t_2 \in I := [0, 1] \subset \mathbb{R}$, $x_0 + t_1 h_0 + t_2 l_0 \in D \cap \mathcal{B}_{\mathcal{X}}(x_0, \delta_1)$.

In summary, $\exists h_0, l_0 \in \mathcal{X} \setminus \{\vartheta_{\mathcal{X}}\}$ such that $\epsilon_0 := \big\| f^{(2)}(x_0)(h_0)(l_0) - f^{(2)}(x_0)(l_0)(h_0) \big\| / (\|h_0\| \|l_0\|) \in (0, \infty) \subset \mathbb{R}$, $\forall t_1, t_2 \in I$, f is twice differentiable at $x_0 + t_1 h_0 + t_2 l_0 \in D \cap \mathcal{B}_{\mathcal{X}}(x_0, \delta_0)$, and $\big\| f^{(2)}(x_0 + t_1 h_0 + t_2 l_0) - f^{(2)}(x_0) \big\| < \epsilon_0/2$.

Let $\bar{D} := I$, if $\mathbb{K} = \mathbb{R}$; or $\bar{D} := \{a + i0 \,|\, a \in I\} \subset \mathbb{C}$, if $\mathbb{K} = \mathbb{C}$. $\forall t_1 \in I$, define $\psi_{t_1} : \bar{D} \to B(\mathcal{X}, \mathcal{Y})$ by $\psi_{t_1}(t_2) = f^{(1)}(x_0 + t_1 h_0 + t_2 l_0) - f^{(1)}(x_0 + t_1 h_0) - f^{(2)}(x_0)(t_2 l_0)$, $\forall t_2 \in \bar{D}$. By Propositions 9.19, 9.16, and 9.15 and chain rule, each term in the definition of ψ_{t_1} is Fréchet differentiable. Then, by Proposition 9.15, ψ_{t_1} is Fréchet differentiable at t_2, $\forall t_2 \in \bar{D}$. By mean value theorem, $\exists \bar{t}_2 \in I^\circ$ such that

$$\begin{aligned}
&\left\| f^{(1)}(x_0 + t_1 h_0 + l_0) - f^{(1)}(x_0 + t_1 h_0) - f^{(2)}(x_0)(l_0) \right\| \\
&= \left\| \psi_{t_1}(1) - \psi_{t_1}(0) \right\| \le \left\| D\psi_{t_1}(\bar{t}_2) \right\| \\
&= \left\| f^{(2)}(x_0 + t_1 h_0 + \bar{t}_2 l_0)(l_0) - f^{(2)}(x_0)(l_0) \right\| \\
&\le \left\| f^{(2)}(x_0 + t_1 h_0 + \bar{t}_2 l_0) - f^{(2)}(x_0) \right\| \|l_0\| < \epsilon_0 \|l_0\|/2
\end{aligned}$$

where the second inequality follows from Proposition 7.64.

Define $\gamma : \bar{D} \to \mathcal{Y}$ by $\gamma(t_1) = f(x_0 + l_0 + t_1 h_0) - f(x_0 + t_1 h_0) - f^{(2)}(x_0)(l_0)(t_1 h_0)$, $\forall t_1 \in \bar{D}$. By Propositions 9.19, 9.16, and 9.15 and chain rule, each term in the definition of γ is Fréchet differentiable. Then, by Proposition 9.15, γ is Fréchet differentiable at t_1, $\forall t_1 \in \bar{D}$. By mean value theorem, $\exists \bar{t}_1 \in I^\circ$ such that

$$\begin{aligned}
&\left\| f(x_0 + l_0 + h_0) - f(x_0 + h_0) - f(x_0 + l_0) + f(x_0) - f^{(2)}(x_0)(l_0)(h_0) \right\| \\
&= \|\gamma(1) - \gamma(0)\| \le \left\| D\gamma(\bar{t}_1) \right\| = \left\| f^{(1)}(x_0 + l_0 + \bar{t}_1 h_0)(h_0) \right. \\
&\quad \left. - f^{(1)}(x_0 + \bar{t}_1 h_0)(h_0) - f^{(2)}(x_0)(l_0)(h_0) \right\| \\
&\le \left\| f^{(1)}(x_0 + l_0 + \bar{t}_1 h_0) - f^{(1)}(x_0 + \bar{t}_1 h_0) - f^{(2)}(x_0)(l_0) \right\| \|h_0\| \\
&= \left\| \psi_{\bar{t}_1}(1) - \psi_{\bar{t}_1}(0) \right\| \|h_0\| < \epsilon_0 \|l_0\| \|h_0\|/2
\end{aligned}$$

where the second inequality follows from Proposition 7.64.

By symmetry, we have $\| f(x_0 + h_0 + l_0) - f(x_0 + l_0) - f(x_0 + h_0) + f(x_0) - f^{(2)}(x_0)(h_0)(l_0) \| < \epsilon_0 \|h_0\| \|l_0\|/2$. Then, $\| f^{(2)}(x_0)(h_0)(l_0) - f^{(2)}(x_0)(l_0) \cdot (h_0) \| < \epsilon_0 \|h_0\| \|l_0\|$. This leads to the contradiction $\epsilon_0 := \| f^{(2)}(x_0)(h_0)(l_0) - f^{(2)}(x_0)(l_0)(h_0) \| / (\|h_0\| \|l_0\|) < \epsilon_0$.

Hence, $f^{(2)}(x_0)$ must be symmetric and $f^{(2)}(x_0) \in B_{S\,2}(\mathcal{X}, \mathcal{Y})$.

2° Assume that the result holds $\forall n \le k$, $k \in \{2, 3, \ldots\}$.

3° Consider the case $n = k + 1$. $\forall (h_1, \ldots, h_{k+1}) \in \mathcal{X}^{k+1}$, let $(v_1, \ldots, v_{k+1})$ be a permutation of $(h_1, \ldots, h_{k+1})$. We need to show that

$$f^{(k+1)}(x_0)(h_{k+1}) \cdots (h_1) = f^{(k+1)}(x_0)(v_{k+1}) \cdots (v_1)$$

Since any permutation can be arrived at in a finite number of steps by interchanging two consecutive elements, then all we need to show is that, $\forall i = 1, \ldots, k$,

$$\begin{aligned} & f^{(k+1)}(x_0)(h_{k+1}) \cdots (h_{i+1})(h_i) \cdots (h_1) \\ & = f^{(k+1)}(x_0)(h_{k+1}) \cdots (h_i)(h_{i+1}) \cdots (h_1) \end{aligned}$$

We will distinguish two exhaustive and mutually exclusive cases: Case 1: $1 \leq i < k$; Case 2: $i = k$. Case 1: $1 \leq i < k$. Define $g : \operatorname{dom}\left(f^{(k)}\right) \to \mathcal{Y}$ by $g(x) = f^{(k)}(x)(h_k) \cdots (h_{i+1})(h_i) \cdots (h_1)$, $\forall x \in \operatorname{dom}\left(f^{(k)}\right)$. $\forall x \in D \cap \mathcal{B}_{\mathcal{X}}(x_0, \delta_0) = \operatorname{dom}\left(f^{(k+1)}\right) \cap \mathcal{B}_{\mathcal{X}}(x_0, \delta_0)$, f is $(k+1)$-times differentiable at x, and $\operatorname{dom}\left(f^{(k)}\right) \supseteq D \cap \mathcal{B}_{\mathcal{X}}(x, \delta_x)$, where $\delta_x := \delta_0 - \|x - x_0\| \in (0, \infty) \subset \mathbb{R}$. Then, f is $\mathcal{C}_k$ at x. Clearly, $D \cap \mathcal{B}_{\mathcal{X}}(x, \delta_x) = (D \cap \mathcal{B}_{\mathcal{X}}(x_0, \delta_0)) \cap \mathcal{B}_{\mathcal{X}}(x, \delta_x)$ is convex. By inductive assumption, $g(x) = f^{(k)}(x)(h_k) \cdots (h_i)(h_{i+1}) \cdots (h_1)$. By Proposition 9.27, g is $\mathcal{C}_1$ at x_0 and $f^{(k+1)}(x_0)(h_{k+1}) \cdots (h_{i+1})(h_i) \cdots (h_1) = g^{(1)}(x_0)(h_{k+1}) = f^{(k+1)}(x_0)(h_{k+1}) \cdots \cdot (h_i)(h_{i+1}) \cdots (h_1)$. This case is proved.

Case 2: $i = k$. Define $g : \operatorname{dom}\left(f^{(k-1)}\right) \to \mathcal{Y}$ by, $\forall x \in \operatorname{dom}\left(f^{(k-1)}\right)$, $g(x) = f^{(k-1)}(x)(h_{k-1}) \cdots (h_1)$. Then, by Proposition 9.27, g is $\mathcal{C}_2$ at x_0 and $g^{(2)}(x_0)(u)(v) = f^{(k+1)}(x_0)(u)(v)(h_{k-1}) \cdots (h_1)$, $\forall u, v \in \mathcal{X}$. Note that $\operatorname{dom}\left(f^{(k-1)}\right) \cap \mathcal{B}_{\mathcal{X}}(x_0, \delta_0) = D \cap \mathcal{B}_{\mathcal{X}}(x_0, \delta_0)$ is convex. By the case of $n = 2$, we have $f^{(k+1)}(x_0)(h_{k+1})(h_k) \cdots (h_1) = g^{(2)}(x_0)(h_{k+1})(h_k) = g^{(2)}(x_0)(h_k)(h_{k+1}) = f^{(k+1)}(x_0)(h_k)(h_{k+1}) \cdots (h_1)$. This case is proved.

Hence, $f^{(k+1)}(x_0) \in \mathrm{B}_{\mathrm{S}\,k+1}(\mathcal{X}, \mathcal{Y})$.

This completes the induction process and the proof of the proposition. □

Definition 9.29 Let $\mathcal{X}_1, \ldots, \mathcal{X}_p$, and $\mathcal{Y}$ be normed linear spaces over $\mathbb{K}$, where $p \in \{2, 3, \ldots\}$, $D \subseteq \mathcal{X} := \prod_{i=1}^{p} \mathcal{X}_i$, $f : D \to \mathcal{Y}$, and $x_o := (x_{1o}, \ldots, x_{po}) \in D$. Assume that $\frac{\partial f}{\partial x_{i_1}} : \operatorname{dom}\left(\frac{\partial f}{\partial x_{i_1}}\right) \to \mathrm{B}(\mathcal{X}_{i_1}, \mathcal{Y})$ is partial differentiable with respect to x_{i_2} at $x_o \in \operatorname{dom}\left(\frac{\partial f}{\partial x_{i_1}}\right)$, where $i_1, i_2 \in \{1, \ldots, p\}$. Then, this partial derivative is denoted by $\frac{\partial^2 f}{\partial x_{i_2} \partial x_{i_1}}(x_o) \in \mathrm{B}(\mathcal{X}_{i_2}, \mathrm{B}(\mathcal{X}_{i_1}, \mathcal{Y}))$, which is one of the *second order partial derivatives of* f. We will let $\frac{\partial^2 f}{\partial x_{i_2} \partial x_{i_1}}$ denote the $\mathrm{B}(\mathcal{X}_{i_2}, \mathrm{B}(\mathcal{X}_{i_1}, \mathcal{Y}))$-valued function whose domain of definition is $\operatorname{dom}\left(\frac{\partial^2 f}{\partial x_{i_2} \partial x_{i_1}}\right) := \left\{x \in \operatorname{dom}\left(\frac{\partial f}{\partial x_{i_1}}\right) \,\middle|\, \frac{\partial^2 f}{\partial x_{i_2} \partial x_{i_1}}(x) \in \mathrm{B}(\mathcal{X}_{i_2}, \mathrm{B}(\mathcal{X}_{i_1}, \mathcal{Y})) \text{ exists}\right\}$. There are a total of p^2 second order partial derivatives of f. Recursively, we may define *kth order partial derivatives*, $k \in \{3, 4, \ldots\}$. There are a total of p^k k order partial derivatives. A typical such derivative is denoted by $\frac{\partial^k f}{\partial x_{i_k} \cdots \partial x_{i_1}}$, where $i_1, \ldots, i_k \in \{1, \ldots, p\}$. ◇

Let $\mathcal{X}$ be a normed linear space over $\mathbb{K}$ and $D \subseteq \mathcal{X}$. Define $\operatorname{Sm}(D) := \bigcup_{\alpha \in \mathbb{K}} \alpha D$. Clearly, $\operatorname{Sm}(D) \subseteq \operatorname{span}(D)$.

Proposition 9.30 *Let $\mathcal{X}$ be a normed linear space over $\mathbb{K}$, $D \subseteq \mathcal{X}$, and $x_0 \in \overline{D}$. Then,* $\overline{\operatorname{span}(\mathrm{A}_D(x_0))} \subseteq \overline{\operatorname{span}((D \cap \mathcal{B}(x_0,\delta)) - x_0)}$ *and* $\mathrm{A}_D(x_0) \subseteq \overline{\operatorname{Sm}((D \cap \mathcal{B}(x_0,\delta)) - x_0)}$, $\forall \delta \in (0,\infty) \subset \mathbb{R}$.

Proof $\forall u \in \mathrm{A}_D(x_0)$, we will show that $u \in \overline{\operatorname{Sm}(D - x_0)}$. $\forall \epsilon \in (0,\infty) \subset \mathbb{R}$, by $u \in \mathrm{A}_D(x_0)$, we have $\exists \bar{x} := x_0 + r\bar{u} \in D$ such that $0 < r < \epsilon$ and $\bar{u} \in \mathcal{B}(u,\epsilon)$. Then, $\bar{u} \in \operatorname{Sm}(D - x_0) \cap \mathcal{B}(u,\epsilon) \neq \emptyset$. Hence, by Proposition 3.3, $u \in \overline{\operatorname{Sm}(D - x_0)}$. By the arbitrariness of u, we have $\mathrm{A}_D(x_0) \subseteq \overline{\operatorname{Sm}(D - x_0)} \subseteq \overline{\operatorname{span}(D - x_0)}$. By Proposition 7.17, $\operatorname{span}(\mathrm{A}_D(x_0)) \subseteq \overline{\operatorname{span}(D - x_0)}$. By Definition 3.2, $\overline{\operatorname{span}(\mathrm{A}_D(x_0))} \subseteq \overline{\operatorname{span}(D - x_0)}$. $\forall \delta \in (0,\infty) \subset \mathbb{R}$, we have $x_0 \in \overline{D \cap \mathcal{B}(x_0,\delta)}$ and $\mathrm{A}_D(x_0) = \mathrm{A}_{D\cap\mathcal{B}(x_0,\delta)}(x_0)$. Then, $\mathrm{A}_D(x_0) \subseteq \overline{\operatorname{Sm}((D \cap \mathcal{B}(x_0,\delta)) - x_0)}$ and $\overline{\operatorname{span}((D \cap \mathcal{B}(x_0,\delta)) - x_0)} \supseteq \overline{\operatorname{span}(\mathrm{A}_D(x_0))}$. This completes the proof of the proposition. □

9.4.2 *Interchange Order of Differentiation*

Next, we present two results on the interchangeability of order of differentiation. The first result does not assume the existence of the derivative after the interchange, which then requires a stronger assumption on the set $D = \operatorname{dom}(f)$. The second result assumes the existence of the derivative after the interchange, where the stronger assumption on D can be removed.

Proposition 9.31 *Let $\mathcal{X}$, $\mathcal{Y}$, and $\mathcal{Z}$ be normed linear spaces over $\mathbb{K}$, $D \subseteq \mathcal{X} \times \mathcal{Y}$, $f : D \to \mathcal{Z}$, and $(x_0, y_0) \in D$. Assume that the following conditions are satisfied:*

(i) *D is locally convex at (x_0, y_0), that is, $\exists \delta_0 \in (0,\infty) \subset \mathbb{R}$ such that $D \cap \mathcal{B}_{\mathcal{X}\times\mathcal{Y}}((x_0,y_0),\delta_0) =: \tilde{D}$ is convex. Furthermore, $\forall (x,y) \in \tilde{D}$, we have $(x,y_0), (x_0,y) \in \tilde{D}$.*
(ii) *$\frac{\partial f}{\partial x}(x,y)$, $\frac{\partial f}{\partial y}(x,y)$, and $\frac{\partial^2 f}{\partial x \partial y}(x,y)$ exist, $\forall (x,y) \in \tilde{D}$, and $\frac{\partial^2 f}{\partial x \partial y}$ is continuous at (x_0, y_0).*

Then, $\forall \epsilon \in (0,\infty) \subset \mathbb{R}$, $\exists \delta \in (0,\delta_0] \subset \mathbb{R}$ such that $\forall (x,y) \in D \cap \mathcal{B}_{\mathcal{X}\times\mathcal{Y}}((x_0,y_0),\delta) =: \tilde{D}_\delta$, we have $\Big\| f(x,y) - f(x,y_0) - f(x_0,y) + f(x_0,y_0) - \frac{\partial^2 f}{\partial x \partial y}(x_0,y_0)(x - x_0)(y - y_0) \Big\| \leq \epsilon \|x - x_0\| \|y - y_0\|$.

Furthermore, if, in addition, the following condition is satisfied:

(iii) *$\exists M \in [0,\infty) \subset \mathbb{R}$, $\forall h \in \mathcal{X}$, $\forall \epsilon \in (0,\infty) \subset \mathbb{R}$, $\forall (x,y) \in \tilde{D}$, $\forall \bar{\delta} \in (0,\infty) \subset \mathbb{R}$, $\exists h_1, h_2 \in \operatorname{Sm}\big((D_y \cap \mathcal{B}_{\mathcal{X}}(x_0,\bar{\delta})) - x_0\big)$ such that $\|h - h_1 + h_2\| \leq \epsilon \|h\|$, $\|h_1\| \leq M\|h\|$, and $\|h_2\| \leq M\|h\|$, where $D_y := \{x \in \mathcal{X} \mid (x,y) \in D\}$, $\forall y \in \mathcal{Y}$.*

Then, $\frac{\partial^2 f}{\partial y \partial x}(x_0, y_0)$ *exists and is given by*

$$\frac{\partial^2 f}{\partial y \partial x}(x_0, y_0)(k)(h) = \frac{\partial^2 f}{\partial x \partial y}(x_0, y_0)(h)(k); \quad \forall h \in \mathcal{X}, \forall k \in \mathcal{Y}$$

Proof $\forall \epsilon \in (0, \infty) \subset \mathbb{R}$, by the continuity of $\frac{\partial^2 f}{\partial x \partial y}$ at (x_0, y_0), $\exists \delta \in (0, \delta_0] \subset \mathbb{R}$ such that $\forall (x, y) \in D \cap \mathcal{B}_{\mathcal{X} \times \mathcal{Y}}((x_0, y_0), \delta) =: \tilde{D}_\delta \subseteq \tilde{D}$, we have $\left\| \frac{\partial^2 f}{\partial x \partial y}(x, y) - \frac{\partial^2 f}{\partial x \partial y}(x_0, y_0) \right\| < \epsilon/3$. Note that, by (i), $\tilde{D}_\delta$ is convex and, $\forall (x, y) \in \tilde{D}_\delta$, we have $(x_0, y), (x, y_0) \in \tilde{D}_\delta$.

Claim 9.31.1 $\forall (x, y) \in \tilde{D}_\delta$, $\Big\| f(x, y) - f(x, y_0) - f(x_0, y) + f(x_0, y_0) - \frac{\partial^2 f}{\partial x \partial y}(x_0, y_0)(x - x_0)(y - y_0) \Big\| \leq \epsilon/3 \|x - x_0\| \|y - y_0\|$.

Proof of Claim Fix $(x, y) \in \tilde{D}_\delta$. Let $I := [0, 1] \subset \mathbb{R}$. Define $\bar{D} := I$ if $\mathbb{K} = \mathbb{R}$, or $\bar{D} := \{a + \mathrm{i}0 \mid a \in I\} \subseteq \mathbb{C}$ if $\mathbb{K} = \mathbb{C}$.

$\forall t_2 \in I$, define $\psi_{t_2} : \bar{D} \to \mathrm{B}(\mathcal{Y}, \mathcal{Z})$ by $\psi_{t_2}(t_1) = \frac{\partial f}{\partial y}(t_1 x + (1 - t_1)x_0, t_2 y + (1 - t_2)y_0) - \frac{\partial f}{\partial y}(x_0, t_2 y + (1 - t_2)y_0) - \frac{\partial^2 f}{\partial x \partial y}(x_0, y_0)(t_1 (x - x_0))$, $\forall t_1 \in \bar{D}$. By Propositions 9.19, 9.16, and 9.15 and chain rule, each term in the definition of ψ_{t_2} is differentiable. By Proposition 9.15, ψ_{t_2} is differentiable. By mean value theorem, $\exists \bar{t}_1 \in I^\circ$ such that

$$\begin{aligned}
&\Big\| \frac{\partial f}{\partial y}(x, t_2 y + (1 - t_2)y_0) - \frac{\partial f}{\partial y}(x_0, t_2 y + (1 - t_2)y_0) \\
&\quad - \frac{\partial^2 f}{\partial x \partial y}(x_0, y_0)(x - x_0) \Big\| \\
&= \left\| \psi_{t_2}(1) - \psi_{t_2}(0) \right\| \leq \left\| \mathrm{D}\psi_{t_2}(\bar{t}_1) \right\| \\
&= \Big\| \frac{\partial^2 f}{\partial x \partial y}(\bar{t}_1 x + (1 - \bar{t}_1)x_0, t_2 y + (1 - t_2)y_0)(x - x_0) \\
&\quad - \frac{\partial^2 f}{\partial x \partial y}(x_0, y_0)(x - x_0) \Big\| \leq \epsilon/3 \|x - x_0\|
\end{aligned}$$

where the last inequality follows from Proposition 7.64.

Define $\gamma : \bar{D} \to \mathcal{Z}$ by $\gamma(t_2) = f(x, t_2 y + (1 - t_2)y_0) - f(x_0, t_2 y + (1 - t_2)y_0) - \frac{\partial^2 f}{\partial x \partial y}(x_0, y_0)(x - x_0)(t_2 (y - y_0))$, $\forall t_2 \in \bar{D}$. By Propositions 9.19, 9.16, and 9.15 and chain rule, each term in the definition of γ is differentiable. By Proposition 9.15, γ is differentiable. By mean value theorem, $\exists \bar{t}_2 \in I^\circ$ such that

$$\begin{aligned}
&\left\| f(x, y) - f(x, y_0) - f(x_0, y) + f(x_0, y_0) - \frac{\partial^2 f}{\partial x \partial y}(x_0, y_0)(x - x_0)(y - y_0) \right\| \\
&= \|\gamma(1) - \gamma(0)\| \leq \left\| \mathrm{D}\gamma(\bar{t}_2) \right\|
\end{aligned}$$

$$= \left\| \frac{\partial f}{\partial y}(x, \bar{t}_2 y + (1 - \bar{t}_2) y_0)(y - y_0) - \frac{\partial f}{\partial y}(x_0, \bar{t}_2 y + (1 - \bar{t}_2) y_0)(y - y_0) \right.$$
$$\left. - \frac{\partial^2 f}{\partial x \partial y}(x_0, y_0)(x - x_0)(y - y_0) \right\|$$
$$\leq \left\| \frac{\partial f}{\partial y}(x, \bar{t}_2 y + (1 - \bar{t}_2) y_0) - \frac{\partial f}{\partial y}(x_0, \bar{t}_2 y + (1 - \bar{t}_2) y_0) \right.$$
$$\left. - \frac{\partial^2 f}{\partial x \partial y}(x_0, y_0)(x - x_0) \right\| \|y - y_0\|$$
$$= \left\| \psi_{\bar{t}_2}(1) - \psi_{\bar{t}_2}(0) \right\| \|y - y_0\| \leq \epsilon/3 \|x - x_0\| \|y - y_0\|$$

where the second inequality follows from Proposition 7.64. This completes the proof of the claim. □

Define $D_{x_0} := \{y \in \mathcal{Y} \mid (x_0, y) \in D\}$ and $\hat{D} := \left\{ y \in \mathcal{Y} \;\middle|\; (x_0, y) \in \operatorname{dom}\left(\frac{\partial f}{\partial x}\right) \right\}$. By (ii), we have $\overline{\operatorname{span}\left(\mathrm{A}_{D_{x_0}}(y_0)\right)} = \mathcal{Y}$ and $\overline{\operatorname{span}\left(\mathrm{A}_{D_y}(x)\right)} = \mathcal{X}$, $\forall (x, y) \in \tilde{D}$. Clearly, we have $\hat{D} \cap \mathcal{B}_{\mathcal{Y}}(y_0, \delta_0) = D_{x_0} \cap \mathcal{B}_{\mathcal{Y}}(y_0, \delta_0)$. This implies that $\overline{\operatorname{span}\left(\mathrm{A}_{\hat{D}}(y_0)\right)} = \mathcal{Y}$.

$\forall k \in \mathcal{Y}$, define $L_k : \mathcal{X} \to \mathcal{Z}$ by $L_k(h) = \frac{\partial^2 f}{\partial x \partial y}(x_0, y_0)(h)(k)$, $\forall h \in \mathcal{X}$. Clearly, L_k is a linear operator. Note that $\|L_k\| = \sup_{h \in \mathcal{X}, \|h\| \leq 1} \|L_k(h)\| \leq \left\| \frac{\partial^2 f}{\partial x \partial y}(x_0, y_0) \right\| \|k\| < +\infty$, where the first inequality follows from Proposition 7.64. Hence, $L_k \in \mathrm{B}(\mathcal{X}, \mathcal{Z})$.

Define $L : \mathcal{Y} \to \mathrm{B}(\mathcal{X}, \mathcal{Z})$ by $L(k) = L_k$, $\forall k \in \mathcal{Y}$. Clearly, L is a linear operator. Note that $\|L\| = \sup_{k \in \mathcal{Y}, \|k\| \leq 1} \|L(k)\| \leq \left\| \frac{\partial^2 f}{\partial x \partial y}(x_0, y_0) \right\| < +\infty$. Hence, $L \in \mathrm{B}(\mathcal{Y}, \mathrm{B}(\mathcal{X}, \mathcal{Z}))$.

Now, fix any $y \in \hat{D} \cap \mathcal{B}_{\mathcal{Y}}\left(y_0, \delta/\sqrt{2}\right)$; we will show that $\left\| \Delta_y \right\| \leq (2M + 1)\epsilon \|y - y_0\|$, where $\Delta_y := \frac{\partial f}{\partial x}(x_0, y) - \frac{\partial f}{\partial x}(x_0, y_0) - L(y - y_0) \in \mathrm{B}(\mathcal{X}, \mathcal{Z})$. This immediately implies that $\frac{\partial^2 f}{\partial y \partial x}(x_0, y_0) = L$ and completes the proof of the proposition.

We will distinguish two exhaustive and mutually exclusive cases: Case 1: $y = y_0$; Case 2: $y \neq y_0$. Case 1: $y = y_0$. The result is immediate. Case 2: $y \neq y_0$. By the existence of $\frac{\partial f}{\partial x}(x_0, y)$, $\exists \delta_{y1} \in (0, \delta/\sqrt{2}] \subset \mathbb{R}$ such that $\forall x \in D_y \cap \mathcal{B}_{\mathcal{X}}\left(x_0, \delta_{y1}\right)$, we have $\left\| f(x, y) - f(x_0, y) - \frac{\partial f}{\partial x}(x_0, y)(x - x_0) \right\| \leq \epsilon/3 \|y - y_0\| \|x - x_0\|$. By the existence of $\frac{\partial f}{\partial x}(x_0, y_0)$, $\exists \delta_{y2} \in (0, \delta/\sqrt{2}] \subset \mathbb{R}$ such that $\forall x \in D_{y_0} \cap \mathcal{B}_{\mathcal{X}}\left(x_0, \delta_{y2}\right)$, we have $\left\| f(x, y_0) - f(x_0, y_0) - \frac{\partial f}{\partial x}(x_0, y_0)(x - x_0) \right\| \leq \epsilon/3 \|y - y_0\| \|x - x_0\|$. Let $\delta_y = \min\left\{\delta_{y1}, \delta_{y2}\right\} \in (0, \delta/\sqrt{2}] \subset \mathbb{R}$. $\forall x \in D_y \cap \mathcal{B}_{\mathcal{X}}\left(x_0, \delta_y\right)$, we have $(x, y), (x, y_0), (x_0, y) \in \tilde{D}_\delta$. Then, $x \in D_{y_0} \cap \mathcal{B}_{\mathcal{X}}\left(x_0, \delta_y\right)$. By Claim 9.31.1 and the preceding argument, we have

$$\begin{aligned}
&\big\|\Delta_y(x-x_0)\big\| \\
&= \Big\|\frac{\partial f}{\partial x}(x_0,y)(x-x_0) - \frac{\partial f}{\partial x}(x_0,y_0)(x-x_0) - L(y-y_0)(x-x_0)\Big\| \\
&\le \Big\|\frac{\partial f}{\partial x}(x_0,y)(x-x_0) - f(x,y) + f(x_0,y)\Big\| + \Big\|f(x,y_0) - f(x_0,y_0) \\
&\quad -\frac{\partial f}{\partial x}(x_0,y_0)(x-x_0)\Big\| + \Big\|f(x,y) - f(x,y_0) - f(x_0,y) + f(x_0,y_0) \\
&\quad -\frac{\partial^2 f}{\partial x \partial y}(x_0,y_0)(x-x_0)(y-y_0)\Big\| \le \epsilon\|x-x_0\|\|y-y_0\|
\end{aligned}$$

$\forall h \in \mathcal{X}$, by (iii) and the continuity of Δ_y, $\exists h_1, h_2 \in \mathrm{Sm}\big((D_y \cap \mathcal{B}_{\mathcal{X}}(x_0,\delta_y)) - x_0\big)$, such that $\big\|\Delta_y(h - h_1 + h_2)\big\| \le \epsilon\|h\|\|y-y_0\|$, $\|h_1\| \le M\|h\|$, and $\|h_2\| \le M\|h\|$. Then, $\exists \alpha_1, \alpha_2 \in \mathbb{K}$ and $\exists x_1, x_2 \in D_y \cap \mathcal{B}_{\mathcal{X}}(x_0,\delta_y)$ such that $h_i = \alpha_i\,(x_i - x_0)$, $i = 1, 2$. Then, we have

$$\begin{aligned}
\big\|\Delta_y(h)\big\| &\le \big\|\Delta_y(h_1 - h_2)\big\| + \epsilon\|h\|\|y-y_0\| \\
&= \big\|\alpha_1\Delta_y(x_1 - x_0) - \alpha_2\Delta_y(x_2 - x_0)\big\| + \epsilon\|h\|\|y-y_0\| \\
&\le |\alpha_1|\big\|\Delta_y(x_1 - x_0)\big\| + |\alpha_2|\big\|\Delta_y(x_2 - x_0)\big\| + \epsilon\|h\|\|y-y_0\| \\
&\le (|\alpha_1|\|x_1 - x_0\| + |\alpha_2|\|x_2 - x_0\| + \|h\|)\epsilon\|y-y_0\| \\
&= (\|h_1\| + \|h_2\| + \|h\|)\epsilon\|y-y_0\| \le (2M+1)\epsilon\|y-y_0\|\|h\|
\end{aligned}$$

Then, we have $\big\|\Delta_y\big\| \le (2M+1)\epsilon\|y-y_0\|$. This case is proved.

This completes the proof of the proposition. □

In the preceding proposition, conditions (i) and (iii) are assumptions on the set D. It is clear that (i) and (iii) are satisfied if $(x_0, y_0) \in D^\circ$.

Proposition 9.32 *Let $\mathcal{X}$, $\mathcal{Y}$, and $\mathcal{Z}$ be normed linear spaces over $\mathbb{K}$, $D \subseteq \mathcal{X} \times \mathcal{Y}$, $f : D \to \mathcal{Z}$, and $(x_0, y_0) \in D$. Assume that the following conditions are satisfied:*

(i) *D is locally convex at (x_0, y_0), that is, $\exists \delta_0 \in (0, \infty) \subset \mathbb{R}$ such that $D \cap \mathcal{B}_{\mathcal{X}\times\mathcal{Y}}((x_0, y_0), \delta_0) =: \tilde{D}$ is convex. Furthermore, $\forall (x, y) \in \tilde{D}$, we have $(x, y_0), (x_0, y) \in \tilde{D}$.*

(ii) *$\frac{\partial f}{\partial x}(x, y)$, $\frac{\partial f}{\partial y}(x, y)$, $\frac{\partial^2 f}{\partial x \partial y}(x, y)$, and $\frac{\partial^2 f}{\partial y \partial x}(x, y)$ exist, $\forall (x, y) \in \tilde{D}$, and $\frac{\partial^2 f}{\partial x \partial y}$ and $\frac{\partial^2 f}{\partial y \partial x}$ are continuous at (x_0, y_0).*

Then, $\frac{\partial^2 f}{\partial y \partial x}(x_0, y_0)(k)(h) = \frac{\partial^2 f}{\partial x \partial y}(x_0, y_0)(h)(k)$, $\forall h \in \mathcal{X}$, $\forall k \in \mathcal{Y}$.

Proof Define $D_{x_0} := \{y \in \mathcal{Y} \mid (x_0, y) \in D\}$ and $D_{y_0} := \{x \in \mathcal{X} \mid (x, y_0) \in D\}$. By (ii), we have $\overline{\text{span}\left(\text{A}_{D_{x_0}}(y_0)\right)} = \mathcal{Y}$ and $\overline{\text{span}\left(\text{A}_{D_{y_0}}(x_0)\right)} = \mathcal{X}$.

$\forall \epsilon \in (0, \infty) \subset \mathbb{R}$, by Proposition 9.31, $\exists \delta_1 \in (0, \delta_0] \subset \mathbb{R}$ such that $\forall (x, y) \in D \cap \mathcal{B}_{\mathcal{X}\times\mathcal{Y}}((x_0, y_0), \delta_1) =: \tilde{D}_{\delta_1} \subseteq \tilde{D}$, we have $\Big\| f(x, y) - f(x, y_0) - f(x_0, y) + f(x_0, y_0) - \frac{\partial^2 f}{\partial x \partial y}(x_0, y_0)(x - x_0)(y - y_0)\Big\| \leq \epsilon/2\|x - x_0\|\|y - y_0\|$.

By symmetry, $\exists \delta_2 \in (0, \delta_0] \subset \mathbb{R}$ such that $\forall (x, y) \in \mathcal{B}_{\mathcal{X}\times\mathcal{Y}}((x_0, y_0), \delta_2) \cap D =: \tilde{D}_{\delta_2} \subseteq \tilde{D}$, we have $\Big\| f(x, y) - f(x, y_0) - f(x_0, y) + f(x_0, y_0) - \frac{\partial^2 f}{\partial y \partial x}(x_0, y_0) (y - y_0)(x - x_0)\Big\| \leq \epsilon/2\|x - x_0\|\|y - y_0\|$.

Therefore, let $\delta := \min\{\delta_1, \delta_2\} \in (0, \delta_0] \subset \mathbb{R}$, and $\forall (x, y) \in D \cap \mathcal{B}_{\mathcal{X}\times\mathcal{Y}}((x_0, y_0), \delta) =: \tilde{D}_\delta = \tilde{D}_{\delta_1} \cap \tilde{D}_{\delta_2} \subseteq \tilde{D}$, we have $\Big\| \frac{\partial^2 f}{\partial y \partial x}(x_0, y_0)(y - y_0) (x - x_0) - \frac{\partial^2 f}{\partial x \partial y}(x_0, y_0)(x - x_0)(y - y_0)\Big\| \leq \epsilon\|x - x_0\|\|y - y_0\|$. By the arbitrariness of ϵ, we have $\frac{\partial^2 f}{\partial y \partial x}(x_0, y_0)(y - y_0)(x - x_0) = \frac{\partial^2 f}{\partial x \partial y}(x_0, y_0)(x - x_0)(y - y_0)$.

Clearly, by (i), $\tilde{D}_\delta$ is convex.

Suppose that the result of the proposition is not true; then $\exists h_0 \in \mathcal{X}$ and $\exists k_0 \in \mathcal{Y}$ such that $\frac{\partial^2 f}{\partial y \partial x}(x_0, y_0)(k_0)(h_0) \neq \frac{\partial^2 f}{\partial x \partial y}(x_0, y_0)(h_0)(k_0)$. By the fact that $\overline{\text{span}\left(\text{A}_{D_{x_0}}(y_0)\right)} = \mathcal{Y}$ and $\overline{\text{span}\left(\text{A}_{D_{y_0}}(x_0)\right)} = \mathcal{X}$ and the continuity of $\frac{\partial^2 f}{\partial y \partial x}(x_0, y_0)$ and $\frac{\partial^2 f}{\partial x \partial y}(x_0, y_0)$, $\exists \bar{h}_0 \in \text{span}\left(\text{A}_{D_{y_0}}(x_0)\right)$ and $\exists \bar{k}_0 \in \text{span}\left(\text{A}_{D_{x_0}}(y_0)\right)$ such that $\frac{\partial^2 f}{\partial y \partial x}(x_0, y_0)(\bar{k}_0)(\bar{h}_0) \neq \frac{\partial^2 f}{\partial x \partial y}(x_0, y_0)(\bar{h}_0)(\bar{k}_0)$. By the multi-linearity of these two second order partial derivatives, $\exists \hat{h}_0 \in \text{A}_{D_{y_0}}(x_0)$ and $\exists \hat{k}_0 \in \text{A}_{D_{x_0}}(y_0)$ such that $\frac{\partial^2 f}{\partial y \partial x}(x_0, y_0)(\hat{k}_0)(\hat{h}_0) \neq \frac{\partial^2 f}{\partial x \partial y}(x_0, y_0)(\hat{h}_0)(\hat{k}_0)$. By Proposition 9.30, we have

$$\text{A}_{D_{y_0}}(x_0) \subseteq \overline{\text{Sm}\left((D_{y_0} \cap \mathcal{B}_{\mathcal{X}}(x_0, \delta)) - x_0\right)}$$

$$\text{A}_{D_{x_0}}(y_0) \subseteq \overline{\text{Sm}\left((D_{x_0} \cap \mathcal{B}_{\mathcal{Y}}(y_0, \delta)) - y_0\right)}$$

This implies that, by the continuity of these two second order partial derivatives, $\exists \check{h}_0 \in \text{Sm}\left((D_{y_0} \cap \mathcal{B}_{\mathcal{X}}(x_0, \delta)) - x_0\right)$ and $\exists \check{k}_0 \in \text{Sm}\left((D_{x_0} \cap \mathcal{B}_{\mathcal{Y}}(y_0, \delta)) - y_0\right)$ such that $\frac{\partial^2 f}{\partial y \partial x}(x_0, y_0)(\check{k}_0)(\check{h}_0) \neq \frac{\partial^2 f}{\partial x \partial y}(x_0, y_0)(\check{h}_0)(\check{k}_0)$. Then, $\exists \bar{x} \in D_{y_0} \cap \mathcal{B}_{\mathcal{X}}(x_0, \delta)$, $\exists \bar{y} \in D_{x_0} \cap \mathcal{B}_{\mathcal{Y}}(y_0, \delta)$, and $\exists \alpha_h, \alpha_k \in \mathbb{K}$ such that $\check{h}_0 = \alpha_h(\bar{x} - x_0)$ and $\check{k}_0 = \alpha_k(\bar{y} - y_0)$. By the multi-linearity of these two second order partial derivatives, $\frac{\partial^2 f}{\partial y \partial x}(x_0, y_0)(\bar{y} - y_0)(\bar{x} - x_0) \neq \frac{\partial^2 f}{\partial x \partial y}(x_0, y_0)(\bar{x} - x_0)(\bar{y} - y_0)$. Clearly, $(\bar{x}, y_0), (x_0, \bar{y}) \in \tilde{D}_\delta$. By the convexity of $\tilde{D}_\delta$, we have $(\hat{x}, \hat{y}) := 0.5\,(\bar{x}, y_0) + 0.5\,(x_0, \bar{y}) \in \tilde{D}_\delta$. By the multi-linearity of these two second order partial derivatives, we have $\frac{\partial^2 f}{\partial y \partial x}(x_0, y_0)(\hat{y} - y_0)(\hat{x} - x_0) \neq \frac{\partial^2 f}{\partial x \partial y}(x_0, y_0)(\hat{x} - x_0)(\hat{y} - y_0)$. This is a contradiction. Hence, the result of the proposition must hold.

This completes the proof of the proposition. $\square$

9.4.3 High Order Derivatives of Some Common Functions

Proposition 9.33 *Let $\mathcal{X}$ and $\mathcal{Y}$ be normed linear spaces over $\mathbb{K}$, $D \subseteq \mathcal{X}$, $f : D \to \mathcal{Y}$, and $x_0 \in D$. Assume that $\exists \delta_0 \in (0, \infty) \subset \mathbb{R}$ and $\exists y_0 \in \mathcal{Y}$ such that $f(x) = y_0$ and $\overline{\operatorname{span}(A_D(x))} = \mathcal{X}$, $\forall x \in D \cap \mathcal{B}_{\mathcal{X}}(x_0, \delta_0)$. Then, f is $\mathcal{C}_\infty$ at x_0 and $f^{(i)}(x) = \vartheta_{B_{S_i}(\mathcal{X},\mathcal{Y})}$, $\forall x \in D \cap \mathcal{B}_{\mathcal{X}}(x_0, \delta_0)$, $\forall i \in \mathbb{N}$.*

Proof $\forall x \in D \cap \mathcal{B}_{\mathcal{X}}(x_0, \delta_0)$, let $\delta_x := \delta_0 - \|x - x_0\| > 0$. Then, $\forall \bar{x} \in D \cap \mathcal{B}_{\mathcal{X}}(x, \delta_x) \subseteq D \cap \mathcal{B}_{\mathcal{X}}(x_0, \delta_0)$, we have $f(\bar{x}) = y_0$. Note that $\overline{\operatorname{span}(A_D(x))} = \mathcal{X}$. By Proposition 9.10, $f^{(1)}(x) = \vartheta_{B(\mathcal{X},\mathcal{Y})}$. Let $D_1 := \operatorname{dom}\left(f^{(1)}\right)$. Then, $D_1 \cap \mathcal{B}_{\mathcal{X}}(x_0, \delta_0) = D \cap \mathcal{B}_{\mathcal{X}}(x_0, \delta_0)$. Then, $\overline{\operatorname{span}\left(A_{D_1}(x)\right)} = \mathcal{X}$, $\forall x \in D_1 \cap \mathcal{B}_{\mathcal{X}}(x_0, \delta_0)$. By recursively applying the above argument, we have $f^{(i)}(x) = \vartheta_{B_i(\mathcal{X},\mathcal{Y})} = \vartheta_{B_{S_i}(\mathcal{X},\mathcal{Y})}$, $\forall x \in D \cap \mathcal{B}_{\mathcal{X}}(x_0, \delta_0)$, $\forall i \in \mathbb{N}$. This completes the proof of the proposition. □

Proposition 9.34 *Let $\mathcal{X}$ and $\mathcal{Y}$ be normed linear spaces over $\mathbb{K}$, $D_2 \subseteq D_1 \subseteq \mathcal{X}$, $x_0 \in D_2$, $f : D_1 \to \mathcal{Y}$, $g := f|_{D_2}$, $k \in \mathbb{N}$, and $n \in \mathbb{N} \cup \{\infty\}$. Then, the following statements hold:*

(i) *If, $\forall x \in D_2$, $f^{(k)}(x)$ exists and $\overline{\operatorname{span}\left(A_{D_2}(x)\right)} = \mathcal{X}$, then g is k-times Fréchet differentiable and $g^{(i)}(x) = f^{(i)}(x)$, $\forall x \in D_2$, $\forall i \in \{1, \ldots, k\}$.*
(ii) *If, $\forall x \in D_2$, $g^{(k)}(x)$ exists and $\exists \delta_x \in (0, \infty) \subset \mathbb{R}$ such that $D_1 \cap \mathcal{B}_{\mathcal{X}}(x, \delta_x) = D_2 \cap \mathcal{B}_{\mathcal{X}}(x, \delta_x)$, then $f^{(k)}(x)$ exists and $f^{(i)}(x) = g^{(i)}(x)$, $\forall x \in D_2$, $\forall i \in \{1, \ldots, k\}$.*
(iii) *If f is $\mathcal{C}_n$ at x_0 and $\exists \delta \in (0, \infty) \subset \mathbb{R}$ such that $\overline{\operatorname{span}\left(A_{D_2}(x)\right)} = \mathcal{X}$, $\forall x \in D_2 \cap \mathcal{B}_{\mathcal{X}}(x_0, \delta)$, then, g is $\mathcal{C}_n$ at x_0.*
(iv) *If g is $\mathcal{C}_n$ at x_0 and $\exists \delta \in (0, \infty) \subset \mathbb{R}$ such that $D_1 \cap \mathcal{B}_{\mathcal{X}}(x_0, \delta) = D_2 \cap \mathcal{B}_{\mathcal{X}}(x_0, \delta)$, then f is $\mathcal{C}_n$ at x_0.*

Proof (i) We will use mathematical induction on k to prove this statement.
1° $k = 1$. $\forall x \in D_2$, by Proposition 9.11, we have $g^{(1)}(x)$ exists and $g^{(1)}(x) = f^{(1)}(x)$. Hence, the result holds in this case.
2° Assume that the result holds for $k \leq \bar{k} \in \mathbb{N}$.
3° Consider the case $k = \bar{k} + 1$. By inductive assumption, $g^{(\bar{k})}(x)$ exists and $g^{(i)}(x) = f^{(i)}(x)$, $\forall x \in D_2$, $\forall i \in \left\{1, \ldots, \bar{k}\right\}$. Then, $\operatorname{dom}\left(g^{(\bar{k})}\right) = D_2 \subseteq \operatorname{dom}\left(f^{(\bar{k})}\right) =: \bar{D}_1$. $\forall x \in D_2$, by the assumption, $f^{(\bar{k})}$ is differentiable at x and $\overline{\operatorname{span}\left(A_{D_2}(x)\right)} = \mathcal{X}$. By Proposition 9.11, $g^{(\bar{k})}$ is Fréchet differentiable at x and $g^{(\bar{k}+1)}(x) = \mathrm{D}g^{(\bar{k})}(x) = \mathrm{D}f^{(\bar{k})}(x) = f^{(\bar{k}+1)}(x)$. This completes the induction process.

(ii) We will use mathematical induction on k to prove this statement.
1° $k = 1$. $\forall x \in D_2$, by Proposition 9.11, $f^{(1)}(x)$ exists and $f^{(1)}(x) = g^{(1)}(x)$. Hence, the result holds.
2° Assume that the result holds for $k \leq \bar{k} \in \mathbb{N}$.

3° Consider the case $k = \bar{k} + 1$. By inductive assumption, $f^{(\bar{k})}(x)$ exists and $f^{(i)}(x) = g^{(i)}(x)$, $\forall x \in D_2$, $\forall i \in \{1, \ldots, \bar{k}\}$. Then, $\operatorname{dom}\left(g^{(\bar{k})}\right) = D_2 \subseteq \operatorname{dom}\left(f^{(\bar{k})}\right) =: \bar{D}_1 \subseteq D_1$. $\forall x \in D_2$, by the assumption, $g^{(\bar{k})}$ is differentiable at x and $\exists \delta_x \in (0, \infty) \subset \mathbb{R}$ such that $\bar{D}_1 \cap \mathcal{B}_{\mathcal{X}}(x, \delta_x) = D_2 \cap \mathcal{B}_{\mathcal{X}}(x, \delta_x)$. Then, by Proposition 9.11, $f^{(\bar{k})}$ is Fréchet differentiable at x and $f^{(\bar{k}+1)}(x) = \mathrm{D}f^{(\bar{k})}(x) = \mathrm{D}g^{(\bar{k})}(x) = g^{(\bar{k}+1)}(x)$. This completes the induction process.

(iii) We will distinguish two exhaustive and mutually exclusive cases: Case 1: $n \in \mathbb{N}$; Case 2: $n = \infty$.

Case 1: $n \in \mathbb{N}$. Without loss of generality, assume f is n-times differentiable at x, $\forall x \in D_1 \cap \mathcal{B}_{\mathcal{X}}(x_0, \delta)$. Let $\bar{D}_2 := D_2 \cap \mathcal{B}_{\mathcal{X}}(x_0, \delta)$ and $\bar{g} := f|_{\bar{D}_2}$. $\forall x \in \bar{D}_2$, $f^{(n)}(x)$ exists. Let $\delta_x := \delta - \|x - x_0\| \in (0, \infty) \subset \mathbb{R}$. Then, $\bar{D}_2 \cap \mathcal{B}_{\mathcal{X}}(x, \delta_x) = D_2 \cap \mathcal{B}_{\mathcal{X}}(x, \delta_x)$. Hence, $\overline{\operatorname{span}(\mathrm{A}_{\bar{D}_2}(x))} = \overline{\operatorname{span}\left(\mathrm{A}_{D_2}(x)\right)} = \mathcal{X}$. Then, by (i), $\bar{g}$ is n-times differentiable and $\bar{g}^{(i)}(x) = f^{(i)}(x)$, $\forall x \in \bar{D}_2$, $\forall i \in \{1, \ldots, n\}$. By (ii), $g^{(n)}(x)$ exists and $g^{(i)}(x) = \bar{g}^{(i)}(x) = f^{(i)}(x)$, $\forall x \in \bar{D}_2$, $\forall i \in \{1, \ldots, n\}$. By the continuity of $f^{(n)}$ at x_0, $\forall \epsilon \in (0, \infty) \subset \mathbb{R}$, $\exists \bar{\delta} \in (0, \delta] \subset \mathbb{R}$, and $\forall x \in \operatorname{dom}\left(f^{(n)}\right) \cap \mathcal{B}_{\mathcal{X}}\left(x_0, \bar{\delta}\right) = D_1 \cap \mathcal{B}_{\mathcal{X}}\left(x_0, \bar{\delta}\right)$, we have $\left\|f^{(n)}(x) - f^{(n)}(x_0)\right\| < \epsilon$. $\forall x \in \operatorname{dom}\left(g^{(n)}\right) \cap \mathcal{B}_{\mathcal{X}}\left(x_0, \bar{\delta}\right) = D_2 \cap \mathcal{B}_{\mathcal{X}}\left(x_0, \bar{\delta}\right)$, we have $\left\|g^{(n)}(x) - g^{(n)}(x_0)\right\| = \left\|f^{(n)}(x) - f^{(n)}(x_0)\right\| < \epsilon$. Hence, $g^{(n)}$ is continuous at x_0. Therefore, g is $\mathcal{C}_n$ at x_0.

Case 2: $n = \infty$. By the assumption, $\exists \delta \in (0, \infty) \subset \mathbb{R}$ such that f is $\mathcal{C}_i$ at x, $\forall x \in D \cap \mathcal{B}_{\mathcal{X}}(x_0, \delta)$, $\forall i \in \mathbb{N}$. By Case 1, g is $\mathcal{C}_i$ at x, $\forall x \in D \cap \mathcal{B}_{\mathcal{X}}(x_0, \delta)$, $\forall i \in \mathbb{N}$. Hence, g is $\mathcal{C}_\infty$ at x_0.

(iv) We will distinguish two exhaustive and mutually exclusive cases: Case 1: $n \in \mathbb{N}$; Case 2: $n = \infty$.

Case 1: $n \in \mathbb{N}$. Without loss of generality, assume g is n-times differentiable at x, $\forall x \in D_2 \cap \mathcal{B}_{\mathcal{X}}(x_0, \delta) =: \bar{D}_2$. Let $\bar{g} := g|_{\bar{D}_2}$. $\forall x \in \bar{D}_2$, $g^{(n)}(x)$ exists. Let $\delta_x := \delta - \|x - x_0\| \in (0, \infty) \subset \mathbb{R}$. Then, $\bar{D}_2 \cap \mathcal{B}_{\mathcal{X}}(x, \delta_x) = D_2 \cap \mathcal{B}_{\mathcal{X}}(x, \delta_x) = D_1 \cap \mathcal{B}_{\mathcal{X}}(x, \delta_x)$. Hence, $\overline{\operatorname{span}(\mathrm{A}_{\bar{D}_2}(x))} = \overline{\operatorname{span}\left(\mathrm{A}_{D_2}(x)\right)} = \mathcal{X}$. Then, by (i), $\bar{g}$ is n-times differentiable and $\bar{g}^{(i)}(x) = g^{(i)}(x)$, $\forall x \in \bar{D}_2$, $\forall i \in \{1, \ldots, n\}$. By (ii), $f^{(n)}(x)$ exists and $f^{(i)}(x) = \bar{g}^{(i)}(x) = g^{(i)}(x)$, $\forall x \in \bar{D}_2$, $\forall i \in \{1, \ldots, n\}$. By the continuity of $g^{(n)}$ at x_0, $\forall \epsilon \in (0, \infty) \subset \mathbb{R}$, $\exists \bar{\delta} \in (0, \delta] \subset \mathbb{R}$, and $\forall x \in \operatorname{dom}\left(g^{(n)}\right) \cap \mathcal{B}_{\mathcal{X}}\left(x_0, \bar{\delta}\right) = D_2 \cap \mathcal{B}_{\mathcal{X}}\left(x_0, \bar{\delta}\right) = D_1 \cap \mathcal{B}_{\mathcal{X}}\left(x_0, \bar{\delta}\right)$, we have $\left\|g^{(n)}(x) - g^{(n)}(x_0)\right\| < \epsilon$. $\forall x \in \operatorname{dom}\left(f^{(n)}\right) \cap \mathcal{B}_{\mathcal{X}}\left(x_0, \bar{\delta}\right) = D_1 \cap \mathcal{B}_{\mathcal{X}}\left(x_0, \bar{\delta}\right)$, we have $\left\|f^{(n)}(x) - f^{(n)}(x_0)\right\| = \left\|g^{(n)}(x) - g^{(n)}(x_0)\right\| < \epsilon$. Hence, $f^{(n)}$ is continuous at x_0. Therefore, f is $\mathcal{C}_n$ at x_0.

Case 2: $n = \infty$. By the assumption, g is $\mathcal{C}_i$ at x, $\forall x \in \bar{D}_2$, $\forall i \in \mathbb{N}$. By Case 1, f is $\mathcal{C}_i$ at x, $\forall x \in \bar{D}_2$, $\forall i \in \mathbb{N}$. Hence, f is $\mathcal{C}_\infty$ at x_0.

This completes the proof of the proposition. □

Proposition 9.35 *Let $\mathcal{X}$ and $\mathcal{Y}$ be normed linear spaces over $\mathbb{K}$, $D \subseteq \mathcal{X}$, $f : D \to \mathcal{Y}$. Assume that there exist open sets $O_1, O_2 \subseteq \mathcal{X}$ such that $D_1 := D \cap O_1$ and $D_2 := D \cap O_2$ satisfy $D_1 \cup D_2 = D$ and $f|_{D_1}$ and $f|_{D_2}$ are $\mathcal{C}_k$, where $k \in \mathbb{N} \cup \{\infty\}$. Then, f is $\mathcal{C}_k$.*

Proof $\forall x \in D$, without loss of generality, assume that $x \in D_1$. Then, $\exists \delta \in (0, \infty) \subset \mathbb{R}$ such that $\mathcal{B}_{\mathcal{X}}(x, \delta) \subseteq O_1$, which implies that $D_1 \cap \mathcal{B}_{\mathcal{X}}(x, \delta) = D \cap \mathcal{B}_{\mathcal{X}}(x, \delta)$. Since $f|_{D_1}$ is $\mathcal{C}_k$, then, by Proposition 9.34, $\forall \bar{x} \in D_1$, $f^{(i)}(\bar{x}) = f|_{D_1}^{(i)}(\bar{x})$, $\forall i \in \mathbb{N}$ with $i \leq k$. Then, f is $\mathcal{C}_k$ at x. By the arbitrariness of x, f is $\mathcal{C}_k$. This completes the proof of the proposition. □

Note that D_1 and D_2 are open sets in the subset topology of D. This result should be compared with Theorem 3.11, where continuity on D can be concluded when D_1 and D_2 are relatively open or are relatively closed. For continuous differentiability, D_1 and D_2 must be relatively open for the conclusion to hold.

Proposition 9.36 *Let $\mathcal{X}$ be a normed linear space over $\mathbb{K}$ and $f : \mathcal{X} \to \mathcal{X}$ be given by $f = \mathrm{id}_{\mathcal{X}}$, that is, $f(x) = x$, $\forall x \in \mathcal{X}$. Then, f is $\mathcal{C}_\infty$, $f^{(1)}(x) = \mathrm{id}_{\mathcal{X}}$, and $f^{(i+1)}(x) = \vartheta_{\mathrm{B}_{\mathrm{S}\,i+1}(\mathcal{X},\mathcal{X})}$, $\forall x \in \mathcal{X}$, $\forall i \in \mathbb{N}$.*

Proof This is straightforward and is therefore omitted. □

Proposition 9.37 *Let $\mathcal{X}$ and $\mathcal{Y}$ be normed linear spaces over $\mathbb{K}$ and $f : \mathcal{X} \times \mathcal{Y} \to \mathcal{X}$ be given by $f = \pi_{\mathcal{X}}$, that is, $f(x, y) = x$, $\forall (x, y) \in \mathcal{X} \times \mathcal{Y}$. Then, f is $\mathcal{C}_\infty$, $f^{(1)}(x, y) = \begin{bmatrix} \mathrm{id}_{\mathcal{X}} & \vartheta_{\mathrm{B}(\mathcal{Y},\mathcal{X})} \end{bmatrix}$, and $f^{(i+1)}(x, y) = \vartheta_{\mathrm{B}_{\mathrm{S}\,i+1}(\mathcal{X}\times\mathcal{Y},\mathcal{X})}$, $\forall (x, y) \in \mathcal{X} \times \mathcal{Y}$, $\forall i \in \mathbb{N}$.*

Proof This is straightforward and is therefore omitted. □

Proposition 9.38 *Let $\mathcal{X}$ be a normed linear space over $\mathbb{K}$ and $f : \mathcal{X} \times \mathcal{X} \to \mathcal{X}$ be given by $f(x_1, x_2) = x_1 + x_2$, $\forall (x_1, x_2) \in \mathcal{X} \times \mathcal{X}$. Then, f is $\mathcal{C}_\infty$, $f^{(1)}(x_1, x_2) = \begin{bmatrix} \mathrm{id}_{\mathcal{X}} & \mathrm{id}_{\mathcal{X}} \end{bmatrix}$, and $f^{(i+1)}(x_1, x_2) = \vartheta_{\mathrm{B}_{\mathrm{S}\,i+1}(\mathcal{X}\times\mathcal{X},\mathcal{X})}$, $\forall (x_1, x_2) \in \mathcal{X} \times \mathcal{X}$, $\forall i \in \mathbb{N}$.*

Proof This is straightforward and is therefore omitted. □

Proposition 9.39 *Let $\mathcal{X}$ be a normed linear space over $\mathbb{K}$ and $f : \mathbb{K} \times \mathcal{X} \to \mathcal{X}$ be given by $f(\alpha, x) = \alpha x$, $\forall (\alpha, x) \in \mathbb{K} \times \mathcal{X}$. Then, f is $\mathcal{C}_\infty$, $f^{(1)}(\alpha, x) = \begin{bmatrix} x & \alpha\,\mathrm{id}_{\mathcal{X}} \end{bmatrix}$, $f^{(2)}(\alpha, x)(d_2, h_2)(d_1, h_1) = d_1 h_2 + d_2 h_1$, and $f^{(i+2)}(\alpha, x) = \vartheta_{\mathrm{B}_{\mathrm{S}\,i+2}(\mathbb{K}\times\mathcal{X},\mathcal{X})}$, $\forall (\alpha, x) \in \mathbb{K} \times \mathcal{X}$, $\forall i \in \mathbb{N}$, $\forall (d_1, h_1) \in \mathbb{K} \times \mathcal{X}$, $\forall (d_2, h_2) \in \mathbb{K} \times \mathcal{X}$.*

Proof This is straightforward and is therefore omitted. □

Proposition 9.40 *Let $\mathcal{X}$ and $\mathcal{Y}$ be normed linear spaces over $\mathbb{K}$, $D \subseteq \mathcal{X}$, $f_1 : D \to \mathcal{Y}$, $f_2 : D \to \mathcal{Y}$, $x_0 \in D$, $\alpha_1, \alpha_2 \in \mathbb{K}$, $n \in \mathbb{N}$, $k \in \mathbb{N} \cup \{\infty\}$, and $g : D \to \mathcal{Y}$ be given by $g(x) = \alpha_1 f_1(x) + \alpha_2 f_2(x)$, $\forall x \in D$. If f_1 and f_2 are n-times differentiable, then g is n-times differentiable and $g^{(i)}(x) = \alpha_1 f_1^{(i)}(x) + \alpha_2 f_2^{(i)}(x)$, $\forall x \in D$, $\forall i \in \{1, \dots, n\}$. If f_1 and f_2 are $\mathcal{C}_k$ at x_0, then g is $\mathcal{C}_k$ at x_0 and $g^{(i)}(x) = \alpha_1 f_1^{(i)}(x) + \alpha_2 f_2^{(i)}(x)$, $\forall x \in D \cap \mathcal{B}_{\mathcal{X}}(x_0, \delta_0)$, $\forall i \in \mathbb{N}$ with $i \leq k$, for some $\delta_0 \in (0, \infty) \subset \mathbb{R}$.*

Proof This is straightforward and is therefore omitted. □

Proposition 9.41 *Let $\mathcal{X}$ and $\mathcal{Y}$ be normed linear spaces over $\mathbb{K}$ and $f : \mathrm{B}(\mathcal{X}, \mathcal{Y}) \times \mathcal{X} \to \mathcal{Y}$ be given by $f(A, x) = Ax$, $\forall (A, x) \in \mathrm{B}(\mathcal{X}, \mathcal{Y}) \times \mathcal{X}$. Then, f is $\mathcal{C}_\infty$, $f^{(1)}(A, x) = \begin{bmatrix} \mathrm{ro}(x) & A \end{bmatrix}$, $f^{(2)}(A, x)(\Delta_2, h_2)(\Delta_1, h_1) = \Delta_1 h_2 + \Delta_2 h_1$, and*

$f^{(i+2)}(A,x) = \vartheta_{\mathrm{BS}_{i+2}(\mathrm{B}(\mathcal{X},\mathcal{Y})\times\mathcal{X},\mathcal{Y})}$, $\forall (A,x) \in \mathrm{B}(\mathcal{X},\mathcal{Y}) \times \mathcal{X}$, $\forall i \in \mathbb{N}$, $\forall (\Delta_1, h_1) \in \mathrm{B}(\mathcal{X},\mathcal{Y}) \times \mathcal{X}$, $\forall (\Delta_2, h_2) \in \mathrm{B}(\mathcal{X},\mathcal{Y}) \times \mathcal{X}$.

Proof This is straightforward and is therefore omitted. □

Proposition 9.42 *Let $\mathcal{X}$, $\mathcal{Y}$, and $\mathcal{Z}$ be normed linear spaces over $\mathbb{K}$ and $f : \mathrm{B}(\mathcal{Y},\mathcal{Z}) \times \mathrm{B}(\mathcal{X},\mathcal{Y}) \to \mathrm{B}(\mathcal{X},\mathcal{Z})$ be given by $f(A_{yz}, A_{xy}) = A_{yz}A_{xy}$, $\forall (A_{yz}, A_{xy}) \in \mathrm{B}(\mathcal{Y},\mathcal{Z}) \times \mathrm{B}(\mathcal{X},\mathcal{Y})$. Then, f is $\mathcal{C}_\infty$, $f^{(1)}(A_{yz}, A_{xy}) = \left[\mathrm{ro}(A_{xy})\ A_{yz}\right]$, $f^{(2)}(A_{yz}, A_{xy})(\Delta_{yz2}, \Delta_{xy2})(\Delta_{yz1}, \Delta_{xy1}) = \Delta_{yz1}\Delta_{xy2} + \Delta_{yz2}\Delta_{xy1}$, and $f^{(i+2)}(A_{yz}, A_{xy}) = \vartheta_{\mathrm{BS}_{i+2}(\mathrm{B}(\mathcal{Y},\mathcal{Z})\times\mathrm{B}(\mathcal{X},\mathcal{Y}),\mathrm{B}(\mathcal{X},\mathcal{Z}))}$, $\forall (A_{yz}, A_{xy}) \in \mathrm{B}(\mathcal{Y},\mathcal{Z}) \times \mathrm{B}(\mathcal{X},\mathcal{Y})$, $\forall i \in \mathbb{N}$, $\forall (\Delta_{yz1}, \Delta_{xy1}) \in \mathrm{B}(\mathcal{Y},\mathcal{Z}) \times \mathrm{B}(\mathcal{X},\mathcal{Y})$, $\forall (\Delta_{yz2}, \Delta_{xy2}) \in \mathrm{B}(\mathcal{Y},\mathcal{Z}) \times \mathrm{B}(\mathcal{X},\mathcal{Y})$.*

Proof This is straightforward and is therefore omitted. □

Proposition 9.43 *Let $\mathcal{X}_1, \ldots, \mathcal{X}_p$ and $\mathcal{Y}_1, \ldots, \mathcal{Y}_m$ be normed linear spaces over $\mathbb{K}$, where $p, m \in \mathbb{N}$, $\mathcal{Z}_{ji} := \mathrm{B}(\mathcal{X}_i, \mathcal{Y}_j)$, $i = 1, \ldots, p$, $j = 1, \ldots, m$, $\mathcal{Z} := \prod_{j=1}^{m} \prod_{i=1}^{p} \mathcal{Z}_{ji}$, and $f : \mathcal{Z} \to \mathrm{B}\left(\prod_{i=1}^{p} \mathcal{X}_i, \prod_{j=1}^{m} \mathcal{Y}_j\right)$ be given by*

$$f(A_{11}, \ldots, A_{mp}) = \begin{bmatrix} A_{11} & \cdots & A_{1p} \\ \vdots & & \vdots \\ A_{m1} & \cdots & A_{mp} \end{bmatrix}; \qquad \forall (A_{11}, \ldots, A_{mp}) \in \mathcal{Z}$$

Then, f is $\mathcal{C}_\infty$ and

$$f^{(1)}(A_{11}, \ldots, A_{mp})(\Delta_{11}, \ldots, \Delta_{mp}) = \begin{bmatrix} \Delta_{11} & \cdots & \Delta_{1p} \\ \vdots & & \vdots \\ \Delta_{m1} & \cdots & \Delta_{mp} \end{bmatrix}$$

$$f^{(l+1)}(A_{11}, \ldots, A_{mp}) = \vartheta_{\mathrm{BS}_{l+1}\left(\mathcal{Z}, \mathrm{B}\left(\prod_{i=1}^{p} \mathcal{X}_i, \prod_{j=1}^{m} \mathcal{Y}_j\right)\right)}$$

$\forall (A_{11}, \ldots, A_{mp}) \in \mathcal{Z}$, $\forall l \in \mathbb{N}$, $\forall (\Delta_{11}, \ldots, \Delta_{mp}) \in \mathcal{Z}$.

Proof This is straightforward and is therefore omitted. □

9.4.4 Properties of High Order Derivatives

Proposition 9.44 *Let $\mathcal{X}$, $\mathcal{Y}$, and $\mathcal{Z}$ be normed linear spaces over $\mathbb{K}$, $D \subseteq \mathcal{X}$, $f_1 : D \to \mathcal{Y}$, $f_2 : D \to \mathcal{Z}$, $x_0 \in D$, $k \in \mathbb{N}$, and $g : D \to \mathcal{Y} \times \mathcal{Z}$ be given by $g(x) = (f_1(x), f_2(x))$, $\forall x \in D$. Then, the following statements hold:*

(i) *$\exists \delta_0 \in (0, \infty) \subset \mathbb{R}$ such that $f_1^{(k)}(x)$ and $f_2^{(k)}(x)$ exist, $\forall x \in D \cap \mathcal{B}_{\mathcal{X}}(x_0, \delta_0)$ if, and only if, $\exists \delta_0 \in (0, \infty) \subset \mathbb{R}$ such that $g^{(k)}(x)$ exists, $\forall x \in D \cap \mathcal{B}_{\mathcal{X}}(x_0, \delta_0)$. In this case, $g^{(i)}(x) = \begin{bmatrix} f_1^{(i)}(x) \\ f_2^{(i)}(x) \end{bmatrix}$, $\forall x \in D \cap \mathcal{B}_{\mathcal{X}}(x_0, \delta_0)$, $\forall i \in \{1, \ldots, k\}$.*

(ii) *Let $n \in \mathbb{N} \cup \{\infty\}$. Then, f_1 and f_2 are $\mathcal{C}_n$ at x_0 if, and only if, g is $\mathcal{C}_n$ at x_0.*

Proof (i) We will use mathematical induction on k to prove this statement.

1° $k = 1$. The statement holds by Proposition 9.19.

2° Assume that the result holds for $k = \bar{k} \in \mathbb{N}$.

3° Consider the case $k = \bar{k} + 1$. "Necessity" Let $f_1^{(\bar{k}+1)}(x)$ and $f_2^{(\bar{k}+1)}(x)$ exist, $\forall x \in D \cap \mathcal{B}_{\mathcal{X}}(x_0, \delta_0)$. By inductive assumption, $g^{(\bar{k})}(x)$ exists and $g^{(i)}(x) = \begin{bmatrix} f_1^{(i)}(x) \\ f_2^{(i)}(x) \end{bmatrix}$, $\forall x \in D \cap \mathcal{B}_{\mathcal{X}}(x_0, \delta_0)$, $\forall i \in \{1, \ldots, \bar{k}\}$. Let $\bar{D}_1 := \operatorname{dom}\left(f_1^{(\bar{k})}\right)$, $\bar{D}_2 := \operatorname{dom}\left(f_2^{(\bar{k})}\right)$, and $\bar{D} := \operatorname{dom}\left(g^{(\bar{k})}\right)$. Then, $\bar{D}_1 \subseteq D$, $\bar{D}_2 \subseteq D$, $\bar{D} \subseteq D$, $\bar{D} \supseteq D \cap \mathcal{B}_{\mathcal{X}}(x_0, \delta_0)$, and $\bar{D} \cap \mathcal{B}_{\mathcal{X}}(x_0, \delta_0) \subseteq \bar{D}_1 \cap \bar{D}_2$. This implies that $\bar{D} \cap \mathcal{B}_{\mathcal{X}}(x_0, \delta_0) = D \cap \mathcal{B}_{\mathcal{X}}(x_0, \delta_0) =: \hat{D}$ and $\overline{\operatorname{span}\left(A_{\hat{D}}(x)\right)} = \overline{\operatorname{span}\left(A_D(x)\right)} = \mathcal{X}$, $\forall x \in \hat{D}$. By the assumption, $f_1^{(\bar{k})}$ and $f_2^{(\bar{k})}$ are differentiable at x, $\forall x \in \hat{D}$. By Propositions 9.19 and 9.11, $\forall x \in \hat{D}$, $g^{(\bar{k})}$ is differentiable at x and $g^{(\bar{k}+1)}(x) = \mathrm{D}\, g^{(\bar{k})}\big|_{\hat{D}}(x) = \begin{bmatrix} \mathrm{D}\, f_1^{(\bar{k})}\big|_{\hat{D}}(x) \\ \mathrm{D}\, f_2^{(\bar{k})}\big|_{\hat{D}}(x) \end{bmatrix} = \begin{bmatrix} f_1^{(\bar{k}+1)}(x) \\ f_2^{(\bar{k}+1)}(x) \end{bmatrix}$.

"Sufficiency" Let $g^{(\bar{k}+1)}(x)$ exist, $\forall x \in D \cap \mathcal{B}_{\mathcal{X}}(x_0, \delta_0)$. By inductive assumption, $f_1^{(\bar{k})}(x)$ and $f_2^{(\bar{k})}(x)$ exist and $g^{(i)}(x) = \begin{bmatrix} f_1^{(i)}(x) \\ f_2^{(i)}(x) \end{bmatrix}$, $\forall x \in D \cap \mathcal{B}_{\mathcal{X}}(x_0, \delta_0)$, $\forall i \in \{1, \ldots, \bar{k}\}$. Let $\bar{D}_1 := \operatorname{dom}\left(f_1^{(\bar{k})}\right)$, $\bar{D}_2 := \operatorname{dom}\left(f_2^{(\bar{k})}\right)$, and $\bar{D} := \operatorname{dom}\left(g^{(\bar{k})}\right)$. Then, $\bar{D}_1 \subseteq D$, $\bar{D}_2 \subseteq D$, $\bar{D} \subseteq D$, $\bar{D} \supseteq D \cap \mathcal{B}_{\mathcal{X}}(x_0, \delta_0)$, and $D \cap \mathcal{B}_{\mathcal{X}}(x_0, \delta_0) \subseteq \bar{D}_1 \cap \bar{D}_2$. This implies that $\bar{D} \cap \mathcal{B}_{\mathcal{X}}(x_0, \delta_0) = D \cap \mathcal{B}_{\mathcal{X}}(x_0, \delta_0) = \bar{D}_1 \cap \mathcal{B}_{\mathcal{X}}(x_0, \delta_0) = \bar{D}_2 \cap \mathcal{B}_{\mathcal{X}}(x_0, \delta_0) =: \hat{D}$. Then, $\overline{\operatorname{span}\left(A_{\hat{D}}(x)\right)} = \overline{\operatorname{span}\left(A_D(x)\right)} = \mathcal{X}$, $\forall x \in \hat{D}$. By Propositions 9.11 and 9.19, we have, $\forall x \in \hat{D}$, $f_1^{(\bar{k})}\big|_{\hat{D}}$ and $f_2^{(\bar{k})}\big|_{\hat{D}}$ are differentiable at x and $g^{(\bar{k}+1)}(x) = \mathrm{D}\, g^{(\bar{k})}\big|_{\hat{D}}(x) = \begin{bmatrix} \mathrm{D}\, f_1^{(\bar{k})}\big|_{\hat{D}}(x) \\ \mathrm{D}\, f_2^{(\bar{k})}\big|_{\hat{D}}(x) \end{bmatrix}$. By Proposition 9.11, we have $\mathrm{D}\, f_1^{(\bar{k})}\big|_{\hat{D}}(x) = \mathrm{D} f_1^{(\bar{k})}(x) = f_1^{(\bar{k}+1)}(x)$ and $\mathrm{D}\, f_2^{(\bar{k})}\big|_{\hat{D}}(x) = \mathrm{D} f_2^{(\bar{k})}(x) = f_2^{(\bar{k}+1)}(x)$, $\forall x \in \hat{D}$. Then, $g^{(\bar{k}+1)}(x) = \begin{bmatrix} f_1^{(\bar{k}+1)}(x) \\ f_2^{(\bar{k}+1)}(x) \end{bmatrix}$, $\forall x \in \hat{D}$.

This completes the induction process.

(ii) We will distinguish two exhaustive and mutually exclusive cases: Case 1: $n \in \mathbb{N}$; Case 2: $n = \infty$.

Case 1: $n \in \mathbb{N}$. "Sufficiency" Let g be $\mathcal{C}_n$ at x_0. Then, $\exists \delta_0 \in (0, \infty) \subset \mathbb{R}$ such that g is n-times differentiable at x, $\forall x \in D \cap \mathcal{B}_{\mathcal{X}}(x_0, \delta_0) =: \bar{D}$. By (i), f_1 and f_2 are n-times differentiable at x, $\forall x \in \bar{D}$ and $g^{(n)}(x) = \begin{bmatrix} f_1^{(n)}(x) \\ f_2^{(n)}(x) \end{bmatrix}$, $\forall x \in \bar{D}$. By the continuity of $g^{(n)}$ at x_0 and Proposition 3.32, $f_1^{(n)}$ and $f_2^{(n)}$ are continuous at x_0. Then, f_1 and f_2 are $\mathcal{C}_n$ at x_0.

"Necessity" Let f_1 and f_2 be $\mathcal{C}_n$ at x_0. Then, $\exists \delta_0 \in (0, \infty) \subset \mathbb{R}$ such that f_1 and f_2 are n-times differentiable at x, $\forall x \in D \cap \mathcal{B}_{\mathcal{X}}(x_0, \delta_0) =: \bar{D}$. By (i), g is n-times differentiable at x, $\forall x \in \bar{D}$, and $g^{(n)}(x) = \begin{bmatrix} f_1^{(n)}(x) \\ f_2^{(n)}(x) \end{bmatrix}$, $\forall x \in \bar{D}$. By the continuity of $f_1^{(n)}$ and $f_2^{(n)}$ at x_0 and Proposition 3.32, $g^{(n)}$ is continuous at x_0. Then, g is $\mathcal{C}_n$ at x_0.

Case 2: $n = \infty$. "Sufficiency" Let g be $\mathcal{C}_\infty$ at x_0. Then, $\exists \delta \in (0, \infty) \subset \mathbb{R}$ such that g is $\mathcal{C}_i$ at x, $\forall x \in D \cap \mathcal{B}_{\mathcal{X}}(x_0, \delta)$, $\forall i \in \mathbb{N}$. By Case 1, f_1 and f_2 are $\mathcal{C}_i$ at x, $\forall x \in D \cap \mathcal{B}_{\mathcal{X}}(x_0, \delta)$, $\forall i \in \mathbb{N}$. Then, f_1 and f_2 are $\mathcal{C}_\infty$ at x_0. "Necessity" Let f_1 and f_2 be $\mathcal{C}_\infty$ at x_0. Then, $\exists \delta \in (0, \infty) \subset \mathbb{R}$ such that f_1 and f_2 are $\mathcal{C}_i$ at x, $\forall x \in D \cap \mathcal{B}_{\mathcal{X}}(x_0, \delta)$, $\forall i \in \mathbb{N}$. By Case 1, g is $\mathcal{C}_i$ at x, $\forall x \in D \cap \mathcal{B}_{\mathcal{X}}(x_0, \delta)$, $\forall i \in \mathbb{N}$. Then, g is $\mathcal{C}_\infty$ at x_0.

This completes the proof of the proposition. □

Proposition 9.45 *Let $\mathcal{X}$, $\mathcal{Y}$, and $\mathcal{Z}$ be normed linear spaces over $\mathbb{K}$, $D_1 \subseteq \mathcal{X}$, $D_2 \subseteq \mathcal{Y}$, $f : D_1 \to D_2$, $g : D_2 \to \mathcal{Z}$, $x_0 \in D_1$, and $y_0 := f(x_0) \in D_2$. Then, the following statements hold:*

(i) *Assume that f is $\mathcal{C}_k$ at x_0 and g is $\mathcal{C}_k$ at y_0, for some $k \in \mathbb{N} \cup \{\infty\}$. Then, $h := g \circ f$ is $\mathcal{C}_k$ at x_0.*
(ii) *Let $k \in \mathbb{N}$. Assume that f is k-times differentiable and g is k-times differentiable. Then, h is k-times differentiable.*

Proof (i) We will first use mathematical induction on k to show that the result holds if $k \in \mathbb{N}$.

1° $k = 1$. By g being $\mathcal{C}_1$ at y_0, then $\exists \delta_1 \in (0, \infty) \subset \mathbb{R}$ such that $g^{(1)}(y)$ exists, $\forall y \in D_2 \cap \mathcal{B}_{\mathcal{Y}}(y_0, \delta_1)$, and $g^{(1)}$ is continuous at y_0. By f being $\mathcal{C}_1$ at x_0, then, by Proposition 9.7, $\exists \delta \in (0, \infty) \subset \mathbb{R}$ such that $f(x) \in D_2 \cap \mathcal{B}_{\mathcal{Y}}(y_0, \delta_1)$ and $f^{(1)}(x)$ exists, $\forall x \in D_1 \cap \mathcal{B}_{\mathcal{X}}(x_0, \delta)$, and $f^{(1)}$ is continuous at x_0. $\forall x \in D_1 \cap \mathcal{B}_{\mathcal{X}}(x_0, \delta)$, by chain rule, $h^{(1)}(x)$ exists and $h^{(1)}(x) = g^{(1)}(f(x))f^{(1)}(x)$. By Propositions 3.12, 9.7, 3.32, and 9.42, $h^{(1)}$ is continuous at x_0. Hence, h is $\mathcal{C}_1$ at x_0.

2° Assume that the result holds for $k \leq \bar{k} \in \mathbb{N}$.

3° Consider the case $k = \bar{k} + 1$. By g being $\mathcal{C}_{\bar{k}+1}$ at y_0, then $\exists \delta_1 \in (0, \infty) \subset \mathbb{R}$ such that $g^{(1)}(y)$ exists, $\forall y \in D_2 \cap \mathcal{B}_{\mathcal{Y}}(y_0, \delta_1) =: \bar{D}$, and $g^{(1)}$ is $\mathcal{C}_{\bar{k}}$ at y_0. By f being $\mathcal{C}_{\bar{k}+1}$ at x_0, then, by Proposition 9.7, $\exists \delta \in (0, \infty) \subset \mathbb{R}$ such that $f(x) \in \bar{D}$ and $f^{(1)}(x)$ exists, $\forall x \in D_1 \cap \mathcal{B}_{\mathcal{X}}(x_0, \delta) =: \hat{D}$, and $f^{(1)}$ is $\mathcal{C}_{\bar{k}}$ at x_0. $\forall x \in \hat{D}$, by chain rule, $h^{(1)}(x)$ exists and $h^{(1)}(x) = g^{(1)}(f(x))f^{(1)}(x)$. Then,

$h^{(1)}|_{\hat{D}} : \hat{D} \to \mathrm{B}(\mathcal{X}, \mathcal{Z})$ is given by $h^{(1)}|_{\hat{D}}(x) = g^{(1)}|_{\bar{D}}(f|_{\hat{D}}(x))\, f^{(1)}|_{\hat{D}}(x)$, $\forall x \in \hat{D}$. By Proposition 9.34, $g^{(1)}|_{\bar{D}}$ is $\mathcal{C}_{\bar{k}}$ at y_0 and $f|_{\hat{D}}$ and $f^{(1)}|_{\hat{D}}$ are $\mathcal{C}_{\bar{k}}$ at x_0. By inductive assumption and Propositions 9.44, 9.42, and 3.32, $h^{(1)}|_{\hat{D}}$ is $\mathcal{C}_{\bar{k}}$ at x_0. By Proposition 9.34, $h^{(1)}$ is $\mathcal{C}_{\bar{k}}$ at x_0. Hence, h is $\mathcal{C}_{\bar{k}+1}$ at x_0.

This completes the induction process.

When $k = \infty$, then $\exists \delta \in (0, \infty) \subset \mathbb{R}$ such that g is $\mathcal{C}_i$ at y, $\forall y \in D_2 \cap \mathcal{B}_{\mathcal{Y}}(y_0, \delta)$, and f is $\mathcal{C}_i$ at x, $\forall x \in D_1 \cap \mathcal{B}_{\mathcal{X}}(x_0, \delta)$, $\forall i \in \mathbb{N}$, which further implies by the induction conclusion, h is $\mathcal{C}_i$ at x, $\forall x \in D_1 \cap \mathcal{B}_{\mathcal{X}}(x_0, \delta)$, $\forall i \in \mathbb{N}$. Hence, h is $\mathcal{C}_\infty$ at x_0.

(ii) We will first use mathematical induction on k to show that the result holds.

1° $k = 1$. By chain rule, $h^{(1)}(x)$ exists and $h^{(1)}(x) = g^{(1)}(f(x))f^{(1)}(x)$, $\forall x \in D_1$. Hence, the result holds.

2° Assume that the result holds for $k \leq \bar{k} \in \mathbb{N}$.

3° Consider the case $k = \bar{k} + 1$. By chain rule, $h^{(1)}(x)$ exists and $h^{(1)}(x) = g^{(1)}(f(x))f^{(1)}(x)$, $\forall x \in D_1$. By inductive assumption and Propositions 9.44 and 9.42, $h^{(1)}$ is $\bar{k}$-times differentiable. Hence, h is $(\bar{k} + 1)$-times differentiable. This completes the induction process and the proof of the proposition. □

Proposition 9.46 *Let $\mathcal{X}$, $\mathcal{Y}$, and $\mathcal{Z}$ be normed linear spaces over $\mathbb{K}$, $D \subseteq \mathcal{X} \times \mathcal{Y}$, and $f : D \to \mathcal{Z}$ be partial differentiable with respect to x and partial differentiable with respect to y. Then, the following statements hold:*

(i) *If f is $(n + 1)$-times differentiable, where $n \in \mathbb{N}$, then, $\frac{\partial f}{\partial x}$ and $\frac{\partial f}{\partial y}$ are n-times differentiable.*
(ii) *If f is $\mathcal{C}_1$ at $(x_0, y_0) \in D$, then $\frac{\partial f}{\partial x}$ and $\frac{\partial f}{\partial y}$ are continuous at (x_0, y_0).*
(iii) *If f is $\mathcal{C}_n$ at $(x_0, y_0) \in D$, where $n \in \{2, 3, \ldots\} \cup \{\infty\}$, then $\frac{\partial f}{\partial x}$ and $\frac{\partial f}{\partial y}$ are $\mathcal{C}_{n-1}$ at (x_0, y_0).*

Proof By Proposition 9.9, $f^{(1)}(x, y) = \begin{bmatrix} \frac{\partial f}{\partial x}(x, y) & \frac{\partial f}{\partial y}(x, y) \end{bmatrix}$, $\forall (x, y) \in D$. Define $g : \mathcal{X} \times \mathcal{Y} \to \mathrm{B}(\mathcal{X}, \mathcal{X} \times \mathcal{Y})$ by $g(x, y) = \begin{bmatrix} \mathrm{id}_{\mathcal{X}} \\ \vartheta_{\mathrm{B}(\mathcal{X}, \mathcal{Y})} \end{bmatrix}$. By Proposition 9.33, g is $\mathcal{C}_\infty$. It is clear that $\frac{\partial f}{\partial x}(x, y) = f^{(1)}(x, y)g(x, y)$, $\forall (x, y) \in D$:

(i) Since f is $(n + 1)$-times differentiable, then $f^{(1)}$ is n-times differentiable. By Propositions 9.42, 9.44, and 9.45, we have $\frac{\partial f}{\partial x}$ is n-times differentiable. By symmetry, $\frac{\partial f}{\partial y}$ is n-times differentiable.

(ii) Since f is $\mathcal{C}_1$ at (x_0, y_0), then $f^{(1)}$ is continuous at (x_0, y_0). By Propositions 9.42, 9.7, 3.32, and 3.12, we have $\frac{\partial f}{\partial x}$ is continuous at (x_0, y_0). By symmetry, $\frac{\partial f}{\partial y}$ is continuous at (x_0, y_0).

(iii) Since f is $\mathcal{C}_n$ at (x_0, y_0), then $f^{(1)}$ is $\mathcal{C}_{n-1}$ at (x_0, y_0). By Propositions 9.42, 9.44, and 9.45, we have $\frac{\partial f}{\partial x}$ is $\mathcal{C}_{n-1}$ at (x_0, y_0). By symmetry, $\frac{\partial f}{\partial y}$ is $\mathcal{C}_{n-1}$ at (x_0, y_0).

This completes the proof of the proposition. □

Proposition 9.47 *Let $\mathcal{X}_1, \ldots, \mathcal{X}_p$ and $\mathcal{Y}$ be normed linear spaces over $\mathbb{K}$, where $p \in \{2, 3, \ldots\}$, $D \subseteq \mathcal{X} := \prod_{i=1}^{p} \mathcal{X}_i$, $f : D \to \mathcal{Y}$, $x_o \in D^\circ$, and $k \in \mathbb{N}$. Assume that $\exists \delta_0 \in (0, \infty) \subset \mathbb{R}$ such that all partial derivatives of f up to kth order exist and are continuous at x, $\forall x \in \tilde{D} := \mathcal{B}_{\mathcal{X}}(x_o, \delta_0) \subseteq D$. Then, f is $\mathcal{C}_k$ at x, $\forall x \in \tilde{D}$.*

Proof We will prove this using mathematical induction on k.
1° $k = 1$. $\forall x \in \tilde{D}$, by repeated application of Proposition 9.24, we have $f^{(1)}(x)$ exists and $f^{(1)}(x) = \left[\frac{\partial f}{\partial x_1}(x) \cdots \frac{\partial f}{\partial x_p}(x) \right]$.
2° Assume that the result holds for $k = \bar{k} \in \mathbb{N}$.
3° Consider the case $k = \bar{k} + 1$. $\forall x \in \tilde{D}$, by the case $k = 1$, $f^{(1)}(x)$ exists and $f^{(1)}(x) = \left[\frac{\partial f}{\partial x_1}(x) \cdots \frac{\partial f}{\partial x_p}(x) \right]$. $\forall i \in \{1, \ldots, p\}$, by the assumption, all partial derivatives of $\frac{\partial f}{\partial x_i}$ up to $\bar{k}$th order exist and are continuous at x. Then, by inductive assumption, $\frac{\partial f}{\partial x_i}$ is $\mathcal{C}_{\bar{k}}$ at x. Define the function $g : \hat{D} \to \prod_{i=1}^{p} \mathrm{B}(\mathcal{X}_i, \mathcal{Y})$ by $g(x) = (\frac{\partial f}{\partial x_1}(x), \ldots, \frac{\partial f}{\partial x_p}(x))$, $\forall x \in \hat{D} := \bigcap_{i=1}^{p} \mathrm{dom}\left(\frac{\partial f}{\partial x_i}\right)$. Clearly, $\tilde{D} \subseteq \hat{D} \subseteq D$. Then, by Proposition 9.44, g is $\bar{k}$-times differentiable at x and $g^{(j)}(x) = \begin{bmatrix} \mathrm{D}^j \frac{\partial f}{\partial x_1}(x) \\ \vdots \\ \mathrm{D}^j \frac{\partial f}{\partial x_p}(x) \end{bmatrix}$, $\forall j = 1, \ldots, \bar{k}$. By Proposition 3.32, g is $\mathcal{C}_{\bar{k}}$ at x. By Propositions 9.45 and 9.43, $f^{(1)}$ is $\mathcal{C}_{\bar{k}}$ at x. Therefore, f is $\mathcal{C}_{\bar{k}+1}$ at x. This completes the induction process and the proof of the proposition. □

Theorem 9.48 (Taylor's Theorem) *Let $\mathcal{X}$ and $\mathcal{Y}$ be normed linear spaces over $\mathbb{K}$, $D \subseteq \mathcal{X}$, $f : D \to \mathcal{Y}$, $x_0, x_1 \in D$, and $n \in \mathbb{Z}_+$. Let $\bar{D} := I := [0, 1] \subset \mathbb{R}$ and $\bar{\bar{D}} := I^\circ$ if $\mathbb{K} = \mathbb{R}$ or $\bar{D} := \{a + \mathrm{i}0 \mid a \in I\} \subset \mathbb{C}$ and $\bar{\bar{D}} := \{a + \mathrm{i}0 \mid a \in I^\circ\} \subset \mathbb{C}$ if $\mathbb{K} = \mathbb{C}$. Let $\varphi : \bar{D} \to D$ be given by $\varphi(t) = t x_1 + (1 - t) x_0$, $\forall t \in \bar{D}$. Assume that $\mathrm{dom}\left(f^{(n)}\right) \supseteq \varphi(\bar{D})$, $\mathrm{dom}\left(f^{(n+1)}\right) \supseteq \varphi(\bar{\bar{D}})$, and $f^{(n)}$ is continuous at $x = \varphi(t)$, $\forall t \in \bar{D}$. Let $R_n \in \mathcal{Y}$ be given by*

$$R_n := f(x_1) - \Bigg(f(x_0) + \frac{1}{1!} f^{(1)}(x_0)(x_1 - x_0) + \cdots + \frac{1}{n!} f^{(n)}(x_0) \underbrace{(x_1 - x_0) \cdots (x_1 - x_0)}_{n\text{-times}} \Bigg)$$

Then, the following statements hold:

(i) *If $\mathcal{Y} = \mathbb{R}$ and $\mathbb{K} = \mathbb{R}$, then $\exists \bar{t}_0 \in I^\circ$ such that*

$$R_n = \frac{1}{(n+1)!} f^{(n+1)}(\varphi(\bar{t}_0)) \underbrace{(x_1 - x_0) \cdots (x_1 - x_0)}_{(n+1)\text{-times}}$$

(ii) $\exists \bar{t}_0 \in I^\circ$ *such that*

$$\|R_n\| \leq \frac{1}{(n+1)!} \left\| f^{(n+1)}(\varphi(\bar{t}_0)) \right\| \|x_1 - x_0\|^{n+1}$$

Proof (i) Let $\mathcal{Y} = \mathbb{R}$ and $\mathbb{K} = \mathbb{R}$. Define $F : I \to \mathbb{R}$ by

$$F(t) = f(\varphi(1)) - \Bigg(f(\varphi(1-t)) + \frac{t}{1!} f^{(1)}(\varphi(1-t))(x_1 - x_0) + \cdots$$
$$+ \frac{t^n}{n!} f^{(n)}(\varphi(1-t)) \underbrace{(x_1 - x_0) \cdots (x_1 - x_0)}_{n\text{-times}} + R_n t^{n+1} \Bigg); \quad \forall t \in I$$

By Propositions 3.12, 3.32, 9.7, 7.23, and 7.65, F is continuous. Clearly, φ is differentiable. By chain rule and Propositions 9.10, 9.15–9.17, and 9.19, F is differentiable at t, $\forall t \in I^\circ$. Clearly, $F(0) = F(1) = 0$. By mean value theorem 9.20, $\exists t_0 \in I^\circ$ such that $0 = F(1) - F(0) = \mathrm{D}F(t_0)$. Then, we have

$$0 = -\Bigg(- f^{(1)}(\varphi(1-t_0))(x_1 - x_0) + f^{(1)}(\varphi(1-t_0))(x_1 - x_0)$$
$$- \frac{t_0}{1!} f^{(2)}(\varphi(1-t_0))(x_1 - x_0)(x_1 - x_0) + \cdots$$
$$+ \frac{t_0^{n-1}}{(n-1)!} f^{(n)}(\varphi(1-t_0)) \underbrace{(x_1 - x_0) \cdots (x_1 - x_0)}_{n\text{-times}}$$
$$- \frac{t_0^n}{n!} f^{(n+1)}(\varphi(1-t_0)) \underbrace{(x_1 - x_0) \cdots (x_1 - x_0)}_{(n+1)\text{-times}} + (n+1) R_n t_0^n \Bigg)$$
$$= \frac{t_0^n}{n!} f^{(n+1)}(\varphi(1-t_0)) \underbrace{(x_1 - x_0) \cdots (x_1 - x_0)}_{(n+1)\text{-times}} - (n+1) R_n t_0^n$$

Hence,

$$R_n = \frac{1}{(n+1)!} f^{(n+1)}(\varphi(\bar{t}_0)) \underbrace{(x_1 - x_0) \cdots (x_1 - x_0)}_{(n+1)\text{-times}}$$

where $\bar{t}_0 = 1 - t_0 \in I^\circ$.

(ii) By Proposition 7.85, $\exists y_* \in \mathcal{Y}^*$ with $\|y_*\| \leq 1$ such that $\|R_n\| = \langle\langle y_*, R_n\rangle\rangle$. Define $G : \bar{D} \to \mathbb{K}$ by

$$G(t) = \left\langle\!\!\left\langle y_*, f(x_1) - f(\varphi(1-t)) - \sum_{i=1}^{n} \frac{t^i}{i!} f^{(i)}(\varphi(1-t)) \underbrace{(x_1 - x_0)\cdots(x_1 - x_0)}_{i\text{-times}} \right\rangle\!\!\right\rangle - \|R_n\| t^{n+1}$$

$\forall t \in \bar{D}$. By Propositions 3.12, 3.32, 9.7, 7.23, and 7.65, G is continuous. Clearly, φ is differentiable. By chain rule and Propositions 9.10, 9.15–9.17, and 9.19, G is differentiable at t, $\forall t \in \tilde{\bar{D}}$. Clearly, $G(0) = G(1) = 0$.

We will distinguish two exhaustive and mutually exclusive cases: Case 1: $\mathbb{K} = \mathbb{R}$; Case 2: $\mathbb{K} = \mathbb{C}$.

Case 1: $\mathbb{K} = \mathbb{R}$. By mean value theorem 9.20, $\exists t_0 \in I^\circ$ such that $0 = G(1) - G(0) = \mathrm{D}G(t_0)$. Then, we have

$$\begin{aligned} 0 &= \left\langle\!\!\left\langle y_*, f^{(1)}(\varphi(1-t_0))(x_1 - x_0) - \sum_{i=1}^{n} \frac{t_0^{i-1}}{(i-1)!} f^{(i)}(\varphi(1-t_0)) \underbrace{(x_1 - x_0)\cdots(x_1 - x_0)}_{i\text{-times}} \right.\right. \\ &\quad \left.\left. + \sum_{i=1}^{n} \frac{t_0^i}{i!} f^{(i+1)}(\varphi(1-t_0)) \underbrace{(x_1 - x_0)\cdots(x_1 - x_0)}_{(i+1)\text{-times}} \right\rangle\!\!\right\rangle - (n+1)\|R_n\| t_0^n \\ &= \left\langle\!\!\left\langle y_*, \frac{t_0^n}{n!} f^{(n+1)}(\varphi(1-t_0)) \underbrace{(x_1 - x_0)\cdots(x_1 - x_0)}_{(n+1)\text{-times}} \right\rangle\!\!\right\rangle - (n+1)\|R_n\| t_0^n \end{aligned}$$

Then, we have

$$\begin{aligned} \|R_n\| &= \frac{1}{(n+1)!} \left\langle\!\!\left\langle y_*, f^{(n+1)}(\varphi(1-t_0)) \underbrace{(x_1 - x_0)\cdots(x_1 - x_0)}_{(n+1)\text{-times}} \right\rangle\!\!\right\rangle \\ &\leq \frac{1}{(n+1)!} \|y_*\| \left\| f^{(n+1)}(\varphi(1-t_0)) \underbrace{(x_1 - x_0)\cdots(x_1 - x_0)}_{(n+1)\text{-times}} \right\| \\ &\leq \frac{1}{(n+1)!} \left\| f^{(n+1)}(\varphi(\bar{t}_0)) \right\| \|x_1 - x_0\|^{n+1} \end{aligned}$$

where the last two inequalities follow from Proposition 7.64 and $\bar{t}_0 = 1 - t_0 \in I^\circ$.

Case 2: $\mathbb{K} = \mathbb{C}$. By Lemma 9.22, $\exists t_0 \in I^\circ$ such that $\mathrm{Re}\,(G(1) - G(0)) = \mathrm{Re}\,(\mathrm{D}G(t_0))$. Then, we have

$$\begin{aligned}
0 &= \mathrm{Re}\left(\langle\langle y_*, f^{(1)}(\varphi(1-t_0))(x_1 - x_0)\rangle\rangle\right) \\
&\quad - \mathrm{Re}\left(\left\langle\left\langle y_*, \sum_{i=1}^{n} \frac{t_0^{i-1}}{(i-1)!} f^{(i)}(\varphi(1-t_0)) \underbrace{(x_1 - x_0)\cdots(x_1 - x_0)}_{i\text{-times}}\right\rangle\right\rangle\right) \\
&\quad + \mathrm{Re}\left(\left\langle\left\langle y_*, \sum_{i=1}^{n} \frac{t_0^{i}}{i!} f^{(i+1)}(\varphi(1-t_0)) \underbrace{(x_1 - x_0)\cdots(x_1 - x_0)}_{(i+1)\text{-times}}\right\rangle\right\rangle\right) \\
&\quad - (n+1)\|R_n\| t_0^n \\
&= \mathrm{Re}\left(\left\langle\left\langle y_*, \frac{t_0^n}{n!} f^{(n+1)}(\varphi(1-t_0)) \underbrace{(x_1 - x_0)\cdots(x_1 - x_0)}_{(n+1)\text{-times}}\right\rangle\right\rangle\right) \\
&\quad - (n+1)\|R_n\| t_0^n
\end{aligned}$$

Hence,

$$\begin{aligned}
\|R_n\| &= \frac{1}{(n+1)!}\mathrm{Re}\left(\left\langle\left\langle y_*, f^{(n+1)}(\varphi(1-t_0)) \underbrace{(x_1 - x_0)\cdots(x_1 - x_0)}_{(n+1)\text{-times}}\right\rangle\right\rangle\right) \\
&\le \frac{1}{(n+1)!}\|y_*\| \left\| f^{(n+1)}(\varphi(1-t_0)) \underbrace{(x_1 - x_0)\cdots(x_1 - x_0)}_{(n+1)\text{-times}} \right\| \\
&\le \frac{1}{(n+1)!}\left\| f^{(n+1)}(\varphi(\bar{t}_0))\right\| \|x_1 - x_0\|^{n+1}
\end{aligned}$$

where the last two inequalities follow from Proposition 7.64 and $\bar{t}_0 = 1 - t_0 \in I^\circ$.

This completes the proof of the theorem. □

9.5 Mapping Theorems

Definition 9.49 Let $\mathcal{X} := (X, \rho)$ be a metric space, $S \subseteq X$, and $T : S \to X$. T is said to be a *contraction mapping on* S if $T(S) \subseteq S$ and $\exists \alpha \in [0, 1) \subset \mathbb{R}$ such that $\forall x_1, x_2 \in S$, we have $\rho(T(x_1), T(x_2)) \le \alpha\rho(x_1, x_2)$. Then, α is called a *contraction index* for T. ◇

Theorem 9.50 (Contraction Mapping Theorem) *Let $S \ne \emptyset$ be a closed subset of a complete metric space $\mathcal{X} := (X, \rho)$ and T be a contraction mapping on S with contraction index $\alpha \in [0, 1) \subset \mathbb{R}$. Then, the following statements hold:*

(i) $\exists!\, x_0 \in S$ *such that* $x_0 = T(x_0)$.
(ii) $\forall x_1 \in S$, *recursively define* $x_{n+1} = T(x_n)$, $\forall n \in \mathbb{N}$. *Then,* $\lim_{n\in\mathbb{N}} x_n = x_0$.
(iii) $\rho(x_n, x_0) \le \frac{\alpha^{n-1}}{1-\alpha}\rho(x_2, x_1)$, $\rho(x_n, x_0) \le \frac{\alpha}{1-\alpha}\rho(x_n, x_{n-1})$, *and* $\rho(x_n, x_0) \le \alpha\rho(x_{n-1}, x_0)$, $\forall n \in \{2, 3, \ldots\}$.

Proof Fix any $x_1 \in S \neq \emptyset$. Recursively define $x_{n+1} = T(x_n) \in S$, $\forall n \in \mathbb{N}$. Then, $\rho(x_n, x_{n-1}) = \rho(T(x_{n-1}), T(x_{n-2})) \le \alpha\rho(x_{n-1}, x_{n-2}) \le \cdots \le \alpha^{n-2}\rho(x_2, x_1)$, $\forall n \in \{3, 4, \ldots\}$. Therefore, $(x_n)_{n=1}^{\infty} \subseteq S$ is a Cauchy sequence and converges to $x_0 \in S$ by $\mathcal{X}$ being complete, S being closed, and Proposition 4.39. Clearly, T is continuous. Then, by Proposition 3.66, $x_0 = T(x_0)$. $\forall \bar{x} \in S$ such that $\bar{x} = T(\bar{x})$. Then, $\rho(x_0, \bar{x}) \le \alpha\rho(x_0, \bar{x})$ and $\rho(x_0, \bar{x}) = 0$. Hence, $\bar{x} = x_0$. Hence, the statements (i) and (ii) are true.

Note that, by Propositions 3.66 and 4.30, $\forall n \in \{2, 3, \ldots\}$,

$$\rho(x_n, x_0) = \lim_{m\in\mathbb{N}} \rho(x_n, x_m) \le \sum_{i=n}^{\infty} \rho(x_{i+1}, x_i) \le \sum_{i=n+1}^{\infty} \alpha^{i-2}\rho(x_2, x_1)$$
$$= \frac{\alpha^{n-1}}{1-\alpha}\rho(x_2, x_1)$$
$$\rho(x_n, x_0) \le \sum_{i=n}^{\infty} \rho(x_{i+1}, x_i) \le \sum_{i=1}^{\infty} \alpha^i \rho(x_n, x_{n-1}) = \frac{\alpha}{1-\alpha}\rho(x_n, x_{n-1})$$
$$\rho(x_n, x_0) = \rho(T(x_{n-1}), T(x_0)) \le \alpha\rho(x_{n-1}, x_0)$$

This completes the proof of the theorem. □

Lemma 9.51 *Let* $\mathcal{X}$ *and* $\mathcal{Y}$ *be normed linear spaces over* $\mathbb{K}$, $D \subseteq \mathcal{X}$, $x_0 \in D$, *and* $f : D \to \mathcal{Y}$ *be* $\mathcal{C}_1$ *at* x_0. *Assume that* D *is locally convex at* x_0. *Then,* $\forall \epsilon \in (0, \infty) \subset \mathbb{R}$, $\exists \delta \in (0, \infty) \subset \mathbb{R}$ *such that,* $\forall x_1, x_2 \in D \cap \mathcal{B}_{\mathcal{X}}(x_0, \delta)$, *we have* $\left\| f(x_1) - f(x_2) - f^{(1)}(x_0)(x_1 - x_2)\right\| \le \epsilon\|x_1 - x_2\|$.

Proof By the assumption, $\exists \delta_0 \in (0, \infty) \subset \mathbb{R}$ such that $D \cap \mathcal{B}_{\mathcal{X}}(x_0, \delta_0) =: \tilde{D}$ is convex. $\forall \epsilon \in (0, \infty) \subset \mathbb{R}$, by f being $\mathcal{C}_1$ at x_0, $\exists \delta \in (0, \delta_0] \subset \mathbb{R}$ such that $\forall x \in D \cap \mathcal{B}_{\mathcal{X}}(x_0, \delta) =: \bar{D}$, we have $\left\| f^{(1)}(x) - f^{(1)}(x_0)\right\| < \epsilon$. Define $\gamma : D \to \mathcal{Y}$ by $\gamma(x) = f(x) - f^{(1)}(x_0)(x - x_0)$, $\forall x \in D$. By Propositions 9.45, 9.38, 9.41, and 9.44, γ is $\mathcal{C}_1$ at x_0 and $\gamma^{(1)}(x) = f^{(1)}(x) - f^{(1)}(x_0)$, $\forall x \in \bar{D}$. Clearly, $\bar{D}$ is convex. Then, $\forall x_1, x_2 \in \bar{D}$, by mean value theorem 9.23 and Proposition 7.64, we have $\left\| f(x_1) - f(x_2) - f^{(1)}(x_0)(x_1 - x_2)\right\| = \|\gamma(x_1) - \gamma(x_2)\| \le \left\|\gamma^{(1)}(t_0 x_1 + (1 - t_0)x_2)(x_1 - x_2)\right\| \le \epsilon\|x_1 - x_2\|$, for some $t_0 \in (0, 1) \subset \mathbb{R}$. This completes the proof of the lemma. □

Theorem 9.52 (Injective Mapping Theorem) *Let* $\mathcal{X}$ *and* $\mathcal{Y}$ *be Banach spaces over* $\mathbb{K}$, $D \subseteq \mathcal{X}$, *and* $F : D \to \mathcal{Y}$ *be* $\mathcal{C}_1$ *at* $x_0 \in D^\circ$. *Assume that* $F^{(1)}(x_0) \in \mathrm{B}(\mathcal{X}, \mathcal{Y})$ *is injective and* $M := \mathcal{R}\big(F^{(1)}(x_0)\big) \subseteq \mathcal{Y}$ *is closed. Then,* $\exists \delta \in (0, \infty) \subset \mathbb{R}$ *with*

$U := \mathcal{B}_{\mathcal{X}}(x_0, \delta) \subseteq D$ such that $F|_U : U \to F(U)$ is bijective and admits a continuous inverse $F_i : F(U) \to U$.

Proof By Proposition 7.13 and Proposition 4.39, $M \subseteq \mathcal{Y}$ is a Banach space. Then, $F^{(1)}(x_0) : \mathcal{X} \to M$ is bijective. By Open Mapping Theorem 7.103, the inverse A of $F^{(1)}(x_0) : \mathcal{X} \to M$ belongs to $\mathrm{B}(M, \mathcal{X})$. $\forall h \subset \mathcal{X}$, we have $\|h\| = \left\|AF^{(1)}(x_0)h\right\| \leq \|A\|\left\|F^{(1)}(x_0)h\right\|$. Then, $\exists r \in (0, \infty) \subset \mathbb{R}$ such that $r\|A\| \leq 1$ and $r\|h\| \leq \left\|F^{(1)}(x_0)h\right\|$, $\forall h \in \mathcal{X}$.

By Lemma 9.51, $\exists \delta \in (0, \infty) \subset \mathbb{R}$ with $U := \mathcal{B}_{\mathcal{X}}(x_0, \delta) \subseteq D$ such that $\forall x_1, x_2 \in U$, we have $\left\|F(x_1) - F(x_2) - F^{(1)}(x_0)(x_1 - x_2)\right\| \leq r\|x_1 - x_2\|/2$. $\forall x_1, x_2 \in U$, $r\|x_1 - x_2\| \leq \left\|F^{(1)}(x_0)(x_1 - x_2)\right\| \leq \|F(x_1) - F(x_2)\| + \left\|F(x_1) - F(x_2) - F^{(1)}(x_0)(x_1 - x_2)\right\| \leq \|F(x_1) - F(x_2)\| + r\|x_1 - x_2\|/2$. This implies that $\|F(x_1) - F(x_2)\| \geq r\|x_1 - x_2\|/2$. Hence, $F|_U : U \to F(U)$ is injective and surjective. Then, $F|_U : U \to F(U)$ is bijective and admits an inverse $F_i : F(U) \to U$. $\forall y_1, y_2 \in F(U)$, we have $\|F_i(y_1) - F_i(y_2)\| \leq 2/r\|F(F_i(y_1)) - F(F_i(y_2))\| = 2/r\|y_1 - y_2\|$. Hence, F_i is uniformly continuous. This completes the proof of the theorem. □

Theorem 9.53 (Surjective Mapping Theorem) *Let $\mathcal{X}$ and $\mathcal{Y}$ be Banach spaces over $\mathbb{K}$, $D \subseteq \mathcal{X}$, $F : D \to \mathcal{Y}$, $x_0 \in D^\circ$, and $y_0 := F(x_0) \in \mathcal{Y}$. Assume that F is $\mathcal{C}_1$ at x_0, and $F^{(1)}(x_0) \in \mathrm{B}(\mathcal{X}, \mathcal{Y})$ is surjective. Then, $\exists r \in (0, \infty) \subset \mathbb{R}$, $\exists \delta \in (0, \infty) \subset \mathbb{R}$, and $\exists c_1 \in [0, \infty) \subset \mathbb{R}$ with $c_1\delta \leq r$ such that $\forall \bar{y} \in \mathcal{B}_{\mathcal{Y}}(y_0, \delta/2)$, $\forall \bar{x} \in \mathcal{B}_{\mathcal{X}}(x_0, r/2)$ with $\bar{y} = F(\bar{x})$, $\forall y \in \mathcal{B}_{\mathcal{Y}}(\bar{y}, \delta/2)$, $\exists x \in \mathcal{B}_{\mathcal{X}}(x_0, r) \subseteq D$ with $\|x - \bar{x}\| \leq c_1\|y - \bar{y}\|$, we have $y = F(x)$.*

Proof Let $M := \mathcal{N}\big(F^{(1)}(x_0)\big)$, which is a closed subspace by Proposition 7.68. By Proposition 7.45, the quotient space $\mathcal{X}/M$ is a Banach space. By Proposition 7.70, $F^{(1)}(x_0) = A \circ \phi$, where $\phi : \mathcal{X} \to \mathcal{X}/M$ is the natural homomorphism, and $A \in \mathrm{B}(\mathcal{X}/M, \mathcal{Y})$ is injective. Since $F^{(1)}(x_0)$ is surjective, then A is bijective and, by Open Mapping Theorem 7.103, $A^{-1} \in \mathrm{B}(\mathcal{Y}, \mathcal{X}/M)$. Let $c_1 := 4\left\|A^{-1}\right\| \in [0, \infty) \subset \mathbb{R}$.

Define $\gamma : D \to \mathcal{X}/M$ by $\gamma(x) = A^{-1}(F(x) - F^{(1)}(x_0)(x - x_0))$, $\forall x \in D$. By Propositions 9.45, 9.38, 9.41, 9.34, and 9.44, γ is $\mathcal{C}_1$ at x_0 and $\gamma^{(1)}(x) = A^{-1}\left(F^{(1)}(x) - F^{(1)}(x_0)\right)$, $\forall x \in \mathcal{B}_{\mathcal{X}}(x_0, r_0) \subseteq D$, for some $r_0 \in (0, \infty) \subset \mathbb{R}$. Clearly, $\gamma^{(1)}(x_0) = \vartheta_{\mathrm{B}(\mathcal{X}, \mathcal{X}/M)}$. Then, by Lemma 9.51, $\exists r \in (0, r_0] \subset \mathbb{R}$ such that $\forall x_1, x_2 \in \mathcal{B}_{\mathcal{X}}(x_0, r) \subseteq D$, we have $\|\gamma(x_1) - \gamma(x_2)\| \leq \|x_1 - x_2\|/4$.

Let $\delta \in (0, \infty) \subset \mathbb{R}$ be such that $c_1\delta \leq r$. Fix any $y \in \mathcal{B}_{\mathcal{Y}}(y_0, \delta/2)$ and any $\bar{x} \in \mathcal{B}_{\mathcal{X}}(x_0, r/2)$ with $\bar{y} = F(\bar{x})$. Fix any $y \in \mathcal{B}_{\mathcal{Y}}(\bar{y}, \delta/2)$. Recursively define $x_1 := \bar{x}$, $[x_{k+1}] = [x_k] + A^{-1}(y - F(x_k))$ and select $x_{k+1} \in [x_{k+1}]$ such that $\|x_{k+1} - x_k\| \leq 2\|[x_{k+1}] - [x_k]\|$, $\forall k \in \mathbb{N}$, where $[x_k] = \phi(x_k) = x_k + M$ is the coset containing x_k. Clearly, $x_1 \in \mathcal{B}_{\mathcal{X}}(x_0, r)$. Note that $\|x_2 - x_1\| \leq 2\|[x_2] - [x_1]\| = 2\left\|A^{-1}(y - \bar{y})\right\| \leq 2\left\|A^{-1}\right\|\|y - \bar{y}\| = c_1\|y - \bar{y}\|/2 < r/4$, where the second inequality follows from Proposition 7.64. Then, $x_2 \in \mathcal{B}_{\mathcal{X}}(x_0, r)$. Assume that $x_1, \ldots, x_k \in \mathcal{B}_{\mathcal{X}}(x_0, r)$ for some $k \in \{2, 3, \ldots\}$. Note that $\forall i \in \{2, \ldots, k\}$, $\|x_{i+1} - x_i\| \leq 2\|[x_{i+1}] - [x_i]\| = 2\left\|A^{-1}(y - F(x_i) + A\phi(x_i)) - A^{-1}(y - F(x_{i-1}) + A\phi(x_{i-1}))\right\| = 2\|\gamma(x_i) - \gamma(x_{i-1})\| \leq \|x_i - x_{i-1}\|/2$. Then, $\|x_{i+1} - x_i\| \leq \|x_2 - x_1\|/2^{i-1}$. Hence,

$\|x_{k+1} - \bar{x}\| \leq \sum_{i=1}^{k} \|x_{i+1} - x_i\| \leq \sum_{i=1}^{k} \|x_2 - x_1\|/2^{i-1} = (2 - 1/2^{k-1}) \|x_2 - x_1\| < r/2$. This implies that $x_{k+1} \in \mathcal{B}_{\mathcal{X}}(x_0, r)$. Inductively, we have $(x_k)_{k=0}^{\infty} \subseteq \mathcal{B}_{\mathcal{X}}(x_0, r)$ and $\|x_{k+1} - x_k\| \leq \|x_2 - x_1\|/2^{k-1}$, $\forall k \in \mathbb{N}$. Hence, $(x_k)_{k=0}^{\infty}$ is a Cauchy sequence. By the completeness of $\mathcal{X}$, we have $\lim_{k\in\mathbb{N}} x_k = x \in \mathcal{X}$. Note that, by Propositions 3.66 and 4.30, $\|x - \bar{x}\| = \lim_{k\in\mathbb{N}} \|x_k - x_1\| \leq \lim_{k\in\mathbb{N}}(2 - 1/2^{k-2})\|x_2 - x_1\| = 2\|x_2 - x_1\| < r/2$. Hence, $x \in \mathcal{B}_{\mathcal{X}}(x_0, r)$. By the differentiability of F and Propositions 9.7, 3.66, and 7.69, we have $\phi(x) = \lim_{k\in\mathbb{N}} \phi(x_{k+1}) = \lim_{k\in\mathbb{N}}(A^{-1}(y - F(x_k)) + \phi(x_k)) = \phi(x) + A^{-1}(y - F(x))$. This implies that $y = F(x)$. Note that $\|x - \bar{x}\| \leq 2\|x_2 - x_1\| \leq c_1\|y - \bar{y}\|$. This completes the proof of the theorem. □

Theorem 9.54 (Open Mapping Theorem) *Let $\mathcal{X}$ and $\mathcal{Y}$ be Banach spaces over $\mathbb{K}$, $D \subseteq \mathcal{X}$ be open, and $F : D \to \mathcal{Y}$ be $\mathcal{C}_1$. Assume that $F^{(1)}(x) \in \mathrm{B}(\mathcal{X}, \mathcal{Y})$ is surjective, $\forall x \in D$. Then, F is an open mapping.*

Proof Fix any open subset $U \subseteq D$, where U is open in the subset topology of D. Since D is open, then U is open in $\mathcal{X}$. We will show that $F(U)$ is an open set in $\mathcal{Y}$. Fix any $y_0 \in F(U)$; there exists $x_0 \in U$ such that $y_0 = F(x_0)$. Then, $\exists r \in (0, \infty) \subset \mathbb{R}$ such that $\mathcal{B}_{\mathcal{X}}(x_0, r) \subseteq U$. It is easy to check that all assumptions of Surjective Mapping Theorem are satisfied at x_0. Then, there exist an open set $V \subseteq \mathcal{Y}$ with $y_0 \in V$ and $c_1 \in [0, \infty) \subset \mathbb{R}$ such that $\forall \bar{y} \in V$, $\exists \bar{x} \in D$ with $\|\bar{x} - x_0\| \leq c_1\|\bar{y} - y_0\|$, we have $\bar{y} = F(\bar{x})$. Take $\delta \in (0, \infty) \subset \mathbb{R}$ such that $c_1\delta \leq r$ and $\mathcal{B}_{\mathcal{Y}}(y_0, \delta) \subseteq V$. Then, $\forall \bar{y} \in \mathcal{B}_{\mathcal{Y}}(y_0, \delta)$, $\|\bar{x} - x_0\| \leq c_1\|\bar{y} - y_0\| < r$. Then, $\bar{x} \in \mathcal{B}_{\mathcal{X}}(x_0, r) \subseteq U$ and $\bar{y} \in F(U)$. Hence, $\mathcal{B}_{\mathcal{Y}}(y_0, \delta) \subseteq F(U)$. Therefore, $y_0 \in (F(U))^{\circ}$. By the arbitrariness of y_0, we have $F(U)$ is open in $\mathcal{Y}$. By the arbitrariness of U, F is an open mapping. This completes the proof of the theorem. □

Proposition 9.55 *Let $\mathcal{X}$ and $\mathcal{Y}$ be Banach spaces over $\mathbb{K}$ and $A \in \mathrm{B}(\mathcal{X}, \mathcal{Y})$ be bijective. Then, $\forall T \in \mathrm{B}(\mathcal{X}, \mathcal{Y})$ with $\|T - A\|\|A^{-1}\| < 1$, we have T is bijective and $\|T^{-1} - A^{-1}\| \leq \frac{\|A^{-1}\|^2\|T-A\|}{1-\|A^{-1}\|\|T-A\|}$.*

Proof By Open Mapping Theorem 7.103, $A^{-1} \in \mathrm{B}(\mathcal{Y}, \mathcal{X})$. We will first prove the result for the special case $\mathcal{Y} = \mathcal{X}$ and $A = \mathrm{id}_{\mathcal{X}}$. We will distinguish two exhaustive and mutually exclusive cases: Case 1: $\mathcal{X}$ is a singleton set; Case 2: $\exists \bar{x} \in \mathcal{X}$ such that $\bar{x} \neq \vartheta_{\mathcal{X}}$.

Case 1: $\mathcal{X}$ is a singleton set. Then, $\|A^{-1}\| = \|A\| = 0$. $\forall T \in \mathrm{B}(\mathcal{X}, \mathcal{X})$, we have $T = \mathrm{id}_{\mathcal{X}}$. Then, T is bijective and $\|T^{-1} - \mathrm{id}_{\mathcal{X}}\| = 0$. The result holds for this case.

Case 2: $\exists \bar{x} \in \mathcal{X}$ such that $\bar{x} \neq \vartheta_{\mathcal{X}}$. Then, $\|A\| = \|A^{-1}\| = 1$. $\forall T \in \mathrm{B}(\mathcal{X}, \mathcal{X})$ with $\|T - \mathrm{id}_{\mathcal{X}}\| < 1$, let $\Delta := T - \mathrm{id}_{\mathcal{X}}$. We will show that T is bijective. $\forall x_1, x_2 \in \mathcal{X}$ with $T(x_1) = T(x_2)$, we have $x_1 + \Delta(x_1) = x_2 + \Delta(x_2)$, which implies that $\Delta(x_1 - x_2) = x_2 - x_1$. By Proposition 7.64, we have $\|\Delta\|\|x_1 - x_2\| \geq \|x_1 - x_2\|$. Since $\|\Delta\| < 1$, then $\|x_1 - x_2\| = 0$ and $x_1 = x_2$. Therefore, T is injective. $\forall x_0 \in \mathcal{X}$, define $\phi : \mathcal{X} \to \mathcal{X}$ by $\phi(x) = x_0 - \Delta(x)$, $\forall x \in \mathcal{X}$. Clearly, ϕ is a contraction mapping on $\mathcal{X}$ with contraction index $\|\Delta\|$. By Contraction Mapping Theorem, there exists a unique $\bar{x} \in \mathcal{X}$ such that $\bar{x} = \phi(\bar{x})$. Then, $x_0 = \bar{x} + \Delta(\bar{x}) = T(\bar{x})$. Hence, T is surjective. Then, T is bijective. By Open Mapping Theorem 7.103, $T^{-1} \in B(\mathcal{X}, \mathcal{X})$.

$\forall y \in \mathcal{Y}$, let $x = T^{-1}y$. Then, $y = Tx = x + \Delta x$ and $x = y - \Delta x$. By Proposition 7.64, we have $\|x\| \le \|y\| + \|\Delta\| \|x\|$ and $\|x\| \le \frac{\|y\|}{1-\|\Delta\|}$. By the arbitrariness of y, we have $\|T^{-1}\| \le 1/(1 - \|\Delta\|)$. This further implies that $\|T^{-1} - \mathrm{id}_{\mathcal{X}}\| = \|T^{-1}(\mathrm{id}_{\mathcal{X}} - T)\| \le \|T^{-1}\| \|\Delta\| \le \|\Delta\|/(1 - \|\Delta\|)$. The result holds in this case.

Hence, the result holds for the special case $\mathcal{Y} = \mathcal{X}$ and $A = \mathrm{id}_{\mathcal{X}}$. Now consider the general case. $\forall T \in B(\mathcal{X}, \mathcal{Y})$ with $\|T - A\| \|A^{-1}\| < 1$, we have $\bar{T} := A^{-1}T \in B(\mathcal{X}, \mathcal{X})$. Note that $\|\bar{T} - \mathrm{id}_{\mathcal{X}}\| = \|A^{-1}(T - A)\| \le \|A^{-1}\| \|T - A\| < 1$. Then, $\|\bar{T} - \mathrm{id}_{\mathcal{X}}\| \|\mathrm{id}_{\mathcal{X}}\| < 1$. By the special case, we have $\bar{T}$ is bijective and $\|\bar{T}^{-1} - \mathrm{id}_{\mathcal{X}}\| \le \frac{\|\mathrm{id}_{\mathcal{X}}\|^2 \|\bar{T} - \mathrm{id}_{\mathcal{X}}\|}{1 - \|\mathrm{id}_{\mathcal{X}}\| \|\bar{T} - \mathrm{id}_{\mathcal{X}}\|}$. Then, T is bijective and, by Proposition 7.64,

$$\|\bar{T} - \mathrm{id}_{\mathcal{X}}\| = \|A^{-1}(T - A)\| \le \|A^{-1}\| \|T - A\|$$

$$\|A^{-1}\| \|\mathrm{id}_{\mathcal{X}}\| = \|A^{-1}\|$$

$$\|T^{-1} - A^{-1}\| = \|(\bar{T}^{-1} - \mathrm{id}_{\mathcal{X}})A^{-1}\| \le \|A^{-1}\| \frac{\|\mathrm{id}_{\mathcal{X}}\|^2 \|\bar{T} - \mathrm{id}_{\mathcal{X}}\|}{1 - \|\mathrm{id}_{\mathcal{X}}\| \|\bar{T} - \mathrm{id}_{\mathcal{X}}\|}$$

$$\le \frac{\|A^{-1}\|^2 \|T - A\|}{1 - \|A^{-1}\| \|T - A\|}$$

This completes the proof of the proposition. □

Proposition 9.56 *Let $\mathcal{X}$ and $\mathcal{Y}$ be Banach spaces over $\mathbb{K}$, $D := \{L \in B(\mathcal{X}, \mathcal{Y}) \mid L$ is bijective$\}$, and $f : D \to B(\mathcal{Y}, \mathcal{X})$ be given by $f(A) = A^{-1}$, $\forall A \in D$. Then, D is open in $B(\mathcal{X}, \mathcal{Y})$, f is $\mathcal{C}_\infty$, and $f^{(1)}(A)(\Delta) = -A^{-1}\Delta A^{-1}$, $\forall A \in D$, $\forall \Delta \in B(\mathcal{X}, \mathcal{Y})$.*

Proof By Proposition 9.55 and Open Mapping Theorem 7.103, D is open and f is continuous. $\forall A \in D$, $\overline{\mathrm{span}\,(A_D(A))} = B(\mathcal{X}, \mathcal{Y})$ since D is open and $A \in D^{\circ}$. Define $L : B(\mathcal{X}, \mathcal{Y}) \to B(\mathcal{Y}, \mathcal{X})$ by $L(\Delta) = -A^{-1}\Delta A^{-1}$, $\forall \Delta \in B(\mathcal{X}, \mathcal{Y})$. Clearly, L is a linear operator. Note that

$$\|L\| = \sup_{\Delta \in B(\mathcal{X}, \mathcal{Y}),\ \|\Delta\| \le 1} \|L(\Delta)\| \le \|A^{-1}\|^2$$

where the inequality follows from Proposition 7.64. Hence, L is a bounded linear operator.

$\forall \epsilon \in (0, \infty) \subset \mathbb{R}$, by the continuity of f, $\exists \delta \in (0, \infty) \subset \mathbb{R}$ such that $\| f(\bar{A}) - f(A) \| < \epsilon$, $\forall \bar{A} \in \mathcal{B}_{\mathrm{B}(\mathcal{X},\mathcal{Y})}(A, \delta)$. Then, $\forall \bar{A} \in \mathcal{B}_{\mathrm{B}(\mathcal{X},\mathcal{Y})}(A, \delta)$, we have

$$\begin{aligned} &\| f(\bar{A}) - f(A) - L(\bar{A} - A) \| = \| \bar{A}^{-1} - A^{-1} + A^{-1}(\bar{A} - A)A^{-1} \| \\ &= \| - A^{-1}(\bar{A} - A)\bar{A}^{-1} + A^{-1}(\bar{A} - A)A^{-1} \| \\ &\leq \| A^{-1} \| \| \bar{A} - A \| \| f(\bar{A}) - f(A) \| \leq \epsilon \| A^{-1} \| \| \bar{A} - A \| \end{aligned}$$

where the first inequality follows from Proposition 7.64. Hence, we have $f^{(1)}(A) = L$. Then, f is differentiable. Note that $f^{(1)}(A) = -f(A)\mathrm{ro}(f(A))$, $\forall A \in D$. By Propositions 9.42, 9.7, 3.12, and 3.32, $f^{(1)}$ is continuous. Hence, f is $\mathcal{C}_1$.

Assume that f is $\mathcal{C}_k$, for some $k \in \mathbb{N}$. We will show that f is $\mathcal{C}_{k+1}$. Then, f is $\mathcal{C}_\infty$. Note that $f^{(1)}(A) = -f(A)\mathrm{ro}(f(A))$, $\forall A \in D$. By Propositions 9.45, 9.42, and 9.44, $f^{(1)}$ is $\mathcal{C}_k$. Then, f is $\mathcal{C}_{k+1}$. This completes the proof of the proposition. □

Theorem 9.57 (Inverse Function Theorem) *Let $\mathcal{X}$ and $\mathcal{Y}$ be Banach spaces over $\mathbb{K}$, $D \subseteq \mathcal{X}$, and $F : D \to \mathcal{Y}$ be $\mathcal{C}_1$ at $x_0 \in D^\circ$. Assume that $F^{(1)}(x_0) \in \mathrm{B}(\mathcal{X}, \mathcal{Y})$ is bijective. Then, $\exists$ open set $U \subseteq D$ with $x_0 \in U$ and $\exists$ open set $V \subseteq \mathcal{Y}$ with $y_0 := F(x_0) \in V$ such that:*

(i) *$F|_U : U \to V$ is bijective.*
(ii) *The inverse mapping $F_i : V \to U$ of $F|_U$ is differentiable, $F_i^{(1)} : V \to \mathrm{B}(\mathcal{Y}, \mathcal{X})$ is given by $F_i^{(1)}(y) = (F^{(1)}(F_i(y)))^{-1}$, $\forall y \in V$, and $F_i^{(1)}$ is continuous at y_0.*
(iii) *If F is k-times differentiable for some $k \in \mathbb{N}$, then F_i is k-times differentiable.*
(iv) *If F is $\mathcal{C}_k$ at x_0 for some $k \in \mathbb{N} \cup \{\infty\}$, then F_i is $\mathcal{C}_k$ at y_0.*

Proof By Open Mapping Theorem 7.103, $(F^{(1)}(x_0))^{-1} \in \mathrm{B}(\mathcal{Y}, \mathcal{X})$. By F being $\mathcal{C}_1$ at $x_0 \in D^\circ$, $\exists \bar{r} \in (0, \infty) \subset \mathbb{R}$ such that $F^{(1)}(x)$ exists, $\forall x \in \mathcal{B}_\mathcal{X}(x_0, \bar{r}) \subseteq D$. Define $T : \mathcal{B}_\mathcal{X}(\vartheta_\mathcal{X}, \bar{r}) \to \mathcal{X}$ by $T(x) = (F^{(1)}(x_0))^{-1}(F(x+x_0) - y_0)$, $\forall x \in \mathcal{B}_\mathcal{X}(\vartheta_\mathcal{X}, \bar{r}) \subseteq D - x_0$. Clearly, $T(\vartheta_\mathcal{X}) = \vartheta_\mathcal{X}$. By Propositions 9.45, 9.38, 9.34, and 9.44, T is $\mathcal{C}_1$ at $\vartheta_\mathcal{X}$, T is differentiable, and $T^{(1)}(x) = (F^{(1)}(x_0))^{-1} F^{(1)}(x + x_0)$, $\forall x \in \mathcal{B}_\mathcal{X}(\vartheta_\mathcal{X}, \bar{r})$. Clearly, $T^{(1)}(\vartheta_\mathcal{X}) = \mathrm{id}_\mathcal{X}$.

Define $\psi : \mathcal{B}_\mathcal{X}(\vartheta_\mathcal{X}, \bar{r}) \to \mathcal{X}$ by $\psi(x) = T(x) - x$. Then, by Propositions 9.38, 9.45, and 9.44, ψ is differentiable, ψ is $\mathcal{C}_1$ at $\vartheta_\mathcal{X}$, and $\psi^{(1)}(x) = T^{(1)}(x) - \mathrm{id}_\mathcal{X}$, $\forall x \in \mathcal{B}_\mathcal{X}(\vartheta_\mathcal{X}, \bar{r})$. Clearly, $\psi(\vartheta_\mathcal{X}) = \vartheta_\mathcal{X}$ and $\psi^{(1)}(\vartheta_\mathcal{X}) = \vartheta_{\mathrm{B}(\mathcal{X},\mathcal{X})}$. Fix any $\alpha \in (0, 1) \subset \mathbb{R}$. Then, $\exists r_1 \in (0, \bar{r}) \subset \mathbb{R}$ such that $\overline{\mathcal{B}}_\mathcal{X}(\vartheta_\mathcal{X}, r_1) \subseteq D - x_0$ and $\| \psi^{(1)}(x) \| \leq \alpha$, $\forall x \in \overline{\mathcal{B}}_\mathcal{X}(\vartheta_\mathcal{X}, r_1)$. $\forall x_1, x_2 \in \overline{\mathcal{B}}_\mathcal{X}(\vartheta_\mathcal{X}, r_1)$, by mean value theorem, $\| \psi(x_1) - \psi(x_2) \| \leq \sup_{t_0 \in (0,1) \subset \mathbb{R}} \| \psi^{(1)}(t_0 x_1 + (1 - t_0)x_2)(x_1 - x_2) \| \leq \alpha \| x_1 - x_2 \|$, where the last inequality follows from Proposition 7.64.

$\forall \bar{x} \in \mathcal{B}_\mathcal{X}(\vartheta_\mathcal{X}, (1 - \alpha)r_1)$, define $\phi : \overline{\mathcal{B}}_\mathcal{X}(\vartheta_\mathcal{X}, r_1) \to \mathcal{X}$ by $\phi(x) = \bar{x} - \psi(x)$, $\forall x \in \overline{\mathcal{B}}_\mathcal{X}(\vartheta_\mathcal{X}, r_1)$. $\forall x \in \overline{\mathcal{B}}_\mathcal{X}(\vartheta_\mathcal{X}, r_1)$, $\| \phi(x) \| \leq \| \bar{x} \| + \| \psi(x) \| < (1 - \alpha)r_1 + \| \psi(x) - \psi(\vartheta_\mathcal{X}) \| \leq (1 - \alpha)r_1 + \alpha \| x \| \leq r_1$. Hence, $\phi : \overline{\mathcal{B}}_\mathcal{X}(\vartheta_\mathcal{X}, r_1) \to \mathcal{B}_\mathcal{X}(\vartheta_\mathcal{X}, r_1) \subseteq \overline{\mathcal{B}}_\mathcal{X}(\vartheta_\mathcal{X}, r_1)$. It is easy to see that ϕ is a contraction mapping with contraction index α. By Contraction Mapping Theorem, $\exists!\ \hat{x} \in \overline{\mathcal{B}}_{\mathcal{X}}(\vartheta_{\mathcal{X}}, r_1)$ such that $\hat{x} = \phi(\hat{x}) \in \mathcal{B}_{\mathcal{X}}(\vartheta_{\mathcal{X}}, r_1)$, which is equivalent to $\bar{x} = T(\hat{x})$. Let $\bar{U} := T_{\text{inv}}(\mathcal{B}_{\mathcal{X}}(\vartheta_{\mathcal{X}}, (1-\alpha)r_1)) \cap \mathcal{B}_{\mathcal{X}}(\vartheta_{\mathcal{X}}, r_1)$ and $\bar{V} := \mathcal{B}_{\mathcal{X}}(\vartheta_{\mathcal{X}}, (1-\alpha)r_1)$. Note that $\bar{U}$ and $\bar{V}$ are open sets in $\mathcal{X}$ since T is continuous by Proposition 9.7. Then, $T|_{\bar{U}} : \bar{U} \to \bar{V}$ is bijective. Since $T(\vartheta_{\mathcal{X}}) = \vartheta_{\mathcal{X}}$, then $\vartheta_{\mathcal{X}} \in \bar{U}$.

Hence, there exists an inverse mapping $T_i : \bar{V} \to \bar{U}$ of $T|_{\bar{U}}$. Let $U := \bar{U} + x_0$ and $V := F^{(1)}(x_0)(\bar{V}) + y_0$. Clearly, U and V are open sets in $\mathcal{X}$ and $\mathcal{Y}$, respectively. Note that $F|_U(x) = F^{(1)}(x_0)\ T|_{\bar{U}}(x - x_0) + y_0$, $\forall x \in U$. Then, $F|_U : U \to V$ is bijective, whose inverse function is $F_i : V \to U$. The inverse function F_i is given by $F_i(y) = T_i((F^{(1)}(x_0))^{-1}(y - y_0)) + x_0$, $\forall y \in V$. Hence, the statement (i) holds.

Next, we will show that $F_i : V \to U$ is differentiable. Note that $\forall x \in \bar{V}$, $x = T(T_i(x)) = T_i(x) + \psi(T_i(x))$ and $T_i(x) = x - \psi(T_i(x))$. $\forall x_1, x_2 \in \bar{V}$, we have $T_i(x_1), T_i(x_2) \in \bar{U} \subseteq \mathcal{B}_{\mathcal{X}}(\vartheta_{\mathcal{X}}, r_1)$ and

$$\begin{aligned}
\|T_i(x_1) - T_i(x_2)\| &= \|x_1 - x_2 - \psi(T_i(x_1)) + \psi(T_i(x_2))\| \\
&\le \|x_1 - x_2\| + \|\psi(T_i(x_1)) - \psi(T_i(x_2))\| \\
&\le \|x_1 - x_2\| + \alpha\|T_i(x_1) - T_i(x_2)\|
\end{aligned}$$

which further implies that $\|T_i(x_1) - T_i(x_2)\| \le \|x_1 - x_2\|/(1-\alpha)$. Therefore, T_i is continuous. By Propositions 3.12, 7.23, and 3.32, F_i is continuous. We need the following intermediate result.

Claim 9.57.1 $\forall x \in U$, let $y = F(x)$. Then, F_i is differentiable at $F(x)$ and $F_i^{(1)}(y) = (F^{(1)}(x))^{-1}$.

Proof of Claim Fix any $x \in U$ and let $y = F(x)$. Then, $x - x_0 \in \bar{U} \subseteq \mathcal{B}_{\mathcal{X}}(\vartheta_{\mathcal{X}}, r_1)$ and $\left\|\psi^{(1)}(x - x_0)\right\| \le \alpha$. Note that $T = \text{id}_{\mathcal{X}} + \psi$ and $T^{(1)}(x - x_0) = \text{id}_{\mathcal{X}} + \psi^{(1)}(x - x_0)$. By Proposition 9.55, $T^{(1)}(x - x_0)$ is bijective and $\left\|(T^{(1)}(x - x_0))^{-1}\right\| < \infty$. Note that $F^{(1)}(x) = F^{(1)}(x_0)T^{(1)}(x - x_0)$. Then, $F^{(1)}(x)$ is bijective and $c_1 := \left\|(F^{(1)}(x))^{-1}\right\| < \infty$.

$\forall \epsilon \in (0, \infty) \subset \mathbb{R}$ with $\epsilon c_1 < 1$, by the differentiability of F at x, $\exists \delta_1 \in (0, \infty) \subset \mathbb{R}$ such that $\forall h \in \mathcal{X}$ with $\|h\| < \delta_1$, we have $\left\|F(x + h) - F(x) - F^{(1)}(x)(h)\right\| \le \epsilon\|h\|$. By the continuity of F_i, $\exists \delta \in (0, \infty) \subset \mathbb{R}$ such that $\forall u \in \mathcal{Y}$ with $\|u\| < \delta$, we have $y + u \in V$ and $\|F_i(y + u) - F_i(y)\| < \delta_1$. $\forall u \in \mathcal{Y}$ with $\|u\| < \delta$, let $\beta := \left\|F_i(y + u) - F_i(y) - (F^{(1)}(x))^{-1}u\right\| \ge 0$. Let $h := F_i(y + u) - x \in \mathcal{X}$. Then, $\|h\| = \|F_i(y + u) - F_i(y)\| < \delta_1$. Note that

$$\begin{aligned}
\beta &= \left\|x + h - x - (F^{(1)}(x))^{-1}u\right\| = \left\|(F^{(1)}(x))^{-1}(u - F^{(1)}(x)h)\right\| \\
&\le \left\|(F^{(1)}(x))^{-1}\right\|\left\|u - F^{(1)}(x)h\right\| \\
&= c_1\left\|F(x + h) - F(x) - F^{(1)}(x)h\right\| \\
&\le c_1\epsilon\|h\| = \epsilon c_1\|F_i(y + u) - F_i(y)\|
\end{aligned}$$

$$\leq \epsilon c_1 (\| F_i(y+u) - F_i(y) - (F^{(1)}(x))^{-1} u\| + \|(F^{(1)}(x))^{-1} u\|)$$
$$\leq \epsilon c_1 (\beta + c_1 \|u\|)$$

Then, $\beta \leq \frac{\epsilon c_1^2}{1-\epsilon c_1} \|u\|$. Hence, F_i is differentiable at y and $F_i^{(1)}(y) = (F^{(1)}(x))^{-1}$. This completes the proof of the claim. □

Then, $\forall y \in V$, $F_i^{(1)}(y) = (F^{(1)}(F_i(y)))^{-1}$. By Propositions 9.56 and 3.12, the continuity of F_i, and the continuity of $F^{(1)}$ at x_0, we have $F_i^{(1)}$ is continuous at y_0. Then, the statement (ii) holds.

For (iii), we will use mathematical induction on k.

1° $k = 1$. The result holds by (ii).

2° Assume that the result holds for $k = \bar{k} \in \mathbb{N}$.

3° Consider the case $k = \bar{k} + 1$. By inductive assumption, F_i is $\bar{k}$-times differentiable. Clearly, $F^{(1)}$ is $\bar{k}$-times differentiable. By (ii) and Propositions 9.45 and 9.56, $F_i^{(1)}$ is $\bar{k}$-times differentiable. Then, F_i is $(\bar{k}+1)$-times differentiable. This completes the induction process and the proof of the statement (iii).

For (iv), we will use mathematical induction on k to show that the result holds when $k \in \mathbb{N}$.

1° $k = 1$. The result holds by (ii).

2° Assume that the result holds for $k = \bar{k} \in \mathbb{N}$.

3° Consider the case $k = \bar{k} + 1$. Clearly, $F^{(1)}$ is $\mathcal{C}_{\bar{k}}$ at x_0. By the inductive assumption, F_i is $\mathcal{C}_{\bar{k}}$ at y_0. By (ii) and Propositions 9.45 and 9.56, $F_i^{(1)}$ is $\mathcal{C}_{\bar{k}}$ at y_0. Hence, F_i is $\mathcal{C}_{\bar{k}+1}$ at y_0. This completes the induction process.

When $k = \infty$. $\exists \delta \in (0, \infty) \subset \mathbb{R}$ such that F is $\mathcal{C}_i$ at x, $\forall x \in \mathcal{B}_{\mathcal{X}}(x_0, \delta) \subseteq D$, $\forall i \in \mathbb{N}$. By taking $r_1 < \delta$ in the proof of (i), we have $U \subseteq \mathcal{B}_{\mathcal{X}}(x_0, \delta)$. Then, by the induction conclusion, F_i is $\mathcal{C}_i$ at y, $\forall y \in F(U) = V$, $\forall i \in \mathbb{N}$. Hence, F_i is $\mathcal{C}_\infty$ at y_0.

This completes the proof of the theorem. □

Theorem 9.58 (Implicit Function Theorem) *Let $\mathcal{X} := (X, \mathcal{O})$ be a topological space, $\mathcal{Y}$ and $\mathcal{Z}$ be Banach spaces over $\mathbb{K}$, $D \subseteq \mathcal{X} \times \mathcal{Y}$, and $F : D \to \mathcal{Z}$ be continuous. Assume that F is partial differentiable with respect to y and $\frac{\partial F}{\partial y}$ is continuous at $(x_0, y_0) \in D^\circ$, $F(x_0, y_0) = \vartheta_{\mathcal{Z}}$, and $\frac{\partial F}{\partial y}(x_0, y_0) \in \mathrm{B}(\mathcal{Y}, \mathcal{Z})$ is bijective. Then, the following statements hold:*

(i) *There exists an open set $U_0 \in \mathcal{O}$ with $x_0 \in U_0$ and $r_1 \in (0, \infty) \subset \mathbb{R}$ such that $U_0 \times \mathcal{B}_{\mathcal{Y}}(y_0, r_1) \subseteq D$ and $\forall x \in U_0$, $\exists! y \in \mathcal{B}_{\mathcal{Y}}(y_0, r_1)$ satisfying $F(x, y) = \vartheta_{\mathcal{Z}}$. This defines a function $\phi : U_0 \to \mathcal{B}_{\mathcal{Y}}(y_0, r_1)$ by $\phi(x) = y$, $\forall x \in U_0$. Then, ϕ is continuous.*

(ii) *$\forall (x, y) \in U_0 \times \mathcal{B}_{\mathcal{Y}}(y_0, r_1)$, $\frac{\partial F}{\partial y}(x, y)$ is bijective and $\left\| \left(\frac{\partial F}{\partial y}(x, y) \right)^{-1} \right\| < +\infty$.*

Proof By Open Mapping Theorem 7.103, $\left(\frac{\partial F}{\partial y}(x_0, y_0)\right)^{-1} \in B(\mathcal{Z}, \mathcal{Y})$. Define a mapping $\psi : D \to \mathcal{Y}$ by

$$\psi(x, y) = y - \left(\frac{\partial F}{\partial y}(x_0, y_0)\right)^{-1} F(x, y); \qquad \forall (x, y) \in D$$

Note that $\psi(x_0, y_0) = y_0$. Then, by Propositions 7.23, 3.12, 3.27, and 3.32, ψ is continuous. By the partial differentiability of F with respect to y, chain rule, and Propositions 9.41, 9.15, and 9.19, ψ is partial differentiable with respect to y and

$$\frac{\partial \psi}{\partial y}(x, y) = \mathrm{id}_{\mathcal{Y}} - \left(\frac{\partial F}{\partial y}(x_0, y_0)\right)^{-1} \frac{\partial F}{\partial y}(x, y); \qquad \forall (x, y) \in D$$

By the continuity of $\frac{\partial F}{\partial y}$ at (x_0, y_0), then $\frac{\partial \psi}{\partial y}$ is continuous at $(x_0, y_0) \in D^\circ$ and $\frac{\partial \psi}{\partial y}(x_0, y_0) = \vartheta_{B(\mathcal{Y},\mathcal{Y})}$. Fix any $\alpha \in (0, 1) \subset \mathbb{R}$. Then, $\exists U_1 \in \mathcal{O}$ with $x_0 \in U_1$ and $\exists r_1 \in (0, \infty) \subset \mathbb{R}$ such that $U_1 \times \overline{\mathcal{B}}_{\mathcal{Y}}(y_0, r_1) \subseteq D$ and $\left\|\frac{\partial \psi}{\partial y}(x, y)\right\| \leq \alpha$, $\forall (x, y) \in U_1 \times \overline{\mathcal{B}}_{\mathcal{Y}}(y_0, r_1)$. By the continuity of ψ, $\exists U_0 \in \mathcal{O}$ with $x_0 \in U_0 \subseteq U_1$ such that $\|\psi(x, y_0) - y_0\| = \|\psi(x, y_0) - \psi(x_0, y_0)\| < (1 - \alpha)r_1$, $\forall x \in U_0$.

Fix any $x \in U_0$, define mapping $\gamma_x : \overline{\mathcal{B}}_{\mathcal{Y}}(y_0, r_1) \to \mathcal{Y}$ by $\gamma_x(y) = \psi(x, y)$, $\forall y \in \overline{\mathcal{B}}_{\mathcal{Y}}(y_0, r_1)$. We will show that γ_x is a contraction mapping with contraction index α. $\forall y \in \overline{\mathcal{B}}_{\mathcal{Y}}(y_0, r_1)$, we have

$$\begin{aligned}
\|\gamma_x(y) - y_0\| &= \|\psi(x, y) - \psi(x_0, y_0)\| \\
&\leq \|\psi(x, y) - \psi(x, y_0)\| + \|\psi(x, y_0) - \psi(x_0, y_0)\| \\
&< \sup_{t \in (0,1) \subset \mathbb{R}} \left\|\frac{\partial \psi}{\partial y}(x, ty + (1 - t)y_0)(y - y_0)\right\| + (1 - \alpha)r_1 \\
&\leq \sup_{t \in (0,1) \subset \mathbb{R}} \left\|\frac{\partial \psi}{\partial y}(x, ty + (1 - t)y_0)\right\| \|y - y_0\| + (1 - \alpha)r_1 \\
&\leq \alpha r_1 + (1 - \alpha)r_1 = r_1
\end{aligned}$$

where the second inequality follows from mean value theorem 9.23 and the third inequality follows from Proposition 7.64. Then, $\gamma_x : \overline{\mathcal{B}}_{\mathcal{Y}}(y_0, r_1) \to \mathcal{B}_{\mathcal{Y}}(y_0, r_1) \subseteq \overline{\mathcal{B}}_{\mathcal{Y}}(y_0, r_1)$. $\forall y_1, y_2 \in \overline{\mathcal{B}}_{\mathcal{Y}}(y_0, r_1)$, we have

$$\begin{aligned}
\|\gamma_x(y_1) - \gamma_x(y_2)\| &= \|\psi(x, y_1) - \psi(x, y_2)\| \\
&\leq \sup_{t \in (0,1) \subset \mathbb{R}} \left\|\frac{\partial \psi}{\partial y}(x, ty_1 + (1 - t)y_2)(y_1 - y_2)\right\|
\end{aligned}$$

$$\le \sup_{t\in(0,1)\subset\mathbb{R}} \left\| \frac{\partial\psi}{\partial y}(x, ty_1 + (1-t)y_2) \right\| \|y_1 - y_2\|$$
$$\le \alpha \|y_1 - y_2\|$$

where the first inequality follows from mean value theorem 9.23 and the second inequality follows from Proposition 7.64. Hence, γ_x is a contraction mapping with contraction index α.

By Contraction Mapping Theorem, $\exists!\ y \in \overline{\mathcal{B}}_{\mathcal{Y}}(y_0, r_1)$ such that $y = \gamma_x(y) \in \mathcal{B}_{\mathcal{Y}}(y_0, r_1)$, $y = \lim_{n\in\mathbb{N}} \gamma_{x,n}(y_0)$, where $\gamma_{x,n}(y_0)$ is recursively defined by $\gamma_{x,1}(y_0) = y_0$ and $\gamma_{x,k+1}(y_0) = \gamma_x(\gamma_{x,k}(y_0))$, $\forall k \in \mathbb{N}$, and $\left\|\gamma_{x,n}(y_0) - y\right\| \le \frac{\alpha^{n-1}}{1-\alpha}\|\gamma_x(y_0) - y_0\| < r_1\alpha^{n-1}$, $\forall n \in \{2, 3, \ldots\}$. By $\gamma_x(y) = y$, we conclude that $F(x, y) = \vartheta_{\mathcal{Z}}$. Hence, $\forall x \in U_0$, $\exists!\ y \in \mathcal{B}_{\mathcal{Y}}(y_0, r_1)$ such that $F(x, y) = \vartheta_{\mathcal{Z}}$, since $F(x, y) = \vartheta_{\mathcal{Z}} \Leftrightarrow y = \gamma_x(y)$. Then, we may define $\phi : U_0 \to \mathcal{B}_{\mathcal{Y}}(y_0, r_1)$ by $\phi(x) = y = \lim_{n\in\mathbb{N}} \gamma_{x,n}(y_0)$, $\forall x \in U_0$. Hence, $F(x, \phi(x)) = \vartheta_{\mathcal{Z}}$, $\forall x \in U_0$.

Next, we show that ϕ is continuous. Fix any $\bar{x} \in U_0$. $\forall \epsilon \in (0, \infty) \subset \mathbb{R}$, $\exists n_0 \in \mathbb{N}$ with $n_0 > 1$ such that $\alpha^{n_0-1}r_1 < \epsilon/3$. By the continuity of ψ, we have that $\gamma_{x,n_0}(y_0)$ is continuous with respect to x, that is, $\exists \bar{U} \in \mathcal{O}$ with $\bar{x} \in \bar{U} \subseteq U_0$ such that $\forall x_1 \in \bar{U}$, we have $\left\|\gamma_{x_1,n_0}(y_0) - \gamma_{\bar{x},n_0}(y_0)\right\| < \epsilon/3$. Then,

$$\begin{aligned}\|\phi(x_1) - \phi(\bar{x})\| &\le \left\|\phi(x_1) - \gamma_{x_1,n_0}(y_0)\right\| + \left\|\gamma_{x_1,n_0}(y_0) - \gamma_{\bar{x},n_0}(y_0)\right\| \\ &\quad + \left\|\gamma_{\bar{x},n_0}(y_0) - \phi(\bar{x})\right\| \le r_1\alpha^{n_0-1} + \epsilon/3 + r_1\alpha^{n_0-1} < \epsilon\end{aligned}$$

Hence, ϕ is continuous. Thus, the statement (i) is proved.

(ii) Note that, $\forall (x, y) \in U_1 \times \overline{\mathcal{B}}_{\mathcal{Y}}(y_0, r_1)$, we have $\left\|\frac{\partial\psi}{\partial y}(x, y)\right\| \le \alpha$. By Proposition 9.55, $\mathrm{id}_{\mathcal{Y}} - \frac{\partial\psi}{\partial y}(x, y)$ is bijective with continuous inverse. Note that $\mathrm{id}_{\mathcal{Y}} - \frac{\partial\psi}{\partial y}(x, y) = \left(\frac{\partial F}{\partial y}(x_0, y_0)\right)^{-1}\frac{\partial F}{\partial y}(x, y)$. Therefore, $\frac{\partial F}{\partial y}(x, y)$ is bijective with continuous inverse.

This completes the proof of the theorem. □

Theorem 9.59 (Implicit Function Theorem) *Let $\mathcal{X}$ be normed linear space over $\mathbb{K}$, $\mathcal{Y}$ and $\mathcal{Z}$ be Banach spaces over $\mathbb{K}$, $D \subseteq \mathcal{X} \times \mathcal{Y}$, and $F : D \to \mathcal{Z}$ be continuous. Assume that F is partial differentiable with respect to y and $\frac{\partial F}{\partial y}$ is continuous at $(x_0, y_0) \in D^\circ$, $F(x_0, y_0) = \vartheta_{\mathcal{Z}}$, and $\frac{\partial F}{\partial y}(x_0, y_0) \in \mathrm{B}(\mathcal{Y}, \mathcal{Z})$ is bijective. Then, the following statements hold:*

(i) *There exist $r_0, r_1 \in (0, \infty) \subset \mathbb{R}$ such that $\mathcal{B}_{\mathcal{X}}(x_0, r_0) \times \mathcal{B}_{\mathcal{Y}}(y_0, r_1) \subseteq D$ and $\forall x \in \mathcal{B}_{\mathcal{X}}(x_0, r_0)$, $\exists!\ y \in \mathcal{B}_{\mathcal{Y}}(y_0, r_1)$ satisfying $F(x, y) = \vartheta_{\mathcal{Z}}$. This defines a function $\phi : \mathcal{B}_{\mathcal{X}}(x_0, r_0) \to \mathcal{B}_{\mathcal{Y}}(y_0, r_1)$ by $\phi(x) = y$, $\forall x \in \mathcal{B}_{\mathcal{X}}(x_0, r_0)$. Then, ϕ is continuous.*

(ii) *If F is Fréchet differentiable at $(x, \phi(x)) \in \mathcal{B}_{\mathcal{X}}(x_0, r_0) \times \mathcal{B}_{\mathcal{Y}}(y_0, r_1)$ for some $x \in \mathcal{B}_{\mathcal{X}}(x_0, r_0)$, then ϕ is Fréchet differentiable at x and*

$$\phi^{(1)}(x) = -\left(\frac{\partial F}{\partial y}(x, \phi(x))\right)^{-1} \frac{\partial F}{\partial x}(x, \phi(x))$$

(iii) *Let $n \in \mathbb{N}$. If F is n-times Fréchet differentiable, then ϕ is n-times Fréchet differentiable.*

(iv) *Let $n \in \mathbb{N} \cup \{\infty\}$ and $\bar{x} \in \mathcal{B}_{\mathcal{X}}(x_0, r_0)$. If F is $\mathcal{C}_n$ at $(\bar{x}, \phi(\bar{x}))$, then ϕ is $\mathcal{C}_n$ at $\bar{x}$.*

Proof By Implicit Function Theorem 9.58, the statement (i) holds. Furthermore, $\forall (x, y) \in \mathcal{B}_{\mathcal{X}}(x_0, r_0) \times \mathcal{B}_{\mathcal{Y}}(y_0, r_1)$, $\frac{\partial F}{\partial y}(x, y)$ is bijective and $\left\| \left(\frac{\partial F}{\partial y}(x, y)\right)^{-1} \right\| < +\infty$.

(ii) Fix some $x \in \mathcal{B}_{\mathcal{X}}(x_0, r_0)$ such that F is differentiable at $(x, \phi(x))$. Let $y := \phi(x) \in \mathcal{B}_{\mathcal{Y}}(y_0, r_1)$. By Proposition 9.9, $\left[\frac{\partial F}{\partial x}(x, y) \; \frac{\partial F}{\partial y}(x, y) \right] = F^{(1)}(x, y)$. Let $L := -\left(\frac{\partial F}{\partial y}(x, \phi(x))\right)^{-1} \frac{\partial F}{\partial x}(x, \phi(x)) \in \mathrm{B}(\mathcal{X}, \mathcal{Y})$ and $c_1 := \left\| \left(\frac{\partial F}{\partial y}(x, y)\right)^{-1} \right\| < +\infty$. $\forall \epsilon \in (0, \infty) \subset \mathbb{R}$ with $\epsilon c_1 < 1$, by the differentiability of F at (x, y), $\exists \delta_1 \in (0, \infty) \subset \mathbb{R}$ such that $\forall (h, k) \in \mathcal{B}_{\mathcal{X} \times \mathcal{Y}}((\vartheta_{\mathcal{X}}, \vartheta_{\mathcal{Y}}), \delta_1)$, we have $(x + h, y + k) \in D$ and $\left\| F(x + h, y + k) - F(x, y) - \frac{\partial F}{\partial x}(x, y)h - \frac{\partial F}{\partial y}(x, y)k \right\| \leq \epsilon \|(h, k)\|$. By the continuity of ϕ at x, $\exists \delta \in (0, \min\{r_0 - \|x - x_0\|, \delta_1/\sqrt{2}\}] \subset \mathbb{R}$ such that $\forall h \in \mathcal{B}_{\mathcal{X}}(\vartheta_{\mathcal{X}}, \delta)$, we have $\|\phi(x + h) - \phi(x)\| = \|\phi(x + h) - y\| < \delta_1/\sqrt{2}$.

$\forall h \in \mathcal{B}_{\mathcal{X}}(\vartheta_{\mathcal{X}}, \delta)$, let $\beta := \|\phi(x + h) - \phi(x) - Lh\| \geq 0$. Then, $\|(h, \phi(x + h) - y)\| < \delta_1$. Note that, by Proposition 7.64,

$$\begin{aligned}
\beta &= \left\| \left(\frac{\partial F}{\partial y}(x, y)\right)^{-1} \left(\frac{\partial F}{\partial y}(x, y)(\phi(x + h) - \phi(x)) + \frac{\partial F}{\partial x}(x, y)h\right) \right\| \\
&\leq c_1 \Big\| F(x + h, \phi(x + h)) - F(x, y) - \frac{\partial F}{\partial x}(x, y)h \\
&\quad - \frac{\partial F}{\partial y}(x, y)(\phi(x + h) - y) \Big\| \\
&\leq c_1 \epsilon \|(h, \phi(x + h) - y)\| \leq c_1 \epsilon \left(\|h\| + \|\phi(x + h) - y\|\right) \\
&\leq c_1 \epsilon \left(\|h\| + \|\phi(x + h) - \phi(x) - Lh\| + \|Lh\|\right) \\
&\leq c_1 \epsilon \left(\beta + (1 + \|L\|)\|h\|\right)
\end{aligned}$$

Then, we have $\beta \leq \frac{c_1 \epsilon (1 + \|L\|)}{1 - c_1 \epsilon} \|h\|$. Hence, $\phi^{(1)}(x) = L$. Then, the statement (ii) holds.

For (iii), we will use mathematical induction on n to prove this result.

1° $n = 1$. The result follows from (ii).
2° Assume that the result holds for $n = \bar{n} \in \mathbb{N}$.
3° Consider the case $n = \bar{n} + 1$. By (ii), we have ϕ is Fréchet differentiable and $\phi^{(1)}(x) = -\left(\frac{\partial F}{\partial y}(x, \phi(x))\right)^{-1} \frac{\partial F}{\partial x}(x, \phi(x))$, $\forall x \in \mathcal{B}_{\mathcal{X}}(x_0, r_0)$. By inductive assumption, ϕ is $\bar{n}$-times differentiable. By Proposition 9.46, $\frac{\partial F}{\partial x}$ and $\frac{\partial F}{\partial y}$ are $\bar{n}$-times differentiable. By Propositions 9.45, 9.56, 9.44, and 9.42, $\phi^{(1)}$ is $\bar{n}$-times differentiable. Then, ϕ is $(\bar{n} + 1)$-times differentiable. This completes the induction process.

For (iv), let $\bar{y} = \phi(\bar{x})$. We will first use mathematical induction on n to prove the result for $n \in \mathbb{N}$.
1° $n = 1$. By F being $\mathcal{C}_1$ at $(\bar{x}, \bar{y})$, then $\exists \bar{r} \in (0, \infty) \subset \mathbb{R}$ such that F is differentiable at (x, y), $\forall (x, y) \in \bar{D} := \mathcal{B}_{\mathcal{X}\times\mathcal{Y}}((\bar{x}, \bar{y}), \bar{r}) \subseteq D$. By Propositions 9.34 and 9.9, $F|_{\bar{D}}^{(1)}(x, y) = \left[\frac{\partial F}{\partial x}(x, y)\ \frac{\partial F}{\partial y}(x, y)\right]$, $\forall (x, y) \in \bar{D}$. By Proposition 9.46, $\left.\frac{\partial F}{\partial x}\right|_{\bar{D}}$ and $\left.\frac{\partial F}{\partial y}\right|_{\bar{D}}$ are continuous at $(\bar{x}, \bar{y})$. Since $\bar{D}$ is open, then $\frac{\partial F}{\partial x}$ and $\frac{\partial F}{\partial y}$ are continuous at $(\bar{x}, \bar{y})$. By the continuity of ϕ, $\exists \bar{\delta} \in (0, \min\{r_0 - \|\bar{x} - x_0\|, \bar{r}\}] \subset \mathbb{R}$ such that F is differentiable at $(x, \phi(x))$, $\forall x \in \mathcal{B}_{\mathcal{X}}(\bar{x}, \bar{\delta})$. By (ii), ϕ is differentiable at x and $\phi^{(1)}(x) = -\left(\frac{\partial F}{\partial y}(x, \phi(x))\right)^{-1} \frac{\partial F}{\partial x}(x, \phi(x))$, $\forall x \in \mathcal{B}_{\mathcal{X}}(\bar{x}, \bar{\delta})$. By Propositions 3.12, 3.32, 9.56, 9.42, and 9.7, $\phi^{(1)}$ is continuous at $\bar{x}$. Hence, ϕ is $\mathcal{C}_1$ at $\bar{x}$.
2° Assume that the result holds for $n = \bar{n} \in \mathbb{N}$.
3° Consider the case $n = \bar{n}+1$. By F being $\mathcal{C}_{\bar{n}+1}$ at $(\bar{x}, \bar{y})$, then $\exists \bar{r} \in (0, \infty) \subset \mathbb{R}$ such that F is differentiable at (x, y), $\forall (x, y) \in \bar{D} := \mathcal{B}_{\mathcal{X}\times\mathcal{Y}}((\bar{x}, \bar{y}), \bar{r}) \subseteq D$. By Propositions 9.34 and 9.9, $F|_{\bar{D}}^{(1)}(x, y) = \left[\frac{\partial F}{\partial x}(x, y)\ \frac{\partial F}{\partial y}(x, y)\right]$, $\forall (x, y) \in \bar{D}$. By Proposition 9.46, $\left.\frac{\partial F}{\partial x}\right|_{\bar{D}}$ and $\left.\frac{\partial F}{\partial y}\right|_{\bar{D}}$ are $\mathcal{C}_{\bar{n}}$ at $(\bar{x}, \bar{y})$. Since $\bar{D}$ is open, then, by Proposition 9.34, $\frac{\partial F}{\partial x}$ and $\frac{\partial F}{\partial y}$ are $\mathcal{C}_{\bar{n}}$ at $(\bar{x}, \bar{y})$. By the inductive assumption, ϕ is $\mathcal{C}_{\bar{n}}$ at $\bar{x}$. By the continuity of ϕ, $\exists \bar{\delta} \in (0, \min\{r_0 - \|\bar{x} - x_0\|, \bar{r}\}] \subset \mathbb{R}$ such that F is differentiable at $(x, \phi(x))$, $\forall x \in \mathcal{B}_{\mathcal{X}}(\bar{x}, \bar{\delta})$. By (ii), ϕ is differentiable at x and $\phi^{(1)}(x) = -\left(\frac{\partial F}{\partial y}(x, \phi(x))\right)^{-1} \frac{\partial F}{\partial x}(x, \phi(x))$, $\forall x \in \mathcal{B}_{\mathcal{X}}(\bar{x}, \bar{\delta})$. By Propositions 9.45, 9.44, 9.56, and 9.42, $\phi^{(1)}$ is $\mathcal{C}_{\bar{n}}$ at $\bar{x}$. Hence, ϕ is $\mathcal{C}_{\bar{n}+1}$ at $\bar{x}$.

This completes the induction process. Hence, (iv) holds when $n \in \mathbb{N}$.

When $n = \infty$. There exists $\delta \in (0, \infty) \subset \mathbb{R}$, such that F is $\mathcal{C}_i$ at $(\hat{x}, \hat{y})$, $\forall (\hat{x}, \hat{y}) \in \mathcal{B}_{\mathcal{X}\times\mathcal{Y}}((\bar{x}, \bar{y}), \delta) \subseteq \mathcal{B}_{\mathcal{X}}(x_0, r_0) \times \mathcal{B}_{\mathcal{Y}}(y_0, r_0)$, $\forall i \in \mathbb{N}$. By the induction result, we have ϕ is $\mathcal{C}_i$ at $\hat{x}$, $\forall \hat{x} \in \mathcal{B}_{\mathcal{X}}\left(\bar{x}, \delta/\sqrt{2}\right) \cap \phi_{\text{inv}}(\mathcal{B}_{\mathcal{Y}}\left(\bar{y}, \delta/\sqrt{2}\right))$. Hence, ϕ is $\mathcal{C}_\infty$ at $\bar{x}$. Hence, (iv) holds.

This completes the proof of the proposition. □

Proposition 9.60 *Let $\mathcal{X} := (X, \mathcal{O})$ be a topological space, $\mathcal{Y}$ and $\mathcal{Z}$ be normed linear spaces over $\mathbb{K}$, $D \subseteq \mathcal{X} \times \mathcal{Y}$, $f : D \to \mathcal{Z}$ be partial differentiable with respect to y, and $\frac{\partial f}{\partial y}(x, y) = \vartheta_{\mathrm{B}(\mathcal{Y},\mathcal{Z})}$, $\forall (x, y) \in D$. Let $\bar{D} := \pi_{\mathcal{X}}(D)$, where $\pi_{\mathcal{X}}$*

is the projection function of $\mathcal{X} \times \mathfrak{Y}$ to $\mathcal{X}$. Assume that $\forall x \in \bar{D}$, the set $D_x := \{y \in \mathfrak{Y} \mid (x, y) \in D\} \subseteq \mathfrak{Y}$ is convex. Then, there exists a function $\phi : \bar{D} \to \mathfrak{Z}$ such that $f(x, y) = \phi(x)$, $\forall (x, y) \in D$. Furthermore, the following statements hold:

(i) *If f is continuous at $(x_0, y_0) \in D^\circ$, then ϕ is continuous at x_0.*
(ii) *If $\mathcal{X}$ is a normed linear space $\mathfrak{X}$ over $\mathbb{K}$ and f is $\mathcal{C}_k$ at $(x_0, y_0) \in D^\circ$, where $k \in \mathbb{N} \cup \{\infty\}$, then ϕ is $\mathcal{C}_k$ at x_0.*

Proof $\forall x \in \bar{D} = \pi_{\mathcal{X}}(D)$, $D_x \neq \emptyset$. By axiom of choice, $\exists g : \bar{D} \to \mathfrak{Y}$ such that $g(x) \in D_x$, $\forall x \in \bar{D}$. Define $\phi : \bar{D} \to \mathfrak{X}$ by $\phi(x) = f(x, g(x))$, $\forall x \in \bar{D}$. $\forall (x, y) \in D$, we have $x \in \bar{D}$ and $y, g(x) \in D_x$. By the convexity of D_x, the line segment connecting y and $g(x)$ is contained in D_x. By mean value theorem 9.23, $\exists t_0 \in (0, 1) \subset \mathbb{R}$ such that $\|f(x, y) - \phi(x)\| = \|f(x, y) - f(x, g(x))\| \leq \left\| \frac{\partial f}{\partial y}(x, t_0 y + (1 - t_0) g(x))(y - g(x)) \right\| = 0$. Hence, $f(x, y) = \phi(x)$.

(i) Let f be continuous at $(x_0, y_0) \in D^\circ$. Then, $\exists U \in \mathcal{O}$ with $x_0 \in U$ and $\exists \delta \in (0, \infty) \subset \mathbb{R}$ such that $U \times \mathcal{B}_{\mathfrak{Y}}(y_0, \delta) \subseteq D$. $\forall x \in U$, we have $\phi(x) = f(x, y_0)$. Hence, ϕ is continuous at x_0.

(ii) Let $\mathcal{X}$ be a normed linear space $\mathfrak{X}$ over $\mathbb{K}$ and f be $\mathcal{C}_k$ at $(x_0, y_0) \in D^\circ$, where $k \in \mathbb{N} \cup \{\infty\}$. Then, $\exists \delta_x, \delta_y \in (0, \infty) \subset \mathbb{R}$ such that $\mathcal{B}_{\mathfrak{X}}(x_0, \delta_x) \times \mathcal{B}_{\mathfrak{Y}}(y_0, \delta_y) \subseteq D$. $\forall x \in \mathcal{B}_{\mathfrak{X}}(x_0, \delta_x)$, we have $\phi(x) = f(x, y_0)$. By Proposition 9.45, $\phi|_{\mathcal{B}_{\mathfrak{X}}(x_0, \delta_x)}$ is $\mathcal{C}_k$ at x_0. By Proposition 9.34, ϕ is $\mathcal{C}_k$ at x_0.

This completes the proof of the proposition. □

9.6 Global Inverse Function Theorem

Definition 9.61 Let $\mathcal{X}$, $\mathcal{Y}$, and $\mathcal{Z}$ be topological spaces, $F : \mathcal{X} \to \mathcal{Y}$, and $\sigma : \mathcal{Z} \to \mathcal{Y}$. We will say $\theta : \mathcal{Z} \to \mathcal{X}$ *inverts F along σ* if $\sigma = F \circ \theta$. ◇

Lemma 9.62 *Let $\mathcal{X}$ and $\mathcal{Y}$ be Hausdorff topological spaces, $F : \mathcal{X} \to \mathcal{Y}$ be continuous and countably proper, $x_0 \in \mathcal{X}$, and $y_0 := F(x_0) \in \mathcal{Y}$. Assume that $\forall x \in \mathcal{X}$, $\exists U \in \mathcal{O}_{\mathcal{X}}$ with $x \in U$ and $\exists V \in \mathcal{O}_{\mathcal{Y}}$ with $F(x) \in V$ such that $F|_U : U \to V$ is a homeomorphism. Then, given any continuous mapping $\sigma : [a, b] \to \mathcal{Y}$ with $\sigma(t_0) = y_0$, where $a, t_0, b \in \mathbb{R}$ and $a \leq t_0 \leq b$, there exists a unique continuous mapping $\theta : [a, b] \to \mathcal{X}$ with $\theta(t_0) = x_0$ that inverts F along σ.*

Proof We will distinguish three exhaustive and mutually exclusive cases: Case 1: $a = b = t_0$; Case 2: $a = t_0 < b$; Case 3: $a < t_0 \leq b$.

Case 1: $a = b = t_0$. Clearly, θ exists and is unique. This case is proved.

Case 2: $a = t_0 < b$. "Uniqueness" Let $\theta_1 : [a, b] \to \mathcal{X}$ and $\theta_2 : [a, b] \to \mathcal{X}$ be continuous mappings that invert F along σ with $\theta_1(a) = \theta_2(a) = x_0$. Let $S := \{s \in [a, b] \subset \mathbb{R} \mid \theta_1(t) = \theta_2(t)\ \forall t \in [a, s] \subset \mathbb{R}\}$ and $\xi := \sup S$. Clearly, $a \in S$ and $a \leq \xi \leq b$. It is easy to show that $\theta_1(t) = \theta_2(t)$, $\forall t \in \mathbb{R}$ with $a \leq t < \xi$. There exists $(t_n)_{n=1}^{\infty} \subseteq S$ such that $\lim_{n \in \mathbb{N}} t_n = \xi$. By Proposition 3.66 and the continuity of θ_1 and θ_2, we have $\theta_1(\xi) = \lim_{n\in\mathbb{N}} \theta_1(t_n) = \lim_{n\in\mathbb{N}} \theta_2(t_n) = \theta_2(t_n)$, where the limit operator makes sense since $\mathcal{X}$ is Hausdorff. Then, $\xi \in S$.

We will next show that $\xi = b$ by an argument of contradiction. Suppose that $\xi < b$. Let $x := \theta_1(\xi) = \theta_2(\xi)$ and $y := F(x)$. Then, $\exists U \in \mathcal{O}_{\mathcal{X}}$ with $x \in U$ and $\exists V \in \mathcal{O}_{\mathcal{Y}}$ with $y \in V$ such that $F|_U : U \to V$ is a homeomorphism. By the continuity of θ_1 and θ_2, $\exists \bar{\xi} \in (\xi, b]$ such that $\theta_1(t), \theta_2(t) \in U$, $\forall t \in [\xi, \bar{\xi}] \subset \mathbb{R}$. Then, $\sigma(t) = F(\theta_1(t)) = F(\theta_2(t)) \in V$, $\forall t \in [\xi, \bar{\xi}] \subset \mathbb{R}$. Since $F|_U : U \to V$ is a homeomorphism, then $\theta_1(t) = \theta_2(t)$, $\forall t \in [\xi, \bar{\xi}] \subset \mathbb{R}$. Then, $\bar{\xi} \in S$ and $\xi < \bar{\xi} \leq \sup S = \xi$. This is a contradiction. Therefore, we must have $\xi = b$.

Therefore, $\theta_1(t) = \theta_2(t)$, $\forall t \in [a, \xi] = [a, b] \subset \mathbb{R}$, since $\xi \in S$. This shows that $\theta_1 = \theta_2$. Hence, if θ exists, then it must be unique.

"Existence" Let $S := \{s \in [a, b] \subset \mathbb{R} \mid$ there exists a continuous $\theta : [a, s] \to \mathcal{X}$ that inverts F along $\sigma|_{[a,s]}$ with $\theta(a) = x_0\} \subset \mathbb{R}$ and $\xi := \sup S$. Clearly, $a \in S$ and $a \leq \xi \leq b$.

We will show that $\xi \in S$ by an argument of contradiction. Suppose $\xi \notin S$. Then, $a < \xi \leq b$ and $\exists (t_n)_{n=1}^{\infty} \subseteq S$, which is nondecreasing, such that $\lim_{n\in\mathbb{N}} t_n = \xi$. $\forall n \in \mathbb{N}$, there exists a continuous $\theta_n : [a, t_n] \to \mathcal{X}$ that inverts F along $\sigma|_{[a,t_n]}$ with $\theta_n(a) = x_0$. By the uniqueness property that we have shown, we have $\theta_n = \theta_{n+1}|_{[a,t_n]}$. Hence, we may define $\theta : [a, \xi) \to \mathcal{X}$ such that $\theta(t) = \theta_n(t)$, $\forall t \in [a, t_n] \subset \mathbb{R}$, $\forall n \in \mathbb{N}$. Then, θ is continuous and inverts F along $\sigma|_{[a,\xi)}$ with $\theta(a) = x_0$. Note that $\sigma(t_n) = F(\theta(t_n))$, $\forall n \in \mathbb{N}$. By continuity of σ and Proposition 3.66, we have $\lim_{n\in\mathbb{N}} \sigma(t_n) = \sigma(\xi) \in \mathcal{Y}$, where the limit operator makes sense since $\mathcal{Y}$ is Hausdorff. Then, $(\theta(t_n))_{n=1}^{\infty} \subseteq F_{\text{inv}}(M)$, where $M := \{\sigma(t_n) \in \mathcal{Y} \mid n \in \mathbb{N}\} \cup \{\sigma(\xi)\} \subseteq \mathcal{Y}$. Clearly, M is compact in $\mathcal{Y}$. Since F is countably proper, then $F_{\text{inv}}(M)$ is countably compact. By Proposition 5.26, $F_{\text{inv}}(M)$ has the Bolzano–Weierstrass property, and $(\theta(t_n))_{n=1}^{\infty}$ admits a cluster point $x \in F_{\text{inv}}(M)$. By the continuity of F and Proposition 3.66, $(F(\theta(t_n)))_{n=1}^{\infty} = (\sigma(t_n))_{n=1}^{\infty}$ admits a cluster point $F(x)$. Since $\mathcal{Y}$ is Hausdorff, then $F(x) = \sigma(\xi) =: y$, by Proposition 3.65.

By the assumption of the lemma, $\exists U \in \mathcal{O}_{\mathcal{X}}$ with $x \in U$ and $\exists V \in \mathcal{O}_{\mathcal{Y}}$ with $y \in V$ such that $F|_U : U \to V$ is a homeomorphism. Since σ is continuous, then $\sigma_{\text{inv}}(V)$ is open in $[a, b] \subset \mathbb{R}$. Since $\xi \in \sigma_{\text{inv}}(V)$, then $\exists \hat{\xi} \in [a, \xi) \subset \mathbb{R}$ such that $[\hat{\xi}, \xi] \subseteq \sigma_{\text{inv}}(V)$. By $\lim_{n\in\mathbb{N}} t_n = \xi$, $\exists N \in \mathbb{N}$ such that $\forall n \in \mathbb{N}$ with $n \geq N$, $t_n \in [\hat{\xi}, \xi] \subset \mathbb{R}$. Since $(\theta(t_n))_{n=1}^{\infty}$ admits a cluster point $x \in U$, then $\exists n_0 \in \mathbb{N}$ with $n_0 \geq N$ such that $\theta(t_{n_0}) \in U$. Clearly, $\sigma([t_{n_0}, \xi]) \subseteq V$. Define $\theta_1 := (F|_U)_{\text{inv}} \circ \sigma|_{[t_{n_0},\xi]} : [t_{n_0}, \xi] \to U$. Clearly, θ_1 is continuous and inverts F along $\sigma|_{[t_{n_0},\xi]}$ with $\theta_1(t_{n_0}) = \theta(t_{n_0})$. Define $\bar{\theta} : [a, \xi] \to \mathcal{X}$ by $\bar{\theta}(t) = \theta(t)$, $\forall t \in [a, t_{n_0}] \subset \mathbb{R}$, $\bar{\theta}(t) = \theta_1(t)$, $\forall t \in [t_{n_0}, \xi]$. By Theorem 3.11, $\bar{\theta}$ is continuous. Clearly, $\bar{\theta}$ inverts F along $\sigma|_{[a,\xi]}$ with $\bar{\theta}(a) = \theta(a) = x_0$. Hence, $\xi \in S$. This is a contradiction. Therefore, we must have $\xi \in S$.

Next, we will show that $\xi = b$ by an argument of contradiction. Suppose $\xi < b$. Since $\xi \in S$, then $\exists$ a continuous function $\theta : [a, \xi] \to \mathcal{X}$ that inverts F along $\sigma|_{[a,\xi]}$ with $\theta(a) = x_0$. Let $x := \theta(\xi) \in \mathcal{X}$ and $y := F(x) = \sigma(\xi) \in \mathcal{Y}$. Then, $\exists U \in \mathcal{O}_{\mathcal{X}}$ with $x \in U$ and $\exists V \in \mathcal{O}_{\mathcal{Y}}$ with $y \in V$ such that $F|_U : U \to V$ is a homeomorphism. By the continuity of σ, $\exists \bar{\xi} \in (\xi, b] \subset \mathbb{R}$ such that $\sigma(t) \in V$,

$\forall t \in [\xi, \bar{\xi}] \subset \mathbb{R}$. Define $\bar{\theta} : [a, \bar{\xi}] \to \mathcal{X}$ by $\bar{\theta}(t) = \theta(t)$, $\forall t \in [a, \xi]$, $\bar{\theta}(t) = (F|_U)_{\text{inv}}(\sigma(t))$, $\forall t \in [\xi, \bar{\xi}] \subset \mathbb{R}$. By Theorem 3.11, $\bar{\theta}$ is continuous. It is clear that $\bar{\theta}$ inverts F along $\sigma|_{[a,\bar{\xi}]}$ with $\bar{\theta}(a) = x_0$. Then, $\bar{\xi} \in S$. This implies that $\xi < \bar{\xi} \leq \sup S = \xi$, which is a contradiction. Hence, $\xi = b \in S$.

This case is proved.

Case 3: $a < t_0 \leq b$. By Cases 1 and 2, $\exists!\, \theta_1 : [t_0, b] \to \mathcal{X}$ that is continuous and inverts F along $\sigma|_{[t_0,b]}$ with $\theta_1(t_0) = x_0$. Define $\bar{\sigma} : [t_0, 2t_0 - a] \to \mathcal{Y}$ by $\bar{\sigma}(t) = \sigma(2t_0 - t)$, $\forall t \in [t_0, 2t_0 - a] \subset \mathbb{R}$. By Proposition 3.12, $\bar{\sigma}$ is continuous and $\bar{\sigma}(t_0) = y_0$. By Case 2, $\exists!\, \theta_2 : [t_0, 2t_0 - a] \to \mathcal{X}$ that is continuous and inverts F along $\bar{\sigma}$ with $\theta_2(t_0) = x_0$. Define $\bar{\theta} : [a, b] \to \mathcal{X}$ by $\bar{\theta}(t) = \theta_1(t)$, $\forall t \in [t_0, b] \subset \mathbb{R}$, $\bar{\theta}(t) = \theta_2(2t_0 - t)$, $\forall t \in [a, t_0] \subset \mathbb{R}$. By Theorem 3.11, $\bar{\theta}$ is continuous and inverts F along σ with $\bar{\theta}(t_0) = x_0$. The uniqueness of $\bar{\theta}$ follows from the uniqueness of θ_1 and θ_2. This case is proved.

This completes the proof of the lemma. □

Lemma 9.63 *Let $\mathcal{X}$ and $\mathcal{Y}$ be Hausdorff topological spaces, $F : \mathcal{X} \to \mathcal{Y}$ be continuous and countably proper, $x_0 \in \mathcal{X}$, and $y_0 = F(x_0) \in \mathcal{Y}$. Assume that $\forall x \in \mathcal{X}$, $\exists U \in \mathcal{O}_{\mathcal{X}}$ with $x \in U$ and $\exists V \in \mathcal{O}_{\mathcal{Y}}$ with $F(x) \in V$ such that $F|_U : U \to V$ is a homeomorphism. Then, given any continuous mapping $\sigma : [a, b] \times [c, d] \to \mathcal{Y}$ with $\sigma(a, c) = y_0$, where $a, b, c, d \in \mathbb{R}$, $a \leq b$, and $c \leq d$, there exists a unique continuous mapping $\theta : [a, b] \times [c, d] \to \mathcal{X}$ with $\theta(a, c) = x_0$ that inverts F along σ.*

Proof We will distinguish two exhaustive and mutually exclusive cases: Case 1: $a = b$; Case 2: $a < b$. Case 1: $a = b$. The result holds by Lemma 9.62.

Case 2: $a < b$. "Uniqueness" Let $\theta_1 : [a, b] \times [c, d] \to \mathcal{X}$ and $\theta_2 : [a, b] \times [c, d] \to \mathcal{X}$ be continuous mappings that invert F along σ with $\theta_1(a, c) = \theta_2(a, c) = x_0$. Fix any $(t, r) \in [a, b] \times [c, d] \subset \mathbb{R}^2$. Let $\bar{\sigma} : [0, 1] \to \mathcal{Y}$, $\bar{\theta}_1 : [0, 1] \to \mathcal{X}$, and $\bar{\theta}_2 : [0, 1] \to \mathcal{X}$ be defined by, $\forall \lambda \in [0, 1] \subset \mathbb{R}$,

$$\bar{\sigma}(\lambda) = \sigma(\lambda t + (1 - \lambda)a, \lambda r + (1 - \lambda)c)$$

$$\bar{\theta}_1(\lambda) = \theta_1(\lambda t + (1 - \lambda)a, \lambda r + (1 - \lambda)c)$$

$$\bar{\theta}_2(\lambda) = \theta_2(\lambda t + (1 - \lambda)a, \lambda r + (1 - \lambda)c)$$

Since $\sigma = F \circ \theta_1$ and $\sigma = F \circ \theta_2$, then $\bar{\sigma} = F \circ \bar{\theta}_1$ and $\bar{\sigma} = F \circ \bar{\theta}_2$. By Proposition 3.12, $\bar{\sigma}$, $\bar{\theta}_1$, and $\bar{\theta}_2$ are continuous. This implies that $\bar{\theta}_1$ inverts F along $\bar{\sigma}$ with $\bar{\theta}_1(0) = \theta_1(a, c) = x_0$ and $\bar{\theta}_2$ inverts F along $\bar{\sigma}$ with $\bar{\theta}_2(0) = x_0$. By Lemma 9.62, we have $\bar{\theta}_1 = \bar{\theta}_2$. Then, $\theta_1(t, r) = \bar{\theta}_1(1) = \bar{\theta}_2(1) = \theta_2(t, r)$. Hence, $\theta_1 = \theta_2$. This shows that $\theta : [a, b] \times [c, d] \to \mathcal{X}$ is unique when it exists.

"Existence" Define $\sigma_{\cdot c} : [a, b] \to \mathcal{Y}$ by $\sigma_{\cdot c}(t) = \sigma(t, c)$, $\forall t \in [a, b] \subset \mathbb{R}$. By Proposition 3.12, $\sigma_{\cdot c}$ is continuous with $\sigma_{\cdot c}(a) = y_0$. By Lemma 9.62, there exists a unique continuous mapping $\theta_{\cdot c} : [a, b] \to \mathcal{X}$ that inverts F along $\sigma_{\cdot c}$ with $\theta_{\cdot c}(a) = x_0$. Fix any $t \in [a, b] \subset \mathbb{R}$. Define $\sigma_t : [c, d] \to \mathcal{Y}$ by $\sigma_t(r) = \sigma(t, r)$, $\forall r \in [c, d] \subset \mathbb{R}$. By Proposition 3.12, σ_t is continuous with $\sigma_t(c) = \sigma(t, c) =$

$\sigma_{,c}(t) = F(\theta_{,c}(t))$. By Lemma 9.62, there exists a unique continuous mapping $\theta_t : [c,d] \to \mathcal{X}$ that inverts F along σ_t with $\theta_t(c) = \theta_{,c}(t)$.

Define $\theta : [a,b] \times [c,d] \to \mathcal{X}$ by $\theta(t,r) = \theta_t(r)$, $\forall (t,r) \in [a,b] \times [c,d]$. Clearly, $\forall (t,r) \in [a,b] \times [c,d]$, $\sigma(t,r) = \sigma_t(r) = F(\theta_t(r)) = F(\theta(t,r))$. Hence, θ inverts F along σ. $\theta(a,c) = \theta_a(c) = \theta_{,c}(a) = x_0$. All we need to show is that θ is continuous to complete the proof of the lemma. Define $S := \{r \in [c,d] \subset \mathbb{R} \mid \theta|_{[a,b]\times[c,r]}$ is continuous$\} \subset \mathbb{R}$ and $\xi = \sup S$. Clearly, $c \in S$ and $c \le \xi \le d$.

We will show that $\xi \in S$ by an argument of contradiction. Suppose $\xi \notin S$. Then, $c < \xi \le d$ and $\exists (r_n)_{n=1}^{\infty} \subseteq S$, which is nondecreasing, such that $\lim_{n\in\mathbb{N}} r_n = \xi$. $\forall (t,r) \in [a,b] \times [c,\xi) \subset \mathbb{R}^2$, there exists $n_0 \in \mathbb{N}$ such that $\forall n \ge n_0$, we have $r_n > r$. Then, $\theta|_{[a,b]\times[c,r_n]}$ is continuous implies that θ is continuous at (t,r). Hence, $\theta|_{[a,b]\times[c,\xi)}$ is continuous.

Fix any $t \in [a,b]$, and let $x = \theta(t,\xi) \in \mathcal{X}$, and $y = F(x) = \sigma(t,\xi) \in \mathcal{Y}$. Then, $\exists U \in \mathcal{O}_{\mathcal{X}}$ with $x \in U$ and $\exists V \in \mathcal{O}_{\mathcal{Y}}$ with $y \in V$ such that $F|_U : U \to V$ is a homeomorphism. Since σ is continuous, then $\exists a_t, b_t, c_t, d_t \in \mathbb{R}$ with $a_t < t < b_t$ and $c \le c_t < \xi < d_t$ such that $\sigma(D_t) \subseteq V$, where $D_t := ((a_t,b_t) \times [c_t,d_t)) \cap ([a,b]\times[c,d]) \subset \mathbb{R}^2$. Define $\bar{\theta} : D_t \to U$ by $\bar{\theta}(\bar{t},\bar{r}) = (F|_U)_{\text{inv}}(\sigma(\bar{t},\bar{r}))$, $\forall (\bar{t},\bar{r}) \in D_t$. By Proposition 3.12, $\bar{\theta}$ is continuous.

Claim 9.63.1 $\theta(\bar{t},\bar{r}) = \bar{\theta}(\bar{t},\bar{r})$, $\forall (\bar{t},\bar{r}) \in D_t$.

Proof of Claim Fix any $(\bar{t},\bar{r}) \in D_t$. Note that $\bar{\theta}$ inverts F along $\sigma|_{D_t}$. Define $\bar{\theta}_t : D_{t,2} \to \mathcal{X}$ by $\bar{\theta}_t(\hat{r}) = \bar{\theta}(t,\hat{r})$, $\forall \hat{r} \in D_{t,2} := [c_t,d_t) \cap [c,d] \subset \mathbb{R}$. Then, $\bar{\theta}_t$ is continuous and inverts F along $\sigma_t|_{D_{t,2}}$ with $\bar{\theta}_t(\xi) = x$. Note that $\theta_t|_{D_{t,2}}$ is also continuous and inverts F along $\sigma_t|_{D_{t,2}}$ with $\theta_t(\xi) = x$. By Lemma 9.62, we have $\bar{\theta}_t = \theta_t|_{D_{t,2}}$ and, in particular, $\bar{\theta}(t,c_t) = \bar{\theta}_t(c_t) = \theta_t(c_t) = \theta(t,c_t)$. Define $\bar{\theta}_{,c_t} : D_{t,1} \to \mathcal{X}$ by $\bar{\theta}_{,c_t}(\hat{t}) = \bar{\theta}(\hat{t},c_t)$, $\forall \hat{t} \in D_{t,1} := (a_t,b_t) \cap [a,b] \subset \mathbb{R}$. Define $\sigma_{,c_t} : [a,b] \to \mathcal{Y}$ by $\sigma_{,c_t}(\hat{t}) = \sigma(\hat{t},c_t)$, $\forall \hat{t} \in [a,b]$. Define $\theta_{,c_t} : [a,b] \to \mathcal{X}$ by $\theta_{,c_t}(\hat{t}) = \theta(\hat{t},c_t)$, $\forall \hat{t} \in [a,b]$. Then, $\bar{\theta}_{,c_t}$ is continuous and inverts F along $\sigma_{,c_t}\big|_{D_{t,1}}$ with $\bar{\theta}_{,c_t}(t) = \theta(t,c_t)$. Since $c_t \in [c,\xi) \subset \mathbb{R}$, then $\theta_{,c_t}\big|_{D_{t,1}}$ is continuous and inverts F along $\sigma_{,c_t}\big|_{D_{t,1}}$ with $\theta_{,c_t}(t) = \theta(t,c_t)$. By Lemma 9.62, we have $\theta_{,c_t}\big|_{D_{t,1}} = \bar{\theta}_{,c_t}$ and, in particular, $\theta(\bar{t},c_t) = \theta_{,c_t}(\bar{t}) = \bar{\theta}_{,c_t}(\bar{t}) = \bar{\theta}(\bar{t},c_t)$. Define $\bar{\theta}_{\bar{t}} : D_{t,2} \to \mathcal{X}$ by $\bar{\theta}_{\bar{t}}(\hat{r}) = \bar{\theta}(\bar{t},\hat{r})$, $\forall \hat{r} \in D_{t,2}$. Then, $\bar{\theta}_{\bar{t}}$ is continuous and inverts F along $\sigma_{\bar{t}}|_{D_{t,2}}$ with $\bar{\theta}_{\bar{t}}(c_t) = \bar{\theta}(\bar{t},c_t) = \theta(\bar{t},c_t)$. Note that $\theta_{\bar{t}}|_{D_{t,2}}$ is also continuous and inverts F along $\sigma_{\bar{t}}|_{D_{t,2}}$ with $\theta_{\bar{t}}(c_t) = \theta(\bar{t},c_t)$. By Lemma 9.62, we have $\bar{\theta}_{\bar{t}} = \theta_{\bar{t}}|_{D_{t,2}}$ and, in particular, $\bar{\theta}(\bar{t},\bar{r}) = \bar{\theta}_{\bar{t}}(\bar{r}) = \theta_{\bar{t}}(\bar{r}) = \theta(\bar{t},\bar{r})$. This completes the proof of the claim. □

Then, $\theta|_{D_t}$ is continuous. Then, θ is continuous at (t,ξ). Then, $\theta|_{[a,b]\times[c,\xi]}$ is continuous and $\xi \in S$. This contradicts with the hypothesis $\xi \notin S$. Therefore, $\xi \in S$.

Now, we show $\xi = d$ by an argument of contradiction. Suppose $\xi < d$. Fix any $t \in [a,b]$, and let $x = \theta(t,\xi) \in \mathcal{X}$ and $y = F(x) = \sigma(t,\xi) \in \mathcal{Y}$. Then, $\exists U \in \mathcal{O}_{\mathcal{X}}$ with $x \in U$ and $\exists V \in \mathcal{O}_{\mathcal{Y}}$ with $y \in V$ such that $F|_U : U \to V$ is a homeomorphism. Since σ is continuous, then $\exists a_t, b_t, c_t, d_t \in \mathbb{R}$ with $a_t < t < b_t$

and $c_t < \xi < d_t \leq d$ such that $\sigma(D_t) \subseteq V$, where $D_t := ((a_t, b_t) \times (c_t, d_t)) \cap ([a, b] \times [c, d]) \subset \mathbb{R}^2$. Define $\bar{\theta} : D_t \to U$ by $\bar{\theta}(\bar{t}, \bar{r}) = (F|_U)_{\text{inv}}(\sigma(\bar{t}, \bar{r}))$, $\forall(\bar{t}, \bar{r}) \in D_t$. By Proposition 3.12, $\bar{\theta}$ is continuous.

Claim 9.63.2 $\theta(\bar{t}, \bar{r}) = \bar{\theta}(\bar{t}, \bar{r})$, $\forall(\bar{t}, \bar{r}) \in D_t$.

Proof of Claim Fix any $(\bar{t}, \bar{r}) \in D_t$. Note that $\bar{\theta}$ inverts F along $\sigma|_{D_t}$. Define $\bar{\theta}_{\cdot\xi} : D_{t,1} \to \mathcal{X}$ by $\bar{\theta}_{\cdot\xi}(\hat{t}) = \bar{\theta}(\hat{t}, \xi)$, $\forall \hat{t} \in D_{t,1} := (a_t, b_t) \cap [a, b] \subset \mathbb{R}$. Define $\sigma_{\cdot\xi} : [a, b] \to \mathcal{Y}$ by $\sigma_{\cdot\xi}(\hat{t}) = \sigma(\hat{t}, \xi)$, $\forall \hat{t} \in [a, b]$. Define $\theta_{\cdot\xi} : [a, b] \to \mathcal{X}$ by $\theta_{\cdot\xi}(\hat{t}) = \theta(\hat{t}, \xi)$, $\forall \hat{t} \in [a, b]$. Then, $\bar{\theta}_{\cdot\xi}$ is continuous and inverts F along $\sigma_{\cdot\xi}|_{D_{t,1}}$ with $\bar{\theta}_{\cdot\xi}(t) = x$. Since $\xi \in S$, then $\theta_{\cdot\xi}|_{D_{t,1}}$ is continuous and inverts F along $\sigma_{\cdot\xi}|_{D_{t,1}}$ with $\theta_{\cdot\xi}(t) = \theta(t, \xi) = x$. By Lemma 9.62, we have $\theta_{\cdot\xi}|_{D_{t,1}} = \bar{\theta}_{\cdot\xi}$ and, in particular, $\theta(\bar{t}, \xi) = \theta_{\cdot\xi}(\bar{t}) = \bar{\theta}_{\cdot\xi}(\bar{t}) = \bar{\theta}(\bar{t}, \xi)$. Define $\bar{\theta}_{\bar{t}} : D_{t,2} \to \mathcal{X}$ by $\bar{\theta}_{\bar{t}}(\hat{r}) = \bar{\theta}(\bar{t}, \hat{r})$, $\forall \hat{r} \in D_{t,2} := (c_t, d_t) \cap [c, d] \subset \mathbb{R}$. Then, $\bar{\theta}_{\bar{t}}$ is continuous and inverts F along $\sigma_{\bar{t}}|_{D_{t,2}}$ with $\bar{\theta}_{\bar{t}}(\xi) = \bar{\theta}(\bar{t}, \xi) = \theta(\bar{t}, \xi)$. Note that $\theta_{\bar{t}}|_{D_{t,2}}$ is also continuous and inverts F along $\sigma_{\bar{t}}|_{D_{t,2}}$ with $\theta_{\bar{t}}(\xi) = \theta(\bar{t}, \xi)$. By Lemma 9.62, we have $\bar{\theta}_{\bar{t}} = \theta_{\bar{t}}|_{D_{t,2}}$ and, in particular, $\bar{\theta}(\bar{t}, \bar{r}) = \bar{\theta}_{\bar{t}}(\bar{r}) = \theta_{\bar{t}}(\bar{r}) = \theta(\bar{t}, \bar{r})$. This completes the proof of the claim. □

Then, $\theta|_{D_t}$ is continuous. Note that $[a, b] \subseteq \bigcup_{t \in [a,b] \subset \mathbb{R}} (a_t, b_t)$. By the compactness of $[a, b] \subseteq \mathbb{R}$, there exists a finite set $T_N \subseteq [a, b] \subset \mathbb{R}$ such that $[a, b] \subseteq \bigcup_{t \in T_N} (a_t, b_t)$. Note that $\theta|_{D_t}$, $\forall t \in T_N$, and $\theta|_{[a,b] \times [c,\xi)}$ are continuous. Then, by Theorem 3.11, $\theta|_D$ is continuous, where $D := ([a, b] \times [c, \xi)) \cup (\bigcup_{t \in T_N} D_t) \subseteq [a, b] \times [c, d] \subseteq \mathbb{R}^2$ and all of the sets are relatively open. Set $\bar{d} = (\min_{t \in T_N} d_t + \xi)/2 \in (\xi, d]$. It is clear that $[a, b] \times [c, \bar{d}] \subseteq D \subseteq \mathbb{R}^2$. Then, $\theta|_{[a,b] \times [c,\bar{d}]}$ is continuous and $\bar{d} \in S$. This leads to the contradiction $\xi < \bar{d} \leq \sup S = \xi$. Therefore, $\xi = d$. Then, θ is continuous. This completes the proof of Case 2.

This completes the proof of the lemma. □

Theorem 9.64 (Global Inverse Function Theorem) *Let $\mathcal{X}$ and $\mathcal{Y}$ be Hausdorff topological spaces, $\mathcal{X} \neq \emptyset$, and $F : \mathcal{X} \to \mathcal{Y}$ be continuous and countably proper. Assume that $\forall x \in \mathcal{X}$, $\exists U \in \mathcal{O}_{\mathcal{X}}$ with $x \in U$ and $\exists V \in \mathcal{O}_{\mathcal{Y}}$ with $F(x) \in V$ such that $F|_U : U \to V$ is a homeomorphism, $\mathcal{X}$ is arcwise connected, and $\mathcal{Y}$ is simply connected. Then, $F : \mathcal{X} \to \mathcal{Y}$ is a homeomorphism.*

Proof Fix $x_0 \in \mathcal{X} \neq \emptyset$, and let $y_0 = F(x_0) \in \mathcal{Y}$. $\forall y \in \mathcal{Y}$, since $\mathcal{Y}$ is simply connected, then $\mathcal{Y}$ is arcwise connected by Definition 3.53. Then, there exists a curve $\sigma : I \to \mathcal{Y}$, where $I := [0, 1] \subset \mathbb{R}$, such that $\sigma(0) = y_0$ and $\sigma(1) = y$. By Lemma 9.62, there exists a continuous mapping $\theta : I \to \mathcal{X}$ that inverts F along σ with $\theta(0) = x_0$. Then, $y = \sigma(1) = F(\theta(1))$. Hence, F is surjective.

Fix $x_1, x_2 \in \mathcal{X}$ such that $F(x_1) = F(x_2) = y$. Since $\mathcal{X}$ is arcwise connected, then there exists a curve $\delta : I \to \mathcal{X}$ such that $\delta(0) = x_1$ and $\delta(1) = x_2$. Consider the curve $\eta := F \circ \delta$, which is continuous by Proposition 3.12. η is a closed curve since $\eta(0) = F(x_1) = y = F(x_2) = \eta(1)$. Since $\mathcal{Y}$ is simply connected, then η is homotopic to a single point $\bar{y} \in \mathcal{Y}$. Then, there exists a continuous mapping

$\gamma : I \times I \to \mathcal{Y}$ such that $\gamma(t, 0) = \eta(t)$, $\gamma(t, 1) = \bar{y}$, and $\gamma(0, t) = \gamma(1, t)$, $\forall t \in I$. By Lemma 9.63, there exists a continuous function $\zeta : I \times I \to \mathcal{X}$ that inverts F along γ with $\zeta(0, 0) = x_1$. Define $\gamma_0 : I \to \mathcal{Y}$ by $\gamma_0(t) = \gamma(0, t)$, $\gamma_1 : I \to \mathcal{Y}$ by $\gamma_1(t) = \gamma(1, t)$, $\gamma_{,0} : I \to \mathcal{Y}$ by $\gamma_{,0}(t) = \gamma(t, 0)$, and $\gamma_{,1} : I \to \mathcal{Y}$ by $\gamma_{,1}(t) = \gamma(t, 1)$, $\forall t \in I$. Then, $\gamma_0 = \gamma_1$. Define $\zeta_0 : I \to \mathcal{X}$ by $\zeta_0(t) = \zeta(0, t)$, $\zeta_1 : I \to \mathcal{X}$ by $\zeta_1(t) = \zeta(1, t)$, $\zeta_{,0} : I \to \mathcal{X}$ by $\zeta_{,0}(t) = \zeta(t, 0)$, and $\zeta_{,1} : I \to \mathcal{X}$ by $\zeta_{,1}(t) = \zeta(t, 1)$, $\forall t \in I$. Then, ζ_0 is continuous and inverts F along γ_0. Set $\bar{x} = \zeta_0(1) = \zeta(0, 1) \in \mathcal{X}$, and then, $\bar{y} = F(\bar{x})$. $\zeta_{,1}$ is continuous and inverts F along $\gamma_{,1}$ with $\zeta_{,1}(0) = \zeta(0, 1) = \bar{x}$. Since $\gamma_{,1}$ is a constant function with value $\bar{y}$, then the constant mapping $\lambda : I \to \mathcal{X}$ given by $\lambda(t) = \bar{x}$, $\forall t \in I$, is continuous and inverts F along $\gamma_{,1}$ with $\lambda(0) = \bar{x}$. By Lemma 9.62, we have $\lambda = \zeta_{,1}$ and, in particular, $\bar{x} = \lambda(1) = \zeta_{,1}(1) = \zeta(1, 1)$. Note that ζ_1 is continuous and inverts F along γ_1 with $\zeta_1(1) = \zeta(1, 1) = \bar{x}$. Since $\gamma_1 = \gamma_0$, then ζ_0 is continuous and inverts F along γ_1 with $\zeta_0(1) = \bar{x}$. By Lemma 9.62, we have $\zeta_0 = \zeta_1$ and in particular $\zeta(1, 0) = \zeta_1(0) = \zeta_0(0) = \zeta(0, 0) = x_1$. Note that $\zeta_{,0}$ is continuous and inverts F along $\gamma_{,0} = \eta$ with $\zeta_{,0}(0) = \zeta(0, 0) = x_1$. By construction, δ is continuous and inverts F along η with $\delta(0) = x_1$. By Lemma 9.62, $\delta = \zeta_{,0}$ and in particular, $x_2 = \delta(1) = \zeta_{,0}(1) = \zeta(1, 0) = x_1$. Hence, $x_1 = x_2$. Therefore, F is injective.

Hence, F is bijective and admits inverse $F_i : \mathcal{Y} \to \mathcal{X}$. $\forall y \in \mathcal{Y}$, let $x = F_i(y) \in \mathcal{X}$. Then, $\exists U \in \mathcal{O}_{\mathcal{X}}$ with $x \in U$ and $\exists V \in \mathcal{O}_{\mathcal{Y}}$ with $y \in V$ such that $F|_U : U \to V$ is a homeomorphism. Then, $F_i|_V$ is the inverse of $F|_U$ and is continuous. Then, F_i is continuous at y since V is open. By the arbitrariness of y, F_i is continuous. Hence, F is a homeomorphism.

This completes the proof of the theorem. □

Theorem 9.65 (Global Inverse Function Theorem) *Let $\mathcal{X}$ and $\mathcal{Y}$ be Hausdorff topological spaces and $F : \mathcal{X} \to \mathcal{Y}$ be continuous and countably proper. Let $H := \{x \in \mathcal{X} \mid \exists U \in \mathcal{O}_{\mathcal{X}}$ with $x \in U \ni \cdot\ F|_U : U \to F(U) \in \mathcal{O}_{\mathcal{Y}}$ is a homeomorphism$\} \subseteq \mathcal{X}$, $\Sigma := \mathcal{X} \setminus H$, $\Sigma_0 := F_{\text{inv}}(F(\Sigma))$, $X_0 := \mathcal{X} \setminus \Sigma_0$, and $Y_0 := \mathcal{Y} \setminus F(\Sigma)$. Assume that $X_0 \neq \emptyset$ is arcwise connected and Y_0 is simply connected. Then, $G := F|_{X_0} : X_0 \to Y_0$ is a homeomorphism.*

Proof Let $\mathcal{O}_{X_0}$ and $\mathcal{O}_{Y_0}$ be the subset topology on X_0 and Y_0, respectively.

Claim 9.65.1 $\Sigma \subseteq \mathcal{X}$ is closed.

Proof of Claim $\forall x \in H, \exists U_x \in \mathcal{O}_{\mathcal{X}}$ with $x \in U_x$ such that $F|_{U_x} : U_x \to F(U_x) \in \mathcal{O}_{\mathcal{Y}}$ is a homeomorphism. $\forall \bar{x} \in U_x$, $\bar{x} \in H$. Then, $U_x \subseteq H$ and $H = \bigcup_{x \in H} U_x$. Hence, $H \in \mathcal{O}_{\mathcal{X}}$. Then, $\Sigma := \mathcal{X} \setminus H$ is closed. □

Claim 9.65.2 $F_{\text{inv}}(Y_0) = X_0$. $G : X_0 \to Y_0$ is continuous and countably proper.

Proof of Claim By Proposition 2.5, $F_{\text{inv}}(Y_0) = F_{\text{inv}}(\mathcal{Y} \setminus F(\Sigma)) = F_{\text{inv}}(\mathcal{Y}) \setminus F_{\text{inv}}(F(\Sigma)) = \mathcal{X} \setminus \Sigma_0 = X_0$ and $F(X_0) = F(F_{\text{inv}}(Y_0)) \subseteq Y_0$. Hence, G is a function of X_0 to Y_0.

Fix any $K \subseteq Y_0$ such that K is compact in $\mathcal{O}_{Y_0}$. Then K is a compact set in $\mathcal{O}_{\mathcal{Y}}$. Then, $G_{\text{inv}}(K) = F_{\text{inv}}(K) \subseteq X_0$. By the countable properness of F, we have

$F_{\text{inv}}(K) \subseteq \mathcal{X}$ is countably compact in $\mathcal{O}_{\mathcal{X}}$. Then, it is easy to show that $G_{\text{inv}}(K)$ is countably compact in $\mathcal{O}_{X_0}$. Hence, G is countably proper.

$\forall V_0 \in \mathcal{O}_{Y_0}$, $V_0 = Y_0 \cap V$, where $V \in \mathcal{O}_{\mathcal{Y}}$. $G_{\text{inv}}(V_0) = F_{\text{inv}}(V_0) = F_{\text{inv}}(Y_0) \cap F_{\text{inv}}(V) = X_0 \cap F_{\text{inv}}(V)$. Since F is continuous, $F_{\text{inv}}(V) \in \mathcal{O}_{\mathcal{X}}$. Thus, $G_{\text{inv}}(V_0) \in \mathcal{O}_{X_0}$. Hence, G is continuous. □

By Proposition 2.5, $\Sigma_0 = F_{\text{inv}}(F(\Sigma)) \supseteq \Sigma \cap \text{dom}(F) = \Sigma$. Then, $X_0 = \mathcal{X} \setminus \Sigma_0 \subseteq \mathcal{X} \setminus \Sigma = H$. $\forall x \in X_0 \subseteq H$, $\exists U \in \mathcal{O}_{\mathcal{X}}$ with $x \in U$ such that $F|_U : U \to F(U) \in \mathcal{O}_{\mathcal{Y}}$ is a homeomorphism. Let $U_0 := X_0 \cap U \in \mathcal{O}_{X_0}$, $V := F(U) \in \mathcal{O}_{\mathcal{Y}}$, and $V_0 := F(U) \cap Y_0 \in \mathcal{O}_{Y_0}$. Clearly, $x \in U_0$. By Proposition 2.5, $G(U_0) = F(U_0) \subseteq F(X_0) \cap F(U) \subseteq Y_0 \cap V = V_0$. Then, $G|_{U_0} : U_0 \to V_0$. Note that $G|_{U_0} = (F|_U)\big|_{U_0}$. Since $F|_U$ is injective, then $G|_{U_0}$ is injective. $\forall \bar{y} \in V_0$, $\bar{y} \in V$ and then $\exists \bar{x} \in U$ such that $\bar{y} = F(\bar{x})$. Note that $\bar{y} \in Y_0$ and $F_{\text{inv}}(Y_0) = X_0$, and then $\bar{x} \in X_0$. Hence, $\bar{x} \in U_0$ and $G(\bar{x}) = F(\bar{x}) = \bar{y}$. Then, $G|_{U_0} : U_0 \to V_0$ is surjective. Hence, $G|_{U_0}$ is bijective with inverse $G_i = F_i|_{V_0}$, where F_i is the inverse of $F|_U : U \to V$. Since $F|_U$ is homeomorphism, then $F|_U$ and F_i are continuous. Then, $G|_{U_0}$ and G_i are continuous. This shows that $G|_{U_0} : U_0 \to V_0$ is a homeomorphism.

By Global Inverse Function Theorem 9.64, $G : X_0 \to Y_0$ is a homeomorphism. This completes the proof of the theorem. □

Proposition 9.66 *Let $\mathcal{X} := (X, \mathcal{O})$ be a topological space and $A_\alpha \subseteq X$ be arcwise connected (in subset topology), $\forall \alpha \in \Lambda$, where Λ is an index set. Assume that $A_{\alpha_1} \cap A_{\alpha_2} \neq \emptyset$, $\forall \alpha_1, \alpha_2 \in \Lambda$. Then, $A := \bigcup_{\alpha \in \Lambda} A_\alpha$ is arcwise connected (in subset topology).*

Proof $\forall x_1, x_2 \in A$, $\exists \alpha_1, \alpha_2 \in \Lambda$ such that $x_i \in A_{\alpha_i}$, $i = 1, 2$. By the assumption, let $x_0 \in A_{\alpha_1} \cap A_{\alpha_2} \neq \emptyset$. $\forall i \in \{1, 2\}$, since A_{α_i} is arcwise connected, $\exists$ curve $\gamma_i : [0, 1] \subset \mathbb{R} \to A_{\alpha_i}$ such that $\gamma_i(0) = x_0$ and $\gamma_i(1) = x_i$. Define $\gamma : [0, 1] \subset \mathbb{R} \to A$ by $\gamma(t) = \begin{cases} \gamma_1(1-2t) & 0 \le t \le 1/2 \\ \gamma_2(2t-1) & 1/2 < t \le 1 \end{cases}$, $\forall t \in [0, 1] \subset \mathbb{R}$. Then $\gamma(0) = \gamma_1(1) = x_1$ and $\gamma(1) = \gamma_2(1) = x_2$. Note that $\gamma|_{[0,1/2]}(t) = \gamma_1(1-2t)$, $\forall t \in [0, 1/2] \subset \mathbb{R}$, and $\gamma|_{[1/2,1]}(t) = \gamma_2(2t-1)$, $\forall t \in [1/2, 1] \subset \mathbb{R}$, are continuous functions. By Theorem 3.11, γ is continuous. Therefore, γ is a curve connecting x_1 and x_2. By the arbitrariness of x_1 and x_2, we have A is arcwise connected. This completes the proof of the proposition. □

Proposition 9.67 *Let $\mathfrak{X}$ be a normed linear space and $O \subseteq \mathfrak{X}$ be open and connected. Then, O is arcwise connected.*

Proof The result is trivial if $O = \emptyset$. Let $x_0 \in O \neq \emptyset$. Let $\mathcal{M} := \{A \subseteq O \mid x_0 \in A \text{ and } A \text{ is arcwise connected}\}$ and $A_0 := \bigcup_{A \in \mathcal{M}} A$. Then, $x_0 \in A_0 \subseteq O$ and, by Proposition 9.66, A_0 is arcwise connected. $\forall x \in A_0 \subseteq O$, $\exists \delta \in (0, \infty) \subset \mathbb{R}$ such that $\mathcal{B}_{\mathfrak{X}}(x, \delta) \subseteq O$. Clearly, $\mathcal{B}_{\mathfrak{X}}(x, \delta)$ is arcwise connected. Then, by Proposition 9.66, $A_0 \cup \mathcal{B}_{\mathfrak{X}}(x, \delta)$ is arcwise connected. This implies that $A_0 \cup \mathcal{B}_{\mathfrak{X}}(x, \delta) \in \mathcal{M}$. Then, $\mathcal{B}_{\mathfrak{X}}(x, \delta) \subseteq A_0$. By the arbitrariness of x, we have A_0

is open in $\mathcal{X}$. We will show that $A_0 = O$ by an argument of contradiction. Suppose $A_0 \subset O$. Let $E := O \setminus A_0 \neq \emptyset$. Let ∂E be the boundary of E in $\mathcal{X}$.

Claim 9.67.1 $\partial E \cap E \neq \emptyset$.

Proof of Claim Suppose $\partial E \cap E = \emptyset$. By Proposition 3.3, $E = \overline{E} \cap E = (\partial E \cup E^\circ) \cap E = E^\circ \cap E = E^\circ$. Hence, E is open in $\mathcal{X}$. Then, A_0 and E form a separation of O. This contradicts with the assumption that O is connected. Hence, the claim holds. □

Let $x_1 \in \partial E \cap E \subseteq O$. Then, $\exists \delta_1 \in (0, \infty) \subset \mathbb{R}$ such that $\mathcal{B}_{\mathcal{X}}(x_1, \delta_1) \subseteq O$. $x_1 \in \partial E$ implies that $\mathcal{B}_{\mathcal{X}}(x_1, \delta_1) \cap \tilde{E} \neq \emptyset$. This implies that $\exists x_2 \in \mathcal{B}_{\mathcal{X}}(x_1, \delta_1) \cap (O \setminus E) = \mathcal{B}_{\mathcal{X}}(x_1, \delta_1) \cap A_0$. By Proposition 9.66, $A_0 \cup \mathcal{B}_{\mathcal{X}}(x_1, \delta_1)$ is arcwise connected. Then, by the definition of A_0, we have $\mathcal{B}_{\mathcal{X}}(x_1, \delta_1) \subseteq A_0$ and $x_1 \in A_0$. This contradicts with the fact that $x_1 \in E = O \setminus A_0$. Therefore, $A_0 = O$ and O is arcwise connected. This completes the proof of the proposition. □

9.7 Interchange Differentiation and Limit

Proposition 9.68 *Let $\mathcal{X}$ and $\mathcal{Y}$ be normed linear spaces over $\mathbb{K}$, $D \subseteq \mathcal{X}$, $x_0 \in D$, and $F_n : D \to \mathcal{Y}$, $\forall n \in \mathbb{N}$. Assume that:*

(i) *$\exists \delta_0 \in (0, \infty) \subset \mathbb{R}$ such that $\bar{D} := D \cap \mathcal{B}_{\mathcal{X}}(x_0, \delta_0) - x_0$ is a conic segment.*
(ii) *F_n is differentiable, $\forall n \in \mathbb{N}$.*
(iii) *$\left(F_n^{(1)}\right)_{n=1}^{\infty}$ converges uniformly to $G : D \to \mathrm{B}(\mathcal{X}, \mathcal{Y})$.*
(iv) *$\forall x \in D$, $\lim_{n \in \mathbb{N}} F_n(x) = F(x)$, where $F : D \to \mathcal{Y}$.*

Then, F is differentiable at x_0 and $F^{(1)}(x_0) = G(x_0) = \lim_{n \in \mathbb{N}} F_n^{(1)}(x_0)$.

Proof By the differentiability of F_1, we have $\overline{\operatorname{span}(\mathrm{A}_D(x))} = \mathcal{X}$, $\forall x \in D$. $\forall \epsilon \in (0, \infty) \subset \mathbb{R}$, by (iii), $\exists n_0 \in \mathbb{N}$ such that $\forall n \in \mathbb{N}$ with $n \geq n_0$, $\left\| F_n^{(1)}(x) - G(x) \right\| < \epsilon/4$, $\forall x \in D$. $\forall n, m \in \mathbb{N}$ with $n \geq n_0$ and $m \geq n_0$, by Proposition 9.15, $g_{n,m} : D \to \mathcal{Y}$, defined by $g_{n,m}(x) = F_n(x) - F_m(x)$, $\forall x \in D$, is differentiable and $g_{n,m}^{(1)}(x) = F_n^{(1)}(x) - F_m^{(1)}(x)$, $\forall x \in D$. By mean value theorem 9.23, $\forall x \in \bar{D} + x_0$ and $\forall n, m \in \mathbb{N}$ with $n \geq n_0$ and $m \geq n_0$, $\|F_n(x) - F_m(x) - F_n(x_0) + F_m(x_0)\| = \left\| g_{n,m}(x) - g_{n,m}(x_0) \right\| \leq \left\| g_{n,m}^{(1)}(\bar{x})(x - x_0) \right\| \leq \left\| F_n^{(1)}(\bar{x}) - F_m^{(1)}(\bar{x}) \right\| \|x - x_0\| \leq \epsilon \|x - x_0\|/2$, where $\bar{x} = t_0 x + (1 - t_0)x_0 \in \bar{D} + x_0$ and $t_0 \in (0, 1) \subset \mathbb{R}$. By $F_{n_0}^{(1)}(x_0) \in \mathrm{B}(\mathcal{X}, \mathcal{Y})$, $\exists \delta \in (0, \delta_0] \subset \mathbb{R}$ such that, $\forall x \in \hat{D} := D \cap \mathcal{B}_{\mathcal{X}}(x_0, \delta) \subseteq \bar{D} + x_0$, we have $\left\| F_{n_0}(x) - F_{n_0}(x_0) - F_{n_0}^{(1)}(x_0)(x - x_0) \right\| \leq \epsilon \|x - x_0\|/4$.

Fix any $x \in \hat{D}$. $\forall m \in \mathbb{N}$ with $m \geq n_0$, we have $\|F_m(x) - F_m(x_0) - G(x_0)(x - x_0)\| \leq \left\| F_m(x) - F_m(x_0) - F_{n_0}(x) + F_{n_0}(x_0) \right\| + \left\| F_{n_0}(x) - F_{n_0}(x_0) - F_{n_0}^{(1)}(x_0)(x - x_0) \right\| + \left\| F_{n_0}^{(1)}(x_0)(x - x_0) - G(x_0)(x - x_0) \right\| \leq \epsilon \|x - x_0\|/2 + \epsilon \|x - x_0\|/4 + \epsilon \|x - x_0\|/4 = \epsilon \|x - x_0\|$. Take limit as $m \to \infty$, we have $\|F(x) - F(x_0) - G(x_0)(x - x_0)\| \leq \lim_{m\in\mathbb{N}} \|F_m(x) - F_m(x_0) - G(x_0)(x - x_0)\| \leq \epsilon\|x - x_0\|$. By the arbitrariness of x, F is differentiable at x_0 and $F^{(1)}(x_0) = G(x_0) = \lim_{n\in\mathbb{N}} \mathrm{D}F_n(x_0)$. This completes the proof of the proposition. □

Example 9.69 Let $\mathcal{X}$ and $\mathcal{Y}$ be normed linear spaces over $\mathbb{K}$, $\Omega \subseteq \mathcal{X}$ be a compact set, which satisfies $\overline{\operatorname{span}(A_\Omega(x))} = \mathcal{X}$, $\forall x \in \Omega$, $k \in \mathbb{N}$, and $\mathcal{W} := \mathcal{C}(\Omega, \mathcal{Y})$ be the normed linear space defined in Example 7.31. Define $Z := \{f \in \mathcal{C}(\Omega, \mathcal{Y}) \mid f \text{ is } \mathcal{C}_k\}$. By Proposition 9.40, Z is a subspace of $\mathcal{C}(\Omega, \mathcal{Y})$. Then, $\mathcal{Z} := (Z, \oplus_{\mathcal{W}}, \otimes_{\mathcal{W}}, \vartheta_{\mathcal{W}})$ is the vector space over the field $\mathbb{K}$. $\forall f \in Z$, $\forall i \in \{0, \ldots, k\}$, $f^{(i)} \in \mathcal{C}(\Omega, \mathrm{B}_i(\mathcal{X}, \mathcal{Y}))$ with norm $\|f^{(i)}\|_{\mathcal{C}} \in \mathbb{R}$. Now, define a norm on Z by $\|f\|_{\mathcal{C}_k} := \left(\sum_{i=0}^{k} \|f^{(i)}\|_{\mathcal{C}}^2\right)^{1/2} \in [0, \infty) \subset \mathbb{R}$, $\forall f \in Z$. If $f = \vartheta_{\mathcal{W}}$, by Proposition 9.33, f is $\mathcal{C}_\infty$ and $f^{(i)}(x) = \vartheta_{\mathrm{B}_i(\mathcal{X},\mathcal{Y})}$, $\forall x \in \Omega$, $\forall i \in \mathbb{N}$, then $\|f\|_{\mathcal{C}_k} = 0$. On the other hand, if $\|f\|_{\mathcal{C}_k} = 0$, then $f = \vartheta_{\mathcal{W}}$. $\forall f_1, f_2 \in Z$, by Proposition 9.40, $\|f_1 + f_2\|_{\mathcal{C}_k} = \left(\sum_{i=0}^{k} \left\|f_1^{(i)} + f_2^{(i)}\right\|_{\mathcal{C}}^2\right)^{1/2} \leq \left(\sum_{i=0}^{k} \left\|f_1^{(i)}\right\|_{\mathcal{C}}^2\right)^{1/2} + \left(\sum_{i=0}^{k} \left\|f_2^{(i)}\right\|_{\mathcal{C}}^2\right)^{1/2} = \|f_1\|_{\mathcal{C}_k} + \|f_2\|_{\mathcal{C}_k}$, where the inequality follows from Minkowski's inequality, Theorem 7.9. $\forall \alpha \in \mathbb{K}$, $\forall f \in Z$, by Proposition 9.40, $\|\alpha f\|_{\mathcal{C}_k} = \left(\sum_{i=0}^{k} \|\alpha f^{(i)}\|_{\mathcal{C}}^2\right)^{1/2} = \left(\sum_{i=0}^{k} |\alpha|^2 \|f^{(i)}\|_{\mathcal{C}}^2\right)^{1/2} = |\alpha| \|f\|_{\mathcal{C}_k}$. This shows that $(\mathcal{Z}, \mathbb{K}, \|\cdot\|_{\mathcal{C}_k})$ is a normed linear space, which will be denoted $\mathcal{C}_k(\Omega, \mathcal{Y})$. ◇

Example 9.70 Let $\mathcal{X}$ be a normed linear space over $\mathbb{K}$ and $\mathcal{Y}$ be a Banach space over $\mathbb{K}$, $\Omega \subseteq \mathcal{X}$ be a compact set, and $k \in \mathbb{N}$. Assume that, $\forall x \in \Omega$, $\exists \delta_x \in (0, \infty) \subset \mathbb{R}$ such that $\Omega \cap \mathcal{B}_{\mathcal{X}}(x, \delta_x) - x$ is a conic segment and $\overline{\operatorname{span}(A_\Omega(x))} = \mathcal{X}$. Let $\mathcal{C}_k(\Omega, \mathcal{Y})$ be the normed linear space defined in Example 9.69. We will show that $\mathcal{C}_k(\Omega, \mathcal{Y})$ is also a Banach space over $\mathbb{K}$.

Fix any Cauchy sequence $(f_n)_{n=1}^{\infty} \subseteq \mathcal{C}_k(\Omega, \mathcal{Y})$. By the definition of the norm $\|\cdot\|_{\mathcal{C}_k}$, $\left(f_n^{(i)}\right)_{n=1}^{\infty} \subseteq \mathcal{C}(\Omega, \mathrm{B}_i(\mathcal{X}, \mathcal{Y})) =: \mathcal{W}_i$ is a Cauchy sequence, $\forall i \in \{0, \ldots, k\}$. $\forall i \in \{0, \ldots, k\}$, by Example 7.32 and Proposition 7.66, $\mathcal{C}(\Omega, \mathrm{B}_i(\mathcal{X}, \mathcal{Y}))$ is a Banach space. Then, $\exists g_i \in \mathcal{C}(\Omega, \mathrm{B}_i(\mathcal{X}, \mathcal{Y}))$ such that $\lim_{n\in\mathbb{N}} f_n^{(i)} = g_i$ in $\mathcal{C}(\Omega, \mathrm{B}_i(\mathcal{X}, \mathcal{Y}))$. Then, $\left(f_n^{(i)}\right)_{n=1}^{\infty}$ converges uniformly to g_i and g_i is continuous. By Proposition 9.68, we have $g_i^{(1)} = g_{i+1}$, $\forall i \in \{0, \ldots, k-1\}$. Then, we have $g_0^{(i)} = g_i$, $\forall i \in \{0, \ldots, k\}$, and $g_0 \in \mathcal{C}_k(\Omega, \mathcal{Y})$. Then, $\|f_n - g_0\|_{\mathcal{C}_k} = \left(\sum_{i=0}^{k} \left\|f_n^{(i)} - g_0^{(i)}\right\|_{\mathcal{C}}^2\right)^{1/2} = \left(\sum_{i=0}^{k} \left\|f_n^{(i)} - g_i\right\|_{\mathcal{C}}^2\right)^{1/2} \to 0$ as $n \to \infty$, where the first equality follows from Proposition 9.40. Then, $\lim_{n\in\mathbb{N}} f_n = g_0$ in $\mathcal{C}_k(\Omega, \mathcal{Y})$. Hence, $\mathcal{C}_k(\Omega, \mathcal{Y})$ is complete and therefore a Banach space. ◇

Sometimes, we need to consider a normed linear space of continuous functions on a topological space that is not necessarily compact. This leads us to the following examples.

Example 9.71 Let $\mathcal{X} := (X, \mathcal{O})$ be a topological space, $\mathcal{Y}$ be a normed linear space over the field $\mathbb{K}$, and $\mathcal{C}_v(\mathcal{X}, \mathcal{Y})$ be the vector space of all continuous functions of $\mathcal{X}$ to $\mathcal{Y}$ as defined in Example 7.50 with null vector ϑ. Define a function $\|\cdot\| : \mathcal{C}_v(\mathcal{X}, \mathcal{Y}) \to \mathbb{R}_e$ by $\|f\| = \max\left\{\sup_{x\in\mathcal{X}} \|f(x)\|_\mathcal{Y}, 0\right\}$, $\forall f \in \mathcal{C}_v(\mathcal{X}, \mathcal{Y})$. Consider the set $\mathcal{M} := \{f \in \mathcal{C}_v(\mathcal{X}, \mathcal{Y}) \mid \|f\| < +\infty\}$. Clearly, $\vartheta \in \mathcal{M}$. $\forall f_1, f_2 \in \mathcal{M}$, $\forall \alpha, \beta \in \mathbb{K}$, $\|\alpha f_1 + \beta f_2\| = \max\left\{\sup_{x\in\mathcal{X}} \|\alpha f_1(x) + \beta f_2(x)\|_\mathcal{Y}, 0\right\} \le \max\left\{\sup_{x\in\mathcal{X}}(|\alpha|\|f_1(x)\|_\mathcal{Y} + |\beta|\|f_2(x)\|_\mathcal{Y}), 0\right\} < +\infty$. Then, $\alpha f_1 + \beta f_2 \in \mathcal{M}$. Hence, $\mathcal{M}$ is a subspace of $\mathcal{C}_v(\mathcal{X}, \mathcal{Y})$.

Clearly, $\forall f \in \mathcal{M}$, $\|f\| \in [0, \infty) \subset \mathbb{R}$ and $\|f\| = 0 \Leftrightarrow f = \vartheta$. $\forall f_1, f_2 \in \mathcal{M}$, $\forall \alpha \in \mathbb{K}$, $\|f_1 + f_2\| = \max\left\{\sup_{x\in\mathcal{X}} \|f_1(x) + f_2(x)\|_\mathcal{Y}, 0\right\} \le \max\left\{\sup_{x\in\mathcal{X}} \|f_1(x)\|_\mathcal{Y} + \sup_{x\in\mathcal{X}} \|f_2(x)\|_\mathcal{Y}, 0\right\} = \|f_1\| + \|f_2\|$, where the first inequality follows from Proposition 3.81.

$$\begin{aligned}
\|\alpha f_1\| &= \max\{\sup_{x\in\mathcal{X}} \|\alpha f_1(x)\|_\mathcal{Y}, 0\} \\
&= \max\{\sup_{x\in\mathcal{X}} |\alpha|\|f_1(x)\|_\mathcal{Y}, 0\} = \begin{cases} \max\left\{|\alpha| \sup_{x\in\mathcal{X}} \|f_1(x)\|_\mathcal{Y}, 0\right\} & \alpha \neq 0 \\ 0 & \alpha = 0 \end{cases} \\
&= |\alpha|\|f_1\|
\end{aligned}$$

where the third equality follows from Proposition 3.81. Hence, $(\mathcal{M}, \mathbb{K}, \|\cdot\|)$ is a normed linear space, which will be denoted by $\mathcal{C}_b(\mathcal{X}, \mathcal{Y})$. ◇

Example 9.72 Let $\mathcal{X} := (X, \mathcal{O})$ be a topological space and $\mathcal{Y}$ be a Banach space over the field $\mathbb{K}$ (with norm $\|\cdot\|_\mathcal{Y}$). Consider the normed linear space $\mathcal{C}_b(\mathcal{X}, \mathcal{Y})$ (with norm $\|\cdot\|$) defined in Example 9.71. We will show that this space is a Banach space. We will distinguish two exhaustive and mutually exclusive cases: Case 1: $\mathcal{X} = \emptyset$; Case 2: $\mathcal{X} \neq \emptyset$. Case 1: $\mathcal{X} = \emptyset$. Then, $\mathcal{C}_b(\mathcal{X}, \mathcal{Y})$ is a singleton set. Hence, any Cauchy sequence in $\mathcal{C}_b(\mathcal{X}, \mathcal{Y})$ must converge. Thus, $\mathcal{C}_b(\mathcal{X}, \mathcal{Y})$ is a Banach space. Case 2: $\mathcal{X} \neq \emptyset$. Take a Cauchy sequence $(f_n)_{n=1}^\infty \subseteq \mathcal{C}_b(\mathcal{X}, \mathcal{Y})$. $\forall \epsilon \in (0, \infty) \subset \mathbb{R}$, $\exists N \in \mathbb{N}$ such that $\forall n, m \ge N$, $0 \le \|f_n(x) - f_m(x)\|_\mathcal{Y} \le \|f_n - f_m\| < \epsilon, \forall x \in \mathcal{X}$. This shows that, $\forall x \in \mathcal{X}$, $(f_n(x))_{n=1}^\infty \subseteq \mathcal{Y}$ is a Cauchy sequence, which converges to $f(x) \in \mathcal{Y}$ since $\mathcal{Y}$ is complete. This defines a function $f : \mathcal{X} \to \mathcal{Y}$. It is easy to show that $(f_n)_{n=1}^\infty$, viewed as a sequence of functions of $\mathcal{X}$ to $\mathcal{Y}$, converges uniformly to f. By Proposition 4.26, f is continuous. $\forall x \in \mathcal{X}$, $\|f(x)\|_\mathcal{Y} \le \|f_N(x) - f(x)\|_\mathcal{Y} + \|f_N(x)\|_\mathcal{Y} = \lim_{m\in\mathbb{N}} \|f_N(x) - f_m(x)\|_\mathcal{Y} + \|f_N(x)\|_\mathcal{Y} \le \epsilon + \|f_N\|$. Hence, $\|f\| \le \|f_N\| + \epsilon$. Then, $f \in \mathcal{C}_b(\mathcal{X}, \mathcal{Y})$. It is easy to show that $\lim_{n\in\mathbb{N}} \|f_n - f\| = 0$. Hence, $\lim_{n\in\mathbb{N}} f_n = f$ in $\mathcal{C}_b(\mathcal{X}, \mathcal{Y})$. Hence, $\mathcal{C}_b(\mathcal{X}, \mathcal{Y})$ is a Banach space. In both cases, we have shown that $\mathcal{C}_b(\mathcal{X}, \mathcal{Y})$ is a Banach space when $\mathcal{Y}$ is a Banach space. ◇

Example 9.73 Let $\mathfrak{X}$ and $\mathcal{Y}$ be normed linear spaces over $\mathbb{K}$, $\Omega \subseteq \mathfrak{X}$ be endowed with the subset topology, which satisfies $\overline{\operatorname{span}(\mathrm{A}_\Omega(x))} = \mathfrak{X}$, $\forall x \in \Omega$,

$k \in \mathbb{N}$, and $\mathcal{W} := \mathcal{C}_b(\Omega, \mathcal{Y})$ be the normed linear space defined in Example 9.71 with null vector $\vartheta_{\mathcal{W}}$. Define $Z := \{f \in \mathcal{C}_b(\Omega, \mathcal{Y}) \mid f \text{ is } \mathcal{C}_k \text{ and } f^{(i)} \in \mathcal{C}_b(\Omega, \mathrm{B}_i(\mathcal{X}, \mathcal{Y})),\ i = 1, \ldots, k\}$. By Proposition 9.40, Z is a subspace of $\mathcal{W}$. Then, $\mathcal{Z} := (Z, \oplus_{\mathcal{W}}, \otimes_{\mathcal{W}}, \vartheta_{\mathcal{W}})$ is vector space over the field $\mathbb{K}$. $\forall f \in Z$, $\forall i \in \{0, \ldots, k\}$, $f^{(i)} \in \mathcal{C}_b(\Omega, \mathrm{B}_i(\mathcal{X}, \mathcal{Y}))$ with norm $\|f^{(i)}\|_{\mathcal{C}_b} \in \mathbb{R}$. Now, define a norm on Z by $\|f\|_{\mathcal{C}_{bk}} := \left(\sum_{i=0}^{k} \|f^{(i)}\|_{\mathcal{C}_b}^2\right)^{1/2} \in [0, \infty) \subset \mathbb{R}$, $\forall f \in Z$. If $f = \vartheta_{\mathcal{W}}$, by Proposition 9.33, f is $\mathcal{C}_\infty$ and $f^{(i)}(x) = \vartheta_{\mathrm{B}_i(\mathcal{X}, \mathcal{Y})}$, $\forall x \in \Omega$, $\forall i \in \mathbb{N}$, then $\|f\|_{\mathcal{C}_{bk}} = 0$. On the other hand, if $\|f\|_{\mathcal{C}_{bk}} = 0$, then $f = \vartheta_{\mathcal{W}}$. $\forall f_1, f_2 \in Z$, by Proposition 9.40, $\|f_1 + f_2\|_{\mathcal{C}_{bk}} = \left(\sum_{i=0}^{k} \left\|f_1^{(i)} + f_2^{(i)}\right\|_{\mathcal{C}_b}^2\right)^{1/2} \le \left(\sum_{i=0}^{k} \left\|f_1^{(i)}\right\|_{\mathcal{C}_b}^2\right)^{1/2} + \left(\sum_{i=0}^{k} \left\|f_2^{(i)}\right\|_{\mathcal{C}_b}^2\right)^{1/2} = \|f_1\|_{\mathcal{C}_{bk}} + \|f_2\|_{\mathcal{C}_{bk}}$, where the inequality follows from Minkowski's Inequality, Theorem 7.9. $\forall \alpha \in \mathbb{K}$, $\forall f \in Z$, by Proposition 9.40, $\|\alpha f\|_{\mathcal{C}_{bk}} = \left(\sum_{i=0}^{k} \|\alpha f^{(i)}\|_{\mathcal{C}_b}^2\right)^{1/2} = \left(\sum_{i=0}^{k} |\alpha|^2 \|f^{(i)}\|_{\mathcal{C}_b}^2\right)^{1/2} = |\alpha| \|f\|_{\mathcal{C}_{bk}}$. This shows that $(\mathcal{Z}, \mathbb{K}, \|\cdot\|_{\mathcal{C}_{bk}})$ is a normed linear space, which will be denoted $\mathcal{C}_{bk}(\Omega, \mathcal{Y})$. ◇

Example 9.74 Let $\mathcal{X}$ be a normed linear space over $\mathbb{K}$ and $\mathcal{Y}$ be a Banach space over $\mathbb{K}$, $\Omega \subseteq \mathcal{X}$ be endowed with the subset topology, and $k \in \mathbb{N}$. Assume that, $\forall x \in \Omega$, $\exists \delta_x \in (0, \infty) \subset \mathbb{R}$ such that $\Omega \cap \mathcal{B}_{\mathcal{X}}(x, \delta_x) - x$ is a conic segment and $\overline{\operatorname{span}(\mathrm{A}_\Omega(x))} = \mathcal{X}$. Let $\mathcal{C}_{bk}(\Omega, \mathcal{Y})$ be the normed linear space defined in Example 9.73. We will show that $\mathcal{C}_{bk}(\Omega, \mathcal{Y})$ is also a Banach space over $\mathbb{K}$.

Fix any Cauchy sequence $(f_n)_{n=1}^{\infty} \subseteq \mathcal{C}_{bk}(\Omega, \mathcal{Y})$. By the definition of the norm $\|\cdot\|_{\mathcal{C}_{bk}}$, $\left(f_n^{(i)}\right)_{n=1}^{\infty} \subseteq \mathcal{C}_b(\Omega, \mathrm{B}_i(\mathcal{X}, \mathcal{Y})) =: \mathcal{W}_i$ is a Cauchy sequence, $\forall i \in \{0, \ldots, k\}$. $\forall i \in \{0, \ldots, k\}$, by Example 9.72 and Proposition 7.66, $\mathcal{C}_b(\Omega, \mathrm{B}_i(\mathcal{X}, \mathcal{Y}))$ is a Banach space. Then, $\exists g_i \in \mathcal{C}_b(\Omega, \mathrm{B}_i(\mathcal{X}, \mathcal{Y}))$ such that $\lim_{n\in\mathbb{N}} f_n^{(i)} = g_i$ in $\mathcal{C}_b(\Omega, \mathrm{B}_i(\mathcal{X}, \mathcal{Y}))$. Then, $\left(f_n^{(i)}\right)_{n=1}^{\infty}$ converges uniformly to g_i and g_i is continuous. By Proposition 9.68, we have $g_i^{(1)} = g_{i+1}$, $\forall i \in \{0, \ldots, k-1\}$. Then, we have $g_0^{(i)} = g_i$, $\forall i \in \{0, \ldots, k\}$, and $g_0 \in \mathcal{C}_{bk}(\Omega, \mathcal{Y})$. Furthermore, $\|f_n - g_0\|_{\mathcal{C}_{bk}} = \left(\sum_{i=0}^{k} \left\|f_n^{(i)} - g_0^{(i)}\right\|_{\mathcal{C}_b}^2\right)^{1/2} = \left(\sum_{i=0}^{k} \left\|f_n^{(i)} - g_i\right\|_{\mathcal{C}_b}^2\right)^{1/2} \to 0$ as $n \to \infty$, where the first equality follows from Proposition 9.40. Then, $\lim_{n\in\mathbb{N}} f_n = g_0$ in $\mathcal{C}_{bk}(\Omega, \mathcal{Y})$. Hence, $\mathcal{C}_{bk}(\Omega, \mathcal{Y})$ is complete and therefore a Banach space. ◇

9.8 Tensor Algebra

Definition 9.75 Let $m \in \mathbb{Z}_+$, $\mathcal{X}_i$ be a normed linear space over $\mathbb{K}$, $i = 1, \ldots, m$, and $\mathcal{Z}$ be a normed linear space over $\mathbb{K}$. A bounded linear operator $A \in \mathrm{B}(\mathcal{X}_m, \mathrm{B}(\mathcal{X}_{m-1}, \ldots, \mathrm{B}(\mathcal{X}_1, \mathcal{Z}) \cdots))$ is said to be an *mth order $\mathcal{Z}$-valued tensor*. Let $B \in \mathrm{B}(\mathcal{Y}_n, \mathrm{B}(\mathcal{Y}_{n-1}, \ldots, \mathrm{B}(\mathcal{Y}_1, \mathcal{X}_m) \cdots))$ be another nth order $\mathcal{X}_m$-valued tensor. We define

$$AB := A \cdot B \in \mathrm{B}(\mathcal{Y}_n, \ldots, \mathrm{B}(\mathcal{Y}_1, \mathrm{B}(\mathcal{X}_{m-1}, \ldots, \mathrm{B}(\mathcal{X}_1, \mathcal{Z}) \cdots)) \cdots)$$

to be an $(n + m - 1)$th order $\mathcal{Z}$-valued tensor such that

$$(AB)(y_n) \cdots (y_1)(x_{m-1}) \cdots (x_1) = A(B(y_n) \cdots (y_1))(x_{m-1}) \cdots (x_1) \in \mathcal{Z}$$

$\forall y_i \in \mathcal{Y}_i$, $i = 1, \ldots, n$, $\forall x_j \in \mathcal{X}_j$, $j = 1, \ldots, m-1$. Let $(n_1, \ldots, n_m)$ be any permutation of $(1, \ldots, m)$. Then, we may define *the transpose of tensor A with permutation $(n_1, \ldots, n_m)$* to be $A^{T_{n_1,\ldots,n_m}} \in \mathrm{B}\big(\mathcal{X}_{n_m}, \ldots, \mathrm{B}\big(\mathcal{X}_{n_1}, \mathcal{Z}\big) \cdots\big)$ such that

$$A^{T_{n_1,\ldots,n_m}}(x_{n_m}) \cdots (x_{n_1}) = A(x_m) \cdots (x_1) \qquad \forall x_i \in \mathcal{X}_i,\ i = 1, \ldots, m \qquad \diamond$$

Proposition 9.76 *Let $m, n \in \mathbb{Z}_+$, $\mathcal{X}_i$, $i = 1, \ldots, m$, $\mathcal{Y}_j$, $j = 1, \ldots, n$, $\mathcal{Z}$ be normed linear spaces over $\mathbb{K}$, $A, A_k \in \mathcal{W}_1 := \mathrm{B}(\mathcal{X}_m, \ldots, \mathrm{B}(\mathcal{X}_1, \mathcal{Z}) \cdots)$, $k = 1, 2$, be mth order $\mathcal{Z}$-valued tensors, $B, B_l \in \mathcal{W}_2 := \mathrm{B}(\mathcal{Y}_n, \ldots, \mathrm{B}(\mathcal{Y}_1, \mathcal{X}_m) \cdots)$ be nth order $\mathcal{X}_m$-valued tensors, $l = 1, 2$, and*

$$\mathcal{W}_3 := \mathrm{B}(\mathcal{Y}_n, \ldots, \mathrm{B}(\mathcal{Y}_1, \mathrm{B}(\mathcal{X}_{m-1}, \ldots, \mathrm{B}(\mathcal{X}_1, \mathcal{Z}) \cdots)) \cdots)$$

Then, the following statements hold:

(i) $\|AB\| \le \|A\| \|B\|$.
(ii) $\forall \alpha_k, \beta_l \in \mathbb{K}$, $k = 1, 2$, $l = 1, 2$, *we have* $(\alpha_1 A_1 + \alpha_2 A_2)(\beta_1 B_1 + \beta_2 B_2) = \alpha_1 \beta_1 A_1 B_1 + \alpha_1 \beta_2 A_1 B_2 + \alpha_2 \beta_1 A_2 B_1 + \alpha_2 \beta_2 A_2 B_2$.
(iii) *Let $A \in \mathrm{B}(\mathcal{X}_2, \mathrm{B}(\mathcal{X}_1, \mathbb{K}))$ be a second order $\mathbb{K}$-valued tensor. Then, $A^{T_{2,1}} = A' \phi_{\mathcal{X}_1}$, where $\phi_{\mathcal{X}_1} : \mathcal{X}_1 \to \mathcal{X}_1^{**}$ is the natural mapping as defined in Remark 7.88.*
(iv) *Let $f : \mathcal{W}_1 \times \mathcal{W}_2 \to \mathcal{W}_3$ be defined by $f(A, B) = AB \in \mathcal{W}_3$, $\forall A \in \mathcal{W}_1$, $\forall B \in \mathcal{W}_2$. Then, f is $\mathcal{C}_\infty$, $f^{(1)}(A_0, B_0)(\Delta_{1,1}, \Delta_{2,1}) = A_0 \Delta_{2,1} + \Delta_{1,1} B_0$, $f^{(2)}(A_0, B_0)(\Delta_{1,1}, \Delta_{2,1})(\Delta_{1,2}, \Delta_{2,2}) = \Delta_{1,2}\Delta_{2,1} + \Delta_{1,1}\Delta_{2,2}$, and $f^{(i+2)}(A_0, B_0) = \vartheta_{\mathrm{B_S}_{i+2}(\mathcal{W}_1 \times \mathcal{W}_2, \mathcal{W}_3)}$, $\forall (A_0, B_0) \in \mathcal{W}_1 \times \mathcal{W}_2$, $\forall (\Delta_{1,1}, \Delta_{2,1}) \in \mathcal{W}_1 \times \mathcal{W}_2$, $\forall (\Delta_{1,2}, \Delta_{2,2}) \in \mathcal{W}_1 \times \mathcal{W}_2$, $\forall i \in \mathbb{N}$.*

(v) *Let* $\mathcal{X}$ *be a normed linear space over* $\mathbb{K}$, $x_0 \in D \subseteq \mathcal{X}$, $A : D \to \mathcal{W}_1$, *and* $B : D \to \mathcal{W}_2$ *be tensor-valued functions that are Fréchet differentiable at* x_0, *and* $C : D \to \mathcal{W}_3$ *be defined by* $C(x) = A(x)B(x)$, $\forall x \in D$. *Then,*

$$C^{(1)}(x_0) = \Big((A^{(1)}(x_0))^{T_{1,\dots,m-1,m+1,m}} B(x_0) \Big)^{T_{1,\dots,m-1,m+1,\dots,n+m,m}}$$
$$+ A(x_0)B^{(1)}(x_0)$$

(vi) *Let* $\mathcal{X}$ *be a normed linear space over* $\mathbb{K}$, $x_0 \in D \subseteq \mathcal{X}$, $A : D \to \mathcal{W}_1$ *be a tensor-valued function that is Fréchet differentiable at* x_0, $(n_1, \dots, n_m)$ *be a permutation of* $(1, \dots, m)$, *and* $C : D \to \mathcal{W}_p := \mathrm{B}(\mathcal{X}_{n_m}, \dots, \mathrm{B}(\mathcal{X}_{n_1}, \mathcal{Z}) \cdots)$ *be defined by* $C(x) = (A(x))^{T_{n_1,\dots,n_m}}$, $\forall x \in D$. *Then, we have*

$$C^{(1)}(x_0) = (A^{(1)}(x_0))^{T_{n_1,\dots,n_m,m+1}}$$

Proof (i) and (ii) These are straightforward and are therefore omitted.

(iii) Note that $A \in \mathrm{B}(\mathcal{X}_2, \mathcal{X}_1^*)$. Then, $A' \in \mathrm{B}(\mathcal{X}_1^{**}, \mathcal{X}_2^*)$ and $A^{T_{2,1}} \in \mathrm{B}(\mathcal{X}_1, \mathrm{B}(\mathcal{X}_2, \mathbb{K})) = \mathrm{B}(\mathcal{X}_1, \mathcal{X}_2^*)$. Then, $A^{T_{2,1}}(x_1)(x_2) = A(x_2)(x_1) = \langle\langle A(x_2), x_1 \rangle\rangle = \langle\langle \phi_{\mathcal{X}_1}(x_1), A(x_2) \rangle\rangle = \langle\langle A'(\phi_{\mathcal{X}_1}(x_1)), x_2 \rangle\rangle = A'(\phi_{\mathcal{X}_1}(x_1))(x_2)$, $\forall x_i \in \mathcal{X}_i$, $i = 1, 2$. Then, we have $A^{T_{2,1}} = A'\phi_{\mathcal{X}_1}$.

(iv) is straightforward and is therefore omitted.

(v) follows directly from (iv), the chain rule, and Proposition 9.19.

(vi) is straightforward and is therefore omitted. □

Definition 9.77 Let $\mathcal{X}_i$, $i = 1, \dots, m$, $\mathcal{Z}$, $\mathcal{Y}_j$, $j = 1, \dots, n$, be normed linear spaces over $\mathbb{K}$, $A \in \mathrm{B}(\mathcal{X}_m, \dots, \mathrm{B}(\mathcal{X}_1, \mathcal{Z}) \cdots)$ be an mth order $\mathcal{Z}$-valued tensor, and $B \in \mathrm{B}(\mathcal{Y}_n, \dots, \mathrm{B}(\mathcal{Y}_1, \mathbb{K}) \cdots)$ be an nth order $\mathbb{K}$-valued tensor. Define the *outer product of* A *and* B to be an $(n+m)$th order $\mathcal{Z}$-valued tensor $C := A \otimes B \in \mathrm{B}(\mathcal{Y}_n, \dots, \mathrm{B}(\mathcal{Y}_1, \mathrm{B}(\mathcal{X}_m, \dots, \mathrm{B}(\mathcal{X}_1, \mathcal{Z}) \cdots)) \cdots)$ such that $C(y_n) \cdots (y_1)(x_m) \cdots (x_1) = B(y_n) \cdots (y_1) A(x_m) \cdots (x_1) \in \mathcal{Z}$, $\forall x_i \in \mathcal{X}_i$, $i = 1, \dots, m$, $\forall y_j \in \mathcal{Y}_j$, $j = 1, \dots, n$. Similarly, we may define the *outer product of* B *and* A to be an $(n+m)$th order $\mathcal{Z}$-valued tensor $\bar{C} := B \otimes A \in \mathrm{B}(\mathcal{X}_m, \dots, \mathrm{B}(\mathcal{X}_1, \mathrm{B}(\mathcal{Y}_n, \dots, \mathrm{B}(\mathcal{Y}_1, \mathcal{Z}) \cdots)) \cdots)$ such that $\bar{C}(x_m) \cdots (x_1)(y_n) \cdots (y_1) = B(y_n) \cdots (y_1) A(x_m) \cdots (x_1) \in \mathcal{Z}$, $\forall x_i \in \mathcal{X}_i$, $i = 1, \dots, m$, $\forall y_j \in \mathcal{Y}_j$, $j = 1, \dots, n$. ◇

Proposition 9.78 *Let* $m, n \in \mathbb{Z}_+$, $\mathcal{X}_i$, $i = 1, \dots, m$, $\mathcal{Y}_j$, $j = 1, \dots, n$, $\mathcal{Z}_\tau$, $\tau = 1, 2$, *be normed linear spaces over* $\mathbb{K}$ *with* $\mathcal{Z}_1 = \mathbb{K}$ *or* $\mathcal{Z}_2 = \mathbb{K}$, $\mathcal{Z} = \mathcal{Z}_1$ *if* $\mathcal{Z}_2 = \mathbb{K}$, $\mathcal{Z} = \mathcal{Z}_2$ *if* $\mathcal{Z}_1 = \mathbb{K}$, $A, A_k \in \mathcal{W}_1 := \mathrm{B}(\mathcal{X}_m, \dots, \mathrm{B}(\mathcal{X}_1, \mathcal{Z}_1) \cdots)$, $k = 1, 2$, *be* m*th order* $\mathcal{Z}_1$*-valued tensors,* $B, B_l \in \mathcal{W}_2 := \mathrm{B}(\mathcal{Y}_n, \dots, \mathrm{B}(\mathcal{Y}_1, \mathcal{Z}_2) \cdots)$ *be* n*th order* $\mathcal{Z}_2$*-valued tensors,* $l = 1, 2$, *and*

$$\mathcal{W}_3 := \mathrm{B}(\mathcal{Y}_n, \dots, \mathrm{B}(\mathcal{Y}_1, \mathrm{B}(\mathcal{X}_m, \dots, \mathrm{B}(\mathcal{X}_1, \mathcal{Z}) \cdots)) \cdots)$$

Then, the following statements hold:

(i) $\|A \otimes B\| \leq \|A\|\|B\|$.
(ii) $\forall \alpha_k, \beta_l \in \mathbb{K}$, $k = 1, 2$, $l = 1, 2$, *we have* $(\alpha_1 A_1 + \alpha_2 A_2) \otimes (\beta_1 B_1 + \beta_2 B_2) = \alpha_1\beta_1 A_1 \otimes B_1 + \alpha_1\beta_2 A_1 \otimes B_2 + \alpha_2\beta_1 A_2 \otimes B_1 + \alpha_2\beta_2 A_2 \otimes B_2$.
(iii) *Let* $f : \mathcal{W}_1 \times \mathcal{W}_2 \to \mathcal{W}_3$ *be defined by* $f(A, B) = A \otimes B \in \mathcal{W}_3$, $\forall A \in \mathcal{W}_1$, $\forall B \in \mathcal{W}_2$. *Then,* f *is* $\mathcal{C}_\infty$, $f^{(1)}(A_0, B_0)(\Delta_{1,1}, \Delta_{2,1}) = A_0 \otimes \Delta_{2,1} + \Delta_{1,1} \otimes B_0$, $f^{(2)}(A_0, B_0)(\Delta_{1,1}, \Delta_{2,1})(\Delta_{1,2}, \Delta_{2,2}) = \Delta_{1,2} \otimes \Delta_{2,1} + \Delta_{1,1} \otimes \Delta_{2,2}$, *and* $f^{(i+2)}(A_0, B_0) = \vartheta_{\mathrm{BS}_{i+2}(\mathcal{W}_1 \times \mathcal{W}_2, \mathcal{W}_3)}$, $\forall (A_0, B_0) \in \mathcal{W}_1 \times \mathcal{W}_2$, $\forall (\Delta_{1,1}, \Delta_{2,1}) \in \mathcal{W}_1 \times \mathcal{W}_2$, $\forall (\Delta_{1,2}, \Delta_{2,2}) \in \mathcal{W}_1 \times \mathcal{W}_2$, $\forall i \in \mathbb{N}$.
(iv) *Let* $\mathcal{X}$ *be a normed linear space over* $\mathbb{K}$, $x_0 \in D \subseteq \mathcal{X}$, $A : D \to \mathcal{W}_1$ *and* $B : D \to \mathcal{W}_2$ *be tensor-valued functions that are Fréchet differentiable at* x_0, *and* $C : D \to \mathcal{W}_3$ *be defined by* $C(x) = A(x) \otimes B(x)$, $\forall x \in D$. *Then,*

$$C^{(1)}(x_0) = \left(A^{(1)}(x_0) \otimes B(x_0)\right)^{T_{1,\ldots,m,m+2,\ldots,n+m+1,m+1}} + A(x_0) \otimes B^{(1)}(x_0)$$

Proof These are straightforward and are therefore omitted. □

9.9 Analytic Functions

Definition 9.79 Let $\mathcal{X}$ and $\mathcal{Y}$ be normed linear spaces over $\mathbb{K}$, $D \subseteq \mathcal{X}$, $f : D \to \mathcal{Y}$, and $x_0 \in D^\circ$. f is said to be *analytic at* x_0 if f is $\mathcal{C}_\infty$ at x_0, and $\exists \delta \in (0, \infty) \subset \mathbb{R}$, $\exists c \in [0, \infty) \subset \mathbb{R}$, and $\exists M \in (0, \infty) \subset \mathbb{R}$, such that $\left\| f^{(n)}(x) \right\| \leq c n! M^n$, $\forall n \in \mathbb{Z}_+$, $\forall x \in \mathcal{B}_{\mathcal{X}}(x_0, \delta) \subseteq D$. In this case, $\forall \bar{\delta} \in (0, \delta] \subset \mathbb{R}$ with $M\bar{\delta} < 1$ is called an *analytic radius* of f at x_0. If $D \supseteq D_1 \in \mathcal{O}_{\mathcal{X}}$ and f is analytic at x, $\forall x \in D_1$, then we say that f is *analytic on* D_1. When f is analytic on D, then we say that f is *analytic* or an *analytic function*. ◇

Theorem 9.80 (Taylor Series) *Let* $\mathcal{X}$ *and* $\mathcal{Y}$ *be normed linear spaces over* $\mathbb{K}$, $D \subseteq \mathcal{X}$, $f : D \to \mathcal{Y}$, *and* $x_0 \in D^\circ$. *Assume that* f *is analytic at* x_0 *with an analytic radius* $\delta \in (0, \infty) \subset \mathbb{R}$. *Then,* $\forall x \in \mathcal{B}_{\mathcal{X}}(x_0, \delta)$,

$$f(x) = \sum_{n=0}^{\infty} \frac{1}{n!} f^{(n)}(x_0) \underbrace{(x - x_0) \cdots (x - x_0)}_{n-times}$$

Proof Since f is analytic at $x_0 \in D^\circ$ with an analytic radius $\delta \in (0, \infty) \subset \mathbb{R}$, then $\exists c \in [0, \infty) \subset \mathbb{R}$ and $\exists M \in (0, \infty) \subset \mathbb{R}$ such that $\left\| f^{(n)}(x) \right\| \leq c n! M^n$, $\forall n \in \mathbb{Z}_+$,

$\forall x \in \mathcal{B}_{\mathcal{X}}(x_0, \delta) \subseteq D$, and $M\delta < 1$. Then, $\forall x \in \mathcal{B}_{\mathcal{X}}(x_0, \delta)$, $\forall n \in \mathbb{Z}_+$, let

$$R_n(x, x_0; f) := f(x) - \sum_{i=0}^{n} \frac{1}{i!} f^{(i)}(x_0) \underbrace{(x - x_0) \cdots (x - x_0)}_{i-\text{times}}$$

By Taylor Theorem 9.48, $\exists t_0 \in (0, 1) \subset \mathbb{R}$ such that $\|R_n(x, x_0; f)\| \leq \frac{1}{(n+1)!} \left\| f^{(n+1)}(t_0 x + (1 - t_0)x_0) \right\| \|x - x_0\|^{n+1} < c(M\delta)^{n+1}$. Clearly, $\lim_{n \in \mathbb{N}} \|R_n(x, x_0; f)\| = 0$. Hence, the desired equality holds. This completes the proof of the theorem. □

Next, we establish a result that is needed in the further analysis of analytic functions.

Proposition 9.81 *Let $\mathcal{X}$, $\mathcal{Y}$, $\mathcal{Z}$, and $\mathcal{W}$ be normed linear spaces over $\mathbb{K}$, $D_1 \subseteq \mathcal{X}$, $f : D_1 \to \mathrm{B}(\mathcal{Y}, \mathcal{Z})$, $g : D_1 \to \mathrm{B}(\mathcal{Z}, \mathcal{W})$, $n \in \mathbb{Z}_+$, and $x_0 \in D_1^\circ$. Assume that f and g are $\mathcal{C}_n$ at x_0. Define $h : D_1 \to \mathrm{B}(\mathcal{Y}, \mathcal{W})$ by $h(x) = g(x) f(x)$, $\forall x \in D_1$. Then, h is $\mathcal{C}_n$ at x_0 such that*

$$h^{(n)}(x_0)(\xi_{x_n}) \cdots (\xi_{x_1})(\xi_y) = \tag{9.1}$$

$$\sum_{i=0}^{n} \sum_{\substack{\Lambda_1 := \{j_1, \ldots, j_i\} \text{ is any } i\text{-element subset of} \\ \Lambda := \{1, \ldots, n\};\ \Lambda_2 := \{j_{i+1}, \ldots, j_n\} := \Lambda \setminus \Lambda_1}} g^{(i)}(x_0)(\xi_{x_{j_1}}) \cdots (\xi_{x_{j_i}})$$

$$(f^{(n-i)}(x_0)(\xi_{x_{j_{i+1}}}) \cdots (\xi_{x_{j_n}})(\xi_y))$$

$\forall \xi_{x_l} \in \mathcal{X}$, $l = 1, \ldots, n$, $\forall \xi_y \in \mathcal{Y}$.

Proof By Propositions 9.45, 9.44, and 9.42, h is $\mathcal{C}_n$ at x_0. Since f and g are $\mathcal{C}_n$ at x_0, by Proposition 9.28, then $\exists \delta \in (0, \infty) \subset \mathbb{R}$ such that $f^{(n)}(x_0) \in \mathrm{B_S}_n(\mathcal{X}, \mathrm{B}(\mathcal{Y}, \mathcal{Z}))$ and $g^{(n)}(x_0) \in \mathrm{B_S}_n(\mathcal{X}, \mathrm{B}(\mathcal{Z}, \mathcal{W}))$, and $f^{(i)}(x) \in \mathrm{B_S}_i(\mathcal{X}, \mathrm{B}(\mathcal{Y}, \mathcal{Z}))$ and $g^{(i)}(x) \in \mathrm{B_S}_i(\mathcal{X}, \mathrm{B}(\mathcal{Z}, \mathcal{W}))$, $i = 1, \ldots, n-1$, $\forall x \in \mathcal{B}_{\mathcal{X}}(x_0, \delta) \subseteq D_1$. We will use mathematical induction on n to prove the result.

1° $n = 0$. We have, $\forall \xi_y \in \mathcal{Y}$,

$$\text{LHS} = h^{(0)}(x_0)(\xi_y) = h(x_0)(\xi_y) = g(x_0) f(x_0) \xi_y = g(x_0)(f(x_0)(\xi_y)) = \text{RHS}$$

This case is proved.

2° Assume that the result holds for $n \leq l \in \mathbb{Z}_+$.

3° Consider the case when $n = l + 1$. Since f and g are $\mathcal{C}_{l+1}$ at x_0, then $\exists \delta \in (0, \infty) \subset \mathbb{R}$ such that f and g are $\mathcal{C}_l$ at x, $\forall x \in \mathcal{B}_{\mathcal{X}}(x_0, \delta) \subseteq D_1$. By the inductive assumption, $\forall x \in \mathcal{B}_{\mathcal{X}}(x_0, \delta)$, $\forall \xi_{x_s} \in \mathcal{X}$, $s = 1, \ldots, l$, $\forall \xi_y \in \mathcal{Y}$,

$$h^{(l)}(x)(\xi_{x_l}) \cdots (\xi_{x_1})(\xi_y) =$$

$$\sum_{i=0}^{l} \sum_{\substack{\Lambda_1 := \{j_1, \ldots, j_i\} \text{ is any } i\text{-element subset of} \\ \Lambda := \{1, \ldots, l\};\ \Lambda_2 := \{j_{i+1}, \ldots, j_l\} := \Lambda \setminus \Lambda_1}} g^{(i)}(x)(\xi_{x_{j_1}}) \cdots (\xi_{x_{j_i}})$$

$$(f^{(l-i)}(x)(\xi_{x_{j_{i+1}}}) \cdots (\xi_{x_{j_l}})(\xi_y))$$

Then, by Proposition 9.27, we have, $\forall \xi_{x_s} \in \mathcal{X}$, $s = 1, \ldots, l+1$, $\forall \xi_y \in \mathcal{Y}$,

$$h^{(l+1)}(x_0)(\xi_{x_{l+1}}) \cdots (\xi_{x_1}) = \bar{h}^{(1)}(x_0)(\xi_{x_{l+1}})$$

where $\bar{h} : \mathcal{B}_{\mathcal{X}}(x_0, \delta) \to \mathrm{B}(\mathcal{Y}, \mathcal{W})$ is defined by $\bar{h}(x) = h^{(l)}(x)(\xi_{x_l}) \cdots (\xi_{x_1})$, $\forall x \in \mathcal{B}_{\mathcal{X}}(x_0, \delta)$. This implies that, by the chain rule (Theorem 9.18) and Proposition 9.42,

$$h^{(l+1)}(x_0)(\xi_{x_{l+1}}) \cdots (\xi_{x_1})(\xi_y)$$

$$= \sum_{i=0}^{l} \sum_{\substack{\Lambda_1 := \{j_1, \ldots, j_i\} \text{ is any } i\text{-element subset of} \\ \Lambda := \{1, \ldots, l\};\ \Lambda_2 := \{j_{i+1}, \ldots, j_l\} := \Lambda \setminus \Lambda_1}} \Big(g^{(i+1)}(x_0)(\xi_{x_{l+1}})(\xi_{x_{j_1}}) \cdots (\xi_{x_{j_i}})$$

$$(f^{(l-i)}(x_0)(\xi_{x_{j_{i+1}}}) \cdots (\xi_{x_{j_l}})(\xi_y)) + g^{(i)}(x_0)(\xi_{x_{j_1}}) \cdots (\xi_{x_{j_i}})$$

$$(f^{(l-i+1)}(x_0)(\xi_{x_{l+1}})(\xi_{x_{j_{i+1}}}) \cdots (\xi_{x_{j_l}})(\xi_y)) \Big)$$

$$= \sum_{i=0}^{l+1} \Bigg(\sum_{\substack{\Lambda_1 := \{j_1, \ldots, j_i\} \text{ is any } i\text{-element subset of} \\ \Lambda := \{1, \ldots, l\};\ \Lambda_2 := \{j_{i+1}, \ldots, j_l\} := \Lambda \setminus \Lambda_1}} g^{(i)}(x_0)(\xi_{x_{j_1}}) \cdots (\xi_{x_{j_i}})$$

$$(f^{(l+1-i)}(x_0)(\xi_{x_{l+1}})(\xi_{x_{j_{i+1}}}) \cdots (\xi_{x_{j_l}})(\xi_y))$$

$$+ \sum_{\substack{\Lambda_1 := \{j_1, \ldots, j_{i-1}\} \text{ is any } (i-1)\text{-element subset of} \\ \Lambda := \{1, \ldots, l\};\ \Lambda_2 := \{j_i, \ldots, j_l\} := \Lambda \setminus \Lambda_1}} g^{(i)}(x_0)(\xi_{x_{l+1}})(\xi_{x_{j_1}}) \cdots (\xi_{x_{j_{i-1}}})$$

$$(f^{(l+1-i)}(x_0)(\xi_{x_{j_i}}) \cdots (\xi_{x_{j_l}})(\xi_y)) \Bigg)$$

$$= \sum_{i=0}^{l+1} \sum_{\substack{\Lambda_1 := \{j_1, \ldots, j_i\} \text{ is any } i\text{-element subset of} \\ \Lambda := \{1, \ldots, l+1\};\ \Lambda_2 := \{j_{i+1}, \ldots, j_{l+1}\} := \Lambda \setminus \Lambda_1}} g^{(i)}(x_0)(\xi_{x_{j_1}}) \cdots (\xi_{x_{j_i}})$$

$$(f^{(l+1-i)}(x_0)(\xi_{x_{j_{i+1}}}) \cdots (\xi_{x_{j_{l+1}}})(\xi_y))$$

where the last equality follows since all i-element subsets of $\{1, \ldots, l+1\}$ are divided into two disjoint classes: one with $l+1$ as an element; and the other without $l+1$ as an element. This completes the induction process.

This completes the proof of the proposition. □

Now, we establish the important result that the composition of analytic functions is an analytic function.

Theorem 9.82 *Let $\mathcal{X}$, $\mathcal{Y}$, and $\mathcal{Z}$ be normed linear spaces over $\mathbb{K}$, $D_1 \subseteq \mathcal{X}$, $D_2 \subseteq \mathcal{Y}$, $f : D_1 \to D_2$, $g : D_2 \to \mathcal{Z}$, $x_0 \in D_1^\circ$, and $y_0 := f(x_0) \in D_2^\circ$. Assume that f and g are analytic at x_0 and y_0, respectively. Then, $h := g \circ f : D_1 \to \mathcal{Z}$ is analytic at x_0.*

Proof Since g is analytic at y_0, then $\exists \delta_g \in (0, \infty) \subset \mathbb{R}$, $\exists c_g \in [0, \infty) \subset \mathbb{R}$, and $\exists M_g \in (0, \infty) \subset \mathbb{R}$ such that $\|g^{(n)}(y)\| \leq c_g n! M_g^n$, $\forall n \in \mathbb{Z}_+$, $\forall y \in \mathcal{B}_{\mathcal{Y}}(y_0, \delta_g) \subseteq D_2$. Since f is analytic at x_0, then $\exists \delta_f \in (0, \infty) \subset \mathbb{R}$, $\exists c_f \in [0, \infty) \subset \mathbb{R}$, and $\exists M_f \in (0, \infty) \subset \mathbb{R}$ such that $\|f^{(n)}(x)\| \leq c_f n! M_f^n$ and $f(x) \in \mathcal{B}_{\mathcal{Y}}(y_0, \delta_g)$, $\forall n \in \mathbb{Z}_+$, $\forall x \in \mathcal{B}_{\mathcal{X}}(x_0, \delta_f)$. By Proposition 9.45, h is $\mathcal{C}_\infty$ at $\forall x \in \mathcal{B}_{\mathcal{X}}(x_0, \delta_f)$.

Define nonnegative constants $K_{n,i} := \sup_{x \in \mathcal{B}_{\mathcal{X}}(x_0, \delta_f)} \|\mathrm{D}^n(g^{(i)} \circ f)(x)\|$, $\forall i, n \in \mathbb{Z}_+$. Clearly, $\forall i \in \mathbb{Z}_+$,

$$K_{0,i} = \sup_{x \in \mathcal{B}_{\mathcal{X}}(x_0, \delta_f)} \|g^{(i)} \circ f(x)\| \leq \sup_{y \in \mathcal{B}_{\mathcal{Y}}(y_0, \delta_g)} \|g^{(i)}(y)\| \leq c_g i! M_g^i$$

Note that, $\forall i, n \in \mathbb{Z}_+$, by Chain Rule (Theorem 9.18),

$$\begin{aligned} K_{n+1,i} &= \sup_{x \in \mathcal{B}_{\mathcal{X}}(x_0, \delta_f)} \|\mathrm{D}^{n+1}(g^{(i)} \circ f)(x)\| \\ &= \sup_{x \in \mathcal{B}_{\mathcal{X}}(x_0, \delta_f)} \|\mathrm{D}^n((g^{(i+1)} \circ f)(f^{(1)}))(x)\| \end{aligned}$$

Define $l_{i+1} : \mathcal{B}_{\mathcal{X}}(x_0, \delta_f) \to \mathrm{B}_{\mathrm{S}\,i+1}(\mathcal{Y}, \mathcal{Z})$ by $l_{i+1}(x) := g^{(i+1)}(f(x))$, $\forall x \in \mathcal{B}_{\mathcal{X}}(x_0, \delta_f)$, $\forall i \in \mathbb{Z}_+$. Then, $\forall i, n \in \mathbb{Z}_+$,

$$\begin{aligned} K_{n+1,i} &= \sup_{x \in \mathcal{B}_{\mathcal{X}}(x_0, \delta_f)} \|\mathrm{D}^n((g^{(i+1)} \circ f)(f^{(1)}))(x)\| \\ &= \sup_{x \in \mathcal{B}_{\mathcal{X}}(x_0, \delta_f)} \|\mathrm{D}^n(l_{i+1} f^{(1)})(x)\| \\ &= \sup_{x \in \mathcal{B}_{\mathcal{X}}(x_0, \delta_f)} \sup_{\substack{\xi_{x_s} \in \mathcal{X}, s=1,\ldots,n, \xi_x \in \mathcal{X}, \\ \|\xi_{x_s}\| \leq 1, \|\xi_x\| \leq 1}} \|\mathrm{D}^n(l_{i+1} f^{(1)})(x)(\xi_{x_n}) \cdots (\xi_{x_1})(\xi_x)\| \\ &= \sup_{x \in \mathcal{B}_{\mathcal{X}}(x_0, \delta_f)} \sup_{\substack{\xi_{x_s} \in \mathcal{X}, s=1,\ldots,n, \xi_x \in \mathcal{X}, \\ \|\xi_{x_s}\| \leq 1, \|\xi_x\| \leq 1}} \Bigg\| \sum_{k=0}^{n} \end{aligned}$$

$$\sum_{\substack{\Lambda_1:=\{j_1,\ldots,j_k\}\text{ is any }k\text{-element subset of}\\ \Lambda:=\{1,\ldots,n\};\Lambda_2:=\{j_{k+1},\ldots,j_n\}:=\Lambda\setminus\Lambda_1}} l_{i+1}^{(k)}(x)(\xi_{x j_1})\cdots(\xi_{x j_k})$$

$$(f^{(n-k+1)}(x)(\xi_{x j_{k+1}})\cdots(\xi_{x j_n})(\xi_x))\Big\|$$

$$\leq \sup_{x\in\mathcal{B}_{\mathcal{X}}(x_0,\delta_f)} \sum_{k=0}^{n} \sum_{\substack{\Lambda_1:=\{j_1,\ldots,j_k\}\text{ is any }k\text{-element subset of}\\ \Lambda:=\{1,\ldots,n\};\Lambda_2:=\{j_{k+1},\ldots,j_n\}:=\Lambda\setminus\Lambda_1}} \left\|l_{i+1}^{(k)}(x)\right\| \cdot$$

$$\left\|f^{(n-k+1)}(x)\right\|$$

$$= \sup_{x\in\mathcal{B}_{\mathcal{X}}(x_0,\delta_f)} \sum_{k=0}^{n} \binom{n}{k} \left\|\mathrm{D}^k(g^{(i+1)}\circ f)(x)\right\| \left\|f^{(n-k+1)}(x)\right\|$$

$$\leq \sum_{k=0}^{n} \binom{n}{k} K_{k,i+1} c_f (n+1-k)! M_f^{n+1-k}$$

where the fourth equality follows from Proposition 9.81. Hence, $0 \leq K_{n+1,i} \leq \sum_{k=0}^{n} \binom{n}{k} K_{k,i+1} c_f (n-k+1)! M_f^{n-k+1}$, $\forall i, n \in \mathbb{Z}_+$.

We will use mathematical induction on $n \in \mathbb{N}$ to show that $\forall n \in \mathbb{N}$, $\forall i \in \mathbb{Z}_+$,

$$0 \leq K_{n,i} \leq (n+i)! c_g M_g^{i+1} M_f^n \sum_{j=0}^{n-1} \binom{n}{j+1}\binom{n-1}{j}\binom{n+i}{n-j-1}^{-1} c_f^{j+1} M_g^j$$

1° $n = 1$. By the recursive formula and the inequality for $K_{0,i}$, $\forall i \in \mathbb{Z}_+$,

$$K_{1,i} \leq K_{0,i+1} c_f M_f \leq (i+1)! c_f M_f c_g M_g^{i+1} = \text{RHS}$$

This case is proved.

2° Assume that the result holds for $n = 1, \ldots, l \in \mathbb{N}$.

3° Consider the case $n = l + 1$. $\forall i \in \mathbb{Z}_+$, by the recursive formula,

$$K_{l+1,i} \leq \sum_{k=0}^{l} \binom{l}{k} K_{k,i+1} c_f (l-k+1)! M_f^{l-k+1}$$

$$\leq c_f (l+1)! M_f^{l+1} c_g (i+1)! M_g^{i+1} + \sum_{k=1}^{l} \binom{l}{k} c_f (l-k+1)! M_f^{l-k+1}$$

$$\cdot (k+i+1)! c_g M_g^{i+2} M_f^k \sum_{j=0}^{k-1} \binom{k}{j+1}\binom{k-1}{j}\binom{k+i+1}{k-j-1}^{-1} c_f^{j+1} M_g^j$$

$$= c_f c_g M_g^{i+1} M_f^{l+1} (l+i+1)!(l+1) \binom{l+i+1}{l}^{-1}$$

$$+ \sum_{k=1}^{l} \sum_{j=0}^{k-1} \frac{l!(l-k+1)!k!(k-1)!(i+j+2)!}{k!(l-k)!j!(j+1)!(k-j-1)!} c_f^{j+2} c_g M_f^{l+1} M_g^{j+i+2}$$

$$= c_f c_g M_g^{i+1} M_f^{l+1} (l+i+1)!(l+1) \binom{l+i+1}{l}^{-1}$$

$$+ \sum_{j=0}^{l-1} \sum_{k=j+1}^{l} \frac{l!(l-k+1)!(k-1)!(i+j+2)!}{(l-k)!j!(j+1)!(k-j-1)!} c_f^{j+2} c_g M_f^{l+1} M_g^{j+i+2}$$

$$= (l+i+1)! c_g M_g^{i+1} M_f^{l+1} \Bigg(c_f (l+1) \binom{l+i+1}{l}^{-1}$$

$$+ \sum_{j=0}^{l-1} \sum_{k=j+1}^{l} \frac{(l-k+1)l!(k-1)!(i+j+2)!}{j!(j+1)!(k-j-1)!(l+i+1)!} c_f^{j+2} M_g^{j+1} \Bigg)$$

$$= (l+i+1)! c_g M_g^{i+1} M_f^{l+1} \Bigg(c_f (l+1) \binom{l+i+1}{l}^{-1}$$

$$+ \sum_{j=0}^{l-1} \sum_{k=j+1}^{l} (l-k+1) \binom{k-1}{j} \binom{l}{j+1} \binom{l+i+1}{l-j-1}^{-1} c_f^{j+2} M_g^{j+1} \Bigg)$$

$$= (l+i+1)! c_g M_g^{i+1} M_f^{l+1} \Bigg(c_f (l+1) \binom{l+i+1}{l}^{-1}$$

$$+ \sum_{j=1}^{l} \binom{l}{j} \binom{l+i+1}{l-j}^{-1} c_f^{j+1} M_g^{j} \sum_{k=j}^{l} (l-k+1) \binom{k-1}{j-1} \Bigg)$$

where the second inequality follows from the inductive assumption and the bound for $K_{0,i}$. Note that $\sum_{k=j}^{l}(l-k+1)\binom{k-1}{j-1} = \sum_{k=j}^{l}(l-k+1)(\binom{k}{j} - \binom{k-1}{j}) = \sum_{k=j}^{l}(l-k+1)\binom{k}{j} - \sum_{k=j-1}^{l-1}(l-k)\binom{k}{j}) = \sum_{k=j}^{l}\binom{k}{j} = \sum_{k=j}^{l}(\binom{k+1}{j+1} - \binom{k}{j+1}) -$

$\binom{l+1}{j+1}$, $\forall 1 \le j \le l$. This leads to

$$K_{l+1,i} \le (l+i+1)!c_g M_g^{i+1} M_f^{l+1} \Big(c_f(l+1) \binom{l+i+1}{l}^{-1} + \sum_{j=1}^{l} \binom{l}{j} \binom{l+1}{j+1} \binom{l+i+1}{l-j}^{-1} c_f^{j+1} M_g^j \Big)$$

$$= (l+i+1)!c_g M_g^{i+1} M_f^{l+1} \sum_{j=0}^{l} \binom{l}{j} \binom{l+1}{j+1} \binom{l+i+1}{l-j}^{-1} c_f^{j+1} M_g^j$$

$$= \text{RHS}$$

This completes the induction process.

Hence, $\forall n \in \mathbb{N}$,

$$\sup_{x \in \mathcal{B}_{\mathcal{X}}(x_0, \delta_f)} \left\| h^{(n)}(x) \right\| = \sup_{x \in \mathcal{B}_{\mathcal{X}}(x_0, \delta_f)} \left\| \mathrm{D}^n (g \circ f)(x) \right\| = K_{n,0}$$

$$\le n!c_g M_g M_f^n \sum_{j=0}^{n-1} \binom{n}{j+1} \binom{n-1}{j} \binom{n}{n-j-1}^{-1} c_f^{j+1} M_g^j$$

$$= n!c_g M_g M_f^n \sum_{j=0}^{n-1} \binom{n-1}{j} c_f^{j+1} M_g^j = n!c_f c_g M_g M_f^n (1 + c_f M_g)^{n-1}$$

$$= n! \frac{c_f c_g M_g}{1 + c_f M_g} (M_f (1 + c_f M_g))^n$$

Hence, h is analytic at x_0. This completes the proof of the theorem. □

Proposition 9.83 *The constant mapping as defined in Proposition* 9.33, *the identity mapping as delineated in Proposition* 9.36, *the projection mapping as delineated in Proposition* 9.37, *the vector addition, the scalar multiplication, the operation of a linear operator on a vector as delineated in Proposition* 9.41, *the operation of composition of two linear operators as delineated in Proposition* 9.42, *and the building up of linear operator on the product space using linear operators on individual spaces as delineated in Proposition* 9.43 *are analytic on the interior of the domain of the mapping.*

Proof This is immediate from Definition 9.79. □

Proposition 9.84 *Let $\mathcal{X}$ and $\mathcal{Y}$ be Banach spaces over $\mathbb{K}$, $D := \{L \in \mathrm{B}(\mathcal{X}, \mathcal{Y}) \mid L$ is bijective.$\}$, and $f : D \to \mathrm{B}(\mathcal{Y}, \mathcal{X})$ be defined by $f(A) = A^{-1}$, $\forall A \in D \subseteq \mathrm{B}(\mathcal{X}, \mathcal{Y})$. Then, f is analytic.*

Proof By Proposition 9.56, D is open in $\mathrm{B}(\mathcal{X}, \mathcal{Y})$, f is $\mathcal{C}_\infty$, and $f^{(1)}(A)(\Delta) = -f(A)\Delta f(A)$, $\forall A \in D$, $\forall \Delta \in \mathrm{B}(\mathcal{X}, \mathcal{Y})$. Fix any $A \in D$. Since D is open, then $\exists \delta \in (0, \infty) \subset \mathbb{R}$ such that $U := \mathcal{B}_{\mathrm{B}(\mathcal{X}, \mathcal{Y})}(A, \delta) \subseteq D$. Without loss of generality, assume that $\delta \|f(A)\| < 1$. By Proposition 9.55, $\forall T \in U$, we have $\|f(T) - f(A)\| \le \frac{\|f(A)\|^2 \|T-A\|}{1-\|f(A)\|\|T-A\|} \le \|f(A)\|^2 \delta (1 - \|f(A)\|\delta)^{-1}$. This implies that $\|f(T)\| \le \|f(A)\| + \|f(A)\|^2 \delta (1 - \|f(A)\|\delta)^{-1} = \|f(A)\| (1 - \delta \|f(A)\|)^{-1} =: \alpha \in [0, \infty) \subset \mathbb{R}$. This leads to

$$\left\| f^{(1)}(T) \right\| = \sup_{\Delta \in \mathrm{B}(\mathcal{X}, \mathcal{Y}), \|\Delta\| \le 1} \left\| f^{(1)}(T)(\Delta) \right\| \le \|f(T)\|^2 \le \alpha^2$$

We will use mathematical induction on $n \in \mathbb{Z}_+$ to show that $\left\| f^{(n)}(T) \right\| \le n!\alpha^{n+1}$, $\forall n \in \mathbb{Z}_+$, $\forall T \in U$.

1° Clearly, $n = 0$ and $n = 1$ cases are proved.

2° Assume that the result holds for $n = 0, \ldots, k \in \mathbb{N}$.

3° Consider the case when $n = k + 1$. $\forall T \in U$, $\forall \Delta_1 \in \mathrm{B}(\mathcal{X}, \mathcal{Y})$, $f^{(1)}(T)(\Delta_1) = -f(T)\Delta_1 f(T) =: g_{\Delta_1}(T)$, where $g_{\Delta_1} : U \to \mathrm{B}(\mathcal{Y}, \mathcal{X})$. By Proposition 9.27, we have, $\forall \Delta_i \in \mathrm{B}(\mathcal{X}, \mathcal{Y})$, $i = 2, \ldots, k + 1$,

$$g_{\Delta_1}^{(k)}(T)(\Delta_{k+1}) \cdots (\Delta_2) = f^{(k+1)}(T)(\Delta_{k+1}) \cdots (\Delta_2)(\Delta_1)$$

This implies that

$$\begin{aligned} \left\| f^{(k+1)}(T) \right\| &= \sup_{\Delta_i \in \mathrm{B}(\mathcal{X}, \mathcal{Y}), \|\Delta_i\| \le 1, i=1,\ldots,k+1} \left\| f^{(k+1)}(T)(\Delta_{k+1}) \cdots (\Delta_1) \right\| \\ &= \sup_{\Delta_i \in \mathrm{B}(\mathcal{X}, \mathcal{Y}), \|\Delta_i\| \le 1, i=1,\ldots,k+1} \left\| g_{\Delta_1}^{(k)}(T)(\Delta_{k+1}) \cdots (\Delta_2) \right\| \end{aligned}$$

By Proposition 9.81, we have the following equality

$$\begin{aligned} & g_{\Delta_1}^{(k)}(T)(\Delta_{k+1}) \cdots (\Delta_2) = \\ & -\sum_{i=0}^{k} \sum_{\substack{\Lambda_1 := \{j_1, \ldots, j_i\} \text{ is any } i\text{-element subset of} \\ \Lambda := \{2, \ldots, k+1\};\, \Lambda_2 := \{j_{i+1}, \ldots, j_k\} := \Lambda \setminus \Lambda_1}} f^{(i)}(T)(\Delta_{j_1}) \cdots (\Delta_{j_i}) \\ & (\Delta_1 f^{(k-i)}(T)(\Delta_{j_{i+1}}) \cdots (\Delta_{j_k})) \end{aligned}$$

Then,

$$\begin{aligned}\left\|f^{(k+1)}(T)\right\| &\leq \sum_{i=0}^{k} \sum_{\substack{\Lambda_1:=\{j_1,\ldots,j_i\} \text{ is any } i\text{-element subset of} \\ \Lambda:=\{2,\ldots,k+1\};\, \Lambda_2:=\{j_{i+1},\ldots,j_k\}:=\Lambda\setminus\Lambda_1}} \left\|f^{(i)}(T)\right\| \left\|f^{(k-i)}(T)\right\| \\ &= \sum_{i=0}^{k} \binom{k}{i} \left\|f^{(i)}(T)\right\| \left\|f^{(k-i)}(T)\right\| \\ &\leq \sum_{i=0}^{k} \binom{k}{i} i!\alpha^{i+1}(k-i)!\alpha^{k+1-i} = \sum_{i=0}^{k} k!\alpha^{k+2} = (k+1)!\alpha^{k+2}\end{aligned}$$

where the last inequality follows from the inductive assumption. This completes the induction process.

Hence, f is analytic at T. By the arbitrariness of T, f is an analytic function. This completes the proof of the proposition. □

Finally, we establish the analytic versions of the Inverse Function Theorem and the Implicit Function Theorem.

Theorem 9.85 (Inverse Function Theorem) *Let $\mathcal{X}$ and $\mathcal{Y}$ be Banach spaces over $\mathbb{K}$, $D \subseteq \mathcal{X}$, $F : D \to \mathcal{Y}$ be analytic at $x_0 \in D^\circ$, and $y_0 = F(x_0) \in \mathcal{Y}$. Assume that $F^{(1)}(x_0) \in \mathrm{B}(\mathcal{X}, \mathcal{Y})$ is bijective. Then, $\exists \delta \in (0, \infty) \subset \mathbb{R}$, $\exists U \subseteq D$ with $x_0 \in U \in \mathcal{O}_{\mathcal{X}}$ such that:*

(i) *$F|_U : U \to \mathcal{B}_{\mathcal{Y}}(y_0, \delta) =: V \subseteq \mathcal{Y}$ is bijective.*
(ii) *The inverse mapping $F_i : V \to U$ is $\mathcal{C}_\infty$, and $F_i^{(1)} : V \to \mathrm{B}(\mathcal{Y}, \mathcal{X})$ is given by $F_i^{(1)}(y) = (F^{(1)}(F_i(y)))^{-1}$, $\forall y \in V$.*
(iii) *The inverse mapping F_i is analytic at $y_0 \in V$.*

Proof The result (i) and (ii) are implied by Inverse Function Theorem 9.57. All we need to show is (iii). Since F is analytic at x_0, then $\exists \delta_1 \in (0, \infty) \subset \mathbb{R}$, $\exists c \in [0, \infty) \subset \mathbb{R}$, and $\exists M \in (0, \infty) \subset \mathbb{R}$ such that $\bar{U} := \mathcal{B}_{\mathcal{X}}(x_0, \delta_1) \subseteq U$, $\left\|F^{(n)}(x)\right\| \leq cn!M^n$, $\forall n \in \mathbb{Z}_+$, $\forall x \in \bar{U}$. Then, $\exists \delta_2 \in (0, \delta] \subset \mathbb{R}$ such that $F_i(\mathcal{B}_{\mathcal{Y}}(y_0, \delta_2)) \subseteq \bar{U}$.

Define $h : \bar{U} \to \mathrm{B}(\mathcal{Y}, \mathcal{X})$ by $h(x) = (F^{(1)}(x))^{-1}$, $\forall x \in \bar{U}$. Since $F|_{\bar{U}}$ is an analytic function, then $F^{(1)}|_{\bar{U}}$ is an analytic function. By Proposition 9.84 and Theorem 9.82, h is an analytic function. Then, $\exists \bar{\delta}_1 \in (0, \delta_1] \subset \mathbb{R}$, $\exists c_h \in [0, \infty) \subset \mathbb{R}$, and $\exists M_h \in (0, \infty) \subset \mathbb{R}$ such that $\left\|h^{(n)}(x)\right\| \leq c_h n! M_h^n$, $\forall n \in \mathbb{Z}_+$, $\forall x \in \mathcal{B}_{\mathcal{X}}\big(x_0, \bar{\delta}_1\big)$. By choosing δ_1 sufficiently small, without loss of generality, we may take $\bar{\delta}_1 = \delta_1$.

Note that $F_i^{(1)} \circ F(x) = h(x)$, $\forall x \in \bar{U}$. Then, $h^{(1)}(x) = \mathrm{D}(F_i^{(1)} \circ F)(x) = (F_i^{(2)} \circ F(x))F^{(1)}(x)$ and $F_i^{(2)} \circ F(x) = (h^{(1)}(x))h(x)$, $\forall x \in \bar{U}$. Define $h_k : \bar{U} \to \mathrm{B}_{S^k}(\mathcal{Y}, \mathcal{X})$ by $h_k(x) = F_i^{(k)} \circ F(x)$, $\forall x \in \bar{U}$, $\forall 2 \leq k \in \mathbb{N}$. This leads to the recursive relationship

$$(\mathrm{D}h_k(x))h(x) = F_i^{(k+1)} \circ F(x) = h_{k+1}(x), \quad \forall x \in \bar{U}, \ \forall 2 \le k \in \mathbb{N}$$

with the initial condition $h_2(x) = h^{(1)}(x)h(x)$, $\forall x \in \bar{U}$. Define the nonnegative constants $K_{k,j} := \sup_{x\in\bar{U}} \left\|\mathrm{D}^j h_k(x)\right\|$, $\forall j \in \mathbb{Z}_+$, $\forall 2 \le k \in \mathbb{N}$. We will use mathematical induction on $2 \le k \in \mathbb{N}$ to show that

$$0 \le K_{k,j} \le \frac{(j+2k-2)!}{(k-1)!2^{k-1}} c_h^k M_h^{j+k-1}, \qquad \forall j \in \mathbb{Z}_+, \ \forall 2 \le k \in \mathbb{N}$$

1° $k = 2$. $\forall j \in \mathbb{Z}_+$, by the initial condition of the recursion,

$$\begin{aligned}
K_{2,j} &= \sup_{x\in\bar{U}} \left\|\mathrm{D}^j h_2(x)\right\| = \sup_{x\in\bar{U}} \left\|\mathrm{D}^j (h^{(1)}h)(x)\right\| \\
&= \sup_{x\in\bar{U}} \sup_{\substack{\xi_{x_l}\in\mathcal{X}, \|\xi_{x_l}\|\le 1, l=1,\ldots,j \\ \xi_y\in\mathcal{Y}, \|\xi_y\|\le 1}} \left\|\mathrm{D}^j (h^{(1)}h)(x)(\xi_{x_j})\cdots(\xi_{x_1})(\xi_y)\right\| \\
&= \sup_{x\in\bar{U}} \sup_{\substack{\xi_{x_l}\in\mathcal{X}, \|\xi_{x_l}\|\le 1, l=1,\ldots,j \\ \xi_y\in\mathcal{Y}, \|\xi_y\|\le 1}} \Bigg\| \sum_{l=0}^{j} \\
&\qquad \sum_{\substack{\Lambda_1:=\{s_1,\ldots,s_l\} \text{ is any } l\text{-element subset of} \\ \Lambda:=\{1,\ldots,j\};\ \Lambda_2:=\{s_{l+1},\ldots,s_j\}:=\Lambda\setminus\Lambda_1}} h^{(l+1)}(x)(\xi_{x_{s_1}})\cdots(\xi_{x_{s_l}}) \\
&\qquad (h^{(j-l)}(x)(\xi_{x_{s_{l+1}}})\cdots(\xi_{x_{s_j}})(\xi_y))\Bigg\| \\
&\le \sup_{x\in\bar{U}} \sum_{l=0}^{j} \sum_{\substack{\Lambda_1:=\{s_1,\ldots,s_l\} \text{ is any } l\text{-element subset of} \\ \Lambda:=\{1,\ldots,j\};\ \Lambda_2:=\{s_{l+1},\ldots,s_j\}:=\Lambda\setminus\Lambda_1}} \left\|h^{(l+1)}(x)\right\| \left\|h^{(j-l)}(x)\right\| \\
&= \sup_{x\in\bar{U}} \sum_{l=0}^{j} \binom{j}{l} \left\|h^{(l+1)}(x)\right\| \left\|h^{(j-l)}(x)\right\| \\
&\le \sum_{l=0}^{j} \frac{j!}{l!(j-l)!} c_h (l+1)! M_h^{l+1} c_h (j-l)! M_h^{j-l} \\
&= j! c_h^2 M_h^{j+1} \sum_{l=0}^{j} (l+1) = \frac{(j+2)!}{2} c_h^2 M_h^{j+1} = \text{RHS}
\end{aligned}$$

where the fourth equality follows from Proposition 9.81. This case is proved.

2° Assume that the result holds for $k = n \geq 2$ with $n \in \mathbb{N}$.

3° Consider the case when $k = n + 1$. $\forall j \in \mathbb{Z}_+$, by the recursive relationship,

$$K_{n+1,j} = \sup_{x\in\bar{U}} \left\| \mathrm{D}^j h_{n+1}(x) \right\| = \sup_{x\in\bar{U}} \left\| \mathrm{D}^j (h_n^{(1)} h)(x) \right\|$$

$$= \sup_{x\in\bar{U}} \sup_{\substack{\xi_{x_l}\in\mathcal{X}, \|\xi_{x_l}\|\leq 1, l=1,\ldots,j \\ \xi_y\in\mathcal{Y}, \|\xi_y\|\leq 1}} \left\| \mathrm{D}^j (h_n^{(1)} h)(x)(\xi_{x_j}) \cdots (\xi_{x_1})(\xi_y) \right\|$$

$$= \sup_{x\in\bar{U}} \sup_{\substack{\xi_{x_l}\in\mathcal{X}, \|\xi_{x_l}\|\leq 1, l=1,\ldots,j \\ \xi_y\in\mathcal{Y}, \|\xi_y\|\leq 1}} \left\| \sum_{l=0}^{j} \right.$$

$$\sum_{\substack{\Lambda_1:=\{s_1,\ldots,s_l\} \text{ is any } l\text{-element subset of} \\ \Lambda:=\{1,\ldots,j\};\, \Lambda_2:=\{s_{l+1},\ldots,s_j\}:=\Lambda\setminus\Lambda_1}} h_n^{(l+1)}(x)(\xi_{x_{s_1}}) \cdots (\xi_{x_{s_l}})$$

$$\left. (h^{(j-l)}(x)(\xi_{x_{s_{l+1}}}) \cdots (\xi_{x_{s_j}})(\xi_y)) \right\|$$

$$\leq \sup_{x\in\bar{U}} \sum_{l=0}^{j} \sum_{\substack{\Lambda_1:=\{s_1,\ldots,s_l\} \text{ is any } l\text{-element subset of} \\ \Lambda:=\{1,\ldots,j\};\, \Lambda_2:=\{s_{l+1},\ldots,s_j\}:=\Lambda\setminus\Lambda_1}} \left\| h_n^{(l+1)}(x) \right\| \left\| h^{(j-l)}(x) \right\|$$

$$= \sup_{x\in\bar{U}} \sum_{l=0}^{j} \binom{j}{l} \left\| h_n^{(l+1)}(x) \right\| \left\| h^{(j-l)}(x) \right\|$$

$$\leq \sup_{x\in\bar{U}} \sum_{l=0}^{j} \binom{j}{l} K_{n,l+1} \left\| h^{(j-l)}(x) \right\|$$

$$\leq \sum_{l=0}^{j} \frac{j!}{l!(j-l)!} \frac{(l+2n-1)!}{(n-1)!2^{n-1}} c_h^n M_h^{l+n} c_h (j-l)! M_h^{j-l}$$

$$= c_h^{n+1} M_h^{j+n} \frac{j!}{(n-1)!2^{n-1}} \sum_{l=0}^{j} \prod_{s=1}^{2n-1} (l+s)$$

$$= c_h^{n+1} M_h^{j+n} \frac{j!}{(n-1)!2^{n-1}} \frac{1}{2n} \frac{(j+2n)!}{j!} = \frac{(j+2n)!}{n!2^n} c_h^{n+1} M_h^{j+n} = \mathrm{RHS}$$

where the fourth equality follows from Proposition 9.81; the second inequality follows the inductive assumption; and the seventh equality follows from Proposition A.1. This completes the induction process.

Based on this inequality, we have the following bounds for the high order derivatives of F_i around y_0. $\forall y \in \mathcal{B}_{\mathcal{Y}}(y_0, \delta_2)$, $\forall 2 \leq n \in \mathbb{N}$,

$$\left\| F_i^{(n)}(y) \right\| \leq \sup_{x \in \bar{U}} \left\| F_i^{(n)} \circ F(x) \right\| = \sup_{x \in \bar{U}} \|h_n(x)\| \leq K_{n,0}$$

$$\leq \frac{(2n-2)!}{(n-1)!2^{n-1}} c_h^n M_h^{n-1} \leq (n-1)!2^{n-1} c_h^n M_h^{n-1} = \frac{1}{2M_h n} n!(2c_h M_h)^n$$

Hence, F_i is analytic at y_0. This completes the proof of the theorem. □

Theorem 9.86 (Implicit Function Theorem) *Let $\mathcal{X}$, $\mathcal{Y}$, and $\mathcal{Z}$ be Banach spaces over $\mathbb{K}$, $D \subseteq \mathcal{X} \times \mathcal{Y}$, $F : D \to \mathcal{Z}$, and $(x_0, y_0) \in D^\circ$. Assume that F is analytic at (x_0, y_0), $F(x_0, y_0) = \vartheta_{\mathcal{Z}}$, and $\frac{\partial F}{\partial y}(x_0, y_0) \in \mathrm{B}(\mathcal{Y}, \mathcal{Z})$ is bijective. Then, the following statements hold:*

(i) *There exists $r_0, r_1 \in (0, \infty) \subset \mathbb{R}$ such that $U \times V := \mathcal{B}_{\mathcal{X}}(x_0, r_0) \times \mathcal{B}_{\mathcal{Y}}(y_0, r_1) \subseteq D$ and $\forall x \in U$, $\exists!\ y \in V$ satisfying $F(x, y) = \vartheta_{\mathcal{Z}}$. This defines a function $\phi : U \to V$ by $\phi(x) = y$, $\forall x \in U$. Then, ϕ is $\mathcal{C}_\infty$.*
(ii) $\phi^{(1)}(x) = -(\frac{\partial F}{\partial y}(x, \phi(x)))^{-1} \frac{\partial F}{\partial x}(x, \phi(x))$, $\forall x \in U$.
(iii) *ϕ is analytic at x_0.*

Proof The results (i) and (ii) follow immediately from Implicit Function Theorem 9.59. All we need to show is (iii). Since F is analytic at (x_0, y_0), we may choose r_0 and r_1 to be sufficiently small such that $\exists c \in [0, \infty) \subset \mathbb{R}$ and $\exists M \in (0, \infty) \subset \mathbb{R}$, $\left\| F^{(n)}(x, y) \right\| \leq cn!M^n$, $\forall n \in \mathbb{Z}_+$, $\forall (x, y) \in U \times V$.

Define mapping $\bar{F} : D \to \mathcal{X} \times \mathcal{Z}$ by $\bar{F}(x, y) = (x, F(x, y))$, $\forall (x, y) \in U \times V$. By Proposition 9.83 and Theorem 9.82, $\bar{F}$ is analytic at $(x_0, y_0) \in D^\circ$. By Propositions 9.44, 9.13, and 9.24,

$$\bar{F}^{(1)}(x_0, y_0) = \begin{bmatrix} \mathrm{id}_{\mathcal{X}} & \vartheta_{\mathrm{B}(\mathcal{Y}, \mathcal{X})} \\ \frac{\partial F}{\partial x}(x_0, y_0) & \frac{\partial F}{\partial y}(x_0, y_0) \end{bmatrix}$$

Clearly, $\bar{F}^{(1)}(x_0, y_0)$ is bijective. By Proposition 4.31, $\mathcal{X} \times \mathcal{Y}$ and $\mathcal{X} \times \mathcal{Z}$ are Banach spaces over $\mathbb{K}$. By the analytic version of the Inverse Function Theorem 9.85, $\exists \bar{U} \subseteq D$ with $(x_0, y_0) \in \bar{U} \in \mathcal{O}_{\mathcal{X} \times \mathcal{Y}}$, $\exists \delta \in (0, \infty) \subset \mathbb{R}$ such that $\bar{F}|_{\bar{U}} : \bar{U} \to \mathcal{B}_{\mathcal{X} \times \mathcal{Z}}((x_0, \vartheta_{\mathcal{Z}}), \delta)$ is bijective, whose inverse function $\bar{F}_i : \mathcal{B}_{\mathcal{X} \times \mathcal{Z}}((x_0, \vartheta_{\mathcal{Z}}), \delta) \to \bar{U}$ is analytic at $(x_0, \vartheta_{\mathcal{Z}})$. By taking r_0 and r_1 sufficiently small, we may assume that $U \times V \subseteq \bar{U}$. Then, $\exists \delta_1 \in (0, \min\{\delta, r_0\}] \subset \mathbb{R}$, $\exists W \subseteq U \times V$ such that $\bar{F}|_W \to \mathcal{B}_{\mathcal{X} \times \mathcal{Z}}((x_0, \vartheta_{\mathcal{Z}}), \delta_1)$ is bijective. By the uniqueness of $y \in V$ that solves $F(x, y) = \vartheta_{\mathcal{Z}}$, we have $\phi(x) = \pi_{\mathcal{Y}} \circ \bar{F}_i(x, \vartheta_{\mathcal{Z}})$, $\forall x \in \mathcal{B}_{\mathcal{X}}(x_0, \delta_1)$. By Theorem 9.82 and Proposition 9.83, ϕ is analytic at x_0. This completes the proof of the theorem. □

Definition 9.87 Let $\mathcal{X}$ be a Banach space over $\mathbb{K}$ and $A \in \mathrm{B}(\mathcal{X}, \mathcal{X})$. We will define $\exp(A) \in \mathrm{B}(\mathcal{X}, \mathcal{X})$ by $\exp(A) := \sum_{n=0}^{\infty} \frac{1}{n!} A^n$, where $A^0 := \mathrm{id}_{\mathcal{X}}$ by notation. ◇

This function is well-defined since $\mathrm{B}(\mathcal{X}, \mathcal{X})$ is a Banach space over $\mathbb{K}$ by Proposition 7.66, and the series is absolutely summable since $\|A\|_{\mathrm{B}(\mathcal{X},\mathcal{X})} < \infty$. Thus, the series converges in $\mathrm{B}(\mathcal{X}, \mathcal{X})$ by Proposition 7.27. The following proposition lists a number of properties for the exponential function.

Proposition 9.88 *Let $\mathcal{X}$ be a Banach space over $\mathbb{K}$, and $A \in \mathrm{B}(\mathcal{X}, \mathcal{X})$. $\forall t, \tau \in \mathbb{K}$, we have:*

(i) $h : \mathbb{K} \to \mathrm{B}(\mathcal{X}, \mathcal{X})$ *defined by* $h(t) = \exp(At)$, $\forall t \in \mathbb{K}$, *is an analytic function.* $\forall t \in \mathbb{K}$, *h admits arbitrarily large analytic radius at t.*
(ii) $\exp(A(t+\tau)) = \exp(At)\exp(A\tau)$ *and* $\exp(\vartheta_{\mathrm{B}(\mathcal{X},\mathcal{X})}) = \mathrm{id}_{\mathcal{X}}$.
(iii) $\frac{\mathrm{d}}{\mathrm{d}t}\exp(At) = A\exp(At) = \exp(At)A$.
(iv) *If* $B \in \mathrm{B}(\mathcal{X}, \mathcal{X})$ *and* $AB = BA$, *then* $\exp(At)B = B\exp(At)$, *and* $\exp(A)\exp(B) = \exp(B)\exp(A) = \exp(A+B)$.
(v) $(\exp(A))^{-1} = \exp(-A)$ *and* $(\exp(A))' = \exp(A')$.
(vi) $\|\exp(A)\|_{\mathrm{B}(\mathcal{X},\mathcal{X})} \leq \exp(\|A\|_{\mathrm{B}(\mathcal{X},\mathcal{X})}) < \infty$.
(vii) *The function* $\exp : \mathrm{B}(\mathcal{X}, \mathcal{X}) \to \mathrm{B}(\mathcal{X}, \mathcal{X})$ *is Fréchet differentiable with* $\|\mathrm{D}\exp(A)\|_{\mathrm{B}(\mathrm{B}(\mathcal{X},\mathcal{X}),\mathrm{B}(\mathcal{X},\mathcal{X}))} \leq \exp(\|A\|_{\mathrm{B}(\mathcal{X},\mathcal{X})}) < \infty$, $\forall A \in \mathrm{B}(\mathcal{X}, \mathcal{X})$.

Proof (iii) $\exp(At) = \sum_{n=0}^{\infty} \frac{1}{n!} A^n t^n$. By Propositions 9.16 and 9.68, it can be differentiated term by term and $\frac{\mathrm{d}}{\mathrm{d}t}\exp(At) = \frac{\mathrm{d}}{\mathrm{d}t}\sum_{n=0}^{\infty}\frac{1}{n!}A^n t^n = \sum_{n=1}^{\infty}\frac{1}{(n-1)!}A^n t^{n-1} = A\exp(At) = \exp(At)A$.

(i) By repeated application of (iii), we have $h^{(n)}(t) = \frac{\mathrm{d}^n}{\mathrm{d}t^n}\exp(At) = A^n\exp(At)$, $\forall n \in \mathbb{N}$. Then, h is an analytic function by Definition 9.79. Furthermore, $\forall t \in \mathbb{R}$, h admits arbitrarily large analytic radius at t.

(iv) $\exp(At)B = (\sum_{n=0}^{\infty}\frac{1}{n!}A^n t^n)B = \sum_{n=0}^{\infty}\frac{1}{n!}A^n B t^n = \sum_{n=0}^{\infty}\frac{1}{n!}BA^n \cdot t^n = B\exp(At)$, where the third equality follows from $AB = BA$. Furthermore, $\exp(A)\exp(B) = \exp(A)(\sum_{n=0}^{\infty}\frac{1}{n!}B^n) = \sum_{n=0}^{\infty}\frac{1}{n!}\exp(A)B^n = \sum_{n=0}^{\infty}\frac{1}{n!}B^n\exp(A) = \exp(B)\exp(A)$, where the third equality follows from $\exp(A)B = B\exp(A)$. $\exp(A+B) = \sum_{n=0}^{\infty}\frac{1}{n!}(A+B)^n = \sum_{n=0}^{\infty}\frac{1}{n!}\cdot(\sum_{i=0}^{n}\binom{n}{i}A^i B^{n-i}) = \sum_{n=0}^{\infty}\sum_{i=0}^{n}\frac{1}{i!(n-i)!}A^i B^{n-i} = \sum_{i=0}^{\infty}\sum_{n=i}^{\infty}\frac{1}{i!(n-i)!}A^i \cdot B^{n-i} = \sum_{i=0}^{\infty}\frac{1}{i!}A^i\sum_{n=i}^{\infty}\frac{1}{(n-i)!}B^{n-i} = \sum_{i=0}^{\infty}\frac{1}{i!}A^i\sum_{n=0}^{\infty}\frac{1}{n!}B^n = \exp(A)\cdot\exp(B)$, where the second equality follows from $AB = BA$; and the third through sixth equalities follow from the absolute summability of the series involved.

(ii) This directly follows from (iv) and Definition 9.87.

(v) The inverse relationship follows from (ii). $(\exp(A))' = (\sum_{n=0}^{\infty}\frac{1}{n!}\cdot A^n)' = \sum_{n=0}^{\infty}\frac{1}{n!}(A^n)' = \sum_{n=0}^{\infty}\frac{1}{n!}(A')^n = \exp(A')$, where the second and third equalities follow from Proposition 7.110.

(vi) $\|\exp(A)\|_{\mathrm{B}(\mathcal{X},\mathcal{X})} = \|\sum_{n=0}^{\infty}\frac{1}{n!}A^n\|_{\mathrm{B}(\mathcal{X},\mathcal{X})} \leq \sum_{n=0}^{\infty}\frac{1}{n!}\|A\|_{\mathrm{B}(\mathcal{X},\mathcal{X})}^n = \exp(\|A\|_{\mathrm{B}(\mathcal{X},\mathcal{X})}) < \infty$, where the first equality follows from Definition 9.87; and the first inequality follows from Definition 7.1 and Proposition 7.64; and the second equality follows from standard algebra.

(vii) Fix any $A_0 \in \mathrm{B}(\mathcal{X}, \mathcal{X})$, and we will show that the function $\exp : \mathrm{B}(\mathcal{X}, \mathcal{X}) \to \mathrm{B}(\mathcal{X}, \mathcal{X})$ is Fréchet differentiable at A_0. Note that $\exp(A_0) =$

$\lim_{n\to\infty}\sum_{i=0}^{n}\frac{1}{i!}A_0^i =: \lim_{n\to\infty}F_n(A_0)$. Then, $\mathrm{D}F_n(A)(B) = B + \sum_{i=2}^{n}\frac{1}{i!}\cdot(\sum_{j=0}^{i-1}A^jBA^{i-j-1})$, $\forall A, B \in \mathrm{B}(\mathcal{X},\mathcal{X})$.

Define $G : \mathrm{B}(\mathcal{X},\mathcal{X}) \to \mathrm{B}(\mathrm{B}(\mathcal{X},\mathcal{X}),\mathrm{B}(\mathcal{X},\mathcal{X}))$ by $G(A)(B) = B + \lim_{n\to\infty}\sum_{i=2}^{n}\frac{1}{i!}(\sum_{j=0}^{i-1}A^jBA^{i-j-1})$, $\forall A, B \in \mathrm{B}(\mathcal{X},\mathcal{X})$. $\forall A \in \mathrm{B}(\mathcal{X},\mathcal{X})$, $G(A)$ is a linear function of B with norm $\|G(A)\|_{\mathrm{B}(\mathrm{B}(\mathcal{X},\mathcal{X}),\mathrm{B}(\mathcal{X},\mathcal{X}))} \leq 1 + \lim_{n\to\infty}\sum_{i=2}^{n}\frac{1}{i!}i\cdot\|A\|_{\mathrm{B}(\mathcal{X},\mathcal{X})}^{i-1} = \exp(\|A\|_{\mathrm{B}(\mathcal{X},\mathcal{X})}) < \infty$. Hence, $G(A) \in \mathrm{B}(\mathrm{B}(\mathcal{X},\mathcal{X}),\mathrm{B}(\mathcal{X},\mathcal{X}))$. Then, G is well-defined. Clearly, $\mathrm{D}F_n$ converges uniformly to G as $n\to\infty$ over the open set $\mathcal{B}_{\mathrm{B}(\mathcal{X},\mathcal{X})}(A_0, 1)$. Then, by Proposition 9.68, we have $\mathrm{D}\exp(A_0) = G(A_0)$. Clearly,

$$\begin{aligned}\|\mathrm{D}\exp(A_0)\|_{\mathrm{B}(\mathrm{B}(\mathcal{X},\mathcal{X}),\mathrm{B}(\mathcal{X},\mathcal{X}))} &= \|G(A_0)\|_{\mathrm{B}(\mathrm{B}(\mathcal{X},\mathcal{X}),\mathrm{B}(\mathcal{X},\mathcal{X}))}\\ &\leq \exp(\|A_0\|_{\mathrm{B}(\mathcal{X},\mathcal{X})}) < \infty\end{aligned}$$

This completes the proof of the proposition. □

9.10 Newton's Method

Let $\mathcal{X}$ and $\mathcal{Y}$ be Banach spaces and $F : \mathcal{X} \to \mathcal{Y}$. To solve the equation $F(x) = \vartheta_{\mathcal{Y}}$, we may use Newton's method. Assume that F is twice Fréchet differentiable, and we set $x_{n+1} = x_n - (F^{(1)}(x_n))^{-1}F(x_n)$, $\forall n \in \mathbb{N}$. Then, when x_1 is sufficiently close to a solution, the sequence $(x_n)_{n=1}^{\infty}$ converges to a solution x_{opt} with $F(x_{\mathrm{opt}}) = \vartheta_{\mathcal{Y}}$. This result is formalized in the following proposition.

Theorem 9.89 (Newton's Method) *Let $\mathcal{X}$ and $\mathcal{Y}$ be Banach spaces over $\mathbb{K}$, $D_1 \subseteq \mathcal{X}$, $F : D_1 \to \mathcal{Y}$, and $x_1 \in \mathcal{X}$. Assume that $\exists \beta_1, \eta_1, h_1, K \in [0,\infty) \subset \mathbb{R}$ such that:*

(i) $h_1 := \beta_1\eta_1 K \leq \frac{1}{2}$.
(ii) $\mathrm{dom}\left(F^{(2)}\right) \supseteq \overline{\mathcal{B}}_{\mathcal{X}}\left(x_1, \frac{2-2h_1}{2-3h_1}\eta_1\right) =: D \subseteq \overline{\mathcal{B}}_{\mathcal{X}}(x_1, 2\eta_1)$ *and,* $\forall x \in D$, $\left\|F^{(2)}(x)\right\| \leq K$.
(iii) $F^{(1)}(x_1)$ *is bijective,* $\left\|(F^{(1)}(x_1))^{-1}\right\| \leq \beta_1$, *and* $\left\|(F^{(1)}(x_1))^{-1}F(x_1)\right\| \leq \eta_1$.

Then, $\exists (x_n)_{n=1}^{\infty} \subseteq D$, defined by $x_{n+1} = x_n - (F^{(1)}(x_n))^{-1}F(x_n)$, $\forall n \in \mathbb{N}$, that converges to $x_{\mathrm{opt}} \in D$ with $F(x_{\mathrm{opt}}) = \vartheta_{\mathcal{Y}}$.

Proof We will use mathematical induction to prove the following claim.

Claim 9.89.1 $\forall n \in \mathbb{N}$, we have:

(a) $F^{(1)}(x_n)$ is bijective and $\left\|(F^{(1)}(x_n))^{-1}\right\| \leq \beta_n := \frac{\beta_{n-1}}{1-h_{n-1}} \in [0,\infty) \subset \mathbb{R}$.
(b) $\left\|(F^{(1)}(x_n))^{-1}F(x_n)\right\| \leq \eta_n := \frac{h_{n-1}\eta_{n-1}}{2(1-h_{n-1})} \in [0,\infty) \subset \mathbb{R}$.

(c) $x_{n+1} \in \overline{\mathcal{B}}_{\mathcal{Y}}\left(x_1, \frac{2-2h_1}{2-3h_1}\left(1 - \left(\frac{h_1}{2(1-h_1)}\right)^n\right)\eta_1\right)$.
(d) $0 \le h_n := \beta_n \eta_n K \le h_{n-1} \le \frac{1}{2}$.

Proof of Claim 1° Clearly, (a)–(d) are satisfied for $n = 1$.
2° Assume that (a)–(d) are satisfied for $n \le k \in \mathbb{N}$.
3° Consider the case of $n = k+1$. By mean value theorem 9.23, $\exists t_0 \in (0, 1) \subset \mathbb{R}$ such that $\left\| F^{(1)}(x_{k+1}) - F^{(1)}(x_k) \right\| \le \left\| F^{(2)}(t_0 x_{k+1} + (1-t_0)x_k)(x_{k+1} - x_k) \right\| \le \left\| F^{(2)}(t_0 x_{k+1} + (1-t_0)x_k) \right\| \|x_{k+1} - x_k\| \le K\eta_k$. Then, $\left\| (F^{(1)}(x_k))^{-1} \right\| \left\| F^{(1)}(x_{k+1}) - F^{(1)}(x_k) \right\| \le \beta_k \eta_k K = h_k < 1$. By Proposition 9.55, $F^{(1)}(x_{k+1})$ is bijective and

$$\begin{aligned}
&\left\| (F^{(1)}(x_{k+1}))^{-1} - (F^{(1)}(x_k))^{-1} \right\| \\
&\le \frac{\left\| (F^{(1)}(x_k))^{-1} \right\|^2 \left\| F^{(1)}(x_{k+1}) - F^{(1)}(x_k) \right\|}{1 - \left\| (F^{(1)}(x_k))^{-1} \right\| \left\| F^{(1)}(x_{k+1}) - F^{(1)}(x_k) \right\|} \\
&\le \frac{\beta_k^2 K \eta_k}{1 - \beta_k \eta_k K} = \frac{h_k \beta_k}{1 - h_k}
\end{aligned}$$

This leads to $\left\| (F^{(1)}(x_{k+1}))^{-1} \right\| \le \left\| (F^{(1)}(x_k))^{-1} \right\| + \left\| (F^{(1)}(x_{k+1}))^{-1} - (F^{(1)}(x_k))^{-1} \right\| \le \beta_k + h_k \beta_k / (1 - h_k) = \beta_k / (1 - h_k) = \beta_{k+1}$. Hence, (a) holds.

Note that $F(x_{k+1}) = F(x_{k+1}) - F(x_k) - F^{(1)}(x_k)(x_{k+1} - x_k)$. By Taylor's Theorem 9.48, $\exists t_1 \in (0, 1) \subset \mathbb{R}$ such that $\|F(x_{k+1})\| \le \frac{1}{2} \left\| F^{(2)}(t_1 x_{k+1} + (1-t_1)x_k) \right\| \|x_{k+1} - x_k\|^2 \le K\eta_k^2/2$. Then, $\left\| (F^{(1)}(x_{k+1}))^{-1} F(x_{k+1}) \right\| \le \beta_k K \eta_k^2 / (2 - 2h_k) = \frac{h_k \eta_k}{2(1-h_k)} = \eta_{k+1}$. Hence, (b) holds.

Note also that $\|x_{k+2} - x_{k+1}\| = \left\| (F^{(1)}(x_{k+1}))^{-1} F(x_{k+1}) \right\| \le \eta_{k+1} = \frac{h_k}{2(1-h_k)} \eta_k \le \frac{h_1}{2(1-h_1)} \eta_k \le \left(\frac{h_1}{2(1-h_1)}\right)^k \eta_1$. Then, $\|x_{k+2} - x_1\| \le \|x_{k+2} - x_{k+1}\| + \|x_{k+1} - x_1\| \le \left(\frac{h_1}{2(1-h_1)}\right)^k \eta_1 + \frac{2-2h_1}{2-3h_1}\left(1 - \left(\frac{h_1}{2(1-h_1)}\right)^k\right)\eta_1 = \frac{2-2h_1}{2-3h_1}\left(1 - \left(\frac{h_1}{2(1-h_1)}\right)^{k+1}\right)\eta_1$. Hence, (c) holds.

Note that $h_{k+1} = \beta_{k+1}\eta_{k+1} K = \frac{h_k^2}{2(1-h_k)^2}$. Since $0 \le h_k \le 1/2$, then $0 \le \frac{h_k}{2(1-h_k)^2} \le 1$. This implies that $0 \le h_{k+1} \le h_k \le 1/2$. Hence, (d) holds.

This completes the induction process and the proof of the claim. □

Then, the sequence $(x_n)_{n=1}^\infty \subseteq D$ is well-defined. $\forall n \in \mathbb{N}$, $\|x_{n+2} - x_{n+1}\| = \left\| (F^{(1)}(x_{n+1}))^{-1} F(x_{n+1}) \right\| \le \eta_{n+1} \le \left(\frac{h_1}{2(1-h_1)}\right)^n \eta_1$. Thus, $(x_n)_{n=1}^\infty$ is a Cauchy sequence since $0 \le \frac{h_1}{2(1-h_1)} \le 1/2 < 1$. It must converge to $x_{\text{opt}} \in D$ since D is closed and $\mathcal{X}$ is complete. Note that $F(x_n) = F^{(1)}(x_n)(x_n - x_{n+1})$, $\forall n \in \mathbb{N}$. By Proposition 9.7, we have F and $F^{(1)}$ are continuous at x_{opt}. This leads to $F(x_{\text{opt}}) =$

$\lim_{n\in\mathbb{N}} F(x_n) = \lim_{n\in\mathbb{N}} F^{(1)}(x_n)(x_n - x_{n+1}) = F^{(1)}(x_{\text{opt}})(x_{\text{opt}} - x_{\text{opt}}) = \vartheta_{\mathcal{Y}}$, where the first equality follows from Proposition 3.66 and the third equality follows from Propositions 3.66, 3.67, 7.23, and 7.65. This completes the proof of the theorem. □

It is numerically expensive to use Newton's method due to the necessity of inverting $F^{(1)}(x_n)$ at each step of the iteration. But the benefit of this method is the quadratic convergence of $(x_n)_{n=1}^{\infty}$ to x_{opt}, which is summarized in the following proposition.

Proposition 9.90 *Let $\mathcal{X}$ and $\mathcal{Y}$ be Banach spaces over $\mathbb{K}$, $D_1 \subseteq \mathcal{X}$, $F : D_1 \to \mathcal{Y}$, and $x_1 \in \mathcal{X}$. Assume that $\exists \beta_1, \eta_1, h_1, K, M_1, M_2, M_4 \in [0, \infty) \subset \mathbb{R}$ such that:*

(i) $h_1 := \beta_1 \eta_1 K \leq \frac{1}{2}$.

(ii) $\operatorname{dom}\left(F^{(3)}\right) \supseteq D_2 \supseteq \overline{\mathcal{B}}_{\mathcal{X}}\left(x_1, \frac{2-2h_1}{2-3h_1}\eta_1\right) =: D$, *where $D_2 \subseteq D_1$ is an open set in $\mathcal{X}$, $\forall x \in D_2$, $F^{(1)}(x)$ is bijective, and, $\forall x \in D$, $\left\|F^{(2)}(x)\right\| \leq K$, $\left\|(F^{(1)}(x))^{-1}\right\| \leq M_1$, $\left\|F^{(3)}(x)\right\| \leq M_2$, and $\|F(x)\| \leq M_4$.*

(iii) $\left\|(F^{(1)}(x_1))^{-1}\right\| \leq \beta_1$ *and* $\left\|(F^{(1)}(x_1))^{-1} F(x_1)\right\| \leq \eta_1$.

Then, $\exists (x_n)_{n=1}^{\infty} \subseteq D$, defined by $x_{n+1} = x_n - (F^{(1)}(x_n))^{-1} F(x_n)$, $\forall n \in \mathbb{N}$, that converges to $x_{\text{opt}} \in D$ with $F(x_{\text{opt}}) = \vartheta_{\mathcal{Y}}$. Furthermore, $\forall n \in \mathbb{N}$, $\left\|x_{n+1} - x_{\text{opt}}\right\| \leq c\left\|x_n - x_{\text{opt}}\right\|^2$, where $c := (M_1 K + 2M_1^3 K^2 M_4 + M_1^2 M_2 M_4)/2$.

Proof By Newton's method (Theorem 9.89), $(x_n)_{n=1}^{\infty} \subseteq D$ is well-defined and converges to $x_{\text{opt}} \in D$ with $F(x_{\text{opt}}) = \vartheta_{\mathcal{Y}}$. Define $T : D_2 \to \mathcal{X}$ by $T(x) = x - (F^{(1)}(x))^{-1} F(x)$, $\forall x \in D_2$. Then, by Propositions 9.34, 9.44, 9.45, 9.41, 9.40, and 9.55, T is twice Fréchet differentiable. This leads to, $\forall x \in D_2$, $\forall h_1 \in \mathcal{X}$

$$\begin{aligned}
T^{(1)}(x)(h_1) &= h_1 - (F^{(1)}(x))^{-1} F^{(1)}(x)(h_1) \\
&\quad + (F^{(1)}(x))^{-1} F^{(2)}(x)(h_1)((F^{(1)}(x))^{-1} F(x)) \\
&= (F^{(1)}(x))^{-1} F^{(2)}(x)(h_1)((F^{(1)}(x))^{-1} F(x))
\end{aligned}$$

Then, $T^{(1)}(x_{\text{opt}})(h_1) = \vartheta_{\mathcal{X}}$, $\forall h_1 \in \mathcal{X}$, since $F(x_{\text{opt}}) = \vartheta_{\mathcal{Y}}$. This implies that $T^{(1)}(x_{\text{opt}}) = \vartheta_{\mathrm{B}(\mathcal{X},\mathcal{X})}$. The second order derivative of T is given by, $\forall x \in D_2$, $\forall h_1, h_2 \in \mathcal{X}$,

$$\begin{aligned}
&T^{(2)}(x)(h_2)(h_1) \\
&= -(F^{(1)}(x))^{-1} F^{(2)}(x)(h_2)((F^{(1)}(x))^{-1} F^{(2)}(x)(h_1)((F^{(1)}(x))^{-1} F(x))) \\
&\quad + (F^{(1)}(x))^{-1} F^{(3)}(x)(h_2)(h_1)((F^{(1)}(x))^{-1} F(x)) \\
&\quad + (F^{(1)}(x))^{-1} F^{(2)}(x)(h_1)\Big(- (F^{(1)}(x))^{-1} F^{(2)}(x)(h_2)((F^{(1)}(x))^{-1} \\
&\quad \cdot F(x)) + (F^{(1)}(x))^{-1} F^{(1)}(x)(h_2)\Big)
\end{aligned}$$

$$
\begin{aligned}
&= -(F^{(1)}(x))^{-1}F^{(2)}(x)(h_2)((F^{(1)}(x))^{-1}F^{(2)}(x)(h_1)((F^{(1)}(x))^{-1}F(x))) \\
&\quad +(F^{(1)}(x))^{-1}F^{(3)}(x)(h_2)(h_1)((F^{(1)}(x))^{-1}F(x)) \\
&\quad -(F^{(1)}(x))^{-1}F^{(2)}(x)(h_1)((F^{(1)}(x))^{-1}F^{(2)}(x)(h_2)((F^{(1)}(x))^{-1}F(x))) \\
&\quad +(F^{(1)}(x))^{-1}F^{(2)}(x)(h_1)(h_2)
\end{aligned}
$$

Then, $\forall x \in D$, $\left\|T^{(2)}(x)\right\| = \sup_{h_1,h_2\in\mathcal{X},\ \|h_1\|\leq 1,\ \|h_2\|\leq 1}\left\|T^{(2)}(x)(h_2)(h_1)\right\| \leq 2M_1^3K^2M_4 + M_1^2M_2M_4 + M_1K = 2c$. $\forall n \in \mathbb{N}$, by Taylor's Theorem 9.48, $\exists t_0 \in (0,1) \subset \mathbb{R}$ such that $\left\|x_{n+1} - x_{\mathrm{opt}}\right\| = \left\|T(x_n) - T(x_{\mathrm{opt}})\right\| = \left\|T(x_n) - T(x_{\mathrm{opt}}) - T^{(1)}(x_{\mathrm{opt}})(x_n - x_{\mathrm{opt}})\right\| \leq \frac{1}{2}\left\|T^{(2)}(t_0x_n + (1-t_0)x_{\mathrm{opt}})\right\| \left\|x_n - x_{\mathrm{opt}}\right\|^2 \leq c\left\|x_n - x_{\mathrm{opt}}\right\|^2$. This completes the proof of the proposition. □

Chapter 10
Local Theory of Optimization

In this chapter, we will develop a number of tools for optimization of sufficiently many times differentiable functions. As in Chap. 8, we will be mainly concerned with real spaces, rather than complex ones.

10.1 Basic Notion

Definition 10.1 Let $\mathcal{X} := (X, \mathcal{O})$ be a topological space, $f : \mathcal{X} \to \mathbb{R}$, and $x_0 \in \mathcal{X}$. x_0 is said to be *a point of minimum* for f if $f(x_0) \leq f(x)$, $\forall x \in \mathcal{X}$. It is said to be *the point of strict minimum* for f if $f(x_0) < f(x)$, $\forall x \in \mathcal{X} \setminus \{x_0\}$. It is said to be *a point of relative minimum* for f if $\exists O \in \mathcal{O}$ with $x_0 \in O$ such that $f(x_0) \leq f(x)$, $\forall x \in O$. It is said to be *a point of relative strict minimum* for f if $\exists O \in \mathcal{O}$ with $x_0 \in O$ such that $f(x_0) < f(x)$, $\forall x \in O \setminus \{x_0\}$. Similar definitions for *points of maxima*. Moreover, x_0 is said to be *a point of relative extremum* if it is a point of relative minimum or relative maximum. It is said to be *a point of relative strict extremum* if it is a point of relative strict minimum or relative strict maximum. ◇

Proposition 10.2 *Let $\mathfrak{X}$ be a real normed linear space, $D \subseteq \mathfrak{X}$, $x_0 \in D$, $f : D \to \mathbb{R}$, $u \in \mathrm{A}_D(x_0)$. Assume that the directional derivative of f at x_0 along u exists and x_0 is a point of relative minimum for f. Then, $\mathrm{D}f(x_0; u) \geq 0$.*

Proof This is immediate from Definition 9.4. □

Definition 10.3 Let $\mathfrak{X}$ be a real normed linear space and $A \in \mathrm{B_S}_2(\mathfrak{X}, \mathbb{R})$. A is said to be *positive definite* if $\exists m \in (0, \infty) \subset \mathbb{R}$ such that $A(x)(x) = \langle\langle Ax, x\rangle\rangle \geq m\|x\|^2$, $\forall x \in \mathfrak{X}$. Let the set of all such positive definite operators be denoted by $\mathcal{S}_{+\,\mathfrak{X}}$. A is said to be *positive semi-definite* if $A(x)(x) \geq 0$, $\forall x \in \mathfrak{X}$. Let the set of all such positive semi-definite operators be denoted by $\mathcal{S}_{\mathrm{psd}\,\mathfrak{X}}$. A is said to be *negative definite* if $\exists m \in (0, \infty) \subset \mathbb{R}$ such that $A(x)(x) \leq -m\|x\|^2$, $\forall x \in \mathfrak{X}$. Let the set of all such negative definite operators be denoted by $\mathcal{S}_{-\,\mathfrak{X}}$. A is said to be

Z. Pan, *Measure-Theoretic Calculus in Abstract Spaces*,
https://doi.org/10.1007/978-3-031-21912-2_10

negative semi-definite if $A(x)(x) \leq 0$, $\forall x \in \mathcal{X}$. Let the set of all such negative semi-definite operators be denoted by $\mathcal{S}_{\mathrm{nsd}\,\mathcal{X}}$. We will denote $\mathrm{B}_{\mathrm{S}\,2}(\mathcal{X}, \mathbb{R})$ by $\mathcal{S}_{\mathcal{X}}$, which is clearly a Banach subspace of $\mathrm{B}_2(\mathcal{X}, \mathbb{R})$. ◇

Proposition 10.4 *Let $\mathcal{X}$ be a real normed linear space. Then,*

(i) $\mathcal{S}_{-\mathcal{X}} = -\mathcal{S}_{+\mathcal{X}}$ *and* $\mathcal{S}_{\mathrm{nsd}\,\mathcal{X}} = -\mathcal{S}_{\mathrm{psd}\,\mathcal{X}}$.
(ii) $\mathcal{S}_{+\mathcal{X}}$ *and* $\mathcal{S}_{-\mathcal{X}}$ *are open sets in* $\mathrm{B}_{\mathrm{S}\,2}(\mathcal{X}, \mathbb{R}) = \mathcal{S}_{\mathcal{X}}$.
(iii) $\mathcal{S}_{\mathrm{psd}\,\mathcal{X}}$ *and* $\mathcal{S}_{\mathrm{nsd}\,\mathcal{X}}$ *are closed convex cones in* $\mathcal{S}_{\mathcal{X}}$.
(iv) $\mathcal{S}_{+\mathcal{X}} \subseteq \mathcal{S}^{\circ}_{\mathrm{psd}\,\mathcal{X}}$ *and* $\mathcal{S}_{-\mathcal{X}} \subseteq \mathcal{S}^{\circ}_{\mathrm{nsd}\,\mathcal{X}}$.

Proof (i) This is clear. (ii) Fix $A \in \mathcal{S}_{+\mathcal{X}}$. Then, $\exists m \in (0, \infty) \subset \mathbb{R}$ such that $A(x)(x) \geq m\|x\|^2$. $\forall B \in \mathrm{B}_{\mathrm{S}\,2}(\mathcal{X}, \mathbb{R})$ with $\|B - A\| < m/2$, $\forall x \in \mathcal{X}$, we have $B(x)(x) = A(x)(x) + (B - A)(x)(x) \geq m\|x\|^2 - \|B - A\|\|x\|^2 \geq m\|x\|^2/2$, where the first inequality follows from Proposition 7.64. This implies that $B \in \mathcal{S}_{+\mathcal{X}}$. Then, $A \in \mathcal{S}^{\circ}_{+\mathcal{X}}$. Hence, $\mathcal{S}_{+\mathcal{X}}$ is open in $\mathcal{S}_{\mathcal{X}}$. Therefore, $\mathcal{S}_{-\mathcal{X}} = -\mathcal{S}_{+\mathcal{X}}$ is open in $\mathcal{S}_{\mathcal{X}}$.

(iii) Clearly, $\vartheta_{\mathrm{B}_{\mathrm{S}\,2}(\mathcal{X},\mathbb{R})} \in \mathcal{S}_{\mathrm{psd}\,\mathcal{X}}$. $\forall A \in \mathcal{S}_{\mathrm{psd}\,\mathcal{X}}$, $\forall \alpha \in [0, \infty) \subset \mathbb{R}$, $\alpha A \in \mathcal{S}_{\mathrm{psd}\,\mathcal{X}}$. Hence, $\mathcal{S}_{\mathrm{psd}\,\mathcal{X}}$ is a cone. $\forall A, B \in \mathcal{S}_{\mathrm{psd}\,\mathcal{X}}$, $\forall x \in \mathcal{X}$, $(A + B)(x)(x) = A(x)(x) + B(x)(x) \geq 0$. Hence, $A + B \in \mathcal{S}_{\mathrm{psd}\,\mathcal{X}}$. Then, $\mathcal{S}_{\mathrm{psd}\,\mathcal{X}}$ is a convex cone.

Let $M := \mathcal{S}_{\mathcal{X}} \setminus \mathcal{S}_{\mathrm{psd}\,\mathcal{X}}$. $\forall A \in M$, $\exists x_0 \in \mathcal{X}$ such that $A(x_0)(x_0) < 0$. Then, there exists $m \in (0, \infty) \subset \mathbb{R}$ such that $A(x_0)(x_0) < -m\|x_0\|^2 < 0$. $\forall B \in \mathrm{B}_{\mathrm{S}\,2}(\mathcal{X}, \mathbb{R})$ with $\|B - A\| < m/2$, we have $B(x_0)(x_0) = A(x_0)(x_0) + (B - A)(x_0)(x_0) < -m\|x_0\|^2 + \|B - A\|\|x_0\|^2 \leq -m\|x_0\|^2/2 < 0$, where the first inequality follows from Proposition 7.64. This implies that $B \in M$ and $A \in M^{\circ}$. Then, $M = M^{\circ}$ is open. Therefore, $\mathcal{S}_{\mathrm{psd}\,\mathcal{X}}$ is closed.

$\mathcal{S}_{\mathrm{nsd}\,\mathcal{X}} = -\mathcal{S}_{\mathrm{psd}\,\mathcal{X}}$ is clearly also a closed convex cone.

(iv) Clearly, $\mathcal{S}_{+\mathcal{X}} \subseteq \mathcal{S}_{\mathrm{psd}\,\mathcal{X}}$. Then, $\mathcal{S}_{+\mathcal{X}} \subseteq \mathcal{S}^{\circ}_{\mathrm{psd}\,\mathcal{X}}$. This further implies that $\mathcal{S}_{-\mathcal{X}} \subseteq \mathcal{S}^{\circ}_{\mathrm{nsd}\,\mathcal{X}}$. This completes the proof of the proposition. □

Definition 10.5 Let $(\mathcal{X}, \mathbb{K})$ be a vector space and $K \subseteq \mathcal{X}$ be convex. $M \subseteq K$ is said to be an *extreme subset* of K if $M \neq \emptyset$ and $\forall x_1, x_2 \in K$, $\forall \alpha \in (0, 1) \subset \mathbb{R}$, $\alpha x_1 + (1 - \alpha)x_2 \in M$ implies that $x_1, x_2 \in M$. If a singleton set $\{x_0\} \subseteq K$ is an extreme subset, then x_0 is called an *extreme point* of K. ◇

Proposition 10.6 *Let $\mathcal{X}$ be a real normed linear space, $K \subseteq \mathcal{X}$ be a nonempty convex set, and $H := \{x \in \mathcal{X} \mid \langle\langle x_{*0}, x\rangle\rangle = c\}$ be a supporting hyperplane of K, where $x_{*0} \in \mathcal{X}^*$ with $x_{*0} \neq \vartheta_{\mathcal{X}^*}$ and $c \in \mathbb{R}$. Then, any extreme subset of $K_1 := K \cap H$ is also an extreme subset of K.*

Proof Without loss of generality, assume that $\inf_{k \in K} \langle\langle x_{*0}, x\rangle\rangle = c$. Let $M \subseteq K_1$ be an extreme subset of K_1. Then, $M \neq \emptyset$. $\forall x_1, x_2 \in \mathbb{K}$, $\forall \alpha \in (0, 1) \subset \mathbb{R}$, let $\bar{x} := \alpha x_1 + (1 - \alpha)x_2 \in M \subseteq H$. Then, $\langle\langle x_{*0}, x_1\rangle\rangle \geq c$ and $\langle\langle x_{*0}, x_2\rangle\rangle \geq c$ and $\langle\langle x_{*0}, \bar{x}\rangle\rangle = c$. This implies that $\alpha\,(\langle\langle x_{*0}, x_1\rangle\rangle - c) + (1 - \alpha)\,(\langle\langle x_{*0}, x_2\rangle\rangle - c) = 0$ and $\langle\langle x_{*0}, x_1\rangle\rangle = c = \langle\langle x_{*0}, x_2\rangle\rangle$. Hence, $x_1, x_2 \in K_1$. Since, M is an extreme subset of K_1, then $x_1, x_2 \in M$. Therefore, M is an extreme subset of K. This completes the proof of the proposition. □

Proposition 10.7 (Krein–Milman) *Let $\mathcal{X}$ be a real reflexive Banach space, $K \subseteq \mathcal{X}$ be a nonempty bounded closed convex set, and $M \subseteq K$ be a weakly compact extreme subset of K. Then, M contains at least one extreme point of K.*

Proof By Proposition 8.11, K is compact in $\mathcal{X}_{\text{weak}}$. Let $\mathcal{M} = \{E \subseteq M \mid E$ is a weakly compact extreme subset of $K\}$. Clearly, $M \in \mathcal{M} \neq \emptyset$. Clearly, $\supseteq$ defines an antisymmetric partial ordering on $\mathcal{M}$, where smaller sets are further down the stream. Next, we will use Zorn's Lemma to show that $\mathcal{M}$ admits a maximal element.

Let $\mathcal{E} \subseteq \mathcal{M}$ be a nonempty totally ordered (by $\supseteq$) subcollection. Let $E_0 := \bigcap_{E \in \mathcal{E}} E$. $\forall E \in \mathcal{E}$, $E \subseteq M$ is weakly compact extreme subset of K. Then, by Propositions 7.116, 5.5, and 3.61, E is weakly closed. By Proposition 5.5, E_0 is weakly compact. $\forall x_1, x_2 \in K$, $\forall \alpha \in (0, 1) \subset \mathbb{R}$, let $\alpha x_1 + (1 - \alpha)x_2 \in E_0$. $\forall E \in \mathcal{E}$, $\alpha x_1 + (1-\alpha)x_2 \in E$. Since E is an extreme subset of K, then $x_1, x_2 \in E$. By the arbitrariness of E, we have $x_1, x_2 \in E_0$. Hence, E_0 is an extreme subset of K if $E_0 \neq \emptyset$. $\forall E \in \mathcal{E}$, E is nonempty. Since $\mathcal{E}$ is totally ordered by $\supseteq$, then the intersection of finite number of sets in $\mathcal{E}$ is again in $\mathcal{E}$, and hence nonempty. By Proposition 5.12, $E_0 \neq \emptyset$. Then, E_0 is a weakly compact extreme subset of K and $E_0 \in \mathcal{M}$. Clearly, E_0 is an upper bound of $\mathcal{E}$ (in terms of $\supseteq$). Then, by Zorn's Lemma, $\mathcal{M}$ admits a maximal element E_M. Then, $E_M \subseteq M$ is a weakly compact extreme subset of K and $E_M \neq \emptyset$.

We will show that E_M is a singleton set, which then proves that M contains an extreme point of K. Suppose that $\exists x_1, x_2 \in E_M$ with $x_1 \neq x_2$. Let $N := \operatorname{span}(\{x_2 - x_1\})$ and define a functional $f : N \to \mathbb{R}$ by $f(\alpha(x_2 - x_1)) = \alpha$, $\forall \alpha \in \mathbb{R}$. Clearly, f is a linear functional on N, and $\|f\|_N = 1/\|x_2 - x_1\| < \infty$. By Hahn–Banach Theorem 7.83, there exists $x_{*0} \in \mathcal{X}^*$ with $\|x_{*0}\| = 1/\|x_2 - x_1\|$ such that $\langle\langle x_{*0}, \alpha(x_2 - x_1)\rangle\rangle = \alpha$, $\forall \alpha \in \mathbb{R}$. Clearly, $\langle\langle x_{*0}, x_1\rangle\rangle \neq \langle\langle x_{*0}, x_2\rangle\rangle$. Note that E_M is nonempty and compact in $\mathcal{X}_{\text{weak}}$ and x_{*0} is weakly continuous. By Proposition 5.29, $c := \langle\langle x_{*0}, x_0\rangle\rangle = \inf_{x \in E_M} \langle\langle x_{*0}, x\rangle\rangle \in \mathbb{R}$ for some $x_0 \in E_M$. Define $H := \{x \in \mathcal{X} \mid \langle\langle x_{*0}, x\rangle\rangle = c\}$. Let $E_m := E_M \cap H$. Then, at least one of x_1 and x_2 is not in E_m. Hence, $E_M \supset E_m$. Clearly, $E_m \ni x_0$ is nonempty. Note that H is weakly closed. Then, by Proposition 5.5, $E_m \subseteq M$ is weakly compact. $\forall \bar{x}_1, \bar{x}_2 \in K$, $\forall \bar{\alpha} \in (0, 1) \subset \mathbb{R}$, let $\bar{\alpha}\bar{x}_1 + (1 - \bar{\alpha})\bar{x}_2 \in E_m \subseteq E_M$. Since E_M is an extreme subset of K, then we have $\bar{x}_1, \bar{x}_2 \in E_M$. This further implies that $\langle\langle x_{*0}, \bar{x}_1\rangle\rangle \geq c$ and $\langle\langle x_{*0}, \bar{x}_2\rangle\rangle \geq c$. Note that $\langle\langle x_{*0}, \bar{\alpha}\bar{x}_1 + (1 - \bar{\alpha})\bar{x}_2\rangle\rangle = c$. Then, we must have $\langle\langle x_{*0}, \bar{x}_1\rangle\rangle = c = \langle\langle x_{*0}, \bar{x}_2\rangle\rangle$ and $\bar{x}_1, \bar{x}_2 \in E_m$. This shows that E_m is an extreme subset of K. Then $E_m \in \mathcal{M}$. This contradicts with the fact that E_M is maximal with respect to $\supseteq$. Therefore, E_M is a singleton set.

This completes the proof of the proposition. □

Proposition 10.8 *Let $\mathcal{X}$ be a real reflexive Banach space, $K \subseteq \mathcal{X}$ be a nonempty bounded closed convex set, and E be the set of extreme points of K. Then, $K = \overline{\operatorname{co}(E)}$.*

Proof Let $C := \overline{\operatorname{co}(E)}$. By Proposition 7.15, C is closed and convex and $C \subseteq K$. We will prove the result by an argument of contradiction. Suppose $K \supset C$.

Then, $\exists x_0 \in K \setminus C$. By Proposition 8.10, there exists $x_{*0} \in \mathcal{X}^*$ such that $\langle\langle x_{*0}, x_0\rangle\rangle < \inf_{x\in C} \langle\langle x_{*0}, x\rangle\rangle$. By Proposition 8.11, K is compact in the weak topology and x_{*0} is continuous in the weak topology. Then, by Proposition 5.29, $c_0 := \langle\langle x_{*0}, x_1\rangle\rangle = \inf_{x\in K} \langle\langle x_{*0}, x\rangle\rangle \in \mathbb{R}$ for some $x_1 \in K$. Clearly $c_0 \leq \langle\langle x_{*0}, x_0\rangle\rangle < \inf_{x\in C} \langle\langle x_{*0}, x\rangle\rangle$. We will distinguish two exhaustive and mutually exclusive cases: Case 1: $x_{*0} = \vartheta_{\mathcal{X}^*}$; Case 2: $x_{*0} \neq \vartheta_{\mathcal{X}^*}$. Case 1: $x_{*0} = \vartheta_{\mathcal{X}^*}$. Then, $c_0 = 0$ and $C = \emptyset$. By Proposition 10.7, $E \neq \emptyset$. Then, $C \neq \emptyset$. This is a contradiction. Case 2: $x_{*0} \neq \vartheta_{\mathcal{X}^*}$. Let $H := \{x \in \mathcal{X} \mid \langle\langle x_{*0}, x\rangle\rangle = c_0\}$. H is a supporting hyperplane of K and $H \cap C = \emptyset$. Let $C_m := K \cap H \neq \emptyset$. Clearly, C_m is bounded closed and convex. By Proposition 10.7, there is an extreme point $x_m \in C_m$ of C_m. Then, $\{x_m\} \subseteq C_m$ is an extreme subset of C_m. By Proposition 10.6, $\{x_m\}$ is an extreme subset of K. Then, x_m is an extreme point of K and $x_m \in E$. This leads to the contradiction $x_m \in C_m \subseteq H$, $x_m \in E \subseteq C$ and $C \cap H = \emptyset$. Thus, in both cases, we have arrived at a contradiction. Then, the hypothesis must be false. Hence, $K = \overline{\text{co}(E)}$. This completes the proof of the proposition. □

Proposition 10.9 *Let $\mathcal{X}$ be a real vector space, $\Omega \subseteq \mathcal{X}$ be a convex set, and $f_1 : \Omega \to \mathbb{R}$ and $f_2 : \Omega \to \mathbb{R}$ be convex functionals. Then, the following statements hold.*

(i) $\forall n \in \mathbb{N}$, $\forall x_1, \ldots, x_n \in \Omega$, $\forall \alpha_1, \ldots, \alpha_n \in [0, 1] \subset \mathbb{R}$ *with* $\sum_{i=1}^n \alpha_i = 1$, *we have* $f_1\big(\sum_{i=1}^n \alpha_i x_i\big) \leq \sum_{i=1}^n \alpha_i f_1(x_i)$. *If, in addition, f_1 is strictly convex, $x_1, \ldots, x_n$ are distinct, and $\alpha_1, \ldots, \alpha_n \in (0, 1) \subset \mathbb{R}$, then* $f_1\big(\sum_{i=1}^n \alpha_i x_i\big) < \sum_{i=1}^n \alpha_i f_1(x_i)$.

(ii) $\forall \alpha_1, \alpha_2 \in [0, \infty) \subset \mathbb{R}$, $\alpha_1 f_1 + \alpha_2 f_2$ *is convex. If, in addition, f_1 is strictly convex and $\alpha_1 \in (0, \infty) \subset \mathbb{R}$, then $\alpha_1 f_1 + \alpha_2 f_2$ is strictly convex.*

(iii) $\forall c \in \mathbb{R}$, $\{x \in \Omega \mid f_1(x) \leq c\}$ *is convex.*

(iv) *Let $\mathcal{Y}$ be a real vector space, $A : \mathcal{Y} \to \mathcal{X}$ be an affine operator, $D \subseteq \mathcal{Y}$ be convex, and $A(D) \subseteq \Omega$. Then, $f_1 \circ A|_D : D \to \mathbb{R}$ is convex.*

Proof This is straightforward and is therefore omitted. □

Proposition 10.10 *Let $f : [a, b] \to \mathbb{R}$, where $a, b \in \mathbb{R}$ with $a < b$. Then, the following statements hold.*

(i) *f is convex if, and only if,* $\forall x_1, x_2, x_3 \in [a, b] \subset \mathbb{R}$ *with* $x_1 < x_2 < x_3$, *we have*

$$\frac{f(x_2) - f(x_1)}{x_2 - x_1} \leq \frac{f(x_3) - f(x_2)}{x_3 - x_2}$$

if, and only if, $\forall x_1, x_2, x_3 \in [a, b] \subset \mathbb{R}$ *with* $x_1 < x_2 < x_3$, *we have*

$$\frac{f(x_2) - f(x_1)}{x_2 - x_1} \leq \frac{f(x_3) - f(x_1)}{x_3 - x_1}$$

(ii) *f is strictly convex if, and only if, $\forall x_1, x_2, x_3 \in [a, b] \subset \mathbb{R}$ with $x_1 < x_2 < x_3$, we have*

$$\frac{f(x_2) - f(x_1)}{x_2 - x_1} < \frac{f(x_3) - f(x_2)}{x_3 - x_2}$$

if, and only if, $\forall x_1, x_2, x_3 \in [a, b] \subset \mathbb{R}$ with $x_1 < x_2 < x_3$, we have

$$\frac{f(x_2) - f(x_1)}{x_2 - x_1} < \frac{f(x_3) - f(x_1)}{x_3 - x_1}$$

(iii) *If f is convex and $c \in (a, b) \subset \mathbb{R}$, then, the one-sided derivatives*

$$\lim_{x \to c^+} \frac{f(x) - f(c)}{x - c}; \qquad \lim_{x \to c^-} \frac{f(x) - f(c)}{x - c}$$

exist.

Proof This is straightforward and is therefore omitted. □

Proposition 10.11 *Let $\mathcal{X}$ be a real normed linear space, $\Omega \subseteq \mathcal{X}$ be convex, and $f : \Omega \to \mathbb{R}$ be differentiable. Then, the following statements hold.*

(i) *f is convex if, and only if, $\forall x, y \in \Omega$, we have $f(y) \geq f(x) + f^{(1)}(x)(y - x)$.*
(ii) *f is strictly convex if, and only if, $\forall x, y \in \Omega$ with $x \neq y$, we have $f(y) > f(x) + f^{(1)}(x)(y - x)$.*

Proof (i) "Necessity" Let f be convex. $\forall x, y \in \Omega$, by the convexity of Ω, we have $y - x \in \mathrm{A}_\Omega(x)$. $\forall \alpha \in (0, 1] \subset \mathbb{R}$, we have $f(\alpha y + (1-\alpha)x) \leq \alpha f(y) + (1-\alpha) f(x)$, which further implies that $f(y) - f(x) \geq (f(x + \alpha(y - x)) - f(x))/\alpha$. Then, we have

$$\begin{aligned} f(y) - f(x) &\geq \lim_{\alpha \to 0^+} (f(x + \alpha(y - x)) - f(x))/\alpha = \mathrm{D}f(x; y - x) \\ &= f^{(1)}(x)(y - x) \end{aligned}$$

where the first equality follows from Definition 9.4 and the last equality follows from Proposition 9.5.

"Sufficiency" Let $f(y) \geq f(x) + f^{(1)}(x)(y - x)$, $\forall x, y \in \Omega$. $\forall x_1, x_2 \in \Omega$, $\forall \alpha \in [0, 1] \subset \mathbb{R}$, let $x := \alpha x_1 + (1 - \alpha)x_2 \in \Omega$. Then, $f(x_2) = f(x + \alpha(x_2 - x_1)) \geq f(x) + \alpha f^{(1)}(x)(x_2 - x_1)$. Note also, $f(x_1) = f(x + (1 - \alpha)(x_1 - x_2)) \geq f(x) + (1 - \alpha) f^{(1)}(x)(x_1 - x_2)$. Then, $\alpha f(x_1) + (1 - \alpha) f(x_2) \geq f(x)$. Hence, f is convex.

(ii) "Necessity" Let f be strictly convex. $\forall x, y \in \Omega$ with $x \neq y$, by the convexity of Ω, we have $y - x \in \mathrm{A}_\Omega(x)$. $\forall \alpha \in (0, 1) \subset \mathbb{R}$, we have $f(\alpha y + (1 - \alpha)x) < \alpha f(y) + (1 - \alpha) f(x)$, which further implies that $f(y) - f(x) > (f(x + \alpha(y -$

$x)) - f(x))/\alpha$. Define $A : \mathbb{R} \to \mathcal{X}$ by $A(\beta) = x + \beta(y - x)$, $\forall \beta \in \mathbb{R}$. Clearly, A is a affine operator and $A(I) \subseteq \Omega$, where $I := [0, 1] \subset \mathbb{R}$. By Proposition 10.9, $g := f \circ A|_I$ is convex. Then, we have

$$\begin{aligned} f(y) - f(x) &> 2(f(x + 0.5(y - x)) - f(x)) = \frac{g(0.5) - g(0)}{0.5} \\ &\geq \lim_{\alpha \to 0^+} \frac{g(\alpha) - g(0)}{\alpha} = \lim_{\alpha \to 0^+} \frac{f(x + \alpha(y - x)) - f(x)}{\alpha} \\ &= \mathrm{D}f(x; y - x) = f^{(1)}(x)(y - x) \end{aligned}$$

where the second inequality follows from Proposition 10.10 and the third equality follows from Definition 9.4 and the last equality follows from Proposition 9.5.

"Sufficiency" Let $f(y) > f(x) + f^{(1)}(x)(y - x)$, $\forall x, y \in \Omega$ with $x \neq y$. $\forall x_1, x_2 \in \Omega$ with $x_1 \neq x_2$, $\forall \alpha \in (0, 1) \subset \mathbb{R}$, let $x := \alpha x_1 + (1 - \alpha)x_2 \in \Omega$. Then, $f(x_2) = f(x + \alpha(x_2 - x_1)) > f(x) + \alpha f^{(1)}(x)(x_2 - x_1)$. Note also, $f(x_1) = f(x + (1 - \alpha)(x_1 - x_2)) > f(x) + (1 - \alpha) f^{(1)}(x)(x_1 - x_2)$. Then, $\alpha f(x_1) + (1 - \alpha) f(x_2) > f(x)$. Hence, f is strictly convex.

This completes the proof of the proposition. □

Proposition 10.12 *Let $\mathcal{X}$ be a real normed linear space, $\Omega \subseteq \mathcal{X}$ be convex, and $f : \Omega \to \mathbb{R}$ be twice differentiable. Then, the following statements hold.*

(i) *If $f^{(2)}(x)$ is positive semi-definite, $\forall x \in \Omega$, then f is convex.*
(ii) *If f is convex and $\mathcal{C}_2$, then $f^{(2)}(x)$ is positive semi-definite, $\forall x \in \overline{\Omega^\circ} \cap \Omega$.*
(iii) *If f is convex and $f^{(2)}(x)$ is positive definite for all $x \in \Omega \setminus E$, where $E \subseteq \Omega$ does not contain any line segment, then f is strictly convex.*

Proof (i) $\forall x, y \in \Omega$, by Taylor's Theorem, $\exists t_0 \in (0, 1) \subset \mathbb{R}$ we have

$$f(y) = f(x) + f^{(1)}(x)(y - x) + \frac{1}{2} f^{(2)}(t_0 y + (1 - t_0)x)(y - x)(y - x)$$

By the assumption, $f^{(2)}(t_0 y + (1 - t_0)x)(y - x)(y - x) \geq 0$. Then, $f(y) \geq f(x) + f^{(1)}(x)(y - x)$. By the arbitrariness of x, y and Proposition 10.11, f is convex.

(ii) We will prove this statement by an argument of contradiction. Suppose $\exists x_0 \in \overline{\Omega^\circ} \cap \Omega$ such that $f^{(2)}(x_0)$ is not positive semi-definite. Then, $\exists h \in \mathcal{X}$ such that $f^{(2)}(x_0)(h)(h) < 0$. We will distinguish two exhaustive and mutually exclusive cases: Case 1: $x_0 \in \Omega^\circ$; Case 2: $x_0 \notin \Omega^\circ$. Case 1: $x_0 \in \Omega^\circ$. By the continuity of $f^{(2)}$ and $x_0 \in \Omega^\circ$, $\exists \delta \in (0, \infty) \subset \mathbb{R}$ such that $x_1 := x_0 + \delta h \in \Omega$ and

$f^{(2)}(x_0 + \alpha\delta h)(h)(h) < 0, \forall \alpha \in (0, 1) \subset \mathbb{R}$. By Taylor's Theorem, $\exists t_0 \in (0, 1) \subset \mathbb{R}$ we have

$$f(x_1) = f(x_0) + f^{(1)}(x_0)(x_1 - x_0) + \frac{\delta^2}{2} f^{(2)}(x_0 + t_0\delta h)(h)(h)$$

Then, $f(x_1) < f(x_0) + f^{(1)}(x_0)(x_1 - x_0)$. This contradicts with the fact that f is convex and Proposition 10.11.

Case 2: $x_0 \notin \Omega^\circ$. Then, $x_0 \in (\overline{\Omega^\circ} \cap \Omega) \setminus \Omega^\circ$. Then, by Proposition 4.13, there exists $(x_n)_{n=1}^\infty \subseteq \Omega^\circ$ such that $\lim_{n\in\mathbb{N}} x_n = x_0$. By the continuity of $f^{(2)}$, we have $\exists n_0 \in \mathbb{N}$ such that $f^{(2)}(x_{n_0})(h)(h) < 0$. Hence, $f^{(2)}(x_{n_0})$ is not positive semi-definite and $x_{n_0} \in \Omega^\circ$. By Case 1, there is a contradiction. Hence, the statement must be true.

(iii) We will prove this statement by an argument of contradiction. Suppose f is not strictly convex. By Proposition 10.11, $\exists x, y \in \Omega$ with $x \neq y$ such that $f(y) = f(x) + f^{(1)}(x)(y - x)$. By the convexity of f and Proposition 10.11, $\forall \alpha \in [0, 1] \subset \mathbb{R}$, we have $f(x + \alpha(y - x)) \geq f(x) + \alpha f^{(1)}(x)(y - x) = f(x) + \alpha(f(y) - f(x))$. By the convexity of f, we have $f(x + \alpha(y - x)) = f(x) + \alpha(f(y) - f(x))$. Define $g : [0, 1] \to \mathbb{R}$ by $g(\alpha) = f(\alpha y + (1 - \alpha)x) - \alpha f(y) - (1 - \alpha)f(x), \forall \alpha \in I := [0, 1] \subset \mathbb{R}$. Then, $g(\alpha) = 0, \forall \alpha \in I$. This implies that $g^{(2)}(\alpha) = 0, \forall \alpha \in I$, and $f^{(2)}(\alpha y + (1 - \alpha)x)(y - x)(y - x) = 0, \forall \alpha \in I$. Hence, $f^{(2)}(\alpha y + (1 - \alpha)x)$ is not positive definite, $\forall \alpha \in I$. Hence, E contains the line segment connecting x and y. This contradicts with the assumption. Therefore, the statement must be true.

This completes the proof of the proposition. □

10.2 Unconstrained Optimization

The basic problem to be considered in this section is

$$\mu_0 := \inf_{x\in\Omega} f(x) \tag{10.1}$$

where $\mathcal{X}$ is a real normed linear space, $\Omega \subseteq \mathcal{X}$ is a set, and $f : \Omega \to \mathbb{R}$ is a functional.

Proposition 10.13 *Let $\mathcal{X}$ be a real normed linear space, $\Omega \subseteq \mathcal{X}$ be convex, $f : \Omega \to \mathbb{R}$ be a convex functional, and $\mu_0 := \inf_{x\in\Omega} f(x) \in \mathbb{R}$. Then, the following statements hold.*

(i) *The set of all points of minimum for f, which is given by $\{x \in \Omega \mid f(x) = \mu_0\}$, is convex.*
(ii) *Any point of relative minimum for f is a point of minimum for f.*
(iii) *Any point of relative strict minimum for f is the point of strict minimum for f.*

Proof (i) Note that $\{x \in \Omega \mid f(x) = \mu_0\} = \{x \in \Omega \mid f(x) \leq \mu_0\}$, which is convex by Proposition 10.9.

(ii) Fix any $x_0 \in \Omega$ that is a point of relative minimum for f. Then, $\exists \epsilon \in (0, \infty) \subset \mathbb{R}$ such that $f(x) \geq f(x_0)$, $\forall x \in \Omega \cap \mathcal{B}_{\mathcal{X}}(x_0, \epsilon)$. $\forall y \in \Omega$, $\exists \alpha \in (0, 1) \subset \mathbb{R}$ such that $x_0 + \alpha(y - x_0) \in \mathcal{B}_{\mathcal{X}}(x_0, \epsilon)$. Then, $f(x_0) \leq f(x_0 + \alpha(y - x_0)) = f(\alpha y + (1 - \alpha)x_0) \leq \alpha f(y) + (1 - \alpha) f(x_0)$. This implies that $f(x_0) \leq f(y)$. Hence, x_0 is a point of minimum for f.

(iii) Fix any $x_0 \in \Omega$ that is a point of relative strict minimum for f. Then, $\exists \epsilon \in (0, \infty) \subset \mathbb{R}$ such that $f(x) > f(x_0)$, $\forall x \in (\Omega \cap \mathcal{B}_{\mathcal{X}}(x_0, \epsilon)) \setminus \{x_0\}$. $\forall y \in \Omega$ with $y \neq x_0$, $\exists \alpha \in (0, 1) \subset \mathbb{R}$ such that $x_0 + \alpha(y - x_0) \in \mathcal{B}_{\mathcal{X}}(x_0, \epsilon)$. Then, $f(x_0) < f(x_0 + \alpha(y - x_0)) = f(\alpha y + (1 - \alpha)x_0) \leq \alpha f(y) + (1 - \alpha) f(x_0)$. This implies that $f(x_0) < f(y)$. Hence, x_0 is a point of strict minimum for f. This completes the proof of the proposition. □

Proposition 10.14 *Let $\mathcal{X}$ be a real reflexive Banach space, $\Omega \subseteq \mathcal{X}$ be a nonempty bounded closed convex set, and $f : \Omega \to \mathbb{R}$ be a weakly upper semicontinuous convex functional. Then, there exists an extreme point $x_0 \in \Omega$ of Ω such that x_0 is a point of maximum for f.*

Proof By Proposition 8.11, Ω is weakly compact. Let $\mathcal{O}_{\text{weak}}(\mathcal{X})$ be the weak topology on $\mathcal{X}$ and $\mathcal{O}_{\text{weak},\Omega}$ be the subset topology on Ω with respect to $\mathcal{O}_{\text{weak}}(\mathcal{X})$. By Proposition 5.30, there exists $x_1 \in \Omega$ that is a point of maximum for f. Let $M := \{x \in \Omega \mid f(x) \geq f(x_1)\}$. Then $x_1 \in M \neq \emptyset$. Note that, by Proposition 2.5, $M = f_{\text{inv}}(\mathbb{R} \setminus I) = \Omega \setminus f_{\text{inv}}(I)$, where $I = (-\infty, f(x_1)) \subset \mathbb{R}$. Since I is open in $\mathbb{R}$ and f is weakly upper semicontinuous, then, $f_{\text{inv}}(I) \in \mathcal{O}_{\text{weak},\Omega}$. Then, M is closed in $\mathcal{O}_{\text{weak},\Omega}$. By Proposition 5.5, M is weakly compact. Since $f(x_1) = \max_{x \in \Omega} f(x)$, then M is an extreme subset of Ω. By Proposition 10.7, there exists $x_0 \in M$ that is an extreme point of Ω. This completes the proof of the proposition. □

Proposition 10.15 *Let $\mathcal{X}$ be a real normed linear space, $\Omega \subseteq \mathcal{X}$, $x_0 \in \Omega^\circ$, and $f : \Omega \to \mathbb{R}$ be differentiable at x_0. Assume that x_0 is a point of relative extremum of f, then $f^{(1)}(x_0) = \vartheta_{\mathcal{X}^*}$.*

Proof Without loss of generality, assume that x_0 is a point of relative minimum for f. The case when x_0 is a point of relative maximum for f can be proved similarly. Since $x_0 \in \Omega^\circ$, then $\mathrm{A}_\Omega(x_0) = \mathcal{X}$. $\forall u \in \mathcal{X}$, by Propositions 10.2 and 9.5, $\mathrm{D}f(x_0; u) = f^{(1)}(x_0)u \geq 0$ and $\mathrm{D}f(x_0; -u) = -f^{(1)}(x_0)u \geq 0$. Then, $f^{(1)}(x_0)u = 0$. By the arbitrariness of u, we have $f^{(1)}(x_0) = \vartheta_{\mathcal{X}^*}$. This completes the proof of the proposition. □

Proposition 10.16 *Let $\mathcal{X}$ be a real normed linear space, $\Omega \subseteq \mathcal{X}$, $x_0 \in \Omega^\circ$, and $f : \Omega \to \mathbb{R}$ be $\mathcal{C}_2$ at x_0. Assume that x_0 is a point of relative minimum for f, then $f^{(1)}(x_0) = \vartheta_{\mathcal{X}^*}$ and $f^{(2)}(x_0)$ is positive semi-definite.*

Proof By Proposition 10.15, $f^{(1)}(x_0) = \vartheta_{\mathcal{X}^*}$. We will show that $f^{(2)}(x_0)$ is positive semi-definite by an argument of contradiction. Suppose $f^{(2)}(x_0)$ is not positive semi-definite. Then, $\exists h \in \mathcal{X}$ such that $f^{(2)}(x_0)(h)(h) < 0$. By the continuity of $f^{(2)}$ at x_0, there exists $\delta \in (0, \infty) \subset \mathbb{R}$ with $D := \mathcal{B}_{\mathcal{X}}(x_0, \delta) \subseteq \Omega$ such that $f^{(2)}(x)(h)(h) < 0$, $\forall x \in D$. Let $x_1 := x_0 + \alpha h \in D$, where $\alpha \in (0, \infty) \subset \mathbb{R}$ is an arbitrary constant. By Taylor's Theorem 9.48, there exists $t_0 \in (0, 1) \subset \mathbb{R}$ such that

$$
\begin{aligned}
f(x_1) &= f(x_0) + \frac{1}{2} f^{(2)}(x_0 + t_0 \alpha h)(x_1 - x_0)(x_1 - x_0) \\
&= f(x_0) + \frac{\alpha^2}{2} f^{(2)}(x_0 + t_0 \alpha h)(h)(h) < f(x_0)
\end{aligned}
$$

This contradicts with the fact that x_0 is a point of relative minimum for f. Hence, the result must be true. This completes the proof of the proposition. □

Proposition 10.17 *Let $\mathcal{X}$ be a real normed linear space, $\Omega \subseteq \mathcal{X}$, $x_0 \in \Omega^\circ$, and $f : \Omega \to \mathbb{R}$ be twice differentiable at x, $\forall x \in D := \mathcal{B}_{\mathcal{X}}(x_0, \delta_0) \subseteq \Omega$, where $\delta_0 \in (0, \infty) \subset \mathbb{R}$ is some constant. Assume that $f^{(1)}(x_0) = \vartheta_{\mathcal{X}^*}$ and $f^{(2)}(x)$ is positive semi-definite, $\forall x \in D$. Then, x_0 is a point of relative minimum for f.*

Proof $\forall x \in D$, by Taylor's Theorem 9.48, there exists $t_0 \in (0, 1) \subset \mathbb{R}$ such that

$$
f(x) = f(x_0) + \frac{1}{2} f^{(2)}(x_0 + t_0 (x - x_0))(x - x_0)(x - x_0) \geq f(x_0)
$$

Hence, x_0 is a point of relative minimum for f. This completes the proof of the proposition. □

Proposition 10.18 *Let $\mathcal{X}$ be a real normed linear space, $\Omega \subseteq \mathcal{X}$, $x_0 \in \Omega^\circ$, and $f : \Omega \to \mathbb{R}$ be $\mathcal{C}_2$ at x_0. Assume that $f^{(1)}(x_0) = \vartheta_{\mathcal{X}^*}$ and $f^{(2)}(x_0)$ is positive definite. Then, x_0 is a point of relative strict minimum for f.*

Proof By Proposition 10.4 and the continuity of $f^{(2)}$ at x_0, $\exists \delta \in (0, \infty) \subset \mathbb{R}$ such that $f^{(2)}(x)$ is positive definite, $\forall x \in D := \mathcal{B}_{\mathcal{X}}(x_0, \delta) \subseteq \Omega$. $\forall x \in D \setminus \{x_0\}$, by Taylor's Theorem, there exists $t_0 \in (0, 1) \subset \mathbb{R}$ such that

$$
f(x) = f(x_0) + \frac{1}{2} f^{(2)}(x_0 + t_0 (x - x_0))(x - x_0)(x - x_0) > f(x_0)
$$

Hence, x_0 is a point of relative strict minimum for f. This completes the proof of the proposition. □

10.3 Optimization with Equality Constraints

The basic problem to be considered in this section is

$$\begin{aligned} \mu_0 := \quad \inf \quad & f(x) \\ \text{subject to} \quad & x \in \Omega \\ & H(x) = \vartheta_{\mathcal{Y}} \end{aligned} \tag{10.2}$$

where $\mathcal{X}$ and $\mathcal{Y}$ are real normed linear spaces, $\Omega \subseteq \mathcal{X}$ is a set, $f : \Omega \to \mathbb{R}$ is a functional, and $H : \Omega \to \mathcal{Y}$ is a function.

Definition 10.19 Let $\mathcal{X}$ and $\mathcal{Y}$ be real normed linear spaces, $\Omega \subseteq \mathcal{X}$, $H : \Omega \to \mathcal{Y}$ be $\mathcal{C}_1$ at $x_0 \in \Omega^\circ$. x_0 is said to be a *regular point* of H if $H^{(1)}(x_0) \in \mathrm{B}(\mathcal{X}, \mathcal{Y})$ is surjective. $\diamond$

Lemma 10.20 *Let $\mathcal{X}$ and $\mathcal{Y}$ be real Banach spaces, $\Omega \subseteq \mathcal{X}$, $f : \Omega \to \mathbb{R}$ be $\mathcal{C}_1$ at $x_0 \in \Omega^\circ$, and $H : \Omega \to \mathcal{Y}$ be $\mathcal{C}_1$ at x_0. Consider the optimization problem* (10.2). *Assume that x_0 is a point of relative minimum for f on the set $\Omega_c := \{x \in \Omega \mid H(x) = \vartheta_{\mathcal{Y}}\}$; and x_0 is a regular point of H. Then, $\forall u \in \mathcal{X}$ with $H^{(1)}(x_0)u = \vartheta_{\mathcal{Y}}$, we have $f^{(1)}(x_0)u = 0$.*

Proof Define $T : \Omega \to \mathbb{R} \times \mathcal{Y}$ by $T(x) = (f(x), H(x))$, $\forall x \in \Omega$. By Proposition 9.44, T is $\mathcal{C}_1$ at x_0 and $T^{(1)}(x_0) = \begin{bmatrix} f^{(1)}(x_0) \\ H^{(1)}(x_0) \end{bmatrix}$.

We will prove the result using an argument of contradiction. Suppose the result is not true. Then, $\exists u_0 \in \mathcal{X}$ with $H^{(1)}(x_0)u_0 = \vartheta_{\mathcal{Y}}$, we have $f^{(1)}(x_0)u_0 \neq 0$. We will show that $T^{(1)}(x_0)$ is surjective. $\forall (r, y) \in \mathbb{R} \times \mathcal{Y}$, by x_0 being a regular point of H, $\exists u_1 \in \mathcal{X}$ such that $H^{(1)}(x_0)u_1 = y$. Let $u = \frac{r - f^{(1)}(x_0)u_1}{f^{(1)}(x_0)u_0} u_0 + u_1$. Then, $T^{(1)}(x_0)u = (r, y)$. Hence, $T^{(1)}(x_0)$ is surjective.

Note that $T(x_0) = (f(x_0), \vartheta_{\mathcal{Y}})$ and $\mathbb{R} \times \mathcal{Y}$ is a real Banach space, by Propositions 7.22 and 4.31. By Surjective Mapping Theorem 9.53, $\exists \delta_r \in (0, \infty) \subset \mathbb{R}$, $\exists \delta \in (0, \infty) \subset \mathbb{R}$, and $c_1 \in [0, \infty) \subset \mathbb{R}$ with $c_1\delta \leq \delta_r$ such that $\forall (r, y) \in \mathcal{B}_{\mathbb{R}\times\mathcal{Y}}(T(x_0), \delta/2)$, $\exists x \in \mathcal{B}_{\mathcal{X}}(x_0, \delta_r) \subseteq \Omega$ with $\|x - x_0\| \leq c_1\|(r, y) - T(x_0)\|$, we have $T(x) = (r, y)$. Then, $\forall \bar{\delta}_r \in (0, \delta/2) \subset \mathbb{R}$, let $r_1 = f(x_0) - \bar{\delta}_r \in \mathbb{R}$. $(r_1, \vartheta_{\mathcal{Y}}) \in \mathcal{B}_{\mathbb{R}\times\mathcal{Y}}(T(x_0), \delta/2)$. Then, $\exists x_1 \in \Omega$ with $\|x_1 - x_0\| \leq c_1\|(r_1, \vartheta_{\mathcal{Y}}) - T(x_0)\| = c_1|r_1 - f(x_0)| = c_1\bar{\delta}_r < (1 + c_1)\bar{\delta}_r$ such that $T(x_1) = (r_1, \vartheta_{\mathcal{Y}})$. Then, $H(x_1) = \vartheta_{\mathcal{Y}}$ and $x_1 \in \Omega_c \cap \mathcal{B}_{\mathcal{X}}(x_0, (1 + c_1)\bar{\delta}_r)$. Furthermore, $f(x_1) = r_1 < f(x_0)$. This contradicts with the assumption that x_0 is a point of relative minimum for f on Ω_c. Therefore, the result must be true.

This completes the proof of the lemma. $\square$

Proposition 10.21 (Lagrange Multiplier) *Let $\mathcal{X}$ and $\mathcal{Y}$ be real Banach spaces, $\Omega \subseteq \mathcal{X}$, $f : \Omega \to \mathbb{R}$ be $\mathcal{C}_1$ at $x_0 \in \Omega^\circ$, and $H : \Omega \to \mathcal{Y}$ be $\mathcal{C}_1$ at x_0. Consider the optimization problem* (10.2)*. Assume that x_0 is a point of relative minimum for f on the set $\Omega_c := \{x \in \Omega \mid H(x) = \vartheta_{\mathcal{Y}}\}$; and x_0 is a regular point of H. Then, there exists a Lagrange multiplier $y_{*0} \in \mathcal{Y}^*$ such that the Lagrangian $L : \Omega \times \mathcal{Y}^* \to \mathbb{R}$ defined by $L(x, y_*) = f(x) + \langle\langle y_*, H(x)\rangle\rangle$, $\forall (x, y_*) \in \Omega \times \mathcal{Y}^*$, is* stationary *at (x_0, y_{*0}), that is $L^{(1)}(x_0, y_{*0}) = \vartheta_{\mathrm{B}(\mathcal{X}\times\mathcal{Y}^*,\mathbb{R})}$.*

Proof By Lemma 10.20, $\forall u \in \mathcal{X}$ with $H^{(1)}(x_0)u = \vartheta_{\mathcal{Y}}$, we have $f^{(1)}(x_0)u = 0$. Then, $f^{(1)}(x_0) \in \big(\mathcal{N}\big(H^{(1)}(x_0)\big)\big)^\perp$. Since x_0 is a regular point of H, then $\mathcal{R}\big(H^{(1)}(x_0)\big) = \mathcal{Y}$ is closed. By Proposition 7.114, we have $\big(\mathcal{N}\big(H^{(1)}(x_0)\big)\big)^\perp = \mathcal{R}\Big(\big(H^{(1)}(x_0)\big)'\Big)$. Then, there exists $y_{*0} \in \mathcal{Y}^*$ such that $f^{(1)}(x_0) = -\big(H^{(1)}(x_0)\big)' y_{*0}$. By Propositions 9.34, 9.41, 9.38, 9.37, 9.44, and 9.45, the Lagrangian L is $\mathcal{C}_1$ at (x_0, y_{*0}) and, $\forall (u, v_*) \in \mathcal{X} \times \mathcal{Y}^*$,

$$
\begin{aligned}
L^{(1)}(x_0, y_{*0})(u, v_*) &= f^{(1)}(x_0)u + \langle\langle v_*, H(x_0)\rangle\rangle + \big\langle\big\langle y_{*0}, H^{(1)}(x_0)u\big\rangle\big\rangle \\
&= f^{(1)}(x_0)u + \Big\langle\Big\langle \Big(H^{(1)}(x_0)\Big)' y_{*0}, u\Big\rangle\Big\rangle = 0
\end{aligned}
$$

where the second equality follows from that fact that $x_0 \in \Omega_c$. Hence, $L^{(1)}(x_0, y_{*0}) = \vartheta_{\mathrm{B}(\mathcal{X}\times\mathcal{Y}^*,\mathbb{R})}$. This completes the proof of the proposition. □

Proposition 10.22 (Generalized Lagrange Multiplier) *Let $\mathcal{X}$ and $\mathcal{Y}$ be real Banach spaces, $\Omega \subseteq \mathcal{X}$, $f : \Omega \to \mathbb{R}$ be $\mathcal{C}_1$ at $x_0 \in \Omega^\circ$, and $H : \Omega \to \mathcal{Y}$ be $\mathcal{C}_1$ at x_0. Consider the optimization problem* (10.2)*. Assume that x_0 is a point of relative minimum for f on the set $\Omega_c := \{x \in \Omega \mid H(x) = \vartheta_{\mathcal{Y}}\}$; and $\mathcal{R}\big(H^{(1)}(x_0)\big) \subseteq \mathcal{Y}$ is closed. Then, there exists a Lagrange multiplier $(r_0, y_{*0}) \in \mathbb{R} \times \mathcal{Y}^*$ with $(r_0, y_{*0}) \neq (0, \vartheta_{\mathcal{Y}^*})$ such that the Lagrangian $L : \Omega \times \mathcal{Y}^* \to \mathbb{R}$ defined by $L(x, y_*) = r_0 f(x) + \langle\langle y_*, H(x)\rangle\rangle$, $\forall (x, y_*) \in \Omega \times \mathcal{Y}^*$, is stationary at (x_0, y_{*0}), that is $L^{(1)}(x_0, y_{*0}) = \vartheta_{\mathrm{B}(\mathcal{X}\times\mathcal{Y}^*,\mathbb{R})}$.*

Proof We will distinguish two exhaustive and mutually exclusive cases: Case 1: $\mathcal{R}\big(H^{(1)}(x_0)\big) = \mathcal{Y}$; Case 2: $\mathcal{R}\big(H^{(1)}(x_0)\big) \subset \mathcal{Y}$.

Case 1: $\mathcal{R}\big(H^{(1)}(x_0)\big) = \mathcal{Y}$. Take $r_0 = 1$. Clearly, x_0 is a regular point of H. The result follows from Proposition 10.21.

Case 2: $\mathcal{R}\big(H^{(1)}(x_0)\big) \subset \mathcal{Y}$. Take $r_0 = 0$. Clearly, x_0 is not a regular point of H. Then, $\exists y_0 \in \mathcal{Y} \setminus \mathcal{R}\big(H^{(1)}(x_0)\big)$. Let $M = \mathcal{R}\big(H^{(1)}(x_0)\big)$, which is a closed subspace of $\mathcal{Y}$ by the assumption. Then, by Proposition 4.10, $\delta := \inf_{m\in M} \|y_0 - m\| > 0$. By Proposition 7.97, we have $\delta = \max_{y_*\in M^\perp,\ \|y_*\|\leq 1} \langle\langle y_*, y_0\rangle\rangle$, where the maximum is achieved at $y_{*0} \in M^\perp$. Then, $y_{*0} \neq \vartheta_{\mathcal{Y}^*}$ and, by Proposition 7.112, $y_{*0} \in \mathcal{N}\Big(\big(H^{(1)}(x_0)\big)'\Big)$. The Lagrangian L is given by $L(x, y_*) = \langle\langle y_*, H(x)\rangle\rangle$, $\forall (x, y_*) \in \Omega \times \mathcal{Y}^*$. By Propositions 9.34, 9.41, 9.37, 9.44, and 9.45, L is $\mathcal{C}_1$ at

(x_0, y_{*0}) and, $\forall (u, v_*) \in \mathcal{X} \times \mathcal{Y}^*$,

$$L^{(1)}(x_0, y_{*0})(u, v_*) = \langle\langle v_*, H(x_0)\rangle\rangle + \left\langle\left\langle y_{*0}, H^{(1)}(x_0)u\right\rangle\right\rangle$$
$$= \left\langle\left\langle \left(H^{(1)}(x_0)\right)' y_{*0}, u\right\rangle\right\rangle = \langle\langle \vartheta_{\mathcal{X}^*}, u\rangle\rangle = 0$$

where the second equality follows from that fact that $x_0 \in \Omega_c$. Hence, $L^{(1)}(x_0, y_{*0}) = \vartheta_{\mathrm{B}(\mathcal{X}\times\mathcal{Y}^*,\mathbb{R})}$. This case is proved.

This completes the proof of the proposition. □

Proposition 10.23 *Let $\mathcal{X}$ and $\mathcal{Y}$ be real Banach spaces, $\Omega \subseteq \mathcal{X}$, $f : \Omega \to \mathbb{R}$ be $\mathcal{C}_2$ at $x_0 \in \Omega^\circ$, $H : \Omega \to \mathcal{Y}$ be $\mathcal{C}_2$ at x_0. Consider the optimization problem* (10.2)*. Assume that*

(i) *$H(x_0) = \vartheta_{\mathcal{Y}}$ and $\mathcal{R}\left(H^{(1)}(x_0)\right) \subseteq \mathcal{Y}$ is closed.*
(ii) *the Lagrangian $L : \Omega \times \mathcal{Y}^* \to \mathbb{R}$ defined by $L(x, y_*) = f(x) + \langle\langle y_*, H(x)\rangle\rangle$, $\forall (x, y_*) \in \Omega \times \mathcal{Y}^*$, is stationary at (x_0, y_{*0}), where $y_{*0} \in \mathcal{Y}^*$ is a Lagrange multiplier.*
(iii) *$\frac{\partial^2 L}{\partial x^2}(x_0, y_{*0})$ is positive definite on the subspace $M := \mathcal{N}\left(H^{(1)}(x_0)\right)$, that is, $\exists m \in (0, \infty) \subset \mathbb{R}$ such that $\frac{\partial^2 L}{\partial x^2}(x_0, y_{*0})(h)(h) \geq m\|h\|^2$, $\forall h \in \mathcal{N}\left(H^{(1)}(x_0)\right)$.*

Then, x_0 is a point of relative strict minimum for f on the set $\Omega_c := \{x \in \Omega \mid H(x) = \vartheta_{\mathcal{Y}}\}$.

Proof By Propositions 9.34, 9.37, 9.38, 9.41, 9.44, and 9.45, L is $\mathcal{C}_2$ at (x_0, y_{*0}). By Proposition 9.9 and (ii),

$$L^{(1)}(x_0, y_{*0}) = \left[\ \frac{\partial L}{\partial x}(x_0, y_{*0}) \ \ \frac{\partial L}{\partial y_*}(x_0, y_{*0})\ \right] = \vartheta_{\mathrm{B}(\mathcal{X}\times\mathcal{Y}^*,\mathbb{R})}$$

Then, $\exists \delta_0 \in (0, \infty) \subset \mathbb{R}$ such that $f^{(2)}(x)$, $H^{(2)}(x)$, and $\frac{\partial^2 L}{\partial x^2}(x, y_*)$ exists, $\forall x \in \mathcal{B}_{\mathcal{X}}(x_0, \delta_0) \subseteq \Omega$ and $\forall y_* \in \mathcal{B}_{\mathcal{Y}^*}(y_{*0}, \delta_0)$, and, by Proposition 9.46, $\frac{\partial^2 L}{\partial x^2}$ is continuous at (x_0, y_{*0}). Then, $\exists \delta_1 \in (0, \delta_0] \subset \mathbb{R}$ and $\exists c_1 \in [0, \infty) \subset \mathbb{R}$ such that $\left\|\frac{\partial^2 L}{\partial x^2}(x, y_{*0}) - \frac{\partial^2 L}{\partial x^2}(x_0, y_{*0})\right\| < m/5$ and $\left\|H^{(2)}(x)\right\| \leq c_1$, $\forall x \in D_1 := \mathcal{B}_{\mathcal{X}}(x_0, \delta_1)$. By Propositions 7.114 and 7.98 and (i), $M^\perp = \mathcal{R}\left(\left(H^{(1)}(x_0)\right)'\right)$ is closed. Then, by Proposition 7.113, $\exists c_2 \in [0, \infty) \subset \mathbb{R}$ such that, $\forall x_* \in \mathcal{R}\left(\left(H^{(1)}(x_0)\right)'\right)$, there exists $y_* \in \mathcal{Y}^*$ such that $x_* = \left(H^{(1)}(x_0)\right)' y_*$ and $\|y_*\| \leq c_2\|x_*\|$.

Claim 10.23.1 $\forall x \in \Omega_c \cap D_1$, $\exists h_0 \in M$ such that $\|x - x_0 - h_0\| \leq c_1 c_2 \|x - x_0\|^2$, $\|x - x_0\| - c_1 c_2 \|x - x_0\|^2 \leq \|h_0\| \leq \|x - x_0\| + c_1 c_2 \|x - x_0\|^2$.

Proof of Claim Fix any $x \in \Omega_c \cap D_1$. Then, $H(x) = H(x_0) = \vartheta_{\mathcal{Y}}$. By Taylor's Theorem 9.48, $\exists t_0 \in (0,1) \subset \mathbb{R}$ such that $\left\| H^{(1)}(x_0)(x-x_0) \right\| = \left\| H(x) - H(x_0) - H^{(1)}(x_0)(x-x_0) \right\| \leq \frac{1}{2}\left\| H^{(2)}(t_0 x + (1-t_0)x_0) \right\| \|x-x_0\|^2 \leq c_1 \|x-x_0\|^2/2$. By Proposition 7.97, we have $\inf_{h\in M} \|x-x_0-h\| = \max_{x_*\in M^\perp,\, \|x_*\|\leq 1} \langle\langle x_*, x-x_0\rangle\rangle$, where the maximum is achieved at $x_{*1} \in M^\perp = \mathcal{R}\Big(\big(H^{(1)}(x_0)\big)'\Big)$ with $\|x_{*1}\| \leq 1$. Then, $\exists y_{*1} \in \mathcal{Y}^*$ such that $x_{*1} = \big(H^{(1)}(x_0)\big)' y_{*1}$ and $\|y_{*1}\| \leq c_2$. By Proposition 7.68, M is closed. Then, $\exists h_0 \in M$ such that $\|x-x_0-h_0\| \leq 2\inf_{h\in M}\|x-x_0-h\| = 2\langle\langle x_{*1}, x-x_0\rangle\rangle = 2\big\langle\big\langle y_{*1}, H^{(1)}(x_0)(x-x_0)\big\rangle\big\rangle \leq 2\|y_{*1}\| \left\| H^{(1)}(x_0)(x-x_0)\right\| \leq c_1 c_2 \|x-x_0\|^2$.

Then, $\|h_0\| \geq \|x-x_0\| - \|x-x_0-h_0\| \geq \|x-x_0\| - c_1c_2\|x-x_0\|^2$; and $\|h_0\| \leq \|x-x_0\| + \|x-x_0-h_0\| \leq \|x-x_0\| + c_1c_2\|x-x_0\|^2$. This completes the proof of the claim. □

Let $c_3 = \left\| \frac{\partial^2 L}{\partial x^2}(x_0, y_{*0}) \right\|$. Let $\delta \in (0, \delta_1] \subset \mathbb{R}$ such that $c_1 c_2 \delta \leq 1/4$, $c_1^2 c_2^2 (c_3 + m/5)\delta^2 \leq m/5$, and $5c_1c_2(c_3+m/5)\delta/2 \leq m/5$. $\forall x \in \Omega_c \cap \mathcal{B}_{\mathcal{X}}(x_0, \delta)$, by Claim 10.23.1, $\exists h_0 \in M$ such that $\|x-x_0-h_0\| \leq c_1c_2\|x-x_0\|^2 \leq \|x-x_0\|/4$, $3\|x-x_0\|/4 \leq \|h_0\| \leq 5\|x-x_0\|/4$. By Taylor's Theorem 9.48, $\exists t_1 \in (0,1) \subset \mathbb{R}$ such that

$$
\begin{aligned}
f(x) - f(x_0) &= L(x, y_{*0}) - L(x_0, y_{*0}) - \frac{\partial L}{\partial x}(x_0, y_{*0})(x-x_0) \\
&= \frac{1}{2}\frac{\partial^2 L}{\partial x^2}(\underbrace{t_1 x + (1-t_1)x_0}_{\bar{x}}, y_{*0})(x-x_0)(x-x_0) \\
&= \frac{1}{2}\Big(\frac{\partial^2 L}{\partial x^2}(\bar{x}, y_{*0})(h_0)(h_0) + \frac{\partial^2 L}{\partial x^2}(\bar{x}, y_{*0})(x-x_0-h_0)(h_0) \\
&\quad + \frac{\partial^2 L}{\partial x^2}(\bar{x}, y_{*0})(h_0)(x-x_0-h_0) \\
&\quad + \frac{\partial^2 L}{\partial x^2}(\bar{x}, y_{*0})(x-x_0-h_0)(x-x_0-h_0)\Big) \\
&\geq \frac{1}{2}\Big(\frac{\partial^2 L}{\partial x^2}(x_0, y_{*0})(h_0)(h_0) + \Big(\frac{\partial^2 L}{\partial x^2}(\bar{x}, y_{*0}) - \frac{\partial^2 L}{\partial x^2}(x_0, y_{*0})\Big)(h_0)(h_0) \\
&\quad - \left\|\frac{\partial^2 L}{\partial x^2}(\bar{x}, y_{*0})\right\|\big(2\|x-x_0-h_0\|\|h_0\| + \|x-x_0-h_0\|^2\big)\Big) \\
&\geq \frac{1}{2}\Big(m\|h_0\|^2 - m\|h_0\|^2/5 - \Big(\left\|\frac{\partial^2 L}{\partial x^2}(\bar{x}, y_{*0}) - \frac{\partial^2 L}{\partial x^2}(x_0, y_{*0})\right\| \\
&\quad + \left\|\frac{\partial^2 L}{\partial x^2}(x_0, y_{*0})\right\|\Big)\big(2\|x-x_0-h_0\|\|h_0\| + \|x-x_0-h_0\|^2\big)\Big)
\end{aligned}
$$

$$\geq \frac{1}{2}\Big(\frac{4m}{5}\|h_0\|^2 - (c_3 + m/5)\,(\frac{5}{2}c_1c_2\|x - x_0\|^3 + c_1^2c_2^2\|x - x_0\|^4)\Big)$$

$$\geq \frac{1}{2}(\frac{4m}{5}\frac{9}{16} - \frac{m}{5} - \frac{m}{5})\|x - x_0\|^2 = \frac{m}{40}\|x - x_0\|^2$$

Hence, x_0 is a point of relative strict minimum for f on the set Ω_c. This completes the proof of the proposition. □

Proposition 10.24 *Let $\mathcal{X}$ and $\mathcal{Y}$ be real Banach spaces, $\Omega \subseteq \mathcal{X}$, $f : \Omega \to \mathbb{R}$ be $\mathcal{C}_2$ at $x_0 \in \Omega^\circ$, $H : \Omega \to \mathcal{Y}$ be $\mathcal{C}_2$ at x_0. Consider the optimization problem* (10.2). *Assume that x_0 is a regular point of H and is a point of relative minimum for f on the set $\Omega_c := \{x \in \Omega \mid H(x) = \vartheta_{\mathcal{Y}}\}$. Then, there exists a Lagrange multiplier $y_{*0} \in \mathcal{Y}^*$ such that*

(i) *The Lagrangian $L : \Omega \times \mathcal{Y}^* \to \mathbb{R}$ defined by $L(x, y_*) = f(x) + \langle\langle y_*, H(x)\rangle\rangle$, $\forall (x, y_*) \in \Omega \times \mathcal{Y}^*$, is stationary at (x_0, y_{*0}).*
(ii) *$\frac{\partial^2 L}{\partial x^2}(x_0, y_{*0})$ is positive semi-definite on the subspace $\mathcal{N}\big(H^{(1)}(x_0)\big) =: M$, that is, $\frac{\partial^2 L}{\partial x^2}(x_0, y_{*0})(h)(h) \geq 0$, $\forall h \in \mathcal{N}\big(H^{(1)}(x_0)\big)$.*

Proof Under the assumption of the proposition, by Proposition 10.21, there exists a Lagrange multiplier $y_{*0} \in \mathcal{Y}^*$ such that (i) holds. We will show that (ii) also holds by an argument of contradiction. Suppose (ii) does not hold. Then, $\exists h_0 \in M$ with $\|h_0\| = 1$ such that $\frac{\partial^2 L}{\partial x^2}(x_0, y_{*0})(h_0)(h_0) < -m < 0$ for some $m \in (0, \infty) \subset \mathbb{R}$. By Surjective Mapping Theorem 9.53, $\exists r_1 \in (0, \infty) \subset \mathbb{R}$, $\exists \delta_1 \in (0, \infty) \subset \mathbb{R}$, and $\exists c_1 \in [0, \infty) \subset \mathbb{R}$ with $c_1\delta_1 \leq r_1$ such that $\forall \bar{y} \in \mathcal{B}_{\mathcal{Y}}(\vartheta_{\mathcal{Y}}, \delta_1/2)$, $\forall \bar{x} \in \mathcal{B}_{\mathcal{X}}(x_0, r_1/2)$ with $\bar{y} = H(\bar{x})$, $\forall y \in \mathcal{B}_{\mathcal{Y}}(\bar{y}, \delta_1/2)$, $\exists x \in \mathcal{B}_{\mathcal{X}}(x_0, r_1) \subseteq \Omega$ with $\|x - \bar{x}\| \leq c_1\|y - \bar{y}\|$, we have $y = H(x)$. By Propositions 9.34, 9.37, 9.38, 9.41, 9.44, and 9.45, L is $\mathcal{C}_2$ at (x_0, y_{*0}). By Proposition 9.9 and (i),

$$L^{(1)}(x_0, y_{*0}) = \Big[\, \frac{\partial L}{\partial x}(x_0, y_{*0}) \;\; \frac{\partial L}{\partial y_*}(x_0, y_{*0}) \,\Big] = \vartheta_{\mathrm{B}(\mathcal{X}\times\mathcal{Y}^*, \mathbb{R})}$$

By Proposition 9.46, $\exists r_2 \in (0, r_1] \subset \mathbb{R}$ such that $\frac{\partial^2 L}{\partial x^2}(x, y_{*0})$ exists and $\Big\|\frac{\partial^2 L}{\partial x^2}(x, y_{*0}) - \frac{\partial^2 L}{\partial x^2}(x_0, y_{*0})\Big\| < m/5$, $\forall x \in \mathcal{B}_{\mathcal{X}}(x_0, r_2)$. Since H is $\mathcal{C}_2$ at x_0, then $\exists c_2 \in [0, \infty) \subset \mathbb{R}$ and $\exists r_3 \in (0, r_2] \subset \mathbb{R}$ such that $\big\|H^{(2)}(x)\big\| \leq c_2$, $\forall x \in \mathcal{B}_{\mathcal{X}}(x_0, r_3)$. Let $c_3 := \Big\|\frac{\partial^2 L}{\partial x^2}(x_0, y_{*0})\Big\|$.

$\forall \delta \in (0, r_3/2) \subset \mathbb{R}$ such that $c_2\delta^2 < \delta_1$, $c_1c_2\delta/2 \leq 1/4$, $c_1^2c_2^2\,(c_3 + m/5)\delta^2/4 \leq m/5$, and $c_1c_2\,(c_3 + m/5)\delta \leq m/5$. By Taylor's Theorem 9.48, $\exists t_0 \in (0, 1) \subset \mathbb{R}$ such that

$$\|H(x_0 + \delta h_0)\| = \big\|H(x_0 + \delta h_0) - H(x_0) - \delta H^{(1)}(x_0)h_0\big\|$$
$$\leq \frac{1}{2}\big\|H^{(2)}(x_0 + t_0\delta h_0)\big\|\delta^2 \leq c_2\delta^2/2 < \delta_1/2$$

Let $\bar{y}_\delta := H(x_0 + \delta h_0) \in \mathcal{B}_{\mathcal{Y}}(\vartheta_{\mathcal{Y}}, \delta_1/2)$ and $\bar{x}_\delta := x_0 + \delta h_0 \in \mathcal{B}_{\mathcal{X}}(x_0, r_1/2)$. Note that $\bar{y}_\delta = H(\bar{x}_\delta)$ and $\vartheta_{\mathcal{Y}} \in \mathcal{B}_{\mathcal{Y}}(\bar{y}_\delta, \delta_1/2)$. Then, $\exists x_\delta \in \mathcal{B}_{\mathcal{X}}(x_0, r_1)$ with $\|x_\delta - \bar{x}_\delta\| \leq c_1\|\bar{y}_\delta\| \leq c_1c_2\delta^2/2$ such that $H(x_\delta) = \vartheta_{\mathcal{Y}}$. Then, we have $\|x_\delta - x_0\| \geq \|\delta h_0\| - \|x_\delta - \bar{x}_\delta\| \geq \delta - c_1c_2\delta^2/2 \geq 3\delta/4$; and $\|x_\delta - x_0\| \leq \|\delta h_0\| + \|x_\delta - \bar{x}_\delta\| \leq \delta + c_1c_2\delta^2/2 \leq 5\delta/4 < r_3$. Hence, $x_\delta \in \Omega_c \cap \mathcal{B}_{\mathcal{X}}(x_0, r_3)$. By Taylor's Theorem 9.48, $\exists t_1 \in (0, 1) \subset \mathbb{R}$ such that

$$
\begin{aligned}
f(x_\delta) - f(x_0) &= L(x_\delta, y_{*0}) - L(x_0, y_{*0}) - \frac{\partial L}{\partial x}(x_0, y_{*0})(x_\delta - x_0) \\
&= \frac{1}{2}\frac{\partial^2 L}{\partial x^2}(\underbrace{t_1x_\delta + (1-t_1)x_0}_{\hat{x}}, y_{*0})(x_\delta - x_0)(x_\delta - x_0) \\
&= \frac{1}{2}\Big(\frac{\partial^2 L}{\partial x^2}(\hat{x}, y_{*0})(\bar{x}_\delta - x_0)(\bar{x}_\delta - x_0) \\
&\quad + \frac{\partial^2 L}{\partial x^2}(\hat{x}, y_{*0})(x_\delta - \bar{x}_\delta)(\bar{x}_\delta - x_0) + \frac{\partial^2 L}{\partial x^2}(\hat{x}, y_{*0})(\bar{x}_\delta - x_0)(x_\delta - \bar{x}_\delta) \\
&\quad + \frac{\partial^2 L}{\partial x^2}(\hat{x}, y_{*0})(x_\delta - \bar{x}_\delta)(x_\delta - \bar{x}_\delta)\Big) \\
&\leq \frac{1}{2}\Big(\delta^2\frac{\partial^2 L}{\partial x^2}(x_0, y_{*0})(h_0)(h_0) + \delta^2\Big(\frac{\partial^2 L}{\partial x^2}(\hat{x}, y_{*0}) \\
&\quad - \frac{\partial^2 L}{\partial x^2}(x_0, y_{*0})\Big)(h_0)(h_0) + \Big\|\frac{\partial^2 L}{\partial x^2}(\hat{x}, y_{*0})\Big\|(2\|x_\delta - \bar{x}_\delta\|\delta\|h_0\| \\
&\quad + \|x_\delta - \bar{x}_\delta\|^2)\Big) \\
&< \frac{1}{2}\Big(-m\delta^2 + m\delta^2/5 + \Big(\Big\|\frac{\partial^2 L}{\partial x^2}(\hat{x}, y_{*0}) - \frac{\partial^2 L}{\partial x^2}(x_0, y_{*0})\Big\| \\
&\quad + \Big\|\frac{\partial^2 L}{\partial x^2}(x_0, y_{*0})\Big\|\Big)(c_1c_2\delta^3 + c_1^2c_2^2\delta^4/4)\Big) \\
&\leq \frac{1}{2}\Big(-\frac{4m}{5}\delta^2 + (c_3 + m/5)(c_1c_2\delta^3 + c_1^2c_2^2\delta^4/4)\Big) \\
&\leq \frac{1}{2}\Big(-\frac{4m}{5}\delta^2 + \frac{m}{5}\delta^2 + \frac{m}{5}\delta^2\Big) = -\frac{m}{5}\delta^2 < 0
\end{aligned}
$$

Thus, $\lim_{\delta\to 0^+} x_\delta = x_0$, $x_\delta \in \Omega_c$, and $f(x_\delta) < f(x_0)$, for sufficiently small δ. Then, x_0 is not a point of relative minimum for f on the set Ω_c. This contradicts with the assumption. Hence, (ii) must hold. This completes the proof of the proposition. □

10.4 Inequality Constraints

The basic problem to be considered in this section is

$$\begin{aligned} \mu_0 := \quad & \inf \quad f(x) \\ & \text{subject to } x \in \Omega \\ & G(x) \leqq \vartheta_{\mathcal{Z}} \end{aligned} \tag{10.3}$$

where $\mathcal{X}$ is a real normed linear spaces, $\Omega \subseteq \mathcal{X}$ is a set, $f : \Omega \to \mathbb{R}$ is a functional, $\mathcal{Z}$ is a real normed linear space with a positive cone $P \subseteq \mathcal{Z}$, and $G : \Omega \to \mathcal{Z}$ is a function.

Theorem 10.25 (Generalized Kuhn–Tucker Theorem) *Let $\mathcal{X}$ be a real Banach space, $\Omega \subseteq \mathcal{X}$, $x_0 \in \Omega^\circ$, $\mathcal{Z}$ be a real Banach space with a positive cone $P \subseteq \mathcal{Z}$, and $f : \Omega \to \mathbb{R}$ and $G : \Omega \to \mathcal{Z}$ be $\mathcal{C}_1$ at x_0. Consider the optimization problem* (10.3). *Assume that x_0 is a regular point of G and is a point of relative minimum for f on the set $\Omega_c := \left\{x \in \Omega \,\middle|\, G(x) \leqq \vartheta_{\mathcal{Z}}\right\}$. Then, there exists a Lagrange multiplier $z_{*0} \in \mathcal{Z}^*$ with $z_{*0} \geqq \vartheta_{\mathcal{Z}^*}$ such that the Lagrangian $L : \Omega \to \mathbb{R}$ defined by $L(x) = f(x) + \langle\langle z_{*0}, G(x)\rangle\rangle$, $\forall x \in \Omega$, is stationary at x_0 and $\langle\langle z_{*0}, G(x_0)\rangle\rangle = 0$.*

Proof Consider the small variation equivalent problem of (10.3) at x_0:

$$\begin{aligned} \bar{\mu}_0 := \quad & \inf \quad f^{(1)}(x_0)u \\ & \text{subject to } u \in \mathcal{X} \\ & G(x_0) + G^{(1)}(x_0)u \leqq \vartheta_{\mathcal{Z}} \end{aligned} \tag{10.4}$$

This problem is convex and belongs to the class of problems studied in Sect. 8.8. The primal function for (10.4) is $\omega : \bar{\Gamma} \to \mathbb{R}_e$, where

$$\bar{\Gamma} := \left\{\bar{z} \in \mathcal{Z} \,\middle|\, \exists u \in \mathcal{X} \ni \cdot\, G(x_0) + G^{(1)}(x_0)u \leqq \bar{z}\right\} \tag{10.5a}$$

$$\omega(\bar{z}) = \inf_{u \in \mathcal{X},\, G(x_0)+G^{(1)}(x_0)u \leqq \bar{z}} f^{(1)}(x_0)u; \qquad \forall \bar{z} \in \bar{\Gamma} \tag{10.5b}$$

Since $G^{(1)}(x_0) \in B(\mathcal{X}, \mathcal{Z})$ is surjective, then $\bar{\Gamma} = \mathcal{Z}$. Note that $\bar{\mu}_0 = \omega(\vartheta_{\mathcal{Z}})$. Since $x_0 \in \Omega_c$, then $z_0 := G(x_0) \leqq \vartheta_{\mathcal{Z}}$. Hence, $\bar{\mu}_0 \leq 0$.

Claim 10.25.1 $\bar{\mu}_0 = 0$.

Proof of Claim We will prove this using an argument of contradiction. Suppose $\bar{\mu}_0 < 0$. Then, $\exists u_0 \in \mathcal{X}$ with $\|u_0\| > 0$ such that $f^{(1)}(x_0)u_0 = -m\|u_0\| < 0$ and $G(x_0) + G^{(1)}(x_0)u_0 \leqq \vartheta_{\mathcal{Z}}$. By Surjective Mapping Theorem 9.53, $\exists r_1 \in (0, \infty) \subset \mathbb{R}$, $\exists \delta_1 \in (0, \infty) \subset \mathbb{R}$, and $\exists c_1 \in [0, \infty) \subset \mathbb{R}$ with $c_1\delta_1 \leq r_1$ such that $\forall \bar{z} \in \mathcal{B}_{\mathcal{Z}}(z_0, \delta_1/2)$, $\forall \bar{x} \in \mathcal{B}_{\mathcal{X}}(x_0, r_1/2)$ with $\bar{z} = G(\bar{x})$, $\forall z \in \mathcal{B}_{\mathcal{Z}}(\bar{z}, \delta_1/2)$,

$\exists x \in \mathcal{B}_{\mathcal{X}}(x_0, r_1) \subseteq \Omega$ with $\|x - \bar{x}\| \leq c_1 \|z - \bar{z}\|$, we have $z = G(x)$. Let $c_2 := \left\| G^{(1)}(x_0) \right\|$ and $c_3 := \left\| f^{(1)}(x_0) \right\|$. By f and G being $\mathcal{C}_1$ at x_0, $\exists r_2 \in (0, r_1] \subset \mathbb{R}$ such that $\forall x \in \mathcal{B}_{\mathcal{X}}(x_0, r_2)$, $\left\| f^{(1)}(x) - f^{(1)}(x_0) \right\| < m/3$ and $\left\| G^{(1)}(x) - G^{(1)}(x_0) \right\| < \frac{m}{3(1+c_1+c_1(c_3+m/3))}$.

$\forall \delta \in (0,1) \subset \mathbb{R}$ such that $\delta \|u_0\| < r_2/2$, $(c_2 + m/3)\delta\|u_0\| < \delta_1/2$, and $\delta m \|u_0\|/3 < r_2/2$. By Mean Value Theorem 9.23, $\exists t_0 \in (0,1) \subset \mathbb{R}$ such that $\left\| G(x_0 + \delta u_0) - G(x_0) - \delta G^{(1)}(x_0) u_0 \right\| \leq \left\| (G^{(1)}(x_0 + t_0 \delta u_0) - G^{(1)}(x_0)) (\delta u_0) \right\| < \frac{m}{3(1+c_1+c_1(c_3+m/3))} \delta \|u_0\| < \delta_1/2$. Let $\bar{x}_\delta := x_0 + \delta u_0 \in \mathcal{B}_{\mathcal{X}}(x_0, r_2/2)$, $\bar{z}_\delta := G(\bar{x}_\delta)$, and $z_\delta := G(x_0) + \delta G^{(1)}(x_0) u_0 \in \mathcal{B}_{\mathcal{Z}}(\bar{z}_\delta, \delta_1/2)$. Note that $\|\bar{z}_\delta - z_0\| \leq \left\| G(x_0 + \delta u_0) - G(x_0) - \delta G^{(1)}(x_0) u_0 \right\| + \left\| \delta G^{(1)}(x_0) u_0 \right\| < c_2 \delta \|u_0\| + \frac{m}{3(1+c_1+c_1(c_3+m/3))} \delta \|u_0\| < \delta_1/2$ and $\bar{z}_\delta \in \mathcal{B}_{\mathcal{Z}}(z_0, \delta_1/2)$. Then, $\exists x_\delta \in \mathcal{B}_{\mathcal{X}}(x_0, r_1)$ with $\|x_\delta - \bar{x}_\delta\| \leq c_1 \|z_\delta - \bar{z}_\delta\| \leq \frac{c_1 m}{3(1+c_1+c_1(c_3+m/3))} \delta \|u_0\| < r_2/2$ such that $z_\delta = G(x_\delta)$. Note that $G(x_\delta) = z_\delta = (1-\delta) G(x_0) + \delta (G(x_0) + G^{(1)}(x_0) u_0) \stackrel{\leq}{=} \vartheta_{\mathcal{Z}}$. Then, $x_\delta \in \Omega_c$. Note also that $x_\delta \in \mathcal{B}_{\mathcal{X}}(x_0, r_2)$. By Mean Value Theorem 9.20, $\exists t_1, t_2 \in (0,1) \subset \mathbb{R}$ such that

$$
\begin{aligned}
f(x_\delta) - f(x_0) &= f(x_\delta) - f(\bar{x}_\delta) + f(\bar{x}_\delta) - f(x_0) \\
&= f^{(1)}(t_1 x_\delta + (1-t_1)\bar{x}_\delta)(x_\delta - \bar{x}_\delta) + \delta f^{(1)}(x_0 + t_2 \delta u_0) u_0 \\
&\leq \left(\left\| f^{(1)}(x_0) \right\| + \left\| f^{(1)}(t_1 x_\delta + (1-t_1)\bar{x}_\delta) - f^{(1)}(x_0) \right\| \right) \|x_\delta - \bar{x}_\delta\| \\
&\quad + \delta f^{(1)}(x_0) u_0 + \delta \left\| f^{(1)}(x_0 + t_2 \delta u_0) - f^{(1)}(x_0) \right\| \|u_0\| \\
&\leq -m\delta \|u_0\| + (c_3 + m/3) \frac{c_1 m}{3(1+c_1+c_1(c_3+m/3))} \delta \|u_0\| \\
&\quad + m\delta \|u_0\|/3 < -m\delta \|u_0\|/3 < 0
\end{aligned}
$$

Note that $\|x_\delta - x_0\| \leq \|x_\delta - \bar{x}_\delta\| + \delta \|u_0\| < (1 + m/3)\delta \|u_0\|$. Then, we have shown that $f(x_\delta) < f(x_0)$, $x_\delta \in \Omega_c$, and $\lim_{\delta \to 0^+} x_\delta = x_0$. This contradicts with the assumption that x_0 is a point of relative minimum for f on Ω_c. Hence, $\bar{\mu}_0 = 0$.

This completes the proof of the claim. □

Then, $\omega(\vartheta_{\mathcal{Z}}) = 0$. By Fact 8.52, $\omega : \mathcal{Z} \to \mathbb{R}$ is real-valued and convex.

Next, we will show that ω is continuous at $\vartheta_{\mathcal{Z}}$ by Proposition 8.22. By Proposition 7.113, $\exists c_4 \in [0, \infty) \subset \mathbb{R}$, $\forall \bar{z} \in \mathcal{Z}$, $\exists \bar{u} \subset \mathcal{X}$ such that $\bar{z} = G^{(1)}(x_0)\bar{u}$ and $\|\bar{u}\| \leq c_4 \|\bar{z}\|$. Then, $G(x_0) + G^{(1)}(x_0)\bar{u} = G(x_0) + \bar{z} \stackrel{\leq}{=} \bar{z}$ and $f^{(1)}(x_0)\bar{u} \leq \left\| f^{(1)}(x_0) \right\| \|\bar{u}\| \leq c_4 \left\| f^{(1)}(x_0) \right\| \|\bar{z}\|$. This implies that $\forall \bar{z} \in \mathcal{B}_{\mathcal{Z}}(\vartheta_{\mathcal{Z}}, 1/2)$, $\omega(\bar{z}) \leq c_4 \left\| f^{(1)}(x_0) \right\| /2 =: r_0 - 1/2$. Then, $\mathcal{B}_{\mathbb{R} \times \mathcal{Z}}((r_0, \vartheta_{\mathcal{Z}}), 1/2) \subseteq [\omega, \mathcal{Z}]$. Hence, $(r_0, \vartheta_{\mathcal{Z}}) \in \left[\omega, \bar{\Gamma}\right]^\circ$. By Proposition 8.22, ω is continuous at $\vartheta_{\mathcal{Z}}$. By Proposition 8.23, ω is continuous.

By Proposition 8.34, we have

$$0 = \bar{\mu}_0 = \omega(\vartheta_{\mathcal{Z}}) = \max_{z_* \in \bar{\Gamma}_{\text{conj}}} (-\omega_{\text{conj}}(z_*))$$

where $\bar{\Gamma}_{\text{conj}}$ and $\omega_{\text{conj}} : \bar{\Gamma}_{\text{conj}} \to \mathbb{R}$ are the conjugate set and conjugate functional of ω. By Fact 8.53, $\bar{\Gamma}_{\text{conj}} \subseteq P^{\ominus}$. Let the maximum be achieved at $-z_{*0} \in \bar{\Gamma}_{\text{conj}}$, where $z_{*0} \stackrel{\geq}{=} \vartheta_{\mathcal{Z}^*}$. Define $\bar{\omega} : P^{\ominus} \to \mathbb{R}_e$ by $\bar{\omega}(z_*) = \sup_{z \in \bar{\Gamma}}(\langle\langle z_*, z \rangle\rangle - \omega(z))$, $\forall z_* \in P^{\ominus}$. By Fact 8.54, we have $\max_{z_* \in \bar{\Gamma}_{\text{conj}}}(-\omega_{\text{conj}}(z_*)) = \max_{z_* \in P^{\ominus}}(-\bar{\omega}(z_*))$, where both maximums are achieved at $-z_{*0}$, and $\forall z_* \in P^{\oplus}$,

$$-\bar{\omega}(-z_*) = \inf_{u \in \mathcal{X}} \left(f^{(1)}(x_0)u + \left\langle\!\left\langle z_*, G(x_0) + G^{(1)}(x_0)u \right\rangle\!\right\rangle \right)$$

Then, we have

$$0 = \bar{\mu}_0 = \max_{z_* \stackrel{\geq}{=} \vartheta_{\mathcal{Z}^*}} \inf_{u \in \mathcal{X}} \left(f^{(1)}(x_0)u + \left\langle\!\left\langle z_*, G(x_0) + G^{(1)}(x_0)u \right\rangle\!\right\rangle \right)$$

where the maximum is achieved at z_{*0}. This implies that $\langle\langle z_{*0}, G(x_0) \rangle\rangle + \inf_{u \in \mathcal{X}} \left\langle\!\left\langle f^{(1)}(x_0) + \left(G^{(1)}(x_0)\right)' z_{*0}, u \right\rangle\!\right\rangle = 0$ by Proposition 8.37. For the infimum to be finite, we must have $f^{(1)}(x_0) + \left(G^{(1)}(x_0)\right)' z_{*0} = \vartheta_{\mathcal{X}^*}$. Hence, the Lagrangian L is stationary at x_0. Then, $0 = \langle\langle z_{*0}, G(x_0) \rangle\rangle$.

This completes the proof of the proposition. □

Proposition 10.26 *Let $\mathcal{X}$ be a real Banach space, $\mathcal{Y}$ be a real Banach space with a positive cone $P \subseteq \mathcal{Y}$, and $A \in \mathrm{B}(\mathcal{X}, \mathcal{Y})$ be surjective. Then,*

$$(A_{\text{inv}}(P))^{\oplus} = A'(P^{\oplus})$$

Proof $\forall x_* \in A'(P^{\oplus})$, $\exists y_* \in P^{\oplus}$ such that $x_* = A'y_*$. $\forall x \in A_{\text{inv}}(P)$, we have $Ax \in P$. Then, $\langle\langle x_*, x \rangle\rangle = \left\langle\!\left\langle A'y_*, x \right\rangle\!\right\rangle = \langle\langle y_*, Ax \rangle\rangle \geq 0$. By the arbitrariness of x, $x_* \in (A_{\text{inv}}(P))^{\oplus}$. By the arbitrariness of x_*, we have $A'(P^{\oplus}) \subseteq (A_{\text{inv}}(P))^{\oplus}$.

On the other hand, fix any $x_* \in (A_{\text{inv}}(P))^{\oplus}$. $\forall x \in \mathcal{N}(A)$, $x \in A_{\text{inv}}(P)$, since $\vartheta_{\mathcal{Y}} \in P$. Then, $\langle\langle x_*, x \rangle\rangle \geq 0$. Since $-x \in \mathcal{N}(A)$ as well, then $\langle\langle x_*, x \rangle\rangle = 0$. Hence, $x_* \in (\mathcal{N}(A))^{\perp}$. By Proposition 7.114, $(\mathcal{N}(A))^{\perp} = \mathcal{R}(A')$. Then, $\exists y_* \in \mathcal{Y}^*$ such that $x_* = A'y_*$. $\forall y \in P$, since A is surjective, then $\exists x \in A_{\text{inv}}(P)$ such that $y = Ax$. Then, we have $\langle\langle y_*, y \rangle\rangle = \langle\langle y_*, Ax \rangle\rangle = \left\langle\!\left\langle A'y_*, x \right\rangle\!\right\rangle = \langle\langle x_*, x \rangle\rangle \geq 0$. By the arbitrariness of y, we have $y_* \in P^{\oplus}$. Then, $x_* \in A'(P^{\oplus})$. By the arbitrariness of x_*, we have $(A_{\text{inv}}(P))^{\oplus} \subseteq A'(P^{\oplus})$.

Therefore, we have $A'(P^{\oplus}) = (A_{\text{inv}}(P))^{\oplus}$. This completes the proof of the proposition. □

Next, we present the second-order sufficient condition for a relative strict minimum point in the optimization problem (10.3) with inequality constraints.

Proposition 10.27 *Let $\mathcal{X}$ be a real Banach space, $\Omega \subseteq \mathcal{X}$, $x_0 \in \Omega^\circ$, $\mathcal{Z}$ be a real Banach space with a positive cone $P \subseteq \mathcal{Z}$, and $f : \Omega \to \mathbb{R}$ and $G : \Omega \to \mathcal{Z}$ be $\mathcal{C}_2$ at x_0. Consider the optimization problem (10.3). Assume that*

(i) $G(x_0) \stackrel{\leq}{=} \vartheta_{\mathcal{Z}}$ *and x_0 is regular point of G.*
(ii) $z_{*0} \in P^\oplus$ *is the Lagrange multiplier, the Lagrangian $L : \Omega \to \mathbb{R}$ defined by $L(x) = f(x) + \langle\langle z_{*0}, G(x)\rangle\rangle$, $\forall x \in \Omega$, is stationary at x_0.*
(iii) $\langle\langle z_{*0}, G(x_0)\rangle\rangle = 0$.
(iv) $\exists m \in (0, \infty) \subset \mathbb{R}$ *such that* $L^{(2)}(x_0)(u)(u) \geq m\|u\|^2$, $\forall u \in M_c := \left\{u \in \mathcal{X} \mid G(x_0) + G^{(1)}(x_0)u \stackrel{\leq}{=} \vartheta_{\mathcal{Z}}\right\}$, *that is, $L^{(2)}(x_0)$ is positive definite on the set M_c.*

Then, x_0 is a point of relative strict minimum of f on the set $\Omega_c := \left\{x \in \mathcal{X} \mid G(x) \stackrel{\leq}{=} \vartheta_{\mathcal{Z}}\right\}$.

Proof By Propositions 9.34, 9.37, 9.40, 9.41, 9.44, and 9.45, L is $\mathcal{C}_2$ at x_0. By (ii), $L^{(1)}(x_0) = \vartheta_{\mathrm{B}(\mathcal{X},\mathbb{R})} = \vartheta_{\mathcal{X}^*}$. Then, $\exists \delta_0 \in (0, \infty) \subset \mathbb{R}$ such that $f^{(2)}(x)$, $G^{(2)}(x)$, and $L^{(2)}(x)$ exist, $\forall x \in \mathcal{B}_{\mathcal{X}}(x_0, \delta_0) \subseteq \Omega$ and $L^{(2)}$ is continuous at x_0. Then, $\exists \delta_1 \in (0, \delta_0] \subset \mathbb{R}$ and $\exists c_1 \in [0, \infty) \subset \mathbb{R}$ such that $\left\|L^{(2)}(x) - L^{(2)}(x_0)\right\| < m/5$ and $\left\|G^{(2)}(x)\right\| \leq c_1$, $\forall x \in D_1 := \mathcal{B}_{\mathcal{X}}(x_0, \delta_1)$. By Propositions 7.114 and 7.98 and (i), $(\mathcal{N}(G^{(1)}(x_0)))^\perp = \mathcal{R}\Big(\big(G^{(1)}(x_0)\big)'\Big)$ is closed. Then, by Proposition 7.113, $\exists c_2 \in [0, \infty) \subset \mathbb{R}$ such that, $\forall x_* \in \mathcal{R}\Big(\big(G^{(1)}(x_0)\big)'\Big)$, there exists $z_* \in \mathcal{Z}^*$ such that $x_* = \big(G^{(1)}(x_0)\big)' z_*$ and $\|z_*\| \leq c_2\|x_*\|$.

Claim 10.27.1 $\forall x \in \Omega_c \cap D_1, \exists h_0 \in M_c$ such that $\|x - x_0 - h_0\| \leq c_1 c_2 \|x - x_0\|^2$ and $\|x - x_0\| - c_1 c_2 \|x - x_0\|^2 \leq \|h_0\| \leq \|x - x_0\| + c_1 c_2 \|x - x_0\|^2$.

Proof of Claim Fix any $x \in \Omega_c \cap D_1$. Then, $G(x) \stackrel{\leq}{=} \vartheta_{\mathcal{Z}}$. It is easy to see that $M_c \ni \vartheta_{\mathcal{X}}$ is a nonempty closed convex set. Then, by Proposition 8.15, we have $0 \leq \bar{\delta} := \inf_{h \in M_c} \|x - x_0 - h\| = \max_{x_* \in M_{c\mathrm{supp}}, \|x_*\| \leq 1} (\langle\langle x_*, x - x_0\rangle\rangle - g(x_*)) \geq 0$, where $g : M_{c\mathrm{supp}} \to \mathbb{R}$ is the support functional of M_c, $M_{c\mathrm{supp}} := \{x_* \in \mathcal{X}^* \mid \sup_{u \in M_c} \langle\langle x_*, u\rangle\rangle < +\infty\}$, and $g(x_*) = \sup_{u \in M_c} \langle\langle x_*, u\rangle\rangle$, $\forall x_* \in M_{c\mathrm{supp}}$.

We will show that $M_{c\mathrm{supp}} \subseteq ((G^{(1)}(x_0))_{\mathrm{inv}}(P))^\oplus$. Fix any $x_* \in M_{c\mathrm{supp}}$. $\forall u \in (G^{(1)}(x_0))_{\mathrm{inv}}(P)$, we have $G^{(1)}(x_0)u \stackrel{\geq}{=} \vartheta_{\mathcal{Z}}$. $\forall \alpha \in [0, \infty) \subset \mathbb{R}$, it is easy to show that $-\alpha u \in M_c$. Suppose $\langle\langle x_*, u\rangle\rangle < 0$. Then, $\sup_{\bar{u} \in M_c} \langle\langle x_*, \bar{u}\rangle\rangle \geq \sup_{\alpha \in [0,\infty) \subset \mathbb{R}} \langle\langle x_*, -\alpha u\rangle\rangle = +\infty$. This implies that $x_* \notin M_{c\mathrm{supp}}$, which is a contradiction. Hence, we must have $\langle\langle x_*, u\rangle\rangle \geq 0$. By the arbitrariness of u, we have $x_* \in ((G^{(1)}(x_0))_{\mathrm{inv}}(P))^\oplus$. By the arbitrariness of x_*, we have $M_{c\mathrm{supp}} \subseteq ((G^{(1)}(x_0))_{\mathrm{inv}}(P))^\oplus$.

Then, $\bar{\delta} = \langle\langle x_{*0}, x - x_0\rangle\rangle - g(x_{*0})$ for some $x_{*0} \in M_{c\,\mathrm{supp}} \subseteq ((G^{(1)}(x_0))_{\mathrm{inv}}(P))^{\oplus}$ with $\|x_{*0}\| \leq 1$. By (i) and Proposition 10.26, we have $x_{*0} \in (G^{(1)}(x_0))'(P^{\oplus}) \subseteq \mathcal{R}\big((G^{(1)}(x_0))'\big)$. Then, $\exists z_{*1} \in \mathcal{Z}^*$ such that $x_{*0} = (G^{(1)}(x_0))'z_{*1}$ and $\|z_{*1}\| \leq c_2\|x_{*0}\| \leq c_2$. This further implies that

$$\begin{aligned}\bar{\delta} &= \langle\!\langle z_{*1}, G^{(1)}(x_0)(x - x_0)\rangle\!\rangle - \sup_{u\in M_c} \langle\!\langle z_{*1}, G^{(1)}(x_0)u\rangle\!\rangle \\ &= \langle\!\langle z_{*1}, -G(x) + G(x_0) + G^{(1)}(x_0)(x - x_0)\rangle\!\rangle \\ &\quad - \sup_{u\in M_c} \langle\!\langle z_{*1}, -G(x) + G(x_0) + G^{(1)}(x_0)u\rangle\!\rangle\end{aligned}$$

where the second equality follows from Proposition 8.37. By (i), we have $\mathcal{R}\big(G^{(1)}(x_0)\big) = \mathcal{Z}$ and $\exists u_1 \in \mathcal{X}$ such that $G^{(1)}(x_0)u_1 = G(x) - G(x_0)$. It is easy to see that $u_1 \in M_c$ since $x \in \Omega_c$ and $G(x) \stackrel{\leq}{=} \vartheta_{\mathcal{Z}}$. Then, $\bar{\delta} \leq \langle\!\langle z_{*1}, -G(x) + G(x_0) + G^{(1)}(x_0)(x - x_0)\rangle\!\rangle$.

By Taylor's Theorem 9.48, $\exists t_0 \in (0, 1) \subset \mathbb{R}$ such that $\big\|G(x) - G(x_0) - G^{(1)}(x_0)(x - x_0)\big\| \leq \frac{1}{2}\big\|G^{(2)}(t_0x + (1 - t_0)x_0)\big\|\,\|x - x_0\|^2 \leq c_1\|x - x_0\|^2/2$. Then, we have $\bar{\delta} \leq \|z_{*1}\|\,\big\|G(x) - G(x_0) - G^{(1)}(x_0)(x - x_0)\big\| \leq c_1c_2\|x - x_0\|^2/2$. Since M_c is closed, then $\exists h_0 \in M_c$ such that $\|x - x_0 - h_0\| \leq 2\bar{\delta} \leq c_1c_2\|x - x_0\|^2$.

Then, $\|h_0\| \geq \|x - x_0\| - \|x - x_0 - h_0\| \geq \|x - x_0\| - c_1c_2\|x - x_0\|^2$; and $\|h_0\| \leq \|x - x_0\| + \|x - x_0 - h_0\| \leq \|x - x_0\| + c_1c_2\|x - x_0\|^2$. This completes the proof of the claim. □

Let $c_3 = \big\|L^{(2)}(x_0)\big\|$. Let $\delta \in (0, \delta_1] \subset \mathbb{R}$ such that $c_1c_2\delta \leq 1/4$, $c_1^2c_2^2(c_3 + m/5)\delta^2 \leq m/5$, and $5c_1c_2(c_3 + m/5)\delta/2 \leq m/5$. $\forall x \in \Omega_c \cap \mathcal{B}_{\mathcal{X}}(x_0, \delta)$, by Claim 10.27.1, $\exists h_0 \in M_c$ such that $\|x - x_0 - h_0\| \leq c_1c_2\|x - x_0\|^2 \leq \|x - x_0\|/4$ and $3\|x - x_0\|/4 \leq \|h_0\| \leq 5\|x - x_0\|/4$. By Taylor's Theorem 9.48, $\exists t_1 \in (0, 1) \subset \mathbb{R}$ such that

$$\begin{aligned}&f(x) - f(x_0) \geq L(x) - L(x_0) - L^{(1)}(x_0)(x - x_0) \\ &= \frac{1}{2}L^{(2)}(\underbrace{t_1x + (1 - t_1)x_0}_{\bar{x}})(x - x_0)(x - x_0) \\ &= \frac{1}{2}\big(L^{(2)}(\bar{x})(h_0)(h_0) + L^{(2)}(\bar{x})(x - x_0 - h_0)(h_0) \\ &\quad + L^{(2)}(\bar{x})(h_0)(x - x_0 - h_0) + L^{(2)}(\bar{x})(x - x_0 - h_0)(x - x_0 - h_0)\big) \\ &\geq \frac{1}{2}\big(L^{(2)}(x_0)(h_0)(h_0) + (L^{(2)}(\bar{x}) - L^{(2)}(x_0))(h_0)(h_0) \\ &\quad - \big\|L^{(2)}(\bar{x})\big\|(2\|x - x_0 - h_0\|\,\|h_0\| + \|x - x_0 - h_0\|^2)\big)\end{aligned}$$

$$\geq \frac{1}{2}\Big(m\|h_0\|^2 - m\|h_0\|^2/5 - \big(\big\|L^{(2)}(\bar{x}) - L^{(2)}(x_0)\big\| + \big\|L^{(2)}(x_0)\big\|\big)$$
$$\cdot(2\|x - x_0 - h_0\|\,\|h_0\| + \|x - x_0 - h_0\|^2)\Big)$$
$$\geq \frac{1}{2}\Big(\frac{4m}{5}\|h_0\|^2 - (c_3 + m/5)\,(\frac{5}{2}c_1c_2\|x - x_0\|^3 + c_1^2c_2^2\|x - x_0\|^4)\Big)$$
$$\geq \frac{1}{2}(\frac{4m}{5}\frac{9}{16} - \frac{m}{5} - \frac{m}{5})\|x - x_0\|^2 = \frac{m}{40}\|x - x_0\|^2$$

Hence, x_0 is a point of relative strict minimum for f on the set Ω_c. This completes the proof of the proposition. □

Proposition 10.28 *Let $\mathcal{X}$ be a real Banach space, $\Omega \subseteq \mathcal{X}$, $x_0 \in \Omega^\circ$, $\mathcal{Z}$ be a real Banach space with a positive cone $P \subseteq \mathcal{Z}$, and $f : \Omega \to \mathbb{R}$ and $G : \Omega \to \mathcal{Z}$ be $\mathcal{C}_2$ at x_0. Consider the optimization problem (10.3). Assume that x_0 is a regular point of G and a point of relative minimum for f on the set $\Omega_c := \{x \in \mathcal{X} \mid G(x) \stackrel{\leq}{=} \vartheta_{\mathcal{Z}}\}$. Then, there exists a Lagrange multiplier $z_{*0} \in P^{\oplus}$ such that*

(i) *The Lagrangian $L : \Omega \to \mathbb{R}$ defined by $L(x) = f(x) + \langle\langle z_{*0}, G(x)\rangle\rangle$, $\forall x \in \Omega$, is stationary at x_0 and $\langle\langle z_{*0}, G(x_0)\rangle\rangle = 0$.*
(ii) *$L^{(2)}(x_0)$ is positive semi-definite on the subspace $\mathcal{N}\big(G^{(1)}(x_0)\big) =: M$, that is, $L^{(2)}(x_0)(h)(h) \geq 0$, $\forall h \in \mathcal{N}\big(G^{(1)}(x_0)\big)$.*

Proof By the Generalized Kuhn–Tucker Theorem 10.25, there exists a Lagrange multiplier $z_{*0} \in \mathcal{Z}^*$ with $z_{*0} \stackrel{\geq}{=} \vartheta_{\mathcal{Z}^*}$ such that (i) holds. We will show that (ii) also holds by an argument of contradiction. Suppose (ii) does not hold. Then, $\exists h_0 \in M$ with $\|h_0\| = 1$ such that $L^{(2)}(x_0)(h_0)(h_0) < -m < 0$ for some $m \in (0, \infty) \subset \mathbb{R}$. Let $z_0 := G(x_0) \in \mathcal{Z}$. By Surjective Mapping Theorem 9.53, $\exists r_1 \in (0, \infty) \subset \mathbb{R}$, $\exists \delta_1 \in (0, \infty) \subset \mathbb{R}$, and $\exists c_1 \in [0, \infty) \subset \mathbb{R}$ with $c_1\delta_1 \leq r_1$ such that $\forall \bar{z} \in \mathcal{B}_{\mathcal{Z}}(z_0, \delta_1/2)$, $\forall \bar{x} \in \mathcal{B}_{\mathcal{X}}(x_0, r_1/2)$ with $\bar{z} = G(\bar{x})$, $\forall z \in \mathcal{B}_{\mathcal{Z}}(\bar{z}, \delta_1/2)$, $\exists x \in \mathcal{B}_{\mathcal{X}}(x_0, r_1) \subseteq \Omega$ with $\|x - \bar{x}\| \leq c_1\|z - \bar{z}\|$, we have $z = G(x)$. By Propositions 9.34, 9.37, 9.40, 9.41, 9.44, and 9.45, L is $\mathcal{C}_2$ at x_0. By (i), $L^{(1)}(x_0) = \vartheta_{\mathrm{B}(\mathcal{X},\mathbb{R})} = \vartheta_{\mathcal{X}^*}$. Then, $\exists r_2 \in (0, r_1] \subset \mathbb{R}$ such that $L^{(2)}(x)$ exists and $\big\|L^{(2)}(x) - L^{(2)}(x_0)\big\| < m/5$, $\forall x \in \mathcal{B}_{\mathcal{X}}(x_0, r_2)$. Since G is $\mathcal{C}_2$ at x_0, then $\exists c_2 \in [0, \infty) \subset \mathbb{R}$ and $\exists r_3 \in (0, r_2] \subset \mathbb{R}$ such that $\big\|G^{(2)}(x)\big\| \leq c_2$, $\forall x \in \mathcal{B}_{\mathcal{X}}(x_0, r_3)$. Let $c_3 := \big\|L^{(2)}(x_0)\big\|$.

$\forall \delta \in (0, r_3/2) \subset \mathbb{R}$ such that $c_2\delta^2 < \delta_1$, $c_1c_2\delta/2 \leq 1/4$, $c_1^2c_2^2\,(c_3 + m/5)\delta^2/4 \leq m/5$, and $c_1c_2\,(c_3 + m/5)\delta \leq m/5$. By Taylor's Theorem 9.48, $\exists t_0 \in (0, 1) \subset \mathbb{R}$ such that

$$\|G(x_0 + \delta h_0) - z_0\| = \big\|G(x_0 + \delta h_0) - G(x_0) - \delta G^{(1)}(x_0)h_0\big\|$$
$$\leq \frac{1}{2}\big\|G^{(2)}(x_0 + t_0\delta h_0)\big\|\delta^2 \leq c_2\delta^2/2 < \delta_1/2$$

Let $\bar{z}_\delta := G(x_0 + \delta h_0) \in \mathcal{B}_{\mathcal{Z}}(z_0, \delta_1/2)$ and $\bar{x}_\delta := x_0 + \delta h_0 \in \mathcal{B}_{\mathcal{X}}(x_0, r_1/2)$. Note that $\bar{z}_\delta = G(\bar{x}_\delta)$ and $z_0 \in \mathcal{B}_{\mathcal{Z}}(\bar{z}_\delta, \delta_1/2)$. Then, $\exists x_\delta \in \mathcal{B}_{\mathcal{X}}(x_0, r_1)$ with $\|x_\delta - \bar{x}_\delta\| \leq c_1\|z_0 - \bar{z}_\delta\| \leq c_1 c_2 \delta^2/2$ such that $G(x_\delta) = z_0$. Then, we have $\|x_\delta - x_0\| \geq \|\delta h_0\| - \|x_\delta - \bar{x}_\delta\| \geq \delta - c_1 c_2 \delta^2/2 \geq 3\delta/4$; and $\|x_\delta - x_0\| \leq \|\delta h_0\| + \|x_\delta - \bar{x}_\delta\| \leq \delta + c_1 c_2 \delta^2/2 \leq 5\delta/4 < r_3$. Hence, $x_\delta \in \Omega_c \cap \mathcal{B}_{\mathcal{X}}(x_0, r_3)$. By Taylor's Theorem 9.48, $\exists t_1 \in (0, 1) \subset \mathbb{R}$ such that

$$
\begin{aligned}
f(x_\delta) - f(x_0) &= L(x_\delta) - L(x_0) - L^{(1)}(x_0)(x_\delta - x_0) - \langle\langle z_{*0}, G(x_\delta)\rangle\rangle \\
&\quad + \langle\langle z_{*0}, G(x_0)\rangle\rangle = L(x_\delta) - L(x_0) - L^{(1)}(x_0)(x_\delta - x_0) \\
&= \frac{1}{2} L^{(2)}(\underbrace{t_1 x_\delta + (1 - t_1) x_0}_{\hat{x}})(x_\delta - x_0)(x_\delta - x_0) \\
&= \frac{1}{2}\big(L^{(2)}(\hat{x})(\bar{x}_\delta - x_0)(\bar{x}_\delta - x_0) + L^{(2)}(\hat{x})(x_\delta - \bar{x}_\delta)(\bar{x}_\delta - x_0) \\
&\quad + L^{(2)}(\hat{x})(\bar{x}_\delta - x_0)(x_\delta - \bar{x}_\delta) + L^{(2)}(\hat{x})(x_\delta - \bar{x}_\delta)(x_\delta - \bar{x}_\delta)\big) \\
&\leq \frac{1}{2}\big(\delta^2 L^{(2)}(x_0)(h_0)(h_0) + \delta^2 (L^{(2)}(\hat{x}) - L^{(2)}(x_0))(h_0)(h_0) \\
&\quad + \big\|L^{(2)}(\hat{x})\big\|(2\|x_\delta - \bar{x}_\delta\|\delta\|h_0\| + \|x_\delta - \bar{x}_\delta\|^2)\big) \\
&< \frac{1}{2}\big(- m\delta^2 + m\delta^2/5 + \big(\big\|L^{(2)}(\hat{x}) - L^{(2)}(x_0)\big\| + \big\|L^{(2)}(x_0)\big\|\big) \\
&\quad \cdot (c_1 c_2 \delta^3 + c_1^2 c_2^2 \delta^4/4)\big) \\
&\leq \frac{1}{2}\big(- \frac{4m}{5}\delta^2 + (c_3 + m/5)\,(c_1 c_2 \delta^3 + c_1^2 c_2^2 \delta^4/4)\big) \\
&\leq \frac{1}{2}\big(- \frac{4m}{5}\delta^2 + \frac{m}{5}\delta^2 + \frac{m}{5}\delta^2\big) = -\frac{m}{5}\delta^2 < 0
\end{aligned}
$$

Thus, $\lim_{\delta \to 0^+} x_\delta = x_0$, $x_\delta \in \Omega_c$, and $f(x_\delta) < f(x_0)$, for sufficiently small δ. Then, x_0 is not a point of relative minimum for f on the set Ω_c. This contradicts with the assumption. Hence, (ii) must hold. This completes the proof of the proposition. □

Proposition 10.29 *Let $\mathcal{X}$ be a real reflexive Banach space and $L \in \mathcal{S}_{\mathcal{X}}$. Then, the following statements holds.*

(i) $L' = L$.
(ii) *If $L \in \mathcal{S}_{+\mathcal{X}}$, that is, $\exists \delta \in (0, \infty) \subset \mathbb{R}$ such that $L(x)(x) \geq \delta\|x\|_{\mathcal{X}}^2$, $\forall x \in \mathcal{X}$, then, $L \in \mathrm{B}(\mathcal{X}, \mathcal{X}^*)$ is bijective, $L^{-1} \in \mathcal{S}_{+\mathcal{X}^*}$, that is, $L^{-1}(x_*)(x_*) \geq \frac{\delta}{\epsilon + \|L\|_{\mathrm{B}(\mathcal{X},\mathcal{X}^*)}^2}\|x_*\|_{\mathcal{X}^*}^2$, $\forall x_* \in \mathcal{X}^*$, $\forall \epsilon \in (0, \infty) \subset \mathbb{R}$.*

Proof (i) $\forall x, y \in \mathcal{X}$, we have

$$L(x)(y) = L(y)(x) = \langle\langle L(x), y\rangle\rangle = \langle\langle \phi_{\mathcal{X}}(y), L(x)\rangle\rangle = \left\langle\left\langle L'(\phi_{\mathcal{X}}(y)), x\right\rangle\right\rangle$$

where the first equality follows from the fact that $L \in \mathcal{S}_{\mathcal{X}} = \mathrm{B}_{\mathrm{S}2}(\mathcal{X}, \mathbb{R})$; and $\phi_{\mathcal{X}} : \mathcal{X} \to \mathcal{X}^{**}$ is the natural mapping introduced in Remark 7.88. Since $\mathcal{X}$ is reflexive, then $\mathcal{X}^{**}$ and $\mathcal{X}$ is isometrically isomorphic and $\phi_{\mathcal{X}} = \mathrm{id}_{\mathcal{X}}$. This implies that

$$L(x)(y) = \langle\langle L'(y), x\rangle\rangle = L(y)(x) = \langle\langle L(y), x\rangle\rangle$$

By the arbitrariness of x and y, we have $L = L'$.

(ii) By the assumption, $L \in \mathrm{B}_{\mathrm{S}2}(\mathcal{X}, \mathbb{R}) \subseteq \mathrm{B}(\mathcal{X}, \mathcal{X}^*)$. $\forall x \in \mathcal{X}$ with $x \neq \vartheta_{\mathcal{X}}$, $L(x) \neq \vartheta_{\mathcal{X}^*}$ since $L(x)(x) > 0$. Hence, $\mathcal{N}(L) = \{\vartheta_{\mathcal{X}}\}$ and L is injective. By (i), $L = L'$. Then, $\mathcal{N}(L') = \{\vartheta_{\mathcal{X}}\}$. By Proposition 7.112, $(\mathcal{R}(L))^{\perp} = \mathcal{N}(L')$. This shows that ${}^{\perp}((\mathcal{R}(L))^{\perp}) = \mathcal{X}^*$. By Proposition 7.98, $\overline{\mathcal{R}(L)} = \mathcal{X}^*$. Hence, $\mathcal{R}(L)$ is dense in $\mathcal{X}^*$. Clearly, $\vartheta_{\mathcal{X}^*} \in \mathcal{R}(L)$. $\forall x_* \in \mathcal{X}^*$ with $x_* \neq \vartheta_{\mathcal{X}^*}$, there exists $(x_{*i})_{i=1}^{\infty} \subseteq \mathcal{R}(L)$ such that $x_* = \lim_{i\in\mathbb{N}} x_{*i}$. Without loss of generality, we may assume that $x_{*i} \neq \vartheta_{\mathcal{X}^*}$, $\forall i \in \mathbb{N}$. Let $x_{*i} = L(x_i)$, $\forall i \in \mathbb{N}$, where $x_i \in \mathcal{X}$. Clearly, $x_i \neq \vartheta_{\mathcal{X}}$ since $x_{*i} \neq \vartheta_{\mathcal{X}^*}$, $\forall i \in \mathbb{N}$. Then, we have

$$\|x_{*i}\|_{\mathcal{X}^*}\|x_i\|_{\mathcal{X}} \geq \langle\langle x_{*i}, x_i\rangle\rangle = L(x_i)(x_i) \geq \delta\|x_i\|_{\mathcal{X}}^2; \quad \forall i \in \mathbb{N}$$

This implies that $\|x_{*i}\|_{\mathcal{X}^*} \geq \delta\|x_i\|_{\mathcal{X}}$ and $\|x_i\|_{\mathcal{X}} \leq \|x_{*i}\|_{\mathcal{X}^*}/\delta$, $\forall i \in \mathbb{N}$. Since $(x_{*i})_{i=1}^{\infty}$ is convergent, then there exists $c \in (0, \infty) \subset \mathbb{R}$ such that $\|x_{*i}\|_{\mathcal{X}^*} \leq c$, $\forall i \in \mathbb{N}$. Then, $\|x_i\|_{\mathcal{X}} \leq c/\delta =: c_1 \in (0, \infty) \subset \mathbb{R}$, $\forall i \in \mathbb{N}$. This shows that $(x_i)_{i=1}^{\infty} \subseteq \overline{\mathcal{B}}_{\mathcal{X}}(\vartheta_{\mathcal{X}}, c_1) =: S_1 \subseteq \mathcal{X} = \mathcal{X}^{**}$. By Alaoglu Theorem 7.122, $S_1 \subseteq \mathcal{X}^{**}$ is weak* compact. By Propositions 5.22 and 5.26, $(x_i)_{i=1}^{\infty}$ has a cluster point $x \in \mathcal{X}^{**} = \mathcal{X}$ in weak* topology. Since $\mathcal{X}$ is reflexive, the weak* topology on $\mathcal{X}^{**} = \mathcal{X}$ is identical to the weak topology on $\mathcal{X}$, and the weak* topology on $\mathcal{X}^*$ is identical to the weak topology on $\mathcal{X}^*$. By Proposition 7.123, $L \in \mathrm{B}(\mathcal{X}, \mathcal{X}^*)$ is a continuous as a mapping $L : \mathcal{X}_{\text{weak}} \to \mathcal{X}^*_{\text{weak}^*}$. By Proposition 3.66, $(x_{*i})_{i=1}^{\infty} = (L(x_i))_{i=1}^{\infty}$ has a cluster point $L(x) \in \mathcal{X}^*_{\text{weak}^*}$ in weak* topology. By Proposition 7.120, $\mathcal{X}^*_{\text{weak}^*}$ is completely regular, then it is Hausdorff by Proposition 3.61. By Proposition 3.65, $L(x) = x_*$, since $x_* = \lim_{i\in\mathbb{N}} x_{*i}$ and $x_* = \lim_{i\in\mathbb{N}} x_{*i}$ weak*. This shows that $x_* \in \mathcal{R}(L)$. By the arbitrariness of x_*, we have $\mathcal{R}(L) = \mathcal{X}^*$ and L is surjective.

Hence, L is bijective. By Open Mapping Theorem 7.103, $L^{-1} \in \mathrm{B}(\mathcal{X}^*, \mathcal{X}) = \mathrm{B}(\mathcal{X}^*, \mathcal{X}^{**}) = \mathrm{B}(\mathcal{X}^*, \mathrm{B}(\mathcal{X}^*, \mathbb{R}))$.

$\forall x_*, y_* \in \mathcal{X}^*$, let $x = L^{-1}(x_*)$ and $y = L^{-1}(y_*)$. We have

$$\begin{aligned} L^{-1}(x_*)(y_*) &= \left\langle\left\langle \phi_{\mathcal{X}}(L^{-1}(x_*)), y_*\right\rangle\right\rangle = \left\langle\left\langle y_*, L^{-1}(x_*)\right\rangle\right\rangle = \langle\langle L(y), x\rangle\rangle \\ &= L(y)(x) = L(x)(y) = \langle\langle L(x), y\rangle\rangle = \left\langle\left\langle x_*, L^{-1}(y_*)\right\rangle\right\rangle = L^{-1}(y_*)(x_*) \end{aligned}$$

where $\phi_{\mathcal{X}} : \mathcal{X} \to \mathcal{X}^{**}$ is the natural mapping introduced in Remark 7.88; and the fifth equality follows from the fact that $L \in \mathcal{S}_{\mathcal{X}}$. By the arbitrariness of x_* and y_*, we have $L^{-1} \in \mathcal{S}_{\mathcal{X}^*}$.

$\forall x_* \in \mathcal{X}^*$, let $x = L^{-1}(x_*)$. Then, $x_* = L(x)$ and $\|x_*\|_{\mathcal{X}^*} \leq \|L\|_{\mathrm{B}(\mathcal{X},\mathcal{X}^*)} \cdot \|x\|_{\mathcal{X}}$. We will distinguish two exhaustive and mutually exclusive cases: Case 1: $\|L\|_{\mathrm{B}(\mathcal{X},\mathcal{X}^*)} = 0$; Case 2: $\|L\|_{\mathrm{B}(\mathcal{X},\mathcal{X}^*)} > 0$.

Case 1: $\|L\|_{\mathrm{B}(\mathcal{X},\mathcal{X}^*)} = 0$, we have $\|x_*\|_{\mathcal{X}^*} = 0$, $\forall x_* \in \mathcal{X}^*$. Hence, $\mathcal{X}^* = \{\vartheta_{\mathcal{X}^*}\}$. Then, $L^{-1}(x_*)(x_*) = 0 \geq \frac{\delta}{\epsilon + \|L\|^2_{\mathrm{B}(\mathcal{X},\mathcal{X}^*)}} \|x_*\|^2_{\mathcal{X}^*} = 0$, $\forall \epsilon \in (0, \infty) \subset \mathbb{R}$.

Case 2: $\|L\|_{\mathrm{B}(\mathcal{X},\mathcal{X}^*)} > 0$. Then, we have

$$
\begin{aligned}
L^{-1}(x_*)(x_*) = L(x)(x) \geq \delta \|x\|^2_{\mathcal{X}} &\geq \frac{\delta}{\|L\|^2_{\mathrm{B}(\mathcal{X},\mathcal{X}^*)}} \|x_*\|^2_{\mathcal{X}^*} \\
&\geq \frac{\delta}{\epsilon + \|L\|^2_{\mathrm{B}(\mathcal{X},\mathcal{X}^*)}} \|x_*\|^2_{\mathcal{X}^*}, \quad \forall \epsilon \in (0, \infty) \subset \mathbb{R}
\end{aligned}
$$

Hence, we have shown, in both cases, that $L^{-1}(x_*)(x_*) \geq \frac{\delta}{\epsilon + \|L\|^2_{\mathrm{B}(\mathcal{X},\mathcal{X}^*)}} \cdot \|x_*\|^2_{\mathcal{X}^*}$, $\forall x_* \in \mathcal{X}^*$ and $\forall \epsilon \in (0, \infty) \subset \mathbb{R}$. This proves that $L^{-1} \in \mathcal{S}_{+\,\mathcal{X}^*}$.

This completes the proof of the proposition. □

Chapter 11
General Measure and Integration

11.1 Measure Spaces

Definition 11.1 A *measurable space* $(X, \mathcal{B})$ is a set X and a σ-algebra $\mathcal{B}$ (of subsets of X) on X. A subset $A \subseteq X$ is said to be $\mathcal{B}$-*measurable* if $A \in \mathcal{B}$, or simply *measurable* when no confusion may arise from the context. ◇

Definition 11.2 Let $(X, \mathcal{B})$ be a measurable space. A *measure* μ on $(X, \mathcal{B})$ is a nonnegative extended real-valued function on $\mathcal{B}$, i.e., a mapping $\mu : \mathcal{B} \to [0, \infty] \subseteq \mathbb{R}_e$, such that:

(i) $\mu(\emptyset) = 0$.

(ii) $\mu\big(\bigcup_{i=1}^{\infty} E_i\big) = \sum_{i=1}^{\infty} \mu(E_i)$, where the sequence $(E_i)_{i=1}^{\infty} \subseteq \mathcal{B}$ is disjoint.

Then, the triple $(X, \mathcal{B}, \mu)$ is called a *measure space*. ◇

Fact 11.3 *Let $(X, \mathcal{B}, \mu)$ be a measure space. Then, μ is finitely additive, i.e., $\mu\big(\bigcup_{i=1}^{n} E_i\big) = \sum_{i=1}^{n} \mu(E_i)$, where $n \in \mathbb{Z}_+$ and $(E_i)_{i=1}^{n} \subseteq \mathcal{B}$ is pairwise disjoint.*

Proof Consider the sequence of measurable sets $(E_i)_{i=1}^{\infty} \subseteq \mathcal{B}$, with $E_i = \emptyset$, $i = n+1, n+2, \ldots$. Clearly, the sets in $(E_i)_{i=1}^{\infty}$ are pairwise disjoint. Therefore, $\mu\big(\bigcup_{i=1}^{n} E_i\big) = \mu\big(\bigcup_{i=1}^{\infty} E_i\big) = \sum_{i=1}^{\infty} \mu(E_i) = \sum_{i=1}^{n} \mu(E_i)$. This completes the proof of the fact. □

Proposition 11.4 *Let $(X, \mathcal{B}, \mu)$ be a measure space. $\forall A, B \in \mathcal{B}$ such that $A \subseteq B$, then $\mu(A) \le \mu(B)$.*

Proof The set $B \cap \widetilde{A} \in \mathcal{B}$ since $\mathcal{B}$ is a σ-algebra. Note that $A \cap \big(B \cap \widetilde{A}\big) = \emptyset$, and then, $\mu(B) = \mu\big(A \cup \big(B \cap \widetilde{A}\big)\big) = \mu(A) + \mu\big(B \cap \widetilde{A}\big) \ge \mu(A)$, by Fact 11.3. This completes the proof of the proposition. □

Z. Pan, *Measure-Theoretic Calculus in Abstract Spaces*,
https://doi.org/10.1007/978-3-031-21912-2_11

Proposition 11.5 *Let $(X, \mathcal{B}, \mu)$ be a measure space. A sequence $(E_n)_{n=1}^{\infty} \subseteq \mathcal{B}$ is such that $E_n \supseteq E_{n+1}$, $n = 1, 2, \ldots$. If $\mu(E_1) < +\infty$, then $\lim_{n\in\mathbb{N}} \mu(E_n) = \mu\left(\bigcap_{n=1}^{\infty} E_n\right)$.*

Proof Let $E = \bigcap_{n=1}^{\infty} E_n \in \mathcal{B}$. Then, $\widetilde{E} = \bigcup_{n=1}^{\infty} \widetilde{E_n} \in \mathcal{B}$. Note that $E_1 = E \cup \left(\bigcup_{n=1}^{\infty} \left(E_n \cap \widetilde{E_{n+1}}\right)\right)$. Clearly, the right-hand side of the above equality is a disjoint union of measurable sets. Therefore,

$$\mu(E_1) = \mu(E) + \sum_{n=1}^{\infty} \mu\left(E_n \cap \widetilde{E_{n+1}}\right) \tag{11.1}$$

On the other hand, note that $E_n = E_{n+1} \cup \left(E_n \cap \widetilde{E_{n+1}}\right)$, $n = 1, 2, \ldots$, which implies that $\mu(E_n) = \mu(E_{n+1}) + \mu\left(E_n \cap \widetilde{E_{n+1}}\right)$, since the two sets on the right-hand side of the above equality are disjoint, and all three numbers are finite by Proposition 11.4.

Then, substituting the above equality into (11.1) yields $\mu(E_1) = \mu(E) + \sum_{n=1}^{\infty} (\mu(E_n) - \mu(E_{n+1})) = \mu(E) + \mu(E_1) - \lim_{n\in\mathbb{N}} \mu(E_n)$. Since $\mu(E_1) < +\infty$, then $\mu(E) < +\infty$ by Proposition 11.4 and $E \subseteq E_1$. Therefore, we have $\lim_{n\in\mathbb{N}} \mu(E_n) = \mu(E)$.

This completes the proof of the proposition. □

Proposition 11.6 *Let $(X, \mathcal{B}, \mu)$ be a measure space. A sequence $(E_n)_{n=1}^{\infty} \subseteq \mathcal{B}$. Then, $\mu\left(\bigcup_{n=1}^{\infty} E_n\right) \leq \sum_{n=1}^{\infty} \mu(E_n)$.*

Proof Let $G_1 = E_1$, $G_n = E_n \setminus \left(\bigcup_{i=1}^{n-1} E_i\right)$, $n = 2, 3, \ldots$. Then, $(G_n)_{n=1}^{\infty}$ is a sequence of disjoint and measurable subsets of X. Therefore, $\mu\left(\bigcup_{i=1}^{\infty} G_i\right) = \sum_{i=1}^{\infty} \mu(G_i)$. Note that $\bigcup_{i=1}^{n} G_i = \bigcup_{i=1}^{n} E_i$, $n = 1, 2, \ldots$. Therefore, $\bigcup_{i=1}^{\infty} G_i = \bigcup_{i=1}^{\infty} E_i$. Thus,

$$\mu\left(\bigcup_{i=1}^{\infty} E_i\right) = \sum_{i=1}^{\infty} \mu(G_i) \tag{11.2}$$

By Proposition 11.4, $G_n \subseteq E_n$ implies that $\mu(G_n) \leq \mu(E_n)$, $n = 1, 2, \ldots$. Applying these inequalities in the equality (11.2) yields $\mu\left(\bigcup_{i=1}^{\infty} E_i\right) \leq \sum_{i=1}^{\infty} \mu(E_i)$.

This completes the proof of the proposition. □

Proposition 11.7 *Let $(X, \mathcal{B}, \mu)$ be a measure space and $(A_i)_{i=1}^{\infty} \subseteq \mathcal{B}$ with $A_i \subseteq A_{i+1}$, $\forall i \in \mathbb{N}$. Then, $\mu(\bigcup_{i=1}^{\infty} A_i) = \lim_{i\in\mathbb{N}} \mu(A_i) \in \mathbb{R}_e$.*

Proof Let $A := \bigcup_{i=1}^{\infty} A_i \in \mathcal{B}$. Note that $A = A_1 \cup (\bigcup_{i=1}^{\infty}(A_{i+1} \setminus A_i))$, where the sets in the union are pairwise disjoint. Then, by countable additivity of measure, $\mu(A) = \mu(A_1) + \sum_{i=1}^{\infty} \mu(A_{i+1} \setminus A_i) = \lim_{n\in\mathbb{N}}(\mu(A_1) + \sum_{i=1}^{n} \mu(A_{i+1} \setminus A_i)) = \lim_{n\in\mathbb{N}} \mu(A_{n+1}) = \lim_{i\in\mathbb{N}} \mu(A_i)$, where the third equality follows from Fact 11.3. This completes the proof of the proposition. □

Proposition 11.8 *Let $(X, \mathcal{B}, \mu)$ be a measure space. Then, $\forall E_1, E_2 \in \mathcal{B}$ with $\mu(E_1 \triangle E_2) = 0$, we have $\mu(E_1) = \mu(E_2)$.*

Proof $0 = \mu(E_1 \triangle E_2) = \mu(E_1 \setminus E_2) + \mu(E_2 \setminus E_1)$. Then, $\mu(E_1 \setminus E_2) = \mu(E_2 \setminus E_1) = 0$. Hence, we have $\mu(E_1) = \mu(E_1 \cap E_2) + \mu(E_1 \cap \widetilde{E_2}) = \mu(E_1 \cap E_2) + \mu(E_2 \cap \widetilde{E_1}) = \mu(E_2)$. This completes the proof of the proposition. □

Lemma 11.9 (Borel–Cantelli) *Let $(X, \mathcal{B}, \mu)$ be a measure space and $(E_n)_{n=1}^{\infty} \subseteq \mathcal{B}$. Assume that $\sum_{n=1}^{\infty} \mu(E_n) < \infty$. Then, $\mu(G) := \mu(\bigcap_{n\in\mathbb{N}} \bigcup_{m=n}^{\infty} E_m) = 0$.*

Proof $\forall \epsilon \in (0, \infty) \subset \mathbb{R}$, $\exists N \in \mathbb{N}$ such that $\sum_{n=N}^{\infty} \mu(E_n) < \epsilon$. Then, $0 \leq \mu(G) \leq \mu(\bigcup_{m=N}^{\infty} E_m) \leq \sum_{m=N}^{\infty} \mu(E_m) < \epsilon$, where the second inequality follows from Proposition 11.4; and the third inequality follows from Proposition 11.6. By the arbitrariness of ϵ, we have $\mu(G) = 0$. This completes the proof of the lemma. □

The set $\bigcap_{n\in\mathbb{N}} \bigcup_{m=n}^{\infty} E_m$ is equal to the set of elements that is in E_n, $n \in \mathbb{N}$, infinitely often. The book Williams (1991) uses the notation $\limsup_{n\in\mathbb{N}} E_n$ to denote this set. The book Williams (1991) also uses the notation $\liminf_{n\in\mathbb{N}} E_n$ to denote the set $\bigcup_{n\in\mathbb{N}} \bigcap_{m=n}^{\infty} E_m$, which is equal to the set of elements that is in all of E_m, $m \in \mathbb{N}$ with $n \leq m$, eventually. We will not formally use these notations. But, these sets are very important in the study of probability measure spaces.

Definition 11.10 Let $(X, \mathcal{B}, \mu)$ be a measure space. μ is said to be *finite* if $\mu(X) < +\infty$. μ is said to be *σ-finite* if there exists a sequence of measurable subsets of X, $(X_n)_{n=1}^{\infty}$, such that $\bigcup_{n=1}^{\infty} X_n = X$ and $\mu(X_n) < +\infty$, $n = 1, 2, \ldots$.

A set $E \subseteq X$ is said to be *of finite measure*, if $E \in \mathcal{B}$, and $\mu(E) < +\infty$. A set $E \subseteq X$ is said to be *of σ-finite measure*, if there exists a sequence of measurable subsets of X, $(E_n)_{n=1}^{\infty}$, such that $\bigcup_{n=1}^{\infty} E_n = E$ and $\mu(E_n) < +\infty$, $n = 1, 2, \ldots$. ◇

Definition 11.11 Let $(X, \mathcal{B}, \mu)$ be a measure space. It is said to be *complete* if $\forall A, E \subseteq X$, with $A \subseteq E$, $E \in \mathcal{B}$, and $\mu(E) = 0$, then $A \in \mathcal{B}$. ◇

Proposition 11.12 *Let $(X, \mathcal{B}, \mu)$ be a measure space. Then, there is a unique complete measure space $(X, \mathcal{B}_0, \mu_0)$ such that:*

(i) $\mathcal{B} \subseteq \mathcal{B}_0$.
(ii) $\forall E \in \mathcal{B}$, $\mu(E) = \mu_0(E)$.
(iii) *$\forall E \subseteq X$, $E \subset \mathcal{B}_0$ if, and only if, $E = A \cup B$, where $A, B \subseteq X$ with $B \in \mathcal{B}$, and there exists a $C \in \mathcal{B}$, such that $A \subseteq C$ and $\mu(C) = 0$.*

Then, $(X, \mathcal{B}_0, \mu_0)$ is said to be the completion of $(X, \mathcal{B}, \mu)$.

Furthermore, let $(X, \mathcal{B}_1, \mu_1)$ be another complete measure space that satisfies (i) *and* (ii), *and then, $\mathcal{B}_0 \subseteq \mathcal{B}_1$ and $\mu_0(E) = \mu_1(E)$, $\forall E \in \mathcal{B}_0$.*

Proof Define $\mathcal{B}_0$ to be the collection of subsets of X as the following.

$$\mathcal{B}_0 = \{E \subseteq X \mid \text{there exists } A, B, C \subseteq X \text{ such that } E = A \cup B, \\ B, C \in \mathcal{B},\ \mu(C) = 0, \text{ and } A \subseteq C\}$$

$\forall E \in \mathcal{B}$, let $A = \emptyset$, $B = E$, and $C = \emptyset$. Then, $E = A \cup B$, $B, C \in \mathcal{B}$, $\mu(C) = 0$, and $A \subseteq C$. Therefore, $E \in \mathcal{B}_0$. Hence, $\mathcal{B} \subseteq \mathcal{B}_0$.

The collection $\mathcal{B}_0$ forms a σ-algebra on X, which is proved in the following.

$\forall E_1, E_2 \in \mathcal{B}_0$, there exist $A_1, B_1, C_1, A_2, B_2, C_2 \subseteq X$ such that $E_1 = A_1 \cup B_1$, $E_2 = A_2 \cup B_2$, $B_1, C_1, B_2, C_2 \in \mathcal{B}$, $\mu(C_1) = \mu(C_2) = 0$, and $A_1 \subseteq C_1$, $A_2 \subseteq C_2$. Then, $E_1 \cup E_2 = (A_1 \cup A_2) \cup (B_1 \cup B_2)$. The sets $B_1 \cup B_2, C_1 \cup C_2 \in \mathcal{B}$, and $A_1 \cup A_2 \subseteq C_1 \cup C_2$. The sequence of inequalities holds, $0 \leq \mu(C_1 \cup C_2) \leq \mu(C_1) + \mu(C_2) \leq 0$, where the first one follows from Proposition 11.4 and the second one follows from Proposition 11.6. This further implies that $\mu(C_1 \cup C_2) = 0$. By the definition of $\mathcal{B}_0$, $E_1 \cup E_2 \in \mathcal{B}_0$.

Note that $\widetilde{E_1} = \widetilde{B_1} \cap \widetilde{A_1} = \widetilde{B_1} \cap \left(\left(\widetilde{A_1} \setminus \widetilde{C_1}\right) \cup \widetilde{C_1}\right)$, since $\widetilde{C_1} \subseteq \widetilde{A_1}$. This implies that $\widetilde{E_1} = \left(\widetilde{B_1} \cap \left(\widetilde{A_1} \setminus \widetilde{C_1}\right)\right) \cup \left(\widetilde{B_1} \cap \widetilde{C_1}\right)$. Since $\widetilde{B_1} \cap \widetilde{C_1} \in \mathcal{B}$ and $\widetilde{B_1} \cap \left(\widetilde{A_1} \setminus \widetilde{C_1}\right) = \widetilde{B_1} \cap \widetilde{A_1} \cap C_1 \subseteq C_1$ and $\mu(C_1) = 0$, then $\widetilde{E_1} \in \mathcal{B}_0$.

Therefore, $\mathcal{B}_0$ forms an algebra on the set X.

Consider an arbitrary sequence $(E_n)_{n=1}^{\infty} \subseteq \mathcal{B}_0$. Then, there exists $A_n, B_n, C_n \subseteq X$, such that $E_n = A_n \cup B_n$, $B_n, C_n \in \mathcal{B}$, $A_n \subseteq C_n$, and $\mu(C_n) = 0$, $n = 1, 2, \ldots$. Then, $\bigcup_{n=1}^{\infty} E_n = \left(\bigcup_{n=1}^{\infty} A_n\right) \cup \left(\bigcup_{n=1}^{\infty} B_n\right)$. Since $\mathcal{B}$ is a σ-algebra, then $\bigcup_{n=1}^{\infty} B_n \in \mathcal{B}$ and $\bigcup_{n=1}^{\infty} C_n \in \mathcal{B}$. Furthermore, $\bigcup_{n=1}^{\infty} A_n \subseteq \bigcup_{n=1}^{\infty} C_n$ and $0 \leq \mu\left(\bigcup_{n=1}^{\infty} C_n\right) \leq \sum_{n=1}^{\infty} \mu(C_n) = 0$. Then, $\bigcup_{n=1}^{\infty} E_n \in \mathcal{B}_0$.

This proves that $\mathcal{B}_0$ is a σ-algebra.

Define a mapping $\mu_0 : \mathcal{B}_0 \to [0, \infty] \subseteq \mathbb{R}_e$ as following. $\forall E \in \mathcal{B}_0$, there exists $A, B, C \subseteq X$ such that $E = A \cup B$, $B, C \in \mathcal{B}$, $A \subseteq C$, and $\mu(C) = 0$. $\mu_0(E) := \mu(B)$.

The mapping μ_0 is well-defined because of the following. $\forall E \in \mathcal{B}_0$, let $A_1, A_2, B_1, B_2, C_1, C_2 \subseteq X$ be such that

$$E = A_1 \cup B_1 = A_2 \cup B_2$$
$$B_1, C_1, B_2, C_2 \in \mathcal{B}$$
$$\mu(C_1) = \mu(C_2) = 0 \qquad A_1 \subseteq C_1 \qquad A_2 \subseteq C_2$$

Then, $\mu(B_1) = \mu\left(\left(B_1 \cap \widetilde{B_2}\right) \cup (B_1 \cap B_2)\right) = \mu\left(B_1 \cap \widetilde{B_2}\right) + \mu(B_1 \cap B_2) \leq \mu\left(B_1 \cap \widetilde{B_2}\right) + \mu(B_2)$, where the last inequality follows from Proposition 11.4.

Since $A_1 \cup B_1 = A_2 \cup B_2 = E$ implies that $\widetilde{B_2} \cap (A_1 \cup B_1) = \widetilde{B_2} \cap (A_2 \cup B_2) = \widetilde{B_2} \cap A_2 \subseteq C_2$, then $\widetilde{B_2} \cap B_1 \subseteq \left(\widetilde{B_2} \cap B_1\right) \cup \left(\widetilde{B_2} \cap A_1\right) \subseteq C_2$. By Proposition 11.4, this implies that $\mu\left(\widetilde{B_2} \cap B_1\right) = 0$. Therefore, $\mu(B_1) \leq \mu(B_2)$.

By an argument similar to the above, it can be concluded that $\mu(B_2) \leq \mu(B_1)$. Combining the two inequalities yields $\mu(B_2) = \mu(B_1)$.

This shows that $\mu_0(E)$ is well-defined.

$\forall E \in \mathcal{B} \subseteq \mathcal{B}_0$, $E = E \cup \emptyset$, then $\mu_0(E) = \mu(E)$. This proves that μ_0 agrees with μ on $\mathcal{B}$.

In the following, it will be shown that the nonnegative, extended real-valued function μ_0 defines a measure on the measurable space $(X, \mathcal{B}_0)$. Since $\emptyset \in \mathcal{B}$, then $\mu_0(\emptyset) = \mu(\emptyset) = 0$.

Consider any sequence of subsets $(E_n)_{n=1}^\infty \subseteq \mathcal{B}_0$, which are pairwise disjoint. There exists $A_n, B_n, C_n \subseteq X$, such that $E_n = A_n \cup B_n$, $B_n, C_n \in \mathcal{B}$, $A_n \subseteq C_n$, and $\mu(C_n) = 0$, $n = 1, 2, \ldots$. Since $B_n \subseteq E_n$, $n = 1, 2, \ldots$, then $(B_n)_{n=1}^\infty$ are pairwise disjoint. Hence, $\mu\big(\bigcup_{n=1}^\infty B_n\big) = \sum_{n=1}^\infty \mu(B_n)$. Since $\bigcup_{n=1}^\infty E_n = \big(\bigcup_{n=1}^\infty A_n\big) \cup \big(\bigcup_{n=1}^\infty B_n\big)$, $\bigcup_{n=1}^\infty A_n \subseteq \bigcup_{n=1}^\infty C_n$, $0 \leq \mu\big(\bigcup_{n=1}^\infty C_n\big) \leq \sum_{n=1}^\infty \mu(C_n) = 0$, and $\bigcup_{n=1}^\infty B_n, \bigcup_{n=1}^\infty C_n \in \mathcal{B}$, then $\mu_0\big(\bigcup_{n=1}^\infty E_n\big) = \mu\big(\bigcup_{n=1}^\infty B_n\big) = \sum_{n=1}^\infty \mu(B_n) = \sum_{n=1}^\infty \mu_0(E_n)$. This proves that μ_0 is countably additive for disjoint $\mathcal{B}_0$-measurable sets.

Hence, μ_0 is a measure on $(X, \mathcal{B}_0)$.

In the following, it is proved that $(X, \mathcal{B}_0, \mu_0)$ is a complete measure space.

$\forall E_0 \subseteq X$, such that there exists $E \in \mathcal{B}_0$ with $E_0 \subseteq E$ and $\mu_0(E) = 0$. Then, there exists $A, B, C \subseteq X$ such that $E = A \cup B$, $B, C \in \mathcal{B}$, $A \subseteq C$, and $\mu(C) = 0$, by the definition of $\mathcal{B}_0$ and $E \in \mathcal{B}_0$. By the definition of μ_0, we have $\mu_0(E) = \mu(B) = 0$. Hence, $E_0 = E_0 \cup \emptyset$ and $E_0 \subseteq A \cup B \subseteq C \cup B$. Since $\emptyset, C \cup B \in \mathcal{B}$, and $0 \leq \mu(C \cup B) \leq \mu(C) + \mu(B) = 0$, then $E_0 \in \mathcal{B}_0$. This proves that $(X, \mathcal{B}_0, \mu_0)$ is complete.

Let $(X, \mathcal{B}_1, \mu_1)$ be any complete measure space that satisfies the (i) and (ii). It is shown in the following that $\mathcal{B}_0 \subseteq \mathcal{B}_1$ by an argument of contradiction. Suppose there exists a set $E \in \mathcal{B}_0 \setminus \mathcal{B}_1$, and then, $E \notin \mathcal{B}$. Since $E \in \mathcal{B}_0$, then there exists $A, B, C \subseteq X$, such that $E = A \cup B$, $A \subseteq C$, $B, C \in \mathcal{B}$, and $\mu(C) = 0$. This implies $B, C \in \mathcal{B}_1$ and $\mu_1(C) = \mu(C) = 0$. Hence, $A \notin \mathcal{B}_1$; otherwise, we have $E \in \mathcal{B}_1$. But, $A \subseteq C$, with $\mu_1(C) = 0$, which contradicts the fact that $(X, \mathcal{B}_1, \mu_1)$ being complete.

Therefore, we have $\mathcal{B}_0 \subseteq \mathcal{B}_1$.

$\forall E \in \mathcal{B}_0$, there exists $A, B, C \subseteq X$, such that $E = A \cup B$, $A \subseteq C$, $B, C \in \mathcal{B}$, and $\mu(C) = 0$. Then, $\mu_0(E) = \mu(B) = \mu_1(B) \leq \mu_1(E) \leq \mu_1(B \cup C) \leq \mu_1(B) + \mu_1(C) = \mu(B)$, where Propositions 11.4 and 11.6 have been applied. Therefore, μ_0 and μ_1 agree on the σ-algebra $\mathcal{B}_0$.

Hence, $(X, \mathcal{B}_0, \mu_0)$ is a complete measure space that satisfies (i), (ii), and (iii). It is the unique such space since $\mathcal{B}_0$ is unique by (iii); then μ_0 is also unique since any complete measure space $(X, \mathcal{B}_0, \mu_1)$ satisfying (i) and (ii) will be such that $\mu_1(E) = \mu_0(E)$, $\forall E \subset \mathcal{B}_0$.

This completes the proof of the proposition. □

Proposition 11.13 *Let $\mathcal{X} := (X, \mathcal{B}, \mu)$ be a measure space, $A \in \mathcal{B}$, $\hat{\mathcal{B}} \subseteq \mathcal{B}$ be a σ-algebra on X, and $\mathcal{B}_A := \{C \subseteq A \mid C \in \mathcal{B}\}$. Then, the following statements hold:*

(i) *$(X, \hat{\mathcal{B}}, \mu|_{\hat{\mathcal{B}}})$ and $\mathcal{A} := (A, \mathcal{B}_A, \mu|_{\mathcal{B}_A})$ are measure spaces. The measure space $\mathcal{A}$ is said to be the* subspace *of the measure space $\mathcal{X}$.*

(ii) *If, in addition, $\mathcal{X}$ is complete, then $\mathcal{A}$ is also complete.*

(iii) *Let* $\bar{\mathcal{X}} := (X, \bar{\mathcal{B}}, \bar{\mu})$ *be the completion of* $\mathcal{X}$ *as defined in Proposition* 11.12, *and* $\bar{\mathcal{A}} := (A, \bar{\mathcal{B}}_A, \bar{\mu}_A)$ *be the measure subspace of* $\bar{\mathcal{X}}$. *Then,* $\bar{\mathcal{A}}$ *is the completion of* $\mathcal{A}$.

Proof (i) Since $\hat{\mathcal{B}}$ is a σ-algebra on X, then $(X, \hat{\mathcal{B}})$ is a measurable space. $\forall B \in \hat{\mathcal{B}} \subseteq \mathcal{B}$, $0 \leq \mu(B) \leq +\infty$ is well-defined. Then, $\mu|_{\hat{\mathcal{B}}} : \hat{\mathcal{B}} \to [0, \infty] \subseteq \mathbb{R}_e$. Clearly, $\mu|_{\hat{\mathcal{B}}}(\emptyset) = 0$. $\forall$ pairwise disjoint sequence $(E_i)_{i=1}^{\infty} \subseteq \hat{\mathcal{B}} \subseteq \mathcal{B}$, we have $E := \bigcup_{i=1}^{\infty} E_i \in \hat{\mathcal{B}}$ and $\mu|_{\hat{\mathcal{B}}}(E) = \mu(E) = \sum_{i=1}^{\infty} \mu(E_i) = \sum_{i=1}^{\infty} \mu|_{\hat{\mathcal{B}}}(E_i)$. Hence, $(X, \hat{\mathcal{B}}, \mu|_{\hat{\mathcal{B}}})$ is a measure space.

Note that $\emptyset \in \mathcal{B}_A$ and $A \in \mathcal{B}_A$. $\forall E_A \in \mathcal{B}_A$, then $A \supseteq A \setminus E_A \in \mathcal{B}$ and $A \setminus E_A \in \mathcal{B}_A$. $\forall (E_{Ai})_{i=1}^{\infty} \subseteq \mathcal{B}_A$, $\forall i \in \mathbb{N}$, $A \supseteq E_{Ai} \in \mathcal{B}$. Then, $A \supseteq \bigcup_{i=1}^{\infty} E_{Ai} \in \mathcal{B}$ and $\bigcup_{i=1}^{\infty} E_{Ai} \in \mathcal{B}_A$. Hence, $\mathcal{B}_A$ is a σ-algebra on A. It is straightforward to show that $\mu|_{\mathcal{B}_A}$ is a measure on the measurable space $(A, \mathcal{B}_A)$. Hence, $(A, \mathcal{B}_A, \mu|_{\mathcal{B}_A})$ is a measure space.

(ii) If, in addition, $(X, \mathcal{B}, \mu)$ is complete, $\forall E_A \subseteq A$ such that there exists $B \in \mathcal{B}_A$ with $E_A \subseteq B$ and $\mu|_{\mathcal{B}_A}(B) = 0$, Then $B \in \mathcal{B}$ and $\mu(B) = 0$. Then, $E_A \in \mathcal{B}$ by the completeness of $(X, \mathcal{B}, \mu)$. Thus, $E_A \in \mathcal{B}_A$. By the arbitrariness of E_A, $(A, \mathcal{B}_A, \mu|_{\mathcal{B}_A})$ is complete.

(iii) Note that $\bar{\mathcal{B}}_A = \{C \subseteq A \mid C \in \bar{\mathcal{B}}\}$ and $\bar{\mu}_A = \bar{\mu}|_{\bar{\mathcal{B}}_A}$. By (ii), $\bar{\mathcal{A}}$ is a complete measure space. Since $\mathcal{B} \subseteq \bar{\mathcal{B}}$, then $\mathcal{B}_A \subseteq \bar{\mathcal{B}}_A$. $\forall E \in \mathcal{B}_A$, $\mu|_{\mathcal{B}_A}(E) = \mu(E) = \bar{\mu}(E) = \bar{\mu}_A(E)$. $\forall E \subseteq A$, assume that $E = E_A \cup E_B$ with $E_B, E_C \in \mathcal{B}_A$, $E_A \subseteq E_C$, and $\mu|_{\mathcal{B}_A}(E_C) = 0$. Then, $\bar{\mu}_A(E_C) = 0$ and $E_A \in \bar{\mathcal{B}}_A$ since $\bar{\mathcal{A}}$ is complete. Then, $E \in \bar{\mathcal{B}}_A$. $\forall E \subseteq A$, assume that $E \in \bar{\mathcal{B}}_A$. Then, $E \subseteq A$ and $E \in \bar{\mathcal{B}}$. By Proposition 11.12, $E = E_A \cup E_B$ with $E_B, E_C \in \mathcal{B}$, $E_A \subseteq E_C$, and $\mu(E_C) = 0$. Then, $E_B, E_C \cap A \in \mathcal{B}_A$, $E_A \subseteq E_C \cap A$, and $0 \leq \mu|_{\mathcal{B}_A}(E_C \cap A) = \mu(E_C \cap A) \leq \mu(E_C) = 0$. Hence, (i)–(iii) of Proposition 11.12 are satisfied, and $\bar{\mathcal{A}}$ is the completion of $\mathcal{A}$.

This completes the proof of the proposition. □

11.2 Outer Measure and the Extension Theorem

Definition 11.14 Let X be a set. An *outer measure* is a function $\mu_o : {}^{X}2 \to [0, \infty] \subset \mathbb{R}_e$ satisfying:

(i) $\mu_o(\emptyset) = 0$.
(ii) If $A \subseteq B \subseteq X$, then $\mu_o(A) \leq \mu_o(B)$ (monotonicity).
(iii) If $E \subseteq \bigcup_{i=1}^{\infty} E_i \subseteq X$, then $\mu_o(E) \leq \sum_{i=1}^{\infty} \mu_o(E_i)$ (countable subadditivity). ◇

Note that (i) and (iii) above imply finite subadditivity.

Definition 11.15 Let X be a set and $\mu_o : {}^X2 \to [0, \infty] \subset \mathbb{R}_e$ be an outer measure. $E \subseteq X$ is said to be *measurable with respect to* μ_o if $\forall A \subseteq X$, we have $\mu_o(A) = \mu_o(A \cap E) + \mu_o(A \cap \widetilde{E})$. ◇

Lemma 11.16 *Let X be a set and $\mu_o : {}^X2 \to [0, \infty] \subset \mathbb{R}_e$ be an outer measure. $\forall E \subseteq X$ with $\mu_o(E) = 0$, then E is measurable with respect to μ_o.*

Proof Fix any $E \subseteq X$ with $\mu_o(E) = 0$. $\forall A \subseteq X$, $0 \le \mu_o(A \cap E) \le \mu_o(E) = 0$ and $\mu_o(A \cap \widetilde{E}) \le \mu_o(A)$. Then, we have $\mu_o(A) \le \mu_o(A \cap E) + \mu_o(A \cap \widetilde{E}) \le \mu_o(A)$. Hence, E is measurable with respect to μ_o. This completes the proof of the lemma. □

Theorem 11.17 *Let X be a set, $\mu_o : {}^X2 \to [0, \infty] \subset \mathbb{R}_e$ be an outer measure, $\mathcal{B} := \{E \subseteq X \mid E$ is measurable with respect to $\mu_o\}$, and $\bar{\mu} := \mu_o|_{\mathcal{B}} : \mathcal{B} \to [0, \infty] \subset \mathbb{R}_e$. Then, $\mathcal{B}$ is a σ-algebra on X and $(X, \mathcal{B}, \bar{\mu})$ is a complete measure space.*

Proof Clearly, $\emptyset, X \subseteq \mathcal{B}$. $\forall E \in \mathcal{B}$, we have $\widetilde{E} \in \mathcal{B}$. $\forall E_1, E_2 \in \mathcal{B}$, $\forall A \subseteq X$, by the measurability of E_1 and E_2, we have $\mu_o(A) = \mu_o(A \cap E_2) + \mu_o(A \cap \widetilde{E_2}) = \mu_o(A \cap E_2) + \mu_o(A \cap \widetilde{E_2} \cap E_1) + \mu_o(A \cap \widetilde{E_2} \cap \widetilde{E_1})$. Note that $A \cap (E_1 \cup E_2) = A \cap (E_2 \cup (E_1 \setminus E_2)) = (A \cap E_2) \cup (A \cap E_1 \cap \widetilde{E_2})$. Then, we have $\mu_o(A \cap (E_1 \cup E_2)) \le \mu_o(A \cap E_2) + \mu_o(A \cap E_1 \cap \widetilde{E_2})$. Then, we have $\mu_o(A) \ge \mu_o(A \cap (E_1 \cup E_2)) + \mu_o(A \cap \widetilde{E_1} \cap \widetilde{E_2}) = \mu_o(A \cap (E_1 \cup E_2)) + \mu_o(A \cap \widetilde{E_1 \cup E_2}) \ge \mu_o(A)$. Hence, $E_1 \cup E_2 \in \mathcal{B}$. Thus, $\mathcal{B}$ is an algebra on X.

$\forall (E_i)_{i=1}^\infty \subseteq \mathcal{B}$. Let $G_n := \bigcup_{i=1}^n E_i \in \mathcal{B}$ and $\bar{E}_1 := G_1 \in \mathcal{B}$, $\bar{E}_{n+1} := G_{n+1} \setminus G_n \in \mathcal{B}$, $\forall n \in \mathbb{N}$. Then, $\bar{E}_1, \bar{E}_2, \ldots$ are pairwise disjoint. Let $E := \bigcup_{i=1}^\infty E_i = \bigcup_{i=1}^\infty \bar{E}_i$. $\forall A \subseteq X$, we have $\mu_o(A) = \mu_o(A \cap G_n) + \mu_o(A \cap \widetilde{G_n}) \ge \mu_o(A \cap G_n) + \mu_o(A \cap \widetilde{E})$ and $\mu_o(A \cap G_{n+1}) = \mu_o(A \cap G_{n+1} \cap \bar{E}_{n+1}) + \mu_o(A \cap G_{n+1} \cap \widetilde{\bar{E}_{n+1}}) = \mu_o(A \cap \bar{E}_{n+1}) + \mu_o(A \cap G_n)$, $\forall n \in \mathbb{N}$. Hence, we have $\mu_o(A \cap G_n) = \sum_{i=1}^n \mu_o(A \cap \bar{E}_i)$ and $\mu_o(A) \ge \sum_{i=1}^n \mu_o(A \cap \bar{E}_i) + \mu_o(A \cap \widetilde{E})$, $\forall n \in \mathbb{N}$. Therefore, $\mu_o(A) \ge \sum_{i=1}^\infty \mu_o(A \cap \bar{E}_i) + \mu_o(A \cap \widetilde{E})$. Note that $A \cap E = \bigcup_{i=1}^\infty (A \cap \bar{E}_i)$, which implies that $\mu_o(A \cap E) \le \sum_{i=1}^\infty \mu_o(A \cap \bar{E}_i)$, by countable subadditivity of outer measures. Then, we have $\mu_o(A) \ge \sum_{i=1}^\infty \mu_o(A \cap \bar{E}_i) + \mu_o(A \cap \widetilde{E}) \ge \mu_o(A \cap E) + \mu_o(A \cap \widetilde{E}) \ge \mu_o(A)$. Hence, by the arbitrariness of A, $E \in \mathcal{B}$. This implies that $\mathcal{B}$ is a σ-algebra on X.

Set $A = E$ in the above, and we have $\bar{\mu}(E) = \mu_o(E) = \sum_{i=1}^\infty \mu_o(E \cap \bar{E}_i) = \sum_{i=1}^\infty \bar{\mu}(\bar{E}_i)$. Hence, $(X, \mathcal{B}, \bar{\mu})$ is a measure space.

Fix any $A, E \subseteq X$ with $A \subseteq E$, $E \in \mathcal{B}$, and $\bar{\mu}(E) = 0$. Then, $0 \le \mu_o(A) \le \mu_o(E) = \bar{\mu}(E) = 0$. By Lemma 11.16, $A \in \mathcal{B}$. Therefore, $(X, \mathcal{B}, \bar{\mu})$ is complete.

This completes the proof of the theorem. □

Definition 11.18 Let X be a set and $\mathcal{A} \subseteq {}^X2$ be an algebra on X. A *measure on algebra* $\mathcal{A}$ is a function $\mu : \mathcal{A} \to [0, \infty] \subset \mathbb{R}_e$ that satisfies:

(i) $\mu(\emptyset) = 0$.
(ii) $\forall (A_i)_{i=1}^\infty \subseteq \mathcal{A}$ with $A := \bigcup_{i=1}^\infty A_i \in \mathcal{A}$ and $A_1, A_2, \ldots$ being pairwise disjoint, then $\mu(A) = \sum_{i=1}^\infty \mu(A_i)$.

μ is said to be *finite* if $\mu(X) < +\infty$; and it is said to be σ-*finite* if $\exists (X_i)_{i=1}^{\infty} \subseteq \mathcal{A}$ such that $X = \bigcup_{i=1}^{\infty} X_i$ and $\mu(X_i) < \infty$, $\forall i \in \mathbb{N}$. ◇

Clearly, a measure on an algebra is monotonic.

Theorem 11.19 (Carathéodory Extension Theorem) *Let X be a set, $\mathcal{A} \subseteq {}^X2$ be an algebra on X, and $\mu : \mathcal{A} \to [0, \infty] \subset \mathbb{R}_e$ be a measure on algebra $\mathcal{A}$. Define $\mu_o : {}^X2 \to [0, \infty] \subset \mathbb{R}_e$ by*

$$\mu_o(A) = \inf_{(A_i)_{i=1}^{\infty} \subseteq \mathcal{A},\ A \subseteq \bigcup_{i=1}^{\infty} A_i} \sum_{i=1}^{\infty} \mu(A_i), \qquad \forall A \subseteq X$$

Then, the following statements hold:

(i) *μ_o is an outer measure on X and is said to be* the outer measure induced by μ.
(ii) *Let $\mathcal{B} := \{E \subseteq X \mid E$ is measurable with respect to $\mu_o\}$ and $\bar{\mu} := \mu_o|_{\mathcal{B}} : \mathcal{B} \to [0, \infty] \subset \mathbb{R}_e$. Then, $(X, \mathcal{B}, \bar{\mu})$ is a complete measure space and $\bar{\mu}(A) = \mu(A)$, $\forall A \in \mathcal{A}$. Hence, $\bar{\mu}$ is an extension of μ to the σ-algebra $\mathcal{B} \supseteq \mathcal{A}$.*
(iii) *$\bar{\mu}$ is finite if μ is finite; and $\bar{\mu}$ is σ-finite if μ is σ-finite.*
(iv) *Let $\mathcal{B}_a$ be the σ-algebra on X generated by $\mathcal{A}$. If μ is σ-finite, then $\bar{\mu}|_{\mathcal{B}_a}$ is the unique measure on the measurable space $(X, \mathcal{B}_a)$ that is an extension of μ.*
(v) *If μ is σ-finite, then the measure space $(X, \mathcal{B}, \bar{\mu})$ is the completion of the measure space $(X, \mathcal{B}_a, \bar{\mu}|_{\mathcal{B}_a})$.*

Proof We need the following intermediate results.

Claim 11.19.1 Let $A \in \mathcal{A}$, $(A_i)_{i=1}^{\infty} \subseteq \mathcal{A}$, and $A \subseteq \bigcup_{i=1}^{\infty} A_i$. Then, $\mu(A) \leq \sum_{i=1}^{\infty} \mu(A_i)$.

Proof of Claim Let $G_1 := A \cap A_1 \in \mathcal{A}$ and $G_{n+1} := A \cap A_{n+1} \cap \left(\bigcap_{i=1}^{n} \widetilde{A_i} \right) \in \mathcal{A}$, $\forall n \in \mathbb{N}$. Then, $A = \bigcup_{i=1}^{\infty} G_i$ and $G_1, G_2, \ldots$ are pairwise disjoint. This implies that $\mu(A) = \sum_{i=1}^{\infty} \mu(G_i) \leq \sum_{i=1}^{\infty} \mu(A_i)$, where the inequality follows from the monotonicity of μ. This completes the proof of the claim. □

Hence, $\forall A \in \mathcal{A}$, we have $\mu_o(A) = \mu(A)$.

Claim 11.19.2 μ_o is an outer measure.

Proof of Claim Clearly, $\mu_o(\emptyset) = \mu(\emptyset) = 0$ since $\emptyset \in \mathcal{A}$. $\forall A_1 \subseteq A_2 \subseteq X$, by the definition of μ_o, we have $\mu_o(A_1) \leq \mu_o(A_2)$. We only need to show the countable subadditivity for μ_o. Fix any $E \subseteq \bigcup_{i=1}^{\infty} E_i \subseteq X$. We will distinguish two exhaustive and mutually exclusive cases: Case 1: $\mu_o(E_i) < \infty$, $\forall i \in \mathbb{N}$; Case 2: $\exists i_0 \in \mathbb{N}$ such that $\mu_o(E_{i_0}) = \infty$.

Case 1: $\mu_o(E_i) < \infty$, $\forall i \in \mathbb{N}$. $\forall \epsilon \in (0, \infty) \subset \mathbb{R}$, $\forall i \in \mathbb{N}$, $\exists \left(A_{i,j}\right)_{j=1}^{\infty} \subseteq \mathcal{A}$ such that $E_i \subseteq \bigcup_{j=1}^{\infty} A_{i,j}$ and $\mu_o(E_i) \leq \sum_{j=1}^{\infty} \mu(A_{i,j}) < \mu_o(E_i) + \epsilon/2^i$. Then, $E \subseteq \bigcup_{i=1}^{\infty} E_i \subseteq \bigcup_{i=1}^{\infty} \bigcup_{j=1}^{\infty} A_{i,j}$. By the definition of μ_o, we have

$\mu_o(E) \leq \sum_{i=1}^{\infty} \sum_{j=1}^{\infty} \mu(A_{i,j}) \leq \sum_{i=1}^{\infty} \mu_o(E_i) + \epsilon$. By the arbitrariness of ϵ, we have $\mu_o(E) \leq \sum_{i=1}^{\infty} \mu_o(E_i)$.

Case 2: $\exists i_0 \in \mathbb{N}$ such that $\mu_o(E_{i_0}) = \infty$. Then, $\mu_o(E) \leq \infty = \sum_{i=1}^{\infty} \mu_o(E_i)$.

In both cases, we have shown that $\mu_o(E) \leq \sum_{i=1}^{\infty} \mu_o(E_i)$. Hence, μ_o is an outer measure. This completes the proof of the claim. □

Hence, the statement (i) holds.

(ii) By Theorem 11.17, $(X, \mathcal{B}, \bar{\mu})$ is a complete measure space.

Claim 11.19.3 $\mathcal{A} \subseteq \mathcal{B}$.

Proof of Claim Fix any $A \in \mathcal{A}$. $\forall \bar{A} \subseteq X$, we will show that $\mu_o(\bar{A}) = \mu_o(\bar{A} \cap A) + \mu_o(\bar{A} \cap \widetilde{A})$ by distinguishing two exhaustive and mutually exclusive cases: Case 1: $\mu_o(\bar{A}) = \infty$; Case 2: $\mu_o(\bar{A}) < \infty$. Case 1: $\mu_o(\bar{A}) = \infty$. Then, $\mu_o(\bar{A}) \geq \mu_o(\bar{A} \cap A) + \mu_o(\bar{A} \cap \widetilde{A}) \geq \mu_o(\bar{A})$, where the second inequality follows from the countable subadditivity of an outer measure. This case is proved.

Case 2: $\mu_o(\bar{A}) < \infty$. $\forall \epsilon \in (0, \infty) \subset \mathbb{R}$, $\exists (A_i)_{i=1}^{\infty} \subseteq \mathcal{A}$ such that $\bar{A} \subseteq \bigcup_{i=1}^{\infty} A_i$ and $\mu_o(\bar{A}) \leq \sum_{i=1}^{\infty} \mu(A_i) < \mu_o(\bar{A}) + \epsilon$. By the countable additivity of μ, we have $\mu(A_i) = \mu(A_i \cap A) + \mu(A_i \cap \widetilde{A})$, $\forall i \in \mathbb{N}$. Then, $\mu_o(\bar{A}) + \epsilon > \sum_{i=1}^{\infty}(\mu(A_i \cap A) + \mu(A_i \cap \widetilde{A})) \geq \mu_o(\bar{A} \cap A) + \mu_o(\bar{A} \cap \widetilde{A}) \geq \mu_o(\bar{A})$, where the second inequality follows from the definition of μ_o and the third inequality follows from the countable subadditivity of μ_o. By the arbitrariness of ϵ, we have $\mu_o(\bar{A}) = \mu_o(\bar{A} \cap A) + \mu_o(\bar{A} \cap \widetilde{A})$. This case is proved.

Hence, in both cases, we have $\mu_o(\bar{A}) = \mu_o(\bar{A} \cap A) + \mu_o(\bar{A} \cap \widetilde{A})$. By the arbitrariness of $\bar{A}$, we have A is measurable with respect to μ_o. Hence, $A \in \mathcal{B}$ and $\mathcal{A} \subseteq \mathcal{B}$. This completes the proof of the claim. □

$\forall A \in \mathcal{A}$, we have $\bar{\mu}(A) = \mu_o(A) = \mu(A)$. Hence, the statement (ii) holds.

(iii) Note that $X \in \mathcal{A}$. If μ is finite, then $\bar{\mu}(X) = \mu(X) < \infty$. Hence, $\bar{\mu}$ is finite.

If μ is σ-finite, then $\exists (X_i)_{i=1}^{\infty} \subseteq \mathcal{A}$ with $X = \bigcup_{i=1}^{\infty} X_i$ such that $\mu(X_i) < +\infty$, $\forall i \in \mathbb{N}$. Then, $\bar{\mu}(X_i) = \mu(X_i) < +\infty$ and $(X_i)_{i=1}^{\infty} \subseteq \mathcal{B}$. Hence, $\bar{\mu}$ is σ-finite.

(iv) Let μ be σ-finite and $\tilde{\mu} : \mathcal{B}_a \to [0, \infty] \subset \mathbb{R}_e$ be any measure that is an extension of μ, that is, $\tilde{\mu}(A) = \mu(A)$, $\forall A \in \mathcal{A}$. Clearly, $\mathcal{B}_a \subseteq \mathcal{B}$. Then, by Proposition 11.13, $\bar{\mu}|_{\mathcal{B}_a}$ is a measure on $(X, \mathcal{B}_a)$, and it is an extension of μ. Let $\mathcal{A}_\sigma := \{E \subseteq X \mid \exists (A_i)_{i=1}^{\infty} \subseteq \mathcal{A} \ni \cdot E = \bigcup_{i=1}^{\infty} A_i\}$. Since $\mathcal{B} \supseteq \mathcal{A}$ and is a σ-algebra on X, then $\mathcal{A}_\sigma \subseteq \mathcal{B}_a \subseteq \mathcal{B}$.

Claim 11.19.4 $\forall E \in \mathcal{A}_\sigma$, $\tilde{\mu}(E) = \bar{\mu}(E)$.

Proof of Claim Fix any $E \in \mathcal{A}_\sigma$. Then, $\exists (A_i)_{i=1}^{\infty} \subseteq \mathcal{A} \ni \cdot E = \bigcup_{i=1}^{\infty} A_i$. Let $\bar{A}_1 := A_1$ and $\bar{A}_{n+1} = (\bigcup_{i=1}^{n+1} A_i) \setminus (\bigcup_{i=1}^{n} A_i)$, $\forall n \in \mathbb{N}$. Then, $\left(\bar{A}_i\right)_{i=1}^{\infty} \subseteq \mathcal{A}$, $E = \bigcup_{i=1}^{\infty} \bar{A}_i$, and $\bar{A}_1, \bar{A}_2, \ldots$ are pairwise disjoint. Then, we have $\tilde{\mu}(E) = \sum_{i=1}^{\infty} \tilde{\mu}(\bar{A}_i) = \sum_{i=1}^{\infty} \mu(\bar{A}_i) = \sum_{i=1}^{\infty} \bar{\mu}(\bar{A}_i) = \bar{\mu}(E)$, where the first and last equalities follow from the countable additivity of measures. This completes the proof of the claim. □

Claim 11.19.5 $\forall B \in \mathcal{B}_a$ with $\bar{\mu}(B) < \infty$, we have $\bar{\mu}(B) = \tilde{\mu}(B)$.

Proof of Claim Fix any $B \in \mathcal{B}_a$ with $\bar{\mu}(B) < \infty$. Then, $\mu_o(B) = \bar{\mu}(B) < \infty$. $\forall \epsilon \in (0, \infty) \subset \mathbb{R}$, by the definition of μ_o, $\exists (A_i)_{i=1}^{\infty} \subseteq \mathcal{A}$ such that $B \subseteq \bigcup_{i=1}^{\infty} A_i =: E \in \mathcal{A}_\sigma \subseteq \mathcal{B}_a$ and $\bar{\mu}(B) = \mu_o(B) \leq \sum_{i=1}^{\infty} \mu(A_i) < \bar{\mu}(B) + \epsilon$. Then, $\tilde{\mu}(B) \leq \tilde{\mu}(E) = \bar{\mu}(E) \leq \sum_{i=1}^{\infty} \mu(A_i) < \bar{\mu}(B) + \epsilon$, where the first inequality follows from the monotonicity of measure, Proposition 11.4, the first equality follows from Claim 11.19.4, and the second inequality follows from the countable subadditivity of measure, Proposition 11.6. By the arbitrariness of ϵ, we have $\tilde{\mu}(B) \leq \bar{\mu}(B)$.

On the other hand, $\bar{\mu}(E) = \bar{\mu}(B) + \bar{\mu}(E \setminus B) < \bar{\mu}(B) + \epsilon$. Since $\bar{\mu}(B) < \infty$, then $\bar{\mu}(E \setminus B) < \epsilon$. Note that $E \setminus B \in \mathcal{B}_a$. This implies that $\bar{\mu}(B) \leq \bar{\mu}(E) = \tilde{\mu}(E) = \tilde{\mu}(B) + \tilde{\mu}(E \setminus B) \leq \tilde{\mu}(B) + \bar{\mu}(E \setminus B) < \tilde{\mu}(B) + \epsilon$, where the second inequality follows from the result of the previous paragraph. By the arbitrariness of ϵ, we have $\bar{\mu}(B) \leq \tilde{\mu}(B)$.

Hence, $\bar{\mu}(B) = \tilde{\mu}(B)$. This completes the proof of the claim. □

Since μ is σ-finite, then $\exists (X_i)_{i=1}^{\infty} \subseteq \mathcal{A}$ such that $X = \bigcup_{i=1}^{\infty} X_i$ and $\mu(X_i) < \infty$, $\forall i \in \mathbb{N}$. Without loss of generality, we may assume that $X_1, X_2, \ldots$ are pairwise disjoint. $\forall B \in \mathcal{B}_a$, we have $B = \bigcup_{i=1}^{\infty}(B \cap X_i)$, $B \cap X_1, B \cap X_2, \ldots$ are pairwise disjoint, and $\bar{\mu}(B \cap X_i) \leq \bar{\mu}(X_i) < \infty$ with $B \cap X_i \in \mathcal{B}_a$, $\forall i \in \mathbb{N}$. This implies that $\tilde{\mu}(B) = \sum_{i=1}^{\infty} \tilde{\mu}(B \cap X_i) = \sum_{i=1}^{\infty} \bar{\mu}(B \cap X_i) = \bar{\mu}(B)$, where the first and last equalities follow from the countable additivity of measures and the second equality follows from Claim 11.19.5. Hence, $\bar{\mu}|_{\mathcal{B}_a} = \tilde{\mu}$. Therefore, $\bar{\mu}|_{\mathcal{B}_a}$ is the unique measure on $(X, \mathcal{B}_a)$ that is an extension of μ.

(v) Let μ be σ-finite. Consider the measure space $(X, \mathcal{B}_a, \bar{\mu}|_{\mathcal{B}_a})$. By Proposition 11.12, this measure space admits the completion $(X, \hat{\mathcal{B}}, \hat{\mu})$. Since $(X, \mathcal{B}, \bar{\mu})$ is a complete measure space that agrees with $\bar{\mu}|_{\mathcal{B}_a}$ on $\mathcal{B}_a$, by Proposition 11.12, we have $\hat{\mathcal{B}} \subseteq \mathcal{B}$ and $\hat{\mu} = \bar{\mu}|_{\hat{\mathcal{B}}}$.

Claim 11.19.6 $\forall B \in \mathcal{B}, \exists U \in \mathcal{B}_a$ with $B \subseteq U$ such that $\bar{\mu}(U \setminus B) = 0$.

Proof of Claim $\forall B \in \mathcal{B}$, $\bar{\mu}(B) = \inf_{(A_i)_{i=1}^{\infty} \subseteq \mathcal{A}, B \subseteq \bigcup_{i=1}^{\infty} A_i} \sum_{i=1}^{\infty} \mu(A_i)$. We first prove the special case $\bar{\mu}(B) < +\infty$. $\forall i \in \mathbb{N}$, $\exists \big(U_{i,j}\big)_{j=1}^{\infty} \subseteq \mathcal{A}$ with $B \subseteq \bigcup_{j=1}^{\infty} U_{i,j} =: U_i \in \mathcal{B}_a$ such that $\bar{\mu}(B) \leq \bar{\mu}(U_i) = \hat{\mu}(U_i) \leq \sum_{j=1}^{\infty} \mu(U_{i,j}) < \bar{\mu}(B) + 2^{-i} < +\infty$. Then, $\bar{\mu}(U_i) = \bar{\mu}(B) + \bar{\mu}(U_i \setminus B) < \bar{\mu}(B) + 2^{-i} < +\infty$. This implies that $\bar{\mu}(U_i \setminus B) < 2^{-i}$. Let $U := \bigcap_{i=1}^{\infty} U_i \in \mathcal{B}_a$. Clearly, $B \subseteq U$ and $\bar{\mu}(U \setminus B) \leq \bar{\mu}(U_i \setminus B) < 2^{-i}$, $\forall i \in \mathbb{N}$. Hence, $\bar{\mu}(U \setminus B) = 0$.

Now, we prove the general case $\bar{\mu}(B) \in [0, \infty] \subset \mathbb{R}_e$. $B = \bigcup_{i=1}^{\infty}(X_i \cap B) =: \bigcup_{i=1}^{\infty} B_i$. $\forall i \in \mathbb{N}$, $\bar{\mu}(B_i) \leq \bar{\mu}(X_i) = \mu(X_i) < +\infty$. By the special case, $\exists U_i \in \mathcal{B}_a$ with $B_i \subseteq U_i$ such that $\bar{\mu}(U_i \setminus B_i) = 0$. Then, $B \subseteq \bigcup_{i=1}^{\infty} U_i =: U \in \mathcal{B}_a$ and $0 \leq \bar{\mu}(U \setminus B) \leq \sum_{i=1}^{\infty} \bar{\mu}(U_i \setminus B) \leq \sum_{i=1}^{\infty} \bar{\mu}(U_i \setminus B_i) = 0$, where the second inequality follows from the countable subadditivity of measure. This completes the proof of the claim. □

$\forall B \in \mathcal{B}$, by Claim 11.19.6, $\exists U, \bar{U} \in \mathcal{B}_a$ with $B \subseteq U$ and $\widetilde{B} \subseteq \bar{U}$ such that $\bar{\mu}(U \setminus B) = \bar{\mu}(\bar{U} \setminus \widetilde{B}) = 0$. Let $F := \widetilde{\bar{U}} \in \mathcal{B}_a$. Then, $F \subseteq B$ and $\bar{\mu}(B \setminus F) =$

$\bar{\mu}(\bar{U} \setminus \widetilde{B}) = 0$. Then, $B = F \cup (B \setminus F)$, $B \setminus F \subseteq U \setminus F$, $F, U \setminus F \in \mathcal{B}_a$, and $\hat{\mu}(U \setminus F) = \bar{\mu}|_{\mathcal{B}_a}(U \setminus F) = \bar{\mu}(U \setminus F) = \bar{\mu}(U \setminus B) + \bar{\mu}(B \setminus F) = 0$. This shows that $B \in \hat{\mathcal{B}}$, by the completeness of $(X, \hat{\mathcal{B}}, \hat{\mu})$. By the arbitrariness of B, we have $\mathcal{B} \subseteq \hat{\mathcal{B}}$. Hence, $\mathcal{B} = \hat{\mathcal{B}}$ and $\bar{\mu} = \hat{\mu}$. Hence, $(X, \mathcal{B}, \bar{\mu})$ is the completion of $(X, \mathcal{B}_a, \bar{\mu}|_{\mathcal{B}_a})$.

This completes the proof of the theorem. □

Definition 11.20 An *interval* I in $\mathbb{R}$ is any of the following sets: (i) $\emptyset$; (ii) $(a, b) \subseteq \mathbb{R}$ with $a, b \in \mathbb{R}_e$ and $a < b$; (iii) $[a, b) \subset \mathbb{R}$ with $a \in \mathbb{R}$, $b \in \mathbb{R}_e$, and $a < b$; (iv) $(a, b] \subset \mathbb{R}$ with $a \in \mathbb{R}_e$, $b \in \mathbb{R}$, and $a < b$; and (v) $[a, b] \subset \mathbb{R}$ with $a, b \in \mathbb{R}$ and $a \leq b$. ◇

Example 11.21 Let $X = \mathbb{R}$,

$$\mathcal{A} := \{E \subseteq \mathbb{R} \mid E \text{ is the union of finitely many intervals in } \mathbb{R}.\}$$

It is clear that $\mathcal{A}$ is an algebra on $\mathbb{R}$. $\forall E \in \mathcal{A}$, E may be written as the union of finitely many pairwise disjoint intervals: $E = \bigcup_{i=1}^{n} I_i$, for some $n \in \mathbb{N}$. Define $\mu(E) = \sum_{i=1}^{n} \mu(I_i)$, where

$$\mu(I) = \begin{cases} 0 & \text{if } I = \emptyset \\ b - a & \text{if } I \text{ is any other types of interval} \end{cases}$$

Clearly, $\mu(\emptyset) = 0$; $\mu(I_1 \cup I_2) = \mu(I_1) + \mu(I_2)$, $\forall$ intervals $I_1, I_2 \subseteq \mathbb{R}$ with $I_1 \cap I_2 = \emptyset$. Then, $\mu : \mathcal{A} \to [0, \infty] \subset \mathbb{R}_e$ is well-defined and $\mu(E_1 \cup E_2) = \mu(E_1) + \mu(E_2)$, $\forall E_1, E_2 \in \mathcal{A}$ with $E_1 \cap E_2 = \emptyset$. Hence, μ is finitely additive.

Fix any interval $I = \bigcup_{i=1}^{\infty} E_i$, where $(E_i)_{i=1}^{\infty} \subseteq \mathcal{A}$ and $E_1, E_2, \ldots$ are pairwise disjoint. We will show that $\mu(I) = \sum_{i=1}^{\infty} \mu(E_i)$. $\forall n \in \mathbb{N}$, $I \supseteq \bigcup_{i=1}^{n} E_i$. Then, by finite additivity of μ, $\mu(I) \geq \sum_{i=1}^{n} \mu(E_i)$. Hence, $\mu(I) \geq \sum_{i=1}^{\infty} \mu(E_i)$. We will distinguish three exhaustive and mutually exclusive cases: Case 1: $\sum_{i=1}^{\infty} \mu(E_i) = \infty$; Case 2: $I = \emptyset$; Case 3: $I \neq \emptyset$ and $\sum_{i=1}^{\infty} \mu(E_i) < \infty$. Case 1: $\sum_{i=1}^{\infty} \mu(E_i) = \infty$. Then, $\infty \geq \mu(I) \geq \sum_{i=1}^{\infty} \mu(E_i) = \infty$. Hence, $\mu(I) = \sum_{i=1}^{\infty} \mu(E_i) = \infty$. Case 2: $I = \emptyset$. Then, $0 = \mu(I) \leq \sum_{i=1}^{\infty} \mu(E_i)$. Hence, we have $0 = \mu(I) = \sum_{i=1}^{\infty} \mu(E_i)$. Case 3: $I \neq \emptyset$ and $\sum_{i=1}^{\infty} \mu(E_i) < \infty$. $\forall i \in \mathbb{N}$, $E_i = \bigcup_{j=1}^{n_i} I_{i,j}$, where $n_i \in \mathbb{N}$ and $I_{i,1}, \ldots, I_{i,n_i}$ are pairwise disjoint intervals. Clearly, $\mu(E_i) = \sum_{j=1}^{n_i} \mu(I_{i,j})$. Then, $I = \bigcup_{i=1}^{\infty} E_i = \bigcup_{i=1}^{\infty} \bigcup_{j=1}^{n_i} I_{i,j}$. Then, we have $\sum_{i=1}^{\infty} \mu(E_i) = \sum_{i=1}^{\infty} \sum_{j=1}^{n_i} \mu(I_{i,j}) < \infty$. Fix any closed interval $[a, b] := \bar{I} \subseteq I$, where $a, b \in \mathbb{R}$ and $a \leq b$. Such $a, b \in \mathbb{R}$ exists since $I \neq \emptyset$. $\forall \epsilon \in (0, \infty) \subset \mathbb{R}$, $\forall i \in \mathbb{N}$, $\forall j \in \{1, \ldots, n_i\}$, we may enlarge $I_{i,j}$ to an open interval $\bar{I}_{i,j} \supseteq I_{i,j}$ with $\mu(\bar{I}_{i,j}) \leq \mu(I_{i,j}) + \frac{2^{-i}\epsilon}{n_i}$. Then, $\bar{I} \subseteq \bigcup_{i=1}^{\infty} \bigcup_{j=1}^{n_i} \bar{I}_{i,j}$. By the compactness of $\bar{I}$, $\exists N \in \mathbb{N}$ such that $\bar{I} \subseteq \bigcup_{i=1}^{N} \bigcup_{j=1}^{n_i} \bar{I}_{i,j}$. Clearly, we have $b - a = \mu(\bar{I}) \leq \sum_{i=1}^{N} \sum_{j=1}^{n_i} \mu(\bar{I}_{i,j}) \leq \sum_{i=1}^{N} \sum_{j=1}^{n_i} (\mu(I_{i,j}) + \frac{2^{-i}\epsilon}{n_i}) \leq \epsilon + \sum_{i=1}^{\infty} \sum_{j=1}^{n_i} \mu(I_{i,j})$. By the arbitrariness of ϵ, a, b, we have $\mu(I) \leq \sum_{i=1}^{\infty} \sum_{j=1}^{n_i} \mu(I_{i,j})$. Then, $\mu(I) \leq$

$\sum_{i=1}^{\infty} \mu(E_i)$. Thus, we have $\mu(I) = \sum_{i=1}^{\infty} \mu(E_i)$. Hence, in all three cases, we have $\mu(I) = \sum_{i=1}^{\infty} \mu(E_i)$.

$\forall E \in \mathcal{A}$, $\forall (E_i)_{i=1}^{\infty} \subseteq \mathcal{A}$ with $E = \bigcup_{i=1}^{\infty} E_i$ and $E_1, E_2, \ldots$ being pairwise disjoint, we have $E = \bigcup_{i=1}^{n} I_i$, where $n \in \mathbb{N}$ and $I_1, \ldots, I_n$ are pairwise disjoint intervals. Then, $I_j = \bigcup_{i=1}^{\infty} (I_j \cap E_i)$, $\forall j \in \{1, \ldots, n\}$. By the above argument, we have $\mu(I_j) = \sum_{i=1}^{\infty} \mu(I_j \cap E_i)$, $\forall j \in \{1, \ldots, n\}$. Then, $\mu(E) = \sum_{j=1}^{n} \mu(I_j) = \sum_{j=1}^{n} \sum_{i=1}^{\infty} \mu(I_j \cap E_i) = \sum_{i=1}^{\infty} \sum_{j=1}^{n} \mu(I_j \cap E_i) = \sum_{i=1}^{\infty} \mu(E_i)$, where the last equality follows from the finite additivity of μ.

Hence, μ is a measure on algebra $\mathcal{A}$ and is σ-finite. By Carathéodory Extension Theorem 11.19, μ extends uniquely to a σ-finite complete measure space $(X, \mathcal{B}, \bar{\mu})$. This measure space is called the *Lebesgue measure space*. Since $\mathcal{B}$ is a σ-algebra on $\mathbb{R}$ and the topology $\mathcal{O}_{\mathbb{R}}$ on $\mathbb{R}$ is second countable, then $\mathcal{B} \supseteq \mathcal{O}_{\mathbb{R}}$. We will denote the Lebesgue measure space by $(\mathbb{R}, \mathcal{B}_{\mathrm{L}}, \mu_{\mathrm{L}})$, where $\mathcal{B}_{\mathrm{L}}$ is the collection of *Lebesgue measurable sets* in $\mathbb{R}$ and μ_{L} is the *Lebesgue measure*.

Let $\mu_{\mathrm{L}o}$ be the *Lebesgue outer measure* induced by μ as defined in Carathéodory Extension Theorem 11.19. We claim that

$$\mu_{\mathrm{L}o}(E) = \inf_{(I_i)_{i=1}^{\infty} \text{ are open intervals, } E \subseteq \bigcup_{i=1}^{\infty} I_i} \sum_{i=1}^{\infty} \mu(I_i); \quad \forall E \subseteq \mathbb{R} \tag{11.3}$$

To prove this, we note that $\mu_{\mathrm{L}o}(E) = \inf_{(E_i)_{i=1}^{\infty} \subseteq \mathcal{A},\ E \subseteq \bigcup_{i=1}^{\infty} E_i} \sum_{i=1}^{\infty} \mu(E_i)$. Then, clearly, we have

$$\mu_{\mathrm{L}o}(E) \leq \inf_{(I_i)_{i=1}^{\infty} \text{ are open intervals, } E \subseteq \bigcup_{i=1}^{\infty} I_i} \sum_{i=1}^{\infty} \mu(I_i)$$

On the other hand, $\forall (E_i)_{i=1}^{\infty} \subseteq \mathcal{A}$ with $E \subseteq \bigcup_{i=1}^{\infty} E_i$, $\forall i \in \mathbb{N}$, $E_i = \bigcup_{j=1}^{n_i} I_{i,j}$, where $n_i \in \mathbb{N}$ and $I_{i,1}, \ldots, I_{i,n_i}$ are pairwise disjoint intervals, and $\mu(E_i) = \sum_{j=1}^{n_i} \mu(I_{i,j})$. $\forall \epsilon \in (0, \infty) \subset \mathbb{R}$, $\forall j \in \{1, \ldots, n_i\}$, we may enlarge $I_{i,j}$ to an open interval $\bar{I}_{i,j} \supseteq I_{i,j}$ with $\mu(\bar{I}_{i,j}) \leq \mu(I_{i,j}) + \frac{2^{-i}\epsilon}{n_i}$. Then, we have $E \subseteq \bigcup_{i=1}^{\infty} \bigcup_{j=1}^{n_i} I_{i,j} \subseteq \bigcup_{i=1}^{\infty} \bigcup_{j=1}^{n_i} \bar{I}_{i,j}$ and $\sum_{i=1}^{\infty} \mu(E_i) = \sum_{i=1}^{\infty} \sum_{j=1}^{n_i} \mu(I_{i,j}) \geq \sum_{i=1}^{\infty} \sum_{j=1}^{n_i} (\mu(\bar{I}_{i,j}) - \frac{2^{-i}\epsilon}{n_i}) = \sum_{i=1}^{\infty} \sum_{j=1}^{n_i} \mu(\bar{I}_{i,j}) - \epsilon \geq -\epsilon + \inf_{(I_i)_{i=1}^{\infty} \text{ are open intervals, } E \subseteq \bigcup_{i=1}^{\infty} I_i} \sum_{i=1}^{\infty} \mu(I_i)$. This implies that $\mu_{\mathrm{L}o}(E) \geq -\epsilon + \inf_{(I_i)_{i=1}^{\infty} \text{ are open intervals, } E \subseteq \bigcup_{i=1}^{\infty} I_i} \sum_{i=1}^{\infty} \mu(I_i)$. Hence, by the arbitrariness of ϵ, we have

$$\mu_{\mathrm{L}o}(E) \geq \inf_{(I_i)_{i=1}^{\infty} \text{ are open intervals, } E \subseteq \bigcup_{i=1}^{\infty} I_i} \sum_{i=1}^{\infty} \mu(I_i)$$

Therefore, (11.3) is true.

By Carathéodory Extension Theorem 11.19, $\mu_L = \mu_{Lo}|_{\mathcal{B}_L}$ and $(\mathbb{R}, \mathcal{B}_L, \mu_L)$ is the completion of $(\mathbb{R}, \mathcal{B}_a, \mu_L|_{\mathcal{B}_a})$, where $\mathcal{B}_a$ is the σ-algebra on $\mathbb{R}$ generated by $\mathcal{A}$. Clearly, $\mathcal{O}_\mathbb{R} \subseteq \mathcal{B}_a$ and $\mathcal{B}_a$ is the σ-algebra on $\mathbb{R}$ generated by $\mathcal{O}_\mathbb{R}$. $\diamond$

Definition 11.22 Let $\mathcal{X} := (X, \mathcal{O})$ be a topological space. The collection of *Borel sets* on $\mathcal{X}$ is the smallest σ-algebra on X that contains $\mathcal{O}$, which will be denoted by $\mathcal{B}_B(\mathcal{X})$. The smallest algebra on X that contains $\mathcal{O}$ will be denoted by $\mathcal{A}(\mathcal{X})$. $\diamond$

Clearly, we have $\mathcal{B}_B(\mathbb{R}) \subseteq \mathcal{B}_L$.

Proposition 11.23 *Let $E \subseteq \mathbb{R}$. Then, the following statements are equivalent:*

(i) $E \in \mathcal{B}_L$.
(ii) $\forall \epsilon \in (0, \infty) \subset \mathbb{R}$, $\exists O \in \mathcal{O}_\mathbb{R}$ *with* $E \subseteq O$ *such that* $\mu_{Lo}(O \setminus E) < \epsilon$.
(iii) $\forall \epsilon \in (0, \infty) \subset \mathbb{R}$, $\exists F \subseteq \mathbb{R}$ *with* $\widetilde{F} \in \mathcal{O}_\mathbb{R}$ *and* $F \subseteq E$ *such that* $\mu_{Lo}(E \setminus F) < \epsilon$.

If $\mu_{Lo}(E) < +\infty$, *then (i)–(iii) are also equivalent to:*

(iv) $\forall \epsilon \in (0, \infty) \subset \mathbb{R}$, $\exists U \in \mathcal{O}_\mathbb{R}$, *which is finite union of open intervals, such that* $\mu_{Lo}(U \triangle E) < \epsilon$.

Hence, the Lebesgue measure space $(\mathbb{R}, \mathcal{B}_L, \mu_L)$ *is the completion of the Borel measure space* $(\mathbb{R}, \mathcal{B}_B(\mathbb{R}), \mu_B := \mu_L|_{\mathcal{B}_B(\mathbb{R})})$.

Proof Fix $E \subseteq \mathbb{R}$.

(i) $\Rightarrow$ (ii). First, we show the special case that $\mu_{Lo}(E) < +\infty$. Let $E \in \mathcal{B}_L$. By Example 11.21,

$$\mu_{Lo}(E) = \mu_L(E) = \inf_{(I_i)_{i=1}^\infty \text{ are open intervals, } E \subseteq \bigcup_{i=1}^\infty I_i} \sum_{i=1}^\infty \mu_L(I_i)$$

Then, $\forall \epsilon \in (0, \infty) \subset \mathbb{R}$, $\exists$ a sequence of open intervals $(I_i)_{i=1}^\infty \subseteq \mathcal{O}_\mathbb{R}$ with $E \subseteq \bigcup_{i=1}^\infty I_i =: O \in \mathcal{O}_\mathbb{R}$ such that $\mu_L(E) \le \mu_L(O) \le \sum_{i=1}^\infty \mu_L(I_i) < \mu_L(E) + \epsilon < +\infty$. Then, $\epsilon > \mu_L(O) - \mu_L(E) = \mu_L(O \setminus E) = \mu_{Lo}(O \setminus E)$, where the first equality follows from the countable additivity of μ_L. Hence, (ii) holds.

Next, we show the general case $\mu_{Lo}(E) \in [0, \infty] \subset \mathbb{R}_e$. Let $E \in \mathcal{B}_L$ and $I_1 := (-1, 1] \subset \mathbb{R}$, $I_n := (-n, -n+1] \cup (n-1, n] \subset \mathbb{R}$, $\forall n = 2, 3, \ldots$. Clearly, $E = \bigcup_{i=1}^\infty (I_i \cap E) =: \bigcup_{i=1}^\infty E_i$. Fix any $\epsilon \in (0, \infty) \subset \mathbb{R}$. $\forall i \in \mathbb{N}$, $I_i \in \mathcal{B}_L$ and $E_i \in \mathcal{B}_L$ with $\mu_L(E_i) \le \mu_L(I_i) = 2$. By the special case, we have just shown, $\exists O_i \in \mathcal{O}_\mathbb{R} \subseteq \mathcal{B}_L$ with $E_i \subseteq O_i$ such that $\mu_{Lo}(O_i \setminus E_i) = \mu_L(O_i \setminus E_i) < 2^{-i}\epsilon$. Let $O := \bigcup_{i=1}^\infty O_i \in \mathcal{O}_\mathbb{R}$. Then, $\mu_{Lo}(O \setminus E) = \mu_L(O \setminus E) = \mu_L((\bigcup_{i=1}^\infty O_i) \cap \widetilde{\bigcup_{i=1}^\infty E_i}) = \mu_L((\bigcup_{i=1}^\infty O_i) \cap (\bigcap_{i=1}^\infty \widetilde{E_i})) = \mu_L(\bigcup_{i=1}^\infty (O_i \cap (\bigcap_{j=1}^\infty \widetilde{E_j}))) \le \sum_{i=1}^\infty \mu_L(O_i \cap (\bigcap_{j=1}^\infty \widetilde{E_j})) \le \sum_{i=1}^\infty \mu_L(O_i \cap \widetilde{E_i}) = \sum_{i=1}^\infty \mu_L(O_i \setminus E_i) < \epsilon$. Hence, (ii) holds.

(ii) $\Rightarrow$ (i). Fix any $A \subseteq \mathbb{R}$. $\forall \epsilon \in (0, \infty) \subset \mathbb{R}$, by (ii), $\exists O \in \mathcal{O}_\mathbb{R}$ with $E \subseteq O$ such that $\mu_{Lo}(O \setminus E) < \epsilon$. Note that $A \cap \widetilde{E} = (A \cap \widetilde{E} \cap O) \cup (A \cap \widetilde{E} \cap \widetilde{O}) \subseteq (O \setminus E) \cup (A \cap \widetilde{O})$ and $A \cap E \subseteq A \cap O$. By Definition 11.14, we have $\mu_{Lo}(A) \le \mu_{Lo}(A \cap E) + \mu_{Lo}(A \cap \widetilde{E}) \le \mu_{Lo}(A \cap O) + \mu_{Lo}(O \setminus E) + \mu_{Lo}(A \cap \widetilde{O}) \le \mu_{Lo}(A) + \epsilon$, where the third inequality follows from the fact that $O \in \mathcal{O}_{\mathbb{R}} \subseteq \mathcal{B}_{\mathrm{L}}$ and Definition 11.15. By the arbitrariness of ϵ, we have $\mu_{\mathrm{L}o}(A) = \mu_{\mathrm{L}o}(A \cap E) + \mu_{\mathrm{L}o}(A \cap \widetilde{E})$. By the arbitrariness of A, we have $E \in \mathcal{B}_{\mathrm{L}}$.

(i) $\Leftrightarrow$ (iii). $E \in \mathcal{B}_{\mathrm{L}} \Leftrightarrow \widetilde{E} \in \mathcal{B}_{\mathrm{L}} \Leftrightarrow \forall \epsilon \in (0, \infty) \subset \mathbb{R}, \exists O \in \mathcal{O}_{\mathbb{R}}$ with $\widetilde{E} \subseteq O$ such that $\mu_{\mathrm{L}o}(O \setminus \widetilde{E}) < \epsilon$, where the last $\Leftrightarrow$ follows from (i) $\Leftrightarrow$ (ii). Let $F := \widetilde{O}$; we have $F \subseteq E$ and $\mu_{\mathrm{L}o}(E \setminus F) = \mu_{\mathrm{L}o}(O \setminus \widetilde{E}) < \epsilon$.

Next, we show (i)–(iii) are equivalent to (iv) under the additional assumption that $\mu_{\mathrm{L}o}(E) < +\infty$.

(i) $\Rightarrow$ (iv). Let $E \in \mathcal{B}_{\mathrm{L}}$. $\forall \epsilon \in (0, \infty) \subset \mathbb{R}$, by Example 11.21, $\exists$ a sequence of open intervals $(I_i)_{i=1}^{\infty}$ with $E \subseteq \bigcup_{i=1}^{\infty} I_i =: O \in \mathcal{O}_{\mathbb{R}}$ such that $\mu_{\mathrm{L}o}(E) = \mu_{\mathrm{L}}(E) \leq \sum_{i=1}^{\infty} \mu_{\mathrm{L}}(I_i) \leq \mu_{\mathrm{L}}(E) + \epsilon/2 < +\infty$. Then, $\exists N \in \mathbb{N}$ such that $\sum_{i=N+1}^{\infty} \mu_{\mathrm{L}}(I_i) < \epsilon/2$. Let $U := \bigcup_{i=1}^{N} I_i \in \mathcal{O}_{\mathbb{R}}$. We have $\mu_{\mathrm{L}o}(U \Delta E) = \mu_{\mathrm{L}}(U \Delta E) = \mu_{\mathrm{L}}(U \setminus E) + \mu_{\mathrm{L}}(E \setminus U) \leq \mu_{\mathrm{L}}(O \setminus E) + \mu_{\mathrm{L}}(E \cap \widetilde{U})$. Note that $\mu_{\mathrm{L}}(O) = \mu_{\mathrm{L}}(E) + \mu_{\mathrm{L}}(O \setminus E) \leq \sum_{i=1}^{\infty} \mu_{\mathrm{L}}(I_i) \leq \mu_{\mathrm{L}}(E) + \epsilon/2 < +\infty$ and $E \cap \widetilde{U} = (E \cap \widetilde{U} \cap O) \cup (E \cap \widetilde{U} \cap \widetilde{O}) = E \cap \widetilde{U} \cap O \subseteq O \setminus U \subseteq \bigcup_{i=N+1}^{\infty} I_i$. Then, $\mu_{\mathrm{L}o}(U \Delta E) \leq \epsilon/2 + \mu_{\mathrm{L}}(\bigcup_{i=N+1}^{\infty} I_i) \leq \epsilon/2 + \sum_{i=N+1}^{\infty} \mu_{\mathrm{L}}(I_i) < \epsilon$. Hence, (iv) holds.

(iv) $\Rightarrow$ (i). Fix any set $A \subseteq \mathbb{R}$. By (iv), $\forall \epsilon \in (0, \infty) \subset \mathbb{R}, \exists U \in \mathcal{O}_{\mathbb{R}}$, which is finite union of open intervals, such that $\mu_{\mathrm{L}o}(U \Delta E) < \epsilon/2$. Note that $A \cap E = (A \cap U \cap E) \cup (A \cap E \cap \widetilde{U}) \subseteq (A \cap U) \cup (E \setminus U)$ and $A \cap \widetilde{E} = (A \cap \widetilde{E} \cap U) \cup (A \cap \widetilde{E} \cap \widetilde{U}) \subseteq (U \setminus E) \cup (A \cap \widetilde{U})$. By Definition 11.14, $\mu_{\mathrm{L}o}(A) \leq \mu_{\mathrm{L}o}(A \cap E) + \mu_{\mathrm{L}o}(A \cap \widetilde{E}) \leq \mu_{\mathrm{L}o}((A \cap U) \cup (E \setminus U)) + \mu_{\mathrm{L}o}((U \setminus E) \cup (A \cap \widetilde{U})) \leq \mu_{\mathrm{L}o}(A \cap U) + \mu_{\mathrm{L}o}(E \setminus U) + \mu_{\mathrm{L}o}(U \setminus E) + \mu_{\mathrm{L}o}(A \cap \widetilde{U}) = \mu_{\mathrm{L}o}(A) + \mu_{\mathrm{L}o}(E \setminus U) + \mu_{\mathrm{L}o}(U \setminus E) \leq \mu_{\mathrm{L}o}(A) + 2\mu_{\mathrm{L}o}(E \Delta U) \leq \mu_{\mathrm{L}o}(A) + \epsilon$, where the equality follows from the fact that $U \in \mathcal{O}_{\mathbb{R}} \subseteq \mathcal{B}_{\mathrm{L}}$ and Definition 11.15. By the arbitrariness of ϵ, we have $\mu_{\mathrm{L}o}(A) = \mu_{\mathrm{L}o}(A \cap E) + \mu_{\mathrm{L}o}(A \cap \widetilde{E})$. By the arbitrariness of A, we have $E \in \mathcal{B}_{\mathrm{L}}$. Hence, (i) holds.

Finally, by Example 11.21, the Lebesgue measure space is the completion of the Borel measure space $(\mathbb{R}, \mathcal{B}_{\mathrm{B}}(\mathbb{R}), \mu_{\mathrm{L}}|_{\mathcal{B}_{\mathrm{B}}(\mathbb{R})})$. This completes the proof of the proposition. □

Proposition 11.24 *Let $\mathcal{X}_i := (X_i, \mathcal{O}_i)$ be a second countable topological space, $i = 1, 2$, $\mathcal{X} := \mathcal{X}_1 \times \mathcal{X}_2 =: (X_1 \times X_2, \mathcal{O})$ be the product topological space, $\mathcal{E} := \{B_1 \times B_2 \subseteq X_1 \times X_2 \mid B_i \in \mathcal{B}_{\mathrm{B}}(\mathcal{X}_i),\ i = 1, 2\}$, and $\mathcal{B}$ be the σ-algebra on $X_1 \times X_2$ generated by $\mathcal{E}$. Then, $\mathcal{B} = \mathcal{B}_{\mathrm{B}}(\mathcal{X})$.*

Proof Let $\mathcal{O}_{Bi} \subseteq \mathcal{O}_i$ be a countable basis for $\mathcal{X}_i$, $i = 1, 2$. Without loss of generality, assume that $X_i \in \mathcal{O}_{Bi}$, $i = 1, 2$. By Proposition 3.28, $\mathcal{X}$ is second countable with a countable basis $\mathcal{O}_B := \{O_{B1} \times O_{B2} \subseteq X_1 \times X_2 \mid O_{Bi} \in \mathcal{O}_{Bi},\ i = 1, 2\} \subseteq \mathcal{O}$. Clearly, $\mathcal{O}_{Bi} \subseteq \mathcal{O}_i \subseteq \mathcal{B}_{\mathrm{B}}(\mathcal{X}_i)$, $i = 1, 2$. Then, $\mathcal{O}_B \subseteq \mathcal{E}$. $\forall O \in \mathcal{O}$, $O = \bigcup_{i=1}^{\infty} O_i$, where $O_i \in \mathcal{O}_B$, $\forall i \in \mathbb{N}$. Hence, $O \in \mathcal{B}$. This implies that $\mathcal{O} \subseteq \mathcal{B}$. Since $\mathcal{B}$ is a σ-algebra, then $\mathcal{O} \subseteq \mathcal{B}_{\mathrm{B}}(\mathcal{X}) \subseteq \mathcal{B}$.

On the other hand, note that $\mathcal{B}_{\mathrm{B}}(\mathcal{X})$ is the σ-algebra generated by $\mathcal{O}$. Then, $\forall O_1 \in \mathcal{O}_1, \forall O_2 \in \mathcal{O}_2$, we have $O_1 \times O_2 \in \mathcal{O} \subseteq \mathcal{B}_{\mathrm{B}}(\mathcal{X})$.

Claim 11.24.1 $\forall O_1 \in \mathcal{O}_1, \forall B_2 \in \mathcal{B}_{\mathrm{B}}(\mathcal{X}_2)$, we have $O_1 \times B_2 \in \mathcal{B}_{\mathrm{B}}(\mathcal{X})$.

Proof of Claim Fix any $O_1 \in \mathcal{O}_1$. Define $\mathcal{F} := \{E \subseteq X_2 \mid O_1 \times E \in \mathcal{B}_B(\mathcal{X})\}$. Clearly, $\mathcal{O}_2 \subseteq \mathcal{F}$. Then, $\emptyset, X_2 \in \mathcal{O}_2 \subseteq \mathcal{F}$. $\forall E \in \mathcal{F}$, we have $O_1 \times E \in \mathcal{B}_B(\mathcal{X})$. Note that $O_1 \times X_2 \in \mathcal{B}_B(\mathcal{X})$. This leads to $O_1 \times (X_2 \setminus E) = (O_1 \times X_2) \setminus (O_1 \times E) \in \mathcal{B}_B(\mathcal{X})$. Thus, $X_2 \setminus E \in \mathcal{F}$. $\forall (E_i)_{i=1}^{\infty} \subseteq \mathcal{F}$, we have $O_1 \times E_i \in \mathcal{B}_B(\mathcal{X})$, $\forall i \in \mathbb{N}$. This implies that $O_1 \times (\bigcup_{i=1}^{\infty} E_i) = \bigcup_{i=1}^{\infty} (O_1 \times E_i) \in \mathcal{B}_B(\mathcal{X})$ and $\bigcup_{i=1}^{\infty} E_i \in \mathcal{F}$. The above shows that $\mathcal{F}$ is a σ-algebra on X_2. Then, $\mathcal{O}_2 \subseteq \mathcal{B}_B(\mathcal{X}_2) \subseteq \mathcal{F}$. Hence, $\forall B_2 \in \mathcal{B}_B(\mathcal{X}_2)$, we have $B_2 \in \mathcal{F}$ and $O_1 \times B_2 \in \mathcal{B}_B(\mathcal{X})$. This completes the proof of the claim. □

Claim 11.24.2 $\forall B_1 \in \mathcal{B}_B(\mathcal{X}_1)$, $\forall B_2 \in \mathcal{B}_B(\mathcal{X}_2)$, we have $B_1 \times B_2 \in \mathcal{B}_B(\mathcal{X})$.

Proof of Claim Fix any $B_2 \in \mathcal{B}_B(\mathcal{X}_2)$. Define $\mathcal{F} := \{E \subseteq X_1 \mid E \times B_2 \in \mathcal{B}_B(\mathcal{X})\}$. By Claim 11.24.1, $\mathcal{O}_1 \subseteq \mathcal{F}$. Then, $\emptyset, X_1 \in \mathcal{O}_1 \subseteq \mathcal{F}$. $\forall E \in \mathcal{F}$, we have $E \times B_2 \in \mathcal{B}_B(\mathcal{X})$. Note that $X_1 \times B_2 \in \mathcal{B}_B(\mathcal{X})$. This leads to $(X_1 \setminus E) \times B_2 = (X_1 \times B_2) \setminus (E \times B_2) \in \mathcal{B}_B(\mathcal{X})$. Thus, $X_1 \setminus E \in \mathcal{F}$. $\forall (E_i)_{i=1}^{\infty} \subseteq \mathcal{F}$, we have $E_i \times B_2 \in \mathcal{B}_B(\mathcal{X})$, $\forall i \in \mathbb{N}$. This implies that $(\bigcup_{i=1}^{\infty} E_i) \times B_2 = \bigcup_{i=1}^{\infty} (E_i \times B_2) \in \mathcal{B}_B(\mathcal{X})$ and $\bigcup_{i=1}^{\infty} E_i \in \mathcal{F}$. The above shows that $\mathcal{F}$ is a σ-algebra on X_1. Then, $\mathcal{O}_1 \subseteq \mathcal{B}_B(\mathcal{X}_1) \subseteq \mathcal{F}$. Hence, $\forall B_1 \in \mathcal{B}_B(\mathcal{X}_1)$, we have $B_1 \in \mathcal{F}$ and $B_1 \times B_2 \in \mathcal{B}_B(\mathcal{X})$. This completes the proof of the claim. □

By Claim 11.24.2, $\mathcal{E} \subseteq \mathcal{B}_B(\mathcal{X})$. By the definition of $\mathcal{B}$, we have $\mathcal{B} \subseteq \mathcal{B}_B(\mathcal{X})$. Hence, $\mathcal{B} = \mathcal{B}_B(\mathcal{X})$. This completes the proof of the proposition. □

Proposition 11.25 *Let $\mathcal{X} := (X, \mathcal{O})$ be a topological space, $E \in \mathcal{B}_B(\mathcal{X})$, $\mathcal{E} := (E, \mathcal{O}_E)$ be the topological space with the subset topology of $\mathcal{X}$, and $\mathcal{B}_E := \{C \subseteq E \mid C \in \mathcal{B}_B(\mathcal{X})\}$. Then, $\mathcal{B}_E = \mathcal{B}_B(\mathcal{E})$.*

Proof Clearly, $\emptyset, E \in \mathcal{B}_E$. $\forall A \in \mathcal{B}_E$, we have $A \subseteq E$ and $A \in \mathcal{B}_B(\mathcal{X})$. Then, $E \supseteq E \setminus A \in \mathcal{B}_B(\mathcal{X})$ and $E \setminus A \in \mathcal{B}_E$. $\forall (A_i)_{i=1}^{\infty} \subseteq \mathcal{B}_E$, we have $E \supseteq A_i \in \mathcal{B}_B(\mathcal{X})$, $\forall i \in \mathbb{N}$. Then, $E \supseteq \bigcup_{i=1}^{\infty} A_i \in \mathcal{B}_B(\mathcal{X})$ and $\bigcup_{i=1}^{\infty} A_i \in \mathcal{B}_E$. This shows that $\mathcal{B}_E$ is a σ-algebra on E.

$\forall O_E \in \mathcal{O}_E$, $\exists O \in \mathcal{O} \subseteq \mathcal{B}_B(\mathcal{X})$ such that $O_E = O \cap E$. Then, $O_E \in \mathcal{B}_E$ and $\mathcal{O}_E \subseteq \mathcal{B}_E$. Hence, $\mathcal{O}_E \subseteq \mathcal{B}_B(\mathcal{E}) \subseteq \mathcal{B}_E$.

On the other hand, let $\tilde{\mathcal{E}} := (X \setminus E, \mathcal{O}_{X \setminus E})$ be the topological space with the subset topology of $\mathcal{X}$.

Claim 11.25.1 $\forall C \in \mathcal{B}_B(\mathcal{X})$, we have $C_1 := C \cap E \in \mathcal{B}_B(\mathcal{E})$ and $C_2 := C \cap (X \setminus E) \in \mathcal{B}_B\left(\tilde{\mathcal{E}}\right)$.

Proof of Claim Define $\mathcal{F} := \left\{ A_1 \cup A_2 \subseteq X \,\middle|\, A_1 \in \mathcal{B}_B(\mathcal{E}),\ A_2 \in \mathcal{B}_B\left(\tilde{\mathcal{E}}\right) \right\}$. $\forall O \in \mathcal{O}$, $O \cap E \in \mathcal{O}_E \subseteq \mathcal{B}_B(\mathcal{E})$ and $O \cap (X \setminus E) \in \mathcal{O}_{X \setminus E} \subseteq \mathcal{B}_B\left(\tilde{\mathcal{E}}\right)$. Then, $O = (O \cap E) \cup (O \cap (X \setminus E)) \in \mathcal{F}$ and $\mathcal{O} \subseteq \mathcal{F}$.

Clearly, $\emptyset, E \in \mathcal{B}_B(\mathcal{E})$ and $\emptyset, X \setminus E \in \mathcal{B}_B\left(\tilde{\mathcal{E}}\right)$. Then, $\emptyset = \emptyset \cup \emptyset \in \mathcal{F}$ and $X = E \cup (X \setminus E) \in \mathcal{F}$. $\forall A \in \mathcal{F}$, $\exists A_1 \in \mathcal{B}_B(\mathcal{E})$ and $\exists A_2 \in \mathcal{B}_B\left(\tilde{\mathcal{E}}\right)$ such that $A = A_1 \cup A_2$. Then, $A_1 = A \cap E$ and $A_2 = A \cap (X \setminus E)$. Note that $X \setminus A =$

$(E \cup (X \setminus E)) \setminus A = (E \setminus A) \cup ((X \setminus E) \setminus A) = (E \setminus A_1) \cup ((X \setminus E) \setminus A_2)$, $E \setminus A_1 \in \mathcal{B}_B(\mathcal{E})$, and $(X \setminus E) \setminus A_2 \in \mathcal{B}_B\left(\tilde{\mathcal{E}}\right)$. This implies that $X \setminus A \in \mathcal{F}$. $\forall (A_i)_{i=1}^{\infty} \subseteq \mathcal{F}$, let $A_{i,1} := E \cap A_i \in \mathcal{B}_B(\mathcal{E})$ and $A_{i,2} := (X \setminus E) \cap A_i \in \mathcal{B}_B\left(\tilde{\mathcal{E}}\right)$, $\forall i \in \mathbb{N}$. Note that $\bigcup_{i=1}^{\infty} A_i = \bigcup_{i=1}^{\infty}(A_{i,1} \cup A_{i,2}) = (\bigcup_{i=1}^{\infty} A_{i,1}) \cup (\bigcup_{i=1}^{\infty} A_{i,2})$, $\bigcup_{i=1}^{\infty} A_{i,1} \in \mathcal{B}_B(\mathcal{E})$, and $\bigcup_{i=1}^{\infty} A_{i,2} \in \mathcal{B}_B\left(\tilde{\mathcal{E}}\right)$. This implies that $\bigcup_{i=1}^{\infty} A_i \in \mathcal{F}$. The above shows that $\mathcal{F}$ is a σ-algebra on X. Then, $\mathcal{O} \subseteq \mathcal{B}_B(\mathcal{X}) \subseteq \mathcal{F}$. This completes the proof of the claim. □

$\forall C \in \mathcal{B}_E$, we have $E \supseteq C \in \mathcal{B}_B(\mathcal{X})$. By Claim 11.25.1, $C = C \cap E \in \mathcal{B}_B(\mathcal{E})$. Then, $\mathcal{B}_E \subseteq \mathcal{B}_B(\mathcal{E})$. Therefore, $\mathcal{B}_E = \mathcal{B}_B(\mathcal{E})$. This completes the proof of the proposition. □

Definition 11.26 Let $\mathcal{X} := (X, \mathcal{O})$ be a topological space and $(X, \mathcal{B}, \mu)$ be a measure space on the same set X. The triple $\mathbb{X} := (\mathcal{X}, \mathcal{B}, \mu)$ is said to be a *topological measure space* if $\mathcal{B} = \mathcal{B}_B(\mathcal{X})$ and $\forall E \in \mathcal{B}_B(\mathcal{X})$, $\forall \epsilon \in (0, \infty) \subset \mathbb{R}$, $\exists U \in \mathcal{O}$ with $E \subseteq U$ such that $\mu(U \setminus E) < \epsilon$. We will say that $\mathbb{X}$ is *finite* or *σ-finite* if the underlying measure space is so. We will say that $\mathbb{X}$ is *Tychonoff, Hausdorff, regular, completely regular,* or *normal* if $\mathcal{X}$ is so. We will say that $\mathbb{X}$ is *first countable* or *second countable* if $\mathcal{X}$ is so. We will say that $\mathbb{X}$ is *separable, second category everywhere, connected,* or *locally connected* if $\mathcal{X}$ is so. We will say that $\mathbb{X}$ is *compact, countably compact, sequentially compact, locally compact, σ-compact,* or *paracompact* if $\mathcal{X}$ is so. We will say that $\mathbb{X}$ is *locally finite* if $\forall$ compact $K \subseteq \mathcal{X}$, $\mu(K) < \infty$.

Let $\mathcal{X} := (X, \rho)$ be a metric space, $\mathcal{O}$ be the natural topology on $\mathcal{X}$ generated by the metric ρ, and $(X, \mathcal{B}, \mu)$ be a measure space on the same set X. The triple $\mathbb{X} := (\mathcal{X}, \mathcal{B}, \mu)$ is said to be a *metric measure space* if $((X, \mathcal{O}), \mathcal{B}, \mu)$ is a topological measure space. $\mathbb{X}$ is said to be a *complete metric measure space*, if $\mathcal{X}$ is a complete metric space. $\mathbb{X}$ is said to be *totally bounded* if $\mathcal{X}$ is so.

Let $\mathfrak{X} := (\mathcal{X}, \mathbb{K}, \|\cdot\|)$ be a normed linear space over the field $\mathbb{K}$, $\mathcal{O}$ be the natural topology on $\mathfrak{X}$ generated by the norm $\|\cdot\|$, and $(X, \mathcal{B}, \mu)$ be a measure space on the same set X. The triple $\mathbb{X} := (\mathfrak{X}, \mathcal{B}, \mu)$ is said to be a *normed linear measure space* if $((X, \mathcal{O}), \mathcal{B}, \mu)$ is a topological measure space. Furthermore, if $\mathfrak{X}$ is a Banach space, then $\mathbb{X}$ is said to be a *Banach measure space*. Depending on whether $\mathbb{K} = \mathbb{R}$ or $\mathbb{K} = \mathbb{C}$, we will say that $\mathbb{X}$ is a *real* or *complex Banach measure space*. ◇

Proposition 11.27 *Let $\mathcal{X} := (X, \mathcal{O})$ be a topological space, $(X, \mathcal{B}, \mu)$ be a measure space on the same set X, and $\mathcal{B} = \mathcal{B}_B(\mathcal{X})$. The triple $\mathbb{X} := (\mathcal{X}, \mathcal{B}, \mu)$ is topological measure space if, and only if, $\forall E \in \mathcal{B}_B(\mathcal{X})$, $\forall \epsilon \in (0, \infty) \subset \mathbb{R}$, $\exists X \setminus F \in \mathcal{O}$ with $F \subseteq E$ such that $\mu(E \setminus F) < \epsilon$.*

Proof "Necessity" $\forall E \in \mathcal{B}, \forall \epsilon \in (0, \infty) \subset \mathbb{R}$, $X \setminus E \in \mathcal{B}$. By $\mathbb{X}$ being a topological measure space, $\exists U \in \mathcal{O}$ with $X \setminus E \subseteq U$ such that $\mu(U \setminus (X \setminus E)) < \epsilon$. Then, $X \setminus U \subseteq E$ and $\mu(E \setminus (X \setminus U)) = \mu(E \cap U) = \mu(U \setminus (X \setminus E)) < \epsilon$. Hence, $F := X \setminus U$ is the set we seek.

"Sufficiency" $\forall E \in \mathcal{B}$, $\forall \epsilon \in (0, \infty) \subset \mathbb{R}$, $X \setminus E \in \mathcal{B}$. By the assumption, $\exists X \setminus F \in \mathcal{O}$ with $F \subseteq X \setminus E$ such that $\mu((X \setminus E) \setminus F) < \epsilon$. Let $U := X \setminus F$. Then, $E \subseteq U$ and $\mu(U \setminus E) = \mu((X \setminus F) \setminus E) = \mu((X \setminus E) \setminus F) < \epsilon$. Hence, $\mathbb{X}$ is a topological measure space. □

Example 11.28 The real line $\mathbb{R}$ is a real Banach space with norm defined to be the absolute value. Then, by Proposition 11.23 and Example 11.21, $\mathbb{R} := ((\mathbb{R}, \mathbb{R}, |\cdot|), \mathcal{B}_B(\mathbb{R}), \mu_B)$ is a σ-finite real Banach measure space. ◇

Proposition 11.29 *Let $\mathbb{X} := (\mathcal{X}, \mathcal{B}, \mu)$ be a topological measure space, where $\mathcal{X} := (X, \mathcal{O})$ is a topological space, $E \in \mathcal{B}$, $\mathcal{E} := (E, \mathcal{O}_E)$ be the topological space with the subset topology of $\mathcal{X}$, and $(E, \mathcal{B}_E, \mu_E)$ be the measure subspace of $\mathbb{X}$ as defined in Proposition* 11.13. *Then, $\mathbb{E} := (\mathcal{E}, \mathcal{B}_E, \mu_E)$ is a topological measure space and is said to be the* topological measure subspace *of* $\mathbb{X}$.

Proof By Definition 11.26, $\mathcal{B} = \mathcal{B}_B(\mathcal{X})$. By Proposition 11.13, $\mathcal{B}_E = \{C \subseteq E \mid C \in \mathcal{B}\}$ and $\mu_E = \mu|_{\mathcal{B}_E}$. By Proposition 11.25, we have $\mathcal{B}_E = \mathcal{B}_B(\mathcal{E})$.

$\forall A \in \mathcal{B}_E$, $\forall \epsilon \in (0, \infty) \subset \mathbb{R}$, $E \supseteq A \in \mathcal{B}$. By $\mathbb{X}$ being a topological measure space, $\exists U \in \mathcal{O}$ with $A \subseteq U$ such that $\mu(U \setminus A) < \epsilon$. Let $\bar{U} := U \cap E \in \mathcal{O}_E \subseteq \mathcal{B}_E$. Clearly, $A \subseteq \bar{U}$ and $0 \le \mu_E(\bar{U} \setminus A) = \mu((U \cap E) \setminus A) \le \mu(U \setminus A) < \epsilon$. Hence, $\mathbb{E}$ is a topological measure space. This completes the proof of the proposition. □

Definition 11.30 Let X be a set and $\mathcal{C} \subseteq {}^X2$ be a nonempty collection of subsets of X. $\mathcal{C}$ is said to be a *semialgebra* on X if $\forall C_1, C_2 \in \mathcal{C}$, $C_1 \cap C_2 \in \mathcal{C}$ and $\widetilde{C_1}$ is a finite disjoint union of sets in $\mathcal{C}$. ◇

Proposition 11.31 *Let X be a set and $\mathcal{C} \subseteq {}^X2$ be a semialgebra on X. $\mathcal{A} := \{A \subseteq X \mid A$ is the finite disjoint union of sets in $\mathcal{C}\}$. Then, $\mathcal{A}$ is the algebra on X generated by $\mathcal{C}$.*

Proof Clearly, $\emptyset \in \mathcal{A}$. Fix $C_1 \in \mathcal{C} \neq \emptyset$. Then, $\widetilde{C_1} = \bigcup_{i=2}^n C_i$, where $n \in \mathbb{N}$, $(C_i)_{i=2}^n \subseteq \mathcal{C}$ is pairwise disjoint. Then, $X = C_1 \cup \widetilde{C_1} = \bigcup_{i=1}^n C_i$, where the sets in the union are pairwise disjoint. Hence, $X \in \mathcal{A}$. $\forall A_1, A_2 \in \mathcal{A}$, then $A_i = \bigcup_{j=1}^{n_i} C_{j,i}$, where $n_i \in \mathbb{Z}_+$ and $\left(C_{j,i}\right)_{j=1}^{n_i} \subseteq \mathcal{C}$ is pairwise disjoint, $i = 1, 2$. Note that $\widetilde{A_1} = \bigcap_{j=1}^{n_1} \widetilde{C_{j,1}}$. $\forall j \in \{1, \ldots, n_1\}$, $\widetilde{C_{j,1}} = \bigcup_{l=1}^{m_j} C_{l,j,1}$, where $m_j \in \mathbb{Z}_+$ and $\left(C_{l,j,1}\right)_{l=1}^{m_j} \subseteq \mathcal{C}$ is pairwise disjoint. Then,

$$\widetilde{A_1} = \bigcap_{j=1}^{n_1} \bigcup_{l=1}^{m_j} C_{l,j,1} = \bigcup_{l_1=1}^{m_1} \Big(C_{l_1,1,1} \cap \Big(\bigcap_{j=2}^{n_1} \bigcup_{l=1}^{m_j} C_{l,j,1}\Big)\Big)$$

$$= \bigcup_{l_1=1}^{m_1} \Big(C_{l_1,1,1} \cap \Big(\bigcup_{l_2=1}^{m_2} \Big(C_{l_2,2,1} \cap \Big(\bigcap_{j=3}^{n_1} \bigcup_{l=1}^{m_j} C_{l,j,1}\Big)\Big)\Big)\Big)$$

$$= \bigcup_{l_1=1}^{m_1} \bigcup_{l_2=1}^{m_2} (C_{l_1,1,1} \cap C_{l_2,2,1} \cap (\bigcap_{j=3}^{n_1} \bigcup_{l=1}^{m_j} C_{l,j,1})))$$

$$= \bigcup_{l_1=1}^{m_1} \cdots \bigcup_{l_{n_1}=1}^{m_{n_1}} \bigcap_{j=1}^{n_1} C_{l_j,j,1}$$

When $n_1 = 0$, clearly $\widetilde{A_1} = X \in \mathcal{A}$. When $n_1 \geq 1$, $\bigcap_{j=1}^{n_1} C_{l_j,j,1} \in \mathcal{C}$ for any admissible choice of indices $l_1, \ldots, l_{n_1}$. For different admissible choices of indices $l_1, \ldots, l_{n_1}$ and $\bar{l}_1, \ldots, \bar{l}_{n_1}$, $\left(\bigcap_{j=1}^{n_1} C_{l_j,j,1}\right) \cap \left(\bigcap_{j=1}^{n_1} C_{\bar{l}_j,j,1}\right) = \emptyset$. Hence, $\widetilde{A_1} \in \mathcal{A}$. Note that $A_1 \cap A_2 = \bigcup_{j_1=1}^{n_1} \bigcup_{j_2=1}^{n_2} \left(C_{j_1,1} \cap C_{j_2,2}\right)$. $C_{j_1,1} \cap C_{j_2,2} \in \mathcal{C}$ for any admissible choice of indices j_1, j_2. For different admissible choices of indices j_1, j_2 and $\bar{j}_1, \bar{j}_2$, $\left(C_{j_1,1} \cap C_{j_2,2}\right) \cap \left(C_{\bar{j}_1,1} \cap C_{\bar{j}_2,2}\right) = \emptyset$. Hence, $A_1 \cap A_2 \in \mathcal{A}$.

Therefore, $\mathcal{A}$ is an algebra on X. Clearly, $\mathcal{A}$ is the smallest algebra on X containing $\mathcal{C}$. Hence, $\mathcal{A}$ is the algebra on X generated by $\mathcal{C}$. This completes the proof of the proposition. □

Proposition 11.32 *Let X be a set, $\mathcal{C}$ be a semialgebra on X, and $\mu : \mathcal{C} \to [0, +\infty] \subset \mathbb{R}_e$. Assume that:*

(i) $\forall C \in \mathcal{C}$, $\forall n \in \mathbb{Z}_+$, $\forall$ *pairwise disjoint* $\left(C_i\right)_{i=1}^n \subseteq \mathcal{C}$ *with* $C = \bigcup_{i=1}^n C_i$, *then* $\mu\left(C\right) = \sum_{i=1}^n \mu\left(C_i\right)$.

(ii) $\forall C \in \mathcal{C}$, $\forall$ *pairwise disjoint* $\left(C_i\right)_{i=1}^\infty \subseteq \mathcal{C}$ *with* $C = \bigcup_{i=1}^\infty C_i$, *then* $\mu\left(C\right) \leq \sum_{i=1}^\infty \mu\left(C_i\right)$.

Then, μ admits a unique extension to a measure $\bar{\mu}$ on the algebra on X, $\mathcal{A}$, generated by $\mathcal{C}$.

Proof By Proposition 11.31,

$$\mathcal{A} = \{A \subseteq X \mid A \text{ is the finite disjoint union of sets in } \mathcal{C}\}$$

Define $\bar{\mu} : \mathcal{A} \to [0, \infty] \subset \mathbb{R}_e$ by $\bar{\mu}(A) = \sum_{i=1}^n \mu(C_i)$, $\forall A = \bigcup_{i=1}^n C_i \in \mathcal{A}$, where $n \in \mathbb{Z}_+$ and $(C_i)_{i=1}^n \subseteq \mathcal{C}$ is pairwise disjoint. By (i), $\bar{\mu}$ is well-defined. Clearly, $\bar{\mu}(\emptyset) = 0$. $\forall A_1, A_2 \in \mathcal{A}$ with $A_1 \cap A_2 = \emptyset$, $A_l = \bigcup_{i=1}^{n_l} C_{i,l}$, where $n_l \in \mathbb{Z}_+$ and $\left(C_{i,l}\right)_{i=1}^{n_l} \subseteq \mathcal{C}$ is pairwise disjoint, $l = 1, 2$. $A_1 \cup A_2 = \bigcup_{l=1}^2 \bigcup_{i=1}^{n_l} C_{i,l}$, where $C_{i,l}$, $l = 1, 2$, $i = 1, \ldots, n_l$, are pairwise disjoint. Then, $\bar{\mu}(A_1 \cup A_2) = \sum_{l=1}^2 \sum_{i=1}^{n_l} \mu(C_{i,l}) = \bar{\mu}(A_1) + \bar{\mu}(A_2)$. Hence, $\bar{\mu}$ is finitely additive.

$\forall (A_i)_{i=1}^\infty \subseteq \mathcal{A}$ with $A = \bigcup_{i=1}^\infty A_i \in \mathcal{A}$ and $A_1, A_2, \ldots$ being pairwise disjoint. By finite additivity $\bar{\mu}(A) = \bar{\mu}((\bigcup_{i=1}^n A_i) \cup (A \setminus \bigcup_{i=1}^n A_i)) \geq \sum_{i=1}^n \bar{\mu}(A_i)$, $\forall n \in \mathbb{N}$. Then, $\bar{\mu}(A) \geq \sum_{i=1}^\infty \bar{\mu}(A_i)$. On the other hand, since $A \in \mathcal{A}$, then $A = \bigcup_{i=1}^{\bar{n}} C_i$, where $\bar{n} \in \mathbb{Z}_+$ and $(C_i)_{i=1}^{\bar{n}} \subseteq \mathcal{C}$ is pairwise disjoint. $\forall i \in \mathbb{N}$, $A_i = \bigcup_{j=1}^{n_i} C_{j,i}$, where $n_i \in \mathbb{Z}_+$ and $\left(C_{j,i}\right)_{j=1}^{n_i} \subseteq \mathcal{C}$ is pairwise disjoint. Then, $\forall i \in \{1, \ldots, \bar{n}\}$, $C_i = C_i \cap (\bigcup_{l=1}^\infty A_l) = \bigcup_{l=1}^\infty \bigcup_{j=1}^{n_l} (C_{j,l} \cap$

$C_i)$. By (ii), we have $\mu(C_i) \leq \sum_{l=1}^{\infty} \sum_{j=1}^{n_l} \mu(C_{j,l} \cap C_i)$. Therefore, $\bar{\mu}(A) = \sum_{i=1}^{\bar{n}} \mu(C_i) \leq \sum_{i=1}^{\bar{n}} \sum_{l=1}^{\infty} \sum_{j=1}^{n_l} \mu(C_{j,l} \cap C_i) = \sum_{l=1}^{\infty} \sum_{j=1}^{n_l} \sum_{i=1}^{\bar{n}} \mu(C_{j,l} \cap C_i) = \sum_{l=1}^{\infty} \sum_{j=1}^{n_l} \mu(C_{j,l}) = \sum_{l=1}^{\infty} \bar{\mu}(A_l)$, where the third equality follows from (i). Hence, $\bar{\mu}(A) = \sum_{l=1}^{\infty} \bar{\mu}(A_l)$.

This implies that $\bar{\mu}$ is a measure on algebra $\mathcal{A}$. Clearly, it is an extension of μ to $\mathcal{A}$ and it is the unique extension. This completes the proof of the proposition. □

11.3 Measurable Functions

Definition 11.33 Let $(X, \mathcal{B})$ be a measurable space, $\mathcal{Y} := (Y, \mathcal{O}_Y)$ be a topological space, $D \in \mathcal{B}$, and $f : D \to \mathcal{Y}$. We will say that f is *$\mathcal{B}$-measurable*, if $\forall O \in \mathcal{O}_Y$, $f_{\text{inv}}(O) \in \mathcal{B}$. When it is clear from the context, we will simply say that f is measurable. ◇

Proposition 11.34 *Let $(X, \mathcal{B})$ be a measurable space, $\mathcal{Y} := (Y, \mathcal{O}_Y)$ be a topological space, $D \in \mathcal{B}$, and $f : D \to \mathcal{Y}$. Then:*

(i) *f is $\mathcal{B}$-measurable if, and only if, $\forall E \in \mathcal{B}_B(\mathcal{Y})$, $f_{\text{inv}}(E) \in \mathcal{B}$.*
(ii) *If $\mathcal{Y}$ is second countable with a countable basis $\bar{\mathcal{B}}$, then f is $\mathcal{B}$-measurable if, and only if, $\forall B_Y \in \bar{\mathcal{B}}$, $f_{\text{inv}}(B_Y) \in \mathcal{B}$.*

Proof (i) "Sufficiency" is obvious. "Necessity" Let f be $\mathcal{B}$-measurable. Define $\mathcal{B}_Y := \{E \subseteq Y \mid f_{\text{inv}}(E) \in \mathcal{B}\}$. Clearly, $\mathcal{O}_Y \subseteq \mathcal{B}_Y$. We will show that $\mathcal{B}_Y$ is a σ-algebra on Y. Then, $\mathcal{B}_B(\mathcal{Y}) \subseteq \mathcal{B}_Y$ and the result is established.

Clearly, $f_{\text{inv}}(\emptyset) = \emptyset \in \mathcal{B}$ and $f_{\text{inv}}(Y) = D \in \mathcal{B}$; then $\emptyset, Y \in \mathcal{B}_Y$. $\forall E \in \mathcal{B}_Y$, by Proposition 2.5, $f_{\text{inv}}(Y \setminus E) = f_{\text{inv}}(Y) \setminus f_{\text{inv}}(E) = D \setminus f_{\text{inv}}(E) \in \mathcal{B}$. Hence, $Y \setminus E \in \mathcal{B}_Y$. $\forall (E_i)_{i=1}^{\infty} \subseteq \mathcal{B}_Y$, we have $f_{\text{inv}}(E_i) \in \mathcal{B}$, $\forall i \in \mathbb{N}$. Then, by Proposition 2.5, $f_{\text{inv}}(\bigcup_{i=1}^{\infty} E_i) = \bigcup_{i=1}^{\infty} f_{\text{inv}}(E_i) \in \mathcal{B}$. Then, $\bigcup_{i=1}^{\infty} E_i \in \mathcal{B}_Y$. This shows $\mathcal{B}_Y$ is a σ-algebra on Y.

(ii) "Necessity" is obvious. "Sufficiency" $\forall O \in \mathcal{O}_Y$, since $\mathcal{Y}$ is second countable, then $\exists (B_i)_{i=1}^{\infty} \subseteq \bar{\mathcal{B}}$ such that $O = \bigcup_{i=1}^{\infty} B_i$. This implies that $f_{\text{inv}}(O) = \bigcup_{i=1}^{\infty} f_{\text{inv}}(B_i) \in \mathcal{B}$, where the equality follows from Proposition 2.5. Hence, f is $\mathcal{B}$-measurable.

This completes the proof of the proposition. □

Proposition 11.35 *Let $(X, \mathcal{B})$ be a measurable space, $D \in \mathcal{B}$, $f : D \to \mathbb{R}_e$. Then, the following statements are equivalent:*

(i) *f is $\mathcal{B}$-measurable.*
(ii) $\forall \alpha \in \mathbb{R}$, *the set* $\{x \in D \mid f(x) < \alpha\} \in \mathcal{B}$.
(iii) $\forall \alpha \in \mathbb{R}$, *the set* $\{x \in D \mid f(x) \geq \alpha\} \in \mathcal{B}$.
(iv) $\forall \alpha \in \mathbb{R}$, *the set* $\{x \in D \mid f(x) \leq \alpha\} \in \mathcal{B}$.
(v) $\forall \alpha \in \mathbb{R}$, *the set* $\{x \in D \mid f(x) > \alpha\} \in \mathcal{B}$.

These statements imply:

(vi) $\forall \alpha \in \mathbb{R}_e$, *the set* $\{x \in D \mid f(x) = \alpha\} \in \mathcal{B}$.

Proof The proof is straightforward and is therefore omitted. □

Proposition 11.36 *Let* $\mathcal{X} := (X, \mathcal{B})$ *be a measurable space,* $D \in \mathcal{B}$, $f_i : D \to [0, \infty] \subset \mathbb{R}_e$ *be* $\mathcal{B}$*-measurable,* $i = 1, 2$, *and* $c \in (0, \infty) \subset \mathbb{R}$. *Then,* cf_1 *and* $f_1 + f_2$ *are* $\mathcal{B}$*-measurable.*

Proof $\forall \alpha \in \mathbb{R}$, $\{x \in D \mid cf_1(x) < \alpha\} = \{x \in D \mid f_1(x) < \alpha/c\} \in \mathcal{B}$ by Proposition 11.35. Then, by Proposition 11.35, cf_1 is $\mathcal{B}$-measurable. $\{x \in D \mid f_1(x) + f_2(x) < \alpha\} = \bigcup_{r \in \mathbb{Q}}(\{x \in D \mid f_1(x) < r\} \cap \{x \in D \mid f_2(x) < \alpha - r\}) \in \mathcal{B}$ by Proposition 11.35. Then, by Proposition 11.35, $f_1 + f_2$ is $\mathcal{B}$-measurable. This completes the proof of the proposition. □

Proposition 11.37 *Let* $\mathbb{X} := (\mathcal{X}, \mathcal{B}, \mu)$ *be a topological measure space,* $\mathcal{Y}$ *be a topological space,* $D \in \mathcal{B}$, *and* $f : D \to \mathcal{Y}$ *be continuous. Then,* f *is* $\mathcal{B}$*-measurable.*

Proof The proof is straightforward and is therefore omitted. □

Proposition 11.38 *Let* $(X, \mathcal{B})$ *be a measurable space,* $\mathcal{Y}$, *and* $\mathcal{Z}$ *be topological spaces,* $D \in \mathcal{B}$, $f : D \to \mathcal{Y}$ *be* $\mathcal{B}$*-measurable, and* $g : \mathcal{Y} \to \mathcal{Z}$ *be continuous. Then,* $h := g \circ f : D \to \mathcal{Z}$ *is* $\mathcal{B}$*-measurable.*

Proof $\forall O \in \mathcal{O}_{\mathcal{Z}}$, $g_{\text{inv}}(O) \in \mathcal{O}_{\mathcal{Y}}$ since g is continuous. $h_{\text{inv}}(O) = f_{\text{inv}}(g_{\text{inv}}(O)) \in \mathcal{B}$. Hence, h is $\mathcal{B}$-measurable. □

Proposition 11.39 *Let* $(X, \mathcal{B})$ *be a measurable space,* $\mathcal{Y}_i := (Y_i, \mathcal{O}_i)$ *be a second countable topological space,* $D \in \mathcal{B}$, $f_i : D \to \mathcal{Y}_i$, $\forall i \in N \subseteq \mathbb{N}$, N *is a countable index set,* $\mathcal{Y} := (Y, \mathcal{O}) := \prod_{i \in N} \mathcal{Y}_i$ *be the product topological space, and* $f : D \to \mathcal{Y}$ *be given by* $\pi_i(f(x)) = f_i(x)$, $\forall x \in D$, $\forall i \in N$. *Then,* f *is* $\mathcal{B}$*-measurable if, and only if,* f_i *is* $\mathcal{B}$*-measurable,* $\forall i \in N$.

Proof "Sufficiency" Let f_i's be $\mathcal{B}$-measurable. $\forall i \in N$, let $\mathcal{B}_{Y i}$ be a countable basis for $\mathcal{Y}_i$. Without loss of generality, assume $\emptyset, Y_i \in \mathcal{B}_{Y i}$. Then, $\mathcal{Y}$ is second countable with a countable basis $\mathcal{B}_Y$ as defined in Proposition 3.28. $\forall O \in \mathcal{O}$, $O = \bigcup_{l=1}^{\infty} O_{Bl}$, where $O_{Bl} \in \mathcal{B}_Y$, $\forall l \in \mathbb{N}$. $\forall l \in \mathbb{N}$, $O_{Bl} = \prod_{i \in N} B_{li}$, where $B_{li} \in \mathcal{B}_{Y i}$, $\forall i \in N$, $B_{li} = Y_i$, $\forall i \in N \setminus N_l$, and $N_l \subseteq N$ is a finite set. Then, $f_{\text{inv}}(O_{Bl}) = \bigcap_{i \in N} f_{i\,\text{inv}}(B_{li}) \in \mathcal{B}$ since f_i's are $\mathcal{B}$-measurable and $\mathcal{B}$ is a σ-algebra. Then, $f_{\text{inv}}(O) = f_{\text{inv}}(\bigcup_{l=1}^{\infty} O_{Bl}) = \bigcup_{l=1}^{\infty} f_{\text{inv}}(O_{Bl}) \in \mathcal{B}$. Hence, f is $\mathcal{B}$-measurable.

"Necessity" Let f be $\mathcal{B}$-measurable. $\forall i_0 \in N$, $\forall O_{i_0} \in \mathcal{O}_{i_0}$, let $O := \prod_{i \in N} O_i \in \mathcal{O}$, where $O_i = Y_i$, $\forall i \in N$ with $i \neq i_0$. By the measurability of f, we have $f_{i_0\,\text{inv}}(O_{i_0}) = f_{\text{inv}}(O) \in \mathcal{B}$. Hence, f_{i_0} is $\mathcal{B}$-measurable.

This completes the proof of the proposition. □

Proposition 11.40 *Let* $(X, \mathcal{B})$ *be a measurable space,* $D \in \mathcal{B}$, $f_n : D \to \mathbb{R}_e$ *be* $\mathcal{B}$*-measurable,* $\forall n \in \mathbb{N}$, $\bar{f}_S := \limsup_{n \in \mathbb{N}} f_n : D \to \mathbb{R}_e$ *defined by* $\bar{f}_S(x) =$

$\limsup_{n\in\mathbb{N}} f_n(x)$, $\forall x \in D$, *and* $\bar{f}_I := \liminf_{n\in\mathbb{N}} f_n : D \to \mathbb{R}_e$ *defined by* $\bar{f}_I(x) = \liminf_{n\in\mathbb{N}} f_n(x)$, $\forall x \in D$. *Then,* $\forall n \in \mathbb{Z}_+$, $\bigvee_{i=1}^n f_i$, $\bigwedge_{i=1}^n f_i$, $\sup_{i\in\mathbb{N}} f_i$, $\inf_{i\in\mathbb{N}} f_i$, $\limsup_{n\in\mathbb{N}} f_n = \bar{f}_S$, *and* $\liminf_{n\in\mathbb{N}} f_n = \bar{f}_I$ *are* $\mathcal{B}$*-measurable.*

Proof $\forall \alpha \in \mathbb{R}$, $\{x \in D \mid \bigvee_{i=1}^n f_i(x) > \alpha\} = \bigcup_{i=1}^n \{x \in D \mid f_i(x) > \alpha\} \in \mathcal{B}$. Hence, by Proposition 11.35, $\bigvee_{i=1}^n f_i$ is $\mathcal{B}$-measurable.

$\forall \alpha \in \mathbb{R}$, $\{x \in D \mid \sup_{i\in\mathbb{N}} f_i(x) > \alpha\} = \bigcup_{i=1}^\infty \{x \in D \mid f_i(x) > \alpha\} \in \mathcal{B}$. Hence, by Proposition 11.35, $\sup_{i\in\mathbb{N}} f_i$ is $\mathcal{B}$-measurable.

By arguments that are similar to the above, $\bigwedge_{i=1}^n f_i$ and $\inf_{i\in\mathbb{N}} f_i$ are $\mathcal{B}$-measurable.

By Definition 3.82, $\bar{f}_S = \inf_{i\in\mathbb{N}} \sup_{j\in\mathbb{N}, j\geq i} f_j$ and $\bar{f}_I = \sup_{i\in\mathbb{N}} \inf_{j\in\mathbb{N}, j\geq i} f_j$. Then, $\bar{f}_S$ and $\bar{f}_I$ are $\mathcal{B}$-measurable. This completes the proof of the proposition. □

Proposition 11.41 *Let* $(X, \mathcal{B})$ *be a measurable space,* $(E_i)_{i=1}^\infty \subseteq \mathcal{B}$ *be such that* $\bigcup_{i=1}^\infty E_i = X$, $\mathcal{Y} := (Y, \mathcal{O})$ *be a topological space,* $D \in \mathcal{B}$, $\mathcal{B}_{E_i} := \{B \in \mathcal{B} \mid B \subseteq E_i\}$, $\forall i \in \mathbb{N}$, *and* $f : D \to \mathcal{Y}$. *Then,* $f|_{E_i}$ *is* $\mathcal{B}_{E_i}$*-measurable,* $\forall i \in \mathbb{N}$, *if, and only if,* $f|_{E_i}$ *is* $\mathcal{B}$*-measurable,* $\forall i \in \mathbb{N}$, *if, and only if,* f *is* $\mathcal{B}$*-measurable.*

Proof (i) ⇒ (ii). Assume that $f|_{E_i}$ is $\mathcal{B}_{E_i}$-measurable, $\forall i \in \mathbb{N}$. $\forall O \in \mathcal{O}$, by the assumption, we have $(f|_{E_i})_{\text{inv}}(O) \in \mathcal{B}_{E_i} \subseteq \mathcal{B}$, $\forall i \in \mathbb{N}$. Then, $f|_{E_i}$ is $\mathcal{B}$-measurable.

(ii) ⇒ (iii). Assume that $f|_{E_i}$ is $\mathcal{B}$-measurable, $\forall i \in \mathbb{N}$. $\forall O \in \mathcal{O}$, by the assumption, we have $(f|_{E_i})_{\text{inv}}(O) \in \mathcal{B}$, $\forall i \in \mathbb{N}$. Then, $f_{\text{inv}}(O) = \bigcup_{i=1}^\infty (f|_{E_i})_{\text{inv}}(O) \in \mathcal{B}$. Hence, f is $\mathcal{B}$-measurable.

(iii) ⇒ (i). Assume that f is $\mathcal{B}$-measurable. $\forall O \in \mathcal{O}$, by the assumption, $f_{\text{inv}}(O) \in \mathcal{B}$. Then, $(f|_{E_i})_{\text{inv}}(O) = f_{\text{inv}}(O) \cap E_i \in \mathcal{B}_{E_i}$, $\forall i \in \mathbb{N}$. Hence, $f|_{E_i}$ is $\mathcal{B}_{E_i}$-measurable, $\forall i \in \mathbb{N}$. This completes the proof of the proposition. □

Definition 11.42 Let $\mathcal{X} := (X, \mathcal{B}, \mu)$ be a measure space. A property P is said to hold *almost everywhere in* $\mathcal{X}$ (abbreviated a.e.) if the set of points where it fails to hold or does not make sense belongs to $\mathcal{B}$ and have measure 0. We will write P a.e. in $\mathcal{X}$ or $P(x)$ a.e. $x \in \mathcal{X}$. ◇

Lemma 11.43 *Let* $\mathcal{X} := (X, \mathcal{B})$ *be a measurable space,* $\mathcal{Y} := (Y, \rho)$ *be a separable metric space,* $D_1, D_2 \in \mathcal{B}$, *and* $f : D_1 \to \mathcal{Y}$ *and* $g : D_2 \to \mathcal{Y}$ *be* $\mathcal{B}$*-measurable. Then, the function* $h : D_1 \cap D_2 \to \mathbb{R}$, *defined by* $h(x) = \rho(f(x), g(x))$, $\forall x \in D_1 \cap D_2$, *is* $\mathcal{B}$*-measurable.*

Proof By Proposition 4.4, $\mathcal{Y}$ is second countable. By Propositions 11.39 and 11.41, the function $h_1 : D_1 \cap D_2 \to \mathcal{Y} \times \mathcal{Y}$, defined by $h_1(x) = (f(x), g(x))$, $\forall x \in D_1 \cap D_2$, is $\mathcal{B}$-measurable. By Propositions 11.38 and 4.30, the function h is $\mathcal{B}$-measurable. □

Lemma 11.44 *Let* $\mathcal{X} := (X, \mathcal{B}, \mu)$ *be a measure space,* $\mathcal{Y} := (Y, \rho)$ *be a separable metric space,* $D_i \in \mathcal{B}$, *and* $f_i : D_i \to \mathcal{Y}$ *be* $\mathcal{B}$*-measurable,* $i = 1, 2, 3$. *Assume that* $f_1 = f_2$ *a.e. in* $\mathcal{X}$ *and* $f_2 = f_3$ *a.e. in* $\mathcal{X}$. *Then,* $f_1 = f_3$ *a.e. in* $\mathcal{X}$.

Proof Let $E_1 := \widetilde{D_1} \cup \widetilde{D_2} \cup \{x \in D_1 \cap D_2 \mid f_1(x) \neq f_2(x)\}$, $E_2 := \widetilde{D_2} \cup \widetilde{D_3} \cup \{x \in D_2 \cap D_3 \mid f_2(x) \neq f_3(x)\}$, and $E_3 := \widetilde{D_1} \cup \widetilde{D_3} \cup \{x \in D_1 \cap D_3 \mid f_3(x) \neq f_1(x)\}$. By the assumption, we have $E_1, E_2 \in \mathcal{B}$ and $\mu(E_1) = \mu(E_2) = 0$. Clearly, $E_3 \subseteq E_1 \cup E_2$. Note that $E_3 = \widetilde{D_1} \cup \widetilde{D_3} \cup \{x \in D_1 \cap D_3 \mid \rho(f_1(x), f_3(x)) > 0\} \in \mathcal{B}$ by Lemma 11.43. Then, we have $0 \leq \mu(E_3) \leq \mu(E_1) + \mu(E_2) = 0$. Hence, $f_1 = f_3$ a.e. in $\mathcal{X}$. This completes the proof of the lemma. □

Lemma 11.45 *Let $\mathcal{X} := (X, \mathcal{B}, \mu)$ be a measure space, $\mathcal{Y}_i := (Y_i, \mathcal{O}_i)$ be a second countable metrizable topological space, $\forall i \in N \subseteq \mathbb{N}$, where N is a countable index set, $\mathcal{Y} := \prod_{i \in N} \mathcal{Y}_i$ be the product space, $D_1, D_2 \in \mathcal{B}$, $f_i : D_1 \to \mathcal{Y}_i$ and $g_i : D_2 \to \mathcal{Y}_i$ be $\mathcal{B}$-measurable, $\forall i \in N$, $f : D_1 \to \mathcal{Y}$ be defined by $\pi_i(f(x)) = f_i(x)$, $\forall i \in N$, $\forall x \in D_1$, and $g : D_2 \to \mathcal{Y}$ be defined by $\pi_i(g(x)) = g_i(x)$, $\forall i \in N$, $\forall x \in D_2$. Then, $f = g$ a.e. in $\mathcal{X}$ if, and only if, $f_i = g_i$ a.e. in $\mathcal{X}$, $\forall i \in N$.*

Proof Define $E_i := \widetilde{D_1} \cup \widetilde{D_2} \cup \{x \in D_1 \cap D_2 \mid f_i(x) \neq g_i(x)\}$, $\forall i \in N$, and $E := \widetilde{D_1} \cup \widetilde{D_2} \cup \{x \in D_1 \cap D_2 \mid f(x) \neq g(x)\}$.

"Sufficiency" Let $f_i = g_i$ a.e. in $\mathcal{X}$, $\forall i \in N$. Then, $E_i \in \mathcal{B}$ and $\mu(E_i) = 0$, $\forall i \in N$. Note that $E = \bigcup_{i \in N} E_i \in \mathcal{B}$ and $\mu(E) = 0$. Hence, $f = g$ a.e. in $\mathcal{X}$.

"Necessity" Let $f = g$ a.e. in $\mathcal{X}$. Then, $E \in \mathcal{B}$ and $\mu(E) = 0$. $\forall i \in N$, $E_i \subseteq E$. Since $\mathcal{Y}_i$ is a second countable metrizable topological space, let $\rho_i : Y_i \times Y_i \to [0, \infty) \subset \mathbb{R}$ be the metric on $\mathcal{Y}_i$ whose natural topology is $\mathcal{O}_i$. Then, (Y_i, ρ_i) is a separable metric space, by Proposition 4.4. By Lemma 11.43, we have $E_i = \widetilde{D_1} \cup \widetilde{D_2} \cup \{x \in D_1 \cap D_2 \mid \rho_i(f_i(x), g_i(x)) > 0\} \in \mathcal{B}$. Then, $0 \leq \mu(E_i) \leq \mu(E) = 0$. Hence, $f_i = g_i$ a.e. in $\mathcal{X}$.

This completes the proof of the lemma. □

Lemma 11.46 *Let $\mathcal{X} := (X, \mathcal{B}, \mu)$ be a measure space, $\mathcal{Y}$ be a topological space, $\mathcal{Z} := (Z, \rho)$ be a separable metric space, $D_1, D_2 \in \mathcal{B}$, $f_i : D_i \to \mathcal{Y}$ be $\mathcal{B}$-measurable, $i = 1, 2$, and $g : \mathcal{Y} \to \mathcal{Z}$ be continuous. Assume that $f_1 = f_2$ a.e. in $\mathcal{X}$. Then, $g \circ f_1 = g \circ f_2$ a.e. in $\mathcal{X}$.*

Proof By the assumption, $E := \widetilde{D_1} \cup \widetilde{D_2} \cup \{x \in D_1 \cap D_2 \mid f_1(x) \neq f_2(x)\} \in \mathcal{B}$ and $\mu(E) = 0$. By Proposition 11.38, we have $g \circ f_i$ is $\mathcal{B}$-measurable, $i = 1, 2$. By Lemma 11.43, $E \supseteq \bar{E} := \widetilde{D_1} \cup \widetilde{D_2} \cup \{x \in D_1 \cap D_2 \mid g(f_1(x)) \neq g(f_2(x))\} = \widetilde{D_1} \cup \widetilde{D_2} \cup \{x \in D_1 \cap D_2 \mid \rho(g(f_1(x)), g(f_2(x))) > 0\} \in \mathcal{B}$. Then, $0 \leq \mu(\bar{E}) \leq \mu(E) = 0$. Hence, $g \circ f_1 = g \circ f_2$ a.e. in $\mathcal{X}$. This completes the proof of the lemma. □

Proposition 11.47 *Let $\mathcal{X} := (X, \mathcal{B}, \mu)$ be a complete measure space, $\mathcal{Y}$ be a topological space, $D_1, D_2 \in {}^X2$, $f : D_1 \to \mathcal{Y}$ be $\mathcal{B}$-measurable, $g : D_2 \to \mathcal{Y}$, and $g = f$ a.e. in $\mathcal{X}$. Then, g is $\mathcal{B}$-measurable.*

Proof Let $E := \widetilde{D_1} \cup \widetilde{D_2} \cup \{x \in D_1 \cap D_2 \mid f(x) \neq g(x)\}$. Then, $E \in \mathcal{B}$ and $\mu(E) = 0$. $\forall O \in \mathcal{O}_{\mathcal{Y}}$, $g_{\text{inv}}(O) = g_{\text{inv}}(O) \cap (E \cup (X \setminus E)) = (g_{\text{inv}}(O) \cap E) \cup (g_{\text{inv}}(O) \cap (X \setminus E)) = (g_{\text{inv}}(O) \cap E) \cup (f_{\text{inv}}(O) \cap (X \setminus E))$. Note that $X \setminus E, f_{\text{inv}}(O) \in \mathcal{B}$. By the completeness of $\mathcal{X}$, $E \supseteq g_{\text{inv}}(O) \cap E \in \mathcal{B}$. Then, $g_{\text{inv}}(O) \in \mathcal{B}$. Hence, g is $\mathcal{B}$-measurable. □

Proposition 11.48 *Let $\mathcal{X} := (X, \mathcal{B}, \mu)$ be a measure space, $\mathcal{Y} := (Y, \rho)$ be a metric space, $D \in \mathcal{B}$, $f_n : D \to \mathcal{Y}$ be $\mathcal{B}$-measurable, $\forall n \in \mathbb{N}$, and $f : D \to \mathcal{Y}$. Assume that* $\lim_{n\in\mathbb{N}} f_n(x) = f(x)$, $\forall x \in D$. *Then, f is $\mathcal{B}$-measurable.*

Proof Fix any open set $O \subseteq \mathcal{Y}$. $\forall n \in \mathbb{N}$, define $O_n := \{y \in O \mid \mathcal{B}_{\mathcal{Y}}(y, 1/n) \subseteq O\}$. Then, $O_n = O \cap \left(\bigcap_{y\in\mathcal{Y}\setminus O}(\mathcal{Y} \setminus \mathcal{B}_{\mathcal{Y}}(y, 1/n))\right)$. Note that $\bigcap_{y\in\mathcal{Y}\setminus O}(\mathcal{Y}\setminus\mathcal{B}_{\mathcal{Y}}(y, 1/n))$ is a closed set in $\mathcal{Y}$. Then, $O_n \in \mathcal{B}_{\mathrm{B}}(\mathcal{Y})$.

Claim 11.48.1 $f_{\mathrm{inv}}(O) = \bigcup_{l=1}^{\infty} \bigcap_{k=l}^{\infty} \bigcup_{n=1}^{\infty} \bigcap_{j=n}^{\infty} f_{j\,\mathrm{inv}}(O_k) =: U$.

Proof of Claim $\forall x \in f_{\mathrm{inv}}(O) \subseteq D$, then $y := f(x) \in O$. Since O is open, then $\exists L \in \mathbb{N}$ such that $\mathcal{B}_{\mathcal{Y}}(y, 1/L) \subseteq O$. Then, $\forall l \geq L + 1$, $\forall \bar{y} \in \mathcal{B}_{\mathcal{Y}}\left(y, \frac{1}{L(L+1)}\right)$, $\forall \tilde{y} \in \mathcal{B}_{\mathcal{Y}}(\bar{y}, 1/l)$, we have $\rho(y, \tilde{y}) \leq \rho(y, \bar{y}) + \rho(\bar{y}, \tilde{y}) < \frac{1}{L(L+1)} + \frac{1}{l} \leq \frac{1}{L}$. This implies that $\tilde{y} \in \mathcal{B}_{\mathcal{Y}}(y, 1/L) \subseteq O$. Then, $\mathcal{B}_{\mathcal{Y}}(\bar{y}, 1/l) \subseteq O$. Thus, $\bar{y} \in O_l$ and $\mathcal{B}_{\mathcal{Y}}\left(y, \frac{1}{L(L+1)}\right) \subseteq O_l$. Since $\lim_{n\in\mathbb{N}} f_n(x) = f(x) = y$, then $\exists N \in \mathbb{N}$ such that $\forall n \geq N$, $\rho(f_n(x), y) < \frac{1}{L(L+1)}$. Then, $f_n(x) \in \mathcal{B}_{\mathcal{Y}}\left(y, \frac{1}{L(L+1)}\right) \subseteq O_l$ and $x \in f_{n\,\mathrm{inv}}(O_l)$. By arbitrariness of n, $x \in \bigcap_{j=N}^{\infty} f_{j\,\mathrm{inv}}(O_l) \subseteq \bigcup_{n=1}^{\infty} \bigcap_{j=n}^{\infty} f_{j\,\mathrm{inv}}(O_l)$. By arbitrariness of l, we have $x \in \bigcap_{k=L+1}^{\infty} \bigcup_{n=1}^{\infty} \bigcap_{j=n}^{\infty} f_{j\,\mathrm{inv}}(O_k) \subseteq U$. Hence, by the arbitrariness of x, $f_{\mathrm{inv}}(O) \subseteq U$.

On the other hand, $\forall x \in U$, $\exists L \in \mathbb{N}$ such that $x \in \bigcup_{n=1}^{\infty} \bigcap_{j=n}^{\infty} f_{j\,\mathrm{inv}}(O_L)$. Then, $\exists N \in \mathbb{N}$ such that $\forall n \geq N$, $x \in f_{n\,\mathrm{inv}}(O_L)$ and $f_n(x) \in O_L$. Since $\lim_{n\in\mathbb{N}} f_n(x) = f(x)$, then $\exists \bar{N} \in \mathbb{N}$ with $\bar{N} \geq N$ such that $\rho(f_{\bar{N}}(x), f(x)) < 1/L$. Then, $f(x) \in \mathcal{B}_{\mathcal{Y}}(f_{\bar{N}}(x), 1/L)$. Note that $f_{\bar{N}}(x) \in O_L$ and $\mathcal{B}_{\mathcal{Y}}(f_{\bar{N}}(x), 1/L) \subseteq O$. Then, $f(x) \in O$ and $x \in f_{\mathrm{inv}}(O)$. By the arbitrariness of x, $U \subseteq f_{\mathrm{inv}}(O)$.

Hence, $f_{\mathrm{inv}}(O) = U$. This completes the proof of the claim. □

$\forall k \in \mathbb{N}$, $\forall j \in \mathbb{N}$, by Proposition 11.34, $f_{j\,\mathrm{inv}}(O_k) \in \mathcal{B}$. By Claim 11.48.1, $f_{\mathrm{inv}}(O) \in \mathcal{B}$. Hence, f is $\mathcal{B}$-measurable. This completes the proof of the proposition. □

Proposition 11.49 *Let $\mathcal{X} := (X, \mathcal{B}, \mu)$ be a complete measure space, $\mathcal{Y} := (Y, \rho)$ be a metric space, $D_1, D_2 \in {}^{X}2$, $f_n : D_1 \to \mathcal{Y}$ be $\mathcal{B}$-measurable, $\forall n \in \mathbb{N}$, and $f : D_2 \to \mathcal{Y}$. Assume that* $\lim_{n\in\mathbb{N}} f_n = f$ *a.e. in $\mathcal{X}$. Then, f is $\mathcal{B}$-measurable.*

Proof Let $E := \{x \in D_1 \cap D_2 \mid (f_n(x))_{n=1}^{\infty}$ does not converge to $f(x)\} \cup \widetilde{D_1} \cup \widetilde{D_2}$. By the assumption, $E \in \mathcal{B}$ and $\mu(E) = 0$. Let $g := f|_{X\setminus E}$ and $g_n := f_n|_{X\setminus E}$, $\forall n \in \mathbb{N}$. Then, $\lim_{n\in\mathbb{N}} g_n(x) = g(x)$, $\forall x \in X \setminus E \in \mathcal{B}$. By Proposition 11.48, g is $\mathcal{B}$-measurable. $\forall$ open set $O \subseteq \mathcal{Y}$, $f_{\mathrm{inv}}(O) = g_{\mathrm{inv}}(O) \cup (f_{\mathrm{inv}}(O) \cap E)$. By the completeness of $\mathcal{X}$, we have $E \supseteq f_{\mathrm{inv}}(O) \cap E \in \mathcal{B}$. Then, $f_{\mathrm{inv}}(O) \in \mathcal{B}$. Hence, f is $\mathcal{B}$-measurable. This completes the proof of the proposition. □

Proposition 11.50 *Let $(X, \mathcal{B})$ be a measurable space, $\mathcal{Y} := (Y, \rho)$ be a separable metric space, $D_1, D_2 \in \mathcal{B}$, $f_n : D_1 \to \mathcal{Y}$ be $\mathcal{B}$-measurable, $\forall n \in \mathbb{N}$, and $f : D_2 \to \mathcal{Y}$ be $\mathcal{B}$-measurable. Then, the set*

$$E := \left\{x \in D_1 \cap D_2 \,\middle|\, \left(f_n(x)\right)_{n=1}^{\infty} \text{ converges to } f(x)\right\} \in \mathcal{B}.$$

Proof We will show that

$$E = \bigcap_{m=1}^{\infty} \bigcup_{i=1}^{\infty} \bigcap_{k=i}^{\infty} E_{k,m} =: \bar{E}$$

where $E_{k,m} := \{x \in D_1 \cap D_2 \mid \rho(f_k(x), f(x)) < 1/m\} \in \mathcal{B}$, $\forall m \in \mathbb{N}$, $\forall k \in \mathbb{N}$, by Lemma 11.43. This further implies that $\bar{E} \in \mathcal{B}$.

$\forall x \in E$, $\forall m \in \mathbb{N}$, $\exists k_0 \in \mathbb{N}$, $\forall k \in \mathbb{N}$ with $k \geq k_0$, $x \in D_1 \cap D_2$, and $\rho(f_k(x), f(x)) < 1/m$. Then, $x \in E_{k,m}$ and $x \in \bigcap_{k=k_0}^{\infty} E_{k,m} \subseteq \bigcup_{i=1}^{\infty} \bigcap_{k=i}^{\infty} E_{k,m}$. By the arbitrariness of m, we have $x \in \bar{E}$. This implies that $E \subseteq \bar{E}$. On the other hand, $\forall x \in \bar{E}$, $\forall m \in \mathbb{N}$, $x \in \bigcup_{i=1}^{\infty} \bigcap_{k=i}^{\infty} E_{k,m}$. Then, $\exists k_0 \in \mathbb{N}$, such that $x \in \bigcap_{k=k_0}^{\infty} E_{k,m}$. $\forall k \in \mathbb{N}$ with $k \geq k_0$, $x \in E_{k,m}$. Then, $x \in D_1 \cap D_2$ and $\rho(f_k(x), f(x)) < 1/m$. This leads to $\lim_{n \in \mathbb{N}} f_n(x) = f(x)$ and $x \in E$. Then, we have $\bar{E} \subseteq E$. Hence, $E = \bar{E} \in \mathcal{B}$. This completes the proof of the proposition. □

Proposition 11.51 *Let $\mathcal{X} := (X, \mathcal{B}, \mu)$ be a measure space, $\mathcal{Y} := (Y, \rho)$ be a separable complete metric space, $D \in \mathcal{B}$, $f_i : D \to \mathcal{Y}$ be $\mathcal{B}$-measurable, $\forall i \in \mathbb{N}$, $E := \{x \in D \mid (f_i(x))_{i=1}^{\infty}$ converges in $\mathcal{Y}\}$, and $f : E \to \mathcal{Y}$ defined by $f(x) = \lim_{i \in \mathbb{N}} f_i(x)$, $\forall x \in E$. Then, $E = \bigcap_{j=1}^{\infty} \bigcup_{m=1}^{\infty} \bigcap_{l=m}^{\infty} \bigcap_{n=m}^{\infty} E_{j,l,n} =: \bar{E} \in \mathcal{B}$, where $E_{j,l,n} := \{x \in D \mid \rho(f_l(x), f_n(x)) < 1/j\}$, $\forall j, l, n \in \mathbb{N}$, and f is $\mathcal{B}$-measurable.*

Proof $\forall j, l, n \in \mathbb{N}$, by Lemma 11.43 and the separability of $\mathcal{Y}$, $E_{j,l,n} \in \mathcal{B}$. Then, $\bar{E} \in \mathcal{B}$.

$\forall x \in \bar{E}$, $\forall j \in \mathbb{N}$, $\exists m_0 \in \mathbb{N}$, $\forall l, n \in \mathbb{N}$ with $l \geq m_0$ and $n \geq m_0$, we have $x \in D$ and $\rho(f_l(x), f_n(x)) < 1/j$. Then, $(f_i(x))_{i=1}^{\infty} \subseteq \mathcal{Y}$ is a Cauchy sequence. By the completeness of $\mathcal{Y}$, $(f_i(x))_{i=1}^{\infty}$ converges in $\mathcal{Y}$. Hence, $x \in E$. By the arbitrariness of x, we have $\bar{E} \subseteq E$.

On the other hand, $\forall x \in E$, since $(f_i(x))_{i=1}^{\infty}$ converges in $\mathcal{Y}$, then it is a Cauchy sequence. $\forall j \in \mathbb{N}$, $\exists m_0 \in \mathbb{N}$, $\forall l, n \in \mathbb{N}$ with $l \geq m_0$ and $n \geq m_0$, we have $\rho(f_l(x), f_n(x)) < 1/j$. Then, $x \in E_{j,l,n}$. Then, $x \in \bar{E}$. By the arbitrariness of x, we have $E \subseteq \bar{E}$. Therefore, $E = \bar{E} \in \mathcal{B}$.

Note that $f(x) = \lim_{i \in \mathbb{N}} f_i|_E(x)$, $\forall x \in E$. By Proposition 11.41, $f_i|_E$ is $\mathcal{B}$-measurable, $\forall i \in \mathbb{N}$. By Proposition 11.48, f is $\mathcal{B}$-measurable.

This completes the proof of the proposition. □

Proposition 11.52 *Let $\mathcal{X} := (X, \mathcal{B}, \mu)$ be a measure space, $\mathcal{Y} := (Y, \mathcal{O})$ be a Hausdorff topological space, $\mathcal{Z} := (Z, \rho)$ be a separable metric space, $D_1, D_2 \in \mathcal{B}$, $f_n : D_1 \to \mathcal{Y}$ be $\mathcal{B}$-measurable, $\forall n \in \mathbb{N}$, $f : D_2 \to \mathcal{Y}$ be $\mathcal{B}$-measurable, and $g : \mathcal{Y} \to \mathcal{Z}$ be continuous. Assume that $\lim_{n \in \mathbb{N}} f_n = f$ a.e. in $\mathcal{X}$. Then, $\lim_{n \in \mathbb{N}} g \circ f_n = g \circ f$ a.e. in $\mathcal{X}$.*

Proof Let $E := \{x \in D_1 \cap D_2 \mid \lim_{n \in \mathbb{N}} f_n(x) = f(x)\}$ and $\bar{E} := \{x \in D_1 \cap D_2 \mid \lim_{n \in \mathbb{N}} g(f_n(x)) = g(f(x))\}$. By Propositions 3.65 and 3.66, we have $E \subseteq \bar{E}$. By the assumption, $E \in \mathcal{B}$ and $\mu(\widetilde{E}) = 0$. By Proposition 11.38, $g \circ f_n$ and $g \circ f$ are $\mathcal{B}$-measurable, $\forall n \in \mathbb{N}$. By Proposition 11.50, $\bar{E} \in \mathcal{B}$. Then, $0 \leq \mu(\tilde{\bar{E}}) \leq$

$\mu(\widetilde{E}) = 0$. Hence, $\lim_{n\in\mathbb{N}} g \circ f_n = g \circ f$ a.e. in $\mathcal{X}$. This completes the proof of the proposition. □

Proposition 11.53 *Let $\mathcal{X} := (X, \mathcal{B}, \mu)$ be a measure space, $D_1, D_2 \in \mathcal{B}$, $\mathcal{Y}_i := (Y_i, \mathcal{O}_i)$ be a second countable metrizable topological space, $f_i : D_1 \to \mathcal{Y}_i$ be $\mathcal{B}$-measurable, $g_{n,i} : D_2 \to \mathcal{Y}_i$ be $\mathcal{B}$-measurable, $\forall n \in \mathbb{N}$, $\forall i \in N \subseteq \mathbb{N}$, N is a countable index set, $\mathcal{Y} := (Y, \mathcal{O}) := \prod_{i\in N} \mathcal{Y}_i$ be the product topological space, $f : D_1 \to \mathcal{Y}$ be given by $\pi_i(f(x)) = f_i(x)$, $\forall x \in D_1$, $\forall i \in N$, and $g_n : D_2 \to \mathcal{Y}$ be given by $\pi_i(g_n(x)) = g_{n,i}(x)$, $\forall x \in D_2$, $\forall i \in N$, $\forall n \in \mathbb{N}$. Then, $\lim_{n\in\mathbb{N}} g_n = f$ a.e. in $\mathcal{X}$ if, and only if, $\lim_{n\in\mathbb{N}} g_{n,i} = f_i$ a.e. in $\mathcal{X}$, $\forall i \in N$.*

Proof By Proposition 11.39, f and g_n are $\mathcal{B}$-measurable, $\forall n \in \mathbb{N}$. "Sufficiency" Assume $\lim_{n\in\mathbb{N}} g_{n,i} = f_i$ a.e. in $\mathcal{X}$, $\forall i \in N$. $\forall i \in N$, let $E_i := \widetilde{D_1} \cup \widetilde{D_2} \cup \{x \in D_1 \cap D_2 \mid \left(g_{n,i}(x)\right)_{n=1}^{\infty}$ does not converge to $f_i(x)\}$. Then, $E_i \in \mathcal{B}$ and $\mu(E_i) = 0$. Let $E := \widetilde{D_1} \cup \widetilde{D_2} \cup \{x \in D_1 \cap D_2 \mid (g_n(x))_{n=1}^{\infty}$ does notconverge to $f(x)\}$. By Proposition 3.67, $E = \bigcup_{i\in N} E_i \in \mathcal{B}$. Then, $0 \leq \mu(E) \leq \sum_{i\in N} \mu(E_i) = 0$. Hence, $\lim_{n\in\mathbb{N}} g_n = f$ a.e. in $\mathcal{X}$. "Necessity" Assume $\lim_{n\in\mathbb{N}} g_n = f$ a.e. in $\mathcal{X}$. Let E and E_i be as defined above, $\forall i \in \mathbb{N}$. Then, $E \in \mathcal{B}$ and $\mu(E) = 0$. $\forall i \in N$, by Proposition 3.67, we have $E_i \subseteq E$. Since $\mathcal{Y}_i$ is second countable metrizable topological space, let $\rho_i : Y_i \times Y_i \to [0, \infty) \subset \mathbb{R}$ be the metric on $\mathcal{Y}_i$ whose natural topology is $\mathcal{O}_i$. Then, by Proposition 4.4, (Y_i, ρ_i) is a separable metric space. By Proposition 11.50, we have $E_i \in \mathcal{B}$. Then, $0 \leq \mu(E_i) \leq \mu(E) = 0$. Hence, $\lim_{n\in\mathbb{N}} g_{n,i} = f_i$ a.e. in $\mathcal{X}$. This completes the proof of the proposition. □

Proposition 11.54 *Let $\mathcal{X} := (X, \mathcal{B}, \mu)$ be a measure space, $\mathcal{Y}$ be a separable metric space, $D_i \in \mathcal{B}$, $i = 1, 2, 3, 4$, $f_n : D_1 \to \mathcal{Y}$ and $g_n : D_2 \to \mathcal{Y}$ be $\mathcal{B}$-measurable, $\forall n \in \mathbb{N}$, and $f : D_3 \to \mathcal{Y}$ and $g : D_4 \to \mathcal{Y}$ be $\mathcal{B}$-measurable. Assume that $\lim_{n\in\mathbb{N}} f_n = f$ a.e. in $\mathcal{X}$, $f_n = g_n$ a.e. in $\mathcal{X}$, $\forall n \in \mathbb{N}$, and $f = g$ a.e. in $\mathcal{X}$. Then, $\lim_{n\in\mathbb{N}} g_n = g$ a.e. in $\mathcal{X}$.*

Proof By the assumption, we have $E_1 := \widetilde{D_1} \cup \widetilde{D_3} \cup \{x \in D_1 \cap D_3 \mid (f_n(x))_{n=1}^{\infty}$ does not converge to $f(x)\} \in \mathcal{B}$; $\bar{E}_n := \widetilde{D_1} \cup \widetilde{D_2} \cup \{x \in D_1 \cap D_2 \mid f_n(x) \neq g_n(x)\} \in \mathcal{B}$, $\forall n \in \mathbb{N}$; $\bar{E} := \widetilde{D_3} \cup \widetilde{D_4} \cup \{x \in D_3 \cap D_4 \mid f(x) \neq g(x)\} \in \mathcal{B}$ and $\mu(E_1) = \mu(\bar{E}_n) = \mu(\bar{E}) = 0$, $\forall n \in \mathbb{N}$. Then, $E := E_1 \cup (\bigcup_{n=1}^{\infty} \bar{E}_n) \cup \bar{E} \in \mathcal{B}$ and $\mu(E) = 0$. Let $E_2 := \widetilde{D_2} \cup \widetilde{D_4} \cup \{x \in D_2 \cap D_4 \mid (g_n(x))_{n=1}^{\infty}$ does not converge to $g(x)\}$. By Proposition 11.50, we have $E_2 \in \mathcal{B}$. Clearly, $E \subseteq E_2$ and $E_2 \subseteq E$. Then, $0 \leq \mu(E_2) \leq \mu(E) = 0$. This shows that $\lim_{n\in\mathbb{N}} g_n = g$ a.e. in $\mathcal{X}$. This completes the proof of the proposition. □

Theorem 11.55 (Egoroff's Theorem) *Let $\mathcal{X} := (X, \mathcal{B}, \mu)$ be a finite measure space, $\mathcal{Y} := (Y, \rho)$ be a separable metric space, $D_1, D_2 \in \mathcal{B}$, $f_n : D_1 \to \mathcal{Y}$ be $\mathcal{B}$-measurable, $\forall n \in \mathbb{N}$, and $f : D_2 \to \mathcal{Y}$ be $\mathcal{B}$-measurable. Assume that $\lim_{n\in\mathbb{N}} f_n = f$ a.e. in $\mathcal{X}$. Then, $\forall \eta \in (0, \infty) \subset \mathbb{R}$, $\exists A \in \mathcal{B}$ with $\mu(A) < \eta$ and $A \supseteq \widetilde{D_1} \cup \widetilde{D_2}$ such that $\left(f_n|_{X\setminus A}\right)_{n=1}^{\infty}$ converges uniformly to $f|_{X\setminus A}$.*

Proof Let $E := \{x \in D_1 \cap D_2 \mid (f_n(x))_{n=1}^{\infty}$ does not converge to $f(x)\} \cup \widetilde{D_1} \cup \widetilde{D_2}$. By the assumption, $E \in \mathcal{B}$ and $\mu(E) = 0$.

$\forall \eta \in (0, \infty) \subset \mathbb{R}$, $\forall n \in \mathbb{N}$, $\forall m \in \mathbb{N}$, let

$$G_{m,n} := \{x \in X \setminus E \mid \rho(f_m(x), f(x)) \geq 2^{-n}\}$$

By Lemma 11.43, $G_{m,n} \in \mathcal{B}$. Let

$$H_{m,n} := \bigcup_{j=m}^{\infty} G_{j,n} = \{x \in X \setminus E \mid \rho(f_i(x), f(x)) \geq 2^{-n} \text{ for some } i \geq m\}$$

Then, $H_{m+1,n} \subseteq H_{m,n} \in \mathcal{B}$. $\forall x \in X \setminus E$, we have $\lim_{j \in \mathbb{N}} f_j(x) = f(x)$; then $\exists m_x \in \mathbb{N}$ such that $x \notin H_{m_x,n}$. Hence, $\bigcap_{m=1}^{\infty} H_{m,n} = \emptyset$. Note that $\mu(H_{1,n}) \leq \mu(X) < \infty$. By Proposition 11.5, $\lim_{m \in \mathbb{N}} \mu(H_{m,n}) = \mu(\bigcap_{m=1}^{\infty} H_{m,n}) = 0$. Then, $\exists m_n \in \mathbb{N}$ such that $\mu(H_{m_n,n}) < 2^{-n}\eta$. Let $A_n := H_{m_n,n} \in \mathcal{B}$. Then, $\mu(A_n) < 2^{-n}\eta$, and $\forall i \geq m_n$, $\forall x \in (X \setminus E) \setminus A_n$, we have $\rho(f_i(x), f(x)) < 2^{-n}$.

Let $A := E \cup (\bigcup_{n=1}^{\infty} A_n) \in \mathcal{B}$. Clearly, $A \supseteq \widetilde{D_1} \cup \widetilde{D_2}$. By countable subadditivity of measure, we have $\mu(A) \leq \mu(E) + \sum_{n=1}^{\infty} \mu(A_n) < \eta$. $\forall \epsilon \in (0, \infty) \subset \mathbb{R}$, $\exists n_0 \in \mathbb{N}$ with $0 < 2^{-n_0} \leq \epsilon$. $\forall j \in \mathbb{N}$ with $m_{n_0} \leq j$, $\forall x \in X \setminus A \subseteq (X \setminus E) \setminus A_{n_0}$, we have $\rho(f_j(x), f(x)) < 2^{-n_0} \leq \epsilon$. Hence, $\left(f_n|_{X \setminus A}\right)_{n=1}^{\infty}$ converges uniformly to $f|_{X \setminus A}$. This completes the proof of the proposition. □

Definition 11.56 Let $\mathcal{X} := (X, \mathcal{B}, \mu)$ be a measure space, $\mathcal{Y} := (Y, \rho)$ be a separable metric space, $D_n \in \mathcal{B}$, $f_n : D_n \to \mathcal{Y}$ be $\mathcal{B}$-measurable, $\forall n \in \mathbb{N}$, $D \in \mathcal{B}$, and $f : D \to \mathcal{Y}$ be $\mathcal{B}$-measurable. We will say that $(f_n)_{n=1}^{\infty}$ *converges to* f *in measure in* $\mathcal{X}$ if $\forall \epsilon \in (0, \infty) \subset \mathbb{R}$, $\exists N \in \mathbb{N}$ such that $\forall n \in \mathbb{N}$ with $n \geq N$, we have $\mu(A_{n,\epsilon}) := \mu(\widetilde{D_n} \cup \widetilde{D} \cup \{x \in D_n \cap D \mid \rho(f_n(x), f(x)) \geq \epsilon\}) < \epsilon$. In this case, we will write $\lim_{n \in \mathbb{N}} f_n = f$ in measure in $\mathcal{X}$. ◇

Note that in the above definition, $\forall \epsilon \in (0, \infty) \subset \mathbb{R}$, $\forall n \in \mathbb{N}$, $A_{n,\epsilon} \in \mathcal{B}$ by Lemma 11.43. Hence, the definition is well-defined.

Proposition 11.57 *Let $\mathcal{X} := (X, \mathcal{B}, \mu)$ be a measure space, $\mathcal{Y} := (Y, \rho)$ be a separable metric space, $D_1, D_2 \in \mathcal{B}$, $f_n : D_1 \to \mathcal{Y}$ be $\mathcal{B}$-measurable, $\forall n \in \mathbb{N}$, and $f : D_2 \to \mathcal{Y}$ be $\mathcal{B}$-measurable. Assume that* $\lim_{n \in \mathbb{N}} f_n = f$ *in measure in $\mathcal{X}$. Then, there exists a subsequence $\left(f_{n_k}\right)_{k=1}^{\infty}$ of $(f_n)_{n=1}^{\infty}$ such that* $\lim_{k \in \mathbb{N}} f_{n_k} = f$ *a.e. in $\mathcal{X}$.*

Proof Define $A_{n,k} := \widetilde{D_1} \cup \widetilde{D_2} \cup \{x \in D_1 \cap D_2 \mid \rho(f_n(x), f(x)) \geq 2^{-k}\}$, $\forall n, k \in \mathbb{N}$. By Lemma 11.43, $A_{n,k} \in \mathcal{B}$. By the assumption, $\exists n_1 \in \mathbb{N}$ such that $\forall n \in \mathbb{N}$ with $n \geq n_1$, we have $\mu(A_{n,1}) < 2^{-1}$. $\forall k \in \mathbb{N}$ with $k > 1$, by the assumption, $\exists n_k \in \mathbb{N}$ with $n_k > n_{k-1}$ such that $\forall n \in \mathbb{N}$ with $n \geq n_k$, we have $\mu(A_{n,k}) < 2^{-k}$. This defines a subsequence $\left(f_{n_k}\right)_{k=1}^{\infty}$.

Let $A := \bigcap_{i=1}^{\infty} \bigcup_{k=i}^{\infty} A_{n_k,k} \in \mathcal{B}$. Then, $\mu(A) \leq \mu(\bigcup_{k=i}^{\infty} A_{n_k,k}) \leq \sum_{k=i}^{\infty} \mu(A_{n_k,k}) < 2^{-i+1}$, $\forall i \in \mathbb{N}$. Hence, $\mu(A) = 0$. $\forall x \in \widetilde{X \setminus A} = \bigcup_{i=1}^{\infty} \bigcap_{k=i}^{\infty} \widetilde{A_{n_k,k}}$, $\exists i_0 \in \mathbb{N}$ such that $\forall k \in \mathbb{N}$ with $k \geq i_0$, $x \in \widetilde{A_{n_k,k}}$. Then, we have $x \in D_1 \cap D_2$ and $\rho(f_{n_k}(x), f(x)) < 2^{-k}$. Hence, $\lim_{k \in \mathbb{N}} f_{n_k}(x) = f(x)$. Then, $E := \{x \in D_1 \cap D_2 \mid \left(f_{n_k}(x)\right)_{k=1}^{\infty}$ does not converge to $f(x)\} \cup \widetilde{D_1} \cup \widetilde{D_2} \subseteq A$.

By Proposition 11.50, we have $E \in \mathcal{B}$ and $\mu(E) = 0$. Therefore, $\lim_{k\in\mathbb{N}} f_{n_k} = f$ a.e. in $\mathcal{X}$. This completes the proof of the proposition. □

Proposition 11.58 *Let $\mathcal{X} := (X, \mathcal{B}, \mu)$ be a finite measure space, $\mathcal{Y} := (Y, \rho)$ be a separable metric space, $D_1, D_2 \in \mathcal{B}$, $f_n : D_1 \to \mathcal{Y}$ be $\mathcal{B}$-measurable, $\forall n \in \mathbb{N}$, and $f : D_2 \to \mathcal{Y}$ be $\mathcal{B}$-measurable. Assume that $\lim_{n\in\mathbb{N}} f_n = f$ a.e. in $\mathcal{X}$. Then, $\lim_{n\in\mathbb{N}} f_n = f$ in measure in $\mathcal{X}$.*

Proof $\forall \epsilon \in (0, \infty) \subset \mathbb{R}$, by Egoroff's Theorem 11.55, $\exists A \in \mathcal{B}$ with $\mu(A) < \epsilon$ and $A \supseteq \widetilde{D_1} \cup \widetilde{D_2}$ such that $\left(f_n|_{X\setminus A}\right)_{n=1}^{\infty}$ converges uniformly to $f|_{X\setminus A}$. Then, $\exists n_0 \in \mathbb{N}$, $\forall n \in \mathbb{N}$ with $n \geq n_0$, we have $\rho(f_n(x), f(x)) < \epsilon$, $\forall x \in X \setminus A$. Then, $A_{n,\epsilon} := \widetilde{D_1} \cup \widetilde{D_2} \cup \{x \in D_1 \cap D_2 \mid \rho(f_n(x), f(x)) \geq \epsilon\} \subseteq A$. By Lemma 11.43, $A_{n,\epsilon} \in \mathcal{B}$ and $\mu(A_{n,\epsilon}) \leq \mu(A) < \epsilon$. This shows that $\lim_{n\in\mathbb{N}} f_n = f$ in measure in $\mathcal{X}$. This completes the proof of the proposition. □

Proposition 11.59 *Let $\mathcal{X} := (X, \mathcal{B}, \mu)$ be a measure space, $\mathcal{Y} := (Y, \rho)$ be a separable metric space, $D_n \in \mathcal{B}$, $f_n : D_n \to \mathcal{Y}$ be $\mathcal{B}$-measurable, $\forall n \in \mathbb{N}$, $D \in \mathcal{B}$, and $f : D \to \mathcal{Y}$ be $\mathcal{B}$-measurable. Then, $\lim_{n\in\mathbb{N}} f_n = f$ in measure in $\mathcal{X}$ if, and only if, every subsequence $\left(f_{n_k}\right)_{k=1}^{\infty}$ of $(f_n)_{n=1}^{\infty}$ admits a subsequence $\left(f_{n_{k_j}}\right)_{j=1}^{\infty}$ that converges to f in measure.*

Proof "Necessity" Let $(f_n)_{n=1}^{\infty}$ converges to f in measure and $\left(f_{n_k}\right)_{k=1}^{\infty}$ be a subsequence of $(f_n)_{n=1}^{\infty}$. Clearly, $\left(f_{n_k}\right)_{k=1}^{\infty}$ converges to f in measure, and it is a subsequence of itself. Hence, the result holds.

"Sufficiency" Assume that every subsequence $\left(f_{n_k}\right)_{k=1}^{\infty}$ of $(f_n)_{n=1}^{\infty}$ admits a subsequence $\left(f_{n_{k_j}}\right)_{j=1}^{\infty}$ that converges to f in measure. We will prove the result using an argument of contradiction. Suppose $(f_n)_{n=1}^{\infty}$ does not converge to f in measure. Then, $\exists \epsilon_0 \in (0, \infty) \subset \mathbb{R}$, $\forall n_0 \in \mathbb{N}$, $\exists n \in \mathbb{N}$ with $n \geq n_0$ such that $\mu(A_n) := \mu(\widetilde{D_n} \cup \widetilde{D} \cup \{x \in D_n \cap D \mid \rho(f_n(x), f(x)) \geq \epsilon_0\}) \geq \epsilon_0$, where $A_n \in \mathcal{B}$ by Lemma 11.43. Then, $\exists n_1 \in \mathbb{N}$ such that $\mu(A_{n_1}) \geq \epsilon_0$. $\forall k \in \mathbb{N}$, $\exists n_{k+1} \in \mathbb{N}$ with $n_{k+1} > n_k$ such that $\mu(A_{n_{k+1}}) \geq \epsilon_0$. This defines a subsequence $\left(f_{n_k}\right)_{k=1}^{\infty}$, which satisfies that $\forall k \in \mathbb{N}$, $\mu(A_{n_k}) \geq \epsilon_0$. Clearly, there does not exist a subsequence of $\left(f_{n_k}\right)_{k=1}^{\infty}$ that converges to f in measure. This contradicts with the assumption. Therefore, $(f_n)_{n=1}^{\infty}$ converges to f in measure.

This completes the proof of the proposition. □

Proposition 11.60 *Let $\mathcal{X} := (X, \mathcal{B}, \mu)$ be a finite measure space, $\mathcal{Y} := (Y, \rho)$ be a separable metric space, $D_1, D_2 \in \mathcal{B}$, $f_n : D_1 \to \mathcal{Y}$ be $\mathcal{B}$-measurable, $\forall n \in \mathbb{N}$, and $f : D_2 \to \mathcal{Y}$ be $\mathcal{B}$-measurable. Then, $\lim_{n\in\mathbb{N}} f_n = f$ in measure in $\mathcal{X}$ if, and only if, every subsequence $\left(f_{n_k}\right)_{k=1}^{\infty}$ of $(f_n)_{n=1}^{\infty}$ admits a subsequence $\left(f_{n_{k_j}}\right)_{j=1}^{\infty}$ that converges to f a.e. in $\mathcal{X}$.*

Proof "Necessity" Let $(f_n)_{n=1}^{\infty}$ converge to f in measure. Let $\left(f_{n_k}\right)_{k=1}^{\infty}$ be a subsequence of $(f_n)_{n=1}^{\infty}$. Clearly, $\left(f_{n_k}\right)_{k=1}^{\infty}$ converges to f in measure. By Proposition 11.57, there exists a subsequence $\left(f_{n_{k_j}}\right)_{j=1}^{\infty}$ of $\left(f_{n_k}\right)_{k=1}^{\infty}$ that converges to f a.e. in $\mathcal{X}$.

"Sufficiency" Assume that every subsequence $\left(f_{n_k}\right)_{k=1}^{\infty}$ of $(f_n)_{n=1}^{\infty}$ admits a subsequence $\left(f_{n_{k_j}}\right)_{j=1}^{\infty}$ that converges to f a.e. in $\mathcal{X}$. By Proposition 11.58, $\left(f_{n_{k_j}}\right)_{j=1}^{\infty}$ converges to f in measure. By Proposition 11.59, we have $(f_n)_{n=1}^{\infty}$ converges to f in measure in $\mathcal{X}$.

This completes the proof of the proposition. □

Definition 11.61 Let $\mathcal{X} := (X, \mathcal{B}, \mu)$ be a measure space, $\mathcal{Y} := (Y, \rho)$ be a separable metric space, $D_n \in \mathcal{B}$, and $f_n : D_n \to \mathcal{Y}$ be $\mathcal{B}$-measurable, $\forall n \in \mathbb{N}$. We will say that $(f_n)_{n=1}^{\infty}$ is *Cauchy* in measure in $\mathcal{X}$ if $\forall \epsilon \in (0, \infty) \subset \mathbb{R}$, $\exists n_0 \in \mathbb{N}$, $\forall n, m \in \mathbb{N}$ with $n \geq n_0$ and $m \geq n_0$, we have $\mu(\widetilde{D_n} \cup \widetilde{D_m} \cup \{x \in D_n \cap D_m \mid \rho(f_n(x), f_m(x)) \geq \epsilon\}) < \epsilon$. ◇

Clearly, a sequence of measurable functions that converges in measure to a measurable function is Cauchy in measure. The converse is established in the following proposition.

Proposition 11.62 *Let $\mathcal{X} := (X, \mathcal{B}, \mu)$ be a measure space, $\mathcal{Y} := (Y, \rho)$ be a separable complete metric space, $D_n \in \mathcal{B}$, and $f_n : D_n \to \mathcal{Y}$ be $\mathcal{B}$-measurable, $\forall n \in \mathbb{N}$. Assume that $(f_n)_{n=1}^{\infty}$ is Cauchy in measure in $\mathcal{X}$. Then, $\exists f : D \to \mathcal{Y}$, where $D := \bigcup_{n=1}^{\infty} D_n \in \mathcal{B}$ such that f is $\mathcal{B}$-measurable and $\lim_{n\in\mathbb{N}} f_n = f$ in measure in $\mathcal{X}$.*

Proof Let $n_0 := 0$. $\forall i \in \mathbb{N}$, by the assumption, $\exists n_i \in \mathbb{N}$ with $n_i > n_{i-1}$ such that $\forall n, m \in \mathbb{N}$ with $n \geq n_i$ and $m \geq n_i$, we have $\mu(\widetilde{D_n} \cup \widetilde{D_m} \cup \{x \in D_n \cap D_m \mid \rho(f_n(x), f_m(x)) \geq 2^{-i}\}) < 2^{-i}$. This defines a subsequence $\left(f_{n_i}\right)_{i=1}^{\infty}$ of $(f_n)_{n=1}^{\infty}$. $\forall i \in \mathbb{N}$, let $A_i := \widetilde{D_{n_i}} \cup \widetilde{D_{n_{i+1}}} \cup \{x \in D_{n_i} \cap D_{n_{i+1}} \mid \rho(f_{n_i}(x), f_{n_{i+1}}(x)) \geq 2^{-i}\}$. Then, $A_i \in \mathcal{B}$ and $\mu(A_i) < 2^{-i}$. Let $A := \bigcap_{i=1}^{\infty} \bigcup_{j=i}^{\infty} A_j \in \mathcal{B}$. Then, $\mu(A) \leq \mu(\bigcup_{j=i}^{\infty} A_j) \leq \sum_{j=i}^{\infty} \mu(A_j) < 2^{-i+1}$, $\forall i \in \mathbb{N}$. Hence, $\mu(A) = 0$. $\forall x \in X \setminus A = \bigcup_{i=1}^{\infty} \bigcap_{j=i}^{\infty} \widetilde{A_j} \subseteq D$, $\exists i_0 \in \mathbb{N}$ such that $x \in \widetilde{A_i}$ and $\rho(f_{n_i}(x), f_{n_{i+1}}(x)) < 2^{-i}$, $\forall i \in \mathbb{N}$ with $i \geq i_0$. Then, $\left(f_{n_i}(x)\right)_{i=i_0}^{\infty} \subseteq \mathcal{Y}$ is a Cauchy sequence. By the completeness of $\mathcal{Y}$, $\exists!\, f(x) \in \mathcal{Y}$ such that $\lim_{i\in\mathbb{N}} f_{n_i}(x) = f(x)$. $\forall x \in A \cap D$, we assign $f(x) := f_{\bar{n}}(\bar{x}) \in \mathcal{Y}$, where $\bar{n} \in \mathbb{N}$ and $\bar{x} \in D_{\bar{n}} \subseteq D$ are fixed independent of x. This defines a function $f : D \to \mathcal{Y}$ that satisfies $\lim_{i\in\mathbb{N}} f_{n_i}(x) = f(x)$, $\forall x \in X \setminus A$. $\forall i \in \mathbb{N}$, define $\bar{f}_{n_i} : D \to \mathcal{Y}$ by $\bar{f}_{n_i}(x) = \begin{cases} f_{n_i}(x) & x \in D_{n_i} \setminus A \\ f_{\bar{n}}(\bar{x}) & x \in D \setminus (D_{n_i} \setminus A) \end{cases}$, $\forall x \in D$. By Proposition 11.41, $\bar{f}_{n_i}$ is $\mathcal{B}$-measurable. Clearly, $\lim_{i\in\mathbb{N}} \bar{f}_{n_i}(x) = f(x)$, $\forall x \in D$. By Proposition 11.48, f is $\mathcal{B}$-measurable.

$\forall i \in \mathbb{N}$, $\forall x \in \widetilde{\bigcup_{j=i}^{\infty} A_j} = \bigcap_{j=i}^{\infty} \widetilde{A_j} \subseteq X \setminus A$, we have $\rho(f_{n_j}(x), f_{n_{j+1}}(x)) < 2^{-j}$, $\forall j \geq i$. Then, by Propositions 4.30, 3.66, and 3.67, $\rho(f_{n_i}(x), f(x)) = \lim_{l\in\mathbb{N}} \rho(f_{n_i}(x), f_{n_l}(x)) \leq \lim_{l\in\mathbb{N}} \sum_{j=i}^{l-1} \rho(f_{n_j}(x), f_{n_{j+1}}(x)) < 2^{-i+1}$. Then,

$E_i := \widetilde{D_{n_i}} \cup \{x \in D_{n_i} \mid \rho(f_{n_i}(x), f(x)) \geq 2^{-i+1}\} \subseteq \bigcup_{j=i}^{\infty} A_j$. By Lemma 11.43, $E_i \in \mathcal{B}$ and $\mu(E_i) \leq \sum_{j=i}^{\infty} \mu(A_j) < 2^{-i+1}$.

$\forall \epsilon \in (0, \infty) \subset \mathbb{R}$, $\exists i_0 \in \mathbb{N}$ such that $2^{-i_0+1} < \epsilon/2$. $\forall n \in \mathbb{N}$ with $n \geq n_{i_0}$, we have $\mu(\widetilde{D_n} \cup \{x \in D_n \mid \rho(f_n(x), f(x)) \geq \epsilon\}) \leq \mu(\widetilde{D_n} \cup \widetilde{D_{n_{i_0}}} \cup \{x \in D_n \cap D_{n_{i_0}} \mid \rho(f_n(x), f_{n_{i_0}}(x)) + \rho(f_{n_{i_0}}(x), f(x)) \geq \epsilon\}) \leq \mu(\widetilde{D_n} \cup \widetilde{D_{n_{i_0}}} \cup \{x \in D_n \cap D_{n_{i_0}} \mid \rho(f_n(x), f_{n_{i_0}}(x)) \geq \epsilon/2\}) + \mu(\widetilde{D_{n_{i_0}}} \cup \{x \in D_{n_{i_0}} \mid \rho(f_{n_{i_0}}(x), f(x)) \geq \epsilon/2\}) < 2^{-i_0} + 2^{-i_0+1} < \epsilon$. Hence, $\lim_{n\in\mathbb{N}} f_n = f$ in measure in $\mathcal{X}$. This completes the proof of the proposition. □

Definition 11.63 Let X be a set and $A \subseteq X$. The *indicator function* $\chi_{A,X} : X \to \{0, 1\} \subset \mathbb{R}$ is defined by $\chi_{A,X}(x) = \begin{cases} 1 & x \in A \\ 0 & x \in X \setminus A \end{cases}$, $\forall x \in X$. ◇

Definition 11.64 Let $\mathcal{X} := (X, \mathcal{B}, \mu)$ be a measure space and $\mathcal{Y}$ be a normed linear space. $\phi : X \to \mathcal{Y}$ is said to be a *simple function* if $\exists n \in \mathbb{Z}_+$, $\exists y_1, \ldots, y_n \in \mathcal{Y}$, and $\exists A_1, \ldots, A_n \in \mathcal{B}$ with $\mu(A_i) < +\infty$, $i = 1, \ldots, n$, such that $\phi(x) = \sum_{i=1}^{n} y_i \chi_{A_i,X}(x)$, $\forall x \in X$. We will say that a simple function ϕ is in *canonical representation* if $y_1, \ldots, y_n$ are distinct and none equals to $\vartheta_{\mathcal{Y}}$, and $A_1, \ldots, A_n$ are nonempty and pairwise disjoint. ◇

Clearly, every simple function admits a unique canonical representation.

Proposition 11.65 *Let $\mathcal{X} := (X, \mathcal{B}, \mu)$ be a finite measure space, $\mathcal{Y} := (\mathcal{Y}, \mathbb{K}, \|\cdot\|)$ be a separable normed linear space, and $f : X \to U \subseteq \mathcal{Y}$ be $\mathcal{B}$-measurable. Then:*

(i) *$\forall \epsilon \in (0, \infty) \subset \mathbb{R}$, $\exists$ a simple function $\phi_\epsilon : X \to U$ such that $\mu(\{x \in X \mid \|f(x) - \phi_\epsilon(x)\| \geq \epsilon\}) < \epsilon$.*
(ii) *There exists a sequence of simple functions $(\varphi_n)_{n=1}^{\infty}$, $\varphi_n : X \to U$, $\forall n \in \mathbb{N}$, that converges to f in measure.*

Proof (i) By Proposition 4.38, U, considered as a subspace of the metric space $\mathcal{Y}$, is separable. We will distinguish two exhaustive and mutually exclusive cases: Case 1: $U = \emptyset$; Case 2: $U \neq \emptyset$. Case 1: $U = \emptyset$. Then, $X = \emptyset$. $\forall \epsilon \in (0, \infty) \subset \mathbb{R}$, choose $\phi_\epsilon : X \to U$ to be the simple function with $n = 0$ as in Definition 11.64. Then, $\mu(\{x \in X \mid \|f(x) - \phi_\epsilon(x)\| \geq \epsilon\}) = \mu(\emptyset) = 0 < \epsilon$. Hence, (i) holds. Case 2: $U \neq \emptyset$. Then, $\exists (y_i)_{i=1}^{\infty} \subseteq U$ such that $\forall y \in U$, $\forall \delta \in (0, \infty) \subset \mathbb{R}$, $\exists l \in \mathbb{N} \ni \cdot \|y - y_l\| < \delta$. $\forall \epsilon \in (0, \infty) \subset \mathbb{R}$, we have $U \subseteq \bigcup_{i=1}^{\infty} \mathcal{B}_{\mathcal{Y}}(y_i, \epsilon) =: \bigcup_{i=1}^{\infty} \bar{V}_i$. Define $V_1 := \bar{V}_1 \in \mathcal{B}_B(\mathcal{Y})$, $V_{i+1} := \bar{V}_{i+1} \setminus (\bigcup_{j=1}^{i} \bar{V}_j) \in \mathcal{B}_B(\mathcal{Y})$, $\forall i \in \mathbb{N}$. Clearly, $U \subseteq \bigcup_{i=1}^{\infty} V_i$ and $V_1, V_2, \ldots$ are pairwise disjoint. Then, by Propositions 2.5 and 11.34, $X = f_{\text{inv}}(U) = \bigcup_{i=1}^{\infty} f_{\text{inv}}(V_i)$ and $(f_{\text{inv}}(V_i))_{i=1}^{\infty} \subseteq \mathcal{B}$ and are pairwise disjoint. Then, $\mu(X) = \sum_{i=1}^{\infty} \mu(f_{\text{inv}}(V_i)) < +\infty$. This implies that $\exists n_0 \in \mathbb{N}$ such that $\sum_{i=n_0+1}^{\infty} \mu(f_{\text{inv}}(V_i)) < \epsilon$. Let $A_i := f_{\text{inv}}(V_i) \in \mathcal{B}$, $i = 1, \ldots, n_0$, and $A_{n_0+1} := X \setminus (\bigcup_{i=1}^{n_0} A_i) \in \mathcal{B}$. Define the simple function $\phi_\epsilon : X \to U$ by $\phi_\epsilon(x) = \sum_{i=1}^{n_0+1} y_i \chi_{A_i,X}(x)$, $\forall x \in X$. $\forall x \in X$, $\exists! \, i_x \in \{1, \ldots, n_0 + 1\}$ such that $x \in A_{i_x}$. Then, $\phi_\epsilon(x) = y_{i_x} \in U$. $\forall x \in X \setminus A_{n_0+1}$, $i_x \in \{1, \ldots, n_0\}$

and $x \in A_{i_x} = f_{\text{inv}}(V_{i_x}) \subseteq f_{\text{inv}}(\bar{V}_{i_x})$. Then, $\phi_\epsilon(x) = y_{i_x}$ and $f(x) \in \bar{V}_{i_x} = \mathcal{B}_{\mathcal{Y}}(y_{i_x}, \epsilon)$. Hence, $\|f(x) - \phi_\epsilon(x)\| < \epsilon$. Then, $A := \{x \in X \mid \|f(x) - \phi_\epsilon(x)\| \geq \epsilon\} \subseteq A_{n_0+1} = \bigcup_{i=n_0+1}^{\infty} f_{\text{inv}}(V_i)$. By Lemma 11.43, $A \in \mathcal{B}$. Then, $\mu(A) \leq \sum_{i=n_0+1}^{\infty} \mu(f_{\text{inv}}(V_i)) < \epsilon$. Hence, (i) holds. In both cases, we have proved (i).

(ii) $\forall n \in \mathbb{N}$, let $\varphi_n := \phi_{2^{-n}}$. Then, $(\varphi_n)_{n=1}^{\infty}$ converges to f in measure. This completes the proof of the proposition. □

Proposition 11.66 *Let $\mathcal{X} := (X, \mathcal{B}, \mu)$ be a σ-finite measure space, $\mathcal{Y}$ be a separable normed linear space, $U \subseteq \mathcal{Y}$ be a conic segment, and $f : X \to U \subseteq \mathcal{Y}$ be $\mathcal{B}$-measurable. Then, there exists a sequence of simple functions $(\varphi_n)_{n=1}^{\infty}$, $\varphi_n : X \to U$, $\forall n \in \mathbb{N}$, such that $\lim_{n\in\mathbb{N}} \varphi_n = f$ a.e. in $\mathcal{X}$ and $\|\varphi_n(x)\| \leq \|f(x)\|$, $\forall x \in X$, $\forall n \in \mathbb{N}$.*

Proof We will first consider the special case $\mathcal{Y} = \mathbb{R}$ and $\mathcal{X}$ is finite. $\forall \epsilon \in (0, \infty) \subset \mathbb{R}$, by Proposition 11.5, $\exists M_\epsilon \in \mathbb{N}$ such that $\mu(\{x \in X \mid |f(x)| \geq M_\epsilon \epsilon\}) =: \mu(A_\epsilon) < \epsilon$. Let $I_i := [-(M_\epsilon - i + 1)\epsilon, -(M_\epsilon - i)\epsilon) \subset \mathbb{R}$, $i = 1, \ldots, 2M_\epsilon$ and $A_i := f_{\text{inv}}(I_i) \in \mathcal{B}$. Note that $\mu(A_i) \leq \mu(X) < +\infty$, $i = 1, \ldots, 2M_\epsilon$, and A_i's are pairwise disjoint. Define $\phi_\epsilon := \sum_{i=1}^{M_\epsilon} -(M_\epsilon - i)\epsilon\chi_{A_i,X} + \sum_{i=M_\epsilon+1}^{2M_\epsilon} -(M_\epsilon - i + 1)\epsilon\chi_{A_i,X}$. Clearly, we have $\phi_\epsilon : X \to U$, $|\phi_\epsilon(x)| \leq |f(x)|$, $\forall x \in X$, and $\mu(\{x \in X \mid |f(x) - \phi_\epsilon(x)| > \epsilon\}) \leq \mu(A_\epsilon) < \epsilon$. Then, the sequence of simple functions $(\phi_{2^{-n}})_{n=1}^{\infty}$ converges to f in measure in $\mathcal{X}$. By Proposition 11.57, there exists a subsequence $(\varphi_k)_{k=1}^{\infty} = (\phi_{2^{-n_k}})_{k=1}^{\infty}$ such that $\lim_{k\in\mathbb{N}} \varphi_k = f$ a.e. in $\mathcal{X}$. The result holds in this special case.

Next, we consider the special case $\mathcal{X}$ is finite. By Proposition 11.65, there exists a sequence of simple functions $(\bar{\psi}_i)_{i=1}^{\infty}$, $\bar{\psi}_i : X \to U$, $\forall i \in \mathbb{N}$, that converges to f in measure. By Proposition 11.57, there exists a subsequence $(\psi_n)_{n=1}^{\infty} = (\bar{\psi}_{i_n})_{n=1}^{\infty}$, such that $\lim_{n\in\mathbb{N}} \psi_n = f$ a.e. in $\mathcal{X}$. By Propositions 7.21 and 11.38, $\mathcal{P} \circ f$ is $\mathcal{B}$-measurable, where $\mathcal{P} \circ f : X \to [0, \infty) \subset \mathbb{R}$ is defined by $\mathcal{P} \circ f(x) = \|f(x)\|$, $\forall x \in X$.[1] By the previous special case, there exists a sequence of simple functions $(\phi_n)_{n=1}^{\infty}$, $\phi_n : X \to [0, \infty) \subset \mathbb{R}$, $\forall n \in \mathbb{N}$, such that $\lim_{n\in\mathbb{N}} \phi_n = \mathcal{P} \circ f$ a.e. in $\mathcal{X}$ and $0 \leq \phi_n(x) \leq \mathcal{P} \circ f(x)$, $\forall x \in X$, $\forall n \in \mathbb{N}$. Let $E_1 := \{x \in X \mid (\psi_n(x))_{n=1}^{\infty}$ does not converge to $f(x)\}$ and $E_2 := \{x \in X \mid (\phi_n(x))_{n=1}^{\infty}$ does not converge to $\|f(x)\|\}$. Then, $E_1, E_2 \in \mathcal{B}$ and $\mu(E_1) = \mu(E_2) = 0$. Fix any $n \in \mathbb{N}$. Let ψ_n admit the canonical representation $\psi_n = \sum_{j=1}^{\bar{n}} y_j\chi_{A_j,X}$, where $\bar{n} \in \mathbb{Z}_+$, $y_1, \ldots, y_{\bar{n}} \in U$ are distinct, and none equals to $\vartheta_{\mathcal{Y}}$, $A_1, \ldots, A_{\bar{n}} \in \mathcal{B}$ are pairwise disjoint, nonempty, and of finite measure. Let $y_{\bar{n}+1} := \vartheta_{\mathcal{Y}} \in U$ and $A_{\bar{n}+1} := X \setminus (\bigcup_{j=1}^{\bar{n}} A_j) \in \mathcal{B}$. Then, $\psi_n = \sum_{j=1}^{\bar{n}+1} y_j\chi_{A_j,X}$. Let ϕ_n admit the canonical representation $\sum_{j=1}^{\bar{\bar{n}}} a_j\chi_{\bar{A}_j,X}$, where $\bar{\bar{n}} \in \mathbb{Z}_+$, $a_1, \ldots, a_{\bar{\bar{n}}} \in [0, \infty) \subset \mathbb{R}$ are distinct and none equals to 0, $\bar{A}_1, \ldots, \bar{A}_{\bar{\bar{n}}} \in \mathcal{B}$ are pairwise disjoint, nonempty, and of finite measure. Let $a_{\bar{\bar{n}}+1} := 0$ and $\bar{A}_{\bar{\bar{n}}+1} := X \setminus (\bigcup_{j=1}^{\bar{\bar{n}}} \bar{A}_j) \in \mathcal{B}$. Then, $\phi_n = \sum_{j=1}^{\bar{\bar{n}}+1} a_j\chi_{\bar{A}_j,X}$. Define a simple

[1] This notation will be used throughout the rest of the book. With this notation, we distinguish that $\|f\|$ is a nonnegative real number and $\mathcal{P} \circ f$ is a nonnegative real-valued function.

function $\varphi_n : X \to U$ by $\varphi_n(x) = \begin{cases} y_j & \text{if } \|y_j\| \le a_l \\ (a_l/\|y_j\|)y_j & \text{if } \|y_j\| > a_l \end{cases}$, $\forall x \in A_j \cap \bar{A}_l$, $j = 1, \ldots, \tilde{n}+1$, $l = 1, \ldots, \bar{n}+1$. Clearly, $\varphi_n(x) \in U$, $\forall x \in X$. Alternatively, we have $\varphi_n(x) = \begin{cases} \psi_n(x) & \text{if } \|\psi_n(x)\| \le \phi_n(x) \\ (\phi_n(x)/\|\psi_n(x)\|)\psi_n(x) & \text{if } \|\psi_n(x)\| > \phi_n(x) \end{cases}$, $\forall x \in X$. Then, we have $\|\varphi_n(x)\| \le \phi_n(x) \le \mathcal{P} \circ f(x)$, $\forall x \in X$, $\forall n \in \mathbb{N}$, and $\lim_{n\in\mathbb{N}} \varphi_n(x) = f(x)$, $\forall x \in X \setminus (E_1 \cup E_2)$. Let $E := \{x \in X \mid (\varphi_n(x))_{n=1}^\infty \text{ does not converge to } f(x)\}$. Then, $E \subseteq E_1 \cup E_2$. By Proposition 11.50, $E \in \mathcal{B}$. This implies that $0 \le \mu(E) \le \mu(E_1) + \mu(E_2) = 0$. Therefore, $\lim_{n\in\mathbb{N}} \varphi_n = f$ a.e. in $\mathcal{X}$. Hence, the result holds in this special case.

Finally, we consider the general case. Since $\mathcal{X}$ is σ-finite, then $\exists (X_i)_{i=1}^\infty \subseteq \mathcal{B}$ such that $X = \bigcup_{i=1}^\infty X_i$ and $\mu(X_i) < +\infty$, $\forall i \in \mathbb{N}$. Without loss of generality, we may assume that $X_i \subseteq X_{i+1}$, $\forall i \in \mathbb{N}$. Fix any $i \in \mathbb{N}$; let $\mathcal{X}_i := (X_i, \mathcal{B}_i, \mu_i)$ be the finite measure subspace of $\mathcal{X}$ as defined in Proposition 11.13. Then, $f|_{X_i} : X_i \to U$ is $\mathcal{B}_i$-measurable. By the second special case, there exists a sequence of simple functions $(\phi_{i,n})_{n=1}^\infty$, $\phi_{i,n} : X_i \to U$, $\forall n \in \mathbb{N}$, such that $\|\phi_{i,n}(x)\| \le \|f(x)\|$, $\forall x \in X_i$, $\forall n \in \mathbb{N}$, and $\lim_{n\in\mathbb{N}} \phi_{i,n} = f|_{X_i}$ a.e. in $\mathcal{X}_i$. By Proposition 11.58, $\lim_{n\in\mathbb{N}} \phi_{i,n} = f|_{X_i}$ in measure in $\mathcal{X}_i$. Then, $\exists n_i \in \mathbb{N}$ such that $\mu_i(E_i) := \mu_i(\{x \in X_i \mid \|f|_{X_i}(x) - \phi_{i,n_i}(x)\| \ge 2^{-i}\}) < 2^{-i}$. Let ϕ_{i,n_i} admit the canonical representation $\phi_{i,n_i} = \sum_{j=1}^{\bar{n}_i} y_{i,j}\chi_{A_{i,j},X_i}$, where $\bar{n}_i \in \mathbb{Z}_+$, $y_{i,1}, \ldots, y_{i,\bar{n}_i} \in U$ are distinct, and none equals to $\vartheta_{\mathcal{Y}}$, and $A_{i,1}, \ldots, A_{i,\bar{n}_i} \in \mathcal{B}_i \subseteq \mathcal{B}$ are pairwise disjoint, nonempty, and of finite measure. Then, we may define a simple function $\varphi_i : X \to U \cup \{\vartheta_{\mathcal{Y}}\} = U$ by $\varphi_i = \sum_{j=1}^{\bar{n}_i} y_{i,j}\chi_{A_{i,j},X}$. Note that $\varphi_i(x) = \phi_{i,n_i}(x)$ and $\|\varphi_i(x)\| \le \|f(x)\|$, $\forall x \in X_i$, $\varphi_i(x) = \vartheta_{\mathcal{Y}} \in U$ and $\|\varphi_i(x)\| = 0 \le \|f(x)\|$, $\forall x \in X \setminus X_i$. Hence, $\|\varphi_i(x)\| \le \|f(x)\|$, $\forall x \in X$. Then, $E_i = \{x \in X \mid \|f(x) - \varphi_i(x)\| \ge 2^{-i}\} \cap X_i \in \mathcal{B}$ and $\mu(E_i) = \mu_i(E_i) < 2^{-i}$ by Proposition 11.13.

Let $E := \bigcap_{i=1}^\infty \bigcup_{j=i}^\infty E_j \in \mathcal{B}$. $\mu(E) \le \mu(\bigcup_{j=i}^\infty E_j) < 2^{-i+1}$, $\forall i \in \mathbb{N}$. Then, $\mu(E) = 0$. $\forall x \in X \setminus E$, $\exists i_0 \in \mathbb{N}$ such that $x \in X_{i_0}$. Then, $x \in X_j$, $\forall j \in \mathbb{N}$ with $j \ge i_0$. $\exists i_1 \in \mathbb{N}$ such that $x \in X \setminus E_j$, $\forall j \in \mathbb{N}$ with $j \ge i_1$. Let $k_0 = \max\{i_0, i_1\}$. $\forall k \ge k_0$, $x \in X_k \setminus E_k$ and $\|\varphi_k(x) - f(x)\| < 2^{-k}$. This implies that $\lim_{k\in\mathbb{N}} \varphi_k(x) = f(x)$. Then, $\bar{E} := \{x \in X \mid (\varphi_k(x))_{k=1}^\infty \text{ does not converge to } f(x)\} \subseteq E$. By Proposition 11.50, we have $\bar{E} \in \mathcal{B}$ and $\mu(\bar{E}) = 0$. Hence, $\lim_{k\in\mathbb{N}} \varphi_k = f$ a.e. in $\mathcal{X}$. This completes the proof of the proposition. □

Proposition 11.67 *Let $\mathcal{X} := (X, \mathcal{B}, \mu)$ be a σ-finite measure space and $f : X \to [0, \infty] \subset \mathbb{R}_e$ be $\mathcal{B}$-measurable. Then, there exists a sequence of simple functions $(\varphi_n)_{n=1}^\infty$, $\varphi_n : X \to [0, \infty) \subset \mathbb{R}$, $\forall n \in \mathbb{N}$, such that $0 \le \varphi_n(x) \le \varphi_{n+1}(x) \le f(x)$, $\forall x \in X$, $\forall n \in \mathbb{N}$, and $\lim_{n\in\mathbb{N}} \varphi_n(x) = f(x)$, $\forall x \in X$.*

Proof $\exists (X_n)_{n=1}^\infty \subseteq \mathcal{B}$ such that $X = \bigcup_{n=1}^\infty X_n$ and $\mu(X_n) < \infty$, $\forall n \in \mathbb{N}$. Without loss of generality, we may assume that $X_n \subseteq X_{n+1}$, $\forall n \in \mathbb{N}$. Fix any $n \in \mathbb{N}$. Let

$I_{i,n} := [(i-1)/n, i/n) \subseteq \mathbb{R}$, $E_{i,n} := f_{\text{inv}}(I_{i,n}) \cap X_n \in \mathcal{B}$, $i = 1, \ldots, n^2$, and $E_{n^2+1,n} := X_n \setminus (\bigcup_{i=1}^{n^2} E_{i,n})$. Note that $E_{i,n}$'s are pairwise disjoint and $\mu(E_{i,n}) \leq \mu(X_n) < \infty$, $\forall i \in \{1, \ldots, n^2+1\}$. Define $\phi_n := \sum_{i=1}^{n^2+1} (i-1)/n \chi_{E_{i,n},X}$. Then, $\phi_n : X \to [0, \infty) \subset \mathbb{R}$, $0 \leq \phi_n(x) \leq f(x)$, $\forall x \in X$, ϕ_n is a simple function, and $0 \leq f(x) - \phi_n(x) \leq 1/n$, $\forall x \in f_{\text{inv}}([0, n)) \cap X_n$. Define $\varphi_n := \bigvee_{i=1}^{n} \phi_i$. Then, $\varphi_n : X \to [0, \infty) \subset \mathbb{R}$, $0 \leq \varphi_n(x) \leq \varphi_{n+1}(x) \leq f(x)$, $\forall x \in X$, φ_n is a simple function, and $0 \leq f(x) - \varphi_n(x) \leq 1/n$, $\forall x \in f_{\text{inv}}([0, n)) \cap X_n$.

Fix any $x \in X$. If $f(x) = \infty$, then $\exists n_0 \in \mathbb{N}$ such that $x \in X_{n_0}$. $\forall n \in \mathbb{N}$ with $n_0 \leq n$, $x \in X_n$, $x \in E_{n^2+1,n}$, and $\phi_n(x) = n$. This further implies that $\varphi_n(x) = n$, $\forall n \geq n_0$. Hence, $\infty = f(x) = \lim_{n\in\mathbb{N}} \varphi_n(x)$. On the other hand, if $f(x) < \infty$, then $\exists n_0 \in \mathbb{N}$ such that $x \in X_{n_0}$ and $0 \leq f(x) < n_0$. Then, $\forall n \in \mathbb{N}$ with $n_0 \leq n$, $|f(x) - \varphi_n(x)| \leq 1/n$. This implies that $\lim_{n\in\mathbb{N}} \varphi_n(x) = f(x)$.

Hence, we have $f(x) = \lim_{n\in\mathbb{N}} \varphi_n(x)$, $\forall x \in X$. This completes the proof of the proposition. □

Lemma 11.68 *Let $\mathcal{X} := (X, \mathcal{B}, \mu)$ be a σ-finite measure space, $\mathcal{Y}$ be a normed linear space over $\mathbb{K}$, $U \subseteq \mathcal{Y}$ be a σ-compact conic segment, and $f : X \to U$ be $\mathcal{B}$-measurable. Then, there exists a sequence of simple functions $(\varphi_n)_{n=1}^{\infty}$, $\varphi_n : X \to U$, $\forall n \in \mathbb{N}$ such that $\lim_{n\in\mathbb{N}} \varphi_n(x) = f(x)$, $\forall x \in X$, and $\|\varphi_n(x)\| \leq \|f(x)\|$, $\forall x \in X$, $\forall n \in \mathbb{N}$.*

Proof Since U is σ-compact, then $U = \bigcup_{n=1}^{\infty} K_n$, where K_n is compact, $\forall n \in \mathbb{N}$. Since $\mathcal{X}$ is σ-finite, then $\exists (X_n)_{n=1}^{\infty} \subseteq \mathcal{B}$ such that $X = \bigcup_{n=1}^{\infty} X_n$, and $\mu(X_n) < +\infty$, $\forall n \in \mathbb{N}$. Without loss of generality, we may assume that $K_n \subseteq K_{n+1}$ and $X_n \subseteq X_{n+1}$, $\forall n \in \mathbb{N}$.

Fix any $n \in \mathbb{N}$. $K_n \subseteq \bigcup_{y\in K_n} \mathcal{B}_{\mathcal{Y}}(y, 1/n)$. By the compactness of K_n, $\exists D_n \subseteq K_n$, where D_n is a finite set, such that $K_n \subseteq \bigcup_{y\in D_n} \mathcal{B}_{\mathcal{Y}}(y, 1/n)$. Without loss of generality, assume that $D_n = \{y_{n,1}, \ldots, y_{n,m_n}\}$, where $m_n \in \mathbb{Z}_+$. Let $E_{n,1} := \mathcal{B}_{\mathcal{Y}}(y_{n,1}, 1/n)$, and $E_{n,i} := \mathcal{B}_{\mathcal{Y}}(y_{n,i}, 1/n) \setminus \left(\bigcup_{j=1}^{i-1} E_{n,j}\right)$, $\forall i = 2, \ldots, m_n$. Then, $K_n \subseteq \bigcup_{i=1}^{m_n} E_{n,i}$ and the sets in the union are pairwise disjoint. Let $A_{n,i} := f_{\text{inv}}(E_{n,i}) \cap X_n \in \mathcal{B}$, $i = 1, \ldots, m_n$, since $E_{n,i} \in \mathcal{B}_{\text{B}}(\mathcal{Y})$ and f is $\mathcal{B}$-measurable. Note that $\mu(A_{n,i}) \leq \mu(X_n) < +\infty$, $i = 1, \ldots, m_n$, and $(A_{n,i})_{i=1}^{m_n}$ is pairwise disjoint. Define $\psi_n : X \to K_n \subseteq U$ by $\psi_n := \sum_{i=1}^{m_n} y_{n,i} \chi_{A_{n,i},X}$. Clearly, ψ_n is a simple function. $\forall x \in X$, $\exists N \in \mathbb{N}$ such that $x \in X_N$ and $f(x) \in K_N$. Then, $\|f(x) - \psi_n(x)\| < 1/n$, $\forall n \in \mathbb{N}$ with $n \geq N$. Hence, $\lim_{n\in\mathbb{N}} \psi_n(x) = f(x)$, $\forall x \in X$.

By Propositions 7.21 and 11.38, $\mathcal{P} \circ f : X \to [0, \infty) \subset \mathbb{R}$ is $\mathcal{B}$-measurable. By Proposition 11.67, there exists a sequence of simple functions $(\phi_n)_{n=1}^{\infty}$ such that $0 \leq \phi_n(x) \leq \phi_{n+1}(x) \leq \mathcal{P} \circ f(x) = \|f(x)\|$, $\forall x \in X$, $\forall n \in \mathbb{N}$, and $\lim_{n\in\mathbb{N}} \phi_n(x) = \mathcal{P} \circ f(x)$, $\forall x \in X$.

Fix any $n \in \mathbb{N}$; let ϕ_n admit the canonical representation $\phi_n = \sum_{i=1}^{\bar{m}_n} a_j \cdot \chi_{\bar{A}_{n,j},X}$, where $\bar{m}_n \in \mathbb{Z}_+$, $a_1, \ldots, a_{\bar{m}_n} \in [0, \infty) \subset \mathbb{R}$ are distinct and none equals to 0, $\bar{A}_{n,1}, \ldots, \bar{A}_{n,\bar{m}_n} \in \mathcal{B}$ are pairwise disjoint, nonempty, and of finite measure. Define $\varphi_n : X \to U$ by

$\varphi_n(x) =$

$$\begin{cases} y_{n,i} & \text{if } x \in A_{n,i} \cap \bar{A}_{n,j}, \|y_{n,i}\| \le a_j, i = 1, \ldots, m_n, j = 1, \ldots, \bar{m}_n \\ \frac{a_j}{\|y_{n,i}\|} y_{n,i} & \text{if } x \in A_{n,i} \cap \bar{A}_{n,j}, \|y_{n,i}\| > a_j, i = 1, \ldots, m_n, j = 1, \ldots, \bar{m}_n \\ \vartheta_{\mathcal{Y}} & x \in X \setminus \left(\left(\bigcup_{i=1}^{m_n} A_{n,i} \right) \cap \left(\bigcup_{j=1}^{\bar{m}_n} \bar{A}_{n,j} \right) \right) \end{cases}$$

Clearly, $\varphi_n(x) \in U$, $\forall x \in X$, since U is a conic segment. An equivalent way to express φ_n is that $\varphi_n(x) = \begin{cases} \psi_n(x) & \text{if } \|\psi_n(x)\| \le \phi_n(x) \\ \frac{\phi_n(x)}{\|\psi_n(x)\|} \psi_n(x) & \text{if } \|\psi_n(x)\| > \phi_n(x) \end{cases}$, $\forall x \in X$. Then, we have $\|\varphi_n(x)\| \le \phi_n(x) \le \mathcal{P} \circ f(x)$, $\forall x \in X$. Furthermore, $\lim_{n \in \mathbb{N}} \varphi_n(x) = f(x)$, $\forall x \in X$.

This completes the proof of the lemma. □

Thus, we have generalized Littlewood's three principles: "every Lebesgue measurable set in $\mathbb{R}$ is almost a finite union of open intervals" (Proposition 11.23); "every measurable function taking value in a separable normed linear spaces is almost a simple function" (Propositions 11.65, 11.66, and 11.67 and Lemma 11.68); "every convergent sequence of measurable functions taking value in a separable metric space is nearly uniformly convergent" (Egoroff's Theorem 11.55).

11.4 Integration

Proposition 11.69 *Let $\mathcal{X}$ be a normed linear space and $\mathcal{B}_B(\mathcal{X})$ be the collection of Borel sets on $\mathcal{X}$. A* representation *of $\mathcal{X}$ is the collection $R := \{(x_\alpha, U_\alpha) \mid \alpha \in \Lambda\}$, where Λ is a finite index set, $(U_\alpha)_{\alpha \in \Lambda} \subseteq \mathcal{B}_B(\mathcal{X})$ are nonempty and pairwise disjoint, $\bigcup_{\alpha \in \Lambda} U_\alpha = \mathcal{X}$, $x_\alpha \in U_\alpha$, and $\|x_\alpha\| < \inf_{x \in U_\alpha} \|x\| + 1$, $\forall \alpha \in \Lambda$. The set of all representations of $\mathcal{X}$ is denoted $\mathcal{R}(\mathcal{X})$. Introduce a relation $\preceq$ on $\mathcal{R}(\mathcal{X})$ by $\forall R_1 := \{(x_\alpha, U_\alpha) \mid \alpha \in \Lambda\}$, $R_2 := \{(y_\beta, V_\beta) \mid \beta \in \Gamma\} \in \mathcal{R}(\mathcal{X})$, we will say $R_1 \preceq R_2$ if, $\forall \alpha \in \Lambda$, $\exists \beta \in \Gamma$ such that $x_\alpha = y_\beta$, and, $\forall \beta \in \Gamma$, $\exists \alpha \in \Lambda$ such that $V_\beta \subseteq U_\alpha$. Then, $\mathcal{I}(\mathcal{X}) := (\mathcal{R}(\mathcal{X}), \preceq)$ is a directed system, and $\preceq$ is an antisymmetric partial ordering on $\mathcal{R}(\mathcal{X})$. $\mathcal{I}(\mathcal{X})$ is said to be the* integration system *on $\mathcal{X}$.*

Proof Clearly, $R_0 := \{(\vartheta_{\mathcal{X}}, \mathcal{X})\} \in \mathcal{R}(\mathcal{X}) \neq \emptyset$. Clearly, $\preceq$ is reflexive and transitive. $\forall R_1 := \{(x_\alpha, U_\alpha) \mid \alpha \in \Lambda\}$, $R_2 := \{(y_\beta, V_\beta) \mid \beta \in \Gamma\} \in \mathcal{R}(\mathcal{X})$ with $R_1 \preceq R_2$ and $R_2 \preceq R_1$, $\forall \alpha \in \Lambda$, $\exists \beta \in \Gamma$ such that $U_\alpha \subseteq V_\beta$, by $R_2 \preceq R_1$. By $R_1 \preceq R_2$, $\exists \bar{\alpha} \in \Lambda$ such that $V_\beta \subseteq U_{\bar{\alpha}}$. Then, we have $U_\alpha \subseteq V_\beta \subseteq U_{\bar{\alpha}}$. By $R_1 \in \mathcal{R}(\mathcal{X})$, $(U_{\tilde{\alpha}})_{\tilde{\alpha} \in \Lambda}$ are nonempty and pairwise disjoint. Then, we must have $\alpha = \bar{\alpha}$ and $U_\alpha = V_\beta$. By $R_1 \preceq R_2$, $\exists \bar{\beta} \in \Gamma$ such that $x_\alpha = y_{\bar{\beta}}$ Then, $U_\alpha = V_\beta \ni x_\alpha = y_{\bar{\beta}} \in V_{\bar{\beta}}$. By $R_2 \in \mathcal{R}(\mathcal{X})$, $\left(V_{\tilde{\beta}}\right)_{\tilde{\beta} \in \Gamma}$ are nonempty and pairwise disjoint. Then, we must have $\bar{\beta} = \beta$ and $x_\alpha = y_\beta$. Then, $(x_\alpha, U_\alpha) \in R_2$. Hence, by the arbitrariness of α, we have $R_1 \subseteq R_2$. By an argument that is similar to the above, we have $R_2 \subseteq R_1$. Therefore, $R_1 = R_2$. This shows that $\preceq$ is an antisymmetric partial ordering on $\mathcal{R}(\mathcal{X})$.

$\forall R_1 := \{(x_\alpha, U_\alpha) \mid \alpha \in \Lambda\}, R_2 := \{(y_\beta, V_\beta) \mid \beta \in \Gamma\} \in \mathcal{R}(\mathcal{X})$, define $R_3 \in \mathcal{R}(\mathcal{X})$ as follows. $\forall \alpha \in \Lambda$, $\forall \beta \in \Gamma$, let $W_{\alpha,\beta} := U_\alpha \cap V_\beta \in \mathcal{B}_{\rm B}(\mathcal{X})$. We will distinguish six exhaustive and mutually exclusive cases.

Case 1: $W_{\alpha,\beta} = \emptyset$. Define $W^{\langle 1\rangle}_{\alpha,\beta} = W^{\langle 2\rangle}_{\alpha,\beta} = \emptyset$.

Case 2: $W_{\alpha,\beta} \neq \emptyset$, $x_\alpha \notin W_{\alpha,\beta}$, and $y_\beta \notin W_{\alpha,\beta}$. Define $W^{\langle 1\rangle}_{\alpha,\beta} = W_{\alpha,\beta} \in \mathcal{B}_{\rm B}(\mathcal{X})$, $z^{\langle 1\rangle}_{\alpha,\beta} \in W^{\langle 1\rangle}_{\alpha,\beta}$ such that $\left\| z^{\langle 1\rangle}_{\alpha,\beta} \right\| < \inf_{x \in W^{\langle 1\rangle}_{\alpha,\beta}} \|x\| + 1$, and $W^{\langle 2\rangle}_{\alpha,\beta} = \emptyset$.

Case 3: $x_\alpha \in W_{\alpha,\beta}$ and $y_\beta \notin W_{\alpha,\beta}$. Define $W^{\langle 1\rangle}_{\alpha,\beta} = W_{\alpha,\beta} \in \mathcal{B}_{\rm B}(\mathcal{X})$, $z^{\langle 1\rangle}_{\alpha,\beta} = x_\alpha \in W^{\langle 1\rangle}_{\alpha,\beta}$, and $W^{\langle 2\rangle}_{\alpha,\beta} = \emptyset$. Then, $\left\| z^{\langle 1\rangle}_{\alpha,\beta} \right\| = \|x_\alpha\| < \inf_{x \in U_\alpha} \|x\| + 1 \le \inf_{x \in W^{\langle 1\rangle}_{\alpha,\beta}} \|x\| + 1$.

Case 4: $x_\alpha \notin W_{\alpha,\beta}$ and $y_\beta \in W_{\alpha,\beta}$. Define $W^{\langle 1\rangle}_{\alpha,\beta} = W_{\alpha,\beta} \in \mathcal{B}_{\rm B}(\mathcal{X})$, $z^{\langle 1\rangle}_{\alpha,\beta} = y_\beta \in W^{\langle 1\rangle}_{\alpha,\beta}$, and $W^{\langle 2\rangle}_{\alpha,\beta} = \emptyset$. Then, $\left\| z^{\langle 1\rangle}_{\alpha,\beta} \right\| = \left\| y_\beta \right\| < \inf_{x \in V_\beta} \|x\| + 1 \le \inf_{x \in W^{\langle 1\rangle}_{\alpha,\beta}} \|x\| + 1$.

Case 5: $x_\alpha = y_\beta \in W_{\alpha,\beta}$. Define $W^{\langle 1\rangle}_{\alpha,\beta} = W_{\alpha,\beta} \in \mathcal{B}_{\rm B}(\mathcal{X})$, $z^{\langle 1\rangle}_{\alpha,\beta} = x_\alpha \in W^{\langle 1\rangle}_{\alpha,\beta}$, and $W^{\langle 2\rangle}_{\alpha,\beta} = \emptyset$. Then, $\left\| z^{\langle 1\rangle}_{\alpha,\beta} \right\| = \|x_\alpha\| < \inf_{x \in U_\alpha} \|x\| + 1 \le \inf_{x \in W^{\langle 1\rangle}_{\alpha,\beta}} \|x\| + 1$.

Case 6: $x_\alpha \in W_{\alpha,\beta}$, $y_\beta \in W_{\alpha,\beta}$, and $x_\alpha \neq y_\beta$. Let $\delta = \left\| x_\alpha - y_\beta \right\| > 0$. Define $W^{\langle 1\rangle}_{\alpha,\beta} = W_{\alpha,\beta} \cap B_{\mathcal{X}}(x_\alpha, \delta/2) \in \mathcal{B}_{\rm B}(\mathcal{X})$, $z^{\langle 1\rangle}_{\alpha,\beta} = x_\alpha \in W^{\langle 1\rangle}_{\alpha,\beta}$, $W^{\langle 2\rangle}_{\alpha,\beta} = W_{\alpha,\beta} \cap \widetilde{B_{\mathcal{X}}(x_\alpha, \delta/2)} \in \mathcal{B}_{\rm B}(\mathcal{X})$, and $z^{\langle 2\rangle}_{\alpha,\beta} = y_\beta \in W^{\langle 2\rangle}_{\alpha,\beta}$. Then, $\left\| z^{\langle 1\rangle}_{\alpha,\beta} \right\| = \|x_\alpha\| < \inf_{x \in U_\alpha} \|x\| + 1 \le \inf_{x \in W^{\langle 1\rangle}_{\alpha,\beta}} \|x\| + 1$ and $\left\| z^{\langle 2\rangle}_{\alpha,\beta} \right\| = \left\| y_\beta \right\| < \inf_{x \in V_\beta} \|x\| + 1 \le \inf_{x \in W^{\langle 2\rangle}_{\alpha,\beta}} \|x\| + 1$.

Define $R_3 := \left\{ (z^{\langle i\rangle}_{\alpha,\beta}, W^{\langle i\rangle}_{\alpha,\beta}) \,\middle|\, \alpha \in \Lambda,\ \beta \in \Gamma,\ i = 1, 2,\ W^{\langle i\rangle}_{\alpha,\beta} \neq \emptyset \right\}$. Clearly, $R_3 \in \mathcal{R}(\mathcal{X})$, $R_1 \preceq R_3$, and $R_2 \preceq R_3$. Therefore, $\mathcal{I}(\mathcal{X})$ is a directed system. This completes the proof of the proposition. □

Definition 11.70 Let $\mathcal{X} := (X, \mathcal{B}, \mu)$ be a finite measure space, $\mathcal{Y}$ be a normed linear space with the integration system $\mathcal{I}(\mathcal{Y}) := (\mathcal{R}(\mathcal{Y}), \preceq)$, and $f : X \to \mathcal{Y}$ be $\mathcal{B}$-measurable. $\forall R = \{(y_\alpha, U_\alpha) \mid \alpha \in \Lambda\} \in \mathcal{R}(\mathcal{Y})$, define $F_R := \sum_{\alpha \in \Lambda} y_\alpha \mu(f_{\rm inv}(U_\alpha)) \in \mathcal{Y}$. This defines a net $(F_R)_{R \in \mathcal{I}(\mathcal{Y})}$. f is said to be *integrable* if the net admits a limit in $\mathcal{Y}$. In this case, $\lim_{R \in \mathcal{I}(\mathcal{Y})} F_R \in \mathcal{Y}$ is said to be the *integral* of f over $\mathcal{X}$ and denoted by $\int_X f \, {\rm d}\mu$ or $\int_X f(x) \, {\rm d}\mu(x)$. When $\mathcal{Y} = \mathbb{R}$, we will denote $\int_X f \, {\rm d}\mu := \int_X f(x) \, {\rm d}\mu(x) := \lim_{R \in \mathcal{I}(\mathbb{R})} F_R \in \mathbb{R}_e$ whenever the limit exists, which is said to be the integral of f over $\mathcal{X}$. ◇

Definition 11.71 Let $\mathcal{X} := (X, \mathcal{B}, \mu)$ be a measure space with $\mu(X) = \infty$, $\mathcal{Y}$ be a normed linear space, and $f : X \to \mathcal{Y}$ be $\mathcal{B}$-measurable. Define the set $\mathcal{M}(\mathcal{X}) := \{A \in \mathcal{B} \mid \mu(A) < +\infty\}$. Clearly, $\mathcal{M}(\mathcal{X}) := (\mathcal{M}(\mathcal{X}), \subseteq)$ is a directed system. $\forall A \in \mathcal{M}(\mathcal{X})$, let $(A, \mathcal{B}_A, \mu_A)$ be the finite measure subspace of $\mathcal{X}$ as defined in Proposition 11.13. Define a net $(F_A)_{A \in \mathcal{M}(\mathcal{X})}$ by $F_A := \int_A f|_A \, {\rm d}\mu_A \in \mathcal{Y}$, $\forall A \in \mathcal{M}(\mathcal{X})$. f is said to be *integrable* if the net is well-defined and admits a limit in $\mathcal{Y}$. In this case, $\lim_{A \in \mathcal{M}(\mathcal{X})} F_A \in \mathcal{Y}$ is said to be the *integral* of f over $\mathcal{X}$ and denoted by $\int_X f \, {\rm d}\mu$ or $\int_X f(x) \, {\rm d}\mu(x)$. When $\mathcal{Y} = \mathbb{R}$, we will allow the net

$(F_A)_{A\in\mathcal{M}(\mathcal{X})} \subseteq \mathbb{R}_e$ and denote $\int_X f\,\mathrm{d}\mu := \int_X f(x)\,\mathrm{d}\mu(x) := \lim_{A\in\mathcal{M}(\mathcal{X})} F_A \in \mathbb{R}_e$ whenever the limit exists, which is said to be the integral of f over $\mathcal{X}$. ◇

Proposition 11.72 *Let $\mathcal{X} := (X, \mathcal{B}, \mu)$ be a finite measure space, $\mathcal{Y}$ be a normed linear space, $f : X \to \mathcal{Y}$ be $\mathcal{B}$-measurable, and $\bar{\mathcal{X}} := (X, \bar{\mathcal{B}}, \bar{\mu})$ be a finite measure space that is an extension of $\mathcal{X}$, i.e., $\mathcal{B} \subseteq \bar{\mathcal{B}}$ and $\mu = \bar{\mu}|_{\mathcal{B}}$. Then, $\int_X f\,\mathrm{d}\mu = \int_X f\,\mathrm{d}\bar{\mu}$, whenever one of the integrals exists.*

Proof Let $\mathcal{I}(\mathcal{Y})$ be the integration system on $\mathcal{Y}$ as defined in Proposition 11.69, $(F_R)_{R\in\mathcal{I}(\mathcal{Y})}$ be the net for $\int_X f\,\mathrm{d}\mu$, and $(\bar{F}_R)_{R\in\mathcal{I}(\mathcal{Y})}$ be the net for $\int_X f\,\mathrm{d}\bar{\mu}$, as defined in Definition 11.70. $\forall R := \{(y_\alpha, U_\alpha) \mid \alpha \in \Lambda\} \in \mathcal{I}(\mathcal{Y})$, $\forall \alpha \in \Lambda$, we have $f_{\mathrm{inv}}(U_\alpha) \in \mathcal{B} \subseteq \bar{\mathcal{B}}$ since f is $\mathcal{B}$-measurable, and $\mu(f_{\mathrm{inv}}(U_\alpha)) = \bar{\mu}(f_{\mathrm{inv}}(U_\alpha)) \in \mathbb{R}$ since $\mu = \bar{\mu}|_{\mathcal{B}}$. Then, $F_R = \sum_{\alpha\in\Lambda} y_\alpha \cdot \mu(f_{\mathrm{inv}}(U_\alpha)) = \sum_{\alpha\in\Lambda} y_\alpha \bar{\mu}(f_{\mathrm{inv}}(U_\alpha)) = \bar{F}_R \in \mathcal{Y}$. Hence, $\int_X f\,\mathrm{d}\mu = \lim_{R\in\mathcal{I}(\mathcal{Y})} F_R = \lim_{R\in\mathcal{I}(\mathcal{Y})} \bar{F}_R = \int_X f\,\mathrm{d}\bar{\mu}$, whenever one of the integrals exists. This completes the proof of the proposition. □

Lemma 11.73 *Let $\mathcal{X} := (X, \mathcal{B}, \mu)$ be a finite measure space, $\mathcal{Y}$ be a normed linear space, $f : X \to \mathcal{Y}$ be $\mathcal{B}$-measurable, $\hat{X} \in \mathcal{B}$, and $\hat{\mathcal{X}} := (\hat{X}, \hat{\mathcal{B}}, \hat{\mu})$ be the finite measure subspace of $\mathcal{X}$ as defined in Proposition 11.13. Assume that $\mu(X \setminus \hat{X}) = 0$. Then, $\int_X f\,\mathrm{d}\mu = \int_{\hat{X}} f|_{\hat{X}}\,\mathrm{d}\hat{\mu}$, whenever one of the integrals exists.*

Proof Let $\mathcal{I}(\mathcal{Y})$ be the integration system on $\mathcal{Y}$ as defined in Proposition 11.69, $(F_R)_{R\in\mathcal{I}(\mathcal{Y})}$ be the net for $\int_X f\,\mathrm{d}\mu$, and $\left(\hat{F}_R\right)_{R\in\mathcal{I}(\mathcal{Y})}$ be the net for $\int_{\hat{X}} f|_{\hat{X}}\,\mathrm{d}\hat{\mu}$, as defined in Definition 11.70. $\forall R := \{(y_\alpha, U_\alpha) \mid \alpha \in \Lambda\} \in \mathcal{I}(\mathcal{Y})$, $\forall \alpha \in \Lambda$, we have $f_{\mathrm{inv}}(U_\alpha) \cap \hat{X} = f|_{\hat{X}\,\mathrm{inv}}(U_\alpha) \in \hat{\mathcal{B}} \subseteq \mathcal{B}$ since f is $\mathcal{B}$-measurable, and $0 \le \mu(f_{\mathrm{inv}}(U_\alpha) \cap (X \setminus \hat{X})) \le \mu(X \setminus \hat{X}) = 0$. Then, $F_R = \sum_{\alpha\in\Lambda} y_\alpha \cdot \mu(f_{\mathrm{inv}}(U_\alpha)) = \sum_{\alpha\in\Lambda} y_\alpha (\mu(f_{\mathrm{inv}}(U_\alpha) \cap \hat{X}) + \mu(f_{\mathrm{inv}}(U_\alpha) \cap (X \setminus \hat{X}))) = \sum_{\alpha\in\Lambda} y_\alpha \cdot \mu(f|_{\hat{X}\,\mathrm{inv}}(U_\alpha)) = \sum_{\alpha\in\Lambda} y_\alpha \hat{\mu}(f|_{\hat{X}\,\mathrm{inv}}(U_\alpha)) = \hat{F}_R \in \mathcal{Y}$. Hence, $\int_X f\,\mathrm{d}\mu = \lim_{R\in\mathcal{I}(\mathcal{Y})} F_R = \lim_{R\in\mathcal{I}(\mathcal{Y})} \hat{F}_R = \int_{\hat{X}} f|_{\hat{X}}\,\mathrm{d}\hat{\mu}$, whenever one of the integrals exists. This completes the proof of the proposition. □

Proposition 11.74 *Let $\mathcal{X} := (X, \mathcal{B}, \mu)$ be a measure space with $\mu(X) = +\infty$, $\mathcal{Y}$ be a normed linear space, $f : X \to \mathcal{Y}$ be $\mathcal{B}$-measurable, and $\bar{\mathcal{X}} := (X, \bar{\mathcal{B}}, \bar{\mu})$ be the completion of $\mathcal{X}$. Then, $\int_X f\,\mathrm{d}\mu = \int_X f\,\mathrm{d}\bar{\mu}$, whenever one of the integrals exists.*

Proof Let $(F_A)_{A\in\mathcal{M}(\mathcal{X})}$ be the net for $\int_X f\,\mathrm{d}\mu$ and $(\bar{F}_{\bar{A}})_{\bar{A}\in\mathcal{M}(\bar{\mathcal{X}})}$ be the net for $\int_X f\,\mathrm{d}\bar{\mu}$ as defined in Definition 11.71. Clearly, $\mathcal{M}(\mathcal{X}) \subseteq \mathcal{M}(\bar{\mathcal{X}})$, since $\bar{\mathcal{X}}$ is the completion of $\mathcal{X}$. We will distinguish two exhaustive cases: Case 1: $\int_X f\,\mathrm{d}\bar{\mu}$ exists; Case 2: $\int_X f\,\mathrm{d}\mu$ exists.

Case 1: $\int_X f\,\mathrm{d}\bar{\mu}$ exists. Then, the net $(\bar{F}_{\bar{A}})_{\bar{A}\in\mathcal{M}(\bar{\mathcal{X}})}$ is well-defined. $\forall A \in \mathcal{M}(\mathcal{X})$, $A \in \mathcal{M}(\bar{\mathcal{X}})$. By Proposition 11.72, $F_A = \int_A f|_A\,\mathrm{d}\mu_A = \int_A f|_A\,\mathrm{d}\bar{\mu}_A = \bar{F}_A$, where $(A, \mathcal{B}_A, \mu_A)$ is the finite measure subspace of $\mathcal{X}$ and $(A, \bar{\mathcal{B}}_A, \bar{\mu}_A)$ is the finite complete measure subspace of $\bar{\mathcal{X}}$. Then, $(F_A)_{A\in\mathcal{M}(\mathcal{X})}$ is well-defined. $\forall \bar{A} \in \mathcal{M}(\bar{\mathcal{X}})$, we have $\bar{A} \in \bar{\mathcal{B}}$ with $\bar{\mu}(\bar{A}) < +\infty$. By Proposition 11.12 and its proof, $\exists \tilde{A}, \tilde{B}, \tilde{C} \subseteq$

X with $\bar{A} = \tilde{A} \cup \tilde{B}$, $\tilde{B}, \tilde{C} \in \mathcal{B}$, $\tilde{A} \subseteq \tilde{C}$, $\mu(\tilde{C}) = 0$, and $\bar{\mu}(\bar{A}) = \mu(\tilde{B})$. Then, $\bar{A} \subseteq \tilde{B} \cup \tilde{C} =: A \in \mathcal{B}$ and $\mu(A) \leq \mu(\tilde{B}) + \mu(\tilde{C}) = \mu(\tilde{B}) = \bar{\mu}(\bar{A}) < +\infty$. This shows that $A \in \mathcal{M}(\mathcal{X})$ and $\bar{A} \subseteq A$. Hence, $(F_A)_{A\in\mathcal{M}(\mathcal{X})}$ is a subnet of $\left(\bar{F}_{\bar{A}}\right)_{\bar{A}\in\mathcal{M}(\bar{\mathcal{X}})}$. By Proposition 3.70, $\int_X f \,d\mu = \lim_{A\in\mathcal{M}(\mathcal{X})} F_A = \lim_{\bar{A}\in\mathcal{M}(\bar{\mathcal{X}})} \bar{F}_{\bar{A}} = \int_X f \,d\bar{\mu}$.

Case 2: $\int_X f \,d\mu$ exists. Then, the net $(F_A)_{A\in\mathcal{M}(\mathcal{X})}$ is well-defined. $\forall \bar{A} \in \mathcal{M}(\bar{\mathcal{X}})$, we have $\bar{A} \in \bar{\mathcal{B}}$ with $\bar{\mu}(\bar{A}) < +\infty$. By Proposition 11.12 and its proof, $\exists \tilde{A}, \tilde{B}, \tilde{C} \subseteq X$ with $\bar{A} = \tilde{A} \cup \tilde{B}$, $\tilde{B}, \tilde{C} \in \mathcal{B}$, $\tilde{A} \subseteq \tilde{C}$, $\mu(\tilde{C}) = 0$, and $\bar{\mu}(\bar{A}) = \mu(\tilde{B})$. Let $\tilde{B} =: A \in \mathcal{B}$. Then, $0 \leq \bar{\mu}(\bar{A} \setminus A) = \bar{\mu}(\tilde{A} \setminus \tilde{B}) \leq \bar{\mu}(\tilde{C}) = \mu(\tilde{C}) = 0$ and $\mu(A) = \bar{\mu}(\bar{A}) < +\infty$. This shows that $A \in \mathcal{M}(\mathcal{X})$ and $A \subseteq \bar{A}$. Let $\mathcal{A} := (A, \mathcal{B}_A, \mu_A)$ be the finite measure subspace of $\mathcal{X}$, $\bar{\mathcal{A}} := (\bar{A}, \bar{\mathcal{B}}_{\bar{A}}, \bar{\mu}_{\bar{A}})$ be the finite complete measure subspace of $\bar{\mathcal{X}}$, and $\hat{\mathcal{A}} := (A, \bar{\mathcal{B}}_A, \bar{\mu}_A)$ be the finite complete measure subspace of $\bar{\mathcal{A}}$. By Lemma 11.73, we have $\bar{F}_{\bar{A}} = \int_{\bar{A}} f|_{\bar{A}} \,d\bar{\mu}_{\bar{A}} = \int_A f|_A \,d\bar{\mu}_A$, whenever one of them exists. Since $\bar{\mathcal{X}}$ is the completion of $\mathcal{X}$, then $\hat{\mathcal{A}}$ is an extension of $\mathcal{A}$. By Proposition 11.72, we have $\int_A f|_A \,d\bar{\mu}_A = \int_A f|_A \,d\mu_A = F_A$. Hence, $\bar{F}_{\bar{A}} = F_A$. Then, the net $\left(\bar{F}_{\bar{A}}\right)_{\bar{A}\in\mathcal{M}(\bar{\mathcal{X}})}$ is well-defined.

Fix any open set U ($U \subseteq \mathbb{R}_e$ if $\mathcal{Y} = \mathbb{R}$ or $U \subseteq \mathcal{Y}$ if $\mathcal{Y} \neq \mathbb{R}$) with $\int_X f \,d\mu \in U$. Since $\lim_{A\in\mathcal{M}(\mathcal{X})} F_A = \int_X f \,d\mu \in U$, then $\exists A_0 \in \mathcal{M}(\mathcal{X})$ such that $\forall A \in \mathcal{M}(\mathcal{X})$ with $A_0 \subseteq A$, we have $F_A \in U$. Note that $A_0 \in \mathcal{B} \subseteq \bar{\mathcal{B}}$ and $\mu(A_0) = \bar{\mu}(A_0) < +\infty$. Then, $A_0 \in \mathcal{M}(\bar{\mathcal{X}})$. $\forall \bar{A} \in \mathcal{M}(\bar{\mathcal{X}})$ with $A_0 \subseteq \bar{A}$, then $\bar{A} \in \bar{\mathcal{B}}$ and $\bar{\mu}(\bar{A}) < +\infty$. By Proposition 11.12 and its proof, $\exists \tilde{A}, \tilde{B}, \tilde{C} \subseteq X$ with $\bar{A} = \tilde{A} \cup \tilde{B}$, $\tilde{B}, \tilde{C} \in \mathcal{B}$, $\tilde{A} \subseteq \tilde{C}$, $\mu(\tilde{C}) = 0$, and $\bar{\mu}(\bar{A}) = \mu(\tilde{B})$. Let $A := A_0 \cup \tilde{B} \in \mathcal{B}$. Then, $0 \leq \bar{\mu}(\bar{A} \setminus A) = \bar{\mu}((\tilde{A} \cup \tilde{B}) \setminus (A_0 \cup \tilde{B})) \leq \bar{\mu}((\tilde{A} \cup \tilde{B}) \setminus \tilde{B}) \leq \bar{\mu}(\tilde{A} \setminus \tilde{B}) \leq \bar{\mu}(\tilde{C}) = \mu(\tilde{C}) = 0$, $A \subseteq \bar{A}$, and $\mu(A) \leq \mu(\tilde{B}) + \mu(A_0) = \bar{\mu}(\bar{A}) + \mu(A_0) < +\infty$. This shows that $A \in \mathcal{M}(\mathcal{X})$ and $A_0 \subseteq A \subseteq \bar{A}$. Let $\mathcal{A} := (A, \mathcal{B}_A, \mu_A)$ be the finite measure subspace of $\mathcal{X}$, $\bar{\mathcal{A}} := (\bar{A}, \bar{\mathcal{B}}_{\bar{A}}, \bar{\mu}_{\bar{A}})$ be the finite complete measure subspace of $\bar{\mathcal{X}}$, and $\hat{\mathcal{A}} := (A, \bar{\mathcal{B}}_A, \bar{\mu}_A)$ be the finite complete measure subspace of $\bar{\mathcal{A}}$. By Lemma 11.73, we have $\bar{F}_{\bar{A}} = \int_{\bar{A}} f|_{\bar{A}} \,d\bar{\mu}_{\bar{A}} = \int_A f|_A \,d\bar{\mu}_A$, whenever one of them exists. Since $\bar{\mathcal{X}}$ is the completion of $\mathcal{X}$, then $\hat{\mathcal{A}}$ is an extension of $\mathcal{A}$. By Proposition 11.72, we have $\int_A f|_A \,d\bar{\mu}_A = \int_A f|_A \,d\mu_A = F_A$. Hence, $\bar{F}_{\bar{A}} = F_A \in U$. Therefore, we have $\int_X f \,d\bar{\mu} = \lim_{\bar{A}\in\mathcal{M}(\bar{\mathcal{X}})} \bar{F}_{\bar{A}} = \int_X f \,d\mu$.

Hence, in both cases, we have $\int_X f \,d\mu = \int_X f \,d\bar{\mu}$. This completes the proof of the proposition. □

Proposition 11.75 *Let $\mathcal{X} := (X, \mathcal{B}, \mu)$ be a measure space, $\mathcal{Y}$ be a normed linear space, $\phi : X \to \mathcal{Y}$ be a simple function in canonical representation, i.e., $\exists n \in \mathbb{Z}_+$, $\exists y_1, \ldots, y_n \in \mathcal{Y}$, which are distinct and none equals to $\vartheta_{\mathcal{Y}}$, $\exists A_1, \ldots, A_n \in \mathcal{B}$, which are nonempty, pairwise disjoint, and of finite measure, such that $\phi(x) = \sum_{i=1}^{n} y_i \chi_{A_i,X}(x)$, $\forall x \in X$. Then, $\int_X \phi \,d\mu = \sum_{i=1}^{n} y_i \mu(A_i) =: I \in \mathcal{Y}$.*

Proof We will distinguish two exhaustive and mutually exclusive cases: Case 1: $\mu(X) < +\infty$; Case 2: $\mu(X) = +\infty$.

Case 1: $\mu(X) < +\infty$. Let $y_{n+1} := \vartheta_{\mathcal{Y}}$ and $A_{n+1} := X \setminus (\bigcup_{i=1}^{n} A_i) \in \mathcal{B}$. Let $\epsilon_0 := \min\{1, \min_{1\leq i<j\leq n+1} \|y_i - y_j\|\} > 0$, by the assumption. Let $(\Phi_R)_{R\in\mathfrak{I}(\mathcal{Y})}$ be the net as defined in Definition 11.70 for the simple function ϕ.

$\forall \epsilon \in (0, \epsilon_0/2) \subset \mathbb{R}$, let $U_i := \mathcal{B}_{\mathcal{Y}}(y_i, \epsilon) \in \mathcal{B}_{\mathrm{B}}(\mathcal{Y})$, $i = 1, \ldots, n+1$, which are clearly pairwise disjoint and nonempty. We will distinguish two exhaustive and mutually exclusive subcases: Case 1a: $\mathcal{Y} = \bigcup_{i=1}^{n+1} U_i$; Case 1b: $\mathcal{Y} \supset \bigcup_{i=1}^{n+1} U_i$.

Case 1a: $\mathcal{Y} = \bigcup_{i=1}^{n+1} U_i$. Then, $\exists \delta \in (0, \infty) \subset \mathbb{R}$ such that $\mathcal{Y} = \mathcal{B}_{\mathcal{Y}}(\vartheta_{\mathcal{Y}}, \delta)$. Then, $\mathcal{Y}$ must be the trivial normed linear space, i.e., $\mathcal{Y}$ is a singleton set containing $\vartheta_{\mathcal{Y}}$. There is a single representation of $\mathcal{Y}$, which is $R_0 := \{(\vartheta_{\mathcal{Y}}, \mathcal{Y})\}$. Then, $\Phi_{R_0} = \vartheta_{\mathcal{Y}}$. Clearly, we must have $n = 0$, since $y_1, \ldots, y_n \in \mathcal{Y}$ are distinct and none equals to $\vartheta_{\mathcal{Y}}$. Then, $I = \vartheta_{\mathcal{Y}}$. Hence, $\forall R \in \mathfrak{I}(\mathcal{Y})$ with $R_0 \preceq R$, $\|\Phi_R - I\| = \|\Phi_{R_0} - I\| = 0$.

Case 1b: $\mathcal{Y} \supset \bigcup_{i=1}^{n+1} U_i$. Let $U_{n+2} = \bigcap_{i=1}^{n+1} \widetilde{U}_i \in \mathcal{B}_{\mathrm{B}}(\mathcal{Y})$, which is nonempty. Let $y_{n+2} \in U_{n+2}$ be such that $\|y_{n+2}\| < \inf_{y\in U_{n+2}} \|y\| + 1$. Let $R_0 := \left\{(y_i, U_i) \,\middle|\, i = 1, \ldots, n+2\right\}$. It is easy to check that $R_0 \in \mathfrak{I}(\mathcal{Y})$. $\forall \tilde{R} \in \mathfrak{I}(\mathcal{Y})$ with $R_0 \preceq \tilde{R}$, then $\tilde{R} = \left\{(\tilde{y}_\alpha, \tilde{U}_\alpha) \,\middle|\, \alpha \in \Lambda\right\}$, where Λ is a finite index set, $\left(\tilde{U}_\alpha\right)_{\alpha\in\Lambda} \subseteq \mathcal{B}_{\mathrm{B}}(\mathcal{Y})$ are nonempty and pairwise disjoint, $\bigcup_{\alpha\in\Lambda} \tilde{U}_\alpha = \mathcal{Y}$, $\tilde{y}_\alpha \in \tilde{U}_\alpha$, and $\|\tilde{y}_\alpha\| < \inf_{y\in\tilde{U}_\alpha} \|y\| + 1$, $\forall \alpha \in \Lambda$. $\forall i \in \{1, \ldots, n+2\}$, by $R_0 \preceq \tilde{R}$, $\exists!\, \alpha_i \in \Lambda$ such that $y_i = \tilde{y}_{\alpha_i}$ and, by $\tilde{R} \in \mathfrak{I}(\mathcal{Y})$, $\exists! \bar{\alpha}_i \in \Lambda$ such that $y_i \in \tilde{U}_{\bar{\alpha}_i}$. Since $\left(\tilde{U}_\alpha\right)_{\alpha\in\Lambda} \subseteq \mathcal{B}_{\mathrm{B}}(\mathcal{Y})$ are nonempty and pairwise disjoint and $\tilde{y}_\alpha \in \tilde{U}_\alpha$, $\forall \alpha \in \Lambda$, then $\alpha_i = \bar{\alpha}_i$. Again by $R_0 \preceq \tilde{R}$, we have $\tilde{U}_{\alpha_i} \subseteq U_i$. Hence, $\alpha_1, \ldots, \alpha_{n+2}$ are distinct. Let $\bar{\Lambda} := \{\alpha_1, \ldots, \alpha_{n+2}\}$. $\forall \alpha \in \Lambda \setminus \bar{\Lambda}$, $\phi_{\mathrm{inv}}(\tilde{U}_\alpha) = \emptyset$. Then, $\Phi_{\tilde{R}} = \sum_{\alpha\in\Lambda} \tilde{y}_\alpha \mu(\phi_{\mathrm{inv}}(\tilde{U}_\alpha)) = \sum_{i=1}^{n+2} \tilde{y}_{\alpha_i} \mu(\phi_{\mathrm{inv}}(\tilde{U}_{\alpha_i})) = \sum_{i=1}^{n+2} y_i \mu(\phi_{\mathrm{inv}}(\tilde{U}_{\alpha_i}))$. Note that $\forall i \in \{1, \ldots, n+1\}$, $\phi_{\mathrm{inv}}(\tilde{U}_{\alpha_i}) = A_i$ and $\mu(A_i) \leq \mu(X) < +\infty$. Furthermore, $\phi_{\mathrm{inv}}(\tilde{U}_{\alpha_{n+2}}) = \emptyset$. Hence, $\Phi_{\tilde{R}} = \sum_{i=1}^{n} y_i \mu(A_i) = I$. Then, we have $\|\Phi_{\tilde{R}} - I\| = 0$.

In both subcases, we have $\exists R_0 \in \mathfrak{I}(\mathcal{Y})$, $\forall R \in \mathfrak{I}(\mathcal{Y})$ with $R_0 \preceq R$, we have $\|\Phi_R - I\| = 0 < \epsilon$. Hence, $\int_X \phi \, d\mu = \lim_{R\in\mathfrak{I}(\mathcal{Y})} \Phi_R = I$.

Case 2: $\mu(X) = +\infty$. $\forall A \in \mathcal{M}(\mathcal{X})$, $\mu(A) < +\infty$. $\phi|_A : A \to \mathcal{Y}$ is a simple function. By Case 1, $\Phi_A = \int_A \phi|_A \, d\mu_A \in \mathcal{Y}$ is well-defined, where $(A, \mathcal{B}_A, \mu_A)$ is the finite measure subspace of $\mathcal{X}$. Hence, the net $(\Phi_A)_{A\in\mathcal{M}(\mathcal{X})}$ is well-defined. Take $A_0 := \bigcup_{i=1}^{n} A_i \in \mathcal{B}$. Then, $\mu(A_0) = \sum_{i=1}^{n} \mu(A_i) < +\infty$. Therefore, $A_0 \in \mathcal{M}(\mathcal{X})$. $\forall A \in \mathcal{M}(\mathcal{X})$ with $A_0 \subseteq A$, by Case 1, $\Phi_A = \sum_{i=1}^{n} y_i \mu(A_i) = I$. Then, $\int_X \phi \, d\mu = \lim_{A\in\mathcal{M}(\mathcal{X})} \Phi_A = I$.

This completes the proof of the proposition. □

Clearly, the above result holds for simple functions given in any form, not necessarily in canonical representation. It also shows that integrals of simple functions are linear, that is, $\forall$ simple functions ϕ_1 and ϕ_2, $\forall c \in \mathbb{K}$, we have $\int_X (\phi_1 + \phi_2) \, d\mu = \int_X \phi_1 \, d\mu + \int_X \phi_2 \, d\mu$ and $\int_X (c\phi_1) \, d\mu = c \int_X \phi_1 \, d\mu$.

Lemma 11.76 *Let $\mathcal{X} := (X, \mathcal{B}, \mu)$ be a finite measure space, $\mathcal{Y}$ be a separable Banach space, $\phi_n : X \to \mathcal{Y}$ be a simple function in canonical representation, $\forall n \in \mathbb{N}$, and $f : X \to \mathcal{Y}$ be $\mathcal{B}$-measurable. Assume that $\lim_{n\in\mathbb{N}} \phi_n = f$ a.e. in $\mathcal{X}$ and*

$\exists M \in [0, \infty) \subset \mathbb{R}$ such that $\|\phi_n(x)\| \leq M$, $\forall x \in X$, $\forall n \in \mathbb{N}$. Then, f is integrable over X and $\int_X f \,\mathrm{d}\mu = \lim_{n\in\mathbb{N}} \int_X \phi_n \,\mathrm{d}\mu \in \mathcal{Y}$.

Proof Let $E := \{x \in X \mid (\phi_n(x))_{n=1}^{\infty}$ does not converge to $f(x)\}$. Then, $E \in \mathcal{B}$ and $\mu(E) = 0$. Let $(F_R)_{R\in\mathfrak{I}(\mathcal{Y})}$ be the net for $\int_X f \,\mathrm{d}\mu$ as defined in Definition 11.70.

$\forall \epsilon \in (0, 1) \subset \mathbb{R}$, by Egoroff's Theorem 11.55, $\exists E_1 \in \mathcal{B}$ with $\mu(E_1) < \frac{\epsilon}{4M+2}$ such that $\left(\phi_n|_{X\setminus E_1}\right)_{n=1}^{\infty}$ converges uniformly to $f|_{X\setminus E_1}$. Clearly, $E \subseteq E_1$. Then, $\exists n_0 \in \mathbb{N}$, $\forall n \in \mathbb{N}$ with $n \geq n_0$, $\forall x \in X \setminus E_1$, $\|\phi_n(x) - f(x)\| < \frac{\epsilon}{6\mu(X)+1} =: \bar{\epsilon}$. Fix any $n \geq n_0$. Let ϕ_n admit the canonical representation $\phi_n = \sum_{i=1}^{\bar{n}} y_i \chi_{A_i,X}$, where $\bar{n} \in \mathbb{Z}_+$, $y_1, \ldots, y_{\bar{n}} \in \mathcal{Y}$ are distinct and none equals to $\vartheta_{\mathcal{Y}}$, and $A_1, \ldots, A_{\bar{n}} \in \mathcal{B}$ are nonempty, pairwise disjoint, and of finite measure. Let $y_{\bar{n}+1} := \vartheta_{\mathcal{Y}}$ and $A_{\bar{n}+1} := X \setminus (\bigcup_{i=1}^{\bar{n}} A_i) \in \mathcal{B}$. Then, $X = \bigcup_{i=1}^{\bar{n}+1} A_i$ and the sets in the union are pairwise disjoint and of finite measure. Define $\bar{V}_i := \mathcal{B}_{\mathcal{Y}}(y_i, \bar{\epsilon})$, $i = 1, \ldots, \bar{n}+1$. Define $V_1 := \bar{V}_1 \in \mathcal{B}_{\mathrm{B}}(\mathcal{Y})$, $V_{i+1} := \bar{V}_{i+1} \setminus (\bigcup_{j=1}^{i} \bar{V}_j) \in \mathcal{B}_{\mathrm{B}}(\mathcal{Y})$, $i = 1, \ldots, \bar{n}$. Let $V_{\bar{n}+2} := \mathcal{Y} \setminus (\bigcup_{j=1}^{\bar{n}+1} \bar{V}_j)$. Clearly, we may form a representation $\bar{R} := \{(\bar{y}_i, V_i) \mid i = 1, \ldots, \bar{n}+2, V_i \neq \emptyset\} \in \mathfrak{I}(\mathcal{Y})$, where $\bar{y}_i \in V_i$, $i = 1, \ldots, \bar{n}+2$, are any vectors that satisfy the assumption of Proposition 11.69. $\forall R \in \mathfrak{I}(\mathcal{Y})$ with $\bar{R} \preceq R$. Let $R := \left\{(\hat{y}_\gamma, \hat{U}_\gamma) \,\middle|\, \gamma \in \Gamma\right\}$. $\forall \gamma \in \Gamma$, by $\bar{R} \preceq R$, $\exists! i_\gamma \in \{1, \ldots, \bar{n}+2\}$ such that $\hat{U}_\gamma \subseteq V_{i_\gamma}$. Define $\bar{\Gamma} := \{\gamma \in \Gamma \mid i_\gamma = \bar{n}+2\}$. Let $A_{\gamma,i} := A_i \cap f_{\mathrm{inv}}(\hat{U}_\gamma) \cap (X \setminus E_1) \in \mathcal{B}$, $i = 1, \ldots, \bar{n}+1$.

Claim 11.76.1 $\forall \gamma \in \bar{\Gamma}$, $f_{\mathrm{inv}}(\hat{U}_\gamma) \subseteq E_1$ and $\left\|\hat{y}_\gamma \mu(f_{\mathrm{inv}}(\hat{U}_\gamma) \cap E_1)\right\| \leq (M+1)\mu(f_{\mathrm{inv}}(\hat{U}_\gamma) \cap E_1) = (M+1)\mu(f_{\mathrm{inv}}(\hat{U}_\gamma))$.

Proof of Claim Fix any $\gamma \in \bar{\Gamma}$. $\hat{U}_\gamma \subseteq V_{\bar{n}+2} = \mathcal{Y} \setminus (\bigcup_{j=1}^{\bar{n}+1} \bar{V}_j)$. Then, $\mathcal{B} \ni f_{\mathrm{inv}}(\hat{U}_\gamma) \subseteq E_1$. We will distinguish two exhaustive and mutually exclusive cases: Case 1: $f_{\mathrm{inv}}(\hat{U}_\gamma) \subseteq E$; Case 2: $f_{\mathrm{inv}}(\hat{U}_\gamma) \cap (E_1 \setminus E) \neq \emptyset$. Case 1: $f_{\mathrm{inv}}(\hat{U}_\gamma) \subseteq E$. Then, $\mu(f_{\mathrm{inv}}(\hat{U}_\gamma) \cap E_1) = 0$ and $\left\|\hat{y}_\gamma \mu(f_{\mathrm{inv}}(\hat{U}_\gamma) \cap E_1)\right\| = 0 = (M+1)\mu(f_{\mathrm{inv}}(\hat{U}_\gamma) \cap E_1)$. Case 2: $f_{\mathrm{inv}}(\hat{U}_\gamma) \cap (E_1 \setminus E) \neq \emptyset$. Then, $\exists x \in E_1 \setminus E$ such that $f(x) \in \hat{U}_\gamma$. Note that $\lim_{m\in\mathbb{N}} \phi_m(x) = f(x)$ and $\|\phi_m(x)\| \leq M$, $\forall m \in \mathbb{N}$. Then, by Propositions 7.21 and 3.66, we have $\|f(x)\| = \lim_{m\in\mathbb{N}} \|\phi_m(x)\| \leq M$. Then, by $R \in \mathfrak{I}(\mathcal{Y})$, $\left\|\hat{y}_\gamma\right\| < \inf_{y\in\hat{U}_\gamma} \|y\| + 1 \leq M+1$. Therefore, $\left\|\hat{y}_\gamma \mu(f_{\mathrm{inv}}(\hat{U}_\gamma) \cap E_1)\right\| \leq (M+1)\mu(f_{\mathrm{inv}}(\hat{U}_\gamma) \cap E_1)$. This completes the proof of the claim. □

$\forall \gamma \in \Gamma \setminus \bar{\Gamma}$, $\hat{U}_\gamma \subseteq V_{i_\gamma}$, where $i_\gamma \in \{1, \ldots, \bar{n}+1\}$. $\forall i \in \{1, \ldots, \bar{n}+1\}$ with $A_{\gamma,i} \neq \emptyset$, then $\exists x \in A_{\gamma,i}$, $\phi_n(x) = y_i$, and $f(x) \in \mathcal{B}_{\mathcal{Y}}(y_i, \bar{\epsilon}) \cap \hat{U}_\gamma \subseteq \mathcal{B}_{\mathcal{Y}}(y_i, \bar{\epsilon}) \cap V_{i_\gamma} \subseteq \mathcal{B}_{\mathcal{Y}}(y_i, \bar{\epsilon}) \cap \bar{V}_{i_\gamma} \subseteq \mathcal{B}_{\mathcal{Y}}(y_i, \bar{\epsilon}) \cap \mathcal{B}_{\mathcal{Y}}\left(y_{i_\gamma}, \bar{\epsilon}\right)$. Note that $\hat{y}_\gamma \in \hat{U}_\gamma \subseteq \mathcal{B}_{\mathcal{Y}}\left(y_{i_\gamma}, \bar{\epsilon}\right)$. Therefore, we have $\left\|\hat{y}_\gamma - y_i\right\| \leq \left\|\hat{y}_\gamma - y_{i_\gamma}\right\| + \left\|y_{i_\gamma} - f(x)\right\| + \|f(x) - y_i\| < 3\bar{\epsilon}$.

Note that $F_R = \sum_{\gamma\in\Gamma} \hat{y}_\gamma \mu(f_{\mathrm{inv}}(\hat{U}_\gamma))$ and $\int_X \phi_n \, d\mu = \sum_{i=1}^{\bar{n}} y_i \mu(A_i)$, by Proposition 11.75. Then,

$$\begin{aligned}
\Big\| F_R - \int_X \phi_n \, d\mu \Big\| &= \Big\| \sum_{\gamma\in\Gamma} \hat{y}_\gamma \mu(f_{\mathrm{inv}}(\hat{U}_\gamma)) - \sum_{i=1}^{\bar{n}} y_i \mu(A_i) \Big\| \\
&= \Big\| \sum_{\gamma\in\Gamma} \hat{y}_\gamma (\mu(f_{\mathrm{inv}}(\hat{U}_\gamma) \cap (X \setminus E_1)) + \mu(f_{\mathrm{inv}}(\hat{U}_\gamma) \cap E_1)) \\
&\quad - \sum_{i=1}^{\bar{n}+1} y_i (\mu(A_i \cap E_1) + \mu(A_i \cap (X \setminus E_1))) \Big\| \\
&= \Big\| \sum_{\gamma\in\Gamma} \hat{y}_\gamma (\sum_{i=1}^{\bar{n}+1} \mu(A_{\gamma,i}) + \mu(f_{\mathrm{inv}}(\hat{U}_\gamma) \cap E_1)) - \sum_{i=1}^{\bar{n}+1} y_i (\mu(A_i \cap E_1) \\
&\quad + \sum_{\gamma\in\Gamma} \mu(A_{\gamma,i})) \Big\| \\
&= \Big\| \sum_{\gamma\in\Gamma} \sum_{i=1}^{\bar{n}+1} \hat{y}_\gamma \mu(A_{\gamma,i}) + \sum_{\gamma\in\Gamma} \hat{y}_\gamma \mu(f_{\mathrm{inv}}(\hat{U}_\gamma) \cap E_1) - \sum_{i=1}^{\bar{n}+1} y_i \mu(A_i \cap E_1) \\
&\quad - \sum_{\gamma\in\Gamma} \sum_{i=1}^{\bar{n}+1} y_i \mu(A_{\gamma,i}) \Big\| \\
&= \Big\| \sum_{\gamma\in\Gamma} \sum_{i=1}^{\bar{n}+1} (\hat{y}_\gamma - y_i) \mu(A_{\gamma,i}) + \sum_{\gamma\in\Gamma} \hat{y}_\gamma \mu(f_{\mathrm{inv}}(\hat{U}_\gamma) \cap E_1) \\
&\quad - \sum_{i=1}^{\bar{n}+1} y_i \mu(A_i \cap E_1) \Big\| \\
&= \Big\| \sum_{\gamma\in\Gamma\setminus\bar{\Gamma}} \sum_{i=1}^{\bar{n}+1} (\hat{y}_\gamma - y_i) \mu(A_{\gamma,i}) + \sum_{\gamma\in\bar{\Gamma}} \sum_{i=1}^{\bar{n}+1} (\hat{y}_\gamma - y_i) \mu(A_{\gamma,i}) \\
&\quad + \sum_{\gamma\in\Gamma} \hat{y}_\gamma \mu(f_{\mathrm{inv}}(\hat{U}_\gamma) \cap E_1) - \sum_{i=1}^{\bar{n}+1} y_i \mu(A_i \cap E_1) \Big\| \\
&= \Big\| \sum_{\gamma\in\Gamma\setminus\bar{\Gamma}} \sum_{i=1}^{\bar{n}+1} (\hat{y}_\gamma - y_i) \mu(A_{\gamma,i}) + \sum_{\gamma\in\Gamma} \hat{y}_\gamma \mu(f_{\mathrm{inv}}(\hat{U}_\gamma) \cap E_1) \\
&\quad - \sum_{i=1}^{\bar{n}+1} y_i \mu(A_i \cap E_1) \Big\|
\end{aligned}$$

$$\leq \sum_{\gamma\in\Gamma\setminus\bar{\Gamma}} \sum_{i=1}^{\bar{n}+1} \|\hat{y}_\gamma - y_i\|\mu(A_{\gamma,i}) + \sum_{\gamma\in\Gamma} \left\| \hat{y}_\gamma \mu(f_{\text{inv}}(\hat{U}_\gamma) \cap E_1) \right\|$$

$$+ \sum_{i=1}^{\bar{n}+1} \|y_i\|\mu(A_i \cap E_1)$$

$$\leq 3\bar{\epsilon} \sum_{\gamma\in\Gamma\setminus\bar{\Gamma}} \sum_{i=1}^{\bar{n}+1} \mu(A_{\gamma,i}) + \sum_{\gamma\in\Gamma\setminus\bar{\Gamma}} \left\| \hat{y}_\gamma \mu(f_{\text{inv}}(\hat{U}_\gamma) \cap E_1) \right\|$$

$$+ \sum_{\gamma\in\bar{\Gamma}} \left\| \hat{y}_\gamma \mu(f_{\text{inv}}(\hat{U}_\gamma) \cap E_1) \right\| + \sum_{i=1}^{\bar{n}+1} \|y_i\|\mu(A_i \cap E_1)$$

$$\leq \frac{3\mu(X)\epsilon}{6\mu(X)+1} + \sum_{\gamma\in\Gamma\setminus\bar{\Gamma}} (M+1)\mu(f_{\text{inv}}(\hat{U}_\gamma) \cap E_1))$$

$$+ \sum_{\gamma\in\bar{\Gamma}} (M+1)\mu(f_{\text{inv}}(\hat{U}_\gamma) \cap E_1)) + \sum_{i=1}^{\bar{n}+1} M\mu(A_i \cap E_1)$$

$$< \epsilon/2 + \sum_{\gamma\in\Gamma} (M+1)\mu(f_{\text{inv}}(\hat{U}_\gamma) \cap E_1)) + \sum_{i=1}^{\bar{n}+1} M\mu(A_i \cap E_1)$$

$$= \epsilon/2 + (2M+1)\mu(E_1) < \epsilon$$

where the seventh equality follows from the fact that $\forall\gamma \in \bar{\Gamma}$, $A_{\gamma,i} \subseteq f_{\text{inv}}(\hat{U}_\gamma) \cap (X \setminus E_1) = \emptyset$ and the third inequality follows from the fact that $\forall\gamma \in \Gamma \setminus \bar{\Gamma}$, $\hat{y}_\gamma \in \hat{U}_\gamma \subseteq V_{i_\gamma} \subseteq \bar{V}_{i_\gamma} = \mathcal{B}_{\mathcal{Y}}(y_{i_\gamma}, \bar{\epsilon})$ and $\|\hat{y}_\gamma\| \leq \|y_{i_\gamma}\| + \|\hat{y}_\gamma - y_{i_\gamma}\| < M + \bar{\epsilon} < M + 1$.

In summary, we have shown that $\forall\epsilon \in (0, 1) \subset \mathbb{R}$, $\exists n_0 \in \mathbb{N}$, $\forall n \in \mathbb{N}$ with $n \geq n_0$, $\exists\bar{R} \in \mathcal{I}(\mathcal{Y})$ such that $\forall R \in \mathcal{I}(\mathcal{Y})$ with $\bar{R} \preceq R$, we have $\left\| F_R - \int_X \phi_n \, d\mu \right\| < \epsilon$. Then, the net $(F_R)_{R\in\mathcal{I}(\mathcal{Y})}$ is a Cauchy net. By Proposition 4.44, the net admits a limit $\int_X f \, d\mu \in \mathcal{Y}$. By Propositions 4.30, 3.66, and 3.67, we have $\left\| \int_X f \, d\mu - \int_X \phi_n \, d\mu \right\| = \lim_{R\in\mathcal{I}(\mathcal{Y})} \left\| F_R - \int_X \phi_n \, d\mu \right\| \leq \epsilon$, $\forall n \geq n_0$. Hence, we have $\int_X f \, d\mu = \lim_{n\in\mathbb{N}} \int_X \phi_n \, d\mu$. This completes the proof of the lemma. □

Theorem 11.77 (Bounded Convergence Theorem) *Let* $(X, \mathcal{B}, \mu) =: \mathcal{X}$ *be a finite measure space,* $\mathcal{Y}$ *be a separable Banach space,* $f_n : X \to \mathcal{Y}$ *be* $\mathcal{B}$*-measurable,* $\forall n \in \mathbb{N}$*, and* $f : X \to \mathcal{Y}$ *be* $\mathcal{B}$*-measurable. Assume that* $\lim_{n\in\mathbb{N}} f_n = f$ *a.e. in* $\mathcal{X}$ *and* $\exists M \in [0, \infty) \subset \mathbb{R}$ *such that* $\|f_n(x)\| \leq M$, $\forall x \in X$, $\forall n \in \mathbb{N}$. *Then,* f *is integrable over* X *and* $\int_X f \, d\mu = \lim_{n\in\mathbb{N}} \int_X f_n \, d\mu \in \mathcal{Y}$.

Proof Let $U := \overline{\mathcal{B}}_{\mathcal{Y}}(\vartheta_{\mathcal{Y}}, M)$, which is a conic segment. Then, $\forall n \in \mathbb{N}$, $f_n : X \to U \subseteq \mathcal{Y}$. By Proposition 11.66, there exists a sequence of simple functions $(\phi_{n,i})_{i=1}^\infty$, $\phi_{n,i} : X \to U$, $\forall i \in \mathbb{N}$, that converges to f_n a.e. in $\mathcal{X}$ and $\|\phi_{n,i}(x)\| \leq \|f_n(x)\| \leq$

M, $\forall x \in X$, $\forall i \in \mathbb{N}$. By Lemma 11.76, $\int_X f_n \, d\mu = \lim_{i \in \mathbb{N}} \int_X \phi_{n,i} \, d\mu \in \mathcal{Y}$. By Proposition 11.58, we have $\lim_{i \in \mathbb{N}} \phi_{n,i} = f_n$ in measure in $\mathcal{X}$. Then, $\exists i_n \in \mathbb{N}$ such that $\left\| \int_X f_n \, d\mu - \int_X \phi_{n,i_n} \, d\mu \right\| < 2^{-n}$ and $\mu(E_n) := \mu(\{x \in X \mid \left\| f_n(x) - \phi_{n,i_n}(x) \right\| \geq 2^{-n}\}) < 2^{-n}$. Denote $\bar{\phi}_n := \phi_{n,i_n}$.

By Proposition 11.58, $\lim_{n \in \mathbb{N}} f_n = f$ in measure in $\mathcal{X}$. $\forall \epsilon \in (0, \infty) \subset \mathbb{R}$, $\exists n_0 \in \mathbb{N}$ such that $\forall n \in \mathbb{N}$ with $n \geq n_0$, we have $2^{-n} < \epsilon/2$ and $\mu(\bar{E}_n) := \mu(\{x \in X \mid \| f_n(x) - f(x) \| \geq \epsilon/2\}) < \epsilon/2$. Note that, by Lemma 11.43, $\mathcal{B} \ni F_n := \{x \in X \mid \left\| \bar{\phi}_n(x) - f(x) \right\| \geq \epsilon\} \subseteq \{x \in X \mid \left\| \bar{\phi}_n(x) - f_n(x) \right\| + \| f_n(x) - f(x) \| \geq \epsilon\} \subseteq E_n \cup \bar{E}_n$. Then, we have $\mu(F_n) \leq \mu(E_n) + \mu(\bar{E}_n) < \epsilon$. This implies that $\lim_{n \in \mathbb{N}} \bar{\phi}_n = f$ in measure in $\mathcal{X}$. By Proposition 11.57, there exists a subsequence $\left(\bar{\phi}_{n_k}\right)_{k=1}^{\infty}$ of $\left(\bar{\phi}_n\right)_{n=1}^{\infty}$ that converges to f a.e. in $\mathcal{X}$. By Lemma 11.76, f is integrable over X and $\int_X f \, d\mu = \lim_{k \in \mathbb{N}} \int_X \bar{\phi}_{n_k} \, d\mu \in \mathcal{Y}$. By the fact that $\left\| \int_X f_{n_k} \, d\mu - \int_X \bar{\phi}_{n_k} \, d\mu \right\| < 2^{-n_k}$, $\forall k \in \mathbb{N}$, then we have $\lim_{k \in \mathbb{N}} \int_X f_{n_k} \, d\mu = \lim_{k \in \mathbb{N}} \int_X \bar{\phi}_{n_k} \, d\mu = \int_X f \, d\mu \in \mathcal{Y}$. So far, we have shown that for any uniformly bounded $\mathcal{B}$-measurable sequence of functions $(f_n)_{n=1}^{\infty}$ that converges to f a.e. in $\mathcal{X}$, there exists a subsequence $\left(f_{n_k}\right)_{k=1}^{\infty}$ such that $\lim_{k \in \mathbb{N}} \int_X f_{n_k} \, d\mu = \int_X f \, d\mu \in \mathcal{Y}$. This then holds for any subsequence of $(f_n)_{n=1}^{\infty}$. By Proposition 3.71, we have $\lim_{n \in \mathbb{N}} \int_X f_n \, d\mu = \int_X f \, d\mu \in \mathcal{Y}$ and f is integrable over X.

This completes the proof of the theorem. □

Proposition 11.78 *Let $\mathcal{X} := (X, \mathcal{B}, \mu)$ be a measure space and $f : X \to [0, \infty) \subset \mathbb{R}$ be $\mathcal{B}$-measurable. Then,*

$$\int_X f \, d\mu = \sup_{0 \leq \phi \leq f} \int_X \phi \, d\mu =: S \in \mathbb{R}_e$$

where the supremum is over all simple functions $\phi : X \to [0, \infty) \subset \mathbb{R}$ that satisfy $0 \leq \phi(x) \leq f(x)$, $\forall x \in X$.

Proof We will distinguish two exhaustive and mutually exclusive cases: Case 1: $\mu(X) < +\infty$; Case 2: $\mu(X) = +\infty$.

Case 1: $\mu(X) < +\infty$. Let $(F_R)_{R \in \mathcal{J}(\mathbb{R})}$ be the net as defined in Definition 11.70 for $\int_X f \, d\mu$. Clearly, $S \geq 0$. Fix any simple function $\phi : X \to [0, \infty) \subset \mathbb{R}$ such that $0 \leq \phi(x) \leq f(x)$, $\forall x \in X$. Without loss of generality, we may assume that ϕ is in canonical representation. Then, $\exists n \in \mathbb{Z}_+$, $\exists y_1, \ldots, y_n \in [0, \infty) \subset \mathbb{R}$, which are distinct and none equals to 0, and $\exists A_1, \ldots, A_n \in \mathcal{B}$, which are nonempty, pairwise disjoint, and $\mu(A_i) < \infty$, $i = 1, \ldots, n$, such that $\phi = \sum_{i=1}^{n} y_i \chi_{A_i,X}$. Let $y_{n+1} := 0$ and $A_{n+1} := X \setminus (\bigcup_{i=1}^{n} A_i) \in \mathcal{B}$. Then, $X = \bigcup_{i=1}^{n+1} A_i$, and the sets in the union are pairwise disjoint and of finite measure. Without loss of generality, assume that $0 = y_{n+1} < y_n < \cdots < y_1 < +\infty$. Let $I_1 := [y_1, \infty) \subset \mathbb{R}$, $I_i := [y_i, y_{i-1}) \subset \mathbb{R}$, $i = 2, \ldots, n+1$, $I_{n+2} := (-\infty, 0) \subset \mathbb{R}$, and $y_{n+2} = -\frac{1}{2}$. Then, $R_0 := \{(y_i, I_i) \mid i = 1, \ldots, n+2\} \in \mathcal{J}(\mathbb{R})$. $\forall R \in \mathcal{J}(\mathbb{R})$ with $R_0 \preceq R$, let $R = \{(\bar{y}_\alpha, \bar{U}_\alpha) \mid \alpha \in \Lambda\}$. $\forall \alpha \in \Lambda$, by $R_0 \preceq R$, $\exists!\, i_\alpha \in \{1, \ldots, n+2\}$ such that $\bar{U}_\alpha \subseteq I_{i_\alpha}$. Define $\bar{\Lambda} := \{\alpha \in \Lambda \mid i_\alpha = n+2\}$. Note that $\forall \alpha \in \bar{\Lambda}$, $\bar{U}_\alpha \subseteq I_{n+2}$ and

$f_{\text{inv}}(\bar{U}_\alpha) = \emptyset$. Then,

$$F_R = \sum_{\alpha \in \Lambda} \bar{y}_\alpha \mu(f_{\text{inv}}(\bar{U}_\alpha)) = \sum_{\alpha \in \Lambda} \bar{y}_\alpha \sum_{i=1}^{n+1} \mu(A_i \cap f_{\text{inv}}(\bar{U}_\alpha))$$

$$=: \sum_{\alpha \in \Lambda} \bar{y}_\alpha \sum_{i=1}^{n+1} \mu(E_{i,\alpha}) = \sum_{\alpha \in \Lambda \setminus \bar{\Lambda}} \bar{y}_\alpha \sum_{i=1}^{n+1} \mu(E_{i,\alpha})$$

$\forall \alpha \in \Lambda \setminus \bar{\Lambda}$, $\forall i \in \{1, \ldots, n+1\}$, if $E_{i,\alpha} \neq \emptyset$, then $\exists x \in E_{i,\alpha}$, $y_i = \phi(x) \leq f(x) \in \bar{U}_\alpha \subseteq I_{i_\alpha}$, which further implies that $y_i \leq y_{i_\alpha}$. Note that $\bar{y}_\alpha \in \bar{U}_\alpha \subseteq I_{i_\alpha}$ and $\bar{y}_\alpha \geq y_{i_\alpha}$. Then, $\bar{y}_\alpha \mu(E_{i,\alpha}) \geq y_i \mu(E_{i,\alpha})$. If $E_{i,\alpha} = \emptyset$, clearly, $0 = \bar{y}_\alpha \mu(E_{i,\alpha}) = y_i \mu(E_{i,\alpha}) = 0$. Therefore, we have $\bar{y}_\alpha \mu(E_{i,\alpha}) \geq y_i \mu(E_{i,\alpha})$, $\forall \alpha \in \Lambda \setminus \bar{\Lambda}$, $\forall i \in \{1, \ldots, n+1\}$. This leads to

$$F_R = \sum_{\alpha \in \Lambda \setminus \bar{\Lambda}} \sum_{i=1}^{n+1} \bar{y}_\alpha \mu(E_{i,\alpha}) \geq \sum_{\alpha \in \Lambda \setminus \bar{\Lambda}} \sum_{i=1}^{n+1} y_i \mu(E_{i,\alpha}) = \sum_{i=1}^{n+1} \sum_{\alpha \in \Lambda \setminus \bar{\Lambda}} y_i \mu(E_{i,\alpha})$$

$$= \sum_{i=1}^{n+1} \sum_{\alpha \in \Lambda} y_i \mu(E_{i,\alpha}) = \sum_{i=1}^{n+1} y_i \mu(A_i) = \int_X \phi \, d\mu$$

where the last equality follows from Proposition 11.75. In summary, we have shown that $\forall R \in \mathcal{J}(\mathbb{R})$ with $R_0 \preceq R$, we have $F_R \geq \int_X \phi \, d\mu$. Then, $\liminf_{R \in \mathcal{J}(\mathbb{R})} F_R \geq \int_X \phi \, d\mu$. By the arbitrariness of ϕ, we have $\liminf_{R \in \mathcal{J}(\mathbb{R})} F_R \geq S$.

We will show that $\limsup_{R \in \mathcal{J}(\mathbb{R})} F_R \leq S$ by an argument of contradiction. Suppose $\bar{S} := \limsup_{R \in \mathcal{J}(\mathbb{R})} F_R > S$. Then, $\exists \epsilon_0 \in (0, \infty) \subset \mathbb{R}$, $\forall R_0 \in \mathcal{J}(\mathbb{R})$, $\exists R \in \mathcal{J}(\mathbb{R})$ with $R_0 \preceq R$, such that $F_R > S + \epsilon_0$. Let $\epsilon_1 := \min\{1, \frac{\epsilon_0}{2+2\mu(X)}\} > 0$. By Proposition 11.5, $\exists E_1 \in \mathcal{B}$ with $\mu(E_1) < \epsilon_1$ and $\exists M \in \mathbb{N}$ such that $0 \leq f(x) \leq M - 1$, $\forall x \in X \setminus E_1$. Then, $\exists n \in \mathbb{N}$ such that $M/n < \epsilon_1$. Let $U_i := [(i-1)M/n, iM/n) \subset \mathbb{R}$, $y_i := (i-1)M/n$, $i = 1, \ldots, n$, $U_{n+1} := [M, \infty) \subset \mathbb{R}$, $y_{n+1} := M$, $U_{n+2} := (-\infty, 0) \subset \mathbb{R}$, and $y_{n+2} = -\frac{1}{2}$. Then, $R_0 := \{(y_i, U_i) \mid i = 1, \ldots, n+2\} \in \mathcal{J}(\mathbb{R})$. Then, $\exists R \in \mathcal{J}(\mathbb{R})$ with $R_0 \preceq R$ such that $F_R > S + \epsilon_0$. Let $R = \{(\bar{y}_\alpha, \bar{U}_\alpha) \mid \alpha \in \Lambda\}$. $\forall \alpha \in \Lambda$, $\exists! i_\alpha \in \{1, \ldots, n+2\}$ such that $\bar{U}_\alpha \subseteq U_{i_\alpha}$. Define $\bar{\Lambda} := \{\alpha \in \Lambda \mid i_\alpha = n+2\}$, $\Lambda_1 := \{\alpha \in \Lambda \mid i_\alpha = 1, \ldots, n\}$, and $\Lambda_2 := \{\alpha \in \Lambda \mid i_\alpha = n+1\}$. Clearly, $\Lambda = \Lambda_1 \cup \Lambda_2 \cup \bar{\Lambda}$, and the sets in the union are pairwise disjoint. $\forall \alpha \in \bar{\Lambda}$, $\bar{U}_\alpha \subseteq U_{n+2} = (-\infty, 0) \subset \mathbb{R}$. Then, $f_{\text{inv}}(\bar{U}_\alpha) = \emptyset$. $\forall \alpha \in \Lambda_1$, note that $\bar{y}_\alpha \in \bar{U}_\alpha \subseteq U_{i_\alpha}$, and then $y_{i_\alpha} \leq \bar{y}_\alpha < y_{i_\alpha+1}$. Let $\tilde{y}_\alpha := \inf_{y \in \bar{U}_\alpha} y \in \overline{\bar{U}_\alpha} \subseteq \overline{U_{i_\alpha}}$. Then, $y_{i_\alpha} \leq \tilde{y}_\alpha \leq y_{i_\alpha+1}$. Let $A_\alpha := f_{\text{inv}}(\bar{U}_\alpha)$. $\forall \alpha \in \Lambda_2$, note that $\bar{y}_\alpha \in \bar{U}_\alpha \subseteq U_{n+1} = [M, \infty) \subset \mathbb{R}$ and $\bar{y}_\alpha = |\bar{y}_\alpha| < \inf_{y \in \bar{U}_\alpha} |y| + 1 = \inf_{y \in \bar{U}_\alpha} y + 1$. Let $\tilde{y}_\alpha := \inf_{y \in \bar{U}_\alpha} y$. Then, $\tilde{y}_\alpha \leq \bar{y}_\alpha < \tilde{y}_\alpha + 1$. Let $A_\alpha := f_{\text{inv}}(\bar{U}_\alpha)$. Since $\bar{U}_\alpha \subseteq U_{n+1}$, then $f_{\text{inv}}(\bar{U}_\alpha) \subseteq E_1$.

Define a simple function $\phi : X \to [0,\infty) \subset \mathbb{R}$ by $\phi := \sum_{\alpha \in \Lambda_1 \cup \Lambda_2} \tilde{y}_\alpha \chi_{A_\alpha,X}$. Clearly, $0 \leq \phi(x) \leq f(x)$, $\forall x \in X$. Then, by Proposition 11.75,

$$\begin{aligned}
\int_X \phi \,\mathrm{d}\mu &= \sum_{\alpha \in \Lambda_1 \cup \Lambda_2} \tilde{y}_\alpha \mu(A_\alpha) = \sum_{\alpha \in \Lambda_1 \cup \Lambda_2} \tilde{y}_\alpha \mu(f_{\mathrm{inv}}(\bar{U}_\alpha)) - F_R + F_R \\
&> S + \epsilon_0 + \sum_{\alpha \in \Lambda_1 \cup \Lambda_2} \tilde{y}_\alpha \mu(f_{\mathrm{inv}}(\bar{U}_\alpha)) - F_R \\
&= S + \epsilon_0 + \sum_{\alpha \in \Lambda_1 \cup \Lambda_2} (\tilde{y}_\alpha - \bar{y}_\alpha) \mu(f_{\mathrm{inv}}(\bar{U}_\alpha)) \\
&\geq S + \epsilon_0 - \sum_{\alpha \in \Lambda_1} |\tilde{y}_\alpha - \bar{y}_\alpha| \mu(f_{\mathrm{inv}}(\bar{U}_\alpha)) - \sum_{\alpha \in \Lambda_2} |\tilde{y}_\alpha - \bar{y}_\alpha| \mu(f_{\mathrm{inv}}(\bar{U}_\alpha)) \\
&\geq S + \epsilon_0 - M/n \sum_{\alpha \in \Lambda_1} \mu(f_{\mathrm{inv}}(\bar{U}_\alpha)) - \sum_{\alpha \in \Lambda_2} \mu(f_{\mathrm{inv}}(\bar{U}_\alpha)) \\
&\geq S + \epsilon_0 - \epsilon_1 \mu(X) - \mu(E_1) > S
\end{aligned}$$

This contradicts with the definition of S. Therefore, $\limsup_{R \in \mathcal{J}(\mathbb{R})} F_R \leq S$.

Then, we have $S \leq \liminf_{R \in \mathcal{J}(\mathbb{R})} F_R \leq \limsup_{R \in \mathcal{J}(\mathbb{R})} F_R \leq S$. By Proposition 3.83, $S = \lim_{R \in \mathcal{J}(\mathbb{R})} F_R = \int_X f \,\mathrm{d}\mu$.

Case 2: $\mu(X) = +\infty$. Let $(F_A)_{A \in \mathcal{M}(\mathcal{X})}$ be the net as defined in Definition 11.71. $\forall A \in \mathcal{M}(\mathcal{X})$, $A \in \mathcal{B}$ and $\mu(A) < \infty$. Let $(A, \mathcal{B}_A, \mu_A)$ be the measure subspace of $\mathcal{X}$ as defined in Proposition 11.13. By Case 1, $F_A = \int_A f|_A \,\mathrm{d}\mu_A = \sup_{0 \leq \phi \leq f|_A} \int_A \phi \,\mathrm{d}\mu_A \in \mathbb{R}_e$, where supremum is over all simple functions $\phi : A \to [0,\infty) \subset \mathbb{R}$ with $0 \leq \phi(x) \leq f(x)$, $\forall x \in A$. Then, the net $(F_A)_{A \in \mathcal{M}(\mathcal{X})} \subseteq \mathbb{R}_e$ is well-defined.

Fix any simple function $\phi : X \to [0,\infty) \subset \mathbb{R}$ with $0 \leq \phi(x) \leq f(x)$, $\forall x \in X$. Then, $\exists n \in \mathbb{Z}_+$, $\exists y_1, \ldots, y_n \in [0,\infty) \subset \mathbb{R}$, and $\exists A_1, \ldots, A_n \in \mathcal{B}$, which are of finite measure, such that $\phi = \sum_{i=1}^n y_i \chi_{A_i,X}$. Take $A_0 = \bigcup_{i=1}^n A_i \in \mathcal{B}$. Then, $\mu(A_0) \leq \sum_{i=1}^n \mu(A_i) < +\infty$. Then, $A_0 \in \mathcal{M}(\mathcal{X})$. $\forall A \in \mathcal{M}(\mathcal{X})$ with $A_0 \subseteq A$, we have $F_A = \sup_{0 \leq \varphi \leq f|_A} \int_A \varphi \,\mathrm{d}\mu_A \geq \int_A \phi|_A \,\mathrm{d}\mu_A = \sum_{i=1}^n y_i \mu(A_i) = \int_X \phi \,\mathrm{d}\mu$, where the last two equalities follow from Proposition 11.75. Then, $\liminf_{A \in \mathcal{M}(\mathcal{X})} F_A \geq \int_X \phi \,\mathrm{d}\mu$. By the arbitrariness of ϕ, we have $\liminf_{A \in \mathcal{M}(\mathcal{X})} F_A \geq S$.

On the other hand, $\sup_{A \in \mathcal{M}(\mathcal{X})} F_A =: \bar{S} \geq \liminf_{A \in \mathcal{M}(\mathcal{X})} F_A \geq S$. Suppose $\bar{S} > S$. Then, $\exists A \in \mathcal{M}(\mathcal{X})$ such that $S < F_A = \int_A f|_A \,\mathrm{d}\mu_A = \sup_{0 \leq \phi \leq f|_A} \int_A \phi \,\mathrm{d}\mu_A$. This implies that $\exists$ a simple function $\phi : A \to [0,\infty) \subset \mathbb{R}$ with $0 \leq \phi(x) \leq f(x)$, $\forall x \in A$, such that $\int_A \phi \,\mathrm{d}\mu_A > S$. We may extend this simple function on A to a simple function on X that is zero on $X \setminus A$. Let the extended simple function be $\bar{\phi} : X \to [0,\infty) \subset \mathbb{R}$. Then, clearly, $0 \leq \bar{\phi}(x) \leq f(x)$, $\forall x \in X$, and, by Proposition 11.75, $\int_X \bar{\phi} \,\mathrm{d}\mu = \int_A \phi \,\mathrm{d}\mu_A > S$. This contradicts with definition of S. Therefore, we have $\bar{S} \leq S$.

In summary, we have $\underline{S} \le \liminf_{A \in \mathcal{M}(\mathcal{X})} F_A \le \limsup_{A \in \mathcal{M}(\mathcal{X})} F_A \le \sup_{A \in \mathcal{M}(\mathcal{X})} F_A = \bar{S} \le S$. Then, by Proposition 3.83, we have $\int_X f \,\mathrm{d}\mu = \lim_{A \in \mathcal{M}(\mathcal{X})} F_A = S \in \mathbb{R}_e$.

This completes the proof of the proposition. □

Definition 11.79 Let $\mathcal{X} := (X, \mathcal{B}, \mu)$ be a measure space, $f : X \to [0, \infty] \subset \mathbb{R}_e$ be $\mathcal{B}$-measurable. We will define the *integral of* f *over* $\mathcal{X}$ by $\int_X f \,\mathrm{d}\mu := \sup_{0 \le \phi \le f} \int_X \phi \,\mathrm{d}\mu =: S \in [0, \infty] \subset \mathbb{R}_e$, where the supremum is over all simple functions $\phi : X \to [0, \infty) \subset \mathbb{R}$ that satisfy $0 \le \phi(x) \le f(x)$, $\forall x \in X$. We will say that f is integrable if $S \in \mathbb{R}$. ◇

This definition generalizes the definition of integral to nonnegative extended real-valued functions. By Proposition 11.78, it is consistent with our earlier definitions of integral. The reason that we introduce this generalization is that this leads to considerable simplification in the integrals of nonnegative functions over product measure spaces.

Theorem 11.80 (Fatou's Lemma) *Let $\mathcal{X} := (X, \mathcal{B}, \mu)$ be a measure space, $f_n : X \to [0, \infty] \subset \mathbb{R}_e$ be $\mathcal{B}$-measurable, $\forall n \in \mathbb{N}$, and $f : X \to [0, \infty] \subset \mathbb{R}_e$ be $\mathcal{B}$-measurable. Assume that $\lim_{n \in \mathbb{N}} f_n = f$ a.e. in $\mathcal{X}$. Then, $\int_X f \,\mathrm{d}\mu \le \liminf_{n \in \mathbb{N}} \int_X f_n \,\mathrm{d}\mu$.*

Proof By Definition 11.79, $\int_X f \,\mathrm{d}\mu = \sup_{0 \le \phi \le f} \int_X \phi \,\mathrm{d}\mu$, where the supremum is over all simple functions $\phi : X \to [0, \infty) \subset \mathbb{R}$ that satisfy $0 \le \phi(x) \le f(x)$, $\forall x \in X$. Fix any such simple function ϕ. Let $E := \{x \in X \mid \phi(x) > 0\}$. We must have $E \in \mathcal{B}$ and $\mu(E) < +\infty$. Let $M \in [0, \infty) \subset \mathbb{R}$ be such that $0 \le \phi(x) \le M$, $\forall x \in X$. Define $g_n := f_n \wedge \phi$, $\forall n \in \mathbb{N}$. By Proposition 11.40, g_n's are $\mathcal{B}$-measurable. Since $\lim_{n \in \mathbb{N}} f_n = f$ a.e. in $\mathcal{X}$ and $0 \le \phi(x) \le f(x)$, $\forall x \in X$, then $\lim_{n \in \mathbb{N}} g_n = \phi$ a.e. in $\mathcal{X}$. By Bounded Convergence Theorem 11.77 and Proposition 11.75, we have $\int_X \phi \,\mathrm{d}\mu = \int_E \phi|_E \,\mathrm{d}\mu_E = \lim_{n \in \mathbb{N}} \int_E g_n|_E \,\mathrm{d}\mu_E$, where $(E, \mathcal{B}_E, \mu_E)$ is the finite measure subspace of $\mathcal{X}$ as defined in Proposition 11.13. By Definition 11.79, we have $\int_E g_n|_E \,\mathrm{d}\mu_E \le \int_X g_n \,\mathrm{d}\mu \le \int_X f_n \,\mathrm{d}\mu$. Then, $\int_X \phi \,\mathrm{d}\mu \le \liminf_{n \in \mathbb{N}} \int_X f_n \,\mathrm{d}\mu$. By the arbitrariness of ϕ, we have $\int_X f \,\mathrm{d}\mu \le \liminf_{n \in \mathbb{N}} \int_X f_n \,\mathrm{d}\mu$. This completes the proof of the theorem. □

Theorem 11.81 (Monotone Convergence Theorem) *Let $(X, \mathcal{B}, \mu) =: \mathcal{X}$ be a measure space, $f_n : X \to [0, \infty] \subset \mathbb{R}_e$ be $\mathcal{B}$-measurable, $\forall n \in \mathbb{N}$, and $f : X \to [0, \infty] \subset \mathbb{R}_e$ be $\mathcal{B}$-measurable. Assume that $\lim_{n \in \mathbb{N}} f_n = f$ a.e. in $\mathcal{X}$ and $f_n(x) \le f_{n+1}(x)$, $\forall x \in X$, $\forall n \in \mathbb{N}$. Then, $\int_X f \,\mathrm{d}\mu = \lim_{n \in \mathbb{N}} \int_X f_n \,\mathrm{d}\mu$.*

Proof By the assumption, $f_n \le f$ a.e. in $\mathcal{X}$, $\forall n \in \mathbb{N}$. By Definition 11.79 and the fact that any two simple functions that equal to each other almost everywhere have the same integral, we have $\mathbb{R}_e \ni \int_X f \,\mathrm{d}\mu \ge \int_X f_n \,\mathrm{d}\mu \in \mathbb{R}_e$, $\forall n \in \mathbb{N}$. By Fatou's Lemma 11.80, we have $\int_X f \,\mathrm{d}\mu \le \liminf_{n \in \mathbb{N}} \int_X f_n \,\mathrm{d}\mu \le \limsup_{n \in \mathbb{N}} \int_X f_n \,\mathrm{d}\mu \le \int_X f \,\mathrm{d}\mu$. Hence, by Proposition 3.83, we have $\int_X f \,\mathrm{d}\mu = \lim_{n \in \mathbb{N}} \int_X f_n \,\mathrm{d}\mu \in \mathbb{R}_e$. This completes the proof of the proposition. □

Proposition 11.82 *Let $\mathcal{X} := (X, \mathcal{B}, \mu)$ be a σ-finite measure space and $g : X \to [0, \infty] \subset \mathbb{R}_e$ be $\mathcal{B}$-measurable. Assume that $\int_X g \, d\mu =: \lambda < \infty$. Then, define $f : X \to [0, \infty) \subset \mathbb{R}$ by $f(x) = \begin{cases} g(x) & g(x) < \infty \\ 0 & g(x) = \infty \end{cases}$, $\forall x \in X$. Then, f is $\mathcal{B}$-measurable, $f = g$ a.e. in $\mathcal{X}$, and $\int_X f \, d\mu = \int_X g \, d\mu \in \mathbb{R}$.*

Proof Let $E := \{x \in X \mid g(x) < \infty\} \in \mathcal{B}$. We will show that $\mu(X \setminus E) = 0$ by an argument of contradiction. Suppose $\mu(X \setminus E) > 0$. Since $\mathcal{X}$ is σ-finite, $\exists (X_n)_{n=1}^{\infty} \subseteq \mathcal{B}$ such that $X = \bigcup_{n=1}^{\infty} X_n$ and $\mu(X_n) < +\infty$, $\forall n \in \mathbb{N}$. Without loss of generality, we may assume $X_n \subseteq X_{n+1}$, $\forall n \in \mathbb{N}$. Let $B_n := X_n \setminus E \in \mathcal{B}$, $\forall n \in \mathbb{N}$. By Proposition 11.7, we have $\lim_{n \in \mathbb{N}} \mu(B_n) = \mu(X \setminus E) > 0$. Then, $\exists n \in \mathbb{N}$ such that $\mu(B_n) \in (0, +\infty) \subset \mathbb{R}$. Clearly, $\lambda \in [0, \infty) \subset \mathbb{R}$. Let $M := \frac{\lambda+1}{\mu(B_n)} \in (0, \infty) \subset \mathbb{R}$. Let $\tilde{g} := g \wedge M$. By Proposition 11.40, $\tilde{g}$ is $\mathcal{B}$-measurable. Then, $\tilde{g} : X \to [0, M] \subset \mathbb{R}$ and $\lambda = \int_X g \, d\mu \geq \int_X \tilde{g} \, d\mu \geq M\mu(B_n) = \lambda + 1 > \lambda$, where the first inequality follows from Definition 11.79 and the second inequality follows from Definition 11.79 and Proposition 11.75. This is a contradiction. Hence, $\mu(X \setminus E) = 0$.

Clearly, $f : X \to [0, \infty) \subset \mathbb{R}$ satisfies $f(x) = \begin{cases} g(x) & x \in E \\ 0 & x \in X \setminus E \end{cases}$. By Proposition 11.41, f is $\mathcal{B}$-measurable and $f = g$ a.e. in $\mathcal{X}$. By Monotone Convergence Theorem 11.81, $\int_X f \, d\mu = \int_X g \, d\mu = \lambda \in \mathbb{R}$. This completes the proof of the proposition. □

The above proposition says that for a measurable extended real-valued function, if its integral is finite, then it can be modified to be a real-valued function by replacing its ∞ value by zero on a set of measure zero, and the integral remains the same for modified and unmodified functions.

Proposition 11.83 *Let $\mathcal{X} := (X, \mathcal{B}, \mu)$ be a measure space, $f_i : X \to [0, \infty] \subset \mathbb{R}_e$ be $\mathcal{B}$-measurable, $i = 1, 2$, $g_n : X \to [0, \infty] \subset \mathbb{R}_e$ be $\mathcal{B}$-measurable, $\forall n \in \mathbb{N}$, and $g : X \to [0, \infty] \subset \mathbb{R}_e$ be $\mathcal{B}$-measurable. Then, the following statements hold:*

(i) *$\forall c \in (0, \infty) \subset \mathbb{R}$, $\int_X (cg) \, d\mu = c \int_X g \, d\mu$; $\forall c \in [0, \infty) \subset \mathbb{R}$, if $g : X \to [0, \infty)$ and $\int_X g \, d\mu < \infty$, then $\int_X (cg) \, d\mu = c \int_X g \, d\mu$.*
(ii) *If $f_1 \leq f_2$ a.e. in $\mathcal{X}$, then $\int_X f_1 \, d\mu \leq \int_X f_2 \, d\mu$.*
(iii) *$\forall E \in \mathcal{B}$, we have $\int_E g|_E \, d\mu_E =: \int_X (g\chi_{E,X}) \, d\mu =: \int_E g \, d\mu \leq \int_X g \, d\mu$, where $\mathcal{E} := (E, \mathcal{B}_E, \mu_E)$ is the measure subspace of $\mathcal{X}$.*
(iv) *$\int_X (f_1 + f_2) \, d\mu = \int_X f_1 \, d\mu + \int_X f_2 \, d\mu \in \mathbb{R}_e$.*
(v) *$\forall E_1, E_2 \in \mathcal{B}$ with $E_1 \cap E_2 = \emptyset$, we have $\int_{E_1 \cup E_2} g \, d\mu = \int_{E_1} g \, d\mu + \int_{E_2} g \, d\mu \in \mathbb{R}_e$.*
(vi) *$\forall$ pairwise disjoint $(E_n)_{n=1}^{\infty} \subseteq \mathcal{B}$, we have $\int_{\bigcup_{n=1}^{\infty} E_n} g \, d\mu = \sum_{n=1}^{\infty} \int_{E_n} g \, d\mu \in \mathbb{R}_e$.*
(vii) *If g and g_n's are integrable over $\mathcal{X}$, $\lim_{n \in \mathbb{N}} g_n = g$ a.e. in $\mathcal{X}$, and $\lim_{n \in \mathbb{N}} \int_X g_n \, d\mu = \int_X g \, d\mu \in \mathbb{R}$. Then, $\forall E \in \mathcal{B}$, we have $\int_E g \, d\mu = \lim_{n \in \mathbb{N}} \int_E g_n \, d\mu \in \mathbb{R}$.*

Proof (i) and (ii) follow directly from Propositions 11.75 and 3.81 and Definition 11.79.

(iii) Note that $g|_E$ is $\mathcal{B}_E$-measurable by Proposition 11.41. Thus, $g|_E$ is $\mathcal{B}$-measurable. This implies that $g\chi_{E,X}$ is $\mathcal{B}$-measurable. Then, the result follows directly from Definition 11.79 and Proposition 11.75.

(iv) By Propositions 3.89, 11.38, and 11.39 and Example 5.84, $f_1 + f_2$ is $\mathcal{B}$-measurable. By Definition 11.79, we have $\int_X f_1 \, d\mu = \sup_{0 \le \phi \le f_1} \int_X \phi \, d\mu$, $\int_X f_2 \, d\mu = \sup_{0 \le \phi \le f_2} \int_X \phi \, d\mu$, and $\int_X (f_1 + f_2) \, d\mu = \sup_{0 \le \phi \le f_1 + f_2} \int_X \phi \, d\mu$, where the suprema are over all simple functions $\phi : X \to [0, \infty) \subset \mathbb{R}$ satisfying the stated inequalities. Any simple function $\phi_1 : X \to [0, \infty) \subset \mathbb{R}$ and $\phi_2 : X \to [0, \infty) \subset \mathbb{R}$ satisfying $0 \le \phi_1 \le f_1$ and $0 \le \phi_2 \le f_2$, we have $\phi_1 + \phi_2$ is a simple function satisfying $0 \le \phi_1 + \phi_2 \le f_1 + f_2$. Then, $\int_X \phi_1 \, d\mu + \int_X \phi_2 \, d\mu = \int_X (\phi_1 + \phi_2) \, d\mu \le \int_X (f_1 + f_2) \, d\mu$, where the equality follows from Proposition 11.75. Hence, we have $\int_X f_1 \, d\mu + \int_X f_2 \, d\mu \le \int_X (f_1 + f_2) \, d\mu$.

On the other hand, fix any simple function ϕ satisfying $0 \le \phi \le f_1 + f_2$. Let $\bar{f}_1 := \phi \wedge f_1$ and $\bar{f}_2 := \phi - \bar{f}_1$. By Propositions 11.40, 11.38, 11.39, and 7.23, $\bar{f}_1$ and $\bar{f}_2$ are $\mathcal{B}$-measurable. Then, we have $0 \le \bar{f}_1 \le f_1$, $0 \le \bar{f}_2 \le f_2$, $\bar{f}_1(x) = \bar{f}_2(x) = 0$, $\forall x \in X \setminus E$, where $E := \{x \in X \mid \phi(x) > 0\} \in \mathcal{B}$, and $0 \le \bar{f}_1(x) \le \phi(x) \le M$ and $0 \le \bar{f}_2(x) \le \phi(x) \le M$, $\forall x \in X$, for some $M \in [0, \infty) \subset \mathbb{R}$. Note that $\mu(E) < +\infty$ since ϕ is a simple function. Let $\mathcal{E} := (E, \mathcal{B}_E, \mu_E)$ be the finite measure subspace of $\mathcal{X}$ as defined in Proposition 11.13. By Proposition 11.66, there exist sequences of simple functions $(\phi_{1,n})_{n=1}^{\infty}$ and $(\phi_{2,n})_{n=1}^{\infty}$, $\phi_{i,n} : E \to [0, M] \subset \mathbb{R}$, $\forall i \in \{1, 2\}$, $\forall n \in \mathbb{N}$, that converge to $\bar{f}_1\big|_E$ and $\bar{f}_2\big|_E$ a.e. in $\mathcal{E}$, respectively. By Bounded Convergence Theorem 11.77, we have $\int_E \bar{f}_1\big|_E \, d\mu_E = \lim_{n \in \mathbb{N}} \int_E \phi_{1,n} \, d\mu_E$ and $\int_E \bar{f}_2\big|_E \, d\mu_E = \lim_{n \in \mathbb{N}} \int_E \phi_{2,n} \, d\mu_E$. By Propositions 11.52, 7.23, 11.53, $\lim_{n \in \mathbb{N}} (\phi_{1,n} + \phi_{2,n}) = \bar{f}_1\big|_E + \bar{f}_2\big|_E = \phi|_E$ a.e. in $\mathcal{E}$. Again, by Bounded Convergence Theorem 11.77 and Proposition 11.75, we have $\int_E \phi|_E \, d\mu_E = \lim_{n \in \mathbb{N}} \int_E (\phi_{1,n} + \phi_{2,n}) \, d\mu_E = \lim_{n \in \mathbb{N}} (\int_E \phi_{1,n} \, d\mu_E + \int_E \phi_{2,n} \, d\mu_E) = \int_E \bar{f}_1\big|_E \, d\mu_E + \int_E \bar{f}_2\big|_E \, d\mu_E$. By Proposition 11.75, (ii), and (iii), we have $\int_X \phi \, d\mu = \int_E \phi|_E \, d\mu_E = \int_E \bar{f}_1\big|_E \, d\mu_E + \int_E \bar{f}_2\big|_E \, d\mu_E \le \int_X \bar{f}_1 \, d\mu + \int_X \bar{f}_2 \, d\mu \le \int_X f_1 \, d\mu + \int_X f_2 \, d\mu$. By the arbitrariness of ϕ, we have $\int_X (f_1 + f_2) \, d\mu \le \int_X f_1 \, d\mu + \int_X f_2 \, d\mu$.

Therefore, we have $\int_X (f_1 + f_2) \, d\mu = \int_X f_1 \, d\mu + \int_X f_2 \, d\mu$.

(v) By (iii) and (iv), we have $\int_{E_1 \cup E_2} g \, d\mu = \int_X (g\chi_{E_1 \cup E_2, X}) \, d\mu = \int_X (g\chi_{E_1, X} + g\chi_{E_2, X}) \, d\mu = \int_X (g\chi_{E_1, X}) \, d\mu + \int_X (g\chi_{E_2, X}) \, d\mu = \int_{E_1} g \, d\mu + \int_{E_2} g \, d\mu \in \mathbb{R}_e$.

(vi) Fix any pairwise disjoint $(E_n)_{n=1}^{\infty} \subseteq \mathcal{B}$. Let $E := \bigcup_{n=1}^{\infty} E_n \in \mathcal{B}$ and $\bar{E}_n := \bigcup_{i=1}^{n} E_i \in \mathcal{B}$, $\forall n \in \mathbb{N}$. By (iii), $g\chi_{\bar{E}_n, X}$ and $g\chi_{E,X}$ are $\mathcal{B}$-measurable, $\forall n \in \mathbb{N}$. Clearly, $\lim_{n \in \mathbb{N}} g(x)\chi_{\bar{E}_n, X}(x) = g(x)\chi_{E,X}(x)$, $\forall x \in X$, and $0 \le g(x)\chi_{\bar{E}_n, X}(x) \le g(x)\chi_{\bar{E}_{n+1}, X}(x)$, $\forall x \in X$, $\forall n \in \mathbb{N}$. By Monotone Convergence Theorem 11.81, $\int_E g \, d\mu = \int_X (g\chi_{E,X}) \, d\mu = \lim_{n \in \mathbb{N}} \int_X (g\chi_{\bar{E}_n, X}) \, d\mu = \lim_{n \in \mathbb{N}} \int_{\bar{E}_n} g \, d\mu = \lim_{n \in \mathbb{N}} \sum_{i=1}^{n} \int_{E_i} g \, d\mu = \sum_{n=1}^{\infty} \int_{E_n} g \, d\mu$, where the fourth equality follows from (v).

(vii) By (iii), we have $0 \leq \int_E g \, d\mu \leq \int_X g \, d\mu < +\infty$, $0 \leq \int_{X \setminus E} g \, d\mu \leq \int_X g \, d\mu < +\infty$, $0 \leq \int_E g_n \, d\mu \leq \int_X g_n \, d\mu < +\infty$, and $0 \leq \int_{X \setminus E} g_n \, d\mu \leq \int_X g_n \, d\mu < +\infty$, $\forall n \in \mathbb{N}$. By Fatou's Lemma 11.80, $\int_E g \, d\mu = \int_E g|_E \, d\mu_E \leq \liminf_{n \in \mathbb{N}} \int_E g_n|_E \, d\mu_E = \liminf_{n \in \mathbb{N}} \int_E g_n \, d\mu$. By a similar argument, $\int_{X \setminus E} g \, d\mu \leq \liminf_{n \in \mathbb{N}} \int_{X \setminus E} g_n \, d\mu$. By (v) and Proposition 3.83, we have

$$\begin{aligned}
\int_E g \, d\mu &\leq \liminf_{n \in \mathbb{N}} \int_E g_n \, d\mu \leq \limsup_{n \in \mathbb{N}} \int_E g_n \, d\mu = \limsup_{n \in \mathbb{N}} \Big(\int_X g_n \, d\mu \\
&\quad - \int_{X \setminus E} g_n \, d\mu \Big) = \lim_{n \in \mathbb{N}} \int_X g_n \, d\mu - \liminf_{n \in \mathbb{N}} \int_{X \setminus E} g_n \, d\mu \\
&\leq \int_X g \, d\mu - \int_{X \setminus E} g \, d\mu = \int_E g \, d\mu
\end{aligned}$$

By Proposition 3.83, we have $\int_E g \, d\mu = \lim_{n \in \mathbb{N}} \int_E g_n \, d\mu \in \mathbb{R}$.

This completes the proof of the proposition. □

Proposition 11.84 *Let $\mathcal{X} := (X, \mathcal{B}, \mu)$ be a measure space and $f : X \to [0, \infty] \subset \mathbb{R}_e$ be $\mathcal{B}$-measurable. Assume that f is integrable over $\mathcal{X}$, that is, $\int_X f \, d\mu \in \mathbb{R}$. Then, $\forall \epsilon \in (0, \infty) \subset \mathbb{R}$, $\exists \delta \in (0, \infty) \subset \mathbb{R}$ such that $\forall A \in \mathcal{B}$ with $\mu(A) < \delta$, we have $\int_A f|_A \, d\mu_A < \epsilon$, where $\mathcal{A} := (A, \mathcal{B}_A, \mu_A)$ is the measure subspace of $\mathcal{X}$ as defined in Proposition 11.13.*

Proof By Proposition 11.82, we may, without loss of generality, consider $f : X \to [0, \infty) \subset \mathbb{R}$. $\forall n \in \mathbb{N}$, let $f_n := f \wedge n : X \to [0, n] \subset \mathbb{R}$. Then, $f_n(x) \leq f_{n+1}(x)$, $\forall x \in X$ and $\lim_{n \in \mathbb{N}} f_n(x) = f(x)$, $\forall x \in X$. By Proposition 11.40, f_n is $\mathcal{B}$-measurable. By Propositions 7.23, 11.38, and 11.39, $f - f_n$ is $\mathcal{B}$-measurable. By monotone convergence theorem 11.81, $\int_X f \, d\mu = \lim_{n \in \mathbb{N}} \int_X f_n \, d\mu$. $\forall \epsilon \in (0, \infty) \subset \mathbb{R}$, then $\exists n_0 \in \mathbb{N}$ such that $\left| \int_X f \, d\mu - \int_X f_{n_0} \, d\mu \right| < \epsilon/2$. Choose $\delta := \frac{\epsilon}{2n_0} \in (0, \infty) \subset \mathbb{R}$. $\forall A \in \mathcal{B}$ with $\mu(A) < \delta$, by Proposition 11.83 and Definition 11.79, we have

$$\begin{aligned}
&\int_A f|_A \, d\mu_A = \int_A (f - f_{n_0})\big|_A \, d\mu_A + \int_A f_{n_0}\big|_A \, d\mu_A \\
&\leq \int_X (f - f_{n_0}) \, d\mu + n_0 \mu(A) = \int_X f \, d\mu - \int_X f_{n_0} \, d\mu + n_0 \mu(A) \\
&< \epsilon/2 + \epsilon/2 = \epsilon
\end{aligned}$$

This completes the proof of the proposition. □

11.5 General Convergence Theorems

Lemma 11.85 *Let $\mathcal{X} := (X, \mathcal{B}, \mu)$ be a measure space, $f_n : X \to [0, \infty] \subset \mathbb{R}_e$ be $\mathcal{B}$-measurable, and $f : X \to [0, \infty] \subset \mathbb{R}_e$ be $\mathcal{B}$-measurable. Assume that $\lim_{n\in\mathbb{N}} f_n = f$ a.e. in $\mathcal{X}$ and $\int_X f \,\mathrm{d}\mu = \lim_{n\in\mathbb{N}} \int_X f_n \,\mathrm{d}\mu \in \mathbb{R}$. Then, $\forall \epsilon \in (0, \infty) \subset \mathbb{R}$, $\exists \delta \in (0, \infty) \subset \mathbb{R}$ and $\exists n_0 \in \mathbb{N}$ such that $\forall A \in \mathcal{B}$ with $\mu(A) < \delta$, $\forall n \in \mathbb{N}$ with $n \geq n_0$, we have $0 \leq \int_A f|_A \,\mathrm{d}\mu_A < \epsilon$ and $0 \leq \int_A f_n|_A \,\mathrm{d}\mu_A < \epsilon$, where $(A, \mathcal{B}_A, \mu_A)$ is the finite measure subspace of $\mathcal{X}$ as defined in Proposition 11.13.*

Proof Let $\bar{f}_n := f_n \wedge f$, $\forall n \in \mathbb{N}$. By Proposition 11.40, $\bar{f}_n$ is $\mathcal{B}$-measurable, $\forall n \in \mathbb{N}$. Then, $0 \leq \bar{f}_n(x) \leq f(x)$, $\forall x \in X$, $\forall n \in \mathbb{N}$, and $\int_X \bar{f}_n \,\mathrm{d}\mu \leq \int_X f \,\mathrm{d}\mu$, $\forall n \in \mathbb{N}$, by Definition 11.79. By Proposition 11.50, $\lim_{n\in\mathbb{N}} \bar{f}_n = f$ a.e. in $\mathcal{X}$. By Fatou's Lemma 11.80, $\int_X f \,\mathrm{d}\mu \leq \liminf_{n\in\mathbb{N}} \int_X \bar{f}_n \,\mathrm{d}\mu \leq \limsup_{n\in\mathbb{N}} \int_X \bar{f}_n \,\mathrm{d}\mu \leq \int_X f \,\mathrm{d}\mu$, where the last two inequalities follow from Proposition 3.83 and Definition 3.82, respectively. By Proposition 3.83, $\int_X f \,\mathrm{d}\mu = \lim_{n\in\mathbb{N}} \int_X \bar{f}_n \,\mathrm{d}\mu \in \mathbb{R}$.

$\forall \epsilon \in (0, \infty) \subset \mathbb{R}$, by Proposition 11.84, $\exists \delta \in (0, \infty) \subset \mathbb{R}$ and $\exists n_0 \in \mathbb{N}$ such that $\forall A \in \mathcal{B}$ with $\mu(A) < \delta$, $\forall n \in \mathbb{N}$ with $n \geq n_0$, we have $0 \leq \int_A f|_A \,\mathrm{d}\mu_A < \epsilon/3$, $\left|\int_X f \,\mathrm{d}\mu - \int_X \bar{f}_n \,\mathrm{d}\mu\right| < \epsilon/3$, and $\left|\int_X f \,\mathrm{d}\mu - \int_X f_n \,\mathrm{d}\mu\right| < \epsilon/3$. Then,

$$\begin{aligned}
0 &\leq \int_A f_n|_A \,\mathrm{d}\mu_A = \int_A (f_n - \bar{f}_n)\big|_A \,\mathrm{d}\mu_A + \int_A \bar{f}_n\big|_A \,\mathrm{d}\mu_A \\
&\leq \int_A (f_n - \bar{f}_n)\big|_A \,\mathrm{d}\mu_A + \int_A f|_A \,\mathrm{d}\mu_A < \int_X (f_n - \bar{f}_n) \,\mathrm{d}\mu + \epsilon/3 \\
&= \int_X f_n \,\mathrm{d}\mu - \int_X \bar{f}_n \,\mathrm{d}\mu + \epsilon/3 \\
&\leq \epsilon/3 + \left|\int_X f \,\mathrm{d}\mu - \int_X \bar{f}_n \,\mathrm{d}\mu\right| + \left|\int_X f \,\mathrm{d}\mu - \int_X f_n \,\mathrm{d}\mu\right| < \epsilon
\end{aligned}$$

where the first inequality follows from Definition 11.79, the first equality, the second inequality, and the third inequality follow from Proposition 11.83, and the second equality follows from Proposition 11.83 and the fact that $\int_X f_n \,\mathrm{d}\mu < +\infty$. This completes the proof of the lemma. □

Lemma 11.86 *Let $\mathcal{X} := (X, \mathcal{B}, \mu)$ be a finite measure space, $\mathcal{Y}$ be a Banach space, $\mathcal{W}$ be a separable subspace of $\mathcal{Y}$, $\phi_n : X \to \mathcal{W}$ be a simple function, $\forall n \in \mathbb{N}$, $f : X \to \mathcal{W}$ be $\mathcal{B}$-measurable, $g_n : X \to [0, \infty) \subset \mathbb{R}$ be $\mathcal{B}$-measurable, $\forall n \in \mathbb{N}$, and $g : X \to [0, \infty) \subset \mathbb{R}$ be $\mathcal{B}$-measurable. Assume that:*

(i) $\lim_{n\in\mathbb{N}} \phi_n = f$ *a.e. in* $\mathcal{X}$ *and* $\lim_{n\in\mathbb{N}} g_n = g$ *a.e. in* $\mathcal{X}$.
(ii) $\|\phi_n(x)\| \leq g_n(x)$, $\forall x \in X$, $\forall n \in \mathbb{N}$.
(iii) g_n*'s and* g *are integrable over* X *and* $\lim_{n\in\mathbb{N}} \int_X g_n \,\mathrm{d}\mu = \int_X g \,\mathrm{d}\mu \in \mathbb{R}$.

Then, f is integrable over X and $\int_X f \,\mathrm{d}\mu = \lim_{n\in\mathbb{N}} \int_X \phi_n \,\mathrm{d}\mu \in \mathcal{Y}$.

Proof Let

$$E := \Big\{x \in X \;\Big|\; (\phi_n(x))_{n=1}^{\infty} \text{ does not converge to } f(x) \text{ or } (g_n(x))_{n=1}^{\infty} \text{ does not converge to } g(x)\Big\}$$

Then, by (i), $E \in \mathcal{B}$ and $\mu(E) = 0$. Let $(F_R)_{R\in\mathfrak{I}(\mathcal{Y})}$ be the net for $\int_X f\,\mathrm{d}\mu$ as defined in Definition 11.70.

$\forall \epsilon \in (0, \infty) \subset \mathbb{R}$, by (i), (iii), and Lemma 11.85, $\exists \delta \in (0, \epsilon/6] \subset \mathbb{R}$ and $\exists n_0 \in \mathbb{N}$ such that $\forall A \in \mathcal{B}$ with $\mu(A) < \delta$, $\forall n \in \mathbb{N}$ with $n \geq n_0$, we have $0 \leq \int_A g|_A \,\mathrm{d}\mu_A < \epsilon/6$ and $0 \leq \int_A g_n|_A \,\mathrm{d}\mu_A < \epsilon/6$, where $(A, \mathcal{B}_A, \mu_A)$ is the finite measure subspace of $\mathcal{X}$. By Egoroff's Theorem 11.55, $\exists \bar{E}_1 \in \mathcal{B}$ with $\mu(\bar{E}_1) < \delta$ such that $\left(\phi_n|_{X\setminus\bar{E}_1}\right)_{n=1}^{\infty}$ converges uniformly to $f|_{X\setminus\bar{E}_1}$. Let $E_1 := E \cup \bar{E}_1$. Then, $E_1 \in \mathcal{B}$ with $\mu(E_1) = \mu(\bar{E}_1) < \delta$ such that $\left(\phi_n|_{X\setminus E_1}\right)_{n=1}^{\infty}$ converges uniformly to $f|_{X\setminus E_1}$. Then, $\exists n_1 \in \mathbb{N}$ with $n_1 \geq n_0$, $\forall n \in \mathbb{N}$ with $n \geq n_1$, $\forall x \in X \setminus E_1$, $\|\phi_n(x) - f(x)\| < \frac{\epsilon}{6\mu(X)+1} =: \bar{\epsilon}$. Fix any $n \geq n_1$. Let ϕ_n admit the canonical representation $\phi_n = \sum_{i=1}^{\bar{n}} w_i \chi_{A_i,X}$, where $\bar{n} \in \mathbb{Z}_+$, $w_1, \ldots, w_{\bar{n}} \in \mathcal{W}$ are distinct and none equals to $\vartheta_{\mathcal{Y}}$, and $A_1, \ldots, A_{\bar{n}} \in \mathcal{B}$ are nonempty, pairwise disjoint, and of finite measure. Let $w_{\bar{n}+1} := \vartheta_{\mathcal{Y}}$ and $A_{\bar{n}+1} := X \setminus (\bigcup_{i=1}^{\bar{n}} A_i) \in \mathcal{B}$. Then, $X = \bigcup_{i=1}^{\bar{n}+1} A_i$, and the sets in the union are pairwise disjoint and of finite measure. Define $\bar{V}_i := \mathcal{B}_{\mathcal{Y}}(w_i, \bar{\epsilon})$, $i = 1, \ldots, \bar{n}+1$. Define $V_1 := \bar{V}_1 \in \mathcal{B}_{\mathrm{B}}(\mathcal{Y})$, $V_{i+1} := \bar{V}_{i+1} \setminus (\bigcup_{j=1}^{i} \bar{V}_j) \in \mathcal{B}_{\mathrm{B}}(\mathcal{Y})$, $i = 1, \ldots, \bar{n}$. Let $V_{\bar{n}+2} := \mathcal{Y} \setminus (\bigcup_{j=1}^{\bar{n}+1} \bar{V}_j) \in \mathcal{B}_{\mathrm{B}}(\mathcal{Y})$. Clearly, we may form a representation $\bar{R} := \{(\bar{y}_i, V_i) \mid i = 1, \ldots, \bar{n}+2, V_i \neq \emptyset\} \in \mathfrak{I}(\mathcal{Y})$, where $\bar{y}_i \in V_i$, $i = 1, \ldots, \bar{n}+2$, are any vectors that satisfy the assumption of Proposition 11.69. $\forall R \in \mathfrak{I}(\mathcal{Y})$ with $\bar{R} \preceq R$. Let $R := \{(\hat{y}_\gamma, \hat{U}_\gamma) \mid \gamma \in \Gamma\}$. $\forall \gamma \in \Gamma$, by $\bar{R} \preceq R$, $\exists!\, i_\gamma \in \{1, \ldots, \bar{n}+2\}$ such that $\hat{U}_\gamma \subseteq V_{i_\gamma}$. Define $\bar{\Gamma} := \{\gamma \in \Gamma \mid i_\gamma = \bar{n}+2\}$. Let $A_{\gamma,i} := A_i \cap f_{\mathrm{inv}}(\hat{U}_\gamma) \cap (X \setminus E_1) \in \mathcal{B}$, $i = 1, \ldots, \bar{n}+1$. Let $\bar{A}_\gamma := f_{\mathrm{inv}}(\hat{U}_\gamma) \cap (E_1 \setminus E) \in \mathcal{B}$.

Claim 11.86.1 $\forall \gamma \in \bar{\Gamma}$, $f_{\mathrm{inv}}(\hat{U}_\gamma) \subseteq E_1$; $\forall \gamma \in \Gamma_s := \{\gamma \in \Gamma \mid \bar{A}_\gamma \neq \emptyset\}$, $\left\|\hat{y}_\gamma \mu(\bar{A}_\gamma)\right\| \leq (1 + \inf_{x\in\bar{A}_\gamma} g(x))\mu(\bar{A}_\gamma)$; and $\left\|\sum_{\gamma\in\Gamma} \hat{y}_\gamma \mu(f_{\mathrm{inv}}(\hat{U}_\gamma) \cap E_1)\right\| < \epsilon/3$.

Proof of Claim Fix any $\gamma \in \bar{\Gamma}$. $\hat{U}_\gamma \subseteq V_{\bar{n}+2} = \mathcal{Y} \setminus (\bigcup_{j=1}^{\bar{n}+1} \bar{V}_j)$. Then, $\mathcal{B} \ni f_{\mathrm{inv}}(\hat{U}_\gamma) \subseteq E_1$.

$\forall \gamma \in \Gamma_s$, $\bar{A}_\gamma = f_{\mathrm{inv}}(\hat{U}_\gamma) \cap (E_1 \setminus E) \neq \emptyset$. $\forall x \in \bar{A}_\gamma$, we have $f(x) \in \hat{U}_\gamma$, $\lim_{m\in\mathbb{N}} \phi_m(x) = f(x)$, $\|\phi_m(x)\| \leq g_m(x)$, $\forall m \in \mathbb{N}$, and $\lim_{m\in\mathbb{N}} g_m(x) = g(x)$. Then, by Propositions 4.30, 3.66, and 3.67, we have $\|f(x)\| = \lim_{m\in\mathbb{N}} \|\phi_m(x)\| \leq \lim_{m\in\mathbb{N}} g_m(x) = g(x)$. Then, by $R \in \mathfrak{I}(\mathcal{Y})$, $\left\|\hat{y}_\gamma\right\| < \inf_{y\in\hat{U}_\gamma} \|y\| + 1 \leq \|f(x)\| + 1 \leq g(x)+1$. By the arbitrariness of x, we have $\left\|\hat{y}_\gamma\right\| \leq \inf_{x\in\bar{A}_\gamma} g(x)+1$. Therefore, $\left\|\hat{y}_\gamma \mu(\bar{A}_\gamma)\right\| \leq (1 + \inf_{x\in\bar{A}_\gamma} g(x))\mu(\bar{A}_\gamma)$.

Note that

$$\Big\| \sum_{\gamma\in\Gamma} \hat{y}_\gamma \mu(f_{\rm inv}(\hat{U}_\gamma) \cap E_1)\Big\| = \Big\| \sum_{\gamma\in\Gamma} \hat{y}_\gamma (\mu(\bar{A}_\gamma) + \mu(f_{\rm inv}(\hat{U}_\gamma) \cap E_1 \cap E))\Big\|$$

$$= \Big\| \sum_{\gamma\in\Gamma} \hat{y}_\gamma \mu(\bar{A}_\gamma)\Big\| = \Big\| \sum_{\gamma\in\Gamma_s} \hat{y}_\gamma \mu(\bar{A}_\gamma)\Big\| \le \sum_{\gamma\in\Gamma_s} \Big\|\hat{y}_\gamma \mu(\bar{A}_\gamma)\Big\|$$

$$\le \sum_{\gamma\in\Gamma_s} (1 + \inf_{x\in\bar{A}_\gamma} g(x))\mu(\bar{A}_\gamma) = \sum_{\gamma\in\Gamma_s} \mu(\bar{A}_\gamma) + \sum_{\gamma\in\Gamma_s} (\inf_{x\in\bar{A}_\gamma} g(x))\mu(\bar{A}_\gamma)$$

The simple function $\psi : X \to [0,\infty) \subset \mathbb{R}$, defined by $\psi(x) = \sum_{\gamma\in\Gamma_s}(\inf_{\bar{x}\in\bar{A}_\gamma} g(\bar{x}))\chi_{\bar{A}_\gamma,X}(x)$, $\forall x \in X$, satisfies $0 \le \psi(x) \le g(x)$, $\forall x \in X$ and $\psi(x) = 0$, $\forall x \in X \setminus E_1$. Then, $\sum_{\gamma\in\Gamma_s}(\inf_{x\in\bar{A}_\gamma} g(x))\mu(\bar{A}_\gamma) = \int_X \psi\, \mathrm{d}\mu = \int_{E_1} \psi|_{E_1}\, \mathrm{d}\mu_{E_1} \le \int_{E_1} g|_{E_1}\, \mathrm{d}\mu_{E_1} < \epsilon/6$, where the equalities follow from Proposition 11.75 and the first inequality follows from Proposition 11.83. Then, $\Big\|\sum_{\gamma\in\Gamma} \hat{y}_\gamma \mu(f_{\rm inv}(\hat{U}_\gamma) \cap E_1)\Big\| < \mu(E_1) + \epsilon/6 < \delta + \epsilon/6 \le \epsilon/3$.

This completes the proof of the claim. □

Claim 11.86.2 $\Big\|\sum_{i=1}^{\bar{n}+1} w_i \mu(A_i \cap E_1)\Big\| < \epsilon/6$.

Proof of Claim Since $\|\phi_n(x)\| \le g_n(x)$, $\forall x \in X$, then $\|w_i\| \le g_n(x)$, $\forall x \in A_i$, $i = 1, \ldots, \bar{n}+1$. Then, $\Big\|\sum_{i=1}^{\bar{n}+1} w_i \mu(A_i \cap E_1)\Big\| \le \sum_{i=1}^{\bar{n}+1} \|w_i\|\mu(A_i \cap E_1) \le \int_{E_1} g_n|_{E_1}\, \mathrm{d}\mu_{E_1} < \epsilon/6$, where the second inequality follows from Definition 11.79. □

Claim 11.86.3 $\forall \gamma \in \Gamma \setminus \bar{\Gamma}$, $\forall i \in \{1, \ldots, \bar{n}+1\}$, we have $\big\|\hat{y}_\gamma - w_i\big\|\mu(A_{\gamma,i}) \le 3\bar{\epsilon}\mu(A_{\gamma,i})$.

Proof of Claim $\forall \gamma \in \Gamma \setminus \bar{\Gamma}$, $\forall i \in \{1, \ldots, \bar{n}+1\}$, $i_\gamma \in \{1, \ldots, \bar{n}+1\}$ and $\hat{U}_\gamma \subseteq V_{i_\gamma}$. The result is trivial if $A_{\gamma,i} = \emptyset$. If $A_{\gamma,i} \neq \emptyset$, then $\exists x \in A_{\gamma,i} = A_i \cap f_{\rm inv}(\hat{U}_\gamma) \cap (X \setminus E_1)$. This implies that $\phi_n(x) = w_i$ and $f(x) \in \mathcal{B}_\mathcal{Y}(w_i, \bar{\epsilon}) \cap \hat{U}_\gamma \subseteq \mathcal{B}_\mathcal{Y}(w_i, \bar{\epsilon}) \cap V_{i_\gamma} \subseteq \mathcal{B}_\mathcal{Y}(w_i, \bar{\epsilon}) \cap \bar{V}_{i_\gamma} = \mathcal{B}_\mathcal{Y}(w_i, \bar{\epsilon}) \cap \mathcal{B}_\mathcal{Y}\big(w_{i_\gamma}, \bar{\epsilon}\big)$. Note that $\hat{y}_\gamma \in \hat{U}_\gamma \subseteq \mathcal{B}_\mathcal{Y}\big(w_{i_\gamma}, \bar{\epsilon}\big)$. Therefore, we have $\big\|\hat{y}_\gamma - w_i\big\| \le \big\|\hat{y}_\gamma - w_{i_\gamma}\big\| + \big\|w_{i_\gamma} - f(x)\big\| + \|f(x) - w_i\| < 3\bar{\epsilon}$. Hence, the result holds. □

Note that $F_R = \sum_{\gamma\in\Gamma} \hat{y}_\gamma \mu(f_{\rm inv}(\hat{U}_\gamma))$ and $\int_X \phi_n\, \mathrm{d}\mu = \sum_{i=1}^{\bar{n}} w_i \mu(A_i)$, by Proposition 11.75. Then,

$$\Big\|F_R - \int_X \phi_n\, \mathrm{d}\mu\Big\| = \Big\| \sum_{\gamma\in\Gamma} \hat{y}_\gamma \mu(f_{\rm inv}(\hat{U}_\gamma)) - \sum_{i=1}^{\bar{n}} w_i \mu(A_i)\Big\|$$

$$= \Big\| \sum_{\gamma\in\Gamma} \hat{y}_\gamma (\mu(f_{\rm inv}(\hat{U}_\gamma) \cap (X \setminus E_1)) + \mu(f_{\rm inv}(\hat{U}_\gamma) \cap E_1))$$

$$
- \sum_{i=1}^{\bar{n}+1} w_i (\mu(A_i \cap E_1) + \mu(A_i \cap (X \setminus E_1))) \Big\|
$$

$$
= \Big\| \sum_{\gamma \in \Gamma} \hat{y}_\gamma \Big(\sum_{i=1}^{\bar{n}+1} \mu(A_{\gamma,i}) + \mu(f_{\text{inv}}(\hat{U}_\gamma) \cap E_1) \Big) - \sum_{i=1}^{\bar{n}+1} w_i (\mu(A_i \cap E_1)
$$

$$
+ \sum_{\gamma \in \Gamma} \mu(A_{\gamma,i})) \Big\|
$$

$$
= \Big\| \sum_{\gamma \in \Gamma} \sum_{i=1}^{\bar{n}+1} \hat{y}_\gamma \mu(A_{\gamma,i}) + \sum_{\gamma \in \Gamma} \hat{y}_\gamma \mu(f_{\text{inv}}(\hat{U}_\gamma) \cap E_1) - \sum_{i=1}^{\bar{n}+1} w_i \mu(A_i \cap E_1)
$$

$$
- \sum_{\gamma \in \Gamma} \sum_{i=1}^{\bar{n}+1} w_i \mu(A_{\gamma,i}) \Big\|
$$

$$
= \Big\| \sum_{\gamma \in \Gamma} \sum_{i=1}^{\bar{n}+1} (\hat{y}_\gamma - w_i) \mu(A_{\gamma,i}) + \sum_{\gamma \in \Gamma} \hat{y}_\gamma \mu(f_{\text{inv}}(\hat{U}_\gamma) \cap E_1)
$$

$$
- \sum_{i=1}^{\bar{n}+1} w_i \mu(A_i \cap E_1) \Big\|
$$

$$
< \Big\| \sum_{\gamma \in \Gamma} \sum_{i=1}^{\bar{n}+1} (\hat{y}_\gamma - w_i) \mu(A_{\gamma,i}) \Big\| + \frac{\epsilon}{3} + \frac{\epsilon}{6}
$$

$$
= \frac{\epsilon}{2} + \Big\| \sum_{\gamma \in \Gamma \setminus \bar{\Gamma}} \sum_{i=1}^{\bar{n}+1} (\hat{y}_\gamma - w_i) \mu(A_{\gamma,i}) + \sum_{\gamma \in \bar{\Gamma}} \sum_{i=1}^{\bar{n}+1} (\hat{y}_\gamma - w_i) \mu(A_{\gamma,i}) \Big\|
$$

$$
= \frac{\epsilon}{2} + \Big\| \sum_{\gamma \in \Gamma \setminus \bar{\Gamma}} \sum_{i=1}^{\bar{n}+1} (\hat{y}_\gamma - w_i) \mu(A_{\gamma,i}) \Big\| \leq \frac{\epsilon}{2} + \sum_{\gamma \in \Gamma \setminus \bar{\Gamma}} \sum_{i=1}^{\bar{n}+1} \| \hat{y}_\gamma - w_i \| \mu(A_{\gamma,i})
$$

$$
\leq \frac{\epsilon}{2} + \sum_{\gamma \in \Gamma \setminus \bar{\Gamma}} \sum_{i=1}^{\bar{n}+1} 3\bar{\epsilon} \mu(A_{\gamma,i}) \leq \frac{\epsilon}{2} + \frac{3\mu(X)\epsilon}{6\mu(X)+1} < \epsilon
$$

where the first inequality follows from Claims 11.86.1 and 11.86.2, the seventh equality follows from the fact that $A_{\gamma,i} = \emptyset$, $\forall \gamma \in \bar{\Gamma}$, $\forall i \in \{1, \ldots, \bar{n}+1\}$, and the third inequality follows from Claim 11.86.3.

In summary, we have shown that $\forall \epsilon \in (0, \infty) \subset \mathbb{R}$, $\exists n_1 \in \mathbb{N}$, $\forall n \in \mathbb{N}$ with $n \geq n_1$, $\exists \bar{R} \in \mathcal{I}(\mathcal{Y})$ such that $\forall R \in \mathcal{I}(\mathcal{Y})$ with $\bar{R} \preceq R$, we have $\| F_R - \int_X \phi_n \, d\mu \| < \epsilon$. Then, the net $(F_R)_{R \in \mathcal{I}(\mathcal{Y})}$ is a Cauchy net. By Proposition 4.44, the net admits a limit $\int_X f \, d\mu \in \mathcal{Y}$. By Propositions 4.30, 3.66, and 3.67, we have $\| \int_X f \, d\mu - \int_X \phi_n \, d\mu \| = \lim_{R \in \mathcal{I}(\mathcal{Y})} \| F_R - \int_X \phi_n \, d\mu \| \leq \epsilon$, $\forall n \geq n_1$. Hence, we have $\int_X f \,\mathrm{d}\mu = \lim_{n\in\mathbb{N}} \int_X \phi_n \,\mathrm{d}\mu \in \mathcal{Y}$ and f is integrable over X. This completes the proof of the lemma. □

Lemma 11.87 *Let $\mathcal{X} := (X, \mathcal{B}, \mu)$ be a finite measure space, $\mathcal{Y}$ be a Banach space, $\mathcal{W}$ be a separable subspace of $\mathcal{Y}$, $U \subseteq \mathcal{W}$ be a conic segment, $f : X \to U \subseteq \mathcal{W}$ be $\mathcal{B}$-measurable, and $g : X \to [0, \infty) \subset \mathbb{R}$ be $\mathcal{B}$-measurable. Assume that $\|f(x)\| \leq g(x)$, $\forall x \in X$, and g is integrable over X. Then, there exists a sequence of simple functions $(\varphi_k)_{k=1}^{\infty}$, $\varphi_k : X \to U$, $\forall k \in \mathbb{N}$, such that $\lim_{k\in\mathbb{N}} \varphi_k = f$ a.e. in $\mathcal{X}$, $\|\varphi_k(x)\| \leq \|f(x)\|$, $\forall x \in X$, $\forall k \in \mathbb{N}$, and f is integrable over X with $\int_X f \,\mathrm{d}\mu = \lim_{k\in\mathbb{N}} \int_X \varphi_k \,\mathrm{d}\mu \in \mathcal{Y}$.*

Furthermore, if $\mathcal{Y}$ admits a positive cone P and $U \subseteq P \subseteq \mathcal{Y}$, then $\int_X f \,\mathrm{d}\mu \in P$.

Proof By Proposition 11.66, there exists a sequence of simple functions $(\varphi_k)_{k=1}^{\infty}$, $\varphi_k : X \to U$, $\forall k \in \mathbb{N}$, such that $\|\varphi_k(x)\| \leq \|f(x)\| \leq g(x)$, $\forall x \in X$, $\forall k \in \mathbb{N}$, and $\lim_{k\in\mathbb{N}} \varphi_k = f$ a.e. in $\mathcal{X}$. By Lemma 11.86, f is integrable over X and $\int_X f \,\mathrm{d}\mu = \lim_{k\in\mathbb{N}} \int_X \varphi_k \,\mathrm{d}\mu \in \mathcal{Y}$.

Furthermore, if $\mathcal{Y}$ admits a positive cone P and $U \subseteq P \subseteq \mathcal{Y}$, then, by Propositions 8.41 and 11.75, $\varphi_k : X \to U \subseteq P \subseteq \mathcal{Y}$, $\forall k \in \mathbb{N}$, and $\int_X \varphi_k \,\mathrm{d}\mu \in P$. Then, by Proposition 4.13, $\int_X f \,\mathrm{d}\mu \in \overline{P} = P$, since P is closed. This completes the proof of the lemma. □

Theorem 11.88 (Lebesgue Dominated Convergence) *Let $(X, \mathcal{B}, \mu) =: \mathcal{X}$ be a finite measure space, $\mathcal{Y}$ be a Banach space, $\mathcal{W}$ be a separable subspace of $\mathcal{Y}$, $f_n : X \to \mathcal{W}$ be $\mathcal{B}$-measurable, $\forall n \in \mathbb{N}$, $f : X \to \mathcal{W}$ be $\mathcal{B}$-measurable, $g_n : X \to [0, \infty) \subset \mathbb{R}$ be $\mathcal{B}$-measurable, $\forall n \in \mathbb{N}$, and $g : X \to [0, \infty) \subset \mathbb{R}$ be $\mathcal{B}$-measurable. Assume that:*

(i) *$\lim_{n\in\mathbb{N}} f_n = f$ a.e. in $\mathcal{X}$, $\lim_{n\in\mathbb{N}} g_n = g$ a.e. in $\mathcal{X}$.*
(ii) *$\|f_n(x)\| \leq g_n(x)$, $\forall x \in X$, $\forall n \in \mathbb{N}$.*
(iii) *g_n's and g are integrable over X and $\int_X g \,\mathrm{d}\mu = \lim_{n\in\mathbb{N}} \int_X g_n \,\mathrm{d}\mu \in \mathbb{R}$.*

Then, f is integrable over X and $\int_X f \,\mathrm{d}\mu = \lim_{n\in\mathbb{N}} \int_X f_n \,\mathrm{d}\mu \in \mathcal{Y}$.

Proof $\forall n \in \mathbb{N}$, by Lemma 11.87, there exists a sequence of simple functions $(\bar{\varphi}_{n,k})_{k=1}^{\infty}$, $\bar{\varphi}_{n,k} : X \to \mathcal{W}$, $\forall k \in \mathbb{N}$, such that $\lim_{k\in\mathbb{N}} \bar{\varphi}_{n,k} = f_n$ a.e. in $\mathcal{X}$, $\|\bar{\varphi}_{n,k}(x)\| \leq \|f_n(x)\|$, $\forall x \in X$, $\forall k \in \mathbb{N}$, f_n is integrable over X, and $\int_X f_n \,\mathrm{d}\mu = \lim_{k\in\mathbb{N}} \int_X \bar{\varphi}_{n,k} \,\mathrm{d}\mu \in \mathcal{Y}$. By Proposition 11.58, $\lim_{k\in\mathbb{N}} \bar{\varphi}_{n,k} = f_n$ in measure in $\mathcal{X}$. Then, $\exists k_n \in \mathbb{N}$ such that $\mu(E_n) := \mu(\{x \in X \mid \|\bar{\varphi}_{n,k_n}(x) - f_n(x)\| \geq 2^{-n}\}) < 2^{-n}$ and $\left\|\int_X f_n \,\mathrm{d}\mu - \int_X \bar{\varphi}_{n,k_n} \,\mathrm{d}\mu\right\| < 2^{-n}$. Denote $\psi_n := \bar{\varphi}_{n,k_n}$.

By Proposition 11.58, $\lim_{n\in\mathbb{N}} f_n = f$ in measure in $\mathcal{X}$. $\forall \epsilon \in (0, \infty) \subset \mathbb{R}$, $\exists n_0 \in \mathbb{N}$ such that $\forall n \in \mathbb{N}$ with $n \geq n_0$, we have $2^{-n} < \epsilon/2$ and $\mu(\bar{E}_n) := \mu(\{x \in X \mid \|f_n(x) - f(x)\| \geq \epsilon/2\}) < \epsilon/2$. Note that, by Lemma 11.43, $\mathcal{B} \ni F_n := \{x \in X \mid \|\psi_n(x) - f(x)\| \geq \epsilon\} \subseteq \{x \in X \mid \|\psi_n(x) - f_n(x)\| + \|f_n(x) - f(x)\| \geq \epsilon\} \subseteq E_n \cup \bar{E}_n$. Then, we have $\mu(F_n) \leq \mu(E_n) + \mu(\bar{E}_n) < \epsilon$. This implies that $\lim_{n\in\mathbb{N}} \psi_n = f$ in measure in $\mathcal{X}$. By Proposition 11.57, there exists a subsequence $(\psi_{n_i})_{i=1}^{\infty}$ of $(\psi_n)_{n=1}^{\infty}$ that converges to f a.e. in $\mathcal{X}$. By Lemma 11.86, f is integrable over X and $\int_X f \,\mathrm{d}\mu = \lim_{i\in\mathbb{N}} \int_X \psi_{n_i} \,\mathrm{d}\mu \in \mathcal{Y}$.

By the fact that $\left\| \int_X f_{n_i} \,d\mu - \int_X \psi_{n_i} \,d\mu \right\| < 2^{-n_i}$, $\forall i \in \mathbb{N}$, then we have $\lim_{i\in\mathbb{N}} \int_X f_{n_i} \,d\mu = \lim_{i\in\mathbb{N}} \int_X \psi_{n_i} \,d\mu = \int_X f \,d\mu \in \mathcal{Y}$. So far, we have shown that, for the sequence $(f_n)_{n=1}^{\infty}$, there exists a subsequence $\left(f_{n_i}\right)_{i=1}^{\infty}$ such that $\lim_{i\in\mathbb{N}} \int_X f_{n_i} \,d\mu = \int_X f \,d\mu \in \mathcal{Y}$. This then holds for any subsequence of $(f_n)_{n=1}^{\infty}$. By Proposition 3.71, we have $\lim_{n\in\mathbb{N}} \int_X f_n \,d\mu = \int_X f \,d\mu$.

This completes the proof of the theorem. □

Proposition 11.89 *Let $\mathcal{X} := (X, \mathcal{B}, \mu)$ be a finite measure space, $\mathcal{Y}$ be a Banach space over $\mathbb{K}$, $\mathcal{W}$ be a separable subspace of $\mathcal{Y}$, $\mathcal{Z}$ be a Banach space over $\mathbb{K}$, $f_i : X \to \mathcal{W}$ be $\mathcal{B}$-measurable, $g_i : X \to [0, \infty) \subset \mathbb{R}$ be integrable over $\mathcal{X}$, $i = 1, 2$. Assume that $\|f_i(x)\| \leq g_i(x)$, $\forall x \in X$, $i = 1, 2$. Then, the following statements hold:*

(i) *f_1, f_2, and $f_1 + f_2$ are integrable over $\mathcal{X}$ and $\int_X (f_1 + f_2) \,d\mu = \int_X f_1 \,d\mu + \int_X f_2 \,d\mu \in \mathcal{Y}$.*
(ii) *$\forall A \in B(\mathcal{Y}, \mathcal{Z})$, Af_1 is integrable over $\mathcal{X}$ and $\int_X (Af_1) \,d\mu = A \int_X f_1 \,d\mu \in \mathcal{Z}$.*
(iii) *$\forall c \in \mathbb{K}$, cf_1 is integrable over $\mathcal{X}$ and $\int_X (cf_1) \,d\mu = c \int_X f_1 \,d\mu \in \mathcal{Y}$.*
(iv) *$\forall E \in \mathcal{B}$, $f_1|_E$ is integrable over $\mathcal{E}$ and $\int_X (f_1 \chi_{E,X}) \,d\mu = \int_E f_1|_E \,d\mu_E \in \mathcal{Y}$, where $\mathcal{E} := (E, \mathcal{B}_E, \mu_E)$ is the finite measure subspace of $\mathcal{X}$ as defined in Proposition 11.13. We will henceforth denote $\int_E f_1|_E \,d\mu_E$ by $\int_E f_1 \,d\mu$.*
(v) *If $f_1 = f_2$ a.e. in $\mathcal{X}$, then $\int_X f_1 \,d\mu = \int_X f_2 \,d\mu \in \mathcal{Y}$.*
(vi) *$\forall$ pairwise disjoint $(E_i)_{i=1}^{\infty} \subseteq \mathcal{B}$, $\sum_{i=1}^{\infty} \int_{E_i} f_1 \,d\mu = \int_{\bigcup_{i=1}^{\infty} E_i} f_1 \,d\mu \in \mathcal{Y}$.*
(vii) *$0 \leq \left\| \int_X f_1 \,d\mu \right\| \leq \int_X \mathcal{P} \circ f_1 \,d\mu \leq \int_X g_1 \,d\mu < +\infty$, where $\mathcal{P} \circ f_1 : X \to [0, \infty) \subset \mathbb{R}$ is as defined on page 388.*
(viii) *If $\mathcal{Y}$ admits a positive cone P and $f_1 \leqq f_2$ a.e. in $\mathcal{X}$, then $\int_X f_1 \,d\mu \leqq \int_X f_2 \,d\mu$.*

Proof By Lemma 11.87, there exist sequences of simple functions $(\phi_{i,n})_{n=1}^{\infty}$, $\phi_{i,n} : X \to \mathcal{W}$, $\forall n \in \mathbb{N}$, $i = 1, 2$ such that $i = 1, 2$, $\lim_{n\in\mathbb{N}} \phi_{i,n} = f_i$ a.e. in $\mathcal{X}$, $\left\|\phi_{i,n}(x)\right\| \leq \|f_i(x)\|$, $\forall x \in X$, $\forall n \in \mathbb{N}$, f_i is integrable over $\mathcal{X}$, and $\int_X f_i \,d\mu = \lim_{n\in\mathbb{N}} \int_X \phi_{i,n} \,d\mu \in \mathcal{Y}$.

(i) By Propositions 7.23, 11.38, and 11.39, $f_1 + f_2$, $g_1 + g_2$, and $\phi_{1,n} + \phi_{2,n}$ are $\mathcal{B}$-measurable, $\forall n \in \mathbb{N}$. By Proposition 11.83, $0 \leq \int_X (g_1 + g_2) \,d\mu = \int_X g_1 \,d\mu + \int_X g_2 \,d\mu < +\infty$. Note that, by Propositions 7.23, 11.52, and 11.53, $\lim_{n\in\mathbb{N}} (\phi_{1,n} + \phi_{2,n}) = f_1 + f_2$ a.e. in $\mathcal{X}$ and $\left\|\phi_{1,n}(x) + \phi_{2,n}(x)\right\| \leq \left\|\phi_{1,n}(x)\right\| + \left\|\phi_{2,n}(x)\right\| \leq g_1(x) + g_2(x)$, $\forall x \in X$, $\forall n \in \mathbb{N}$. Clearly, $\phi_{1,n} + \phi_{2,n} : X \to \mathcal{W}$ and $f_1 + f_2 : X \to \mathcal{W}$, $\forall n \in \mathbb{N}$. By Lebesgue Dominated Convergence Theorem 11.88, $f_1 + f_2$ is integrable over $\mathcal{X}$ and $\int_X (f_1 + f_2) \,d\mu = \lim_{n\in\mathbb{N}} \int_X (\phi_{1,n} + \phi_{2,n}) \,d\mu = \lim_{n\in\mathbb{N}} \left(\int_X \phi_{1,n} \,d\mu + \int_X \phi_{2,n} \,d\mu \right) = \int_X f_1 \,d\mu + \int_X f_2 \,d\mu \in \mathcal{Y}$, where the second equality follows from Proposition 11.75 and the third equality follows from Propositions 3.66, 3.67, and 7.23.

(ii) $\forall A \in B(\mathcal{Y}, \mathcal{Z})$, by Proposition 11.83 and the fact that $\int_X g_1 \,d\mu \in \mathbb{R}$, we have $\int_X (\|A\| g_1) \,d\mu = \|A\| \int_X g_1 \,d\mu \in \mathbb{R}$. By Propositions 7.62 and 11.38, Af_1 and $A\phi_{1,n}$ are $\mathcal{B}$-measurable, $\forall n \in \mathbb{N}$. Clearly, $A\phi_{1,n} : X \to \hat{\mathcal{W}}$ and $Af_1 : X \to \hat{\mathcal{W}}$, $\forall n \in \mathbb{N}$, where $\hat{\mathcal{W}} := \{z \in \mathcal{Z} \,|\, z = Aw,\ w \in \mathcal{W}\}$ is a separable subspace of $\mathcal{Z}$, by Proposition 7.126. By Propositions 11.52 and 7.62, $\lim_{n\in\mathbb{N}} A\phi_{1,n} = Af_1$ a.e. in $\mathcal{X}$. By Proposition 11.75, we have $\int_X (A\phi_{1,n})\,d\mu = A\int_X \phi_{1,n}\,d\mu \in \mathcal{Z}$. Note that $\|A\phi_{1,n}(x)\| \le \|A\|\cdot\|\phi_{1,n}(x)\| \le \|A\|g_1(x)$, $\forall x\in X$, $\forall n\in\mathbb{N}$. By Lebesgue Dominated Convergence Theorem 11.88, Af_1 is integrable over $\mathcal{X}$ and $\int_X (Af_1)\,d\mu = \lim_{n\in\mathbb{N}}\int_X (A\phi_{1,n})\,d\mu = \lim_{n\in\mathbb{N}} A\int_X \phi_{1,n}\,d\mu = A\int_X f_1\,d\mu \in \mathcal{Z}$, where the last equality follows from Propositions 3.66 and 7.62.

(iii) This follows immediately from (ii).

(iv) $\left(\phi_{1,n}|_E\right)_{n=1}^\infty$ is a sequence of simple functions that converges to $f_1|_E$ a.e. in $\mathcal{E}$. Note that $\left\|\phi_{1,n}|_E(x)\right\| \le g_1|_E(x)$, $\forall x\in E$, $\forall n\in\mathbb{N}$. By Proposition 11.83, $0 \le \int_E g_1|_E\,d\mu_E \le \int_X g_1\,d\mu < +\infty$. By Lebesgue Dominated Convergence Theorem 11.88, we have $\int_E f_1|_E\,d\mu_E = \lim_{n\in\mathbb{N}}\int_E \phi_{1,n}|_E\,d\mu_E \in \mathcal{Y}$. Note that $\left(\phi_{1,n}\chi_{E,X}\right)_{n=1}^\infty$ is a sequence of simple functions that converges to $f_1\chi_{E,X}$ a.e. in $\mathcal{X}$, and $\|\phi_{1,n}(x)\chi_{E,X}(x)\| \le g_1(x)$, $\forall x\in X$, $\forall n\in\mathbb{N}$. By Propositions 7.23, 11.38, and 11.39, $\phi_{1,n}\chi_{E,X}$ and $f_1\chi_{E,X}$ are $\mathcal{B}$-measurable. Again by Lebesgue Dominated Convergence Theorem 11.88, we have $\int_X (f_1\chi_{E,X})\,d\mu = \lim_{n\in\mathbb{N}}\int_X (\phi_{1,n}\chi_{E,X})\,d\mu = \lim_{n\in\mathbb{N}}\int_E \phi_{1,n}|_E\,d\mu_E = \int_E f_1|_E\,d\mu_E \in \mathcal{Y}$, where the second equality follows from Proposition 11.75.

(v) If $f_1 = f_2$ a.e. in $\mathcal{X}$, then $\lim_{n\in\mathbb{N}}\phi_{1,n} = f_2$ a.e. in $\mathcal{X}$ by Proposition 11.54. By Lebesgue Dominated Convergence Theorem 11.88, we have $\int_X f_2\,d\mu = \lim_{n\in\mathbb{N}}\int_X \phi_{1,n}\,d\mu = \int_X f_1\,d\mu$.

(vi) $\forall$ pairwise disjoint $(E_i)_{i=1}^\infty \subseteq \mathcal{B}$, let $E := \bigcup_{i=1}^\infty E_i \in \mathcal{B}$ and $\bar{E}_n := \bigcup_{i=1}^n E_i \in \mathcal{B}$, $\forall n\in\mathbb{N}$. Then, $\lim_{n\in\mathbb{N}} f_1(x)\chi_{\bar{E}_n,X}(x) = f_1(x)\chi_{E,X}(x)$, $\forall x\in X$. Then, we have $\sum_{i=1}^\infty \int_{E_i} f_1\,d\mu = \lim_{n\in\mathbb{N}}\sum_{i=1}^n \int_X (f_1\chi_{E_i,X})\,d\mu = \lim_{n\in\mathbb{N}}\int_X (f_1\chi_{\bar{E}_n,X})\,d\mu = \int_X (f_1\chi_{E,X})\,d\mu = \int_E f_1\,d\mu \in \mathcal{Y}$, where the first equality follows from (iv), the second equality follows from (i), the third equality follows from Lebesgue Dominated Convergence Theorem 11.88, the last equality follows from (iv), and the last step follows from (iv).

(vii) Note that, by Proposition 11.75, $\left\|\int_X \phi_{1,n}\,d\mu\right\| \le \int_X \mathcal{P}\circ\phi_{1,n}\,d\mu \in \mathbb{R}$, $\forall n\in\mathbb{N}$. By Propositions 7.21 and 11.38, $\mathcal{P}\circ\phi_{1,n}$ and $\mathcal{P}\circ f_1$ are $\mathcal{B}$-measurable, $\forall n\in\mathbb{N}$. By Propositions 7.21 and 11.52, we have $\lim_{n\in\mathbb{N}}\mathcal{P}\circ\phi_{1,n} = \mathcal{P}\circ f_1$ a.e. in $\mathcal{X}$. Note that $0 \le \mathcal{P}\circ\phi_{1,n}(x) \le g_1(x)$, $\forall x\in X$, $\forall n\in\mathbb{N}$. Then, $0 \le \left\|\int_X f_1\,d\mu\right\| = \left\|\lim_{n\in\mathbb{N}}\int_X \phi_{1,n}\,d\mu\right\| = \lim_{n\in\mathbb{N}}\left\|\int_X \phi_{1,n}\,d\mu\right\| \le \lim_{n\in\mathbb{N}}\int_X \mathcal{P}\circ\phi_{1,n}\,d\mu = \int_X \mathcal{P}\circ f_1\,d\mu \le \int_X g_1\,d\mu < +\infty$, where the second equality follows from Propositions 7.21 and 3.66, the third equality follows from Lebesgue Dominated Convergence Theorem 11.88, and the third inequality follows from Proposition 11.83.

(viii) Let $f := f_2 - f_1$. Then, $f \gtrapprox \vartheta_{\mathcal{Y}}$ a.e. in $\mathcal{X}$, $f : X \to \mathcal{W}$, and f is $\mathcal{B}$-measurable by Propositions 11.38, 11.39, and 7.23. Note that P is a closed set, $P \in \mathcal{B}_B(\mathcal{Y})$, and $f_{\mathrm{inv}}(P) \in \mathcal{B}$. Define $\bar{f} : X \to \mathcal{W}\cap P \subseteq \mathcal{Y}$ by $\bar{f}(x) = \begin{cases} f(x) & x\in f_{\mathrm{inv}}(P) \\ \vartheta_{\mathcal{Y}} & x\in X\setminus f_{\mathrm{inv}}(P)\end{cases}$. Then, $f = \bar{f}$ a.e. in $\mathcal{X}$, $\bar{f} : X \to \mathcal{W}\cap P$, and $\|\bar{f}(x)\| \le \|f(x)\| \le \|f_1(x)\| + \|f_2(x)\| \le g_1(x) + g_2(x)$, $\forall x\in X$. By Proposition 11.41, $\bar{f}$ is $\mathcal{B}$-measurable. By Proposition 11.83, g_1+g_2 is integrable over $\mathcal{X}$. By Lemma 11.87,

$\int_X \bar{f}\, d\mu \in P$. By (i) and (v), we have $\int_X f_2\, d\mu - \int_X f_1\, d\mu = \int_X f\, d\mu = \int_X \bar{f}\, d\mu \stackrel{\geq}{=} \vartheta_{\mathcal{Y}}$. Then, $\int_X f_1\, d\mu \stackrel{\leq}{=} \int_X f_2\, d\mu$.

This completes the proof of the proposition. □

Definition 11.90 Let $\mathcal{X} := (X, \mathcal{B}, \mu)$ be a measure space, $\mathcal{Y}$ be a normed linear space, and $f : X \to \mathcal{Y}$ be $\mathcal{B}$-measurable. f is said to be *absolutely integrable over* $\mathcal{X}$ if $\mathcal{P} \circ f$ is integrable over $\mathcal{X}$, that is, $0 \leq \int_X \mathcal{P} \circ f\, d\mu < +\infty$. ◇

Note that $\mathcal{P} \circ f$ is $\mathcal{B}$-measurable by Propositions 7.21 and 11.38. We will show that the desired properties of integration are valid if the integrand is absolutely integrable.

Theorem 11.91 (Lebesgue Dominated Convergence) *Let $(X, \mathcal{B}, \mu) =: \mathcal{X}$ be a measure space, $\mathcal{Y}$ be a Banach space, $\mathcal{W}$ be a separable subspace of $\mathcal{Y}$, $f_n : X \to \mathcal{W}$ be $\mathcal{B}$-measurable, $\forall n \in \mathbb{N}$, $f : X \to \mathcal{W}$ be $\mathcal{B}$-measurable, $g_n : X \to [0, \infty) \subset \mathbb{R}$ be $\mathcal{B}$-measurable, $\forall n \in \mathbb{N}$, and $g : X \to [0, \infty) \subset \mathbb{R}$ be $\mathcal{B}$-measurable. Assume that:*

(i) $\lim_{n\in\mathbb{N}} f_n = f$ *a.e. in* $\mathcal{X}$; $\lim_{n\in\mathbb{N}} g_n = g$ *a.e. in* $\mathcal{X}$.
(ii) $\|f_n(x)\| \leq g_n(x)$, $\forall x \in X$, $\forall n \in \mathbb{N}$.
(iii) g_n*'s and* g *are integrable over* $\mathcal{X}$ *and* $\int_X g\, d\mu = \lim_{n\in\mathbb{N}} \int_X g_n\, d\mu \in \mathbb{R}$.

Then, f is integrable over $\mathcal{X}$ and $\int_X f\, d\mu = \lim_{n\in\mathbb{N}} \int_X f_n\, d\mu \in \mathcal{Y}$. Furthermore, f is absolutely integrable over $\mathcal{X}$ and $0 \leq \int_X (\mathcal{P} \circ f)\, d\mu = \lim_{n\in\mathbb{N}} \int_X (\mathcal{P} \circ f_n)\, d\mu \leq \int_X g\, d\mu < \infty$.

Proof We will show that f is integrable over $\mathcal{X}$ and $\int_X f\, d\mu = \lim_{n\in\mathbb{N}} \int_X f_n\, d\mu \in \mathcal{Y}$ by distinguishing two exhaustive and mutually exclusive cases: Case 1: $\mu(X) < +\infty$; Case 2: $\mu(X) = +\infty$.

Case 1: $\mu(X) < +\infty$. The result follows immediately from Lebesgue Dominated Convergence Theorem 11.88.

Case 2: $\mu(X) = +\infty$. Let $(F_A)_{A\in\mathcal{M}(\mathcal{X})}$ and $\left(F_{n,A}\right)_{A\in\mathcal{M}(\mathcal{X})}$ be the nets for the functions f and f_n as defined in Definition 11.71, respectively, $\forall n \in \mathbb{N}$. $\forall A \in \mathcal{M}(\mathcal{X})$, we have $A \in \mathcal{B}$ and $\mu(A) < \infty$. Let $\mathcal{A} := (A, \mathcal{B}_A, \mu_A)$ be the finite measure subspace of $\mathcal{X}$ as defined in Proposition 11.13. By Proposition 11.83, $g|_A$ and $g_n|_A$'s are integrable over $\mathcal{A}$, $\lim_{n\in\mathbb{N}} g_n|_A = g|_A$ a.e. in $\mathcal{A}$, and $\int_A g|_A\, d\mu_A = \lim_{n\in\mathbb{N}} \int_A g_n|_A\, d\mu_A$. By Lebesgue Dominated Convergence Theorem 11.88, we have $f|_A$ and $f_n|_A$'s are integrable over $\mathcal{A}$ and $\int_A f|_A\, d\mu_A = \lim_{n\in\mathbb{N}} \int_A f_n|_A\, d\mu_A \in \mathcal{Y}$. Note that $F_A = \int_A f|_A\, d\mu_A \in \mathcal{Y}$ and $F_{n,A} = \int_A f_n|_A\, d\mu_A \in \mathcal{Y}$, $\forall n \in \mathbb{N}$. Then, the nets $(F_A)_{A\in\mathcal{M}(\mathcal{X})}$ and $\left(F_{n,A}\right)_{A\in\mathcal{M}(\mathcal{X})}$'s are well-defined.

$\forall \epsilon \in (0, \infty) \subset \mathbb{R}$, by Definition 11.79, $\exists$ a simple function $\phi : X \to [0, \infty) \subset \mathbb{R}$ such that $0 \leq \phi(x) \leq g(x)$, $\forall x \in X$, and $\int_X g\, d\mu - \epsilon/5 < \int_X \phi\, d\mu \leq \int_X g\, d\mu < +\infty$. Let $A_0 := \{x \in X \mid \phi(x) > 0\}$. Then, $A_0 \in \mathcal{B}$, $\mu(A_0) < +\infty$, and $A_0 \in \mathcal{M}(\mathcal{X})$. $\forall A \in \mathcal{M}(\mathcal{X})$ with $A_0 \subseteq A$, we have $0 \leq \int_{A\setminus A_0} g|_{A\setminus A_0}\, d\mu_{A\setminus A_0} = \int_A g|_A\, d\mu_A - \int_{A_0} g|_{A_0}\, d\mu_{A_0} \leq \int_X g\, d\mu - \int_{A_0} \phi|_{A_0}\, d\mu_{A_0} = \int_X g\, d\mu - \int_X \phi\, d\mu < \epsilon/5$, where the first equality and the second inequality follow from Proposition 11.83, and the second equality follows from Proposition 11.75. Note that $\|F_A - F_{A_0}\| = \left\| \int_A f|_A \, d\mu_A - \int_{A_0} f|_{A_0} \, d\mu_{A_0} \right\| = \left\| \int_{A \setminus A_0} f|_{A \setminus A_0} \, d\mu_{A \setminus A_0} \right\| = \lim_{n \in \mathbb{N}} \left\| \int_{A \setminus A_0} f_n|_{A \setminus A_0} \cdot d\mu_{A \setminus A_0} \right\| \leq \lim_{n \in \mathbb{N}} \int_{A \setminus A_0} g_n|_{A \setminus A_0} \, d\mu_{A \setminus A_0} = \int_{A \setminus A_0} g|_{A \setminus A_0} \, d\mu_{A \setminus A_0} < \epsilon/5$, where the second equality follows form Proposition 11.89, the third equality follows from the preceding paragraph, the first inequality follows from Proposition 11.89, and the fourth equality follows from the preceding paragraph. Then, the net $(F_A)_{A \in \mathcal{M}(\mathcal{X})}$ is a Cauchy net, which admits a limit, by Proposition 4.44, $\int_X f \, d\mu = \lim_{A \in \mathcal{M}(\mathcal{X})} F_A \in \mathcal{Y}$. Hence, f is integrable over $\mathcal{X}$.

$\forall n \in \mathbb{N}$, by Definition 11.79, $\exists$ a simple function $\phi_n : X \to [0, \infty) \subset \mathbb{R}$ such that $0 \leq \phi_n(x) \leq g_n(x)$, $\forall x \in X$, and $\int_X g_n \, d\mu - \epsilon/5 < \int_X \phi_n \, d\mu \leq \int_X g_n \, d\mu < +\infty$. Let $A_n := \{x \in X \mid \phi_n(x) > 0\}$. Then, $A_n \in \mathcal{B}$, $\mu(A_n) < +\infty$, and $A_n \in \mathcal{M}(\mathcal{X})$. $\forall A \in \mathcal{M}(\mathcal{X})$ with $A_n \subseteq A$, we have $0 \leq \int_{A \setminus A_n} g_n|_{A \setminus A_n} \, d\mu_{A \setminus A_n} = \int_A g_n|_A \, d\mu_A - \int_{A_n} g_n|_{A_n} \, d\mu_{A_n} \leq \int_X g_n \, d\mu - \int_{A_n} \phi_n|_{A_n} \, d\mu_{A_n} = \int_X g_n \, d\mu - \int_X \phi_n \, d\mu < \epsilon/5$, where the first equality and the second inequality follow from Proposition 11.83 and the second equality follows from Proposition 11.75. Note that $\|F_{n,A} - F_{n,A_n}\| = \left\| \int_A f_n|_A \, d\mu_A - \int_{A_n} f_n|_{A_n} \, d\mu_{A_n} \right\| = \left\| \int_{A \setminus A_n} f_n|_{A \setminus A_n} \, d\mu_{A \setminus A_n} \right\| \leq \int_{A \setminus A_n} g_n|_{A \setminus A_n} \, d\mu_{A \setminus A_n} < \epsilon/5$, where the second equality and the first inequality follow from Proposition 11.89. Then, the net $(F_{n,A})_{A \in \mathcal{M}(\mathcal{X})}$ is a Cauchy net, which admits a limit, by Proposition 4.44, $\int_X f_n \, d\mu = \lim_{A \in \mathcal{M}(\mathcal{X})} F_{n,A} \in \mathcal{Y}$.

Let $\bar{g}_n := g_n \wedge g$, $\forall n \in \mathbb{N}$. Then, $0 \leq \bar{g}_n(x) \leq g(x)$, $\forall x \in X$, and $\bar{g}_n$ is $\mathcal{B}$-measurable by Proposition 11.40, $\forall n \in \mathbb{N}$. Then, $\lim_{n \in \mathbb{N}} \bar{g}_n = g$ a.e. in $\mathcal{X}$ by Propositions 11.52 and 11.53. By Fatou's Lemma 11.80, $\int_X g \, d\mu \leq \liminf_{n \in \mathbb{N}} \int_X \bar{g}_n \, d\mu$. By Proposition 11.83, $0 \leq \int_X \bar{g}_n \, d\mu \leq \int_X g \, d\mu < +\infty$, $\forall n \in \mathbb{N}$. Then, we have $\int_X g \, d\mu \leq \liminf_{n \in \mathbb{N}} \int_X \bar{g}_n \, d\mu \leq \limsup_{n \in \mathbb{N}} \int_X \bar{g}_n \, d\mu \leq \int_X g \, d\mu$. Therefore, by Proposition 3.83, $\int_X g \, d\mu = \lim_{n \in \mathbb{N}} \int_X \bar{g}_n \, d\mu \in \mathbb{R}$. This coupled with (iii) and the fact that $F_{A_0} = \lim_{n \in \mathbb{N}} F_{n,A_0}$ implies that $\exists n_0 \in \mathbb{N}$, $\forall n \in \mathbb{N}$ with $n \geq n_0$, we have $0 \leq \int_X g \, d\mu - \int_X \bar{g}_n \, d\mu < \epsilon/5$, $\left| \int_X g \, d\mu - \int_X g_n \, d\mu \right| < \epsilon/5$, and $\|F_{A_0} - F_{n,A_0}\| < \epsilon/5$. $\forall A \in \mathcal{M}(\mathcal{X})$ with $A_0 \subseteq A$, we have $0 \leq \int_{A \setminus A_0} g_n|_{A \setminus A_0} \, d\mu_{A \setminus A_0} = \int_{A \setminus A_0} (g_n - \bar{g}_n)|_{A \setminus A_0} \, d\mu_{A \setminus A_0} + \int_{A \setminus A_0} \bar{g}_n|_{A \setminus A_0} \, d\mu_{A \setminus A_0} \leq \int_X (g_n - \bar{g}_n) \, d\mu + \int_{A \setminus A_0} g|_{A \setminus A_0} \, d\mu_{A \setminus A_0} < \int_X g_n \, d\mu - \int_X \bar{g}_n \, d\mu + \epsilon/5 \leq \left| \int_X g_n \, d\mu - \int_X g \, d\mu \right| + \int_X g \, d\mu - \int_X \bar{g}_n \, d\mu + \epsilon/5 < 3\epsilon/5$, where the first equality, the second inequality, and the third inequality follow from Proposition 11.83. Furthermore, $\|F_{n,A} - F_{n,A_0}\| = \left\| \int_A f_n|_A \, d\mu_A - \int_{A_0} f_n|_{A_0} \, d\mu_{A_0} \right\| = \left\| \int_{A \setminus A_0} f_n|_{A \setminus A_0} \cdot d\mu_{A \setminus A_0} \right\| \leq \int_{A \setminus A_0} g_n|_{A \setminus A_0} \, d\mu_{A \setminus A_0} < 3\epsilon/5$, where the second equality and the first inequality follow from Proposition 11.89. This implies that $\left\| \int_X f \, d\mu - \int_X f_n \, d\mu \right\| \leq \left\| \int_X f \, d\mu - F_{A_0} \right\| + \|F_{A_0} - F_{n,A_0}\| + \left\| F_{n,A_0} - \int_X f_n \, d\mu \right\| < \lim_{A \in \mathcal{M}(\mathcal{X})} \|F_A - F_{A_0}\| + \epsilon/5 + \lim_{A \in \mathcal{M}(\mathcal{X})} \|F_{n,A_0} - F_{n,A}\| \leq \epsilon/5 + \epsilon/5 + 3\epsilon/5 = \epsilon$, where the second inequality follows from Propositions 7.21 and 3.66. Hence, $\int_X f \, d\mu = \lim_{n \in \mathbb{N}} \int_X f_n \, d\mu \in \mathcal{Y}$.

Then, in both cases, we have shown that f is integrable over $\mathcal{X}$ and $\int_X f \, d\mu = \lim_{n \in \mathbb{N}} \int_X f_n \, d\mu \in \mathcal{Y}$.

Note that $0 \le \mathcal{P} \circ f_n(x) \le g_n(x)$, $\forall x \in X$, $\forall n \in \mathbb{N}$, $\lim_{n \in \mathbb{N}} \mathcal{P} \circ f_n = \mathcal{P} \circ f$ a.e. in $\mathcal{X}$ by Propositions 7.21 and 11.52, and $\mathcal{P} \circ f$ and $\mathcal{P} \circ f_n$'s are $\mathcal{B}$-measurable by Propositions 7.21 and 11.38. Then, by what we have proved so far, $\int_X \mathcal{P} \circ f \, d\mu = \lim_{n \in \mathbb{N}} \int_X \mathcal{P} \circ f_n \, d\mu \le \lim_{n \in \mathbb{N}} \int_X g_n \, d\mu = \int_X g \, d\mu < \infty$. Hence, f is absolutely integrable over $\mathcal{X}$.

This completes the proof of the theorem. □

Proposition 11.92 *Let $\mathcal{X} := (X, \mathcal{B}, \mu)$ be a measure space, $\mathcal{Y}$ be a Banach space over $\mathbb{K}$, $\mathcal{W}$ be a separable subspace of $\mathcal{Y}$, $\mathcal{Z}$ be a Banach space over $\mathbb{K}$, and $f_i : X \to \mathcal{W}$ be absolutely integrable over $\mathcal{X}$, $i = 1, 2$. Then, the following statements hold:*

(i) *f_i is integrable over $\mathcal{X}$ and $\int_X f_i \, d\mu \in \mathcal{Y}$, $i = 1, 2$.*
(ii) *$f_1 + f_2$ is absolutely integrable over $\mathcal{X}$ and $\int_X (f_1 + f_2) \, d\mu = \int_X f_1 \, d\mu + \int_X f_2 \, d\mu \in \mathcal{Y}$.*
(iii) *$\forall A \in B(\mathcal{Y}, \mathcal{Z})$, $A f_1$ is absolutely integrable over $\mathcal{X}$ and $\int_X (A f_1) \, d\mu = A \int_X f_1 \, d\mu \in \mathcal{Z}$.*
(iv) *$\forall c \in \mathbb{K}$, $c f_1$ is absolutely integrable over $\mathcal{X}$ and $\int_X (c f_1) \, d\mu = c \int_X f_1 \, d\mu \in \mathcal{Y}$.*
(v) *$\forall H \in \mathcal{B}$, $f_1|_H$ is absolutely integrable over $\mathcal{H}$ and $\int_X (f_1 \chi_{H,X}) \, d\mu = \int_H f_1|_H \, d\mu_H \in \mathcal{Y}$, where $\mathcal{H} := (H, \mathcal{B}_H, \mu_H)$ is the measure subspace of $\mathcal{X}$ as defined in Proposition 11.13. We will henceforth denote $\int_H f_1|_H \, d\mu_H$ by $\int_H f_1 \, d\mu$.*
(vi) *If $f_1 = f_2$ a.e. in $\mathcal{X}$, then $\int_X f_1 \, d\mu = \int_X f_2 \, d\mu \in \mathcal{Y}$.*
(vii) *$\forall$ pairwise disjoint $(E_i)_{i=1}^{\infty} \subseteq \mathcal{B}$, $\sum_{i=1}^{\infty} \int_{E_i} f_1 \, d\mu = \int_{\bigcup_{i=1}^{\infty} E_i} f_1 \, d\mu \in \mathcal{Y}$.*
(viii) *$0 \le \left\| \int_X f_1 \, d\mu \right\| \le \int_X \mathcal{P} \circ f_1 \, d\mu < +\infty$.*
(ix) *If $\mathcal{Y}$ admits a positive cone P and $f_1 \leqq f_2$ a.e. in $\mathcal{X}$, then $\int_X f_1 \, d\mu \leqq \int_X f_2 \, d\mu$.*

Proof We will distinguish two exhaustive and mutually exclusive cases: Case 1: $\mu(X) < +\infty$; Case 2: $\mu(X) = +\infty$. Case 1: $\mu(X) < +\infty$. The results follow immediately from Proposition 11.89.

Case 2: $\mu(X) = +\infty$. f_1, f_2, $\mathcal{P} \circ f_1$, and $\mathcal{P} \circ f_2$ are $\mathcal{B}$-measurable by Propositions 7.21 and 11.38.

(i) By Lebesgue Dominated Convergence Theorem 11.91, we have that f_i is integrable over $\mathcal{X}$ and $\int_X f_i \, d\mu \in \mathcal{Y}$, $i = 1, 2$. Let $(F_{i,E})_{E \in \mathcal{M}(\mathcal{X})}$ be the net for $\int_X f_i \, d\mu$ as defined in Definition 11.71, $i = 1, 2$. $\forall E \in \mathcal{M}(\mathcal{X})$, $F_{i,E} = \int_E f_i|_E \, d\mu_E$, $i = 1, 2$. Then, by Definition 11.71, we have the net $(F_{i,E})_{E \in \mathcal{M}(\mathcal{X})}$ is well-defined, $i = 1, 2$, and $\lim_{E \in \mathcal{M}(\mathcal{X})} F_{i,E} = \int_X f_i \, d\mu \in \mathcal{Y}$, $i = 1, 2$.

(ii) By Propositions 11.38, 11.39, 7.21, and 7.23, $f_1 + f_2$ and $\mathcal{P} \circ (f_1 + f_2)$ are $\mathcal{B}$-measurable. Note that $\mathcal{P} \circ (f_1 + f_2)(x) = \|f_1(x) + f_2(x)\| \le \|f_1(x)\| + \|f_2(x)\| = (\mathcal{P} \circ f_1 + \mathcal{P} \circ f_2)(x)$, $\forall x \in X$. By Proposition 11.83, $0 \le \int_X \mathcal{P} \circ (f_1 + f_2) \, d\mu \le \int_X (\mathcal{P} \circ f_1 + \mathcal{P} \circ f_2) \, d\mu = \int_X \mathcal{P} \circ f_1 \, d\mu + \int_X \mathcal{P} \circ f_2 \, d\mu <$

$+\infty$. Hence, $f_1 + f_2$ is absolutely integrable over $\mathcal{X}$. Let $\left(\bar{F}_E\right)_{E\in\mathcal{M}(\mathcal{X})}$ be the net for $\int_X (f_1 + f_2)\,\mathrm{d}\mu$ as defined in Definition 11.71. $\forall E \in \mathcal{M}(\mathcal{X})$, $\bar{F}_E = \int_E (f_1 + f_2)|_E \,\mathrm{d}\mu_E = \int_E f_1|_E \,\mathrm{d}\mu_E + \int_E f_2|_E \,\mathrm{d}\mu_E = F_{1,E} + F_{2,E}$, where the second equality follows from Proposition 11.89. Then, $\int_X (f_1 + f_2)\,\mathrm{d}\mu = \lim_{E\in\mathcal{M}(\mathcal{X})} \bar{F}_E = \lim_{E\in\mathcal{M}(\mathcal{X})}(F_{1,E} + F_{2,E}) = \int_X f_1\,\mathrm{d}\mu + \int_X f_2\,\mathrm{d}\mu \in \mathcal{Y}$, where the third equality follows from Propositions 3.66, 3.67, and 7.23. Hence, $f_1 + f_2$ is integrable over $\mathcal{X}$.

(iii) By Propositions 11.38, 7.21, and 7.62, Af_1 and $\mathcal{P}\circ(Af_1)$ are $\mathcal{B}$-measurable. Note that $\mathcal{P}\circ(Af_1)(x) = \|Af_1(x)\| \le \|A\|\|f_1(x)\| = \|A\|\mathcal{P}\circ f_1(x)$, $\forall x \in X$. By Proposition 11.83 and the fact that $\int_X \mathcal{P}\circ f_1\,\mathrm{d}\mu \in \mathbb{R}$, we have $0 \le \int_X \mathcal{P}\circ(Af_1)\,\mathrm{d}\mu \le \int_X (\|A\|\mathcal{P}\circ f_1)\,\mathrm{d}\mu = \|A\|\int_X \mathcal{P}\circ f_1\,\mathrm{d}\mu < +\infty$. Hence, Af_1 is absolutely integrable over $\mathcal{X}$. Let $\left(\bar{F}_E\right)_{E\in\mathcal{M}(\mathcal{X})}$ be the net for $\int_X (Af_1)\,\mathrm{d}\mu$ as defined in Definition 11.71. $\forall E \in \mathcal{M}(\mathcal{X})$, $\bar{F}_E = \int_E (Af_1)|_E \,\mathrm{d}\mu_E = A\int_E f_1|_E\,\mathrm{d}\mu_E = AF_{1,E}$, where the second equality follows from Proposition 11.89. Then, by Propositions 3.66 and 7.62, we have $\int_X (Af_1)\,\mathrm{d}\mu = \lim_{E\in\mathcal{M}(\mathcal{X})}\bar{F}_E = \lim_{E\in\mathcal{M}(\mathcal{X})} AF_{1,E} = A\int_X f_1\,\mathrm{d}\mu \in \mathcal{Z}$. Hence, Af_1 is integrable over $\mathcal{X}$.

(iv) This follows immediately from (iii).

(v) Fix any $H \in \mathcal{B}$. By Proposition 11.41, $f_1|_H$ and $\mathcal{P}\circ(f_1|_H) = (\mathcal{P}\circ f_1)|_H$ are $\mathcal{B}_H$-measurable. By Proposition 11.83, $f_1|_H$ is absolutely integrable over $\mathcal{H}$. Then, by (i), $f_1|_H$ is integrable over $\mathcal{H}$. Again, by (i), $f_1\chi_{H,X}$ is integrable over $\mathcal{X}$. $\forall \epsilon \in (0,\infty) \subset \mathbb{R}$, $\exists A_0 \in \mathcal{M}(\mathcal{X})$ such that $\forall A \in \mathcal{M}(\mathcal{X})$ with $A_0 \subseteq A$, we have $\left\|\int_A (f_1\chi_{H,X})\big|_A \,\mathrm{d}\mu_A - \int_X (f_1\chi_{H,X})\,\mathrm{d}\mu\right\| < \epsilon/2$. We will distinguish two exhaustive and mutually exclusive subcases: Case 2a: $\mu(H) = +\infty$; Case 2b: $\mu(H) < +\infty$.

Case 2a: $\mu(H) = +\infty$. $\exists E_0 \in \mathcal{M}(\mathcal{H})$ such that $\forall E \in \mathcal{M}(\mathcal{H})$ with $E_0 \subseteq E$, we have $\left\|\int_E f_1|_E\,\mathrm{d}\mu_E - \int_H f_1|_H\,\mathrm{d}\mu_H\right\| < \epsilon/2$. Let $A_1 := A_0 \cup E_0 \in \mathcal{M}(\mathcal{X})$ and $E_1 := A_1 \cap H \in \mathcal{M}(\mathcal{H})$. By Proposition 11.89, $\int_{A_1} (f_1\chi_{H,X})\big|_{A_1}\,\mathrm{d}\mu_{A_1} = \int_{H\cap A_1} f_1|_{H\cap A_1}\,\mathrm{d}\mu_{H\cap A_1} = \int_{E_1} f_1|_{E_1}\,\mathrm{d}\mu_{E_1}$. Then, we have $\left\|\int_X (f_1\chi_{H,X})\,\mathrm{d}\mu - \int_H f_1|_H\,\mathrm{d}\mu_H\right\| \le \left\|\int_X (f_1\chi_{H,X})\,\mathrm{d}\mu - \int_{A_1} (f_1\chi_{H,X})\big|_{A_1}\cdot \mathrm{d}\mu_{A_1}\right\| + \left\|\int_{E_1} f_1|_{E_1}\cdot\mathrm{d}\mu_{E_1} - \int_H f_1|_H\,\mathrm{d}\mu_H\right\| < \epsilon/2+\epsilon/2 = \epsilon$. By the arbitrariness of ϵ, we have $\int_X (f_1\chi_{H,X})\,\mathrm{d}\mu = \int_H f_1|_H\,\mathrm{d}\mu_H \in \mathcal{Y}$.

Case 2b: $\mu(H) < +\infty$. Let $A_1 := A_0 \cup H \in \mathcal{M}(\mathcal{X})$. By Proposition 11.89, $\int_{A_1} (f_1\chi_{H,X})\big|_{A_1}\,\mathrm{d}\mu_{A_1} = \int_H f_1|_H\,\mathrm{d}\mu_H$. Then, we have $\left\|\int_X (f_1\chi_{H,X})\,\mathrm{d}\mu - \int_H f_1|_H\,\mathrm{d}\mu_H\right\| = \left\|\int_X (f_1\chi_{H,X})\,\mathrm{d}\mu - \int_{A_1} (f_1\chi_{H,X})\big|_{A_1}\,\mathrm{d}\mu_{A_1}\right\| < \epsilon/2$. By the arbitrariness of ϵ, we have $\int_X (f_1\chi_{H,X})\,\mathrm{d}\mu = \int_H f_1|_H\,\mathrm{d}\mu_H \in \mathcal{Y}$.

Hence, in both subcases, $\int_X (f_1\chi_{H,X})\,\mathrm{d}\mu = \int_H f_1|_H\,\mathrm{d}\mu_H \in \mathcal{Y}$.

(vi) This follows immediately from Lebesgue Dominated Convergence Theorem 11.91.

(vii) $\forall$ pairwise disjoint $(E_i)_{i=1}^{\infty} \subseteq \mathcal{B}$, let $E := \bigcup_{i=1}^{\infty} E_i \in \mathcal{B}$ and $\bar{E}_n := \bigcup_{i=1}^{n} E_i \in \mathcal{B}$, $\forall n \in \mathbb{N}$. Then, $\lim_{n\in\mathbb{N}} f_1(x)\chi_{\bar{E}_n,X}(x) = f_1(x)\chi_{E,X}(x)$, $\forall x \in X$. Then, we have $\sum_{i=1}^{\infty} \int_{E_i} f_1 \,\mathrm{d}\mu = \lim_{n\in\mathbb{N}} \sum_{i=1}^{n} \int_X (f_1\chi_{E_i,X}) \,\mathrm{d}\mu = \lim_{n\in\mathbb{N}} \int_X (f_1\chi_{\bar{E}_n,X}) \,\mathrm{d}\mu = \int_X (f_1\chi_{E,X}) \,\mathrm{d}\mu = \int_E f_1 \,\mathrm{d}\mu \in \mathcal{Y}$, where the first equality follows from (v), the second equality follows from (ii), the third equality follows from Lebesgue Dominated Convergence Theorem 11.91, the last equality follows from (v), and the last set membership follows from (v).

(viii) By (i), $\mathcal{P} \circ f_1$ is integrable over $\mathcal{X}$. $\forall E \in \mathcal{M}(\mathcal{X})$, by Proposition 11.89, $\|F_{1,E}\| \leq \int_E \mathcal{P} \circ f_1 \,\mathrm{d}\mu \leq \int_X \mathcal{P} \circ f_1 \,\mathrm{d}\mu < +\infty$, where the second inequality follows from Proposition 11.83. Then, $0 \leq \left\|\int_X f_1 \,\mathrm{d}\mu\right\| = \lim_{E\in\mathcal{M}(\mathcal{X})} \|F_{1,E}\| \leq \int_X \mathcal{P} \circ f_1 \,\mathrm{d}\mu < +\infty$, where the equality follows from Propositions 3.66 and 7.21.

(ix) $\forall E \in \mathcal{M}(\mathcal{X})$, by Proposition 11.89 and (v), we have $\int_E (f_2 - f_1)|_E \cdot \mathrm{d}\mu_E = \int_E f_2|_E \,\mathrm{d}\mu_E - \int_E f_1|_E \,\mathrm{d}\mu_E \in P$. Note that, by (i) and (iv), $\int_X f_2 \,\mathrm{d}\mu - \int_X f_1 \,\mathrm{d}\mu = \int_X (f_2 - f_1) \,\mathrm{d}\mu = \lim_{E\in\mathcal{M}(\mathcal{X})} \int_E (f_2 - f_1)|_E \,\mathrm{d}\mu_E \in \overline{P} = P$, where the second to last step follows from Proposition 3.68 and the last step follows from the fact that P is closed. Therefore, we have $\int_X f_1 \,\mathrm{d}\mu \stackrel{\leq}{=} \int_X f_2 \,\mathrm{d}\mu$.

This completes the proof of the proposition. □

Proposition 11.93 *Let $\mathcal{X} := (X, \mathcal{B}, \mu)$ be a measure space, $\mathcal{Y}_i$ be a Banach space over $\mathbb{K}$, $\mathcal{W}_i \subseteq \mathcal{Y}_i$ be a separable subspace, $f_i : X \to \mathcal{W}_i$, $i = 1, 2$, and $h : X \to \mathcal{W}_1 \times \mathcal{W}_2$ be defined by $h(x) = (f_1(x), f_2(x))$, $\forall x \in X$. Then, h is absolutely integrable over $\mathcal{X}$ if, and only if, f_1 and f_2 are absolutely integrable over $\mathcal{X}$. In this case, we have $\int_X h \,\mathrm{d}\mu = (\int_X f_1 \,\mathrm{d}\mu, \int_X f_2 \,\mathrm{d}\mu) \in \mathcal{Y}_1 \times \mathcal{Y}_2$.*

Proof Let f_1 and f_2 be absolutely integrable over $\mathcal{X}$. By Proposition 11.39, h is $\mathcal{B}$-measurable. By Propositions 7.21 and 11.38, $\mathcal{P} \circ h$ is $\mathcal{B}$-measurable. $\forall x \in X$, $\mathcal{P} \circ h(x) = \|(f_1(x), f_2(x))\| = (\|f_1(x)\|^2 + \|f_2(x)\|^2)^{1/2} \leq \mathcal{P} \circ f_1(x) + \mathcal{P} \circ f_2(x)$. By Proposition 11.83, $0 \leq \int_X (\mathcal{P} \circ h) \,\mathrm{d}\mu \leq \int_X (\mathcal{P} \circ f_1 + \mathcal{P} \circ f_2) \,\mathrm{d}\mu = \int_X (\mathcal{P} \circ f_1) \,\mathrm{d}\mu + \int_X (\mathcal{P} \circ f_2) \,\mathrm{d}\mu < \infty$. Hence, h is absolutely integrable over $\mathcal{X}$. By Propositions 4.31 and 7.22, $\mathcal{Y}_1 \times \mathcal{Y}_2$ is a Banach space over $\mathbb{K}$. By Propositions 7.22, 3.28, and 4.4, $\mathcal{W}_1 \times \mathcal{W}_2 \subseteq \mathcal{Y}_1 \times \mathcal{Y}_2$ is a separable subspace. By Proposition 11.92, h, f_1, and f_2 are integrable over $\mathcal{X}$ and $\int_X h \,\mathrm{d}\mu \in \mathcal{Y}_1 \times \mathcal{Y}_2$.

On the other hand, let h be absolutely integrable over $\mathcal{X}$. Then, by Proposition 11.39, f_1 and f_2 are $\mathcal{B}$-measurable. Fix any $i \in \{1, 2\}$. By Propositions 7.21 and 11.38, $\mathcal{P} \circ f_i$ is $\mathcal{B}$-measurable. $\forall x \in X$, $\mathcal{P} \circ f_i(x) \leq \mathcal{P} \circ h(x)$. Then, by Proposition 11.83, $0 \leq \int_X \mathcal{P} \circ f_i \,\mathrm{d}\mu \leq \int_X \mathcal{P} \circ h \,\mathrm{d}\mu < \infty$. Hence, f_i is absolutely integrable over $\mathcal{X}$. By Proposition 11.92, f_i is integrable over $\mathcal{X}$ and $\int_X f_i \,\mathrm{d}\mu \in \mathcal{Y}_i$.

Hence, h is absolutely integrable over $\mathcal{X}$ if, and only if, f_1 and f_2 are absolutely integrable over $\mathcal{X}$.

Let h, f_1, and f_2 be absolutely integrable over $\mathcal{X}$. We will show that $\int_X h \,\mathrm{d}\mu = (\int_X f_1 \,\mathrm{d}\mu, \int_X f_2 \,\mathrm{d}\mu) \in \mathcal{Y}_1 \times \mathcal{Y}_2$ by distinguishing two exhaustive and mutually exclusives cases: Case 1: $\mu(X) < \infty$; Case 2: $\mu(X) = \infty$.

Case 1: $\mu(X) < \infty$. By Lemma 11.87, $\forall i \in \{1, 2\}$, $\exists$ a sequence of simple functions $(\varphi_{i,n})_{n=1}^{\infty}$, $\varphi_{i,n} : X \to \mathcal{W}_i$, $\forall n \in \mathbb{N}$, such that $\lim_{n\in\mathbb{N}} \varphi_{i,n} = f_i$ a.e. in $\mathcal{X}$, $\|\varphi_{i,n}(x)\| \leq \|f_i(x)\|$, $\forall x \in X$, $\forall n \in \mathbb{N}$, and $\int_X f_i \,\mathrm{d}\mu = \lim_{n\in\mathbb{N}} \int_X \varphi_{i,n} \,\mathrm{d}\mu \in \mathcal{Y}_i$.

By Proposition 11.53, we have $\lim_{n\in\mathbb{N}} \psi_n = h$ a.e. in $\mathcal{X}$, where $\psi_n : X \to \mathcal{W}_1 \times \mathcal{W}_2$ is defined by $\psi_n(x) = (\varphi_{1,n}(x), \varphi_{2,n}(x))$, $\forall x \in X$, $\forall n \in \mathbb{N}$. Clearly, ψ_n is a simple function, $\forall n \in \mathbb{N}$. By Lebesgue Dominated Convergence Theorem 11.91, $\int_X h \,d\mu = \lim_{n\in\mathbb{N}} \int_X \psi_n \cdot d\mu = \lim_{n\in\mathbb{N}} (\int_X \varphi_{1,n} \,d\mu, \int_X \varphi_{2,n} \,d\mu) = (\int_X f_1 \,d\mu, \int_X f_2 \,d\mu) \in \mathcal{Y}_1 \times \mathcal{Y}_2$, where the second equality follows from Proposition 11.75 and the third equality follows from Proposition 3.67.

Case 2: $\mu(X) = \infty$. Let $(H_E)_{E\in\mathcal{M}(\mathcal{X})}$ and $\left(F_{i,E}\right)_{E\in\mathcal{M}(\mathcal{X})}$ be the nets for $\int_X h \,d\mu$ and $\int_X f_i \,d\mu$ as defined in Definition 11.71, respectively, $i = 1, 2$. $\forall E \in \mathcal{M}(\mathcal{X})$, by Case 1,

$$H_E = \int_E h|_E \,d\mu_E = (\int_E f_1|_E \,d\mu_E, \int_E f_2|_E \,d\mu_E) = (F_{1,E}, F_{2,E})$$

where $\mathcal{E} := (E, \mathcal{B}_E, \mu_E)$ is the finite measure subspace of $\mathcal{X}$. Since h, f_1, and f_2 are integrable over $\mathcal{X}$, we have $\int_X h \,d\mu = \lim_{E\in\mathcal{M}(\mathcal{X})} H_E = \lim_{E\in\mathcal{M}(\mathcal{X})} (F_{1,E}, F_{2,E}) = (\int_X f_1 \,d\mu, \int_X f_2 \,d\mu) \in \mathcal{Y}_1 \times \mathcal{Y}_2$, where the last equality follows from Proposition 3.67.

Hence, in both cases, we have $\int_X h \,d\mu = (\int_X f_1 \,d\mu, \int_X f_2 \,d\mu) \in \mathcal{Y}_1 \times \mathcal{Y}_2$. This completes the proof of the proposition. □

Proposition 11.94 *Let $\mathcal{X} := (X, \mathcal{B}, \mu)$ be a σ-finite measure space, $\mathcal{Y}$ be a Banach space, $\mathcal{W}$ be a separable subspace of $\mathcal{Y}$, $U \subseteq \mathcal{W}$ be a conic segment, and $f : X \to U$ be absolutely integrable over $\mathcal{X}$. Then, there exists a sequence of simple functions $(\varphi_n)_{n=1}^{\infty}$, $\varphi_n : X \to U$, $\forall n \in \mathbb{N}$, such that $\|\varphi_n(x)\| \leq \mathcal{P} \circ f(x)$, $\forall x \in X$, $\forall n \in \mathbb{N}$, $\lim_{n\in\mathbb{N}} \varphi_n = f$ a.e. in $\mathcal{X}$, $\lim_{n\in\mathbb{N}} \int_X \varphi_n \,d\mu = \int_X f \,d\mu \in \mathcal{Y}$, $\lim_{n\in\mathbb{N}} \int_X \mathcal{P}\circ\varphi_n \,d\mu = \int_X \mathcal{P} \circ f \,d\mu \in [0, \infty) \subset \mathbb{R}$, and $\lim_{n\in\mathbb{N}} \int_X \mathcal{P} \circ (\varphi_n - f) \,d\mu = 0$.*

Proof By Proposition 11.66, there exists a sequence of simple functions $(\varphi_n)_{n=1}^{\infty}$, $\varphi_n : X \to U$, $\forall n \in \mathbb{N}$, such that $\|\varphi_n(x)\| \leq \mathcal{P} \circ f(x)$, $\forall x \in X$, $\forall n \in \mathbb{N}$, and $\lim_{n\in\mathbb{N}} \varphi_n = f$ a.e. in $\mathcal{X}$. By Lebesgue Dominated Convergence Theorem 11.91, $\lim_{n\in\mathbb{N}} \int_X \varphi_n \,d\mu = \int_X f \,d\mu \in \mathcal{Y}$. Note that $\lim_{n\in\mathbb{N}} \mathcal{P} \circ \varphi_n = \mathcal{P} \circ f$ a.e. in $\mathcal{X}$, by Propositions 7.21 and 11.52. Again by Lebesgue Dominated Convergence Theorem 11.91, we have $\lim_{n\in\mathbb{N}} \int_X (\mathcal{P} \circ \varphi_n) \,d\mu = \int_X (\mathcal{P} \circ f) \,d\mu \in [0, \infty) \subset \mathbb{R}$. By Lemma 11.43, $\mathcal{P} \circ (\varphi_n - f)$ is $\mathcal{B}$-measurable. By Propositions 7.21, 7.23, 11.53, and 11.52, we have $\lim_{n\in\mathbb{N}} \mathcal{P} \circ (\varphi_n - f) = 0$ a.e. in $\mathcal{X}$. Note that $0 \leq \mathcal{P} \circ (\varphi_n - f)(x) = \|\varphi_n(x) - f(x)\| \leq \|\varphi_n(x)\| + \|f(x)\| \leq 2\mathcal{P} \circ f(x)$, $\forall x \in X$, $\forall n \in \mathbb{N}$. By Lebesgue Dominated Convergence Theorem 11.91 and Proposition 11.75, we have $\lim_{n\in\mathbb{N}} \int_X \mathcal{P} \circ (\varphi_n - f) \,d\mu = 0$. This completes the proof of the proposition. □

Proposition 11.95 *Let $\mathcal{X} := (X, \mathcal{B}, \mu)$ be a measure space, $\mathcal{Y}$ be a Banach space, $\mathcal{W}$ be a separable subspace of $\mathcal{Y}$, $U \subseteq \mathcal{W}$ be a conic segment, and $f : X \to U$ be absolutely integrable over $\mathcal{X}$. Then, there exists a sequence of simple functions $(\phi_n)_{n=1}^{\infty}$, $\phi_n : X \to U$, $\forall n \in \mathbb{N}$, such that $\|\phi_n(x)\| \leq \mathcal{P} \circ f(x)$, $\forall x \in X$, $\forall n \in \mathbb{N}$, $\lim_{n\in\mathbb{N}} \int_X \phi_n \,d\mu = \int_X f \,d\mu \in \mathcal{Y}$, $\lim_{n\in\mathbb{N}} \int_X \mathcal{P} \circ \phi_n \,d\mu = \int_X \mathcal{P} \circ f \,d\mu \in [0, \infty) \subset \mathbb{R}$, and $\lim_{n\in\mathbb{N}} \int_X \mathcal{P} \circ (\phi_n - f) \,d\mu = 0$.*

Proof We will distinguish two exhaustive and mutually exclusive cases: Case 1: $\mu(X) < +\infty$; Case 2: $\mu(X) = +\infty$.

Case 1: $\mu(X) < +\infty$. Then, $\mathcal{X}$ is a finite measure space. The result follows immediately from Proposition 11.94.

Case 2: $\mu(X) = +\infty$. Since $\int_X \mathcal{P} \circ f \, d\mu < +\infty$, then, by Definition 11.71, $\forall n \in \mathbb{N}$, $\exists E_n \in \mathcal{B}$ with $\mu(E_n) < +\infty$ such that $\int_X \mathcal{P} \circ f \, d\mu - 2^{-n-1} < \int_{E_n} \mathcal{P} \circ f \, d\mu \leq \int_X \mathcal{P} \circ f \, d\mu$. By Case 1, $\exists$ a simple function $\bar{\phi}_n : E_n \to U$ such that $\|\bar{\phi}_n(x)\| \leq \mathcal{P} \circ f(x)$, $\forall x \in E_n$, $\left\| \int_{E_n} \bar{\phi}_n \, d\mu_{E_n} - \int_{E_n} f \, d\mu \right\| < 2^{-n-1}$, $\left| \int_{E_n} \mathcal{P} \circ \bar{\phi}_n \, d\mu_{E_n} - \int_{E_n} \mathcal{P} \circ f \, d\mu \right| < 2^{-n-1}$, and $0 \leq \int_{E_n} \mathcal{P} \circ (\bar{\phi}_n - f|_{E_n}) \, d\mu_{E_n} < 2^{-n-1}$, where $(E_n, \mathcal{B}_{E_n}, \mu_{E_n})$ is the finite measure subspace of $\mathcal{X}$. This simple function $\bar{\phi}_n$ may be extended to a simple function $\phi_n : X \to U$ such that $\phi_n|_{E_n} = \bar{\phi}_n$ and $\phi_n(x) = \vartheta_{\mathcal{Y}}$, $\forall x \in X \setminus E_n$. Then, we have $\|\phi_n(x)\| \leq \mathcal{P} \circ f(x)$, $\forall x \in X$, and $\left\| \int_X \phi_n \, d\mu - \int_X f \, d\mu \right\| = \left\| \int_X (\phi_n \chi_{E_n,X}) \, d\mu - \int_{E_n} f \, d\mu - \int_{X \setminus E_n} f \, d\mu \right\| = \left\| \int_{E_n} \bar{\phi}_n \, d\mu_{E_n} - \int_{E_n} f \, d\mu - \int_{X \setminus E_n} f \, d\mu \right\| \leq \left\| \int_{E_n} \bar{\phi}_n \, d\mu_{E_n} - \int_{E_n} f \, d\mu \right\| + \left\| \int_{X \setminus E_n} f \, d\mu \right\| < 2^{-n-1} + \int_{X \setminus E_n} \mathcal{P} \circ f \, d\mu = 2^{-n-1} + \int_X \mathcal{P} \circ f \, d\mu - \int_{E_n} \mathcal{P} \circ f \, d\mu < 2^{-n}$, where the first and second equalities follow from Proposition 11.92, the second inequality follows from Proposition 11.92, and the third equality follows from Proposition 11.83. Next, we have $\left| \int_X \mathcal{P} \circ \phi_n \, d\mu - \int_X \mathcal{P} \circ f \, d\mu \right| \leq \left| \int_X ((\mathcal{P} \circ \phi_n) \chi_{E_n,X}) \, d\mu - \int_{E_n} \mathcal{P} \circ f \, d\mu \right| + \left| \int_X \mathcal{P} \circ f \, d\mu - \int_{E_n} \mathcal{P} \circ f \, d\mu \right| < \left| \int_{E_n} \mathcal{P} \circ \bar{\phi}_n \, d\mu_{E_n} - \int_{E_n} \mathcal{P} \circ f \, d\mu \right| + 2^{-n-1} < 2^{-n}$, where the second inequality follows from Proposition 11.92. Finally, we have $0 \leq \int_X \mathcal{P} \circ (\phi_n - f) \, d\mu = \int_{E_n} \mathcal{P} \circ ((\phi_n - f)|_{E_n}) \, d\mu_{E_n} + \int_{X \setminus E_n} \mathcal{P} \circ ((\phi_n - f)|_{X \setminus E_n}) \, d\mu_{X \setminus E_n} = \int_{E_n} \mathcal{P} \circ (\bar{\phi}_n - f|_{E_n}) \, d\mu_{E_n} + \int_{X \setminus E_n} \mathcal{P} \circ (f|_{X \setminus E_n}) \, d\mu_{X \setminus E_n} < 2^{-n-1} + \int_{X \setminus E_n} \mathcal{P} \circ f \, d\mu < 2^{-n}$, where the first equality follows from Propositions 11.83. Hence, $(\phi_n)_{n=1}^{\infty}$ is the sequence we seek.

This completes the proof of the proposition. □

Proposition 11.96 *Let $\mathcal{X} := (X, \mathcal{B}, \mu)$ be a σ-finite measure space, and $f_1 : X \to \mathbb{R}$ and $f_2 : X \to \mathbb{R}$ be $\mathcal{B}$-measurable. Assume that $\int_E f_1 \, d\mu \leq \int_E f_2 \, d\mu$, $\forall E \in \mathcal{B}$ with $\int_E \mathcal{P} \circ f_i \, d\mu < \infty$, $i = 1, 2$. Then, $f_1 \leq f_2$ a.e. in $\mathcal{X}$.*

Proof By Propositions 7.21, 7.23, 11.38, and 11.39, $f_1 - f_2$, $\mathcal{P} \circ f_1$, and $\mathcal{P} \circ f_2$ are $\mathcal{B}$-measurable. Let $E := \{x \in X \mid f_2(x) - f_1(x) < 0\} \in \mathcal{B}$. Suppose $\mu(E) > 0$. $E = \bigcup_{n=1}^{\infty} E_n := \bigcup_{n=1}^{\infty} \{x \in X \mid f_2(x) - f_1(x) \leq -1/n, \ |f_1(x)| \leq n, \ |f_2(x)| \leq n\}$. By Proposition 11.7, $\mu(E) = \lim_{n \in \mathbb{N}} \mu(E_n)$ and $\exists n \in \mathbb{N}$ such that $\mu(E_n) > 0$. Since $\mathcal{X}$ is σ-finite, then $\exists (X_m)_{m=1}^{\infty} \subseteq \mathcal{B}$ such that $X = \bigcup_{m=1}^{\infty} X_m$ and $\mu(X_m) < +\infty$, $\forall m \in \mathbb{N}$. Without loss of generality, we may assume that $X_m \subseteq X_{m+1}$, $\forall m \in \mathbb{N}$. Then, by Proposition 11.7, $\mu(E_n) = \lim_{m \in \mathbb{N}} \mu(E_n \cap X_m)$. Then, $\exists m \in \mathbb{N}$ such that $\mu(\bar{E}) := \mu(E_n \cap X_m) \in (0, +\infty) \subset \mathbb{R}$. By Definition 11.79, $\int_{\bar{E}} \mathcal{P} \circ f_i \, d\mu \leq n\mu(\bar{E}) < \infty$, $i = 1, 2$. Then, by Propositions 11.75 and 11.92, $\mathbb{R} \ni$

$\int_{\bar{E}} f_1 \, \mathrm{d}\mu - \int_{\bar{E}} f_2 \, \mathrm{d}\mu - \mu(\bar{E})/n = \int_{\bar{E}} (f_1 - f_2 - 1/n) \, \mathrm{d}\mu \geq 0$. This contradicts with the assumption. Hence, $\mu(E) = 0$ and $f_1 \leq f_2$ a.e. in $\mathcal{X}$. □

Proposition 11.97 *Let $\mathcal{X} := (X, \mathcal{B}, \mu)$ be a σ-finite measure space and $f_i : X \to [0, \infty) \subset \mathbb{R}$ be $\mathcal{B}$-measurable, $i = 1, 2$. Assume that $f_1 \leq f_2$ a.e. in $\mathcal{X}$, $\mu(\{x \in X \mid f_1(x) < f_2(x)\}) > 0$, and $0 \leq \int_X f_1 \, \mathrm{d}\mu < +\infty$. Then, $\int_X f_1 \, \mathrm{d}\mu < \int_X f_2 \, \mathrm{d}\mu$.*

Proof By Proposition 11.83, $\int_X f_1 \, \mathrm{d}\mu \leq \int_X f_2 \, \mathrm{d}\mu$. Suppose $\int_X f_2 \, \mathrm{d}\mu = \int_X f_1 \, \mathrm{d}\mu \in \mathbb{R}$. Then, f_1 and f_2 are absolutely integrable over $\mathcal{X}$. Let $g := f_2 - f_1$. Then, by Propositions 7.23, 11.38, and 11.39, we have $g : X \to \mathbb{R}$ is $\mathcal{B}$-measurable and $g \geq 0$ a.e. in $\mathcal{X}$. By Proposition 11.92, g is absolutely integrable over $\mathcal{X}$ and $\int_X g \, \mathrm{d}\mu = 0$. $\forall E \in \mathcal{B}$, by Proposition 11.92, we have $\int_E g \, \mathrm{d}\mu + \int_{X \setminus E} g \, \mathrm{d}\mu = \int_X g \, \mathrm{d}\mu = 0$, and both of the summands on the left-hand side are nonnegative. Then, $\int_E g \, \mathrm{d}\mu = 0$. By Proposition 11.96, we have $g = 0$ a.e. in $\mathcal{X}$. This contradicts with the fact that $\mu(\{x \in X \mid g(x) > 0\}) = \mu(\{x \in X \mid f_1(x) < f_2(x)\}) > 0$. Therefore, we must have $\int_X f_1 \, \mathrm{d}\mu < \int_X f_2 \, \mathrm{d}\mu$. This completes the proof of the proposition. □

Theorem 11.98 (Jensen's Inequality) *Let $\mathcal{X} := (X, \mathcal{B}, \mu)$ be a finite measure space with $\mu(X) = 1$, $\mathcal{Y}$ be a real Banach space, $\mathcal{W}$ be a separable subspace of $\mathcal{Y}$, $\Omega \subseteq \mathcal{Y}$ be a nonempty closed convex set, $f : X \to \Omega \cap \mathcal{W}$ be absolutely integrable over $\mathcal{X}$, and $G : \Omega \to \mathbb{R}$ be a convex functional. Assume that $G \circ f$ is absolutely integrable over $\mathcal{X}$ and the epigraph $[G, \Omega]$ is closed. Then, $\int_X f \, \mathrm{d}\mu \in \Omega$ and $G(\int_X f \, \mathrm{d}\mu) \leq \int_X (G \circ f) \, \mathrm{d}\mu \in \mathbb{R}$.*

Proof By Proposition 11.92, $y_0 := \int_X f \, \mathrm{d}\mu \in \mathcal{Y}$. We will first show that $y_0 \in \Omega$ by distinguishing two exhaustive and mutually exclusive cases: Case 1: $\vartheta_{\mathcal{Y}} \in \Omega$; Case 2: $\vartheta_{\mathcal{Y}} \notin \Omega$.

Case 1: $\vartheta_{\mathcal{Y}} \in \Omega$. Then, $\Omega \cap \mathcal{W}$ is a conic segment. By Lemma 11.87, there exists a sequence of simple functions $(\varphi_n)_{n=1}^{\infty}$, $\varphi_n : \mathcal{X} \to \Omega \cap \mathcal{W}$, $\forall n \in \mathbb{N}$, such that $\lim_{n \in \mathbb{N}} \varphi_n = f$ a.e. in $\mathcal{X}$, $\|\varphi_n(x)\| \leq \mathcal{P} \circ f(x)$, $\forall x \in X$, $\forall n \in \mathbb{N}$, and $y_0 = \int_X f \, \mathrm{d}\mu = \lim_{n \in \mathbb{N}} \int_X \varphi_n \, \mathrm{d}\mu \in \mathcal{Y}$. Fix any $n \in \mathbb{N}$. Let φ_n admit the canonical representation $\varphi_n = \sum_{i=1}^{\bar{n}} y_i \chi_{A_i, X}$. Let $y_{\bar{n}+1} := \vartheta_{\mathcal{Y}} \in \Omega \cap \mathcal{W}$ and $A_{\bar{n}+1} := X \setminus (\bigcup_{i=1}^{\bar{n}} A_i) \in \mathcal{B}$. Then, by Proposition 11.75, $\int_X \varphi_n \, \mathrm{d}\mu = \sum_{i=1}^{\bar{n}+1} y_i \mu(A_i)$. Note that $1 = \mu(X) = \sum_{i=1}^{\bar{n}+1} \mu(A_i)$ and the summands are nonnegative. Since Ω is convex, then $\int_X \varphi_n \, \mathrm{d}\mu \in \Omega$. Since Ω is closed, then, by Proposition 4.13, $y_0 = \lim_{n \in \mathbb{N}} \int_X \varphi_n \, \mathrm{d}\mu \in \Omega$.

Case 2: $\vartheta_{\mathcal{Y}} \notin \Omega$. Note that $\mu(X) = 1$ implies that $X \neq \emptyset$. Then, $\Omega \cap \mathcal{W} \neq \emptyset$. Let $\bar{y} \in \Omega \cap \mathcal{W}$ and $\bar{\Omega} := \Omega - \bar{y}$. Then, $\vartheta_{\mathcal{Y}} \in \bar{\Omega}$ and, by Propositions 7.16 and 6.39, $\bar{\Omega}$ is a closed convex set. Let $\bar{f} := f - \bar{y}$. Then, we have $\bar{f} : X \to \bar{\Omega} \cap \mathcal{W}$. By Propositions 11.38, 11.39, and 7.23, $\bar{f}$ is $\mathcal{B}$-measurable. Note that $\mathcal{P} \circ \bar{f}(x) = \|f(x) - \bar{y}\| \leq \mathcal{P} \circ f(x) + \|\bar{y}\|$, $\forall x \in X$. Then, by Propositions 11.83 and 11.75, $0 \leq \int_X (\mathcal{P} \circ \bar{f}) \, \mathrm{d}\mu \leq \int_X (\mathcal{P} \circ f + \|\bar{y}\|) \, \mathrm{d}\mu = \int_X (\mathcal{P} \circ f) \, \mathrm{d}\mu + \int_X \|\bar{y}\| \, \mathrm{d}\mu = \int_X (\mathcal{P} \circ f) \, \mathrm{d}\mu + \|\bar{y}\| < +\infty$. Hence, $\bar{f}$ is absolutely integrable over $\mathcal{X}$. By Case 1, we have $\int_X \bar{f}\,\mathrm{d}\mu \in \bar{\Omega}$. By Propositions 11.92 and 11.75, we have $\int_X f\,\mathrm{d}\mu = \int_X \bar{f}\,\mathrm{d}\mu + \bar{y} \in \Omega$.

Hence, in both cases, we have $y_0 \in \Omega$. By Proposition 8.32, we have $G(y_0) = \sup_{y_* \in \Omega_{\mathrm{conj}}}(\langle\langle y_*, y_0\rangle\rangle - G_{\mathrm{conj}}(y_*))$, where $G_{\mathrm{conj}} : \Omega_{\mathrm{conj}} \to \mathbb{R}$ is the conjugate functional to G. $\forall \epsilon \in (0, \infty) \subset \mathbb{R}$, $\exists y_{*\epsilon} \in \Omega_{\mathrm{conj}}$, such that $G(y_0) - \epsilon \leq \langle\langle y_{*\epsilon}, y_0\rangle\rangle - G_{\mathrm{conj}}(y_{*\epsilon}) \leq \langle\langle y_{*\epsilon}, y_0\rangle\rangle - \langle\langle y_{*\epsilon}, y\rangle\rangle + G(y) = G(y) - \langle\langle y_{*\epsilon}, y - y_0\rangle\rangle$, $\forall y \in \Omega$. Then, $\forall x \in X$, we have $G(y_0) - \epsilon \leq G(f(x)) - \langle\langle y_{*\epsilon}, f(x) - y_0\rangle\rangle$. By Proposition 11.92, we have

$$\begin{aligned} G(y_0) - \epsilon &= \int_X (G(y_0) - \epsilon)\,\mathrm{d}\mu \leq \int_X (G \circ f - \langle\langle y_{*\epsilon}, f - y_0\rangle\rangle)\,\mathrm{d}\mu \\ &= \int_X (G \circ f)\,\mathrm{d}\mu - \int_X \langle\langle y_{*\epsilon}, f\rangle\rangle\,\mathrm{d}\mu + \int_X \langle\langle y_{*\epsilon}, y_0\rangle\rangle\,\mathrm{d}\mu \\ &= \int_X (G \circ f)\,\mathrm{d}\mu - \langle\langle y_{*\epsilon}, \int_X f\,\mathrm{d}\mu\rangle\rangle + \langle\langle y_{*\epsilon}, y_0\rangle\rangle = \int_X (G \circ f)\,\mathrm{d}\mu \in \mathbb{R} \end{aligned}$$

By the arbitrariness of ϵ, we have $\mathbb{R} \ni G(\int_X f\,\mathrm{d}\mu) \leq \int_X (G \circ f)\,\mathrm{d}\mu \in \mathbb{R}$. This completes the proof of the theorem. □

11.6 Banach Space Valued Measures

Definition 11.99 Let $(X, \mathcal{B})$ be a measurable space and $\mathcal{Y}$ be a normed linear space. A *$\mathcal{Y}$-valued pre-measure* μ on $(X, \mathcal{B})$ is a function $\mu : \mathcal{B} \to \mathcal{Y}$ such that:

(i) $\mu(\emptyset) = \vartheta_{\mathcal{Y}}$.
(ii) $\forall (E_i)_{i=1}^\infty \subseteq \mathcal{B}$, which is pairwise disjoint, we have $\sum_{i=1}^\infty \|\mu(E_i)\| < +\infty$ and $\mu(\bigcup_{i=1}^\infty E_i) = \sum_{i=1}^\infty \mu(E_i)$.

Then, the triple $(X, \mathcal{B}, \mu)$ is said to be a *$\mathcal{Y}$-valued pre-measure space*.

Define $\mathcal{P} \circ \mu : \mathcal{B} \to [0, \infty] \subset \mathbb{R}_e$ by, $\forall E \in \mathcal{B}$, $\mathcal{P} \circ \mu(E) := \sup_{n \in \mathbb{Z}_+, (E_i)_{i=1}^n \subseteq \mathcal{B}, \bigcup_{i=1}^n E_i = E, E_i \cap E_j = \emptyset, \forall 1 \leq i < j \leq n} \sum_{i=1}^n \|\mu(E_i)\|$. $\mathcal{P} \circ \mu$ is said to be the *total variation* of μ. ◇

Proposition 11.100 *Let $\mathcal{X} := (X, \mathcal{B}, \mu)$ be a $\mathcal{Y}$-valued pre-measure space, where $\mathcal{Y}$ is a normed linear space. Then, $\mathcal{P} \circ \mu$ defines a measure on the measurable space $(X, \mathcal{B})$.*

Proof Clearly, $\mathcal{P} \circ \mu(\emptyset) = 0$. $\forall$ pairwise disjoint $(E_i)_{i=1}^\infty \subseteq \mathcal{B}$, let $E := \bigcup_{i=1}^\infty E_i$. $\forall i \in \mathbb{N}$ with $\mathcal{P} \circ \mu(E_i) > 0$, $\forall 0 \leq t_i < \mathcal{P} \circ \mu(E_i)$, $\exists n_i \in \mathbb{Z}_+$ and $\exists$ pairwise disjoint $(E_{i,j})_{j=1}^{n_i} \subseteq \mathcal{B}$ with $E_i = \bigcup_{j=1}^{n_i} E_{i,j}$ such that $t_i < \sum_{j=1}^{n_i} \|\mu(E_{i,j})\| \leq \mathcal{P} \circ \mu(E_i)$. $\forall i \in \mathbb{N}$ with $\mathcal{P} \circ \mu(E_i) = 0$, let $t_i = 0$, $n_i = 1$, $E_{i,1} = E_i$. Then, $0 = t_i \leq \|\mu(E_{i,1})\| \leq \mathcal{P} \circ \mu(E_i) = 0$.

Claim 11.100.1 $0 \leq t := \sum_{i=1}^\infty t_i \leq \mathcal{P} \circ \mu(E)$.

Proof of Claim Clearly, $t \geq 0$ since $t_i \geq 0$, $\forall i \in \mathbb{N}$. We will distinguish two exhaustive and mutually exclusive cases: Case 1: $t < +\infty$; Case 2: $t = +\infty$.

Case 1: $t < +\infty$. Then, $t \leq \sum_{i=1}^{\infty} \sum_{j=1}^{n_i} \|\mu(E_{i,j})\| \geq 0$. Therefore, $\forall \epsilon \in (0, \infty) \subset \mathbb{R}$, $\exists N \in \mathbb{N}$ such that $t - \epsilon < \sum_{i=1}^{N} \sum_{j=1}^{n_i} \|\mu(E_{i,j})\| \leq \sum_{i=1}^{N} \sum_{j=1}^{n_i} \|\mu(E_{i,j})\| + \left\|\mu(E \setminus \left(\bigcup_{i=1}^{N} \bigcup_{j=1}^{n_i} E_{i,j}\right))\right\| \leq \mathcal{P} \circ \mu(E)$. By the arbitrariness of ϵ, we have $t \leq \mathcal{P} \circ \mu(E)$.

Case 2: $t = +\infty$. Then $t = \sum_{i=1}^{\infty} \sum_{j=1}^{n_i} \|\mu(E_{i,j})\| = \infty$. $\forall M \in [0, \infty) \subset \mathbb{R}$, $\exists N \in \mathbb{N}$ such that

$$
\begin{aligned}
M &\leq \sum_{i=1}^{N} \sum_{j=1}^{n_i} \|\mu(E_{i,j})\| \\
&\leq \sum_{i=1}^{N} \sum_{j=1}^{n_i} \|\mu(E_{i,j})\| + \left\|\mu(E \setminus \left(\bigcup_{i=1}^{N} \bigcup_{j=1}^{n_i} E_{i,j}\right))\right\| \leq \mathcal{P} \circ \mu(E)
\end{aligned}
$$

This implies that $\mathcal{P} \circ \mu(E) = +\infty = t$. This completes the proof of the claim. □

Hence, $\sum_{i=1}^{\infty} t_i \leq \mathcal{P} \circ \mu(E)$. By the arbitrariness of t_i's, we have $\sum_{i=1}^{\infty} \mathcal{P} \circ \mu(E_i) \leq \mathcal{P} \circ \mu(E)$.

On the other hand, $\forall n \in \mathbb{Z}_+$, $\forall$ pairwise disjoint $\left(A_j\right)_{j=1}^{n} \subseteq \mathcal{B}$ with $E = \bigcup_{j=1}^{n} A_j$, we have

$$
\begin{aligned}
\sum_{j=1}^{n} \|\mu(A_j)\| &= \sum_{j=1}^{n} \|\mu(\bigcup_{i=1}^{\infty}(E_i \cap A_j))\| = \sum_{j=1}^{n} \|\sum_{i=1}^{\infty} \mu(E_i \cap A_j)\| \\
&\leq \sum_{j=1}^{n} \sum_{i=1}^{\infty} \|\mu(E_i \cap A_j)\| = \sum_{i=1}^{\infty} \sum_{j=1}^{n} \|\mu(E_i \cap A_j)\| \leq \sum_{i=1}^{\infty} \mathcal{P} \circ \mu(E_i)
\end{aligned}
$$

By the arbitrariness of n and $\left(A_j\right)_{j=1}^{n}$, we have $\mathcal{P} \circ \mu(E) \leq \sum_{i=1}^{\infty} \mathcal{P} \circ \mu(E_i)$.

Hence, $\mathcal{P} \circ \mu(E) = \sum_{i=1}^{\infty} \mathcal{P} \circ \mu(E_i)$. This implies that $\mathcal{P} \circ \mu$ is a measure on $(X, \mathcal{B})$. This completes the proof of the proposition. □

Definition 11.101 Let $\mathcal{X} := (X, \mathcal{B}, \mu)$ be a $\mathcal{Y}$-valued pre-measure space, where $\mathcal{Y}$ is a normed linear space. We will say $\mathcal{X}$ is *finite* if $\mathcal{P} \circ \mu(X) < +\infty$. We will say a property holds *almost everywhere in* $\mathcal{X}$ if it holds almost everywhere in $\bar{\mathcal{X}} := (X, \mathcal{B}, \mathcal{P} \circ \mu)$. We will say that $\mathcal{X}$ is *complete* if the measure space $\bar{\mathcal{X}}$ is complete. ◇

Fact 11.102 *Let $(X, \mathcal{B}, \mu)$ be a $\mathcal{Y}$-valued pre-measure space, where $\mathcal{Y}$ is a normed linear space. Then, μ is finitely additive, i.e., $\mu\left(\bigcup_{i=1}^{n} E_i\right) = \sum_{i=1}^{n} \mu(E_i)$, where $n \in \mathbb{Z}_+$ and $(E_i)_{i=1}^{n} \subseteq \mathcal{B}$ is pairwise disjoint.*

Proof Let $E_{n+i} = \emptyset$, $\forall i \in \mathbb{N}$. Clearly, $(E_i)_{i=1}^{\infty} \subseteq \mathcal{B}$ is pairwise disjoint. Therefore, $\mu\big(\bigcup_{i=1}^{n} E_i\big) = \mu\big(\bigcup_{i=1}^{\infty} E_i\big) = \sum_{i=1}^{\infty} \mu(E_i) = \sum_{i=1}^{n} \mu(E_i)$. This completes the proof of the fact. □

Proposition 11.103 *Let $(X, \mathcal{B}, \mu)$ be a $\mathcal{Y}$-valued pre-measure space, where $\mathcal{Y}$ is a normed linear space, and $(E_n)_{n=1}^{\infty} \subseteq \mathcal{B}$ be such that $E_n \supseteq E_{n+1}$, $\forall n \in \mathbb{N}$. Then, $\lim_{n\in\mathbb{N}} \mu(E_n) = \mu\big(\bigcap_{n=1}^{\infty} E_n\big) \in \mathcal{Y}$.*

Proof Let $E := \bigcap_{n=1}^{\infty} E_n \in \mathcal{B}$. Note that $E_1 = E \cup \big(\bigcup_{n=1}^{\infty} (E_n \setminus E_{n+1})\big)$, and the sets in the union are pairwise disjoint and measurable. Therefore, $\mu(E_1) = \mu(E) + \sum_{n=1}^{\infty} \mu(E_n \setminus E_{n+1})$ and $\|\mu(E)\| + \sum_{n=1}^{\infty} \|\mu(E_n \setminus E_{n+1})\| < +\infty$. Note that $E_n = E_{n+1} \cup (E_n \setminus E_{n+1})$, $n = 1, 2, \ldots$, which implies that $\mu(E_n) = \mu(E_{n+1}) + \mu(E_n \setminus E_{n+1})$, by Fact 11.102. This leads to $\mu(E_1) = \mu(E) + \sum_{n=1}^{\infty} (\mu(E_n) - \mu(E_{n+1})) = \mu(E) + \mu(E_1) - \lim_{n\in\mathbb{N}} \mu(E_n) \in \mathcal{Y}$. Hence, we have $\lim_{n\in\mathbb{N}} \mu(E_n) = \mu(E) \in \mathcal{Y}$. This completes the proof of the proposition. □

Proposition 11.104 *Let $(X, \mathcal{B}, \mu)$ be a $\mathcal{Y}$-valued pre-measure space, where $\mathcal{Y}$ is a normed linear space, and $(A_i)_{i=1}^{\infty} \subseteq \mathcal{B}$ with $A_i \subseteq A_{i+1}$, $\forall i \in \mathbb{N}$. Then, $\mu(\bigcup_{i=1}^{\infty} A_i) = \lim_{i\in\mathbb{N}} \mu(A_i) \in \mathcal{Y}$.*

Proof Let $A := \bigcup_{i=1}^{\infty} A_i \in \mathcal{B}$. Note that $A = A_1 \cup (\bigcup_{i=1}^{\infty} (A_{i+1} \setminus A_i))$, where the sets in the union are pairwise disjoint and measurable. Then, by countable additivity, $\mu(A) = \mu(A_1) + \sum_{i=1}^{\infty} \mu(A_{i+1} \setminus A_i) = \lim_{n\in\mathbb{N}} (\mu(A_1) + \sum_{i=1}^{n} \mu(A_{i+1} \setminus A_i)) = \lim_{n\in\mathbb{N}} \mu(A_{n+1}) = \lim_{i\in\mathbb{N}} \mu(A_i) \in \mathcal{Y}$, where the third equality follows from Fact 11.102. This completes the proof of the proposition. □

Proposition 11.105 *Let $(X, \mathcal{B}, \mu)$ be a $\mathcal{Y}$-valued pre-measure space, where $\mathcal{Y}$ is a normed linear space. Then, $\forall E_1, E_2 \in \mathcal{B}$ with $\mathcal{P} \circ \mu(E_1 \Delta E_2) = 0$, we have $\mu(E_1) = \mu(E_2) \in \mathcal{Y}$.*

Proof By Proposition 11.100, $0 = \mathcal{P}\circ\mu(E_1 \Delta E_2) = \mathcal{P}\circ\mu(E_1 \setminus E_2) + \mathcal{P}\circ\mu(E_2 \setminus E_1)$. Then, $\mathcal{P} \circ \mu(E_1 \setminus E_2) = \mathcal{P} \circ \mu(E_2 \setminus E_1) = 0$ and $\mu(E_1 \setminus E_2) = \vartheta_{\mathcal{Y}} = \mu(E_2 \setminus E_1)$. Hence, we have $\mu(E_1) = \mu(E_1 \cap E_2) + \mu(E_1 \setminus E_2) = \mu(E_1 \cap E_2) + \mu(E_2 \setminus E_1) = \mu(E_2)$, where the first and last equalities follow from Fact 11.102. This completes the proof of the proposition. □

Proposition 11.106 *Let $(X, \mathcal{B}, \mu)$ be a $\mathcal{Y}$-valued pre-measure space, where $\mathcal{Y}$ is a normed linear space. Then, there is a unique complete $\mathcal{Y}$-valued pre-measure space $(X, \mathcal{B}_0, \mu_0)$ such that:*

(i) $\mathcal{B} \subseteq \mathcal{B}_0$.
(ii) $\forall E \in \mathcal{B}$, $\mu(E) = \mu_0(E)$ *and* $\mathcal{P} \circ \mu(E) = \mathcal{P} \circ \mu_0(E)$.
(iii) $\forall E \subseteq X$, $E \in \mathcal{B}_0$ *if, and only if,* $E = A \cup B$, *where* $A, B \subseteq X$ *with* $B \in \mathcal{B}$, *and there exists a* $C \in \mathcal{B}$, *such that* $A \subseteq C$ *and* $\mathcal{P} \circ \mu(C) = 0$.

Then, $(X, \mathcal{B}_0, \mu_0)$ is said to be the completion of $(X, \mathcal{B}, \mu)$. *Then, $(X, \mathcal{B}_0, \mathcal{P} \circ \mu_0)$ is the completion of $(X, \mathcal{B}, \mathcal{P} \circ \mu)$.*

Furthermore, let $(X, \mathcal{B}_1, \mu_1)$ *be another complete* $\mathcal{Y}$*-valued pre-measure space that satisfies* (i) *and* (ii)*; then,* $\mathcal{B}_0 \subseteq \mathcal{B}_1$, $\mu_0(E) = \mu_1(E)$*, and* $\mathcal{P} \circ \mu_0(E) = \mathcal{P} \circ \mu_1(E)$, $\forall E \in \mathcal{B}_0$.

Proof By Proposition 11.100, $(X, \mathcal{B}, \mathcal{P} \circ \mu)$ is a measure space. By Proposition 11.12, it admits the completion $(X, \mathcal{B}_0, \nu_0)$, where $\mathcal{B}_0$ satisfies (i) and (iii) and $\nu_0(E) = \mathcal{P} \circ \mu(E)$, $\forall E \in \mathcal{B}$. Define $\mu_0 : \mathcal{B}_0 \to \mathcal{Y}$ as following. $\forall E \in \mathcal{B}_0$, there exists $A, B, C \subseteq X$ such that $E = A \cup B$, $B, C \in \mathcal{B}$, $A \subseteq C$, and $\mathcal{P} \circ \mu(C) = 0$. $\mu_0(E) := \mu(B) \in \mathcal{Y}$.

The mapping μ_0 is well-defined because of the following. $\forall E \in \mathcal{B}_0$, let $A_1, A_2, B_1, B_2, C_1, C_2 \subseteq X$ be such that

$$E = A_1 \cup B_1 = A_2 \cup B_2$$

$$B_1, C_1, B_2, C_2 \in \mathcal{B}$$

$$\mathcal{P} \circ \mu(C_1) = \mathcal{P} \circ \mu(C_2) = 0; \qquad A_1 \subseteq C_1; \qquad A_2 \subseteq C_2$$

Then, $\mathcal{B} \ni B_1 \setminus B_2 \subseteq E \setminus B_2 = A_2 \setminus B_2 \subseteq C_2$ and $\mathcal{B} \ni B_2 \setminus B_1 \subseteq E \setminus B_1 = A_1 \setminus B_1 \subseteq C_1$. This implies, by Proposition 11.4, $0 \leq \mathcal{P} \circ \mu(B_1 \triangle B_2) = \mathcal{P} \circ \mu(B_1 \setminus B_2) + \mathcal{P} \circ \mu(B_2 \setminus B_1) \leq \mathcal{P} \circ \mu(C_2) + \mathcal{P} \circ \mu(C_1) = 0$. By Proposition 11.105, we have $\mu(B_1) = \mu(B_2)$. This shows that μ_0 is well-defined.

$\forall E \in \mathcal{B} \subseteq \mathcal{B}_0$, $E = E \cup \emptyset$; then, $\mu_0(E) = \mu(E)$. This proves that μ_0 agrees with μ on $\mathcal{B}$.

Next, we will show that μ_0 is a $\mathcal{Y}$-valued pre-measure on the measurable space $(X, \mathcal{B}_0)$. Since $\emptyset \in \mathcal{B}$, then $\mu_0(\emptyset) = \mu(\emptyset) = \vartheta_{\mathcal{Y}}$. Consider any sequence of pairwise disjoint subsets $(E_n)_{n=1}^{\infty} \subseteq \mathcal{B}_0$. $\forall n \in \mathbb{N}$, there exists $A_n, B_n, C_n \subseteq X$ such that $E_n = A_n \cup B_n$, $B_n, C_n \in \mathcal{B}$, $A_n \subseteq C_n$, and $\mathcal{P} \circ \mu(C_n) = 0$. Since $B_n \subseteq E_n$, $\forall n \in \mathbb{N}$, then, $(B_n)_{n=1}^{\infty} \subseteq \mathcal{B}$ is pairwise disjoint. Hence, $\mu\big(\bigcup_{n=1}^{\infty} B_n\big) = \sum_{n=1}^{\infty} \mu(B_n)$ and $\sum_{n=1}^{\infty} \|\mu(B_n)\| < +\infty$. Since $\bigcup_{n=1}^{\infty} E_n = \big(\bigcup_{n=1}^{\infty} A_n\big) \cup \big(\bigcup_{n=1}^{\infty} B_n\big)$, $\bigcup_{n=1}^{\infty} A_n \subseteq \bigcup_{n=1}^{\infty} C_n$, $0 \leq \mathcal{P} \circ \mu\big(\bigcup_{n=1}^{\infty} C_n\big) \leq \sum_{n=1}^{\infty} \mathcal{P} \circ \mu(C_n) = 0$, and $\bigcup_{n=1}^{\infty} B_n, \bigcup_{n=1}^{\infty} C_n \in \mathcal{B}$, then $\mu_0\big(\bigcup_{n=1}^{\infty} E_n\big) = \mu\big(\bigcup_{n=1}^{\infty} B_n\big) = \sum_{n=1}^{\infty} \mu(B_n) = \sum_{n=1}^{\infty} \mu_0(E_n)$. Also, $\sum_{n=1}^{\infty} \|\mu_0(E_n)\| = \sum_{n=1}^{\infty} \|\mu(B_n)\| < +\infty$. This proves that μ_0 is countably additive. Hence, $(X, \mathcal{B}_0, \mu_0)$ is a $\mathcal{Y}$-valued pre-measure space.

Next, we will show that $\mathcal{P} \circ \mu_0 = \nu_0$. $\forall E \in \mathcal{B}_0$, $\forall n \in \mathbb{Z}_+$, $\forall$ pairwise disjoint $(E_i)_{i=1}^{n} \subseteq \mathcal{B}_0$ with $E = \bigcup_{i=1}^{n} E_i$, $\forall i \in \{1, \ldots, n\}$, $E_i = A_i \cup B_i$ such that $A_i \subseteq C_i$, $B_i, C_i \in \mathcal{B}$, and $\mathcal{P} \circ \mu(C_i) = 0$. Then, $(B_i)_{i=1}^{n} \subseteq \mathcal{B}$ is pairwise disjoint, $E = (\bigcup_{i=1}^{n} A_i) \cup (\bigcup_{i=1}^{n} B_i) =: A \cup B$, $A \subseteq \bigcup_{i=1}^{n} C_i =: C \in \mathcal{B}$, $B \in \mathcal{B}$, and $0 \leq \mathcal{P} \circ \mu(C) \leq \sum_{i=1}^{n} \mathcal{P} \circ \mu(C_i) = 0$, where the last inequality follows from Proposition 11.6. By Proposition 11.12 and its proof, $\nu_0(E) = \mathcal{P} \circ \mu(B) = \sum_{i=1}^{n} \mathcal{P} \circ \mu(B_i) \geq \sum_{i=1}^{n} \|\mu(B_i)\| = \sum_{i=1}^{n} \|\mu_0(E_i)\|$. Hence, $\nu_0(E) \geq \mathcal{P} \circ \mu_0(E)$.

On the other hand, $\nu_0(E) = \mathcal{P} \circ \mu(B) \leq \mathcal{P} \circ \mu_0(B) \leq \mathcal{P} \circ \mu_0(E)$, where the first inequality follows from Definition 11.99 and the fact that $\mu = \mu_0|_{\mathcal{B}}$ and the second inequality follows from Proposition 11.4. Therefore, $\mathcal{P} \circ \mu_0 = \nu_0$. $\forall E \in \mathcal{B}$, $\mathcal{P} \circ \mu_0(E) = \nu_0(E) = \mathcal{P} \circ \mu(E)$. This shows that (ii) is satisfied. This proves that

$(X, \mathcal{B}_0, \mu_0)$ is a complete $\mathcal{Y}$-valued pre-measure space satisfying (i), (ii), and (iii) and $(X, \mathcal{B}_0, \mathcal{P} \circ \mu_0)$ is the completion of $(X, \mathcal{B}, \mathcal{P} \circ \mu)$.

Let $(X, \mathcal{B}_1, \mu_1)$ be any complete $\mathcal{Y}$-valued pre-measure space that satisfies the (i) and (ii). By Proposition 11.100, $(X, \mathcal{B}_1, \mathcal{P} \circ \mu_1)$ is a complete measure space that extends $(X, \mathcal{B}, \mathcal{P} \circ \mu)$. By Proposition 11.12 and the fact that $(X, \mathcal{B}_0, \mathcal{P} \circ \mu_0)$ is the completion of $(X, \mathcal{B}, \mathcal{P} \circ \mu)$, then $\mathcal{B}_0 \subseteq \mathcal{B}_1$ and $(\mathcal{P} \circ \mu_1)|_{\mathcal{B}_0} = \mathcal{P} \circ \mu_0$. $\forall E \in \mathcal{B}_0$, there exists $A, B, C \subseteq X$ such that $E = A \cup B$, $A \subseteq C$, $B, C \in \mathcal{B}$, and $\mathcal{P} \circ \mu(C) = 0$. Then, $\mathcal{P} \circ \mu_1(C) = 0$, $E = B \cup (E \setminus B)$, $\mathcal{B}_0 \ni E \setminus B = A \setminus B \subseteq C$, and $0 \le \mathcal{P} \circ \mu_1(E \setminus B) \le \mathcal{P} \circ \mu_1(C) = 0$. This leads to $\mu_1(E \setminus B) = \vartheta_{\mathcal{Y}}$ and $\mu_1(E) = \mu_1(B) + \mu_1(E \setminus B) = \mu(B) = \mu_0(E)$. Therefore, $\mu_0 = \mu_1|_{\mathcal{B}_0}$.

Therefore, $(X, \mathcal{B}_0, \mu_0)$ is a complete $\mathcal{Y}$-valued pre-measure space that satisfies (i), (ii), and (iii). It is the unique such space since $\mathcal{B}_0$ is unique by (iii), and μ_0 is also unique since any complete $\mathcal{Y}$-valued pre-measure space $(X, \mathcal{B}_0, \mu_1)$ satisfying (i) and (ii) will be such that $\mu_1 = \mu_0$ and $\mathcal{P} \circ \mu_0 = \mathcal{P} \circ \mu_1$.

This completes the proof of the proposition. □

Proposition 11.107 *Let $\mathcal{X} := (X, \mathcal{B}, \mu)$ be a $\mathcal{Y}$-valued pre-measure space, where $\mathcal{Y}$ is a normed linear space, $A \in \mathcal{B}$, $\hat{\mathcal{B}} \subseteq \mathcal{B}$ be a σ-algebra on X, and $\mathcal{B}_A := \{C \subseteq A \mid C \in \mathcal{B}\}$. Then, the following statements hold:*

(i) *$(X, \hat{\mathcal{B}}, \mu|_{\hat{\mathcal{B}}})$ and $\mathcal{A} := (A, \mathcal{B}_A, \mu_A := \mu|_{\mathcal{B}_A})$ are $\mathcal{Y}$-valued pre-measure spaces, and $(\mathcal{P} \circ \mu)|_{\mathcal{B}_A} = \mathcal{P} \circ (\mu|_{\mathcal{B}_A}) = \mathcal{P} \circ \mu_A$. The $\mathcal{Y}$-valued pre-measure space $\mathcal{A}$ is said to be the* subspace *of the $\mathcal{Y}$-valued pre-measure space $\mathcal{X}$.*
(ii) *If, in addition, $\mathcal{X}$ is complete, then $\mathcal{A}$ is also complete.*
(iii) *Let $\bar{\mathcal{X}} := (X, \bar{\mathcal{B}}, \bar{\mu})$ be the completion of $\mathcal{X}$ as defined in Proposition 11.106, and $\bar{\mathcal{A}} := (A, \bar{\mathcal{B}}_A, \bar{\mu}_A)$ be the $\mathcal{Y}$-valued pre-measure subspace of $\bar{\mathcal{X}}$. Then, $\bar{\mathcal{A}}$ is the completion of $\mathcal{A}$.*

Proof (i) Since $\hat{\mathcal{B}}$ is a σ-algebra on X, then $(X, \hat{\mathcal{B}})$ is a measurable space. $\forall B \in \hat{\mathcal{B}} \subseteq \mathcal{B}$, $\mu|_{\hat{\mathcal{B}}}(B) = \mu(B) \in \mathcal{Y}$. Then, $\mu|_{\hat{\mathcal{B}}} : \hat{\mathcal{B}} \to \mathcal{Y}$. Clearly, $\mu|_{\hat{\mathcal{B}}}(\emptyset) = \mu(\emptyset) = \vartheta_{\mathcal{Y}}$. $\forall$ pairwise disjoint sequence $(E_i)_{i=1}^{\infty} \subseteq \hat{\mathcal{B}} \subseteq \mathcal{B}$, we have $E := \bigcup_{i=1}^{\infty} E_i \in \hat{\mathcal{B}}$, $\sum_{i=1}^{\infty} \left\| \mu|_{\hat{\mathcal{B}}}(E_i) \right\| = \sum_{i=1}^{\infty} \|\mu(E_i)\| < +\infty$, and $\mu|_{\hat{\mathcal{B}}}(E) = \mu(E) = \sum_{i=1}^{\infty} \mu(E_i) = \sum_{i=1}^{\infty} \mu|_{\hat{\mathcal{B}}}(E_i)$. Hence, $(X, \hat{\mathcal{B}}, \mu|_{\hat{\mathcal{B}}})$ is a $\mathcal{Y}$-valued pre-measure space.

Clearly, $\mathcal{B}_A$ is a σ-algebra on A as we have shown in Proposition 11.13. $\emptyset \in \mathcal{B}_A$ and $\mu|_{\mathcal{B}_A}(\emptyset) = \mu(\emptyset) = \vartheta_{\mathcal{Y}}$. $\forall$ pairwise disjoint sequence $(E_i)_{i=1}^{\infty} \subseteq \mathcal{B}_A \subseteq \mathcal{B}$, we have $E := \bigcup_{i=1}^{\infty} E_i \in \mathcal{B}_A$, $\sum_{i=1}^{\infty} \left\| \mu|_{\mathcal{B}_A}(E_i) \right\| = \sum_{i=1}^{\infty} \|\mu(E_i)\| < +\infty$, and $\mu|_{\mathcal{B}_A}(E) = \mu(E) = \sum_{i=1}^{\infty} \mu(E_i) = \sum_{i=1}^{\infty} \mu|_{\mathcal{B}_A}(E_i)$. Hence, $\mathcal{A} = (A, \mathcal{B}_A, \mu|_{\mathcal{B}_A})$ is a $\mathcal{Y}$-valued pre-measure space. By Definition 11.99, $(\mathcal{P} \circ \mu)|_{\mathcal{B}_A} = \mathcal{P} \circ (\mu|_{\mathcal{B}_A}) = \mathcal{P} \circ \mu_A$.

(ii) If, in addition, $\mathcal{X}$ is complete, then $(X, \mathcal{B}, \mathcal{P} \circ \mu)$ is a complete measure space. $\forall E_A \subseteq A$ such that there exists $B \in \mathcal{B}_A$ with $E_A \subseteq B$ and $\mathcal{P} \circ (\mu|_{\mathcal{B}_A})(B) = 0$, we have $B \in \mathcal{B}$ and $\mathcal{P} \circ \mu(B) = (\mathcal{P} \circ \mu)|_{\mathcal{B}_A}(B) = \mathcal{P} \circ (\mu|_{\mathcal{B}_A})(B) = \mathcal{P} \circ \mu_A(B) = 0$. Then, $E_A \in \mathcal{B}$ by the completeness of $(X, \mathcal{B}, \mathcal{P} \circ \mu)$. Thus, $E_A \in \mathcal{B}_A$. By the arbitrariness of E_A, $(A, \mathcal{B}_A, \mathcal{P} \circ \mu_A)$ is complete. Hence, $\mathcal{A}$ is complete.

(iii) Note that $\bar{\mathcal{B}}_A = \{C \subseteq A \mid C \in \bar{\mathcal{B}}\}$, $\bar{\mu}_A = \bar{\mu}|_{\bar{\mathcal{B}}_A}$, and $\mathcal{P} \circ \bar{\mu}_A = (\mathcal{P} \circ \bar{\mu})|_{\bar{\mathcal{B}}_A}$. By (ii), $\bar{\mathcal{A}}$ is a complete $\mathcal{Y}$-valued pre-measure space. Since $\mathcal{B} \subseteq \bar{\mathcal{B}}$, then $\mathcal{B}_A \subseteq \bar{\mathcal{B}}_A$. $\forall E \in \mathcal{B}_A$, $\mu_A(E) = \mu|_{\mathcal{B}_A}(E) = \mu(E) = \bar{\mu}(E) = \bar{\mu}_A(E)$ and, by (i), $\mathcal{P}\circ\mu_A(E) = (\mathcal{P} \circ \mu)|_{\mathcal{B}_A}(E) = \mathcal{P} \circ \mu(E) = \mathcal{P} \circ \bar{\mu}(E) = (\mathcal{P} \circ \bar{\mu})|_{\bar{\mathcal{B}}_A}(E) = \mathcal{P} \circ \bar{\mu}_A(E)$, where the third equality follows from Proposition 11.106. $\forall E \subseteq A$, assume that $E = E_A \cup E_B$ with $E_B, E_C \in \mathcal{B}_A$, $E_A \subseteq E_C$, and $\mathcal{P} \circ \mu_A(E_C) = 0$. Then, $\mathcal{P} \circ \bar{\mu}_A(E_C) = 0$ and $E_A \in \bar{\mathcal{B}}_A$ since $\bar{\mathcal{A}}$ is complete. Then, $E \in \bar{\mathcal{B}}_A$. $\forall E \subseteq A$, assume that $E \in \bar{\mathcal{B}}_A$. Then, $E \subseteq A$ and $E \in \bar{\mathcal{B}}$. By Proposition 11.106, $E = E_A \cup E_B$ with $E_B, E_C \in \mathcal{B}$, $E_A \subseteq E_C$, and $\mathcal{P} \circ \mu(E_C) = 0$. Then, $E_B, E_C \cap A \in \mathcal{B}_A$, $E_A \subseteq E_C \cap A$, and $0 \leq \mathcal{P}\circ\mu_A(E_C \cap A) = \mathcal{P}\circ\mu(E_C \cap A) \leq \mathcal{P}\circ\mu(E_C) = 0$, where the last inequality follows form Proposition 11.4. Hence, (i)–(iii) of Proposition 11.106 are satisfied and $\bar{\mathcal{A}}$ is the completion of $\mathcal{A}$.

This completes the proof of the proposition. □

Definition 11.108 Let $(X, \mathcal{B}, \nu)$ be a measure space and $\mathcal{Y}$ be a normed linear space. A $\mathcal{Y}$-*valued measure* μ on $(X, \mathcal{B})$ is a function from $\mathcal{B}$ to $\mathcal{Y}$ that satisfies:

(i) $\mu(\emptyset) = \vartheta_{\mathcal{Y}}$.
(ii) $\forall E \in \mathcal{B}$ with $\nu(E) = +\infty$, $\mu(E)$ is undefined.
(iii) $\forall E \in \mathcal{B}$ with $\nu(E) < +\infty$, $\mu(E) \in \mathcal{Y}$ and $\forall$ pairwise disjoint $(E_i)_{i=1}^{\infty} \subseteq \mathcal{B}$ with $E = \bigcup_{i=1}^{\infty} E_i$, we have $\sum_{i=1}^{\infty} \|\mu(E_i)\| < +\infty$ and $\mu(E) = \sum_{i=1}^{\infty} \mu(E_i)$ (countable additivity).
(iv) $\forall E \in \mathcal{B}$ with $\nu(E) < +\infty$, we have

$$\nu(E) = \sup_{n\in\mathbb{Z}_+,\ (E_i)_{i=1}^n \subseteq \mathcal{B},\ \bigcup_{i=1}^n E_i = E,\ E_i \cap E_j = \emptyset,\ \forall 1 \leq i < j \leq n} \sum_{i=1}^{n} \|\mu(E_i)\|$$

Then, $(X, \mathcal{B}, \mu)$ is said to be a $\mathcal{Y}$-*valued measure space*, and ν is said to be the *total variation* of μ, denoted by $\mathcal{P} \circ \mu$. ◇

Definition 11.109 Let $\mathcal{X} := (X, \mathcal{B}, \mu)$ be a $\mathcal{Y}$-valued measure space, where $\mathcal{Y}$ is a normed linear space. We will say $\mathcal{X}$ is *finite* if $\mathcal{P} \circ \mu(X) < +\infty$. We will say $\mathcal{X}$ is σ-*finite* if $\bar{\mathcal{X}} := (X, \mathcal{B}, \mathcal{P} \circ \mu)$ is σ-finite. We will say a property holds *almost everywhere in* $\mathcal{X}$ if it holds almost everywhere in $\bar{\mathcal{X}}$. We will say that $\mathcal{X}$ is *complete* if the measure space $\bar{\mathcal{X}}$ is complete. We will say that a sequence of measurable functions on $\mathcal{X}$ *converges in measure in* $\mathcal{X}$ if it converges in measure in $\bar{\mathcal{X}}$. ◇

Clearly, $(X, \mathcal{B}, \mu)$ is a finite $\mathcal{Y}$-valued measure space if, and only if, it is a finite $\mathcal{Y}$-valued pre-measure space. In this case, the definitions of total variations of μ coincide considering μ as a $\mathcal{Y}$-valued measure or as $\mathcal{Y}$-valued pre-measure. Hence, a $\mathcal{Y}$-valued pre-measure space becomes a finite $\mathcal{Y}$-valued measure space when we show that its total variation is finite.

Fact 11.110 *Let $(X, \mathcal{B}, \mu)$ be a $\mathcal{Y}$-valued measure space, where $\mathcal{Y}$ is a normed linear space. Then, μ is finitely additive, i.e., $\forall n \in \mathbb{Z}_+$ and $\forall$ pairwise disjoint*

$(E_i)_{i=1}^n \subseteq \mathcal{B}$, $\mu\left(\bigcup_{i=1}^n E_i\right) = \sum_{i=1}^n \mu(E_i)$, whenever the right-hand side or the left-hand side makes sense.

Proof Let $E := \bigcup_{i=1}^n E_i \in \mathcal{B}$. Clearly, $\mathcal{P} \circ \mu(E) = \sum_{i=1}^n \mathcal{P} \circ \mu(E_i)$. Then, $E \in \operatorname{dom}(\mu)$ whenever $(E_i)_{i=1}^n \subseteq \operatorname{dom}(\mu)$. Let $E_{n+i} = \emptyset$, $\forall i \in \mathbb{N}$. Clearly, $(E_i)_{i=1}^\infty \subseteq \mathcal{B}$ is pairwise disjoint. Therefore, $\mu(E) = \mu\left(\bigcup_{i=1}^\infty E_i\right) = \sum_{i=1}^\infty \mu(E_i) = \sum_{i=1}^n \mu(E_i) \in \mathcal{Y}$, when $E \in \operatorname{dom}(\mu)$. This completes the proof of the fact. □

Proposition 11.111 *Let $(X, \mathcal{B}, \mu)$ be a $\mathcal{Y}$-valued measure space, where $\mathcal{Y}$ is a normed linear space, and $(E_n)_{n=1}^\infty \subseteq \mathcal{B}$ be such that $\mathcal{P} \circ \mu(E_1) < +\infty$ and $E_n \supseteq E_{n+1}$, $\forall n \in \mathbb{N}$. Then, $\lim_{n\in\mathbb{N}} \mu(E_n) = \mu\left(\bigcap_{n=1}^\infty E_n\right) \in \mathcal{Y}$.*

Proof Clearly, $E_1 \in \operatorname{dom}(\mu)$ and then $E_n \in \operatorname{dom}(\mu)$, $\forall n \in \mathbb{N}$. Let $E := \bigcap_{n=1}^\infty E_n \in \mathcal{B}$. Note that $E_1 = E \cup \left(\bigcup_{n=1}^\infty (E_n \setminus E_{n+1})\right)$, and the sets in the union are pairwise disjoint and measurable. Therefore, $\mu(E_1) = \mu(E) + \sum_{n=1}^\infty \mu(E_n \setminus E_{n+1}) \in \mathcal{Y}$ and $\|\mu(E)\| + \sum_{n=1}^\infty \|\mu(E_n \setminus E_{n+1})\| < \infty$. Note that $E_n = E_{n+1} \cup (E_n \setminus E_{n+1})$, $n = 1, 2, \ldots$, which implies that $\mu(E_n) = \mu(E_{n+1}) + \mu(E_n \setminus E_{n+1}) \in \mathcal{Y}$, by Fact 11.110. This leads to $\mu(E_1) = \mu(E) + \sum_{n=1}^\infty (\mu(E_n) - \mu(E_{n+1})) = \mu(E) + \mu(E_1) - \lim_{n\in\mathbb{N}} \mu(E_n) \in \mathcal{Y}$. Hence, we have $\lim_{n\in\mathbb{N}} \mu(E_n) = \mu(E) \in \mathcal{Y}$. This completes the proof of the proposition. □

Proposition 11.112 *Let $(X, \mathcal{B}, \mu)$ be a $\mathcal{Y}$-valued measure space, where $\mathcal{Y}$ is a normed linear space, and $(A_i)_{i=1}^\infty \subseteq \mathcal{B}$ be such that $A_i \subseteq A_{i+1}$, $\forall i \in \mathbb{N}$, and $A := \bigcup_{i=1}^\infty A_i \in \operatorname{dom}(\mu)$. Then, $\mu(A) = \lim_{i\in\mathbb{N}} \mu(A_i) \in \mathcal{Y}$.*

Proof $A \in \operatorname{dom}(\mu)$ implies that $\mathcal{P} \circ \mu(A) < +\infty$. Then, $\mathcal{P} \circ \mu(A_i) \leq \mathcal{P} \circ \mu(A) < +\infty$ and $A_i \in \operatorname{dom}(\mu)$, $\forall i \in \mathbb{N}$. Note that $A = A_1 \cup (\bigcup_{i=1}^\infty (A_{i+1} \setminus A_i))$, where the sets in the union are pairwise disjoint and measurable. Then, by countable additivity, $\mu(A) = \mu(A_1) + \sum_{i=1}^\infty \mu(A_{i+1} \setminus A_i) = \lim_{n\in\mathbb{N}} (\mu(A_1) + \sum_{i=1}^n \mu(A_{i+1} \setminus A_i)) = \lim_{n\in\mathbb{N}} \mu(A_{n+1}) = \lim_{i\in\mathbb{N}} \mu(A_i) \in \mathcal{Y}$, where the third equality follows from Fact 11.110. This completes the proof of the proposition. □

Proposition 11.113 *Let $(X, \mathcal{B}, \mu)$ be a $\mathcal{Y}$-valued measure space, where $\mathcal{Y}$ is a normed linear space. Then, $\forall E_1, E_2 \in \mathcal{B}$ with $\mathcal{P} \circ \mu(E_1 \Delta E_2) = 0$, we have $\mu(E_1) = \mu(E_2)$ whenever one of them makes sense.*

Proof By Fact 11.3, $0 = \mathcal{P} \circ \mu(E_1 \Delta E_2) = \mathcal{P} \circ \mu(E_1 \setminus E_2) + \mathcal{P} \circ \mu(E_2 \setminus E_1)$. Then, $\mathcal{P} \circ \mu(E_1 \setminus E_2) = \mathcal{P} \circ \mu(E_2 \setminus E_1) = 0$ and $\mu(E_1 \setminus E_2) = \vartheta_{\mathcal{Y}} = \mu(E_2 \setminus E_1)$. Without loss of generality, assume $E_1 \in \operatorname{dom}(\mu)$. Then, $\mathcal{P} \circ \mu(E_1) < +\infty$. This implies that $\mathcal{P} \circ \mu(E_2) = \mathcal{P} \circ \mu(E_1 \cap E_2) + \mathcal{P} \circ \mu(E_2 \setminus E_1) = \mathcal{P} \circ \mu(E_1 \cap E_2) + \mathcal{P} \circ \mu(E_1 \setminus E_2) = \mathcal{P} \circ \mu(E_1) < +\infty$ and $E_2 \in \operatorname{dom}(\mu)$. By Fact 11.110, we have $\mu(E_1) = \mu(E_1 \cap E_2) + \mu(E_1 \setminus E_2) = \mu(E_1 \cap E_2) + \mu(E_2 \setminus E_1) = \mu(E_2)$. This completes the proof of the proposition. □

Proposition 11.114 *Let $(X, \mathcal{B}, \mu)$ be a $\mathcal{Y}$-valued measure space, where $\mathcal{Y}$ is a normed linear space. Then, there is a unique complete $\mathcal{Y}$-valued measure space*

$(X, \mathcal{B}_0, \mu_0)$ such that:

(i) $\mathcal{B} \subseteq \mathcal{B}_0$.
(ii) $(\mathcal{P} \circ \mu_0)|_{\mathcal{B}} = \mathcal{P} \circ \mu$ *and* $\mu_0|_{\mathcal{B}} = \mu$.
(iii) $\forall E \subseteq X$, $E \in \mathcal{B}_0$ *if, and only if,* $E = A \cup B$, *where* $A, B \subseteq X$ *with* $B \in \mathcal{B}$, *and there exists a* $C \in \mathcal{B}$, *such that* $A \subseteq C$ *and* $\mathcal{P} \circ \mu(C) = 0$.

Then, $(X, \mathcal{B}_0, \mu_0)$ *is said to be the* completion of $(X, \mathcal{B}, \mu)$. *Then,* $(X, \mathcal{B}_0, \mathcal{P} \circ \mu_0)$ *is the completion of* $(X, \mathcal{B}, \mathcal{P} \circ \mu)$.

Furthermore, let $(X, \mathcal{B}_1, \mu_1)$ *be another complete* $\mathcal{Y}$*-valued measure space that satisfies* (i) *and* (ii)*; then,* $\mathcal{B}_0 \subseteq \mathcal{B}_1$, $\mathcal{P} \circ \mu_0 = (\mathcal{P} \circ \mu_1)|_{\mathcal{B}_0}$, *and* $\mu_0 = \mu_1|_{\mathcal{B}_0}$.

Proof By Definition 11.108, $(X, \mathcal{B}, \mathcal{P} \circ \mu)$ is a measure space. By Proposition 11.12, it admits the completion $(X, \mathcal{B}_0, \nu_0)$, where $\mathcal{B}_0$ satisfies (i) and (iii) and $\nu_0(E) = \mathcal{P} \circ \mu(E)$, $\forall E \in \mathcal{B}$. Define a function μ_0 from $\mathcal{B}_0$ to $\mathcal{Y}$ as following. $\forall E \in \mathcal{B}_0$ with $\nu_0(E) = +\infty$, $\mu_0(E)$ is undefined. $\forall E \in \mathcal{B}_0$ with $\nu_0(E) < +\infty$, there exists $A, B, C \subseteq X$ such that $E = A \cup B$, $B, C \in \mathcal{B}$, $A \subseteq C$, and $\mathcal{P} \circ \mu(C) = 0$. By Proposition 11.12 and its proof, $\nu_0(E) = \mathcal{P} \circ \mu(B) < +\infty$ and $B \in \operatorname{dom}(\mu)$. We will set $\mu_0(E) := \mu(B) \in \mathcal{Y}$.

The mapping μ_0 is well-defined because of the following. $\forall E \in \mathcal{B}_0$ with $\nu_0(E) < +\infty$, let $A_1, A_2, B_1, B_2, C_1, C_2 \subseteq X$ be such that

$$E = A_1 \cup B_1 = A_2 \cup B_2$$
$$B_1, C_1, B_2, C_2 \in \mathcal{B}$$
$$\mathcal{P} \circ \mu(C_1) = \mathcal{P} \circ \mu(C_2) = 0; \qquad A_1 \subseteq C_1; \qquad A_2 \subseteq C_2$$

Then, $\mathcal{B} \ni B_1 \setminus B_2 \subseteq E \setminus B_2 = A_2 \setminus B_2 \subseteq C_2$ and $\mathcal{B} \ni B_2 \setminus B_1 \subseteq E \setminus B_1 = A_1 \setminus B_1 \subseteq C_1$. This implies, by Proposition 11.4, $0 \leq \mathcal{P} \circ \mu(B_1 \triangle B_2) = \mathcal{P} \circ \mu(B_1 \setminus B_2) + \mathcal{P} \circ \mu(B_2 \setminus B_1) \leq \mathcal{P} \circ \mu(C_2) + \mathcal{P} \circ \mu(C_1) = 0$. By Proposition 11.113, we have $\mu(B_1) = \mu(B_2) \in \mathcal{Y}$. This shows that μ_0 is well-defined.

$\forall E \in \mathcal{B} \subseteq \mathcal{B}_0$, $E = E \cup \emptyset$; then, $\nu_0(E) = \mathcal{P} \circ \mu(E)$. If $\mathcal{P} \circ \mu(E) = +\infty$, then $\mu(E)$ is undefined and $\mu_0(E)$ is also undefined. If $\mathcal{P} \circ \mu(E) < +\infty$, then $\mu_0(E) = \mu(E) \in \mathcal{Y}$. This proves that $\mu_0|_{\mathcal{B}} = \mu$.

Next, we will show that $(X, \mathcal{B}_0, \mu_0)$ is a $\mathcal{Y}$-valued measure space with $\mathcal{P} \circ \mu_0 = \nu_0$. Since $\emptyset \in \mathcal{B}$, then $\mu_0(\emptyset) = \mu(\emptyset) = \vartheta_{\mathcal{Y}}$. Hence, (i) of Definition 11.108 is satisfied. (ii) of Definition 11.108 is satisfied by the definition of μ_0. Consider any $E \in \mathcal{B}_0$ with $\nu_0(E) < +\infty$; then $\mu_0(E) \in \mathcal{Y}$. $\forall$ pairwise disjoint $(E_n)_{n=1}^{\infty} \subseteq \mathcal{B}_0$ with $E = \bigcup_{n=1}^{\infty} E_n$, $\forall n \in \mathbb{N}$, there exists $A_n, B_n, C_n \subseteq X$ such that $E_n = A_n \cup B_n$, $B_n, C_n \in \mathcal{B}$, $A_n \subseteq C_n$, and $\mathcal{P} \circ \mu(C_n) = 0$. Since $B_n \subseteq E_n$, $\forall n \in \mathbb{N}$, then $(B_n)_{n=1}^{\infty} \subseteq \mathcal{B}$ is pairwise disjoint. Let $A := \bigcup_{n=1}^{\infty} A_n$, $B := \bigcup_{n=1}^{\infty} B_n \in \mathcal{B}$, and $C := \bigcup_{n=1}^{\infty} C_n \in \mathcal{B}$. Then, $E = A \cup B$, $A \subseteq C$, and $\mathcal{P} \circ \mu(C) = 0$. This implies that $\mathcal{P} \circ \mu(B) = \nu_0(E) < +\infty$, $\mu(B) = \sum_{n=1}^{\infty} \mu(B_n) \in \mathcal{Y}$, and $\sum_{n=1}^{\infty} \|\mu(B_n)\| < +\infty$. Hence, $\mu_0(E) = \mu(B) = \sum_{n=1}^{\infty} \mu(B_n) = \sum_{n=1}^{\infty} \mu_0(E_n) \in \mathcal{Y}$. Also, $\sum_{n=1}^{\infty} \|\mu_0(E_n)\| = \sum_{n=1}^{\infty} \|\mu(B_n)\| < +\infty$. This proves that (iii) of Definition 11.108 is satisfied and μ_0 is countably additive.

$\forall E \in \mathcal{B}_0$ with $\nu_0(E) < +\infty$, $\forall n \in \mathbb{Z}_+$, $\forall$ pairwise disjoint $(E_i)_{i=1}^n \subseteq \mathcal{B}_0$ with $E = \bigcup_{i=1}^n E_i$, $\forall i \in \{1, \ldots, n\}$, $\exists A_i, B_i, C_i \subseteq X$ such that $E_i = A_i \cup B_i$, $B_i, C_i \in \mathcal{B}$, $A_i \subseteq C_i$, and $\mathcal{P} \circ \mu(C_i) = 0$. Then, $(B_i)_{i=1}^n \subseteq \mathcal{B}$ is pairwise disjoint. Let $A := \bigcup_{i=1}^n A_i$, $B := \bigcup_{i=1}^n B_i \in \mathcal{B}$, and $C := \bigcup_{i=1}^n C_i \in \mathcal{B}$; we have $E = A \cup B$, $A \subseteq C$, and $\mathcal{P} \circ \mu(C) = 0$. By Proposition 11.12 and its proof, $+\infty > \nu_0(E) = \mathcal{P} \circ \mu(B) = \sum_{i=1}^n \mathcal{P} \circ \mu(B_i) \geq \sum_{i=1}^n \|\mu(B_i)\| = \sum_{i=1}^n \|\mu_0(E_i)\|$. Hence,

$$\nu_0(E) \geq \sup_{n\in\mathbb{Z}_+,\, (E_i)_{i=1}^n\subseteq\mathcal{B}_0,\, \bigcup_{i=1}^n E_i=E,\, E_i\cap E_j=\emptyset,\, \forall 1\leq i<j\leq n} \sum_{i=1}^n \|\mu_0(E_i)\| =: s_E$$

On the other hand,

$$\begin{aligned} +\infty > \nu_0(E) &= \mathcal{P} \circ \mu(B) \\ &= \sup_{n\in\mathbb{Z}_+,\, (B_i)_{i=1}^n\subseteq\mathcal{B},\, \bigcup_{i=1}^n B_i=B,\, B_i\cap B_j=\emptyset,\, \forall 1\leq i<j\leq n} \sum_{i=1}^n \|\mu(B_i)\| \\ &\leq \sup_{n\in\mathbb{Z}_+,\, (E_i)_{i=1}^n\subseteq\mathcal{B}_0,\, \bigcup_{i=1}^n E_i=B,\, E_i\cap E_j=\emptyset,\, \forall 1\leq i<j\leq n} \sum_{i=1}^n \|\mu_0(E_i)\| \leq s_E \end{aligned}$$

Hence, we have $\nu_0(E) = s_E$ and (iv) of Definition 11.108 is satisfied. Hence, $(X, \mathcal{B}_0, \mu_0)$ is a $\mathcal{Y}$-valued measure space with total variation $\mathcal{P} \circ \mu_0 = \nu_0$. Since ν_0 is complete, then $(X, \mathcal{B}_0, \mu_0)$ is a complete $\mathcal{Y}$-valued measure space. Since $\nu_0|_{\mathcal{B}} = \mathcal{P} \circ \mu$, then $(\mathcal{P} \circ \mu_0)|_{\mathcal{B}} = \mathcal{P} \circ \mu$. Hence, (i)–(iii) are satisfied by $(X, \mathcal{B}_0, \mu_0)$. Clearly, $(X, \mathcal{B}_0, \mathcal{P} \circ \mu_0)$ is the completion of $(X, \mathcal{B}, \mathcal{P} \circ \mu)$.

Let $(X, \mathcal{B}_1, \mu_1)$ be any complete $\mathcal{Y}$-valued measure space that satisfies (i) and (ii). $(X, \mathcal{B}_1, \mathcal{P} \circ \mu_1)$ is a complete measure space that extends $(X, \mathcal{B}, \mathcal{P} \circ \mu)$. By Proposition 11.12 and the fact that $(X, \mathcal{B}_0, \mathcal{P}\circ\mu_0)$ is the completion of $(X, \mathcal{B}, \mathcal{P}\circ\mu)$, then $\mathcal{B}_0 \subseteq \mathcal{B}_1$ and $(\mathcal{P} \circ \mu_1)|_{\mathcal{B}_0} = \mathcal{P} \circ \mu_0$. $\forall E \in \mathcal{B}_0$ with $\mathcal{P} \circ \mu_0(E) = +\infty$, we have $\mathcal{P} \circ \mu_1(E) = +\infty$ and $\mu_0(E)$ and $\mu_1(E)$ are both undefined. $\forall E \in \mathcal{B}_0$ with $\mathcal{P} \circ \mu_0(E) < +\infty$, by (iii), there exists $A, B, C \subseteq X$ such that $E = A \cup B$, $A \subseteq C$, $B, C \in \mathcal{B}$, and $\mathcal{P} \circ \mu(C) = 0$. Then, $\mu_0(E) = \mu(B) = \mu_1(B) \in \mathcal{Y}$ and $\mathcal{P} \circ \mu_1(E) = \mathcal{P} \circ \mu_0(E) < +\infty$. Note that $\mathcal{P} \circ \mu_1(C) = \mathcal{P} \circ \mu(C) = 0$, $E = B \cup (E \setminus B)$, $\mathcal{B}_0 \ni E \setminus B = A \setminus B \subseteq C$, and $0 \leq \mathcal{P}\circ\mu_1(E \setminus B) \leq \mathcal{P}\circ\mu_1(C) = 0$. This leads to $\mu_1(E \setminus B) = \vartheta_{\mathcal{Y}}$ and, by Fact 11.110, $\mu_1(E) = \mu_1(B) + \mu_1(E \setminus B) = \mu(B) = \mu_0(E) \in \mathcal{Y}$. Therefore, $\mu_0 = \mu_1|_{\mathcal{B}_0}$.

Therefore, $(X, \mathcal{B}_0, \mu_0)$ is a complete $\mathcal{Y}$-valued measure space that satisfies (i), (ii), and (iii). It is the unique such space since $\mathcal{B}_0$ is unique by (iii), and μ_0 is also unique since any complete $\mathcal{Y}$-valued measure space $(X, \mathcal{B}_0, \mu_1)$ satisfying (i) and (ii) will be such that $\mu_1 = \mu_0$ and $\mathcal{P} \circ \mu_0 = \mathcal{P} \circ \mu_1$.

This completes the proof of the proposition. □

Proposition 11.115 *Let $\mathcal{X} := (X, \mathcal{B}, \mu)$ be a $\mathcal{Y}$-valued measure space, where $\mathcal{Y}$ is a normed linear space, $A \in \mathcal{B}$, and $\mathcal{B}_A := \{C \subseteq A \mid C \in \mathcal{B}\}$. Then, the following statements hold:*

(i) *$\mathcal{A} := (A, \mathcal{B}_A, \mu_A := \mu|_{\mathcal{B}_A})$ is a $\mathcal{Y}$-valued measure space, which is said to be the* subspace *of $\mathcal{X}$, and $(\mathcal{P} \circ \mu)|_{\mathcal{B}_A} = \mathcal{P} \circ \mu_A$.*
(ii) *If, in addition, $\mathcal{X}$ is complete, then $\mathcal{A}$ is also complete.*
(iii) *Let $\bar{\mathcal{X}} := (X, \bar{\mathcal{B}}, \bar{\mu})$ be the completion of $\mathcal{X}$ as defined in Proposition 11.114, and $\bar{\mathcal{A}} := (A, \bar{\mathcal{B}}_A, \bar{\mu}_A)$ be the $\mathcal{Y}$-valued measure subspace of $\bar{\mathcal{X}}$. Then, $\bar{\mathcal{A}}$ is the completion of $\mathcal{A}$.*

Proof (i) Clearly, $\mathcal{B}_A$ is a σ-algebra on A as we have shown in Proposition 11.13. By Proposition 11.13, $\nu_A := (\mathcal{P} \circ \mu)|_{\mathcal{B}_A} : \mathcal{B}_A \to [0, +\infty] \subset \mathbb{R}_e$ is a measure on $(A, \mathcal{B}_A)$. Clearly, $\emptyset \in \mathcal{B}_A$ and $\mu_A(\emptyset) = \mu|_{\mathcal{B}_A}(\emptyset) = \mu(\emptyset) = \vartheta_{\mathcal{Y}}$. $\forall E \in \mathcal{B}_A$ with $\nu_A(E) = +\infty$, we have $\mathcal{P} \circ \mu(E) = +\infty$ and $\mu(E)$ is undefined. Then, $\mu_A(E)$ is also undefined. $\forall E \in \mathcal{B}_A$ with $\nu_A(E) < +\infty$, we have $\mathcal{P} \circ \mu(E) = \nu_A(E) < +\infty$. Then, $\mu_A(E) = \mu(E) \in \mathcal{Y}$. $\forall$ pairwise disjoint $(E_i)_{i=1}^{\infty} \subseteq \mathcal{B}_A \subseteq \mathcal{B}$ with $E = \bigcup_{i=1}^{\infty} E_i$, we have $\sum_{i=1}^{\infty} \|\mu_A(E_i)\| = \sum_{i=1}^{\infty} \|\mu(E_i)\| < +\infty$, and $\mu_A(E) = \mu(E) = \sum_{i=1}^{\infty} \mu(E_i) = \sum_{i=1}^{\infty} \mu_A(E_i) \in \mathcal{Y}$. $\forall E \in \mathcal{B}_A$ with $\nu_A(E) < +\infty$, we have $\mathcal{P} \circ \mu(E) = \nu_A(E) < +\infty$. Then,

$$\begin{aligned}\nu_A(E) &= \mathcal{P} \circ \mu(E) \\ &= \sup_{n \in \mathbb{Z}_+,\ (E_i)_{i=1}^n \subseteq \mathcal{B},\ \bigcup_{i=1}^n E_i = E,\ E_i \cap E_j = \emptyset,\ \forall 1 \le i < j \le n} \sum_{i=1}^{n} \|\mu(E_i)\| \\ &= \sup_{n \in \mathbb{Z}_+,\ (E_i)_{i=1}^n \subseteq \mathcal{B}_A,\ \bigcup_{i=1}^n E_i = E,\ E_i \cap E_j = \emptyset,\ \forall 1 \le i < j \le n} \sum_{i=1}^{n} \|\mu_A(E_i)\|\end{aligned}$$

Hence, $\mathcal{A} = (A, \mathcal{B}_A, \mu_A)$ is a $\mathcal{Y}$-valued measure space with $\mathcal{P} \circ \mu_A = \nu_A = (\mathcal{P} \circ \mu)|_{\mathcal{B}_A}$.

(ii) If, in addition, $\mathcal{X}$ is complete, then $(X, \mathcal{B}, \mathcal{P} \circ \mu)$ is a complete measure space. Since $(A, \mathcal{B}_A, \mathcal{P} \circ \mu_A)$ is a subspace of $(X, \mathcal{B}, \mathcal{P} \circ \mu)$, then it is complete by Proposition 11.13. Hence, $\mathcal{A}$ is complete.

(iii) Note that $\bar{\mathcal{B}}_A = \{C \subseteq A \mid C \in \bar{\mathcal{B}}\}$, $\bar{\mu}_A = \bar{\mu}|_{\bar{\mathcal{B}}_A}$, and $\mathcal{P} \circ \bar{\mu}_A = (\mathcal{P} \circ \bar{\mu})|_{\bar{\mathcal{B}}_A}$. By (ii), $\bar{\mathcal{A}}$ is a complete $\mathcal{Y}$-valued measure space. Since $\mathcal{B} \subseteq \bar{\mathcal{B}}$, then $\mathcal{B}_A \subseteq \bar{\mathcal{B}}_A$. Note that $\mathcal{P} \circ \mu_A = (\mathcal{P} \circ \mu)|_{\mathcal{B}_A} = ((\mathcal{P} \circ \bar{\mu})|_{\mathcal{B}})\big|_{\mathcal{B}_A} = (\mathcal{P} \circ \bar{\mu})|_{\mathcal{B}_A} = ((\mathcal{P} \circ \bar{\mu})|_{\bar{\mathcal{B}}_A})\Big|_{\mathcal{B}_A} = (\mathcal{P} \circ \bar{\mu}_A)|_{\mathcal{B}_A}$, where the first equality follows from (i), the second equality follows from Proposition 11.114, the third and fourth equalities follow from the fact that $\mathcal{B}_A \subseteq \mathcal{B}$ and $\mathcal{B}_A \subseteq \bar{\mathcal{B}}_A$, and the last equality follows from (i). Note also that $\mu_A = \mu|_{\mathcal{B}_A} = (\bar{\mu}|_{\mathcal{B}})\big|_{\mathcal{B}_A} = (\bar{\mu}|_{\bar{\mathcal{B}}_A})\Big|_{\mathcal{B}_A} = \bar{\mu}_A|_{\mathcal{B}_A}$, where the first equality follows from (i), the second equality follows from Proposition 11.114; the third equality follows from the fact that $\mathcal{B}_A \subseteq \mathcal{B}$ and $\mathcal{B}_A \subseteq \bar{\mathcal{B}}_A$, and the last equality follows from (i). $\forall E \subseteq A$, assume that $E = E_A \cup E_B$ with $E_B, E_C \in \mathcal{B}_A$,

$E_A \subseteq E_C$, and $\mathcal{P} \circ \mu_A(E_C) = 0$. Then, $\mathcal{P} \circ \bar{\mu}_A(E_C) = 0$ and $E_A \in \bar{\mathcal{B}}_A$ since $\bar{\mathcal{A}}$ is complete. Then, $E \in \bar{\mathcal{B}}_A$. On the other hand, $\forall E \subseteq A$, assume that $E \in \bar{\mathcal{B}}_A$. Then, $E \subseteq A$ and $E \in \bar{\mathcal{B}}$. By Proposition 11.114, $E = E_A \cup E_B$ with $E_B, E_C \in \mathcal{B}$, $E_A \subseteq E_C$, and $\mathcal{P} \circ \mu(E_C) = 0$. Then, $E_B, E_C \cap A \in \mathcal{B}_A$, $E_A \subseteq E_C \cap A$, and $0 \le \mathcal{P} \circ \mu_A(E_C \cap A) = \mathcal{P} \circ \mu(E_C \cap A) \le \mathcal{P} \circ \mu(E_C) = 0$, where the last inequality follows form Proposition 11.4. Hence, (i)–(iii) of Proposition 11.114 are satisfied, and $\bar{\mathcal{A}}$ is the completion of $\mathcal{A}$.

This completes the proof of the proposition. □

Proposition 11.116 *Let $\mathcal{X} := (X, \mathcal{B}, \mu)$ be a measure space, $\mathcal{Y}$ be a Banach space, $\mathcal{W}$ be a separable subspace of $\mathcal{Y}$, and $f : X \to \mathcal{W}$ be $\mathcal{B}$-measurable. Define $\bar{\nu} : \mathcal{B} \to [0, +\infty] \subset \mathbb{R}_e$ by $\bar{\nu}(E) = \int_E \mathcal{P} \circ f \, d\mu$, $\forall E \in \mathcal{B}$. Then, $\bar{\nu}$ is a measure on $(X, \mathcal{B})$. Define a function ν from $\mathcal{B}$ to $\mathcal{Y}$ by $\nu(E)$ is undefined, $\forall E \in \mathcal{B}$ with $\bar{\nu}(E) = +\infty$, and $\nu(E) = \int_E f \, d\mu$, $\forall E \in \mathcal{B}$ with $\bar{\nu}(E) < +\infty$. Then, $(X, \mathcal{B}, \nu)$ is a $\mathcal{Y}$-valued measure space with $\mathcal{P} \circ \nu = \bar{\nu}$. We will call ν the* $\mathcal{Y}$-valued measure with kernel f over $\mathcal{X}$. *Furthermore, the following statements hold:*

(i) *If $\mathcal{X}$ is σ-finite, then ν is also σ-finite.*
(ii) *f is absolutely integrable over $\mathcal{X}$ if, and only if, ν is finite.*
(iii) *If $\mathcal{P} \circ f > 0$ a.e. in $\mathcal{X}$ and $\mathcal{X}$ is σ-finite and complete, then $(X, \mathcal{B}, \nu)$ is a σ-finite complete $\mathcal{Y}$-valued measure space.*

Proof Since f is $\mathcal{B}$-measurable, by Propositions 7.21 and 11.38, $\mathcal{P} \circ f$ is $\mathcal{B}$-measurable. Then, $\bar{\nu} : \mathcal{B} \to [0, +\infty] \subset \mathbb{R}_e$ is well-defined by Definition 11.79. By Proposition 11.75, $\bar{\nu}(\emptyset) = \int_\emptyset (\mathcal{P} \circ f) \, d\mu = 0$. $\forall$ pairwise disjoint $(E_i)_{i=1}^\infty \subseteq \mathcal{B}$, by Proposition 11.83, we have $\bar{\nu}(\bigcup_{i=1}^\infty E_i) = \int_{\bigcup_{i=1}^\infty E_i} (\mathcal{P} \circ f) \, d\mu = \sum_{i=1}^\infty \int_{E_i} (\mathcal{P} \circ f) \, d\mu = \sum_{i=1}^\infty \bar{\nu}(E_i)$. Hence, $\bar{\nu}$ is a measure on $(X, \mathcal{B})$. By Proposition 11.75, $\nu(\emptyset) = \int_\emptyset f \, d\mu = \vartheta_{\mathcal{Y}}$. $\forall E \in \mathcal{B}$ with $\bar{\nu}(E) = +\infty$, $\nu(E)$ is undefined. $\forall E \in \mathcal{B}$ with $\bar{\nu}(E) < +\infty$, by Proposition 11.92, $\nu(E) = \int_E f \, d\mu = \int_E f|_E \, d\mu_E \in \mathcal{Y}$, where $\mathcal{E} := (E, \mathcal{B}_E, \mu_E)$ is the measure subspace of $\mathcal{X}$ as defined in Proposition 11.13. $\forall$ pairwise disjoint $(E_i)_{i=1}^\infty \subseteq \mathcal{B}$ with $\bigcup_{i=1}^\infty E_i = E$, we have $\sum_{i=1}^\infty \|\nu(E_i)\| = \sum_{i=1}^\infty \left\| \int_{E_i} f \, d\mu \right\| \le \sum_{i=1}^\infty \int_{E_i} (\mathcal{P} \circ f) \, d\mu = \int_E (\mathcal{P} \circ f) \, d\mu = \bar{\nu}(E) < +\infty$, where the first equality follows from the fact that $E_i \subseteq E$ and $0 \le \bar{\nu}(E_i) \le \bar{\nu}(E) < +\infty$, the first inequality follows from Proposition 11.92, and the second equality follows from Proposition 11.83. Then, $\sum_{i=1}^\infty \nu(E_i) = \sum_{i=1}^\infty \int_{E_i} f \, d\mu = \sum_{i=1}^\infty \int_{E_i} f|_E \, d\mu_E = \int_E f|_E \, d\mu_E = \int_E f \, d\mu = \nu(E)$, where the second, the third, and the fourth equalities follow from Proposition 11.92.

$\forall E \in \mathcal{B}$ with $\bar{\nu}(E) < +\infty$, $\forall n \in \mathbb{Z}_+$, $\forall$ pairwise disjoint $(E_i)_{i=1}^n \subseteq \mathcal{B}$ with $E = \bigcup_{i=1}^n E_i$, $f|_E$ is absolutely integrable over $\mathcal{E}$. $\sum_{i=1}^n \|\nu(E_i)\| = \sum_{i=1}^n \left\| \int_{E_i} f \, d\mu \right\| \le \sum_{i=1}^n \int_{E_i} \mathcal{P} \circ f \, d\mu = \int_E \mathcal{P} \circ f \, d\mu = \bar{\nu}(E) < +\infty$, where the first equality follows from the fact that $E_i \subseteq E$ and $0 \le \bar{\nu}(E_i) \le \bar{\nu}(E) < +\infty$, the first inequality follows from Proposition 11.92, and the second equality follows from Proposition 11.83. Hence, $s_E := \sup_{n \in \mathbb{Z}_+, (E_i)_{i=1}^n \subseteq \mathcal{B}, E = \bigcup_{i=1}^n E_i, E_i \cap E_j = \emptyset, \forall 1 \le i < j \le n} \sum_{i=1}^n \|\nu(E_i)\| \le \bar{\nu}(E)$.

On the other hand, $\forall \epsilon \in (0, \infty) \subset \mathbb{R}$, by Proposition 11.95, $\exists$ a simple function $\phi : \mathcal{E} \to \mathcal{W}$ such that $0 \le \left| \int_E (\mathcal{P} \circ f)|_E \, d\mu_E - \int_E (\mathcal{P} \circ \phi) \, d\mu_E \right| < \epsilon/2$ and $0 \le \int_E \mathcal{P} \circ (\phi - f|_E) \, d\mu_E < \epsilon/2$. Let ϕ admit canonical representation $\phi = \sum_{i=1}^{n} w_i \chi_{E_i, E}$, where $n \in \mathbb{Z}_+$, $w_1, \ldots, w_n \in \mathcal{W}$ are distinct and none equals to $\vartheta_{\mathcal{Y}}$, $E_1, \ldots, E_n \in \mathcal{B}_E$ are nonempty, pairwise disjoint, and $\mu_E(E_i) < +\infty$, $\forall i \in \{1, \ldots, n\}$. Then, we have

$$\begin{aligned}
\bar{\nu}(E) &= \int_E (\mathcal{P} \circ f) \, d\mu = \int_E (\mathcal{P} \circ f)|_E \, d\mu_E < \int_E (\mathcal{P} \circ \phi) \, d\mu_E + \epsilon/2 \\
&= \sum_{i=1}^{n} \|\nu(E_i)\| + \sum_{i=1}^{n} \|w_i\| \mu_E(E_i) + \epsilon/2 - \sum_{i=1}^{n} \|\nu(E_i)\| \\
&= \sum_{i=1}^{n} \|\nu(E_i)\| + \epsilon/2 + \sum_{i=1}^{n} \Big\| \int_{E_i} \phi \, d\mu_E \Big\| - \sum_{i=1}^{n} \Big\| \int_{E_i} f|_E \, d\mu_E \Big\| \\
&\le s_E + \epsilon/2 + \sum_{i=1}^{n} \Big\| \int_{E_i} \phi \, d\mu_E - \int_{E_i} f|_E \, d\mu_E \Big\| \\
&= s_E + \epsilon/2 + \sum_{i=1}^{n} \Big\| \int_{E_i} (\phi - f|_E) \, d\mu_E \Big\| \\
&\le s_E + \epsilon/2 + \sum_{i=1}^{n} \int_{E_i} \mathcal{P} \circ (\phi - f|_E) \, d\mu_E \\
&= s_E + \epsilon/2 + \int_{\bigcup_{i=1}^{n} E_i} \mathcal{P} \circ (\phi - f|_E) \, d\mu_E < s_E + \epsilon
\end{aligned}$$

where the second equality follows from Proposition 11.92, the third equality follows from Proposition 11.75, the fourth equality follows from Proposition 11.75 and the fact that $0 \le \bar{\nu}(E_i) \le \bar{\nu}(E) < +\infty$, the fifth equality and the third inequality follow from Proposition 11.92, and the last equality and the last inequality follow from Proposition 11.83. Hence, by the arbitrariness of ϵ, we have $\bar{\nu}(E) \le s_E$. Therefore, $\bar{\nu}(E) = s_E$.

The above shows that $(X, \mathcal{B}, \nu)$ is a $\mathcal{Y}$-valued measure space with $\mathcal{P} \circ \nu = \bar{\nu}$.

(i) Let $\mathcal{X}$ be σ-finite. Then, $\exists (X_i)_{i=1}^{\infty} \subseteq \mathcal{B}$ such that $X = \bigcup_{i=1}^{\infty} X_i$ and $\mu(X_i) < +\infty$, $\forall i \in \mathbb{N}$. $\forall i \in \mathbb{N}$, $\forall m \in \mathbb{N}$, let $X_{i,m} := X_i \cap \{x \in X \mid \|f(x)\| \le m\} \in \mathcal{B}$. Clearly, $X = \bigcup_{i=1}^{\infty} X_i = \bigcup_{i=1}^{\infty} \bigcup_{m=1}^{\infty} X_{i,m}$. By Bounded Convergence Theorem 11.77, we have $\bar{\nu}(X_{i,m}) = \int_{X_{i,m}} (\mathcal{P} \circ f) \, d\mu \le m\mu(X_{i,m}) \le m\mu(X_i) < +\infty$. Then, $\bar{\nu}$ is σ-finite. Hence, $(X, \mathcal{B}, \nu)$ is a σ-finite $\mathcal{Y}$-valued measure space.

(ii) Let f be absolutely integrable over $\mathcal{X}$. Then, $\bar{\nu}(X) = \int_X (\mathcal{P} \circ f) \, d\mu < +\infty$. Hence, $(X, \mathcal{B}, \nu)$ is a finite $\mathcal{Y}$-valued (pre-)measure space. On the other hand, let ν be finite. Then, $\bar{\nu}(X) = \int_X (\mathcal{P} \circ f) \, d\mu < +\infty$. Hence, f is absolutely integrable over $\mathcal{X}$.

(iii) By (i), $(X, \mathcal{B}, \nu)$ is σ-finite. $\forall E \in \mathcal{B}$ with $\mathcal{P} \circ \nu(E) = 0$, then $\int_E (\mathcal{P} \circ f)\, d\mu = 0$. By Proposition 11.96, $\mathcal{P} \circ f = 0$ a.e. in $\mathcal{E}$. Since $\mathcal{P} \circ f > 0$ a.e. in $\mathcal{X}$, then $\mu(E) = 0$. $\forall A \subseteq E$, by the completeness of $\mathcal{X}$, $A \in \mathcal{B}$. Hence, $(X, \mathcal{B}, \nu)$ is a σ-finite complete $\mathcal{Y}$-valued measure space.

This completes the proof of the proposition. □

Proposition 11.117 *Let $\mathcal{X} := (X, \mathcal{B}, \mu)$ be a σ-finite $\mathcal{Y}$-valued measure space, where $\mathcal{Y}$ is a normed linear space, $(X_n)_{n=1}^{\infty} \subseteq \mathcal{B}$ be such that $X = \bigcup_{n=1}^{\infty} X_n$ and $\mathcal{P} \circ \mu(X_n) < +\infty$, $\forall n \in \mathbb{N}$, and $\mathcal{X}_n := (X_n, \mathcal{B}_n, \mu_n)$ be the finite $\mathcal{Y}$-valued measure subspace of $\mathcal{X}$. Then, any $\mathcal{Y}$-valued measure space $\bar{\mathcal{X}} := (X, \bar{\mathcal{B}}, \bar{\mu})$ with $\mathcal{X}_n$ as its finite $\mathcal{Y}$-valued measure subspace, $\forall n \in \mathbb{N}$, must be identical to $\mathcal{X}$.*

Proof $\forall E \in \mathcal{B}$, $E \cap X_n \in \mathcal{B}_n \subseteq \bar{\mathcal{B}}$, $\forall n \in \mathbb{N}$. Then, $E = \bigcup_{n=1}^{\infty}(E \cap X_n) \in \bar{\mathcal{B}}$. Then, $\mathcal{B} \subseteq \bar{\mathcal{B}}$. On the other hand, $\forall E \in \bar{\mathcal{B}}$, $E \cap X_n \in \mathcal{B}_n \subseteq \mathcal{B}$, $\forall n \in \mathbb{N}$. Then, $E = \bigcup_{n=1}^{\infty}(E \cap X_n) \in \mathcal{B}$. Then, $\bar{\mathcal{B}} \subseteq \mathcal{B}$. Hence, $\bar{\mathcal{B}} = \mathcal{B}$.

$\forall E \in \mathcal{B}$ with $E \in \operatorname{dom}(\mu)$, we have $\mathcal{P} \circ \mu(E) < \infty$ and $\mu(E) \in \mathcal{Y}$. Let $E_n := E \cap (X_n \setminus \bigcup_{i=1}^{n-1} X_i) \in \mathcal{B}_n$, $\forall n \in \mathbb{N}$. Then, $\mu(E_n) = \mu_n(E_n) = \bar{\mu}(E_n)$, $\forall n \in \mathbb{N}$. Note that $\mathcal{P} \circ \mu(E) = \sum_{n=1}^{\infty} \mathcal{P} \circ \mu(E_n) = \sum_{n=1}^{\infty} \mathcal{P} \circ \mu_n(E_n) = \sum_{n=1}^{\infty} \mathcal{P} \circ \bar{\mu}(E_n) = \mathcal{P} \circ \bar{\mu}(E) < +\infty$, where the second and third equalities follow from Proposition 11.115. Then, $E \in \operatorname{dom}(\bar{\mu})$ and $\mu(E) = \sum_{n=1}^{\infty} \mu(E_n) = \sum_{n=1}^{\infty} \bar{\mu}(E_n) = \bar{\mu}(E) \in \mathcal{Y}$.

On the other hand, $\forall E \in \bar{\mathcal{B}} = \mathcal{B}$ with $E \in \operatorname{dom}(\bar{\mu})$, we have $\mathcal{P} \circ \bar{\mu}(E) < \infty$ and $\bar{\mu}(E) \in \mathcal{Y}$. Let $E_n := E \cap (X_n \setminus \bigcup_{i=1}^{n-1} X_i) \in \mathcal{B}_n$, $\forall n \in \mathbb{N}$. Then, $\mu(E_n) = \mu_n(E_n) = \bar{\mu}(E_n)$, $\forall n \in \mathbb{N}$. Note that $\mathcal{P} \circ \mu(E) = \sum_{n=1}^{\infty} \mathcal{P} \circ \mu(E_n) = \sum_{n=1}^{\infty} \mathcal{P} \circ \mu_n(E_n) = \sum_{n=1}^{\infty} \mathcal{P} \circ \bar{\mu}(E_n) = \mathcal{P} \circ \bar{\mu}(E) < +\infty$, where the second and third equalities follow from Proposition 11.115. Then, $E \in \operatorname{dom}(\mu)$ and $\mu(E) = \sum_{n=1}^{\infty} \mu(E_n) = \sum_{n=1}^{\infty} \bar{\mu}(E_n) = \bar{\mu}(E) \in \mathcal{Y}$.

Therefore, $\operatorname{dom}(\mu) = \operatorname{dom}(\bar{\mu})$ and $\mu = \bar{\mu}$. Hence, $\mathcal{X} = \bar{\mathcal{X}}$. This completes the proof of the proposition. □

Proposition 11.118 *Let $\mathcal{X}_n := (X_n, \mathcal{B}_n, \mu_n)$ be a finite $\mathcal{Y}$-valued measure space, where $\mathcal{Y}$ is a Banach space, $\forall n \in \mathbb{N}$. $\forall n, m \in \mathbb{N}$, let $(X_n \cap X_m, \mathcal{B}_{n,m}, \mu_{n,m})$ be the finite $\mathcal{Y}$-valued measure subspace of $\mathcal{X}_n$. Assume that $(\mathcal{X}_n)_{n=1}^{\infty}$ is consistent, that is $\mathcal{B}_{n,m} = \mathcal{B}_{m,n}$ and $\mu_{n,m} = \mu_{m,n}$, $\forall n, m \in \mathbb{N}$. Let $X := \bigcup_{n=1}^{\infty} X_n$, $\mathcal{B} := \{B \subseteq X \mid B \cap X_n \in \mathcal{B}_n, \forall n \in \mathbb{N}\}$, $\nu : \mathcal{B} \to [0, \infty] \subset \mathbb{R}_e$ be defined by $\nu(E) := \sum_{n=1}^{\infty} \mathcal{P} \circ \mu_n(E_n)$, where $E_1 := E \cap X_1$ and $E_{n+1} := E \cap (X_{n+1} \setminus \bigcup_{i=1}^{n} X_i)$, $\forall n \in \mathbb{N}$, and μ be a function from $\mathcal{B}$ to $\mathcal{Y}$ defined by $\mu(E)$ is undefined, $\forall E \in \mathcal{B}$ with $\nu(E) = +\infty$, and $\mu(E) := \sum_{n=1}^{\infty} \mu_n(E_n) \in \mathcal{Y}$, $\forall E \in \mathcal{B}$ with $\nu(E) < +\infty$, where $(E_n)_{n=1}^{\infty}$ is defined as above. Then, $\mathcal{X} := (X, \mathcal{B}, \mu)$ is a σ-finite $\mathcal{Y}$-valued measure space with $\mathcal{P} \circ \mu = \nu$ and $\mathcal{X}_n$ is the finite $\mathcal{Y}$-valued measure subspace of $\mathcal{X}$, $\forall n \in \mathbb{N}$, and $\mathcal{X}$ is unique among any $\mathcal{Y}$-valued measure spaces on X. The above process is said to be the* generation process *on $(\mathcal{X}_n)_{n=1}^{\infty}$ for σ-finite $\mathcal{Y}$-valued measure space $\mathcal{X}$.*

Proof We will first show that $\mathcal{B}$ is a σ-algebra on X. Clearly, $\emptyset \in \mathcal{B}$ and $X \in \mathcal{B}$. $\forall (E_i)_{i=1}^{\infty} \subseteq \mathcal{B}$, let $E := \bigcup_{i=1}^{\infty} E_i$. $\forall n \in \mathbb{N}$, $E \cap X_n = \bigcup_{i=1}^{\infty}(E_i \cap X_n)$. Note that $\forall i \in \mathbb{N}$, $E_i \in \mathcal{B}$ implies $E_i \cap X_n \in \mathcal{B}_n$. Since $\mathcal{B}_n$ is a σ-algebra, then $E \cap X_n \in \mathcal{B}_n$. By the arbitrariness of n, we have $E \in \mathcal{B}$. $\forall E \in \mathcal{B}$, $E \cap X_n \in \mathcal{B}_n$, $\forall n \in \mathbb{N}$.

Then, $(X \setminus E) \cap X_n = X_n \setminus E = X_n \setminus (X_n \cap E) \in \mathcal{B}_n$. By the arbitrariness of n, we have $X \setminus E \in \mathcal{B}$. Hence, $\mathcal{B}$ is a σ-algebra on X. $\forall n \in \mathbb{N}$, $\forall m \in \mathbb{N}$, $X_n \cap X_m \in \mathcal{B}_{n,m} = \mathcal{B}_{m,n} \subseteq \mathcal{B}_m$. By the arbitrariness of m, we have $X_n \in \mathcal{B}$.

Next, we will show that ν is a σ-finite measure on $(X, \mathcal{B})$. Clearly, $\nu(\emptyset) = 0$. $\forall$ pairwise disjoint $(E_i)_{i=1}^{\infty} \subseteq \mathcal{B}$, let $E := \bigcup_{i=1}^{\infty} E_i \in \mathcal{B}$. $\forall i \in \mathbb{N}$, let $E_{i,1} := E_i \cap X_1 \in \mathcal{B}_1$ and $E_{i,n+1} := E_i \cap (X_{n+1} \setminus \bigcup_{j=1}^{n} X_j) \in \mathcal{B}_{n+1}$, $\forall n \in \mathbb{N}$. Let $B_1 := E \cap X_1 = \bigcup_{i=1}^{\infty} E_{i,1} \in \mathcal{B}_1$ and $B_{n+1} := E \cap (X_{n+1} \setminus \bigcup_{j=1}^{n} X_j) = \bigcup_{i=1}^{\infty} E_{i,n+1} \in \mathcal{B}_{n+1}$, $\forall n \in \mathbb{N}$. Then, $\nu(E) = \sum_{n=1}^{\infty} \mathcal{P} \circ \mu_n(B_n) = \sum_{n=1}^{\infty} \sum_{i=1}^{\infty} \mathcal{P} \circ \mu_n(E_{i,n}) = \sum_{i=1}^{\infty} \sum_{n=1}^{\infty} \mathcal{P} \circ \mu_n(E_{i,n}) = \sum_{i=1}^{\infty} \nu(E_i)$, where the first equality follows from the definition of ν, the second equality follows from the fact that $\mathcal{P} \circ \mu_n$ is a measure on $(X_n, \mathcal{B}_n)$, the third equality follows from the fact that all summands are nonnegative, and the fourth equality follows from the definition of ν. Hence, ν is a measure on $(X, \mathcal{B})$. $\forall n \in \mathbb{N}$, let $X_{n,1} := X_n \cap X_1$ and $X_{n,i+1} := X_n \cap (X_{i+1} \setminus \bigcup_{j=1}^{i} X_j)$, $\forall i \in \mathbb{N}$. Then, $\nu(X_n) = \sum_{i=1}^{\infty} \mathcal{P} \circ \mu_i(X_{n,i}) = \sum_{i=1}^{\infty} \mathcal{P} \circ \mu_{i+1}((X_n \cap X_{i+1}) \setminus \bigcup_{j=1}^{i} X_j) + \mathcal{P} \circ \mu_1(X_n \cap X_1) = \sum_{i=1}^{\infty} \mathcal{P} \circ \mu_{i+1,n}((X_n \cap X_{i+1}) \setminus \bigcup_{j=1}^{i} X_j) + \mathcal{P} \circ \mu_{1,n}(X_n \cap X_1) = \sum_{i=1}^{\infty} \mathcal{P} \circ \mu_{n,i+1}((X_n \cap X_{i+1}) \setminus \bigcup_{j=1}^{i} X_j) + \mathcal{P} \circ \mu_{n,1}(X_n \cap X_1) = \sum_{i=1}^{\infty} \mathcal{P} \circ \mu_n((X_n \cap X_{i+1}) \setminus \bigcup_{j=1}^{i} X_j) + \mathcal{P} \circ \mu_n(X_n \cap X_1) = \mathcal{P} \circ \mu_n(X_n) < +\infty$, where the first equality follows from the definition of ν, the third equality follows from Proposition 11.115, the fourth equality follows from the consistency assumption, the fifth equality follows from Proposition 11.115, and the last equality follows from the fact that $\mathcal{P} \circ \mu_n$ is a measure on $(X_n, \mathcal{B}_n)$. Hence, ν is σ-finite.

Next, we will show that $(X, \mathcal{B}, \mu)$ is a σ-finite $\mathcal{Y}$-valued measure space with $\mathcal{P} \circ \mu = \nu$. Clearly, $\mu(\emptyset) = \vartheta_{\mathcal{Y}}$. $\forall E \in \mathcal{B}$ with $\nu(E) = +\infty$, $\mu(E)$ is undefined. $\forall E \in \mathcal{B}$ with $\nu(E) < +\infty$, let $(B_n)_{n=1}^{\infty}$ be as defined in the previous paragraph. Then, $+\infty > \nu(E) = \sum_{n=1}^{\infty} \mathcal{P} \circ \mu_n(B_n) \geq \sum_{n=1}^{\infty} \|\mu_n(B_n)\|$. This implies that $\mu(E) = \sum_{n=1}^{\infty} \mu_n(B_n) \in \mathcal{Y}$ by Proposition 7.27. $\forall$ pairwise disjoint $(E_i)_{i=1}^{\infty} \subseteq \mathcal{B}$ with $E = \bigcup_{i=1}^{\infty} E_i$, $\forall i \in \mathbb{N}$, let $\left(E_{i,n}\right)_{n=1}^{\infty}$ be as defined in the previous paragraph. Then, we have $\sum_{i=1}^{\infty} \|\mu(E_i)\| = \sum_{i=1}^{\infty} \left\|\sum_{n=1}^{\infty} \mu_n(E_{i,n})\right\| \leq \sum_{i=1}^{\infty} \sum_{n=1}^{\infty} \left\|\mu_n(E_{i,n})\right\| \leq \sum_{i=1}^{\infty} \sum_{n=1}^{\infty} \mathcal{P} \circ \mu_n(E_{i,n}) = \sum_{i=1}^{\infty} \nu(E_i) = \nu(E) < +\infty$, where the first equality follows from the definition of μ, the second inequality follows from the fact that $\mathcal{X}_n$ is a finite $\mathcal{Y}$-valued measure space, the second equality follows from the definition of ν, and the third equality follows from the fact that ν is a measure. We also have $\sum_{i=1}^{\infty} \mu(E_i) = \sum_{i=1}^{\infty} \sum_{n=1}^{\infty} \mu_n(E_{i,n}) = \sum_{n=1}^{\infty} \sum_{i=1}^{\infty} \mu_n(E_{i,n}) = \sum_{n=1}^{\infty} \mu_n(\bigcup_{i=1}^{\infty} E_{i,n}) = \sum_{n=1}^{\infty} \mu_n(B_n) = \mu(E)$, where the first equality follows from the definition of μ, the second equality follows from the fact that the double summation is absolutely summable, the third equality follows from the fact that μ_n is a finite $\mathcal{Y}$-valued measure, and the last equality follows from the definition of μ. Hence, (i)–(iii) of Definition 11.108 are satisfied.

$\forall E \in \mathcal{B}$ with $\nu(E) < +\infty$, let $(B_n)_{n=1}^{\infty}$ be as defined in the previous paragraph. $\forall m \in \mathbb{Z}_+$, $\forall$ pairwise disjoint $(E_i)_{i=1}^{m} \subseteq \mathcal{B}$ with $E = \bigcup_{i=1}^{m} E_i$, $\forall i \in \{1, \ldots, m\}$, let $\left(E_{i,n}\right)_{n=1}^{\infty}$ be as defined in the previous paragraph. Then, $\sum_{i=1}^{m} \|\mu(E_i)\| = \sum_{i=1}^{m} \left\|\sum_{n=1}^{\infty} \mu_n(E_{i,n})\right\| \leq \sum_{i=1}^{m} \sum_{n=1}^{\infty} \left\|\mu_n(E_{i,n})\right\| \leq \sum_{i=1}^{m} \sum_{n=1}^{\infty} \mathcal{P} \circ \mu_n(E_{i,n}) = \sum_{i=1}^{m} \nu(E_i) = \nu(E) < \infty$. Hence,

$$s_E := \sup_{m\in\mathbb{Z}_+,\ (E_i)_{i=1}^m\subseteq\mathcal{B},\ E=\bigcup_{i=1}^m E_i,\ E_i\cap E_j=\emptyset,\ \forall 1\le i<j\le m} \sum_{i=1}^{m} \|\mu(E_i)\| \le \nu(E)$$

On the other hand, $\nu(E) = \sum_{n=1}^{\infty} \mathcal{P} \circ \mu_n(B_n) < +\infty$, where $B_1, B_2, \ldots$ are pairwise disjoint. Since $(X_n, \mathcal{B}_n, \mu_n)$ is a finite $\mathcal{Y}$-valued measure space, then, $\forall \epsilon \in (0, +\infty) \subset \mathbb{R}$, $\forall n \in \mathbb{N}$, $\exists m_n \in \mathbb{Z}_+$, $\exists$ pairwise disjoint $\left(B_{n,i}\right)_{i=1}^{m_n} \subseteq \mathcal{B}_n$ with $B_n = \bigcup_{i=1}^{m_n} B_{n,i}$ such that $\mathcal{P} \circ \mu_n(B_n) - 2^{-n}\epsilon < \sum_{i=1}^{m_n} \left\|\mu_n(B_{n,i})\right\| \le \mathcal{P} \circ \mu_n(B_n)$. It is easy to see that $\mu_n(B_{n,i}) = \mu(B_{n,i})$, $\forall i \in \{1, \ldots, m_n\}$. Then, $\nu(E) - \epsilon = \sum_{n=1}^{\infty}(\mathcal{P}\circ\mu_n(B_n) - 2^{-n}\epsilon) < \sum_{n=1}^{\infty} \sum_{i=1}^{m_n} \left\|\mu_n(B_{n,i})\right\| = \sum_{n=1}^{\infty} \sum_{i=1}^{m_n} \left\|\mu(B_{n,i})\right\| \le \nu(E) < +\infty$. Then, $\exists N \in \mathbb{N}$ such that $\nu(E) \le \sum_{n=1}^{N} \sum_{i=1}^{m_n} \left\|\mu(B_{n,i})\right\| + \epsilon \le s_E + \epsilon$. By the arbitrariness of ϵ, we have $\nu(E) \le s_E$. Therefore, $\nu(E) = s_E$ and (iv) of Definition 11.108 is satisfied. This shows that $(X, \mathcal{B}, \mu)$ is a σ-finite $\mathcal{Y}$-valued measure space with $\mathcal{P} \circ \mu = \nu$.

Finally, we will show that $\mathcal{X}_n$ is the finite $\mathcal{Y}$-valued measure subspace of $\mathcal{X}$, $\forall n \in \mathbb{N}$. Fix any $n \in \mathbb{N}$. $\forall E \in \mathcal{B}$ with $E \subseteq X_n$, we have $E = E \cap X_n \in \mathcal{B}_n$. On the other hand, $\forall E \in \mathcal{B}_n$, we have $E \subseteq X_n$ and $\forall i \in \mathbb{N}$, $E \cap X_i = E \cap X_i \cap X_n \in \mathcal{B}_{n,i} = \mathcal{B}_{i,n} \subseteq \mathcal{B}_i$. Then, $E \in \mathcal{B}$. Hence, $\mathcal{B}_n = \{E \in \mathcal{B} \,|\, E \subseteq X_n\}$. $\forall E \in \mathcal{B}_n$, let $B_1 := E \cap X_1 \in \mathcal{B}_1$ and $B_{i+1} := E \cap (X_{i+1} \setminus \bigcup_{j=1}^{i} X_j) \in \mathcal{B}_{i+1}$, $\forall i \in \mathbb{N}$. Then, $\nu(E) = \sum_{i=1}^{\infty} \mathcal{P} \circ \mu_i(B_i)$. Note that $B_i \in \mathcal{B}_i$ and $B_i \subseteq X_n$, $\forall i \in \mathbb{N}$, and this implies that $B_i \in \mathcal{B}_{i,n} = \mathcal{B}_{n,i} \subseteq \mathcal{B}_n$. Then, we have $\nu(E) = \sum_{i=1}^{\infty} \mathcal{P} \circ \mu_{i,n}(B_i) = \sum_{i=1}^{\infty} \mathcal{P} \circ \mu_{n,i}(B_i) = \sum_{i=1}^{\infty} \mathcal{P} \circ \mu_n(B_i) = \mathcal{P} \circ \mu_n(E) \le \mathcal{P} \circ \mu_n(X_n) < +\infty$, where the first equality follows from the fact that $\mu_{i,n} = \mu_i|_{\mathcal{B}_{i,n}}$, the second equality follows from the consistency assumption, the third equality follows from the fact that $\mu_{n,i} = \mu_n|_{\mathcal{B}_{n,i}}$, the fourth equality follows from countable additivity of $\mathcal{P} \circ \mu_n$, and the first inequality follows from Proposition 11.4. Then, $\mathcal{Y} \ni \mu(E) = \sum_{i=1}^{\infty} \mu_i(B_i) = \sum_{i=1}^{\infty} \mu_{i,n}(B_i) = \sum_{i=1}^{\infty} \mu_{n,i}(B_i) = \sum_{i=1}^{\infty} \mu_n(B_i) = \mu_n(E)$, where the first equality follows from the definition of μ, the second equality follows from the fact that $\mu_{i,n} = \mu_i|_{\mathcal{B}_{i,n}}$, the third equality follows from the consistency assumption, the fourth equality follows from the fact that $\mu_{n,i} = \mu_n|_{\mathcal{B}_{n,i}}$, and the last equality follows from the fact that μ_n is a finite $\mathcal{Y}$-valued measure. Hence, $\mu_n = \mu|_{\mathcal{B}_n}$. This shows that $\mathcal{X}_n$ is the subspace of $\mathcal{X}$. By Proposition 11.117, $\mathcal{X}$ is the unique $\mathcal{Y}$-valued measure space on the set X such that $\mathcal{X}_n$ is the finite $\mathcal{Y}$-valued measure subspace of $\mathcal{X}$, $\forall n \in \mathbb{N}$.

This completes the proof of the proposition. □

In the application of Proposition 11.118, if the sequence $(X_n)_{n=1}^{\infty}$ is pairwise disjoint, then the consistency requirement is automatically satisfied as long as $\mathcal{X}_n$ is a $\mathcal{Y}$-valued measure space, $\forall n \in \mathbb{N}$.

Definition 11.119 Let $\mathcal{X} := (X, \mathcal{B}, \mu)$ be a finite $\mathcal{Y}$-valued measure space, where $\mathcal{Y}$ is a normed linear space over $\mathbb{K}$, $\mathcal{Z}$ be a normed linear space over $\mathbb{K}$, $\mathcal{W} \subseteq \mathrm{B}(\mathcal{Y}, \mathcal{Z})$ be a subspace, $f : X \to \mathcal{W}$ be $\mathcal{B}$-measurable, and $\mathcal{I}(\mathcal{W}) := (\mathcal{R}(\mathcal{W}), \preceq)$ be the integration system on $\mathcal{W}$. $\forall R = \{(w_\alpha, U_\alpha) \,|\, \alpha \in \Lambda\} \in \mathcal{R}(\mathcal{W})$, define $F_R := \sum_{\alpha\in\Lambda} w_\alpha \mu(f_{\mathrm{inv}}(U_\alpha)) \in \mathcal{Z}$. This defines a net $(F_R)_{R\in\mathcal{I}(\mathcal{W})}$. f is said to be *integrable over* $\mathcal{X}$ if the net admits a limit in $\mathcal{Z}$. In this case, $\lim_{R\in\mathcal{I}(\mathcal{W})} F_R \in \mathcal{Z}$ is said to be the *integral* of f over $\mathcal{X}$ and denoted by $\int_X f\,\mathrm{d}\mu$ or $\int_X f(x)\,\mathrm{d}\mu(x)$. When $\mathcal{Z} = \mathbb{R}$, we will denote $\int_X f\,\mathrm{d}\mu := \int_X f(x)\,\mathrm{d}\mu(x) := \lim_{R\in\mathcal{I}(\mathcal{W})} F_R \in \mathbb{R}_e$, whenever the limit exists. ◇

Definition 11.120 Let $\mathcal{X} := (X, \mathcal{B}, \mu)$ be a $\mathcal{Y}$-valued measure space with $\mathcal{P} \circ \mu(X) = +\infty$, where $\mathcal{Y}$ is a normed linear space over $\mathbb{K}$, $\mathcal{Z}$ be a normed linear space over $\mathbb{K}$, $\bar{\mathcal{X}} := (X, \mathcal{B}, \mathcal{P} \circ \mu)$ be the total variation, and $f : X \to \mathcal{W} := \mathrm{B}(\mathcal{Y}, \mathcal{Z})$ be $\mathcal{B}$-measurable. Define $\mathcal{M}(\mathcal{X})$ to be the directed system, $\mathcal{M}(\bar{\mathcal{X}})$, of subsets of X as defined in Definition 11.71. $\forall A \in \mathcal{M}(\mathcal{X})$, let $(A, \mathcal{B}_A, \mu_A)$ be the finite $\mathcal{Y}$-valued measure subspace of $\mathcal{X}$ and define $F_A := \int_A f|_A\,\mathrm{d}\mu_A \in \mathcal{Z}$, if the integral exists. This defines a net $(F_A)_{A\in\mathcal{M}(\mathcal{X})}$ when all of the integrals involved exist. f is said to be *integrable over* $\mathcal{X}$ if the net is well-defined (i.e., all integrals involved exist) and admits a limit in $\mathcal{Z}$. In this case, $\lim_{A\in\mathcal{M}(\mathcal{X})} F_A \in \mathcal{Z}$ is said to be the *integral* of f over $\mathcal{X}$ and denoted by $\int_X f\,\mathrm{d}\mu$ or $\int_X f(x)\,\mathrm{d}\mu(x)$. When $\mathcal{Z} = \mathbb{R}$, we will allow the net $(F_A)_{A\in\mathcal{M}(\mathcal{X})} \subseteq \mathbb{R}_e$ and denote $\int_X f\,\mathrm{d}\mu := \int_X f(x)\,\mathrm{d}\mu(x) := \lim_{A\in\mathcal{M}(\mathcal{X})} F_A \in \mathbb{R}_e$, whenever the limit exists. ◇

Proposition 11.121 *Let $\mathcal{X} := (X, \mathcal{B}, \mu)$ be a finite $\mathcal{Y}$-valued measure space, where $\mathcal{Y}$ is a normed linear space over $\mathbb{K}$, $\mathcal{Z}$ be a normed linear space over $\mathbb{K}$, $f : X \to \mathcal{W} := \mathrm{B}(\mathcal{Y}, \mathcal{Z})$ be $\mathcal{B}$-measurable, and $\bar{\mathcal{X}} := (X, \bar{\mathcal{B}}, \bar{\mu})$ be a finite $\mathcal{Y}$-valued measure space that satisfies $\mathcal{B} \subseteq \bar{\mathcal{B}}$ and $\mu = \bar{\mu}|_{\mathcal{B}}$. Then, $\int_X f\,\mathrm{d}\mu = \int_X f\,\mathrm{d}\bar{\mu}$, whenever one of the integrals exists.*

Proof Let $\mathcal{I}(\mathcal{W})$ be the integration system on $\mathcal{W}$ as defined in Proposition 11.69, $(F_R)_{R\in\mathcal{I}(\mathcal{W})}$ be the net for $\int_X f\,\mathrm{d}\mu$, and $\left(\bar{F}_R\right)_{R\in\mathcal{I}(\mathcal{W})}$ be the net for $\int_X f\,\mathrm{d}\bar{\mu}$, as defined in Definition 11.119. $\forall R := \{(w_\alpha, U_\alpha) \mid \alpha \in \Lambda\} \in \mathcal{I}(\mathcal{W})$, $\forall \alpha \in \Lambda$, we have $f_{\mathrm{inv}}(U_\alpha) \in \mathcal{B} \subseteq \bar{\mathcal{B}}$ since f is $\mathcal{B}$-measurable, and $\mu(f_{\mathrm{inv}}(U_\alpha)) = \bar{\mu}(f_{\mathrm{inv}}(U_\alpha)) \in \mathcal{Y}$ since $\mu = \bar{\mu}|_{\mathcal{B}}$. Then, $F_R = \sum_{\alpha\in\Lambda} w_\alpha \mu(f_{\mathrm{inv}}(U_\alpha)) = \sum_{\alpha\in\Lambda} w_\alpha \bar{\mu}(f_{\mathrm{inv}}(U_\alpha)) = \bar{F}_R \in \mathcal{Z}$. Hence, $\int_X f\,\mathrm{d}\mu = \lim_{R\in\mathcal{I}(\mathcal{W})} F_R = \lim_{R\in\mathcal{I}(\mathcal{W})} \bar{F}_R = \int_X f\,\mathrm{d}\bar{\mu}$, whenever one of the integrals exists. This completes the proof of the proposition. □

Lemma 11.122 *Let $\mathcal{X} := (X, \mathcal{B}, \mu)$ be a finite $\mathcal{Y}$-valued measure space, where $\mathcal{Y}$ is a normed linear space over $\mathbb{K}$, $\mathcal{Z}$ be a normed linear space over $\mathbb{K}$, $f : X \to \mathcal{W} := \mathrm{B}(\mathcal{Y}, \mathcal{Z})$ be $\mathcal{B}$-measurable, $\hat{X} \in \mathcal{B}$, and $\hat{\mathcal{X}} := (\hat{X}, \hat{\mathcal{B}}, \hat{\mu})$ be the finite $\mathcal{Y}$-valued measure subspace of $\mathcal{X}$ as defined in Proposition 11.115. Assume that $\mathcal{P} \circ \mu(X \setminus \hat{X}) = 0$. Then, $\int_X f\,\mathrm{d}\mu = \int_{\hat{X}} f|_{\hat{X}}\,\mathrm{d}\hat{\mu}$, whenever one of the integrals exists.*

Proof Let $\mathcal{I}(\mathcal{W})$ be the integration system on $\mathcal{W}$ as defined in Proposition 11.69, $(F_R)_{R\in\mathcal{I}(\mathcal{W})}$ be the net for $\int_X f\,\mathrm{d}\mu$, and $\left(\hat{F}_R\right)_{R\in\mathcal{I}(\mathcal{W})}$ be the net for $\int_{\hat{X}} f|_{\hat{X}}\,\mathrm{d}\hat{\mu}$, as defined in Definition 11.119. $\forall R := \{(w_\alpha, U_\alpha) \mid \alpha \in \Lambda\} \in \mathcal{I}(\mathcal{W})$, $\forall \alpha \in \Lambda$, we have $f_{\mathrm{inv}}(U_\alpha) \cap \hat{X} = f|_{\hat{X}\,\mathrm{inv}}(U_\alpha) \in \hat{\mathcal{B}} \subseteq \mathcal{B}$ since f is $\mathcal{B}$-measurable, and $0 \le \mathcal{P}\circ\mu(f_{\mathrm{inv}}(U_\alpha)\cap(X\setminus\hat{X})) \le \mathcal{P}\circ\mu(X\setminus\hat{X}) = 0$. This implies that $\mu(f_{\mathrm{inv}}(U_\alpha)) = \mu(f_{\mathrm{inv}}(U_\alpha) \cap \hat{X}) + \mu(f_{\mathrm{inv}}(U_\alpha) \cap (X \setminus \hat{X})) = \mu(f|_{\hat{X}\,\mathrm{inv}}(U_\alpha)) = \hat{\mu}(f|_{\hat{X}\,\mathrm{inv}}(U_\alpha))$ by Fact 11.110. Then, $F_R = \sum_{\alpha\in\Lambda} w_\alpha \mu(f_{\mathrm{inv}}(U_\alpha)) = \sum_{\alpha\in\Lambda} w_\alpha \hat{\mu}(f|_{\hat{X}\,\mathrm{inv}}(U_\alpha)) = \hat{F}_R \in \mathcal{Z}$. Hence, $\int_X f\,\mathrm{d}\mu = \lim_{R\in\mathcal{I}(\mathcal{W})} F_R = \lim_{R\in\mathcal{I}(\mathcal{W})} \hat{F}_R = \int_{\hat{X}} f|_{\hat{X}}\,\mathrm{d}\hat{\mu}$, whenever one of the integrals exists. This completes the proof of the proposition. □

Proposition 11.123 *Let $\mathcal{X} := (X, \mathcal{B}, \mu)$ be a $\mathcal{Y}$-valued measure space with $\mathcal{P} \circ \mu(X) = +\infty$, where $\mathcal{Y}$ is a normed linear space over $\mathbb{K}$, $\mathcal{Z}$ be a normed linear space over $\mathbb{K}$, $f : X \to \mathcal{W} := \mathrm{B}(\mathcal{Y}, \mathcal{Z})$ be $\mathcal{B}$-measurable, and $\bar{\mathcal{X}} := (X, \bar{\mathcal{B}}, \bar{\mu})$ be the completion of $\mathcal{X}$. Then, $\int_X f \, \mathrm{d}\mu = \int_X f \, \mathrm{d}\bar{\mu}$, whenever one of the integrals exists.*

Proof Let $\hat{\mathcal{X}} := (X, \mathcal{B}, \mathcal{P} \circ \mu =: \nu)$ and $\hat{\bar{\mathcal{X}}} := (X, \bar{\mathcal{B}}, \mathcal{P} \circ \bar{\mu} =: \bar{\nu})$. Then, $\hat{\mathcal{X}}$ and $\hat{\bar{\mathcal{X}}}$ are measure spaces and $\hat{\bar{\mathcal{X}}}$ is the completion of $\hat{\mathcal{X}}$ by Proposition 11.114. Let $(F_A)_{A \in \mathcal{M}(\mathcal{X})}$ be the net for $\int_X f \, \mathrm{d}\mu$ and $\left(\bar{F}_{\bar{A}}\right)_{\bar{A} \in \mathcal{M}\left(\bar{\mathcal{X}}\right)}$ be the net for $\int_X f \, \mathrm{d}\bar{\mu}$ as defined in Definition 11.120. Clearly, $\mathcal{M}(\mathcal{X}) \subseteq \mathcal{M}\left(\bar{\mathcal{X}}\right)$, since $\hat{\bar{\mathcal{X}}}$ is the completion of $\hat{\mathcal{X}}$. We will distinguish two exhaustive cases: Case 1: $\int_X f \, \mathrm{d}\bar{\mu}$ exists; Case 2: $\int_X f \, \mathrm{d}\mu$ exists.

Case 1: $\int_X f \, \mathrm{d}\bar{\mu}$ exists. Then, the net $\left(\bar{F}_{\bar{A}}\right)_{\bar{A} \in \mathcal{M}\left(\bar{\mathcal{X}}\right)}$ is well-defined. $\forall A \in \mathcal{M}(\mathcal{X})$, we have $A \in \mathcal{B}$ and $\nu(A) < +\infty$. Then, $A \in \mathcal{M}\left(\bar{\mathcal{X}}\right)$. By Propositions 11.121 and 11.114, $F_A = \int_A f|_A \, \mathrm{d}\mu_A = \int_A f|_A \, \mathrm{d}\bar{\mu}_A = \bar{F}_A$, where $(A, \mathcal{B}_A, \mu_A)$ is the finite $\mathcal{Y}$-valued measure subspace of $\mathcal{X}$ and $(A, \bar{\mathcal{B}}_A, \bar{\mu}_A)$ is the finite complete $\mathcal{Y}$-valued measure subspace of $\bar{\mathcal{X}}$. Then, $(F_A)_{A \in \mathcal{M}(\mathcal{X})}$ is well-defined. $\forall \bar{A} \in \mathcal{M}\left(\bar{\mathcal{X}}\right)$, we have $\bar{A} \in \bar{\mathcal{B}}$ with $\bar{\nu}(\bar{A}) < +\infty$. By Proposition 11.114 and its proof, $\exists \tilde{A}, \tilde{B}, \tilde{C} \subseteq X$ with $\bar{A} = \tilde{A} \cup \tilde{B}$, $\tilde{B}, \tilde{C} \in \mathcal{B}$, $\tilde{A} \subseteq \tilde{C}$, $\nu(\tilde{C}) = 0$, and $\bar{\nu}(\bar{A}) = \nu(\tilde{B})$. Then, $\bar{A} \subseteq \tilde{B} \cup \tilde{C} =: A \in \mathcal{B}$ and $\nu(A) \leq \nu(\tilde{B}) + \nu(\tilde{C}) = \nu(\tilde{B}) = \bar{\nu}(\bar{A}) < +\infty$. This shows that $A \in \mathcal{M}(\mathcal{X})$ and $\bar{A} \subseteq A$. Hence, $(F_A)_{A \in \mathcal{M}(\mathcal{X})}$ is a subnet of $\left(\bar{F}_{\bar{A}}\right)_{\bar{A} \in \mathcal{M}\left(\bar{\mathcal{X}}\right)}$. By Proposition 3.70, $\int_X f \, \mathrm{d}\mu = \lim_{A \in \mathcal{M}(\mathcal{X})} F_A = \lim_{\bar{A} \in \mathcal{M}\left(\bar{\mathcal{X}}\right)} \bar{F}_{\bar{A}} = \int_X f \, \mathrm{d}\bar{\mu}$.

Case 2: $\int_X f \, \mathrm{d}\mu$ exists. Then, the net $(F_A)_{A \in \mathcal{M}(\mathcal{X})}$ is well-defined. $\forall \bar{A} \in \mathcal{M}\left(\bar{\mathcal{X}}\right)$, we have $\bar{A} \in \bar{\mathcal{B}}$ with $\bar{\nu}(\bar{A}) < +\infty$. By Proposition 11.114 and its proof, $\exists \tilde{A}, \tilde{B}, \tilde{C} \subseteq X$ with $\bar{A} = \tilde{A} \cup \tilde{B}$, $\tilde{B}, \tilde{C} \in \mathcal{B}$, $\tilde{A} \subseteq \tilde{C}$, $\nu(\tilde{C}) = 0$, and $\bar{\nu}(\bar{A}) = \nu(\tilde{B})$. Let $\tilde{B} =: A \in \mathcal{B}$. Then, $0 \leq \bar{\nu}(\bar{A} \setminus A) = \bar{\nu}(\bar{A} \setminus \tilde{B}) \leq \bar{\nu}(\tilde{C}) = \nu(\tilde{C}) = 0$ and $\nu(A) = \bar{\nu}(\bar{A}) < +\infty$. This shows that $A \in \mathcal{M}(\mathcal{X})$ and $A \subseteq \bar{A}$. Let $\mathcal{A} := (A, \mathcal{B}_A, \mu_A)$ be the finite $\mathcal{Y}$-valued measure subspace of $\mathcal{X}$, $\bar{\mathcal{A}} := (\bar{A}, \bar{\mathcal{B}}_{\bar{A}}, \bar{\mu}_{\bar{A}})$ be the finite complete $\mathcal{Y}$-valued measure subspace of $\bar{\mathcal{X}}$, and $\hat{\mathcal{A}} := (A, \bar{\mathcal{B}}_A, \bar{\mu}_A)$ be the finite complete $\mathcal{Y}$-valued measure subspace of $\bar{\mathcal{A}}$. Since $\bar{\mathcal{X}}$ is the completion of $\mathcal{X}$, then, by Proposition 11.115, $\hat{\mathcal{A}}$ is the completion of $\mathcal{A}$. By Proposition 11.121, we have $\int_A f|_A \, \mathrm{d}\bar{\mu}_A = \int_A f|_A \, \mathrm{d}\mu_A = F_A$. By Lemma 11.122, we have $\bar{F}_{\bar{A}} = \int_{\bar{A}} f|_{\bar{A}} \, \mathrm{d}\bar{\mu}_{\bar{A}} = \int_A f|_A \, \mathrm{d}\bar{\mu}_A = F_A$. Hence, $\bar{F}_{\bar{A}} = F_A$. Then, the net $\left(\bar{F}_{\bar{A}}\right)_{\bar{A} \in \mathcal{M}\left(\bar{\mathcal{X}}\right)}$ is well-defined.

Fix any open set U ($U \subseteq \mathbb{R}_e$ if $\mathcal{Z} = \mathbb{R}$ or $U \subseteq \mathcal{Z}$ if $\mathcal{Z} \neq \mathbb{R}$) with $\int_X f \, \mathrm{d}\mu \in U$. Since $\lim_{A \in \mathcal{M}(\mathcal{X})} F_A = \int_X f \, \mathrm{d}\mu \in U$, then $\exists A_0 \in \mathcal{M}(\mathcal{X})$ such that $\forall A \in \mathcal{M}(\mathcal{X})$ with $A_0 \subseteq A$, we have $F_A \in U$. Note that $A_0 \in \mathcal{B} \subseteq \bar{\mathcal{B}}$ and $\nu(A_0) = \bar{\nu}(A_0) < +\infty$. Then, $A_0 \in \mathcal{M}\left(\bar{\mathcal{X}}\right)$. $\forall \bar{A} \in \mathcal{M}\left(\bar{\mathcal{X}}\right)$ with $A_0 \subseteq \bar{A}$, then $\bar{A} \in \bar{\mathcal{B}}$ and $\bar{\nu}(\bar{A}) < +\infty$.

By Proposition 11.114 and its proof, $\exists \tilde{A}, \tilde{B}, \tilde{C} \subseteq X$ with $\bar{A} = \tilde{A} \cup \tilde{B}$, $\tilde{B}, \tilde{C} \in \mathcal{B}$, $\tilde{A} \subseteq \tilde{C}$, $\nu(\tilde{C}) = 0$, and $\bar{\nu}(\bar{A}) = \nu(\tilde{B})$. Let $A := A_0 \cup \tilde{B} \in \mathcal{B}$. Then, $A_0 \subseteq A \subseteq \bar{A}$, $0 \leq \bar{\nu}(\bar{A} \setminus A) = \bar{\nu}((\tilde{A} \cup \tilde{B}) \setminus (A_0 \cup \tilde{B})) \leq \bar{\nu}((\tilde{A} \cup \tilde{B}) \setminus \tilde{B}) \leq \bar{\nu}(\tilde{A} \setminus \tilde{B}) \leq \bar{\nu}(\tilde{C}) = \nu(\tilde{C}) = 0$, and $\nu(A) \leq \nu(\tilde{B}) + \nu(A_0) = \bar{\nu}(\bar{A}) + \nu(A_0) < +\infty$. This shows that $A \in \mathcal{M}(\mathcal{X})$. Let $\mathcal{A} := (A, \mathcal{B}_A, \mu_A)$ be the finite $\mathcal{Y}$-valued measure subspace of $\mathcal{X}$, $\bar{\mathcal{A}} := (\bar{A}, \bar{\mathcal{B}}_{\bar{A}}, \bar{\mu}_{\bar{A}})$ be the finite complete $\mathcal{Y}$-valued measure subspace of $\bar{\mathcal{X}}$, and $\hat{\mathcal{A}} := (A, \bar{\mathcal{B}}_A, \bar{\mu}_A)$ be the finite complete $\mathcal{Y}$-valued measure subspace of $\bar{\mathcal{A}}$. Since $\bar{\mathcal{X}}$ is the completion of $\mathcal{X}$, then, by Proposition 11.115, $\hat{\mathcal{A}}$ is the completion of $\mathcal{A}$. By Proposition 11.121, we have $\int_A f|_A \, d\bar{\mu}_A = \int_A f|_A \, d\mu_A = F_A$. By Lemma 11.122, we have $\bar{F}_{\bar{A}} = \int_{\bar{A}} f|_{\bar{A}} \, d\bar{\mu}_{\bar{A}} = \int_A f|_A \, d\bar{\mu}_A = F_A \in U$. Therefore, we have $\int_X f \, d\bar{\mu} = \lim_{\bar{A} \in \mathcal{M}(\bar{\mathcal{X}})} \bar{F}_{\bar{A}} = \int_X f \, d\mu$.

Hence, in both cases, we have $\int_X f \, d\mu = \int_X f \, d\bar{\mu}$. This completes the proof of the proposition. □

Definition 11.124 Let $\mathcal{X} := (X, \mathcal{B}, \mu)$ be a $\mathcal{Y}$-valued measure space, where $\mathcal{Y}$ is a normed linear space over $\mathbb{K}$, and $\mathcal{Z}$ be a normed linear space over $\mathbb{K}$. $\phi : X \to \mathcal{Z}$ is said to be a *simple function* if $\exists n \in \mathbb{Z}_+$, $\exists z_1, \ldots, z_n \in \mathcal{Z}$, and $\exists A_1, \ldots, A_n \in \mathcal{B}$ with $\mathcal{P} \circ \mu(A_i) < +\infty$, $i = 1, \ldots, n$, such that $\phi(x) = \sum_{i=1}^{n} z_i \chi_{A_i,X}(x)$, $\forall x \in X$. We will say that a simple function ϕ is in *canonical representation* if $z_1, \ldots, z_n$ are distinct and none equals to $\vartheta_{\mathcal{Z}}$, and $A_1, \ldots, A_n$ are nonempty and pairwise disjoint. ◇

Clearly, every simple function admits a unique canonical representation.

Proposition 11.125 *Let $\mathcal{X} := (X, \mathcal{B}, \mu)$ be a $\mathcal{Y}$-valued measure space, where $\mathcal{Y}$ is a normed linear space over $\mathbb{K}$, $\mathcal{Z}$ be a normed linear space over $\mathbb{K}$, $\phi : X \to \mathrm{B}(\mathcal{Y}, \mathcal{Z}) =: \mathcal{W}$ be a simple function in canonical representation, i.e., $\exists n \in \mathbb{Z}_+$, $\exists w_1, \ldots, w_n \in \mathcal{W}$, which are distinct and none equals to $\vartheta_{\mathcal{W}}$, $\exists A_1, \ldots, A_n \in \mathcal{B}$, which are nonempty, pairwise disjoint, and $\mathcal{P} \circ \mu(A_i) < +\infty$, $i = 1, \ldots, n$, such that $\phi(x) = \sum_{i=1}^{n} w_i \chi_{A_i,X}(x)$, $\forall x \in X$. Then, $\int_X \phi \, d\mu = \sum_{i=1}^{n} w_i \mu(A_i) =: I \in \mathcal{Z}$ and $0 \leq \left\| \int_X \phi \, d\mu \right\| \leq \int_X \mathcal{P} \circ \phi \, d\mathcal{P} \circ \mu < +\infty$.*

Proof We will distinguish two exhaustive and mutually exclusive cases: Case 1: $\mathcal{P} \circ \mu(X) < +\infty$; Case 2: $\mathcal{P} \circ \mu(X) = +\infty$.

Case 1: $\mathcal{P} \circ \mu(X) < +\infty$. Let $w_{n+1} := \vartheta_{\mathcal{W}}$ and $A_{n+1} := X \setminus (\bigcup_{i=1}^{n} A_i) \in \mathcal{B}$. Let $\epsilon_0 := \min\{1, \min_{1 \leq i < j \leq n+1} \left\| w_i - w_j \right\|\} > 0$, by the assumption. Let $(\Phi_R)_{R \in \mathcal{I}(\mathcal{W})}$ be the net as defined in Definition 11.119 for $\int_X \phi \, d\mu$.

$\forall \epsilon \in (0, \epsilon_0/2) \subset \mathbb{R}$, let $U_i := \mathcal{B}_{\mathcal{W}}(w_i, \epsilon) \in \mathcal{B}_{\mathrm{B}}(\mathcal{W})$, $i = 1, \ldots, n+1$, which are clearly pairwise disjoint and nonempty. We will distinguish two exhaustive and mutually exclusive subcases: Case 1a: $\mathcal{W} = \bigcup_{i=1}^{n+1} U_i$; Case 1b: $\mathcal{W} \supset \bigcup_{i=1}^{n+1} U_i$.

Case 1a: $\mathcal{W} = \bigcup_{i=1}^{n+1} U_i$. Then, $\exists \delta \in (0, \infty) \subset \mathbb{R}$ such that $\mathcal{W} = \mathcal{B}_{\mathcal{W}}(\vartheta_{\mathcal{W}}, \delta)$. Then, $\mathcal{W}$ must be the trivial normed linear space, i.e., $\mathcal{W}$ is a singleton set containing $\vartheta_{\mathcal{W}}$. There is a single representation of $\mathcal{W}$, which is $R_0 := \{(\vartheta_{\mathcal{W}}, \mathcal{W})\}$. Then, $\Phi_{R_0} = \vartheta_{\mathcal{Z}}$. Clearly, we must have $n = 0$, since $w_1, \ldots, w_n \in \mathcal{Y}$ are distinct and none equals to $\vartheta_{\mathcal{W}}$. Then, $I = \vartheta_{\mathcal{Z}}$. Hence, $\forall R \in \mathcal{J}(\mathcal{W})$ with $R_0 \preceq R$, $\|\Phi_R - I\| = \|\Phi_{R_0} - I\| = 0$.

Case 1b: $\mathcal{W} \supset \bigcup_{i=1}^{n+1} U_i$. Let $U_{n+2} = \mathcal{W} \setminus (\bigcup_{i=1}^{n+1} U_i) \in \mathcal{B}_{\mathrm{B}}(\mathcal{W})$, which is nonempty. Let $w_{n+2} \in U_{n+2}$ be such that $\|w_{n+2}\| < \inf_{w \in U_{n+2}} \|w\| + 1$. Let $R_0 := \{(w_i, U_i) \mid i = 1, \ldots, n+2\}$. It is easy to check that $R_0 \in \mathcal{J}(\mathcal{W})$. $\forall \tilde{R} \in \mathcal{J}(\mathcal{W})$ with $R_0 \preceq \tilde{R}$, then $\tilde{R} = \{(\tilde{w}_\alpha, \tilde{U}_\alpha) \mid \alpha \in \Lambda\}$, where Λ is a finite index set, $\left(\tilde{U}_\alpha\right)_{\alpha \in \Lambda} \subseteq \mathcal{B}_{\mathrm{B}}(\mathcal{W})$ are nonempty and pairwise disjoint, $\bigcup_{\alpha \in \Lambda} \tilde{U}_\alpha = \mathcal{W}$, $\tilde{w}_\alpha \in \tilde{U}_\alpha$, and $\|\tilde{w}_\alpha\| < \inf_{w \in \tilde{U}_\alpha} \|w\| + 1$, $\forall \alpha \in \Lambda$. $\forall i \in \{1, \ldots, n+2\}$, by $R_0 \preceq \tilde{R}$, $\exists!\, \alpha_i \in \Lambda$ such that $w_i = \tilde{w}_{\alpha_i}$ and, by $\tilde{R} \in \mathcal{J}(\mathcal{W})$, $\exists!\, \bar{\alpha}_i \in \Lambda$ such that $w_i \in \tilde{U}_{\bar{\alpha}_i}$. Since $\left(\tilde{U}_\alpha\right)_{\alpha \in \Lambda} \subseteq \mathcal{B}_{\mathrm{B}}(\mathcal{W})$ are nonempty and pairwise disjoint and $\tilde{w}_\alpha \in \tilde{U}_\alpha$, $\forall \alpha \in \Lambda$, then $\alpha_i = \bar{\alpha}_i$. Again by $R_0 \preceq \tilde{R}$, we have $\tilde{U}_{\alpha_i} \subseteq U_i$. Hence, $\alpha_1, \ldots, \alpha_{n+2}$ are distinct. Let $\bar{\Lambda} := \{\alpha_1, \ldots, \alpha_{n+2}\}$. $\forall \alpha \in \Lambda \setminus \bar{\Lambda}$, $\phi_{\mathrm{inv}}(\tilde{U}_\alpha) = \emptyset$. Then, $\Phi_{\tilde{R}} = \sum_{\alpha \in \Lambda} \tilde{w}_\alpha \mu(\phi_{\mathrm{inv}}(\tilde{U}_\alpha)) = \sum_{i=1}^{n+2} \tilde{w}_{\alpha_i} \mu(\phi_{\mathrm{inv}}(\tilde{U}_{\alpha_i})) = \sum_{i=1}^{n+2} w_i \mu(\phi_{\mathrm{inv}}(\tilde{U}_{\alpha_i}))$. Note that $\forall i \in \{1, \ldots, n+1\}$, $\phi_{\mathrm{inv}}(\tilde{U}_{\alpha_i}) = A_i$ and $\mu(A_i) \in \mathcal{Y}$. Furthermore, $\phi_{\mathrm{inv}}(\tilde{U}_{\alpha_{n+2}}) = \emptyset$. Hence, $\Phi_{\tilde{R}} = \sum_{i=1}^{n} w_i \mu(A_i) = I \in \mathcal{Z}$. Then, we have $\left\|\Phi_{\tilde{R}} - I\right\| = 0$.

In both subcases, we have $\exists R_0 \in \mathcal{J}(\mathcal{W})$, $\forall R \in \mathcal{J}(\mathcal{W})$ with $R_0 \preceq R$, we have $\|\Phi_R - I\| = 0 < \epsilon$. Hence, $\int_X \phi \,\mathrm{d}\mu = \lim_{R \in \mathcal{J}(\mathcal{W})} \Phi_R = I$. Note that $0 \leq \left\|\int_X \phi \,\mathrm{d}\mu\right\| \leq \sum_{i=1}^{n} \|w_i\| \|\mu(A_i)\| \leq \sum_{i=1}^{n} \|w_i\| \mathcal{P} \circ \mu(A_i) = \int_X \mathcal{P} \circ \phi \,\mathrm{d}\mathcal{P} \circ \mu < +\infty$, where the second inequality follows from Proposition 7.64, the third inequality follows from Definition 11.108, the equality follows from what we have proved in this proposition, and the last inequality follows from Definition 11.124.

Case 2: $\mathcal{P} \circ \mu(X) = +\infty$. $\forall A \in \mathcal{M}(\mathcal{X})$, $\mathcal{P} \circ \mu(A) < +\infty$. $\phi|_A : A \to \mathcal{W}$ is a simple function. By Case 1, $\Phi_A = \int_A \phi|_A \,\mathrm{d}\mu_A \in \mathcal{Z}$ is well-defined, where $(A, \mathcal{B}_A, \mu_A)$ is the finite $\mathcal{Y}$-valued measure subspace of $\mathcal{X}$. Hence, the net $(\Phi_A)_{A \in \mathcal{M}(\mathcal{X})}$ is well-defined. Take $A_0 := \bigcup_{i=1}^{n} A_i \in \mathcal{B}$. Then, $\mathcal{P} \circ \mu(A_0) = \sum_{i=1}^{n} \mathcal{P} \circ \mu(A_i) < +\infty$. Therefore, $A_0 \in \mathcal{M}(\mathcal{X})$. $\forall A \in \mathcal{M}(\mathcal{X})$ with $A_0 \subseteq A$, by Case 1, $\Phi_A = \sum_{i=1}^{n} w_i \mu(A_i) = I \in \mathcal{Z}$. Then, $\int_X \phi \,\mathrm{d}\mu = \lim_{A \in \mathcal{M}(\mathcal{X})} \Phi_A = I \in \mathcal{Z}$. Note that $0 \leq \left\|\int_X \phi \,\mathrm{d}\mu\right\| \leq \sum_{i=1}^{n} \|w_i\| \|\mu(A_i)\| \leq \sum_{i=1}^{n} \|w_i\| \mathcal{P} \circ \mu(A_i) = \int_X \mathcal{P} \circ \phi \,\mathrm{d}\mathcal{P} \circ \mu < +\infty$, where the second inequality follows from Proposition 7.64, the third inequality follows from Definition 11.108, the equality follows from what we have proved in this proposition, and the last inequality follows from Definition 11.124.

This completes the proof of the proposition. □

Clearly, the above result holds for simple functions given in any form, not necessarily in the canonical representation. It also shows that the integrals of simple functions are linear.

Lemma 11.126 *Let $\mathcal{X} := (X, \mathcal{B}, \mu)$ be a finite $\mathcal{Y}$-valued measure space, where $\mathcal{Y}$ is a normed linear space over $\mathbb{K}$, $\mathcal{Z}$ be a Banach space over $\mathbb{K}$, $\mathcal{W}$ be a separable subspace of $\mathrm{B}(\mathcal{Y}, \mathcal{Z})$, $\phi_n : X \to \mathcal{W}$ be a simple function, $\forall n \in \mathbb{N}$, $f : X \to \mathcal{W}$ be $\mathcal{B}$-measurable, $g_n : X \to [0, \infty) \subset \mathbb{R}$ be $\mathcal{B}$-measurable, $\forall n \in \mathbb{N}$, and $g : X \to [0, \infty) \subset \mathbb{R}$ be $\mathcal{B}$-measurable, $\bar{\mathcal{W}} :=$*

$\overline{\text{span}(\{z \in \mathcal{Z} \mid \exists y \in \mathcal{Y}, \exists w \in \mathcal{W} \ni \cdot z = wy\})} \subseteq \mathcal{Z}$ *be the Banach subspace of* $\mathcal{Z}$. *Assume that:*

(i) $\lim_{n\in\mathbb{N}} \phi_n = f$ *a.e. in* $\mathcal{X}$ *and* $\lim_{n\in\mathbb{N}} g_n = g$ *a.e. in* $\mathcal{X}$.
(ii) $\|\phi_n(x)\| \leq g_n(x)$, $\forall x \in X$, $\forall n \in \mathbb{N}$.
(iii) $\int_X g_n \, d\mathcal{P} \circ \mu \in \mathbb{R}$, $\forall n \in \mathbb{N}$, *and* $\lim_{n\in\mathbb{N}} \int_X g_n \, d\mathcal{P} \circ \mu = \int_X g \, d\mathcal{P} \circ \mu \in \mathbb{R}$.

Then, f *is integrable over* $\mathcal{X}$, $\int_X f \, d\mu = \lim_{n\in\mathbb{N}} \int_X \phi_n \, d\mu \in \bar{\mathcal{W}} \subseteq \mathcal{Z}$, *and* $0 \leq \left\|\int_X f \, d\mu\right\| \leq \int_X g \, d\mathcal{P} \circ \mu < +\infty$.

Proof Let $\bar{\mathcal{X}} := (X, \mathcal{B}, \mathcal{P} \circ \mu)$ be the finite measure space as defined in Definition 11.108. Let

$$E := \big\{x \in X \mid (\phi_n(x))_{n=1}^{\infty} \text{ does not converge to } f(x) \text{ or } (g_n(x))_{n=1}^{\infty} \text{ does not converge to } g(x)\big\}$$

Then, by (i), $E \in \mathcal{B}$ and $\mathcal{P} \circ \mu(E) = 0$. Let $(F_R)_{R\in\mathcal{I}(\mathcal{W})}$ be the net for $\int_X f \, d\mu$ as defined in Definition 11.119.

$\forall \epsilon \in (0, \infty) \subset \mathbb{R}$, by (i), (iii), and Lemma 11.85, $\exists \delta \in (0, \epsilon/6] \subset \mathbb{R}$ and $\exists n_0 \in \mathbb{N}$ such that $\forall A \in \mathcal{B}$ with $\mathcal{P}\circ\mu(A) < \delta$, $\forall n \in \mathbb{N}$ with $n \geq n_0$, we have $0 \leq \int_A g \, d\mathcal{P}\circ\mu < \epsilon/6$ and $0 \leq \int_A g_n \, d\mathcal{P} \circ \mu < \epsilon/6$. By Egoroff's Theorem 11.55, $\exists \bar{E}_1 \in \mathcal{B}$ with $\mathcal{P} \circ \mu(\bar{E}_1) < \delta$ such that $\left(\phi_n|_{X\setminus\bar{E}_1}\right)_{n=1}^{\infty}$ converges uniformly to $f|_{X\setminus\bar{E}_1}$. Let $E_1 := E \cup \bar{E}_1$. Then, $E_1 \in \mathcal{B}$ with $\mathcal{P} \circ \mu(E_1) = \mathcal{P} \circ \mu(\bar{E}_1) < \delta$ such that $\left(\phi_n|_{X\setminus E_1}\right)_{n=1}^{\infty}$ converges uniformly to $f|_{X\setminus E_1}$. Then, $\exists n_1 \in \mathbb{N}$ with $n_1 \geq n_0$, $\forall n \in \mathbb{N}$ with $n \geq n_1$, $\forall x \in X \setminus E_1$, $\|\phi_n(x) - f(x)\| < \frac{\epsilon}{6\mathcal{P}\circ\mu(X)+1} =: \bar{\epsilon} \in (0, \infty) \subset \mathbb{R}$. Fix any $n \geq n_1$. Let ϕ_n admit the canonical representation $\phi_n = \sum_{i=1}^{\bar{n}} w_i \chi_{A_i,X}$, where $\bar{n} \in \mathbb{Z}_+$, $w_1, \ldots, w_{\bar{n}} \in \mathcal{W}$ are distinct and none equals to $\vartheta_{\mathcal{W}}$, and $A_1, \ldots, A_{\bar{n}} \in \mathcal{B}$ are nonempty, pairwise disjoint, and $\mathcal{P} \circ \mu(A_i) < +\infty$, $i = 1, \ldots, \bar{n}$. Let $w_{\bar{n}+1} := \vartheta_{\mathcal{W}}$ and $A_{\bar{n}+1} := X \setminus (\bigcup_{i=1}^{\bar{n}} A_i) \in \mathcal{B}$. Then, $X = \bigcup_{i=1}^{\bar{n}+1} A_i$, and the sets in the union are pairwise disjoint and of finite total variation. Define $\bar{V}_i := \mathcal{B}_{\mathcal{W}}(w_i, \bar{\epsilon})$, $i = 1, \ldots, \bar{n} + 1$. Define $V_1 := \bar{V}_1 \in \mathcal{B}_{\mathrm{B}}(\mathcal{W})$, $V_{i+1} := \bar{V}_{i+1} \setminus (\bigcup_{j=1}^{i} \bar{V}_j) \in \mathcal{B}_{\mathrm{B}}(\mathcal{W})$, $i = 1, \ldots, \bar{n}$. Let $V_{\bar{n}+2} := \mathcal{W} \setminus (\bigcup_{j=1}^{\bar{n}+1} \bar{V}_j) \in \mathcal{B}_{\mathrm{B}}(\mathcal{W})$. Clearly, we may form a representation $\bar{R} := \big\{(\bar{w}_i, V_i) \,\big|\, i = 1, \ldots, \bar{n} + 2, V_i \neq \emptyset\big\} \in \mathcal{I}(\mathcal{W})$, where $\bar{w}_i \in V_i$ are any vectors that satisfy the assumption of Proposition 11.69. $\forall R \in \mathcal{I}(\mathcal{W})$ with $\bar{R} \preceq R$. Let $R := \{(\hat{w}_\gamma, \hat{U}_\gamma) \mid \gamma \in \Gamma\}$. $\forall \gamma \in \Gamma$, by $\bar{R} \preceq R$, $\exists! \, i_\gamma \in \{1, \ldots, \bar{n} + 2\}$ such that $\hat{U}_\gamma \subseteq V_{i_\gamma}$. Define $\bar{\Gamma} := \{\gamma \in \Gamma \mid i_\gamma = \bar{n}+2\}$. Let $A_{\gamma,i} := A_i \cap f_{\text{inv}}(\hat{U}_\gamma) \cap (X \setminus E_1) \in \mathcal{B}$, $i = 1, \ldots, \bar{n} + 1$. Let $\bar{A}_\gamma := f_{\text{inv}}(\hat{U}_\gamma) \cap E_1 \setminus E$.

Claim 11.126.1 $\forall \gamma \in \bar{\Gamma}$, $f_{\text{inv}}(\hat{U}_\gamma) \subseteq E_1$; $\forall \gamma \in \Gamma_s := \{\gamma \in \Gamma \mid \bar{A}_\gamma \neq \emptyset\}$, $\left\|\hat{w}_\gamma \mu(\bar{A}_\gamma)\right\| \leq (1+\inf_{x\in\bar{A}_\gamma} g(x))\mathcal{P}\circ\mu(\bar{A}_\gamma)$; and $\left\|\sum_{\gamma\in\Gamma} \hat{w}_\gamma \mu(f_{\text{inv}}(\hat{U}_\gamma) \cap E_1)\right\| < \epsilon/3$.

Proof of Claim Fix any $\gamma \in \bar{\Gamma}$. $\hat{U}_\gamma \subseteq V_{\bar{n}+2} = \mathcal{W} \setminus (\bigcup_{j=1}^{\bar{n}+1} \bar{V}_j)$. Then, $\mathcal{B} \ni f_{\text{inv}}(\hat{U}_\gamma) \subseteq E_1$.

$\forall \gamma \in \Gamma_s$, $\bar{A}_\gamma = f_{\text{inv}}(\hat{U}_\gamma) \cap (E_1 \setminus E) \neq \emptyset$. $\forall x \in \bar{A}_\gamma$, we have $f(x) \in \hat{U}_\gamma$, $\lim_{m\in\mathbb{N}} \phi_m(x) = f(x)$, $\|\phi_m(x)\| \leq g_m(x)$, $\forall m \in \mathbb{N}$, and $\lim_{m\in\mathbb{N}} g_m(x) = g(x)$. Then, by Propositions 7.21 and 3.66, we have $\|f(x)\| = \lim_{m\in\mathbb{N}} \|\phi_m(x)\| < \lim_{m\in\mathbb{N}} g_m(x) = g(x)$. Then, by $R \in \mathfrak{I}(\mathcal{W})$, $\left\|\hat{w}_\gamma\right\| < \inf_{w\in\hat{U}_\gamma} \|w\| + 1 \leq \|f(x)\|+1 \leq g(x)+1$. By the arbitrariness of x, we have $\left\|\hat{w}_\gamma\right\| \leq \inf_{x\in\bar{A}_\gamma} g(x)+1$. Therefore, $\left\|\hat{w}_\gamma \mu(\bar{A}_\gamma)\right\| \leq \left\|\mu(\bar{A}_\gamma)\right\|(\inf_{x\in\bar{A}_\gamma} g(x) + 1) \leq (1 + \inf_{x\in\bar{A}_\gamma} g(x))\mathcal{P} \circ \mu(\bar{A}_\gamma)$.

Note that

$$\begin{aligned}
&\left\| \sum_{\gamma\in\Gamma} \hat{w}_\gamma \mu(f_{\text{inv}}(\hat{U}_\gamma) \cap E_1)\right\| = \left\| \sum_{\gamma\in\Gamma} \hat{w}_\gamma (\mu(\bar{A}_\gamma) + \mu(f_{\text{inv}}(\hat{U}_\gamma) \cap E_1 \cap E))\right\| \\
&\leq \left\| \sum_{\gamma\in\Gamma} \hat{w}_\gamma \mu(\bar{A}_\gamma)\right\| + \sum_{\gamma\in\Gamma} \|\hat{w}_\gamma\| \left\|\mu(f_{\text{inv}}(\hat{U}_\gamma) \cap E_1 \cap E)\right\| \\
&\leq \left\| \sum_{\gamma\in\Gamma_s} \hat{w}_\gamma \mu(\bar{A}_\gamma)\right\| + \sum_{\gamma\in\Gamma} \left\|\hat{w}_\gamma\right\| \mathcal{P} \circ \mu(f_{\text{inv}}(\hat{U}_\gamma) \cap E_1 \cap E) \\
&\leq \sum_{\gamma\in\Gamma_s} (1 + \inf_{x\in\bar{A}_\gamma} g(x))\mathcal{P} \circ \mu(\bar{A}_\gamma) \\
&= \sum_{\gamma\in\Gamma_s} \mathcal{P} \circ \mu(\bar{A}_\gamma) + \sum_{\gamma\in\Gamma_s} (\inf_{x\in\bar{A}_\gamma} g(x))\mathcal{P} \circ \mu(\bar{A}_\gamma)
\end{aligned}$$

The simple function $\psi : X \to [0, \infty) \subset \mathbb{R}$, defined by $\psi(x) = \sum_{\gamma\in\Gamma_s} (\inf_{\bar{x}\in\bar{A}_\gamma} g(\bar{x})) \cdot \chi_{\bar{A}_\gamma,X}(x)$, $\forall x \in X$, satisfies $0 \leq \psi(x) \leq g(x)$, $\forall x \in X$ and $\psi(x) = 0$, $\forall x \in X \setminus E_1$. Then, $\sum_{\gamma\in\Gamma_s} (\inf_{x\in\bar{A}_\gamma} g(x))\mathcal{P} \circ \mu(\bar{A}_\gamma) = \int_X \psi \, d\mathcal{P} \circ \mu = \int_{E_1} \psi \, d\mathcal{P} \circ \mu \leq \int_{E_1} g \, d\mathcal{P} \circ \mu < \epsilon/6$, where the equalities follow from Proposition 11.75 and the first inequality follows from Proposition 11.83. Then, $\left\|\sum_{\gamma\in\Gamma} \hat{w}_\gamma \mu(f_{\text{inv}}(\hat{U}_\gamma) \cap E_1)\right\| < \mathcal{P} \circ \mu(E_1) + \epsilon/6 < \delta + \epsilon/6 \leq \epsilon/3$.

This completes the proof of the claim. □

Claim 11.126.2 $\left\|\sum_{i=1}^{\bar{n}+1} w_i \mu(A_i \cap E_1)\right\| < \epsilon/6$.

Proof of Claim Since $\|\phi_n(x)\| \leq g_n(x)$, $\forall x \in X$, then $\|w_i\| \leq g_n(x)$, $\forall x \in A_i$, $i = 1, \ldots, \bar{n} + 1$. Then, by Definition 11.79, $\left\|\sum_{i=1}^{\bar{n}+1} w_i \mu(A_i \cap E_1)\right\| = \left\|\int_{E_1} \phi \, d\mu\right\| \leq \int_{E_1} \|\phi_n(x)\| \, d\mathcal{P} \circ \mu(x) \leq \int_{E_1} g_n \, d\mathcal{P} \circ \mu < \epsilon/6$. □

Claim 11.126.3 $\forall \gamma \in \Gamma \setminus \bar{\Gamma}$, $\forall i \in \{1, \ldots, \bar{n} + 1\}$, we have $\left\|\hat{w}_\gamma - w_i\right\| \cdot \left\|\mu(A_{\gamma,i})\right\| \leq \frac{3\epsilon}{6\mathcal{P}\circ\mu(X)+1} \mathcal{P} \circ \mu(A_{\gamma,i})$.

Proof of Claim $\forall \gamma \in \Gamma \setminus \bar{\Gamma}$, $\forall i \in \{1, \ldots, \bar{n}+1\}$, $\hat{U}_\gamma \subseteq V_{i_\gamma}$, where $i_\gamma \in \{1, \ldots, \bar{n}+1\}$. The result is trivial if $A_{\gamma,i} = \emptyset$. If $A_{\gamma,i} \neq \emptyset$, then $\exists x \in A_{\gamma,i} = A_i \cap f_{\text{inv}}(\hat{U}_\gamma) \cap (X \setminus E_1)$. This implies that $\phi_n(x) = w_i$ and $f(x) \in \mathcal{B}_{\mathcal{W}}(w_i, \bar{\epsilon}) \cap \hat{U}_\gamma \subseteq \mathcal{B}_{\mathcal{W}}(w_i, \bar{\epsilon}) \cap V_{i_\gamma} \subseteq \mathcal{B}_{\mathcal{W}}(w_i, \bar{\epsilon}) \cap \bar{V}_{i_\gamma} = \mathcal{B}_{\mathcal{W}}(w_i, \bar{\epsilon}) \cap \mathcal{B}_{\mathcal{W}}\big(w_{i_\gamma}, \bar{\epsilon}\big)$. Note that $\hat{w}_\gamma \in \hat{U}_\gamma \subseteq \mathcal{B}_{\mathcal{W}}\big(w_{i_\gamma}, \bar{\epsilon}\big)$. Therefore, we have $\big\| \hat{w}_\gamma - w_i \big\| \leq \big\| \hat{w}_\gamma - w_{i_\gamma} \big\| + \big\| w_{i_\gamma} - f(x) \big\| + \| f(x) - w_i \| < 3\bar{\epsilon} = \frac{3\epsilon}{6\mathcal{P} \circ \mu(X)+1}$. Hence, the result holds. □

Note that $F_R = \sum_{\gamma \in \Gamma} \hat{w}_\gamma \mu(f_{\text{inv}}(\hat{U}_\gamma))$ and $\int_X \phi_n \, \mathrm{d}\mu = \sum_{i=1}^{\bar{n}} w_i \mu(A_i) = \sum_{i=1}^{\bar{n}+1} w_i \mu(A_i)$, by Proposition 11.125. Then,

$$
\begin{aligned}
&\Big\| F_R - \int_X \phi_n \, \mathrm{d}\mu \Big\| = \Big\| \sum_{\gamma \in \Gamma} \hat{w}_\gamma \mu(f_{\text{inv}}(\hat{U}_\gamma)) - \sum_{i=1}^{\bar{n}} w_i \mu(A_i) \Big\| \\
&= \Big\| \sum_{\gamma \in \Gamma} \hat{w}_\gamma (\mu(f_{\text{inv}}(\hat{U}_\gamma) \cap (X \setminus E_1)) + \mu(f_{\text{inv}}(\hat{U}_\gamma) \cap E_1)) \\
&\quad - \sum_{i=1}^{\bar{n}+1} w_i (\mu(A_i \cap E_1) + \mu(A_i \cap (X \setminus E_1))) \Big\| \\
&= \Big\| \sum_{\gamma \in \Gamma} \hat{w}_\gamma (\sum_{i=1}^{\bar{n}+1} \mu(A_{\gamma,i}) + \mu(f_{\text{inv}}(\hat{U}_\gamma) \cap E_1)) - \sum_{i=1}^{\bar{n}+1} w_i (\mu(A_i \cap E_1) \\
&\quad + \sum_{\gamma \in \Gamma} \mu(A_{\gamma,i})) \Big\| \\
&= \Big\| \sum_{\gamma \in \Gamma} \sum_{i=1}^{\bar{n}+1} \hat{w}_\gamma \mu(A_{\gamma,i}) + \sum_{\gamma \in \Gamma} \hat{w}_\gamma \mu(f_{\text{inv}}(\hat{U}_\gamma) \cap E_1) \\
&\quad - \sum_{i=1}^{\bar{n}+1} w_i \mu(A_i \cap E_1) - \sum_{\gamma \in \Gamma} \sum_{i=1}^{\bar{n}+1} w_i \mu(A_{\gamma,i}) \Big\| \\
&= \Big\| \sum_{\gamma \in \Gamma} \sum_{i=1}^{\bar{n}+1} (\hat{w}_\gamma - w_i) \mu(A_{\gamma,i}) + \sum_{\gamma \in \Gamma} \hat{w}_\gamma \mu(f_{\text{inv}}(\hat{U}_\gamma) \cap E_1) \\
&\quad - \sum_{i=1}^{\bar{n}+1} w_i \mu(A_i \cap E_1) \Big\| \\
&< \Big\| \sum_{\gamma \in \Gamma} \sum_{i=1}^{\bar{n}+1} (\hat{w}_\gamma - w_i) \mu(A_{\gamma,i}) \Big\| + \frac{\epsilon}{3} + \frac{\epsilon}{6}
\end{aligned}
$$

$$= \frac{\epsilon}{2} + \Big\| \sum_{\gamma \in \Gamma \setminus \bar{\Gamma}} \sum_{i=1}^{\bar{n}+1} (\hat{w}_\gamma - w_i)\mu(A_{\gamma,i}) + \sum_{\gamma \in \bar{\Gamma}} \sum_{i=1}^{\bar{n}+1} (\hat{w}_\gamma - w_i)\mu(A_{\gamma,i}) \Big\|$$

$$= \frac{\epsilon}{2} + \Big\| \sum_{\gamma \in \Gamma \setminus \bar{\Gamma}} \sum_{i=1}^{\bar{n}+1} (\hat{w}_\gamma - w_i)\mu(A_{\gamma,i}) \Big\|$$

$$\leq \frac{\epsilon}{2} + \sum_{\gamma \in \Gamma \setminus \bar{\Gamma}} \sum_{i=1}^{\bar{n}+1} \|\hat{w}_\gamma - w_i\| \, \|\mu(A_{\gamma,i})\|$$

$$\leq \frac{\epsilon}{2} + \sum_{\gamma \in \Gamma \setminus \bar{\Gamma}} \sum_{i=1}^{\bar{n}+1} \frac{3\epsilon}{6\mathcal{P} \circ \mu(X) + 1} \mathcal{P} \circ \mu(A_{\gamma,i}) \leq \frac{\epsilon}{2} + \frac{3\mathcal{P} \circ \mu(X)\epsilon}{6\mathcal{P} \circ \mu(X) + 1} < \epsilon$$

where the first inequality follows from Claims 11.126.1 and 11.126.2, the seventh equality follows from Claim 11.126.1 and the fact that $A_{\gamma,i} = \emptyset$, $\forall \gamma \in \bar{\Gamma}$, $\forall i \in \{1, \ldots, \bar{n} + 1\}$, and the third inequality follows from Claim 11.126.3.

In summary, we have shown that $\forall \epsilon \in (0, \infty) \subset \mathbb{R}$, $\exists n_1 \in \mathbb{N}$, $\forall n \in \mathbb{N}$ with $n \geq n_1$, $\exists \bar{R} \in \mathcal{I}(\mathcal{W})$ such that $\forall R \in \mathcal{I}(\mathcal{W})$ with $\bar{R} \preceq R$, we have $\big\| F_R - \int_X \phi_n \, \mathrm{d}\mu \big\| < \epsilon$. Then, the net $(F_R)_{R \in \mathcal{I}(\mathcal{W})} \subseteq \mathcal{Z}$ is a Cauchy net. By Proposition 4.44, the net admits a limit $\int_X f \, \mathrm{d}\mu \in \mathcal{Z}$. By Propositions 7.21, 7.23, 3.67, and 3.66, we have $\big\| \int_X f \, \mathrm{d}\mu - \int_X \phi_n \, \mathrm{d}\mu \big\| = \lim_{R \in \mathcal{I}(\mathcal{W})} \big\| F_R - \int_X \phi_n \, \mathrm{d}\mu \big\| \leq \epsilon$, $\forall n \geq n_1$. Hence, we have $\int_X f \, \mathrm{d}\mu = \lim_{n \in \mathbb{N}} \int_X \phi_n \, \mathrm{d}\mu$ and f is integrable over $\mathcal{X}$. By Proposition 11.125, $\int_X \phi_n \, \mathrm{d}\mu \in \bar{\mathcal{W}}$, $\forall n \in \mathbb{N}$. Then, by the completeness of $\bar{\mathcal{W}}$, we have $\int_X f \, \mathrm{d}\mu \in \bar{\mathcal{W}}$.

Note that $0 \leq \big\| \int_X f \, \mathrm{d}\mu \big\| = \lim_{n \in \mathbb{N}} \big\| \int_X \phi_n \, \mathrm{d}\mu \big\| \leq \limsup_{n \in \mathbb{N}} \int_X \mathcal{P} \circ \phi_n \, \mathrm{d}\mathcal{P} \circ \mu \leq \lim_{n \in \mathbb{N}} \int_X g_n \, \mathrm{d}\mathcal{P} \circ \mu = \int_X g \, \mathrm{d}\mathcal{P} \circ \mu < +\infty$, where the first equality follows from Propositions 7.21 and 3.66, the second inequality follows from Proposition 11.125, and the third inequality follows from Proposition 11.83. This completes the proof of the lemma. □

Lemma 11.127 *Let $\mathcal{X} := (X, \mathcal{B}, \mu)$ be a finite $\mathcal{Y}$-valued measure space, where $\mathcal{Y}$ is a normed linear space over $\mathbb{K}$, $\mathcal{Z}$ be a Banach space over $\mathbb{K}$, $\mathcal{W}$ be a separable subspace of* $\mathrm{B}(\mathcal{Y}, \mathcal{Z})$, *$f : X \to \mathcal{W}$ be $\mathcal{B}$-measurable, and $g : X \to [0, \infty) \subset \mathbb{R}$ be $\mathcal{B}$-measurable. Assume that $\|f(x)\| \leq g(x)$, $\forall x \in X$, and g is integrable over $\bar{\mathcal{X}} := (X, \mathcal{B}, \mathcal{P} \circ \mu)$. Then, there exists a sequence of simple functions $(\varphi_k)_{k=1}^{\infty}$, $\varphi_k : X \to \mathcal{W}$, $\forall k \in \mathbb{N}$, such that $\lim_{k \in \mathbb{N}} \varphi_k = f$ a.e. in $\mathcal{X}$, $\|\varphi_k(x)\| \leq g(x)$, $\forall x \in X$, $\forall k \in \mathbb{N}$, f is integrable over $\mathcal{X}$ with $\int_X f \, \mathrm{d}\mu = \lim_{k \in \mathbb{N}} \int_X \varphi_k \, \mathrm{d}\mu \in \mathcal{Z}$, $\lim_{k \in \mathbb{N}} \int_X \mathcal{P} \circ \varphi_k \, \mathrm{d}\mathcal{P} \circ \mu = \int_X \mathcal{P} \circ f \, \mathrm{d}\mathcal{P} \circ \mu \in [0, \infty) \subset \mathbb{R}$, and $\lim_{k \in \mathbb{N}} \int_X \mathcal{P} \circ (\varphi_k - f) \, \mathrm{d}\mathcal{P} \circ \mu = 0$.*

Proof By Proposition 11.66, there exists a sequence of simple functions $(\varphi_k)_{k=1}^{\infty}$, $\varphi_k : X \to \mathcal{W}$, $\forall k \in \mathbb{N}$, such that $\|\varphi_k(x)\| \leq \|f(x)\| \leq g(x)$, $\forall x \in X$, $\forall k \in \mathbb{N}$, and $\lim_{k \in \mathbb{N}} \varphi_k = f$ a.e. in $\mathcal{X}$. By Lemma 11.126, f is integrable over $\mathcal{X}$ and $\int_X f \, \mathrm{d}\mu = \lim_{k \in \mathbb{N}} \int_X \varphi_k \, \mathrm{d}\mu \in \mathcal{Z}$.

Note that $0 \leq \mathcal{P} \circ \varphi_k(x) \leq g(x)$, $\forall x \in X$, $\forall k \in \mathbb{N}$, $\mathcal{P} \circ f$ is $\mathcal{B}$-measurable, and $\lim_{k \in \mathbb{N}} \mathcal{P} \circ \varphi_k = \mathcal{P} \circ f$ a.e. in $\mathcal{X}$, by Propositions 7.21 and 11.52. Then, by Lebesgue Dominated Convergence Theorem 11.91, we have $0 \leq \lim_{k \in \mathbb{N}} \int_X \mathcal{P} \circ \varphi_k \, d\mathcal{P} \circ \mu = \int_X \mathcal{P} \circ f \, d\mathcal{P} \circ \mu \leq \int_X g \, d\mathcal{P} \circ \mu < +\infty$, where the second inequality follows from Proposition 11.83. By Lemma 11.43, $\mathcal{P} \circ (\varphi_k - f)$ is $\mathcal{B}$-measurable. By Propositions 7.21, 7.23, 11.52, and 11.53, $\lim_{k \in \mathbb{N}} \mathcal{P} \circ (\varphi_k - f) = 0$ a.e. in $\mathcal{X}$. Note that $0 \leq \mathcal{P} \circ (\varphi_k - f)(x) \leq 2g(x)$, $\forall x \in X$, $\forall k \in \mathbb{N}$. By Lebesgue Dominated Convergence Theorem 11.91 and Proposition 11.75, we have $\lim_{k \in \mathbb{N}} \int_X \mathcal{P} \circ (\varphi_k - f) \, d\mathcal{P} \circ \mu = 0$. This completes the proof of the lemma. □

Theorem 11.128 (Lebesgue Dominated Convergence) *Let $(X, \mathcal{B}, \mu) =: \mathcal{X}$ be a finite $\mathcal{Y}$-valued measure space, where $\mathcal{Y}$ is a normed linear space over $\mathbb{K}$, $\mathcal{Z}$ be a Banach space over $\mathbb{K}$, $\mathcal{W}$ be a separable subspace of $B(\mathcal{Y}, \mathcal{Z})$, $\bar{\mathcal{W}} := \operatorname{span}(\{z \in \mathcal{Z} \mid \exists y \in \mathcal{Y}, \exists w \in \mathcal{W} \ni \cdot z = wy\}) \subseteq \mathcal{Z}$ be the Banach subspace of $\mathcal{Z}$, $f_n : X \to \mathcal{W}$ be $\mathcal{B}$-measurable, $\forall n \in \mathbb{N}$, $f : X \to \mathcal{W}$ be $\mathcal{B}$-measurable, $g_n : X \to [0, \infty) \subset \mathbb{R}$ be $\mathcal{B}$-measurable, $\forall n \in \mathbb{N}$, and $g : X \to [0, \infty) \subset \mathbb{R}$ be $\mathcal{B}$-measurable. Assume that:*

(i) *$\lim_{n \in \mathbb{N}} f_n = f$ a.e. in $\mathcal{X}$, $\lim_{n \in \mathbb{N}} g_n = g$ a.e. in $\mathcal{X}$.*
(ii) *$\|f_n(x)\| \leq g_n(x)$, $\forall x \in X$, $\forall n \in \mathbb{N}$.*
(iii) *$\int_X g_n \, d\mathcal{P} \circ \mu \in \mathbb{R}$, $\forall n \in \mathbb{N}$, and $\int_X g \, d\mathcal{P} \circ \mu = \lim_{n \in \mathbb{N}} \int_X g_n \, d\mathcal{P} \circ \mu \in \mathbb{R}$.*

Then, f is integrable over $\mathcal{X}$, $\int_X f \, d\mu = \lim_{n \in \mathbb{N}} \int_X f_n \, d\mu \in \bar{\mathcal{W}} \subseteq \mathcal{Z}$, and $0 \leq \left\| \int_X f \, d\mu \right\| \leq \int_X g \, d\mathcal{P} \circ \mu < +\infty$.

Proof $\forall n \in \mathbb{N}$, by Lemma 11.127, there exists a sequence of simple functions $\left(\bar{\varphi}_{n,k}\right)_{k=1}^{\infty}$, $\bar{\varphi}_{n,k} : X \to \mathcal{W}$, $\forall k \in \mathbb{N}$, such that $\lim_{k \in \mathbb{N}} \bar{\varphi}_{n,k} = f_n$ a.e. in $\mathcal{X}$, $\left\|\bar{\varphi}_{n,k}(x)\right\| \leq g_n(x)$, $\forall x \in X$, $\forall k \in \mathbb{N}$, f_n is integrable over $\mathcal{X}$, and $\int_X f_n \, d\mu = \lim_{k \in \mathbb{N}} \int_X \bar{\varphi}_{n,k} \, d\mu \in \mathcal{Z}$. By Lemma 11.126, $\int_X f_n \, d\mu \in \bar{\mathcal{W}}$, $\forall n \in \mathbb{N}$. By Proposition 11.58, $\lim_{k \in \mathbb{N}} \bar{\varphi}_{n,k} = f_n$ in measure in $\mathcal{X}$. Then, $\exists k_n \in \mathbb{N}$ such that $\mathcal{P} \circ \mu(E_n) := \mathcal{P} \circ \mu(\{x \in X \mid \left\|\bar{\varphi}_{n,k_n}(x) - f_n(x)\right\| \geq 2^{-n}\}) < 2^{-n}$ and $\left\| \int_X f_n \, d\mu - \int_X \bar{\varphi}_{n,k_n} \, d\mu \right\| < 2^{-n}$. Denote $\psi_n := \bar{\varphi}_{n,k_n}$.

By Proposition 11.58, $\lim_{n \in \mathbb{N}} f_n = f$ in measure in $\mathcal{X}$. $\forall \epsilon \in (0, \infty) \subset \mathbb{R}$, $\exists n_0 \in \mathbb{N}$ such that $\forall n \in \mathbb{N}$ with $n \geq n_0$, we have $2^{-n} < \epsilon/2$ and $\mathcal{P} \circ \mu(\bar{E}_n) := \mathcal{P} \circ \mu(\{x \in X \mid \|f_n(x) - f(x)\| \geq \epsilon/2\}) < \epsilon/2$. Note that, by Lemma 11.43, $\mathcal{B} \ni F_n := \{x \in X \mid \|\psi_n(x) - f(x)\| \geq \epsilon\} \subseteq \{x \in X \mid \|\psi_n(x) - f_n(x)\| + \|f_n(x) - f(x)\| \geq \epsilon\} \subseteq E_n \cup \bar{E}_n$. Then, we have $\mathcal{P} \circ \mu(F_n) \leq \mathcal{P} \circ \mu(E_n) + \mathcal{P} \circ \mu(\bar{E}_n) < \epsilon$. This implies that $\lim_{n \in \mathbb{N}} \psi_n = f$ in measure in $\mathcal{X}$. By Proposition 11.57, there exists a subsequence $\left(\psi_{n_i}\right)_{i=1}^{\infty}$ of $(\psi_n)_{n=1}^{\infty}$ that converges to f a.e. in $\mathcal{X}$. By Lemma 11.126, f is integrable over $\mathcal{X}$, $\int_X f \, d\mu = \lim_{i \in \mathbb{N}} \int_X \psi_{n_i} \, d\mu \in \mathcal{Z}$, and $0 \leq \left\| \int_X f \, d\mu \right\| \leq \int_X g \, d\mathcal{P} \circ \mu < +\infty$. By the fact that $\left\| \int_X f_{n_i} \, d\mu - \int_X \psi_{n_i} \, d\mu \right\| < 2^{-n_i}$, $\forall i \in \mathbb{N}$, then we have $\lim_{i \in \mathbb{N}} \int_X f_{n_i} \, d\mu = \lim_{i \in \mathbb{N}} \int_X \psi_{n_i} \, d\mu = \int_X f \, d\mu \in \mathcal{Z}$. So far, we have shown that, for the sequence $(f_n)_{n=1}^{\infty}$, there exists a subsequence $\left(f_{n_i}\right)_{i=1}^{\infty}$ such that $\lim_{i \in \mathbb{N}} \int_X f_{n_i} \, d\mu = \int_X f \, d\mu \in \mathcal{Z}$. This then holds for any subsequence of $(f_n)_{n=1}^{\infty}$.

By Proposition 3.71, we have $\lim_{n\in\mathbb{N}} \int_X f_n \, d\mu = \int_X f \, d\mu$. By the completeness of $\bar{\mathcal{W}}$, we have $\int_X f \, d\mu \in \bar{\mathcal{W}} \subseteq \mathcal{Z}$.

This completes the proof of the theorem. □

Definition 11.129 Let $\mathcal{X} := (X, \mathcal{B}, \mu)$ be a $\mathcal{Y}$-valued measure space, where $\mathcal{Y}$ is a normed linear space over $\mathbb{K}$, $\mathcal{Z}$ be a normed linear space over $\mathbb{K}$, $\mathcal{W} := B(\mathcal{Y}, \mathcal{Z})$, and $f : X \to \mathcal{W}$ be $\mathcal{B}$-measurable. f is said to be *absolutely integrable over* $\mathcal{X}$ if $\mathcal{P} \circ f$ is integrable over $\bar{\mathcal{X}} := (X, \mathcal{B}, \mathcal{P} \circ \mu)$, that is, $0 \le \int_X \mathcal{P} \circ f \, d\mathcal{P} \circ \mu < +\infty$. ◇

Proposition 11.130 *Let $\mathcal{X} := (X, \mathcal{B}, \mu)$ be a finite $\mathcal{Y}$-valued measure space, where $\mathcal{Y}$ is a normed linear space over $\mathbb{K}$, $\mathcal{Z}$ be a Banach space over $\mathbb{K}$, $\mathcal{W}$ be a separable subspace of $B(\mathcal{Y}, \mathcal{Z})$, $\mathcal{U}$ be a Banach space over $\mathbb{K}$, $\bar{\mathcal{W}} := \overline{\operatorname{span}(\{z \in \mathcal{Z} \mid \exists y \in \mathcal{Y}, \exists w \in \mathcal{W} \ni \cdot z = wy\})} \subseteq \mathcal{Z}$ be the Banach subspace of $\mathcal{Z}$, and $f_i : X \to \mathcal{W}$ be absolutely integrable over $\mathcal{X}$, $i = 1, 2$. Then, the following statements hold:*

(i) *f_i is integrable over $\mathcal{X}$ and $\int_X f_i \, d\mu \in \bar{\mathcal{W}} \subseteq \mathcal{Z}$, $i = 1, 2$.*
(ii) *$f_1 + f_2$ is absolutely integrable over $\mathcal{X}$ and $\int_X (f_1 + f_2) \, d\mu = \int_X f_1 \, d\mu + \int_X f_2 \, d\mu \in \bar{\mathcal{W}}$.*
(iii) *$\forall A \in B(\mathcal{Z}, \mathcal{U})$, Af_1 is absolutely integrable over $\mathcal{X}$ and $\int_X (Af_1) \, d\mu = A \int_X f_1 \, d\mu \in \mathcal{U}$.*
(iv) *$\forall c \in \mathbb{K}$, cf_1 is absolutely integrable over $\mathcal{X}$ and $\int_X (cf_1) \, d\mu = c \int_X f_1 \, d\mu \in \bar{\mathcal{W}}$.*
(v) *$\forall H \in \mathcal{B}$, let $\mathcal{H} := (H, \mathcal{B}_H, \mu_H)$ be the finite $\mathcal{Y}$-valued measure subspace of $\mathcal{X}$ as defined in Proposition 11.115. Then, $f_1|_H$ is absolutely integrable over $\mathcal{H}$ and $\int_X (f_1 \chi_{H,X}) \, d\mu = \int_H f_1|_H \, d\mu_H \in \bar{\mathcal{W}}$. We will henceforth denote $\int_H f_1|_H \, d\mu_H$ by $\int_H f_1 \, d\mu$.*
(vi) *If $f_1 = f_2$ a.e. in $\mathcal{X}$, then $\int_X f_1 \, d\mu = \int_X f_2 \, d\mu \in \bar{\mathcal{W}}$.*
(vii) *$\forall$ pairwise disjoint $(E_i)_{i=1}^{\infty} \subseteq \mathcal{B}$, $\sum_{i=1}^{\infty} \int_{E_i} f_1 \, d\mu = \int_{\bigcup_{i=1}^{\infty} E_i} f_1 \, d\mu \in \bar{\mathcal{W}}$.*
(viii) *$0 \le \left\| \int_X f_1 \, d\mu \right\| \le \int_X \mathcal{P} \circ f_1 \, d\mathcal{P} \circ \mu < +\infty$.*

Proof (i) $\forall i \in \{1, 2\}$, by Lemma 11.127, there exists a sequence of simple functions $(\phi_{i,n})_{n=1}^{\infty}$, $\phi_{i,n} : X \to \mathcal{W}$, $\forall n \in \mathbb{N}$, such that $\lim_{n\in\mathbb{N}} \phi_{i,n} = f_i$ a.e. in $\mathcal{X}$, $\|\phi_{i,n}(x)\| \le \mathcal{P} \circ f_i(x)$, $\forall x \in X$, $\forall n \in \mathbb{N}$, f_i is integrable over $\mathcal{X}$, and $\int_X f_i \, d\mu = \lim_{n\in\mathbb{N}} \int_X \phi_{i,n} \, d\mu \in \mathcal{Z}$. By Lebesgue Dominated Convergence Theorem 11.128, $\int_X f_i \, d\mu \in \bar{\mathcal{W}}$, $i = 1, 2$.

(ii) By Propositions 11.38, 11.39, 7.21, and 7.23, $f_1 + f_2$ and $\mathcal{P} \circ (f_1 + f_2)$ are $\mathcal{B}$-measurable. Note that $\mathcal{P} \circ (f_1 + f_2)(x) = \|f_1(x) + f_2(x)\| \le \|f_1(x)\| + \|f_2(x)\| = (\mathcal{P} \circ f_1 + \mathcal{P} \circ f_2)(x)$, $\forall x \in X$. By Proposition 11.83, we have $0 \le \int_X \mathcal{P} \circ (f_1 + f_2) \, d\mathcal{P} \circ \mu \le \int_X (\mathcal{P} \circ f_1 + \mathcal{P} \circ f_2) \, d\mathcal{P} \circ \mu = \int_X \mathcal{P} \circ f_1 \, d\mathcal{P} \circ \mu + \int_X \mathcal{P} \circ f_2 \, d\mathcal{P} \circ \mu < +\infty$. Hence, $f_1 + f_2$ is absolutely integrable over $\mathcal{X}$.

Note that $\phi_{1,n} + \phi_{2,n} : X \to \mathcal{W}$, $f_1 + f_2 : X \to \mathcal{W}$, and $\|\phi_{1,n}(x) + \phi_{2,n}(x)\| \le \|\phi_{1,n}(x)\| + \|\phi_{2,n}(x)\| \le (\mathcal{P} \circ f_1 + \mathcal{P} \circ f_2)(x)$, $\forall x \in X$, $\forall n \in \mathbb{N}$. By Propositions 7.23, 11.52, and 11.53, $\lim_{n\in\mathbb{N}} (\phi_{1,n} + \phi_{2,n}) = f_1 + f_2$ a.e. in $\mathcal{X}$. By Lebesgue

Dominated Convergence Theorem 11.128, f_1+f_2 is integrable over $\mathcal{X}$ and $\int_X (f_1+f_2)\,d\mu = \lim_{n\in\mathbb{N}} \int_X (\phi_{1,n}+\phi_{2,n})\,d\mu = \lim_{n\in\mathbb{N}} \left(\int_X \phi_{1,n}\,d\mu + \int_X \phi_{2,n}\,d\mu\right) = \int_X f_1\,d\mu + \int_X f_2\,d\mu \in \bar{\mathcal{W}}$, where the second equality follows from Proposition 11.125 and the third equality follows from Propositions 3.66, 3.67, and 7.23.

(iii) Fix any $A \in B(\mathcal{Z},\mathcal{U})$. By Propositions 11.38, 7.21, and 7.62, $A\phi_{1,n}$, Af_1, and $\mathcal{P}\circ(Af_1)$ are $\mathcal{B}$-measurable, $\forall n\in\mathbb{N}$. Note that $\mathcal{P}\circ(Af_1)(x) = \|Af_1(x)\| \leq \|A\|\|f_1(x)\| = \|A\|\mathcal{P}\circ f_1(x)$, $\forall x\in X$. By Proposition 11.83 and the fact that $0\leq \int_X \mathcal{P}\circ f_1\,d\mathcal{P}\circ\mu < +\infty$, we have $0 \leq \int_X \mathcal{P}\circ(Af_1)\,d\mathcal{P}\circ\mu \leq \int_X (\|A\|\mathcal{P}\circ f_1)\,d\mathcal{P}\circ\mu = \|A\|\int_X \mathcal{P}\circ f_1\,d\mathcal{P}\circ\mu < +\infty$. Hence, Af_1 is absolutely integrable over $\mathcal{X}$.

Note that $Af_1 : X \to \hat{\mathcal{W}}$ and $A\phi_{1,n} : X \to \hat{\mathcal{W}}$, $\forall n\in\mathbb{N}$, where $\hat{\mathcal{W}} := \{F \in B(\mathcal{Y},\mathcal{U}) \mid F = AF_1,\ F_1\in\mathcal{W}\}$ is a separable subspace of $B(\mathcal{Y},\mathcal{U})$, by Proposition 7.126. By Propositions 11.52 and 7.62, $\lim_{n\in\mathbb{N}} A\phi_{1,n} = Af_1$ a.e. in $\mathcal{X}$. By Proposition 11.125, we have $\int_X (A\phi_{1,n})\,d\mu = A\int_X \phi_{1,n}\,d\mu \in \mathcal{U}$, $\forall n\in\mathbb{N}$. Note also that $\|A\phi_{1,n}(x)\| \leq \|A\|\|\phi_{1,n}(x)\| \leq \|A\|\mathcal{P}\circ f_1(x)$, $\forall x\in X$, $\forall n\in\mathbb{N}$. By Lebesgue Dominated Convergence Theorem 11.128, Af_1 is integrable over $\mathcal{X}$ and $\int_X (Af_1)\,d\mu = \lim_{n\in\mathbb{N}} \int_X (A\phi_{1,n})\,d\mu = \lim_{n\in\mathbb{N}} A\int_X \phi_{1,n}\,d\mu = A\int_X f_1\,d\mu \in \mathcal{U}$, where the last equality follows from Propositions 3.66 and 7.62.

(iv) This follows immediately from (iii).

(v) Fix any $H\in\mathcal{B}$. Note that $(\mathcal{P}\circ f_1)|_H = \mathcal{P}\circ(f_1|_H)$. Then, by Proposition 11.115, we have $0 \leq \int_H \mathcal{P}\circ(f_1|_H)\,d\mathcal{P}\circ\mu_H = \int_H (\mathcal{P}\circ f_1)|_H\,d(\mathcal{P}\circ\mu)|_{\mathcal{B}_H} = \int_H \mathcal{P}\circ f_1\,d\mathcal{P}\circ\mu \leq \int_X \mathcal{P}\circ f_1\,d\mathcal{P}\circ\mu < +\infty$, where the second equality and the second inequality follow from Proposition 11.83. Hence, $f_1|_H$ is absolutely integrable over $\mathcal{H}$.

$\left(\phi_{1,n}|_H\right)_{n=1}^{\infty}$ is a sequence of simple functions that converges to $f_1|_H$ a.e. in $\mathcal{H}$. Note that $\left\|\phi_{1,n}|_H(x)\right\| \leq (\mathcal{P}\circ f_1)|_H(x) = \mathcal{P}\circ(f_1|_H)(x)$, $\forall x\in H$, $\forall n\in\mathbb{N}$. By Lebesgue Dominated Convergence Theorem 11.128, we have $\int_H f_1|_H\,d\mu_H = \lim_{n\in\mathbb{N}} \int_H \phi_{1,n}|_H\,d\mu_H \in \bar{\mathcal{W}}$. Note that $(\phi_{1,n}\chi_{H,X})_{n=1}^{\infty}$ is a sequence of simple functions that converges to $f_1\chi_{H,X}$ a.e. in $\mathcal{X}$, and $\left\|\phi_{1,n}(x)\chi_{H,X}(x)\right\| \leq \mathcal{P}\circ f_1(x)$, $\forall x\in X$, $\forall n\in\mathbb{N}$. Again by Lebesgue Dominated Convergence Theorem 11.128, we have $\int_X (f_1\chi_{H,X})\,d\mu = \lim_{n\in\mathbb{N}} \int_X (\phi_{1,n}\chi_{H,X})\,d\mu = \lim_{n\in\mathbb{N}} \int_H \phi_{1,n}|_H\,d\mu_H = \int_H f_1|_H\,d\mu_H \in \bar{\mathcal{W}}$, where the second equality follows from Proposition 11.125.

(vi) If $f_1 = f_2$ a.e. in $\mathcal{X}$, then, by Proposition 11.54, $\lim_{n\in\mathbb{N}} \phi_{1,n} = f_2$ a.e. in $\mathcal{X}$. By Lebesgue Dominated Convergence Theorem 11.128, we have $\int_X f_2\,d\mu = \lim_{n\in\mathbb{N}} \int_X \phi_{1,n}\,d\mu = \int_X f_1\,d\mu$.

(vii) $\forall$ pairwise disjoint $(E_i)_{i=1}^{\infty} \subseteq \mathcal{B}$, let $E := \bigcup_{i=1}^{\infty} E_i \in \mathcal{B}$ and $\bar{E}_n := \bigcup_{i=1}^{n} E_i \in \mathcal{B}$, $\forall n\in\mathbb{N}$. Then, $\lim_{n\in\mathbb{N}} f_1(x)\chi_{\bar{E}_n,X}(x) = f_1(x)\chi_{E,X}(x)$, $\forall x\in X$. Then, we have $\sum_{i=1}^{\infty} \int_{E_i} f_1\,d\mu = \lim_{n\in\mathbb{N}} \sum_{i=1}^{n} \int_X (f_1\chi_{E_i,X})\,d\mu = \lim_{n\in\mathbb{N}} \int_X (f_1\chi_{\bar{E}_n,X})\,d\mu = \int_X (f_1\chi_{E,X})\,d\mu = \int_E f_1\,d\mu \in \bar{\mathcal{W}}$, where the first equality follows from (v), the second equality follows from (ii), the third equality follows from Lebesgue Dominated Convergence Theorem 11.128, the last equality follows from (v), and the last set membership follows from (v).

(viii) The result follows immediately from Lebesgue Dominated Convergence Theorem 11.128.

This completes the proof of the proposition. □

Theorem 11.131 (Lebesgue Dominated Convergence) *Let* $(X, \mathcal{B}, \mu) =: \mathcal{X}$ *be a* $\mathcal{Y}$*-valued measure space, where* $\mathcal{Y}$ *is a normed linear space over* $\mathbb{K}$*,* $\mathcal{Z}$ *be a Banach space over* $\mathbb{K}$*,* $\mathcal{W}$ *be a separable subspace of* $\mathrm{B}(\mathcal{Y}, \mathcal{Z})$*,* $\bar{\mathcal{W}} := \overline{\mathrm{span}\,(\{z \in \mathcal{Z} \mid \exists y \in \mathcal{Y}, \exists w \in \mathcal{W} \ni \cdot z = wy\})} \subseteq \mathcal{Z}$ *be the Banach subspace of* $\mathcal{Z}$*,* $f_n : X \to \mathcal{W}$ *be* $\mathcal{B}$*-measurable,* $\forall n \in \mathbb{N}$*,* $f : X \to \mathcal{W}$ *be* $\mathcal{B}$*-measurable,* $g_n : X \to [0, \infty) \subset \mathbb{R}$ *be* $\mathcal{B}$*-measurable,* $\forall n \in \mathbb{N}$*, and* $g : X \to [0, \infty) \subset \mathbb{R}$ *be* $\mathcal{B}$*-measurable. Assume that:*

(i) $\lim_{n \in \mathbb{N}} f_n = f$ *a.e. in* $\mathcal{X}$*,* $\lim_{n \in \mathbb{N}} g_n = g$ *a.e. in* $\mathcal{X}$*.*
(ii) $\|f_n(x)\| \le g_n(x)$*,* $\forall x \in X$*,* $\forall n \in \mathbb{N}$*.*
(iii) $\int_X g_n \,\mathrm{d}\mathcal{P} \circ \mu \in \mathbb{R}$*,* $\forall n \in \mathbb{N}$ *and* $\int_X g \,\mathrm{d}\mathcal{P} \circ \mu = \lim_{n \in \mathbb{N}} \int_X g_n \,\mathrm{d}\mathcal{P} \circ \mu \in \mathbb{R}$*.*

Then, f *is integrable over* $\mathcal{X}$*,* $\int_X f \,\mathrm{d}\mu = \lim_{n \in \mathbb{N}} \int_X f_n \,\mathrm{d}\mu \in \bar{\mathcal{W}} \subseteq \mathcal{Z}$*, and* $0 \le \left\| \int_X f \,\mathrm{d}\mu \right\| \le \int_X g \,\mathrm{d}\mathcal{P} \circ \mu < +\infty$*. Furthermore,* f *is absolutely integrable over* $\mathcal{X}$*.*

Proof We will show that f is integrable over $\mathcal{X}$, $\int_X f \,\mathrm{d}\mu = \lim_{n \in \mathbb{N}} \int_X f_n \,\mathrm{d}\mu \in \bar{\mathcal{W}}$, and $0 \le \left\| \int_X f \,\mathrm{d}\mu \right\| \le \int_X g \,\mathrm{d}\mathcal{P} \circ \mu < +\infty$ by distinguishing two exhaustive and mutually exclusive cases: Case 1: $\mathcal{P} \circ \mu(X) < +\infty$; Case 2: $\mathcal{P} \circ \mu(X) = +\infty$.

Case 1: $\mathcal{P} \circ \mu(X) < +\infty$. Then, $\mathcal{X}$ is a finite $\mathcal{Y}$-valued measure space. By Lebesgue Dominated Convergence Theorem 11.128, f is integrable over $\mathcal{X}$, $\int_X f \,\mathrm{d}\mu = \lim_{n \in \mathbb{N}} \int_X f_n \,\mathrm{d}\mu \in \bar{\mathcal{W}}$, and $0 \le \left\| \int_X f \,\mathrm{d}\mu \right\| \le \int_X g \,\mathrm{d}\mathcal{P} \circ \mu < +\infty$.

Case 2: $\mathcal{P} \circ \mu(X) = +\infty$. Let $(F_A)_{A \in \mathcal{M}(\mathcal{X})}$ and $\left(F_{n,A}\right)_{A \in \mathcal{M}(\mathcal{X})}$ be the nets for the integrals $\int_X f \,\mathrm{d}\mu$ and $\int_X f_n \,\mathrm{d}\mu$ as defined in Definition 11.120, respectively, $\forall n \in \mathbb{N}$. $\forall A \in \mathcal{M}(\mathcal{X})$, we have $A \in \mathcal{B}$ and $\mathcal{P} \circ \mu(A) < \infty$. Let $\mathcal{A} := (A, \mathcal{B}_A, \mu_A)$ be the finite $\mathcal{Y}$-valued measure subspace of $\mathcal{X}$ as defined in Proposition 11.115. By Proposition 11.83, $g|_A$ and $g_n|_A$'s are integrable over $\bar{\mathcal{A}} := (A, \mathcal{B}_A, \mathcal{P} \circ \mu_A)$, which is the finite measure subspace of $\bar{\mathcal{X}} := (X, \mathcal{B}, \mathcal{P} \circ \mu)$ as shown in Proposition 11.115. By Proposition 11.83, $\lim_{n \in \mathbb{N}} g_n|_A = g|_A$ a.e. in $\mathcal{A}$ and $\int_A g|_A \,\mathrm{d}\mathcal{P} \circ \mu_A = \lim_{n \in \mathbb{N}} \int_A g_n|_A \,\mathrm{d}\mathcal{P} \circ \mu_A \in \mathbb{R}$. By Lebesgue Dominated Convergence Theorem 11.128, we have $f|_A$ and $f_n|_A$'s are integrable over $\mathcal{A}$, $\int_A f|_A \,\mathrm{d}\mu_A = \lim_{n \in \mathbb{N}} \int_A f_n|_A \,\mathrm{d}\mu_A \in \bar{\mathcal{W}}$, and $0 \le \left\| \int_A f|_A \,\mathrm{d}\mu_A \right\| \le \int_A g|_A \,\mathrm{d}\mathcal{P} \circ \mu_A < +\infty$. Note that $F_A = \int_A f|_A \,\mathrm{d}\mu_A \in \bar{\mathcal{W}}$ and $F_{n,A} = \int_A f_n|_A \,\mathrm{d}\mu_A \in \bar{\mathcal{W}}$, $\forall n \in \mathbb{N}$. Then, the nets $(F_A)_{A \in \mathcal{M}(\mathcal{X})}$ and $\left(F_{n,A}\right)_{A \in \mathcal{M}(\mathcal{X})}$'s are well-defined.

$\forall \epsilon \in (0, \infty) \subset \mathbb{R}$, by Definition 11.79, $\exists$ a simple function $\phi : X \to [0, \infty) \subset \mathbb{R}$ such that $0 \le \phi(x) \le g(x)$, $\forall x \in X$, and $\int_X g \,\mathrm{d}\mathcal{P} \circ \mu - \epsilon/5 < \int_X \phi \,\mathrm{d}\mathcal{P} \circ \mu \le \int_X g \,\mathrm{d}\mathcal{P} \circ \mu < +\infty$. Let $A_0 := \{x \in X \mid \phi(x) > 0\}$. Then, $A_0 \in \mathcal{B}$, $\mathcal{P} \circ \mu(A_0) < +\infty$, and $A_0 \in \mathcal{M}(\mathcal{X})$. $\forall A \in \mathcal{M}(\mathcal{X})$ with $A_0 \subseteq A$, we have $0 \le \int_{A \setminus A_0} g|_{A \setminus A_0} \,\mathrm{d}\mathcal{P} \circ \mu_{A \setminus A_0} = \int_A g|_A \,\mathrm{d}\mathcal{P} \circ \mu_A - \int_{A_0} g|_{A_0} \,\mathrm{d}\mathcal{P} \circ \mu_{A_0} \le \int_X g \,\mathrm{d}\mathcal{P} \circ \mu - \int_{A_0} \phi|_{A_0} \,\mathrm{d}\mathcal{P} \circ \mu_{A_0} = \int_X g \,\mathrm{d}\mathcal{P} \circ \mu - \int_X \phi \,\mathrm{d}\mathcal{P} \circ \mu < \epsilon/5$, where the first equality and the second inequality follow from Proposition 11.83 and the second equality follows from Proposition 11.75. Note that $\left\| F_A - F_{A_0} \right\| = \left\| \int_A f|_A \,\mathrm{d}\mu_A - \int_{A_0} f|_{A_0} \,\mathrm{d}\mu_{A_0} \right\| =$

$\left\|\int_{A\setminus A_0} f|_{A\setminus A_0}\, \mathrm{d}\mu_{A\setminus A_0}\right\| \leq \int_{A\setminus A_0} g|_{A\setminus A_0}\, \mathrm{d}\mathcal{P}\circ\mu_{A\setminus A_0} < \epsilon/5$, where the second equality and the first inequality follow from Proposition 11.130. Then, the net $(F_A)_{A\in\mathcal{M}(\mathcal{X})}$ is a Cauchy net, which admits a limit, by Proposition 4.44, $\int_X f\, \mathrm{d}\mu = \lim_{A\in\mathcal{M}(\mathcal{X})} F_A \in \bar{\mathcal{W}}$. Hence, f is integrable over $\mathcal{X}$. By Propositions 7.21 and 3.66, $0 \leq \left\|\int_X f\, \mathrm{d}\mu\right\| = \lim_{A\in\mathcal{M}(\mathcal{X})} \|F_A\| \leq \lim_{A\in\mathcal{M}(\mathcal{X})} \int_A g|_A\, \mathrm{d}\mathcal{P}\circ\mu_A = \int_X g\, \mathrm{d}\mathcal{P}\circ\mu < +\infty$.

$\forall n \in \mathbb{N}$, by Definition 11.79, $\exists$ a simple function $\phi_n : X \to [0,\infty) \subset \mathbb{R}$ such that $0 \leq \phi_n(x) \leq g_n(x)$, $\forall x \in X$, and $\int_X g_n\, \mathrm{d}\mathcal{P}\circ\mu - \epsilon/5 < \int_X \phi_n\, \mathrm{d}\mathcal{P}\circ\mu \leq \int_X g_n\, \mathrm{d}\mathcal{P}\circ\mu < +\infty$. Let $A_n := \{x \in X \mid \phi_n(x) > 0\}$. Then, $A_n \in \mathcal{B}$, $\mathcal{P}\circ\mu(A_n) < +\infty$, and $A_n \in \mathcal{M}(\mathcal{X})$. $\forall A \in \mathcal{M}(\mathcal{X})$ with $A_n \subseteq A$, we have $0 \leq \int_{A\setminus A_n} g_n|_{A\setminus A_n}\, \mathrm{d}\mathcal{P}\circ\mu_{A\setminus A_n} = \int_A g_n|_A\, \mathrm{d}\mathcal{P}\circ\mu_A - \int_{A_n} g_n|_{A_n}\, \mathrm{d}\mathcal{P}\circ\mu_{A_n} \leq \int_X g_n\, \mathrm{d}\mathcal{P}\circ\mu - \int_{A_n} \phi_n|_{A_n}\, \mathrm{d}\mathcal{P}\circ\mu_{A_n} = \int_X g_n\, \mathrm{d}\mathcal{P}\circ\mu - \int_X \phi_n\, \mathrm{d}\mathcal{P}\circ\mu < \epsilon/5$, where the first equality and the second inequality follow from Proposition 11.83 and the second equality follows from Proposition 11.75. Note that $\|F_{n,A} - F_{n,A_n}\| = \left\|\int_A f_n|_A\, \mathrm{d}\mu_A - \int_{A_n} f_n|_{A_n}\, \mathrm{d}\mu_{A_n}\right\| = \left\|\int_{A\setminus A_n} f_n|_{A\setminus A_n}\, \mathrm{d}\mu_{A\setminus A_n}\right\| \leq \int_{A\setminus A_n} g_n|_{A\setminus A_n}\, \mathrm{d}\mathcal{P}\circ\mu_{A\setminus A_n} < \epsilon/5$, where the second equality and the first inequality follow from Proposition 11.130. Then, the net $\left(F_{n,A}\right)_{A\in\mathcal{M}(\mathcal{X})}$ is a Cauchy net, which admits a limit, by Proposition 4.44, $\int_X f_n\, \mathrm{d}\mu = \lim_{A\in\mathcal{M}(\mathcal{X})} F_{n,A} \in \bar{\mathcal{W}}$.

Let $\bar{g}_n := g_n \wedge g$, $\forall n \in \mathbb{N}$. Then, $0 \leq \bar{g}_n(x) \leq g(x)$, $\forall x \in X$, and $\bar{g}_n$ is $\mathcal{B}$-measurable by Proposition 11.40, $\forall n \in \mathbb{N}$. Then, $\lim_{n\in\mathbb{N}} \bar{g}_n = g$ a.e. in $\mathcal{X}$ by Proposition 11.50. By Fatou's Lemma 11.80, $\int_X g\, \mathrm{d}\mathcal{P}\circ\mu \leq \liminf_{n\in\mathbb{N}} \int_X \bar{g}_n\, \mathrm{d}\mathcal{P}\circ\mu$. By Definition 11.79, $0 \leq \int_X \bar{g}_n\, \mathrm{d}\mathcal{P}\circ\mu \leq \int_X g\, \mathrm{d}\mathcal{P}\circ\mu < +\infty$, $\forall n \in \mathbb{N}$. Then, we have $\int_X g\, \mathrm{d}\mathcal{P}\circ\mu \leq \liminf_{n\in\mathbb{N}} \int_X \bar{g}_n\, \mathrm{d}\mathcal{P}\circ\mu \leq \limsup_{n\in\mathbb{N}} \int_X \bar{g}_n\, \mathrm{d}\mathcal{P}\circ\mu \leq \int_X g\, \mathrm{d}\mathcal{P}\circ\mu < +\infty$. Therefore, by Proposition 3.83, $\int_X g\, \mathrm{d}\mathcal{P}\circ\mu = \lim_{n\in\mathbb{N}} \int_X \bar{g}_n\, \mathrm{d}\mathcal{P}\circ\mu \in \mathbb{R}$. This coupled with (iii) and the fact that $F_{A_0} = \lim_{n\in\mathbb{N}} F_{n,A_0}$ implies that $\exists n_0 \in \mathbb{N}$, $\forall n \in \mathbb{N}$ with $n_0 \leq n$, we have $0 \leq \int_X g\, \mathrm{d}\mathcal{P}\circ\mu - \int_X \bar{g}_n\, \mathrm{d}\mathcal{P}\circ\mu < \epsilon/5$, $\left|\int_X g\, \mathrm{d}\mathcal{P}\circ\mu - \int_X g_n\, \mathrm{d}\mathcal{P}\circ\mu\right| < \epsilon/5$, and $\left\|F_{A_0} - F_{n,A_0}\right\| < \epsilon/5$. $\forall A \in \mathcal{M}(\mathcal{X})$ with $A_0 \subseteq A$, we have $0 \leq \int_{A\setminus A_0} g_n|_{A\setminus A_0}\, \mathrm{d}\mathcal{P}\circ\mu_{A\setminus A_0} = \int_{A\setminus A_0} (g_n - \bar{g}_n)|_{A\setminus A_0}\, \mathrm{d}\mathcal{P}\circ\mu_{A\setminus A_0} + \int_{A\setminus A_0} \bar{g}_n|_{A\setminus A_0}\, \mathrm{d}\mathcal{P}\circ\mu_{A\setminus A_0} \leq \int_X (g_n - \bar{g}_n)\, \mathrm{d}\mathcal{P}\circ\mu + \int_{A\setminus A_0} g|_{A\setminus A_0}\, \mathrm{d}\mathcal{P}\circ\mu_{A\setminus A_0} < \int_X g_n\, \mathrm{d}\mathcal{P}\circ\mu - \int_X \bar{g}_n\, \mathrm{d}\mathcal{P}\circ\mu + \epsilon/5 \leq \left|\int_X g_n\, \mathrm{d}\mathcal{P}\circ\mu - \int_X g\, \mathrm{d}\mathcal{P}\circ\mu\right| + \int_X g\, \mathrm{d}\mathcal{P}\circ\mu - \int_X \bar{g}_n\, \mathrm{d}\mathcal{P}\circ\mu + \epsilon/5 < 3\epsilon/5$, where the first equality, the second inequality, and the third inequality follow from Proposition 11.83. Furthermore, $\left\|F_{n,A} - F_{n,A_0}\right\| = \left\|\int_A f_n|_A\, \mathrm{d}\mu_A - \int_{A_0} f_n|_{A_0}\, \mathrm{d}\mu_{A_0}\right\| = \left\|\int_{A\setminus A_0} f_n|_{A\setminus A_0}\, \mathrm{d}\mu_{A\setminus A_0}\right\| \leq \int_{A\setminus A_0} g_n|_{A\setminus A_0}\, \mathrm{d}\mathcal{P}\circ\mu_{A\setminus A_0} < 3\epsilon/5$, where the second equality and the first inequality follow from Proposition 11.130. This implies that $\left\|\int_X f\, \mathrm{d}\mu - \int_X f_n\, \mathrm{d}\mu\right\| \leq \left\|\int_X f\, \mathrm{d}\mu - F_{A_0}\right\| + \left\|F_{A_0} - F_{n,A_0}\right\| + \left\|F_{n,A_0} - \int_X f_n\, \mathrm{d}\mu\right\| < \lim_{A\in\mathcal{M}(\mathcal{X})} \left\|F_A - F_{A_0}\right\| + \epsilon/5 + \lim_{A\in\mathcal{M}(\mathcal{X})} \left\|F_{n,A_0} - F_{n,A}\right\| \leq \epsilon/5 + \epsilon/5 + 3\epsilon/5 = \epsilon$, where the second inequality follows from Propositions 7.21 and 3.66. Hence, $\int_X f\, \mathrm{d}\mu = \lim_{n\in\mathbb{N}} \int_X f_n\, \mathrm{d}\mu \in \bar{\mathcal{W}}$.

Then, in both cases, we have shown that f is integrable over $\mathcal{X}$, $\int_X f \, d\mu = \lim_{n\in\mathbb{N}} \int_X f_n \, d\mu \in \bar{\mathcal{W}}$, and $0 \leq \left\| \int_X f \, d\mu \right\| \leq \int_X g \, d\mathcal{P} \circ \mu < +\infty$.

Note that $0 \leq \mathcal{P} \circ f_n(x) \leq g_n(x)$, $\forall x \in X$, $\forall n \in \mathbb{N}$, $\lim_{n\in\mathbb{N}} \mathcal{P} \circ f_n = \mathcal{P} \circ f$ a.e. in $\mathcal{X}$ by Propositions 7.21 and 11.52, and $\mathcal{P} \circ f$ and $\mathcal{P} \circ f_n$'s are $\mathcal{B}$-measurable by Propositions 7.21 and 11.38. Then, by Lebesgue Dominated Convergence Theorem 11.91, $\int_X \mathcal{P} \circ f \, d\mathcal{P} \circ \mu = \lim_{n\in\mathbb{N}} \int_X \mathcal{P} \circ f_n \, d\mathcal{P} \circ \mu \in \mathbb{R}$. Hence, f is absolutely integrable over $\mathcal{X}$.

This completes the proof of the theorem. □

Proposition 11.132 *Let $\mathcal{X} := (X, \mathcal{B}, \mu)$ be a $\mathcal{Y}$-valued measure space, where $\mathcal{Y}$ is a normed linear space over $\mathbb{K}$, $\mathcal{Z}$ be a Banach space over $\mathbb{K}$, $\mathcal{W}$ be a separable subspace of $\mathrm{B}(\mathcal{Y}, \mathcal{Z})$, $\mathcal{U}$ be a Banach space over $\mathbb{K}$, $\bar{\mathcal{W}} := \overline{\operatorname{span}\left(\{z \in \mathcal{Z} \mid \exists y \in \mathcal{Y}, \exists w \in \mathcal{W} \ni \cdot z = wy\}\right)} \subseteq \mathcal{Z}$ be the Banach subspace of $\mathcal{Z}$, $f_i : X \to \mathcal{W}$ be absolutely integrable over $\mathcal{X}$, $i = 1, 2$. Then, the following statements hold:*

(i) *f_i is integrable over $\mathcal{X}$ and $\int_X f_i \, d\mu \in \bar{\mathcal{W}} \subseteq \mathcal{Z}$, $i = 1, 2$.*
(ii) *$f_1 + f_2$ is absolutely integrable over $\mathcal{X}$ and $\int_X (f_1 + f_2) \, d\mu = \int_X f_1 \, d\mu + \int_X f_2 \, d\mu \in \bar{\mathcal{W}}$.*
(iii) *$\forall A \in \mathrm{B}(\mathcal{Z}, \mathcal{U})$, Af_1 is absolutely integrable over $\mathcal{X}$ and $\int_X (Af_1) \, d\mu = A \int_X f_1 \, d\mu \in \mathcal{U}$.*
(iv) *$\forall c \in \mathbb{K}$, cf_1 is absolutely integrable over $\mathcal{X}$ and $\int_X (cf_1) \, d\mu = c \int_X f_1 \, d\mu \in \bar{\mathcal{W}}$.*
(v) *$\forall H \in \mathcal{B}$, let $\mathcal{H} := (H, \mathcal{B}_H, \mu_H)$ be the $\mathcal{Y}$-valued measure subspace of $\mathcal{X}$ as defined in Proposition 11.115. Then, $f_1|_H$ is absolutely integrable over $\mathcal{H}$ and $\int_X (f_1 \chi_{H,X}) \, d\mu = \int_H f_1|_H \, d\mu_H \in \bar{\mathcal{W}}$. We will henceforth denote $\int_H f_1|_H \, d\mu_H$ by $\int_H f_1 \, d\mu$.*
(vi) *If $f_1 = f_2$ a.e. in $\mathcal{X}$, then $\int_X f_1 \, d\mu = \int_X f_2 \, d\mu \in \bar{\mathcal{W}}$.*
(vii) *$\forall$ pairwise disjoint $(E_i)_{i=1}^\infty \subseteq \mathcal{B}$, $\sum_{i=1}^\infty \int_{E_i} f_1 \, d\mu = \int_{\bigcup_{i=1}^\infty E_i} f_1 \, d\mu \in \bar{\mathcal{W}}$.*
(viii) $0 \leq \left\| \int_X f_1 \, d\mu \right\| \leq \int_X \mathcal{P} \circ f_1 \, d\mathcal{P} \circ \mu < +\infty$.

Proof We will distinguish two exhaustive and mutually exclusive cases: Case 1: $\mathcal{P} \circ \mu(X) < +\infty$; Case 2: $\mathcal{P} \circ \mu(X) = +\infty$. Case 1: $\mathcal{P} \circ \mu(X) < +\infty$. The results follow immediately from Proposition 11.130.

Case 2: $\mathcal{P} \circ \mu(X) = +\infty$. f_1, f_2, $\mathcal{P} \circ f_1$, and $\mathcal{P} \circ f_2$ are $\mathcal{B}$-measurable by Propositions 7.21 and 11.38. Let $\left(F_{i,E}\right)_{E\in\mathcal{M}(\mathcal{X})}$ be the net for $\int_X f_i \, d\mu$ as defined in Definition 11.120, $i = 1, 2$.

(i) $\forall E \in \mathcal{M}(\mathcal{X})$, by Proposition 11.130, $F_{i,E} = \int_E f_i|_E \, d\mu_E \in \bar{\mathcal{W}}$, $i = 1, 2$. Then, the net $\left(F_{i,E}\right)_{E\in\mathcal{M}(\mathcal{X})}$ is well-defined, $i = 1, 2$. By Lebesgue Dominated Convergence Theorem 11.131, f_i is integrable over $\mathcal{X}$ and $\int_X f_i \, d\mu \in \bar{\mathcal{W}}$, $i = 1, 2$. Then, by Definition 11.120, we have $\lim_{E\in\mathcal{M}(\mathcal{X})} F_{i,E} = \int_X f_i \, d\mu \in \bar{\mathcal{W}}$, $i = 1, 2$.

(ii) By Propositions 11.38, 11.39, 7.21, and 7.23, $f_1 + f_2$ and $\mathcal{P} \circ (f_1 + f_2)$ are $\mathcal{B}$-measurable. Note that $\mathcal{P} \circ (f_1 + f_2)(x) = \|f_1(x) + f_2(x)\| \leq \|f_1(x)\| + \|f_2(x)\| = (\mathcal{P} \circ f_1 + \mathcal{P} \circ f_2)(x)$, $\forall x \in X$. By Proposition 11.83, $0 \leq \int_X \mathcal{P} \circ (f_1 + f_2) \, d\mathcal{P} \circ \mu \leq \int_X (\mathcal{P} \circ f_1 + \mathcal{P} \circ f_2) \, d\mathcal{P} \circ \mu = \int_X \mathcal{P} \circ f_1 \, d\mathcal{P} \circ \mu + \int_X \mathcal{P} \circ f_2 \, d\mathcal{P} \circ \mu <$

$+\infty$. Hence, $f_1 + f_2$ is absolutely integrable over $\mathcal{X}$. Let $(\bar{F}_E)_{E\in\mathcal{M}(\mathcal{X})}$ be the net for $\int_X (f_1 + f_2)\,d\mu$ as defined in Definition 11.120. $\forall E \in \mathcal{M}(\mathcal{X})$, $\bar{F}_E = \int_E (f_1+f_2)|_E \,d\mu_E = \int_E f_1|_E \,d\mu_E + \int_E f_2|_E \,d\mu_E = F_{1,E} + F_{2,E} \in \bar{\mathcal{W}}$, where the second equality follows from Proposition 11.130. Then, $\int_X (f_1 + f_2)\,d\mu = \lim_{E\in\mathcal{M}(\mathcal{X})} \bar{F}_E = \lim_{E\in\mathcal{M}(\mathcal{X})}(F_{1,E} + F_{2,E}) = \int_X f_1\,d\mu + \int_X f_2\,d\mu \in \bar{\mathcal{W}}$, where the third equality follows from Propositions 3.66, 3.67, and 7.23. Hence, $f_1 + f_2$ is integrable over $\mathcal{X}$.

(iii) By Propositions 11.38, 7.21, and 7.62, Af_1 and $\mathcal{P}\circ(Af_1)$ are $\mathcal{B}$-measurable. Note that $\mathcal{P}\circ(Af_1)(x) = \|Af_1(x)\| \leq \|A\|\|f_1(x)\| = \|A\|\mathcal{P}\circ f_1(x)$, $\forall x \in X$. By Proposition 11.83 and the fact that $\int_X \mathcal{P}\circ f_1\,d\mathcal{P}\circ\mu \in \mathbb{R}$, we have $0 \leq \int_X \mathcal{P}\circ(Af_1)\,d\mathcal{P}\circ\mu \leq \int_X (\|A\|\mathcal{P}\circ f_1)\,d\mathcal{P}\circ\mu = \|A\|\int_X \mathcal{P}\circ f_1\,d\mathcal{P}\circ\mu < +\infty$. Hence, Af_1 is absolutely integrable over $\mathcal{X}$. Note also that $Af_1 : X \to \hat{\mathcal{W}}$, where $\hat{\mathcal{W}} := \{F \in \mathrm{B}(\mathcal{Y},\mathcal{U}) \mid F = AF_1,\ F_1 \in \mathcal{W}\}$ is a separable subspace of $\mathrm{B}(\mathcal{Y},\mathcal{U})$ by Proposition 7.126. By (i), $\int_X (Af_1)\,d\mu \in \mathcal{U}$. Let $(\bar{F}_E)_{E\in\mathcal{M}(\mathcal{X})}$ be the net for $\int_X (Af_1)\,d\mu$ as defined in Definition 11.120. $\forall E \in \mathcal{M}(\mathcal{X})$, $\bar{F}_E = \int_E (Af_1)|_E \,d\mu_E = A\int_E f_1|_E \,d\mu_E = AF_{1,E} \in \mathcal{U}$, where the second equality follows from Proposition 11.130. Then, by Propositions 3.66 and 7.62, we have $\int_X (Af_1)\,d\mu = \lim_{E\in\mathcal{M}(\mathcal{X})} \bar{F}_E = \lim_{E\in\mathcal{M}(\mathcal{X})} AF_{1,E} = A\int_X f_1\,d\mu \in \mathcal{U}$. Hence, Af_1 is integrable over $\mathcal{X}$.

(iv) This follows immediately from (iii).

(v) Fix any $H \in \mathcal{B}$. By Proposition 11.41, $f_1|_H$ and $\mathcal{P}\circ(f_1|_H) = (\mathcal{P}\circ f_1)|_H$ are $\mathcal{B}_H$-measurable. By Proposition 11.83, $0 \leq \int_H \mathcal{P}\circ(f_1|_H)\,d\mathcal{P}\circ\mu_H \leq \int_X (\mathcal{P}\circ f_1)\,d\mathcal{P}\circ\mu < +\infty$. Hence, $f_1|_H$ is absolutely integrable over $\mathcal{H}$. By (i), $f_1|_H$ is integrable over $\mathcal{H}$. By Propositions 11.38, 11.39, and 7.23, $f_1\chi_{H,X}$ is $\mathcal{B}$-measurable. By Proposition 11.83, $f_1\chi_{H,X}$ is absolutely integrable over $\mathcal{X}$. Again, by (i), $f_1\chi_{H,X}$ is integrable over $\mathcal{X}$. $\forall\epsilon \in (0,\infty) \subset \mathbb{R}$, $\exists A_0 \in \mathcal{M}(\mathcal{X})$ such that $\forall A \in \mathcal{M}(\mathcal{X})$ with $A_0 \subseteq A$, we have $\left\|\int_A (f_1\chi_{H,X})\big|_A \,d\mu_A - \int_X (f_1\chi_{H,X})\,d\mu\right\| < \epsilon/2$. We will distinguish two exhaustive and mutually exclusive subcases: Case 2a: $\mathcal{P}\circ\mu(H) = +\infty$; Case 2b: $\mathcal{P}\circ\mu(H) < +\infty$.

Case 2a: $\mathcal{P}\circ\mu(H) = +\infty$. $\exists E_0 \in \mathcal{M}(\mathcal{H})$ such that $\forall E \in \mathcal{M}(\mathcal{H})$ with $E_0 \subseteq E$, we have $\left\|\int_E f_1|_E \,d\mu_E - \int_H f_1|_H \,d\mu_H\right\| < \epsilon/2$. Let $A_1 := A_0 \cup E_0 \in \mathcal{M}(\mathcal{X})$ and $E_1 := A_1 \cap H \in \mathcal{M}(\mathcal{H})$. By Proposition 11.130, $\int_{A_1} (f_1\chi_{H,X})\big|_{A_1} \,d\mu_{A_1} = \int_{H\cap A_1} f_1|_{H\cap A_1} \,d\mu_{H\cap A_1} = \int_{E_1} f_1|_{E_1} \,d\mu_{E_1}$. Then, we have $\left\|\int_X (f_1\chi_{H,X})\,d\mu - \int_H f_1|_H \,d\mu_H\right\| \leq \left\|\int_X (f_1\chi_{H,X})\,d\mu - \int_{A_1} (f_1\chi_{H,X})\big|_{A_1} \,d\mu_{A_1}\right\| + \left\|\int_{E_1} f_1|_{E_1} \cdot d\mu_{E_1} - \int_H f_1|_H \,d\mu_H\right\| < \epsilon/2 + \epsilon/2 = \epsilon$. By the arbitrariness of ϵ, we have $\int_X (f_1\chi_{H,X})\,d\mu = \int_H f_1|_H \,d\mu_H \in \bar{\mathcal{W}}$.

Case 2b: $\mathcal{P}\circ\mu(H) < +\infty$. Let $A_1 := A_0 \cup H \in \mathcal{M}(\mathcal{X})$. By Proposition 11.130, $\int_{A_1} (f_1\chi_{H,X})\big|_{A_1} \,d\mu_{A_1} = \int_H f_1|_H \,d\mu_H$. Then, we have $\left\|\int_X (f_1 \cdot \chi_{H,X})d\mu - \int_H f_1|_H \,d\mu_H\right\| = \left\|\int_X (f_1\chi_{H,X})d\mu - \int_{A_1} (f_1\chi_{H,X})\big|_{A_1} d\mu_{A_1}\right\| < \epsilon/2$. By the arbitrariness of ϵ, we have $\int_X (f_1\chi_{H,X})\,d\mu = \int_H f_1|_H \,d\mu_H \in \bar{\mathcal{W}}$.

Hence, in both subcases, $\int_X (f_1 \chi_{H,X})\, d\mu = \int_H f_1|_H \, d\mu_H \in \bar{\mathcal{W}}$.

(vi) This follows immediately from Lebesgue Dominated Convergence Theorem 11.131.

(vii) $\forall$ pairwise disjoint $(E_i)_{i=1}^{\infty} \subseteq \mathcal{B}$, let $E := \bigcup_{i=1}^{\infty} E_i \in \mathcal{B}$ and $\bar{E}_n := \bigcup_{i=1}^{n} E_i \in \mathcal{B}$, $\forall n \in \mathbb{N}$. Then, $\lim_{n\in\mathbb{N}} f_1(x)\chi_{\bar{E}_n,X}(x) = f_1(x)\chi_{E,X}(x)$, $\forall x \in X$. Then, we have $\sum_{i=1}^{\infty} \int_{E_i} f_1 \, d\mu = \lim_{n\in\mathbb{N}} \sum_{i=1}^{n} \int_X (f_1 \chi_{E_i,X})\, d\mu = \lim_{n\in\mathbb{N}} \int_X (f_1 \chi_{\bar{E}_n,X})\, d\mu = \int_X (f_1 \chi_{E,X})\, d\mu = \int_E f_1 \, d\mu \in \bar{\mathcal{W}}$, where the first equality follows from (v), the second equality follows from (ii), the third equality follows from Lebesgue Dominated Convergence Theorem 11.131, the last equality follows from (v), and the last set membership follows from (v).

(viii) This follows immediately from Lebesgue Dominated Convergence Theorem 11.131.

This completes the proof of the proposition. □

Proposition 11.133 *Let $\mathcal{X} := (X, \mathcal{B}, \mu)$ be a σ-finite $\mathcal{Y}$-valued measure space, where $\mathcal{Y}$ is a normed linear space over $\mathbb{K}$, $\mathcal{Z}$ be a Banach space over $\mathbb{K}$, $\mathcal{W}$ be a separable subspace of* $B(\mathcal{Y}, \mathcal{Z})$, *and $f : X \to \mathcal{W}$ be absolutely integrable over $\mathcal{X}$. Then, there exists a sequence of simple functions $(\varphi_n)_{n=1}^{\infty}$, $\varphi_n : X \to \mathcal{W}$, $\forall n \in \mathbb{N}$, such that $\|\varphi_n(x)\| \le \mathcal{P} \circ f(x)$, $\forall x \in X$, $\forall n \in \mathbb{N}$, $\lim_{n\in\mathbb{N}} \varphi_n = f$ a.e. in $\mathcal{X}$, $\lim_{n\in\mathbb{N}} \int_X \varphi_n \, d\mu = \int_X f \, d\mu \in \mathcal{Z}$, $\lim_{n\in\mathbb{N}} \int_X \mathcal{P} \circ \varphi_n \, d\mathcal{P} \circ \mu = \int_X \mathcal{P} \circ f \, d\mathcal{P} \circ \mu \in [0, \infty) \subset \mathbb{R}$, and $\lim_{n\in\mathbb{N}} \int_X \mathcal{P} \circ (\varphi_n - f)\, d\mathcal{P} \circ \mu = 0$.*

Proof By Proposition 11.66, there exists a sequence of simple functions $(\varphi_n)_{n=1}^{\infty}$, $\varphi_n : X \to \mathcal{W}$, $\forall n \in \mathbb{N}$, such that $\|\varphi_n(x)\| \le \mathcal{P} \circ f(x)$, $\forall x \in X$, $\forall n \in \mathbb{N}$, and $\lim_{n\in\mathbb{N}} \varphi_n = f$ a.e. in $\mathcal{X}$. By Lebesgue Dominated Convergence Theorem 11.131, $\lim_{n\in\mathbb{N}} \int_X \varphi_n \, d\mu = \int_X f \, d\mu \in \mathcal{Z}$. Note that $\lim_{n\in\mathbb{N}} \mathcal{P} \circ \varphi_n = \mathcal{P} \circ f$ a.e. in $\mathcal{X}$, by Propositions 7.21 and 11.52. By Lebesgue Dominated Convergence Theorem 11.91, we have $\lim_{n\in\mathbb{N}} \int_X (\mathcal{P} \circ \varphi_n)\, d\mathcal{P} \circ \mu = \int_X (\mathcal{P} \circ f)\, d\mathcal{P} \circ \mu \in [0, \infty) \subset \mathbb{R}$. By Lemma 11.43, $\mathcal{P} \circ (\varphi_n - f)$ is $\mathcal{B}$-measurable. By Propositions 7.21, 7.23, 11.53, and 11.52, we have $\lim_{n\in\mathbb{N}} \mathcal{P} \circ (\varphi_n - f) = 0$ a.e. in $\mathcal{X}$. Note that $0 \le \mathcal{P} \circ (\varphi_n - f)(x) = \|\varphi_n(x) - f(x)\| \le \|\varphi_n(x)\| + \|f(x)\| \le 2\mathcal{P} \circ f(x)$, $\forall x \in X$, $\forall n \in \mathbb{N}$. By Lebesgue Dominated Convergence Theorem 11.91 and Proposition 11.75, we have $\lim_{n\in\mathbb{N}} \int_X \mathcal{P} \circ (\varphi_n - f)\, d\mathcal{P} \circ \mu = 0$. This completes the proof of the proposition. □

Proposition 11.134 *Let $\mathcal{X} := (X, \mathcal{B}, \mu)$ be a $\mathcal{Y}$-valued measure space, where $\mathcal{Y}$ is a normed linear space over $\mathbb{K}$, $\mathcal{Z}$ be a Banach space over $\mathbb{K}$, $\mathcal{W}$ be a separable subspace of* $B(\mathcal{Y}, \mathcal{Z})$, *and $f : X \to \mathcal{W}$ be absolutely integrable over $\mathcal{X}$. Then, there exists a sequence of simple functions $(\phi_n)_{n=1}^{\infty}$, $\phi_n : X \to \mathcal{W}$, $\forall n \in \mathbb{N}$, such that $\|\phi_n(x)\| \le \mathcal{P} \circ f(x)$, $\forall x \in X$, $\forall n \in \mathbb{N}$, $\lim_{n\in\mathbb{N}} \int_X \phi_n \, d\mu = \int_X f \, d\mu \in \mathcal{Z}$, $\lim_{n\in\mathbb{N}} \int_X \mathcal{P} \circ \phi_n \, d\mathcal{P} \circ \mu = \int_X \mathcal{P} \circ f \, d\mathcal{P} \circ \mu \in [0, \infty) \subset \mathbb{R}$, and $\lim_{n\in\mathbb{N}} \int_X \mathcal{P} \circ (\phi_n - f)\, d\mathcal{P} \circ \mu = 0$.*

Proof Let $\bar{\mathcal{X}} := (X, \mathcal{B}, \mathcal{P} \circ \mu)$, which is a measure space. By the assumption, f is absolutely integrable over $\bar{\mathcal{X}}$. By Proposition 11.95, there exists a sequence of simple functions $(\phi_n)_{n=1}^{\infty}$, $\phi_n : X \to \mathcal{W}$, $\forall n \in \mathbb{N}$, such that $\|\phi_n(x)\| \le \mathcal{P} \circ f(x)$,

$\forall x \in X$, $\forall n \in \mathbb{N}$, $\lim_{n\in\mathbb{N}} \int_X \mathcal{P} \circ \phi_n \, \mathrm{d}\mathcal{P} \circ \mu = \int_X \mathcal{P} \circ f \, \mathrm{d}\mathcal{P} \circ \mu \in [0, \infty) \subset \mathbb{R}$, and $\lim_{n\in\mathbb{N}} \int_X \mathcal{P} \circ (\phi_n - f) \, \mathrm{d}\mathcal{P} \circ \mu = 0$. By Propositions 11.132 and 11.125, we have $\int_X f \, \mathrm{d}\mu \in \mathcal{Z}$ and $\int_X \phi_n \, \mathrm{d}\mu \in \mathcal{Z}$, $\forall n \in \mathbb{N}$. By Proposition 11.132, $\lim_{n\in\mathbb{N}} \left\| \int_X \phi_n \, \mathrm{d}\mu - \int_X f \, \mathrm{d}\mu \right\| = \lim_{n\in\mathbb{N}} \left\| \int_X (\phi_n - f) \, \mathrm{d}\mu \right\| \leq \lim_{n\in\mathbb{N}} \int_X \mathcal{P} \circ (\phi_n - f) \, \mathrm{d}\mathcal{P} \circ \mu = 0$. Then, $\lim_{n\in\mathbb{N}} \int_X \phi_n \, \mathrm{d}\mu = \int_X f \, \mathrm{d}\mu \in \mathcal{Z}$. This completes the proof of the proposition. □

11.7 Calculation with Measures

It is easy to see that a σ-finite measure space $(X, \mathcal{B}, \mu)$ may be identified with a σ-finite $\mathbb{R}$-valued measure space $(X, \mathcal{B}, \bar{\mu})$ as follows: $\bar{\mu}(E) = \mu(E) \in [0, +\infty) \subset \mathbb{R}$, $\forall E \in \mathcal{B}$ with $\mu(E) < +\infty$; and $\bar{\mu}(E)$ is undefined, $\forall E \in \mathcal{B}$ with $\mu(E) = +\infty$. Then, $\mathcal{P} \circ \bar{\mu} = \mu$. The converse of this identification is proved in the following proposition. Thus, a σ-finite measure on $(X, \mathcal{B})$ is a special kind of σ-finite $\mathbb{R}$-valued measure on $(X, \mathcal{B})$. We will make use of this identification to imply definitions and results for σ-finite measures from definitions and results for σ-finite $\mathbb{R}$-valued measures.

Proposition 11.135 *Let $\mathcal{X} := (X, \mathcal{B}, \mu)$ be a σ-finite $\mathbb{R}$-valued measure space with $\nu := \mathcal{P} \circ \mu$ and $\mu(E) \geq 0$, $\forall E \in \operatorname{dom}(\mu)$. Then, any function $\bar{\nu} : \mathcal{B} \to \mathbb{R}_e$ satisfying $\bar{\nu}(E) = \sum_{i=1}^{\infty} \bar{\nu}(E_i)$, $\forall$ pairwise disjoint $(E_i)_{i=1}^{\infty} \subseteq \mathcal{B}$; and $\bar{\nu}(E) = \mu(E)$, $\forall E \in \operatorname{dom}(\mu)$, must be such that $\bar{\nu} = \nu$.*

Proof $\forall E \in \mathcal{B}$, we will show that $\bar{\nu}(E) = \nu(E)$ by distinguishing two exhaustive and mutually exclusive cases: Case 1: $E \in \operatorname{dom}(\mu)$; Case 2: $E \in \mathcal{B} \setminus \operatorname{dom}(\mu)$.

Case 1: $E \in \operatorname{dom}(\mu)$. Then, $\bar{\nu}(E) = \mu(E) \in [0, \infty) \subset \mathbb{R}$. By Definition 11.108, $\mu(E) \leq \nu(E) < \infty$ and

$$\nu(E) = \sup_{\substack{n\in\mathbb{Z}_+,\, (E_i)_{i=1}^n \subseteq \mathcal{B},\, \bigcup_{i=1}^n E_i = E \\ E_i \cap E_j = \emptyset,\, \forall 1 \leq i < j \leq n}} \sum_{i=1}^{n} |\mu(E_i)| = \mu(E) = \bar{\nu}(E)$$

Case 2: $E \in \mathcal{B} \setminus \operatorname{dom}(\mu)$. Then, $\nu(E) = \infty$. Since $\mathcal{X}$ is σ-finite, then $\exists (X_n)_{n=1}^{\infty} \subseteq \mathcal{B}$ with $\nu(X_n) < \infty$, $\forall n \in \mathbb{N}$ and $X = \bigcup_{n=1}^{\infty} X_n$. Without loss of generality, we may assume that $(X_n)_{n=1}^{\infty}$ is pairwise disjoint. Then, $E = \bigcup_{n=1}^{\infty} (E \cap X_n)$, $0 \leq \nu(E \cap X_n) \leq \nu(X_n) < \infty$, and $E \cap X_n \in \operatorname{dom}(\mu)$, $\forall n \in \mathbb{N}$. Then, $\bar{\nu}(E) = \sum_{n=1}^{\infty} \bar{\nu}(E \cap X_n) = \sum_{n=1}^{\infty} \nu(E \cap X_n) = \nu(E) = \infty$, where the second equality follows from Case 1.

Hence, $\bar{\nu}(E) = \nu(E)$ in both cases. This completes the proof of the proposition. □

Proposition 11.136 *Let $(X, \mathcal{B})$ be a measurable space and $\mathcal{Y}$ and $\mathcal{Z}$ be normed linear spaces over $\mathbb{K}$. Then, the following statements hold:*

(i) *Let μ_1 and μ_2 be measures on $(X, \mathcal{B})$, $a \in (0, \infty) \subset \mathbb{R}$. Define $\mu := \mu_1 + \mu_2 : \mathcal{B} \to [0, \infty] \subset \mathbb{R}_e$ by $\mu(E) = \mu_1(E) + \mu_2(E)$, $\forall E \in \mathcal{B}$; then μ is a measure on $(X, \mathcal{B})$. μ is σ-finite if μ_1 and μ_2 are σ-finite. Define $\bar{\mu} := a\mu_1 : \mathcal{B} \to [0, \infty] \subset \mathbb{R}_e$ by $\bar{\mu}(E) = a\mu_1(E)$, $\forall E \in \mathcal{B}$, and then $\bar{\mu}$ is a measure on $(X, \mathcal{B})$. $\bar{\mu}$ is σ-finite if μ_1 is σ-finite.*

(ii) *Let μ be a finite measure on $(X, \mathcal{B})$ and $a \in [0, \infty) \subset \mathbb{R}$. Define $\bar{\mu} := a\mu : \mathcal{B} \to [0, \infty) \subset \mathbb{R}$ by $\bar{\mu}(E) = a\mu(E)$, $\forall E \in \mathcal{B}$. Then, $\bar{\mu}$ is a finite measure on $(X, \mathcal{B})$.*

(iii) *Let ν_1 and ν_2 be finite $\mathcal{Y}$-valued measures on $(X, \mathcal{B})$ and $\alpha \in \mathbb{K}$. Define $\nu := \nu_1 + \nu_2 : \mathcal{B} \to \mathcal{Y}$ by $\nu(E) = \nu_1(E) + \nu_2(E)$, $\forall E \in \mathcal{B}$. Then, ν is a finite $\mathcal{Y}$-valued measure on $(X, \mathcal{B})$ with $\mathcal{P} \circ \nu \leq \mathcal{P} \circ \nu_1 + \mathcal{P} \circ \nu_2$. Define $\bar{\nu} := \alpha\nu_1 : \mathcal{B} \to \mathcal{Y}$ by $\bar{\nu}(E) = \alpha\nu_1(E)$, $\forall E \in \mathcal{B}$. Then, $\bar{\nu}$ is a finite $\mathcal{Y}$-valued measure on $(X, \mathcal{B})$ with $\mathcal{P} \circ \bar{\nu} = |\alpha| \mathcal{P} \circ \nu_1$.*

(iv) *Let μ be a finite $\mathbb{K}$-valued measure on $(X, \mathcal{B})$ and $y \in \mathcal{Y}$. Define $\nu := \mu y : \mathcal{B} \to \mathcal{Y}$ by $\nu(E) = \mu(E)y$, $\forall E \in \mathcal{B}$. Then, ν is a finite $\mathcal{Y}$ valued measure on $(X, \mathcal{B})$ with $\mathcal{P} \circ \nu = \|y\| \mathcal{P} \circ \mu$.*

(v) *Let μ be a finite $\mathcal{Y}$-valued measure on $(X, \mathcal{B})$ and $A \in \mathrm{B}(\mathcal{Y}, \mathcal{Z})$. Define $\bar{\mu} := A\mu : \mathcal{B} \to \mathcal{Z}$ by $\bar{\mu}(E) = A\mu(E)$, $\forall E \in \mathcal{B}$. Then, $\bar{\mu}$ is a finite $\mathcal{Z}$-valued measure on $(X, \mathcal{B})$ with $\mathcal{P} \circ \bar{\mu} \leq \|A\| \mathcal{P} \circ \mu$.*

Proof (i) Clearly, $\mu(\emptyset) = \mu_1(\emptyset) + \mu_2(\emptyset) = 0$ and $\bar{\mu}(\emptyset) = a\mu(\emptyset) = 0$. $\forall$ pairwise disjoint $(E_i)_{i=1}^{\infty} \subseteq \mathcal{B}$, let $E := \bigcup_{i=1}^{\infty} E_i \in \mathcal{B}$. Then, $\sum_{i=1}^{\infty} (\mu_1 + \mu_2)(E_i) = \sum_{i=1}^{\infty} (\mu_1(E_i) + \mu_2(E_i)) = \sum_{i=1}^{\infty} \mu_1(E_i) + \sum_{i=1}^{\infty} \mu_2(E_i) = \mu_1(E) + \mu_2(E) = (\mu_1 + \mu_2)(E)$ and $\sum_{i=1}^{\infty} (a\mu_1)(E_i) = \sum_{i=1}^{\infty} a\mu_1(E_i) = a \sum_{i=1}^{\infty} \mu_1(E_i) = a\mu_1(E)$. Hence, $\mu_1 + \mu_2$ and $a\mu_1$ are measures on $(X, \mathcal{B})$.

When μ_1 is σ-finite, there exists $(X_n)_{n=1}^{\infty} \subseteq \mathcal{B}$ such that $X = \bigcup_{n=1}^{\infty} X_n$ and $\mu_1(X_n) < \infty$, $\forall n \in \mathbb{N}$. Then, $(a\mu_1)(X_n) = a\mu_1(X_n) < \infty$. Hence, $a\mu_1$ is σ-finite.

When μ_2 is also σ-finite, there exists $\left(\bar{X}_n\right)_{n=1}^{\infty} \subseteq \mathcal{B}$ such that $X = \bigcup_{n=1}^{\infty} \bar{X}_n$ and $\mu_2(\bar{X}_n) < \infty$, $\forall n \in \mathbb{N}$. Then, $X = \bigcup_{n=1}^{\infty} \bigcup_{m=1}^{\infty} (X_n \cap \bar{X}_m)$, $X_n \cap \bar{X}_m \in \mathcal{B}$, and $\mu(X_n \cap \bar{X}_m) = \mu_1(X_n \cap \bar{X}_m) + \mu_2(X_n \cap \bar{X}_m) \leq \mu_1(X_n) + \mu_2(\bar{X}_m) < \infty$. Hence, μ is also σ-finite.

(ii) Clearly, $\bar{\mu}(\emptyset) = a\mu(\emptyset) = 0$. $\forall$ pairwise disjoint $(E_i)_{i=1}^{\infty} \subseteq \mathcal{B}$, let $E := \bigcup_{i=1}^{\infty} E_i \in \mathcal{B}$. Then, $\sum_{i=1}^{\infty} (a\mu)(E_i) = \sum_{i=1}^{\infty} a\mu(E_i) = a \sum_{i=1}^{\infty} \mu(E_i) = a\mu(E) = (a\mu)(E)$, where the second equality follows from the fact that $\sum_{i=1}^{\infty} \mu(E_i) = \mu(E) \leq \mu(X) < +\infty$. Hence, $a\mu$ is a finite measure on $(X, \mathcal{B})$.

(iii) Clearly, $(\nu_1 + \nu_2)(\emptyset) = \nu_1(\emptyset) + \nu_2(\emptyset) = \vartheta_{\mathcal{Y}}$ and $(\alpha\nu_1)(\emptyset) = \alpha\nu_1(\emptyset) = \vartheta_{\mathcal{Y}}$. $\forall$ pairwise disjoint $(E_i)_{i=1}^{\infty} \subseteq \mathcal{B}$, let $E := \bigcup_{i=1}^{\infty} E_i \in \mathcal{B}$. Then, $\sum_{i=1}^{\infty} \|(\nu_1 + \nu_2)(E_i)\| = \sum_{i=1}^{\infty} \|\nu_1(E_i) + \nu_2(E_i)\| \leq \sum_{i=1}^{\infty} (\|\nu_1(E_i)\| + \|\nu_2(E_i)\|) = \sum_{i=1}^{\infty} \|\nu_1(E_i)\| + \sum_{i=1}^{\infty} \|\nu_2(E_i)\| < +\infty$, where the last inequality follows from the fact that ν_1 and ν_2 are finite $\mathcal{Y}$-valued measures, and $\sum_{i=1}^{\infty} \|(\alpha\nu_1)(E_i)\| = \sum_{i=1}^{\infty} |\alpha| \|\nu_1(E_i)\| = |\alpha| \sum_{i=1}^{\infty} \|\nu_1(E_i)\| < +\infty$.

Note that $\sum_{i=1}^{\infty}(\nu_1+\nu_2)(E_i) = \sum_{i=1}^{\infty}(\nu_1(E_i)+\nu_2(E_i)) = \sum_{i=1}^{\infty}\nu_1(E_i) + \sum_{i=1}^{\infty}\nu_2(E_i) = \nu_1(E)+\nu_2(E) = (\nu_1+\nu_2)(E) \in \mathcal{Y}$, where the second equality follows from Propositions 7.23, 3.66, and 3.67, and $\sum_{i=1}^{\infty}(\alpha\nu_1)(E_i) = \sum_{i=1}^{\infty}\alpha\nu_1(E_i) = \alpha\nu_1(E) = (\alpha\nu_1)(E) \in \mathcal{Y}$, where the second equality follows from Propositions 7.23, 3.66, and 3.67. Hence, $\nu_1+\nu_2$ and $\alpha\nu_1$ are $\mathcal{Y}$-valued pre-measures on $(X, \mathcal{B})$.

$\forall E \in \mathcal{B}$, $\forall n \in \mathbb{Z}_+$, $\forall$ pairwise disjoint $(E_i)_{i=1}^{n} \subseteq \mathcal{B}$ with $E = \bigcup_{i=1}^{n} E_i$, we have $\sum_{i=1}^{n}\|(\nu_1+\nu_2)(E_i)\| \le \sum_{i=1}^{n}\|\nu_1(E_i)\| + \sum_{i=1}^{n}\|\nu_2(E_i)\| \le \mathcal{P}\circ\nu_1(E)+\mathcal{P}\circ\nu_2(E)$. Hence, $\mathcal{P}\circ(\nu_1+\nu_2)(E) \le \mathcal{P}\circ\nu_1(E)+\mathcal{P}\circ\nu_2(E)$. Then, $\mathcal{P}\circ\nu \le \mathcal{P}\circ\nu_1+\mathcal{P}\circ\nu_2$. Since ν_1 and ν_2 are finite, then $\mathcal{P}\circ\nu(X) \le \mathcal{P}\circ\nu_1(X)+\mathcal{P}\circ\nu_2(X) < +\infty$. Therefore, $\nu_1+\nu_2$ is a finite $\mathcal{Y}$-valued measure on $(X, \mathcal{B})$.

$\forall E \in \mathcal{B}$, we have

$$\mathcal{P}\circ(\alpha\nu_1)(E) = \sup_{\substack{n\in\mathbb{Z}_+,\,(E_i)_{i=1}^{n}\subseteq\mathcal{B},\,\bigcup_{i=1}^{n}E_i=E\\ E_i\cap E_j=\emptyset,\,\forall 1\le i<j\le n}} \sum_{i=1}^{n}\|(\alpha\nu_1)(E_i)\|$$

$$= \sup_{\substack{n\in\mathbb{Z}_+,\,(E_i)_{i=1}^{n}\subseteq\mathcal{B},\,\bigcup_{i=1}^{n}E_i=E\\ E_i\cap E_j=\emptyset,\,\forall 1\le i<j\le n}} |\alpha|\sum_{i=1}^{n}\|\nu_1(E_i)\| = |\alpha|\mathcal{P}\circ\nu_1(E)$$

where the last equality follows from Proposition 3.81 and the fact that $0 \le \mathcal{P}\circ\nu_1(E) \le \mathcal{P}\circ\nu_1(X) < +\infty$. Hence, $\mathcal{P}\circ(\alpha\nu_1) = |\alpha|\mathcal{P}\circ\nu_1$. Therefore, $\alpha\nu_1$ is a finite $\mathcal{Y}$-valued measure on $(X, \mathcal{B})$.

(iv) Clearly, $(\mu y)(\emptyset) = \mu(\emptyset)y = \vartheta_{\mathcal{Y}}$. $\forall$ pairwise disjoint $(E_i)_{i=1}^{\infty} \subseteq \mathcal{B}$, let $E := \bigcup_{i=1}^{\infty} E_i \in \mathcal{B}$. Then, $\sum_{i=1}^{\infty}\|(\mu y)(E_i)\| = \sum_{i=1}^{\infty}\|y\||\mu(E_i)| = \|y\|\sum_{i=1}^{\infty}|\mu(E_i)| < +\infty$. Note that $\sum_{i=1}^{\infty}(\mu y)(E_i) = \sum_{i=1}^{\infty}\mu(E_i)y = \mu(E)y = (\mu y)(E) \in \mathcal{Y}$, where the second equality follows from Propositions 7.23, 3.66, and 3.67. Hence, μy is a $\mathcal{Y}$-valued pre-measure on $(X, \mathcal{B})$. $\forall E \in \mathcal{B}$,

$$\mathcal{P}\circ(\mu y)(E) = \sup_{\substack{n\in\mathbb{Z}_+,\,(E_i)_{i=1}^{n}\subseteq\mathcal{B},\,\bigcup_{i=1}^{n}E_i=E\\ E_i\cap E_j=\emptyset,\,\forall 1\le i<j\le n}} \sum_{i=1}^{n}\|(\mu y)(E_i)\|$$

$$= \sup_{\substack{n\in\mathbb{Z}_+,\,(E_i)_{i=1}^{n}\subseteq\mathcal{B},\,\bigcup_{i=1}^{n}E_i=E\\ E_i\cap E_j=\emptyset,\,\forall 1\le i<j\le n}} \|y\|\sum_{i=1}^{n}|\mu(E_i)| = \|y\|\mathcal{P}\circ\mu(E)$$

where the last equality follows from Proposition 3.81 and the fact that $0 \le \mathcal{P}\circ\mu(E) \le \mathcal{P}\circ\mu(X) < +\infty$. Hence, $\mathcal{P}\circ(\mu y) = \|y\|\mathcal{P}\circ\mu$. Therefore, μy is a finite $\mathcal{Y}$-valued measure on $(X, \mathcal{B})$.

(v) Clearly, $\bar{\mu}(\emptyset) = A\mu(\emptyset) = \vartheta_{\mathcal{Z}}$. $\forall$ pairwise disjoint $(E_i)_{i=1}^{\infty} \subseteq \mathcal{B}$, let $E := \bigcup_{i=1}^{\infty} E_i \in \mathcal{B}$. Then, $\mu(E) = \sum_{i=1}^{\infty}\mu(E_i) \in \mathcal{Y}$, and $\sum_{i=1}^{\infty}\|\mu(E_i)\| < \infty$. This implies that $\bar{\mu}(E) = A\sum_{i=1}^{\infty}\mu(E_i) = \sum_{i=1}^{\infty}A\mu(E_i) = \sum_{i=1}^{\infty}\bar{\mu}(E_i) \in \mathcal{Z}$, where the second equality follows from Proposition 3.66, and $\sum_{i=1}^{\infty}\|\bar{\mu}(E_i)\| \le$

$\sum_{i=1}^{\infty} \|A\| \|\mu(E_i)\| = \|A\| \sum_{i=1}^{\infty} \|\mu(E_i)\| < \infty$. Hence, $\bar{\mu}$ is a $\mathcal{Z}$-valued pre-measure on $(X, \mathcal{B})$. $\forall E \in \mathcal{B}$,

$$\begin{aligned}
\mathcal{P} \circ \bar{\mu}(E) &= \sup_{n\in\mathbb{Z}_+,\, (E_i)_{i=1}^n \subseteq \mathcal{B},\, E=\bigcup_{i=1}^n E_i,\, E_i \cap E_j = \emptyset,\, \forall 1\le i<j\le n} \sum_{i=1}^{n} \|\bar{\mu}(E_i)\| \\
&\le \sup_{n\in\mathbb{Z}_+,\, (E_i)_{i=1}^n \subseteq \mathcal{B},\, E=\bigcup_{i=1}^n E_i,\, E_i \cap E_j = \emptyset,\, \forall 1\le i<j\le n} \sum_{i=1}^{n} \|A\| \|\mu(E_i)\| \\
&= \|A\| \sup_{n\in\mathbb{Z}_+,\, (E_i)_{i=1}^n \subseteq \mathcal{B},\, E=\bigcup_{i=1}^n E_i,\, E_i \cap E_j = \emptyset,\, \forall 1\le i<j\le n} \sum_{i=1}^{n} \|\mu(E_i)\| \\
&= \|A\| \mathcal{P} \circ \mu(E) \le \|A\| \mathcal{P} \circ \mu(X) < \infty
\end{aligned}$$

where the second equality follows from Proposition 3.81. Hence, $\mathcal{P} \circ \bar{\mu} \le \|A\| \mathcal{P} \circ \mu$ and $\bar{\mu}$ is a finite $\mathcal{Z}$-valued pre-measure on $(X, \mathcal{B})$. Then, $\bar{\mu}$ is a finite $\mathcal{Z}$-valued measure on $(X, \mathcal{B})$.

This completes the proof of the proposition. □

Proposition 11.137 *Let $\mathcal{X} := (X, \mathcal{B}, \mu)$ be a σ-finite measure space, $\mathcal{Y}$ be a normed linear space over $\mathbb{K}$, and ν_1 and ν_2 be $\mathcal{Y}$-valued measures on $(X, \mathcal{B})$. Assume that, $\forall E \in \mathcal{B}$ with $\mu(E) < \infty$, we have $\nu_1(E) = \nu_2(E) \in \mathcal{Y}$. Then, $\nu_1 = \nu_2$ is σ-finite.*

Proof Since μ is σ-finite, then $\exists (X_n)_{n=1}^{\infty} \subseteq \mathcal{B}$ such that $X = \bigcup_{n=1}^{\infty} X_n$ and $\mu(X_n) < \infty$, $\forall n \in \mathbb{N}$. Fix any $n \in \mathbb{N}$. By the assumption, $\mu(X_n) < \infty$ implies that $X_n \in \operatorname{dom}(\nu_1) \cap \operatorname{dom}(\nu_2)$ and $\mathcal{P} \circ \nu_i(X_n) < \infty$, $i = 1, 2$. Hence, ν_1 and ν_2 are σ-finite. Let $\mathcal{X}_n := (X_n, \mathcal{B}_n, \mu_n)$ be the finite measure subspace of $\mathcal{X}$, and $(X_n, \mathcal{B}_n, \nu_{n,i})$ be the finite $\mathcal{Y}$-valued measure subspace of $(X, \mathcal{B}, \nu_i)$, $i = 1, 2$. $\forall E \in \mathcal{B}_n$, we have $\mu(E) \le \mu(X_n) < \infty$. Then, $\nu_{n,1}(E) = \nu_1(E) = \nu_2(E) = \nu_{n,2}(E)$. Thus, $\nu_{n,1} = \nu_{n,2}$. By Proposition 11.117, we have $\nu_1 = \nu_2$. □

Proposition 11.138 *Let $(X, \mathcal{B})$ be a measurable space, $\mathcal{Y}$ be a normed linear space over $\mathbb{K}$, and $\mathcal{Z}$ be a Banach space over $\mathbb{K}$. Then, the following statements hold.*

(i) *Let μ_1 and μ_2 be σ-finite $\mathcal{Z}$-valued measures on $(X, \mathcal{B})$. Then, there exists a unique $\mathcal{Z}$-valued measure μ on $(X, \mathcal{B})$ such that $\mu(E) = \mu_1(E) + \mu_2(E) \in \mathcal{Z}$, $\forall E \in \operatorname{dom}(\mu_1) \cap \operatorname{dom}(\mu_2)$. Furthermore, μ is σ-finite. We will denote this σ-finite $\mathcal{Z}$-valued measure by $\mu := \mu_1 + \mu_2$. Then, $\mathcal{P} \circ \mu \le \mathcal{P} \circ \mu_1 + \mathcal{P} \circ \mu_2$.*
(ii) *Let μ be a σ-finite measure on $(X, \mathcal{B})$ and $\alpha \in [0, \infty) \subset \mathbb{R}$. Then, there exists a unique measure $\bar{\mu}$ on $(X, \mathcal{B})$ such that $\bar{\mu}(E) = \alpha\mu(E)$, $\forall E \in \mathcal{B}$ with $\mu(E) < \infty$. Furthermore, $\bar{\mu}$ is σ-finite. We will denote this σ-finite measure by $\bar{\mu} := \alpha\mu$. The σ-finite measure $\bar{\mu}$ satisfies: if $\alpha > 0$, then $\bar{\mu}(E) = \alpha\mu(E)$, $\forall E \in \mathcal{B}$; if $\alpha = 0$, then $\bar{\mu}(E) = 0$, $\forall E \in \mathcal{B}$.*
(iii) *Let μ be a σ-finite $\mathcal{Y}$-valued measure on $(X, \mathcal{B})$ and $\alpha \in \mathbb{K}$. Then, there exists a unique $\mathcal{Y}$-valued measure $\bar{\mu}$ on $(X, \mathcal{B})$ such that $\bar{\mu}(E) = \alpha\mu(E)$, $\forall E \in$*

dom (μ). *Furthermore, $\bar{\mu}$ is σ-finite. We will denote this σ-finite $\mathcal{Y}$-valued measure by $\bar{\mu} := \alpha\mu$. The σ-finite $\mathcal{Y}$-valued measure $\bar{\mu}$ satisfies: if $\alpha \neq 0$, then $\bar{\mu}(E) = \alpha\mu(E)$, $\forall E \in$ dom (μ), and $\bar{\mu}(E)$ is undefined, $\forall E \in \mathcal{B} \setminus$ dom (μ); if $\alpha = 0$, then $\bar{\mu}(E) = \vartheta_{\mathcal{Y}}$, $\forall E \in \mathcal{B}$; and $\mathcal{P} \circ (\alpha\mu) = |\alpha|\mathcal{P} \circ \mu$.*

(iv) *Let μ be a σ-finite $\mathbb{K}$-valued measure on $(X, \mathcal{B})$ and $y \in \mathcal{Y}$. Then, there exists a unique $\mathcal{Y}$-valued measure $\bar{\mu}$ on $(X, \mathcal{B})$ such that $\bar{\mu}(E) = \mu(E)y$, $\forall E \in$ dom (μ). Furthermore, $\bar{\mu}$ is σ-finite. We will denote this σ-finite $\mathcal{Y}$-valued measure by $\bar{\mu} := \mu y$. The σ-finite $\mathcal{Y}$-valued measure $\bar{\mu}$ satisfies: if $y \neq \vartheta_{\mathcal{Y}}$, then $\bar{\mu}(E) = \mu(E)y$, $\forall E \in$ dom (μ), and $\bar{\mu}(E)$ is undefined, $\forall E \in \mathcal{B} \setminus$ dom (μ); if $y = \vartheta_{\mathcal{Y}}$, then $\bar{\mu}(E) = \vartheta_{\mathcal{Y}}$, $\forall E \in \mathcal{B}$; and $\mathcal{P} \circ (\mu y) = \|y\|\mathcal{P} \circ \mu$.*

(v) *Let μ be a σ-finite $\mathcal{Y}$-valued measure on $(X, \mathcal{B})$ and $A \in \mathrm{B}(\mathcal{Y}, \mathcal{Z})$. Then, there is a unique $\mathcal{Z}$-valued measure $\bar{\mu}$ on $(X, \mathcal{B})$ such that $\bar{\mu}(E) = A\mu(E)$, $\forall E \in$ dom (μ). Furthermore, $\bar{\mu}$ is σ-finite and $\mathcal{P} \circ \bar{\mu} \leq \|A\|\mathcal{P} \circ \mu$. We will denote this σ-finite $\mathcal{Z}$-valued measure by $\bar{\mu} := A\mu$.*

Proof (i) Since μ_1 and μ_2 are σ-finite, then there exists $(X_n)_{n=1}^{\infty} \subseteq \mathcal{B}$ such that $X = \bigcup_{n=1}^{\infty} X_n$ and $\mathcal{P} \circ \mu_i(X_n) < \infty$, $\forall n \in \mathbb{N}$, $i = 1, 2$. Without loss of generality, we may assume that $(X_n)_{n=1}^{\infty}$ is pairwise disjoint. Let $(X_n, \mathcal{B}_n, \mu_{i,n})$ be the finite $\mathcal{Z}$-valued measure subspace of $(X, \mathcal{B}, \mu_i)$, $\forall n \in \mathbb{N}$, $i = 1, 2$. By Proposition 11.136, $\mathcal{X}_n := (X_n, \mathcal{B}_n, \mu_{1,n} + \mu_{2,n})$ is a finite $\mathcal{Z}$-valued measure space, $\forall n \in \mathbb{N}$. By Proposition 11.118, the generation process on $(\mathcal{X}_n)_{n=1}^{\infty}$ yields a unique σ-finite $\mathcal{Z}$-valued measure space $\mathcal{X} := (X, \mathcal{B}, \mu)$ on the set X.

We will show that $\mathcal{P} \circ \mu \leq \mathcal{P} \circ \mu_1 + \mathcal{P} \circ \mu_2$. $\forall E \in \mathcal{B}$, $\mathcal{P} \circ \mu(E) \in [0, \infty] \subset \mathbb{R}_e$. We will show that $\mathcal{P} \circ \mu(E) \leq \mathcal{P} \circ \mu_1(E) + \mathcal{P} \circ \mu_2(E)$ by distinguishing two exhaustive and mutually exclusive cases: Case 1: $\mathcal{P} \circ \mu_1(E) + \mathcal{P} \circ \mu_2(E) = \infty$; Case 2: $\mathcal{P} \circ \mu_1(E) + \mathcal{P} \circ \mu_2(E) < \infty$. Case 1: $\mathcal{P} \circ \mu_1(E) + \mathcal{P} \circ \mu_2(E) = \infty$. Then, it is trivially true that $\mathcal{P} \circ \mu(E) \leq \mathcal{P} \circ \mu_1(E) + \mathcal{P} \circ \mu_2(E) = \infty$. Case 2: $\mathcal{P} \circ \mu_1(E) + \mathcal{P} \circ \mu_2(E) < \infty$. Then, $\mathcal{P} \circ \mu_i(E) \in [0, \infty) \subset \mathbb{R}$, $i = 1, 2$. Hence, $E \in$ dom $(\mu_1) \cap$ dom (μ_2). Let $E_n := E \cap X_n \in \mathcal{B}_n$, $\forall n \in \mathbb{N}$. Then, $\mathcal{P} \circ \mu(E) = \sum_{n=1}^{\infty} \mathcal{P} \circ \mu(E_n) = \sum_{n=1}^{\infty} \mathcal{P} \circ (\mu_{1,n} + \mu_{2,n})(E_n) \leq \sum_{n=1}^{\infty} (\mathcal{P} \circ \mu_{1,n}(E_n) + \mathcal{P} \circ \mu_{2,n}(E_n)) = \sum_{n=1}^{\infty} \mathcal{P} \circ \mu_1(E_n) + \sum_{n=1}^{\infty} \mathcal{P} \circ \mu_2(E_n) = \mathcal{P} \circ \mu_1(E) + \mathcal{P} \circ \mu_2(E) < \infty$, where the second equality follows from Propositions 11.118 and 11.115, the first inequality follows Proposition 11.136, and the third equality follows from Proposition 11.115. Hence, $\mathcal{P} \circ \mu \leq \mathcal{P} \circ \mu_1 + \mathcal{P} \circ \mu_2$.

$\forall E \in$ dom $(\mu_1) \cap$ dom (μ_2). Then, $\mathcal{P} \circ \mu_i(E) < \infty$ and $\mu_i(E) \in \mathcal{Z}$, $i = 1, 2$. This implies that $\mathcal{P} \circ \mu(E) \leq \mathcal{P} \circ \mu_1(E) + \mathcal{P} \circ \mu_2(E) < \infty$, $E \in$ dom (μ), and $\mu(E) \in \mathcal{Z}$. Let $E_n := E \cap X_n \in \mathcal{B}_n$, $\forall n \in \mathbb{N}$. Then, $\mu(E) = \sum_{n=1}^{\infty} \mu(E_n) = \sum_{n=1}^{\infty} (\mu_{1,n} + \mu_{2,n})(E_n) = \sum_{n=1}^{\infty} (\mu_1(E_n) + \mu_2(E_n)) = \sum_{n=1}^{\infty} \mu_1(E_n) + \sum_{n=1}^{\infty} \mu_2(E_n) = \mu_1(E) + \mu_2(E) \in \mathcal{Z}$, where the first equality follows from Definition 11.108, the second equality follows from Proposition 11.118, the third equality follows from Propositions 11.115 and 11.136, the fourth equality follows from Propositions 7.23, 3.66, and 3.67, and the last equality follows from Definition 11.108. Hence, μ is the σ-finite $\mathcal{Z}$-valued measure we seek.

(ii) Define $\bar{\mu} : \mathcal{B} \to [0, \infty] \subset \mathbb{R}_e$ by $\bar{\mu}(E) = \alpha\mu(E)$, $\forall E \in \mathcal{B}$, if $\alpha > 0$, and $\bar{\mu}(E) = 0$, $\forall E \in \mathcal{B}$, if $\alpha = 0$. When $\alpha \in (0, \infty) \subset \mathbb{R}$, $\bar{\mu}$ is a σ-finite measure on $(X, \mathcal{B})$ by Proposition 11.136. When $\alpha = 0$, clearly $\bar{\mu}$ is a finite measure on $(X, \mathcal{B})$. Hence, $\bar{\mu}$ is the σ-finite measure we seek. Let $\hat{\mu}$ be any measure on $(X, \mathcal{B})$ such that $\hat{\mu}(E) = \alpha\mu(E)$, $\forall E \in \mathcal{B}$ with $\mu(E) < \infty$. Since μ is σ-finite, then there exists $(X_n)_{n=1}^{\infty} \subseteq \mathcal{B}$ such that $X = \bigcup_{n=1}^{\infty} X_n$ and $\mu(X_n) < \infty$, $\forall n \in \mathbb{N}$. Without loss of generality, we may assume that $(X_n)_{n=1}^{\infty}$ is pairwise disjoint. $\forall E \in \mathcal{B}$, let $E_n := E \cap X_n$, $\forall n \in \mathbb{N}$. Then, $\mu(E_n) < \infty$, $\forall n \in \mathbb{N}$. This implies that $\hat{\mu}(E) = \sum_{n=1}^{\infty} \hat{\mu}(E_n) = \sum_{n=1}^{\infty} \alpha\mu(E_n) = \sum_{n=1}^{\infty} \bar{\mu}(E_n) = \bar{\mu}(E) \in [0, \infty] \subset \mathbb{R}_e$. Hence, $\bar{\mu}$ is unique.

(iii) Define $\nu := |\alpha|\mathcal{P} \circ \mu$ as prescribed in (ii), which is a σ-finite measure on $(X, \mathcal{B})$. Define $\bar{\mu}$ to be a function from $\mathcal{B}$ to $\mathcal{Y}$ by, if $\alpha \neq 0$, then $\bar{\mu}(E) = \alpha\mu(E) \in \mathcal{Y}$, $\forall E \in \operatorname{dom}(\mu)$, and $\bar{\mu}(E)$ is undefined, $\forall E \in \mathcal{B} \setminus \operatorname{dom}(\mu)$; if $\alpha = 0$, then $\bar{\mu}(E) = \vartheta_{\mathcal{Y}}$, $\forall E \in \mathcal{B}$. We will show that $\bar{\mu}$ is a σ-finite $\mathcal{Y}$-valued measure on $(X, \mathcal{B})$ satisfying: $\bar{\mu}(E) = \alpha\mu(E)$, $\forall E \in \operatorname{dom}(\mu)$, and $\mathcal{P} \circ \bar{\mu} = \nu$ by distinguishing two exhaustive and mutually exclusive cases: Case 1: $\alpha = 0$; Case 2: $\alpha \neq 0$.

Case 1: $\alpha = 0$. Then, $\bar{\mu}(E) = \vartheta_{\mathcal{Y}}$ and $\nu(E) = 0$, $\forall E \in \mathcal{B}$. It is easy to show that $\bar{\mu}$ is a finite $\mathcal{Y}$-valued measure on $(X, \mathcal{B})$ with $\mathcal{P} \circ \bar{\mu} = \nu$. It is straightforward to check that $\bar{\mu}(E) = 0\mu(E) = \vartheta_{\mathcal{Y}}$, $\forall E \in \operatorname{dom}(\mu)$. Hence, this case is proved.

Case 2: $\alpha \neq 0$. $\forall E \in \mathcal{B}$ with $\nu(E) = \infty$, then $\mathcal{P} \circ \mu(E) = \infty$. This leads to that $\mu(E)$ is undefined and $\bar{\mu}(E)$ is also undefined. $\forall E \in \mathcal{B}$ with $\nu(E) < \infty$, then $\mathcal{P} \circ \mu(E) = \nu(E)/|\alpha| < \infty$ and $E \in \operatorname{dom}(\mu)$. Then, $\bar{\mu}(E) = \alpha\mu(E) \in \mathcal{Y}$. $\forall$ pairwise disjoint $(E_i)_{i=1}^{\infty} \subseteq \mathcal{B}$ with $E = \bigcup_{i=1}^{\infty} E_i$, we have $\bar{\mu}(E) = \alpha\mu(E) = \alpha\sum_{i=1}^{\infty} \mu(E_i) = \sum_{i=1}^{\infty} \alpha\mu(E_i) = \sum_{i=1}^{\infty} \bar{\mu}(E_i) \in \mathcal{Y}$ and $\sum_{i=1}^{\infty} \|\bar{\mu}(E_i)\| = \sum_{i=1}^{\infty} |\alpha|\|\mu(E_i)\| \le \sum_{i=1}^{\infty} |\alpha|\mathcal{P} \circ \mu(E_i) = |\alpha|\mathcal{P} \circ \mu(E) = \nu(E) < \infty$. $\forall E \in \mathcal{B}$ with $\nu(E) < \infty$,

$$\begin{aligned}
\nu(E) &= |\alpha|\mathcal{P} \circ \mu(E) \\
&= |\alpha| \sup_{n\in\mathbb{Z}_+,\ (E_i)_{i=1}^n\subseteq\mathcal{B},\ E=\bigcup_{i=1}^n E_i,\ E_i\cap E_j=\emptyset,\ \forall 1\le i<j\le n} \sum_{i=1}^{n} \|\mu(E_i)\| \\
&= \sup_{n\in\mathbb{Z}_+,\ (E_i)_{i=1}^n\subseteq\mathcal{B},\ E=\bigcup_{i=1}^n E_i,\ E_i\cap E_j=\emptyset,\ \forall 1\le i<j\le n} \sum_{i=1}^{n} |\alpha|\|\mu(E_i)\| \\
&= \sup_{n\in\mathbb{Z}_+,\ (E_i)_{i=1}^n\subseteq\mathcal{B},\ E=\bigcup_{i=1}^n E_i,\ E_i\cap E_j=\emptyset,\ \forall 1\le i<j\le n} \sum_{i=1}^{n} \|\bar{\mu}(E_i)\|
\end{aligned}$$

where the third equality follows from Proposition 3.81. Hence, $\bar{\mu}$ is a σ-finite $\mathcal{Y}$-valued measure on $(X, \mathcal{B})$ with $\mathcal{P} \circ \bar{\mu} = \nu = |\alpha|\mathcal{P} \circ \mu$. This case is proved.

Hence, $\bar{\mu}$ is the σ-finite $\mathcal{Y}$-valued measure on $(X, \mathcal{B})$ we seek. Let $\hat{\mu}$ be any $\mathcal{Y}$-valued measure on $(X, \mathcal{B})$ such that $\hat{\mu}(E) = \alpha\mu(E)$, $\forall E \in \operatorname{dom}(\mu)$. $\forall E \in \mathcal{B}$ with $\mathcal{P} \circ \mu(E) < \infty$, then $E \in \operatorname{dom}(\mu)$ and $\hat{\mu}(E) = \bar{\mu}(E) \in \mathcal{Y}$. By Proposition 11.137, $\bar{\mu} = \hat{\mu}$. Hence, $\bar{\mu}$ is unique.

(iv) Define $\nu := \|y\|\mathcal{P} \circ \mu$ as prescribed in (ii), which is a σ-finite measure on $(X, \mathcal{B})$. Define $\bar{\mu}$ to be a function from $\mathcal{B}$ to $\mathcal{Y}$ by, if $y \neq \vartheta_{\mathcal{Y}}$, then $\bar{\mu}(E) = \mu(E)y \in \mathcal{Y}$, $\forall E \in \operatorname{dom}(\mu)$, and $\bar{\mu}(E)$ is undefined, $\forall E \in \mathcal{B} \setminus \operatorname{dom}(\mu)$; if $y = \vartheta_{\mathcal{Y}}$, then $\bar{\mu}(E) = \vartheta_{\mathcal{Y}}, \forall E \in \mathcal{B}$. We will show that $\bar{\mu}$ is a σ-finite $\mathcal{Y}$-valued measure on $(X, \mathcal{B})$ satisfying: $\bar{\mu}(E) = \mu(E)y$, $\forall E \in \operatorname{dom}(\mu)$, and $\mathcal{P} \circ \bar{\mu} = \nu$ by distinguishing two exhaustive and mutually exclusive cases: Case 1: $y = \vartheta_{\mathcal{Y}}$; Case 2: $y \neq \vartheta_{\mathcal{Y}}$.

Case 1: $y = \vartheta_{\mathcal{Y}}$. Then, $\bar{\mu}(E) = \vartheta_{\mathcal{Y}}$ and $\nu(E) = 0$, $\forall E \in \mathcal{B}$. It is easy to show that $\bar{\mu}$ is a finite $\mathcal{Y}$-valued measure on $(X, \mathcal{B})$ with $\mathcal{P} \circ \bar{\mu} = \nu$. It is straightforward to check that $\bar{\mu}(E) = \mu(E)y = \vartheta_{\mathcal{Y}}$, $\forall E \in \operatorname{dom}(\mu)$. Hence, this case is proved.

Case 2: $y \neq \vartheta_{\mathcal{Y}}$. $\forall E \in \mathcal{B}$ with $\nu(E) = \infty$, then $\mathcal{P} \circ \mu(E) = \infty$. This leads to that $\mu(E)$ is undefined and $\bar{\mu}(E)$ is also undefined. $\forall E \in \mathcal{B}$ with $\nu(E) < \infty$, then $\mathcal{P} \circ \mu(E) = \nu(E)/\|y\| < \infty$ and $E \in \operatorname{dom}(\mu)$. Then, $\bar{\mu}(E) = \mu(E)y \in \mathcal{Y}$. $\forall$ pairwise disjoint $(E_i)_{i=1}^{\infty} \subseteq \mathcal{B}$ with $E = \bigcup_{i=1}^{\infty} E_i$, we have $\bar{\mu}(E) = \mu(E)y = (\sum_{i=1}^{\infty} \mu(E_i))y = \sum_{i=1}^{\infty} \mu(E_i)y = \sum_{i=1}^{\infty} \bar{\mu}(E_i) \in \mathcal{Y}$ and $\sum_{i=1}^{\infty} \|\bar{\mu}(E_i)\| = \sum_{i=1}^{\infty} \|y\||\mu(E_i)| \leq \sum_{i=1}^{\infty} \|y\|\mathcal{P} \circ \mu(E_i) = \|y\|\mathcal{P} \circ \mu(E) = \nu(E) < \infty$. $\forall E \in \mathcal{B}$ with $\nu(E) < \infty$,

$$
\begin{aligned}
\nu(E) &= \|y\|\mathcal{P} \circ \mu(E) \\
&= \|y\| \sup_{n\in\mathbb{Z}_+,\ (E_i)_{i=1}^n\subseteq\mathcal{B},\ E=\bigcup_{i=1}^n E_i,\ E_i\cap E_j=\emptyset,\ \forall 1\leq i<j\leq n} \sum_{i=1}^{n} |\mu(E_i)| \\
&= \sup_{n\in\mathbb{Z}_+,\ (E_i)_{i=1}^n\subseteq\mathcal{B},\ E=\bigcup_{i=1}^n E_i,\ E_i\cap E_j=\emptyset,\ \forall 1\leq i<j\leq n} \sum_{i=1}^{n} \|y\||\mu(E_i)| \\
&= \sup_{n\in\mathbb{Z}_+,\ (E_i)_{i=1}^n\subseteq\mathcal{B},\ E=\bigcup_{i=1}^n E_i,\ E_i\cap E_j=\emptyset,\ \forall 1\leq i<j\leq n} \sum_{i=1}^{n} \|\bar{\mu}(E_i)\|
\end{aligned}
$$

where the third equality follows from Proposition 3.81. Hence, $\bar{\mu}$ is a σ-finite $\mathcal{Y}$-valued measure on $(X, \mathcal{B})$ with $\mathcal{P} \circ \bar{\mu} = \nu = \|y\|\mathcal{P} \circ \mu$. This case is proved.

Hence, $\bar{\mu}$ is the σ-finite $\mathcal{Y}$-valued measure on $(X, \mathcal{B})$ we seek. Let $\hat{\mu}$ be any $\mathcal{Y}$-valued measure on $(X, \mathcal{B})$ such that $\hat{\mu}(E) = \mu(E)y$, $\forall E \in \operatorname{dom}(\mu)$. $\forall E \in \mathcal{B}$ with $\mathcal{P} \circ \mu(E) < \infty$, then $E \in \operatorname{dom}(\mu)$ and $\hat{\mu}(E) = \bar{\mu}(E) \in \mathcal{Y}$. By Proposition 11.137, $\bar{\mu} = \hat{\mu}$. Hence, $\bar{\mu}$ is unique.

(v) Since μ is σ-finite, then $\exists (X_n)_{n=1}^{\infty} \subseteq \mathcal{B}$ such that $X = \bigcup_{n=1}^{\infty} X_n$ and $\mathcal{P} \circ \mu(X_n) < \infty$, $\forall n \in \mathbb{N}$. Without loss of generality, we may assume that $(X_n)_{n=1}^{\infty}$ is pairwise disjoint. Fix any $n \in \mathbb{N}$. Let $\mathcal{X}_n := (X_n, \mathcal{B}_n, \mu_n)$ be the finite $\mathcal{Y}$-valued measure subspace of $\mathcal{X} := (X, \mathcal{B}, \mu)$. By Proposition 11.136, $\bar{\mathcal{X}}_n := (X_n, \mathcal{B}_n, \bar{\mu}_n := A\mu_n)$ is a finite $\mathcal{Z}$-valued measure space with $\mathcal{P} \circ \bar{\mu}_n \leq \|A\|\mathcal{P} \circ \mu_n$. By Proposition 11.118, the generation process on $(\bar{\mathcal{X}}_n)_{n=1}^{\infty}$ yields a unique σ-finite $\mathcal{Z}$-valued measure space $\bar{\mathcal{X}} := (X, \mathcal{B}, \bar{\mu})$ on the set X.

$\forall E \in \mathcal{B}$, $\mathcal{P} \circ \bar{\mu}(E) = \sum_{n=1}^{\infty} \mathcal{P} \circ \bar{\mu}(E \cap X_n) = \sum_{n=1}^{\infty} \mathcal{P} \circ \bar{\mu}_n(E \cap X_n) \leq \sum_{n=1}^{\infty} \|A\| \mathcal{P} \circ \mu_n(E \cap X_n)$. If $\|A\| = 0$, then $\mathcal{P} \circ \bar{\mu}(E) \leq \sum_{n=1}^{\infty} 0 = 0 = (\|A\|\mathcal{P} \circ \mu)(E)$. If $\|A\| > 0$, then $\mathcal{P} \circ \bar{\mu}(E) \leq \sum_{n=1}^{\infty} \|A\|\mathcal{P} \circ \mu_n(E \cap X_n) = \|A\| \sum_{n=1}^{\infty} \mathcal{P} \circ \mu_n(E \cap X_n) = \|A\|\mathcal{P} \circ \mu(E) = (\|A\|\mathcal{P} \circ \mu)(E)$. Hence, we have $\mathcal{P} \circ \bar{\mu} \leq \|A\|\mathcal{P} \circ \mu$.

$\forall E \in \operatorname{dom}(\mu)$, $\mathcal{P} \circ \mu(E) < \infty$. Then, $\mathcal{P} \circ \bar{\mu}(E) \leq (\|A\|\mathcal{P} \circ \mu)(E) < \infty$ and $E \in \operatorname{dom}(\bar{\mu})$. This leads to $\bar{\mu}(E) = \sum_{n=1}^{\infty} \bar{\mu}(E \cap X_n) = \sum_{n=1}^{\infty} \mu_n(E \cap X_n) = \sum_{n=1}^{\infty} A\mu_n(E \cap X_n) = \sum_{n=1}^{\infty} A\mu(E \cap X_n) = A \sum_{n=1}^{\infty} \mu(E \cap X_n) = A\mu(E)$, where the first equality follows from Definition 11.108, the second and fourth equalities follow from Propositions 11.115 and 11.136, the fifth equality follows from Proposition 3.66, and the last equality follows from Definition 11.108. Hence, $\bar{\mu}$ is the σ-finite $\mathcal{Z}$-valued measure we seek.

This completes the proof of the proposition. □

Note that, for σ-finite measures μ_1 and μ_2, by Proposition 11.135, they are identified with σ-finite $\mathbb{R}$-valued measures $\bar{\mu}_1$ and $\bar{\mu}_2$, respectively. Then, the σ-finite measure $\mu_1 + \mu_2$ defined in Proposition 11.136 is identified with the σ-finite $\mathbb{R}$-valued measure $\bar{\mu}_1 + \bar{\mu}_2$ defined in Proposition 11.138. Similar statement can be made for $\alpha\mu_1$ and $\alpha\bar{\mu}_1$, where $\alpha \in [0, \infty) \subset \mathbb{R}$. Hence, the definitions for $\mu_1 + \mu_2$ and $\alpha\mu_1$ in Propositions 11.136 and 11.138 are unambiguous.

Proposition 11.139 *Let $m, n \in \mathbb{Z}_+$, $\mathcal{Y}_i$ be a normed linear space over $\mathbb{K}$, $i = 1, \ldots, m$, $\mathcal{Z}_j$ be a normed linear space over $\mathbb{K}$, $j = 1, \ldots, n$, $\mathcal{Y} := \prod_{i=1}^{m} \mathcal{Y}_i$, $\mathcal{Z} := \prod_{j=1}^{n} \mathcal{Z}_j$, $h : \prod_{i=1}^{m} \prod_{j=1}^{n} \mathrm{B}(\mathcal{Y}_i, \mathcal{Z}_j) \to \mathrm{B}(\mathcal{Y}, \mathcal{Z})$ be defined by*

$$h(A_{1,1}, A_{2,1}, \ldots, A_{n,1}, \ldots, A_{1,m}, \ldots, A_{n,m}) = \begin{bmatrix} A_{1,1} & \cdots & A_{1,m} \\ \vdots & & \vdots \\ A_{n,1} & \cdots & A_{n,m} \end{bmatrix}$$

$\forall A_{j,i} \in \mathrm{B}(\mathcal{Y}_i, \mathcal{Z}_j)$, $i = 1, \ldots, m$, $j = 1, \ldots, n$, and $p_{j,i} : \mathrm{B}(\mathcal{Y}, \mathcal{Z}) \to \mathrm{B}(\mathcal{Y}_i, \mathcal{Z}_j)$ defined by

$$p_{j,i}\left(\begin{bmatrix} A_{1,1} & \cdots & A_{1,m} \\ \vdots & & \vdots \\ A_{n,1} & \cdots & A_{n,m} \end{bmatrix}\right) = A_{j,i}$$

$\forall \begin{bmatrix} A_{1,1} & \cdots & A_{1,m} \\ \vdots & & \vdots \\ A_{n,1} & \cdots & A_{n,m} \end{bmatrix} \in \mathrm{B}(\mathcal{Y}, \mathcal{Z})$, *$i = 1, \ldots, m$, $j = 1, \ldots, n$. Then, h and $p_{j,i}$, $i = 1, \ldots, m$, $j = 1, \ldots, n$, are bound linear functions with norm less than or equal to 1.*

Proof It is easy to show that h is a linear function. Then, we have

$$\begin{aligned}
\|h\| &= \sup_{\sum_{i=1}^{m}\sum_{j=1}^{n}\|A_{j,i}\|^2\leq 1} \left\| \begin{bmatrix} A_{1,1} & \cdots & A_{1,m} \\ \vdots & & \vdots \\ A_{n,1} & \cdots & A_{n,m} \end{bmatrix} \right\| \\
&= \sup_{\sum_{i=1}^{m}\sum_{j=1}^{n}\|A_{j,i}\|^2\leq 1} \; \sup_{\sum_{i=1}^{m}\|y_i\|^2\leq 1} \left\| \begin{bmatrix} A_{1,1} & \cdots & A_{1,m} \\ \vdots & & \vdots \\ A_{n,1} & \cdots & A_{n,m} \end{bmatrix} \begin{bmatrix} y_1 \\ \vdots \\ y_m \end{bmatrix} \right\| \\
&= \sup_{\sum_{i=1}^{m}\sum_{j=1}^{n}\|A_{j,i}\|^2\leq 1} \; \sup_{\sum_{i=1}^{m}\|y_i\|^2\leq 1} \Big(\sum_{j=1}^{n} \Big\| \sum_{i=1}^{m} A_{j,i} y_i \Big\|^2 \Big)^{1/2} \\
&\leq \sup_{\sum_{i=1}^{m}\sum_{j=1}^{n}\|A_{j,i}\|^2\leq 1} \; \sup_{\sum_{i=1}^{m}\|y_i\|^2\leq 1} \Big(\sum_{j=1}^{n} \big(\sum_{i=1}^{m} \|A_{j,i}\| \|y_i\| \big)^2 \Big)^{1/2} \\
&\leq \sup_{\sum_{i=1}^{m}\sum_{j=1}^{n}\|A_{j,i}\|^2\leq 1} \; \sup_{\sum_{i=1}^{m}\|y_i\|^2\leq 1} \Big(\sum_{j=1}^{n} \Big(\sum_{i=1}^{m} \|A_{j,i}\|^2 \Big) \Big(\sum_{i=1}^{m} \|y_i\|^2 \Big) \Big)^{1/2} \\
&\leq \sup_{\sum_{i=1}^{m}\sum_{j=1}^{n}\|A_{j,i}\|^2\leq 1} \Big(\sum_{j=1}^{n}\sum_{i=1}^{m} \|A_{j,i}\|^2 \Big)^{1/2} \leq 1
\end{aligned}$$

where the second inequality follows from the Cauchy–Schwarz Inequality.

Fix any $i \in \{1, \ldots, m\}$ and any $j \in \{1, \ldots, n\}$. It is easy to show that $p_{j,i}$ is a linear function. Then, we have

$$\begin{aligned}
\|p_{j,i}\| &= \sup_{\left\| \begin{bmatrix} A_{1,1} & \cdots & A_{1,m} \\ \vdots & & \vdots \\ A_{n,1} & \cdots & A_{n,m} \end{bmatrix} \right\| \leq 1} \|A_{j,i}\| \\
&= \sup_{\left\| \begin{bmatrix} A_{1,1} & \cdots & A_{1,m} \\ \vdots & & \vdots \\ A_{n,1} & \cdots & A_{n,m} \end{bmatrix} \right\| \leq 1} \; \sup_{\|y_i\|\leq 1} \|A_{j,i} y_i\|
\end{aligned}$$

$$\leq \sup_{\left\|\begin{bmatrix} A_{1,1} & \cdots & A_{1,m} \\ \vdots & & \vdots \\ A_{n,1} & \cdots & A_{n,m} \end{bmatrix}\right\| \leq 1} \sup_{\|y_i\| \leq 1} \left\| \begin{bmatrix} A_{1,1} & \cdots & A_{1,m} \\ \vdots & & \vdots \\ A_{n,1} & \cdots & A_{n,m} \end{bmatrix} \begin{bmatrix} \vartheta_{\mathcal{Y}_1} \\ \vdots \\ y_i \\ \vdots \\ \vartheta_{\mathcal{Y}_m} \end{bmatrix} \right\|$$

$$\leq \sup_{\left\|\begin{bmatrix} A_{1,1} & \cdots & A_{1,m} \\ \vdots & & \vdots \\ A_{n,1} & \cdots & A_{n,m} \end{bmatrix}\right\| \leq 1} \sup_{\|y_i\| \leq 1} \left\| \begin{bmatrix} A_{1,1} & \cdots & A_{1,m} \\ \vdots & & \vdots \\ A_{n,1} & \cdots & A_{n,m} \end{bmatrix} \right\| \|y_i\|$$

$$\leq 1$$

This completes the proof of the proposition. □

Proposition 11.140 *Let $\mathcal{X} := (X, \mathcal{B}, \mu)$ be a $\mathbb{K}$-valued measure space, $\mathcal{Y}_i$ be a normed linear space over $\mathbb{K}$, $\mathcal{Z}_j$ be a Banach space over $\mathbb{K}$, $\mathcal{W}_{j,i} \subseteq \mathrm{B}(\mathcal{Y}_i, \mathcal{Z}_j)$ be a separable subspace, $A_{j,i} : X \to \mathcal{W}_{j,i}$, $i = 1, 2$, $j = 1, 2$, and $A : X \to \mathrm{B}(\mathcal{Y}_1 \times \mathcal{Y}_2, \mathcal{Z}_1 \times \mathcal{Z}_2)$ be defined by $A(x) = \begin{bmatrix} A_{1,1}(x) & A_{1,2}(x) \\ A_{2,1}(x) & A_{2,2}(x) \end{bmatrix}$, $\forall x \in X$. Then, A is absolutely integrable over $\mathcal{X}$ if, and only if, $A_{j,i}$, $i = 1, 2$, $j = 1, 2$, are absolutely integrable over $\mathcal{X}$. In this case, we have $\int_X A \,\mathrm{d}\mu = \begin{bmatrix} \int_X A_{1,1} \,\mathrm{d}\mu & \int_X A_{1,2} \,\mathrm{d}\mu \\ \int_X A_{2,1} \,\mathrm{d}\mu & \int_X A_{2,2} \,\mathrm{d}\mu \end{bmatrix} \in \mathrm{B}(\mathcal{Y}_1 \times \mathcal{Y}_2, \mathcal{Z}_1 \times \mathcal{Z}_2)$.*

Proof Let $A_{j,i}$, $i = 1, 2$, $j = 1, 2$, be absolutely integrable over $\mathcal{X}$. By Proposition 11.39, $a : X \to \mathcal{W}_{1,1} \times \mathcal{W}_{2,1} \times \mathcal{W}_{1,2} \times \mathcal{W}_{2,2}$ defined by $a(x) = (A_{1,1}(x), A_{2,1}(x), A_{1,2}(x), A_{2,2}(x))$, $\forall x \in X$, is $\mathcal{B}$-measurable. By Propositions 11.139 and 11.38, $A = ha$ is $\mathcal{B}$-measurable, where h is the bounded linear function defined in Proposition 11.139 for $m = n = 2$. By Propositions 11.38 and 7.21, $\mathcal{P} \circ A$ is $\mathcal{B}$-measurable. $\forall x \in X$, $\mathcal{P} \circ A(x) = \|ha(x)\| \leq \|h\| \|a(x)\| \leq \|A_{1,1}(x)\| + \|A_{2,1}(x)\| + \|A_{1,2}(x)\| + \|A_{2,2}(x)\| = \sum_{i=1}^{2} \sum_{j=1}^{2} \mathcal{P} \circ A_{j,i}(x)$. By Proposition 11.83, $0 \leq \int_X (\mathcal{P} \circ A) \,\mathrm{d}\mathcal{P} \circ \mu \leq \int_X (\sum_{i=1}^{2} \sum_{j=1}^{2} \mathcal{P} \circ A_{j,i}) \,\mathrm{d}\mathcal{P} \circ \mu = \sum_{i=1}^{2} \sum_{j=1}^{2} \int_X (\mathcal{P} \circ A_{j,i}) \,\mathrm{d}\mathcal{P} \circ \mu < \infty$. Hence, A is absolutely integrable over $\mathcal{X}$. By Propositions 4.31, 7.66, and 7.22, $\mathrm{B}(\mathcal{Y}_i, \mathcal{Z}_j)$, $i = 1, 2$, $j = 1, 2$, $\prod_{i=1}^{2} \prod_{j=1}^{2} \mathrm{B}(\mathcal{Y}_i, \mathcal{Z}_j)$, and $\mathrm{B}(\mathcal{Y}_1 \times \mathcal{Y}_2, \mathcal{Z}_1 \times \mathcal{Z}_2)$ are Banach spaces over $\mathbb{K}$. By Propositions 7.22, 3.28, and 4.4, $\prod_{i=1}^{2} \prod_{j=1}^{2} \mathcal{W}_{j,i} \subseteq \prod_{i=1}^{2} \prod_{j=1}^{2} \mathrm{B}(\mathcal{Y}_i, \mathcal{Z}_j)$ is a separable subspace. By Proposition 11.132, $A_{j,i}$, $i = 1, 2$, $j = 1, 2$, a, and A are integrable over $\mathcal{X}$ and $\int_X A \,\mathrm{d}\mu = h \int_X a \,\mathrm{d}\mu \in \mathrm{B}(\mathcal{Y}_1 \times \mathcal{Y}_2, \mathcal{Z}_1 \times \mathcal{Z}_2)$.

On the other hand, let A be absolutely integrable over $\mathcal{X}$. Fix any $i \in \{1, 2\}$ and any $j \in \{1, 2\}$. Then, by Propositions 11.139, 11.38, $A_{j,i} = p_{j,i} A$ is $\mathcal{B}$-measurable, where $p_{j,i}$ is the bounded linear function as defined in Proposition 11.139. By Propositions 7.21 and 11.38, $\mathcal{P} \circ A_{j,i}$ is $\mathcal{B}$-measurable. $\forall x \in X$, $\mathcal{P} \circ A_{j,i}(x) \leq \| p_{j,i} \| \mathcal{P} \circ A(x) \leq \mathcal{P} \circ A(x)$. Then, by Proposition 11.83, $0 \leq \int_X \mathcal{P} \circ A_{j,i} \, d\mathcal{P} \circ \mu \leq \int_X \mathcal{P} \circ A \, d\mathcal{P} \circ \mu < \infty$. Hence, $A_{j,i}$ is absolutely integrable over $\mathcal{X}$. By Proposition 11.132, $A_{j,i}$ is integrable over $\mathcal{X}$ and $\int_X A_{j,i} \, d\mu \in \mathrm{B}(\mathcal{Y}_i, \mathcal{Z}_j)$.

Hence, A is absolutely integrable over $\mathcal{X}$ if, and only if, $A_{j,i}$, $i = 1, 2$, $j = 1, 2$, are absolutely integrable over $\mathcal{X}$.

Let A, $A_{j,i}$, $i = 1, 2$, $j = 1, 2$, be absolutely integrable over $\mathcal{X}$. We will show that $\int_X A \, d\mu = \begin{bmatrix} \int_X A_{1,1} \, d\mu & \int_X A_{1,2} \, d\mu \\ \int_X A_{2,1} \, d\mu & \int_X A_{2,2} \, d\mu \end{bmatrix} \in \mathrm{B}(\mathcal{Y}_1 \times \mathcal{Y}_2, \mathcal{Z}_1 \times \mathcal{Z}_2)$ by distinguishing two exhaustive and mutually exclusives cases: Case 1: $\mathcal{P} \circ \mu(X) < \infty$; Case 2: $\mathcal{P} \circ \mu(X) = \infty$.

Case 1: $\mathcal{P} \circ \mu(X) < \infty$. By Proposition 11.133, $\forall i \in \{1, 2\}$, $\forall j \in \{1, 2\}$, $\exists$ a sequence of simple functions $(\varphi_{j,i,n})_{n=1}^{\infty}$, $\varphi_{j,i,n} : X \to \mathcal{W}_{j,i}$, $\forall n \in \mathbb{N}$, such that $\lim_{n \in \mathbb{N}} \varphi_{j,i,n} = A_{j,i}$ a.e. in $\mathcal{X}$, $\| \varphi_{j,i,n}(x) \| \leq \mathcal{P} \circ A_{j,i}(x)$, $\forall x \in X$, $\forall n \in \mathbb{N}$, and $\int_X A_{j,i} \, d\mu = \lim_{n \in \mathbb{N}} \int_X \varphi_{j,i,n} \, d\mu \in \mathrm{B}(\mathcal{Y}_i, \mathcal{Z}_j)$. By Proposition 11.53, we have $\lim_{n \in \mathbb{N}} \psi_n = a$ a.e. in $\mathcal{X}$, where $\psi_n : X \to \prod_{i=1}^{2} \prod_{j=1}^{2} \mathcal{W}_{j,i}$ is defined by $\psi_n(x) = (\varphi_{1,1,n}(x), \varphi_{2,1,n}(x), \varphi_{1,2,n}(x), \varphi_{2,2,n}(x))$, $\forall x \in X$, $\forall n \in \mathbb{N}$. Clearly, ψ_n is a simple function, $\forall n \in \mathbb{N}$. By Lebesgue Dominated Convergence Theorem 11.131, $\int_X a \, d\mu = \lim_{n \in \mathbb{N}} \int_X \psi_n \, d\mu = \lim_{n \in \mathbb{N}} (\int_X \varphi_{1,1,n} \, d\mu, \int_X \varphi_{2,1,n} \, d\mu, \int_X \varphi_{1,2,n} \, d\mu, \int_X \varphi_{2,2,n} \, d\mu) = (\int_X A_{1,1} \, d\mu, \int_X A_{2,1} \, d\mu, \int_X A_{1,2} \, d\mu, \int_X A_{2,2} \, d\mu) \in \prod_{i=1}^{2} \prod_{j=1}^{2} \mathrm{B}(\mathcal{Y}_i, \mathcal{Z}_j)$, where the second equality follows from Proposition 11.75 and the third equality follows from Proposition 3.67. Then, by Proposition 11.132, $\int_X A \, d\mu = h \int_X a \, d\mu = \begin{bmatrix} \int_X A_{1,1} \, d\mu & \int_X A_{1,2} \, d\mu \\ \int_X A_{2,1} \, d\mu & \int_X A_{2,2} \, d\mu \end{bmatrix} \in \mathrm{B}(\mathcal{Y}_1 \times \mathcal{Y}_2, \mathcal{Z}_1 \times \mathcal{Z}_2)$.

Case 2: $\mathcal{P} \circ \mu(X) = \infty$. Let $(A_E)_{E \in \mathcal{M}(\mathcal{X})}$ and $(A_{j,i,E})_{E \in \mathcal{M}(\mathcal{X})}$ be the nets for $\int_X A \, d\mu$ and $\int_X A_{j,i} \, d\mu$ as defined in Definition 11.71, respectively, $i = 1, 2$, $j = 1, 2$. $\forall E \in \mathcal{M}(\mathcal{X})$, by Case 1, $A_E = \int_E A|_E \, d\mu_E = \begin{bmatrix} \int_E A_{1,1}|_E \, d\mu_E & \int_E A_{1,2}|_E \, d\mu_E \\ \int_E A_{2,1}|_E \, d\mu_E & \int_E A_{2,2}|_E \, d\mu_E \end{bmatrix} = \begin{bmatrix} A_{1,1,E} & A_{1,2,E} \\ A_{2,1,E} & A_{2,2,E} \end{bmatrix} \in \mathrm{B}(\mathcal{Y}_1 \times \mathcal{Y}_2, \mathcal{Z}_1 \times \mathcal{Z}_2)$, where $\mathcal{E} := (E, \mathcal{B}_E, \mu_E)$ is the finite $\mathbb{K}$-valued measure subspace of $\mathcal{X}$. Since A, $A_{j,i}$, $i = 1, 2$, $j = 1, 2$, are integrable over $\mathcal{X}$, we have

$$\int_X A \, d\mu = \lim_{E \in \mathcal{M}(\mathcal{X})} A_E = \lim_{E \in \mathcal{M}(\mathcal{X})} \begin{bmatrix} A_{1,1,E} & A_{1,2,E} \\ A_{2,1,E} & A_{2,2,E} \end{bmatrix}$$
$$= h \lim_{E \in \mathcal{M}(\mathcal{X})} (A_{1,1,E}, A_{2,1,E}, A_{1,2,E}, A_{2,2,E})$$

$$= h\left(\int_X A_{1,1}\,d\mu, \int_X A_{2,1}\,d\mu, \int_X A_{1,2}\,d\mu, \int_X A_{2,2}\,d\mu\right)$$

$$= \begin{bmatrix} \int_X A_{1,1}\,d\mu & \int_X A_{1,2}\,d\mu \\ \int_X A_{2,1}\,d\mu & \int_X A_{2,2}\,d\mu \end{bmatrix} \in B(\mathcal{Y}_1 \times \mathcal{Y}_2, \mathcal{Z}_1 \times \mathcal{Z}_2)$$

where the first equality follows from Definition 11.119, the third equality follows from Propositions 11.139 and 3.66, and the fourth equality follows from Proposition 3.67 and Definition 11.119.

Hence, in both cases, we have $\int_X A\,d\mu = \begin{bmatrix} \int_X A_{1,1}\,d\mu & \int_X A_{1,2}\,d\mu \\ \int_X A_{2,1}\,d\mu & \int_X A_{2,2}\,d\mu \end{bmatrix} \in B(\mathcal{Y}_1 \times \mathcal{Y}_2, \mathcal{Z}_1 \times \mathcal{Z}_2)$. This completes the proof of the proposition. □

Proposition 11.141 *Let $(X, \mathcal{B})$ be a measurable space, $\mathcal{Y}$ be a Banach space over $\mathbb{K}$, and Z be the set of σ-finite $\mathcal{Y}$-valued measures on $(X, \mathcal{B})$. Define vector addition and scalar multiplication on Z according to Proposition 11.138 and $\vartheta \in Z$ by $\vartheta(E) = \vartheta_{\mathcal{Y}}$, $\forall E \in \mathcal{B}$. Then, $(Z, +, \cdot, \vartheta) =: \mathcal{M}_\sigma(X, \mathcal{B}, \mathcal{Y})$ is a vector space over $\mathbb{K}$.*

Proof We will show that $\mathcal{M}_\sigma(X, \mathcal{B}, \mathcal{Y})$ is a vector space over $\mathbb{K}$. Fix any $\mu_1, \mu_2, \mu_3 \in Z$ and fix any $\alpha, \beta \in \mathbb{K}$.

(i) $\forall E \in \mathcal{B}$ with $(\mathcal{P} \circ \mu_1 + \mathcal{P} \circ \mu_2)(E) < \infty$, then $E \in \text{dom}(\mu_1) \cap \text{dom}(\mu_2)$. Then, we have $(\mu_1 + \mu_2)(E) = \mu_1(E) + \mu_2(E) = (\mu_2 + \mu_1)(E) \in \mathcal{Y}$. By Proposition 11.137, $\mu_1 + \mu_2 = \mu_2 + \mu_1$.

(ii) $\forall E \in \mathcal{B}$ with $(\mathcal{P} \circ \mu_1 + \mathcal{P} \circ \mu_2 + \mathcal{P} \circ \mu_3)(E) < \infty$, then $E \in \text{dom}(\mu_1) \cap \text{dom}(\mu_2) \cap \text{dom}(\mu_3)$ and, by Proposition 11.138, $E \in \text{dom}(\mu_1 + \mu_2) \cap \text{dom}(\mu_2 + \mu_3)$. Then, we have $((\mu_1 + \mu_2) + \mu_3)(E) = (\mu_1 + \mu_2)(E) + \mu_3(E) = (\mu_1(E) + \mu_2(E)) + \mu_3(E) = \mu_1(E) + (\mu_2(E) + \mu_3(E)) = \mu_1(E) + (\mu_2 + \mu_3)(E) = (\mu_1 + (\mu_2 + \mu_3))(E) \in \mathcal{Y}$. By Proposition 11.137, $(\mu_1 + \mu_2) + \mu_3 = \mu_1 + (\mu_2 + \mu_3)$.

(iii) $\forall E \in \mathcal{B}$ with $\mathcal{P} \circ \mu_1(E) < \infty$, we have $(\mu_1 + \vartheta)(E) = \mu_1(E) + \vartheta(E) = \mu_1(E) \in \mathcal{Y}$. By Proposition 11.137, $\mu_1 + \vartheta = \mu_1$.

(iv) $\forall E \in \mathcal{B}$ with $(\mathcal{P} \circ \mu_1 + \mathcal{P} \circ \mu_2)(E) < \infty$, we have $E \in \text{dom}(\mu_1) \cap \text{dom}(\mu_2)$ and $E \in \text{dom}(\mu_1 + \mu_2) \cap \text{dom}(\alpha\mu_1) \cap \text{dom}(\alpha\mu_2)$. Then, $(\alpha(\mu_1 + \mu_2))(E) = \alpha(\mu_1 + \mu_2)(E) = \alpha(\mu_1(E) + \mu_2(E)) = \alpha\mu_1(E) + \alpha\mu_2(E) = (\alpha\mu_1)(E) + (\alpha\mu_2)(E) = (\alpha\mu_1 + \alpha\mu_2)(E) \in \mathcal{Y}$. By Proposition 11.137, $\alpha(\mu_1 + \mu_2) = \alpha\mu_1 + \alpha\mu_2$.

(v) $\forall E \in \mathcal{B}$ with $\mathcal{P} \circ \mu_1(E) < \infty$, we have $((\alpha + \beta)\mu_1)(E) = (\alpha + \beta)\mu_1(E) = \alpha\mu_1(E) + \beta\mu_1(E) = (\alpha\mu_1)(E) + (\beta\mu_1)(E) = (\alpha\mu_1 + \beta\mu_1)(E) \in \mathcal{Y}$. By Proposition 11.137, $(\alpha + \beta)\mu_1 = \alpha\mu_1 + \beta\mu_1$.

(vi) $\forall E \in \mathcal{B}$ with $\mathcal{P} \circ \mu_1(E) < \infty$, we have $((\alpha\beta)\mu_1)(E) = (\alpha\beta)\mu_1(E) = \alpha(\beta\mu_1(E)) = \alpha(\beta\mu_1)(E) = (\alpha(\beta\mu_1))(E) \in \mathcal{Y}$. By Proposition 11.137, $(\alpha\beta)\mu_1 = \alpha(\beta\mu_1)$.

(vii) $\forall E \in \mathcal{B}$ with $\mathcal{P} \circ \mu_1(E) < \infty$, we have $(0\mu_1)(E) = 0\mu_1(E) = \vartheta_{\mathcal{Y}} = \vartheta(E) \in \mathcal{Y}$ and $(1\mu_1)(E) = 1\mu_1(E) = \mu_1(E) \in \mathcal{Y}$. By Proposition 11.137, $0\mu_1 = \vartheta$ and $1\mu_1 = \mu_1$.

Therefore, $\mathcal{M}_\sigma(X, \mathcal{B}, \mathcal{Y}) = (Z, +, \cdot, \vartheta)$ is a vector space over $\mathbb{K}$. This completes the proof of the proposition. □

Proposition 11.142 *Let $(X, \mathcal{B})$ be a measurable space, $\mathcal{Y}$ be a normed linear space over $\mathbb{K}$, and Z be the set of finite $\mathcal{Y}$-valued measures on $(X, \mathcal{B})$. Define vector addition and scalar multiplication on Z according to Proposition 11.136 and $\vartheta \in Z$ by $\vartheta(E) = \vartheta_{\mathcal{Y}}$, $\forall E \in \mathcal{B}$. Then, $\mathcal{Z} := (Z, +, \cdot, \vartheta)$ is a vector space over $\mathbb{K}$. Define $\|\cdot\| : Z \to [0, \infty) \subset \mathbb{R}$ by $\|\mu\| = \mathcal{P} \circ \mu(X)$, $\forall \mu \in Z$. Then, $\mathcal{M}_f(X, \mathcal{B}, \mathcal{Y}) := (\mathcal{Z}, \mathbb{K}, \|\cdot\|)$ is a normed linear space.*

Furthermore, if $\mathcal{Y}$ is a Banach space, then $\mathcal{M}_f(X, \mathcal{B}, \mathcal{Y})$ is a subspace of $\mathcal{M}_\sigma(X, \mathcal{B}, \mathcal{Y})$ and is a Banach space over $\mathbb{K}$.

Proof We will first show that $\mathcal{Z}$ is a vector space over $\mathbb{K}$. Fix any $\mu_1, \mu_2, \mu_3 \in Z$ and fix any $\alpha, \beta \in \mathbb{K}$.

(i) By Proposition 11.136, $\mu_1 + \mu_2, \mu_2 + \mu_1 \in Z$ and, $\forall E \in \mathcal{B}$, $(\mu_1 + \mu_2)(E) = \mu_1(E) + \mu_2(E) = (\mu_2 + \mu_1)(E) \in \mathcal{Y}$. Hence, $\mu_1 + \mu_2 = \mu_2 + \mu_1$.

(ii) By Proposition 11.136, $(\mu_1 + \mu_2) + \mu_3, \mu_1 + (\mu_2 + \mu_3) \in Z$ and, $\forall E \in \mathcal{B}$, $((\mu_1 + \mu_2) + \mu_3)(E) = (\mu_1 + \mu_2)(E) + \mu_3(E) = (\mu_1(E) + \mu_2(E)) + \mu_3(E) = \mu_1(E) + (\mu_2(E) + \mu_3(E)) = \mu_1(E) + (\mu_2 + \mu_3)(E) = (\mu_1 + (\mu_2 + \mu_3))(E) \in \mathcal{Y}$. Hence, $(\mu_1 + \mu_2) + \mu_3 = \mu_1 + (\mu_2 + \mu_3)$.

(iii) By Proposition 11.136, $\mu_1 + \vartheta \in Z$ and, $\forall E \in \mathcal{B}$, $(\mu_1 + \vartheta)(E) = \mu_1(E) + \vartheta(E) = \mu_1(E) \in \mathcal{Y}$. Hence, $\mu_1 + \vartheta = \mu_1$.

(iv) By Proposition 11.136, $\alpha(\mu_1 + \mu_2), \alpha\mu_1 + \alpha\mu_2 \in Z$ and, $\forall E \in \mathcal{B}$, $(\alpha(\mu_1 + \mu_2))(E) = \alpha(\mu_1 + \mu_2)(E) = \alpha(\mu_1(E) + \mu_2(E)) = \alpha\mu_1(E) + \alpha\mu_2(E) = (\alpha\mu_1)(E) + (\alpha\mu_2)(E) = (\alpha\mu_1 + \alpha\mu_2)(E) \in \mathcal{Y}$. Hence, $\alpha(\mu_1 + \mu_2) = \alpha\mu_1 + \alpha\mu_2$.

(v) By Proposition 11.136, $(\alpha + \beta)\mu_1, \alpha\mu_1 + \beta\mu_1 \in Z$ and, $\forall E \in \mathcal{B}$, $((\alpha + \beta)\mu_1)(E) = (\alpha + \beta)\mu_1(E) = \alpha\mu_1(E) + \beta\mu_1(E) = (\alpha\mu_1)(E) + (\beta\mu_1)(E) = (\alpha\mu_1 + \beta\mu_1)(E) \in \mathcal{Y}$. Hence, $(\alpha + \beta)\mu_1 = \alpha\mu_1 + \beta\mu_1$.

(vi) By Proposition 11.136, $(\alpha\beta)\mu_1, \alpha(\beta\mu_1) \in Z$ and, $\forall E \in \mathcal{B}$, $((\alpha\beta)\mu_1)(E) = (\alpha\beta)\mu_1(E) = \alpha(\beta\mu_1(E)) = \alpha(\beta\mu_1)(E) = (\alpha(\beta\mu_1))(E) \in \mathcal{Y}$. Then, $(\alpha\beta)\mu_1 = \alpha(\beta\mu_1)$.

(vii) By Proposition 11.136, $0\mu_1, 1\mu_1 \in Z$ and, $\forall E \in \mathcal{B}$, $(0\mu_1)(E) = 0\mu_1(E) = \vartheta_{\mathcal{Y}} = \vartheta(E) \in \mathcal{Y}$ and $(1\mu_1)(E) = 1\mu_1(E) = \mu_1(E) \in \mathcal{Y}$. Hence, $0\mu_1 = \vartheta$ and $1\mu_1 = \mu_1$.

Therefore, $\mathcal{Z}$ is a vector space over $\mathbb{K}$. In the following, we show that $\|\cdot\|$ defines a norm on Z. Fix any $\mu_1, \mu_2 \in Z$ and fix any $\alpha \in \mathbb{K}$.

(i) $\|\mu_1\| = \mathcal{P} \circ \mu_1(X) \in [0, \infty) \subset \mathbb{R}$ since $\mu_1 \in Z$. Clearly, $\|\vartheta\| = \mathcal{P} \circ \vartheta(X) = 0$. If $\|\mu_1\| = 0$, we have, $\forall E \in \mathcal{B}$, $0 \leq \|\mu_1(E)\| \leq \mathcal{P} \circ \mu_1(E) \leq \mathcal{P} \circ \mu_1(X) = \|\mu_1\| = 0$, which implies that $\mu_1(E) = \vartheta_{\mathcal{Y}}$, $\forall E \in \mathcal{B}$. Then, $\mu_1 = \vartheta$. This shows that $\|\mu_1\| = 0 \Leftrightarrow \mu_1 = \vartheta$.

(ii) $\|\mu_1 + \mu_2\| = \mathcal{P} \circ (\mu_1 + \mu_2)(X) \leq (\mathcal{P} \circ \mu_1 + \mathcal{P} \circ \mu_2)(X) = \mathcal{P} \circ \mu_1(X) + \mathcal{P} \circ \mu_2(X) = \|\mu_1\| + \|\mu_2\|$, where the inequality follows from Proposition 11.136.

(iii) By Proposition 11.136, $\|\alpha\mu_1\| = \mathcal{P} \circ (\alpha\mu_1)(X) = (|\alpha|\mathcal{P} \circ \mu_1)(X) = |\alpha|\mathcal{P} \circ \mu_1(X) = |\alpha|\|\mu_1\|$.

Hence, $\mathcal{M}_f(X, \mathcal{B}, \mathcal{Y})$ is a normed linear space.

Let $\mathcal{Y}$ be a Banach space. Clearly, $\mathcal{M}_f(X, \mathcal{B}, \mathcal{Y}) \subseteq \mathcal{M}_\sigma(X, \mathcal{B}, \mathcal{Y})$. Therefore, $\mathcal{M}_f(X, \mathcal{B}, \mathcal{Y})$ is a subspace of $\mathcal{M}_\sigma(X, \mathcal{B}, \mathcal{Y})$. Finally, we show that $\mathcal{M}_f(X, \mathcal{B}, \mathcal{Y})$ is a Banach space when $\mathcal{Y}$ is a Banach space. Let $(\mu_n)_{n=1}^\infty \subseteq \mathcal{M}_f(X, \mathcal{B}, \mathcal{Y})$ be a Cauchy sequence. $\forall \epsilon \in (0, +\infty) \subset \mathbb{R}$, $\exists N \in \mathbb{N}$, $\forall m, n \in \mathbb{N}$ with $N \leq m$ and $N \leq n$, we have $\|\mu_n - \mu_m\| = \mathcal{P} \circ (\mu_n - \mu_m)(X) < \epsilon$. $\forall E \in \mathcal{B}$, $\|\mu_n(E) - \mu_m(E)\| = \|(\mu_n - \mu_m)(E)\| \leq \mathcal{P} \circ (\mu_n - \mu_m)(E) \leq \mathcal{P} \circ (\mu_n - \mu_m)(X) < \epsilon$. Hence, $(\mu_n(E))_{n=1}^{\infty}$ is a uniform Cauchy sequence uniform in $E \in \mathcal{B}$. By $\mathcal{Y}$ being complete, $\forall E \in \mathcal{B}$, there exists a unique $\mu(E) \in \mathcal{Y}$ such that $\lim_{n\in\mathbb{N}} \mu_n(E) = \mu(E)$. This defines a function $\mu : \mathcal{B} \to \mathcal{Y}$. Clearly, $\mu(\emptyset) = \lim_{n\in\mathbb{N}} \mu_n(\emptyset) = \vartheta_{\mathcal{Y}}$. $\forall$ pairwise disjoint $(E_i)_{i=1}^{\infty} \subseteq \mathcal{B}$, let $E := \bigcup_{i=1}^{\infty} E_i \in \mathcal{B}$. Then, $\sum_{i=1}^{\infty} \|\mu(E_i)\| = \sum_{i=1}^{\infty} \lim_{n\in\mathbb{N}} \|\mu_n(E_i)\| \leq \liminf_{n\in\mathbb{N}} \sum_{i=1}^{\infty} \|\mu_n(E_i)\| \leq \liminf_{n\in\mathbb{N}} \sum_{i=1}^{\infty} \mathcal{P} \circ \mu_n(E_i) = \liminf_{n\in\mathbb{N}} \mathcal{P} \circ \mu_n(E) \leq \liminf_{n\in\mathbb{N}} \mathcal{P} \circ \mu_n(X) = \lim_{n\in\mathbb{N}} \|\mu_n\| \in \mathbb{R}$, where the first equality follows from Propositions 7.21 and 3.66, the first inequality follows from Fatou's Lemma 11.80 (with respect to counting measure on $\mathbb{N}$), the second inequality follows from Definition 11.108, and the last step follows from Propositions 4.23 and 7.21 and the fact that $(\mu_n)_{n=1}^{\infty}$ is a Cauchy sequence. This implies that

$$\sum_{i=1}^{\infty} \mu(E_i) = \sum_{i=1}^{\infty} \lim_{n\in\mathbb{N}} \mu_n(E_i) = \lim_{m\in\mathbb{N}} \lim_{n\in\mathbb{N}} \sum_{i=1}^{m} \mu_n(E_i) = \lim_{m\in\mathbb{N}} \lim_{n\in\mathbb{N}} \mu_n(\bigcup_{i=1}^{m} E_i)$$

$$= \lim_{n\in\mathbb{N}} \lim_{m\in\mathbb{N}} \mu_n(\bigcup_{i=1}^{m} E_i) = \lim_{n\in\mathbb{N}} \mu_n(E) = \mu(E) \in \mathcal{Y}$$

where fourth equality follows from Iterated Limit Theorem 4.56. Hence, μ is a $\mathcal{Y}$-valued pre-measure on $(X, \mathcal{B})$.

Note that

$$\begin{aligned}
\mathcal{P} \circ \mu(X) &= \sup_{n\in\mathbb{Z}_+,\ (E_i)_{i=1}^{n}\subseteq\mathcal{B},\ X=\bigcup_{i=1}^{n} E_i,\ E_i\cap E_j=\emptyset,\ \forall 1\leq i<j\leq n} \sum_{i=1}^{n} \|\mu(E_i)\| \\
&= \sup_{n\in\mathbb{Z}_+,\ (E_i)_{i=1}^{n}\subseteq\mathcal{B},\ X=\bigcup_{i=1}^{n} E_i,\ E_i\cap E_j=\emptyset,\ \forall 1\leq i<j\leq n} \lim_{m\in\mathbb{N}} \sum_{i=1}^{n} \|\mu_m(E_i)\| \\
&\leq \sup_{n\in\mathbb{Z}_+,\ (E_i)_{i=1}^{n}\subseteq\mathcal{B},\ X=\bigcup_{i=1}^{n} E_i,\ E_i\cap E_j=\emptyset,\ \forall 1\leq i<j\leq n} \lim_{m\in\mathbb{N}} \sum_{i=1}^{n} \mathcal{P} \circ \mu_m(E_i) \\
&= \lim_{m\in\mathbb{N}} \mathcal{P} \circ \mu_m(X) = \lim_{m\in\mathbb{N}} \|\mu_m\| \in \mathbb{R}
\end{aligned}$$

where the second equality follows from Propositions 7.21 and 3.66 and the last step follows from Propositions 4.23 and 7.21 and the fact that $(\mu_n)_{n=1}^{\infty}$ is a Cauchy sequence. Hence, μ is a finite $\mathcal{Y}$-valued measure on $(X, \mathcal{B})$ and $\mu \in \mathcal{M}_f(X, \mathcal{B}, \mathcal{Y})$.

Note that, $\forall n \in \mathbb{N}$ with $N \leq n$,

$$\begin{aligned}
\|\mu_n - \mu\| &= \mathcal{P} \circ (\mu_n - \mu)(X) \\
&= \sup_{m\in\mathbb{Z}_+,\ (E_i)_{i=1}^{m}\subseteq\mathcal{B},\ X=\bigcup_{i=1}^{m} E_i,\ E_i\cap E_j=\emptyset,\ \forall 1\leq i<j\leq m} \sum_{i=1}^{m} \|(\mu_n - \mu)(E_i)\|
\end{aligned}$$

$$= \sup_{m\in\mathbb{Z}_+,\,(E_i)_{i=1}^m\subseteq\mathcal{B},\,X=\bigcup_{i=1}^m E_i,\,E_i\cap E_j=\emptyset,\,\forall 1\le i<j\le m} \sum_{i=1}^m \|\mu_n(E_i)-\mu(E_i)\|$$

$$= \sup_{\substack{m\in\mathbb{Z}_+,\,(E_i)_{i=1}^m\subseteq\mathcal{B},\,X=\bigcup_{i=1}^m E_i\\ E_i\cap E_j=\emptyset,\,\forall 1\le i<j\le m}} \sum_{i=1}^m \lim_{k\in\mathbb{N}} \|\mu_n(E_i)-\mu_k(E_i)\|$$

$$= \sup_{\substack{m\in\mathbb{Z}_+,\,(E_i)_{i=1}^m\subseteq\mathcal{B},\,X=\bigcup_{i=1}^m E_i\\ E_i\cap E_j=\emptyset,\,\forall 1\le i<j\le m}} \lim_{k\in\mathbb{N}} \sum_{i=1}^m \|(\mu_n-\mu_k)(E_i)\|$$

$$\le \sup_{\substack{m\in\mathbb{Z}_+,\,(E_i)_{i=1}^m\subseteq\mathcal{B},\,X=\bigcup_{i=1}^m E_i\\ E_i\cap E_j=\emptyset,\,\forall 1\le i<j\le m}} \liminf_{k\in\mathbb{N}} \sum_{i=1}^m \mathcal{P}\circ(\mu_n-\mu_k)(E_i)$$

$$= \liminf_{k\in\mathbb{N}} \mathcal{P}\circ(\mu_n-\mu_k)(X) = \liminf_{k\in\mathbb{N}} \|\mu_n-\mu_k\| \le \epsilon$$

where the fourth equality follows from Propositions 7.21, 7.23, 3.66, and 3.67. Hence, $\lim_{n\in\mathbb{N}} \|\mu_n-\mu\| = 0$ and $\lim_{n\in\mathbb{N}} \mu_n = \mu \in \mathcal{M}_f(X,\mathcal{B},\mathcal{Y})$. Thus, we have shown that every Cauchy sequence in $\mathcal{M}_f(X,\mathcal{B},\mathcal{Y})$ converges in $\mathcal{M}_f(X,\mathcal{B},\mathcal{Y})$. Hence, $\mathcal{M}_f(X,\mathcal{B},\mathcal{Y})$ is a Banach space.

This completes the proof of the proposition. □

In the above proposition, $\mathcal{M}_f(X,\mathcal{B},\mathcal{Y})$ admits a topology $\mathcal{O}_1$ that is induced by the norm. We may define a topology on the vector space $\mathcal{M}_\sigma(X,\mathcal{B},\mathcal{Y})$ when $\mathcal{Y}$ is a Banach space, whose subset topology on $\mathcal{M}_f(X,\mathcal{B},\mathcal{Y})$ is exactly $\mathcal{O}_1$. Let $\mathcal{O}_B$ be a collection of subsets of $\mathcal{M}_\sigma(X,\mathcal{B},\mathcal{Y})$ such that, $\forall B\in\mathcal{O}_B$, $\exists E\in\mathcal{B}$, let $\mathcal{B}_E := \{A\in\mathcal{B} \mid A\subseteq E\}$, $\exists\epsilon\in(0,\infty)\subset\mathbb{R}$, $\exists \nu_E\in\mathcal{M}_f(E,\mathcal{B}_E,\mathcal{Y})$ such that

$$B = \Big\{\mu\in\mathcal{M}_\sigma(X,\mathcal{B},\mathcal{Y}) \;\Big|\; (E,\mathcal{B}_E,\mu_E) \text{ is the finite } \mathcal{Y}\text{-valued measure subspace of } (X,\mathcal{B},\mu) \text{ and } \|\mu_E-\nu_E\|_{\mathcal{M}_f(E,\mathcal{B}_E,\mathcal{Y})} < \epsilon\Big\}$$

It is easy to show that $\mathcal{O}_B$ is a basis for a topology $\mathcal{O}$ on $\mathcal{M}_\sigma(X,\mathcal{B},\mathcal{Y})$ by Proposition 3.18. We will abuse the notation and denote the topological space $(\mathcal{M}_\sigma(X,\mathcal{B},\mathcal{Y}),\mathcal{O})$ by $\mathcal{M}_\sigma(X,\mathcal{B},\mathcal{Y})$. With this topology in place, we may define convergence for nets in $\mathcal{M}_\sigma(X,\mathcal{B},\mathcal{Y})$ according to Sect. 3.9. For clarity, we summarize the concept of convergence for sequences in $\mathcal{M}_\sigma(X,\mathcal{B},\mathcal{Y})$ in the following definition.

Definition 11.143 Let $(X,\mathcal{B})$ be a measurable space, $\mathcal{Y}$ be a Banach space over $\mathbb{K}$, $(\mu_n)_{n=1}^\infty \subseteq \mathcal{M}_\sigma(X,\mathcal{B},\mathcal{Y})$, and $\nu\in\mathcal{M}_\sigma(X,\mathcal{B},\mathcal{Y})$. We will say the sequence $(\mu_n)_{n=1}^\infty$ converges to ν and write $\lim_{n\to\mathbb{N}} \mu_n = \nu$ if, $\forall E\in\operatorname{dom}(\nu)$, $\lim_{n\in\mathbb{N}} \mathcal{P}\circ(\mu_n-\nu)(E) = 0$. ◇

This definition means that if $(\mu_n)_{n=1}^\infty$ converges to ν, $\forall E\in\operatorname{dom}(\nu)$, the subspace $(E,\mathcal{B}_E,\mu_{n,E})$ of $(X,\mathcal{B},\mu_n)$ is finite for sufficiently large $n\in\mathbb{N}$ and

$\lim_{n\in\mathbb{N}} \mu_{n,E} = \nu_E$ in $\mathcal{M}_f(E, \mathcal{B}_E, \mathcal{Y})$, where $(E, \mathcal{B}_E, \nu_E)$ is the finite $\mathcal{Y}$-valued measure subspace of $(X, \mathcal{B}, \nu)$. Furthermore, by Proposition 11.142 and its proof and Propositions 7.21 and 3.66, we have $\lim_{n\in\mathbb{N}} \mu_n(E) = \nu(E)$ and $\lim_{n\in\mathbb{N}} \mathcal{P} \circ \mu_n(E) = \mathcal{P} \circ \nu(E)$, $\forall E \in \text{dom}(\nu)$.

Proposition 11.144 *Let $(X, \mathcal{B})$ be a measurable space, μ_1 and μ_2 be σ-finite measures on $(X, \mathcal{B})$, $a \in [0, \infty) \subset \mathbb{R}$, $f : X \to [0, \infty) \subset \mathbb{R}$ be $\mathcal{B}$-measurable, and $\mathcal{X}_i := (X, \mathcal{B}, \mu_i)$, $i = 1, 2$. Then, the following statements hold:*

(i) *If $\mu_1 \le \mu_2$ (that is $\mu_1(E) \le \mu_2(E)$, $\forall E \in \mathcal{B}$), then $0 \le \int_X f \, d\mu_1 \le \int_X f \, d\mu_2 \le \infty$.*
(ii) *Let $\mu := \mu_1 + \mu_2$ (as defined in Proposition 11.136); then $0 \le \int_X f \, d\mu = \int_X f \, d\mu_1 + \int_X f \, d\mu_2 \le \infty$.*
(iii) *Let $\nu := a\mu_1$ (as defined in Proposition 11.138); then $0 \le \int_X f \, d\nu = \int_X (af) \, d\mu_1 \le \infty$.*

Proof (i) We will distinguish two exhaustive and mutually exclusives cases: Case 1: $\int_X f \, d\mu_2 = \infty$; Case 2: $\int_X f \, d\mu_2 < \infty$. Case 1: $\int_X f \, d\mu_2 = \infty$. The result holds trivially in this case. Case 2: $\int_X f \, d\mu_2 < \infty$. By Definition 11.79, $0 \le \int_X f \, d\mu_1 = \sup_{0 \le \phi \le f} \int_X \phi \, d\mu_1 \le \infty$, where the supremum is over all simple functions $\phi : \mathcal{X}_1 \to [0, \infty) \subset \mathbb{R}$ such that $0 \le \phi(x) \le f(x)$, $\forall x \in X$. Fix any such simple function ϕ. Let ϕ admit the canonical representation $\phi = \sum_{i=1}^n a_i \chi_{A_i,X}$, where $n \in \mathbb{Z}_+$, $a_1, \ldots, a_n \in (0, \infty) \subset \mathbb{R}$ are distinct, and $A_1, \ldots, A_n \in \mathcal{B}$ are nonempty, pairwise disjoint, and $\mu_1(A_i) < \infty$, $i = 1, \ldots, n$. Then, by Proposition 11.75, $\int_X \phi \, d\mu_1 = \sum_{i=1}^n a_i \mu_1(A_i) \in [0, +\infty) \subset \mathbb{R}$. By Proposition 11.83, $\int_X \phi \, d\mu_2 \le \int_X f \, d\mu_2 < \infty$.

Claim 11.144.1 $\mu_2(A_i) < \infty$, $i = 1, \ldots, n$.

Proof of Claim Suppose $\exists i_0 \in \{1, \ldots, n\}$ such that $\mu_2(A_{i_0}) = \infty$. Since $\mathcal{X}_2$ is σ-finite, then, by Proposition 11.7, $\exists E \in \mathcal{B}$ with $E \subseteq A_{i_0}$ such that $\frac{1}{a_{i_0}} \int_X f \, d\mu_2 < \mu_2(E) < \infty$. Choose $\varphi : \mathcal{X}_2 \to [0, \infty) \subset \mathbb{R}$ by $\varphi = a_{i_0} \chi_{E,X}$. Clearly, φ is a simple function of $\mathcal{X}_2$ and $0 \le \varphi(x) \le f(x)$, $\forall x \in X$. By Proposition 11.75, $\int_X \varphi \, d\mu_2 = a_{i_0} \mu_2(E) > \int_X f \, d\mu_2$, which contradicts Definition 11.79. Hence, the claim is true. □

By the claim, ϕ is a simple function of $\mathcal{X}_2$. By Proposition 11.75, $\int_X \phi \, d\mu_2 = \sum_{i=1}^n a_i \mu_2(A_i) \ge \sum_{i=1}^n a_i \mu_1(A_i) = \int_X \phi \, d\mu_1$. This implies that $\int_X f \, d\mu_2 \ge \int_X \phi \, d\mu_2 \ge \int_X \phi \, d\mu_1$. By the arbitrariness of ϕ, we have $\int_X f \, d\mu_2 \ge \int_X f \, d\mu_1$. Hence, in both cases, we have $0 \le \int_X f \, d\mu_1 \le \int_X f \, d\mu_2 \le \infty$.

(ii) By Proposition 11.136, μ is a σ-finite measure on $(X, \mathcal{B})$. By Proposition 11.66, there exists a sequence of simple functions $(\phi_n)_{n=1}^\infty$, $\phi_n : \mathcal{X} \to [0, \infty) \subset \mathbb{R}$, $\forall n \in \mathbb{N}$, such that $0 \le \phi_n(x) \le f(x) < \infty$, $\forall x \in X$, $\forall n \in \mathbb{N}$, and $\lim_{n\in\mathbb{N}} \phi_n = f$ a.e. in $\mathcal{X}$, where $\mathcal{X} := (X, \mathcal{B}, \mu)$. Since $\mu = \mu_1 + \mu_2$, then $\mu_i \le \mu$, $i = 1, 2$. This implies that $(\phi_n)_{n=1}^\infty$ is a sequence of simple functions of $\mathcal{X}_i$ and $\lim_{n\in\mathbb{N}} \phi_n = f$ a.e. in $\mathcal{X}_i$, $i = 1, 2$. By Fatou's Lemma 11.80 and Propositions 3.83 and 11.83, we have $0 \le \int_X f \, d\mu \le \liminf_{n\in\mathbb{N}} \int_X \phi_n \, d\mu \le$

$\limsup_{n\in\mathbb{N}} \int_X \phi_n \,\mathrm{d}\mu \le \sup_{n\in\mathbb{N}} \int_X \phi_n \,\mathrm{d}\mu \le \int_X f \,\mathrm{d}\mu \le \infty$. Then, $\int_X f \,\mathrm{d}\mu = \lim_{n\in\mathbb{N}} \int_X \phi_n \,\mathrm{d}\mu \in [0, \infty] \subset \mathbb{R}_e$. By an argument that is similar to the above, we have $\int_X f \,\mathrm{d}\mu_i = \lim_{n\in\mathbb{N}} \int_X \phi_n \,\mathrm{d}\mu_i \in [0, \infty] \subset \mathbb{R}_e$, $i = 1, 2$.

Fix any $n \in \mathbb{N}$. Let ϕ_n admit the canonical representation $\phi_n = \sum_{i=1}^{m_n} a_{i,n} \cdot \chi_{A_{i,n},X}$. By Proposition 11.75, $\int_X \phi_n \,\mathrm{d}\mu = \sum_{i=1}^{m_n} a_{i,n}\mu(A_{i,n}) = \sum_{i=1}^{m_n} a_{i,n} \cdot \mu_1(A_{i,n}) + \sum_{i=1}^{m_n} a_{i,n}\mu_2(A_{i,n}) = \int_X \phi_n \,\mathrm{d}\mu_1 + \int_X \phi_n \,\mathrm{d}\mu_2 \in [0, \infty) \subset \mathbb{R}$. Then, $\int_X f \,\mathrm{d}\mu = \lim_{n\in\mathbb{N}} \int_X \phi_n \,\mathrm{d}\mu = \lim_{n\in\mathbb{N}}(\int_X \phi_n \,\mathrm{d}\mu_1 + \int_X \phi_n \,\mathrm{d}\mu_2) = \int_X f \,\mathrm{d}\mu_1 + \int_X f \,\mathrm{d}\mu_2 \in [0, \infty] \subset \mathbb{R}_e$.

(iii) We will distinguish two exhaustive and mutually exclusives cases: Case 1: $a = 0$; Case 2: $a > 0$. Case 1: $a = 0$. Then, by Proposition 11.138, $\nu(E) = 0$, $\forall E \in \mathcal{B}$. This implies that, by Definition 11.70 Proposition 11.75, $0 = \int_X f \,\mathrm{d}\nu = \int_X (af) \,\mathrm{d}\mu_1$. Hence, the result holds. Case 2: $a > 0$. By Definition 11.79, $\int_X (af) \,\mathrm{d}\mu_1 = \sup_{0\le\phi\le af} \int_X \phi \,\mathrm{d}\mu_1$, where the supremum is over all simple functions $\phi : \mathcal{X}_1 \to [0, \infty) \subset \mathbb{R}$ such that $0 \le \phi(x) \le af(x) < \infty$, $\forall x \in X$. Clearly, ϕ is such a simple function if, and only if, $\phi/a : \bar{\mathcal{X}} \to [0, \infty) \subset \mathbb{R}$ is a simple function of $\bar{\mathcal{X}} := (X, \mathcal{B}, \nu)$ with $0 \le \phi(x)/a \le f(x)$, $\forall x \in X$. Then, $\int_X (af) \,\mathrm{d}\mu_1 = \sup_{0\le\phi\le af} \int_X \phi \,\mathrm{d}\mu_1 = \sup_{0\le\phi/a\le f} \int_X \phi \,\mathrm{d}\mu_1 = \sup_{0\le\phi\le f} \int_X (a\phi) \,\mathrm{d}\mu_1 = \sup_{0\le\phi\le f} \int_X \phi \,\mathrm{d}\nu = \int_X f \,\mathrm{d}\nu \in [0, \infty] \subset \mathbb{R}_e$, where the fourth equality follows from Proposition 11.75 and the fifth equality follows from Definition 11.79.

This completes the proof of the proposition. □

Proposition 11.145 *Let $(X, \mathcal{B})$ be a measurable space, $\mathcal{Y}$ be a Banach space over $\mathbb{K}$, $\mu_1, \mu_2 \in \mathcal{M}_\sigma(X, \mathcal{B}, \mathcal{Y})$, $\mathcal{Z}$ be a Banach space over $\mathbb{K}$, $\mathcal{W}$ be a separable subspace of $\mathrm{B}(\mathcal{Y}, \mathcal{Z})$, $f : X \to \mathcal{W}$ be $\mathcal{B}$-measurable, and $\mu := \mu_1 + \mu_2 \in \mathcal{M}_\sigma(X, \mathcal{B}, \mathcal{Y})$. Assume that f is absolutely integrable over $\mathcal{X}_i := (X, \mathcal{B}, \mu_i)$, $i = 1, 2$. Then, f is absolutely integrable over $\mathcal{X} := (X, \mathcal{B}, \mu)$ and $\int_X f \,\mathrm{d}\mu = \int_X f \,\mathrm{d}\mu_1 + \int_X f \,\mathrm{d}\mu_2 \in \mathcal{Z}$.*

Proof By Propositions 11.138 and 11.136, $\mathcal{P} \circ \mu \le \mathcal{P} \circ \mu_1 + \mathcal{P} \circ \mu_2 =: \nu$ and ν is σ-finite. By Proposition 11.144, $0 \le \int_X \mathcal{P} \circ f \,\mathrm{d}\mathcal{P} \circ \mu \le \int_X \mathcal{P} \circ f \,\mathrm{d}\nu = \int_X \mathcal{P} \circ f \,\mathrm{d}\mathcal{P} \circ \mu_1 + \int_X \mathcal{P} \circ f \,\mathrm{d}\mathcal{P} \circ \mu_2 < \infty$. Hence, f is absolutely integrable over $\mathcal{X}$.

By Proposition 11.66, there exists a sequence of simple functions $(\varphi_n)_{n=1}^\infty$, $\varphi_n : \hat{\mathcal{X}} \to \mathcal{W}$, $\forall n \in \mathbb{N}$, such that $\|\varphi_n(x)\| \le \|f(x)\|$, $\forall x \in X$, $\forall n \in \mathbb{N}$, and $\lim_{n\in\mathbb{N}} \varphi_n = f$ a.e. in $\hat{\mathcal{X}}$, where $\hat{\mathcal{X}} := (X, \mathcal{B}, \nu)$. Since $\mathcal{P} \circ \mu \le \nu$ and $\mathcal{P} \circ \mu_i \le \nu$, $i = 1, 2$, then $\lim_{n\in\mathbb{N}} \varphi_n = f$ a.e. in $\mathcal{X}$, $\lim_{n\in\mathbb{N}} \varphi_n = f$ a.e. in $\mathcal{X}_i$, $i = 1, 2$, and φ_n is a simple function of $\mathcal{X}$, $\mathcal{X}_1$, or $\mathcal{X}_2$, $\forall n \in \mathbb{N}$. By Lebesgue Dominated Convergence Theorem 11.131, we have $\int_X f \,\mathrm{d}\mu = \lim_{n\in\mathbb{N}} \int_X \varphi_n \,\mathrm{d}\mu \in \mathcal{Z}$ and $\int_X f \,\mathrm{d}\mu_i = \lim_{n\in\mathbb{N}} \int_X \varphi_n \,\mathrm{d}\mu_i \in \mathcal{Z}$, $i = 1, 2$.

$\forall n \in \mathbb{N}$, let φ_n admit canonical representation $\varphi_n = \sum_{i=1}^{m_n} w_{i,n}\chi_{A_{i,n},X}$. By Proposition 11.125, we have $\int_X \varphi_n \,\mathrm{d}\mu = \sum_{i=1}^{m_n} w_{i,n}\mu(A_{i,n}) = \sum_{i=1}^{m_n} w_{i,n} \cdot (\mu_1(A_{i,n}) + \mu_2(A_{i,n})) = \sum_{i=1}^{m_n} w_{i,n}\mu_1(A_{i,n}) + \sum_{i=1}^{m_n} w_{i,n}\mu_2(A_{i,n}) = \int_X \varphi_n \,\mathrm{d}\mu_1 + \int_X \varphi_n \,\mathrm{d}\mu_2 \in \mathcal{Z}$. Then, by Propositions 7.23, 3.66, and 3.67, $\int_X f \,\mathrm{d}\mu = \lim_{n\in\mathbb{N}} \int_X \varphi_n \,\mathrm{d}\mu = \lim_{n\in\mathbb{N}}(\int_X \varphi_n \,\mathrm{d}\mu_1 + \int_X \varphi_n \,\mathrm{d}\mu_2) = \lim_{n\in\mathbb{N}} \int_X \varphi_n \,\mathrm{d}\mu_1 +$

$\lim_{n\in\mathbb{N}} \int_X \varphi_n \, d\mu_2 = \int_X f \, d\mu_1 + \int_X f \, d\mu_2 \in \mathcal{Z}$. This completes the proof of the proposition. □

Proposition 11.146 *Let $\mathcal{X} := (X, \mathcal{B}, \mu)$ be a σ-finite $\mathcal{Y}_1$-valued measure space, where $\mathcal{Y}_1$ is a normed linear space over $\mathbb{K}$, $\mathcal{Y}_2$ and $\mathcal{Z}$ be Banach spaces over $\mathbb{K}$, $A \in \mathrm{B}(\mathcal{Y}_1, \mathcal{Y}_2)$, $\nu := A\mu \in \mathcal{M}_\sigma(X, \mathcal{B}, \mathcal{Y}_2)$ as defined in Proposition 11.138, $\mathcal{W}_2$ be a separable subspace of $\mathrm{B}(\mathcal{Y}_2, \mathcal{Z})$, and $f : X \to \mathcal{W}_2$ be $\mathcal{B}$-measurable. Assume that $\|A\|\mathcal{P} \circ f$ is integrable over $(X, \mathcal{B}, \mathcal{P} \circ \mu)$. Then, f is absolutely integrable over $\bar{\mathcal{X}} := (X, \mathcal{B}, \nu)$, fA is absolutely integrable over $\mathcal{X}$, and $\int_X f \, d\nu = \int_X (fA) \, d\mu \in \mathcal{Z}$.*

Proof By Propositions 11.38, 7.64, and 7.21, fA and $\mathcal{P} \circ (fA)$ are $\mathcal{B}$-measurable. By Proposition 11.83, $0 \le \int_X \mathcal{P} \circ (fA) \, d\mathcal{P} \circ \mu \le \int_X (\|A\|\mathcal{P} \circ f) \, d\mathcal{P} \circ \mu < \infty$. Thus, fA is absolutely integrable over $\mathcal{X}$. By Proposition 11.138, $\mathcal{P} \circ \nu \le \|A\|\mathcal{P} \circ \mu$. By Proposition 11.144, $0 \le \int_X \mathcal{P} \circ f \, d\mathcal{P} \circ \nu \le \int_X \mathcal{P} \circ f \, d(\|A\|\mathcal{P} \circ \mu) = \int_X (\|A\|\mathcal{P} \circ f) \, d\mathcal{P} \circ \mu < \infty$. Hence, f is absolutely integrable over $\bar{\mathcal{X}}$.

By Proposition 11.66, there exists a sequence of simple functions $(\varphi_n)_{n=1}^{\infty}$, $\varphi_n : \mathcal{X} \to \mathcal{W}_2$, $\forall n \in \mathbb{N}$, such that $\|\varphi_n(x)\| \le \|f(x)\|$, $\forall x \in X$, $\forall n \in \mathbb{N}$, and $\lim_{n\in\mathbb{N}} \varphi_n = f$ a.e. in $\mathcal{X}$. Since $\mathcal{P} \circ \nu \le \|A\|\mathcal{P} \circ \mu$, then $\lim_{n\in\mathbb{N}} \varphi_n = f$ a.e. in $\bar{\mathcal{X}}$ and φ_n is a simple function of $\bar{\mathcal{X}}$, $\forall n \in \mathbb{N}$. By Lebesgue Dominated Convergence Theorem 11.131, we have $\int_X f \, d\nu = \lim_{n\in\mathbb{N}} \int_X \varphi_n \, d\nu \in \mathcal{Z}$. Clearly, $\|\varphi_n(x)A\| \le \|A\|\|f(x)\|$, $\forall x \in X$, $\forall n \in \mathbb{N}$. Note that $fA : X \to \mathcal{W}_1$ and $\varphi_n A : X \to \mathcal{W}_1$, $\forall n \in \mathbb{N}$, where $\mathcal{W}_1 := \{F \in \mathrm{B}(\mathcal{Y}_1, \mathcal{Z}) \mid F = F_1 A, \ F_1 \in \mathcal{W}_2\}$ is a separable subspace of $\mathrm{B}(\mathcal{Y}_1, \mathcal{Z})$ by Proposition 7.126. By Proposition 11.52, we have $\lim_{n\in\mathbb{N}} \varphi_n A = fA$ a.e. in $\mathcal{X}$. By Lebesgue Dominated Convergence Theorem 11.131, $\int_X (fA) \, d\mu = \lim_{n\in\mathbb{N}} \int_X (\varphi_n A) \, d\mu \in \mathcal{Z}$. By Proposition 11.125, we have $\int_X (\varphi_n A) \, d\mu = \int_X \varphi_n \, d\nu \in \mathcal{Z}$, $\forall n \in \mathbb{N}$. Then, $\int_X (fA) \, d\mu = \int_X f \, d\nu \in \mathcal{Z}$. This completes the proof of the proposition. □

Proposition 11.147 *Let $n, m \in \mathbb{Z}_+$, $(X, \mathcal{B})$ be a measurable space, and $\mu_{j,i}$ be a σ-finite $\mathrm{B}(\mathcal{Y}_i, \mathcal{Z}_j)$-valued measure on $(X, \mathcal{B})$, where $\mathcal{Y}_i$ is a normed linear space over $\mathbb{K}$ and $\mathcal{Z}_j$ is a Banach space over $\mathbb{K}$, $i = 1, \ldots, m$, $j = 1, \ldots, n$. Then, there exists a unique σ-finite $\mathrm{B}(\mathcal{Y}, \mathcal{Z})$-valued measure μ on $(X, \mathcal{B})$ such that*

$$\mu(E) = \begin{bmatrix} \mu_{1,1}(E) & \cdots & \mu_{1,m}(E) \\ \vdots & & \vdots \\ \mu_{n,1}(E) & \cdots & \mu_{n,m}(E) \end{bmatrix} \in \mathrm{B}(\mathcal{Y}, \mathcal{Z}), \ \forall E \in \bigcap_{i=1}^{m} \bigcap_{j=1}^{n} \operatorname{dom}(\mu_{j,i}),$$

where $\mathcal{Y} := \prod_{i=1}^{m} \mathcal{Y}_i$ and $\mathcal{Z} := \prod_{j=1}^{n} \mathcal{Z}_j$. Furthermore, $\max_{\substack{i\in\{1,\ldots,m\} \\ j\in\{1,\ldots,n\}}} \mathcal{P} \circ \mu_{j,i} \le \mathcal{P} \circ \mu \le \sum_{i=1}^{m} \sum_{j=1}^{n} \mathcal{P} \circ \mu_{j,i}$. μ will be called the vector measure *and denoted by*

$$\begin{bmatrix} \mu_{1,1} & \cdots & \mu_{1,m} \\ \vdots & & \vdots \\ \mu_{n,1} & \cdots & \mu_{n,m} \end{bmatrix}.$$

Proof Consider the special case where $\mu_{j,i}$'s are finite. Then, μ is uniquely defined since $\operatorname{dom}(\mu_{j,i}) = \mathcal{B}$, $i = 1, \ldots, m$, $j = 1, \ldots, n$. This implies that

$$\mu(\emptyset) = \begin{bmatrix} \mu_{1,1}(\emptyset) & \cdots & \mu_{1,m}(\emptyset) \\ \vdots & & \vdots \\ \mu_{n,1}(\emptyset) & \cdots & \mu_{n,m}(\emptyset) \end{bmatrix} = \begin{bmatrix} \vartheta_{\mathrm{B}(\mathcal{Y}_1,\mathcal{Z}_1)} & \cdots & \vartheta_{\mathrm{B}(\mathcal{Y}_m,\mathcal{Z}_1)} \\ \vdots & & \vdots \\ \vartheta_{\mathrm{B}(\mathcal{Y}_1,\mathcal{Z}_n)} & \cdots & \vartheta_{\mathrm{B}(\mathcal{Y}_m,\mathcal{Z}_n)} \end{bmatrix} = \vartheta_{\mathrm{B}(\mathcal{Y},\mathcal{Z})}$$

$\forall$ pairwise disjoint $(E_k)_{k=1}^{\infty} \subseteq \mathcal{B}$, we have

$$\begin{aligned}\sum_{k=1}^{\infty} \|\mu(E_k)\| &\le \sum_{k=1}^{\infty} \|h\| (\sum_{i=1}^{m} \sum_{j=1}^{n} \|\mu_{j,i}(E_k)\|^2)^{\frac{1}{2}} \\ &\le \sum_{k=1}^{\infty} \sum_{i=1}^{m} \sum_{j=1}^{n} \|\mu_{j,i}(E_k)\| = \sum_{i=1}^{m} \sum_{j=1}^{n} \sum_{k=1}^{\infty} \|\mu_{j,i}(E_k)\| < \infty\end{aligned}$$

where the first inequality follows from Proposition 11.139 with h being the bounded linear function defined in Proposition 11.139 and the last inequality follows from the fact that $\mu_{j,i}$ is a finite $\mathrm{B}(\mathcal{Y}_i, \mathcal{Z}_j)$-valued measure on $(X, \mathcal{B})$ and therefore is a finite $\mathrm{B}(\mathcal{Y}_i, \mathcal{Z}_j)$-valued pre-measure on $(X, \mathcal{B})$, $i = 1, \ldots, m$, $j = 1, \ldots, n$. We also have $\sum_{k=1}^{\infty} \mu(E_k) = \sum_{k=1}^{\infty} \begin{bmatrix} \mu_{1,1}(E_k) & \cdots & \mu_{1,m}(E_k) \\ \vdots & & \vdots \\ \mu_{n,1}(E_k) & \cdots & \mu_{n,m}(E_k) \end{bmatrix} = \begin{bmatrix} \sum_{k=1}^{\infty} \mu_{1,1}(E_k) & \cdots & \sum_{k=1}^{\infty} \mu_{1,m}(E_k) \\ \vdots & & \vdots \\ \sum_{k=1}^{\infty} \mu_{n,1}(E_k) & \cdots & \sum_{k=1}^{\infty} \mu_{n,m}(E_k) \end{bmatrix} = \mu(E) \in \mathrm{B}(\mathcal{Y}, \mathcal{Z})$, where $E := \bigcup_{k=1}^{\infty} E_k \in \mathcal{B}$; the second equality follows from Propositions 3.66, 3.67, and Proposition 11.139; and the third equality follows from the fact that $\mu_{j,i}$ is a finite $\mathrm{B}(\mathcal{Y}_i, \mathcal{Z}_j)$-valued pre-measure on $(X, \mathcal{B})$, $i = 1, \ldots, m$ $j = 1, \ldots, n$. This shows that μ is a $\mathcal{Y}$-valued pre-measure on $(X, \mathcal{B})$. Observe that, $\forall s \in \{1, \ldots, m\}$, $\forall t \in \{1, \ldots, n\}$, $\forall E \in \mathcal{B}$,

$$\begin{aligned}\mathcal{P} \circ \mu_{t,s}(E) &= \sup_{k \in \mathbb{Z}_+,\, (E_i)_{i=1}^{k} \subseteq \mathcal{B},\, E = \bigcup_{i=1}^{k} E_i,\, E_i \cap E_j = \emptyset,\, \forall 1 \le i < j \le k} \sum_{i=1}^{k} \|\mu_{t,s}(E_i)\| \\ &= \sup_{k \in \mathbb{Z}_+,\, (E_i)_{i=1}^{k} \subseteq \mathcal{B},\, E = \bigcup_{i=1}^{k} E_i,\, E_i \cap E_j = \emptyset,\, \forall 1 \le i < j \le k} \sum_{i=1}^{k} \|p_{t,s}\mu(E_i)\| \\ &\le \sup_{k \in \mathbb{Z}_+,\, (E_i)_{i=1}^{k} \subseteq \mathcal{B},\, E = \bigcup_{i=1}^{k} E_i,\, E_i \cap E_j = \emptyset,\, \forall 1 \le i < j \le k} \sum_{i=1}^{k} \|\mu(E_i)\| \\ &= \mathcal{P} \circ \mu(E)\end{aligned}$$

where the first equality follows from Definition 11.108, the second equality and the first inequality follow from Propositions 11.139 and 7.64, and the last equality follows from Definition 11.99. Observe also that, $\forall E \in \mathcal{B}$,

$$\begin{aligned}
\mathcal{P} \circ \mu(E) &= \sup_{k\in\mathbb{Z}_+,\ (E_i)_{i=1}^k \subseteq \mathcal{B},\ E=\bigcup_{i=1}^k E_i,\ E_i\cap E_j=\emptyset,\ \forall 1\leq i<j\leq k} \sum_{i=1}^{k} \|\mu(E_i)\| \\
&\leq \sup_{k\in\mathbb{Z}_+,\ (E_i)_{i=1}^k \subseteq \mathcal{B},\ E=\bigcup_{i=1}^k E_i,\ E_i\cap E_j=\emptyset,\ \forall 1\leq i<j\leq k} \sum_{i=1}^{k}\sum_{s=1}^{m}\sum_{t=1}^{n} \|\mu_{t,s}(E_i)\| \\
&\leq \sum_{s=1}^{m}\sum_{t=1}^{n} \sup_{k\in\mathbb{Z}_+,\ (E_i)_{i=1}^k \subseteq \mathcal{B},\ E=\bigcup_{i=1}^k E_i,\ E_i\cap E_j=\emptyset,\ \forall 1\leq i<j\leq k} \sum_{i=1}^{k} \|\mu_{t,s}(E_i)\| \\
&= \sum_{s=1}^{m}\sum_{t=1}^{n} \mathcal{P} \circ \mu_{t,s}(E) \leq \sum_{s=1}^{m}\sum_{t=1}^{n} \mathcal{P} \circ \mu_{t,s}(X) < \infty
\end{aligned}$$

where the first equality follows from Definition 11.99, the first inequality follows from Propositions 11.139, 7.64, and 7.22, the second inequality follows from Proposition 3.81, and the second equality follows from Definition 11.108. Then, μ is a finite $\mathcal{Y}$-valued pre-measure on $(X, \mathcal{B})$ and $\max_{\substack{i\in\{1,\ldots,m\}\\ j\in\{1,\ldots,n\}}} \mathcal{P} \circ \mu_{j,i} \leq \mathcal{P} \circ \mu \leq \sum_{i=1}^{m}\sum_{j=1}^{n} \mathcal{P}\circ\mu_{j,i}$. Then, $(X, \mathcal{B}, \mu)$ is a finite $\mathrm{B}(\mathcal{Y}, \mathcal{Z})$-valued measure space. This completes the proof for the special case.

Consider the general case where $\mu_{j,i}$'s are σ-finite. Then, $\exists (X_k)_{k=1}^{\infty} \subseteq \mathcal{B}$ such that $X = \bigcup_{k=1}^{\infty} X_k$ and $\mathcal{P} \circ \mu_{j,i}(X_k) < \infty$, $\forall i \in \{1, \ldots, m\}$, $\forall j \in \{1, \ldots, n\}$, $\forall k \in \mathbb{N}$. Without loss of generality, assume that $(X_k)_{k=1}^{\infty}$ is pairwise disjoint. Let $\mathcal{X}_{j,i,k} := (X_k, \mathcal{B}_k, \mu_{j,i,k})$ be the finite $\mathrm{B}(\mathcal{Y}_i, \mathcal{Z}_j)$-valued measure subspace of $(X, \mathcal{B}, \mu_{j,i})$, $\forall k \in \mathbb{N}$, $\forall i \in \{1, \ldots, m\}$, $\forall j \in \{1, \ldots, n\}$. Then, by the special case, there exists a unique finite $\mathrm{B}(\mathcal{Y}, \mathcal{Z})$-valued measure $\bar{\mu}_k$ on $(X_k, \mathcal{B}_k)$ such that

$$\bar{\mu}_k(E) = \begin{bmatrix} \mu_{1,1,k}(E) & \cdots & \mu_{1,m,k}(E) \\ \vdots & & \vdots \\ \mu_{n,1,k}(E) & \cdots & \mu_{n,m,k}(E) \end{bmatrix} = \begin{bmatrix} \mu_{1,1}(E) & \cdots & \mu_{1,m}(E) \\ \vdots & & \vdots \\ \mu_{n,1}(E) & \cdots & \mu_{n,m}(E) \end{bmatrix} \in \mathrm{B}(\mathcal{Y}, \mathcal{Z}),$$

$\forall E \in \mathcal{B}_k$. Let $\bar{\mathcal{X}}_k := (X_k, \mathcal{B}_k, \bar{\mu}_k)$ be the finite $\mathrm{B}(\mathcal{Y}, \mathcal{Z})$-valued measure space, $\forall k \in \mathbb{N}$. By Propositions 4.31, 7.22, and 7.66, $\mathrm{B}(\mathcal{Y}, \mathcal{Z})$ is a Banach space over $\mathbb{K}$. By Proposition 11.118, there is a unique σ-finite $\mathrm{B}(\mathcal{Y}, \mathcal{Z})$-valued measure space $\mathcal{X} := (X, \mathcal{B}, \mu)$ on X such that $\bar{\mathcal{X}}_k$ is the finite $\mathrm{B}(\mathcal{Y}, \mathcal{Z})$-valued measure subspace of $\mathcal{X}$, $\forall k \in \mathbb{N}$. $\forall s \in \{1, \ldots, m\}$, $\forall t \in \{1, \ldots, n\}$, $\forall E \in \mathcal{B}$, we have $\mathcal{P} \circ \mu_{t,s}(E) = \sum_{k=1}^{\infty} \mathcal{P} \circ \mu_{t,s}(E \cap X_k) = \sum_{k=1}^{\infty} \mathcal{P} \circ \mu_{t,s,k}(E \cap X_k) \leq \sum_{k=1}^{\infty} \mathcal{P}\circ\bar{\mu}_k(E\cap X_k) = \sum_{k=1}^{\infty} \mathcal{P}\circ\mu(E\cap X_k) = \mathcal{P}\circ\mu(E) \leq \sum_{k=1}^{\infty}\sum_{i=1}^{m}\sum_{j=1}^{n} \mathcal{P}\circ \mu_{j,i,k}(E\cap X_k) = \sum_{k=1}^{\infty}\sum_{i=1}^{m}\sum_{j=1}^{n} \mathcal{P}\circ\mu_{j,i}(E\cap X_k) = \sum_{i=1}^{m}\sum_{j=1}^{n} \mathcal{P}\circ\mu_{j,i}(E)$, where the second equality follows from Proposition 11.115, the first inequality follows from the special case, the third equality follows from Proposition 11.115, the second inequality follows from the special case, and the fifth equality follows from

Proposition 11.115. Then, $\max_{\substack{i\in\{1,\ldots,m\}\\ j\in\{1,\ldots,n\}}} \mathcal{P}\circ\mu_{j,i} \le \mathcal{P}\circ\mu \le \sum_{i=1}^m\sum_{j=1}^n \mathcal{P}\circ\mu_{j,i}$. $\forall E \in \bigcap_{i=1}^m\bigcap_{j=1}^n \operatorname{dom}(\mu_{j,i})$, $\mathcal{P}\circ\mu(E) \le \sum_{i=1}^m\sum_{j=1}^n \mathcal{P}\circ\mu_{j,i}(E) < \infty$ and $\mu(E)\in \mathrm{B}(\mathcal{Y},\mathcal{Z})$. This implies that

$$\begin{aligned}
\mu(E) &= \sum_{k=1}^{\infty}\mu(E\cap X_k) = \sum_{k=1}^{\infty}\bar{\mu}_k(E\cap X_k)\\
&= \sum_{k=1}^{\infty}\begin{bmatrix}\mu_{1,1}(E\cap X_k) & \cdots & \mu_{1,m}(E\cap X_k)\\ \vdots & & \vdots\\ \mu_{n,1}(E\cap X_k) & \cdots & \mu_{n,m}(E\cap X_k)\end{bmatrix}\\
&= \begin{bmatrix}\sum_{k=1}^{\infty}\mu_{1,1}(E\cap X_k) & \cdots & \sum_{k=1}^{\infty}\mu_{1,m}(E\cap X_k)\\ \vdots & & \vdots\\ \sum_{k=1}^{\infty}\mu_{n,1}(E\cap X_k) & \cdots & \sum_{k=1}^{\infty}\mu_{n,m}(E\cap X_k)\end{bmatrix}\\
&= \begin{bmatrix}\mu_{1,1}(E) & \cdots & \mu_{1,m}(E)\\ \vdots & & \vdots\\ \mu_{n,1}(E) & \cdots & \mu_{n,m}(E)\end{bmatrix}
\end{aligned}$$

where the first equality follows from Definition 11.108, the second equality follows from Proposition 11.115, the third equality follows from the special case, the fourth equality follows from Propositions 11.139, 3.66, and 3.67, and the fifth equality follows from Definition 11.108. This completes the proof of the proposition. □

Proposition 11.148 *Let $\mathcal{X}_i := (X,\mathcal{B},\mu_i)$ be a σ-finite $\mathcal{Y}_i$-valued measure space, where $\mathcal{Y}_i$ is a Banach space over $\mathbb{K}$, $\mathcal{Z}_j$ be a Banach space over $\mathbb{K}$, $\mathcal{W}_{j,i} \subseteq \mathrm{B}(\mathcal{Y}_i,\mathcal{Z}_j)$ be a separable subspace, $f_{j,i} : X \to \mathcal{W}_{j,i}$, $i = 1,2$, $j = 1,2$, $\mu := \begin{bmatrix}\mu_1\\ \mu_2\end{bmatrix}$ be the vector measure as defined in Proposition 11.147, $\mathcal{X} := (X,\mathcal{B},\mu)$ be the σ-finite $\mathcal{Y}_1\times\mathcal{Y}_2$-valued measure space, and $F : X \to \mathrm{B}(\mathcal{Y}_1\times\mathcal{Y}_2,\mathcal{Z}_1\times\mathcal{Z}_2)$ be defined by $F(x) = \begin{bmatrix}f_{1,1}(x) & f_{1,2}(x)\\ f_{2,1}(x) & f_{2,2}(x)\end{bmatrix}$, $\forall x\in X$. Assume that F is absolutely integrable over $\mathcal{X}$. Then, $f_{j,i}$, $j = 1,2$, are absolutely integrable over $\mathcal{X}_i$, $i = 1,2$. In this case, we have $\int_X F\,\mathrm{d}\mu = \begin{bmatrix}\int_X f_{1,1}\,\mathrm{d}\mu_1 + \int_X f_{1,2}\,\mathrm{d}\mu_2\\ \int_X f_{2,1}\,\mathrm{d}\mu_1 + \int_X f_{2,2}\,\mathrm{d}\mu_2\end{bmatrix} \in \mathcal{Z}_1\times\mathcal{Z}_2$.*

Proof Let F be absolutely integrable over $\mathcal{X}$. Fix any $i \in \{1,2\}$ and any $j \in \{1,2\}$. Then, by Propositions 11.139 and 11.38, $f_{j,i} = p_{j,i}F$ is $\mathcal{B}$-measurable, where $p_{j,i}$ is the bounded linear function as defined in Proposition 11.139. By Propositions 7.21 and 11.38, $\mathcal{P}\circ f_{j,i}$ is $\mathcal{B}$-measurable. $\forall x \in X$, $\mathcal{P}\circ f_{j,i}(x) \le \|p_{j,i}\|\mathcal{P}\circ F(x) \le \mathcal{P}\circ F(x)$. Then, by Propositions 11.83, 11.147, and 11.144, $0 \le \int_X \mathcal{P}\circ f_{j,i}\,\mathrm{d}\mathcal{P}\circ\mu_i \le \int_X \mathcal{P}\circ f_{j,i}\,\mathrm{d}\mathcal{P}\circ\mu \le \int_X \mathcal{P}\circ F\,\mathrm{d}\mathcal{P}\circ\mu < \infty$. Hence,

$f_{j,i}$ is absolutely integrable over $\mathcal{X}_i$. By Proposition 11.132, $f_{j,i}$ is integrable over $\mathcal{X}_i$ and $\int_X f_{j,i}\, d\mu_i \in B(\mathcal{Y}_i, \mathcal{Z}_j)$.

By Propositions 7.22, 4.4, and 3.28, $\prod_{i=1}^{2}\prod_{j=1}^{2} \mathcal{W}_{j,i} \subseteq \prod_{i=1}^{2}\prod_{j=1}^{2} B(\mathcal{Y}_i, \mathcal{Z}_j)$ is a separable subspace. Let h be the bounded linear function as defined in Proposition 11.139 for $n = m = 2$. Then, $\mathcal{W} := h(\prod_{i=1}^{2}\prod_{j=1}^{2} \mathcal{W}_{j,i}) \subseteq B(\mathcal{Y}_1 \times \mathcal{Y}_2, \mathcal{Z}_1 \times \mathcal{Z}_2)$ is a separable subspace. By Proposition 11.66, $\exists$ a sequence of simple functions $(\psi_n)_{n=1}^{\infty}$, $\psi_n : \mathcal{X} \to \mathcal{W}$, $\forall n \in \mathbb{N}$, such that $\lim_{n\in\mathbb{N}} \psi_n = F$ a.e. in $\mathcal{X}$, $\|\psi_n(x)\| \leq \|F(x)\|$, $\forall x \in X$, $\forall n \in \mathbb{N}$. Fix any $i \in \{1, 2\}$ and any $j \in \{1, 2\}$. Let $\varphi_{j,i,n} := p_{j,i}\psi_n$. Then, $\varphi_{j,i,n} : \mathcal{X} \to p_{j,i}(\mathcal{W}) \subseteq B(\mathcal{Y}_i, \mathcal{Z}_j)$ is a simple function. By Proposition 11.52, $\lim_{n\in\mathbb{N}} \varphi_{j,i,n} = p_{j,i}F = f_{j,i}$ a.e. in $\mathcal{X}$. Since $\mathcal{P} \circ \mu \geq \mathcal{P} \circ \mu_i$, we have $\varphi_{j,i,n} : \mathcal{X}_i \to p_{j,i}(\mathcal{W}) \subseteq B(\mathcal{Y}_i, \mathcal{Z}_j)$ is a simple function, $\forall n \in \mathbb{N}$, $\lim_{n\in\mathbb{N}} \varphi_{j,i,n} = f_{j,i}$ a.e. in $\mathcal{X}_i$, and F is absolutely integrable over $\mathcal{X}_i$, by Proposition 11.144. Note that $\|\varphi_{j,i,n}(x)\| \leq \|p_{j,i}\|\|\psi_n(x)\| \leq \|F(x)\|$, $\forall x \in X$, $\forall n \in \mathbb{N}$. By Lebesgue Dominated Convergence Theorem 11.131, $\int_X f_{j,i}\, d\mu_i = \lim_{n\in\mathbb{N}} \int_X \varphi_{j,i,n}\, d\mu_i \in \mathcal{Z}_j$. By Lebesgue Dominated Convergence Theorem 11.131,

$$\int_X F\, d\mu = \lim_{n\in\mathbb{N}} \int_X \psi_n\, d\mu = \lim_{n\in\mathbb{N}} \begin{bmatrix} \int_X \varphi_{1,1,n}\, d\mu_1 + \int_X \varphi_{1,2,n}\, d\mu_2 \\ \int_X \varphi_{2,1,n}\, d\mu_1 + \int_X \varphi_{2,2,n}\, d\mu_2 \end{bmatrix}$$

$$= \begin{bmatrix} \int_X f_{1,1}\, d\mu_1 + \int_X f_{1,2}\, d\mu_2 \\ \int_X f_{2,1}\, d\mu_1 + \int_X f_{2,2}\, d\mu_2 \end{bmatrix} \in \mathcal{Z}_1 \times \mathcal{Z}_2$$

where the second equality follows from Proposition 11.125, and the third equality follows from Propositions 3.67, 3.66, and 7.23.

This completes the proof of the proposition. □

Theorem 11.149 (Fatou's Lemma) *Let $(X, \mathcal{B})$ be a measurable space, $(\mu_n)_{n=1}^{\infty}$ be a sequence of measures on $(X, \mathcal{B})$, μ be a measure on $(X, \mathcal{B})$, $f_n : X \to [0, \infty] \subset \mathbb{R}_e$ be $\mathcal{B}$-measurable, $\forall n \in \mathbb{N}$, $f : X \to [0, \infty] \subset \mathbb{R}_e$ be $\mathcal{B}$-measurable. Assume that $\lim_{n\in\mathbb{N}} f_n = f$ a.e. in $\mathcal{X}$, where $\mathcal{X} := (X, \mathcal{B}, \mu)$, and $\lim_{n\in\mathbb{N}} \mu_n(E) = \mu(E)$, $\forall E \in \mathcal{B}$ with $\mu(E) < \infty$. Then, $0 \leq \int_X f\, d\mu \leq \liminf_{n\in\mathbb{N}} \int_X f_n\, d\mu_n \leq \infty$.*

Proof By Definition 11.79, $\int_X f\, d\mu = \sup_{0<\phi\leq f} \int_X \phi\, d\mu$, where the supremum is over all simple functions $\phi : \mathcal{X} \to [0, \infty) \subset \mathbb{R}$ that satisfies $0 \leq \phi(x) \leq f(x)$, $\forall x \in X$. Fix any such simple function ϕ. Then, $\exists M \in [0, \infty) \subset \mathbb{R}$ such that $\phi(x) \leq M$, $\forall x \in X$ and $\mu(\bar{E}) := \mu(\{x \in X \mid \phi(x) > 0\}) < \infty$. By $\lim_{n\in\mathbb{N}} \mu_n(\bar{E}) = \mu(\bar{E}) < \infty$, $\exists n_0 \in \mathbb{N}$ such that $\mu_n(\bar{E}) < \infty$, $\forall n \in \mathbb{N}$ with $n_0 \leq n$. Thus, $\forall n \in \mathbb{N}$ with $n_0 \leq n$, ϕ is a simple function of $\mathcal{X}_n := (X, \mathcal{B}, \mu_n)$. By Proposition 11.75, $\int_X \phi\, d\mu =: c_1 \in [0, \infty) \subset \mathbb{R}$. Let $\hat{E} := \{x \in X \mid (f_n(x))_{n=1}^{\infty}$ does not converge to $f(x)\}$. By the assumption, $\hat{E} \in \mathcal{B}$ and $\mu(\hat{E}) = 0$.

Fix any $\epsilon \in (0,1) \subset \mathbb{R}$. Define $E_n := \{x \in X \setminus \hat{E} \mid f_k(x) \geq (1 - \frac{\epsilon}{2+2c_1})\phi(x), \forall k \geq n, k \in \mathbb{N}\}$, $\forall n \in \mathbb{N}$. By Propositions 7.23, 11.38, and 11.39, $E_n \in \mathcal{B}, \forall n \in \mathbb{N}$. Clearly, $\forall x \in X \setminus (\hat{E} \cup \bar{E})$, $x \in E_n, \forall n \in \mathbb{N}$. Then, $X \setminus E_n \subseteq \hat{E} \cup \bar{E}$, $\forall n \in \mathbb{N}$. $\mu(X \setminus E_n) \leq \mu(\hat{E} \cup \bar{E}) \leq \mu(\bar{E}) < \infty$, $\forall n \in \mathbb{N}$. $\forall x \in \bar{E} \setminus \hat{E}$, then $\lim_{n \in \mathbb{N}} f_n(x) = f(x) \geq \phi(x) > 0$, and there exists $n_x \in \mathbb{N}$ such that $x \in E_{n_x}$. Then, $\bigcup_{n=1}^{\infty} E_n = X \setminus \hat{E}$. Clearly, $E_n \subseteq E_{n+1}, \forall n \in \mathbb{N}$. Then, $X \setminus E_n \supseteq X \setminus E_{n+1}$, $\forall n \in \mathbb{N}$, and $\bigcap_{n=1}^{\infty}(X \setminus E_n) = X \setminus (\bigcup_{n=1}^{\infty} E_n) = \hat{E}$. By Proposition 11.5, we have $\lim_{n \in \mathbb{N}} \mu(X \setminus E_n) = \mu(\hat{E}) = 0$. Then, $\exists n_1 \in \mathbb{N}$ with $n_0 \leq n_1$ such that $\mu(X \setminus E_{n_1}) < \frac{\epsilon}{4+4M}$. By the assumption that $\lim_{i \in \mathbb{N}} \mu_i(X \setminus E_{n_1}) = \mu(X \setminus E_{n_1}) < \frac{\epsilon}{4+4M}$. Then, there exists $n_2 \in \mathbb{N}$ with $n_1 \leq n_2$, $\forall n \in \mathbb{N}$ with $n_2 \leq n$, we have $\mu_n(X \setminus E_{n_1}) < \frac{\epsilon}{2+2M}$. Fix any $n \in \mathbb{N}$ with $n_2 \leq n$, $X \setminus E_{n_1} \supseteq X \setminus E_n$, which implies that $0 \leq \mu_n(X \setminus E_n) \leq \mu_n(X \setminus E_{n_1}) < \frac{\epsilon}{2+2M}$. Thus, by Proposition 11.83, $\int_X f_n \,\mathrm{d}\mu_n \geq \int_{E_n} f_n \,\mathrm{d}\mu_n \geq \int_{E_n} (1 - \frac{\epsilon}{2+2c_1})\phi \,\mathrm{d}\mu_n = (1 - \frac{\epsilon}{2+2c_1})(\int_X \phi \,\mathrm{d}\mu_n - \int_{X \setminus E_n} \phi \,\mathrm{d}\mu_n) \geq (1 - \frac{\epsilon}{2+2c_1})(\int_X \phi \,\mathrm{d}\mu_n - M\mu_n(X \setminus E_n)) > (1 - \frac{\epsilon}{2+2c_1}) \int_X \phi \,\mathrm{d}\mu_n - \epsilon/2$. By the assumption that $\lim_{n \in \mathbb{N}} \mu_n(E) = \mu(E), \forall E \in \mathcal{B}$ with $\mu(E) < \infty$, and Proposition 11.75, we have $\lim_{n \in \mathbb{N}} \int_X \phi \,\mathrm{d}\mu_n = \int_X \phi \,\mathrm{d}\mu = c_1$. Then, $\liminf_{n \in \mathbb{N}} \int_X f_n \,\mathrm{d}\mu_n \geq (1 - \frac{\epsilon}{2+2c_1})c_1 - \epsilon/2 > c_1 - \epsilon = \int_X \phi \,\mathrm{d}\mu - \epsilon$. By the arbitrariness of ϵ, we have $\liminf_{n \in \mathbb{N}} \int_X f_n \,\mathrm{d}\mu_n \geq \int_X \phi \,\mathrm{d}\mu$. By the arbitrariness of ϕ, we have $\liminf_{n \in \mathbb{N}} \int_X f_n \,\mathrm{d}\mu_n \geq \int_X f \,\mathrm{d}\mu$.

This completes the proof of the lemma. □

Theorem 11.150 (Lebesgue Dominated Convergence) *Let $(X, \mathcal{B})$ be a measurable space, $(\mu_n)_{n=1}^{\infty}$ be a sequence of measures on $(X, \mathcal{B})$, μ be a measure on $(X, \mathcal{B})$, $f_n : X \to \mathbb{R}$ be $\mathcal{B}$-measurable, $\forall n \in \mathbb{N}$, $f : X \to \mathbb{R}$ be $\mathcal{B}$-measurable, $g_n : X \to [0, \infty) \subset \mathbb{R}$ be $\mathcal{B}$-measurable, $\forall n \in \mathbb{N}$, $g : X \to [0, \infty) \subset \mathbb{R}$ be $\mathcal{B}$-measurable. Assume that:*

(i) $\lim_{n \in \mathbb{N}} f_n = f$ *a.e. in* $\mathcal{X}$ *and* $\lim_{n \in \mathbb{N}} g_n = g$ *a.e. in* $\mathcal{X}$, *where* $\mathcal{X} := (X, \mathcal{B}, \mu)$.
(ii) $|f_n(x)| \leq g_n(x)$, $\forall x \in X$, $\forall n \in \mathbb{N}$.
(iii) $\int_X g_n \,\mathrm{d}\mu_n \in \mathbb{R}$, $\forall n \in \mathbb{N}$, *and* $\int_X g \,\mathrm{d}\mu = \lim_{n \in \mathbb{N}} \int_X g_n \,\mathrm{d}\mu_n \in \mathbb{R}$.
(iv) $\lim_{n \in \mathbb{N}} \mu_n(E) = \mu(E)$, $\forall E \in \mathcal{B}$ *with* $\mu(E) < \infty$.

Then, f is absolutely integrable over $\mathcal{X}$, $\int_X f \,\mathrm{d}\mu = \lim_{n \in \mathbb{N}} \int_X f_n \,\mathrm{d}\mu_n \in \mathbb{R}$, and $0 \leq \left|\int_X f \,\mathrm{d}\mu\right| \leq \int_X g \,\mathrm{d}\mu < \infty$.

Proof By Proposition 11.92, f_n is absolutely integrable over $\mathcal{X}_n := (X, \mathcal{B}, \mu_n)$ and $\int_X f_n \,\mathrm{d}\mu_n \in \mathbb{R}$. Clearly, $|f(x)| \leq g(x)$ a.e. in $\mathcal{X}$. By Proposition 11.83, f is absolutely integrable over $\mathcal{X}$. By Proposition 11.92, $\int_X f \,\mathrm{d}\mu \in \mathbb{R}$ and $0 \leq \left|\int_X f \,\mathrm{d}\mu\right| \leq \int_X \mathcal{P} \circ f \,\mathrm{d}\mu \leq \int_X g \,\mathrm{d}\mu < \infty$.

By Propositions 7.23, 11.52, and 11.53, $\lim_{n \in \mathbb{N}}(g_n - f_n) = g - f$ a.e. in $\mathcal{X}$ and $\lim_{n \in \mathbb{N}}(g_n + f_n) = g + f$ a.e. in $\mathcal{X}$. Clearly, $(g_n - f_n)(x) \geq 0$ and $(g_n + f_n)(x) \geq 0$, $\forall x \in X$, $\forall n \in \mathbb{N}$. By Propositions 7.23, 11.52, and 11.53, $\lim_{n \in \mathbb{N}}(g_n - f_n) = \lim_{n \in \mathbb{N}}(g_n - f_n) \vee 0 = (g - f) \vee 0 = g - f$ a.e. in $\mathcal{X}$ and $\lim_{n \in \mathbb{N}}(g_n + f_n) = \lim_{n \in \mathbb{N}}(g_n + f_n) \vee 0 = (g + f) \vee 0 = g + f$ a.e. in $\mathcal{X}$. By Fatou's Lemma 11.149 and Propositions 11.92 and 3.83, we have $0 \leq \int_X ((g - f) \vee 0)\, d\mu = \int_X (g - f)\, d\mu = \int_X g\, d\mu - \int_X f\, d\mu \leq \liminf_{n\in\mathbb{N}} \int_X (g_n - f_n)\, d\mu_n = \liminf_{n\in\mathbb{N}} (\int_X g_n\, d\mu_n - \int_X f_n\, d\mu_n) = \int_X g\, d\mu - \limsup_{n\in\mathbb{N}} \int_X f_n\, d\mu_n$ and $0 \leq \int_X ((g + f) \vee 0)\, d\mu = \int_X (g + f)\, d\mu = \int_X g\, d\mu + \int_X f\, d\mu \leq \liminf_{n\in\mathbb{N}} \int_X (g_n + f_n)\, d\mu_n = \liminf_{n\in\mathbb{N}} (\int_X g_n\, d\mu_n + \int_X f_n\, d\mu_n) = \int_X g\, d\mu + \liminf_{n\in\mathbb{N}} \int_X f_n\, d\mu_n$. Then, by Proposition 3.83, we have $\int_X f\, d\mu \leq \liminf_{n\in\mathbb{N}} \int_X f_n\, d\mu_n \leq \limsup_{n\in\mathbb{N}} \int_X f_n\, d\mu_n \leq \int_X f\, d\mu \in \mathbb{R}$. This further implies that $\int_X f\, d\mu = \lim_{n\in\mathbb{N}} \int_X f_n\, d\mu_n \in \mathbb{R}$. This completes the proof of the theorem. □

In the preceding two results, we work with measure spaces that may not be σ-finite, and we assume a mode of convergence for the measures to be set-wise convergence rather than the stronger "norm" convergence alluded to in Definition 11.143. The next result shows that Lebesgue Dominated convergence Theorem holds in general under the "norm" convergence assumption on the Banach space valued measures.

Theorem 11.151 (Lebesgue Dominated Convergence) *Let $(X, \mathcal{B})$ be a measurable space, $\mathcal{Y}$ be a Banach space over $\mathbb{K}$, $\mu \in \mathcal{M}_\sigma(X, \mathcal{B}, \mathcal{Y})$, $(\mu_n)_{n=1}^\infty \subseteq \mathcal{M}_\sigma(X, \mathcal{B}, \mathcal{Y})$, $\mathcal{Z}$ be a Banach space over $\mathbb{K}$, $\mathcal{W}$ be a separable subspace of $\mathrm{B}(\mathcal{Y}, \mathcal{Z})$, $\bar{\mathcal{W}} := \overline{\mathrm{span}\,(\{z \in \mathcal{Z} \mid \exists y \in \mathcal{Y}, \exists w \in \mathcal{W} \ni \cdot\, z = wy\})} \subseteq \mathcal{Z}$ be the Banach subspace of $\mathcal{Z}$, $f_n : X \to \mathcal{W}$ be $\mathcal{B}$-measurable, $\forall n \in \mathbb{N}$, $f : X \to \mathcal{W}$ be $\mathcal{B}$-measurable, $g_n : X \to [0, \infty) \subset \mathbb{R}$ be $\mathcal{B}$-measurable, $\forall n \in \mathbb{N}$, $g : X \to [0, \infty) \subset \mathbb{R}$ be $\mathcal{B}$-measurable. Assume that:*

(i) $\lim_{n\in\mathbb{N}} f_n = f$ *a.e. in $\mathcal{X}$ and* $\lim_{n\in\mathbb{N}} g_n = g$ *a.e. in $\mathcal{X}$, where $\mathcal{X} := (X, \mathcal{B}, \mu)$.*
(ii) $\| f_n(x) \| \leq g_n(x)$, $\forall x \in X$, $\forall n \in \mathbb{N}$.
(iii) $\int_X g_n\, d\mathcal{P} \circ \mu_n \in \mathbb{R}$, $\forall n \in \mathbb{N}$, *and* $\int_X g\, d\mathcal{P} \circ \mu = \lim_{n\in\mathbb{N}} \int_X g_n\, d\mathcal{P} \circ \mu_n \in \mathbb{R}$.
(iv) $\lim_{n\in\mathbb{N}} \mu_n = \mu$ *as defined in Definition* 11.143.

Then, f is absolutely integrable over $\mathcal{X}$, $\int_X f\, d\mu = \lim_{n\in\mathbb{N}} \int_X f_n\, d\mu_n \in \bar{\mathcal{W}} \subseteq \mathcal{Z}$, and $0 \leq \| \int_X f\, d\mu \| \leq \int_X g\, d\mathcal{P} \circ \mu < \infty$.

Proof By Proposition 11.132, f_n is absolutely integrable over $(X, \mathcal{B}, \mu_n) =: \mathcal{X}_n$ and $\int_X f_n\, d\mu_n \in \bar{\mathcal{W}}$. Clearly, $\| f(x) \| \leq g(x)$ a.e. in $\mathcal{X}$. By Proposition 11.83, f is absolutely integrable over $\mathcal{X}$. By Proposition 11.132, $\int_X f\, d\mu \in \bar{\mathcal{W}}$ and $0 \leq \| \int_X f\, d\mu \| \leq \int_X \mathcal{P} \circ f\, d\mathcal{P} \circ \mu \leq \int_X g\, d\mathcal{P} \circ \mu < \infty$.

We will distinguish two exhaustive and mutually exclusives cases: Case 1: $\mathcal{P} \circ \mu(X) < \infty$; Case 2: $\mathcal{P} \circ \mu(X) = \infty$.

Case 1: $\mathcal{P} \circ \mu(X) < \infty$. By $\lim_{n\in\mathbb{N}} \mu_n = \mu$, we have $\lim_{n\in\mathbb{N}} \mathcal{P} \circ \mu_n(X) = \mathcal{P} \circ \mu(X) < \infty$. Without loss of generality, we may assume that $\mathcal{P} \circ \mu_n(X) < \infty$, $\forall n \in \mathbb{N}$. Fix any $\epsilon \in (0, \infty) \subset \mathbb{R}$. By Proposition 11.84, $\exists \delta \in (0, \epsilon/6] \subset \mathbb{R}$ such that $\forall A \in \mathcal{B}$ with $\mathcal{P} \circ \mu(A) < \delta$, we have $\int_A g\, d\mathcal{P} \circ \mu < \epsilon/6$. By Egoroff's Theorem 11.55, $\exists \bar{E} \in \mathcal{B}$ with $\mathcal{P} \circ \mu(\bar{E}) < \delta/2$ such that $\left(f_n|_{X\setminus\bar{E}} \right)_{n=1}^\infty$ converges uniformly to $f|_{X\setminus\bar{E}}$. Then, $\exists n_0 \in \mathbb{N}$, $\forall n \in \mathbb{N}$ with $n_0 \leq n$, $\| f_n(x) - f(x) \| < \frac{\epsilon}{4\mathcal{P}\circ\mu(X)+1}$, $\forall x \in X \setminus \bar{E}$. By Proposition 11.5, there exists $M \in \mathbb{N}$ such that $E_1 := \{x \in X \mid g(x) \geq M\} \in \mathcal{B}$ and $\mathcal{P} \circ \mu(E_1) < \delta/2$. Let $E := \bar{E} \cup E_1 \in \mathcal{B}$. Then,

$\mathcal{P} \circ \mu(E) < \delta$. Hence, $\int_E g \, \mathrm{d}\mathcal{P} \circ \mu < \epsilon/6$. By Lebesgue Dominated Convergence Theorem 11.150, we have $\lim_{n \in \mathbb{N}} \int_E g_n \, \mathrm{d}\mathcal{P} \circ \mu_n = \lim_{n \in \mathbb{N}} \int_X (g_n \chi_{E,X}) \, \mathrm{d}\mathcal{P} \circ \mu_n = \int_X (g \chi_{E,X}) \, \mathrm{d}\mathcal{P} \circ \mu = \int_E g \, \mathrm{d}\mathcal{P} \circ \mu < \epsilon/6$. Then, $\exists n_1 \in \mathbb{N}$ with $n_0 \leq n_1$, $\forall n \in \mathbb{N}$ with $n_1 \leq n$, we have $0 \leq \int_E g_n \, \mathrm{d}\mathcal{P} \circ \mu_n < \epsilon/3$. By the assumption (iv), $\lim_{n \in \mathbb{N}} \mathcal{P} \circ (\mu_n - \mu)(X) = 0$. This implies that $\exists n_2 \in \mathbb{N}$ with $n_1 \leq n_2$, $\forall n \in \mathbb{N}$ with $n_2 \leq n$, we have $\mathcal{P} \circ (\mu_n - \mu)(X) < \frac{\epsilon}{4M}$ and $\mathcal{P} \circ \mu_n(X) < \mathcal{P} \circ \mu(X) + 1/4$.

Fix any $n \in \mathbb{N}$ with $n_2 \leq n$. Clearly, $f|_{X \setminus E}$ is absolutely integrable over $\mathcal{E}_n$ and $\int_{X \setminus E} f \, \mathrm{d}\mu_n \in \mathcal{Z}$, since $\|f(x)\| \leq M$ a.e. in $\mathcal{E}_n$, where $\mathcal{E}_n := (X \setminus E, \mathcal{B}_{X \setminus E}, \mu_{n,X \setminus E})$ is the finite $\mathcal{Y}$-valued measure subspace of $\mathcal{X}_n$. We have

$$
\begin{aligned}
&\Big\| \int_X f_n \, \mathrm{d}\mu_n - \int_X f \, \mathrm{d}\mu \Big\| \\
&= \Big\| \int_E f_n \, \mathrm{d}\mu_n + \int_{X \setminus E} f_n \, \mathrm{d}\mu_n - \int_{X \setminus E} f \, \mathrm{d}\mu_n + \int_{X \setminus E} f \, \mathrm{d}\mu_n \\
&\quad - \int_{X \setminus E} f \, \mathrm{d}\mu - \int_E f \, \mathrm{d}\mu \Big\| \\
&\leq \Big\| \int_E f_n \, \mathrm{d}\mu_n \Big\| + \Big\| \int_E f \, \mathrm{d}\mu \Big\| + \Big\| \int_{X \setminus E} (f_n - f) \, \mathrm{d}\mu_n \Big\| \\
&\quad + \Big\| \int_{X \setminus E} f \, \mathrm{d}(\mu_n - \mu) \Big\| \\
&\leq \int_E g_n \, \mathrm{d}\mathcal{P} \circ \mu_n + \int_E g \, \mathrm{d}\mathcal{P} \circ \mu + \int_{X \setminus E} \mathcal{P} \circ (f_n - f) \, \mathrm{d}\mathcal{P} \circ \mu_n \\
&\quad + \int_{X \setminus E} \mathcal{P} \circ f \, \mathrm{d}\mathcal{P} \circ (\mu_n - \mu) \\
&< \epsilon/3 + \epsilon/6 + \frac{\mathcal{P} \circ \mu_n(X \setminus E)\epsilon}{4\mathcal{P} \circ \mu(X) + 1} + M\mathcal{P} \circ (\mu_n - \mu)(X \setminus E) < \epsilon
\end{aligned}
$$

where the first equality follows from Proposition 11.132, the first inequality follows from Propositions 11.132, 11.145, and 11.146, the second inequality follows from Proposition 11.132, and the third inequality follows from Proposition 11.83. Hence, we have $\lim_{n \in \mathbb{N}} \int_X f_n \, \mathrm{d}\mu_n = \int_X f \, \mathrm{d}\mu \in \bar{\mathcal{W}}$. This case is proved.

Case 2: $\mathcal{P} \circ \mu(X) = \infty$. Fix any $\epsilon \in (0, \infty) \subset \mathbb{R}$. By Definition 11.79, $\exists$ a simple function $\phi : \mathcal{X} \to [0, \infty) \subset \mathbb{R}$ such that $0 \leq \phi(x) \leq g(x)$, $\forall x \in X$ and $\int_X g \, \mathrm{d}\mathcal{P} \circ \mu - \epsilon/4 < \int_X \phi \, \mathrm{d}\mathcal{P} \circ \mu \leq \int_X g \, \mathrm{d}\mathcal{P} \circ \mu < \infty$. Let $A_0 := \{x \in X \mid \phi(x) > 0\}$. Then, $A_0 \in \mathcal{B}$ and $\mathcal{P} \circ \mu(A_0) < \infty$. Note that, by Propositions 11.83 and 11.75, $0 \leq \int_{X \setminus A_0} g \, \mathrm{d}\mathcal{P} \circ \mu = \int_X g \, \mathrm{d}\mathcal{P} \circ \mu - \int_{A_0} g \, \mathrm{d}\mathcal{P} \circ \mu \leq \int_X g \, \mathrm{d}\mathcal{P} \circ \mu - \int_{A_0} \phi \, \mathrm{d}\mathcal{P} \circ \mu = \int_X g \, \mathrm{d}\mathcal{P} \circ \mu - \int_X \phi \, \mathrm{d}\mathcal{P} \circ \mu < \epsilon/4$. This implies that $0 \leq \Big\| \int_X f \, \mathrm{d}\mu - \int_{A_0} f \, \mathrm{d}\mu \Big\| = \Big\| \int_{X \setminus A_0} f \, \mathrm{d}\mu \Big\| \leq \int_{X \setminus A_0} \mathcal{P} \circ f \, \mathrm{d}\mathcal{P} \circ \mu \leq \int_{X \setminus A_0} g \, \mathrm{d}\mathcal{P} \circ \mu < \epsilon/4$, where the first equality and the second inequality follow from Proposition 11.132, and the third inequality follows from Proposition 11.83. By Lebesgue Dominated Convergence

Theorem 11.150, we have $\int_{X\setminus A_0} g \,\mathrm{d}\mathcal{P}\circ\mu = \lim_{n\in\mathbb{N}} \int_{X\setminus A_0} g_n \,\mathrm{d}\mathcal{P}\circ\mu_n$. By Case 1, we have $\int_{A_0} f \,\mathrm{d}\mu = \lim_{n\in\mathbb{N}} \int_{A_0} f_n \,\mathrm{d}\mu_n \in \bar{\mathcal{W}}$. Then, $\exists n_0 \in \mathbb{N}$, $\forall n \in \mathbb{N}$ with $n_0 \leq n$, we have $\left\| \int_{A_0} f \,\mathrm{d}\mu - \int_{A_0} f_n \,\mathrm{d}\mu_n \right\| < \epsilon/4$ and $\int_{X\setminus A_0} g_n \,\mathrm{d}\mathcal{P}\circ\mu_n < \epsilon/2$. This leads to $\left\| \int_X f \,\mathrm{d}\mu - \int_X f_n \,\mathrm{d}\mu_n \right\| \leq \left\| \int_X f \,\mathrm{d}\mu - \int_{A_0} f \,\mathrm{d}\mu \right\| + \left\| \int_{A_0} f \,\mathrm{d}\mu - \int_{A_0} f_n \,\mathrm{d}\mu_n \right\| + \left\| \int_{A_0} f_n \,\mathrm{d}\mu_n - \int_X f_n \,\mathrm{d}\mu_n \right\| < \epsilon/4 + \epsilon/4 + \left\| \int_{X\setminus A_0} f_n \,\mathrm{d}\mu_n \right\| \leq \epsilon/2 + \int_{X\setminus A_0} g_n \,\mathrm{d}\mathcal{P}\circ\mu_n < \epsilon$, where the second and third inequalities follow from Proposition 11.132. Hence, $\int_X f \,\mathrm{d}\mu = \lim_{n\in\mathbb{N}} \int_X f_n \,\mathrm{d}\mu_n \in \bar{\mathcal{W}}$. This case is proved.

This completes the proof of the theorem. □

Proposition 11.152 *Let $\mathcal{X} := (X, \mathcal{B}, \mu)$ be a $\mathcal{Y}$-valued measure space, where $\mathcal{Y}$ is a normed linear space over $\mathbb{K}$, $\mathcal{Z}$ be a Banach space over $\mathbb{K}$, $\mathcal{W}$ be a separable subspace of* $\mathrm{B}(\mathcal{Y}, \mathcal{Z})$*, and $f : X \to \mathcal{W}$ be absolutely integrable over $\mathcal{X}$. Define $\nu : \mathcal{B} \to \mathcal{Z}$ by $\nu(E) = \int_E f \,\mathrm{d}\mu$, $\forall E \in \mathcal{B}$. Then, $\bar{\mathcal{X}} := (X, \mathcal{B}, \nu)$ is a finite $\mathcal{Z}$-valued measure space and $0 \leq \mathcal{P}\circ\nu(E) \leq \int_E \mathcal{P}\circ f \,\mathrm{d}\mathcal{P}\circ\mu \leq \int_X \mathcal{P}\circ f \,\mathrm{d}\mathcal{P}\circ\mu < \infty$, $\forall E \in \mathcal{B}$.*

Proof By Proposition 11.125, $\nu(\emptyset) = \int_\emptyset f \,\mathrm{d}\mu = \vartheta_{\mathcal{Z}}$. $\forall E \in \mathcal{B}$, by Proposition 11.132, $\nu(E) = \int_E f \,\mathrm{d}\mu \in \mathcal{Z}$ and $0 \leq \|\nu(E)\| \leq \int_E \mathcal{P}\circ f \,\mathrm{d}\mathcal{P}\circ\mu \leq \int_X \mathcal{P}\circ f \,\mathrm{d}\mathcal{P}\circ\mu < \infty$. $\forall$ pairwise disjoint $(E_n)_{n=1}^\infty \subseteq \mathcal{B}$, let $E := \bigcup_{n=1}^\infty E_n \in \mathcal{B}$. Then, $\sum_{n=1}^\infty \|\nu(E_n)\| \leq \sum_{n=1}^\infty \int_{E_n} \mathcal{P}\circ f \,\mathrm{d}\mathcal{P}\circ\mu = \int_E \mathcal{P}\circ f \,\mathrm{d}\mathcal{P}\circ\mu < \infty$, where the equality follows from Proposition 11.83. By Proposition 11.132, we have $\nu(E) = \int_E f \,\mathrm{d}\mu = \sum_{n=1}^\infty \int_{E_n} f \,\mathrm{d}\mu = \sum_{n=1}^\infty \nu(E_n) \in \mathcal{Z}$. Hence, ν is a $\mathcal{Z}$-valued pre-measure on $(X, \mathcal{B})$. $\forall E \in \mathcal{B}$,

$$\begin{aligned}
\mathcal{P}\circ\nu(E) &= \sup_{n\in\mathbb{Z}_+,\ (E_i)_{i=1}^n\subseteq\mathcal{B},\ E=\bigcup_{i=1}^n E_i,\ E_i\cap E_j=\emptyset,\ \forall 1\leq i<j\leq n} \sum_{i=1}^n \|\nu(E_i)\| \\
&\leq \sup_{n\in\mathbb{Z}_+,\ (E_i)_{i=1}^n\subseteq\mathcal{B},\ E=\bigcup_{i=1}^n E_i,\ E_i\cap E_j=\emptyset,\ \forall 1\leq i<j\leq n} \sum_{i=1}^n \int_{E_i} \mathcal{P}\circ f \,\mathrm{d}\mathcal{P}\circ\mu \\
&= \int_E \mathcal{P}\circ f \,\mathrm{d}\mathcal{P}\circ\mu \leq \int_X \mathcal{P}\circ f \,\mathrm{d}\mathcal{P}\circ\mu < \infty
\end{aligned}$$

Thus, ν is a finite $\mathcal{Z}$-valued pre-measure, and hence, a finite $\mathcal{Z}$-valued measure on $(X, \mathcal{B})$. This completes the proof of the proposition. □

Proposition 11.153 *Let $\mathcal{X} := (X, \mathcal{B}, \mu)$ be a σ-finite $\mathcal{Y}$-valued measure space, where $\mathcal{Y}$ is a normed linear space over $\mathbb{K}$, $\mathcal{Z}$ be a Banach space over $\mathbb{K}$, $\mathcal{W}$ be a separable subspace of* $\mathrm{B}(\mathcal{Y}, \mathcal{Z})$*, and $f : X \to \mathcal{W}$ be $\mathcal{B}$-measurable. Then, there exists a unique $\mathcal{Z}$-valued measure ν on $(X, \mathcal{B})$ such that $\nu(E) = \int_E f \,\mathrm{d}\mu \in \mathcal{Z}$, $\forall E \in \mathcal{B}$ with $\int_E \mathcal{P}\circ f \,\mathrm{d}\mathcal{P}\circ\mu < \infty$. Furthermore, ν is σ-finite and $0 \leq \mathcal{P}\circ\nu(E) \leq \int_E \mathcal{P}\circ f \,\mathrm{d}\mathcal{P}\circ\mu$, $\forall E \in \mathcal{B}$.*

Proof Since $\mathcal{X}$ is σ-finite, then $\exists (X_n)_{n=1}^{\infty} \subseteq \mathcal{B}$ such that $X = \bigcup_{n=1}^{\infty} X_n$ and $\mathcal{P} \circ \mu(X_n) < \infty$, $\forall n \in \mathbb{N}$. Without loss of generality, we may assume that $(X_n)_{n=1}^{\infty}$ is pairwise disjoint. $\forall m \in \mathbb{N}$, let $\bar{X}_m := \{x \in X \mid m-1 \leq \|f(x)\| < m\} \in \mathcal{B}$. Clearly, $(\bar{X}_m)_{m=1}^{\infty} \subseteq \mathcal{B}$ is pairwise disjoint and $X = \bigcup_{m=1}^{\infty} \bar{X}_m$. $\forall m, n \in \mathbb{N}$, let $\mathcal{X}_{m,n} := (\bar{X}_m \cap X_n, \mathcal{B}_{m,n}, \mu_{m,n})$ be the finite $\mathcal{Y}$-valued measure subspace of $\mathcal{X}$. Then, $f|_{\bar{X}_m \cap X_n}$ is absolutely integrable over $\mathcal{X}_{m,n}$. By Propositions 11.152 and 11.132, we may define a finite $\mathcal{Z}$-valued measure space $\hat{\mathcal{X}}_{m,n} := (\bar{X}_m \cap X_n, \mathcal{B}_{m,n}, \nu_{m,n})$ by $\nu_{m,n}(E) = \int_E f \, d\mu \in \mathcal{Z}$, $\forall E \in \mathcal{B}_{m,n}$. Clearly, the sets $\bar{X}_m \cap X_n$, $m \in \mathbb{N}$, $n \in \mathbb{N}$, are pairwise disjoint. Then, by Proposition 11.118, the generation process on $\left(\hat{\mathcal{X}}_{m,n}\right)_{m \in \mathbb{N},\, n \in \mathbb{N}}$ yields the unique σ-finite $\mathcal{Z}$-valued measure ν on $(X, \mathcal{B})$ such that $\hat{\mathcal{X}}_{m,n}$ is a finite $\mathcal{Z}$-valued measure subspace of $\hat{\mathcal{X}} := (X, \mathcal{B}, \nu)$, $\forall m, n \in \mathbb{N}$.

Fix any $E \in \mathcal{B}$ with $\int_E \mathcal{P} \circ f \, d\mathcal{P} \circ \mu < \infty$. $\forall m, n \in \mathbb{N}$, let $E_{m,n} := E \cap \bar{X}_m \cap X_n \in \mathcal{B}$. Then, $\mathcal{P} \circ \nu(E) = \sum_{m=1}^{\infty} \sum_{n=1}^{\infty} \mathcal{P} \circ \nu_{m,n}(E_{m,n}) \leq \sum_{m=1}^{\infty} \sum_{n=1}^{\infty} \int_{E_{m,n}} \mathcal{P} \circ f \, d\mathcal{P} \circ \mu = \int_E \mathcal{P} \circ f \, d\mathcal{P} \circ \mu < \infty$, where the first equality follows from Proposition 11.118, the first inequality follows from Proposition 11.152, and the second equality follows from Proposition 11.83. This implies that $E \in \operatorname{dom}(\nu)$ and $\nu(E) = \sum_{m=1}^{\infty} \sum_{n=1}^{\infty} \nu_{m,n}(E_{m,n}) = \sum_{m=1}^{\infty} \sum_{n=1}^{\infty} \int_{E_{m,n}} f \, d\mu = \int_E f \, d\mu \in \mathcal{Z}$, where the first equality follows from Proposition 11.118, the second equality follows from Proposition 11.152, and the third equality follows from Proposition 11.132. Hence, ν is the σ-finite $\mathcal{Z}$-valued measure we seek.

This completes the proof of the proposition. □

Proposition 11.154 *Let $\mathcal{X} := (X, \mathcal{B}, \mu)$ be a σ-finite $\mathbb{K}$-valued measure space, $\mathcal{Z}$ be a Banach space over $\mathbb{K}$, $\mathcal{W}$ be a separable subspace of* $\mathrm{B}(\mathbb{K}, \mathcal{Z}) = \mathcal{Z}$*, and $f : X \to \mathcal{W}$ be $\mathcal{B}$-measurable. Then, there exists a unique $\mathcal{Z}$-valued measure ν on $(X, \mathcal{B})$ such that $\mathcal{P} \circ \nu(E) = \int_E \mathcal{P} \circ f \, d\mathcal{P} \circ \mu \in [0, \infty] \subset \mathbb{R}_e$, $\forall E \in \mathcal{B}$, and $\nu(E) = \int_E f \, d\mu \in \mathcal{Z}$, $\forall E \in \mathcal{B}$ with $\mathcal{P} \circ \nu(E) < \infty$. Furthermore, ν is σ-finite.*

Proof By Proposition 11.153, there exists a unique $\mathcal{Z}$-valued measure ν on $(X, \mathcal{B})$ such that $\nu(E) = \int_E f \, d\mu \in \mathcal{Z}$, $\forall E \in \mathcal{B}$ with $\int_E \mathcal{P} \circ f \, d\mathcal{P} \circ \mu < \infty$. Furthermore, ν is σ-finite and $0 \leq \mathcal{P} \circ \nu(E) \leq \int_E \mathcal{P} \circ f \, d\mathcal{P} \circ \mu$, $\forall E \in \mathcal{B}$. All we need to show is that $\mathcal{P} \circ \nu(E) \geq \int_E \mathcal{P} \circ f \, d\mathcal{P} \circ \mu$, $\forall E \in \mathcal{B}$.

Define $\bar{\nu} : \mathcal{B} \to [0, \infty] \subset \mathbb{R}_e$ by $\bar{\nu}(E) = \int_E \mathcal{P} \circ f \, d\mathcal{P} \circ \mu$, $\forall E \in \mathcal{B}$. By Proposition 11.116, $\bar{\nu}$ is a σ-finite measure on $(X, \mathcal{B})$. Fix any $E \in \mathcal{B}$. We will distinguish two exhaustive and mutually exclusives cases: Case 1: $E \in \mathcal{B} \setminus \operatorname{dom}(\nu)$; Case 2: $E \in \operatorname{dom}(\nu)$. Case 1: $E \in \mathcal{B} \setminus \operatorname{dom}(\nu)$. Clearly, we have $\mathcal{P} \circ \nu(E) = \infty \geq \bar{\nu}(E)$. This case is proved. Case 2: $E \in \operatorname{dom}(\nu)$. Since $\bar{\nu}$ is σ-finite, then $\exists (X_n)_{n=1}^{\infty} \subseteq \mathcal{B}$ such that $X = \bigcup_{n=1}^{\infty} X_n$ and $\bar{\nu}(X_n) < \infty$, $\forall n \in \mathbb{N}$. Without loss of generality, we may assume that $(X_n)_{n=1}^{\infty}$ is pairwise disjoint. Let $E_n := E \cap X_n \in \mathcal{B}$, $\forall n \in \mathbb{N}$. Then, $\forall n \in \mathbb{N}$, we have $\int_{E_n} \mathcal{P} \circ f \, d\mathcal{P} \circ \mu = \bar{\nu}(E_n) \leq \bar{\nu}(X_n) < \infty$. $\forall n \in \mathbb{N}$, $\forall F \in \mathcal{B}$ with $F \subseteq E_n$, we have $0 \leq \mathcal{P} \circ \nu(F) \leq \int_F \mathcal{P} \circ f \, d\mathcal{P} \circ \mu = \bar{\nu}(F) \leq \bar{\nu}(E_n) < \infty$ and $\nu(F) = \int_F f \, d\mu \in \mathcal{Z}$. Fix any $\epsilon \in (0, \infty) \subset \mathbb{R}$. Fix any $n \in \mathbb{N}$. Let $\mathcal{E}_n := (E_n, \mathcal{B}_{E_n}, \mu_{E_n})$ be the σ-finite $\mathbb{K}$-valued measure subspace of $\mathcal{X}$. By Proposition 11.94, $\exists$ simple function $\varphi_n : \mathcal{E}_n \to \mathcal{W}$ such that $\|\varphi_n(x)\| \leq$

$\mathcal{P} \circ f(x)$, $\forall x \in E_n$, $\left| \int_{E_n} \mathcal{P} \circ \varphi_n \,\mathrm{d}\mathcal{P} \circ \mu - \int_{E_n} \mathcal{P} \circ f \,\mathrm{d}\mathcal{P} \circ \mu \right| < 2^{-n-2}\epsilon$, and $\int_{E_n} \mathcal{P} \circ (f|_{E_n} - \varphi_n) \,\mathrm{d}\mathcal{P} \circ \mu < 2^{-n-1}\epsilon$. Let φ_n admit the canonical representation $\varphi_n = \sum_{i=1}^{m_n} w_{i,n} \chi_{E_{i,n},E_n}$, where $m_n \in \mathbb{Z}_+$, $w_{1,n}, \ldots, w_{m_n,n} \in \mathcal{W}$ are distinct and none equals to $\vartheta_{\mathcal{Z}}$, and $E_{1,n}, \ldots, E_{m_n,n} \in \mathcal{B}$ are subsets of E_n, nonempty, pairwise disjoint, and $\mathcal{P} \circ \mu(E_{i,n}) < \infty$, $i = 1, \ldots, m_n$. $\forall i \in \{1, \ldots, m_n\}$, $\exists m_{i,n} \in \mathbb{Z}_+$, $\exists$ pairwise disjoint $\left(E_{j,i,n}\right)_{j=1}^{m_{i,n}} \subseteq \mathcal{B}$ with $E_{i,n} = \bigcup_{j=1}^{m_{i,n}} E_{j,i,n}$ such that $\mathcal{P} \circ \mu(E_{i,n}) < \sum_{j=1}^{m_{i,n}} \left|\mu(E_{j,i,n})\right| + \frac{2^{-n-2}\epsilon}{(1+m_n)\|w_{i,n}\|}$. This leads to

$$
\begin{aligned}
\bar{\nu}(E_n) &= \int_{E_n} \mathcal{P} \circ f \,\mathrm{d}\mathcal{P} \circ \mu < \int_{E_n} \mathcal{P} \circ \varphi_n \,\mathrm{d}\mathcal{P} \circ \mu + 2^{-n-2}\epsilon \\
&= \sum_{i=1}^{m_n} \sum_{j=1}^{m_{i,n}} \left\| \nu(E_{j,i,n}) \right\| + \sum_{i=1}^{m_n} \left\| w_{i,n} \right\| \mathcal{P} \circ \mu(E_{i,n}) + 2^{-n-2}\epsilon \\
&\quad - \sum_{i=1}^{m_n} \sum_{j=1}^{m_{i,n}} \left\| \nu(E_{j,i,n}) \right\| \\
&\le \mathcal{P} \circ \nu(E_n) + \sum_{i=1}^{m_n} \left\| w_{i,n} \right\| \Big(\sum_{j=1}^{m_{i,n}} \left|\mu(E_{j,i,n})\right| + \frac{2^{-n-2}\epsilon}{(1+m_n)\left\|w_{i,n}\right\|} \Big) \\
&\quad + 2^{-n-2}\epsilon - \sum_{i=1}^{m_n} \sum_{j=1}^{m_{i,n}} \left\| \nu(E_{j,i,n}) \right\| \\
&< \mathcal{P} \circ \nu(E_n) + \sum_{i=1}^{m_n} \sum_{j=1}^{m_{i,n}} \left\| w_{i,n} \mu(E_{j,i,n}) \right\| + 2^{-n-1}\epsilon - \sum_{i=1}^{m_n} \sum_{j=1}^{m_{i,n}} \left\| \nu(E_{j,i,n}) \right\| \\
&= \mathcal{P} \circ \nu(E_n) + 2^{-n-1}\epsilon + \sum_{i=1}^{m_n} \sum_{j=1}^{m_{i,n}} \Big(\Big\| \int_{E_{j,i,n}} \varphi_n \,\mathrm{d}\mu \Big\| - \Big\| \int_{E_{j,i,n}} f \,\mathrm{d}\mu \Big\| \Big) \\
&< \mathcal{P} \circ \nu(E_n) + 2^{-n-1}\epsilon + \sum_{i=1}^{m_n} \sum_{j=1}^{m_{i,n}} \Big\| \int_{E_{j,i,n}} (\varphi_n - f|_{E_n}) \,\mathrm{d}\mu \Big\| \\
&\le \mathcal{P} \circ \nu(E_n) + 2^{-n-1}\epsilon + \sum_{i=1}^{m_n} \sum_{j=1}^{m_{i,n}} \int_{E_{j,i,n}} \mathcal{P} \circ (\varphi_n - f|_{E_n}) \,\mathrm{d}\mathcal{P} \circ \mu \\
&= \mathcal{P} \circ \nu(E_n) + 2^{-n-1}\epsilon + \int_{\bigcup_{i=1}^{m_n} \bigcup_{j=1}^{m_{i,n}} E_{j,i,n}} \mathcal{P} \circ (\varphi_n - f|_{E_n}) \,\mathrm{d}\mathcal{P} \circ \mu \\
&\le \mathcal{P} \circ \nu(E_n) + 2^{-n-1}\epsilon + \int_{E_n} \mathcal{P} \circ (\varphi_n - f|_{E_n}) \,\mathrm{d}\mathcal{P} \circ \mu \\
&< \mathcal{P} \circ \nu(E_n) + 2^{-n}\epsilon
\end{aligned}
$$

where the second equality follows from Proposition 11.75, the third equality follows from Proposition 11.125, the fourth and the fifth inequalities follow from Proposition 11.132, and the fourth equality and the sixth inequality follow from Proposition 11.83. Then, $\bar{\nu}(E) = \sum_{n=1}^{\infty} \bar{\nu}(E_n) < \sum_{n=1}^{\infty} (\mathcal{P} \circ \nu(E_n) + 2^{-n}\epsilon) = \mathcal{P} \circ \nu(E) + \epsilon$. By the arbitrariness of ϵ, we have $\bar{\nu}(E) \leq \mathcal{P} \circ \nu(E)$. This case is proved.

Hence, $\mathcal{P} \circ \nu(E) = \bar{\nu}(E) = \int_E \mathcal{P} \circ f \, d\mathcal{P} \circ \mu$, $\forall E \in \mathcal{B}$. This completes the proof of the proposition. □

Definition 11.155 Let $(X, \mathcal{B})$ be a measurable space, μ_1 and μ_2 be measures on $(X, \mathcal{B})$. We will say that μ_1 is *absolutely continuous* with respect to μ_2 if $\forall A \in \mathcal{B}$ with $\mu_2(A) = 0$, we have $\mu_1(A) = 0$. In this case, we will denote $\mu_1 \ll \mu_2$. μ_1 and μ_2 are said to be *mutually singular* if $\exists A, B \in \mathcal{B}$ with $A \cap B = \emptyset$ and $X = A \cup B$, such that $\mu_1(A) = \mu_2(B) = 0$. In this case, we will denote $\mu_1 \perp \mu_2$. ◇

Proposition 11.156 *Let μ, λ_1, and λ_2 be measures on a measurable space $(X, \mathcal{B})$, $\mathcal{Y}$ be a normed linear space over $\mathbb{K}$, ν_1 and ν_2 be σ-finite $\mathcal{Y}$-valued measures on $(X, \mathcal{B})$, $a, b \in (0, \infty) \subset \mathbb{R}$, and $\alpha, \beta \in \mathbb{K}$. Then, the following statements hold:*

(i) *$\lambda_1 \ll \mu$ and $\lambda_2 \ll \mu$ if, and only if, $a\lambda_1 + b\lambda_2 \ll \mu$.*
(ii) *$\lambda_1 \perp \mu$ and $\lambda_2 \perp \mu$ if, and only if, $a\lambda_1 + b\lambda_2 \perp \mu$.*
(iii) *If $\lambda_1 \ll \mu$ and $\lambda_2 \perp \mu$, then $\lambda_1 \perp \lambda_2$.*
(iv) *If $\lambda_1 \perp \lambda_1$, then $\lambda_1(E) = 0$, $\forall E \in \mathcal{B}$.*
(v) *If $\mathcal{P} \circ \nu_1 \perp \mathcal{P} \circ \nu_2$, then $\mathcal{P} \circ (\alpha\nu_1) \perp \mathcal{P} \circ (\beta\nu_2)$.*
(vi) *If $\mathcal{P} \circ \nu_1 \perp \mathcal{P} \circ \nu_1$, then $\nu_1(E) = \vartheta_{\mathcal{Y}}$, $\forall E \in \mathcal{B}$.*

If, in addition, $\mathcal{Y}$ is complete, then the following statements hold:

(vii) *If $\mathcal{P} \circ \nu_1 \perp \mathcal{P} \circ \nu_2$, then $\mathcal{P} \circ (\nu_1 + \nu_2) = \mathcal{P} \circ \nu_1 + \mathcal{P} \circ \nu_2$.*
(viii) *If $\mathcal{P} \circ \nu_1 \perp \mathcal{P} \circ \nu_2$, then $\mathcal{P} \circ (\alpha\nu_1 + \beta\nu_2) = |\alpha|\mathcal{P} \circ \nu_1 + |\beta|\mathcal{P} \circ \nu_2$.*
(ix) *If $\mathcal{P} \circ \nu_1 \ll \mu$ and $\mathcal{P} \circ \nu_2 \ll \mu$, then $\mathcal{P} \circ (\alpha\nu_1 + \beta\nu_2) \ll \mu$.*
(x) *If $\mathcal{P} \circ \nu_1 \perp \mu$ and $\mathcal{P} \circ \nu_2 \perp \mu$, then $\mathcal{P} \circ (\alpha\nu_1 + \beta\nu_2) \perp \mu$.*

Proof (i) By Proposition 11.136, $a\lambda_1 + b\lambda_2$ is a measure on $(X, \mathcal{B})$. "Sufficiency" $\forall E \in \mathcal{B}$ with $\mu(E) = 0$, by $a\lambda_1 + b\lambda_2 \ll \mu$, we have $a\lambda_1(E) + b\lambda_2(E) = 0$. Then, $\lambda_1(E) = \lambda_2(E) = 0$. Hence, $\lambda_1 \ll \mu$ and $\lambda_2 \ll \mu$. "Necessity" $\forall E \in \mathcal{B}$ with $\mu(E) = 0$, by $\lambda_1 \ll \mu$ and $\lambda_2 \ll \mu$, we have $\lambda_1(E) = \lambda_2(E) = 0$. Then, $a\lambda_1(E) + b\lambda_2(E) = 0$. Hence, $a\lambda_1 + b\lambda_2 \ll \mu$.

(ii) By Proposition 11.136, $a\lambda_1 + b\lambda_2$ is a measure on $(X, \mathcal{B})$. "Sufficiency" Let $a\lambda_1 + b\lambda_2 \perp \mu$. Then, $\exists A, B \in \mathcal{B}$ with $A = X \setminus B$ such that $(a\lambda_1 + b\lambda_2)(A) = 0 = \mu(B)$. Then, $\lambda_1(A) = \lambda_2(A) = \mu(B) = 0$. Hence, $\lambda_1 \perp \mu$ and $\lambda_2 \perp \mu$. "Necessity" Let $\lambda_1 \perp \mu$ and $\lambda_2 \perp \mu$. Then, $\exists A_i, B_i \in \mathcal{B}$ with $A_i = X \setminus B_i$ such that $\lambda_i(A_i) = \mu(B_i) = 0$, $i = 1, 2$. Let $A := A_1 \cap A_2 \in \mathcal{B}$ and $B := B_1 \cup B_2 \in \mathcal{B}$. Then, $A = X \setminus B$ and $0 \leq a\lambda_1(A) + b\lambda_2(A) = (a\lambda_1 + b\lambda_2)(A) \leq a\lambda_1(A_1) + b\lambda_2(A_2) = 0 \leq \mu(B) \leq \mu(B_1) + \mu(B_2) = 0$. Hence, $a\lambda_1 + b\lambda_2 \perp \mu$.

(iii) By $\lambda_2 \perp \mu$, $\exists A, B \in \mathcal{B}$ with $A = X \setminus B$ such that $\lambda_2(A) = 0 = \mu(B)$. By $\lambda_1 \ll \mu$, we have $\lambda_1(B) = 0$. Hence, $\lambda_1 \perp \lambda_2$.

(iv) By $\lambda_1 \perp \lambda_1$, $\exists A, B \in \mathcal{B}$ with $A = X \setminus B$ such that $\lambda_1(A) = \lambda_1(B) = 0$. Then, $\lambda_1(X) = \lambda_1(A) + \lambda_1(B) = 0$. Hence, the result holds.

(v) By Proposition 11.138, $\alpha\nu_1$ and $\beta\nu_2$ are σ-finite $\mathcal{Y}$-valued measures on $(X, \mathcal{B})$ with $\mathcal{P}\circ(\alpha\nu_1) = |\alpha|\mathcal{P}\circ\nu_1$ and $\mathcal{P}\circ(\beta\nu_2) = |\beta|\mathcal{P}\circ\nu_2$. By $\mathcal{P}\circ\nu_1 \perp \mathcal{P}\circ\nu_2$, $\exists A, B \in \mathcal{B}$ with $A = X \setminus B$ such that $\mathcal{P}\circ\nu_1(A) = 0 = \mathcal{P}\circ\nu_2(B)$. Then, by Proposition 11.138, $\mathcal{P} \circ (\alpha\nu_1)(A) = |\alpha|\mathcal{P} \circ \nu_1(A) = 0 = |\beta|\mathcal{P} \circ \nu_2(B) = \mathcal{P} \circ (\beta\nu_2)(B)$. Hence, $\mathcal{P} \circ (\alpha\nu_1) \perp \mathcal{P} \circ (\beta\nu_2)$.

(vi) By (iv), $\mathcal{P} \circ \nu_1(E) = 0$, $\forall E \in \mathcal{B}$. Then, $\nu_1(E) = \vartheta_{\mathcal{Y}}$, $\forall E \in \mathcal{B}$.

Let $\mathcal{Y}$ be a Banach space over $\mathbb{K}$; then $\mathcal{M}_\sigma(X, \mathcal{B}, \mathcal{Y})$ is well-defined by Proposition 11.141.

(vii) By Proposition 11.138, $\nu_1 + \nu_2 \in \mathcal{M}_\sigma(X, \mathcal{B}, \mathcal{Y})$ and $\mathcal{P} \circ (\nu_1 + \nu_2) \leq \mathcal{P} \circ \nu_1 + \mathcal{P}\circ\nu_2$. By $\mathcal{P}\circ\nu_1 \perp \mathcal{P}\circ\nu_2$, $\exists A, B \in \mathcal{B}$ with $A = X \setminus B$ such that $\mathcal{P}\circ\nu_1(A) = 0 = \mathcal{P}\circ\nu_2(B)$. Fix any $E \in \mathcal{B}$. We will show that $\mathcal{P}\circ(\nu_1 + \nu_2)(E) \geq \mathcal{P}\circ\nu_1(E) + \mathcal{P} \circ \nu_2(E)$ by distinguishing two exhaustive and mutually exclusive cases: Case 1: $\mathcal{P} \circ \nu_1(E) + \mathcal{P} \circ \nu_2(E) < \infty$; Case 2: $\mathcal{P} \circ \nu_1(E) + \mathcal{P} \circ \nu_2(E) = \infty$.

Case 1: $\mathcal{P} \circ \nu_1(E) + \mathcal{P} \circ \nu_2(E) < \infty$. Then, $\mathcal{P} \circ \nu_i(E) < \infty$ and $E \in \operatorname{dom}(\nu_i)$, $i = 1, 2$. $\forall \epsilon \in (0, \infty) \subset \mathbb{R}$, $\exists n_1 \in \mathbb{Z}_+$, $\exists$ pairwise disjoint $(E_i)_{i=1}^{n_1} \subseteq \mathcal{B}$ with $E = \bigcup_{i=1}^{n_1} E_i$ such that $\mathcal{P}\circ\nu_1(E) < \sum_{i=1}^{n_1} \|\nu_1(E_i)\| + \epsilon/2 < +\infty$. $\exists n_2 \in \mathbb{Z}_+$, $\exists$ pairwise disjoint $(F_i)_{i=1}^{n_2} \subseteq \mathcal{B}$ with $E = \bigcup_{i=1}^{n_2} F_i$ such that $\mathcal{P} \circ \nu_2(E) < \sum_{i=1}^{n_2} \|\nu_2(F_i)\| + \epsilon/2 < +\infty$. Then,

$$\begin{aligned}
\mathcal{P} \circ \nu_1(E) + \mathcal{P} \circ \nu_2(E) &< \sum_{i=1}^{n_1} \|\nu_1(E_i)\| + \sum_{i=1}^{n_2} \|\nu_2(F_i)\| + \epsilon \\
&= \sum_{i=1}^{n_1} \|\nu_1(E_i \cap A) + \nu_1(E_i \cap B)\| + \sum_{i=1}^{n_2} \|\nu_2(F_i \cap A) + \nu_2(F_i \cap B)\| + \epsilon \\
&= \sum_{i=1}^{n_1} \|\nu_2(E_i \cap B) + \nu_1(E_i \cap B)\| + \sum_{i=1}^{n_2} \|\nu_2(F_i \cap A) + \nu_1(F_i \cap A)\| + \epsilon \\
&= \sum_{i=1}^{n_1} \|(\nu_1 + \nu_2)(E_i \cap B)\| + \sum_{i=1}^{n_2} \|(\nu_1 + \nu_2)(F_i \cap A)\| + \epsilon \\
&\leq \sum_{i=1}^{n_1} \mathcal{P} \circ (\nu_1 + \nu_2)(E_i \cap B) + \sum_{i=1}^{n_2} \mathcal{P} \circ (\nu_1 + \nu_2)(F_i \cap A) + \epsilon \\
&= \mathcal{P} \circ (\nu_1 + \nu_2)(E \cap B) + \mathcal{P} \circ (\nu_1 + \nu_2)(E \cap A) + \epsilon \\
&= \mathcal{P} \circ (\nu_1 + \nu_2)(E) + \epsilon
\end{aligned}$$

where the second equality follows from the fact that $\forall \bar{A} \in \mathcal{B}$ with $\bar{A} \subseteq A$, we have $\mathcal{P} \circ \nu_1(\bar{A}) = 0$ and $\nu_1(\bar{A}) = \vartheta_{\mathcal{Y}}$, and the fact that $\forall \bar{B} \in \mathcal{B}$ with $\bar{B} \subseteq B$, we have $\mathcal{P}\circ\nu_2(\bar{B}) = 0$ and $\nu_2(\bar{B}) = \vartheta_{\mathcal{Y}}$; the third equality follows from Proposition 11.138; and the second inequality and the last two equalities follow from Definition 11.108.

By the arbitrariness of ϵ, we have $\mathcal{P} \circ \nu_1(E) + \mathcal{P} \circ \nu_2(E) \leq \mathcal{P} \circ (\nu_1 + \nu_2)(E)$. This case is proved.

Case 2: $\mathcal{P} \circ \nu_1(E) + \mathcal{P} \circ \nu_2(E) = \infty$. Without loss of generality, assume $\mathcal{P} \circ \nu_1(E) = \infty$. Since ν_1 and ν_2 are σ-finite, then $\exists (X_n)_{n=1}^{\infty} \subseteq \mathcal{B}$ such that $X = \bigcup_{n=1}^{\infty} X_n$ and $\mathcal{P} \circ \nu_i(X_n) < \infty, \forall n \in \mathbb{N}, i = 1, 2$. Without loss of generality, we may assume that $X_n \subseteq X_{n+1}, \forall n \in \mathbb{N}$. Let $E_n := E \cap X_n, \forall n \in \mathbb{N}$. Then, $E = \bigcup_{n=1}^{\infty} E_n$ and $E_n \subseteq E_{n+1}, \forall n \in \mathbb{N}$. By Proposition 11.7, we have $\lim_{n \in \mathbb{N}} \mathcal{P} \circ \nu_1(E_n) = \mathcal{P} \circ \nu_1(E) = \infty$. $\forall M \in (0, \infty) \subset \mathbb{R}$, $\exists n_0 \in \mathbb{N}$ such that $\mathcal{P} \circ \nu_1(E_{n_0}) > M + 1$. Clearly, $\mathcal{P} \circ \nu_i(E_{n_0}) \leq \mathcal{P} \circ \nu_i(X_{n_0}) < \infty$, $i = 1, 2$. Then, $\exists n_1 \in \mathbb{Z}_+$, $\exists$ pairwise disjoint $(\bar{E}_i)_{i=1}^{n_1} \subseteq \mathcal{B}$ with $E_{n_0} = \bigcup_{i=1}^{n_1} \bar{E}_i$ such that $\mathcal{P} \circ \nu_1(E_{n_0}) < \sum_{i=1}^{n_1} \|\nu_1(\bar{E}_i)\| + 1$. Then,

$$\begin{aligned}
M &< \sum_{i=1}^{n_1} \|\nu_1(\bar{E}_i)\| = \sum_{i=1}^{n_1} \|\nu_1(\bar{E}_i \cap A) + \nu_1(\bar{E}_i \cap B)\| \\
&= \sum_{i=1}^{n_1} \|\nu_2(\bar{E}_i \cap B) + \nu_1(\bar{E}_i \cap B)\| \\
&= \sum_{i=1}^{n_1} \|(\nu_1 + \nu_2)(\bar{E}_i \cap B)\| \leq \sum_{i=1}^{n_1} \mathcal{P} \circ (\nu_1 + \nu_2)(\bar{E}_i \cap B) \\
&= \mathcal{P} \circ (\nu_1 + \nu_2)(E_{n_0} \cap B) \leq \mathcal{P} \circ (\nu_1 + \nu_2)(E)
\end{aligned}$$

where the first equality follows from Definition 11.108; the second equality follows from the fact that $\forall \bar{A} \in \mathcal{B}$ with $\bar{A} \subseteq A$, we have $\mathcal{P} \circ \nu_1(\bar{A}) = 0$ and $\nu_1(\bar{A}) = \vartheta_{\mathcal{Y}}$, and the fact that $\forall \bar{B} \in \mathcal{B}$ with $\bar{B} \subseteq B$, we have $\mathcal{P} \circ \nu_2(\bar{B}) = 0$ and $\nu_2(\bar{B}) = \vartheta_{\mathcal{Y}}$; the third equality follows from Proposition 11.138; and the second inequality follows from Definition 11.108. By the arbitrariness of M, we have $\mathcal{P} \circ (\nu_1 + \nu_2)(E) = \infty$. Then, $\mathcal{P} \circ \nu_1(E) + \mathcal{P} \circ \nu_2(E) = \mathcal{P} \circ (\nu_1 + \nu_2)(E) = \infty$. This case is proved.

In both cases, we have $\mathcal{P} \circ \nu_1(E) + \mathcal{P} \circ \nu_2(E) \leq \mathcal{P} \circ (\nu_1 + \nu_2)(E)$. By the arbitrariness of E, we have $\mathcal{P} \circ \nu_1 + \mathcal{P} \circ \nu_2 \leq \mathcal{P} \circ (\nu_1 + \nu_2)$. Hence, $\mathcal{P} \circ (\nu_1 + \nu_2) = \mathcal{P} \circ \nu_1 + \mathcal{P} \circ \nu_2$.

(viii) By Proposition 11.138, $\alpha\nu_1, \beta\nu_2, \alpha\nu_1 + \beta\nu_2 \in \mathcal{M}_\sigma(X, \mathcal{B}, \mathcal{Y})$. By (v), we have $\mathcal{P} \circ (\alpha\nu_1) \perp \mathcal{P} \circ (\beta\nu_2)$. By (vii) and Proposition 11.138, we have $\mathcal{P} \circ (\alpha\nu_1 + \beta\nu_2) = \mathcal{P} \circ (\alpha\nu_1) + \mathcal{P} \circ (\beta\nu_2) = |\alpha|\mathcal{P} \circ \nu_1 + |\beta|\mathcal{P} \circ \nu_2$.

(ix) By Proposition 11.138, $\alpha\nu_1 + \beta\nu_2 \in \mathcal{M}_\sigma(X, \mathcal{B}, \mathcal{Y})$. $\forall E \in \mathcal{B}$ with $\mu(E) = 0$, by $\mathcal{P} \circ \nu_i \ll \mu$, we have $\mathcal{P} \circ \nu_i(E) = 0$, $i = 1, 2$. Then, $0 \leq \mathcal{P} \circ (\alpha\nu_1 + \beta\nu_2)(E) \leq |\alpha|\mathcal{P} \circ \nu_1(E) + |\beta|\mathcal{P} \circ \nu_2(E) = 0$, where the second inequality follows from Proposition 11.138. Hence, $\mathcal{P} \circ (\alpha\nu_1 + \beta\nu_2) \ll \mu$.

(x) By Proposition 11.138, $\alpha\nu_1 + \beta\nu_2 \in \mathcal{M}_\sigma(X, \mathcal{B}, \mathcal{Y})$. By $\mathcal{P} \circ \nu_i \perp \mu$, $\exists A_i, B_i \in \mathcal{B}$ with $A_i = X \setminus B_i$ such that $\mathcal{P} \circ \nu_i(A_i) = \mu(B_i) = 0$, $i = 1, 2$. Let $A := A_1 \cap A_2 \in \mathcal{B}$ and $B := B_1 \cup B_2 \in \mathcal{B}$. Then, $A = X \setminus B$ and $0 \leq \mathcal{P} \circ (\alpha\nu_1 + \beta\nu_2)(A) \leq (|\alpha|\mathcal{P} \circ \nu_1 + |\beta|\mathcal{P} \circ \nu_2)(A) \leq |\alpha|\mathcal{P} \circ \nu_1(A_1) + |\beta|\mathcal{P} \circ \nu_2(A_2) = 0 \leq \mu(B) \leq$

$\mu(B_1) + \mu(B_2) = 0$, where the second inequality follows from Proposition 11.138. Hence, $\mathcal{P} \circ (\alpha\nu_1 + \beta\nu_2) \perp \mu$.

This completes the proof of the proposition. □

Proposition 11.157 *Let $(X, \mathcal{B}, \mu)$ be a $\mathcal{Y}$-valued measure space, where $\mathcal{Y}$ is a normed linear space, $A \in \mathcal{B}$, and $B := X \setminus A$. Define μ_A to be a function from $\mathcal{B}$ to $\mathcal{Y}$ by: $\forall E \in \mathcal{B}$ with $E \cap A \in \operatorname{dom}(\mu)$, $\mu_A(E) := \mu(E \cap A) \in \mathcal{Y}$; $\forall E \in \mathcal{B}$ with $E \cap A \in \mathcal{B} \setminus \operatorname{dom}(\mu)$, $\mu_A(E)$ is undefined. Define μ_B to be a function from $\mathcal{B}$ to $\mathcal{Y}$ similarly as μ_A with A replaced with B. Then, the following statements hold:*

(i) *μ_A and μ_B are $\mathcal{Y}$-valued measures on $(X, \mathcal{B})$ with $\mathcal{P} \circ \mu_A(E) = \mathcal{P} \circ \mu(E \cap A)$ and $\mathcal{P} \circ \mu_B(E) = \mathcal{P} \circ \mu(E \cap B)$, $\forall E \in \mathcal{B}$.*
(ii) *$\mathcal{P} \circ \mu_A \perp \mathcal{P} \circ \mu_B$ and $\mathcal{P} \circ \mu = \mathcal{P} \circ \mu_A + \mathcal{P} \circ \mu_B$.*
(iii) *If μ is σ-finite and $\mathcal{Y}$ is a Banach space, then μ_A and μ_B are σ-finite and $\mu = \mu_A + \mu_B$.*

Proof (i) Define $\nu_A : \mathcal{B} \to [0, \infty] \subset \mathbb{R}_e$ by $\nu_A(E) = \mathcal{P} \circ \mu(E \cap A)$, $\forall E \in \mathcal{B}$. It is straightforward to check that ν_A is a measure on $(X, \mathcal{B})$. $\forall E \in \mathcal{B}$ with $\nu_A(E) = \infty$, then $\mathcal{P} \circ \mu(E \cap A) = \infty$, $E \cap A \in \mathcal{B} \setminus \operatorname{dom}(\mu)$, and $\mu_A(E)$ is undefined. $\forall E \in \mathcal{B}$ with $\nu_A(E) < \infty$, then $\mathcal{P} \circ \mu(E \cap A) < \infty$, $E \cap A \in \operatorname{dom}(\mu)$, and $\mu_A(E) = \mu(E \cap A) \in \mathcal{Y}$. $\forall$ pairwise disjoint $(E_n)_{n=1}^{\infty} \subseteq \mathcal{B}$ with $E = \bigcup_{n=1}^{\infty} E_n$, clearly, $0 \leq \mathcal{P} \circ \mu(E_n \cap A) \leq \mathcal{P} \circ \mu(E \cap A) < \infty$, $E_n \cap A \in \operatorname{dom}(\mu)$, and $\mu_A(E_n) = \mu(E_n \cap A) \in \mathcal{Y}$, $\forall n \in \mathbb{N}$. By Definition 11.108, $\mu_A(E) = \mu(E \cap A) = \sum_{n=1}^{\infty} \mu(E_n \cap A) = \sum_{n=1}^{\infty} \mu_A(E_n)$ and $\sum_{n=1}^{\infty} \|\mu_A(E_n)\| = \sum_{n=1}^{\infty} \|\mu(E_n \cap A)\| < \infty$. $\forall E \in \mathcal{B}$ with $\nu_A(E) < \infty$,

$$\begin{aligned}
\nu_A(E) &= \mathcal{P} \circ \mu(E \cap A) \\
&= \sup_{n \in \mathbb{Z}_+,\ (E_i)_{i=1}^n \subseteq \mathcal{B},\ E \cap A = \bigcup_{i=1}^n E_i,\ E_i \cap E_j = \emptyset,\ \forall 1 \leq i < j \leq n} \sum_{i=1}^{n} \|\mu(E_i)\| \\
&= \sup_{n \in \mathbb{Z}_+,\ (E_i)_{i=1}^n \subseteq \mathcal{B},\ E \cap A = \bigcup_{i=1}^n E_i,\ E_i \cap E_j = \emptyset,\ \forall 1 \leq i < j \leq n} \sum_{i=1}^{n} \|\mu_A(E_i)\| \\
&= \sup_{n \in \mathbb{Z}_+,\ (E_i)_{i=1}^n \subseteq \mathcal{B},\ E = \bigcup_{i=1}^n E_i,\ E_i \cap E_j = \emptyset,\ \forall 1 \leq i < j \leq n} \sum_{i=1}^{n} \|\mu_A(E_i)\|
\end{aligned}$$

Hence, μ_A is a $\mathcal{Y}$-valued measure on $(X, \mathcal{B})$ with $\mathcal{P} \circ \mu_A = \nu_A$. By symmetry, μ_B is a $\mathcal{Y}$-valued measure on $(X, \mathcal{B})$ with $\mathcal{P} \circ \mu_B(E) = \mathcal{P} \circ \mu(E \cap B)$, $\forall E \in \mathcal{B}$.

(ii) Clearly, $\mathcal{P} \circ \mu_A(B) = 0 = \mathcal{P} \circ \mu_B(A)$. Then, $\mathcal{P} \circ \mu_A \perp \mathcal{P} \circ \mu_B$. $\forall E \in \mathcal{B}$, $\mathcal{P} \circ \mu_A(E) + \mathcal{P} \circ \mu_B(E) = \mathcal{P} \circ \mu(E \cap A) + \mathcal{P} \circ \mu(E \cap B) = \mathcal{P} \circ \mu(E)$. Hence, $\mathcal{P} \circ \mu = \mathcal{P} \circ \mu_A + \mathcal{P} \circ \mu_B$.

(iii) If μ is σ-finite and $\mathcal{Y}$ is a Banach space, then $\mathcal{P} \circ \mu$ is σ-finite and $\mathcal{P} \circ \mu_A$ and $\mathcal{P} \circ \mu_B$ are σ-finite. Therefore, μ_A and μ_B are σ-finite. Then, $\mu_A + \mu_B$ is well-defined as in Proposition 11.138. $\forall E \in \operatorname{dom}(\mu_A) \cap \operatorname{dom}(\mu_B)$, we have $E \cap A \in \operatorname{dom}(\mu)$ and $E \cap B \in \operatorname{dom}(\mu)$. Then, $\mathcal{P} \circ \mu(E) = \mathcal{P} \circ \mu(E \cap A) + \mathcal{P} \circ \mu(E \cap B) < \infty$

and $E \in \text{dom}(\mu)$. Hence, $\mu(E) = \mu(E \cap A) + \mu(E \cap B) = \mu_A(E) + \mu_B(E)$. By Proposition 11.138, $\mu = \mu_A + \mu_B$.

This completes the proof of the proposition. □

11.8 The Radon–Nikodym Theorem

Definition 11.158 Let $(X, \mathcal{B}, \mu)$ be an $\mathbb{R}$-valued pre-measure space (or a σ-finite $\mathbb{R}$-valued measure space). $A \in \mathcal{B}$ is said to be a *positive set* with respect to μ if $\forall E \in \text{dom}(\mu)$ with $E \subseteq A$, we have $\mu(E) \in [0, \infty) \subset \mathbb{R}$. $B \in \mathcal{B}$ is said to be a *negative set* with respect to μ if $\forall E \in \text{dom}(\mu)$ with $E \subseteq B$, we have $\mu(E) \in (-\infty, 0] \subset \mathbb{R}$. ◇

Lemma 11.159 *Let $(X, \mathcal{B}, \mu)$ be an $\mathbb{R}$-valued pre-measure space (or a σ-finite $\mathbb{R}$-valued measure space). Then, every measurable subset of a positive set is a positive set; the union of countably many positive sets is a positive set. Similar statements hold for negative sets.*

Proof The first statement follows directly from Definition 11.158. For the second statement, let $(A_n)_{n=1}^{\infty} \subseteq \mathcal{B}$ be a sequence of positive sets and $A := \bigcup_{n=1}^{\infty} A_n$. $\forall E \in \text{dom}(\mu)$ with $E \subseteq A$, let $E_n := E \cap A_n \setminus (\bigcup_{i=1}^{n-1} A_i) \in \mathcal{B}$, $\forall n \in \mathbb{N}$. Then, $E = \bigcup_{n=1}^{\infty} E_n$, $(E_n)_{n=1}^{\infty}$ is pairwise disjoint, $E_n \in \text{dom}(\mu)$, $E_n \subseteq A_n$, and $\mu(E_n) \in [0, \infty) \subset \mathbb{R}$, $\forall n \in \mathbb{N}$. This leads to $\mu(E) = \sum_{i=1}^{\infty} \mu(E_n) \in [0, \infty) \subset \mathbb{R}$. Hence, A is a positive set. For the case of finitely many positive sets, we may form a sequence of positive sets by appending $\emptyset$, which is clearly a positive set, to the collection. This completes the proof of the lemma. □

Lemma 11.160 *Let $(X, \mathcal{B}, \mu)$ be an $\mathbb{R}$-valued pre-measure space (or a σ-finite $\mathbb{R}$-valued measure space) and $E \in \mathcal{B}$. Assume that $E \in \text{dom}(\mu)$ and $\mu(E) \in (0, +\infty) \subset \mathbb{R}$. Then, there exists a positive set $A \subseteq E$ with $\mu(A) \in (0, +\infty) \subset \mathbb{R}$. Similar statement holds for negative sets.*

Proof Note that, $\forall \bar{E} \in \mathcal{B}$ with $\bar{E} \subseteq E$, we have $\bar{E} \in \text{dom}(\mu)$ and $\mu(\bar{E}) \in \mathbb{R}$. Either E is a positive set, in which case the result holds, or E contains a subset of negative measures. In the latter case, let $n_1 \in \mathbb{N}$ be the smallest natural number such that there is a measurable set $E_1 \subseteq E$ with $\mu(E_1) < -1/n_1$. Then, $\mu(E \setminus E_1) = \mu(E) - \mu(E_1) \in (0, +\infty) \subset \mathbb{R}$. Inductively, assume that we have obtained $n_j \in \mathbb{N}$, $E_j \in \mathcal{B}$, $\forall j \in \{1, \ldots, k\} \subseteq \mathbb{N}$, such that $E_j \subseteq E \setminus (\bigcup_{l=1}^{j-1} E_l)$, $\mu(E_j) < -1/n_j$, and $\mu(E \setminus (\bigcup_{l=1}^{j} E_l)) \in (0, +\infty) \subset \mathbb{R}$, $\forall j \in \{1, \ldots, k\}$. Either $E \setminus (\bigcup_{j=1}^{k} E_j)$ is a positive set, in which case the result holds, or it contains a subset of negative measures. In the latter case, let $n_{k+1} \in \mathbb{N}$ be the smallest natural number such that there is a measurable set $E_{k+1} \subseteq E \setminus (\bigcup_{j=1}^{k} E_j)$ with $\mu(E_{k+1}) < -1/n_{k+1}$. Then, $\mu(E \setminus (\bigcup_{j=1}^{k+1} E_j)) = \mu(E \setminus (\bigcup_{j=1}^{k} E_j)) - \mu(E_{k+1}) \in (0, +\infty) \subset \mathbb{R}$. Thus, either the result holds or we have $(n_k)_{k=1}^{\infty} \subseteq \mathbb{N}$ and $(E_k)_{k=1}^{\infty} \subseteq \mathcal{B}$. Let $A := E \setminus (\bigcup_{k=1}^{\infty} E_k) \in \mathcal{B}$. Then, $E = A \cup (\bigcup_{k=1}^{\infty} E_k)$, where the sets in the union are pairwise disjoint. Then, $\mu(E) = \mu(A) + \sum_{k=1}^{\infty} \mu(E_k)$ and $|\mu(A)| - \sum_{k=1}^{\infty} \mu(E_k) < +\infty$. Then, $\sum_{k=1}^{\infty} 1/n_k < +\infty$, $\lim_{k\in\mathbb{N}} n_k = \infty$, and $\mu(A) \in (0, +\infty) \subset \mathbb{R}$. $\forall \epsilon \in (0, \infty) \subset \mathbb{R}$, $\exists k_0 \in \mathbb{N}$ such that $(n_{k_0} - 1)^{-1} < \epsilon$. Since $A \subseteq E \setminus (\bigcup_{j=1}^{k_0-1} E_j)$, then A cannot contain any measurable subset with measure less than $-1/(n_{k_0} - 1) > -\epsilon$. Thus, A contains no measurable subset with measure less than $-\epsilon$. By the arbitrariness of ϵ, A is a positive set. This completes the proof of the lemma. □

Theorem 11.161 (Hahn Decomposition Theorem) *Let $(X, \mathcal{B}, \mu) =: \mathcal{X}$ be an $\mathbb{R}$-valued pre-measure space (or a σ-finite $\mathbb{R}$-valued measure space). Then, there is a positive set $A \in \mathcal{B}$ and a negative set $B \in \mathcal{B}$ such that $X = A \cup B$ and $A \cap B = \emptyset$.*

Proof We will distinguish two exhaustive and mutually exclusives cases: Case 1: $\mathcal{X}$ is an $\mathbb{R}$-valued pre-measure space; Case 2: $\mathcal{X}$ is a σ-finite $\mathbb{R}$-valued measure space.

Case 1: $\mathcal{X}$ is an $\mathbb{R}$-valued pre-measure space. Let $\lambda := \sup\{\mu(A) \in \mathbb{R} \mid A \text{ is a positive set}\} \in \mathbb{R}_e$. Since $\emptyset$ is a positive set, then $\lambda \geq 0$. Let $(A_i)_{i=1}^{\infty} \subseteq \mathcal{B}$ be a sequence of positive sets such that $\lim_{i\in\mathbb{N}} \mu(A_i) = \lambda$. Let $A := \bigcup_{i=1}^{\infty} A_i \in \mathcal{B}$. By Lemma 11.159, A is a positive set. This implies that $\mu(A) \leq \lambda$. Note that $\mu(A) = \mu(A_i) + \mu(A \setminus A_i) \geq \mu(A_i)$, $\forall i \in \mathbb{N}$. Then, $\lambda = \mu(A) \in \mathbb{R}$. Let $B := X \setminus A$. Suppose that there exists $E \subseteq B$ such that E is a positive set with $\mu(E) > 0$. Then $A \cup E$ is a positive set by Lemma 11.159 and $\mu(A \cup E) = \mu(A) + \mu(E) > \lambda$. This contradicts with the definition of λ. Hence, B does not contain any subset that is a positive set with positive measure. Then, by Lemma 11.160, B does not contain any subset with positive measure. Hence, B is a negative set. This case is proved.

Case 2: $\mathcal{X}$ is a σ-finite $\mathbb{R}$-valued measure space. Then, $\exists (X_n)_{n=1}^{\infty} \subseteq \mathcal{B}$ such that $X = \bigcup_{n=1}^{\infty} X_n$ and $\mathcal{P} \circ \mu(X_n) < \infty$, $\forall n \in \mathbb{N}$. Without loss of generality, we may assume that $(X_n)_{n=1}^{\infty}$ is pairwise disjoint. Let $\mathcal{X}_n := (X_n, \mathcal{B}_n, \mu_n)$ be the finite $\mathbb{R}$-valued measure subspace of $\mathcal{X}$, $\forall n \in \mathbb{N}$. Fix any $n \in \mathbb{N}$. $\mathcal{X}_n$ is a finite $\mathbb{R}$-valued pre-measure space. By Case 1, there exists $A_n, B_n \in \mathcal{B}_n \subseteq \mathcal{B}$ such that $A_n = X_n \setminus B_n$ and A_n is a positive set with respect to μ_n and B_n is a negative set with respect to μ_n. It is straightforward to see that A_n is a positive set with respect to μ and B_n is a negative set with respect to μ. Let $A := \bigcup_{n=1}^{\infty} A_n \in \mathcal{B}$ and $B := \bigcup_{n=1}^{\infty} B_n \in \mathcal{B}$. Then, by Lemma 11.159, A is a positive set with respect to μ and B is a negative set with respect to μ. $A \cup B = (\bigcup_{n=1}^{\infty} A_n) \cup (\bigcup_{n=1}^{\infty} B_n) = \bigcup_{n=1}^{\infty} (A_n \cup B_n) = \bigcup_{n=1}^{\infty} X_n = X$ and $A \cap B = (\bigcup_{n=1}^{\infty} A_n) \cap (\bigcup_{n=1}^{\infty} B_n) = \bigcup_{n=1}^{\infty} \bigcup_{m=1}^{\infty} (A_n \cap B_m) = \emptyset$. This case is proved.

This completes the proof of the theorem. □

Theorem 11.162 (Jordan Decomposition Theorem) *Let $(X, \mathcal{B}, \mu)$ be an $\mathbb{R}$-valued pre-measure space. Then, there exists a unique pair of mutually singular finite measures μ_+ and μ_- on $(X, \mathcal{B})$ such that $\mu = \mu_+ - \mu_-$. Furthermore, $\mathcal{P} \circ \mu = \mu_+ + \mu_-$ and $(X, \mathcal{B}, \mu)$ is a finite $\mathbb{R}$-valued measure space.*

Proof By Hahn Decomposition Theorem 11.161, $\exists A, B \in \mathcal{B}$ with $A \cap B = \emptyset$ and $X = A \cup B$ such that A is a positive set with respect to μ and B is a negative set with respect to μ. Define $\mu_+ : \mathcal{B} \to [0, \infty) \subset \mathbb{R}$ by $\mu_+(E) = \mu(A \cap E)$, $\forall E \in \mathcal{B}$ and $\mu_- : \mathcal{B} \to [0, \infty) \subset \mathbb{R}$ by $\mu_-(E) = -\mu(B \cap E)$, $\forall E \in \mathcal{B}$. It is straightforward to show that μ_+ and μ_- are finite measures on $(X, \mathcal{B})$. Clearly, $\mu_+(B) = \mu_-(A) = 0$. Then, $\mu_+ \perp \mu_-$. It is straightforward to show that $\mu = \mu_+ - \mu_-$. Then, by Proposition 11.136, μ is a finite $\mathbb{R}$-valued measure on $(X, \mathcal{B})$. By Proposition 11.156, $\mathcal{P} \circ \mu = \mathcal{P} \circ \mu_+ + \mathcal{P} \circ \mu_- = \mu_+ + \mu_-$.

Let $\hat{\mu}_+$ and $\hat{\mu}_-$ be any pair of mutually singular finite measures on $(X, \mathcal{B})$ such that $\mu = \hat{\mu}_+ - \hat{\mu}_-$. Then, $\exists \hat{A}, \hat{B} \in \mathcal{B}$ with $\hat{A} \cap \hat{B} = \emptyset$ and $\hat{A} \cup \hat{B} = X$ such that $\hat{\mu}_+(\hat{B}) = \hat{\mu}_-(\hat{A}) = 0$. Note that $\mu_+(A \cap \hat{A}) = \mu_+(A) - \mu_+(A \setminus \hat{A}) \leq \mu_+(A) = \mu(A) = \hat{\mu}_+(A) - \hat{\mu}_-(A) = \hat{\mu}_+(A \cap \hat{A}) + \hat{\mu}_+(A \setminus \hat{A}) - \hat{\mu}_-(A \cap \hat{A}) - \hat{\mu}_-(A \setminus \hat{A}) = \hat{\mu}_+(A \cap \hat{A}) - \hat{\mu}_-(A \setminus \hat{A}) \leq \hat{\mu}_+(A \cap \hat{A}) = \hat{\mu}_+(\hat{A}) - \hat{\mu}_+(\hat{A} \setminus A) \leq \hat{\mu}_+(\hat{A}) = \mu(\hat{A}) = \mu_+(\hat{A}) - \mu_-(\hat{A}) = \mu_+(A \cap \hat{A}) + \mu_+(\hat{A} \setminus A) - \mu_-(A \cap \hat{A}) - \mu_-(\hat{A} \setminus A) = \mu_+(A \cap \hat{A}) - \mu_-(\hat{A} \setminus A) \leq \mu_+(A \cap \hat{A}) \in \mathbb{R}$. This implies that $\mu_+(A \cap \hat{A}) = \mu_+(A) = \mu(A) = \hat{\mu}_+(A \cap \hat{A}) = \hat{\mu}_+(\hat{A}) = \mu(\hat{A})$ and $\mu_+(A \setminus \hat{A}) = \hat{\mu}_-(A \setminus \hat{A}) = \hat{\mu}_+(\hat{A} \setminus A) = \mu_-(\hat{A} \setminus A) = 0$. $\forall E \in \mathcal{B}$, we have $\mu_+(E) = \mu_+(E \cap A) = \mu_+((E \cap A) \setminus \hat{A}) + \mu_+(E \cap A \cap \hat{A}) = \mu_+(E \cap A \cap \hat{A}) = \mu(E \cap A \cap \hat{A}) = \hat{\mu}_+(E \cap A \cap \hat{A}) = \hat{\mu}_+(E \cap \hat{A}) - \hat{\mu}_+((E \cap \hat{A}) \setminus A) = \hat{\mu}_+(E \cap \hat{A}) = \hat{\mu}_+(E) - \hat{\mu}_+(E \setminus \hat{A}) = \hat{\mu}_+(E) \in \mathbb{R}$ and $\mu_-(E) = \mu_+(E) - \mu(E) = \hat{\mu}_+(E) - \mu(E) = \hat{\mu}_-(E) \in \mathbb{R}$. Hence, $\mu_+ = \hat{\mu}_+$ and $\mu_- = \hat{\mu}_-$. This shows that the pair μ_+ and μ_- is unique.

This completes the proof of the theorem. □

Theorem 11.163 (Jordan Decomposition Theorem) *Let $(X, \mathcal{B}, \mu) =: \mathcal{X}$ be an σ-finite $\mathbb{R}$-valued measure space. Then, there exists a unique pair of mutually singular σ-finite measures μ_+ and μ_- on $(X, \mathcal{B})$ such that $\mu = \mu_+ - \mu_-$. Furthermore, $\mathcal{P} \circ \mu = \mu_+ + \mu_-$.*

Proof By Hahn Decomposition Theorem 11.161, there exists $A, B \in \mathcal{B}$ such that $A = X \setminus B$, A is a positive set with respect to μ, and B is a negative set with respect to μ. Define μ_A to be a function from $\mathcal{B}$ to $\mathbb{R}$ by: $\mu_A(E)$ is undefined if $E \in \mathcal{B}$ and $E \cap A \in \mathcal{B} \setminus \operatorname{dom}(\mu)$; $\mu_A(E) = \mu(E \cap A) \in [0, \infty) \subset \mathbb{R}$ if $E \in \mathcal{B}$ and $E \cap A \in \operatorname{dom}(\mu)$. Define μ_B to be a function from $\mathcal{B}$ to $\mathbb{R}$ by: $\mu_B(E)$ is undefined if $E \in \mathcal{B}$ and $E \cap B \in \mathcal{B} \setminus \operatorname{dom}(\mu)$; $\mu_B(E) = \mu(E \cap B) \in (-\infty, 0] \subset \mathbb{R}$ if $E \in \mathcal{B}$ and $E \cap B \in \operatorname{dom}(\mu)$. Then, by Proposition 11.157, μ_A and μ_B are σ-finite $\mathbb{R}$-valued measures on $(X, \mathcal{B})$ such that $\mu = \mu_A + \mu_B$, $\mathcal{P} \circ \mu = \mathcal{P} \circ \mu_A + \mathcal{P} \circ \mu_B$, and $\mathcal{P} \circ \mu_A \perp \mathcal{P} \circ \mu_B$. Since $\mu_A(E) \in [0, \infty) \subset \mathbb{R}$, $\forall E \in \operatorname{dom}(\mu_A)$, then μ_A is identified with a σ-finite measure μ_+ on $(X, \mathcal{B})$ according to Proposition 11.135. By Proposition 11.138, $-\mu_B$ is a σ-finite $\mathbb{R}$-valued measure on $(X, \mathcal{B})$. Since $-\mu_B(E) \in [0, \infty) \subset \mathbb{R}$, $\forall E \in \operatorname{dom}(\mu_B)$, then $-\mu_B$ is identified with a σ-finite measure μ_- on $(X, \mathcal{B})$ according to Proposition 11.135. Then, $\mathcal{P} \circ \mu = \mathcal{P} \circ \mu_A + \mathcal{P} \circ (-\mu_B) = \mu_+ + \mu_-$. Since $\mathcal{P} \circ \mu_A \perp \mathcal{P} \circ \mu_B$, then, $\mu_+ \perp \mu_-$. $\forall E \in \mathcal{B}$ with $\mathcal{P} \circ \mu(E) = \mu_+(E) + \mu_-(E) < \infty$, we have $\mu(E) = (\mu_A + \mu_B)(E) = \mu_A(E) + \mu_B(E) = \mu_+(E) - \mu_-(E) = (\mu_+ - \mu_-)(E) \in \mathbb{R}$. By Propositions 11.137 and 11.138, we have $\mu = \mu_+ - \mu_-$. Hence, the pair μ_+ and μ_- is the pair of mutually singular σ-finite measures on $(X, \mathcal{B})$ that we seek.

Next, we will show that the pair is unique. Let $\hat{\mu}_+$ and $\hat{\mu}_-$ be another pair of mutually singular σ-finite measures on $(X, \mathcal{B})$ such that $\mu = \hat{\mu}_+ - \hat{\mu}_-$ and $\mathcal{P} \circ \mu = \hat{\mu}_+ + \hat{\mu}_-$. Since the four measures μ_+, μ_-, $\hat{\mu}_+$, and $\hat{\mu}_-$ are σ-finite, then $\exists (X_n)_{n=1}^\infty \subseteq \mathcal{B}$ such that $X = \bigcup_{n=1}^\infty X_n$ and $\mu_+(X_n) \in \mathbb{R}$, $\mu_-(X_n) \in \mathbb{R}$, $\hat{\mu}_+(X_n) \in \mathbb{R}$, and $\hat{\mu}_-(X_n) \in \mathbb{R}$, $\forall n \in \mathbb{N}$. Then, $\mathcal{P} \circ \mu(X_n) < \infty$, $\forall n \in \mathbb{N}$. Fix any $n \in \mathbb{N}$. Let $\mathcal{X}_n := (X_n, \mathcal{B}_n, \mu_n)$ be the finite $\mathbb{R}$-valued measure subspace of $\mathcal{X}$, $(X_n, \mathcal{B}_n, \mu_{n,+})$ be the finite measure subspace of $(X, \mathcal{B}, \mu_+)$, $(X_n, \mathcal{B}_n, \mu_{n,-})$ be the finite measure subspace of $(X, \mathcal{B}, \mu_-)$, $(X_n, \mathcal{B}_n, \hat{\mu}_{n,+})$ be the finite measure subspace of $(X, \mathcal{B}, \hat{\mu}_+)$, and $(X_n, \mathcal{B}_n, \hat{\mu}_{n,-})$ be the finite measure subspace of $(X, \mathcal{B}, \hat{\mu}_-)$. Then, it is easy to see that $\mu_n = \mu_{n,+} - \mu_{n,-} = \hat{\mu}_{n,+} - \hat{\mu}_{n,-}$, $\mu_{n,+} \perp \mu_{n,-}$, $\hat{\mu}_{n,+} \perp \hat{\mu}_{n,-}$, and $\mathcal{P} \circ \mu_n = \mu_{n,+} + \mu_{n,-} = \hat{\mu}_{n,+} + \hat{\mu}_{n,-}$. By Jordan Decomposition Theorem 11.162, we have $\mu_{n,+} = \hat{\mu}_{n,+}$ and $\mu_{n,-} = \hat{\mu}_{n,-}$. By Proposition 11.117, we have $\mu_+ = \hat{\mu}_+$ and $\mu_- = \hat{\mu}_-$. Hence, the pair is unique.

This completes the proof of the theorem. □

Proposition 11.164 *Let $\mathcal{X} := (X, \mathcal{B}, \mu)$ be a $\mathbb{C}$-valued pre-measure space. Then, $\mu = \mu_r + \mathrm{i}\mu_i$, where $\mu_r : \mathcal{B} \to \mathbb{R}$ and $\mu_i : \mathcal{B} \to \mathbb{R}$ are finite $\mathbb{R}$-valued measures on $(X, \mathcal{B})$, and $\max\{\mathcal{P} \circ \mu_r(E), \mathcal{P} \circ \mu_i(E)\} \leq \mathcal{P} \circ \mu(E) \leq \mathcal{P} \circ \mu_r(E) + \mathcal{P} \circ \mu_i(E)$, $\forall E \in \mathcal{B}$. Then, $\mathcal{X}$ is a finite $\mathbb{C}$-valued measure space.*

Proof Define $\mu_r : \mathcal{B} \to \mathbb{R}$ and $\mu_i : \mathcal{B} \to \mathbb{R}$ by $\mu_r(E) = \mathrm{Re}\,(\mu(E))$ and $\mu_i(E) = \mathrm{Im}\,(\mu(E))$, $\forall E \in \mathcal{B}$. Clearly, $\mu = \mu_r + \mathrm{i}\mu_i$, $\mu_r(\emptyset) = 0$, and $\mu_i(\emptyset) = 0$. $\forall$ pairwise disjoint sequence $(E_n)_{n=1}^\infty \subseteq \mathcal{B}$, $\sum_{n=1}^\infty |\mu_r(E_n)| \leq \sum_{n=1}^\infty |\mu(E_n)| < +\infty$, where the last inequality follows from the fact that μ is a $\mathbb{C}$-value pre-measure. We also have $\sum_{n=1}^\infty \mu_r(E_n) = \sum_{n=1}^\infty \mathrm{Re}\,(\mu(E_n)) = \mathrm{Re}\left(\sum_{n=1}^\infty \mu(E_n)\right) = \mathrm{Re}\left(\mu(\bigcup_{n=1}^\infty E_n)\right) = \mu_r(\bigcup_{n=1}^\infty E_n)$. Hence, μ_r is an $\mathbb{R}$-valued pre-measure on $(X, \mathcal{B})$. Note that $\sum_{n=1}^\infty |\mu_i(E_n)| \leq \sum_{n=1}^\infty |\mu(E_n)| < +\infty$ and $\sum_{n=1}^\infty \mu_i(E_n) = \sum_{n=1}^\infty \mathrm{Im}\,(\mu(E_n)) = \mathrm{Im}\left(\sum_{n=1}^\infty \mu(E_n)\right) = \mathrm{Im}\left(\mu(\bigcup_{n=1}^\infty E_n)\right) = \mu_i(\bigcup_{n=1}^\infty E_n)$. Hence, μ_i is an $\mathbb{R}$-valued pre-measure on $(X, \mathcal{B})$. By Jordan Decomposition Theorem 11.162, μ_r and μ_i are finite $\mathbb{R}$-valued measures on $(X, \mathcal{B})$.

$\forall E \in \mathcal{B}$, $\forall n \in \mathbb{Z}_+$, $\forall$ pairwise disjoint $(E_k)_{k=1}^n \subseteq \mathcal{B}$ with $E = \bigcup_{k=1}^n E_k$, $\sum_{k=1}^n |\mu_r(E_k)| \leq \sum_{k=1}^n |\mu(E_k)| \leq \mathcal{P} \circ \mu(E)$. Hence, $\mathcal{P} \circ \mu_r(E) \leq \mathcal{P} \circ \mu(E)$.

$\forall E \in \mathcal{B}$, $\forall n \in \mathbb{Z}_+$, $\forall$ pairwise disjoint $(E_k)_{k=1}^n \subseteq \mathcal{B}$ with $E = \bigcup_{k=1}^n E_k$, $\sum_{k=1}^n |\mu_i(E_k)| \leq \sum_{k=1}^n |\mu(E_k)| \leq \mathcal{P} \circ \mu(E)$. Hence, $\mathcal{P} \circ \mu_i(E) \leq \mathcal{P} \circ \mu(E)$.

$\forall E \in \mathcal{B}$, $\forall n \in \mathbb{Z}_+$, $\forall$ pairwise disjoint $(E_k)_{k=1}^n \subseteq \mathcal{B}$ with $E = \bigcup_{k=1}^n E_k$, $\sum_{k=1}^n |\mu(E_k)| = \sum_{k=1}^n \sqrt{(\mu_r(E_k))^2 + (\mu_i(E_k))^2} \leq \sum_{k=1}^n (|\mu_r(E_k)| + |\mu_i(E_k)|) \leq \mathcal{P} \circ \mu_r(E) + \mathcal{P} \circ \mu_i(E)$. Hence, $\mathcal{P} \circ \mu(E) \leq \mathcal{P} \circ \mu_r(E) + \mathcal{P} \circ \mu_i(E)$. Then, $\mathcal{P} \circ \mu(X) < +\infty$. Hence, $\mathcal{X}$ is a finite $\mathbb{C}$-valued measure space.

This completes the proof of the proposition. □

Proposition 11.165 *Let $\mathcal{X} := (X, \mathcal{B}, \mu)$ be a σ-finite $\mathbb{C}$-valued measure space. Then, there exists a pair of σ-finite $\mathbb{R}$-valued measures μ_r and μ_i on $(X, \mathcal{B})$ such that $\mu = \mu_r + \mathrm{i}\mu_i$ and $0 \leq \max\{\mathcal{P} \circ \mu_r(E), \mathcal{P} \circ \mu_i(E)\} \leq \mathcal{P} \circ \mu(E) \leq \mathcal{P} \circ \mu_r(E) + \mathcal{P} \circ \mu_i(E)$, $\forall E \in \mathcal{B}$.*

Proof Since $\mathcal{X}$ is σ-finite, then $\exists (X_n)_{n=1}^\infty \subseteq \mathcal{B}$ such that $X = \bigcup_{n=1}^\infty X_n$ and $\mathcal{P} \circ \mu(X_n) < \infty$, $\forall n \in \mathbb{N}$. Without loss of generality, we may assume that $(X_n)_{n=1}^\infty$ is pairwise disjoint. Fix any $n \in \mathbb{N}$. Let $\mathcal{X}_n := (X_n, \mathcal{B}_n, \mu_n)$ be the finite $\mathbb{C}$-valued measure subspace of $\mathcal{X}$. Then, $\mathcal{X}_n$ is a finite $\mathbb{C}$-valued pre-measure space. By

Proposition 11.164, there exist finite $\mathbb{R}$-valued measures $\mu_{n,r}$ and $\mu_{n,i}$ on $(X_n, \mathcal{B}_n)$ such that $\mu_n = \mu_{n,r} + \mathrm{i}\mu_{n,i}$ and $0 \le \max\{\mathcal{P}\circ\mu_{n,r}(E), \mathcal{P}\circ\mu_{n,i}(E)\} \le \mathcal{P}\circ\mu_n(E) \le \mathcal{P}\circ\mu_{n,r}(E) + \mathcal{P}\circ\mu_{n,i}(E) < \infty$, $\forall E \in \mathcal{B}_n$. By Proposition 11.118, the generation process on $\left((X_n, \mathcal{B}_n, \mu_{n,r})\right)_{n=1}^{\infty}$ yields a unique σ-finite $\mathbb{R}$-valued measure space $(X, \mathcal{B}, \mu_r)$ on X, and the generation process on $\left((X_n, \mathcal{B}_n, \mu_{n,i})\right)_{n=1}^{\infty}$ yields a unique σ-finite $\mathbb{R}$-valued measure space $(X, \mathcal{B}, \mu_i)$ on X.

$\forall E \in \mathcal{B}$, let $E_n := E \cap X_n \in \mathcal{B}_n$, $\forall n \in \mathbb{N}$. Then, $\mathcal{P}\circ\mu(E) = \sum_{n=1}^{\infty} \mathcal{P}\circ\mu(E_n) = \sum_{n=1}^{\infty} \mathcal{P}\circ\mu_n(E_n)$, where the first equality follows from the fact that $\mathcal{P}\circ\mu$ is a measure and the second equality follows from Propositions 11.118 and 11.115. By Propositions 11.118 and 11.115, we have $\mathcal{P}\circ\mu(E) \ge \sum_{n=1}^{\infty} \mathcal{P}\circ\mu_{n,r}(E_n) = \sum_{n=1}^{\infty} \mathcal{P}\circ\mu_r(E_n) = \mathcal{P}\circ\mu_r(E)$. Similarly, we have $\mathcal{P}\circ\mu(E) \ge \mathcal{P}\circ\mu_i(E)$. Again, by Propositions 11.118 and 11.115, $\mathcal{P}\circ\mu(E) \le \sum_{n=1}^{\infty}(\mathcal{P}\circ\mu_{n,r}(E_n) + \mathcal{P}\circ\mu_{n,i}(E_n)) = \mathcal{P}\circ\mu_r(E) + \mathcal{P}\circ\mu_i(E)$. Hence, we have $0 \le \max\{\mathcal{P}\circ\mu_r(E), \mathcal{P}\circ\mu_i(E)\} \le \mathcal{P}\circ\mu(E) \le \mathcal{P}\circ\mu_r(E) + \mathcal{P}\circ\mu_i(E) \le \infty$, $\forall E \in \mathcal{B}$.

By Proposition 11.138, $\mu_r + \mathrm{i}\mu_i$ is a σ-finite $\mathbb{C}$-valued measure on $(X, \mathcal{B})$. $\forall n \in \mathbb{N}$, $\forall E \in \mathcal{B}_n$, $\mathcal{P}\circ\mu(E) \le \mathcal{P}\circ\mu(X_n) < \infty$ and $E \in \operatorname{dom}(\mu)$. This implies that $\max\{\mathcal{P}\circ\mu_r(E), \mathcal{P}\circ\mu_i(E)\} \le \mathcal{P}\circ\mu(E) < \infty$ and $E \in \operatorname{dom}(\mu_r) \cap \operatorname{dom}(\mu_i)$. Then, $(\mu_r + \mathrm{i}\mu_i)(E) = \mu_r(E) + \mathrm{i}\mu_i(E) = \mu_{n,r}(E) + \mathrm{i}\mu_{n,i}(E) = (\mu_{n,r} + \mathrm{i}\mu_{n,i})(E) = \mu_n(E) = \mu(E)$, where the first equality follows from Proposition 11.138, the second equality follows from Proposition 11.115, the third equality follows from Proposition 11.138, and the last equality follows from Proposition 11.115. Hence, $\mathcal{X}_n$ is the finite $\mathbb{C}$-valued measure subspace of $(X, \mathcal{B}, \mu_r + \mathrm{i}\mu_i)$, $\forall n \in \mathbb{N}$. By Proposition 11.117, we have $\mu = \mu_r + \mathrm{i}\mu_i$.

This completes the proof of the proposition. □

Definition 11.166 Let $(X, \mathcal{B})$ be a measurable space, λ_1 and λ_2 be measures on $(X, \mathcal{B})$, and $f : X \to [0, \infty) \subset \mathbb{R}$ be $\mathcal{B}$-measurable. We will say that f is a *Radon–Nikodym derivative of* λ_1 *with respect to* λ_2 if $\lambda_1(E) = \int_E f \,\mathrm{d}\lambda_2$, $\forall E \in \mathcal{B}$. Let μ be a $\mathbb{K}$-valued measure on $(X, \mathcal{B})$, $\mathcal{Y}$ be a normed linear space over $\mathbb{K}$, ν be a $\mathcal{Y}$-valued measure on $(X, \mathcal{B})$, and $\bar{f} : X \to \mathcal{Y}$ be $\mathcal{B}$-measurable. We will say that $\bar{f}$ is a *Radon–Nikodym derivative of* ν *with respect to* μ if $\mathcal{P}\circ\bar{f}$ is a Radon–Nikodym derivative of $\mathcal{P}\circ\nu$ with respect to $\mathcal{P}\circ\mu$ (that is, $\mathcal{P}\circ\nu(E) = \int_E \mathcal{P}\circ\bar{f}\,\mathrm{d}\mathcal{P}\circ\mu \in [0, \infty] \subset \mathbb{R}_e$, $\forall E \in \mathcal{B}$) and $\nu(E) = \int_E \bar{f}\,\mathrm{d}\mu \in \mathcal{Y}$, $\forall E \in \operatorname{dom}(\nu)$. ◇

It is easy to see that the definition above is compatible with the identification of σ-finite $\mathbb{R}$-valued measures and σ-finite measures eluded to in Section 11.7. It is straightforward to show that if $\bar{f} : X \to \mathcal{Y}$ is a Radon–Nikodym derivative of ν with respect to μ, then $\mathcal{P}\circ\nu \ll \mathcal{P}\circ\mu$. When μ is σ-finite, $\mathcal{Y}$ is a Banach space, and $\mathcal{W}$ is a separable subspace of $\mathcal{Y}$, then Proposition 11.154 guarantees that any measurable function $\bar{f} : X \to \mathcal{W}$ is a Radon–Nikodym derivative of some σ-finite $\mathcal{Y}$-valued measure ν with respect to μ. The next result gives the sufficient condition under which the Radon–Nikodym derivative is unique. The Radon–Nikodym theorems then give sufficient conditions on ν and μ under which the Radon–Nikodym derivative exists.

Proposition 11.167 *Let $\mathcal{X} := (X, \mathcal{B}, \mu)$ be a σ-finite $\mathbb{K}$-valued measure space, $\mathcal{Y}$ be a separable Banach space over $\mathbb{K}$, ν be a σ-finite $\mathcal{Y}$-valued measure on $(X, \mathcal{B})$, and $f : X \to \mathcal{Y}$ be a Radon–Nikodym derivative of the ν with respect to μ, and $g : X \to \mathcal{Y}$ be $\mathcal{B}$-measurable. Then, g is a Radon–Nikodym derivative of ν with respect to μ if, and only if, $g = f$ a.e. in $\mathcal{X}$. In this case, we will write $\frac{\mathrm{d}\nu}{\mathrm{d}\mu} = f$ a.e. in $\mathcal{X}$.*

Proof "Sufficiency" Let $g = f$ a.e. in $\mathcal{X}$. Then, by Propositions 7.21, 7.23, 11.38, and 11.39, $\mathcal{P} \circ g = \mathcal{P} \circ f$ a.e. in $\mathcal{X}$. $\forall E \in \mathcal{B}$, $\mathcal{P} \circ \nu(E) = \int_E \mathcal{P} \circ f \, \mathrm{d}\mathcal{P} \circ \mu = \int_E \mathcal{P} \circ g \, \mathrm{d}\mathcal{P} \circ \mu$, where the second equality follows from Proposition 11.83. $\forall E \in \operatorname{dom}(\nu)$, $\infty > \mathcal{P} \circ \nu(E) = \int_E \mathcal{P} \circ f \, \mathrm{d}\mathcal{P} \circ \mu = \int_E \mathcal{P} \circ g \, \mathrm{d}\mathcal{P} \circ \mu$. Then, $f|_E$ and $g|_E$ are absolutely integrable over $\mathcal{E} := (E, \mathcal{B}_E, \mu_E)$, which is the $\mathbb{K}$-valued measure subspace of $\mathcal{X}$. This implies that, by Proposition 11.132, $\nu(E) = \int_E f \, \mathrm{d}\mu = \int_E g \, \mathrm{d}\mu \in \mathcal{Y}$. Hence, g is a Radon–Nikodym derivative of ν with respect to μ.

"Necessity" Let $g : X \to \mathcal{Y}$ be another Radon–Nikodym derivative of ν with respect to μ. Define $\lambda := \nu - \nu$, which is a σ-finite $\mathcal{Y}$-valued measure on $(X, \mathcal{B})$, by Proposition 11.138. Then, by Proposition 11.137, $\lambda(E) = \vartheta_{\mathcal{Y}}$ and $\mathcal{P} \circ \lambda(E) = 0$, $\forall E \in \mathcal{B}$. By Proposition 11.154, we may define a σ-finite $\mathcal{Y}$-valued measure $\bar{\lambda}$ on $(X, \mathcal{B})$ such that $\mathcal{P} \circ \bar{\lambda}(E) = \int_E \mathcal{P} \circ (f - g) \, \mathrm{d}\mathcal{P} \circ \mu$, $\forall E \in \mathcal{B}$, and $\bar{\lambda}(E) = \int_E (f - g) \, \mathrm{d}\mu$, $\forall E \in \mathcal{B}$ with $\int_E \mathcal{P} \circ (f - g) \, \mathrm{d}\mathcal{P} \circ \mu < \infty$. $\forall E \in \mathcal{B}$ with $\mathcal{P} \circ \nu(E) < \infty$, then $f|_E$ and $g|_E$ are absolutely integrable over $\mathcal{E}$. By Proposition 11.132, $(f - g)|_E$ is absolutely integrable over $\mathcal{E}$. Then, we have $\vartheta_{\mathcal{Y}} = \lambda(E) = \nu(E) - \nu(E) = \int_E f \, \mathrm{d}\mu - \int_E g \, \mathrm{d}\mu = \int_E (f - g) \, \mathrm{d}\mu = \bar{\lambda}(E)$. By Proposition 11.137, we have $\lambda = \bar{\lambda}$. Then, $\forall E \in \mathcal{B}$, $E \in \operatorname{dom}(\lambda)$, $0 = \mathcal{P} \circ \lambda(E) = \mathcal{P} \circ \bar{\lambda}(E) = \int_E \mathcal{P} \circ (f - g) \, \mathrm{d}\mathcal{P} \circ \mu$. By Proposition 11.96, $\mathcal{P} \circ (f - g) = 0$ a.e. in $\mathcal{X}$ and $f = g$ a.e. in $\mathcal{X}$.

This completes the proof of the proposition. □

Proposition 11.168 *Let $(X, \mathcal{B})$ be a measurable space, λ and λ_1 be σ-finite measures on $(X, \mathcal{B})$, $\hat{\mathcal{X}} := (X, \mathcal{B}, \lambda)$, $\hat{\mathcal{X}}_1 := (X, \mathcal{B}, \lambda_1)$, $\mu, \mu_1 \in \mathcal{M}_\sigma(X, \mathcal{B}, \mathbb{K})$, $\mathcal{X} := (X, \mathcal{B}, \mu)$, $\mathcal{X}_1 := (X, \mathcal{B}, \mu_1)$, $\mathcal{Y}$ be a separable Banach space over $\mathbb{K}$, $\mathcal{Z}$ be a Banach space over $\mathbb{K}$, $\mathcal{W}$ be a separable subspace of $\mathrm{B}(\mathcal{Y}, \mathcal{Z})$, $\nu_1, \nu_2 \in \mathcal{M}_\sigma(X, \mathcal{B}, \mathcal{Y})$, $\bar{\mathcal{X}}_1 := (X, \mathcal{B}, \nu_1)$, $\frac{\mathrm{d}\nu_i}{\mathrm{d}\mu} =: \bar{f}_i : X \to \mathcal{Y}$ a.e. in $\mathcal{X}$, $i = 1, 2$, $\frac{\mathrm{d}\lambda}{\mathrm{d}\lambda_1} =: \hat{f} : X \to [0, \infty) \subset \mathbb{R}$ a.e. in $\hat{\mathcal{X}}_1$, $\frac{\mathrm{d}\mu}{\mathrm{d}\mu_1} =: f : X \to \mathbb{K}$ a.e. in $\mathcal{X}_1$, $g : X \to \mathcal{W}$ be $\mathcal{B}$-measurable, $\bar{f} : X \to [0, \infty) \subset \mathbb{R}$ be $\mathcal{B}$-measurable, $\hat{\bar{f}} : X \to \mathbb{K}$ be $\mathcal{B}$-measurable, $\alpha_0 \in \mathbb{K}$, $A \in \mathrm{B}(\mathcal{Y}, \mathcal{Z})$, and $y_0 \in \mathcal{Y}$. Then, the following statements hold:*

(i) $\int_X \bar{f} \, \mathrm{d}\lambda = \int_X (\bar{f} \hat{f}) \, \mathrm{d}\lambda_1 \in [0, \infty] \subset \mathbb{R}_e$.
(ii) *g is absolutely integrable over $\bar{\mathcal{X}}_1$ if, and only if, $(\mathcal{P} \circ g)(\mathcal{P} \circ \bar{f}_1)$ is integrable over $(X, \mathcal{B}, \mathcal{P} \circ \mu)$. In which case, $\int_X g \, \mathrm{d}\nu_1 = \int_X (g \bar{f}_1) \, \mathrm{d}\mu \in \mathcal{Z}$.*
(iii) $\frac{\mathrm{d}(\nu_1 + \nu_2)}{\mathrm{d}\mu} = \bar{f}_1 + \bar{f}_2 = \frac{\mathrm{d}\nu_1}{\mathrm{d}\mu} + \frac{\mathrm{d}\nu_2}{\mathrm{d}\mu}$ *a.e. in $\mathcal{X}$.*
(iv) *$A \bar{f}_1$ is a Radon–Nikodym derivative of $A \nu_1$ with respect to μ.*
(v) $\frac{\mathrm{d}(\alpha_0 \nu_1)}{\mathrm{d}\mu} = \alpha_0 \bar{f}_1 = \alpha_0 \frac{\mathrm{d}\nu_1}{\mathrm{d}\mu}$ *a.e. in $\mathcal{X}$.*
(vi) $\frac{\mathrm{d}(\mu y_0)}{\mathrm{d}\mu_1} = f y_0 = \left(\frac{\mathrm{d}\mu}{\mathrm{d}\mu_1} \right) y_0$ *a.e. in $\mathcal{X}_1$.*

(vii) $\frac{\mathrm{d}\nu_1}{\mathrm{d}\mu_1} = \bar{f}_1 f = \left(\frac{\mathrm{d}\nu_1}{\mathrm{d}\mu}\right)\left(\frac{\mathrm{d}\mu}{\mathrm{d}\mu_1}\right)$ *a.e. in* $\mathcal{X}_1$.

(viii) $\frac{\mathrm{d}\mu_1}{\mathrm{d}\mu} = \hat{\bar{f}}$ *a.e. in* $\mathcal{X}$ *if, and only if,* $\hat{\bar{f}} f = 1$ *a.e. in* $\mathcal{X}_1$.

Proof (i) By Definition 11.166, $\hat{f}$ is $\mathcal{B}$-measurable and $\lambda(E) = \int_E \hat{f}\,\mathrm{d}\lambda_1 \in \mathbb{R}_e$, $\forall E \in \mathcal{B}$. By Propositions 7.23, 11.38, and 11.39, $\bar{f}\hat{f}$ is $\mathcal{B}$-measurable. By Proposition 11.66, there exists a sequence of simple functions $(\bar{\phi}_n)_{n=1}^{\infty}$, $\bar{\phi}_n : \tilde{\mathcal{X}} \to [0, \infty) \subset \mathbb{R}$, $\forall n \in \mathbb{N}$, such that $0 \leq \bar{\phi}_n(x) \leq \bar{f}(x)$, $\forall x \in X$, $\forall n \in \mathbb{N}$, and $\lim_{n\in\mathbb{N}} \bar{\phi}_n = \bar{f}$ a.e. in $\tilde{\mathcal{X}}$, where $\tilde{\mathcal{X}} := (X, \mathcal{B}, \lambda + \lambda_1)$ is a σ-finite measure space by Proposition 11.136. $\forall n \in \mathbb{N}$, define $\phi_n := \bigvee_{i=1}^{n} \bar{\phi}_i$. Then, ϕ_n is a simple function of $\tilde{\mathcal{X}}$ to $[0, \infty) \subset \mathbb{R}$, $0 \leq \phi_n(x) \leq \phi_{n+1}(x) \leq \bar{f}(x)$, $\forall x \in X$, $\forall n \in \mathbb{N}$. By Proposition 11.50, $\lim_{n\in\mathbb{N}} \phi_n = \bar{f}$ a.e. in $\tilde{\mathcal{X}}$. Since $\lambda \leq \lambda + \lambda_1$, then ϕ_n is a simple function of $\hat{\mathcal{X}}$, $\forall n \in \mathbb{N}$, and $\lim_{n\in\mathbb{N}} \phi_n = \bar{f}$ a.e. in $\hat{\mathcal{X}}$. By Monotone Convergence Theorem 11.81, $\int_X \bar{f}\,\mathrm{d}\lambda = \lim_{n\in\mathbb{N}} \int_X \phi_n\,\mathrm{d}\lambda \in \mathbb{R}_e$. Fix any $n \in \mathbb{N}$. Let ϕ_n admit the canonical representation $\phi_n = \sum_{i=1}^{\bar{n}} a_i \chi_{A_i,X}$. Then, by Propositions 11.75 and 11.83, $\int_X \phi_n\,\mathrm{d}\lambda = \sum_{i=1}^{\bar{n}} a_i \lambda(A_i) = \sum_{i=1}^{\bar{n}} a_i \int_{A_i} \hat{f}\,\mathrm{d}\lambda_1 = \sum_{i=1}^{\bar{n}} \int_X (a_i \chi_{A_i,X} \hat{f})\,\mathrm{d}\lambda_1 = \int_X (\sum_{i=1}^{\bar{n}} a_i \chi_{A_i,X}) \hat{f}\,\mathrm{d}\lambda_1 = \int_X (\phi_n \hat{f})\,\mathrm{d}\lambda_1$. Note that $0 \leq \phi_n(x)\hat{f}(x) \leq \phi_{n+1}(x)\hat{f}(x) \leq \bar{f}(x)\hat{f}(x)$, $\forall x \in X$, $\forall n \in \mathbb{N}$. By Proposition 11.50, $\lim_{n\in\mathbb{N}} \phi_n \hat{f} = \bar{f}\hat{f}$ a.e. in $\tilde{\mathcal{X}}$. Since $\lambda_1 \leq \lambda + \lambda_1$, then $\lim_{n\in\mathbb{N}} \phi_n \hat{f} = \bar{f}\hat{f}$ a.e. in $\hat{\mathcal{X}}_1$. By Monotone Convergence Theorem 11.81, we have

$$\int_X (\bar{f}\hat{f})\,\mathrm{d}\lambda_1 = \lim_{n\in\mathbb{N}} \int_X (\phi_n \hat{f})\,\mathrm{d}\lambda_1 = \lim_{n\in\mathbb{N}} \int_X \phi_n\,\mathrm{d}\lambda = \int_X \bar{f}\,\mathrm{d}\lambda \in \mathbb{R}_e$$

(ii) Since $\frac{\mathrm{d}\nu_1}{\mathrm{d}\mu} = \bar{f}_1$ a.e. in $\mathcal{X}$, then $\frac{\mathrm{d}\mathcal{P}\circ\nu_1}{\mathrm{d}\mathcal{P}\circ\mu} = \mathcal{P} \circ \bar{f}_1$ a.e. in $\mathcal{X}$. By (i), $\int_X \mathcal{P}\circ g\,\mathrm{d}\mathcal{P}\circ\nu_1 = \int_X (\mathcal{P}\circ g)(\mathcal{P}\circ \bar{f}_1)\,\mathrm{d}\mathcal{P}\circ\mu \in \mathbb{R}_e$. Hence, g is absolutely integrable over $\bar{\mathcal{X}}_1$ if, and only if, $(\mathcal{P} \circ g)(\mathcal{P} \circ \bar{f}_1)$ is integrable over $(X, \mathcal{B}, \mathcal{P} \circ \mu)$. Let g be absolutely integrable over $\bar{\mathcal{X}}_1$, then $\int_X \mathcal{P}\circ g\,\mathrm{d}\mathcal{P}\circ\nu_1 = \int_X (\mathcal{P}\circ g)(\mathcal{P}\circ \bar{f}_1)\,\mathrm{d}\mathcal{P}\circ\mu < \infty$. By Proposition 11.66, there exists a sequence of simple functions $(\phi_n)_{n=1}^{\infty}$, $\phi_n : \tilde{\mathcal{X}}_1 \to \mathcal{W}$, $\forall n \in \mathbb{N}$, such that $\|\phi_n(x)\| \leq \|g(x)\|$, $\forall x \in X$, $\forall n \in \mathbb{N}$, and $\lim_{n\in\mathbb{N}} \phi_n = g$ a.e. in $\tilde{\mathcal{X}}_1$, where $\tilde{\mathcal{X}}_1 := (X, \mathcal{B}, \mathcal{P} \circ \mu + \mathcal{P} \circ \nu_1)$ is a σ-finite measure space by Proposition 11.136. Since $\mathcal{P} \circ \nu_1 \leq \mathcal{P} \circ \mu + \mathcal{P} \circ \nu_1$, then ϕ_n is a simple function of $\bar{\mathcal{X}}_1$, $\forall n \in \mathbb{N}$, and $\lim_{n\in\mathbb{N}} \phi_n = g$ a.e. in $\bar{\mathcal{X}}_1$. By Lebesgue Dominated Convergence Theorem 11.131, $\int_X g\,\mathrm{d}\nu_1 = \lim_{n\in\mathbb{N}} \int_X \phi_n\,\mathrm{d}\nu_1 \in \mathcal{Z}$. Fix any $n \in \mathbb{N}$. Let ϕ_n admit the canonical representation $\phi_n = \sum_{i=1}^{\bar{n}} w_i \chi_{A_i,X}$. Note that $\hat{\mathcal{W}} := \mathrm{span}(\{z \in \mathcal{Z} \mid z = wy,\ w \in \mathcal{W},\ y \in \mathcal{Y}\})$ is a separable subspace of $\mathcal{Z}$ by Propositions 7.126, 7.65, 3.28, and 4.4. Then, by Propositions 11.125 and 11.132, we have $\int_X \phi_n\,\mathrm{d}\nu_1 = \sum_{i=1}^{\bar{n}} w_i \nu_1(A_i) = \sum_{i=1}^{\bar{n}} w_i \int_{A_i} \bar{f}_1\,\mathrm{d}\mu = \sum_{i=1}^{\bar{n}} \int_X (\chi_{A_i,X} w_i \bar{f}_1)\,\mathrm{d}\mu = \int_X (\sum_{i=1}^{\bar{n}} \chi_{A_i,X} w_i \bar{f}_1)\,\mathrm{d}\mu = \int_X (\phi_n \bar{f}_1)\,\mathrm{d}\mu$. Note that $\mathcal{P} \circ (\phi_n \bar{f}_1)(x) = \|\phi_n(x)\bar{f}_1(x)\| \leq \|\phi_n(x)\|\|\bar{f}_1(x)\| \leq (\mathcal{P} \circ g(x))(\mathcal{P} \circ \bar{f}_1(x))$, $\forall x \in X$, $\forall n \in \mathbb{N}$. Note also that $\phi_n \bar{f}_1 : X \to \hat{\mathcal{W}} \subseteq \mathcal{Z}$ and $g\bar{f}_1 : X \to \hat{\mathcal{W}}$, $\forall n \in \mathbb{N}$. By Propositions 7.65, 11.38, and 11.39, $\phi_n \bar{f}_1$ and $g\bar{f}_1$ are $\mathcal{B}$-

measurable, $\forall n \in \mathbb{N}$. By Proposition 11.50, $\lim_{n\in\mathbb{N}} \phi_n \bar{f}_1 = g\bar{f}_1$ a.e. in $\tilde{\mathcal{X}}_1$. Since $\mathcal{P}\circ\mu \leq \mathcal{P}\circ\mu + \mathcal{P}\circ\nu_1$, then $\lim_{n\in\mathbb{N}} \phi_n \bar{f}_1 = g\bar{f}_1$ a.e. in $\mathcal{X}$. By Lebesgue Dominated Convergence Theorem 11.131, $\int_X (g\bar{f}_1)\,\mathrm{d}\mu = \lim_{n\in\mathbb{N}} \int_X (\phi_n \bar{f}_1)\,\mathrm{d}\mu = \lim_{n\in\mathbb{N}} \int_X \phi_n\,\mathrm{d}\nu_1 = \int_X g\,\mathrm{d}\nu_1 \in \mathcal{Z}$.

(iii) By Proposition 11.154, $\bar{f}_1 + \bar{f}_2$ is a Radon–Nikodym derivative of a σ-finite $\mathcal{Y}$-valued measure $\bar{\nu}$ on $(X, \mathcal{B})$ with respect to μ. $\forall E \in \mathcal{B}$ with $(\mathcal{P}\circ\nu_1 + \mathcal{P}\circ\nu_2 + \mathcal{P}\circ\bar{\nu})(E) < \infty$, then $\mathcal{P}\circ\nu_i(E) = \int_E \mathcal{P}\circ\bar{f}_i\,\mathrm{d}\mathcal{P}\circ\mu < \infty$, $i = 1, 2$. This implies that $\bar{f}_i\big|_E$, $i = 1, 2$, are absolutely integrable over $\mathcal{E} := (E, \mathcal{B}_E, \mu_E)$, which is the $\mathbb{K}$-valued measure subspace of $\mathcal{X}$. Then, $\bar{\nu}(E) = \int_E (\bar{f}_1 + \bar{f}_2)\,\mathrm{d}\mu = \int_E \bar{f}_1\,\mathrm{d}\mu + \int_E \bar{f}_2\,\mathrm{d}\mu = \nu_1(E) + \nu_2(E) = (\nu_1 + \nu_2)(E) \in \mathcal{Y}$, where the second equality follows from Proposition 11.132 and the last equality follows from Proposition 11.138. Note that $\mathcal{P}\circ\nu_1 + \mathcal{P}\circ\nu_2 + \mathcal{P}\circ\bar{\nu}$ is a σ-finite measure on $(X, \mathcal{B})$ by Proposition 11.136. By Proposition 11.137, we have $\bar{\nu} = \nu_1 + \nu_2$. Hence, by Proposition 11.167, $\frac{\mathrm{d}(\nu_1+\nu_2)}{\mathrm{d}\mu} = \bar{f}_1 + \bar{f}_2$ a.e. in $\mathcal{X}$.

(iv) Note that $A\bar{f}_1 : X \to \mathcal{R}(A) \subseteq \mathcal{Z}$, where $\mathcal{R}(A)$ is a separable subspace of $\mathcal{Z}$ since $\mathcal{Y}$ is separable. By Proposition 11.154, $A\bar{f}_1$ is a Radon–Nikodym derivative of a σ-finite $\mathcal{Z}$-valued measure $\bar{\nu}$ on $(X, \mathcal{B})$ with respect to μ. $\forall E \in \mathcal{B}$ with $(\mathcal{P}\circ\nu_1 + \mathcal{P}\circ\bar{\nu})(E) < \infty$, then $\mathcal{P}\circ\nu_1(E) = \int_E \mathcal{P}\circ\bar{f}_1\,\mathrm{d}\mathcal{P}\circ\mu < \infty$. This implies that $\bar{f}_1\big|_E$ is absolutely integrable over $\mathcal{E} := (E, \mathcal{B}_E, \mu_E)$, which is the $\mathbb{K}$-valued measure subspace of $\mathcal{X}$. Then, $\bar{\nu}(E) = \int_E (A\bar{f}_1)\,\mathrm{d}\mu = A\int_E \bar{f}_1\,\mathrm{d}\mu = A\nu_1(E) = (A\nu_1)(E) \in \mathcal{Z}$, where the second equality follows from Proposition 11.132 and the last equality follows from Proposition 11.138. Note that $\mathcal{P}\circ\nu_1 + \mathcal{P}\circ\bar{\nu}$ is a σ-finite measure on $(X, \mathcal{B})$ by Proposition 11.136. By Proposition 11.137, we have $\bar{\nu} = A\nu_1$. Hence, $A\bar{f}_1$ is a Radon–Nikodym derivative of $A\nu_1$ with respect to μ.

(v) and (vi) follow immediately from (iv) and Proposition 11.167.

(vii) By Proposition 11.154, $\bar{f}_1 f$ is a Radon–Nikodym derivative of a σ-finite $\mathcal{Y}$-valued measure $\bar{\nu}$ on $(X, \mathcal{B})$ with respect to μ_1. $\forall E \in \mathcal{B}$ with $(\mathcal{P}\circ\mu + \mathcal{P}\circ\mu_1 + \mathcal{P}\circ\nu_1 + \mathcal{P}\circ\bar{\nu})(E) < \infty$, then $\mathcal{P}\circ\nu_1(E) = \int_E \mathcal{P}\circ\bar{f}_1\,\mathrm{d}\mathcal{P}\circ\mu < \infty$, $\mathcal{P}\circ\bar{\nu}(E) = \int_E \mathcal{P}\circ(\bar{f}_1 f)\,\mathrm{d}\mathcal{P}\circ\mu_1 = \int_E (\mathcal{P}\circ\bar{f}_1)(\mathcal{P}\circ f)\,\mathrm{d}\mathcal{P}\circ\mu_1 < \infty$, and $\bar{\nu}(E) = \int_E (\bar{f}_1 f)\,\mathrm{d}\mu_1 = \int_E \bar{f}_1\,\mathrm{d}\mu = \nu_1(E) \in \mathcal{Y}$, where the second equality follows from (ii). Note that $\mathcal{P}\circ\mu + \mathcal{P}\circ\mu_1 + \mathcal{P}\circ\nu_1 + \mathcal{P}\circ\bar{\nu}$ is a σ-finite measure on $(X, \mathcal{B})$ by Proposition 11.136. By Proposition 11.137, we have $\bar{\nu} = \nu_1$. Hence, by Proposition 11.167, $\frac{\mathrm{d}\nu_1}{\mathrm{d}\mu_1} = \bar{f}_1 f$ a.e. in $\mathcal{X}_1$.

(viii) "Necessity" Let $\frac{\mathrm{d}\mu_1}{\mathrm{d}\mu} = \hat{\bar{f}}$ a.e. in $\mathcal{X}$. By (vii), $\frac{\mathrm{d}\mu_1}{\mathrm{d}\mu_1} = \hat{\bar{f}} f$ a.e. in $\mathcal{X}_1$. Clearly, the constant function 1 of X is a Radon–Nikodym derivative of μ_1 with respect to μ_1. By Proposition 11.167, we have $\hat{\bar{f}} f = 1$ a.e. in $\mathcal{X}_1$.

"Sufficiency" Let $\hat{\bar{f}}$ be such that $\hat{\bar{f}} f = 1$ a.e. in $\mathcal{X}_1$. By Proposition 11.154, $\hat{\bar{f}}$ is a Radon–Nikodym derivative of a σ-finite $\mathbb{K}$-valued measure $\bar{\mu}$ on $(X, \mathcal{B})$ with respect to μ. $\forall E \in \mathcal{B}$ with $(\mathcal{P}\circ\mu_1 + \mathcal{P}\circ\bar{\mu})(E) < \infty$, $\bar{\mu}(E) = \int_E \hat{\bar{f}}\,\mathrm{d}\mu = \int_E (\hat{\bar{f}} f)\,\mathrm{d}\mu_1 = \int_E 1\,\mathrm{d}\mu_1 = \mu_1(E) \in \mathbb{K}$, where the second equality follows from (ii), the third equality follows from Proposition 11.132, and the last equality follows from Proposition 11.125. By Proposition 11.137, we have $\bar{\mu} = \mu_1$. Then, by Proposition 11.167, $\frac{\mathrm{d}\mu_1}{\mathrm{d}\mu} = \hat{\bar{f}}$ a.e. in $\mathcal{X}$.

This completes the proof of the proposition. □

Theorem 11.169 (Radon–Nikodym) *Let $\mathcal{X} := (X, \mathcal{B}, \mu)$ be a σ-finite measure space and ν be a σ-finite $\mathbb{R}$-valued measure on $(X, \mathcal{B})$. Assume that $\mathcal{P} \circ \nu \ll \mu$. Then, there exists $f : X \to \mathbb{R}$ such that f is the Radon–Nikodym derivative of ν with respect to μ. The function f is unique in the sense that $g : X \to \mathbb{R}$ is another function with this property if, and only if, g is $\mathcal{B}$-measurable and $f = g$ a.e. in $\mathcal{X}$. Hence, $\frac{\mathrm{d}\nu}{\mathrm{d}\mu} = f$ a.e. in $\mathcal{X}$. Furthermore, when ν is a σ-finite measure on $(X, \mathcal{B})$, then we may take $f : X \to [0, \infty) \subset \mathbb{R}$.*

Proof We will show the existence of f in three steps. First, consider the special case that $\mathcal{X}$ is a finite measure space and ν is a finite measure on $(X, \mathcal{B})$. Define $\mathcal{F} := \{f : X \to [0, \infty) \subset \mathbb{R} \mid f \text{ is } \mathcal{B}\text{-measurable, and } \forall A \in \mathcal{B}, \int_A f \,\mathrm{d}\mu \leq \nu(A)\}$. Clearly, the identically 0 function belongs to $\mathcal{F} \neq \emptyset$. Let $\lambda := \sup_{f \in \mathcal{F}} \int_X f \,\mathrm{d}\mu \in \mathbb{R}_e$. Then, $\lambda \geq 0$. Clearly, $\lambda \leq \nu(X) < +\infty$, by the definition of $\mathcal{F}$. $\forall f_1, f_2 \in \mathcal{F}$, by Propositions 11.38, 11.39, 7.23, $f_1 - f_2$ is $\mathcal{B}$-measurable. Then, $B := \{x \in X \mid f_1(x) - f_2(x) > 0\} \in \mathcal{B}$ and $C := \{x \in X \mid f_1(x) - f_2(x) \leq 0\} \in \mathcal{B}$. Clearly, $B = X \setminus C$. By Proposition 11.40, $f_1 \vee f_2$ is $\mathcal{B}$-measurable. $\forall A \in \mathcal{B}$, by Proposition 11.83, $\int_A (f_1 \vee f_2) \,\mathrm{d}\mu = \int_{A \cap B} (f_1 \vee f_2) \,\mathrm{d}\mu + \int_{A \cap C} (f_1 \vee f_2) \,\mathrm{d}\mu = \int_{A \cap B} f_1 \,\mathrm{d}\mu + \int_{A \cap C} f_2 \,\mathrm{d}\mu \leq \nu(A \cap B) + \nu(A \cap C) = \nu(A)$. Therefore, $f_1 \vee f_2 \in \mathcal{F}$. By the definition of λ, there exists $\left(\bar{f}_n\right)_{n=1}^{\infty} \subseteq \mathcal{F}$ such that $\lambda = \lim_{n \in \mathbb{N}} \int_X \bar{f}_n \,\mathrm{d}\mu$. $\forall n \in \mathbb{N}$, let $\tilde{f}_n := \bigvee_{i=1}^{n} \bar{f}_i \in \mathcal{F}$. Then, $\tilde{f}_n(x) \leq \tilde{f}_{n+1}(x)$, $\forall x \in X$, $\forall n \in \mathbb{N}$. By Proposition 11.83 and the definition of λ, we have $\int_X \bar{f}_n \,\mathrm{d}\mu \leq \int_X \tilde{f}_n \,\mathrm{d}\mu \leq \lambda$. Then, $\lambda = \lim_{n \in \mathbb{N}} \int_X \tilde{f}_n \,\mathrm{d}\mu \in [0, \nu(X)] \subset \mathbb{R}$. By Proposition 11.82 and Monotone Convergence Theorem 11.81, $\exists f : X \to [0, \infty) \subset \mathbb{R}$, which is $\mathcal{B}$-measurable, such that $\lim_{n \in \mathbb{N}} \tilde{f}_n = f$ a.e. in $\mathcal{X}$ and $\int_X f \,\mathrm{d}\mu = \lim_{n \in \mathbb{N}} \int_X \tilde{f}_n \,\mathrm{d}\mu = \lambda$. $\forall A \in \mathcal{B}$, by Monotone Convergence Theorem 11.81, we have $\int_A f \,\mathrm{d}\mu = \int_X (f \chi_{A,X}) \,\mathrm{d}\mu = \lim_{n \in \mathbb{N}} \int_X (\tilde{f}_n \chi_{A,X}) \,\mathrm{d}\mu = \lim_{n \in \mathbb{N}} \int_A \tilde{f}_n \,\mathrm{d}\mu \leq \nu(A)$. Then, $f \in \mathcal{F}$. Clearly, f is integrable over $\mathcal{X}$.

Claim 11.169.1 $\nu(E) = \int_E f \,\mathrm{d}\mu$, $\forall E \in \mathcal{B}$.

Proof of Claim Suppose that $\exists E_0 \in \mathcal{B}$ such that $\nu(E_0) > \int_{E_0} f \,\mathrm{d}\mu \geq 0$. Then, $\exists \epsilon_0 > 0$ such that $\nu(E_0) > \epsilon_0 \mu(E_0) + \int_{E_0} f \,\mathrm{d}\mu$. Define $\bar{\nu} : \mathcal{B} \to \mathbb{R}$ by $\bar{\nu}(E) = \nu(E) - \epsilon_0 \mu(E) - \int_E f \,\mathrm{d}\mu$. By Propositions 11.116 and 11.136, $\bar{\nu}$ is a finite $\mathbb{R}$-valued measure with $\bar{\nu}(E_0) \in (0, \infty) \subset \mathbb{R}$. By Lemma 11.160, $\exists A_0 \in \mathcal{B}$ with $A_0 \subseteq E_0$ such that A_0 is a positive set for $\bar{\nu}$ with $\bar{\nu}(A_0) \in (0, \infty) \subset \mathbb{R}$. Clearly, $0 < \bar{\nu}(A_0) \leq \nu(A_0)$. By $\nu \ll \mu$ and μ is finite, we have $\mu(A_0) \in (0, \infty) \subset \mathbb{R}$. $\forall E \in \mathcal{B}$, $\nu(E) = \nu(E \cap A_0) + \nu(E \setminus A_0) \geq \bar{\nu}(E \cap A_0) + \epsilon_0 \mu(E \cap A_0) + \int_{E \cap A_0} f \,\mathrm{d}\mu + \int_{E \setminus A_0} f \,\mathrm{d}\mu \geq \epsilon_0 \mu(E \cap A_0) + \int_E f \,\mathrm{d}\mu$, where the last inequality follows from Proposition 11.83. Let $f_1 := \epsilon_0 \chi_{A_0,X}$. By Propositions 11.38, 11.39, and 7.23, $f + f_1$ is $\mathcal{B}$-measurable. $\forall E \in \mathcal{B}$, $\int_E (f + f_1) \,\mathrm{d}\mu = \int_E f \,\mathrm{d}\mu + \int_E f_1 \,\mathrm{d}\mu = \int_E f \,\mathrm{d}\mu + \epsilon_0 \mu(E \cap A_0) \leq \nu(E)$, where the first equality follows from Proposition 11.83 and the second equality follows from Proposition 11.75. Hence, $f + f_1 \in \mathcal{F}$. Then, $\int_X (f + f_1)\, d\mu = \int_X f\, d\mu + \int_X f_1\, d\mu = \lambda + \epsilon_0 \mu(A_0) > \lambda$. This contradicts with the definition of λ. Therefore, the claim holds. □

Hence, f is a Radon–Nikodym derivative of ν with respect to μ. This special case is proved.

Next step, we consider the case when $\mathcal{X}$ is a σ-finite measure space and ν is a σ-finite measure on $(X, \mathcal{B})$. Then, $X = \bigcup_{n=1}^{\infty} X_n$ where $(X_n)_{n=1}^{\infty} \subseteq \mathcal{B}$, $\nu(X_n) < +\infty$, and $\mu(X_n) < +\infty$, $\forall n \in \mathbb{N}$. Without loss of generality, we may assume that X_n's are pairwise disjoint. $\forall n \in \mathbb{N}$, let $(X_n, \mathcal{B}_n, \mu_n)$ be the finite measure subspace of $\mathcal{X}$ and $(X_n, \mathcal{B}_n, \nu_n)$ be the finite measure subspace of $(X, \mathcal{B}, \nu)$ as defined in Proposition 11.13. By $\nu \ll \mu$, we have $\nu_n \ll \mu_n$. Then, by the first step, $\exists \bar{f}_n : X_n \to [0, \infty) \subset \mathbb{R}$ such that $\bar{f}_n$ is $\mathcal{B}_n$-measurable and $\nu_n(E_n) = \int_{E_n} \bar{f}_n\, d\mu_n$, $\forall E_n \in \mathcal{B}_n$. Define $f : X \to [0, \infty) \subset \mathbb{R}$ by $f(x) = \bar{f}_n(x)$, $\forall x \in X_n$, $\forall n \in \mathbb{N}$. By Proposition 11.41, f is $\mathcal{B}$-measurable. $\forall E \in \mathcal{B}$, $\nu(E) = \sum_{n=1}^{\infty} \nu(E \cap X_n) = \sum_{n=1}^{\infty} \nu_n(E \cap X_n) = \sum_{n=1}^{\infty} \int_{E\cap X_n} \bar{f}_n\, d\mu_n = \sum_{n=1}^{\infty} \int_{E\cap X_n} f|_{X_n}\, d\mu_n = \sum_{n=1}^{\infty} \int_{E\cap X_n} f\, d\mu = \int_E f\, d\mu$, where the fifth and sixth equalities follow from Proposition 11.83. Hence, f is a Radon–Nikodym derivative of ν with respect to μ.

The third step, we consider the case when $\mathcal{X}$ is a σ-finite measure space and ν is an σ-finite $\mathbb{R}$-valued measure on $(X, \mathcal{B})$. By Jordan Decomposition Theorem 11.163, there exists a pair of mutually singular σ-finite measures ν_+ and ν_- on $(X, \mathcal{B})$ such that $\nu = \nu_+ - \nu_-$ and $\mathcal{P} \circ \nu = \nu_+ + \nu_-$. By $\mathcal{P} \circ \nu \ll \mu$, we have $\nu_+ \ll \mu$ and $\nu_- \ll \mu$. By the second step, $\exists f_+ : X \to [0, \infty) \subset \mathbb{R}$ and $\exists f_- : X \to [0, \infty) \subset \mathbb{R}$ such that f_+ and f_- are Radon–Nikodym derivatives of ν_+ and ν_- with respect to μ, respectively. Let $f := f_+ - f_-$. $\forall E \in \operatorname{dom}(\nu)$, we have $\mathcal{P} \circ \nu(E) = \nu_+(E) + \nu_-(E) = \int_E f_+\, d\mu + \int_E f_-\, d\mu < \infty$. This leads to $\int_E f_+\, d\mu < \infty$ and $\int_E f_-\, d\mu < \infty$. Then, $f_+|_E$ and $f_-|_E$ are (absolutely) integrable over $\mathcal{E} := (E, \mathcal{B}_E, \mu_E)$, which is the σ-finite measure subspace of $\mathcal{X}$. Then, $\nu(E) = \nu_+(E) - \nu_-(E) = \int_E f_+\, d\mu - \int_E f_-\, d\mu = \int_E f\, d\mu \in \mathbb{R}$, where the third equality follows from Proposition 11.92. By Proposition 11.116, f is a Radon–Nikodym derivative of a σ-finite $\mathbb{R}$-valued measure $\bar{\nu}$ on $(X, \mathcal{B})$ with respect to μ. The above shows that $\bar{\nu}(E) = \nu(E) \in \mathbb{R}$, $\forall E \in \operatorname{dom}(\nu)$. By Proposition 11.137, we have $\bar{\nu} = \nu$. Hence, f is a Radon–Nikodym derivative of ν with respect to μ.

This completes the three steps to show the existence of f. The uniqueness of f follows directly from Proposition 11.167. This completes the proof of the theorem. □

Proposition 11.170 *Let $\mathcal{X} := (X, \mathcal{B})$ be a measurable space, $\mathcal{Y}$ be a separable normed linear space over $\mathbb{K}$, and $f : X \to \mathcal{Y}$. Assume that $\mathcal{Y}^*$ is separable, $D_* \subseteq \mathcal{Y}^*$ is a countable dense set, and, $\forall y_* \in D_*$, $f_{y_*} : X \to \mathbb{K}$ defined by $f_{y_*}(x) = \langle\langle y_*, f(x)\rangle\rangle$, $\forall x \in X$, is $\mathcal{B}$-measurable. Then, f is $\mathcal{B}$-measurable.*

Proof Let $D \subseteq \mathcal{Y}$ be a countable dense set. By Proposition 4.4, $\mathcal{Y}$ is second countable with a countable basis $\mathcal{E} := \{\mathcal{B}_{\mathcal{Y}}(y_0, r) \mid y_0 \in D,\ r \in \mathbb{Q},\ r > 0\}$. By Proposition 11.34, all we need to show is that $\forall \mathcal{B}_{\mathcal{Y}}(y_0, r) \in \mathcal{E}$, $V :=$

$f_{\text{inv}}(\mathcal{B}_{\mathcal{Y}}(y_0, r)) \in \mathcal{B}$. $\exists N \in \mathbb{N}$ such that $r > 1/N$. We will show that $V = U := \bigcup_{n=N}^{\infty} \bigcap_{y_* \in D_*,\ y_* \neq \vartheta_{\mathcal{Y}^*}} f_{y_* \text{inv}}(\mathcal{B}_{\mathbb{K}}(\langle\langle y_*, y_0 \rangle\rangle, \|y_*\| (r - 1/n))) \in \mathcal{B}$. $\forall n \in \mathbb{N}$ with $n \geq N$, $\forall y_* \in D_*$ with $y_* \neq \vartheta_{\mathcal{Y}^*}$, f_{y_*} is $\mathcal{B}$-measurable implies that $f_{y_* \text{inv}}(\mathcal{B}_{\mathbb{K}}(\langle\langle y_*, y_0 \rangle\rangle, \|y_*\| (r - 1/n))) \in \mathcal{B}$. Hence, $U \in \mathcal{B}$.

$\forall x \in V$, we have $\|f(x) - y_0\| < r$. Then, $\exists n \in \mathbb{N}$ with $n \geq N$ such that $\|f(x) - y_0\| < r - 1/n$. $\forall y_* \in D_*$ with $y_* \neq \vartheta_{\mathcal{Y}^*}$, we have $\left|f_{y_*}(x) - \langle\langle y_*, y_0 \rangle\rangle\right| = |\langle\langle y_*, f(x) - y_0 \rangle\rangle| \leq \|y_*\| \|f(x) - y_0\| < \|y_*\| (r - 1/n)$. Hence, $x \in f_{y_* \text{inv}}(\mathcal{B}_{\mathbb{K}}(\langle\langle y_*, y_0 \rangle\rangle, \|y_*\| (r - 1/n)))$. Then, $x \in U$. By the arbitrariness of x, we have $V \subseteq U$.

On the other hand, $\forall x \in U$, $\exists n \in \mathbb{N}$ with $n \geq N$, $\forall y_* \in D_*$ with $y_* \neq \vartheta_{\mathcal{Y}^*}$, we have $\left|f_{y_*}(x) - \langle\langle y_*, y_0 \rangle\rangle\right| < \|y_*\| (r - 1/n)$. Then, $|\langle\langle y_*, f(x) - y_0 \rangle\rangle| < \|y_*\| (r - 1/n)$. If $f(x) = y_0$, then $x \in V$. If $f(x) \neq y_0$, by Proposition 7.85, $\exists y_{*0} \in \mathcal{Y}^*$ with $\|y_{*0}\| = 1$ such that $\|f(x) - y_0\| = \langle\langle y_{*0}, f(x) - y_0 \rangle\rangle$. Since D_* is dense in $\mathcal{Y}^*$, then, by Proposition 4.13, $\exists (y_{*i})_{i=1}^{\infty} \subseteq D_*$ that converges to y_{*0}. Without loss of generality, assume $y_{*i} \neq \vartheta_{\mathcal{Y}^*}$, $\forall i \in \mathbb{N}$. By Propositions 3.66, 3.67, 7.21, and 7.72, we have $\|f(x) - y_0\| = \lim_{i \in \mathbb{N}} |\langle\langle y_{*i}, f(x) - y_0 \rangle\rangle| \leq \lim_{i \in \mathbb{N}} \|y_{*i}\| (r - 1/n) = r - 1/n < r$. Hence, $x \in V$. This implies that $U \subseteq V$.

Hence, $V = U \in \mathcal{B}$. This completes the proof of the proposition. □

Theorem 11.171 (Radon–Nikodym) *Let $\mathcal{X} := (X, \mathcal{B}, \mu)$ be a σ-finite $\mathbb{K}$-valued measure space, $\mathcal{Y}$ be a separable reflexive Banach space over $\mathbb{K}$, ν be a σ-finite $\mathcal{Y}$-valued measure on $(X, \mathcal{B})$. Assume that $\mathcal{P} \circ \nu \ll \mathcal{P} \circ \mu$ and $\mathcal{Y}^*$ is separable. Then, there exists $f : X \to \mathcal{Y}$ such that f is the Radon–Nikodym derivative of ν with respect to μ. The function f is unique in the sense that $g : X \to \mathcal{Y}$ is another function with this property if, and only if, g is $\mathcal{B}$-measurable and $f = g$ a.e. in $\mathcal{X}$. Hence, $\frac{\mathrm{d}\nu}{\mathrm{d}\mu} = f$ a.e. in $\mathcal{X}$.*

Proof First, we consider the special case when $\mathcal{X}$ is a σ-finite measure space and $\mathcal{Y} = \mathbb{C}$. By Proposition 11.165, $\nu = \nu_r + \mathrm{i}\nu_i$, where ν_r and ν_i are σ-finite $\mathbb{R}$-valued measures on $(X, \mathcal{B})$, and $\max\{\mathcal{P} \circ \nu_r(E), \mathcal{P} \circ \nu_i(E)\} \leq \mathcal{P} \circ \nu(E) \leq \mathcal{P} \circ \nu_r(E) + \mathcal{P} \circ \nu_i(E)$, $\forall E \in \mathcal{B}$. Then, $\mathcal{P} \circ \nu \ll \mu$ implies that $\mathcal{P} \circ \nu_r \ll \mu$ and $\mathcal{P} \circ \nu_i \ll \mu$. By Radon–Nikodym Theorem 11.169, there exist $f_r : X \to \mathbb{R}$ and $f_i : X \to \mathbb{R}$ such that $\frac{\mathrm{d}\nu_r}{\mathrm{d}\mu} = f_r$ a.e. in $\mathcal{X}$ and $\frac{\mathrm{d}\nu_i}{\mathrm{d}\mu} = f_i$ a.e. in $\mathcal{X}$. By Proposition 11.168, we have $\frac{\mathrm{d}\nu}{\mathrm{d}\mu} = \frac{\mathrm{d}(\nu_r + \mathrm{i}\nu_i)}{\mathrm{d}\mu} = \frac{\mathrm{d}\nu_r}{\mathrm{d}\mu} + \mathrm{i}\frac{\mathrm{d}\nu_i}{\mathrm{d}\mu} = f_r + \mathrm{i} f_i =: f : X \to \mathbb{C}$ a.e. in $\mathcal{X}$. This completes the proof of the special case.

Next, we consider the special case when $\mathcal{X}$ is a σ-finite measure space. $\forall y_* \in \mathcal{Y}^*$, define ν_{y_*} to be the σ-finite $\mathbb{K}$-valued measure on $(X, \mathcal{B})$ such that $\nu_{y_*}(E) = \langle\langle y_*, \nu(E) \rangle\rangle$, $\forall E \in \operatorname{dom}(\nu)$, as shown in Proposition 11.138. Then, by Proposition 11.138, $\mathcal{P} \circ \nu_{y_*} \leq \|y_*\| \mathcal{P} \circ \nu$. Then, $\mathcal{P} \circ \nu_{y_*} \ll \mu$ since $\mathcal{P} \circ \nu \ll \mu$. By Radon–Nikodym Theorem 11.169 and the special case, $\exists!\, f_{y_*} : X \to \mathbb{K}$ such that $\frac{\mathrm{d}\nu_{y_*}}{\mathrm{d}\mu} = f_{y_*}$ a.e. in $\mathcal{X}$. $\forall y_{*1}, y_{*2} \in \mathcal{Y}^*$, $\forall \alpha, \beta \in \mathbb{K}$, $\forall E \in \operatorname{dom}(\nu)$, we have $E \in \operatorname{dom}(\nu_{y_{*1}}) \cap \operatorname{dom}(\nu_{y_{*2}}) \cap \operatorname{dom}(\nu_{\alpha y_{*1} + \beta y_{*2}})$. Then, $\mathbb{K} \ni \nu_{\alpha y_{*1} + \beta y_{*2}}(E) = \int_E f_{\alpha y_{*1} + \beta y_{*2}} \,\mathrm{d}\mu = \langle\langle \alpha y_{*1} + \beta y_{*2}, \nu(E) \rangle\rangle = \alpha \langle\langle y_{*1}, \nu(E) \rangle\rangle + \beta \langle\langle y_{*2}, \nu(E) \rangle\rangle = \alpha \nu_{y_{*1}}(E) + \beta \nu_{y_{*2}}(E) = \alpha \int_E f_{y_{*1}} \,\mathrm{d}\mu + \beta \int_E f_{y_{*2}} \,\mathrm{d}\mu$, $\int_E \mathcal{P} \circ f_{\alpha y_{*1} + \beta y_{*2}} \,\mathrm{d}\mu < \infty$, $\int_E \mathcal{P} \circ f_{y_{*1}} \,\mathrm{d}\mu < \infty$, and $\int_E \mathcal{P} \circ f_{y_{*2}} \,\mathrm{d}\mu < \infty$. Then, $f_{y_{*1}}\big|_E$ and $f_{y_{*2}}\big|_E$

are absolutely integrable over $\mathcal{E} := (E, \mathcal{B}_E, \mu_E)$, which is the σ-finite measure subspace of $\mathcal{X}$. By Proposition 11.92, we have $\nu_{\alpha y_{*1}+\beta y_{*2}}(E) = \int_E f_{\alpha y_{*1}+\beta y_{*2}}\, d\mu = \int_E (\alpha f_{y_{*1}} + \beta f_{y_{*2}})\, d\mu \in \mathbb{K}$, $\forall E \in \mathcal{B}$ with $\mathcal{P} \circ \nu(E) < \infty$. By Propositions 11.116 and 11.137, $\alpha f_{y_{*1}} + \beta f_{y_{*2}}$ is a Radon–Nikodym derivative of $\nu_{\alpha y_{*1}+\beta y_{*2}}$ with respect to μ. Then, by Proposition 11.167,

$$f_{\alpha y_{*1}+\beta y_{*2}} = \alpha f_{y_{*1}} + \beta f_{y_{*2}} \text{ a.e. in } \mathcal{X}; \quad \forall y_{*1}, y_{*2} \in \mathcal{Y}^*, \forall \alpha, \beta \in \mathbb{K}$$

Note that $\mathcal{P} \circ \nu$ is a σ-finite measure on $(X, \mathcal{B})$ and $\mathcal{P} \circ \nu \ll \mu$. Again, by Radon–Nikodym Theorem 11.169, $\exists!\ \hat{f} : X \to [0, \infty) \subset \mathbb{R}$ such that $\frac{d\mathcal{P}\circ\nu}{d\mu} = \hat{f}$ a.e. in $\mathcal{X}$. This yields $\mathcal{P} \circ \nu(E) = \int_E \hat{f}\, d\mu$, $\forall E \in \mathcal{B}$. $\forall y_* \in \mathcal{Y}^*$, by Definition 11.166, $\mathcal{P} \circ \nu_{y_*}(E) = \int_E (\mathcal{P} \circ f_{y_*})\, d\mu \leq (\|y_*\|\mathcal{P} \circ \nu)(E)$, $\forall E \in \mathcal{B}$. When $\|y_*\| > 0$, $\mathcal{P} \circ \nu_{y_*}(E) \leq (\|y_*\|\mathcal{P} \circ \nu)(E) = \|y_*\|\mathcal{P} \circ \nu(E) = \|y_*\| \int_E \hat{f}\, d\mu = \int_E (\|y_*\|\hat{f})\, d\mu$, where the first equality follows from Proposition 11.136 and the third equality follows from Proposition 11.83. When $\|y_*\| = 0$, $\mathcal{P} \circ \nu_{y_*}(E) \leq 0 = \int_E (\|y_*\|\hat{f})\, d\mu$, where the inequality follows from Proposition 11.138 and the equality follows from Proposition 11.75. Hence, $\mathcal{P} \circ \nu_{y_*}(E) = \int_E (\mathcal{P} \circ f_{y_*})\, d\mu \leq \int_E (\|y_*\|\hat{f})\, d\mu$, $\forall E \in \mathcal{B}$, $\forall y_* \in \mathcal{Y}^*$. Then, by Proposition 11.96,

$$\mathcal{P} \circ f_{y_*} \leq \|y_*\|\hat{f} \text{ a.e. in } \mathcal{X}; \quad \forall y_* \in \mathcal{Y}^*$$

Let $\mathbb{K}_\mathbb{Q} := \mathbb{Q}$ if $\mathbb{K} = \mathbb{R}$ and $\mathbb{K}_\mathbb{Q} := \{\alpha + i\beta \in \mathbb{C} \mid \alpha, \beta \in \mathbb{Q}\}$ if $\mathbb{K} = \mathbb{C}$. Clearly, $\mathbb{K}_\mathbb{Q}$ is a countable dense set in $\mathbb{K}$.

Let $D \subseteq \mathcal{Y}^*$ be a countable dense set and $\hat{D} := \{\sum_{i=1}^n \alpha_i y_{*i} \in \mathcal{Y}^* \mid n \in \mathbb{Z}_+,\ \alpha_1, \ldots, \alpha_n \in \mathbb{K}_\mathbb{Q},\ y_{*1}, \ldots, y_{*n} \in D\}$. Then, $\hat{D} \subseteq \mathcal{Y}^*$ is also a countable dense set. Note that, $\forall y_{*1}, y_{*2} \in \hat{D}$, $\forall \alpha, \beta \in \mathbb{K}_\mathbb{Q}$, we have $\alpha y_{*1} + \beta y_{*2} \in \hat{D}$.

Define

$$E_0 := \Big\{x \in X \ \Big|\ \exists y_{*1}, y_{*2} \in \hat{D},\ \exists \alpha, \beta \in \mathbb{K}_\mathbb{Q} \ni \cdot\ f_{\alpha y_{*1}+\beta y_{*2}}(x) \neq \alpha f_{y_{*1}}(x) + \beta f_{y_{*2}}(x) \text{ or } \big|f_{y_{*1}}(x)\big| > \|y_{*1}\|\hat{f}(x)\Big\}$$

Then, $E_0 \in \mathcal{B}$ and $\mu(E_0) = 0$.

We have the following results:

$$f_{\alpha y_{*1}+\beta y_{*2}}(x) = \alpha f_{y_{*1}}(x) + \beta f_{y_{*2}}(x); \qquad \forall x \in X \setminus E_0, \forall y_{*1}, y_{*2} \in \hat{D}, \forall \alpha, \beta \in \mathbb{K}_\mathbb{Q} \tag{11.4}$$

and

$$\left|f_{y_*}(x)\right| \leq \|y_*\| \hat{f}(x); \quad \forall x \in X \setminus E_0, \forall y_* \in \hat{D} \tag{11.5}$$

$\forall x \in X \setminus E_0$, $\forall y_* \in \mathcal{Y}^*$, by Proposition 4.13, $\exists (y_{*i})_{i=1}^\infty \subseteq \hat{D}$ that converges to y_*. Then, this sequence is a Cauchy sequence. By (11.4) and (11.5), $\left(f_{y_{*i}}(x)\right)_{i=1}^\infty \subseteq \mathbb{K}$ is a Cauchy sequence and admits limit $F(x, y_*) \in \mathbb{K}$. It should be clear that $F(x, y_*)$ is independent of the choice of the sequence $(y_{*i})_{i=1}^\infty$, since let $(\bar{y}_{*i})_{i=1}^\infty \subseteq \hat{D}$ be any sequence converging to y_*, the combined sequence $(y_{*1}, \bar{y}_{*1}, y_{*2}, \bar{y}_{*2}, \ldots) \subseteq \hat{D}$ converges to y_*. Therefore, the sequence $(f_{y_{*1}}(x), f_{\bar{y}_{*1}}(x), f_{y_{*2}}(x), f_{\bar{y}_{*2}}(x), \ldots) \subseteq \mathbb{K}$ is a Cauchy sequence and admits a limit, which has to be equal to $F(x, y_*)$. Then, $\forall x \in X \setminus E_0$, $\forall y_* \in \mathcal{Y}^*$, $F(x, y_*) \in \mathbb{K}$. $\forall x \in X \setminus E_0$, $\forall \bar{y}_* \in \hat{D}$, $F(x, \bar{y}_*) = f_{\bar{y}_*}(x)$.

$\forall x \in X \setminus E_0$, $\forall y_{*1}, y_{*2} \in \mathcal{Y}^*$, $\forall \alpha, \beta \in \mathbb{K}$, $\exists \left(\bar{y}_{*1,i}\right)_{i=1}^\infty \subseteq \hat{D}$ that converges to y_{*1}, $\exists \left(\bar{y}_{*2,i}\right)_{i=1}^\infty \subseteq \hat{D}$ that converges to y_{*2}, $\exists (\alpha_i)_{i=1}^\infty \subseteq \mathbb{K}_{\mathbb{Q}}$ that converges to α, and $\exists (\beta_i)_{i=1}^\infty \subseteq \mathbb{K}_{\mathbb{Q}}$ that converges to β. By Propositions 7.23, 3.66, and 3.67, we have $\lim_{i \in \mathbb{N}} (\alpha_i \bar{y}_{*1,i} + \beta_i \bar{y}_{*2,i}) = \alpha y_{*1} + \beta y_{*2}$, where the sequence on the left-hand side is contained in $\hat{D}$. This implies that $F(x, \alpha y_{*1} + \beta y_{*2}) = \lim_{i \in \mathbb{N}} f_{\alpha_i \bar{y}_{*1,i} + \beta_i \bar{y}_{*2,i}}(x) = \lim_{i \in \mathbb{N}} (\alpha_i f_{\bar{y}_{*1,i}}(x) + \beta_i f_{\bar{y}_{*2,i}}(x)) = \alpha F(x, y_{*1}) + \beta F(x, y_{*2})$, where the second equality follows from (11.4) and the last equality follows from Propositions 7.23, 3.66, and 3.67. Furthermore, $|F(x, y_{*1})| = \lim_{i \in \mathbb{N}} \left|f_{\bar{y}_{*1,i}}(x)\right| \leq \lim_{i \in \mathbb{N}} \left\|\bar{y}_{*1,i}\right\| \hat{f}(x) = \|y_{*1}\| \hat{f}(x)$, where the inequality follows from (11.5) and the last equality follows from Propositions 3.66 and 7.21. Hence, $\forall x \in X \setminus E_0$, $F(x, \cdot) : \mathcal{Y}^* \to \mathbb{K}$ is a bounded linear functional on $\mathcal{Y}^*$. Since $\mathcal{Y}$ is reflexive, then $\exists f(x) \in \mathcal{Y}$ such that $F(x, y_*) = \langle\langle y_*, f(x) \rangle\rangle$, $\forall y_* \in \mathcal{Y}^*$ and $\|f(x)\| \leq \hat{f}(x)$. Therefore, we may define $f : X \to \mathcal{Y}$ by assigning $f(x) = \vartheta_{\mathcal{Y}}$, $\forall x \in E_0$.

$\forall y_* \in \mathcal{Y}^*$, define $F_{y_*} : X \to \mathbb{K}$ by $F_{y_*}(x) = \langle\langle y_*, f(x) \rangle\rangle$, $\forall x \in X$. Fix any $\bar{y}_* \in \hat{D}$. Then, $\forall x \in X \setminus E_0$, $F_{\bar{y}_*}(x) = \langle\langle \bar{y}_*, f(x) \rangle\rangle = F(x, \bar{y}_*) = f_{\bar{y}_*}(x)$ and, $\forall x \in E_0$, $F_{\bar{y}_*}(x) = \langle\langle \bar{y}_*, f(x) \rangle\rangle = 0$. By Proposition 11.41, $F_{\bar{y}_*}$ is $\mathcal{B}$-measurable. By the arbitrariness of $\bar{y}_*$ and Proposition 11.170, f is $\mathcal{B}$-measurable.

Claim 11.171.1 $\forall y_{*0} \in \mathcal{Y}^*$, $f_{y_{*0}}(x) = \langle\langle y_{*0}, f(x) \rangle\rangle$ a.e. $x \in \mathcal{X}$.

Proof of Claim Let $\bar{D} := D \cup \{y_{*0}\}$ and $\hat{\hat{D}} := \{\sum_{i=1}^n \alpha_i y_{*i} \in \mathcal{Y}^* \mid n \in \mathbb{Z}_+,\ \alpha_1, \ldots, \alpha_n \in \mathbb{K}_{\mathbb{Q}},\ y_{*1}, \ldots, y_{*n} \in \bar{D}\}$. Then, $\hat{\hat{D}} \subseteq \mathcal{Y}^*$ is also a countable dense set and $\hat{D} \subseteq \hat{\hat{D}}$. Define

$$E_1 := \{x \in X \mid \exists y_{*1}, y_{*2} \in \hat{\hat{D}},\ \exists \alpha, \beta \in \mathbb{K}_{\mathbb{Q}} \ni \cdot\ f_{\alpha y_{*1} + \beta y_{*2}}(x) \neq \alpha f_{y_{*1}}(x) + \beta f_{y_{*2}}(x) \text{ or } \left|f_{y_{*1}}(x)\right| > \|y_{*1}\| \hat{f}(x)\}$$

Then, $E_1 \in \mathcal{B}$, $\mu(E_1) = 0$, and $E_0 \subseteq E_1$.

$\forall x \in X \setminus E_1$, $\forall y_{*1}, y_{*2} \in \hat{\hat{D}}$, $\forall \alpha, \beta \in \mathbb{K}_{\mathbb{Q}}$, we have $f_{\alpha y_{*1}+\beta y_{*2}}(x) = \alpha f_{y_{*1}}(x) + \beta f_{y_{*2}}(x)$ and $\left|f_{y_{*1}}(x)\right| \leq \|y_{*1}\| \hat{f}(x)$. By Proposition 4.13, $\exists (y_{*i})_{i=1}^{\infty} \subseteq \hat{\hat{D}}$ that converges to y_{*0}. Then, this sequence is a Cauchy sequence. By the above, $\left(f_{y_{*i}}(x)\right)_{i=1}^{\infty} \subseteq \mathbb{K}$ is a Cauchy sequence and admits limit $\bar{F}(x, y_{*0}) \in \mathbb{K}$. It should be clear that $\bar{F}(x, y_{*0})$ is independent of the choice of the sequence $(y_{*i})_{i=1}^{\infty}$. Two particular sequences $(\bar{y}_{*i})_{i=1}^{\infty} \subseteq \hat{D}$ converging to y_{*0} and $(y_{*0}, y_{*0}, \ldots) \subseteq \hat{\hat{D}}$ lead to $f_{y_{*0}}(x) = \bar{F}(x, y_{*0}) = F(x, y_{*0})$. Therefore, $f_{y_{*0}}(x) = F(x, y_{*0}) = \langle\langle y_{*0}, f(x)\rangle\rangle = F_{y_{*0}}(x)$, $\forall x \in X \setminus E_1$. Note that $f_{y_{*0}}$ is $\mathcal{B}$-measurable and, by Propositions 7.72 and 11.38, $F_{y_{*0}}$ is $\mathcal{B}$-measurable. Then, $E_1 \supseteq \{x \in X \mid f_{y_{*0}}(x) \neq F_{y_{*0}}(x)\} = \{x \in X \mid \left|f_{y_{*0}}(x) - F_{y_{*0}}(x)\right| > 0\} \in \mathcal{B}$. Hence, $f_{y_{*0}} = F_{y_{*0}}$ a.e. in $\mathcal{X}$. Therefore, the claim holds. □

Note that $\mathcal{P} \circ f(x) \leq \hat{f}(x)$, $\forall x \in X$, and $\mathcal{P} \circ f$ is $\mathcal{B}$-measurable, by Propositions 7.21 and 11.38. $\forall E \in \mathrm{dom}(\nu)$, we have $\mathcal{P} \circ \nu(E) = \int_E \hat{f} \, \mathrm{d}\mu < \infty$. This implies that $\int_E \mathcal{P} \circ f \, \mathrm{d}\mu < \infty$. By Proposition 11.92, we have $\int_E f \, \mathrm{d}\mu \in \mathcal{Y}$. $\forall y_* \in \mathcal{Y}^*$, $\mathcal{P} \circ \nu_{y_*}(E) \leq (\|y_*\| \mathcal{P} \circ \nu)(E) < \infty$, $E \in \mathrm{dom}\left(\nu_{y_*}\right)$, and $\langle\langle y_*, \nu(E)\rangle\rangle = \nu_{y_*}(E) = \int_E f_{y_*} \, \mathrm{d}\mu = \int_E F_{y_*} \, \mathrm{d}\mu = \int_E \langle\langle y_*, f(x)\rangle\rangle \, \mathrm{d}\mu(x) = \left\langle\!\left\langle y_*, \int_E f \, \mathrm{d}\mu\right\rangle\!\right\rangle$, where the last three equalities follow from Proposition 11.92. Hence, by Proposition 7.85, $\nu(E) = \int_E f \, \mathrm{d}\mu \in \mathcal{Y}$, $\forall E \in \mathrm{dom}(\nu)$. Hence, by Propositions 11.116 and 11.137, f is a Radon–Nikodym derivative of ν with respect to μ. By Proposition 11.167, f is unique as desired and $\frac{\mathrm{d}\nu}{\mathrm{d}\mu} = f$ a.e. in $\mathcal{X}$.

Finally, we consider the general case. Let $\bar{\mathcal{X}} := (X, \mathcal{B}, \mathcal{P} \circ \mu)$, which is a σ-finite measure space. By the second special case, $\exists!\ f_1 : X \to \mathcal{Y}$ such that $\frac{\mathrm{d}\nu}{\mathrm{d}\mathcal{P}\circ\mu} = f_1$ a.e. in $\bar{\mathcal{X}}$. Again by the second special case, $\exists!\ f_2 : X \to \mathbb{K}$ such that $\frac{\mathrm{d}\mu}{\mathrm{d}\mathcal{P}\circ\mu} = f_2$ a.e. in $\bar{\mathcal{X}}$. By Definition 11.166, $\frac{\mathrm{d}\mathcal{P}\circ\mu}{\mathrm{d}\mathcal{P}\circ\mu} = \mathcal{P} \circ f_2$ a.e. in $\bar{\mathcal{X}}$. Clearly, the constant function 1 is the Radon–Nikodym derivative of $\mathcal{P} \circ \mu$ with respect to $\mathcal{P} \circ \mu$. Then, $\mathcal{P} \circ f_2 = 1$ a.e. in $\bar{\mathcal{X}}$ and $\exists!\ f_3 : X \to \mathbb{K}$ such that $f_3 f_2 = 1$ a.e. in $\bar{\mathcal{X}}$. By Proposition 11.168, $\frac{\mathrm{d}\mathcal{P}\circ\mu}{\mathrm{d}\mu} = f_3$ a.e. in $\mathcal{X}$ and $\frac{\mathrm{d}\nu}{\mathrm{d}\mu} = \left(\frac{\mathrm{d}\nu}{\mathrm{d}\mathcal{P}\circ\mu}\right)\left(\frac{\mathrm{d}\mathcal{P}\circ\mu}{\mathrm{d}\mu}\right) = f_1 f_3 =: f : X \to \mathcal{Y}$ a.e. in $\mathcal{X}$.

This completes the proof of the theorem. □

Theorem 11.172 (Lebesgue Decomposition) *Let $\mathcal{X} := (X, \mathcal{B}, \mu)$ be a σ-finite measure space, $\mathcal{Y}$ be a Banach space, and ν be a σ-finite $\mathcal{Y}$-valued measure on $(X, \mathcal{B})$. Then, there exists a unique pair of σ-finite $\mathcal{Y}$-valued measures ν_0 and ν_1 on $(X, \mathcal{B})$ such that $\mathcal{P} \circ \nu_0 \perp \mu$, $\mathcal{P} \circ \nu_1 \ll \mu$, $\nu = \nu_0 + \nu_1$, and $\mathcal{P} \circ \nu = \mathcal{P} \circ \nu_0 + \mathcal{P} \circ \nu_1$. Furthermore, the following statements hold:*

(i) *If ν is a finite $\mathcal{Y}$-valued measure on $(X, \mathcal{B})$, then ν_0 and ν_1 are finite $\mathcal{Y}$-valued measures on $(X, \mathcal{B})$.*
(ii) *If ν is a σ-finite measure on $(X, \mathcal{B})$, then ν_0 and ν_1 are σ-finite measures on $(X, \mathcal{B})$.*
(iii) *If ν is a finite measure on $(X, \mathcal{B})$, then ν_0 and ν_1 are finite measures on $(X, \mathcal{B})$.*

Proof Let $\lambda := \mu + \mathcal{P} \circ \nu$. By Proposition 11.136, λ is a σ-finite measure on $(X, \mathcal{B})$. Let $\bar{\mathcal{X}} := (X, \mathcal{B}, \lambda)$. Clearly, we have $\mu \ll \lambda$ and $\mathcal{P} \circ \nu \ll \lambda$. By Radon–Nikodym Theorem 11.169, $\exists f : X \to [0, \infty) \subset \mathbb{R}$ such that f is the Radon–Nikodym derivative of μ with respect to λ. Then, $\mu(E) = \int_E f \, d\lambda$, $\forall E \in \mathcal{B}$. Let $A := \{x \in X \mid f(x) > 0\} \in \mathcal{B}$ and $B := \{x \in X \mid f(x) = 0\} \in \mathcal{B}$. Then, $A \cap B = \emptyset$ and $A \cup B = X$. By Proposition 11.75, $\mu(B) = \int_B f \, d\lambda = 0$. Define ν_0 to be a function from $\mathcal{B}$ to $\mathcal{Y}$ by: $\nu_0(E)$ is undefined, $\forall E \in \mathcal{B}$ with $E \cap B \in \mathcal{B} \setminus \text{dom}(\nu)$; and $\nu_0(E) = \nu(E \cap B) \in \mathcal{Y}$, $\forall E \in \mathcal{B}$ with $E \cap B \in \text{dom}(\nu)$. Define ν_1 to be a function from $\mathcal{B}$ to $\mathcal{Y}$ by: $\nu_1(E)$ is undefined, $\forall E \in \mathcal{B}$ with $E \cap A \in \mathcal{B} \setminus \text{dom}(\nu)$; and $\nu_1(E) = \nu(E \cap A) \in \mathcal{Y}$, $\forall E \in \mathcal{B}$ with $E \cap A \in \text{dom}(\nu)$. By Proposition 11.157, ν_0 and ν_1 are σ-finite $\mathcal{Y}$-valued measures on $(X, \mathcal{B})$ such that $\nu = \nu_0 + \nu_1$, $\mathcal{P} \circ \nu = \mathcal{P} \circ \nu_0 + \mathcal{P} \circ \nu_1$, $\mathcal{P} \circ \nu_0(E) = \mathcal{P} \circ \nu(E \cap B)$, and $\mathcal{P} \circ \nu_1(E) = \mathcal{P} \circ \nu(E \cap A)$, $\forall E \in \mathcal{B}$. This implies that $\mathcal{P} \circ \nu_0(A) = \mathcal{P} \circ \nu(A \cap B) = 0$. Then, $\mathcal{P} \circ \nu_0 \perp \mu$.

$\forall E \in \mathcal{B}$ with $\mu(E) = 0$, we have $\int_E f \, d\lambda = 0$. Let $\bar{\mathcal{E}} := (E, \mathcal{B}_E, \lambda_E)$ be the σ-finite measure subspace of $\bar{\mathcal{X}}$ as defined in Proposition 11.13. Then, by Proposition 11.96, we have $f = 0$ a.e. in $\bar{\mathcal{E}}$. Then, $\lambda(E \cap A) = 0$ and $\mathcal{P} \circ \nu_1(E) = \mathcal{P} \circ \nu(E \cap A) = 0$, since $\mathcal{P} \circ \nu \ll \lambda$. Hence, $\mathcal{P} \circ \nu_1 \ll \mu$. This shows that ν_0 and ν_1 is the pair of σ-finite $\mathcal{Y}$-valued measures on $(X, \mathcal{B})$ that we seek.

Next, we show the uniqueness of the pair ν_0 and ν_1. Let $\hat{\nu}_0$ and $\hat{\nu}_1$ be another pair of σ-finite $\mathcal{Y}$-valued measures on $(X, \mathcal{B})$ such that $\mathcal{P} \circ \hat{\nu}_0 \perp \mu$, $\mathcal{P} \circ \hat{\nu}_1 \ll \mu$, $\nu = \hat{\nu}_0 + \hat{\nu}_1$, and $\mathcal{P} \circ \nu = \mathcal{P} \circ \hat{\nu}_0 + \mathcal{P} \circ \hat{\nu}_1$. Then, $\exists \hat{A}, \hat{B} \in \mathcal{B}$ with $\hat{A} = X \setminus \hat{B}$ such that $\mathcal{P} \circ \hat{\nu}_0(\hat{A}) = \mu(\hat{B}) = 0$. $\forall E \in \mathcal{B}$ with $\lambda(E) < \infty$, we have $\mu(E) < \infty$, $\mathcal{P} \circ \nu(E) = \mathcal{P} \circ \nu_0(E) + \mathcal{P} \circ \nu_1(E) = \mathcal{P} \circ \hat{\nu}_0(E) + \mathcal{P} \circ \hat{\nu}_1(E) < \infty$. Then, $\hat{\nu}_0(E) = \hat{\nu}_0(E \cap \hat{A}) + \hat{\nu}_0(E \cap \hat{B}) = \hat{\nu}_0(E \cap \hat{B}) = \hat{\nu}_0(E \cap \hat{B}) + \hat{\nu}_1(E \cap \hat{B}) = \nu(E \cap \hat{B}) = \nu_0(E \cap \hat{B}) + \nu_1(E \cap \hat{B}) + \hat{\nu}_0(E \cap \hat{A} \cap B) + \hat{\nu}_1(E \cap \hat{A} \cap B) = \nu_0(E \cap \hat{B}) + \nu(E \cap \hat{A} \cap B) = \nu_0(E \cap \hat{B}) + \nu_0(E \cap \hat{A}) = \nu_0(E) \in \mathcal{Y}$, where the second equality follows from the fact that $\mathcal{P} \circ \hat{\nu}_0(\hat{A}) = 0$, the third equality follows from $\mu(\hat{B}) = 0$ and $\mathcal{P} \circ \hat{\nu}_1 \ll \mu$, the fifth equality follows from the fact that $\mathcal{P} \circ \hat{\nu}_0(\hat{A}) = 0$, $\mu(B) = 0$, and $\mathcal{P} \circ \hat{\nu}_1 \ll \mu$, and the sixth equality follows from the fact that $\mu(\hat{B}) = 0$ and $\mathcal{P} \circ \nu_1 \ll \mu$. We also have $\hat{\nu}_1(E) = \hat{\nu}_1(E \cap \hat{A}) + \hat{\nu}_1(E \cap \hat{B}) = \hat{\nu}_1(E \cap \hat{A}) = \hat{\nu}_1(E \cap \hat{A}) + \hat{\nu}_0(E \cap \hat{A}) = \nu(E \cap \hat{A}) = \nu_0(E \cap \hat{A}) + \nu_1(E \cap \hat{A}) = \nu(E \cap \hat{A} \cap B) + \nu_1(E \cap \hat{A}) = \hat{\nu}_0(E \cap \hat{A} \cap B) + \hat{\nu}_1(E \cap \hat{A} \cap B) + \nu_1(E \cap \hat{A}) = \nu_1(E \cap \hat{A}) = \nu_1(E \cap \hat{A}) + \nu_1(E \cap \hat{B}) = \nu_1(E) \in \mathcal{Y}$, where the second equality follows from the fact that $\mu(\hat{B}) = 0$ and $\mathcal{P} \circ \hat{\nu}_1 \ll \mu$; the third equality follows from $\mathcal{P} \circ \hat{\nu}_0(\hat{A}) = 0$; the eighth equality follows from the fact that $\mathcal{P} \circ \hat{\nu}_0(\hat{A}) = 0$, $\mu(B) = 0$, and $\mathcal{P} \circ \hat{\nu}_1 \ll \mu$; and the ninth equality follows from the fact that $\mu(\hat{B}) = 0$ and $\mathcal{P} \circ \nu_1 \ll \mu$. By Proposition 11.137, we have $\hat{\nu}_0 = \nu_0$ and $\hat{\nu}_1 = \nu_1$. Hence, the pair ν_0 and ν_1 is unique.

(i) If ν is finite, then $\mathcal{P} \circ \nu(X) = \mathcal{P} \circ \nu_0(X) + \mathcal{P} \circ \nu_1(X) < \infty$. Then, ν_0 and ν_1 are finite.

(ii) If ν is a σ-finite measure on $(X, \mathcal{B})$, by Proposition 11.135, it is identified with a σ-finite $\mathbb{R}$-valued measure $\bar{\nu}$ on $(X, \mathcal{B})$ such that $\bar{\nu}(E) \in [0, \infty) \subset \mathbb{R}$, $\forall E \in \text{dom}(\bar{\nu})$. By the general case, there exists a pair of σ-finite $\mathbb{R}$-valued measures

$\bar{\nu}_0$ and $\bar{\nu}_1$ on $(X, \mathcal{B})$ that satisfies the desired properties. By the proof of the general case, $\bar{\nu}_0(E) \in [0, \infty) \subset \mathbb{R}$, $\forall E \in \text{dom}(\bar{\nu}_0)$, and $\bar{\nu}_1(E) \in [0, \infty) \subset \mathbb{R}$, $\forall E \in \text{dom}(\bar{\nu}_1)$. By Proposition 11.135, $\bar{\nu}_0$ and $\bar{\nu}_1$ are identified with σ-finite measures ν_0 and ν_1, respectively. Then, $\nu_0 = \mathcal{P} \circ \bar{\nu}_0 \perp \mu$, $\nu_1 = \mathcal{P} \circ \bar{\nu}_1 \ll \mu$, and $\nu = \mathcal{P} \circ \bar{\nu} = \mathcal{P} \circ \bar{\nu}_0 + \mathcal{P} \circ \bar{\nu}_1 = \nu_0 + \nu_1$. Then, ν_0 and ν_1 are σ-finite measures on $(X, \mathcal{B})$.

(iii) If ν is a finite measure on $(X, \mathcal{B})$, then, by (i) and (ii), ν_0 and ν_1 are finite measures on $(X, \mathcal{B})$.

This completes the proof of the theorem. □

11.9 L_p Spaces

Example 11.173 Let $p \in [1, \infty) \subset \mathbb{R}$, $\mathcal{X} := (X, \mathcal{B}, \mu)$ be a σ-finite measure space, $\mathcal{Y}$ be a separable normed linear space over $\mathbb{K}$, and $(\mathcal{M}(\mathcal{X}, \mathcal{Y}), \mathbb{K})$ be the vector space of functions of $\mathcal{X}$ to $\mathcal{Y}$ as defined in Example 6.20 with the usual vector addition, scalar multiplication, and the null vector ϑ. Define $l : [0, \infty) \subset \mathbb{R} \to [0, \infty) \subset \mathbb{R}$ by $l(t) = t^p$, $\forall t \in [0, \infty) \subset \mathbb{R}$. We will introduce the notation $\mathcal{P}_p \circ f$ to denote $l \circ \mathcal{P} \circ f$. Let $Z_p := \{f \in \mathcal{M}(\mathcal{X}, \mathcal{Y}) \mid f \text{ is } \mathcal{B}\text{-measurable and } \mathcal{P}_p \circ f \text{ is integrable over } \mathcal{X}\}$. Define $\|\cdot\|_p : Z_p \to [0, \infty) \subset \mathbb{R}$ by $\|f\|_p = \left(\int_X (\mathcal{P}_p \circ f)\, d\mu\right)^{1/p}$, $\forall f \in Z_p$. We will next show that Z_p is a subspace of $(\mathcal{M}(\mathcal{X}, \mathcal{Y}), \mathbb{K})$ and $\|\cdot\|_p$ defines a pseudo-norm on Z_p.

Clearly, $\vartheta \in Z_p \neq \emptyset$. $\forall f, g \in Z_p$, $\forall \alpha \in \mathbb{K}$, by Propositions 7.23, 11.38, and 11.39, $f + g$ and αf are $\mathcal{B}$-measurable. By Minkowski's Inequality 11.174, $f + g \in Z_p$ and $\|f + g\|_p \leq \|f\|_p + \|g\|_p$. $\|\alpha f\|_p = \left(\int_X (\mathcal{P}_p \circ (\alpha f))\, d\mu\right)^{1/p} = \left(|\alpha|^p \int_X (\mathcal{P}_p \circ f)\, d\mu\right)^{1/p} = |\alpha| \left(\int_X (\mathcal{P}_p \circ f)\, d\mu\right)^{1/p} = |\alpha| \|f\|_p \in \mathbb{R}$, where the second equality follows from Proposition 11.92. Hence, $\alpha f \in Z_p$.

The above shows that Z_p is a subspace of $(\mathcal{M}(\mathcal{X}, \mathcal{Y}), \mathbb{K})$. Hence, $(Z_p, \mathbb{K})$ is a vector space. Clearly, $\|\vartheta\|_p = 0$. Then, combined with the above, we have $\|\cdot\|_p$ defines a pseudo-norm on $(Z_p, \mathbb{K})$. By Proposition 7.47, the quotient space of $(Z_p, \mathbb{K})$ modulo $\|\cdot\|_p$ is a normed linear space, to be denoted $L_p(\mathcal{X}, \mathcal{Y})$. We will denote the vector space $(Z_p, \mathbb{K})$ with the pseudo-norm $\|\cdot\|_p$ by $\bar{L}_p(\mathcal{X}, \mathcal{Y})$.

$\forall f \in Z_p$ with $\|f\|_p = 0$, we have $\int_X \mathcal{P}_p \circ f\, d\mu = 0$. By Proposition 11.96, we have $\mathcal{P}_p \circ f = 0$ a.e. in $\mathcal{X}$ and $f = \vartheta_{\mathcal{Y}}$ a.e. in $\mathcal{X}$. On the other hand, $\forall f \in \mathcal{M}(\mathcal{X}, \mathcal{Y})$ with $f = \vartheta_{\mathcal{Y}}$ a.e. in $\mathcal{X}$ and f being $\mathcal{B}$-measurable, then, by Proposition 11.83, $\|f\|_p = 0$ and $f \in Z_p$. Hence, $\forall f \in \mathcal{M}(\mathcal{X}, \mathcal{Y})$, $f \in Z_p$ and $\|f\|_p = 0$ if, and only if, $f = \vartheta_{\mathcal{Y}}$ a.e. in $\mathcal{X}$ and f is $\mathcal{B}$-measurable. We will denote the norm in $L_p(\mathcal{X}, \mathcal{Y})$ by $\|\cdot\|_p$ and elements in $L_p(\mathcal{X}, \mathcal{Y})$ by $[f]$, where $f \in \bar{L}_p(\mathcal{X}, \mathcal{Y})$. ◇

Theorem 11.174 (Minkowski's Inequality) *Let $p \in [1, \infty) \subset \mathbb{R}$, $\mathcal{X} := (X, \mathcal{B}, \mu)$ be a σ-finite measure space, $\mathcal{Y}$ be a separable normed linear space over $\mathbb{K}$, and $\bar{L}_p(\mathcal{X}, \mathcal{Y})$ be the vector space over $\mathbb{K}$ with the pseudo-norm $\|\cdot\|_p$ as defined* *in Example 11.173.* $\forall f, g \in \bar{\mathrm{L}}_p(\mathcal{X}, \mathcal{Y})$*, then,* $f + g \in \bar{\mathrm{L}}_p(\mathcal{X}, \mathcal{Y})$ *and*

$$\|f + g\|_p \leq \|f\|_p + \|g\|_p$$

When $1 < p < \infty$*, equality holds if, and only if,* $\exists \alpha, \beta \in [0, \infty) \subset \mathbb{R}$*, which are not both zeros, such that* $\alpha \mathcal{P} \circ f = \beta \mathcal{P} \circ g$ *a.e. in* $\mathcal{X}$ *and* $\mathcal{P} \circ (f + g) = \mathcal{P} \circ f + \mathcal{P} \circ g$ *a.e. in* $\mathcal{X}$*.*

Proof Clearly, f and g are $\mathcal{B}$-measurable. By Propositions 7.23, 11.38, and 11.39, $f + g$ is $\mathcal{B}$-measurable. We will distinguish two exhaustive and mutually exclusive cases: Case 1: $\|f\|_p \|g\|_p = 0$; Case 2: $\|f\|_p \|g\|_p > 0$. Case 1: $\|f\|_p \|g\|_p = 0$. Without loss of generality, assume that $\|g\|_p = 0$. Then, $\int_X (\mathcal{P}_p \circ g) \,\mathrm{d}\mu = 0$. By Proposition 11.96, we have $\mathcal{P}_p \circ g = 0$ a.e. in $\mathcal{X}$. Thus, $g = \vartheta_{\mathcal{Y}}$ a.e. in $\mathcal{X}$ and $f = f + g$ a.e. in $\mathcal{X}$. By Proposition 11.83, the integral $\int_X (\mathcal{P}_p \circ (f + g)) \,\mathrm{d}\mu = \int_X (\mathcal{P}_p \circ f) \,\mathrm{d}\mu \in \mathbb{R}$. Hence, $\|f + g\|_p = \|f\|_p = \|f\|_p + \|g\|_p \in \mathbb{R}$ and $f + g \in \bar{\mathrm{L}}_p(\mathcal{X}, \mathcal{Y})$. Clearly, equality holds. Let $\alpha = 0$, $\beta = 1$; we have $\alpha \mathcal{P} \circ f = \beta \mathcal{P} \circ g$ a.e. in $\mathcal{X}$ and $\mathcal{P} \circ (f + g) = \mathcal{P} \circ f + \mathcal{P} \circ g$ a.e. in $\mathcal{X}$.

Case 2: $\|f\|_p \|g\|_p > 0$. Let $\lambda := \frac{\|f\|_p}{\|f\|_p + \|g\|_p} \in (0, 1) \subset \mathbb{R}$. $\forall x \in X$, we have

$$\frac{\|f(x) + g(x)\|^p}{(\|f\|_p + \|g\|_p)^p} \leq \Big(\frac{\|f(x)\| + \|g(x)\|}{\|f\|_p + \|g\|_p}\Big)^p$$
$$= \Big(\lambda \frac{\|f(x)\|}{\|f\|_p} + (1 - \lambda) \frac{\|g(x)\|}{\|g\|_p}\Big)^p \leq \lambda \frac{\|f(x)\|^p}{\|f\|_p^p} + (1 - \lambda) \frac{\|g(x)\|^p}{\|g\|_p^p}$$

where the second inequality follows from the convexity of the function t^p on $[0, \infty) \subset \mathbb{R}$. In the above, when $p > 1$, equality holds if, and only if, $\|f(x) + g(x)\| = \|f(x)\| + \|g(x)\|$ and $\frac{\|f(x)\|}{\|f\|_p} = \frac{\|g(x)\|}{\|g\|_p}$.

By Proposition 11.83, we have $\int_X \frac{(\mathcal{P}_p \circ (f+g))}{(\|f\|_p + \|g\|_p)^p} \,\mathrm{d}\mu \leq \int_X \Big(\lambda \frac{(\mathcal{P}_p \circ f)}{\|f\|_p^p} + (1 - \lambda) \frac{(\mathcal{P}_p \circ g)}{\|g\|_p^p}\Big) \,\mathrm{d}\mu = \frac{\lambda}{\|f\|_p^p} \int_X (\mathcal{P}_p \circ f) \,\mathrm{d}\mu + \frac{1-\lambda}{\|g\|_p^p} \int_X (\mathcal{P}_p \circ g) \,\mathrm{d}\mu = 1$. This implies that $\int_X (\mathcal{P}_p \circ (f + g)) \,\mathrm{d}\mu \leq (\|f\|_p + \|g\|_p)^p < +\infty$ and $\|f + g\|_p \leq \|f\|_p + \|g\|_p$. Therefore, $f + g \in \bar{\mathrm{L}}_p(\mathcal{X}, \mathcal{Y})$.

Fix any $p \in (1, +\infty) \subset \mathbb{R}$. By Proposition 11.97, equality holds if, and only if, $\mathcal{P} \circ (f + g) = \mathcal{P} \circ f + \mathcal{P} \circ g$ a.e. in $\mathcal{X}$ and $\frac{\mathcal{P} \circ f}{\|f\|_p} = \frac{\mathcal{P} \circ g}{\|g\|_p}$ a.e. in $\mathcal{X}$. Equality implies that $\alpha = 1/\|f\|_p$ and $\beta = 1/\|g\|_p$ such that $\alpha \mathcal{P} \circ f = \beta \mathcal{P} \circ g$ a.e. in $\mathcal{X}$ and $\mathcal{P} \circ (f + g) = \mathcal{P} \circ f + \mathcal{P} \circ g$ a.e. in $\mathcal{X}$. On the other hand, if $\exists \alpha, \beta \in [0, \infty) \subset \mathbb{R}$, which are not both zeros, such that $\alpha \mathcal{P} \circ f = \beta \mathcal{P} \circ g$ a.e. in $\mathcal{X}$ and $\mathcal{P} \circ (f + g) = \mathcal{P} \circ f + \mathcal{P} \circ g$ a.e. in $\mathcal{X}$. Without loss of generality, assume that $\beta \neq 0$. Then, $\mathcal{P} \circ g = \alpha_1 \mathcal{P} \circ f$ a.e. in $\mathcal{X}$ with $\alpha_1 := \alpha/\beta$. It is easy to show that $\|g\|_p = \alpha_1 \|f\|_p$. This implies that $\alpha_1 = \|g\|_p / \|f\|_p$. Then, equality holds. Hence, equality holds, if, and only if, $\exists \alpha, \beta \in [0, \infty) \subset \mathbb{R}$, which are not both zeros, such that $\alpha \mathcal{P} \circ f = \beta \mathcal{P} \circ g$ a.e. in $\mathcal{X}$ and $\mathcal{P} \circ (f + g) = \mathcal{P} \circ f + \mathcal{P} \circ g$ a.e. in $\mathcal{X}$.

Hence, the result holds in both cases. This completes the proof of the theorem. □

Definition 11.175 Let $\mathcal{X} := (X, \mathcal{B}, \mu)$ be a measure space and $f : X \to \mathbb{R}$ be $\mathcal{B}$-measurable. The *essential supremum* of f is

$$\operatorname*{ess\,sup}_{x \in \mathcal{X}} f(x) := \inf\{M \in \mathbb{R} \mid \mu(\{x \in X \mid f(x) > M\}) = 0\} \in \mathbb{R}_e$$

◇

Proposition 11.176 *Let $\mathcal{X} := (X, \mathcal{B}, \mu)$ be a measure space and $f : X \to \mathbb{R}$ and $g : X \to \mathbb{R}$ be $\mathcal{B}$-measurable. Then:*

(i) *$\text{ess sup}_{x\in\mathcal{X}}(f(x) + g(x)) \leq \text{ess sup}_{x\in\mathcal{X}} f(x) + \text{ess sup}_{x\in\mathcal{X}} g(x)$.*
(ii) *If $f \leq g$ a.e. in $\mathcal{X}$, then $\text{ess sup}_{x\in\mathcal{X}} f(x) \leq \text{ess sup}_{x\in\mathcal{X}} g(x)$.*
(iii) *$\forall \alpha \in (0, \infty) \subset \mathbb{R}$, $\text{ess sup}_{x\in\mathcal{X}}(\alpha f(x)) = \alpha\, \text{ess sup}_{x\in\mathcal{X}} f(x)$; and $\forall \alpha \in [0, \infty) \subset \mathbb{R}$, $\text{ess sup}_{x\in\mathcal{X}}(\alpha f(x)) = \alpha\, \text{ess sup}_{x\in\mathcal{X}} f(x)$ if $\text{ess sup}_{x\in\mathcal{X}} f(x) \in \mathbb{R}$.*
(iv) *$\lambda := \text{ess sup}_{x\in\mathcal{X}} f(x) \in \mathbb{R}_e$. Then, $f(x) \leq \lambda$ a.e. $x \in \mathcal{X}$.*

Proof (i) We will distinguish two exhaustive and mutually exclusive cases: Case 1: $\mu(X) = 0$; Case 2: $\mu(X) > 0$.

Case 1: $\mu(X) = 0$. Then, $\forall M \in \mathbb{R}$, $0 \leq \mu(\{x \in X \mid f(x) > M\}) \leq \mu(X) = 0$. Then, $\text{ess sup}_{x\in\mathcal{X}} f(x) = -\infty$. Similarly, $\text{ess sup}_{x\in\mathcal{X}} g(x) = \text{ess sup}_{x\in\mathcal{X}}(f(x) + g(x)) = -\infty$. Then, the (i) holds.

Case 2: $\mu(X) > 0$. Note that $\bigcup_{n=1}^{\infty} A_n := \bigcup_{n=1}^{\infty} \{x \in X \mid f(x) > -n\} = X$. By Proposition 11.7, $\exists n \in \mathbb{N}$ such that $\mu(A_n) > 0$. Then, $\text{ess sup}_{x\in\mathcal{X}} f(x) \geq -n > -\infty$. Similarly, $\text{ess sup}_{x\in\mathcal{X}} g(x) > -\infty$ and

$$\operatorname*{ess\,sup}_{x \in \mathcal{X}} (f(x) + g(x)) > -\infty$$

Then, $\lambda := \text{ess sup}_{x\in\mathcal{X}} f(x) + \text{ess sup}_{x\in\mathcal{X}} g(x) \in (-\infty, +\infty] \subset \mathbb{R}_e$. If $\lambda = +\infty$, then (i) holds. On the other hand, if $\lambda \in \mathbb{R}$, $\forall M \in \mathbb{R}$ with $\lambda < M$, $\exists M_1, M_2 \in \mathbb{R}$ with $\text{ess sup}_{x\in\mathcal{X}} f(x) < M_1$ and $\text{ess sup}_{x\in\mathcal{X}} g(x) < M_2$ and $M_1 + M_2 = M$. Then, $\mu(\{x \in X \mid f(x) > M_1\}) = 0 = \mu(\{x \in X \mid g(x) > M_2\})$. By Propositions 7.23, 11.38, and 11.39, $f + g$ is $\mathcal{B}$-measurable. Hence, $\mu(\{x \in X \mid f(x)+g(x) > M\}) = 0$ and $\text{ess sup}_{x\in\mathcal{X}}(f(x)+g(x)) \leq M$. By the arbitrariness of M, (i) holds.

(ii) Let $\lambda := \text{ess sup}_{x\in\mathcal{X}} g(x) \in \mathbb{R}_e$. We will distinguish two exhaustive and mutually exclusive cases: Case 1: $\lambda = +\infty$; Case 2: $\lambda < +\infty$. Case 1: $\lambda = +\infty$. The result holds. Case 2: $\lambda < +\infty$. $\forall M \in \mathbb{R}$ with $\lambda < M$, $\mu(\{x \in X \mid g(x) > M\}) = 0$. Then, $\{x \in X \mid f(x) > M\} \subseteq \{x \in X \mid g(x) > M\} \cup \{x \in X \mid f(x) - g(x) > 0\}$. By Propositions 7.23, 11.38, and 11.39, $f - g$ is $\mathcal{B}$-measurable. Then, $0 \leq \mu(\{x \in X \mid f(x) > M\}) \leq \mu(\{x \in X \mid g(x) > M\}) + \mu(\{x \in X \mid f(x) - g(x) > 0\}) = 0$. Hence, $\text{ess sup}_{x\in\mathcal{X}} f(x) \leq M$. By the arbitrariness of M, the result holds.

(iii) Let $\alpha \in (0, +\infty) \subset \mathbb{R}$. Then, $\text{ess sup}_{x\in\mathcal{X}}(\alpha f(x)) = \inf\{M \in \mathbb{R} \mid \mu(\{x \in X \mid \alpha f(x) > M\}) = 0\} = \inf\{\alpha M \in \mathbb{R} \mid \mu(\{x \in X \mid \alpha f(x) > \alpha M\}) = 0\} = \alpha \inf\{M \in \mathbb{R} \mid \mu(\{x \in X \mid f(x) > M\}) = 0\} = \alpha\, \text{ess sup}_{x\in\mathcal{X}} f(x)$, where the third equality follows from Proposition 3.81.

Let $\alpha = 0$ and $\operatorname{ess\,sup}_{x \in \mathcal{X}} f(x) \in \mathbb{R}$. Then, $\mu(X) > 0$. Thus,

$$\operatorname*{ess\,sup}_{x \in \mathcal{X}} (\alpha f(x)) = \operatorname*{ess\,sup}_{x \in \mathcal{X}} 0 = 0 = \alpha \operatorname*{ess\,sup}_{x \in \mathcal{X}} f(x)$$

(iv) We will distinguish three exhaustive and mutually exclusive cases: Case 1: $\mu(X) = 0$; Case 2: $\mu(X) > 0$ and $\lambda < +\infty$; Case 3: $\mu(X) > 0$ and $\lambda = +\infty$. Case 1: $\mu(X) = 0$. Then, $\lambda = -\infty$. Clearly, the result holds. Case 2: $\mu(X) > 0$ and $\lambda < +\infty$. Then, $\lambda \in \mathbb{R}$. $\forall n \in \mathbb{N}$, $\mu(E_n) := \mu(\{x \in X \mid f(x) > \lambda + 1/n\}) = 0$. By Proposition 11.7, $\mu(E) := \mu(\{x \in X \mid f(x) > \lambda\}) = \mu(\bigcup_{n=1}^{\infty} E_n) = \lim_{n \in \mathbb{N}} \mu(E_n) = 0$. Hence, $f(x) \leq \lambda$ a.e. $x \in \mathcal{X}$. Case 3: $\mu(X) > 0$ and $\lambda = +\infty$. Clearly, the result holds. Hence, (iv) holds in all three cases.

This completes the proof of the proposition. □

Example 11.177 Let $\mathcal{X} := (X, \mathcal{B}, \mu)$ be a measure space, $\mathcal{Y}$ be a separable normed linear space over $\mathbb{K}$, and $(\mathcal{M}(\mathcal{X}, \mathcal{Y}), \mathbb{K})$ be the vector space of functions of $\mathcal{X}$ to $\mathcal{Y}$ as defined in Example 6.20 with the usual vector addition, scalar multiplication, and the null vector ϑ. Let $Z_\infty := \{f \in \mathcal{M}(\mathcal{X}, \mathcal{Y}) \mid f$ is $\mathcal{B}$-measurable and $\operatorname{ess\,sup}_{x \in \mathcal{X}} \|f(x)\| < +\infty\}$. Define $\|\cdot\|_\infty : Z_\infty \to [0, \infty) \subset \mathbb{R}$ by $\|f\|_\infty = \max\{\operatorname{ess\,sup}_{x \in \mathcal{X}} \|f(x)\|, 0\}$, $\forall f \in Z_\infty$. We will next show that Z_∞ is a subspace of $(\mathcal{M}(\mathcal{X}, \mathcal{Y}), \mathbb{K})$ and $\|\cdot\|_\infty$ defines a pseudo-norm on Z_∞. Clearly, $\vartheta \in Z_\infty \neq \emptyset$. $\forall f, g \in Z_\infty$, $\forall \alpha \in \mathbb{K}$, by Propositions 7.23, 11.38, and 11.39, $f + g$ and αf are $\mathcal{B}$-measurable. By Proposition 11.176,

$$\begin{aligned}
\|f + g\|_\infty &= \max\{\operatorname*{ess\,sup}_{x \in \mathcal{X}} \|f(x) + g(x)\|, 0\} \\
&\leq \max\{\operatorname*{ess\,sup}_{x \in \mathcal{X}} (\|f(x)\| + \|g(x)\|), 0\} \\
&\leq \max\{\operatorname*{ess\,sup}_{x \in \mathcal{X}} \|f(x)\| + \operatorname*{ess\,sup}_{x \in \mathcal{X}} \|g(x)\|, 0\} \\
&= \begin{cases} \operatorname{ess\,sup}_{x \in \mathcal{X}} \|f(x)\| + \operatorname{ess\,sup}_{x \in \mathcal{X}} \|g(x)\| & \text{if } \mu(X) > 0 \\ 0 & \text{if } \mu(X) = 0 \end{cases} \\
&\leq \|f\|_\infty + \|g\|_\infty < +\infty
\end{aligned}$$

and

$$\begin{aligned}
\|\alpha f\|_\infty &= \max\{\operatorname*{ess\,sup}_{x \in \mathcal{X}} \|\alpha f(x)\|, 0\} = \max\{\operatorname*{ess\,sup}_{x \in \mathcal{X}} |\alpha| \|f(x)\|, 0\} \\
&= \begin{cases} |\alpha| \operatorname{ess\,sup}_{x \in \mathcal{X}} \|f(x)\| & \text{if } \mu(X) > 0 \\ 0 & \text{if } \mu(X) = 0 \end{cases} = |\alpha| \|f\|_\infty < +\infty
\end{aligned}$$

Then, $f + g, \alpha f \in Z_\infty$. Hence, Z_∞ is a subspace of $(\mathcal{M}(\mathcal{X}, \mathcal{Y}), \mathbb{K})$. This implies that $(Z_\infty, \mathbb{K})$ is a vector space. Clearly, $\|\vartheta\|_\infty = 0$. Therefore, $\|\cdot\|_\infty$ defines a pseudo-norm on $(Z_\infty, \mathbb{K})$. By Proposition 7.47, the quotient space of $(Z_\infty, \mathbb{K})$

modulo $\|\cdot\|_\infty$ is a normed linear space, to be denoted $\mathrm{L}_\infty(\mathcal{X}, \mathcal{Y})$. We will denote the vector space $(Z_\infty, \mathbb{K})$ with the pseudo-norm $\|\cdot\|_\infty$ by $\bar{\mathrm{L}}_\infty(\mathcal{X}, \mathcal{Y})$.

$\forall f \in Z_\infty$ with $\|f\|_\infty = 0$, we have $\operatorname{ess\,sup}_{x\in\mathcal{X}} \|f(x)\| \leq 0$. Then, by Proposition 11.176, $\mathcal{P} \circ f = 0$ a.e. in $\mathcal{X}$. Hence, $f = \vartheta_\mathcal{Y}$ a.e. in $\mathcal{X}$. On the other hand, $\forall f \in \mathcal{M}(\mathcal{X}, \mathcal{Y})$ with $f = \vartheta_\mathcal{Y}$ a.e. in $\mathcal{X}$ and f being $\mathcal{B}$-measurable, we have $\|f\|_\infty = 0$ and $f \in Z_\infty$. Hence, $\forall f \in \mathcal{M}(\mathcal{X}, \mathcal{Y})$, $f \in Z_\infty$ and $\|f\|_\infty = 0$ if, and only if, $f = \vartheta_\mathcal{Y}$ a.e. in $\mathcal{X}$ and f is $\mathcal{B}$-measurable. We will denote the norm in $\mathrm{L}_\infty(\mathcal{X}, \mathcal{Y})$ by $\|\cdot\|_\infty$ and elements in $\mathrm{L}_\infty(\mathcal{X}, \mathcal{Y})$ by $[f]$, where $f \in \bar{\mathrm{L}}_\infty(\mathcal{X}, \mathcal{Y})$. ◇

In the following, we will write $\lim_{n\in\mathbb{N}} z_n \doteq z$ in $\bar{\mathrm{L}}_p(\mathcal{X}, \mathcal{Y})$, when the sequence $(z_n)_{n=1}^\infty \subseteq \bar{\mathrm{L}}_p(\mathcal{X}, \mathcal{Y})$ converges to $z \in \bar{\mathrm{L}}_p(\mathcal{X}, \mathcal{Y})$ in $\bar{\mathrm{L}}_p(\mathcal{X}, \mathcal{Y})$ pseudo-norm. We will simply write $\lim_{n\in\mathbb{N}} z_n \doteq z$ if there is no confusion in which pseudo-norm convergence occurs. For $z \in \bar{\mathrm{L}}_p(\mathcal{X}, \mathcal{Y})$, we will denote the corresponding equivalence class in $\mathrm{L}_p(\mathcal{X}, \mathcal{Y})$ by $[z]$. Then, for $(z_n)_{n=1}^\infty \subseteq \bar{\mathrm{L}}_p(\mathcal{X}, \mathcal{Y})$, $\lim_{n\in\mathbb{N}} z_n \doteq z$ in $\bar{\mathrm{L}}_p(\mathcal{X}, \mathcal{Y})$ if, and only if, $\lim_{n\in\mathbb{N}} [z_n] = [z]$ in $\mathrm{L}_p(\mathcal{X}, \mathcal{Y})$ (or simply $\lim_{n\in\mathbb{N}} [z_n] = [z]$ when there is no confusion in which norm convergence occurs).

Theorem 11.178 (Hölder's Inequality) *Let $p \in [1, +\infty) \subset \mathbb{R}$ and $q \in (1, +\infty] \subset \mathbb{R}_e$ with $1/p + 1/q = 1$, $\mathcal{X} := (X, \mathcal{B}, \mu)$ be a σ-finite measure space, and $\mathcal{Y}$ be a separable normed linear space over $\mathbb{K}$ with $\mathcal{Y}^*$ being separable. Then, $\forall f \in \bar{\mathrm{L}}_p(\mathcal{X}, \mathcal{Y})$, $\forall g = \bar{\mathrm{L}}_q(\mathcal{X}, \mathcal{Y}^*)$, the function $r : X \to \mathbb{K}$, defined by $r(x) = \langle\langle g(x), f(x)\rangle\rangle$, $\forall x \in X$, is absolutely integrable over $\mathcal{X}$ and*

$$\int_X |\langle\langle g(x), f(x)\rangle\rangle| \, d\mu(x) \leq \|f\|_p \|g\|_q$$

When $q < \infty$, equality holds if, and only if, $|\langle\langle g(x), f(x)\rangle\rangle| = \|f(x)\| \cdot \|g(x)\|$ a.e. $x \in \mathcal{X}$ and $\exists \alpha, \beta \in \mathbb{R}$, which are not both zeros, such that $\alpha \mathcal{P}_p \circ f = \beta \mathcal{P}_q \circ g$ a.e. in $\mathcal{X}$.

Proof By Propositions 7.65, 11.38, 11.39, and 7.21, r and $\mathcal{P} \circ r$ are $\mathcal{B}$-measurable. We will distinguish two exhaustive and mutually exclusive cases: Case 1: $q = \infty$; Case 2: $1 < q < +\infty$. Case 1: $q = \infty$. Then, $p = 1$. By Proposition 11.176, $\|g(x)\| \leq \|g\|_\infty$ a.e. $x \in \mathcal{X}$. Note that $|\langle\langle g(x), f(x)\rangle\rangle| \leq \|g(x)\|\|f(x)\| \leq \|g\|_\infty \|f(x)\|$ a.e. $x \in \mathcal{X}$. Then, by Propositions 11.83 and 11.92,

$$\int_X |\langle\langle g(x), f(x)\rangle\rangle| \, d\mu(x) \leq \|g\|_\infty \int_X \mathcal{P} \circ f \, d\mu = \|f\|_1 \|g\|_\infty$$

Case 2: $1 < q < +\infty$. Then, $1 < p < +\infty$. We will further distinguish two exhaustive and mutually exclusive cases: Case 2a: $\|f\|_p \|g\|_q = 0$; Case 2b: $\|f\|_p \|g\|_q > 0$. Case 2a: $\|f\|_p \|g\|_q = 0$. Without loss of generality, assume $\|g\|_q = 0$. Then, $g = \vartheta_{\mathcal{Y}^*}$ a.e. in $\mathcal{X}$. This implies that $|\langle\langle g(x), f(x)\rangle\rangle| = 0$ a.e. $x \in$

$\mathcal{X}$ and, by Propositions 11.83 and 11.75,

$$\int_X |\langle\langle g(x), f(x)\rangle\rangle| \,d\mu(x) = 0 = \|f\|_p \|g\|_q$$

Equality holds $\Rightarrow \alpha = 0$, $\beta = 1$, $\alpha\mathcal{P}_p \circ f = 0 = \beta\mathcal{P}_q \circ g$ a.e. in $\mathcal{X}$, and $|\langle\langle g(x), f(x)\rangle\rangle| = 0 = \|f(x)\|\|g(x)\|$ a.e. $x \in \mathcal{X}$. This subcase is proved. Case 2b: $\|f\|_p\|g\|_q > 0$. Then, $\|f\|_p > 0$ and $\|g\|_q > 0$. $\forall x \in X$, by Lemma 7.7 with $a = \left(\frac{\|f(x)\|}{\|f\|_p}\right)^p$, $b = \left(\frac{\|g(x)\|}{\|g\|_q}\right)^q$, and $\lambda = 1/p$, we have

$$\frac{|\langle\langle g(x), f(x)\rangle\rangle|}{\|f\|_p \|g\|_q} \le \frac{\|g(x)\| \, \|f(x)\|}{\|g\|_q \, \|f\|_p} \le \frac{1}{p}\frac{\|f(x)\|^p}{\|f\|_p^p} + \frac{1}{q}\frac{\|g(x)\|^q}{\|g\|_q^q}$$

with equality holding if, and only if, $|\langle\langle g(x), f(x)\rangle\rangle| = \|g(x)\|\|f(x)\|$ and $\frac{\|f(x)\|^p}{\|f\|_p^p} = \frac{\|g(x)\|^q}{\|g\|_q^q}$. Integrating the above inequality over $\mathcal{X}$, we have, by Propositions 11.83 and 11.97,

$$\frac{\int_X |\langle\langle g(x), f(x)\rangle\rangle| \,d\mu(x)}{\|f\|_p \, \|g\|_q} \le \frac{1}{p} + \frac{1}{q} = 1$$

with equality holding if, and only if, $|\langle\langle g(x), f(x)\rangle\rangle| = \|g(x)\|\|f(x)\|$ a.e. $x \in \mathcal{X}$ and $\frac{\mathcal{P}_p \circ f}{\|f\|_p^p} = \frac{\mathcal{P}_q \circ g}{\|g\|_q^q}$ a.e. in $\mathcal{X}$. Equality $\Rightarrow \alpha = 1/\|f\|_p^p$, $\beta = 1/\|g\|_q^q$, $\alpha\mathcal{P}_p \circ f = \beta\mathcal{P}_q \circ g$ a.e. in $\mathcal{X}$ and $|\langle\langle g(x), f(x)\rangle\rangle| = \|g(x)\|\|f(x)\|$ a.e. $x \in \mathcal{X}$. On the other hand, if $|\langle\langle g(x), f(x)\rangle\rangle| = \|g(x)\|\|f(x)\|$ a.e. $x \in \mathcal{X}$ and $\exists \alpha, \beta \in \mathbb{R}$, which are not both zeros, such that $\alpha\mathcal{P}_p \circ f = \beta\mathcal{P}_q \circ g$ a.e. in $\mathcal{X}$, then, without loss of generality, assume $\beta \neq 0$. Let $\alpha_1 = \alpha/\beta$. Then, $\alpha_1\mathcal{P}_p \circ f = \mathcal{P}_q \circ g$ a.e. in $\mathcal{X}$. Hence, $\alpha_1\|f\|_p^p = \|g\|_q^q$, which further implies that $\alpha_1 = \|g\|_q^q/\|f\|_p^p$. Hence, $\frac{\mathcal{P}_p \circ f}{\|f\|_p^p} = \frac{\mathcal{P}_q \circ g}{\|g\|_q^q}$ a.e. in $\mathcal{X}$. This implies equality. Therefore, equality holds if, and only if, $|\langle\langle g(x), f(x)\rangle\rangle| = \|g(x)\|\|f(x)\|$ a.e. $x \in \mathcal{X}$ and $\exists \alpha, \beta \in \mathbb{R}$, which are not both zeros, such that $\alpha\mathcal{P}_p \circ f = \beta\mathcal{P}_q \circ g$ a.e. in $\mathcal{X}$. This subcase is proved.

This completes the proof of the theorem. □

When $p = 2 = q$, the Hölder's inequality becomes the well-known *Cauchy–Schwarz inequality*:

$$\int_X |\langle\langle g(x), f(x)\rangle\rangle| \,d\mu(x) \le \Big(\int_X \mathcal{P}_2 \circ f \,d\mu\Big)^{1/2} \Big(\int_X \mathcal{P}_2 \circ g \,d\mu\Big)^{1/2}$$

Example 11.179 Let $p \in [1, \infty) \subset \mathbb{R}$, $\mathcal{X} := (X, \mathcal{B}, \mu)$ be a σ-finite measure space, $\mathcal{Y}$ be a separable Banach space over $\mathbb{K}$, and $\mathrm{L}_p(\mathcal{X}, \mathcal{Y})$ be the normed linear space over $\mathbb{K}$ as defined in Example 11.173. We will show that $\mathrm{L}_p(\mathcal{X}, \mathcal{Y})$ is a Banach space by Proposition 7.27. Define $l : [0, \infty) \subset \mathbb{R} \to [0, \infty) \subset \mathbb{R}$ by $l(t) = t^p$, $\forall t \in [0, \infty) \subset \mathbb{R}$. Fix any $([f_n])_{n=1}^{\infty} \subseteq \mathrm{L}_p(\mathcal{X}, \mathcal{Y})$ with $f_n \in \bar{\mathrm{L}}_p(\mathcal{X}, \mathcal{Y})$, $\forall n \in \mathbb{N}$, and $\sum_{n=1}^{\infty} \|[f_n]\|_p =: M \in [0, \infty) \subset \mathbb{R}$. $\forall n \in \mathbb{N}$, define $g_n : X \to [0, \infty) \subset \mathbb{R}$ by $g_n(x) = \sum_{i=1}^{n} \|f_i(x)\|$, $\forall x \in X$. By Propositions 7.23, 7.21, 11.38, and 11.39, g_n is $\mathcal{B}$-measurable. Note that $\mathcal{P} \circ f_n \in \bar{\mathrm{L}}_p(\mathcal{X}, \mathbb{R})$. Then, $g_n \in \bar{\mathrm{L}}_p(\mathcal{X}, \mathbb{R})$ and $\|g_n\|_p \leq \sum_{i=1}^{n} \|\mathcal{P} \circ f_i\|_p = \sum_{i=1}^{n} \|f_i\|_p = \sum_{i=1}^{n} \|[f_i]\|_p \leq M$. This implies that $\int_X (l \circ g_n)\, d\mu \leq M^p$. Clearly, $(g_n(x))^p \leq (g_{n+1}(x))^p$, $\forall x \in X$, $\forall n \in \mathbb{N}$. By Proposition 11.82 and Monotone Convergence Theorem 11.81, $\exists g : X \to [0, \infty) \subset \mathbb{R}$, which is $\mathcal{B}$-measurable, such that $\lim_{n\in\mathbb{N}} l \circ g_n = l \circ g$ a.e. in $\mathcal{X}$ and $\int_X (l \circ g)\, d\mu = \lim_{n\in\mathbb{N}} \int_X (l \circ g_n)\, d\mu \leq M^p$.

Let $E := \{x \in X \mid (g_n(x))_{n=1}^{\infty}$ does not converge to $g(x)\}$. Then, $E \in \mathcal{B}$ and $\mu(E) = 0$. $\forall x \in X \setminus E$, $\lim_{n\in\mathbb{N}} g_n(x) = g(x) \in \mathbb{R}$ and $\lim_{n\in\mathbb{N}} \sum_{i=1}^{n} \|f_i(x)\| = g(x) \in \mathbb{R}$. By Proposition 7.27 and the completeness of $\mathcal{Y}$, we have $\lim_{n\in\mathbb{N}} \sum_{i=1}^{n} f_i(x) =: \lim_{n\in\mathbb{N}} s_n(x) =: s(x) \in \mathcal{Y}$. Define $s(x) = s_n(x) = \vartheta_{\mathcal{Y}}$, $\forall x \in E$ and $\forall n \in \mathbb{N}$. Then, by Proposition 11.41, s_n is $\mathcal{B}$-measurable, $\forall n \in \mathbb{N}$. Clearly, $\lim_{n\in\mathbb{N}} s_n(x) = s(x)$, $\forall x \in X$. By Proposition 11.48, s is $\mathcal{B}$-measurable. By Lemma 11.43, $s_n \doteq \sum_{i=1}^{n} f_i$ a.e. in $\mathcal{X}$. This yields that $[s_n] = \left[\sum_{i=1}^{n} f_i\right] = \sum_{i=1}^{n} [f_i]$ and $s_n \in \bar{\mathrm{L}}_p(\mathcal{X}, \mathcal{Y})$. Then, by Proposition 11.50, $\lim_{n\in\mathbb{N}} \sum_{i=1}^{n} f_i = s$ a.e. in $\mathcal{X}$. Note that $\lim_{n\in\mathbb{N}} \mathcal{P}_p \circ (s_n(x) - s(x)) = 0$, $\forall x \in X$. Note also $\|s_n(x) - s(x)\|^p \leq (\|s_n(x)\| + \|s(x)\|)^p \leq (g_n(x) + g(x))^p \leq 2^p (g(x))^p$, $\forall x \in X \setminus E$, $\forall n \in \mathbb{N}$, and $\|s_n(x) - s(x)\|^p = 0 \leq 2^p (g(x))^p$, $\forall x \in E$, $\forall n \in \mathbb{N}$. Hence, $\|s_n(x) - s(x)\|^p \leq 2^p (g(x))^p$, $\forall x \in X$, $\forall n \in \mathbb{N}$. By Lebesgue Dominated Convergence Theorem 11.91, $\lim_{n\in\mathbb{N}} \int_X (\mathcal{P}_p \circ (s_n - s))\, d\mu = 0$. This implies that $\lim_{n\in\mathbb{N}} \|s_n - s\|_p = 0$. Hence, $s \in \bar{\mathrm{L}}_p(\mathcal{X}, \mathcal{Y})$ and $\lim_{n\in\mathbb{N}} \sum_{i=1}^{n} f_i \doteq \lim_{n\in\mathbb{N}} s_n \doteq s$. Hence, $([f_n])_{n=1}^{\infty}$ is summable in $\mathrm{L}_p(\mathcal{X}, \mathcal{Y})$. By Proposition 7.27 and the arbitrariness of $([f_n])_{n=1}^{\infty}$, we have $\mathrm{L}_p(\mathcal{X}, \mathcal{Y})$ is complete. This shows that $\mathrm{L}_p(\mathcal{X}, \mathcal{Y})$ is a Banach space when $\mathcal{X}$ is a σ-finite measure space and $\mathcal{Y}$ is a separable Banach space. ◇

Example 11.180 Let $\mathcal{X} := (X, \mathcal{B}, \mu)$ be a measure space, $\mathcal{Y}$ be a separable Banach space over $\mathbb{K}$, and $\mathrm{L}_\infty(\mathcal{X}, \mathcal{Y})$ be the normed linear space over $\mathbb{K}$ as defined in Example 11.177. We will show that $\mathrm{L}_\infty(\mathcal{X}, \mathcal{Y})$ is a Banach space. Fix any Cauchy sequence $([f_n])_{n=1}^{\infty} \subseteq \mathrm{L}_\infty(\mathcal{X}, \mathcal{Y})$ with $f_n \in \bar{\mathrm{L}}_\infty(\mathcal{X}, \mathcal{Y})$, $\forall n \in \mathbb{N}$. $\forall k \in \mathbb{N}$, $\exists N_k \in \mathbb{N}$, $\forall n, m \in \mathbb{N}$ with $n \geq N_k$ and $m \geq N_k$, we have $\|f_n - f_m\|_\infty < 1/k$. Then, by Proposition 11.176, $A_{n,m,k} := \{x \in X \mid \|f_n(x) - f_m(x)\| \geq 1/k\} \in \mathcal{B}$ and $\mu(A_{n,m,k}) = 0$. Let $A := \bigcup_{k=1}^{\infty} \bigcup_{n=N_k}^{\infty} \bigcup_{m=N_k}^{\infty} A_{n,m,k} \in \mathcal{B}$. Clearly, $\mu(A) = 0$. $\forall x \in X \setminus A$, $(f_n(x))_{n=1}^{\infty} \subseteq \mathcal{Y}$ is a Cauchy sequence, which converges to $f(x) \in \mathcal{Y}$ by the completeness of $\mathcal{Y}$. Define $f(x) = \vartheta_{\mathcal{Y}}$, $\forall x \in A$. Then, $f : X \to \mathcal{Y}$ is well-defined, and $\lim_{n\in\mathbb{N}} f_n(x) = f(x)$, $\forall x \in X \setminus A$. By Propositions 11.48 and 11.41, f is $\mathcal{B}$-measurable. $\forall k \in \mathbb{N}$, $\forall n \in \mathbb{N}$ with $n \geq N_k$, $\forall x \in X \setminus A$, by Propositions 3.66, 3.67, 7.21, and 7.23, we have $\|f_n(x) - f(x)\| = \lim_{m\in\mathbb{N}} \|f_n(x) - f_m(x)\| \leq 1/k$. Then, $0 \leq \mu(\{x \in X \mid \|f(x) - f_n(x)\| > 1/k\}) \leq \mu(A) = 0$. This shows that $\|f_n - f\|_\infty \leq 1/k$. Then, $\lim_{n\in\mathbb{N}} \|f_n - f\|_\infty = 0$, $\lim_{n\in\mathbb{N}} f_n \doteq f$, and $f \in \bar{\mathrm{L}}_\infty(\mathcal{X}, \mathcal{Y})$. Hence,

$\lim_{n\in\mathbb{N}}[f_n] = [f]$ in $L_\infty(\mathcal{X}, \mathcal{Y})$. Therefore, $L_\infty(\mathcal{X}, \mathcal{Y})$ is a Banach space when $\mathcal{X}$ is a measure space and $\mathcal{Y}$ is a separable Banach space. ◇

Proposition 11.181 *Let $p \in [1, \infty) \subset \mathbb{R}$, $\mathcal{X} := (X, \mathcal{B}, \mu)$ be a σ-finite measure space, $\mathcal{Y}$ be a separable normed linear space over $\mathbb{K}$, $L_p(\mathcal{X}, \mathcal{Y})$ be the normed linear space over $\mathbb{K}$ as defined in Example 11.173, and $f \in \bar{L}_p(\mathcal{X}, \mathcal{Y})$. Then, there exists a sequence of simple functions $(\phi_i)_{i=1}^\infty$, $\phi_i : \mathcal{X} \to \mathcal{Y}$, $\forall i \in \mathbb{N}$, such that $\lim_{i\in\mathbb{N}} \phi_i = f$ a.e. in $\mathcal{X}$, $\|\phi_i(x)\| \leq \|f(x)\|$, $\forall x \in X$, $\forall i \in \mathbb{N}$, $\lim_{i\in\mathbb{N}} \int_X \mathcal{P}_p \circ (\phi_i - f)\, d\mu = 0$, and $\lim_{i\in\mathbb{N}} \phi_i \doteq f$ in $\bar{L}_p(\mathcal{X}, \mathcal{Y})$.*

Proof Since $f \in \bar{L}_p(\mathcal{X}, \mathcal{Y})$, then f is $\mathcal{B}$-measurable and $\mathcal{P}_p \circ f$ is integrable over $\mathcal{X}$. By Proposition 11.66, there exists a sequence of simple functions $(\phi_i)_{i=1}^\infty$, $\phi_i : \mathcal{X} \to \mathcal{Y}$, $\forall i \in \mathbb{N}$, such that $\|\phi_i(x)\| \leq \|f(x)\|$, $\forall x \in X$, $\forall i \in \mathbb{N}$, and $\lim_{i\in\mathbb{N}} \phi_i = f$ a.e. in $\mathcal{X}$. By Propositions 7.23, 7.21, 11.38, and 11.39, $\mathcal{P}_p \circ (\phi_i - f)$ is $\mathcal{B}$-measurable, $\forall i \in \mathbb{N}$. Note that, by Propositions 7.23, 7.21, 11.52, and 11.53, $\lim_{i\in\mathbb{N}} \mathcal{P}_p \circ (\phi_i - f) = 0$ a.e. in $\mathcal{X}$ and $\mathcal{P}_p \circ (\phi_i - f)(x) \leq 2^p \mathcal{P}_p \circ f(x)$, $\forall x \in X$, $\forall i \in \mathbb{N}$. By Lebesgue Dominated Convergence Theorem 11.91, we have $\lim_{i\in\mathbb{N}} \int_X \mathcal{P}_p \circ (\phi_i - f)\, d\mu = 0$. Hence, $\lim_{i\in\mathbb{N}} \phi_i \doteq f$. This completes the proof of the proposition. □

Proposition 11.182 *Let $p \in [1, \infty) \subset \mathbb{R}$, $\mathbb{X} := (\mathcal{X}, \mathcal{B}, \mu)$ be a σ-finite normal topological measure space, $\mathcal{Y}$ be a separable normed linear space over $\mathbb{K}$, and $f \in \bar{L}_p(\mathbb{X}, \mathcal{Y})$. Then, $\forall \epsilon \in (0, \infty) \subset \mathbb{R}$, $\exists$ a continuous function $g : \mathbb{X} \to \mathcal{Y}$ such that $g \in \bar{L}_p(\mathbb{X}, \mathcal{Y})$ and $\|g - f\|_p < \epsilon$.*

Proof Fix $\epsilon \in (0, \infty) \subset \mathbb{R}$. By Proposition 11.181, there exists a simple function $\phi : \mathbb{X} \to \mathcal{Y}$ such that $\|\phi(x)\| \leq \|f(x)\|$, $\forall x \in \mathbb{X}$, and $\|\phi - f\|_p < \epsilon/2$. Let ϕ admit the canonical representation $\phi = \sum_{i=1}^n y_i \chi_{A_i,\mathbb{X}}$, where $n \in \mathbb{Z}_+$, $y_1, \ldots, y_n \in \mathcal{Y}$ are distinct, and none equals to $\vartheta_\mathcal{Y}$, $A_1, \ldots, A_n \in \mathcal{B}$ are pairwise disjoint, nonempty, and of finite measure. $\forall i \in \{1, \ldots, n\}$, by $\mathbb{X}$ being a topological measure space and Proposition 11.27, $\exists U_i, \mathbb{X} \setminus F_i \in \mathcal{O}_\mathcal{X}$ such that $F_i \subseteq A_i \subseteq U_i$ and $\mu(U_i \setminus F_i) = \mu(U_i \setminus A_i) + \mu(A_i \setminus F_i) < \frac{\epsilon^p}{2^p (n+1)^p \|y_i\|^p}$. By $\mathcal{X}$ being a normal topological space and Urysohn's Lemma 3.55, there exists a continuous function $g_i : \mathbb{X} \to [0, 1] \subset \mathbb{R}$ such that $g_i(x) = 1$, $\forall x \in F_i$ and $g_i(x) = 0$, $\forall x \in \mathbb{X} \setminus U_i$. By Proposition 11.37, g_i is $\mathcal{B}$-measurable. By Proposition 11.83, $\left\|g_i - \chi_{A_i,\mathbb{X}}\right\|_p \leq \frac{\epsilon}{2(n+1)\|y_i\|}$.

Define $g : \mathbb{X} \to \mathcal{Y}$ by $g(x) = \sum_{i=1}^n y_i g_i(x)$, $\forall x \in \mathbb{X}$. By Propositions 7.23, 3.12, and 3.32, g is continuous. By Proposition 11.37, g is $\mathcal{B}$-measurable. Then, $\|g - f\|_p \leq \|g - \phi\|_p + \|\phi - f\|_p \leq \sum_{i=1}^n \left\|y_i(g_i - \chi_{A_i,\mathbb{X}})\right\|_p + \epsilon/2 = \sum_{i=1}^n \|y_i\| \left\|g_i - \chi_{A_i,\mathbb{X}}\right\|_p + \epsilon/2 \leq \sum_{i=1}^n \frac{\|y_i\|\epsilon}{2(n+1)\|y_i\|} + \epsilon/2 < \epsilon$. Hence, $g \in \bar{L}_p(\mathbb{X}, \mathcal{Y})$.

This completes the proof of the proposition. □

Proposition 11.183 *Let $\mathcal{X} := (X, \mathcal{B}, \mu)$ be a σ-finite measure space, $\mathcal{Y}$ be a separable Banach space, $f : X \to \mathcal{Y}$ be $\mathcal{B}$-measurable, $\forall E \in \mathcal{B}$ with $\mu(E) < \infty$, $f|_E$ be absolutely integrable over $\mathcal{E} := (E, \mathcal{B}_E, \mu_E)$, which is the finite measure* *subspace of $\mathcal{X}$, and $M \in [0, \infty) \subset \mathbb{R}$. Assume that, $\forall E \in \mathcal{B}$ with $0 < \mu(E) < \infty$, we have $\left\| \frac{1}{\mu(E)} \int_E f \, d\mu \right\| \leq M$. Then, $\mathcal{P} \circ f \leq M$ a.e. in $\mathcal{X}$.*

Proof Consider the open set $O := \{y \in \mathcal{Y} \mid \|y\| > M\} \subseteq \mathcal{Y}$. Since $\mathcal{Y}$ is separable, by Propositions 4.38 and 4.4, O is second countable and separable. Let $D \subseteq O$ be a countable dense set in O (the relative closure of D with respect to O equals to O). It is easy to show that $\mathcal{M} := \{\mathcal{B}_{\mathcal{Y}}(y, r) \subseteq O \mid y \in D,\ r \in \mathbb{Q},\ r > 0\}$ is a countable basis for O. Let $E := f_{\text{inv}}(O) = f_{\text{inv}}(\bigcup_{\mathcal{B}_{\mathcal{Y}}(y,r) \in \mathcal{M}} \mathcal{B}_{\mathcal{Y}}(y, r)) = \bigcup_{\mathcal{B}_{\mathcal{Y}}(y,r) \in \mathcal{M}} f_{\text{inv}}(\mathcal{B}_{\mathcal{Y}}(y, r))$.

We will show that $\mu(f_{\text{inv}}(\mathcal{B}_{\mathcal{Y}}(y, r))) = 0$, $\forall \mathcal{B}_{\mathcal{Y}}(y, r) \in \mathcal{M}$, by an argument of contradiction. Suppose that $\exists y \in D$, $\exists r \in \mathbb{Q}$ with $r > 0$ such that $\mathcal{B}_{\mathcal{Y}}(y, r) \subseteq O$, and $\mu(\bar{E}) := \mu(f_{\text{inv}}(\mathcal{B}_{\mathcal{Y}}(y, r))) > 0$. Since $\mathcal{X}$ is σ-finite, then $\exists \hat{E} \in \mathcal{B}$ with $\hat{E} \subseteq \bar{E}$ such that $0 < \mu(\hat{E}) < +\infty$. $\forall x \in \hat{E}$, we have $x \in \bar{E}$ and $\|f(x) - y\| < r$. Then, $\left\| \frac{1}{\mu(\hat{E})} \int_{\hat{E}} f \, d\mu - y \right\| = \left\| \frac{1}{\mu(\hat{E})} \int_{\hat{E}} (f - y) \, d\mu \right\| = \frac{1}{\mu(\hat{E})} \left\| \int_{\hat{E}} (f(x) - y) \, d\mu(x) \right\| \leq \frac{1}{\mu(\hat{E})} \int_{\hat{E}} \|f(x) - y\| \, d\mu(x) < r$, where the first equality follows from Propositions 11.92 and 11.75, the first inequality follows from Proposition 11.92, and the second inequality follows from Proposition 11.97. This implies that $\frac{1}{\mu(\hat{E})} \int_{\hat{E}} f \, d\mu \in \mathcal{B}_{\mathcal{Y}}(y, r) \subseteq O$ and $\left\| \frac{1}{\mu(\hat{E})} \int_{\hat{E}} f \, d\mu \right\| > M$. This contradicts the assumption. Hence, $\mu(f_{\text{inv}}(\mathcal{B}_{\mathcal{Y}}(y, r))) = 0$, $\forall \mathcal{B}_{\mathcal{Y}}(y, r) \in \mathcal{M}$.

Then, $0 \leq \mu(E) \leq \sum_{\mathcal{B}_{\mathcal{Y}}(y,r) \in \mathcal{M}} \mu(f_{\text{inv}}(\mathcal{B}_{\mathcal{Y}}(y, r))) = 0$. Hence, $\mathcal{P} \circ f \leq M$ a.e. in $\mathcal{X}$. □

Lemma 11.184 *Let $p, q \in (1, \infty) \subset \mathbb{R}$ with $1/p + 1/q = 1$, $\mathcal{X} := (X, \mathcal{B}, \mu)$ be a σ-finite measure space, $\mathcal{Y}$ be a separable normed linear space over $\mathbb{K}$ with $\mathcal{Y}^*$ being separable, $\mathcal{Z} := L_p(\mathcal{X}, \mathcal{Y})$ be the normed linear space over $\mathbb{K}$ as defined in Example 11.173, and $g : X \to \mathcal{Y}^*$ be $\mathcal{B}$-measurable. Assume that:*

(i) *$\forall E \in \mathcal{B}$ with $\mu(E) < +\infty$, g is absolutely integrable over $\mathcal{E} := (E, \mathcal{B}_E, \mu_E)$, which is the finite measure subspace of $\mathcal{X}$.*
(ii) *$\exists M \in [0, \infty) \subset \mathbb{R}$ such that $\forall$ simple function $\phi : \mathcal{X} \to \mathcal{Y}$ ($\phi \in \bar{L}_p(\mathcal{X}, \mathcal{Y})$), the function $\langle\langle g(\cdot), \phi(\cdot)\rangle\rangle : X \to \mathbb{K}$ is absolutely integrable over $\mathcal{X}$ and $\left| \int_X \langle\langle g(x), \phi(x)\rangle\rangle \, d\mu(x) \right| \leq M \|\phi\|_p$.*

Then, $g \in \bar{L}_q(\mathcal{X}, \mathcal{Y}^*)$ and $\|g\|_q \leq M$.

Proof By Proposition 11.116, define a σ-finite $\mathcal{Y}^*$-valued measure ν on $(X, \mathcal{B})$ by: $\nu(E)$ is undefined, $\forall E \in \mathcal{B}$ with $\int_E \mathcal{P} \circ g \, d\mu = \infty$; $\nu(E) = \int_E g \, d\mu \in \mathcal{Y}^*$, $\forall E \in \mathcal{B}$ with $\int_E \mathcal{P} \circ g \, d\mu < \infty$. Then, $\mathcal{P} \circ \nu(E) = \int_E \mathcal{P} \circ g \, d\mu$, $\forall E \in \mathcal{B}$. By (i), $\mathcal{P} \circ \nu(E) < \infty$, $\forall E \in \mathcal{B}$ with $\mu(E) < \infty$. By Proposition 11.92, $\nu(E) = \vartheta_{\mathcal{Y}^*}$, $\forall E \in \mathcal{B}$ with $\mu(E) = 0$. Fix any $E \in \mathcal{B}$ with $\mu(E) < \infty$. Define $\mathcal{P}_q \circ \nu(E) \in [0, \infty] \subset \mathbb{R}_e$ by

$$\mathcal{P}_q \circ \nu(E) := \sup_{\substack{n \in \mathbb{Z}_+,\ (E_i)_{i=1}^n \subseteq \mathcal{B},\ \bigcup_{i=1}^n E_i = E \\ E_i \cap E_j = \emptyset,\ \forall 1 \leq i < j \leq n}} \sum_{\substack{i=1 \\ \mu(E_i) > 0}}^{n} \frac{\|\nu(E_i)\|^q}{(\mu(E_i))^q} \mu(E_i)$$

$\forall n \in \mathbb{Z}_+$, $\forall$ pairwise disjoint $(E_i)_{i=1}^n \subseteq \mathcal{B}$ with $E = \bigcup_{i=1}^n E_i$, $\forall \epsilon \in (0,1) \subset \mathbb{R}$, $\forall i \in \{1,\ldots,n\}$ with $\mu(E_i) > 0$, by Lemma 7.75, $\exists y_i \in \mathcal{Y}$ such that $\|y_i\| = \|\nu(E_i)\|^{q-1}/(\mu(E_i))^{q-1}$ and $\langle\langle \nu(E_i), y_i\rangle\rangle \geq (1-\epsilon)\|\nu(E_i)\|\|y_i\| = (1-\epsilon)\|\nu(E_i)\|^q/(\mu(E_i))^{q-1}$. Define a simple function $\phi = \sum_{i=1\ \mu(E_i)>0}^{n} y_i \chi_{E_i,X}$. Then,

$$\|\phi\|_p = \Big(\sum_{\substack{i=1\\ \mu(E_i)>0}}^{n} \|\nu(E_i)\|^{(q-1)p} (\mu(E_i))^{(1-q)p} \mu(E_i) \Big)^{1/p}$$
$$= \Big(\sum_{\substack{i=1\\ \mu(E_i)>0}}^{n} \|\nu(E_i)\|^{q} (\mu(E_i))^{-q} \mu(E_i) \Big)^{1/p}$$

By (ii), we have

$$\sum_{\substack{i=1\\ \mu(E_i)>0}}^{n} (1-\epsilon)\|\nu(E_i)\|^q/(\mu(E_i))^{q-1} \leq \Big| \sum_{\substack{i=1\\ \mu(E_i)>0}}^{n} \langle\langle \nu(E_i), y_i\rangle\rangle \Big|$$
$$= \Big| \sum_{\substack{i=1\\ \mu(E_i)>0}}^{n} \Big\langle\Big\langle \int_{E_i} g \, \mathrm{d}\mu, y_i \Big\rangle\Big\rangle \Big| = \Big| \sum_{\substack{i=1\\ \mu(E_i)>0}}^{n} \int_{E_i} \langle\langle g(x), y_i\rangle\rangle \, \mathrm{d}\mu(x) \Big|$$
$$= \Big| \int_X \langle\langle g(x), \phi(x)\rangle\rangle \, \mathrm{d}\mu(x) \Big| \leq M \Big(\sum_{\substack{i=1\\ \mu(E_i)>0}}^{n} \|\nu(E_i)\|^q (\mu(E_i))^{1-q} \Big)^{1/p}$$

where the second and third equalities follow from Proposition 11.92. Hence, we have $\sum_{i=1\ \mu(E_i)>0}^{n} \|\nu(E_i)\|^q (\mu(E_i))^{-q} \mu(E_i) \leq M^q/(1-\epsilon)^q$. By the arbitrariness of ϵ, we have $\sum_{i=1\ \mu(E_i)>0}^{n} \|\nu(E_i)\|^q (\mu(E_i))^{-q} \mu(E_i) \leq M^q$. Then, $\mathcal{P}_q \circ \nu(E) \leq M^q$.

Since $\mathcal{X}$ is σ-finite, then $\exists (X_n)_{n=1}^\infty \subseteq \mathcal{B}$ such that $X = \bigcup_{n=1}^\infty X_n$ and $\mu(X_n) < \infty$, $\forall n \in \mathbb{N}$. Without loss of generality, we may assume that $X_n \subseteq X_{n+1}$, $\forall n \in \mathbb{N}$. By Proposition 11.66, there exists a sequence of simple functions $(\psi_i)_{i=1}^\infty$, $\psi_i : \mathcal{X} \to \mathcal{Y}^*$, $\forall i \in \mathbb{N}$, such that $\|\psi_i(x)\| \leq \|g(x)\|$, $\forall x \in X$, $\forall i \in \mathbb{N}$, and $\lim_{i\in\mathbb{N}} \psi_i = g$ a.e. in $\mathcal{X}$. Fix any $n \in \mathbb{N}$; let $E_n := \{x \in X_n \mid \|g(x)\| \leq n\} \in \mathcal{B}$ and $\mathcal{E}_n := (E_n, \mathcal{B}_{E_n}, \mu_{E_n})$ be the finite measure subspace of $\mathcal{X}$. Then, $(\mathcal{P} \circ g)|_{E_n}$ and $(\mathcal{P}_q \circ g)\big|_{E_n}$ are integrable over $\mathcal{E}_n$. By Propositions 7.21, 7.23, 11.52, and 11.53, $\lim_{i\in\mathbb{N}} (\mathcal{P} \circ (\psi_i - g))|_{E_n} = 0$ a.e. in $\mathcal{E}_n$. Note that $(\mathcal{P} \circ (\psi_i - g))|_{E_n}(x) \leq 2\,(\mathcal{P} \circ g)|_{E_n}(x)$, $\forall x \in E_n$, $\forall i \in \mathbb{N}$. By Lebesgue Dominated Convergence Theorem 11.91, we have $\lim_{i\in\mathbb{N}} \int_{E_n} \mathcal{P} \circ (\psi_i - g)\, \mathrm{d}\mu = 0$. Note also that $(\mathcal{P}_q \circ \psi_i)\big|_{E_n}(x) \leq (\mathcal{P}_q \circ g)\big|_{E_n}(x)$, $\forall x \in E_n$, $\forall i \in \mathbb{N}$. By Propositions 7.21 and 11.52, $\lim_{i\in\mathbb{N}} (\mathcal{P}_q \circ \psi_i)\big|_{E_n} = (\mathcal{P}_q \circ g)\big|_{E_n}$ a.e. in $\mathcal{E}_n$. Again, by Lebesgue

Dominated Convergence Theorem 11.91, we have $\lim_{i\in\mathbb{N}} \int_{E_n} \mathcal{P}_q \circ \psi_i \,\mathrm{d}\mu = \int_{E_n} \mathcal{P}_q \circ g \,\mathrm{d}\mu$. $\forall \epsilon \in (0,\infty) \subset \mathbb{R}$, $\exists i_0 \in \mathbb{N}$ such that $0 \le \int_{E_n} \mathcal{P} \circ (\psi_{i_0} - g)\,\mathrm{d}\mu < \frac{\epsilon}{2qn^{q-1}}$ and $\left|\int_{E_n} \mathcal{P}_q \circ \psi_{i_0}\,\mathrm{d}\mu - \int_{E_n} \mathcal{P}_q \circ g\,\mathrm{d}\mu\right| < \epsilon/2$. Let ψ_{i_0} admit the canonical representation $\psi_{i_0} = \sum_{j=1}^{\bar{n}} y_{*j}\chi_{A_j,X}$. Then,

$$
\begin{aligned}
\int_{E_n} \mathcal{P}_q \circ g \,\mathrm{d}\mu &< \int_{E_n} \mathcal{P}_q \circ \psi_{i_0}\,\mathrm{d}\mu + \epsilon/2 \\
&= \sum_{\substack{j=1\\ \mu(A_j\cap E_n)>0}}^{\bar{n}} \|\nu(A_j\cap E_n)\|^q (\mu(A_j\cap E_n))^{1-q} + \sum_{j=1}^{\bar{n}} \|y_{*j}\|^q \mu(A_j\cap E_n) \\
&\quad - \sum_{\substack{j=1\\ \mu(A_j\cap E_n)>0}}^{\bar{n}} \|\nu(A_j\cap E_n)\|^q (\mu(A_j\cap E_n))^{1-q} + \epsilon/2
\end{aligned}
$$

Let $\bar{A}_j = A_j \cap E_n$, $j = 1, \ldots, \bar{n}$. Then,

$$
\begin{aligned}
\int_{E_n} \mathcal{P}_q \circ g \,\mathrm{d}\mu &< \sum_{\substack{j=1\\ \mu(\bar{A}_j)>0}}^{\bar{n}} \|\nu(\bar{A}_j)\|^q (\mu(\bar{A}_j))^{1-q} \\
&\quad - \sum_{\substack{j=1\\ \mu(\bar{A}_j)>0}}^{\bar{n}} \|\nu(\bar{A}_j)\|^q (\mu(\bar{A}_j))^{1-q} + \sum_{\substack{j=1\\ \mu(\bar{A}_j)>0}}^{\bar{n}} \|y_{*j}\|^q \mu(\bar{A}_j) + \epsilon/2 \\
&\le \mathcal{P}_q \circ \nu(E_n) - \sum_{\substack{j=1\\ \mu(\bar{A}_j)>0}}^{\bar{n}} \|\nu(\bar{A}_j)\|^q (\mu(\bar{A}_j))^{1-q} \\
&\quad + \sum_{\substack{j=1\\ \mu(\bar{A}_j)>0}}^{\bar{n}} \left\|\int_{\bar{A}_j} \psi_{i_0}\,\mathrm{d}\mu\right\|^q (\mu(\bar{A}_j))^{1-q} + \epsilon/2 \\
&\le M^q + \sum_{\substack{j=1\\ \mu(\bar{A}_j)>0}}^{\bar{n}} (\mu(\bar{A}_j))^{1-q} \left(\left\|\int_{\bar{A}_j} \psi_{i_0}\,\mathrm{d}\mu\right\|^q - \left\|\int_{\bar{A}_j} g\,\mathrm{d}\mu\right\|^q\right) + \epsilon/2 \\
&= M^q + \sum_{\substack{j=1\\ \mu(\bar{A}_j)>0}}^{\bar{n}} (\mu(\bar{A}_j))^{1-q} \left(\left\|\int_{\bar{A}_j} \psi_{i_0}\,\mathrm{d}\mu\right\| - \left\|\int_{\bar{A}_j} g\,\mathrm{d}\mu\right\|\right) q
\end{aligned}
$$

$$\cdot\left(t_j\left\|\int_{\bar{A}_j}\psi_{i_0}\,\mathrm{d}\mu\right\|+(1-t_j)\left\|\int_{\bar{A}_j}g\,\mathrm{d}\mu\right\|\right)^{q-1}+\epsilon/2$$

$$\leq M^q+\sum_{\substack{j=1\\ \mu(\bar{A}_j)>0}}^{\bar{n}}(\mu(\bar{A}_j))^{1-q}\left\|\int_{\bar{A}_j}\psi_{i_0}\,\mathrm{d}\mu-\int_{\bar{A}_j}g\,\mathrm{d}\mu\right\|q$$

$$\cdot\left(t_jn\mu(\bar{A}_j)+(1-t_j)n\mu(\bar{A}_j)\right)^{q-1}+\epsilon/2$$

$$=M^q+\sum_{\substack{j=1\\ \mu(\bar{A}_j)>0}}^{\bar{n}}qn^{q-1}\left\|\int_{\bar{A}_j}(\psi_{i_0}-g)\,\mathrm{d}\mu\right\|+\epsilon/2$$

$$\leq M^q+\sum_{\substack{j=1\\ \mu(\bar{A}_j)>0}}^{\bar{n}}qn^{q-1}\int_{\bar{A}_j}\mathcal{P}\circ(\psi_{i_0}-g)\,\mathrm{d}\mu+\epsilon/2$$

$$\leq M^q+qn^{q-1}\int_{E_n}\mathcal{P}\circ(\psi_{i_0}-g)\,\mathrm{d}\mu+\epsilon/2<M^q+\epsilon$$

where the first equality follows from the Mean Value Theorem 9.20 and $t_j \in (0,1) \subset \mathbb{R}$, $j = 1,\ldots,\bar{n}$; the fourth inequality follows from Proposition 11.92 and the fact that $\left\|\psi_{i_0}(x)\right\| \leq \|g(x)\| \leq n$, $\forall x \in E_n$; the second equality and the fifth inequality follow from Proposition 11.92; and the sixth inequality follows from Proposition 11.83. By arbitrariness of ϵ, we have $\int_{E_n}\mathcal{P}_q\circ g\,\mathrm{d}\mu \leq M^q$. Clearly, we have $E_n \subseteq E_{n+1}$, $\forall n \in \mathbb{N}$, and $\bigcup_{n=1}^{\infty}E_n = X$. Then, by Monotone Convergence Theorem 11.81, we have $\int_X \mathcal{P}_q\circ g\,\mathrm{d}\mu = \lim_{n\in\mathbb{N}}\int_X(\mathcal{P}_q\circ g)\chi_{E_n,X}\,\mathrm{d}\mu = \lim_{n\in\mathbb{N}}\int_{E_n}\mathcal{P}_q\circ g\,\mathrm{d}\mu \leq M^q$. Hence, $\|g\|_q \leq M$ and $g \in \bar{\mathrm{L}}_q(\mathcal{X},\mathcal{Y}^*)$.

This completes the proof of the lemma. □

Lemma 11.185 *Let $p \in [1,\infty) \subset \mathbb{R}$, $\mathcal{X} := (X,\mathcal{B},\mu)$ be a σ-finite measure space, $\mathcal{Y}$ be a separable reflexive Banach space over $\mathbb{K}$ with $\mathcal{Y}^*$ being separable, and $\mathcal{Z} := \mathrm{L}_p(\mathcal{X},\mathcal{Y})$ be the Banach space over $\mathbb{K}$ as defined in Example 11.179. Then, $\forall f \in \mathcal{Z}^*$, $\exists g : X \to \mathcal{Y}^*$, which is $\mathcal{B}$-measurable, such that:*

(i) *$\forall E \in \mathcal{B}$ with $\mu(E) < +\infty$, g is absolutely integrable over $\mathcal{E} := (E,\mathcal{B}_E,\mu_E)$, which is the finite measure subspace of $\mathcal{X}$.*
(ii) *$\forall$ simple function $\phi : \mathcal{X} \to \mathcal{Y}$ ($\phi \in \bar{\mathcal{Z}} := \bar{\mathrm{L}}_p(\mathcal{X},\mathcal{Y})$), the function $\langle\langle g(\cdot),\phi(\cdot)\rangle\rangle : X \to \mathbb{K}$ is absolutely integrable over $\mathcal{X}$, $f([\phi]) = \int_X\langle\langle g(x),\phi(x)\rangle\rangle\,\mathrm{d}\mu(x)$, and $\left|\int_X\langle\langle g(x),\phi(x)\rangle\rangle\,\mathrm{d}\mu(x)\right| \leq \|f\|\|\phi\|_p$.*

Furthermore, g is unique in the sense that $\tilde{g} : X \to \mathcal{Y}^*$ is another function with the above properties if, and only if, $\tilde{g}$ is $\mathcal{B}$-measurable and $g = \tilde{g}$ a.e. in $\mathcal{X}$.

Proof Fix any $f \in \mathcal{Z}^*$. $\forall E \in \mathcal{B}$ with $\mu(E) < \infty$, $\forall y \in \mathcal{Y}$, let $z_{E,y} := y\chi_{E,X} \in \bar{\mathcal{Z}}$. Define $f_E : \mathcal{Y} \to \mathbb{K}$ by $f_E(y) = f([z_{E,y}])$, $\forall y \in \mathcal{Y}$. Since $f \in \mathcal{Z}^*$, then f_E is linear and continuous and $f_E = y_{*E} \in \mathcal{Y}^*$. Note that $[z_{\emptyset,y}] = \vartheta_{\mathcal{Z}}$, $\forall y \in \mathcal{Y}$, and $f([z_{\emptyset,y}]) = 0$. Then, $f_\emptyset = \vartheta_{\mathcal{Y}^*}$ and $y_{*\emptyset} = \vartheta_{\mathcal{Y}^*}$.

Claim 11.185.1 $\forall E \in \mathcal{B}$ with $\mu(E) < \infty$, $\forall$ pairwise disjoint $(E_i)_{i=1}^\infty \subseteq \mathcal{B}$ with $E = \bigcup_{i=1}^\infty E_i$. Then, $\sum_{i=1}^\infty \|y_{*E_i}\| \leq \|f\|(\mu(E))^{1/p} < \infty$.

Proof of Claim $\sum_{i=1}^\infty \|y_{*E_i}\| = \sum_{i=1}^\infty \sup_{y\in\mathcal{Y}, \|y\|\leq 1} |f_{E_i}(y)|$. $\forall i \in \mathbb{N}$, by Propositions 7.85 and 7.90, $\exists y_i \in \mathcal{Y}$ with $\|y_i\| \leq 1$ such that $\|y_{*E_i}\| = f_{E_i}(y_i) = f([z_{E_i,y_i}])$. Then, $\forall n \in \mathbb{N}$, $\sum_{i=1}^n \|y_{*E_i}\| = \sum_{i=1}^n f([z_{E_i,y_i}]) = f([\sum_{i=1}^n z_{E_i,y_i}]) \leq \|f\| \|[\sum_{i=1}^n z_{E_i,y_i}]\|_p = \|f\| \|\sum_{i=1}^n z_{E_i,y_i}\|_p \leq \|f\| \cdot (\mu(E))^{1/p} < \infty$, where the first inequality follows from Proposition 7.72 and the second inequality follows from Propositions 11.83 and 11.75. By the arbitrariness of n, we have $\sum_{n=1}^\infty \|y_{*E_n}\| \leq \|f\|(\mu(E))^{1/p}$. This completes the proof of the claim. □

Claim 11.185.2 $\forall E \in \mathcal{B}$ with $\mu(E) < \infty$, $\forall$ pairwise disjoint $(E_n)_{n=1}^\infty \subseteq \mathcal{B}$ with $E = \bigcup_{n=1}^\infty E_n$, we have $y_{*E} = \sum_{n=1}^\infty y_{*E_n} \in \mathcal{Y}^*$.

Proof of Claim By Claim 11.185.1 and Propositions 7.27 and 7.72, $\sum_{n=1}^\infty y_{*E_n} \in \mathcal{Y}^*$. $\forall y \in \mathcal{Y}$,

$$\langle\langle \sum_{n=1}^\infty y_{*E_n}, y \rangle\rangle = \lim_{n\in\mathbb{N}} \langle\langle \sum_{i=1}^n y_{*E_i}, y \rangle\rangle = \lim_{n\in\mathbb{N}} \sum_{i=1}^n \langle\langle y_{*E_i}, y \rangle\rangle$$

$$= \lim_{n\in\mathbb{N}} \sum_{i=1}^n f([z_{E_i,y}]) = \lim_{n\in\mathbb{N}} f([\sum_{i=1}^n z_{E_i,y}])$$

where the first equality follows from Propositions 7.72 and 3.66 and the last equality follows from the linearity of f. Note that $\lim_{n\in\mathbb{N}} \sum_{i=1}^n z_{E_i,y}(x) = z_{E,y}(x)$, $\forall x \in X$, and $\|\sum_{i=1}^n z_{E_i,y}(x) - z_{E,y}(x)\|^p \leq \|y\|^p \chi_{E,X}(x)$, $\forall x \in X$, $\forall n \in \mathbb{N}$. Then, by Lebesgue Dominated Convergence Theorem 11.91, $\lim_{n\in\mathbb{N}} \int_X \mathcal{P}_p \circ (\sum_{i=1}^n z_{E_i,y} - z_{E,y})\, d\mu = 0$ and $\lim_{n\in\mathbb{N}} \sum_{i=1}^n z_{E_i,y} \doteq z_{E,y}$ in $\bar{\mathcal{Z}}$. By Propositions 7.72 and 3.66, $\lim_{n\in\mathbb{N}} f([\sum_{i=1}^n z_{E_i,y}]) = f([z_{E,y}]) = \langle\langle y_{*E}, y \rangle\rangle$. Hence, $\langle\langle y_{*E}, y \rangle\rangle = \langle\langle \sum_{n=1}^\infty y_{*E_n}, y \rangle\rangle$, $\forall y \in \mathcal{Y}$. This implies that, by Proposition 7.85, $y_{*E} = \sum_{n=1}^\infty y_{*E_n}$. □

Since $\mathcal{X}$ is σ-finite, then $\exists (X_n)_{n=1}^\infty \subseteq \mathcal{B}$ such that $X = \bigcup_{n=1}^\infty X_n$ and $\mu(X_n) < \infty$, $\forall n \in \mathbb{N}$. Without loss of generality, we may assume that $(X_n)_{n=1}^\infty$ is pairwise disjoint. Fix any $n \in \mathbb{N}$; let $\mathcal{X}_n := (X_n, \mathcal{B}_n, \mu_n)$ be the finite measure subspace of $\mathcal{X}$. We may define a function $\nu_n : \mathcal{B}_n \to \mathcal{Y}^*$ by $\nu_n(E) = f_E = y_{*E}$, $\forall E \in \mathcal{B}_n$. Clearly, $\nu_n(\emptyset) = y_{*\emptyset} = \vartheta_{\mathcal{Y}^*}$. $\forall$ pairwise disjoint $(E_i)_{i=1}^\infty \subseteq \mathcal{B}_n$, let $E := \bigcup_{i=1}^\infty E_i \in \mathcal{B}_n$. By Claim 11.185.1, $\sum_{i=1}^\infty \|\nu_n(E_i)\| \leq \|f\|(\mu(E))^{1/p} \leq \|f\|(\mu(X_n))^{1/p} < \infty$. By Claim 11.185.2, $\nu_n(E) = \sum_{i=1}^\infty \nu_n(E_i) \in \mathcal{Y}^*$. This shows that ν_n is a $\mathcal{Y}^*$-valued pre-measure on $(X_n, \mathcal{B}_n)$. By Claim 11.185.1, $\mathcal{P} \circ \nu_n(X_n) \leq \|f\|(\mu(X_n))^{1/p} < +\infty$. Then, ν_n is finite. Hence, $(X_n, \mathcal{B}_n, \nu_n)$ is a finite $\mathcal{Y}^*$-valued measure space.

By Proposition 11.118, the generation process on $((X_n, \mathcal{B}_n, \nu_n))_{n=1}^{\infty}$ yields a unique σ-finite $\mathcal{Y}^*$-valued measure space $(X, \mathcal{B}, \nu)$ on X.

Next, we will show that $\mathcal{P} \circ \nu(E) \leq \|f\|(\mu(E))^{1/p} < \infty$ and $\nu(E) = y_{*E}$, $\forall E \in \mathcal{B}$ with $\mu(E) < \infty$. Fix any $E \in \mathcal{B}$ with $\mu(E) < \infty$. Let $E_n := X_n \cap E \in \mathcal{B}_n$, $\forall n \in \mathbb{N}$. By Proposition 11.118, $\mathcal{P} \circ \nu(E) = \sum_{n=1}^{\infty} \mathcal{P} \circ \nu_n(E_n)$. $\forall \epsilon \in (0, +\infty) \subset \mathbb{R}$, $\forall n \in \mathbb{N}$, $\exists m_n \in \mathbb{Z}_+$, $\exists$ pairwise disjoint $(E_{n,i})_{i=1}^{m_n} \subseteq \mathcal{B}_n$ with $E_n = \bigcup_{i=1}^{m_n} E_{n,i}$, such that $\mathcal{P} \circ \nu_n(E_n) < \sum_{i=1}^{m_n} \|\nu_n(E_{n,i})\| + 2^{-n}\epsilon < \infty$. Then,

$$\begin{aligned}\mathcal{P} \circ \nu(E) &< \sum_{n=1}^{\infty}(\sum_{i=1}^{m_n} \|\nu_n(E_{n,i})\| + 2^{-n}\epsilon) \\ &\leq \sum_{n=1}^{\infty}\sum_{i=1}^{m_n} \|y_{*E_{n,i}}\| + \epsilon \leq \|f\|(\mu(E))^{1/p} + \epsilon < \infty\end{aligned}$$

where the third inequality follows from Claim 11.185.1. By the arbitrariness of ϵ, we have $\mathcal{P} \circ \nu(E) \leq \|f\|(\mu(E))^{1/p} < \infty$. Then, $E \in \operatorname{dom}(\nu)$ and $\nu(E) = \sum_{n=1}^{\infty} \nu_n(E_n) \in \mathcal{Y}^*$. By Claim 11.185.2, we have $\nu(E) = \sum_{n=1}^{\infty} y_{*E_n} = y_{*E}$. Thus, by Proposition 11.137, ν is uniquely defined independent of the choice of $(X_n)_{n=1}^{\infty}$.

$\forall E \in \mathcal{B}$ with $\mu(E) = 0$, $\mathcal{P} \circ \nu(E) \leq 0$. Thus, $\mathcal{P} \circ \nu \ll \mu$. By Radon–Nikodym Theorem 11.171 and Proposition 7.90, $\exists!\ g : \mathcal{X} \to \mathcal{Y}^*$ such that g is the Radon–Nikodym derivative of ν with respect to μ. $\forall E \in \mathcal{B}$ with $\mu(E) < \infty$, we have $\mathcal{P} \circ \nu(E) = \int_E \mathcal{P} \circ g \, d\mu < \infty$. Thus, (i) holds. $\forall$ simple function $\phi : \mathcal{X} \to \mathcal{Y}$, $\exists M \in [0, \infty) \subset \mathbb{R}$ such that $\|\phi(x)\| \leq M$, $\forall x \in X$, and $\mu(\hat{E}) := \mu(\{x \in X \mid \phi(x) \neq \vartheta_{\mathcal{Y}}\}) < \infty$. Then, by Proposition 7.72, $|\langle\langle g(x), \phi(x)\rangle\rangle| \leq M\|g(x)\|\chi_{\hat{E},X}(x) = M\mathcal{P} \circ g(x)\chi_{\hat{E},X}(x)$, $\forall x \in X$. By Propositions 7.72, 11.38, and 11.39, the function $\langle\langle g(\cdot), \phi(\cdot)\rangle\rangle : X \to \mathbb{K}$ is $\mathcal{B}$-measurable. Hence, the function $\langle\langle g(\cdot), \phi(\cdot)\rangle\rangle : X \to \mathbb{K}$ is absolutely integrable over $\mathcal{X}$. Let ϕ admit the canonical representation $\phi = \sum_{i=1}^{n} y_i \chi_{E_i,X}$. Then,

$$\begin{aligned}f([\phi]) &= f([\sum_{i=1}^{n} y_i \chi_{E_i,X}]) = \sum_{i=1}^{n} f([y_i \chi_{E_i,X}]) = \sum_{i=1}^{n} f([z_{E_i,y_i}]) \\ &= \sum_{i=1}^{n} \langle\langle y_{*E_i}, y_i\rangle\rangle = \sum_{i=1}^{n} \langle\langle \nu(E_i), y_i\rangle\rangle = \sum_{i=1}^{n} \left\langle\left\langle \int_{E_i} g \, d\mu, y_i\right\rangle\right\rangle \\ &= \sum_{i=1}^{n} \int_{E_i} \langle\langle g(x), y_i\rangle\rangle \, d\mu(x) = \int_X \langle\langle g(x), \phi(x)\rangle\rangle \, d\mu(x) \in \mathbb{K}\end{aligned}$$

where the last two equalities follow from Proposition 11.92. Then,

$$\Big|\int_X \langle\langle g(x), \phi(x)\rangle\rangle \, d\mu(x)\Big| = |f([\phi])| \leq \|f\| \cdot \|[\phi]\|_p = \|f\|\|\phi\|_p$$

Thus, (ii) holds. Hence, g is the function we seek. Since ν is uniquely defined and g is also uniquely defined, then g is unique as desired.

This completes the proof of the lemma. □

Theorem 11.186 (Riesz Representation Theorem) *Let $p \in [1, \infty) \subset \mathbb{R}$, $q \in (1, \infty] \subset \mathbb{R}_e$ with $1/p + 1/q = 1$, $\mathcal{X} := (X, \mathcal{B}, \mu)$ be a σ-finite measure space, $\mathcal{Y}$ be a separable reflexive Banach space over $\mathbb{K}$ with $\mathcal{Y}^*$ being separable, and $\mathcal{Z} := \mathrm{L}_p(\mathcal{X}, \mathcal{Y})$ be the Banach space over $\mathbb{K}$ as defined in Example 11.179. Then, $\mathcal{Z}^* = \mathrm{L}_q(\mathcal{X}, \mathcal{Y}^*)$ isometrically isomorphically. In particular, $\forall f \in \mathcal{Z}^*$, $\exists! [g] \in \mathrm{L}_q(\mathcal{X}, \mathcal{Y}^*)$ with $g \in \bar{\mathrm{L}}_q(\mathcal{X}, \mathcal{Y}^*)$ such that $f([z]) = \langle\langle [g], [z] \rangle\rangle := \int_X \langle\langle g(x), z(x) \rangle\rangle \, \mathrm{d}\mu(x)$ and the function $\langle\langle g(\cdot), z(\cdot) \rangle\rangle : X \to \mathbb{K}$ is absolutely integrable over $\mathcal{X}$, $\forall z \in \bar{\mathcal{Z}} := \bar{\mathrm{L}}_p(\mathcal{X}, \mathcal{Y})$, and $\|f\| = \|g\|_q$.*

Proof $\forall f \in \mathcal{Z}^*$, by Lemma 11.185, $\exists g : X \to \mathcal{Y}^*$, which is $\mathcal{B}$-measurable, such that:

(i) $\forall E \in \mathcal{B}$ with $\mu(E) < +\infty$, g is absolutely integrable over $\mathcal{E} := (E, \mathcal{B}_E, \mu_E)$, which is the finite measure subspace of $\mathcal{X}$.
(ii) $\forall$ simple function $\phi : \mathcal{X} \to \mathcal{Y}$ ($\phi \in \bar{\mathrm{L}}_p(\mathcal{X}, \mathcal{Y})$), the function $\langle\langle g(\cdot), \phi(\cdot) \rangle\rangle : X \to \mathbb{K}$ is absolutely integrable over $\mathcal{X}$, $f([\phi]) = \int_X \langle\langle g(x), \phi(x) \rangle\rangle \, \mathrm{d}\mu(x)$ and $\left| \int_X \langle\langle g(x), \phi(x) \rangle\rangle \, \mathrm{d}\mu(x) \right| \leq \|f\| \|\phi\|_p$.

Furthermore, g is unique in the sense that $\tilde{g} : X \to \mathcal{Y}^*$ is another function with the above properties if, and only if, $\tilde{g}$ is $\mathcal{B}$-measurable and $g = \tilde{g}$ a.e. in $\mathcal{X}$.

We will show that $g \in \bar{\mathrm{L}}_q(\mathcal{X}, \mathcal{Y}^*)$ and $\|g\|_q \leq \|f\| < +\infty$ by distinguishing two exhaustive and mutually exclusive cases: Case 1: $1 < p < +\infty$; Case 2: $p = 1$. Case 1: $1 < p < +\infty$. By Lemma 11.184, we have $g \in \bar{\mathrm{L}}_q(\mathcal{X}, \mathcal{Y}^*)$ and $\|g\|_q \leq \|f\| < +\infty$. Case 2: $p = 1$. $\forall E \in \mathcal{B}$ with $0 < \mu(E) < \infty$, by (i) and Proposition 11.92, $y_{*E} := \int_E g \, \mathrm{d}\mu \in \mathcal{Y}^*$ and, $\forall y \in \mathcal{Y}$, $|\langle\langle y_{*E}, y \rangle\rangle| = \left| \int_E \langle\langle g(x), y \rangle\rangle \, \mathrm{d}\mu(x) \right| = \left| \int_X \langle\langle g(x), y\chi_{E,X}(x) \rangle\rangle \, \mathrm{d}\mu(x) \right| \leq \|f\| \left\| y\chi_{E,X} \right\|_1 = \|f\| \|y\| \mu(E)$, where the first two equalities follow from Proposition 11.92 and the inequality follows from (ii). Then, $\|y_{*E}\| / \mu(E) \leq \|f\|$. By Proposition 11.183, we have $\mathcal{P} \circ g \leq \|f\|$ a.e. in $\mathcal{X}$. Hence, $\|g\|_\infty \leq \|f\|$ and $g \in \bar{\mathrm{L}}_\infty(\mathcal{X}, \mathcal{Y}^*)$.

Hence, in both cases, we have $g \in \bar{\mathrm{L}}_q(\mathcal{X}, \mathcal{Y}^*)$ and $\|g\|_q \leq \|f\| < +\infty$. $\forall z \in \bar{\mathcal{Z}}$, by Proposition 11.181, there exists a sequence of simple functions $(\varphi_i)_{i=1}^\infty$, $\varphi_i : \mathcal{X} \to \mathcal{Y}$, $\forall i \in \mathbb{N}$, such that $\lim_{i \in \mathbb{N}} \varphi_i = z$ a.e. in $\mathcal{X}$, $\|\varphi_i(x)\| \leq \|z(x)\|$, $\forall x \in X$, $\forall i \in \mathbb{N}$ and $\lim_{i \in \mathbb{N}} \varphi_i \doteq z$ in $\bar{\mathcal{Z}}$. By Propositions 7.72 and 3.66, $f([z]) = \lim_{i \in \mathbb{N}} f([\varphi_i]) = \lim_{i \in \mathbb{N}} \int_X \langle\langle g(x), \varphi_i(x) \rangle\rangle \, \mathrm{d}\mu(x)$. Note that, by Propositions 7.72, 11.52, and 11.53, $\lim_{i \in \mathbb{N}} \langle\langle g(x), \varphi_i(x) \rangle\rangle = \langle\langle g(x), z(x) \rangle\rangle$ a.e. $x \in \mathcal{X}$ and $|\langle\langle g(x), \varphi_i(x) \rangle\rangle| \leq \|g(x)\| \|\varphi_i(x)\| \leq \|g(x)\| \|z(x)\|$, $\forall x \in X$, $\forall i \in \mathbb{N}$. By Hölder's Inequality 11.178, $\int_X \|g(x)\| \cdot \|z(x)\| \, \mathrm{d}\mu(x) \leq \|g\|_q \|z\|_p < \infty$. By Propositions 7.72, 11.38, and 11.39, $\langle\langle g(\cdot), \varphi_i(\cdot) \rangle\rangle : X \to \mathbb{K}$ and $\langle\langle g(\cdot), z(\cdot) \rangle\rangle : X \to \mathbb{K}$ are $\mathcal{B}$-measurable, $\forall i \in \mathbb{N}$. By Lebesgue Dominated Convergence Theorem 11.91, $f([z]) = \lim_{i \in \mathbb{N}} \int_X \langle\langle g(x), \varphi_i(x) \rangle\rangle \, \mathrm{d}\mu(x) = \int_X \langle\langle g(x), z(x) \rangle\rangle \, \mathrm{d}\mu(x)$ and the function $\langle\langle g(\cdot), z(\cdot) \rangle\rangle : X \to \mathbb{K}$ is absolutely integrable over $\mathcal{X}$. Note that

$$\|f\| = \sup_{[z]\in\mathcal{Z},\ \|[z]\|_p\le 1} |f([z])| = \sup_{z\in\bar{\mathcal{Z}},\ \|z\|_p\le 1} \left| \int_X \langle\langle g(x), z(x)\rangle\rangle \, \mathrm{d}\mu(x) \right|$$

$$\le \sup_{z\in\bar{\mathcal{Z}},\ \|z\|_p\le 1} \int_X |\langle\langle g(x), z(x)\rangle\rangle| \, \mathrm{d}\mu(x)$$

$$\le \sup_{z\in\bar{\mathcal{Z}},\ \|z\|_p\le 1} \|g\|_q \|z\|_p \le \|g\|_q$$

where the first inequality follows from Proposition 11.92 and the second inequality follows from Hölder's inequality 11.178. Hence, we have $\|f\| = \|g\|_q$.

Thus, we may define a function $\Phi : \mathcal{Z}^* \to \mathbf{L}_q(\mathcal{X}, \mathcal{Y}^*)$ by $\Phi(f) = [g]$, $\forall f \in \mathcal{Z}^*$. Clearly, Φ is well-defined and $\|\Phi(f)\|_q = \|f\|$, $\forall f \in \mathcal{Z}^*$. $\forall f_1, f_2 \in \mathcal{Z}^*$, $\forall \alpha, \beta \in \mathbb{K}$, let $[g_i] = \Phi(f_i)$, $i = 1, 2$. $\forall [z] \in \mathcal{Z}$,

$$(\alpha f_1 + \beta f_2)([z]) = \alpha f_1([z]) + \beta f_2([z])$$

$$= \alpha \int_X \langle\langle g_1(x), z(x)\rangle\rangle \, \mathrm{d}\mu(x) + \beta \int_X \langle\langle g_2(x), z(x)\rangle\rangle \, \mathrm{d}\mu(x)$$

$$= \int_X \langle\langle \alpha g_1(x) + \beta g_2(x), z(x)\rangle\rangle \, \mathrm{d}\mu(x)$$

where the last equality follows from Proposition 11.92. By the uniqueness of g, we have $\Phi(\alpha f_1 + \beta f_2) = [\alpha g_1 + \beta g_2] = \alpha[g_1] + \beta[g_2] = \alpha\Phi(f_1) + \beta\Phi(f_2)$. Hence, Φ is linear. $\forall [g] \in \mathbf{L}_q(\mathcal{X}, \mathcal{Y}^*)$, we may define $f : \mathcal{Z} \to \mathbb{K}$ by $f([z]) = \int_X \langle\langle g(x), z(x)\rangle\rangle \, \mathrm{d}\mu(x)$, $\forall z \in \bar{\mathcal{Z}}$. By Hölder's Inequality 11.178 and Proposition 11.92, we have $f \in \mathcal{Z}^*$. Then, $\Phi(f) = [g]$ and Φ is surjective. $\forall f_1, f_2 \in \mathcal{Z}^*$ with $\Phi(f_1) = \Phi(f_2)$, we have $0 = \|\Phi(f_1) - \Phi(f_2)\|_q = \|\Phi(f_1 - f_2)\|_q = \|f_1 - f_2\|$ and $f_1 = f_2$. Hence, Φ is injective. Therefore, $\Phi : \mathcal{Z}^* \to \mathbf{L}_q(\mathcal{X}, \mathcal{Y}^*)$ is an isometrical isomorphism. This completes the proof of the theorem. □

11.10 Dual of $\mathcal{C}(\mathcal{X}, \mathcal{Y})$ and $\mathcal{C}_c(\mathcal{X}, \mathcal{Y})$

Proposition 11.187 *Let $\mathcal{X} := (X, \mathcal{O})$ be a Hausdorff topological space and $\mathbb{X} := (\mathcal{X}, \mathcal{B}_B(\mathcal{X}), \mu)$ be a topological measure space. Then, the following statements hold:*

(i) *$\forall$ compact subset $K \subseteq \mathcal{X}$, we have $K \in \mathcal{B}_B(\mathcal{X})$.*
(ii) *If $\mathbb{X}$ is locally finite and σ-compact, then it is σ-finite.*
(iii) *If $\mathbb{X}$ is finite, then it is locally finite.*

Proof (i) By Proposition 5.5, K is closed in $\mathcal{X}$ and $\widetilde{K} \in \mathcal{O}$. Thus, we have $K \in \mathcal{B}_B(\mathcal{X})$.

(ii) By $\mathbb{X}$ being σ-compact, there exists $(K_n)_{n=1}^{\infty} \subseteq \mathcal{X}$ such that K_n is compact, $\forall n \in \mathbb{N}$. By (i), $K_n \in \mathcal{B}_B(\mathcal{X})$ and, by $\mathbb{X}$ being locally finite, $\mu(K_n) < \infty$. Then, $\mathbb{X}$ is σ-finite.

(iii) $\forall$ compact subset $K \subseteq \mathcal{X}$, by (i), $K \in \mathcal{B}_B(\mathcal{X})$. Since $\mathbb{X}$ is finite, then $\mu(K) \leq \mu(X) < \infty$. Then, $\mathbb{X}$ is locally finite.

This completes the proof of the proposition. □

Proposition 11.188 *Let $\mathcal{X} := (X, \mathcal{O})$ be a topological space, $(X_n)_{n=1}^{\infty} \subseteq \mathcal{O}$, $\mathcal{X}_n := (X_n, \mathcal{O}_n)$ be a topological space with the subset topology $\mathcal{O}_n$, and $(\mathcal{X}_n, \mathcal{B}_B(\mathcal{X}_n), \mu_n) =: \mathbb{X}_n$ be a finite topological measure space, $\forall n \in \mathbb{N}$. Assume that $(\mathbb{X}_n)_{n=1}^{\infty}$ satisfies the assumptions of Proposition 11.118. By Proposition 11.118, the generation process on $(\mathbb{X}_n)_{n=1}^{\infty}$ yields a unique σ-finite measure space $\bar{\mathcal{X}} := (X, \mathcal{B}, \mu)$ such that $\mathbb{X}_n$ is the finite measure subspace of $\bar{\mathcal{X}}$, $\forall n \in \mathbb{N}$. Then, $\mathbb{X} := (\mathcal{X}, \mathcal{B}_B(\mathcal{X}), \mu)$ is the unique σ-finite topological measure space on $\mathcal{X}$ such that $\mathbb{X}_n$ is the finite topological measure subspace of $\mathbb{X}$, $\forall n \in \mathbb{N}$.*

Proof $\forall E \in \mathcal{B}_B(\mathcal{X})$, by Proposition 11.25, we have $E \cap X_n \in \mathcal{B}_B(\mathcal{X}_n)$, $\forall n \in \mathbb{N}$. Define $E_{\langle 1\rangle} := E \cap X_1$ and $E_{\langle n+1\rangle} := (E \cap X_{n+1}) \setminus (\bigcup_{i=1}^{n} E_{\langle i\rangle})$, $\forall n \in \mathbb{N}$. Then, $\left(E_{\langle n\rangle}\right)_{n=1}^{\infty}$ is pairwise disjoint, $\bigcup_{n=1}^{\infty} E_{\langle n\rangle} = E$, and $E_{\langle n\rangle} \in \mathcal{B}_B(\mathcal{X}_n) \subseteq \mathcal{B}_B(\mathcal{X})$, $\forall n \in \mathbb{N}$. Define $\bar{\mu} : \mathcal{B}_B(\mathcal{X}) \to [0, \infty] \subset \mathbb{R}_e$ by $\bar{\mu}(E) = \sum_{n=1}^{\infty} \mu_n(E_{\langle n\rangle}) \in [0, \infty] \subset \mathbb{R}_e$, $\forall E \in \mathcal{B}_B(\mathcal{X})$.

We will show that $(X, \mathcal{B}_B(\mathcal{X}), \bar{\mu})$ is a σ-finite measure space and $\mathbb{X}_n$ is the finite measure subspace of it, $\forall n \in \mathbb{N}$. It is easy to show that $\bar{\mu}(\emptyset) = 0$. $\forall$pairwise disjoint $(E_i)_{i=1}^{\infty} \subseteq \mathcal{B}_B(\mathcal{X})$, let $E := \bigcup_{i=1}^{\infty} E_i \in \mathcal{B}_B(\mathcal{X})$. Note that $E_{\langle 1\rangle} = E \cap X_1 = \bigcup_{i=1}^{\infty}(E_i \cap X_1) = \bigcup_{i=1}^{\infty} E_{i,\langle 1\rangle}$. Inductively, $E_{\langle n+1\rangle} = (E \cap X_{n+1}) \setminus (\bigcup_{i=1}^{n} E_{\langle i\rangle}) = (\bigcup_{j=1}^{\infty}(E_j \cap X_{n+1})) \setminus (\bigcup_{i=1}^{n} \bigcup_{j=1}^{\infty} E_{j,\langle i\rangle}) = \bigcup_{j=1}^{\infty}((E_j \cap X_{n+1}) \setminus (\bigcup_{i=1}^{n} E_{j,\langle i\rangle})) = \bigcup_{j=1}^{\infty} E_{j,\langle n+1\rangle}$, $\forall n \in \mathbb{N}$, where the second equality follows from the inductive assumption and the third equality follows from the fact that $\left(E_j\right)_{j=1}^{\infty}$ is pairwise disjoint and $E_{j,\langle i\rangle} \subseteq E_j$, $\forall i \in \mathbb{N}$, $\forall j \in \mathbb{N}$. Then,

$$\bar{\mu}(E) = \sum_{n=1}^{\infty} \mu_n(E_{\langle n\rangle}) = \sum_{n=1}^{\infty} \mu_n(\bigcup_{j=1}^{\infty} E_{j,\langle n\rangle}) = \sum_{n=1}^{\infty}\sum_{j=1}^{\infty} \mu_n(E_{j,\langle n\rangle})$$
$$= \sum_{j=1}^{\infty}\sum_{n=1}^{\infty} \mu_n(E_{j,\langle n\rangle}) = \sum_{j=1}^{\infty} \bar{\mu}(E_j)$$

where the fourth equality follows from the fact that all summands are nonnegative real numbers. Hence, $(X, \mathcal{B}_B(\mathcal{X}), \bar{\mu})$ is a measure space. It is σ-finite since $\bar{\mu}(X_n) = \sum_{i=1}^{\infty} \mu_i(X_{n,\langle i\rangle}) = \sum_{i=1}^{\infty} \mu_n(X_{n,\langle i\rangle}) = \mu_n(X_n) < \infty$, $\forall n \in \mathbb{N}$, where the second equality follows from the consistency assumption of Proposition 11.118 on $(\mathbb{X}_n)_{n=1}^{\infty}$.

Fix any $n \in \mathbb{N}$. $\forall E \in \mathcal{B}_B(\mathcal{X}_n) \subseteq \mathcal{B}_B(\mathcal{X})$, $\bar{\mu}(E) = \sum_{i=1}^{\infty} \mu_i(E_{\langle i\rangle}) = \sum_{i=1}^{\infty} \mu_n(E_{\langle i\rangle}) = \mu_n(\bigcup_{i=1}^{\infty} E_{\langle i\rangle}) = \mu_n(E)$, where the second equality follows from the consistency assumption of Proposition 11.118 on $(\mathbb{X}_n)_{n=1}^{\infty}$. By Propositions 11.25 and 11.13, we have $\mathbb{X}_n$ is the finite measure subspace of $(X, \mathcal{B}_B(\mathcal{X}), \bar{\mu})$, $\forall n \in \mathbb{N}$. By Proposition 11.118, we have $\mathcal{B} = \mathcal{B}_B(\mathcal{X})$ and $\mu = \bar{\mu}$. Then, $\mathbb{X} = (\mathcal{X}, \mathcal{B}_B(\mathcal{X}), \bar{\mu})$.

Finally, we will show that $\mathbb{X}$ is a topological measure space. Fix any $E \in \mathcal{B}_B(\mathcal{X})$. Then, $E = \bigcup_{n=1}^{\infty} E_{\langle n \rangle}$. $\forall \epsilon \in (0, \infty) \subset \mathbb{R}$, $\forall n \in \mathbb{N}$, since $\mathbb{X}_n$ is a topological measure space and $E_{\langle n \rangle} \in \mathcal{B}_B(\mathcal{X}_n)$, $\exists O_n \in \mathcal{O}_n$ with $E_{\langle n \rangle} \subseteq O_n$ such that $\mu_n(O_n \setminus E_{\langle n \rangle}) < 2^{-n}\epsilon$. Since $\mathcal{X}_n$ is a topological subspace of $\mathcal{X}$ and $X_n \in \mathcal{O}$, then $O_n \in \mathcal{O}_n \subseteq \mathcal{O} \subseteq \mathcal{B}_B(\mathcal{X})$. Let $O := \bigcup_{n=1}^{\infty} O_n \in \mathcal{O}$. Then, $E = \bigcup_{n=1}^{\infty} E_{\langle n \rangle} \subseteq \bigcup_{n=1}^{\infty} O_n = O$ and

$$\begin{aligned}
\mu(O \setminus E) &= \bar{\mu}(O \setminus E) = \bar{\mu}\Big(\big(\bigcup_{n=1}^{\infty} O_n\big) \setminus \big(\bigcup_{n=1}^{\infty} E_{\langle n \rangle}\big)\Big) \\
&= \bar{\mu}\Big(\bigcup_{n=1}^{\infty}\big(O_n \cap \big(\bigcap_{j=1}^{\infty} \widetilde{E_{\langle j \rangle}}\big)\big)\Big) \le \bar{\mu}\Big(\bigcup_{n=1}^{\infty}\big(O_n \cap \widetilde{E_{\langle n \rangle}}\big)\Big) \\
&= \bar{\mu}\Big(\bigcup_{n=1}^{\infty}\big(O_n \setminus E_{\langle n \rangle}\big)\Big) \le \sum_{n=1}^{\infty} \bar{\mu}(O_n \setminus E_{\langle n \rangle}) < \epsilon
\end{aligned}$$

Hence, $\mathbb{X}$ is a topological measure space. By Proposition 11.29, $\mathbb{X}_n$ is the finite topological measure subspace of $\mathbb{X}$, $\forall n \in \mathbb{N}$. This completes the proof of the proposition. □

Proposition 11.189 *Let $\mathcal{X} := (X, \mathcal{O})$ be a Hausdorff topological space, $(X_n)_{n=1}^{\infty} \subseteq \mathcal{O}$, $\mathcal{X}_n := (X_n, \mathcal{O}_n)$ be the topological space with the subset topology $\mathcal{O}_n$, and $\mathbb{X}_n := (\mathcal{X}_n, \mathcal{B}_B(\mathcal{X}_n), \mu_n)$ be a finite topological measure space, $\forall n \in \mathbb{N}$. Assume that $(\mathbb{X}_n)_{n=1}^{\infty}$ satisfies the assumptions of Proposition 11.118. By Proposition 11.188, the generation process on $(\mathbb{X}_n)_{n=1}^{\infty}$ yields a unique σ-finite topological measure space $\mathbb{X} := (\mathcal{X}, \mathcal{B}_B(\mathcal{X}), \mu)$ such that $\mathbb{X}_n$ is the finite topological measure subspace of $\mathbb{X}$, $\forall n \in \mathbb{N}$. Then, $\mathbb{X}$ is locally finite.*

Proof Fix any compact subset $K \subseteq \mathcal{X}$. By Proposition 11.187, $K \in \mathcal{B}_B(\mathcal{X})$. Clearly, $K \subseteq \bigcup_{n=1}^{\infty} X_n = X$. By the compactness of K, $\exists N \in \mathbb{N}$ such that $K \subseteq \bigcup_{n=1}^{N} X_n$. Then, $\mu(K) \le \sum_{n=1}^{N} \mu(X_n) = \sum_{n=1}^{N} \mu_n(X_n) < \infty$, where the first inequality follows from Proposition 11.6 and the equality follows from Proposition 11.13. Hence, $\mathbb{X}$ is locally finite. □

Proposition 11.190 *Let $p \in [1, \infty) \subset \mathbb{R}$, $\mathfrak{X}_1$ be a finite-dimensional Banach space, $\mathcal{X}_2$ be a σ-compact metric space, $\mathcal{Y}_1$ be a finite-dimensional Banach space over $\mathbb{K}$, $\mathcal{Y}_2$ be a separable normed linear space over $\mathbb{K}$, $X \subseteq \mathfrak{X}_1$ be a subset with subset topology $\mathcal{O}$, $\mathcal{X} := (X, \mathcal{O})$, and $\mathbb{X} := (\mathcal{X}, \mathcal{B}_B(\mathcal{X}), \mu)$ be a metric measure space. Then, the following statements hold:*

(i) *$\mathcal{Y}_1$ is separable.*
(ii) *$\mathbf{l}_p(\mathcal{Y}_2) =: \mathcal{Z}_2$ is a separable normed linear space.*
(iii) *If $\mathcal{X}$ is compact, then $\mathcal{C}(\mathcal{X}, \mathcal{Y}_1)$ is a separable Banach space.*

(iv) *If $\mathbb{X}$ is a locally finite σ-compact metric measure space, then* $\mathrm{L}_p(\mathbb{X}, \mathcal{Y}_1)$ *is a separable Banach space.*
(v) $\mathcal{X}_2$ *is separable.*

Proof Let $\mathcal{Y}_1$ be m-dimensional, where $m \in \mathbb{Z}_+$, and $\{e_1, \ldots, e_m\} \subseteq \mathcal{Y}_1$ be a set of basis vectors. Then, $\mathcal{Y}_1$ is isomorphic to $\mathbb{K}^m$. The norm $\|\cdot\|_{\mathcal{Y}_1}$ induces a norm $\|\cdot\|_1$ on $\mathbb{K}^m$ as defined by, $\forall a := (\alpha_1, \ldots, \alpha_m) \in \mathbb{K}^m$, $\|a\|_1 := \left\|\sum_{i=1}^m \alpha_i e_i\right\|_{\mathcal{Y}_1} \in \mathbb{R}$. Then, $\mathcal{Y}_1$ is isometrically isomorphic to the normed linear space $(\mathbb{K}^m, \mathbb{K}, \|\cdot\|_1)$, which is a Banach space by Theorem 7.36. By Theorem 7.38, $\|\cdot\|_1$ is equivalent to the Euclidean norm $|\cdot|$ on $\mathbb{K}^m$.

(i) Let $\mathbb{K}_{\mathbb{Q}} := \mathbb{Q}$ if $\mathbb{K} = \mathbb{R}$; and $\mathbb{K}_{\mathbb{Q}} := \{a + ib \in \mathbb{C} \mid a, b \in \mathbb{Q}\}$, if $\mathbb{K} = \mathbb{C}$. Clearly, $\mathbb{K}_{\mathbb{Q}}$ is a countable dense subset of $\mathbb{K}$. Then, $\mathbb{K}_{\mathbb{Q}}^m$ is a countable dense subset of $(\mathbb{K}^m, \mathbb{K}, |\cdot|)$. Since $\|\cdot\|_1$ is equivalent to $|\cdot|$, then $\mathbb{K}_{\mathbb{Q}}^m$ is a countable dense subset of $(\mathbb{K}^m, \mathbb{K}, \|\cdot\|_1)$. Hence, $\mathcal{Y}_1$ is separable, since $(\mathbb{K}^m, \mathbb{K}, \|\cdot\|_1)$ is separable.

(ii) Let $D \subseteq \mathcal{Y}_2$ be a countable dense subset and, $\forall i \in \mathbb{N}$, $D_i := \{(y_1, \ldots, y_i, \vartheta_{\mathcal{Y}_2}, \vartheta_{\mathcal{Y}_2}, \ldots) \in \mathcal{Z}_2 \mid y_j \in D, 1 \leq j \leq i\}$. Then, D_i is countable, $\forall i \in \mathbb{N}$. Let $\bar{D} := \bigcup_{i=1}^\infty D_i \subseteq \mathcal{Z}_2$, which is also countable. We will show that $\bar{D}$ is dense in $\mathcal{Z}_2 = \mathrm{l}_p(\mathcal{Y}_2)$. $\forall z := (y_1, y_2, \ldots) \in \mathcal{Z}_2$, $\forall \epsilon \in (0, +\infty) \subset \mathbb{R}$, we have $\sum_{i=1}^\infty \|y_i\|_{\mathcal{Y}_2}^p < \infty$. Then, $\exists N \in \mathbb{N}$ such that $\sum_{i=N+1}^\infty \|y_i\|_{\mathcal{Y}_2}^p < \epsilon^p/2$. $\forall i \in \{1, \ldots, N\}$, $\exists \bar{y}_i \in D$ such that $\|y_i - \bar{y}_i\|_{\mathcal{Y}_2} < \frac{\epsilon}{2^{1/p} N^{1/p}}$. Let $\bar{z} := (\bar{y}_1, \ldots, \bar{y}_N, \vartheta_{\mathcal{Y}_2}, \vartheta_{\mathcal{Y}_2}, \ldots) \in \bar{D}$. Then, $\|z - \bar{z}\|_p = \left(\sum_{i=1}^N \|y_i - \bar{y}_i\|_{\mathcal{Y}_2}^p + \sum_{i=N+1}^\infty \|y_i\|_{\mathcal{Y}_2}^p\right)^{1/p} < \left(\sum_{i=1}^N \frac{\epsilon^p}{2N} + \epsilon^p/2\right)^{1/p} = \epsilon$. Hence, $\bar{D}$ is dense in $\mathcal{Z}_2$. Therefore, $\mathcal{Z}_2$ is separable.

(iii) This result is standard, which follows directly from Corollary 7.57 of Stone–Weierstrass Theorem, and therefore, the proof is omitted.

(iv) First, consider the special case where $\mathbb{X}$ is a finite compact metric measure space. By (iii), let $D \subseteq \mathcal{C}(\mathcal{X}, \mathcal{Y}_1)$ be a countable dense subset. By Proposition 4.11, $\mathcal{X}$ is a normal topological space. Then, by Proposition 11.182 and (i), $\forall f \in \bar{\mathrm{L}}_p(\mathbb{X}, \mathcal{Y}_1)$, $\forall \epsilon \in (0, +\infty) \subset \mathbb{R}$, $\exists g \in \bar{\mathrm{L}}_p(\mathbb{X}, \mathcal{Y}_1)$ such that g is continuous and $\|g - f\|_p < \epsilon/2$. Clearly, $g \in \mathcal{C}(\mathcal{X}, \mathcal{Y}_1)$. Then, $\exists h \in D$ such that $\|h - g\|_{\mathcal{C}(\mathcal{X}, \mathcal{Y}_1)} < \frac{\epsilon}{2(\mu(X)+1)^{1/p}}$. Since X is compact and $h \in \mathcal{C}(\mathcal{X}, \mathcal{Y}_1)$, then $\exists M \in [0, \infty) \subset \mathbb{R}$ such that $\mathcal{P} \circ h(x) \leq M < \infty$, $\forall x \in X$. By Proposition 11.37, h is $\mathcal{B}$-measurable. Hence, $h \in \bar{\mathrm{L}}_p(\mathbb{X}, \mathcal{Y}_1)$. Then, $\|f - h\|_p \leq \|f - g\|_p + \|g - h\|_p < \epsilon/2 + \left(\int_X \mathcal{P}_p \circ (g - h) \, d\mu\right)^{1/p} \leq \epsilon/2 + \left(\|g - h\|_{\mathcal{C}(\mathcal{X}, \mathcal{Y}_1)}^p \mu(X)\right)^{1/p} < \epsilon$. Hence, $D \subseteq \bar{\mathrm{L}}_p(\mathbb{X}, \mathcal{Y}_1)$ is dense. Therefore, $\mathrm{L}_p(\mathbb{X}, \mathcal{Y}_1)$ is separable. By Example 11.179, $\mathrm{L}_p(\mathbb{X}, \mathcal{Y}_1)$ is a Banach space. Hence, $\mathrm{L}_p(\mathbb{X}, \mathcal{Y}_1)$ is a separable Banach space.

Now, consider the general case where $\mathbb{X}$ is a locally finite σ-compact metric measure space. Then, $\exists (X_n)_{n=1}^\infty \subseteq \mathbb{X}$ such that $\mathbb{X} = \bigcup_{n=1}^\infty X_n$ and X_n is compact, $\forall n \in \mathbb{N}$. Without loss of generality, we may assume that $X_n \subseteq X_{n+1}$, $\forall n \in \mathbb{N}$. By Proposition 11.187, $X_n \in \mathcal{B}_{\mathrm{B}}(\mathcal{X})$ and, since $\mathbb{X}$ is locally finite, then $\mu(X_n) < \infty$, $\forall n \in \mathbb{N}$. Then, $\mathbb{X}$ is σ-finite. Fix any $n \in \mathbb{N}$. By Proposition 4.37, let $\mathcal{X}_n$ be the metric subspace of $\mathcal{X}$. Let $\mathbb{X}_n := (X_n, \mathcal{B}_n, \mu_n)$ be the finite measure subspace of $\mathbb{X}$. By Proposition 11.29, $\mathcal{B}_n = \mathcal{B}_{\mathrm{B}}(\mathcal{X}_n)$ and $\mathbb{X}_n$ is a finite compact metric measure space. By the special case, $\mathrm{L}_p(\mathbb{X}_n, \mathcal{Y}_1)$ is a separable Banach space. Let $D_n \subseteq \mathrm{L}_p(\mathbb{X}_n, \mathcal{Y}_1)$ be a countable dense subset. $\forall f \in \bar{\mathrm{L}}_p(\mathbb{X}_n, \mathcal{Y}_1)$, define $\bar{f} \in \bar{\mathrm{L}}_p(\mathbb{X}, \mathcal{Y}_1)$ by $\bar{f}(x) = \begin{cases} f(x) & \forall x \in X_n \\ \vartheta_{\mathcal{Y}_1} & \forall x \in X \setminus X_n \end{cases}$, $\forall x \in \mathbb{X}$. Let $D := \{[\bar{f}] \in \mathrm{L}_p(\mathbb{X}, \mathcal{Y}_1) \mid \exists n \in \mathbb{N}, [f] \in D_n\}$. Clearly, $D \subseteq \mathrm{L}_p(\mathbb{X}, \mathcal{Y}_1)$ is countable.

We will show that D is dense in $\mathrm{L}_p(\mathbb{X}, \mathcal{Y}_1)$. Fix any $f \in \bar{\mathrm{L}}_p(\mathbb{X}, \mathcal{Y}_1)$. We have $\int_X (\mathcal{P}_p \circ f)\, \mathrm{d}\mu < \infty$. By Monotone Convergence Theorem 11.81, $\int_X (\mathcal{P}_p \circ f)\, \mathrm{d}\mu = \lim_{n\in\mathbb{N}} \int_X \mathcal{P}_p \circ (f\chi_{X_n,X})\, \mathrm{d}\mu$. Let $f_n := f\chi_{X_n,X} \in \bar{\mathrm{L}}_p(\mathbb{X}, \mathcal{Y}_1)$, $\forall n \in \mathbb{N}$. Then,

$$\begin{aligned}
\lim_{n\in\mathbb{N}} \|f_n - f\|_p &= \lim_{n\in\mathbb{N}} \Big(\int_X \mathcal{P}_p \circ (f_n - f)\, \mathrm{d}\mu \Big)^{1/p} \\
&= \lim_{n\in\mathbb{N}} \Big(\int_X \mathcal{P}_p \circ (f\chi_{X\setminus X_n,X})\, \mathrm{d}\mu \Big)^{1/p} \\
&= \lim_{n\in\mathbb{N}} \Big(\int_X (\mathcal{P}_p \circ f - \mathcal{P}_p \circ f_n)\, \mathrm{d}\mu \Big)^{1/p} \\
&= \lim_{n\in\mathbb{N}} \Big(\int_X \mathcal{P}_p \circ f\, \mathrm{d}\mu - \int_X \mathcal{P}_p \circ f_n\, \mathrm{d}\mu \Big)^{1/p} = 0
\end{aligned}$$

where the fourth equality follows from Proposition 11.83. Hence, $\lim_{n\in\mathbb{N}} f_n \doteq f$ in $\bar{\mathrm{L}}_p(\mathbb{X}, \mathcal{Y}_1)$. $\forall \epsilon \in (0, \infty) \subset \mathbb{R}$, $\exists n_0 \in \mathbb{N}$ such that $\|f_{n_0} - f\|_p < \epsilon/2$.

Let $\hat{f}_{n_0} := f_{n_0}|_{X_{n_0}}$. Then, $\hat{f}_{n_0} \in \bar{\mathrm{L}}_p(\mathbb{X}_{n_0}, \mathcal{Y}_1)$. This implies that $\exists [g] \in D_{n_0}$ with $g \in \bar{\mathrm{L}}_p(\mathbb{X}_{n_0}, \mathcal{Y}_1)$ such that $\big\|\hat{f}_{n_0} - g\big\|_p < \epsilon/2$. Let $\bar{g} \in \bar{\mathrm{L}}_p(\mathbb{X}, \mathcal{Y}_1)$ be defined as in the second paragraph from last. Then, $[\bar{g}] \in D$ and

$$\begin{aligned}
\|f - \bar{g}\|_p &\le \|f - f_{n_0}\|_p + \|f_{n_0} - \bar{g}\|_p \\
&< \epsilon/2 + \Big(\int_X \mathcal{P}_p \circ (f_{n_0} - \bar{g})\, \mathrm{d}\mu \Big)^{1/p} \\
&= \epsilon/2 + \Big(\int_X \mathcal{P}_p \circ ((f_{n_0} - \bar{g})\chi_{X_{n_0},X})\, \mathrm{d}\mu \Big)^{1/p} \\
&= \epsilon/2 + \Big(\int_{X_{n_0}} \mathcal{P}_p \circ (\hat{f}_{n_0} - g)\, \mathrm{d}\mu_{n_0} \Big)^{1/p} \\
&= \epsilon/2 + \big\|\hat{f}_{n_0} - g\big\|_p < \epsilon
\end{aligned}$$

where the second equality follows from Proposition 11.83. Hence, D is dense in $\mathrm{L}_p(\mathbb{X}, \mathcal{Y}_1)$. Then, by Example 11.179, $\mathrm{L}_p(\mathbb{X}, \mathcal{Y}_1)$ is a separable Banach space.

(v) Since $\mathcal{X}_2$ is σ-compact, then $\exists$ compact sets $(K_n)_{n=1}^\infty$ such that $\mathcal{X}_2 = \bigcup_{n=1}^\infty K_n$. $\forall n \in \mathbb{N}$, $\forall i \in \mathbb{N}$, $K_n \subseteq \bigcup_{x\in K_n} \mathcal{B}_{\mathcal{X}_2}(x, 1/i)$. By the compactness of K_n, $\exists$ finite subset $D_{n,i} \subseteq K_n$ such that $K_n \subseteq \bigcup_{x \in D_{n,i}} \mathcal{B}_{\mathcal{X}_2}(x, 1/i)$. Then, $D := \bigcup_{n=1}^{\infty} \bigcup_{i=1}^{\infty} D_{n,i} \subseteq \mathcal{X}_2$ is a countable dense set. Hence, $\mathcal{X}_2$ is separable.

This completes the proof of the proposition. □

Definition 11.191 Let $\mathcal{X} := (X, \mathcal{O})$ be a topological space, $\mathcal{Y}$ be a normed linear space, and $(X, \mathcal{B}, \mu)$ be a $\mathcal{Y}$-valued measure space on the same set X. The triple $\mathbb{X} := (\mathcal{X}, \mathcal{B}, \mu)$ is said to be a *$\mathcal{Y}$-valued topological measure space* if $\mathcal{B} = \mathcal{B}_B(\mathcal{X})$ and $\bar{\mathbb{X}} := (\mathcal{X}, \mathcal{B}_B(\mathcal{X}), \mathcal{P} \circ \mu)$ is a topological measure space. We will say that $\mathbb{X}$ is *finite*, *σ-finite*, or *locally finite* if $\bar{\mathbb{X}}$ is so. We will say that $\mathbb{X}$ is *Tychonoff*, *Hausdorff*, *regular*, *completely regular*, *normal*, *first countable*, *second countable*, *separable*, *second category everywhere*, *connected*, *locally connected*, *compact*, *countably compact*, *sequentially compact*, *locally compact*, *σ-compact*, or *paracompact* if $\mathcal{X}$ is so.

Let $\mathcal{X} := (X, \rho)$ be a metric space with the natural topology $\mathcal{O}$, $\mathcal{Y}$ be a normed linear space, and $(X, \mathcal{B}_B(\mathcal{X}), \mu)$ be a $\mathcal{Y}$-valued measure space on the same set X. The triple $\mathbb{X} := (\mathcal{X}, \mathcal{B}_B(\mathcal{X}), \mu)$ is said to be a *$\mathcal{Y}$-valued metric measure space* if $((X, \mathcal{O}), \mathcal{B}_B(\mathcal{X}), \mu)$ is a $\mathcal{Y}$-valued topological measure space. $\mathbb{X}$ is said to be *complete* or *totally bounded* if $\mathcal{X}$ is so.

Let $\mathfrak{X} := (\mathcal{X}, \mathbb{K}, \|\cdot\|)$ be a normed linear space over the field $\mathbb{K}$, $\mathcal{O}$ be the natural topology on $\mathfrak{X}$ generated by the norm $\|\cdot\|$, $\mathcal{Y}$ be a normed linear space, and $(X, \mathcal{B}_B(\mathfrak{X}), \mu)$ be a $\mathcal{Y}$-valued measure space on the same set X. The triple $\mathbb{X} := (\mathfrak{X}, \mathcal{B}_B(\mathfrak{X}), \mu)$ is said to be a *$\mathcal{Y}$-valued normed linear measure space* if $((X, \mathcal{O}), \mathcal{B}_B(\mathfrak{X}), \mu)$ is a $\mathcal{Y}$-valued topological measure space. When $\mathfrak{X}$ is a Banach space, then $\mathbb{X}$ is said to be a *$\mathcal{Y}$-valued Banach measure space*. Depending on whether $\mathbb{K} = \mathbb{R}$ or $\mathbb{K} = \mathbb{C}$, we will say that $\mathbb{X}$ is a $\mathcal{Y}$-valued *real* or *complex Banach measure space*. ◇

Proposition 11.192 *Let $\mathcal{X} := (X, \mathcal{O})$ be a topological space and $\mathcal{Y}$ be a normed linear space over $\mathbb{K}$. Define $\bar{\mathcal{Z}} := \{\mu \in \mathcal{M}_f(X, \mathcal{B}_B(\mathcal{X}), \mathcal{Y}) \mid (\mathcal{X}, \mathcal{B}_B(\mathcal{X}), \mu)$ is a finite $\mathcal{Y}$-valued topological measure space$\}$. Then, $\bar{\mathcal{Z}}$ is a closed subspace of $\mathcal{M}_f(X, \mathcal{B}_B(\mathcal{X}), \mathcal{Y})$ and $(\bar{\mathcal{Z}}, \mathbb{K}, \|\cdot\|_{\mathcal{M}_f(X, \mathcal{B}_B(\mathcal{X}), \mathcal{Y})}) =: \mathcal{M}_{ft}(\mathcal{X}, \mathcal{Y})$ is a normed linear space.*

If, in addition, $\mathcal{Y}$ is a Banach space, then $\mathcal{M}_{ft}(\mathcal{X}, \mathcal{Y})$ is a Banach space. Furthermore, define $\mathcal{Z} := \{\mu \in \mathcal{M}_\sigma(X, \mathcal{B}_B(\mathcal{X}), \mathcal{Y}) \mid (\mathcal{X}, \mathcal{B}_B(\mathcal{X}), \mu)$ is a σ-finite $\mathcal{Y}$-valued topological measure space$\} \subseteq \mathcal{M}_\sigma(X, \mathcal{B}_B(\mathcal{X}), \mathcal{Y})$. Let $\mathcal{Z}$ admit the subset topology $\mathcal{O}_{\mathcal{Z}}$. Then, $\mathcal{Z} =: \mathcal{M}_{\sigma t}(\mathcal{X}, \mathcal{Y})$ is a subspace of $\mathcal{M}_\sigma(X, \mathcal{B}_B(\mathcal{X}), \mathcal{Y})$. We will abuse the notation and denote the topological space $(\mathcal{Z}, \mathcal{O}_{\mathcal{Z}})$ by $\mathcal{M}_{\sigma t}(\mathcal{X}, \mathcal{Y})$.

Proof Let $\mathcal{Y}$ be a normed linear space. We will show that $\bar{\mathcal{Z}}$ is a subspace of $\mathcal{M}_f(X, \mathcal{B}_B(\mathcal{X}), \mathcal{Y})$. $\forall \alpha_1, \alpha_2 \in \mathbb{K}$, $\forall \mu_1, \mu_2 \in \mathcal{Z}$, by Proposition 11.136, $\mu := \alpha_1 \mu_1 + \alpha_2 \mu_2 \in \mathcal{M}_f(X, \mathcal{B}_B(\mathcal{X}), \mathcal{Y})$. $\forall E \in \mathcal{B}_B(\mathcal{X})$, $\forall \epsilon \in (0, +\infty) \subset \mathbb{R}$, $\forall i = 1, 2$, by $\mu_i \in \mathcal{Z}$, $\exists O_i \in \mathcal{O}$ with $E \subseteq O_i$ such that $\mathcal{P} \circ \mu_i(O_i \setminus E) < \epsilon/(1 + 2|\alpha_i|)$. Let $O := O_1 \cap O_2 \in \mathcal{O}$. Clearly, $E \subseteq O$ and $\mathcal{P} \circ \mu(O \setminus E) \le (\sum_{i=1}^{2} |\alpha_i| \mathcal{P} \circ \mu_i)(O \setminus E) = \sum_{i=1}^{2} |\alpha_i| \mathcal{P} \circ \mu_i(O \setminus E) < \sum_{i=1}^{2} |\alpha_i| \mathcal{P} \circ \mu_i(O_i \setminus E) < \epsilon$, where the first inequality and the first equality follow from Proposition 11.136. This shows that $(\mathcal{X}, \mathcal{B}_B(\mathcal{X}), \mathcal{P} \circ \mu)$ is a topological measure space. Then, $(\mathcal{X}, \mathcal{B}_B(\mathcal{X}), \mu)$ is a finite

$\mathcal{Y}$-valued topological measure space and $\mu \in \bar{\mathcal{Z}}$. Clearly, $\vartheta_{\mathcal{M}_f(X,\mathcal{B}_B(\mathcal{X}),\mathcal{Y})} \in \bar{\mathcal{Z}} \neq \emptyset$. Hence, $\bar{\mathcal{Z}}$ is a subspace of $\mathcal{M}_f(X, \mathcal{B}_B(\mathcal{X}), \mathcal{Y})$. Then, $\mathcal{M}_{ft}(\mathcal{X}, \mathcal{Y})$ is a normed linear space since $\mathcal{M}_f(X, \mathcal{B}_B(\mathcal{X}), \mathcal{Y})$ is a normed linear space.

Next, we will show that $\bar{\mathcal{Z}}$ is closed. $\forall \mu \in \overline{\bar{\mathcal{Z}}} \subseteq \mathcal{M}_f(X, \mathcal{B}_B(\mathcal{X}), \mathcal{Y})$, by Proposition 4.13, $\exists (\mu_n)_{n=1}^{\infty} \subseteq \bar{\mathcal{Z}}$ such that $\lim_{n\in\mathbb{N}} \mu_n = \mu$. Then, $\lim_{n\in\mathbb{N}} \|\mu_n - \mu\|_{\mathcal{M}_f(X,\mathcal{B}_B(\mathcal{X}),\mathcal{Y})} = \lim_{n\in\mathbb{N}} \mathcal{P} \circ (\mu_n - \mu)(X) = 0$. $\forall E \in \mathcal{B}_B(\mathcal{X})$, $\forall \epsilon \in (0, +\infty) \subset \mathbb{R}$, $\exists n \in \mathbb{N}$ such that $\mathcal{P} \circ (\mu_n - \mu)(X) < \epsilon/2$. By $\mu_n \in \bar{\mathcal{Z}}$, $\exists O \in \mathcal{O}$ with $E \subseteq O$ such that $\mathcal{P} \circ \mu_n(O \setminus E) < \epsilon/2$. Then, $\mathcal{P} \circ \mu(O \setminus E) = \mathcal{P} \circ (\mu_n - \mu_n + \mu)(O \setminus E) \leq (\mathcal{P} \circ \mu_n + \mathcal{P} \circ (\mu_n - \mu))(O \setminus E) = \mathcal{P} \circ \mu_n(O \setminus E) + \mathcal{P} \circ (\mu_n - \mu)(O \setminus E) < \epsilon/2 + \mathcal{P} \circ (\mu_n - \mu)(X) < \epsilon$, where the first equality and the first inequality follow from Proposition 11.136. Hence, $\mu \in \bar{\mathcal{Z}}$. By the arbitrariness of μ, we have $\overline{\bar{\mathcal{Z}}} = \bar{\mathcal{Z}}$ and $\bar{\mathcal{Z}}$ is closed.

Let $\mathcal{Y}$ be a Banach space. By Proposition 11.142, $\mathcal{M}_f(X, \mathcal{B}_B(\mathcal{X}), \mathcal{Y})$ is a Banach space. By Proposition 4.39, $\mathcal{M}_{ft}(\mathcal{X}, \mathcal{Y})$ is a Banach space.

Finally, we will show that $\mathcal{Z}$ is a subspace of $\mathcal{M}_\sigma(X, \mathcal{B}_B(\mathcal{X}), \mathcal{Y})$. $\forall \alpha_1, \alpha_2 \in \mathbb{K}$, $\forall \mu_1, \mu_2 \in \mathcal{Z}$, by Proposition 11.138, $\mu := \alpha_1\mu_1 + \alpha_2\mu_2 \in \mathcal{M}_\sigma(X, \mathcal{B}_B(\mathcal{X}), \mathcal{Y})$. $\forall E \in \mathcal{B}_B(\mathcal{X})$, $\forall \epsilon \in (0, +\infty) \subset \mathbb{R}$, $\forall i = 1, 2$, by $\mu_i \in \mathcal{Z}$, $\exists O_i \in \mathcal{O}$ with $E \subseteq O_i$ such that $\mathcal{P} \circ \mu_i(O_i \setminus E) < \epsilon/(1 + 2|\alpha_i|)$. Let $O := O_1 \cap O_2 \in \mathcal{O}$. Clearly, $E \subseteq O$ and $\mathcal{P} \circ \mu(O \setminus E) \leq (\sum_{i=1}^{2} |\alpha_i|\mathcal{P} \circ \mu_i)(O \setminus E) = \sum_{i=1}^{2} |\alpha_i|\mathcal{P} \circ \mu_i(O \setminus E) \leq \sum_{i=1}^{2} |\alpha_i|\mathcal{P} \circ \mu_i(O_i \setminus E) < \epsilon$, where the first inequality and the first equality follow from Proposition 11.138. This shows that $(\mathcal{X}, \mathcal{B}_B(\mathcal{X}), \mathcal{P} \circ \mu)$ is a topological measure space. Then, $(\mathcal{X}, \mathcal{B}_B(\mathcal{X}), \mu)$ is a σ-finite $\mathcal{Y}$-valued topological measure space and $\mu \in \mathcal{Z}$. Clearly, $\vartheta_{\mathcal{M}_\sigma(X,\mathcal{B}_B(\mathcal{X}),\mathcal{Y})} \in \mathcal{Z} \neq \emptyset$. Hence, $\mathcal{Z}$ is a subspace of $\mathcal{M}_\sigma(X, \mathcal{B}_B(\mathcal{X}), \mathcal{Y})$.

This completes the proof of the proposition. □

A bit of notation to simplify our presentation. Let $\mathcal{M}_\sigma(X, \mathcal{B})$ denote the set of σ-finite measures on the measurable space $(X, \mathcal{B})$; $\mathcal{M}_f(X, \mathcal{B})$ denote the set of finite measures on the measurable space $(X, \mathcal{B})$; $\mathcal{M}_{\sigma t}(\mathcal{X})$ denote the set of σ-finite topological measures on the topological space $\mathcal{X}$; and $\mathcal{M}_{ft}(\mathcal{X})$ denote the set of finite topological measures on the topological space $\mathcal{X}$.

Proposition 11.193 *Let $\mathcal{X} := (X, \mathcal{O})$ be a topological space and $\mu_o : \mathcal{O} \to [0, \infty) \subset \mathbb{R}$. Assume that:*

(i) $\mu_o(\emptyset) = 0$.
(ii) $\mu_o(O_1) \leq \mu_o(O_2)$, $\forall O_1, O_2 \in \mathcal{O}$ *with* $O_1 \subseteq O_2$.
(iii) $\mu_o(\bigcup_{i=1}^{\infty} O_i) \leq \sum_{i=1}^{\infty} \mu_o(O_i)$, $\forall (O_i)_{i=1}^{\infty} \subseteq \mathcal{O}$.
(iv) $\mu_o(O_1 \cup O_2) = \mu_o(O_1) + \mu_o(O_2)$, $\forall O_1, O_2 \in \mathcal{O}$ *with* $O_1 \cap O_2 = \emptyset$.
(v) $\mu_o(O) = \sup_{U\in\mathcal{O},\, U\subseteq\overline{U}\subseteq O} \mu_o(U)$, $\forall O \in \mathcal{O}$.

Define $\bar{\mu}_o : {}^{X}2 \to [0, \infty) \subset \mathbb{R}$ by $\bar{\mu}_o(E) = \inf_{O\in\mathcal{O},\, E\subseteq O} \mu_o(O)$, $\forall E \subseteq X$. Then, the following statements hold:

1. *$\bar{\mu}_o$ is an outer measure. It induces a finite complete measure space $(X, \bar{\mathcal{B}}, \bar{\mu})$, where $\bar{\mathcal{B}} := \{E \subseteq X \mid E$ is measurable with respect to $\bar{\mu}_o\}$ and $\bar{\mu} := \bar{\mu}_o|_{\bar{\mathcal{B}}}$.*

2. *$\mathcal{B}_B(\mathcal{X}) \subseteq \bar{\mathcal{B}}$, and the triple $\mathbb{X} := (\mathcal{X}, \mathcal{B}_B(\mathcal{X}), \mu := \bar{\mu}|_{\mathcal{B}_B(\mathcal{X})})$ is a finite topological measure space with $\mu(O) = \mu_o(O)$, $\forall O \in \mathcal{O}$.*
3. *The measure μ is unique in the sense that if $\hat{\mu}$ be another measure on $(X, \mathcal{B}_B(\mathcal{X}))$ satisfying $\hat{\mu}(O) = \mu_o(O) = \mu(O)$, $\forall O \in \mathcal{O}$, then $\hat{\mu} = \mu$.*

Proof 1. By (i), we have $\bar{\mu}_o(\emptyset) = 0$. $\forall A \subseteq B \subseteq X$, we have $0 \leq \bar{\mu}_o(A) = \inf_{O \in \mathcal{O}, A \subseteq O} \mu_o(O) \leq \inf_{O \in \mathcal{O}, B \subseteq O} \mu_o(O) = \bar{\mu}_o(B) \leq \mu_o(X) < \infty$. $\forall E \subseteq \bigcup_{i=1}^{\infty} E_i \subseteq X$, $\forall \epsilon \in (0, +\infty) \subset \mathbb{R}$, $\forall i \in \mathbb{N}$, $\exists O_i \in \mathcal{O}$ with $E_i \subseteq O_i$ such that $\mu_o(O_i) < \bar{\mu}_o(E_i) + 2^{-i}\epsilon$. Then, $\bar{\mu}_o(E) = \inf_{O \in \mathcal{O}, E \subseteq O} \mu_o(O) \leq \mu_o(\bigcup_{i=1}^{\infty} O_i) \leq \sum_{i=1}^{\infty} \mu_o(O_i) < \sum_{i=1}^{\infty} \bar{\mu}_o(E_i) + \epsilon$, where the second inequality follows from (iii). By the arbitrariness of ϵ, we have $\bar{\mu}_o(E) \leq \sum_{i=1}^{\infty} \bar{\mu}_o(E_i)$. Hence, $\bar{\mu}_o : {}^X2 \to [0, \infty) \subset \mathbb{R}$ is an outer measure. It is easy to see that $\bar{\mu}_o(O) = \mu_o(O)$, $\forall O \in \mathcal{O}$. By Theorem 11.17, $(X, \bar{\mathcal{B}}, \bar{\mu})$ is a finite complete measure space.

2. $\forall O \in \mathcal{O}$, $\forall E \subseteq X$, $\forall \epsilon \in (0, +\infty) \subset \mathbb{R}$, $\exists O_1 \in \mathcal{O}$ with $E \subseteq O_1$ such that $\mu_o(O_1) < \bar{\mu}_o(E) + \epsilon/2$. By (v), $\exists U \in \mathcal{O}$ with $U \subseteq \overline{U} \subseteq O_1 \cap O$ such that $\mu_o(O \cap O_1) < \mu_o(U) + \epsilon/2$. Then, $\bar{\mu}_o(E) > \mu_o(O_1) - \epsilon/2 \geq \mu_o((O_1 \setminus \overline{U}) \cup U) - \epsilon/2 = \mu_o(O_1 \setminus \overline{U}) + \mu_o(U) - \epsilon/2 = \bar{\mu}_o(O_1 \setminus \overline{U}) + \mu_o(U) - \epsilon/2 > \bar{\mu}_o(O_1 \setminus (O_1 \cap O)) + \mu_o(O \cap O_1) - \epsilon = \bar{\mu}_o(O_1 \setminus O) + \bar{\mu}_o(O \cap O_1) - \epsilon \geq \bar{\mu}_o(E \setminus O) + \bar{\mu}_o(E \cap O) - \epsilon \geq \bar{\mu}_o(E) - \epsilon$, where the second inequality follows from (ii), the first equality follows from (iv), and the third, the fourth, and the last inequalities follow from the fact that $\bar{\mu}_o$ is an outer measure. By the arbitrariness of ϵ, we have $\bar{\mu}_o(E) = \bar{\mu}_o(E \setminus O) + \bar{\mu}_o(E \cap O)$. By the arbitrariness of E, O is measurable with respect to $\bar{\mu}_o$ and $O \in \bar{\mathcal{B}}$. By the arbitrariness of O, we have $\mathcal{O} \subseteq \bar{\mathcal{B}}$. Since $\bar{\mathcal{B}}$ is a σ-algebra on X, then $\mathcal{B}_B(\mathcal{X}) \subseteq \bar{\mathcal{B}}$. By Proposition 11.13, $(X, \mathcal{B}_B(\mathcal{X}), \mu)$ is a measure space. Clearly, $\mu(O) = \bar{\mu}(O) = \bar{\mu}_o(O) = \mu_o(O)$, $\forall O \in \mathcal{O}$. This coupled with the definition of $\bar{\mu}_o$ leads to the conclusion that $\mathbb{X}$ is a topological measure space. Clearly, $\mu(X) = \mu_o(X) < \infty$. Hence, $\mathbb{X}$ is a finite topological measure space.

3. Let $\hat{\mu}$ be another measure on $(X, \mathcal{B}_B(\mathcal{X}))$ satisfying $\hat{\mu}(O) = \mu_o(O) = \mu(O)$, $\forall O \in \mathcal{O}$. Then, $\hat{\mu}$ is finite. Suppose $\mu \neq \hat{\mu}$. Then, $\exists E \in \mathcal{B}_B(\mathcal{X})$ such that $\hat{\mu}(E) \neq \mu(E) = \inf_{O \in \mathcal{O}, E \subseteq O} \mu_o(O) = \inf_{O \in \mathcal{O}, E \subseteq O} \mu(O) = \inf_{O \in \mathcal{O}, E \subseteq O} \hat{\mu}(O)$. Since $\hat{\mu}(E) \leq \hat{\mu}(O)$, $\forall O \in \mathcal{O}$ with $E \subseteq O$, then we must have $\hat{\mu}(E) < \mu(E)$. Since $\mathbb{X}$ is a topological measure space, then, by Proposition 11.27, $\exists X \setminus F \in \mathcal{O}$ with $F \subseteq E$ such that $\mu(E \setminus F) < (\mu(E) - \hat{\mu}(E))/2$. This implies that $\mu(F) = \mu(E) - \mu(E \setminus F) > \mu(E)/2 + \hat{\mu}(E)/2 > \hat{\mu}(E) \geq \hat{\mu}(F)$, where the equality follows from μ being a finite measure and the last inequality follows from $E \supseteq F$ and $\hat{\mu}$ being a measure. Therefore, $\mu_o(X \setminus F) = \mu(X \setminus F) = \mu(X) - \mu(F) < \hat{\mu}(X) - \hat{\mu}(F) = \hat{\mu}(X \setminus F) = \mu_o(X \setminus F)$, where the first equality follows from $X \setminus F \in \mathcal{O}$, the second equality follows from μ being a finite measure, the inequality follows from $X \in \mathcal{O}$, the third equality follows from $\hat{\mu}$ being a measure, and the last equality follows from $X \setminus F \in \mathcal{O}$. This is a contradiction. Hence, we have $\hat{\mu} = \mu$ and μ is unique.

This completes the proof of the proposition. □

Proposition 11.194 *Let $\mathcal{X} := (X, \mathcal{O})$ be a compact Hausdorff topological space, $\mathcal{Y}$ be a Banach space, $\bar{\mu}$ be a function that assigns a vector $\bar{\mu}(F) \in \mathcal{Y}$ for each* *closed subset $F \subseteq X$, and $\mathbb{X} := (\mathcal{X}, \mathcal{B}_B(\mathcal{X}), \nu)$ be a finite topological measure space. Assume that:*

(i) $\bar{\mu}(F_1 \cup F_2) = \bar{\mu}(F_1) + \bar{\mu}(F_2)$, $\forall X \setminus F_1, X \setminus F_2 \in \mathcal{O}$ *with* $F_1 \cap F_2 = \emptyset$.
(ii) $\|\bar{\mu}(F_1) - \bar{\mu}(F_2)\| \leq \nu(O)$, $\forall X \setminus F_1, X \setminus F_2, O \in \mathcal{O}$ *with* $F_1 \Delta F_2 \subseteq O$.

Then, there exists a unique $\mu \in \mathcal{M}_{f1}(\mathcal{X}, \mathcal{Y})$ such that $\mu(F) = \bar{\mu}(F)$, $\forall X \setminus F \in \mathcal{O}$. Furthermore, $\mathcal{P} \circ \mu \leq \nu$.

Proof Fix any $E \in \mathcal{B}_B(\mathcal{X})$. Let $\hat{\mathcal{A}}_E := \{F \subseteq E \mid X \setminus F \in \mathcal{O}\}$ and $\bar{\hat{\mathcal{A}}}_E := (\hat{\mathcal{A}}_E, \subseteq)$. Clearly, $\bar{\hat{\mathcal{A}}}_E$ is a directed system.

Claim 11.194.1 The net $(\bar{\mu}(F))_{F \in \bar{\hat{\mathcal{A}}}_E}$ is Cauchy.

Proof of Claim $\forall \epsilon \in (0, +\infty) \subset \mathbb{R}$, by $\mathbb{X}$ being a topological measure space, $\exists V \in \mathcal{O}$ with $E \subseteq V$ such that $\nu(V \setminus E) < \epsilon/2$. By Proposition 11.27, $\exists X \setminus F \in \mathcal{O}$ with $F \subseteq E$ such that $\nu(E \setminus F) < \epsilon/2$. $\forall F_1 \in \bar{\hat{\mathcal{A}}}_E$ with $F \subseteq F_1$, we have $F \subseteq F_1 \subseteq E \subseteq V$ and $F \Delta F_1 = F_1 \setminus F \subseteq V \setminus F \in \mathcal{O}$. By (ii), $\|\bar{\mu}(F_1) - \bar{\mu}(F)\| \leq \nu(V \setminus F) = \nu(V \setminus E) + \nu(E \setminus F) < \epsilon$. Hence, the net is Cauchy. This completes the proof of the claim. □

By Proposition 4.44, we may define $\mu(E) = \lim_{F \in \bar{\hat{\mathcal{A}}}_E} \bar{\mu}(F) \in \mathcal{Y}$. Thus, we have defined a function $\mu : \mathcal{B}_B(\mathcal{X}) \to \mathcal{Y}$. We will show that μ is the $\mathcal{Y}$-valued measure we seek. Clearly, $\mu(F) = \bar{\mu}(F)$, $\forall X \setminus F \in \mathcal{O}$. By (i), $\bar{\mu}(\emptyset) + \bar{\mu}(\emptyset) = \bar{\mu}(\emptyset)$ and then $\mu(\emptyset) = \bar{\mu}(\emptyset) = \vartheta_{\mathcal{Y}}$.

$\forall E \in \mathcal{B}_B(\mathcal{X})$, $\forall \epsilon \in (0, +\infty) \subset \mathbb{R}$, $\exists V \in \mathcal{O}$ with $E \subseteq V$ such that $\nu(V \setminus E) < \epsilon/2$. By $\mu(E) = \lim_{F \in \bar{\hat{\mathcal{A}}}_E} \bar{\mu}(F)$, $\exists F \in \bar{\hat{\mathcal{A}}}_E$ such that $\|\mu(E) - \bar{\mu}(F)\| < \epsilon/2$. Clearly, $F \subseteq E \subseteq V$. Then, $\|\mu(E)\| < \|\bar{\mu}(F)\| + \epsilon/2 \leq \nu(V) + \epsilon/2 = \nu(E) + \nu(V \setminus E) + \epsilon/2 < \nu(E) + \epsilon$, where the second inequality follows from (ii). By the arbitrariness of ϵ, we have $\|\mu(E)\| \leq \nu(E)$.

Fix any pairwise disjoint $(E_n)_{n=1}^{\infty} \subseteq \mathcal{B}_B(\mathcal{X})$; let $E := \bigcup_{n=1}^{\infty} E_n \in \mathcal{B}_B(\mathcal{X})$. $\forall \epsilon \in (0, +\infty) \subset \mathbb{R}$, $\forall n \in \mathbb{N}$, $\exists V_n \in \mathcal{O}$ with $E_n \subseteq V_n$ such that $\nu(V_n \setminus E_n) < 2^{-n-1}\epsilon/5$. By Proposition 11.27, $\exists X \setminus F_n \in \mathcal{O}$ with $F_n \subseteq E_n$ such that $\nu(E_n \setminus F_n) < 2^{-n-1}\epsilon/5$. By $\mu(E_n) = \lim_{F \in \bar{\hat{\mathcal{A}}}_{E_n}} \bar{\mu}(F)$, $\exists \hat{F}_n \in \bar{\hat{\mathcal{A}}}_{E_n}$ with $F_n \subseteq \hat{F}_n$ such that $\left\|\mu(E_n) - \bar{\mu}(\hat{F}_n)\right\| < 2^{-n}\epsilon/5$. Clearly, $F_n \subseteq \hat{F}_n \subseteq E_n \subseteq V_n$. $\exists V \in \mathcal{O}$ with $E \subseteq V$ such that $\nu(V \setminus E) < \epsilon/5$. By Proposition 11.27, $\exists X \setminus F \in \mathcal{O}$ with $F \subseteq E$ such that $\nu(E \setminus F) < \epsilon/5$. By $\mu(E) = \lim_{F \in \bar{\hat{\mathcal{A}}}_E} \bar{\mu}(F)$, $\exists \hat{F} \in \bar{\hat{\mathcal{A}}}_E$ with $F \subseteq \hat{F}$ such that $\left\|\mu(E) - \bar{\mu}(\hat{F})\right\| < \epsilon/5$. Clearly, $F \subseteq \hat{F} \subseteq E \subseteq V$. By Proposition 5.5, $\hat{F}$ is compact. Since $\hat{F} \subseteq E = \bigcup_{n=1}^{\infty} E_n \subseteq \bigcup_{n=1}^{\infty} V_n$, then $\exists n_0 \in \mathbb{N}$ such that $\hat{F} \subseteq \bigcup_{i=1}^{n_0} V_i$. $\forall n \in \mathbb{N}$ with $n_0 \leq n$, $\left\|\mu(E) - \sum_{i=1}^{n} \mu(E_i)\right\| \leq \left\|\mu(E) - \bar{\mu}(\hat{F})\right\| + \left\|\bar{\mu}(\hat{F}) - \sum_{i=1}^{n} \bar{\mu}(\hat{F}_i)\right\| + \sum_{i=1}^{n} \left\|\bar{\mu}(\hat{F}_i) - \mu(E_i)\right\| < \epsilon/5 + \left\|\bar{\mu}(\hat{F}) - \bar{\mu}(\bigcup_{i=1}^{n} \hat{F}_i)\right\| + \epsilon/5$, where the second inequality follows from (i). Clearly,

$\hat{F}$ and $\bigcup_{i=1}^n \hat{F}_i$ are closed sets. $\hat{F}\triangle(\bigcup_{i=1}^n \hat{F}_i) = (\hat{F}\setminus(\bigcup_{i=1}^n \hat{F}_i))\cup((\bigcup_{i=1}^n \hat{F}_i)\setminus\hat{F}) \subseteq ((\bigcup_{i=1}^n V_i) \setminus (\bigcup_{i=1}^n \hat{F}_i)) \cup (E \setminus \hat{F}) \subseteq (\bigcup_{i=1}^n (V_i \setminus F_i)) \cup (V \setminus F) \in \mathcal{O}$. By (ii), $\left\|\mu(E) - \sum_{i=1}^n \mu(E_i)\right\| < 2\epsilon/5 + \nu((\bigcup_{i=1}^n (V_i \setminus F_i)) \cup (V \setminus F)) \leq 2\epsilon/5 + \sum_{i=1}^n \nu(V_i \setminus F_i) + \nu(V \setminus F) = 2\epsilon/5 + \sum_{i=1}^n (\nu(V_i \setminus E_i) + \nu(E_i \setminus F_i)) + \nu(V \setminus E) + \nu(E \setminus F) < \epsilon$. Then, $\mu(E) = \sum_{n=1}^\infty \mu(E_n) \in \mathcal{Y}$.

$\forall$ pairwise disjoint $(E_n)_{n=1}^\infty \subseteq \mathcal{B}_B(\mathcal{X})$, $\sum_{n=1}^\infty \|\mu(E_n)\| \leq \sum_{n=1}^\infty \nu(E_n) = \nu(\bigcup_{n=1}^\infty E_n) \leq \nu(X) < \infty$. This shows that μ is a $\mathcal{Y}$-valued pre-measure on $(X, \mathcal{B}_B(\mathcal{X}))$.

$\forall E \in \mathcal{B}_B(\mathcal{X})$, $\forall n \in \mathbb{Z}_+$, $\forall$ pairwise disjoint $(E_i)_{i=1}^n \subseteq \mathcal{B}_B(\mathcal{X})$ with $E = \bigcup_{i=1}^n E_i$, $\sum_{i=1}^n \|\mu(E_i)\| \leq \sum_{i=1}^n \nu(E_i) = \nu(E)$. Hence, $\mathcal{P} \circ \mu(E) \leq \nu(E)$. By the arbitrariness of E, we have $\mathcal{P} \circ \mu \leq \nu$. Hence, μ is a finite $\mathcal{Y}$-valued measure on $(X, \mathcal{B}_B(\mathcal{X}))$. It is easy to show that $(\mathcal{X}, \mathcal{B}_B(\mathcal{X}), \mu)$ is a $\mathcal{Y}$-valued topological measure space. Then, $\mu \in \mathcal{M}_{ft}(\mathcal{X}, \mathcal{Y})$.

Finally, we need to show that $\mu \in \mathcal{M}_{ft}(\mathcal{X}, \mathcal{Y})$ is unique. Let $\hat{\mu} \in \mathcal{M}_{ft}(\mathcal{X}, \mathcal{Y})$ be such that: $\hat{\mu}(F) = \bar{\mu}(F) = \mu(F)$, $\forall X \setminus F \in \mathcal{O}$. $\forall E \in \mathcal{B}_B(\mathcal{X})$, $\forall \epsilon \in (0, +\infty) \subset \mathbb{R}$, by Proposition 11.27, $\exists X \setminus F \in \mathcal{O}$ with $F \subseteq E$ such that $\mathcal{P} \circ \mu(E \setminus F) < \epsilon/2$. Again by Proposition 11.27, $\exists X \setminus F_1 \in \mathcal{O}$ with $F_1 \subseteq E$ such that $\mathcal{P} \circ \hat{\mu}(E \setminus F_1) < \epsilon/2$. Let $\bar{F} := F \cup F_1 \subseteq E$, which is clearly closed. Then, $\left\|\mu(E) - \hat{\mu}(E)\right\| \leq \left\|\mu(E) - \mu(\bar{F})\right\| + \left\|\mu(\bar{F}) - \hat{\mu}(\bar{F})\right\| + \left\|\hat{\mu}(\bar{F}) - \hat{\mu}(E)\right\| = \left\|\mu(E \setminus \bar{F})\right\| + 0 + \left\|\hat{\mu}(E \setminus \bar{F})\right\| \leq \mathcal{P} \circ \mu(E \setminus \bar{F}) + \mathcal{P} \circ \hat{\mu}(E \setminus \bar{F}) \leq \mathcal{P} \circ \mu(E \setminus F) + \mathcal{P} \circ \hat{\mu}(E \setminus F_1) < \epsilon$. By the arbitrariness of ϵ, we have $\mu(E) = \hat{\mu}(E)$. Hence, $\mu = \hat{\mu}$.

This completes the proof of the proposition. □

Definition 11.195 Let $\mathcal{X} := (X, \mathcal{O})$ be a topological space. $E \subseteq X$ is said to be a G_δ if $\exists (O_i)_{i=1}^\infty \subseteq \mathcal{O}$ such that $E = \bigcap_{i=1}^\infty O_i$. $\bar{E} \subseteq X$ is said to be an F_σ if exists $\left(\widetilde{F}_i\right)_{i=1}^\infty \subseteq \mathcal{O}$ such that $\bar{E} = \bigcup_{i=1}^\infty F_i$. ◇

Proposition 11.196 *Let $\mathcal{X} := (X, \rho)$ be a locally compact separable metric space with the natural topology $\mathcal{O}$, $\mathcal{E} := \{E \subseteq X \mid E$ is a compact $G_\delta\}$, and $\mathcal{B}_a(\mathcal{X})$ be the σ-algebra generated by $\mathcal{E}$. Then, $\mathcal{B}_a(\mathcal{X}) = \mathcal{B}_B(\mathcal{X})$.*

Proof $\forall E \in \mathcal{E}$, by Proposition 5.5, E is closed. Then, $E \in \mathcal{B}_B(\mathcal{X})$. Hence, $\mathcal{E} \subseteq \mathcal{B}_B(\mathcal{X})$. Since $\mathcal{B}_B(\mathcal{X})$ is a σ-algebra, then $\mathcal{B}_a(\mathcal{X}) \subseteq \mathcal{B}_B(\mathcal{X})$.

$\forall x \in X$, by Definition 5.49, $\exists O_x \in \mathcal{O}$ such that $x \in O_x$ and $\overline{O_x}$ is compact. Then, $X = \bigcup_{x \in X} O_x$. By Propositions 4.4 and 3.24, $\exists$ a countable set $D \subseteq X$ such that $X = \bigcup_{x \in D} O_x = \bigcup_{x \in D} \overline{O_x}$.

$\forall \widetilde{F} \in \mathcal{O}$, $\forall x \in D$, $F_x := F \cap \overline{O_x}$ is compact and closed by Propositions 5.5 and 3.5. Then, $F_x = \bigcap_{n=1}^\infty \bigcup_{\bar{x} \in F_x} B_{\mathcal{X}}(\bar{x}, 1/n)$, by Proposition 4.10. This implies that F_x is a G_δ in addition to being compact. Then, $F_x \in \mathcal{E} \subseteq \mathcal{B}_a(\mathcal{X})$. This implies that $F = \bigcup_{x \in D} F_x \in \mathcal{B}_a(\mathcal{X})$. Hence, $\widetilde{F} \in \mathcal{B}_a(\mathcal{X})$. By the arbitrariness of $\widetilde{F}$, we have $\mathcal{O} \subseteq \mathcal{B}_a(\mathcal{X})$. By $\mathcal{B}_a(\mathcal{X})$ being a σ-algebra, we have $\mathcal{B}_B(\mathcal{X}) \subseteq \mathcal{B}_a(\mathcal{X})$.

Therefore, $\mathcal{B}_a(\mathcal{X}) = \mathcal{B}_B(\mathcal{X})$. This completes the proof of the proposition. □

Lemma 11.197 *Let $\mathcal{X} := (X, \mathcal{O})$ be a normal topological space. Then, $\forall \widetilde{E} \in \mathcal{O}$ with E being a G_δ, there exists a continuous function $\phi : X \to [0, 1] \subset \mathbb{R}$ such that $E = \{x \in X \mid \phi(x) = 1\}$.*

Proof Since E is a G_δ, then $\exists (O_i)_{i=1}^\infty \subseteq \mathcal{O}$ such that $E = \bigcap_{i=1}^\infty O_i$. $\forall i \in \mathbb{N}$, by Urysohn's Lemma 3.55, there exists a continuous function $\phi_i : X \to [0, 1] \subset \mathbb{R}$ such that $\phi_i|_E = 1$ and $\phi_i|_{\widetilde{O_i}} = 0$. Let $\phi := \sum_{i=1}^\infty 2^{-i}\phi_i$. By Proposition 4.26, $\phi : X \to [0, 1] \subset \mathbb{R}$ is continuous. Clearly, $\phi|_E = 1$ and $\phi(x) < 1$, $\forall x \in \bigcup_{i=1}^\infty \widetilde{O_i} = \widetilde{E}$. Hence, the result holds. This completes the proof of the lemma. □

Theorem 11.198 *Let $\mathcal{X} := (X, \rho)$ be a locally compact separable metric space with the natural topology $\mathcal{O}$ and μ be a finite measure on $(X, \mathcal{B}_B(\mathcal{X}))$. Then, $\mathbb{X} := (\mathcal{X}, \mathcal{B}_B(\mathcal{X}), \mu)$ is a finite metric measure space. Thus, $\mathcal{M}_{ft}(\mathcal{X}) = \mathcal{M}_f(X, \mathcal{B}_B(\mathcal{X}))$. As a consequence, $\mathcal{M}_f(X, \mathcal{B}_B(\mathcal{X}), \mathcal{Y}) = \mathcal{M}_{ft}(\mathcal{X}, \mathcal{Y})$, where $\mathcal{Y}$ is any normed linear space.*

Proof By Propositions 4.4, 3.24, and 5.72, $\mathcal{X}$ is σ-compact. Let $\mathcal{R} \subseteq \mathcal{B}_B(\mathcal{X})$ be such that $E \in \mathcal{R}$ if $\forall \epsilon \in (0, +\infty) \subset \mathbb{R}$, we have:

(i) $\exists O \in \mathcal{O}$ with O being σ-compact and $E \subseteq O$ such that $\mu(O \setminus E) < \epsilon$.
(ii) $\exists \widetilde{F} \in \mathcal{O}$ with F being a compact G_δ and $F \subseteq E$ such that $\mu(E \setminus F) < \epsilon$.
We will show that $\mathcal{R} = \mathcal{B}_B(\mathcal{X})$.

Claim 11.198.1 $\forall \widetilde{F} \in \mathcal{O}$, $F = \bigcap_{n=1}^\infty \bigcup_{x\in F} \mathcal{B}_{\mathcal{X}}(x, 1/n) =: \bar{F}$ and is a G_δ.

Proof of Claim Clearly, $F \subseteq \bar{F}$. $\forall x_0 \in \widetilde{F}$, by Proposition 4.10, $\epsilon_0 := \operatorname{dist}(x_0, F) > 0$. Then, $\forall n \in \mathbb{N}$ with $1/n < \epsilon_0$, $x_0 \notin \bigcup_{x\in F} \mathcal{B}_{\mathcal{X}}(x, 1/n)$ and $x_0 \in \widetilde{\bar{F}}$. Hence, $\widetilde{F} \subseteq \widetilde{\bar{F}}$ and $F = \bar{F}$. Clearly, $\bar{F}$ is a G_δ. This completes the proof of the claim. □

Claim 11.198.2 $\forall O \in \mathcal{O}$, O is σ-compact.

Proof of Claim By Claim 11.198.1, $\widetilde{O}$ is a G_δ. Then, $\exists (O_i)_{i=1}^\infty \subseteq \mathcal{O}$ such that $\widetilde{O} = \bigcap_{i=1}^\infty O_i$ and $O = \bigcup_{i=1}^\infty \widetilde{O_i}$. By $\mathcal{X}$ being σ-compact, $\exists$ compact sets $(K_i)_{i=1}^\infty$ such that $X = \bigcup_{i=1}^\infty K_i$. Then, $O = O \cap X = \bigcup_{i=1}^\infty \bigcup_{j=1}^\infty (\widetilde{O_i} \cap K_j)$. $\forall i, j \in \mathbb{N}$, by Propositions 3.5 and 5.5, $\widetilde{O_i} \cap K_j$ is compact. Hence, O is σ-compact. □

$\forall (E_i)_{i=1}^\infty \subseteq \mathcal{R}$, let $E := \bigcup_{i=1}^\infty E_i \in \mathcal{B}_B(\mathcal{X})$. $\forall \epsilon \in (0, +\infty) \subset \mathbb{R}$, $\forall i \in \mathbb{N}$, $\exists O_i \in \mathcal{O}$ with O_i being σ-compact and $E_i \subseteq O_i$ such that $\mu(O_i \setminus E_i) < 2^{-i}\epsilon$; $\exists \widetilde{F_i} \in \mathcal{O}$ with F_i being a compact G_δ and $F_i \subseteq E_i$ such that $\mu(E_i \setminus F_i) < 2^{-i-1}\epsilon$. Let $O := \bigcup_{i=1}^\infty O_i \in \mathcal{O}$, which is σ-compact by Claim 11.198.2. Then, $E \subseteq O$ and $\mu(O \setminus E) = \mu((\bigcup_{i=1}^\infty O_i) \setminus (\bigcup_{i=1}^\infty E_i)) \leq \mu(\bigcup_{i=1}^\infty (O_i \setminus E_i)) \leq \sum_{i=1}^\infty \mu(O_i \setminus E_i) < \epsilon$. Hence, E satisfies (i). By Proposition 11.7, $\mu(E) = \lim_{n\in\mathbb{N}} \mu(\bigcup_{i=1}^n E_i)$. Then, $\exists n_0 \in \mathbb{N}$ such that $\mu(E) < \mu(\bigcup_{i=1}^{n_0} E_i) + \epsilon/2$. Let $F := \bigcup_{i=1}^{n_0} F_i$. Clearly, F is compact and closed. By Claim 11.198.1, F is a G_δ. Then, $F \subseteq E$ and $\mu(E \setminus F) = \mu(E) - \mu(F) < \mu(\bigcup_{i=1}^{n_0} E_i) - \mu(\bigcup_{i=1}^{n_0} F_i) + \epsilon/2 = \mu((\bigcup_{i=1}^{n_0} E_i) \setminus (\bigcup_{i=1}^{n_0} F_i)) + \epsilon/2 \leq \mu(\bigcup_{i=1}^{n_0} (E_i \setminus F_i)) + \epsilon/2 \leq \sum_{i=1}^{n_0} \mu(E_i \setminus F_i) + \epsilon/2 < \epsilon$. Hence, E satisfies (ii). Then, $E \in \mathcal{R}$. Thus, $\mathcal{R}$ is closed under countable unions.

$\forall E \in \mathcal{R}$, $\forall \epsilon \in (0, +\infty) \subset \mathbb{R}$, $\exists O \in \mathcal{O}$ with O being σ-compact and $E \subseteq O$ such that $\mu(O \setminus E) < \epsilon$; $\exists \widetilde{F} \in \mathcal{O}$ with F being a compact G_δ and $F \subseteq E$ such that $\mu(E \setminus F) < \epsilon$. Then, $\widetilde{O}$ is a closed set, $\widetilde{O} \subseteq \widetilde{E}$, and $\mu(\widetilde{E} \setminus \widetilde{O}) = \mu(O \setminus E) < \epsilon$. By $\mathcal{X}$ being σ-compact, $\exists$ compact sets $(K_i)_{i=1}^\infty$ such that $X = \bigcup_{i=1}^\infty K_i$. Without loss of generality, we may assume that $K_i \subseteq K_{i+1}$, $\forall i \in \mathbb{N}$. Note that $\epsilon > \mu(\widetilde{E} \setminus \widetilde{O}) = \mu(\widetilde{E} \setminus (\widetilde{O} \cap \bigcup_{i=1}^{\infty} K_i)) = \mu(\widetilde{E} \setminus (\bigcup_{i=1}^{\infty} (\widetilde{O} \cap K_i))) = \mu(\bigcap_{i=1}^{\infty} (\widetilde{E} \setminus (\widetilde{O} \cap K_i))) = \lim_{i \in \mathbb{N}} \mu(\widetilde{E} \setminus (\widetilde{O} \cap K_i))$, where the last equality follows from Proposition 11.5. Then, $\exists n \in \mathbb{N}$ such that $\mu(\widetilde{E} \setminus (\widetilde{O} \cap K_n)) < \epsilon$. By Propositions 3.5 and 5.5, $\widetilde{O} \cap K_n$ is closed and compact; then it is a G_δ by Claim 11.198.1. Clearly, $\widetilde{O} \cap K_n \subseteq \widetilde{E}$. Hence, $\widetilde{E}$ satisfies (ii). $\widetilde{F} \in \mathcal{O}$ and is σ-compact by Claim 11.198.2. Note that $\widetilde{E} \subseteq \widetilde{F}$ and $\mu(\widetilde{F} \setminus \widetilde{E}) = \mu(E \setminus F) < \epsilon$. Then, $\widetilde{E}$ satisfies (i). Hence, $\widetilde{E} \in \mathcal{R}$. $\mathcal{R}$ is closed under set complements. Clearly, $\emptyset \in \mathcal{R}$. This proves that $\mathcal{R}$ is a σ-algebra.

$\forall E \in \mathcal{O}$ with E being a compact G_δ, $\forall \epsilon \in (0, +\infty) \subset \mathbb{R}$, then E satisfies (ii) trivially. By Lemma 11.197, there exists a continuous function $\phi : X \to [0, 1] \subset \mathbb{R}$ such that $E = \{x \in X \mid \phi(x) = 1\}$. $\forall i \in \mathbb{N}$, let $O_i := \{x \in X \mid \phi(x) > 1 - 1/i\} \in \mathcal{O}$. Clearly, $E = \bigcap_{i=1}^{\infty} O_i$ and $O_{i+1} \subseteq O_i$, $\forall i \in \mathbb{N}$. By Proposition 11.5, $\mu(E) = \lim_{i \in \mathbb{N}} \mu(O_i)$. $\exists n \in \mathbb{N}$ such that $\mu(O_n) - \epsilon \leq \mu(E) \leq \mu(O_n)$. Clearly, we have $E \subseteq O_n$. Then, $\mu(O_n \setminus E) = \mu(O_n) - \mu(E) < \epsilon$. By Claim 11.198.2, O_n is σ-compact. Then, E satisfies (i) and $E \in \mathcal{R}$.

Thus, $\mathcal{R}$ is a σ-algebra and $\mathcal{E} := \{E \in \mathcal{B}_B(\mathcal{X}) \mid E \text{ is a compact } G_\delta\} \subseteq \mathcal{R}$. By Proposition 11.196, $\mathcal{R} = \mathcal{B}_B(\mathcal{X})$. Then, $\mathbb{X}$ is a finite metric measure space.

Thus, we have $\mathcal{M}_f(X, \mathcal{B}_B(\mathcal{X})) \subseteq \mathcal{M}_{ft}(\mathcal{X})$. Clearly, $\mathcal{M}_f(X, \mathcal{B}_B(\mathcal{X})) \supseteq \mathcal{M}_{ft}(\mathcal{X})$. Hence, $\mathcal{M}_f(X, \mathcal{B}_B(\mathcal{X})) = \mathcal{M}_{ft}(\mathcal{X})$.

$\forall \bar{\mu} \in \mathcal{M}_f(X, \mathcal{B}_B(\mathcal{X}), \mathcal{Y})$, then $\mathcal{P} \circ \bar{\mu} \in \mathcal{M}_f(X, \mathcal{B}_B(\mathcal{X})) = \mathcal{M}_{ft}(\mathcal{X})$. This implies that $\bar{\mu} \in \mathcal{M}_{ft}(\mathcal{X}, \mathcal{Y})$. Hence, $\mathcal{M}_f(X, \mathcal{B}_B(\mathcal{X}), \mathcal{Y}) \subseteq \mathcal{M}_{ft}(\mathcal{X}, \mathcal{Y})$. Clearly, $\mathcal{M}_f(X, \mathcal{B}_B(\mathcal{X}), \mathcal{Y}) \supseteq \mathcal{M}_{ft}(\mathcal{X}, \mathcal{Y})$. Then, $\mathcal{M}_f(X, \mathcal{B}_B(\mathcal{X}), \mathcal{Y}) = \mathcal{M}_{ft}(\mathcal{X}, \mathcal{Y})$.

This completes the proof of the theorem. □

Definition 11.199 Let $\mathcal{X} := (X, \mathcal{O})$ be a topological space, $\mathcal{Y}$ be a normed linear space, and $f : X \to \mathcal{Y}$. The *support of* f is the set $\text{supp}(f) := \overline{\{x \in X \mid f(x) \neq \vartheta_{\mathcal{Y}}\}}$. ◇

Lemma 11.200 *Let $\mathcal{X} := (X, \mathcal{O})$ be compact Hausdorff topological space, $\mathcal{Z} := \mathcal{C}(\mathcal{X}, \mathbb{R})$, and $f \in \mathcal{Z}^*$. $\bar{f} \in \mathcal{Z}^*$ is said to be a* positive linear functional *if $\bar{f}(z) \geq 0$, $\forall z \in P := \{h \in \mathcal{Z} \mid h : X \to [0, \infty) \subset \mathbb{R}\}$. Then, $f = f_+ - f_-$, where $f_+, f_- \in \mathcal{Z}^*$ are positive linear functionals.*

Proof $\forall z \in P$, define $f_+(z) := \sup_{\phi \in \mathcal{Z},\, 0 \leq \phi(x) \leq z(x),\, \forall x \in X} f(\phi)$. Then, $0 \leq f_+(z) \leq \|f\| \|z\| < \infty$ and $f_+(z) \geq f(z)$, $\forall z \in P$. Clearly:

(i) $f_+(\alpha z) = \alpha f_+(z)$, $\forall z \in P$ and $\forall \alpha \in [0, \infty) \subset \mathbb{R}$.

$\forall z_1, z_2 \in P$, $\forall i \in \{1, 2\}$, $\forall \phi_i \in \mathcal{Z}$ with $0 \leq \phi_i(x) \leq z_i(x)$, $\forall x \in X$, we have $f(\phi_1) + f(\phi_2) = f(\phi_1 + \phi_2) \leq f_+(z_1 + z_2)$. Then, $f_+(z_1) + f_+(z_2) \leq f_+(z_1 + z_2)$. On the other hand, $\forall \phi \in \mathcal{Z}$ with $0 \leq \phi(x) \leq z_1(x) + z_2(x)$, $\forall x \in X$, we have $f(\phi) = f(\phi \wedge z_1) + f(\phi - \phi \wedge z_1) \leq f_+(z_1) + f_+(z_2)$. Then, $f_+(z_1 + z_2) \leq f_+(z_1) + f_+(z_2)$. Therefore:

(ii) $f_+(z_1) + f_+(z_2) = f_+(z_1 + z_2)$, $\forall z_1, z_2 \in P$.

$\forall z \in \mathcal{Z}$, define $z_+ := z \vee 0 \in P$ and $z_- := (-z) \vee 0 \in P$. Clearly, $z = z_+ - z_-$. Define $f_+(z) := f_+(z_+) - f_+(z_-) \in \mathbb{R}$. Then, $f_+ : \mathcal{Z} \to \mathbb{R}$ is well-defined. We will show that $f_+ \in \mathcal{Z}^*$.

$\forall z \in \mathcal{Z}$, $|f_+(z)| = |f_+(z_+) - f_+(z_-)| \leq f_+(z_+) + f_+(z_-) = f_+(z_+ + z_-) \leq \|f\|\|z_+ + z_-\| = \|f\|\|z\|$.

$\forall z_1, z_2 \in \mathcal{Z}$, let $z := z_1 + z_2$. Then, $f_+(z_1) + f_+(z_2) = f_+(z_{1+}) - f_+(z_{1-}) + f_+(z_{2+}) - f_+(z_{2-}) = f_+(z_{1+} + z_{2+}) - f_+(z_{1-} + z_{2-}) = f_+(z_+) + f_+(z_{1+} + z_{2+} - z_+) - f_+(z_-) - f_+(z_{1-} + z_{2-} - z_-) = f_+(z) + f_+(z_{1+} + z_{2+} - z_+) - f_+(z_{1-} + z_{2-} - z_-)$, where the first equality follows from the definition of f_+, the second equality follows from (ii), the third equality follows from (ii), and the last equality follows from the definition of f_+. Note that, $\forall x \in X$,

$$\begin{aligned}
(z_{1+} + z_{2+} - z_+)(x) &= z_1(x) \vee 0 + z_2(x) \vee 0 - (z_1(x) + z_2(x)) \vee 0 \\
&= \begin{cases} 0 & z_1(x) \geq 0 \text{ and } z_2(x) \geq 0 \\ -z_2(x) & z_1(x) \geq 0 > z_2(x) \geq -z_1(x) \\ z_1(x) & z_1(x) \geq 0 \geq -z_1(x) > z_2(x) \\ -z_1(x) & z_2(x) \geq 0 > z_1(x) \geq -z_2(x) \\ z_2(x) & z_2(x) \geq 0 \geq -z_2(x) > z_1(x) \\ 0 & z_1(x) < 0 \text{ and } z_2(x) < 0 \end{cases} \\
&= (-z_1(x)) \vee 0 + (-z_2(x)) \vee 0 - (-z_1(x) - z_2(x)) \vee 0) \\
&= z_{1-}(x) + z_{2-}(x) - z_-(x)
\end{aligned}$$

Hence, (iii) $f_+(z_1) + f_+(z_2) = f_+(z) = f_+(z_1 + z_2)$, $\forall z_1, z_2 \in \mathcal{Z}$.

$\forall z \in \mathcal{Z}$, $\forall \alpha \in \mathbb{R}$. If $\alpha = 0$, then $f_+(\alpha z) = 0 = \alpha f_+(z)$. If $\alpha > 0$, then $f_+(\alpha z) = f_+(\alpha z_+) - f_+(\alpha z_-) = \alpha f_+(z_+) - \alpha f_+(z_-) = \alpha f_+(z)$, where the first equality follows from the definition of f_+, the second equality follows from (i), and the third equality follows from the definition of f_+. If $\alpha < 0$, then $f_+(\alpha z) = f_+(-\alpha z_-) - f_+(-\alpha z_+) = -\alpha f_+(z_-) + \alpha f_+(z_+) = \alpha f_+(z)$, where the first equality follows from the definition of f_+, the second equality follows from (i), and the third equality follows from the definition of f_+. Hence, we have:
(iv) $f_+(\alpha z) = \alpha f_+(z)$, $\forall z \in \mathcal{Z}$ and $\forall \alpha \in \mathbb{R}$.

Hence, $f_+ \in \mathcal{Z}^*$ and is a positive linear functional.

Let $f_- := f_+ - f \in \mathcal{Z}^*$. Clearly, $f_-(z) \geq 0$, $\forall z \in P$. Hence, f_- is also a positive linear functional.

This completes the proof of the lemma. □

Theorem 11.201 (Riesz Representation Theorem) *Let $\mathcal{X} := (X, \mathcal{O})$ be a compact Hausdorff topological space, $\mathcal{Y}$ be a normed linear space over $\mathbb{K}$, $\mathcal{Z} := \mathcal{C}(\mathcal{X}, \mathcal{Y})$, and $Z := \{z \in \mathcal{Z} \mid z = hy,\ h \in \mathcal{C}(\mathcal{X}, \mathbb{R}),\ y \in \mathcal{Y}\}$. Assume that $\overline{\text{span}\,(Z)} = \mathcal{Z}$. Then, $\forall f \in \mathcal{Z}^*$, $\exists!\ \mu \in \mathcal{M}_{ft}(\mathcal{X}, \mathcal{Y}^*)$ such that $f(z) = \int_X \langle\langle \mathrm{d}\mu(x), z(x)\rangle\rangle =: \langle\langle \mu, z\rangle\rangle$, $\forall z \in \mathcal{Z}$. Furthermore, $\mathcal{Z}^* = \mathcal{M}_{ft}(\mathcal{X}, \mathcal{Y}^*)$ isometrically isomorphically.*

Proof By Proposition 5.14, $\mathcal{X}$ is normal. By Example 7.31, $\mathcal{Z}$ is a normed linear space over $\mathbb{K}$. Fix any $f \in \mathcal{Z}^*$. Define $\nu_o : \mathcal{O} \to [0, \|f\|] \subset \mathbb{R}$ by $\nu_o(O) := \sup_{z \in \mathcal{Z}, \|z\| \leq 1, \text{supp}(z) \subseteq O} |f(z)|$, $\forall O \in \mathcal{O}$. Clearly, we have:
(i) $\nu_o(\emptyset) = 0$.
(ii) $0 \leq \nu_o(O) \leq \|f\| < \infty$, $\forall O \in \mathcal{O}$.

(iii) $\nu_o(O_1) \leq \nu_o(O_2)$, $\forall O_1, O_2 \in \mathcal{O}$ with $O_1 \subseteq O_2$.

$\forall (O_i)_{i=1}^{\infty} \subseteq \mathcal{O}$, let $O := \bigcup_{i=1}^{\infty} O_i \in \mathcal{O}$. $\forall z \in \mathcal{Z}$ with $K := \text{supp}(z) \subseteq O$ and $\|z\| \leq 1$, by Proposition 5.5, K is compact. By Corollary 5.65 of Partition of Unity, $\exists m \in \mathbb{Z}_+$, $\exists n_1, \ldots, n_m \in \mathbb{N}$, which may be taken to be distinct, $\exists (\varphi_{n_i})_{i=1}^{m} \subseteq \mathcal{C}(\mathcal{X}, \mathbb{R})$ such that $\varphi_{n_i} : X \to [0, 1] \subset \mathbb{R}$, $\text{supp}(\varphi_{n_i}) \subseteq O_{n_i}$, $i = 1, \ldots, m$, $\sum_{i=1}^{m} \varphi_{n_i}(x) = 1$, $\forall x \in K$, and $0 \leq \sum_{i=1}^{m} \varphi_{n_i}(x) \leq 1$, $\forall x \in X$. Then, $z = \sum_{i=1}^{m} z\varphi_{n_i}$. $\forall i \in \{1, \ldots, m\}$, $z\varphi_{n_i} \in \mathcal{Z}$, $\|z\varphi_{n_i}\| \leq 1$, and $\text{supp}(z\varphi_{n_i}) \subseteq \overline{\text{supp}(z) \cap \text{supp}(\varphi_{n_i})} \subseteq \text{supp}(\varphi_{n_i}) \subseteq O_{n_i}$. Then, $|f(z)| = \left|\sum_{i=1}^{m} f(z\varphi_{n_i})\right| \leq \sum_{i=1}^{m} \left|f(z\varphi_{n_i})\right| \leq \sum_{i=1}^{m} \nu_o(O_{n_i}) \leq \sum_{i=1}^{\infty} \nu_o(O_i)$. By the arbitrariness of z, we have:

(iv) $\nu_o(O) = \nu_o(\bigcup_{i=1}^{\infty} O_i) \leq \sum_{i=1}^{\infty} \nu_o(O_i)$, $\forall (O_i)_{i=1}^{\infty} \subseteq \mathcal{O}$.

$\forall O_1, O_2 \in \mathcal{O}$ with $O_1 \cap O_2 = \emptyset$, $\forall \epsilon \in (0, +\infty) \subset \mathbb{R}$, $\exists \bar{z}_i \in \mathcal{Z}$ with $\text{supp}(\bar{z}_i) \subseteq O_i$ and $\|\bar{z}_i\| \leq 1$, such that $|f(\bar{z}_i)| > \nu_o(O_i) - \epsilon/2$, $i = 1, 2$. $\forall i \in \{1, 2\}$, when $f(\bar{z}_i) \neq 0$, take $z_i := \frac{|f(\bar{z}_i)|}{f(\bar{z}_i)} \bar{z}_i \in \mathcal{Z}$; and when $f(\bar{z}_i) = 0$, take $z_i := \bar{z}_i \in \mathcal{Z}$. Then, $z_i \in \mathcal{Z}$, $\|z_i\| = \|\bar{z}_i\| \leq 1$, $\text{supp}(z_i) = \text{supp}(\bar{z}_i) \subseteq O_i$, and $f(z_i) = |f(\bar{z}_i)| > \nu_o(O_i) - \epsilon/2$. Let $z := z_1 + z_2 \in \mathcal{Z}$. Clearly, $\|z\| \leq 1$ since $O_1 \cap O_2 = \emptyset$. Note that $\text{supp}(z) \subseteq \overline{\text{supp}(z_1) \cup \text{supp}(z_2)} = \text{supp}(z_1) \cup \text{supp}(z_2) \subseteq O_1 \cup O_2$, where the equality follows from Proposition 3.3. Then, $\nu_o(O_1) + \nu_o(O_2) - \epsilon < f(z_1) + f(z_2) = f(z) \leq \nu_o(O_1 \cup O_2) \leq \nu_o(O_1) + \nu_o(O_2)$, where the last inequality follows from (iv). By the arbitrariness of ϵ, we have:

(v) $\nu_o(O_1 \cup O_2) = \nu_o(O_1) + \nu_o(O_2)$, $\forall O_1, O_2 \in \mathcal{O}$ with $O_1 \cap O_2 = \emptyset$.

$\forall O \in \mathcal{O}$, $\forall \epsilon \in (0, +\infty) \subset \mathbb{R}$, $\exists z \in \mathcal{Z}$ with $K := \text{supp}(z) \subseteq O$ and $\|z\| \leq 1$, such that $|f(z)| > \nu_o(O) - \epsilon$. By Proposition 5.5, K is compact. By Proposition 3.35, $\exists U \in \mathcal{O}$, such that $K \subseteq U \subseteq \overline{U} \subseteq O$. Then, $\nu_o(O) \geq \nu_o(U) \geq |f(z)| > \nu_o(O) - \epsilon$, where the first inequality follows from (iii). Hence, we have:

(vi) $\nu_o(O) = \sup_{U \in \mathcal{O},\, U \subseteq \overline{U} \subseteq O} \nu_o(U)$, $\forall O \in \mathcal{O}$.

By Proposition 11.193, there exists a finite topological measure space $\bar{\mathbb{X}} := (\mathcal{X}, \mathcal{B}_{\mathrm{B}}(\mathcal{X}), \nu)$ such that $\nu(O) = \nu_o(O)$, $\forall O \in \mathcal{O}$ and $\nu(E) = \inf_{O \in \mathcal{O},\, E \subseteq O} \nu_o(O)$, $\forall E \in \mathcal{B}_{\mathrm{B}}(\mathcal{X})$. Furthermore, ν is unique among measures $\hat{\nu}$ on $(\mathcal{X}, \mathcal{B}_{\mathrm{B}}(\mathcal{X}))$ with $\hat{\nu}|_{\mathcal{O}} = \nu_o$. Clearly, $\nu \in \mathcal{M}_{ft}(\mathcal{X})$.

Fix any $\widetilde{F} \in \mathcal{O}$. Let $\mathcal{A}_F := \{(U, V, h) \in \mathcal{O} \times \mathcal{O} \times \mathcal{C}(\mathcal{X}, \mathbb{R}) \mid F \subseteq U \subseteq \overline{U} \subseteq V,\ h : X \to [0, 1] \subset \mathbb{R},\ h|_{\overline{U}} = 1,\ \text{supp}(h) \subseteq V\}$. Define a relation $\prec$ on $\mathcal{A}_F$ by $(U_1, V_1, h_1) \prec (U_2, V_2, h_2)$ if $V_1 \supseteq V_2$. By Proposition 3.35 and Urysohn's Lemma 3.55, $\bar{\mathcal{A}}_F := (\mathcal{A}_F, \prec)$ is a directed system. $\forall (U, V, h) \in \bar{\mathcal{A}}_F$, the function $f_h : \mathcal{Y} \to \mathbb{K}$, defined by $f_h(y) = f(hy)$, $\forall y \in \mathcal{Y}$, is a bounded linear functional since $f \in \mathcal{Z}^*$. Then, $f_h \in \mathcal{Y}^*$. This defines a net $(f_h)_{(U,V,h) \in \bar{\mathcal{A}}_F} \subseteq \mathcal{Y}^*$.

Claim 11.201.1 The net $(f_h)_{(U,V,h) \in \bar{\mathcal{A}}_F} \subseteq \mathcal{Y}^*$ is Cauchy.

Proof of Claim $\forall \epsilon \in (0, +\infty) \subset \mathbb{R}$, $\exists V \in \mathcal{O}$ with $F \subseteq V$ such that $\nu(V \setminus F) < \epsilon$. By Proposition 3.35, $\exists U, \check{U} \in \mathcal{O}$ such that $F \subseteq U \subseteq \overline{U} \subseteq \check{U} \subseteq \overline{\check{U}} \subseteq V$. By Urysohn's Lemma 3.55, $\exists h \in \mathcal{C}(\mathcal{X}, \mathbb{R})$ with $h : X \to [0, 1] \subset \mathbb{R}$ such that $h|_{\overline{U}} = 1$ and $h|_{\widetilde{\check{U}}} = 0$. Then, $\text{supp}(h) \subseteq \overline{\check{U}} \subseteq V$ and $(U, V, h) \in \bar{\mathcal{A}}_F$. $\forall (U_1, V_1, h_1) \in \mathcal{A}_F$ with $(U, V, h) \prec (U_1, V_1, h_1)$, we have $V_1 \subseteq V$ and $\left\| f_h - f_{h_1} \right\| = \sup_{y \in \mathcal{Y},\, \|y\| \le 1} \left| \langle\langle f_h, y \rangle\rangle - \langle\langle f_{h_1}, y \rangle\rangle \right| = \sup_{y \in \mathcal{Y},\, \|y\| \le 1} |f(hy) - f(h_1 y)| = \sup_{y \in \mathcal{Y},\, \|y\| \le 1} |f((h - h_1)y)|$. Note that, $\forall y \in \mathcal{Y}$ with $\|y\| \le 1$, $(h - h_1)y \in \mathcal{Z}$, $\|(h - h_1)y\| \le 1$, and

$$
\begin{aligned}
&\operatorname{supp}((h - h_1)y) \subseteq \operatorname{supp}(h - h_1) \\
&\subseteq \overline{(\operatorname{supp}(h) \cup \operatorname{supp}(h_1)) \setminus \{x \in X \mid h(x) = 1 = h_1(x)\}} \\
&\subseteq \overline{(\operatorname{supp}(h) \cup \operatorname{supp}(h_1)) \setminus (\overline{U} \cap \overline{U_1})} \\
&= \overline{\operatorname{supp}(h) \setminus (\overline{U} \cap \overline{U_1})} \cup \overline{\operatorname{supp}(h_1) \setminus (\overline{U} \cap \overline{U_1})} \\
&\subseteq \overline{\operatorname{supp}(h) \setminus (U \cap U_1)} \cup \overline{\operatorname{supp}(h_1) \setminus (U \cap U_1)} \\
&= (\operatorname{supp}(h) \setminus (U \cap U_1)) \cup (\operatorname{supp}(h_1) \setminus (U \cap U_1)) \\
&\subseteq (V \setminus F) \cup (V_1 \setminus F) \subseteq V \setminus F \in \mathcal{O}
\end{aligned}
$$

where the third containment follows from the fact that $h|_{\overline{U}} = 1$ and $h_1|_{\overline{U_1}} = 1$ and the first equality follows from Proposition 3.3. Then, $\left\| f_h - f_{h_1} \right\| \le \nu_o(V \setminus F) = \nu(V \setminus F) < \epsilon$. Hence, the net is Cauchy. This completes the proof of the claim. □

By Propositions 4.44 and 7.72:
(vii) $\lim_{(U,V,h) \in \bar{\mathcal{A}}_F} f_h =: \bar{\mu}(F) \in \mathcal{Y}^*$, $\forall \widetilde{F} \in \mathcal{O}$.
Thus, we have defined a function $\bar{\mu}$ that assigns a vector $\bar{\mu}(F) \in \mathcal{Y}^*$ to each closed subset $F \subseteq X$. We will next show that $\bar{\mu}$ satisfies the assumption of Proposition 11.194.

Fix any $\widetilde{F_1}, \widetilde{F_2} \in \mathcal{O}$ with $F_1 \cap F_2 = \emptyset$. Fix any $\epsilon \in (0, \infty) \subset \mathbb{R}$. By the normality of $\mathcal{X}$, $\exists V_1, V_2 \in \mathcal{O}$ with $V_1 \cap V_2 = \emptyset$ such that $F_i \subseteq V_i$, $i = 1, 2$. Without loss of generality, we may assume that $\nu(V_i \setminus F_i) < \epsilon/4$, $i = 1, 2$. Fix any $i \in \{1, 2\}$. By $\lim_{(U,V,h) \in \bar{\mathcal{A}}_{F_i}} f_h = \bar{\mu}(F_i)$, $\exists (\hat{U}_i, \hat{V}_i, \hat{h}_i) \in \bar{\mathcal{A}}_{F_i}$ with $\hat{V}_i \subseteq V_i$ such that $\left\| \bar{\mu}(F_i) - f_{\hat{h}_i} \right\| < \epsilon/6$. By $\lim_{(U,V,h) \in \bar{\mathcal{A}}_{F_1 \cup F_2}} f_h = \bar{\mu}(F_1 \cup F_2)$, $\exists (\hat{U}, \hat{V}, \hat{h}) \in \bar{\mathcal{A}}_{F_1 \cup F_2}$ with $\hat{V} \subseteq V_1 \cup V_2$ such that $\left\| \bar{\mu}(F_1 \cup F_2) - f_{\hat{h}} \right\| < \epsilon/6$. This implies that $\|\bar{\mu}(F_1 \cup F_2) - \bar{\mu}(F_1) - \bar{\mu}(F_2)\| \le \left\| \bar{\mu}(F_1 \cup F_2) - f_{\hat{h}} \right\| + \left\| f_{\hat{h}} - f_{\hat{h}_1} - f_{\hat{h}_2} \right\| + \left\| f_{\hat{h}_1} - \bar{\mu}(F_1) \right\| + \left\| f_{\hat{h}_2} - \bar{\mu}(F_2) \right\| < \epsilon/2 + \sup_{y \in \mathcal{Y},\, \|y\| \le 1} \left| \langle\langle f_{\hat{h}}, y \rangle\rangle - \langle\langle f_{\hat{h}_1}, y \rangle\rangle - \langle\langle f_{\hat{h}_2}, y \rangle\rangle \right| = \epsilon/2 + \sup_{y \in \mathcal{Y},\, \|y\| \le 1} \left| f(\hat{h} y) - f(\hat{h}_1 y) - f(\hat{h}_2 y) \right| = \epsilon/2 + \sup_{y \in \mathcal{Y},\, \|y\| \le 1} \left| f((\hat{h} - \hat{h}_1 - \hat{h}_2)y) \right|$. Note that, $\forall y \in \mathcal{Y}$ with $\|y\| \le 1$, $\hat{z} := (\hat{h} - \hat{h}_1 - \hat{h}_2)y \in \mathcal{Z}$, $\|\hat{z}\| \le 1$, since $\operatorname{supp}(\hat{h}_1) \cap \operatorname{supp}(\hat{h}_2) \subseteq \hat{V}_1 \cap \hat{V}_2 = \emptyset$, and $\operatorname{supp}(\hat{z}) \subseteq \overline{(\operatorname{supp}(\hat{h}) \cup \operatorname{supp}(\hat{h}_1) \cup \operatorname{supp}(\hat{h}_2)) \setminus ((\overline{\hat{U}} \cap \overline{\hat{U}_1}) \cup (\overline{\hat{U}} \cap \overline{\hat{U}_2}))}$. Then, $\operatorname{supp}(\hat{z}) \subseteq \overline{\operatorname{supp}(\hat{h}) \cup \operatorname{supp}(\hat{h}_1) \cup \operatorname{supp}(\hat{h}_2)} = \operatorname{supp}(\hat{h}) \cup \operatorname{supp}(\hat{h}_1) \cup \operatorname{supp}(\hat{h}_2) \subseteq \hat{V} \cup \hat{V}_1 \cup \hat{V}_2 \subseteq V_1 \cup V_2$. In addition, $\operatorname{supp}(\hat{z}) \subseteq \overline{\left((\overline{\hat{U}} \cap \overline{\hat{U}_1}) \cup (\overline{\hat{U}} \cap \overline{\hat{U}_2})\right)^{\sim}} \subseteq \overline{\left((\hat{U} \cap \hat{U}_1) \cup (\hat{U} \cap \hat{U}_2)\right)^{\sim}} = \left(\hat{U} \cap (\hat{U}_1 \cup \hat{U}_2)\right)^{\sim} \subseteq \widetilde{F_1 \cup F_2}$. This yields

$\text{supp}(\hat{z}) \subseteq (V_1 \cup V_2) \cap \widetilde{F_1 \cup F_2} = (V_1 \setminus F_1) \cup (V_2 \setminus F_2) =: \hat{O} \in \mathcal{O}$. Then, we have $\|\bar{\mu}(F_1 \cup F_2) - \bar{\mu}(F_1) - \bar{\mu}(F_2)\| < \epsilon/2 + \nu_o(\hat{O}) = \epsilon/2 + \nu(\hat{O}) = \epsilon/2 + \nu(V_1 \setminus F_1) + \nu(V_2 \setminus F_2) < \epsilon$. By the arbitrariness of ϵ, we have:
(viii) $\bar{\mu}(F_1) + \bar{\mu}(F_2) = \bar{\mu}(F_1 \cup F_2)$, $\forall \widetilde{F_1}, \widetilde{F_2} \in \mathcal{O}$ with $F_1 \cap F_2 = \emptyset$.

Fix any $\widetilde{F_1}, \widetilde{F_2}, O \in \mathcal{O}$ with $F_1 \Delta F_2 \subseteq O$. $\forall \epsilon \in (0, +\infty) \subset \mathbb{R}$, $\forall i \in \{1, 2\}$, $\exists V_i \in \mathcal{O}$ with $F_i \subseteq V_i$ such that $\nu(V_i \setminus F_i) < \epsilon/4$. By $\lim_{(U,V,h)\in\bar{\mathcal{A}}_{F_i}} f_h = \bar{\mu}(F_i)$, $\exists(\hat{U}_i, \hat{V}_i, \hat{h}_i) \in \bar{\mathcal{A}}_{F_i}$ with $\hat{V}_i \subseteq V_i$ such that $\left\|\bar{\mu}(F_i) - f_{\hat{h}_i}\right\| < \epsilon/4$. This leads to $\|\bar{\mu}(F_1) - \bar{\mu}(F_2)\| \leq \left\|\bar{\mu}(F_1) - f_{\hat{h}_1}\right\| + \left\|f_{\hat{h}_1} - f_{\hat{h}_2}\right\| + \left\|f_{\hat{h}_2} - \bar{\mu}(F_2)\right\| < \epsilon/2 + \sup_{y\in\mathcal{Y}, \|y\|\leq 1} \left|\langle\langle f_{\hat{h}_1}, y\rangle\rangle - \langle\langle f_{\hat{h}_2}, y\rangle\rangle\right| = \epsilon/2 + \sup_{y\in\mathcal{Y}, \|y\|\leq 1} \left|f((\hat{h}_1 - \hat{h}_2)y)\right|$. Note that, $\forall y \in \mathcal{Y}$ with $\|y\| \leq 1$, $\hat{z} := (\hat{h}_1 - \hat{h}_2)y \in \mathcal{Z}$, $\|\hat{z}\| \leq 1$, and $\text{supp}(\hat{z}) \subseteq \overline{(\text{supp}(\hat{h}_1) \cup \text{supp}(\hat{h}_2)) \setminus (\hat{U}_1 \cap \hat{U}_2)} \subseteq \overline{(\text{supp}(\hat{h}_1) \cup \text{supp}(\hat{h}_2)) \setminus (\hat{U}_1 \cap \hat{U}_2)} = (\text{supp}(\hat{h}_1) \cup \text{supp}(\hat{h}_2)) \setminus (\hat{U}_1 \cap \hat{U}_2) \subseteq (V_1 \cup V_2) \setminus (F_1 \cap F_2) = (V_1 \setminus (F_1 \cap F_2)) \cup (V_2 \setminus (F_1 \cap F_2)) \subseteq (V_1 \setminus F_1) \cup (F_1 \setminus F_2) \cup (F_2 \setminus F_1) \cup (V_2 \setminus F_2) = (V_1 \setminus F_1) \cup (V_2 \setminus F_2) \cup (F_1 \Delta F_2) \subseteq (V_1 \setminus F_1) \cup (V_2 \setminus F_2) \cup O =: \hat{O} \in \mathcal{O}$. Thus, $\|\bar{\mu}(F_1) - \bar{\mu}(F_2)\| < \epsilon/2 + \nu_o(\hat{O}) = \epsilon/2 + \nu(\hat{O}) \leq \epsilon/2 + \nu(V_1 \setminus F_1) + \nu(V_2 \setminus F_2) + \nu(O) < \epsilon + \nu(O)$. By the arbitrariness of ϵ, we have:
(ix) $\|\bar{\mu}(F_1) - \bar{\mu}(F_2)\| \leq \nu(O)$, $\forall \widetilde{F_1}, \widetilde{F_2}, O \in \mathcal{O}$ with $F_1 \Delta F_2 \subseteq O$.

By Proposition 11.194, $\exists!\, \mu \in \mathcal{M}_{ft}(\mathcal{X}, \mathcal{Y}^*)$ such that $\mu(F) = \bar{\mu}(F)$, $\forall \widetilde{F} \in \mathcal{O}$. Furthermore, $\mathcal{P} \circ \mu \leq \nu$. Then, $\|\mu\| = \mathcal{P} \circ \mu(X) \leq \nu(X) \leq \|f\| < \infty$. This defines a mapping $\Phi_{\mathcal{Z}} : \mathcal{Z}^* \to \mathcal{M}_{ft}(\mathcal{X}, \mathcal{Y}^*)$ by $\Phi_{\mathcal{Z}}(f) = \mu$, $\forall f \in \mathcal{Z}^*$. The above can be expressed as:
(x) $\|\Phi_{\mathcal{Z}}(f)\| \leq \|f\|$, $\forall f \in \mathcal{Z}^*$.
$\forall f_1, f_2 \in \mathcal{Z}^*$, $\forall \alpha \in \mathbb{K}$, let $\mu_i := \Phi_{\mathcal{Z}}(f_i) \in \mathcal{M}_{ft}(\mathcal{X}, \mathcal{Y}^*)$, $i = 1, 2$, $f := f_1 + f_2 \in \mathcal{Z}^*$, and $\mu := \Phi_{\mathcal{Z}}(f) \in \mathcal{M}_{ft}(\mathcal{X}, \mathcal{Y}^*)$. $\forall \widetilde{F} \in \mathcal{O}$, $\mu(F) = \bar{\mu}(F) = \lim_{(U,V,h)\in\bar{\mathcal{A}}_F} f_h = \lim_{(U,V,h)\in\bar{\mathcal{A}}_F} (f_{1h} + f_{2h}) = \lim_{(U,V,h)\in\bar{\mathcal{A}}_F} f_{1h} + \lim_{(U,V,h)\in\bar{\mathcal{A}}_F} f_{2h} = \bar{\mu}_1(F) + \bar{\mu}_2(F) = \mu_1(F) + \mu_2(F) = (\mu_1 + \mu_2)(F)$. By Proposition 11.192, $\mu_1 + \mu_2 \in \mathcal{M}_{ft}(\mathcal{X}, \mathcal{Y}^*)$. Then, by Proposition 11.194, $\Phi_{\mathcal{Z}}(f_1 + f_2) = \mu = \mu_1 + \mu_2 = \Phi_{\mathcal{Z}}(f_1) + \Phi_{\mathcal{Z}}(f_2)$. Let $\hat{\mu} := \Phi_{\mathcal{Z}}(\alpha f_1) \in \mathcal{M}_{ft}(\mathcal{X}, \mathcal{Y}^*)$. $\forall \widetilde{F} \in \mathcal{O}$, we have $\hat{\mu}(F) = \bar{\hat{\mu}}(F) = \lim_{(U,V,h)\in\bar{\mathcal{A}}_F} (\alpha f_1)_h = \lim_{(U,V,h)\in\bar{\mathcal{A}}_F} \alpha f_{1h} = \alpha\bar{\mu}_1(F) = (\alpha\mu_1)(F)$. By Proposition 11.192, $\alpha\mu_1 \in \mathcal{M}_{ft}(\mathcal{X}, \mathcal{Y}^*)$. Then, by Proposition 11.194, $\Phi_{\mathcal{Z}}(\alpha f_1) = \hat{\mu} = \alpha\mu_1 = \alpha\Phi_{\mathcal{Z}}(f_1)$. This shows that $\Phi_{\mathcal{Z}}$ is a bound linear operator.

Fix any $\mu \in \mathcal{M}_{ft}(\mathcal{X}, \mathcal{Y}^*)$. $\forall z \in \mathcal{Z}$, by Proposition 11.37, z is $\mathcal{B}_B(\mathcal{X})$-measurable. Clearly, $\mathcal{P} \circ z(x) \leq \|z\|$, $\forall x \in X$. Then, $\int_X \mathcal{P} \circ z \, d\mathcal{P} \circ \mu \leq \|z\| \mathcal{P} \circ \mu(X) = \|z\| \|\mu\| < \infty$. Hence, z is absolutely integrable over $\mathbb{X} := (\mathcal{X}, \mathcal{B}_B(\mathcal{X}), \mu)$. By Proposition 5.7, $z(X) \subseteq \mathcal{Y}$ is compact. Then, by Propositions 11.190, 7.35, and 7.17, $H := \overline{\text{span}\,(z(X))} \subseteq \mathcal{Y} \subseteq \mathcal{Y}^{**}$ is a separable subspace. By Proposition 11.132,

$\langle\langle\mu, z\rangle\rangle = \int_X \langle\langle d\mu(x), z(x)\rangle\rangle \in \mathbb{K}$ and $|\langle\langle\mu, z\rangle\rangle| = \left|\int_X \langle\langle d\mu(x), z(x)\rangle\rangle\right| \le \|z\|\|\mu\| < \infty$. Hence, we have:
(xi) $\forall\mu \in \mathcal{M}_{ft}(\mathcal{X}, \mathcal{Y}^*)$, $\forall z \in \mathcal{Z}$, we have z is absolutely integrable and integrable over $(\mathcal{X}, \mathcal{B}_B(\mathcal{X}), \mu)$, and $|\langle\langle\mu, z\rangle\rangle| \le \|z\|\|\mu\| < \infty$.

We will show that $f(z) = \int_X \langle\langle d\mu(x), z(x)\rangle\rangle = \langle\langle\mu, z\rangle\rangle$, $\forall z \in \mathcal{Z}$, $\forall f \in \mathcal{Z}^*$, where $\mu = \Phi_{\mathcal{Z}}(f)$, in four steps. In the first step, we consider the special case: $\mathcal{Y} = \mathbb{R}$ and f is a positive linear functional. $\forall \widetilde{F} \in \mathcal{O}$, $\forall(U, V, h) \in \bar{\mathcal{A}}_F$, $f_h = f(h) \in [0, \|f\|] \subset \mathbb{R}$, since $h \in P := \{z \in \mathcal{C}(\mathcal{X}, \mathbb{R}) \mid z : X \to [0, \infty) \subset \mathbb{R}\} \subseteq \mathcal{C}(\mathcal{X}, \mathbb{R}) = \mathcal{Z}$. Then, $\bar{\mu}(F) = \lim_{(U,V,h)\in\bar{\mathcal{A}}_F} f_h \in [0, \|f\|] \subset \mathbb{R}$. Hence, by Proposition 11.194 and its proof, $\mu : \mathcal{B}_B(\mathcal{X}) \to [0, \|f\|] \subset \mathbb{R}$ and $\mu \in \mathcal{M}_{ft}(\mathcal{X})$. Fix any $z \in \mathcal{Z}$. We will distinguish three exhaustive cases: Case 1: $z = \vartheta_{\mathcal{Z}}$; Case 2: $z \ne \vartheta_{\mathcal{Z}}$ and $z \in P$; Case 3: $z \in \mathcal{Z}$. Case 1: $z = \vartheta_{\mathcal{Z}}$. Then, $f(z) = 0 = \int_X z(x) d\mu(x) = \langle\langle\mu, z\rangle\rangle$.

Case 2: $z \ne \vartheta_{\mathcal{Z}}$ and $z \in P$. Let $\bar{z} := z/\|z\| \in P$. Then, $\bar{z} : X \to [0, 1] \subset \mathbb{R}$. Fix any $n \in \mathbb{N}$ and fix any $i \in \{1, \ldots, n\}$. Define $\phi_i := ((n\bar{z} - i + 1) \vee 0) \wedge 1 \in P$, $O_i := \{x \in X \mid n\bar{z}(x) > i - 1\}$, and $F_i := \{x \in X \mid n\bar{z}(x) \ge i - 1\}$. Let $O_0 := \{x \in X \mid n\bar{z}(x) > -1\} = X$, $O_{n+1} := \{x \in X \mid n\bar{z}(x) > n\} = \emptyset$, $F_{n+1} := \{x \in X \mid n\bar{z}(x) \ge n\}$, and $F_{n+2} := \{x \in X \mid n\bar{z}(x) \ge n + 1\} = \emptyset$. Clearly, $F_{i+2} \subseteq O_{i+1} \subseteq \overline{O_{i+1}} \subseteq F_{i+1} \subseteq O_i \subseteq \overline{O_i} \subseteq F_i \subseteq O_{i-1}$, $\phi_i : X \to [0, 1] \subset \mathbb{R}$, $\phi_i|_{F_{i+1}} = 1$, and $\operatorname{supp}(\phi_i) \subseteq \overline{O_i} \subseteq F_i \subseteq O_{i-1}$. Clearly, $O_i \in \mathcal{O}$ and $\widetilde{F_{i+1}} \in \mathcal{O}$, $i = 0, \ldots, n + 1$, and $\bar{z} = \frac{1}{n}\sum_{i=1}^n \phi_i$. Fix any $i \in \{1, \ldots, n\}$. By the definition of $\bar{\mu}$ and the fact that f is a positive linear functional, $\forall(U, V, h) \in \bar{\mathcal{A}}_{F_{i+2}}$ with $V \subseteq O_{i+1}$, we have $h(x) \le \chi_{O_{i+1},X}(x) \le \chi_{F_{i+1},X}(x) \le \phi_i(x)$, $\forall x \in X$, and $f_h = f(h) \le f(\phi_i)$. This implies that $\mu(F_{i+2}) \le f(\phi_i)$. $\forall(U, V, h) \in \bar{\mathcal{A}}_{F_i}$, we have $\phi_i(x) \le \chi_{F_i,X}(x) \le h(x)$, $\forall x \in X$, and $f_h = f(h) \ge f(\phi_i)$. This leads to $f(\phi_i) \le \mu(F_i)$. By Propositions 11.75 and 11.83, we have $\mu(F_{i+2}) \le \mu(F_{i+1}) = \int_X \chi_{F_{i+1},X}\, d\mu \le \int_X \phi_i\, d\mu \le \int_X \chi_{F_i,X}\, d\mu = \mu(F_i)$. Then, $\left|f(\phi_i) - \int_X \phi_i\, d\mu\right| \le \mu(F_i) - \mu(F_{i+2})$. Summing both sides from 1 to n, we have $\left|f(\bar{z}) - \int_X \bar{z}\, d\mu\right| = \left|\frac{1}{n}\sum_{i=1}^n (f(\phi_i) - \int_X \phi_i\, d\mu)\right| \le \frac{1}{n}\sum_{i=1}^n \left|f(\phi_i) - \int_X \phi_i\, d\mu\right| \le \frac{1}{n}(\mu(F_1) + \mu(F_2)) \le 2\mu(X)/n = 2\|\mu\|/n \le 2\|f\|/n$, where the first equality follows from (xi) and Proposition 11.92 and the last inequality follows from (x). By the arbitrariness of n, we have $f(\bar{z}) = \int_X \bar{z}\, d\mu$. By (xi) and Proposition 11.92, $f(z) = f(\|z\|\bar{z}) = \|z\| f(\bar{z}) = \|z\| \int_X \bar{z}\, d\mu = \int_X z\, d\mu = \langle\langle\mu, z\rangle\rangle$.

Case 3: $z \in \mathcal{Z}$. Let $z_+ := z \vee 0 \in P$ and $z_- := (-z) \vee 0 \in P$. Clearly, $z = z_+ - z_-$. By Cases 1 and 2, $f(z) = f(z_+) - f(z_-) = \int_X z_+\, d\mu - \int_X z_-\, d\mu = \int_X z\, d\mu = \langle\langle\mu, z\rangle\rangle$, where the third equality follows from (xi) and Proposition 11.92.

Hence, $f(z) = \int_X z\, d\mu = \langle\langle\mu, z\rangle\rangle$, $\forall z \in \mathcal{Z}$. This completes the first step.

In the second step, we consider the special case: $\mathcal{Y} = \mathbb{R}$. By Lemma 11.200, $f = f_+ - f_-$, where $f_+, f_- \in \mathcal{Z}^*$ are positive linear functionals. Let $\mu_+ := \Phi_{\mathcal{Z}}(f_+)$ and $\mu_- := \Phi_{\mathcal{Z}}(f_-)$. By the first step, $f_+(z) = \int_X z\, d\mu_+$ and $f_-(z) = \int_X z\, d\mu_-$, $\forall z \in \mathcal{Z}$. Then, by linearity of $\Phi_{\mathcal{Z}}$, $\mu = \Phi_{\mathcal{Z}}(f) = \Phi_{\mathcal{Z}}(f_+ - f_-) = \Phi_{\mathcal{Z}}(f_+) - \Phi_{\mathcal{Z}}(f_-) = \mu_+ - \mu_-$. By Propositions 11.145 and 11.146, $f(z) = f_+(z) - f_-(z) = \int_X z\, d\mu_+ - \int_X z\, d\mu_- = \int_X z\, d\mu = \langle\langle\mu, z\rangle\rangle$. This completes the second step.

In the third step, we consider the special case when $\mathcal{Y}$ is a real normed linear space. $\forall y \in \mathcal{Y}$, define $\mu_y : \mathcal{B}_B(\mathcal{X}) \to \mathbb{R}$ by $\mu_y(E) = \langle\langle \mu(E), y \rangle\rangle$, $\forall E \in \mathcal{B}_B(\mathcal{X})$. Then, $\mu_y \in \mathcal{M}_{ft}(\mathcal{X}, \mathbb{R})$ by Proposition 11.136. Define $f_y : \mathcal{C}(\mathcal{X}, \mathbb{R}) \to \mathbb{R}$ by $f_y(h) = f(hy)$, $\forall h \in \mathcal{C}(\mathcal{X}, \mathbb{R})$. By $f \in \mathcal{Z}^*$, we have $f_y \in (\mathcal{C}(\mathcal{X}, \mathbb{R}))^*$. $\forall \widetilde{F} \in \mathcal{O}$, $\mu_y(F) = \langle\langle \mu(F), y \rangle\rangle = \left\langle\left\langle \lim_{(U,V,h) \in \bar{\mathcal{A}}_F} f_h, y \right\rangle\right\rangle = \lim_{(U,V,h) \in \bar{\mathcal{A}}_F} \langle\langle f_h, y \rangle\rangle = \lim_{(U,V,h) \in \bar{\mathcal{A}}_F} f(hy) = \lim_{(U,V,h) \in \bar{\mathcal{A}}_F} f_y(h) = \Phi_{\mathcal{C}(\mathcal{X},\mathbb{R})}(f_y)(F)$. By Proposition 11.194, we have $\mu_y = \Phi_{\mathcal{C}(\mathcal{X},\mathbb{R})}(f_y)$. By the second step, we have $f(hy) = f_y(h) = \int_X h \, d\mu_y = \int_X \langle\langle d\mu(x), h(x)y \rangle\rangle = \langle\langle \mu, hy \rangle\rangle$, $\forall h \in \mathcal{C}(\mathcal{X}, \mathbb{R})$, where the third equality follows from Proposition 11.146. $\forall z \in \mathcal{Z}$, $\forall \epsilon \in (0, +\infty) \subset \mathbb{R}$, $\exists \bar{z} \in \operatorname{span}(Z)$ such that $\|z - \bar{z}\| < \epsilon/(2\|f\| + 1)$. Then, $\bar{z} = \sum_{i=1}^n h_i y_i$, where $n \in \mathbb{Z}_+$, $h_1, \ldots, h_n \in \mathcal{C}(\mathcal{X}, \mathbb{R})$, and $y_1, \ldots, y_n \in \mathcal{Y}$. Then, $|f(z) - \langle\langle \mu, z \rangle\rangle| \le |f(z) - f(\bar{z})| + |f(\bar{z}) - \langle\langle \mu, \bar{z} \rangle\rangle| + |\langle\langle \mu, \bar{z} \rangle\rangle - \langle\langle \mu, z \rangle\rangle| \le \|f\|\|z - \bar{z}\| + \left|\sum_{i=1}^n f(h_i y_i) - \int_X \left\langle\left\langle d\mu(x), \sum_{i=1}^n h_i(x) y_i \right\rangle\right\rangle\right| + \left|\int_X \langle\langle d\mu(x), \bar{z}(x) \rangle\rangle - \int_X \langle\langle d\mu(x), z(x) \rangle\rangle\right| \le \epsilon/2 + \left|\sum_{i=1}^n (f(h_i y) - \int_X \langle\langle d\mu(x), h_i(x) y \rangle\rangle)\right| + \left|\int_X \langle\langle d\mu(x), \bar{z}(x) - z(x) \rangle\rangle\right| \le \epsilon/2 + 0 + \int_X \mathcal{P} \circ (\bar{z} - z)(x) \, d\mathcal{P} \circ \mu(x) \le \epsilon/2 + \|z - \bar{z}\| \mathcal{P} \circ \mu(X) \le \epsilon/2 + \frac{\epsilon}{2\|f\|+1} \|\mu\| \le \epsilon/2 + \frac{\epsilon}{2\|f\|+1} \|f\| < \epsilon$, where the second inequality follows from Proposition 7.72, the third inequality follows from (xi) and Proposition 11.132, the fourth inequality follows from Proposition 11.132, the fifth inequality follows from Propositions 11.75 and 11.83, and the seventh inequality follows from (x). By the arbitrariness of ϵ, we have $f(z) = \langle\langle \mu, z \rangle\rangle = \int_X \langle\langle d\mu(x), z(x) \rangle\rangle$. This completes the third step.

In the fourth step, we consider the special case when $\mathcal{Y}$ is a complex normed linear space. Let $g := \operatorname{Re} \circ f$. By Lemmas 7.40 and 7.81, g is a bounded linear functional of the real normed linear space $\mathcal{Z}_\mathbb{R} := (\mathcal{C}(\mathcal{X}, \mathcal{Y}_\mathbb{R}), \mathbb{R}, \|\cdot\|_{\mathcal{C}(\mathcal{X},\mathcal{Y}_\mathbb{R})})$ and $f(z) = g(z) - \mathrm{i} g(\mathrm{i}z)$, $\forall z \in \mathcal{Z}_\mathbb{R}$. Let $\bar{g} \in \mathcal{Z}_\mathbb{R}^*$ be defined by $\bar{g}(z) = g(\mathrm{i}z)$, $\forall z \in \mathcal{Z}_\mathbb{R}$, $\mu_{gr} := \Phi_{\mathcal{Z}_\mathbb{R}}(g) \in \mathcal{M}_{ft}(\mathcal{X}, \mathcal{Y}_\mathbb{R}^*)$, and $\mu_{gi} := \Phi_{\mathcal{Z}_\mathbb{R}}(\bar{g}) \in \mathcal{M}_{ft}(\mathcal{X}, \mathcal{Y}_\mathbb{R}^*)$. $\forall \widetilde{F} \in \mathcal{O}$, $\mu(F) = \lim_{(U,V,h) \in \bar{\mathcal{A}}_F} f_h = \lim_{(U,V,h) \in \bar{\mathcal{A}}_F} (g_h - \mathrm{i}\bar{g}_h) = \lim_{(U,V,h) \in \bar{\mathcal{A}}_F} g_h - \mathrm{i} \lim_{(U,V,h) \in \bar{\mathcal{A}}_F} \bar{g}_h = \Phi_{\mathcal{Z}_\mathbb{R}}(g)(F) - \mathrm{i}\Phi_{\mathcal{Z}_\mathbb{R}}(\bar{g})(F) = \mu_{gr}(F) - \mathrm{i}\mu_{gi}(F) = (\mu_{gr} - \mathrm{i}\mu_{gi})(F)$. By the definition of $\bar{g}$, we have $\mu_{gi}(E)(y) = \mu_{gr}(E)(\mathrm{i}y) \in \mathbb{C}$, $\forall E \in \mathcal{B}_B(\mathcal{X})$, $\forall y \in \mathcal{Y}_\mathbb{R}$. Then, by Lemma 7.81, $\mu_{gr}(E) - \mathrm{i}\mu_{gi}(E) \in \mathcal{Y}^*$, $\forall E \in \mathcal{B}_B(\mathcal{X})$. By Proposition 11.136, $\mu_{gr} - \mathrm{i}\mu_{gi} \in \mathcal{M}_{ft}(\mathcal{X}, \mathcal{Y}^*)$. By Proposition 11.194, we have $\mu = \mu_{gr} - \mathrm{i}\mu_{gi}$. By the third step, $g(z) = \int_X \left\langle\left\langle d\mu_{gr}(x), z(x) \right\rangle\right\rangle$ and $\bar{g}(z) = \int_X \left\langle\left\langle d\mu_{gi}(x), z(x) \right\rangle\right\rangle$, $\forall z \in \mathcal{Z}_\mathbb{R}$. This implies, $\forall z \in \mathcal{Z}$, $f(z) = g(z) - \mathrm{i}g(\mathrm{i}z) = g(z) - \mathrm{i}\bar{g}(z) = \int_X \left\langle\left\langle d\mu_{gr}(x), z(x) \right\rangle\right\rangle - \mathrm{i}\int_X \left\langle\left\langle d\mu_{gi}(x), z(x) \right\rangle\right\rangle = \int_X \left\langle\left\langle d(\mu_{gr} - \mathrm{i}\mu_{gi})(x), z(x) \right\rangle\right\rangle = \int_X \langle\langle d\mu(x), z(x) \rangle\rangle = \langle\langle \mu, z \rangle\rangle$, where the fourth equality follows from Propositions 11.145 and 11.146. This completes the fourth step.

Thus, we have the representation:

(xii) $f(z) = \int_X \langle\langle d\mu(x), z(x) \rangle\rangle = \langle\langle \mu, z \rangle\rangle$, $\forall z \in \mathcal{Z}$, $\forall f \in \mathcal{Z}^*$, where $\mu = \Phi_\mathcal{Z}(f) \in \mathcal{M}_{ft}(\mathcal{X}, \mathcal{Y}^*)$.

Fix any $\mu \in \mathcal{M}_{ft}(\mathcal{X}, \mathcal{Y}^*)$. Define $f : \mathcal{Z} \to \mathbb{K}$ by $f(z) = \langle\langle \mu, z \rangle\rangle = \int_X \langle\langle d\mu(x), z(x) \rangle\rangle$, $\forall z \in \mathcal{Z}$. We will show that $f \in \mathcal{Z}^*$, $\|f\| \le \|\mu\|$, and $\Phi_\mathcal{Z}(f) = \mu$. $\forall z \in \mathcal{Z}$. By Proposition 5.7, $z(X) \subseteq \mathcal{Y}$ is compact. Then, by Propositions 11.190, 7.35, and 7.17, $H := \overline{\operatorname{span}(z(X))} \subseteq \mathcal{Y}$ is a separable subspace

By (xi) and Proposition 11.132, $f(z) = \langle\langle \mu, z\rangle\rangle = \int_X \langle\langle \mathrm{d}\mu(x), z(x)\rangle\rangle \in \mathbb{K}$ is well-defined. By Proposition 11.132, $f \in \mathcal{Z}^*$ and $\|f\| = \sup_{z\in\mathcal{Z},\, \|z\|\leq 1} |\langle\langle \mu, z\rangle\rangle| = \sup_{z\in\mathcal{Z},\, \|z\|\leq 1} \left|\int_X \langle\langle \mathrm{d}\mu(x), z(x)\rangle\rangle\right| \leq \sup_{z\in\mathcal{Z},\, \|z\|\leq 1} \|z\|\|\mu\| \leq \|\mu\|$.

Let $\hat{\mu} := \Phi_{\mathcal{Z}}(f) \in \mathcal{M}_{ft}(\mathcal{X}, \mathcal{Y}^*)$. $\forall F \in \widetilde{\mathcal{O}}$, $\forall y \in \mathcal{Y}$, $\left\langle\!\left\langle \hat{\mu}(F), y\right\rangle\!\right\rangle = \left\langle\!\left\langle \lim_{(U,V,h)\in\bar{\mathcal{A}}_F} f_h, y\right\rangle\!\right\rangle = \lim_{(U,V,h)\in\bar{\mathcal{A}}_F} \left\langle\!\left\langle f_h, y\right\rangle\!\right\rangle = \lim_{(U,V,h)\in\bar{\mathcal{A}}_F} f(hy) = \lim_{(U,V,h)\in\bar{\mathcal{A}}_F} \int_X \left\langle\!\left\langle \mathrm{d}\mu(x), h(x)y\right\rangle\!\right\rangle$. $\forall \epsilon \in (0, +\infty) \subset \mathbb{R}$, $\exists O \in \mathcal{O}$ with $F \subseteq O$ such that $\mathcal{P} \circ \mu(O \setminus F) < \epsilon/(1 + \|y\|)$. $\forall (U, V, h) \in \bar{\mathcal{A}}_F$ with $V \subseteq O$, $\left|\int_X \langle\langle \mathrm{d}\mu(x), h(x)y\rangle\rangle - \langle\langle \mu(F), y\rangle\rangle\right| = \Big|\int_F \langle\langle \mathrm{d}\mu(x), h(x)y\rangle\rangle - \langle\langle \mu(F), y\rangle\rangle + \int_{O\setminus F} \langle\langle \mathrm{d}\mu(x), h(x)y\rangle\rangle + \int_{\widetilde{O}} \langle\langle \mathrm{d}\mu(x), h(x)y\rangle\rangle\Big| = \Big|\langle\langle \mu(F), y\rangle\rangle - \langle\langle \mu(F), y\rangle\rangle + \int_{O\setminus F} \langle\langle \mathrm{d}\mu(x), h(x)y\rangle\rangle + 0\Big| \leq \int_{O\setminus F} (\|y\|\mathcal{P} \circ h)\, \mathrm{d}\mathcal{P} \circ \mu \leq \|y\|\mathcal{P} \circ \mu(O \setminus F) < \epsilon$, where the first equality follows from (xi) and Proposition 11.132, the second equality follows from Proposition 11.125, the first inequality follows from Proposition 11.132, and the second inequality follows from Proposition 11.83. This implies that $\left|\left\langle\!\left\langle \hat{\mu}(F), y\right\rangle\!\right\rangle - \langle\langle \mu(F), y\rangle\rangle\right| = \lim_{(U,V,h)\in\bar{\mathcal{A}}_F} \left|\int_X \langle\langle \mathrm{d}\mu(x), h(x)y\rangle\rangle - \langle\langle \mu(F), y\rangle\rangle\right| \leq \epsilon$. By the arbitrariness of ϵ and y, we have $\hat{\mu}(F) = \mu(F)$. By Proposition 11.194, $\hat{\mu} = \mu$. This shows that $\Phi_{\mathcal{Z}}$ is surjective.

$\forall f \in \mathcal{Z}^*$, let $\mu := \Phi_{\mathcal{Z}}(f) \in \mathcal{M}_{ft}(\mathcal{X}, \mathcal{Y}^*)$. Then, by the above and (x), we have $\|\mu\| = \|\Phi_{\mathcal{Z}}(f)\| \leq \|f\| \leq \|\mu\|$. Hence, $\|\Phi_{\mathcal{Z}}(f)\| = \|f\|$. Then, $\Phi_{\mathcal{Z}}$ is injective. This shows that $\Phi_{\mathcal{Z}}$ is bijective, continuous, linear, and norm preserving. Hence, $\Phi_{\mathcal{Z}} : \mathcal{Z}^* \to \mathcal{M}_{ft}(\mathcal{X}, \mathcal{Y}^*)$ is an isometrical isomorphism.

This completes the proof of the theorem. □

Definition 11.202 Let $m \in \mathbb{Z}_+$, $I := [0, 1] \subset \mathbb{R}$, $\mathcal{Y}$ be a normed linear space, $f : I^m \to \mathcal{Y}$, and $n \in \mathbb{N}$. The *nth Bernsteĭn function for* f, $B_n : I^m \to \mathcal{Y}$, is defined by, $\forall x := (x_1, \ldots, x_m) \in I^m$,

$$B_n(x) = B_n(x; f) = \sum_{k_1=0}^{n} \cdots \sum_{k_m=0}^{n} f(\frac{k_1}{n}, \ldots, \frac{k_m}{n}) \prod_{i=1}^{m} \binom{n}{k_i} x_i^{k_i} (1 - x_i)^{n-k_i}$$

where, for notational consistency, we have made the arbitrary assignment of $0^0 = 1$ in this definition as well as the following theorem. ◇

Theorem 11.203 (Bernsteĭn Approximation Theorem) *Let* $m \in \mathbb{N}$, $I := [0, 1] \subset \mathbb{R}$, $\mathcal{Y}$ *be a normed linear space,* $f : I^m \to \mathcal{Y}$ *be continuous, and* $(B_n)_{n=1}^{\infty}$ *be the sequence of Bernsteĭn functions for* f. *Then,* $\lim_{n\in\mathbb{N}} \|f - B_n\|_{\mathcal{C}(I^m,\mathcal{Y})} = 0$.

Proof $\forall n \in \mathbb{Z}_+$, $\forall y \in I$, we have

$$1 = (y + 1 - y)^n = \sum_{k=0}^{n} \binom{n}{k} y^k (1-y)^{n-k} \tag{11.6}$$

Then, $\forall n \in \mathbb{N}$, $\forall y \in I$, we have

$$\begin{aligned} y &= y \cdot 1 = y\Big(\sum_{k=0}^{n-1} \binom{n-1}{k} y^k (1-y)^{n-1-k}\Big) \\ &= \sum_{k=0}^{n-1} \binom{n-1}{k} y^{k+1} (1-y)^{n-1-k} = \sum_{k=1}^{n} \binom{n-1}{k-1} y^k (1-y)^{n-k} \\ &= \sum_{k=1}^{n} \frac{k}{n} \binom{n}{k} y^k (1-y)^{n-k} = \sum_{k=0}^{n} \frac{k}{n} \binom{n}{k} y^k (1-y)^{n-k} \end{aligned} \tag{11.7}$$

where the second equality follows from (11.6). Similarly, $\forall n - 1 \in \mathbb{N}$, $\forall y \in I$,

$$\begin{aligned} y^2 &= y\Big(\sum_{k=0}^{n-1} \frac{k}{n-1} \binom{n-1}{k} y^k (1-y)^{n-1-k}\Big) \\ &= \sum_{k=0}^{n-1} \frac{k}{n-1} \binom{n-1}{k} y^{k+1} (1-y)^{n-1-k} \\ &= \sum_{k=1}^{n} \frac{k-1}{n-1} \binom{n-1}{k-1} y^k (1-y)^{n-k} = \sum_{k=1}^{n} \frac{k(k-1)}{n(n-1)} \binom{n}{k} y^k (1-y)^{n-k} \\ &= \sum_{k=0}^{n} \frac{k(k-1)}{n(n-1)} \binom{n}{k} y^k (1-y)^{n-k} \end{aligned}$$

where the first equality follows from (11.7). This implies that

$$\begin{aligned} (1 - \frac{1}{n}) y^2 &= \sum_{k=0}^{n} \frac{k^2 - k}{n^2} \binom{n}{k} y^k (1-y)^{n-k} \\ &= \sum_{k=0}^{n} \frac{k^2}{n^2} \binom{n}{k} y^k (1-y)^{n-k} - \frac{1}{n} \sum_{k=0}^{n} \frac{k}{n} \binom{n}{k} y^k (1-y)^{n-k} \\ &= \sum_{k=0}^{n} \frac{k^2}{n^2} \binom{n}{k} y^k (1-y)^{n-k} - \frac{1}{n} y \end{aligned}$$

where the last equality follows from (11.7). Rearranging terms in the above yields, $\forall y \in I, \forall n - 1 \in \mathbb{N}$,

$$
\begin{aligned}
\frac{1}{n} y(1-y) &= \sum_{k=0}^{n} \frac{k^2}{n^2} \binom{n}{k} y^k (1-y)^{n-k} + y^2 - 2y^2 \\
&= \sum_{k=0}^{n} \frac{k^2}{n^2} \binom{n}{k} y^k (1-y)^{n-k} + y^2 \sum_{k=0}^{n} \binom{n}{k} y^k (1-y)^{n-k} \\
&\quad -2y \sum_{k=0}^{n} \frac{k}{n} \binom{n}{k} y^k (1-y)^{n-k} \\
&= \sum_{k=0}^{n} (y^2 - 2y\frac{k}{n} + \frac{k^2}{n^2}) \binom{n}{k} y^k (1-y)^{n-k} \\
&= \sum_{k=0}^{n} (y - \frac{k}{n})^2 \binom{n}{k} y^k (1-y)^{n-k} \qquad (11.8)
\end{aligned}
$$

where the second equality follows from (11.6) and (11.7). $\forall n \in \mathbb{Z}_+$, $\forall x := (x_1, \ldots, x_m) \in I^m$, we have

$$
\begin{aligned}
1 &= \prod_{i=1}^{m} (x_i + 1 - x_i)^n = \prod_{i=1}^{m} \Big(\sum_{k=0}^{n} \binom{n}{k} x_i^k (1-x_i)^{n-k} \Big) \\
&= \sum_{k_1=0}^{n} \cdots \sum_{k_m=0}^{n} \prod_{i=1}^{m} \binom{n}{k_i} x_i^{k_i} (1-x_i)^{n-k_i} \qquad (11.9)
\end{aligned}
$$

Then, $\forall x := (x_1, \ldots, x_m) \in I^m$, $\forall n - 1 \in \mathbb{N}$, we have

$$
\begin{aligned}
\frac{1}{n} \sum_{i=1}^{m} x_i (1-x_i) &= \sum_{i=1}^{m} \sum_{k_i=0}^{n} (x_i - \frac{k_i}{n})^2 \binom{n}{k_i} x_i^{k_i} (1-x_i)^{n-k_i} \\
&= \sum_{i=1}^{m} \Big(\sum_{k_i=0}^{n} (x_i - \frac{k_i}{n})^2 \binom{n}{k_i} x_i^{k_i} (1-x_i)^{n-k_i} \Big) \\
&\quad \cdot \prod_{\substack{j=1 \\ j \neq i}}^{m} \Big(\sum_{k_j=0}^{n} \binom{n}{k_j} x_j^{k_j} (1-x_j)^{n-k_j} \Big) \\
&= \sum_{i=1}^{m} \sum_{k_1=0}^{n} \cdots \sum_{k_m=0}^{n} (x_i - \frac{k_i}{n})^2 \prod_{j=1}^{m} \binom{n}{k_j} x_j^{k_j} (1-x_j)^{n-k_j}
\end{aligned}
$$

$$= \sum_{k_1=0}^{n} \cdots \sum_{k_m=0}^{n} \Big(\prod_{j=1}^{m} \binom{n}{k_j} x_j^{k_j} (1 - x_j)^{n-k_j} \Big) \Big(\sum_{i=1}^{m} (x_i - \frac{k_i}{n})^2 \Big)$$

$$= \sum_{k_1=0}^{n} \cdots \sum_{k_m=0}^{n} \Big| x - (\frac{k_1}{n}, \ldots, \frac{k_m}{n}) \Big|^2 \prod_{j=1}^{m} \binom{n}{k_j} x_j^{k_j} (1 - x_j)^{n-k_j} \qquad (11.10)$$

where the first equality follows from (11.8) and the second equality follows from (11.9).

Note that $I^m \subseteq \mathbb{R}^m$ is a compact metric space. By Propositions 5.22, 5.29, 7.21, 3.12, and 3.9, $\exists M \in [0, \infty) \subset \mathbb{R}$ such that $\| f(x) \| \leq M$, $\forall x \in I^m$. By Proposition 5.39, f is uniformly continuous. $\forall \epsilon \in (0, \infty) \subset \mathbb{R}$, $\exists \delta \in (0, \infty) \subset \mathbb{R}$, $\forall x, \bar{x} \in I^m$ with $|x - \bar{x}| < \delta$, we have $\| f(x) - f(\bar{x}) \| < \epsilon/2$. Let $n_0 := \lceil \max\{\delta^{-4}, m^2 M^2/\epsilon^2, 2\} \rceil \in \mathbb{N}$. $\forall n \in \mathbb{N}$ with $n_0 \leq n$, $\forall x := (x_1, \ldots, x_m) \in I^m$, we have

$$\begin{aligned}
&\| f(x) - B_n(x) \| \\
&= \Big\| f(x) \sum_{k_1=0}^{n} \cdots \sum_{k_m=0}^{n} \prod_{i=1}^{m} \binom{n}{k_i} x_i^{k_i} (1 - x_i)^{n-k_i} \\
&\quad - \sum_{k_1=0}^{n} \cdots \sum_{k_m=0}^{n} f(\frac{k_1}{n}, \ldots, \frac{k_m}{n}) \prod_{i=1}^{m} \binom{n}{k_i} x_i^{k_i} (1 - x_i)^{n-k_i} \Big\| \\
&= \Big\| \sum_{k_1=0}^{n} \cdots \sum_{k_m=0}^{n} (f(x) - f(\frac{k_1}{n}, \ldots, \frac{k_m}{n})) \prod_{i=1}^{m} \binom{n}{k_i} x_i^{k_i} (1 - x_i)^{n-k_i} \Big\| \\
&\leq \sum_{k_1=0}^{n} \cdots \sum_{k_m=0}^{n} \| f(x) - f(\frac{k_1}{n}, \ldots, \frac{k_m}{n}) \| \prod_{i=1}^{m} \binom{n}{k_i} x_i^{k_i} (1 - x_i)^{n-k_i}
\end{aligned}$$

where the first equality follows from (11.9) and Definition 11.202. Let $J := \{0, \ldots, n\}^m$ and $J_x := \{(k_1, \ldots, k_m) \in J \mid \Big| x - (\frac{k_1}{n}, \ldots, \frac{k_m}{n}) \Big| < n^{-1/4}\}$. Then,

$$\begin{aligned}
&\| f(x) - B_n(x) \| \\
&\leq \sum_{(k_1, \ldots, k_m) \in J_x} \| f(x) - f(\frac{k_1}{n}, \ldots, \frac{k_m}{n}) \| \prod_{i=1}^{m} \binom{n}{k_i} x_i^{k_i} (1 - x_i)^{n-k_i} \\
&\quad + \sum_{(k_1, \ldots, k_m) \in J \setminus J_x} \| f(x) - f(\frac{k_1}{n}, \ldots, \frac{k_m}{n}) \| \prod_{i=1}^{m} \binom{n}{k_i} x_i^{k_i} (1 - x_i)^{n-k_i}
\end{aligned}$$

Note that, $\forall (k_1, \ldots, k_m) \in J_x$, $\left|(x_1, \ldots, x_m) - (\frac{k_1}{n}, \ldots, \frac{k_m}{n})\right| < n^{-1/4} \leq \delta$, then $\left\| f(x) - f(\frac{k_1}{n}, \ldots, \frac{k_m}{n})\right\| < \epsilon/2$. This implies that

$$\sum_{(k_1,\ldots,k_m)\in J_x} \left\| f(x) - f(\frac{k_1}{n}, \ldots, \frac{k_m}{n})\right\| \prod_{i=1}^{m} \binom{n}{k_i} x_i^{k_i} (1 - x_i)^{n-k_i}$$
$$< \frac{\epsilon}{2} \sum_{(k_1,\ldots,k_m)\in J_x} \prod_{i=1}^{m} \binom{n}{k_i} x_i^{k_i} (1 - x_i)^{n-k_i}$$
$$\leq \frac{\epsilon}{2} \sum_{(k_1,\ldots,k_m)\in J} \prod_{i=1}^{m} \binom{n}{k_i} x_i^{k_i} (1 - x_i)^{n-k_i} = \frac{\epsilon}{2}$$

where the equality follows from (11.9). Note also that, $\forall (k_1, \ldots, k_m) \in J \setminus J_x$, $\left\| f(x) - f(\frac{k_1}{n}, \ldots, \frac{k_m}{n})\right\| \leq 2M = 2M \frac{\left|x-(\frac{k_1}{n},\ldots,\frac{k_m}{n})\right|^2}{\left|x-(\frac{k_1}{n},\ldots,\frac{k_m}{n})\right|^2} \leq 2M\sqrt{n}\left|x - (\frac{k_1}{n}, \ldots, \frac{k_m}{n})\right|^2$.

Then,

$$\sum_{(k_1,\ldots,k_m)\in J\setminus J_x} \| f(x) - f(\frac{k_1}{n}, \ldots, \frac{k_m}{n})\| \prod_{i=1}^{m} \binom{n}{k_i} x_i^{k_i} (1 - x_i)^{n-k_i}$$
$$\leq \sum_{(k_1,\ldots,k_m)\in J\setminus J_x} 2M\sqrt{n}|x - (\frac{k_1}{n}, \ldots, \frac{k_m}{n})|^2 \prod_{i=1}^{m} \binom{n}{k_i} x_i^{k_i} (1 - x_i)^{n-k_i}$$
$$\leq 2M\sqrt{n} \sum_{(k_1,\ldots,k_m)\in J} |x - (\frac{k_1}{n}, \ldots, \frac{k_m}{n})|^2 \prod_{i=1}^{m} \binom{n}{k_i} x_i^{k_i} (1 - x_i)^{n-k_i}$$
$$= 2M\sqrt{n}\frac{1}{n}\sum_{i=1}^{m} x_i (1 - x_i) \leq \frac{Mm}{2\sqrt{n}} \leq \epsilon/2$$

Hence, we have $\|f(x) - B_n(x)\| < \epsilon$. By the arbitrariness of $x \in I^m$, we have $\|f - B_n\|_{\mathcal{C}(I^m,\mathcal{Y})} < \epsilon$. Therefore, $\lim_{n\in\mathbb{N}} \|f - B_n\|_{\mathcal{C}(I^m,\mathcal{Y})} = 0$.

This completes the proof of the theorem. □

Theorem 11.204 (Riesz Representation Theorem) *Let $m \in \mathbb{Z}_+$, $I_1, \ldots, I_m \subset \mathbb{R}$ be compact intervals, $X := \prod_{i=1}^{m} I_i \subset \mathbb{R}^m$ with subset topology $\mathcal{O}$, $\mathcal{X} := (X, \mathcal{O})$, $\mathcal{Y}$ be a normed linear space, and $\mathcal{Z} := \mathcal{C}(\mathcal{X}, \mathcal{Y})$. Then, $\mathcal{Z}^* = \mathcal{M}_{ft}(\mathcal{X}, \mathcal{Y}^*) = \mathcal{M}_f(X, \mathcal{B}_B(\mathcal{X}), \mathcal{Y}^*)$.*

Proof By Tychonoff Theorem 5.47 and Proposition 4.37, $\mathcal{X}$ is a compact metric space. By Proposition 5.5, X is closed in $\mathbb{R}^m$. By Proposition 4.39, $\mathcal{X}$ is a complete metric space. Since $\mathbb{R}^m$ is separable, by Proposition 4.38, $\mathcal{X}$ is separable. Hence, $\mathcal{X}$ is a separable compact complete metric space.

Let $Z := \{z \in \mathcal{Z} \mid z = hy, h \in \mathcal{C}(\mathcal{X}, \mathbb{R}), y \in \mathcal{Y}\}$. We will show that $\mathcal{Z} = \overline{\text{span}(Z)}$ by distinguishing two exhaustive and mutually exclusive cases: Case 1: $\exists i_0 \in \{1, \dots, m\}$ such that $I_{i_0} = \emptyset$; Case 2: $I_1, \dots, I_m$ are nonempty. Case 1: $\exists i_0 \in \{1, \dots, m\}$ such that $I_{i_0} = \emptyset$. Then, $X = \emptyset$ and $\mathcal{Z}$ is the trivial Banach space with a singleton element. Clearly, $Z = \mathcal{Z}$. Hence, $\mathcal{Z} = \overline{\text{span}(Z)}$.

Case 2: $I_1, \dots, I_m$ are nonempty. Without loss of generality, assume $I_i = [a_i, b_i] \subset \mathbb{R}$ with $a_i, b_i \in \mathbb{R}$ and $a_i \le b_i$, $i = 1, \dots, m$, and $\exists \bar{m} \in \mathbb{Z}_+$ with $\bar{m} \le m$ such that $a_i < b_i$, $i = 1, \dots, \bar{m}$, and $a_i = b_i$, $i = \bar{m} + 1, \dots, m$. Let $I := [0, 1] \subset \mathbb{R}$. Then, we may define a homeomorphism $\psi : \mathcal{X} \to I^{\bar{m}}$ by $\psi(x) = (\frac{x_1 - a_1}{b_1 - a_1}, \dots, \frac{x_{\bar{m}} - a_{\bar{m}}}{b_{\bar{m}} - a_{\bar{m}}})$, $\forall x := (x_1, \dots, x_m) \in X$. $\forall f \in \mathcal{Z}$, let $\bar{f} : I^{\bar{m}} \to \mathcal{Y}$ be defined by $\bar{f} := f \circ \psi_{\text{inv}}$. By Proposition 3.12, $\bar{f} \in \mathcal{C}(I^{\bar{m}}, \mathcal{Y})$. By Bernsteĭn Approximation Theorem 11.203, $\forall \epsilon \in (0, \infty) \subset \mathbb{R}$, $\exists n \in \mathbb{N}$ such that $\|\bar{f} - B_n\|_{\mathcal{C}(I^{\bar{m}}, \mathcal{Y})} < \epsilon$, where $B_n : I^{\bar{m}} \to \mathcal{Y}$ is the nth Bernsteĭn function for $\bar{f}$. Then, $\|f - B_n \circ \psi\|_{\mathcal{Z}} = \|(\bar{f} - B_n) \circ \psi\|_{\mathcal{Z}} = \|\bar{f} - B_n\|_{\mathcal{C}(I^{\bar{m}}, \mathcal{Y})} < \epsilon$. Clearly, $B_n \circ \psi \in \text{span}(Z)$. By the arbitrariness of ϵ, we have $f \in \overline{\text{span}(Z)}$. By the arbitrariness of f, we have $\mathcal{Z} = \overline{\text{span}(Z)}$.

Hence, in both cases, we have $\mathcal{Z} = \overline{\text{span}(Z)}$. By Riesz Representation Theorem 11.201, we have $\mathcal{Z}^* = \mathcal{M}_{ft}(\mathcal{X}, \mathcal{Y}^*)$. By Theorem 11.198, $\mathcal{Z}^* = \mathcal{M}_{ft}(\mathcal{X}, \mathcal{Y}^*) = \mathcal{M}_f(X, \mathcal{B}_B(\mathcal{X}), \mathcal{Y}^*)$. This completes the proof of the theorem. □

Theorem 11.205 (Riesz Representation Theorem) *$\mathcal{X} := (X, \mathcal{O})$ be a compact Hausdorff topological space, $\mathcal{Y}$ be a finite-dimensional Banach space over $\mathbb{K}$, and $\mathcal{Z} := \mathcal{C}(\mathcal{X}, \mathcal{Y})$. Then, $\mathcal{Z}^* = \mathcal{M}_{ft}(\mathcal{X}, \mathcal{Y}^*)$.*

Proof Let $Z := \{z \in \mathcal{Z} \mid z = hy, h \in \mathcal{C}(\mathcal{X}, \mathbb{R}), y \in \mathcal{Y}\}$. We will show that $\mathcal{Z} = \text{span}(Z)$. Then, the theorem is a direct consequence of Riesz Representation Theorem 11.201.

Let the dimension of $\mathcal{Y}$ be $m \in \mathbb{Z}_+$ and $\{y_1, \dots, y_m\}$ be a basis of $\mathcal{Y}$. Then, there is an invertible bounded linear mapping $\phi : \mathcal{Y} \to \mathbb{K}^m$ such that $y = \sum_{i=1}^m (\pi_i \circ \phi(y)) y_i$, $\forall y \in \mathcal{Y}$, where $\pi_i : \mathbb{K}^m \to \mathbb{K}$ is the ith coordinate projection function, $i = 1, \dots, m$. We will distinguish two exhaustive and mutually exclusives cases: Case 1: $\mathbb{K} = \mathbb{R}$; Case 2: $\mathbb{K} = \mathbb{C}$.

Case 1: $\mathbb{K} = \mathbb{R}$. $\forall f \in \mathcal{Z}$, let $g_i := \pi_i \circ \phi \circ f \in \mathcal{C}(\mathcal{X}, \mathbb{R})$, $i = 1, \dots, m$. Then, $f(x) = \sum_{i=1}^m (\pi_i \circ \phi(f(x))) y_i = \sum_{i=1}^m g_i(x) y_i$, $\forall x \in \mathcal{X}$. Hence, $f \in \text{span}(Z)$. This case is proved.

Case 2: $\mathbb{K} = \mathbb{C}$. $\forall f \in \mathcal{Z}$, let $g_i := \operatorname{Re} \circ \pi_i \circ \phi \circ f \in \mathcal{C}(\mathcal{X}, \mathbb{R})$ and $h_i := \operatorname{Im} \circ \pi_i \circ \phi \circ f \in \mathcal{C}(\mathcal{X}, \mathbb{R})$, $i = 1, \dots, m$. Then, $f(x) = \sum_{i=1}^m (\pi_i \circ \phi(f(x))) y_i = \sum_{i=1}^m (\operatorname{Re} \circ \pi_i \circ \phi \circ f(x) + \mathrm{i} \operatorname{Im} \circ \pi_i \circ \phi \circ f(x)) y_i = \sum_{i=1}^m (g_i(x) y_i + h_i(x) \mathrm{i} y_i)$, $\forall x \in \mathcal{X}$. Hence, $f \in \text{span}(Z)$. This case is also proved.

This completes the proof of the theorem. □

Proposition 11.206 *Let $m \in \mathbb{Z}_+$, $I_1, \dots, I_m \subset \mathbb{R}$ be compact intervals, $X := \prod_{i=1}^m I_i \subset \mathbb{R}^m$ with subset topology $\mathcal{O}$, $\mathcal{X} := (X, \mathcal{O})$, and $\mathcal{Y}$ be a separable normed linear space. Then, $\mathcal{Z} := \mathcal{C}(\mathcal{X}, \mathcal{Y})$ is separable.*

Proof Let $Z := \{z \in \mathcal{Z} \mid z = hy, h \in \mathcal{C}(\mathcal{X}, \mathbb{R}), y \in \mathcal{Y}\}$. By Riesz Representation Theorem 11.204 and its proof, we have $\mathcal{Z} = \overline{\text{span}(Z)}$. By Proposition 11.190, $\mathcal{C}(\mathcal{X}, \mathbb{R})$ is separable. Let $D_1 \subseteq \mathcal{C}(\mathcal{X}, \mathbb{R})$ be a countable dense subset. Let $D_2 \subseteq \mathcal{Y}$ be a countable dense subset. Then, $Z_0 := \{z \in Z \mid z = h_1 y_1, h_1 \in D_1, y_1 \in D_2\}$ is a countable set. Clearly, $\overline{\text{span}(Z_0)} \supseteq Z$. Then, Z is separable. By Proposition 7.35, $\mathcal{Z}$ is separable. This completes the proof of the proposition. □

Theorem 11.207 *Let $\mathcal{X}$ be a finite-dimensional normed linear space over $\mathbb{K}$. Then, $\mathcal{X}$ is a second countable locally compact Hausdorff space.*

Proof Clearly, $\mathcal{X}$ is Hausdorff. By Propositions 7.42, $\overline{\mathcal{B}}_{\mathcal{X}}(x_0, r)$ is compact, $\forall x_0 \in \mathcal{X}$ and $\forall r \in \mathbb{R}_+$. By Theorem 7.36, $\mathcal{X}$ is a Banach space. By Proposition 11.190, $\mathcal{X}$ is separable. Let $D \subseteq \mathcal{X}$ be a countable dense set. Clearly, $\mathcal{M} := \{\mathcal{B}_{\mathcal{X}}(x_0, r) \subseteq \mathcal{X} \mid x_0 \in D, r \in \mathbb{Q} \cap \mathbb{R}_+\}$ is a basis for the topology of $\mathcal{X}$, which is countable. Then, $\mathcal{X}$ is second countable. By Proposition 5.51, $\mathcal{X}$ is a locally compact space. Therefore, $\mathcal{X}$ is a second countable locally compact Hausdorff space. □

Proposition 11.208 *Let $m \in \mathbb{N}$, $\mathcal{X}$ be a m-dimensional Banach space over $\mathbb{K}_1$, $\mathcal{X}_c := (X_c, \mathcal{O}_c)$ be the Alexandroff one-point compactification of $\mathcal{X}$, $\mathcal{Y}$ be a normed linear space over $\mathbb{K}_2$, $Z_c := \{\bar{z} \in \mathcal{C}(\mathcal{X}_c, \mathcal{Y}) \mid \bar{z}(\omega) = \vartheta_{\mathcal{Y}}\}$, where $\omega \in \mathcal{X}_c$ is the point at infinity in $\mathcal{X}_c$, $Z := \{z \in \mathcal{C}_b(\mathcal{X}, \mathcal{Y}) \mid \exists \bar{z} \in Z_c \ni \cdot z = \bar{z}|_{\mathcal{X}}\}$. Then, $Z = \{z \in \mathcal{C}_b(\mathcal{X}, \mathcal{Y}) \mid \lim_{\|x\|_{\mathcal{X}} \to \infty} z(x) = \vartheta_{\mathcal{Y}}\} =: \bar{Z}$, and $(Z, \mathbb{K}_2, \|\cdot\|)$ is a normed linear space over $\mathbb{K}_2$, where $\|\cdot\|$ is the norm as in $\mathcal{C}_b(\mathcal{X}, \mathcal{Y})$. We will denote $(Z, \mathbb{K}_2, \|\cdot\|) =: \mathcal{C}_c(\mathcal{X}, \mathcal{Y})$.*

Furthermore, if $\mathcal{Y}$ is a Banach space, then $\mathcal{C}_c(\mathcal{X}, \mathcal{Y})$ is also a Banach space.

Proof By Theorem 11.207, $\mathcal{X}$ is a second countable locally compact Hausdorff space. Then, by Alexandroff One-Point Compactification Theorem 5.66, $\mathcal{X}_c$ is well-defined. By Example 9.71, $\mathcal{C}_b(\mathcal{X}, \mathcal{Y})$ is a normed linear space over $\mathbb{K}_2$. By Example 7.31 and Proposition 5.22, $\mathcal{C}(\mathcal{X}_c, \mathcal{Y})$ is a normed linear space over $\mathbb{K}_2$.

Fix any $z \in Z$. $z \in \mathcal{C}_b(\mathcal{X}, \mathcal{Y})$ implies that $\|z\|_{\mathcal{C}_b(\mathcal{X},\mathcal{Y})} \in \overline{\mathbb{R}}_+$. By the definition of Z, $\exists \bar{z} \in Z_c$ such that $z = \bar{z}|_{\mathcal{X}}$. Then, $\bar{z}(\omega) = \vartheta_{\mathcal{Y}}$. $\forall \epsilon \in \mathbb{R}_+$, $O := \bar{z}_{\text{inv}}(\mathcal{B}_{\mathcal{Y}}(\vartheta_{\mathcal{Y}}, \epsilon)) \in \mathcal{O}_c$, since $\bar{z}$ is continuous. Clearly, $\omega \in O$. Then, $\mathcal{X}_c \setminus O \subseteq \mathcal{X}$ is a compact set by Alexandroff One-Point Compactification Theorem 5.66. Thus, by Proposition 7.42, $\exists M \in \overline{\mathbb{R}}_+$ such that $\mathcal{X}_c \setminus O \subseteq \mathcal{B}_{\mathcal{X}}(\vartheta_{\mathcal{X}}, M)$. Hence, $\forall x \in \mathcal{X}$ with $\|x\|_{\mathcal{X}} \geq M$, we have $x \in O \cap \mathcal{X}$, and then $\|z(x)\|_{\mathcal{Y}} = \|\bar{z}(x)\|_{\mathcal{Y}} < \epsilon$. This shows that $\lim_{\|x\|_{\mathcal{X}} \to \infty} z(x) = \vartheta_{\mathcal{Y}}$. Therefore, $z \in \bar{Z}$. By the arbitrariness of z, we have $Z \subseteq \bar{Z}$.

On the other hand, fix any $z \in \bar{Z}$. Define $\bar{z} : \mathcal{X}_c \to \mathcal{Y}$ by $\bar{z}(x) = \begin{cases} z(x) & x \in \mathcal{X} \\ \vartheta_{\mathcal{Y}} & x = \omega \end{cases}$, $\forall x \in \mathcal{X}_c$. $\forall x_0 \in \mathcal{X}_c$, if $x_0 \in \mathcal{X}$, we have $\bar{z}$ is continuous at x_0. If $x_0 = \omega$, then, by the definition of $\bar{Z}$, $\forall \epsilon \in \mathbb{R}_+$, $\exists M \in \overline{\mathbb{R}}_+$ such that $\forall x \in \mathcal{X}$ with $\|x\|_{\mathcal{X}} \geq M$, we have $\|z(x)\|_{\mathcal{Y}} < \epsilon$. Thus, let $O := (\mathcal{X} \setminus \overline{\mathcal{B}}_{\mathcal{X}}(\vartheta_{\mathcal{X}}, M)) \cup \{\omega\} \subseteq \mathcal{X}_c$. By Alexandroff One-Point Compactification Theorem 5.66, $O \in \mathcal{O}_c$. Clearly, $\forall x \in O$, we have $\|\bar{z}(x)\|_{\mathcal{Y}} < \epsilon$. This proves that $\omega \in O \subseteq \bar{z}_{\text{inv}}(\mathcal{B}_{\mathcal{Y}}(\vartheta_{\mathcal{Y}}, \epsilon))$. Hence, $\bar{z}$ is continuous at $x_0 = \omega$. Thus, $\bar{z} \in Z_c$. Therefore, $z \in Z$. By the arbitrariness of z, we have $\bar{Z} \subseteq Z$.

Combining the above two paragraphs, we have $Z = \bar{Z}$.

Clearly, $\vartheta_{\mathcal{C}_b(\mathcal{X},\mathcal{Y})} \in \bar{Z} = Z \neq \emptyset$. $\forall z_1, z_2 \in Z$, $\forall \alpha \in \mathbb{K}_2$, we have $z_1 + z_2 \in \mathcal{C}_b(\mathcal{X}, \mathcal{Y})$, since $z_1, z_2 \in \bar{Z} \subseteq \mathcal{C}_b(\mathcal{X}, \mathcal{Y})$ and $\lim_{\|x\|_{\mathcal{X}}\to\infty}(z_1(x) + z_2(x)) = \lim_{\|x\|_{\mathcal{X}}\to\infty} z_1(x) + \lim_{\|x\|_{\mathcal{X}}\to\infty} z_2(x) = \vartheta_{\mathcal{Y}}$ by the definition of $\bar{Z}$ and $Z = \bar{Z}$. This shows that $z_1 + z_2 \in Z$. Furthermore, $\alpha z_1 \in \mathcal{C}_b(\mathcal{X}, \mathcal{Y})$, and $\lim_{\|x\|_{\mathcal{X}}\to\infty} \alpha z_1(x) = \alpha\vartheta_{\mathcal{Y}} = \vartheta_{\mathcal{Y}}$. This shows that $\alpha z_1 \in Z$. Hence, $Z \subseteq \mathcal{C}_b(\mathcal{X}, \mathcal{Y})$ is a subspace, and therefore, $(Z, \mathbb{K}_2, \|\cdot\|)$ is a normed linear subspace of $\mathcal{C}_b(\mathcal{X}, \mathcal{Y})$.

Consider the case that $\mathcal{Y}$ is a Banach space. By Example 9.72, $\mathcal{C}_b(\mathcal{X}, \mathcal{Y})$ is a Banach space. We will show that Z is a complete subspace of $\mathcal{C}_b(\mathcal{X}, \mathcal{Y})$. Then, $(Z, \mathbb{K}_2, \|\cdot\|)$ is a Banach subspace of $\mathcal{C}_b(\mathcal{X}, \mathcal{Y})$.

Fix any Cauchy sequence $(z_n)_{n=1}^{\infty} \subseteq Z$. Then, $\lim_{n\in\mathbb{N}} z_n = z_0 \in \mathcal{C}_b(\mathcal{X}, \mathcal{Y})$ since $Z \subseteq \mathcal{C}_b(\mathcal{X}, \mathcal{Y})$ and $\mathcal{C}_b(\mathcal{X}, \mathcal{Y})$ is a Banach space. $\forall \epsilon \in \mathbb{R}_+$, $\exists N \in \mathbb{N}$, $\forall n, m \in \mathbb{N}$ with $n, m \geq N$, we have $\|z_n - z_m\|_{\mathcal{C}_b(\mathcal{X},\mathcal{Y})} < \epsilon/2$. By $z_N \in Z$, we have $\lim_{\|x\|_{\mathcal{X}}\to\infty} z_N(x) = \vartheta_{\mathcal{Y}}$. Then, $\exists M \in \mathbb{R}_+$, $\forall x \in \mathcal{X}$ with $\|x\|_{\mathcal{X}} > M$, we have $\|z_N(x)\|_{\mathcal{Y}} < \epsilon/2$. Then, $\|z_0(x)\|_{\mathcal{Y}} \leq \|z_0(x) - z_N(x)\|_{\mathcal{Y}} + \|z_N(x)\|_{\mathcal{Y}} < \|z_0 - z_N\|_{\mathcal{C}_b(\mathcal{X},\mathcal{Y})} + \epsilon/2 \leq \lim_{n\in\mathbb{N}} \|z_n - z_N\|_{\mathcal{C}_b(\mathcal{X},\mathcal{Y})} + \epsilon/2 \leq \epsilon$. Hence, $\lim_{\|x\|_{\mathcal{X}}\to\infty} z_0(x) = \vartheta_{\mathcal{Y}}$ and $z_0 \in Z$. By the arbitrariness of $(z_n)_{n=1}^{\infty}$, we have Z is complete. Then, $\mathcal{C}_c(\mathcal{X}, \mathcal{Y})$ is a Banach space. □

Theorem 11.209 (Riesz Representation Theorem) *Let $m \in \mathbb{N}$ and $n \in \mathbb{Z}_+$, $\mathcal{X}$ be an m-dimensional Banach space over $\mathbb{K}_1$, and $\mathcal{Y}$ be an n-dimensional Banach space over $\mathbb{K}_2$. Then, the following statements hold:*

(i) $(\mathcal{C}_c(\mathcal{X}, \mathcal{Y}))^* = \mathcal{M}_{ft}(\mathcal{X}, \mathcal{Y}^*) = \mathcal{M}_f(\mathcal{X}, \mathcal{B}_B(\mathcal{X}), \mathcal{Y}^*)$ *isometrically isomorphically.*
(ii) $\mathcal{C}_c(\mathcal{X}, \mathcal{Y})$ *is separable.*

Proof (i) By Theorems 11.198 and 11.207, $\mathcal{M}_f(\mathcal{X}, \mathcal{B}_B(\mathcal{X}), \mathcal{Y}^*) = \mathcal{M}_{ft}(\mathcal{X}, \mathcal{Y}^*)$. By Proposition 11.192, $\mathcal{M}_{ft}(\mathcal{X}, \mathcal{Y}^*)$ is a Banach space over $\mathbb{K}_2$.

Let $\mathcal{X}_c := (X_c, \mathcal{O}_c)$ be the Alexandroff one-point compactification of $\mathcal{X}$ as defined in Theorem 5.66. $\mathcal{X}_c$ is well-defined since $\mathcal{X}$ is a second countable locally compact Hausdorff space, by Theorem 11.207. By Riesz Representation Theorem 11.205, we have $(\mathcal{C}(\mathcal{X}_c, \mathcal{Y}))^* = \mathcal{M}_{ft}(\mathcal{X}_c, \mathcal{Y}^*)$. Let $\mathcal{F} := \{\emptyset, \{\omega\}\}$, where $\omega \in \mathcal{X}_c$ is the point at infinity. Define $\mu_1 : \mathcal{F} \to \mathcal{Y}^*$ by $\mu_1(\emptyset) = \vartheta_{\mathcal{Y}^*} = \mu_1(\{\omega\})$. It is easy to see that μ_1 is a finite $\mathcal{Y}^*$-valued measure on the measurable space $(\{\omega\}, \mathcal{F})$. Now, we need the following intermediate result.

Claim 11.209.1 $\forall \mu \in \mathcal{M}_{ft}(\mathcal{X}, \mathcal{Y}^*)$, we may define a unique finite $\mathcal{Y}^*$-valued measure space $(\mathcal{X}_c, \mathcal{B}_B(\mathcal{X}_c), \bar{\mu})$ such that $(\mathcal{X}, \mathcal{B}_B(\mathcal{X}), \mu)$ and $(\{\omega\}, \mathcal{F}, \mu_1)$ are finite $\mathcal{Y}^*$-valued measure subspaces of $(\mathcal{X}_c, \mathcal{B}_B(\mathcal{X}_c), \bar{\mu})$. Furthermore, $\bar{\mu} \in \mathcal{M}_{ft}(\mathcal{X}_c, \mathcal{Y}^*)$.

Proof of Claim By Proposition 11.118, there exists a unique σ-finite $\mathcal{Y}^*$-valued measure space $(\mathcal{X}_c, \mathcal{B}_B(\mathcal{X}_c), \bar{\mu})$ such that $(\mathcal{X}, \mathcal{B}_B(\mathcal{X}), \mu)$ and $(\{\omega\}, \mathcal{F}, \mu_1)$ are finite $\mathcal{Y}^*$-valued measure subspaces of $(\mathcal{X}_c, \mathcal{B}_B(\mathcal{X}_c), \bar{\mu})$. This implies that $\mathcal{P}\circ\bar{\mu}(\mathcal{X}_c) = \mathcal{P}\circ\bar{\mu}(\mathcal{X}) + \mathcal{P}\circ\bar{\mu}(\{\omega\}) = \mathcal{P}\circ\mu(\mathcal{X}) + \mathcal{P}\circ\mu_1(\{\omega\}) = \mathcal{P}\circ\mu(\mathcal{X}) + 0 = \mathcal{P}\circ\mu(\mathcal{X}) < \infty$, where the first equality follows from the fact that $\mathcal{P} \circ \bar{\mu}$ is a measure, the second equality follows from Proposition 11.115, the third equality follows from the definition of μ_1, and the inequality follows from μ is finite. Hence, $\bar{\mu}$ is finite.

Thus, $\bar{\mu} \in \mathcal{M}_f(\mathcal{X}_c, \mathcal{B}_B(\mathcal{X}_c), \mathcal{Y}^*)$. By Alexandroff One-Point Compactification Theorem 5.66, $\mathcal{X}_c$ is a compact Hausdorff space. By Proposition 5.14, $\mathcal{X}_c$ is normal. Clearly, $\mathcal{X}_c$ is second countable. By Urysohn Metrization Theorem 4.53, $\mathcal{X}_c$ is metrizable. Hence, $\mathcal{X}_c$ is a compact Hausdorff metrizable topological space satisfying the second axiom of countability. By Theorem 11.198, $\mathcal{M}_f(\mathcal{X}_c, \mathcal{B}_B(\mathcal{X}_c), \mathcal{Y}^*) = \mathcal{M}_{ft}(\mathcal{X}_c, \mathcal{Y}^*)$. This completes the proof of the claim. □

Fix any $\mu \in \mathcal{M}_{ft}(\mathcal{X}, \mathcal{Y}^*)$. By Claim 11.209.1, let $\bar{\mu}$ be the finite $\mathcal{Y}^*$-valued measure on the measurable space $(\mathcal{X}_c, \mathcal{B}_B(\mathcal{X}_c))$ such that $(\mathcal{X}, \mathcal{B}_B(\mathcal{X}), \mu)$ and $(\{\omega\}, \mathcal{F}, \mu_1)$ are its finite $\mathcal{Y}^*$-valued measure subspaces. Then, $\bar{\mu} \in \mathcal{M}_{ft}(\mathcal{X}_c, \mathcal{Y}^*) = (\mathcal{C}(\mathcal{X}_c, \mathcal{Y}))^*$. Thus, by Riesz Representation Theorem 11.205, we may define $\bar{f} : \mathcal{C}(\mathcal{X}_c, \mathcal{Y}) \to \mathbb{K}_2$ by $\bar{f}(\bar{z}) = \int_{\mathcal{X}_c} \langle\langle d\bar{\mu}(x), \bar{z}(x)\rangle\rangle$, $\forall \bar{z} \in \mathcal{C}(\mathcal{X}_c, \mathcal{Y})$. Then, $\bar{f}$ is a bounded linear functional on $\mathcal{C}(\mathcal{X}_c, \mathcal{Y})$, and $\|\bar{f}\|_{(\mathcal{C}(\mathcal{X}_c,\mathcal{Y}))^*} = \|\bar{\mu}\|_{\mathcal{M}_{ft}(\mathcal{X}_c,\mathcal{Y}^*)}$. This implies that $\int_{\mathcal{X}_c} \langle\langle d\bar{\mu}(x), \bar{z}(x)\rangle\rangle = \bar{f}(\bar{z}) = \int_{\mathcal{X}} \langle\langle d\bar{\mu}(x), \bar{z}(x)\rangle\rangle + \int_{\{\omega\}} \langle\langle d\bar{\mu}(x), \bar{z}(x)\rangle\rangle = \int_{\mathcal{X}} \langle\langle d\mu(x), \bar{z}(x)\rangle\rangle + \int_{\{\omega\}} \langle\langle d\mu_1(x), \bar{z}(x)\rangle\rangle = \int_{\mathcal{X}} \langle\langle d\mu(x), \bar{z}(x)\rangle\rangle + \langle\langle \mu_1(\{\omega\}), \bar{z}(\omega)\rangle\rangle = \int_{\mathcal{X}} \langle\langle d\mu(x), \bar{z}(x)\rangle\rangle$, $\forall \bar{z} \in \mathcal{C}(\mathcal{X}_c, \mathcal{Y})$, where the second equality follows from Proposition 11.130, the third equality follows from Proposition 11.115, the fourth equality follows from Proposition 11.125, and the fifth equality follows from the definition of μ_1. Let Z_c be defined as in Proposition 11.208. We may define $f : \mathcal{C}_c(\mathcal{X}, \mathcal{Y}) \to \mathbb{K}_2$ by $f(z) = \bar{f}(\bar{z}) = \int_{\mathcal{X}} \langle\langle d\mu(x), \bar{z}(x)\rangle\rangle = \int_{\mathcal{X}} \langle\langle d\mu(x), z(x)\rangle\rangle$, $\forall z \in \mathcal{C}_c(\mathcal{X}, \mathcal{Y})$ and $\bar{z} \in Z_c$ is such that $z = \bar{z}|_{\mathcal{X}}$. Thus, f is a bounded linear functional on $\mathcal{C}_c(\mathcal{X}, \mathcal{Y})$, since $\bar{f}$ is a bounded linear functional on $\mathcal{C}(\mathcal{X}_c, \mathcal{Y})$ and therefore on Z_c, which is a subspace of $\mathcal{C}(\mathcal{X}_c, \mathcal{Y})$. Hence, μ defines a bounded linear functional on $\mathcal{C}_c(\mathcal{X}, \mathcal{Y})$ by $f := \Phi(\mu)$, where $f(z) = \int_{\mathcal{X}} \langle\langle d\mu(x), z(x)\rangle\rangle$, $\forall z \in \mathcal{C}_c(\mathcal{X}, \mathcal{Y})$. Therefore, $\|f\|_{(\mathcal{C}_c(\mathcal{X},\mathcal{Y}))^*} = \sup_{\substack{z \in \mathcal{C}_c(\mathcal{X},\mathcal{Y}) \\ \|z\|_{\mathcal{C}_c(\mathcal{X},\mathcal{Y})} \leq 1}} \left|\int_{\mathcal{X}} \langle\langle d\mu(x), z(x)\rangle\rangle\right| \leq \sup_{\substack{z \in \mathcal{C}_c(\mathcal{X},\mathcal{Y}) \\ \|z\|_{\mathcal{C}_c(\mathcal{X},\mathcal{Y})} \leq 1}} \int_{\mathcal{X}} \|z(x)\|_{\mathcal{Y}} \, d\mathcal{P} \circ \mu(x) \leq \mathcal{P} \circ \mu(\mathcal{X}) = \|\mu\|_{\mathcal{M}_{ft}(\mathcal{X},\mathcal{Y}^*)} < \infty$, where the first equality follows from Proposition 7.63, the first inequality follows from Proposition 11.130, the second inequality follows from Proposition 11.83, and the second equality follows from Proposition 11.142. Thus, we may define $\Phi : \mathcal{M}_{ft}(\mathcal{X}, \mathcal{Y}^*) \to (\mathcal{C}_c(\mathcal{X}, \mathcal{Y}))^*$ by $f = \Phi(\mu)$, $\forall \mu \in \mathcal{M}_{ft}(\mathcal{X}, \mathcal{Y}^*)$. Clearly, Φ is a bounded linear operator.

On the other hand, fix any $f \in (\mathcal{C}_c(\mathcal{X}, \mathcal{Y}))^*$. We may define a functional $\bar{f} : Z_c \to \mathbb{K}_2$ as follows. $\forall \bar{z} \in Z_c$, let $z : \mathcal{X} \to \mathcal{Y}$ be defined by $z := \bar{z}|_{\mathcal{X}}$. Then, $z \in \mathcal{C}_c(\mathcal{X}, \mathcal{Y})$. We define $\bar{f}(\bar{z}) := f(z)$. It is easy to show that $\bar{f}$ is a bounded linear functional on Z_c with $\|\bar{f}\|_{Z_c^*} = \|f\|_{(\mathcal{C}_c(\mathcal{X},\mathcal{Y}))^*}$. Clearly, Z_c is a subspace of $\mathcal{C}(\mathcal{X}_c, \mathcal{Y})$. By Hahn–Banach Theorem 7.83, $\exists \tilde{f} \in (\mathcal{C}(\mathcal{X}_c, \mathcal{Y}))^* = \mathcal{M}_{ft}(\mathcal{X}_c, \mathcal{Y}^*)$ such that $\left\|\tilde{f}\right\|_{(\mathcal{C}(\mathcal{X}_c,\mathcal{Y}))^*} = \|\bar{f}\|_{Z_c^*} = \|f\|_{(\mathcal{C}_c(\mathcal{X},\mathcal{Y}))^*}$ and $\tilde{f}(\bar{z}) = \bar{f}(\bar{z}) = f(z)$, $\forall \bar{z} \in Z_c$ and $z = \bar{z}|_{\mathcal{X}}$. By Riesz Representation Theorem 11.205 (and Theorem 11.201), $\exists!\, \bar{\mu} \in \mathcal{M}_{ft}(\mathcal{X}_c, \mathcal{Y}^*)$ such that $\tilde{f}(\bar{z}) = \int_{\mathcal{X}_c} \langle\langle d\bar{\mu}(x), \bar{z}(x)\rangle\rangle$, $\forall \bar{z} \in \mathcal{C}(\mathcal{X}_c, \mathcal{Y})$ and $\|\bar{\mu}\|_{\mathcal{M}_{ft}(\mathcal{X}_c,\mathcal{Y}^*)} = \mathcal{P} \circ \bar{\mu}(\mathcal{X}_c) = \left\|\tilde{f}\right\|_{(\mathcal{C}(\mathcal{X}_c,\mathcal{Y}))^*} = \|f\|_{(\mathcal{C}_c(\mathcal{X},\mathcal{Y}))^*}$. $\forall z \in \mathcal{C}_c(\mathcal{X}, \mathcal{Y})$, let

$\bar{z} \in Z_c$ with $z = \bar{z}|_{\mathcal{X}}$. We have

$$\tilde{f}(\bar{z}) = f(z) = \int_{\mathcal{X}_c} \langle\langle \mathrm{d}\bar{\mu}(x), \bar{z}(x)\rangle\rangle = \int_{\mathcal{X}} \langle\langle \mathrm{d}\bar{\mu}(x), \bar{z}(x)\rangle\rangle$$
$$+ \int_{\{\omega\}} \langle\langle \mathrm{d}\bar{\mu}(x), \bar{z}(x)\rangle\rangle = \int_{\mathcal{X}} \langle\langle \mathrm{d}\bar{\mu}(x), z(x)\rangle\rangle + \langle\langle \bar{\mu}(\{\omega\}), \bar{z}(\omega)\rangle\rangle$$
$$= \int_{\mathcal{X}} \langle\langle \mathrm{d}\bar{\mu}(x), z(x)\rangle\rangle =: \int_{\mathcal{X}} \langle\langle \mathrm{d}\mu(x), z(x)\rangle\rangle$$

where the first and second equalities follow from the preceding discussion of this paragraph, the third equality follows from Proposition 11.130, the fourth equality follows from Proposition 11.125, the fifth equality follows from the fact $\bar{z}(\omega) = \vartheta_{\mathcal{Y}}$ since $\bar{z} \in Z_c$, and the assignment is such that $(\mathcal{X}, \mathcal{B}_B(\mathcal{X}), \mu)$ is the $\mathcal{Y}^*$-valued topological measure subspace of $(\mathcal{X}_c, \mathcal{B}_B(\mathcal{X}_c), \bar{\mu})$, by Proposition 11.115. Hence, $\mu \in \mathcal{M}_{ft}(\mathcal{X}, \mathcal{Y}^*)$ and $f(z) = \int_{\mathcal{X}} \langle\langle \mathrm{d}\mu(x), z(x)\rangle\rangle$, $\forall z \in \mathcal{C}_c(\mathcal{X}, \mathcal{Y})$. This implies that $\|\mu\|_{\mathcal{M}_{ft}(\mathcal{X},\mathcal{Y}^*)} \leq \|\bar{\mu}\|_{\mathcal{M}_{ft}(\mathcal{X}_c,\mathcal{Y}^*)} = \|f\|_{(\mathcal{C}_c(\mathcal{X},\mathcal{Y}))^*}$. This shows that $f \in \Phi(\mathcal{M}_{ft}(\mathcal{X}, \mathcal{Y}^*))$. By the arbitrariness of f, we have Φ is surjective.

Now, we will show that the mapping Φ is injective. Then, we have Φ is bijective. By the preceding two paragraphs, $\forall \mu \in \mathcal{M}_{ft}(\mathcal{X}, \mathcal{Y}^*)$, let $f = \Phi(\mu)$; then we have $\|f\|_{(\mathcal{C}_c(\mathcal{X},\mathcal{Y}))^*} \leq \|\mu\|_{\mathcal{M}_{ft}(\mathcal{X},\mathcal{Y}^*)} \leq \|f\|_{(\mathcal{C}_c(\mathcal{X},\mathcal{Y}))^*}$. Hence, Φ is an isometry. By the expression for Φ, we can easily see that Φ is linear. Then, Φ is an isometrical isomorphism. This proves that $\mathcal{M}_{ft}(\mathcal{X}, \mathcal{Y}^*) = (\mathcal{C}_c(\mathcal{X}, \mathcal{Y}))^*$.

We will show that Φ is injective by an argument of contradiction. Suppose Φ is not injective. Then, $\exists \mu, \hat{\mu} \in \mathcal{M}_{ft}(\mathcal{X}, \mathcal{Y}^*)$ such that $f := \Phi(\mu) = \Phi(\hat{\mu})$ and $\mu \neq \hat{\mu}$. Thus, $\exists E \in \mathcal{B}_B(\mathcal{X})$ such that $\mu(E) \neq \hat{\mu}(E)$. Let $\bar{\mu} := \hat{\mu} - \mu \in \mathcal{M}_{ft}(\mathcal{X}, \mathcal{Y}^*)$. Then, by linearity of Φ, $\vartheta_{(\mathcal{C}_c(\mathcal{X},\mathcal{Y}))^*} = \Phi(\bar{\mu})$, but $\bar{\mu}(E) \neq \vartheta_{\mathcal{Y}^*}$. By Claim 11.209.1, we let $\tilde{\mu}$ be the finite $\mathcal{Y}^*$-valued measure on $(\mathcal{X}_c, \mathcal{B}_B(\mathcal{X}_c))$ such that $(\mathcal{X}, \mathcal{B}_B(\mathcal{X}), \bar{\mu})$ and $(\{\omega\}, \mathcal{F}, \mu_1)$ are its finite $\mathcal{Y}^*$-valued measure subspaces. Therefore, $\tilde{\mu} \in \mathcal{M}_{ft}(\mathcal{X}_c, \mathcal{Y}^*) = (\mathcal{C}(\mathcal{X}_c, \mathcal{Y}))^*$. Since $\bar{\mu}(E) \neq \vartheta_{\mathcal{Y}^*}$, then $\tilde{\mu}(E) = \bar{\mu}(E) \neq \vartheta_{\mathcal{Y}^*}$ and $\tilde{\mu} \neq \vartheta_{\mathcal{M}_{ft}(\mathcal{X}_c,\mathcal{Y}^*)}$. This implies that $\exists \bar{z} \in \mathcal{C}(\mathcal{X}_c, \mathcal{Y})$ such that $\int_{\mathcal{X}_c} \langle\langle \mathrm{d}\tilde{\mu}(x), \bar{z}(x)\rangle\rangle \neq 0$. Note that $\int_{\mathcal{X}_c} \langle\langle \mathrm{d}\tilde{\mu}(x), \bar{z}(x)\rangle\rangle = \int_{\mathcal{X}} \langle\langle \mathrm{d}\tilde{\mu}(x), \bar{z}(x)\rangle\rangle + \int_{\{\omega\}} \langle\langle \mathrm{d}\tilde{\mu}(x), \bar{z}(x)\rangle\rangle = \int_{\mathcal{X}} \langle\langle \mathrm{d}\bar{\mu}(x), \bar{z}(x)\rangle\rangle + \langle\langle \tilde{\mu}(\{\omega\}), \bar{z}(\omega)\rangle\rangle = \int_{\mathcal{X}} \langle\langle \mathrm{d}\bar{\mu}(x), \bar{z}(x)\rangle\rangle$, where the first equality follows from Proposition 11.130, the second equality follows from Proposition 11.125, and the last equality follows from $\tilde{\mu}(\{\omega\}) = \mu_1(\{\omega\}) = \vartheta_{\mathcal{Y}^*}$. Let $z := \bar{z}|_{\mathcal{X}}$. Then, $z \in \mathcal{C}_b(\mathcal{X}, \mathcal{Y})$ and $\int_{\mathcal{X}} \langle\langle \mathrm{d}\bar{\mu}(x), z(x)\rangle\rangle = \int_{\mathcal{X}} \langle\langle \mathrm{d}\bar{\mu}(x), \bar{z}(x)\rangle\rangle = \int_{\mathcal{X}_c} \langle\langle \mathrm{d}\tilde{\mu}(x), \bar{z}(x)\rangle\rangle \neq 0$. By Theorem 11.207, $\mathcal{X}$ is a second countable locally compact Hausdorff space. $\forall n \in \mathbb{N}$, by Proposition 5.62, there exists $h_n : \mathcal{X} \to [0, 1] \subset \mathbb{R}$ such that $h_n(x) = 1$, $\forall x \in E_{n,1} := \overline{\mathcal{B}}_{\mathcal{X}}(\vartheta_{\mathcal{X}}, n)$ and $h_n(x) = 0$, $\forall x \in E_{n,2} := \mathcal{X} \setminus \mathcal{B}_{\mathcal{X}}(\vartheta_{\mathcal{X}}, n + 1)$. Clearly, $\lim_{n\in\mathbb{N}} z(x)h_n(x) = z(x)$, $\forall x \in \mathcal{X}$. By Lebesgue Dominated Convergence Theorem 11.128, we have $\int_{\mathcal{X}} \langle\langle \mathrm{d}\bar{\mu}(x), z(x)\rangle\rangle = \lim_{n\in\mathbb{N}} \int_{\mathcal{X}} \langle\langle \mathrm{d}\bar{\mu}(x), z(x)h_n(x)\rangle\rangle \neq 0$. Then, $\exists n \in \mathbb{N}$ such that $\Phi(\bar{\mu})(zh_n) = \int_{\mathcal{X}} \langle\langle \mathrm{d}\bar{\mu}(x), z(x)h_n(x)\rangle\rangle \neq 0$, since $zh_n \in \mathcal{C}_c(\mathcal{X}, \mathcal{Y})$. This is a contradiction to $\Phi(\bar{\mu}) = \vartheta_{(\mathcal{C}_c(\mathcal{X},\mathcal{Y}))^*}$. This shows that hypothesis is impossible. Hence, Φ is injective.

(ii) Let $K_n := \overline{\mathcal{B}}_{\mathcal{X}}(\vartheta_{\mathcal{X}}, n) \subseteq \mathcal{X}$, $\forall n \in \mathbb{N}$. Then, by Proposition 7.42, we have $K_n \subseteq K_{n+1} \subseteq \mathcal{X}$ are compact. By Proposition 5.62 and Theorem 11.207, $\exists h_n : \mathcal{X} \to [0, 1] \subset \mathbb{R}$ such that $h_n(x) = 1$, $\forall x \in K_n$; $h_n(x) = 0$, $\forall x \in \mathcal{X} \setminus K_{n+1}^\circ$, $\forall n \in \mathbb{N}$. Fix any $n \in \mathbb{N}$. By Proposition 11.190, $\mathcal{C}(K_n, \mathcal{Y})$ is separable. Let $D_n \subseteq \mathcal{C}(K_n, \mathcal{Y})$ be a countable dense subset. Define $\bar{D}_n \subseteq \mathcal{C}_c(\mathcal{X}, \mathcal{Y})$ by $\bar{D}_n := \left\{ f : \mathcal{X} \to \mathcal{Y} \;\middle|\; \exists g \in D_n \ni \cdot f(x) = \begin{cases} h_n(x)g(x) & x \in K_n \\ h_n(x)g(\frac{nx}{\|x\|_{\mathcal{X}}}) & x \in \mathcal{X} \setminus K_n \end{cases} \right\}$. Clearly, $\bar{D}_n$ is countable. Define $\bar{D} := \bigcup_{n\in\mathbb{N}} \bar{D}_n \subseteq \mathcal{C}_c(\mathcal{X}, \mathcal{Y})$. We will show that $\bar{D}$ is dense in $\mathcal{C}_c(\mathcal{X}, \mathcal{Y})$.

Fix any $z \in \mathcal{C}_c(\mathcal{X}, \mathcal{Y})$. Fix any $\epsilon \in \mathbb{R}_+$. Then, $\exists N \in \mathbb{N}$ such that $\|z(x)\|_{\mathcal{Y}} < \frac{\epsilon}{3}$, $\forall x \in \mathcal{X} \setminus K_N^\circ$. Since $z|_{K_N} \in \mathcal{C}(K_N, \mathcal{Y})$, then $\exists g \in D_N$ such that $\left\| z|_{K_N} - g \right\|_{\mathcal{C}(K_N,\mathcal{Y})} < \frac{\epsilon}{3}$. Let $f \in \bar{D}_N \subseteq \bar{D}$ be such that $f(x) = \begin{cases} h_N(x)g(x) & x \in K_N \\ h_N(x)g(\frac{Nx}{\|x\|_{\mathcal{X}}}) & x \in \mathcal{X} \setminus K_N \end{cases}$. Then, $\forall x \in K_N$, $\|f(x) - z(x)\|_{\mathcal{Y}} = \|g(x) - z(x)\|_{\mathcal{Y}} \leq \left\| z|_{K_N} - g \right\|_{\mathcal{C}(K_N,\mathcal{Y})} < \frac{\epsilon}{3}$. $\forall x \in \mathcal{X} \setminus K_N$, $\|f(x) - z(x)\|_{\mathcal{Y}} \leq \|f(x)\|_{\mathcal{Y}} + \|z(x)\|_{\mathcal{Y}} = \left\| h_N(x)g(\frac{Nx}{\|x\|_{\mathcal{X}}}) \right\|_{\mathcal{Y}} + \frac{\epsilon}{3} \leq \left\| g(\frac{Nx}{\|x\|_{\mathcal{X}}}) \right\|_{\mathcal{Y}} + \frac{\epsilon}{3} \leq \left\| g(\frac{Nx}{\|x\|_{\mathcal{X}}}) - z(\frac{Nx}{\|x\|_{\mathcal{X}}}) \right\|_{\mathcal{Y}} + \left\| z(\frac{Nx}{\|x\|_{\mathcal{X}}}) \right\|_{\mathcal{Y}} + \frac{\epsilon}{3} \leq \left\| z|_{K_N} - g \right\|_{\mathcal{C}(K_N,\mathcal{Y})} + \frac{\epsilon}{3} + \frac{\epsilon}{3} < \epsilon$. Hence, $\|f - z\|_{\mathcal{C}_c(\mathcal{X},\mathcal{Y})} \leq \epsilon$. This shows that $\bar{D}$ is dense in $\mathcal{C}_c(\mathcal{X}, \mathcal{Y})$. Therefore, $\mathcal{C}_c(\mathcal{X}, \mathcal{Y})$ is separable.

This completes the proof of the theorem. □

$\mathcal{C}_c(\mathbb{R}^m, \mathbb{R})$ is the continuous domain analogy of c_0.

Definition 11.210 Let $m \in \mathbb{N}$, $n \in \mathbb{Z}_+$, $\mathcal{X}$ be an m-dimensional Banach space over $\mathbb{K}_1$, $\mathcal{Y}$ be an n-dimensional Banach space over $\mathbb{K}_2$, and $(\mu_n)_{n=1}^\infty \subseteq \mathcal{M}_{ft}(\mathcal{X}, \mathcal{Y}^*) = \mathcal{M}_f(\mathcal{X}, \mathcal{B}_B(\mathcal{X}), \mathcal{Y}^*)$. We will say that $(\mu_n)_{n=1}^\infty$ *converges to* $\mu_0 \in \mathcal{M}_{ft}(\mathcal{X}, \mathcal{Y}^*)$ *in distribution as* $n \to \infty$, if $\lim_{n\in\mathbb{N}} \mu_n = \mu_0$ weak* (see Proposition 7.120 for details). In this case, we will denote $\mu_n \looparrowright \mu_0$ as $n \to \infty$. ◇

Proposition 11.211 *Let* $\boldsymbol{\Omega} := (\Omega, \mathcal{B}, \mu)$ *be a* σ*-finite measure space,* $\mathcal{X}$ *be a separable Banach space over* $\mathbb{K}$, $(x_n)_{n=0}^\infty$ *be a sequence of* $\mathcal{X}$*-valued functions of* $\boldsymbol{\Omega}$ *that is* $\mathcal{B}$*-measurable, and* $p \in [1, \infty] \subset \mathbb{R}_e$. *Assume that* $\lim_{n\in\mathbb{N}} \|x_n - x_0\|_{\bar{L}_p(\boldsymbol{\Omega},\mathcal{X})} = 0$. *Then,* $\lim_{n\in\mathbb{N}} x_n = x_0$ *in measure in* $\boldsymbol{\Omega}$.

Proof We will distinguish two exhaustive and mutually exclusive cases: Case 1: $p = \infty$; and Case 2: $p \in [1, \infty) \subset \mathbb{R}$.

Case 1: $p = \infty$. Then, $\lim_{n\in\mathbb{N}} \|x_n - x_0\|_{\bar{L}_\infty(\boldsymbol{\Omega},\mathcal{X})} = 0$ implies that $\lim_{n\in\mathbb{N}} \max \left\{ \operatorname{ess\,sup}_{\omega\in\Omega} \|x_n(\omega) - x_0(\omega)\|_{\mathcal{X}}, 0 \right\} = 0$. Thus, $\forall \epsilon \in (0, \infty) \subset \mathbb{R}$, $\exists n_0 \in \mathbb{N}$ such that $\max \left\{ \operatorname{ess\,sup}_{\omega\in\Omega} \|x_n(\omega) - x_0(\omega)\|_{\mathcal{X}}, 0 \right\} < \epsilon$, $\forall n \in \mathbb{N}$ with $n \geq n_0$. Then, $\mu(\{\omega \in \Omega \mid \|x_n(\omega) - x_0(\omega)\|_{\mathcal{X}} \geq \epsilon\} = 0 < \epsilon$. Hence, $\lim_{n\in\mathbb{N}} x_n = x_0$ in measure in $\boldsymbol{\Omega}$.

Case 2: $p \in [1, \infty) \subset \mathbb{R}$. Suppose that $(x_n)_{n=1}^{\infty}$ does not converge to x_0 in measure. Then, $\exists \epsilon_0 \in (0, \infty) \subset \mathbb{R}$ such that $\forall n \in \mathbb{N}$, there exists $n_1 \in \mathbb{N}$ with $n \leq n_1$ such that $\mu(\{\omega \in \Omega \mid \|x_{n_1}(\omega) - x_0(\omega)\|_{\mathcal{X}} \geq \epsilon_0\}) =: \mu(A) \geq \epsilon_0$. Then, $\|x_{n_1} - x_0\|_{\bar{\mathrm{L}}_p(\boldsymbol{\Omega}, \mathcal{X})} = (\int_{\Omega} \|x_{n_1}(\omega) - x_0(\omega)\|_{\mathcal{X}}^p \, \mathrm{d}\mu(\omega))^{1/p} \geq (\int_A \epsilon_0^p \, \mathrm{d}\mu(\omega))^{1/p} \geq \epsilon_0^{\frac{1+p}{p}} =: \delta_0 \in (0, \infty) \subset \mathbb{R}$, where the first inequality follows from Proposition 11.83 and the second inequality follows from Proposition 11.75. Hence, $\limsup_{n \in \mathbb{N}} \|x_n - x_0\|_{\bar{\mathrm{L}}_p(\boldsymbol{\Omega}, \mathcal{X})} \geq \delta_0$. This contradicts with the assumption. Hence, the hypothesis is false. We must have $\lim_{n \in \mathbb{N}} x_n = x_0$ in measure in $\boldsymbol{\Omega}$. □

Proposition 11.212 *Let $\boldsymbol{\Omega} := (\Omega, \mathcal{B}, \mu)$ be a measure space, $\mathcal{X}$ and $\mathcal{Y}$ be separable metric spaces, $(x_n)_{n=0}^{\infty}$ be a sequence of $\mathcal{X}$-valued functions of $\boldsymbol{\Omega}$ that are $\mathcal{B}$-measurable, $(y_n)_{n=0}^{\infty}$ be a sequence of $\mathcal{Y}$-valued functions of $\boldsymbol{\Omega}$ that are $\mathcal{B}$-measurable, and $(z_n)_{n=0}^{\infty}$ is a sequence of $\mathcal{Z} := \mathcal{X} \times \mathcal{Y}$-valued functions of $\boldsymbol{\Omega}$, defined by $z_n(\omega) := (x_n(\omega), y_n(\omega))$, $\forall \omega \in \boldsymbol{\Omega}$, $\forall n \in \mathbb{N}$. Then, $\lim_{n \in \mathbb{N}} z_n = z_0$ in measure in $\boldsymbol{\Omega}$ if, and only if, $\lim_{n \in \mathbb{N}} x_n = x_0$ in measure in $\boldsymbol{\Omega}$ and $\lim_{n \in \mathbb{N}} y_n = y_0$ in measure in $\boldsymbol{\Omega}$.*

Proof By Propositions 3.28 and 4.4, $\mathcal{Z}$ is separable. By Proposition 11.39, z_n is $\mathcal{B}$-measurable, $\forall n \in \mathbb{N}$. "Sufficiency" Let $\lim_{n \in \mathbb{N}} x_n = x_0$ in measure in $\boldsymbol{\Omega}$ and $\lim_{n \in \mathbb{N}} y_n = y_0$ in measure in $\boldsymbol{\Omega}$. Fix any $\epsilon \in (0, \infty) \subset \mathbb{R}$. There exists $n_1(\epsilon) \in \mathbb{N}$ such that $\forall n \in \mathbb{N}$ with $n_1(\epsilon) \leq n$, we have $\mu(\{\omega \in \boldsymbol{\Omega} \mid \rho_{\mathcal{X}}(x_n(\omega), x_0(\omega)) \geq \epsilon\}) =: \mu(E_{n,\epsilon}) < \epsilon$. There exists $n_2(\epsilon) \in \mathbb{N}$ such that $\forall n \in \mathbb{N}$ with $n_2(\epsilon) \leq n$, we have $\mu(\{\omega \in \boldsymbol{\Omega} \mid \rho_{\mathcal{Y}}(y_n(\omega), y_0(\omega)) \geq \epsilon\}) =: \mu(F_{n,\epsilon}) < \epsilon$. Let $n_0(\epsilon) := \max\left\{n_1(\frac{\epsilon}{2}), n_2(\frac{\epsilon}{2})\right\} \in \mathbb{N}$. $\forall n \in \mathbb{N}$ with $n_0(\epsilon) \leq n$, we have $A_{n,\epsilon} := \{\omega \in \boldsymbol{\Omega} \mid \rho_{\mathcal{Z}}(z_n(\omega), z_0(\omega)) \geq \epsilon\} \subseteq \{\omega \in \boldsymbol{\Omega} \mid \rho_{\mathcal{X}}(x_n(\omega), x_0(\omega)) \geq \frac{\epsilon}{\sqrt{2}}\} \cup \{\omega \in \boldsymbol{\Omega} \mid \rho_{\mathcal{Y}}(y_n(\omega), y_0(\omega)) \geq \frac{\epsilon}{\sqrt{2}}\} \subseteq \{\omega \in \boldsymbol{\Omega} \mid \rho_{\mathcal{X}}(x_n(\omega), x_0(\omega)) \geq \frac{\epsilon}{2}\} \cup \{\omega \in \boldsymbol{\Omega} \mid \rho_{\mathcal{Y}}(y_n(\omega), y_0(\omega)) \geq \frac{\epsilon}{2}\} = E_{n,\frac{\epsilon}{2}} \cup F_{n,\frac{\epsilon}{2}}$ and $\mu(A_{n,\epsilon}) \leq \mu(E_{n,\frac{\epsilon}{2}}) + \mu(F_{n,\frac{\epsilon}{2}}) < \epsilon$. Hence, we have $\lim_{n \in \mathbb{N}} z_n = z_0$ in measure in $\boldsymbol{\Omega}$.

"Necessity" Let $\lim_{n \in \mathbb{N}} z_n = z_0$ in measure in $\boldsymbol{\Omega}$. Fix any $\epsilon \in (0, \infty) \subset \mathbb{R}$. There exists $n_0(\epsilon) \in \mathbb{N}$ such that $\forall n \in \mathbb{N}$ with $n_0(\epsilon) \leq n$, we have $\mu(A_{n,\epsilon}) < \epsilon$. $\forall \omega \in E_{n,\epsilon}$, we have $\rho_{\mathcal{Z}}(z_n(\omega), z_0(\omega)) \geq \epsilon$. Hence, $E_{n,\epsilon} \subseteq A_{n,\epsilon}$. Then, by Lemma 11.43, we have $\mu(E_{n,\epsilon}) \leq \mu(A_{n,\epsilon}) < \epsilon$. Hence, $\lim_{n \in \mathbb{N}} x_n = x_0$ in measure in $\boldsymbol{\Omega}$. By symmetry, $\lim_{n \in \mathbb{N}} y_n = y_0$ in measure in $\boldsymbol{\Omega}$.

This completes the proof of the proposition. □

Here, we will summarize different modes of convergence for functions of a measure space. Let $\mathcal{X} := (X, \mathcal{B}, \mu)$ be a measure space, $\mathcal{Y}$ be a separable metric space, and $(f_0)_{0=1}^{\infty}$ be a $\mathcal{Y}$-valued functions of $\mathcal{X}$ that are $\mathcal{B}$-measurable. Then, we have the diagram in Fig. 11.1.

Proposition 11.213 *Let $\mathcal{X} := (X, \mathcal{B}, \mu)$ be a finite measure space, $\mathcal{Y}$ be a separable Banach space over $\mathbb{K}$, $p \in [1, \infty] \subset \mathbb{R}_e$, $(f_n)_{n=0}^{\infty} \subseteq \bar{\mathrm{L}}_p(\mathcal{X}, \mathcal{Y})$, and $\lim_{n \in \mathbb{N}} f_n \doteq f_0$ in $\bar{\mathrm{L}}_p(\mathcal{X}, \mathcal{Y})$. Then, we have f_n are absolutely integrable over $\mathcal{X}$, $\forall n \in \mathbb{Z}_+$, $\lim_{n \in \mathbb{N}} \int_X \mathcal{P} \circ (f_n - f_0) \, \mathrm{d}\mu = 0$, and $\lim_{n \in \mathbb{N}} \int_X f_n \, \mathrm{d}\mu = \int_X f_0 \, \mathrm{d}\mu \in \mathcal{Y}$.*

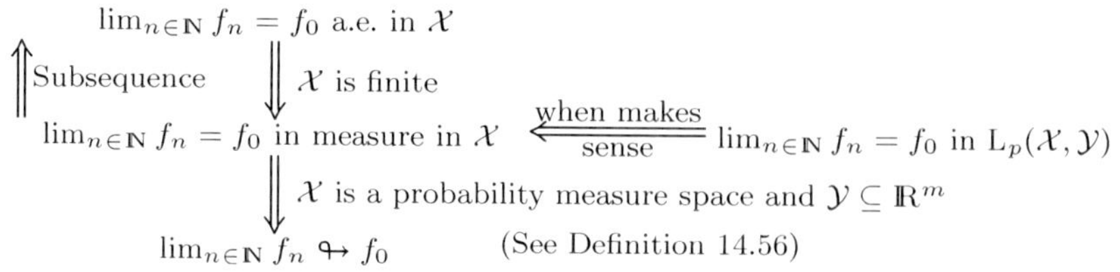

Fig. 11.1 Modes of convergence for functions from $\mathcal{X}$

Proof We will distinguish two exhaustive and mutually exclusive cases: Case 1: $p = \infty$; Case 2: $p \in [1, \infty) \subset \mathbb{R}_+$.

Case 1: $p = \infty$. The result is a direct consequence of Bounded Convergence Theorem 11.77.

Case 2: $p \in [1, \infty) \subset \mathbb{R}_+$. $\forall n \in \mathbb{Z}_+$, $f_n \in \bar{\mathrm{L}}_p(\mathcal{X}, \mathcal{Y})$ implies that $\int_X \mathcal{P}_p \circ f_n \, d\mu < \infty$. Since $\mathcal{P} \circ f_n(x) \leq \mathcal{P}_p \circ f_n(x) \vee 1$, $\forall x \in X$, then, by Proposition 11.83, we have $0 \leq \int_X \mathcal{P} \circ f \, d\mu \leq \int_X (\mathcal{P}_p \circ f \vee 1) \, d\mu \leq \int_X \mathcal{P}_p \circ f \, d\mu + \mu(X) < \infty$. Therefore, f_n is absolutely integrable over $\mathcal{X}$. By Proposition 11.92, we have $\int_X f_n \, d\mu \in \mathcal{Y}$, $\forall n \in \mathbb{Z}_+$.

Next, we will show that $\lim_{n\in\mathbb{N}} \int_X f_n \, d\mu = \int_X f_0 \, d\mu \in \mathcal{Y}$ by further distinguishing two exhaustive and mutually exclusive subcases: Case 2a: $\mu(X) = 0$; Case 2b: $\mu(X) \in \mathbb{R}_+$.

Case 2a: $\mu(X) = 0$. Then, by Proposition 11.92, we have $\vartheta_{\mathcal{Y}} = \int_X f_n \, d\mu$ and $\int_X \mathcal{P} \circ (f_n - f_0) \, d\mu = 0$, $\forall n \in \mathbb{Z}_+$. This subcase is proved.

Case 2b: $\mu(X) \in \mathbb{R}_+$. We have the following line of arguments.

$$0 \leq \left\| \frac{1}{\mu(X)} \int_X (f_n - f_0) \, d\mu \right\|^p \leq \left(\frac{1}{\mu(X)} \int_X \mathcal{P} \circ (f_n - f_0) \, d\mu \right)^p$$
$$\leq \frac{1}{\mu(X)} \int_X \mathcal{P}_p \circ (f_n - f_0) \, d\mu \to 0 \text{ as } n \to \infty$$

where the second inequality follows from Proposition 11.92, the third inequality follows from Jensen's Inequality and Theorem 11.98, and the convergence statement follows from the assumption of the proposition. Hence, we have $\lim_{n\in\mathbb{N}} \int_X \mathcal{P} \circ (f_n - f_0) \, d\mu = 0$, $\lim_{n\in\mathbb{N}} \int_X (f_n - f_0) \, d\mu = \vartheta_{\mathcal{Y}}$, and $\lim_{n\in\mathbb{N}} \inf_X f_n \, d\mu = \int_X f_0 \, d\mu \in \mathcal{Y}$. This subcase is proved.

In both subcases, we have $\lim_{n\in\mathbb{N}} \int_X \mathcal{P} \circ (f_n - f_0) \, d\mu = 0$ and $\lim_{n\in\mathbb{N}} \int_X f_n \cdot d\mu = \int_X f_0 \, d\mu \in \mathcal{Y}$. This completes the proof of Case 2.

Hence, the result holds in both cases, and this completes the proof of the proposition. □

Chapter 12
Differentiation and Integration

12.1 Carathéodory Extension Theorem

Definition 12.1 Let X be a set, $\mathcal{A} \subseteq {}^X2$ be an algebra on X, $\mathcal{Y}$ be a normed linear space over $\mathbb{K}$, and $\nu : \mathcal{A} \to [0, \infty] \subset \mathbb{R}_e$ be a measure on the algebra $\mathcal{A}$. A *$\mathcal{Y}$-valued measure on the algebra $\mathcal{A}$* is a function μ from $\mathcal{A}$ to $\mathcal{Y}$ that satisfies:

(i) $\mu(\emptyset) = \vartheta_{\mathcal{Y}}$.
(ii) $\forall A \in \mathcal{A}$ with $\nu(A) = \infty$, $\mu(A)$ is undefined.
(iii) $\forall A \in \mathcal{A}$ with $\nu(A) < \infty$, $\mu(A) \in \mathcal{Y}$ and $\forall$ pairwise disjoint $(A_i)_{i=1}^{\infty} \subseteq \mathcal{A}$ with $A = \bigcup_{i=1}^{\infty} A_i$, we have $\sum_{i=1}^{\infty} \|\mu(A_i)\| < +\infty$ and $\mu(A) = \sum_{i=1}^{\infty} \mu(A_i)$.
(iv) $\forall A \in \mathcal{A}$ with $\nu(A) < \infty$, we have

$$\nu(A) = \sup_{n \in \mathbb{Z}_+,\ (A_i)_{i=1}^{n} \subseteq \mathcal{A},\ A = \bigcup_{i=1}^{n} A_i,\ A_i \cap A_j = \emptyset,\ \forall 1 \le i < j \le n} \sum_{i=1}^{n} \|\mu(A_i)\|.$$

Then, ν is said to be the *total variation* of μ and denoted by $\mathcal{P} \circ \mu$. μ is said to be finite if ν is finite, and μ is said to be σ-finite if ν is σ-finite. ◇

Proposition 12.2 *Let X be a set, $\mathcal{A} \subseteq {}^X2$ be an algebra on X, $\mathcal{Y}$ be a normed linear space over $\mathbb{K}$, μ be a $\mathcal{Y}$-valued measure on the algebra $\mathcal{A}$, and $E \in \mathcal{A}$. Then, $\mathcal{A}_E := \{C \in \mathcal{A} \mid C \subseteq E\}$ is an algebra on E, $\mu_E := \mu|_{\mathcal{A}_E}$ is a $\mathcal{Y}$-valued measure on the algebra $\mathcal{A}_E$, and $\mathcal{P} \circ \mu_E = (\mathcal{P} \circ \mu)|_{\mathcal{A}_E}$.*

Proof It is easy to show that $\mathcal{A}_E$ is an algebra on E. Note that $\mathcal{P} \circ \mu$ is a measure on the algebra $\mathcal{A}$. Define $\nu_E := (\mathcal{P} \circ \mu)|_{\mathcal{A}_E}$. Then, $\nu_E(\emptyset) = \mathcal{P} \circ \mu(\emptyset) = 0$. $\forall$ pairwise disjoint $(A_i)_{i=1}^{\infty} \subseteq \mathcal{A}_E$ with $A := \bigcup_{i=1}^{\infty} A_i \in \mathcal{A}_E$, we have $\nu_E(A) = \mathcal{P} \circ \mu(A) = \sum_{i=1}^{\infty} \mathcal{P} \circ \mu(A_i) = \sum_{i=1}^{\infty} \nu_E(A_i)$. Then, ν_E is a measure on the algebra $\mathcal{A}_E$.

$\mu_E(\emptyset) = \mu(\emptyset) = \vartheta_{\mathcal{Y}}$. $\forall A \in \mathcal{A}_E$ with $\nu_E(A) = \infty$, we have $\mathcal{P} \circ \mu(A) = \infty$, $\mu(A)$ is undefined, and therefore, $\mu_E(A)$ is undefined. $\forall A \in \mathcal{A}_E$ with $\nu_E(A) < \infty$,

Z. Pan, *Measure-Theoretic Calculus in Abstract Spaces*,
https://doi.org/10.1007/978-3-031-21912-2_12

we have $\mathcal{P} \circ \mu(A) < \infty$ and $\mu_E(A) = \mu(A) \in \mathcal{Y}$. $\forall$ pairwise disjoint $(A_i)_{i=1}^{\infty} \subseteq \mathcal{A}_E$ with $A = \bigcup_{i=1}^{\infty} A_i$, we have $\sum_{i=1}^{\infty} \|\mu_E(A_i)\| = \sum_{i=1}^{\infty} \|\mu(A_i)\| < \infty$ and $\sum_{i=1}^{\infty} \mu_E(A_i) = \sum_{i=1}^{\infty} \mu(A_i) = \mu(A) = \mu_E(A)$. $\forall A \in \mathcal{A}_E$ with $\nu_E(A) < \infty$, we have

$$\begin{aligned} \nu_E(A) &= \mathcal{P} \circ \mu(A) \\ &= \sup_{n \in \mathbb{Z}_+, (A_i)_{i=1}^n \subseteq \mathcal{A}, A = \bigcup_{i=1}^n A_i, A_i \cap A_j = \emptyset, \forall 1 \le i < j \le n} \sum_{i=1}^n \|\mu(A_i)\| \\ &= \sup_{n \in \mathbb{Z}_+, (A_i)_{i=1}^n \subseteq \mathcal{A}_E, A = \bigcup_{i=1}^n A_i, A_i \cap A_j = \emptyset, \forall 1 \le i < j \le n} \sum_{i=1}^n \|\mu_E(A_i)\| \end{aligned}$$

Hence, μ_E is a $\mathcal{Y}$-valued measure on the algebra $\mathcal{A}_E$ with $\mathcal{P} \circ \mu_E = \mathcal{P} \circ (\mu|_{\mathcal{A}_E}) = \nu_E = (\mathcal{P} \circ \mu)|_{\mathcal{A}_E}$.

This completes the proof of the proposition. □

Proposition 12.3 *Let $\mathcal{X} := (X, \mathcal{B}, \mu)$ be a finite measure space. Define $\rho : \mathcal{B} \times \mathcal{B} \to [0, \infty) \subset \mathbb{R}$ by $\rho(E_1, E_2) = \mu(E_1 \triangle E_2)$, $\forall E_1, E_2 \in \mathcal{B}$. Then, ρ defines a pseudo-metric on $\mathcal{B}$.*

Proof $\forall E_1, E_2, E_3 \in \mathcal{B}$, $\rho(E_1, E_2) = \mu(E_1 \triangle E_2) \in [0, +\infty) \subset \mathbb{R}$ since μ is a finite measure on $(X, \mathcal{B})$. Clearly, $\rho(E_1, E_1) = \mu(E_1 \triangle E_1) = \mu(\emptyset) = 0$ and $\rho(E_1, E_2) = \mu(E_1 \triangle E_2) = \mu(E_2 \triangle E_1) = \rho(E_2, E_1)$. Note that $\rho(E_1, E_2) + \rho(E_2, E_3) = \mu(E_1 \triangle E_2) + \mu(E_2 \triangle E_3) = \mu(E_1 \setminus E_2) + \mu(E_2 \setminus E_1) + \mu(E_2 \setminus E_3) + \mu(E_3 \setminus E_2) \ge \mu((E_1 \setminus E_2) \cup (E_2 \setminus E_3)) + \mu((E_2 \setminus E_1) \cup (E_3 \setminus E_2))$. Clearly, we have $(E_1 \setminus E_2) \cup (E_2 \setminus E_3) \supseteq E_1 \setminus E_3$ and $(E_2 \setminus E_1) \cup (E_3 \setminus E_2) \supseteq E_3 \setminus E_1$. Then, $\rho(E_1, E_2) + \rho(E_2, E_3) \ge \mu(E_1 \setminus E_3) + \mu(E_3 \setminus E_1) = \mu(E_1 \triangle E_3) = \rho(E_1, E_3)$. Hence, ρ defines a pseudo-metric on $\mathcal{B}$. This completes the proof of the proposition.

□

Theorem 12.4 (Carathéodory Extension Theorem) *Let $\mathcal{A}$ be an algebra on a set X, $\mathcal{Y}$ be a Banach space over $\mathbb{K}$, μ be a σ-finite $\mathcal{Y}$-valued measure on the algebra $\mathcal{A}$, and $\mathcal{B}$ be the σ-algebra on X generated by $\mathcal{A}$. Then, there is a unique $\mathcal{Y}$-valued measure $\bar{\mu}$ on $(X, \mathcal{B})$ such that $\bar{\mu}|_{\mathcal{A}} = \mu$ and $(\mathcal{P} \circ \bar{\mu})|_{\mathcal{A}} = \mathcal{P} \circ \mu$. Furthermore, $\bar{\mu}$ is σ-finite.*

Proof We first consider the special case where μ is finite. Then, $\mu : \mathcal{A} \to \mathcal{Y}$ and $\mathcal{P} \circ \mu : \mathcal{A} \to [0, \mathcal{P} \circ \mu(X)] \subset \mathbb{R}$. By Carathéodory Extension Theorem 11.19, there is a unique finite measure $\nu : \mathcal{B} \to [0, \mathcal{P} \circ \mu(X)] \subset \mathbb{R}$ on the measurable space $(X, \mathcal{B})$ such that $\nu(A) = \mathcal{P} \circ \mu(A)$, $\forall A \in \mathcal{A}$, and $\nu(E) = \inf_{(A_i)_{i=1}^{\infty} \subseteq \mathcal{A}, \bigcup_{i=1}^{\infty} A_i \supseteq E} \sum_{i=1}^{\infty} \mathcal{P} \circ \mu(A_i)$, $\forall E \in \mathcal{B}$. Define $\rho : \mathcal{B} \times \mathcal{B} \to [0, \mathcal{P} \circ \mu(X)] \subset \mathbb{R}$ by $\rho(E_1, E_2) = \nu(E_1 \triangle E_2)$, $\forall E_1, E_2 \in \mathcal{B}$. By Proposition 12.3, ρ defines a pseudo-metric on $\mathcal{B}$. Define an equivalence relation $\equiv$ on $\mathcal{B}$ by $E_1 \equiv E_2$ if $\rho(E_1, E_2) = \nu(E_1 \triangle E_2) = 0$. By Lemma 4.48, $(\mathcal{B}/\equiv, \bar{\rho})$ is a metric space, where $\bar{\rho} : (\mathcal{B}/\equiv) \times (\mathcal{B}/\equiv) \to$

$[0, \mathcal{P} \circ \mu(X)] \subset \mathbb{R}$ is defined by $\bar{\rho}([E_1], [E_2]) = \rho(E_1, E_2)$, $\forall [E_1], [E_2] \in \mathcal{B}/\equiv$ with $E_1, E_2 \in \mathcal{B}$. Let $\mathcal{A}_\equiv := \{[A] \in \mathcal{B}/\equiv \mid A \in \mathcal{A}\}$.

Claim 12.4.1 $\mathcal{A}_\equiv$ is dense in $\mathcal{B}/\equiv$, that is, $\overline{\mathcal{A}_\equiv} = \mathcal{B}/\equiv$.

Proof of Claim $\forall [E] \in \mathcal{B}/\equiv$ with $E \in \mathcal{B}$, $\forall \epsilon \in (0, \infty) \subset \mathbb{R}$, $\exists (A_i)_{i=1}^\infty \subseteq \mathcal{A}$ with $E \subseteq \bigcup_{i=1}^\infty A_i$ such that $\nu(E) \le \sum_{i=1}^\infty \mathcal{P} \circ \mu(A_i) < \nu(E) + \epsilon/2 < +\infty$. Then, $\exists n_0 \in \mathbb{N}$ such that $\sum_{i=n_0+1}^\infty \mathcal{P} \circ \mu(A_i) < \epsilon/2$. Let $A := \bigcup_{i=1}^{n_0} A_i \in \mathcal{A}$. Note that $E \setminus A \subseteq (\bigcup_{i=1}^\infty A_i) \setminus A \subseteq \bigcup_{i=n_0+1}^\infty A_i$. This implies that $0 \le \nu(E \setminus A) \le \nu(\bigcup_{i=n_0+1}^\infty A_i) \le \sum_{i=n_0+1}^\infty \nu(A_i) = \sum_{i=n_0+1}^\infty \mathcal{P} \circ \mu(A_i) < \epsilon/2$. Note also that $A \setminus E \subseteq (\bigcup_{i=1}^\infty A_i) \setminus E$. This further implies that $0 \le \nu(A \setminus E) \le \nu((\bigcup_{i=1}^\infty A_i) \setminus E) = \nu(\bigcup_{i=1}^\infty A_i) - \nu(E) \le \sum_{i=1}^\infty \nu(A_i) - \nu(E) = \sum_{i=1}^\infty \mathcal{P} \circ \mu(A_i) - \nu(E) < \epsilon/2$. Hence, $\bar{\rho}([E], [A]) = \rho(E, A) = \nu(E \Delta A) = \nu(E \setminus A) + \nu(A \setminus E) < \epsilon$. Hence, $\overline{\mathcal{A}_\equiv} = \mathcal{B}/\equiv$. This completes the proof of the claim. □

Define $g : \mathcal{A}_\equiv \to \mathcal{Y}$ by $g([A]) = \mu(A)$, $\forall [A] \in \mathcal{A}_\equiv$ with $A \in \mathcal{A}$. We will show that g is well-defined. $\forall A_1, A_2 \in \mathcal{A}$ with $A_1 \equiv A_2$, we have $\rho(A_1, A_2) = \nu(A_1 \Delta A_2) = \mathcal{P} \circ \mu(A_1 \Delta A_2) = 0$. Then, $\mathcal{P} \circ \mu(A_1 \setminus A_2) = 0 = \mathcal{P} \circ \mu(A_2 \setminus A_1)$. This implies that $\mu(A_1 \setminus A_2) = \vartheta_\mathcal{Y} = \mu(A_2 \setminus A_1)$ and $\mu(A_1) = \mu(A_1 \cap A_2) + \mu(A_1 \setminus A_2) = \mu(A_1 \cap A_2) + \mu(A_2 \setminus A_1) = \mu(A_2) \in \mathcal{Y}$. Hence, g is well-defined.

Claim 12.4.2 g is uniformly continuous.

Proof of Claim $\forall \epsilon \in (0, \infty) \subset \mathbb{R}$, let $\delta = \epsilon$. $\forall [A_1], [A_2] \in \mathcal{A}_\equiv$ with $A_1, A_2 \in \mathcal{A}$ and $\bar{\rho}([A_1], [A_2]) < \delta$, we have $\|g([A_1]) - g([A_2])\| = \|\mu(A_1) - \mu(A_2)\| = \|\mu(A_1 \cap A_2) + \mu(A_1 \setminus A_2) - \mu(A_2 \cap A_1) - \mu(A_2 \setminus A_1)\| = \|\mu(A_1 \setminus A_2) - \mu(A_2 \setminus A_1)\| \le \|\mu(A_1 \setminus A_2)\| + \|\mu(A_2 \setminus A_1)\| \le \mathcal{P} \circ \mu(A_1 \setminus A_2) + \mathcal{P} \circ \mu(A_2 \setminus A_1) = \mathcal{P} \circ \mu(A_1 \Delta A_2) = \nu(A_1 \Delta A_2) = \rho(A_1, A_2) = \bar{\rho}([A_1], [A_2]) < \delta = \epsilon$. Hence, g is uniformly continuous. This completes the proof of the claim. □

By Proposition 4.46 and the completeness of $\mathcal{Y}$, there exists a unique continuous function $\bar{g} : \mathcal{B}/\equiv \to \mathcal{Y}$ such that $\bar{g}|_{\mathcal{A}_\equiv} = g$, and $\bar{g}$ is uniformly continuous. Define $\bar{\mu} : \mathcal{B} \to \mathcal{Y}$ by $\bar{\mu}(E) = \bar{g}([E])$, $\forall E \in \mathcal{B}$. We will show that $\bar{\mu}$ is the $\mathcal{Y}$-valued measure we seek. Clearly, $\bar{\mu}(A) = \bar{g}([A]) = g([A]) = \mu(A)$, $\forall A \in \mathcal{A}$. Then, $\bar{\mu}(\emptyset) = \mu(\emptyset) = \vartheta_\mathcal{Y}$. $\forall \epsilon \in (0, \infty) \subset \mathbb{R}$, by the uniform continuity of $\bar{g}$, $\exists \delta(\epsilon) \in (0, \epsilon] \subset \mathbb{R}$ such that, $\forall [E_1], [E_2] \in \mathcal{B}/\equiv$ with $E_1, E_2 \in \mathcal{B}$ and $\bar{\rho}([E_1], [E_2]) = \rho(E_1, E_2) = \nu(E_1 \Delta E_2) < \delta(\epsilon)$, we have $\|\bar{\mu}(E_1) - \bar{\mu}(E_2)\| = \|\bar{g}([E_1]) - \bar{g}([E_2])\| < \epsilon$.

Claim 12.4.3 $\bar{\mu}$ is finitely additive, that is, $\bar{\mu}(E_1 \cup E_2) = \bar{\mu}(E_1) + \bar{\mu}(E_2)$, $\forall E_1, E_2 \in \mathcal{B}$ with $E_1 \cap E_2 = \emptyset$.

Proof of Claim $\forall E_1, E_2 \in \mathcal{B}$ with $E_1 \cap E_2 = \emptyset$, $\forall \epsilon \in (0, \infty) \subset \mathbb{R}$, by Claim 12.4.1, $\exists A_1, A_2 \in \mathcal{A}$ such that $\bar{\rho}([A_i], [E_i]) = \rho(A_i, E_i) = \nu(A_i \Delta E_i) < \delta(\epsilon/4)/2$, $i = 1, 2$. Then, $\|\bar{\mu}(E_i) - \bar{\mu}(A_i)\| < \epsilon/4$, $i = 1, 2$. Note that $(E_1 \cup E_2) \setminus (A_1 \cup A_2) = (E_1 \cup E_2) \cap \widetilde{A_1} \cap \widetilde{A_2} = (E_1 \cap \widetilde{A_1} \cap \widetilde{A_2}) \cup (E_2 \cap \widetilde{A_1} \cap \widetilde{A_2}) \subseteq (E_1 \setminus A_1) \cup (E_2 \setminus A_2)$ and $(A_1 \cup A_2) \setminus (E_1 \cup E_2) \subseteq (A_1 \setminus E_1) \cup (A_2 \setminus E_2)$. This leads to $(E_1 \cup E_2) \Delta (A_1 \cup A_2) \subseteq (E_1 \setminus A_1) \cup (E_2 \setminus A_2) \cup (A_1 \setminus E_1) \cup (A_2 \setminus E_2) = (E_1 \Delta A_1) \cup$

$(E_2 \triangle A_2)$. This implies that $\bar{\rho}([E_1 \cup E_2], [A_1 \cup A_2]) = \nu((E_1 \cup E_2) \triangle (A_1 \cup A_2)) \leq \nu(E_1 \triangle A_1) + \nu(E_2 \triangle A_2) < \delta(\epsilon/4)$. Hence, $\|\bar{\mu}(E_1 \cup E_2) - \bar{\mu}(A_1 \cup A_2)\| < \epsilon/4$. Note also that $(A_1 \cap A_2) \triangle (E_1 \cap E_2) = ((A_1 \cap A_2) \setminus (E_1 \cap E_2)) \cup ((E_1 \cap E_2) \setminus (A_1 \cap A_2)) = (A_1 \cap A_2 \cap \widetilde{E_1}) \cup (A_1 \cap A_2 \cap \widetilde{E_2}) \cup (E_1 \cap E_2 \cap \widetilde{A_1}) \cup (E_1 \cap E_2 \cap \widetilde{A_2}) \subseteq (A_1 \setminus E_1) \cup (A_2 \setminus E_2) \cup (E_1 \setminus A_1) \cup (E_2 \setminus A_2) = (E_1 \triangle A_1) \cup (E_2 \triangle A_2)$. This leads to $\bar{\rho}([E_1 \cap E_2], [A_1 \cap A_2]) = \bar{\rho}([\emptyset], [A_1 \cap A_2]) = \nu((E_1 \cap E_2) \triangle (A_1 \cap A_2)) \leq \nu((E_1 \triangle A_1) \cup (E_2 \triangle A_2)) \leq \nu(E_1 \triangle A_1) + \nu(E_2 \triangle A_2) < \delta(\epsilon/4)$. This implies that $\|\bar{\mu}(A_1 \cup A_2) - \bar{\mu}(A_1) - \bar{\mu}(A_2)\| = \|\mu(A_1 \cup A_2) - \mu(A_1) - \mu(A_2)\| = \|\mu(A_1) + \mu(A_2 \setminus A_1) - \mu(A_1) - \mu(A_2)\| = \|\mu(A_1 \cap A_2) - \mu(\emptyset)\| = \|\bar{\mu}(A_1 \cap A_2) - \bar{\mu}(\emptyset)\| < \epsilon/4$. Therefore, we have $\|\bar{\mu}(E_1 \cup E_2) - \bar{\mu}(E_1) - \bar{\mu}(E_2)\| \leq \|\bar{\mu}(E_1 \cup E_2) - \bar{\mu}(A_1 \cup A_2)\| + \|\bar{\mu}(A_1 \cup A_2) - \bar{\mu}(A_1) - \bar{\mu}(A_2)\| + \|\bar{\mu}(A_1) - \bar{\mu}(E_1)\| + \|\bar{\mu}(A_2) - \bar{\mu}(E_2)\| < \epsilon$. By the arbitrariness of ϵ, we have $\bar{\mu}(E_1 \cup E_2) = \bar{\mu}(E_1) + \bar{\mu}(E_2)$. Hence, $\bar{\mu}$ is finitely additive. This completes the proof of the claim. □

Claim 12.4.4 $\forall E \in \mathcal{B}$, $\|\bar{\mu}(E)\| \leq \nu(E)$.

Proof of Claim $\forall E \in \mathcal{B}$, $\forall \epsilon \in (0, \infty) \subset \mathbb{R}$, by Claim 12.4.1, $\exists A \in \mathcal{A}$ such that $\bar{\rho}([E], [A]) = \nu(E \triangle A) < \delta(\epsilon/2) \leq \epsilon/2$. Then, $\|\bar{\mu}(E) - \bar{\mu}(A)\| < \epsilon/2$. Note that $\|\bar{\mu}(A)\| = \|\mu(A)\| \leq \mathcal{P} \circ \mu(A) = \nu(A)$. Then, $\|\bar{\mu}(E)\| \leq \|\bar{\mu}(E) - \bar{\mu}(A)\| + \|\bar{\mu}(A)\| < \epsilon/2 + \nu(A) \leq \epsilon/2 + \nu(A \cap E) + \nu(A \setminus E) \leq \epsilon/2 + \nu(E) + \nu(A \triangle E) < \nu(E) + \epsilon$. By the arbitrariness of ϵ, we have $\|\bar{\mu}(E)\| \leq \nu(E)$. This completes the proof of the claim. □

$\forall$ pairwise disjoint $(E_n)_{n=1}^\infty \subseteq \mathcal{B}$, let $E := \bigcup_{n=1}^\infty E_n \in \mathcal{B}$. By Claim 12.4.4, we have $\sum_{n=1}^\infty \|\bar{\mu}(E_n)\| \leq \sum_{n=1}^\infty \nu(E_n) = \nu(E) < +\infty$. Let $\bar{E}_n := \bigcup_{i=1}^n E_i \in \mathcal{B}$, $\forall n \in \mathbb{N}$. By Claim 12.4.3, $\bar{\mu}(\bar{E}_n) = \sum_{i=1}^n \bar{\mu}(E_i)$. We will show that $\lim_{n \in \mathbb{N}} \bar{\mu}(\bar{E}_n) = \bar{\mu}(E)$, which then implies that $\bar{\mu}$ is countably additive. $\forall \epsilon \in (0, \infty) \subset \mathbb{R}$, $\exists n_0 \in \mathbb{N}$ such that $\forall n \in \mathbb{N}$ with $n \geq n_0$, we have $0 \leq \sum_{i=n+1}^\infty \nu(E_i) < \delta(\epsilon)$. Then, $\bar{\rho}\left(\left[\bar{E}_n\right], \left[E\right]\right) = \rho(\bar{E}_n, E) = \nu(\bar{E}_n \triangle E) = \nu(E \setminus \bar{E}_n) = \sum_{i=n+1}^\infty \nu(E_i) < \delta(\epsilon)$. This implies that $\left\|\bar{\mu}(E) - \bar{\mu}(\bar{E}_n)\right\| < \epsilon$. Hence, $\bar{\mu}(E) = \lim_{n \in \mathbb{N}} \bar{\mu}(\bar{E}_n) = \sum_{n=1}^\infty \bar{\mu}(E_n)$. Therefore, $\bar{\mu}$ is countably additive. This shows that $\bar{\mu}$ is a $\mathcal{Y}$-valued pre-measure on $(X, \mathcal{B})$.

$\forall E \in \mathcal{B}$, $\forall n \in \mathbb{Z}_+$, $\forall$ pairwise disjoint $\left(E_i\right)_{i=1}^n \subseteq \mathcal{B}$ with $E = \bigcup_{i=1}^n E_i$, by Claim 12.4.4, we have $\sum_{i=1}^n \|\bar{\mu}(E_i)\| \leq \sum_{i=1}^n \nu(E_i) = \nu(E)$. Therefore, $\mathcal{P} \circ \bar{\mu}(E) \leq \nu(E)$. By Proposition 11.100, $\mathcal{P} \circ \bar{\mu}$ is a measure on $(X, \mathcal{B})$. By the above argument, $\mathcal{P} \circ \bar{\mu}(X) \leq \nu(X) = \mathcal{P} \circ \mu(X) < +\infty$. Hence, $\mathcal{P} \circ \bar{\mu}$ is a finite measure on $(X, \mathcal{B})$. $\forall A \in \mathcal{A}$, by Definitions 11.108 and 12.1, $\nu(A) = \mathcal{P} \circ \mu(A) \leq \mathcal{P} \circ \bar{\mu}(A) \leq \nu(A)$. Then, $\mathcal{P} \circ \bar{\mu}(A) = \nu(A) = \mathcal{P} \circ \mu(A)$, $\forall A \in \mathcal{A}$. By Carathéodory Extension Theorem 11.19, $\mathcal{P} \circ \bar{\mu} = \nu$.

To show the uniqueness of $\bar{\mu}$, let $\tilde{\mu}$ be another $\mathcal{Y}$-valued measure on $(X, \mathcal{B})$ such that $\tilde{\mu}|_{\mathcal{A}} = \mu$ and $(\mathcal{P} \circ \tilde{\mu})|_{\mathcal{A}} = \mathcal{P} \circ \mu$. By Carathéodory Extension Theorem 11.19, $\mathcal{P} \circ \tilde{\mu} = \nu$. Hence, $\tilde{\mu}$ is a finite $\mathcal{Y}$-valued measure and $\tilde{\mu} : \mathcal{B} \to \mathcal{Y}$. Define $\tilde{g} : \mathcal{B}/\equiv \to \mathcal{Y}$ by $\tilde{g}([E]) = \tilde{\mu}(E)$, $\forall [E] \in \mathcal{B}/\equiv$ with $E \in \mathcal{B}$. We need to show that $\tilde{g}$ is well-defined. $\forall E_1, E_2 \in \mathcal{B}$ with $E_1 \equiv E_2$, we have $\rho(E_1, E_2) = \nu(E_1 \triangle E_2) = 0$. This leads to $\nu(E_1 \setminus E_2) = 0 = \nu(E_2 \setminus E_1)$ and $\tilde{\mu}(E_1 \setminus E_2) = \vartheta_{\mathcal{Y}} = \tilde{\mu}(E_2 \setminus E_1)$. Then,

$\tilde{\mu}(E_1) = \tilde{\mu}(E_1 \cap E_2) + \tilde{\mu}(E_1 \setminus E_2) = \tilde{\mu}(E_1 \cap E_2) + \tilde{\mu}(E_2 \setminus E_1) = \tilde{\mu}(E_2)$. Hence, $\tilde{g}$ is well-defined. Clearly, $\tilde{g}([A]) = \tilde{\mu}(A) = \mu(A) = g([A])$, $\forall [A] \in \mathcal{A}_{\equiv}$ with $A \in \mathcal{A}$. Hence, $\tilde{g}|_{\mathcal{A}_{\equiv}} = g$. $\forall \epsilon \in (0, \infty) \subset \mathbb{R}$, let $\delta = \epsilon$. $\forall [E_1], [E_2] \in \mathcal{B}/\equiv$ with $E_1, E_2 \in \mathcal{B}$ and $\bar{\rho}([E_1], [E_2]) = \nu(E_1 \Delta E_2) < \delta$, we have $\|\tilde{g}([E_1]) - \tilde{g}([E_2])\| = \|\tilde{\mu}(E_1) - \tilde{\mu}(E_2)\| = \|\tilde{\mu}(E_1 \cap E_2) + \tilde{\mu}(E_1 \setminus E_2) - \tilde{\mu}(E_2 \cap E_1) - \tilde{\mu}(E_2 \setminus E_1)\| \leq \|\tilde{\mu}(E_1 \setminus E_2)\| + \|\tilde{\mu}(E_2 \setminus E_1)\| \leq \mathcal{P} \circ \tilde{\mu}(E_1 \setminus E_2) + \mathcal{P} \circ \tilde{\mu}(E_2 \setminus E_1) = \mathcal{P} \circ \tilde{\mu}(E_1 \Delta E_2) = \nu(E_1 \Delta E_2) < \delta = \epsilon$. Hence, $\tilde{g}$ is uniformly continuous. By Proposition 4.46, $\tilde{g} = \bar{g}$. Then, $\tilde{\mu}(E) = \tilde{g}([E]) = \bar{g}([E]) = \bar{\mu}(E)$, $\forall E \in \mathcal{B}$. Hence, $\tilde{\mu} = \bar{\mu}$. This establishes the uniqueness and completes the proof for the special case.

Next, we consider the general case where μ is σ-finite. Then, $\exists (X_n)_{n=1}^{\infty} \subseteq \mathcal{A}$ such that $X = \bigcup_{n=1}^{\infty} X_n$ and $\mathcal{P} \circ \mu(X_n) < \infty$, $\forall n \in \mathbb{N}$. Without loss of generality, we may assume that $(X_n)_{n=1}^{\infty}$ is pairwise disjoint. Fix any $n \in \mathbb{N}$. Let $\mathcal{A}_n := \{E \in \mathcal{A} \mid E \subseteq X_n\}$ and $\mu_n := \mu|_{\mathcal{A}_n}$. By Proposition 12.2, μ_n is a finite $\mathcal{Y}$-valued measure on the algebra $\mathcal{A}_n$ and $\mathcal{P} \circ \mu_n = (\mathcal{P} \circ \mu)|_{\mathcal{A}_n}$. By the special case, there exists a unique $\mathcal{Y}$-valued measure $\bar{\mu}_n$ on $(X_n, \mathcal{B}_n)$, where $\mathcal{B}_n$ is the σ-algebra generated by $\mathcal{A}_n$, such that $\bar{\mu}_n(A) = \mu_n(A)$ and $\mathcal{P} \circ \bar{\mu}_n(A) = \mathcal{P} \circ \mu_n(A)$, $\forall A \in \mathcal{A}_n$. Furthermore, $\bar{\mu}_n$ is finite. By Proposition 11.118, the generation process on $(\mathcal{X}_n := (X_n, \mathcal{B}_n, \bar{\mu}_n))_{n=1}^{\infty}$ yields a unique σ-finite $\mathcal{Y}$-valued measure space $\mathcal{X} := (X, \bar{\mathcal{B}}, \bar{\mu})$ on X, where $\bar{\mathcal{B}} := \{E \subseteq X \mid E \cap X_n \in \mathcal{B}_n, \forall n \in \mathbb{N}\}$, $\forall E \in \bar{\mathcal{B}}$, $\mathcal{P} \circ \bar{\mu}(E) = \sum_{n=1}^{\infty} \mathcal{P} \circ \bar{\mu}_n(E \cap X_n)$, $\bar{\mu}(E)$ is undefined if $\mathcal{P} \circ \bar{\mu}(E) = \infty$, $\bar{\mu}(E) = \sum_{n=1}^{\infty} \bar{\mu}_n(E \cap X_n) \in \mathcal{Y}$ if $\mathcal{P} \circ \bar{\mu}(E) < \infty$, and $\mathcal{X}_n$ is the finite $\mathcal{Y}$-valued measure subspace of $\mathcal{X}$, $\forall n \in \mathbb{N}$.

Claim 12.4.5 $\mathcal{B} = \bar{\mathcal{B}}$.

Proof of Claim $\forall A \in \mathcal{A}$, $\forall n \in \mathbb{N}$, $A \cap X_n \in \mathcal{A}_n \subseteq \mathcal{B}_n$. This implies that $A \in \bar{\mathcal{B}}$. Then, $\mathcal{A} \subseteq \bar{\mathcal{B}}$. Since $\bar{\mathcal{B}}$ is a σ-algebra, then $\mathcal{B} \subseteq \bar{\mathcal{B}}$. On the other hand, $\forall E \in \bar{\mathcal{B}}$, $\forall n \in \mathbb{N}$, we have $E \cap X_n \in \mathcal{B}_n \subseteq \mathcal{B}$. Then, $E = \bigcup_{n=1}^{\infty}(E \cap X_n) \in \mathcal{B}$ and, hence, $\bar{\mathcal{B}} \subseteq \mathcal{B}$. This shows that $\mathcal{B} = \bar{\mathcal{B}}$ and completes the proof of the claim. □

$\forall A \in \mathcal{A}$, $\forall n \in \mathbb{N}$, we have $A \cap X_n \in \mathcal{A}_n \subseteq \mathcal{B}_n$ and $\mathcal{P} \circ \bar{\mu}(A) = \sum_{n=1}^{\infty} \mathcal{P} \circ \bar{\mu}_n(A \cap X_n) = \sum_{n=1}^{\infty} \mathcal{P} \circ \mu_n(A \cap X_n) = \sum_{n=1}^{\infty} \mathcal{P} \circ \mu(A \cap X_n) = \mathcal{P} \circ \mu(A)$. Hence, $(\mathcal{P} \circ \bar{\mu})|_{\mathcal{A}} = \mathcal{P} \circ \mu$. $\forall A \in \mathcal{A}$ with $\mathcal{P} \circ \bar{\mu}(A) = \infty$, then $\bar{\mu}(A)$ is undefined and $\mathcal{P} \circ \mu(A) = \infty$, which implies that $\mu(A)$ is undefined. $\forall A \in \mathcal{A}$ with $\mathcal{P} \circ \bar{\mu}(A) < \infty$, then $\mathcal{P} \circ \mu(A) < \infty$ and $\bar{\mu}(A) = \sum_{n=1}^{\infty} \bar{\mu}_n(A \cap X_n) = \sum_{n=1}^{\infty} \mu_n(A \cap X_n) = \sum_{n=1}^{\infty} \mu(A \cap X_n) = \mu(A) \in \mathcal{Y}$. Hence, $\bar{\mu}|_{\mathcal{A}} = \mu$. This shows that $\bar{\mu}$ is the σ-finite $\mathcal{Y}$-valued measure on $(X, \mathcal{B})$ that we seek.

This completes the proof of the theorem. □

Definition 12.5 Let X be a set, $\mathcal{A} \subseteq {}^X2$ be an algebra on X, $\mathcal{Y}$ be a normed linear space over $\mathbb{K}$, and $\mu : \mathcal{A} \to \mathcal{Y}$. μ is said to be a *$\mathcal{Y}$-valued pre-measure on the algebra $\mathcal{A}$* if:

(i) $\mu(\emptyset) = \vartheta_{\mathcal{Y}}$.
(ii) $\forall$ pairwise disjoint $(A_i)_{i=1}^{\infty} \subseteq \mathcal{A}$ with $A := \bigcup_{i=1}^{\infty} A_i \in \mathcal{A}$, we have $\sum_{i=1}^{\infty} \|\mu(A_i)\| < \infty$ and $\mu(A) = \sum_{i=1}^{\infty} \mu(A_i) \in \mathcal{Y}$.

Define $\mathcal{P} \circ \mu : \mathcal{A} \to [0, \infty] \subset \mathbb{R}_e$ by, $\forall A \in \mathcal{A}$,

$$\mathcal{P} \circ \mu(A) := \sup_{n\in\mathbb{Z}_+,\ (A_i)_{i=1}^n \subseteq \mathcal{A},\ A=\bigcup_{i=1}^n A_i,\ A_i\cap A_j=\emptyset,\ \forall 1\le i<j\le n} \sum_{i=1}^{n} \|\mu(A_i)\|$$

$\mathcal{P} \circ \mu$ is said to be the *total variation of* μ.

μ is said to be *finite* if $\mathcal{P} \circ \mu(X) < \infty$. ◇

Proposition 12.6 *Let X be a set, $\mathcal{A} \subseteq {}^X2$ be an algebra on X, $\mathcal{Y}$ be a normed linear space over $\mathbb{K}$, and $\mu : \mathcal{A} \to \mathcal{Y}$ be a $\mathcal{Y}$-valued pre-measure on the algebra $\mathcal{A}$. Then, $\mathcal{P} \circ \mu : \mathcal{A} \to [0, \infty] \subset \mathbb{R}_e$ defines a measure on the algebra $\mathcal{A}$.*

Proof Clearly, $\mathcal{P} \circ \mu(\emptyset) = 0$. $\forall$ pairwise disjoint $(A_i)_{i=1}^\infty \subseteq \mathcal{A}$ with $A := \bigcup_{i=1}^\infty A_i \in \mathcal{A}$, $\forall i \in \mathbb{N}$ with $\mathcal{P} \circ \mu(A_i) > 0$, $\forall 0 \le t_i < \mathcal{P} \circ \mu(A_i)$, $\exists n_i \in \mathbb{Z}_+$ and $\exists$ pairwise disjoint $(A_{i,j})_{j=1}^{n_i} \subseteq \mathcal{A}$ with $A_i = \bigcup_{j=1}^{n_i} A_{i,j}$ such that $t_i < \sum_{j=1}^{n_i} \|\mu(A_{i,j})\| \le \mathcal{P} \circ \mu(A_i)$. $\forall i \in \mathbb{N}$ with $\mathcal{P} \circ \mu(A_i) = 0$, let $t_i = 0$, $n_i = 1$, $A_{i,1} = A_i$. Then, $0 = t_i \le \|\mu(A_{i,1})\| \le \mathcal{P} \circ \mu(A_i) = 0$.

Claim 12.6.1 $0 \le t := \sum_{i=1}^\infty t_i \le \mathcal{P} \circ \mu(A)$.

Proof of Claim Clearly, $t \ge 0$ since $t_i \ge 0$, $\forall i \in \mathbb{N}$. We will distinguish two exhaustive and mutually exclusive cases: Case 1: $t < +\infty$; Case 2: $t = +\infty$.

Case 1: $t < +\infty$. Then, $t \le \sum_{i=1}^\infty \sum_{j=1}^{n_i} \|\mu(A_{i,j})\| \ge 0$. Therefore, $\forall \epsilon \in (0, \infty) \subset \mathbb{R}$, $\exists N \in \mathbb{N}$ such that $t - \epsilon < \sum_{i=1}^N \sum_{j=1}^{n_i} \|\mu(A_{i,j})\| \le \sum_{i=1}^N \sum_{j=1}^{n_i} \|\mu(A_{i,j})\| + \left\|\mu(A \setminus \left(\bigcup_{i=1}^N \bigcup_{j=1}^{n_i} A_{i,j}\right))\right\| \le \mathcal{P} \circ \mu(A)$. By the arbitrariness of ϵ, we have $t \le \mathcal{P} \circ \mu(A)$.

Case 2: $t = +\infty$. Then $t = \sum_{i=1}^\infty \sum_{j=1}^{n_i} \|\mu(A_{i,j})\| = \infty$. $\forall M \in [0, \infty) \subset \mathbb{R}$, $\exists N \in \mathbb{N}$ such that

$$\begin{aligned} M &\le \sum_{i=1}^{N} \sum_{j=1}^{n_i} \|\mu(A_{i,j})\| \\ &\le \sum_{i=1}^{N} \sum_{j=1}^{n_i} \|\mu(A_{i,j})\| + \left\|\mu(A \setminus \left(\bigcup_{i=1}^{N} \bigcup_{j=1}^{n_i} A_{i,j}\right))\right\| \le \mathcal{P} \circ \mu(A) \end{aligned}$$

This implies that $\mathcal{P} \circ \mu(A) = +\infty = t$.

This completes the proof of the claim. □

Hence, $\sum_{i=1}^\infty t_i \le \mathcal{P} \circ \mu(A)$. By the arbitrariness of t_i's, we have $\sum_{i=1}^\infty \mathcal{P} \circ \mu(A_i) \le \mathcal{P} \circ \mu(A)$.

On the other hand, $\forall n \in \mathbb{Z}_+$, $\forall$ pairwise disjoint $(E_j)_{j=1}^n \subseteq \mathcal{A}$ with $A = \bigcup_{j=1}^n E_j$, we have

$$\sum_{j=1}^n \|\mu(E_j)\| = \sum_{j=1}^n \|\mu(\bigcup_{i=1}^\infty (A_i \cap E_j))\| = \sum_{j=1}^n \|\sum_{i=1}^\infty \mu(A_i \cap E_j)\|$$

$$\le \sum_{j=1}^n \sum_{i=1}^\infty \|\mu(A_i \cap E_j)\| = \sum_{i=1}^\infty \sum_{j=1}^n \|\mu(A_i \cap E_j)\| \le \sum_{i=1}^\infty \mathcal{P} \circ \mu(A_i)$$

By the arbitrariness of n and $(E_j)_{j=1}^n$, we have $\mathcal{P} \circ \mu(A) \le \sum_{i=1}^\infty \mathcal{P} \circ \mu(A_i)$.

Hence, $\mathcal{P} \circ \mu(A) = \sum_{i=1}^\infty \mathcal{P} \circ \mu(A_i)$. This implies that $\mathcal{P} \circ \mu$ is a measure on the algebra $\mathcal{A}$. This completes the proof of the proposition. □

Clearly, $\mu : \mathcal{A} \to \mathcal{Y}$ is a finite $\mathcal{Y}$-valued pre-measure on the algebra $\mathcal{A}$ if, and only if, μ is a finite $\mathcal{Y}$-valued measure on the algebra $\mathcal{A}$. In this case, the definitions of total variations of μ coincide considering μ as a $\mathcal{Y}$-valued pre-measure on the algebra $\mathcal{A}$ or as a $\mathcal{Y}$-valued measure on the algebra $\mathcal{A}$.

12.2 Change of Variable

Definition 12.7 Let $\mathcal{X} := (X, \mathcal{B}, \mu)$ be a ($\mathcal{Y}$-valued) measure space, $\bar{\mathcal{X}} := (\bar{X}, \bar{\mathcal{B}}, \nu)$ be a ($\mathcal{Y}$-valued) measure space, ($\mathcal{Y}$ be a normed linear space,) and $g : X \to \bar{X}$. g is said to be an *isomeasure* between $\mathcal{X}$ and $\bar{\mathcal{X}}$ if g is bijective; $\forall E \in \mathcal{B}$, $g(E) \in \bar{\mathcal{B}}$; $\forall \bar{E} \in \bar{\mathcal{B}}$, $g_{\text{inv}}(\bar{E}) \in \mathcal{B}$; ($\forall \bar{E} \in \bar{\mathcal{B}}$, $\mathcal{P} \circ \nu(\bar{E}) = \mathcal{P} \circ \mu(g_{\text{inv}}(\bar{E}))$;) $\forall E \in \text{dom}(\mu)$, we have $g(E) \in \text{dom}(\nu)$ and $\nu(g(E)) = \mu(E)$; and $\forall \bar{E} \in \text{dom}(\nu)$, we have $g_{\text{inv}}(\bar{E}) \in \text{dom}(\mu)$ and $\nu(\bar{E}) = \mu(g_{\text{inv}}(\bar{E}))$. We will say $\mathcal{X}$ and $\bar{\mathcal{X}}$ are *isomeasuric* if there is an isomeasure between $\mathcal{X}$ and $\bar{\mathcal{X}}$. ◇

Two ($\mathcal{Y}$-valued) measure spaces are isomeasuric, and then they are equivalent up to relabeling of elements. Thus, isomeasure preserves the completeness, finiteness, and σ-finiteness of the ($\mathcal{Y}$-valued) measure spaces.

Definition 12.8 Let $\mathbb{X} := (\mathcal{X}, \mathcal{B}, \mu)$ be a ($\mathcal{Y}$-valued) topological measure space, $\bar{\mathbb{X}} := (\bar{\mathcal{X}}, \bar{\mathcal{B}}, \nu)$ be a ($\mathcal{Y}$-valued) topological measure space, ($\mathcal{Y}$ be a normed linear space,) and $g : \mathbb{X} \to \bar{\mathbb{X}}$. g is said to be an *homeomorphical isomeasure* between $\mathbb{X}$ and $\bar{\mathbb{X}}$ if $g : \mathcal{X} \to \bar{\mathcal{X}}$ is a homeomorphism and g is an isomeasure between $\mathbb{X}$ and $\bar{\mathbb{X}}$. We will say $\mathbb{X}$ and $\bar{\mathbb{X}}$ are *homeomorphically isomeasuric* if there is an homeomorphic isomeasure between $\mathbb{X}$ and $\bar{\mathbb{X}}$. ◇

If two ($\mathcal{Y}$-valued) topological measure spaces are homeomorphically isomeasuric, then they are equivalent up to relabeling of elements.

Proposition 12.9 *Let $\mathcal{X} := (X, \mathcal{B}, \mu)$ be a measure space, $(\bar{X}, \bar{\mathcal{B}})$ be a measurable space, and $g : X \to \bar{X}$ be bijective and such that $\forall \bar{E} \in \bar{\mathcal{B}}$, $g_{\text{inv}}(\bar{E}) \in \mathcal{B}$ and, $\forall E \in$*

$\mathcal{B}$, $g(E) \in \bar{\mathcal{B}}$. Then, we may define the induced measure $g_*\mu : \bar{\mathcal{B}} \to [0, \infty] \subset \mathbb{R}_e$ *by $g_*\mu(\bar{E}) = \mu(g_{\text{inv}}(\bar{E}))$, $\forall \bar{E} \in \bar{\mathcal{B}}$. Then, $\bar{\mathcal{X}} := (\bar{X}, \bar{\mathcal{B}}, g_*\mu)$ is a measure space and g is an isomeasure between $\mathcal{X}$ and $\bar{\mathcal{X}}$.*

Proof This is straightforward and therefore omitted. □

Proposition 12.10 *Let $\mathcal{Y}$ be a normed linear space, $\mathcal{X} := (X, \mathcal{B}, \mu)$ be a $\mathcal{Y}$-valued measure space, $(\bar{X}, \bar{\mathcal{B}})$ be a measurable space, and $g : X \to \bar{X}$ be bijective and such that $\forall \bar{E} \in \bar{\mathcal{B}}$, $g_{\text{inv}}(\bar{E}) \in \mathcal{B}$ and, $\forall E \in \mathcal{B}$, $g(E) \in \bar{\mathcal{B}}$. Then, we may define the* $\mathcal{Y}$-valued induced measure *$g_*\mu$ from $\bar{\mathcal{B}}$ to $\mathcal{Y}$ by, $\forall \bar{E} \in \bar{\mathcal{B}}$ with $g_{\text{inv}}(\bar{E}) \in$ dom (μ), $g_*\mu(\bar{E}) = \mu(g_{\text{inv}}(\bar{E})) \in \mathcal{Y}$; and $\forall \bar{E} \in \bar{\mathcal{B}}$ with $g_{\text{inv}}(\bar{E}) \in \mathcal{B} \setminus$ dom (μ), $g_*\mu(\bar{E})$ is undefined. Then, $\bar{\mathcal{X}} := (\bar{X}, \bar{\mathcal{B}}, g_*\mu)$ is a $\mathcal{Y}$-valued measure space, $\mathcal{P} \circ (g_*\mu) = g_*(\mathcal{P} \circ \mu)$, and g is an isomeasure between $\mathcal{X}$ and $\bar{\mathcal{X}}$.*

Proof Since $\mathcal{X}$ is a $\mathcal{Y}$-valued measure space, then $(X, \mathcal{B}, \mathcal{P} \circ \mu)$ is a measure space. By Proposition 12.9, $(\bar{X}, \bar{\mathcal{B}}, g_*(\mathcal{P} \circ \mu))$ is a measure space, and g is an isomeasure between $(X, \mathcal{B}, \mathcal{P} \circ \mu)$ and $(\bar{X}, \bar{\mathcal{B}}, g_*(\mathcal{P} \circ \mu))$:

(i) $g_*\mu(\emptyset) = \mu(g_{\text{inv}}(\emptyset)) = \mu(\emptyset) = \vartheta_{\mathcal{Y}} \in \mathcal{Y}$.

(ii) $\forall \bar{E} \in \bar{\mathcal{B}}$ with $g_*(\mathcal{P} \circ \mu)(\bar{E}) = +\infty$, we have $\mathcal{P} \circ \mu(g_{\text{inv}}(\bar{E})) = +\infty$ and $g_{\text{inv}}(\bar{E}) \in \mathcal{B} \setminus$ dom (μ). This implies that $g_*\mu(\bar{E})$ is undefined.

(iii) $\forall \bar{E} \in \bar{\mathcal{B}}$ with $g_*(\mathcal{P} \circ \mu)(\bar{E}) < \infty$, we have $\mathcal{P} \circ \mu(g_{\text{inv}}(\bar{E})) < \infty$ and $g_{\text{inv}}(\bar{E}) \in$ dom (μ). This implies that $g_*\mu(\bar{E}) = \mu(g_{\text{inv}}(\bar{E})) \in \mathcal{Y}$. $\forall$ pairwise disjoint $(\bar{E}_i)_{i=1}^{\infty} \subseteq \bar{\mathcal{B}}$ with $\bar{E} = \bigcup_{i=1}^{\infty} \bar{E}_i$, $\forall i \in \mathbb{N}$, $\bar{E}_i \subseteq \bar{E}$, $g_{\text{inv}}(\bar{E}_i) \subseteq g_{\text{inv}}(\bar{E})$, $\mathcal{P} \circ \mu(g_{\text{inv}}(\bar{E}_i)) \leq \mathcal{P} \circ \mu(g_{\text{inv}}(\bar{E})) < \infty$, and $g_{\text{inv}}(\bar{E}_i) \in$ dom (μ). Then, $\sum_{i=1}^{\infty} \|g_*\mu(\bar{E}_i)\| = \sum_{i=1}^{\infty} \|\mu(g_{\text{inv}}(\bar{E}_i))\| \leq \sum_{i=1}^{\infty} \mathcal{P} \circ \mu(g_{\text{inv}}(\bar{E}_i)) = \mathcal{P} \circ \mu(\bigcup_{i=1}^{\infty} g_{\text{inv}}(\bar{E}_i)) = \mathcal{P} \circ \mu(g_{\text{inv}}(\bigcup_{i=1}^{\infty} \bar{E}_i)) = \mathcal{P} \circ \mu(g_{\text{inv}}(\bar{E})) < \infty$, where the first inequality follows from Definition 11.108, the second equality follows from the fact that $\mathcal{P} \circ \mu$ is a measure, and the third equality follows from Proposition 2.5. Furthermore, $\sum_{i=1}^{\infty} g_*\mu(\bar{E}_i) = \sum_{i=1}^{\infty} \mu(g_{\text{inv}}(\bar{E}_i)) = \mu(\bigcup_{i=1}^{\infty} g_{\text{inv}}(\bar{E}_i)) = \mu(g_{\text{inv}}(\bigcup_{i=1}^{\infty} \bar{E}_i)) = \mu(g_{\text{inv}}(\bar{E})) = g_*\mu(\bar{E}) \in \mathcal{Y}$, where the second equality follows from Definition 11.108 and the third equality follows from Proposition 2.5.

(iv) $\forall \bar{E} \in \bar{\mathcal{B}}$ with $g_*(\mathcal{P} \circ \mu)(\bar{E}) < \infty$, by the previous paragraph, we have $\sup_{n \in \mathbb{Z}_+, (\bar{E}_i)_{i=1}^n \subseteq \bar{\mathcal{B}}, \bar{E} = \bigcup_{i=1}^n \bar{E}_i, \bar{E}_i \cap \bar{E}_j = \emptyset, \forall 1 \leq i < j \leq n} \sum_{i=1}^n \|g_*\mu(\bar{E}_i)\| =: s_{\bar{E}} \leq \mathcal{P} \circ \mu(g_{\text{inv}}(\bar{E})) = g_*(\mathcal{P} \circ \mu)(\bar{E})$. $\forall \epsilon \in (0, \infty) \subset \mathbb{R}$, $\exists n \in \mathbb{Z}_+$, $\exists$ pairwise disjoint $(E_i)_{i=1}^n \subseteq \mathcal{B}$ with $g_{\text{inv}}(\bar{E}) = \bigcup_{i=1}^n E_i$, such that $\mathcal{P} \circ \mu(g_{\text{inv}}(\bar{E})) - \epsilon < \sum_{i=1}^n \|\mu(E_i)\| \leq \mathcal{P} \circ \mu(g_{\text{inv}}(\bar{E})) = g_*(\mathcal{P} \circ \mu)(\bar{E})$. Then, $(\bar{E}_i)_{i=1}^n := (g(E_i))_{i=1}^{\infty} \subseteq \bar{\mathcal{B}}$ is pairwise disjoint, $\bigcup_{i=1}^n \bar{E}_i = g(\bigcup_{i=1}^n E_i) = g(g_{\text{inv}}(\bar{E})) = \bar{E}$, and $\sum_{i=1}^n \|g_*\mu(\bar{E}_i)\| = \sum_{i=1}^n \|\mu(g_{\text{inv}}(\bar{E}_i))\| = \sum_{i=1}^n \|\mu(E_i)\| > g_*(\mathcal{P} \circ \mu)(\bar{E}) - \epsilon$. Hence, $s_{\bar{E}} > g_*(\mathcal{P} \circ \mu)(\bar{E}) - \epsilon$. By the arbitrariness of ϵ, we have $s_{\bar{E}} \geq g_*(\mathcal{P} \circ \mu)(\bar{E})$ and $s_{\bar{E}} = g_*(\mathcal{P} \circ \mu)(\bar{E})$.

Therefore, by Definition 11.108, $(\bar{X}, \bar{\mathcal{B}}, g_*\mu)$ is a $\mathcal{Y}$-valued measure space with $\mathcal{P} \circ (g_*\mu) = g_*(\mathcal{P} \circ \mu)$. It is straightforward to show that g is an isomeasure between $\mathcal{X}$ and $\bar{\mathcal{X}}$. This completes the proof of the proposition. □

Proposition 12.11 *Let $\mathcal{X} := (X, \mathcal{O})$ and $\bar{\mathcal{X}} := (\bar{X}, \bar{\mathcal{O}})$ be topological spaces, $g : \mathcal{X} \to \bar{\mathcal{X}}$ be a homeomorphism, ($\mathcal{Y}$ be a normed linear space,) and $\mathbb{X} := (\mathcal{X}, \mathcal{B}_B(\mathcal{X}), \mu)$ be a ($\mathcal{Y}$-valued) topological measure space. Then, $\bar{\mathbb{X}} := (\bar{\mathcal{X}}, \mathcal{B}_B(\bar{\mathcal{X}}), g_*\mu)$ is a ($\mathcal{Y}$-valued) topological measure space and g is a homeomorphical isomeasure between $\mathbb{X}$ and $\bar{\mathbb{X}}$.*

Proof Clearly, g is bijective and g and g_{inv} are continuous. By Proposition 11.34, $\forall E \in \mathcal{B}_B(\mathcal{X})$, we have $g(E) \in \mathcal{B}_B(\bar{\mathcal{X}})$, and $\forall \bar{E} \in \mathcal{B}_B(\bar{\mathcal{X}})$, we have $g_{\text{inv}}(\bar{E}) \in \mathcal{B}_B(\mathcal{X})$. Then, by Propositions 12.9 and 12.10, the ($\mathcal{Y}$-valued) induced measure $g_*\mu$ is well-defined, $(\bar{X}, \mathcal{B}_B(\bar{\mathcal{X}}), g_*\mu)$ is a ($\mathcal{Y}$-valued) measure space, and g is an isomeasure between $\mathbb{X}$ and $\bar{\mathbb{X}}$.

We will show that $\bar{\mathbb{X}}$ is a ($\mathcal{Y}$-valued) topological measure space. $\forall \bar{E} \in \mathcal{B}_B(\bar{\mathcal{X}})$, $\forall \epsilon \in (0, \infty) \subset \mathbb{R}$, we have $g_{\text{inv}}(\bar{E}) \in \mathcal{B}_B(\mathcal{X})$. Since $\mathbb{X}$ is a ($\mathcal{Y}$-valued) topological measure space, then $\exists U \in \mathcal{O}$ with $g_{\text{inv}}(\bar{E}) \subseteq U$ such that $\mu(U \setminus g_{\text{inv}}(\bar{E})) < \epsilon$ ($\mathcal{P} \circ \mu(U \setminus g_{\text{inv}}(\bar{E})) < \epsilon$). Let $\bar{U} := g(U) \in \bar{\mathcal{O}}$. Then, $\bar{E} \subseteq g(U) = \bar{U}$ and $g_*\mu(\bar{U} \setminus \bar{E}) = \mu(g_{\text{inv}}(\bar{U} \setminus \bar{E})) = \mu(g_{\text{inv}}(\bar{U}) \setminus g_{\text{inv}}(\bar{E})) = \mu(U \setminus g_{\text{inv}}(\bar{E})) < \epsilon$ (in the case when μ is a $\mathcal{Y}$-valued measure, we have $\mathcal{P} \circ (g_*\mu)(\bar{U} \setminus \bar{E}) = g_*(\mathcal{P} \circ \mu)(\bar{U} \setminus \bar{E}) = \mathcal{P} \circ \mu(g_{\text{inv}}(\bar{U} \setminus \bar{E})) = \mathcal{P} \circ \mu(g_{\text{inv}}(\bar{U}) \setminus g_{\text{inv}}(\bar{E})) = \mathcal{P} \circ \mu(U \setminus g_{\text{inv}}(\bar{E})) < \epsilon$, where the first equality follows from Proposition 12.10). Hence, $\bar{\mathbb{X}}$ is a ($\mathcal{Y}$-valued) topological measure space. Then, g is a homeomorphical isomeasure between $\mathbb{X}$ and $\bar{\mathbb{X}}$. □

Proposition 12.12 *Let $\mathcal{X} := (X, \mathcal{B}, \mu)$ be a measure space, $\bar{\mathcal{X}} := (\bar{X}, \bar{\mathcal{B}}, \nu)$ be a measure space, $\mathcal{Y}$ be a normed linear space, $f : \bar{X} \to \mathcal{Y}$ be $\bar{\mathcal{B}}$-measurable, and $g : X \to \bar{X}$ be an isomeasure between $\mathcal{X}$ and $\bar{\mathcal{X}}$. Then, $\int_X (f \circ g)\, d\mu = \int_{\bar{X}} f\, d\nu$, whenever one of the integrals exists.*

Proof Clearly, $f \circ g$ is $\mathcal{B}$-measurable. We will distinguish two exhaustive and mutually exclusive cases: Case 1: $\mu(X) < +\infty$; Case 2: $\mu(X) = +\infty$.

Case 1: $\mu(X) < +\infty$. $\nu(\bar{X}) = \mu(g_{\text{inv}}(\bar{X})) = \mu(X) < +\infty$. Let $\mathcal{I}(\mathcal{Y})$ be the integration system on $\mathcal{Y}$. $\forall R = \{(y_\alpha, U_\alpha) \mid \alpha \in \Lambda\} \in \mathcal{I}(\mathcal{Y})$, let $\bar{F}_R := \sum_{\alpha \in \Lambda} y_\alpha \nu(f_{\text{inv}}(U_\alpha)) \in \mathcal{Y}$ and $F_R := \sum_{\alpha \in \Lambda} y_\alpha \mu((f \circ g)_{\text{inv}}(U_\alpha)) \in \mathcal{Y}$. Note that $\bar{F}_R = \sum_{\alpha \in \Lambda} y_\alpha \mu(g_{\text{inv}}(f_{\text{inv}}(U_\alpha))) = F_R$. Then, by Definition 11.70, we have $\int_{\bar{X}} f\, d\nu = \lim_{R \in \mathcal{I}(\mathcal{Y})} \bar{F}_R = \lim_{R \in \mathcal{I}(\mathcal{Y})} F_R = \int_X (f \circ g)\, d\mu$, whenever one of the integrals exists.

Case 2: $\mu(X) = +\infty$. $\nu(\bar{X}) = \mu(g_{\text{inv}}(\bar{X})) = \mu(X) = +\infty$. Let $(\bar{F}_{\bar{A}})_{\bar{A} \in \mathcal{M}(\bar{\mathcal{X}})}$ be the net for $\int_{\bar{X}} f\, d\nu$ and $(F_A)_{A \in \mathcal{M}(\mathcal{X})}$ be the net for $\int_X (f \circ g)\, d\mu$ as defined in Definition 11.71. Without loss of generality, assume that $\int_{\bar{X}} f\, d\nu$ exists. Then,

$$\begin{aligned}
\int_{\bar{X}} f\, d\nu &= \lim_{\bar{A} \in \mathcal{M}(\bar{\mathcal{X}})} \bar{F}_{\bar{A}} = \lim_{\bar{A} \in \mathcal{M}(\bar{\mathcal{X}})} \int_{\bar{A}} f|_{\bar{A}}\, d\nu_{\bar{A}} \\
&= \lim_{\bar{A} \in \mathcal{M}(\bar{\mathcal{X}})} \int_{g_{\text{inv}}(\bar{A})} (f \circ g)|_{g_{\text{inv}}(\bar{A})}\, d\mu_{g_{\text{inv}}(\bar{A})}
\end{aligned}$$

where the first two equalities follow from Definition 11.71 and the third equality follows from Case 1. Fix any open set U ($U \subseteq \mathbb{R}_e$ if $\mathcal{Y} = \mathbb{R}$ or $U \subseteq \mathcal{Y}$ if $\mathcal{Y} \neq \mathbb{R}$) with $\int_{\bar{X}} f \, d\nu \in U$, $\exists \bar{A}_0 \in \mathcal{M}(\bar{\mathcal{X}})$, $\forall \bar{A} \in \mathcal{M}(\bar{\mathcal{X}})$ with $\bar{A}_0 \subseteq \bar{A}$, $\bar{F}_{\bar{A}} = \int_{g_{\text{inv}}(\bar{A})} (f \circ g)|_{g_{\text{inv}}(\bar{A})} \, d\mu_{g_{\text{inv}}(\bar{A})} \in U$. $\forall A \in \mathcal{M}(\mathcal{X})$ with $g_{\text{inv}}(\bar{A}_0) \subseteq A$, we have $A \in \mathcal{B}$ and $\mu(A) < +\infty$. Then, $g(A) \in \bar{\mathcal{B}}$ and $\mu(A) = \nu(g(A)) < +\infty$. Hence, $g(A) \in \mathcal{M}(\bar{\mathcal{X}})$ and $g(A) \supseteq g(g_{\text{inv}}(\bar{A}_0)) = \bar{A}_0$. Then, $F_A = \int_A (f \circ g)|_A \, d\mu_A = \int_{g_{\text{inv}}(g(A))} (f \circ g)|_{g_{\text{inv}}(g(A))} \, d\mu_{g_{\text{inv}}(g(A))} \in U$. Hence, $\int_X (f \circ g) \, d\mu = \lim_{A \in \mathcal{M}(\mathcal{X})} F_A = \int_{\bar{X}} f \, d\nu$.

This completes the proof of the proposition. □

Proposition 12.13 *Let $\mathcal{X} := (X, \mathcal{B}, \mu)$ be a measure space, $\bar{\mathcal{X}} := (\bar{X}, \bar{\mathcal{B}}, \nu)$ be a measure space, $\mathcal{Y}$ be a Banach space, $\mathcal{W}$ be a separable subspace of $\mathcal{Y}$, $f : \bar{X} \to \mathcal{W}$ be absolutely integrable over $\bar{\mathcal{X}}$, and $g : X \to \bar{X}$ be an isomeasure between $\mathcal{X}$ and $\bar{\mathcal{X}}$. Then, $f \circ g$ is absolutely integrable over $\mathcal{X}$ and $\int_X (f \circ g) \, d\mu = \int_{\bar{X}} f \, d\nu \in \mathcal{Y}$.*

Proof Clearly, $f \circ g$ is $\mathcal{B}$-measurable. By the assumption, $0 \leq \int_{\bar{X}} (\mathcal{P} \circ f) \, d\nu < +\infty$. By Proposition 12.12, $\int_X (\mathcal{P} \circ f \circ g) \, d\mu = \int_{\bar{X}} (\mathcal{P} \circ f) \, d\nu \in [0, \infty) \subset \mathbb{R}$. Hence, $f \circ g$ is absolutely integrable over $\mathcal{X}$. By Proposition 11.92, f is integrable over $\bar{\mathcal{X}}$. By Proposition 12.12, $\int_X (f \circ g) \, d\mu = \int_{\bar{X}} f \, d\nu \in \mathcal{Y}$. This completes the proof of the proposition. □

Proposition 12.14 *Let $\mathcal{X} := (X, \mathcal{B}, \mu)$ be a $\mathcal{Y}$-valued measure space, where $\mathcal{Y}$ is a normed linear space over $\mathbb{K}$, $\bar{\mathcal{X}} := (\bar{X}, \bar{\mathcal{B}}, \nu)$ be a $\mathcal{Y}$-valued measure space, $\mathcal{Z}$ be a normed linear space over $\mathbb{K}$, $\mathcal{W} := \mathrm{B}(\mathcal{Y}, \mathcal{Z})$, $f : \bar{X} \to \mathcal{W}$ be $\bar{\mathcal{B}}$-measurable, and $g : X \to \bar{X}$ be an isomeasure between $\mathcal{X}$ and $\bar{\mathcal{X}}$. Then, $\int_{\bar{X}} f \, d\nu = \int_X (f \circ g) \, d\mu$, whenever one of the integrals exists.*

Proof Clearly, $f \circ g$ is $\mathcal{B}$-measurable. We will distinguish two exhaustive and mutually exclusive cases: Case 1: $\mathcal{P} \circ \mu(X) < +\infty$; Case 2: $\mathcal{P} \circ \mu(X) = +\infty$.

Case 1: $\mathcal{P} \circ \mu(X) < +\infty$. $\mathcal{P} \circ \nu(\bar{X}) = \mathcal{P} \circ \mu(g_{\text{inv}}(\bar{X})) = \mathcal{P} \circ \mu(X) < +\infty$. Let $\mathcal{I}(\mathcal{W})$ be the integration system on $\mathcal{W}$. $\forall R = \{(w_\alpha, U_\alpha) \mid \alpha \in \Lambda\} \in \mathcal{I}(\mathcal{W})$, let $\bar{F}_R := \sum_{\alpha \in \Lambda} w_\alpha \nu(f_{\text{inv}}(U_\alpha)) \in \mathcal{Z}$ and $F_R := \sum_{\alpha \in \Lambda} w_\alpha \mu((f \circ g)_{\text{inv}}(U_\alpha)) \in \mathcal{Z}$. Then, $\bar{F}_R = \sum_{\alpha \in \Lambda} w_\alpha \mu(g_{\text{inv}}(f_{\text{inv}}(U_\alpha))) = F_R$. Then, by Definition 11.119, we have $\int_{\bar{X}} f \, d\nu = \lim_{R \in \mathcal{I}(\mathcal{W})} \bar{F}_R = \lim_{R \in \mathcal{I}(\mathcal{W})} F_R = \int_X (f \circ g) \, d\mu$, whenever one of the integrals exists.

Case 2: $\mathcal{P} \circ \mu(X) = +\infty$. $\mathcal{P} \circ \nu(\bar{X}) = \mathcal{P} \circ \mu(g_{\text{inv}}(\bar{X})) = \mathcal{P} \circ \mu(X) = +\infty$. Let $(\bar{F}_{\bar{A}})_{\bar{A} \in \mathcal{M}(\bar{\mathcal{X}})}$ be the net for $\int_{\bar{X}} f \, d\nu$ and $(F_A)_{A \in \mathcal{M}(\mathcal{X})}$ be the net for $\int_X (f \circ g) \, d\mu$ as defined in Definition 11.120. Without loss of generality, assume that $\int_{\bar{X}} f \, d\nu$ exists. Then,

$$\int_{\bar{X}} f \, d\nu = \lim_{\bar{A} \in \mathcal{M}(\bar{\mathcal{X}})} \bar{F}_{\bar{A}} = \lim_{\bar{A} \in \mathcal{M}(\bar{\mathcal{X}})} \int_{\bar{A}} f|_{\bar{A}} \, d\nu_{\bar{A}}$$

$$= \lim_{\bar{A} \in \mathcal{M}(\bar{\mathcal{X}})} \int_{g_{\text{inv}}(\bar{A})} (f \circ g)|_{g_{\text{inv}}(\bar{A})} \, d\mu_{g_{\text{inv}}(\bar{A})}$$

where the first two equalities follow from Definition 11.120 and the third equality follows from Case 1. Fix any open set U ($U \subseteq \mathbb{R}_e$ if $\mathcal{Z} = \mathbb{R}$ or $U \subseteq \mathcal{Z}$ if $\mathcal{Z} \neq \mathbb{R}$) with $\int_{\bar{X}} f \, d\nu \in U$, $\exists \bar{A}_0 \in \mathcal{M}(\bar{\mathcal{X}})$, $\forall \bar{A} \in \mathcal{M}(\bar{\mathcal{X}})$ with $\bar{A}_0 \subseteq \bar{A}$, $\bar{F}_{\bar{A}} = \int_{g_{\text{inv}}(\bar{A})} (f \circ g)|_{g_{\text{inv}}(\bar{A})} \, d\mu_{g_{\text{inv}}(\bar{A})} \in U$. $\forall A \in \mathcal{M}(\mathcal{X})$ with $g_{\text{inv}}(\bar{A}_0) \subseteq A$, we have $A \in \mathcal{B}$ and $\mathcal{P} \circ \mu(A) < +\infty$. Then, $g(A) \in \bar{\mathcal{B}}$ and $\mathcal{P} \circ \mu(A) = \mathcal{P} \circ \nu(g(A)) < +\infty$. Hence, $g(A) \in \mathcal{M}(\bar{\mathcal{X}})$ and $g(A) \supseteq g(g_{\text{inv}}(\bar{A}_0)) = \bar{A}_0$. Then, $F_A = \int_A (f \circ g)|_A \, d\mu_A = \int_{g_{\text{inv}}(g(A))} (f \circ g)|_{g_{\text{inv}}(g(A))} \, d\mu_{g_{\text{inv}}(g(A))} \in U$. Hence, $\int_X (f \circ g) \, d\mu = \lim_{A \in \mathcal{M}(\mathcal{X})} F_A = \int_{\bar{X}} f \, d\nu$.

This completes the proof of the proposition. □

Proposition 12.15 *Let $\mathcal{X} := (X, \mathcal{B}, \mu)$ be a $\mathcal{Y}$-valued measure space, where $\mathcal{Y}$ is a normed linear space over $\mathbb{K}$, $\bar{\mathcal{X}} := (\bar{X}, \bar{\mathcal{B}}, \nu)$ be a $\mathcal{Y}$-valued measure space, $\mathcal{Z}$ be a Banach space over $\mathbb{K}$, $\mathcal{W}$ be a separable subspace of $\mathrm{B}(\mathcal{Y}, \mathcal{Z})$, $f : \bar{X} \to \mathcal{W}$ be absolutely integrable over $\bar{\mathcal{X}}$, and $g : X \to \bar{X}$ be an isomeasure between $\mathcal{X}$ and $\bar{\mathcal{X}}$. Then, $f \circ g$ is absolutely integrable over $\mathcal{X}$, and $\int_{\bar{X}} f \, d\nu = \int_X (f \circ g) \, d\mu \in \mathcal{Z}$.*

Proof Clearly, $f \circ g$ is $\mathcal{B}$-measurable. By Definition 12.7, g is an isomeasure between $(X, \mathcal{B}, \mathcal{P} \circ \mu)$ and $(\bar{X}, \bar{\mathcal{B}}, \mathcal{P} \circ \nu)$. By Proposition 12.12, $\int_X (\mathcal{P} \circ f \circ g) \, d\mathcal{P} \circ \mu = \int_{\bar{X}} (\mathcal{P} \circ f) \, d\mathcal{P} \circ \nu \in \mathbb{R}$, where the last step follows from the assumption that f is absolutely integrable over $\bar{\mathcal{X}}$. Hence, $f \circ g$ is absolutely integrable over $\mathcal{X}$. By Proposition 11.132, f is integrable over $\bar{\mathcal{X}}$. By Proposition 12.14, $\int_X (f \circ g) \, d\mu = \int_{\bar{X}} f \, d\nu \in \mathcal{Z}$. □

Theorem 12.16 (Change of Variable) *Let $\mathcal{X} := (X, \mathcal{B}, \mu)$ be a σ-finite $\mathbb{K}$-valued measure space, $\bar{\mathcal{X}} := (\bar{X}, \bar{\mathcal{B}}, \nu)$ be a σ-finite $\mathcal{Y}$-valued measure space, where $\mathcal{Y}$ is a separable Banach space over $\mathbb{K}$, $g : \bar{X} \to X$ be an isomeasure between $\bar{\mathcal{X}}$ and $\hat{\mathcal{X}} := (X, \mathcal{B}, g_* \nu)$, $\mathcal{Z}$ be a Banach space over $\mathbb{K}$, $\mathcal{W}$ be a separable subspace of $\mathrm{B}(\mathcal{Y}, \mathcal{Z})$, $f : \bar{X} \to \mathcal{W}$ be $\bar{\mathcal{B}}$-measurable, and $\frac{d(g_* \nu)}{d\mu} =: h : X \to \mathcal{Y}$. Then, f is absolutely integrable over $\bar{\mathcal{X}}$ if, and only if, $(\mathcal{P} \circ f \circ g_{\text{inv}})(\mathcal{P} \circ h)$ is integrable over $(X, \mathcal{B}, \mathcal{P} \circ \mu)$. In which case, $(f \circ g_{\text{inv}})h$ is absolutely integrable over $\mathcal{X}$ and $\int_{\bar{X}} f \, d\nu = \int_X (f \circ g_{\text{inv}}) h \, d\mu \in \mathcal{Z}$.*

Proof By Proposition 12.15, f is absolutely integrable over $\bar{\mathcal{X}}$ if, and only if, $f \circ g_{\text{inv}}$ is absolutely integrable over $\hat{\mathcal{X}}$. By (ii) of Proposition 11.168, $f \circ g_{\text{inv}}$ is absolutely integrable over $\hat{\mathcal{X}}$ if, and only if, $(\mathcal{P} \circ f \circ g_{\text{inv}})(\mathcal{P} \circ h)$ is integrable over $(X, \mathcal{B}, \mathcal{P} \circ \mu)$. Hence, f is absolutely integrable over $\bar{\mathcal{X}}$ if, and only if, $(\mathcal{P} \circ f \circ g_{\text{inv}})(\mathcal{P} \circ h)$ is integrable over $(X, \mathcal{B}, \mathcal{P} \circ \mu)$. In this case, $(f \circ g_{\text{inv}})h$ is absolutely integrable over $\mathcal{X}$ and, by Proposition 12.15 and (ii) of Proposition 11.168, $\int_{\bar{X}} f \, d\nu = \int_X (f \circ g_{\text{inv}}) \, d(g_* \nu) = \int_X (f \circ g_{\text{inv}}) h \, d\mu \in \mathcal{Z}$. This completes the proof of the theorem. □

12.3 Product Measure

Definition 12.17 Let X be a set. $\mathcal{A} \subseteq {}^{X}2$ is said to be a *π-system on X* if: (i) $\emptyset, X \in \mathcal{A}$; and (ii) $\forall A, B \in \mathcal{A}$, we have $A \cap B \in \mathcal{A}$. $\mathcal{D} \subseteq {}^{X}2$ is said to be a *monotone class on X* if:

(i) $\emptyset, X \in \mathcal{D}$.
(ii) $\forall A, B \in \mathcal{D}$ with $A \subseteq B$, we have $B \setminus A \in \mathcal{D}$.
(iii) $\forall (A_i)_{i=1}^{\infty} \subseteq \mathcal{D}$ with $A_i \subseteq A_{i+1}$, $\forall i \in \mathbb{N}$, we have $\bigcup_{i=1}^{\infty} A_i \in \mathcal{D}$. ◇

Let X be a set and $\mathcal{A} \subseteq {}^{X}2$. Then, there exists the smallest monotone class on X that contains $\mathcal{A}$, which is the intersection of all monotone classes on X that contains $\mathcal{A}$.

Proposition 12.18 *Let X be a set. $\mathcal{B} \subseteq {}^{X}2$ is a σ-algebra on X if, and only if, $\mathcal{B}$ is a π-system on X and is a monotone class on X.*

Proof "Necessity" Let $\mathcal{B}$ be a σ-algebra on X. Then, $\emptyset, X \in \mathcal{B}$. $\forall A, B \in \mathcal{B}$, we have $A \cap B \in \mathcal{B}$. Hence, $\mathcal{B}$ is a π-system on X. $\forall A, B \in \mathcal{B}$ with $A \subseteq B$, $B \setminus A \in \mathcal{B}$. $\forall (A_i)_{i=1}^{\infty} \subseteq \mathcal{B}$ with $A_i \subseteq A_{i+1}$, $\forall i \in \mathbb{N}$, we have $\bigcup_{i=1}^{\infty} A_i \in \mathcal{B}$. Hence, $\mathcal{B}$ is a monotone class on X.

"Sufficiency" Let $\mathcal{B}$ be a monotone class on X and a π-system on X. Then, $\emptyset, X \in \mathcal{B}$. $\forall A \subseteq \mathcal{B}$, we have $A \subseteq X$ and $X \setminus A = \widetilde{A} \in \mathcal{B}$, since $\mathcal{B}$ is a monotone class on X. $\forall A_1, A_2 \in \mathcal{B}$, we have $A_1 \cup A_2 = \widetilde{\widetilde{A_1} \cap \widetilde{A_2}}$. By the above, $\widetilde{A_1}, \widetilde{A_2} \in \mathcal{B}$. Then, $\widetilde{A_1} \cap \widetilde{A_2} \in \mathcal{B}$ since $\mathcal{B}$ is a π-system on X. This further implies that $A_1 \cup A_2 \in \mathcal{B}$. Hence, $\mathcal{B}$ is an algebra on X. $\forall (E_i)_{i=1}^{\infty} \subseteq \mathcal{B}$, $\forall n \in \mathbb{N}$, let $A_n := \bigcup_{i=1}^{n} E_i \in \mathcal{B}$, since $\mathcal{B}$ is an algebra. Then, $A_n \subseteq A_{n+1}$, $\forall n \in \mathbb{N}$. By $\mathcal{B}$ being a monotone class on X, we have $\bigcup_{n=1}^{\infty} A_n = \bigcup_{i=1}^{\infty} E_i \in \mathcal{B}$. This shows that $\mathcal{B}$ is a σ-algebra.

This completes the proof of the proposition. □

Lemma 12.19 (Monotone Class Lemma) *Let X be a set, $\mathcal{A} \subseteq {}^{X}2$ be a π-system on X, $\mathcal{B}$ be the σ-algebra on X generated by $\mathcal{A}$, and $\mathcal{D}$ be the smallest monotone class on X that contains $\mathcal{A}$. Then, $\mathcal{B} = \mathcal{D}$.*

Proof By Proposition 12.18, $\mathcal{B}$ is a monotone class on X. Hence, $\mathcal{D} \subseteq \mathcal{B}$.

Define $\mathcal{D}_1 := \{E \in \mathcal{D} \mid E \cap A \in \mathcal{D},\ \forall A \in \mathcal{A}\}$. Since $\mathcal{A}$ is a π-system on X, then $\mathcal{A} \subseteq \mathcal{D}_1$. Clearly, $\mathcal{D}_1 \subseteq \mathcal{D}$. We will show that $\mathcal{D}_1$ is a monotone class on X. Then, $\mathcal{D} \subseteq \mathcal{D}_1$ and $\mathcal{D} = \mathcal{D}_1$.

Clearly, $\emptyset \in \mathcal{A} \subseteq \mathcal{D}_1$ and $X \in \mathcal{A} \subseteq \mathcal{D}_1$. $\forall E_1, E_2 \in \mathcal{D}_1$ with $E_1 \subseteq E_2$, we have $E_1, E_2 \in \mathcal{D}$ and $E_2 \setminus E_1 \in \mathcal{D}$. $\forall A \in \mathcal{A}$, we have $A \cap E_1 \in \mathcal{D}$ and $A \cap E_2 \in \mathcal{D}$. Clearly, $A \cap E_1 \subseteq A \cap E_2$. Since $\mathcal{D}$ is a monotone class on X, we have $\mathcal{D} \ni (A \cap E_2) \setminus (A \cap E_1) = A \cap (E_2 \setminus E_1)$. By the arbitrariness of A, we have $E_2 \setminus E_1 \in \mathcal{D}_1$. $\forall (E_i)_{i=1}^{\infty} \subseteq \mathcal{D}_1$ with $E_i \subseteq E_{i+1}$, $\forall i \in \mathbb{N}$, we have $\bigcup_{i=1}^{\infty} E_i \in \mathcal{D}$ since $\mathcal{D}$ is a monotone class on X. $\forall A \in \mathcal{A}$, $\forall i \in \mathbb{N}$, $E_i \in \mathcal{D}_1$ implies that $A \cap E_i \in \mathcal{D}$. Clearly, $A \cap E_i \subseteq A \cap E_{i+1}$, $\forall i \in \mathbb{N}$. Then, $\mathcal{D} \ni \bigcup_{i=1}^{\infty}(A \cap E_i) = A \cap (\bigcup_{i=1}^{\infty} E_i)$, since $\mathcal{D}$ is a monotone class on X. By the arbitrariness of A, we have $\bigcup_{i=1}^{\infty} E_i \in \mathcal{D}_1$. This shows that $\mathcal{D}_1$ is a monotone class on X.

Define $\mathcal{D}_2 := \{E \in \mathcal{D} \mid E \cap B \in \mathcal{D},\ \forall B \in \mathcal{D}\}$. Since $\mathcal{D}_1 = \mathcal{D}$, then $\mathcal{A} \subseteq \mathcal{D}_2$. Clearly, $\mathcal{D}_2 \subseteq \mathcal{D}$. We will show that $\mathcal{D}_2$ is a monotone class on X. Then, $\mathcal{D} \subseteq \mathcal{D}_2$ and $\mathcal{D} = \mathcal{D}_2$.

Clearly, $\emptyset \in \mathcal{A} \subseteq \mathcal{D}_2$ and $X \in \mathcal{A} \subseteq \mathcal{D}_2$. $\forall E_1, E_2 \in \mathcal{D}_2$ with $E_1 \subseteq E_2$, we have $E_2 \setminus E_1 \in \mathcal{D}$ since $\mathcal{D}$ is a monotone class on X. $\forall B \in \mathcal{D}$, we have $E_1 \cap B \in \mathcal{D}$ and $E_2 \cap B \subset \mathcal{D}$ since $E_1, E_2 \in \mathcal{D}_2$. Clearly, $E_1 \cap B \subseteq E_2 \cap B$. Then, $\mathcal{D} \ni (E_2 \cap B) \setminus (E_1 \cap B) = (E_2 \setminus E_1) \cap B$, since $\mathcal{D}$ is a monotone class on X. By the arbitrariness of B, we have $E_2 \setminus E_1 \in \mathcal{D}_2$. $\forall (E_i)_{i=1}^\infty \subseteq \mathcal{D}_2$ with $E_i \subseteq E_{i+1}$, $\forall i \in \mathbb{N}$, we have $\bigcup_{i=1}^\infty E_i \in \mathcal{D}$ since $\mathcal{D}$ is a monotone class on X. $\forall B \in \mathcal{D}$, $\forall i \in \mathbb{N}$, $E_i \in \mathcal{D}_2$ implies that $B \cap E_i \in \mathcal{D}$. Clearly, $B \cap E_i \subseteq B \cap E_{i+1}$, $\forall i \in \mathbb{N}$. Then, $\mathcal{D} \ni \bigcup_{i=1}^\infty (B \cap E_i) = B \cap (\bigcup_{i=1}^\infty E_i)$, since $\mathcal{D}$ is a monotone class on X. By the arbitrariness of B, we have $\bigcup_{i=1}^\infty E_i \in \mathcal{D}_2$. This shows that $\mathcal{D}_2$ is a monotone class on X.

Therefore, $\mathcal{D}_2 = \mathcal{D}$. This implies, by the definition of $\mathcal{D}_2$, that $\mathcal{D}$ is a π-system on X. By Proposition 12.18, we have $\mathcal{D}$ is σ-algebra. Then, $\mathcal{B} \subseteq \mathcal{D}$. Hence, $\mathcal{B} = \mathcal{D}$. This completes the proof of the lemma. □

Proposition 12.20 *Let X be a set, $\mathcal{C}$ be a semialgebra on X, $\mathcal{Y}$ be a normed linear space, $\mathcal{A}$ be the algebra on X generated by $\mathcal{C}$, $\nu : \mathcal{C} \to [0, \infty] \subset \mathbb{R}_e$, and μ be a function from $\mathcal{C}$ to $\mathcal{Y}$. Assume that:*

(i) $\mu(\emptyset) = \vartheta_{\mathcal{Y}}$ *and* $\nu(\emptyset) = 0$.
(ii) $\forall C \in \mathcal{C}$, $\forall n \in \mathbb{Z}_+$, $\forall$ *pairwise disjoint* $(C_i)_{i=1}^n \subseteq \mathcal{C}$ *with* $C = \bigcup_{i=1}^n C_i$, *we have* $\nu(C) = \sum_{i=1}^n \nu(C_i)$.
(iii) $\forall C \in \mathcal{C}$, $\forall$ *pairwise disjoint* $(C_i)_{i=1}^\infty \subseteq \mathcal{C}$ *with* $C = \bigcup_{i=1}^\infty C_i$, *we have* $\nu(C) \le \sum_{i=1}^\infty \nu(C_i)$.
(iv) $\forall C \in \mathcal{C}$ *with* $\nu(C) = \infty$, $\mu(C)$ *is undefined.*
(v) $\forall C \in \mathcal{C}$ *with* $\nu(C) < \infty$, $\mu(C) \in \mathcal{Y}$ *and* $\forall n \in \mathbb{Z}_+$, $\forall$ *pairwise disjoint* $(C_i)_{i=1}^n \subseteq \mathcal{C}$ *with* $C = \bigcup_{i=1}^n C_i$, *we have* $\mu(C) = \sum_{i=1}^n \mu(C_i)$.
(vi) $\forall C \in \mathcal{C}$ *with* $\nu(C) < \infty$, *we have*

$$\nu(C) = \sup_{n\in\mathbb{Z}_+,\ (C_i)_{i=1}^n\subseteq\mathcal{C},\ C=\bigcup_{i=1}^n C_i,\ C_i\cap C_j=\emptyset,\ \forall 1\le i<j\le n} \sum_{i=1}^n \|\mu(C_i)\|$$

Then, ν admits a unique extension to a measure $\bar{\nu}$ on the algebra $\mathcal{A}$, and μ admits a unique extension to a $\mathcal{Y}$-valued measure $\bar{\mu}$ on the algebra $\mathcal{A}$ with $(\mathcal{P} \circ \bar{\mu})|_{\mathcal{C}} = \nu$. Furthermore, $\mathcal{P} \circ \bar{\mu} = \bar{\nu}$.

Proof By Proposition 11.31,

$$\mathcal{A} = \{A \subseteq X \mid A \text{ is the finite disjoint union of sets in } \mathcal{C}\}$$

Define $\bar{\nu} : \mathcal{A} \to [0, \infty] \subset \mathbb{R}_e$ by $\bar{\nu}(A) = \sum_{i=1}^n \nu(C_i)$, $\forall A = \bigcup_{i=1}^n C_i \in \mathcal{A}$, where $n \in \mathbb{Z}_+$ and $(C_i)_{i-1}^n \subseteq \mathcal{C}$ is pairwise disjoint. By (ii), (iii), and Proposition 11.32 and its proof, $\bar{\nu}$ is the unique measure on the algebra $\mathcal{A}$ that is an extension of ν.

Hence, $\bar{\nu}$ is the measure on the algebra $\mathcal{A}$ that we seek. Define $\bar{\mu}$ to be a function from $\mathcal{A}$ to $\mathcal{Y}$ by: $\bar{\mu}(A)$ is undefined if $\bar{\nu}(A) = \infty$; and $\bar{\mu}(A) = \sum_{i=1}^{n} \mu(C_i) \in \mathcal{Y}$ if $\bar{\nu}(A) < \infty$, $A = \bigcup_{i=1}^{n} C_i$, $n \in \mathbb{Z}_+$, and $(C_i)_{i=1}^{n} \subseteq \mathcal{C}$ is pairwise disjoint. By (i) and (v), $\bar{\mu}$ is well-defined. Clearly, $\bar{\mu}|_{\mathcal{C}} = \mu$ and, hence, $\bar{\mu}$ is an extension of μ.

We will show that $\bar{\mu}$ is a $\mathcal{Y}$-valued measure on the algebra $\mathcal{A}$ with $\mathcal{P} \circ \bar{\mu} = \bar{\nu}$. Then, $\bar{\mu}$ is the $\mathcal{Y}$-valued measure on the algebra $\mathcal{A}$ that we seek. By (i), $\bar{\mu}(\emptyset) = \vartheta_{\mathcal{Y}}$. $\forall A \in \mathcal{A}$ with $\bar{\nu}(A) = \infty$, $\bar{\mu}(A)$ is undefined. $\forall A \in \mathcal{A}$ with $\bar{\nu}(A) < \infty$, $\forall$ pairwise disjoint $(A_i)_{i=1}^{\infty} \subseteq \mathcal{A}$ with $A = \bigcup_{i=1}^{\infty} A_i$, we have $\bar{\mu}(A) \in \mathcal{Y}$, $\bar{\nu}(A_i) < \infty$, and $\bar{\mu}(A_i) \in \mathcal{Y}$, $\forall i \in \mathbb{N}$. Since $A \in \mathcal{A}$, then $A = \bigcup_{i=1}^{n_0} C_i$, where $n_0 \in \mathbb{Z}_+$ and $(C_i)_{i=1}^{n_0} \subseteq \mathcal{C}$ is pairwise disjoint. Fix any $i \in \mathbb{N}$, $A_i \in \mathcal{A}$ implies that $A_i = \bigcup_{j=1}^{n_i} C_{i,j}$, where $n_i \in \mathbb{Z}_+$ and $(C_{i,j})_{j=1}^{n_i} \subseteq \mathcal{C}$ is pairwise disjoint. This leads to $\sum_{i=1}^{\infty} \|\bar{\mu}(A_i)\| = \sum_{i=1}^{\infty} \left\| \sum_{j=1}^{n_i} \mu(C_{i,j}) \right\| \leq \sum_{i=1}^{\infty} \sum_{j=1}^{n_i} \|\mu(C_{i,j})\| \leq \sum_{i=1}^{\infty} \sum_{j=1}^{n_i} \nu(C_{i,j}) = \sum_{i=1}^{\infty} \sum_{j=1}^{n_i} \bar{\nu}(C_{i,j}) = \bar{\nu}(\bigcup_{i=1}^{\infty} \bigcup_{j=1}^{n_i} C_{i,j}) = \bar{\nu}(A) < \infty$, where the first equality follows from the definition of $\bar{\mu}$, the second inequality follows from (vi), the second equality follows from $\bar{\nu}|_{\mathcal{C}} = \nu$, and the third equality follows from the fact that $\bar{\nu}$ is a measure on the algebra $\mathcal{A}$. This implies that $\forall \epsilon \in (0, +\infty) \subset \mathbb{R}$, $\exists N \in \mathbb{N}$ such that $\sum_{i=N+1}^{\infty} \sum_{j=1}^{n_i} \bar{\nu}(C_{i,j}) < \epsilon$. $\forall n \in \mathbb{N}$ with $N \leq n$, let $\bar{A} := A \setminus (\bigcup_{i=1}^{n} \bigcup_{j=1}^{n_i} C_{i,j}) \in \mathcal{A}$. Clearly, $\bar{A} = \bigcup_{i=n+1}^{\infty} \bigcup_{j=1}^{n_i} C_{i,j}$ and $\bar{\nu}(\bar{A}) = \sum_{i=n+1}^{\infty} \sum_{j=1}^{n_i} \bar{\nu}(C_{i,j}) < \epsilon$. By $\bar{A} \in \mathcal{A}$, we have $\bar{A} = \bigcup_{i=1}^{\bar{n}} \bar{C}_i$, where $\bar{n} \in \mathbb{Z}_+$ and $\left(\bar{C}_i\right)_{i=1}^{\bar{n}} \subseteq \mathcal{C}$ is pairwise disjoint. Then,

$$
\begin{aligned}
&\left\|\bar{\mu}(A) - \sum_{i=1}^{n} \bar{\mu}(A_i)\right\| = \left\| \sum_{k=1}^{n_0} \mu(C_k) - \sum_{i=1}^{n} \sum_{j=1}^{n_i} \mu(C_{i,j}) \right\| \\
&= \left\| \sum_{k=1}^{n_0} \mu(C_k \cap ((\bigcup_{i=1}^{n} \bigcup_{j=1}^{n_i} C_{i,j}) \cup \bar{A})) - \sum_{i=1}^{n} \sum_{j=1}^{n_i} \mu(C_{i,j} \cap (\bigcup_{k=1}^{n_0} C_k)) \right\| \\
&= \left\| \sum_{k=1}^{n_0} \mu((\bigcup_{i=1}^{n} \bigcup_{j=1}^{n_i} (C_k \cap C_{i,j})) \cup (\bigcup_{i=1}^{\bar{n}} (C_k \cap \bar{C}_i))) - \sum_{i=1}^{n} \sum_{j=1}^{n_i} \mu(\bigcup_{k=1}^{n_0} (C_{i,j} \cap C_k)) \right\| \\
&= \left\| \sum_{k=1}^{n_0} \sum_{i=1}^{n} \sum_{j=1}^{n_i} \mu(C_k \cap C_{i,j}) + \sum_{k=1}^{n_0} \sum_{i=1}^{\bar{n}} \mu(C_k \cap \bar{C}_i) - \sum_{i=1}^{n} \sum_{j=1}^{n_i} \sum_{k=1}^{n_0} \mu(C_{i,j} \cap C_k) \right\| \\
&= \left\| \sum_{i=1}^{\bar{n}} \sum_{k=1}^{n_0} \mu(C_k \cap \bar{C}_i) \right\| = \left\| \sum_{i=1}^{\bar{n}} \mu(\bar{C}_i) \right\| \leq \sum_{i=1}^{\bar{n}} \left\| \mu(\bar{C}_i) \right\| \leq \sum_{i=1}^{\bar{n}} \nu(\bar{C}_i) \\
&= \sum_{i=1}^{\bar{n}} \bar{\nu}(\bar{C}_i) = \bar{\nu}(\bigcup_{i=1}^{\bar{n}} \bar{C}_i) = \bar{\nu}(\bar{A}) < \epsilon
\end{aligned}
$$

where the first equality follows from the definition of $\bar{\mu}$, the fourth equality follows from (v), the sixth equality follows from (v), the second inequality follows from (vi), the seventh equality follows from $\bar{\nu}|_{\mathcal{C}} = \nu$, and the eighth equality follows from the fact that $\bar{\nu}$ is a measure on the algebra $\mathcal{A}$. Hence, $\bar{\mu}(A) = \sum_{i=1}^{\infty} \bar{\mu}(A_i)$.

$\forall A \in \mathcal{A}$ with $\bar{\nu}(A) < \infty$, then $A = \bigcup_{i=1}^{n_0} C_i$, where $n_0 \in \mathbb{Z}_+$ and $(C_i)_{i=1}^{n_0} \subseteq \mathcal{C}$ is pairwise disjoint. $\forall n \in \mathbb{Z}_+$, $\forall$ pairwise disjoint $(A_i)_{i=1}^{n} \subseteq \mathcal{A}$ with $A = \bigcup_{i=1}^{n} A_i$, $\forall i \in \{1, \ldots, n\}$, $A_i \in \mathcal{A}$ implies that $A_i = \bigcup_{j=1}^{n_i} C_{i,j}$, where $n_i \in \mathbb{Z}_+$ and $(C_{i,j})_{j=1}^{n_i} \subseteq \mathcal{C}$ is pairwise disjoint. Then, $\sum_{i=1}^{n} \|\bar{\mu}(A_i)\| = \sum_{i=1}^{n} \left\| \sum_{j=1}^{n_i} \mu(C_{i,j}) \right\| \leq \sum_{i=1}^{n} \sum_{j=1}^{n_i} \|\mu(C_{i,j})\| \leq \sum_{i=1}^{n} \sum_{j=1}^{n_i} \nu(C_{i,j}) = \sum_{i=1}^{n} \sum_{j=1}^{n_i} \bar{\nu}(C_{i,j}) = \sum_{i=1}^{n} \bar{\nu}(A_i) = \bar{\nu}(A)$, where the first equality follows from the definition of $\bar{\mu}$, the second inequality follows from (vi), the second equality follows from $\bar{\nu}|_{\mathcal{C}} = \nu$, and the last two equalities follow from the fact that $\bar{\nu}$ is a measure on the algebra $\mathcal{A}$. Hence, $s_A := \sup_{n \in \mathbb{Z}_+, (A_i)_{i=1}^{n} \subseteq \mathcal{A}, A = \bigcup_{i=1}^{n} A_i, A_i \cap A_j = \emptyset, \forall 1 \leq i < j \leq n} \sum_{i=1}^{n} \|\bar{\mu}(A_i)\| \leq \bar{\nu}(A)$. On the other hand, $\forall \epsilon \in (0, +\infty) \subset \mathbb{R}$, $\forall i \in \{1, \ldots, n_0\}$, by (vi), $\exists n_i \in \mathbb{Z}_+$, $\exists$ pairwise disjoint $(C_{i,j})_{j=1}^{n_i} \subseteq \mathcal{C}$ with $C_i = \bigcup_{j=1}^{n_i} C_{i,j}$ such that $\bar{\nu}(C_i) = \nu(C_i) < \sum_{j=1}^{n_i} \|\mu(C_{i,j})\| + \epsilon/(1 + n_0) = \sum_{j=1}^{n_i} \|\bar{\mu}(C_{i,j})\| + \epsilon/(1 + n_0)$. Then, $\bar{\nu}(A) = \sum_{i=1}^{n_0} \bar{\nu}(C_i) < \sum_{i=1}^{n_0} \sum_{j=1}^{n_i} \|\bar{\mu}(C_{i,j})\| + \epsilon \leq s_A + \epsilon$. By the arbitrariness of ϵ, we have $\bar{\nu}(A) \leq s_A$. Hence, $s_A = \bar{\nu}(A)$. This shows that $\bar{\mu}$ is a $\mathcal{Y}$-valued measure on the algebra $\mathcal{A}$ with $\mathcal{P} \circ \bar{\mu} = \bar{\nu}$.

To show the uniqueness of $\bar{\mu}$, let $\hat{\mu}$ be another $\mathcal{Y}$-valued measure on the algebra $\mathcal{A}$ with $\hat{\mu}|_{\mathcal{C}} = \mu$ and $(\mathcal{P} \circ \hat{\mu})|_{\mathcal{C}} = \nu$. Since $\mathcal{P} \circ \hat{\mu}$ is a measure on the algebra $\mathcal{A}$ that is an extension of ν, then $\mathcal{P} \circ \hat{\mu} = \bar{\nu} = \mathcal{P} \circ \bar{\mu}$ by Proposition 11.32. $\forall A \in \mathcal{A}$ with $\bar{\nu}(A) = \infty$, then $\hat{\mu}(A)$ and $\bar{\mu}(A)$ are both undefined. $\forall A \in \mathcal{A}$ with $\bar{\nu}(A) < \infty$, then $A = \bigcup_{i=1}^{n} C_i$, where $n \in \mathbb{Z}_+$ and $(C_i)_{i=1}^{n} \subseteq \mathcal{C}$ is pairwise disjoint. Then, $\hat{\mu}(A) = \sum_{i=1}^{n} \hat{\mu}(C_i) = \sum_{i=1}^{n} \mu(C_i) = \bar{\mu}(A)$, where the first equality follows from the fact that $\hat{\mu}$ is a $\mathcal{Y}$-valued measure on the algebra $\mathcal{A}$, the second equality follows from $\hat{\mu}|_{\mathcal{C}} = \mu$, and the last equality follows from the definition of $\bar{\mu}$. Hence, $\hat{\mu} = \bar{\mu}$. Therefore, $\bar{\mu}$ is unique.

This completes the proof of the proposition. □

Proposition 12.21 *Let $m \in \mathbb{N}$, $\mathcal{X}_j := (X_j, \mathcal{B}_j, \mu_j)$ be a finite $\mathbb{K}$-valued measure space, $\forall j \in \{1, \ldots, m\}$, $\mathcal{C}_m := \{\prod_{j=1}^{m} E_j \subseteq \prod_{j=1}^{m} X_j \mid E_j \in \mathcal{B}_j, \forall j \in \{1, \ldots, m\}\}$, which is the set of measurable rectangles in $\prod_{j=1}^{m} X_j$, $\bar{\mathcal{B}}_m$ be the σ-algebra on $\prod_{j=1}^{m} X_j$ generated by $\mathcal{C}_m$, $\mu : \mathcal{C}_m \to \mathbb{K}$ be defined by $\mu\left(\prod_{j=1}^{m} E_j\right) = \prod_{j=1}^{m} \mu_j(E_j)$, $\forall \prod_{j=1}^{m} E_j \in \mathcal{C}_m$, and $\nu : \mathcal{C}_m \to [0, \infty) \subset \mathbb{R}$ be defined by $\nu\left(\prod_{j=1}^{m} E_j\right) = \prod_{j=1}^{m} \mathcal{P} \circ \mu_j(E_j)$, $\forall \prod_{j=1}^{m} E_j \in \mathcal{C}_m$. Then, $\mathcal{C}_m$ is a semialgebra on $\prod_{j=1}^{m} X_j$, and μ and ν satisfy the assumptions of Proposition 12.20. Therefore, there exists a unique $\mathbb{K}$-valued measure $\lambda := \prod_{j=1}^{m} \mu_j$ on the measurable space $\left(\prod_{j=1}^{m} X_j, \bar{\mathcal{B}}_m\right)$ such that $\lambda\left(\prod_{j=1}^{m} E_j\right) = \prod_{j=1}^{m} \mu_j(E_j)$ and $\mathcal{P} \circ \lambda\left(\prod_{j=1}^{m} E_j\right) = \prod_{j=1}^{m} \mathcal{P} \circ \mu_j(E_j)$, $\forall \prod_{j=1}^{m} E_j \in \mathcal{C}_m$. Furthermore, λ is finite. The $\mathbb{K}$-valued measure space $\mathcal{X} := \left(\prod_{j=1}^{m} X_j, \bar{\mathcal{B}}_m, \lambda\right)$ is said to be the* product $\mathbb{K}$-valued measure space *of*

$\mathcal{X}_1, \ldots, \mathcal{X}_m$ *and denoted by* $\prod_{j=1}^m \mathcal{X}_j$. λ *is said to be the* product $\mathbb{K}$-valued measure *of* $\mu_1, \ldots, \mu_m$ *and denoted by* $\prod_{j=1}^m \mu_j$.

Proof We will prove the result using mathematical induction on m:
1° $m = 1$. The result is trivial and $\prod_{j=1}^1 \mathcal{X}_1 = \mathcal{X}_1$ and $\prod_{j=1}^1 \mu_1 = \mu_1$.
2° Assume that the result holds $\forall m \in \{1, \ldots, k-1\}$ with $k \in \mathbb{N}$ and $k \geq 2$.
3° Consider the case where $m = k$. Let $\bar{\lambda} := \prod_{j=1}^{k-1} \mu_j$ by the inductive assumption.

$\forall \prod_{j=1}^k E_{j,1}, \prod_{j=1}^k E_{j,2} \in \mathcal{C}_k$, we have $\big(\prod_{j=1}^k E_{j,1}\big) \cap \big(\prod_{j=1}^k E_{j,2}\big) = \prod_{j=1}^k (E_{j,1} \cap E_{j,2}) \in \mathcal{C}_m$. Note that $\Big(\prod_{j=1}^k E_{j,1}\Big)^{\sim} = \Big(\Big(\prod_{j=1}^{k-1} E_{j,1}\Big)^{\sim} \times E_{k,1}\Big) \cup \Big(\Big(\prod_{j=1}^{k-1} E_{j,1}\Big)^{\sim} \times \widetilde{E_{k,1}}\Big) \cup \Big(\Big(\prod_{j=1}^{k-1} E_{j,1}\Big) \times \widetilde{E_{k,1}}\Big)$. By inductive assumption, $\prod_{j=1}^{k-1} E_{j,1} \in \mathcal{C}_{k-1}$ and $\mathcal{C}_{k-1}$ is a semialgebra on $\prod_{j=1}^{k-1} X_j$; then $\Big(\prod_{j=1}^{k-1} E_{j,1}\Big)^{\sim}$ is a finite disjoint union of sets in $\mathcal{C}_{k-1}$. Thus, $\Big(\prod_{j=1}^k E_{j,1}\Big)^{\sim}$ is a finite disjoint union of sets in $\mathcal{C}_k$. Clearly, $\emptyset = \prod_{j=1}^k \emptyset \in \mathcal{C}_k \neq \emptyset$. Hence, $\mathcal{C}_k$ is a semialgebra on $X_1 \times X_2$.

Next, we will show that μ and ν satisfy (i)–(vi) of Proposition 12.20. $\mu(\emptyset) = \mu\big(\prod_{j=1}^k \emptyset\big) = \prod_{j=1}^k \mu_j(\emptyset) = 0$ and $\nu(\emptyset) = \nu\big(\prod_{j=1}^k \emptyset\big) = \prod_{j=1}^k \mathcal{P} \circ \emptyset) = 0$. Hence, (i) of Proposition 12.20 holds.

(iv) of Proposition 12.20 holds trivially since $\nu\big(\prod_{j=1}^k E_j\big) = \prod_{j=1}^k \mathcal{P} \circ \mu_j(E_j) \leq \prod_{j=1}^k \mathcal{P} \circ \mu_j(X_j) < \infty$, $\forall \prod_{j=1}^k E_j \in \mathcal{C}_k$.

$\forall \prod_{j=1}^k E_j \in \mathcal{C}_k$, $\forall$ pairwise disjoint $\Big(\prod_{j=1}^k E_{j,i}\Big)_{i=1}^{\infty} \subseteq \mathcal{C}_k$ with $\prod_{j=1}^k E_j = \bigcup_{i=1}^{\infty} (\prod_{j=1}^k E_{j,i})$. Fix any $x_k \in E_k$. $\forall (x_1, \ldots, x_{k-1}) \in \prod_{j=1}^{k-1} E_j$, $\exists!\ i \in \mathbb{N}$ such that $(x_1, \ldots, x_k) \in \prod_{j=1}^k E_{j,i}$. Then, $\prod_{j=1}^{k-1} E_j = \bigcup_{i \in \mathbb{N},\, x_k \in E_{k,i}} \prod_{j=1}^{k-1} E_{j,i}$ and the sets in the union are pairwise disjoint. Then, $\sum_{i \in \mathbb{N}} \chi_{E_{k,i}, X_k}(x_k) \prod_{j=1}^{k-1} \mathcal{P} \circ \mu_j(E_{j,i}) = \sum_{i \in \mathbb{N},\, x_k \in E_{k,i}} \prod_{j=1}^{k-1} \mathcal{P} \circ \mu_j(E_{j,i}) = \sum_{i \in \mathbb{N},\, x_k \in E_{k,i}} \mathcal{P} \circ \bar{\lambda}\big(\prod_{j=1}^{k-1} E_{j,i}\big) = \mathcal{P} \circ \bar{\lambda}\Big(\bigcup_{i \in \mathbb{N},\, x_k \in E_{k,i}} \prod_{j=1}^{k-1} E_{j,i}\Big) = \mathcal{P} \circ \bar{\lambda}\Big(\prod_{j=1}^{k-1} E_j\Big) = \chi_{E_k, X_k}(x_k) \prod_{j=1}^{k-1} \mathcal{P} \circ \mu_j(E_j) \in [0, \infty) \subset \mathbb{R}$, where the last four equalities follow from the inductive assumption. $\forall x_k \in X_k \setminus E_k$, we have $\chi_{E_k, X_k}(x_k) \cdot \prod_{j=1}^{k-1} \mathcal{P} \circ \mu_j(E_j) = 0 = \sum_{i \in \mathbb{N}} \chi_{E_{k,i}, X_k}(x_k) \prod_{j=1}^{k-1} \mathcal{P} \circ \mu_j(E_{j,i})$. Then, $\forall x_k \in X_k$, $\chi_{E_k, X_k}(x_k) \prod_{j=1}^{k-1} \mathcal{P} \circ \mu_j(E_j) = \sum_{i \in \mathbb{N}} \chi_{E_{k,i}, X_k}(x_k) \prod_{j=1}^{k-1} \mathcal{P} \circ \mu_j(E_{j,i})$. By Monotone Convergence Theorem 11.81 and Proposition 11.75, $\nu\big(\prod_{j=1}^k E_j\big) = \prod_{j=1}^k \mathcal{P} \circ \mu_j(E_j) = \int_{X_k} \chi_{E_k, X_k} \prod_{j=1}^{k-1} \mathcal{P} \circ \mu_j(E_j)\, d\mathcal{P} \circ \mu_k = \int_{X_k} \sum_{i \in \mathbb{N}} \chi_{E_{k,i}, X_k} \prod_{j=1}^{k-1} \mathcal{P} \circ \mu_j(E_{j,i}) \cdot d\mathcal{P} \circ \mu_k = \sum_{i=1}^{\infty} \big(\prod_{j=1}^{k-1} \mathcal{P} \circ \mu_j(E_{j,i})\big) \mathcal{P} \circ \mu_k(E_{k,i}) = \sum_{i=1}^{\infty} \prod_{j=1}^k \mathcal{P} \circ \mu_j(E_{j,i}) = \sum_{i=1}^{\infty} \nu\big(\prod_{j=1}^k E_{j,i}\big)$. Hence, (ii) and (iii) of Proposition 12.20 hold.

$\forall \prod_{j=1}^{k} E_j \in \mathcal{C}_k$, $\forall n \in \mathbb{Z}_+$, $\forall$ pairwise disjoint $\left(\prod_{j=1}^{k} E_{j,i}\right)_{i=1}^{n} \subseteq \mathcal{C}_k$ with $\prod_{j=1}^{k} E_j = \bigcup_{i=1}^{n} \prod_{j=1}^{k} E_{j,i}$. Fix any $x_k \in E_k$. $\forall (x_1, \ldots, x_{k-1}) \in \prod_{j=1}^{k-1} E_j$, $\exists!\ i \in N := \{1, \ldots, n\}$ such that $(x_1, \ldots, x_k) \in \prod_{j=1}^{k} E_{j,i}$. Thus, $\prod_{j=1}^{k-1} E_j = \bigcup_{i \in N,\, x_k \in E_{k,i}} \prod_{j=1}^{k-1} E_{j,i}$ and the sets in the union are pairwise disjoint. Then, $\sum_{i \in N} \chi_{E_{k,i},X_k}(x_k) \prod_{j=1}^{k-1} \mu_j(E_{j,i}) = \sum_{i \in N,\, x_k \in E_{k,i}} \prod_{j=1}^{k-1} \mu_j(E_{j,i}) = \sum_{i \in N,\, x_k \in E_{k,i}} \bar{\lambda}\Big(\prod_{j=1}^{k-1} E_{j,i}\Big) = \bar{\lambda}\Big(\bigcup_{i \in N,\, x_k \in E_{k,i}} \prod_{j=1}^{k-1} E_{j,i}\Big) = \bar{\lambda}\Big(\prod_{j=1}^{k-1} E_j\Big) = \chi_{E_k,X_k}(x_k) \prod_{j=1}^{k-1} \mu_j(E_j) \in \mathbb{K}$, where the last four equalities follow from the inductive assumption. $\forall x_k \in X_k \setminus E_k$, we have $\chi_{E_k,X_k}(x_k) \prod_{j=1}^{k-1} \mu_j(E_j) = 0 = \sum_{i \in N} \chi_{E_{k,i},X_k}(x_k) \prod_{j=1}^{k-1} \mu_j(E_{j,i})$. Then, $\forall x_1 \in X_1$, $\chi_{E_k,X_k}(x_k) \cdot \prod_{j=1}^{k-1} \mu_j(E_j) = \sum_{i \in N} \chi_{E_{k,i},X_k}(x_k) \prod_{j=1}^{k-1} \mu_j(E_{j,i})$. By Propositions 11.125 and 11.132, we have $\mu\big(\prod_{j=1}^{k} E_j\big) = \prod_{j=1}^{k} \mu_j(E_j) = \int_{X_k} \chi_{E_k,X_k} \prod_{j=1}^{k-1} \mu_j(E_j)\, d\mu_k = \int_{X_k} \big(\sum_{i \in N} \chi_{E_{k,i},X_k} \prod_{j=1}^{k-1} \mu_j(E_{j,i})\big)\, d\mu_k = \sum_{i=1}^{n} \prod_{j=1}^{k} \mu_j(E_{j,i}) = \sum_{i=1}^{n} \mu\big(\prod_{j=1}^{k} E_{j,i}\big)$. Hence, (v) of Proposition 12.20 holds.

$\forall \prod_{j=1}^{k} E_j \in \mathcal{C}_k$, $\forall n \in \mathbb{Z}_+$, $\forall$ pairwise disjoint $\left(\prod_{j=1}^{k} E_{j,i}\right)_{i=1}^{n} \subseteq \mathcal{C}_k$ with $\prod_{j=1}^{k} E_j = \bigcup_{i=1}^{n} \prod_{j=1}^{k} E_{j,i}$, we have $\sum_{i=1}^{n} \left|\mu\big(\prod_{j=1}^{k} E_{j,i}\big)\right| = \sum_{i=1}^{n} \prod_{j=1}^{k} |\mu_j(E_{j,i})| \leq \sum_{i=1}^{n} \prod_{j=1}^{k} \mathcal{P} \circ \mu_j(E_{j,i}) = \sum_{i=1}^{n} \nu\big(\prod_{j=1}^{k} E_{j,i}\big) = \nu\big(\prod_{j=1}^{k} E_j\big)$, where the inequality follows from Definition 11.108 and the last equality follows from (ii). Then, $\nu\big(\prod_{j=1}^{k} E_j\big) \geq s_{\prod_{j=1}^{k} E_j} := \sup_{\substack{n \in \mathbb{Z}_+,\, \left(\prod_{j=1}^{k} E_{j,i}\right)_{i=1}^{n} \subseteq \mathcal{C}_k,\, \prod_{j=1}^{k} E_j = \bigcup_{i=1}^{n} \prod_{j=1}^{k} E_{j,i} \\ \left(\prod_{j=1}^{k} E_{j,i}\right) \cap \left(\prod_{j=1}^{k} E_{j,p}\right) = \emptyset,\, \forall 1 \leq i < p \leq n}} \sum_{i=1}^{n} \left|\mu(\prod_{j=1}^{k} E_{j,i})\right|$. On the other hand, $\forall \epsilon \in (0, 1] \subset \mathbb{R}$, $\forall j \in \{1, \ldots, k\}$, $\exists n_j \in \mathbb{Z}_+$, $\exists$ pairwise disjoint $(E_{j,i})_{i-1}^{n_j} \subseteq \mathcal{B}_j$ with $E_j = \bigcup_{i=1}^{n_j} E_{j,i}$ such that $\sum_{i=1}^{n_j} |\mu_j(E_{j,i})| \geq \mathcal{P} \circ \mu_j(E_j)(1 - \frac{\epsilon}{1+\nu(\prod_{j=1}^{k} E_j)})^{1/k}$. Thus, $\prod_{j=1}^{k} E_j = \bigcup_{i_1=1}^{n_1} \cdots \bigcup_{i_k=1}^{n_k} (\prod_{j=1}^{k} E_{j,i_j})$ and the sets in the union are pairwise disjoint and in $\mathcal{C}_k$. This implies that $\sum_{i_1=1}^{n_1} \cdots \sum_{i_k=1}^{n_k} \left|\mu(\prod_{j=1}^{k} E_{j,i_j})\right| = \sum_{i_1=1}^{n_1} \cdots \sum_{i_k=1}^{n_k} \prod_{j=1}^{k} |\mu_j(E_{j,i_j})| = \prod_{j=1}^{k} (\sum_{i_j=1}^{n_j} |\mu_j(E_{j,i_j})|) \geq (1 - \frac{\epsilon}{1+\nu(\prod_{j=1}^{k} E_j)}) \prod_{j=1}^{k} \mathcal{P} \circ \mu_j(E_j) > \nu(\prod_{j=1}^{k} E_j) - \epsilon$. By the arbitrariness of ϵ, we have $s_{\prod_{j=1}^{k} E_j} \geq \nu(\prod_{j=1}^{k} E_j)$. Hence, (vi) of Proposition 12.20 holds.

Let $\mathcal{A}_k$ be the algebra on $\prod_{j=1}^{k} X_j$ generated by $\mathcal{C}_k$. By Proposition 12.20, ν admits a unique extension to a measure $\bar{\nu}$ on the algebra $\mathcal{A}_k$, and μ admits a unique extension to a $\mathbb{K}$-valued measure $\bar{\mu}$ on the algebra $\mathcal{A}_k$ with $(\mathcal{P} \circ \bar{\mu})|_{\mathcal{C}_k} = \nu$. Furthermore, $\mathcal{P} \circ \bar{\mu} = \bar{\nu}$. Clearly, $\bar{\mu}$ is finite. By Carathéodory Extension Theorem 12.4, there is a unique $\mathbb{K}$-valued measure λ on the measurable space $(\prod_{j=1}^k X_j, \bar{\mathcal{B}}_k)$ such that $\lambda|_{\mathcal{A}} = \bar{\mu}$ and $(\mathcal{P} \circ \lambda)|_{\mathcal{A}} = \mathcal{P} \circ \bar{\mu} = \bar{\nu}$. Then, $\lambda|_{\mathcal{C}_k} = \mu$ and $(\mathcal{P} \circ \lambda)|_{\mathcal{C}_k} = \nu$. This implies that $\mathcal{P} \circ \lambda(\prod_{j=1}^k X_j) = \nu(\prod_{j=1}^k X_j) = \prod_{j=1}^k \mathcal{P} \circ \mu_j(X_j) < \infty$. Hence, λ is finite.

This completes the induction process.

This completes the proof of the proposition. □

Proposition 12.22 *Let $k \in \mathbb{N}$, $\mathcal{X}_j := (X_j, \mathcal{B}_j, \mu_j)$ be a σ-finite $\mathbb{K}$-valued measure space, $\forall j \in \{1, \ldots, k\}$, $\mathcal{C} := \left\{\prod_{j=1}^k E_j \subseteq \prod_{j=1}^k X_j \;\middle|\; E_j \in \mathcal{B}_j,\ \forall j \in \{1, \ldots, k\}\right\}$ be the set of measurable rectangles, and $\mathcal{B}$ be the σ-algebra on $\prod_{j=1}^k X_j$ generated by $\mathcal{C}$. Then, there is a unique $\mathbb{K}$-valued measure $\lambda := \prod_{j=1}^k \mu_j$ on the measurable space $(\prod_{j=1}^k X_j, \mathcal{B})$ such that $\lambda(\prod_{j=1}^k E_j) = \prod_{j=1}^k \mu_j(E_j)$ and $\mathcal{P} \circ \lambda(\prod_{j=1}^k E_j) = \prod_{j=1}^k \mathcal{P} \circ \mu_j(E_j)$, $\forall E_j \in \operatorname{dom}(\mu_j)$, $\forall j \in \{1, \ldots, k\}$. Furthermore, λ is σ-finite. The $\mathbb{K}$-valued measure space $\mathcal{X} := (\prod_{j=1}^k X_j, \mathcal{B}, \lambda)$ is said to be the* product $\mathbb{K}$-valued measure space *of $\mathcal{X}_1, \ldots, \mathcal{X}_k$ and denoted by $\prod_{j=1}^k \mathcal{X}_j$. λ is said to be the product* $\mathbb{K}$-valued measure *of $\mu_1, \ldots, \mu_k$ and denoted by $\prod_{j=1}^k \mu_j$.*

Proof Fix any $j \in \{1, \ldots, k\}$. Since $\mathcal{X}_j$ is σ-finite, then $\exists (X_{j,n})_{n=1}^\infty \subseteq \mathcal{B}_j$ such that $X_j = \bigcup_{n=1}^\infty X_{j,n}$ and $\mathcal{P} \circ \mu_j(X_{j,n}) < \infty$, $\forall n \in \mathbb{N}$. Without loss of generality, we may assume that $(X_{j,n})_{n=1}^\infty$ is pairwise disjoint. Let $\mathcal{X}_{j,n} := (X_{j,n}, \mathcal{B}_{j,n}, \mu_{j,n})$ be the finite $\mathbb{K}$-valued measure subspace of $\mathcal{X}_j$, $\forall n \in \mathbb{N}$. Fix any $n_1, \ldots, n_k \in \mathbb{N}$. Let $\mathcal{C}_{n_1,\ldots,n_k} := \left\{\prod_{j=1}^k E_j \subseteq \prod_{j=1}^k X_{j,n_j} \;\middle|\; E_j \in \mathcal{B}_{j,n_j},\ j = 1, \ldots, k\right\}$ and $\bar{\mathcal{B}}_{n_1,\ldots,n_k}$ be the σ-algebra on $\prod_{j=1}^k X_{j,n_j}$ generated by $\mathcal{C}_{n_1,\ldots,n_k}$. By Proposition 12.21, there is a unique $\mathbb{K}$-valued measure $\lambda_{n_1,\ldots,n_k} = \prod_{j=1}^k \mu_{j,n_j}$ on $(\prod_{j=1}^k X_{j,n_j}, \bar{\mathcal{B}}_{n_1,\ldots,n_k})$ such that $\lambda_{n_1,\ldots,n_k}(\prod_{j=1}^k E_j) = \prod_{j=1}^k \mu_{j,n_j}(E_j)$ and $\mathcal{P} \circ \lambda_{n_1,\ldots,n_k}(\prod_{j=1}^k E_j) = \prod_{j=1}^k \mathcal{P} \circ \mu_{j,n_j}(E_j)$, $\forall \prod_{j=1}^k E_j \in \mathcal{C}_{n_1,\ldots,n_k}$. Furthermore, $\lambda_{n_1,\ldots,n_k}$ is finite. Denote the finite $\mathbb{K}$-valued measure space $(\prod_{j=1}^k X_{j,n_j}, \bar{\mathcal{B}}_{n_1,\ldots,n_k}, \lambda_{n_1,\ldots,n_k}) =: \bar{\mathcal{X}}_{n_1,\ldots,n_k}$. By Proposition 11.118, the generation process on $\left(\bar{\mathcal{X}}_{n_1,\ldots,n_k}\right)_{n_j \in \mathbb{N},\ j \in \{1,\ldots,k\}}$ yields a unique σ-finite $\mathbb{K}$-valued measure space $\mathcal{X} := (\prod_{j=1}^k X_j, \bar{\mathcal{B}}, \lambda)$ on $\prod_{j=1}^k X_j$ such that $\bar{\mathcal{X}}_{n_1,\ldots,n_k}$ is the finite $\mathbb{K}$-valued measure subspace of $\mathcal{X}$, $\forall n_1, \ldots, n_k \in \mathbb{N}$, where $\bar{\mathcal{B}} := \left\{E \subseteq \prod_{j=1}^k X_j \;\middle|\; E \cap (\prod_{j=1}^k X_{j,n_j}) \in \bar{\mathcal{B}}_{n_1,\ldots,n_k}, \forall n_j \in \mathbb{N},\ \forall j \in \{1, \ldots, k\}\right\}$, $\mathcal{P} \circ \lambda(E) = \sum_{n_1=1}^\infty \cdots \sum_{n_k=1}^\infty \mathcal{P} \circ \lambda_{n_1,\ldots,n_k}(E \cap (\prod_{j=1}^k X_{j,n_j}))$, $\forall E \in \bar{\mathcal{B}}$, and $\lambda(E) = \sum_{n_1=1}^\infty \cdots \sum_{n_k=1}^\infty \lambda_{n_1,\ldots,n_k}(E \cap (\prod_{j=1}^k X_{j,n_j}))$, $\forall E \in \bar{\mathcal{B}}$ with $\mathcal{P} \circ \lambda(E) < \infty$.

Claim 12.22.1 $\mathcal{B} = \bar{\mathcal{B}}$.

Proof of Claim $\forall \prod_{j=1}^{k} E_j \in \mathcal{C}$, $\forall n_1, \ldots, n_k \in \mathbb{N}$, $E_j \cap X_{j,n_j} \in \mathcal{B}_{j,n_j}$, $\forall j \in \{1, \ldots, k\}$, by Proposition 11.115. Then, $(\prod_{j=1}^{k} E_j) \cap (\prod_{j=1}^{k} X_{j,n_j}) \in \mathcal{C}_{n_1,\ldots,n_k} \subseteq \bar{\mathcal{B}}_{n_1,\ldots,n_k}$. This implies that $\prod_{j=1}^{k} E_j \in \bar{\mathcal{B}}$. Then, $\mathcal{C} \subseteq \bar{\mathcal{B}}$. Since $\bar{\mathcal{B}}$ is a σ-algebra, then $\mathcal{B} \subseteq \bar{\mathcal{B}}$.

On the other hand, $\forall n_1, \ldots, n_k \in \mathbb{N}$, $\forall \prod_{j=1}^{k} E_j \in \mathcal{C}_{n_1,\ldots,n_k}$, we have $E_j \in \mathcal{B}_{j,n_j} \subseteq \mathcal{B}_j$, $\forall j \in \{1, \ldots, k\}$. Thus, $\prod_{j=1}^{k} E_j \in \mathcal{C} \subseteq \mathcal{B}$. Then, $\mathcal{C}_{n_1,\ldots,n_k} \subseteq \mathcal{B}$ and $\bar{\mathcal{B}}_{n_1,\ldots,n_k} \subseteq \mathcal{B}$. This further implies that $\bar{\mathcal{B}} \subseteq \mathcal{B}$. Hence, $\mathcal{B} = \bar{\mathcal{B}}$. This completes the proof of the claim. □

$\forall j \in \{1, \ldots, k\}$, $\forall E_j \in \operatorname{dom}(\mu_j)$, we have $\infty > \mathcal{P} \circ \mu_j(E_j) = \sum_{n_j=1}^{\infty} \mathcal{P} \circ \mu_j(E_j \cap X_{j,n_j}) = \sum_{n_j=1}^{\infty} \mathcal{P} \circ \mu_{j,n_j}(E_j \cap X_{j,n_j})$, $\mathbb{K} \ni \mu_j(E_j) = \sum_{n_j=1}^{\infty} \mu_j(E_j \cap X_{j,n_j}) = \sum_{n_j=1}^{\infty} \mu_{j,n_j}(E_j \cap X_{j,n_j})$. Then,

$$\begin{aligned}
\mathcal{P} \circ \lambda\Big(\prod_{j=1}^{k} E_j\Big) &= \sum_{n_1=1}^{\infty} \cdots \sum_{n_k=1}^{\infty} \mathcal{P} \circ \lambda_{n_1,\ldots,n_k}\Big(\big(\prod_{j=1}^{k} E_j\big) \cap \big(\prod_{j=1}^{k} X_{j,n_j}\big)\Big) \\
&= \sum_{n_1=1}^{\infty} \cdots \sum_{n_k=1}^{\infty} \mathcal{P} \circ \lambda_{n_1,\ldots,n_k}\Big(\prod_{j=1}^{k} (E_j \cap X_{j,n_j})\Big) \\
&= \sum_{n_1=1}^{\infty} \cdots \sum_{n_k=1}^{\infty} \prod_{j=1}^{k} \mathcal{P} \circ \mu_{j,n_j}(E_j \cap X_{j,n_j}) \\
&= \prod_{j=1}^{k} \Big(\sum_{n_j=1}^{\infty} \mathcal{P} \circ \mu_{j,n_j}(E_j \cap X_{j,n_j})\Big) = \prod_{j=1}^{k} \mathcal{P} \circ \mu_j(E_j) \\
&< \infty
\end{aligned}$$

where the first equality follows from Proposition 11.118 and the third equality follows from Proposition 12.21. We also have

$$\begin{aligned}
\lambda\Big(\prod_{j=1}^{k} E_j\Big) &= \sum_{n_1=1}^{\infty} \cdots \sum_{n_k=1}^{\infty} \lambda_{n_1,\ldots,n_k}\Big(\big(\prod_{j=1}^{k} E_j\big) \cap \big(\prod_{j=1}^{k} X_{j,n_j}\big)\Big) \\
&= \sum_{n_1=1}^{\infty} \cdots \sum_{n_k=1}^{\infty} \lambda_{n_1,\ldots,n_k}\Big(\prod_{j=1}^{k} (E_j \cap X_{j,n_j})\Big)
\end{aligned}$$

$$= \sum_{n_1=1}^{\infty} \cdots \sum_{n_k=1}^{\infty} \prod_{j=1}^{k} \mu_{j,n_j}(E_j \cap X_{j,n_j})$$

$$= \prod_{j=1}^{k} \Big(\sum_{n_j=1}^{\infty} \mu_{j,n_j}(E_j \cap X_{j,n_j}) \Big) = \prod_{j=1}^{k} \mu_j(E_j) \in \mathbb{K}$$

where the first equality follows from Proposition 11.118 and the third equality follows from Proposition 12.21. Hence, $\mathcal{X}$ is the σ-finite $\mathbb{K}$-valued measure space that we seek.

This completes the proof of the proposition. □

Proposition 12.23 *Let $m \in \mathbb{N}$, $\mathcal{X}_i := (X_i, \mathcal{B}_i, \mu_i)$ be a σ-finite $\mathbb{K}$-valued measure space, $\bar{X}_i \in \mathcal{B}_i$, $\bar{\mathcal{X}}_i := (\bar{X}_i, \bar{\mathcal{B}}_i, \bar{\mu}_i)$ be the σ-finite $\mathbb{K}$-valued measure subspace of $\mathcal{X}_i$, $i = 1, \ldots, m$, and $\mathcal{X} := (\prod_{i=1}^{m} X_i, \mathcal{B}, \lambda) := \prod_{i=1}^{m} \mathcal{X}_i$ be the σ-finite product $\mathbb{K}$-valued measure space of $\mathcal{X}_1, \ldots, \mathcal{X}_m$. Then, $\bar{\mathcal{X}} := (\prod_{i=1}^{m} \bar{X}_i, \bar{\mathcal{B}}, \bar{\lambda})$ is the $\mathbb{K}$-valued measure subspace of $\mathcal{X}$ if, and only if, $\bar{\mathcal{X}} = \prod_{i=1}^{m} \bar{\mathcal{X}}_i$.*

Proof Let $\bar{\mathcal{X}} := (\prod_{i=1}^{m} \bar{X}_i, \bar{\mathcal{B}}, \bar{\lambda})$ be the σ-finite $\mathbb{K}$-valued measure subspace of $\mathcal{X}$ and $\tilde{\mathcal{X}} := \prod_{i=1}^{m} \bar{\mathcal{X}}_i := (\prod_{i=1}^{m} \bar{X}_i, \tilde{\mathcal{B}}, \tilde{\lambda})$ be the σ-finite product $\mathbb{K}$-valued measure space of $\bar{\mathcal{X}}_1, \ldots, \bar{\mathcal{X}}_m$. All we need to show is that $\bar{\mathcal{X}} = \tilde{\mathcal{X}}$.

Let $\bar{X}_{i,1} := \bar{X}_i$ and $\bar{X}_{i,2} := X_i \setminus \bar{X}_i$, $i = 1, \ldots, m$, $\bar{\mathcal{C}}_{n_1,\ldots,n_m} := \big\{\prod_{i=1}^{m} B_i \subseteq \prod_{i=1}^{m} \bar{X}_{i,n_i} \mid B_i \in \mathcal{B}_i,\ \forall i \in \{1, \ldots, m\}\big\}$ be the set of measurable rectangles on $\prod_{i=1}^{m} \bar{X}_{i,n_i}$, $\tilde{\mathcal{B}}_{n_1,\ldots,n_m}$ be the σ-algebra generated by $\bar{\mathcal{C}}_{n_1,\ldots,n_m}$ on $\prod_{i=1}^{m} \bar{X}_{i,n_i}$, $\forall n_1, \ldots, n_m \in \{1, 2\}$, and $\mathcal{C} := \big\{\prod_{i=1}^{m} B_i \subseteq \prod_{i=1}^{m} X_i \,\big|\, B_i \in \mathcal{B}_i,\ \forall i \in \{1, \ldots, m\}\big\}$ be the set of measurable rectangles on $\prod_{i=1}^{m} X_i$. Then, by Proposition 12.22, $\mathcal{B}$ is the σ-algebra generated by $\mathcal{C}$ on $\prod_{i=1}^{m} X_i$.

Claim 12.23.1 $\mathcal{B} = \Big\{E \subseteq \prod_{i=1}^{m} X_i \,\Big|\, E \cap (\prod_{i=1}^{m} \bar{X}_{i,n_i}) \in \tilde{\mathcal{B}}_{n_1,\ldots,n_m},\ \forall n_i \in \{1, 2\},\ \forall i \in \{1, \ldots, m\}\Big\} =: \hat{\mathcal{B}}$.

Proof of Claim $\forall \prod_{i=1}^{m} B_i \in \mathcal{C}$, we have $B_i \in \mathcal{B}_i$, $i = 1, \ldots, m$, which implies that $\mathcal{B}_i \ni B_i \cap \bar{X}_{i,n_i} \subseteq \bar{X}_{i,n_i}$, $n_i = 1, 2$, $i = 1, \ldots, m$. Hence, $\prod_{i=1}^{m} B_i \in \hat{\mathcal{B}}$. By the arbitrariness of $\prod_{i=1}^{m} B_i$, we have $\mathcal{C} \subseteq \hat{\mathcal{B}}$.

Now, we will show that $\hat{\mathcal{B}}$ is a σ-algebra on $\prod_{i=1}^{m} X_i$. Clearly, $\emptyset = \prod_{i=1}^{m} \emptyset \in \mathcal{C} \subseteq \hat{\mathcal{B}}$ and $\prod_{i=1}^{m} X_i \in \mathcal{C} \subseteq \hat{\mathcal{B}}$. $\forall (E_n)_{n=1}^{\infty} \subseteq \hat{\mathcal{B}}$, $\forall n_i = 1, 2$, $\forall i = 1, \ldots, m$, $\forall n \in \mathbb{N}$, by the definition of $\hat{\mathcal{B}}$, $E_n \cap (\prod_{i=1}^{m} \bar{X}_{i,n_i}) \in \tilde{\mathcal{B}}_{n_1,\ldots,n_m}$, which implies $(\bigcup_{n=1}^{\infty} E_n) \cap (\prod_{i=1}^{m} \bar{X}_{i,n_i}) = \bigcup_{n=1}^{\infty} (E_n \cap (\prod_{i=1}^{m} \bar{X}_{i,n_i})) \in \tilde{\mathcal{B}}_{n_1,\ldots,n_m}$, since $\tilde{\mathcal{B}}_{n_1,\ldots,n_m}$ is a σ-algebra. Hence, $\bigcup_{n=1}^{\infty} E_n \in \hat{\mathcal{B}}$. $\forall E \in \hat{\mathcal{B}}$, $\forall n_i = 1, 2$, $\forall i = 1, \ldots, m$, $E \cap (\prod_{i=1}^{m} \bar{X}_{i,n_i}) \in \tilde{\mathcal{B}}_{n_1,\ldots,n_m}$ and then $((\prod_{i=1}^{m} X_i) \setminus E) \cap (\prod_{i=1}^{m} \bar{X}_{i,n_i}) = (\prod_{i=1}^{m} \bar{X}_{i,n_i}) \setminus E = (\prod_{i=1}^{m} \bar{X}_{i,n_i}) \setminus (E \cap (\prod_{i=1}^{m} \bar{X}_{i,n_i})) \in \tilde{\mathcal{B}}_{n_1,\ldots,n_m}$, since $\tilde{\mathcal{B}}_{n_1,\ldots,n_m}$ is a σ-algebra on $\prod_{i=1}^{m} \bar{X}_{i,n_i}$. This implies that $(\prod_{i=1}^{m} X_i) \setminus E \in \hat{\mathcal{B}}$. Hence, $\hat{\mathcal{B}}$ is a σ-algebra on $\prod_{i=1}^{m} X_i$.

Since $\mathcal{B}$ is the σ-algebra on $\prod_{i=1}^{m} X_i$ generated by $\mathcal{C}$ and $\mathcal{C} \subseteq \hat{\mathcal{B}}$, then $\mathcal{B} \subseteq \hat{\mathcal{B}}$.

On the other hand, $\bar{\mathcal{C}}_{n_1,\ldots,n_m} \subseteq \mathcal{C}$, $n_i = 1, 2$, $i = 1, \ldots, m$, implies that $\tilde{\mathcal{B}}_{n_1,\ldots,n_m} \subseteq \mathcal{B}$. $\forall E \in \hat{\mathcal{B}}$, we have $E \cap (\prod_{i=1}^{m} \bar{X}_{i,n_i}) \in \tilde{\mathcal{B}}_{n_1,\ldots,n_m} \subseteq \mathcal{B}$, $n_i = 1, 2$, $i = 1, \ldots, m$, which implies that $E = \bigcup_{n_1=1}^{2} \cdots \bigcup_{n_m=1}^{2} (E \cap (\prod_{i=1}^{m} \bar{X}_{i,n_i})) \in \mathcal{B}$. Hence, $\hat{\mathcal{B}} \subseteq \mathcal{B}$. Therefore, $\mathcal{B} = \hat{\mathcal{B}}$. This completes the proof of the claim. □

Note that $\tilde{\mathcal{B}} = \tilde{\mathcal{B}}_{1,\ldots,1} = \left\{ E \subseteq \prod_{i=1}^{m} \bar{X}_i \;\middle|\; E \in \hat{\mathcal{B}} \right\} = \left\{ E \subseteq \prod_{i=1}^{m} \bar{X}_i \mid E \in \mathcal{B} \right\} = \bar{\mathcal{B}}$, where the first equality follows from Proposition 12.22, the second equality follows from the definition of $\hat{\mathcal{B}}$, the third equality follows from Claim 12.23.1, and the last equality follows from Proposition 11.115.

Finally, we will show that $\bar{\lambda} = \tilde{\lambda}$ by Propositions 12.22 and 11.115. By Proposition 11.115 and Claim 12.23.1, $\bar{\lambda}$ is a $\mathbb{K}$-valued measure on $(\prod_{i=1}^{m} \bar{X}_i, \bar{\mathcal{B}})$, $\bar{\lambda} = \lambda|_{\bar{\mathcal{B}}} = \lambda|_{\tilde{\mathcal{B}}}$, and $\mathcal{P} \circ \bar{\lambda} = (\mathcal{P} \circ \lambda)|_{\bar{\mathcal{B}}}$. By Proposition 12.22, λ is the unique $\mathbb{K}$-valued measure on $(\prod_{i=1}^{m} X_i, \mathcal{B})$ such that $\lambda(\prod_{i=1}^{m} E_i) = \prod_{i=1}^{m} \mu_i(E_i)$ and $\mathcal{P} \circ \lambda(\prod_{i=1}^{m} E_i) = \prod_{i=1}^{m} \mathcal{P} \circ \mu_i(E_i)$, $\forall E_i \in \operatorname{dom}(\mu_i)$, $\forall i \in \{1, \ldots, m\}$. Then, $\forall i \in \{1, \ldots, m\}$, $\forall E_i \in \operatorname{dom}(\bar{\mu}_i)$, by Proposition 11.115, $\bar{\mu}_i = \mu_i|_{\bar{\mathcal{B}}_i}$ and $\mathcal{P} \circ \bar{\mu}_i = (\mathcal{P} \circ \mu_i)|_{\bar{\mathcal{B}}_i}$. Hence, $E_i \in \operatorname{dom}(\mu_i)$. This implies that $\bar{\lambda}(\prod_{i=1}^{m} E_i) = \lambda(\prod_{i=1}^{m} E_i) = \prod_{i=1}^{m} \mu_i(E_i) = \prod_{i=1}^{m} \bar{\mu}_i(E_i)$ and $\mathcal{P} \circ \bar{\lambda}(\prod_{i=1}^{m} E_i) = \mathcal{P} \circ \lambda(\prod_{i=1}^{m} E_i) = \prod_{i=1}^{m} \mathcal{P} \circ \mu_i(E_i) = \prod_{i=1}^{m} \mathcal{P} \circ \bar{\mu}_i(E_i)$. By Proposition 12.22, $\tilde{\lambda}$ is the unique $\mathbb{K}$-valued measure on $(\prod_{i=1}^{m} \bar{X}_i, \bar{\mathcal{B}})$ such that $\tilde{\lambda}(\prod_{i=1}^{m} E_i) = \prod_{i=1}^{m} \bar{\mu}_i(E_i)$ and $\mathcal{P} \circ \tilde{\lambda}(\prod_{i=1}^{m} E_i) = \prod_{i=1}^{m} \mathcal{P} \circ \bar{\mu}_i(E_i)$, $\forall E_i \in \operatorname{dom}(\bar{\mu}_i)$, $\forall i \in \{1, \ldots, m\}$. Hence, we have $\bar{\lambda} = \tilde{\lambda}$.

Hence, $\bar{\mathcal{X}} = \tilde{\mathcal{X}}$. This completes the proof of the proposition. □

Proposition 12.24 *Let Γ be a finite nonempty index set, Λ_β be a finite nonempty index set, $\forall \beta \in \Gamma$, with Λ_β's being pairwise disjoint, $\Lambda := \bigcup_{\beta \in \Gamma} \Lambda_\beta$, $\mathcal{X}_\alpha := (X_\alpha, \mathcal{B}_\alpha, \mu_\alpha)$ be a σ-finite $\mathbb{K}$-valued measure space, $\forall \alpha \in \Lambda$, $\hat{\mathcal{X}}_\beta := \prod_{\alpha \in \Lambda_\beta} \mathcal{X}_\alpha := (\prod_{\alpha \in \Lambda_\beta} X_\alpha, \hat{\mathcal{B}}_\beta, \hat{\mu}_\beta)$ be the σ-finite product $\mathbb{K}$-valued measure space, $\forall \beta \in \Gamma$, $\hat{\mathcal{X}} := \prod_{\beta \in \Gamma} \hat{\mathcal{X}}_\beta := (\prod_{\beta \in \Gamma} \prod_{\alpha \in \Lambda_\beta} X_\alpha, \hat{\mathcal{B}}, \hat{\mu})$ be the σ-finite product $\mathbb{K}$-valued measure space, and $\bar{\mathcal{X}} := \prod_{\alpha \in \Lambda} \mathcal{X}_\alpha := (\prod_{\alpha \in \Lambda} X_\alpha, \bar{\mathcal{B}}, \bar{\mu})$ be the σ-finite product $\mathbb{K}$-valued measure space. Then, $\bar{\mathcal{X}}$ and $\hat{\mathcal{X}}$ are isomeasuric.*

Proof Define $\Psi : \hat{\mathcal{X}} \to \bar{\mathcal{X}}$ by, $\forall x \in \hat{\mathcal{X}}$, $\forall \alpha \in \Lambda$, $\exists!\, \beta_\alpha \in \Gamma \ni \cdot\, \alpha \in \Lambda_{\beta_\alpha}$, $\pi_\alpha(\Psi(x)) = \pi_\alpha^{\langle \beta_\alpha \rangle}(\pi_{\beta_\alpha}^{\langle \Gamma \rangle}(x))$. By the proof of Proposition 3.30, Ψ is bijective with inverse $\Psi_{\text{inv}} : \bar{\mathcal{X}} \to \hat{\mathcal{X}}$.

Let $\hat{\mathcal{C}}_\beta := \left\{ \prod_{\alpha \in \Lambda_\beta} E_\alpha \subseteq \prod_{\alpha \in \Lambda_\beta} X_\alpha \;\middle|\; E_\alpha \in \mathcal{B}_\alpha,\ \forall \alpha \in \Lambda_\beta \right\}$, $\forall \beta \in \Gamma$, $\hat{\mathcal{C}} := \left\{ \prod_{\beta \in \Gamma} E_\beta \subseteq \prod_{\beta \in \Gamma} \prod_{\alpha \in \Lambda_\beta} X_\alpha \;\middle|\; E_\beta \in \hat{\mathcal{B}}_\beta,\ \forall \beta \in \Gamma \right\}$, $\bar{\hat{\mathcal{C}}} := \{ \prod_{\beta \in \Gamma} \prod_{\alpha \in \Lambda_\beta} E_\alpha \subseteq \prod_{\beta \in \Gamma} \prod_{\alpha \in \Lambda_\beta} X_\alpha \mid E_\alpha \in \mathcal{B}_\alpha,\ \forall \alpha \in \Lambda \}$, and $\bar{\mathcal{C}} := \{ \prod_{\alpha \in \Lambda} E_\alpha \subseteq \prod_{\alpha \in \Lambda} X_\alpha \mid E_\alpha \in \mathcal{B}_\alpha,\ \forall \alpha \in \Lambda \}$. By Proposition 12.22, $\hat{\mathcal{B}}_\beta$ is the σ-algebra generated by $\hat{\mathcal{C}}_\beta$ on

$\prod_{\alpha\in\Lambda_\beta} X_\alpha$, $\forall\beta\in\Gamma$, $\hat{\mathcal{B}}$ is the σ-algebra generated by $\hat{\mathcal{C}}$ on $\prod_{\beta\in\Gamma}\prod_{\alpha\in\Lambda_\beta} X_\alpha$, and $\bar{\mathcal{B}}$ is the σ-algebra generated by $\bar{\mathcal{C}}$ on $\prod_{\alpha\in\Lambda} X_\alpha$. Clearly, $\hat{\bar{\mathcal{C}}}\subseteq\hat{\mathcal{C}}$.

Fix any $\prod_{\beta\in\Gamma}\prod_{\alpha\in\Lambda_\beta} E_\alpha\in\hat{\bar{\mathcal{C}}}$, and we will show that $\Psi(\prod_{\beta\in\Gamma}\prod_{\alpha\in\Lambda_\beta} E_\alpha)=\prod_{\alpha\in\Lambda} E_\alpha$. $\forall x\in\prod_{\beta\in\Gamma}\prod_{\alpha\in\Lambda_\beta} E_\alpha$, $\forall\alpha\in\Lambda$, we have $\pi_\alpha(\Psi(x))=\pi_\alpha^{\langle\beta_\alpha\rangle}(\pi_{\beta_\alpha}^{\langle\Gamma\rangle}(x))\in E_\alpha$. Hence, $\Psi(x)\in\prod_{\alpha\in\Lambda} E_\alpha$. This implies that $\Psi(\prod_{\beta\in\Gamma}\prod_{\alpha\in\Lambda_\beta} E_\alpha)\subseteq\prod_{\alpha\in\Lambda} E_\alpha$. On the other hand, $\forall y\in\prod_{\alpha\in\Lambda} E_\alpha$, $\forall\beta\in\Gamma$, $\forall\alpha\in\Lambda_\beta$, we have $\pi_\alpha^{\langle\beta\rangle}(\pi_\beta^{\langle\Gamma\rangle}(\Psi_{\text{inv}}(y)))=\pi_\alpha(y)\in E_\alpha$. Then, $\Psi_{\text{inv}}(y)\in\prod_{\beta\in\Gamma}\prod_{\alpha\in\Lambda_\beta} E_\alpha$. This leads to $\Psi_{\text{inv}}(\prod_{\alpha\in\Lambda} E_\alpha)\subseteq\prod_{\beta\in\Gamma}\prod_{\alpha\in\Lambda_\beta} E_\alpha$. This is equivalent to $\prod_{\alpha\in\Lambda} E_\alpha\subseteq\Psi(\prod_{\beta\in\Gamma}\prod_{\alpha\in\Lambda_\beta} E_\alpha)$. Therefore, $\Psi(\prod_{\beta\in\Gamma}\prod_{\alpha\in\Lambda_\beta} E_\alpha)=\prod_{\alpha\in\Lambda} E_\alpha\in\bar{\mathcal{C}}$. Conversely, $\forall\prod_{\alpha\in\Lambda} E_\alpha\in\bar{\mathcal{C}}$, we have $\Psi_{\text{inv}}(\prod_{\alpha\in\Lambda} E_\alpha)=\prod_{\beta\in\Gamma}\prod_{\alpha\in\Lambda_\beta} E_\alpha\in\hat{\bar{\mathcal{C}}}$.

Let $\hat{\bar{\mathcal{B}}}$ be the σ-algebra on $\prod_{\beta\in\Gamma}\prod_{\alpha\in\Lambda_\beta} X_\alpha$ generated by $\hat{\bar{\mathcal{C}}}$. We will show that $\hat{\bar{\mathcal{B}}}=\left\{E\subseteq\prod_{\beta\in\Gamma}\prod_{\alpha\in\Lambda_\beta} X_\alpha\;\middle|\;\Psi(E)\in\bar{\mathcal{B}}\right\}=:\hat{\hat{\mathcal{B}}}$. Since $\bar{\mathcal{C}}\subseteq\bar{\mathcal{B}}$, by the previous paragraph, we have $\hat{\bar{\mathcal{C}}}\subseteq\hat{\hat{\mathcal{B}}}$. It is easy to show that $\hat{\hat{\mathcal{B}}}$ is a σ-algebra on $\prod_{\beta\in\Gamma}\prod_{\alpha\in\Lambda_\beta} X_\alpha$ since $\bar{\mathcal{B}}$ is a σ-algebra. Then, $\hat{\bar{\mathcal{B}}}\subseteq\hat{\hat{\mathcal{B}}}$. On the other hand, let $\tilde{\mathcal{B}}:=\left\{E\subseteq\prod_{\alpha\in\Lambda} X_\alpha\;\middle|\;\Psi_{\text{inv}}(E)\in\hat{\bar{\mathcal{B}}}\right\}$. Since $\hat{\bar{\mathcal{B}}}$ is a σ-algebra on $\prod_{\beta\in\Gamma}\prod_{\alpha\in\Lambda_\beta} X_\alpha$, then $\tilde{\mathcal{B}}$ is a σ-algebra on $\prod_{\alpha\in\Lambda} X_\alpha$. Since $\hat{\bar{\mathcal{C}}}\subseteq\hat{\bar{\mathcal{B}}}$, by the previous paragraph, then $\bar{\mathcal{C}}\subseteq\tilde{\mathcal{B}}$. By the definition of $\bar{\mathcal{B}}$, we have $\bar{\mathcal{B}}\subseteq\tilde{\mathcal{B}}$. Since $\hat{\bar{\mathcal{B}}}\subseteq\hat{\hat{\mathcal{B}}}$, then, by the definitions of $\tilde{\mathcal{B}}$ and $\hat{\hat{\mathcal{B}}}$, $\tilde{\mathcal{B}}\subseteq\bar{\mathcal{B}}$. Thus, we have $\tilde{\mathcal{B}}=\bar{\mathcal{B}}$ and $\hat{\bar{\mathcal{B}}}=\hat{\hat{\mathcal{B}}}$. We can now prove the following result.

Claim 12.24.1 $\hat{\mathcal{C}}\subseteq\hat{\bar{\mathcal{B}}}$.

Proof of Claim Without loss of generality, assume that $\Gamma=\{\beta_1,\ldots,\beta_n\}$ for some $n\in\mathbb{N}$. We will prove the following statement:

- $\forall i\in\{1,\ldots,n+1\}$, $\forall E_{\beta_j}:=\prod_{\alpha\in\Lambda_{\beta_j}} E_\alpha\in\hat{\mathcal{C}}_{\beta_j}$, $j=i,\ldots,n$, $\forall E_{\beta_j}\in\hat{\mathcal{B}}_{\beta_j}$, $j=1,\ldots,i-1$, we have $\prod_{j=1}^n E_{\beta_j}\in\hat{\bar{\mathcal{B}}}$

by mathematical induction on i.

1° $i=1$. $\forall E_{\beta_j}:=\prod_{\alpha\in\Lambda_{\beta_j}} E_\alpha\in\hat{\mathcal{C}}_{\beta_j}$, $j=1,\ldots,n$, we have $\prod_{j=1}^n E_{\beta_j}=\prod_{j=1}^n\prod_{\alpha\in\Lambda_{\beta_j}} E_\alpha\in\hat{\bar{\mathcal{C}}}\subseteq\hat{\bar{\mathcal{B}}}$. This case is proved.

2° Assume that the statement holds for $i=k\in\{1,\ldots,n\}$.

3° Consider the case $i=k+1$. Fix any $E_{\beta_j}:=\prod_{\alpha\in\Lambda_{\beta_j}} E_\alpha\in\hat{\mathcal{C}}_{\beta_j}$, $j=k+1,\ldots,n$, and any $E_{\beta_j}\in\hat{\mathcal{B}}_{\beta_j}$, $j=1,\ldots,k-1$. Define $\mathcal{F}:=\left\{E_{\beta_k}\in\hat{\mathcal{B}}_{\beta_k}\;\middle|\;\prod_{j=1}^n E_{\beta_j}\in\hat{\bar{\mathcal{B}}}\right\}$. By the inductive assumption, we have $\hat{\mathcal{C}}_{\beta_k}\subseteq\mathcal{F}$. Then, $\emptyset=\prod_{\alpha\in\Lambda_{\beta_k}}\emptyset\in\hat{\mathcal{C}}_{\beta_k}\subseteq\mathcal{F}$ and $X_{\beta_k}:=\prod_{\alpha\in\Lambda_{\beta_k}} X_\alpha\in$

$\hat{\mathcal{C}}_{\beta_k} \subseteq \mathcal{F}$. $\forall E_{\beta_k} \in \mathcal{F}$, we have $E_{\beta_k} \in \hat{\mathcal{B}}_{\beta_k}$ and $\prod_{j=1}^{n} E_{\beta_j} \in \hat{\bar{\mathcal{B}}}$. By $X_{\beta_k} \in \hat{\mathcal{C}}_{\beta_k} \subseteq \mathcal{F}$, we have $(\prod_{j=1}^{k-1} E_{\beta_j}) \times X_{\beta_k} \times (\prod_{j=k+1}^{n} E_{\beta_j}) \in \hat{\bar{\mathcal{B}}}$. This implies that $\left((\prod_{j=1}^{k-1} E_{\beta_j}) \times X_{\beta_k} \times (\prod_{j=k+1}^{n} E_{\beta_j})\right) \setminus \left(\prod_{j=1}^{n} E_{\beta_j}\right) = (\prod_{j=1}^{k-1} E_{\beta_j}) \times (X_{\beta_k} \setminus E_{\beta_k}) \times (\prod_{j=k+1}^{n} E_{\beta_j}) \in \hat{\bar{\mathcal{B}}}$. Then, $X_{\beta_k} \setminus E_{\beta_k} = (\prod_{\alpha \in \Lambda_{\beta_k}} X_\alpha) \setminus E_{\beta_k} \in \mathcal{F}$. $\forall (E_{\beta_k,i})_{i=1}^{\infty} \subseteq \mathcal{F}$, we have $E_{\beta_k,i} \in \hat{\mathcal{B}}_{\beta_k}$ and $(\prod_{j=1}^{k-1} E_{\beta_j}) \times E_{\beta_k,i} \times (\prod_{j=k+1}^{n} E_{\beta_j}) \in \hat{\bar{\mathcal{B}}}$, $\forall i \in \mathbb{N}$. Then, $\bigcup_{i=1}^{\infty} E_{\beta_k,i} \in \hat{\mathcal{B}}_{\beta_k}$ and $\bigcup_{i=1}^{\infty} \left((\prod_{j=1}^{k-1} E_{\beta_j}) \times E_{\beta_k,i} \times (\prod_{j=k+1}^{n} E_{\beta_j})\right) = (\prod_{j=1}^{k-1} E_{\beta_j}) \times (\bigcup_{i=1}^{\infty} E_{\beta_k,i}) \times (\prod_{j=k+1}^{n} E_{\beta_j}) \in \hat{\bar{\mathcal{B}}}$. This leads to $\bigcup_{i=1}^{\infty} E_{\beta_k,i} \in \mathcal{F}$. Hence, $\mathcal{F}$ is a σ-algebra on $\prod_{\alpha \in \Lambda_{\beta_k}} X_\alpha$. Since $\hat{\mathcal{B}}_{\beta_k} \supseteq \mathcal{F} \supseteq \hat{\mathcal{C}}_{\beta_k}$, then $\mathcal{F} = \hat{\mathcal{B}}_{\beta_k}$. Therefore, the statement holds for $i = k + 1$.

This completes the induction process and the statement holds for $i = 1, \ldots, n + 1$.

By the above statement for $i = n + 1$ case, we have $\forall E_{\beta_j} \in \hat{\mathcal{B}}_{\beta_j}$, $j = 1, \ldots, n$, $\prod_{j=1}^{n} E_{\beta_j} \in \hat{\bar{\mathcal{B}}}$. Hence, $\hat{\mathcal{C}} \subseteq \hat{\bar{\mathcal{B}}}$. This completes the proof of the claim. □

By Claim 12.24.1 and the definition of $\hat{\mathcal{B}}$, we have $\hat{\mathcal{B}} \subseteq \hat{\bar{\mathcal{B}}}$. Since $\hat{\bar{\mathcal{C}}} \subseteq \hat{\mathcal{C}}$, we have $\hat{\bar{\mathcal{B}}} \subseteq \hat{\mathcal{B}}$. Therefore, $\hat{\mathcal{B}} = \hat{\bar{\mathcal{B}}}$. This further implies that, $\forall E \in \bar{\mathcal{B}}$, $\Psi_{\text{inv}}(E) \in \hat{\bar{\hat{\mathcal{B}}}} = \hat{\bar{\mathcal{B}}} = \hat{\mathcal{B}}$ and, $\forall E \in \hat{\mathcal{B}} = \hat{\bar{\hat{\mathcal{B}}}}$, $\Psi(E) \in \bar{\mathcal{B}}$.

By Proposition 12.10, Ψ is an isomeasure between $\hat{\mathcal{X}}$ and $(\prod_{\alpha \in \Lambda} X_\alpha, \bar{\mathcal{B}}, \Psi_* \hat{\mu})$. $\forall \alpha \in \Lambda$, $\forall E_\alpha \in \operatorname{dom}(\mu_\alpha)$, we have

$$\mathcal{P} \circ (\Psi_* \hat{\mu})\Big(\prod_{\alpha \in \Lambda} E_\alpha\Big) = \Psi_* (\mathcal{P} \circ \hat{\mu})\Big(\prod_{\alpha \in \Lambda} E_\alpha\Big) = \mathcal{P} \circ \hat{\mu}\Big(\Psi_{\text{inv}}\Big(\prod_{\alpha \in \Lambda} E_\alpha\Big)\Big)$$

$$= \mathcal{P} \circ \hat{\mu}\Big(\prod_{\beta \in \Gamma} \prod_{\alpha \in \Lambda_\beta} E_\alpha\Big) = \prod_{\beta \in \Gamma} \mathcal{P} \circ \hat{\mu}_\beta\Big(\prod_{\alpha \in \Lambda_\beta} E_\alpha\Big)$$

$$= \prod_{\beta \in \Gamma} \prod_{\alpha \in \Lambda_\beta} \mathcal{P} \circ \mu_\alpha(E_\alpha) = \prod_{\alpha \in \Lambda} \mathcal{P} \circ \mu_\alpha(E_\alpha) \in [0, \infty) \subset \mathbb{R}$$

and

$$\Psi_* \hat{\mu}\Big(\prod_{\alpha \in \Lambda} E_\alpha\Big) = \hat{\mu}\Big(\Psi_{\text{inv}}\Big(\prod_{\alpha \in \Lambda} E_\alpha\Big)\Big) = \hat{\mu}\Big(\prod_{\beta \in \Gamma} \prod_{\alpha \in \Lambda_\beta} E_\alpha\Big)$$

$$= \prod_{\beta \in \Gamma} \hat{\mu}_\beta\Big(\prod_{\alpha \in \Lambda_\beta} E_\alpha\Big) = \prod_{\beta \in \Gamma} \prod_{\alpha \in \Lambda_\beta} \mu_\alpha(E_\alpha) = \prod_{\alpha \in \Lambda} \mu_\alpha(E_\alpha) \in \mathbb{K}$$

By Proposition 12.22, $\bar{\mu}$ is the unique $\mathbb{K}$-valued measure on $(\prod_{\alpha\in\Lambda} X_\alpha, \bar{\mathcal{B}})$ such that $\bar{\mu}(\prod_{\alpha\in\Lambda} E_\alpha) = \prod_{\alpha\in\Lambda} \mu_\alpha(E_\alpha)$ and $\mathcal{P}\circ\bar{\mu}(\prod_{\alpha\in\Lambda} E_\alpha) = \prod_{\alpha\in\Lambda} \mathcal{P}\circ\mu_\alpha(E_\alpha)$, $\forall E_\alpha \in \operatorname{dom}(\mu_\alpha)$, $\forall \alpha \in \Lambda$. Therefore, $\bar{\mu} = \Psi_* \hat{\mu}$. This implies that Ψ is an isomeasure between $\hat{\mathcal{X}}$ and $\bar{\mathcal{X}}$. This completes the proof of the proposition. □

Proposition 12.25 *Let $\mathcal{X}_i := (X_i, \mathcal{B}_i, \mu_i)$ be a σ-finite measure space, $i = 1, 2$, and $\mathcal{X} := (X_1 \times X_2, \mathcal{B}, \mu) := \mathcal{X}_1 \times \mathcal{X}_2$ be the product measure space. Then, $\forall E \in \mathcal{B}$, $\forall x_1 \in X_1$, $E_{x_1} := \{\bar{x}_2 \in X_2 \mid (x_1, \bar{x}_2) \in E\} \in \mathcal{B}_2$ and, $\forall x_2 \in X_2$, $E_{\langle x_2\rangle} := \{\bar{x}_1 \in X_1 \mid (\bar{x}_1, x_2) \in E\} \in \mathcal{B}_1$.*

Proof Let $\mathcal{E} := \{E \in \mathcal{B} \mid E_{x_1} \in \mathcal{B}_2,\ \forall x_1 \in X_1\} \subseteq \mathcal{B}$. $\forall E \in \mathcal{C} := \{E_1 \times E_2 \subseteq X_1 \times X_2 \mid E_1 \in \mathcal{B}_1,\ E_2 \in \mathcal{B}_2\}$, we have $E \in \mathcal{E}$. Then, $\mathcal{C} \subseteq \mathcal{E}$. We will show that $\mathcal{E}$ is a σ-algebra on $X_1 \times X_2$. Then, by Proposition 12.22, we have $\mathcal{B} \subseteq \mathcal{E}$ and $\mathcal{B} = \mathcal{E}$.

Clearly, $\emptyset \in \mathcal{C} \subseteq \mathcal{E}$ and $X_1 \times X_2 \in \mathcal{C} \subseteq \mathcal{E}$. $\forall E \in \mathcal{E}$, $\forall x_1 \in X_1$, we have $((X_1 \times X_2) \setminus E)_{x_1} = X_2 \setminus E_{x_1} \in \mathcal{B}_2$ since $\mathcal{B}_2$ is a σ-algebra. By the arbitrariness of x_1, we have $(X_1 \times X_2) \setminus E \in \mathcal{E}$. $\forall (E_i)_{i=1}^\infty \subseteq \mathcal{E}$, let $E := \bigcup_{i=1}^\infty E_i \in \mathcal{B}$. $\forall x_1 \in X_1$, we have $E_{x_1} = \bigcup_{i=1}^\infty E_{i\,x_1} \in \mathcal{B}_2$ since $\mathcal{B}_2$ is a σ-algebra. Then, $E \in \mathcal{E}$. Therefore, $\mathcal{E}$ is a σ-algebra and $\mathcal{B} = \mathcal{E}$. By the definition of $\mathcal{E}$, we have $\forall E \in \mathcal{B} = \mathcal{E}$, $\forall x_1 \in X_1$, $E_{x_1} \in \mathcal{B}_2$. By symmetry, $\forall E \in \mathcal{B}$, $\forall x_2 \in X_2$, $E_{\langle x_2\rangle} \in \mathcal{B}_1$. This completes the proof of the proposition. □

Proposition 12.26 *Let $\mathcal{X}_i := (X_i, \mathcal{B}_i, \mu_i)$ be a σ-finite measure space, $i = 1, 2$, and $\mathcal{X} := (X_1 \times X_2, \mathcal{B}, \mu) := \mathcal{X}_1 \times \mathcal{X}_2$ be the product measure space. Then, $\forall E \in \mathcal{B}$, the function $g_E : X_1 \to [0, \infty] \subseteq \mathbb{R}_e$, defined by $g_E(x_1) = \mu_2(E_{x_1})$, $\forall x_1 \in X_1$, is $\mathcal{B}_1$-measurable, where E_{x_1} is as defined in Proposition 12.25. By symmetry, $\forall E \in \mathcal{B}$, the function $h_E : X_2 \to [0, \infty] \subseteq \mathbb{R}_e$, defined by $h_E(x_2) = \mu_1(E_{\langle x_2\rangle})$, $\forall x_2 \in X_2$, is $\mathcal{B}_2$-measurable, where $E_{\langle x_2\rangle}$ is as defined in Proposition 12.25.*

Proof By Proposition 12.25, $\forall x_1 \in X_1$, $E_{x_1} \in \mathcal{B}_2$. Then, $g_E(x_1) \in [0, \infty] \subset \mathbb{R}_e$. Hence, the function g_E is well-defined.

First, consider the special case when $\mathcal{X}_1$ and $\mathcal{X}_2$ are finite. Let $\mathcal{E} := \{E \in \mathcal{B} \mid g_E$ is $\mathcal{B}_1$-measurable$\}$, and $\mathcal{C}$ be the set of measurable rectangles in $X_1 \times X_2$. $\forall E := E_1 \times E_2 \in \mathcal{C}$, $g_E(x_1) = \mu_2(E_2)\chi_{E_1, X_1}(x_1)$, $\forall x_1 \in X_1$. Clearly, g_E is $\mathcal{B}_1$-measurable. Then, $E \in \mathcal{E}$ and $\mathcal{C} \subseteq \mathcal{E}$.

Clearly, $\mathcal{C}$ is a π-system on $X_1 \times X_2$. We will show that $\mathcal{E}$ is a monotone class on $X_1 \times X_2$. Then, by Monotone Class Lemma 12.19, we have $\mathcal{E} = \mathcal{B}$.

Clearly, $\emptyset = \emptyset \times \emptyset \in \mathcal{C} \subseteq \mathcal{E}$ and $X_1 \times X_2 \in \mathcal{C} \subseteq \mathcal{E}$. $\forall E_1, E_2 \in \mathcal{E}$ with $E_1 \subseteq E_2$, then $g_{E_i} : X_1 \to [0, \mu_2(X_2)] \subset \mathbb{R}$ is $\mathcal{B}_1$-measurable, $i = 1, 2$. Let $E := E_2 \setminus E_1$. $\forall x_1 \in X_1$, we have $g_E(x_1) = \mu_2(E_{2x_1} \setminus E_{1x_1}) = \mu_2(E_{2x_1}) - \mu_2(E_{1x_1}) = g_{E_2}(x_1) - g_{E_1}(x_1) \in [0, \mu_2(X_2)] \subset \mathbb{R}$. By Propositions 7.23, 11.39, and 11.38, g_E is $\mathcal{B}_1$-measurable. Then, $E = E_2 \setminus E_1 \in \mathcal{E}$. $\forall (E_i)_{i=1}^\infty \subseteq \mathcal{E}$ with $E_i \subseteq E_{i+1}$, $\forall i \in \mathbb{N}$, then $g_{E_i} : X_1 \to [0, \mu_2(X_2)] \subset \mathbb{R}$ is $\mathcal{B}_1$-measurable, $\forall i \in \mathbb{N}$. Let $E := \bigcup_{i=1}^\infty E_i \in \mathcal{B}$. $\forall x_1 \in X_1$, $g_E(x_1) = \mu_2(\bigcup_{i=1}^\infty E_{i\,x_1}) = \lim_{i\in\mathbb{N}} \mu_2(E_{i\,x_1}) = \lim_{i\in\mathbb{N}} g_{E_i}(x_1)$, where the first equality follows from the definition of g_E, the second equality follows from Proposition 11.7, and the third equality follows from the definition of g_{E_i}'s. Hence, by Proposition 11.48, g_E is $\mathcal{B}_1$-measurable. Then, $E = \bigcup_{i=1}^{\infty} E_i \in \mathcal{E}$. This shows that $\mathcal{E}$ is a monotone class on $X_1 \times X_2$.

Then, $\mathcal{E} = \mathcal{B}$ and, $\forall E \in \mathcal{B}$, g_E is $\mathcal{B}_1$-measurable. By symmetry, $\forall E \in \mathcal{B}$, h_E is $\mathcal{B}_2$-measurable. The proposition holds for the special case.

Next, consider the general case when $\mathcal{X}_1$ and $\mathcal{X}_2$ are σ-finite. Since $\mathcal{X}_1$ is σ-finite, then $\exists (X_{1,n})_{n=1}^{\infty} \subseteq \mathcal{B}_1$ such that $X_1 = \bigcup_{n=1}^{\infty} X_{1,n}$ and $\mu_1(X_{1,n}) < \infty$, $\forall n \in \mathbb{N}$. Without loss of generality, we may assume that $X_{1,n} \subseteq X_{1,n+1}$, $\forall n \in \mathbb{N}$. Since $\mathcal{X}_2$ is σ-finite, then $\exists (X_{2,m})_{m=1}^{\infty} \subseteq \mathcal{B}_2$ such that $X_2 = \bigcup_{m=1}^{\infty} X_{2,m}$ and $\mu_2(X_{2,m}) < \infty$, $\forall m \in \mathbb{N}$. Without loss of generality, we may assume that $X_{2,m} \subseteq X_{2,m+1}$, $\forall m \in \mathbb{N}$. $\forall n, m \in \mathbb{N}$, let $\mathcal{X}_{1,n} := (X_{1,n}, \mathcal{B}_{1,n}, \mu_{1,n})$ be the finite measure subspace of $\mathcal{X}_1$, $\mathcal{X}_{2,m} := (X_{2,m}, \mathcal{B}_{2,m}, \mu_{2,m})$ be the finite measure subspace of $\mathcal{X}_2$, $(X_{1,n} \times X_{2,m}, \bar{\mathcal{B}}_{n,m}, \lambda_{n,m}) := \mathcal{X}_{1,n} \times \mathcal{X}_{2,m}$, which is the finite measure subspace of $\mathcal{X}$ by Proposition 12.23. $\forall E \in \mathcal{B}$, $\forall n, m \in \mathbb{N}$, let $E_{n,m} := E \cap (X_{1,n} \times X_{2,m}) \in \bar{\mathcal{B}}_{n,m}$. By the special case, the function $g_{n,m} : X_{1,n} \to [0, \infty) \subset \mathbb{R}$, defined by $g_{n,m}(x_1) = \mu_{2,m}((E_{n,m})_{x_1} \cap X_{2,m})$, $\forall x_1 \in X_{1,n}$, is $\mathcal{B}_{1,n}$-measurable. Define $\bar{g}_{n,m} : X_1 \to [0, \infty) \subset \mathbb{R}$ by $\bar{g}_{n,m}(x_1) = \mu_2((E_{n,m})_{x_1})$, $\forall x_1 \in X_1$. Then, $\bar{g}_{n,m}(x_1) = \begin{cases} g_{n,m}(x_1) & \forall x_1 \in X_{1,n} \\ 0 & \forall x_1 \in X_1 \setminus X_{1,n} \end{cases}$, $\forall x_1 \in X_1$. By Proposition 11.41, $\bar{g}_{n,m}$ is $\mathcal{B}_1$-measurable. Note that $g_E(x_1) = \mu_2(E_{x_1}) = \mu_2(\bigcup_{n=1}^{\infty} (E_{n,n})_{x_1}) = \lim_{n \in \mathbb{N}} \mu_2((E_{n,n})_{x_1}) = \lim_{n \in \mathbb{N}} \bar{g}_{n,n}(x_1)$, $\forall x_1 \in X_1$, where the second equality follows from the fact that $E = \bigcup_{n=1}^{\infty} E_{n,n}$ and the third equality follows from Proposition 11.7. By Proposition 11.48 and the fact that $[0, \infty] \subset \mathbb{R}_e$ is metrizable, g_E is $\mathcal{B}_1$-measurable. By symmetry, $\forall E \in \mathcal{B}$, h_E is $\mathcal{B}_2$-measurable.

This completes the proof of the proposition. □

Lemma 12.27 *Let $\mathcal{X}_i := (X_i, \mathcal{B}_i, \mu_i)$ be a σ-finite measure space, $i = 1, 2$, and $\mathcal{X} := (X_1 \times X_2, \mathcal{B}, \mu) := \mathcal{X}_1 \times \mathcal{X}_2$ be the product measure space. Then, $\forall E \in \mathcal{B}$, we have $\mu(E) = \int_{X_1} g_E \, d\mu_1 = \int_{X_2} h_E \, d\mu_2$, where the function $g_E : X_1 \to [0, \infty] \subseteq \mathbb{R}_e$, as defined in Proposition 12.26, is $\mathcal{B}_1$-measurable and the function $h_E : X_2 \to [0, \infty] \subseteq \mathbb{R}_e$, as defined in Proposition 12.26, is $\mathcal{B}_2$-measurable.*

Proof First, consider the special case when $\mathcal{X}_1$ and $\mathcal{X}_2$ are finite. Then $\mathcal{X}$ is finite. $\forall E \in \mathcal{B}$, $0 \le \mu(E) \le \mu(X_1 \times X_2) = \mu_1(X_1)\mu_2(X_2) < \infty$. By Proposition 12.26, the function $g_E : X_1 \to [0, \mu_2(X_2)] \subset \mathbb{R}$ is $\mathcal{B}_1$-measurable. Let $\mathcal{E} := \left\{ E \in \mathcal{B} \,\middle|\, \mu(E) = \int_{X_1} g_E \, d\mu_1 \right\} \subseteq \mathcal{B}$. Let $\mathcal{C}$ be the set of measurable rectangles in $X_1 \times X_2$, which is clearly a π-system. $\forall E := E_1 \times E_2 \in \mathcal{C}$, $g_E = \mu_2(E_2)\chi_{E_1, X_1}$. Then, by Proposition 11.75, $\mu(E) = \mu_1(E_1)\mu_2(E_2) = \int_{X_1} g_E \, d\mu_1$. This shows that $E \in \mathcal{E}$. Then $\mathcal{C} \subseteq \mathcal{E}$. We will show that $\mathcal{E}$ is a monotone class on $X_1 \times X_2$. By Monotone Class Lemma 12.19, we have $\mathcal{E} = \mathcal{B}$.

Clearly, $\emptyset \in \mathcal{C} \subseteq \mathcal{E}$ and $X_1 \times X_2 \in \mathcal{C} \subseteq \mathcal{E}$. $\forall E_1, E_2 \in \mathcal{E}$ with $E_1 \subseteq E_2$, let $E := E_2 \setminus E_1 \in \mathcal{B}$. By Proposition 12.26, g_{E_1}, g_{E_2}, and g_E are $\mathcal{B}_1$-measurable. As shown in the proof of Proposition 12.26, $g_E = g_{E_2} - g_{E_1}$. Then, $\mu(E) = \mu(E_2) - \mu(E_1) = \int_{X_1} g_{E_2} \, d\mu_1 - \int_{X_1} g_{E_1} \, d\mu_1 = \int_{X_1} g_E \, d\mu_1$, where the first equality follows from the fact that μ is a measure, the second equality follows from the definition of $\mathcal{E}$, the last equality follows from Proposition 11.83, and all equalities follow from the fact all quantities involved in the equalities are finite. This shows that $E = E_2 \setminus E_1 \in \mathcal{E}$.

$\forall (E_i)_{i=1}^\infty \subseteq \mathcal{E}$ with $E_i \subseteq E_{i+1}$, $\forall i \in \mathbb{N}$, let $E := \bigcup_{i=1}^\infty E_i \in \mathcal{B}$. $\forall i \in \mathbb{N}$, by Proposition 12.26, g_{E_i} is $\mathcal{B}_1$-measurable. $\forall x_1 \in X_1$, we have $g_{E_i}(x_1) \leq g_{E_{i+1}}(x_1)$ since $E_i \subseteq E_{i+1}$. By Proposition 12.26, g_E is $\mathcal{B}_1$-measurable. As shown in the proof of Proposition 12.26, $g_E(x_1) = \lim_{i\in\mathbb{N}} g_{E_i}(x_1)$, $\forall x_1 \in X_1$. This implies that $\mu(E) = \lim_{i\in\mathbb{N}} \mu(E_i) = \lim_{i\in\mathbb{N}} \int_{X_1} g_{E_i}\, d\mu_1 = \int_{X_1} g_E\, d\mu_1$, where the first equality follows from Proposition 11.7, the second equality follows from the definition of $\mathcal{E}$, and the last equality follows from Monotone Convergence Theorem 11.81. Hence, $E = \bigcup_{i=1}^\infty E_i \in \mathcal{E}$.

Hence, $\mathcal{E}$ is a monotone class on $X_1 \times X_2$ and $\mathcal{E} = \mathcal{B}$. Therefore, $\mu(E) = \int_{X_1} g_E\, d\mu_1$, $\forall E \in \mathcal{B}$. By symmetry, $\mu(E) = \int_{X_2} h_E\, d\mu_2$, $\forall E \in \mathcal{B}$. This completes the proof for the special case.

Next, consider the general case when $\mathcal{X}_1$ and $\mathcal{X}_2$ are σ-finite. Since $\mathcal{X}_1$ is σ-finite, then $\exists (X_{1,n})_{n=1}^\infty \subseteq \mathcal{B}_1$ such that $X_1 = \bigcup_{n=1}^\infty X_{1,n}$ and $\mu_1(X_{1,n}) < \infty$, $\forall n \in \mathbb{N}$. Without loss of generality, we may assume that $X_{1,n} \subseteq X_{1,n+1}$, $\forall n \in \mathbb{N}$. Since $\mathcal{X}_2$ is σ-finite, then $\exists (X_{2,m})_{m=1}^\infty \subseteq \mathcal{B}_2$ such that $X_2 = \bigcup_{m=1}^\infty X_{2,m}$ and $\mu_2(X_{2,m}) < \infty$, $\forall m \in \mathbb{N}$. Without loss of generality, we may assume that $X_{2,m} \subseteq X_{2,m+1}$, $\forall m \in \mathbb{N}$. $\forall n, m \in \mathbb{N}$, let $\mathcal{X}_{1,n} := (X_{1,n}, \mathcal{B}_{1,n}, \mu_{1,n})$ be the finite measure subspace of $\mathcal{X}_1$, $\mathcal{X}_{2,m} := (X_{2,m}, \mathcal{B}_{2,m}, \mu_{2,m})$ be the finite measure subspace of $\mathcal{X}_2$, $(X_{1,n} \times X_{2,m}, \bar{\mathcal{B}}_{n,m}, \lambda_{n,m}) := \mathcal{X}_{1,n} \times \mathcal{X}_{2,m}$, which is the finite measure subspace of $\mathcal{X}$ by Proposition 12.23. $\forall E \in \mathcal{B}$, $\forall n, m \in \mathbb{N}$, let $E_{n,m} := E \cap (X_{1,n} \times X_{2,m}) \in \bar{\mathcal{B}}_{n,m}$. By the special case, the function $g_{n,m} : X_{1,n} \to [0, \infty) \subset \mathbb{R}$, defined by $g_{n,m}(x_1) = \mu_{2,m}((E_{n,m})_{x_1} \cap X_{2,m})$, $\forall x_1 \in X_{1,n}$, is $\mathcal{B}_{1,n}$-measurable and $\mu(E_{n,m}) = \lambda_{n,m}(E_{n,m}) = \int_{X_{1,n}} g_{n,m}\, d\mu_{1,n}$. Define $\bar{g}_{n,m} : X_1 \to [0, \infty) \subset \mathbb{R}$ by $\bar{g}_{n,m}(x_1) = \mu_2((E_{n,m})_{x_1})$, $\forall x_1 \in X_1$. Then, $\bar{g}_{n,m}(x_1) = \begin{cases} g_{n,m}(x_1) & \forall x_1 \in X_{1,n} \\ 0 & \forall x_1 \in X_1 \setminus X_{1,n} \end{cases}$, $\forall x_1 \in X_1$. By Proposition 11.41, $\bar{g}_{n,m}$ is $\mathcal{B}_1$-measurable. By Definition 11.79, $\mu(E_{n,m}) = \int_{X_1} \bar{g}_{n,m}\, d\mu_1$. By Proposition 12.26 and its proof, we have $g_E(x_1) = \lim_{n\in\mathbb{N}} \bar{g}_{n,n}(x_1)$, $\forall x_1 \in X_1$, and g_E is $\mathcal{B}_1$-measurable. Clearly, $\bar{g}_{n,n}(x_1) \leq \bar{g}_{n+1,n+1}(x_1)$, $\forall x_1 \in X_1$, $\forall n \in \mathbb{N}$, since $E_{n,n} \subseteq E_{n+1,n+1}$. Then, $\mu(E) = \lim_{n\in\mathbb{N}} \mu(E_{n,n}) = \lim_{n\in\mathbb{N}} \int_{X_1} \bar{g}_{n,n}\, d\mu_1 = \int_{X_1} g_E\, d\mu_1$, where the first equality follows from Proposition 11.7 and the last equality follows from Monotone Convergence Theorem 11.81.

By symmetry, $\mu(E) = \int_{X_2} h_E\, d\mu_2$, $\forall E \in \mathcal{B}$.

This completes the proof of the proposition. □

Proposition 12.28 *Let $\mathcal{X}_i := (X_i, \mathcal{B}_i, \mu_i)$ be a σ-finite measure space, $i = 1, 2$, $\mathcal{X} := (X_1 \times X_2, \mathcal{B}, \mu) := \mathcal{X}_1 \times \mathcal{X}_2$ be the product measure space, $\mathcal{Y} := (Y, \mathcal{O})$ be a topological space, and $f : X_1 \times X_2 \to \mathcal{Y}$ be $\mathcal{B}$-measurable. Then:*

(i) *$\forall x_1 \in X_1$, $f_{x_1} : X_2 \to \mathcal{Y}$, defined by $f_{x_1}(x_2) = f(x_1, x_2)$, $\forall x_2 \in X_2$, is $\mathcal{B}_2$-measurable.*

(ii) *$\forall x_2 \in X_2$, $f_{\langle x_2 \rangle} : X_1 \to \mathcal{Y}$, defined by $f_{\langle x_2 \rangle}(x_1) = f(x_1, x_2)$, $\forall x_1 \in X_1$, is $\mathcal{B}_1$-measurable.*

Proof (i) $\forall$ open set $O \in \mathcal{O}$, $f_{\text{inv}}(O) \in \mathcal{B}$ since f is $\mathcal{B}$-measurable. $\forall x_1 \in X_1$, by Proposition 12.25, $(f_{\text{inv}}(O))_{x_1} := \{x_2 \in X_2 \,|\, (x_1, x_2) \in f_{\text{inv}}(O)\} \in \mathcal{B}_2$. Clearly, $f_{x_1\text{inv}}(O) = (f_{\text{inv}}(O))_{x_1} \in \mathcal{B}_2$. Then, f_{x_1} is $\mathcal{B}_2$-measurable.

(ii) By symmetry, (ii) holds. This completes the proof of the proposition. □

Theorem 12.29 (Tonelli) *Let $\mathcal{X}_i := (X_i, \mathcal{B}_i, \mu_i)$ be a σ-finite measure space, $i = 1, 2$, $\mathcal{X} := (X_1 \times X_2, \mathcal{B}, \mu) := \mathcal{X}_1 \times \mathcal{X}_2$ be the product measure space, and $f : X_1 \times X_2 \to [0, \infty] \subset \mathbb{R}_e$ be $\mathcal{B}$-measurable. Define $f_{x_1} : X_2 \to [0, \infty] \subset \mathbb{R}_e$ and $f_{\langle x_2 \rangle} : X_1 \to [0, \infty] \subset \mathbb{R}_e$ as in Proposition 12.28. Then, the following statements hold:*

(i) *$\int_{X_1 \times X_2} f \, d\mu = \int_{X_1} p \, d\mu_1$, where $p : X_1 \to [0, \infty] \subset \mathbb{R}_e$, defined by $p(x_1) = \int_{X_2} f_{x_1} \, d\mu_2$, $\forall x_1 \in X_1$, is $\mathcal{B}_1$-measurable.*

(ii) *$\int_{X_1 \times X_2} f \, d\mu = \int_{X_2} q \, d\mu_2$, where $q : X_2 \to [0, \infty] \subset \mathbb{R}_e$, defined by $q(x_2) = \int_{X_1} f_{\langle x_2 \rangle} \, d\mu_1$, $\forall x_2 \in X_2$, is $\mathcal{B}_2$-measurable.*

Proof (i) By Proposition 11.67, there exists a sequence of simple functions $(\varphi_n)_{n=1}^{\infty}$, $\varphi_n : \mathcal{X} \to [0, \infty) \subset \mathbb{R}$, $\forall n \in \mathbb{N}$, such that $0 \le \varphi_n(x_1, x_2) \le \varphi_{n+1}(x_1, x_2) \le f(x_1, x_2)$, $\forall (x_1, x_2) \in X_1 \times X_2$, $\forall n \in \mathbb{N}$, and $\lim_{n \in \mathbb{N}} \varphi_n(x_1, x_2) = f(x_1, x_2)$, $\forall (x_1, x_2) \in X_1 \times X_2$. By Proposition 12.28 and Definition 11.79, p is well-defined. Define $p_n : X_1 \to [0, \infty] \subset \mathbb{R}_e$ by $p_n(x_1) = \int_{X_2} \varphi_n(x_1, x_2) \cdot d\mu_2(x_2)$, $\forall x_1 \in X_1$, $\forall n \in \mathbb{N}$. By Proposition 12.28, p_n's are well-defined. By Proposition 11.83, $p_n(x_1) \le p_{n+1}(x_1) \le p(x_1)$, $\forall x_1 \in X_1$, $\forall n \in \mathbb{N}$. By Monotone Convergence Theorem 11.81, $p(x_1) = \lim_{n \in \mathbb{N}} p_n(x_1)$, $\forall x_1 \in X_1$. Fix any $n \in \mathbb{N}$. Let φ_n admit the canonical representation $\varphi_n = \sum_{i=1}^{m_n} a_i \chi_{E_i, X_1 \times X_2}$, where $m_n \in \mathbb{Z}_+$, $a_1, \ldots, a_{m_n} \in (0, \infty) \subset \mathbb{R}$ are distinct, $E_1, \ldots, E_{m_n} \in \mathcal{B}$ are nonempty, pairwise disjoint, and $\mu(E_i) < \infty$, $\forall i \in \{1, \ldots, m_n\}$. Fix any $i \in \{1, \ldots, m_n\}$, and define $g_i : X_1 \to [0, \infty] \subset \mathbb{R}_e$ by $g_i(x_1) = \mu_2((E_i)_{x_1})$, $\forall x_1 \in X_1$, where $(E_i)_{x_1} := \{x_2 \in X_2 \,|\, (x_1, x_2) \in E_i\}$. By Proposition 12.26, g_i is $\mathcal{B}_1$-measurable. By Lemma 12.27, $\mu(E_i) = \int_{X_1} g_i \, d\mu_1$. By Proposition 11.75, $\int_{X_1 \times X_2} \varphi_n \, d\mu = \sum_{i=1}^{m_n} a_i \mu(E_i) = \sum_{i=1}^{m_n} a_i \int_{X_1} g_i \, d\mu_1 = \int_{X_1} (\sum_{i=1}^{m_n} a_i g_i) \, d\mu_1 =: \int_{X_1} \bar{p}_n \, d\mu_1$, where the third equality follows from Proposition 11.83. By Propositions 3.89, 11.38, and 11.39, $\bar{p}_n : X_1 \to [0, \infty] \subset \mathbb{R}_e$ is $\mathcal{B}_1$-measurable. Note that $p_n(x_1) = \int_{X_2} \varphi_n(x_1, x_2) \, d\mu_2(x_2) = \int_{X_2} (\sum_{i=1}^{m_n} a_i \chi_{E_i, X_1 \times X_2}(x_1, x_2)) \, d\mu_2(x_2) = \sum_{i=1}^{m_n} a_i \int_{X_2} \chi_{(E_i)_{x_1}, X_2} \, d\mu_2 = \sum_{i=1}^{m_n} a_i \mu_2(E_{i\,x_1}) = \sum_{i=1}^{m_n} a_i g_i(x_1) = \bar{p}_n(x_1)$, $\forall x_1 \in X_1$, where the first equality follows from the definition of p_n, the second equality follows from the canonical representation of φ_n, the third equality follows from Proposition 11.83, the fourth equality follows from the fact that $\mathcal{X}_2$ is σ-finite, Propositions 11.75 and 11.7, and the Monotone Convergence Theorem 11.81, the fifth equality follows from the definition of g_i's, and the last equality follows from the definition of $\bar{p}_n$. Then, p_n is $\mathcal{B}_1$-measurable. By Proposition 11.48 and the fact that $[0, \infty] \subset \mathbb{R}_e$ is metrizable, p is $\mathcal{B}_1$-measurable. Then, $\int_{X_1 \times X_2} \varphi_n \, d\mu = \int_{X_1} \bar{p}_n \, d\mu_1 = \int_{X_1} p_n \, d\mu_1 \in [0, \infty] \subset \mathbb{R}_e$. By Monotone Convergence Theorem 11.81, $\int_{X_1\times X_2} f\,d\mu = \lim_{n\in\mathbb{N}} \int_{X_1\times X_2} \varphi_n\,d\mu = \lim_{n\in\mathbb{N}} \int_{X_1} p_n\,d\mu_1 = \int_{X_1} p\,d\mu_1$.

Hence, (i) holds. By symmetry, (ii) holds. This completes the proof of the theorem. □

Theorem 12.30 (Fubini) *Let $\mathcal{X}_i := (X_i, \mathcal{B}_i, \mu_i)$ be a σ-finite measure space, $i = 1, 2$, $\mathcal{X} := (X_1 \times X_2, \mathcal{B}, \mu) := \mathcal{X}_1 \times \mathcal{X}_2$ be the product measure space, $\mathcal{Y}$ be a separable Banach space, and $f : X_1 \times X_2 \to \mathcal{Y}$ be absolutely integrable over $\mathcal{X}$. Define $f_{x_1} : X_2 \to \mathcal{Y}$ and $f_{\langle x_2\rangle} : X_1 \to \mathcal{Y}$ as in Proposition 12.28. Then, the following statements hold:*

(i) *There exists $\check{p} : X_1 \to \mathcal{Y}$ and exists $\check{U} \in \mathcal{B}_1$ such that $\mu_1(X_1 \setminus \check{U}) = 0$; $\forall x_1 \in \check{U}$, f_{x_1} is absolutely integrable over $\mathcal{X}_2$ and $\check{p}(x_1) = \int_{X_2} f_{x_1}\,d\mu_2 \in \mathcal{Y}$; $\forall x_1 \in X_1 \setminus \check{U}$, $\check{p}(x_1) = \vartheta_{\mathcal{Y}}$; $\check{p}$ is absolutely integrable over $\mathcal{X}_1$; and $\int_{X_1\times X_2} f\,d\mu = \int_{X_1} \check{p}\,d\mu_1 \in \mathcal{Y}$.*

(ii) *There exists $\check{q} : X_2 \to \mathcal{Y}$ and exists $\check{V} \in \mathcal{B}_2$ such that $\mu_2(X_2 \setminus \check{V}) = 0$; $\forall x_2 \in \check{V}$, $f_{\langle x_2\rangle}$ is absolutely integrable over $\mathcal{X}_1$ and $\check{q}(x_2) = \int_{X_1} f_{\langle x_2\rangle}\,d\mu_1 \in \mathcal{Y}$; $\forall x_2 \in X_2 \setminus \check{V}$, $\check{q}(x_2) = \vartheta_{\mathcal{Y}}$; $\check{q}$ is absolutely integrable over $\mathcal{X}_2$; and $\int_{X_1\times X_2} f\,d\mu = \int_{X_2} \check{q}\,d\mu_2 \in \mathcal{Y}$.*

(iii) *$\forall p : X_1 \to \mathcal{Y}$ satisfying: p is $\mathcal{B}_1$-measurable; and $\exists U \in \mathcal{B}_1$ with $\mu_1(X_1 \setminus U) = 0$ such that, $\forall x_1 \in U$, f_{x_1} is absolutely integrable over $\mathcal{X}_2$ and $p(x_1) = \int_{X_2} f_{x_1}\,d\mu_2 \in \mathcal{Y}$, we have p is absolutely integrable over $\mathcal{X}_1$ and $\int_{X_1\times X_2} f\,d\mu = \int_{X_1} p\,d\mu_1 \in \mathcal{Y}$.*

(iv) *$\forall q : X_2 \to \mathcal{Y}$ satisfying: q is $\mathcal{B}_2$-measurable; and $\exists V \in \mathcal{B}_2$ with $\mu_2(X_2 \setminus V) = 0$ such that, $\forall x_2 \in V$, $f_{\langle x_2\rangle}$ is absolutely integrable over $\mathcal{X}_1$ and $q(x_2) = \int_{X_1} f_{\langle x_2\rangle}\,d\mu_1 \in \mathcal{Y}$, we have q is absolutely integrable over $\mathcal{X}_2$ and $\int_{X_1\times X_2} f\,d\mu = \int_{X_2} q\,d\mu_2 \in \mathcal{Y}$.*

Proof (i) Since f is absolutely integrable over $\mathcal{X}$, then $\mathcal{P} \circ f$ is $\mathcal{B}$-measurable and $\int_{X_1\times X_2} \mathcal{P} \circ f\,d\mu < \infty$. By Tonelli Theorem 12.29, $g : X_1 \to [0, \infty] \subset \mathbb{R}_e$, defined by $g(x_1) = \int_{X_2} \mathcal{P} \circ f(x_1, x_2)\,d\mu_2(x_2)$, $\forall x_1 \in X_1$, is $\mathcal{B}_1$-measurable, and $\int_{X_1\times X_2} \mathcal{P} \circ f\,d\mu = \int_{X_1} g\,d\mu_1$. By Proposition 11.82, the function $\tilde{g} : X_1 \to [0, \infty) \subset \mathbb{R}$, defined by $\tilde{g}(x_1) = \begin{cases} g(x_1) & g(x_1) < \infty \\ 0 & g(x_1) = \infty \end{cases}$, $\forall x_1 \in X_1$, is $\mathcal{B}_1$-measurable, $g = \tilde{g}$ a.e. in $\mathcal{X}_1$, and $\int_{X_1\times X_2} \mathcal{P} \circ f\,d\mu = \int_{X_1} \tilde{g}\,d\mu_1 < \infty$. Let $U := \{x_1 \in X_1 \mid g(x_1) < \infty\}$. Then, $U \in \mathcal{B}_1$ and $\mu_1(X_1 \setminus U) = 0$. $\forall x_1 \in U$, we have $g(x_1) = \int_{X_2} \mathcal{P} \circ f(x_1, x_2)\,d\mu_2(x_2) = \int_{X_2} \mathcal{P} \circ f_{x_1}\,d\mu_2 < \infty$. Then, f_{x_1} is absolutely integrable over $\mathcal{X}_2$. By Propositions 12.28 and 11.92, $\int_{X_2} f_{x_1}\,d\mu_2 \in \mathcal{Y}$, $\forall x_1 \in U$.

Define $\tilde{p} : X_1 \to \mathcal{Y}$ by $\tilde{p}(x_1) = \begin{cases} \int_{X_2} f_{x_1}\,d\mu_2 & x_1 \in U \\ \vartheta_{\mathcal{Y}} & x_1 \in X_1 \setminus U \end{cases}$, $\forall x_1 \in X_1$.

By Proposition 11.66, $\exists$ a sequence of simple functions $(\varphi_n)_{n=1}^{\infty}$, $\varphi_n : \mathcal{X} \to \mathcal{Y}$, $\forall n \in \mathbb{N}$, such that $\|\varphi_n(x_1, x_2)\| \le \|f(x_1, x_2)\|$, $\forall (x_1, x_2) \in X_1 \times X_2$, $\forall n \in \mathbb{N}$, and $\lim_{n\in\mathbb{N}} \varphi_n = f$ a.e. in $\mathcal{X}$. Let $A := \{(x_1, x_2) \in X_1 \times$

$X_2 \,|\, (\varphi_n(x_1, x_2))_{n=1}^{\infty}$ do not converge to $f(x_1, x_2)\}$. Then, $A \in \mathcal{B}$ and $\mu(A) = 0$. Define $g_A : X_1 \to [0, \infty] \subset \mathbb{R}_e$ by $g_A(x_1) = \mu_2(A_{x_1})$, $\forall x_1 \in X_1$, where $A_{x_1} := \{\bar{x}_2 \in X_2 \,|\, (x_1, \bar{x}_2) \in A\}$ is as defined in Proposition 12.25, and $\hat{g}_A : X_1 \to [0, \infty) \subset \mathbb{R}$ by $\hat{g}_A(x_1) = \begin{cases} g_A(x_1) & g_A(x_1) < \infty \\ 0 & g_A(x_1) = \infty \end{cases}$, $\forall x_1 \in X_1$. Then, by Lemma 12.27 and Proposition 11.82, g_A and $\hat{g}_A$ are well-defined and $\mathcal{B}_1$-measurable, $g_A = \hat{g}_A$ a.e. in $\mathcal{X}_1$, and $\mu(A) = \int_{X_1} g_A \, d\mu_1 = \int_{X_1} \hat{g}_A \, d\mu_1 = 0$. Since $\hat{g}_A(x_1) \geq 0$, $\forall x_1 \in X_1$, then, by Proposition 11.96, we have $\hat{g}_A = 0$ a.e. in $\mathcal{X}_1$. Then, by Lemma 11.44 and the fact that $[0, \infty] \subset \mathbb{R}_e$ is second countable and metrizable, $g_A = 0$ a.e. in $\mathcal{X}_1$. Let $A_1 := \{x_1 \in X_1 \,|\, g_A(x_1) \neq 0\}$. Then, $A_1 \in \mathcal{B}_1$ and $\mu_1(A_1) = 0$. Fix any $x_1 \in U \setminus A_1$. $g_A(x_1) = 0$ implies that $\mu_2(A_{x_1}) = 0$, which further implies that $\lim_{n\in\mathbb{N}} \varphi_n(x_1, x_2) = f(x_1, x_2)$ a.e. $x_2 \in \mathcal{X}_2$. Then, by Lebesgue Dominated Convergence Theorem 11.91, $\tilde{p}(x_1) = \lim_{n\in\mathbb{N}} \int_{X_2} \varphi_n(x_1, x_2) \, d\mu_2(x_2) \in \mathcal{Y}$, $\forall x_1 \in U \setminus A_1$.

Fix any $n \in \mathbb{N}$. Let φ_n admit the canonical representation $\varphi_n = \sum_{i=1}^{m} y_i \chi_{E_i, X_1 \times X_2}$, where $m \in \mathbb{Z}_+$, $y_1, \ldots, y_m \in \mathcal{Y}$ are distinct and none equals to $\vartheta_{\mathcal{Y}}$, $E_1, \ldots, E_m \in \mathcal{B}$ are nonempty, pairwise disjoint, and $\mu(E_i) < \infty$, $i = 1, \ldots, m$. Fix any $i \in \{1, \ldots, m\}$. Define $g_i : X_1 \to [0, \infty] \subset \mathbb{R}_e$ by $g_i(x_1) = \mu_2(E_{i\,x_1})$, $\forall x_1 \in X_1$, where $E_{i\,x_1} := \{\bar{x}_2 \in X_2 \,|\, (x_1, \bar{x}_2) \in E_i\}$. By Lemma 12.27, g_i is $\mathcal{B}_1$-measurable and $\mu(E_i) = \int_{X_1} g_i \, d\mu_1$. Define $\hat{g}_i : X_1 \to [0, \infty) \subset \mathbb{R}$ by $\hat{g}_i(x_1) = \begin{cases} g_i(x_1) & g_i(x_1) < \infty \\ 0 & g_i(x_1) = \infty \end{cases}$, $\forall x_1 \in X_1$. By Proposition 11.82, $\hat{g}_i$ is $\mathcal{B}_1$-measurable, $g_i = \hat{g}_i$ a.e. in $\mathcal{X}_1$, and $\mu(E_i) = \int_{X_1} \hat{g}_i \, d\mu_1$. Then, $\int_{X_1 \times X_2} \varphi_n \, d\mu = \sum_{i=1}^{m} y_i \mu(E_i) = \sum_{i=1}^{m} y_i \int_{X_1} \hat{g}_i \, d\mu_1 = \int_{X_1} (\sum_{i=1}^{m} y_i \hat{g}_i) \, d\mu_1 =: \int_{X_1} \bar{p}_n \, d\mu_1$, where the first equality follows from Proposition 11.75 and the third equality follows from Proposition 11.92. By Propositions 7.23, 11.38, and 11.39, $\bar{p}_n : X_1 \to \mathcal{Y}$ is $\mathcal{B}_1$-measurable and, by Proposition 11.92, $\bar{p}_n$ is absolutely integrable over $\mathcal{X}_1$. Let $\bar{A}_n := \bigcup_{i=1}^{m} \{x_1 \in X_1 \,|\, g_i(x_1) = \infty\}$. Then, $\bar{A}_n \in \mathcal{B}_1$ and $\mu_1(\bar{A}_n) = 0$. $\forall x_1 \in X_1 \setminus \bar{A}_n$, $\bar{p}_n(x_1) = \sum_{i=1}^{m} y_i \hat{g}_i(x_1) = \sum_{i=1}^{m} y_i g_i(x_1) = \sum_{i=1}^{m} y_i \mu_2(E_{i\,x_1}) = \sum_{i=1}^{m} y_i \int_{X_2} \chi_{E_{i\,x_1}, X_2} \, d\mu_2 = \int_{X_2} (\sum_{i=1}^{m} y_i \chi_{E_{i\,x_1}, X_2}) \, d\mu_2 = \int_{X_2} \varphi_n(x_1, x_2) \cdot d\mu_2(x_2) \in \mathcal{Y}$, where the first equality follows from the definition of $\bar{p}_n$, the second equality follows from the definitions of $\hat{g}_i$'s, the third equality follows from the definitions of g_i's, the fourth equality follows from Proposition 11.75, and the fifth equality follows from Proposition 11.92. Let $A_2 := \bigcup_{n=1}^{\infty} \bar{A}_n$. Then, $A_2 \in \mathcal{B}_1$ and $\mu_1(A_2) = 0$. $\forall x_1 \in U \setminus (A_1 \cup A_2) =: \check{U}$, we have $\tilde{p}(x_1) = \lim_{n\in\mathbb{N}} \int_{X_2} \varphi_n(x_1, x_2) \, d\mu_2(x_2) = \lim_{n\in\mathbb{N}} \bar{p}_n(x_1) \in \mathcal{Y}$. Clearly, $\check{U} \in \mathcal{B}_1$ and $0 \leq \mu_1(X_1 \setminus \check{U}) = \mu_1((X_1 \setminus U) \cup A_1 \cup A_2) \leq \mu_1(X_1 \setminus U) + \mu_1(A_1) + \mu_1(A_2) = 0$. Fix any $n \in \mathbb{N}$. Define $\check{p}_n : X_1 \to \mathcal{Y}$ by $\check{p}_n(x_1) = \begin{cases} \bar{p}_n(x_1) & x_1 \in \check{U} \\ \vartheta_{\mathcal{Y}} & x_1 \in X_1 \setminus \check{U} \end{cases}$, $\forall x_1 \in X_1$. By Proposition 11.41, $\check{p}_n$ is $\mathcal{B}_1$-measurable. By Lemma 11.43, $\check{p}_n = \bar{p}_n$ a.e. in $\mathcal{X}_1$. Define $\check{p} : X_1 \to \mathcal{Y}$ by $\check{p}(x_1) = \begin{cases} \tilde{p}(x_1) & x_1 \in \check{U} \\ \vartheta_{\mathcal{Y}} & x_1 \in X_1 \setminus \check{U} \end{cases}$, $\forall x_1 \in X_1$. Then, $\check{p}(x_1) = \lim_{n\in\mathbb{N}} \check{p}_n(x_1)$, $\forall x_1 \in X_1$. By Proposition 11.48,

$\check{p}$ is $\mathcal{B}_1$-measurable. $\forall x_1 \in \check{U}$, $\check{p}(x_1) = \tilde{p}(x_1) = \int_{X_2} f_{x_1} \, d\mu_2$ and f_{x_1} is absolutely integrable over $\mathcal{X}_2$. Note that $\int_{X_1 \times X_2} f \, d\mu = \lim_{n \in \mathbb{N}} \int_{X_1 \times X_2} \varphi_n \, d\mu = \lim_{n \in \mathbb{N}} \int_{X_1} \bar{p}_n \, d\mu_1 = \lim_{n \in \mathbb{N}} \int_{X_1} \check{p}_n \, d\mu_1$, where the first equality follows from Lebesgue Dominated Convergence Theorem 11.91, the second equality follows from the definition of $\bar{p}_n$, and the third equality follows from Proposition 11.92. $\forall n \in \mathbb{N}$, $\forall x_1 \in \check{U}$, $\|\check{p}_n(x_1)\| = \|\bar{p}_n(x_1)\| = \left\| \int_{X_2} \varphi_n(x_1, x_2) \, d\mu_2(x_2) \right\| \leq \int_{X_2} \mathcal{P} \circ \varphi_n(x_1, x_2) \, d\mu_2(x_2) \leq \int_{X_2} \mathcal{P} \circ f(x_1, x_2) \cdot d\mu_2(x_2) = g(x_1) = \tilde{g}(x_1)$. $\forall x_1 \in X_1 \setminus \check{U}$, $\|\check{p}_n(x_1)\| = 0 \leq \tilde{g}(x_1)$. Hence, we have $\|\check{p}_n(x_1)\| \leq \tilde{g}(x_1)$, $\forall x_1 \in X_1$, $\forall n \in \mathbb{N}$. By Lebesgue Dominated Convergence Theorem 11.91, we have $\lim_{n \in \mathbb{N}} \int_{X_1} \check{p}_n \, d\mu_1 = \int_{X_1} \check{p} \, d\mu_1$ and $\check{p}$ is absolutely integrable over $\mathcal{X}_1$. Then, $\int_{X_1 \times X_2} f \, d\mu = \int_{X_1} \check{p} \, d\mu_1$. Hence, (i) holds.

By symmetry, (ii) also holds.

(iii) Fix any $p : X_1 \to \mathcal{Y}$ satisfying: p is $\mathcal{B}_1$-measurable; and $\exists U \in \mathcal{B}_1$ with $\mu_1(X_1 \setminus U) = 0$ such that, $\forall x_1 \in U$, f_{x_1} is absolutely integrable over $\mathcal{X}_2$ and $p(x_1) = \int_{X_2} f_{x_1} \, d\mu_2$. By (i), we have $U_0 := \check{U} \cap U \in \mathcal{B}_1$ and $\mu_1(X_1 \setminus U_0) = \mu_1((X_1 \setminus \check{U}) \cup (X_1 \setminus U)) = 0$. $\forall x_1 \in U_0$, $p(x_1) = \int_{X_2} f_{x_1} \, d\mu_2 = \check{p}(x_1)$. By Lemma 11.43, $p = \check{p}$ a.e. in $\mathcal{X}_1$. By Proposition 11.83, we have p is absolutely integrable over $\mathcal{X}_1$ since $\check{p}$ is so. By Proposition 11.92, $\int_{X_1} p \, d\mu_1 = \int_{X_1} \check{p} \, d\mu_1 = \int_{X_1 \times X_2} f \, d\mu \in \mathcal{Y}$. Hence, (iii) holds.

By symmetry, (iv) holds. This completes the proof of the theorem. □

When the range of the function f is a σ-compact subset of a separable Banach space, then the set $\check{U}$ and $\check{V}$ of the previous theorem takes a simpler form. This is established in the following theorem.

Theorem 12.31 (Fubini) *Let $\mathcal{X}_i := (X_i, \mathcal{B}_i, \mu_i)$ be a σ-finite measure space, $i = 1, 2$, $\mathcal{X} := (X_1 \times X_2, \mathcal{B}, \mu) := \mathcal{X}_1 \times \mathcal{X}_2$ be the product measure space, $\mathcal{Y}$ be a Banach space, $W \subseteq \mathcal{Y}$ be a σ-compact conic segment, and $f : X_1 \times X_2 \to W$ be absolutely integrable over $\mathcal{X}$. Define $f_{x_1} : X_2 \to W \subseteq \bar{\mathcal{Y}} := \overline{\text{span}(W)} \subseteq \mathcal{Y}$ and $f_{\langle x_2 \rangle} : X_1 \to W \subseteq \bar{\mathcal{Y}}$ as in Proposition 12.28. Then, the following statements hold:*

(i) *Define $U := \{x_1 \in X_1 \mid f_{x_1}$ is absolutely integrable over $\mathcal{X}_2\}$, and $p : X_1 \to \bar{\mathcal{Y}}$ by $p(x_1) = \begin{cases} \int_{X_2} f_{x_1} \, d\mu_2 & x_1 \in U \\ \vartheta_{\mathcal{Y}} & x_1 \in X_1 \setminus U \end{cases}$, $\forall x_1 \in X_1$. Then, $U \in \mathcal{B}_1$ with $\mu_1(X_1 \setminus U) = 0$, p is $\mathcal{B}_1$-measurable and is absolutely integrable over $\mathcal{X}_1$, and $\int_{X_1 \times X_2} f \, d\mu = \int_{X_1} p \, d\mu_1 \in \bar{\mathcal{Y}}$.*

(ii) *Define $V := \{x_2 \in X_2 \mid f_{\langle x_2 \rangle}$ is absolutely integrable over $\mathcal{X}_1\}$, and $q : X_2 \to \bar{\mathcal{Y}}$ by $q(x_2) = \begin{cases} \int_{X_1} f_{\langle x_2 \rangle} \, d\mu_1 & x_2 \in V \\ \vartheta_{\mathcal{Y}} & x_2 \in X_2 \setminus V \end{cases}$, $\forall x_2 \in X_2$. Then, $V \in \mathcal{B}_2$ with $\mu_2(X_2 \setminus V) = 0$, q is $\mathcal{B}_2$-measurable and is absolutely integrable over $\mathcal{X}_2$, and $\int_{X_1 \times X_2} f \, d\mu = \int_{X_2} q \, d\mu_2 \in \bar{\mathcal{Y}}$.*

Proof (i) By Proposition 11.190, W is a separable. Then, by Propositions 7.35 and 4.39, $\bar{\mathcal{Y}}$ is a separable Banach subspace of $\mathcal{Y}$. Since f is absolutely integrable over $\mathcal{X}$, then f and $\mathcal{P} \circ f$ are $\mathcal{B}$-measurable and $\int_{X_1 \times X_2} \mathcal{P} \circ f \,\mathrm{d}\mu < \infty$. By Tonelli Theorem 12.29, $g : X_1 \to [0, \infty] \subset \mathbb{R}_e$, defined by $g(x_1) = \int_{X_2} \mathcal{P} \circ f(x_1, x_2) \,\mathrm{d}\mu_2(x_2)$, $\forall x_1 \in X_1$, is $\mathcal{B}_1$-measurable and $\int_{X_1 \times X_2} \mathcal{P} \circ f \,\mathrm{d}\mu = \int_{X_1} g \,\mathrm{d}\mu_1$. By Proposition 11.82, the function $\tilde{g} : X_1 \to [0, \infty) \subset \mathbb{R}$, defined by $\tilde{g}(x_1) = \begin{cases} g(x_1) & g(x_1) < \infty \\ 0 & g(x_1) = \infty \end{cases}$, $\forall x_1 \in X_1$, is $\mathcal{B}_1$-measurable, $g = \tilde{g}$ a.e. in $\mathcal{X}_1$, and $\int_{X_1 \times X_2} \mathcal{P} \circ f \,\mathrm{d}\mu = \int_{X_1} \tilde{g} \,\mathrm{d}\mu_1$. Let $\bar{U} := \{x_1 \in X_1 \mid g(x_1) < \infty\}$. Then, $\bar{U} \in \mathcal{B}_1$, and $\mu_1(X_1 \setminus \bar{U}) = 0$. $\forall x_1 \in \bar{U}$, we have $g(x_1) = \int_{X_2} \mathcal{P} \circ f(x_1, x_2) \,\mathrm{d}\mu_2(x_2) = \int_{X_2} \mathcal{P} \circ f_{x_1} \,\mathrm{d}\mu_2 < \infty$. Then, f_{x_1} is absolutely integrable over $\mathcal{X}_2$, $\forall x_1 \in \bar{U}$. By Propositions 12.28 and 11.92, $\int_{X_2} f_{x_1} \,\mathrm{d}\mu_2 \in \bar{\mathcal{Y}}$, $\forall x_1 \in \bar{U}$. $\forall x_1 \in X_1 \setminus \bar{U}$, $\int_{X_2} \mathcal{P} \circ f_{x_1} \,\mathrm{d}\mu_2 = \infty$ and f_{x_1} is not absolutely integrable over $\mathcal{X}_2$. Hence, $U = \bar{U}$.

By Lemma 11.68, $\exists$ a sequence of simple functions $(\varphi_n)_{n=1}^\infty$, $\varphi_n : \mathcal{X} \to W$, $\forall n \in \mathbb{N}$, such that $\|\varphi_n(x_1, x_2)\| \leq \|f(x_1, x_2)\|$, $\forall (x_1, x_2) \in X_1 \times X_2$, $\forall n \in \mathbb{N}$, and $\lim_{n \in \mathbb{N}} \varphi_n(x_1, x_2) = f(x_1, x_2)$, $\forall (x_1, x_2) \in X_1 \times X_2$. Fix any $x_1 \in X_1$, and we have $\lim_{n \in \mathbb{N}} \varphi_n(x_1, x_2) = f_{x_1}(x_2)$, $\forall x_2 \in X_2$. Then, by Lebesgue Dominated Convergence Theorem 11.91, $p(x_1) = \lim_{n \in \mathbb{N}} \int_{X_2} \varphi_n(x_1, x_2) \,\mathrm{d}\mu_2(x_2)$, $\forall x_1 \in U$.

Fix any $n \in \mathbb{N}$. Let φ_n admit the canonical representation $\varphi_n = \sum_{i=1}^m y_i \chi_{E_i, X_1 \times X_2}$, where $m \in \mathbb{Z}_+$, $y_1, \ldots, y_m \in W$ are distinct and none equals to $\vartheta_{\mathcal{Y}}$, $E_1, \ldots, E_m \in \mathcal{B}$ are nonempty, pairwise disjoint, and $\mu(E_i) < \infty$, $i = 1, \ldots, m$. Fix any $i \in \{1, \ldots, m\}$. Define $g_i : X_1 \to [0, \infty] \subset \mathbb{R}_e$ by $g_i(x_1) = \mu_2(E_{i\,x_1})$, $\forall x_1 \in X_1$, where $E_{i\,x_1} := \{\bar{x}_2 \in X_2 \mid (x_1, \bar{x}_2) \in E_i\}$. By Lemma 12.27, g_i is $\mathcal{B}_1$-measurable and $\mu(E_i) = \int_{X_1} g_i \,\mathrm{d}\mu_1$. Define $\hat{g}_i : X_1 \to [0, \infty) \subset \mathbb{R}$ by $\hat{g}_i(x_1) = \begin{cases} g_i(x_1) & g_i(x_1) < \infty \\ 0 & g_i(x_1) = \infty \end{cases}$, $\forall x_1 \in X_1$. By Proposition 11.82, $\hat{g}_i$ is $\mathcal{B}_1$-measurable, $g_i = \hat{g}_i$ a.e. in $\mathcal{X}_1$, and $\mu(E_i) = \int_{X_1} \hat{g}_i \,\mathrm{d}\mu_1$. Then, $\int_{X_1 \times X_2} \varphi_n \,\mathrm{d}\mu = \sum_{i=1}^m y_i \mu(E_i) = \sum_{i=1}^m y_i \int_{X_1} \hat{g}_i \,\mathrm{d}\mu_1 = \int_{X_1} (\sum_{i=1}^m y_i \hat{g}_i) \,\mathrm{d}\mu_1 =: \int_{X_1} \bar{p}_n \,\mathrm{d}\mu_1$, where the first equality follows from Proposition 11.75 and the third equality follows from Proposition 11.92. By Propositions 7.23, 11.38, and 11.39, $\bar{p}_n : X_1 \to \bar{\mathcal{Y}}$ is $\mathcal{B}_1$-measurable and, by Proposition 11.92, $\bar{p}_n$ is absolutely integrable over $\mathcal{X}_1$ $\forall x_1 \in U$, $g(x_1) < \infty$, then f_{x_1} is absolutely integrable over $\mathcal{X}_2$. Since $\|\varphi_n(x_1, x_2)\| \leq \|f(x_1, x_2)\|$, $\forall (x_1, x_2) \in \mathcal{X}$, $\forall n \in \mathbb{N}$, then, $\forall x_1 \in U$, $\|\varphi_n(x_1, x_2)\|$, as a function of x_2, is absolutely integrable over $\mathcal{X}_2$. Then, $\hat{g}_i(x_1) = g_i(x_1)$, $i = 1, \ldots, m$, $\forall x_1 \in U$. Hence, $\bar{p}_n(x_1) = \sum_{i=1}^m y_i \hat{g}_i(x_1) = \sum_{i=1}^m y_i g_i(x_1) = \sum_{i=1}^m y_i \mu_2(E_{i\,x_1}) = \sum_{i=1}^m y_i \int_{X_2} \chi_{E_{i\,x_1}, X_2} \,\mathrm{d}\mu_2 = \int_{X_2} (\sum_{i=1}^m y_i \chi_{E_{i\,x_1}, X_2}) \,\mathrm{d}\mu_2 = \int_{X_2} \varphi_n(x_1, x_2) \,\mathrm{d}\mu_2(x_2) \in \bar{\mathcal{Y}}$, $\forall x_1 \in U$, where the first equality follows from the definition of $\bar{p}_n$, the second equality follows from the preceding discussion, the third equality follows from the definitions of g_i's, the fourth equality follows from Proposition 11.75, and the fifth equality follows from Proposition 11.92.

Define $\hat{p}_n : X_1 \to \bar{\mathcal{Y}}$ by $\hat{p}_n(x_1) = \begin{cases} \bar{p}_n(x_1) & x_1 \in U \\ \vartheta_{\mathcal{Y}} & x_1 \in X_1 \setminus U \end{cases}$, $\forall x_1 \in X_1$. By Proposition 11.41, $\hat{p}_n$ is $\mathcal{B}_1$-measurable. Then, $\bar{p}_n = \hat{p}_n$ a.e. in $\mathcal{X}_1$.

Note that, $\forall x_1 \in U$, we have

$$\|\hat{p}_n(x_1)\| = \|\bar{p}_n(x_1)\| = \left\| \int_{X_2} \varphi_n(x_1, x_2)\, \mathrm{d}\mu_2(x_2) \right\|$$

$$\leq \int_{X_2} \mathcal{P} \circ \varphi_n(x_1, x_2)\, \mathrm{d}\mu_2(x_2) \leq \int_{X_2} \mathcal{P} \circ f(x_1, x_2)\, \mathrm{d}\mu_2(x_2) = \tilde{g}(x_1) < \infty$$

where the second equality follows from the previous paragraph, the first inequality follows from Proposition 11.92, the second inequality follows from the property of φ_n, and the last equality follows from the definition of U. Note also that, $\forall x_1 \in X_1 \setminus U$, we have $\|\hat{p}_n(x_1)\| = 0 = \tilde{g}(x_1)$. Then, $\|\hat{p}_n(x_1)\| \leq \tilde{g}(x_1)$, $\forall x_1 \in X_1$. Hence, $\hat{p}_n$ is absolutely integrable over $\mathcal{X}_1$. By Proposition 11.92, $\int_{X_1 \times X_2} \varphi_n \, \mathrm{d}\mu = \int_{X_1} \bar{p}_n \, \mathrm{d}\mu_1 = \int_{X_1} \hat{p}_n \, \mathrm{d}\mu_1$, where the first equality follows from the definition of $\bar{p}_n$.

By the result of the third preceding paragraph, $\forall x_1 \in U$,

$$p(x_1) = \lim_{n \in \mathbb{N}} \int_{X_2} \varphi_n(x_1, x_2)\, \mathrm{d}\mu_2(x_2) = \lim_{n \in \mathbb{N}} \bar{p}_n(x_1) = \lim_{n \in \mathbb{N}} \hat{p}_n(x_1)$$

where the second equality follows from the result of the second preceding paragraph and the last equality follows from the definition of $\hat{p}_n$. Note that $p(x_1) = \vartheta_{\mathcal{Y}} = \lim_{n \in \mathbb{N}} \hat{p}_n(x_1)$, $\forall x_1 \in X_1 \setminus U$. Then, $p(x_1) = \lim_{n \in \mathbb{N}} \hat{p}_n(x_1)$, $\forall x_1 \in X_1$. This implies that, by Proposition 11.51, p is $\mathcal{B}_1$-measurable. Then, we have $\int_{X_1 \times X_2} f \, \mathrm{d}\mu = \lim_{n \in \mathbb{N}} \int_{X_1 \times X_2} \varphi_n \, \mathrm{d}\mu = \lim_{n \in \mathbb{N}} \int_{X_1} \hat{p}_n \, \mathrm{d}\mu_1 = \int_{X_1} p \, \mathrm{d}\mu_1$, where the first equality follows from Lebesgue Dominated Convergence Theorem 11.91, the second equality follows from the result of the previous paragraph, and the last equality follows from the Lebesgue Dominated Convergence Theorem 11.91. Hence, (i) holds.

By symmetry, (ii) also holds. This completes the proof of the theorem. □

Proposition 12.32 *Let $m \in \mathbb{N}$, $\mathcal{X}_i := (X_i, \mathcal{O}_i)$ be a second countable topological space, $\mathbb{X}_i := (\mathcal{X}_i, \mathcal{B}_i, \mu_i)$ be a second countable σ-finite topological measure space, $i = 1, \ldots, m$, $\mathcal{X} := (\prod_{i=1}^{m} X_i, \mathcal{O}) := \prod_{i=1}^{m} \mathcal{X}_i$ be the product topological space, and $(\prod_{i=1}^{m} X_i, \mathcal{B}, \mu) := \prod_{i=1}^{m} (X_i, \mathcal{B}_i, \mu_i)$ be the product measure space. Then, $\mathbb{X} := (\mathcal{X}, \mathcal{B}, \mu)$ is a second countable σ-finite topological measure space, which is said to be the product topological measure space of $\mathbb{X}_1, \ldots, \mathbb{X}_m$, denoted by $\mathbb{X} = \prod_{i=1}^{m} \mathbb{X}_i$.*

Proof By Proposition 12.22, $\mathcal{B}$ is the σ-algebra on $\prod_{i=1}^{m} X_i$ generated by $\mathcal{C} := \{\prod_{i=1}^{m} B_i \subseteq \prod_{i=1}^{m} X_i \mid B_i \in \mathcal{B}_i, \ \forall i \in \{1, \ldots, m\}\}$. By $\mathbb{X}_i$ being a topological measure space, we have $\mathcal{B}_i = \mathcal{B}_{\mathrm{B}}(\mathcal{X}_i)$, $i = 1, \ldots, m$. Let $\bar{\mathcal{X}}_1 := \mathcal{X}_1$ and $\bar{\mathcal{X}}_i := \bar{\mathcal{X}}_{i-1} \times \mathcal{X}_i := ((\cdots(X_1 \times X_2) \times \cdots) \times X_i, \bar{\mathcal{O}}_i)$, $i = 2, \ldots, m$, $(X_1, \bar{\mathcal{B}}_1, \bar{\mu}_1) := (X_1, \mathcal{B}_1, \mu_1)$ and $((\cdots(X_1 \times X_2) \times \cdots) \times X_i, \bar{\mathcal{B}}_i, \bar{\mu}_i) := ((\cdots(X_1 \times X_2) \times$

$\cdots) \times X_{i-1}, \bar{\mathcal{B}}_{i-1}, \bar{\mu}_{i-1}) \times (X_i, \mathcal{B}_i, \mu_i)$, $i = 2, \ldots, m$. By Proposition 3.28, $\bar{\mathcal{X}}_i$, $i = 1, \ldots, m$, are second countable topological spaces, and by Proposition 11.24, $\bar{\mathcal{B}}_i = \mathcal{B}_{\mathrm{B}}(\bar{\mathcal{X}}_i)$, $i = 1, \ldots, m$. By Propositions 3.30 and 12.24, $\bar{\mathcal{X}}_m$ is homeomorphic to $\mathcal{X}$ and $((\cdots(X_1 \times X_2) \times \cdots) \times X_m, \bar{\mathcal{B}}_m, \bar{\mu}_m)$ is isomeasuric to $(\prod_{i=1}^m X_i, \mathcal{B}, \mu)$ with the same bijective mapping. Then, we have $\mathcal{B} = \mathcal{B}_{\mathrm{B}}(\mathcal{X})$. By Proposition 3.28, $\mathcal{X}$ is second countable.

First, consider the special case when $\mathbb{X}_i$, $i = 1, \ldots, m$, are finite. Let $\mathcal{E} := \{E \in \mathcal{B} \mid \forall \epsilon \in (0, \infty) \subset \mathbb{R}$, (i) $\exists U \in \mathcal{O}$ with $E \subseteq U$ such that $\mu(U \setminus E) < \epsilon$; (ii) $\exists \widetilde{F} \in \mathcal{O}$ with $F \subseteq E$ such that $\mu(E \setminus F) < \epsilon\}$. We will show that $\mathcal{E}$ is a σ-algebra on $\prod_{i=1}^m X_i$ and contains $\mathcal{C}$. Then, $\mathcal{E} = \mathcal{B}$ and $\mathbb{X}$ is a second countable finite topological measure space.

$\forall E := \prod_{i=1}^m B_i \in \mathcal{C}$, $\forall \epsilon \in (0, \infty) \subset \mathbb{R}$, fix any $i \in \{1, \ldots, m\}$, and then $B_i \in \mathcal{B}_i$. Since $\mathbb{X}_i$ is a topological measure space, then $\exists O_i \in \mathcal{O}_i$ with $B_i \subseteq O_i$ such that $\mu_i(O_i \setminus B_i) < \frac{\epsilon \wedge 1}{1 + m \prod_{j=1, j \neq i}^m (1 + \mu_j(B_j))}$. By Proposition 11.27, $\exists X_i \setminus F_i \in \mathcal{O}_i$ with $F_i \subseteq B_i$ such that $\mu_i(B_i \setminus F_i) < \frac{\epsilon}{1 + m \prod_{j=1, j \neq i}^m \mu_j(B_j)}$. Let $U := \prod_{i=1}^m O_i \in \mathcal{O}$ and $F := \prod_{i=1}^m F_i$. Clearly, $(\prod_{i=1}^m X_i) \setminus F = \bigcup_{i=1}^m ((\prod_{j=1}^{i-1} X_j) \times (X_i \setminus F_i) \times (\prod_{j=i+1}^m X_j)) \in \mathcal{O}$. Then, $F \subseteq E \subseteq U$,

$$\begin{aligned}
\mu(U \setminus E) &= \mu((\prod_{i=1}^m O_i) \setminus (\prod_{i=1}^m B_i)) \\
&= \mu\Big(\bigcup_{i=1}^m \Big(\Big(\prod_{j=1}^{i-1} O_j\Big) \times (O_i \setminus B_i) \times \Big(\prod_{j=i+1}^m O_j\Big)\Big)\Big) \\
&\leq \sum_{i=1}^m \mu\Big(\Big(\prod_{j=1}^{i-1} O_j\Big) \times (O_i \setminus B_i) \times \Big(\prod_{j=i+1}^m O_j\Big)\Big) \\
&= \sum_{i=1}^m \mu_i(O_i \setminus B_i) \prod_{j=1, j \neq i}^m \mu_j(O_j) \\
&\leq \sum_{i=1}^m \frac{\epsilon \wedge 1}{1 + m \prod_{j=1, j \neq i}^m (1 + \mu_j(B_j))} \prod_{j=1, j \neq i}^m (1 + \mu_j(B_j)) < \epsilon
\end{aligned}$$

where the first inequality follows from Proposition 11.6 and the third equality follows from Proposition 12.21, and

$$\begin{aligned}
\mu(E \setminus F) &= \mu((\prod_{i=1}^m B_i) \setminus (\prod_{i=1}^m F_i)) \\
&= \mu\Big(\bigcup_{i=1}^m \Big(\Big(\prod_{j=1}^{i-1} B_j\Big) \times (B_i \setminus F_i) \times \Big(\prod_{j=i+1}^m B_j\Big)\Big)\Big)
\end{aligned}$$

$$\le \sum_{i=1}^{m} \mu\Big(\Big(\prod_{j=1}^{i-1} B_j\Big) \times (B_i \setminus F_i) \times \Big(\prod_{j=i+1}^{m} B_j\Big)\Big)$$

$$= \sum_{i=1}^{m} \mu_i(B_i \setminus F_i) \prod_{j=1, j\neq i}^{m} \mu_j(B_j)$$

$$\le \sum_{i=1}^{m} \frac{\epsilon}{1 + m \prod_{j=1, j\neq i}^{m} \mu_j(B_j)} \prod_{j=1, j\neq i}^{m} \mu_j(B_j) < \epsilon$$

where the first inequality follows from Proposition 11.6 and the third equality follows from Proposition 12.21. Hence, $E \in \mathcal{E}$. This shows that $\mathcal{C} \subseteq \mathcal{E}$.

$\forall E \in \mathcal{E}$, $\forall \epsilon \in (0, \infty) \subset \mathbb{R}$, $\exists U \in \mathcal{O}$ with $E \subseteq U$ such that $\mu(U \setminus E) < \epsilon$, $\exists \widetilde{F} \in \mathcal{O}$ with $F \subseteq E$ such that $\mu(E \setminus F) < \epsilon$. Then, $\mu(\widetilde{E} \setminus \widetilde{U}) = \mu(U \setminus E) < \epsilon$ and $\mu(\widetilde{F} \setminus \widetilde{E}) = \mu(E \setminus F) < \epsilon$. Hence, $\widetilde{E} \in \mathcal{E}$.

$\forall (E_i)_{i=1}^{\infty} \subseteq \mathcal{E}$, let $E := \bigcup_{i=1}^{\infty} E_i \in \mathcal{B}$. $\forall \epsilon \in (0, \infty) \subset \mathbb{R}$, $\forall i \in \mathbb{N}$, $\exists U_i \in \mathcal{O}$ with $E_i \subseteq U_i$ such that $\mu(U_i \setminus E_i) < 2^{-i}\epsilon$; $\exists \widetilde{F_i} \in \mathcal{O}$ with $F_i \subseteq E_i$ such that $\mu(E_i \setminus F_i) < 2^{-i-1}\epsilon$. Let $U := \bigcup_{i=1}^{\infty} U_i \in \mathcal{O}$. Then, $E = \bigcup_{i=1}^{\infty} E_i \subseteq \bigcup_{i=1}^{\infty} U_i = U$ and $\mu(U \setminus E) = \mu(\bigcup_{i=1}^{\infty}(U_i \setminus E)) \le \sum_{i=1}^{\infty} \mu(U_i \setminus E) \le \sum_{i=1}^{\infty} \mu(U_i \setminus E_i) < \epsilon$, where the first inequality follows from Proposition 11.6 and the second inequality follows from Proposition 11.4. Hence, E satisfies (i). By Proposition 11.7, $\mu(E) = \mu(\bigcup_{i=1}^{\infty} E_i) = \lim_{n\in\mathbb{N}} \mu(\bigcup_{i=1}^{n} E_i)$. Then, by $\mu(E) \le \mu(\prod_{i=1}^{m} X_i) = \prod_{i=1}^{m} \mu_i(X_i) < \infty$, $\exists n_0 \in \mathbb{N}$ such that $\mu(\bigcup_{i=1}^{n_0} E_i) \le \mu(E) < \mu(\bigcup_{i=1}^{n_0} E_i) + \epsilon/2$. Let $F := \bigcup_{i=1}^{n_0} F_i$. Clearly, $\widetilde{F} = \bigcap_{i=1}^{n_0} \widetilde{F_i} \in \mathcal{O}$. Then, $F \subseteq E$ and $\mu(E \setminus F) = \mu(E \setminus (\bigcup_{i=1}^{n_0} E_i)) + \mu((\bigcup_{i=1}^{n_0} E_i) \setminus F) = \mu(E) - \mu(\bigcup_{i=1}^{n_0} E_i) + \mu(\bigcup_{i=1}^{n_0}(E_i \setminus F)) < \epsilon/2 + \sum_{i=1}^{n_0} \mu(E_i \setminus F) \le \epsilon/2 + \sum_{i=1}^{n_0} \mu(E_i \setminus F_i) < \epsilon$, where the first and second equalities follow from Fact 11.3, the first inequality follows from Proposition 11.6, and the second inequality follows from Proposition 11.4. Hence, E satisfies (ii). Then, $E = \bigcup_{i=1}^{\infty} E_i \in \mathcal{E}$.

Clearly, $\emptyset = \prod_{i=1}^{m} \emptyset \in \mathcal{C} \subseteq \mathcal{E}$ and $\prod_{i=1}^{m} X_i \in \mathcal{C} \subseteq \mathcal{E}$. This shows that $\mathcal{E}$ is a σ-algebra on $\prod_{i=1}^{m} X_i$ and contains $\mathcal{C}$. This completes the proof of the special case.

Next, consider the general case when $\mathbb{X}_i$, $i = 1, \ldots, m$, are σ-finite. Fix any $i = 1, \ldots, m$. By $\mathbb{X}_i$ being σ-finite, $\exists (X_{i,n})_{n=1}^{\infty} \subseteq \mathcal{B}_i$ such that $X_i = \bigcup_{n=1}^{\infty} X_{i,n}$ and $\mu_i(X_{i,n}) < \infty$, $\forall n \in \mathbb{N}$. By $\mathbb{X}_i$ being a topological measure space, without loss of generality, we may assume that $X_{i,n} \subseteq X_{i,n+1}$ and $X_{i,n} \in \mathcal{O}_i$, $\forall n \in \mathbb{N}$. $\forall n \in \mathbb{N}$, by Proposition 11.29, let $\mathbb{X}_{i,n} := ((X_{i,n}, \mathcal{O}_{i,n}), \mathcal{B}_{i,n}, \mu_{i,n})$ be the finite topological measure subspace of $\mathbb{X}_i$. $\forall n \in \mathbb{N}$, by the special case, let $\hat{\mathbb{X}}_n := \prod_{i=1}^{m} \mathbb{X}_{i,n} := ((\prod_{i=1}^{m} X_{i,n}, \hat{\mathcal{O}}_n), \hat{\mathcal{B}}_n, \hat{\mu}_n)$ be the second countable finite product topological measure space. By Proposition 12.23, $(\prod_{i=1}^{m} X_{i,n}, \hat{\mathcal{B}}_n, \hat{\mu}_n)$ is the finite measure subspace of $(\prod_{i=1}^{m} X_i, \mathcal{B}, \mu)$. $\forall E \in \mathcal{B}$, $\forall \epsilon \in (0, \infty) \subset \mathbb{R}$, $\forall n \in \mathbb{N}$, let

$E_n := E \cap (\prod_{i=1}^m X_{i,n}) \in \hat{\mathcal{B}}_n$. Since $\hat{\mathbb{X}}_n$ is a topological measure space, $\exists U_n \in \hat{\mathcal{O}}_n$ with $E_n \subseteq U_n$ such that $\hat{\mu}_n(U_n \setminus E_n) < 2^{-n}\epsilon$. Since $X_{i,n} \in \mathcal{O}_i$, $i = 1, \dots, m$, then $\prod_{i=1}^m X_{i,n} \in \mathcal{O}$, which further implies that $U_n \in \mathcal{O}$. Let $U := \bigcup_{n=1}^\infty U_n \in \mathcal{O}$. Then, $E = \bigcup_{n=1}^\infty E_n \subseteq \bigcup_{n=1}^\infty U_n = U$ and $\mu(U \setminus E) = \mu(\bigcup_{n=1}^\infty (U_n \setminus E)) \le \sum_{n=1}^\infty \mu(U_n \setminus E) \le \sum_{n=1}^\infty \mu(U_n \setminus E_n) = \sum_{n=1}^\infty \hat{\mu}_n(U_n \setminus E_n) < \epsilon$, where the first inequality follows from Proposition 11.6, the second inequality follows from Proposition 11.4, and the second equality follows from Proposition 11.13. By the arbitrariness of E, we have $\mathbb{X}$ is a σ-finite topological measure space. It is second countable since $\mathcal{X}$ is second countable. This completes the proof of the proposition. □

Proposition 12.33 *Let $m \in \mathbb{N}$, $\mathcal{X}_i := (X_i, \mathcal{O}_i)$ be a second countable topological space, $\mathbb{X}_i := (\mathcal{X}_i, \mathcal{B}_i, \mu_i)$ be a second countable σ-finite $\mathbb{K}$-valued topological measure space, $i = 1, \dots, m$, $\mathcal{X} := (\prod_{i=1}^m X_i, \mathcal{O}) := \prod_{i=1}^m \mathcal{X}_i$ be the product topological space, and $(\prod_{i=1}^m X_i, \mathcal{B}, \mu) := \prod_{i=1}^m (X_i, \mathcal{B}_i, \mu_i)$ be the product $\mathbb{K}$-valued measure space. Then, $\mathbb{X} := (\mathcal{X}, \mathcal{B}, \mu)$ is a second countable σ-finite $\mathbb{K}$-valued topological measure space, which is said to be the product* $\mathbb{K}$-valued topological measure space of $\mathbb{X}_1, \dots, \mathbb{X}_m$, *denoted by $\mathbb{X} = \prod_{i=1}^m \mathbb{X}_i$.*

Proof Let $\hat{\mathcal{X}}_i := (X_i, \mathcal{B}_i, \mathcal{P} \circ \mu_i =: \nu_i)$, $i = 1, \dots, m$, and $\hat{\mathcal{X}} := (\prod_{i=1}^m X_i, \mathcal{B}, \nu := \mathcal{P} \circ \mu)$. By Proposition 12.22, $\hat{\mathcal{X}}_i$, $i = 1, \dots, m$, and $\hat{\mathcal{X}}$ are σ-finite measure spaces and $\hat{\mathcal{X}} = \prod_{i=1}^m \hat{\mathcal{X}}_i$. Since $\mathbb{X}_i$ is a second countable σ-finite $\mathbb{K}$-valued topological measure space, then, by Definition 11.191, $\hat{\mathbb{X}}_i := (\mathcal{X}_i, \mathcal{B}_i, \nu_i)$ is a second countable σ-finite topological measure space, $i = 1, \dots, m$. By Proposition 12.32, $\hat{\mathbb{X}} := (\mathcal{X}, \mathcal{B}, \nu)$ is a second countable σ-finite topological measure space. Then, $\mathbb{X}$ is a second countable σ-finite $\mathbb{K}$-valued topological measure space. This completes the proof of the proposition. □

Proposition 12.34 *Let $m \in \mathbb{N}$, $\mathcal{X}_i := (X_i, \mathcal{B}_i, \mu_i)$ be a σ-finite measure space, ν_i be a σ-finite $\mathbb{K}$-valued measure on $(X_i, \mathcal{B}_i)$ with $\frac{d\nu_i}{d\mu_i} = f_i : X_i \to \mathbb{K}$ a.e. in $\mathcal{X}_i$, $i = 1, \dots, m$, and $f : \prod_{i=1}^m X_i \to \mathbb{K}$ be defined by $f(x_1, \dots, x_m) = \prod_{i=1}^m f_i(x_i)$, $\forall (x_1, \dots, x_m) \in \prod_{i=1}^m X_i$. Then, $\frac{d\nu}{d\mu} = f$ a.e. in $\mathcal{X}$, where $\nu := \prod_{i=1}^m \nu_i$, $\mu := \prod_{i=1}^m \mu_i$, and $\mathcal{X} := (\prod_{i=1}^m X_i, \mathcal{B}, \mu) := \prod_{i=1}^m \mathcal{X}_i$.*

Proof By Proposition 12.22, $\mathcal{X}$ is a σ-finite measure space. By Propositions 7.23, 11.38, and 11.39, f is $\mathcal{B}$-measurable. By Proposition 11.116 and Definition 11.166, f is a Radon–Nikodym derivative of a σ-finite $\mathbb{K}$-valued measure $\bar{\nu}$ on $(\prod_{i=1}^m X_i, \mathcal{B})$ with respect to μ. By Proposition 11.167, $\frac{d\bar{\nu}}{d\mu} = f$ a.e. in $\mathcal{X}$.

First, consider the special case when ν_i, $i = 1, \dots, m$, are finite $\mathbb{K}$-valued measures. Let $\mathcal{E} := \{E \in \mathcal{B} \mid \nu(E) = \bar{\nu}(E)\}$ and $\mathcal{C} := \{\prod_{i=1}^m B_i \subseteq \prod_{i=1}^m X_i \mid B_i \in \mathcal{B}_i, \ \forall i \in \{1, \dots, m\}\}$, where $\mathcal{C}$ is the set of measurable rectangles in $\prod_{i=1}^m X_i$. $\forall E := \prod_{i=1}^m B_i \in \mathcal{C}$, we have $\nu(E) = \prod_{i=1}^m \nu_i(B_i)$ by Proposition 12.21. By repeated application of Tonelli Theorem 12.29 and Proposition 12.24, we have

$$\int_{\prod_{i=1}^{m} X_i} \mathcal{P} \circ f \, \mathrm{d}\mu = \Big(\int_{\prod_{i=1}^{m-1} X_i} \prod_{i=1}^{m-1} \mathcal{P} \circ f_i \, \mathrm{d}(\prod_{i=1}^{m-1} \mu_i) \Big) \Big(\int_{X_m} \mathcal{P} \circ f_m \, \mathrm{d}\mu_m \Big)$$

$$= \Big(\int_{\prod_{i=1}^{m-1} X_i} \prod_{i=1}^{m-1} \mathcal{P} \circ f_i \, \mathrm{d}(\prod_{i=1}^{m-1} \mu_i) \Big) \mathcal{P} \circ \nu_m(X_m) = \cdots = \prod_{i=1}^{m} \mathcal{P} \circ \nu_i(X_i)$$

$$< \infty$$

where the first equality follows from Tonelli Theorem 12.29 and the second equality follows from Definition 11.166. Hence, f is absolutely integrable over $\mathcal{X}$ and $\bar{\nu}$ is a finite $\mathbb{K}$-valued measure by Proposition 11.116. By Proposition 11.83, $f \chi_{\prod_{i=1}^{m} B_i, \prod_{i=1}^{m} X_i}$ is absolutely integrable over $\mathcal{X}$. By repeated application of Fubini Theorem 12.31,

$$\bar{\nu}(E) = \int_{\prod_{i=1}^{m} B_i} f \, \mathrm{d}\mu = \int_{\prod_{i=1}^{m} X_i} f \chi_{\prod_{i=1}^{m} B_i, \prod_{i=1}^{m} X_i} \, \mathrm{d}(\prod_{i=1}^{m} \mu_i)$$

$$= \Big(\int_{\prod_{i=1}^{m-1} X_i} \Big(\prod_{i=1}^{m-1} f_i \Big) \chi_{\prod_{i=1}^{m-1} B_i, \prod_{i=1}^{m-1} X_i} \, \mathrm{d}(\prod_{i=1}^{m-1} \mu_i) \Big) \Big(\int_{B_m} f_m \, \mathrm{d}\mu_m \Big)$$

$$= \Big(\int_{\prod_{i=1}^{m-1} X_i} \Big(\prod_{i=1}^{m-1} f_i \Big) \chi_{\prod_{i=1}^{m-1} B_i, \prod_{i=1}^{m-1} X_i} \, \mathrm{d}(\prod_{i=1}^{m-1} \mu_i) \Big) \nu_m(B_m)$$

$$= \cdots = \prod_{i=1}^{m} \nu_i(B_i) = \nu(E)$$

where the first equality follows from Definition 11.166, the second equality follows from Proposition 11.92, the third equality follows from Fubini Theorem, and the fourth equality follows from Definition 11.166. This implies that $E \in \mathcal{E}$. By the arbitrariness of E, we have $\mathcal{C} \subseteq \mathcal{E}$.

Clearly, $\mathcal{C}$ is a π-system on $\prod_{i=1}^{m} X_i$. We will show that $\mathcal{E}$ is a monotone class on $\prod_{i=1}^{m} X_i$. Then, by Monotone Class Lemma 12.19, we have $\mathcal{E} = \mathcal{B}$. This further implies that $\nu = \bar{\nu}$ and $\frac{\mathrm{d}\nu}{\mathrm{d}\mu} = f$ a.e. in $\mathcal{X}$, which completes the proof of the special case.

Note that $\emptyset = \prod_{i=1}^{m} \emptyset \in \mathcal{C} \subseteq \mathcal{E}$ and $\prod_{i=1}^{m} X_i \in \mathcal{C} \subseteq \mathcal{E}$. $\forall E_1, E_2 \in \mathcal{E}$ with $E_1 \subseteq E_2$, we have $\nu(E_2 \setminus E_1) = \nu(E_2) - \nu(E_1) = \bar{\nu}(E_2) - \bar{\nu}(E_1) = \bar{\nu}(E_2 \setminus E_1)$, where the first equality follows from the fact that ν is a finite $\mathbb{K}$-valued measure, the second equality follows from the fact that $E_1, E_2 \in \mathcal{E}$, and the third equality follows from fact that $\bar{\nu}$ is a finite $\mathbb{K}$-valued measure. Hence, $E_2 \setminus E_1 \in \mathcal{E}$.

$\forall (E_i)_{i=1}^{\infty} \subseteq \mathcal{E}$ with $E_i \subseteq E_{i+1}$, $\forall i \in \mathbb{N}$, let $E := \bigcup_{i=1}^{\infty} E_i \in \mathcal{B}$. Then, $\nu(E) = \lim_{i\in\mathbb{N}} \nu(E_i) = \lim_{i\in\mathbb{N}} \bar{\nu}(E_i) = \bar{\nu}(E)$, where the first equality follows from Proposition 11.112 and the fact that ν is finite, the second equality follows from the fact that $E_i \in \mathcal{E}$, $\forall i \in \mathbb{N}$, and the third equality follows from Proposition 11.112 and the fact that $\bar{\nu}$ is finite. Then, $E = \bigcup_{i=1}^{\infty} E_i \in \mathcal{E}$.

Hence, $\mathcal{E}$ is a monotone class on $\prod_{i=1}^{m} X_i$. This completes the proof of the special case.

Next, consider the general case when ν_i, $i = 1, \ldots, m$, are σ-finite. Then, there exists $(X_{i,n})_{n=1}^{\infty} \subseteq \mathcal{B}_i$ such that $X_i = \bigcup_{n=1}^{\infty} X_{i,n}$ and $\mathcal{P} \circ \nu_i(X_{i,n}) < \infty$, $\forall n \in \mathbb{N}$, $i = 1, \ldots, m$. Without loss of generality, we may assume that $(X_{i,n})_{n=1}^{\infty}$ is pairwise disjoint, $i = 1, \ldots, m$. For any $i = 1, \ldots, m$ and $\forall n \in \mathbb{N}$, let $\mathcal{X}_{i,n} := (X_{i,n}, \mathcal{B}_{i,n}, \mu_{i,n})$ be the σ-finite measure subspace of $\mathcal{X}_i$ and $\hat{\mathcal{X}}_{i,n} := (X_{i,n}, \mathcal{B}_{i,n}, \nu_{i,n})$ be the finite $\mathbb{K}$-valued measure subspace of $\hat{\mathcal{X}}_i := (X_i, \mathcal{B}_i, \nu_i)$. Fix any $n_1, \ldots, n_m \in \mathbb{N}$. Let $\bar{\mathcal{X}}_{n_1,\ldots,n_m} := (\prod_{i=1}^{m} X_{i,n_i}, \bar{\mathcal{B}}_{n_1,\ldots,n_m}, \bar{\mu}_{n_1,\ldots,n_m}) := \prod_{i=1}^{m} \mathcal{X}_{i,n_i}$ be the σ-finite product measure space, and $\tilde{\mathcal{X}}_{n_1,\ldots,n_m} := (\prod_{i=1}^{m} X_{i,n_i}, \bar{\mathcal{B}}_{n_1,\ldots,n_m}, \tilde{\nu}_{n_1,\ldots,n_m}) := \prod_{i=1}^{m} \hat{\mathcal{X}}_{i,n_i}$ be the finite product $\mathbb{K}$-valued measure space. By Proposition 12.23, $\bar{\mathcal{X}}_{n_1,\ldots,n_m}$ is the measure subspace of $\mathcal{X}$ and $\tilde{\mathcal{X}}_{n_1,\ldots,n_m}$ is the finite $\mathbb{K}$-valued measure subspace of $\tilde{\mathcal{X}} := (\prod_{i=1}^{m} X_i, \mathcal{B}, \nu)$. By Definition 11.166, we have $\frac{d\nu_{i,n_i}}{d\mu_{i,n_i}} = f_i|_{X_{i,n_i}}$ a.e. in $\mathcal{X}_{i,n_i}$, $i = 1, \ldots, m$. By the special case, we have $\frac{d\tilde{\nu}_{n_1,\ldots,n_m}}{d\bar{\mu}_{n_1,\ldots,n_m}} = f|_{\prod_{i=1}^{m} X_{i,n_i}}$ a.e. in $\bar{\mathcal{X}}_{n_1,\ldots,n_m}$. $\forall E \in \mathcal{B}$ with $\mathcal{P} \circ \nu(E) < \infty$, let $E_{n_1,\ldots,n_m} := E \cap (\prod_{i=1}^{m} X_{i,n_i}) \in \bar{\mathcal{B}}_{n_1,\ldots,n_m}$. Then, we have

$$
\begin{aligned}
\mathcal{P} \circ \nu(E) &= \sum_{n_1=1}^{\infty} \cdots \sum_{n_m=1}^{\infty} \mathcal{P} \circ \nu(E_{n_1,\ldots,n_m}) \\
&= \sum_{n_1=1}^{\infty} \cdots \sum_{n_m=1}^{\infty} \mathcal{P} \circ \tilde{\nu}_{n_1,\ldots,n_m}(E_{n_1,\ldots,n_m}) \\
&= \sum_{n_1=1}^{\infty} \cdots \sum_{n_m=1}^{\infty} \int_{E_{n_1,\ldots,n_m}} \mathcal{P} \circ (f|_{\prod_{i=1}^{m} X_{i,n_i}})\, d\bar{\mu}_{n_1,\ldots,n_m} \\
&= \sum_{n_1=1}^{\infty} \cdots \sum_{n_m=1}^{\infty} \int_{E_{n_1,\ldots,n_m}} \mathcal{P} \circ f\, d\mu = \int_E \mathcal{P} \circ f\, d\mu < \infty
\end{aligned}
$$

where the first equality follows from the fact that $\mathcal{P} \circ \nu$ is a measure, the second equality follows from Proposition 11.115, the third equality follows from the fact that $\frac{d\tilde{\nu}_{n_1,\ldots,n_m}}{d\bar{\mu}_{n_1,\ldots,n_m}} = f|_{\prod_{i=1}^{m} X_{i,n_i}}$ a.e. in $\bar{\mathcal{X}}_{n_1,\ldots,n_m}$, and the last two equalities follow from Proposition 11.83. We also have

$$\begin{aligned}
\nu(E) &= \sum_{n_1=1}^{\infty} \cdots \sum_{n_m=1}^{\infty} \nu(E_{n_1,\ldots,n_m}) = \sum_{n_1=1}^{\infty} \cdots \sum_{n_m=1}^{\infty} \tilde{\nu}_{n_1,\ldots,n_m}(E_{n_1,\ldots,n_m}) \\
&= \sum_{n_1=1}^{\infty} \cdots \sum_{n_m=1}^{\infty} \int_{E_{n_1,\ldots,n_m}} f|_{\prod_{i=1}^m X_{i,n_i}} \, \mathrm{d}\bar{\mu}_{n_1,\ldots,n_m} \\
&= \sum_{n_1=1}^{\infty} \cdots \sum_{n_m=1}^{\infty} \int_{E_{n_1,\ldots,n_m}} f \, \mathrm{d}\mu = \int_E f \, \mathrm{d}\mu = \bar{\nu}(E)
\end{aligned}$$

where the first equality follows from Definition 11.108, the second equality follows from Proposition 11.115, the third equality follows from the fact that $\frac{\mathrm{d}\tilde{\nu}_{n_1,\ldots,n_m}}{\mathrm{d}\bar{\mu}_{n_1,\ldots,n_m}} = f|_{\prod_{i=1}^m X_{i,n_i}}$ a.e. in $\bar{\mathcal{X}}_{n_1,\ldots,n_m}$, the fourth equality follows from Proposition 11.92, the fifth equality follows from Proposition 11.92 and the fact that $\int_E \mathcal{P} \circ f \, \mathrm{d}\mu < \infty$, and the last equality follows from the fact that $\frac{\mathrm{d}\bar{\nu}}{\mathrm{d}\mu} = f$ a.e. in $\mathcal{X}$. Therefore, by Proposition 11.137, we have $\bar{\nu} = \nu$ and $\frac{\mathrm{d}\nu}{\mathrm{d}\mu} = f$ a.e. in $\mathcal{X}$.

This completes the proof of the proposition. □

Theorem 12.35 (Fubini) *Let $\mathcal{X}_i := (X_i, \mathcal{B}_i, \mu_i)$ be a σ-finite $\mathbb{K}$-valued measure space, $i = 1, 2$, $\mathcal{X} := (X_1 \times X_2, \mathcal{B}, \mu) := \mathcal{X}_1 \times \mathcal{X}_2$ be the product $\mathbb{K}$-valued measure space, $\mathcal{Y}$ be a separable Banach space over $\mathbb{K}$, and $f : X_1 \times X_2 \to \mathcal{Y}$ be absolutely integrable over $\mathcal{X}$. Define $f_{x_1} : X_2 \to \mathcal{Y}$ and $f_{\langle x_2 \rangle} : X_1 \to \mathcal{Y}$ as in Proposition 12.28. Then, the following statements hold:*

(i) *There exists $\check{p} : X_1 \to \mathcal{Y}$ and exists $\check{U} \in \mathcal{B}_1$ such that $\mathcal{P} \circ \mu_1(X_1 \setminus \check{U}) = 0$; $\forall x_1 \in \check{U}$, f_{x_1} is absolutely integrable over $\mathcal{X}_2$ and $\check{p}(x_1) = \int_{X_2} f_{x_1} \, \mathrm{d}\mu_2 \in \mathcal{Y}$; $\forall x_1 \in X_1 \setminus \check{U}$, $\check{p}(x_1) = \vartheta_{\mathcal{Y}}$; $\check{p}$ is absolutely integrable over $\mathcal{X}_1$; and $\int_{X_1 \times X_2} f \, \mathrm{d}\mu = \int_{X_1} \check{p} \, \mathrm{d}\mu_1 \in \mathcal{Y}$.*

(ii) *There exists $\check{q} : X_2 \to \mathcal{Y}$ and exists $\check{V} \in \mathcal{B}_2$ such that $\mathcal{P} \circ \mu_2(X_2 \setminus \check{V}) = 0$; $\forall x_2 \in \check{V}$, $f_{\langle x_2 \rangle}$ is absolutely integrable over $\mathcal{X}_1$ and $\check{q}(x_2) = \int_{X_1} f_{\langle x_2 \rangle} \, \mathrm{d}\mu_1 \in \mathcal{Y}$; $\forall x_2 \in X_2 \setminus \check{V}$, $\check{q}(x_2) = \vartheta_{\mathcal{Y}}$; $\check{q}$ is absolutely integrable over $\mathcal{X}_2$; and $\int_{X_1 \times X_2} f \, \mathrm{d}\mu = \int_{X_2} \check{q} \, \mathrm{d}\mu_2 \in \mathcal{Y}$.*

(iii) *$\forall p : X_1 \to \mathcal{Y}$ satisfying: p is $\mathcal{B}_1$-measurable; and $\exists U \in \mathcal{B}_1$ with $\mathcal{P} \circ \mu_1(X_1 \setminus U) = 0$ such that, $\forall x_1 \in U$, f_{x_1} is absolutely integrable over $\mathcal{X}_2$ and $p(x_1) = \int_{X_2} f_{x_1} \, \mathrm{d}\mu_2 \in \mathcal{Y}$, and we have p is absolutely integrable over $\mathcal{X}_1$ and $\int_{X_1 \times X_2} f \, \mathrm{d}\mu = \int_{X_1} p \, \mathrm{d}\mu_1 \in \mathcal{Y}$.*

(iv) *$\forall q : X_2 \to \mathcal{Y}$ satisfying: q is $\mathcal{B}_2$-measurable; and $\exists V \in \mathcal{B}_2$ with $\mathcal{P} \circ \mu_2(X_2 \setminus V) = 0$ such that, $\forall x_2 \in V$, $f_{\langle x_2 \rangle}$ is absolutely integrable over $\mathcal{X}_1$ and $q(x_2) = \int_{X_1} f_{\langle x_2 \rangle} \, \mathrm{d}\mu_1 \in \mathcal{Y}$, and we have q is absolutely integrable over $\mathcal{X}_2$ and $\int_{X_1 \times X_2} f \, \mathrm{d}\mu = \int_{X_2} q \, \mathrm{d}\mu_2 \in \mathcal{Y}$.*

Proof (i) Let $\hat{\mathcal{X}}_i := (X_i, \mathcal{B}_i, \mathcal{P} \circ \mu_i =: \nu_i)$, $i = 1, 2$, and $\hat{\mathcal{X}} := (X_1 \times X_2, \mathcal{B}, \mathcal{P} \circ \mu =: \nu)$. By Proposition 12.22, $\hat{\mathcal{X}}_1$, $\hat{\mathcal{X}}_2$, and $\hat{\mathcal{X}}$ are σ-finite measure spaces and $\hat{\mathcal{X}} = \hat{\mathcal{X}}_1 \times \hat{\mathcal{X}}_2$. Fix any $i = 1, 2$. By Radon–Nikodym Theorem 11.171, we may define functions $\frac{d\mu_i}{d\nu_i} =: g_i : X_i \to \mathbb{K}$. $\forall E \in \mathcal{B}_i$, $\nu_i(E) = \mathcal{P} \circ \mu_i(E) = \int_E \mathcal{P} \circ g_i \, d\nu_i$. By Proposition 11.167, we have $\mathcal{P} \circ g_i = 1$ a.e. in $\hat{\mathcal{X}}_i$. By Proposition 12.34, we have $\frac{d\mu}{d\nu} = g_1 g_2$ a.e. in $\hat{\mathcal{X}}$.

Since f is absolutely integrable over $\mathcal{X}$, then, by Proposition 11.168, $\int_{X_1 \times X_2} f \, d\mu = \int_{X_1 \times X_2} (f g_1 g_2) \, d\nu$ and $(\mathcal{P} \circ f)(\mathcal{P} \circ (g_1 g_2)) = (\mathcal{P} \circ f)(\mathcal{P} \circ g_1)(\mathcal{P} \circ g_2) = \mathcal{P} \circ (f g_1 g_2)$ is integrable over $\hat{\mathcal{X}}$. Define $\bar{f} := f g_1 g_2 : X_1 \times X_2 \to \mathcal{Y}$. By Fubini Theorem 12.30, there exists $\bar{p} : X_1 \to \mathcal{Y}$ and exists $\bar{U} \in \mathcal{B}_1$ such that $\nu_1(X_1 \setminus \bar{U}) = 0$; $\forall x_1 \in \bar{U}$, $\bar{f}_{x_1}$ is absolutely integrable over $\hat{\mathcal{X}}_2$ and $\bar{p}(x_1) = \int_{X_2} \bar{f}_{x_1} \, d\nu_2$; $\forall x_1 \in X_1 \setminus \bar{U}$, $\bar{p}(x_1) = \vartheta_{\mathcal{Y}}$; $\bar{p}$ is absolutely integrable over $\hat{\mathcal{X}}_1$; and $\int_{X_1 \times X_2} \bar{f} \, d\nu = \int_{X_1} \bar{p} \, d\nu_1$. Let $A_1 := \{x_1 \in X_1 \mid g_1(x_1) = 0\}$. Clearly, $A_1 \in \mathcal{B}_1$ and $\nu_1(A_1) = 0$. Let $\check{U} := \bar{U} \setminus A_1 \in \mathcal{B}_1$. Then, $\mathcal{P} \circ \mu_1(X_1 \setminus \check{U}) = \nu_1(X_1 \setminus \check{U}) = \nu_1((X_1 \setminus \bar{U}) \cup A_1) = 0$. $\forall x_1 \in \check{U}$, $\bar{f}_{x_1}$ is absolutely integrable over $\hat{\mathcal{X}}_2$ implies that $f_{x_1} g_1(x_1) g_2$ is absolutely integrable over $\hat{\mathcal{X}}_2$, which further implies that $(\mathcal{P} \circ f_{x_1})(\mathcal{P} \circ g_2)$ is integrable over $\hat{\mathcal{X}}_2$. By Proposition 11.168, f_{x_1} is absolutely integrable over $\mathcal{X}_2$ and $\int_{X_2} f_{x_1} \, d\mu_2 = \int_{X_2} (f_{x_1} g_2) \, d\nu_2$. Then, $\bar{p}(x_1) = g_1(x_1) \int_{X_2} f_{x_1} \, d\mu_2$. Define $\check{p} : X_1 \to \mathcal{Y}$ by $\check{p}(x_1) = \begin{cases} \int_{X_2} f_{x_1} \, d\mu_2 & x_1 \in \check{U} \\ \vartheta_{\mathcal{Y}} & x_1 \in X_1 \setminus \check{U} \end{cases}$, $\forall x_1 \in X_1$. Then, $\check{p}(x_1) = \begin{cases} \bar{p}(x_1)/g_1(x_1) & x_1 \in \check{U} \\ \vartheta_{\mathcal{Y}} & x_1 \in X_1 \setminus \check{U} \end{cases}$. By Propositions 11.41, 11.38, and 11.39, $\check{p}$ is $\mathcal{B}_1$-measurable. Since $\bar{p}$ and $g_1 \check{p}$ are $\mathcal{B}_1$-measurable, then, by Lemma 11.43, $\bar{p} = g_1 \check{p}$ a.e. in $\mathcal{X}_1$. Since $\bar{p}$ is absolutely integrable over $\hat{\mathcal{X}}_1$, then $\mathcal{P} \circ (g_1 \check{p}) = (\mathcal{P} \circ \check{p})(\mathcal{P} \circ g_1)$ is integrable over $\hat{\mathcal{X}}_1$. By Proposition 11.168, $\check{p}$ is absolutely integrable over $\mathcal{X}_1$ and $\int_{X_1} (\check{p} g_1) \, d\nu_1 = \int_{X_1} \check{p} \, d\mu_1$. By Proposition 11.92, we have $\int_{X_1 \times X_2} f \, d\mu = \int_{X_1 \times X_2} \bar{f} \, d\nu = \int_{X_1} \bar{p} \, d\nu_1 = \int_{X_1} (\check{p} g_1) \, d\nu_1 = \int_{X_1} \check{p} \, d\mu_1$. Hence, (i) holds.

By symmetry, (ii) holds.

(iii) Fix any $p : X_1 \to \mathcal{Y}$ satisfying: p is $\mathcal{B}_1$-measurable; and $\exists U \in \mathcal{B}_1$ with $\mathcal{P} \circ \mu_1(X_1 \setminus U) = 0$ such that, $\forall x_1 \in U$, f_{x_1} is absolutely integrable over $\mathcal{X}_2$ and $p(x_1) = \int_{X_2} f_{x_1} \, d\mu_2 \in \mathcal{Y}$. By (i), we have $U_0 := \check{U} \cap U \in \mathcal{B}_1$ and $\mathcal{P} \circ \mu_1(X_1 \setminus U_0) = \mathcal{P} \circ \mu_1((X_1 \setminus \check{U}) \cup (X_1 \setminus U)) = 0$. $\forall x_1 \in U_0$, $p(x_1) = \int_{X_2} f_{x_1} \, d\mu_2 = \check{p}(x_1) \in \mathcal{Y}$. By Lemma 11.43, $p = \check{p}$ a.e. in $\mathcal{X}_1$. By Proposition 11.83, we have p is absolutely integrable over $\mathcal{X}_1$ since $\check{p}$ is so. By Proposition 11.92, $\int_{X_1} p \, d\mu_1 = \int_{X_1} \check{p} \, d\mu_1 = \int_{X_1 \times X_2} f \, d\mu \in \mathcal{Y}$. Hence, (iii) holds.

By symmetry, (iv) holds. This completes the proof of the theorem. □

12.4 Functions of Bounded Variation

Definition 12.36 Let $m \in \mathbb{N}$, $\mathbb{R}^m$ be endowed with the usual positive cone, $\mathcal{Y}$ be a normed linear space over $\mathbb{K}$, $D \subseteq \mathbb{R}^m$, and $F : D \to \mathcal{Y}$. F is said to be *continuous on the right at* $x_0 \in D$ if $\forall \epsilon \in (0, \infty) \subset \mathbb{R}$, $\exists \delta \in (0, \infty) \subset \mathbb{R}$, $\forall x \in \mathcal{B}_{\mathbb{R}^m}(x_0, \delta) \cap D$ with $x_0 \leqq x$, we have $\|F(x) - F(x_0)\| < \epsilon$. F is said to be *continuous on the right* if it is continuous on the right at x, $\forall x \in D$. F is said to be *continuous on the right on* $E \subseteq D$ if it is continuous on the right at x, $\forall x \in E$. ◇

Proposition 12.37 *Let $m \in \mathbb{N}$, $\mathbb{R}^m$ be endowed with the usual positive cone P, $\mathcal{Y}$ be a normed linear space over $\mathbb{K}$, $x_0 \in \Omega \subseteq \mathbb{R}^m$, and $F : \Omega \to \mathcal{Y}$. Then, the following statements are equivalent:*

(i) *F is continuous on the right at x_0.*

(ii) *If x_0 is an accumulation point of $\Omega \cap (x_0 + P)$, then $\lim_{\substack{h \to \mathbf{0}_m \\ h \geqq \mathbf{0}_m}} F(x_0 + h) = F(x_0)$.*

(iii) *If x_0 is an accumulation point of $\Omega \cap (x_0 + P)$, then $\forall (h_n)_{n=1}^{\infty} \subseteq (P \cap (\Omega - x_0)) \setminus \{\mathbf{0}_m\}$ with $\lim_{n \in \mathbb{N}} h_n = \mathbf{0}_m$, we have $\lim_{n \in \mathbb{N}} F(x_0 + h_n) = F(x_0)$.*

Proof (i) ⇒ (iii). Let F be continuous on the right at x_0 and x_0 be an accumulation point of $\Omega \cap (x_0 + P)$. By (i), $\forall \epsilon \in (0, \infty) \subset \mathbb{R}$, $\exists \delta \in (0, \infty) \subset \mathbb{R}$, $\forall x \in \mathcal{B}_{\mathbb{R}^m}(x_0, \delta) \cap \Omega$ with $x_0 \leqq x$, we have $\|F(x) - F(x_0)\| < \epsilon$. Then, $\forall (h_n)_{n=1}^{\infty} \subseteq (P \cap (\Omega - x_0)) \setminus \{\mathbf{0}_m\}$ with $\lim_{n \in \mathbb{N}} h_n = \mathbf{0}_m$, $\exists N \in \mathbb{N}$ such that $\forall n \in \mathbb{N}$ with $N \leq n$, $h_n \in \mathcal{B}_{\mathbb{R}^m}(\mathbf{0}_m, \delta)$. This implies that $x_0 + h_n \in \mathcal{B}_{\mathbb{R}^m}(x_0, \delta) \cap \Omega$, $x_0 + h_n \geqq x_0$, and $\|F(x_0 + h_n) - F(x_0)\| < \epsilon$. Hence, $\lim_{n \in \mathbb{N}} F(x_0 + h_n) = F(x_0)$.

(iii) ⇒ (ii). Let $D := P \cap (\Omega - x_0)$. Define $\bar{F} : D \to \mathcal{Y}$ by $\bar{F}(h) = F(x_0 + h)$, $\forall h \in D$. By (iii), we have $\forall (h_n)_{n=1}^{\infty} \subseteq D \setminus \{\mathbf{0}_m\}$ with $\lim_{n \in \mathbb{N}} h_n = \mathbf{0}_m$, we have $\lim_{n \in \mathbb{N}} \bar{F}(h_n) = F(x_0)$. By Proposition 4.16, we have $\lim_{h \to \mathbf{0}_m} \bar{F}(h) = F(x_0)$. Hence, $\lim_{\substack{h \to \mathbf{0}_m \\ h \geqq \mathbf{0}_m}} F(x_0 + h) = F(x_0)$.

(ii) ⇒ (i). We will distinguish two exhaustive and mutually exclusive cases: Case 1: x_0 is not an accumulation point of $\Omega \cap (x_0 + P)$; Case 2: x_0 is an accumulation point of $\Omega \cap (x_0 + P)$. Case 1: x_0 is not an accumulation point of $\Omega \cap (x_0 + P)$. $\exists \delta \in (0, \infty) \subset \mathbb{R}$ such that $\mathcal{B}_{\mathbb{R}^m}(x_0, \delta) \cap ((\Omega \cap (x_0 + P)) \setminus \{x_0\}) = \emptyset$ and $\mathcal{B}_{\mathbb{R}^m}(x_0, \delta) \cap \Omega \cap (x_0 + P) = \{x_0\}$. $\forall \epsilon \in (0, \infty) \subset \mathbb{R}$, $\forall x \in \mathcal{B}_{\mathbb{R}^m}(x_0, \delta) \cap \Omega$ with $x_0 \leqq x$, we have $x = x_0$ and $\|F(x) - F(x_0)\| = 0 < \epsilon$. Hence, F is continuous on the right at x_0.

Case 2: x_0 is an accumulation point of $\Omega \cap (x_0 + P)$. By (ii), $\lim_{\substack{h \to \mathbf{0}_m \\ h \geqq \mathbf{0}_m}} F(x_0 + h) = F(x_0)$. $\forall \epsilon \in (0, \infty) \subset \mathbb{R}$, $\exists \delta \in (0, \infty) \subset \mathbb{R}$, $\forall h \in (\mathcal{B}_{\mathbb{R}^m}(\mathbf{0}_m, \delta) \cap P \cap (\Omega - x_0)) \setminus \{\mathbf{0}_m\}$, we have $\|F(x_0 + h) - F(x_0)\| < \epsilon$. Then, $\forall x \in \mathcal{B}_{\mathbb{R}^m}(x_0, \delta) \cap \Omega \cap (x_0 + P)$, we have $x - x_0 \in \mathcal{B}_{\mathbb{R}^m}(\mathbf{0}_m, \delta) \cap (\Omega - x_0) \cap P$ and $\|F(x) - F(x_0)\| < \epsilon$. Hence, F is continuous on the right at x_0.

In both cases, F is continuous on the right at x_0.

This completes the proof of the proposition. □

Definition 12.38 Let $m \in \mathbb{N}$, $\mathbb{R}^m$ be endowed with the usual positive cone. $\forall x_1, x_2 \in \mathbb{R}^m$ with $x_1 \stackrel{\lessdot}{=} x_2$, the *semi-open rectangle with corners* x_1 *and* x_2 is $\mathrm{r}_{x_1,x_2} := \left\{x \in \mathbb{R}^m \;\middle|\; x_1 \lessdot x \stackrel{\lessdot}{=} x_2\right\}$. The *closed rectangle with corners* x_1 *and* x_2 is $\bar{\mathrm{r}}_{x_1,x_2} := \left\{x \in \mathbb{R}^m \;\middle|\; x_1 \stackrel{\lessdot}{=} x \stackrel{\lessdot}{=} x_2\right\}$. The *open rectangle with corners* x_1 *and* x_2 is $\mathrm{r}^\circ_{x_1,x_2} := \{x \in \mathbb{R}^m \mid x_1 \lessdot x \lessdot x_2\}$. $\Omega \subseteq \mathbb{R}^m$ is said to be a *region* if $\exists N \subseteq \mathbb{N}$ and $(\bar{x}_i)_{i\in N}, (\tilde{x}_i)_{i\in N} \subseteq \Omega$ with $\bar{x}_i \stackrel{\lessdot}{=} \tilde{x}_i$ and $\bar{\mathrm{r}}_{\bar{x}_i,\tilde{x}_i} \subseteq \Omega$, $\forall i \in N$, such that $\Omega = \bigcup_{i\in N} \bar{\mathrm{r}}_{\bar{x}_i,\tilde{x}_i}$, $(\mathrm{r}_{\bar{x}_i,\tilde{x}_i})_{i\in N}$ is pairwise disjoint, $\mathrm{P}(\Omega) := \bigcup_{i\in N} \mathrm{r}_{\bar{x}_i,\tilde{x}_i}$, and $\forall x_1, x_2 \in \Omega$ with $x_1 \stackrel{\lessdot}{=} x_2$ and $\bar{\mathrm{r}}_{x_1,x_2} \subseteq \Omega$, we have $\mathrm{r}_{x_1,x_2} \subseteq \mathrm{P}(\Omega)$. $\mathrm{P}(\Omega)$ is uniquely defined for each region Ω (see below) and is called the *principal* of Ω. ◇

Clearly, any region $\Omega \subseteq \mathbb{R}^m$ we have $\Omega \in \mathcal{B}_{\mathrm{B}}(\mathbb{R}^m)$ and $\mathrm{P}(\Omega) \in \mathcal{B}_{\mathrm{B}}(\mathbb{R}^m)$.

Proposition 12.39 *Let $m \in \mathbb{N}$, $\mathbb{R}^m$ be endowed with the usual positive cone. Then, the following holds:*

(i) *If $\Omega \subseteq \mathbb{R}^m$ is a region, then $\mathrm{P}(\Omega) \subseteq \Omega$ is uniquely defined, $\Omega \in \mathcal{B}_{\mathrm{B}}(\mathbb{R}^m)$, and $\mathrm{P}(\Omega) \in \mathcal{B}_{\mathrm{B}}(\mathbb{R}^m)$.*
(ii) *If $\Omega_1, \Omega_2 \subseteq \mathbb{R}^m$ are regions with $\Omega_1 \subseteq \Omega_2$, then $\mathrm{P}(\Omega_1) \subseteq \mathrm{P}(\Omega_2)$.*
(iii) *If $\Omega_1, \Omega_2, \Omega_1 \cup \Omega_2 \subseteq \mathbb{R}^m$ are regions, then $\mathrm{P}(\Omega_1) \cup \mathrm{P}(\Omega_2) \subseteq \mathrm{P}(\Omega_1 \cup \Omega_2)$.*
(iv) *If $\Omega_1, \Omega_2, \Omega_1 \cap \Omega_2 \subseteq \mathbb{R}^m$ are regions, then $\mathrm{P}(\Omega_1) \cap \mathrm{P}(\Omega_2) \supseteq \mathrm{P}(\Omega_1 \cap \Omega_2)$.*
(v) *Let $\Omega \in \mathcal{B}_{\mathrm{B}}(\mathbb{R}^m)$ be a rectangle, that is, $\exists$ intervals $I_1, \ldots, I_m$ such that $\Omega = \prod_{i=1}^m I_i$. $\forall i \in \{1, \ldots, m\} =: \bar{J}$, if $I_i = [a_i, b_i] \subset \mathbb{R}$ with $a_i, b_i \in \mathbb{R}$ and $a_i < b_i$, let $\bar{I}_i := (a_i, b_i] \subset \mathbb{R}$; if $I_i = [a_i, a_i] \subset \mathbb{R}$ with $a_i \in \mathbb{R}$, let $\bar{I}_i := \emptyset$; if $I_i = [a_i, b_i) \subset \mathbb{R}$ with $a_i, b_i \in \mathbb{R}$ and $a_i < b_i$, let $\bar{I}_i := (a_i, b_i) \subset \mathbb{R}$; if $I_i = [a_i, \infty) \subset \mathbb{R}$ with $a_i \in \mathbb{R}$, let $\bar{I}_i := (a_i, \infty) \subset \mathbb{R}$; in all other cases, let $\bar{I}_i := I_i$. Then, Ω is a region and $\mathrm{P}(\Omega) = \prod_{i=1}^m \bar{I}_i$ is a rectangle.*
(vi) *$\forall x_1, x_2 \in \mathbb{R}^m$ with $x_1 \stackrel{\lessdot}{=} x_2$, we have $\bar{\mathrm{r}}_{x_1,x_2}$ and r_{x_1,x_2} are regions with $\mathrm{P}(\bar{\mathrm{r}}_{x_1,x_2}) = \mathrm{P}(\mathrm{r}_{x_1,x_2}) = \mathrm{r}_{x_1,x_2}$.*
(vii) *Let $M \subseteq \mathbb{N}$ and $(\check{x}_j)_{j\in M}, (\hat{x}_j)_{j\in M} \subseteq \mathbb{R}^m$ with $\check{x}_j \stackrel{\lessdot}{=} \hat{x}_j$, $\forall j \in M$. Then, $\Omega := \bigcup_{j\in M} \mathrm{r}_{\check{x}_j,\hat{x}_j} \subseteq \mathbb{R}^m$ is a region and $\mathrm{P}(\Omega) = \Omega$.*
(viii) *If $\Omega_1 \subseteq \mathbb{R}^l$ and $\Omega_2 \subseteq \mathbb{R}^{m-l}$ are regions, where $l \in \{1, \ldots, m-1\}$, then $\Omega_1 \times \Omega_2 \subseteq \mathbb{R}^m$ is a region with $\mathrm{P}(\Omega_1 \times \Omega_2) = \mathrm{P}(\Omega_1) \times \mathrm{P}(\Omega_2)$.*

Proof (i) Let $j = 1, 2$, and $\exists N_j \subseteq \mathbb{N}$ and $\left(\bar{x}_{j,i}\right)_{i\in N_j}, \left(\tilde{x}_{j,i}\right)_{i\in N_j} \subseteq \Omega$ with $\bar{x}_{j,i} \stackrel{\lessdot}{=} \tilde{x}_{j,i}$ and $\bar{\mathrm{r}}_{\bar{x}_{j,i},\tilde{x}_{j,i}} \subseteq \Omega$, $\forall i \in N_j$, such that $\Omega = \bigcup_{i\in N_j} \bar{\mathrm{r}}_{\bar{x}_{j,i},\tilde{x}_{j,i}}$, $\left(\mathrm{r}_{\bar{x}_{j,i},\tilde{x}_{j,i}}\right)_{i\in N_j}$ is pairwise disjoint, $\Omega_j := \bigcup_{i\in N_j} \mathrm{r}_{\bar{x}_{j,i},\tilde{x}_{j,i}}$, and $\forall x_1, x_2 \in \Omega$ with $x_1 \stackrel{\lessdot}{=} x_2$ and $\bar{\mathrm{r}}_{x_1,x_2} \subseteq \Omega$, we have $\mathrm{r}_{x_1,x_2} \subseteq \Omega_j$. We will show that $\Omega_1 = \Omega_2$. Then, the result follows.

$\forall i \in N_1$, we have $\bar{x}_{1,i} \stackrel{\lessdot}{=} \tilde{x}_{1,i}$ and $\bar{\mathrm{r}}_{\bar{x}_{1,i},\tilde{x}_{1,i}} \subseteq \Omega$. This implies that $\mathrm{r}_{\bar{x}_{1,i},\tilde{x}_{1,i}} \subseteq \Omega_2$. Hence, $\Omega_1 = \bigcup_{i\in N_1} \mathrm{r}_{\bar{x}_{1,i},\tilde{x}_{1,i}} \subseteq \Omega_2$. By symmetry, we have $\Omega_2 \subseteq \Omega_1$.

(ii) By Ω_1 being a region, $\exists N_1 \subseteq \mathbb{N}$ and $(\bar{x}_{1,i})_{i\in N_1}, (\tilde{x}_{1,i})_{i\in N_1} \subseteq \Omega_1 \subseteq \Omega_2$ with $\bar{x}_{1,i} \stackrel{\leq}{=} \tilde{x}_{1,i}$ and $\bar{\mathrm{r}}_{\bar{x}_{1,i},\tilde{x}_{1,i}} \subseteq \Omega_1 \subseteq \Omega_2$, $\forall i \in N_1$, such that $\Omega_1 = \bigcup_{i\in N_1} \bar{\mathrm{r}}_{\bar{x}_{1,i},\tilde{x}_{1,i}}$ and $\mathrm{P}(\Omega_1) = \bigcup_{i\in N_1} \mathrm{r}_{\bar{x}_{1,i},\tilde{x}_{1,i}}$. By Ω_2 being a region, we have $\mathrm{r}_{\bar{x}_{1,i},\tilde{x}_{1,i}} \subseteq \mathrm{P}(\Omega_2)$, $\forall i \in N_1$. Hence, $\mathrm{P}(\Omega_1) \subseteq \mathrm{P}(\Omega_2)$.

(iii) By (ii), $\mathrm{P}(\Omega_1 \cup \Omega_2) \supseteq \mathrm{P}(\Omega_i)$, $i = 1, 2$. Then, the result follows.

(iv) By (ii), $\mathrm{P}(\Omega_1 \cap \Omega_2) \subseteq \mathrm{P}(\Omega_i)$, $i = 1, 2$. Then, the result follows.

(v) This is straightforward and is therefore omitted.

(vi) This follows immediately from (v).

(vii) We will first prove the result under the additional assumption that $(\mathrm{r}_{\check{x}_j,\hat{x}_j})_{j\in M} =: (\Omega_j)_{j\in M}$ is pairwise disjoint. $\forall j \in M$, by (vi), Ω_j is a region and $\mathrm{P}(\Omega_j) = \Omega_j$. Then, $\exists N_j \subseteq \mathbb{N}$ and $(\bar{x}_{j,i})_{i\in N_j}, (\tilde{x}_{j,i})_{i\in N_j} \subseteq \Omega_j$ with $\bar{x}_{j,i} \stackrel{\leq}{=} \tilde{x}_{j,i}$, $\forall i \in N_j$, such that $\Omega_j = \bigcup_{i\in N_j} \bar{\mathrm{r}}_{\bar{x}_{j,i},\tilde{x}_{j,i}}$, $(\mathrm{r}_{\bar{x}_{j,i},\tilde{x}_{j,i}})_{i\in N_j}$ is pairwise disjoint, $\mathrm{P}(\Omega_j) = \Omega_j = \bigcup_{i\in N_j} \mathrm{r}_{\bar{x}_{j,i},\tilde{x}_{j,i}}$, and $\forall x_1, x_2 \in \Omega_j$ with $x_1 \stackrel{\leq}{=} x_2$ and $\bar{\mathrm{r}}_{x_1,x_2} \subseteq \Omega_j$, we have $\mathrm{r}_{x_1,x_2} \subseteq \mathrm{P}(\Omega_j) = \Omega_j$. For $\Omega := \bigcup_{j\in M} \Omega_j$, consider $(\bar{x}_{j,i})_{j\in M,\, i\in N_j}, (\tilde{x}_{j,i})_{j\in M,\, i\in N_j} \subseteq \Omega$. We have $\bar{x}_{j,i} \stackrel{\leq}{=} \tilde{x}_{j,i}$, $\forall j \in M$, $\forall i \in N_j$, $(\mathrm{r}_{\bar{x}_{j,i},\tilde{x}_{j,i}})_{j\in M,\, i\in N_j}$ is pairwise disjoint, $\Omega = \bigcup_{j\in M} \bigcup_{i\in N_j} \bar{\mathrm{r}}_{\bar{x}_{j,i},\tilde{x}_{j,i}}$, $\mathrm{P}(\Omega) = \bigcup_{j\in M} \bigcup_{i\in N_j} \mathrm{r}_{\bar{x}_{j,i},\tilde{x}_{j,i}} = \bigcup_{j\in M} \Omega_j = \Omega$, and $\forall x_1, x_2 \in \Omega$ with $x_1 \stackrel{\leq}{=} x_2$ and $\bar{\mathrm{r}}_{x_1,x_2} \subseteq \Omega$, we have $\mathrm{r}_{x_1,x_2} \subseteq \Omega = \mathrm{P}(\Omega)$. Hence, Ω is a region with $\mathrm{P}(\Omega) = \Omega$.

Next, we prove the general case when $(\Omega_j)_{j\in M}$ is not necessarily pairwise disjoint. Without loss of generality, take $M = \mathbb{N}$. Then, $\Omega = \bigcup_{n=1}^{\infty} \left(\Omega_n \setminus (\bigcup_{i=1}^{n-1} \Omega_i)\right)$. Since for semi-open rectangles A and B, $A \setminus B$ can be expressed as a finite union of pairwise disjoint semi-open rectangles. Then, $\Omega = \bigcup_{n=1}^{\infty} \bar{\Omega}_n$, where $\bar{\Omega}_n = \mathrm{r}_{\check{x}_n,\hat{x}_n}$ for some $\check{x}_n, \hat{x}_n \in \mathbb{R}^m$ with $\check{x}_n \stackrel{\leq}{=} \hat{x}_n$, $\forall n \in \mathbb{N}$, and the sets in the union are pairwise disjoint. By the special case, we have Ω is a region and $\mathrm{P}(\Omega) = \Omega$.

(viii) By Ω_j being a region, $\exists N_j \subseteq \mathbb{N}$ and $(\bar{x}_{j,i})_{i\in N_j}, (\tilde{x}_{j,i})_{i\in N_j} \subseteq \Omega_j$ with $\bar{x}_{j,i} \stackrel{\leq}{=} \tilde{x}_{j,i}$ and $\bar{\mathrm{r}}_{\bar{x}_{j,i},\tilde{x}_{j,i}} \subseteq \Omega_j$, $\forall i \in N_j$, such that $\Omega_j = \bigcup_{i\in N_j} \bar{\mathrm{r}}_{\bar{x}_{j,i},\tilde{x}_{j,i}}$, $(\mathrm{r}_{\bar{x}_{j,i},\tilde{x}_{j,i}})_{i\in N_j}$ is pairwise disjoint, $\mathrm{P}(\Omega_j) := \bigcup_{i\in N_j} \mathrm{r}_{\bar{x}_{j,i},\tilde{x}_{j,i}}$, and $\forall x_1, x_2 \in \Omega$ with $x_1 \stackrel{\leq}{=} x_2$ and $\bar{\mathrm{r}}_{x_1,x_2} \subseteq \Omega_j$, we have $\mathrm{r}_{x_1,x_2} \subseteq \mathrm{P}(\Omega_j)$, $j = 1, 2$.

Consider $\left((\bar{x}_{1,i}, \bar{x}_{2,j})\right)_{i\in N_1,\, j\in N_2}, \left((\tilde{x}_{1,i}, \tilde{x}_{2,j})\right)_{i\in N_1,\, j\in N_2} \subseteq \Omega_1 \times \Omega_2$. We have $(\bar{x}_{1,i}, \bar{x}_{2,j}) \stackrel{\leq}{=} (\tilde{x}_{1,i}, \tilde{x}_{2,j})$ and $\bar{\mathrm{r}}_{(\bar{x}_{1,i},\bar{x}_{2,j}),(\tilde{x}_{1,i},\tilde{x}_{2,j})} = \bar{\mathrm{r}}_{\bar{x}_{1,i},\tilde{x}_{1,i}} \times \bar{\mathrm{r}}_{\bar{x}_{2,j},\tilde{x}_{2,j}} \subseteq \Omega_1 \times \Omega_2$, $\forall i \in N_1$, $\forall j \in N_2$. Furthermore, $\Omega_1 \times \Omega_2 = \bigcup_{i\in N_1} \bigcup_{j\in N_2} \bar{\mathrm{r}}_{\bar{x}_{1,i},\tilde{x}_{1,i}} \times \bar{\mathrm{r}}_{\bar{x}_{2,j},\tilde{x}_{2,j}} = \bigcup_{i\in N_1} \bigcup_{j\in N_2} \bar{\mathrm{r}}_{(\bar{x}_{1,i},\bar{x}_{2,j}),(\tilde{x}_{1,i},\tilde{x}_{2,j})}$, $\left(\mathrm{r}_{(\bar{x}_{1,i},\bar{x}_{2,j}),(\tilde{x}_{1,i},\tilde{x}_{2,j})}\right)_{\substack{i\in N_1\\ j\in N_2}} = \left(\mathrm{r}_{\bar{x}_{1,i},\tilde{x}_{1,i}} \times \mathrm{r}_{\bar{x}_{2,j},\tilde{x}_{2,j}}\right)_{\substack{i\in N_1\\ j\in N_2}}$ are pairwise disjoint. $\bigcup_{i\in N_1} \bigcup_{j\in N_2} \mathrm{r}_{\bar{x}_{1,i},\tilde{x}_{1,i}} \times \mathrm{r}_{\bar{x}_{2,j},\tilde{x}_{2,j}} = \mathrm{P}(\Omega_1) \times \mathrm{P}(\Omega_2) = \bigcup_{i\in N_1} \bigcup_{j\in N_2} \mathrm{r}_{(\bar{x}_{1,i},\bar{x}_{2,j}),(\tilde{x}_{1,i},\tilde{x}_{2,j})}$. $\forall (\bar{x}, \bar{y}), (\tilde{x}, \tilde{y}) \in \Omega_1 \times \Omega_2$ with $(\bar{x}, \bar{y}) \stackrel{\leq}{=} (\tilde{x}, \tilde{y})$ and $\bar{\mathrm{r}}_{(\bar{x},\bar{y}),(\tilde{x},\tilde{y})} \subseteq \Omega_1 \times \Omega_2$, we have $\bar{x} \stackrel{\leq}{=} \tilde{x}$ and $\bar{y} \stackrel{\leq}{=} \tilde{y}$, $\bar{\mathrm{r}}_{(\bar{x},\bar{y}),(\tilde{x},\tilde{y})} = \bar{\mathrm{r}}_{\bar{x},\tilde{x}} \times \bar{\mathrm{r}}_{\bar{y},\tilde{y}} \subseteq \Omega_1 \times \Omega_2$, $\bar{\mathrm{r}}_{\bar{x},\tilde{x}} \subseteq \Omega_1$, $\bar{\mathrm{r}}_{\bar{y},\tilde{y}} \subseteq \Omega_2$, which further implies that

$\mathrm{r}_{\bar{x},\tilde{x}} \subseteq \mathrm{P}(\Omega_1)$, $\mathrm{r}_{\bar{y},\tilde{y}} \subseteq \mathrm{P}(\Omega_2)$, and $\mathrm{r}_{(\bar{x},\bar{y}),(\tilde{x},\tilde{y})} = \mathrm{r}_{\bar{x},\tilde{x}} \times \mathrm{r}_{\bar{y},\tilde{y}} \subseteq \mathrm{P}(\Omega_1) \times \mathrm{P}(\Omega_2)$. Hence, $\Omega_1 \times \Omega_2$ is a region with $\mathrm{P}(\Omega_1 \times \Omega_2) = \mathrm{P}(\Omega_1) \times \mathrm{P}(\Omega_2)$.

This completes the proof of the proposition. □

Proposition 12.40 *Let $m \in \mathbb{N}$, $\mathbb{R}^m$ be endowed with the usual positive cone, $\Omega \in \mathcal{O}_{\mathbb{R}^m}$ with the subset topology $\mathcal{O}$. Then, Ω is a region with $\mathrm{P}(\Omega) = \Omega$.*

Proof By $\mathbb{R}$ being second countable and Proposition 3.28, $\mathbb{R}^m$ is second countable. By Proposition 4.4, $\mathbb{R}^m$ is separable, which implies that, by Proposition 4.38, $(\Omega, \mathcal{O})$ is separable and second countable. Let $D \subseteq \Omega$ be a countable dense subset and define $\mathcal{B} := \{\mathrm{r}^{\circ}_{x-\delta\mathbf{1}_m, x+\delta\mathbf{1}_m} \subseteq \Omega \mid x \in D, \delta \in \mathbb{Q}, \delta > 0, \bar{\mathrm{r}}_{x-\delta\mathbf{1}_m, x+\delta\mathbf{1}_m} \subseteq \Omega\}$. It is easy to show, by Definition 3.17, that $\mathcal{B}$ is a countable basis for $\mathcal{O}$. Then, $\Omega = \bigcup_{B\in\mathcal{B}} B =: \bigcup_{i\in N} B_i = \bigcup_{i\in N} \bar{\mathrm{r}}_{x_i-\delta_i\mathbf{1}_m, x_i+\delta_i\mathbf{1}_m} = \bigcup_{i\in N} \mathrm{r}_{x_i-\delta_i\mathbf{1}_m, x_i+\delta_i\mathbf{1}_m}$, where $N \subseteq \mathbb{N}$. By (vii) of Proposition 12.39, Ω is a region with $\mathrm{P}(\Omega) = \Omega$. This completes the proof of the proposition. □

Definition 12.41 Let $m \in \mathbb{N}$, $\mathbb{R}^m$ be endowed with the usual positive cone, $\mathcal{Y}$ be a normed linear space over $\mathbb{K}$, $\Omega \in \mathcal{B}_{\mathrm{B}}(\mathbb{R}^m)$ be a region, and $F : \Omega \to \mathcal{Y}$. $\forall x_1, x_2 \in \mathbb{R}^m$ with $x_1 \stackrel{\lessdot}{=} x_2$, $\forall J \subseteq \bar{J} := \{1, \ldots, m\}$, the vertex of r_{x_1,x_2} with any of J coordinates equal to that of x_1 and the rest of coordinates equal to that of x_2 is $x_J \in \bar{\mathrm{r}}_{x_1,x_2} \subseteq \mathbb{R}^m$ such that $\pi_i(x_J) = \begin{cases} \pi_i(x_1) & \forall i \in J \\ \pi_i(x_2) & \forall i \in \bar{J} \setminus J \end{cases}$. We define the set of vertexes of r_{x_1,x_2} with i coordinates equal to that of x_1 to be $\mathrm{VRect}_{i,x_1,x_2} := \{x_J \in \mathbb{R}^m \mid J \subseteq \bar{J}, \mathrm{card}(J) = i\}$, if $x_1 \lessdot x_2$. When $\bar{\mathrm{r}}_{x_1,x_2} \subseteq \Omega$, *the increment of F on r_{x_1,x_2} is* $\Delta_F(\mathrm{r}_{x_1,x_2}) := \begin{cases} \sum_{i=0}^{m} \sum_{\bar{x}\in\mathrm{VRect}_{i,x_1,x_2}} (-1)^i F(\bar{x}) & x_1 \lessdot x_2 \\ \vartheta_{\mathcal{Y}} & x_1 \not\lessdot x_2 \end{cases} = \sum_{J\subseteq\bar{J}} (-1)^{\mathrm{card}(J)} F(x_J)$. The *total variation of F on the semi-open rectangle* r_{x_1,x_2} is

$$T_F\big(\mathrm{r}_{x_1,x_2}\big) := \sup_{\substack{n\in\mathbb{Z}_+,\, \check{x}_1,\hat{x}_1,\ldots,\check{x}_n,\hat{x}_n\in\bar{\mathrm{r}}_{x_1,x_2},\, \check{x}_i \stackrel{\lessdot}{=} \hat{x}_i,\, \forall i\in\{1,\ldots,n\} \\ \bigcup_{i=1}^{n} \mathrm{r}_{\check{x}_i,\hat{x}_i} = \mathrm{r}_{x_1,x_2},\, \mathrm{r}_{\check{x}_i,\hat{x}_i}\cap\mathrm{r}_{\check{x}_j,\hat{x}_j}=\emptyset,\, \forall 1\le i<j\le n}} \sum_{i=1}^{n} \big\| \Delta_F(\mathrm{r}_{\check{x}_i,\hat{x}_i}) \big\|$$

where $T_F(\mathrm{r}_{x_1,x_2}) \in [0, \infty] \subset \mathbb{R}_e$. F is said to be *of locally bounded variation* if:

(i) F is continuous on the right.
(ii) $\forall x_1, x_2 \in \Omega$ with $x_1 \stackrel{\lessdot}{=} x_2$ and $\bar{\mathrm{r}}_{x_1,x_2} \subseteq \Omega$, we have $T_F(\mathrm{r}_{x_1,x_2}) < \infty$.
(iii) $\forall x_1, x_2 \in \Omega$ with $x_1 \lessdot x_2$ and $\bar{\mathrm{r}}_{x_1,x_2} \subseteq \Omega$, $\bar{G} : \bar{\mathrm{r}}_{x_1,x_2} \to [0, \infty) \subset \mathbb{R}$ defined by $\bar{G}(x) = T_F\big(\mathrm{r}_{x_1,x}\big)$, $\forall x \in \bar{\mathrm{r}}_{x_1,x_2}$, is continuous on the right.

The *total variation of F* is defined by $T_F := \sum_{i\in N} T_F(\mathrm{r}_{\bar{x}_i,\tilde{x}_i})$ if the sum is independent of the choice of $(\bar{x}_i)_{i\in N}$ and $(\tilde{x}_i)_{i\in N}$ as prescribed in Definition 12.38. F is said to be *of bounded variation* if F is of locally bounded variation and $T_F < \infty$. ◇

Definition 12.42 Let $m \in \mathbb{N}$, $\mathbb{R}^m$ be endowed with the usual positive cone, $\mathcal{Y}$ be a normed linear space, $\Omega \in \mathcal{B}_B(\mathbb{R}^m)$ be a region with the subset topology, μ be a σ-finite $\mathcal{Y}$-valued measure on $(\mathrm{P}(\Omega), \mathcal{B}_B(\mathrm{P}(\Omega)))$, and $F : \Omega \to \mathcal{Y}$. F is said to be a *cumulative distribution function of* μ if $\forall x_1, x_2 \in \Omega$ with $\bar{\mathrm{r}}_{x_1,x_2} \subseteq \Omega$ and $x_1 \leqq x_2$, we have $\mu(\mathrm{r}_{x_1,x_2}) = \Delta_F(\mathrm{r}_{x_1,x_2})$. ◇

Note that, for σ-finite $\mathcal{Y}$-valued measures, the cumulative distribution function is not unique and may not exist. For finite $\mathcal{Y}$-valued measures, the function $F : \Omega \to \mathcal{Y}$ defined by $F(x) = \mu(\mathrm{P}(\Omega) \cap \{\bar{x} \in \mathbb{R}^m \mid \bar{x} \leqq x\})$, $\forall x \in \Omega$, is a cumulative distribution function of μ.

Proposition 12.43 *Let $m \in \mathbb{N}$, $\forall i \in \{1, \ldots, m\}$, X_i be a set, $\mathcal{C}_i$ be a semialgebra on X_i, $\mathcal{B}_i$ be the σ-algebra on X_i generated by $\mathcal{C}_i$, $\mathcal{C} := \{\prod_{i=1}^m C_i \subseteq \prod_{i=1}^m X_i \mid C_i \in \mathcal{C}_i,\ i = 1, \ldots, m\}$, $\mathcal{B}$ be the σ-algebra on $X := \prod_{i=1}^m X_i$ generated by $\mathcal{C}$, and $\mathcal{M} := \{\prod_{i=1}^m B_i \subseteq X \mid B_i \in \mathcal{B}_i,\ i = 1, \ldots, m\}$. Then, $\mathcal{M} \subseteq \mathcal{B}$ and $\mathcal{B}$ is the σ-algebra on X generated by $\mathcal{M}$.*

Proof First, we will prove the following intermediate result.

Claim 12.43.1 $\forall C_i \in \mathcal{C}_i$, $i = 1, \ldots, m-1$, $\forall B_m \in \mathcal{B}_m$, we have $(\prod_{i=1}^{m-1} C_i) \times B_m \in \mathcal{B}$.

Proof of Claim Fix any $C_i \in \mathcal{C}_i$, $i = 1, \ldots, m-1$. Define $\mathcal{F} := \left\{E \subseteq X_m \,\middle|\, (\prod_{i=1}^{m-1} C_i) \times E \in \mathcal{B}\right\}$. By the definition of $\mathcal{B}$, we have $\mathcal{C}_m \subseteq \mathcal{F}$. Clearly, $\emptyset \in \mathcal{C}_m \subseteq \mathcal{F}$. Then, by $\mathcal{C}_m$ being a semialgebra on X_m, we have $X_m = X_m \setminus \emptyset = \bigcup_{i=1}^n C_{m,i}$, where $n \in \mathbb{Z}_+$ and $(C_{m,i})_{i=1}^n \subseteq \mathcal{C}_m$ is pairwise disjoint. Then, $(\prod_{i=1}^{m-1} C_i) \times C_{m,i} \in \mathcal{B}$, $i = 1, \ldots, n$, and hence, $(\prod_{i=1}^{m-1} C_i) \times X_m = \bigcup_{i=1}^n ((\prod_{i=1}^{m-1} C_i) \times C_{m,i}) \in \mathcal{B}$. This implies that $X_m \in \mathcal{F}$. $\forall E \in \mathcal{F}$, we have $(\prod_{i=1}^{m-1} C_i) \times E \in \mathcal{B}$. Then, by $\mathcal{B}$ being a σ-algebra and $(\prod_{i=1}^{m-1} C_i) \times X_m \in \mathcal{B}$, we have $(\prod_{i=1}^{m-1} C_i) \times (X_m \setminus E) = ((\prod_{i=1}^{m-1} C_i) \times X_m) \setminus ((\prod_{i=1}^{m-1} C_i) \times E) \in \mathcal{B}$. Thus, $X_m \setminus E \in \mathcal{F}$. $\forall (E_j)_{j=1}^\infty \in \mathcal{F}$, we have $(\prod_{i=1}^{m-1} C_i) \times E_j \in \mathcal{B}$, $\forall j \in \mathbb{N}$. Then, by $\mathcal{B}$ being a σ-algebra, we have $(\prod_{i=1}^{m-1} C_i) \times (\bigcup_{j=1}^\infty E_j) = \bigcup_{j=1}^\infty ((\prod_{i=1}^{m-1} C_i) \times E_j) \in \mathcal{B}$. Thus, $\bigcup_{j=1}^\infty E_j \in \mathcal{F}$. This shows that $\mathcal{F}$ is a σ-algebra. Hence, $\mathcal{B}_m \subseteq \mathcal{F}$ since $\mathcal{B}_m$ is the σ-algebra generated by $\mathcal{C}_m$. This completes the proof of the claim. □

Fix any $B_i \in \mathcal{B}_i$, $i = 1, \ldots, m$. Clearly, $\emptyset \in \mathcal{C}_i$, $i = 1, \ldots, m-1$. Fix any $i = 1, \ldots, m-1$. Since $\mathcal{C}_i$ is a semialgebra on X_i, we have $X_i = X_i \setminus \emptyset = \bigcup_{j=1}^{n_i} C_{i,j}$, where $n_i \in \mathbb{Z}_+$ and $(C_{i,j})_{j=1}^{n_i} \subseteq \mathcal{C}_i$ is pairwise disjoint. By Claim 12.43.1, $(\prod_{i=1}^{m-1} C_{i,j_i}) \times B_m \in \mathcal{B}$, $j_i = 1, \ldots, n_i$, $i = 1, \ldots, m-1$. Then, $(\prod_{i=1}^{m-1} X_i) \times B_m = (\bigcup_{j_1=1}^{n_1} \cdots \bigcup_{j_{m-1}=1}^{n_{m-1}} \prod_{i=1}^{m-1} C_{i,j_i}) \times B_m = \bigcup_{j_1=1}^{n_1} \cdots \bigcup_{j_{m-1}=1}^{n_{m-1}} ((\prod_{i=1}^{m-1} C_{i,j_i}) \times B_m) \in \mathcal{B}$. By symmetry, we have $(\prod_{i=1}^{k-1} X_i) \times B_k \times (\prod_{i=k+1}^m X_i) \in \mathcal{B}$, $k = 1, \ldots, m$. Then, $\prod_{i=1}^m B_i = \bigcap_{k=1}^m \left((\prod_{i=1}^{k-1} X_i) \times B_k \times (\prod_{i=k+1}^m X_i)\right) \in \mathcal{B}$. By the arbitrariness of B_i's, we have $\mathcal{C} \subseteq \mathcal{M} \subseteq \mathcal{B}$. Since $\mathcal{B}$ is the σ-algebra on X generated by $\mathcal{C}$, then $\mathcal{B}$ is the σ-algebra on X generated by $\mathcal{M}$. $\square$

Proposition 12.44 *Let $m \in \mathbb{N}$, $\mathbb{R}^m$ be endowed with the usual positive cone, $\bar{x}, \tilde{x} \in \mathbb{R}^m$ with $\bar{x} \stackrel{<}{=} \tilde{x}$, $X := \mathrm{r}_{\bar{x},\tilde{x}} \subseteq \mathbb{R}^m$ with subset topology $\mathcal{O}$, $\mathcal{X} := (X, \mathcal{O})$ be the topological space, $\mathcal{C} := \left\{\mathrm{r}_{x_1,x_2} \,\middle|\, \bar{x} \stackrel{<}{=} x_1 \stackrel{<}{=} x_2 \stackrel{<}{=} \tilde{x}\right\}$, and $\mathcal{B}$ be the σ-algebra on X generated by $\mathcal{C}$. Then, $\mathcal{B} = \mathcal{B}_{\mathrm{B}}(\mathcal{X})$.*

Proof Let $\bar{x} = (\bar{x}_1, \ldots, \bar{x}_m)$ and $\tilde{x} = (\tilde{x}_1, \ldots, \tilde{x}_m)$. $\forall i \in \{1, \ldots, m\}$, let $I_i := (\bar{x}_i, \tilde{x}_i] \subset \mathbb{R}$ with the subset topology $\mathcal{O}_i$, and $\mathcal{C}_i := \{(a, b] \subset \mathbb{R} \mid \bar{x}_i \le a \le b \le \tilde{x}_i\}$. Clearly, $\mathcal{C}_i$ is a semialgebra on I_i and the σ-algebra $\mathcal{B}_i$ on I_i generated by $\mathcal{C}_i$ is equal to $\mathcal{B}_{\mathrm{B}}(I_i)$. Then, $\mathcal{C} = \{\prod_{i=1}^{m} C_i \mid C_i \in \mathcal{C}_i,\ i = 1, \ldots, m\}$. Then, by Proposition 12.43, $\mathcal{B}$ equals to the σ-algebra generated by $\mathcal{M} := \{\prod_{i=1}^{m} B_i \mid B_i \in \mathcal{B}_i,\ i = 1, \ldots, m\}$. By Propositions 12.22 and 12.32, we have $\mathcal{B} = \mathcal{B}_{\mathrm{B}}(\mathcal{X})$. This completes the proof of the proposition. $\square$

Proposition 12.45 *Let $\mathcal{X} := (X, \mathcal{O})$ be a topological space, $(X_i)_{i=1}^{\infty} \subseteq \mathcal{B}_{\mathrm{B}}(\mathcal{X})$ be pairwise disjoint with $X = \bigcup_{i=1}^{\infty} X_i$, $\mathcal{X}_i := (X_i, \mathcal{O}_i)$ be the topological subspace of $\mathcal{X}$, $\forall i \in \mathbb{N}$, and $\mathcal{B} := \{E \subseteq X \mid E \cap X_i \in \mathcal{B}_{\mathrm{B}}(\mathcal{X}_i), \forall i \in \mathbb{N}\}$. Then, $\mathcal{B} = \mathcal{B}_{\mathrm{B}}(\mathcal{X})$.*

Proof $\forall E \in \mathcal{B}$, $\forall i \in \mathbb{N}$, $E \cap X_i \in \mathcal{B}_{\mathrm{B}}(\mathcal{X}_i)$. By Proposition 11.25, $\mathcal{B}_{\mathrm{B}}(\mathcal{X}_i) = \{C \subseteq X_i \mid C \in \mathcal{B}_{\mathrm{B}}(\mathcal{X})\}$, $\forall i \in \mathbb{N}$. Then, $E \cap X_i \in \mathcal{B}_{\mathrm{B}}(\mathcal{X})$, $\forall i \in \mathbb{N}$. Then, $E = \bigcup_{i=1}^{\infty}(E \cap X_i) \in \mathcal{B}_{\mathrm{B}}(\mathcal{X})$. Hence, $\mathcal{B} \subseteq \mathcal{B}_{\mathrm{B}}(\mathcal{X})$. On the other hand, $\forall E \in \mathcal{B}_{\mathrm{B}}(\mathcal{X})$, $E \cap X_i \in \mathcal{B}_{\mathrm{B}}(\mathcal{X})$, $\forall i \in \mathbb{N}$. By Proposition 11.25, $E \cap X_i \in \mathcal{B}_{\mathrm{B}}(\mathcal{X}_i)$, $\forall i \in \mathbb{N}$. Hence, $E \in \mathcal{B}$. This implies that $\mathcal{B}_{\mathrm{B}}(\mathcal{X}) \subseteq \mathcal{B}$. This completes the proof of the proposition. $\square$

Proposition 12.46 *Let $m \in \mathbb{N}$, $\mathbb{R}^m$ be endowed with the usual positive cone, $\Omega \subseteq \mathbb{R}^m$, $\mathcal{Y}$ be a normed linear space, and $F : \Omega \to \mathcal{Y}$. Then, $\forall \bar{x}, \tilde{x} \in \Omega$ with $\bar{x} \stackrel{<}{=} \tilde{x}$ and $\bar{\mathrm{r}}_{\bar{x},\tilde{x}} \subseteq \Omega$, $\forall n \in \mathbb{Z}_+$, $\forall \big(\check{x}_i\big)_{i=1}^{n}, \big(\hat{x}_i\big)_{i=1}^{n} \subseteq \Omega$ with $\check{x}_i \stackrel{<}{=} \hat{x}_i$, $\bar{\mathrm{r}}_{\check{x}_i,\hat{x}_i} \subseteq \Omega$, $\forall i \in \{1, \ldots, n\}$, $\mathrm{r}_{\bar{x},\tilde{x}} = \bigcup_{i=1}^{n} \mathrm{r}_{\check{x}_i,\hat{x}_i}$, and the sets in the union being pairwise disjoint, we have $\Delta_F(\mathrm{r}_{\bar{x},\tilde{x}}) = \sum_{i=1}^{n} \Delta_F(\mathrm{r}_{\check{x}_i,\hat{x}_i})$.*

Proof We will prove the result by mathematical induction on n.
1° $n = 0$. Clearly, $\mathrm{r}_{\bar{x},\tilde{x}} = \emptyset$ and $\bar{x} \not< \tilde{x}$. Then, $\vartheta_{\mathcal{Y}} = \sum_{i=1}^{0} \Delta_F(\mathrm{r}_{\check{x}_i,\hat{x}_i}) = \Delta_F(\mathrm{r}_{\bar{x},\tilde{x}})$. This case is proved.

$n = 1$. This case is trivial and proved.

$n = 2$. Without loss of generality, assume that $\bar{x} < \tilde{x}$ and $\check{x}_i < \hat{x}_i$, $i = 1, 2$. Otherwise, the result follows from $n = 1$ or $n = 0$ cases, since $\Delta_F(\mathrm{r}_{x_1,x_2}) = \vartheta_{\mathcal{Y}}$ if $x_1 \stackrel{<}{=} x_2$ and $x_1 \not< x_2$. Then, we have $(\check{x}_i)_{i=1}^{n}, (\hat{x}_i)_{i=1}^{n} \subseteq \bar{\mathrm{r}}_{\bar{x},\tilde{x}}$. Without loss of generality, assume $\tilde{x} \in \mathrm{r}_{\check{x}_2,\hat{x}_2}$. Then, $\tilde{x} = \hat{x}_2$ and $\bar{x} \stackrel{<}{=} \check{x}_2 < \hat{x}_2 = \tilde{x}$. Clearly, $\bar{x} \neq \check{x}_2$ since $\mathrm{r}_{\bar{x},\tilde{x}} = \bigcup_{i=1}^{2} \mathrm{r}_{\check{x}_i,\hat{x}_i}$ and the sets in the union are nonempty and pairwise disjoint. Then, $\exists j_0 \in \{1, \ldots, m\}$ such that $\pi_{j_0}(\tilde{x}) > \pi_{j_0}(\check{x}_2) > \pi_{j_0}(\bar{x})$.

Without loss of generality, assume that $j_0 = m$. Define $x_C \in \bar{\text{r}}_{\bar{x},\tilde{x}}$ be such that $\pi_j(x_C) = \begin{cases} \pi_j(\bar{x}) & j = 1, \ldots, m-1 \\ \pi_m(\check{x}_2) & j = m \end{cases}$ and $x_D \in \bar{\text{r}}_{\bar{x},\tilde{x}}$ be such that $\pi_j(x_D) = \begin{cases} \pi_j(\tilde{x}) & j = 1, \ldots, m-1 \\ \pi_m(\check{x}_2) & j = m \end{cases}$. Clearly, we have $x_D \in \text{r}_{\check{x}_1,\hat{x}_1} \setminus \text{r}_{\check{x}_2,\hat{x}_2}$. Then, $\check{x}_1 = \bar{x}$, $\hat{x}_1 = x_D$, $\check{x}_2 = x_C$, and $\hat{x}_2 = \tilde{x}$. By Definition 12.41, we have

$$\sum_{i=1}^{2} \Delta_F(\text{r}_{\check{x}_i,\hat{x}_i}) = \sum_{i=0}^{m} \sum_{s \in \text{VRect}_{i,\bar{x},x_D}} (-1)^i F(s) + \sum_{i=0}^{m} \sum_{s \in \text{VRect}_{i,x_C,\tilde{x}}} (-1)^i F(s)$$

$$= \sum_{i=0}^{m} \sum_{\substack{s \in \text{VRect}_{i,\bar{x},x_D} \\ \pi_m(s)=\pi_m(\bar{x})}} (-1)^i F(s) + \sum_{i=0}^{m} \sum_{\substack{s \in \text{VRect}_{i,\bar{x},x_D} \\ \pi_m(s)=\pi_m(x_D)}} (-1)^i F(s)$$

$$+ \sum_{i=0}^{m} \sum_{\substack{s \in \text{VRect}_{i,x_C,\tilde{x}} \\ \pi_m(s)=\pi_m(\tilde{x})}} (-1)^i F(s) + \sum_{i=0}^{m} \sum_{\substack{s \in \text{VRect}_{i,x_C,\tilde{x}} \\ \pi_m(s)=\pi_m(x_C)}} (-1)^i F(s)$$

$$= \sum_{i=0}^{m} \sum_{\substack{s \in \text{VRect}_{i,\bar{x},\tilde{x}} \\ \pi_m(s)=\pi_m(\bar{x})}} (-1)^i F(s) + \sum_{i=0}^{m} \sum_{\substack{s \in \text{VRect}_{i+1,x_C,\tilde{x}} \\ \pi_m(s)=\pi_m(x_C)}} (-1)^i F(s)$$

$$+ \sum_{i=0}^{m} \sum_{\substack{s \in \text{VRect}_{i,\bar{x},\tilde{x}} \\ \pi_m(s)=\pi_m(\tilde{x})}} (-1)^i F(s) + \sum_{i=0}^{m} \sum_{\substack{s \in \text{VRect}_{i,x_C,\tilde{x}} \\ \pi_m(s)=\pi_m(x_C)}} (-1)^i F(s)$$

$$= \sum_{i=0}^{m} \sum_{s \in \text{VRect}_{i,\bar{x},\tilde{x}}} (-1)^i F(s) = \Delta_F(\text{r}_{\bar{x},\tilde{x}})$$

where the first two equalities follow from Definition 12.41, the third equality follows from the fact that, $\forall i \in \{0, \ldots, m\}$, $\forall s \in \text{VRect}_{i,\bar{x},x_D}$ with $\pi_m(s) = \pi_m(x_D)$ if, and only if, $s \in \text{VRect}_{i+1,x_C,\tilde{x}}$ with $\pi_m(s) = \pi_m(x_C)$, $\forall s \in \text{VRect}_{i,\bar{x},x_D}$ with $\pi_m(s) = \pi_m(\bar{x})$ if, and only if, $s \in \text{VRect}_{i,\bar{x},\tilde{x}}$ with $\pi_m(s) = \pi_m(\bar{x})$, and $\forall s \in \text{VRect}_{i,x_C,\tilde{x}}$ with $\pi_m(s) = \pi_m(\tilde{x})$ if, and only if, $s \in \text{VRect}_{i,\bar{x},\tilde{x}}$ with $\pi_m(s) = \pi_m(\tilde{x})$, and the fourth equality follows from the fact that the second and fourth summations cancel each other out. This case is proved.

2° Assume that the result holds for $n \leq k-1$, where $k \in \mathbb{N}$ and $k \geq 3$.

3° Consider the case when $n = k \in \{3, 4, \ldots\}$. Without loss of generality, assume that $\bar{x} < \tilde{x}$ and $\check{x}_i < \hat{x}_i$, $i = 1, \ldots, n$. Otherwise, the result follows from 1° or 2°, since $\Delta_F(\text{r}_{x_1,x_2}) = \vartheta_{\mathcal{Y}}$ if $x_1 \leqq x_2$ and $x_1 \not< x_2$. Then, we have $(\check{x}_i)_{i=1}^n, (\hat{x}_i)_{i=1}^n \subseteq \bar{\text{r}}_{\bar{x},\tilde{x}}$. Without loss of generality, assume $\tilde{x} \in \text{r}_{\check{x}_n,\hat{x}_n}$. Then, $\tilde{x} = \hat{x}_n$ and $\bar{x} \leqq \check{x}_n < \hat{x}_n = \tilde{x}$. Clearly, $\bar{x} \neq \check{x}_n$ since $\text{r}_{\bar{x},\tilde{x}} = \bigcup_{i=1}^n \text{r}_{\check{x}_i,\hat{x}_i}$ and the sets in the union are nonempty and pairwise disjoint. Then, $\exists j_0 \in \{1, \ldots, m\}$ such that

$\pi_{j_0}(\tilde{x}) > \pi_{j_0}(\check{x}_n) > \pi_{j_0}(\bar{x})$. Without loss of generality, assume that $j_0 = m$. Define $x_C \in \bar{r}_{\bar{x},\tilde{x}}$ be such that $\pi_j(x_C) = \begin{cases} \pi_j(\bar{x}) & j = 1, \ldots, m-1 \\ \pi_m(\check{x}_n) & j = m \end{cases}$ and $x_D \in \bar{r}_{\bar{x},\tilde{x}}$ be such that $\pi_j(x_D) = \begin{cases} \pi_j(\tilde{x}) & j = 1, \ldots, m-1 \\ \pi_m(\check{x}_n) & j = m \end{cases}$. Then, $r_{\bar{x},\tilde{x}} = r_{\bar{x},x_D} \cup r_{x_C,\tilde{x}} = (\bigcup_{i=1}^{n}(r_{\bar{x},x_D} \cap r_{\check{x}_i,\hat{x}_i})) \cup (\bigcup_{i=1}^{n}(r_{x_C,\tilde{x}} \cap r_{\check{x}_i,\hat{x}_i}))$. Clearly, $r_{\bar{x},x_D} \cap r_{\check{x}_n,\hat{x}_n} = \emptyset$. Note that $x_D \in r_{\bar{x},\tilde{x}} \setminus r_{\check{x}_n,\hat{x}_n}$. Without loss of generality, assume that $x_D \in r_{\check{x}_{n-1},\hat{x}_{n-1}}$. Then, $\bar{x} \leqq \check{x}_{n-1} \lessdot x_D \leqq \hat{x}_{n-1} \leqq \tilde{x}$. Then, we must have $x_D = \hat{x}_{n-1}$ since $r_{\check{x}_{n-1},\hat{x}_{n-1}} \cap r_{\check{x}_n,\hat{x}_n} = \emptyset$. Thus, $r_{x_C,\tilde{x}} \cap r_{\check{x}_{n-1},\hat{x}_{n-1}} = \emptyset$. Then, $r_{\bar{x},x_D} = \bigcup_{i=1}^{n-1}(r_{\bar{x},x_D} \cap r_{\check{x}_i,\hat{x}_i})$ and $r_{x_C,\tilde{x}} = \bigcup_{\substack{i=1 \\ i\neq n-1}}^{n} (r_{x_C,\tilde{x}} \cap r_{\check{x}_i,\hat{x}_i})$. By the inductive assumption, we have $\Delta_F(r_{\bar{x},\tilde{x}}) = \Delta_F(r_{\bar{x},x_D}) + \Delta_F(r_{x_C,\tilde{x}}) = \sum_{i=1}^{n-1} \Delta_F(r_{\bar{x},x_D} \cap r_{\check{x}_i,\hat{x}_i}) + \sum_{\substack{i=1 \\ i\neq n-1}}^{n} \Delta_F(r_{x_C,\tilde{x}} \cap r_{\check{x}_i,\hat{x}_i}) = \sum_{i=1}^{n-2}(\Delta_F(r_{\bar{x},x_D} \cap r_{\check{x}_i,\hat{x}_i}) + \Delta_F(r_{x_C,\tilde{x}} \cap r_{\check{x}_i,\hat{x}_i})) + \Delta_F(r_{\check{x}_{n-1},\hat{x}_{n-1}}) + \Delta_F(r_{\check{x}_n,\hat{x}_n}) = \sum_{i=1}^{n} \Delta_F(r_{\check{x}_i,\hat{x}_i})$. This case is proved.

Therefore, the result holds for any $n \in \mathbb{N}$. This completes the proof of the proposition. □

Proposition 12.47 *Let $m \in \mathbb{N}$, $\mathbb{R}^m$ be endowed with the usual positive cone, $\mathcal{Y}$ be a normed linear space, $\Omega \subseteq \mathbb{R}^m$, and $F : \Omega \to \mathcal{Y}$. Assume that F satisfies (i) and (ii) of Definition 12.41. Then, $\forall \bar{x}, \tilde{x} \in \Omega$ with $\bar{x} \leqq \tilde{x}$ and $\bar{r}_{\bar{x},\tilde{x}} \subseteq \Omega$, $\forall n \in \mathbb{Z}_+$, $\forall (\bar{x}_i)_{i=1}^n, (\tilde{x}_i)_{i=1}^n \subseteq \bar{r}_{\bar{x},\tilde{x}}$ with $\bar{x}_i \leqq \tilde{x}_i$, $\forall i \in \{1, \ldots, n\}$, $r_{\bar{x},\tilde{x}} = \bigcup_{i=1}^n r_{\bar{x}_i,\tilde{x}_i}$, and the sets in the union being pairwise disjoint, we have $T_F(r_{\bar{x},\tilde{x}}) = \sum_{i=1}^n T_F(r_{\bar{x}_i,\tilde{x}_i})$.*

Proof $\forall \epsilon \in (0,\infty) \subset \mathbb{R}$, $\forall i \in \{1, \ldots, n\}$, $\exists n_i \in \mathbb{Z}_+$, $\exists (\check{x}_{i,j})_{j=1}^{n_i}, (\hat{x}_{i,j})_{j=1}^{n_i} \subseteq \bar{r}_{\bar{x}_i,\tilde{x}_i}$ with $\check{x}_{i,j} \leqq \hat{x}_{i,j}$, $\forall j \in \{1, \ldots, n_i\}$, $r_{\bar{x}_i,\tilde{x}_i} = \bigcup_{j=1}^{n_i} r_{\check{x}_{i,j},\hat{x}_{i,j}}$, and sets in the union being pairwise disjoint, such that $\sum_{j=1}^{n_i} \left\| \Delta_F(r_{\check{x}_{i,j},\hat{x}_{i,j}}) \right\| > T_F(r_{\bar{x}_i,\tilde{x}_i}) - \epsilon/(n+1)$. Then, $\sum_{i=1}^n T_F(r_{\bar{x}_i,\tilde{x}_i}) - \epsilon < \sum_{i=1}^n \sum_{j=1}^{n_i} \left\| \Delta_F(r_{\check{x}_{i,j},\hat{x}_{i,j}}) \right\| \leq T_F(r_{\bar{x},\tilde{x}})$. By the arbitrariness of ϵ, we have $\sum_{i=1}^n T_F(r_{\bar{x}_i,\tilde{x}_i}) \leq T_F(r_{\bar{x},\tilde{x}})$.

On the other hand, $\forall \epsilon \in (0,\infty) \subset \mathbb{R}$, $\exists \bar{n} \in \mathbb{Z}_+$, $\exists (\check{x}_i)_{i=1}^{\bar{n}}, (\hat{x}_i)_{i=1}^{\bar{n}} \subseteq \bar{r}_{\bar{x},\tilde{x}}$ with $\check{x}_i \leqq \hat{x}_i$, $\forall i \in \{1, \ldots, \bar{n}\}$, $r_{\bar{x},\tilde{x}} = \bigcup_{i=1}^{\bar{n}} r_{\check{x}_i,\hat{x}_i}$, and the sets in the union being pairwise disjoint, such that $\sum_{i=1}^{\bar{n}} \left\| \Delta_F(r_{\check{x}_i,\hat{x}_i}) \right\| > T_F(r_{\bar{x},\tilde{x}}) - \epsilon$. Then, $T_F(r_{\bar{x},\tilde{x}}) - \epsilon < \sum_{i=1}^{\bar{n}} \left\| \Delta_F(r_{\check{x}_i,\hat{x}_i}) \right\| = \sum_{i=1}^{\bar{n}} \left\| \Delta_F(\bigcup_{j=1}^n (r_{\check{x}_i,\hat{x}_i} \cap r_{\bar{x}_j,\tilde{x}_j})) \right\| = \sum_{i=1}^{\bar{n}} \left\| \sum_{j=1}^n \Delta_F(r_{\check{x}_i,\hat{x}_i} \cap r_{\bar{x}_j,\tilde{x}_j}) \right\| \leq \sum_{i=1}^{\bar{n}} \sum_{j=1}^n \left\| \Delta_F(r_{\check{x}_i,\hat{x}_i} \cap r_{\bar{x}_j,\tilde{x}_j}) \right\| = \sum_{j=1}^n \sum_{i=1}^{\bar{n}} \left\| \Delta_F(r_{\check{x}_i,\hat{x}_i} \cap r_{\bar{x}_j,\tilde{x}_j}) \right\| \leq \sum_{j=1}^n T_F(r_{\bar{x}_j,\tilde{x}_j})$, where the second equality follows from Proposition 12.46. By the arbitrariness of ϵ, we have $T_F(r_{\bar{x},\tilde{x}}) \leq \sum_{i=1}^n T_F(r_{\bar{x}_i,\tilde{x}_i})$. This completes the proof of the proposition. □

Proposition 12.48 *Let $m \in \mathbb{N}$, $\mathbb{R}^m$ be endowed with the usual positive cone, $\Omega \in \mathcal{B}_B(\mathbb{R}^m)$ be a region with the subset topology $\mathcal{O}$, and $F : \Omega \to \mathbb{R}$ be of locally bounded variation. Assume that $\Delta_F(r_{x_1,x_2}) \geq 0$, $\forall x_1, x_2 \in \Omega$ with $\bar{r}_{x_1,x_2} \subseteq \Omega$*

and $x_1 \stackrel{\lessdot}{=} x_2$. *Then, there exists a unique* σ*-finite measure* $\hat{\mu}$ *on the measurable space* $(\mathrm{P}(\Omega), \mathcal{B}_{\mathrm{B}}(\mathrm{P}(\Omega)))$ *such that* F *is a cumulative distribution function of* $\hat{\mu}$. *Furthermore,* $T_F(\mathrm{r}_{x_1,x_2}) = \Delta_F(\mathrm{r}_{x_1,x_2}) = \hat{\mu}(\mathrm{r}_{x_1,x_2})$, $\forall x_1, x_2 \in \Omega$ *with* $x_1 \stackrel{\lessdot}{=} x_2$ *and* $\bar{\mathrm{r}}_{x_1,x_2} \subseteq \Omega$, *and* $T_F = \hat{\mu}(\mathrm{P}(\Omega)) \in [0, \infty] \subset \mathbb{R}_e$.

Proof Since Ω is a region, then $\exists N \subseteq \mathbb{N}$, $\exists (\bar{x}_i)_{i\in N}, (\tilde{x}_i)_{i\in N} \subseteq \Omega$ with $\bar{x}_i \stackrel{\lessdot}{=} \tilde{x}_i$ and $\bar{\mathrm{r}}_{\bar{x}_i,\tilde{x}_i} \subseteq \Omega$, $\forall i \in N$, such that $\Omega = \bigcup_{i\in N} \bar{\mathrm{r}}_{\bar{x}_i,\tilde{x}_i}$, $\left(\mathrm{r}_{\bar{x}_i,\tilde{x}_i}\right)_{i\in N}$ is pairwise disjoint, $\mathrm{P}(\Omega) = \bigcup_{i\in N} \mathrm{r}_{\bar{x}_i,\tilde{x}_i}$, and $\forall x_1, x_2 \in \Omega$ with $x_1 \stackrel{\lessdot}{=} x_2$ and $\bar{\mathrm{r}}_{x_1,x_2} \subseteq \Omega$, we have $\mathrm{r}_{x_1,x_2} \subseteq \mathrm{P}(\Omega)$.

Claim 12.48.1 $\forall \bar{x}, \tilde{x} \in \Omega$ with $\bar{\mathrm{r}}_{\bar{x},\tilde{x}} \subseteq \Omega$ and $\bar{x} \stackrel{\lessdot}{=} \tilde{x}$, let $\mathcal{C} := \left\{\mathrm{r}_{x_1,x_2} \,\middle|\, \bar{x} \stackrel{\lessdot}{=} x_1 \stackrel{\lessdot}{=} x_2 \stackrel{\lessdot}{=} \tilde{x}\right\}$, which is the collection of semi-open rectangles contained in $\mathrm{r}_{\bar{x},\tilde{x}}$. Clearly, $\mathcal{C}$ is semialgebra on $\mathrm{r}_{\bar{x},\tilde{x}}$. Define $\mu : \mathcal{C} \to [0, \infty) \subset \mathbb{R}$ by $\mu(\mathrm{r}_{x_1,x_2}) = \Delta_F(\mathrm{r}_{x_1,x_2})$, $\forall \mathrm{r}_{x_1,x_2} \in \mathcal{C}$. Then, μ admits a unique extension to a measure $\bar{\mu}$ on the algebra on $\mathrm{r}_{\bar{x},\tilde{x}}$, $\mathcal{A}$, generated by $\mathcal{C}$.

Proof of Claim We will prove the claim by Proposition 11.32. By Proposition 12.46, (i) of Proposition 11.32 is satisfied. $\forall C := \mathrm{r}_{\check{x},\hat{x}} \in \mathcal{C}$, $\forall$ pairwise disjoint $\left(C_j := \mathrm{r}_{\check{x}_j,\hat{x}_j}\right)_{j=1}^{\infty} \subseteq \mathcal{C}$ with $C = \bigcup_{j=1}^{\infty} C_j$, we will show that $\mu(C) \le \sum_{j=1}^{\infty} \mu(C_j)$ by distinguishing two exhaustive and mutually exclusive cases: Case 1: $C = \emptyset$; Case 2: $C \neq \emptyset$.

Case 1: $C = \emptyset$. Then, $\mu(C) = 0$. Clearly, $C_j = \emptyset$, $\forall j \in \mathbb{N}$. Then, $\mu(C) = 0 = \sum_{j=1}^{\infty} \mu(C_j)$. This case is proved.

Case 2: $C \neq \emptyset$. Then, $\check{x} \lessdot \hat{x}$. By (i) of Definition 12.41, $\forall \epsilon \in (0, \infty) \subset \mathbb{R}$, $\exists \check{y} \in \mathrm{r}_{\check{x},\hat{x}}$ such that $\Delta_F(\mathrm{r}_{\check{y},\hat{x}}) > \Delta_F(\mathrm{r}_{\check{x},\hat{x}}) - \epsilon/2$. Since $\mathrm{r}_{\check{y},\hat{x}} \subseteq \mathrm{r}_{\check{x},\hat{x}}$, by the assumption and (i) of Proposition 11.32, we have $\Delta_F(\mathrm{r}_{\check{y},\hat{x}}) \le \Delta_F(\mathrm{r}_{\check{x},\hat{x}})$.

Fix any $j \in \mathbb{N}$. $C_j = \mathrm{r}_{\check{x}_j,\hat{x}_j} \subseteq \overline{C_j} \subseteq \overline{C} \subseteq \Omega$. By (i) of Definition 12.41, $\exists \hat{y}_j \in \mathbb{R}^m$ with $\hat{x}_j \lessdot \hat{y}_j$ such that $\Delta_F(\mathrm{r}_{\check{x}_j,\hat{y}_j} \cap \mathrm{r}_{\bar{x},\tilde{x}}) < \Delta_F(\mathrm{r}_{\check{x}_j,\hat{x}_j}) + 2^{-j-1}\epsilon$. Since $C_j \subseteq \mathrm{r}_{\check{x}_j,\hat{y}_j} \cap \mathrm{r}_{\bar{x},\tilde{x}}$, by the assumption and (i) of Proposition 11.32, we have $\Delta_F(\mathrm{r}_{\check{x}_j,\hat{x}_j}) \le \Delta_F(\mathrm{r}_{\check{x}_j,\hat{y}_j} \cap \mathrm{r}_{\bar{x},\tilde{x}})$.

Obviously, $\bar{\mathrm{r}}_{\check{y},\hat{x}} \subseteq \mathrm{r}_{\check{x},\hat{x}} = C = \bigcup_{j=1}^{\infty} C_j = \bigcup_{j=1}^{\infty} \mathrm{r}_{\check{x}_j,\hat{x}_j} \subseteq \bigcup_{j=1}^{\infty} (\mathrm{r}^{\circ}_{\check{x}_j,\hat{y}_j} \cap \mathrm{r}_{\bar{x},\tilde{x}})$. Since $\bar{\mathrm{r}}_{\check{y},\hat{x}}$ is compact, then $\exists \bar{N} \in \mathbb{N}$ such that $\mathrm{r}_{\check{y},\hat{x}} \subseteq \bar{\mathrm{r}}_{\check{y},\hat{x}} \subseteq \bigcup_{j=1}^{\bar{N}} (\mathrm{r}^{\circ}_{\check{x}_j,\hat{y}_j} \cap \mathrm{r}_{\bar{x},\tilde{x}}) \subseteq \bigcup_{j=1}^{\bar{N}} (\mathrm{r}_{\check{x}_j,\hat{y}_j} \cap \mathrm{r}_{\bar{x},\tilde{x}})$. By (i) of Proposition 11.32, $\mu(C) = \Delta_F(\mathrm{r}_{\check{x},\hat{x}}) < \Delta_F(\mathrm{r}_{\check{y},\hat{x}}) + \epsilon/2 < \sum_{j=1}^{\bar{N}} \Delta_F(\mathrm{r}_{\check{x}_j,\hat{y}_j} \cap \mathrm{r}_{\bar{x},\tilde{x}}) + \epsilon/2 < \sum_{j=1}^{\bar{N}} (\Delta_F(\mathrm{r}_{\check{x}_j,\hat{x}_j}) + 2^{-j-1}\epsilon) + \epsilon/2 < \sum_{j=1}^{\bar{N}} \mu(C_j) + \epsilon \le \sum_{j=1}^{\infty} \mu(C_j) + \epsilon$. By the arbitrariness of ϵ, we have $\mu(C) \le \sum_{j=1}^{\infty} \mu(C_j)$. This case is also proved.

In both cases, we have $\mu(C) \le \sum_{j=1}^{\infty} \mu(C_j)$. Then, (ii) of Proposition 11.32 holds. Then, by Proposition 11.32, μ admits a unique extension to a measure $\bar{\mu}$ on the algebra $\mathcal{A}$. This completes the proof of the claim. □

Fix any $i \in N$. By Claim 12.48.1 and Carathéodory Extension Theorem 11.19, there exists a unique finite measure $\hat{\mu}_i$ on the measurable space $(\mathrm{r}_{\bar{x}_i,\tilde{x}_i}, \mathcal{B}_i)$ that is an extension of $\bar{\mu}$, where $\mathcal{B}_i$ is the σ-algebra on $\mathrm{r}_{\bar{x}_i,\tilde{x}_i}$ generated by $\mathcal{C}_i := \left\{\mathrm{r}_{x_1,x_2} \,\middle|\, \bar{x}_i \stackrel{\lessdot}{=} x_1 \stackrel{\lessdot}{=} x_2 \stackrel{\lessdot}{=} \tilde{x}_i\right\}$. By Proposition 12.44, $\mathcal{B}_i = \mathcal{B}_{\mathrm{B}}(\mathrm{r}_{\bar{x}_i,\tilde{x}_i})$. Let $\mathcal{X}_i := (\mathrm{r}_{\bar{x}_i,\tilde{x}_i}, \mathcal{B}_i, \hat{\mu}_i)$ be the finite measure space. By Proposition 11.118, the generation process on $(\mathcal{X}_i)_{i=1}^{\infty}$ yields a unique σ-finite measure space $\mathcal{X} := (\mathrm{P}(\Omega), \mathcal{B}, \hat{\mu})$ such that $\mathcal{X}_i$ is the finite measure subspace of $\mathcal{X}$, $\forall i \in N$.

By Proposition 12.45, $\mathcal{B} = \mathcal{B}_{\mathrm{B}}(\mathrm{P}(\Omega))$. Then, $\mathcal{X} = (\mathrm{P}(\Omega), \mathcal{B}_{\mathrm{B}}(\mathrm{P}(\Omega)), \hat{\mu})$. $\forall x_1, x_2 \in \Omega$ with $x_1 \stackrel{\lessdot}{=} x_2$ and $\bar{\mathrm{r}}_{x_1,x_2} \subseteq \Omega$, $\hat{\mu}(\mathrm{r}_{x_1,x_2}) = \hat{\mu}(\bigcup_{i\in N}(\mathrm{r}_{x_1,x_2} \cap \mathrm{r}_{\bar{x}_i,\tilde{x}_i})) = \sum_{i\in N} \hat{\mu}(\mathrm{r}_{x_1,x_2} \cap \mathrm{r}_{\bar{x}_i,\tilde{x}_i}) = \sum_{i\in N} \hat{\mu}_i(\mathrm{r}_{x_1,x_2} \cap \mathrm{r}_{\bar{x}_i,\tilde{x}_i}) = \sum_{i\in N} \Delta_F(\mathrm{r}_{x_1,x_2} \cap \mathrm{r}_{\bar{x}_i,\tilde{x}_i}) = \Delta_F(\bigcup_{i\in N}(\mathrm{r}_{x_1,x_2} \cap \mathrm{r}_{\bar{x}_i,\tilde{x}_i})) = \Delta_F(\mathrm{r}_{x_1,x_2}) = T_F(\mathrm{r}_{x_1,x_2})$, where the second equality follows from the fact that $\hat{\mu}$ is a measure; the third equality follows from Proposition 11.13, the fourth and fifth equalities follow from Claim 12.48.1, and the last equality follows from Definition 12.41 and the assumption. Hence, F is a cumulative distribution function of $\hat{\mu}$.

Finally, we will show that $T_F = \hat{\mu}(\mathrm{P}(\Omega)) \in [0, \infty] \subset \mathbb{R}_e$. Note that $T_F = \sum_{i\in N} T_F\big(\mathrm{r}_{\bar{x}_i,\tilde{x}_i}\big) = \sum_{i\in\mathbb{N}} \Delta_F(\mathrm{r}_{\bar{x}_i,\tilde{x}_i}) = \sum_{i\in\mathbb{N}} \hat{\mu}(\mathrm{r}_{\bar{x}_i,\tilde{x}_i}) = \hat{\mu}(\mathrm{P}(\Omega)) \in [0, \infty] \subset \mathbb{R}_e$, where the first equality follows from Definition 12.41, the second equality follows from the assumption of the proposition and Definition 12.41, the third equality follows from the fact that F is a cumulative distribution function of $\hat{\mu}$, and the fourth equality follows from the fact that $\hat{\mu}$ is a measure. Clearly, T_F is well-defined since it is independent of $(\bar{x}_i)_{i\in N}$ and $(\tilde{x}_i)_{i\in N}$, by the uniqueness of $\hat{\mu}$. This completes the proof of the proposition. □

Proposition 12.49 *Let $m \in \mathbb{N}$, $\mathbb{R}^m$ be endowed with the usual positive cone, $\Omega \in \mathcal{B}_{\mathrm{B}}(\mathbb{R}^m)$ be a region with the subset topology $\mathcal{O}$, and $F : \Omega \to \mathbb{R}$ satisfy:*

(i) *F is continuous on the right on Ω.*
(ii) *$\forall x_1, x_2 \in \Omega$ with $x_1 \stackrel{\lessdot}{=} x_2$ and $\bar{\mathrm{r}}_{x_1,x_2} \subseteq \Omega$, we have $\Delta_F(\mathrm{r}_{x_1,x_2}) \geq 0$.*

Then:

1. *F is of locally bounded variation.*
2. *There exists a unique σ-finite measure $\hat{\mu}$ on the measurable space $(\mathrm{P}(\Omega), \mathcal{B}_{\mathrm{B}}(\mathrm{P}(\Omega)))$ such that F is a cumulative distribution function of $\hat{\mu}$.*
3. *$T_F(\mathrm{r}_{x_1,x_2}) = \Delta_F(\mathrm{r}_{x_1,x_2}) = \hat{\mu}(\mathrm{r}_{x_1,x_2})$, $\forall x_1, x_2 \in \Omega$ with $x_1 \stackrel{\lessdot}{=} x_2$ and $\bar{\mathrm{r}}_{x_1,x_2} \subseteq \Omega$, and $T_F = \hat{\mu}(\mathrm{P}(\Omega)) \in [0, \infty] \subset \mathbb{R}_e$.*

Proof By (i), (i) of Definition 12.41 is satisfied. $\forall \bar{x}, \tilde{x} \in \Omega$ with $\bar{x} \stackrel{\lessdot}{=} \tilde{x}$ and $\bar{\mathrm{r}}_{\bar{x},\tilde{x}} \subseteq \Omega$, $\forall n \in \mathbb{Z}_+$, $\forall \check{x}_1, \hat{x}_1, \ldots, \check{x}_n, \hat{x}_n \in \bar{\mathrm{r}}_{\bar{x},\tilde{x}}$ with $\check{x}_i \stackrel{\lessdot}{=} \hat{x}_i$, $\forall i \in \{1, \ldots, n\}$, $\mathrm{r}_{\bar{x},\tilde{x}} = \bigcup_{i=1}^{n} \mathrm{r}_{\check{x}_i,\hat{x}_i}$, and the sets in the union being pairwise disjoint, we have $\sum_{i=1}^{n} \left|\Delta_F(\mathrm{r}_{\check{x}_i,\hat{x}_i})\right| = \sum_{i=1}^{n} \Delta_F(\mathrm{r}_{\check{x}_i,\hat{x}_i}) = \Delta_F(\mathrm{r}_{\bar{x},\tilde{x}}) \in [0, \infty) \subset \mathbb{R}$, where the first equality follows from (ii) and the second equality follows from Proposition 12.46.

Then, $T_F(\mathrm{r}_{\bar{x},\tilde{x}}) = \Delta_F(\mathrm{r}_{\bar{x},\tilde{x}}) \in [0,\infty) \subset \mathbb{R}$. Thus, (ii) of Definition 12.41 is satisfied.

$\forall x_1, x_2 \in \Omega$ with $x_1 \lessdot x_2$ and $\bar{\mathrm{r}}_{x_1,x_2} \subseteq \Omega$, define $\bar{G} : \bar{\mathrm{r}}_{x_1,x_2} \to [0,\infty) \subset \mathbb{R}$ by $\bar{G}(x) = T_F(\mathrm{r}_{x_1,x})$, $\forall x \in \bar{\mathrm{r}}_{x_1,x_2}$. By the previous paragraph, $\bar{G}(x) = \Delta_F(\mathrm{r}_{x_1,x})$, $\forall x \in \bar{\mathrm{r}}_{x_1,x_2}$. Then, $\forall x \in \bar{\mathrm{r}}_{x_1,x_2}$ with $x \neq x_2$, $\bar{G}(x) = \Delta_F(\mathrm{r}_{x_1,x}) = \lim_{\substack{h \to \mathbf{0}_m \\ h \geqq \mathbf{0}_m}} \Delta_F(\mathrm{r}_{x_1,x+h}) = \lim_{\substack{h \to \mathbf{0}_m \\ h \geqq \mathbf{0}_m}} \bar{G}(x+h)$, where the second equality follows from Definition 12.41 and (i). Hence, by Proposition 12.37, (iii) of Definition 12.41 is satisfied.

This shows that F is of locally bounded variation. Hence, the first statement holds. The second and third statements of the proposition follow directly from Proposition 12.48. This completes the proof of the proposition. □

Theorem 12.50 *Let $m \in \mathbb{N}$, $\mathbb{R}^m$ be endowed with the usual positive cone, $\Omega \in \mathcal{B}_{\mathrm{B}}(\mathbb{R}^m)$ be a region with the subset topology $\mathcal{O}$, $\mathcal{Y}$ be a Banach space, and $F : \Omega \to \mathcal{Y}$ be of locally bounded variation. Then, there exists a unique σ-finite $\mathcal{Y}$-valued measure $\hat{\mu}$ on the measurable space $(\mathrm{P}(\Omega), \mathcal{B}_{\mathrm{B}}(\mathrm{P}(\Omega)))$ such that F is a cumulative distribution function of $\hat{\mu}$. Furthermore, $\mathcal{P} \circ \hat{\mu}(\mathrm{r}_{x_1,x_2}) = T_F(\mathrm{r}_{x_1,x_2})$, $\forall x_1, x_2 \in \Omega$ with $x_1 \leqq x_2$ and $\bar{\mathrm{r}}_{x_1,x_2} \subseteq \Omega$, and $\mathcal{P} \circ \hat{\mu}(\mathrm{P}(\Omega)) = T_F$.*

Proof Since $\Omega \in \mathcal{B}_{\mathrm{B}}(\mathbb{R}^m)$ is a region, then $\exists N \subseteq \mathbb{N}$ and $(\bar{x}_i)_{i\in N}, (\tilde{x}_i)_{i\in N} \subseteq \Omega$ with $\bar{x}_i \leqq \tilde{x}_i$ and $\bar{\mathrm{r}}_{\bar{x}_i,\tilde{x}_i} \subseteq \Omega$, $\forall i \in N$, such that $\Omega = \bigcup_{i\in N} \bar{\mathrm{r}}_{\bar{x}_i,\tilde{x}_i}$, $\left(\mathrm{r}_{\bar{x}_i,\tilde{x}_i}\right)_{i\in N}$ is pairwise disjoint, $\mathrm{P}(\Omega) := \bigcup_{i\in N} \mathrm{r}_{\bar{x}_i,\tilde{x}_i}$, and $\forall x_1, x_2 \in \Omega$ with $x_1 \leqq x_2$ and $\bar{\mathrm{r}}_{x_1,x_2} \subseteq \Omega$, we have $\mathrm{r}_{x_1,x_2} \subseteq \mathrm{P}(\Omega)$. We need the following result.

Claim 12.50.1 Fix any $\bar{x}, \tilde{x} \in \Omega$ with $\bar{x} \leqq \tilde{x}$ and $\bar{\mathrm{r}}_{\bar{x},\tilde{x}} \subseteq \Omega$, let $\hat{\Omega} := \mathrm{r}_{\bar{x},\tilde{x}} \subseteq \Omega$ be endowed with subset topology, and then there exists a unique finite $\mathcal{Y}$-valued measure $\check{\mu}$ on the measurable space $(\hat{\Omega}, \mathcal{B}_{\mathrm{B}}(\hat{\Omega}))$ such that $\check{\mu}(\mathrm{r}_{x_1,x_2}) = \Delta_F(\mathrm{r}_{x_1,x_2})$ and $\mathcal{P} \circ \check{\mu}(\mathrm{r}_{x_1,x_2}) = T_F(\mathrm{r}_{x_1,x_2})$, $\forall \bar{x} \leqq x_1 \leqq x_2 \leqq \tilde{x}$. Furthermore, $\check{\mu} \in \mathcal{M}_{ft}(\hat{\Omega}, \mathcal{Y})$.

Proof of Claim Let $\bar{\Omega} := \bar{\mathrm{r}}_{\bar{x},\tilde{x}}$, $\bar{F} := F|_{\bar{\Omega}} : \bar{\Omega} \to \mathcal{Y}$, and $\bar{G} : \bar{\Omega} \to [0,\infty) \subset \mathbb{R}$ be defined by $\bar{G}(x) = T_F(\mathrm{r}_{\bar{x},x})$, $\forall x \in \bar{\Omega}$. Note that $\bar{G}$ is real-valued since F is of locally bounded variation. By (iii) of Definition 12.41, $\bar{G}$ is continuous on the right. By Proposition 12.47, $\Delta_{\bar{G}}(\mathrm{r}_{x_1,x_2}) = T_F(\mathrm{r}_{x_1,x_2}) \geq 0$, $\forall x_1, x_2 \in \bar{\Omega}$ with $x_1 \leqq x_2$. By Proposition 12.49, $\bar{G}$ is of locally bounded variation, there exists a unique σ-finite measure $\hat{\nu}$ on the measurable space $(\hat{\Omega}, \mathcal{B}_{\mathrm{B}}(\hat{\Omega}))$ such that $\bar{G}$ is a cumulative distribution function of $\hat{\nu}$, and $\hat{\nu}(\hat{\Omega}) = T_{\bar{G}}(\hat{\Omega}) < \infty$. Hence, $\hat{\nu}$ is finite.

Let $\mathcal{C} := \left\{\mathrm{r}_{x_1,x_2} \,\middle|\, \bar{x} \leqq x_1 \leqq x_2 \leqq \tilde{x}\right\}$ be the collection of semi-open rectangles contained in $\hat{\Omega}$. Clearly, $\mathcal{C}$ is a semialgebra on $\hat{\Omega}$. Define $\nu : \mathcal{C} \to [0,\infty) \subset \mathbb{R}$ by $\nu := \hat{\nu}|_{\mathcal{C}}$ and $\mu : \mathcal{C} \to \mathcal{Y}$ by $\mu(\mathrm{r}_{x_1,x_2}) = \Delta_F(\mathrm{r}_{x_1,x_2}) \in \mathcal{Y}$, $\forall \mathrm{r}_{x_1,x_2} \in \mathcal{C}$. It is easy to show that (i)–(v) of Proposition 12.20 are satisfied by μ and ν. $\forall \mathrm{r}_{x_1,x_2} \in \mathcal{C}$, we have

$$\nu(\mathrm{r}_{x_1,x_2}) = \hat{\nu}(\mathrm{r}_{x_1,x_2}) = T_F\big(\mathrm{r}_{x_1,x_2}\big)$$

$$= \sup_{\substack{n\in\mathbb{Z}_+,\ \check{x}_1,\hat{x}_1,\ldots,\check{x}_n,\hat{x}_n\in\bar{\mathrm{r}}_{x_1,x_2},\ \check{x}_i \stackrel{\lessdot}{=} \hat{x}_i,\ \forall i\in\{1,\ldots,n\} \\ \bigcup_{i=1}^n \mathrm{r}_{\check{x}_i,\hat{x}_i}=\mathrm{r}_{x_1,x_2},\ \mathrm{r}_{\check{x}_i,\hat{x}_i}\cap \mathrm{r}_{\check{x}_j,\hat{x}_j}=\emptyset,\ \forall 1\leq i<j\leq n}} \sum_{i=1}^n \big\|\Delta_F(\mathrm{r}_{\check{x}_i,\hat{x}_i})\big\|$$

$$= \sup_{n\in\mathbb{Z}_+,\ (C_i)_{i=1}^n\subseteq\mathcal{C},\ \mathrm{r}_{x_1,x_2}=\bigcup_{i=1}^n C_i,\ C_i\cap C_j=\emptyset,\ \forall 1\leq i<j\leq n} \sum_{i=1}^n \|\mu(C_i)\|$$

Hence, (vi) of Proposition 12.20 is satisfied. By Proposition 12.20, ν admits a unique extension to a measure $\tilde{\nu}$ on the algebra on $\hat{\Omega}$, $\mathcal{A}$, generated by $\mathcal{C}$, and μ admits a unique extension to a $\mathcal{Y}$-valued measure $\tilde{\mu}$ on the algebra $\mathcal{A}$ with $\mathcal{P} \circ \tilde{\mu} = \tilde{\nu}$ and $\mathcal{P} \circ \tilde{\mu}|_{\mathcal{C}} = \nu$. Clearly, $\tilde{\nu}$ is finite. By Carathéodory Extension Theorem 12.4, there is a unique $\mathcal{Y}$-valued measure $\check{\mu}$ on $(\hat{\Omega}, \bar{\mathcal{B}})$ such that $\check{\mu}\big|_{\mathcal{A}} = \tilde{\mu}$ and $(\mathcal{P} \circ \check{\mu})\big|_{\mathcal{A}} = \tilde{\nu}$, where $\bar{\mathcal{B}}$ is the σ-algebra on $\hat{\Omega}$ generated by $\mathcal{A}$. Then, $\bar{\mathcal{B}}$ equals to the σ-algebra on $\hat{\Omega}$ generated by $\mathcal{C}$. By Proposition 12.44, $\bar{\mathcal{B}} = \mathcal{B}_{\mathrm{B}}(\hat{\Omega})$. By Carathéodory Extension Theorem 11.19, we have $\mathcal{P} \circ \check{\mu} = \hat{\nu}$. Then, $\check{\mu}$ is a finite $\mathcal{Y}$-valued measure on $(\hat{\Omega}, \mathcal{B}_{\mathrm{B}}(\hat{\Omega}))$. By Theorem 11.198, $\check{\mu} \in \mathcal{M}_{ft}(\hat{\Omega}, \mathcal{Y})$. This completes the proof of the claim. □

$\forall i \in N$, let $\mathbb{X}_i := (\mathrm{r}_{\bar{x}_i,\tilde{x}_i}, \mathcal{B}_{\mathrm{B}}(\mathrm{r}_{\bar{x}_i,\tilde{x}_i}), \hat{\mu}_i)$ be the unique finite $\mathcal{Y}$-valued metric measure space as defined in Claim 12.50.1. By Proposition 11.118, the generation process on $(\mathbb{X}_i)_{i\in N}$ yields a unique σ-finite $\mathcal{Y}$-valued measure space $\mathcal{X} := (\mathrm{P}(\Omega), \mathcal{B}, \hat{\mu})$ such that $\mathbb{X}_i$ is the finite $\mathcal{Y}$-valued measure subspace of $\mathcal{X}$, $\forall i \in N$. By Proposition 12.45, we have $\mathcal{B} = \mathcal{B}_{\mathrm{B}}(\mathrm{P}(\Omega))$.

Next, we will show that F is a cumulative distribution function of $\hat{\mu}$. Fix any $x_1, x_2 \in \Omega$ with $x_1 \stackrel{\lessdot}{=} x_2$ and $\bar{\mathrm{r}}_{x_1,x_2} \subseteq \Omega$. Let $(\mathrm{r}_{x_1,x_2}, \mathcal{B}_{\mathrm{B}}(\mathrm{r}_{x_1,x_2}), \check{\mu})$ be the finite $\mathcal{Y}$-valued metric measure space as prescribed in Claim 12.50.1. Note that $\mathcal{P} \circ \hat{\mu}(\mathrm{r}_{x_1,x_2}) = \sum_{i\in N} \mathcal{P} \circ \hat{\mu}(\mathrm{r}_{x_1,x_2} \cap \mathrm{r}_{\bar{x}_i,\tilde{x}_i}) = \sum_{i\in N} \mathcal{P} \circ \hat{\mu}_i(\mathrm{r}_{x_1,x_2} \cap \mathrm{r}_{\bar{x}_i,\tilde{x}_i}) = \sum_{i\in N} T_F(\mathrm{r}_{x_1,x_2} \cap \mathrm{r}_{\bar{x}_i,\tilde{x}_i}) = \sum_{i\in N} \mathcal{P} \circ \check{\mu}(\mathrm{r}_{x_1,x_2} \cap \mathrm{r}_{\bar{x}_i,\tilde{x}_i}) = \mathcal{P} \circ \check{\mu}(\mathrm{r}_{x_1,x_2}) = T_F(\mathrm{r}_{x_1,x_2}) < \infty$, where the first equality follows from the fact $\mathcal{P} \circ \hat{\mu}$ is a measure, the second equality follows from Proposition 11.115, the third equality follows from Claim 12.50.1, the fourth equality follows from Claim 12.50.1, the fifth equality follows from the fact that $\mathcal{P} \circ \check{\mu}$ is a measure, and the sixth equality and the inequality follow from Claim 12.50.1. This implies that $\hat{\mu}(\mathrm{r}_{x_1,x_2}) = \sum_{i\in N} \hat{\mu}(\mathrm{r}_{x_1,x_2} \cap \mathrm{r}_{\bar{x}_i,\tilde{x}_i}) = \sum_{i\in N} \hat{\mu}_i(\mathrm{r}_{x_1,x_2} \cap \mathrm{r}_{\bar{x}_i,\tilde{x}_i}) = \sum_{i\in N} \Delta_F(\mathrm{r}_{x_1,x_2} \cap \mathrm{r}_{\bar{x}_i,\tilde{x}_i}) = \sum_{i\in N} \check{\mu}(\mathrm{r}_{x_1,x_2} \cap \mathrm{r}_{\bar{x}_i,\tilde{x}_i}) = \check{\mu}(\mathrm{r}_{x_1,x_2}) = \Delta_F(\mathrm{r}_{x_1,x_2}) \in \mathcal{Y}$, where the first equality follows from the fact that $\hat{\mu}$ is a $\mathcal{Y}$-valued measure, the second equality follows from Proposition 11.115, the third equality follows from Claim 12.50.1, the fourth equality follows from Claim 12.50.1, the fifth equality follows from the fact that $\check{\mu}$ is a finite $\mathcal{Y}$-valued measure, and the last equality follows from Claim 12.50.1. By the arbitrariness of x_1 and x_2, F is a cumulative distribution function of $\hat{\mu}$.

Clearly, $T_F = \sum_{i\in N} T_F(\mathrm{r}_{\bar{x}_i,\tilde{x}_i}) = \sum_{i\in N} \mathcal{P}\circ\hat{\mu}(\mathrm{r}_{\bar{x}_i,\tilde{x}_i}) = \mathcal{P}\circ\hat{\mu}(\mathrm{P}(\Omega))$, where the first equality follows from Definition 12.41, the second equality follows from the preceding argument, and the last equality follows from the fact $\mathcal{P}\circ\hat{\mu}$ is a measure. T_F is well-defined since it is independent of $(\bar{x}_i)_{i\in N}$ and $(\tilde{x}_i)_{i\in N}$, by the uniqueness of $\hat{\mu}$.

This completes the proof of the theorem. □

Proposition 12.51 *Let $m \in \mathbb{N}$, $\mathbb{R}^m$ be endowed with the usual positive cone P, $\Omega \in \mathcal{B}_\mathrm{B}(\mathbb{R}^m)$ be a region with the subset topology $\mathcal{O}$, $\mathcal{Y}$ be a normed linear space over $\mathbb{K}$, and $\mu \in \mathcal{M}_f(\mathrm{P}(\Omega), \mathcal{B}_\mathrm{B}(\mathrm{P}(\Omega)), \mathcal{Y})$. Define $F : \Omega \to \mathcal{Y}$ and $G : \Omega \to [0,\infty) \subset \mathbb{R}$ by $F(x) = \mu\left(\left\{\bar{x}\in\mathrm{P}(\Omega) \,\middle|\, \bar{x} \leqq x\right\}\right)$ and $G(x) = \mathcal{P}\circ\mu\left(\left\{\bar{x}\in\mathrm{P}(\Omega) \,\middle|\, \bar{x} \leqq x\right\}\right)$, $\forall x \in \Omega$. Then:*

(i) *F is a cumulative distribution function of μ, F is of locally bounded variation, $T_F(\mathrm{r}_{x_1,x_2}) \le \mathcal{P}\circ\mu(\mathrm{r}_{x_1,x_2})$, $\forall x_1, x_2 \in \Omega$ with $x_1 \leqq x_2$ and $\bar{\mathrm{r}}_{x_1,x_2} \subseteq \Omega$, and $T_F \le \mathcal{P}\circ\mu(\mathrm{P}(\Omega))$ if it is well-defined.*
(ii) *G is a cumulative distribution function of $\mathcal{P}\circ\mu$ and G is of bounded variation.*
(iii) *If, in addition, $\mathcal{Y}$ is a Banach space, then $T_F(\mathrm{r}_{x_1,x_2}) = \mathcal{P}\circ\mu(\mathrm{r}_{x_1,x_2}) = \Delta_G(\mathrm{r}_{x_1,x_2})$, $\forall x_1, x_2 \in \Omega$ with $x_1 \leqq x_2$ and $\bar{\mathrm{r}}_{x_1,x_2} \subseteq \Omega$, and $T_F = \mathcal{P}\circ\mu(\mathrm{P}(\Omega))$. Therefore, F is of bounded variation.*

Proof Since Ω is a region, then $\exists N \subseteq \mathbb{N}$ and $(\bar{x}_i)_{i\in N}, (\tilde{x}_i)_{i\in N} \subseteq \Omega$ with $\bar{x}_i \leqq \tilde{x}_i$ and $\bar{\mathrm{r}}_{\bar{x}_i,\tilde{x}_i} \subseteq \Omega$, $\forall i \in N$, such that $\Omega = \bigcup_{i\in N} \bar{\mathrm{r}}_{\bar{x}_i,\tilde{x}_i}$, $\left(\mathrm{r}_{\bar{x}_i,\tilde{x}_i}\right)_{i\in N}$ is pairwise disjoint, $\mathrm{P}(\Omega) := \bigcup_{i\in N} \mathrm{r}_{\bar{x}_i,\tilde{x}_i}$, and $\forall x_1, x_2 \in \Omega$ with $x_1 \leqq x_2$ and $\bar{\mathrm{r}}_{x_1,x_2} \subseteq \Omega$, we have $\mathrm{r}_{x_1,x_2} \subseteq \mathrm{P}(\Omega)$.

$\forall x_1, x_2 \in \Omega$ with $x_1 \leqq x_2$ and $\bar{\mathrm{r}}_{x_1,x_2} \subseteq \Omega$, we have

$$\begin{aligned}
\Delta_F(\mathrm{r}_{x_1,x_2}) &= \begin{cases} \sum_{i=0}^{m} \sum_{\hat{x}\in\mathrm{VRect}_{i,x_1,x_2}} (-1)^i F(\hat{x}) & x_1 < x_2 \\ \vartheta_\mathcal{Y} & x_1 \nless x_2 \end{cases} \\
&= \begin{cases} \sum_{i=0}^{m} \sum_{\hat{x}\in\mathrm{VRect}_{i,x_1,x_2}} (-1)^i \mu\left(\left\{\bar{x}\in\mathrm{P}(\Omega) \mid \bar{x} \leqq \hat{x}\right\}\right) & x_1 < x_2 \\ \vartheta_\mathcal{Y} & x_1 \nless x_2 \end{cases} \\
&= \mu(\mathrm{P}(\Omega)\cap \mathrm{r}_{x_1,x_2}) = \mu(\mathrm{r}_{x_1,x_2})
\end{aligned}$$

Hence, F is a cumulative distribution function of μ.

By Proposition 11.4, $G(x_1) \le G(x_2)$, $\forall x_1, x_2 \in \Omega$ with $x_1 \leqq x_2$. By an argument that is similar to the previous paragraph, we have G is a cumulative distribution function of $\mathcal{P}\circ\mu$ and $\Delta_G(\mathrm{r}_{x_1,x_2}) = \mathcal{P}\circ\mu(\mathrm{r}_{x_1,x_2}) \ge 0$, $\forall x_1, x_2 \in \Omega$ with $x_1 \leqq x_2$ and $\bar{\mathrm{r}}_{x_1,x_2} \subseteq \Omega$. Since μ is finite, by Proposition 11.5, we have, $\forall x \in \Omega$, $G(x) = \mathcal{P}\circ\mu\left(\left\{\bar{x}\in\mathrm{P}(\Omega) \,\middle|\, \bar{x} \leqq x\right\}\right) = \mathcal{P}\circ\mu\left(\bigcap_{i=1}^{\infty} \left\{\bar{x}\in\mathrm{P}(\Omega) \,\middle|\, \bar{x} \leqq x + \mathbf{1}_m/i\right\}\right) = \lim_{i\in\mathbb{N}} \mathcal{P}\circ\mu\left(\left\{\bar{x}\in\mathrm{P}(\Omega) \,\middle|\, \bar{x} \leqq x + \mathbf{1}_m/i\right\}\right)$. $\forall \epsilon \in (0,\infty) \subset \mathbb{R}$, $\exists n_0 \in \mathbb{N}$ such that

$\mathcal{P} \circ \mu\left(\left\{\bar{x} \in \mathrm{P}(\Omega) \,\middle|\, \bar{x} \leqq x + \mathbf{1}_m/n_0\right\}\right) < G(x)+\epsilon$. Let $\delta := 1/n_0 \in (0, \infty) \subset \mathbb{R}$. $\forall \tilde{x} \in \mathcal{B}_{\mathbb{R}^m}(x, \delta) \cap \Omega$ with $x \leqq \tilde{x}$, we have $G(x) = \mathcal{P} \circ \mu\left(\left\{\bar{x} \in \mathrm{P}(\Omega) \,\middle|\, \bar{x} \leqq x\right\}\right) \leq \mathcal{P} \circ \mu\left(\left\{\bar{x} \in \mathrm{P}(\Omega) \,\middle|\, \bar{x} \leqq \tilde{x}\right\}\right) = G(\tilde{x}) \leq \mathcal{P} \circ \mu\left(\left\{\bar{x} \in \mathrm{P}(\Omega) \,\middle|\, \bar{x} \leqq x + \delta\mathbf{1}_m\right\}\right) < G(x) + \epsilon$. Hence, G is continuous on the right. By Proposition 12.49, G is of locally bounded variation. Then, $T_G = \sum_{i\in N} T_G\left(\mathrm{r}_{\bar{x}_i,\tilde{x}_i}\right) = \sum_{i\in N} \Delta_G(\mathrm{r}_{\bar{x}_i,\tilde{x}_i}) = \sum_{i\in N} \mathcal{P} \circ \mu(\mathrm{r}_{\bar{x}_i,\tilde{x}_i}) = \mathcal{P} \circ \mu(\mathrm{P}(\Omega)) < \infty$, where the first equality follows from Definition 12.41, the second equality follows from the fact that $\Delta_G(\mathrm{r}_{x_1,x_2}) = \mathcal{P} \circ \mu(\mathrm{r}_{x_1,x_2}) \geq 0$, $\forall x_1, x_2 \in \Omega$ with $x_1 \leqq x_2$ and $\bar{\mathrm{r}}_{x_1,x_2} \subseteq \Omega$, the third equality follows from the fact that G is the cumulative distribution function of $\mathcal{P} \circ \mu$, and the last two steps follow from the fact that $\mathcal{P} \circ \mu$ is a finite measure. Clearly, T_G is well-defined since it is independent of $(\bar{x}_i)_{i\in N}$ and $(\tilde{x}_i)_{i\in N}$. Hence, G is of bounded variation. Thus, (ii) holds.

$\forall x \in \Omega$, $\forall \epsilon \in (0, \infty) \subset \mathbb{R}$, by G being continuous on the right, $\exists \delta \in (0, \infty) \subset \mathbb{R}$, $\forall \mathbf{0}_m \leqq h < \delta\mathbf{1}_m$ with $x + h \in \Omega$, we have $G(x) \leq G(x + h) < G(x) + \epsilon$. Then, $\|F(x + h) - F(x)\| = \left\|\mu\left(\left\{\bar{x} \in \mathrm{P}(\Omega) \,\middle|\, \bar{x} \leqq x + h\right\}\right) - \mu\left(\left\{\bar{x} \in \mathrm{P}(\Omega) \,\middle|\, \bar{x} \leqq x\right\}\right)\right\| = \left\|\mu\left(\left\{\bar{x} \in \mathrm{P}(\Omega) \,\middle|\, \bar{x} \leqq x + h\right\} \setminus \left\{\bar{x} \in \mathrm{P}(\Omega) \,\middle|\, \bar{x} \leqq x\right\}\right)\right\| \leq \mathcal{P} \circ \mu\left(\left\{\bar{x} \in \mathrm{P}(\Omega) \,\middle|\, \bar{x} \leqq x + h\right\} \setminus \left\{\bar{x} \in \mathrm{P}(\Omega) \,\middle|\, \bar{x} \leqq x\right\}\right) = \mathcal{P} \circ \mu\left(\left\{\bar{x} \in \mathrm{P}(\Omega) \,\middle|\, \bar{x} \leqq x + h\right\}\right) - \mathcal{P} \circ \mu\left(\left\{\bar{x} \in \mathrm{P}(\Omega) \,\middle|\, \bar{x} \leqq x\right\}\right) = G(x + h) - G(x) < \epsilon$, where the second equality and the first inequality follow from the fact that μ is a finite $\mathcal{Y}$-valued measure and the third equality follows from Fact 11.3. Hence, F is continuous on the right. Then, F satisfies (i) of Definition 12.41.

$\forall x_1, x_2 \in \Omega$ with $x_1 \leqq x_2$ and $\bar{\mathrm{r}}_{x_1,x_2} \subseteq \Omega$, we have

$$
\begin{aligned}
T_F\left(\mathrm{r}_{x_1,x_2}\right) &= \sup_{\substack{n\in\mathbb{Z}_+,\, \check{x}_1,\hat{x}_1,\ldots,\check{x}_n,\hat{x}_n\in\bar{\mathrm{r}}_{x_1,x_2},\, \check{x}_i\leqq\hat{x}_i,\, \forall i\in\{1,\ldots,n\} \\ \bigcup_{i=1}^n \mathrm{r}_{\check{x}_i,\hat{x}_i}=\mathrm{r}_{x_1,x_2},\, \mathrm{r}_{\check{x}_i,\hat{x}_i}\cap\mathrm{r}_{\check{x}_j,\hat{x}_j}=\emptyset,\, \forall 1\leq i<j\leq n}} \sum_{i=1}^n \left\|\Delta_F(\mathrm{r}_{\check{x}_i,\hat{x}_i})\right\| \\
&= \sup_{\substack{n\in\mathbb{Z}_+,\, \check{x}_1,\hat{x}_1,\ldots,\check{x}_n,\hat{x}_n\in\bar{\mathrm{r}}_{x_1,x_2},\, \check{x}_i\leqq\hat{x}_i,\, \forall i\in\{1,\ldots,n\} \\ \bigcup_{i=1}^n \mathrm{r}_{\check{x}_i,\hat{x}_i}=\mathrm{r}_{x_1,x_2},\, \mathrm{r}_{\check{x}_i,\hat{x}_i}\cap\mathrm{r}_{\check{x}_j,\hat{x}_j}=\emptyset,\, \forall 1\leq i<j\leq n}} \sum_{i=1}^n \left\|\mu(\mathrm{r}_{\check{x}_i,\hat{x}_i})\right\| \\
&\leq \sup_{\substack{n\in\mathbb{Z}_+,\, \check{x}_1,\hat{x}_1,\ldots,\check{x}_n,\hat{x}_n\in\bar{\mathrm{r}}_{x_1,x_2},\, \check{x}_i\leqq\hat{x}_i,\, \forall i\in\{1,\ldots,n\} \\ \bigcup_{i=1}^n \mathrm{r}_{\check{x}_i,\hat{x}_i}=\mathrm{r}_{x_1,x_2},\, \mathrm{r}_{\check{x}_i,\hat{x}_i}\cap\mathrm{r}_{\check{x}_j,\hat{x}_j}=\emptyset,\, \forall 1\leq i<j\leq n}} \sum_{i=1}^n \mathcal{P} \circ \mu(\mathrm{r}_{\check{x}_i,\hat{x}_i}) \\
&= \mathcal{P} \circ \mu(\mathrm{r}_{x_1,x_2}) < \mathcal{P} \circ \mu(\Omega) < \infty
\end{aligned}
$$

where the second equality follows from the fact that F is a cumulative distribution function of μ, the first inequality follows from Definition 11.108, the third equality follows from the fact that $\mathcal{P} \circ \mu$ is a measure, the second inequality follows from Proposition 11.4, and the last inequality follows from the fact that μ is finite. Then, F satisfies (ii) of Definition 12.41.

$\forall x_1, x_2 \in \Omega$ with $x_1 \lessdot x_2$ and $\bar{\mathrm{r}}_{x_1,x_2} \subseteq \Omega$, define $\bar{G} : \bar{\mathrm{r}}_{x_1,x_2} \to [0, \infty) \subset \mathbb{R}$ by $\bar{G}(x) = T_F(\mathrm{r}_{x_1,x})$, $\forall x \in \bar{\mathrm{r}}_{x_1,x_2}$. $\forall x \in \bar{\mathrm{r}}_{x_1,x_2}$, by G being continuous on the right, $\forall \epsilon \in (0, \infty) \subset \mathbb{R}$, $\exists \delta \in (0, \infty) \subset \mathbb{R}$, and $\forall \mathbf{0}_m \stackrel{\lessdot}{=} h \lessdot \delta \mathbf{1}_m$ with $x + h \in \Omega$, we have $G(x) \leq G(x + h) < G(x) + \epsilon$. Then, $\forall \mathbf{0}_m \stackrel{\lessdot}{=} h \lessdot \delta \mathbf{1}_m$ with $x + h \in \bar{\mathrm{r}}_{x_1,x_2}$, $0 \leq \bar{G}(x + h) - \bar{G}(x) = T_F(\mathrm{r}_{x_1,x+h}) - T_F(\mathrm{r}_{x_1,x}) = \sum_{(\check{x},\hat{x})\in M} T_F(\mathrm{r}_{\check{x},\hat{x}}) \leq \sum_{(\check{x},\hat{x})\in M} \mathcal{P} \circ \mu(\mathrm{r}_{\check{x},\hat{x}}) = \mathcal{P} \circ \mu(\mathrm{r}_{x_1,x+h} \setminus \mathrm{r}_{x_1,x}) \leq \mathcal{P} \circ \mu\left(\left\{\bar{x} \in \mathrm{P}(\Omega) \,\middle|\, \bar{x} \stackrel{\lessdot}{=} x + h\right\} \setminus \{\bar{x} \in \mathrm{P}(\Omega) \mid \bar{x} \stackrel{\lessdot}{=} x\right\}\right) = \mathcal{P} \circ \mu\left(\left\{\bar{x} \in \mathrm{P}(\Omega) \,\middle|\, \bar{x} \stackrel{\lessdot}{=} x + h\right\}\right) - \mathcal{P} \circ \mu\left(\left\{\bar{x} \in \mathrm{P}(\Omega) \,\middle|\, \bar{x} \stackrel{\lessdot}{=} x\right\}\right) = G(x + h) - G(x) < \epsilon$, where the first inequality follows from the monotonicity of $\bar{G}$; $M := \left\{(\check{x}, \hat{x}) \in \bar{\mathrm{r}}_{x_1,x} \times \bar{\mathrm{r}}_{x,x+h} \,\middle|\, i \in \{0, \ldots, m-1\}, J \subseteq \bar{J}_m := \{1, \ldots, m\},\right.$ $\mathrm{card}(J) = i, \pi_k(\check{x}) = \pi_k(x_1), \forall k \in J, \pi_k(\check{x}) = \pi_k(x), \forall k \in \bar{J}_m \setminus J, \pi_k(\hat{x}) = \pi_k(x), \forall k \in J, \pi_k(\hat{x}) = \pi_k(x + h), \forall k \in \bar{J}_m \setminus J\big\}$; the second equality follows from Proposition 12.47; the second inequality follows from the last paragraph; the third equality follows from the fact that $\mathcal{P} \circ \mu$ is a measure; the third inequality follows from Proposition 11.4; and the fourth equality follows from the fact that $\mathcal{P} \circ \mu$ is a finite measure. Hence, $\bar{G}$ is continuous on the right. Thus, F satisfies (iii) of Definition 12.41.

Therefore, F is of locally bounded variation. When T_F is well-defined, we have $T_F = \sum_{i\in N} T_F(\mathrm{r}_{\bar{x}_i,\tilde{x}_i}) \leq \sum_{i\in N} \mathcal{P} \circ \mu(\mathrm{r}_{\bar{x}_i,\tilde{x}_i}) = \mathcal{P} \circ \mu(\mathrm{P}(\Omega)) < \infty$, where the first equality follows from Definition 12.41, the first inequality follows from the second to last paragraph, and the second equality and the second inequality follow from the fact that $\mathcal{P} \circ \mu$ is a finite measure. Thus, (i) holds.

(iii) Let $\mathcal{Y}$ be a Banach space. By Theorem 12.50 and (i), μ is the unique σ-finite $\mathcal{Y}$-valued measure on the measurable space $(\mathrm{P}(\Omega), \mathcal{B}_{\mathrm{B}}(\mathrm{P}(\Omega)))$ such that F is a cumulative distribution function of μ. Theorem 12.50, coupled with (ii), further concludes that $T_F(\mathrm{r}_{x_1,x_2}) = \mathcal{P} \circ \mu(\mathrm{r}_{x_1,x_2}) = \Delta_G(\mathrm{r}_{x_1,x_2})$, $\forall x_1, x_2 \in \Omega$ with $x_1 \stackrel{\lessdot}{=} x_2$ and $\bar{\mathrm{r}}_{x_1,x_2} \subseteq \Omega$, and $T_F = \mathcal{P} \circ \mu(\mathrm{P}(\Omega)) < \infty$. Since T_F is well-defined and is finite, then F is of bounded variation.

This completes the proof of the proposition. □

Proposition 12.52 *Let $m \in \mathbb{N}$, $\mathbb{R}^m$ be endowed with the usual positive cone P, $\Omega \in \mathcal{B}_{\mathrm{B}}(\mathbb{R}^m)$ be a rectangle with the subset topology $\mathcal{O}$, $\mathcal{Y}$ be a normed linear space over $\mathbb{K}$, and μ be a σ-finite $\mathcal{Y}$-valued measure on $(\mathrm{P}(\Omega), \mathcal{B}_{\mathrm{B}}(\mathrm{P}(\Omega)))$ such that $\mathcal{P} \circ \mu(\mathrm{r}_{x_1,x_2}) < \infty$, $\forall x_1, x_2 \in \Omega$ with $x_1 \stackrel{\lessdot}{=} x_2$. By Proposition 12.39, $\mathrm{P}(\Omega) = \prod_{i=1}^{m} \bar{I}_i$ and Ω is a region.*

Define $F : \Omega \to \mathcal{Y}$ by, if $\Omega \neq \emptyset$, then fix $x_0 \in \Omega$, $F(x) = \vartheta_{\mathcal{Y}}$, $\forall x \in \Omega$ with $\pi_{i_0}(x) = \pi_{i_0}(x_0)$ for some $i_0 \in \bar{J} := \{1, \ldots, m\}$; and $\forall x \in \Omega$ with $\pi_i(x) \neq \pi_i(x_0)$, $\forall i \in \bar{J}$, $F(x)$ is such that $\Delta_F(\mathrm{r}_{\check{x},\hat{x}}) = \mu(\mathrm{r}_{\check{x},\hat{x}})$, where $\check{x} := x \wedge x_0$ and $\hat{x} := x \vee x_0$. Then, F is a cumulative distribution function of μ, which is called the cumulative *distribution function of μ with origin x_0. Define $G : \Omega \to \mathbb{R}$ to be the cumulative distribution function of $\mathcal{P} \circ \mu$ with origin x_0. Then, the following statements hold:*

(i) *F and G are of locally bounded variation.*
(ii) *If, in addition, $\mathcal{Y}$ is a Banach space, then $T_F(\mathrm{r}_{x_1,x_2}) = \mathcal{P} \circ \mu(\mathrm{r}_{x_1,x_2}) = \Delta_G(\mathrm{r}_{x_1,x_2})$, $\forall x_1, x_2 \in \Omega$ with $x_1 \stackrel{\lessdot}{=} x_2$ and $\bar{\mathrm{r}}_{x_1,x_2} \subseteq \Omega$, and $T_F = \mathcal{P} \circ \mu(\mathrm{P}(\Omega))$.*

Proof (i) By the assumption on μ, we have $\mathcal{P} \circ \mu(\mathrm{r}_{x_1,x_2}) < \infty$, $\forall x_1, x_2 \in \Omega$ with $x_1 \stackrel{\lessdot}{=} x_2$. Then, $\mathrm{r}_{x_1,x_2} \in \mathrm{dom}(\mu)$. This implies that $F(x) \in \mathcal{Y}$ and $G(x) \in \mathbb{R}$, $\forall x \in \Omega$. Hence, F and G are well-defined.

If $\Omega = \emptyset$, then the result holds trivially. This case is proved. Consider the case $\Omega \neq \emptyset$.

Claim 12.52.1 F is a cumulative distribution function of μ.

Proof of Claim $\forall x_1, x_2 \in \Omega$ with $x_1 \stackrel{\lessdot}{=} x_2$. If $x_1 \not\lessdot x_2$, then by Definition 12.41, we have $\mu(\mathrm{r}_{x_1,x_2}) = \mu(\emptyset) = \vartheta_{\mathcal{Y}} = \Delta_F(\mathrm{r}_{x_1,x_2})$. On the other hand, if $x_1 \lessdot x_2$, then let $n := \mathrm{card}\left(\left\{i \in \bar{J} \,\middle|\, (\pi_i(x_1) - \pi_i(x_0))\,(\pi_i(x_2) - \pi_i(x_0)) < 0\right\}\right) \in \{0, \ldots, m\}$. We will show that $\mu(\mathrm{r}_{x_1,x_2}) = \Delta_F(\mathrm{r}_{x_1,x_2})$ by mathematical induction on n.
1° Consider the case $n = 0$. Define $J := \left\{i \in \bar{J} \,\middle|\, \pi_i(x_1) < \pi_i(x_0) \text{ or } \pi_i(x_2) < \pi_i(x_0)\right\}$ and $\bar{n} := \mathrm{card}\,(J)$.

$\forall j \in \{0, \ldots, m\}$, $\forall x \in \mathrm{VRect}_{j,x_1,x_2}$, if $\pi_i(x) \neq \pi_i(x_0)$, $\forall i \in \bar{J}$, then, $\bar{n} = \mathrm{card}\left(\left\{i \in \bar{J} \,\middle|\, \pi_i(x) < \pi_i(x_0)\right\}\right)$, by the definition of F and Definition 12.41, we have $\Delta_F(\mathrm{r}_{\check{x},\hat{x}}) = \mu(\mathrm{r}_{\check{x},\hat{x}}) = (-1)^{\bar{n}} F(x)$. On the other hand, if $\pi_{i_0}(x) = \pi_{i_0}(x_0)$ for some $i_0 \in \bar{J}$, then, by the definition of F and Definition 12.41, $\Delta_F(\mathrm{r}_{\check{x},\hat{x}}) = \vartheta_{\mathcal{Y}} = \mu(\emptyset) = \mu(\mathrm{r}_{\check{x},\hat{x}}) = F(x) = (-1)^{\bar{n}} F(x)$. Hence, we have $\mu(\mathrm{r}_{\check{x},\hat{x}}) = (-1)^{\bar{n}} F(x)$, $\forall x \in \mathrm{VRect}_{j,x_1,x_2}$, $\forall j \in \{0, \ldots, m\}$.

Then, we have

$$\begin{aligned}
\mu(\mathrm{r}_{x_1,x_2}) &= \mu\left(\left\{x \in \mathrm{P}(\Omega) \,\middle|\, x_1 \lessdot x \stackrel{\lessdot}{=} x_2\right\}\right) \\
&= \mu\big(\big\{x \in \mathrm{P}(\Omega) \,\big|\, \pi_i(x_1) < \pi_i(x) \leq \pi_i(x_2) \leq \pi_i(x_0),\ \forall i \in J; \\
&\quad \pi_i(x_0) \leq \pi_i(x_1) < \pi_i(x) \leq \pi_i(x_2),\ \forall i \in \bar{J} \setminus J\big\}\big) \\
&= \mu\big(\big\{x \in \mathrm{P}(\Omega) \,\big|\, \pi_i(x_1) < \pi_i(x) \leq \pi_i(x_0),\ \forall i \in J; \\
&\quad \pi_i(x_0) < \pi_i(x) \leq \pi_i(x_2),\ \forall i \in J \setminus J\big\}\big) \\
&\quad -\mu\Big(\bigcup_{i_0=1}^{m} \{x \in \mathrm{P}(\Omega) \,\big|\, \pi_i(x_1) < \pi_i(x) \leq \pi_i(x_0),\ \forall i \in J \setminus \{i_0\}; \\
&\quad \pi_i(x_0) < \pi_i(x) \leq \pi_i(x_2),\ \forall i \in (\bar{J} \setminus J) \setminus \{i_0\}; \\
&\quad \pi_i(x_2) < \pi_i(x) \leq \pi_i(x_0),\ \forall i \in J \cap \{i_0\}; \\
&\quad \pi_i(x_0) < \pi_i(x) \leq \pi_i(x_1),\ \forall i \in (\bar{J} \setminus J) \cap \{i_0\}\}\Big)
\end{aligned}$$

$$= \sum_{j=0}^{m} (-1)^j \sum_{\hat{J} \subseteq \bar{J}, \mathrm{card}(\hat{J})=j} \mu\Big(\Big\{x \in \mathrm{P}(\Omega) \;\Big|\; \pi_i(x_1) < \pi_i(x) \le \pi_i(x_0),$$

$$\forall i \in J \setminus \hat{J};\; \pi_i(x_0) < \pi_i(x) \le \pi_i(x_2),\ \forall i \in (\bar{J} \setminus J) \setminus \hat{J};$$

$$\pi_i(x_2) < \pi_i(x) \le \pi_i(x_0),\ \forall i \in J \cap \hat{J};$$

$$\pi_i(x_0) < \pi_i(x) \le \pi_i(x_1),\ \forall i \in (\bar{J} \setminus J) \cap \hat{J}\Big\}\Big)$$

where the equalities follow from the assumption on μ. $\forall \hat{J} \subseteq \bar{J}$ with $\mathrm{card}\left(\hat{J}\right) = j \in \{0, \ldots, m\}$, define $x_{\hat{J}} \in \Omega$ by, $\forall i \in \bar{J}$,

$$\pi_i(x_{\hat{J}}) = \begin{cases} \pi_i(x_1)\ \forall i \in (J \setminus \hat{J}) \cup ((\bar{J} \setminus J) \cap \hat{J}) = J \triangle \hat{J} \\ \pi_i(x_2)\ \forall i \in ((\bar{J} \setminus J) \setminus \hat{J}) \cup (\hat{J} \cap J) = \bar{J} \setminus (\hat{J} \triangle J) \end{cases}$$

Then, $\Big\{x \in \mathrm{P}(\Omega) \;\Big|\; \pi_i(x_1) < \pi_i(x) \le \pi_i(x_0),\ \forall i \in J \setminus \hat{J};\ \pi_i(x_0) < \pi_i(x) \le \pi_i(x_2),\ \forall i \in (\bar{J} \setminus J) \setminus \hat{J};\ \pi_i(x_2) < \pi_i(x) \le \pi_i(x_0),\ \forall i \in J \cap \hat{J};\ \pi_i(x_0) < \pi_i(x) \le \pi_i(x_1),\ \forall i \in (\bar{J} \setminus J) \cap \hat{J}\Big\} = \mathrm{r}_{\check{x}_{\hat{J}}, \hat{x}_{\hat{J}}}$. By the definition of $x_{\hat{J}}$, we have $x_{\hat{J}} \in \mathrm{VRect}_{\mathrm{card}\left(J \triangle \hat{J}\right), x_1, x_2}$ and $\mathrm{card}\left(J \triangle \hat{J}\right) \equiv \mathrm{card}(J) + \mathrm{card}\left(\hat{J}\right) = \bar{n} + j \pmod 2$, $\forall \hat{J} \subseteq \bar{J}$ with $\mathrm{card}\left(\hat{J}\right) = j \in \{0, \ldots, m\}$. This leads to the conclusion that

$$\begin{aligned} \mu(\mathrm{r}_{x_1, x_2}) &= \sum_{j=0}^{m} (-1)^j \sum_{\hat{J} \subseteq \bar{J}, \mathrm{card}(\hat{J})=j} \mu(\mathrm{r}_{\check{x}_{\hat{J}}, \hat{x}_{\hat{J}}}) \\ &= \sum_{j=0}^{m} (-1)^j \sum_{\hat{J} \subseteq \bar{J}, \mathrm{card}(\hat{J})=j} (-1)^{\bar{n}} F(x_{\hat{J}}) \\ &= \sum_{j=0}^{m} \sum_{\hat{J} \subseteq \bar{J}, \mathrm{card}(\hat{J})=j} (-1)^{\mathrm{card}(J \triangle \hat{J})} F(x_{\hat{J}}) \\ &= \sum_{i=0}^{m} \sum_{\bar{x} \in \mathrm{VRect}_{i, x_1, x_2}} (-1)^i F(\bar{x}) = \Delta_F(\mathrm{r}_{x_1, x_2}) \end{aligned}$$

where the second equality follows from the previous paragraph. This case is proved.

2° Assume that $\mu(\mathrm{r}_{x_1,x_2}) = \Delta_F(\mathrm{r}_{x_1,x_2})$, $\forall n \le k-1 \in \{0,\ldots,m-1\}$.
3° Consider the case when $n = k \in \bar{J}$. Since $n > 0$, then $\exists i_0 \in \bar{J}$ such that $\pi_{i_0}(x_1) < \pi_{i_0}(x_0) < \pi_{i_0}(x_2)$. Without loss of generality, assume that $i_0 = 1$. Then, we have $\mu(\mathrm{r}_{x_1,x_2}) = \mu\left(\left\{x \in \mathrm{P}(\Omega) \,\middle|\, \pi_i(x_1) < \pi_i(x) \le \pi_i(x_2),\ \forall i \in \bar{J}\right\}\right) = \mu\left(\left\{x \in \mathrm{P}(\Omega) \,\middle|\, \pi_1(x_1) < \pi_1(x) \le \pi_1(x_0);\ \pi_i(x_1) < \pi_i(x) \le \pi_i(x_2),\ \forall i \in \bar{J} \setminus \{1\}\right\}\right) + \mu\left(\left\{x \in \mathrm{P}(\Omega) \,\middle|\, \pi_1(x_0) < \pi_1(x) \le \pi_1(x_2);\ \pi_i(x_1) < \pi_i(x) \le \pi_i(x_2),\ \forall i \in \bar{J} \setminus \{1\}\right\}\right) = \mu(\mathrm{r}_{x_1,x_3}) + \mu(\mathrm{r}_{x_4,x_2})$, where $x_3, x_4 \in \Omega$ are defined by $\pi_i(x_3) = \begin{cases} \pi_i(x_0) \text{ if } i = 1 \\ \pi_i(x_2)\ \forall i \in \bar{J} \setminus \{1\} \end{cases}$ and $\pi_i(x_4) = \begin{cases} \pi_i(x_0) \text{ if } i = 1 \\ \pi_i(x_1)\ \forall i \in \bar{J} \setminus \{1\} \end{cases}$, and all equalities follows from the assumption on μ. By the inductive assumption, we have $\mu(\mathrm{r}_{x_1,x_2}) = \mu(\mathrm{r}_{x_1,x_3}) + \mu(\mathrm{r}_{x_4,x_2}) = \Delta_F(\mathrm{r}_{x_1,x_3}) + \Delta_F(\mathrm{r}_{x_4,x_2}) = \Delta_F(\mathrm{r}_{x_1,x_2})$, where the last equality follows from Proposition 12.46. This completes the induction process.

Therefore, F is a cumulative distribution function of μ. □

By Claim 12.52.1, F and G are cumulative distribution functions of μ and $\mathcal{P} \circ \mu$, respectively. We need the following intermediate result.

Claim 12.52.2 $\forall x_1, x_2 \in \Omega$ with $x_1 \stackrel{\lessdot}{=} x_2$, if x_2 is an accumulation point of $\Omega \cap (x_2 + P)$, then $\lim_{\substack{h \to \mathbf{0}_m \\ h \stackrel{\gtrdot}{=} \mathbf{0}_m}} \mathcal{P} \circ \mu(\mathrm{r}_{x_1,x_2+h} \setminus \mathrm{r}_{x_1,x_2}) = 0$. $\forall x_1, x_2 \in \Omega$ with $x_1 \stackrel{\lessdot}{=} x_2$ and $x_1 \ne x_2$, then $\lim_{\substack{h \to \mathbf{0}_m,\ h \stackrel{\gtrdot}{=} \mathbf{0}_m \\ x_1 + h \stackrel{\lessdot}{=} x_2}} \mathcal{P} \circ \mu(\mathrm{r}_{x_1,x_2} \setminus \mathrm{r}_{x_1+h,x_2}) = 0$.

Proof of Claim $\forall (h_n)_{n=1}^\infty \subseteq P \cap (\Omega - x_2) \setminus \{\mathbf{0}_m\}$ with $\lim_{n\in\mathbb{N}} h_n = \mathbf{0}_m$, we have $\lim_{n\in\mathbb{N}} \bar{h}_n := \lim_{n\in\mathbb{N}} \bigvee_{i=n}^\infty h_i = \mathbf{0}_m$, $\mathbf{0}_m \stackrel{\lessdot}{=} \bar{h}_{n+1} \stackrel{\lessdot}{=} \bar{h}_n \ne \mathbf{0}_m$, and $\bar{h}_n \in P \cap (\Omega - x_2) \setminus \{\mathbf{0}_m\}$, $\forall n \in \mathbb{N}$. Then, $0 \le \lim_{n\in\mathbb{N}} \mathcal{P} \circ \mu(\mathrm{r}_{x_1,x_2+h_n} \setminus \mathrm{r}_{x_1,x_2}) \le \lim_{n\in\mathbb{N}} \mathcal{P} \circ \mu(\mathrm{r}_{x_1,x_2+\bar{h}_n} \setminus \mathrm{r}_{x_1,x_2}) = \mathcal{P} \circ \mu\left(\bigcap_{n=1}^\infty (\mathrm{r}_{x_1,x_2+\bar{h}_n} \setminus \mathrm{r}_{x_1,x_2})\right) = \mathcal{P} \circ \mu(\emptyset) = 0$, where the first inequality follows from the fact that $\mathcal{P} \circ \mu$ is a measure, the second inequality follows from Proposition 11.4, and the first equality follows from Proposition 11.5. Then, $\lim_{n\in\mathbb{N}} \mathcal{P} \circ \mu(\mathrm{r}_{x_1,x_2+h_n} \setminus \mathrm{r}_{x_1,x_2}) = 0$. By Proposition 4.16, we have $\lim_{\substack{h \to \mathbf{0}_m \\ h \stackrel{\gtrdot}{=} \mathbf{0}_m}} \mathcal{P} \circ \mu(\mathrm{r}_{x_1,x_2+h} \setminus \mathrm{r}_{x_1,x_2}) = 0$. This proves the first statement.

$\forall (h_n)_{n=1}^\infty \subseteq \left\{h \in \mathbb{R}^m \,\middle|\, x_1 \stackrel{\lessdot}{=} x_1 + h \stackrel{\lessdot}{=} x_2,\ h \ne \mathbf{0}_m\right\}$ with $\lim_{n\in\mathbb{N}} h_n = \mathbf{0}_m$, we have $\lim_{n\in\mathbb{N}} \bar{h}_n := \lim_{n\in\mathbb{N}} \bigvee_{i=n}^\infty h_i = \mathbf{0}_m$ and $\mathbf{0}_m \stackrel{\lessdot}{=} \bar{h}_{n+1} \stackrel{\lessdot}{=} \bar{h}_n \ne \mathbf{0}_m$, $\forall n \in \mathbb{N}$. Then, $0 \le \lim_{n\in\mathbb{N}} \mathcal{P} \circ \mu(\mathrm{r}_{x_1,x_2} \setminus \mathrm{r}_{x_1+h_n,x_2}) \le \lim_{n\in\mathbb{N}} \mathcal{P} \circ \mu(\mathrm{r}_{x_1,x_2} \setminus \mathrm{r}_{x_1+\bar{h}_n,x_2}) = \mathcal{P} \circ \mu\left(\bigcap_{n=1}^\infty \left(\mathrm{r}_{x_1,x_2} \setminus \mathrm{r}_{x_1+\bar{h}_n,x_2}\right)\right) = \mathcal{P} \circ \mu(\emptyset) = 0$, where the first inequality follows from the fact that $\mathcal{P} \circ \mu$ is a measure, the second inequality follows from Proposition 11.4, and the first equality follows from Proposition 11.5. Then, $\lim_{n\in\mathbb{N}} \mathcal{P} \circ \mu(\mathrm{r}_{x_1,x_2} \setminus \mathrm{r}_{x_1+h_n,x_2}) = 0$. By Proposition 4.16, we have $\lim_{\substack{h \to \mathbf{0}_m,\ h \stackrel{\gtrdot}{=} \mathbf{0}_m \\ x_1 + h \stackrel{\lessdot}{=} x_2}} \mathcal{P} \circ \mu(\mathrm{r}_{x_1,x_2} \setminus \mathrm{r}_{x_1+h,x_2}) = 0$.

This completes the proof of the claim. □

$\forall x \in \Omega$, let $J := \left\{ i \in \bar{J} \,\middle|\, \pi_i(x) < \pi_i(x_0) \right\}$ and $n := \mathrm{card}(J)$. Let x be an accumulation point of $\Omega \cap (x + P)$. Then, $\exists \delta \in (0, \infty) \subset \mathbb{R}$ such that, $\forall h \in P \cap (\Omega - x)$ with $h \neq \mathbf{0}_m$ and $h \lessdot \delta \mathbf{1}_m$, we have $x + h \in \Omega$ and $\left\{ i \in \bar{J} \,\middle|\, \pi_i(x + h) < \pi_i(x_0) \right\} = J$. Then, $0 \leq |G(x + h) - G(x)| = |(-1)^n (\mathcal{P} \circ \mu(\mathrm{r}_{\check{x}_h, \hat{x}_h}) - \mathcal{P} \circ \mu(\mathrm{r}_{\check{x}, \hat{x}}))| \leq |\mathcal{P} \circ \mu(\mathrm{r}_{\check{x}_h, \hat{x}_h}) - \mathcal{P} \circ \mu(\mathrm{r}_{\check{x}_h, \hat{x}})| + |\mathcal{P} \circ \mu(\mathrm{r}_{\check{x}_h, \hat{x}}) - \mathcal{P} \circ \mu(\mathrm{r}_{\check{x}, \hat{x}})| = \mathcal{P} \circ \mu(\mathrm{r}_{\check{x}_h, \hat{x}_h} \setminus \mathrm{r}_{\check{x}_h, \hat{x}}) + \mathcal{P} \circ \mu(\mathrm{r}_{\check{x}, \hat{x}} \setminus \mathrm{r}_{\check{x}_h, \hat{x}}) \leq \mathcal{P} \circ \mu(\mathrm{r}_{\check{x}, \hat{x}_h} \setminus \mathrm{r}_{\check{x}, \hat{x}}) + \mathcal{P} \circ \mu(\mathrm{r}_{\check{x}, \hat{x}} \setminus \mathrm{r}_{\check{x}_h, \hat{x}})$, where $x_h := x + h$, the first equality follows from the definition of G; the second inequality follows from the fact that $x \stackrel{\lessdot}{=} x_h$ and $\check{x} \stackrel{\lessdot}{=} \check{x}_h \stackrel{\lessdot}{=} \hat{x} \stackrel{\lessdot}{=} \hat{x}_h$, the second equality follows from Fact 11.3, and the last inequality follows from Proposition 11.4. By Claim 12.52.2, we have $0 \leq \lim_{\substack{h \to \mathbf{0}_m \\ h \stackrel{\gtrdot}{=} \mathbf{0}_m}} |G(x + h) - G(x)| \leq \lim_{\substack{h \to \mathbf{0}_m \\ h \stackrel{\gtrdot}{=} \mathbf{0}_m}} \mathcal{P} \circ \mu(\mathrm{r}_{\check{x}, \hat{x}_h} \setminus \mathrm{r}_{\check{x}, \hat{x}}) + \lim_{\substack{h \to \mathbf{0}_m \\ h \stackrel{\gtrdot}{=} \mathbf{0}_m}} \mathcal{P} \circ \mu(\mathrm{r}_{\check{x}, \hat{x}} \setminus \mathrm{r}_{\check{x}_h, \hat{x}}) = 0$. Hence, by Proposition 12.37, G is continuous on the right. By Proposition 12.49, G is of locally bounded variation.

$\forall x \in \Omega$, let x be an accumulation point of $\Omega \cap (x + P)$, and let J, n, δ, and x_h be as defined in the previous paragraph. $\forall h \in P \cap (\Omega - x)$ with $h \neq \mathbf{0}_m$ and $h \lessdot \delta \mathbf{1}_m$, we have $0 \leq \| F(x + h) - F(x) \| = \left\| (-1)^n (\mu(\mathrm{r}_{\check{x}_h, \hat{x}_h}) - \mu(\mathrm{r}_{\check{x}, \hat{x}})) \right\| \leq \left\| \mu(\mathrm{r}_{\check{x}_h, \hat{x}_h}) - \mu(\mathrm{r}_{\check{x}_h, \hat{x}}) \right\| + \left\| \mu(\mathrm{r}_{\check{x}_h, \hat{x}}) - \mu(\mathrm{r}_{\check{x}, \hat{x}}) \right\| = \left\| \mu(\mathrm{r}_{\check{x}_h, \hat{x}_h} \setminus \mathrm{r}_{\check{x}_h, \hat{x}}) \right\| + \left\| \mu(\mathrm{r}_{\check{x}, \hat{x}} \setminus \mathrm{r}_{\check{x}_h, \hat{x}}) \right\| \leq \mathcal{P} \circ \mu(\mathrm{r}_{\check{x}_h, \hat{x}_h} \setminus \mathrm{r}_{\check{x}_h, \hat{x}}) + \mathcal{P} \circ \mu(\mathrm{r}_{\check{x}, \hat{x}} \setminus \mathrm{r}_{\check{x}_h, \hat{x}}) \leq \mathcal{P} \circ \mu(\mathrm{r}_{\check{x}, \hat{x}_h} \setminus \mathrm{r}_{\check{x}, \hat{x}}) + \mathcal{P} \circ \mu(\mathrm{r}_{\check{x}, \hat{x}} \setminus \mathrm{r}_{\check{x}_h, \hat{x}})$, where the first equality follows from the definition of F, the second inequality follows from the fact that $x \stackrel{\lessdot}{=} x_h$ and $\check{x} \stackrel{\lessdot}{=} \check{x}_h \stackrel{\lessdot}{=} \hat{x} \stackrel{\lessdot}{=} \hat{x}_h$, the second equality and the third inequality follow from the assumption on μ, and the last inequality follows from Proposition 11.4. By Claim 12.52.2, we have $0 \leq \lim_{\substack{h \to \mathbf{0}_m \\ h \stackrel{\gtrdot}{=} \mathbf{0}_m}} \| F(x + h) - F(x) \| \leq \lim_{\substack{h \to \mathbf{0}_m \\ h \stackrel{\gtrdot}{=} \mathbf{0}_m}} \mathcal{P} \circ \mu(\mathrm{r}_{\check{x}, \hat{x}_h} \setminus \mathrm{r}_{\check{x}, \hat{x}}) + \lim_{\substack{h \to \mathbf{0}_m \\ h \stackrel{\gtrdot}{=} \mathbf{0}_m}} \mathcal{P} \circ \mu(\mathrm{r}_{\check{x}, \hat{x}} \setminus \mathrm{r}_{\check{x}_h, \hat{x}}) = 0$. Hence, by Proposition 12.37, F is continuous on the right and F satisfies (i) of Definition 12.41.

$\forall x_1, x_2 \in \Omega$ with $x_1 \stackrel{\lessdot}{=} x_2$, we have

$$\begin{aligned} T_F\big(\mathrm{r}_{x_1, x_2}\big) &= \sup_{\substack{n \in \mathbb{Z}_+,\ \check{x}_1, \hat{x}_1, \ldots, \check{x}_n, \hat{x}_n \in \bar{\mathrm{r}}_{x_1, x_2},\ \check{x}_i \stackrel{\lessdot}{=} \hat{x}_i,\ \forall i \in \{1, \ldots, n\} \\ \bigcup_{i=1}^n \mathrm{r}_{\check{x}_i, \hat{x}_i} = \mathrm{r}_{x_1, x_2},\ \mathrm{r}_{\check{x}_i, \hat{x}_i} \cap \mathrm{r}_{\check{x}_j, \hat{x}_j} = \emptyset,\ \forall 1 \leq i < j \leq n}} \sum_{i=1}^n \left\| \Delta_F(\mathrm{r}_{\check{x}_i, \hat{x}_i}) \right\| \\ &= \sup_{\substack{n \in \mathbb{Z}_+,\ \check{x}_1, \hat{x}_1, \ldots, \check{x}_n, \hat{x}_n \in \bar{\mathrm{r}}_{x_1, x_2},\ \check{x}_i \stackrel{\lessdot}{=} \hat{x}_i,\ \forall i \in \{1, \ldots, n\} \\ \bigcup_{i=1}^n \mathrm{r}_{\check{x}_i, \hat{x}_i} = \mathrm{r}_{x_1, x_2},\ \mathrm{r}_{\check{x}_i, \hat{x}_i} \cap \mathrm{r}_{\check{x}_j, \hat{x}_j} = \emptyset,\ \forall 1 \leq i < j \leq n}} \sum_{i=1}^n \left\| \mu(\mathrm{r}_{\check{x}_i, \hat{x}_i}) \right\| \\ &\leq \sup_{\substack{n \in \mathbb{Z}_+,\ \check{x}_1, \hat{x}_1, \ldots, \check{x}_n, \hat{x}_n \in \bar{\mathrm{r}}_{x_1, x_2},\ \check{x}_i \stackrel{\lessdot}{=} \hat{x}_i,\ \forall i \in \{1, \ldots, n\} \\ \bigcup_{i=1}^n \mathrm{r}_{\check{x}_i, \hat{x}_i} = \mathrm{r}_{x_1, x_2},\ \mathrm{r}_{\check{x}_i, \hat{x}_i} \cap \mathrm{r}_{\check{x}_j, \hat{x}_j} = \emptyset,\ \forall 1 \leq i < j \leq n}} \sum_{i=1}^n \mathcal{P} \circ \mu(\mathrm{r}_{\check{x}_i, \hat{x}_i}) \\ &= \mathcal{P} \circ \mu(\mathrm{r}_{x_1, x_2}) < \infty \end{aligned}$$

where the second equality follows from the fact that F is a cumulative distribution function of μ, the first inequality follows from Definition 11.108, the third equality follows from the fact that $\mathcal{P} \circ \mu$ is a measure, and the last inequality follows from the assumption on μ. Then, F satisfies (ii) of Definition 12.41.

$\forall x_1, x_2 \subset \Omega$ with $x_1 \lessdot x_2$ and $\bar{\mathrm{r}}_{x_1,x_2} \subseteq \Omega$, define $\bar{G} : \bar{\mathrm{r}}_{x_1,x_2} \to [0,\infty) \subset \mathbb{R}$ by $\bar{G}(x) = T_F(\mathrm{r}_{x_1,x})$, $\forall x \in \bar{\mathrm{r}}_{x_1,x_2}$. $\forall x \in \bar{\mathrm{r}}_{x_1,x_2}$ with $x \neq x_2$, we have

$$\begin{aligned}
0 &\leq \lim_{\substack{h\to \mathbf{0}_m \\ h \geqq \mathbf{0}_m}} (\bar{G}(x+h) - \bar{G}(x)) = \lim_{\substack{h\to \mathbf{0}_m \\ h \geqq \mathbf{0}_m}} \left(T_F\left(\mathrm{r}_{x_1,x+h}\right) - T_F\left(\mathrm{r}_{x_1,x}\right)\right) \\
&= \lim_{\substack{h\to \mathbf{0}_m \\ h \geqq \mathbf{0}_m}} \sum_{(\bar{x},\tilde{x})\in M} T_F\left(\mathrm{r}_{\bar{x},\tilde{x}}\right) \leq \lim_{\substack{h\to \mathbf{0}_m \\ h \geqq \mathbf{0}_m}} \sum_{(\bar{x},\tilde{x})\in M} \mathcal{P} \circ \mu(\mathrm{r}_{\bar{x},\tilde{x}}) \\
&= \lim_{\substack{h\to \mathbf{0}_m \\ h \geqq \mathbf{0}_m}} \mathcal{P} \circ \mu(\mathrm{r}_{x_1,x+h} \setminus \mathrm{r}_{x_1,x}) = 0
\end{aligned}$$

where the first inequality follows from the monotonicity of $\bar{G}$; the first equality follows from the definition of $\bar{G}$; $M := \{(\bar{x}, \tilde{x}) \in \bar{\mathrm{r}}_{x_1,x} \times \bar{\mathrm{r}}_{x,x+h} \mid i \in \{0,\ldots,m-1\}, J \subseteq \bar{J}, \mathrm{card}\,(J) = i, \pi_k(\bar{x}) = \pi_k(x_1), \forall k \in J, \pi_k(\bar{x}) = \pi_k(x), \forall k \in \bar{J} \setminus J, \pi_k(\tilde{x}) = \pi_k(x), \forall k \in J, \pi_k(\tilde{x}) = \pi_k(x+h), \forall k \in \bar{J} \setminus J\}$; the second equality follows from Proposition 12.47; the second inequality follows from the last paragraph; the third equality follows from the fact that $\mathcal{P} \circ \mu$ is a measure; and the last equality follows from Claim 12.52.2. Hence, by Proposition 12.37, $\bar{G}$ is continuous on the right. Thus, F satisfies (iii) of Definition 12.41.

Therefore, F is of locally bounded variation. This proves (i).

(ii) Let $\mathcal{Y}$ be a Banach space. By Theorem 12.50 and (i), μ is the unique σ-finite $\mathcal{Y}$-valued measure on the measurable space $(\mathrm{P}(\Omega), \mathcal{B}_\mathrm{B}(\mathrm{P}(\Omega)))$ such that F is a cumulative distribution function of μ. Theorem 12.50, coupled with (i), further concludes that $T_F(\mathrm{r}_{x_1,x_2}) = \mathcal{P} \circ \mu(\mathrm{r}_{x_1,x_2}) = \Delta_G(\mathrm{r}_{x_1,x_2})$, $\forall x_1, x_2 \in \Omega$ with $x_1 \leqq x_2$ and $\bar{\mathrm{r}}_{x_1,x_2} \subseteq \Omega$, and $T_F = \mathcal{P} \circ \mu(\mathrm{P}(\Omega))$.

This completes the proof of the proposition. □

Lemma 12.53 *Let $m \in \mathbb{N}$, $\mathbb{R}^m$ be endowed with the usual positive cone P, $\Omega \in \mathcal{B}_\mathrm{B}(\mathbb{R}^m)$ be a region with the subset topology $\mathcal{O}$, and $F : \Omega \to \mathbb{R}$. Assume that the following holds:*

(i) *F is continuous on the right.*
(ii) *F is monotonically nondecreasing, i. e., $F(x_1) \leq F(x_2)$, $\forall x_1, x_2 \in \Omega$ with $x_1 \leqq x_2$.*

Then, F is $\mathcal{B}_\mathrm{B}(\mathbb{R}^m)$-measurable.

Proof Consider the special case when Ω is a rectangle. $\forall \alpha \in \mathbb{R}$, let $B := \{\bar{x} \in \Omega \mid F(\bar{x}) < \alpha\}$. We will show that $B \in \mathcal{O} \subseteq \mathcal{B}_\mathrm{B}(\Omega) \subseteq \mathcal{B}_\mathrm{B}(\mathbb{R}^m)$. Then, by Proposition 11.35, F is $\mathcal{B}_\mathrm{B}(\mathbb{R}^m)$-measurable.

$\forall x \in B$, $F(x) < \alpha$. Since F is continuous on the right, then $\exists \delta \in (0, \infty) \subset \mathbb{R}$ such that $F(\bar{x}) < \alpha$, $\forall \bar{x} \in \Omega \cap (x + P)$ with $\bar{x} \lessdot x + \delta \mathbf{1}_m$. $\forall \bar{x} \in \{\tilde{x} \in \Omega \mid \tilde{x} \lessdot x + \delta \mathbf{1}_m\} =: O \in \mathcal{O}$, since $\bar{x}, x \in \Omega$ and Ω is a rectangle, we have $x \vee \bar{x} \in \Omega \cap (x + P)$ with $x \vee \bar{x} \lessdot x + \delta \mathbf{1}_m$. Then, $F(\bar{x}) \leq F(x \vee \bar{x}) < \alpha$. This implies that $\bar{x} \in B$. By the arbitrariness of $\bar{x}$, we have $O \subseteq B$. By the arbitrariness of x, we have $B \in \mathcal{O} \subseteq \mathcal{B}_B(\Omega) \subseteq \mathcal{B}_B(\mathbb{R}^m)$.

Now, consider the general case when Ω is a region. Then, $\exists N \subseteq \mathbb{N}$, $\exists (\bar{x}_i)_{i \in N}, (\tilde{x}_i)_{i \in N} \subseteq \Omega$ with $\bar{x}_i \stackrel{\lessdot}{=} \tilde{x}_i$ and $\bar{r}_{\bar{x}_i, \tilde{x}_i} \subseteq \Omega$, $\forall i \in N$, such that $\Omega = \bigcup_{i \in N} \bar{r}_{\bar{x}_i, \tilde{x}_i}$. $\forall \alpha \in \mathbb{R}$, let $B := \{\hat{x} \in \Omega \mid F(\hat{x}) < \alpha\} = \bigcup_{i \in N} (B \cap \bar{r}_{\bar{x}_i, \tilde{x}_i}) =: \bigcup_{i \in N} B_i$. By the special case, $B_i \in \mathcal{B}_B(\mathbb{R}^m)$, $\forall i \in N$. Then, $B \in \mathcal{B}_B(\mathbb{R}^m)$. By Proposition 11.35, F is $\mathcal{B}_B(\mathbb{R}^m)$-measurable.

This completes the proof of the lemma. □

Proposition 12.54 *Let $m \in \mathbb{N}$, $\mathbb{R}^m$ be endowed with the usual positive cone, $\Omega \in \mathcal{B}_B(\mathbb{R}^m)$ be a region with the subset topology $\mathcal{O}$, $\mathcal{Y}$ be a separable normed linear space over $\mathbb{K}$ with $\mathcal{Y}^*$ being separable, $\mu \in \mathcal{M}_f(\mathrm{P}(\Omega), \mathcal{B}_B(\mathrm{P}(\Omega)), \mathcal{Y})$, and $F : \Omega \to \mathcal{Y}$ be the cumulative distribution function of μ as defined by $F(x) = \mu\left(\left\{\bar{x} \in \mathrm{P}(\Omega) \,\middle|\, \bar{x} \stackrel{\lessdot}{=} x\right\}\right)$, $\forall x \in \Omega$, as shown in Proposition 12.51. Then, F is $\mathcal{B}_B(\mathbb{R}^m)$-measurable.*

Proof We will prove the result in three steps. In the first step, we consider the special case $\mathcal{Y} = \mathbb{R}$. By Jordan Decomposition Theorem 11.162, there exists a unique pair of mutually singular finite measures μ_+ and μ_- on the measurable space $(\mathrm{P}(\Omega), \mathcal{B}_B(\mathrm{P}(\Omega)))$ such that $\mu = \mu_+ - \mu_-$. Furthermore, $\mathcal{P} \circ \mu = \mu_+ + \mu_-$. Let $F_+ : \Omega \to [0, \infty) \subset \mathbb{R}$ and $F_- : \Omega \to [0, \infty) \subset \mathbb{R}$ be the cumulative distribution functions of μ_+ and μ_-, respectively, as defined by $F_+(x) = \mu_+\left(\left\{\bar{x} \in \mathrm{P}(\Omega) \,\middle|\, \bar{x} \stackrel{\lessdot}{=} x\right\}\right)$ and $F_-(x) = \mu_-\left(\left\{\bar{x} \in \mathrm{P}(\Omega) \,\middle|\, \bar{x} \stackrel{\lessdot}{=} x\right\}\right)$, $\forall x \in \Omega$. By Proposition 12.51, F_+ and F_- are of locally bounded variation and, hence, are continuous on the right. By their definitions, $F_+(x_1) \leq F_+(x_2)$ and $F_-(x_1) \leq F_-(x_2)$, $\forall x_1, x_2 \in \Omega$ with $x_1 \stackrel{\lessdot}{=} x_2$. Hence, F_+ and F_- are monotonically nondecreasing. By Lemma 12.53, F_+ and F_- are $\mathcal{B}_B(\mathbb{R}^m)$-measurable. Clearly, $F(x) = \mu\left(\left\{\bar{x} \in \mathrm{P}(\Omega) \,\middle|\, \bar{x} \stackrel{\lessdot}{=} x\right\}\right) = \mu_+\left(\left\{\bar{x} \in \mathrm{P}(\Omega) \,\middle|\, \bar{x} \stackrel{\lessdot}{=} x\right\}\right) - \mu_-\left(\left\{\bar{x} \in \mathrm{P}(\Omega) \,\middle|\, \bar{x} \stackrel{\lessdot}{=} x\right\}\right) = F_+(x) - F_-(x)$, $\forall x \in \Omega$. By Propositions 11.38, 11.39, and 7.23, F is $\mathcal{B}_B(\mathbb{R}^m)$-measurable. This completes the first step.

In the second step, we consider the special case $\mathcal{Y} = \mathbb{C}$. By Proposition 11.164, $\mu = \mu_r + \mathrm{i}\mu_i$, where μ_r and μ_i are finite $\mathbb{R}$-valued measures on $(\mathrm{P}(\Omega), \mathcal{B}_B(\mathrm{P}(\Omega)))$. Let $F_r : \Omega \to \mathbb{R}$ and $F_i : \Omega \to \mathbb{R}$ be defined by $F_r(x) = \mu_r\left(\left\{\bar{x} \in \mathrm{P}(\Omega) \,\middle|\, \bar{x} \stackrel{\lessdot}{=} x\right\}\right)$ and $F_i(x) = \mu_i\left(\left\{\bar{x} \in \mathrm{P}(\Omega) \,\middle|\, \bar{x} \stackrel{\lessdot}{=} x\right\}\right)$, $\forall x \in \Omega$. By Proposition 12.51, F_r and F_i are cumulative distribution functions of μ_r and μ_i, respectively. By the first step, F_r and F_i are $\mathcal{B}_B(\mathbb{R}^m)$-measurable. Clearly, $F(x) = \mu\left(\left\{\bar{x} \in \mathrm{P}(\Omega) \,\middle|\, \bar{x} \stackrel{\lessdot}{=} x\right\}\right) = \mu_r\left(\left\{\bar{x} \in \mathrm{P}(\Omega) \,\middle|\, \bar{x} \stackrel{\lessdot}{=} x\right\}\right) +$

$\mathrm{i}\mu_i\left(\left\{\bar{x} \in \mathrm{P}(\Omega) \;\middle|\; \bar{x} \stackrel{\leq}{=} x\right\}\right) = F_r(x) + \mathrm{i}F_i(x)$, $\forall x \in \Omega$. By Propositions 11.38, 11.39, and 7.23, F is $\mathcal{B}_{\mathrm{B}}(\mathbb{R}^m)$-measurable. This completes the second step.

In the last step, we consider the general case where $\mathcal{Y}$ is a separable normed linear space over $\mathbb{K}$ with $\mathcal{Y}^*$ being separable. Fix any $y_* \in \mathcal{Y}^*$. Define $\mu_{y_*} : \mathcal{B}_{\mathrm{B}}(\mathrm{P}(\Omega)) \to \mathbb{K}$ by $\mu_{y_*}(E) = \langle\langle y_*, \mu(E)\rangle\rangle$, $\forall E \in \mathcal{B}_{\mathrm{B}}(\mathrm{P}(\Omega))$. By Proposition 11.136, μ_{y_*} is a finite $\mathbb{K}$-valued measure on $(\mathrm{P}(\Omega), \mathcal{B}_{\mathrm{B}}(\mathrm{P}(\Omega)))$ and $\mathcal{P} \circ \mu_{y_*} \leq \|y_*\|\mathcal{P} \circ \mu$. Define $F_{y_*} : \Omega \to \mathbb{K}$ by $F_{y_*}(x) = \langle\langle y_*, F(x)\rangle\rangle$, $\forall x \in \Omega$. Clearly, $F_{y_*}(x) = \langle\langle y_*, F(x)\rangle\rangle = \left\langle\left\langle y_*, \mu\left(\left\{\bar{x} \in \mathrm{P}(\Omega) \;\middle|\; \bar{x} \stackrel{\leq}{=} x\right\}\right)\right\rangle\right\rangle = \mu_{y_*}\left(\left\{\bar{x} \in \mathrm{P}(\Omega) \;\middle|\; \bar{x} \stackrel{\leq}{=} x\right\}\right)$, $\forall x \in \Omega$. Then, F_{y_*} is a cumulative distribution function of μ_{y_*}. By the first and second steps, we have F_{y_*} is $\mathcal{B}_{\mathrm{B}}(\mathbb{R}^m)$-measurable. By Proposition 11.170, F is $\mathcal{B}_{\mathrm{B}}(\mathbb{R}^m)$-measurable. This completes the last step and, hence, the proof of the proposition. □

Proposition 12.55 *Let $m \in \mathbb{N}$, $\mathbb{R}^m$ be endowed with the usual positive cone, $\Omega \in \mathcal{B}_{\mathrm{B}}(\mathbb{R}^m)$ be a rectangle with the subset topology $\mathcal{O}$, $\mathcal{Y}$ be a separable normed linear space over $\mathbb{K}$ with $\mathcal{Y}^*$ being separable, μ be a σ-finite $\mathcal{Y}$-valued measure on $(\mathrm{P}(\Omega), \mathcal{B}_{\mathrm{B}}(\mathrm{P}(\Omega)))$ such that $\mathcal{P} \circ \mu(\mathrm{r}_{x_1,x_2}) < \infty$, $\forall x_1, x_2 \in \Omega$ with $x_1 \stackrel{\leq}{=} x_2$, and $F : \Omega \to \mathcal{Y}$ be the cumulative distribution function of μ with origin $x_0 \in \Omega$ when $\Omega \neq \emptyset$. Then, F is $\mathcal{B}_{\mathrm{B}}(\mathbb{R}^m)$-measurable.*

Proof The result is trivial if $\Omega = \emptyset$. Consider the case $\Omega \neq \emptyset$. Since Ω is a rectangle, then, by Proposition 12.52, $\mathrm{P}(\Omega)$ is also a rectangle and Ω is a region. Then, $\exists (\bar{x}_i)_{i=1}^{\infty}, (\tilde{x}_i)_{i=1}^{\infty} \subseteq \Omega$ such that $\bar{x}_i \stackrel{\leq}{=} x_0 \stackrel{\leq}{=} \tilde{x}_i$, $\forall i \in \mathbb{N}$, $\bigcup_{i=1}^{\infty} \bar{\mathrm{r}}_{\bar{x}_i,\tilde{x}_i} = \Omega$, and $\mathrm{P}(\Omega) = \bigcup_{i=1}^{\infty} \mathrm{r}_{\bar{x}_i,\tilde{x}_i}$.

Fix any $i \in \mathbb{N}$, let $\bar{E}_i := \bar{\mathrm{r}}_{\bar{x}_i,\tilde{x}_i} \in \mathcal{B}_{\mathrm{B}}(\mathbb{R}^m)$ and $E_i := \mathrm{r}_{\bar{x}_i,\tilde{x}_i} \in \mathcal{B}_{\mathrm{B}}(\mathbb{R}^m)$. Note that $\mathcal{P} \circ \mu(E_i) < \infty$, by the assumption on μ. Let $(E_i, \mathcal{B}_{E_i}, \mu_{E_i})$ be the finite $\mathcal{Y}$-valued measure subspace of $(\mathrm{P}(\Omega), \mathcal{B}_{\mathrm{B}}(\mathrm{P}(\Omega)), \mu)$, $\mathcal{O}_{\bar{E}_i}$ be the subset topology on $\bar{E}_i \subseteq \Omega$, and $\bar{\mathcal{E}}_i := (\bar{E}_i, \mathcal{O}_{\bar{E}_i})$. Define $\bar{F} : \bar{E}_i \to \mathcal{Y}$ by $\bar{F}(x) = \mu_{E_i}\left(\left\{\bar{x} \in E_i \;\middle|\; \bar{x} \stackrel{\leq}{=} x\right\}\right)$, $\forall x \in \bar{E}_i$. By Proposition 12.51, $\bar{F}$ is a cumulative distribution function of μ_{E_i} and is of locally bounded variation. By Proposition 12.54, $\bar{F}$ is $\mathcal{B}_{\mathrm{B}}(\mathbb{R}^m)$-measurable. $\forall x \in \bar{E}_i$, let $\check{x} := x \wedge x_0 \in \bar{E}_i$ and $\hat{x} := x \vee x_0 \in \bar{E}_i$. Then, $\Delta_F(\mathrm{r}_{\check{x},\hat{x}}) = \mu(\mathrm{r}_{\check{x},\hat{x}}) = \mu_{E_i}(\mathrm{r}_{\check{x},\hat{x}}) = \Delta_{\bar{F}}(\mathrm{r}_{\check{x},\hat{x}})$, where the first equality follows from the fact that F is a cumulative distribution function of μ, the second equality follows from Proposition 11.115, and the third equality follows from the fact that $\bar{F}$ is a cumulative distribution function of μ_{E_i}. If $\exists k_0 \in \bar{J} := \{1, \ldots, m\}$ such that $\pi_{k_0}(x) = \pi_{k_0}(x_0)$, then $F(x) = \vartheta_{\mathcal{Y}} = \Delta_F(\mathrm{r}_{\check{x},\hat{x}})$. Otherwise, we have $(-1)^{n(x)}F(x) = \Delta_F(\mathrm{r}_{\check{x},\hat{x}})$, where $n(x) = \mathrm{card}\left(\left\{k \in \bar{J} \;\middle|\; \pi_k(x) < \pi_k(x_0)\right\}\right)$. Hence, we have $F(x) = (-1)^{n(x)}\Delta_F(\mathrm{r}_{\check{x},\hat{x}}) = (-1)^{n(x)}\Delta_{\bar{F}}(\mathrm{r}_{\check{x},\hat{x}}) = (-1)^{n(x)}\sum_{\hat{J} \subseteq \bar{J}}(-1)^{\mathrm{card}(\hat{J})}\bar{F}(\bar{x}_{\hat{J}})$, where $\bar{x}_{\hat{J}} \in \bar{E}_i$ is defined by $\pi_k(\bar{x}_{\hat{J}}) = \begin{cases} \pi_k(\check{x}) \ \forall k \in \hat{J} \\ \pi_k(\hat{x}) \ \forall k \in \bar{J} \setminus \hat{J} \end{cases}$.

$\forall j \in \{0, \ldots, 2^m - 1\}$, let $\underline{j_1 \ldots j_m}$ be the representation of j in base 2, $J_j := \left\{k \in \bar{J} \mid j_k = 0\right\}$, $n_j := \mathrm{card}(J_j)$. Define $O_j := \left\{x \in \bar{E}_i \;\middle|\; \pi_k(x) > \pi_k(x_0),\ \forall k \in \bar{J} \setminus\right.$

J_j, $\pi_k(x) < \pi_k(x_0)$, $\forall k \in J_j\big\} \in \mathcal{O}_{\bar{E}_i}$. Let $D := \big\{x \in \bar{E}_i \mid \pi_{i_0}(x) = \pi_{i_0}(x_0)$ for some $i_0 \in \bar{J}\big\} \in \mathcal{B}_{\mathrm{B}}(\mathbb{R}^m)$. Clearly, $\bar{E}_i = (\bigcup_{j=0}^{2^m-1} O_j) \cup D$. $F|_D(x) = \vartheta_{\mathcal{Y}}$, $\forall x \in D$. Then, $F|_D$ is $\mathcal{B}_{\mathrm{B}}(\mathbb{R}^m)$-measurable. Fix any $j \in \{0, \ldots, 2^m - 1\}$. Denote $J_j =: J$ and $n_j =: n$. Then, $F|_{O_j}(x) = (-1)^n \sum_{\hat{J} \subseteq \bar{J}} (-1)^{\mathrm{card}(\hat{J})} \bar{F}(\bar{x}_{\hat{J}})$, $\forall x \in O_j$. Note that $\pi_k(\bar{x}_{\hat{J}}) = \begin{cases} \pi_k(\check{x}) \ \forall k \in \hat{J} \\ \pi_k(\hat{x}) \ \forall k \in \bar{J} \setminus \hat{J} \end{cases} = \begin{cases} \pi_k(x) \ \ \forall k \in \bar{J} \setminus (J \triangle \hat{J}) \\ \pi_k(x_0) \ \forall k \in J \triangle \hat{J} \end{cases}$, $\forall k \in \bar{J}$, $\forall \hat{J} \subseteq \bar{J}$. We may define function $\tilde{F}_{\hat{J}} : O_j \to \mathcal{Y}$ by $\tilde{F}_{\hat{J}}(x) = \bar{F}(\bar{x}_{\hat{J}})$, $\forall x \in O_j$. Since $\bar{E}_i$ is a closed rectangle, it can be viewed as product of a closed set $E_{i,\delta}$ in the $J \triangle \hat{J}$ collection of coordinates and another closed set $E_{i,\sigma}$ in the $\bar{J} \setminus (J \triangle \hat{J})$ collection of coordinates. $\forall x \in \bar{E}_i$, $\pi_\delta(x) \in E_{i,\delta}$ gives the one coordinate and $\pi_\sigma(x) \in E_{i,\sigma}$ gives the other coordinate. Then, $x = (\pi_\delta(x), \pi_\sigma(x))$. For any open set $O_Y \in \mathcal{O}_{\mathcal{Y}}$, $\bar{F}_{\mathrm{inv}}(O_Y) \in \mathcal{B}_{\mathrm{B}}(\bar{\mathcal{E}}_i) \subseteq \mathcal{B}_{\mathrm{B}}(\mathbb{R}^m)$, since $\bar{F}$ is $\mathcal{B}_{\mathrm{B}}(\mathbb{R}^m)$-measurable. Define $A_1 := \bar{F}_{\mathrm{inv}}(O_Y)_{\pi_\delta(x_0)} := \big\{s \in E_{i,\sigma} \mid (\pi_\delta(x_0), s) \in \bar{F}_{\mathrm{inv}}(O_Y)\big\}$. By Proposition 12.25, $A_1 \in \mathcal{B}_{\mathrm{B}}(E_{i,\sigma})$. By Proposition 11.24, $E_{i,\delta} \times A_1 \in \mathcal{B}_{\mathrm{B}}(\bar{\mathcal{E}}_i)$. Then, $(\tilde{F}_{\hat{J}})_{\mathrm{inv}}(O_Y) = O_j \cap (E_{i,\delta} \times A_1) \in \mathcal{B}_{\mathrm{B}}(\bar{\mathcal{E}}_i) \subseteq \mathcal{B}_{\mathrm{B}}(\mathbb{R}^m)$. Hence, $\tilde{F}_{\hat{J}}$ is $\mathcal{B}_{\mathrm{B}}(\mathbb{R}^m)$-measurable. Clearly, $F|_{O_j} = (-1)^n \sum_{\hat{J} \subseteq \bar{J}} (-1)^{\mathrm{card}(\hat{J})} \tilde{F}_{\hat{J}}$. By Propositions 7.23, 11.39, and 11.38, $F|_{O_j}$ is $\mathcal{B}_{\mathrm{B}}(\mathbb{R}^m)$-measurable. By the arbitrariness of j and Proposition 11.41, $F|_{\bar{E}_i}$ is $\mathcal{B}_{\mathrm{B}}(\mathbb{R}^m)$-measurable. By the arbitrariness of i and Proposition 11.41, F is $\mathcal{B}_{\mathrm{B}}(\mathbb{R}^m)$-measurable. This completes the proof of the proposition. □

Example 12.56 Let $m \in \mathbb{N}$, $\mathbb{R}$ be the second countable σ-finite real Banach measure space as defined in Example 11.28, and $\mathbb{R}^m$ be the second countable σ-finite product topological measure space according to Proposition 12.32. Since $\mathbb{R}^m$ is a real Banach space, then $\mathbb{R}^m =: ((\mathbb{R}^m, \mathbb{R}, |\cdot|), \mathcal{B}_{\mathrm{B}}(\mathbb{R}^m), \mu_{\mathrm{B}m})$ is a second countable σ-finite real Banach measure space. $\mu_{\mathrm{B}m}$ is the *m-dimensional Borel measure*. By Carathéodory Extension Theorem 11.19, $\mu_{\mathrm{B}m}$ induces the *m-dimensional Lebesgue outer measure* $\mu_{\mathrm{L}mo} : {}^{\mathbb{R}^m}2 \to [0, \infty] \subset \mathbb{R}_e$, which is defined by $\mu_{\mathrm{L}mo}(E) = \inf_{(A_i)_{i=1}^\infty \subseteq \mathcal{B}_{\mathrm{B}}(\mathbb{R}^m), E \subseteq \bigcup_{i=1}^\infty A_i} \sum_{i=1}^\infty \mu_{\mathrm{B}m}(A_i)$, $\forall E \subseteq \mathbb{R}^m$. $\mathcal{B}_{\mathrm{L}m} := \{E \subseteq \mathbb{R}^m \mid E \text{ is measurable with respect to } \mu_{\mathrm{L}mo}\}$ is the collection of *m-dimensional Lebesgue measurable sets*, and $\mu_{\mathrm{L}m} := \mu_{\mathrm{L}mo}|_{\mathcal{B}_{\mathrm{L}m}}$ is the *m-dimensional Lebesgue measure*. By Carathéodory Extension Theorem 11.19, $(\mathbb{R}^m, \mathcal{B}_{\mathrm{L}m}, \mu_{\mathrm{L}m})$ is the completion of the m-dimensional Borel measure space $(\mathbb{R}^m, \mathcal{B}_{\mathrm{B}}(\mathbb{R}^m), \mu_{\mathrm{B}m})$ and is said to be the *m-dimensional Lebesgue measure space*. When $m = 1$, $\mu_{\mathrm{B}m}$, $\mu_{\mathrm{L}mo}$, $\mu_{\mathrm{L}m}$, and $\mathcal{B}_{\mathrm{L}m}$ are also denoted by μ_{B}, $\mu_{\mathrm{L}o}$, μ_{L}, and $\mathcal{B}_{\mathrm{L}}$, respectively.

$\forall E \in {}^{\mathbb{R}^m}2$, we will show that

$$\mu_{\mathrm{L}mo}(E) = \inf_{(O_i)_{i=1}^\infty \subseteq \mathcal{O}_{\mathbb{R}^m}, E \subseteq \bigcup_{i=1}^\infty O_i} \sum_{i=1}^\infty \mu_{\mathrm{B}m}(O_i) =: \bar{\mu}_o(E)$$

Since $\mathcal{O}_{\mathbb{R}^m} \subseteq \mathcal{B}_{\mathrm{B}}(\mathbb{R}^m)$, then $\mu_{\mathrm{L}mo}(E) \leq \bar{\mu}_o(E)$. On the other hand, if $\mu_{\mathrm{L}mo}(E) = \infty$, then $\infty = \mu_{\mathrm{L}mo}(E) \leq \bar{\mu}_o(E) \leq \infty$ and $\mu_{\mathrm{L}mo}(E) = \bar{\mu}_o(E)$. If $\mu_{\mathrm{L}mo}(E) < \infty$,

$\forall \epsilon \in (0, \infty) \subset \mathbb{R}$, $\exists (A_i)_{i=1}^{\infty} \subseteq \mathcal{B}_{\mathrm{B}}(\mathbb{R}^m)$ such that $E \subseteq \bigcup_{i=1}^{\infty} A_i$ and $\mu_{\mathrm{L}mo}(E) \leq \sum_{i=1}^{\infty} \mu_{\mathrm{B}m}(A_i) < \mu_{\mathrm{L}mo}(E) + \epsilon/2$. $\forall i \in \mathbb{N}$, since $\mathbb{R}^m$ is a topological measure space, then $\exists O_i \in \mathcal{O}_{\mathbb{R}^m}$ with $A_i \subseteq O_i$ such that $\mu_{\mathrm{B}m}(O_i \setminus A_i) < 2^{-i-1}\epsilon$. Then, $E \subseteq \bigcup_{i=1}^{\infty} O_i$ and $\bar{\mu}_o(E) \leq \sum_{i=1}^{\infty} \mu_{\mathrm{B}m}(O_i) = \sum_{i=1}^{\infty}(\mu_{\mathrm{B}m}(A_i) + \mu_{\mathrm{B}m}(O_i \setminus A_i)) < \mu_{\mathrm{L}mo}(E) + \epsilon/2 + \epsilon/2 = \mu_{\mathrm{L}mo}(E) + \epsilon$. By the arbitrariness of ϵ, we have $\bar{\mu}_o(E) \leq \mu_{\mathrm{L}mo}(E)$. Hence, $\mu_{\mathrm{L}mo}(E) = \bar{\mu}_o(E)$.

By the translational invariance of $\mathbb{R}$, we have $\mathbb{R}^m$ is translational invariant. Then, by the above, $\mu_{\mathrm{L}mo}$ and $\mu_{\mathrm{L}m}$ are translational invariant. ◇

Definition 12.57 (Lebesgue–Stieltjes Integral) Let $m \in \mathbb{N}$, $\Omega \in \mathcal{B}_{\mathrm{B}}(\mathbb{R}^m)$ be a region with the subset topology, $\mathcal{Y}$ be a Banach space over $\mathbb{K}$, $F : \Omega \to \mathcal{Y}$ be of locally bounded variation, $(\mathrm{P}(\Omega), \mathcal{B}_{\mathrm{B}}(\mathrm{P}(\Omega)), \mu)$ be the unique σ-finite $\mathcal{Y}$-valued measure space that admits F as a cumulative distribution function as prescribed in Theorem 12.50, $\mathcal{Z}$ be a normed linear space over $\mathbb{K}$, $U \in \mathcal{B}_{\mathrm{B}}(\mathrm{P}(\Omega))$, and $g : U \to \mathrm{B}(\mathcal{Y}, \mathcal{Z})$ be $\mathcal{B}_{\mathrm{B}}(\mathrm{P}(\Omega))$-measurable (or $\mathcal{B}_{\mathrm{B}}(\mathbb{R}^m)$-measurable). We will denote the integral $\int_U g(x)\,\mathrm{d}\mu(x)$ by $\int_U g\,\mathrm{d}F$ or $\int_U g(x)\,\mathrm{d}F(x)$, whenever the first integral makes sense. ◇

12.5 Absolute and Lipschitz Continuity

Definition 12.58 Let $I \subseteq \mathbb{R}$ be an interval and $\mathcal{Y}$ be a normed linear space. A function $f : I \to \mathcal{Y}$ is said to be *absolutely continuous at* $x_0 \in I$ if $\exists a \in (0, \infty) \subset \mathbb{R}$, $\forall \epsilon \in (0, \infty) \subset \mathbb{R}$, $\exists \delta(\epsilon) \in (0, a] \subset \mathbb{R}$, $\forall n \in \mathbb{N}$, $\forall x_0 - a \leq x_1 \leq \bar{x}_1 \leq x_2 \leq \bar{x}_2 \leq \cdots \leq x_n \leq \bar{x}_n \leq x_0 + a$ with $x_i, \bar{x}_i \in I$, $i = 1, \ldots, n$, and $\sum_{i=1}^{n} \mu_{\mathrm{B}}(\mathrm{r}_{x_i, \bar{x}_i}) = \sum_{i=1}^{n}(\bar{x}_i - x_i) < \delta(\epsilon)$, and we have $\sum_{i=1}^{n} \left\| \Delta_F(\mathrm{r}_{x_i, \bar{x}_i}) \right\| = \sum_{i=1}^{n} \| f(\bar{x}_i) - f(x_i) \| < \epsilon$. We will say that f is *absolutely continuous* if it is absolutely continuous at any $x_0 \in I$. We will say f is *absolutely continuous on* $E \subseteq I$ if it is absolutely continuous at any $x_0 \in E$.

Let $m \in \mathbb{N}$, $\mathbb{R}^m$ be endowed with the usual positive cone, $\Omega \in \mathcal{B}_{\mathrm{B}}(\mathbb{R}^m)$ be a rectangle with subset topology $\mathcal{O}$, and $\mathcal{Y}$ be a normed linear space over $\mathbb{K}$. $F : \Omega \to \mathcal{Y}$ is said to be *absolutely continuous at* $x_0 \in \Omega$ if the following conditions hold:

(i) When $m > 1$, $\forall i_0 \in \bar{J} := \{1, \ldots, m\}$, let $D_{i_0} := \left\{x \in \mathbb{R}^{m-1} \,\middle|\, \exists \bar{x} \in \Omega \ni \cdot\ \pi_{i_0}(\bar{x}) = \pi_{i_0}(x_0) \text{ and } \pi_{\widehat{i_0}}(\bar{x}) = x\right\}$, where $\pi_{\widehat{i_0}} : \mathbb{R}^m \to \mathbb{R}^{m-1}$ is defined by $\pi_{\widehat{i_0}}(x) = (\pi_1(x), \ldots, \pi_{i_0-1}(x), \pi_{i_0+1}(x), \ldots, \pi_m(x)) \in \mathbb{R}^{m-1}$, $\forall x \in \mathbb{R}^m$, and $F_{i_0} : D_{i_0} \to \mathcal{Y}$ be defined by $F_{i_0}(x) = F(\bar{x})$, $\forall x \in D_{i_0}$, $\forall \bar{x} \in \Omega$ with $\pi_{i_0}(\bar{x}) = \pi_{i_0}(x_0)$ and $\pi_{\widehat{i_0}}(\bar{x}) = x$. The function F_{i_0} is absolutely continuous at $\pi_{\widehat{i_0}}(x_0)$.

(ii) $\exists a \in (0, \infty) \subset \mathbb{R}$, $\forall \epsilon \in (0, \infty) \subset \mathbb{R}$, $\exists \delta(\epsilon) \in (0, \infty) \subset \mathbb{R}$, $\forall n \in \mathbb{N}$, $\forall (\bar{x}_i)_{i=1}^{n}, (\tilde{x}_i)_{i=1}^{n} \subset \bar{\mathrm{r}}_{x_0 - a\mathbf{1}_m, x_0 + a\mathbf{1}_m} \cap \Omega$ with $\bar{x}_i \overset{\lessdot}{=} \tilde{x}_i$, $\forall i \in \{1, \ldots, n\}$,

$\left(\mathrm{r}_{\bar{x}_i,\tilde{x}_i}\right)_{i=1}^n$ being pairwise disjoint, and $\sum_{i=1}^n \mu_{\mathrm{B}m}(\mathrm{r}_{\bar{x}_i,\tilde{x}_i}) < \delta(\epsilon)$, we have $\sum_{i=1}^n \left\| \Delta_F(\mathrm{r}_{x_i,\tilde{x}_i}) \right\| < \epsilon$.

F is said to be *absolutely continuous* if it is absolutely continuous at any $x_0 \in \Omega$. F is said to be *absolutely continuous on* $E \subseteq \Omega$ if it is absolutely continuous at any $x_0 \in E$. ◇

In the above definition, the definition for general $\mathbb{R}^m$ case subsumes the definition for $\mathbb{R}$ case. We explicitly specify the $\mathbb{R}$ case for its direct implications.

Proposition 12.59 *Let $m \in \mathbb{N}$, $\mathbb{R}^m$ be endowed with the usual positive cone, $\Omega \in \mathcal{B}_{\mathrm{B}}(\mathbb{R}^m)$ be a rectangle with subset topology $\mathcal{O}$, $\mathcal{Y}$ be a normed linear space over $\mathbb{K}$, and $F : \Omega \to \mathcal{Y}$ be absolutely continuous at $x_0 \in \Omega$. Then, F is continuous at x_0.*

Proof We will prove the result using mathematical induction on m:
1° $m = 1$. By Definition 12.58, F is continuous at x_0. This case is proved.
2° Assume that the result holds for $m \le k - 1 \in \mathbb{N}$.
3° Consider the case when $m = k \in \{2, 3, \ldots\}$. By (i) of Definition 12.58, $\forall i_0 \in \bar{J} := \{1, \ldots, m\}$, $F_{i_0} : D_{i_0} \to \mathcal{Y}$ is absolutely continuous at $\pi_{\widehat{i_0}}(x_0)$, where F_{i_0} and D_{i_0} are as defined in Definition 12.58. Clearly, $D_{i_0} \in \mathcal{B}_{\mathrm{B}}\left(\mathbb{R}^{m-1}\right)$ is a rectangle since Ω is a rectangle. By the inductive assumption, F_{i_0} is continuous at $\pi_{\widehat{i_0}}(x_0)$. $\forall \epsilon \in (0, \infty) \subset \mathbb{R}$, $\exists \delta_{i_0}(\epsilon) \in (0, \infty) \subset \mathbb{R}$ such that $\left\| F_{i_0}(x) - F_{i_0}(\pi_{\widehat{i_0}}(x_0)) \right\| < \epsilon$, $\forall x \in \mathcal{B}_{\mathbb{R}^{m-1}}\left(\pi_{\widehat{i_0}}(x_0), \delta_{i_0}(\epsilon)\right) \cap D_{i_0}$.

By (ii) of Definition 12.58, $\exists a \in (0, \infty) \subset \mathbb{R}$, $\forall \epsilon \in (0, \infty) \subset \mathbb{R}$, $\exists \bar{\delta}(\epsilon) \in (0, \infty) \subset \mathbb{R}$, $\forall n \in \mathbb{N}$, $\forall (\bar{x}_i)_{i=1}^n, (\tilde{x}_i)_{i=1}^n \subseteq \bar{\mathrm{r}}_{x_0 - a\mathbf{1}_m, x_0 + a\mathbf{1}_m} \cap \Omega$ with $\bar{x}_i \stackrel{\le}{=} \tilde{x}_i$, $\forall i \in \{1, \ldots, n\}$, $(\mathrm{r}_{\bar{x}_i,\tilde{x}_i})_{i=1}^n$ being pairwise disjoint, and $\sum_{i=1}^n \mu_{\mathrm{B}m}(\mathrm{r}_{\bar{x}_i,\tilde{x}_i}) < \bar{\delta}(\epsilon)$, and we have $\sum_{i=1}^n \left\| \Delta_F(\mathrm{r}_{\bar{x}_i,\tilde{x}_i}) \right\| < \epsilon$. Take $\delta(\epsilon) := \min\{a, \sqrt[m]{\bar{\delta}(2^{-m}\epsilon)}, \delta_1(2^{-m}\epsilon), \ldots, \delta_m(2^{-m}\epsilon)\} \in \mathbb{R}_+$. $\forall x \in \mathcal{B}_{\mathbb{R}^m}(x_0, \delta(\epsilon)) \cap \Omega$, we have $\|F(x) - F(x_0)\| \le \left\| \Delta_G(\mathrm{r}_{\check{x},\hat{x}}) \right\| + \sum_{J \subset \bar{J}} \|F(x_J) - F(x_0)\|$, where $\check{x} := x \wedge x_0$ and $\hat{x} := x \vee x_0$, $G : \Omega \to \mathcal{Y}$ is defined by $G(x) = F(x) - F(x_0)$, $\forall x \in \Omega$, and $x_J \in \bar{\mathrm{r}}_{\check{x},\hat{x}}$ is defined by $\pi_i(x_J) = \begin{cases} \pi_i(x) & \forall i \in J \\ \pi_i(x_0) & \forall i \in \bar{J} \setminus J \end{cases}$, $\forall J \subset \bar{J}$.

Clearly, $\check{x}, \hat{x} \in \bar{\mathrm{r}}_{x_0 - a\mathbf{1}_m, x_0 + a\mathbf{1}_m} \cap \Omega$ since $a > |x - x_0| \ge \max\{|\check{x} - x_0|, |\hat{x} - x_0|\} \ge 0$. Since $\mu_{\mathrm{B}m}(\mathrm{r}_{\check{x},\hat{x}}) \le |x - x_0|^m < \bar{\delta}(2^{-m}\epsilon)$, then $\|\Delta_F(\mathrm{r}_{\check{x},\hat{x}})\| = \|\Delta_G(\mathrm{r}_{\check{x},\hat{x}})\| < 2^{-m}\epsilon$. $\forall J \subset \bar{J}$, fix any $i_0 \in \bar{J} \setminus J$, and we have $\|F(x_J) - F(x_0)\| = \|F_{i_0}(\pi_{\widehat{i_0}}(x_J)) - F_{i_0}(\pi_{\widehat{i_0}}(x_0))\|$. Clearly, $\pi_{\widehat{i_0}}(x_J) \in D_{i_0}$ and $\left|\pi_{\widehat{i_0}}(x_J) - \pi_{\widehat{i_0}}(x_0)\right| = |x_J - x_0| \le |x - x_0| < \delta_{i_0}(2^{-m}\epsilon)$. Then, $\|F(x_J) - F(x_0)\| = \left\| F_{i_0}(\pi_{\widehat{i_0}}(x_J)) - F_{i_0}(\pi_{\widehat{i_0}}(x_0)) \right\| < 2^{-m}\epsilon$. Hence, $\|F(x) - F(x_0)\| < \epsilon$. This shows that F is continuous at x_0.

This completes the induction process.

This completes the proof of the proposition. □

Proposition 12.60 *Let $m \in \mathbb{N}$, $\mathbb{R}^m$ be endowed with the usual positive cone, $\Omega \in \mathcal{B}_{\mathrm{B}}(\mathbb{R}^m)$ be a rectangle with subset topology $\mathcal{O}$, $\mathcal{Y}$ be a normed linear space over $\mathbb{K}$, $F : \Omega \to \mathcal{Y}$, and $x_0 \in \Omega$. Then, F is absolutely continuous at x_0 if, and only if:*

(a) *$\forall i \in \{0, \ldots, m-1\}$, $\forall J \subset \bar{J} := \{1, \ldots, m\}$ with $\operatorname{card}(J) = i$, let $\pi_J : \mathbb{R}^m \to \mathbb{R}^i$ be such that $\pi_J(\bar{x}) = (\pi_l(\bar{x}))_{l \in J}$, $\forall \bar{x} \in \mathbb{R}^m$, $\pi_{\bar{J} \setminus J} : \mathbb{R}^m \to \mathbb{R}^{m-i}$ be such that $\pi_{\bar{J} \setminus J}(\bar{x}) = (\pi_l(\bar{x}))_{l \in \bar{J} \setminus J}$, $\forall \bar{x} \in \mathbb{R}^m$, $M : \mathbb{R}^i \times \mathbb{R}^{m-i} \to \mathbb{R}^m$ be such that $M(x_a, x_b) = \bar{x}$ with $x_a = \pi_J(\bar{x})$ and $x_b = \pi_{\bar{J} \setminus J}(\bar{x})$, $\forall x_a \in \mathbb{R}^i$ and $\forall x_b \in \mathbb{R}^{m-i}$, $x_f := \pi_J(x_0)$, $\tilde{x}_f := \pi_{\bar{J} \setminus J}(x_0)$, $\tilde{\Omega}_J := \pi_{\bar{J} \setminus J}(\Omega)$, and $F_J : \tilde{\Omega}_J \to \mathcal{Y}$ be defined by $F_J(\bar{x}) = F(M(x_f, \bar{x}))$, $\forall \bar{x} \in \tilde{\Omega}_J$, $\exists a_J \in (0, \infty) \subset \mathbb{R}$, $\forall \epsilon \in (0, \infty) \subset \mathbb{R}$, $\exists \delta_J(\epsilon) \in (0, \infty) \subset \mathbb{R}$, $\forall n \in \mathbb{N}$, $\forall \left(\bar{x}_j\right)_{j=1}^n, \left(\tilde{x}_j\right)_{j=1}^n \subseteq \bar{\mathrm{r}}_{\tilde{x}_f - a_J \mathbf{1}_{m-i}, \tilde{x}_f + a_J \mathbf{1}_{m-i}} \cap \tilde{\Omega}_J$ with $\bar{x}_j \stackrel{\lessdot}{=} \tilde{x}_j$, $\forall j \in \{1, \ldots, n\}$, $(\mathrm{r}_{\bar{x}_j, \tilde{x}_j})_{j=1}^n$ being pairwise disjoint, and $\sum_{j=1}^n \mu_{\mathrm{B}(m-i)}(\mathrm{r}_{\bar{x}_j, \tilde{x}_j}) < \delta_J(\epsilon)$, and we have $\sum_{j=1}^n \left\| \Delta_{F_J}(\mathrm{r}_{\bar{x}_j, \tilde{x}_j}) \right\| < \epsilon$.*

Proof We will prove the result using mathematical induction on m:
1° The result is clearly a restatement of Definition 12.58. Hence, the case is proved.
2° Assume that the result holds for $m \le k - 1 \in \mathbb{N}$.
3° Consider the case when $m = k \in \{2, 3, \ldots\}$.

"Sufficiency" We need to show that F is absolutely continuous at x_0 assuming (a). Let $i = 0$, and (a) implies (ii) of Definition 12.58. $\forall i_0 \in \bar{J}$, let π_{i_0}, $\pi_{\widehat{i_0}}$, D_{i_0}, and F_{i_0} be as defined in (i) of Definition 12.58. By (a) holds for any $i \in \{1, \ldots, m-1\}$ and any $J \subset \bar{J}$ with $\operatorname{card}(J) = i$ such that $i_0 \in J$, we have (a) holds for F_{i_0} at $\pi_{\widehat{i_0}}(x_0)$. By the inductive assumption, F_{i_0} is absolutely continuous at $\pi_{\widehat{i_0}}(x_0)$. Hence, (i) of Definition 12.58 holds for F at x_0. Then, F is absolutely continuous at x_0.

"Necessity" We need to show that (a) holds assuming that F is absolutely continuous at x_0. By (ii) of Definition 12.58, (a) holds when $i = 0$. By (i) of Definition 12.58, $\forall i_0 \in \bar{J}$, let π_{i_0}, $\pi_{\widehat{i_0}}$, D_{i_0}, and F_{i_0} be as defined in (i) of Definition 12.58, and we have F_{i_0} is absolutely continuous at $\pi_{\widehat{i_0}}(x_0)$. By the inductive assumption, (a) holds for F_{i_0}. Then, (a) holds for F, $\forall i \in \{1, \ldots, m-1\}$, $\forall J \subset \bar{J}$ with $\operatorname{card}(J) = i$ and $i_0 \in J$. Hence, (a) holds for F. This completes the proof for the necessity.

This completes the induction process.

This completes the proof of the proposition. □

Proposition 12.61 *Let $m \in \mathbb{N}$, $\mathbb{R}^m$ be endowed with the usual positive cone, $\Omega \in \mathcal{B}_{\mathrm{B}}(\mathbb{R}^m)$ be a rectangle with subset topology $\mathcal{O}$, $\mathcal{Y}$ and $\mathcal{Z}$ be normed linear spaces over $\mathbb{K}$, $f : \Omega \to \mathcal{Y}$, $g : \Omega \to \mathcal{Z}$, $x_0 \in \Omega$, and $h : \Omega \to \mathcal{Y} \times \mathcal{Z}$ be defined by $h(x) = (f(x), g(x))$, $\forall x \in \Omega$. Then, h is absolutely continuous at x_0 if, and only if, f and g are absolutely continuous at x_0.*

Proof We will prove the result using mathematical induction on m:
$1°$ $m = 1$. "Sufficiency" By f being absolutely continuous at x_0, $\exists a_f \in (0, \infty) \subset \mathbb{R}$, $\forall \epsilon \in (0, \infty) \subset \mathbb{R}$, $\exists \delta_f(\epsilon) \in (0, a_f] \subset \mathbb{R}$, $\forall n \in \mathbb{N}$, $\forall x_0 - a_f \leq x_1 \leq \bar{x}_1 \leq x_2 \leq \bar{x}_2 \leq \cdots \leq x_n \leq \bar{x}_n \leq x_0 + a_f$ with $x_i, \bar{x}_i \in I$, $i = 1, \ldots, n$, and $\sum_{i=1}^{n} (\bar{x}_i - x_i) < \delta_f(\epsilon)$, and we have $\sum_{i=1}^{n} \| f(\bar{x}_i) - f(x_i) \| < \epsilon$. By g being absolutely continuous at x_0, $\exists a_g \in (0, \infty) \subset \mathbb{R}$, $\forall \epsilon \in (0, \infty) \subset \mathbb{R}$, $\exists \delta_g(\epsilon) \in (0, a_g] \subset \mathbb{R}$, $\forall n \in \mathbb{N}$, $\forall x_0 - a_g \leq x_1 \leq \bar{x}_1 \leq x_2 \leq \bar{x}_2 \leq \cdots \leq x_n \leq \bar{x}_n \leq x_0 + a_g$ with $x_i, \bar{x}_i \in I$, $i = 1, \ldots, n$, and $\sum_{i=1}^{n} (\bar{x}_i - x_i) < \delta_g(\epsilon)$, and we have $\sum_{i=1}^{n} \| g(\bar{x}_i) - g(x_i) \| < \epsilon$. Let $a := \min\{a_f, a_g\} \in (0, +\infty) \subset \mathbb{R}$. $\forall \epsilon \in (0, \infty) \subset \mathbb{R}$, let $\delta(\epsilon) := \min\{\delta_f(\epsilon/2), \delta_g(\epsilon/2)\} \in (0, a] \subset \mathbb{R}$. $\forall n \in \mathbb{N}$, $\forall x_0 - a \leq x_1 \leq \bar{x}_1 \leq x_2 \leq \bar{x}_2 \leq \cdots \leq x_n \leq \bar{x}_n \leq x_0 + a$ with $x_i, \bar{x}_i \in I$, $i = 1, \ldots, n$, and $\sum_{i=1}^{n} (\bar{x}_i - x_i) < \delta(\epsilon)$, we have $\sum_{i=1}^{n} \| h(\bar{x}_i) - h(x_i) \| = \sum_{i=1}^{n} \| (f(\bar{x}_i) - f(x_i), g(\bar{x}_i) - g(x_i)) \| = \sum_{i=1}^{n} (\| f(\bar{x}_i) - f(x_i) \|^2 + \| g(\bar{x}_i) - g(x_i) \|^2)^{1/2} \leq \sum_{i=1}^{n} (\| f(\bar{x}_i) - f(x_i) \| + \| g(\bar{x}_i) - g(x_i) \|) < \epsilon$. Hence, h is absolutely continuous at x_0.

"Necessity" By h being absolutely continuous at x_0, $\exists a \in (0, \infty) \subset \mathbb{R}$, $\forall \epsilon \in (0, \infty) \subset \mathbb{R}$, $\exists \delta(\epsilon) \in (0, a] \subset \mathbb{R}$, $\forall n \in \mathbb{N}$, $\forall x_0 - a \leq x_1 \leq \bar{x}_1 \leq x_2 \leq \bar{x}_2 \leq \cdots \leq x_n \leq \bar{x}_n \leq x_0 + a$ with $x_i, \bar{x}_i \in I$, $i = 1, \ldots, n$, and $\sum_{i=1}^{n} (\bar{x}_i - x_i) < \delta(\epsilon)$, and we have $\epsilon > \sum_{i=1}^{n} \| h(\bar{x}_i) - h(x_i) \| = \sum_{i=1}^{n} (\| f(\bar{x}_i) - f(x_i) \|^2 + \| g(\bar{x}_i) - g(x_i) \|^2)^{1/2} \geq \sum_{i=1}^{n} \| f(\bar{x}_i) - f(x_i) \|$. Hence, f is absolutely continuous at x_0. By symmetry, g is absolutely continuous at x_0. This case is proved.
$2°$ Assume that the result holds for $m \leq k - 1 \in \mathbb{N}$.
$3°$ Consider the case $m = k \in \{2, 3, \ldots\}$.

"Sufficiency" Assume that f and g are absolutely continuous at x_0. $\forall i_0 \in \bar{J} := \{1, \ldots, m\}$, let $D_{i_0} \subseteq \mathbb{R}^{m-1}$, $\pi_{\widehat{i_0}} : \mathbb{R}^m \to \mathbb{R}^{m-1}$, $f_{i_0} : D_{i_0} \to \mathcal{Y}$, $g_{i_0} : D_{i_0} \to \mathcal{Z}$, and $h_{i_0} : D_{i_0} \to \mathcal{Y} \times \mathcal{Z}$ be defined as in Definition 12.58. By (i) of Definition 12.58, f_{i_0} and g_{i_0} are absolutely continuous at $\pi_{\widehat{i_0}}(x_0)$. By the inductive assumption, h_{i_0} is absolutely continuous at $\pi_{\widehat{i_0}}(x_0)$. Then, h satisfies (i) of Definition 12.58.

By (ii) of Definition 12.58, $\exists a_f \in (0, \infty) \subset \mathbb{R}$, $\forall \epsilon \in (0, \infty) \subset \mathbb{R}$, $\exists \delta_f(\epsilon) \in (0, \infty) \subset \mathbb{R}$, $\forall n \in \mathbb{N}$, $\forall (\bar{x}_i)_{i=1}^{n}, (\tilde{x}_i)_{i=1}^{n} \subseteq \bar{\mathrm{r}}_{x_0 - a_f \mathbf{1}_m, x_0 + a_f \mathbf{1}_m} \cap \Omega$ with $\bar{x}_i \leqq \tilde{x}_i$, $\forall i \in \{1, \ldots, n\}$, $(\mathrm{r}_{\bar{x}_i, \tilde{x}_i})_{i=1}^{n}$ being pairwise disjoint, and $\sum_{i=1}^{n} \mu_{\mathrm{B}m}(\mathrm{r}_{\bar{x}_i, \tilde{x}_i}) < \delta_f(\epsilon)$, we have $\sum_{i=1}^{n} \| \Delta_f(\mathrm{r}_{\bar{x}_i, \tilde{x}_i}) \| < \epsilon$; $\exists a_g \in (0, \infty) \subset \mathbb{R}$, $\forall \epsilon \in (0, \infty) \subset \mathbb{R}$, $\exists \delta_g(\epsilon) \in (0, \infty) \subset \mathbb{R}$, $\forall n \in \mathbb{N}$, $\forall (\bar{x}_i)_{i=1}^{n}, (\tilde{x}_i)_{i=1}^{n} \subseteq \bar{\mathrm{r}}_{x_0 - a_g \mathbf{1}_m, x_0 + a_g \mathbf{1}_m} \cap \Omega$ with $\bar{x}_i \leqq \tilde{x}_i$, $\forall i \in \{1, \ldots, n\}$, $(\mathrm{r}_{\bar{x}_i, \tilde{x}_i})_{i=1}^{n}$ being pairwise disjoint, and $\sum_{i=1}^{n} \mu_{\mathrm{B}m}(\mathrm{r}_{\bar{x}_i, \tilde{x}_i}) < \delta_g(\epsilon)$, and we have $\sum_{i=1}^{n} \| \Delta_g(\mathrm{r}_{\bar{x}_i, \tilde{x}_i}) \| < \epsilon$. Let $a := a_f \wedge a_g \in (0, \infty) \subset \mathbb{R}$. $\forall \epsilon \in (0, \infty) \subset \mathbb{R}$, let $\delta(\epsilon) := \min\{\delta_f(\epsilon/2), \delta_g(\epsilon/2)\} \in (0, \infty) \subset \mathbb{R}$. $\forall n \in \mathbb{N}$, $\forall (\bar{x}_i)_{i=1}^{n}, (\tilde{x}_i)_{i=1}^{n} \subseteq \bar{\mathrm{r}}_{x_0 - a \mathbf{1}_m, x_0 + a \mathbf{1}_m} \cap \Omega$ with $\bar{x}_i \leqq \tilde{x}_i$, $\forall i \in \{1, \ldots, n\}$, $(\mathrm{r}_{\bar{x}_i, \tilde{x}_i})_{i=1}^{n}$ being pairwise disjoint, and $\sum_{i=1}^{n} \mu_{\mathrm{B}m}(\mathrm{r}_{\bar{x}_i, \tilde{x}_i}) < \delta(\epsilon)$, we have $\sum_{i=1}^{n} \| \Delta_h(\mathrm{r}_{\bar{x}_i, \tilde{x}_i}) \| = \sum_{i=1}^{n} \| (\Delta_f(\mathrm{r}_{\bar{x}_i, \tilde{x}_i}), \Delta_g(\mathrm{r}_{\bar{x}_i, \tilde{x}_i})) \| = \sum_{i=1}^{n} (\| \Delta_f(\mathrm{r}_{\bar{x}_i, \tilde{x}_i}) \|^2 + \| \Delta_g(\mathrm{r}_{\bar{x}_i, \tilde{x}_i}) \|^2)^{1/2} \leq \sum_{i=1}^{n} \| \Delta_f(\mathrm{r}_{\bar{x}_i, \tilde{x}_i}) \| + \sum_{i=1}^{n} \| \Delta_g(\mathrm{r}_{\bar{x}_i, \tilde{x}_i}) \| < \epsilon$. Hence, h satisfies (ii) of Definition 12.58. Then, h is absolutely continuous at x_0.

"Necessity" Let h be absolutely continuous at x_0. $\forall i_0 \in \bar{J}$, let D_{i_0}, $\pi_{\widehat{i_0}}$, f_{i_0}, g_{i_0}, and h_{i_0} be defined as in Definition 12.58. By (i) of Definition 12.58, h_{i_0} is absolutely continuous at $\pi_{\widehat{i_0}}(x_0)$. By the inductive assumption, f_{i_0} and g_{i_0} are absolutely continuous at x_0. Hence, f and g satisfy (i) of Definition 12.58.

By (ii) of Definition 12.58, $\exists a \in (0, \infty) \subset \mathbb{R}$, $\forall \epsilon \in (0, \infty) \subset \mathbb{R}$, $\exists \delta(\epsilon) \in (0, \infty) \subset \mathbb{R}$, $\forall n \in \mathbb{N}$, $\forall (\bar{x}_i)_{i=1}^n, (\tilde{x}_i)_{i=1}^n \subseteq \bar{\mathrm{r}}_{x_0 - a\mathbf{1}_m, x_0 + a\mathbf{1}_m} \cap \Omega$ with $\bar{x}_i \leqq \tilde{x}_i$, $\forall i \in \{1, \ldots, n\}$, $(\mathrm{r}_{\bar{x}_i, \tilde{x}_i})_{i=1}^n$ being pairwise disjoint, and $\sum_{i=1}^n \mu_{\mathrm{B}m}(\mathrm{r}_{\bar{x}_i, \tilde{x}_i}) < \delta(\epsilon)$, we have $\epsilon > \sum_{i=1}^n \|\Delta_h(\mathrm{r}_{\bar{x}_i, \tilde{x}_i})\| \geq \sum_{i=1}^n \|\Delta_f(\mathrm{r}_{\bar{x}_i, \tilde{x}_i})\|$. Hence, f satisfies (ii) of Definition 12.58. Then, f is absolutely continuous at x_0. By symmetry, g is absolutely continuous at x_0.

This completes the induction process.

This completes the proof of the proposition. □

Proposition 12.62 *Let $m \in \mathbb{N}$, $\mathbb{R}^m$ be endowed with the usual positive cone, $\Omega \in \mathcal{B}_{\mathrm{B}}(\mathbb{R}^m)$ be a rectangle with subset topology $\mathcal{O}$, $\mathcal{Y}$ be a normed linear space over $\mathbb{K}$, $F : \Omega \to \mathcal{Y}$, and $\Omega_1, \Omega_2 \subseteq \Omega$ be rectangles in $\mathbb{R}^m$ such that $\Omega_1 \cup \Omega_2 = \Omega$. Then, the following statements hold:*

(i) *If F is absolutely continuous on $E \subseteq \Omega$, then $F|_{\Omega_1}$ is absolutely continuous on $E \cap \Omega_1$.*

(ii) *If $F|_{\Omega_1}$ and $F|_{\Omega_2}$ are absolutely continuous and Ω_1 and Ω_2 are both relatively open or both relatively closed (with respect to the subset topology $\mathcal{O}$), then F is absolutely continuous.*

Proof (i) $\forall x_0 \in E \cap \Omega_1$, since F is absolutely continuous at x_0, then $F|_{\Omega_1}$ is absolutely continuous at x_0. By the arbitrariness of x_0, $F|_{\Omega_1}$ is absolutely continuous on $E \cap \Omega_1$.

(ii) We will distinguish two exhaustive cases: Case 1: Ω_1 and Ω_2 are relatively open in Ω; Case 2: Ω_1 and Ω_2 are relatively closed in Ω.

Case 1: Ω_1 and Ω_2 are relatively open in Ω. Fix any $x_0 \in \Omega$. We will prove the statement that "if $F|_{\Omega_i}$ is absolutely continuous at x_0 when $x_0 \in \Omega_i$, and Ω_1 and Ω_2 are relatively open in Ω, then F is absolutely continuous at x_0" by mathematical induction on m:

1° $m = 1$. Without loss of generality, assume $x_0 \in \Omega_1$. Since Ω_1 is open relative to Ω, then $\exists a \in (0, \infty) \subset \mathbb{R}$ such that $(x_0 - a, x_0 + a) \cap \Omega_1 = (x_0 - a, x_0 + a) \cap \Omega$. By $F|_{\Omega_1}$ being absolutely continuous at x_0, we can easily show that that F is absolutely continuous at x_0. This completes the first step in the induction process.

2° Assume that the result holds for $m \leq k - 1 \in \mathbb{N}$.

3° Consider the case when $m = k \in \{2, 3, \ldots\}$. Without loss of generality, assume $x_0 \in \Omega_1$. $\forall i_0 \in \bar{J} := \{1, \ldots, m\}$, let D_{i_0} and $F_{i_0} : D_{i_0} \to \mathcal{Y}$ be as defined in Definition 12.58. Let $D_{i_0,j} := \left\{x \in \mathbb{R}^{m-1} \mid \exists \bar{x} \in \Omega_j \ni \cdot \pi_{i_0}(\bar{x}) = \pi_{i_0}(x_0) \text{ and } \pi_{\widehat{i_0}}(\bar{x}) = x\right\}$ and $F_{i_0,j} : D_{i_0,j} \to \mathcal{Y}$ be defined by $F_{i_0,j}(x) = F|_{\Omega_j}(\bar{x})$, $\forall x \in D_{i_0,j}$, where $\bar{x} \in \Omega_j$ is such that $\pi_{i_0}(\bar{x}) = \pi_{i_0}(x_0)$ and $\pi_{\widehat{i_0}}(\bar{x}) = x$, $j = 1, 2$. Fix any $j \in \{1, 2\}$. Since $\Omega_j \subseteq \Omega$ is a relatively open rectangle, then $D_{i_0,j} \subseteq D_{i_0}$ is a relatively open rectangle. By $F|_{\Omega_j}$ being absolutely continuous at x_0 when $x_0 \in \Omega_j$, we have $F_{i_0,j}$ is absolutely continuous at $\pi_{\widehat{i_0}}(x_0)$ when $\pi_{\widehat{i_0}}(x_0) \in D_{i_0,j}$. It is easy to see that $F_{i_0,j} = F_{i_0}|_{D_{i_0,j}}$ and $D_{i_0,1} \cup D_{i_0,2} = D_{i_0}$. By the inductive assumption, F_{i_0} is absolutely continuous at $\pi_{\widehat{i_0}}(x_0) \in D_{i_0}$. Hence, F satisfies (i) of Definition 12.58.

Since Ω_1 is relative open in Ω, then $\exists a \in (0, \infty) \subset \mathbb{R}$ such that $\mathrm{r}^\circ_{x_0-a\mathbf{1}_m, x_0+a\mathbf{1}_m} \cap \Omega_1 = \mathrm{r}^\circ_{x_0-a\mathbf{1}_m, x_0+a\mathbf{1}_m} \cap \Omega$. By $F|_{\Omega_1}$ being absolutely continuous at x_0 and (ii) of Definition 12.58, $\exists a_F \in (0, a) \subset \mathbb{R}$, $\forall \epsilon \in (0, \infty) \subset \mathbb{R}$, $\exists \delta(\epsilon) \in (0, \infty) \subset \mathbb{R}$, $\forall n \in \mathbb{N}$, $\forall (\bar{x}_i)_{i=1}^n, (\tilde{x}_i)_{i=1}^n \subseteq \bar{\mathrm{r}}_{x_0-a_F\mathbf{1}_m, x_0+a_F\mathbf{1}_m} \cap \Omega_1 = \bar{\mathrm{r}}_{x_0-a_F\mathbf{1}_m, x_0+a_F\mathbf{1}_m} \cap \Omega$ with $\bar{x}_i \leqq \tilde{x}_i$, $\forall i \in \{1, \ldots, n\}$, $(\mathrm{r}_{\bar{x}_i, \tilde{x}_i})_{i=1}^n$ being pairwise disjoint, and $\sum_{i=1}^n \mu_{\mathrm{B}m}(\mathrm{r}_{\bar{x}_i, \tilde{x}_i}) < \delta(\epsilon)$, and we have $\sum_{i=1}^n \left\| \Delta_F(\mathrm{r}_{\bar{x}_i, \tilde{x}_i}) \right\| < \epsilon$. Hence, F satisfies (ii) of Definition 12.58.

Hence, F is absolutely continuous at x_0. This completes the induction process.

By the arbitrariness of x_0, F is absolutely continuous. This completes the proof for Case 1.

Case 2: Ω_1 and Ω_2 are relatively closed in Ω. Fix any $x_0 \in \Omega$. We will prove the statement that "if $F|_{\Omega_i}$ is absolutely continuous at x_0 when $x_0 \in \Omega_i$, and Ω_1 and Ω_2 are relatively closed in Ω, then F is absolutely continuous at x_0" by mathematical induction on m.

1° $m = 1$. Without loss of generality, assume $x_0 \in \Omega_1$. We will further distinguish two exhaustive and mutually exclusive cases: Case 2a: $x_0 \notin \Omega_2$; Case 2b: $x_0 \in \Omega_2$. Case 2a: $x_0 \notin \Omega_2$. Since Ω_2 is relatively closed in Ω and $x_0 \in \Omega \setminus \Omega_2$, then, by Proposition 4.10, $\exists a \in (0, \infty) \subset \mathbb{R}$ such that $(x_0 - a, x_0 + a) \cap \Omega \subseteq \Omega \setminus \Omega_2 \subseteq \Omega_1$. $F|_{\Omega_1}$ is absolutely continuous at x_0 implies that F is absolutely continuous at x_0. This subcase is proved. Case 2b: $x_0 \in \Omega_2$. Then, $x_0 \in \Omega_1 \cap \Omega_2$. By $F|_{\Omega_i}$ being absolutely continuous at x_0, $\exists a_i \in (0, \infty) \subset \mathbb{R}$, $\forall \epsilon \in (0, \infty) \subset \mathbb{R}$, $\exists \delta_i(\epsilon) \in (0, a_i] \subset \mathbb{R}$, $\forall n \in \mathbb{N}$, $\forall x_0 - a_i \le x_1 \le \bar{x}_1 \le \cdots \le x_n \le \bar{x}_n \le x_0 + a_i$ with $x_j, \bar{x}_j \in \Omega_i$, $j = 1, \ldots, n$, and $\sum_{j=1}^n (\bar{x}_j - x_j) < \delta_i(\epsilon)$, and we have $\sum_{j=1}^n \|F(\bar{x}_j) - F(x_j)\| < \epsilon$, $i = 1, 2$. Take $a := \min\{a_1, a_2\} \in (0, \infty) \subset \mathbb{R}$, $\forall \epsilon \in (0, \infty) \subset \mathbb{R}$, and let $\delta(\epsilon) := \min\{\delta_1(\epsilon/2), \delta_2(\epsilon/2)\} \in (0, a] \subset \mathbb{R}$. $\forall n \in \mathbb{N}$, $\forall x_0 - a \le x_1 \le \bar{x}_1 \le \cdots \le x_n \le \bar{x}_n \le x_0 + a$ with $x_j, \bar{x}_j \in \Omega$, $j = 1, \ldots, n$, and $\sum_{j=1}^n (\bar{x}_j - x_j) < \delta(\epsilon)$. Without loss of generality, assume $\exists j_0 \in \{1, \ldots, n-1\}$ such that $\bar{x}_{j_0} \le x_0 \le x_{j_0+1}$. (In case, j_0 does not exist, and add a pair of x_0's into the collection of x_j's and $\bar{x}_j$'s.) Note that $x_1 \in \Omega = \Omega_1 \cup \Omega_2$. Without loss of generality, assume $x_1 \in \Omega_1$. Then $x_1, \bar{x}_1, \ldots, x_{j_0}, \bar{x}_{j_0} \in \Omega_1$, since Ω_1 is a rectangle (interval, since $m = 1$), and $\sum_{j=1}^{j_0} (\bar{x}_j - x_j) < \delta(\epsilon) \le \delta_1(\epsilon/2)$. Thus, $\sum_{j=1}^{j_0} \|F(\bar{x}_j) - F(x_j)\| < \epsilon/2$. By a similar argument, we have $\sum_{j=j_0+1}^n \|F(\bar{x}_j) - F(x_j)\| < \epsilon/2$. This implies that $\sum_{j=1}^n \|F(\bar{x}_j) - F(x_j)\| < \epsilon$. Hence, F is absolutely continuous at x_0. This subcase is proved.

Hence, in both subcases, F is absolutely continuous at x_0. This completes the proof for the first step of the induction process.

2° Assume that the result holds for $m \le k - 1 \in \mathbb{N}$.

3° Consider the case when $m = k \in \{2, 3, \ldots\}$. Without loss of generality, assume $x_0 \in \Omega_1$. $\forall i_0 \in \bar{J}$, let D_{i_0} and $F_{i_0} : D_{i_0} \to \mathcal{Y}$ be as defined in Definition 12.58. Let $D_{i_0,j} := \left\{ x \in \mathbb{R}^{m-1} \,\middle|\, \exists \bar{x} \in \Omega_j \ni \cdot \pi_{i_0}(\bar{x}) = \pi_{i_0}(x_0) \text{ and } \pi_{\widehat{i_0}}(\bar{x}) = x \right\}$ and $F_{i_0,j}$:

$D_{i_0,j} \to \mathcal{Y}$ be defined by $F_{i_0,j}(x) = F|_{\Omega_j}(\bar{x})$, $\forall x \in D_{i_0,j}$, where $\bar{x} \in \Omega_j$ is such that $\pi_{i_0}(\bar{x}) = \pi_{i_0}(x_0)$ and $\pi_{\widehat{i_0}}(\bar{x}) = x$, $j = 1, 2$. Fix any $j \in \{1, 2\}$. Since $\Omega_j \subseteq \Omega$ is a relatively closed rectangle, then $D_{i_0,j} \subseteq D_{i_0}$ is a relatively closed rectangle. By $F|_{\Omega_j}$ being absolutely continuous at x_0 when $x_0 \in \Omega_j$, we have $F_{i_0,j}$ is absolutely continuous at $\pi_{\widehat{i_0}}(x_0)$ when $\pi_{\widehat{i_0}}(x_0) \in D_{i_0,j}$. It is easy to see that $F_{i_0,j} = F_{i_0}|_{D_{i_0,j}}$ and $D_{i_0,1} \cup D_{i_0,2} = D_{i_0}$. By the inductive assumption, F_{i_0} is absolutely continuous at $\pi_{\widehat{i_0}}(x_0) \in D_{i_0}$. Hence, F satisfies (i) of Definition 12.58.

We will further distinguish two exhaustive and mutually exclusives subcases: Case 2α: $x_0 \notin \Omega_2$; Case 2β: $x_0 \in \Omega_2$.

Case 2α: $x_0 \notin \Omega_2$. Then, $x_0 \in \Omega \setminus \Omega_2$. Since Ω_2 is relatively closed in Ω, by Proposition 4.10, $\exists a \in (0, \infty) \subset \mathbb{R}$ such that $\mathrm{r}^\circ_{x_0 - a\mathbf{1}_m, x_0 + a\mathbf{1}_m} \cap \Omega \subseteq \Omega \setminus \Omega_2 \subseteq \Omega_1$. By $F|_{\Omega_1}$ being absolutely continuous at x_0, $\exists \bar{a} \in (0, a) \subset \mathbb{R}$, $\forall \epsilon \in (0, \infty) \subset \mathbb{R}$, $\exists \delta(\epsilon) \in (0, \infty) \subset \mathbb{R}$, $\forall n \in \mathbb{N}$, $\forall (\bar{x}_i)_{i=1}^n, (\tilde{x}_i)_{i=1}^n \subseteq \bar{\mathrm{r}}_{x_0 - \bar{a}\mathbf{1}_m, x_0 + \bar{a}\mathbf{1}_m} \cap \Omega_1 = \bar{\mathrm{r}}_{x_0 - \bar{a}\mathbf{1}_m, x_0 + \bar{a}\mathbf{1}_m} \cap \Omega$ with $\bar{x}_i \leqq \tilde{x}_i$, $\forall i \in \{1, \ldots, n\}$, and $(\mathrm{r}_{\bar{x}_i, \tilde{x}_i})_{i=1}^n$ being pairwise disjoint, and $\sum_{i=1}^n \mu_{\mathrm{B}m}(\mathrm{r}_{\bar{x}_i, \tilde{x}_i}) < \delta(\epsilon)$, and we have $\sum_{i=1}^n \|\Delta_F(\mathrm{r}_{\bar{x}_i, \tilde{x}_i})\| < \epsilon$. Hence, F satisfies (ii) of Definition 12.58. Then, F is absolutely continuous at x_0. This subcase is proved.

Case 2β: $x_0 \in \Omega_2$. Then, $x_0 \in \Omega_1 \cap \Omega_2$. Fix any $i \in \{1, 2\}$. By $F|_{\Omega_i}$ being absolutely continuous at x_0 and (ii) of Definition 12.58, $\exists a_i \in (0, \infty) \subset \mathbb{R}$, $\forall \epsilon \in (0, \infty) \subset \mathbb{R}$, $\exists \delta_i(\epsilon) \in (0, \infty) \subset \mathbb{R}$, $\forall n \in \mathbb{N}$, $\forall (\bar{x}_j)_{j=1}^n, \left(\tilde{x}_j\right)_{j=1}^n \subseteq \bar{\mathrm{r}}_{x_0 - a_i\mathbf{1}_m, x_0 + a_i\mathbf{1}_m} \cap \Omega_i$ with $\bar{x}_j \leqq \tilde{x}_j$, $\forall j \in \{1, \ldots, n\}$, $\left(\mathrm{r}_{\bar{x}_j, \tilde{x}_j}\right)_{j=1}^n$ being pairwise disjoint, and $\sum_{j=1}^n \mu_{\mathrm{B}m}(\mathrm{r}_{\bar{x}_j, \tilde{x}_j}) < \delta_i(\epsilon)$, we have $\sum_{j=1}^n \left\|\Delta_F(\mathrm{r}_{\bar{x}_j, \tilde{x}_j})\right\| < \epsilon$.

Take $a := \min\{a_1, a_2\} \in (0, \infty) \subset \mathbb{R}$, $\forall \epsilon \in (0, \infty) \subset \mathbb{R}$, let $\delta(\epsilon) := \min\{\delta_1(\epsilon/2), \delta_2(\epsilon/2)\} \in \mathbb{R}_+$, $\forall n \in \mathbb{N}$, $\forall (\bar{x}_j)_{j=1}^n, \left(\tilde{x}_j\right)_{j=1}^n \subseteq \bar{\mathrm{r}}_{x_0 - a\mathbf{1}_m, x_0 + a\mathbf{1}_m} \cap \Omega$ with $\bar{x}_j \leqq \tilde{x}_j$, $\forall j \in \{1, \ldots, n\}$, $(\mathrm{r}_{\bar{x}_j, \tilde{x}_j})_{j=1}^n$ being pairwise disjoint, and $\sum_{j=1}^n \mu_{\mathrm{B}m}(\mathrm{r}_{\bar{x}_j, \tilde{x}_j}) < \delta(\epsilon)$. $\forall j \in \{1, \ldots, n\}$, since $\Omega = \Omega_1 \cup \Omega_2$ and Ω_1 and Ω_2 are relatively closed rectangles in the rectangle Ω, then $\mathrm{r}_{\bar{x}_j, \tilde{x}_j} = \mathrm{r}_{\bar{x}_{j,1}, \tilde{x}_{j,1}} \cup \mathrm{r}_{\bar{x}_{j,2}, \tilde{x}_{j,2}}$, where $\bar{x}_{j,k}, \tilde{x}_{j,k} \in \bar{\mathrm{r}}_{x_0 - a\mathbf{1}_m, x_0 + a\mathbf{1}_m} \cap \Omega_k$, $\bar{x}_{j,k} \leqq \tilde{x}_{j,k}$, $k = 1, 2$, and the two sets in the union are disjoint. To see the above, note that $\mathrm{r}_{\bar{x}_j, \tilde{x}_j} \cap \Omega_k$ is a rectangle, $k = 1, 2$. If $\bar{\mathrm{r}}_{\bar{x}_j, \tilde{x}_j} \subseteq \Omega_l$, for some $l \in \{1, 2\}$, then let $\bar{x}_{j,l} = \bar{x}_j$, $\tilde{x}_{j,l} = \tilde{x}_j$, $\bar{x}_{j,3-l} = x_0$, and $\tilde{x}_{j,3-l} = x_0$. The desired result holds. Consider the remaining possibility, we must have $\bar{x}_j \in \Omega_l \setminus \Omega_{3-l}$ and $\tilde{x}_j \in \Omega_{3-l} \setminus \Omega_l$, for some $l \in \{1, 2\}$. Without loss of generality, assume $l = 1$. Ω_1 and Ω_2 are m-dimensional rectangles and none contains the other, their union is a m-dimensional rectangle Ω, in which both Ω_1 and Ω_2 are relatively closed. The only possibility for this to happen is that $\pi_i(\Omega_1) = \pi_i(\Omega_2) = \pi_i(\Omega)$, $\forall i \in \bar{J} \setminus \{i_0\}$, for some $i_0 \in \bar{J}$, and $\pi_{i_0}(\Omega_1)$ and $\pi_{i_0}(\Omega_2)$ are overlapping relatively closed intervals in $\pi_{i_0}(\Omega)$. Let $\bar{x}_{j,1} = \bar{x}_j \in \Omega_1 \setminus \Omega_2$. Since $\bar{\mathrm{r}}_{\bar{x}_j, \tilde{x}_j} \subseteq \Omega$, then $\bar{\mathrm{r}}_{\bar{x}_j, \tilde{x}_j} \cap \Omega_1 \subseteq \Omega$ is a relatively closed rectangle with respect to $\bar{\mathrm{r}}_{\bar{x}_j, \tilde{x}_j}$. Then, $\emptyset \neq \bar{\mathrm{r}}_{\bar{x}_j, \tilde{x}_j} \cap \Omega_1 =: \bar{\mathrm{r}}_{\bar{x}_{j,1}, \tilde{x}_{j,1}} \subseteq \bar{\mathrm{r}}_{\bar{x}_j, \tilde{x}_j} \subseteq \Omega$. This implies that $\bar{x}_{j,1}, \tilde{x}_{j,1} \in \bar{\mathrm{r}}_{x_0 - a\mathbf{1}_m, x_0 + a\mathbf{1}_m} \cap \Omega_1$ and $\bar{x}_{j,1} \leqq \tilde{x}_{j,1}$. Since $\tilde{x}_j \in \Omega_2 \setminus \Omega_1$, then $\emptyset \neq \bar{\mathrm{r}}_{\bar{x}_j, \tilde{x}_j} \cap \Omega_2 =: \bar{\mathrm{r}}_{\hat{x}_j, \tilde{x}_j} \subseteq \bar{\mathrm{r}}_{\bar{x}_j, \tilde{x}_j} \subseteq \Omega$, where $\hat{x}_j \in \bar{\mathrm{r}}_{x_0 - a\mathbf{1}_m, x_0 + a\mathbf{1}_m} \cap \Omega_2$

and $\hat{x}_j \stackrel{\lessdot}{=} \tilde{x}_j$. Clearly, $\tilde{x}_{j,1} \in \Omega_1 \cap \Omega_2$ since Ω_2 is relatively closed in Ω and $\hat{x}_j \in \Omega_1 \cap \Omega_2$ since Ω_1 is relatively closed in Ω. Then, $\hat{x}_j \in \bar{\mathrm{r}}_{\bar{x}_j,\tilde{x}_j} \cap \Omega_1 = \bar{\mathrm{r}}_{\bar{x}_{j,1},\tilde{x}_{j,1}}$ and $\hat{x}_j \stackrel{\lessdot}{=} \tilde{x}_{j,1}$. This implies that $\bar{\mathrm{r}}_{\hat{x}_j,\tilde{x}_{j,1}} \subseteq \Omega_1 \cap \Omega_2$. Let $\tilde{x}_{j,2} = \tilde{x}_j$ and $\bar{x}_{j,2} \in \bar{\mathrm{r}}_{\hat{x}_j,\tilde{x}_{j,1}}$ be such that $\pi_l(\bar{x}_{j,2}) = \begin{cases} \pi_l(\hat{x}_j) & \text{if } \pi_l(\tilde{x}_{j,1}) = \pi_l(\tilde{x}_j) \\ \pi_l(\tilde{x}_{j,1}) & \text{if } \pi_l(\tilde{x}_{j,1}) < \pi_l(\tilde{x}_j) \end{cases}$, $\forall l \in \bar{J}$. Then, $\bar{x}_{j,2} \stackrel{\lessdot}{=} \tilde{x}_{j,2}$ and $\bar{x}_{j,2}, \tilde{x}_{j,2} \in \Omega_2 \cap \bar{\mathrm{r}}_{x_0 - a\mathbf{1}_m, x_0 + a\mathbf{1}_m}$. Clearly, $\mathrm{r}_{\bar{x}_{j,1},\tilde{x}_{j,1}} \cap \mathrm{r}_{\bar{x}_{j,2},\tilde{x}_{j,2}} = \emptyset$ and $\mathrm{r}_{\bar{x}_{j,1},\tilde{x}_{j,1}} \cup \mathrm{r}_{\bar{x}_{j,2},\tilde{x}_{j,2}} = \mathrm{r}_{\bar{x}_j,\tilde{x}_j}$ since $\bar{\mathrm{r}}_{\bar{x}_{j,1},\tilde{x}_{j,1}} \cup \bar{\mathrm{r}}_{\hat{x}_j,\tilde{x}_j} = \bar{\mathrm{r}}_{\bar{x}_j,\tilde{x}_j}$. These $\bar{x}_{j,1}, \bar{x}_{j,2}, \tilde{x}_{j,1}$, and $\tilde{x}_{j,2}$ satisfy the conditions specified above.

Then, $\sum_{j=1}^{n} \mu_{\mathrm{B}m}(\mathrm{r}_{\bar{x}_j,\tilde{x}_j}) = \sum_{k=1}^{2} \sum_{j=1}^{m} \mu_{\mathrm{B}m}(\mathrm{r}_{\bar{x}_{j,k},\tilde{x}_{j,k}}) < \delta(\epsilon)$ and $\sum_{j=1}^{m} \mu_{\mathrm{B}m}(\mathrm{r}_{\bar{x}_{j,k},\tilde{x}_{j,k}}) < \delta_k(\epsilon/2)$, $k = 1, 2$. This implies that $\sum_{j=1}^{n} \left\| \Delta_F(\mathrm{r}_{\bar{x}_j,\tilde{x}_j}) \right\| = \sum_{j=1}^{n} \left\| \Delta_F(\mathrm{r}_{\bar{x}_{j,1},\tilde{x}_{j,1}}) + \Delta_F(\mathrm{r}_{\bar{x}_{j,2},\tilde{x}_{j,2}}) \right\| \leq \sum_{k=1}^{2} \sum_{j=1}^{n} \left\| \Delta_F(\mathrm{r}_{\bar{x}_{j,k},\tilde{x}_{j,k}}) \right\| < \epsilon$, where the first equality follows from Proposition 12.46. Hence, F satisfies (ii) of Definition 12.58. Therefore, F is absolutely continuous at x_0. This subcase is proved.

In both subcases, F is absolutely continuous at x_0. This completes the induction process.

By the arbitrariness of x_0, F is absolutely continuous. This completes the proof for Case 2.

In both cases, F is absolutely continuous. This completes the proof of the proposition. □

Definition 12.63 Let $\mathcal{X} := (X, \rho_X)$ and $\mathcal{Y} := (Y, \rho_Y)$ be metric spaces, $x_0 \in \mathcal{X}$, and $f : \mathcal{X} \to \mathcal{Y}$. f is said to be *locally Lipschitz continuous at* x_0 *(with Lipschitz constant* $L_0 \in \overline{\mathbb{R}_+}$*)* if $\exists \delta \in \mathbb{R}_+$, $\forall x_1, x_2 \in \mathcal{B}_{\mathcal{X}}(x_0, \delta)$, we have $\rho_Y(f(x_1), f(x_2)) \leq L_0 \rho_X(x_1, x_2)$. f is said to be *locally Lipschitz continuous on* $\mathcal{X}$ if it is locally Lipschitz continuous at x, $\forall x \in \mathcal{X}$. f is said to be *Lipschitz continuous on* $\mathcal{X}$ *(with Lipschitz constant* $L \in \overline{\mathbb{R}_+}$*)* if $\forall x_1, x_2 \in \mathcal{X}$, we have $\rho_Y(f(x_1), f(x_2)) \leq L \rho_X(x_1, x_2)$. ◇

Clearly, if f is locally Lipschitz continuous at x_0, then f is continuous at x_0; if f is Lipschitz continuous on $\mathcal{X}$, then it is uniformly continuous; if f is Lipschitz continuous on $\mathcal{X}$, then it is locally Lipschitz continuous on $\mathcal{X}$.

Definition 12.64 Let $\mathcal{X} := (X, \rho_X)$ and $\mathcal{Z} := (Z, \rho_Z)$ be metric spaces, Y be a set, $f : \mathcal{X} \times Y \to \mathcal{Z}$, and $x_0 \in \mathcal{X}$. f is said to be *locally Lipschitz continuous at* x_0 *(with Lipschitz constant* $L_0 \in \overline{\mathbb{R}_+}$*) uniformly over* Y if $\exists \delta \in \mathbb{R}_+$, $\forall x_1, x_2 \in \mathcal{B}_{\mathcal{X}}(x_0, \delta)$, $\forall y \in Y$, we have $\rho_Z(f(x_1, y), f(x_2, y)) \leq L_0 \rho_X(x_1, x_2)$. f is said to be *locally Lipschitz continuous on* $\mathcal{X}$ *uniformly over* Y if it is locally Lipschitz at x uniformly over Y, $\forall x \in \mathcal{X}$. f is said to be *Lipschitz continuous on* $\mathcal{X}$ *(with Lipschitz constant* $L \in \overline{\mathbb{R}_+}$*) uniformly over* Y if $\forall x_1, x_2 \in \mathcal{X}$, $\forall y \in Y$, we have $\rho_Z(f(x_1, y), f(x_2, y)) \leq L \rho_X(x_1, x_2)$. ◇

Clearly, if f is Lipschitz continuous on $\mathcal{X}$ uniformly over Y, then it is locally Lipschitz continuous on $\mathcal{X}$ uniformly over Y.

Proposition 12.65 *Let $\mathcal{X} := (X, \rho_X)$ be a compact metric space, Y be a set, $y_0 \in Y$, $\mathcal{Z} := (Z, \rho_Z)$ be a metric space, and $f : \mathcal{X} \times Y \to \mathcal{Z}$ be locally Lipschitz continuous on $\mathcal{X}$ uniformly over Y. Assume that $\forall x \in \mathcal{X}$, $0 \leq \sup_{y \in Y} \rho_Z(f(x, y), f(x, y_0)) < +\infty$. Then, f is Lipschitz continuous on $\mathcal{X}$ uniformly over Y.*

Proof $\forall x \in \mathcal{X}$, by the local Lipschitz continuity of f, $\exists L_x \in \overline{\mathbb{R}_+}$, $\exists \delta_x \in \mathbb{R}_+$, $\forall x_1, x_2 \in O_x := \mathcal{B}_{\mathcal{X}}(x, \delta_x)$, $\forall y \in Y$, we have $\rho_Z(f(x_1, y), f(x_2, y)) \leq L_x \rho_X(x_1, x_2)$. Let $\hat{O}_x := \mathcal{B}_{\mathcal{X}}(x, \delta_x/2)$. Then, $\mathcal{X} \subseteq \bigcup_{x \in \mathcal{X}} \hat{O}_x$, which is an open covering of $\mathcal{X}$. By the compactness of $\mathcal{X}$, there exists a finite set $X_N \subseteq \mathcal{X}$ such that $\mathcal{X} \subseteq \bigcup_{x \in X_N} \hat{O}_x$. Let $L_M := \max\{0, \max_{x \in X_N} L_x, \frac{2}{\min_{x \in X_N} \delta_x} \cdot \max_{\hat{x}_1, \hat{x}_2 \in X_N} (\sup_{y \in Y} \rho_Z(f(\hat{x}_1, y), f(\hat{x}_1, y_0)) + \rho_Z(f(\hat{x}_1, y_0), f(\hat{x}_2, y_0)) + \sup_{y \in Y} \rho_Z(f(\hat{x}_2, y), f(\hat{x}_2, y_0)) + \frac{L_{\hat{x}_1} \delta_{\hat{x}_1}}{2} + \frac{L_{\hat{x}_2} \delta_{\hat{x}_2}}{2})\} \in \overline{\mathbb{R}_+}$. $\forall x_1, x_2 \in \mathcal{X}$, $\forall y \in Y$, $\exists \hat{x}_1 \in X_N$ such that $x_1 \in \hat{O}_{\hat{x}_1}$. We will distinguish two exhaustive and mutually exclusive cases: Case 1: $x_2 \in O_{\hat{x}_1}$; Case 2: $x_2 \notin O_{\hat{x}_1}$. Case 1: $x_2 \in O_{\hat{x}_1}$. Thus, $x_1, x_2 \in O_{\hat{x}_1}$ and $\rho_Z(f(x_1, y), f(x_2, y)) \leq L_{\hat{x}_1} \rho_X(x_1, x_2) \leq L_M \rho_X(x_1, x_2)$. Case 2: $x_2 \notin O_{\hat{x}_1}$. Thus, $\rho_X(x_2, \hat{x}_1) \geq \delta_{\hat{x}_1}$ and $\rho_X(x_2, x_1) \geq \rho_X(x_2, \hat{x}_1) - \rho_X(x_1, \hat{x}_1) > \delta_{\hat{x}_1}/2 \geq \frac{\min_{x \in X_N} \delta_x}{2} > 0$. $\exists \hat{x}_2 \in X_N$ such that $x_2 \in \hat{O}_{\hat{x}_2}$. This implies that $\rho_Z(f(x_1, y), f(x_2, y)) \leq \rho_Z(f(x_1, y), f(\hat{x}_1, y)) + \rho_Z(f(\hat{x}_1, y), f(\hat{x}_1, y_0)) + \rho_Z(f(\hat{x}_1, y_0), f(\hat{x}_2, y_0)) + \rho_Z(f(\hat{x}_2, y_0), f(\hat{x}_2, y)) + \rho_Z(f(\hat{x}_2, y), f(x_2, y)) \leq L_{\hat{x}_1} \rho_X(x_1, \hat{x}_1) + \sup_{y \in Y} \rho_Z(f(\hat{x}_1, y), f(\hat{x}_1, y_0)) + \rho_Z(f(\hat{x}_1, y_0), f(\hat{x}_2, y_0)) + \sup_{y \in Y} \rho_Z(f(\hat{x}_2, y), f(\hat{x}_2, y_0)) + L_{\hat{x}_2} \rho_X(x_2, \hat{x}_2) \leq \sup_{y \in Y} \rho_Z(f(\hat{x}_1, y), f(\hat{x}_1, y_0)) + \rho_Z(f(\hat{x}_1, y_0), f(\hat{x}_2, y_0)) + \frac{L_{\hat{x}_1} \delta_{\hat{x}_1}}{2} + \frac{L_{\hat{x}_2} \delta_{\hat{x}_2}}{2} + \sup_{y \in Y} \rho_Z(f(\hat{x}_2, y), f(\hat{x}_2, y_0)) \leq L_M \frac{\min_{x \in X_N} \delta_x}{2} \leq L_M \rho_X(x_1, x_2)$. Hence, in both cases, we have $\rho_Z(f(x_1, y), f(x_2, y)) \leq L_M \rho_X(x_1, x_2)$. Therefore, f is Lipschitz continuous on $\mathcal{X}$ uniformly over Y. This completes the proof of the proposition. □

Proposition 12.66 *Let $\mathcal{X}$ be a normed linear space over $\mathbb{K}$, Y be a set, $\mathcal{Z}$ be a normed linear space over $\mathbb{K}$, $U \subseteq \mathcal{X}$ be convex, and $f : U \times Y \to \mathcal{Z}$ be partial differentiable with respect to x. Assume that $\exists L \in [0, \infty) \subset \mathbb{R}$ such that $\left\| \frac{\partial f}{\partial x}(x, y) \right\| \leq L$, $\forall (x, y) \in U \times Y$. Then, f is Lipschitz continuous on U with Lipschitz constant L uniformly over Y.*

Proof $\forall x_1, x_2 \in U$, $\forall y \in Y$, by Mean Value Theorem 9.23, $\exists t \in (0, 1) \subset \mathbb{R}$ such that $\|f(x_1, y) - f(x_2, y)\| \leq \left\| \frac{\partial f}{\partial x}(t x_1 + (1 - t) x_2, y) \right\| \|x_1 - x_2\| \leq L \|x_1 - x_2\|$. Hence, f is Lipschitz continuous on U with Lipschitz constant L uniformly over Y. This completes the proof of the proposition. □

Proposition 12.67 *Let $I \subseteq \mathbb{R}$ be an interval, $\mathcal{Y}$ be a normed linear space over $\mathbb{K}_1$, $\mathcal{Z}$ be a normed linear space over $\mathbb{K}_2$, $U \subseteq \mathcal{Y}$, $f : I \to U$ be absolutely continuous* *at $x_0 \in I$, and $g : U \to \mathcal{Z}$ be locally Lipschitz continuous at $y_0 := f(x_0) \in U$. Then, $g \circ f$ is absolutely continuous at x_0.*

Proof By g being locally Lipschitz at y_0, then $\exists L_0 \in [0, \infty) \subset \mathbb{R}$, $\exists \delta_g \in (0, \infty) \subset \mathbb{R}$, $\forall y_1, y_2 \in U \cap \mathcal{B}_{\mathcal{Y}}(y_0, \delta_g)$, we have $\|g(y_1) - g(y_2)\| \leq L_0 \|y_1 - y_2\|$. Since f is absolutely continuous at x_0, then it is continuous at x_0 by Proposition 12.59. Then, $\exists a_0 \in (0, \infty) \subset \mathbb{R}$ such that $\forall x \in \mathcal{B}_{\mathbb{R}}(x_0, a_0) \cap I$, we have $f(x) \in U \cap \mathcal{B}_{\mathcal{Y}}(y_0, \delta_g)$. By the absolute continuity of f at x_0, $\exists a_f \in (0, a_0) \subset \mathbb{R}$, $\forall \epsilon \in (0, \infty) \subset \mathbb{R}$, $\exists \delta_f(\epsilon) \in (0, a_f] \subset \mathbb{R}$, $\forall n \in \mathbb{N}$, $\forall x_0 - a_f \leq x_1 \leq \bar{x}_1 \leq x_2 \leq \bar{x}_2 \leq \cdots \leq x_n \leq \bar{x}_n \leq x_0 + a_f$ with $x_i, \bar{x}_i \in I$, $i = 1, \ldots, n$, and $\sum_{i=1}^n (\bar{x}_i - x_i) < \delta_f(\epsilon)$, we have $\sum_{i=1}^n \|f(\bar{x}_i) - f(x_i)\| < \epsilon$. For the function $g \circ f$, take $a = a_f \in (0, \infty) \subset \mathbb{R}$, and $\forall \epsilon \in (0, \infty) \subset \mathbb{R}$, let $\delta(\epsilon) = \delta_f(\epsilon/(1 + L_0)) \in (0, a] \subset \mathbb{R}$. $\forall n \in \mathbb{N}$, $\forall x_0 - a \leq x_1 \leq \bar{x}_1 \leq x_2 \leq \bar{x}_2 \leq \cdots \leq x_n \leq \bar{x}_n \leq x_0 + a$ with $x_i, \bar{x}_i \in I$, $i = 1, \ldots, n$, and $\sum_{i=1}^n (\bar{x}_i - x_i) < \delta(\epsilon)$, we have $f(\bar{x}_i), f(x_i) \in U \cap \mathcal{B}_{\mathcal{Y}}(y_0, \delta_g)$, and $\sum_{i=1}^n \|g \circ f(\bar{x}_i) - g \circ f(x_i)\| \leq \sum_{i=1}^n L_0 \|f(\bar{x}_i) - f(x_i)\| \leq L_0 \frac{\epsilon}{1+L_0} < \epsilon$. Hence, $g \circ f$ is absolutely continuous at x_0. This completes the proof of the proposition. □

Proposition 12.68 *Let $I \subseteq \mathbb{R}$ be an interval, $\mathcal{Y}$ be a normed linear space over $\mathbb{K}$, $\mathcal{Z}$ be a normed linear space over $\mathbb{K}$, $f : I \to \mathcal{Y}$ and $g : I \to \mathcal{Y}$ be absolutely continuous at $x_0 \in I$, $A : I \to \mathrm{B}(\mathcal{Y}, \mathcal{Z})$ be absolutely continuous at x_0, and $h : I \to \mathcal{Z}$ be defined by $h(x) = A(x) f(x)$, $\forall x \in I$. Then, $f + g$ and h are absolutely continuous at x_0.*

Proof By Propositions 12.61, 9.14, 12.67, and 12.66, we have that $f + g$ is absolutely continuous at x_0.

By Propositions 12.61, 9.17, 12.67, and 12.66, we have that h is absolutely continuous at x_0.

This completes the proof of the proposition. □

Proposition 12.69 *Let $m \in \mathbb{N}$, $\mathbb{R}^m$ be endowed with the usual positive cone, $\Omega \in \mathcal{B}_{\mathrm{B}}(\mathbb{R}^m)$ be a rectangle, $\mathcal{Y}$ be a normed linear space over $\mathbb{K}$, $\mathcal{Z}$ be a normed linear space over $\mathbb{K}$, $F : \Omega \to \mathcal{Y}$ be absolutely continuous at $x_0 \in \Omega$, and $A \in \mathrm{B}(\mathcal{Y}, \mathcal{Z})$. Then, $h : \Omega \to \mathcal{Z}$, defined by $h(x) = Af(x)$, $\forall x \in \Omega$, is absolutely continuous at x_0.*

Proof We will prove this result using mathematical induction on m:

1° $m = 1$. Clearly, the result follows from Proposition 12.68.

2° Assume that the result holds for $m \leq k \in \mathbb{N}$.

3° Consider the case when $m = k + 1$. We will prove that h is absolutely continuous at x_0 by Definition 12.58. Fix any $i_0 \in \bar{J} := \{1, \ldots, m\}$, and let D_{i_0}, $\pi_{\widehat{i_0}}$, and F_{i_0} be defined as in (i) of Definition 12.58. Let $h_{i_0} : D_{i_0} \to \mathcal{Z}$ be defined by $h_{i_0}(x) = h(\bar{x}) = AF(\bar{x}) = AF_{i_0}(x)$, $\forall x \in D_{i_0} \subseteq \mathbb{R}^k$, $\forall \bar{x} \in \Omega$ with $\pi_{i_0}(\bar{x}) = \pi_{i_0}(x_0)$ and $\pi_{\widehat{i_0}}(\bar{x}) = x$. By the assumption, F is absolutely continuous at x_0, and then F_{i_0} is absolutely continuous at $\pi_{\widehat{i_0}}(x_0)$. By inductive assumption, h_{i_0} is absolutely continuous at $\pi_{\widehat{i_0}}(x_0)$. Thus, (i) of Definition 12.58 holds.

Since F is absolutely continuous at x_0, by (ii) of Definition 12.58, $\exists a \in (0, \infty) \subset \mathbb{R}$, $\forall \epsilon \in (0, \infty) \subset \mathbb{R}$, $\exists \delta(\epsilon) \in \mathbb{R}_+$, $\forall n \in \mathbb{N}$, $\forall (\bar{x}_i)_{i=1}^n, (\tilde{x}_i)_{i=1}^n \subseteq \bar{\mathrm{r}}_{x_0 - a\mathbf{1}_m, x_0 + a\mathbf{1}_m} \cap \Omega$ with $\bar{x}_i \lneqq \tilde{x}_i$, $\forall i \in \{1, \ldots, n\}$, $(\mathrm{r}_{\bar{x}_i, \tilde{x}_i})_{i=1}^n$ being pairwise disjoint, and $\sum_{i=1}^n \mu_{\mathrm{B}m}(\mathrm{r}_{\bar{x}_i, \tilde{x}_i}) < \delta(\epsilon)$, we have $\sum_{i=1}^n \|\Delta_F(\mathrm{r}_{\bar{x}_i, \tilde{x}_i})\| < \frac{\epsilon}{1+\|A\|}$. It is easy to see that $\Delta_h(\mathrm{r}_{\bar{x}_i, \tilde{x}_i}) = A\Delta_F(\mathrm{r}_{\bar{x}_i, \tilde{x}_i})$, $\forall i \in \{1, \ldots, n\}$. Then, $\sum_{i=1}^n \|\Delta_h(\mathrm{r}_{\bar{x}_i, \tilde{x}_i})\| \leq \|A\| \sum_{i=1}^n \left\|\Delta_F(\mathrm{r}_{\bar{x}_i, \tilde{x}_i})\right\| < \epsilon$. Hence, (ii) of Definition 12.58 holds.

Thus, h is absolutely continuous at x_0.

This completes the induction process and therefore the proof of the proposition. □

Proposition 12.70 *Let $\mathcal{X}$ be a normed linear space over $\mathbb{K}_1$, Y be a set, $\mathcal{Z}$ be a normed linear space over $\mathbb{K}_2$, $U \subseteq \mathcal{X}$ be endowed with subset topology $\mathcal{O}$, $f : U \times Y \to \mathcal{Z}$, $g : U \to \mathbb{R}$ be continuous, $(c_i)_{i \in \mathbb{Z}} \subset \mathbb{R}$ be a sequence of strictly increasing real numbers with $\lim_{i \to -\infty} c_i = -\infty$ and $\lim_{i \to \infty} c_i = \infty$, and $U_i := \{x \in U \mid c_i \leq g(x) \leq c_{i+1}\}$, $\forall i \in \mathbb{Z}$. Assume that:*

(i) *U is locally convex according to Definition 7.124.*
(ii) *$f|_{U_i \times Y}$ is locally Lipschitz continuous on U_i uniformly over Y, $\forall i \in \mathbb{Z}$.*

Then, f is locally Lipschitz continuous on U uniformly over Y.

Proof Fix any $x_0 \in U = \bigcup_{i \in \mathbb{Z}} U_i$. Then, $g(x_0) \in \mathbb{R}$. We will distinguish two exhaustive and mutually exclusives cases: Case 1: $\exists i \in \mathbb{Z}$ such that $c_i < g(x_0) < c_{i+1}$; Case 2: $\exists i \in \mathbb{Z}$ such that $g(x_0) = c_i$. Case 1: $\exists i \in \mathbb{Z}$ such that $c_i < g(x_0) < c_{i+1}$. Then, $x_0 \in U_i$. Since g is continuous, then $\exists \delta_1 \in (0, \infty) \subset \mathbb{R}$ such that $\mathcal{B}_{\mathcal{X}}(x_0, \delta_1) \cap U \subseteq U_i$. By $f|_{U_i \times Y}$ being locally Lipschitz continuous on U_i uniformly over Y, we have $f|_{U_i \times Y}$ is locally Lipschitz continuous at x_0 uniformly over Y. By Definition 12.64, $\exists L_1 \in [0, \infty) \subset \mathbb{R}$, $\exists \delta \in (0, \delta_1] \subset \mathbb{R}$, $\forall x_1, x_2 \in \mathcal{B}_{\mathcal{X}}(x_0, \delta) \cap U_i = \mathcal{B}_{\mathcal{X}}(x_0, \delta) \cap U$, $\forall y \in Y$, we have $\|f(x_1, y) - f(x_2, y)\| \leq L_1 \|x_1 - x_2\|$. Hence, f is locally Lipschitz continuous at x_0 uniformly over Y.

Case 2: $\exists i \in \mathbb{Z}$ such that $g(x_0) = c_i$. Then, $x_0 \in U_{i-1} \cap U_i$. By (i) and the continuity of g, $\exists \delta_{x_0} \in (0, \infty) \subset \mathbb{R}$ such that $U \cap \mathcal{B}_{\mathcal{X}}(x_0, \delta_{x_0}) \subseteq U_{i-1} \cup U_i$ is convex. By $f|_{U_j \times Y}$ being locally Lipschitz continuous on U_j uniformly over Y, we have $f|_{U_j \times Y}$ is locally Lipschitz continuous at x_0 uniformly over Y, $j = i-1, i$. By Definition 12.64, $\exists L_j \in [0, \infty) \subset \mathbb{R}$, $\exists \delta_j \in (0, \delta_{x_0}] \subset \mathbb{R}$, $\forall x_1, x_2 \in \mathcal{B}_{\mathcal{X}}(x_0, \delta_j) \cap U_j$, $\forall y \in Y$, we have $\|f(x_1, y) - f(x_2, y)\| \leq L_j \|x_1 - x_2\|$, $j = i - 1, i$. Take $L := L_{i-1} \vee L_i \in [0, \infty) \subset \mathbb{R}$ and $\delta := \delta_{i-1} \wedge \delta_i \in (0, \delta_{x_0}] \subset \mathbb{R}$. $\forall x_1, x_2 \in \mathcal{B}_{\mathcal{X}}(x_0, \delta) \cap U =: D$, $\forall y \in Y$, by Proposition 6.40, D is convex. If $\exists j \in \{i-1, i\}$ such that $x_1, x_2 \in U_j$, then $\|f(x_1, y) - f(x_2, y)\| \leq L_j \|x_1 - x_2\| \leq L\|x_1 - x_2\|$. On the other hand, without loss of generality, assume that $x_1 \in U_{i-1}$ and $x_2 \in U_i$. Consider the line segment connecting x_1 and x_2.

Claim 12.70.1 $\exists t_0 \in [0, 1] \subset \mathbb{R}$ such that $x_{t_0} := t_0 x_1 + (1 - t_0) x_2 \in U_{i-1} \cap U_i \cap D$.

Proof of Claim Since D is convex, then $x_t := tx_1 + (1-t)x_2 \in D \subseteq U_{i-1} \cup U_i$, $\forall t \in [0, 1] \subset \mathbb{R}$. Define

$$t_0 := \sup\{t \in [0, 1] \subset \mathbb{R} \mid x_t \in U_i\} =: \sup A$$

Clearly, $0 \in A$ and $0 \leq t_0 \leq 1$. Suppose $t_0 \notin A$. This leads to $t_0 \in (0, 1] \subset \mathbb{R}$. Then, $\exists (t_n)_{n=1}^{\infty} \subseteq [0, t_0) \cap A$ such that $\lim_{n\in\mathbb{N}} t_n = t_0$. $\forall n \in \mathbb{N}$, by $t_n \in A$, we have $g(x_{t_n}) \geq c_i$. Then, $g(x_{t_0}) = \lim_{n\in\mathbb{N}} g(x_{t_n}) \geq c_i$, $x_{t_0} \in U_i$, and $t_0 \in A$. This is a contradiction. Hence, $t_0 \in A$ and $x_{t_0} \in U_2$. If $t_0 = 1$, then $x_{t_0} = x_1 \in U_{i-1}$ and $x_{t_0} \in U_{i-1} \cap U_i$. Hence, the claim holds. On the other hand, if $t_0 < 1$, then $\forall t \in (t_0, 1] \subset \mathbb{R}$, $t \notin A$ and $x_t \in U_{i-1}$. This implies that $g(x_t) \leq c_i$, $\forall t \in (t_0, 1] \subset \mathbb{R}$. Then, $g(x_{t_0}) = \lim_{t\to t_0^+} g(x_t) \leq c_i$ and $x_{t_0} \in U_{i-1}$. Hence, $x_{t_0} \in U_{i-1} \cap U_i$. The claim holds as well. This completes the proof of the claim. □

By Claim 12.70.1, $\|f(x_1, y) - f(x_2, y)\| \leq \|f(x_1, y) - f(x_{t_0}, y)\| + \|f(x_{t_0}, y) - f(x_2, y)\| \leq L_{i-1}\|x_1 - x_{t_0}\| + L_i\|x_2 - x_{t_0}\| = L_{i-1}(1 - t_0)\|x_1 - x_2\| + L_i t_0\|x_1 - x_2\| \leq L\|x_1 - x_2\|$.

Hence, we have that $\|f(x_1, y) - f(x_2, y)\| \leq L\|x_1 - x_2\|$ in either situations. Then, f is locally Lipschitz continuous at x_0 uniformly over Y.

In both cases, we have that f is locally Lipschitz continuous at x_0 uniformly over Y. By the arbitrariness of x_0, f is locally Lipschitz continuous on U uniformly over Y. This completes the proof of the proposition. □

Definition 12.71 Let $a, b \in \mathbb{R}$, $I := r_{a\wedge b, a\vee b} \subset \mathbb{R}$ be the semi-open interval with a and b as end points, $\mathbb{I} := ((I, |\cdot|), \mathcal{B}, \mu)$ be the finite metric measure subspace of $\mathbb{R}$, $\mathcal{Y}$ be a normed linear space, and $f : I \to \mathcal{Y}$ be $\mathcal{B}$-measurable. We will write

$$\int_a^b f(x)\,dx := \int_a^b f\,d\mu_B := \begin{cases} \int_I f\,d\mu_B & \text{if } b \geq a \\ -\int_I f\,d\mu_B & \text{if } b < a \end{cases}$$

whenever the right-hand side makes sense. ◇

By Lemma 11.73, the interval I may be $\bar{r}_{a\wedge b, a\vee b}$, or $r_{a\wedge b, a\vee b}$, or $r^{\circ}_{a\wedge b, a\vee b}$, which will not change the value of the integral.

Fact 12.72 *Let $a, b, c \in \mathbb{R}$, $I := r_{a\wedge b\wedge c, a\vee b\vee c} \subset \mathbb{R}$, $\mathbb{I} := ((I, |\cdot|), \mathcal{B}, \mu)$ be the finite metric measure subspace of $\mathbb{R}$, $\mathcal{Y}$ be a separable Banach space, and $f : I \to \mathcal{Y}$ be absolutely integrable over $\mathbb{I}$. Then, $\int_a^c f(x)\,dx + \int_c^b f(x)\,dx = \int_a^b f(x)\,dx \in \mathcal{Y}$.*

Proof We will distinguish six exhaustive and mutually exclusives cases: Case 1: $a \leq c \leq b$; Case 2: $a \leq b \leq c$; Case 3: $b \leq a \leq c$; Case 4: $b \leq c \leq a$; Case 5: $c \leq a \leq b$; Case 6: $c \leq b \leq a$. Case 1: $a \leq c \leq b$. Let $J_1 := r_{a,c} \subset \mathbb{R}$, $J_2 := r_{c,b} \subset \mathbb{R}$. By Proposition 11.92, $\int_a^b f(x)\,dx = \int_I f\,d\mu_B = \int_{J_1} f\,d\mu_B + \int_{J_2} f\,d\mu_B = \int_a^c f(x)\,dx + \int_c^b f(x)\,dx \in \mathcal{Y}$. Then, $\int_a^c f(x)\,dx + \int_c^b f(x)\,dx = \int_a^b f(x)\,dx \in \mathcal{Y}$.

Case 2: $a \leq b \leq c$. By Case 1, $\int_a^c f(x)\,dx = \int_a^b f(x)\,dx + \int_b^c f(x)\,dx$. Then, $\int_a^b f(x)\,dx = \int_a^c f(x)\,dx - \int_b^c f(x)\,dx = \int_a^c f(x)\,dx + \int_c^b f(x)\,dx$.

Case 3: $b \leq a \leq c$. By Case 1, $\int_b^c f(x)\,dx = \int_b^a f(x)\,dx + \int_a^c f(x)\,dx$. Then, $\int_a^b f(x)\,dx = -\int_b^a f(x)\,dx = \int_a^c f(x)\,dx - \int_b^c f(x)\,dx = \int_a^c f(x)\,dx + \int_c^b f(x)\,dx$.

Case 4: $b \leq c \leq a$. By Case 1, $\int_b^a f(x)\,dx = \int_b^c f(x)\,dx + \int_c^a f(x)\,dx$. Then, $\int_a^b f(x)\,dx = -\int_b^a f(x)\,dx = -\int_b^c f(x)\,dx - \int_c^a f(x)\,dx = \int_a^c f(x)\,dx + \int_c^b f(x)\,dx$.

Case 5: $c \leq a \leq b$. By Case 1, $\int_c^b f(x)\,dx = \int_c^a f(x)\,dx + \int_a^b f(x)\,dx$. Then, $\int_a^b f(x)\,dx = \int_c^b f(x)\,dx - \int_c^a f(x)\,dx = \int_a^c f(x)\,dx + \int_c^b f(x)\,dx$.

Case 6: $c \leq b \leq a$. By Case 1, $\int_c^a f(x)\,dx = \int_c^b f(x)\,dx + \int_b^a f(x)\,dx$. Then, $\int_a^b f(x)\,dx = -\int_b^a f(x)\,dx = \int_c^b f(x)\,dx - \int_c^a f(x)\,dx = \int_a^c f(x)\,dx + \int_c^b f(x)\,dx$.

This completes the proof of the fact. □

Proposition 12.73 *Let $m \in \mathbb{N}$, $\mathbb{R}^m$ be endowed with the usual positive cone, $\Omega \in \mathcal{B}_B(\mathbb{R}^m)$ be a rectangle with the subset topology $\mathcal{O}$, $((\mathrm{P}(\Omega), |\cdot|), \mathcal{B}, \mu)$ be the σ-finite metric measure subspace of $\mathbb{R}^m$, $\mathcal{Y}$ be a normed linear space over $\mathbb{K}$, and $F : \Omega \to \mathcal{Y}$ be absolutely continuous. Then, the following statements hold:*

(i) *F is of locally bounded variation.*
(ii) *If, in addition, $\mathcal{Y}$ is a Banach space over $\mathbb{K}$, then there exists a unique σ-finite $\mathcal{Y}$-valued measure space $(\mathrm{P}(\Omega), \mathcal{B}, \nu)$ such that ν admits F as a cumulative distribution function. Furthermore, $\mathcal{P} \circ \nu(\mathrm{r}_{x_1,x_2}) = T_F(\mathrm{r}_{x_1,x_2})$, $\forall x_1, x_2 \in \Omega$ with $x_1 \stackrel{<}{=} x_2$, and $\mathcal{P} \circ \nu(\mathrm{P}(\Omega)) = T_F$ and $\mathcal{P} \circ \nu \ll \mu$.*

Proof (i) By Proposition 12.52, Ω is a region and $\mathrm{P}(\Omega) \in \mathcal{B}_B(\mathbb{R}^m)$ is a rectangle. By Proposition 12.59, F is continuous. Then, (i) of Definition 12.41 holds.

$\forall x_1, x_2 \in \Omega$ with $x_1 \stackrel{<}{=} x_2$, consider the closed rectangle $\bar{\mathrm{r}}_{x_1,x_2} \subseteq \Omega$. $\forall x_0 \in \bar{\mathrm{r}}_{x_1,x_2}$, F is absolutely continuous at x_0. By (ii) of Definition 12.58, $\exists a_{x_0} \in (0, \infty) \subset \mathbb{R}$, $\forall \epsilon \in (0, \infty) \subset \mathbb{R}$, $\exists \delta_{x_0}(\epsilon) \in (0, \infty) \subset \mathbb{R}$, $\forall n \in \mathbb{N}$, $\forall (\bar{x}_i)_{i=1}^n, (\tilde{x}_i)_{i=1}^n \subseteq \bar{\mathrm{r}}_{x_0 - a_{x_0}\mathbf{1}_m, x_0 + a_{x_0}\mathbf{1}_m} \cap \Omega$ with $\bar{x}_i \stackrel{<}{=} \tilde{x}_i$, $\forall i \in \{1, \ldots, n\}$, $(\mathrm{r}_{\bar{x}_i,\tilde{x}_i})_{i=1}^n$ being pairwise disjoint, and $\sum_{i=1}^n \mu_{Bm}(\mathrm{r}_{\bar{x}_i,\tilde{x}_i}) < \delta_{x_0}(\epsilon)$, we have $\sum_{i=1}^n \|\Delta_F(\mathrm{r}_{\bar{x}_i,\tilde{x}_i})\| < \epsilon$. Then, $\bar{\mathrm{r}}_{x_1,x_2} \subseteq \bigcup_{x \in \bar{\mathrm{r}}_{x_1,x_2}} \mathrm{r}^\circ_{x - a_x\mathbf{1}_m, x + a_x\mathbf{1}_m}$. By the compactness of $\bar{\mathrm{r}}_{x_1,x_2}$, $\exists$ a finite set $N \subseteq \bar{\mathrm{r}}_{x_1,x_2}$ such that $\mathrm{r}_{x_1,x_2} \subseteq \bigcup_{x \in N} \mathrm{r}^\circ_{x - a_x\mathbf{1}_m, x + a_x\mathbf{1}_m}$. Let $\epsilon_0 := 1 \in \mathbb{R}$, $a := \max_{x \in N} a_x \in (0, \infty) \subset \mathbb{R}$, and $\delta := \min_{x \in N} \delta_x(\epsilon_0) \in (0, \infty) \subset \mathbb{R}$. $\forall n \in \mathbb{Z}_+$, $\forall (\check{x}_i)_{i=1}^n, (\hat{x}_i)_{i=1}^n \subseteq \bar{\mathrm{r}}_{x_1,x_2}$ with $\check{x}_i \stackrel{<}{=} \hat{x}_i$, $\forall i \in \{1, \ldots, n\}$, $\bigcup_{i=1}^n \mathrm{r}_{\check{x}_i,\hat{x}_i} = \mathrm{r}_{x_1,x_2}$, and sets in the union being pairwise disjoint, we have, by Proposition 12.46,

$$\sum_{i=1}^n \left\|\Delta_F(\mathrm{r}_{\check{x}_i,\hat{x}_i})\right\| = \sum_{i=1}^n \left\| \sum_{x \in N} \sum_{j=1}^{n_x} \Delta_F(\mathrm{r}_{\check{x}_{i,x,j},\hat{x}_{i,x,j}}) \right\|$$
$$\leq \sum_{x \in N} \sum_{i=1}^n \sum_{j=1}^{n_x} \left\|\Delta_F(\mathrm{r}_{\check{x}_{i,x,j},\hat{x}_{i,x,j}})\right\|$$

where $n_x \in \mathbb{Z}_+$, $\forall x \in N$, $(\check{x}_{i,x,j})_{x\in N,\ j\in\{1,\ldots,n_x\}}, (\hat{x}_{i,x,j})_{x\in N,\ j\in\{1,\ldots,n_x\}} \subseteq \bar{\mathrm{r}}_{\check{x}_i,\hat{x}_i}$, $\check{x}_{i,x,j} \stackrel{\lessdot}{=} \hat{x}_{i,x,j}$, $\forall x \in N$, $\forall j \in \{1,\ldots,n_x\}$, $\bigcup_{x\in N}\bigcup_{j=1}^{n_x} \mathrm{r}_{\check{x}_{i,x,j},\hat{x}_{i,x,j}} = \mathrm{r}_{\check{x}_i,\hat{x}_i}$, the sets in the union are pairwise disjoint, and $\bigcup_{j=1}^{n_x} \mathrm{r}_{\check{x}_{i,x,j},\hat{x}_{i,x,j}} \subseteq \mathrm{r}_{x-a_x\mathbf{1}_m,x+a_x\mathbf{1}_m}$, $\forall x \in N$, $i = 1,\ldots,n$. $\forall x \in N$, by further partition each $\mathrm{r}_{\check{x}_{i,x,j},\hat{x}_{i,x,j}}$ into $\bar{n}_x$, where $\bar{n}_x = 1 + \lfloor 2^m a_x^m/\delta_x(\epsilon_0)\rfloor$, equal measure and pairwise disjoint rectangles $\mathrm{r}_{\check{x}_{i,x,j},\hat{x}_{i,x,j}} = \bigcup_{k=1}^{\bar{n}_x} \mathrm{r}_{\check{x}_{i,x,j,k},\hat{x}_{i,x,j,k}}$, we have

$$\begin{aligned}\sum_{i=1}^{n}\left\|\Delta_F(\mathrm{r}_{\check{x}_i,\hat{x}_i})\right\| &\le \sum_{x\in N}\sum_{i=1}^{n}\sum_{j=1}^{n_x}\left\|\Delta_F(\mathrm{r}_{\check{x}_{i,x,j},\hat{x}_{i,x,j}})\right\| \\ &= \sum_{x\in N}\sum_{i=1}^{n}\sum_{j=1}^{n_x}\left\|\sum_{k=1}^{\bar{n}_x}\Delta_F(\mathrm{r}_{\check{x}_{i,x,j,k},\hat{x}_{i,x,j,k}})\right\| \\ &\le \sum_{x\in N}\sum_{i=1}^{n}\sum_{j=1}^{n_x}\sum_{k=1}^{\bar{n}_x}\left\|\Delta_F(\mathrm{r}_{\check{x}_{i,x,j,k},\hat{x}_{i,x,j,k}})\right\| \\ &= \sum_{x\in N}\sum_{k=1}^{\bar{n}_x}\sum_{i=1}^{n}\sum_{j=1}^{n_x}\left\|\Delta_F(\mathrm{r}_{\check{x}_{i,x,j,k},\hat{x}_{i,x,j,k}})\right\|\end{aligned}$$

Note that, $\forall k = 1,\ldots,\bar{n}_x$,

$$\sum_{i=1}^{n}\sum_{j=1}^{n_x}\mu_{\mathrm{B}m}(\mathrm{r}_{\check{x}_{i,x,j,k},\hat{x}_{i,x,j,k}}) < \frac{1}{\bar{n}_x}\sum_{i=1}^{n}\sum_{j=1}^{n_x}\mu_{\mathrm{B}m}(\mathrm{r}_{\check{x}_{i,x,j},\hat{x}_{i,x,j}}) < \delta_x(\epsilon_0)$$

Then, $\sum_{i=1}^{n}\sum_{j=1}^{n_x}\|\Delta_F(\mathrm{r}_{\check{x}_{i,x,j,k},\hat{x}_{i,x,j,k}})\| < 1$. This implies that $\sum_{i=1}^{n}\|\Delta_F(\mathrm{r}_{\check{x}_i,\hat{x}_i})\| \le \sum_{x\in N}\sum_{k=1}^{\bar{n}_x}\sum_{i=1}^{n}\sum_{j=1}^{n_x}\|\Delta_F(\mathrm{r}_{\check{x}_{i,x,j,k},\hat{x}_{i,x,j,k}})\| < \sum_{x\in N}\sum_{k=1}^{\bar{n}_x} 1 = \mathrm{card}(N)(1 + \lfloor 2^m a^m/\delta\rfloor)$. Hence, $T_F(\mathrm{r}_{x_1,x_2}) \le \mathrm{card}(N)(1+\lfloor 2^m a^m/\delta\rfloor) < \infty$. Hence, F satisfies (ii) of Definition 12.41.

$\forall x_1, x_2 \in \Omega$ with $x_1 \lessdot x_2$, define $\bar{G} : \bar{\mathrm{r}}_{x_1,x_2} \to [0,\infty) \subset \mathbb{R}$ by $\bar{G}(x) = T_F(\mathrm{r}_{x_1,x})$, $\forall x \in \bar{\mathrm{r}}_{x_1,x_2}$. We will show that $\bar{G}$ is continuous on the right. Fix any $x_0 \in \bar{\mathrm{r}}_{x_1,x_2}$. Consider the sets $S_J := \{x \in \bar{\mathrm{r}}_{x_1,x_2} \mid \pi_i(x) = \pi_i(x_0), \forall i \in J\}$, $\forall J \subseteq \bar{J} := \{1,\ldots,m\}$ with $J \neq \emptyset$. Fix any $J \subseteq \bar{J}$ with $J \neq \emptyset$. Clearly, S_J is compact. $\forall x \in S_J$, by (ii) of Definition 12.58, $\exists a_x \in (0,\infty) \subset \mathbb{R}$, $\forall \epsilon \in (0,\infty) \subset \mathbb{R}$, $\exists \delta_x(\epsilon) \in (0,\infty) \subset \mathbb{R}$, $\forall n \in \mathbb{N}$, $\forall (\bar{x}_i)_{i=1}^n, (\tilde{x}_i)_{i=1}^n \subseteq \bar{\mathrm{r}}_{x-a_x\mathbf{1}_m,x+a_x\mathbf{1}_m} \cap \bar{\mathrm{r}}_{x_1,x_2}$ with $\bar{x}_i \stackrel{\lessdot}{=} \tilde{x}_i$, $\forall i \in \{1,\ldots,n\}$, $\left(\mathrm{r}_{\bar{x}_i,\tilde{x}_i}\right)_{i=1}^n$ being pairwise disjoint, and $\sum_{i=1}^{n}\mu_{\mathrm{B}m}(\mathrm{r}_{\bar{x}_i,\tilde{x}_i}) < \delta_x(\epsilon)$, we have $\sum_{i=1}^{n}\left\|\Delta_F(\mathrm{r}_{\bar{x}_i,\tilde{x}_i})\right\| < \epsilon$. Then, $S_J \subseteq \bigcup_{x\in S_J}\mathrm{r}^{\circ}_{x-a_x\mathbf{1}_m,x+a_x\mathbf{1}_m}$. By the compactness of S_J, $\exists$ a finite set $N_J \subseteq S_J$ such that $S_J \subseteq \bigcup_{x\in N_J}\mathrm{r}^{\circ}_{x-a_x\mathbf{1}_m,x+a_x\mathbf{1}_m}$. Let $N := \bigcup_{J\subseteq\bar{J},\ J\neq\emptyset} N_J$. Let $\bar{a} := \max_{x\in N} a_x \in$

$(0, \infty) \subset \mathbb{R}$ and $\tilde{a} := \min_{x \in N} a_x \in (0, \infty) \subset \mathbb{R}$. $\forall \epsilon \in (0, \infty) \subset \mathbb{R}$, let $\delta := \min_{x \in N} \delta_x(\epsilon/(1 + \mathrm{card}(N))) \in (0, \infty) \subset \mathbb{R}$. Let $\tilde{\delta}$ be the minimum positive real root of $\prod_{i=1}^{m}(\pi_i(x_0) - \pi_i(x_1) + \tilde{\delta}) - \prod_{i=1}^{m}(\pi_i(x_0) - \pi_i(x_1)) = \delta/2$ and $\bar{\delta} := \tilde{\delta} \wedge \tilde{a} \in (0, \infty) \subset \mathbb{R}$. $\forall \bar{x} \in \mathcal{B}_{\mathbb{R}^m}(x_0, \bar{\delta}) \cap \bar{\mathrm{r}}_{x_1,x_2}$ with $x_0 \stackrel{\leq}{=} \bar{x}$, we have $|\bar{G}(\bar{x}) - \bar{G}(x_0)| = |T_F(\mathrm{r}_{x_1,\bar{x}}) - T_F(\mathrm{r}_{x_1,x_0})| = \sum_{J \subseteq \bar{J},\, J \neq \emptyset} T_F(\mathrm{r}_{\check{x}_J,\hat{x}_J})$, where $\check{x}_J \in \bar{\mathrm{r}}_{x_1,x_0}$ is such that $\pi_i(\check{x}_J) = \pi_i(x_0)$, $\forall i \in J$, $\pi_i(\check{x}_J) = \pi_i(x_1)$, $\forall i \in \bar{J} \setminus J$; and $\hat{x}_J \in \bar{\mathrm{r}}_{x_0,\bar{x}}$ is such that $\pi_i(\hat{x}_J) = \pi_i(\bar{x})$, $\forall i \in J$, $\pi_i(\hat{x}_J) = \pi_i(x_0)$, $\forall i \in \bar{J} \setminus J$; and the second equality follows from Proposition 12.47. $\forall J \subseteq \bar{J}$ with $J \neq \emptyset$, $\mathrm{r}_{\check{x}_J,\hat{x}_J} \subseteq \bigcup_{x \in N_J} \mathrm{r}^{\circ}_{x-a_x\mathbf{1}_m,x+a_x\mathbf{1}_m}$ implies that $\mathrm{r}_{\check{x}_J,\hat{x}_J} = \bigcup_{x \in N_J}(\mathrm{r}^{\circ}_{x-a_x\mathbf{1}_m,x+a_x\mathbf{1}_m} \cap \mathrm{r}_{\check{x}_J,\hat{x}_J})$. Then, we have $|\bar{G}(\bar{x}) - \bar{G}(x_0)| \leq \sum_{x \in N} \sum_{J \subseteq \bar{J},\, J \neq \emptyset} T_F(\mathrm{r}_{\check{x}_J,\hat{x}_J} \cap \mathrm{r}_{x-a_x\mathbf{1}_m,x+a_x\mathbf{1}_m})$. Note that, $\forall x \in N$, $\sum_{J \subseteq \bar{J},\, J \neq \emptyset} \mu_{\mathrm{B}m}(\mathrm{r}_{\check{x}_J,\hat{x}_J} \cap \mathrm{r}_{x-a_x\mathbf{1}_m,x+a_x\mathbf{1}_m}) \leq \sum_{J \subseteq \bar{J},\, J \neq \emptyset} \mu_{\mathrm{B}m}(\mathrm{r}_{\check{x}_J,\hat{x}_J}) \leq \prod_{i=1}^{m}(\pi_i(x_0) - \pi_i(x_1) + \bar{\delta}) - \prod_{i=1}^{m}(\pi_i(x_0) - \pi_i(x_1)) \leq \delta/2 < \delta$. Then, $\sum_{J \subseteq \bar{J},\, J \neq \emptyset} T_F(\mathrm{r}_{\check{x}_J,\hat{x}_J} \cap \mathrm{r}_{x-a_x\mathbf{1}_m,x+a_x\mathbf{1}_m}) < \epsilon/(1 + \mathrm{card}(N))$. Hence, $|\bar{G}(\bar{x}) - \bar{G}(x_0)| < \sum_{x \in N} \epsilon/(1 + \mathrm{card}(N)) < \epsilon$. This shows that $\bar{G}$ is continuous on the right at x_0. By the arbitrariness of x_0, we have $\bar{G}$ is continuous on the right. Hence, (iii) of Definition 12.41 holds. Therefore, F is of locally bounded variation. This completes the proof of (i).

(ii) By Theorem 12.50, there exists a unique σ-finite $\mathcal{Y}$-valued measure space $(\mathrm{P}(\Omega), \mathcal{B}, \nu)$ such that F is a cumulative distribution function of ν, $\mathcal{P} \circ \nu(\mathrm{r}_{x_1,x_2}) = T_F(\mathrm{r}_{x_1,x_2})$, $\forall x_1, x_2 \in \Omega$ with $x_1 \stackrel{\leq}{=} x_2$, and $\mathcal{P} \circ \nu(\mathrm{P}(\Omega)) = T_F$.

All we need to show is $\mathcal{P} \circ \nu \ll \mu$. First, we consider the special case $\Omega = \bar{\mathrm{r}}_{x_1,x_2} \subseteq \mathbb{R}^m$, where $x_1, x_2 \in \mathbb{R}^m$ with $x_1 \stackrel{\leq}{=} x_2$. $\forall x \in \bar{\mathrm{r}}_{x_1,x_2}$, F being absolutely continuous at x implies that $\exists \lambda_x \in (0, \infty) \subset \mathbb{R}$, $\forall \epsilon \in (0, \infty) \subset \mathbb{R}$, $\exists \delta_x(\epsilon) \in (0, \lambda_x] \subset \mathbb{R}$, $\forall n \in \mathbb{N}$, $\forall (\hat{x}_i)_{i=1}^n, (\bar{x}_i)_{i=1}^n \subseteq \Omega \cap \bar{\mathrm{r}}_{x-\lambda_x\mathbf{1}_m,x+\lambda_x\mathbf{1}_m}$ with $\hat{x}_i \stackrel{\leq}{=} \bar{x}_i$, $\forall i = 1, \ldots, n$, $(\mathrm{r}_{\hat{x}_i,\bar{x}_i})_{i=1}^n$ being pairwise disjoint, and $\sum_{i=1}^{n} \mu(\mathrm{r}_{\hat{x}_i,\bar{x}_i}) < \delta_x(\epsilon)$, we have $\sum_{i=1}^{n} \|\Delta_F(\mathrm{r}_{\hat{x}_i,\bar{x}_i})\| < \epsilon$. Then, $\bar{\mathrm{r}}_{x_1,x_2} \subseteq \bigcup_{x \in \bar{\mathrm{r}}_{x_1,x_2}} \mathrm{r}^{\circ}_{x-\lambda_x\mathbf{1}_m,x+\lambda_x\mathbf{1}_m} \subset \mathbb{R}^m$. By the compactness of $\bar{\mathrm{r}}_{x_1,x_2}$, there exists a finite set $X_N := \{\tilde{x}_1, \ldots, \tilde{x}_p\} \subseteq \bar{\mathrm{r}}_{x_1,x_2}$, where $p \in \mathbb{N}$, such that $\bar{\mathrm{r}}_{x_1,x_2} \subseteq \bigcup_{i=1}^{p} \mathrm{r}^{\circ}_{\tilde{x}_i-\lambda_{\tilde{x}_i}\mathbf{1}_m,\tilde{x}_i+\lambda_{\tilde{x}_i}\mathbf{1}_m} \subset \mathbb{R}^m$.

$\forall E \in \mathcal{B}$ with $\mu(E) = 0$, $\forall \epsilon \in (0, \infty) \subset \mathbb{R}$, let $\delta(\epsilon) := \min_{i \in \{1,\ldots,p\}} \delta_{\tilde{x}_i}(\frac{\epsilon}{2p}) \in (0, \infty) \subset \mathbb{R}$. By Definition 11.26 and Example 12.56, $\exists O \in \mathcal{O}_{\mathbb{R}^m}$ with $E \subseteq O$ such that $\mu_{\mathrm{B}m}(O \setminus E) < \delta(\epsilon)$. Then, $\mu_{\mathrm{B}m}(O) = \mu_{\mathrm{B}m}(O \setminus E) + \mu_{\mathrm{B}m}(E) = \mu_{\mathrm{B}m}(O \setminus E) < \delta(\epsilon)$. By Proposition 12.40, O is a region with $\mathrm{P}(O) = O$. By Definition 12.38, there exists $N \subseteq \mathbb{N}$ and $\exists (\hat{x}_{i,1})_{i \in N}, (\hat{x}_{i,2})_{i \in N} \subseteq O$ such that $\hat{x}_{i,1} \stackrel{\leq}{=} \hat{x}_{i,2}$, $\forall i \in \mathbb{N}$, and $O = \bigcup_{i \in N} \mathrm{r}_{\hat{x}_{i,1},\hat{x}_{i,2}}$ and the sets in the union are pairwise disjoint. Thus, $\mu_{\mathrm{B}m}(O) = \sum_{i \in N} \mu_{\mathrm{B}m}(\mathrm{r}_{\hat{x}_{i,1},\hat{x}_{i,2}}) < \delta(\epsilon)$ and $\mathcal{P} \circ \nu(E) \leq \mathcal{P} \circ \nu(O \cap \mathrm{r}_{x_1,x_2}) = \sum_{i \in N} \mathcal{P} \circ \nu(\mathrm{r}_{\hat{x}_{i,1},\hat{x}_{i,2}} \cap \mathrm{r}_{x_1,x_2})$. By Definition 12.41, for any fixed $i \in N$, define $\mathrm{r}_{\tilde{x}_{i,1},\tilde{x}_{i,2}} := \mathrm{r}_{\hat{x}_{i,1},\hat{x}_{i,2}} \cap \mathrm{r}_{x_1,x_2}$ with $\tilde{x}_{i,j} \in \bar{\mathrm{r}}_{x_1,x_2}$, $j = 1, 2$, and $\tilde{x}_{i,1} \stackrel{\leq}{=} \tilde{x}_{i,2}$, we have $O \cap \mathrm{r}_{x_1,x_2} = \bigcup_{i \in N} \mathrm{r}_{\tilde{x}_{i,1},\tilde{x}_{i,2}}$. By possibly splitting rectangles into smaller disjoint pieces in the collection, without loss of generality, assume that $\forall i \in N$,

$\exists j \in \{1, \ldots, p\}$ such that $\mathrm{r}_{\tilde{x}_{i,1},\tilde{x}_{i,2}} \subseteq \mathrm{r}_{\tilde{x}_j - \lambda_{\tilde{x}_j}\mathbf{1}_m, \tilde{x}_j + \lambda_{\tilde{x}_j}\mathbf{1}_m}$. This implies that $\forall i \in N$,

$$\begin{aligned}\mathcal{P} \circ \nu(\mathrm{r}_{\tilde{x}_{i,1},\tilde{x}_{i,2}}) &= T_F\big(\mathrm{r}_{\tilde{x}_{i,1},\tilde{x}_{i,2}}\big) \\ &= \sup_{\substack{n\in\mathbb{Z}_+,\, \check{x}_{1,1},\check{x}_{1,2},\ldots,\check{x}_{n,1},\check{x}_{n,2}\in\bar{\mathrm{r}}_{\tilde{x}_{i,1},\tilde{x}_{i,2}},\, \check{x}_{j,1}\stackrel{\lessdot}{=}\check{x}_{j,2},\, \forall j\in\{1,\ldots,n\} \\ \bigcup_{j=1}^n \mathrm{r}_{\check{x}_{j,1},\check{x}_{j,2}}=\mathrm{r}_{\tilde{x}_{i,1},\tilde{x}_{i,2}},\, \mathrm{r}_{\check{x}_{k,1},\check{x}_{k,2}}\cap\mathrm{r}_{\check{x}_{j,1},\check{x}_{j,2}}=\emptyset,\, \forall 1\le k<j\le n}} \sum_{j=1}^{n} \left\| \Delta_F(\mathrm{r}_{\check{x}_{j,1},\check{x}_{j,2}}) \right\|\end{aligned}$$

Then, there exists $n_i \in \mathbb{N}$, $(\check{x}_{i,j,1})_{j=1}^{n_i}, (\check{x}_{i,j,2})_{j=1}^{n_i} \subseteq \bar{\mathrm{r}}_{\tilde{x}_{i,1},\tilde{x}_{i,2}}$ such that $\check{x}_{i,j,1} \stackrel{\lessdot}{=} \check{x}_{i,j,2}$, $\forall j \in \{1, \ldots, n_i\}$, $\bigcup_{j=1}^{n_i} \mathrm{r}_{\check{x}_{i,j,1},\check{x}_{i,j,2}} = \mathrm{r}_{\tilde{x}_{i,1},\tilde{x}_{i,2}}$ and the sets in the union are pairwise disjoint, and $\mathcal{P} \circ \nu(\mathrm{r}_{\tilde{x}_{i,1},\tilde{x}_{i,2}}) < \sum_{j=1}^{n_i} \left\| \Delta_F(\mathrm{r}_{\check{x}_{i,j,1},\check{x}_{i,j,2}}) \right\| + 2^{-i-1}\epsilon$. Fix any finite subset $N_f \subseteq N$, and we have $\sum_{i\in N_f} \mathcal{P} \circ \nu(\mathrm{r}_{\hat{x}_{i,1},\hat{x}_{i,2}} \cap \mathrm{r}_{x_1,x_2}) = \sum_{i\in N_f} \mathcal{P} \circ \nu(\mathrm{r}_{\tilde{x}_{i,1},\tilde{x}_{i,2}}) < \sum_{i\in N_f} \sum_{j=1}^{n_i} \|\Delta_F(\mathrm{r}_{\check{x}_{i,j,1},\check{x}_{i,j,2}})\| + \frac{\epsilon}{2} =: \frac{\epsilon}{2} + \sum_{l=1}^{\tilde{n}} \|\Delta_F(\mathrm{r}_{x_{l,1},x_{l,2}})\|$, where $\tilde{n} \in \mathbb{Z}_+$ and $(x_{l,1})_{l=1}^{\tilde{n}}, (x_{l,2})_{l=1}^{\tilde{n}} \subseteq \bar{\mathrm{r}}_{x_1,x_2}$ with $\sum_{i=1}^{\tilde{n}} \mu(\mathrm{r}_{x_{i,1},x_{i,2}}) = \sum_{i=1}^{\tilde{n}} \mu_{\mathrm{B}m}(\mathrm{r}_{x_{i,1},x_{i,2}}) \le \mu_{\mathrm{B}m}(O \cap \mathrm{r}_{x_1,x_2}) < \delta(\epsilon)$. We may rearrange and partition the collection into p segments: $\mathrm{r}_{x_{1,1},x_{1,2}}, \ldots, \mathrm{r}_{x_{i_1,1},x_{i_1,2}} \subseteq \mathrm{r}_{\tilde{x}_1-\lambda_{\tilde{x}_1}\mathbf{1}_m,\tilde{x}_1+\lambda_{\tilde{x}_1}\mathbf{1}_m}$; $\mathrm{r}_{x_{i_1+1,1},x_{i_1+1,2}}, \ldots, \mathrm{r}_{x_{i_2,1},x_{i_2,2}} \subseteq \mathrm{r}_{\tilde{x}_2-\lambda_{\tilde{x}_2}\mathbf{1}_m,\tilde{x}_2+\lambda_{\tilde{x}_2}\mathbf{1}_m}$; $\ldots$; $\mathrm{r}_{x_{i_{p-1}+1,1},x_{i_{p-1}+1,2}}, \ldots, \mathrm{r}_{x_{i_p,1},x_{i_p,2}} \subseteq \mathrm{r}_{\tilde{x}_p-\lambda_{\tilde{x}_p}\mathbf{1}_m,\tilde{x}_p+\lambda_{\tilde{x}_p}\mathbf{1}_m}$. Note that $\forall j \in \{1, \ldots, p\}$, $\sum_{l=i_{j-1}+1}^{i_j} \mu_{\mathrm{B}m}(\mathrm{r}_{x_{l,1},x_{l,2}}) < \delta(\epsilon) \le \delta_{\tilde{x}_j}(\frac{\epsilon}{2p})$, which implies that $\sum_{l=i_{j-1}+1}^{i_j} \|\Delta_F(\mathrm{r}_{x_{l,1},x_{l,2}})\| < \frac{\epsilon}{2p}$. Then, we have $\sum_{i\in N_f} \mathcal{P} \circ \nu(\mathrm{r}_{\hat{x}_{i,1},\hat{x}_{i,2}} \cap \mathrm{r}_{x_1,x_2}) < \epsilon$. By the arbitrariness of N_f, we have $\mathcal{P} \circ \nu(E) \le \mathcal{P} \circ \nu(O \cap \mathrm{r}_{x_1,x_2}) = \sum_{i\in N} \mathcal{P} \circ \nu(\mathrm{r}_{\hat{x}_{i,1},\hat{x}_{i,2}} \cap \mathrm{r}_{x_1,x_2}) = \sum_{i\in N} \mathcal{P} \circ \nu(\mathrm{r}_{\tilde{x}_{i,1},\tilde{x}_{i,2}}) \le \epsilon$. By the arbitrariness of ϵ, we have $\mathcal{P} \circ \nu(E) = 0$. By the arbitrariness of E, we have $\mathcal{P} \circ \nu \ll \mu$. This proves that $\mathcal{P} \circ \nu \ll \mu$ in the special case.

Consider the general case when Ω is a rectangle. By Proposition 12.39, Ω is a region and $\mathrm{P}(\Omega)$ is a rectangle. By Definition 12.38, $\exists N \subseteq \mathbb{N}$ and $(\bar{x}_i)_{i\in N}, (\tilde{x}_i)_{i\in N} \subseteq \Omega$ with $\bar{x}_i \stackrel{\lessdot}{=} \tilde{x}_i$, $\forall i \in N$, such that $\Omega = \bigcup_{i\in N} \bar{\mathrm{r}}_{\bar{x}_i,\tilde{x}_i}$, $\mathrm{P}(\Omega) = \bigcup_{i\in N} \mathrm{r}_{\bar{x}_i,\tilde{x}_i}$, and the sets in the second union are pairwise disjoint. $\forall E \in \mathcal{B}$ with $\mu(E) = 0$. Then, $\mu(E \cap \mathrm{r}_{\bar{x}_i,\tilde{x}_i}) = 0$, $\forall i \in N$. By the special case, we have $\mathcal{P} \circ \nu(E \cap \mathrm{r}_{\bar{x}_i,\tilde{x}_i}) = 0$, $\forall i \in N$. Then, we have $\mathcal{P} \circ \nu(E) = \sum_{i\in N} \mathcal{P} \circ \nu(E \cap \mathrm{r}_{\bar{x}_i,\tilde{x}_i}) = 0$. By the arbitrariness of E, we have $\mathcal{P} \circ \nu \ll \mu$.

This completes the proof of the proposition. □

Lemma 12.74 *Let $\mathcal{X} := (X, \mathcal{B}, \mu)$ be a measure space and ν be a measure on $(X, \mathcal{B})$. The following statements hold:*

(i) *$\nu \ll \mu$ and ν is finite implies that $\forall \epsilon \in (0, \infty) \subset \mathbb{R}$, $\exists \delta \in (0, \infty) \subset \mathbb{R}$, $\forall E \in \mathcal{B}$ with $\mu(E) < \delta$, we have $\nu(E) < \epsilon$.*
(ii) *$\forall \epsilon \in (0, \infty) \subset \mathbb{R}$, $\exists \delta \in (0, \infty) \subset \mathbb{R}$, $\forall E \in \mathcal{B}$ with $\mu(E) < \delta$, we have $\nu(E) < \epsilon$. Then, $\nu \ll \mu$.*

Proof (i) We will prove the result by an argument of contradiction. Suppose the result does not hold. Then, $\exists \epsilon_0 \in (0, \infty) \subset \mathbb{R}$, $\forall n \in \mathbb{N}$, $\exists E_n \in \mathcal{B}$ with $\mu(E_n) < 2^{-n-1}$ such that $\nu(E_n) > \epsilon_0$. Let $\bar{E}_n := \bigcup_{j=n}^{\infty} E_j$, $\forall n \in \mathbb{N}$. Then, $\bar{E}_n \supseteq \bar{E}_{n+1}$ and $\mu(\bar{E}_n) \leq \sum_{j=n}^{\infty} \mu(E_j) < 2^{-n}$. By Proposition 11.5, we have $0 = \lim_{n\in\mathbb{N}} \mu(\bar{E}_n) = \mu(\bigcap_{n=1}^{\infty} \bar{E}_n)$. Then, by $\nu \ll \mu$ and Proposition 11.5, we have $0 = \nu(\bigcap_{n=1}^{\infty} \bar{E}_n) = \lim_{n\in\mathbb{N}} \nu(\bar{E}_n)$. This contradicts with the fact that $\nu(\bar{E}_n) \geq \nu(E_n) > \epsilon_0$, $\forall n \in \mathbb{N}$. Hence, the result holds.

(ii) $\forall E \in \mathcal{B}$ with $\mu(E) = 0$. By the assumption, $\forall \epsilon \in (0, \infty) \subset \mathbb{R}$, $\exists \delta \in (0, \infty) \subset \mathbb{R}$ such that $\forall \bar{E} \in \mathcal{B}$ with $\mu(\bar{E}) < \delta$, we have $\nu(\bar{E}) < \epsilon$. In particular, $\mu(E) = 0 < \delta$, and this implies that $\nu(E) < \epsilon$. By the arbitrariness of ϵ, we have $\nu(E) = 0$. By the arbitrariness of E, we have $\nu \ll \mu$.

This completes the proof of the lemma. □

Proposition 12.75 *Let $m \in \mathbb{N}$, $\mathbb{R}^m$ be endowed with the usual positive cone, $\Omega \in \mathcal{B}_{\mathrm{B}}(\mathbb{R}^m)$ be a rectangle with subset topology $\mathcal{O}$, $\mathcal{Y}$ be a normed linear space over $\mathbb{K}$, ν be a σ-finite $\mathcal{Y}$-valued measure on the measurable space $(\mathrm{P}(\Omega), \mathcal{B}_{\mathrm{B}}(\mathrm{P}(\Omega)))$, and $\mathbb{X} := (\mathrm{P}(\Omega), \mathcal{B}_{\mathrm{B}}(\mathrm{P}(\Omega)), \mu)$ be the metric measure subspace of $\mathbb{R}^m$. Assume that $\mathcal{P} \circ \nu \ll \mu$ and $\mathcal{P} \circ \nu(\mathrm{r}_{x_1,x_2}) < \infty$, $\forall x_1, x_2 \in \Omega$ with $x_1 \stackrel{<}{=} x_2$. By Proposition 12.52, we may define $F : \Omega \to \mathcal{Y}$ to be the cumulative distribution function of ν with origin x_s, where x_s is an arbitrary fixed vector in Ω if $\Omega \neq \emptyset$. Then, F is of locally bounded variation and is absolutely continuous.*

In particular, if $\mathcal{Y}$ be a Banach space over $\mathbb{K}$, $M \subseteq \mathcal{Y}$ be a separable subspace, and $f : \mathrm{P}(\Omega) \to M$ be $\mathcal{B}$-measurable. We may define ν to be the $\mathcal{Y}$-valued measure with kernel f over $\mathbb{X}$ as outlined in Proposition 11.116. Then, f is the Radon–Nikodym derivative of ν with respect to μ. Assume that, $\forall x_1, x_2 \in \Omega$ with $x_1 \stackrel{<}{=} x_2$, f is absolutely integrable on r_{x_1,x_2}. Then, ν satisfies the assumption of the previous paragraph. Let $F : \Omega \to \mathcal{Y}$ be the cumulative distribution function of ν with origin at x_s as defined in Proposition 12.52. Then, F is absolutely continuous.

Proof We will prove that F is absolutely continuous at x_0, $\forall x_0 \in \Omega$, by Proposition 12.60. Fix any $x_0 \in \Omega$, any $i \in \{0, \ldots, m-1\}$, and any $J \subset \bar{J} := \{1, \ldots, m\}$ with $\mathrm{card}(J) = i$. Let $\pi_J : \mathbb{R}^m \to \mathbb{R}^i$, $\pi_{\bar{J}\setminus J} : \mathbb{R}^m \to \mathbb{R}^{m-i}$, $M : \mathbb{R}^i \times \mathbb{R}^{m-i} \to \mathbb{R}^m$, x_f, $\tilde{x}_f$, $\tilde{\Omega}_J$, and $F_J : \tilde{\Omega}_J \to \mathcal{Y}$ be as defined in (a) Proposition 12.60. Define $\tilde{x} := M(\pi_J(x_s), \pi_{\bar{J}\setminus J}(x_0)) \in \Omega$, $E := \mathrm{r}_{x_f \wedge \pi_J(x_s), x_f \vee \pi_J(x_s)}$, and $E_n := \mathrm{r}_{\tilde{x}_f - \mathbf{1}_{m-i}/n, x_f + \mathbf{1}_{m-i}/n} \cap \tilde{\Omega}_J$, $\forall n \in \mathbb{N}$. Clearly, $V_n := M(E \times E_n) \subseteq \Omega$ and $\mu_{\mathrm{B}m}(\mathrm{P}(V_n)) = \mu_{\mathrm{B}m}(V_n) = \mu_{\mathrm{B}i}(E)\mu_{\mathrm{B}(m-i)}(E_n) \leq (2/n)^{m-i}\mu_{\mathrm{B}i}(E) < \infty$, $\forall n \in \mathbb{N}$. Then, $\lim_{n\in\mathbb{N}} \mu_{\mathrm{B}m}(\mathrm{P}(V_n)) = \mu_{\mathrm{B}m}(\bigcap_{n=1}^{\infty} \mathrm{P}(V_n)) = 0$, by Proposition 11.5. Then, by $\mathcal{P} \circ \nu \ll \mu$, we have $\mathcal{P} \circ \nu(\bigcap_{n=1}^{\infty} \mathrm{P}(V_n)) = 0$. Since Ω is a rectangle, then $\exists n_0 \in \mathbb{N}$ such that $\mathrm{P}(V_{n_0}) = \mathrm{r}_{x_1,x_2}$ for some $x_1, x_2 \in \Omega$ with $x_1 \stackrel{<}{=} x_2$. Thus, $\mathcal{P} \circ \nu(\mathrm{P}(V_{n_0})) = \mathcal{P} \circ \nu(\mathrm{r}_{x_1,x_2}) < \infty$. Let $a_J := 1/(2n_0) \in (0, \infty) \subset \mathbb{R}$. $\forall \epsilon \in (0, \infty) \subset \mathbb{R}$, by Lemma 12.74, $\exists \delta(\epsilon) \in (0, \infty) \subset \mathbb{R}$, $\forall E \in \mathcal{B}_{\mathrm{B}}\big(\mathrm{P}(V_{n_0})\big)$ with $\mu(E) = \mu_{\mathrm{B}m}(E) < \delta(\epsilon)$, we have $\mathcal{P} \circ \nu(E) < \epsilon$. Take $\delta_J(\epsilon) := \delta(\epsilon)/(1 + \mu_{\mathrm{B}i}(E)) \in (0, \infty) \subset \mathbb{R}$.

$\forall n \in \mathbb{N}$, $\forall (\bar{x}_j)_{j=1}^n, (\tilde{x}_j)_{j=1}^n \subseteq \bar{\mathrm{r}}_{\bar{x}_f - a_J \mathbf{1}_{m-i}, \bar{x}_f + a_J \mathbf{1}_{m-i}} \cap \tilde{\Omega}_J$ with $\bar{x}_j \stackrel{\leq}{=} \tilde{x}_j$, $\forall j \in \{1, \ldots, n\}$, $(\mathrm{r}_{\bar{x}_j, \tilde{x}_j})_{j=1}^n$ being pairwise disjoint, and $\sum_{j=1}^n \mu_{\mathrm{B}(m-i)}(\mathrm{r}_{\bar{x}_j,\tilde{x}_j}) < \bar{\delta}_J(\epsilon)$, we have $\sum_{j=1}^n \left\| \Delta_{F_J}(\mathrm{r}_{\bar{x}_j,\tilde{x}_j}) \right\| = \sum_{j=1}^n \left\| \Delta_F(M(E \times \mathrm{r}_{\bar{x}_j,\tilde{x}_j})) \right\| = \sum_{j=1}^n \left\| \nu(M(E \times \mathrm{r}_{\bar{x}_j,\tilde{x}_j})) \right\| \leq \sum_{j=1}^n \mathcal{P} \circ \nu(M(E \times \mathrm{r}_{\bar{x}_j,\tilde{x}_j})) = \mathcal{P} \circ \nu(M(E \times (\bigcup_{j=1}^n \mathrm{r}_{\bar{x}_j,\tilde{x}_j})))$. Since $\bigcup_{j=1}^n \mathrm{r}_{\bar{x}_j,\tilde{x}_j} \subseteq E_{n_0}$ with $\mu_{\mathrm{B}(m-i)}(\bigcup_{j=1}^n \mathrm{r}_{\bar{x}_j,\tilde{x}_j}) < \bar{\delta}_J(\epsilon)$, then $\mu_{\mathrm{B}m}(M(E \times (\bigcup_{j=1}^n \mathrm{r}_{\bar{x}_j,\tilde{x}_j}))) = \mu_{\mathrm{B}i}(E) \mu_{\mathrm{B}m-i}(\bigcup_{j=1}^n \mathrm{r}_{\bar{x}_j,\tilde{x}_j}) < \delta(\epsilon)$ and $M(E \times (\bigcup_{j=1}^n \mathrm{r}_{\bar{x}_j,\tilde{x}_j})) = \bigcup_{j=1}^n M(E \times \mathrm{r}_{\bar{x}_j,\tilde{x}_j}) = \bigcup_{j=1}^n \mathrm{P}(M(E \times \mathrm{r}_{\bar{x}_j,\tilde{x}_j})) \subseteq \mathrm{P}(\bigcup_{j=1}^n M(E \times \mathrm{r}_{\bar{x}_j,\tilde{x}_j})) = \mathrm{P}(M(E \times (\bigcup_{j=1}^n \mathrm{r}_{\bar{x}_j,\tilde{x}_j}))) \subseteq \mathrm{P}(M(E \times E_{n_0})) = \mathrm{P}(V_{n_0})$, where the second equality, the first set containment, the third equality, and the second set containment follow from Proposition 12.39. Hence, $\mathcal{P} \circ \nu(M(E \times (\bigcup_{j=1}^n \mathrm{r}_{\bar{x}_j,\tilde{x}_j}))) < \epsilon$. Therefore, $\sum_{j=1}^n \left\| \Delta_{F_J}(\mathrm{r}_{\bar{x}_j,\tilde{x}_j}) \right\| < \epsilon$.

By Proposition 12.60, F is absolutely continuous at x_0. By the arbitrariness of x_0, F is absolutely continuous. By Proposition 12.73, F is of locally bounded variation. This completes the proof of the first paragraph of the proposition.

By Proposition 11.116, $(\mathrm{P}(\Omega), \mathcal{B}_{\mathrm{B}}(\mathrm{P}(\Omega)), \nu)$ is a σ-finite $\mathcal{Y}$-valued measure space and $\mathcal{P} \circ \nu = \bar{\nu}$, where $\bar{\nu}$ is a σ-finite measure on $(\mathrm{P}(\Omega), \mathcal{B}_{\mathrm{B}}(\mathrm{P}(\Omega)))$ defined by $\bar{\nu}(E) = \int_E \mathcal{P} \circ f \, \mathrm{d}\mu \in [0, \infty] \subset \mathbb{R}_e$, $\forall E \in \mathcal{B}_{\mathrm{B}}(\mathrm{P}(\Omega))$. $\forall E \in \mathcal{B}_{\mathrm{B}}(\mathrm{P}(\Omega))$ with $\mu(E) = 0$, we have $\mathcal{P} \circ \nu(E) = \int_E \mathcal{P} \circ f \, \mathrm{d}\mu_{\mathrm{B}m} = 0$, where the last equality follows from Definition 11.79 and Proposition 11.83. Hence, $\mathcal{P} \circ \nu \ll \mu$. By Definition 11.166, f is the Radon–Nikodym derivative of ν with respect to μ. $\forall x_1, x_2 \in \Omega$ with $x_1 \stackrel{\leq}{=} x_2$, we have $\mathrm{r}_{x_1,x_2} \subseteq \mathrm{P}(\Omega)$. By the assumption, we have $\mathcal{P} \circ \nu(\mathrm{r}_{x_1,x_2}) < \infty$. Thus, ν satisfies all the assumption of the first paragraph of the proposition. This implies that F is absolutely continuous. This completes the proof of the proposition. □

The preceding results state that if a function of a rectangle in $\mathbb{R}^m$ into a normed linear space, then the function being absolutely continuous implies it is of locally bounded variation. If the normed linear space is a Banach space, then there exists a unique Banach space valued measure whose cumulative distribution function is the given function, and this measure is absolutely continuous with respect to the restriction of $\mu_{\mathrm{B}m}$ to the principal. This will further lead to the existence of a unique Radon–Nikodym derivative of the Banach space valued measure with respect to the restriction of $\mu_{\mathrm{B}m}$, if the Banach space is separable and reflexive whose dual is also separable. When the range space is only a Banach space, then the Banach space valued measure may or may not admit a Radon–Nikodym derivative with respect to the restriction of $\mu_{\mathrm{B}m}$ to the principal.

Starting from a measurable function of the principal of a rectangle in $\mathbb{R}^m$ to a separable subspace of a Banach space, we may define a σ-finite Banach space valued measure on the principal. The cumulative distribution function of the Banach space valued measure with origin at some x_0 inside the rectangle exists and is absolutely continuous under the mild condition that the measurable function we started with is absolutely integrable on any semi-open rectangles in $\mathbb{R}^m$ with corners inside the rectangle. In this case, we may further conclude that f is the

Radon–Nikodym derivative of the Banach space valued measure with respect to the restriction of $\mu_{\mathrm{B}m}$ to the principal.

12.6 Fundamental Theorem of Calculus

Definition 12.76 Let $\mathcal{X} := (X, \rho)$ be a metric space and $S \subseteq X$. The *diameter* of S is defined to be $\mathrm{dia}(S) = \max\{\sup_{x_1, x_2 \in S} \rho(x_1, x_2), 0\} \in [0, \infty] \subset \mathbb{R}_e$. $\diamond$

Definition 12.77 Let $m \in \mathbb{N}$, $E \subseteq \mathbb{R}^m$, and $\mathcal{I} \subseteq \mathcal{B}_{\mathrm{B}}(\mathbb{R}^m)$ be a collection of nondegenerate rectangles in $\mathbb{R}^m$ (that is, $\exists$ nondegenerate intervals $I_1, \ldots, I_m$ such that $U = \prod_{i=1}^m I_i$, $\forall U \in \mathcal{I}$), i. e., $\mu_{\mathrm{B}m}(U) > 0$, $\forall U \in \mathcal{I}$. We will say that $\mathcal{I}$ *covers* E *in the sense of Vitali with index* $c \in (0, \frac{1}{\sqrt{m}}] \subset \mathbb{R}$ if $\forall \epsilon \in (0, \infty) \subset \mathbb{R}$, $\forall x \in E$, $\exists U \in \mathcal{I}$ such that $x \in U$, $\mathrm{dia}(U) < \epsilon$, and the shortest side of the rectangle U has length at least $c\,\mathrm{dia}(U)$. $\diamond$

Lemma 12.78 (Vitali) *Let $m \in \mathbb{N}$, $\mathbb{R}^m$ be endowed with the usual positive cone, $E \subseteq \mathbb{R}^m$ with $\mu_{\mathrm{L}mo}(E) < +\infty$, and $\mathcal{I} \subseteq \mathcal{B}_{\mathrm{B}}(\mathbb{R}^m)$ be a collection of nondegenerate rectangles in $\mathbb{R}^m$ that covers E in the sense of Vitali with index $c \in (0, \frac{1}{\sqrt{m}}] \subset \mathbb{R}$. Then, $\forall \epsilon \in (0, \infty) \subset \mathbb{R}$, there exists a finite pairwise disjoint subcollection $\{U_1, \ldots, U_n\} \subseteq \mathcal{I}$, where $n \in \mathbb{Z}_+$, such that $\mu_{\mathrm{L}mo}(E \setminus (\bigcup_{i=1}^n U_i)) < \epsilon$.*

Proof Let $\hat{\mathcal{I}} := \{\overline{U} \subseteq \mathbb{R}^m \mid U \in \mathcal{I}\}$. Then, $\hat{\mathcal{I}}$ is a collection of nondegenerate closed rectangles in $\mathbb{R}^m$ that covers E in the sense of Vitali with index c. We will first show that the result holds for $\hat{\mathcal{I}}$. Fix any $\epsilon \in (0, \infty) \subset \mathbb{R}$. By Example 12.56, $\mu_{\mathrm{L}mo}(E) = \inf_{(O_i)_{i=1}^\infty \in \mathcal{O}_{\mathbb{R}^m},\, E \subseteq \bigcup_{i=1}^\infty O_i} \sum_{i=1}^\infty \mu_{\mathrm{B}m}(O_i) < +\infty$. Then, $\exists O \in \mathcal{O}_{\mathbb{R}^m}$ such that $E \subseteq O$ and $\mu_{\mathrm{B}m}(O) < +\infty$. By neglecting rectangles in $\hat{\mathcal{I}}$, we may, without loss of generality, assume that $U \subseteq O$ and the shortest side of U has length at least $c\,\mathrm{dia}(U)$, $\forall U \in \hat{\mathcal{I}}$. $\forall k \in \mathbb{Z}_+$, assume that pairwise disjoint $U_1, \ldots, U_k \in \hat{\mathcal{I}}$ has already be chosen. We will distinguish two exhaustive and mutually exclusive cases: Case 1: $E \subseteq \bigcup_{i=1}^k U_i$; Case 2: $E \setminus (\bigcup_{i=1}^k U_i) \neq \emptyset$. Case 1: $E \subseteq \bigcup_{i=1}^k U_i$. Then, $\mu_{\mathrm{L}mo}(E \setminus (\bigcup_{i=1}^k U_i)) = 0$. The result holds for $\hat{\mathcal{I}}$. Case 2: $E \setminus (\bigcup_{i=1}^k U_i) \neq \emptyset$. Note that, for any closed rectangle $U \in \hat{\mathcal{I}}$ with $\mathrm{dia}(U) =: \bar{p} > 0$, let $x_0 \in U$ be the center of U, such that $\bar{\mathrm{r}}_{x_0 - \frac{1}{2} c p \mathbf{1}_m, x_0 + \frac{1}{2} c \bar{p} \mathbf{1}_m} \subseteq U \subseteq \bar{\mathrm{r}}_{x_0 - \frac{1}{2} \hat{p} \mathbf{1}_m, x_0 + \frac{1}{2} \hat{p} \mathbf{1}_m}$, where $\hat{p} := \sqrt{1 - (m-1)c^2}\,\bar{p}$ is the upper bound for the longest side of the rectangle U. This implies that $c^m \bar{p}^m \leq \mu_{\mathrm{B}m}(U) \leq \hat{p}^m = (1 - (m-1)c^2)^{m/2} \bar{p}^m$. Let l_k be the supremum of the diameters of rectangles in $\hat{\mathcal{I}}$ that do not intersect $U_1, \ldots, U_k$. Clearly, $l_k \leq \frac{\sqrt[m]{\mu_{\mathrm{B}m}(O)}}{c} < +\infty$. Then, $\exists x_0 \in E \setminus (\bigcup_{i=1}^k U_i)$. Since $\bigcup_{i=1}^k U_i$ is closed, then, by Proposition 4.10, $d := \inf_{x \in \bigcup_{i=1}^k U_i} |x - x_0| > 0$. Then, by the assumption, $\exists \hat{U} \in \hat{\mathcal{I}}$ such that $x_0 \in \hat{U}$ and $\mathrm{dia}(\hat{U}) < d$. Then, $\hat{U} \cap (\bigcup_{i=1}^k U_i) = \emptyset$. This shows that $l_k \geq \mathrm{dia}(\hat{U}) > 0$. Hence, $\exists U_{k+1} \in \hat{\mathcal{I}}$ such that $U_1, \ldots, U_{k+1} \in \hat{\mathcal{I}}$ are pairwise disjoint and $\mathrm{dia}(U_{k+1}) \geq l_k/2$. Inductively, we either have the result holds for $\hat{\mathcal{I}}$ or $\exists (U_k)_{k=1}^\infty \subseteq \hat{\mathcal{I}}$ such that $\mathrm{dia}(U_{k+1}) \geq l_k/2$, $\forall k \in \mathbb{Z}_+$, and the sequence is pairwise disjoint.

In the latter case, we have $O \supseteq \bigcup_{k=1}^{\infty} U_k$ and $+\infty > \mu_{\mathrm{B}m}(O) \geq \mu_{\mathrm{B}m}(\bigcup_{k=1}^{\infty} U_k) = \sum_{k=1}^{\infty} \mu_{\mathrm{B}m}(U_k) \geq \sum_{k=0}^{\infty} 2^{-m} c^m l_k^m$. Note that $(l_k)_{k=0}^{\infty}$ is nonincreasing. Hence, $\lim_{k\in\mathbb{N}} l_k = 0$. Then, $\exists n \in \mathbb{N}$ such that $\sum_{i=n+1}^{\infty} \mu_{\mathrm{B}m}(U_i) < \pi^{-m/2} 2^m 5^{-m} c^m \Gamma(\frac{m}{2}+1)\epsilon$, where we have referred to the special Gamma function. Let $R := E \setminus (\bigcup_{k=1}^{n} U_k)$. Without loss of generality, assume $\mathrm{dia}(U_k) =: p_k \in (0,\infty) \subset \mathbb{R}$ and the center of U_k is x_k, $\forall k \in \mathbb{N}$. $\forall \bar{x} \in R$, since $\bigcup_{k=1}^{n} U_k$ is closed, then, by Proposition 4.10, $d := \inf_{x\in\bigcup_{k=1}^{n} U_k} |x-\bar{x}| > 0$. By the assumption, $\exists U \in \hat{\mathcal{I}}$ such that $\bar{x} \in U$ and $0 < \mathrm{dia}(U) < d$. Then, $U \cap (\bigcup_{k=1}^{n} U_k) = \emptyset$. By the fact that $\lim_{k\in\mathbb{N}} l_k = 0$, we have $\exists i_0 \in \{n+1, n+2, \ldots\}$ such that $U \cap U_{i_0} \neq \emptyset$ and $U \cap U_k = \emptyset$, $\forall k \in \{1,\ldots,i_0-1\}$. Then, $\mathrm{dia}(U) \leq l_{i_0-1} \leq 2\,\mathrm{dia}(U_{i_0}) = 2p_{i_0}$. Let $\hat{x} \in U \cap U_{i_0}$. Then, $|\bar{x}-x_{i_0}| \leq |\bar{x}-\hat{x}| + |\hat{x}-x_{i_0}| \leq \mathrm{dia}(U) + \frac{1}{2}p_{i_0} \leq \frac{5}{2}p_{i_0}$. Then, $\bar{x} \in V_{i_0} := \overline{\mathcal{B}}_{\mathbb{R}^m}(x_{i_0}, \frac{5}{2}p_{i_0}) \subset \mathbb{R}^m$. Then, $\bar{x} \in \bigcup_{k=n+1}^{\infty} V_k := \bigcup_{k=n+1}^{\infty} \overline{\mathcal{B}}_{\mathbb{R}^m}(x_k, \frac{5}{2}p_k) \subset \mathbb{R}^m$. Hence, $R \subseteq \bigcup_{k=n+1}^{\infty} V_k$. Then, $\mu_{\mathrm{L}mo}(R) \leq \mu_{\mathrm{L}mo}(\bigcup_{k=n+1}^{\infty} V_k) \leq \sum_{k=n+1}^{\infty} \mu_{\mathrm{B}m}(V_k) = \sum_{k=n+1}^{\infty} \frac{\pi^{m/2} 2^{-m} 5^m p_k^m}{\Gamma(\frac{m}{2}+1)} = \frac{\pi^{m/2} 2^{-m} 5^m c^{-m}}{\Gamma(\frac{m}{2}+1)} \sum_{k=n+1}^{\infty} c^m p_k^m \leq \frac{\pi^{m/2} 2^{-m} 5^m c^{-m}}{\Gamma(\frac{m}{2}+1)} \cdot \sum_{k=n+1}^{\infty} \mu_{\mathrm{B}m}(U_k) < \epsilon$, where the equality follows from (Mathematics Handbook Editors Group, 1979, pp. 320). Hence, the result holds for $\hat{\mathcal{I}}$.

Then, $\exists n \in \mathbb{Z}_+$, $\exists \{U_1,\ldots,U_n\} \subseteq \mathcal{I}$ such that $\mu_{\mathrm{L}mo}(E \setminus (\bigcup_{i=1}^{n} \overline{U_i})) < \epsilon$. Note that $\mu_{\mathrm{L}mo}(J) := \mu_{\mathrm{L}mo}((\bigcup_{i=1}^{n} \overline{U_i}) \setminus (\bigcup_{i=1}^{n} U_i)) = 0$. Then, $J \in \mathcal{B}_{\mathrm{L}}$ and $\mu_{\mathrm{L}mo}(E \setminus (\bigcup_{i=1}^{n} U_i)) = \mu_{\mathrm{L}mo}((E \setminus (\bigcup_{i=1}^{n} U_i)) \cap J) + \mu_{\mathrm{L}mo}((E \setminus (\bigcup_{i=1}^{n} U_i)) \setminus J) \leq \mu_{\mathrm{L}mo}(E \setminus (\bigcup_{i=1}^{n} \overline{U_i})) < \epsilon$, where the equality follows from Definition 11.15 and the first inequality follows from fact that $\mu_{\mathrm{L}mo}((E \setminus (\bigcup_{i=1}^{n} U_i)) \cap J) \leq \mu_{\mathrm{L}mo}(J) = 0$ and the fact that $(E \setminus (\bigcup_{i=1}^{n} U_i)) \setminus J = E \setminus (\bigcup_{i=1}^{n} \overline{U_i})$.

This completes the proof of the lemma. □

Definition 12.79 Let $m \in \mathbb{N}$, $\mathbb{R}^m$ be endowed with the usual positive cone, $E \in \mathcal{O}_{\mathbb{R}^m}$ with $\mu_{\mathrm{B}m}(E) > 0$, $\mathbb{E} := ((E, |\cdot|), \mathcal{B}, \mu)$ be the σ-finite metric measure subspace of $\mathbb{R}^m$ as defined in Proposition 11.29, $\mathcal{Y}$ be a separable normed linear space over $\mathbb{K}$, $c \in (0, \frac{1}{\sqrt{m}}] \subset \mathbb{R}$, and $f \in \bar{\mathrm{L}}_1(\mathbb{E}, \mathcal{Y})$. A point $x \in E$ is said to be a *rectangular Lebesgue point with regularity* c of f if $\limsup_{r\to 0^+} \sup_{\substack{\mathrm{r}_{x_1,x_2} \subset \mathbb{R}^m \text{ with } \frac{1}{2}(x_1+x_2)=x, \\ |x_2-x_1|=r, \min(x_2-x_1)\geq cr}} \frac{1}{\mu(\mathrm{r}_{x_1,x_2} \cap E)} \int_{\mathrm{r}_{x_1,x_2} \cap E} \|f(y) - f(x)\| \, d\mu(y) = 0$. ◇

Proposition 12.80 *Let $m \in \mathbb{N}$, $\mathbb{R}^m$ be endowed with the usual positive cone, $E \in \mathcal{O}_{\mathbb{R}^m}$ with $\mu_{\mathrm{B}m}(E) > 0$, $\mathbb{E} := ((E, |\cdot|), \mathcal{B}, \mu)$ be the σ-finite metric measure subspace of $\mathbb{R}^m$ as defined in Proposition 11.29, $\mathcal{Y}$ be a separable normed linear space over $\mathbb{K}$, $c \in (0, \frac{1}{\sqrt{m}}] \subset \mathbb{R}$, $f \in \bar{\mathrm{L}}_1(\mathbb{E}, \mathcal{Y})$, and $A := \{x \in E \,|\, x$ is a rectangular Lebesgue point of f with regularity $c\}$. Then, $A \in \mathcal{B}_{\mathrm{L}m}$ and $\mu_{\mathrm{L}m}(E \setminus A) = 0$.*

Proof Since $\mathbb{E}$ is σ-finite, then $\exists (E_n)_{n=1}^{\infty} \subseteq \mathcal{B}$ such that $\bigcup_{n=1}^{\infty} E_n = E$ and $\mu(E_n) < +\infty$, $\forall n \in \mathbb{N}$. $\forall r \in (0, \infty) \subset \mathbb{R}$, define $T_{r,c} f : E \to [0, \infty) \subset \mathbb{R}$ and $M_{r,c} f : E \to [0, \infty) \subset \mathbb{R}$ by, $\forall x \in E$,

$$(T_{r,c} f)(x) = \sup_{\substack{\mathrm{r}_{x_1,x_2} \subset \mathbb{R}^m \text{ with } \frac{1}{2}(x_1+x_2)=x, \\ |x_2-x_1|=r, \min(x_2-x_1) \geq cr}} \frac{1}{\mu(\mathrm{r}_{x_1,x_2} \cap E)} \cdot \int_{\mathrm{r}_{x_1,x_2} \cap E} \|f(y) - f(x)\| \, \mathrm{d}\mu(y)$$

$$(M_{r,c} f)(x) = \sup_{\substack{\mathrm{r}_{x_1,x_2} \subset \mathbb{R}^m \text{ with } \frac{1}{2}(x_1+x_2)=x, \\ |x_2-x_1|=r, \min(x_2-x_1) \geq cr}} \frac{1}{\mu(\mathrm{r}_{x_1,x_2} \cap E)} \int_{\mathrm{r}_{x_1,x_2} \cap E} \mathcal{P} \circ f \, \mathrm{d}\mu$$

$\forall x \in E$, since $E \in \mathcal{O}_{\mathbb{R}^m}$ then $\exists \delta \in (0, \infty) \subset \mathbb{R}$ such that $\mathrm{r}^{\circ}_{x-\delta \mathbf{1}_m, x+\delta \mathbf{1}_m} \subseteq E$. $\forall \mathrm{r}_{x_1,x_2} \subset \mathbb{R}^m$ with $x = \frac{1}{2}(x_1 + x_2)$, $|x_2 - x_1| = r > 0$ and $\min(x_2 - x_1) \geq cr$, we have $\mu(\mathrm{r}_{x_1,x_2} \cap E) = \mu_{\mathrm{B}m}(\mathrm{r}_{x_1,x_2} \cap E) \geq \mu_{\mathrm{B}m}(\mathrm{r}_{x_1,x_2} \cap \mathrm{r}^{\circ}_{x-\delta \mathbf{1}_m, x+\delta \mathbf{1}_m}) \geq ((2\delta) \wedge (cr))^m > 0$. Then, $(M_{r,c} f)(x) \leq ((2\delta) \wedge (cr))^{-m} \|f\|_1 < \infty$ and $(T_{r,c} f)(x) \leq (M_{r,c} f)(x) + \|f(x)\| < \infty$. Hence, $T_{r,c} f$ and $M_{r,c} f$ are well-defined. Define $T_c f : E \to [0, \infty] \subset \mathbb{R}_e$ and $M_c f : E \to [0, \infty] \subset \mathbb{R}_e$ by

$$(T_c f)(x) = \limsup_{r \to 0^+} (T_{r,c} f)(x) \quad (M_c f)(x) = \limsup_{r \to 0^+} (M_{r,c} f)(x); \quad \forall x \in E$$

Clearly, $T_c f$ and $M_c f$ are well-defined. Then, $T_{r,c}$ and $M_{r,c}$ are functions of $\bar{\mathrm{L}}_1(\mathbb{E}, \mathcal{Y})$ to nonnegative real-valued functions of E, and T_c and M_c are functions of $\bar{\mathrm{L}}_1(\mathbb{E}, \mathcal{Y})$ to nonnegative extended real-valued functions of E. Then, $A = \{x \in E \mid (T_c f)(x) = 0\}$ and $E \setminus A = \{x \in E \mid (T_c f)(x) > 0\} = \bigcup_{n=1}^{\infty} \bigcup_{k=1}^{\infty} A_{n,k} := \bigcup_{n=1}^{\infty} \bigcup_{k=1}^{\infty} \{x \in E_n \mid (T_c f)(x) > 1/k\}$.

We will show that $\mu_{\mathrm{L}mo}(A_{n,k}) = 0$, $\forall n \in \mathbb{N}$, $\forall k \in \mathbb{N}$, which implies that $A_{n,k} \in \mathcal{B}_{\mathrm{L}m}$ and $\mu_{\mathrm{L}m}(A_{n,k}) = 0$, $\forall n \in \mathbb{N}$, $\forall k \in \mathbb{N}$. This further implies that $E \setminus A \in \mathcal{B}_{\mathrm{L}m}$, $\mu_{\mathrm{L}m}(E \setminus A) = 0$, and $A \in \mathcal{B}_{\mathrm{L}m}$.

Fix any $n \in \mathbb{N}$ and any $k \in \mathbb{N}$. Suppose $\mu_{\mathrm{L}mo}(A_{n,k}) > 0$. By monotonicity of outer measures, we have $\mu_{\mathrm{L}mo}(A_{n,k}) \leq \mu_{\mathrm{L}mo}(E_n) = \mu_{\mathrm{L}m}(E_n) = \mu(E_n) < +\infty$. Let $\epsilon_0 = \frac{\mu_{\mathrm{L}mo}(A_{n,k})}{5k} \in (0, +\infty) \subset \mathbb{R}$. By Propositions 11.182 and 4.11, there exists a continuous function $g : E \to \mathcal{Y}$ such that $g \in \bar{\mathrm{L}}_1(\mathbb{E}, \mathcal{Y})$ and $\|g - f\|_1 < \epsilon_0$. Let $h := f - g$. Then, $\|h\|_1 < \epsilon_0$. Note that $(T_{r,c} f)(x) \leq (T_{r,c} g)(x) + (T_{r,c} h)(x)$, $\forall r \in (0, \infty) \subset \mathbb{R}$, $\forall x \in E$. Then, by Proposition 3.85, $(T_c f)(x) \leq (T_c g)(x) + (T_c h)(x)$, $\forall x \in E$. Since g is continuous, we have $(T_c g)(x) = 0$, $\forall x \in E$. Then, $(T_c f)(x) \leq (T_c h)(x)$, $\forall x \in E$. Note that

$$(T_{r,c} h)(x) \leq (M_{r,c} h)(x) + \|h(x)\|; \quad \forall r \in (0, \infty) \subset \mathbb{R}, \forall x \in E$$

Then, by Proposition 3.85, we have $(T_c h)(x) \leq (M_c h)(x) + \|h(x)\|$, $\forall x \in E$. Hence, $(T_c f)(x) \leq (M_c h)(x) + \|h(x)\|$, $\forall x \in E$. Then, $A_{n,k} \subseteq \left\{x \in E_n \,\middle|\, (M_c h)(x) > \frac{1}{2k}\right\} \cup \left\{x \in E_n \,\middle|\, \|h(x)\| > \frac{1}{2k}\right\} =: \tilde{A}_{n,k} \cup \bar{A}_{n,k}$. By monotonicity of outer measures, we have $\mu_{\text{L}mo}(\bar{A}_{n,k}) \leq \mu_{\text{L}mo}\left(\left\{x \in E \,\middle|\, \|h(x)\| > \frac{1}{2k}\right\}\right) =: \mu_{\text{L}mo}(\hat{A}_k) = \mu_{\text{B}m}(\hat{A}_k) = \int_E \chi_{\hat{A}_k,E}\,\mathrm{d}\mu \leq \int_E (2k\mathcal{P} \circ h)\,\mathrm{d}\mu = 2k\|h\|_1 < 2k\epsilon_0$, where the second equality follows from the fact that $\hat{A}_k \in \mathcal{B}$ since h is $\mathcal{B}$-measurable. By countable subadditivity of outer measures, we have $\mu_{\text{L}mo}(A_{n,k}) \leq \mu_{\text{L}mo}(\tilde{A}_{n,k}) + \mu_{\text{L}mo}(\bar{A}_{n,k}) < \mu_{\text{L}mo}(\tilde{A}_{n,k}) + 2k\epsilon_0$. Hence, $\mu_{\text{L}mo}(\tilde{A}_{n,k}) > 3k\epsilon_0$. Clearly, $\mu_{\text{L}mo}(\tilde{A}_{n,k}) \leq \mu_{\text{L}mo}(E_n) < +\infty$.

Define $\mathcal{I} := \left\{\mathrm{r}_{x_1,x_2} \subseteq \mathbb{R}^m \,\middle|\, x = \frac{1}{2}(x_1 + x_2) \in E, |x_2 - x_1| =: r > 0, \min(x_2 - x_1) \geq cr, \int_{\mathrm{r}_{x_1,x_2} \cap E} \mathcal{P} \circ h\,\mathrm{d}\mu > \frac{1}{2k}\mu(\mathrm{r}_{x_1,x_2} \cap E)\right\}$. Then, $\mathcal{I}$ covers $\tilde{A}_{n,k}$ in the sense of Vitali with index c. By Vitali's Lemma 12.78, there exists pairwise disjoint rectangles $\left(\mathrm{r}_{x_{i,1},x_{i,2}}\right)_{i=1}^{\bar{m}} \subseteq \mathcal{I}$ with $\bar{m} \in \mathbb{Z}_+$ such that $\mu_{\text{L}mo}(\tilde{A}_{n,k} \setminus (\bigcup_{i=1}^{\bar{m}} \mathrm{r}_{x_{i,1},x_{i,2}})) < \epsilon_0$. Let $B = (\bigcup_{i=1}^{\bar{m}} \mathrm{r}_{x_{i,1},x_{i,2}}) \cap E \in \mathcal{B}$. Then, $\mu(B) = \mu_{\text{L}mo}(B) \geq \mu_{\text{L}mo}(B \cap \tilde{A}_{n,k}) = \mu_{\text{L}mo}(\tilde{A}_{n,k}) - \mu_{\text{L}mo}(\tilde{A}_{n,k} \setminus B) > (3k - 1)\epsilon_0$, where the first inequality follows from the monotonicity of outer measures and the second equality follows from the measurability of B. Then, $\|h\|_1 = \int_E \mathcal{P} \circ h\,\mathrm{d}\mu \geq \int_B \mathcal{P} \circ h\,\mathrm{d}\mu = \sum_{i=1}^{\bar{m}} \int_{\mathrm{r}_{x_{i,1},x_{i,2}} \cap E} \mathcal{P} \circ h\,\mathrm{d}\mu > \sum_{i=1}^{\bar{m}} \frac{1}{2k}\mu(\mathrm{r}_{x_{i,1},x_{i,2}} \cap E) = \frac{1}{2k}\mu(B) > \frac{3k-1}{2k}\epsilon_0 \geq \epsilon_0 > \|h\|_1$. This is a contradiction. Therefore, we must have $\mu_{\text{L}mo}(A_{n,k}) = 0$.

This completes the proof of the proposition. □

In the rest of the book, we will denote the ith unit vector in $\mathbb{R}^m$ by $\boldsymbol{e}_{m,i}$, $\forall m \in \mathbb{N}$, $\forall i \in \{1, \ldots, m\}$.

Proposition 12.81 *Let $m \in \mathbb{N}$, $\mathbb{R}^m$ be endowed with the usual positive cone, $\mathcal{Y}$ be a separable normed linear space over $\mathbb{K}$, $\Omega \in \mathcal{B}_{\text{B}}(\mathbb{R}^m)$ be an open set, and $F : \Omega \to \mathcal{Y}$ be continuous on the right (or continuous on the left) and $\mathcal{B}_{\text{B}}(\mathbb{R}^m)$-measurable. Then, the following statements hold:*

(i) *$\forall i \in \{1, \ldots, m\}$, let $\mathrm{D}_i F : U_i \to \mathrm{B}(\mathbb{K}, \mathcal{Y})$ denote the partial derivative of F with respect to its ith coordinate in its domain and $U_i := \mathrm{dom}(\mathrm{D}_i F)$. Then, $U_i \in \mathcal{B}_{\text{B}}(\mathbb{R}^m)$ and $\mathrm{D}_i F$ is $\mathcal{B}_{\text{B}}(\mathbb{R}^m)$-measurable.*

(ii) *Let $F^{(1)} : U \to \mathrm{B}(\mathbb{K}^m, \mathcal{Y})$ be the Fréchet derivative of F, whose domain is $U := \mathrm{dom}\left(F^{(1)}\right) \subseteq \bigcap_{i=1}^m U_i \subseteq \Omega$. Then, $U \in \mathcal{B}_{\text{B}}(\mathbb{R}^m)$ and $F^{(1)}$ is $\mathcal{B}_{\text{B}}(\mathbb{R}^m)$-measurable.*

Proof (i) Fix any $i \in \{1, \ldots, m\}$. Consider the case that F is continuous on the right. Let $\mathcal{X} := \mathbb{K}$, and $\forall x := (x_1, \ldots, x_m) \in \Omega$, let $D_{i,x} := \{y \in \mathbb{R} \mid (x_1, \ldots, x_{i-1}, y, x_{i+1}, \ldots, x_m) \in \Omega\} \in \mathcal{B}_{\text{B}}(\mathbb{R})$, which is an open set. $\forall s \in D_{i,x}$, clearly $\overline{\mathrm{span}\left(\mathrm{A}_{D_{i,x}}(s)\right)} = \mathbb{K} = \mathcal{X}$. Then, $\mathrm{D}_i F : U_i \to \mathrm{B}(\mathbb{K}, \mathcal{Y})$, where $U_i := \mathrm{dom}(\mathrm{D}_i F) \subseteq \Omega$. Clearly, $\mathrm{B}(\mathbb{K}, \mathcal{Y})$ is isometrically isomorphic to $\mathcal{Y}$. Then, $\mathrm{D}_i F : U_i \to \mathcal{Y}$. Clearly, $\forall x \in U_i$, $\mathrm{D}_i F(x) = \lim_{h \to 0}(F(x + h\boldsymbol{e}_{m,i}) - F(x))/h \in \mathcal{Y}$.

Claim 12.81.1 $U_i = \bigcap_{n=1}^{\infty} \bigcup_{\substack{\delta\in\mathbb{Q}\\ \delta>0}} \bigcap_{\substack{h_1\in\mathbb{Q}\\ 0<|h_1|<\delta}} \bigcap_{\substack{h_2\in\mathbb{Q}\\ 0<|h_2|<\delta}} U_{i,n,h_1,h_2} =: \bar{U}_i \in \mathcal{B}_{\mathrm{B}}(\mathbb{R}^m)$,

where $U_{i,n,h_1,h_2} := \{x \in \Omega \mid \|(F(x+h_1\boldsymbol{e}_{m,i}) - F(x))/h_1 - (F(x+h_2\boldsymbol{e}_{m,i}) - F(x))/h_2\| < 1/n\}$, $\forall n \in \mathbb{N}$, $\forall h_1, h_2 \in \mathbb{Q}$ with $h_1 \neq 0$ and $h_2 \neq 0$.

Proof of Claim $\forall x \in U_i$, $\mathrm{D}_i F(x) = \lim_{h\to 0}(F(x+h\boldsymbol{e}_{m,i}) - F(x))/h \in \mathcal{Y}$. By Proposition 4.45, $\forall n \in \mathbb{N}$, $\exists \delta \in \mathbb{Q}$ with $\delta > 0$, such that $x + h\boldsymbol{e}_{m,i} \in \Omega$, $\forall h \in \mathcal{B}_{\mathbb{R}}(0,\delta)$, and, $\forall \bar{h}_1, \bar{h}_2 \in \mathcal{B}_{\mathbb{R}}(0,\delta) \setminus \{0\}$, we have $\|(F(x+\bar{h}_1\boldsymbol{e}_{m,i}) - F(x))/\bar{h}_1 - (F(x+\bar{h}_2\boldsymbol{e}_{m,i}) - F(x))/\bar{h}_2\| < 1/n$. Then, $\forall h_1, h_2 \in \mathbb{Q}$ with $0 < |h_1| < \delta$ and $0 < |h_2| < \delta$, we have $\|(F(x+h_1\boldsymbol{e}_{m,i}) - F(x))/h_1 - (F(x+h_2\boldsymbol{e}_{m,i}) - F(x))/h_2\| < 1/n$. Thus, $x \in U_{i,n,h_1,h_2}$. Hence, $x \in \bar{U}_i$. By the arbitrariness of x, we have $U_i \subseteq \bar{U}_i$.

On the other hand, $\forall x \in \bar{U}_i$, $\forall n \in \mathbb{N}$, $\exists \delta \in \mathbb{Q}$ with $\delta > 0$, such that, $\forall h \in \mathcal{B}_{\mathbb{R}}(0,\delta)$, $x + h\boldsymbol{e}_{m,i} \in \Omega$, and $\forall h_1, h_2 \in \mathbb{Q}$ with $0 < |h_1| < \delta$ and $0 < |h_2| < \delta$, we have $\|(F(x+h_1\boldsymbol{e}_{m,i}) - F(x))/h_1 - (F(x+h_2\boldsymbol{e}_{m,i}) - F(x))/h_2\| < 1/n$. $\forall \bar{h}_1, \bar{h}_2 \in \mathcal{B}_{\mathbb{R}}(0,\delta) \setminus \{0\}$, $\exists (h_{j,l})_{l=1}^{\infty} \subseteq (\mathbb{Q} \cap \mathcal{B}_{\mathbb{R}}(0,\delta)) \setminus \{0\}$ such that $\bar{h}_j < h_{j,l}$, $\forall l \in \mathbb{N}$, and $\lim_{l\in\mathbb{N}} h_{j,l} = \bar{h}_j$, $j = 1, 2$. $\forall l \in \mathbb{N}$, $\|(F(x+h_{1,l}\boldsymbol{e}_{m,i}) - F(x))/h_{1,l} - (F(x+h_{2,l}\boldsymbol{e}_{m,i}) - F(x))/h_{2,l}\| < 1/n$. Then,

$$\begin{aligned}
&\|(F(x+\bar{h}_1\boldsymbol{e}_{m,i}) - F(x))/\bar{h}_1 - (F(x+\bar{h}_2\boldsymbol{e}_{m,i}) - F(x))/\bar{h}_2\| \\
&= \left\|\lim_{l\in\mathbb{N}} \left(\frac{F(x+h_{1,l}\boldsymbol{e}_{m,i}) - F(x)}{h_{1,l}} - \frac{F(x+h_{2,l}\boldsymbol{e}_{m,i}) - F(x)}{h_{2,l}}\right)\right\| \\
&= \lim_{l\in\mathbb{N}} \|(F(x+h_{1,l}\boldsymbol{e}_{m,i}) - F(x))/h_{1,l} - (F(x+h_{2,l}\boldsymbol{e}_{m,i}) - F(x))/h_{2,l}\| \\
&\leq 1/n
\end{aligned}$$

where the first equality follows from the fact that F is continuous on the right and the second equality follows from Propositions 3.66, 3.67, and 7.23. By Proposition 4.45, $\lim_{h\to 0}(F(x+h\boldsymbol{e}_{m,i}) - F(x))/h \in \mathcal{Y}$. Hence, $x \in \mathrm{dom}(\mathrm{D}_i F) = U_i$. By the arbitrariness of x, we have $\bar{U}_i \subseteq U_i$. Therefore, $U_i = \bar{U}_i$.

By Propositions 7.21, 7.23, 11.38, and 11.39, $U_{i,n,h_1,h_2} \in \mathcal{B}_{\mathrm{B}}(\mathbb{R}^m)$, $\forall n \in \mathbb{N}$, $\forall h_1, h_2 \in \mathbb{Q}$ with $h_1 \neq 0$ and $h_2 \neq 0$. Then, $U_i = \bar{U}_i \in \mathcal{B}_{\mathrm{B}}(\mathbb{R}^m)$. □

$\forall n \in \mathbb{N}$, define $H_n : \Omega \to \mathcal{Y}$ by, $\forall x \in \Omega$,

$$H_n(x) = \begin{cases} n\left(F(x+\frac{1}{n}\boldsymbol{e}_{m,i}) - F(x)\right) & x \in O := \Omega \cap (\Omega - \frac{1}{n}\boldsymbol{e}_{m,i}) \\ \vartheta_{\mathcal{Y}} & x \in \Omega \setminus O \end{cases}$$

By Propositions 7.23, 11.38, 11.39, and 11.41, H_n is $\mathcal{B}_{\mathrm{B}}(\mathbb{R}^m)$-measurable. Then, $H_n|_{U_i}$ is $\mathcal{B}_{\mathrm{B}}(\mathbb{R}^m)$-measurable. Note that, $\forall x \in U_i$, $\mathrm{D}_i F(x) = \lim_{n\in\mathbb{N}} H_n(x) = \lim_{n\in\mathbb{N}} H_n|_{U_i}(x)$. By Proposition 11.48, $\mathrm{D}_i F$ is $\mathcal{B}_{\mathrm{B}}(\mathbb{R}^m)$-measurable.

The case when F is continuous on the left can be proved similarly.

(ii) Consider the case that F is continuous on the right. By Proposition 9.9, $\forall x \in U = \operatorname{dom}\left(F^{(1)}\right)$, we must have $x \in U_i$, $\forall i \in \{1, \ldots, m\}$. Then, $U \subseteq \bigcap_{i=1}^{m} U_i$. $\forall x \in \Omega$, since Ω is open, then $\overline{\operatorname{span}\left(\mathrm{A}_\Omega(x)\right)} = \mathbb{K}^m$.

Claim 12.81.2 $U = \bigcap_{n=1}^{\infty} \bigcup_{\substack{\delta\in\mathbb{Q}\\ \delta>0}} \bigcap_{\substack{h_1\in\mathbb{Q}\\ |h_1|<\delta}} \cdots \bigcap_{\substack{h_m\in\mathbb{Q}\\ |h_m|<\delta}} V_{n,h_1,\ldots,h_m} =: \bar{V} \in \mathcal{B}_{\mathrm{B}}\left(\mathbb{R}^m\right)$, where

$V_{n,h_1,\ldots,h_m} := \left\{x \in \bigcap_{i=1}^{m} U_i \;\middle|\; \left\| F(x + \sum_{i=1}^{m} h_i \boldsymbol{e}_{m,i}) - F(x) - \sum_{i=1}^{m} h_i \mathrm{D}_i F(x)\right\| \leq \frac{1}{n}\sqrt{\sum_{i=1}^{m} h_i^2}\right\}$, $\forall n \in \mathbb{N}, \forall h_1, \ldots, h_m \in \mathbb{Q}$.

Proof of Claim $\forall x \in U$, we have $x \in \bigcap_{i=1}^{m} U_i$. By Definition 9.3 and Proposition 9.9, the candidate for $F^{(1)}(x)$ is $\left[\mathrm{D}_1 F(x) \cdots \mathrm{D}_m F(x)\right] := L \in \mathrm{B}\left(\mathbb{K}^m, \mathcal{Y}\right)$. By $x \in \operatorname{dom}\left(F^{(1)}\right)$, we have $\forall n \in \mathbb{N}$, $\exists \delta \in \mathbb{Q}$ with $\delta > 0$, such that $\mathcal{B}_{\mathbb{R}^m}\left(x, \sqrt{m}\delta\right) \subseteq \Omega$, and $\forall \bar{h} \in \mathcal{B}_{\mathbb{R}^m}\left(\mathbf{0}_m, \sqrt{m}\delta\right)$, we have $\left\|F(x+\bar{h}) - F(x) - L\bar{h}\right\| \leq \frac{1}{n}\left|\bar{h}\right|$. Then, $\forall h := (h_1, \ldots, h_m) \in \mathbb{Q}^m$ with $|h_i| < \delta$, $i = 1, \ldots, m$, we have $\|F(x+h) - F(x) - Lh\| = \left\|F(x + \sum_{i=1}^{m} h_i \boldsymbol{e}_{m,i}) - F(x) - \sum_{i=1}^{m} h_i \mathrm{D}_i F(x)\right\| \leq \frac{1}{n}|h| = \frac{1}{n}\sqrt{\sum_{i=1}^{m} h_i^2}$. Thus, $x \in V_{n,h_1,\ldots,h_m}$. Hence, $x \in \bar{V}$. By the arbitrariness of x, we have $U \subseteq \bar{V}$.

On the other hand, $\forall x \in \bar{V}$. Then, $\forall n \in \mathbb{N}$, $\exists \delta \in \mathbb{Q}$ with $\delta > 0$, such that $\mathcal{B}_{\mathbb{R}^m}(x, \delta) \subseteq \Omega$ and, $\forall h = (h_1, \ldots, h_m) \in \mathbb{Q}^m$ with $|h_i| < \delta$, $i = 1, \ldots, m$, we have $x \in V_{n,h_1,\ldots,h_m} \subseteq \bigcap_{i=1}^{m} U_i$. $\forall \bar{x} \in \mathcal{B}_{\mathbb{K}^m}(x, \delta) \cap \Omega = \mathcal{B}_{\mathbb{R}^m}(x, \delta)$, let $(\bar{h}_1, \ldots, \bar{h}_m) := \bar{x} - x$. Clearly, $\left|\bar{h}_i\right| < \delta$, $i = 1, \ldots, m$. Then, $\exists \left(h_{i,l}\right)_{l=1}^{\infty} \subseteq \mathbb{Q}$ with $\left|h_{i,l}\right| < \delta$ such that $\bar{h}_i < h_{i,l}$, $\forall l \in \mathbb{N}$, and $\lim_{l\in\mathbb{N}} h_{i,l} = \bar{h}_i$, $i = 1, \ldots, m$. By the assumption,

$$
\begin{aligned}
&\|F(\bar{x}) - F(x) - L(\bar{x} - x)\| \\
&= \left\| \lim_{l\in\mathbb{N}} \left(F(x + \sum_{i=1}^{m} h_{i,l} \boldsymbol{e}_{m,i}) - F(x) - \sum_{i=1}^{m} h_{i,l} \mathrm{D}_i F(x) \right) \right\| \\
&= \lim_{l\in\mathbb{N}} \left\| F(x + \sum_{i=1}^{m} h_{i,l} \boldsymbol{e}_{m,i}) - F(x) - \sum_{i=1}^{m} h_{i,l} \mathrm{D}_i F(x) \right\| \\
&\leq \lim_{l\in\mathbb{N}} \frac{1}{n} \left(\sum_{i=1}^{m} h_{i,l}^2 \right)^{1/2} = \frac{1}{n} \left(\sum_{i=1}^{m} \bar{h}_i^2 \right)^{1/2} = \frac{1}{n} |\bar{x} - x|
\end{aligned}
$$

where the first equality follows from the fact that F is continuous on the right, the second equality follows from Propositions 3.66, 3.67, 7.21, and 7.23, and the first inequality follows from the fact that $x \in V_{n,h_{1,l},\ldots,h_{m,l}}$. By Definition 9.3, $F^{(1)}(x) = L$. Hence, $x \in \operatorname{dom}\left(F^{(1)}\right) = U$. By the arbitrariness of x, we have $\bar{V} \subseteq U$. Therefore, $U = \bar{V}$.

By Propositions 7.21, 7.23, 11.38, and 11.39, $V_{n,h_1,\ldots,h_m} \in \mathcal{B}_B(\mathbb{R}^m)$, $\forall n \in \mathbb{N}$, $\forall h_1, \ldots, h_m \in \mathbb{Q}$. Then, $U = \bar{V} \in \mathcal{B}_B(\mathbb{R}^m)$. □

Note that, $\forall x \in U$,

$$F^{(1)}(x) = \begin{bmatrix} D_1 F(x) & \cdots & D_m F(x) \end{bmatrix} = \begin{bmatrix} (D_1 F)|_U (x) & \cdots & (D_m F)|_U (x) \end{bmatrix}$$

By Propositions 11.38, 11.39, 11.41, and 11.139, $F^{(1)}$ is $\mathcal{B}_B(\mathbb{R}^m)$-measurable.

The case when F is continuous on the left can be proved similarly.

This completes the proof of the proposition. □

Theorem 12.82 *Let $I := [a, b] \subset \mathbb{R}$ with $a, b \in \mathbb{R}$ and $a < b$, $\mathcal{Y}$ be a separable Banach space over $\mathbb{K}$, $f : I \to \mathcal{Y}$ be absolutely integrable over I with respect to μ_B, and $F : I \to \mathcal{Y}$ be defined by $F(x) = \int_a^x f(t)\,dt$, $\forall x \in I$. Assume that f is continuous at $x_0 \in I$. Then, F is Fréchet differentiable at x_0 and $F^{(1)}(x_0) = f(x_0)$. (Note that, when $\mathbb{K} = \mathbb{C}$, I is viewed as a subset of $\mathbb{C}$ in calculations of $F^{(1)}$.)*

Proof Clearly, F is well-defined by Proposition 11.92. Note that $\overline{\text{span}\,(A_I(x_0))} = \mathbb{K}$. $\forall \epsilon \in (0, \infty) \subset \mathbb{R}$, by the continuity of f at x_0, $\exists \delta \in (0, \infty) \subset \mathbb{R}$, $\forall x \in I \cap \mathcal{B}_{\mathbb{R}}(x_0, \delta)$, we have $\|f(x) - f(x_0)\| < \epsilon$. Then,

$$\begin{aligned}
&\|F(x) - F(x_0) - f(x_0)(x - x_0)\| \\
&= \left\| \int_a^x f(t)\,dt - \int_a^{x_0} f(t)\,dt - \int_{x_0}^x f(x_0)\,dt \right\| \\
&= \left\| \int_{x_0}^x f(t)\,dt - \int_{x_0}^x f(x_0)\,dt \right\| = \left\| \int_{x_0}^x (f(t) - f(x_0))\,dt \right\| \\
&= \left\| \int_{r_{x_0 \wedge x, x_0 \vee x}} (f - f(x_0))\,d\mu_B \right\| \leq \int_{r_{x_0 \wedge x, x_0 \vee x}} \mathcal{P} \circ (f - f(x_0))\,d\mu_B \\
&\leq \epsilon \mu_B(r_{x_0 \wedge x, x_0 \vee x}) = \epsilon |x - x_0|
\end{aligned}$$

where the second equality follows from Fact 12.72, the third equality and the first inequality follow from Proposition 11.92, and the second inequality follows from Proposition 11.78. Hence, $F^{(1)}(x_0) = f(x_0)$. This completes the proof of the theorem. □

Theorem 12.83 *Let $I := [a, b] \subset \mathbb{R}$ with $a, b \in \mathbb{R}$ and $a < b$, $\mathbb{I} := ((I, |\cdot|), \mathcal{B}, \mu)$ be the finite complete metric measure subspace of $\mathbb{R}$, $\mathcal{Y}$ be a Banach space over $\mathbb{K}$, and $F : I \to \mathcal{Y}$ be $\mathcal{C}_1$. (Note that, when $\mathbb{K} = \mathbb{C}$, I is viewed as a subset of $\mathbb{C}$ in calculations of $F^{(1)}$.) Then, $F^{(1)} : I \to \mathcal{Y}$ is absolutely integrable over $\mathbb{I}$ and $F(b) - F(a) = \int_a^b F^{(1)}(t)\,dt = \int_a^b F^{(1)}\,d\mu_B$.*

Proof Since $F^{(1)} : I \to \mathcal{Y}$ is continuous, then, by Proposition 11.37, $F^{(1)}$ is $\mathcal{B}$-measurable. Since I is compact, then $\exists M \in [0, \infty) \subset \mathbb{R}$ such that $\left\| F^{(1)}(x) \right\| \leq M$, $\forall x \in I$. By Proposition 11.78, we have $F^{(1)}$ is absolutely integrable over $\mathbb{I}$. By Proposition 7.126, $F^{(1)} : I \to \mathcal{N} := \overline{\operatorname{span}\left(F^{(1)}(I)\right)} \subseteq \mathcal{Y}$ and $\mathcal{N}$ is a separable Banach subspace of $\mathcal{Y}$. Define $F_a : I \to \mathcal{N} \subseteq \mathcal{Y}$ by $F_a(x) = \int_a^x F^{(1)}(t)\,\mathrm{d}t = \int_a^x F^{(1)}\,\mathrm{d}\mu_{\mathrm{B}}$, $\forall x \in I$. By Proposition 11.92, F_a is well-defined. By Theorem 12.82, F_a is Fréchet differentiable and $F_a^{(1)} = F^{(1)}$. Then, define $g : I \to \mathcal{Y}$ by $g = F - F_a$. Then, by Proposition 9.15, g is Fréchet differentiable and $g^{(1)}(x) = \vartheta_{\mathcal{Y}}$, $\forall x \in I$. By Mean Value Theorem 9.23, $\|g(x) - g(a)\| \leq 0$, $\forall x \in I$. Hence, $g(x) = g(a) = F(a)$, $\forall x \in I$. Then, $g(b) = F(a) = F(b) - F_a(b) = F(b) - \int_a^b F^{(1)}(t)\,\mathrm{d}t$. Therefore, $F(b) - F(a) = \int_a^b F^{(1)}(t)\,\mathrm{d}t$. This completes the proof of the theorem. □

Proposition 12.84 *Let $m \in \mathbb{N}$, $\mathbb{R}^m$ be endowed with the usual positive cone, $\Omega \in \mathcal{B}_{\mathrm{B}}(\mathbb{R}^m)$ be a region, $\mathbb{X} := ((\mathrm{P}(\Omega), |\cdot|), \mathcal{B}, \mu)$ be the σ-finite metric measure subspace of $\mathbb{R}^m$, $\mathcal{Y}$ be a finite-dimensional Banach space over $\mathbb{K}$, $f : \mathrm{P}(\Omega) \to \mathcal{Y}$ be $\mathcal{B}_{\mathrm{B}}(\mathbb{R}^m)$-measurable, $\bar{J} := \{1, \ldots, m\}$, and $x_0 := (x_{0,1}, \ldots, x_{0,m}) \in \Omega$. Then, we may define the σ-finite $\mathcal{Y}$-valued measure ν with kernel f over $\mathbb{X}$ according to Proposition 11.116. By Proposition 11.167, f is the unique Radon–Nikodym derivative of ν with respect to μ. Assume that, $\forall x_1, x_2 \in \Omega$ with $x_1 \stackrel{\leq}{=} x_2$ and $\bar{\mathrm{r}}_{x_1,x_2} \subseteq \Omega$, we have f is bounded over $\mathrm{r}_{x_1,x_2} \subseteq \mathrm{P}(\Omega)$. Let $F : \Omega \to \mathcal{Y}$ be a cumulative distribution function of ν. Then, $\forall x := (x_1, \ldots, x_m) \in \Omega$ with $\bar{\mathrm{r}}_{\hat{x},\check{x}} \subseteq \Omega$, where $\hat{x} := x_0 \wedge x$ and $\check{x} := x_0 \vee x$, we have*

$$F(x) = \int_{x_{0,1}}^{x_1} \cdots \int_{x_{0,m}}^{x_m} f(s_1, \ldots, s_m)\,\mathrm{d}s_m \cdots \mathrm{d}s_1 - \sum_{J \subseteq \bar{J}, J \neq \emptyset} (-1)^{\operatorname{card}(J)} F(x_J) \tag{12.1}$$

and the order of integration can be arbitrary, where $x_J \in \bar{\mathrm{r}}_{\hat{x},\check{x}} \subseteq \Omega$ is defined by $\pi_i(x_J) = \begin{cases} \pi_i(x_0) = x_{0,i} & i \in J \\ \pi_i(x) = x_i & i \in \bar{J} \setminus J \end{cases}$, $\forall i \in \bar{J}$.

Proof $\mathcal{Y}$ is separable and is σ-compact, since it is finite-dimensional. Fix any $x \in \Omega$ with $\bar{\mathrm{r}}_{\hat{x},\check{x}} \subseteq \Omega$. Then, $\mathrm{r}_{\hat{x},\check{x}} \subseteq \mathrm{P}(\Omega)$. By Definition 12.42 and Definition 11.166, we have $\Delta_F(\mathrm{r}_{\hat{x},\check{x}}) = \nu(\mathrm{r}_{\hat{x},\check{x}}) = \int_{\mathrm{r}_{\hat{x},\check{x}}} f\,\mathrm{d}\mu = \int_{\mathrm{r}_{\hat{x},\check{x}}} f\,\mathrm{d}\mu_{\mathrm{B}m}$. By the assumption and Fubini's Theorem 12.31, we have

$$\int_{\mathrm{r}_{\hat{x},\check{x}}} f\,\mathrm{d}\mu_{\mathrm{B}m} = \int_{x_{0,1}\wedge x_1}^{x_{0,1}\vee x_1} \cdots \int_{x_{0,m}\wedge x_m}^{x_{0,m}\vee x_m} f(s_1, \ldots, s_m)\,\mathrm{d}s_m \cdots \mathrm{d}s_1$$

$$= (-1)^{n(x)} \int_{x_{0,1}}^{x_1} \cdots \int_{x_{0,m}}^{x_m} f(s_1, \ldots, s_m)\,\mathrm{d}s_m \cdots \mathrm{d}s_1 = \Delta_F(\mathrm{r}_{\hat{x},\check{x}})$$

and the order of integration can be arbitrary, where $n(x) = \operatorname{card}\big(\{i \in \bar{J} \mid \pi_i(x_0) > \pi_i(x)\}\big) =: \operatorname{card}(\check{J})$. By Definition 12.41,

$$\Delta_F(\mathrm{r}_{\hat{x},\check{x}}) = \sum_{\hat{J}\subseteq\bar{J}} (-1)^{\operatorname{card}(\hat{J})} F(\bar{x}_{\hat{J}})$$
$$= (-1)^{n(x)} \int_{x_{0,1}}^{x_1} \cdots \int_{x_{0,m}}^{x_m} f(s_1, \ldots, s_m)\, \mathrm{d}s_m \cdots \mathrm{d}s_1$$

where $\bar{x}_{\hat{J}} \in \bar{\mathrm{r}}_{\hat{x},\check{x}} \subseteq \mathbb{R}^m$ is defined by, $\forall i \in \bar{J}$,

$$\pi_i(\bar{x}_{\hat{J}}) = \begin{cases} \pi_i(\hat{x}) \ \forall i \in \hat{J} \\ \pi_i(\check{x}) \ \forall i \in \bar{J} \setminus \hat{J} \end{cases}$$
$$= \begin{cases} \pi_i(x) \ \ \forall i \in (\hat{J} \cap \check{J}) \cup ((\bar{J} \setminus \hat{J}) \cap (\bar{J} \setminus \check{J})) \\ \pi_i(x_0) \text{ otherwise} \end{cases}$$
$$= \begin{cases} \pi_i(x) \ \ \forall i \in \bar{J} \setminus (\hat{J} \triangle \check{J}) \\ \pi_i(x_0) \ \forall i \in \hat{J} \triangle \check{J} \end{cases} = \pi_i(x_{\hat{J}\triangle\check{J}})$$

This leads to

$$\Delta_F(\mathrm{r}_{\hat{x},\check{x}}) = \sum_{\hat{J}\subseteq\bar{J}} (-1)^{\operatorname{card}(\hat{J})} F(x_{\hat{J}\triangle\check{J}})$$
$$= (-1)^{\operatorname{card}(\check{J})} \sum_{\hat{J}\subseteq\bar{J}} (-1)^{\operatorname{card}(\hat{J})+\operatorname{card}(\check{J})} F(x_{\hat{J}\triangle\check{J}})$$
$$= (-1)^{n(x)} \sum_{\hat{J}\subseteq\bar{J}} (-1)^{\operatorname{card}(\hat{J}\triangle\check{J})} F(x_{\hat{J}\triangle\check{J}})$$
$$= (-1)^{n(x)} \int_{x_{0,1}}^{x_1} \cdots \int_{x_{0,m}}^{x_m} f(s_1, \ldots, s_m)\, \mathrm{d}s_m \cdots \mathrm{d}s_1$$

Then, we can conclude

$$\sum_{J\subseteq\bar{J}} (-1)^{\operatorname{card}(J)} F(x_J) = \int_{x_{0,1}}^{x_1} \cdots \int_{x_{0,m}}^{x_m} f(s_1, \ldots, s_m)\, \mathrm{d}s_m \cdots \mathrm{d}s_1$$

The above equation is equivalent to (12.1). This completes the proof of the proposition. □

Proposition 12.85 *Let $m \in \mathbb{N}$, $\mathbb{R}^m$ be endowed with the usual positive cone, $\Omega \in \mathcal{B}_\mathrm{B}(\Omega)$ be a nondegenerate rectangle, $\mathbb{X} := ((\Omega, |\cdot|), \mathcal{B}, \mu)$ be the σ-finite metric measure subspace of $\mathbb{R}^m$, $\mathcal{Y}$ be a separable Banach space, $f : \Omega \to \mathcal{Y}$ be $\mathcal{B}$-*

measurable, $x_0 \in \Omega$*, and* $\bar{J} := \{1, \dots, m\}$*. Assume that* f *is absolutely integrable over* r_{x_a,x_b} *with respect to* μ*,* $\forall x_a, x_b \in \Omega$ *with* $x_a \stackrel{\leq}{=} x_b$*. Define* $\pi_J : \mathbb{R}^m \to \mathbb{R}^{\mathrm{card}(J)}$ *by* $\pi_J(x) = (\pi_i(x))_{i \in J}$*,* $\forall x \in \mathbb{R}^m$*,* $\forall J \subseteq \bar{J}$*;* $(\pi_J(\Omega), \mathcal{B}_J, \mu_J)$ *be the measure subspace of* $\mathbb{R}^{\mathrm{card}(J)}$*,* $\forall J \subseteq \bar{J}$ *with* $J \neq \emptyset$*;* $M_J : \mathbb{R}^{\mathrm{card}(J)} \times \mathbb{R}^{m-\mathrm{card}(J)} \to \mathbb{R}^m$ *by* $M_J(s,t) \in \mathbb{R}^m$ *is such that* $\pi_J(M_J(s,t)) = s$ *and* $\pi_{\bar{J}\setminus J}(M_J(s,t)) = t$*,* $\forall s \in \mathbb{R}^{\mathrm{card}(J)}$*,* $\forall t \in \mathbb{R}^{m-\mathrm{card}(J)}$*,* $\forall J \subseteq \bar{J}$*. Then, the following statements hold:*

(i) $\forall \emptyset \neq J \subseteq \bar{J}$*,* $\exists U_J \in \mathcal{B}$ *with* $\mu(\Omega \setminus U_J) = 0$ *such that* $f(M_J(\cdot, \pi_{\bar{J}\setminus J}(x))) : \pi_J(\Omega) \to \mathcal{Y}$ *is absolutely integrable over* $\pi_J(\mathrm{r}_{\hat{x},\check{x}})$ *with respect to* μ_J*,* $\forall x \in U_J$*; and* $F : \Omega \to \mathcal{Y}$*, defined by,* $\forall x \in \Omega$*,*

$$F(x) = \begin{cases} (-1)^{\mathrm{card}(\check{J}\cap J)} \int_{\pi_J(\mathrm{r}_{\hat{x},\check{x}})} f(M_J(s, \pi_{\bar{J}\setminus J}(x)))\, \mathrm{d}\mu_J(s) & x \in U_J \\ \vartheta_{\mathcal{Y}} & x \in \Omega \setminus U_J \end{cases}$$

where $\hat{x} := x_0 \wedge x$*,* $\check{x} := x_0 \vee x$*, and* $\check{J} := \left\{ i \in \bar{J} \mid \pi_i(x_0) > \pi_i(x) \right\}$*, is* $\mathcal{B}$*-measurable.*

(ii) *Furthermore, if, in addition,* $\mathcal{Y}$ *is finite-dimensional, then* $U_J = \Big\{ x \in \Omega \mid f(M_J(\cdot, \pi_{\bar{J}\setminus J}(x))) : \pi_J(\Omega) \to \mathcal{Y}$ is absolutely integrable over $\pi_J(\mathrm{r}_{\hat{x},\check{x}})$ with respect to $\mu_J \Big\}$.

Proof First consider the special case when $\mu(\Omega) < \infty$ and f is absolutely integrable over $\mathbb{X}$. Since Ω is a nondegenerate rectangle, then Ω is bounded. Define $\bar{f} : \pi_J(\Omega) \times \Omega \to \mathcal{Y}$ by $\bar{f}(s,x) = (-1)^{\mathrm{card}(\check{J}\cap J)} f(M_J(s, \pi_{\bar{J}\setminus J}(x))) \cdot \chi_{\pi_J(\mathrm{r}_{\hat{x},\check{x}}),\pi_J(\Omega)}(s)$, $\forall (s,x) \in \pi_J(\Omega) \times \Omega$. By Propositions 7.23, 11.39, and 11.38, $\bar{f}$ is $\mathcal{B}_{\mathrm{B}}(\mathbb{R}^{\mathrm{card}(J)+m})$-measurable. Note the following derivation,

$$\begin{aligned}
&\int_{\pi_J(\Omega)\times\Omega} \mathcal{P} \circ \bar{f}(s,x)\, \mathrm{d}(\mu_J \times \mu)(s,x) \\
&\leq \int_{\pi_J(\Omega)} \int_{\Omega} \left\| f(M_J(s, \pi_{\bar{J}\setminus J}(x))) \right\| \mathrm{d}\mu(x)\, \mathrm{d}\mu_J(s) \\
&= \int_{\pi_J(\Omega)} \int_{\pi_{\bar{J}\setminus J}(\Omega)} \int_{\pi_J(\Omega)} \| f(M_J(s,t)) \|\, \mathrm{d}\mu_J(\tau)\, \mathrm{d}\mu_{\bar{J}\setminus J}(t)\, \mathrm{d}\mu_J(s) \\
&= \mu_J(\pi_J(\Omega)) \int_{\pi_J(\Omega)} \int_{\pi_{\bar{J}\setminus J}(\Omega)} \| f(x) \|\, \mathrm{d}\mu_{\bar{J}\setminus J}(\pi_{\bar{J}\setminus J}(x))\, \mathrm{d}\mu_J(\pi_J(x)) \\
&= \mu_J(\pi_J(\Omega)) \int_{\Omega} \mathcal{P} \circ f\, \mathrm{d}\mu < \infty
\end{aligned}$$

where the first inequality follows from Proposition 11.83, and the equalities follow from Tonelli's Theorem 12.29. Then, we have that $\bar{f}$ is absolutely integrable over $\pi_J(\Omega) \times \Omega$ with respect to $\mu_J \times \mu$. By Fubini's Theorem 12.30, $\exists p_J : \Omega \to \mathcal{Y}$ and $U_J \in \mathcal{B}$ such that $\mu(\Omega \setminus U_J) = 0$; $\bar{f}(\cdot, x) : \pi_J(\Omega) \to \mathcal{Y}$ is absolutely integrable over $(\pi_J(\Omega), \mathcal{B}_J, \mu_J)$, $\forall x \in U_J$; $p_J : \Omega \to \mathcal{Y}$, defined by, $\forall x \in \Omega$, $p_J(x) = \begin{cases} \int_{\pi_J(\Omega)} \bar{f}(s,x)\,\mathrm{d}\mu_J(s) & x \in U_J \\ \vartheta_{\mathcal{Y}} & x \in \Omega \setminus U_J \end{cases}$, is $\mathcal{B}$-measurable; p_J is absolutely integrable over $\mathbb{X}$, and $\int_{\pi_J(\Omega)\times\Omega} \bar{f}(s,x)\,\mathrm{d}(\mu_J \times \mu)(s,x) = \int_\Omega p_J(x)\,\mathrm{d}\mu(x)$. Clearly, $F(x) = p_J(x)$, $\forall x \in \Omega$. Hence, F is $\mathcal{B}$-measurable. Then, $f(M_J(\cdot, \pi_{\bar{J}\setminus J}(x))) : \pi_J(\Omega) \to \mathcal{Y}$ is absolutely integrable over $\pi_J(\mathrm{r}_{\hat{x},\check{x}}) \subseteq \pi_J(\Omega)$ with respect to μ_J, $\forall x \in U_J$.

If, in addition, $\mathcal{Y}$ is finite-dimensional, let U_J be defined as in the statement (ii). $\forall x \in \Omega$, $f(M_J(\cdot, \pi_{\bar{J}\setminus J}(x))) : \pi_J(\Omega) \to \mathcal{Y}$ is absolutely integrable over $\pi_J(\mathrm{r}_{\hat{x},\check{x}})$ with respect to μ_J if, and only if, $\bar{f}(\cdot, x)$ is absolutely integrable over $(\pi_J(\Omega), \mathcal{B}_J, \mu_J)$. Then, $U_J := \{x \in \Omega \mid \bar{f}(\cdot, x) : \pi_J(\Omega) \to \mathcal{Y}$ is absolutely integrable over $(\pi_J(\Omega), \mathcal{B}_J, \mu_J)\}$. By Fubini's Theorem 12.31, this choice of U_J is admissible for the definition of p_J and p_J is $\mathcal{B}$-measurable. This proves the result in this special case.

Now consider the general case. Since Ω is a nondegenerate rectangle and $x_0 \in \Omega$, then there exists $x_0 \in \Omega_1 \subseteq \Omega_2 \subseteq \cdots \subseteq \Omega_j \subseteq \Omega_{j+1} \subseteq \cdots \subseteq \Omega$ with Ω_j being a nondegenerate closed rectangle, $\mu(\Omega_j) < \infty$, $\forall j \in \mathbb{N}$ such that $\Omega = \bigcup_{j=1}^\infty \Omega_j$. Fix any $j \in \mathbb{N}$. By the assumption of the proposition, we can apply the result of the special case on Ω_j. There exists $U_{J,j} \in \mathcal{B}$ with $\mu(\Omega_j \setminus U_{J,j}) = 0$ such that $f(M_J(\cdot, \pi_{\bar{J}\setminus J}(x))) : \pi_J(\Omega_j) \to \mathcal{Y}$ is absolutely integrable over $\pi_J(\mathrm{r}_{\hat{x},\check{x}})$ with respect to μ_J, $\forall x \in U_{J,j}$; and $F_j : \Omega_j \to \mathcal{Y}$, defined by, $\forall x \in \Omega_j$,

$$F_j(x) = \begin{cases} (-1)^{\mathrm{card}(\check{J}\cap J)} \int_{\pi_J(\mathrm{r}_{\hat{x},\check{x}})} f(M_J(s, \\ \qquad \pi_{\bar{J}\setminus J}(x)))\,\mathrm{d}\mu_J(s) & x \in U_{J,j} \\ \vartheta_{\mathcal{Y}} & x \in \Omega_j \setminus U_{J,j} \end{cases}$$

is $\mathcal{B}$-measurable. Let $U_J := \bigcup_{j=1}^\infty U_{J,j} \in \mathcal{B}$. Then, $0 \le \mu(\Omega \setminus U_J) = \mu(\bigcup_{j=1}^\infty (\Omega_j \setminus U_J)) \le \mu(\bigcup_{j=1}^\infty (\Omega_j \setminus U_{J,j})) \le 0$. Define F according to the statement of the proposition, and we observe that $F(x) = F_j(x)$, $\forall x \in U_{J,j} \setminus (\bigcup_{l=1}^{j-1} U_{J,l})$, if $x \in U_J$; $F(x) = \vartheta_{\mathcal{Y}}$, $\forall x \in \Omega \setminus U_J$. By Proposition 11.41, F is $\mathcal{B}$-measurable. Clearly, $f(M_J(\cdot, \pi_{\bar{J}\setminus J}(x))) : \pi_J(\Omega) \to \mathcal{Y}$ is absolutely integrable over $\pi_J(\mathrm{r}_{\hat{x},\check{x}}) \subseteq \pi_J(\Omega)$ with respect to μ_J, $\forall x \in U_J$.

If, in addition, $\mathcal{Y}$ is finite-dimensional, then, by the special case, $U_{J,j} = \{x \in \Omega_j \mid f(M_J(\cdot, \pi_{\bar{J}\setminus J}(x))) : \pi_J(\Omega_j) \to \mathcal{Y}$ is absolutely integrable over $\pi_J(\mathrm{r}_{\hat{x},\check{x}})$ with respect to $\mu_J\}$, $\forall j \in \mathbb{N}$. Then, $U_J = \bigcup_{j=1}^\infty U_{J,j} = \{x \in \Omega \mid f(M_J(\cdot, \pi_{\bar{J}\setminus J}(x))) : \pi_J(\Omega) \to \mathcal{Y}$ is absolutely integrable over $\pi_J(\mathrm{r}_{\hat{x},\check{x}})$ withrespect to $\mu_J\}$. This completes the proof of the proposition. □

Theorem 12.86 (Fundamental Theorem of Calculus I) *Let $m \in \mathbb{N}$, $\mathbb{R}^m$ be endowed with the usual positive cone, $\Omega \in \mathcal{B}_{\mathrm{B}}(\mathbb{R}^m)$ be an open rectangle, $\mathbb{X} := ((\Omega, |\cdot|), \mathcal{B}, \mu)$ be the σ-finite metric measure subspace of $\mathbb{R}^m$, $\mathcal{Y}$ be a separable Banach space over $\mathbb{K}$, $f : \Omega \to \mathcal{Y}$ be $\mathcal{B}$-measurable, $\bar{J} := \{1, \ldots, m\}$, and $x_0 \in \Omega$. Assume that f is absolutely integrable over r_{x_a,x_b} with respect to μ, $\forall x_a, x_b \in \Omega$ with $x_a \leqq x_b$. Let ν be the σ-finite $\mathcal{Y}$-valued measure with kernel f over $\mathbb{X}$ according* *to Proposition 11.116. Define $F : \Omega \to \mathcal{Y}$ to be the cumulative distribution function of ν with origin at x_0 according to Proposition 12.52. F is absolutely continuous by Proposition 12.75, and $f = \frac{d\nu}{d\mu}$ a.e. in $\mathbb{X}$. Let $D_i F : U_i \to \mathcal{Y}$, $\forall i \in \bar{J}$, where D_i is the derivative of a function with respect to the ith coordinate variable of Ω. $\forall i \in \bar{J}$, define $\pi_{\hat{i}} : \mathbb{R}^m \to \mathbb{R}^{m-1}$ by $\pi_{\hat{i}}(x) = \big(\pi_j(x)\big)_{j \in \bar{J}\setminus\{i\}}$, $\forall x \in \mathbb{R}^m$; also define $M_i : \mathbb{R} \times \mathbb{R}^{m-1} \to \mathbb{R}^m$ by $\pi_i(M_i(x, y)) = x$ and $\pi_{\hat{i}}(M_i(x, y)) = y$, $\forall x \in \mathbb{R}$, $\forall y \in \mathbb{R}^{m-1}$. Then, the following statements hold:*

(i) *$F(x) = (-1)^{\mathrm{card}(\check{J})} \int_{\mathrm{r}_{\hat{x},\check{x}}} f \, d\mu$, $\forall x \in \Omega$, where $\hat{x} := x_0 \wedge x$, $\check{x} := x_0 \vee x$, and $\check{J} := \big\{ i \in \bar{J} \mid \pi_i(x_0) > \pi_i(x) \big\}$.*

(ii) *$U_i \in \mathcal{B}$, $\mu(\Omega \setminus U_i) = 0$, $D_i F$ is $\mathcal{B}$-measurable; $\exists \bar{U}_i \in \mathcal{B}$ with $\mu(\Omega \setminus \bar{U}_i) = 0$, $\forall x \in \bar{U}_i$, $f(M_i(\pi_i(x), \cdot)) : \pi_{\hat{i}}(\Omega) \to \mathcal{Y}$ is absolutely integrable over $\pi_{\hat{i}}(\mathrm{r}_{\hat{x},\check{x}})$ with respect to $\mu_{\hat{i}}$, and $\exists p_i : \Omega \to \mathcal{Y}$ defined by, $\forall x \in \Omega$,*

$$p_i(x) = \begin{cases} (-1)^{\mathrm{card}(\check{J}\setminus\{i\})} \displaystyle\int_{\pi_{\hat{i}}(\mathrm{r}_{\hat{x},\check{x}})} f(M_i(\pi_i(x), s)) \, d\mu_{\hat{i}}(s) & x \in \bar{U}_i \\ \vartheta_{\mathcal{Y}} & x \in \Omega \setminus \bar{U}_i \end{cases}$$

when $m > 1$, $p_i(x) = f(x)$, $\forall x \in \Omega$, $\bar{U}_i = \emptyset$, when $m = 1$; where $(\pi_{\hat{i}}(\Omega), \mathcal{B}_{\hat{i}}, \mu_{\hat{i}})$ is the measure subspace of $\mathbb{R}^{m-1}$, such that p_i is $\mathcal{B}$-measurable, and $D_i F(x) = p_i(x)$ a.e. $x \in \mathbb{X}$, $\forall i \in \bar{J}$.

(iii) *If, in addition, $\mathcal{Y}$ is finite-dimensional,[1] then $\bar{U}_i = \{x \in \Omega \mid f(M_i(\pi_i(x), \cdot)) : \pi_{\hat{i}}(\Omega) \to \mathcal{Y}$ is absolutely integrable over $\pi_{\hat{i}}(\mathrm{r}_{\hat{x},\check{x}})$ withrespect to $\mu_{\hat{i}}\}$.*

Proof By Proposition 12.75, F is absolutely continuous. By Definition 12.41 and Definition 11.166, we have $\Delta_F(\mathrm{r}_{\hat{x},\check{x}}) = \nu(\mathrm{r}_{\hat{x},\check{x}}) = \int_{\mathrm{r}_{\hat{x},\check{x}}} f \, d\mu$. Since F is the cumulative distribution function of ν with origin at x_0, then $\Delta_F(\mathrm{r}_{\hat{x},\check{x}}) = (-1)^{\mathrm{card}(\check{J})} F(x)$. Thus, we have $F(x) = (-1)^{\mathrm{card}(\check{J})} \int_{\mathrm{r}_{\hat{x},\check{x}}} f \, d\mu$, $\forall x \in \Omega$. This establishes (i).

(ii) The result is trivial when $\Omega = \emptyset$. Let $\Omega \neq \emptyset$. By Proposition 12.81, $D_i F$ is $\mathcal{B}$-measurable, $\forall i \in \bar{J}$. We will distinguish two exhaustive and mutually exclusive cases: Case 1: $m = 1$; Case 2: $m > 1$. Case 1: $m = 1$. Then, $i = 1$. $F(x) = (-1)^{\mathrm{card}(\check{J})} \int_{\mathrm{r}_{\hat{x},\check{x}}} f \, d\mu = \int_{x_0}^{x} f \, d\mu_{\mathrm{B}}$, $\forall x \in \Omega$. Since Ω is a nonempty open interval, then $\exists (E_j)_{j=1}^{\infty} := ((a_j, b_j))_{j=1}^{\infty} \subseteq \mathcal{O}_{\mathbb{R}}$ such that $\Omega = \bigcup_{j=1}^{\infty} E_j$ and $-\infty < \cdots < a_n < \cdots < a_2 < a_1 < x_0 < b_1 < b_2 < \cdots < b_n < \cdots < \infty$. Fix any $j \in \mathbb{N}$. Let $\mathcal{E}_j := (E_j, \mathcal{B}_j, \mu_j)$ be the finite measure subspace of $(\Omega, \mathcal{B}, \mu)$. Then, $f \in \bar{L}_1(\mathcal{E}_j, \mathcal{Y})$. Let $c \in (0, 1)$ be a constant. Define $A_j := \{x \in E_j \mid x$ is a rectangular Lebesgue point of $f|_{E_j}$ with regularity $c\}$. By Proposition 12.80, $A_j \in \mathcal{B}_{\mathrm{L}}$ and $\mu_{\mathrm{L}}(E_j \setminus A_j) = 0$. Fix any $x \in A_j$. We have

[1] The finite-dimensionality assumption might possibly be relaxed to σ-compact conic segment.

$$\limsup_{r\to 0^+} \sup_{\substack{\mathrm{r}_{\tilde{x}_1,\tilde{x}_2}\subset\mathbb{R} \text{ with } \frac{1}{2}(\tilde{x}_1+\tilde{x}_2)=x,\\ |\tilde{x}_2-\tilde{x}_1|=r,(\tilde{x}_2-\tilde{x}_1)\geq cr}} \frac{1}{\mu(\mathrm{r}_{\tilde{x}_1,\tilde{x}_2}\cap E_j)} \int_{\mathrm{r}_{\tilde{x}_1,\tilde{x}_2}\cap E_j} \|f(y)-f(x)\|\cdot \mathrm{d}\mu(y) = 0.$$

$\forall \epsilon \in (0,\infty) \subset \mathbb{R}$, $\exists r_0 \in (0,\infty) \subset \mathbb{R}$, $\forall r \in (0,r_0) \subset \mathbb{R}$, we have $0 \leq \sup_{\substack{\mathrm{r}_{\tilde{x}_1,\tilde{x}_2}\subset\mathbb{R} \text{ with } \frac{1}{2}(\tilde{x}_1+\tilde{x}_2)=x,\\ |\tilde{x}_2-\tilde{x}_1|=r,(\tilde{x}_2-\tilde{x}_1)\geq cr}} \frac{1}{\mu(\mathrm{r}_{\tilde{x}_1,\tilde{x}_2}\cap E_j)} \int_{\mathrm{r}_{\tilde{x}_1,\tilde{x}_2}\cap E_j} \|f(y)-f(x)\|\,\mathrm{d}\mu(y) < \epsilon$. Let $\mathrm{r}_{\tilde{x}_1,\tilde{x}_2} \subseteq \mathbb{R}$ be such that $\frac{1}{2}(\tilde{x}_1+\tilde{x}_2) = x$, $\tilde{x}_2 - x < r_0/2$, and $\bar{\mathrm{r}}_{\tilde{x}_1,\tilde{x}_2} \subseteq E_j$. $\forall \bar{x} \in \mathrm{r}_{\tilde{x}_1,\tilde{x}_2}$ with $\bar{x} \neq x$, we have, where $h := \bar{x} - x$,

$$\begin{aligned}
0 &\leq \|F(\bar{x}) - F(x) - f(x)(\bar{x}-x)\| \\
&= \left\| \int_x^{x+h} f(y)\,\mathrm{d}\mu(y) - \int_x^{x+h} f(x)\,\mathrm{d}\mu(y) \right\| \\
&= \left\| \int_x^{x+h} (f(y)-f(x))\,\mathrm{d}\mu(y) \right\| \\
&= \left\| \int_{\mathrm{r}_{x\wedge(x+h),x\vee(x+h)}} (f(y)-f(x))\,\mathrm{d}\mu(y) \right\| \\
&\leq \int_{\mathrm{r}_{x-|h|,x+|h|}} \|f(y)-f(x)\|\,\mathrm{d}\mu(y) \\
&= 2|h| \frac{1}{\mu_{\mathrm{B}}(\mathrm{r}_{x-|h|,x+|h|})} \int_{\mathrm{r}_{x-|h|,x+|h|}} \|f(y)-f(x)\|\,\mathrm{d}\mu_{\mathrm{B}}(y) \\
&\leq 2\epsilon\,|\bar{x}-x|
\end{aligned}$$

where the first equality follows from Fact 12.72 and Proposition 11.75, the second equality follows from Proposition 11.92, the third equality follows from Definition 12.71, the second inequality follows from Proposition 11.92, and the last inequality follows from the fact $x \in A_j$ and the choice of r_0. Then, $x \in \mathrm{dom}(\mathrm{D}_1 F) = U_1$ and $\mathrm{D}_1 F(x) = f(x)$. By the arbitrariness of x, we have $A_j \subseteq U_1$. Clearly, $\bigcup_{j=1}^{\infty} A_j \subseteq U_1 \subseteq \Omega$. This implies that $0 \leq \mu(\Omega \setminus U_1) = \mu_{\mathrm{B}}(\Omega \setminus U_1) = \mu_{\mathrm{L}}(\Omega \setminus U_1) \leq \mu_{\mathrm{L}}(\bigcup_{j=1}^{\infty}(E_j \setminus (\bigcup_{l=1}^{\infty} A_l))) \leq \mu_{\mathrm{L}}(\bigcup_{j=1}^{\infty}(E_j \setminus A_j)) \leq 0$. Hence, $\mu(\Omega \setminus U_1) = 0$. By $\mathrm{D}_1 F$ being $\mathcal{B}$-measurable and f being $\mathcal{B}$-measurable, we have $\mathrm{D}_1 F(x) = f(x)$ a.e. $x \in \mathbb{X}$. This case is proved.

Case 2: $m > 1$. Without loss of generality, consider $i = 1$. Let $\mathbb{X}_1 := (\pi_1(\Omega), \bar{\mathcal{B}}, \bar{\mu})$ be the measure subspace of $\mathbb{R}$. By Proposition 12.85, $\exists \bar{U}_1 \in \mathcal{B}$ with $\mu(\Omega \setminus \bar{U}_1) = 0$, and p_1 as defined in the statement of the theorem, such that p_1 is $\mathcal{B}$-measurable. Let $\bar{E} := \Omega \setminus \bar{U}_1$. By Lemma 12.27, $\exists \bar{U}_{1,\hat{1}} \in \mathcal{B}_{\hat{1}}$ with $\mu_{\hat{1}}(\pi_{\hat{1}}(\Omega) \setminus \bar{U}_{1,\hat{1}}) = 0$ such that $\bar{\mu}(\bar{E}_{\langle \pi_{\hat{1}}(x)\rangle}) = 0$, $\forall \pi_{\hat{1}}(x) \in \bar{U}_{1,\hat{1}}$, where $\bar{E}_{\langle \pi_{\hat{1}}(x)\rangle} := \left\{ \bar{s} \in \pi_1(\Omega) \mid M_1(\bar{s}, \pi_{\hat{1}}(x)) \in \bar{E} \right\}$. By Fubini's Theorem 12.30, we have $F(x) = \int_{\pi_1(x_0)}^{\pi_1(x)} p_1(M_1(s, \pi_{\hat{1}}(x)))\,\mathrm{d}\bar{\mu}(s)$, $\forall x \in \Omega$ with $\pi_{\hat{1}}(x) \in \bar{U}_{1,\hat{1}}$. Then, by Case 1, we have $\mathrm{D}_{\mathrm{I}} F(x) = p_{\mathrm{I}}(x)$ a.e. $\pi_{\mathrm{I}}(x) \in \mathbb{X}_{\mathrm{I}}$, $\forall \pi_{\hat{\mathrm{I}}}(x) \in \bar{U}_{1,\hat{\mathrm{I}}}$. Since $\mathrm{D}_{\mathrm{I}} F$ and p_{I} are $\mathcal{B}$-measurable and $\mathcal{Y}$ is separable, then, by Propositions 7.23, 11.39, and 11.38, $\mathrm{D}_{\mathrm{I}} F - p_{\mathrm{I}}$ is $\mathcal{B}$-measurable. Then, $\hat{U}_1 := \{x \in \Omega \mid \mathrm{D}_{\mathrm{I}} F(x) = p_{\mathrm{I}}(x)\} \in \mathcal{B}$. By Lemma 12.27, $\mu(\Omega \setminus \hat{U}_1) =: \mu(E) = \int_{\pi_{\hat{\mathrm{I}}}(\Omega)} \bar{\mu}(E_{\langle \pi_{\hat{\mathrm{I}}}(x)\rangle}) \,\mathrm{d}\mu_{\hat{\mathrm{I}}}(\pi_{\hat{\mathrm{I}}}(x)) = \int_{\bar{U}_{1,\hat{\mathrm{I}}}} \bar{\mu}(E_{\langle \pi_{\hat{\mathrm{I}}}(x)\rangle}) \,\mathrm{d}\mu_{\hat{\mathrm{I}}}(\pi_{\hat{\mathrm{I}}}(x)) = \int_{\bar{U}_{1,\hat{\mathrm{I}}}} 0 \,\mathrm{d}\mu_{\hat{\mathrm{I}}} = 0$, where $E_{\langle \pi_{\hat{\mathrm{I}}}(x)\rangle} := \{\bar{s} \in \pi_{\mathrm{I}}(\Omega) \mid M_{\mathrm{I}}(\bar{s}, \pi_{\hat{\mathrm{I}}}(x)) \in E\}$; and $\bar{\mu}(E_{\langle \pi_{\hat{\mathrm{I}}}(x)\rangle}) = 0$ follows from $\mathrm{D}_{\mathrm{I}} F(x) = p_{\mathrm{I}}(x)$ a.e. $\pi_{\mathrm{I}}(x) \in \mathbb{X}_{\mathrm{I}}$, $\forall \pi_{\hat{\mathrm{I}}}(x) \in \bar{U}_{1,\hat{\mathrm{I}}}$. Hence, $\mathrm{D}_{\mathrm{I}} F = p_{\mathrm{I}}$ a.e. in $\mathbb{X}$, and $U_1 \in \mathcal{B}$ with $\mu(\Omega \setminus U_1) = 0$. This case is also proved.

In both cases, (ii) holds. Hence, (ii) is true.

(iii) This follows immediately from (ii) and Proposition 12.85.

This completes the proof of the theorem. □

Proposition 12.87 *Let $m \in \mathbb{N}$, $\mathbb{R}^m$ be endowed with the usual positive cone, $\Omega \in \mathcal{B}_{\mathrm{B}}(\mathbb{R}^m)$ be a region, $\mathcal{Z}$ be a normed linear space, $G : \Omega \to \mathcal{Z}$, $\bar{J} := \{1, \ldots, m\}$, and $x_0 \in \Omega$. $\forall J \subseteq \bar{J}$, define $\pi_J : \mathbb{R}^m \to \mathbb{R}^{\mathrm{card}(J)}$ by $\pi_J(x) = (\pi_i(x))_{i \in J}$, $\forall x \in \mathbb{R}^m$; define $M_J : \mathbb{R}^{\mathrm{card}(J)} \times \mathbb{R}^{m-\mathrm{card}(J)} \to \mathbb{R}^m$ by $M_J(s,t) \in \mathbb{R}^m$ such that $\pi_J(M_J(s,t)) = s$ and $\pi_{\bar{J}\setminus J}(M_J(s,t)) = t$, $\forall s \in \mathbb{R}^{\mathrm{card}(J)}$, $\forall t \in \mathbb{R}^{m-\mathrm{card}(J)}$; also define $G_J : \pi_J(\Omega) \to \mathcal{Y}$ by $G_J(s) = G(M_J(s, \pi_{\bar{J}\setminus J}(x_0)))$, $\forall s \in \pi_J(\Omega)$. Then, we have, $\forall x \in \Omega$,*

$$(-1)^{\mathrm{card}(\check{J})} \Delta_G(\mathrm{r}_{\hat{x},\check{x}}) = G(x) - G(x_0) - \sum_{J \subset \bar{J}, J \neq \emptyset} (-1)^{\mathrm{card}(\check{J} \cap J)} \Delta_{G_J}(\pi_J(\mathrm{r}_{\hat{x},\check{x}})) \tag{12.2}$$

where $\hat{x} := x \wedge x_0$, $\check{x} := x \vee x_0$; and $\check{J} := \{i \in \bar{J} \mid \pi_i(x_0) > \pi_i(x)\}$.

Proof We will prove the result using mathematical induction on m:

1° $m = 1$. This is obvious.

2° Assume the claim holds for $m \leq k \in \mathbb{N}$.

3° Consider the case when $m = k + 1 \in \{2, 3, \ldots\}$. Fix any $x \in \Omega$. Define $\bar{G} : \pi_{\bar{J}\setminus\{k+1\}}(\Omega) \to \mathcal{Z}$ by $\bar{G}(s) = G(M_{\bar{J}\setminus\{k+1\}}(s, \pi_{k+1}(x)))$, $\forall s \in \pi_{\bar{J}\setminus\{k+1\}}(\Omega)$. Define $\bar{G}_J : \pi_J(\Omega) \to \mathcal{Z}$ in terms of $\bar{G}$ in a similar manner as G_J in terms of G, $\forall J \subseteq \bar{J} \setminus \{k+1\}$. Then, we have the following sequence of arguments.

$$\begin{aligned}
&(-1)^{\mathrm{card}(\check{J})} \Delta_G(\mathrm{r}_{\hat{x},\check{x}}) \\
&= (-1)^{\mathrm{card}(\check{J}\setminus\{k+1\})} \Big(\Delta_{\bar{G}}(\pi_{\bar{J}\setminus\{k+1\}}(\mathrm{r}_{\hat{x},\check{x}})) - \Delta_G(\pi_{\bar{J}\setminus\{k+1\}}(\mathrm{r}_{\hat{x},\check{x}})) \Big) \\
&= \bar{G}(\pi_{\bar{J}\setminus\{k+1\}}(x)) - \bar{G}(\pi_{\bar{J}\setminus\{k+1\}}(x_0)) \\
&\quad - \sum_{J \subset \bar{J}\setminus\{k+1\}, J \neq \emptyset} (-1)^{\mathrm{card}((\check{J}\setminus\{k+1\}) \cap J)} \Delta_{\bar{G}_J}(\pi_J(\mathrm{r}_{\hat{x},\check{x}})) \\
&\quad - (-1)^{\mathrm{card}(\check{J}\setminus\{k+1\})} \Delta_G(\pi_{\bar{J}\setminus\{k+1\}}(\mathrm{r}_{\hat{x},\check{x}}))
\end{aligned}$$

$$
\begin{aligned}
&= G(x) - G_{\{k+1\}}(\pi_{k+1}(x)) - (-1)^{\operatorname{card}(\check{J}\setminus\{k+1\})} \Delta_G(\pi_{\bar{J}\setminus\{k+1\}}(\mathrm{r}_{\hat{x},\check{x}})) \\
&\quad - \sum_{J\subset\bar{J}\setminus\{k+1\}, J\neq\emptyset} (-1)^{\operatorname{card}((\check{J}\setminus\{k+1\})\cap J)} \Delta_{G_J}(\pi_J(\mathrm{r}_{\hat{x},\check{x}})) \\
&\quad - \sum_{\substack{J\subset\bar{J}\setminus\{k+1\}\\ J\neq\emptyset}} (-1)^{\operatorname{card}((\check{J}\setminus\{k+1\})\cap J)} \Big(\Delta_{\bar{G}_J}(\pi_J(\mathrm{r}_{\hat{x},\check{x}})) - \Delta_{G_J}(\pi_J(\mathrm{r}_{\hat{x},\check{x}})) \Big) \\
&= G(x) - G_{\{k+1\}}(\pi_{k+1}(x)) - \sum_{\substack{J\subseteq\bar{J}\setminus\{k+1\}\\ J\neq\emptyset}} (-1)^{\operatorname{card}(\check{J}\cap J)} \Delta_{G_J}(\pi_J(\mathrm{r}_{\hat{x},\check{x}})) \\
&\quad - \sum_{J\subset\bar{J}, J\supset\{k+1\}} (-1)^{\operatorname{card}(\check{J}\cap J)} \Delta_{G_J}(\pi_J(\mathrm{r}_{\hat{x},\check{x}})) \\
&= G(x) - G_{\{k+1\}}(\pi_{k+1}(x)) - \sum_{J\subseteq\bar{J}, k+1\notin J\neq\emptyset} (-1)^{\operatorname{card}(\check{J}\cap J)} \Delta_{G_J}(\pi_J(\mathrm{r}_{\hat{x},\check{x}})) \\
&\quad - \sum_{J\subset\bar{J}, J\supset\{k+1\}} (-1)^{\operatorname{card}(\check{J}\cap J)} \Delta_{G_J}(\pi_J(\mathrm{r}_{\hat{x},\check{x}})) \\
&= G(x) - G_{\{k+1\}}(\pi_{k+1}(x)) - \sum_{J\subseteq\bar{J}, k+1\notin J\neq\emptyset} (-1)^{\operatorname{card}(\check{J}\cap J)} \Delta_{G_J}(\pi_J(\mathrm{r}_{\hat{x},\check{x}})) \\
&\quad - \sum_{J\subset\bar{J}, J\supseteq\{k+1\}} (-1)^{\operatorname{card}(\check{J}\cap J)} \Delta_{G_J}(\pi_J(\mathrm{r}_{\hat{x},\check{x}})) \\
&\quad + (-1)^{\operatorname{card}(\check{J}\cap\{k+1\})} \Delta_{G_{\{k+1\}}}(\pi_{\{k+1\}}(\mathrm{r}_{\hat{x},\check{x}})) \\
&= G(x) - G_{\{k+1\}}(\pi_{k+1}(x)) - \sum_{J\subset\bar{J}, J\neq\emptyset} (-1)^{\operatorname{card}(\check{J}\cap J)} \Delta_{G_J}(\pi_J(\mathrm{r}_{\hat{x},\check{x}})) \\
&\quad + (-1)^{\operatorname{card}(\check{J}\cap\{k+1\})} \Delta_{G_{\{k+1\}}}(\pi_{\{k+1\}}(\mathrm{r}_{\hat{x},\check{x}})) \\
&= G(x) - G(x_0) - \sum_{J\subset\bar{J}, J\neq\emptyset} (-1)^{\operatorname{card}(\check{J}\cap J)} \Delta_{G_J}(\pi_J(\mathrm{r}_{\hat{x},\check{x}}))
\end{aligned}
$$

where the first equality follows from Definition 12.41, the second equality follows from the inductive assumption 2°, and the fourth equality follows from Definition 12.41. This establishes the claim in this step and completes the induction process.

This completes the proof of the proposition. □

Theorem 12.88 (Fundamental Theorem of Calculus II) *Let $m \in \mathbb{N}$, $\mathbb{R}^m$ be endowed with the usual positive cone, $\Omega \in \mathcal{B}_B(\mathbb{R}^m)$ be an open rectangle, $\mathbb{X} := ((\Omega, |\cdot|), \mathcal{B}, \mu)$ be the σ-finite metric measure subspace of $\mathbb{R}^m$, $\mathcal{Y}$ be a separable reflexive Banach space over $\mathbb{K}$ with $\mathcal{Y}^*$ being separable, $F : \Omega \to \mathcal{Y}$ be absolutely continuous, $\bar{J} := \{1, \ldots, m\}$, $F^{(1)} : U \to B(\mathbb{R}^m, \mathcal{Y})$, $D_i F : U_i \to \mathcal{Y}$, $\forall i \in \bar{J}$, where D_i is the derivative of a function with respect to the ith coordinate variable of Ω, and $x_0 \in \Omega$. $\forall J \subseteq \bar{J}$, define $\pi_J : \mathbb{R}^m \to \mathbb{R}^{\mathrm{card}(J)}$ by $\pi_J(x) = (\pi_i(x))_{i \in J}$, $\forall x \in \mathbb{R}^m$; define $M_J : \mathbb{R}^{\mathrm{card}(J)} \times \mathbb{R}^{m-\mathrm{card}(J)} \to \mathbb{R}^m$ by $M_J(s,t) \in \mathbb{R}^m$ such that $\pi_J(M_J(s,t)) = s$ and $\pi_{\bar{J}\setminus J}(M_J(s,t)) = t$, $\forall s \in \mathbb{R}^{\mathrm{card}(J)}$, $\forall t \in \mathbb{R}^{m-\mathrm{card}(J)}$; also define $F_J : \pi_J(\Omega) \to \mathcal{Y}$ by $F_J(s) = F(M_J(s, \pi_{\bar{J}\setminus J}(x_0)))$, $\forall s \in \pi_J(\Omega)$; and let $\mathbb{X}_J := (\pi_J(\Omega), \mathcal{B}_J, \mu_J)$ be the measure subspace of $\mathbb{R}^{\mathrm{card}(J)}$ when $J \neq \emptyset$. Then, the following statements hold:*

(i) *There exists a set of functions $f_{J,x_0} : \pi_J(\Omega) \to \mathcal{Y}$, which is $\mathcal{B}_J$-measurable, $\forall J \subseteq \bar{J}$ with $J \neq \emptyset$, such that f_{J,x_0}'s are absolutely integrable on $\pi_J(\mathrm{r}_{x_a,x_b})$ with respect to μ_J, $\forall x_a, x_b \in \Omega$ with $x_a \stackrel{\leq}{=} x_b$, and we have, $\forall x \in \Omega$,*

$$F(x) = \sum_{J \subseteq \bar{J}, J \neq \emptyset} (-1)^{\mathrm{card}(\check{J} \cap J)} \int_{\pi_J(\mathrm{r}_{\hat{x},\check{x}})} f_{J,x_0} \,\mathrm{d}\mu_J + F(x_0) \tag{12.3}$$

where $\hat{x} := x \wedge x_0$, $\check{x} := x \vee x_0$; and $\check{J} := \{ i \in \bar{J} \mid \pi_i(x_0) > \pi_i(x) \}$. The set of functions $(f_{J,x_0})_{J \subseteq \bar{J}, J \neq \emptyset}$ is unique in the sense that if $(g_{J,x_0})_{J \subseteq \bar{J}, J \neq \emptyset}$ is another set of such functions satisfying (12.3), then $f_{J,x_0} = g_{J,x_0}$ a.e. in $\mathbb{X}_J$, $\forall J \subseteq \bar{J}$ with $J \neq \emptyset$.

(ii) *If we expand absolutely continuous function F at $\bar{x} \in \Omega$, instead of x_0, then we will have a set of functions $f_{J,\bar{x}} : \pi_J(\Omega) \to \mathcal{Y}$, which is $\mathcal{B}_J$-measurable, $\forall J \subseteq \bar{J}$ with $J \neq \emptyset$, satisfying the conclusion of (i) with x_0 replaced by $\bar{x}$. The $f_{J,\bar{x}}$'s relate to f_{J,x_0}'s according to the following formula, $\forall J \subseteq \bar{J}$ with $J \neq \emptyset$, $\exists \bar{U}_J \in \mathcal{B}_J$ with $\mu_J(\pi_J(\Omega) \setminus \bar{U}_J) = 0$, such that*

$$\begin{aligned} f_{J,\bar{x}}(s) = \sum_{J \subseteq \tilde{J} \subseteq \bar{J}} (-1)^{\mathrm{card}(\check{\bar{J}} \cap \tilde{J} \setminus J)} \int_{\pi_{\tilde{J}\setminus J}(\mathrm{r}_{\hat{\bar{x}},\check{\bar{x}}})} f_{\tilde{J},x_0}(\pi_{\tilde{J}}(M_J(s, \\ \pi_{\tilde{J}\setminus J}(t)))) \,\mathrm{d}\mu_{\tilde{J}\setminus J}(\pi_{\tilde{J}\setminus J}(t)) + f_{J,x_0}(s), \quad \forall s \in \bar{U}_J \end{aligned} \tag{12.4}$$

and $f_{J,\bar{x}}(s) = \vartheta_{\mathcal{Y}}$, $\forall s \in \pi_J(\Omega) \setminus \bar{U}_J$, where $\hat{\bar{x}} := \bar{x} \wedge x_0$, $\check{\bar{x}} := \bar{x} \vee x_0$, and $\check{\bar{J}} := \{ i \in \bar{J} \mid \pi_i(x_0) > \pi_i(\bar{x}) \}$; and all involved integrations are such that the integrands are absolutely integrable over the specified integration domain with respect to the given measures, respectively.

(iii) *$U_i \in \mathcal{B}$, $\mu(\Omega \setminus U_i) = 0$, $D_i F$ is $\mathcal{B}$-measurable, and, $\forall i \in \bar{J}$, $D_i F(x) = f_{\{i\},x_i}(\pi_{\{i\}}(x))$ a.e. $\pi_{\{i\}}(x) \in \mathbb{X}_{\{i\}}$, $\forall \pi_{\bar{J}\setminus\{i\}}(x) \in \mathbb{X}_{\bar{J}\setminus\{i\}}$, where $x_i := M_{\{i\}}(\pi_{\{i\}}(x_0), \pi_{\bar{J}\setminus\{i\}}(x))$; and $F(x) - F(x_i) = \int_{\pi_{\{i\}}(x_0)}^{\pi_{\{i\}}(x)} f_{\{i\},x_i} \,\mathrm{d}\mu_{\{i\}}$, $\forall x \in \mathbb{X}$.*

(iv) $U \subseteq \bigcap_{i=1}^{\infty} U_i$, $U \in \mathcal{B}$, $F^{(1)}$ *is* $\mathcal{B}$*-measurable, and,* $\forall x \in U$, *we have* $F^{(1)}(x) = \left[\mathrm{D}_1 F(x) \cdots \mathrm{D}_m F(x) \right]$. *If, in addition,* $\forall J \subseteq \bar{J}$ *with* $J \neq \emptyset$, $\forall x \in \Omega$, $\exists \delta(J, x) \in (0, \infty) \subset \mathbb{R}$, $\exists c(J, x) \in [0, \infty) \subset \mathbb{R}$ *such that* $\operatorname{ess\,sup}_{y \in \mathcal{B}_{\mathbb{R}^m}(x, \delta(J,x))} \left\| f_{J,x_0}(\pi_J(y)) \right\| \leq c(J, x)$ *and* $\mathcal{B}_{\mathbb{R}^m}(x, \delta(J, x)) \subseteq \Omega$, *then* $\mu(\Omega \setminus U) = 0$.

(v) *If, in addition,* $\mathcal{Y}$ *is finite-dimensional, then* $\bar{U}_J = \left\{ s \in \pi_J(\Omega) \,\middle|\, f_{\tilde{J},x_0} (\pi_{\tilde{J}}(M_J(s, \pi_{\tilde{J} \setminus J}(t)))) \right.$ *is absolutely integrable over* $\pi_{\tilde{J} \setminus J}(\mathrm{r}_{\hat{\tilde{x}}, \check{\tilde{x}}})$ *withrespect to* $\left. \mu_{\tilde{J} \setminus J}(\pi_{\tilde{J} \setminus J}(t)), \forall \tilde{J} \subseteq \bar{J} \text{ with } J \subset \tilde{J} \right\}$, $\forall J \subseteq \bar{J}$ *with* $J \neq \emptyset$.

In the rest of this book, f_{J,x_0} will be called *the stream of* F *in* J *coordinate subspace with respect to* x_0.

Proof (i) By $F = F_{\bar{J}}$ being absolutely continuous and Definition 12.58 and Proposition 12.60, $\forall J \subseteq \bar{J}$ with $J \neq \emptyset$, F_J is absolutely continuous. By Proposition 12.73, there exists a unique σ-finite $\mathcal{Y}$-valued measure ν_J on $(\pi_J(\Omega), \mathcal{B}_J)$ such that F_J is a cumulative distribution function of ν_J. Furthermore, $\mathcal{P} \circ \nu_J \ll \mu_J$. Since $\mathcal{Y}$ is a separable reflexive Banach space with $\mathcal{Y}^*$ being separable, then, by Radon–Nikodym Theorem 11.171, there exists a unique $f_{J,x_0} : \pi_J(\Omega) \to \mathcal{Y}$, which is $\mathcal{B}_J$-measurable, such that $f_{J,x_0} = \frac{\mathrm{d}\nu_J}{\mathrm{d}\mu_J}$ a.e. in $\mathbb{X}_J$. $\forall x_a, x_b \in \Omega$ with $x_a \stackrel{\leq}{=} x_b$, we have $\mathcal{P} \circ \nu_J(\pi_J(\mathrm{r}_{x_a,x_b})) = \int_{\pi_J(\mathrm{r}_{x_a,x_b})} \mathcal{P} \circ f_{J,x_0} \,\mathrm{d}\mu_J = T_{F_J}(\pi_J(\mathrm{r}_{x_a,x_b})) < \infty$, where the first equality follows from Definition 11.166, the second equality follows from Proposition 12.73, and the inequality follows from the fact F_J is absolutely continuous and therefore of locally bounded variation, which implies this according to Definition 12.41. Hence, f_{J,x_0} is absolutely integrable over $\pi_J(\mathrm{r}_{x_a,x_b})$ with respect to μ_J. Then, $\forall x \in \Omega$, we have $\Delta_{F_J}(\pi_J(\mathrm{r}_{\hat{x},\check{x}})) = \nu_J(\pi_J(\mathrm{r}_{\hat{x},\check{x}})) = \int_{\pi_J(\mathrm{r}_{\hat{x},\check{x}})} f_{J,x_0}(s) \,\mathrm{d}\mu_J(s)$. By Proposition 12.87, we have

$$(-1)^{\operatorname{card}(\check{J})} \Delta_F(\mathrm{r}_{\hat{x},\check{x}}) = F(x) - F(x_0) - \sum_{J \subset \bar{J}, J \neq \emptyset} (-1)^{\operatorname{card}(\check{J} \cap J)} \Delta_{F_J}(\pi_J(\mathrm{r}_{\hat{x},\check{x}}))$$

which implies that

$$(-1)^{\operatorname{card}(\check{J})} \int_{\pi_{\bar{J}}(\mathrm{r}_{\hat{x},\check{x}})} f_{\bar{J},x_0}(s) \,\mathrm{d}\mu_{\bar{J}}(s) = F(x) - F(x_0) - \sum_{J \subset \bar{J}, J \neq \emptyset} (-1)^{\operatorname{card}(\check{J} \cap J)} \int_{\pi_J(\mathrm{r}_{\hat{x},\check{x}})} f_{J,x_0}(s) \,\mathrm{d}\mu_J(s)$$

Reorganizing the above formula, we have

$$F(x) = F(x_0) + \sum_{J \subseteq \bar{J}, J \neq \emptyset} (-1)^{\operatorname{card}(\check{J} \cap J)} \int_{\pi_J(\mathrm{r}_{\hat{x},\check{x}})} f_{J,x_0}(s) \,\mathrm{d}\mu_J(s)$$

This establishes (12.3). The uniqueness of the set of functions $\left(f_{J,x_0}\right)_{J\subseteq\bar{J},J\neq\emptyset}$ is clear from the uniqueness of F_J's and the uniqueness of Radon–Nikodym derivatives. Hence, (i) holds.

(ii) By (i), there exists a set of functions $f_{J,\bar{x}} : \pi_J(\Omega) \to \mathcal{Y}$, which is $\mathcal{B}$-measurable, $\forall J \subseteq \bar{J}$ with $J \neq \emptyset$, such that $f_{J,\bar{x}}$'s are absolutely integrable on $\pi_J(\mathrm{r}_{x_a,x_b})$ with respect to μ_J, $\forall x_a, x_b \in \Omega$ with $x_a \leqq x_b$, and we have $F(x) = \sum_{J\subseteq\bar{J},J\neq\emptyset}(-1)^{\mathrm{card}(\check{\bar{J}}\cap J)}\int_{\pi_J(\mathrm{r}_{\hat{\bar{x}},\check{\bar{x}}})} f_{J,\bar{x}}\,\mathrm{d}\mu_J + F(\bar{x})$, where $\hat{\bar{x}} := \bar{x}\wedge x$, $\check{\bar{x}} := \bar{x}\vee x$, and $\check{\bar{J}} := \left\{i \in \bar{J} \,\middle|\, \pi_i(\bar{x}) > \pi_i(x)\right\}$. Now, we let $x = \bar{x} + M_J(s, \mathbf{0}_{m-\mathrm{card}(J)})$, $\forall s \in \mathbb{R}^{\mathrm{card}(J)}$, $\forall J \subseteq \bar{J}$ with $J \neq \emptyset$. By the preceding formula, we have $F(x) = \sum_{\hat{J}\subseteq J,\hat{J}\neq\emptyset}(-1)^{\mathrm{card}(\check{\bar{J}}\cap\hat{J})}\int_{\pi_{\hat{J}}(\mathrm{r}_{\hat{\bar{x}},\check{\bar{x}}})} f_{\hat{J},\bar{x}}\,\mathrm{d}\mu_{\hat{J}} + F(\bar{x})$. This is because $\forall\hat{J} \not\subseteq J$, the domain of integration will be an empty set and that leads to $\vartheta_{\mathcal{Y}}$ for that integral. Now, substitute the formula for $F(x)$ and $F(\bar{x})$ using (12.3), and we have

$$\sum_{\hat{J}\subseteq\bar{J},\hat{J}\neq\emptyset}(-1)^{\mathrm{card}(\check{J}\cap\hat{J})}\int_{\pi_{\hat{J}}(\mathrm{r}_{\hat{x},\check{x}})} f_{\hat{J},x_0}\,\mathrm{d}\mu_{\hat{J}} = \sum_{\hat{J}\subseteq J,\hat{J}\neq\emptyset}(-1)^{\mathrm{card}(\check{\bar{J}}\cap\hat{J})}\int_{\pi_{\hat{J}}(\mathrm{r}_{\hat{\bar{x}},\check{\bar{x}}})}$$

$$f_{\hat{J},\bar{x}}\,\mathrm{d}\mu_{\hat{J}} + \sum_{\hat{J}\subseteq\bar{J},\hat{J}\neq\emptyset}(-1)^{\mathrm{card}(\bar{\check{J}}\cap\hat{J})}\int_{\pi_{\hat{J}}(\mathrm{r}_{\bar{\hat{x}},\bar{\check{x}}})} f_{\hat{J},x_0}\,\mathrm{d}\mu_{\hat{J}}$$

Rearrange terms in the above, we obtain

$$\sum_{\hat{J}\subseteq J,\hat{J}\neq\emptyset}(-1)^{\mathrm{card}(\check{\bar{J}}\cap\hat{J})}\int_{\pi_{\hat{J}}(\mathrm{r}_{\hat{\bar{x}},\check{\bar{x}}})} f_{\hat{J},\bar{x}}\,\mathrm{d}\mu_{\hat{J}} = \sum_{\hat{J}\subseteq\bar{J},\hat{J}\neq\emptyset}(-1)^{\mathrm{card}(\check{J}\cap\hat{J})}\int_{\pi_{\hat{J}}(\mathrm{r}_{\hat{x},\check{x}})}$$

$$f_{\hat{J},x_0}\,\mathrm{d}\mu_{\hat{J}} - \sum_{\hat{J}\subseteq\bar{J},\hat{J}\neq\emptyset}(-1)^{\mathrm{card}(\bar{\check{J}}\cap\hat{J})}\int_{\pi_{\hat{J}}(\mathrm{r}_{\bar{\hat{x}},\bar{\check{x}}})} f_{\hat{J},x_0}\,\mathrm{d}\mu_{\hat{J}}$$

$$= \sum_{\substack{\hat{J}\subseteq\bar{J}\\\hat{J}\neq\emptyset}}\prod_{j\in\hat{J}}\left(\int_{\pi_j(x_0)}^{\pi_j(x)}\right) f_{\hat{J},x_0}(s_1,\ldots,s_{\mathrm{card}(\hat{J})})\,\mathrm{d}\mu_{\mathrm{B}}(s_{\mathrm{card}(\hat{J})})\cdots\mathrm{d}\mu_{\mathrm{B}}(s_1)$$

$$-\sum_{\substack{\hat{J}\subseteq\bar{J}\\\hat{J}\neq\emptyset}}\prod_{j\in\hat{J}}\left(\int_{\pi_j(x_0)}^{\pi_j(\bar{x})}\right) f_{\hat{J},x_0}(s_1,\ldots,s_{\mathrm{card}(\hat{J})})\,\mathrm{d}\mu_{\mathrm{B}}(s_{\mathrm{card}(\hat{J})})\cdots\mathrm{d}\mu_{\mathrm{B}}(s_1)$$

$$= \sum_{\hat{J}\subseteq\bar{J},\hat{J}\neq\emptyset}\prod_{j\in\hat{J}}\left(\int_{\pi_j(x_0)}^{\pi_j(\bar{x})} + \int_{\pi_j(\bar{x})}^{\pi_j(x)}\right) f_{\hat{J},x_0}(s_1,\ldots,s_{\mathrm{card}(\hat{J})})$$

$$\cdot\,\mathrm{d}\mu_{\mathrm{B}}(s_{\mathrm{card}(\hat{J})})\cdots\mathrm{d}\mu_{\mathrm{B}}(s_1) - \sum_{\hat{J}\subseteq\bar{J},\hat{J}\neq\emptyset}\prod_{j\in\hat{J}}\left(\int_{\pi_j(x_0)}^{\pi_j(\bar{x})}\right)$$

$$
f_{\hat{J},x_0}(s_1, \ldots, s_{\mathrm{card}(\hat{J})})\,\mathrm{d}\mu_{\mathrm{B}}(s_{\mathrm{card}(\hat{J})}) \cdots \mathrm{d}\mu_{\mathrm{B}}(s_1)
$$

$$
= \sum_{\hat{J}\subseteq\bar{J},\hat{J}\neq\emptyset} \sum_{\tilde{J}\subseteq\hat{J},\tilde{J}\neq\emptyset} \prod_{j\in\tilde{J}} \left(\int_{\pi_j(\bar{x})}^{\pi_j(x)} \right) \prod_{l\in\hat{J}\setminus\tilde{J}} \left(\int_{\pi_l(x_0)}^{\pi_l(\bar{x})} \right)
$$

$$
f_{\hat{J},x_0}(s_1, \ldots, s_{\mathrm{card}(\hat{J})})\,\mathrm{d}\mu_{\mathrm{B}}(s_{\mathrm{card}(\hat{J})}) \cdots \mathrm{d}\mu_{\mathrm{B}}(s_1)
$$

$$
= \sum_{\substack{\hat{J}\subseteq\bar{J}\\ \hat{J}\neq\emptyset}} \sum_{\substack{\tilde{J}\subseteq\hat{J}\\ \tilde{J}\neq\emptyset}} (-1)^{\mathrm{card}(\check{\bar{J}}\cap\tilde{J})+\mathrm{card}(\bar{\bar{J}}\cap\hat{J}\setminus\tilde{J})} \int_{\pi_{\hat{J}}(\mathrm{r}_{\hat{x}_{\tilde{J},x,\bar{x},x_0},\check{x}_{\tilde{J},x,\bar{x},x_0}})} f_{\hat{J},x_0}\,\mathrm{d}\mu_{\hat{J}}
$$

$$
= \sum_{\hat{J}\subseteq\bar{J},\hat{J}\neq\emptyset} \sum_{\tilde{J}\subseteq\hat{J}\cap J,\tilde{J}\neq\emptyset} (-1)^{\mathrm{card}(\check{\bar{J}}\cap\tilde{J})+\mathrm{card}(\bar{\bar{J}}\cap\hat{J}\setminus\tilde{J})} \int_{\pi_{\hat{J}}(\mathrm{r}_{\hat{x}_{\tilde{J},x,\bar{x},x_0},\check{x}_{\tilde{J},x,\bar{x},x_0}})}
$$

$$
f_{\hat{J},x_0}\,\mathrm{d}\mu_{\hat{J}}
$$

$$
= \sum_{\tilde{J}\subseteq\bar{J},\tilde{J}\neq\emptyset} \sum_{\hat{J}\subseteq\tilde{J}\cap J,\hat{J}\neq\emptyset} (-1)^{\mathrm{card}(\check{\bar{J}}\cap\hat{J})+\mathrm{card}(\bar{\bar{J}}\cap\tilde{J}\setminus\hat{J})} \int_{\pi_{\tilde{J}}(\mathrm{r}_{\hat{x}_{\hat{J},x,\bar{x},x_0},\check{x}_{\hat{J},x,\bar{x},x_0}})}
$$

$$
f_{\tilde{J},x_0}\,\mathrm{d}\mu_{\tilde{J}}
$$

$$
= \sum_{\substack{\hat{J}\subseteq J\\ \hat{J}\neq\emptyset}} \sum_{\substack{\tilde{J}\subseteq\bar{J}\\ \tilde{J}\supseteq\hat{J}}} (-1)^{\mathrm{card}(\check{\bar{J}}\cap\hat{J})+\mathrm{card}(\bar{\bar{J}}\cap\tilde{J}\setminus\hat{J})} \int_{\pi_{\tilde{J}}(\mathrm{r}_{\hat{x}_{\hat{J},x,\bar{x},x_0},\check{x}_{\hat{J},x,\bar{x},x_0}})} f_{\tilde{J},x_0}\,\mathrm{d}\mu_{\tilde{J}}
$$

$$
= \sum_{\hat{J}\subseteq J,\hat{J}\neq\emptyset} \Bigg(\sum_{\tilde{J}\subseteq\bar{J},\tilde{J}\supset\hat{J}} (-1)^{\mathrm{card}(\check{\bar{J}}\cap\hat{J})+\mathrm{card}(\bar{\bar{J}}\cap\tilde{J}\setminus\hat{J})} \int_{\pi_{\hat{J}}(\mathrm{r}_{\hat{\bar{x}},\check{\bar{x}}})} p_{\tilde{J},\hat{J}}\,\mathrm{d}\mu_{\hat{J}}
$$

$$
+(-1)^{\mathrm{card}(\check{\bar{J}}\cap\hat{J})} \int_{\pi_{\hat{J}}(\mathrm{r}_{\hat{\bar{x}},\check{\bar{x}}})} f_{\hat{J},x_0}\,\mathrm{d}\mu_{\hat{J}} \Bigg)
$$

$$
= \sum_{\substack{\hat{J}\subseteq J\\ \hat{J}\neq\emptyset}} (-1)^{\mathrm{card}(\check{\bar{J}}\cap\hat{J})} \int_{\pi_{\hat{J}}(\mathrm{r}_{\hat{\bar{x}},\check{\bar{x}}})} \Bigg(\sum_{\substack{\tilde{J}\subseteq\bar{J}\\ \tilde{J}\supset\hat{J}}} (-1)^{\mathrm{card}(\bar{\bar{J}}\cap\tilde{J}\setminus\hat{J})} p_{\tilde{J},\hat{J}} + f_{\hat{J},x_0} \Bigg)\,\mathrm{d}\mu_{\hat{J}}
$$

where $\hat{x}_{\tilde{J},x,\bar{x},x_0} := x_{1,\tilde{J},x,\bar{x},x_0} \wedge x_{2,\tilde{J},x,\bar{x},x_0}$, $\check{x}_{\tilde{J},x,\bar{x},x_0} := x_{1,\tilde{J},x,\bar{x},x_0} \vee x_{2,\tilde{J},x,\bar{x},x_0}$; $x_{1,\tilde{J},x,\bar{x},x_0} := M_{\tilde{J}}(\pi_{\tilde{J}}(x), \pi_{\bar{J}\setminus\tilde{J}}(\bar{x}))$, $x_{2,\tilde{J},x,\bar{x},x_0} := M_{\tilde{J}}(\pi_{\tilde{J}}(x), \pi_{\bar{J}\setminus J}(x_0))$; the second to fifth equalities follow from intuition; we have not used rigorous notation in these equations in expressing the joint integrals as repeated integrals and the other way around; this should not cause concern since from the first equality to the fifth, all that we have done is breaking down $(-1)^{\mathrm{card}(\check{J}\cap\hat{J})} \int_{\pi_{\hat{J}}(\mathrm{r}_{\hat{x},\check{x}})} f_{\hat{J},x_0}\,\mathrm{d}\mu_{\hat{J}}$ into pieces along the intermediate point $\bar{x}$; the sixth equality follows from the idea that the corresponding integral will be zero if its domain of integration is an empty set, which will happen since $\pi_i(x) = \pi_i(\bar{x})$, $\forall i \in \bar{J} \setminus J$; the seventh equality follows from the interchange of summing variables $\hat{J}$ and $\tilde{J}$;

the ninth equality follows from Fubini's Theorem 12.30, and $\bar{U}_{\bar{J},\hat{J}} \in \mathcal{B}_{\hat{J}}$ with $\mu_{\hat{J}}(\pi_{\hat{J}}(\Omega) \setminus \bar{U}_{\bar{J},\hat{J}}) = 0$, and $p_{\bar{J},\hat{J}} : \pi_{\hat{J}}(\Omega) \to \mathcal{Y}$, $\forall \hat{J} \subset \bar{J}$, defined by $p_{\bar{J},\hat{J}}(s) := \begin{cases} \int_{\pi_{\bar{J}\setminus\hat{J}}(\mathrm{r}_{\bar{\hat{x}},\check{x}})} f_{\bar{J},x_0}(\pi_{\bar{J}}(M_{\hat{J}}(s, \pi_{\bar{J}\setminus\hat{J}}(t)))) \,\mathrm{d}\mu_{\bar{J}\setminus\hat{J}}(\pi_{\bar{J}\setminus\hat{J}}(t)) & s \in \bar{U}_{\bar{J},\hat{J}} \\ \vartheta_{\mathcal{Y}} & s \in \pi_{\hat{J}}(\Omega) \setminus \bar{U}_{\bar{J},\hat{J}} \end{cases}$ is $\mathcal{B}_{\hat{J}}$-measurable; and the tenth equality follows from Proposition 11.92. Let $\bar{U}_{\hat{J}} := \bigcap_{\hat{J}\subset\check{J}\subseteq\bar{J}} \bar{U}_{\check{J},\hat{J}}$, $\forall \hat{J} \subseteq \bar{J}$ with $\hat{J} \neq \emptyset$. Clearly, $\bar{U}_{\hat{J}} \in \mathcal{B}_{\hat{J}}$ and $\mu_{\hat{J}}(\pi_{\hat{J}}(\Omega) \setminus \bar{U}_{\hat{J}}) = 0$. The definition of $f_{J,\bar{x}}$'s (12.4) satisfies the above equalities, $\forall J \subseteq \bar{J}$ with $J \neq \emptyset$. Then, by (i), this set of functions $\left(f_{J,\bar{x}}\right)_{J\subseteq\bar{J},J\neq\emptyset}$ is the unique one we seek.

(iii) Fix any $i \in \bar{J}$. Without loss of generality, let $i = 1$. By Proposition 12.81, $U_1 \in \mathcal{B}$ and $\mathrm{D}_1 F$ is $\mathcal{B}$-measurable. By Fundamental Theorem of Calculus 12.86, every term on the right-hand side of (12.3) is differentiable with respect to the ith coordinate variable of Ω almost everywhere in $\mathbb{X}$. $\forall J \subseteq \bar{J}$ with $1 \in J$, $\exists U_{J,1} \in \mathcal{B}$ with $\mu(\Omega \setminus U_{J,1}) = 0$, $\mathrm{D}_1\Big((-1)^{\mathrm{card}(\check{J}\cap J)} \int_{\pi_J(\mathrm{r}_{\hat{x},\check{x}})} f_{J,x_0}\,\mathrm{d}\mu_J\Big) : U_{J,1} \to \mathcal{Y}$ is $\mathcal{B}$-measurable, $\exists \bar{U}_{J,1} \in \mathcal{B}$ with $\mu(\Omega \setminus \bar{U}_{J,1}) = 0$, and $\exists p_{J,1} : \Omega \to \mathcal{Y}$ defined by $p_{J,1} = \begin{cases} (-1)^{\mathrm{card}(\check{J}\cap J\setminus\{1\})} \int_{\pi_{J\setminus\{1\}}(\mathrm{r}_{\hat{x},\check{x}})} f_{J,x_0}(\pi_J(M_{\{1\}}(\pi_1(x)\\ \quad , \pi_{\bar{J}\setminus\{1\}}(s)))) \,\mathrm{d}\mu_{J\setminus\{1\}}(\pi_{J\setminus\{1\}}(s)) & x \in \bar{U}_{J,1} \\ \vartheta_{\mathcal{Y}} & x \in \Omega \setminus \bar{U}_{J,1} \end{cases}$, $\forall x \in \Omega$, when $J \supset \{1\}$, $p_{J,1} = f_{J,x_0}(\pi_J(x))$, $\forall x \in \Omega$, $\bar{U}_{J,1} = \Omega$, when $J = \{1\}$, such that $p_{J,1}$ is $\mathcal{B}$-measurable and $\mathrm{D}_1\Big((-1)^{\mathrm{card}(\check{J}\cap J)} \int_{\pi_J(\mathrm{r}_{\hat{x},\check{x}})} f_{J,x_0}\,\mathrm{d}\mu_J\Big) = p_{J,1}(x)$ a.e. $x \in \mathbb{X}$. $\forall J \subseteq \bar{J}$ with $1 \notin J$, we have $\mathrm{D}_1\Big((-1)^{\mathrm{card}(\check{J}\cap J)} \cdot \int_{\pi_J(\mathrm{r}_{\hat{x},\check{x}})} f_{J,x_0}\,\mathrm{d}\mu_J\Big) = \vartheta_{\mathcal{Y}}$, $\forall x \in \Omega$. In this case, we have $\bar{U}_{J,1} = \Omega$. Let $\bar{U}_1 := \bigcap_{J\subseteq\bar{J},J\neq\emptyset} \bar{U}_{J,1}$. Then, $\bar{U}_1 \in \mathcal{B}$ and $\mu(\Omega \setminus \bar{U}_1) = 0$. Hence, $\mathrm{D}_1 F(x) = \sum_{J\subseteq\bar{J},J\neq\emptyset} p_{J,1}(x)$, $\forall x \in \bar{U}_1$, and $\mathrm{D}_1 F(x) = \sum_{J\subseteq\bar{J},J\neq\emptyset} p_{J,1}(x)$ a.e. $x \in \mathbb{X}$. This implies that $\mu(\Omega \setminus U_1) = 0$. Now, expand the function F at x_1 (which is as defined in the statement of the theorem.) By (12.3), we have $F(x) = F(x_1) + \int_{\pi_1(x_0)}^{\pi_1(x)} f_{\{1\},x_1}\,\mathrm{d}\mu_{\{1\}}$, $\forall x \in \Omega$. By Fundamental Theorem of Calculus I, Theorem 12.86, we have $\mathrm{D}_1 F(x) = f_{\{1\},x_1}(\pi_{\{1\}}(x))$ a.e. $\pi_{\{1\}}(x) \in \mathbb{X}_{\{1\}}$, $\forall \pi_{\bar{J}\setminus\{1\}}(x) \in \mathbb{X}_{\bar{J}\setminus\{1\}}$. Thus, (iii) holds.

(iv) By Proposition 12.81, $U \subseteq \bigcap_{i=1}^m U_i$, $F^{(1)}$ is $\mathcal{B}$-measurable, and $\forall x \in U$, we have $F^{(1)}(x) = \left[\mathrm{D}_1 F(x) \cdots \mathrm{D}_m F(x)\right]$. All we need to show is that $\mu(\Omega \setminus U) = 0$ under the additional assumption that f_{J,x_0}'s are locally bounded.

$\forall i \in \bar{J}$, $\forall J \subseteq \bar{J}$ with $i \in J$, by the proof of (iii), $\exists \tilde{U}_{J,i} \in \mathcal{B}$ with $\mu(\Omega\setminus\tilde{U}_{J,i}) = 0$, such that $\mathrm{D}_i H_J(x) := \mathrm{D}_i\Big((-1)^{\mathrm{card}(\check{J}\cap J)} \int_{\pi_J(\mathrm{r}_{\hat{x},\check{x}})} f_{J,x_0}\,\mathrm{d}\mu_J\Big)(x) = p_{J,i}(x)$, $\forall x \in \tilde{U}_{J,i}$. Then, let $\hat{U}_{J,i} := U_{J,i} \cap \bar{U}_{J,i} \cap \tilde{U}_{J,i} \in \mathcal{B}$, we have $\mu(\Omega \setminus \hat{U}_{J,i}) = 0$ and $\mathrm{D}_i H_J(x) = (-1)^{\mathrm{card}(\check{J}\cap J\setminus\{i\})} \cdot \int_{\pi_{J\setminus\{i\}}(\mathrm{r}_{\hat{x},\check{x}})} f_{J,x_0}(\pi_J(M_{\{i\}}(\pi_i(x), \pi_{\bar{J}\setminus\{i\}}(s)))) \cdot \mathrm{d}\mu_{J\setminus\{i\}}(\pi_{J\setminus\{i\}}(s)) =: H_{J,i}(x)$, $\forall x \in \hat{U}_{J,i}$.

Let $\check{U} := \left(\bigcap_{i=1}^{m} \bigcap_{i\in J\subseteq \bar{J}} \hat{U}_{J,i}\right) \cap \left(\bigcap_{i=1}^{m} U_i\right) \in \mathcal{B}$. Clearly, $\mu(\Omega \setminus \check{U}) = 0$. Fix any $\bar{x} \in \check{U}$. Then, by (ii), there exists a unique set of functions $\left(f_{J,\bar{x}}\right)_{J\subseteq \bar{J}, J\neq\emptyset}$ such that $F(x) = \sum_{J\subseteq \bar{J}, J\neq\emptyset} (-1)^{\mathrm{card}(\check{\bar{J}}\cap J)} \int_{\pi_J(\mathrm{r}_{\hat{\bar{x}},\check{\bar{x}}})} f_{J,\bar{x}}\, \mathrm{d}\mu_J + F(\bar{x})$, $\forall x \in \Omega$. Furthermore, $f_{J,\bar{x}}$'s are related to f_{J,x_0}'s by (12.4). Fix any $\epsilon \in (0,\infty) \subset \mathbb{R}$. By f_{J,x_0}'s being locally bounded on their domain of definition and (12.4), we can conclude that $f_{J,\bar{x}}$'s are also locally bounded on their domain of definition. $\forall J \subseteq \bar{J}$ with $J \neq \emptyset$, $\forall x \in \Omega$, $\exists \bar{\delta}(J,x) \in (0,\infty) \subset \mathbb{R}$, $\exists \bar{c}(J,x) \in [0,\infty) \subset \mathbb{R}$ such that $\operatorname{ess\,sup}_{y\in\mathcal{B}_{\mathbb{R}^m}(x,\bar{\delta}(J,x))} \|f_{J,\bar{x}}(\pi_J(y))\| \leq \bar{c}(J,x)$ and $\mathcal{B}_{\mathbb{R}^m}\left(x, \bar{\delta}(J,x)\right) \subseteq \Omega$. Let $\bar{\delta}_0 := \min_{J\subseteq \bar{J}, J\neq\emptyset} \bar{\delta}(J,\bar{x}) \in (0,\infty) \subset \mathbb{R}$ and $\bar{c}_0 := \max_{J\subseteq \bar{J}, J\neq\emptyset} \bar{c}(J,\bar{x}) \in [0,\infty) \subset \mathbb{R}$. Then, $\operatorname{ess\,sup}_{y\in\mathcal{B}_{\mathbb{R}^m}(\bar{x},\bar{\delta}_0)} \|f_{J,\bar{x}}(\pi_J(y))\| \leq \bar{c}_0$ and $\mathcal{B}_{\mathbb{R}^m}\left(\bar{x}, \bar{\delta}_0\right) \subseteq \Omega$. $\forall i \in \bar{J}$, we have

$$\begin{aligned}
\mathrm{D}_i F(\bar{x}) &= f_{\{i\},x_0}(\pi_i(\bar{x})) + \sum_{J\subseteq \bar{J}, \{i\}\subset J} (-1)^{\mathrm{card}(\check{\bar{J}}\cap J\setminus\{i\})} \int_{\pi_{J\setminus\{i\}}(\mathrm{r}_{\hat{\bar{x}},\check{\bar{x}}})} \\
&\quad f_{J,x_0}(\pi_J(M_{\{i\}}(\pi_i(\bar{x}), \pi_{\bar{J}\setminus\{i\}}(s))))\, \mathrm{d}\mu_{J\setminus\{i\}}(\pi_{J\setminus\{i\}}(s)) \\
&= \sum_{i\in J\subseteq \bar{J}} H_{J,i}(\bar{x})
\end{aligned}$$

where the equality follows from the preceding discussion. $\forall i \in \bar{J}$, $\forall J \subseteq \bar{J}$ with $i \in J$, $\exists \delta_{J,i}(\bar{x}) \in (0,\infty) \subset \mathbb{R}$ such that $\|H_J(\bar{x} + h_i e_{m,i}) - H_J(\bar{x}) - H_{J,i}(\bar{x}) h_i\| \leq m^{-1} 2^{-m} \epsilon |h_i|$. Let $\delta_1 := \left(\min_{i\in\bar{J}} \min_{i\in J\subseteq\bar{J}} \delta_{J,i}(\bar{x})\right) \wedge \bar{\delta}_0 \wedge \frac{\epsilon}{2(1+\bar{c}_0)(2^m-1-m)} \wedge 1 \in (0,\infty) \subset \mathbb{R}$. This leads to, $\forall x \in \mathcal{B}_{\mathbb{R}^m}(\bar{x}, \delta_1)$,

$$\begin{aligned}
&\left\| F(x) - F(\bar{x}) - \sum_{i=1}^{m} \mathrm{D}_i F(\bar{x}) \pi_i(x - \bar{x}) \right\| \\
&= \left\| \sum_{J\subseteq \bar{J}, J\neq\emptyset} (-1)^{\mathrm{card}(\check{\bar{J}}\cap J)} \int_{\pi_J(\mathrm{r}_{\hat{\bar{x}},\check{\bar{x}}})} f_{J,\bar{x}}\, \mathrm{d}\mu_J - \sum_{i=1}^{m} \mathrm{D}_i F(\bar{x}) \pi_i(x - \bar{x}) \right\| \\
&= \left\| \sum_{J\subseteq \bar{J}, \mathrm{card}(J)\geq 2} (-1)^{\mathrm{card}(\check{\bar{J}}\cap J)} \int_{\pi_J(\mathrm{r}_{\hat{\bar{x}},\check{\bar{x}}})} f_{J,\bar{x}}\, \mathrm{d}\mu_J \right. \\
&\quad + \sum_{i=1}^{m} (-1)^{\mathrm{card}(\check{\bar{J}}\cap\{i\})} \int_{\pi_{\{i\}}(\mathrm{r}_{\hat{\bar{x}},\check{\bar{x}}})} f_{\{i\},\bar{x}}\, \mathrm{d}\mu_{\{i\}} \\
&\quad \left. - \sum_{i=1}^{m} (-1)^{\mathrm{card}(\check{\bar{J}}\cap\{i\})} \int_{\pi_{\{i\}}(\mathrm{r}_{\hat{\bar{x}},\check{\bar{x}}})} \mathrm{D}_i F(\bar{x})\, \mathrm{d}\mu_{\{i\}}(t) \right\|
\end{aligned}$$

$$\leq \sum_{J\subseteq \bar{J},\mathrm{card}(J)\geq 2} \int_{\pi_J(\mathrm{r}_{\hat{x},\check{x}})} \mathcal{P}\circ f_{J,\bar{x}}\, \mathrm{d}\mu_J$$

$$+\left\| \sum_{i=1}^{m} (-1)^{\mathrm{card}(\check{J}\cap\{i\})} \int_{\pi_{\{i\}}(\mathrm{r}_{\hat{x},\check{x}})} \big(f_{\{i\},\bar{x}}(t) - \mathrm{D}_i F(\bar{x})\big)\, \mathrm{d}\mu_{\{i\}}(t) \right\|$$

$$\leq \sum_{J\subseteq \bar{J},\mathrm{card}(J)\geq 2} \int_{\pi_J(\mathrm{r}_{\hat{x},\check{x}})} \mathcal{P}\circ f_{J,\bar{x}}\, \mathrm{d}\mu_J$$

$$+\sum_{i=1}^{m} \sum_{i\in\tilde{J}\subseteq\bar{J}} \left\| H_{\tilde{J}}(\bar{x}+\pi_i(x-\bar{x})e_{m,i}) - H_{\tilde{J}}(\bar{x}) - H_{\tilde{J},i}(\bar{x})\pi_i(x-\bar{x}) \right\|$$

$$\leq \sum_{J\subseteq \bar{J},\mathrm{card}(J)\geq 2} \bar{c}_0 \mu_J(\pi_J(\mathrm{r}_{\hat{x},\check{x}})) + \sum_{i=1}^{m} \sum_{i\in\tilde{J}\subseteq\bar{J}} m^{-1}2^{-m}\epsilon|\pi_i(x-\bar{x})|$$

$$\leq \sum_{J\subseteq \bar{J},\mathrm{card}(J)\geq 2} \bar{c}_0 \prod_{j\in J} \left|\pi_j(x-\bar{x})\right| + \frac{\epsilon}{2}|x-\bar{x}|$$

$$\leq (2^m - 1 - m)\bar{c}_0\delta_1^{\mathrm{card}(J)-1}|x-\bar{x}| + \frac{\epsilon}{2}|x-\bar{x}| \leq \epsilon|x-\bar{x}|$$

where the first equality follows from (12.4); the first inequality follows from Proposition 11.92, the second inequality follows from the fact that

$$\int_{\pi_i(\mathrm{r}_{\hat{x},\check{x}})} f_{\{i\},\bar{x}}\, \mathrm{d}\mu_{\{i\}} = \sum_{J\subseteq \bar{J}, J\neq\emptyset} (H_J(\bar{x}+\pi_i(x-\bar{x})e_{m,i}) - H_J(\bar{x}))$$

and Proposition 11.92, and the third inequality follows from choice of δ_1 and the preceding discussion. Then, $\bar{x} \in \mathrm{dom}\left(F^{(1)}\right)$. By the arbitrariness of $\bar{x}$, we have $\check{U} \subseteq \mathrm{dom}\left(F^{(1)}\right) = U$. Hence, $\mu(\Omega \setminus U) = 0$.

(v) This is a direct consequence of Fubini's Theorem 12.31 and the proof of (ii).

This completes the proof of the theorem. □

Theorem 12.89 (Integration by Parts) *Let $I \subseteq \mathbb{R}$ be a nonempty open or closed interval, $\mathbb{I} := ((\mathrm{P}(I), |\cdot|), \mathcal{B}, \mu)$ be the σ-finite metric measure subspace of $\mathbb{R}$, $\mathcal{Y}$ be a separable Banach space over $\mathbb{K}$, $\mathcal{Z}$ be a separable reflexive Banach space over $\mathbb{K}$ with $\mathcal{Z}^*$ being separable, and $\mathcal{W} \subseteq \mathrm{B}(\mathcal{Y}, \mathcal{Z})$ be a separable Banach subspace, $a : \mathrm{P}(I) \to \mathcal{W}$ be $\mathcal{B}$-measurable and absolutely integrable over $\mathbb{I}$, ν_A be the $\mathcal{W}$-valued measure with kernel a over $\mathbb{I}$, and $A : I \to \mathcal{W}$ be a cumulative distribution function for ν_A, $f : \mathrm{P}(I) \to \mathcal{Y}$ be $\mathcal{B}$-measurable and absolutely integrable over $\mathbb{I}$, ν_F be the $\mathcal{Y}$-valued measure with kernel f over $\mathbb{I}$, and $F : I \to \mathcal{Y}$ be a cumulative distribution function for ν_F. Then, $H : I \to \mathcal{Z}$ defined by $H(x) = A(x)F(x)$, $\forall x \in I$, is absolutely continuous and is a cumulative distribution function of ν_H that is the $\mathcal{Z}$-valued measure with kernel $h : \mathrm{P}(I) \to \mathcal{Z}$ over $\mathbb{I}$, where h is defined* *by* $h(x) = A(x)f(x) + a(x)F(x)$, $\forall x \in \mathrm{P}(I)$, *and* Af, aF, *and* h *are absolutely integrable over* $\mathbb{I}$. *As a consequence,* $\forall b, c \in I$ *with* $b \leq c$, *we have*

$$\int_b^c A(x)f(x)\,\mathrm{d}x + \int_b^c a(x)F(x)\,\mathrm{d}x = A(c)F(c) - A(b)F(b) \tag{12.5}$$

Proof By Proposition 7.66, $\mathrm{B}(\mathcal{Y}, \mathcal{Z})$ is a Banach space. Fix any $x_0 \in I$. Let $F_0 : I \to \mathcal{Y}$ be the cumulative distribution function of ν_F with origin x_0, and $A_0 : I \to \mathcal{W}$ be the cumulative distribution function of ν_A with origin x_0. By Proposition 12.75, F_0 and A_0 are absolutely continuous and therefore continuous. Since f and a are absolutely integrable over $\mathbb{I}$, then $\exists M \in [0, \infty) \subset \mathbb{R}$ such that $\|F_0(x)\|_{\mathcal{Y}} \leq M$ and $\|A_0(x)\|_{\mathcal{W}} \leq M$, $\forall x \in I$. By the assumption, we have $F(x) = F(x_0) + F_0(x)$ and $A(x) = A(x_0) + A_0(x)$, $\forall x \in I$. Then, F and A are absolutely continuous, therefore continuous, and $\exists \bar{M} \in [0, \infty) \subset \mathbb{R}$ such that $\|F(x)\|_{\mathcal{Y}} \leq \bar{M}$ and $\|A(x)\|_{\mathcal{W}} \leq \bar{M}$, $\forall x \in I$. By Proposition 11.37, F and A are $\mathcal{B}$-measurable. By Propositions 7.23, 7.65, 11.38, and 11.39, Af, aF and h are $\mathcal{B}$-measurable. Then, Af, aF are absolutely integrable over $\mathbb{I}$. Hence, h is absolutely integrable over $\mathbb{I}$ by Proposition 11.83. By Proposition 12.68, H is absolutely continuous.

We will first consider the case that I is an open interval. By Fundamental Theorem of Calculus I 12.86, $F^{(1)} = \mathrm{D}_1 F : U_F \to \mathcal{Y}$ and $A^{(1)} = \mathrm{D}_1 A : U_A \to \mathcal{W}$ are such that $U_F, U_A \in \mathcal{B}$, $\mu(I \setminus U_F) = 0 = \mu(I \setminus U_A)$, $F^{(1)}$ and $A^{(1)}$ are $\mathcal{B}$-measurable, $F^{(1)} = f$ a.e. in $\mathbb{I}$, and $A^{(1)} = a$ a.e. in $\mathbb{I}$. Then, $\exists \hat{U}_F, \hat{U}_A \in \mathcal{B}$ such that $\mu(I \setminus \hat{U}_F) = 0 = \mu(I \setminus \hat{U}_A)$ and $F^{(1)}(x) = f(x)$, $\forall x \in \hat{U}_F$, and $A^{(1)}(x) = a(x)$, $\forall x \in \hat{U}_A$.

$\forall x \in \hat{U}_F \cap \hat{U}_A$, by Propositions 9.17 and 9.19 and the Chain Rule, Theorem 9.18, we have $H^{(1)}(x) = A(x)f(x) + a(x)F(x) = h(x)$. By Fundamental Theorem of Calculus II 12.88, $H^{(1)} = \mathrm{D}_1 H : U_H \to \mathcal{Z}$ is such that $U_H \in \mathcal{B}$, $\mu(I \setminus U_H) = 0$, and $H^{(1)}$ is $\mathcal{B}$-measurable. Then, $E := (I \setminus U_H) \cup \left\{x \in U_H \mid H^{(1)}(x) \neq h(x)\right\} \in \mathcal{B}$ and $E \subseteq (I \setminus \hat{U}_F) \cup (I \setminus \hat{U}_A)$. Hence, $\mu(E) = 0$ and $h = H^{(1)}$ a.e. in $\mathbb{I}$. By Fundamental Theorem of Calculus II 12.88, $H(x) = \int_{x_0}^x \bar{h}\,\mathrm{d}\mu + H(x_0)$, $\forall x \in I$, $\bar{h}$ is $\mathcal{B}$-measurable and absolutely integrable over r_{x_a, x_b}, $\forall x_a, x_b \in I$ with $x_a \leq x_b$, and $H^{(1)}(x) = \bar{h}(x)$ a.e. $x \in \mathbb{I}$. Thus, by Lemma 11.44, $h = \bar{h}$ a.e. in $\mathbb{I}$. Then, by Proposition 11.92, we have and $H(x) - H(x_0) = \int_{x_0}^x h(s)\,\mathrm{d}s$, $\forall x \in I$. By Proposition 11.92 and Fact 12.72, $\forall b, c \in I$ with $b \leq c$, we have $A(c)F(c) - A(b)F(b) = H(c) - H(x_0) - (H(b) - H(x_0)) = \int_{x_0}^c h(x)\,\mathrm{d}x - \int_{x_0}^b h(x)\,\mathrm{d}x = \int_b^c (A(x)f(x) + a(x)F(x))\,\mathrm{d}x = \int_b^c A(x)f(x)\,\mathrm{d}x + \int_b^c a(x)F(x)\,\mathrm{d}x$.

Define μ_H to be the $\mathcal{Z}$-valued measure with kernel h over $\mathbb{I}$. By the preceding paragraph, we have H is the cumulative distribution function of μ_H. This completes the proof for this case.

Next, we consider the case when I is a closed interval. We will define $\hat{I}$ to be an open interval such that $I \subset \hat{I}$. Define $\hat{f} : \hat{I} \to \mathcal{Y}$ by $\hat{f}(x) = \begin{cases} f(x) & x \in \mathrm{P}(I) \\ \vartheta_{\mathcal{Y}} & x \in \hat{I} \setminus \mathrm{P}(I) \end{cases}$. Similarly define $\hat{a}$ in terms of a. Define $\hat{F} : \hat{I} \to \mathcal{Y}$ by

$\hat{F}(x) = \begin{cases} F(x) & x \in I \\ F(\min_{s\in I} s) & x < \min_{s\in I} s \\ F(\max_{s\in I} s) & x > \max_{s\in I} s \end{cases}$. Similarly define $\hat{A}$ in terms of A. Then, the result can be applied for $\hat{\mathbb{I}} := ((\hat{I}, |\cdot|), \hat{\mathcal{B}}, \hat{\mu})$ for the functions $\hat{f}$, $\hat{F}$, $\hat{a}$, $\hat{A}$, to yield $\hat{H} : \hat{I} \to \mathcal{Z}$ and $\hat{h} : \hat{I} = \mathrm{P}(\hat{I}) \to \mathcal{Z}$. Restrict these functions to I to yield the desired result.

This completes the proof of the theorem. □

Lemma 12.90 (Vitali) *Let $m \in \mathbb{N}$, $\mathbb{R}^m$ be endowed with the usual positive cone, $E \subseteq \mathbb{R}^m$ with $\mu_{\mathrm{L}mo}(E) < +\infty$, $\mathcal{I} \subseteq \mathcal{B}_{\mathrm{B}}(\mathbb{R}^m)$ be a collection of nondegenerate closed rectangles in $\mathbb{R}^m$, and $\bar{\mathcal{I}} := \big\{V \subseteq \mathbb{R}^m \bigm| V$ is a nondegenerate closed rectangle in $\mathbb{R}^m$ with center $x \in \mathbb{R}^m$, such that the rectangle $U := (\frac{5}{c}(V - x) + x) \in \mathcal{I}\big\}$, and assume that $\bar{\mathcal{I}}$ covers E in the sense of Vitali with index $c \in (0, \frac{1}{\sqrt{m}}] \subset \mathbb{R}$. Then, there exists a countable pairwise disjoint subcollection $(V_i)_{i\in N} \subseteq \bar{\mathcal{I}}$ with $N \subseteq \mathbb{N}$ (each V_i corresponds to $U_i \in \mathcal{I}$ as outlined in the definition of $\bar{\mathcal{I}}$), $\forall \epsilon \in (0, \infty) \subset \mathbb{R}$, $\exists n \in \mathbb{Z}_+$ such that $\{1, \ldots, n\} \subseteq N$, $\{V_1, \ldots, V_n\} \subseteq \bar{\mathcal{I}}$ is pairwise disjoint with $\mu_{\mathrm{L}mo}(E \setminus (\bigcup_{i=1}^n V_i)) < \epsilon$, and $E \subseteq (\bigcup_{i=1}^n V_i) \cup (\bigcup_{i\in N, i>n} U_i)$ with $\sum_{i\in N, i>n} \mu_{\mathrm{B}m}(U_i) \le \epsilon$.*

Proof By Example 12.56,

$$\mu_{\mathrm{L}mo}(E) = \inf_{(O_i)_{i=1}^\infty \in \mathcal{O}_{\mathbb{R}^m},\ E \subseteq \bigcup_{i=1}^\infty O_i} \sum_{i=1}^\infty \mu_{\mathrm{B}m}(O_i) < +\infty$$

Then, $\exists O \in \mathcal{O}_{\mathbb{R}^m}$ such that $E \subseteq O$ and $\mu_{\mathrm{B}m}(O) < +\infty$. By neglecting rectangles in $\bar{\mathcal{I}}$, we may without loss of generality assume that $V \subseteq O$ and the shortest side of V has length at least $c\,\mathrm{dia}(V)$, $\forall V \in \bar{\mathcal{I}}$. $\forall k \in \mathbb{Z}_+$, assume that pairwise disjoint $V_1, \ldots, V_k \in \bar{\mathcal{I}}$ has already be chosen. We will distinguish two exhaustive and mutually exclusive cases: Case 1: $E \subseteq \bigcup_{i=1}^k V_i$; Case 2: $E \setminus (\bigcup_{i=1}^k V_i) \neq \emptyset$. Case 1: $E \subseteq \bigcup_{i=1}^k V_i$. Then, we let $N = \{1, \ldots, n\}$. Clearly, $\mu_{\mathrm{L}mo}(E \setminus (\bigcup_{i=1}^k V_i)) = \mu_{\mathrm{L}mo}(\emptyset) = 0$, $E \subseteq (\bigcup_{i=1}^n V_i)$, and $\sum_{i\in N, i>n} \mu_{\mathrm{B}m}(U_i) = 0$. The procedure terminates. Case 2: $E \setminus (\bigcup_{i=1}^k V_i) \neq \emptyset$. Note that, for any closed rectangle $V \in \bar{\mathcal{I}}$ with $\mathrm{dia}(V) =: \bar{p} > 0$, let $x_0 \in V$ be the center of V, such that $\bar{\mathrm{r}}_{x_0 - \frac{1}{2}c\bar{p}\mathbf{1}_m, x_0 + \frac{1}{2}c\bar{p}\mathbf{1}_m} \subseteq V \subseteq \bar{\mathrm{r}}_{x_0 - \frac{1}{2}\hat{p}\mathbf{1}_m, x_0 + \frac{1}{2}\hat{p}\mathbf{1}_m}$, where $\hat{p} := \sqrt{1 - (m-1)c^2}\,\bar{p}$ is the upper bound for the longest side of the rectangle V. This implies that $c^m \bar{p}^m \le \mu_{\mathrm{B}m}(V) \le \hat{p}^m = (1 - (m-1)c^2)^{m/2} \bar{p}^m$. Let l_k be the supremum of the diameters of rectangles in $\bar{\mathcal{I}}$ that do not intersect $V_1, \ldots, V_k$. Clearly, $l_k \le \frac{\sqrt[m]{\mu_{\mathrm{B}m}(O)}}{c} < +\infty$. Note that $\exists x_0 \in E \setminus (\bigcup_{i=1}^k V_i)$. Since $\bigcup_{i=1}^k V_i$ is closed, then, by Proposition 4.10, $d := \inf_{x \in \bigcup_{i=1}^k V_i} |x - x_0| > 0$. Then, by the assumption, $\exists \hat{V} \in \bar{\mathcal{I}}$ such that $x_0 \in \hat{V}$ and $\mathrm{dia}(\hat{V}) < d$ and the shortest side of the rectangle $\hat{V}$ has length at least $c\,\mathrm{dia}(\hat{V})$. Then, $\hat{V} \cap (\bigcup_{i=1}^k V_i) = \emptyset$. This shows that $l_k \ge \mathrm{dia}(\hat{V}) > 0$. Hence, $\exists V_{k+1} \in \bar{\mathcal{I}}$ such that $V_1, \ldots, V_{k+1} \in \bar{\mathcal{I}}$ are pairwise disjoint and $\mathrm{dia}(V_{k+1}) \ge l_k/2$. Inductively, we either have Case A: $\exists$ a finite pairwise disjoint collection $\{V_1, \ldots, V_{n_0}\} \subseteq \bar{\mathcal{I}}$ with $n_0 \in \mathbb{Z}_+$ and $N = \{1, \ldots, n_0\}$ such that $E \subseteq \bigcup_{i=1}^{n_0} V_i$ or Case B: $\exists (V_k)_{k=1}^{\infty} \subseteq \bar{\mathcal{I}}$ with $N = \mathbb{N}$ such that $\text{dia}(V_{k+1}) \geq l_k/2$, $\forall k \in \mathbb{Z}_+$, and the sets in the sequence are pairwise disjoint.

Fix any $\epsilon \in \mathbb{R}_+$. If Case A holds, then choose $n = n_0 \in \mathbb{Z}_+$ and the result holds. If Case B holds, then $N = \mathbb{N}$ and we have $O \supseteq \bigcup_{k=1}^{\infty} V_k$ and $+\infty > \mu_{\text{B}m}(O) \geq \mu_{\text{B}m}(\bigcup_{k=1}^{\infty} V_k) = \sum_{k=1}^{\infty} \mu_{\text{B}m}(V_k) \geq \sum_{k=0}^{\infty} 2^{-m} c^m l_k^m$. Note that $(l_k)_{k=0}^{\infty}$ is nonincreasing. Hence, $\lim_{k\in\mathbb{N}} l_k = 0$. Thus, $\exists n \in \mathbb{N}$ such that $\sum_{i=n+1}^{\infty} \mu_{\text{B}m}(V_i) < 5^{-m} c^m \epsilon$. Let $R := E \setminus (\bigcup_{k=1}^{n} V_k)$. Without loss of generality, assume $\text{dia}(V_k) =: p_k \in (0, \infty) \subset \mathbb{R}$ and the center of V_k is x_k, $\forall k \in \mathbb{N}$. $\forall \bar{x} \in R$, since $\bigcup_{k=1}^{n} V_k$ is closed, then, by Proposition 4.10, $d := \inf_{x \in \bigcup_{k=1}^{n} V_k} |x - \bar{x}| > 0$. By the assumption, $\exists V \in \bar{\mathcal{I}}$ such that $\bar{x} \in V$ and $0 < \text{dia}(V) < d$ and the shortest side of the rectangle V has length at least $c\,\text{dia}(V)$. Then, $V \cap (\bigcup_{k=1}^{n} V_k) = \emptyset$. By the fact that $\lim_{k\in\mathbb{N}} l_k = 0$, we have $\exists i_0 \in \{n+1, n+2, \ldots\}$ such that $V \cap V_{i_0} \neq \emptyset$ and $V \cap V_k = \emptyset$, $\forall k \in \{1, \ldots, i_0 - 1\}$. Then, $\text{dia}(V) \leq l_{i_0 - 1} \leq 2\,\text{dia}(V_{i_0}) = 2 p_{i_0}$. Let $\hat{x} \in V \cap V_{i_0}$. Then, $\left|\bar{x} - x_{i_0}\right| \leq |\bar{x} - \hat{x}| + \left|\hat{x} - x_{i_0}\right| \leq \text{dia}(V) + \frac{1}{2} p_{i_0} \leq \frac{5}{2} p_{i_0}$. Then, $\bar{x} \in \bar{\mathrm{r}}_{x_{i_0} - \frac{5}{2} p_{i_0} \mathbf{1}_m, x_{i_0} + \frac{5}{2} p_{i_0} \mathbf{1}_m} = \frac{5}{c} \bar{\mathrm{r}}_{-\frac{c}{2} p_{i_0} \mathbf{1}_m, \frac{c}{2} p_{i_0} \mathbf{1}_m} + x_{i_0} \subseteq \frac{5}{c}(V_{i_0} - x_{i_0}) + x_{i_0} = U_{i_0} \in \mathcal{I}$. Then, $\bar{x} \in \bigcup_{k=n+1}^{\infty} U_k$. Hence, $R \subseteq \bigcup_{k=n+1}^{\infty} U_k$. Then, $\mu_{\text{L}mo}(R) \leq \mu_{\text{L}mo}(\bigcup_{k=n+1}^{\infty} U_k) \leq \sum_{k=n+1}^{\infty} \mu_{\text{B}m}(U_k) = \sum_{k=n+1}^{\infty} 5^m c^{-m} \mu_{\text{B}m}(V_k) = 5^m c^{-m} \sum_{k=n+1}^{\infty} \mu_{\text{B}m}(V_k) < \epsilon$. Note that $E \subseteq (\bigcup_{i=1}^{n} V_i) \cup (\bigcup_{i=n+1}^{\infty} U_i)$ and $\sum_{i \in N, i > n} \mu_{\text{B}m}(U_i) < \epsilon$. Hence, the result holds.

This completes the proof of the lemma. □

Theorem 12.91 (Change of Variable) *Let $m \in \mathbb{N}$, $\mathbb{R}^m$ be endowed with the usual positive cone, $\Omega \in \mathcal{O}_{\mathbb{R}^m}$ be an open subset, $\bar{\Omega} \in \mathcal{O}_{\mathbb{R}^m}$ be an open subset, $F : \Omega \to \bar{\Omega}$ be a homeomorphism, $F_i := F_{\text{inv}} : \bar{\Omega} \to \Omega$, $\mathbb{X} := ((\Omega, |\cdot|), \mathcal{B}, \mu)$ be the σ-finite metric measure subspace of $\mathbb{R}^m$, $\bar{\mathbb{X}} := ((\bar{\Omega}, |\cdot|), \bar{\mathcal{B}}, \bar{\mu})$ be the σ-finite metric measure subspace of $\mathbb{R}^m$, and $F_{i*}\bar{\mu}$ be the induced measure on $(\Omega, \mathcal{B})$ and $F_*\mu$ be the induced measure on $(\bar{\Omega}, \bar{\mathcal{B}})$ as defined in Proposition 12.9. Then, $\mathbb{X}$ and $((\bar{\Omega}, |\cdot|), \bar{\mathcal{B}}, F_*\mu)$ is homeomorphically isomeasuric under F, and $\bar{\mathbb{X}}$ and $((\Omega, |\cdot|), \mathcal{B}, F_{i*}\bar{\mu})$ is homeomorphically isomeasuric under F_i. Assume that $F^{(1)} : U \to \mathrm{B}(\mathbb{R}^m, \mathbb{R}^m)$ with $\mu(\Omega \setminus U) = 0$ satisfies, $\forall x_1, x_2 \in \Omega$ with $x_1 \stackrel{\leq}{=} x_2$ and $\bar{\mathrm{r}}_{x_1, x_2} \subseteq \Omega$, we have $\sup_{x \in \bar{\mathrm{r}}_{x_1, x_2} \cap U} \left|\det(F^{(1)}(x))\right| \leq M_{x_1, x_2} \in (0, \infty) \subset \mathbb{R}$; $F_i^{(1)} : \bar{U} \to \mathrm{B}(\mathbb{R}^m, \mathbb{R}^m)$ with $\mu(\bar{\Omega} \setminus \bar{U}) = 0$ satisfies, $\forall \bar{x}_1, \bar{x}_2 \in \Omega$ with $\bar{x}_1 \stackrel{\leq}{=} \bar{x}_2$ and $\bar{\mathrm{r}}_{\bar{x}_1, \bar{x}_2} \subseteq \bar{\Omega}$, we have $\sup_{\bar{x} \in \bar{\mathrm{r}}_{\bar{x}_1, \bar{x}_2} \cap \bar{U}} \left|\det(F_i^{(1)}(\bar{x}))\right| \leq \bar{M}_{\bar{x}_1, \bar{x}_2} \in (0, \infty) \subset \mathbb{R}$; and $\hat{U} := \left\{x \in U \;\middle|\; \det(F^{(1)}(x)) \neq 0\right\}$ and $\tilde{\bar{U}} := \left\{\bar{x} \in \bar{U} \;\middle|\; \det(F_i^{(1)}(\bar{x})) \neq 0\right\}$ are such that $\bar{\mu}(\bar{\Omega} \setminus \tilde{\bar{U}}) = 0 = \mu(\Omega \setminus \tilde{U}) := \mu\left(\Omega \setminus F_i(\tilde{\bar{U}})\right) = \mu(\Omega \setminus \hat{U}) = \bar{\mu}(\bar{\Omega} \setminus F(\hat{U}))$. Then, $F_*\mu \ll \bar{\mu}$ and $\frac{\mathrm{d}F_*\mu}{\mathrm{d}\bar{\mu}} =: g$ a.e. $x \in \bar{\mathbb{X}}$, $F_{i*}\bar{\mu} \ll \mu$ and $\frac{\mathrm{d}F_{i*}\bar{\mu}}{\mathrm{d}\mu} = g_1$ a.e. in $\mathbb{X}$, where $g : \bar{\Omega} \to (0, \infty) \subset \mathbb{R}$ is given by $g(\bar{x}) = \begin{cases} \left|\det(F_i^{(1)}(\bar{x}))\right| & \bar{x} \in \tilde{\bar{U}} \\ 1 & \bar{x} \in \bar{\Omega} \setminus \tilde{\bar{U}} \end{cases}$*

and $g_1(x) = \frac{1}{g(F(x))}$, $\forall x \in \Omega$. *Furthermore, let* $\mathcal{Y}$ *be a separable Banach space,* $\forall f : \Omega \to \mathcal{Y}$ *that is absolutely integrable over* $\mathbb{X}$*, we have*

$$\int_{\Omega} f \, \mathrm{d}\mu = \int_{\bar{\Omega}} f(F_i(\bar{x}))g(\bar{x}) \, \mathrm{d}\bar{\mu}(\bar{x}) \tag{12.6a}$$

and the right-hand side integrand is absolutely integrable over $\bar{\mathbb{X}}$*;* $\forall \bar{f} : \bar{\Omega} \to \mathcal{Y}$ *that is absolutely integrable over* $\bar{\mathbb{X}}$*, we have*

$$\int_{\bar{\Omega}} \bar{f} \, \mathrm{d}\bar{\mu} = \int_{\Omega} \bar{f}(F(x))g_1(x) \, \mathrm{d}\mu(x) \tag{12.6b}$$

and the right-hand side integrand is absolutely integrable over $\mathbb{X}$*.*

Proof Fix $c \in (0, \frac{1}{\sqrt{m}}) \subset \mathbb{R}$. The result is trivial if $\bar{\Omega} = \emptyset$. Consider the case $\bar{\Omega} \neq \emptyset$. Then, $\Omega = F_i(\bar{\Omega}) \neq \emptyset$, $\mu(\Omega) > 0$, and $\bar{\mu}(\bar{\Omega}) > 0$. First, we need the following intermediate result.

Claim 12.91.1 $\forall \bar{x} \in \tilde{\tilde{U}}$, $\forall \epsilon \in (0, \infty) \subset \mathbb{R}$, $\exists \delta \in (0, \infty) \subset \mathbb{R}$ such that, $\forall \bar{h} \in \mathcal{B}_{\mathbb{R}^m}(\mathbf{0}_m, \delta)$ with $\bar{h} \gg \mathbf{0}_m$ and $\min \bar{h} \geq c|\bar{h}|$, we have $\bar{\mathrm{r}}_{\bar{x}-\bar{h},\bar{x}+\bar{h}} \subseteq \bar{\Omega}$ and $F_* \mu(\bar{\mathrm{r}}_{\bar{x}-\bar{h},\bar{x}+\bar{h}}) \leq (1+\epsilon)g(\bar{x})\bar{\mu}(\bar{\mathrm{r}}_{\bar{x}-\bar{h},\bar{x}+\bar{h}})$.

Symmetrically, $\forall x \in \tilde{U} = F_i(\tilde{\tilde{U}})$, $\forall \epsilon \in (0, \infty) \subset \mathbb{R}$, $\exists \delta \in (0, \infty) \subset \mathbb{R}$ such that, $\forall h \in \mathcal{B}_{\mathbb{R}^m}(\mathbf{0}_m, \delta)$ with $h \gg \mathbf{0}_m$ and $\min h \geq c|h|$, we have $\bar{\mathrm{r}}_{x-h,x+h} \subseteq \Omega$, and $F_{i*}\bar{\mu}(\bar{\mathrm{r}}_{x-h,x+h}) \leq (1+\epsilon)\left(g(F(x))\right)^{-1} \mu(\bar{\mathrm{r}}_{x-h,x+h}) = (1+\epsilon)g_1(x)\mu(\bar{\mathrm{r}}_{x-h,x+h})$.

Proof of Claim $\forall \bar{x} \in \tilde{\tilde{U}}$, we have $F_i^{(1)}(\bar{x})$ exists and $\det\left(F_i^{(1)}(\bar{x})\right) \neq 0$. This implies that $F^{(1)}(F_i(\bar{x})) = \left(F_i^{(1)}(\bar{x})\right)^{-1}$ and $\det(F^{(1)}(F_i(\bar{x}))) = \left(\det(F_i^{(1)}(\bar{x}))\right)^{-1}$. By Definition 9.3, $\forall \bar{\epsilon} \in (0, \infty) \subset \mathbb{R}$, $\exists \delta_1 \in (0, \infty) \subset \mathbb{R}$ such that, $\forall \bar{h} \in \mathcal{B}_{\mathbb{R}^m}(\mathbf{0}_m, \delta_1)$, we have $\bar{x} + \bar{h} \in \bar{\Omega}$ and

$$\left|F_i(\bar{x}+\bar{h}) - F_i(\bar{x}) - F_i^{(1)}(\bar{x})\bar{h}\right| \leq \bar{\epsilon}|\bar{h}|$$

Then, $\forall \bar{h} \in \mathcal{B}_{\mathbb{R}^m}(\mathbf{0}_m, \delta_1)$ with $\bar{h} \gg \mathbf{0}_m$ and $\min \bar{h} \geq c|\bar{h}|$, we have $F_i(\bar{\mathrm{r}}_{\bar{x}-\bar{h},\bar{x}+\bar{h}}) \subseteq F_i(\bar{x}) + F_i^{(1)}(\bar{x})(\bar{\mathrm{r}}_{-\bar{h},\bar{h}}) + \overline{\mathcal{B}}_{\mathbb{R}^m}\left(\mathbf{0}_m, \bar{\epsilon}|\bar{h}|\right)$. This implies that $F_*\mu(\bar{\mathrm{r}}_{\bar{x}-\bar{h},\bar{x}+\bar{h}}) = \mu(F_i(\bar{\mathrm{r}}_{\bar{x}-\bar{h},\bar{x}+\bar{h}})) \leq \mu_{\mathrm{B}m}(F_i^{(1)}(\bar{x})(\bar{\mathrm{r}}_{-\bar{h},\bar{h}}) + \overline{\mathcal{B}}_{\mathbb{R}^m}\left(\mathbf{0}_m, \bar{\epsilon}|\bar{h}|\right))$. Clearly, $\mu_{\mathrm{B}m}\left(F_i^{(1)}(\bar{x})(\bar{\mathrm{r}}_{-\bar{h},\bar{h}})\right) = g(\bar{x})\mu_{\mathrm{B}m}(\bar{\mathrm{r}}_{-\bar{h},\bar{h}})$, which can be proved by decomposing the matrix $F_i^{(1)}(\bar{x})$ into basic row operations, and the equality holds for each row operation. Then, for sufficiently small $\bar{\epsilon} > 0$, we have $F_*\mu(\bar{\mathrm{r}}_{\bar{x}-\bar{h},\bar{x}+\bar{h}}) \leq (1+\epsilon)g(\bar{x})\mu_{\mathrm{B}m}(\bar{\mathrm{r}}_{-\bar{h},\bar{h}}) = (1+\epsilon)g(\bar{x})\bar{\mu}(\bar{\mathrm{r}}_{\bar{x}-\bar{h},\bar{x}+\bar{h}})$, since $F_i^{(1)}(\bar{x})(\bar{\mathrm{r}}_{-\bar{h},\bar{h}})$ is a closed parallelogram with all of its sides have length of at least some constant proportion of $|\bar{h}|$. This proves the first paragraph of the claim.

$\forall x \in \tilde{U}$, we have $F(x) \in F(\tilde{U}) = \tilde{\bar{U}}$. Then, $F^{(1)}(x) = \left(F_i^{(1)}(F(x))\right)^{-1}$ and $\left|\det(F^{(1)}(x))\right| = \left|\det(F_i^{(1)}(F(x)))\right|^{-1} = (g(F(x)))^{-1}$. By symmetry, $\forall \epsilon \in (0, \infty) \subset \mathbb{R}$, $\exists \delta \in (0, \infty) \subset \mathbb{R}$, $\forall h \in \mathcal{B}_{\mathbb{R}^m}(\mathbf{0}_m, \delta)$ with $h \gg \mathbf{0}_m$ and $\min h \geq c|h|$, we have $\bar{\mathrm{r}}_{x-h,x+h} \subseteq \Omega$, and $\bar{\mu}(F(\bar{\mathrm{r}}_{x-h,x+h})) = F_{i*}\bar{\mu}(\bar{\mathrm{r}}_{x-h,x+h}) \leq (1+\epsilon)\left|\det(F^{(1)}(x))\right| \mu(\bar{\mathrm{r}}_{x-h,x+h}) = (1+\epsilon)(g(F(x)))^{-1} \cdot \mu(\bar{\mathrm{r}}_{x-h,x+h})$. This proves the second paragraph of the claim.

This completes the proof of the claim. □

We first show that $F_*\mu \ll \bar{\mu}$ under the condition that $\bar{\Omega}$ is a nonempty bounded open set and $\left|\det(F_i^{(1)}(\bar{x}))\right| \leq \bar{M} \in (0, \infty) \subset \mathbb{R}$, $\forall \bar{x} \in \tilde{\bar{U}}$. Then, $0 < \bar{\mu}(\bar{\Omega}) < \infty$. We need the following intermediate result.

Claim 12.91.2 $\forall \bar{O} \in \bar{\mathcal{B}}$ with $\bar{O}$ being open and $\bar{\mu}(\bar{O}) > 0$, we have $\bar{\mu}(\bar{O}) \geq \mu(O)/\bar{M} := \mu(F_i(\bar{O}))/\bar{M}$.

Proof of Claim Fix any $\bar{O} \in \bar{\mathcal{B}}$ with $\bar{O}$ being open and $\bar{\mu}(\bar{O}) > 0$. Then, $O \in \mathcal{B}$ and is open. $\forall \bar{x} \in \tilde{\bar{O}} := \bar{O} \cap \tilde{\bar{U}}$, by Claim 12.91.1, $\forall \epsilon \in (0, \infty) \subset \mathbb{R}$, $\exists \delta(\bar{x}, \epsilon) \in (0, \infty) \subset \mathbb{R}$ such that, $\forall \bar{h} \in \mathcal{B}_{\mathbb{R}^m}(\mathbf{0}_m, \delta(\bar{x}, \epsilon))$ with $\bar{h} \gg \mathbf{0}_m$ and $\min \bar{h} \geq c\left|\bar{h}\right|$, we have $\bar{\mathrm{r}}_{\bar{x}-\bar{h},\bar{x}+\bar{h}} \subseteq \bar{\Omega}$ and $\mu(F_i(\bar{\mathrm{r}}_{\bar{x}-\bar{h},\bar{x}+\bar{h}})) = F_*\mu(\bar{\mathrm{r}}_{\bar{x}-\bar{h},\bar{x}+\bar{h}}) \leq (1+\epsilon)g(\bar{x})\bar{\mu}(\bar{\mathrm{r}}_{\bar{x}-\bar{h},\bar{x}+\bar{h}})$. The collection of such $\bar{\mathrm{r}}_{\bar{x}-\bar{h},\bar{x}+\bar{h}}$ is denoted $\mathcal{I}$. We will restrict $\mathcal{I}$ to consist of all of the above-mentioned $\bar{\mathrm{r}}_{\bar{x}-\bar{h},\bar{x}+\bar{h}}$'s that are subsets of $\bar{O}$. Clearly, $\mathcal{I}$ covers $\tilde{\bar{O}}$ in the sense of Vitali with index c. Define $\bar{\mathcal{I}} := \left\{ V \subseteq \mathbb{R}^m \,\middle|\, V \text{ is a nondegenerate rectangle in } \mathbb{R}^m \text{ with center } x \in \mathbb{R}^m, \text{ such that the rectangle } \left(\frac{5}{c}(V-x)+x\right) \in \mathcal{I}\right\}$. Clearly, $\bar{\mathcal{I}} \subseteq \mathcal{I}$. It is easy to see that $\bar{\mathcal{I}}$ covers $\tilde{\bar{O}}$ in the sense of Vitali with index c. By Vitali's Lemma 12.90, $\exists N \subseteq \mathbb{N}$, $\exists (V_i)_{i\in N} \subseteq \bar{\mathcal{I}}$, $\exists n \in \mathbb{Z}_+$ such that $\{1, \ldots, n\} \subseteq N$, $\{V_1, \ldots, V_n\}$ is pairwise disjoint, $\bar{\mu}\left(\tilde{\bar{O}} \setminus \left(\bigcup_{i=1}^n V_i\right)\right) < \frac{\epsilon}{4}\bar{\mu}(\bar{O})$, $\tilde{\bar{O}} \subseteq \left(\bigcup_{i=1}^n V_i\right) \cup \left(\bigcup_{i\in N, i>n} U_i\right)$, and $\sum_{i\in N, i>n} \bar{\mu}(U_i) \leq \frac{\epsilon}{4}\bar{\mu}(\bar{O})$, where $V_i = \bar{\mathrm{r}}_{\bar{x}_i-\bar{h}_i,\bar{x}_i+\bar{h}_i} \in \bar{\mathcal{I}} \subseteq \mathcal{I}$ and $U_i = \frac{5}{c}(V_i - \bar{x}_i) + \bar{x}_i = \bar{\mathrm{r}}_{\bar{x}_i-\frac{5}{c}\bar{h}_i,\bar{x}_i+\frac{5}{c}\bar{h}_i} \in \mathcal{I}$, $\forall i \in N$. Then, we have the following line of arguments.

$$
\begin{aligned}
\mu(O) &= \mu(\tilde{O} \cup (O \setminus \tilde{O})) = \mu(\tilde{O}) + \mu(O \setminus \tilde{O}) \\
&= \mu(F_i(\tilde{\bar{O}})) \leq \mu\Big(F_i\Big((\bigcup_{i=1}^n V_i) \cup (\bigcup_{i\in N, i>n} U_i)\Big)\Big) \\
&\leq \sum_{i=1}^n \mu(F_i(V_i)) + \sum_{i\in N, i>n} \mu(F_i(U_i)) = \sum_{i=1}^n F_*\mu(V_i) + \sum_{i\in N, i>n} F_*\mu(U_i) \\
&\leq \sum_{i=1}^n (1+\epsilon)g(\bar{x}_i)\bar{\mu}(\bar{\mathrm{r}}_{\bar{x}_i-\bar{h}_i,\bar{x}_i+\bar{h}_i}) + \sum_{\substack{i\in N\\ i>n}} (1+\epsilon)g(\bar{x}_i)\bar{\mu}(\bar{\mathrm{r}}_{\bar{x}_i-\frac{5}{c}\bar{h}_i,\bar{x}_i+\frac{5}{c}\bar{h}_i})
\end{aligned}
$$

$$\leq (1+\epsilon)\bar{M}\sum_{i=1}^{n}\bar{\mu}(\bar{\mathrm{r}}_{\bar{x}_i-\bar{h}_i,\bar{x}_i+\bar{h}_i}) + (1+\epsilon)\bar{M}\sum_{i\in N,i>n}\bar{\mu}(\bar{\mathrm{r}}_{\bar{x}_i-\frac{5}{c}\bar{h}_i,\bar{x}_i+\frac{5}{c}\bar{h}_i})$$

$$= (1+\epsilon)\bar{M}\sum_{i=1}^{n}\bar{\mu}(V_i) + (1+\epsilon)\bar{M}\sum_{i\in N,i>n}\bar{\mu}(U_i)$$

$$\leq (1+\epsilon)\bar{M}\bar{\mu}(\bigcup_{i=1}^{n}V_i) + (1+\epsilon)\bar{M}\frac{\epsilon}{4}\bar{\mu}(\bar{O})$$

$$\leq (1+\epsilon)\bar{M}\bar{\mu}(\bar{O}) + (1+\epsilon)\bar{M}\frac{\epsilon}{4}\bar{\mu}(\bar{O}) = (1+\epsilon)\frac{4+\epsilon}{4}\bar{M}\bar{\mu}(\bar{O})$$

where $\tilde{O} := O \cap \tilde{U} = F_i(\bar{O} \cap \tilde{\bar{U}}) = F_i(\tilde{\bar{O}})$; the third equality follows from the fact that $O \setminus \tilde{O} \subseteq \Omega \setminus \tilde{U}$; the first inequality follows from monotonicity of measure; the second inequality follows from the subadditivity of measure; the fourth equality follows from Proposition 12.9; the third inequality follows from the choice of $\mathcal{I}$ and $\bar{\mathcal{I}}$; the fifth inequality follows from the countable additivity of measure and from the preceding discussion; and the last inequality follows from the monotonicity of measure. Since $\epsilon > 0$ is arbitrary, then we have $\mu(O) \leq \bar{M}\bar{\mu}(\bar{O})$.

This completes the proof of the claim. □

$\forall \bar{E} \in \bar{\mathcal{B}}$ with $\bar{\mu}(\bar{E}) = 0$, $\forall \epsilon > 0$, by the fact that $\bar{\mathbb{X}}$ is a metric measure space, there exists $\bar{O} \in \bar{\mathcal{B}}$ with $\bar{E} \subseteq \bar{O}$ and $\bar{O}$ being open, such that $0 < \bar{\mu}(\bar{O}) < \epsilon/\bar{M}$. Then, $F_*\mu(\bar{E}) = \mu(F_i(\bar{E})) \leq \mu(F_i(\bar{O})) \leq \bar{M}\bar{\mu}(\bar{O}) < \epsilon$. By the arbitrariness of ϵ, we have $F_*\mu(\bar{E}) = 0$. By the arbitrariness of $\bar{E}$, we have $F_*\mu \ll \bar{\mu}$. This proves that $F_*\mu \ll \bar{\mu}$ in the special case.

Now, consider the general case as stipulated in the theorem statement. Since $\bar{\Omega}$ is a nonempty open set and $\mathbb{R}^m$ is second countable, then $\exists (\hat{\bar{x}}_i)_{i=1}^{\infty}, (\check{\bar{x}}_i)_{i=1}^{\infty} \subseteq \bar{\Omega}$ such that $\hat{\bar{x}}_i < \check{\bar{x}}_i$ and $\bar{\mathrm{r}}_{\hat{\bar{x}}_i,\check{\bar{x}}_i} \subseteq \bar{\Omega}$, $\forall i \in \mathbb{N}$, and $\bar{\Omega} = \bigcup_{i=1}^{\infty} \mathrm{r}^{\circ}_{\hat{\bar{x}}_i,\check{\bar{x}}_i}$. $\forall \bar{E} \in \bar{\mathcal{B}}$ with $\bar{\mu}(\bar{E}) = 0$. Then, we have $\bar{\mu}\left(\bar{E} \cap \mathrm{r}^{\circ}_{\hat{\bar{x}}_i,\check{\bar{x}}_i}\right) = 0$, $\forall i \in \mathbb{N}$. By the special case, $F_*\mu(\bar{E} \cap \mathrm{r}^{\circ}_{\hat{\bar{x}}_i,\check{\bar{x}}_i}) = 0$, $\forall i \in \mathbb{N}$. Then, $0 \leq F_*\mu(\bar{E}) \leq \sum_{i=1}^{\infty} F_*\mu(\bar{E} \cap \mathrm{r}^{\circ}_{\hat{\bar{x}}_i,\check{\bar{x}}_i}) = 0$. Hence, we have $F_*\mu \ll \bar{\mu}$.

By symmetry, $F_{i*}\bar{\mu} \ll \mu$.

By Radon–Nikodym Theorem 11.169, $\frac{\mathrm{d}F_*\mu}{\mathrm{d}\bar{\mu}} =: \hat{g}$ exists and is unique almost everywhere on $\bar{\mathbb{X}}$, and $\frac{\mathrm{d}F_{i*}\bar{\mu}}{\mathrm{d}\mu} =: \check{g}$ exists and is unique almost everywhere on $\mathbb{X}$, where $\hat{g} : \bar{\Omega} \to [0, \infty) \subset \mathbb{R}$ is $\bar{\mathcal{B}}$-measurable (or simply $\mathcal{B}_{\mathrm{B}}(\mathbb{R}^m)$-measurable) and $\check{g} : \Omega \to [0, \infty) \subset \mathbb{R}$ is $\mathcal{B}$-measurable (or simply $\mathcal{B}_{\mathrm{B}}(\mathbb{R}^m)$-measurable). By Definitions 12.7 and 12.8 and Proposition 11.168, it is clear that $\check{g}(x) = \frac{1}{\hat{g}(F(x))}$ a.e. $x \in \mathbb{X}$ and $\hat{g}(\bar{x}) = \frac{1}{\check{g}(F_i(\bar{x}))}$ a.e. $\bar{x} \in \bar{\mathbb{X}}$. Then, by changing $\hat{g}$ on a set of measure zero in $\bar{\mu}$ and changing $\check{g}$ on a set of measure zero in μ, we have $\check{g}(x) = \frac{1}{\hat{g}(F(x))}$, $\forall x \in \Omega$.

By Propositions 11.92, 12.11, and 11.168, we have (12.6) holds with g substituted by $\hat{g}$ and g_1 substituted by $\check{g}$. We will only need to show that $\hat{g} = g$ a.e. $\bar{x} \in \bar{\mathbb{X}}$. Then, the theorem is proved.

First, consider the special case when Ω is bounded. Then, $\chi_{\Omega,\Omega}$ is absolutely integrable over $\mathbb{X}$. By the preceding paragraph, we have $\infty > \mu(\Omega) = \int_\Omega \chi_{\Omega,\Omega}\,\mathrm{d}\mu = \int_{\bar{\Omega}} \chi_{\Omega,\Omega}(F_i(\bar{x}))\hat{g}(\bar{x})\,\mathrm{d}\bar{\mu}(\bar{x}) = \int_{\bar{\Omega}} \chi_{\bar{\Omega},\bar{\Omega}}(\bar{x})\hat{g}(\bar{x})\,\mathrm{d}\bar{\mu}(\bar{x})$. Then, we have $\int_{\bar{\Omega}} \hat{g}\,\mathrm{d}\bar{\mu} < \infty$. This implies that $\hat{g} \in \bar{\mathrm{L}}_1(\bar{\Omega}, \mathbb{R})$. Let $\bar{A} := \left\{\bar{x} \in \bar{\Omega} \,\middle|\, \bar{x} \text{ is a rectangular Lebesgue point with regularity } c \text{ of } \hat{g}\right\}$. By Proposition 12.80, $\bar{A} \in \mathcal{B}_{\mathrm{L}m}$ and $\mu_{\mathrm{L}m}(\bar{\Omega} \setminus \bar{A}) = 0$. Define the set $\bar{B} := \{\bar{x} \in \bar{\Omega} \mid \hat{g}(\bar{x}) \leq g(\bar{x})\}$. By Propositions 12.81, 11.38, 11.39, and 11.139, g and g_1 are $\mathcal{B}_{\mathrm{B}}(\mathbb{R}^m)$-measurable. By the measurability of g and $\hat{g}$ and Propositions 7.23, 11.38, 11.39, and 11.35, we have $\bar{B} \in \bar{\mathcal{B}}$. We need the following intermediate result.

Claim 12.91.3 $\bar{B} \supseteq \bar{A} \cap \bar{\bar{U}}$.

Proof of Claim Fix any $\bar{x} \in \bar{A} \cap \bar{\bar{U}}$. Fix any $\epsilon \in (0, \infty) \subset \mathbb{R}$. By Claim 12.91.1, $\exists \delta \in (0, \infty) \subset \mathbb{R}$ such that $\forall \bar{h} \in \mathcal{B}_{\mathbb{R}^m}(\mathbf{0}_m, \delta)$ with $\bar{h} \gg \mathbf{0}_m$ and $\min \bar{h} \geq c\left|\bar{h}\right|$, we have $\bar{\mathrm{r}}_{\bar{x}-\bar{h},\bar{x}+\bar{h}} \subseteq \bar{\Omega}$ and $F_*\mu(\bar{\mathrm{r}}_{\bar{x}-\bar{h},\bar{x}+\bar{h}}) \leq (1+\epsilon)g(\bar{x})\bar{\mu}(\bar{\mathrm{r}}_{\bar{x}-\bar{h},\bar{x}+\bar{h}})$. By Definition 12.79, $\exists r_0 \in (0, \infty) \subset \mathbb{R}$ such that $\forall r \in (0, r_0)$, we have $\mathcal{B}_{\mathbb{R}^m}(\bar{x}, r) \subseteq \bar{\Omega}$ and

$$\sup_{\substack{\bar{h}\in\mathbb{R}^m \text{ with } \bar{h}\gg\mathbf{0}_m \\ \left|\bar{h}\right|=r,\min\bar{h}\geq cr}} \frac{1}{\bar{\mu}(\mathrm{r}_{\bar{x}-\bar{h},\bar{x}+\bar{h}})} \int_{\mathrm{r}_{\bar{x}-\bar{h},\bar{x}+\bar{h}}} \left|\hat{g}(\bar{y}) - \hat{g}(\bar{x})\right| \mathrm{d}\bar{\mu}(\bar{y}) < \epsilon$$

Take $\bar{h} \in \mathbb{R}^m$ with $\left|\bar{h}\right| < r_0 \wedge \delta$, $\bar{h} \gg \mathbf{0}_m$, and $\min \bar{h} \geq c\left|\bar{h}\right|$, we have

$$\begin{aligned}
\hat{g}(\bar{x}) - \epsilon &\leq \hat{g}(\bar{x}) - \frac{1}{\bar{\mu}(\mathrm{r}_{\bar{x}-\bar{h},\bar{x}+\bar{h}})} \int_{\mathrm{r}_{\bar{x}-\bar{h},\bar{x}+\bar{h}}} \left|\hat{g}(\bar{y}) - \hat{g}(\bar{x})\right| \mathrm{d}\bar{\mu}(\bar{y}) \\
&\leq \hat{g}(\bar{x}) - \frac{1}{\bar{\mu}(\mathrm{r}_{\bar{x}-\bar{h},\bar{x}+\bar{h}})} \left| \int_{\mathrm{r}_{\bar{x}-\bar{h},\bar{x}+\bar{h}}} \hat{g}(\bar{y}) - \hat{g}(\bar{x})\,\mathrm{d}\bar{\mu}(\bar{y}) \right| \\
&\leq \frac{1}{\bar{\mu}(\mathrm{r}_{\bar{x}-\bar{h},\bar{x}+\bar{h}})} \int_{\mathrm{r}_{\bar{x}-\bar{h},\bar{x}+\bar{h}}} \hat{g}\,\mathrm{d}\bar{\mu} = \frac{1}{\bar{\mu}(\mathrm{r}_{x-\bar{h},x+\bar{h}})} \int_{F_i(\mathrm{r}_{\bar{x}-\bar{h},\bar{x}+\bar{h}})} 1\,\mathrm{d}\mu \\
&= \frac{\mu(F_i(\mathrm{r}_{\bar{x}-\bar{h},\bar{x}+\bar{h}}))}{\bar{\mu}(\bar{\mathrm{r}}_{\bar{x}-\bar{h},\bar{x}+\bar{h}})} = \frac{F_*\mu(\mathrm{r}_{\bar{x}-\bar{h},\bar{x}+\bar{h}})}{\bar{\mu}(\bar{\mathrm{r}}_{\bar{x}-\bar{h},\bar{x}+\bar{h}})} \leq \frac{F_*\mu(\bar{\mathrm{r}}_{\bar{x}-\bar{h},\bar{x}+\bar{h}})}{\bar{\mu}(\bar{\mathrm{r}}_{\bar{x}-\bar{h},\bar{x}+\bar{h}})} \\
&\leq (1+\epsilon)g(\bar{x})
\end{aligned}$$

where the first inequality follows from the fact $\left|\bar{h}\right| < r_0$; the second inequality follows from Proposition 11.92; the third inequality follows from basic algebra; the first equality follows from (12.6a) and the second to last paragraph preceding the claim; the second equality follows from Proposition 11.75; the third equality follows from Proposition 12.9; the fourth inequality follows from the monotonicity of measure; and the last inequality follows from $|\bar{h}| < \delta$. By the arbitrariness of ϵ, we have $\hat{g}(\bar{x}) \leq g(\bar{x})$ and $\bar{x} \in \bar{B}$. By the arbitrariness of $\bar{x}$, we have $\bar{A} \cap \bar{\tilde{U}} \subseteq \bar{B}$.

This completes the proof of the claim. □

By Claim 12.91.3, we have $0 \leq \bar{\mu}(\bar{\Omega} \setminus \bar{B}) \leq \mu_{\text{L}m}(\bar{\Omega} \setminus (\bar{A} \cap \bar{\tilde{U}})) \leq \mu_{\text{L}m}(\bar{\Omega} \setminus \bar{A}) + \bar{\mu}(\bar{\Omega} \setminus \bar{\tilde{U}}) = 0$. Thus, $\bar{\mu}(\bar{\Omega} \setminus \bar{B}) = 0$.

Define the set $C := \{x \in \Omega \mid \breve{g}(x) \leq g_1(x)\}$. By symmetry, we have $C \in \mathcal{B}$ and $\mu(\Omega \setminus C) = 0$. On the set $\bar{G} := \bar{B} \cap F(C)$, we have $\hat{g}(\bar{x}) \leq g(\bar{x})$ and $\breve{g}(F_i(\bar{x})) \leq g_1(F_i(\bar{x}))$. This implies that $\hat{g}(\bar{x}) \leq g(\bar{x}) = \frac{1}{g_1(F_i(\bar{x}))} \leq \frac{1}{\breve{g}(F_i(\bar{x}))} = \hat{g}(\bar{x})$, $\forall \bar{x} \in \bar{G}$. Note that $0 \leq \bar{\mu}(\bar{\Omega} \setminus \bar{G}) \leq \bar{\mu}(\bar{\Omega} \setminus \bar{B}) + \bar{\mu}(\bar{\Omega} \setminus F(C)) = \bar{\mu}(F(\Omega) \setminus F(C)) = \bar{\mu}(F(\Omega \setminus C)) = F_{i*}\bar{\mu}(\Omega \setminus C)$. By $\mu(\Omega \setminus C) = 0$ and $F_{i*}\bar{\mu} \ll \mu$, we have $F_{i*}\bar{\mu}(\Omega \setminus C) = 0$. Then, we have $\bar{\mu}(\bar{\Omega} \setminus \bar{G}) = 0$. Hence, $g = \hat{g}$ a.e. in $\bar{\mathbb{X}}$. This proves the special case.

Finally, consider the general case. Since $\bar{\Omega}$ is a nonempty open set and $\mathbb{R}^m$ is second countable, then $\exists (\hat{\bar{x}}_i)_{i=1}^{\infty}, (\check{\bar{x}}_i)_{i=1}^{\infty} \subseteq \bar{\Omega}$ such that $\hat{\bar{x}}_i < \check{\bar{x}}_i$ and $\bar{\text{r}}_{\hat{\bar{x}}_i, \check{\bar{x}}_i} \subseteq \bar{\Omega}$, $\forall i \in \mathbb{N}$, and $\bar{\Omega} = \bigcup_{i=1}^{\infty} \text{r}^{\circ}_{\hat{\bar{x}}_i, \check{\bar{x}}_i}$. $\forall i \in \mathbb{N}$, let $\bar{\mathbb{X}}_i := \left(\left(\text{r}^{\circ}_{\hat{\bar{x}}_i, \check{\bar{x}}_i}, |\cdot| \right), \bar{\mathcal{B}}_i, \bar{\mu}_i \right)$ be the finite metric measure subspace of $\bar{\mathbb{X}}$. By the special case, we have $\hat{g}|_{\text{r}^{\circ}_{\hat{\bar{x}}_i, \check{\bar{x}}_i}} = g|_{\text{r}^{\circ}_{\hat{\bar{x}}_i, \check{\bar{x}}_i}}$ a.e. in $\bar{\mathbb{X}}_i$. Let $\bar{D} := \{\bar{x} \in \bar{\Omega} \mid \hat{g}(\bar{x}) \neq g(\bar{x})\}$. By Propositions 7.23, 11.38, 11.39, and 11.35, $\bar{D} \in \bar{\mathcal{B}}$. Then, we have $0 \leq \bar{\mu}(\bar{D}) = \bar{\mu}\big(\bigcup_{i=1}^{\infty} (\bar{D} \cap \text{r}^{\circ}_{\hat{\bar{x}}_i, \check{\bar{x}}_i})\big) \leq \sum_{i=1}^{\infty} \bar{\mu}\big(\bar{D} \cap \text{r}^{\circ}_{\hat{\bar{x}}_i, \check{\bar{x}}_i}\big) = 0$. Hence, $\hat{g} = g$ a.e. in $\bar{\mathbb{X}}$.

This completes the proof of the theorem. □

12.7 Representation of $(\mathcal{C}_k(\Omega, \mathcal{Y}))^*$

Theorem 12.92 (Riesz Representation Theorem) *Let $\mathcal{X}$ be a normed linear space over $\mathbb{K}$, $\mathcal{Y}$ be Banach space over $\mathbb{K}$, $\Omega \subseteq \mathcal{X}$ be compact with the subset topology $\mathcal{O}$, and $k \in \mathbb{N}$. Assume that, $\forall x \in \Omega$, $\exists \delta_x \in (0, \infty) \subset \mathbb{R}$ such that $\Omega \cap \mathcal{B}_{\mathcal{X}}(x, \delta_x) - x$ is a conic segment and* $\text{span}\,(\text{A}_{\Omega}(x)) = \mathcal{X}$. *Let $\mathcal{Z} := \mathcal{C}_k(\Omega, \mathcal{Y})$ be the Banach space over $\mathbb{K}$ as prescribed in Example 9.70. $\forall f \in \mathcal{Z}^*$, $\exists \mu := (\mu_0, \ldots, \mu_k) \in \prod_{i=0}^{k} (\mathcal{C}(\Omega, \text{B}_i(\mathcal{X}, \mathcal{Y})))^*$ such that $f(z) = \sum_{i=0}^{k} \langle\langle \mu_i, z^{(i)} \rangle\rangle$, $\forall z \in \mathcal{Z}$, with $\|\mu\| = \left(\sum_{i=0}^{k} \|\mu_i\|^2 \right)^{1/2} = \|f\|_{\mathcal{Z}^*}$. Furthermore, $\forall \nu := (\nu_0, \ldots, \nu_k) \in \prod_{i=0}^{k} (\mathcal{C}(\Omega, \text{B}_i(\mathcal{X}, \mathcal{Y})))^*$, define $h : \mathcal{Z} \to \mathbb{K}$ by $h(z) = \sum_{i=0}^{k} \langle\langle \nu_i, z^{(i)} \rangle\rangle$, $\forall z \in \mathcal{Z}$. Then, $h \in \mathcal{Z}^*$ and $\|h\|_{\mathcal{Z}^*} \leq \|\nu\|$.*

Proof Let $\mathcal{W} := \prod_{i=0}^{k} \mathcal{C}(\Omega, \text{B}_i(\mathcal{X}, \mathcal{Y})) =: \prod_{i=0}^{k} \mathcal{W}_i$. By Proposition 7.66 and Example 7.32, $\mathcal{W}_i$ is a Banach space over $\mathbb{K}$, $\forall i \in \{0, \ldots, k\}$. Then, by Propositions 7.22 and 4.31, $\mathcal{W}$ is a Banach space over $\mathbb{K}$. $\forall z \in \mathcal{Z}$, $\forall i \in \{0, \ldots, k\}$, we have $z^{(i)} \in \mathcal{W}_i$. Then, we may define a mapping $T : \mathcal{Z} \to \mathcal{W}$ by $T(z) = (z^{(0)}, \ldots, z^{(k)})$,

$\forall z \in \mathcal{Z}$. Clearly, T is linear and norm preserving. Hence, $T \in \mathrm{B}(\mathcal{Z}, \mathcal{W})$ and is injective.

Claim 12.92.1 $\mathcal{R}(T) \subseteq \mathcal{W}$ is closed.

Proof of Claim Clearly, we have $\mathcal{R}(T) \subseteq \overline{\mathcal{R}(T)} \subseteq \mathcal{W}$. $\forall w \in \overline{\mathcal{R}(T)}$, $\exists (z_n)_{n=1}^\infty \subseteq \mathcal{Z}$ such that $w = \lim_{n\in\mathbb{N}} T(z_n)$ in $\mathcal{W}$. Then, $(T(z_n))_{n=1}^\infty \subseteq \mathcal{W}$ is a Cauchy sequence. Since T is norm preserving, then $(z_n)_{n=1}^\infty \subseteq \mathcal{Z}$ is a Cauchy sequence. By the completeness of $\mathcal{Z}$, $\exists z_0 \in \mathcal{Z}$ such that $\lim_{n\in\mathbb{N}} z_n = z_0$ in $\mathcal{Z}$. By the continuity of T, we have $w = \lim_{n\in\mathbb{N}} T(z_n) = T(z_0) \in \mathcal{W}$. Hence, $w \in \mathcal{R}(T)$. By the arbitrariness of w, we have $\overline{\mathcal{R}(T)} \subseteq \mathcal{R}(T)$. Hence, $\mathcal{R}(T)$ is closed. This completes the proof of the claim. □

By Claim 12.92.1 and Proposition 4.39, $\mathcal{R}(T) \subseteq \mathcal{W}$ is a Banach space over $\mathbb{K}$. By Open Mapping Theorem 7.103, $T_{\text{inv}} : \mathcal{R}(T) \to \mathcal{Z}$ and $T_{\text{inv}} \in \mathrm{B}(\mathcal{R}(T), \mathcal{Z})$.

$\forall f \in \mathcal{Z}^*$, define $g : \mathcal{R}(T) \to \mathbb{K}$ by $g(w) = f(T_{\text{inv}}(w))$, $\forall w \in \mathcal{R}(T)$. Clearly, g is a bounded linear functional on $\mathcal{R}(T)$. By Hahn–Banach Extension Theorem 7.83, $\exists \mu \in \mathcal{W}^*$ such that $\mu(w) = g(w)$, $\forall w \in \mathcal{R}(T)$, and $\|\mu\|_{\mathcal{W}^*} = \|g\|_{\mathcal{R}(T)} \leq \|f\|_{\mathcal{Z}^*} \|T_{\text{inv}}\|_{\mathrm{B}(\mathcal{R}(T), \mathcal{Z})} \leq \|f\|_{\mathcal{Z}^*}$, where the last inequality follows from the fact that T is norm preserving and so is T_{inv}. By Example 7.76, $\mathcal{W}^* = \prod_{i=0}^k \mathcal{W}_i^* = \prod_{i=0}^k (\mathcal{C}(\Omega, \mathrm{B}_i(\mathcal{X}, \mathcal{Y})))^*$. Then, $\mu = (\mu_0, \ldots, \mu_k) \in \mathcal{W}^*$ and $f(z) = g(T(z)) = \langle\langle \mu, T(z) \rangle\rangle = \sum_{i=0}^k \langle\langle \mu_i, z^{(i)} \rangle\rangle$, $\forall z \in \mathcal{Z}$.

$\forall \nu := (\nu_0, \ldots, \nu_k) \in \mathcal{W}^*$, define $h : \mathcal{Z} \to \mathbb{K}$ by $h(z) = \sum_{i=0}^k \langle\langle \nu_i, z^{(i)} \rangle\rangle$, $\forall z \in \mathcal{Z}$. Then, $h(z) = \langle\langle \nu, T(z) \rangle\rangle$, $\forall z \in \mathcal{Z}$. Then, $h \in \mathcal{Z}^*$ and $\|h\|_{\mathcal{Z}^*} \leq \|\nu\|_{\mathcal{W}^*} \|T\|_{\mathrm{B}(\mathcal{Z}, \mathcal{W})} \leq \|\nu\|_{\mathcal{W}^*}$, where the last inequality follows from the fact that T is norm preserving.

Then, $\|f\|_{\mathcal{Z}^*} \leq \|\mu\|_{\mathcal{W}^*} \leq \|f\|_{\mathcal{Z}^*}$. Hence, $\|f\|_{\mathcal{Z}^*} = \|\mu\|_{\mathcal{W}^*}$. This completes the proof of the theorem. □

Theorem 12.93 (Riesz Representation Theorem) *Let $m \in \mathbb{Z}_+$, $\mathcal{X} := \mathbb{K}^m$ be the Euclidean space, $I_1, \ldots, I_m \subset \mathbb{R}$ be nondegenerate compact intervals, $\Omega := \prod_{j=1}^m I_j \subset \mathcal{X}$ with the subset topology $\mathcal{O}$, $\mathcal{Y}$ be a Banach space over $\mathbb{K}$, $k \in \mathbb{N}$, and $\mathcal{Z} := \mathcal{C}_k(\Omega, \mathcal{Y})$ be the Banach space over $\mathbb{K}$ as prescribed in Example 9.70. $\forall f \in \mathcal{Z}^*$, $\exists \mu := (\mu_0, \ldots, \mu_k) \in \prod_{i=0}^k \mathcal{M}_{ft}(\Omega, (\mathrm{B}_i(\mathcal{X}, \mathcal{Y}))^*) = \prod_{i=0}^k \mathcal{M}_f(\Omega, \mathcal{B}_\mathrm{B}(\Omega), (\mathrm{B}_i(\mathcal{X}, \mathcal{Y}))^*)$ such that $f(z) = \sum_{i=0}^k \int_\Omega \langle\langle \mathrm{d}\mu_i, z^{(i)} \rangle\rangle$, $\forall z \in \mathcal{Z}$, with $\|\mu\| = \left(\sum_{i=0}^k (\mathcal{P} \circ \mu_i(\Omega))^2\right)^{1/2} = \|f\|_{\mathcal{Z}^*}$. Furthermore, $\forall \nu := (\nu_0, \ldots, \nu_k) \in \prod_{i=0}^k \mathcal{M}_{ft}(\Omega, (\mathrm{B}_i(\mathcal{X}, \mathcal{Y}))^*)$, define $h : \mathcal{Z} \to \mathbb{K}$ by $h(z) = \sum_{i=0}^k \int_\Omega \langle\langle \mathrm{d}\nu_i, z^{(i)} \rangle\rangle$, $\forall z \in \mathcal{Z}$. Then, $h \in \mathcal{Z}^*$ and $\|h\|_{\mathcal{Z}^*} \leq \|\nu\|$.*

Proof Let $\mathcal{W} := \prod_{i=0}^k \mathcal{C}(\Omega, \mathrm{B}_i(\mathcal{X}, \mathcal{Y})) =: \prod_{i=0}^k \mathcal{W}_i$. By Proposition 7.66 and Example 9.70, $\mathcal{W}_i$ is a Banach space over $\mathbb{K}$, $\forall i \in \{0, \ldots, k\}$. It is easy to show that $\Omega \subseteq \mathcal{X}$ satisfies the assumption of Theorem 12.92. By Riesz Representation Theorem 11.204, we have $\mathcal{W}^* = \prod_{i=0}^k (\mathcal{C}(\Omega, \mathrm{B}_i(\mathcal{X}, \mathcal{Y})))^* = \prod_{i=0}^k \mathcal{M}_{ft}(\Omega, (\mathrm{B}_i(\mathcal{X}, \mathcal{Y}))^*) = \prod_{i=0}^k \mathcal{M}_f(\Omega, \mathcal{B}_\mathrm{B}(\Omega), (\mathrm{B}_i(\mathcal{X}, \mathcal{Y}))^*)$. Then, the result follows immediately from Theorem 12.92. This completes the proof of the theorem. □

Theorem 12.94 (Riesz Representation Theorem) *Let $\mathcal{X}$ and $\mathcal{Y}$ be finite-dimensional Banach spaces over $\mathbb{K}$, $\Omega \subseteq \mathcal{X}$ be compact with the subset topology $\mathcal{O}$, and $k \in \mathbb{N}$. Assume that, $\forall x \in \Omega$, $\exists \delta_x \in (0, \infty) \subset \mathbb{R}$ such that $\Omega \cap \mathcal{B}_{\mathcal{X}}(x, \delta_x) - x$ is a conic segment and $\overline{\operatorname{span}(\mathrm{A}_\Omega(x))} = \mathcal{X}$. Let $\mathcal{Z} := \mathcal{C}_k(\Omega, \mathcal{Y})$ be the Banach space over $\mathbb{K}$ as prescribed in Example 9.70. $\forall f \in \mathcal{Z}^*$, $\exists \mu := (\mu_0, \ldots, \mu_k) \in \prod_{i=0}^{k} \mathcal{M}_{ft}(\Omega, (\mathrm{B}_i(\mathcal{X}, \mathcal{Y}))^*) = \prod_{i=0}^{k} \mathcal{M}_f(\Omega, \mathcal{B}_\mathrm{B}(\Omega), (\mathrm{B}_i(\mathcal{X}, \mathcal{Y}))^*)$ such that $f(z) = \sum_{i=0}^{k} \int_\Omega \langle\langle \mathrm{d}\mu_i, z^{(i)} \rangle\rangle$, $\forall z \in \mathcal{Z}$, with $\|\mu\| = \left(\sum_{i=0}^{k} (\mathcal{P} \circ \mu_i(\Omega))^2\right)^{1/2} = \|f\|_{\mathcal{Z}^*}$. Furthermore, $\forall \nu := (\nu_0, \ldots, \nu_k) \in \prod_{i=0}^{k} \mathcal{M}_{ft}(\Omega, (\mathrm{B}_i(\mathcal{X}, \mathcal{Y}))^*)$, define $h : \mathcal{Z} \to \mathbb{R}$ by $h(z) = \sum_{i=0}^{k} \int_\Omega \langle\langle \mathrm{d}\nu_i, z^{(i)} \rangle\rangle$, $\forall z \in \mathcal{Z}$. Then, $h \in \mathcal{Z}^*$ and $\|h\|_{\mathcal{Z}^*} \leq \|\nu\|$.*

Proof Let $\mathcal{W} := \prod_{i=0}^{k} \mathcal{C}(\Omega, \mathrm{B}_i(\mathcal{X}, \mathcal{Y})) =: \prod_{i=0}^{k} \mathcal{W}_i$. The result follows immediately from Theorem 12.92 if we can prove that $\mathcal{W}^* = \prod_{i=0}^{k} \mathcal{M}_{ft}(\Omega, (\mathrm{B}_i(\mathcal{X}, \mathcal{Y}))^*) = \prod_{i=0}^{k} \mathcal{M}_f(\Omega, \mathcal{B}_\mathrm{B}(\Omega), (\mathrm{B}_i(\mathcal{X}, \mathcal{Y}))^*)$. The first equality follows from Example 7.76 and Riesz Representation Theorem 11.205. Since $\Omega \subseteq \mathcal{X}$ is compact, then it is locally compact. Since $\mathcal{X}$ is a finite-dimensional Banach space, then it is separable by Proposition 11.190. By Propositions 4.37 and 4.38, Ω is separable metric space. By Theorem 11.198, we have $\mathcal{M}_{ft}(\Omega, (\mathrm{B}_i(\mathcal{X}, \mathcal{Y}))^*) = \mathcal{M}_f(\Omega, \mathcal{B}_\mathrm{B}(\Omega), (\mathrm{B}_i(\mathcal{X}, \mathcal{Y}))^*)$, $\forall i \in \{0, \ldots, k\}$. This completes the proof of the theorem. □

Example 12.95 Let $\mathbb{R}^2$ be endowed with the usual positive cone, $\Omega := \{(x_1, x_2) \in \mathbb{R}^2 \mid x_1^2 + x_2^2 > 0, \angle(x_1 + \mathrm{i}x_2) \neq 0\}$, $\bar{\Omega} := \{(\rho, \theta) \in \mathbb{R}^2 \mid \rho > 0, \theta \in (0, 2\pi)\}$, and $F : \Omega \to \bar{\Omega}$ be defined by $F(x_1, x_2) := \left(\sqrt{x_1^2 + x_2^2}, \angle(x_1 + \mathrm{i}x_2)\right)$, and $F_i : \bar{\Omega} \to \Omega$ be defined by $F_i(\rho, \theta) = F_{\mathrm{inv}}(\rho, \theta) = (\rho\cos(\theta), \rho\sin(\theta))$. It is easy to check that $F^{(1)} : \Omega \to \mathrm{B}(\mathbb{R}^2, \mathbb{R}^2)$ and $F_i^{(1)} : \bar{\Omega} \to \mathrm{B}(\mathbb{R}^2, \mathbb{R}^2)$. $F_i^{(1)}(\rho, \theta) = \begin{bmatrix} \cos(\theta) & -\rho\sin(\theta) \\ \sin(\theta) & \rho\cos(\theta) \end{bmatrix}$ and $\left|\det(F_i^{(1)}(\rho, \theta))\right| = \rho > 0$, $\forall (\rho, \theta) \in \bar{\Omega}$. Hence, $\bar{\bar{U}} = \bar{\Omega}$ as defined in Change of Variable Theorem 12.91. Then, all assumptions of that theorem are satisfied. Let $\mathcal{Y}$ be a separable Banach space, and $\forall f : \Omega \to \mathcal{Y}$ that is absolutely integrable over Ω, which is the measure subspace of $\mathbb{R}^2$, we have $\int_\Omega f(x_1, x_2)\, \mathrm{d}\mu_{\mathrm{B}2}(x_1, x_2) = \int_{\bar{\Omega}} f(F_i(\rho, \theta))\rho\, \mathrm{d}\mu_{\mathrm{B}2}(\rho, \theta)$. Furthermore, $\forall f : \mathbb{R}^2 \to \mathcal{Y}$ that is absolutely integrable over $\mathbb{R}^2$, we have $\int_{\mathbb{R}^2} f(x_1, x_2)\, \mathrm{d}\mu_{\mathrm{B}2}(x_1, x_2) = \int_\Omega f(x_1, x_2)\, \mathrm{d}\mu_{\mathrm{B}2}(x_1, x_2) = \int_{\bar{\Omega}} f(F_i(\rho, \theta))\rho\, \mathrm{d}\mu_{\mathrm{B}2}(\rho, \theta)$, where the first equality follows from the fact that $\mu_{\mathrm{B}2}(\mathbb{R}^2 \setminus \Omega) = 0$ and Proposition 11.92. $\forall \bar{f} : \bar{\Omega} \to \mathcal{Y}$ that is absolutely integrable over $\bar{\Omega}$, which is the measure subspace of $\mathbb{R}^2$, we have $\int_{\bar{\Omega}} \bar{f}(\rho, \theta)\, \mathrm{d}\mu_{\mathrm{B}2}(\rho, \theta) = \int_\Omega \bar{f}(F(x_1, x_2)) \frac{1}{\sqrt{x_1^2 + x_2^2}}\, \mathrm{d}\mu_{\mathrm{B}2}(x_1, x_2)$. ◇

Example 12.96 Let $m \in \mathbb{N}$ with $m \geq 3$, $\mathbb{R}^m$ be endowed with the usual positive cone, $\Omega := \{(x_1, \ldots, x_m) \in \mathbb{R}^m \mid x_1^2 + x_2^2 > 0, \angle(x_1 + \mathrm{i}x_2) \neq 0\}$, $\bar{\Omega} := \{(\rho, \theta_1, \ldots, \theta_{m-2}, \varphi) \in \mathbb{R}^m \mid \rho > 0, \theta_i \in (0, \pi), i = 1, \ldots, m-2, \varphi \in (0, 2\pi)\}$, and $F : \Omega \to \bar{\Omega}$ be defined by $F(x_1, \ldots, x_m) := \Big(\sqrt{\sum_{i=1}^m x_i^2}, \arccos\big(\frac{x_m}{\sqrt{\sum_{i=1}^m x_i^2}}\big), \ldots, \arccos\big(\frac{x_3}{\sqrt{x_1^2+x_2^2+x_3^2}}\big), \angle(x_1 + \mathrm{i}x_2)\Big)$, and $F_i : \bar{\Omega} \to \Omega$ be defined by $F_i(\rho, \theta_1, \ldots, \theta_{m-2}, \varphi) = F_{\mathrm{inv}}(\rho, \theta_1, \ldots, \theta_{m-2}, \varphi) = (\rho \sin(\theta_1) \cdots \sin(\theta_{m-2}) \cos(\varphi), \rho \sin(\theta_1) \cdots \sin(\theta_{m-2}) \sin(\varphi), \rho \sin(\theta_1) \cdots \sin(\theta_{m-3}) \cos(\theta_{m-2}), \ldots, \rho \sin(\theta_1) \cos(\theta_2), \rho \cos(\theta_1))$. It is easy to check that $F^{(1)} : \Omega \to \mathrm{B}(\mathbb{R}^m, \mathbb{R}^m)$ and $F_i^{(1)} : \bar{\Omega} \to \mathrm{B}(\mathbb{R}^m, \mathbb{R}^m)$, and $\left|\det(F_i^{(1)}(\rho, \theta_1, \ldots, \theta_{m-2}, \varphi))\right| = \rho^{m-1}(\sin(\theta_1))^{m-2}(\sin(\theta_2))^{m-3} \cdots \sin(\theta_{m-2}) > 0$, $\forall(\rho, \theta_1, \ldots, \theta_{m-2}, \varphi) \in \bar{\Omega}$. Hence, $\bar{\bar{U}} = \bar{\Omega}$ as defined in Change of Variable Theorem 12.91. Then, all assumptions of that theorem are satisfied. Let $\mathcal{Y}$ be a separable Banach space, and $\forall f : \Omega \to \mathcal{Y}$ that is absolutely integrable over Ω, which is the measure subspace of $\mathbb{R}^m$, we have $\int_\Omega f(x_1, \ldots, x_m)\, d\mu_{\mathrm{B}m}(x_1, \ldots, x_m) = \int_{\bar{\Omega}} f(F_i(\rho, \theta_1, \ldots, \theta_{m-2}, \varphi))\rho^{m-1}(\sin(\theta_1))^{m-2}(\sin(\theta_2))^{m-3} \cdots \sin(\theta_{m-2}) \cdot d\mu_{\mathrm{B}m}(\rho, \theta_1, \ldots, \theta_{m-2}, \varphi)$. Furthermore, $\forall f : \mathbb{R}^m \to \mathcal{Y}$ that is absolutely integrable over $\mathbb{R}^m$, we have $\int_{\mathbb{R}^m} f(x_1, \ldots, x_m)\, d\mu_{\mathrm{B}m}(x_1, \ldots, x_m) = \int_\Omega f(x_1, \ldots, x_m)\, d\mu_{\mathrm{B}m}(x_1, \ldots, x_m) = \int_{\bar{\Omega}} f(F_i(\rho, \theta_1, \ldots, \theta_{m-2}, \varphi))\rho^{m-1}(\sin(\theta_1))^{m-2} \cdot (\sin(\theta_2))^{m-3} \cdots \sin(\theta_{m-2})\, d\mu_{\mathrm{B}m}(\rho, \theta_1, \ldots, \theta_{m-2}, \varphi)$, where the first equality follows from the fact that $\mu_{\mathrm{B}m}(\mathbb{R}^m \setminus \Omega) = 0$ and Proposition 11.92. $\forall \bar{f} : \bar{\Omega} \to \mathcal{Y}$ that is absolutely integrable over $\bar{\Omega}$, which is the measure subspace of $\mathbb{R}^m$, we have $\int_{\bar{\Omega}} \bar{f}(\rho, \theta_1, \ldots, \theta_{m-2}, \varphi)\, d\mu_{\mathrm{B}m}(\rho, \theta_1, \ldots, \theta_{m-2}, \varphi) = \int_\Omega \bar{f}(F(x_1, \ldots, x_m)) \frac{1}{\sqrt{\sum_{i=1}^m x_i^2}} \frac{1}{\sqrt{\sum_{i=1}^{m-1} x_i^2}} \cdots \frac{1}{\sqrt{\sum_{i=1}^2 x_i^2}}\, d\mu_{\mathrm{B}m}(x_1, \ldots, x_m)$. ◇

Example 12.97 Let $\mathbb{R}^3$ be endowed with the usual positive cone, $\Omega := \{(x_1, x_2, x_3) \in \mathbb{R}^3 \mid x_1^2 + x_2^2 > 0, \angle(x_1 + \mathrm{i}x_2) \neq 0\}$, $\bar{\Omega} := \{(\rho, \varphi, z) \in \mathbb{R}^3 \mid \rho > 0, \varphi \in (0, 2\pi)\}$, and $F : \Omega \to \bar{\Omega}$ be defined by $F(x_1, x_2, x_3) := \big(\sqrt{x_1^2 + x_2^2}, \angle(x_1 + \mathrm{i}x_2), x_3\big)$, and $F_i : \bar{\Omega} \to \Omega$ be defined by $F_i(\rho, \varphi, z) = F_{\mathrm{inv}}(\rho, \varphi, z) = (\rho \cos(\varphi), \rho \sin(\varphi), z)$. It is easy to check that $F^{(1)} : \Omega \to \mathrm{B}(\mathbb{R}^3, \mathbb{R}^3)$ and $F_i^{(1)} : \bar{\Omega} \to \mathrm{B}(\mathbb{R}^3, \mathbb{R}^3)$, $\begin{bmatrix} \cos(\varphi) & -\rho \sin(\varphi) & 0 \\ \sin(\varphi) & \rho \cos(\varphi) & 0 \\ 0 & 0 & 1 \end{bmatrix} = F_i^{(1)}(\rho, \varphi, z)$, and $\left|\det(F_i^{(1)}(\rho, \varphi, z))\right| = \rho > 0$, $\forall(\rho, \varphi, z) \in \bar{\Omega}$. Hence, $\bar{\bar{U}} = \bar{\Omega}$ as defined in Change of Variable Theorem 12.91. Then, all assumptions of that theorem are satisfied. Let $\mathcal{Y}$ be a separable Banach space, and $\forall f : \Omega \to \mathcal{Y}$ that is absolutely integrable over Ω, which is the measure subspace of $\mathbb{R}^3$, we have $\int_\Omega f(x_1, x_2, x_3)\, d\mu_{\mathrm{B}3}(x_1, x_2, x_3) = \int_{\bar{\Omega}} f(F_i(\rho, \varphi, z))\rho\, d\mu_{\mathrm{B}3}(\rho, \varphi, z)$. Furthermore, $\forall f : \mathbb{R}^3 \to \mathcal{Y}$ that is absolutely integrable over $\mathbb{R}^3$, we have $\int_{\mathbb{R}^3} f(x_1, x_2, x_3)\, d\mu_{\mathrm{B}3}(x_1, x_2, x_3) = \int_\Omega f(x_1, x_2, x_3)\, d\mu_{\mathrm{B}3}(x_1, x_2, x_3) = \int_{\bar{\Omega}} f(F_i(\rho, \varphi, z))\rho\, d\mu_{\mathrm{B}3}(\rho, \varphi, z)$, where the first equality follows from the fact that $\mu_{\mathrm{B}3}(\mathbb{R}^3 \setminus \Omega) = 0$ and Proposition 11.92. $\forall \bar{f} : \bar{\Omega} \to \mathcal{Y}$ that is absolutely integrable over $\bar{\Omega}$, which is the measure subspace of $\mathbb{R}^3$, we have $\int_{\bar{\Omega}} \bar{f}(\rho, \varphi, z) \, d\mu_{B3}(\rho, \varphi, z) = \int_{\Omega} \bar{f}(F(x_1, x_2, x_3)) \frac{1}{\sqrt{x_1^2 + x_2^2}} \, d\mu_{B3}(x_1, x_2, x_3)$. ◇

Proposition 12.98 *Let $I \subseteq \mathbb{R}$ be an open interval, $\bar{I} \subseteq \mathbb{R}$ be an open interval, $F : I \to \bar{I}$ be a homeomorphism, $F_i := F_{\text{inv}} : \bar{I} \to I$, and F and F_i are absolutely continuous, $\mathbb{I} := ((I, |\cdot|), \mathcal{B}, \mu)$ be the σ-finite metric measure subspace of $\mathbb{R}$ and $\bar{\mathbb{I}} := ((\bar{I}, |\cdot|), \bar{\mathcal{B}}, \bar{\mu})$ be the σ-finite metric measure subspace of $\mathbb{R}$, and $\mathcal{Y}$ be a separable Banach space. Assume that $F^{(1)} : U \to \mathbb{R}$ satisfies: $\sup_{x \in \bar{r}_{x_1, x_2} \cap U} \left| F^{(1)}(x) \right| < \infty$, $\forall x_1, x_2 \in I$ with $x_1 \leq x_2$; $F_i^{(1)} : \bar{U} \to \mathbb{R}$ satisfies: $\sup_{\bar{x} \in \bar{r}_{\bar{x}_1, \bar{x}_2} \cap \bar{U}} \left| F_i^{(1)}(\bar{x}) \right| < \infty$, $\forall \bar{x}_1, \bar{x}_2 \in \bar{I}$ with $\bar{x}_1 \leq \bar{x}_2$; and $\hat{U} := \{x \in U \mid F^{(1)}(x) \neq 0\}$ and $\bar{\hat{U}} := \left\{ \bar{x} \in \bar{U} \;\middle|\; F_i^{(1)}(\bar{x}) \neq 0 \right\}$ are such that $\bar{\mu}(\bar{I} \setminus \bar{\hat{U}}) = 0 = \mu\big(I \setminus F_i(\bar{\hat{U}})\big) = \mu(I \setminus \hat{U}) = \bar{\mu}\big(\bar{I} \setminus F(\hat{U})\big)$. Then, $\forall f : I \to \mathcal{Y}$ with f being absolutely integrable over $\mathbb{I}$, we have*

$$\int_I f \, d\mu = \int_{F(I)} f(F_i(\bar{x})) g(\bar{x}) \, d\bar{\mu}(\bar{x}) = \int_{\bar{I}} f(F_i(\bar{x})) g(\bar{x}) \, d\bar{\mu}(\bar{x})$$

and $\forall \bar{f} : \bar{I} \to \mathcal{Y}$ with $\bar{f}$ being absolutely integrable over $\bar{\mathbb{I}}$, we have

$$\int_{\bar{I}} \bar{f} \, d\bar{\mu} = \int_{F_i(\bar{I})} \bar{f}(F(x)) g_1(x) \, d\mu(x) = \int_I \bar{f}(F(x)) g_1(x) \, d\mu(x)$$

where $g : \bar{I} \to (0, \infty) \subset \mathbb{R}$ and $g_1 : I \to (0, \infty) \subset \mathbb{R}$ are given by $g(\bar{x}) = \begin{cases} \left| F_i^{(1)}(\bar{x}) \right| & \bar{x} \in \bar{\hat{U}} \\ 1 & \bar{x} \in \bar{I} \setminus \bar{\hat{U}} \end{cases}$ and $g_1(x) = \frac{1}{g(F(x))}$, $\forall x \in I$.

Proof The proposition is a direct consequence of Theorem 12.91. □

Theorem 12.99 (Riemann–Lebesgue) *Let $\mathcal{Y}$ be a separable Banach space over $\mathbb{K}$, $f \in \bar{L}_1(\mathbb{R}, \mathcal{Y})$. Then,*

$$\lim_{|p| \to \infty} \int_{\mathbb{R}} f(t) \sin(pt) \, d\mu_B(t) = \vartheta_{\mathcal{Y}} \tag{12.7}$$

Proof Since $f \in \bar{L}_1(\mathbb{R}, \mathcal{Y})$, then f is absolutely integrable over $\mathbb{R}$. Then, $\bar{f}_p : \mathbb{R} \to \mathcal{Y}$ defined by $\bar{f}_p(t) := f(t) \sin(pt)$, $\forall t \in \mathbb{R}$, is absolutely integrable over $\mathbb{R}$, $\forall p \in \mathbb{R}$.

First, we consider the special case when $f = y\chi_{I,\mathbb{R}}$ with I being an interval and $y \in \mathcal{Y}$ with $y \neq \vartheta_{\mathcal{Y}}$. Then, I must be a finite interval, since $f \in \bar{L}_1(\mathbb{R}, \mathcal{Y})$. The case when $I = \emptyset$ is trivial. Let $\bar{I} := [a, b]$ with $a, b \in \mathbb{R}$ and $a \leq b$. Then, $\lim_{|p| \to \infty} \int_{\mathbb{R}} f(t) \sin(pt) \, d\mu_B(t) = \lim_{|p| \to \infty} \int_a^b y \sin(pt) \, d\mu_B(t) = \lim_{|p| \to \infty} y \frac{1}{p} (\cos(pa) - \cos(pb)) \, d\mu_B(t) = \vartheta_{\mathcal{Y}}$, where the first equality follows from I being an interval and $\overline{I} = [a, b]$ and the second equality follows from Proposition 11.92 and Theorem 12.83. This special case is proved.

Next, we consider the general case when $f \in \bar{L}_1(\mathbb{R}, \mathcal{Y})$. By Proposition 11.66, there exists a sequence of simple functions $(\varphi_n)_{n=1}^{\infty}$, $\varphi_n : \mathbb{R} \to \mathcal{Y}$, $\forall n \in \mathbb{N}$, such that $\lim_{n\in\mathbb{N}} \varphi_n(x) = f(x)$ a.e. $x \in \mathbb{R}$, and $\|\varphi_n(x)\| \leq \|f(x)\|$, $\forall x \in \mathbb{R}$, $\forall n \in \mathbb{N}$. Then, $\lim_{n\in\mathbb{N}} \int_{\mathbb{R}} \mathcal{P} \circ (f - \varphi_n) \, d\mu_B = 0$, by Lebesgue Dominated Convergence Theorem 11.91. $\forall \epsilon \in (0, \infty) \subset \mathbb{R}$, $\exists n_0 \in \mathbb{N}$ such that $\int_{\mathbb{R}} \mathcal{P} \circ (f - \varphi_{n_0}) \, d\mu_B < \frac{\epsilon}{4}$. Let φ_{n_0} admit the canonical representation $\varphi_{n_0} = \sum_{i=1}^{n} y_i \chi_{E_i,\mathbb{R}}$, where $n \in \mathbb{Z}_+$, $y_1, \ldots, y_n \in \mathcal{Y}$ are distinct and none equals to $\vartheta_{\mathcal{Y}}$, and $E_1, \ldots, E_n \in \mathcal{B}_B(\mathbb{R})$ are nonempty, pairwise disjoint, and of finite measure. $\int_{\mathbb{R}} \mathcal{P} \circ \varphi_{n_0} \, d\mu_B = \sum_{i=1}^{n} \|y_i\| \mu_B(E_i)$. By Proposition 11.7, $\exists a, b \in \mathbb{R}$ with $a < b$ such that $\int_{\mathbb{R}\setminus[a,b]} \mathcal{P} \circ \varphi_{n_0} \, d\mu_B < \frac{\epsilon}{4}$. Let $f_1 := \varphi_{n_0} \chi_{[a,b],\mathbb{R}}$. Then, $\int_{\mathbb{R}} \mathcal{P} \circ (f - f_1) \, d\mu_B < \frac{\epsilon}{2}$. By Proposition 11.182, $\exists$ a continuous function $g : [a, b] \to \mathcal{Y}$ such that $g \in \bar{L}_1([a, b], \mathcal{Y})$ and $\left\| g - f_1|_{[a,b]} \right\|_{\bar{L}_1([a,b],\mathcal{Y})} < \frac{\epsilon}{4}$. By Proposition 5.39, g is uniformly continuous. $\exists m \in \mathbb{N}$, $\forall x_1, x_2 \in [a, b]$ with $|x_1 - x_2| \leq \frac{b-a}{m}$, we have $\|g(x_1) - g(x_2)\| < \frac{\epsilon}{4(b-a)}$. Construct a step function $h : \mathbb{R} \to \mathcal{Y}$ by $h(x) = \begin{cases} 0 & x \in \mathbb{R} \setminus [a, b] \\ g(a + \left\lfloor \frac{x-a}{b-a} m \right\rfloor \frac{b-a}{m}) & x \in [a, b] \end{cases}$, $\forall x \in \mathbb{R}$.

Then, we have $\int_{[a,b]} \mathcal{P} \circ (h|_{[a,b]} - g) \, d\mu_B < \frac{\epsilon}{4}$. Hence, $\int_{\mathbb{R}} \mathcal{P} \circ (f - h) \, d\mu_B \leq \int_{\mathbb{R}} \mathcal{P} \circ (f - f_1) \, d\mu_B + \int_{\mathbb{R}} \mathcal{P} \circ (f_1 - h) \, d\mu_B < \frac{\epsilon}{2} + \int_{[a,b]} \mathcal{P} \circ (f_1|_{[a,b]} - h|_{[a,b]}) \, d\mu_B \leq \frac{\epsilon}{2} + \int_{[a,b]} \mathcal{P} \circ (f_1|_{[a,b]} - g) \, d\mu_B + \int_{[a,b]} \mathcal{P} \circ (g - h|_{[a,b]}) \, d\mu_B < \epsilon$. Then, $0 \leq \liminf_{|p|\to\infty} \left\| \int_{\mathbb{R}} \bar{f}_p \, d\mu_B \right\| \leq \limsup_{|p|\to\infty} \left\| \int_{\mathbb{R}} \bar{f}_p \, d\mu_B \right\| \leq \limsup_{|p|\to\infty} \left(\left\| \int_{\mathbb{R}} h(t) \sin(pt) \, d\mu_B(t) \right\| + \left\| \int_{\mathbb{R}} (f(t) - h(t)) \sin(pt) \, d\mu_B(t) \right\| \right) \leq \lim_{|p|\to\infty} \left\| \int_{\mathbb{R}} h(t) \sin(pt) \, d\mu_B(t) \right\| + \limsup_{|p|\to\infty} \int_{\mathbb{R}} \|f(t) - h(t)\| \, d\mu_B(t) < 0 + \epsilon$, where the first three inequalities follow from Proposition 3.83 and Lebesgue Dominated Convergence Theorem 11.91, the fourth inequality follows from Propositions 3.83 and 11.92, and the last inequality follows from the preceding discussion and the special case. Hence, by the arbitrariness of ϵ, we have $\lim_{|p|\to\infty} \left\| \int_{\mathbb{R}} f(t) \sin(pt) \, d\mu_B(t) \right\| = 0$, by Proposition 3.83. Then, $\lim_{|p|\to\infty} \int_{\mathbb{R}} f(t) \sin(pt) \, d\mu_B(t) = \vartheta_{\mathcal{Y}}$.

This completes the proof of the theorem. □

Proposition 12.100 *Let $J := \bar{r}_{a,b} = [a, b] \subset \mathbb{R}$ be a compact interval ($a \leq b$), $g : J \to \mathbb{R}$ be of bounded variation (as defined in Definition 12.41) and is nondecreasing, and $f : J \to \mathbb{R}$ be $\mathcal{B}_B(\mathbb{R})$-measurable. Assume that f is bounded, that is, $\exists M \geq 0$ such that $|f(x)| \leq M$, $\forall x \in J$. If f is Riemann–Stieltjes integrable with respect to g over J (as defined in Definition 29.2 of Bartle (1976), or equivalently in Definition A.4), and let the Riemann–Stieltjes integral be denoted by I, then $I = \int_{\bar{r}_{a,b}} f \, dg$, where the right-hand side is the Lebesgue–Stieltjes integral defined in Definition 12.57.*

Proof By Proposition 12.39, $\mathrm{P}(J) = \mathrm{r}_{a,b}$. By Proposition 12.49, there exists a unique finite measure μ on the measurable space $(\mathrm{P}(J), \mathcal{B}_B(\mathrm{P}(J)))$ such that g is a cumulative distribution function of μ, $\mu(\mathrm{r}_{x_1,x_2}) = g(x_2) - g(x_1)$, $\forall a \leq x_1 \leq x_2 \leq b$, and $\mu(\mathrm{P}(J)) = \mu(\mathrm{r}_{a,b}) = g(b) - g(a) = T_g < \infty$. Since f is bounded, by Bounded Convergence Theorem 11.77, f is absolutely integrable over $\mathrm{P}(J)$ with respect to μ. Then, by Proposition 11.92, $\bar{I} := \int_{\mathrm{r}_{a,b}} f \,\mathrm{d}g \in \mathbb{R}$.

By Riemann Criterion for Integrability (Thoerem 30.1 of Bartle (1976)), $\forall n \in \mathbb{N}$, there exists a partition P_n of J such that, $\forall$ refinement $P := (x_0, \ldots, x_{k_n})$ of P_n, we have

$$\left|\sum_{j=1}^{k_n} M_j(g(x_j) - g(x_{j-1})) - I\right| < 2^{-n}; \ \left|\sum_{j=1}^{k_n} m_j(g(x_j) - g(x_{j-1})) - I\right| < 2^{-n}$$

where $M_j := \sup\{f(x) \in \mathbb{R} \mid x \in [x_{j-1}, x_j]\}$ and $m_j := \inf\{f(x) \in \mathbb{R} \mid x \in [x_{j-1}, x_j]\}$. Without loss of generality, we may assume that P_{n+1} is a refinement of P_n (otherwise, define P_{n+1} to be the partition that contains all of the points in P_n and P_{n+1}). Then,

$$\begin{aligned}
I &= \lim_{n\in\mathbb{N}} \sum_{j=1}^{k_n} M_j(g(x_j) - g(x_{j-1})) = \lim_{n\in\mathbb{N}} \int_{\mathrm{r}_{a,b}} \sum_{j=1}^{k_n} M_j \chi_{\mathrm{r}_{x_{j-1},x_j},J} \,\mathrm{d}\mu \\
&=: \lim_{n\in\mathbb{N}} \int_{\mathrm{r}_{a,b}} \bar{h}_n \,\mathrm{d}\mu = \lim_{n\in\mathbb{N}} \sum_{j=1}^{k_n} m_j(g(x_j) - g(x_{j-1})) \\
&= \lim_{n\in\mathbb{N}} \int_{\mathrm{r}_{a,b}} \sum_{j=1}^{k_n} m_j \chi_{\mathrm{r}_{x_{j-1},x_j},J} \,\mathrm{d}\mu =: \lim_{n\in\mathbb{N}} \int_{\mathrm{r}_{a,b}} h_n \,\mathrm{d}\mu
\end{aligned}$$

where $h_n : \mathrm{r}_{a,b} \to \mathbb{R}$ and $\bar{h}_n : \mathrm{r}_{a,b} \to \mathbb{R}$, $\forall n \in \mathbb{N}$; and the second and fourth equalities follow from Proposition 11.75. Clearly, $-M \le h_n(x) \le h_{n+1}(x) \le f(x) \le \bar{h}_{n+1}(x) \le \bar{h}_n(x) \le M$, $\forall x \in \mathrm{r}_{a,b}$, $\forall n \in \mathbb{N}$. By Proposition 11.48, we have $\lim_{n\in\mathbb{N}} h_n =: h : \mathrm{r}_{a,b} \to \mathbb{R}$ and $\lim_{n\in\mathbb{N}} \bar{h}_n =: \bar{h} : \mathrm{r}_{a,b} \to \mathbb{R}$ are $\mathcal{B}_{\mathrm{B}}(\mathrm{r}_{a,b})$-measurable. By Lebesgue Dominated Convergence Theorem 11.91, we have h and $\bar{h}$ are absolutely integrable over $(\mathrm{r}_{a,b}, \mathcal{B}_{\mathrm{B}}(\mathrm{r}_{a,b}), \mu)$ and $\int_{\mathrm{r}_{a,b}} h \,\mathrm{d}\mu = \lim_{n\in\mathbb{N}} \int_{\mathrm{r}_{a,b}} h_n \,\mathrm{d}\mu = I = \lim_{n\in\mathbb{N}} \int_{\mathrm{r}_{a,b}} \bar{h}_n \,\mathrm{d}\mu = \int_{\mathrm{r}_{a,b}} \bar{h} \,\mathrm{d}\mu$. Note that $h(x) \le f(x) \le \bar{h}(x)$, $\forall x \in \mathrm{r}_{a,b}$, and f is $\mathcal{B}_{\mathrm{B}}(\mathbb{R})$-measurable by assumption, and then, by Proposition 11.92, $I = \int_{\mathrm{r}_{a,b}} h \,\mathrm{d}\mu \le \int_{\mathrm{r}_{a,b}} f \,\mathrm{d}\mu = \bar{I} \le \int_{\mathrm{r}_{a,b}} \bar{h} \,\mathrm{d}\mu = I$.

This completes the proof of the proposition. □

The above can be generalized further. Let $g : J \to \mathbb{R}$ be of bounded variation, but not necessarily nondecreasing and everything else as assumed in the preceding proposition. Then, by Theorem 12.50, there exists a unique finite $\mathbb{R}$-valued measure μ on the measurable space $(\mathrm{P}(J), \mathcal{B}_{\mathrm{B}}(\mathrm{P}(J)))$ such that g is a cumulative distribution function of μ, $\mathcal{P} \circ \mu(\mathrm{r}_{x_1,x_2}) = T_g(\mathrm{r}_{x_1,x_2})$, $\forall a \le x_1 \le x_2 \le b$, and $\mathcal{P} \circ \mu(\mathrm{P}(J)) = \mathcal{P} \circ \mu(\mathrm{r}_{a,b}) = T_g < \infty$. By Jordan Decomposition Theorem 11.162, there exists a unique pair of mutually singular finite measures μ_+ and μ_- on $(\mathrm{r}_{a,b}, \mathcal{B}_{\mathrm{B}}(\mathrm{r}_{a,b}))$, such that $\mu = \mu_+ - \mu_-$ and $\mathcal{P} \circ \mu = \mu_+ + \mu_-$. Let G be a cumulative distribution function for $\mathcal{P} \circ \mu$ as delineated in Proposition 12.51 and G is of bounded variation. We need to assume that f is Riemann–Stieltjes integrable with respect to g and G over J, respectively, and denote these integrals as I and I_G, respectively. It is easy to see that $\frac{1}{2}(g + G)$ and $\frac{1}{2}(G - g)$ are cumulative distribution functions for μ_+ and μ_-, respectively. Then, they are nondecreasing since μ_+ and μ_- are measures. Furthermore, they are continuous on the right since g and G are of bounded variation. Hence, $\frac{1}{2}(g + G)$ and $\frac{1}{2}(G - g)$ are of bounded variation. Thus, by Theorem 29.5 of Bartle (1976) and Proposition 12.100, we have $\frac{1}{2}(I + I_G) = \int_{r_{a,b}} f \, d\mu_+$ and $\frac{1}{2}(I_G - I) = \int_{r_{a,b}} f \, d\mu_-$. This leads to $I = \frac{1}{2}(I + I_G) - \frac{1}{2}(I_G - I) = \int_{r_{a,b}} f \, d\mu_+ - \int_{r_{a,b}} f \, d\mu_- = \int_{r_{a,b}} f \, d\mu = \bar{I}$, where the third equality follows from Propositions 11.145 and 11.146.

Theorem 12.101 (Integration by Parts) *Let $J := \bar{r}_{a,b} \subseteq \mathbb{R}$ be a compact interval, $\mathbb{I} := ((\mathrm{P}(J), |\cdot|), \mathcal{B}, \mu)$ be the finite metric measure subspace of $\mathbb{R}$, $f : J \to R$ be of bounded variation and a cumulative distribution function of the finite $\mathbb{R}$-valued measure space $(\mathrm{P}(J), \mathcal{B}, \nu_f)$, $F : J \to \mathbb{R}$ be the cumulative distribution function for the finite measure space $(\mathrm{P}(J), \mathcal{B}, \mathcal{P} \circ \nu_f)$ as delineated in Proposition 12.51 and is of bounded variation, $g : J \to R$ be of bounded variation and a cumulative distribution function of the finite $\mathbb{R}$-valued measure space $(\mathrm{P}(J), \mathcal{B}, \nu_g)$, and $G : J \to \mathbb{R}$ be the cumulative distribution function for the finite measure space $(\mathrm{P}(J), \mathcal{B}, \mathcal{P} \circ \nu_g)$ as delineated in Proposition 12.51 and is of bounded variation. Assume that f is Riemann–Stieltjes integrable with respect to g and G over J, respectively, and denote the Riemann–Stieltjes integral of f with respect to g over J as I_1; and g is Riemann–Stieltjes integrable with respect to f and F over J, respectively, and denote the Riemann–Stieltjes integral of g with respect to f over J as I_2. Then, we have*

$$\int_{r_{a,b}} f(x) \, dg(x) + \int_{r_{a,b}} g(x) \, df(x) = I_1 + I_2 = f(b)g(b) - f(a)g(a) \tag{12.8}$$

Proof By Proposition 12.54, $f - f(a)$ and $g - g(a)$ are $\mathcal{B}_\mathrm{B}(\mathbb{R})$-measurable. Hence, f and g are $\mathcal{B}_\mathrm{B}(\mathbb{R})$-measurable, by Propositions 11.38, 7.23, and 11.39. Since f and g are of bounded variation, then f and g are bounded. The first equality in the theorem statement follows from Proposition 12.100 and the discussion immediately after it. The second equality in the theorem statement follows from Integration by Parts 29.7 of Bartle (1976) (or Theorem A.8 in the appendix).

This completes the proof of the theorem. □

Proposition 12.102 *Let $I \subseteq \mathbb{R}$ be an open interval, $\mathbb{X} := ((I, |\cdot|), \mathcal{B}, \mu)$ be the σ-finite metric measure subspace of $\mathbb{R}$, $\mathcal{Y}$ be separable reflexive Banach space over $\mathbb{K}$ with $\mathcal{Y}^*$ being separable, $(I, \mathcal{B}, \nu)$ be a σ-finite $\mathcal{Y}$-valued measure space, $F : I \to \mathcal{Y}$ be a cumulative distribution function of ν, and $f : I \to \mathcal{Y}$ be $\mathcal{B}$-measurable. Assume that F is absolutely continuous, and $\mathrm{D}F(x) = f(x)$ a.e. $x \in \mathbb{X}$. Then, $\frac{d\nu}{d\mu} = f$ a.e. in $\mathbb{X}$.*

Proof By Fundamental Theorem of Calculus II, Theorem 12.88, $F(x) = F(a) + \int_a^x f(t)\,\mathrm{d}t$, $\forall a, x \in I$. By Proposition 11.116, we may define $\bar{\nu}$ to be σ-finite $\mathcal{Y}$-valued measure with kernel f over $\mathbb{X}$. Then, by Definition 11.166, f is the Radon–Nikodym derivative of $\bar{\nu}$ with respect to μ. By Definition 12.42, F is a cumulative distribution function of $(I, \mathcal{B}, \bar{\nu})$. By Theorem 12.50 and Proposition 12.73, we have $\nu = \bar{\nu}$. Then, we have $f = \frac{\mathrm{d}\nu}{\mathrm{d}\mu}$ a.e. in $\mathbb{X}$. □

Lemma 12.103 *Let Ω be a set, $\mathcal{S}$ be a π-system on Ω, $\mathcal{B}$ be the σ-algebra on Ω generated by $\mathcal{S}$, $\mathcal{Y}$ be a normed linear space, and μ_1 and μ_2 be finite $\mathcal{Y}$-valued measures on $(\Omega, \mathcal{B})$. Assume that $\mu_1(E) = \mu_2(E)$, $\forall E \in \mathcal{S}$. Then, $\mu_1(E) = \mu_2(E)$, $\forall E \in \mathcal{B}$.*

Proof Define $\mathcal{D} := \{E \in \mathcal{B} \mid \mu_1(E) = \mu_2(E)\}$. Clearly, $\mathcal{S} \subseteq \mathcal{D} \subseteq \mathcal{B}$. We will show that $\mathcal{D}$ is a monotone class on Ω. Then, by Monotone Class Lemma 12.19, we have $\mathcal{D} = \mathcal{B}$. Hence, the result holds.

Clearly, (i) $\emptyset, \Omega \in \mathcal{S} \subseteq \mathcal{D}$. (ii) $\forall E_1, E_2 \in \mathcal{D}$ with $E_1 \subseteq E_2$, we have $\mu_1(E_2 \setminus E_1) = \mu_1(E_2) - \mu_1(E_1) = \mu_2(E_2) - \mu_2(E_1) = \mu_2(E_2 \setminus E_1)$, where the first and last equalities follow from Definitions 11.108 and 11.109 and the second equality follows from the fact $E_1, E_2 \in \mathcal{D}$. This implies that $E_2 \setminus E_1 \in \mathcal{D}$. (iii) $\forall (E_i)_{i=1}^{\infty} \subseteq \mathcal{D}$ with $E_i \subseteq E_{i+1}$, $\forall i \in \mathbb{N}$, we have $\bigcup_{i=1}^{\infty} E_i = \bigcup_{i=1}^{\infty} A_i$, where $A_1 := E_1$, $A_2 := E_2 \setminus E_1$, $A_i := E_i \setminus E_{i-1}$, $\forall i \in \{3, 4, \ldots\}$. Clearly, A_i's are pairwise disjoint. By (ii), we have $A_i \in \mathcal{D}$, $\forall i \in \mathbb{N}$. Then, $\mu_1(\bigcup_{i=1}^{\infty} E_i) = \mu_1(\bigcup_{i=1}^{\infty} A_i) = \sum_{i=1}^{\infty} \mu_1(A_i) = \sum_{i=1}^{\infty} \mu_2(A_i) = \mu_2(\bigcup_{i=1}^{\infty} A_i) = \mu_2(\bigcup_{i=1}^{\infty} E_i)$. Hence, $\bigcup_{i=1}^{\infty} E_i \in \mathcal{D}$. The preceding discussion implies that $\mathcal{D}$ is a monotone class on Ω.

This completes the proof of the lemma. □

Example 12.104 Let $n \in \mathbb{N}$ with $n \geq 2$, and $S_{n-1} := \{x \in \mathbb{R}^n \mid |x| = 1\} \subseteq \mathbb{R}^n$ be the unit sphere (n-dimensional) with the subset topology $\mathcal{O}$. Clearly, $S_{n-1} \in \mathcal{B}_{\mathrm{B}}(\mathbb{R}^n)$. Clearly, $(S_{n-1}, |\cdot|)$ is a metric subspace of $\mathbb{R}^n$. Let $\mathcal{B} := \{E \subseteq S_{n-1} \mid E \in \mathcal{B}_{\mathrm{B}}(\mathbb{R}^n)\}$. Then, $\mathcal{B} = \mathcal{B}_{\mathrm{B}}((S_{n-1}, \mathcal{O}))$ by Proposition 11.25. On S_{n-1}, we introduce function $\mu : \mathcal{O} \to [0, \infty) \subset \mathbb{R}$ by $\mu(O) := n\mu_{\mathrm{B}n}(V_O)$, where $V_O := \{rx \in \mathbb{R}^n \mid x \in O, r \in (0, 1) \subset \mathbb{R}\} \subseteq \mathbb{R}^n$, $\forall O \in \mathcal{O}$. Clearly, $\mu(\emptyset) = 0$, and $\forall O \in \mathcal{O}$, $V_O \in \mathcal{B}_{\mathrm{B}}(\mathbb{R}^n)$ since V_O is an open set in $\mathbb{R}^n$. Hence, $\mu(O)$ is well-defined. Clearly, $\mu(S_{n-1}) = n\mu_{\mathrm{B}n}(\mathcal{B}_{\mathbb{R}^n}(\mathbf{0}_n, 1) \setminus \{\mathbf{0}_m\}) = n\mu_{\mathrm{B}n}(\mathcal{B}_{\mathbb{R}^n}(\mathbf{0}_n, 1)) = \frac{n\pi^{n/2}}{\Gamma(\frac{n}{2}+1)}$, where we have referred to the Γ function, and the formula for the volume of n-dimensional ball in (Mathematics Handbook Editors Group, 1979, pp. 320). Clearly, μ is countably additive on $\mathcal{O}$, since $\mu_{\mathrm{B}n}$ is countably additive. Let $\mathcal{A} \subseteq {}^{S_{n-1}}2$ be the algebra generated by $\mathcal{O}$. By Proposition 2.8, $\forall A \in \mathcal{A}$, $\exists n, m \in \mathbb{N}$, $\forall i_1, \ldots, i_{2n} \in \{1, \ldots, m\}$, $\exists F_{i_1,\ldots,i_{2n}} \subseteq X$ with $F_{i_1,\ldots,i_{2n}} \in \mathcal{O}$ or $S_{n-1} \setminus F_{i_1,\ldots,i_{2n}} \in \mathcal{O}$ such that $A = \bigcup_{i_1=1}^{m} \bigcap_{i_2=1}^{m} \cdots \bigcap_{i_{2n}=1}^{m} F_{i_1,\ldots,i_{2n}}$. Then, $V_A := \{rx \in \mathbb{R}^n \mid x \in A, r \in (0, 1) \subset \mathbb{R}\}$ can be expressed $V_A = \bigcup_{i_1=1}^{m} \bigcap_{i_2=1}^{m} \cdots \bigcap_{i_{2n}=1}^{m} V_{F_{i_1,\ldots,i_{2n}}} \in \mathcal{B}_{\mathrm{B}}(\mathbb{R}^n)$ since $V_{F_{i_1,\ldots,i_{2n}}} \in \mathcal{B}_{\mathrm{B}}(\mathbb{R}^n)$. Thus, we may define $\hat{\mu} : \mathcal{A} \to [0, \infty) \subset \mathbb{R}$ by $\hat{\mu}(A) := n\mu_{\mathrm{B}n}(V_A)$, $\forall A \in \mathcal{A}$. Clearly, $\hat{\mu}$ is an extension of μ to $\mathcal{A}$, and it is a finite measure on the algebra $\mathcal{A}$. By Carathéodory Extension Theorem 11.19, $\hat{\mu}$ admits unique extension to a finite measure $\bar{\mu}$ on the σ-algebra $\mathcal{B}$. Then, $\bar{\mu} : \mathcal{B} \to [0, \infty) \subset \mathbb{R}$ and $\bar{\mu}|_{\mathcal{O}} = \mu$.

Note that $(S_{n-1}, |\cdot|)$ is second countable by Propositions 4.4 and 4.38. By Proposition 5.40, it is compact. Thus, $(S_{n-1}, |\cdot|)$ is a compact (therefore locally compact) separable metric space and $\bar{\mu}$ is a finite measure on $\mathcal{B} = \mathcal{B}_{\mathrm{B}}((S_{n-1}, \mathcal{O}))$. By Theorem 11.198, $\mathbb{S}_{n-1} := ((S_{n-1}, |\cdot|), \mathcal{B}, \bar{\mu})$ is a finite compact separable metric measure space. This is basically the surface area measure space on the n-dimensional unit sphere. ◇

Example 12.105 Let $\mathbb{I} := ((\mathrm{r}_{-\pi,\pi}, |\cdot|), \mathcal{B}, \mu)$ be the finite metric measure subspace of $\mathbb{R}$, and $\mathbb{S}_1 := ((S_1, |\cdot|), \bar{\mathcal{B}}, \bar{\mu})$ be the finite metric measure space as defined in Example 12.104. Define a mapping $g : \mathrm{r}_{-\pi,\pi} \to S_1$ by $g(x) = (\cos(x), \sin(x)) \in S_1, \forall x \in \mathrm{r}_{-\pi,\pi}$. It is easy to see that g is bijective and g is continuous. But, g_{inv} is not continuous and therefore g is not a homeomorphism. $\forall O \in \mathcal{O}_{\mathbb{S}_1}$, we have $g_{\mathrm{inv}}(O) \in \mathcal{O}_{\mathbb{I}}$ since g is continuous. On the other hand, $\forall O \in \mathcal{O}_{\mathbb{I}}$, we will distinguish two exhaustive and mutually exclusive cases: Case 1: $\pi \in O$; and Case 2: $\pi \notin O$. Case 1: $\pi \in O$. Then, $\bar{O} := O \setminus \{\pi\} \in \mathcal{O}_{\mathbb{I}}$, $\bar{O} \in \mathcal{O}_{\mathbb{R}}$, and $g(\bar{O}) \in \mathcal{O}_{\mathbb{S}_1}$. Then, $g(O) = g(\bar{O}) \cup \{g(\pi)\} \in \bar{\mathcal{B}}$. Case 2: $\pi \notin O$. Then, $O \in \mathcal{O}_{\mathbb{R}}$ and $g(O) \in \mathcal{O}_{\mathbb{S}_1} \subseteq \bar{\mathcal{B}}$. Hence, we have shown that $g(O) \in \bar{\mathcal{B}}, \forall O \in \mathcal{O}_{\mathbb{I}}$. By Proposition 11.34, we have $g(B) \in \bar{\mathcal{B}}, \forall B \in \mathcal{B}$ and $g_{\mathrm{inv}}(\bar{B}) \in \mathcal{B}, \forall \bar{B} \in \bar{\mathcal{B}}$. Hence, g_{inv} is an isomeasure between $\mathbb{S}_1$ and $(\mathrm{r}_{-\pi,\pi}, \mathcal{B}, g_{\mathrm{inv}*}\bar{\mu})$.

It is easy to see that $\mu(I_1) = \bar{\mu}(g(I_1)) = g_{\mathrm{inv}*}\bar{\mu}(I_1)$, $\forall$ interval $I_1 \subseteq \mathbb{I}$. Clearly, the collection of all intervals that are subset of $\mathbb{I}$ forms a π-system on $\mathrm{r}_{-\pi,\pi}$. By Lemma 12.103, $\mu = g_{\mathrm{inv}*}\bar{\mu}$. Hence, g is an isomeasure between $\mathbb{I}$ and $\mathbb{S}_1$. $\forall p \in [1, \infty) \subset \mathbb{R}$, let $\mathcal{X}$ be a separable Banach space over $\mathbb{K}$. By Example 11.179, we have $\mathrm{L}_p(\mathbb{I}, \mathcal{X}) \equiv \mathrm{L}_p(\mathbb{S}_1, \mathcal{X})$ in the sense that $\forall z \in \bar{\mathrm{L}}_p(\mathbb{I}, \mathcal{X})$, we have $z \circ g_{\mathrm{inv}} \in \bar{\mathrm{L}}_p(\mathbb{S}_1, \mathcal{X})$ and $\forall \bar{z} \in \bar{\mathrm{L}}_p(\mathbb{S}_1, \mathcal{X})$, we have $\bar{z} \circ g \in \bar{\mathrm{L}}_p(\mathbb{I}, \mathcal{X})$. ◇

12.8 Sobolev Spaces

In this section, we will be concerned with absolutely continuous $\mathcal{Y}$-valued function on an open rectangle Ω of $\mathbb{R}^m$, with $m \in \mathbb{N}$, where $\mathcal{Y}$ is a separable reflexive Banach space over $\mathbb{K}$ with $\mathcal{Y}^*$ being separable. We restrict ourself to consider such functions that further have a finite $\bar{\mathrm{L}}_p(\mathbb{X}, \mathcal{Y})$ pseudo-norm, where $p \in [1, \infty) \subset \mathbb{R}$. We will be using the notations of the Fundamental Theorem of Calculus II, Theorem 12.88, without repeating their definitions. For any such function f, there exists a set of functions $f_{J,x_0} : \pi_J(\Omega) \to \mathcal{Y}$, which are $\mathcal{B}_J$-measurable and absolutely integrable over any bounded rectangles of $\mathbb{X}_J$, $\forall J \subseteq \bar{J} := \{1, \ldots, m\}$ with $J \neq \emptyset$, and any fixed $x_0 \in \Omega$. These are the stream functions of f. They form a unique set of functions, except on a set of measure zero in $\mathbb{X}_J$ for each of the stream functions. They together satisfy (12.3). Theorem 12.88 clearly states that these functions determine the partial derivative functions $\mathrm{D}_i f$, $\forall i \in \bar{J}$. We will further restrict ourself to absolutely continuous functions f such that f_{J,x_0} is in $\bar{\mathrm{L}}_p(\mathbb{X}_J, \mathcal{Y})$,

$\forall J \subseteq \bar{J}$ with $J \neq \emptyset$. It should be clear that $(f+g)_{J,x_0} = f_{J,x_0} + g_{J,x_0}$ a.e. $x \in \mathbb{X}_J$ and $(\alpha f)_{J,x_0} = \alpha f_{J,x_0}$ a.e. in $\mathbb{X}_J$, $\forall J \subseteq \bar{J}$ with $J \neq \emptyset$, $\forall f, g$ under consideration and $\forall \alpha \in \mathbb{K}$. This shows that such functions form a vector space over $\mathbb{K}$. We will define a norm on this space as $\|f\|_{x_0} := (\int_\Omega \mathcal{P}_p \circ f(x)\, d\mu(x) + \sum_{\substack{J\subseteq \bar{J}\\ J\neq\emptyset}} \int_{\pi_J(\Omega)} \mathcal{P}_p \circ f_{J,x_0}(s)\, d\mu_J(s))^{1/p}$. The value of the norm does depends on the point x_0. But the functions included in Z are independent of the choice of x_0 (see Example 12.107). We will show that this defines a Banach space over $\mathbb{K}$, which will be called *the* 1*st order Sobolev space of* $\mathbb{X}$ *to* $\mathcal{Y}$.

Example 12.106 Let $m \in \mathbb{N}$, $\mathbb{R}^m$ be endowed with the usual positive cone, $p \in [1, \infty) \subset \mathbb{R}$, $\Omega \in \mathcal{B}_{\mathrm{B}}(\mathbb{R}^m)$ be a nonempty open rectangle, $\mathbb{X} := ((\Omega, |\cdot|), \mathcal{B}, \mu)$ be the σ-finite metric measure subspace of $\mathbb{R}^m$, and $\mathcal{Y}$ be a separable reflexive Banach space over $\mathbb{K}$ with $\mathcal{Y}^*$ being separable. Fix any $x_0 \in \Omega$. Let $Z := \{f \in \bar{\mathrm{L}}_p(\mathbb{X}, \mathcal{Y}) \mid f$ is absolutely continuous, and the set of stream functions of f with respect to x_0 are in $\bar{\mathrm{L}}_p(\mathbb{X}_J, \mathcal{Y})$, $\forall J \subseteq \bar{J}$ with $J \neq \emptyset$, respectively.$\}$, where $\bar{J} := \{1, \ldots, m\}$, $\mathbb{X}_J := (\pi_J(\Omega), \mathcal{B}_J, \mu_J)$ is as defined in Theorem 12.88. On this set Z, we will define the norm $\|\cdot\|_{Z,x_0} : Z \to [0, \infty) \subset \mathbb{R}$ by

$$\|f\|_{Z,x_0} := \left(\int_\Omega \mathcal{P}_p \circ f(x)\, d\mu(x) + \sum_{\substack{J\subseteq \bar{J}\\ J\neq\emptyset}} \int_{\pi_J(\Omega)} \mathcal{P}_p \circ f_{J,x_0}(s)\, d\mu_J(s) \right)^{1/p} \tag{12.9}$$

We will show that $(Z, \|\cdot\|_{Z,x_0}, \mathbb{K})$ forms a Banach space over $\mathbb{K}$, which will be called the *1st order Sobolev space of* $\mathbb{X}$ *to* $\mathcal{Y}$ *under p-norm with respect to* $x_0 \in \mathbb{X}$ and denoted by $W_{p,1,x_0}(\mathbb{X}, \mathcal{Y})$.

$\forall f_l \in Z$, $l = 1, 2, 3$, we let f_{l,J,x_0} be its stream functions with respect to x_0. Then, $f_l \in \bar{\mathrm{L}}_p(\mathbb{X}, \mathcal{Y})$ and $f_{l,J,x_0} \in \bar{\mathrm{L}}_p(\mathbb{X}_J, \mathcal{Y})$, $\forall J \subseteq \bar{J}$ with $J \neq \emptyset$. Then, $f_1 + f_2 \in \bar{\mathrm{L}}_p(\mathbb{X}, \mathcal{Y})$ and $(f_1 + f_2)_{J,x_0} = f_{1,J,x_0} + f_{2,J,x_0}$ a.e. in $\mathbb{X}_J$, $\forall J \subseteq \bar{J}$ with $J \neq \emptyset$. This implies that $(f_1 + f_2)_{J,x_0} \in \bar{\mathrm{L}}_p(\mathbb{X}_J, \mathcal{Y})$. By Propositions 12.69 and 12.61, $f_1 + f_2$ is absolutely continuous. Hence, $f_1 + f_2 \in Z$. $\forall \alpha \in \mathbb{K}$, by Proposition 12.69, αf_1 is absolutely continuous. It is easy to see that $(\alpha f_1)_{J,x_0} = \alpha f_{1,J,x_0}$ a.e. in $\mathbb{X}_J$. Thus, we have $(\alpha f_1)_{J,x_0} \in \bar{\mathrm{L}}_p(\mathbb{X}_J, \mathcal{Y})$, $\forall J \subseteq \bar{J}$ with $J \neq \emptyset$. Hence, $\alpha f_1 \in Z$. This paragraph then proves that $(Z, \mathbb{K})$ is a subspace of $\bar{\mathrm{L}}_p(\mathbb{X}, \mathcal{Y})$ and hence a vector space over $\mathbb{K}$.

Clearly, $\|\cdot\|_{Z,x_0} : Z \to [0, \infty) \subset \mathbb{R}$. $\|\vartheta_Z\|_{Z,x_0} = 0$. If $\|f_1\|_{Z,x_0} = 0$, we have $f_1(x) = \vartheta_{\mathcal{Y}}$ a.e. $x \in \mathbb{X}$ and $f_{1,J,x_0}(x) = \vartheta_{\mathcal{Y}}$ a.e. $x \in \mathbb{X}_J$. Since f_1 is absolutely continuous, then $f_1(x) = \vartheta_{\mathcal{Y}}$, $\forall x \in \mathbb{X}$. Thus, $f_1 = \vartheta_Z$.

Note that, $\forall \alpha \in \mathbb{K}$,

$$\|\alpha f_1\|_{Z,x_0} = \Big(\int_\Omega \mathcal{P}_p \circ (\alpha f_1)(x)\, \mathrm{d}\mu(x) + \sum_{\substack{J\subseteq \bar{J}\\ J\neq\emptyset}} \int_{\pi_J(\Omega)} \mathcal{P}_p \circ (\alpha f_{1,J,x_0})(s)\, \mathrm{d}\mu_J(s) \Big)^{1/p} = |\alpha| \|f_1\|_{Z,x_0}$$

To prove the triangle inequality, we will distinguish two exhaustive and mutually exclusive cases: Case 1: $\|f_1\|_{Z,x_0}\|f_2\|_{Z,x_0} = 0$; Case 2: $\|f_1\|_{Z,x_0} \cdot \|f_2\|_{Z,x_0} > 0$. Case 1: $\|f_1\|_{Z,x_0}\|f_2\|_{Z,x_0} = 0$. Without loss of generality, assume $\|f_2\|_{Z,x_0} = 0$. Then, $f_2 = \vartheta_Z$ and $f_2(x) = \vartheta_{\mathcal{Y}}$, $\forall x \in \Omega$. Then, we have $\|f_1 + f_2\|_{Z,x_0} = \|f_1\|_{Z,x_0} = \|f_1\|_{Z,x_0} + \|f_2\|_{Z,x_0}$. This case is proved.

Case 2: $\|f_1\|_{Z,x_0}\|f_2\|_{Z,x_0} > 0$. Then, $\|f_1\|_{Z,x_0} > 0$ and $\|f_2\|_{Z,x_0} > 0$. Let $\lambda := \frac{\|f_1\|_{Z,x_0}}{\|f_1\|_{Z,x_0}+\|f_2\|_{Z,x_0}} \in (0,1) \subset \mathbb{R}$. Then, $\forall x \in \mathbb{X}$, we have $\frac{\|f_1(x)+f_2(x)\|_{\mathcal{Y}}^p}{(\|f_1\|_{Z,x_0}+\|f_2\|_{Z,x_0})^p} \leq \left(\frac{\|f_1(x)\|_{\mathcal{Y}}+\|f_2(x)\|_{\mathcal{Y}}}{\|f_1\|_{Z,x_0}+\|f_2\|_{Z,x_0}}\right)^p = \left(\lambda\frac{\|f_1(x)\|_{\mathcal{Y}}}{\|f_1\|_{Z,x_0}} + (1-\lambda)\frac{\|f_2(x)\|_{\mathcal{Y}}}{\|f_2\|_{Z,x_0}}\right)^p \leq \lambda\left(\frac{\|f_1(x)\|_{\mathcal{Y}}}{\|f_1\|_{Z,x_0}}\right)^p + (1-\lambda)\left(\frac{\|f_2(x)\|_{\mathcal{Y}}}{\|f_2\|_{Z,x_0}}\right)^p = \lambda\frac{\|f_1(x)\|_{\mathcal{Y}}^p}{\|f_1\|_{Z,x_0}^p} + (1-\lambda)\frac{\|f_2(x)\|_{\mathcal{Y}}^p}{\|f_2\|_{Z,x_0}^p}$, where the first inequality follows from Definition 7.1 and the second inequality follows from the convexity of the function l^p, $\forall l \in [0,\infty) \subset \mathbb{R}$. By Proposition 11.83, we have $\frac{\int_\Omega \mathcal{P}_p\circ(f_1+f_2)(x)\,\mathrm{d}\mu(x)}{(\|f_1\|_{Z,x_0}+\|f_2\|_{Z,x_0})^p} \leq \lambda\frac{\int_\Omega \mathcal{P}_p\circ f_1(x)\,\mathrm{d}\mu(x)}{\|f_1\|_{Z,x_0}^p} + (1-\lambda)\frac{\int_\Omega \mathcal{P}_p\circ f_2(x)\,\mathrm{d}\mu(x)}{\|f_2\|_{Z,x_0}^p}$.

Fix any $J \subseteq \bar{J}$ with $J \neq \emptyset$, we have, $\forall s \in \pi_J(\Omega)$,

$$\frac{\|f_{1,J,x_0}(s) + f_{2,J,x_0}(s)\|_{\mathcal{Y}}^p}{(\|f_1\|_{Z,x_0} + \|f_2\|_{Z,x_0})^p} \leq \left(\frac{\|f_{1,J,x_0}(s)\|_{\mathcal{Y}} + \|f_{2,J,x_0}(s)\|_{\mathcal{Y}}}{\|f_1\|_{Z,x_0} + \|f_2\|_{Z,x_0}}\right)^p$$
$$= \left(\lambda\frac{\|f_{1,J,x_0}(s)\|_{\mathcal{Y}}}{\|f_1\|_{Z,x_0}} + (1-\lambda)\frac{\|f_{2,J,x_0}(s)\|_{\mathcal{Y}}}{\|f_2\|_{Z,x_0}}\right)^p \leq \lambda\left(\frac{\|f_{1,J,x_0}(s)\|_{\mathcal{Y}}}{\|f_1\|_{Z,x_0}}\right)^p$$
$$+(1-\lambda)\left(\frac{\|f_{2,J,x_0}(s)\|_{\mathcal{Y}}}{\|f_2\|_{Z,x_0}}\right)^p = \lambda\frac{\|f_{1,J,x_0}(s)\|_{\mathcal{Y}}^p}{\|f_1\|_{Z,x_0}^p} + (1-\lambda)\frac{\|f_{2,J,x_0}(s)\|_{\mathcal{Y}}^p}{\|f_2\|_{Z,x_0}^p}$$

where the first inequality follows from Definition 7.1 and the second inequality follows from the convexity of the function l^p, $\forall l \in [0,\infty) \subset \mathbb{R}$. By Proposition 11.83, we have $\frac{\int_{\pi_J(\Omega)} \mathcal{P}_p\circ(f_{1,J,x_0}+f_{2,J,x_0})(s)\,\mathrm{d}\mu_J(s)}{(\|f_1\|_{Z,x_0}+\|f_2\|_{Z,x_0})^p} \leq \lambda\frac{\int_{\pi_J(\Omega)} \mathcal{P}_p\circ f_{1,J,x_0}(s)\,\mathrm{d}\mu_J(s)}{\|f_1\|_{Z,x_0}^p} + (1-\lambda)\frac{\int_{\pi_J(\Omega)} \mathcal{P}_p\circ f_{2,J,x_0}(s)\,\mathrm{d}\mu_J(s)}{\|f_2\|_{Z,x_0}^p}$.

Thus, we have

$$\begin{aligned}
&\frac{\|f_1+f_2\|_{Z,x_0}^p}{(\|f_1\|_{Z,x_0}+\|f_2\|_{Z,x_0})^p}\\
&=\frac{1}{(\|f_1\|_{Z,x_0}+\|f_2\|_{Z,x_0})^p}\bigg(\int_\Omega \mathcal{P}_p\circ(f_1+f_2)(x)\,\mathrm{d}\mu(x)\\
&\quad+\sum_{\substack{J\subseteq\bar{J}\\ J\neq\emptyset}}\int_{\pi_J(\Omega)}\mathcal{P}_p\circ(f_{1,J,x_0}+f_{2,J,x_0})(s)\,\mathrm{d}\mu_J(s)\bigg)\\
&\leq\lambda\frac{\int_\Omega \mathcal{P}_p\circ f_1(x)\,\mathrm{d}\mu(x)}{\|f_1\|_{Z,x_0}^p}+(1-\lambda)\frac{\int_\Omega \mathcal{P}_p\circ f_2(x)\,\mathrm{d}\mu(x)}{\|f_2\|_{Z,x_0}^p}\\
&\quad+\sum_{\substack{J\subseteq\bar{J}\\ J\neq\emptyset}}\bigg(\lambda\frac{\int_{\pi_J(\Omega)}\mathcal{P}_p\circ f_{1,J,x_0}(s)\,\mathrm{d}\mu_J(s)}{\|f_1\|_{Z,x_0}^p}\\
&\quad+(1-\lambda)\frac{\int_{\pi_J(\Omega)}\mathcal{P}_p\circ f_{2,J,x_0}(s)\,\mathrm{d}\mu_J(s)}{\|f_2\|_{Z,x_0}^p}\bigg)\\
&=\frac{\lambda}{\|f_1\|_{Z,x_0}^p}\bigg(\int_\Omega \mathcal{P}_p\circ f_1(x)\,\mathrm{d}\mu(x)+\sum_{\substack{J\subseteq\bar{J}\\ J\neq\emptyset}}\int_{\pi_J(\Omega)}\mathcal{P}_p\circ f_{1,J,x_0}(s)\,\mathrm{d}\mu_J(s)\bigg)\\
&\quad+\frac{1-\lambda}{\|f_2\|_{Z,x_0}^p}\bigg(\int_\Omega \mathcal{P}_p\circ f_2(x)\,\mathrm{d}\mu(x)+\sum_{\substack{J\subseteq\bar{J}\\ J\neq\emptyset}}\int_{\pi_J(\Omega)}\mathcal{P}_p\circ f_{2,J,x_0}(s)\,\mathrm{d}\mu_J(s)\bigg)\\
&=1
\end{aligned}$$

Hence, we have $\|f_1+f_2\|_{Z,x_0}\leq\|f_1\|_{Z,x_0}+\|f_2\|_{Z,x_0}$. This case is proved.

Based on both of the two cases, we have the triangle inequality for $\|\cdot\|_Z$. So far, we have shown that $(Z,\|\cdot\|_{Z,x_0},\mathbb{K})$ is a normed linear space.

Finally, we will show that the space is complete. Fix any Cauchy sequence $(f_l)_{l=1}^\infty\subseteq Z$. By the definition of the norm $\|\cdot\|_{Z,x_0}$, we have $(f_l)_{l=1}^\infty\subseteq\bar{\mathrm{L}}_p(\mathbb{X},\mathcal{Y})$ is a Cauchy sequence, and $(f_{l,J,x_0})_{l=1}^\infty\subseteq\bar{\mathrm{L}}_p(\mathbb{X}_J,\mathcal{Y})$ is a Cauchy sequence, $\forall J\subseteq\bar{J}$ with $J\neq\emptyset$. By the completeness of L_p as proved in Example 11.179, we have $\exists\bar{F}\in\bar{\mathrm{L}}_p(\mathbb{X},\mathcal{Y})$ and $\exists F_{J,x_0}\in\bar{\mathrm{L}}_p(\mathbb{X}_J,\mathcal{Y})$ such that $\lim_{l\in\mathbb{N}}f_l\doteq\bar{F}$ in $\bar{\mathrm{L}}_p(\mathbb{X},\mathcal{Y})$ and $\lim_{l\in\mathbb{N}}f_{l,J,x_0}\doteq F_{J,x_0}$ in $\bar{\mathrm{L}}_p(\mathbb{X}_J,\mathcal{Y})$, $\forall J\subseteq\bar{J}$ with $J\neq\emptyset$. We need to construct a function $F\in Z$ such that its stream functions with respect to x_0 are F_{J,x_0}'s. It should be noted that F_{J,x_0}'s are specified exactly except on a set of measure zero, which are sufficient to serve as stream functions. Note that $\forall l \in \mathbb{N}$, $\forall x \in \Omega$, we have

$$f_l(x) = \sum_{\substack{J \subseteq \bar{J} \\ J \neq \emptyset}} (-1)^{\operatorname{card}(\check{J} \cap J)} \int_{\pi_J(\mathrm{r}_{\hat{x},\check{x}})} f_{l,J,x_0} \, \mathrm{d}\mu_J + f_l(x_0)$$

where $\hat{x} := x \wedge x_0$, $\check{x} := x \vee x_0$, $\check{J} := \{ i \in \bar{J} \mid \pi_i(x_0) > \pi_i(x) \}$. (This is exactly (12.3).) As $l \to \infty$, we have

$$\begin{aligned} \lim_{l \in \mathbb{N}} (f_l(x) - f_l(x_0)) &= \lim_{l \in \mathbb{N}} \sum_{\substack{J \subseteq \bar{J} \\ J \neq \emptyset}} (-1)^{\operatorname{card}(\check{J} \cap J)} \int_{\pi_J(\mathrm{r}_{\hat{x},\check{x}})} f_{l,J,x_0} \, \mathrm{d}\mu_J \\ &= \sum_{\substack{J \subseteq \bar{J} \\ J \neq \emptyset}} (-1)^{\operatorname{card}(\check{J} \cap J)} \int_{\pi_J(\mathrm{r}_{\hat{x},\check{x}})} F_{J,x_0} \, \mathrm{d}\mu_J; \quad \forall x \in \Omega \end{aligned}$$

where the last equality follows from the fact that $\lim_{l \in \mathbb{N}} f_{l,J,x_0} \doteq F_{J,x_0}$ in $\bar{\mathrm{L}}_p(\mathbb{X}_J, \mathcal{Y})$ and Proposition 11.213. This formula determines the function F up to $F(x_0)$. Since $\lim_{l \in \mathbb{N}} f_l \doteq \bar{F}$ in $\bar{\mathrm{L}}_p(\mathbb{X}, \mathcal{Y})$, then by Propositions 11.211 and 11.57, there exists a subsequence $(f_{l_k})_{k \in \mathbb{N}}$ such that $\lim_{k \in \mathbb{N}} f_{l_k} = \bar{F}$ a.e. in $\mathbb{X}$. Then, $\exists x_1 \in \Omega$, (possibly different from x_0) such that $\lim_{k \in \mathbb{N}} f_{l_k}(x_1) = \bar{F}(x_1)$. Then, we may define $F(x_0) := \bar{F}(x_1) - \sum_{\substack{J \subseteq \bar{J} \\ J \neq \emptyset}} (-1)^{\operatorname{card}(\check{J}_1 \cap J)} \int_{\pi_J(\mathrm{r}_{\hat{x}_1,\check{x}_1})} F_{J,x_0} \, \mathrm{d}\mu_J$, where $\hat{x}_1 := x_1 \wedge x_0$, $\check{x}_1 := x_1 \vee x_0$, $\check{J}_1 := \{ i \in \bar{J} \mid \pi_i(x_0) > \pi_i(x_1) \}$. Therefore, we may define $F : \Omega \to \mathcal{Y}$ by, $\forall x \in \Omega$,

$$F(x) = F(x_0) + \sum_{\substack{J \subseteq \bar{J} \\ J \neq \emptyset}} (-1)^{\operatorname{card}(\check{J} \cap J)} \int_{\pi_J(\mathrm{r}_{\hat{x},\check{x}})} F_{J,x_0} \, \mathrm{d}\mu_J$$

By Fundamental Theorem of Calculus I, Theorem 12.86, each term in the above sum is a absolutely continuous function on $\pi_J(\Omega)$ and therefore absolutely continuous on Ω by Definition 12.58. By Propositions 12.69 and 12.61, F is absolutely continuous on Ω. It is easy to see that $F = \bar{F}$ a.e. in $\mathbb{X}$, and F_{J,x_0}'s are the stream functions of F with respect to x_0 by Fundamental Theorem of Calculus II, Theorem 12.88. Hence, $F \in Z$. Then, it is straightforward to check that $\lim_{l \in \mathbb{N}} \| f_l - F \|_{Z,x_0} = 0$ and $\lim_{l \in \mathbb{N}} f_l = F$ in $W_{p,1,x_0}(\mathbb{X}, \mathcal{Y})$.

Hence, any Cauchy sequence in Z converges to an element in Z. This proves that $W_{p,1,x_0}(\mathbb{X}, \mathcal{Y})$ is complete, and therefore, it is a Banach space over $\mathbb{K}$. ◇

We note that the above definition of Sobolev space agrees with that in Zeidler (1995) when $m = 1$. In this case, the norm does not depend on the choice of $x_0 \in \mathbb{X}$.

Example 12.107 We will now show that $\forall f \in W_{p,1,x_0}(\mathbb{X}, \mathcal{Y})$ if, and only if, $f \in W_{p,1,\bar{x}}(\mathbb{X}, \mathcal{Y})$, $\forall \bar{x} \in \mathbb{X}$, under the assumptions of Example 12.106. Furthermore, the norms $\|\cdot\|_{Z,x_0}$ and $\|\cdot\|_{Z,\bar{x}}$ are equivalent to each other.

Fix any $f \in W_{p,1,x_0}(\mathbb{X}, \mathcal{Y})$, we have that $f \in \bar{\mathrm{L}}_p(\mathbb{X}, \mathcal{Y})$ and it is absolutely continuous, and its stream functions $f_{J,x_0} \in \bar{\mathrm{L}}_p(\mathbb{X}_J, \mathcal{Y})$, $\forall J \subseteq \bar{J}$ with $J \neq \emptyset$. Fix any $\bar{x} \in \mathbb{X}$. The stream functions of f with respect to $\bar{x}$ are given by (12.4): $\forall J \subseteq \bar{J}$ with $J \neq \emptyset$,

$$f_{J,\bar{x}}(s) = \begin{cases} \sum_{J \subset \tilde{J} \subseteq \bar{J}} (-1)^{\mathrm{card}(\bar{\check{J}} \cap \tilde{J} \setminus J)} \int_{\pi_{\tilde{J} \setminus J}(\mathrm{r}_{\bar{\hat{x}},\bar{\check{x}}})} f_{\tilde{J},x_0}(\pi_{\tilde{J}}(\\ \quad M_J(s, \pi_{\tilde{J}\setminus J}(t)))) \, \mathrm{d}\mu_{\tilde{J}\setminus J}(\pi_{\tilde{J}\setminus J}(t)) + f_{J,x_0}(s) & s \in \bar{U}_J \\ \vartheta_{\mathcal{Y}} & s \in \pi_J(\Omega) \setminus \bar{U}_J \end{cases}$$

where $\bar{\check{J}} := \{ i \in \bar{J} \mid \pi_i(x_0) > \pi_i(\bar{x}) \}$, $\bar{\hat{x}} := \bar{x} \wedge x_0$, and $\bar{\check{x}} := \bar{x} \vee x_0$; and $\bar{U}_J \in \mathcal{B}_J$ with $\mu_J(\pi_J(\Omega) \setminus \bar{U}_J) = 0$. All we need to do is to show that $f_{J,\bar{x}} \in \bar{\mathrm{L}}_p(\mathbb{X}_J, \mathcal{Y})$, $\forall J \subseteq \bar{J}$ with $J \neq \emptyset$, under the assumption that $f_{J,x_0} \in \bar{\mathrm{L}}_p(\mathbb{X}_J, \mathcal{Y})$, $\forall J \subseteq \bar{J}$ with $J \neq \emptyset$.

Fix any $J \subseteq \bar{J}$ with $J \neq \emptyset$. Then, we have

$$\begin{aligned} \|f_{J,\bar{x}}\|_p &\leq \|f_{J,x_0}\|_p + \sum_{J \subset \tilde{J} \subseteq \bar{J}} \Big(\int_{\bar{U}_J} \Big\| \int_{\pi_{\tilde{J}\setminus J}(\mathrm{r}_{\bar{\hat{x}},\bar{\check{x}}})} f_{\tilde{J},x_0}(\pi_{\tilde{J}}(M_J(s, \\ &\quad \pi_{\tilde{J}\setminus J}(t)))) \, \mathrm{d}\mu_{\tilde{J}\setminus J}(\pi_{\tilde{J}\setminus J}(t)) \Big\|_{\mathcal{Y}}^p \mathrm{d}\mu_J(s) \Big)^{1/p} \\ &\leq \|f_{J,x_0}\|_p + \sum_{J \subset \tilde{J} \subseteq \bar{J}} \Big(\int_{\bar{U}_J} \Big(\int_{\pi_{\tilde{J}\setminus J}(\mathrm{r}_{\bar{\hat{x}},\bar{\check{x}}})} \| f_{\tilde{J},x_0}(\pi_{\tilde{J}}(M_J(s, \\ &\quad \pi_{\tilde{J}\setminus J}(t)))) \|_{\mathcal{Y}} \, \mathrm{d}\mu_{\tilde{J}\setminus J}(\pi_{\tilde{J}\setminus J}(t)) \Big)^p \mathrm{d}\mu_J(s) \Big)^{1/p} \\ &\leq \|f_{J,x_0}\|_p + \sum_{J \subset \tilde{J} \subseteq \bar{J}} \Big(\int_{\bar{U}_J} \Big(\Big(\int_{\pi_{\tilde{J}\setminus J}(\mathrm{r}_{\bar{\hat{x}},\bar{\check{x}}})} 1^q \, \mathrm{d}\mu_{\tilde{J}\setminus J}(\pi_{\tilde{J}\setminus J}(t)) \Big)^{1/q} \\ &\quad \cdot \Big(\int_{\pi_{\tilde{J}\setminus J}(\mathrm{r}_{\bar{\hat{x}},\bar{\check{x}}})} \Big\| f_{\tilde{J},x_0}(\pi_{\tilde{J}}(M_J(s, \pi_{\tilde{J}\setminus J}(t)))) \Big\|_{\mathcal{Y}}^p \mathrm{d}\mu_{\tilde{J}\setminus J}(\pi_{\tilde{J}\setminus J}(t)) \Big)^{1/p} \Big)^p \\ &\quad \mathrm{d}\mu_J(s) \Big)^{1/p} \\ &= \|f_{J,x_0}\|_p + \sum_{J \subset \tilde{J} \subseteq \bar{J}} \Big(\int_{\bar{U}_J} (\mu_{\tilde{J}\setminus J}(\pi_{\tilde{J}\setminus J}(\mathrm{r}_{\bar{\hat{x}},\bar{\check{x}}})))^{p/q} \int_{\pi_{\tilde{J}\setminus J}(\mathrm{r}_{\bar{\hat{x}},\bar{\check{x}}})} \end{aligned}$$

$$\left\| f_{\tilde{J},x_0}(\pi_{\tilde{J}}(M_J(s,\pi_{\tilde{J}\setminus J}(t)))) \right\|_{\mathcal{Y}}^p \mathrm{d}\mu_{\tilde{J}\setminus J}(\pi_{\tilde{J}\setminus J}(t))\,\mathrm{d}\mu_J(s)\Bigg)^{1/p}$$

$$= \|f_{J,x_0}\|_p + \sum_{J\subset \tilde{J}\subseteq \bar{J}} (\mu_{\tilde{J}\setminus J}(\pi_{\tilde{J}\setminus J}(\mathrm{r}_{\hat{\bar{x}},\bar{x}})))^{1/q}\Bigg(\int_{\bar{U}_J}\int_{\pi_{\tilde{J}\setminus J}(\Omega)}$$

$$\left\| f_{\tilde{J},x_0}(\pi_{\tilde{J}}(M_J(s,\pi_{\tilde{J}\setminus J}(t)))) \right\|_{\mathcal{Y}}^p \mathrm{d}\mu_{\tilde{J}\setminus J}(\pi_{\tilde{J}\setminus J}(t))\,\mathrm{d}\mu_J(s)\Bigg)^{1/p}$$

$$= \|f_{J,x_0}\|_p + \sum_{J\subset \tilde{J}\subseteq \bar{J}} (\mu_{\tilde{J}\setminus J}(\pi_{\tilde{J}\setminus J}(\mathrm{r}_{\hat{\bar{x}},\bar{x}})))^{1/q}$$

$$\cdot\Bigg(\int_{\pi_{\tilde{J}}(\Omega)} \left\| f_{\tilde{J},x_0}(s)\right\|_{\mathcal{Y}}^p \mathrm{d}\mu_{\tilde{J}}(s)\Bigg)^{1/p}$$

$$= \|f_{J,x_0}\|_p + \sum_{J\subset \tilde{J}\subseteq \bar{J}} (\mu_{\tilde{J}\setminus J}(\pi_{\tilde{J}\setminus J}(\mathrm{r}_{\hat{\bar{x}},\bar{x}})))^{1/q} \left\| f_{\tilde{J},x_0}\right\|_p < \infty$$

where the first inequality follows from Minkowski's Inequality, Theorem 11.174, the second inequality follows from Proposition 11.92, the third inequality follows from Hölder's Inequality, Theorem 11.178, the first equality follows from Proposition 11.75, the second equality follows from Proposition 11.83, the third equality follows from Tonelli's Theorem 12.29, and the last inequality follows from our assumption. This shows that $f_{J,\bar{x}} \in \bar{\mathrm{L}}_p(\mathbb{X}_J, \mathcal{Y})$. By the arbitrariness of J, we have $\|f\|_{Z,\bar{x}} < \infty$ and $f \in W_{p,1,\bar{x}}(\mathbb{X}, \mathcal{Y})$. By symmetry, $\forall f \in W_{p,1,\bar{x}}(\mathbb{X}, \mathcal{Y})$, we have $f \in W_{p,1,x_0}(\mathbb{X}, \mathcal{Y})$.

Next, we prove that the two norms are equivalent. Note that

$$\|f\|_{Z,\bar{x}} = (\|f\|_p^p + \sum_{\substack{J\subseteq \bar{J}\\ J\neq\emptyset}} \|f_{J,\bar{x}}\|_p^p)^{\frac{1}{p}}$$

$$\leq \Bigg(\|f\|_p^p + \sum_{\substack{J\subseteq \bar{J}\\ J\neq\emptyset}}\Bigg(\|f_{J,x_0}\|_p + \sum_{J\subset \tilde{J}\subseteq \bar{J}} (\mu_{\tilde{J}\setminus J}(\pi_{\tilde{J}\setminus J}(\mathrm{r}_{\hat{\bar{x}},\bar{x}})))^{\frac{1}{q}} \left\| f_{\tilde{J},x_0}\right\|_p\Bigg)^p\Bigg)^{\frac{1}{p}}$$

$$\leq \Bigg(\|f\|_p^p + \sum_{\substack{J\subseteq \bar{J}\\ J\neq\emptyset}}\Bigg(\|f_{J,x_0}\|_p + \sum_{J\subset \tilde{J}\subseteq \bar{J}} l^{\frac{\mathrm{card}(\tilde{J}\setminus J)}{q^2}} l^{\frac{\mathrm{card}(\tilde{J}\setminus J)}{qp}} \left\| f_{\tilde{J},x_0}\right\|_p\Bigg)^p\Bigg)^{\frac{1}{p}}$$

$$\leq \Bigg(\|f\|_p^p + \sum_{\substack{J\subseteq \bar{J}\\ J\neq\emptyset}}\Bigg((1^q + \sum_{J\subset \tilde{J}\subseteq \bar{J}} l^{\frac{\mathrm{card}(\tilde{J}\setminus J)}{q}})^{\frac{1}{q}}$$

$$\cdot \Big(\| f_{J,x_0} \|_p^p + \sum_{J \subset \tilde{J} \subseteq \bar{J}} l^{\frac{\operatorname{card}(\tilde{J} \setminus J)}{q}} \Big\| f_{\tilde{J},x_0} \Big\|_p^p)^{\frac{1}{p}} \Big)^p \Big)^{\frac{1}{p}}$$

$$= \Big(\| f \|_p^p + \sum_{\substack{J \subseteq \bar{J} \\ J \neq \emptyset}} \Big(\sum_{J \subseteq \tilde{J} \subseteq \bar{J}} l^{\frac{\operatorname{card}(\tilde{J} \setminus J)}{q}} \Big)^{\frac{p}{q}} \Big(\sum_{J \subseteq \tilde{J} \subseteq \bar{J}} l^{\frac{\operatorname{card}(\tilde{J} \setminus J)}{q}} \Big\| f_{\tilde{J},x_0} \Big\|_p^p \Big) \Big)^{\frac{1}{p}}$$

$$= \Big(\| f \|_p^p + \sum_{\substack{J \subseteq \bar{J} \\ J \neq \emptyset}} \Big(\sum_{\tilde{J} \subseteq \bar{J} \setminus J} l^{\frac{\operatorname{card}(\tilde{J})}{q}} \Big)^{\frac{p}{q}} \Big(\sum_{J \subseteq \tilde{J} \subseteq \bar{J}} l^{\frac{\operatorname{card}(\tilde{J} \setminus J)}{q}} \Big\| f_{\tilde{J},x_0} \Big\|_p^p \Big) \Big)^{\frac{1}{p}}$$

$$= \Big(\| f \|_p^p + \sum_{\substack{J \subseteq \bar{J} \\ J \neq \emptyset}} \Big(1 + l^{\frac{1}{q}} \Big)^{\frac{p \operatorname{card}(\bar{J} \setminus J)}{q}} \Big(\sum_{J \subseteq \tilde{J} \subseteq \bar{J}} l^{\frac{\operatorname{card}(\tilde{J} \setminus J)}{q}} \Big\| f_{\tilde{J},x_0} \Big\|_p^p \Big) \Big)^{\frac{1}{p}}$$

$$= \Big(\| f \|_p^p + \sum_{\substack{J \subseteq \bar{J} \\ J \neq \emptyset}} \sum_{J \subseteq \tilde{J} \subseteq \bar{J}} \Big(1 + l^{\frac{1}{q}} \Big)^{\frac{p \operatorname{card}(\bar{J} \setminus J)}{q}} l^{\frac{\operatorname{card}(\tilde{J} \setminus J)}{q}} \Big\| f_{\tilde{J},x_0} \Big\|_p^p \Big)^{\frac{1}{p}}$$

$$= \Big(\| f \|_p^p + \sum_{\substack{J \subseteq \bar{J} \\ J \neq \emptyset}} \sum_{\substack{\tilde{J} \subseteq J \\ \tilde{J} \neq \emptyset}} \Big(1 + l^{\frac{1}{q}} \Big)^{\frac{p \operatorname{card}(\bar{J} \setminus \tilde{J})}{q}} l^{\frac{\operatorname{card}(J \setminus \tilde{J})}{q}} \| f_{J,x_0} \|_p^p \Big)^{\frac{1}{p}}$$

$$\leq \Big(\| f \|_p^p + \sum_{\substack{J \subseteq \bar{J} \\ J \neq \emptyset}} \Big(1 + l^{\frac{1}{q}} \Big)^{\frac{pm}{q}} \Bigg(\Bigg(\frac{1 + (1 + l^{\frac{1}{q}})^{\frac{p}{q}} l^{\frac{1}{q}}}{(1 + l^{\frac{1}{q}})^{\frac{p}{q}}} \Bigg)^{\operatorname{card}(J)} - l^{\frac{\operatorname{card}(J)}{q}} \Bigg)$$

$$\cdot \| f_{J,x_0} \|_p^p \Big)^{\frac{1}{p}}$$

$$\leq \Big(1 + l^{\frac{1}{q}} \Big)^{\frac{m-1}{q}} \Big(1 + (1 + l^{\frac{1}{q}})^{\frac{p}{q}} l^{\frac{1}{q}} \Big)^{\frac{m}{p}} \| f \|_{Z,x_0} =: K_{m,p}(l) \| f \|_{Z,x_0}$$

where the first equality follows from (12.9), the first inequality follows from the previous paragraph, the second inequality follows from the definition $l := \max(\bar{\bar{x}} - \bar{\hat{x}}) \in \mathbb{R}_+$, the third inequality follows from Hölder's Inequality, Theorem 7.8, the second through sixth equalities follow from simple algebra, the fourth inequality follows from simple algebra, and the last inequality follows from (12.9) and simple algebra. By symmetry, we have $\| f \|_{Z,x_0} \leq K_{m,p}(l) \| f \|_{Z,\bar{x}}$, where l is the same quantity as before. Hence, the norms are equivalent on $W_{p,1,x_0}(\mathbb{X}, \mathcal{Y})$. $\diamond$

Proposition 12.108 *Let $m \in \mathbb{N}$, $\Omega \in \mathcal{B}_{\mathrm{B}}(\mathbb{R}^m)$ be a nonempty open rectangle, $\mathbb{X} := ((\Omega, |\cdot|), \mathcal{B}, \mu)$ be the σ-finite metric measure subspace of $\mathbb{R}^m$, $\bar{J} := \{1, \ldots, m\}$, and $\mathcal{Y}$ be a separable reflexive Banach space over $\mathbb{K}$ with $\mathcal{Y}^*$ being separable, $x_0 \in$*

Ω, $\mathcal{Z} := W_{p,1,x_0}(\mathbb{X}, \mathcal{Y})$, $f \in \mathcal{Z}$, $\mathrm{D}_i f : U_i \to \mathcal{Y}$ be the partial derivative of f along the ith coordinate direction, $i \in \bar{J}$, and $f_{J,x_0} : \pi_J(\Omega) \to \mathcal{Y}$ be the stream functions of f with respect to x_0, $\forall J \subseteq \bar{J}$ with $J \neq \emptyset$. Then, we have the following results:

(i) $\forall i \in \bar{J}$, $U_i \in \mathcal{B}$ *with* $\mu(\Omega \setminus U_i) = 0$.
(ii) $\forall t \in \mathbb{X}_{\bar{J}\setminus\{i\}}$, $\forall s_1, s_2 \in \mathbb{X}_{\{i\}}$, *we have*

$$\mathrm{D}_i f(M_{\{i\}}(s,t)) = f_{\{i\},M_{\{i\}}(\pi_{\{i\}}(x_0),t)}(s) \text{ a.e. } s \in \mathbb{X}_{\{i\}}$$

$f(M_{\{i\}}(s_1,t)) - f(M_{\{i\}}(s_2,t)) = \int_{s_2}^{s_1} f_{\{i\},M_{\{i\}}(\pi_{\{i\}}(x_0),t)}(s)\,\mathrm{d}s$, *and* $f_{\{i\},M_{\{i\}}(\pi_{\{i\}}(x_0),t)} \in \bar{\mathrm{L}}_p(\mathbb{X}_{\{i\}}, \mathcal{Y})$.

Proof The results are direct consequences of (iii) of Fundamental Theorem of Calculus II, Theorem 12.88, and Example 12.107. □

12.9 Integral Depending on a Parameter

Proposition 12.109 *Let $\mathcal{X} := (X, \rho)$ be a metric space, $D \subseteq \mathcal{X}$ be open, and $K \subseteq D$ be compact. Then, there exists $\delta \in \mathbb{R}_+$ such that $\forall x \in K$, $\mathcal{B}_{\mathcal{X}}(x, \delta) \subseteq D$.*

Proof Since D is open, then $\forall x \in K$, $\exists \delta_x \in \mathbb{R}_+$, such that $\mathcal{B}_{\mathcal{X}}(x, \delta_x) \subseteq D$. This implies $K \subseteq \bigcup_{x\in K} \mathcal{B}_{\mathcal{X}}\big(x, \frac{\delta_x}{2}\big)$. By the compactness of K, there exists $N \in \mathbb{Z}_+$ and $(x_i)_{i=1}^N \subseteq K$ such that $K \subseteq \bigcup_{i=1}^N \mathcal{B}_{\mathcal{X}}\Big(x_i, \frac{\delta_{x_i}}{2}\Big)$. Let $\delta := \min\Big\{1, \bigwedge_{i\in\{1,\dots,N\}} \frac{\delta_{x_i}}{2}\Big\} \in \mathbb{R}_+$. Then, $\forall x \in K$, $\forall y \in \mathcal{B}_{\mathcal{X}}(x, \delta)$, $\exists i_0 \in \{1, \dots, N\}$ such that $x \in \mathcal{B}_{\mathcal{X}}\big(x_{i_0}, \frac{\delta_{x_{i_0}}}{2}\big)$. Thus, $\rho(y, x_{i_0}) \le \rho(y, x) + \rho(x_{i_0}, x) < \delta + \frac{\delta_{x_{i_0}}}{2} \le \delta_{x_{i_0}}$. This leads to that $y \in \mathcal{B}_{\mathcal{X}}(x_{i_0}, \delta_{x_{i_0}}) \subseteq D$. By the arbitrariness of y, we have $\mathcal{B}_{\mathcal{X}}(x, \delta) \subseteq D$. This completes the proof of the proposition. □

Proposition 12.110 *Let $\mathcal{X} := (X, \rho)$ be a metric space, $D \subseteq \mathcal{X}$ be open, $K \subseteq D$ be compact, and $f : D \to \mathcal{Y}$, and $\mathcal{Y}$ be a normed linear space. Assume that f is continuous at x and $\|f(x)\|_{\mathcal{Y}} \le M$, $\forall x \in K$, for some $M \in \overline{\mathbb{R}_+}$. Then, $\forall \epsilon \in \mathbb{R}_+$, $\exists \delta \in \mathbb{R}_+$ such that $\forall x \in \mathcal{X}$ with $\operatorname{dist}(x, K) < \delta$, we have $\|f(x)\|_{\mathcal{Y}} < M + \epsilon$.*

Proof By Proposition 12.109, there exists $\bar{\delta} \in \mathbb{R}_+$ such that $\forall x \in K$, $\mathcal{B}_{\mathcal{X}}(x, \bar{\delta}) \subseteq D$. Fix any $\epsilon \in \mathbb{R}_+$. By the assumptions, $\forall x \in K$, $\exists \delta_x \in \mathrm{r}_{0,\bar{\delta}} \subset \mathbb{R}_+$, such that $\forall y \in \mathcal{B}_{\mathcal{X}}(x, \delta_x)$, we have $\|f(y) - f(x)\|_{\mathcal{Y}} < \epsilon$. This implies that $\|f(y)\|_{\mathcal{Y}} < \epsilon + \|f(x)\|_{\mathcal{Y}} \le M + \epsilon$. Now, $K \subseteq \bigcup_{x\in K} \mathcal{B}_{\mathcal{X}}\big(x, \frac{\delta_x}{2}\big)$. By the compactness of K, there exists $N \in \mathbb{Z}_+$ and $(x_i)_{i=1}^N \subseteq K$ such that $K \subseteq \bigcup_{i=1}^N \mathcal{B}_{\mathcal{X}}\big(x_i, \frac{\delta_{x_i}}{2}\big)$. Let $\delta := \min\Big\{1, \bigwedge_{i\in\{1,\dots,N\}} \frac{\delta_{x_i}}{2}\Big\} \in \mathbb{R}_+$. Then, $\forall x \in \mathcal{X}$ with $\operatorname{dist}(x, K) < \delta$, $\exists y \in K$ such that $x \in \mathcal{B}_{\mathcal{X}}(y, \delta)$. This further implies that $\exists i_0 \in \{1, \dots, N\}$ such that $y \in \mathcal{B}_{\mathcal{X}}\big(x_{i_0}, \frac{\delta_{x_{i_0}}}{2}\big)$. Thus, $\rho(x, x_{i_0}) \le \rho(y, x) + \rho(x_{i_0}, y) < \delta + \frac{\delta_{x_{i_0}}}{2} \le \delta_{x_{i_0}}$. This leads to that $x \in \mathcal{B}_{\mathcal{X}}(x_{i_0}, \delta_{x_{i_0}}) \subseteq D$ and $\|f(x)\|_{\mathcal{Y}} < M + \epsilon$. This completes the proof of the proposition. □

Theorem 12.111 *Let $f : \mathcal{X} \times \mathcal{Y} \to \mathrm{B}(\mathcal{Z}, \mathcal{W})$, where $\mathcal{X} := (X, \rho)$ is a metric space, $\mathcal{Y}$ is a compact metric space, $\mathcal{Z}$ is a normed linear space over $\mathbb{K}$, $\mathcal{W}$ is a Banach space over $\mathbb{K}$, $\mathcal{J} := (J, \mathcal{B}, \mu)$ be a finite $\mathcal{Z}$-valued measure space, and $w : J \to \mathcal{Y}$ be $\mathcal{B}$-measurable. Assume that f is continuous. Define $F : \mathcal{X} \to \mathcal{W}$ by $F(x) := \int_J f(x, w(t))\, \mathrm{d}\mu(t) \in \mathcal{W}$, $\forall x \in \mathcal{X}$. Then, F is continuous, $f(x, w(\cdot)) : \mathcal{J} \to \mathcal{U}_x \subseteq \mathrm{B}(\mathcal{Z}, \mathcal{W})$, where $\mathcal{U}_x$ is a separable Banach subspace of $\mathrm{B}(\mathcal{Z}, \mathcal{W})$, and $F(x) \in \mathcal{W}_x := \overline{\mathrm{span}\,(\{\bar{w} \in \mathcal{W} \mid \exists z \in \mathcal{Z}, \exists A \in \mathcal{U}_x \ni \cdot\ \bar{w} = Az\})}$.*

Proof By Proposition 11.190, $\mathcal{Y}$ is separable. Since f is continuous, by Propositions 7.126 and 5.7, $\forall x \in \mathcal{X}$, the function $h(x) := f(x, \cdot) : \mathcal{Y} \to \mathcal{V}_x \subseteq \mathrm{B}(\mathcal{Z}, \mathcal{W})$ satisfies $\mathcal{V}_x$ is a (separable) compact subset of $\mathrm{B}(\mathcal{Z}, \mathcal{W})$. By Proposition 11.38, $f(x, w(\cdot)) : \mathcal{J} \to \mathcal{V}_x$ is $\mathcal{B}$-measurable. By Bounded Convergence Theorem 11.77, $f(x, w(\cdot)) : \mathcal{J} \to \mathcal{V}_x$ is absolutely integrable over $\mathcal{J}$. By Proposition 7.35, $\mathcal{U}_x := \overline{\mathrm{span}\,(\mathcal{V}_x)}$ is a separable Banach subspace of $\mathrm{B}(\mathcal{Z}, \mathcal{W})$. By Proposition 11.130, $F(x) \in \mathcal{W}_x \subseteq \mathcal{W}$, $\forall x \in \mathcal{X}$, and F is well-defined.

Fix any $x_0 \in \mathcal{X}$. Fix any $\epsilon \in \mathbb{R}_+$. $\forall y \in \mathcal{Y}$, by the continuity of f, $\exists \delta_y \in \mathbb{R}_+$ such that, $\forall (x, \bar{y}) \in \mathcal{B}_{\mathcal{X}\times\mathcal{Y}}\left((x_0, y), \delta_y\right) =: \bar{D}_{x_0,y}$, we have $\|f(x, \bar{y}) - f(x_0, y)\|_{\mathrm{B}(\mathcal{Z},\mathcal{W})} < \epsilon/2$. Now, $\forall \bar{y} \in \mathcal{B}_{\mathcal{Y}}\left(y, \frac{\delta_y}{2}\right)$, $\forall x \in \mathcal{B}_{\mathcal{X}}\left(x_0, \frac{\delta_y}{2}\right)$, we have $(x, \bar{y}), (x_0, \bar{y}) \in \bar{D}_{x_0,y}$ and $\|f(x, \bar{y}) - f(x_0, \bar{y})\|_{\mathrm{B}(\mathcal{Z},\mathcal{W})} \leq \|f(x, \bar{y}) - f(x_0, y)\|_{\mathrm{B}(\mathcal{Z},\mathcal{W})} + \|f(x_0, y) - f(x_0, \bar{y})\|_{\mathrm{B}(\mathcal{Z},\mathcal{W})} < \epsilon$.

Clearly, $\mathcal{Y} \subseteq \bigcup_{y\in\mathcal{Y}} \mathcal{B}_{\mathcal{Y}}\left(y, \frac{\delta_y}{2}\right)$, which is an open covering. By the compactness of $\mathcal{Y}$, $\exists N \in \mathbb{Z}_+$ and $\exists (y_i)_{i=1}^N \subseteq \mathcal{Y}$ such that $\mathcal{Y} \subseteq \bigcup_{i=1}^N \mathcal{B}_{\mathcal{Y}}\left(y_i, \frac{\delta_{y_i}}{2}\right)$. Define $\delta_m := \min\{\min_{i\in\{1,\ldots,N\}} \frac{\delta_{y_i}}{2}, 1\} \in \mathbb{R}_+$. This δ_m has the following property. Fix any $x \in \mathcal{B}_{\mathcal{X}}(x_0, \delta_m)$. $\forall y \in \mathcal{Y}$, $\exists i_0 \in \{1, \ldots, N\}$, such that $y \in \mathcal{B}_{\mathcal{Y}}\left(y_{i_0}, \frac{\delta_{y_{i_0}}}{2}\right)$. Then, $(x, y), (x_0, y) \in \bar{D}_{x_0,y_{i_0}}$ and $\|f(x, y) - f(x_0, y)\|_{\mathrm{B}(\mathcal{Z},\mathcal{W})} < \epsilon$. This further implies that $\|F(x) - F(x_0)\|_{\mathcal{W}} = \left\| \int_J f(x, w(t))\, \mathrm{d}\mu(t) - \int_J f(x_0, w(t))\, \mathrm{d}\mu(t) \right\|_{\mathcal{W}} = \left\| \int_J (f(x, w(t)) - f(x_0, w(t))) \cdot \mathrm{d}\mu(t) \right\|_{\mathcal{W}} \leq \int_J \|f(x, w(t)) - f(x_0, w(t))\|_{\mathrm{B}(\mathcal{Z},\mathcal{W})}\, \mathrm{d}\mathcal{P} \circ \mu(t) \leq \epsilon \mathcal{P} \circ \mu(J)$, where the first equality follows from the definition of F, the second equality follows from Proposition 11.130 and the fact that $\overline{\mathrm{span}\,(\mathcal{U}_{x_0} \cup \mathcal{U}_x)} \subseteq \mathrm{B}(\mathcal{Z}, \mathcal{W})$ is a separable Banach subspace by Proposition 7.35; the first inequality follows from Proposition 11.130; and the last inequality follows from Propositions 11.83 and 11.75.

This proves that F is continuous at x_0. By the arbitrariness of x_0, we have F is continuous. This completes the proof of the theorem. □

This previous theorem is a true generalization of Theorem 31.6 of Bartle (1976).

Theorem 12.112 *Let $f : D \times \mathcal{Y} \to \mathrm{B}(\mathcal{Z}, \mathcal{W})$, where $D \subseteq \mathcal{X}$, $\mathcal{X}$ is a normed linear space over $\mathbb{K}$, $\mathcal{Y}$ is a compact metric space, $\mathcal{Z}$ is a normed linear space over $\mathbb{K}$, $\mathcal{W}$ is a Banach space over $\mathbb{K}$, $x_0 \in D$, and $\mathcal{J} := (J, \mathcal{B}, \mu)$ be a finite $\mathcal{Z}$-valued measure space. Assume that the following conditions hold:*

(i) $\overline{\mathrm{span}\,(\mathrm{A}_D\,(x_0))} = \mathcal{X}$.
(ii) $\exists \delta_{x_0} \in \mathbb{R}_+$, *such that the set* $(D \cap \mathcal{B}_{\mathcal{X}}\left(x_0, \delta_{x_0}\right)) - x_0$ *is a conic segment.*
(iii) $\frac{\partial f}{\partial x} : D \times \mathcal{Y} \to \mathrm{B}(\mathcal{X}, \mathrm{B}(\mathcal{Z}, \mathcal{W}))$ *exists, and* $\frac{\partial f}{\partial x}$ *is continuous at* (x_0, y), $\forall y \in \mathcal{Y}$.
(iv) $w : \mathcal{J} \to \mathcal{Y}$ *is* $\mathcal{B}$*-measurable and* f *is continuous.*

Define $F : D \to \mathcal{W}$ by $F(x) := \int_J f(x, w(t))\,\mathrm{d}\mu(t) \in \mathcal{W}$, $\forall x \in D$. Then, F is Fréchet differentiable at x_0 and $\mathrm{D}F(x_0) = \int_J \left(\frac{\partial f}{\partial x}(x_0, w(t))\right)^{T_{2,1}} \mathrm{d}\mu(t) \in \mathrm{B}(\mathcal{X}, \mathcal{W})$.

Proof By Theorem 12.111, $F(x) \in \mathcal{W}$ and $\int_J \left(\frac{\partial f}{\partial x}(x_0, w(t))\right)^{T_{2,1}} \mathrm{d}\mu(t) \in \mathrm{B}(\mathcal{X}, \mathcal{W})$, $\forall x \in D$, are well-defined.

By (ii), $\exists \bar{\delta}_{x_0} \in \mathbb{R}_+$, such that the set $(D \cap \mathcal{B}_{\mathcal{X}}\left(x_0, \bar{\delta}_{x_0}\right)) - x_0$ is a conic segment. Fix any $\epsilon \in \mathbb{R}_+$. By the continuity of $\frac{\partial f}{\partial x}$, Proposition 12.110, and the compactness of $\{x_0\} \times \mathcal{Y}$, $\exists \delta \in \mathrm{r}_{0,\bar{\delta}_{x_0}} \subset \mathbb{R}_+$, $\forall x \in \mathcal{B}_{\mathcal{X}}(x_0, \delta) \cap D$, $\forall y \in \mathcal{Y}$, we have $\mathrm{dist}((x, y), \{x_0\} \times \mathcal{Y}) < \delta$ and $\left\| \frac{\partial f}{\partial x}(x, y) - \frac{\partial f}{\partial x}(x_0, y) \right\| < \frac{\epsilon}{1+\mathcal{P}\circ\mu(J)}$. Fix any $x \in \mathcal{B}_{\mathcal{X}}(x_0, \delta) \cap D$. $\forall y \in \mathcal{Y}$, by Mean Value Theorem 9.23, $\left\| f(x, y) - f(x_0, y) - \frac{\partial f}{\partial x}(x_0, y)(x - x_0) \right\|_{\mathrm{B}(\mathcal{Z},\mathcal{W})} \leq \left\| \frac{\partial f}{\partial x}(x_0 + \alpha_{x,y}\,(x - x_0), y) - \frac{\partial f}{\partial x}(x_0, y) \right\|_{\mathrm{B}(\mathcal{X},\mathrm{B}(\mathcal{Z},\mathcal{W}))} \|x - x_0\|_{\mathcal{X}} \leq \frac{\epsilon}{1+\mathcal{P}\circ\mu(J)} \|x - x_0\|_{\mathcal{X}}$, where $\alpha_{x,y} \in (0, 1) \subset \mathbb{R}$ and $x_0 + \alpha_{x,y}(x - x_0) \in D \cap \mathcal{B}_{\mathcal{X}}(x_0, \delta)$; and the second inequality follows from the choice of δ. This further yields that $\left\| F(x) - F(x_0) - \left(\int_J \left(\frac{\partial f}{\partial x}(x_0, w(t))\right)^{T_{2,1}} \mathrm{d}\mu(t)\right)(x - x_0) \right\|_{\mathcal{W}} = \left\| \int_J f(x, w(t))\,\mathrm{d}\mu(t) - \int_J f(x_0, w(t))\,\mathrm{d}\mu(t) - \int_J \frac{\partial f}{\partial x}(x_0, w(t))(x - x_0)\,\mathrm{d}\mu(t) \right\|_{\mathcal{W}} = \left\| \int_J (f(x, w(t)) - f(x_0, w(t)) - \frac{\partial f}{\partial x}(x_0, w(t))(x - x_0))\,\mathrm{d}\mu(t) \right\|_{\mathcal{W}} \leq \int_J \left\| f(x, w(t)) - f(x_0, w(t)) - \frac{\partial f}{\partial x}(x_0, w(t))(x - x_0) \right\|_{\mathrm{B}(\mathcal{Z},\mathcal{W})} \mathrm{d}\mathcal{P} \circ \mu(t) \leq \int_J \frac{\epsilon}{1+\mathcal{P}\circ\mu(J)} \|x - x_0\|_{\mathcal{X}}\,\mathrm{d}\mathcal{P} \circ \mu(t) = \frac{\epsilon\mathcal{P}\circ\mu(J)}{1+\mathcal{P}\circ\mu(J)} \|x - x_0\|_{\mathcal{X}} \leq \epsilon \|x - x_0\|_{\mathcal{X}}$, where the first equality follows from the definition of F and Proposition 11.130, the second equality and the first inequality follow from Proposition 11.130, the second inequality follows from Proposition 11.83, and the last equality follows from Proposition 11.75.

This, coupled with (i), proves that F is Fréchet differentiable at x_0 and $\mathrm{D}F(x_0) = \int_J \left(\frac{\partial f}{\partial x}(x_0, w(t))\right)^{T_{2,1}} \mathrm{d}\mu(t)$. This completes the proof of the theorem. □

In the above theorem, the assumptions (i) and (ii) are conditions on the domain D, which are easily satisfied if $x_0 \in D^\circ$. Theorem 12.112 is a true generalization of Theorem 31.7 of Bartle (1976).

Theorem 12.113 (Leibniz's Formula) *Let $\mathcal{X}$ be a real normed linear space, $a, b \in \mathbb{R}$ with $a < b$, and $\mathcal{Z}$ is a real Banach space, $D \subseteq \mathcal{X}$, $x_0 \in D$, $f : D \times \bar{\mathrm{r}}_{a,b} \to \mathcal{Z}$, $\alpha : D \to \bar{\mathrm{r}}_{a,b}$, and $\beta : D \to \bar{\mathrm{r}}_{a,b}$. Assume that the following conditions hold:*

(i) *$\forall x_1 \in D$, we have $\overline{\mathrm{span}\,(\mathrm{A}_D(x_1))} = \mathcal{X}$.*
(ii) *The domain D is locally convex, i. e., $\forall x_1 \in D$, $\exists \delta_{x_1} \in (0, \infty) \subset \mathbb{R}$ such that $D \cap \mathcal{B}_{\mathcal{X}}(x_1, \delta_{x_1})$ is convex.*
(iii) *$\frac{\partial f}{\partial x} : D \times \bar{\mathrm{r}}_{a,b} \to \mathrm{B}(\mathcal{X}, \mathcal{Z})$, $\mathrm{D}\alpha(x_0) \in \mathrm{B}(\mathcal{X}, \mathbb{R})$, and $\mathrm{D}\beta(x_0) \in \mathrm{B}(\mathcal{X}, \mathbb{R})$ exists, and f and $\frac{\partial f}{\partial x}$ are continuous.*

Define $F : D \to \mathcal{Z}$ by $F(x) := \int_{\alpha(x)}^{\beta(x)} f(x,t)\,\mathrm{d}t \in \mathcal{Z}$, $\forall x \in D$. Then, F is Fréchet differentiable at x_0 and

$$\mathrm{D}F(x_0) = f(x_0, \beta(x_0))\mathrm{D}\beta(x_0) - f(x_0, \alpha(x_0))\mathrm{D}\alpha(x_0) + \int_{\alpha(x_0)}^{\beta(x_0)} \frac{\partial f}{\partial x}(x_0, t)\,\mathrm{d}t$$

Proof Let $H : D \times \bar{\mathrm{r}}_{a,b} \times \bar{\mathrm{r}}_{a,b} \to \mathcal{Z}$ be defined by $H(x, u, v) := \int_u^v f(x,t)\,\mathrm{d}t$, $\forall (x, u, v) \in D \times \bar{\mathrm{r}}_{a,b} \times \bar{\mathrm{r}}_{a,b}$. Then, $F(x) = H(x, \alpha(x), \beta(x))$, $\forall x \in D$. Note that, by Theorem 12.112, $\frac{\partial H}{\partial x}(x, u, v) = \int_u^v \frac{\partial f}{\partial x}(x,t)\,\mathrm{d}t$; by Theorem 12.82 and Proposition 7.126, $\frac{\partial H}{\partial u}(x, u, v) = -f(x, u)$; and, by Theorem 12.82 and Proposition 7.126, $\frac{\partial H}{\partial v}(x, u, v) = f(x, v)$; $\forall (x, u, v) \in D \times \bar{\mathrm{r}}_{a,b} \times \bar{\mathrm{r}}_{a,b}$. By Proposition 9.24, we have, $\forall (x, u, v) \in D \times \bar{\mathrm{r}}_{a,b} \times \bar{\mathrm{r}}_{a,b}$,

$$\begin{aligned} \mathrm{D}H(x, u, v) &= \left[\ \frac{\partial H}{\partial x}(x, u, v)\ \ \frac{\partial H}{\partial u}(x, u, v)\ \ \frac{\partial H}{\partial v}(x, u, v)\ \right] \\ &= \left[\ \int_u^v \frac{\partial f}{\partial x}(x,t)\,\mathrm{d}t\ \ -f(x, u)\ \ f(x, v)\ \right] \end{aligned}$$

Now, by Chain Rule (Theorem 9.18) and Proposition 9.19, we have $\mathrm{D}F(x_0) = \mathrm{D}H(x_0, \alpha(x_0), \beta(x_0)) \begin{bmatrix} \mathrm{id}_{\mathcal{X}} \\ \mathrm{D}\alpha(x_0) \\ \mathrm{D}\beta(x_0) \end{bmatrix} = \int_{\alpha(x_0)}^{\beta(x_0)} \frac{\partial f}{\partial x}(x_0, t)\,\mathrm{d}t - f(x_0, \alpha(x_0))\mathrm{D}\alpha(x_0) + f(x_0, \beta(x_0))\mathrm{D}\beta(x_0)$. This completes the proof of the theorem. □

Theorem 12.114 *Let $\mathcal{Z}$ be a normed linear space over $\mathbb{K}$, $\mathcal{W}$ be a Banach space over $\mathbb{K}$, $\mathcal{J} := (J, \mathcal{B}, \mu)$ be a σ-finite $\mathcal{Z}$-valued measure space, $\mathcal{X} := (X, \rho)$ be a* *metric space, $f : \mathcal{X} \times J \to B(\mathcal{Z}, \mathcal{W})$, and $x_0 \in \mathcal{X}$. Assume that the following conditions hold:*

(i) $f(x, \cdot) : J \to \mathcal{U}_x \subseteq B(\mathcal{Z}, \mathcal{W})$ *is $\mathcal{B}$-measurable and $\mathcal{U}_x$ is a separable Banach subspace of* $B(\mathcal{Z}, \mathcal{W})$.
(ii) $f(\cdot, t) : \mathcal{X} \to B(\mathcal{Z}, \mathcal{W})$ *is continuous at x_0, $\forall t \in \mathcal{J}$.*
(iii) *There exists a $\mathcal{B}$-measurable function $g : \mathcal{J} \to \overline{\mathbb{R}_+}$, which is integrable on J with respect to $\mathcal{P} \circ \mu$, such that, $\forall x \in \mathcal{X}$, $\|f(x,t)\|_{B(\mathcal{Z},\mathcal{W})} \le g(t)$, $\forall t \in \mathcal{J}$.*

Define $F : \mathcal{X} \to \mathcal{W}$ by $F(x) := \int_J f(x,t)\,d\mu(t) \in \mathcal{W}$, $\forall x \in \mathcal{X}$. Then, F is continuous at x_0, and $F(x) \in \overline{\text{span}\,(\{w \in \mathcal{W} \mid \exists z \in \mathcal{Z}, \exists A \in \mathcal{U}_x \ni \cdot\, w = Az\})} =: \mathcal{W}_x$.

Proof Fix any $x \in \mathcal{X}$, by Lebesgue Dominated Convergence Theorem 11.131, F is well-defined. By Proposition 11.132, $F(x) \in \mathcal{W}_x \subseteq \mathcal{W}$, $\forall x \in \mathcal{X}$.

We will prove that $\lim_{x\to x_0} F(x) = F(x_0)$ by Proposition 4.15. By the continuity of f, we have $\lim_{x\to x_0} f(x,t) = f(x_0,t)$, $\forall t \in \mathcal{J}$. Fix any sequence $(x_i)_{i=1}^{\infty} \subseteq \mathcal{X}$ with $\lim_{i\in\mathbb{N}} x_i = x_0$. Then, by Proposition 4.15, $\lim_{i\in\mathbb{N}} f(x_i,t) = f(x_0,t)$, $\forall t \in \mathcal{J}$. This implies that $\lim_{i\in\mathbb{N}} F(x_i) = \lim_{i\in\mathbb{N}} \int_J f(x_i,t)\,d\mu(t) = \int_J f(x_0,t)\,d\mu(t) = F(x_0)$, where the second equality follows from Lebesgue Dominated Convergence Theorem 11.131. By Proposition 4.15, we have $\lim_{x\to x_0} F(x) = F(x_0)$. This proves that F is continuous at x_0. This completes the proof of the theorem. □

Theorem 12.115 *Let $\mathcal{X} := (X, \rho)$ be a metric space, $\mathcal{Y}$ be a separable metric space, $\mathcal{Z}$ be a normed linear space over $\mathbb{K}$, $\mathcal{W}$ be a Banach space over $\mathbb{K}$, $\mathcal{J} := (J, \mathcal{B}, \mu)$ be a σ-finite $\mathcal{Z}$-valued measure space, $f : \mathcal{X} \times \mathcal{Y} \to B(\mathcal{Z}, \mathcal{W})$, and $w : \mathcal{J} \to \mathcal{Y}$ be $\mathcal{B}$-measurable. Assume that the following conditions hold:*

(i) *f is continuous.*
(ii) *There exists a $\mathcal{B}$-measurable function $g : \mathcal{J} \to \overline{\mathbb{R}_+}$, which is integrable on J with respect to $\mathcal{P} \circ \mu$, such that, $\forall x \in \mathcal{X}$, $\|f(x, w(t))\|_{B(\mathcal{Z},\mathcal{W})} \le g(t)$, $\forall t \in \mathcal{J}$.*

Define $F : \mathcal{X} \to \mathcal{W}$ by $F(x) := \int_J f(x, w(t))\,d\mu(t) \in \mathcal{W}$, $\forall x \in \mathcal{X}$. Then, F is continuous, $f(x, w(\cdot)) : \mathcal{J} \to \mathcal{U}_x \subseteq B(\mathcal{Z}, \mathcal{W})$, where $\mathcal{U}_x$ is a separable Banach subspace of $B(\mathcal{Z}, \mathcal{W})$, and $F(x) \in \mathcal{W}_x := \overline{\text{span}\,(\{w \in \mathcal{W} \mid \exists z \in \mathcal{Z}, \exists A \in \mathcal{U}_x \ni \cdot\, w = Az\})}$.

Proof Since f is continuous, by Proposition 7.126, $\forall x \in \mathcal{X}$, $h(x) := f(x, \cdot) : \mathcal{Y} \to \mathcal{V}_x \subseteq B(\mathcal{Z}, \mathcal{W})$ satisfies $\mathcal{V}_x$ is a separable subset of $B(\mathcal{Z}, \mathcal{W})$. By Proposition 11.38, $f(x, w(\cdot)) : \mathcal{J} \to \mathcal{V}_x$ is $\mathcal{B}$-measurable. By Lebesgue Dominated Convergence Theorem 11.131, F is well-defined. By Proposition 7.35, $\mathcal{U}_x := \overline{\text{span}\,(\mathcal{V}_x)}$ is a separable Banach subspace of $B(\mathcal{Z}, \mathcal{W})$. By Proposition 11.132, $F(x) \in \mathcal{W}_x \subseteq \mathcal{W}$, $\forall x \in \mathcal{X}$.

Fix any $x_0 \in \mathcal{X}$. We will prove that $\lim_{x\to x_0} F(x) = F(x_0)$ by Proposition 4.15. By the continuity of f, we have $\lim_{x\to x_0} f(x, w(t)) = f(x_0, w(t))$, $\forall t \in \mathcal{J}$. Fix any sequence $(x_i)_{i=1}^{\infty} \subseteq \mathcal{X}$ with $\lim_{i\in\mathbb{N}} x_i = x_0$. Then, by Proposition 4.15, $\lim_{i\in\mathbb{N}} f(x_i, w(t)) = f(x_0, w(t))$, $\forall t \in \mathcal{J}$. This implies that $\lim_{i\in\mathbb{N}} F(x_i) =$

$\lim_{i\in\mathbb{N}} \int_J f(x_i, w(t))\,d\mu(t) = \int_J f(x_0, w(t))\,d\mu(t) = F(x_0)$, where the second equality follows from Lebesgue Dominated Convergence Theorem 11.131. By Proposition 4.15, we have $\lim_{x\to x_0} F(x) = F(x_0)$. This proves that F is continuous at x_0. By the arbitrariness of x_0, F is continuous.

This completes the proof of the theorem. □

Theorem 12.116 *Let $f : D \times \mathcal{Y} \to \mathrm{B}(\mathcal{Z}, \mathcal{W})$, where $D \subseteq \mathcal{X}$, $\mathcal{X}$ is a normed linear space over $\mathbb{K}$, $\mathcal{Y}$ is a separable metric space, $\mathcal{Z}$ is a normed linear space over $\mathbb{K}$, and $\mathcal{W}$ is a Banach space over $\mathbb{K}$, and $\mathcal{J} := (J, \mathcal{B}, \mu)$ be a σ-finite $\mathcal{Z}$-valued measure space. Assume that the following conditions hold:*

(i) $\forall x_0 \in D$, $\overline{\operatorname{span}(\mathrm{A}_D(x_0))} = \mathcal{X}$.
(ii) *$\forall x_0 \in D$, $\exists \delta_{x_0} \in \mathbb{R}_+$, such that the set $(D \cap \mathcal{B}_{\mathcal{X}}(x_0, \delta_{x_0})) - x_0$ is a conic segment.*
(iii) *$w : \mathcal{J} \to \mathcal{Y}$ is $\mathcal{B}$-measurable.*
(iv) *f is continuous, and there exists a measurable function $g : \mathcal{J} \to \overline{\mathbb{R}_+}$, which is integrable over J with respect to $\mathcal{P} \circ \mu$, such that, $\forall x \in D$, $\|f(x, w(t))\|_{\mathrm{B}(\mathcal{Z},\mathcal{W})} \le g(t)$, $\forall t \in J$.*
(v) *$\frac{\partial f}{\partial x}$ exists and is continuous, and there exists a measurable function $\bar{g} : \mathcal{J} \to \overline{\mathbb{R}_+}$, which is integrable over J with respect to $\mathcal{P} \circ \mu$, such that, $\forall x \in D$, $\left\| \frac{\partial f}{\partial x}(x, w(t)) \right\|_{\mathrm{B}(\mathcal{X},\mathrm{B}(\mathcal{Z},\mathcal{W}))} \le \bar{g}(t)$, $\forall t \in J$.*

Define $F : D \to \mathcal{W}$ by $F(x) := \int_J f(x, w(t))\,d\mu(t) \in \mathcal{W}$, $\forall x \in D$. Then, F is continuously Fréchet differentiable and $\mathrm{D}F(x) = \int_J \left(\frac{\partial f}{\partial x}(x, w(t))\right)^{T_{2,1}} d\mu(t) \in \mathrm{B}(\mathcal{X}, \mathcal{W})$, $\forall x \in D$.

Proof By Theorem 12.115, $F(x) \in \mathcal{W}$ and $\int_J \left(\frac{\partial f}{\partial x}(x, w(t))\right)^{T_{2,1}} d\mu(t) \in \mathrm{B}(\mathcal{X}, \mathcal{W})$, $\forall x \in D$, are well-defined and are continuous.

Fix any $x_0 \in D$. Denote $\int_J \left(\frac{\partial f}{\partial x}(x_0, w(t))\right)^{T_{2,1}} d\mu(t) =: L_0 \in \mathrm{B}(\mathcal{X}, \mathcal{W})$. By (ii), $\exists \bar{\delta}_{x_0} \in \mathbb{R}_+$, such that the set $(D \cap \mathcal{B}_{\mathcal{X}}(x_0, \bar{\delta}_{x_0})) - x_0 =: D_0$ is a conic segment. Define function $G : D_0 \setminus \{\vartheta_{\mathcal{X}}\} \to \mathbb{R}$ by $G(h) := \frac{\|F(x_0+h)-F(x_0)-L_0(h)\|_{\mathcal{W}}}{\|h\|_{\mathcal{X}}} \in \mathbb{R}$, $\forall h \in D_0 \setminus \{\vartheta_{\mathcal{X}}\}$. We will show that $\lim_{h\to\vartheta_{\mathcal{X}}} G(h) = 0$ by Proposition 4.16. Fix any sequence $(h_i)_{i=1}^{\infty} \subseteq D_0 \setminus \{\vartheta_{\mathcal{X}}\}$ with $\lim_{i\in\mathbb{N}} h_i = \vartheta_{\mathcal{X}}$. Fix any $i \in \mathbb{N}$. $F(x_0 + h_i) - F(x_0) - L_0(h_i) \in \mathcal{W}$. Then, we have

$$
\begin{aligned}
G(h_i) &= \frac{\|F(x_0 + h_i) - F(x_0) - L_0(h_i)\|_{\mathcal{W}}}{\|h_i\|_{\mathcal{X}}} \\
&= \frac{1}{\|h_i\|_{\mathcal{X}}} \left\| \int_J (f(x_0 + h_i, w(t)) - f(x_0, w(t)) - \frac{\partial f}{\partial x}(x_0, w(t))(h_i))\,d\mu(t) \right\|_{\mathcal{W}} \\
&= \frac{1}{\|h_i\|_{\mathcal{X}}} \left\| \int_J \int_0^1 \left(\frac{\partial f}{\partial x}(x_0 + s h_i, w(t)) - \frac{\partial f}{\partial x}(x_0, w(t))\right)(h_i)\,ds\,d\mu(t) \right\|_{\mathcal{W}}
\end{aligned}
$$

$$\leq \frac{1}{\|h_i\|_{\mathcal{X}}} \int_J \left\| \int_0^1 \left(\frac{\partial f}{\partial x}(x_0 + sh_i, w(t)) - \frac{\partial f}{\partial x}(x_0, w(t)) \right)(h_i) \,ds \right\|_{B(\mathcal{Z},\mathcal{W})}$$
$$d\mathcal{P} \circ \mu(t)$$
$$\leq \frac{1}{\|h_i\|_{\mathcal{X}}} \int_J \int_0^1 \left\| \left(\frac{\partial f}{\partial x}(x_0 + sh_i, w(t)) - \frac{\partial f}{\partial x}(x_0, w(t)) \right)(h_i) \right\|_{B(\mathcal{Z},\mathcal{W})}$$
$$ds\, d\mathcal{P} \circ \mu(t)$$
$$\leq \frac{1}{\|h_i\|_{\mathcal{X}}} \int_J \int_0^1 \left\| \frac{\partial f}{\partial x}(x_0 + sh_i, w(t)) - \frac{\partial f}{\partial x}(x_0, w(t)) \right\|_{B(\mathcal{X},B(\mathcal{Z},\mathcal{W}))}$$
$$\cdot \|h_i\|_{\mathcal{X}}\, ds\, d\mathcal{P} \circ \mu(t)$$
$$= \int_J \int_0^1 \left\| \frac{\partial f}{\partial x}(x_0 + sh_i, w(t)) - \frac{\partial f}{\partial x}(x_0, w(t)) \right\|_{B(\mathcal{X},B(\mathcal{Z},\mathcal{W}))} ds\, d\mathcal{P} \circ \mu(t)$$
$$= \int_0^1 \int_J \left\| \frac{\partial f}{\partial x}(x_0 + sh_i, w(t)) - \frac{\partial f}{\partial x}(x_0, w(t)) \right\|_{B(\mathcal{X},B(\mathcal{Z},\mathcal{W}))} d\mathcal{P} \circ \mu(t)\, ds$$

where the first equality follows from the definitions of G, the second equality follows from the definitions of F and L_0 and Proposition 11.132, the third equality follows from Theorem 12.83, the first inequality follows from Proposition 11.132, the second inequality follows from Proposition 11.92, the third inequality follows from Proposition 7.63, the fourth equality follows from simple algebra, and the last equality follows from Tonelli's Theorem 12.29 and Propositions 11.38 and 11.39. Now, by Lebesgue Dominated Convergence Theorem 11.91, we have $\lim_{i\in\mathbb{N}} G(h_i) = 0$. This further implies that $\lim_{h\to\vartheta_{\mathcal{X}}} G(h) = 0$ by Proposition 4.16.

$\forall \epsilon \in \mathbb{R}_+$, $\exists \delta \in r_{0,\delta_{x_0}}$, $\forall x \in D \cap \mathcal{B}_{\mathcal{X}}(x_0, \delta)$, we have $0 \leq G(x - x_0) < \epsilon$. This is equivalent to the inequality $\|F(x) - F(x_0) - L_0(x - x_0)\|_{\mathcal{Z}} \leq \epsilon \|x - x_0\|_{\mathcal{X}}$. This shows that $DF(x_0) = L_0$. The result follows by the arbitrariness of $x_0 \in D$.

This completes the proof of the theorem. □

Proposition 12.117 *Let $I \subseteq \mathbb{R}$ be an interval, $\mathcal{Y}$ and $\mathcal{Z}$ be normed linear spaces over $\mathbb{K}$, $D \subseteq \mathcal{Y}$, $f : I \to D \subseteq \mathcal{Y}$ be of locally bounded variation, and $g : D \to \mathcal{Z}$ be Lipschitz on D with Lipschitz constant L. Then, the following result holds:*

(i) *$h := g \circ f : I \to \mathcal{Z}$ is of locally bounded variation.*
(ii) *If f is of bounded variation, then h is of bounded variation.*

Proof (i) By Definition 12.41, f is continuous on the right. By Propositions 4.15 and 12.37, h is continuous on the right. $\forall x_1, x_2 \in I$ with $x_1 \leq x_2$. Then, we have

$\bar{\mathrm{r}}_{x_1,x_2} \subseteq I$ since I is an interval. By Definition 12.41, $T_f(\mathrm{r}_{x_1,x_2}) < \infty$. Then,

$$\begin{aligned} T_h\big(\mathrm{r}_{x_1,x_2}\big) &= \sup_{n\in\mathbb{Z}_+,\ x_1=\check{x}_0\leq\check{x}_1\leq\cdots\leq\check{x}_n=x_2} \sum_{i=1}^{n} \big\|\Delta_h(\mathrm{r}_{\check{x}_{i-1},\check{x}_i})\big\|_{\mathcal{Z}} \\ &= \sup_{n\in\mathbb{Z}_+,\ x_1=\check{x}_0\leq\check{x}_1\leq\cdots\leq\check{x}_n=x_2} \sum_{i=1}^{n} \big\|h(\check{x}_i) - h(\check{x}_{i-1})\big\|_{\mathcal{Z}} \\ &= \sup_{n\in\mathbb{Z}_+,\ x_1=\check{x}_0\leq\check{x}_1\leq\cdots\leq\check{x}_n=x_2} \sum_{i=1}^{n} \big\|g(f(\check{x}_i)) - g(f(\check{x}_{i-1}))\big\|_{\mathcal{Z}} \\ &\leq \sup_{n\in\mathbb{Z}_+,\ x_1=\check{x}_0\leq\check{x}_1\leq\cdots\leq\check{x}_n=x_2} \sum_{i=1}^{n} L\big\|f(\check{x}_i) - f(\check{x}_{i-1})\big\|_{\mathcal{Y}} \\ &= L \sup_{n\in\mathbb{Z}_+,\ x_1=\check{x}_0\leq\check{x}_1\leq\cdots\leq\check{x}_n=x_2} \sum_{i=1}^{n} \big\|f(\check{x}_i) - f(\check{x}_{i-1})\big\|_{\mathcal{Y}} \\ &= LT_f\big(\mathrm{r}_{x_1,x_2}\big) < \infty \end{aligned}$$

where the first two equalities follow from Definition 12.41, the third equality follows from the definition of h, the first inequality follows from the Lipschitz continuity of g, the fourth equality follows from Proposition 3.81, and the last equality follows from Definition 12.41. Hence, h satisfies (i) and (ii) of Definition 12.41.

$\forall x_1, x_2 \in I$ with $x_1 < x_2$. Then, $\bar{\mathrm{r}}_{x_1,x_2} \subseteq I$. Define $H : \bar{\mathrm{r}}_{x_1,x_2} \to [0, \infty) \subset \mathbb{R}$ by $H(x) := T_h(\mathrm{r}_{x_1,x})$, $\forall x \in \bar{\mathrm{r}}_{x_1,x_2}$. Fix any $x \in \bar{\mathrm{r}}_{x_1,x_2}$, and let $\delta \in [0, x_2 - x] \subseteq [0, \infty) \subset \mathbb{R}$. Then, we have

$$\begin{aligned} |H(x+\delta) - H(x)| &= \big|T_h\big(\mathrm{r}_{x_1,x+\delta}\big) - T_h\big(\mathrm{r}_{x_1,x}\big)\big| = \big|T_h\big(\mathrm{r}_{x,x+\delta}\big)\big| \\ &\leq LT_f\big(\mathrm{r}_{x,x+\delta}\big) \end{aligned}$$

where the first equality follows from the definition of H, the second equality follows from Proposition 12.47, and the inequality follows from the previous paragraph. By Definition 12.41, we have $\lim_{\delta\to 0^+} T_f(\mathrm{r}_{x,x+\delta}) = \lim_{\delta\to 0^+}(T_f(\mathrm{r}_{x_1,x+\delta}) - T_f(\mathrm{r}_{x_1,x})) = 0$. Then, we have $\lim_{\delta\to 0^+} |H(x+\delta) - H(x)| = 0$. Hence, H is continuous on the right at x. By the arbitrariness of x, H is continuous on the right.

Then, h is of locally bounded variation.

(ii) Clearly, $T_h \leq LT_f < \infty$ if f is of bounded variation. Then, h is of bounded variation. This completes the proof of the proposition. □

Proposition 12.118 *Let $I \subseteq \mathbb{R}$ be an interval, $\mathcal{Y}$ and $\mathcal{Z}$ be Banach space over $\mathbb{K}$, $f : I \to \mathcal{Y}$ and $g : I \to \mathcal{Z}$ be of locally bounded variation, and $h : I \to \mathcal{Y} \times \mathcal{Z}$ be* *defined by* $h(x) = (f(x), g(x)) \in \mathcal{Y} \times \mathcal{Z}$, $\forall x \in I$. *Then, the following statements hold:*

(i) *h is of locally bounded variation, and* $T_h\left(\mathrm{r}_{x_1,x_2}\right) \le T_f(\mathrm{r}_{x_1,x_2}) + T_g(\mathrm{r}_{x_1,x_2}) < \infty$, $\forall x_1, x_2 \in I$ *with* $x_1 \le x_2$.

(ii) *If, in addition, f and g are of bounded variation, then h is of bounded variation.*

Proof (i) By Theorem 12.50, there exists a σ-finite $\mathcal{Y}$-valued measure μ_f on the measurable space $(\mathrm{P}(I), \mathcal{B}_\mathrm{B}(\mathrm{P}(I)))$ such that f is a cumulative distribution function of μ_f; and there exists a σ-finite $\mathcal{Z}$-valued measure μ_g on the measurable space $(\mathrm{P}(I), \mathcal{B}_\mathrm{B}(\mathrm{P}(I)))$ such that g is a cumulative distribution function of μ_g. By Proposition 11.147, there exists a σ-finite $\mathcal{Y} \times \mathcal{Z}$-valued measure $\mu = \begin{bmatrix} \mu_f \\ \mu_g \end{bmatrix}$ such that $\forall E \in \mathrm{dom}\left(\mu_f\right) \cap \mathrm{dom}\left(\mu_g\right)$, we have $\mu(E) = (\mu_f(E), \mu_g(E)) \in \mathcal{Y} \times \mathcal{Z}$, and $\mathcal{P} \circ \mu \le \mathcal{P} \circ \mu_f + \mathcal{P} \circ \mu_g$. Then, $\forall x_1, x_2 \in I$ with $x_1 \le x_2$. Then, $\bar{\mathrm{r}}_{x_1,x_2} \subseteq I$. Then, $T_f(\mathrm{r}_{x_1,x_2}) < \infty$ and $T_g(\mathrm{r}_{x_1,x_2}) < \infty$, by Definition 12.41. By Theorem 12.50, we have $\mathcal{P} \circ \mu_f(\mathrm{r}_{x_1,x_2}) = T_f(\mathrm{r}_{x_1,x_2}) < \infty$ and $\mathcal{P} \circ \mu_g(\mathrm{r}_{x_1,x_2}) = T_g(\mathrm{r}_{x_1,x_2}) < \infty$. This implies that $\mathrm{r}_{x_1,x_2} \in \mathrm{dom}\,(\mu_f) \cap \mathrm{dom}\,(\mu_g)$. Hence, $\mu(\mathrm{r}_{x_1,x_2}) = (\mu_f(\mathrm{r}_{x_1,x_2}), \mu_g(\mathrm{r}_{x_1,x_2})) = (f(x_2) - f(x_1), g(x_2) - g(x_1)) = (f(x_2), g(x_2)) - (f(x_1), g(x_1))$ and $\mathcal{P} \circ \mu(\mathrm{r}_{x_1,x_2}) \le T_f(\mathrm{r}_{x_1,x_2}) + T_g(\mathrm{r}_{x_1,x_2}) < \infty$. This leads to the conclusion that h is a cumulative distribution function of μ. If $I = \emptyset$, then the desired result holds trivially. On the other hand, if $I \neq \emptyset$, then we let $x_0 \in I$. This implies that $\bar{h} := h - h(x_0) : I \to \mathcal{Y} \times \mathcal{Z}$ is a cumulative distribution function for μ with origin at x_0. By Proposition 12.52, $\bar{h}$ is of locally bounded variation and $T_{\bar{h}}(\mathrm{r}_{x_1,x_2}) = \mathcal{P} \circ \mu(\mathrm{r}_{x_1,x_2})$, $\forall x_1, x_2 \in I$ with $x_1 \le x_2$. Then, we have h is of locally bounded variation by Definition 12.41, since $T_{\bar{h}}(\mathrm{r}_{x_1,x_2}) = T_h(\mathrm{r}_{x_1,x_2})$, $\forall x_1, x_2 \in I$ with $x_1 \le x_2$. Furthermore, we have $T_h(\mathrm{r}_{x_1,x_2}) = T_{\bar{h}}(\mathrm{r}_{x_1,x_2}) = \mathcal{P} \circ \mu(\mathrm{r}_{x_1,x_2}) \le T_f(\mathrm{r}_{x_1,x_2}) + T_g(\mathrm{r}_{x_1,x_2}) < \infty$, $\forall x_1, x_2 \in I$ with $x_1 \le x_2$.

(ii) Since f and g are of bounded variation, we have $T_f < \infty$ and $T_g < \infty$. Then, by (i), $T_h \le T_f + T_g < \infty$. Hence, h is of bounded variation.

This completes the proof of the proposition. □

Proposition 12.119 *Let* $I := [a, b] \subset \mathbb{R}$ *be a compact interval with* $a \le b$, $\mathcal{X}$ *and* $\mathcal{Y}$ *be Banach spaces over* $\mathbb{K}$, $p \in [1, \infty) \subset \mathbb{R}$, $f : I \to \mathrm{B}(\mathcal{X}, \mathcal{Y})$ *be of bounded variation, and* $g : I \to \mathcal{X}$ *be continuous. Then, the following result holds:*

(i) $h := \mathcal{P}_p \circ f : I \to [0, \infty) \subset \mathbb{R}$ *is of bounded variation.*

(ii) *h is Riemann–Stieltjes integrable with respect to g on I, g is Riemann–Stieltjes integrable with respect to h on I, and the Riemann–Stieltjes integrals converge in gauge mode.*

(iii) *If, in addition,* $\mathcal{X} = \mathbb{K}$ *and g is of bounded variation and monotonically nondecreasing, then* $\widehat{\int_a^b} h\,\mathrm{d}g = \int_a^b h\,\mathrm{d}g \in \mathbb{R}$, *where the right-hand side is the Lebesgue–Stieltjes integral.*

(iv) *In particular, if* $g = \mathrm{id}_I$, *then* $\widehat{\int_a^b} h\,\mathrm{d}g = \widehat{\int_a^b} \mathcal{P}_p \circ f\,\mathrm{d}t = \int_a^b \mathcal{P}_p \circ f\,\mathrm{d}t = \|f\|^p_{\mathrm{L}_p(\mathrm{r}_{a,b},\mathcal{Y})} = \|f\|^p_{\mathrm{L}_p(\bar{\mathrm{r}}_{a,b},\mathcal{Y})} \in \mathbb{R}$.

Proof (i) Clearly, the norm function $\|\cdot\|_{\mathrm{B}(\mathcal{X},\mathcal{Y})} : \mathrm{B}(\mathcal{X},\mathcal{Y}) \to [0,\infty) \subset \mathbb{R}$ is Lipschitz on $\mathrm{B}(\mathcal{X},\mathcal{Y})$ with Lipschitz constant 1. By Proposition 12.117, the function $\bar{h} := \mathcal{P} \circ f : I \to [0,\infty) \subset \mathbb{R}$ is of bounded variation. Note that, since f is of bounded variation, then f is bounded. Then, $\exists c \in [0,\infty) \subset \mathbb{R}$ such that $\bar{h} : I \to [0,c] \subset \mathbb{R}$. The function $l : \bar{\mathrm{r}}_{0,c} \to \bar{\mathrm{r}}_{0,c^p} \subset \mathbb{R}$, defined by $l(t) = t^p$, $\forall t \in \bar{\mathrm{r}}_{0,c}$, is Lipschitz on $\bar{\mathrm{r}}_{0,c}$ with Lipschitz constant pc^{p-1}. Then, by Proposition 12.117, $h = l \circ \bar{h}$ is of bounded variation.

(ii) By Integrability Theorem A.10, g is Riemann–Stieltjes integrable with respect to h on I, h is Riemann–Stieltjes integrable with respect to g on I, and these Riemann–Stieltjes integrals converge in the gauge mode.

(iii) Let $\mathcal{X} = \mathbb{K}$ and g be of bounded variation and monotonically nondecreasing (hence real-valued). By Proposition 12.55 and Theorem 12.50, h is $\mathcal{B}_{\mathrm{B}}(\mathbb{R})$-measurable. By Proposition 12.100, we have $\widehat{\int}_a^b h\,\mathrm{d}g = \int_a^b h\,\mathrm{d}g \in \mathbb{R}$.

(iv) Let $g = \mathrm{id}_I$, which is continuous, of bounded variation, and monotonically nondecreasing. By (iii), we have $\widehat{\int}_a^b h\,\mathrm{d}g = \widehat{\int}_a^b \mathcal{P}_p \circ f\,\mathrm{d}t = \int_a^b \mathcal{P}_p \circ f\,\mathrm{d}t = \|f\|^p_{\bar{\mathrm{L}}_p(\mathrm{r}_{a,b},\mathcal{Y})} \in \mathbb{R}$. Furthermore, we have $\|f\|^p_{\bar{\mathrm{L}}_p(\mathrm{r}_{a,b},\mathcal{Y})} = \|f\|^p_{\bar{\mathrm{L}}_p(\bar{\mathrm{r}}_{a,b},\mathcal{Y})}$ by Lemma 11.73.

This completes the proof of the proposition. □

Proposition 12.120 *Let $p \in [1,\infty) \subset \mathbb{R}$, $\mathbb{X} := (\bar{\mathrm{r}}_{a,b}, \mathcal{B}, \mu)$ be the finite metric measure subspace of $\mathbb{R}$ with $a \le b$, $\mathcal{Y}$ be a separable normed linear space over $\mathbb{K}$, and $f \in \bar{\mathrm{L}}_p(\mathbb{X},\mathcal{Y})$. Then, $\forall \epsilon \in (0,\infty) \subset \mathbb{R}$, $\exists$ a absolutely continuous function $g : \mathbb{X} \to \mathcal{Y}$ such that $g \in \bar{\mathrm{L}}_p(\mathbb{X},\mathcal{Y})$ and $\|g - f\|_{\bar{\mathrm{L}}_p(\mathbb{X},\mathcal{Y})} < \epsilon$.*

Proof Fix any $\epsilon \in (0,\infty) \subset \mathbb{R}$. By Propositions 4.11 and Proposition 11.182, there exists a continuous function $h \in \bar{\mathrm{L}}_p(\mathbb{X},\mathcal{Y})$ such that $\|h - f\|_{\bar{\mathrm{L}}_p(\mathbb{X},\mathcal{Y})} < \frac{\epsilon}{2}$. Now, by Bernsteĭn Approximation Theorem 11.203, there exists a Bernsteĭn function $g : \mathbb{X} \to \mathcal{Y}$ such that $\|g - h\|_{\mathcal{C}(\mathbb{X},\mathcal{Y})} < \frac{\epsilon}{2(1+b-a)^{1/p}}$. By Definition 11.202, each summand in the function l is absolutely continuous. Note that the vector addition is Lipschitz on the domain $\mathcal{Y} \times \mathcal{Y}$. Then, by Propositions 12.61 and 12.67, we have g is absolutely continuous. By Propositions 12.59 and 11.37, g is $\mathcal{B}$-measurable. Then, $g \in \bar{\mathrm{L}}_p(\mathbb{X},\mathcal{Y})$ since g is then bounded. Clearly, $\|g - f\|_{\bar{\mathrm{L}}_p(\mathbb{X},\mathcal{Y})} \le \|g - h\|_{\bar{\mathrm{L}}_p(\mathbb{X},\mathcal{Y})} + \|h - f\|_{\bar{\mathrm{L}}_p(\mathbb{X},\mathcal{Y})} < (\int_{\mathbb{X}} \mathcal{P}_p \circ (g-h)\,\mathrm{d}\mu)^{1/p} + \frac{\epsilon}{2} \le \left(\int_{\mathbb{X}} \frac{\epsilon^p}{2^p(1+b-a)}\,\mathrm{d}\mu\right)^{1/p} + \frac{\epsilon}{2} < \epsilon$, where the first inequality follows from the triangle inequality, the second inequality follows from the choice of h, the third inequality follows from Proposition 11.83, and the last inequality follows from Proposition 11.75. This completes the proof of the proposition. □

Theorem 12.121 *Let $\mathcal{X}_i := (X_i, \mathcal{B}_i, \mu_i)$ be a finite measure space, $i = 1, 2$, $\mathcal{X} := \mathcal{X}_1 \times \mathcal{X}_2 =: (X_1 \times X_2, \mathcal{B}, \mu)$ be the finite product measure space, $\mathcal{Y}$ be a separable Banach space, $W \subseteq \mathcal{Y}$ be a σ-compact conic segment, $h : X_1 \times X_2 \to W$ be $\mathcal{B}$-measurable, $h_{\langle x_2 \rangle} : X_1 \to W$ be absolutely integrable over $\mathcal{X}_1$, $\forall x_2 \in \mathcal{X}_2$, and $g : X_2 \to \mathcal{Y}$ be defined by $g(x_2) := \int_{X_1} h_{\langle x_2 \rangle}\,\mathrm{d}\mu_1 \in \mathcal{Y}$, $\forall x_2 \in \mathcal{X}_2$. Then, g is $\mathcal{B}_2$-measurable.*

Proof Fix any $n \in \mathbb{N}$. Define $h_n : X_1 \times X_2 \to W$ by $h_n(x_1, x_2) = \begin{cases} h(x_1, x_2) & \text{if } \|h(x_1, x_2)\|_{\mathcal{Y}} \leq n \\ \frac{n}{\|h(x_1,x_2)\|_{\mathcal{Y}}} h(x_1, x_2) & \text{if } \|h(x_1, x_2)\|_{\mathcal{Y}} > n \end{cases}$, $\forall (x_1, x_2) \in X_1 \times X_2$. By Propositions 7.21 and 11.38, we have $\mathcal{P} \circ h : X_1 \times X_2 \to \mathbb{R}$ is $\mathcal{B}$-measurable. By Propositions 11.39 and 11.38, we have h_n is $\mathcal{B}$-measurable. Define $g_n : X_2 \to \mathcal{Y}$ by $g_n(x_2) := \int_{X_1} h_{n\langle x_2 \rangle} \, \mathrm{d}\mu_1 \in \mathcal{Y}$, $\forall x_2 \in \mathcal{X}_2$, which is well-defined by Bounded Convergence Theorem 11.77. Again by Bounded Convergence Theorem 11.77, we have h_n is absolutely integrable over $\mathcal{X}$. By Fubini's Theorem 12.31, we have $\int_{X_1 \times X_2} h_n \, \mathrm{d}\mu = \int_{X_2} g_n \, \mathrm{d}\mu_2$ and g_n is $\mathcal{B}_2$-measurable.

Fix any $x_2 \in X_2$. Note that $\lim_{n \in \mathbb{N}} h_{n\langle x_2 \rangle}(x_1) = h_{\langle x_2 \rangle}(x_1)$, $\forall x_1 \in \mathcal{X}_1$, and $\left\| h_{n\langle x_2 \rangle}(x_1) \right\|_{\mathcal{Y}} \leq \left\| h_{\langle x_2 \rangle}(x_1) \right\|_{\mathcal{Y}}$, $\forall x_1 \in \mathcal{X}_1$. By Lebesgue Dominated Convergence Theorem 11.91, $g(x_2) = \int_{X_1} h_{\langle x_2 \rangle} \, \mathrm{d}\mu_1 = \lim_{n \in \mathbb{N}} \int_{X_1} h_{n\langle x_2 \rangle} \, \mathrm{d}\mu_1 = \lim_{n \in \mathbb{N}} g_n(x_2)$.

By Proposition 11.48, g is $\mathcal{B}_2$-measurable. This completes the proof of the theorem. □

Theorem 12.122 (Taylor's Theorem) *Let $\mathcal{X}$ be a normed linear space over $\mathbb{K}$, $\mathcal{Y}$ be a Banach space over $\mathbb{K}$, $D \subseteq \mathcal{X}$, $f : D \to \mathcal{Y}$, $x_0, x_1 \in D$, and $n \in \mathbb{Z}_+$. Let $\bar{D} := I := [0, 1] \subset \mathbb{R}$ or $\bar{D} := \{a + \mathrm{i}0 \mid a \in I\} \subset \mathbb{C}$ if $\mathbb{K} = \mathbb{C}$. Let $\varphi : \bar{D} \to D$ be given by $\varphi(t) = t x_1 + (1 - t) x_0$, $\forall t \in \bar{D}$. Assume that* $\mathrm{dom}\left(f^{(n+1)}\right) \supseteq \varphi(\bar{D})$, *and $f^{(n+1)}$ is continuous at $x = \varphi(t)$, $\forall t \in \bar{D}$. Let $R_n \in \mathcal{Y}$ be given by*

$$R_n := f(x_1) - \Bigg(f(x_0) + \frac{1}{1!} f^{(1)}(x_0)(x_1 - x_0) + \cdots + \frac{1}{n!} f^{(n)}(x_0) \underbrace{(x_1 - x_0) \cdots (x_1 - x_0)}_{n\text{-times}} \Bigg)$$

Then,

$$R_n = \int_0^1 \frac{s^n}{n!} f^{(n+1)}(\varphi(1 - s)) \underbrace{(x_1 - x_0) \cdots (x_1 - x_0)}_{(n+1)\text{-times}} \, \mathrm{d}s$$

Proof Define $F : I \to \mathcal{Y}$ by

$$F(t) = f(\varphi(1)) - \Bigg(f(\varphi(1 - t)) + \frac{t}{1!} f^{(1)}(\varphi(1 - t))(x_1 - x_0) + \cdots + \frac{t^n}{n!} f^{(n)}(\varphi(1 - t)) \underbrace{(x_1 - x_0) \cdots (x_1 - x_0)}_{n\text{-times}} + R_n t^{n+1} \Bigg); \quad \forall t \in I$$

By Propositions 3.12, 3.32, 9.7, 7.23, and 7.65, F is continuous. Clearly, φ is differentiable. By Chain Rule and Propositions 9.10, 9.15–9.17 and 9.19, F is differentiable at t, $\forall t \in \bar{D}$. By Propositions 3.12, 3.32, 9.7, 7.23, and 7.65, $F^{(1)}$ is continuous. Clearly, $F(0) = F(1) = \vartheta_{\mathcal{Y}}$. By Theorem 12.83, we have

$F(1) - F(0) = \int_0^1 \mathrm{D}F(s)\,\mathrm{d}s$. Then, we have

$$\begin{aligned}
\vartheta_{\mathcal{Y}} &= -\int_0^1 \Big(-f^{(1)}(\varphi(1-s))(x_1-x_0) + f^{(1)}(\varphi(1-s))(x_1-x_0) \\
&\quad -\frac{s}{1!} f^{(2)}(\varphi(1-s))(x_1-x_0)(x_1-x_0) + \cdots \\
&\quad + \frac{s^{n-1}}{(n-1)!} f^{(n)}(\varphi(1-s)) \underbrace{(x_1-x_0)\cdots(x_1-x_0)}_{n\text{-times}} \\
&\quad -\frac{s^n}{n!} f^{(n+1)}(\varphi(1-s)) \underbrace{(x_1-x_0)\cdots(x_1-x_0)}_{(n+1)\text{-times}} + (n+1)R_n s^n \Big)\,\mathrm{d}s \\
&= \int_0^1 \frac{s^n}{n!} f^{(n+1)}(\varphi(1-s)) \underbrace{(x_1-x_0)\cdots(x_1-x_0)}_{(n+1)\text{-times}}\,\mathrm{d}s - R_n
\end{aligned}$$

Hence,

$$R_n = \int_0^1 \frac{s^n}{n!} f^{(n+1)}(\varphi(1-s)) \underbrace{(x_1-x_0)\cdots(x_1-x_0)}_{(n+1)\text{-times}}\,\mathrm{d}s$$

This completes the proof of the theorem. □

12.10 Iterated Integrals

Proposition 12.123 *Let $\mathcal{X}$ be a normed linear space, $S_1, S_2 \subseteq \mathcal{X}$ be compact. Then, $S := \{x \in \mathcal{X} \mid x = \alpha x_1 + (1-\alpha)x_2, x_1 \in S_1, x_2 \in S_2, \alpha \in \bar{\mathrm{r}}_{0,1}\}$ is compact.*

Proof Define a mapping $F : S_1 \times S_2 \times \bar{\mathrm{r}}_{0,1} \to \mathcal{X}$ by $F(x_1, x_2, \alpha) = \alpha x_1 + (1-\alpha)x_2$. By Tychonoff Theorem 5.47, $S_1 \times S_2 \times \bar{\mathrm{r}}_{0,1}$ is compact. Clearly, F is continuous and $F(S_1 \times S_2 \times \bar{\mathrm{r}}_{0,1}) = S$. Then, S is compact by Proposition 5.7. □

Theorem 12.124 *Let $m \in \mathbb{N}$ with $m \geq 2$, $\mathbb{R}^m$ be equipped with the usual positive cone, $a := (a_1, \ldots, a_m), b := (b_1, \ldots, b_m) \in \mathbb{R}^m$ with $a \leqq b$, $\mathcal{Y}$ be a Banach space, and $f : \bar{\mathrm{r}}_{a,b} \to \mathcal{Y}$. Assume that:*

(i) *f is bounded, i. e., $\exists M \in \overline{\mathbb{R}_+}$ such that $\|f(x)\|_{\mathcal{Y}} \leq M$, $\forall x \in \bar{\mathrm{r}}_{a,b}$.*
(ii) *$f(\bar{\mathrm{r}}_{a,b}) \subseteq H \subseteq \mathcal{Y}$, where H is a σ-compact conic segment, and f is $\mathcal{B}_{\mathrm{B}}(\mathbb{R}^m)$-measurable.*

Then,

$$\int_{\mathrm{r}_{a,b}} f(x_1,\ldots,x_m)\,\mathrm{d}\mu_{\mathrm{B}m}(x_1,\ldots,x_m)$$
$$= \int_{a_1}^{b_1}\cdots\int_{a_m}^{b_m} f(x_1,\ldots,x_m)\,\mathrm{d}\mu_{\mathrm{B}}(x_m)\cdots\mathrm{d}\mu_{\mathrm{B}}(x_1)$$

In particular, (ii) holds if the following conditions hold for f:

(a) $\bar{\mathrm{r}}_{a,b} = \bigcup_{i=1}^{\infty} C_i$, *where the sets in the union are $\mathcal{B}_{\mathrm{B}}(\mathbb{R}^m)$-measurable.*
(b) $f|_{C_i}$ *is uniformly continuous,* $\forall i \in \mathbb{N}$.
(c) $\overline{C_i}$ *is compact,* $\forall i \in \mathbb{N}$.

Proof By (i) and (ii) and Bounded Convergence Theorem 11.77, f is absolutely integrable over $\bar{\mathrm{r}}_{a,b}$. By Fubini Theorem 12.31, we have

$$\int_{\mathrm{r}_{a,b}} f(x_1,\ldots,x_m)\,\mathrm{d}\mu_{\mathrm{B}m}(x_1,\ldots,x_m)$$
$$= \int_{a_1}^{b_1}\int_{\mathrm{r}_{\bar{a},\bar{b}}} f(x_1,\ldots,x_m)\,\mathrm{d}\mu_{\mathrm{B}m-1}(x_2,\ldots,x_m)\,\mathrm{d}\mu_{\mathrm{B}}(x_1)$$

where $\bar{a} := (a_2,\ldots,a_m)$, $\bar{b} := (b_2,\ldots,b_m)$.

Fix any $x_1 \in \bar{\mathrm{r}}_{a_1,b_1}$. $f(x_1,\cdot) : \bar{\mathrm{r}}_{\bar{a},\bar{b}} \to H \subseteq \mathcal{Y}$ is absolutely integrable over $\bar{\mathrm{r}}_{\bar{a},\bar{b}}$, since it is $\mathcal{B}_{\mathrm{B}}(\mathbb{R}^{m-1})$-measurable, by Proposition 12.28, and by (i), and Bounded Convergence Theorem 11.77 implies that it is absolutely integrable. It is clear that the range of the function is contained in H, which is a σ-compact conic segment. Then, we can apply Fubini Theorem 12.31 repeatedly to arrive at

$$\int_{\mathrm{r}_{a,b}} f(x_1,\ldots,x_m)\,\mathrm{d}\mu_{\mathrm{B}m}(x_1,\ldots,x_m)$$
$$= \int_{a_1}^{b_1}\int_{\mathrm{r}_{\bar{a},\bar{b}}} f(x_1,\ldots,x_m)\,\mathrm{d}\mu_{\mathrm{B}m-1}(x_2,\ldots,x_m)\,\mathrm{d}\mu_{\mathrm{B}}(x_1)$$
$$= \int_{a_1}^{b_1}\cdots\int_{a_m}^{b_m} f(x_1,\ldots,x_m)\,\mathrm{d}\mu_{\mathrm{B}}(x_m)\cdots\mathrm{d}\mu_{\mathrm{B}}(x_1)$$

Let (a)–(c) hold. Note that $\bar{\mathrm{r}}_{a,b} \subseteq \mathbb{R}^m$ is compact. By (a) and (b) and Proposition 11.37, $f|_{C_i}$ is $\mathcal{B}_{\mathrm{B}}(\mathbb{R}^m)$-measurable, $\forall i \in \mathbb{N}$. By Proposition 11.41, f is $\mathcal{B}_{\mathrm{B}}(\mathbb{R}^m)$-measurable. Fix any $i \in \mathbb{N}$. By (b) and Proposition 4.46, there exists a unique continuous $f_i : \overline{C_i} \to \mathcal{Y}$ such that $f_i|_{C_i} = f|_{C_i}$ and f_i is uniformly continuous. Let $K_i := f_i(\overline{C_i}) \subseteq \mathcal{Y}$. K_i is compact by Proposition 5.7. Then, $f(\bar{\mathrm{r}}_{a,b}) = \bigcup_{i=1}^{\infty} f(C_i) \subseteq \bigcup_{i=1}^{\infty} K_i$. Let $K_0 := \{\vartheta_{\mathcal{Y}}\}$, which is clearly compact. By Proposition 12.123, $\bar{K}_i := \{y \in \mathcal{Y} \mid y = \alpha y_1 + (1-\alpha)y_2,\ y_1 \in K_i,\ y_2 \in K_0\}$ is compact. It is straightforward to show that $\bar{K}_i$ is a conic segment. This implies that

$f(\bar{\mathrm{r}}_{a,b}) \subseteq \bigcup_{i=1}^{\infty} K_i \subseteq \bigcup_{i=1}^{\infty} \bar{K}_i =: H$, where the right-hand side is a σ-compact conic segment. Then, (ii) holds.

This completes the proof of the theorem. □

Theorem 12.125 *Let $\mathbb{R}^2$ be equipped with the usual positive cone, $a := (a_1, a_2), b := (b_1, b_2) \in \mathbb{R}^2$ with $a \leqq b$, $\mathcal{X}$ be a separable Banach space over $\mathbb{K}$, $\mathcal{Y}$ be a Banach space over $\mathbb{K}$, and $\mathcal{Z}$ be a Banach space over $\mathbb{K}$, $x : \bar{\mathrm{r}}_{a_2,b_2} \to \mathcal{X}$ be bounded and $\mathcal{B}_{\mathrm{B}}(\mathbb{R})$-measurable, $A : \bar{\mathrm{r}}_{a,b} \to \mathrm{B}(\mathcal{X}, \mathcal{Y})$ be continuous, and $C : \bar{\mathrm{r}}_{a_1,b_1} \to \mathcal{W} \subseteq \mathrm{B}(\mathcal{Y}, \mathcal{Z})$ be bounded and $\mathcal{B}_{\mathrm{B}}(\mathbb{R})$-measurable. Assume that $\mathcal{W}$ is a separable subspace of $\mathrm{B}(\mathcal{Y}, \mathcal{Z})$. Then,*

$$\int_{\mathrm{r}_{a,b}} C(\tau)A(\tau,s)x(s)\,\mathrm{d}\mu_{\mathrm{B}2}(\tau,s) = \int_{a_2}^{b_2}\int_{a_1}^{b_1} C(\tau)A(\tau,s)x(s)\,\mathrm{d}\tau\,\mathrm{d}s \qquad (12.10)$$

$$= \int_{a_1}^{b_1}\int_{a_2}^{b_2} C(\tau)A(\tau,s)x(s)\,\mathrm{d}s\,\mathrm{d}\tau \in \mathcal{Z}$$

Proof Clearly, $\bar{\mathrm{r}}_{a,b}$ is compact. Since $A : \bar{\mathrm{r}}_{a,b} \to \mathrm{B}(\mathcal{X}, \mathcal{Y})$ is continuous, then A is bounded by Propositions 5.7 and 5.38. A is $\mathcal{B}_{\mathrm{B}}(\mathbb{R}^2)$-measurable by Proposition 11.37. By the boundedness of C, A, and x, and Proposition 7.64, CAx is bounded. By the measurability of C, A, and x, and Propositions 11.38 and 11.39, we have CAx is $\mathcal{B}_{\mathrm{B}}(\mathbb{R}^2)$-measurable. By Proposition 7.126, we have the function $(\tau, s) \mapsto C(\tau)A(\tau, s)x(s)$ has the range contained in a separable Banach subspace of $\mathcal{Z}$. By Bounded Convergence Theorem 11.77, CAx is absolutely integrable over $\bar{\mathrm{r}}_{a,b}$, and therefore, the left-hand side of (12.10) is well-defined and is an element of $\mathcal{Z}$.

Fix any $\tau \in \bar{\mathrm{r}}_{a_1,b_1}$, and it is easy to see that $C(\tau)A(\tau, \cdot)x(\cdot) : \bar{\mathrm{r}}_{a_2,b_2} \to \mathcal{Z}$ is bounded and $\mathcal{B}_{\mathrm{B}}(\mathbb{R})$-measurable. Then, this function is absolutely integrable over r_{a_2,b_2}. Note that $\int_{a_2}^{b_2} C(\tau)A(\tau,s)x(s)\,\mathrm{d}s = C(\tau)\int_{a_2}^{b_2} A(\tau,s)x(s)\,\mathrm{d}s = C(\tau)\int_{a_2}^{b_2} A(\tau,s)\,\mathrm{d}\mu(s)$, where the first equality follows from Proposition 11.92 and the second equality follows from the definition $\mu(E) = \int_E x(s)\,\mathrm{d}s$, $\forall E \in \mathcal{B}_{\mathrm{B}}(\mathbb{R})$ with $E \subseteq \bar{\mathrm{r}}_{a_2,b_2}$, which, by Proposition 11.116, is a finite $\mathcal{X}$-valued measure, and Proposition 11.168. By Theorem 12.111, the mapping $\tau \mapsto \int_{a_2}^{b_2} A(\tau,s)\,\mathrm{d}\mu(s) \in \mathcal{Y}$ is continuous. Then, it is $\mathcal{B}_{\mathrm{B}}(\mathbb{R})$-measurable by Proposition 11.37, and its range is contained in a compact and therefore separable subset of $\mathcal{Y}$. By Propositions 11.38 and 11.39, we have $C(\cdot)\int_{a_2}^{b_2} A(\cdot,s)\,\mathrm{d}\mu(s) : \bar{\mathrm{r}}_{a_1,b_1} \to \mathcal{Z}$ is $\mathcal{B}_{\mathrm{B}}(\mathbb{R})$-measurable. By (iii) of Fubini Theorem 12.30, we have

$$\int_{\mathrm{r}_{a,b}} C(\tau)A(\tau,s)x(s)\,\mathrm{d}\mu_{\mathrm{B}2}(\tau,s) = \int_{a_1}^{b_1}\int_{a_2}^{b_2} C(\tau)A(\tau,s)x(s)\,\mathrm{d}s\,\mathrm{d}\tau \in \mathcal{Z}$$

Fix any $s \in \bar{\mathrm{r}}_{a_2,b_2}$, and it is easy to see that $C(\cdot)A(\cdot, s)x(s) : \bar{\mathrm{r}}_{a_1,b_1} \to \mathcal{Z}$ is bounded and $\mathcal{B}_{\mathrm{B}}(\mathbb{R})$-measurable. Then, this function is absolutely integrable over r_{a_1,b_1}. Note that $\int_{a_1}^{b_1} C(\tau)A(\tau,s)x(s)\,\mathrm{d}\tau = \int_{a_1}^{b_1} C(\tau)A(\tau,s)\,\mathrm{d}\tau \cdot$

$x(s) = (\int_{a_1}^{b_1} \mathrm{d}\bar{\mu}(\tau)\, A(\tau, s))x(s)$, where the first equality follows from Proposition 11.92 and the second equality follows from the definition $\bar{\mu}(E) = \int_E C(\tau)\,\mathrm{d}\tau$, $\forall E \in \mathcal{B}_{\mathrm{B}}(\mathbb{R})$ with $E \subseteq \bar{r}_{a_1,b_1}$, which, by Proposition 11.116, is a finite $\mathrm{B}(\mathcal{Y}, \mathcal{Z})$-valued measure, and Proposition 11.168. By Theorem 12.111, the mapping $s \mapsto \int_{a_1}^{b_1} \mathrm{d}\mu(\tau)\, A(\tau, s) \in \mathrm{B}(\mathcal{X}, \mathcal{Z})$ is continuous. Then, it is $\mathcal{B}_{\mathrm{B}}(\mathbb{R})$-measurable by Proposition 11.37, and its range is contained in a compact and therefore separable subset of $\mathrm{B}(\mathcal{X}, \mathcal{Z})$. By Propositions 11.38 and 11.39, we have $(\int_{a_1}^{b_1} \mathrm{d}\bar{\mu}(\tau)\, A(\tau, \cdot))x(\cdot) : \bar{r}_{a_2,b_2} \to \mathcal{Z}$ is $\mathcal{B}_{\mathrm{B}}(\mathbb{R})$-measurable. By (iv) of Fubini Theorem 12.30, we have

$$\int_{r_{a,b}} C(\tau)A(\tau, s)x(s)\,\mathrm{d}\mu_{\mathrm{B}2}(\tau, s) = \int_{a_2}^{b_2} \int_{a_1}^{b_1} C(\tau)A(\tau, s)x(s)\,\mathrm{d}\tau\,\mathrm{d}s \in \mathcal{Z}$$

This completes the proof of the theorem. □

The preceding theorem can be generalized to m-dimensional case.

Theorem 12.126 *Let $m \in \mathbb{N}$ with $m \geq 2$, $\mathbb{R}^m$ be equipped with the usual positive cone, $a := (a_1, \ldots, a_m)$, $b := (b_1, \ldots, b_m) \in \mathbb{R}^m$ with $a \leqq b$, $\mathcal{X}_i$ be a separable Banach space over $\mathbb{K}$, $i = 1, \ldots, m$, $\mathcal{Y}$ be a Banach space over $\mathbb{K}$, $x_i : \bar{r}_{a_i,b_i} \to \mathcal{X}_i$ be bounded and $\mathcal{B}_{\mathrm{B}}(\mathbb{R})$-measurable, $i = 1, \ldots, m$, and $A : \bar{r}_{a,b} \to \mathrm{B}(\mathcal{X}_m, \mathrm{B}(\mathcal{X}_{m-1}, \cdots, \mathrm{B}(\mathcal{X}_1, \mathcal{Y}) \cdots))$ be continuous. Then,*

$$\int_{r_{a,b}} A(s_1, \ldots, s_m)(x_m(s_m)) \cdots (x_1(s_1))\,\mathrm{d}\mu_{\mathrm{B}m}(s_1, \ldots, s_m) \tag{12.11}$$

$$= \int_{a_1}^{b_1} \cdots \int_{a_m}^{b_m} A(s_1, \ldots, s_m)(x_m(s_m)) \cdots (x_1(s_1))\,\mathrm{d}s_m \cdots \mathrm{d}s_1 \in \mathcal{Y}$$

Furthermore, the order of the iterated integration on the right-hand side of the equality can be arbitrary.

Proof Clearly, $\bar{r}_{a,b}$ is compact. By the continuity of A and Propositions 5.7 and 5.38, $A(\bar{r}_{a,b}) =: H \subseteq \mathrm{B}(\mathcal{X}_m, \mathrm{B}(\mathcal{X}_{m-1}, \cdots, \mathrm{B}(\mathcal{X}_1, \mathcal{Y}) \cdots))$ is a compact set and A is bounded. By Proposition 11.37, A is $\mathcal{B}_{\mathrm{B}}(\mathbb{R}^m)$-measurable. By the boundedness of x_i, $i = 1, \ldots, m$, and A, and Proposition 7.64, we have the f is bounded, where $f : \bar{r}_{a,b} \to \mathcal{Y}$ is defined by $f(s_1, \ldots, s_m) = A(s_1, \ldots, s_m) \cdot (x_m(s_m)) \cdots (x_1(s_1))$, $\forall (s_1, \ldots, s_m) \in \bar{r}_{a,b}$. By Proposition 11.190, H is separable. By the measurability of A, and x_i, $i = 1, \ldots, m$, and Propositions 11.38 and 11.39, we have f is $\mathcal{B}_{\mathrm{B}}(\mathbb{R}^m)$-measurable. By Proposition 7.126, we have the function $f(\bar{r}_{a,b})$ is contained in a separable Banach subspace of $\mathcal{Y}$. By Bounded Convergence Theorem 11.77, f is absolutely integrable over $r_{a,b}$, and therefore, the left-hand side of (12.11) is well-defined and is an element of $\mathcal{Y}$.

We will prove the theorem using mathematical induction on $m \in \mathbb{N}$:

1° $m = 1$. This is trivial.

2° Assume that the result holds for $m = k - 1 \in \mathbb{N}$.

3° Consider the case $m = k \in \{2, 3, \ldots\} \subset \mathbb{N}$. Define $\pi_{\widehat{m}} : \mathbb{R}^m \to \mathbb{R}^{m-1}$ by $\pi_{\widehat{m}}(s_1, \ldots, s_m) = (s_1, \ldots, s_{m-1}) \in \mathbb{R}^{m-1}$, $\forall (s_1, \ldots, s_m) \in \mathbb{R}^m$. Fix any $(s_1, \ldots, s_{m-1}) \in \pi_{\widehat{m}}(\bar{\mathrm{r}}_{a,b})$. It is easy to see that $\int_{a_m}^{b_m} f(s_1, \ldots, s_m)(x_m(s_m)) \cdot \mathrm{d}s_m$ is absolutely integrable and integrable and thus well-defined. Define $f_1 : \pi_{\widehat{m}}(\bar{\mathrm{r}}_{a,b}) \to \mathrm{B}(\mathcal{X}_{m-1}, \mathrm{B}(\mathcal{X}_{m-2}, \cdots, \mathrm{B}(\mathcal{X}_1, \mathcal{Y}) \cdots))$, by $f_1(s_1, \ldots, s_{m-1}) = \int_{a_m}^{b_m} f(s_1, \ldots, s_m)(x_m(s_m))\,\mathrm{d}s_m$, $\forall (s_1, \ldots, s_{m-1}) \in \pi_{\widehat{m}}(\bar{\mathrm{r}}_{a,b})$. Then, f_1 is continuous by Theorem 12.111. By the inductive assumption, $\int_{\pi_{\widehat{m}}(\mathrm{r}_{a,b})} f_1(s_1, \ldots, s_{m-1})(x_{m-1}(s_{m-1})) \cdots (x_2(s_2))(x_1(s_1))\,\mathrm{d}\mu_{\mathrm{B}m-1}(s_1, \ldots, s_{m-1})$ is absolutely integrable and integrable. Then,

$$\begin{aligned}
&\int_{\mathrm{r}_{a,b}} A(s_1, \ldots, s_m)(x_m(s_m)) \cdots (x_1(s_1))\,\mathrm{d}\mu_{\mathrm{B}m}(s_1, \ldots, s_m) \\
&= \int_{\pi_{\widehat{m}}(\mathrm{r}_{a,b})} f_1(s_1, \ldots, s_{m-1})(x_{m-1}(s_{m-1})) \cdots (x_2(s_2))(x_1(s_1)) \\
&\quad \cdot \mathrm{d}\mu_{\mathrm{B}m-1}(s_1, \ldots, s_{m-1}) \\
&= \int_{a_1}^{b_1} \cdots \int_{a_{m-1}}^{b_{m-1}} f_1(s_1, \ldots, s_{m-1})(x_{m-1}(s_{m-1})) \cdots (x_2(s_2))(x_1(s_1)) \\
&\quad \cdot \mathrm{d}s_{m-1} \cdots \mathrm{d}s_1 \\
&= \int_{a_1}^{b_1} \cdots \int_{a_m}^{b_m} f(s_1, \ldots, s_m)(x_m(s_m))\,\mathrm{d}s_m\,(x_{m-1}(s_{m-1})) \cdots (x_1(s_1)) \\
&\quad \cdot \mathrm{d}s_{m-1} \cdots \mathrm{d}s_1 \\
&= \int_{a_1}^{b_1} \cdots \int_{a_m}^{b_m} f(s_1, \ldots, s_m)(x_m(s_m))(x_{m-1}(s_{m-1})) \cdots (x_1(s_1)) \\
&\quad \cdot \mathrm{d}s_m \mathrm{d}s_{m-1} \cdots \mathrm{d}s_1 \in \mathcal{Y}
\end{aligned}$$

where the first equality follows from (iii) of Fubini Theorem 12.30, the second equality follows from the inductive assumption, the third equality follows from the definition of f_1, and the last equality follows from Proposition 11.92. This completes the induction process.

Clearly, the order of the iterated integration can be arbitrary by the symmetry in the result.

This completes the proof of the theorem. □

We will now state and prove the following Iterated Integral Theorem.

Theorem 12.127 (Iterated Integral Theorem) *Let $m \in \mathbb{N}$ with $m \geq 2$, $\mathbb{R}^m$ be equipped with the usual positive cone, $a := (a_1, \ldots, a_m)$, $b := (b_1, \ldots, b_m) \in \mathbb{R}^m$ with $a \leqq b$, $\mathcal{X}_i$ be a separable Banach space over $\mathbb{K}$, $i = 1, \ldots, m$, $\mathcal{Y}$ be a Banach space over $\mathbb{K}$, $x_i : \bar{\mathrm{r}}_{a_i,b_i} \to \mathcal{X}_i$ be bounded and $\mathcal{B}_{\mathrm{B}}(\mathbb{R})$-measurable, $i = 1, \ldots, m$, $\Omega \subseteq \bar{\mathrm{r}}_{a,b}$ be a region, and $A : \Omega \to \mathrm{B}(\mathcal{X}_m, \mathrm{B}(\mathcal{X}_{m-1}, \cdots, \mathrm{B}(\mathcal{X}_1, \mathcal{Y}) \cdots))$ be bounded and continuous. Let $\bar{A} : \bar{\mathrm{r}}_{a,b} \to \mathrm{B}(\mathcal{X}_m, \mathrm{B}(\mathcal{X}_{m-1}, \cdots, \mathrm{B}(\mathcal{X}_1, \mathcal{Y}) \cdots))$ be* *such that* $\bar{A}|_{\mathrm{P}(\Omega)} = A$ *and* $\bar{A}(s) = \vartheta_{\mathrm{B}(\mathcal{X}_m,\mathrm{B}(\mathcal{X}_{m-1},\cdots,\mathrm{B}(\mathcal{X}_1,\mathcal{Y})\cdots))}$, $\forall s \in \bar{\mathrm{r}}_{a,b} \setminus \mathrm{P}(\Omega)$. *Then,*

$$\int_{\mathrm{r}_{a,b}} \bar{A}(s_1,\ldots,s_m)(x_m(s_m))\cdots(x_1(s_1))\,\mathrm{d}\mu_{\mathrm{B}m}(s_1,\ldots,s_m) \tag{12.12}$$

$$= \int_{a_1}^{b_1}\cdots\int_{a_m}^{b_m} \bar{A}(s_1,\ldots,s_m)(x_m(s_m))\cdots(x_1(s_1))\,\mathrm{d}s_m\,\cdots\,\mathrm{d}s_1 \in \mathcal{Y}$$

Furthermore, the order of the iterated integration on the right-hand side of the equality can be arbitrary.

Proof By Definition 12.38, there exists $N \subseteq \mathbb{N}$ and $\left(a^{\langle j\rangle}\right)_{j\in N}, \left(b^{\langle j\rangle}\right)_{j\in N} \subseteq \Omega$ with $a^{\langle j\rangle} \stackrel{\lessdot}{=} b^{\langle j\rangle}$, $\forall j \in N$, such that $\Omega = \bigcup_{j\in N} \bar{\mathrm{r}}_{a^{\langle j\rangle},b^{\langle j\rangle}}$, $\left(\mathrm{r}_{a^{\langle j\rangle},b^{\langle j\rangle}}\right)_{j\in N}$ is pairwise disjoint, $\mathrm{P}(\Omega) = \bigcup_{j\in N} \mathrm{r}_{a^{\langle j\rangle},b^{\langle j\rangle}}$, and $\forall \bar{a},\bar{b} \in \Omega$ with $\bar{a} \stackrel{\lessdot}{=} \bar{b}$ and $\bar{\mathrm{r}}_{\bar{a},\bar{b}} \subseteq \Omega$, we have $\mathrm{r}_{\bar{a},\bar{b}} \subseteq \mathrm{P}(\Omega)$. Note that $\Omega \subseteq \bar{\mathrm{r}}_{a,b}$ implies that, by Proposition 12.39, $\mathrm{P}(\Omega) \subseteq \mathrm{P}(\bar{\mathrm{r}}_{a,b}) = \mathrm{r}_{a,b}$.

Fix any $j \in N$. Define $\bar{A}_j : \bar{\mathrm{r}}_{a,b} \to \mathrm{B}(\mathcal{X}_m, \mathrm{B}(\mathcal{X}_{m-1},\cdots,\mathrm{B}(\mathcal{X}_1,\mathcal{Y})\cdots))$ and $A_j : \bar{\mathrm{r}}_{a,b} \to \mathrm{B}(\mathcal{X}_m, \mathrm{B}(\mathcal{X}_{m-1},\cdots,\mathrm{B}(\mathcal{X}_1,\mathcal{Y})\cdots))$ be such that $\bar{A}_j|_{\bar{\mathrm{r}}_{a^{\langle j\rangle},b^{\langle j\rangle}}} = A|_{\bar{\mathrm{r}}_{a^{\langle j\rangle},b^{\langle j\rangle}}}$ and $A_j|_{\mathrm{r}_{a^{\langle j\rangle},b^{\langle j\rangle}}} = A|_{\mathrm{r}_{a^{\langle j\rangle},b^{\langle j\rangle}}}$; and $\bar{A}_j(s) = \vartheta_{\mathrm{B}(\mathcal{X}_m,\mathrm{B}(\mathcal{X}_{m-1},\cdots,\mathrm{B}(\mathcal{X}_1,\mathcal{Y})\cdots))}$, $\forall s \in \bar{\mathrm{r}}_{a,b} \setminus \bar{\mathrm{r}}_{a^{\langle j\rangle},b^{\langle j\rangle}}$; and $A_j(s) = \vartheta_{\mathrm{B}(\mathcal{X}_m,\mathrm{B}(\mathcal{X}_{m-1},\cdots,\mathrm{B}(\mathcal{X}_1,\mathcal{Y})\cdots))}$, $\forall s \in \bar{\mathrm{r}}_{a,b} \setminus \mathrm{r}_{a^{\langle j\rangle},b^{\langle j\rangle}}$. Then,

$$\int_{\mathrm{r}_{a,b}} A_j(s_1,\ldots,s_m)(x_m(s_m))\cdots(x_1(s_1))\,\mathrm{d}\mu_{\mathrm{B}m}(s_1,\ldots,s_m)$$

$$= \int_{\mathrm{r}_{a^{\langle j\rangle},b^{\langle j\rangle}}} A_j(s_1,\ldots,s_m)(x_m(s_m))\cdots(x_1(s_1))\,\mathrm{d}\mu_{\mathrm{B}m}(s_1,\ldots,s_m)$$

$$= \int_{\mathrm{r}_{a^{\langle j\rangle},b^{\langle j\rangle}}} A(s_1,\ldots,s_m)(x_m(s_m))\cdots(x_1(s_1))\,\mathrm{d}\mu_{\mathrm{B}m}(s_1,\ldots,s_m)$$

$$= \int_{a_1^{\langle j\rangle}}^{b_1^{\langle j\rangle}}\cdots\int_{a_m^{\langle j\rangle}}^{b_m^{\langle j\rangle}} A(s_1,\ldots,s_m)(x_m(s_m))\cdots(x_1(s_1))\,\mathrm{d}s_m\,\cdots\,\mathrm{d}s_1$$

$$= \int_{a_1}^{b_1}\cdots\int_{a_m}^{b_m} A_j(s_1,\ldots,s_m)(x_m(s_m))\cdots(x_1(s_1))\,\mathrm{d}s_m\,\cdots\,\mathrm{d}s_1$$

where the first equality follows from the definition of A_j and Proposition 11.92, the second equality follows from the definition of A_j, the third equality follows from Theorem 12.126, $a^{\langle j\rangle} =: (a_1^{\langle j\rangle},\ldots,a_m^{\langle j\rangle})$ and $b^{\langle j\rangle} =: (b_1^{\langle j\rangle},\ldots,b_m^{\langle j\rangle})$ and the fact that $\bar{A}_j|_{\bar{\mathrm{r}}_{a^{\langle j\rangle},b^{\langle j\rangle}}}$ is continuous, and the fourth equality follows from the definitions of A_j and Definition 12.71. This implies that

$$\begin{aligned}
&\int_{\mathrm{r}_{a,b}} \bar{A}(s_1, \ldots, s_m)(x_m(s_m)) \cdots (x_1(s_1))\, \mathrm{d}\mu_{\mathrm{B}m}(s_1, \ldots, s_m) \\
&= \int_{\mathrm{P}(\Omega)} \bar{A}(s_1, \ldots, s_m)(x_m(s_m)) \cdots (x_1(s_1))\, \mathrm{d}\mu_{\mathrm{B}m}(s_1, \ldots, s_m) \\
&= \int_{\mathrm{P}(\Omega)} A(s_1, \ldots, s_m)(x_m(s_m)) \cdots (x_1(s_1))\, \mathrm{d}\mu_{\mathrm{B}m}(s_1, \ldots, s_m) \\
&= \sum_{j\in N} \int_{\mathrm{r}_{a^{(j)},b^{(j)}}} A(s_1, \ldots, s_m)(x_m(s_m)) \cdots (x_1(s_1))\, \mathrm{d}\mu_{\mathrm{B}m}(s_1, \ldots, s_m) \\
&= \sum_{j\in N} \int_{\mathrm{r}_{a,b}} A_j(s_1, \ldots, s_m)(x_m(s_m)) \cdots (x_1(s_1))\, \mathrm{d}\mu_{\mathrm{B}m}(s_1, \ldots, s_m) \\
&= \sum_{j\in N} \int_{a_1}^{b_1} \cdots \int_{a_m}^{b_m} A_j(s_1, \ldots, s_m)(x_m(s_m)) \cdots (x_1(s_1))\, \mathrm{d}s_m \cdots \mathrm{d}s_1 \\
&= \int_{a_1}^{b_1} \cdots \int_{a_m}^{b_m} \bar{A}(s_1, \ldots, s_m)(x_m(s_m)) \cdots (x_1(s_1))\, \mathrm{d}s_m \cdots \mathrm{d}s_1
\end{aligned}$$

where the first equality follows from $\mathrm{P}(\Omega) \subseteq \mathrm{r}_{a,b}$ and the definition of $\bar{A}$, the second equality follows from the definition of $\bar{A}$, the third equality follows from Proposition 11.92, the fourth and fifth equalities follow from the preceding discussion, and the last equality follows from the repeated application of Proposition 11.92 and the Bounded Convergence Theorem 11.77 and the fact that $\sum_{j\in N} A_j(s_1, \ldots, s_m) = \bar{A}(s_1, \ldots, s_m)$, $\forall (s_1, \ldots, s_m) \in \bar{\mathrm{r}}_{a,b}$.

Clearly, the order of the iterated integration can be arbitrary by the symmetry in the result.

This completes the proof of the theorem. □

12.11 Manifold

12.11.1 Basic Notion

Definition 12.128 Let $\mathcal{X}$ be a nonempty Hausdorff topological space and $\mathcal{Y}$ be a Banach space over $\mathbb{K}$. $\mathcal{X}$ is said to be a *manifold of variant* $\mathcal{Y}$ if $\forall x \in \mathcal{X}$, there exists an open set $U \in \mathcal{O}_{\mathcal{X}}$ with $x \in U$ and an open set $V \subseteq \mathcal{Y}$ such that U and V are homeomorphic. For each such open set U and the homeomorphism $\phi : U \to V \subseteq \mathcal{Y}$, the pair (U, ϕ) is said to be a *chart*. Let (U_1, ϕ_1) and (U_2, ϕ_2) be two charts with $U_3 := U_1 \cap U_2 \neq \emptyset$, then $U_3 \in \mathcal{O}_{\mathcal{X}}$ and $\phi_1(U_3) \subseteq \mathcal{Y}$ and $\phi_2(U_3) \subseteq \mathcal{Y}$ are open subsets of $\mathcal{Y}$. Then, $(\phi_2 \circ (\phi_1)_{\mathrm{inv}})\big|_{\phi_1(U_3)} : \phi_1(U_3) \to \phi_2(U_3)$ is a homeomorphism.

Let $\mathcal{A} := ((U_\alpha, \phi_\alpha))_{\alpha \in A}$ be a collection of charts on the manifold $\mathcal{X}$. $\mathcal{A}$ is said to be an *atlas* if $\bigcup_{\alpha \in A} U_\alpha = \mathcal{X}$. ◇

Definition 12.129 Let $\mathcal{Y}$ be a Banach space over $\mathbb{K}$, $\mathcal{X}$ be a manifold of variant $\mathcal{Y}$, and $\mathcal{A} := ((U_\alpha, \phi_\alpha))_{\alpha \in A}$ be an atlas. $\mathcal{A}$ is said to be a $\mathcal{C}_\infty$ atlas if any two charts (U_1, ϕ_1) and (U_2, ϕ_2) in $\mathcal{A}$ with $U_3 := U_1 \cap U_2 \neq \emptyset$, and we have the $(\phi_2 \circ (\phi_1)_{\text{inv}})\big|_{\phi_1(U_3)} : \phi_1(U_3) \to \phi_2(U_3)$ is a $\mathcal{C}_\infty$ *diffeomorphism*, i. e., the bijective mapping together with its inverse mapping are $\mathcal{C}_\infty$. Any two charts in $\mathcal{A}$ satisfying the above assumption are said to be $\mathcal{C}_\infty$ *compatible*. The $\mathcal{C}_\infty$ atlas $\mathcal{A}$ is said to be *complete*, if it is not a proper subset an another $\mathcal{C}_\infty$ atlas. The manifold $\mathcal{X}$ of variant $\mathcal{Y}$ together with a complete $\mathcal{C}_\infty$ atlas $\mathcal{A}_c$ is said to be a *smooth manifold*. ◇

To specify a $\mathcal{C}_\infty$ manifold, one must specify the topological space $\mathcal{X}$, the Banach space $\mathcal{Y}$, and a $\mathcal{C}_\infty$ atlas.

Definition 12.130 Let $\mathcal{Y}$ be a Banach space over $\mathbb{K}$, $\mathcal{X}$ be a manifold of variant $\mathcal{Y}$, and $\mathcal{A} := ((U_\alpha, \phi_\alpha))_{\alpha \in A}$ be an atlas. $\mathcal{A}$ is said to be an analytic atlas if any two charts (U_1, ϕ_1) and (U_2, ϕ_2) in $\mathcal{A}$ with $U_3 := U_1 \cap U_2 \neq \emptyset$, and we have the $(\phi_2 \circ (\phi_1)_{\text{inv}})\big|_{\phi_1(U_3)} : \phi_1(U_3) \to \phi_2(U_3)$ is an *analytic diffeomorphism*, i. e., the bijective mapping together with its inverse mapping are analytic on their domain of definition. Any two charts in $\mathcal{A}$ satisfying the above assumption are said to be *analytically compatible*. The analytic atlas $\mathcal{A}$ is said to be *complete*, if it is not a proper subset an another analytic atlas. The manifold $\mathcal{X}$ of variant $\mathcal{Y}$ together with a complete analytic atlas $\mathcal{A}_c$ is said to be an *analytic manifold*. ◇

To specify an analytic manifold, one must specify the topological space $\mathcal{X}$, the Banach space $\mathcal{Y}$, and an analytic atlas.

Example 12.131 Let $\mathcal{Y}$ be a Banach space over $\mathbb{K}$ and $\mathcal{X} \subseteq \mathcal{Y}$ be an open subset with the subset topology. Clearly, $\mathcal{X}$ is a manifold of variant $\mathcal{Y}$. It is also a smooth manifold with the $\mathcal{C}_\infty$ atlas that contains the singleton atlas of $\{(\mathcal{X}, \text{id}_\mathcal{X})\}$. It is also an analytic manifold with the analytic atlas that contains the singleton atlas of $\{(\mathcal{X}, \text{id}_\mathcal{X})\}$. ◇

Example 12.132 Let $\mathfrak{X}$ and $\mathcal{Y}$ be Banach spaces over $\mathbb{K}$, $U \subseteq \mathfrak{X} \times \mathcal{Y}$ be an open set, and $F : U \to \mathfrak{X}$ be a $\mathcal{C}_\infty$ (analytic) function with $\frac{\partial F}{\partial x} : U \to \mathrm{B}(\mathfrak{X}, \mathfrak{X})$ being such that $\frac{\partial F}{\partial x}(x, y)$ is bijective, $\forall (x, y) \in M$, where the set $M := \{(x, y) \in U \mid F(x, y) = \vartheta_\mathfrak{X}\}$. Then, the set M together with the subset topology of $\mathfrak{X} \times \mathcal{Y}$ is a smooth (analytic) manifold of variant $\mathcal{Y}$.

Since $\frac{\partial F}{\partial x}(x, y)$ is bijective, then it is invertible and the inverse $(\frac{\partial F}{\partial x}(x, y))^{-1} \in \mathrm{B}(\mathfrak{X}, \mathfrak{X})$, by Open Mapping Theorem 7.103. The rest of the proof of this fact follows immediately from the Implicit Function Theorems 9.59 and 9.86. $\forall (x_0, y_0) \in M$, there exists $U_1 \times V_1 \subseteq \mathfrak{X} \times \mathcal{Y}$, which is open and with $(x_0, y_0) \in U_1 \times V_1$ such that $\forall y \in V_1$, $\exists!\, x(y) \in U_1$ that satisfies $F(x(y), y) = \vartheta_\mathfrak{X}$. Thus, $M \cap (U_1 \times V_1)$ is the open subset on M and the homeomorphism is $\phi(x, y) = y$ that maps to the open set $V_1 \subseteq \mathcal{Y}$. The inverse map $\phi_{\text{inv}}(y) = (x(y), y)$. By the Implicit Function Theorems, $x(y)$ is $\mathcal{C}_\infty$ (analytic). Then, this is one of the chart: $(M \cap (U_1 \times V_1), \phi)$.

Collecting all of these charts for M forms a $\mathcal{C}_\infty$ (analytic) atlas. Therefore, M is a smooth (analytic) manifold of variant $\mathcal{Y}$. ◇

Example 12.133 Let $\mathcal{X}$ be a smooth (analytic) manifold of variant $\mathcal{Y}$, and $E \subseteq \mathcal{X}$ be an open subset. Then, E with the subset topology is a smooth (analytic) manifold of variant $\mathcal{Y}$.

The proof of this fact is as follows. For any chart (U, ϕ) in the complete $\mathcal{C}_\infty$ (analytic) atlas of $\mathcal{X}$, we just take $\check{U} := U \cap E$ and $\check{\phi} := \phi|_{\check{U}}$. Then, take $(\check{U}, \check{\phi})$ as one of the charts for manifold E, if $\check{U} \neq \emptyset$. This defines a $\mathcal{C}_\infty$ (analytic) atlas. Hence, the result follows. ◇

Example 12.134 Let $\mathcal{Y}_i$ be a Banach space over $\mathbb{K}$, $i = 1, 2$, $\mathcal{X}_i$ be a smooth (analytic) manifold of variant $\mathcal{Y}_i$, $i = 1, 2$, and then $\mathcal{X}_1 \times \mathcal{X}_2$ is a smooth (analytic) manifold of variant $\mathcal{Y}_1 \times \mathcal{Y}_2$.

The proof of this fact is as follows. For any chart (U, ϕ) of $\mathcal{X}_1$ and chart (V, ψ) of $\mathcal{X}_2$, then $(U \times V, \text{block diagonal}\,(\phi, \psi))$ is a chart for $\mathcal{X}_1 \times \mathcal{X}_2$. Hence, the result follows. ◇

Definition 12.135 Let $\mathcal{Y}$ and $\mathcal{Z}$ be Banach spaces over $\mathbb{K}$, and M and N be two smooth (analytic) manifolds of variants $\mathcal{Y}$ and $\mathcal{Z}$, respectively, and $F : M \to N$. F is said to be $\mathcal{C}_\infty$*or smooth (analytic)* if $\forall p \in M$, there exists chart of M (among its complete atlas) (U, ϕ) with $p \in U$ and exists chart of N (among its complete atlas) (V, ψ) with $F(p) \in V$, then the expression of F in the local coordinates is $\mathcal{C}_\infty$ (analytic), where the *expression of* F *in local coordinates* is the mapping $\hat{F} := (\psi \circ F \circ \phi_{\text{inv}})|_{\phi(F_{\text{inv}}(V) \cap U)}$. ◇

Clearly, the fact that a mapping F is smooth (analytic) is a property that is independent of the choice of the coordinates charts.

Definition 12.136 Let $\mathcal{Y}$ be a Banach space over $\mathbb{K}$, and M and N be two smooth (analytic) manifolds of variants $\mathcal{Y}$, and $F : M \to N$. F is to be a $\mathcal{C}_\infty$ *(analytic) diffeomorphism* if F is bijective and both F and F_{inv} are $\mathcal{C}_\infty$ (analytic) mappings. Two smooth (analytic) manifolds M and N of the same variant are said to be $\mathcal{C}_\infty$ *(analytically) diffeomorphic* if there exists a $\mathcal{C}_\infty$ (analytic) diffeomorphism $F : M \to N$. ◇

Theorem 12.137 *Let $\mathcal{Y}$ be a Banach space over $\mathbb{K}$, M and N be smooth (analytic) manifolds of variant $\mathcal{Y}$, and $F : M \to N$ be a smooth (analytic) mapping. Then, F is a smooth (analytic) diffeomorphism if, and only if, F is bijective and $\mathrm{D}\hat{F}(\phi(p)) \in \mathrm{B}(\mathcal{Y}, \mathcal{Y})$ is bijective, $\forall p \in M$, where $\hat{F}$ is the expression of F in local coordinates and (U, ϕ) is a chart in the complete $\mathcal{C}_\infty$ (analytic) atlas of M with $p \in U$.*

Proof "Sufficiency" Let $\mathcal{A}_M$ and $\mathcal{A}_N$ be the complete $\mathcal{C}_\infty$ (analytic) atlas for M and N, respectively. Fix any $p \in M$. Then, there exists a chart $(U, \phi) \in \mathcal{A}_M$ with $p \in U$ and another chart $(V, \psi) \in \mathcal{A}_N$ with $F(p) \in V$. The expression of F in local coordinates is $\hat{F} := (\psi \circ F \circ \phi_{\text{inv}})|_{\phi(F_{\text{inv}}(V) \cap U)}$. By the assumption, $\mathrm{D}\hat{F}(\phi(p)) \in \mathrm{B}(\mathcal{Y}, \mathcal{Y})$ is bijective. By Open Mapping Theorem 7.103, $\mathrm{D}\hat{F}(\phi(p))$ is invertible and $(\mathrm{D}\hat{F}(\phi(p)))^{-1} \in \mathrm{B}(\mathcal{Y}, \mathcal{Y})$. By Inverse Function Theorems 9.57

and 9.85, there exists an open subset $\hat{U}_p \subseteq \phi(F_{\text{inv}}(V) \cap U) \subseteq \mathcal{Y}$ with $\phi(p) \in \hat{U}_p$ such that $\hat{F}\big|_{\hat{U}_p}$ is $\mathcal{C}_\infty$ (analytic) diffeomorphism between $\hat{U}_p$ and $\hat{F}(\hat{U}_p)$. Then, $U_p := \phi_{\text{inv}}(\hat{U}_p)$ is an open subset of M and $V_p := \psi_{\text{inv}}(\hat{F}(\hat{U}_p))$ is an open subset of N and $F|_{U_p} : U_p \to V_p$ is bijective and smooth (analytic) with smooth (analytic) inverse. Since F is bijective, then $F_{\text{inv}} : N \to M$ is well-defined. $\forall q \in N$, $(F_{\text{inv}})|_{V_{F_{\text{inv}}(q)}} = (F|_{U_{F_{\text{inv}}(q)}})_{\text{inv}}$ and is smooth (analytic) by our preceding discussion. Hence, F_{inv} is a smooth (analytic) mapping. Then, F is a smooth (analytic) diffeomorphism. This completes the sufficiency part of the proof.

"Necessity" Let F be a smooth (analytic) diffeomorphism between M and N. Then, F is bijective and F_{inv} is smooth (analytic). Then, $\text{id}_M = F_{\text{inv}} \circ F$. $\forall p \in M$, there exists a chart $(U, \phi) \in \mathcal{A}_M$ with $p \in U$, and a chart $(V, \psi) \in \mathcal{A}_N$ with $F(p) \in V$ such that the expression of F in local coordinates $\hat{F} : \phi(U) \to \psi(V)$ is bijective and smooth (analytic) with smooth (analytic) inverse. Then, $\text{id}_{\phi(U)} = \hat{F}_{\text{inv}} \circ \hat{F}$ and $\text{id}_{\psi(V)} = \hat{F} \circ \hat{F}_{\text{inv}}$. By Chain Rule (Theorem 9.18), we have $\text{id}_{\mathcal{Y}} = D(\hat{F}_{\text{inv}})(\hat{F}(y))D\hat{F}(y), \forall y \in \phi(U)$; and $\text{id}_{\mathcal{Y}} = D\hat{F}(\hat{F}_{\text{inv}}(z))D(\hat{F}_{\text{inv}})(z), \forall y \in \psi(V)$. Let $z = \hat{F}(y)$, and by Proposition 2.4, we have $D(\hat{F}_{\text{inv}})(\hat{F}(y)) = (D\hat{F}(y))^{-1}$, $\forall y \in \phi(U)$. Hence, $D\hat{F}(\phi(p)) \in B(\mathcal{Y}, \mathcal{Y})$ and is bijective. This completes the necessity part of the proof.

This completes the proof of the theorem. □

We are particularly interested in smooth (analytic) manifolds of variant $\mathbb{K}^n$.

Theorem 12.138 *Let $n \in \mathbb{Z}_+$, M and N be smooth (analytic) manifolds of variant $\mathbb{K}^n$, and $F : M \to N$ be a smooth (analytic) mapping. Then, F is a smooth (analytic) diffeomorphism if, and only if, F is bijective and* $\text{rank}(D\hat{F}(\phi(p))) = n$, *$\forall p \in M$, where $\hat{F}$ is the expression of F in local coordinates.*

Proof This is immediate from Theorem 12.137. □

Example 12.139 We present here some results on the logarithm on the complex plane. Note that the exponential function (with base e) maps the complex plane $\mathbb{C}$ to $\mathbb{C} \setminus \{0\} =: \mathbb{C}_0$. But, it is not injective, but actually periodic with period $i2\pi$. If we restrict the domain of the exponential function to the strip $E_{\text{exp}} := \{a + ib \in \mathbb{C} \mid -\pi < b \leq \pi\}$, then $\exp|_{E_{\text{exp}}} : E_{\text{exp}} \to \mathbb{C}_0$ is bijective. Define $\text{Ln} : \mathbb{C}_0 \to E_{\text{exp}}$ by $\text{Ln} = (\exp|_{E_{\text{exp}}})_{\text{inv}}$. By the Inverse Function Theorem, Ln is analytic at $\forall z_0 \in \mathbb{C}_0$, and $\text{Ln}^{(1)}(z_0) = \frac{1}{z_0}$. The natural logarithm function $\ln : \mathbb{C}_0 \to \mathbb{C}$ is a multi-valued function that takes value $\text{Ln}(x) + i2k\pi$, $k \in \mathbb{Z}$, depending on the restriction of range of the function. We will use the concept of manifold to resolve the ln function in this example.

Let $\mathcal{X}$ denote the surface of an infinite long cylinder with radius 1. We parameterize $\mathcal{X}$ by the cylindrical polar coordinates: (r, θ, h), where r denotes the radius of projection of the vector to the x-y plane; θ denotes the angle from the positive x-axis direction of the projection of the vector to the x-y plane; and h is the height of the vector from the x-y plane. Then, for the infinite long cylinder, we have $r = 1, \theta \in \mathrm{r}_{-\pi,\pi}$, $h \in \mathbb{R}$. The topology of $\mathcal{X}$ is that of the subset topology it inherits from $\mathbb{R}^3$. On $\mathcal{X}$, we will introduce an analytic atlas that makes it a analytic manifold of variant $\mathbb{C}$. The atlas has two charts: (U_1, ϕ_1) and (U_2, ϕ_2), where $U_1 := \{(1, \theta, h) \in \mathcal{X} \mid h \in \mathbb{R} \text{ and } \theta \in r^{\circ}_{-\pi,\pi}\}$, $\phi_1 : U_1 \to V_1 \subseteq \mathbb{C}$ is defined by $\phi_1(1, \theta, h) := h+i\theta$, $\forall (1, \theta, h) \in \mathcal{X}$, and $V_1 := \{a+ib \in \mathbb{C} \mid a \in \mathbb{R} \text{ and } b \in r^{\circ}_{-\pi,\pi}\}$; and $U_2 := \{(1, \theta, h) \in \mathcal{X} \mid h \in \mathbb{R} \text{ and } \theta \in r_{0,\pi} \cup r^{\circ}_{-\pi,0}\}$, $\phi_2 : U_2 \to V_2 \subseteq \mathbb{C}$ is defined by $\phi_2(1, \theta, h) := \begin{cases} h + i\theta & \theta \geq 0 \\ h + i\theta + i2\pi & \theta < 0 \end{cases}$, $\forall (1, \theta, h) \in \mathcal{X}$, and $V_2 := \{a + ib \in \mathbb{C} \mid a \in \mathbb{R} \text{ and } b \in r^{\circ}_{0,2\pi}\}$. It is easy to check that two charts are analytically compatible. Then, $\mathcal{X}$ together with the atlas $\mathcal{A} := \{(U_1, \phi_1), (U_2, \phi_2)\}$ defines an analytic manifold of variant $\mathbb{C}$. Define $\boxplus : \mathbb{R} \times \mathbb{R} \to r_{-\pi,\pi}$ by $\theta_1 \boxplus \theta_2 = \theta_1 + \theta_2 - 2k\pi$, where $k \in \mathbb{Z}$ is such that $\theta_1 + \theta_2 - 2k\pi \in r_{-\pi,\pi}$. Define on the manifold $\mathcal{X}$ an operation $\oplus$ by $(1, \theta_1, h_1) \oplus (1, \theta_2, h_2) := (1, \theta_1 \boxplus \theta_2, h_1 + h_2)$, $\forall (1, \theta_1, h_1), (1, \theta_2, h_2) \in \mathcal{X}$. Clearly, $\oplus$ is an analytic function on the product manifold $\mathcal{X} \times \mathcal{X}$, and $\oplus$ is commutative and associative. Here, what we have done is just roll the complex plane on to the cylinder with the real axis of the complex plane identifying with the line $\{(1, 0, h) \in \mathcal{X} \mid h \in \mathbb{R}\}$ on the cylinder and identifying whatever overlapped points of the complex plane to the point on the cylinder.

Define the function $\overline{\exp} : \mathcal{X} \to \mathbb{C}_0$ by $\overline{\exp}(1, \theta, h) = \exp(h + i\theta)$, $\forall (1, \theta, h) \in \mathcal{X}$. On the chart (U_1, ϕ_1), the function $\overline{\exp}$ admits expression $\exp : V_1 \to \mathbb{C}_0$. On the chart (U_2, ϕ_2), the function $\overline{\exp}$ admits expression $\exp : V_2 \to \mathbb{C}_0$. Then, $\overline{\exp}$ is an analytic function on the manifold $\mathcal{X}$ of variant $\mathbb{C}$ with atlas $\mathcal{A}$. Clearly, $\overline{\exp} : \mathcal{X} \to \mathbb{C}_0$ is bijective whose derivative (of the expression in local coordinates) is nonzero $\forall p \in \mathcal{X}$. By Theorem 12.138, $\overline{\exp} : \mathcal{X} \to \mathbb{C}_0$ is an analytic diffeomorphism and admits an analytic inverse denoted by $\overline{\ln}$. On chart (U_1, ϕ_1), the function $\overline{\ln}$ admits expression $\ln : \mathbb{C}_0 \setminus \{a + ib \in C \mid a < 0, b = 0\} \to V_1$. On the chart (U_2, ϕ_2), the function $\overline{\ln}$ admits expression $\ln : \mathbb{C}_0 \setminus \{a + ib \in C \mid a > 0, b = 0\} \to V_2$. Here, ln's are the usual multi-valued function, but they are now standard functions since the ranges of the function are limited to sets where it has only one value.

On $\mathcal{X}$, we define the metric $\rho : \mathcal{X} \times \mathcal{X} \to [0, \infty) \subset \mathbb{R}$ by $\rho(p_1, p_2) = \rho((1, \theta_1, h_1), (1, \theta_2, h_2)) = \left((h_1 - h_2)^2 + \min\{(|\theta_1 - \theta_2| - 2\pi)^2, |\theta_1 - \theta_2|^2\}\right)^{\frac{1}{2}}$, $\forall p_1, p_2 \in \mathcal{X}$. It is easy to show that ρ defines a metric on $\mathcal{X}$ and its induced topology is the same as the subset topology inherited from $\mathbb{R}^3$. ◇

Proposition 12.140 *Let $x, y \in \mathbb{C}_0$ and $z_1, z_2, z \in \mathcal{X}$, where $\mathcal{X}$ is the analytic manifold of variant $\mathbb{C}$ as defined in Example 12.139. Using the notations introduced in Example 12.139, we have:*

(i) $\overline{\exp}(\overline{\ln}(x)) = x$.
(ii) $\overline{\ln}(\overline{\exp}(z)) = z$.
(iii) $\overline{\exp}(z_1 \oplus z_2) = \overline{\exp}(z_1)\overline{\exp}(z_2)$.
(iv) $\overline{\ln}(x) \oplus \overline{\ln}(y) = \overline{\ln}(xy)$.
(v) $\rho(z_1 \oplus z_2, p_0) \leq \rho(z_1, p_0) + \rho(z_2, p_0)$, *where* $p_0 := (1, 0, 0) \in \mathcal{X}$.
(vi) $\rho(z_1, z_2) \leq |\phi_1(z_1) - \phi_1(z_2)|$, $\forall z_1, z_2 \in U_1$, *where* (U_1, ϕ_1) *is the chart defined in Example 12.139; and* $\rho(z_1, z_2) \leq |\phi_2(z_1) - \phi_2(z_2)|$, $\forall z_1, z_2 \in U_2$, *where* (U_2, ϕ_2) *is the chart defined in Example 12.139.*
(vii) $\rho(\overline{\ln}(1 + w) \oplus \overline{\ln}(\exp(-w)), p_0) < 2|w|^2$, $\forall w \in \mathcal{B}_{\mathbb{C}}(0, \frac{1}{2})$.

Proof (i) and (ii) These are direct consequences of Example 12.139.

(iii) By Example 12.139, we have $\overline{\exp}(z_1 \oplus z_2) = \overline{\exp}(z_1)\overline{\exp}(z_2)$.

(iv) Let $p_1 := \overline{\ln}(x) \in \mathcal{X}$ and $p_2 := \overline{\ln}(y) \in \mathcal{X}$, and we have, by (i), $\overline{\exp}(p_1) = x$ and $\overline{\exp}(p_2) = y$, which implies that $\overline{\exp}(p_1 \oplus p_2) = \overline{\exp}(p_1)\overline{\exp}(p_2) = xy$, where the first equality follows from (iii). Thus, by (ii), we have $p_1 \oplus p_2 = \overline{\ln}(xy)$.

(v) Let $z_i := (1, \theta_i, h_i)$, $i = 1, 2$, and $\bar{z}_2 := (1, -\theta_2, -h_2) \in \mathcal{X}$. Then, $\rho(z_1 \oplus z_2, p_0) = ((h_1 + h_2)^2 + \min\{(\theta_1 \boxplus \theta_2)^2, (|\theta_1 \boxplus \theta_2| - 2\pi)^2\})^{\frac{1}{2}} = ((h_1 + h_2)^2 + (\theta_1 \boxplus \theta_2)^2)^{\frac{1}{2}} \leq ((h_1 + h_2)^2 + \min\{(\theta_1 + \theta_2)^2, (|\theta_1 + \theta_2| - 2\pi)^2\})^{\frac{1}{2}} = \rho(z_1, \bar{z}_2) \leq \rho(z_1, p_0) + \rho(p_0, \bar{z}_2) = \rho(z_1, p_0) + \rho(p_0, z_2)$, where the second inequality follows from the triangular inequality and the rest of the steps follow from Example 12.139.

(vi) $\forall z_1, z_2 \in U_1$, let $z_i := (1, \theta_i, h_i)$, $i = 1, 2$. $\rho(z_1, z_2) = \big((h_1 - h_2)^2 + \min\{(|\theta_1 - \theta_2| - 2\pi)^2, |\theta_1 - \theta_2|^2\}\big)^{\frac{1}{2}} \leq \big((h_1 - h_2)^2 + |\theta_1 - \theta_2|^2\big)^{\frac{1}{2}} = |h_1 + \mathrm{i}\theta_1 - (h_2 + \mathrm{i}\theta_2)| = |\phi_1(z_1) - \phi_1(z_2)|$.

$\forall z_1, z_2 \in U_2$, let $z_i := (1, \theta_i, h_i)$, $i = 1, 2$. $\rho(z_1, z_2) = \big((h_1 - h_2)^2 + \min\{(|\theta_1 - \theta_2| - 2\pi)^2, |\theta_1 - \theta_2|^2\}\big)^{\frac{1}{2}} \leq |\phi_2(z_1) - \phi_2(z_2)|$.

(vii) $\forall w \in \mathcal{B}_{\mathbb{C}}(0, \frac{1}{2})$, we have $\rho(\overline{\ln}(1 + w) \oplus \overline{\ln}(\exp(-w)), p_0) \leq \big|\phi_1(\overline{\ln}(1 + w) \oplus \overline{\ln}(\exp(-w))) - \phi_1(p_0)\big| = \big|\phi_1(\overline{\ln}((1 + w)\exp(-w))) - 0\big| = \big|\phi_1(\overline{\ln}((1 + w)\exp(-w)))\big| = |\ln((1 + w)\exp(-w))| = |\ln(1 + w) - w| \leq 2|w|^2$, where the first inequality follows from (vi), the first equality follows from (iv), the third equality follows from Example 12.139 and the fact that $\overline{\ln}$ admits the expression ln on the chart (U_1, ϕ_1), the fourth equality follows from the property of ln and the fact that it is single valued on the chart (U_1, ϕ_1), and the last inequality follows from Taylor's Theorem 9.48.

This completes the proof of the proposition. □

12.11.2 Tangent Vectors

Starting in this section, we will not pursue results on analytic manifold. As one can see, the results on smooth manifold have ramifications to analytic manifold, which we shall leave it to the reader to pursue these results.

Definition 12.141 Let $\mathcal{Y}$ be a reflexive Banach space over $\mathbb{K}$, $\mathcal{X}$ be a smooth manifold of variant $\mathcal{Y}$, and $p \in \mathcal{X}$. Let $t := (U, \phi)$ be a chart in the complete $\mathcal{C}_\infty$ atlas of $\mathcal{X}$, denoted by $\mathcal{A}_{\mathcal{X}}$, with $p \in U$. $\lambda : U \to \mathbb{K}$ be a functional such that $\lambda \circ \phi_{\mathrm{inv}} : \phi(U) \to \mathbb{K}$ is $\mathcal{C}_\infty$. Such a functional λ is said to be *smooth in a neighborhood of* p. The collection of all functionals that are smooth in a neighborhood of p is denoted by $\mathcal{C}_\infty(p, \mathbb{K})$. ◇

Note that $\mathcal{C}_\infty(p, \mathbb{K})$ forms a vector space over $\mathbb{K}$ and further satisfies that, any two functionals in $\mathcal{C}_\infty(p, \mathbb{K})$, their product is also a member of $\mathcal{C}_\infty(p, \mathbb{K})$.

Definition 12.142 Let $\mathcal{Y}$ be a reflexive Banach space over $\mathbb{K}$, $\mathcal{X}$ be a smooth manifold of variant $\mathcal{Y}$, and $p \in \mathcal{X}$. A *tangent vector* v *at* p is an assignment of a vector $v_t \in \mathcal{Y}$ to every chart $t := (U, \phi)$ in the complete smooth atlas of $\mathcal{X}$, denoted by $\mathcal{A}_{\mathcal{X}}$, with $p \in U$, such that $\forall h \in \mathcal{C}_\infty(p, \mathbb{K})$, $v(h) \in \mathbb{K}$, that satisfies:

(i) $v(h) = \mathrm{D}(h \circ \phi_{\mathrm{inv}})(\phi(p))(v_t)$.
(ii) $v_{t_1} = \mathrm{D}(\phi_1 \circ (\phi_2)_{\mathrm{inv}})(\phi_2(p))(v_{t_2})$, $\forall t_1 := (U_1, \phi_1), t_2 := (U_2, \phi_2) \in \mathcal{A}_{\mathcal{X}}$ with $p \in U_1 \cap U_2$.
(iii) $v(\alpha h_1 + \beta h_2) = \alpha v(h_1) + \beta v(h_2)$, $\forall \alpha, \beta \in \mathbb{K}$ and $\forall h_1, h_2 \in \mathcal{C}_\infty(\mathcal{X}, \mathbb{K})$.
(iv) $(\alpha v_1 + \beta v_2)(h) = \alpha v_1(h) + \beta v_2(h)$, $\forall \alpha, \beta \in \mathbb{K}$, and for any two tangent vectors v_1 and v_2 at p.
(v) $v(h_1 h_2) = v(h_1) h_2(p) + v(h_2) h_1(p)$, $\forall h_1, h_2 \in \mathcal{C}_\infty(\mathcal{X}, \mathbb{K})$ (Leibniz rule).

The collection of all tangent vectors at $p \in \mathcal{X}$ is the *tangent space to $\mathcal{X}$ at p* and denoted by $\mathsf{T}_p\mathcal{X}$. ◇

Note that $\mathsf{T}_p\mathcal{X}$ is a vector space over $\mathbb{K}$ that is isomorphic to $\mathcal{Y}$.

Definition 12.143 Let $\mathcal{Y}$ and $\mathcal{Z}$ be reflexive Banach spaces over $\mathbb{K}$, $\mathcal{X}_1$ be a smooth manifold of variant $\mathcal{Y}$, and $p \in \mathcal{X}_1$, $\mathcal{X}_2$ be a smooth manifold of variant $\mathcal{Z}$, and $F : \mathcal{X}_1 \to \mathcal{X}_2$ be a smooth mapping. The *differential* of F at p is a mapping $F_{\star p} : \mathsf{T}_p\mathcal{X}_1 \to \mathsf{T}_{F(p)}\mathcal{X}_2$ defined by, $\forall v \in \mathsf{T}_p\mathcal{X}_1$, $\forall \lambda \in \mathcal{C}_\infty(F(p), \mathbb{K})$, we have $(F_{\star p}(v))(\lambda) = v(\lambda \circ F)$. Let $t_1 := (U, \phi) \in \mathcal{A}_{\mathcal{X}_1}$ be a chart at p and $t_2 := (V, \psi) \in \mathcal{A}_{\mathcal{X}_2}$ be a chart at $F(p)$; then, we have the expression of $F_{\star p}$ in local coordinates to be $\mathrm{D}(\psi \circ F \circ \phi_{\mathrm{inv}})(\phi(p))(v_{t_1})$, $\forall v_{t_1} \in \mathcal{Y}$. Then, $F_{\star p}$ is well-defined. ◇

Theorem 12.144 *Let $\mathcal{Y}$ and $\mathcal{Z}$ be reflexive Banach spaces over $\mathbb{K}$, $\mathcal{X}_1$ be a smooth manifold of variant $\mathcal{Y}$, and $p \in \mathcal{X}_1$, $\mathcal{X}_2$ be a smooth manifold of variant $\mathcal{Z}$, and $F : \mathcal{X}_1 \to \mathcal{X}_2$ be a smooth mapping. Then differential of F at p, $F_{\star p} : \mathsf{T}_p\mathcal{X}_1 \to \mathsf{T}_{F(p)}\mathcal{X}_2$ is a linear map.*

Proof $\forall \alpha, \beta \in \mathbb{K}$, $\forall v_1, v_2 \in \mathsf{T}_p\mathcal{X}_1$, $\forall \lambda \in \mathcal{C}_\infty(F(p), \mathbb{K})$, we have $(F_{\star p}(\alpha v_1 + \alpha v_2))(\lambda) = (\alpha v_1 + \beta v_2)(\lambda \circ F) = \alpha v_1(\lambda \circ F) + \beta v_2(\lambda \circ F) = \alpha(F_{\star p}(v_1))(\lambda) + \beta(F_{\star p}(v_2))(\lambda) = (\alpha F_{\star p}(v_1) + \beta F_{\star p}(v_2))(\lambda)$, where the first equality follows from Definition 12.143, the second equality follows from Definition 12.142, the third equality follows from Definition 12.143, and the last equality follows from Definition 12.142. By the arbitrariness of λ, we have $F_{\star p}(\alpha v_1 + \alpha v_2) = \alpha F_{\star p}(v_1) + \beta F_{\star p}(v_2)$. This completes the proof of the theorem. □

Theorem 12.145 (Chain Rule) *Let $\mathcal{Y}_i$ be a reflexive Banach space over $\mathbb{K}$, $\mathcal{X}_i$ be a smooth manifold of variant $\mathcal{Y}_i$, $i = 1, 2, 3$, and $p \in \mathcal{X}_1$, $F : \mathcal{X}_1 \to \mathcal{X}_2$ be a smooth mapping, and $G : \mathcal{X}_2 \to \mathcal{X}_3$ be a smooth mapping. Then we have $(G \circ F)_{\star p} = G_{\star F(p)} \circ F_{\star p}$.*

Proof $\forall v \in \mathsf{T}_p\mathcal{X}_1$, $\forall \lambda \in \mathcal{C}_\infty(G(F(p)), \mathbb{K})$, we have $((G \circ F)_{\star p}(v))(\lambda) = v(\lambda \circ G \circ F) = (F_{\star p}(v))(\lambda \circ G) = (G_{\star F(p)}(F_{\star p}(v)))(\lambda)$. By the arbitrariness of λ, we have $(G \circ F)_{\star p}(v) = G_{\star F(p)}(F_{\star p}(v))$. By the arbitrariness of v, we have $(G \circ F)_{\star p} = G_{\star F(p)} \circ F_{\star p}$. This completes the proof of the theorem. □

Example 12.146 Let $\mathcal{Y}$ be a reflexive Banach space over $\mathbb{K}$, $\Omega \subseteq \mathbb{K}$ be an open set, $\mathcal{X}$ be a smooth manifold of variant $\mathcal{Y}$, $\sigma : \Omega \to \mathcal{X}$ be a smooth mapping, and $t_0 \in \Omega$. Then, $\sigma_{\star t_0}(\frac{\mathrm{d}}{\mathrm{d}t}) \in \mathsf{T}_{\sigma(t_0)}\mathcal{X}$ satisfies, $\forall \lambda \in \mathcal{C}_\infty(\sigma(t_0), \mathbb{K})$, we have $(\sigma_{\star t_0}(\frac{\mathrm{d}}{\mathrm{d}t}))(\lambda) = \frac{\mathrm{d}}{\mathrm{d}t}(\lambda \circ \sigma)(t_0)$. ◇

Definition 12.147 Let $\mathcal{Y}$ be a reflexive Banach space over $\mathbb{K}$, $\mathcal{X}$ be a smooth manifold of variant $\mathcal{Y}$, and $p \in \mathcal{X}$. A *tangent covector* v_* *at* p is an assignment of a vector $v_{*t} \in \mathcal{Y}^*$ to every chart $t := (U, \phi)$ in the complete smooth atlas of $\mathcal{X}$, denoted by $\mathcal{A}_\mathcal{X}$, with $p \in U$, such that $\forall v \in \mathsf{T}_p\mathcal{X}$, $v_*(v) \in \mathbb{K}$, that satisfies:

(i) $v_*(\alpha v_1 + \beta v_2) = \alpha v_*(v_1) + \beta v_*(v_2)$, $\forall \alpha, \beta \in \mathbb{K}$ and $\forall v_1, v_2 \in \mathsf{T}_p\mathcal{X}$.
(ii) $(\alpha v_{*1} + \beta v_{*2})(v) = \alpha v_{*1}(v) + \beta v_{*2}(v)$, $\forall \alpha, \beta \in \mathbb{K}$, and for any two tangent covectors v_{*1} and v_{*2} at p.
(iii) $v_{*t_1} = v_{*t_2} \circ (\mathrm{D}(\phi_2 \circ (\phi_1)_{\mathrm{inv}})(\phi_1(p))) = (\mathrm{D}(\phi_2 \circ (\phi_1)_{\mathrm{inv}})(\phi_1(p)))' v_{*t_2}$, $\forall t_1 := (U_1, \phi_1), t_2 := (U_2, \phi_2) \in \mathcal{A}_\mathcal{X}$ with $p \in U_1 \cap U_2$.

The collection of all tangent covectors at $p \in \mathcal{X}$ is the *cotangent space to* $\mathcal{X}$ *at* p and denoted by $\mathsf{T}_p^*\mathcal{X}$.

If there exists $h \in \mathcal{C}_\infty(p, \mathbb{K})$ such that $v_*(v) = v(h)$, $\forall v \in \mathsf{T}_p\mathcal{X}$, then $v_{*t} := \mathrm{D}(h \circ \phi_{\mathrm{inv}})(\phi(p)) \in \mathcal{Y}^*$, and v_* is denoted by $(\mathrm{d}h)_p$. ◇

Note that $\mathsf{T}_p^*\mathcal{X}$ is a vector space over $\mathbb{K}$ that is isomorphic to $\mathcal{Y}^*$.

12.11.3 Vector Fields

Definition 12.148 Let $\mathcal{Y}$ be a reflexive Banach space over $\mathbb{K}$ and $\mathcal{X}$ be a smooth manifold of variant $\mathcal{Y}$. A *vector field* f is a mapping assigning to each point $p \in \mathcal{X}$ a tangent vector $f(p) \in \mathsf{T}_p\mathcal{X}$. A vector field is said to be *smooth* if, $\forall p \in \mathcal{X}$, there exists a chart $t := (U, \phi) \in \mathcal{A}_\mathcal{X}$ with $p \in U$, such that, on the chart t, the expression of f in local coordinates $\phi(\bar{p}) \in \phi(U) \subseteq \mathcal{Y} \mapsto (f(\bar{p}))_t \in \mathcal{Y}$ is a $\mathcal{C}_\infty$ map. Necessarily, $\forall \lambda \in \mathcal{C}_\infty(\mathcal{X}, \mathbb{K})$, the function $h : \mathcal{X} \to \mathbb{K}$ defined by $h(p) = f(p)(\lambda)$, $\forall p \in \mathcal{X}$, is $\mathcal{C}_\infty$. This function h is said to be the *Lie derivative of* λ *along the vector field* f, and denoted by $\mathsf{L}_f\lambda$. The set of all smooth vector fields on $\mathcal{X}$ is denoted by $\mathsf{V}(\mathcal{X})$. ◇

$\mathsf{V}(\mathcal{X})$ is a vector space over $\mathbb{K}$, since $\forall f, g \in \mathsf{V}(\mathcal{X})$, $\forall \alpha, \beta \in \mathbb{K}$, we have $\alpha f + \beta g$ is the vector field defined by $(\alpha f + \beta g)(p) = \alpha f(p) + \beta g(p)$, $\forall p \in \mathcal{X}$, which is clearly smooth. $\forall a, b \in \mathcal{C}_\infty(\mathcal{X}, \mathbb{K})$, we may still define a linear combination $af + bg$ by $(af + bg)(p) = a(p)f(p) + b(p)g(p)$, $\forall p \in \mathcal{X}$, which is clearly smooth. This makes $\mathsf{V}(\mathcal{X})$ a *module* over the ring $\mathcal{C}_\infty(\mathcal{X}, \mathbb{K})$.

Example 12.149 Let $\mathcal{Y}$ be a reflexive Banach space over $\mathbb{K}$, $\mathcal{X}$ be a smooth manifold of variant $\mathcal{Y}$, $p \in \mathcal{X}$, f be a smooth vector field, and $t_i := (U_i, \phi_i)$ be two charts in $\mathcal{A}_\mathcal{X}$, $i = 1, 2$, with $p \in U_1 \cap U_2$. Note that $f(p) \in \mathsf{T}_p\mathcal{X}$ and its expression in chart t_i is $(f(p))_{t_i} =: f_i(\phi_i(p)) \in \mathcal{Y}$, $i = 1, 2$. Then, by (ii) of Definition 12.142, in chart t_2, $f_2(\phi_2(p)) = \mathrm{D}(\phi_2 \circ (\phi_1)_{\mathrm{inv}})(\phi_1(p))(f_1(\phi_1(p))) =$

$\mathrm{D}(\phi_2 \circ (\phi_1)_{\mathrm{inv}})(\phi_1 \circ (\phi_2)_{\mathrm{inv}}(\phi_2(p)))(f_1(\phi_1 \circ (\phi_2)_{\mathrm{inv}}(\phi_2(p))))$. This is equivalent to $f_2(y_2) = (\mathrm{D}T(y_1) f_1(y_1))|_{y_1 = T_{\mathrm{inv}}(y_2)}$, where $T := \phi_2 \circ (\phi_1)_{\mathrm{inv}}|_{\phi_1(U_1 \cap U_2)}$. We recognize that this is simply the coordinate transformation formula for a differential equation on $\mathcal{Y}$.

Thus, the notion of vector field allows us to define the notion of differential equation on the manifold $\mathcal{X}$ of variant $\mathcal{Y}$. Let f be a vector field on $\mathcal{X}$, and $\sigma : \bar{\mathrm{r}}_{a,b} \to \mathcal{X}$ be a smooth mapping with $a, b \in \mathbb{R}$ and $a < b$. σ is said to be *integral curve* of f if $\sigma_{\star t_0}(\frac{\mathrm{d}}{\mathrm{d}t}) = f(\sigma(t_0))$, $\forall t_0 \in \bar{\mathrm{r}}_{a,b}$. Let $c := (U, \phi) \in \mathcal{A}_{\mathcal{X}}$ with $\sigma(t_0) \in U$. Then, by Example 12.146, we have, $\forall \lambda \in \mathcal{C}_\infty(\mathcal{X}, \mathbb{K})$, $\sigma_{\star t_0}(\frac{\mathrm{d}}{\mathrm{d}t})(\lambda) = f(\sigma(t_0))(\lambda) = \frac{\mathrm{d}}{\mathrm{d}t}(\lambda \circ \sigma)(t_0)$. On the chart c, by (i) of Definition 12.142, we have $\mathrm{D}(\lambda \circ \phi_{\mathrm{inv}})(\phi(\sigma(t_0)))(f(\sigma(t_0))_c) = \frac{\mathrm{d}}{\mathrm{d}t}(\lambda \circ \sigma)(t_0) = \frac{\mathrm{d}}{\mathrm{d}t}(\lambda \circ \phi_{\mathrm{inv}} \circ \phi \circ \sigma)(t_0) = \mathrm{D}(\lambda \circ \phi_{\mathrm{inv}})(\phi(\sigma(t_0))\mathrm{D}(\phi \circ \sigma)(t_0)$. By the arbitrariness of λ, we have $\mathrm{D}(\phi \circ \sigma)(t_0) = f(\sigma(t_0))_c$. This is the desired differential equation formulation: on the left-hand side is the derivative of the expression of σ in local coordinates, and on the right-hand side is the assigned vector $f(\sigma(t_0))_c \in \mathcal{Y}$ for the chart c of the tangent vector $f(\sigma(t_0)) \in \mathsf{T}_{\sigma(t_0)}\mathcal{X}$. Thus, we can say that σ is a solution to the differential equation $\mathrm{D}(\phi \circ \sigma)(t_0) = f(\sigma(t_0))_c$ in a neighborhood of t_0. ◇

Let f and g be smooth vector fields on the manifold $\mathcal{X}$. $\forall \lambda \in \mathcal{C}_\infty(\mathcal{X}, \mathbb{K})$, we may define the following smooth function $\mathsf{L}_f \mathsf{L}_g \lambda - \mathsf{L}_g \mathsf{L}_f \lambda$. (Note that in repeated Lie derivative, we assume that $\mathsf{L}_f \mathsf{L}_g \lambda := \mathsf{L}_f(\mathsf{L}_g \lambda)$.) In a chart $t := (U, \phi) \in \mathcal{A}_{\mathcal{X}}$, we have $\mathsf{L}_g \lambda(p) = \mathrm{D}(\lambda \circ \phi_{\mathrm{inv}})(\phi(p)) g(\phi(p))$, $\forall p \in U$, where g admits expression $g(x)$ in the chart t. Let f admits expression $f(x)$ in the chart t. Then, $\mathsf{L}_f \mathsf{L}_g \lambda(p) = \mathsf{L}_f(\mathsf{L}_g \lambda)(p) = \mathrm{D}(\mathrm{D}(\lambda \circ \phi_{\mathrm{inv}})(x) g(x))(\phi(p)) f(\phi(p)) = (\lambda \circ \phi_{\mathrm{inv}})^{(2)}(\phi(p))(g(\phi(p)))(f(\phi(p))) + \mathrm{D}(\lambda \circ \phi_{\mathrm{inv}})(\phi(p))(\mathrm{D}g(\phi(p))(f(\phi(p))))$. By symmetry, we have $\mathsf{L}_g \mathsf{L}_f \lambda(p) = (\lambda \circ \phi_{\mathrm{inv}})^{(2)}(\phi(p))(f(\phi(p)))(g(\phi(p))) + \mathrm{D}(\lambda \circ \phi_{\mathrm{inv}})(\phi(p))(\mathrm{D}f(\phi(p))(g(\phi(p))))$. Then,

$$
\begin{aligned}
&(\mathsf{L}_f \mathsf{L}_g \lambda - \mathsf{L}_g \mathsf{L}_f \lambda)(p) \\
&= (\lambda \circ \phi_{\mathrm{inv}})^{(2)}(\phi(p))(g(\phi(p)))(f(\phi(p))) \\
&\quad + \mathrm{D}(\lambda \circ \phi_{\mathrm{inv}})(\phi(p))(\mathrm{D}g(\phi(p))(f(\phi(p)))) \\
&\quad - (\lambda \circ \phi_{\mathrm{inv}})^{(2)}(\phi(p))(f(\phi(p)))(g(\phi(p))) \\
&\quad - \mathrm{D}(\lambda \circ \phi_{\mathrm{inv}})(\phi(p))(\mathrm{D}f(\phi(p))(g(\phi(p)))) \\
&= \mathrm{D}(\lambda \circ \phi_{\mathrm{inv}})(\phi(p))(\mathrm{D}g(\phi(p))(f(\phi(p)))) \\
&\quad - \mathrm{D}(\lambda \circ \phi_{\mathrm{inv}})(\phi(p))(\mathrm{D}f(\phi(p))(g(\phi(p)))) \\
&= \mathrm{D}(\lambda \circ \phi_{\mathrm{inv}})(\phi(p))(\mathrm{D}g(\phi(p))(f(\phi(p))) - \mathrm{D}f(\phi(p))(g(\phi(p)))) \\
&=: \mathsf{L}_{[f,g]} \lambda(p)
\end{aligned}
$$

where the first equality follows from the preceding derivation and the second equality follows from Proposition 9.28. Thus, we observe that f and g determine another vector field $[f, g]$, which in chart t admits the expression $\mathrm{D}g(x)(f(x)) - \mathrm{D}f(x)(g(x))$. This is a smooth vector field according to Definition 12.148. We will call $[f, g]$ the *Lie bracket* of the vector fields f and g.

Theorem 12.150 *Let $\mathcal{Y}$ be a reflexive Banach space over $\mathbb{K}$, $\mathcal{X}$ be a smooth manifold of variant $\mathcal{Y}$, $p \in \mathcal{X}$, and f and g be smooth vector fields. Then, there exists a smooth vector field denoted by $[f, g]$ that assigns a tangent vector $[f, g](p) \in \mathsf{T}_p\mathcal{X}$ to every $p \in \mathcal{X}$. In any chart $(U, \phi) \in \mathcal{A}_{\mathcal{X}}$, the expression for $[f, g]$ is given by $\mathrm{D}g(x)(f(x)) - \mathrm{D}f(x)(g(x))$, where $f(x)$ and $g(x)$ are assumed to be the expression of f and g in the chart (U, ϕ). $\forall f, g, h \in \mathsf{V}(\mathcal{X})$, $\forall \alpha, \beta \in \mathbb{K}$, we have:*

(i) $[f, g] = -[g, f]$ *(skew commutative).*
(ii) $[f, \alpha g + \beta h] = \alpha[f, g] + \beta[f, h]$ *(bilinearity).*
(iii) $[\alpha g + \beta h, f] = \alpha[g, f] + \beta[h, f]$ *(bilinearity).*
(iv) $\mathsf{L}_f\mathsf{L}_g\lambda - \mathsf{L}_g\mathsf{L}_f\lambda = \mathsf{L}_{[f,g]}\lambda$, $\forall \lambda \in \mathcal{C}_\infty(\mathcal{X}, \mathbb{K})$.
(v) *It satisfies the Jacobi identity:* $[f, [g, h]] + [g, [h, f]] + [h, [f, g]] = \vartheta_{\mathsf{V}(\mathcal{X})}$.

Proof (i)–(iii) These are immediate from the expression of $[f, g]$ in local coordinates.

(iv) This follows from the derivation preceding the theorem.

(v) Fix any $p \in \mathcal{X}$ and any $\lambda \in \mathcal{C}_\infty(p, \mathbb{K})$. Note that

$$\begin{aligned}\mathsf{L}_{[f,[g,h]]}\lambda &= \mathsf{L}_f(\mathsf{L}_{[g,h]}\lambda) - \mathsf{L}_{[g,h]}(\mathsf{L}_f\lambda)\\ &= \mathsf{L}_f(\mathsf{L}_g\mathsf{L}_h\lambda - \mathsf{L}_h\mathsf{L}_g\lambda) - \mathsf{L}_g\mathsf{L}_h(\mathsf{L}_f\lambda) + \mathsf{L}_h\mathsf{L}_g(\mathsf{L}_f\lambda)\\ &= \mathsf{L}_f\mathsf{L}_g\mathsf{L}_h\lambda + \mathsf{L}_h\mathsf{L}_g\mathsf{L}_f\lambda - \mathsf{L}_f\mathsf{L}_h\mathsf{L}_g\lambda - \mathsf{L}_g\mathsf{L}_h\mathsf{L}_f\lambda\end{aligned}$$

where the first equality follows from (iv), the second equality follows from (iv) and Definition 12.142, and the third equality follows from Definition 12.142. Thus, we have, by symmetry,

$$\begin{aligned}\mathsf{L}_{[f,[g,h]]+[g,[h,f]]+[h,[f,g]]}\lambda &= \mathsf{L}_{[f,[g,h]]}\lambda + \mathsf{L}_{[g,[h,f]]}\lambda + \mathsf{L}_{[h,[f,g]]}\lambda\\ &= \mathsf{L}_f\mathsf{L}_g\mathsf{L}_h\lambda + \mathsf{L}_h\mathsf{L}_g\mathsf{L}_f\lambda - \mathsf{L}_f\mathsf{L}_h\mathsf{L}_g\lambda - \mathsf{L}_g\mathsf{L}_h\mathsf{L}_f\lambda\\ &\quad + \mathsf{L}_g\mathsf{L}_h\mathsf{L}_f\lambda + \mathsf{L}_f\mathsf{L}_h\mathsf{L}_g\lambda - \mathsf{L}_g\mathsf{L}_f\mathsf{L}_h\lambda - \mathsf{L}_h\mathsf{L}_f\mathsf{L}_g\lambda\\ &\quad + \mathsf{L}_h\mathsf{L}_f\mathsf{L}_g\lambda + \mathsf{L}_g\mathsf{L}_f\mathsf{L}_h\lambda - \mathsf{L}_h\mathsf{L}_g\mathsf{L}_f\lambda - \mathsf{L}_f\mathsf{L}_g\mathsf{L}_h\lambda\\ &= 0 \in \mathcal{C}_\infty(p, \mathbb{K})\end{aligned}$$

where the first equality follows from Definition 12.142 and the second equality follows from the preceding discussion. Thus, by the arbitrariness of λ and p, we have $[f, [g, h]] + [g, [h, f]] + [h, [f, g]] = \vartheta_{\mathsf{V}(\mathcal{X})}$.

This completes the proof of the theorem. □

Chapter 13
Hilbert Spaces

Hilbert spaces are vector spaces equipped with inner products. They possess a wealth of structural properties generalizing basic geometrical insights. The concepts of orthonormal basis, Fourier series, and least-square minimization all have natural setting in Hilbert spaces.

13.1 Fundamental Notions

Definition 13.1 A *pre-Hilbert space* $\mathcal{X}$ *over* $\mathbb{K}$ is a vector space $\mathcal{X}$ over $\mathbb{K}$, together with an *inner product* $\langle \cdot, \cdot \rangle : \mathcal{X} \times \mathcal{X} \to \mathbb{K}$ satisfying, $\forall x, y, z \in \mathcal{X}$, $\forall \lambda \in \mathbb{K}$,

(i) $\langle x, y \rangle = \overline{\langle y, x \rangle}$.
(ii) $\langle x + y, z \rangle = \langle x, z \rangle + \langle y, z \rangle$.
(iii) $\langle \lambda x, y \rangle = \lambda \langle x, y \rangle$.
(iv) $\langle x, x \rangle \in [0, \infty) \subset \mathbb{R}$ and $\langle x, x \rangle = 0$ if, and only if, $x = \vartheta_{\mathcal{X}}$. ◇

Proposition 13.2 *Let* $\mathcal{X}$ *be a pre-Hilbert space over* $\mathbb{K}$. *Define* $\|\cdot\| : \mathcal{X} \to [0, \infty) \subset \mathbb{R}$ *by* $\|x\| = \sqrt{\langle x, x \rangle}$, $\forall x \in \mathcal{X}$. *Then,* $\forall x, y, z \in \mathcal{X}$, $\forall \alpha, \beta \in \mathbb{K}$, *we have*

(i) $\langle \alpha x + \beta y, z \rangle = \alpha \langle x, z \rangle + \beta \langle y, z \rangle$.
(ii) $\langle z, \alpha x + \beta y \rangle = \overline{\alpha} \langle z, x \rangle + \overline{\beta} \langle z, y \rangle$.
(iii) $\langle \vartheta_{\mathcal{X}}, x \rangle = 0 = \langle x, \vartheta_{\mathcal{X}} \rangle$.
(iv) $|\langle x, y \rangle| \leq \|x\| \|y\|$, *where equality holds if, and only if,* $y = \vartheta_{\mathcal{X}}$ *or* $x = \lambda y$ *for some* $\lambda \in \mathbb{K}$. *(Cauchy–Schwarz Inequality)*
(v) $\|x + y\| \leq \|x\| + \|y\|$, *where equality holds if, and only if,* $y = \vartheta_{\mathcal{X}}$ *or* $x = \lambda y$, *for some* $\lambda \in [0, \infty) \subset \mathbb{R}$.
(vi) $\|\cdot\|$ *is a norm of* $\mathcal{X}$. *(Induced norm)*

Z. Pan, *Measure-Theoretic Calculus in Abstract Spaces*,
https://doi.org/10.1007/978-3-031-21912-2_13

Proof (i) $\langle \alpha x + \beta y, z\rangle = \langle \alpha x, z\rangle + \langle \beta y, z\rangle = \alpha\langle x, z\rangle + \beta\,\langle y, z\rangle$, where the first equality follows from (ii) of Definition 13.1; and the second equality follows from (iii) of Definition 13.1.

(ii) $\langle z, \alpha x + \beta y\rangle = \overline{\langle \alpha x + \beta y, z\rangle} = \overline{\alpha\langle x, z\rangle + \beta\,\langle y, z\rangle} = \overline{\alpha}\,\overline{\langle x, z\rangle} + \overline{\beta}\,\overline{\langle y, z\rangle} = \overline{\alpha}\langle z, x\rangle + \overline{\beta}\langle z, y\rangle$, where the first equality follows from (i) of Definition 13.1; the second equality follows from (i); and the fourth equality follows from (i) of Definition 13.1.

(iii) $\langle \vartheta_{\mathcal{X}}, x\rangle = \langle y - y, x\rangle = \langle y, x\rangle - \langle y, x\rangle = 0$, where the second equality follows from (i). Then, $\langle x, \vartheta_{\mathcal{X}}\rangle = \overline{\langle \vartheta_{\mathcal{X}}, x\rangle} = 0$.

(iv) We will distinguish two exhaustive and mutually exclusives cases: Case 1: $y = \vartheta_{\mathcal{X}}$; Case 2: $y \neq \vartheta_{\mathcal{X}}$. Case 1: $y = \vartheta_{\mathcal{X}}$. Then, $|\langle x, y\rangle| = 0 = \|x\|\cdot 0 = \|x\|\|y\|$. This case is proved.

Case 2: $y \neq \vartheta_{\mathcal{X}}$. By (iv) of Definition 13.1, we have $\langle y, y\rangle > 0$. $\forall \lambda \in \mathbb{K}$, we have $0 \leq \langle x - \lambda y, x - \lambda y\rangle = \langle x, x - \lambda y\rangle - \lambda\langle y, x - \lambda y\rangle = \langle x, x\rangle - \overline{\lambda}\langle x, y\rangle - \lambda\langle y, x\rangle + \lambda\overline{\lambda}\langle y, y\rangle$, where the inequality follows from (iv) of Definition 13.1; the first equality follows from (i); and the second equality follows from (ii). Take $\lambda := \langle x, y\rangle/\langle y, y\rangle$. Then, $0 \leq \langle x, x\rangle - \overline{\lambda}\lambda\langle y, y\rangle - \lambda\overline{\langle x, y\rangle} + |\lambda|^2\langle y, y\rangle = \langle x, x\rangle - \langle x, y\rangle\overline{\langle x, y\rangle}/\langle y, y\rangle = \langle x, x\rangle - |\langle x, y\rangle|^2/\langle y, y\rangle$, where the equality follows (i) of Definition 13.1. Hence, $|\langle x, y\rangle| \leq \|x\|\|y\|$. Equality holds if, and only if, $x - \lambda y = \vartheta_{\mathcal{X}}$. This case is also proved. Hence, (iv) holds.

(v) Note that $\|x + y\|^2 = \langle x + y, x + y\rangle = \langle x, x\rangle + \langle x, y\rangle + \langle y, x\rangle + \langle y, y\rangle = \|x\|^2 + 2\,\mathrm{Re}\,(\langle x, y\rangle) + \|y\|^2 \leq \|x\|^2 + 2|\langle x, y\rangle| + \|y\|^2 \leq \|x\|^2 + 2\|x\|\|y\| + \|y\|^2 = (\|x\| + \|y\|)^2$, where the first equality follows from (i) and (ii); the second equality follows from (i) of Definition 13.1; the second inequality follows from (iv). Hence, $\|x + y\| \leq \|x\| + \|y\|$. Equality holds if, and only if, $\mathrm{Re}\,(\langle x, y\rangle) = |\langle x, y\rangle| = \|x\|\|y\|$. By (iv), have equality holds if, and only if, $y = \vartheta_{\mathcal{X}}$ or $x = \lambda y$ for some $\lambda \in [0, \infty) \subset \mathbb{R}$.

(vi) Clearly, $\|x\| \in [0, \infty) \subset \mathbb{R}$. $\|x\| = 0 \Leftrightarrow \langle x, x\rangle = 0 \Leftrightarrow x = \vartheta_{\mathcal{X}}$. Hence, (i) of Definition 7.1 holds. By (v), (ii) of Definition 7.1 holds. Note also that $\|\alpha x\| = \sqrt{\langle \alpha x, \alpha x\rangle} = \sqrt{\alpha\overline{\alpha}\langle x, x\rangle} = |\alpha|\|x\|$, where the second equality follows from (i) and (ii); and the third equality follows from the fact that $\alpha\overline{\alpha} = |\alpha|^2$. Hence, (iii) of Definition 7.1 holds. Therefore, $\|\cdot\|$ defines a norm of $\mathcal{X}$. □

Lemma 13.3 (Parallelogram Law) *Let $\mathcal{X}$ be a pre-Hilbert space. $\forall x, y \in \mathcal{X}$, we have $\|x + y\|^2 + \|x - y\|^2 = 2\|x\|^2 + 2\|y\|^2$.*

Proof This is straightforward, and is therefore omitted. □

Pre-Hilbert space is a special kind of normed linear space. The concept of convergence, continuity, topology, and completeness, etc., apply in pre-Hilbert spaces.

Proposition 13.4 *Let $\mathcal{X}$ be a pre-Hilbert space. Then, the inner product is a continuous function of the product space $\mathcal{X} \times \mathcal{X}$.*

Proof We will show that the inner product is continuous at any $(x_0, y_0) \in \mathcal{X} \times \mathcal{X}$. $\forall \epsilon \in (0, \infty) \subset \mathbb{R}$, let $M := \max\{\|x_0\|, \|y_0\|\} + 1 \in (0, \infty) \subset \mathbb{R}$

and $\delta := \min\{\epsilon, 1\}/M \in (0, 1] \subset \mathbb{R}$. Fix any $(x, y) \in \mathcal{B}_{\mathcal{X}\times\mathcal{X}}((x_0, y_0), \delta)$, we have $|\langle x, y\rangle - \langle x_0, y_0\rangle| = |\langle x - x_0, y\rangle + \langle x_0, y - y_0\rangle| \leq |\langle x - x_0, y\rangle| + |\langle x_0, y - y_0\rangle| \leq \|x - x_0\|\|y\| + \|x_0\|\|y - y_0\| \leq \|x - x_0\|(\|y_0\| + \|y - y_0\|) + M\|y - y_0\| \leq M(\|x - x_0\| + \|y - y_0\|) < \sqrt{2}M\delta \leq \sqrt{2}\epsilon$, where the second inequality follows from Proposition 13.2. Hence, the inner product is continuous at (x_0, y_0). By the arbitrariness of (x_0, y_0), we have the inner product is continuous. This completes the proof of the proposition. □

Definition 13.5 A complete pre-Hilbert space is called a *Hilbert space*. ◇

Definition 13.6 Let $\mathcal{X}$ be a pre-Hilbert space, $x, y \in \mathcal{X}$, and $S \subseteq \mathcal{X}$. The vectors x and y are said to be *orthogonal* if $\langle x, y\rangle = 0$. We will then write $x \perp y$. The vector x is said to be *orthogonal* to the set S if $x \perp s$, $\forall s \in S$. We will then write $x \perp S$. ◇

Theorem 13.7 (Pythagorean) *Let $\mathcal{X}$ be a pre-Hilbert space over $\mathbb{K}$ and $x, y \in \mathcal{X}$. If $x \perp y$, then $\|x + y\|^2 = \|x\|^2 + \|y\|^2$. Furthermore, if $\mathbb{K} = \mathbb{R}$ and $\|x + y\|^2 = \|x\|^2 + \|y\|^2$, then $x \perp y$.*

Proof Let $x \perp y$, then $\langle x, y\rangle = 0$. This implies that $\|x + y\|^2 = \langle x + y, x + y\rangle = \langle x, x\rangle + \langle x, y\rangle + \langle y, x\rangle + \langle y, y\rangle = \|x\|^2 + \|y\|^2$, where the second equality follows from Proposition 13.2; and the third equality follows from Definition 13.1.

Let $\mathbb{K} = \mathbb{R}$ and $\|x + y\|^2 = \|x\|^2 + \|y\|^2$. Then, we have $0 = \langle x, y\rangle + \langle y, x\rangle = 2\langle x, y\rangle$, where the first equality follows from Proposition 13.2; and the second equality follows from Definition 13.1 and $\mathbb{K} = \mathbb{R}$. Hence, $x \perp y$. This completes the proof of the theorem. □

Example 13.8 Let $n \in \mathbb{Z}_+$ and $\mathcal{X} = \mathbb{K}^n$. Define the inner product by $\langle x, y\rangle = \sum_{i=1}^{n} \xi_i \overline{\eta_i} \in \mathbb{K}$, $\forall x := (\xi_1, \ldots, \xi_n) \in \mathbb{K}^n$ and $\forall y := (\eta_1, \ldots, \eta_n) \in \mathbb{K}^n$. It is easy to check that the pair $\mathcal{X} := (\mathcal{X}, \langle\cdot,\cdot\rangle)$ is a pre-Hilbert space over $\mathbb{K}$. The induced norm is clearly the Euclidean norm. By Example 7.29, $\mathcal{X}$ is complete. Therefore, $\mathcal{X}$ is a Hilbert space, which is to be denoted simply by $\mathbb{K}^n$. ◇

Example 13.9 Let $\mathcal{X}_i$ be a Hilbert space over $\mathbb{K}$ with inner product $\langle\cdot,\cdot\rangle_i$, $i = 1, 2$. Let $\mathcal{Z} := \mathcal{X}_1 \times \mathcal{X}_2$. Define inner product $\langle\cdot,\cdot\rangle_{\mathcal{Z}} : \mathcal{Z} \times \mathcal{Z} \to \mathbb{K}$ by $\langle(x_1, x_2), (y_1, y_2)\rangle_{\mathcal{Z}} = \langle x_1, y_1\rangle_1 + \langle x_2, y_2\rangle_2$, $\forall (x_1, x_2), (y_1, y_2) \in \mathcal{Z}$. It is easy to show that $\mathcal{Z}$ with this inner product is a pre-Hilbert space over $\mathbb{K}$. It is also easy to see that the induced norm of $\mathcal{Z}$ is equal to the norm of $\mathcal{Z}$ as defined in Proposition 7.22. By Proposition 4.31, $\mathcal{Z}$ is complete. Therefore, $\mathcal{Z}$ is a Hilbert space over $\mathbb{K}$. ◇

The above example can be easily generalized to $\prod_{i=1}^{n} \mathcal{X}_i$ case, where $n \in \mathbb{Z}_+$ and $\mathcal{X}_i$ is a Hilbert space over $\mathbb{K}$, $\forall i \in \{1, \ldots, n\}$.

Example 13.10 Let $\mathcal{X}$ be a Hilbert space over $\mathbb{K}$ with inner product $\langle\cdot,\cdot\rangle_{\mathcal{X}}$ and $\mathcal{Z} := \mathrm{l}_2(\mathcal{X})$. We will show that $\mathcal{Z}$ is a Hilbert space over $\mathbb{K}$ with inner product $\langle\cdot,\cdot\rangle_{\mathcal{Z}}$ defined by $\langle x, y\rangle_{\mathcal{Z}} = \sum_{i=1}^{\infty} \langle\xi_i, \eta_i\rangle$, $\forall x := (\xi_1, \xi_2, \ldots) \in \mathcal{Z}$ and $\forall y := (\eta_1, \eta_2, \ldots) \in \mathcal{Z}$. To see that $\langle x, y\rangle_{\mathcal{Z}} \in \mathbb{K}$, we note that $x \in \mathcal{Z}$ implies that $\sum_{i=1}^{\infty} \|\xi_i\|^2 < \infty$. Fix any $\epsilon \in (0, \infty) \subset \mathbb{R}$. Then, $\exists n_1 \subset \mathbb{N}$ such that

$\sum_{i=n_1}^{\infty} \|\xi_i\|^2 < \epsilon$. By $y \in \mathcal{Z}$, $\exists n_2 \in \mathbb{N}$ such that $\sum_{i=n_2}^{\infty} \|\eta_i\|^2 < \epsilon$. Let $n_0 := \max\{n_1, n_2\} \in \mathbb{N}$. $\forall n, m \in \mathbb{N}$ with $n_0 \leq n \leq m$, we have $\left|\sum_{i=n}^{m} \langle \xi_i, \eta_i \rangle\right| \leq \left(\sum_{i=n}^{m} \|\xi_i\|^2\right)^{1/2} \left(\sum_{i=n}^{m} \|\eta_i\|^2\right)^{1/2} \leq \left(\sum_{i=n_1}^{\infty} \|\xi_i\|^2\right)^{1/2} \left(\sum_{i=n_2}^{\infty} \|\eta_i\|^2\right)^{1/2} < \epsilon$, where the first inequality follows from Hölder's Inequality Theorem 7.8. Hence, $\left(\sum_{i=1}^{n} \langle \xi_i, \eta_i \rangle\right)_{n=1}^{\infty} \subseteq \mathbb{K}$ is a Cauchy sequence, which converges since $\mathbb{K}$ is complete. Hence, $\langle x, y \rangle_{\mathcal{Z}} \in \mathbb{K}$. It is easy to show that (i)–(iv) of Definition 13.1 are satisfied. Hence, $\mathcal{Z}$ is a pre-Hilbert space. It is easy to see that the induced norm is equal to the norm of $\mathcal{Z}$ as defined in Example 7.10. By Example 7.33, we have $\mathcal{Z}$ is complete and therefore a Hilbert space. ◇

Example 13.11 Let $\mathcal{X} := (X, \mathcal{B}, \mu)$ be a σ-finite measure space, $\mathcal{Y}$ be a separable Hilbert space over $\mathbb{K}$ with inner product $\langle \cdot, \cdot \rangle_{\mathcal{Y}}$ and $\mathcal{Z} := \mathrm{L}_2(\mathcal{X}, \mathcal{Y})$. We will show that $\mathcal{Z}$ is a Hilbert space over $\mathbb{K}$ with inner product $\langle \cdot, \cdot \rangle_{\mathcal{Z}}$ defined by $\langle [z_1], [z_2] \rangle_{\mathcal{Z}} = \int_X \langle z_1(x), z_2(x) \rangle_{\mathcal{Y}} \, d\mu(x)$, $\forall z_1, z_2 \in \bar{\mathrm{L}}_2(\mathcal{X}, \mathcal{Y})$.

First, we show that $\langle \cdot, \cdot \rangle_{\mathcal{Z}}$ is well-defined. $\forall z_1, z_2 \in \bar{\mathrm{L}}_2(\mathcal{X}, \mathcal{Y})$, we have z_1 and z_2 are $\mathcal{B}$-measurable. By Propositions 11.39, 11.38, and 13.4, the function $h : \mathcal{X} \to \mathbb{K}$ defined by $h(x) = \langle z_1(x), z_2(x) \rangle_{\mathcal{Y}}$, $\forall x \in \mathcal{X}$, is $\mathcal{B}$-measurable. $\mathcal{P} \circ h(x) = |h(x)| \leq \|z_1(x)\| \|z_2(x)\| = \mathcal{P} \circ z_1(x) \mathcal{P} \circ z_2(x)$, $\forall x \in \mathcal{X}$, where the inequality follows from Proposition 13.2. Then, $\int_X \mathcal{P} \circ h \, d\mu \leq \int_X (\mathcal{P} \circ z_1 \mathcal{P} \circ z_2) \, d\mu \leq (\int_X \mathcal{P} \circ z_1 \, d\mu)^{1/2} (\int_X \mathcal{P} \circ z_2 \, d\mu)^{1/2} = \|z_1\|_2 \|z_2\|_2 < \infty$, where the first inequality follows from Definition 11.79; and the second inequality follows from Hölder's Inequality 11.178. Hence, h is absolute integrable over $\mathcal{X}$. By Proposition 11.92, h is integrable over $\mathcal{X}$ and $\int_X \langle z_1(x), z_2(x) \rangle_{\mathcal{Y}} \, d\mu(x) \in \mathbb{K}$. $\forall \bar{z}_1, \bar{z}_2 \in \bar{\mathrm{L}}_2(\mathcal{X}, \mathcal{Y})$ with $[z_1] = [\bar{z}_1]$ and $[z_2] = [\bar{z}_2]$, by Example 11.173, we have $z_1 = \bar{z}_1$ a.e. in $\mathcal{X}$ and $z_2 = \bar{z}_2$ a.e. in $\mathcal{X}$. Define $\bar{h} : \mathcal{X} \to \mathbb{K}$ by $\bar{h}(x) = \langle \bar{z}_1(x), \bar{z}_2(x) \rangle_{\mathcal{Y}}$, $\forall x \in \mathcal{X}$. By Lemmas 11.45 and 11.46, $h = \bar{h}$ a.e. in $\mathcal{X}$. By Proposition 11.92, $\int_X h \, d\mu = \int_X \bar{h} \, d\mu$. Hence, $\langle [z_1], [z_2] \rangle_{\mathcal{Z}} \in \mathbb{K}$ is well-defined.

Next, we show that $\langle \cdot, \cdot \rangle_{\mathcal{Z}}$ satisfies (i)–(iv) of Definition 13.1. Fix any $z_1, z_2, z_3 \in \bar{\mathrm{L}}_2(\mathcal{X}, \mathcal{Y})$ and any $\lambda \in \mathbb{K}$. (i) $\langle [z_1], [z_2] \rangle_{\mathcal{Z}} = \int_X \langle z_1(x), z_2(x) \rangle_{\mathcal{Y}} d\mu(x) = \int_X \overline{\langle z_2(x), z_1(x) \rangle_{\mathcal{Y}}} \, d\mu(x) = \overline{\int_X \langle z_2(x), z_1(x) \rangle_{\mathcal{Y}} \, d\mu(x)} = \overline{\langle [z_2], [z_1] \rangle_{\mathcal{Z}}}$, where the second equality follows from the fact that $\mathcal{Y}$ is a Hilbert space over $\mathbb{K}$; and the third equality follows from Proposition 11.92. (ii) $\langle [z_1] + [z_2], [z_3] \rangle_{\mathcal{Z}} = \langle [z_1 + z_2], [z_3] \rangle_{\mathcal{Z}} = \int_X \langle z_1(x) + z_2(x), z_3(x) \rangle_{\mathcal{Y}} \cdot d\mu(x) = \int_X (\langle z_1(x), z_3(x) \rangle_{\mathcal{Y}} + \langle z_2(x), z_3(x) \rangle_{\mathcal{Y}}) \, d\mu(x) = \int_X \langle z_1(x), z_3(x) \rangle_{\mathcal{Y}} \cdot d\mu(x) + \int_X \langle z_2(x), z_3(x) \rangle_{\mathcal{Y}} \, d\mu(x) = \langle [z_1], [z_3] \rangle_{\mathcal{Z}} + \langle [z_2], [z_3] \rangle_{\mathcal{Z}}$, where the first equality follows from Proposition 7.43; the third equality follows from the fact that $\mathcal{Y}$ is a Hilbert space; and the fourth equality follows from Proposition 11.92. (iii) $\langle \lambda [z_1], [z_2] \rangle_{\mathcal{Z}} = \langle [\lambda z_1], [z_2] \rangle_{\mathcal{Z}} = \int_X \langle \lambda z_1(x), z_2(x) \rangle_{\mathcal{Y}} \, d\mu(x) = \int_X \lambda \langle z_1(x), z_2(x) \rangle_{\mathcal{Y}} \, d\mu(x) = \lambda \int_X \langle z_1(x), z_2(x) \rangle_{\mathcal{Y}} \, d\mu(x) = \lambda \langle [z_1], [z_2] \rangle_{\mathcal{Z}}$, where the first equality follows from Proposition 7.43; the third equality follows from the fact that $\mathcal{Y}$ is a Hilbert space; and the fourth equality follows from Proposition 11.92. (iv) $\langle [z_1], [z_1] \rangle_{\mathcal{Z}} = \int_X \langle z_1(x), z_1(x) \rangle_{\mathcal{Y}} \, d\mu(x) = \int_X \|z_1(x)\|_{\mathcal{Y}}^2 \, d\mu(x) = \|z_1\|_2^2 \in [0, \infty) \subset \mathbb{R}$. Clearly, $\langle [z_1], [z_1] \rangle_{\mathcal{Z}} = 0 \Leftrightarrow \|z_1\|_2 = 0 \Leftrightarrow [z_1] = \vartheta_{\mathcal{Z}}$. Hence, $\mathcal{Z}$ is a pre-Hilbert space over $\mathbb{K}$.

Note that the induced norm of $\mathcal{Z}$ is equal to the norm of $\mathcal{Z}$ as defined in Example 11.173. By Example 11.179, $\mathcal{Z}$ is complete. Hence, $\mathcal{Z}$ is a Hilbert space over $\mathbb{K}$. ◇

13.2 Projection Theorems

The basic concept of the Projection Theorem is illustrated in Fig. 13.1.

Theorem 13.12 *Let $\mathcal{X}$ be a pre-Hilbert space, $M \subseteq \mathcal{X}$ be a subspace, and $x \in \mathcal{X}$. Consider the problem $\min_{m \in M} \|x - m\|$. If there exists $m_0 \in M$ such that $\|x - m_0\| \leq \|x - m\|$, $\forall m \in M$, then m_0 is the unique vector in M that minimizes $\|x - m\|$. A necessary and sufficient condition for $m_0 \in M$ being the unique minimizing vector is $(x - m_0) \perp M$.*

Proof We first show that if $m_0 \in M$ is a minimizing vector then $(x - m_0) \perp M$ by an argument of contradiction. Suppose that $(x - m_0) \not\perp M$. Then, $\exists m \in M$ such that $\langle x - m_0, m \rangle =: \delta \neq 0$. Without loss of generality, assume $\|m\| = 1$. Define $m_1 := m_0 + \delta m \in M$. Then, by Proposition 13.2, $\|x - m_1\|^2 = \langle x - m_0 - \delta m, x - m_0 - \delta m \rangle = \|x - m_0\|^2 - \langle x - m_0, \delta m \rangle - \langle \delta m, x - m_0 \rangle + \|\delta m\|^2 = \|x - m_0\|^2 - \delta\bar{\delta} - \delta\bar{\delta} + |\delta|^2 \|m\|^2 = \|x - m_0\|^2 - |\delta|^2 < \|x - m_0\|^2$. This contradicts the assumption that m_0 is a minimizing vector. Hence, $(x - m_0) \perp M$.

Next, we show that if $(x - m_0) \perp M$, where $m_0 \in M$, then m_0 is the unique minimizing vector. $\forall m \in M$, by Pythagorean Theorem 13.7, we have $\|x - m\|^2 =$

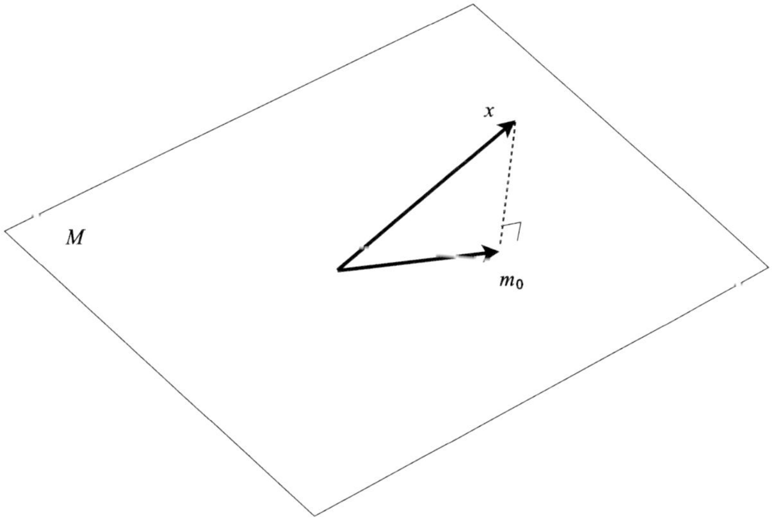

Fig. 13.1 Projection onto a subspace

$\|x - m_0 - (m - m_0)\|^2 = \|x - m_0\|^2 + \|m - m_0\|^2$. Then, $\|x - m\| > \|x - m_0\|$ if $m \neq m_0$. This completes the proof of the theorem. □

Theorem 13.13 (The Classical Projection Theorem) *Let $\mathcal{X}$ be a Hilbert space and $M \subseteq \mathcal{X}$ be a closed subspace. $\forall x \in \mathcal{X}$, there exists a unique vector $m_0 \in M$ such that $\|x - m_0\| = \min_{m \in M} \|x - m\|$. Furthermore, a necessary and sufficient condition for $m_0 \in M$ being the unique minimizing vector is $(x - m_0) \perp M$.*

Proof The uniqueness and orthogonality are immediate consequences of Theorem 13.12. We are only required to establish the existence of m_0. Let $\delta := \inf_{m \in M} \|x - m\| \in [0, \infty) \subset \mathbb{R}$. Then, $\exists (m_n)_{n=1}^{\infty} \subseteq M$ such that $\lim_{n \in \mathbb{N}} \|x - m_n\| = \delta$. $\forall \epsilon \in (0, \infty) \subset \mathbb{R}$, $\exists n_0 \in \mathbb{N}$ such that $\forall n \in \mathbb{N}$ with $n_0 \leq n$, we have $\delta \leq \|x - m_n\| < \sqrt{\delta^2 + \epsilon^2/4}$. By Parallelogram Law 13.3, we have, $\forall i, j \in \mathbb{N}$ with $n_0 \leq i$ and $n_0 \leq j$, $\left\|m_i - x + x - m_j\right\|^2 + \left\|x - m_i + x - m_j\right\|^2 = 2\left\|x - m_j\right\|^2 + 2\|x - m_i\|^2$. Then, $\left\|m_i - m_j\right\|^2 = 2\left\|x - m_j\right\|^2 + 2\|x - m_i\|^2 - 4\left\|x - (m_i + m_j)/2\right\|^2 < 4(\delta^2 + \epsilon^2/4) - 4\delta^2 = \epsilon^2$ and $\left\|m_i - m_j\right\| < \epsilon$. This implies that $(m_n)_{n=1}^{\infty}$ is a Cauchy sequence, which must converge to $m_0 \in M$, since M is a complete subspace by Proposition 4.39. By Proposition 7.21, we have $\|x - m_0\| = \lim_{n \in \mathbb{N}} \|x - m_n\| = \delta = \inf_{m \in M} \|x - m\|$. This completes the proof of the theorem. □

In the above proof, we observe that the key to the existence of the minimizing vector is that M is complete. Hence, we have the following modified version of the projection theorem.

Theorem 13.14 *Let $\mathcal{X}$ be a pre-Hilbert space and $M \subseteq \mathcal{X}$ be a complete subspace. $\forall x \in \mathcal{X}$, there exists a unique vector $m_0 \in M$ such that $\|x - m_0\| = \min_{m \in M} \|x - m\|$. Furthermore, a necessary and sufficient condition for $m_0 \in M$ being the unique minimizing vector is $(x - m_0) \perp M$.*

Proof This is straightforward by the proof of the Projection Theorem 13.13, and is therefore omitted. □

13.3 Dual of Hilbert Spaces

Theorem 13.15 (Riesz-Fréchet) *Let $\mathcal{X}$ be a Hilbert space over $\mathbb{K}$. Then, the following statements hold.*

(i) *$\forall f \in \mathcal{X}^*$, there exists a unique $y_0 \in \mathcal{X}$ such that $f(x) = \langle x, y_0 \rangle$, $\forall x \in \mathcal{X}$, and $\|f\|_{\mathcal{X}^*} = \|y_0\|_{\mathcal{X}}$. Therefore, we may define a mapping $\Phi : \mathcal{X}^* \to \mathcal{X}$ by $\Phi(f) = y_0$.*

(ii) *$\forall y \in \mathcal{X}$, define $g : \mathcal{X} \to \mathbb{K}$ by $g(x) = \langle x, y \rangle$, $\forall x \in \mathcal{X}$, then $g \in \mathcal{X}^*$.*

(iii) *The mapping Φ is bijective, uniformly continuous, norm preserving, and conjugate linear (that is $\Phi(\alpha f_1 + \beta f_2) = \overline{\alpha}\Phi(f_1) + \overline{\beta}\Phi(f_2)$, $\forall f_1, f_2 \in \mathcal{X}^*$, $\forall \alpha, \beta \in \mathbb{K}$).*
(iv) *If $\mathbb{K} = \mathbb{R}$, then Φ is a isometrical isomorphism between $\mathcal{X}^*$ and $\mathcal{X}$.*
(v) *If $\mathbb{K} = \mathbb{C}$, let $\phi : \mathcal{X} \to \mathcal{X}^{**}$ be the natural mapping as defined in Remark 7.88, then ϕ is surjective and $\mathcal{X}$ is reflexive.*
(vi) *If $\mathbb{K} = \mathbb{C}$, then $\mathcal{X}^*$ with the inner product $\langle \cdot, \cdot \rangle_{\mathcal{X}^*}$, defined by $\langle f, g \rangle_{\mathcal{X}^*} := \langle \Phi(g), \Phi(f) \rangle$, $\forall f, g \in \mathcal{X}$, is a Hilbert space, and it is reflexive.*

Henceforth, we will denote $\Phi_{\text{inv}}(x) =: x^*$, $\forall x \in \mathcal{X}$. Furthermore, the following statement hold.

(vii) *When $\mathbb{K} = \mathbb{C}$, let $\Phi_* : \mathcal{X}^{**} = \mathcal{X} \to \mathcal{X}^*$ be the mapping of Φ if $\mathcal{X}$ is replaced by $\mathcal{X}^*$. Then, $\Phi_* = \Phi_{\text{inv}}$. This leads to the identity $(x^*)^* = x$, $\forall x \in \mathcal{X}$.*
(viii) *If $\mathcal{X}$ is separable, then $\mathcal{X}^*$ is separable.*

Proof We will first show that, $\forall y \in \mathcal{X}$, the functional g defined in the statement (ii) of the theorem is a bounded linear functional with $\|g\|_{\mathcal{X}^*} = \|y\|_{\mathcal{X}}$. By Definition 13.1, g is a linear functional. By Proposition 13.2, we have $|g(x)| = |\langle x, y \rangle| \leq \|x\|_{\mathcal{X}} \|y\|_{\mathcal{X}}$. Hence, $g \in \mathcal{X}^*$ and $\|g\|_{\mathcal{X}^*} \leq \|y\|_{\mathcal{X}}$. Note that $|g(y)| = |\langle y, y \rangle| = \|y\|_{\mathcal{X}}^2$. Hence, $\|g\|_{\mathcal{X}^*} \geq \|y\|_{\mathcal{X}}$. Hence, $\|g\|_{\mathcal{X}^*} = \|y\|_{\mathcal{X}}$.

Next, we will show that, for any $f \in \mathcal{X}^*$, there exists a $y_0 \in \mathcal{X}$ such that $f(x) = \langle x, y_0 \rangle$, $\forall x \in \mathcal{X}$. Let $M := \mathcal{N}(f)$. We will distinguish two exhaustive and mutually exclusives cases: Case 1: $M = \mathcal{X}$; Case 2: $M \subset \mathcal{X}$. Case 1: $M = \mathcal{X}$. Take $y = \vartheta_{\mathcal{X}}$. Then, $f(x) = 0 = \langle x, y \rangle$, $\forall x \in \mathcal{X}$. This case is proved.

Case 2: $M \subset \mathcal{X}$. Then, $\exists x_0 \in \mathcal{X} \setminus M$. Since $f \in \mathcal{X}^*$, then M is a closed subspace of $\mathcal{X}$ by Proposition 7.68. By Projection Theorem 13.13, $\exists m_0 \in M$ such that $(x_0 - m_0) \perp M$. Let $z_0 := x_0 - m_0 \in \mathcal{X} \setminus M$. Then, $f(z_0) \neq 0$. Without loss of generality, we may assume that $f(z_0) = 1$. $\forall x \in \mathcal{X}$, by the linearity of f, we have $f(x - f(x)z_0) = f(x) - f(x)f(z_0) = 0$ and $x - f(x)z_0 \in M$. Since $z_0 \perp M$, we have $0 = \langle x - f(x)z_0, z_0 \rangle = \langle x, z_0 \rangle - f(x)\|z_0\|^2$. Then, $f(x) = \langle x, z_0 \rangle / \|z_0\|^2 = \left\langle x, z_0/\|z_0\|^2 \right\rangle$. Let $y_0 := z_0/\|z_0\|^2 \in \mathcal{X}$, we have $f(x) = \langle x, y_0 \rangle$, $\forall x \in \mathcal{X}$. This case is also proved.

Next, we will show that, for any $f \in \mathcal{X}^*$, the vector $y_0 \in \mathcal{X}$ is unique. Let $y \in \mathcal{X}$ be another vector such that $f(x) = \langle x, y \rangle$, $\forall x \in \mathcal{X}$. Then, $0 = f(x) - f(x) = \langle x, y_0 \rangle - \langle x, y \rangle = \langle x, y_0 - y \rangle$, $\forall x \in \mathcal{X}$. Then, $0 = \langle y_0 - y, y_0 - y \rangle = \|y_0 - y\|$. Thus, $y = y_0$. Hence, y_0 is unique.

Based on the above, we have (i) and (ii) holds.

(iii) Note that, $\forall f_1, f_2 \in \mathcal{X}^*$, $\forall \alpha, \beta \in \mathbb{K}$, $(\alpha f_1 + \beta f_2)(x) = \alpha f_1(x) + \beta f_2(x) = \alpha \langle x, \Phi(f_1) \rangle + \beta \langle x, \Phi(f_2) \rangle = \left\langle x, \overline{\alpha}\Phi(f_1) + \overline{\beta}\Phi(f_2) \right\rangle$, $\forall x \in \mathcal{X}$, where the second equality follows from (i); and the third equality follows from Proposition 13.2. Then, by (i), we have $\Phi(\alpha f_1 + \beta f_2) = \overline{\alpha}\Phi(f_1) + \overline{\beta}\Phi(f_2)$. By (i), $\|\Phi(f)\|_{\mathcal{X}} = \|f\|_{\mathcal{X}^*}$, $\forall f \in \mathcal{X}^*$. Therefore, Φ is norm preserving. By (ii), Φ is surjective. $\forall f_1, f_2 \in \mathcal{X}^*$, $\Phi(f_1) = \Phi(f_2)$ implies that $0 = \|\Phi(f_1) - \Phi(f_2)\|_{\mathcal{X}} = \|\Phi(f_1 - f_2)\|_{\mathcal{X}} = \|f_1 - f_2\|_{\mathcal{X}^*}$, which further implies that $f_1 = f_2$. Hence, Φ is injective. Hence, Φ

is bijective. Clearly, Φ is uniformly continuous, which follows from the fact that it is norm preserving and conjugate linear.

(iv) Let $\mathbb{K} = \mathbb{R}$. By (iii), Φ is linear. Then, by Definitions 6.28 and 7.24, Φ is an isometrical isomorphism.

(v) Let $\mathbb{K} = \mathbb{C}$. By Remark 7.88, $\forall x_* \in \mathcal{X}^*$, $\forall x \in \mathcal{X}$, we have $\langle\langle x_*, x\rangle\rangle = \langle\langle \phi(x), x_*\rangle\rangle$. Then, $\forall y \in \mathcal{X}$, we have $\langle x, y\rangle = \langle\langle \Phi_{\rm inv}(y), x\rangle\rangle = \langle\langle \phi(x), \Phi_{\rm inv}(y)\rangle\rangle$. We will show that ϕ is surjective, which then implies that ϕ is isometrical isomorphism between $\mathcal{X}$ and $\mathcal{X}^{**}$, and $\mathcal{X}$ is reflexive. $\forall x_{**} \in \mathcal{X}^{**}$, $\langle\langle x_{**}, x_*\rangle\rangle = \overline{\langle\langle x_{**}, \Phi_{\rm inv}(\Phi(x_*))\rangle\rangle}$, $\forall x_* \in \mathcal{X}^*$. Define $G : \mathcal{X} \to \mathbb{K}$ by $G(y) = \overline{\langle\langle x_{**}, \Phi_{\rm inv}(y)\rangle\rangle}$, $\forall y \in \mathcal{X}$. By (iii), it is easy to show that $G \in \mathcal{X}^*$. By (i), $G(y) = \langle y, \Phi(G)\rangle$, $\forall y \in \mathcal{X}$. Then, $\langle\langle x_{**}, x_*\rangle\rangle = \overline{\langle\langle x_{**}, \Phi_{\rm inv}(\Phi(x_*))\rangle\rangle} = \overline{G(\Phi(x_*))} = \overline{\langle \Phi(x_*), \Phi(G)\rangle} = \langle \Phi(G), \Phi(x_*)\rangle = \langle\langle \Phi_{\rm inv}(\Phi(x_*)), \Phi(G)\rangle\rangle = \langle\langle x_*, \Phi(G)\rangle\rangle = \langle\langle \phi(\Phi(G)), x_*\rangle\rangle$, $\forall x_* \in \mathcal{X}^*$. Thus, we have $x_{**} = \phi(\Phi(G)) \in \mathcal{R}(\phi)$. By the arbitrariness of x_{**}, we have ϕ is surjective.

(vi) We will show that $\langle \cdot, \cdot\rangle_{\mathcal{X}^*}$ is an inner product on $\mathcal{X}^*$. $\forall f, g, h \in \mathcal{X}^*$, $\forall \lambda \in \mathbb{K}$. (a) $\langle g, f\rangle_{\mathcal{X}^*} = \langle \Phi(f), \Phi(g)\rangle = \overline{\langle \Phi(g), \Phi(f)\rangle} = \overline{\langle f, g\rangle_{\mathcal{X}^*}}$, where the second equality follows from (i) of Definition 13.1. (b) $\langle f + g, h\rangle_{\mathcal{X}^*} = \langle \Phi(h), \Phi(f + g)\rangle = \langle \Phi(h), \Phi(f) + \Phi(g)\rangle = \langle \Phi(h), \Phi(f)\rangle + \langle \Phi(h), \Phi(g)\rangle = \langle f, h\rangle_{\mathcal{X}^*} + \langle g, h\rangle_{\mathcal{X}^*}$, where the second equality follows from (iii); and the third equality follows from Definition 13.1. (c) $\langle \lambda f, g\rangle_{\mathcal{X}^*} = \langle \Phi(g), \Phi(\lambda f)\rangle = \big\langle \Phi(g), \bar{\lambda}\Phi(f)\big\rangle = \lambda\langle \Phi(g), \Phi(f)\rangle = \lambda\langle f, g\rangle_{\mathcal{X}^*}$, where the second equality follows from (iii); and the third equality follows from Definition 13.1. (d) $\langle f, f\rangle_{\mathcal{X}^*} = \langle \Phi(f), \Phi(f)\rangle = \|\Phi(f)\|_{\mathcal{X}}^2 = \|f\|_{\mathcal{X}^*}^2 \in [0, \infty) \subset \mathbb{R}$, where the second equality follows from Definition 13.1; and the third equality follows from (iii). Clearly, $\langle f, f\rangle_{\mathcal{X}^*} = 0$ if, and only if, $\|f\|_{\mathcal{X}^*} = 0$ if, and only if, $f = \vartheta_{\mathcal{X}^*}$. Hence, $\mathcal{X}^*$ with the inner product $\langle \cdot, \cdot\rangle_{\mathcal{X}^*}$ form a pre-Hilbert space. The induced norm is equal to the $\|\cdot\|_{\mathcal{X}^*}$. By Proposition 7.72, $\mathcal{X}^*$ is complete and hence a Hilbert space. By Proposition 7.90 and (v), $\mathcal{X}^*$ is reflexive.

(vii) Let $\mathbb{K} = \mathbb{C}$. $\forall x \in \mathcal{X}$. Let $f = \Phi_*(x) = \Phi_*(\phi(x))$. Then, $f(x) = \langle x, \Phi(f)\rangle = (\phi(x))(f) = \langle f, \Phi_*(\phi(x))\rangle_{\mathcal{X}^*} = \langle f, \Phi_*(x)\rangle_{\mathcal{X}^*} = \langle f, f\rangle_{\mathcal{X}^*} = \|f\|_{\mathcal{X}^*}^2$, where the first equality follows from (i); the second equality follows from Remark 7.88; the third equality follows from (i) as applied to $\mathcal{X}^*$; and the fourth equality follows from (v). By (iii), $\|f\|_{\mathcal{X}^*} = \|x\|_{\mathcal{X}}$ and $\|f\|_{\mathcal{X}^*} = \|\Phi(f)\|_{\mathcal{X}}$. This implies that $\langle x, \Phi(f)\rangle = \|x\|_{\mathcal{X}}\|\Phi(f)\|_{\mathcal{X}}$. By (iv) of Proposition 13.2, $\Phi(f) = \vartheta_{\mathcal{X}}$ or $x = \lambda\Phi(f)$, for some $\lambda \in \mathbb{C}$.

If $\Phi(f) = \vartheta_{\mathcal{X}}$, by (iii), $f = \vartheta_{\mathcal{X}^*}$. Thus, by (iii), $x = \vartheta_{\mathcal{X}} = \Phi(f)$ since $f = \Phi_*(x)$.

If $\Phi(f) \neq \vartheta_{\mathcal{X}}$, we must have $x = \lambda\Phi(f)$, for some $\lambda \in \mathbb{C}$. Then, $\|f\|_{\mathcal{X}^*}^2 = \langle x, \Phi(f)\rangle = \lambda\langle \Phi(f), \Phi(f)\rangle = \lambda\|\Phi(f)\|_{\mathcal{X}}^2 = \lambda\|f\|_{\mathcal{X}^*}^2$. Thus, $\lambda = 1$. This implies that $x = \Phi(f)$.

In both cases, we have $x = \Phi(f) = \Phi(\Phi_*(x))$. By the arbitrariness of x, we have $\Phi \circ \Phi_* = \mathrm{id}_{\mathcal{X}}$. By (iii) and Proposition 2.4, we have $\Phi_* = \Phi_{\rm inv}$.

(viii) Let $D \subseteq \mathcal{X}$ be a countable dense set. Then, $\Phi_{\rm inv}(D) \subseteq \mathcal{X}^*$ is a countable dense set by (iii). Hence, $\mathcal{X}^*$ is separable.

This completes the proof of the theorem. □

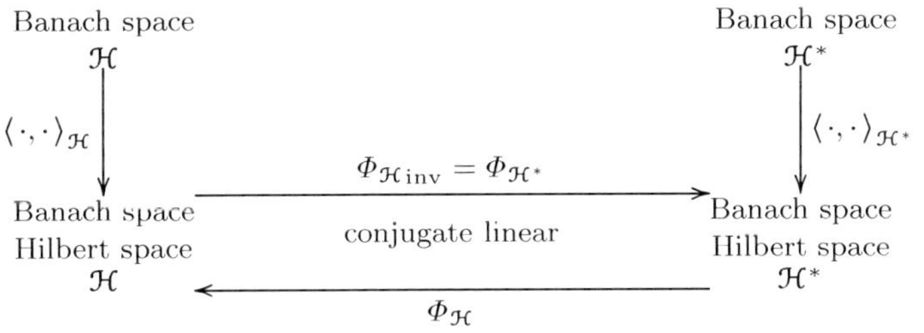

Fig. 13.2 Relationship of a Hilbert space $\mathcal{H}$ and its dual $\mathcal{H}^*$ and $\mathcal{H}$ as Banach space and its dual $\mathcal{H}^*$

Here, we will summarize the relationship between a complex Hilbert space $\mathcal{H}$ and its dual $\mathcal{H}^*$ and the Hilbert space $\mathcal{H}$ as a complex Banach space and its dual $\mathcal{H}^*$. We have the diagram Fig. 13.2.

Definition 13.16 Let $\mathcal{X}$ be a Hilbert space over $\mathbb{K}$, $S \subseteq \mathcal{X}$. By Definition 7.95, $S^{\perp} \subseteq \mathcal{X}^*$ is the set $S^{\perp} = \{x_* \in \mathcal{X}^* \mid \langle\langle x_*, x\rangle\rangle = 0,\ \forall x \in S\}$. We will denote $S^{\perp_h} := \{y \in \mathcal{X} \mid \langle x, y\rangle = \langle\langle y^*, x\rangle\rangle = 0,\ \forall x \in S\}$, which is said to be the *H-orthogonal complement of* S. Note that $S^{\perp_h} = \Phi_{\mathcal{X}}(S^{\perp})$, $S^{\perp} = \Phi_{\mathcal{X}\text{inv}}(S^{\perp_h}) = (\Phi_{\mathcal{X}\text{inv}}(S))^{\perp_h}$, and ${}^{\perp}T = \Phi_{\mathcal{X}}(T^{\perp_h}) = (\Phi_{\mathcal{X}}(T))^{\perp_h}$, $\forall T \subseteq \mathcal{X}^*$. ◇

Proposition 13.17 *Let $\mathcal{X}$ be a Hilbert space over $\mathbb{K}$ and $S, T \subseteq \mathcal{X}$. Then,*

(i) $S^{\perp_h} \subseteq \mathcal{X}$ *is a closed subspace.*
(ii) *If $S \subseteq T$, then $T^{\perp_h} \subseteq S^{\perp_h}$.*
(iii) $S^{\perp_h\perp_h\perp_h} = S^{\perp_h}$.
(iv) $S^{\perp_h\perp_h} = \overline{\text{span}\,(S)}$.

Proof (i) and (ii) follow immediately from Proposition 7.98.
(iv) By Riesz-Fréchet Theorem 13.15, $\mathcal{X}$ is reflexive. Then, $\mathcal{X} = \mathcal{X}^{**}$. Then, by Proposition 7.98, we have $\overline{\text{span}\,(S)} = {}^{\perp}(S^{\perp}) = (\Phi_{\mathcal{X}}(S^{\perp}))^{\perp_h} = S^{\perp_h\perp_h}$.
(iii) By (iv), we have $S \subseteq \overline{\text{span}\,(S)} = S^{\perp_h\perp_h}$. By (ii), we have $S^{\perp_h} \supseteq S^{\perp_h\perp_h\perp_h}$. Again by (iv), we have $S^{\perp_h} \subseteq \overline{\text{span}\left(S^{\perp_h}\right)} = S^{\perp_h\perp_h\perp_h}$. Therefore, $S^{\perp_h} = S^{\perp_h\perp_h\perp_h}$. This completes the proof of the proposition. □

Definition 13.18 Let $\mathcal{X}$ be a vector space over the field $\mathcal{F}$ and $M, N \subseteq \mathcal{X}$ are subspaces. We say that $\mathcal{X}$ is the *direct sum* of M and N if $\forall x \in \mathcal{X}$, $\exists!\, m \in M$ and $\exists!\, n \in N$, such that $x = m + n$. In this case, we will write $\mathcal{X} = M \oplus N$. ◇

Theorem 13.19 *Let M be a closed subspace of a Hilbert space $\mathcal{X}$. Then, $\mathcal{X} = M \oplus M^{\perp_h}$.*

Proof $\forall x \in \mathcal{X}$, by Projection Theorem 13.13, $\exists! m_0 \in M$ such that $n_0 = (x - m_0) \perp M$. Hence, $x = m_0 + n_0$ with $m_0 \in M$ and $n_0 \in M^{\perp_h}$. Clearly, m_0 and n_0 is unique since m_0 is unique. Hence, $\mathcal{X} = M \oplus M^{\perp_h}$. This completes the proof of the theorem. □

Proposition 13.20 *Let $\mathcal{X}$ be a real Hilbert space, $f : \mathcal{X} \times \mathcal{X} \to \mathbb{R}$ be given by $f(x, y) = \langle x, y\rangle$, $\forall (x, y) \in \mathcal{X} \times \mathcal{X}$. Then, f is analytic with arbitrarily large analytic radius, $f^{(1)}(x, y)(h_x, h_y) = \langle h_x, y\rangle + \langle x, h_y\rangle$, $f^{(2)}(x, y)(h_{x2}, h_{y2})(h_{x1}, h_{y1}) = \langle h_{x2}, h_{y1}\rangle + \langle h_{x1}, h_{y2}\rangle$, and $f^{(i+2)}(x, y) = \vartheta_{\mathrm{B_S}_{i+2}(\mathcal{X}\times\mathcal{X},\mathbb{R})}$, $\forall (x, y) \in \mathcal{X}\times\mathcal{X}$, $\forall i \in \mathbb{N}$, $\forall (h_x, h_y) \in \mathcal{X}\times\mathcal{X}$, $\forall (h_{x1}, h_{y1}) \in \mathcal{X}\times\mathcal{X}$, $\forall (h_{x2}, h_{y2}) \in \mathcal{X} \times \mathcal{X}$.*

Proof This follows immediately from Propositions 9.41 and 9.27 and Riesz-Fréchet Theorem 13.15. □

13.4 Hermitian Adjoints

Definition 13.21 Let $\mathcal{X}$ and $\mathcal{Y}$ be Hilbert spaces over $\mathbb{K}$, $A \in \mathrm{B}(\mathcal{X}, \mathcal{Y})$, and $\Phi_{\mathcal{X}} : \mathcal{X}^* \to \mathcal{X}$ and $\Phi_{\mathcal{Y}} : \mathcal{Y}^* \to \mathcal{Y}$ as defined in Riesz-Fréchet Theorem 13.15. Define the *Hermitian adjoint* of A by $A^* := \Phi_{\mathcal{X}} \circ A' \circ \Phi_{\mathcal{Y}\mathrm{inv}} \in \mathrm{B}(\mathcal{Y}, \mathcal{X})$. A is said to be *Hermitian* if $A^* = A$. ◇

Proposition 13.22 *Let $\mathcal{X}$, $\mathcal{Y}$, and $\mathcal{Z}$ be Hilbert spaces over $\mathbb{K}$, $A, B \in \mathrm{B}(\mathcal{X}, \mathcal{Y})$, $C \in \mathrm{B}(\mathcal{Y}, \mathcal{Z})$, $x \in \mathcal{X}$, $y \in \mathcal{Y}$, and $\lambda \in \mathbb{K}$. Then,*

(i) $\mathrm{id}_{\mathcal{X}}^* = \mathrm{id}_{\mathcal{X}}$.
(ii) $(A + B)^* = A^* + B^*$.
(iii) $(\lambda A)^* = \overline{\lambda} A^*$.
(iv) $(CA)^* = A^*C^*$.
(v) *If A is bijective, then* $(A^*)^{-1} = (A^{-1})^* =: A^{-*}$.
(vi) $\langle Ax, y\rangle_{\mathcal{Y}} = \langle x, A^*y\rangle_{\mathcal{X}}$.
(vii) $\langle y, Ax\rangle_{\mathcal{Y}} = \langle A^*y, x\rangle_{\mathcal{X}}$.
(viii) $(A^*)^* = A$;
(ix) $\|A^*\|_{\mathrm{B}(\mathcal{Y},\mathcal{X})} = \|A\|_{\mathrm{B}(\mathcal{X},\mathcal{Y})}$.
(x) $(\mathcal{R}(A))^{\perp_{\mathrm{h}}} = \mathcal{N}(A^*)$ *and* $(\mathcal{R}(A^*))^{\perp_{\mathrm{h}}} = \mathcal{N}(A)$.
(xi) *If $\mathcal{R}(A) \subseteq \mathcal{X}$ is closed, then $\mathcal{R}(A) = (\mathcal{N}(A^*))^{\perp_{\mathrm{h}}}$ and $\mathcal{R}(A^*) = (\mathcal{N}(A))^{\perp_{\mathrm{h}}}$.*
(xii) *Let $(X, \mathcal{B})$ be a measurable space and μ be a $\mathrm{B}(\mathcal{Y}, \mathcal{Z})$-valued measure on $(X, \mathcal{B})$. Define μ^* to be a mapping from $\mathcal{B}$ to $\mathrm{B}(\mathcal{Z}, \mathcal{Y})$ by: $\mu^*(E) = (\mu(E))^*$, if $E \in \mathrm{dom}(\mu)$; and $\mu^*(E)$ is undefined, if $E \in \mathcal{B} \setminus \mathrm{dom}(\mu)$. Then, μ^* is a $\mathrm{B}(\mathcal{Z}, \mathcal{Y})$-valued measure on $(X, \mathcal{B})$ with $\mathcal{P} \circ \mu^* = \mathcal{P} \circ \mu$.*

Proof Let $\Phi_{\mathcal{X}} : \mathcal{X}^* \to \mathcal{X}$, $\Phi_{\mathcal{Y}} : \mathcal{Y}^* \to \mathcal{Y}$, and $\Phi_{\mathcal{Z}} : \mathcal{Z}^* \to \mathcal{Z}$ be as defined in Riesz-Fréchet Theorem 13.15.

(i) $\mathrm{id}_{\mathcal{X}}^* = \Phi_{\mathcal{X}} \circ \mathrm{id}_{\mathcal{X}}' \circ \Phi_{\mathcal{X}\mathrm{inv}} = \Phi_{\mathcal{X}} \circ \mathrm{id}_{\mathcal{X}^*} \circ \Phi_{\mathcal{X}\mathrm{inv}} = \mathrm{id}_{\mathcal{X}}$, where the first equality follows from Definition 13.21; and the second equality follows from Proposition 7.110.

(ii) $(A + B)^* = \Phi_{\mathcal{X}} \circ (A + B)' \circ \Phi_{\mathcal{Y}\mathrm{inv}} = \Phi_{\mathcal{X}} \circ (A' + B') \circ \Phi_{\mathcal{Y}\mathrm{inv}} = \Phi_{\mathcal{X}} \circ A' \circ \Phi_{\mathcal{Y}\mathrm{inv}} + \Phi_{\mathcal{X}} \circ B' \circ \Phi_{\mathcal{Y}\mathrm{inv}} = A^* + B^*$, where the first equality follows from

Definition 13.21; the second equality follows from Proposition 7.110; and the third equality follows from Riesz-Fréchet Theorem 13.15.

(iii) $(\lambda A)^* = \Phi_{\mathcal{X}} \circ (\lambda A)' \circ \Phi_{\mathcal{Y}\text{inv}} = \Phi_{\mathcal{X}} \circ (\lambda A') \circ \Phi_{\mathcal{Y}\text{inv}} = \overline{\lambda} \Phi_{\mathcal{X}} \circ A' \circ \Phi_{\mathcal{Y}\text{inv}} = \overline{\lambda} A^*$, where the first equality follows from Definition 13.21; the second equality follows from Proposition 7.110; and the third equality follows from Riesz-Fréchet Theorem 13.15.

(iv) $(CA)^* = \Phi_{\mathcal{X}} \circ (CA)' \circ \Phi_{\mathcal{Z}\text{inv}} = \Phi_{\mathcal{X}} \circ (A'C') \circ \Phi_{\mathcal{Z}\text{inv}} = \Phi_{\mathcal{X}} \circ A' \circ C' \circ \Phi_{\mathcal{Z}\text{inv}} = \Phi_{\mathcal{X}} \circ A' \circ \Phi_{\mathcal{Y}\text{inv}} \circ \Phi_{\mathcal{Y}} \circ C' \circ \Phi_{\mathcal{Z}\text{inv}} = A^* \circ C^* = A^* C^*$, where the first equality follows from Definition 13.21; and the second equality follows from Proposition 7.110.

(v) By Open Mapping Theorem 7.103, $A^{-1} \in \mathrm{B}(\mathcal{Y}, \mathcal{X})$. Then, we have $(A^{-1})^* = \Phi_{\mathcal{Y}} \circ (A^{-1})' \circ \Phi_{\mathcal{X}\text{inv}} = \Phi_{\mathcal{Y}} \circ (A')^{-1} \circ \Phi_{\mathcal{X}\text{inv}} = (\Phi_{\mathcal{X}} \circ A' \circ \Phi_{\mathcal{Y}\text{inv}})^{-1} = (A^*)^{-1}$, where the first equality follows from Definition 13.21; and the second equality follows from Proposition 7.110.

(vi) $\langle Ax, y \rangle_{\mathcal{Y}} = \langle\langle y^*, Ax \rangle\rangle_{\mathcal{Y}} = \langle\langle A'y^*, x \rangle\rangle_{\mathcal{X}} = \langle x, \Phi_{\mathcal{X}}(A'y^*) \rangle_{\mathcal{X}} = \langle x, \Phi_{\mathcal{X}}(A'(\Phi_{\mathcal{Y}\text{inv}}(y))) \rangle_{\mathcal{X}} = \langle x, A^* y \rangle_{\mathcal{X}}$, where the first equality follows from Riesz-Fréchet Theorem 13.15; the second equality follows from Proposition 7.109; the third and fourth equalities follow from Riesz-Fréchet Theorem 13.15; and the last equality follows from Definition 13.21.

(vii) $\langle y, Ax \rangle_{\mathcal{Y}} = \overline{\langle Ax, y \rangle_{\mathcal{Y}}} = \overline{\langle x, A^* y \rangle_{\mathcal{X}}} = \langle A^* y, x \rangle_{\mathcal{X}}$, where the first equality follows from Definition 13.1; the second equality follows from (vi); and the third equality follows from Definition 13.1.

(viii) $\langle y, Ax \rangle_{\mathcal{Y}} = \langle A^* y, x \rangle_{\mathcal{X}} = \langle y, (A^*)^* x \rangle_{\mathcal{Y}}$, where the first equality follows from (vii); and the second equality follows from (vi). By the arbitrariness of x and y, we have $A = (A^*)^*$.

(ix) $\|A^*\|_{\mathrm{B}(\mathcal{Y},\mathcal{X})} = \sup_{y \in \mathcal{Y},\ \|y\|_{\mathcal{Y}} \leq 1} \|A^* y\|_{\mathcal{X}} = \sup_{y \in \mathcal{Y},\ \|y\|_{\mathcal{Y}} \leq 1} \|\Phi_{\mathcal{X}} \circ A' \circ \Phi_{\mathcal{Y}\text{inv}}(y)\|_{\mathcal{X}} = \|A'\|_{\mathrm{B}(\mathcal{Y}^*, \mathcal{X}^*)} = \|A\|_{\mathrm{B}(\mathcal{X},\mathcal{Y})}$, where the first equality follows from Proposition 7.63; the second equality follows from Definition 13.21; the third equality follows from the fact that Φ is bijective and norm preserving; and the fourth equality follows from Proposition 7.109.

(x) By Proposition 7.112 and Definition 13.16, we have $(\mathcal{R}(A))^{\perp_h} = \Phi_{\mathcal{Y}}(\mathcal{N}(A')) = \mathcal{N}(A^*)$. Then, $(\mathcal{R}(A^*))^{\perp_h} = \mathcal{N}((A^*)^*) = \mathcal{N}(A)$, where the first equality follows from the previous sentence; and the second equality follows from (viii).

(xi) By Propositions 7.112 and 7.114 and Definition 13.16, we have $\mathcal{R}(A) = {}^{\perp}(\mathcal{N}(A')) = {}^{\perp}(\Phi_{\mathcal{Y}\text{inv}}(\Phi_{\mathcal{Y}}(\mathcal{N}(A')))) = (\Phi_{\mathcal{Y}}(\mathcal{N}(A')))^{\perp_h} = (\mathcal{N}(A^*))^{\perp_h}$. Furthermore, $\mathcal{R}(A^*) = (\mathcal{N}((A^*)^*))^{\perp_h} = (\mathcal{N}(A))^{\perp_h}$, where the first equality follows from the previous sentence; the second equality follows from (viii).

(xii) We will prove the result using Definition 11.108. Let $\nu := \mathcal{P} \circ \mu$. Clearly, ν is measure on $(X, \mathcal{B})$. (a) $\nu(\emptyset) = 0$ and $\mu(\emptyset) = \vartheta_{\mathrm{B}(\mathcal{Y},\mathcal{Z})}$. Then, $\mu^*(\emptyset) = (\mu(\emptyset))^* = \vartheta^*_{\mathrm{B}(\mathcal{Y},\mathcal{Z})} = \vartheta_{\mathrm{B}(\mathcal{Z},\mathcal{Y})}$, where the last equality follows from (ii).

(b) $\forall E \in \mathcal{B}$ with $\nu(E) = \infty$, then $\mu(E)$ is undefined and $E \in \mathcal{B} \setminus \mathrm{dom}\,(\mu)$. Then, $\mu^*(E)$ is undefined.

(c) $\forall E \in \mathcal{B}$ with $\nu(E) < \infty$, then $\mu(E) \in B(\mathcal{Y}, \mathcal{Z})$ and $E \in \text{dom}(\mu)$. Then, $\mu^*(E) = (\mu(E))^* \in B(\mathcal{Z}, \mathcal{Y})$. $\forall$ pairwise disjoint $(E_i)_{i=1}^{\infty} \subseteq \mathcal{B}$ with $E = \bigcup_{i=1}^{\infty} E_i$, we have $\sum_{i=1}^{\infty} \|\mu(E_i)\|_{B(\mathcal{Y},\mathcal{Z})} < \infty$ and $\mu(E) = \sum_{i=1}^{\infty} \mu(E_i)$. Then, $\sum_{i=1}^{\infty} \|\mu^*(E_i)\|_{B(\mathcal{Z},\mathcal{Y})} = \sum_{i=1}^{\infty} \|(\mu(E_i))^*\|_{B(\mathcal{Z},\mathcal{Y})} = \sum_{i=1}^{\infty} \|\mu(E_i)\|_{B(\mathcal{Y},\mathcal{Z})} < \infty$, where the second equality follows from (ix). Furthermore, $\sum_{i=1}^{\infty} \mu^*(E_i) = \sum_{i=1}^{\infty} (\mu(E_i))^* = (\sum_{i=1}^{\infty} \mu(E_i))^* = (\mu(E))^* = \mu^*(E)$, where the second equality follows from Propositions 3.66, 7.109, and 7.110, Riesz-Fréchet Theorem 13.15, and Definition 13.21.

(d) $\forall E \in \mathcal{B}$ with $\nu(E) < \infty$, we have

$$\begin{aligned}
\nu(E) &= \sup_{n \in \mathbb{Z}_+,\ (E_i)_{i=1}^n \subseteq \mathcal{B},\ E = \bigcup_{i=1}^n E_i,\ E_i \cap E_j = \emptyset,\ \forall 1 \le i < j \le n} \sum_{i=1}^{n} \|\mu(E_i)\|_{B(\mathcal{Y},\mathcal{Z})} \\
&= \sup_{n \in \mathbb{Z}_+,\ (E_i)_{i=1}^n \subseteq \mathcal{B},\ E = \bigcup_{i=1}^n E_i,\ E_i \cap E_j = \emptyset,\ \forall 1 \le i < j \le n} \sum_{i=1}^{n} \|(\mu(E_i))^*\|_{B(\mathcal{Z},\mathcal{Y})} \\
&= \sup_{n \in \mathbb{Z}_+,\ (E_i)_{i=1}^n \subseteq \mathcal{B},\ E = \bigcup_{i=1}^n E_i,\ E_i \cap E_j = \emptyset,\ \forall 1 \le i < j \le n} \sum_{i=1}^{n} \|\mu^*(E_i)\|_{B(\mathcal{Z},\mathcal{Y})}
\end{aligned}$$

where the first equality follows from Definition 11.108; and the second equality follows from (ix).

This shows that μ^* is a $B(\mathcal{Z}, \mathcal{Y})$-valued measure on $(X, \mathcal{B})$ with $\mathcal{P} \circ \mu^* = \mathcal{P} \circ \mu$. This completes the proof of the proposition. □

13.5 Approximation in Hilbert Spaces

Definition 13.23 Let $\mathcal{X}$ be a pre-Hilbert space and $S \subseteq \mathcal{X}$. We will say that S is an *orthogonal set* if $x \perp y$, $\forall x, y \in S$ with $x \neq y$. The set S is said to be *orthonormal* if S is an orthogonal set and $\|x\| = 1$, $\forall x \in S$. ◇

Proposition 13.24 *Let $\mathcal{X}$ be a pre-Hilbert space over $\mathbb{K}$ and $S \subseteq \mathcal{X}$ be an orthogonal set. Assume that $\vartheta_{\mathcal{X}} \notin S$. Then, S is a linearly independent set.*

Proof Suppose S is not a linearly independent set. Then, $\exists x_0 \in S$ such that $x_0 \in \text{span}(S \setminus \{x_0\})$. Then, $\exists n \in \mathbb{Z}_+$, $\exists x_1, \ldots, x_n \in S \setminus \{x_0\}$ with $x_i \neq x_j$, $\forall 1 \le i < j \le n$, $\exists \alpha_1, \ldots, \alpha_n \in \mathbb{K}$, such that $x_0 = \sum_{i=1}^{n} \alpha_i x_i$. Then, $0 = \|x_0 - \sum_{i=1}^{n} \alpha_i x_i\|^2 = \langle x_0, x_0 \rangle + \sum_{i=1}^{n} \langle \alpha_i x_i, \alpha_i x_i \rangle = \|x_0\|^2 + \sum_{i=1}^{n} |\alpha_i|^2 \|x_i\|^2 > 0$, where the second equality follows from the fact that S is an orthogonal set; and the inequality follows from the fact that $x_0 \neq \vartheta_{\mathcal{X}}$. This is a contradiction. Hence, S is a linearly independent set. This completes the proof of the proposition. □

In Hilbert space, orthonormal sets are greatly favored over other linearly independent sets, and it is reassuring to know that they exist and can be easily constructed. The next theorem presents the Gram–Schmidt orthogonalization procedure.

Theorem 13.25 (Gram–Schmidt Procedure) *Let $n \in \mathbb{N} \cup \{\infty\}$, $\mathcal{X}$ be a pre-Hilbert space over $\mathbb{K}$, and $(x_i)_{i=1}^n \subseteq \mathcal{X}$ be a linearly independent set and vectors in the sequence are distinct. Then, there is an orthonormal sequence $(e_i)_{i=1}^n \subseteq \mathcal{X}$ such that* $\operatorname{span}(\{x_1, \ldots, x_k\}) = \operatorname{span}(\{e_1, \ldots, e_k\})$, $\forall k \in \mathbb{N}$ *with $k \leq n$, and* $\operatorname{span}\left((x_i)_{i=1}^n\right) = \operatorname{span}\left((e_i)_{i=1}^n\right)$.

Proof We construct $(e_i)_{i=1}^n$ recursively as follows.
1° For $i = 1$, clearly $x_1 \neq \vartheta_{\mathcal{X}}$ since $(x_i)_{i=1}^n$ is a linearly independent set. Let $e_1 := x_1/\|x_1\| \in \mathcal{X}$. Then, $\operatorname{span}(\{e_1\}) = \operatorname{span}(\{x_1\})$ and the sequence $(e_i)_{i=1}^1$ is orthonormal. This completes the first step of the construction.
2° Assume that we have constructed $(e_i)_{i=1}^k$, where $k \in \mathbb{N}$ and $k < n$, such that the sequence is orthonormal and $\operatorname{span}(\{e_1, \ldots, e_k\}) = \operatorname{span}(\{x_1, \ldots, x_k\})$.
3° Consider $k + 1$. Let $z_{k+1} := x_{k+1} - \sum_{i=1}^k \langle x_{k+1}, e_i \rangle e_i \in \mathcal{X}$. It is easy to show that $\langle z_{k+1}, e_i \rangle = 0$, $\forall i \in \{1, \ldots, k\}$. Then, $z_{k+1} \perp (e_i)_{i=1}^k$. By the assumption and Lemma 6.49, the vectors $x_1, \ldots, x_{k+1}$ are linearly independent. Then, $z_{k+1} \neq \vartheta_{\mathcal{X}}$. Define $e_{k+1} := z_{k+1}/\|z_{k+1}\| \in \mathcal{X}$. The inductive assumption and the construction imply that $(e_i)_{i=1}^{k+1}$ is orthonormal and $\operatorname{span}(\{e_1, \ldots, e_{k+1}\}) = \operatorname{span}(\{x_1, \ldots, x_{k+1}\})$.

This completes the induction process. Then, there exists an orthonormal $(e_i)_{i=1}^n \subseteq \mathcal{X}$ such that $\operatorname{span}(\{e_1, \ldots, e_k\}) = \operatorname{span}(\{x_1, \ldots, x_k\})$, $\forall k \in \mathbb{N}$ with $k \leq n$.

If $n \in \mathbb{N}$, the theorem is proved. Consider the case when $n = \infty$. We need to show that $\operatorname{span}\left((e_i)_{i=1}^\infty\right) = \operatorname{span}\left((x_i)_{i=1}^\infty\right)$. $\forall x \in \operatorname{span}\left((x_i)_{i=1}^\infty\right)$, $\exists m \in \mathbb{N}$, $\exists \alpha_1, \ldots, \alpha_m \in \mathbb{K}$ such that $x = \sum_{i=1}^m \alpha_i x_i$. Then, $x \in \operatorname{span}(\{x_1, \ldots, x_m\}) = \operatorname{span}(\{e_1, \ldots, e_m\}) \subseteq \operatorname{span}\left((e_i)_{i=1}^\infty\right)$. Hence, we have $\operatorname{span}\left((x_i)_{i=1}^\infty\right) \subseteq \operatorname{span}\left((e_i)_{i=1}^\infty\right)$. By symmetry, we have $\operatorname{span}\left((e_i)_{i=1}^\infty\right) \subseteq \operatorname{span}\left((x_i)_{i=1}^\infty\right)$. Hence, $\operatorname{span}\left((e_i)_{i=1}^\infty\right) = \operatorname{span}\left((x_i)_{i=1}^\infty\right)$. This completes the proof of the theorem. □

Definition 13.26 Let $\mathcal{X}$ be a pre-Hilbert space over $\mathbb{K}$, $n \in \mathbb{N}$, $y_1, \ldots, y_n \in \mathcal{X}$. The *Gram matrix* of $y_1, \ldots, y_n$ is

$$\operatorname{Gram}(y_1, \ldots, y_n) := \begin{bmatrix} \langle y_1, y_1 \rangle & \langle y_1, y_2 \rangle & \cdots & \langle y_1, y_n \rangle \\ \langle y_2, y_1 \rangle & \langle y_2, y_2 \rangle & \cdots & \langle y_2, y_n \rangle \\ \vdots & \vdots & & \vdots \\ \langle y_n, y_1 \rangle & \langle y_n, y_2 \rangle & \cdots & \langle y_n, y_n \rangle \end{bmatrix}$$

The *Gram determinant* of $y_1, \ldots, y_n$ is $\operatorname{gram}(y_1, \ldots, y_n) := \det(\operatorname{Gram}(y_1, \ldots, y_n))$. ◇

Proposition 13.27 *Let $n \in \mathbb{N}$, $\mathcal{X}$ be a pre-Hilbert space over $\mathbb{K}$, $y_1, \ldots, y_n \in \mathcal{X}$. Then,* $\operatorname{gram}(y_1, \ldots, y_n) \neq 0$ *if, and only if, $y_1, \ldots, y_n$ are linearly independent.*

Proof We will prove the equivalent statement that $\text{gram}(y_1, \ldots, y_n) = 0$ if, and only if, $y_1, \ldots, y_n$ are linearly dependent. "Sufficiency" Let $y_1, \ldots, y_n$ be linearly dependent. Then, $\exists \alpha_1, \ldots, \alpha_n \in \mathbb{K}$, which are not all zeros, such that $\sum_{i=1}^{n} \alpha_i y_i = \vartheta$. Then,

$$\begin{aligned}\mathbf{0}_{1\times n} &= \left[\, \langle \sum_{i=1}^{n} \alpha_i y_i, y_1 \rangle \cdots \langle \sum_{i=1}^{n} \alpha_i y_i, y_n \rangle \,\right] \\ &= \left[\, \alpha_1 \cdots \alpha_n \,\right] \text{Gram}(y_1, \ldots, y_n)\end{aligned}$$

This shows that the $\text{rank}(\text{Gram}(y_1, \ldots, y_n)) < n$. Therefore, $\text{gram}(y_1, \ldots, y_n) = 0$.

"Necessity" $\text{gram}(y_1, \ldots, y_n) = 0$ implies that $\text{rank}(\text{Gram}(y_1, \ldots, y_n)) < n$. Then, the row vectors of the Gram matrix is linearly dependent. This implies that $\exists \alpha_1, \ldots, \alpha_n \in \mathbb{K}$, which are not all zeros, such that

$$\begin{aligned}\mathbf{0}_{1\times n} &= \left[\, \alpha_1 \cdots \alpha_n \,\right] \text{Gram}(y_1, \ldots, y_n) \\ &= \left[\, \langle \sum_{i=1}^{n} \alpha_i y_i, y_1 \rangle \cdots \langle \sum_{i=1}^{n} \alpha_i y_i, y_n \rangle \,\right]\end{aligned}$$

Then,

$$\begin{aligned}0 &= \left[\, \langle \sum_{i=1}^{n} \alpha_i y_i, y_1 \rangle \cdots \langle \sum_{i=1}^{n} \alpha_i y_i, y_n \rangle \,\right] \begin{bmatrix} \overline{\alpha_1} \\ \vdots \\ \overline{\alpha_n} \end{bmatrix} \\ &= \sum_{j=1}^{n} \overline{\alpha_j} \langle \sum_{i=1}^{n} \alpha_i y_i, y_j \rangle = \left\langle \sum_{i=1}^{n} \alpha_i y_i, \sum_{j=1}^{n} \alpha_j y_j \right\rangle \\ &= \| \sum_{i=1}^{n} \alpha_i y_i \|^2\end{aligned}$$

Hence, $\sum_{i=1}^{n} \alpha_i y_i = \vartheta$ and $y_1, \ldots, y_n$ are linearly dependent.

This completes the proof of the proposition. □

Theorem 13.28 *Let $\mathcal{X}$ be a pre-Hilbert space over $\mathbb{K}$, $n \in \mathbb{N}$, $y_1, \ldots, y_n \in \mathcal{X}$, $M = \text{span}(\{y_1, \ldots, y_n\})$, $x \in \mathcal{X}$, and $\delta := \min_{m \in M} \|x - m\|$. Then, the following statements hold.*

(i) *There exists a unique $m_0 \in M$ such that $\delta = \|x - m_0\|$ and $m_0 = \sum_{i=1}^{n} \alpha_i y_i$, where $\alpha_1, \ldots, \alpha_n \in \mathbb{K}$ are solutions to the following* normal equation

$$(\text{Gram}(y_1, \ldots, y_n))' \begin{bmatrix} \alpha_1 \\ \vdots \\ \alpha_n \end{bmatrix} = \begin{bmatrix} \langle x, y_1 \rangle \\ \vdots \\ \langle x, y_n \rangle \end{bmatrix} \tag{13.1}$$

(ii) *If the vectors* $y_1, \ldots, y_n$ *are linearly independent, then the Gram matrix* $\text{Gram}(y_1, \ldots, y_n)$ *is invertible and* $\delta^2 = \frac{\text{gram}(y_1,\ldots,y_n,x)}{\text{gram}(y_1,\ldots,y_n)}$.

Proof (i) Clearly, M is a finite-dimensional subspace of $\mathcal{X}$. By Theorem 7.36, M is complete. By Theorem 13.14, $\exists!\, m_0 \in M$ such that $\delta = \|x - m_0\|$ and $(x - m_0) \perp M$. Let $m_0 = \sum_{i=1}^{n} \alpha_i y_i$ with $\alpha_1, \ldots, \alpha_n \in \mathbb{K}$. Then, $\langle x - \sum_{i=1}^{n} \alpha_i y_i, y_j \rangle = 0$, $j = 1, \ldots, n$. By Proposition 13.2, we have $\sum_{i=1}^{n} \alpha_i \langle y_i, y_j \rangle = \langle x, y_j \rangle$, $j = 1, \ldots, n$. Thus, (13.1) holds.

(ii) If vectors $y_1, \ldots, y_n$ are linearly independent, then the Gram determinant $\text{gram}(y_1, \ldots, y_n) \neq 0$, by Proposition 13.27, which further implies that the Gram matrix $\text{Gram}(y_1, \ldots, y_n)$ is invertible. Note that $\delta^2 = \|x - m_0\|^2 = \langle x - m_0, x - m_0 \rangle = \langle x - m_0, x \rangle = \langle x, x \rangle - \sum_{i=1}^{n} \alpha_i \langle y_i, x \rangle$, where the third equality follows from the fact that $(x - m_0) \perp m_0 \in M$. Equivalently, we have $\sum_{i=1}^{n} \alpha_i \langle y_i, x \rangle + \delta^2 = \langle x, x \rangle$. Combining this equation with the normal equation, we have the following matrix equation.

$$\begin{bmatrix} (\text{Gram}(y_1, \ldots, y_n))' & \mathbf{0}_{n\times 1} \\ \langle y_1, x \rangle \cdots \langle y_n, x \rangle & 1 \end{bmatrix} \begin{bmatrix} \alpha_1 \\ \vdots \\ \alpha_n \\ \delta^2 \end{bmatrix} = \begin{bmatrix} \langle x, y_1 \rangle \\ \vdots \\ \langle x, y_n \rangle \\ \langle x, x \rangle \end{bmatrix}$$

By Cramer's Rule, we have $\delta^2 = \frac{\det(G_1)}{\det(G)}$, where

$$G = \begin{bmatrix} \langle y_1, y_1 \rangle & \cdots & \langle y_n, y_1 \rangle & 0 \\ \vdots & & \vdots & \vdots \\ \langle y_1, y_n \rangle & \cdots & \langle y_n, y_n \rangle & 0 \\ \langle y_1, x \rangle & \cdots & \langle y_n, x \rangle & 1 \end{bmatrix}; \qquad G_1 = (\text{Gram}(y_1, \ldots, y_n, x))'$$

By Laplace Expansion Theorem, we have $\det(G) = \det((\text{Gram}(y_1, \ldots, y_n))') = \det(\text{Gram}(y_1, \ldots, y_n)) = \text{gram}(y_1, \ldots, y_n)$. Clearly, $\det(G_1) = \text{gram}(y_1, \ldots, y_n, x)$. Hence, $\delta^2 = \frac{\text{gram}(y_1,\ldots,y_n,x)}{\text{gram}(y_1,\ldots,y_n)}$.

This completes the proof of the theorem. □

In Gram Theorem 13.28, when $y_1, \ldots, y_n$ are orthonormal, we denote y_i by e_i, $i = 1, \ldots, n$. Then, $\alpha_i = \langle x, e_i \rangle$, $i = 1, \ldots, n$, $m_0 = \sum_{i=1}^{n} \langle x, e_i \rangle e_i$, and $\delta = \sqrt{\langle x, x \rangle - \sum_{i=1}^{n} |\langle x, e_i \rangle|^2}$.

Proposition 13.29 *Let* $\mathcal{X}$ *be Hilbert space over* $\mathbb{K}$, $(e_i)_{i=1}^{\infty} \subseteq \mathcal{X}$ *be orthonormal, and* $(\xi_i)_{i=1}^{\infty} \subseteq \mathbb{K}$. *Then,* $x = \sum_{i=1}^{\infty} \xi_i e_i \in \mathcal{X}$ *if, and only if,* $\sum_{i=1}^{\infty} |\xi_i|^2 < \infty$. *In this case,* $\xi_i = \langle x, e_i \rangle$, $\forall i \in \mathbb{N}$.

Proof Define the partial sum : $s_i := \sum_{j=1}^{i} \xi_j e_j \in \mathcal{X}$, $\forall i \in \mathbb{N}$.

"Sufficiency" Let $\sum_{i=1}^{\infty} |\xi_i|^2 < \infty$. $\forall \epsilon \in (0, \infty) \subset \mathbb{R}$, $\exists n_0 \in \mathbb{N}$ such that $\sum_{i=n_0+1}^{\infty} |\xi_i|^2 < \epsilon^2$. $\forall n, m \in \mathbb{N}$ with $n_0 \le n \le m$, $\|s_n - s_m\|^2 = \left\langle \sum_{i=n+1}^{m} \xi_i e_i, \sum_{i=n+1}^{m} \xi_i e_i \right\rangle = \sum_{i=n+1}^{m} |\xi_i|^2 < \epsilon^2$, where the second equality follows from the assumption that $(e_i)_{i=1}^{\infty}$ is orthonormal. Hence, $\|s_n - s_m\| < \epsilon$ and $(s_i)_{i=1}^{\infty} \subseteq \mathcal{X}$ is a Cauchy sequence. By the completeness of $\mathcal{X}$, $\sum_{i=1}^{\infty} \xi_i e_i = \lim_{n \in \mathbb{N}} s_n =: x \in \mathcal{X}$.

"Necessity" Let $x = \sum_{i=1}^{\infty} \xi_i e_i \in \mathcal{X}$. Then, $x = \lim_{n \in \mathbb{N}} s_n$ and $(s_n)_{n=1}^{\infty} \subseteq \mathcal{X}$ is a Cauchy sequence. $\forall \epsilon \in (0, \infty) \subset \mathbb{R}$, $\exists n_0 \in \mathbb{N}$, $\forall n, m \in \mathbb{N}$ with $n_0 \le n \le m$, we have $\|s_n - s_m\| < \sqrt{\epsilon}$. Then, $\sum_{i=n+1}^{m} |\xi_i|^2 = \left\langle \sum_{i=n+1}^{m} \xi_i e_i, \sum_{i=n+1}^{m} \xi_i e_i \right\rangle = \|s_n - s_m\|^2 < \epsilon$. Hence, $\sum_{i=1}^{\infty} |\xi|^2 < \infty$.

When $x = \sum_{i=1}^{\infty} \xi_i e_i \in \mathcal{X}$, by Propositions 13.4 and 3.66, we have $\langle x, e_i \rangle = \lim_{n \in \mathbb{N}} \langle s_n, e_i \rangle = \xi_i$, $\forall i \in \mathbb{N}$. This completes the proof of the proposition. □

Proposition 13.30 (Bessel's Inequality) *Let $\mathcal{X}$ be a pre-Hilbert space over $\mathbb{K}$, $x \in \mathcal{X}$, and $(e_i)_{i=1}^{\infty} \subseteq \mathcal{X}$ be orthonormal. Then, $\sum_{i=1}^{\infty} |\langle x, e_i \rangle|^2 \le \|x\|^2$.*

Proof Let $\alpha_i := \langle x, e_i \rangle \in \mathbb{K}$, $\forall i \in \mathbb{N}$. $\forall n \in \mathbb{N}$, $0 \le \|x - \sum_{i=1}^{n} \alpha_i e_i\|^2 = \|x\|^2 - \sum_{i=1}^{n} \alpha_i \langle e_i, x \rangle - \sum_{i=1}^{n} \overline{\alpha_i} \langle x, e_i \rangle + \sum_{i=1}^{n} |\alpha_i|^2 = \|x\|^2 - \sum_{i=1}^{n} |\alpha_i|^2$. Hence, $\sum_{i=1}^{\infty} |\langle x, e_i \rangle|^2 \le \|x\|^2$. □

Proposition 13.31 *Let $\mathcal{X}$ be a Hilbert space over $\mathbb{K}$, $x \in \mathcal{X}$, and $(e_i)_{i=1}^{\infty} \subseteq \mathcal{X}$ be orthonormal. Then, $m_0 := \sum_{i=1}^{\infty} \langle x, e_i \rangle e_i \in \overline{\operatorname{span}\left((e_i)_{i=1}^{\infty}\right)} =: M$, $(x - m_0) \perp M$, and $\|x - m_0\| = \min_{m \in M} \|x - m\|$.*

Proof By Bessel's Inequality 13.30, $\sum_{i=1}^{\infty} |\langle x, e_i \rangle|^2 \le \|x\|^2 < \infty$. By Proposition 13.29, $m_0 := \sum_{i=1}^{\infty} \langle x, e_i \rangle e_i \in \mathcal{X}$. Since $\sum_{i=1}^{n} \langle x, e_i \rangle e_i \in M$, $\forall n \in \mathbb{N}$, then $m_0 \in M$ by Proposition 4.13. By Propositions 13.4 and 3.66, $\langle x - m_0, e_j \rangle = \lim_{n \in \mathbb{N}} \langle x - \sum_{i=1}^{n} \langle x, e_i \rangle e_i, e_j \rangle = \langle x, e_j \rangle - \langle x, e_j \rangle \|e_j\|^2 = 0$, $\forall j \in \mathbb{N}$. Hence, $(x - m_0) \in ((e_i)_{i=1}^{\infty})^{\perp_h} = ((e_i)_{i=1}^{\infty})^{\perp_h \perp_h \perp_h} = \left(\overline{\operatorname{span}\left((e_i)_{i=1}^{\infty}\right)}\right)^{\perp_h} = M^{\perp_h}$, where the first and the second equalities follow from Proposition 13.17. By Projection Theorem 13.13, $\|x - m_0\| = \min_{m \in M} \|x - m\|$. This completes the proof of the proposition. □

Based on Proposition 13.31, it is apparent that if $M = \mathcal{X}$, then, $\forall x \in \mathcal{X}$ can be expanded as a infinite series of e_i's with coefficients equal to the Fourier coefficients $\langle x, e_i \rangle$. This motivates the following definition and proposition.

Definition 13.32 Let $\mathcal{X}$ be a Hilbert space, $n \in \mathbb{Z}_+ \cup \{\infty\}$, and $(e_i)_{i=1}^{n} \subseteq \mathcal{X}$ be orthonormal. The sequence $(e_i)_{i=1}^{n}$ is said to be *complete* if $\overline{\operatorname{span}\left((e_i)_{i=1}^{n}\right)} = \mathcal{X}$. ◇

Proposition 13.33 *Let $\mathcal{X}$ be a Hilbert space, $n \in \mathbb{Z}_+ \cup \{\infty\}$, and $(e_i)_{i=1}^{n} \subseteq \mathcal{X}$ be orthonormal. The sequence is complete if, and only if, $((e_i)_{i=1}^{n})^{\perp_h} = \{\vartheta\}$.*

Proof By Proposition 13.17, $((e_i)_{i=1}^{n})^{\perp_h} = \{\vartheta\} \Leftrightarrow ((e_i)_{i=1}^{n})^{\perp_h \perp_h \perp_h} = \{\vartheta\} \Leftrightarrow \left(\overline{\operatorname{span}\left((e_i)_{i=1}^{n}\right)}\right)^{\perp_h} = \{\vartheta\} \Leftrightarrow \overline{\operatorname{span}\left((e_i)_{i=1}^{n}\right)} = \mathcal{X}$. This completes the proof of the proposition. □

Example 13.34 Let $I := [-1, 1] \subset \mathbb{R}$, $\mathbb{I} := ((I, |\cdot|), \mathcal{B}, \mu)$ be the complete metric measure subspace of $\mathbb{R}$, $\mathcal{Z} := \mathbf{L}_2(\mathbb{I}, \mathbb{K})$, $z_i \in \bar{\mathbf{L}}_2(\mathbb{I}, \mathbb{K}) =: \bar{\mathcal{Z}}$ be defined by $z_i(t) = t^i$, $\forall t \in I$, $\forall i \in \mathbb{Z}_+$. It is easy to show that $([z_i])_{i=0}^{\infty} \subseteq \mathcal{Z}$ is a linearly independent set. Then, the Gram–Schmidt Procedure 13.25 can be applied to this sequence to produce an orthonormal sequence $([e_i])_{i=0}^{\infty} \subseteq \mathcal{Z}$, where $e_i \in \bar{\mathcal{Z}}$, $\forall i \in \mathbb{Z}_+$, are ith-order polynomials. Let $p_i \in \bar{\mathcal{Z}}$ be defined by $p_i(t) = \frac{(-1)^i}{2^i i!}(1-t^2)^i$, $\forall t \subset I$, $\forall i \in \mathbb{Z}_+$, and $P_i \in \bar{\mathcal{Z}}$ be defined by $P_i(t) = p_i^{(i)}(t)$, $\forall t \in I$, $\forall i \in \mathbb{Z}_+$. Then, we will show that $e_i = \sqrt{\frac{2i+1}{2}} P_i$, $\forall i \in \mathbb{Z}_+$, and $([e_i])_{i=0}^{\infty} \subseteq \mathcal{Z}$ is a complete orthonormal sequence. As defined, $(P_i)_{i=0}^{\infty} \subseteq \bar{\mathcal{Z}}$ is the well-known *Legendre polynomials*.

We will first show that $([P_i])_{i=0}^{\infty}$ is an orthogonal set. $\forall n, m \in \mathbb{Z}_+$, we have

$$
\begin{aligned}
\langle [P_n], [P_m] \rangle &= \int_{-1}^{1} P_n(t)\overline{P_m(t)}\,\mathrm{d}t = \int_{-1}^{1} p_n^{(n)}(t)p_m^{(m)}(t)\,\mathrm{d}t \\
&= p_n^{(n-1)}(t)p_m^{(m)}(t)\Big|_{-1}^{1} - \int_{-1}^{1} p_n^{(n-1)}(t)p_m^{(m+1)}(t)\,\mathrm{d}t \\
&= -\int_{-1}^{1} p_n^{(n-1)}(t)p_m^{(m+1)}(t)\,\mathrm{d}t \\
&\;\vdots \\
&= (-1)^n \int_{-1}^{1} p_n^{(0)}(t)p_m^{(m+n)}(t)\,\mathrm{d}t
\end{aligned}
$$

where the first equality follows from Examples 13.8 and 13.11; the third equality follows from Integration by Parts Theorem 12.89; the fourth equality follows the fact $p_n^{(i)}(x) = 0$, $\forall i \in \mathbb{Z}_+$ with $0 \le i < n$, $x \in \{1, -1\}$; and the last equality follows from the recursive application of the Integration by Parts Theorem 12.89. Thus, if $m \neq n$, we may without loss of generality, assume that $m < n$. Then, we have $p_m^{(m+n)} = \vartheta_{\bar{\mathcal{Z}}}$ since p_m is a polynomial of order $2m < m + n$. Then, $\langle [P_n], [P_m] \rangle = 0$. Hence, $([P_i])_{i=0}^{\infty}$ is an orthogonal set.

Next, we show that $\|[P_i]\|_2 = \sqrt{\frac{2}{2i+1}}$, $\forall i \in \mathbb{Z}_+$. Clearly, $P_0(t) = 1$, $\forall t \in I$ and $\|[P_0]\|_2^2 = \langle [P_0], [P_0] \rangle = 2 = \frac{2}{2\cdot 0+1}$. Recursively, assume that $\|[P_{n-1}]\|_2 = \sqrt{\frac{2}{2n-1}}$, for some $(n-1) \in \mathbb{Z}_+$. Then,

$$
\begin{aligned}
\|[P_n]\|_2^2 &= \langle [P_n], [P_n] \rangle = (-1)^n \int_{-1}^{1} p_n^{(0)}(t)p_n^{(2n)}(t)\,\mathrm{d}t \\
&= \frac{(2n)!}{2^{2n}n!n!}\int_{-1}^{1}(1-t^2)^n\,\mathrm{d}t
\end{aligned}
$$

$$= \frac{(2n)!}{2^{2n}n!n!}(1-t^2)^n t\Big|_{-1}^{1} + \frac{2n(2n)!}{2^{2n}n!n!}\int_{-1}^{1} t^2(1-t^2)^{n-1}\,dt$$

$$= \frac{2n(2n)!}{2^{2n}n!n!}\Big(\int_{-1}^{1}(1-t^2)^{n-1}\,dt - \int_{-1}^{1}(1-t^2)^n\,dt\Big)$$

$$= (2n-1)\|[P_{n-1}]\|_2^2 - 2n\|[P_n]\|_2^2$$

where the second equality follows from the previous paragraph; the third equality follows from the definition of $p_n \in \bar{\mathcal{Z}}$; the fourth equality follows from Integration by Parts Theorem 12.89; the fifth equality follows from Proposition 11.92; and the last equality follows from the third equality. Therefore, we have $\|[P_n]\|_2^2 = \frac{2}{2n+1}$. This completes the induction process.

Clearly, P_i is an ith order polynomial with positive leading coefficient, then e_i is also an ith order polynomial with positive leading coefficient. By the above, we have $([e_i])_{i=0}^{\infty}$ is orthonormal and $\operatorname{span}(\{e_0, \ldots, e_n\}) = \operatorname{span}(\{z_0, \ldots, z_n\})$, $\forall n \in \mathbb{Z}_+$. Then, $([e_i])_{i=0}^{\infty}$ is the sequence generated by the Gram–Schmidt Procedure on $([z_i])_{i=0}^{\infty}$.

Finally, we will show that $([e_i])_{i=0}^{\infty} \subseteq \mathcal{Z}$ is complete by Proposition 13.33. All we need to show is that $\forall z \in \bar{\mathcal{Z}}$ with $\langle [z], [z_i] \rangle = 0$, $\forall i \in \mathbb{Z}_+$, we have $[z] = \vartheta_{\mathcal{Z}}$. Since $\mu(I) = 2 < \infty$, by Cauchy–Schwarz Inequality, we have $\int_{-1}^{1} \mathcal{P} \circ z\,dt \leq (\int_{-1}^{1} \mathcal{P}_2 \circ z\,dt)^{1/2}(\int_{-1}^{1} 1\,dt)^{1/2} = \sqrt{2}\|z\|_2 < \infty$. Then, $z \in \bar{\mathrm{L}}_1(\mathbb{I}, \mathbb{K})$. Define $F : I \to \mathbb{K}$ by $F(t) = \int_{-1}^{t} z(\tau)\,d\tau$, $\forall t \in I$. By Proposition 12.75, F is absolutely continuous and therefore continuous. Since $[z] \perp [z_0]$, then $F(-1) = 0 = F(1) = \langle [z], [z_0] \rangle$. By Proposition 11.37, F is $\mathcal{B}$-measurable. Then, $F \in \bar{\mathcal{Z}}$. $\forall i \in \mathbb{Z}_+$, we have

$$\langle [F], [z_i] \rangle = \int_{-1}^{1} F(t)t^i\,dt = \frac{t^{i+1}}{i+1}F(t)\Big|_{-1}^{1} - \int_{-1}^{1} \frac{t^{i+1}}{i+1} z(t)\,dt = 0$$

where the second equality follows from Integration by Parts Theorem 12.89. Hence, $[F] \perp ([z_i])_{i=0}^{\infty}$. Since F is continuous, by Bernsteĭn Approximation Theorem 11.203, $\forall \epsilon \in (0, \infty) \subset \mathbb{R}$, $\exists$ polynomial $Q : I \to \mathbb{K}$, such that $\|F - Q\|_{\mathcal{C}(I,\mathbb{K})} < \epsilon$. Then,

$$\|[F]\|_2^2 = \int_{-1}^{1} |F(t)|^2\,dt = \int_{-1}^{1} F(t)\overline{(F(t) - Q(t))}\,dt$$
$$\leq \|[F]\|_2 \|[F(t) - Q(t)]\|_2 \leq \|[F]\|_2 \sqrt{2}\epsilon$$

where the second equality follows from the fact that $[F] \perp ([z_i])_{i=0}^{\infty}$; and the first inequality follows from Cauchy–Schwarz Inequality. Then, $\|[F]\|_2 \leq \sqrt{2}\epsilon$. By the arbitrariness of ϵ, we have $\|[F]\|_2 = 0$. Then $[F] = \vartheta_{\mathcal{Z}}$. Since F is continuous, then $F(t) = 0$, $\forall t \in I$. Then, by Fundamental Theorem of Calculus I 12.86, $z = F^{(1)}$ a.e. in $\mathbb{I}$. Then, $[z] = \vartheta_{\mathcal{Z}}$.

Then, by Proposition 13.33, $([e_i])_{i=0}^{\infty}$ is complete. ◇

Proposition 13.35 *Let $S_1 = [0, 2\pi] \subset \mathbb{R}$ and $z \in \mathcal{C}(S_1, \mathbb{C}) =: \mathcal{Z}$. Assume that $z(0) = z(2\pi)$. Let $M = \{\bar{z} \in \mathcal{Z} \mid \bar{z}(x) = \cos(nx), \forall x \in S_1, \forall n \in \mathbb{Z}_+;$ or $\bar{z}(x) = \sin(nx), \forall x \in S_1, \forall n \in \mathbb{N}\} \subseteq \mathcal{Z}$. Let $\mathcal{M} = \operatorname{span}(M) \subseteq \mathcal{Z}$. Then, $z \in \overline{\mathcal{M}}$.*

Proof Let $\mathbb{S}_1 \subset \mathbb{R}^2$ be the unit circle as defined in Example 12.104. Define a mapping $\Psi : S_1 \to \mathbb{S}_1$ by $\Psi(x) = (\cos(x), \sin(x))$, $\forall x \in S_1$. Clearly, Ψ is surjective and continuous. Note that S_1 is compact and $\mathbb{S}_1$ is Hausdorff. It is obvious that we may define a function $\Phi : \mathbb{S}_1 \to \mathbb{C}$ such that $\Phi \circ \Psi = z$. By Proposition 5.18, we have Φ is continuous. Note that $\mathbb{S}_1$ is closed and bounded in $\mathbb{R}^2$, then $\mathbb{S}_1$ is compact by Proposition 5.40. Hence, $\Phi \in \mathcal{C}(\mathbb{S}_1, \mathbb{C})$. Let $\Phi_r := \operatorname{Re} \circ \Phi \in \mathcal{C}(\mathbb{S}_1, \mathbb{R})$ and $\Phi_i := \operatorname{Im} \circ \Phi \in \mathcal{C}(\mathbb{S}_1, \mathbb{R})$. By Tietze Extension Theorem 3.57, there exists $\bar{\Phi}_r \in \mathcal{C}(I^2, \mathbb{R})$ and $\bar{\Phi}_i \in \mathcal{C}(I^2, \mathbb{R})$, where $I = [-1, 1] \subset \mathbb{R}$, such that $\bar{\Phi}_r\big|_{\mathbb{S}_1} = \Phi_r$ and $\bar{\Phi}_i\big|_{\mathbb{S}_1} = \Phi_i$. Let $\bar{\Phi} := \bar{\Phi}_r + \mathrm{i}\bar{\Phi}_i \in \mathcal{C}(I^2, \mathbb{C})$. Then, $\bar{\Phi}\big|_{\mathbb{S}_1} = \Phi$. By Bernsteĭn Approximation Theorem 11.203, $\forall \epsilon \in (0, \infty) \subset \mathbb{R}$, there exists a polynomial $P : I^2 \to \mathbb{C}$ such that $\left|\bar{\Phi}(y_1, y_2) - P(y_1, y_2)\right| < \epsilon$, $\forall y_1, y_2 \in I^2$. Then, $|z(x) - P \circ \Psi(x)| = |\Phi \circ \Psi(x) - P \circ \Psi(x)| < \epsilon, \forall x \in S_1$. Note that $P \circ \Psi \in \mathcal{M}$, since, $\forall \gamma, \theta \in \mathbb{R}$,

$$(\sin(\theta))^2 = \frac{1}{2}(1 - \cos(2\theta)); \qquad (\cos(\theta))^2 = \frac{1}{2}(1 + \cos(2\theta));$$

$$\sin(\gamma)\cos(\theta) = \frac{1}{2}(\sin(\gamma + \theta) + \sin(\gamma - \theta));$$

$$\sin(\gamma)\sin(\theta) = \frac{1}{2}(\cos(\gamma - \theta) - \cos(\gamma + \theta));$$

$$\cos(\gamma)\cos(\theta) = \frac{1}{2}(\cos(\gamma - \theta) + \cos(\gamma + \theta))$$

Hence, $z \in \overline{\mathcal{M}}$, by the arbitrariness of ϵ. This completes the proof of the proposition. □

Example 13.36 Let $\mathbb{I} := ((\mathrm{r}_{-\pi,\pi}, |\cdot|), \mathcal{B}, \mu)$ be the finite metric measure subspace of $\mathbb{R}$; $\mathcal{Z} := \mathrm{L}_2(\mathbb{I}, \sqrt{\pi}\mathbb{K})$ be the Hilbert space as defined in Example 13.11. Then, the inner product on $\mathcal{Z}$, $\langle \cdot, \cdot \rangle_{\mathcal{Z}}$, is defined by $\langle [g], [h] \rangle_{\mathcal{Z}} = \frac{1}{\pi} \int_{\mathbb{I}} g(t)\overline{h(t)}\, \mathrm{d}t$, $\forall g, h \in \bar{\mathcal{Z}}$, where $\bar{\mathcal{Z}} := \bar{\mathrm{L}}_2(\mathbb{I}, \sqrt{\pi}\mathbb{K})$. Let $(e_n)_{n=0}^{\infty} \in \bar{\mathcal{Z}}$ be defined by $e_0(t) = \frac{1}{\sqrt{2}}$, $\forall t \in \mathbb{I}$, $e_{2n-1}(t) = \sin(nt)$, $\forall t \in \mathbb{I}$, $e_{2n}(t) = \cos(nt)$, $\forall t \in \mathbb{I}$, $\forall n \in \mathbb{N}$; and $\forall f \in \bar{\mathcal{Z}}$. We will show that $a_{2n} := \langle [f], [e_{2n}] \rangle_{\mathcal{Z}} \in \mathbb{K}$, $\forall n \in \mathbb{Z}_+$; $a_{2n-1} := \langle [f], [e_{2n-1}] \rangle_{\mathcal{Z}} \in \mathbb{K}$, $\forall n \in \mathbb{N}$, $([e_n])_{n=0}^{\infty} \subseteq \mathcal{Z}$ is a complete orthonormal sequence, and $[f] = \sum_{n=0}^{\infty} a_n [e_n]$ in $\mathcal{Z}$. Then, $\sum_{n=0}^{\infty} a_n e_n$ is called the *Fourier series for f*.

By the finiteness of measure μ, we have $\mathrm{L}_2(\mathbb{I}, \sqrt{\pi}\mathbb{K}) \subseteq \mathrm{L}_1(\mathbb{I}, \sqrt{\pi}\mathbb{K})$. Then, f is absolutely integrable over $\mathbb{I}$. By Propositions 7.23, 11.38, and 11.39, $f e_n$ is $\mathcal{B}$-measurable and therefore absolutely integrable over $\mathbb{I}$, $\forall n \in \mathbb{Z}_+$. Hence, $a_n \in \mathbb{K}$, $\forall n \in \mathbb{Z}_+$. It is straightforward to check that $([e_n])_{n=0}^{\infty}$ is an orthonormal sequence in $\mathcal{Z}$. We will show that it is complete by Proposition 13.33. Fix any

$[z] \in \left(([e_n])_{n=0}^{\infty}\right)^{\perp_h}$, where $z \in \bar{\mathcal{Z}}$. Then, z is $\mathcal{B}$-measurable and absolutely integrable over $\mathbb{I}$. Define $F : \bar{\mathrm{r}}_{-\pi,\pi} \to \mathbb{K}$ by $F(t) = \int_{-\pi}^{t} z(s)\,\mathrm{d}s$, $\forall t \in \bar{\mathrm{r}}_{-\pi,\pi}$. By Proposition 12.75, F is absolutely continuous. Clearly $F(-\pi) = 0$. Since $[z] \perp [e_0]$, then $F(\pi) = 0$. $\forall n \in \mathbb{N}$,

$$\langle [F|_{\mathbb{I}}], [e_{2n-1}] \rangle_{\mathcal{Z}} = \frac{1}{\pi}\int_{-\pi}^{\pi} F(t)\sin(nt)\,\mathrm{d}t$$
$$= -\frac{1}{n\pi}\cos(nt)F(t)|_{t=-\pi}^{\pi} + \frac{1}{n\pi}\int_{-\pi}^{\pi}\cos(nt)z(t)\,\mathrm{d}t = 0$$

where the second equality follows from Integration by Parts Theorem 12.89; and the last equality follows from the preceding discussion and the fact $[z] \perp [e_{2n}]$. $\forall n \in \mathbb{N}$,

$$\langle [F|_{\mathbb{I}}], [e_{2n}] \rangle_{\mathcal{Z}} = \frac{1}{\pi}\int_{-\pi}^{\pi} F(t)\cos(nt)\,\mathrm{d}t$$
$$= \frac{1}{n\pi}\sin(nt)F(t)|_{t=-\pi}^{\pi} - \frac{1}{n\pi}\int_{-\pi}^{\pi}\sin(nt)z(t)\,\mathrm{d}t = 0$$

where the second equality follows from Integration by Parts Theorem 12.89; and the last equality follows from the preceding discussion and the fact $[z] \perp [e_{2n-1}]$. Hence, $\left[F|_{\mathbb{I}}\right] \perp \left(([e_i])_{i=1}^{\infty}\right)$.

Fix any $\epsilon \in (0,\infty) \subset \mathbb{R}$. Since $F \in \mathcal{C}(\bar{\mathrm{r}}_{-\pi,\pi}, \mathbb{K}) =: \mathcal{Y}$ and $F(-\pi) = F(\pi) = 0$, by Proposition 13.35, $\exists Q \in \operatorname{span}\left((\bar{e}_i)_{i=0}^{\infty}\right) \subseteq \mathcal{Y}$ such that $\|F - Q\|_{\mathcal{Y}} < \frac{\epsilon^2}{3\sqrt{2}(1+\|[F|_{\mathbb{I}}]\|_{\mathcal{Z}})}$, where $\bar{e}_i$ is the continuous extension of e_i from $\mathbb{I}$ to $\bar{\mathrm{r}}_{-\pi,\pi}$, $\forall i \in \mathbb{Z}_+$. Let $Q := \sum_{i=0}^{m} \alpha_i \bar{e}_i$, where $\alpha_i \in \mathbb{K}$, $i = 1, \ldots, m$. Then, we have

$$\|[F|_{\mathbb{I}}]\|_{\mathcal{Z}}^2 = \frac{1}{\pi}\int_{-\pi}^{\pi} F(t)\overline{F(t)}\,\mathrm{d}t$$
$$= \frac{1}{\pi}\int_{-\pi}^{\pi} F(t)\overline{(F(t) - Q(t))}\,\mathrm{d}t + \frac{1}{\pi}\int_{-\pi}^{\pi} F(t)\overline{\alpha_0} e_0(t)\,\mathrm{d}t$$
$$= \left|\frac{1}{\pi}\int_{-\pi}^{\pi} F(t)\overline{(F(t) - Q(t))}\,\mathrm{d}t + \frac{1}{\pi}\int_{-\pi}^{\pi} F(t)\overline{\alpha_0} e_0(t)\,\mathrm{d}t\right|$$
$$\leq \left|\frac{1}{\pi}\int_{-\pi}^{\pi} F(t)\overline{(F(t) - Q(t))}\,\mathrm{d}t\right| + \left|\frac{1}{\pi}\int_{-\pi}^{\pi} F(t)\overline{\alpha_0} e_0(t)\,\mathrm{d}t\right|$$
$$\leq \|[F|_{\mathbb{I}}]\|_{\mathcal{Z}} \|[(F - Q)|_{\mathbb{I}}]\|_{\mathcal{Z}} + \left|\frac{\overline{\alpha_0}}{\sqrt{2}\pi} tF(t)|_{t=-\pi}^{\pi} - \frac{\overline{\alpha_0}}{\sqrt{2}\pi}\int_{-\pi}^{\pi} tz(t)\,\mathrm{d}t\right|$$
$$\leq \frac{\epsilon^2}{3} + \frac{|\alpha_0|}{\sqrt{2}\pi}\left|\int_{-\pi}^{\pi} tz(t)\,\mathrm{d}t\right|$$

where the second equality follows from the previous paragraph; the third equality follows from the fact that the left-hand side is a nonnegative real number; the second inequality follows from the Cauchy–Schwarz Inequality and Integration by Parts Theorem 12.89; and the last inequality follows from the preceding discussion.

Define $f_1 : \mathrm{r}_{-\pi,\pi} \to \mathbb{K}$ by $f_1(t) = t$, $\forall t \in \mathrm{r}_{-\pi,\pi}$. Then, $f_1 \in \bar{\mathrm{L}}_2(\mathbb{I}, \mathbb{K})$. Let $g_1 : \mathrm{r}_{-\pi,\pi} \to \mathbb{S}_1$ be defined by $g_1(x) = (\cos(x), \sin(x)) \in \mathbb{S}_1$, $\forall x \in \mathrm{r}_{-\pi,\pi}$, where $\mathbb{S}_1$ is defined in Example 12.104. By Example 12.105, g_1 is an isomeasure between $\mathbb{I}$ and $\mathbb{S}_1$. Then, $f_1 \circ g_{1\,\mathrm{inv}} \in \bar{\mathrm{L}}_2(\mathbb{S}_1, \mathbb{K}) =: \bar{\mathcal{Y}}_1$. By Proposition 11.182, $\exists f_2 \in \bar{\mathrm{L}}_2(\mathbb{S}_1, \mathbb{K})$ such that $f_2 \in \mathcal{C}(\mathbb{S}_1, \mathbb{K}) =: \mathcal{W}$ and $\| f_1 \circ g_{1\,\mathrm{inv}} - f_2 \|_{\bar{\mathcal{Y}}_1} < \frac{\sqrt{2\pi}\epsilon^2}{3(1+|\alpha_0|\|[z]\|_{\mathcal{Z}})}$. Then, by the proof of Proposition 13.35, $\exists h \in \mathcal{W}$ with h being a polynomial in two coordinates, such that $\| f_2 - h \|_{\mathcal{W}} < \frac{\epsilon^2}{3(1+|\alpha_0|\|[z]\|_{\mathcal{Z}})}$. Then $h \circ g_1 \in \mathrm{span}\left((e_i)_{i=0}^{\infty}\right)$. Since $[z] \perp \mathrm{span}\left(([e_i])_{i=0}^{\infty}\right)$, we have

$$
\begin{aligned}
\|[F|_{\mathbb{I}}]\|_{\mathcal{Z}}^2 &\le \frac{\epsilon^2}{3} + \frac{|\alpha_0|}{\sqrt{2\pi}} \left| \int_{-\pi}^{\pi} t z(t)\,\mathrm{d}t \right| \\
&= \frac{\epsilon^2}{3} + \frac{|\alpha_0|}{\sqrt{2\pi}} \left| \int_{-\pi}^{\pi} t z(t)\,\mathrm{d}t - \int_{-\pi}^{\pi} \overline{h(g_1(t))} z(t)\,\mathrm{d}t \right| \\
&= \frac{\epsilon^2}{3} + \frac{|\alpha_0|}{\sqrt{2}} \left| \langle [z], [f_1 - h \circ g_1] \rangle_{\mathcal{Z}} \right| \\
&\le \frac{\epsilon^2}{3} + \frac{|\alpha_0|}{\sqrt{2}} \|[z]\|_{\mathcal{Z}} \|[f_1 - h \circ g_1]\|_{\mathcal{Z}} \\
&\le \frac{\epsilon^2}{3} + \frac{|\alpha_0|}{\sqrt{2}} \|[z]\|_{\mathcal{Z}} \left(\|[f_1 - f_2 \circ g_1]\|_{\mathcal{Z}} + \|[f_2 \circ g_1 - h \circ g_1]\|_{\mathcal{Z}} \right) \\
&= \frac{\epsilon^2}{3} + \frac{|\alpha_0|}{\sqrt{2}} \|[z]\|_{\mathcal{Z}} \frac{1}{\sqrt{\pi}} \left(\| f_1 \circ g_{1\,\mathrm{inv}} - f_2 \|_{\bar{\mathcal{Y}}_1} + \| f_2 - h \|_{\bar{\mathcal{Y}}_1} \right) \\
&< \frac{\epsilon^2}{3} + \frac{\epsilon^2}{3} + \frac{|\alpha_0|}{\sqrt{2}} \|[z]\|_{\mathcal{Z}} \frac{1}{\sqrt{\pi}} \sqrt{2\pi} \| f_2 - h \|_{\mathcal{W}} < \epsilon^2
\end{aligned}
$$

where the first equality follows from the fact that $[z] \perp [e_n]$, $\forall n \in \mathbb{Z}_+$; the second equality follows from Proposition 11.89; the second inequality follows from Cauchy–Schwarz Inequality; the third inequality follows from Definition 7.1; the third equality follows from Proposition 12.12; and the last two inequalities follow from the preceding discussion. By the arbitrariness of ϵ, we have $\left[F|_{\mathbb{I}}\right] = \vartheta_{\mathcal{Z}}$. Then, $F|_{\mathbb{I}} = 0$ a.e. in $\mathbb{I}$. Since F is absolutely continuous, then it is continuous by Proposition 12.59, and then $F(t) = 0$, $\forall t \in \bar{\mathrm{r}}_{-\pi,\pi}$ (otherwise, we can always find a nonempty interval that is a subset of $\mathbb{I}$ such that F is nonzero on the interval.) Then, by Fundamental Theorem of Calculus I, Theorem 12.86, $z(t) = 0$ a.e. $t \in \mathbb{I}$. Hence, $[z] = \vartheta_{\mathcal{Z}}$.

We have shown that $\left(([e_n])_{n=0}^{\infty}\right)^{\perp_h} = \{\vartheta_{\mathcal{Z}}\}$. Then, by Proposition 13.33, $\overline{\text{span}\left(([e_n])_{n=0}^{\infty}\right)} = \mathcal{Z}$. and $([e_n])_{n=0}^{\infty}$ is a complete orthonormal sequence. Then $[f] = \sum_{n=0}^{\infty} a_n[e_n]$ in $\mathcal{Z}$. ◇

Given independent vectors $y_1, \ldots, y_n$ generating a subspace M of a Hilbert space $\mathcal{X}$, we wish to find $m_0 \in M$ that minimizes $\|x - m_0\|$, for some given $x \in \mathcal{X}$. Rather than solving the normal equation (13.1), we may employ the Gram–Schmidt orthogonalization procedure together with Proposition 13.31. First use the Gram–Schmidt Procedure to obtain orthonormal sequence $(e_i)_{i=1}^{n}$. Then, $m_0 = \sum_{i=1}^{n} \langle x, e_i\rangle e_i$. Thus, the optimization problem is easily solved once the vectors y_i's are orthogonalized.

Since solution to the approximation problem is equivalent to solution to the normal equation. It is clear that the Gram–Schmidt procedure can be interpreted as a procedure for inverting the Gram matrix.

It is interesting to note that the Gram–Schmidt procedure may itself be viewed as an approximation problem.

$$e_1 = \frac{y_1}{\|y_1\|}; \qquad e_k = \frac{y_k - \sum_{i=1}^{k-1} \langle y_k, e_i\rangle e_i}{\left\| y_k - \sum_{i=1}^{k-1} \langle y_k, e_i\rangle e_i \right\|}; \qquad \forall k > 1$$

The vector $y_k - \sum_{i=1}^{k-1} \langle y_k, e_i\rangle e_i$ is the optimal error for the problem $\min_{m\in\text{span}(\{y_1,\ldots,y_{k-1}\})} \|y_k - m\|$. Thus, the Gram–Schmidt procedure consists of solving a series of minimum norm approximation problems by use of the Projection Theorem. Alternatively, the minimum norm approximation of x can be found by applying the Gram–Schmidt procedure to the sequence $(y_1, \ldots, y_n, x)$. The optimal error $x - m_0$ is found at the last step.

13.6 Other Minimum Norm Problems

In approximation problems, finite dimensionality of the subspace allows the reduction of the problem into a finite-dimensional normal equation, which leads to a feasible computation procedure. In many important and interesting practical problems, the subspace is not finite-dimensional. In such problems, it is generally not possible to reduce the problem to a finite-dimensional normal equation. However, there is an important class of such problems that can be reduced by the projection theorem to finite-dimensional equations. The motivation for this class of problems is illustrated in Fig. 13.3. In this class of problems, the subspace $M^{\perp_h}$ is finite-dimensional, which leads to feasible computation procedures.

Next, we state a modified version of the Projection Theorem 13.13.

Theorem 13.37 (Restatement of Projection Theorem) *Let $\mathcal{X}$ be a Hilbert space, $M \subseteq \mathcal{X}$ be a closed subspace, $x, y \in \mathcal{X}$, $V = y + M$ be*

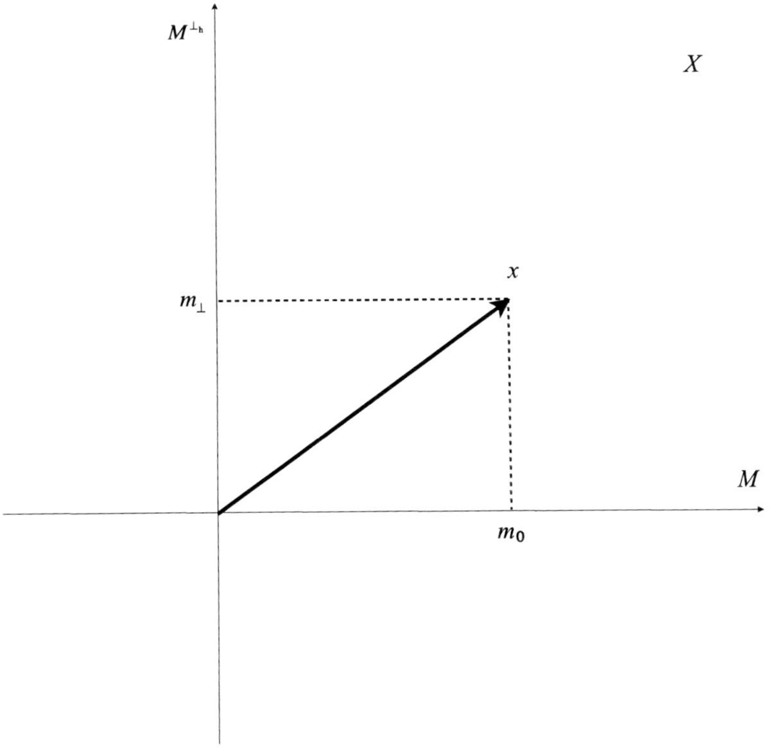

Fig. 13.3 Dual projection problems

a closed linear variety. Then, there is a unique vector $v_0 \in V$ *such that* $\|x - v_0\| = \min_{v \in V} \|x - v\|$. *A necessary and sufficient condition that* $v_0 \in V$ *is the unique minimizing vector is that* $(x - v_0) \perp M$.

Proof This is straightforward, and is therefore omitted. □

A point of caution is necessary here: $(x - v_0) \perp M$ but not $(x - v_0) \perp V$. The concept of the above theorem is illustrated in Fig. 13.4.

A special type of linear variety is $V = \{v \in \mathcal{X} \mid \langle v, y_i \rangle = c_i,\ i = 1, \ldots, n\}$, where $y_1, \ldots, y_n \in \mathcal{X}$ are fixed vectors and $c_1, \ldots, c_n \in \mathbb{K}$ are fixed scalars. Let $M = \text{span}(\{y_1, \ldots, y_n\})$. When $c_1 = \cdots = c_n = 0$, then $V = M^{\perp_h}$. For arbitrary c_i's, V is equal to $y + M^{\perp_h}$ assuming $V \neq \emptyset$. A linear variety of this form is said to be of finite *co-dimension*, since $M^{\perp_h \perp_h} = M$ is finite-dimensional by Proposition 13.17.

Theorem 13.38 *Let* $\mathcal{X}$ *be a Hilbert space over* $\mathbb{K}$, $n \in \mathbb{N}$, $y_1, \ldots, y_n \in \mathcal{X}$, $c_1, \ldots, c_n \in \mathbb{K}$, $V = \{v \in \mathcal{X} \mid \langle v, y_i \rangle = c_i,\ i = 1, \ldots, n\} \neq \emptyset$, $x \in \mathcal{X}$. *Then, the following equation*

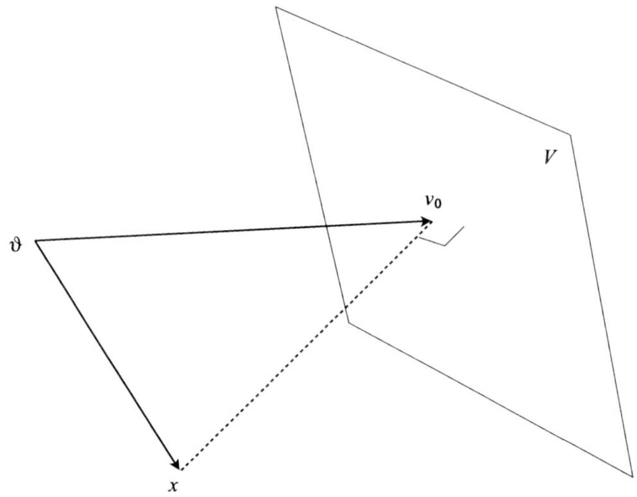

Fig. 13.4 Reformulation of the projection Theorem

$$(\text{Gram}(y_1, \ldots, y_n))' \begin{bmatrix} \beta_1 \\ \vdots \\ \beta_n \end{bmatrix} = \begin{bmatrix} \langle x, y_1 \rangle - c_1 \\ \vdots \\ \langle x, y_n \rangle - c_n \end{bmatrix} \tag{13.2}$$

admits at least one solution $(\beta_1, \ldots, \beta_n) \in \mathbb{K}^n$. *Any such solution implies that* $v_0 = x - \sum_{i=1}^{n} \beta_i y_i \in V$ *is the unique minimizing vector for* $\min_{v \in V} \|x - v\|$.

Proof Let $M := \text{span}(\{y_1, \ldots, y_n\})$. By Theorem 7.36 and Proposition 4.39, M is a closed subspace of $\mathcal{X}$. Then, $M^{\perp_h} = \{\bar{x} \in \mathcal{X} \mid \langle \bar{x}, y_i \rangle = 0,\ i = 1, \ldots, n\}$. Fix any $y_0 \in V \neq \emptyset$, we have $V = y_0 + M^{\perp_h}$. By Theorem 13.37, $\exists!\ v_0 \in V$ such that $\|x - v_0\| = \min_{v \in V} \|x - v\|$ and a necessary and sufficient condition for any $v \in V$ to be v_0 is that $(x - v_0) \perp M^{\perp_h}$. Then, $(x - v_0) \in M^{\perp_h \perp_h} = M$, by Proposition 13.17. Then, $\exists (\beta_1, \ldots, \beta_n) \in \mathbb{K}^n$ such that $x - v_0 = \sum_{i=1}^{n} \beta_i y_i$. Then, $v_0 = x - \sum_{i=1}^{n} \beta_i y_i$. Since $v_0 \in V$, we must have $\langle v_0, y_i \rangle = c_i$, $i = 1, \ldots, n$. This leads to (13.2).

Any solution $(\beta_1, \ldots, \beta_n) \in \mathbb{K}^n$ to (13.2), we have $v_0 \in V$ and $(x - v_0) \in M = M^{\perp_h \perp_h}$. By Theorem 13.37, v_0 is the unique minimizing vector for $\min v \in V \|x - v\|$. This completes the proof of the theorem. □

Much of the discussion in the above can be generalized from linear varieties to convex sets.

Theorem 13.39 *Let* $\mathcal{X}$ *be a real Hilbert space,* $x \in \mathcal{X}$, $K \subseteq \mathcal{X}$ *be a nonempty closed convex subset. Then, there is a unique vector* $k_0 \in K$ *such that* $\|x - k_0\| = \min_{k \in K} \|x - k\|$. *Furthermore, a necessary and sufficient condition for* $k_0 \in K$ *being the unique minimizing vector is that* $\langle x - k_0, k - k_0 \rangle \leq 0$, $\forall k \in K$.

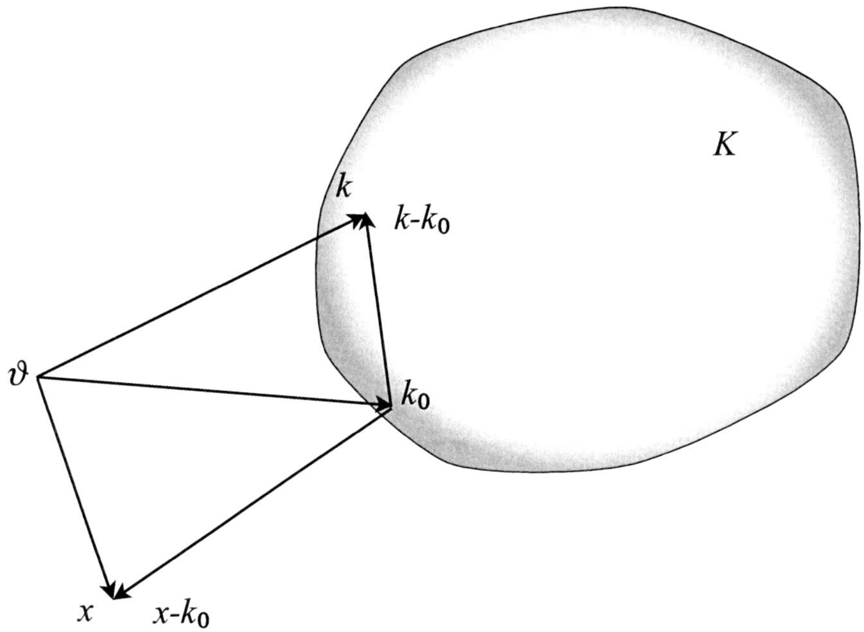

Fig. 13.5 Projection to a convex set

The main idea of this theorem is illustrated in Fig. 13.5, which shows that the angle between $x - k_0$ and $k - k_0$ is greater than or equal to 90°.

Proof First, we show the existence of k_0. Let $\delta := \inf_{k\in K} \|x - k\| \in [0, \infty) \subset \mathbb{R}$ since $K \neq \emptyset$. Then, $\exists\, (k_i)_{i=1}^{\infty} \subseteq K$ such that $\lim_{i\in\mathbb{N}} \|x - k_i\| = \delta$. $\forall \epsilon \in (0, \infty) \subset \mathbb{R}$, $\exists n_0 \in \mathbb{N}$, $\forall i \in \mathbb{N}$ with $n_0 \leq i$, we have $\delta \leq \|x - k_i\| < \sqrt{\delta^2 + \epsilon^2/4}$. $\forall i, j \in \mathbb{N}$ with $n_0 \leq i \leq j$, by Parallelogram Law 13.3, $\|(x - k_i) - (x - k_j)\|^2 + \|(x - k_i) + (x - k_j)\|^2 = 2\|x - k_i\|^2 + 2\|x - k_j\|^2$. This implies that $\|k_i - k_j\|^2 = 2\|x - k_i\|^2 + 2\|x - k_j\|^2 - 4\|x - (k_i + k_j)/2\|^2 < 4(\delta^2 + \epsilon^2/4) - 4\delta^2 = \epsilon^2$, where the inequality follows from the fact that K is convex and $(k_i + k_j)/2 \in K$. Hence, $(k_i)_{i=1}^{\infty} \subseteq K$ is Cauchy sequence. By Proposition 4.39, K is complete and $\lim_{i\in\mathbb{N}} k_i = k_0 \in K$. By Propositions 3.66 and 7.21, $\delta = \lim_{i\in\mathbb{N}} \|x - k_i\| = \|x - k_0\|$.

Next, we show the uniqueness of k_0. Let $\bar{k} \in K$ be such that $\delta = \left\|x - \bar{k}\right\|$. The sequence $(k_i)_{i=1}^{\infty} := (k_0, \bar{k}, k_0, \bar{k}, \ldots) \subseteq K$ satisfies $\lim_{i\in\mathbb{N}} \|x - k_i\| = \delta$. By the proof of the existence, $(k_i)_{i=1}^{\infty}$ is convergent. Then, we must have $\bar{k} = k_0$. Hence, k_0 is unique.

Next, we show that $\langle x - k_0, k - k_0\rangle \leq 0$, $\forall k \in K$ by an argument of contradiction. Suppose this is not true. Then, $\exists k_1 \in K$ such that $\langle x - k_0, k_1 - k_0\rangle =: \lambda > 0$. Consider the vector $k_\alpha := (1 - \alpha)k_0 + \alpha k_1$, where $\alpha \in (0, 1) \subset \mathbb{R}$ is to be determined. Since K is convex, then $k_\alpha \in K$. Note that $\|x - k_\alpha\|^2 = \|(x - k_0) - \alpha\,(k_1 - k_0)\|^2 = \|x - k_0\|^2 - 2\alpha\langle x - k_0, k_1 - k_0\rangle + \alpha^2\|k_1 - k_0\|^2 =$

$\delta^2 - 2\alpha\lambda + \alpha^2\|k_1 - k_0\|^2$. Then, for sufficiently small $\alpha \in (0,1) \subset \mathbb{R}$, we have $\|x - k_\alpha\|^2 < \delta^2$, which contradicts with the definition of δ. Hence, $\langle x - k_0, k - k_0\rangle \le 0$, $\forall k \in K$.

Finally, we show that if $\bar{k} \in K$ satisfies $\langle x - \bar{k}, k - \bar{k}\rangle \le 0$, $\forall k \in K$, then $\|x - \bar{k}\| = \min_{k\in K}\|x - k\|$. $\forall k \in K$, we have $\|x - k\|^2 = \|(x - \bar{k}) - (k - \bar{k})\|^2 = \|x - \bar{k}\|^2 - 2\langle x - \bar{k}, k - \bar{k}\rangle + \|k - \bar{k}\|^2 \ge \|x - \bar{k}\|^2$. Hence, $\|x - \bar{k}\| \le \|x - k\|^2$, $\forall k \in K$.

This completes the proof of the theorem. □

13.7 Positive Definite Operators on Hilbert Spaces

Let $\mathcal{X}$ be a real Hilbert space. In Chap. 10, we defined the symmetric operators $\mathcal{S}_{\mathcal{X}}$ to be $\mathrm{B}_{\mathrm{S}2}(\mathcal{X}, \mathbb{R}) \subseteq \mathrm{B}(\mathcal{X}, \mathcal{X}^*)$. By Riesz-Fréchet Theorem 13.15, $\Phi_{\mathcal{X}}(\mathcal{X}^*) = \mathcal{X}$, where $\Phi_{\mathcal{X}} : \mathcal{X}^* \to \mathcal{X}$ is the isometrical isomorphism. Then, $\bar{\mathcal{S}}_{\mathcal{X}} := \Phi_{\mathcal{X}}(\mathcal{S}_{\mathcal{X}}) \subseteq \mathrm{B}(\mathcal{X}, \mathcal{X})$. Fix any $A = \Phi_{\mathcal{X}}\hat{A} \in \bar{\mathcal{S}}_{\mathcal{X}}$. $\forall x, y \in \mathcal{X}$, we have $\langle y, Ax\rangle = \hat{A}(x)(y) = \hat{A}(y)(x) = \langle x, Ay\rangle = \langle A^*x, y\rangle = \langle y, A^*x\rangle$, where the first equality follows since $A(x)$ is a linear functional on $\mathcal{X}$ and by Riesz-Fréchet Theorem 13.15; the second equality follows since $\hat{A} \in \mathrm{B}_{\mathrm{S}2}(\mathcal{X}, \mathbb{R})$; the third equality follows from the same argument as the first equality; the fourth equality follows from Proposition 13.22; and the last equality follows from Definition 13.1 and $\mathcal{X}$ is a real Hilbert space. By the arbitrariness of x and y, we have $A^* = A$, and A is Hermitian. By Definition 13.21, $A = A^* = \Phi_{\mathcal{X}}A'\Phi_{\mathcal{X}\mathrm{inv}}$ and $\hat{A} = A'\Phi_{\mathcal{X}\mathrm{inv}}$ since $\mathcal{X}$ is a real Hilbert space. Here, we will generalize the definition of $\mathcal{S}_{\mathcal{X}}$, $\mathcal{S}_{+\,\mathcal{X}}$, $\mathcal{S}_{\mathrm{psd}\,\mathcal{X}}$, $\mathcal{S}_{-\,\mathcal{X}}$, and $\mathcal{S}_{\mathrm{nsd}\,\mathcal{X}}$ to the case where $\mathcal{X}$ is a complex Hilbert space.

Definition 13.40 Let $\mathcal{X}$ be a Hilbert space over $\mathbb{K}$ and $A \in \mathrm{B}(\mathcal{X}, \mathcal{X})$. We will write $A \in \bar{\mathcal{S}}_{\mathcal{X}}$, if A is Hermitian. We will write $A \in \bar{\mathcal{S}}_{+\,\mathcal{X}}$ if $A \in \bar{\mathcal{S}}_{\mathcal{X}}$ and $\exists m \in (0,\infty) \subset \mathbb{R}$ such that $\langle x, Ax\rangle \ge m\|x\|^2$, $\forall x \in \mathcal{X}$. We will write $A \in \bar{\mathcal{S}}_{\mathrm{psd}\,\mathcal{X}}$ if $A \in \bar{\mathcal{S}}_{\mathcal{X}}$ and $\exists m \in [0,\infty) \subset \mathbb{R}$ such that $\langle x, Ax\rangle \ge m\|x\|^2$, $\forall x \in \mathcal{X}$. We will write $A \in \bar{\mathcal{S}}_{-\,\mathcal{X}}$ if $A \in \bar{\mathcal{S}}_{\mathcal{X}}$ and $\exists m \in (0,\infty) \subset \mathbb{R}$ such that $\langle x, Ax\rangle \le -m\|x\|^2$, $\forall x \in \mathcal{X}$. We will write $A \in \bar{\mathcal{S}}_{\mathrm{nsd}\,\mathcal{X}}$ if $A \in \bar{\mathcal{S}}_{\mathcal{X}}$ and $\exists m \in [0,\infty) \subset \mathbb{R}$ such that $\langle x, Ax\rangle \le -m\|x\|^2$, $\forall x \in \mathcal{X}$. ◇

Proposition 13.41 *Let $\mathcal{X}$ be a Hilbert space, $A \in \mathrm{B}(\mathcal{X}, \mathcal{X})$ be Hermitian (that is, $A \in \bar{\mathcal{S}}_{\mathcal{X}}$). Assume that $\exists \delta \in (0,\infty) \subset \mathbb{R}$ such that $\langle x, Ax\rangle \ge \delta\|x\|^2$, $\forall x \in \mathcal{X}$. Then, A is bijective and $A^{-1} \in \mathrm{B}(\mathcal{X}, \mathcal{X})$. Furthermore, A^{-1} is Hermitian and $\forall x \in \mathcal{X}$, we have $\langle x, A^{-1}x\rangle \ge \frac{\delta}{\epsilon + \|A\|^2_{\mathrm{B}(\mathcal{X},\mathcal{X})}}\|x\|^2$, $\forall \epsilon \in (0,\infty) \subset \mathbb{R}$.*

Proof By the assumption, $\forall x \in \mathcal{X}$ with $x \neq \vartheta_{\mathcal{X}}$, $A(x) \neq \vartheta_{\mathcal{X}}$ since $\langle x, Ax\rangle \ge \delta\|x\|^2 > 0$. Hence, $\mathcal{N}(A) = \{\vartheta_{\mathcal{X}}\}$ and A is injective. By Proposition 13.22, $(\mathcal{R}(A))^{\perp_{\mathrm{h}}} = \mathcal{N}(A^*) = \mathcal{N}(A) = \{\vartheta_{\mathcal{X}}\}$, where the second equality follows from the fact that A is Hermitian. This shows that $((\mathcal{R}(A))^{\perp_{\mathrm{h}}})^{\perp_{\mathrm{h}}} = \mathcal{X}$. By Proposition 13.17, $\overline{\mathcal{R}(A)} = \mathcal{X}$. Hence, $\mathcal{R}(A)$ is dense in $\mathcal{X}$. Clearly, $\vartheta_{\mathcal{X}} \in \mathcal{R}(A)$. $\forall x \in \mathcal{X}$ with $x \neq \vartheta_{\mathcal{X}}$, there exists $(x_i)_{i=1}^{\infty} \subseteq \mathcal{R}(A)$ such that $x = \lim_{i \in \mathbb{N}} x_i$. Without loss of generality, we may assume that $x_i \neq \vartheta_{\mathcal{X}}$, $\forall i \in \mathbb{N}$. Let $x_i = A(\bar{x}_i)$, $\forall i \in \mathbb{N}$, where $\bar{x}_i \in \mathcal{X}$. Clearly, $\bar{x}_i \neq \vartheta_{\mathcal{X}}$ since $x_i \neq \vartheta_{\mathcal{X}}$, $\forall i \in \mathbb{N}$. Then, we have

$$\|\bar{x}_i\| \|x_i\| \geq |\langle \bar{x}_i, x_i \rangle| = |\langle \bar{x}_i, A\bar{x}_i \rangle| \geq \delta \|\bar{x}_i\|^2; \quad \forall i \in \mathbb{N}$$

where the first inequality follows from the Cauchy–Schwarz Inequality. This implies that $\|x_i\| \geq \delta \|\bar{x}_i\|$ and $\|\bar{x}_i\| \leq \|x_i\|/\delta$, $\forall i \in \mathbb{N}$. Since $(x_i)_{i=1}^{\infty}$ is convergent, then there exists $c \in (0, \infty) \subset \mathbb{R}$ such that $\|x_i\| \leq c$, $\forall i \in \mathbb{N}$. Then, $\|\bar{x}_i\| \leq c/\delta =: c_1 \in (0, \infty) \subset \mathbb{R}$, $\forall i \in \mathbb{N}$. This shows that $(\bar{x}_i)_{i=1}^{\infty} \subseteq \overline{\mathcal{B}}_{\mathcal{X}}(\vartheta_{\mathcal{X}}, c_1) =: S_1 \subseteq \mathcal{X} = \mathcal{X}^{**}$, where the last equality follows from Riesz-Fréchet Theorem 13.15. By Alaoglu Theorem 7.122, $S_1 \subseteq \mathcal{X}^{**}$ is weak* compact. By Propositions 5.22 and 5.26, $(\bar{x}_i)_{i=1}^{\infty}$ has a cluster point $\bar{x} \in \mathcal{X}^{**} = \mathcal{X}$ in weak* topology. Since $\mathcal{X}$ is reflexive, the weak* topology on $\mathcal{X}^{**} = \mathcal{X}$ is identical to the weak topology on $\mathcal{X}$. By Proposition 7.123, $A \in \mathrm{B}(\mathcal{X}, \mathcal{X})$ is continuous as a mapping $A : \mathcal{X}_{\text{weak}} \to \mathcal{X}_{\text{weak}}$. By Proposition 3.66, $(x_i)_{i=1}^{\infty} = (A(\bar{x}_i))_{i=1}^{\infty}$ has a cluster point $A(\bar{x}) \in \mathcal{X}_{\text{weak}}$ in weak topology. By Proposition 7.116, $\mathcal{X}_{\text{weak}}$ is completely regular, then it is Hausdorff by Proposition 3.61. By Proposition 3.65, $A(\bar{x}) = x$, since $x = \lim_{i \in \mathbb{N}} x_i$ and $x = \lim_{i \in \mathbb{N}} x_i$ weakly. This shows that $x \in \mathcal{R}(A)$. By the arbitrariness of x, we have $\mathcal{R}(A) = \mathcal{X}$ and A is surjective.

Hence, A is bijective. By Open Mapping Theorem 7.103, $A^{-1} \in \mathrm{B}(\mathcal{X}, \mathcal{X})$. Then, A^{-1} is Hermitian by Proposition 13.22 and the fact that A is Hermitian.

$\forall x \in \mathcal{X}$, let $\bar{x} = A^{-1}(x)$. We have

$$\langle x, A^{-1}x \rangle = \langle A\bar{x}, \bar{x} \rangle = \overline{\langle \bar{x}, A\bar{x} \rangle} \geq \delta \|\bar{x}\|^2$$

Then, $x = A\bar{x}$ and $\|x\| \leq \|A\|_{\mathrm{B}(\mathcal{X},\mathcal{X})} \|\bar{x}\|$. We will distinguish two exhaustive and mutually exclusive cases: Case 1: $\|A\|_{\mathrm{B}(\mathcal{X},\mathcal{X})} = 0$; Case 2: $\|A\|_{\mathrm{B}(\mathcal{X},\mathcal{X})} > 0$.

Case 1: $\|A\|_{\mathrm{B}(\mathcal{X},\mathcal{X})} = 0$, we have $\|x\| = 0$, $\forall x \in \mathcal{X}$. Hence, $\mathcal{X} = \{\vartheta_{\mathcal{X}}\}$. Then, $\langle x, A^{-1}x \rangle = 0 \geq \frac{\delta}{\epsilon + \|A\|^2_{\mathrm{B}(\mathcal{X},\mathcal{X})}} \|x\|^2 = 0$, $\forall \epsilon \in (0, \infty) \subset \mathbb{R}$.

Case 2: $\|A\|_{\mathrm{B}(\mathcal{X},\mathcal{X})} > 0$. Then, we have

$$\begin{aligned} \langle x, A^{-1}x \rangle &\geq \delta \|\bar{x}\|^2 \geq \frac{\delta}{\|A\|^2_{\mathrm{B}(\mathcal{X},\mathcal{X})}} \|x\|^2 \\ &\geq \frac{\delta}{\epsilon + \|A\|^2_{\mathrm{B}(\mathcal{X},\mathcal{X})}} \|x\|^2, \quad \forall \epsilon \in (0, \infty) \subset \mathbb{R} \end{aligned}$$

Hence, we have shown, in both cases, that $\langle x, A^{-1}x \rangle \geq \frac{\delta}{\epsilon + \|A\|^2_{\mathrm{B}(\mathcal{X},\mathcal{X})}} \|x\|^2$, $\forall x \in \mathcal{X}$ and $\forall \epsilon \in (0, \infty) \subset \mathbb{R}$.

This completes the proof of the proposition. □

Proposition 13.42 *Let $\mathcal{X}$ be a Hilbert space over $\mathbb{K}$ and $A \in \bar{\mathcal{S}}_{\mathcal{X}}$. Then, the following statements hold.*

(i) $\forall x \in \mathcal{X}$, $\langle x, Ax \rangle \in \mathbb{R}$.

(ii) *$\bar{\mathcal{S}}_{\mathcal{X}}$ is a closed subset of* $\mathrm{B}(\mathcal{X}, \mathcal{X})$. $\forall A_1, A_2 \in \bar{\mathcal{S}}_{\mathcal{X}}$, $\forall \lambda \in \mathbb{R}$, *we have* $A_1 + A_2 \in \bar{\mathcal{S}}_{\mathcal{X}}$ *and* $\lambda A_1 \in \bar{\mathcal{S}}_{\mathcal{X}}$. *If* $\mathbb{K} = \mathbb{R}$, *$\bar{\mathcal{S}}_{\mathcal{X}}$ is a closed subspace of* $\mathrm{B}(\mathcal{X}, \mathcal{X})$, *and hence a Banach subspace. If* $\mathbb{K} = \mathbb{C}$, *no further conclusion can be made about* $\bar{\mathcal{S}}_{\mathcal{X}}$ *except that it is a closed convex cone in* $\mathrm{B}(\mathcal{X}, \mathcal{X})$.

(iii) $\bar{\mathcal{S}}_{+\mathcal{X}}$ *and* $\bar{\mathcal{S}}_{-\mathcal{X}}$ *are open subsets of* $\bar{\mathcal{S}}_{\mathcal{X}}$ *(in the subset topology of* $\bar{\mathcal{S}}_{\mathcal{X}} \subseteq \mathrm{B}(\mathcal{X}, \mathcal{X})$*); and* $\bar{\mathcal{S}}_{+\mathcal{X}} = -\bar{\mathcal{S}}_{-\mathcal{X}}$.

(iv) $\bar{\mathcal{S}}_{\mathrm{psd}\,\mathcal{X}}$ *and* $\bar{\mathcal{S}}_{\mathrm{nsd}\,\mathcal{X}}$ *are closed convex cones in* $\bar{\mathcal{S}}_{\mathcal{X}}$ *(in the subset topology of* $\bar{\mathcal{S}}_{\mathcal{X}} \subseteq \mathrm{B}(\mathcal{X}, \mathcal{X})$*); and* $\bar{\mathcal{S}}_{\mathrm{psd}\,\mathcal{X}} = -\bar{\mathcal{S}}_{\mathrm{nsd}\,\mathcal{X}}$.

(v) $A \in \bar{\mathcal{S}}_{+\mathcal{X}}$ *if, and only if,* $A^{-1} \in \bar{\mathcal{S}}_{+\mathcal{X}}$.

(vi) $A \in \bar{\mathcal{S}}_{-\mathcal{X}}$ *if, and only if,* $A^{-1} \in \bar{\mathcal{S}}_{-\mathcal{X}}$.

(vii) *The interior of* $\bar{\mathcal{S}}_{\mathrm{psd}\,\mathcal{X}}$ *relative to* $\bar{\mathcal{S}}_{\mathcal{X}}$ *is* $\bar{\mathcal{S}}_{+\mathcal{X}}$, *and the interior of* $\bar{\mathcal{S}}_{\mathrm{nsd}\,\mathcal{X}}$ *relative to* $\bar{\mathcal{S}}_{\mathcal{X}}$ *is* $\bar{\mathcal{S}}_{-\mathcal{X}}$.

(viii) $A \in \bar{\mathcal{S}}_{+\mathcal{X}}$, $B, C \in \bar{\mathcal{S}}_{\mathrm{psd}\,\mathcal{X}}$, $\alpha \in (0, \infty) \subset \mathbb{R}$, *and* $\beta \in [0, \infty) \subset \mathbb{R}$, *implies that* $A + B \in \bar{\mathcal{S}}_{+\mathcal{X}}$, $B + C \in \bar{\mathcal{S}}_{\mathrm{psd}\,\mathcal{X}}$, $\alpha A \in \bar{\mathcal{S}}_{+\mathcal{X}}$, *and* $\beta B \in \bar{\mathcal{S}}_{\mathrm{psd}\,\mathcal{X}}$.

(ix) *Let* $\mathcal{Y}$ *be a Hilbert space over* $\mathbb{K}$, *and* $B \in \mathrm{B}(\mathcal{X}, \mathcal{Y})$. *Then,* $BB^* \in \bar{\mathcal{S}}_{\mathrm{psd}\,\mathcal{Y}}$ *and* $B^*B \in \bar{\mathcal{S}}_{\mathrm{psd}\,\mathcal{X}}$.

Proof (i) $\forall x \in \mathcal{X}$, $\langle x, Ax \rangle = \langle A^* x, x \rangle = \langle Ax, x \rangle = \overline{\langle x, Ax \rangle}$, where the first equality follows from Proposition 13.22; the second equality follows from $A \in \bar{\mathcal{S}}_{\mathcal{X}}$; and the last equality follows from Definition 13.1. Hence, $\langle x, Ax \rangle \in \mathbb{R}$.

(ii) $\forall A \in \overline{\bar{\mathcal{S}}_{\mathcal{X}}}$, by Proposition 4.13, there exists $(A_i)_{i=1}^{\infty} \subseteq \bar{\mathcal{S}}_{\mathcal{X}}$ such that $\lim_{i \in \mathbb{N}} A_i = A$ in $\mathrm{B}(\mathcal{X}, \mathcal{X})$. Then, $\forall x, y \in \mathcal{X}$, we have $\langle Ax, y \rangle = \lim_{i \in \mathbb{N}} \langle A_i x, y \rangle = \lim_{i \in \mathbb{N}} \langle x, A_i^* y \rangle = \lim_{i \in \mathbb{N}} \langle x, A_i y \rangle = \langle x, Ay \rangle = \langle A^* x, y \rangle$, where the first equality follows from Propositions 7.65, 13.4, 3.12, and 3.66; the second equality follows from Proposition 13.22; the third equality follows from $A_i \in \bar{\mathcal{S}}_{\mathcal{X}}$ and therefore is Hermitian; the fourth equality follows from Propositions 7.65, 13.4, 3.12, and 3.66; and the last equality follows from Proposition 13.22. By the arbitrariness of x and y, we have $A = A^*$ and therefore $A \in \bar{\mathcal{S}}_{\mathcal{X}}$. Thus, we have shown that $\overline{\bar{\mathcal{S}}_{\mathcal{X}}} \subseteq \bar{\mathcal{S}}_{\mathcal{X}} \subseteq \overline{\bar{\mathcal{S}}_{\mathcal{X}}}$. By Proposition 3.3, $\bar{\mathcal{S}}_{\mathcal{X}}$ is a closed subset of $\mathrm{B}(\mathcal{X}, \mathcal{X})$.

$\forall A_1, A_2 \in \bar{\mathcal{S}}_{\mathcal{X}}$ and $\forall \lambda \in \mathbb{R}$, we have $(A_1 + A_2)^* = A_1^* + A_2^* = A_1 + A_2$, where the first equality follows from Proposition 13.22. Then, $A_1 + A_2 \in \bar{\mathcal{S}}_{\mathcal{X}}$. In addition, $(\lambda A_1)^* = \overline{\lambda} A_1^* = \lambda A_1$, where the first equality follows from Proposition 13.22. Hence, $\lambda A_1 \in \bar{\mathcal{S}}_{\mathcal{X}}$.

Note that, the preceding paragraph is enough to conclude that $\bar{\mathcal{S}}_{\mathcal{X}}$ is a subspace if $\mathcal{X}$ is a real Hilbert space. In this case, $\bar{\mathcal{S}}_{\mathcal{X}}$ is a closed subspace and therefore a Banach subspace of $\mathrm{B}(\mathcal{X}, \mathcal{X})$ by Proposition 4.39. When $\mathbb{K} = \mathbb{C}$, we only have $\bar{\mathcal{S}}_{\mathcal{X}}$ is a closed convex cone and hence a complete metric subspace of $\mathrm{B}(\mathcal{X}, \mathcal{X})$ by Proposition 4.39.

(iii) It is clear that $\bar{\mathcal{S}}_{+\mathcal{X}} = -\bar{\mathcal{S}}_{-\mathcal{X}}$. We will show that $\bar{\mathcal{S}}_{+\mathcal{X}}$ is open in $\bar{\mathcal{S}}_{\mathcal{X}}$. Fix any $B \in \bar{\mathcal{S}}_{+\mathcal{X}}$. Then, $\exists m \in (0, \infty) \subset \mathbb{R}$ such that $\langle x, Bx \rangle \geq m\|x\|^2$, $\forall x \in \mathcal{X}$. $\forall C \in \bar{\mathcal{S}}_{\mathcal{X}} \cap \mathcal{B}_{\mathrm{B}(\mathcal{X},\mathcal{X})}\big(B, \frac{1}{2}m\big)$, we have $\langle x, Cx \rangle \in \mathbb{R}$, $\forall x \in \mathcal{X}$, by (i). Then, $\langle x, Cx \rangle \geq \langle x, Bx \rangle - |\langle x, Bx \rangle - \langle x, Cx \rangle| \geq m\|x\|^2 - |\langle x, (B - C)x \rangle| \geq m\|x\|^2 - \|x\|\|(B - C)x\| \geq m\|x\|^2 - \|x\|\|B - C\|_{\mathrm{B}(\mathcal{X},\mathcal{X})}\|x\| \geq m\|x\|^2 - \frac{1}{2}m\|x\|^2 = \frac{1}{2}m\|x\|^2$, $\forall x \in \mathcal{X}$, where the second inequality follows from the previous discussion and Definition 13.1; the third inequality follows from Cauchy–Schwarz Inequality; the fourth inequality follows from Proposition 7.64; and the fifth inequality follows from the fact that $\|B - C\|_{\mathrm{B}(\mathcal{X},\mathcal{X})} < \frac{1}{2}m$. By the arbitrariness of x, $C \in \bar{\mathcal{S}}_{+\mathcal{X}}$. By the arbitrariness of C, we have $\mathcal{B}_{\mathrm{B}(\mathcal{X},\mathcal{X})}\big(B, \frac{1}{2}m\big) \cap \bar{\mathcal{S}}_{\mathcal{X}} \subseteq \bar{\mathcal{S}}_{+\mathcal{X}}$. Hence, $\bar{\mathcal{S}}_{+\mathcal{X}}$ is an open subset of $\bar{\mathcal{S}}_{\mathcal{X}}$. By the relation $\bar{\mathcal{S}}_{+\mathcal{X}} = -\bar{\mathcal{S}}_{-\mathcal{X}}$, we have $\bar{\mathcal{S}}_{-\mathcal{X}}$ is also an open subset of $\bar{\mathcal{S}}_{\mathcal{X}}$.

(iv) It is clear that $\bar{\mathcal{S}}_{\mathrm{psd}\,\mathcal{X}} = -\bar{\mathcal{S}}_{\mathrm{nsd}\,\mathcal{X}}$. We will show that $\bar{\mathcal{S}}_{\mathrm{psd}\,\mathcal{X}}$ is a closed subset of $\bar{\mathcal{S}}_{\mathcal{X}}$. $\forall B \in \overline{\bar{\mathcal{S}}_{\mathrm{psd}\,\mathcal{X}}}$, by Proposition 4.13, there exists $(B_i)_{i=1}^{\infty} \subseteq \bar{\mathcal{S}}_{\mathrm{psd}\,\mathcal{X}}$ such that $\lim_{i\in\mathbb{N}} B_i = B$ in $\mathrm{B}(\mathcal{X}, \mathcal{X})$. By (ii), $B \in \bar{\mathcal{S}}_{\mathcal{X}}$. $\forall x \in \mathcal{X}$, $\langle x, Bx \rangle = \lim_{i\in\mathbb{N}} \langle x, B_i x \rangle \geq 0$, where the equality follows from Propositions 7.65, 13.4, 3.12, and 3.66; and the inequality follows from $B_i \in \bar{\mathcal{S}}_{\mathrm{psd}\,\mathcal{X}}$. By the arbitrariness of x, $B \in \bar{\mathcal{S}}_{\mathrm{psd}\,\mathcal{X}}$. Thus, we have shown that $\overline{\bar{\mathcal{S}}_{\mathrm{psd}\,\mathcal{X}}} \subseteq \bar{\mathcal{S}}_{\mathrm{psd}\,\mathcal{X}} \subseteq \overline{\bar{\mathcal{S}}_{\mathrm{psd}\,\mathcal{X}}}$. By Proposition 3.3, $\bar{\mathcal{S}}_{\mathrm{psd}\,\mathcal{X}}$ is a closed subset of $\bar{\mathcal{S}}_{\mathcal{X}}$.

$\forall A_1, A_2 \in \bar{\mathcal{S}}_{\mathrm{psd}\,\mathcal{X}}$, $\forall \alpha \in [0, 1] \subset \mathbb{R}$. Let $B := \alpha A_1 + (1-\alpha)A_2$. By (ii), $B \in \bar{\mathcal{S}}_{\mathcal{X}}$. $\forall x \in \mathcal{X}$, $\langle x, Bx \rangle = \langle x, \alpha A_1 x + (1-\alpha)A_2 x \rangle = \langle x, \alpha A_1 x \rangle + \langle x, (1-\alpha)A_2 x \rangle = \overline{\alpha}\langle x, A_1 x \rangle + \overline{(1-\alpha)}\langle x, A_2 x \rangle = \alpha\langle x, A_1 x \rangle + (1-\alpha)\langle x, A_2 x \rangle \geq 0 = 0\|x\|^2$, where the second and third equalities follow from Definition 13.1; and the inequality follows from $A_1, A_2 \in \bar{\mathcal{S}}_{\mathrm{psd}\,\mathcal{X}}$. By the arbitrariness of x, we have $B \in \bar{\mathcal{S}}_{\mathrm{psd}\,\mathcal{X}}$. Hence, $\bar{\mathcal{S}}_{\mathrm{psd}\,\mathcal{X}}$ is convex.

Obviously, $\vartheta_{\mathrm{B}(\mathcal{X},\mathcal{X})} \in \bar{\mathcal{S}}_{\mathrm{psd}\,\mathcal{X}}$. $\forall A_1 \in \bar{\mathcal{S}}_{\mathrm{psd}\,\mathcal{X}}$, $\forall \lambda \in [0, \infty) \subset \mathbb{R}$, we have $\lambda A_1 \in \bar{\mathcal{S}}_{\mathcal{X}}$ by (ii). $\forall x \in \mathcal{X}$, $\langle x, \lambda A_1 x \rangle = \lambda\langle x, A_1 x \rangle \geq 0$, where the first equality follows from Definition 13.1 and the fact $\lambda \in \mathbb{R}$; and the inequality follows from $\lambda \geq 0$ and $A_1 \in \bar{\mathcal{S}}_{\mathrm{psd}\,\mathcal{X}}$. By the arbitrariness of x, we have $\lambda A_1 \in \bar{\mathcal{S}}_{\mathrm{psd}\,\mathcal{X}}$. Thus, $\bar{\mathcal{S}}_{\mathrm{psd}\,\mathcal{X}}$ is a cone with vertex at origin in $\bar{\mathcal{S}}_{\mathcal{X}} \subseteq \mathrm{B}(\mathcal{X}, \mathcal{X})$. Hence, $\bar{\mathcal{S}}_{\mathrm{psd}\,\mathcal{X}}$ is a closed convex cone in $\bar{\mathcal{S}}_{\mathcal{X}}$.

By the relation $\bar{\mathcal{S}}_{\mathrm{psd}\,\mathcal{X}} = -\bar{\mathcal{S}}_{\mathrm{nsd}\,\mathcal{X}}$, we have $\bar{\mathcal{S}}_{\mathrm{nsd}\,\mathcal{X}}$ is a closed convex cone in $\bar{\mathcal{S}}_{\mathcal{X}}$.

(v) This follows directly from Proposition 13.41 and Definition 13.40.

(vi) This follows immediately from (v) and (iii).

(vii) Clearly, $\bar{\mathcal{S}}_{+\mathcal{X}}$ is an open subset of $\bar{\mathcal{S}}_{\mathcal{X}}$ and is contained in $\bar{\mathcal{S}}_{\mathrm{psd}\,\mathcal{X}}$, which is a closed subset of $\bar{\mathcal{S}}_{\mathcal{X}}$. Let P be the interior of $\bar{\mathcal{S}}_{\mathrm{psd}\,\mathcal{X}}$ relative $\bar{\mathcal{S}}_{\mathcal{X}}$. Then, $P \supseteq \bar{\mathcal{S}}_{+\mathcal{X}}$. We will show that $P \subseteq \bar{\mathcal{S}}_{+\mathcal{X}}$. This will imply that $P = \bar{\mathcal{S}}_{+\mathcal{X}}$. Suppose $P \not\subseteq \bar{\mathcal{S}}_{+\mathcal{X}}$. Then, there exists $B \in P \setminus \bar{\mathcal{S}}_{+\mathcal{X}}$. $B \in P$ implies that $\exists \delta \in (0, \infty) \subset \mathbb{R}$ such that $\mathcal{B}_{\mathrm{B}(\mathcal{X},\mathcal{X})}(B, \delta) \cap \bar{\mathcal{S}}_{\mathcal{X}} \subseteq P \subseteq \bar{\mathcal{S}}_{\mathrm{psd}\,\mathcal{X}}$. Since $B \notin \bar{\mathcal{S}}_{+\mathcal{X}}$, then $\exists x_0 \in \mathcal{X}$ such that $\langle x_0, Bx_0 \rangle < \frac{\delta}{4}\|x_0\|^2$. Clearly, $x_0 \neq \vartheta_{\mathcal{X}}$. Consider the operator $\hat{B} := B - \frac{\delta}{2}\,\mathrm{id}_{\mathcal{X}}$. Clearly, $\hat{B} \in \mathcal{B}_{\mathrm{B}(\mathcal{X},\mathcal{X})}(B, \delta)$. By Proposition 13.22, $\hat{B}^* = B^* - \frac{\delta}{2}\,\mathrm{id}_{\mathcal{X}}^* = B - \frac{\delta}{2}\,\mathrm{id}_{\mathcal{X}} = \hat{B}$. Hence, $\hat{B} \in \bar{\mathcal{S}}_{\mathcal{X}}$. By our earlier discussion, we must have $\hat{B} \in \bar{\mathcal{S}}_{\mathrm{psd}\,\mathcal{X}}$.

Yet, $\langle x_0, \hat{B}x_0\rangle = \langle x_0, Bx_0\rangle - \overline{(\frac{\delta}{2})}\langle x_0, \mathrm{id}_{\mathcal{X}}\, x_0\rangle < \frac{\delta}{4}\|x_0\|^2 - \frac{\delta}{2}\|x_0\|^2 = -\frac{\delta}{4}\|x_0\|^2 < 0$, where the first equality follows from Definition 13.1; the first inequality follows from our earlier discussion; and the last equality follows from the fact $x_0 \neq \vartheta_{\mathcal{X}}$. This contradicts with the fact that $\hat{B} \in \bar{\mathcal{S}}_{\mathrm{psd}\,\mathcal{X}}$. Hence, the hypothesis does not hold. $P \subseteq \bar{\mathcal{S}}_{+\,\mathcal{X}}$. This proves that $P = \bar{\mathcal{S}}_{+\,\mathcal{X}}$.

The statement that the interior of $\bar{\mathcal{S}}_{\mathrm{nsd}\,\mathcal{X}}$ relative to $\bar{\mathcal{S}}_{\mathcal{X}}$ is $\bar{\mathcal{S}}_{-\,\mathcal{X}}$ can be proved by an argument that is similar to the preceding paragraph.

(viii) This follows immediately from Definition 13.40.

(ix) Clearly, $B^* \in \mathrm{B}(\mathcal{Y}, \mathcal{X})$. Then, $BB^* \in \mathrm{B}(\mathcal{Y}, \mathcal{Y})$. By Proposition 13.22, we have $(BB^*)^* = (B^*)^*B^* = BB^*$. Hence, $BB^* \in \bar{\mathcal{S}}_{\mathcal{Y}}$. $\forall y \in \mathcal{Y}$, $\langle y, BB^*y\rangle_{\mathcal{Y}} = \langle B^*y, B^*y\rangle_{\mathcal{X}} = \|B^*y\|_{\mathcal{X}}^2 \geq 0$, where the first equality follows from Proposition 13.22; and the second equality and the inequality follow from Proposition 13.2. Hence, $BB^* \in \bar{\mathcal{S}}_{\mathrm{psd}\,\mathcal{Y}}$.

By similar arguments, we have $B^*B \in \bar{\mathcal{S}}_{\mathrm{psd}\,\mathcal{X}}$.

This completes the proof of the proposition. □

13.8 Pseudoinverse Operator

Proposition 13.43 *Let $\mathcal{X}$ be a Hilbert space, $M \subseteq \mathcal{X}$ be a closed subspace. Then, we may define a mapping $P : \mathcal{X} \to M$ by $P(x) = x_0$, $\forall x \in \mathcal{X}$, where x_0 is the unique solution to the $\|x - x_0\| = \min_{m\in M} \|x - m\|$, by the Classical Projection Theorem 13.13. P is said to be the* projection operator of $\mathcal{X}$ to M. *Then, we have the following properties for the projection operator.*

(i) $P \in \mathrm{B}(\mathcal{X}, \mathcal{X})$.
(ii) $\|P\|_{\mathrm{B}(\mathcal{X},\mathcal{X})} \leq 1$.
(iii) $P^2 = P$ *(idempotent).*
(iv) $P^* = P$ *(Hermitian).*
(v) $\forall A \in \mathrm{B}(\mathcal{X}, \mathcal{X})$ *that satisfies (iii) and (iv) is a projection operator.*

Proof (i) By Theorem 13.19, $\mathcal{X} = M \oplus M^{\perp_h}$. Then, $\forall x \in \mathcal{X}$, there exists a unique pair $x_1 \in M$ and $x_2 \in M^{\perp_h}$ such that $x_1 + x_2 = x$. Then, $x_1 = Px$. We will show that P is a linear operator. $\forall x, y \in \mathcal{X}$, $\forall \lambda \in \mathbb{K}$, we have $x = x_1 + x_2$ and $y = y_1 + y_2$ with $x_1, y_1 \in M$ and $x_2, y_2 \in M^{\perp_h}$. Then, $x + y = x_1 + y_1 + x_2 + y_2$ with $x_1 + y_1 \in M$ and $x_2 + y_2 \in M^{\perp_h}$. Hence, $P(x + y) = x_1 + y_1 = P(x) + P(y)$. $\lambda x = \lambda x_1 + \lambda x_2$ with $\lambda x_1 \in M$ and $\lambda x_2 \in M^{\perp_h}$. Then, $P(\lambda x) = \lambda x_1 = \lambda P(x)$. Hence, P is linear.

Next, we show that P is bounded. $\|P\|_{\mathrm{B}(\mathcal{X},\mathcal{X})} = \sup_{x\in\mathcal{X},\|x\|\leq 1} \|Px\|$. $\|Px\|^2 = \|x_1\|^2 = \langle x_1, x_1\rangle = \langle x - x_2, x - x_2\rangle = \langle x, x\rangle - \langle x, x_2\rangle - \langle x_2, x\rangle + \langle x_2, x_2\rangle = \|x\|^2 - \langle x_1 + x_2, x_2\rangle - \langle x_2, x_1 + x_2\rangle + \langle x_2, x_2\rangle = \|x\|^2 - \langle x_2, x_2\rangle = \|x\|^2 - \|x_2\|^2 \leq \|x\|^2$, where the first equality follows from the definition of P; the second equality follows from Proposition 13.2; the third through fifth equalities follows from basic algebra; the sixth equality follows from the fact $x_1 \in M$ and $x_2 \in M^{\perp_h}$. Hence, $\|P\|_{B(\mathcal{X},\mathcal{X})} \leq 1$. Hence, P is bounded. Then, $P \in B(\mathcal{X}, \mathcal{X})$.

(ii) This follows from the preceding paragraph.

(iii) This is obvious.

(iv) $\forall x, y \in \mathcal{X}$, $\langle x, Py \rangle = \langle P^* x, y \rangle = \langle x_1 + x_2, y_1 \rangle = \langle x_1, y_1 \rangle = \langle x_1, y_1 + y_2 \rangle = \langle Px, y \rangle$, where the first equality follows from Proposition 13.22; $x = x_1 + x_2$, $y = y_1 + y_2$, $x_1, y_1 \in M$, and $x_2, y_2 \in M^{\perp_h}$. By the arbitrariness of x and y, we have $P^* = P$.

(v) Let $A \in B(\mathcal{X}, \mathcal{X})$ satisfies (iii) and (iv). Let $M := \mathcal{R}(A)$. Then, $M^{\perp_h} = \mathcal{N}(A^*) = \mathcal{N}(A)$, by Proposition 13.22 and (iv). By Theorem 13.19, $\mathcal{X} = M^{\perp_h} \oplus (M^{\perp_h})^{\perp_h}$. By Proposition 13.17, $(M^{\perp_h})^{\perp_h} = \overline{\mathcal{R}(A)}$. Let $P : \mathcal{X} \to \overline{\mathcal{R}(A)}$ be the projection operator. We will show that $P = A$. $\forall x, y \in \mathcal{X}$, we have

$$\begin{aligned}\langle Ax, y - Ay \rangle &= \langle Ax, y \rangle - \langle Ax, Ay \rangle = \langle Ax, y \rangle - \left\langle A^* Ax, y \right\rangle \\ &= \langle Ax, y \rangle - \langle A^2 x, y \rangle = \langle Ax, y \rangle - \langle Ax, y \rangle = 0\end{aligned}$$

where the second equality follows from Proposition 13.22; the third equality follows from (iv); and the fourth equality follows from (iii). Thus, $Ax \perp y - Ay$. By the arbitrariness of x, we have $y - Ay \in M^{\perp_h}$. Note that $Ay \in \overline{\mathcal{R}(A)}$. Then, we have $Ay = Py$. By the arbitrariness of y, we have $A = P$.

This completes the proof of the proposition. □

Proposition 13.44 *Let $\mathcal{X}$ and $\mathcal{Y}$ be Hilbert spaces over $\mathbb{K}$, $A \in B(\mathcal{X}, \mathcal{Y})$, and $y \in \mathcal{Y}$. Assume that $\mathcal{R}(A) \subseteq \mathcal{Y}$ is closed. Then, there exists a unique vector $x_0 \in \mathcal{X}$ such that $x_0 = \arg\min_{x_1} \|x_1\|_{\mathcal{X}}$, where x_1 satisfies $\|Ax_1 - y\|_{\mathcal{Y}} = \min_{x \in \mathcal{X}} \|Ax - y\|_{\mathcal{Y}}$. We will denote $x_0 =: A^\dagger y$, where $A^\dagger : \mathcal{Y} \to \mathcal{X}$ is called the* pseudoinverse of A. *Furthermore, we have the following properties for $A^\dagger$.*

(i) $A^\dagger \in B(\mathcal{Y}, \mathcal{X})$.
(ii) $(A^\dagger)^\dagger = A$.
(iii) $(A^*)^\dagger = (A^\dagger)^*$.
(iv) $A^\dagger A A^\dagger = A^\dagger$.
(v) $A A^\dagger A = A$.
(vi) $(A^\dagger A)^* = A^\dagger A$.
(vii) $(A A^\dagger)^* = A A^\dagger$.
(viii) $A^\dagger = (A^* A)^\dagger A^*$.
(ix) $A^\dagger = A^*(A A^*)^\dagger$.
(x) $A^\dagger = \lim_{\epsilon \to 0^+} A^*(\epsilon\, \mathrm{id}_{\mathcal{Y}} + A A^*)^{-1} = \lim_{\epsilon \to 0^+} (\epsilon\, \mathrm{id}_{\mathcal{X}} + A^* A)^{-1} A^*$.
(xi) *A can be uniquely expressed as $A = P_1 A_r P_2$, where $P_1 : \mathcal{Y} \to \mathcal{R}(A)$ and $P_2 : \mathcal{X} \to (\mathcal{N}(A))^{\perp_h}$ are projection operators, and $A_r \in B\big((\mathcal{N}(A))^{\perp_h}, \mathcal{R}(A)\big)$ is bijective with $A_r^{-1} \in B\big(\mathcal{R}(A), (\mathcal{N}(A))^{\perp_h}\big)$; and $A^\dagger = P_2 A_r^{-1} P_1$.*
(xii) $A A^\dagger = P_1$ *and* $A^\dagger A = P_2$.

Proof Since $\mathcal{R}(A)$ is closed. Then, by the Classical Projection Theorem 13.13, $\exists!\ y_0 \in \mathcal{R}(A)$ such that $\|y - y_0\|_{\mathcal{Y}} = \min_{\bar{y} \in \mathcal{R}(A)} \|y - \bar{y}\|_{\mathcal{Y}}$, furthermore, $y_0 \in \mathcal{R}(A)$ is defined to be the unique vector such that $y - y_0 \in (\mathcal{R}(A))^{\perp_h}$. Then, the set of x_1's defined in the proposition statement is given by $\mathcal{N}(A) + x_1$, where x_1 is any vector such that $Ax_1 = y_0$. Then, by the Classical Projection Theorem 13.13 and Proposition 13.22, $\exists!\ x_0 \in (\mathcal{N}(A))^{\perp_h} = \mathcal{R}(A^*)$ such that $\|x_0\|_{\mathcal{X}} = \min_{x \in \mathcal{N}(A)} \|x_1 - x\|_{\mathcal{X}}$. This proves the existence and uniqueness of $x_0 \in \mathcal{X}$ for any given $y \in \mathcal{Y}$. Thus, $A^\dagger$ is well-defined.

(i) By Theorem 13.19, $\mathcal{X} = \mathcal{N}(A) \oplus (\mathcal{N}(A))^{\perp_h} =: M \oplus M^{\perp_h}$. Define the projection mapping $P_2 : \mathcal{X} \to M^{\perp_h}$. $\forall x \in \mathcal{X}$, there is unique pair $x_1 \in M$ and $x_2 \in M^{\perp_h}$ such that $x = x_1 + x_2 = x_1 + P_2 x$. Then, $Ax = AP_2 x$. This allows us to define a mapping $A_r : M^{\perp_h} \to N := \mathcal{R}(A)$ by $A_r(x_2) = A(x_2)$, $\forall x_2 \in M^{\perp_h}$. Clearly, $A_r \in \mathrm{B}\big(M^{\perp_h}, N\big)$ and A_r is surjective. It is easy to show that $\mathcal{N}(A_r) = \big\{\vartheta_{M^{\perp_h}} = \vartheta_{\mathcal{X}}\big\}$ since the domain of A_r is $M^{\perp_h}$. Then, A_r is bijective. Note that N is a closed subspace of $\mathcal{Y}$ by assumption. Then, by Proposition 4.39, N is a Banach subspace of $\mathcal{Y}$. By Proposition 13.17, $M^{\perp_h}$ is a closed subspace of $\mathcal{X}$. Then, by Proposition 4.39, $M^{\perp_h}$ is a Banach subspace of $\mathcal{X}$. Then, $A_r^{-1} \in \mathrm{B}\left(N, M^{\perp_h}\right)$ by Open Mapping Theorem 7.103. Then, $A = A_r P_2$. By Theorem 13.19, $\mathcal{Y} = N \oplus N^{\perp_h}$. Define $P_1 : \mathcal{Y} \to N$ to be the projection mapping. Clearly, $P_1 \in \mathrm{B}(\mathcal{Y}, N)$. $\forall y \in \mathcal{Y}$, $y_0 = P_1 y$. Let $x_1 := A_r^{-1} y_0 \in M^{\perp_h} \subseteq \mathcal{X}$. Then, $Ax_1 = A_r P_2 A_r^{-1} y_0 = A_r A_r^{-1} y_0 = y_0$. Then, $x_0 = P_2 x_1 = x_1 = A_r^{-1} P_1 y$. Hence, $A^\dagger = A_r^{-1} P_1 = P_2 A_r^{-1} P_1 \in \mathrm{B}\left(\mathcal{Y}, \mathcal{X}\right)$.

(xi) This follows immediately from the preceding discussion in (i).

(ii) For $A \in \mathrm{B}(\mathcal{X}, \mathcal{Y})$, $A_r \in \mathrm{B}\big(M^{\perp_h}, N\big)$, $A = A_r P_2 = P_1 A_r P_2$, and $A^\dagger = A_r^{-1} P_1 = P_2 A_r^{-1} P_1$. Then, $(A^\dagger)^\dagger = P_1 A_r P_2 = A$.

(iii) $(A^\dagger)^* = (A_r^{-1} P_1)^* = (P_2 A_r^{-1} P_1)^* = P_1^* (A_r^{-1})^* P_2^* = P_1 A_r^{-*} P_2$, where the first two equalities follow from (xi); the third equality follows from Proposition 13.22; and the last equality follows from Proposition 13.43. Now, $(\mathcal{N}(A^*))^{\perp_h} = \mathcal{R}(A) = N$ and $\mathcal{R}(A^*) = (\mathcal{N}(A))^{\perp_h} = M^{\perp_h}$, by Proposition 13.22. Therefore, $(A^*)^\dagger = (P_2^* A_r^* P_1^*)^\dagger = (P_2 A_r^* P_1)^\dagger = P_1 (A_r^*)^{-1} P_2 = P_1 A_r^{-*} P_2$, where the first equality follows from (xi); the second equality follows from Proposition 13.43; the third equality follow from (xi); and the last equality follows from Proposition 13.22. Hence, $(A^\dagger)^* = (A^*)^\dagger$.

(xii) By (xi), we have $A^\dagger A = P_2 A_r^{-1} P_1 P_1 A_r P_2 = P_2 A_r^{-1} A_r P_2 = P_2$, where the second and third equalities follow from the fact that P_1 and P_2 are projection operators. Similarly, $AA^\dagger = P_1 A_r P_2 P_2 A_r^{-1} P_1 = P_1 A_r A_r^{-1} P_1 = P_1$, where the second and third equalities follow from the fact that P_1 and P_2 are projection operators.

(iv) By (xi) and (xii), we have $A^\dagger A A^\dagger = P_2 A_r^{-1} P_1 P_1 = P_2 A_r^{-1} P_1 = A^\dagger$, where the second equality follows from Proposition 13.43.

(v) By (xi) and (xiii), we have $AA^\dagger A = P_1 P_1 A_r P_2 = P_1 A_r P_2 = A$, where the second equality follows from Proposition 13.43.

(vi) By (xi) and (xii), we have $(A^\dagger A)^* = P_2^* = P_2 = A^\dagger A$, where the second equality follows from Proposition 13.43.

(vii) By (xi) and (xii), we have $(AA^\dagger)^* = P_1^* = P_1 = AA^\dagger$, where the second equality follows from Proposition 13.43.

(viii) By (xi), we have $A^*A = (P_1 A_r P_2)^* P_1 A_r P_2 = P_2^* A_r^* P_1^* P_1 A_r P_2 = P_2 A_r^* A_r P_2$, where the second equality follows from Proposition 13.22; the third equality follows from Proposition 13.43 and the fact P_1 is a projection operator. Then, we have $(A^*A)^\dagger A^* = P_2(A_r^* A_r)^{-1} P_2 (P_1 A_r P_2)^* = P_2 A_r^{-1} A_r^{-*} P_2 P_2^* A_r^* P_1^* = P_2 A_r^{-1} A_r^{-*} P_2 P_2 A_r^* P_1 = P_2 A_r^{-1} A_r^{-*} A_r^* P_1 = P_2 \cdot A_r^{-1} P_1 = A^\dagger$, where the first equality follows from (xi) applied to the operator A^*A; the second equality follows from Proposition 13.22; the third equality follows from Proposition 13.43; the fourth equality follows from the fact P_2 is a projection operator; and the last equality follows from (xi).

(ix) By (xi), we have $AA^* = P_1 A_r P_2 (P_1 A_r P_2)^* = P_1 A_r P_2 P_2^* A_r^* P_1^* = P_1 A_r A_r^* P_1$, where the second equality follows from Proposition 13.22; the third equality follows from Proposition 13.43 and the fact P_2 is a projection operator. Then, we have $A^*(AA^*)^\dagger = (P_1 A_r P_2)^* P_1 (A_r A_r^*)^{-1} P_1 = P_2^* A_r^* P_1^* P_1 A_r^{-*} A_r^{-1} P_1 = P_2 A_r^* P_1 P_1 A_r^{-*} A_r^{-1} P_1 = P_2 A_r^* A_r^{-*} A_r^{-1} P_1 = P_2 \cdot A_r^{-1} P_1 = A^\dagger$, where the first equality follows from (xi) applied to the operator AA^*; the second equality follows from Proposition 13.22; the third equality follows from Proposition 13.43; the fourth equality follows from the fact P_2 is a projection operator; and the last equality follows from (xi).

(x) Note that $\epsilon\,\mathrm{id}_{\mathcal{X}} + A^*A = \epsilon(P_2 + \bar{P}_2) + P_2 A_r^* A_r P_2 = \epsilon \bar{P}_2\, \mathrm{id}_M\, \bar{P}_2 + P_2(\epsilon\,\mathrm{id}_{M^{\perp_h}} + A_r^* A_r) P_2$, $\forall \epsilon \in (0, \infty) \subset \mathbb{R}$, where $\bar{P}_2 : \mathcal{X} \to M$ is the projection operator; the first equality follows from Theorem 13.19 and the proof of (viii); the second equality follows from Proposition 13.43. Then, $(\epsilon\,\mathrm{id}_{\mathcal{X}} + A^*A)^{-1} = \bar{P}_2 \frac{1}{\epsilon}\, \mathrm{id}_M\, \bar{P}_2 + P_2(\epsilon\,\mathrm{id}_{M^{\perp_h}} + A_r^* A_r)^{-1} P_2$, $\forall \epsilon \in (0, \infty) \subset \mathbb{R}$, where the equality follows from the fact $P_2 \bar{P}_2 = \vartheta_{\mathrm{B}(\mathcal{X},\mathcal{X})}$ and Proposition 13.43; and the invertibility of $\epsilon\,\mathrm{id}_{M^{\perp_h}} + A_r^* A_r$ follows from Proposition 13.41. This yields $(\epsilon\,\mathrm{id}_{\mathcal{X}} + A^*A)^{-1} A^* = (\bar{P}_2 \frac{1}{\epsilon}\, \mathrm{id}_M\, \bar{P}_2 + P_2(\epsilon\,\mathrm{id}_{M^{\perp_h}} + A_r^* A_r)^{-1} P_2) P_2 A_r^* P_1 = P_2(\epsilon\,\mathrm{id}_{M^{\perp_h}} + A_r^* A_r)^{-1} A_r^* P_1$, $\forall \epsilon \in (0, \infty) \subset \mathbb{R}$. Since $A_r^* A_r$ is invertible, by Proposition 9.55, we have $\lim_{\epsilon \to 0^+} (\epsilon\,\mathrm{id}_{M^{\perp_h}} + A_r^* A_r)^{-1} = (A_r^* A_r)^{-1} = A_r^{-1} A_r^{-*}$. Then, $\lim_{\epsilon \to 0^+} (\epsilon\,\mathrm{id}_{\mathcal{X}} + A^*A)^{-1} A^* = \lim_{\epsilon \to 0^+} P_2(\epsilon\,\mathrm{id}_{M^{\perp_h}} + A_r^* A_r)^{-1} A_r^* P_1 = P_2 A_r^{-1} A_r^{-*} A_r^* P_1 = P_2 A_r^{-1} P_1 = A^\dagger$. The other equality in (x) can be proved by symmetry.

This completes the proof of the proposition. □

Proposition 13.45 *Let $\mathcal{X}$ and $\mathcal{Y}$ be Hilbert spaces over $\mathbb{K}$, and $B \in \mathrm{B}(\mathcal{X}, \mathcal{Y})$. If B is surjective, then $BB^* \in \bar{\mathcal{S}}_{+\,\mathcal{Y}}$. If B is injective and $\mathcal{R}(B^*) \subseteq \mathcal{X}$ is closed, then $B^*B \in \bar{\mathcal{S}}_{+\,\mathcal{X}}$.*

Proof If B is surjective, then $\mathcal{R}(B) = \mathcal{Y}$ and it is a closed set in $\mathcal{Y}$. By Proposition 13.44, we have $B = P_1 B_r P_2$, where $P_1 : \mathcal{Y} \to \mathcal{R}(B)$ and $P_2 : \mathcal{X} \to (\mathcal{N}(B))^{\perp_h}$ are projection operators, and $B_r \in \mathrm{B}\big((\mathcal{N}(B))^{\perp_h}, \mathcal{R}(B)\big)$ is bijective. Clearly, $P_1 = \mathrm{id}_{\mathcal{Y}}$. Then, $BB^* = B_r B_r^*$ is bijective, and then $(BB^*)^{-1} \in \mathrm{B}(\mathcal{Y}, \mathcal{Y})$ by Open Mapping Theorem 7.103. Fix any $\forall y \in \mathcal{Y}$. Let $\bar{y} := (BB^*)^{-1} y$. This yields $\langle y, (BB^*)^{-1} y \rangle_{\mathcal{Y}} = \langle BB^* \bar{y}, \bar{y} \rangle_{\mathcal{Y}} = \langle B^* \bar{y}, B^* \bar{y} \rangle_{\mathcal{X}} = \| B^* \bar{y} \|_{\mathcal{X}}^2$. Define

$\bar{x} := B^*\bar{y} = B^*(BB^*)^{-1}y = B^\dagger y \in \mathcal{X}$, where the second equality follows from Proposition 13.44. This leads to $B\bar{x} = BB^\dagger y = P_1 y = y$, where the second equality follows from the proof for (vi) of Proposition 13.44. This yields $\|y\|_{\mathcal{Y}} \leq \|B\|_{B(\mathcal{X},\mathcal{Y})}\|\bar{x}\|_{\mathcal{X}}$. This implies that $\langle y, (BB^*)^{-1}y\rangle_{\mathcal{Y}} = \|\bar{x}\|^2_{\mathcal{X}} \geq \frac{\|B\|^2_{B(\mathcal{X},\mathcal{Y})}}{\epsilon+\|B\|^2_{B(\mathcal{X},\mathcal{Y})}}\|\bar{x}\|^2_{\mathcal{X}} \geq \frac{1}{\epsilon+\|B\|^2_{B(\mathcal{X},\mathcal{Y})}}\|y\|^2_{\mathcal{Y}}$, where the first equality follows from the preceding discussion; the first inequality holds for any $\epsilon \in (0, \infty) \subset \mathbb{R}$; and the last inequality follows the preceding discussion. This yields that $(BB^*)^{-1} \in \bar{\mathcal{S}}_{+\mathcal{Y}}$ (clearly, $(BB^*)^{-1} \in \bar{\mathcal{S}}_{\mathcal{Y}}$). By (v) of Proposition 13.42, $BB^* \in \bar{\mathcal{S}}_{+\mathcal{Y}}$.

If B is injective and $\mathcal{R}(B^*) \subseteq \mathcal{X}$ is closed, then $\mathcal{N}(B) = \{\vartheta_{\mathcal{X}}\}$. By Proposition 13.22, we have $\mathcal{R}(B^*) = (\mathcal{N}(B))^{\perp_h} = \mathcal{X}$. Hence, B^* is surjective. By the preceding paragraph, we have $B^*(B^*)^* = B^*B \in \bar{\mathcal{S}}_{+\mathcal{X}}$, where the equality follows from Proposition 13.22.

This completes the proof of the proposition. □

13.9 Spectral Theory of Linear Operators

Definition 13.46 Let $\mathcal{X}$ be a normed linear space over $\mathbb{K}$, $A \in B(\mathcal{X}, \mathcal{X})$, $\lambda \in \mathbb{K}$, and $x_0 \in \mathcal{X}$ with $x_0 \neq \vartheta_{\mathcal{X}}$. If $\lambda x_0 = Ax_0$, or equivalently $(\lambda\, \mathrm{id}_{\mathcal{X}} - A)x_0 = \vartheta_{\mathcal{X}}$, we will say that λ is an *eigenvalue* of A, x_0 is an *eigenvector* of A associated with the eigenvalue λ. If $(\lambda\, \mathrm{id}_{\mathcal{X}} - A)^{k-1}x_0 \neq \vartheta_{\mathcal{X}}$ and $(\lambda\, \mathrm{id}_{\mathcal{X}} - A)^k x_0 = \vartheta_{\mathcal{X}}$, for some $k \in \mathbb{N}$ with $k \geq 2$, we will say that x_0 is a *generalized eigenvector* of grade k of A associated with the eigenvalue λ. ◇

Proposition 13.47 *Let $\mathcal{X}$ be a Hilbert space over $\mathbb{K}$, $Q \in B(\mathcal{X}, \mathcal{X})$ be Hermitian, and $A \in B(\mathcal{X}, \mathcal{X})$. Then, the following statements hold.*

(i) *If $\lambda \in \mathbb{K}$ is an eigenvalue of Q, then $\lambda \in \mathbb{R}$.*
(ii) *If $S_\gamma \subseteq \mathcal{X}$ be an linearly independent set of eigenvectors of Q associated with eigenvalue $\lambda_\gamma \in \mathbb{R}$, $\forall \gamma \in \Gamma$, and $(\lambda_\gamma)_{\gamma\in\Gamma}$ is pairwise distinct, then the set $S := \bigcup_{\gamma\in\Gamma} S_\gamma$ is a linearly independent set.*
(iii) *If x_i is an eigenvector of Q associated with the eigenvalue $\lambda_i \in \mathbb{K}$, $i = 1, 2$, and $\lambda_1 \neq \lambda_2$, then $\langle x_1, x_2\rangle = 0$ and $x_1 \perp x_2$.*
(iv) *Q has no generalized eigenvectors of grade $k \geq 2$ associated with any eigenvalue of Q, (i. e., Q has only eigenvectors but not generalized eigenvectors).*
(v) *If $\mathcal{R}(Q) \subseteq \mathcal{X}$ is closed, then $0 \in \mathbb{K}$ is not an eigenvalue of Q if, and only if, Q is bijective and $Q^{-1} \in B(\mathcal{X}, \mathcal{X})$.*
(vi) *If $Q \in \bar{\mathcal{S}}_{+\mathcal{X}}$ then $\exists m \in (0, \infty) \subset \mathbb{R}$ such that $\forall \lambda \in \mathbb{R}$ with λ being an eigenvalue of Q, we have $\lambda \geq m$.*
(vii) *If $\lambda \in \mathbb{K}$ is an eigenvalue of A, then $|\lambda| \leq \|A\|_{B(\mathcal{X},\mathcal{X})}\|\mathrm{id}_{\mathcal{X}}\|_{B(\mathcal{X},\mathcal{X})} \leq \|A\|_{B(\mathcal{X},\mathcal{X})}$.*
(viii) *$Q \in \bar{\mathcal{S}}_{\mathrm{psd}\,\mathcal{X}}$ implies that all of the eigenvalues of Q are nonnegative.*

Proof (i) Let $x_0 \in \mathcal{X}$ be the eigenvector of Q associated with the eigenvalue λ. Then, $\lambda x_0 = Qx_0$. This implies that $\overline{\lambda}\|x_0\|^2 = \overline{\lambda}\langle x_0, x_0\rangle = \langle x_0, \lambda x_0\rangle = \langle x_0, Qx_0\rangle = \langle Q^* x_0, x_0\rangle = \langle Qx_0, x_0\rangle = \langle \lambda x_0, x_0\rangle = \lambda\|x_0\|^2$, where the first equality follows from Proposition 13.2; the second equality follows from Definition 13.1; the third equality follows from the fact that x_0 is an eigenvector of Q associated with the eigenvalue λ; the fourth equality follows from Proposition 13.22; the fifth equality follows from the assumption that Q is Hermitian; the sixth equality follows from the fact that x_0 is an eigenvector of Q associated with eigenvalue λ; and the seventh equality follows from Definition 13.1 and Proposition 13.2. Note that $x_0 \neq \vartheta_{\mathcal{X}}$ since it is an eigenvector, then $\|x_0\|^2 > 0$. Thus, the above implies that $\overline{\lambda} = \lambda$. Hence, $\lambda \in \mathbb{R}$.

(ii) We will use mathematical induction on n to show that there does not exist an $n \in \mathbb{Z}_+$ such that $\exists x_0 \in S$ and $\exists x_1, \ldots, x_n \in S \setminus \{x_0\}$, which are distinct, and $\exists \alpha_1, \ldots, \alpha_n \in \mathbb{K} \setminus \{0\}$ such that $x_0 = \sum_{i=1}^n \alpha_i x_i$.

1° $n = 0$. Suppose that the result does not hold, then $\exists x_0 \in S$ such that $x_0 = \vartheta_{\mathcal{X}}$. Note that $x_0 \in S_{\gamma_0}$, for some $\gamma_0 \in \Gamma$. Then, x_0 is an eigenvector of Q associated with eigenvalue λ_{γ_0}. By Definition 13.46, $x_0 \neq \vartheta_{\mathcal{X}}$. This is a contradiction. This case is proved.

2° Assume the claim holds for $n \leq k \in \mathbb{Z}_+$.

3° Consider the case $n = k + 1 \in \mathbb{N}$. Suppose that the result does not hold. Then, $\exists x_0 \in S$ and $\exists x_1, \ldots, x_n \in S \setminus \{x_0\}$, which are distinct, and $\exists \alpha_1, \ldots, \alpha_n \in \mathbb{K} \setminus \{0\}$ such that $x_0 = \sum_{i=1}^n \alpha_i x_i$. We will distinguish two exhaustive and mutually exclusive cases: Case 1: $x_0, \ldots, x_n \in S_{\gamma_0}$ for some $\gamma_0 \in \Lambda$; Case 2: $x_0 \in S_{\gamma_0}$ with $\gamma_0 \in \Gamma$, and $\exists i_1 \in \{1, \ldots, n\}$ such that $x_{i_1} \in S_{\gamma_{i_1}}$ with $\gamma_{i_1} \in \Gamma$ and $\gamma_{i_1} \neq \gamma_0$.

Case 1: $x_0, \ldots, x_n \in S_{\gamma_0}$ for some $\gamma_0 \in \Gamma$. Then, $x_0, \ldots, x_n$ are distinct vectors in S_{γ_0}. By assumption, S_{γ_0} is a linearly independent set. Then, $x_0 = \sum_{i=1}^n \alpha_i x_i$ contradicts with the fact that S_{γ_0} being a linearly independent set.

Case 2: $x_0 \in S_{\gamma_0}$ with $\gamma_0 \in \Gamma$, and $\exists i_1 \in \{1, \ldots, n\}$ such that $x_{i_1} \in S_{\gamma_{i_1}}$ with $\gamma_{i_1} \in \Gamma$ and $\gamma_{i_1} \neq \gamma_0$. Without loss of generality, assume that $x_i \in S_{\gamma_i}, i = 1, \ldots, n$, where $\gamma_i \in \Gamma$. Then, $\vartheta_{\mathcal{X}} = (\lambda_{\gamma_0} \operatorname{id}_{\mathcal{X}} - Q)x_0 = \sum_{i=1}^n (\lambda_{\gamma_0} \operatorname{id}_{\mathcal{X}} - Q)(\alpha_i x_i) = \sum_{i=1}^n \alpha_i(\lambda_{\gamma_0} - \lambda_{\gamma_i})x_i$. By the preceding discussion and (i), $0 \neq \lambda_{\gamma_0} - \lambda_{\gamma_{i_1}} \in \mathbb{R}$. Then, we have $\vartheta_{\mathcal{X}} = \sum_{i=1}^n \frac{(\lambda_{\gamma_0} - \lambda_{\gamma_i})\alpha_i}{(\lambda_{\gamma_0} - \lambda_{\gamma_{i_1}})\alpha_{i_1}} x_i$. Rearranging the equation, we have $x_{i_1} = \sum_{i=1, i \neq i_1}^n -\frac{(\lambda_{\gamma_0} - \lambda_{\gamma_i})\alpha_i}{(\lambda_{\gamma_0} - \lambda_{\gamma_{i_1}})\alpha_{i_1}} x_i$. This contradicts with the inductive assumption. Hence, the result holds in this case as well.

This completes the inductive process. Hence, S is a linearly independent set.

(iii) By (i), $\lambda_1, \lambda_2 \in \mathbb{R}$. We have $\lambda_1\langle x_1, x_2\rangle = \langle \lambda_1 x_1, x_2\rangle = \langle Qx_1, x_2\rangle = \langle x_1, Q^* x_2\rangle = \langle x_1, Qx_2\rangle = \langle x_1, \lambda_2 x_2\rangle = \overline{\lambda_2}\langle x_1, x_2\rangle = \lambda_2\langle x_1, x_2\rangle$, where the first equality follows from Definition 13.1; the second equality follows from the fact that x_1 is an eigenvector of Q associated with the eigenvalue λ_1; the third equality follows from Proposition 13.22; the fourth equality follows from the fact that Q is Hermitian; the fifth equality follows from the fact that x_2 is an eigenvector of Q associated with the eigenvalue λ_2; the six equality follows from Definition 13.1; and the last equality follows from the fact $\lambda_2 \in \mathbb{R}$. Since $\lambda_1 \neq \lambda_2$, then $\langle x_1, x_2\rangle = 0$ and $x_1 \perp x_2$.

(iv) Suppose the result does not hold. Let $x_0 \in \mathcal{X}$ be a generalized eigenvector of grade $k \geq 2$ of Q associated with eigenvalue $\lambda \in \mathbb{R}$. Then, $B^{k-1}x_0 \neq \vartheta_{\mathcal{X}}$ and $B^k x_0 = \vartheta_{\mathcal{X}}$, where $B := \lambda\,\mathrm{id}_{\mathcal{X}} - Q$. Let $x_{k-2} := B^{k-2}x_0$ and $x_{k-1} := B^{k-1}x_0$. Then, $B^2 x_{k-2} = \vartheta_{\mathcal{X}}$. This implies that $0 = \langle x_{k-2}, B^2 x_{k-2}\rangle = \langle B^* x_{k-2}, Bx_{k-2}\rangle = \langle Bx_{k-2}, Bx_{k-2}\rangle = \langle x_{k-1}, x_{k-1}\rangle = \|x_{k-1}\|^2$, where the second equality follows from Proposition 13.22; the third equality follows from the assumption that Q is Hermitian; and the fifth equality follows from Proposition 13.2. Then, $x_{k-1} = \vartheta_{\mathcal{X}}$. This contradicts with $B^{k-1}x_0 \neq \vartheta_{\mathcal{X}}$. Hence, the result holds.

(v) "Necessity" Since 0 is not an eigenvalue of Q, then $\mathcal{N}(Q) = \{\vartheta_{\mathcal{X}}\}$. By Proposition 13.22, $\mathcal{R}(Q) = (\mathcal{N}(Q^*))^{\perp_h} = (\mathcal{N}(Q))^{\perp_h} = \mathcal{X}$. Then, Q is bijective and invertible. By Open Mapping Theorem 7.103, $Q^{-1} \in \mathrm{B}(\mathcal{X}, \mathcal{X})$.

"Sufficiency" Since Q is bijective, then $\mathcal{N}(Q) = \{\vartheta_{\mathcal{X}}\}$. Then, $0 \in \mathbb{K}$ is not an eigenvalue of Q, since there is no eigenvector associated with 0.

(vi) By Definition 13.40, $\exists m \in (0, \infty) \subset \mathbb{R}$ such that $\langle x, Qx\rangle \geq m\|x\|^2$, $\forall x \in \mathcal{X}$. Let $x_0 \in \mathcal{X}$ be an eigenvector of Q associated with the eigenvalue $\lambda \in \mathbb{R}$. Then, we have $\lambda\|x_0\|^2 = \langle x_0, \lambda x_0\rangle = \langle x_0, Qx_0\rangle \geq m\|x_0\|^2$. Since x_0 is an eigenvector, then $x_0 \neq \vartheta_{\mathcal{X}}$ and $\|x_0\| > 0$. Then, we have $\lambda \geq m$.

(vii) Fix any $\lambda \in \mathbb{K}$ with $|\lambda| > \|A\|_{\mathrm{B}(\mathcal{X},\mathcal{X})}\|\mathrm{id}_{\mathcal{X}}\|_{\mathrm{B}(\mathcal{X},\mathcal{X})} \geq 0$. Then, $\|\,\mathrm{id}_{\mathcal{X}} - (\mathrm{id}_{\mathcal{X}} - \frac{1}{\lambda}A)\|_{\mathrm{B}(\mathcal{X},\mathcal{X})}\|\,\mathrm{id}_{\mathcal{X}}^{-1}\,\|_{\mathrm{B}(\mathcal{X},\mathcal{X})} = \frac{1}{|\lambda|}\|A\|_{\mathrm{B}(\mathcal{X},\mathcal{X})}\|\mathrm{id}_{\mathcal{X}}\|_{\mathrm{B}(\mathcal{X},\mathcal{X})} < 1$. By Proposition 9.55, $\mathrm{id}_{\mathcal{X}} - \frac{1}{\lambda}A$ is bijective and admits continuous inverse. Then, $\lambda\,\mathrm{id}_{\mathcal{X}} - A$ is bijective and admits continuous inverse since $\lambda \neq 0$. This implies that $\mathcal{N}(\lambda\,\mathrm{id}_{\mathcal{X}} - A) = \{\vartheta_{\mathcal{X}}\}$. Hence, λ is not an eigenvalue of A. Hence, the result holds noting $\|\mathrm{id}_{\mathcal{X}}\|_{\mathrm{B}(\mathcal{X},\mathcal{X})} \leq 1$.

(viii) Since $Q \in \bar{\mathcal{S}}_{\mathrm{psd}\,\mathcal{X}}$, then $\langle x, Qx\rangle \geq 0$, $\forall x \in \mathcal{X}$. By (i), all eigenvalues of Q are real. Suppose that the result does not hold. Then, $\exists \lambda \in (-\infty, 0) \subset \mathbb{R}$ and $\exists x_0 \in \mathcal{X}$ such that x_0 is an eigenvector of Q associated with the eigenvalue λ. Then, $0 \leq \langle x_0, Qx_0\rangle = \langle x_0, \lambda x_0\rangle = \lambda\|x_0\|^2 < 0$, where the first equality follows from the hypothesis; and the second equality follows from Definition 13.1 and Proposition 13.2; and the last inequality follows from $x_0 \neq \vartheta_{\mathcal{X}}$ since it is an eigenvector. This is a contradiction. Hence, all eigenvalues of Q are nonnegative.

This completes the proof of the proposition. □

Proposition 13.48 *Let $\mathcal{X}$ be a Hilbert space over $\mathbb{K}$, $A \in \mathrm{B}(\mathcal{X}, \mathcal{X})$ be a Hermitian operator. Then,* $\lambda := \sup_{x\in\mathcal{X}, \|x\|\leq 1} |\langle Ax, x\rangle| = \|A\|_{\mathrm{B}(\mathcal{X},\mathcal{X})}$.

Proof Note that $|\langle Ax, x\rangle| \leq \|Ax\|\|x\| \leq \|A\|_{\mathrm{B}(\mathcal{X},\mathcal{X})}\|x\|^2$, $\forall x \in \mathcal{X}$, where the first inequality follows from Cauchy–Schwarz Inequality; and the second inequality follows from Proposition 7.64. Hence, $\lambda \leq \|A\|_{\mathrm{B}(\mathcal{X},\mathcal{X})}$. Fix any $u \in \overline{\mathcal{B}}_{\mathcal{X}}(\vartheta_{\mathcal{X}}, 1)$, let $v_1 := \alpha u + \alpha^{-1}Au$ and $v_2 := \alpha u - \alpha^{-1}Au$, where $\alpha \in (0, \infty) \subset \mathbb{R}$. Note also that

$$\begin{aligned}\|Au\|^2 &= \langle Au, Au\rangle = \langle A^*Au, u\rangle = \langle A^2u, u\rangle = \langle u, A^2u\rangle \\ &= \frac{1}{4}(\langle Av_1, v_1\rangle - \langle Av_2, v_2\rangle) \leq \frac{1}{4}(\lambda\|v_1\|^2 + \lambda\|v_2\|^2) \\ &= \frac{1}{2}\lambda(\alpha^2\|u\|^2 + \alpha^{-2}\|Au\|^2) \leq \frac{1}{2}\lambda(\alpha^2 + \alpha^{-2}\|Au\|^2)\end{aligned}$$

where the first equality follows from Proposition 13.2; the second equality follows from Proposition 13.22; the third equality follows from the fact A is Hermitian; the fourth equality follows from Proposition 13.22 and A being Hermitian; the fifth equality follows from straightforward algebra; the first inequality follows from the definition of λ; the sixth equality follows from straightforward algebra; and the last inequality follows from $u \in \overline{\mathcal{B}}_{\mathcal{X}}(\vartheta_{\mathcal{X}}, 1)$.

If $\|Au\| = 0$, then $\|Au\| \leq \lambda$. On the other hand, if $\|Au\| > 0$, let $\alpha^2 := \|Au\|$, we have $\|Au\|^2 \leq \frac{1}{2}\lambda 2\|Au\|$, which leads to $\|Au\| \leq \lambda$. In both cases, we have $\|Au\| \leq \lambda$, $\forall u \in \overline{\mathcal{B}}_{\mathcal{X}}(\vartheta_{\mathcal{X}}, 1)$. Hence, $\|A\| \leq \lambda$.

This completes the proof of the proposition. □

Definition 13.49 Let $\mathcal{X}$ be a Banach space over $\mathbb{K}_1$ and $\mathcal{Y}$ be a Banach space over $\mathbb{K}_2$, $D \subseteq \mathcal{X}$, $f : D \to \mathcal{Y}$ be continuous. f is said to be a *compact operator* if $\forall r \in [0, \infty) \subset \mathbb{R}$, $\overline{f(\overline{\mathcal{B}}_{\mathcal{X}}(\vartheta_{\mathcal{X}}, r))} \subseteq \mathcal{Y}$ is a compact set. ◇

It is then clear that if $f : \mathcal{X} \to \mathcal{Y}$ is a compact operator, then, $\forall (x_i)_{i=1}^{\infty} \subseteq \mathcal{X}$, which is bounded, then $(f(x_i))_{i=1}^{\infty} \subseteq \mathcal{Y}$ has a subsequence converges to some $y_0 \in \mathcal{Y}$, by Borel-Lebesgue Theorem 5.37.

Proposition 13.50 *Let $\mathcal{X}$ be a Banach space over $\mathbb{K}$ and $\mathcal{Y}$ be a Banach space over $\mathbb{K}$, $A \in \mathrm{B}(\mathcal{X}, \mathcal{Y})$. A is a compact operator if, and only if, $\overline{A(\overline{\mathcal{B}}_{\mathcal{X}}(\vartheta_{\mathcal{X}}, 1))} \subseteq \mathcal{Y}$ is a compact set. We will denote the set of such A's by $\mathrm{K}(\mathcal{X}, \mathcal{Y}) := \{A \in \mathrm{B}(\mathcal{X}, \mathcal{Y}) \mid A$ is a compact operator$\}$.*

Proof "Necessity" is immediate.

"Sufficiency" Let $\Theta := A(\overline{\mathcal{B}}_{\mathcal{X}}(\vartheta_{\mathcal{X}}, 1)) \subseteq \mathcal{Y}$. Then, by A being a compact operator, we have $\overline{\Theta}$ is a compact set. $\forall r \in [0, \infty) \subset \mathbb{R}$, $A(\overline{\mathcal{B}}_{\mathcal{X}}(\vartheta_{\mathcal{X}}, r)) = r\Theta \subseteq \mathcal{Y}$ since A is a linear operator. Then, by Proposition 7.102, $\overline{A(\overline{\mathcal{B}}_{\mathcal{X}}(x_0, r))} = r\overline{\Theta}$ is sequentially compact since $\overline{\Theta}$ is. By Borel-Lebesgue Theorem 5.37, $\overline{A(\overline{\mathcal{B}}_{\mathcal{X}}(x_0, r))}$ is compact. Hence, A is a compact operator.

This completes the proof of the proposition. □

Proposition 13.51 *Let $\mathcal{X}$, $\mathcal{Y}$, and $\mathcal{Z}$ be Banach space over $\mathbb{K}$. Then, the following statements holds.*

(i) *Let $A \in \mathrm{K}(\mathcal{X}, \mathcal{Y})$. $\forall C \in \mathrm{B}(\mathcal{Z}, \mathcal{X})$, $\forall D \in \mathrm{B}(\mathcal{Y}, \mathcal{Z})$, we have $AC \in \mathrm{K}(\mathcal{Z}, \mathcal{Y})$ and $DA \in \mathrm{K}(\mathcal{X}, \mathcal{Z})$.*

(ii) ***(Fredholm Alternative)*** *Let $A \in \mathrm{K}(\mathcal{X}, \mathcal{X})$. $\forall \lambda \in \mathbb{K}$ with $\lambda \neq 0$, we have $E := (\lambda \,\mathrm{id}_{\mathcal{X}} - A)(F) \subseteq \mathcal{X}$ is a closed set, $\forall F \subseteq \mathcal{X}$ that is a closed and bounded set. (The operator $\lambda \,\mathrm{id}_{\mathcal{X}} - A$ is called the* Fredholm alternative *of A.)*

Proof (i) Note that

$$E_1 := \overline{(AC)(\overline{\mathcal{B}}_{\mathcal{Z}}(\vartheta_{\mathcal{Z}}, 1))} \subseteq \overline{A\Big(\overline{\mathcal{B}}_{\mathcal{X}}\Big(\vartheta_{\mathcal{X}}, \|C\|_{\mathrm{B}(\mathcal{Z}, \mathcal{X})}\Big)\Big)} =: F_1 \subseteq \mathcal{Y}$$

where the set containment follows from the fact that $C \in \mathrm{B}(\mathcal{Z}, \mathcal{X})$. By Definition 13.49, F_1 is a compact. Then, E_1 is a compact set by Proposition 5.5. Hence, $AC \in \mathrm{K}(\mathcal{Z}, \mathcal{Y})$ by Proposition 13.50.

Note that $G_1 := \overline{(DA)(\overline{\mathcal{B}}_{\mathcal{X}}(\vartheta_{\mathcal{X}}, 1))} \subseteq \overline{D(\overline{A(\overline{\mathcal{B}}_{\mathcal{X}}(\vartheta_{\mathcal{X}}, 1))})} =: \overline{D(F_2)} \subseteq \mathcal{Z}$. By Proposition 13.50, F_2 is compact. By Proposition 5.7, $D(F_2)$ is compact. By Proposition 5.5, $\overline{D(F_2)} = D(F_2)$ and hence $\overline{D(F_2)}$ is compact. Then, G_1 is compact by Proposition 5.5. Hence, $DA \in \mathrm{K}(\mathcal{X}, \mathcal{Z})$ by Proposition 13.50.

(ii) Fix any $y_0 \in \overline{E}$. By Proposition 4.13, $\exists\, (y_n)_{n=1}^{\infty} \subseteq E$ such that $y_0 = \lim_{n\in\mathbb{N}} y_n$. Then, by the definition of E, there exists $(x_n)_{n=1}^{\infty} \subseteq F \subseteq \overline{\mathcal{B}}_{\mathcal{X}}(\vartheta_{\mathcal{X}}, m)$, for some $m \in \mathbb{N}$ since F is a bounded set, such that $y_n = (\lambda\, \mathrm{id}_{\mathcal{X}} - A)x_n = \lambda x_n - Ax_n$, $\forall n \in \mathbb{N}$. Since A is a compact operator, by Definition 13.49, we have $\overline{A(\overline{\mathcal{B}}_{\mathcal{X}}(\vartheta_{\mathcal{X}}, m))} \subseteq \mathcal{X}$ is a compact set. By Borel-Lebesgue Theorem 5.37, there exists a subsequence $\left(x_{n_k}\right)_{k=1}^{\infty}$ such that $\lim_{k\in\mathbb{N}} Ax_{n_k} = \bar{x} \in \overline{A(\overline{\mathcal{B}}_{\mathcal{X}}(\vartheta_{\mathcal{X}}, m))}$. Thus, we have $\lambda \lim_{k\in\mathbb{N}} x_{n_k} = \lim_{k\in\mathbb{N}} y_{n_k} + \lim_{k\in\mathbb{N}} Ax_{n_k} = y_0 + \bar{x} \in \mathcal{X}$. Since $\lambda \neq 0$, we have $\lim_{k\in\mathbb{N}} x_{n_k} = \lambda^{-1}(y_0 + \bar{x}) =: x_0 \in F$, where the last set membership follows since F is a closed subset of $\mathcal{X}$. Then, we have $(\lambda\, \mathrm{id}_{\mathcal{X}} - A)(x_0) = \lim_{k\in\mathbb{N}}(\lambda\, \mathrm{id}_{\mathcal{X}} - A)(x_{n_k}) = \lim_{k\in\mathbb{N}}(\lambda x_{n_k} - Ax_{n_k}) = \lambda x_0 - \bar{x} = y_0$. Hence, $y_0 \in (\lambda\, \mathrm{id}_{\mathcal{X}} - A)(F) = E$. By the arbitrariness of y_0, we have $\overline{E} \subseteq E \subseteq \overline{E}$. Hence, $E = \overline{E}$ and E is closed by Proposition 3.3. □

Theorem 13.52 (Spectral Theory) *Let $\mathcal{X}$ and $\mathcal{Y}$ be Hilbert space over $\mathbb{K}$, $A \in \mathrm{B}(\mathcal{X}, \mathcal{Y})$, and $Q \in \bar{\mathcal{S}}_{\mathcal{X}}$. Then, the following results hold.*

(i) *Let $A \in \mathrm{K}(\mathcal{X}, \mathcal{Y})$. Then, $A \neq \vartheta_{\mathrm{B}(\mathcal{X},\mathcal{Y})}$ if, and only if, $\|A\|_{\mathrm{B}(\mathcal{X},\mathcal{Y})} =: \sigma > 0$. In this case, $\exists x \in \mathcal{X}$ with $\|x\|_{\mathcal{X}} = 1$ such that $\|Ax\|_{\mathcal{Y}} = \sigma$. Let $y := \frac{1}{\|Ax\|_{\mathcal{Y}}} Ax \in \mathcal{Y}$, then $\|y\|_{\mathcal{Y}} = 1$ and $A = \sigma y x^* + \bar{A}$, where $\bar{A} \in \mathrm{K}(\mathcal{X}, \mathcal{Y})$ and $\bar{A} = \bar{P}_1 A\, \bar{P}_2$, $P_1 : \mathcal{Y} \to \operatorname{span}(\{y\}) =: N$, $P_2 : \mathcal{X} \to \operatorname{span}(\{x\}) =: M$, $\bar{P}_1 = \mathrm{id}_{\mathcal{Y}} - P_1 : \mathcal{Y} \to N^{\perp_h}$, and $\bar{P}_2 = \mathrm{id}_{\mathcal{X}} - P_2 : \mathcal{X} \to M^{\perp_h}$ are projection operators.*

(ii) *$A \in \mathrm{K}(\mathcal{X}, \mathcal{Y})$ if, and only if, $\exists \bar{n} \in \mathbb{Z}_+ \cup \{\infty\}$ and $\exists\, (x_i)_{i=1}^{\bar{n}} \subseteq \mathcal{X}$, which is an orthonormal sequence, and $\exists\, (y_i)_{i=1}^{\bar{n}} \subseteq \mathcal{Y}$, which is an orthonormal sequence, and $\sigma =: \sigma_1 \geq \sigma_2 \geq \cdots \geq \sigma_i \in (0, \infty) \subset \mathbb{R}$, $\forall i \in \mathbb{N}$ with $1 \leq i < \bar{n} + 1$, such that $A = \sum_{i=1}^{\bar{n}} \sigma_i y_i x_i^*$, and in case $\bar{n} = \infty$, we have $\lim_{i\in\mathbb{N}} \sigma_i = 0$.*

(iii) *Let $Q \in \bar{\mathcal{S}}_{\mathcal{X}} \cap \mathrm{K}(\mathcal{X}, \mathcal{X})$. Then, $Q \neq \vartheta_{\mathrm{B}(\mathcal{X},\mathcal{X})}$ if, and only if, Q admits an eigenvector $x_0 \in \mathcal{X}$ with $\|x_0\| = 1$ associated with the eigenvalue $\lambda \in \mathbb{R} \setminus \{0\}$ with $|\lambda| = \|Q\|_{\mathrm{B}(\mathcal{X},\mathcal{X})} \in (0, \infty) \subset \mathbb{R}$. Then, $Q = \lambda x_0 x_0^* + \bar{Q}$, where $\bar{Q} = \bar{P}_1 Q \bar{P}_1 \in \bar{\mathcal{S}}_{\mathcal{X}} \cap \mathrm{K}(\mathcal{X}, \mathcal{X})$, $P_1 : \mathcal{X} \to \operatorname{span}(\{x_0\}) =: M$, and $\bar{P}_1 : \mathcal{X} \to M^{\perp_h}$ are projection operators with $P_1 + \bar{P}_1 = \mathrm{id}_{\mathcal{X}}$.*

(iv) *$Q \in \bar{\mathcal{S}}_{\mathcal{X}} \cap \mathrm{K}(\mathcal{X}, \mathcal{X})$ if, and only if, $\exists n \in \mathbb{Z}_+ \cup \{\infty\}$, Q has n eigenvectors $x_i \in \mathcal{X}$ with $\|x_i\| = 1$, each associated with an eigenvalue $\lambda_i \in \mathbb{R} \setminus \{0\}$, and $\|Q\|_{\mathrm{B}(\mathcal{X},\mathcal{X})} = |\lambda_1| \geq |\lambda_2| \geq \cdots \geq |\lambda_i| \in (0, \infty) \subset \mathbb{R}$, $i \in \mathbb{N}$ with $1 \leq i < n+1$, where $(x_i)_{i=1}^{n} \subseteq \mathcal{X}$ is an orthonormal sequence, and $Q = \sum_{i=1}^{n} \lambda_i x_i x_i^*$; and in case $n = \infty$, we have $\lim_{i\in\mathbb{N}} |\lambda_i| = 0$.*

(v) *Let $Q \in \bar{\mathcal{S}}_{\mathcal{X}} \cap \mathrm{K}(\mathcal{X}, \mathcal{X})$ and the results of (iv) hold. $\forall \lambda \in \mathbb{K}$ with $\lambda \neq 0$ is an eigenvalue of Q if, and only if, $\exists i_0 \in \mathbb{N}$ with $1 \leq i_0 < n + 1$ such that*

$\lambda_{i_0} = \lambda$. *In this case,* $\exists m \in \mathbb{N}$ *with* $m < n+1$ *such that* $\dim(\mathcal{N}(\lambda \, \mathrm{id}_{\mathcal{X}} - A)) = m = \max_{s \in \mathbb{N}} \dim(\mathcal{N}((\lambda \, \mathrm{id}_{\mathcal{X}} - A)^s))$, *which is called the* multiplicity *of the eigenvalue* λ *for* Q, *and there is exactly* m λ_i*'s such that* $\lambda = \lambda_i$.

(vi) *Let* $A \in \mathrm{K}(\mathcal{X}, \mathcal{Y})$ *and the results of* (ii) *hold. Then, the values* $(\sigma_i)_{i=1}^{\bar{n}}$ *is uniquely defined that is independent of the choice of* $(y_i)_{i=1}^{\bar{n}}$ *and* $(x_i)_{i=1}^{\bar{n}}$. *These* $(\sigma_i)_{i=1}^{\bar{n}}$ *is said to be the* singular values *of the operator* A. *The formula* $A = \sum_{i=1}^{\bar{n}} \sigma_i y_i x_i^*$ *is said to be the* singular value decomposition *of the operator* $A \in \mathrm{K}(\mathcal{X}, \mathcal{Y})$.

In the above theorem, we have established that $A \in \mathrm{K}(\mathcal{X}, \mathcal{Y})$ if, and only if, it admits the infinite-dimensional version of the singular value decomposition for matrices. $\sigma_i's$ are the singular values, $1 \leq i < \bar{n}+1$, $(y_i)_{i=1}^{\bar{n}}$ replaces the unitary matrix U; and $(x_i)_{i=1}^{\bar{n}}$ replace the unitary matrix V. For a Hermitian compact operator Q, it can always be diagonalized by its eigenvectors of all of its nonzero eigenvalues. Its singular value decomposition is given in terms of its eigenvectors and eigenvalues, where the $y_i = \mathrm{sgn}(\lambda_i) x_i$ and x_i is the eigenvector associated with eigenvalue $\lambda_i \neq 0$ with $\sigma_i = |\lambda_i| > 0$, $\forall i \in \mathbb{N}$ with $1 \leq i < n+1$.

Proof (i) Let $A \in \mathrm{K}(\mathcal{X}, \mathcal{Y})$. Note that $\|A\|_{\mathrm{B}(\mathcal{X},\mathcal{Y})} = \sup_{\substack{x \in \mathcal{X} \\ \|x\|_{\mathcal{X}} \leq 1}} \|Ax\|_{\mathcal{Y}} = \sigma > 0$ if, and only if, $A \neq \vartheta_{\mathrm{B}(\mathcal{X},\mathcal{Y})}$. In case of $\sigma > 0$, then, $\exists \, (z_n)_{n=1}^{\infty} \subseteq \overline{\mathcal{B}}_{\mathcal{X}}(\vartheta_{\mathcal{X}}, 1)$ such that $\lim_{n \in \mathbb{N}} \|Az_n\|_{\mathcal{Y}} = \sigma$. Then, $(Az_n)_{n=1}^{\infty} \subseteq \overline{A(\overline{\mathcal{B}}_{\mathcal{X}}(\vartheta_{\mathcal{X}}, 1))} =: E \subseteq \mathcal{Y}$. By $A \in \mathrm{K}(\mathcal{X}, \mathcal{Y})$ and Definition 13.49, we have E is compact. Then, E is sequentially compact by Borel-Lebesgue Theorem 5.37. Thus, there exists a subsequence $(Az_{n_k})_{k=1}^{\infty}$ of $(Az_n)_{n=1}^{\infty}$ such that $\lim_{k \in \mathbb{N}} Az_{n_k} = \bar{y} \in E$. Thus, we have $\|\bar{y}\|_{\mathcal{Y}} = \lim_{k \in \mathbb{N}} \|Az_{n_k}\|_{\mathcal{Y}} = \sigma > 0$. By Alaoglu Theorem 7.122 and Riesz-Fréchet Theorem 13.15, $\overline{\mathcal{B}}_{\mathcal{X}}(\vartheta_{\mathcal{X}}, 1) =: F \subseteq \mathcal{X}$ is weakly compact (since $\mathcal{X}$ is a Hilbert space). Then, $(z_{n_k})_{k=1}^{\infty} \subseteq F$ has the Bolzano–Weierstrass property by Propositions 5.22 and 5.26. Then, $\exists \bar{z} \in F$ such that $(z_{n_k})_{k=1}^{\infty}$ has $\bar{z}$ as a weak cluster point. By Proposition 7.123, $A \in \mathrm{B}(\mathcal{X}, \mathcal{Y})$ implies that $A : \mathcal{X}_{\mathrm{weak}} \to \mathcal{Y}_{\mathrm{weak}}$ is continuous. Then, the sequence $(Az_{n_k})_{k=1}^{\infty}$ has a weak cluster point at $A\bar{z} \in \mathcal{Y}_{\mathrm{weak}}$, by Proposition 3.66. By $\lim_{k \in \mathbb{N}} Az_{n_k} = \bar{y}$, we have $\lim_{k \in \mathbb{N}} Az_{n_k} = \bar{y}$ weakly. Then, by Proposition 7.116, $\mathcal{Y}_{\mathrm{weak}}$ is Hausdorff, which further implies, by Proposition 3.65, $\bar{y} = A\bar{z}$. Thus, we have $\|A\bar{z}\|_{\mathcal{Y}} = \sigma > 0$ and $\|\bar{z}\|_{\mathcal{X}} \leq 1$. By the fact that $\|A\|_{\mathrm{B}(\mathcal{X},\mathcal{Y})} = \sigma$, we have $\|\bar{z}\|_{\mathcal{X}} = 1$ (otherwise, $\|A\|_{\mathrm{B}(\mathcal{X},\mathcal{Y})} > \sigma$ and leads to a contradiction).

Let $x := \bar{z}$ and $y := \frac{1}{\sigma} A\bar{z} = \frac{1}{\sigma}\bar{y}$. Then, we have $\|x\|_{\mathcal{X}} = 1 = \|y\|_{\mathcal{Y}}$. Since $\mathcal{X}$ is a Hilbert space, we have $x^* \in \mathcal{X}^*$, $\|x^*\|_{\mathcal{X}^*} = 1$, and $\langle\langle x^*, x \rangle\rangle = \langle x, x \rangle_{\mathcal{X}} = 1$. Let $\bar{A} := A - \sigma y x^*$. By Theorem 13.19, we have $P_1 + \bar{P}_1 = \mathrm{id}_{\mathcal{Y}}$ and $P_2 + \bar{P}_2 = \mathrm{id}_{\mathcal{X}}$. $\forall \tilde{x} \in \mathcal{X}$, we have $\bar{A}\tilde{x} = A\tilde{x} - \sigma y \langle\langle x^*, \tilde{x} \rangle\rangle = A(P_2 + \bar{P}_2)\tilde{x} - \sigma y \langle \tilde{x}, x \rangle_{\mathcal{X}} = AP_2\tilde{x} + A\bar{P}_2\tilde{x} - \sigma y \langle P_2\tilde{x}, x \rangle_{\mathcal{X}} - \sigma y \big\langle \bar{P}_2\tilde{x}, x \big\rangle_{\mathcal{X}}$, where the first equality follows from the definition of $\bar{A}$; and the last two equalities follow from the preceding discussion. Note that $P_2\tilde{x} \in M = \mathrm{span}(\{x\})$ and then $P_2\tilde{x} = \tilde{\alpha}x$, for some $\tilde{\alpha} \in \mathbb{K}$. Then, $\bar{A}\tilde{x} = \tilde{\alpha}Ax + A\bar{P}_2\tilde{x} - \sigma y \langle \tilde{\alpha}x, x \rangle_{\mathcal{X}} - \sigma y \big\langle \bar{P}_2\tilde{x}, x \big\rangle_{\mathcal{X}} = \tilde{\alpha}\sigma y + A\bar{P}_2\tilde{x} - \tilde{\alpha}\sigma y - \sigma y \big\langle \tilde{x}, \bar{P}_2^* x \big\rangle = A\bar{P}_2\tilde{x}$, $\forall \tilde{x} \in \mathcal{X}$, where the second equality follows from the fact $\sigma y = Ax = A\bar{z}$;

and the last equality follows from Propositions 13.22 and 13.43 and the fact that $\bar{P}_2 x = \vartheta_{\mathcal{X}}$. By the arbitrariness of $\tilde{x}$, we have $\bar{A} = A\bar{P}_2$.

Note that $A = \sigma y x^* + \bar{A} = \sigma y x^* + A\bar{P}_2 = \sigma y x^* + P_1 A \bar{P}_2 + \bar{P}_1 A \bar{P}_2$. where the first equality follows from the definition of $\bar{A}$; the second equality follows from the preceding paragraph; and the last equality follows from $P_1 + \bar{P}_1 = \mathrm{id}_{\mathcal{Y}}$. We need the following result.

Claim 13.52.1 $P_1 A \bar{P}_2 = \vartheta_{\mathrm{B}(M^{\perp_h}, N)}$.

Proof of Claim We will prove this by an argument of contradiction. Suppose $P_1 A \bar{P}_2 \neq \vartheta_{\mathrm{B}(M^{\perp_h}, N)}$. Then, $0 < \beta := \left\| P_1 A \bar{P}_2 \right\|_{\mathrm{B}(M^{\perp_h}, N)} = \left\| P_1 A \bar{P}_2 \right\|_{\mathrm{B}(\mathcal{X}, \mathcal{Y})}$. Therefore, $\exists s \in M^{\perp_h}$ with $\|s\|_{\mathcal{X}} = 1$ such that $\left\| P_1 A \bar{P}_2 s \right\|_{\mathcal{Y}} > \frac{\beta}{2} > 0$. Note that $P_1 A \bar{P}_2 s \in N$ and hence $P_1 A \bar{P}_2 s = t y$, for some $t \in \mathbb{K}$ with $t \neq 0$. $\forall \alpha \in \mathbb{K}$, $\|A(x + \alpha s)\|_{\mathcal{Y}}^2 = \|\sigma y + \alpha A s\|_{\mathcal{Y}}^2 = \left\|\sigma y + \alpha P_1 A \bar{P}_2 s + \alpha \bar{P}_1 A \bar{P}_2 s\right\|_{\mathcal{Y}}^2 = \left\|(\sigma + \alpha t) y + \alpha \bar{P}_1 A \bar{P}_2 s\right\|_{\mathcal{Y}}^2$. Choose $\alpha = \bar{t}\delta$, where $\delta = \frac{1}{2\sigma} \in (0, \infty) \subset \mathbb{R}$. Thus, $\|A(x + \alpha s)\|_{\mathcal{Y}}^2 = \left\|(\sigma + |t|^2\delta) y + \bar{t}\delta \bar{P}_1 A \bar{P}_2 s\right\|_{\mathcal{Y}}^2 = (\sigma + |t|^2\delta)^2 + |t|^2\delta^2 \left\| \bar{P}_1 A \bar{P}_2 s\right\|_{\mathcal{Y}}^2 \geq (\sigma + |t|^2\delta)^2 = \sigma^2(1 + \frac{|t|^2\delta}{\sigma})^2 = \sigma^2(1 + \frac{|t|^2}{2\sigma^2})^2 > \sigma^2(1 + \frac{|t|^2}{4\sigma^2})^2 = \sigma^2(1 + |t|^2\delta^2)^2 > \sigma^2(1 + |t|^2\delta^2)$, where the second equality follows from $\bar{P}_1 y = \vartheta_{\mathcal{Y}}$ and Proposition 13.2. Note that $\|x + \alpha s\|_{\mathcal{X}}^2 = \left\|x + \bar{t}\delta s\right\|_{\mathcal{X}}^2 = \|x\|_{\mathcal{X}}^2 + \left\|\bar{t}\delta s\right\|_{\mathcal{X}}^2 = 1 + |t|^2\delta^2$. These implies that $\|A(x + \alpha s)\|_{\mathcal{Y}}^2 > \sigma^2 \|x + \alpha s\|_{\mathcal{X}}^2$ and $\|A\|_{\mathrm{B}(\mathcal{X},\mathcal{Y})} > \sigma$. This is a contradiction. Hence, the hypothesis is invalid. So, $P_1 A \bar{P}_2 = \vartheta_{\mathrm{B}(M^{\perp_h}, N)}$. This completes the proof of the claim. □

Then, we have $A = \sigma y x^* + \bar{P}_1 A \bar{P}_2$ and hence $\bar{A} = \bar{P}_1 A \bar{P}_2$. By Proposition 13.51, $\bar{A} \in \mathrm{K}(\mathcal{X}, \mathcal{Y})$. Furthermore, $\left\|\bar{A}\right\|_{\mathrm{B}(\mathcal{X},\mathcal{Y})} \leq \left\|\bar{P}_1\right\|_{\mathrm{B}(\mathcal{Y},\mathcal{Y})} \|A\|_{\mathrm{B}(\mathcal{X},\mathcal{Y})} \cdot \left\|\bar{P}_2\right\|_{\mathrm{B}(\mathcal{X},\mathcal{X})} \leq \|A\|_{\mathrm{B}(\mathcal{X},\mathcal{Y})} = \sigma$. Clearly, $\bar{A} \in \mathrm{K}(M^{\perp_h}, N^{\perp_h})$. This proves (i).

(ii) "Necessity" Let $A \in \mathrm{K}(\mathcal{X}, \mathcal{Y})$. Then, (i) holds. Repeat (i) for $\bar{A}$. For notational consistency, we let $\sigma_1 = \sigma$, $N_0 := \{\vartheta_{\mathcal{Y}}\}$, $N_1 := N$, $M_0 := \{\vartheta_{\mathcal{X}}\}$, $M_1 := M$, $x_1 := x$, $y_1 := y$, $P_{1,1} := P_1$, $\bar{P}_{1,1} := \bar{P}_1$, $P_{1,2} := P_2$, $\bar{P}_{1,2} := \bar{P}_2$, $A_1 := A$, and $A_2 := \bar{A}$. Recursively, assume that we have completed i steps and arrived at $A_{i+1} = \bar{P}_{i,1} A_i \bar{P}_{i,2}$, $(x_j)_{j=1}^i \subseteq \mathcal{X}$ is an orthonormal sequence, $(y_j)_{j=1}^i \subseteq \mathcal{Y}$ is an orthonormal sequence, $P_{j,1} : N_{j-1}^{\perp_h} \to \mathrm{span}\,(\{y_j\})$ and $P_{j,2} : M_{j-1}^{\perp_h} \to \mathrm{span}\,(\{x_j\})$ are projection operators, $\bar{P}_{j,1} : N_{j-1}^{\perp_h} \to N_j^{\perp_h} := (\mathrm{span}\,(\{y_1, \ldots, y_j\}))^{\perp_h}$ and $\bar{P}_{j,2} : M_{j-1}^{\perp_h} \to M_j^{\perp_h} := (\mathrm{span}\,(\{x_1, \ldots, x_j\}))^{\perp_h}$ are projection operators, $P_{j,1} + \bar{P}_{j,1} = \mathrm{id}_{N_{j-1}^{\perp_h}}$, $P_{j,2} + \bar{P}_{j,2} = \mathrm{id}_{M_{j-1}^{\perp_h}}$, $\sigma_1 \geq \sigma_2 \geq \cdots \geq \sigma_i > 0$, $\sigma_j = \|A_j\|_{\mathrm{B}(\mathcal{X},\mathcal{Y})}$, $j = 1, \ldots, i$, and $A = \sum_{j=1}^i \sigma_j y_j x_j^* + A_{i+1}$. Consider $\sigma_{i+1} := \|A_{i+1}\|_{\mathrm{B}(\mathcal{X},\mathcal{Y})}$. If $\sigma_{i+1} = 0$, then, we have $\bar{n} = i$, $A_{i+1} = \vartheta_{\mathrm{B}(\mathcal{X},\mathcal{Y})}$ and $A = \sum_{j=1}^{\bar{n}} \sigma_j y_j x_j^*$. The desired result holds. In case $\sigma_{i+1} > 0$, we apply (i) to A_{i+1}. By (i), $\exists x_{i+1} \in M_i^{\perp_h}$ and $y_{i+1} := \frac{1}{\sigma_{i+1}} A x_{i+1} \in N_i^{\perp_h}$ with $\|x_{i+1}\|_{\mathcal{X}} = 1 = \|y_{i+1}\|_{\mathcal{Y}}$ such that $A_{i+1} = \sigma_{i+1} y_{i+1} x_{i+1}^* + A_{i+2}$, where $A_{i+2} = \bar{P}_{i+1,1} A_{i+1} \bar{P}_{i+1,2}$,

$P_{i+1,1} : N_i^{\perp_h} \to \operatorname{span}(\{y_{i+1}\})$ and $P_{i+1,2} : M_i^{\perp_h} \to \operatorname{span}(\{x_{i+1}\})$ are projection operators, $\bar{P}_{i+1,1} = \operatorname{id}_{N_i^{\perp_h}} - P_{i+1,1} : N_i^{\perp_h} \to N_{i+1}^{\perp_h} := (\operatorname{span}(\{y_1, \ldots, y_{i+1}\}))^{\perp_h}$ and $\bar{P}_{i+1,2} = \operatorname{id}_{M_i^{\perp_h}} - P_{i+1,2} : M_i^{\perp_h} \to M_{i+1}^{\perp_h} := (\operatorname{span}(\{x_1, \ldots, x_{i+1}\}))^{\perp_h}$ are projection operators. Clearly, $(x_j)_{j=1}^{i+1}$ is orthonormal and so is $(y_j)_{j=1}^{i+1}$. Clearly, $\sigma = \sigma_1 \geq \sigma_2 \geq \cdots \geq \sigma_{i+1} > 0$.

Recursively, either the above process stops in finite number of steps, say $\bar{n} \in \mathbb{N}$, with $\sigma_{\bar{n}+1} = 0$, then, (ii) holds with a finite $\bar{n}$; or the above process continues indefinitely, and we have $\bar{n} = \infty$ and $\sigma_i > 0$, $\forall i \in \mathbb{N}$. In the latter case, clearly $(x_j)_{j=1}^{\infty}$ and $(y_j)_{j=1}^{\infty}$ are orthonormal sequences. We need the following result.

Claim 13.52.2 $\lim_{i\in\mathbb{N}} \sigma_i = 0$.

Proof of Claim Clearly, $\lim_{i\in\mathbb{N}} \sigma_i =: \bar{\sigma} \in [0, \infty) \subset \mathbb{R}$. We will show the claim by an argument of contradiction. Suppose $\lim_{i\in\mathbb{N}} \sigma_i = \bar{\sigma} > 0$. Then, we have $Ax_i = \sigma_i y_i$, $\forall i \in \mathbb{N}$. $(x_i)_{i=1}^{\infty} \subseteq \overline{\mathcal{B}}_{\mathcal{X}}(\vartheta_{\mathcal{X}}, 1)$, but $(\sigma_i y_i)_{i=1}^{\infty} \subseteq E$ does not have any convergent subsequence since $(y_i)_{i=1}^{\infty}$ is orthonormal. This contradicts with the assumption that A is a compact operator and E is sequentially compact. Hence, the hypothesis is invalid. Hence, we must have $\bar{\sigma} = 0 = \lim_{i\in\mathbb{N}} \sigma_i$. This completes the proof of the claim. □

Then, $A_{n+1} = A - \sum_{i=1}^{n} \sigma_i y_i x_i^*$, $\forall n \in \mathbb{N}$. We have $\lim_{n\in\mathbb{N}} \|A_{n+1}\|_{B(\mathcal{X},\mathcal{Y})} = \lim_{n\in\mathbb{N}} \sigma_{n+1} = 0$. This implies that $A = \sum_{i=1}^{\infty} \sigma_i y_i x_i^*$. This completes the necessity part of the proof.

"Sufficiency" Let $\exists \bar{n} \in \mathbb{Z}_+ \cup \{\infty\}$ and $\exists\, (x_i)_{i=1}^{\bar{n}} \subseteq \mathcal{X}$, which is an orthonormal sequence, and $\exists\, (y_i)_{i=1}^{\bar{n}} \subseteq \mathcal{Y}$, which is an orthonormal sequence, and $\sigma =: \sigma_1 \geq \sigma_2 \geq \cdots \geq \sigma_i \in (0, \infty) \subset \mathbb{R}$, $\forall i \in \mathbb{N}$ with $1 \leq i < \bar{n} + 1$, such that $A = \sum_{i=1}^{\bar{n}} \sigma_i y_i x_i^*$, and in case $\bar{n} = \infty$, we have $\lim_{i\in\mathbb{N}} \sigma_i = 0$. Fix any $(\tilde{x}_i)_{i=1}^{\infty} \subseteq F$, we will show that there exists a subsequence $(A\tilde{x}_{i_k})_{k=1}^{\infty} \subseteq E$ that converges to some $y_0 \in \mathcal{Y}$, or equivalently, that the subsequence $(A\tilde{x}_{i_k})_{k=1}^{\infty} \subseteq E$ is a Cauchy sequence. This shows that E is sequentially compact. By Proposition 13.50, $A \in \mathrm{K}(\mathcal{X}, \mathcal{Y})$. This then completes the sufficiency part of the proof. We need the following intermediate result.

Claim 13.52.3 There exists a subsequence $(\tilde{x}_{i_k})_{k=1}^{\infty}$ of $(\tilde{x}_i)_{i=1}^{\infty}$ such that $\lim_{k\in\mathbb{N}} \langle x_{i_k}, x_s\rangle_{\mathcal{X}} = \alpha_s \in \overline{\mathcal{B}}_{\mathbb{K}}(0, 1) =: \bar{F}$, $\forall s \in \mathbb{N}$ with $1 \leq s < \bar{n} + 1$.

Proof of Claim Note that $(\langle \tilde{x}_i, x_1\rangle_{\mathcal{X}})_{i=1}^{\infty} \subseteq \bar{F}$ and $\bar{F} \subseteq \mathbb{K}$ is compact. Then, there exists a subsequence $(\tilde{x}_{i_l})_{l=1}^{\infty}$ such that $\lim_{l\in\mathbb{N}} \langle \tilde{x}_{i_l}, x_1\rangle_{\mathcal{X}} = \alpha_1 \in \bar{F}$. Note that $\left(\langle \tilde{x}_{i_l}, x_2\rangle_{\mathcal{X}}\right)_{l=1}^{\infty} \subseteq \bar{F}$. Then, there exists a subsubsequence $(\tilde{x}_{i_{l_j}})_{j=1}^{\infty}$ such that $\lim_{j\in\mathbb{N}} \langle \tilde{x}_{i_{l_j}}, x_2\rangle_{\mathcal{X}} = \alpha_2 \in \bar{F}$. Clearly, $\lim_{j\in\mathbb{N}} \langle \tilde{x}_{i_{l_j}}, x_1\rangle_{\mathcal{X}} = \alpha_1 \in \bar{F}$. Note that $(\langle \tilde{x}_{i_{l_j}}, x_3\rangle_{\mathcal{X}})_{j=1}^{\infty} \subseteq \bar{F}$. Then, there exists a subsubsubsequence $(\tilde{x}_{i_{l_{j_k}}})_{k=1}^{\infty}$ such that $\lim_{k\in\mathbb{N}} \langle \tilde{x}_{i_{l_{j_k}}}, x_3\rangle_{\mathcal{X}} = \alpha_3 \in \bar{F}$. Clearly, $\lim_{k\in\mathbb{N}} \langle \tilde{x}_{i_{l_{j_k}}}, x_s\rangle_{\mathcal{X}} = \alpha_s \in \bar{F}$, $s = 1, 2, 3$. Recursively, continue this process till $\bar{n}$

If $\bar{n}$ is finite, we may obtain (after finite recursions) a subsequence $(\tilde{x}_{i_k})_{k=1}^{\infty}$ of $(\tilde{x}_i)_{i=1}^{\infty}$ such that $\lim_{k\in\mathbb{N}} \langle \tilde{x}_{i_k}, x_s\rangle_{\mathcal{X}} = \alpha_s \in \bar{F}$, $\forall s \in \mathbb{N}$ with $1 \le s \le \bar{n}$.

If $\bar{n} = \infty$, we may obtain (after countably infinite recursions) infinitely many subsequences by the construction procedure of the second to last paragraph. Then, we take the Cantor's diagonal sequence $(\tilde{x}_{i_1}, \tilde{x}_{i_{l_2}}, \tilde{x}_{i_{l_{j_3}}}, \ldots)$ and denote it as $(\tilde{x}_{i_k})_{k=1}^{\infty}$, which is a subsequence of $(\tilde{x}_i)_{i=1}^{\infty}$. Then, we have $\lim_{k\in\mathbb{N}} \langle \tilde{x}_{i_k}, x_s\rangle_{\mathcal{X}} = \alpha_s \in \bar{F}$, $\forall s \in \mathbb{N}$.

This completes the proof of the claim. □

Then, we have the subsequence $(A\tilde{x}_{i_k})_{k=1}^{\infty} \subseteq E \subseteq \mathcal{Y}$. Note that $A\tilde{x}_{i_k} = \sum_{s=1}^{\bar{n}} \sigma_s y_s \langle \tilde{x}_{i_k}, x_s\rangle_{\mathcal{X}}$. We will show that this is a Cauchy sequence. Fix any $\forall \epsilon \in (0,\infty) \subset \mathbb{R}$. $\exists n_1 \in \mathbb{N}$ with $1 \le n_1 < \bar{n}+1$, such that $\frac{\epsilon}{2\sqrt{2}} \ge \sigma_{n_1+1} \ge \sigma_{n_1+2} \ge \cdots \ge 0$. Then, $\exists n_2 \in \mathbb{N}$ such that $\left|\langle \tilde{x}_{i_k}, x_s\rangle_{\mathcal{X}} - \alpha_s\right| < \frac{\epsilon}{2\sqrt{2n_1}(\sigma_1+1)}$, $\forall s \in \mathbb{N}$ with $1 \le s \le n_1$, $\forall k \in \mathbb{N}$ with $k \ge n_2$. $\forall k, j \in \mathbb{N}$ with $k, j \ge n_2$, we have

$$\begin{aligned}
\|A\tilde{x}_{i_k} - A\tilde{x}_{i_j}\|_{\mathcal{Y}}^2 &= \Big\|\sum_{i=1}^{\bar{n}} \sigma_s y_s \big\langle \tilde{x}_{i_k} - \tilde{x}_{i_j}, x_s\big\rangle_{\mathcal{X}}\Big\|_{\mathcal{Y}}^2 \\
&= \sum_{s=1}^{\bar{n}} \sigma_s^2 \left|\big\langle \tilde{x}_{i_k} - \tilde{x}_{i_j}, x_s\big\rangle_{\mathcal{X}}\right|^2 \\
&= \sum_{s=1}^{n_1} \sigma_s^2 \left|\big\langle \tilde{x}_{i_k} - \tilde{x}_{i_j}, x_s\big\rangle_{\mathcal{X}}\right|^2 + \sum_{s=n_1+1}^{\bar{n}} \sigma_s^2 \left|\big\langle \tilde{x}_{i_k} - \tilde{x}_{i_j}, x_s\big\rangle_{\mathcal{X}}\right|^2 \\
&\le \sum_{s=1}^{n_1} \sigma_1^2 \Big(\frac{\epsilon}{\sqrt{2n_1}(1+\sigma_1)}\Big)^2 + \frac{\epsilon^2}{8} \sum_{s=n_1+1}^{\bar{n}} \left|\big\langle \tilde{x}_{i_k} - \tilde{x}_{i_j}, x_s\big\rangle_{\mathcal{X}}\right|^2 \\
&\le \frac{\epsilon^2}{2} + \frac{\epsilon^2}{8}\|\tilde{x}_{i_k} - \tilde{x}_{i_j}\|_{\mathcal{X}}^2 \le \epsilon^2
\end{aligned}$$

where the second equality follows from the fact that $(y_i)_{i=1}^{\bar{n}}$ is orthonormal; the first inequality follows from the assumption and the preceding discussion; and the second inequality follows from Bessel's Inequality, Proposition 13.30. This proves that $(A\tilde{x}_{i_k})_{k=1}^{\infty} \subseteq E \subseteq \mathcal{Y}$ is a Cauchy sequence. This completes the sufficiency proof.

(iii) Let $Q \in \bar{\mathcal{S}}_{\mathcal{X}} \cap \mathrm{K}(\mathcal{X},\mathcal{X})$. "Necessity" $\|Q\|_{\mathrm{B}(\mathcal{X},\mathcal{X})} > 0$ since $Q \neq \vartheta_{\mathrm{B}(\mathcal{X},\mathcal{X})}$. By Proposition 13.48, $\exists (u_i)_{i=1}^{\infty} \subseteq \mathcal{X}$ with $\|u_i\| = 1$ such that $\lim_{i\in\mathbb{N}} |\langle Q(u_i), u_i\rangle| = \|Q\|_{\mathrm{B}(\mathcal{X},\mathcal{X})} > 0$. Since Q is a compact operator, then $(Q(u_i))_{i=1}^{\infty} \subseteq \overline{Q(\overline{\mathcal{B}}_{\mathcal{X}}(\vartheta_{\mathcal{X}},1))}$, which is a compact set. By Borel-Lebesgue Theorem 5.37, the set $\overline{Q(\overline{\mathcal{B}}_{\mathcal{X}}(\vartheta_{\mathcal{X}},1))} \subseteq \mathcal{X}$ is sequentially compact. Then, there exists a subsequence $(Q(u_{i_k}))_{k=1}^{\infty} \subseteq \overline{Q(\overline{\mathcal{B}}_{\mathcal{X}}(\vartheta_{\mathcal{X}},1))}$ such that $\lim_{k\in\mathbb{N}} Q(u_{i_k}) = u_0$. Then, $\lim_{k\in\mathbb{N}} |\langle Q(u_{i_k}), u_{i_k}\rangle| = \|Q\|_{\mathrm{B}(\mathcal{X},\mathcal{X})}$. Let $\lambda := \lim_{k\in\mathbb{N}} \langle Q(u_{i_k}), u_{i_k}\rangle \in$

$\mathbb{K}$ and we have $|\lambda| = \|Q\|_{B(\mathcal{X},\mathcal{X})}$. Note that $\lambda = \lim_{k\in\mathbb{N}} \langle Q(u_{i_k}), u_{i_k}\rangle = \overline{\lim_{k\in\mathbb{N}} \langle u_{i_k}, Q^*(u_{i_k})\rangle} = \overline{\lim_{k\in\mathbb{N}} \langle u_{i_k}, Q(u_{i_k})\rangle} = \lim_{k\in\mathbb{N}} \overline{\langle Q(u_{i_k}), u_{i_k}\rangle} = \bar{\lambda}$, where the second equality follows from Proposition 13.22; the third equality follows from Q being Hermitian; and the fourth equality follows from Definition 13.1. Hence, we have $\lambda \in \mathbb{R} \setminus \{0\}$. ($\lambda = \|Q\|_{B(\mathcal{X},\mathcal{X})}$ or $\lambda = -\|Q\|_{B(\mathcal{X},\mathcal{X})}$.) Note also that $0 \leq \|Q(u_{i_k}) - \lambda u_{i_k}\|^2 = \langle Q(u_{i_k}) - \lambda u_{i_k}, Q(u_{i_k}) - \lambda u_{i_k}\rangle = \langle Q(u_{i_k}), Q(u_{i_k})\rangle - \lambda\langle Qu_{i_k}, u_{i_k}\rangle - \lambda\langle u_{i_k}, Qu_{i_k}\rangle + \lambda^2\langle u_{i_k}, u_{i_k}\rangle = \|Q(u_{i_k})\|^2 - \lambda^2 - \lambda(\langle Q(u_{i_k}), u_{i_k}\rangle + \langle u_{i_k}, Q(u_{i_k})\rangle - 2\lambda) \leq \|Q\|^2_{B(\mathcal{X},\mathcal{X})}\|u_{i_k}\|^2 - \lambda^2 - \lambda(\langle Q(u_{i_k}), u_{i_k}\rangle + \langle Q^* u_{i_k}, u_{i_k}\rangle - 2\lambda) = -2\lambda(\langle Q(u_{i_k}), u_{i_k}\rangle - \lambda) \to 0$, where the first equality follows from Proposition 13.2; the second equality follows from Definition 13.1; the second inequality follows from simple algebra and Proposition 13.22; and the last equality follows from Q is Hermitian. Then, we have $0 = \lim_{k\in\mathbb{N}} \|Q(u_{i_k}) - \lambda u_{i_k}\| = \lim_{k\in\mathbb{N}} \|u_0 - \lambda u_{i_k}\|$. Hence, $\lim_{k\in\mathbb{N}} u_{i_k} = \frac{1}{\lambda}u_0$. By continuity of Q, we have $Q(\frac{1}{\lambda}u_0) = \lim_{k\in\mathbb{N}} Q(u_{i_k}) = u_0$. Furthermore, $\|\frac{1}{\lambda}u_0\| = \lim_{k\in\mathbb{N}} \|u_{i_k}\| = 1$. Hence, $\frac{1}{\lambda}u_0 =: x_0$ is an eigenvector of Q associated with the eigenvalue λ.

Let span $(\{x_0\}) =: M \subseteq \mathcal{X}$. It is a closed subspace in $\mathcal{X}$ by Theorem 7.36 and Proposition 4.39. By Theorem 13.19, $\mathcal{X} = M \oplus M^{\perp_h}$. Define $P_1 : \mathcal{X} \to M$ and $\bar{P}_1 : \mathcal{X} \to M^{\perp_h}$ be projection operators. We have $P_1 + \bar{P}_1 = \mathrm{id}_{\mathcal{X}}$. This implies that $Q = (P_1 + \bar{P}_1)Q(P_1 + \bar{P}_1) = P_1 Q P_1 + P_1 Q \bar{P}_1 + \bar{P}_1 Q P_1 + \bar{P}_1 Q \bar{P}_1$. Note that $P_1 = x_0 x_0^*$. $\forall x \in \mathcal{X}$, $P_1(x) = x_0\langle\langle x_0^*, x\rangle\rangle = x_0\langle x, x_0\rangle$. This implies that $P_1 Q P_1(x) = \langle x, x_0\rangle P_1 Q(x_0) = \langle x, x_0\rangle P_1(\lambda x_0) = \lambda\langle x, x_0\rangle x_0 = \lambda x_0\langle\langle x_0^*, x\rangle\rangle$ and $\bar{P}_1 Q P_1(x) = \langle x, x_0\rangle \bar{P}_1(\lambda x_0) = \vartheta_{\mathcal{X}}$. Thus, $Q(x) = \lambda x_0 \cdot \langle\langle x_0^*, x\rangle\rangle + P_1 Q \bar{P}_1(x) + \bar{P}_1 Q \bar{P}_1(x)$. Hence, $Q = \lambda x_0 x_0^* + P_1 Q \bar{P}_1 + \bar{P}_1 Q \bar{P}_1$. By Propositions 13.22 and 13.43, we have $Q = Q^* = \lambda x_0 x_0^* + \bar{P}_1 Q^* P_1 + \bar{P}_1 Q^* \bar{P}_1 = \lambda x_0 x_0^* + \bar{P}_1 Q P_1 + \bar{P}_1 Q \bar{P}_1 = \lambda x_0 x_0^* + \bar{P}_1 Q \bar{P}_1$. Hence, we have $Q = \lambda x_0 x_0^* + \bar{P}_1 Q \bar{P}_1$. By Propositions 13.51, 13.43, and 13.22, $\bar{Q} := \bar{P}_1 Q \bar{P}_1 \in \bar{\mathcal{S}}_{\mathcal{X}} \cap \mathrm{K}(\mathcal{X}, \mathcal{X})$.

"Sufficiency" Let Q admit an eigenvector x_0 associated with the eigenvalue $\lambda \in \mathbb{R} \setminus \{0\}$ with $|\lambda| = \|Q\|_{B(\mathcal{X},\mathcal{X})} > 0$. Then, $\langle x_0, Qx_0\rangle = \langle x_0, \lambda x_0\rangle = \lambda\|x_0\|^2 \neq 0$. Hence, $Q \neq \vartheta_{B(\mathcal{X},\mathcal{X})}$.

This proves the (iii).

(iv) "Necessity" If $Q = \vartheta_{B(\mathcal{X},\mathcal{X})}$, the result is trivial with $n = 0$. Consider the case $Q \neq \vartheta_{B(\mathcal{X},\mathcal{X})}$. By (iii), $Q = \lambda x_0 x_0^* + Q_2$, where $Q_2 = \bar{P}_1 Q \bar{P}_1$. Denote $\lambda_1 := \lambda \in \mathbb{R}\setminus\{0\}$ with $|\lambda_1| = \|Q\|_{B(\mathcal{X},\mathcal{X})} > 0$, $x_1 := x_0$, $Q_1 := Q$, and $\bar{P}_{1,1} := \bar{P}_1$. Now, we will repeat (iii) for the operator Q_2. Clearly, $Q_2 \in \bar{\mathcal{S}}_{\mathcal{X}}$. We claim that

Claim 13.52.4 Any eigenvector $x_2 \in \mathcal{X}$ with $\|x_2\| = 1$ of Q_2 associated with an eigenvalue $\lambda_2 \in \mathbb{R}\setminus\{0\}$ must be an eigenvector of Q_1 associated with the eigenvalue λ_2, and $x_2 \perp x_1$.

Proof of Claim $Q_2 x_2 = \lambda_2 x_2$. Since $Q_2 = \bar{P}_1 Q \bar{P}_1$. Then, $\lambda_2\langle x_1, x_2\rangle = \langle x_1, \lambda_2 x_2\rangle = \langle x_1, \bar{P}_1 Q \bar{P}_1 x_2\rangle = \langle \bar{P}_1^*(x_1), Q\bar{P}_1(x_2)\rangle = \langle \bar{P}_1(x_1), Q\bar{P}_1(x_2)\rangle = 0$, where the first equality follows from Definition 13.1 and the fact that $\lambda_2 \in \mathbb{R}$; the second equality follows from our assumption that x_2 is an eigenvector of Q_2 associated with the eigenvalue λ_2; the third equality follows from the expression for Q_2 and Proposition 13.22; the fourth equality follows from Proposition 13.43; and the last equality follows from $\bar{P}_1(x_0) = \vartheta_{\mathcal{X}}$. Since, $\lambda_2 \neq 0$, then $x_1 \perp x_2$. Then, $x_2 \in M^{\perp_h}$. It is then easy to check that $Q(x_2) = Q_1(x_2) = \lambda x_0 \langle\langle x_0^*, x_2 \rangle\rangle + Q_2(x_2) = \lambda x_0 \langle x_2, x_0 \rangle + \lambda_2 x_2 = \lambda x_0 \langle x_2, x_1 \rangle + \lambda_2 x_2 = \lambda_2 x_2$, where the first equality follows from our notation; the second equality follows from the expression for Q; the third equality follows from Riesz-Fréchet Theorem 13.15; the fourth equality follows from our notation; and the last equality follows from the preceding discussion. Hence, x_2 is an eigenvector of Q associated with the eigenvalue λ_2. This completes the proof of the claim. □

Repeat (iii) until $Q_{n+1} = \vartheta_{B(\mathcal{X},\mathcal{X})}$ or indefinitely. After any step $k \in \mathbb{N}$ (with $k < n$), we assume that we have found orthonormal sequence $(x_i)_{i=1}^k \subseteq \mathcal{X}$, x_i is an eigenvector of Q_i (and Q) associated with the eigenvalue $\lambda_i \in \mathbb{R} \setminus \{0\}$ with $|\lambda_i| = \|Q_i\|_{B(\mathcal{X},\mathcal{X})} \in (0, \infty) \subset \mathbb{R}$, $i = 1, \ldots, k$, and $Q = \sum_{j=1}^{i} \lambda_j x_j x_j^* + Q_{i+1}$, $i = 1, \ldots, k$. At step $k + 1$, since $Q_{k+1} \neq \vartheta_{B(\mathcal{X},\mathcal{X})}$, then $\|Q_{k+1}\|_{B(\mathcal{X},\mathcal{X})} > 0$, by (iii), there exists an eigenvector $x_{k+1} \in \mathcal{X}$ with $\|x_{k+1}\| = 1$ of Q_{k+1} associated with the eigenvalue $\lambda_{k+1} \in \mathbb{R} \setminus \{0\}$ with $|\lambda_{k+1}| = \|Q_{k+1}\|_{B(\mathcal{X},\mathcal{X})} > 0$. Furthermore, $Q_{k+1} = \lambda_{k+1} x_{k+1} x_{k+1}^* + \bar{P}_{1,k+1} Q_{k+1} \bar{P}_{1,k+1}$. Now, by Claim 13.52.4, x_{k+1} is an eigenvector of Q_k associated with the eigenvalue λ_{k+1}, and $x_{k+1} \perp x_k$. Recursively, by Claim 13.52.4, x_{k+1} is an eigenvector of Q_i, $i = k - 1, \ldots, 1$, associated with the eigenvalue λ_{k+1}, and $x_{k+1} \perp x_i$. Hence, the sequence $(x_i)_{i=1}^{k+1}$ is an orthonormal sequence. Thus, either the process stops at n, for some $n \in \mathbb{Z}_+$, when $Q_{n+1} = \vartheta_{B(\mathcal{X},\mathcal{X})}$, or we have $n = \infty$ and an orthonormal sequence $(x_i)_{i=1}^{\infty}$. In the latter case, $Q = \sum_{j=1}^{i} \lambda_j x_j x_j^* + Q_{i+1}$, $\forall i \in \mathbb{N}$, where $Q_{i+1} := \bar{P}_{1,i} Q_i \bar{P}_{1,i}$. Clearly, $Q_{i+1} \in \bar{\mathcal{S}}_{\mathcal{X}} \cap K(\mathcal{X}, \mathcal{X})$, $\forall i \in \mathbb{N}$.

$|\lambda_i| = \|Q_i\|_{B(\mathcal{X},\mathcal{X})} \geq \|Q_i\|_{B(\mathcal{X},\mathcal{X})} \left\| \bar{P}_{1,i} \right\|^2_{B(\mathcal{X},\mathcal{X})} \geq \left\| \bar{P}_{1,i} Q_i \bar{P}_{1,i} \right\|_{B(\mathcal{X},\mathcal{X})} = \|Q_{i+1}\|_{B(\mathcal{X},\mathcal{X})} = |\lambda_{i+1}| > 0$, $\forall i \in \mathbb{N}$ with $1 \leq i < n$. In case of $n = \infty$. Suppose that $\lim_{i \in \mathbb{N}} |\lambda_i| > 0$. Then, $Qx_i = \lambda_i x_i$, $\forall i \in \mathbb{N}$. The sequence $(Qx_i)_{i=1}^{\infty} = (\lambda_i x_i)_{i=1}^{\infty}$ does not have a converging subsequence. This contradicts with the assumption that Q is a compact operator. Therefore, $\lim_{i \in \mathbb{N}} |\lambda_i| = 0$. Then $\lim_{k \in \mathbb{N}} \left\| Q - \sum_{i=1}^{k} \lambda_i x_i x_i^* \right\|_{B(\mathcal{X},\mathcal{X})} = \lim_{k \in \mathbb{N}} \|Q_{k+1}\|_{B(\mathcal{X},\mathcal{X})} = \lim_{k \in \mathbb{N}} |\lambda_{k+1}| = 0$, where the first two equalities follow from the recursive construction; and the last equality follows from the preceding discussion. Hence, $Q = \sum_{i=1}^{n} \lambda_i x_i x_i^*$. This completes the necessity part of the proof.

"Sufficiency" By the expression for $Q = \sum_{i=1}^{n} |\lambda_i| (\text{sgn}(\lambda_i) x_i) x_i^*$ and (ii), $Q \in K(\mathcal{X}, \mathcal{X}) \cap \bar{\mathcal{S}}_{\mathcal{X}}$. This completes the sufficiency part of the proof.

(v) Let $Q \in \bar{\mathcal{S}}_{\mathcal{X}} \cap K(\mathcal{X}, \mathcal{X})$. Fix any $\lambda \in \mathbb{K}$ with $\lambda \neq 0$. If there exists an $i_0 \in \mathbb{N}$ with $1 \leq i_0 < n + 1$ such that $\lambda = \lambda_{i_0}$. By (iv), $Q = \sum_{i=1}^{n} \lambda_i x_i x_i^*$. It is easy to show that $Q(x_{i_0}) = \lambda_{i_0} x_{i_0}$. Then, x_{i_0} is an eigenvector of Q associated with the eigenvalue $\lambda = \lambda_{i_0}$, and λ is an eigenvalue of Q.

On the other hand, if λ is an eigenvalue of Q. By Proposition 13.47, $\lambda \in \mathbb{R} \setminus \{0\}$. Then, $M := \mathcal{N}(\lambda \, \text{id}_{\mathcal{X}} - Q) \supset \{\vartheta_{\mathcal{X}}\}$. We need the following intermediate results.

Claim 13.52.5 $M = \mathcal{N}(\lambda \, \text{id}_{\mathcal{X}} - Q)$ is finite-dimensional.

Proof of Claim Suppose M is infinite-dimensional. Then, $\exists\, (\bar{x}_i)_{i=1}^{\infty} \subseteq M$ that is orthonormal. $Q\bar{x}_i = \lambda\bar{x}_i$, $\forall i \in \mathbb{N}$. The sequence $(Q\bar{x}_i)_{i=1}^{\infty} = (\lambda\bar{x}_i)_{i=1}^{\infty}$ does not have a convergent subsequence. This contradicts with the assumption $Q \in \mathrm{K}(\mathcal{X}, \mathcal{X})$. Hence, the hypothesis is invalid. M must be finite-dimensional. □

Claim 13.52.6 $M \supset \{\vartheta_{\mathcal{X}}\}$ implies that $\lambda = \lambda_{i_0}$ for some $i_0 \in \mathbb{N}$ with $1 \leq i_0 < n+1$.

Proof of Claim Let $N := \overline{\mathrm{span}\,((x_i)_{i=1}^{n})}$, and $P : \mathcal{X} \to N$ and $\bar{P} : \mathcal{X} \to N^{\perp_\mathrm{h}}$ be projection operators. By Theorem 13.19, we have $P + \bar{P} = \mathrm{id}_{\mathcal{X}}$. $\forall x \in \mathcal{N}(\lambda\,\mathrm{id}_{\mathcal{X}} - Q) = M$ with $x \neq \vartheta_{\mathcal{X}}$. $x = Px + \bar{P}x =: \bar{x} + \tilde{x}$. $\lambda x = Qx$ implies that $\lambda\bar{x} + \lambda\tilde{x} = \sum_{i=1}^{n} \lambda_i x_i \langle x, x_i\rangle \in N$. This implies that $\lambda\tilde{x} \in N \cap N^{\perp_\mathrm{h}}$ and hence $\tilde{x} = \vartheta_{\mathcal{X}}$. Hence, we have $\lambda\bar{x} = \sum_{i=1}^{n} \lambda_i x_i \langle \bar{x}, x_i\rangle = \lambda \sum_{i=1}^{n} x_i \langle \bar{x}, x_i\rangle$, where the last equality follows from $\bar{x} \in N$ and $(x_i)_{i=1}^{n}$ is orthonormal. Therefore, we must have $\lambda\langle\bar{x}, x_i\rangle = \lambda_i\langle\bar{x}, x_i\rangle$, $\forall i \in \mathbb{N}$ with $1 \leq i < n+1$. Then, $\lambda = \lambda_i$, $\forall i \in \mathbb{N}$ with $1 \leq i < n+1$ and $\langle\bar{x}, x_i\rangle \neq 0$. Since $\vartheta_{\mathcal{X}} \neq x = \bar{x} \in N$, $\exists i_0 \in \mathbb{N}$ with $1 \leq i_0 < n+1$ such that $\langle\bar{x}, x_{i_0}\rangle \neq 0$. Then, $\lambda = \lambda_{i_0}$.

This completes the proof of the claim. □

By the preceding two claims, we have proved that if $\lambda \in \mathbb{K}$ with $\lambda \neq 0$ is an eigenvalue of Q, then $\lambda = \lambda_{i_0}$ for some $i_0 \in \mathbb{N}$ with $1 \leq i_0 < n+1$.

Let $\lambda \in \mathbb{R}\setminus\{0\}$ be an eigenvalue of Q. By Claim 13.52.5, M is finite-dimensional. Let $\dim M =: m \in \mathbb{N}$. We will show that there are exactly m λ_i's such that $\lambda = \lambda_i$.

Let M admit an orthonormal basis $(\bar{x}_i)_{i=1}^{m}$. By Claim 13.52.6 and its proof, $\lambda = \lambda_i$, $\forall i \in \mathbb{N}$ with $1 \leq i < n+1$ and $\langle\bar{x}_j, x_i\rangle \neq 0$, for some $j \in \{1, \ldots, m\}$, and $(\bar{x}_i)_{i=1}^{m} \subseteq N := \overline{\mathrm{span}\,((x_i)_{i=1}^{n})}$. Then, there are at least m λ_i's such that $\lambda = \lambda_i$. Then, there must be exactly m such λ_i's. This is because x_i is an eigenvector of Q associated with λ_i, $\forall i \in \mathbb{N}$ with $1 \leq i < n+1$. Then, $x_i \in M$ whenever $\lambda_i = \lambda$. Those x_i's for such λ_i's form a linearly independent subset of M. M has dimension m, and therefore such λ_i's cannot be more than m. This shows that there are exactly m λ_i's such that $\lambda = \lambda_i$.

We need the following intermediate result.

Claim 13.52.7 Let $\lambda \in \mathbb{R} \setminus \{0\}$ be an eigenvalue of Q. Then,

$$\dim(\mathcal{N}(\lambda\,\mathrm{id}_{\mathcal{X}} - Q)) = \max_{s\in\mathbb{N}} \dim(\mathcal{N}((\lambda\,\mathrm{id}_{\mathcal{X}} - Q)^s)) \in \mathbb{N}$$

Proof of Claim We will show that $\mathcal{N}\big((\lambda\,\mathrm{id}_{\mathcal{X}} - Q)^{s+1}\big) \setminus \mathcal{N}((\lambda\,\mathrm{id}_{\mathcal{X}} - Q)^s) = \emptyset$, $\forall s \in \mathbb{N}$. Suppose we have $x_0 \in \mathcal{N}\big((\lambda\,\mathrm{id}_{\mathcal{X}} - Q)^{s+1}\big) \setminus \mathcal{N}((\lambda\,\mathrm{id}_{\mathcal{X}} - Q)^s)$, for some $s \in \mathbb{N}$. Then, $(\lambda\,\mathrm{id}_{\mathcal{X}} - Q)^{s+1}x_0 = \vartheta_{\mathcal{X}}$ and $(\lambda\,\mathrm{id}_{\mathcal{X}} - Q)^s x_0 =: \xi \neq \vartheta_{\mathcal{X}}$. Clearly, $(\lambda\,\mathrm{id}_{\mathcal{X}} - Q)\xi = \vartheta_{\mathcal{X}}$ and $\xi \in M$. But, $\xi \in \mathcal{R}(\lambda\,\mathrm{id}_{\mathcal{X}} - Q) \subseteq \overline{\mathcal{R}(\lambda\,\mathrm{id}_{\mathcal{X}} - Q)} = ((\mathcal{R}(\lambda\,\mathrm{id}_{\mathcal{X}} - Q))^{\perp_\mathrm{h}})^{\perp_\mathrm{h}} = (\mathcal{N}((\lambda\,\mathrm{id}_{\mathcal{X}} - Q)^*))^{\perp_\mathrm{h}} = M^{\perp_\mathrm{h}}$, where the first equality follows from Proposition 13.17; and the second equality follows from Proposition 13.22. Thus, $\xi \in M \cap M^{\perp_\mathrm{h}}$ and hence $\xi = \vartheta_{\mathcal{X}}$. This contradicts with $\xi \neq \vartheta_{\mathcal{X}}$. Hence, the hypothesis is invalid. Therefore, we must have $\mathcal{N}\big((\lambda\,\mathrm{id}_{\mathcal{X}} - Q)^{s+1}\big) \setminus$

$\mathcal{N}((\lambda \operatorname{id}_{\mathcal{X}} - Q)^s) = \emptyset$, $\forall s \in \mathbb{N}$. Clearly, $\mathcal{N}\big((\lambda \operatorname{id}_{\mathcal{X}} - Q)^{s+1}\big) \supseteq \mathcal{N}((\lambda \operatorname{id}_{\mathcal{X}} - Q)^s)$, $\forall s \in \mathbb{N}$. Then, we must have $\mathcal{N}(\lambda \operatorname{id}_{\mathcal{X}} - Q) = \mathcal{N}((\lambda \operatorname{id}_{\mathcal{X}} - Q)^s)$, $\forall s \in \mathbb{N}$. Then, the result holds, by Claim 13.52.5. This completes the proof of the claim. □

This completes the proof for (v).

(vi) Let $A \in \mathrm{K}(\mathcal{X}, \mathcal{Y})$. Thus, (ii) holds. Then, $A^*A \in \bar{\mathcal{S}}_{\mathrm{psd}\,\mathcal{X}} \cap \mathrm{K}(\mathcal{X}, \mathcal{X})$ by Propositions 13.42 and 13.51. Then, we have $A^*A = \sum_{i=1}^{\bar{n}} \sigma_i^2 x_i x_i^*$. By (v), σ_i^2's are the nonzero eigenvalues of A^*A, and is independent of the choice of $(y_i)_{i=1}^{\bar{n}}$ and $(x_i)_{i=1}^{\bar{n}}$. Hence, $\sigma_i = \sqrt{\lambda_i}$, $\forall i \in \mathbb{N}$ with $1 \leq i < \bar{n} + 1$, where $(\lambda_i)_{i=1}^{\bar{n}}$ are the nonzero eigenvalues of A^*A (including multiplicity), which is independent of the choice of $(y_i)_{i=1}^{\bar{n}}$ and $(x_i)_{i=1}^{\bar{n}}$.

This completes the proof of the theorem. □

Example 13.53 Let $\mathcal{X}$ be a separable Hilbert space over $\mathbb{K}$, $\mathcal{Y}$ be a finite-dimensional Hilbert space over $\mathbb{K}$, $I_i \subseteq \mathbb{R}$ be compact intervals, $\mathcal{I}_i := ((I_i, |\cdot|), \mathcal{B}_i, \mu_i)$ be the compact finite metric measure subspace of $\mathbb{R}$, $i = 1, 2$, $\mathcal{H}_1 := \mathrm{L}_2(\mathcal{I}_1, \mathcal{X})$ and $\mathcal{H}_2 := \mathrm{L}_2(\mathcal{I}_2, \mathcal{Y})$ be Hilbert spaces as defined in Example 13.11, $K : I_2 \times I_1 \to \mathrm{B}(\mathcal{X}, \mathcal{Y})$ be a continuous function. Define the mapping $\bar{T} : \mathcal{H}_1 \to \mathcal{H}_2$ by $\bar{T}([h]) = [T(h)]$ and $T : \bar{\mathrm{L}}_2(\mathcal{I}_1, \mathcal{X}) \to \bar{\mathrm{L}}_2(\mathcal{I}_2, \mathcal{Y})$ defined by

$$T(h)(t) = \int_{\mathcal{I}_1} K(t, s) h(s) \, d\mu_1(s); \quad \forall t \in I_2, \ \forall h \in \bar{\mathrm{L}}_2(\mathcal{I}_1, \mathcal{X})$$

We will show that $\bar{T} \in \mathrm{B}(\mathcal{H}_1, \mathcal{H}_2)$ and is a compact operator.

$\forall [h] \in \mathcal{H}_1$, $\forall t \in I_2$, we have $K(t, \cdot) h(\cdot) : \mathcal{I}_1 \to \mathcal{Y}$ is $\mathcal{B}_1$-measurable, by Propositions 11.37, 11.39, 7.126, and 7.65. By Proposition 5.29, $\exists M \in [0, \infty) \subset \mathbb{R}$ such that $\|K(t, s)\|_{\mathrm{B}(\mathcal{X},\mathcal{Y})} \leq M$, $\forall (t, s) \in I_2 \times I_1$. Then, $\forall [h] \in \mathcal{H}_1$, $\forall t \in I_2$, $\forall s \in I_1$, we have $\int_{\mathcal{I}_1} \|K(t, s) h(s)\|_{\mathcal{Y}} \, d\mu_1(s) \leq \int_{\mathcal{I}_1} M \|h(s)\|_{\mathcal{X}} \, d\mu_1(s) \leq M \sqrt{\mu_1(I_1)} \|[h]\|_{\mathcal{H}_1} < \infty$, where the second inequality follows from Cauchy–Schwarz Inequality. By Proposition 11.89, $T(h)(t) \in \mathcal{Y}$ and $\|T(h)(t)\|_{\mathcal{Y}} \leq M \sqrt{\mu_1(I_1)} \|[h]\|_{\mathcal{H}_1}$, $\forall t \in I_2$, $\forall [h] \in \mathcal{H}_1$.

Next, we will show that $T(h) \in \mathcal{C}(I_2, \mathcal{Y})$. By Proposition 5.39, K is uniformly continuous. $\forall \epsilon \in (0, \infty) \subset \mathbb{R}$, $\exists \delta(\epsilon) \in (0, \infty) \subset \mathbb{R}$ such that $\forall (t_1, s_1), (t_2, s_2) \in I_2 \times I_1$ with $|(t_1, s_1) - (t_2, s_2)| < \delta(\epsilon)$, we have $\|K(t_1, s_1) - K(t_2, s_2)\|_{\mathrm{B}(\mathcal{X},\mathcal{Y})} < \epsilon$. $\forall [h] \in \mathcal{H}_1$, $\forall t_1, t_2 \in I_2$ with $|t_1 - t_2| < \delta\left(\frac{\epsilon}{1 + \sqrt{\mu_1(I_1)} \|[h]\|_{\mathcal{H}_1}}\right)$, we have

$$\begin{aligned}
\|T(h)(t_1) - T(h)(t_2)\|_{\mathcal{Y}} &= \left\| \int_{\mathcal{I}_1} (K(t_1, s) - K(t_2, s)) h(s) \, d\mu_1(s) \right\|_{\mathcal{Y}} \\
&\leq \int_{\mathcal{I}_1} \|(K(t_1, s) - K(t_2, s)) h(s)\|_{\mathcal{Y}} d\mu_1(s) \\
&\leq \int_{\mathcal{I}_1} \|K(t_1, s) - K(t_2, s)\|_{\mathrm{B}(\mathcal{X},\mathcal{Y})} \|h(s)\|_{\mathcal{X}} \, d\mu_1(s)
\end{aligned}$$

$$\leq \int_{\mathcal{I}_1} \frac{\epsilon}{1 + \sqrt{\mu_1(I_1)}\, \|[h]\|_{\mathcal{H}_1}} \|h(s)\|_{\mathcal{X}} \, d\mu_1(s)$$
$$\leq \frac{\epsilon}{1 + \sqrt{\mu_1(I_1)}\, \|[h]\|_{\mathcal{H}_1}} \sqrt{\mu_1(I_1)} \|[h]\|_{\mathcal{H}_1} < \epsilon$$

where the first equality and inequality follow from Proposition 11.89; the third inequality follows from the preceding discussion; and the fourth inequality follows from Cauchy–Schwarz Inequality. Then, by Proposition 11.37, $T(h)$ is $\mathcal{B}_2$-measurable. Furthermore, $T(h) \in \bar{L}_2(\mathcal{I}_2, \mathcal{Y})$ since $\int_{\mathcal{I}_2} \|T(h)(t)\|_{\mathcal{Y}}^2 \, d\mu_2(t) \leq \int_{\mathcal{I}_2} M^2 \mu_1(I_1) \|[h]\|_{\mathcal{H}_1}^2 \, d\mu_2 = M^2 \mu_1(I_1)\mu_2(I_2)\|[h]\|_{\mathcal{H}_1}^2 < \infty$. Then, $\|\bar{T}([h])\|_{\mathcal{H}_2} = \|[T(h)]\|_{\mathcal{H}_2} \leq M\sqrt{\mu_1(I_1)\mu_2(I_2)}\|[h]\|_{\mathcal{H}_1} < \infty$ and $\bar{T}([h]) \in \mathcal{H}_2$. It is clear that $\bar{T}$ is a linear operator. The preceding shows that $\|\bar{T}\|_{B(\mathcal{H}_1, \mathcal{H}_2)} \leq M\sqrt{\mu_1(I_1)\mu_2(I_2)}$. Hence, $\bar{T} \in B(\mathcal{H}_1, \mathcal{H}_2)$.

Finally, we will show that $\bar{T}$ is a compact operator. Let $(y_i)_{i=1}^{\hat{n}} \subseteq \mathcal{Y}$ be a set of orthonormal basis vectors of $\mathcal{Y}$, where $\hat{n} \in \mathbb{Z}_+$ is the dimension of $\mathcal{Y}$. Fix any $([h_i])_{i=1}^{\infty} \subseteq \overline{\mathcal{B}}_{\mathcal{H}_1}(\vartheta_{\mathcal{H}_1}, 1) := F_1$. By Alaoglu Theorem 7.122 and Riesz-Fréchet Theorem 13.15, F_1 is weakly compact. By Propositions 5.22 and 5.26, $\exists [h_0] \in F_1$ such that it is a weak cluster point of $([h_i])_{i=1}^{\infty}$. Let $v_0 := T(h_0) \in \bar{L}_2(\mathcal{I}_2, \mathcal{Y})$. By the preceding paragraph, we have $v_i := T(h_i) \in \mathcal{C}(\mathcal{I}_2, \mathcal{Y})$, $\forall i \in \mathbb{Z}_+$, furthermore, $\forall \epsilon \in (0, \infty) \subset \mathbb{R}$, $\forall t_1, t_2 \in I_2$ with $|t_1 - t_2| < \delta\left(\frac{\epsilon}{1+\sqrt{\mu_1(I_1)}}\right)$, we have $\|T(h_i)(t_1) - T(h_i)(t_2)\|_{\mathcal{Y}} < \epsilon$. Let $i_0 := 0 \in \mathbb{Z}_+$. Fix any $n \in \mathbb{N}$, let $\bar{\delta} := \delta\left(\frac{2^{-n}/3}{1+\sqrt{\mu_1(I_1)}}\right) > 0$. Then, exists $\bar{n} \in \mathbb{N}$ and $(t_i)_{i=1}^{\bar{n}} \subseteq I_2$ with $\min I_2 = t_1 < t_2 < \cdots t_{\bar{n}} = \max I_2$ and $t_j - t_{j-1} < \bar{\delta}$, $\forall j = 2, \ldots, \bar{n}$. Define $x_{*n,j,k} \in \mathcal{H}_1^*$ by $\langle\langle x_{*n,j,k}, [h]\rangle\rangle := \left\langle \int_{\mathcal{I}_1} K(t_j, s)h(s)\, d\mu_1(s), y_k \right\rangle \in \mathbb{K}$, $\forall [h] \in \mathcal{H}_1$, $\forall j = 1, \ldots, \bar{n}$, $\forall k = 1, \ldots, \hat{n}$. It is easy to see that $x_{*n,j,k}$'s are well-defined. Let $O_n := \left\{[h] \in \mathcal{H}_1 \,\middle|\, |\langle\langle x_{*n,j,k}, h - h_0\rangle\rangle| < \frac{2^{-n}}{3\sqrt{\hat{n}}}, j = 1, \ldots, \bar{n}, k = 1, \ldots, \hat{n}.\right\} \subseteq \mathcal{H}_1$. Clearly, $[h_0] \in O_n$ and O_n is a weakly open subset of $\mathcal{H}_1$ by Proposition 7.116. Since $[h_0]$ is a weak cluster point of $([h_i])_{i=1}^{\infty}$, then there exists $i_n \in \mathbb{N}$ with $i_n > i_{n-1}$, such that $[h_{i_n}] \in O_n$. Thus, we have obtained a subsequence $(\bar{T}([h_{i_n}]))_{n=1}^{\infty} \subseteq \bar{T}(F_1) \subset \mathcal{H}_2$. By $[h_{i_n}] \in O_n$, we have $|\langle T(h_{i_n})(t_j) - v_0(t_j), y_k\rangle| = |\langle T(h_{i_n})(t_j) - T(h_0)(t_j), y_k\rangle| = |\langle x_{*n,j,k}, h_{i_n} - h_0\rangle| < \frac{2^{-n}}{3\sqrt{\hat{n}}}$, $\forall j = 1, \ldots, \bar{n}$, $\forall k = 1, \ldots, \hat{n}$. This implies that $\|T(h_{i_n})(t_j) - v_0(t_j)\|_{\mathcal{Y}} = (\sum_{k=1}^{\hat{n}} |\langle T(h_{i_n})(t_j) - v_0(t_j), y_k\rangle|^2)^{1/2} < \frac{2^{-n}}{3}$. Thus, $\forall t \in \mathcal{I}_2$, $\exists j_0 \in 1, \ldots, \bar{n} - 1$ such that $t_{j_0} \leq t \leq t_{j_0+1}$. This yields $\|T(h_{i_n})(t) - v_0(t)\|_{\mathcal{Y}} \leq \|T(h_{i_n})(t) - T(h_{i_n})(t_{j_0})\|_{\mathcal{Y}} + \|T(h_{i_n})(t_{j_0}) - v_0(t_{j_0})\|_{\mathcal{Y}} + \|v_0(t_{j_0}) - v_0(t)\|_{\mathcal{Y}} \leq \frac{2^{-n}}{3} + \frac{2^{-n}}{3} + \frac{2^{-n}}{3} = 2^{-n}$. Thus, we have $\|T(h_{i_n}) - v_0\|_{\mathcal{C}(\mathcal{I}_2, \mathcal{Y})} < 2^{-n}$, and hence $\left\|\bar{T}([h_{i_n}]) - [v_0]\right\|_{\mathcal{H}_2} < 2^{-n}\sqrt{\mu_2(I_2)}$. This shows that $\lim_{n\in\mathbb{N}} \bar{T}([h_{i_n}]) = [v_0]$ in $\mathcal{H}_2$. Thus, we have shown that $\forall (v_i)_{i=1}^{\infty} \subseteq \bar{T}(F_1)$, there exists a subsequence $(v_{i_n})_{n=1}^{\infty}$ that converges to an element of $\bar{T}(F_1)$. This shows that $\bar{T}(F_1)$ is sequentially compact. By Borel-Lebesgue Theorem 5.37, $\bar{T}(F_1)$ is compact and then closed. Hence, $\overline{\bar{T}(F_1)} = \bar{T}(F_1)$ is compact. By Proposition 13.50, $\bar{T}$ is a compact operator.

This completes this example. ◇

Proposition 13.54 *Let $T \in (0, \infty) \subset \mathbb{R}$, $I := [0, T] \subset \mathbb{R}$ with the subset topology $\mathcal{O}$, $\mathcal{I} := (I, \mathcal{O})$, and $z \in \mathcal{C}(\mathcal{I}, \mathbb{K}) =: \mathcal{Z}$. Then, the following statements hold.*

(i) *Assume that $z(0) = z(T)$. Let $M = \{\bar{z} \in \mathcal{Z} \mid \bar{z}(x) = \cos(\frac{2\pi n}{T}x), \forall x \in \mathcal{I}, \forall n \in \mathbb{Z}_+;$ or $\bar{z}(x) = \sin(\frac{2\pi n}{T}x), \forall x \in \mathcal{I}, \forall n \in \mathbb{N}\} \subseteq \mathcal{Z}$. Let $\mathcal{M} = \operatorname{span}(M) \subseteq \mathcal{Z}$. Then, $z \in \overline{\mathcal{M}}$.*

(ii) *Assume that $z(0) = 0$. Let $M = \left\{\bar{z} \in \mathcal{Z} \,\middle|\, \bar{z}(x) = \sin(\frac{(n-\frac{1}{2})\pi}{T}x), \forall x \in \mathcal{I}, \forall n \in \mathbb{N}\right\} \subseteq \mathcal{Z}$. Let $\mathcal{M} = \operatorname{span}(M) \subseteq \mathcal{Z}$. Then, $z \in \overline{\mathcal{M}}$.*

Proof (i) This follows readily from Corollary 7.58 and Proposition 13.35.

(ii) Clearly, we have only to prove this for $T = \frac{\pi}{2}$. For general $T \in (0, \infty) \subset \mathbb{R}$, we may work with $\hat{z} \in \mathcal{C}(\bar{\mathrm{r}}_{0,\frac{\pi}{2}}, \mathbb{K})$ with $\hat{z}(t) := z(\frac{2T}{\pi}t)$, $\forall t \in \bar{\mathrm{r}}_{0,\frac{\pi}{2}}$. Let $T = \frac{\pi}{2}$. Let $\bar{I} := \bar{\mathrm{r}}_{-\pi,\pi} \subset \mathbb{R}$ and $\bar{\mathcal{I}} := (\bar{I}, |\cdot|)$ be the complete metric subspace of $\mathbb{R}$. Define $h \in \mathcal{C}(\bar{\mathcal{I}}, \mathbb{K}) := \mathcal{Y}$ by

$$h(t) = \begin{cases} z(2T - t) & T < t \leq 2T \\ z(t) & 0 \leq t \leq T \\ -z(-t) & -T \leq t < 0 \\ -z(2T + t) & -2T \leq t < -T \end{cases}, \quad \forall t \in \bar{\mathcal{I}}$$

By Theorem 3.11, $h \in \mathcal{Y}$. Clearly, $h(-t) = -h(t)$, $\forall t \in \bar{\mathcal{I}}$. Hence, h is an odd function. Clearly, $h(\pi) = z(0) = 0 = h(-\pi)$. Hence, by (i), $h \in \overline{\bar{\mathcal{M}}}$, where $\bar{\mathcal{M}} := \operatorname{span}(\bar{M})$ and $\bar{M} := \{\bar{z} \in \mathcal{Y} | \bar{z}(x) = \cos(nx), \forall x \in \bar{\mathcal{I}}, \forall n \in \mathbb{Z}_+;$ or $\bar{z}(x) = \sin(nx), \forall x \in \bar{\mathcal{I}}, \forall n \in \mathbb{N}\} \subseteq \mathcal{Y}$. Let $S_n \in \mathcal{C}(\bar{\mathcal{I}}, \mathbb{K})$ be defined by $S_n(t) := \frac{1}{2}a_0 + \sum_{i=1}^{n}(a_i \cos(it) + b_i \sin(it))$, $\forall t \in \bar{\mathcal{I}}$, $\forall n \in \mathbb{Z}_+$, be the partial sum of the Fourier series for h, where $a_i := \frac{1}{\pi}\int_{-\pi}^{\pi} h(t)\cos(it)\,\mathrm{d}t \in \mathbb{K}$, $\forall i \in \mathbb{Z}_+$; and $b_i := \frac{1}{\pi}\int_{-\pi}^{\pi} h(t)\sin(it)\,\mathrm{d}t \in \mathbb{K}$, $\forall i \in \mathbb{N}$. It is straight forward to show that $a_i = 0$, $\forall i \in \mathbb{Z}_+$, since h is an odd function. Now, we will show that $b_{2k} = 0$, $\forall k \in \mathbb{N}$. Fix any $k \in \mathbb{N}$. We have

$$\begin{aligned} b_{2k} &= \frac{1}{\pi}\int_{-\pi}^{\pi} h(t)\sin(2kt)\,\mathrm{d}t \\ &= \frac{1}{\pi}\bigg(\int_{-\pi}^{-\frac{\pi}{2}} -z(\pi + t)\sin(2kt)\,\mathrm{d}t + \int_{-\frac{\pi}{2}}^{0} -z(-t)\sin(2kt)\,\mathrm{d}t \\ &\quad + \int_{0}^{\frac{\pi}{2}} z(t)\sin(2kt)\,\mathrm{d}t + \int_{\frac{\pi}{2}}^{\pi} z(\pi - t)\sin(2kt)\,\mathrm{d}t\bigg) \end{aligned}$$

$$= \frac{1}{\pi}\Big(- \int_0^{\frac{\pi}{2}} z(\bar{t}) \sin(2k\bar{t} - 2k\pi)\, d\bar{t} + \int_{\frac{\pi}{2}}^0 z(\hat{t}) \sin(-2k\hat{t})\, d\hat{t}$$

$$+ \int_0^{\frac{\pi}{2}} z(t) \sin(2kt)\, dt - \int_{\frac{\pi}{2}}^0 z(\tilde{t}) \sin(2k\pi - 2k\tilde{t})\, d\tilde{t}\Big)$$

$$= \frac{1}{\pi}\Big(- \int_0^{\frac{\pi}{2}} z(\bar{t}) \sin(2k\bar{t})\, d\bar{t} + \int_0^{\frac{\pi}{2}} z(\hat{t}) \sin(2k\hat{t})\, d\hat{t}$$

$$+ \int_0^{\frac{\pi}{2}} z(t) \sin(2kt)\, dt - \int_0^{\frac{\pi}{2}} z(\tilde{t}) \sin(2k\tilde{t})\, d\tilde{t}\Big) = 0$$

where the second equality follows from Definition 12.71; the third equality follows from Change of Variable Theorem 12.91 and $\bar{t} = t + \pi$, $\hat{t} = -t$, and $\tilde{t} = \pi - t$; and fourth equality follows from Definition 12.71. Thus, $S_n(t) = \sum_{k=1}^{\lfloor \frac{n+1}{2} \rfloor} b_{2k-1} \sin((2k-1)t)$, $\forall t \in \bar{\mathcal{I}}$, $\forall n \in \mathbb{Z}_+$. By Fejér's Theorem 38.12 of Bartle (1976), the Cesàro means of the Fourier series for h converges to h in $\mathcal{Y}$ (working with the real part and the imaginary part separately when $\mathbb{K} = \mathbb{C}$), that is $\lim_{n\in\mathbb{N}} K_n = h$ in $\mathcal{Y}$, where $K_n \in \mathcal{Y}$ is defined by $K_n = \frac{1}{n}\sum_{i=0}^{n-1} S_i$, $\forall n \in \mathbb{N}$. Since S_i's only involve terms with $\sin((2k-1)t)$, then so is K_n. Hence, we have $h \in \overline{\hat{\mathcal{M}}}$, where $\hat{\mathcal{M}} := \operatorname{span}\left(\hat{M}\right)$ and $\hat{M} := \left\{\bar{z} \in \mathcal{Y} \,\middle|\, \bar{z}(x) = \sin((2k-1)x), \forall x \in \bar{\mathcal{I}}, \forall k \in \mathbb{N}\right\} \subseteq \mathcal{Y}$.

Note that $z = h|_{\mathcal{I}}$. Then, $z \in \overline{\mathcal{M}}$, where $\mathcal{M}$ is as defined in (ii). This completes the proof of the proposition. □

Proposition 13.55 *Let $\mathcal{X}$ and $\mathcal{Y}$ be Hilbert space over $\mathbb{K}$ and $A \in \mathrm{K}(\mathcal{X}, \mathcal{Y})$. Then, the following statements hold.*

(i) *$\forall \lambda \in (0, \infty) \subset \mathbb{R}$, λ is an eigenvalue of AA^* with a corresponding eigenvector y if, and only if, λ is an eigenvalue of A^*A with a corresponding eigenvector A^*y.*

(ii) *$AA^* \in \bar{\mathcal{S}}_{\text{psd}\,\mathcal{Y}}$ and $A^*A \in \bar{\mathcal{S}}_{\text{psd}\,\mathcal{X}}$ are both compact operators.*

(iii) *$A^* \in \mathrm{K}(\mathcal{Y}, \mathcal{X})$.*

(iv) *There exists $\bar{n} \in \mathbb{Z}_+ \cup \{\infty\}$ such that A^*A admits eigenvectors $x_i \in \mathcal{X}$ with $\|x_i\|_{\mathcal{X}} = 1$, each associated with an eigenvalue $\lambda_i \in (0, \infty) \subset \mathbb{R}$, $i = 1, \ldots, \bar{n}$, and $(x_i)_{i=1}^{\bar{n}}$ is an orthonormal sequence, and $A^*A = \sum_{i=1}^{\bar{n}} \lambda_i x_i x_i^*$ and $\lim_{i\in\mathbb{N}} \lambda_i = 0$ if $\bar{n} = \infty$. Then, AA^* admits eigenvectors $y_i := \frac{1}{\|Ax_i\|_{\mathcal{Y}}} Ax_i \in \mathcal{Y}$ with $\|y_i\|_{\mathcal{Y}} = 1$, each associated with an eigenvalue $\lambda_i \in (0, \infty) \subset \mathbb{R}$, $i = 1, \ldots, \bar{n}$, and $(y_i)_{i=1}^{\bar{n}}$ is an orthonormal sequence, and $AA^* = \sum_{i=1}^{\bar{n}} \lambda_i y_i y_i^*$.*

(v) *$\|AA^*\|_{\mathrm{B}(\mathcal{Y},\mathcal{Y})} = \|A^*A\|_{\mathrm{B}(\mathcal{X},\mathcal{X})} = \|A\|^2_{\mathrm{B}(\mathcal{X},\mathcal{Y})}$.*

Proof (i) "Necessity" We have $\lambda y = AA^*y$ and $\|y\| > 0$. Then, $\lambda A^*y = A^*(\lambda y) = A^*AA^*y = (A^*A)(A^*y)$. Now, $y \neq \vartheta_{\mathcal{Y}}$ and $\lambda \neq 0$, implies that $\lambda y \neq \vartheta_{\mathcal{Y}}$ and then

$AA^*y \neq \vartheta_{\mathcal{Y}}$. Hence, $A^*y \neq \vartheta_{\mathcal{X}}$. This shows that A^*y is an eigenvector of A^*A associated with eigenvalue λ.

"Sufficiency" This follows from symmetry and the "Necessity" proof.

(ii) By (ix) of Proposition 13.42, $AA^* \in \bar{\mathcal{S}}_{\text{psd}\,\mathcal{Y}}$ and $A^*A \in \bar{\mathcal{S}}_{\text{psd}\,\mathcal{X}}$. By Proposition 13.51, A^*A and AA^* are compact operators.

(iii) By Spectral Theory Theorem 13.52, $\exists \bar{n} \in \mathbb{Z}_+ \cup \{\infty\}$ and $\exists\,(x_i)_{i=1}^{\bar{n}} \subseteq \mathcal{X}$, which is an orthonormal sequence, and $\exists\,(y_i)_{i=1}^{\bar{n}} \subseteq \mathcal{Y}$, which is an orthonormal sequence, and $\sigma =: \sigma_1 \geq \sigma_2 \geq \cdots \geq \sigma_i \in (0, \infty) \subset \mathbb{R}$, $\forall i \in \mathbb{N}$ with $0 \leq i-1 < \bar{n}$, such that $A = \sum_{i=1}^{\bar{n}} \sigma_i y_i x_i^*$, and in case $\bar{n} = \infty$, we have $\lim_{i\in\mathbb{N}} \sigma_i = 0$. Then, $A^* = \sum_{i=1}^{\bar{n}} \sigma_i x_i y_i^*$. By Spectral Theory Theorem 13.52, $A^* \in \mathrm{K}(\mathcal{Y}, \mathcal{X})$.

(iv) $A^*A = (\sum_{i=1}^{\bar{n}} \sigma_i x_i y_i^*)(\sum_{i=1}^{\bar{n}} \sigma_i y_i x_i^*) = \sum_{i=1}^{\bar{n}} \sigma_i^2 x_i x_i^*$, where the second equality follows from the fact that $(y_i)_{i=1}^{\bar{n}}$ is orthonormal. It is easy to check that $\sigma_i^2 x_i = A^*Ax_i$, $\forall i \in \mathbb{N}$ with $1 \leq i < \bar{n}+1$. Similarly, $AA^* = \sum_{i=1}^{\bar{n}} \sigma_i^2 y_i y_i^*$. By (i) of Spectral Theory Theorem 13.52, we have $y_i = \frac{1}{\sigma_i}Ax_i = \frac{1}{\|Ax_i\|_{\mathcal{Y}}}Ax_i$. Then, (iv) follows.

(v) By Spectral Theory Theorem 13.52, we have $\|AA^*\|_{\mathrm{B}(\mathcal{Y},\mathcal{Y})} = \sigma_1^2 = \|A^*A\|_{\mathrm{B}(\mathcal{X},\mathcal{X})} = \|A\|_{\mathrm{B}(\mathcal{X},\mathcal{Y})}^2$.

This completes the proof of the proposition. □

Proposition 13.56 (Fredholm Alternative) *Let $\mathcal{X}$ be a Hilbert space over $\mathbb{K}$, $A \in \bar{\mathcal{S}}_{\mathcal{X}} \cap \mathrm{K}(\mathcal{X}, \mathcal{X})$, and $\lambda \in \mathbb{R} \setminus \{0\}$. Then, the following statements hold.*

(i) $\mathcal{N}(\lambda\,\mathrm{id}_{\mathcal{X}} - A) = \{\vartheta_{\mathcal{X}}\}$ *if, and only if,* $\mathcal{R}(\lambda\,\mathrm{id}_{\mathcal{X}} - A) = \mathcal{X}$.
(ii) $\mathcal{R}(\lambda\,\mathrm{id}_{\mathcal{X}} - A) = (\mathcal{N}(\lambda\,\mathrm{id}_{\mathcal{X}} - A))^{\perp_{\mathrm{h}}}$.

Proof (i) "Sufficiency" Let $\mathcal{R}(\lambda\,\mathrm{id}_{\mathcal{X}} - A) = \mathcal{X}$. By Proposition 13.22, $\mathcal{N}(\lambda\,\mathrm{id}_{\mathcal{X}} - A) = (\mathcal{R}((\lambda\,\mathrm{id}_{\mathcal{X}} - A)^*))^{\perp_{\mathrm{h}}} = (\mathcal{R}(\lambda\,\mathrm{id}_{\mathcal{X}} - A))^{\perp_{\mathrm{h}}} = \mathcal{X}^{\perp_{\mathrm{h}}} = \{\vartheta_{\mathcal{X}}\}$.

"Necessity" Let $\mathcal{N}(\lambda\,\mathrm{id}_{\mathcal{X}} - A) = \{\vartheta_{\mathcal{X}}\}$. By Proposition 13.22, $\mathcal{N}(\lambda\,\mathrm{id}_{\mathcal{X}} - A) = (\mathcal{R}((\lambda\,\mathrm{id}_{\mathcal{X}} - A)^*))^{\perp_{\mathrm{h}}} = (\mathcal{R}(\lambda\,\mathrm{id}_{\mathcal{X}} - A))^{\perp_{\mathrm{h}}} = \{\vartheta_{\mathcal{X}}\}$. This further implies that $\overline{\mathcal{R}(\lambda\,\mathrm{id}_{\mathcal{X}} - A)} = \left((\mathcal{R}(\lambda\,\mathrm{id}_s\,X - A))^{\perp_{\mathrm{h}}}\right)^{\perp_{\mathrm{h}}} = \{\vartheta_{\mathcal{X}}\}^{\perp_{\mathrm{h}}} = \mathcal{X}$. Fix any $x_0 \in \mathcal{X} = \overline{\mathcal{R}(\lambda\,\mathrm{id}_{\mathcal{X}} - A)}$. Then, there exists $(x_i)_{i=1}^{\infty} \subseteq \mathcal{X}$ such that $\lim_{i\in\mathbb{N}}(\lambda\,\mathrm{id}_{\mathcal{X}} - A)x_i = x_0$. We will distinguish two exhaustive and mutually exclusive cases: Case 1: $\exists r \in [0, \infty) \subset \mathbb{R}$ such that $(x_i)_{i=1}^{\infty} \subseteq \overline{\mathcal{B}}_{\mathcal{X}}(\vartheta_{\mathcal{X}}, r)$; Case 2: $\forall r \in [0, \infty) \subset \mathbb{R}$, $\exists i \in \mathbb{N}$ such that $\|x_i\| > r$.

Case 1: $\exists r \in [0, \infty) \subset \mathbb{R}$ such that $(x_i)_{i=1}^{\infty} \subseteq \overline{\mathcal{B}}_{\mathcal{X}}(\vartheta_{\mathcal{X}}, r)$. By $A \in \mathrm{K}(\mathcal{X}, \mathcal{X})$ and Definition 13.49, there exists a subsequence $(x_{i_k})_{k=1}^{\infty}$ such that $(Ax_{i_k})_{k=1}^{\infty} \subseteq \mathcal{X}$ converges to $\bar{x} \in \mathcal{X}$. Then, we have $\lim_{k\in\mathbb{N}} \lambda x_{i_k} = \lim_{k\in\mathbb{N}}(\lambda\,\mathrm{id}_{\mathcal{X}} - A)(x_{i_k}) + \lim_{k\in\mathbb{N}} Ax_{i_k} = x_0 + \bar{x}$ and $\lim_{k\in\mathbb{N}} x_{i_k} = \frac{1}{\lambda}(x_0 + \bar{x}) =: \bar{x}_0$. Then, $(\lambda\,\mathrm{id}_{\mathcal{X}} - A)(\bar{x}_0) = \lim_{k\in\mathbb{N}}(\lambda\,\mathrm{id}_{\mathcal{X}} - A)(x_{i_k}) = \lim_{k\in\mathbb{N}} \lambda x_{i_k} - \lim_{k\in\mathbb{N}} Ax_{i_k} = \lambda\bar{x}_0 - \bar{x} = x_0$. Hence, $x_0 \in \mathcal{R}(\lambda\,\mathrm{id}_{\mathcal{X}} - A)$.

Case 2: $\forall r \in [0, \infty) \subset \mathbb{R}$, $\exists i \in \mathbb{N}$ such that $\|x_i\| > r$. By possibly further restrict to a subsequence if necessary, we may assume, without loss of generality, $\lim_{i\in\mathbb{N}} \|x_i\| =: \lim_{i\in\mathbb{N}} r_i = \infty$. Then, we have $\lim_{i\in\mathbb{N}} \frac{1}{r_i}(\lambda\,\mathrm{id}_{\mathcal{X}} - A)x_i = \lim_{i\in\mathbb{N}}(\lambda\,\mathrm{id}_{\mathcal{X}} - A)(\frac{1}{r_i}x_i) = \vartheta_{\mathcal{X}}$. Note that $\left\|\frac{1}{r_i}x_i\right\| = 1$, $\forall i \in \mathbb{N}$. By $A \in$

$\mathrm{K}(\mathcal{X}, \mathcal{X})$ and Definition 13.49, there exists a subsequence $\left(\frac{1}{r_{i_k}} x_{i_k}\right)_{k=1}^{\infty}$ such that $\lim_{k\in\mathbb{N}} A(\frac{1}{r_{i_k}} x_{i_k}) = \bar{x} \in \mathcal{X}$. Then, $\lim_{k\in\mathbb{N}} \lambda \frac{1}{r_{i_k}} x_{i_k} = \lim_{k\in\mathbb{N}} (\lambda \operatorname{id}_{\mathcal{X}} - A)(\frac{1}{r_{i_k}} x_{i_k}) + \lim_{k\in\mathbb{N}} A(\frac{1}{r_{i_k}} x_{i_k}) = \bar{x}$. This implies that $\lim_{k\in\mathbb{N}} \frac{1}{r_{i_k}} x_{i_k} = \frac{1}{\lambda}\bar{x}$, and $A(\frac{1}{\lambda}\bar{x}) = \lim_{k\in\mathbb{N}} A(\frac{1}{r_{i_k}} x_{i_k}) = \bar{x}$. We can conclude that $\bar{x} \in \mathcal{N}(\lambda \operatorname{id}_{\mathcal{X}} - A) = \{\vartheta_{\mathcal{X}}\}$. Hence, $\bar{x} = \vartheta_{\mathcal{X}}$. But $\left\| \frac{1}{\lambda}\bar{x} \right\| = \lim_{k\in\mathbb{N}} \left\| \frac{1}{r_{i_k}} x_{i_k} \right\| = 1 > 0$. This is a contradiction! Hence, this case is impossible.

In the above, we have shown that $x_0 \in \mathcal{R}(\lambda \operatorname{id}_{\mathcal{X}} - A)$. By the arbitrariness of x_0, we have $\mathcal{R}(\lambda \operatorname{id}_{\mathcal{X}} - A) = \mathcal{X}$.

This completes the proof of (i).

(ii) Let $\mathcal{N}(\lambda \operatorname{id}_{\mathcal{X}} - A) =: M$. If $M = \{\vartheta_{\mathcal{X}}\}$, then by (i) we have $\mathcal{R}(\lambda \operatorname{id}_{\mathcal{X}} - A) = \mathcal{X} = M^{\perp_{\mathrm{h}}}$. This case is proved.

If $M \supset \{\vartheta_{\mathcal{X}}\}$, then λ is an eigenvalue of A. By Spectral Theory Theorem 13.52, M is finite-dimensional. By Proposition 13.22, we have $(\mathcal{R}(\lambda \operatorname{id}_{\mathcal{X}} - A))^{\perp_{\mathrm{h}}} = \mathcal{N}(\lambda \operatorname{id}_{\mathcal{X}} - A) = M$. Thus, by Proposition 13.17, we have $\overline{\mathcal{R}(\lambda \operatorname{id}_{\mathcal{X}} - A)} = M^{\perp_{\mathrm{h}}}$. Fix any $x_0 \in M^{\perp_{\mathrm{h}}}$. Then, there exists $(\tilde{x}_i)_{i=1}^{\infty} \subseteq \mathcal{X}$ such that $\lim_{i\in\mathbb{N}} (\lambda \operatorname{id}_{\mathcal{X}} - A)\tilde{x}_i = x_0$. Note that M is finite-dimensional implies that M is closed by Theorem 7.36 and Proposition 4.39. Let $P : \mathcal{X} \to M$ and $\bar{P} : \mathcal{X} \to M^{\perp_{\mathrm{h}}}$ be projection operators. By Theorem 13.19, we have $P + \bar{P} = \operatorname{id}_{\mathcal{X}}$. Then, $x_0 = \lim_{i\in\mathbb{N}} (\lambda \operatorname{id}_{\mathcal{X}} - A)\tilde{x}_i = \lim_{i\in\mathbb{N}} (\lambda \operatorname{id}_{\mathcal{X}} - A)(P\tilde{x}_i + \bar{P}\tilde{x}_i) = \lim_{i\in\mathbb{N}} (\lambda \operatorname{id}_{\mathcal{X}} - A)(\bar{P}\tilde{x}_i)$. Hence, the sequence $(\hat{x}_i)_{i=1}^{\infty} := \left(\bar{P}\tilde{x}_i\right)_{i=1}^{\infty} \subseteq M^{\perp_{\mathrm{h}}}$ satisfies $x_0 = \lim_{i\in\mathbb{N}} (\lambda \operatorname{id}_{\mathcal{X}} - A)\hat{x}_i$. We will distinguish two exhaustive and mutually exclusive cases: Case 1: $\exists r \in [0, \infty) \subset \mathbb{R}$ such that $(\hat{x}_i)_{i=1}^{\infty} \subseteq \overline{\mathcal{B}}_{\mathcal{X}}(\vartheta_{\mathcal{X}}, r)$; Case 2: $\forall r \in [0, \infty) \subset \mathbb{R}$, there exists $i_0 \in \mathbb{N}$ such that $\|\hat{x}_{i_0}\| > r$.

Case 1: $\exists r \in [0, \infty) \subset \mathbb{R}$ such that $(\hat{x}_i)_{i=1}^{\infty} \subseteq \overline{\mathcal{B}}_{\mathcal{X}}(\vartheta_{\mathcal{X}}, r)$. By $A \in \mathrm{K}(\mathcal{X}, \mathcal{X})$, there exists a subsequence $(\hat{x}_{i_k})_{k=1}^{\infty}$ such that $\lim_{k\in\mathbb{N}} A\hat{x}_{i_k} = \bar{x} \in \mathcal{X}$. Then, we have $\lim_{k\in\mathbb{N}} \lambda\hat{x}_{i_k} = \lim_{k\in\mathbb{N}} (\lambda \operatorname{id}_{\mathcal{X}} - A)\hat{x}_{i_k} + \lim_{k\in\mathbb{N}} A\hat{x}_{i_k} = x_0 + \bar{x} =: \lambda\hat{x}_0$. This implies that $(\lambda \operatorname{id}_{\mathcal{X}} - A)\hat{x}_0 = \lim_{k\in\mathbb{N}} (\lambda \operatorname{id}_{\mathcal{X}} - A)\hat{x}_{i_k} = x_0 \in \mathcal{R}(\lambda \operatorname{id}_{\mathcal{X}} - A)$.

Case 2: $\forall r \in [0, \infty) \subset \mathbb{R}$, there exists $i_0 \in \mathbb{N}$ such that $\|\hat{x}_{i_0}\| > r$. By possibly further restricting to a subsequence, we may, without loss of generality, assume that $\lim_{i\in\mathbb{N}} \|\hat{x}_i\| =: \lim_{i\in\mathbb{N}} r_i = \infty$. Then, we have $\lim_{i\in\mathbb{N}} (\lambda \operatorname{id}_{\mathcal{X}} - A)(\frac{1}{r_i}\hat{x}_i) = \vartheta_{\mathcal{X}}$. By $A \in \mathrm{K}(\mathcal{X}, \mathcal{X})$ and Definition 13.49, there exists a subsequence $\left(\frac{1}{r_{i_k}}\hat{x}_{i_k}\right)_{k=1}^{\infty} \subseteq M^{\perp_{\mathrm{h}}}$ such that $\lim_{k\in\mathbb{N}} A(\frac{1}{r_{i_k}}\hat{x}_{i_k}) =: \bar{x} \in \mathcal{X}$. This implies that $\lim_{k\in\mathbb{N}} \lambda(\frac{1}{r_{i_k}}\hat{x}_{i_k}) = \lim_{k\in\mathbb{N}} (\lambda \operatorname{id}_{\mathcal{X}} - A)(\frac{1}{r_{i_k}}\hat{x}_{i_k}) + \lim_{k\in\mathbb{N}} A(\frac{1}{r_{i_k}}\hat{x}_{i_k}) = \vartheta_{\mathcal{X}} + \bar{x} = \bar{x}$ and $\lim_{k\in\mathbb{N}} \frac{1}{r_{i_k}}\hat{x}_{i_k} = \frac{1}{\lambda}\bar{x} \in M^{\perp_{\mathrm{h}}}$. Furthermore, $\left\| \frac{1}{\lambda}\bar{x} \right\| = \lim_{k\in\mathbb{N}} \left\| \frac{1}{r_{i_k}}\hat{x}_{i_k} \right\| = 1$ and $\bar{x} \neq \vartheta_{\mathcal{X}}$. Note that $A\bar{x} = \lim_{k\in\mathbb{N}} A(\lambda\frac{1}{r_{i_k}}\hat{x}_{i_k}) = \lambda \lim_{k\in\mathbb{N}} A(\frac{1}{r_{i_k}}\hat{x}_{i_k}) = \lambda\bar{x}$. This implies that $\bar{x}$ is an eigenvector of A associated with the eigenvalue λ, and $\bar{x} \in M$. Thus, we have $\bar{x} \in M \cap M^{\perp_{\mathrm{h}}}$ and hence $\bar{x} = \vartheta_{\mathcal{X}}$. This is a contradiction with $\bar{x} \neq \vartheta_{\mathcal{X}}$. Thus, this case is impossible.

Based on the preceding analysis, we have $x_0 \in \mathcal{R}(\lambda\, \mathrm{id}_{\mathcal{X}} - A)$. By the arbitrariness of x_0, we have $M^{\perp_{\mathrm{h}}} = \mathcal{R}(\lambda\, \mathrm{id}_{\mathcal{X}} - A)$.

This completes the proof of the proposition. □

Example 13.57 Let $m \in \mathbb{N}$, $\mathbb{R}^m$ be endowed with the usual positive cone, $\Omega \in \mathcal{B}_{\mathrm{B}}(\mathbb{R}^m)$ be a nonempty open rectangle, $\mathbb{X} := ((\Omega, |\cdot|), \mathcal{B}, \mu)$ be the σ-finite metric measure subspace of $\mathbb{R}^m$, $\mathcal{Y}$ be a separable Hilbert space over $\mathbb{K}$. (Note that $\mathcal{Y}^*$ is separable by Riesz-Fréchet Theorem 13.15.) Fix $x_0 \in \Omega$. Then, $W_{2,1,x_0}(\mathbb{X}, \mathcal{Y})$ (as defined in Example 12.106) is a Hilbert space over $\mathbb{K}$ with inner product defined by

$$\langle u, v\rangle_{W_{2,1,x_0}(\mathbb{X},\mathcal{Y})}$$
$$:= \int_{\Omega} \langle u(x), v(x)\rangle_{\mathcal{Y}}\, \mathrm{d}\mu(x) + \sum_{\substack{J\subseteq \bar{J}\\ J\neq\emptyset}} \int_{\pi_J(\Omega)} \langle u_{J,x_0}(s), v_{J,x_0}(s)\rangle_{\mathcal{Y}}\, \mathrm{d}\mu_J(s) \in \mathbb{K}$$

$\forall u, v \in W_{2,1,x_0}(\mathbb{X}, \mathcal{Y})$, where u_{J,x_0}'s are the stream functions of u with respect to x_0, v_{J,x_0}'s are the stream functions of v with respect to x_0. (Notations are as defined in Example 12.106.)

It is easy to show that $\langle\cdot,\cdot\rangle_{W_{2,1,x_0}(\mathbb{X},\mathcal{Y})}$ defines an inner product on $W_{2,1,x_0}(\mathbb{X}, \mathcal{Y})$. Then, $W_{2,1,x_0}(\mathbb{X}, \mathcal{Y})$ is a pre-Hilbert space. By Example 12.106, it is a Hilbert space.

Then, by Riesz-Fréchet Theorem 13.15, $W_{2,1,x_0}(\mathbb{X}, \mathcal{Y})$ is reflexive and its dual also a Hilbert space. $\forall f \in (W_{2,1,x_0}(\mathbb{X}, \mathcal{Y}))^*$, there exists a unique $u_0 \in W_{2,1,x_0}(\mathbb{X}, \mathcal{Y})$ such that $f(u) = \langle u, u_0\rangle_{W_{2,1,x_0}(\mathbb{X},\mathcal{Y})}$, $\forall u \in W_{2,1,x_0}(\mathbb{X}, \mathcal{Y})$, and $\|f\|_{(W_{2,1,x_0}(\mathbb{X},\mathcal{Y}))^*} = \|u_0\|_{W_{2,1,x_0}(\mathbb{X},\mathcal{Y})}$. f is denoted to be u_0^*.

The tools of this chapter are applicable to $W_{2,1,x_0}(\mathbb{X}, \mathcal{Y})$. ◇

Chapter 14
Probability Theory

14.1 Fundamental Notions

Definition 14.1 Let $\boldsymbol{\Omega} := (\Omega, \mathcal{B}, \boldsymbol{P})$ be a finite measure space with $\boldsymbol{P}(\Omega) = 1$. Then, it is said to be a *probability measure space*. Let $\mathcal{Y}$ be a topological space. A *$\mathcal{Y}$-valued random variable* is a $\mathcal{B}$-measurable function $x : \Omega \to \mathcal{Y}$. When $\mathcal{Y} \subseteq \mathbb{R}$, we will simply say x is a *random variable*. When $\mathcal{Y} \subseteq \mathcal{Z}$ and $\mathcal{Z}$ is a separable Banach space over $\mathbb{K}$, the integral $\int_{\Omega} x \, d\boldsymbol{P}$ is said to be the *expectation of* x, when it makes sense, which will be denoted by $\boldsymbol{E}(x)$. ◇

Definition 14.2 Let $(\Omega, \mathcal{B})$ be a measurable space, $\mathcal{Y}$ be a topological space, S be a set, $\mathcal{F} \subseteq \mathcal{B}$, $x : \Omega \to \mathcal{Y}$ be $\mathcal{B}$-measurable, and $x_t : \Omega \to \mathcal{Y}$ be $\mathcal{B}$-measurable, $\forall t \in S$. We will denote the σ-algebra generated by $\mathcal{F}$ by $\sigma(\mathcal{F}) \subseteq \mathcal{B}$, denote the σ-algebra generated by $\{E \in \mathcal{B} \,|\, E = x_{\text{inv}}(O), O \in \mathcal{O}_{\mathcal{Y}}\}$ by $\sigma(x) \subseteq \mathcal{B}$, and denote the σ-algebra generated by $\{E \in \mathcal{B} \,|\, E = x_{t\,\text{inv}}(O), O \in \mathcal{O}_{\mathcal{Y}}, t \in S\}$ by $\sigma((x_t)_{t\in S}) \subseteq \mathcal{B}$. Clearly, $\sigma(x)$ is the smallest σ-algebra on which x is measurable, and $\sigma((x_t)_{t\in S})$ is the smallest σ-algebra on which x_t, $\forall t \in S$, are measurable. ◇

Proposition 14.3 *Let $\boldsymbol{\Omega} := (\Omega, \mathcal{B}, \boldsymbol{P})$ be a probability measure space, $n \in \mathbb{N}$, $F_1, \ldots, F_n \in \mathcal{B}$. Then,*

$$\boldsymbol{P}\Big(\bigcup_{i=1}^{n} F_i\Big) = \sum_{l=1}^{n} (-1)^{l-1} \sum_{1 \leq i_1 < \cdots < i_l \leq n} \boldsymbol{P}\Big(\bigcap_{j=1}^{l} F_{i_j}\Big) \tag{14.1}$$

Proof We will prove this using mathematical induction on n.
1° $n = 1$. This is straightforward. $n = 2$. Then, by countable additivity of measure, we have $\boldsymbol{P}(F_1 \cup F_2) = \boldsymbol{P}((F_1 \setminus F_2) \cup F_2) = \boldsymbol{P}(F_1 \setminus F_2) + \boldsymbol{P}(F_2)$. Note that $\boldsymbol{P}(F_1) = \boldsymbol{P}(F_1 \setminus F_2) + \boldsymbol{P}(F_1 \cap F_2)$. This implies that $\boldsymbol{P}(F_1 \cup F_2) = \boldsymbol{P}(F_1) + \boldsymbol{P}(F_2) - \boldsymbol{P}(F_1 \cap F_2)$. This case is proved.
2° Assume the result holds for $n \leq k \in \{2, 3, \ldots\}$.

Z. Pan, *Measure-Theoretic Calculus in Abstract Spaces*,
https://doi.org/10.1007/978-3-031-21912-2_14

3° Consider the case $n = k + 1 \in \{3, 4, \ldots\}$. Then, we have

$$
\begin{aligned}
\boldsymbol{P}(\bigcup_{i=1}^{k+1} F_i) &= \boldsymbol{P}((\bigcup_{i=1}^{k} F_i) \cup F_{k+1}) \\
&= \boldsymbol{P}(\bigcup_{i=1}^{k} F_i) + \boldsymbol{P}(F_{k+1}) - \boldsymbol{P}(\bigcup_{i=1}^{k}(F_i \cap F_{k+1})) \\
&= \sum_{l=1}^{k}(-1)^{l-1} \sum_{1 \le i_1 < \cdots < i_l \le k} \boldsymbol{P}(\bigcap_{j=1}^{l} F_{i_j}) + \boldsymbol{P}(F_{k+1}) \\
&\quad - \sum_{l=1}^{k}(-1)^{l-1} \sum_{1 \le i_1 < \cdots < i_l \le k} \boldsymbol{P}(\bigcap_{j=1}^{l}(F_{i_j} \cap F_{k+1})) \\
&= \sum_{l=1}^{k}(-1)^{l-1} \sum_{1 \le i_1 < \cdots < i_l \le k} \boldsymbol{P}(\bigcap_{j=1}^{l} F_{i_j}) + \boldsymbol{P}(F_{k+1}) \\
&\quad + \sum_{l=1}^{k}(-1)^{l} \sum_{1 \le i_1 < \cdots < i_l \le k} \boldsymbol{P}(\bigcap_{j=1}^{l} F_{i_j} \cap F_{k+1}) \\
&= \sum_{l=1}^{k+1}(-1)^{l-1} \sum_{1 \le i_1 < \cdots < i_l \le k+1} \boldsymbol{P}(\bigcap_{j=1}^{l} F_{i_j})
\end{aligned}
$$

This completes the induction process and therefore the proof of the proposition. □

Lemma 14.4 (Fatou's Lemma) *Let* $\boldsymbol{\Omega} := (\Omega, \mathcal{B}, \boldsymbol{P})$ *be a probability measure space, and* $(E_n)_{n=1}^{\infty} \subseteq \mathcal{B}$. *Define* $F_1 := \bigcap_{n=1}^{\infty} \bigcup_{m=n}^{\infty} E_m = \{\omega \in \Omega \mid \omega \in E_i$ for infinitely many numbers of $i \in \mathbb{N}\}$, *and* $F_2 := \bigcup_{n=1}^{\infty} \bigcap_{m=n}^{\infty} E_m = \{\omega \in \Omega \mid \omega \in E_i$, eventually for all $i \in \mathbb{N}$ with $i \ge n_\omega \in \mathbb{N}\}$. *Then,* $F_2 \subseteq F_1$, *and* $\boldsymbol{P}(F_2) \le \liminf_{n \in \mathbb{N}} \boldsymbol{P}(E_n) \le \limsup_{n \in \mathbb{N}} \boldsymbol{P}(E_n) \le \boldsymbol{P}(F_1)$.

Proof $\forall \omega \in F_2$, $\exists n_\omega \in \mathbb{N}$, $\omega \in \bigcap_{m=n}^{\infty} E_m$, $\forall n \in \mathbb{N}$ with $n_\omega \le n$. Then, $\omega \in \bigcup_{m=n}^{\infty} E_m$, $\forall n \in \mathbb{N}$. Hence, $\omega \in F_1$. By the arbitrariness of ω, we have $F_2 \subseteq F_1$.

Note that $\chi_{F_2,\Omega} = \liminf_{n \in \mathbb{N}} \chi_{E_n,\Omega} \le \limsup_{n \in \mathbb{N}} \chi_{E_n,\Omega} = \chi_{F_1,\Omega}$. Then, by Fatou's Lemma 11.80, we have $\boldsymbol{P}(F_2) = \int_\Omega \chi_{F_2,\Omega} \, \mathrm{d}\boldsymbol{P} \le \liminf_{n \in \mathbb{N}} \int_\Omega \chi_{E_n,\Omega} \, \mathrm{d}\boldsymbol{P} = \liminf_{n \in \mathbb{N}} \boldsymbol{P}(E_n) \le \limsup_{n \in \mathbb{N}} \boldsymbol{P}(E_n)$. Let $G_n := \bigcup_{m=n}^{\infty} E_m$, $\forall n \in \mathbb{N}$. Then, $G_{n+1} \subseteq G_n$, $\forall n \in \mathbb{N}$. By Proposition 11.5, $\boldsymbol{P}(F_1) = \boldsymbol{P}(\bigcap_{n=1}^{\infty} G_n) = \lim_{n \in \mathbb{N}} \boldsymbol{P}(G_n) \ge \limsup_{n \in \mathbb{N}} \boldsymbol{P}(E_n)$. This completes the proof of the lemma. □

Definition 14.5 Let $\boldsymbol{\Omega} := (\Omega, \mathcal{B}, \boldsymbol{P})$ be a probability measure space, Λ be a set, and $\hat{\mathcal{B}}_\lambda \subseteq \mathcal{B}$ be a σ-algebra on Ω, $\forall \lambda \in \Lambda$. We will say that the collection of σ-algebras $(\hat{\mathcal{B}}_\lambda)_{\lambda \in \Lambda}$ is *independent* if, $\forall n \in \mathbb{N}$, $\forall \lambda_1, \ldots, \lambda_n \in \Lambda$, which are distinct, $\forall E_l \in \hat{\mathcal{B}}_{\lambda_l}$, $l = 1, \ldots, n$, we have $\boldsymbol{P}(\bigcap_{l=1}^n E_l) = \prod_{l=1}^n \boldsymbol{P}(E_l)$.

A collection of random variables (possibly Banach space valued) $(x_\lambda)_{\lambda \in \Lambda}$ are said to be *independent* if the collection of σ-algebras $(\sigma(x_\lambda))_{\lambda \in \Lambda}$ is independent. $\diamond$

Lemma 14.6 (Borel–Cantelli) *Let $\boldsymbol{\Omega} := (\Omega, \mathcal{B}, \boldsymbol{P})$ be a probability measure space, $(E_n)_{n=1}^\infty \subseteq \mathcal{B}$. Then, the following statements hold:*

(i) *Assume that $\sum_{n=1}^\infty \boldsymbol{P}(E_n) < \infty$. Then $\boldsymbol{P}(\bigcap_{n=1}^\infty \bigcup_{m=n}^\infty E_m) := \boldsymbol{P}(F_1) = 0$.*
(ii) *Assume that the collection of σ-algebras $(\sigma(\{\emptyset, E_i, \Omega\}))_{i=1}^\infty$ is independent, and $\sum_{n=1}^\infty \boldsymbol{P}(E_n) = \infty$. Then, $\boldsymbol{P}(F_1) = 1$.*

Proof (i) $\boldsymbol{P}(F_1) = \lim_{n \in \mathbb{N}} \boldsymbol{P}(\bigcup_{m=n}^\infty E_m) \leq \lim_{n \in \mathbb{N}} \sum_{m=n}^\infty \boldsymbol{P}(E_m) = 0$, where the first equality follows from the proof of Lemma 14.4, the inequality follows from Proposition 11.6, and the last equality follows from the assumption.

(ii) Note that $\Omega \setminus F_1 = \bigcup_{n=1}^\infty \bigcap_{m=n}^\infty (\Omega \setminus E_m)$. Let $p_k := \boldsymbol{P}(E_k)$, $\forall k \in \mathbb{N}$. Then, $\boldsymbol{P}(\bigcap_{m=n}^\infty (\Omega \setminus E_m)) = \lim_{k \in \mathbb{N}} \boldsymbol{P}(\bigcap_{m=n}^k (\Omega \setminus E_m)) = \lim_{k \in \mathbb{N}} \prod_{m=n}^k \boldsymbol{P}(\Omega \setminus E_m) = \lim_{k \in \mathbb{N}} \prod_{m=n}^k (1 - p_m) = \prod_{m=n}^\infty (1 - p_m)$, $\forall n \in \mathbb{N}$, where the first equality follows from Proposition 11.5, the second equality follows from the independence assumption, and the third equality follows from Definition 11.2. $\forall \alpha \in \mathbb{R}$, $1 + \alpha \leq \exp(\alpha)$. Then, $(1 - p_m) \leq \exp(-p_m)$, $\forall m \in \mathbb{N}$. This implies that $0 \leq \prod_{m=n}^\infty (1 - p_m) \leq \exp(-\sum_{m=n}^\infty p_m) = 0$. Then, we have $\boldsymbol{P}(\bigcap_{m=n}^\infty (\Omega \setminus E_m)) = 0$, $\forall n \in \mathbb{N}$. This implies $0 \leq \boldsymbol{P}(\Omega \setminus F_1) \leq \sum_{n=1}^\infty \boldsymbol{P}(\bigcap_{m=n}^\infty (\Omega \setminus E_m)) = 0$, where the second inequality follows from Proposition 11.6. Hence, $\boldsymbol{P}(F_1) = 1$.

This completes the proof of the lemma. □

Definition 14.7 Let $\boldsymbol{\Omega} := (\Omega, \mathcal{B}, \boldsymbol{P})$ be a probability measure space, $\mathcal{Y}$ be a separable Banach space over $\mathbb{K}$, $x \in \bar{\mathrm{L}}_1(\boldsymbol{\Omega}, \mathcal{Y})$, $\hat{\mathcal{B}} \subseteq \mathcal{B}$ be a σ-algebra on Ω, and $\hat{\boldsymbol{P}} := \boldsymbol{P}|_{\hat{\mathcal{B}}}$. By Proposition 11.13, $\hat{\boldsymbol{\Omega}} := (\Omega, \hat{\mathcal{B}}, \hat{\boldsymbol{P}})$ is a probability measure space. Define $\nu : \mathcal{B} \to \mathcal{Y}$ by $\nu(E) = \int_E x \, \mathrm{d}\boldsymbol{P}$, $\forall E \in \mathcal{B}$. By Proposition 11.116, ν is a finite $\mathcal{Y}$-valued measure on $(\Omega, \mathcal{B})$ with $\mathcal{P} \circ \nu(E) = \int_E \mathcal{P} \circ x \, \mathrm{d}\boldsymbol{P}$, $\forall E \in \mathcal{B}$. By Definition 11.166 and Proposition 11.167, $x = \frac{\mathrm{d}\nu}{\mathrm{d}\boldsymbol{P}}$ a.e. in $\boldsymbol{\Omega}$. By Proposition 11.107, $\hat{\nu} := \nu|_{\hat{\mathcal{B}}}$ is a $\mathcal{Y}$-valued pre-measure on $(\Omega, \hat{\mathcal{B}})$. By Definition 11.99, we have $\mathcal{P} \circ \hat{\nu} \leq (\mathcal{P} \circ \nu)|_{\hat{\mathcal{B}}}$. Then, $(\Omega, \hat{\mathcal{B}}, \hat{\nu})$ is a finite $\mathcal{Y}$-valued measure space. $\forall E \in \hat{\mathcal{B}}$ with $\hat{\boldsymbol{P}}(E) = 0$, we have $E \in \mathcal{B}$, $\boldsymbol{P}(E) = 0$, and $0 \leq \mathcal{P} \circ \hat{\nu}(E) \leq \mathcal{P} \circ \nu(E) = \int_E \mathcal{P} \circ x \, \mathrm{d}\boldsymbol{P} = 0$. Hence, $\mathcal{P} \circ \hat{\nu} \ll \hat{\boldsymbol{P}}$. Let there exists $f : \Omega \to \mathcal{Y}$ that is the Radon–Nikodym derivative of $\hat{\nu}$ with respect to $\hat{\boldsymbol{P}}$, where f is $\hat{\mathcal{B}}$-measurable. The function f is unique as described in Proposition 11.167, $f = \frac{\mathrm{d}\hat{\nu}}{\mathrm{d}\hat{\boldsymbol{P}}}$ a.e. in $\hat{\boldsymbol{\Omega}}$, and $f \in \bar{\mathrm{L}}_1(\hat{\boldsymbol{\Omega}}, \mathcal{Y})$. Then, $[f] \in \mathrm{L}_1(\hat{\boldsymbol{\Omega}}, \mathcal{Y})$ is said to the *conditional expectation of x given $\hat{\mathcal{B}}$* and is denoted by $\boldsymbol{E}(x|\hat{\mathcal{B}})$. $f \in \boldsymbol{E}(x|\hat{\mathcal{B}})$ is a

$\mathcal{Y}$-valued random variable and is said to be a version of the conditional expectation. Then,

$$\int_E x \,\mathrm{d}\boldsymbol{P} = \nu(E) = \hat{\nu}(E) = \int_E f \,\mathrm{d}\hat{\boldsymbol{P}} = \int_E f \,\mathrm{d}\boldsymbol{P}; \quad \forall E \in \hat{\mathcal{B}} \tag{14.2a}$$

$$\int_E \mathcal{P} \circ f \,\mathrm{d}\hat{\boldsymbol{P}} = \mathcal{P} \circ \hat{\nu}(E) = \int_E \mathcal{P} \circ f \,\mathrm{d}\boldsymbol{P} \leq \mathcal{P} \circ \nu(E) = \int_E \mathcal{P} \circ x \,\mathrm{d}\boldsymbol{P} < \infty; \quad \forall E \in \hat{\mathcal{B}} \tag{14.2b}$$

where the last equality in (14.2a) follows from Proposition 11.72, the first equality in (14.2b) follows from Definition 11.166, the second equality in (14.2b) follows from Proposition 11.72, the first inequality in (14.2b) follows from Definition 11.99, and the last equality in (14.2b) follows from Proposition 11.116.

In the above, if $\mathcal{Y}$ is a separable reflexive Banach space with $\mathcal{Y}^*$ being separable, then $\boldsymbol{E}(x|\hat{\mathcal{B}}) \in \mathrm{L}_1(\hat{\boldsymbol{\Omega}}, \mathcal{Y})$ exists by Radon–Nikodym Theorem 11.171. ◇

Proposition 14.8 *Let $\boldsymbol{\Omega} := (\Omega, \mathcal{B}, \boldsymbol{P})$ be a probability measure space, $\mathcal{Y}$ be a separable Banach space over $\mathbb{K}$, $x \in \bar{\mathrm{L}}_1(\boldsymbol{\Omega}, \mathcal{Y})$, $\hat{\mathcal{B}} \subseteq \mathcal{B}$ be a σ-algebra on Ω, $\hat{\boldsymbol{\Omega}} := (\Omega, \hat{\mathcal{B}}, \hat{\boldsymbol{P}} := \boldsymbol{P}|_{\hat{\mathcal{B}}})$, and $f \in \bar{\mathrm{L}}_1(\hat{\boldsymbol{\Omega}}, \mathcal{Y})$. Assume that f satisfies (14.2a), and then $\boldsymbol{E}(x|\hat{\mathcal{B}})$ exists and $f \in \boldsymbol{E}(x|\hat{\mathcal{B}})$.*

Proof Let ν and $\hat{\nu}$ be as defined in Definition 14.7. By Proposition 11.116, we may define a $\mathcal{Y}$-valued measure $\bar{\nu}$ with kernel f on $(\Omega, \hat{\mathcal{B}})$. $\bar{\nu}$ is finite since $f \in \bar{\mathrm{L}}_1(\hat{\boldsymbol{\Omega}}, \mathcal{Y})$. By (14.2a), $\bar{\nu}(E) = \hat{\nu}(E)$, $\forall E \in \hat{\mathcal{B}}$. Therefore, $\bar{\nu} = \hat{\nu}$. By Definition 11.166 and Proposition 11.116, $f = \frac{\mathrm{d}\bar{\nu}}{\mathrm{d}\hat{\boldsymbol{P}}}$ a.e. in $\hat{\boldsymbol{\Omega}}$. Hence, $\boldsymbol{E}(x|\hat{\mathcal{B}})$ exists and $f \in \boldsymbol{E}(x|\hat{\mathcal{B}})$. This completes the proof of the proposition. □

Proposition 14.9 *Let $\boldsymbol{\Omega} := (\Omega, \mathcal{B}, \boldsymbol{P})$ be a probability measure space, $\mathcal{Y}$ be a separable Banach space over $\mathbb{K}$, $\mathcal{Z}$ be a separable Banach space over $\mathbb{K}$, $\mathcal{W} \subseteq \mathrm{B}(\mathcal{Y}, \mathcal{Z})$ be a separable subspace, $\hat{\mathcal{B}} \subseteq \mathcal{B}$ and $\tilde{\mathcal{B}} \subseteq \mathcal{B}$ be independent σ-algebras on Ω, $\hat{\boldsymbol{\Omega}} := (\Omega, \hat{\mathcal{B}}, \hat{\boldsymbol{P}} := \boldsymbol{P}|_{\hat{\mathcal{B}}})$ and $\tilde{\boldsymbol{\Omega}} := (\Omega, \tilde{\mathcal{B}}, \tilde{\boldsymbol{P}} := \boldsymbol{P}|_{\tilde{\mathcal{B}}})$ be probability measure spaces, $x \in \bar{\mathrm{L}}_1(\hat{\boldsymbol{\Omega}}, \mathcal{Y})$ be $\hat{\mathcal{B}}$-measurable, $A \in \bar{\mathrm{L}}_1(\tilde{\boldsymbol{\Omega}}, \mathcal{W})$ be $\tilde{\mathcal{B}}$-measurable. Then, $Ax \in \bar{\mathrm{L}}_1(\boldsymbol{\Omega}, \mathcal{Z})$ and $\boldsymbol{E}(Ax) = \boldsymbol{E}(A)\boldsymbol{E}(x)$.*

Proof First, we will consider the special case where A and x are simple functions. Let A admit the canonical representation $A = \sum_{i=1}^{n} A_i \chi_{E_i,\Omega}$, where $n \in \mathbb{Z}_+$, $A_1, \ldots, A_n \in \mathcal{W}$ are distinct and none equal to $\vartheta_{\mathrm{B}(\mathcal{Y},\mathcal{Z})}$, and $E_1, \ldots, E_n \in \tilde{\mathcal{B}}$ are nonempty and pairwise disjoint. Let x admit the canonical representation $x = \sum_{j=1}^{m} x_j \chi_{F_j,\Omega}$, where $m \in \mathbb{Z}_+$, $x_1, \ldots, x_m \in \mathcal{Y}$ are distinct and none equal to

$\vartheta_{\mathcal{Y}}$ and $F_1, \ldots, F_m \in \hat{\mathcal{B}}$ are nonempty and pairwise disjoint. Then, clearly, $Ax \in \bar{L}_1(\boldsymbol{\Omega}, \mathcal{Z})$ is $\mathcal{B}$-measurable. This leads to

$$E(Ax) = \int_{\Omega} \sum_{i=1}^{n} \sum_{j=1}^{m} A_i x_j \chi_{E_i,\Omega} \chi_{F_j,\Omega} \, \mathrm{d}P = \sum_{i=1}^{n} \sum_{j=1}^{m} A_i x_j \int_{\Omega} \chi_{E_i,\Omega} \chi_{F_j,\Omega} \, \mathrm{d}P$$

$$= \sum_{i=1}^{n} \sum_{j=1}^{m} A_i x_j P(E_i \cap F_j) = \sum_{i=1}^{n} \sum_{j=1}^{m} A_i x_j P(E_i) P(F_j)$$

$$= \sum_{i=1}^{n} A_i P(E_i) \sum_{j=1}^{m} x_j P(F_j) = E(A) E(x)$$

where the second equality follows from Proposition 11.92, the third equality follows from Proposition 11.75, the fourth equality follows from the independence assumption, and the last equality follows from Proposition 11.75. Hence, the result holds in the special case.

Next, we consider the special case that A and x are nonnegative real-valued random variables. Then, $\mathcal{Y} = \mathbb{R} = \mathcal{Z}$. By Proposition 11.67, there exists a sequence of simple functions $(\phi_n)_{n=1}^{\infty}$ with $\phi_n : \Omega \to [0, \infty)$ being $\tilde{\mathcal{B}}$-measurable, $\forall n \in \mathbb{N}$, such that $0 \leq \phi_n(\omega) \leq \phi_{n+1}(\omega) \leq A(\omega)$, $\forall n \in \mathbb{N}$ and $\forall \omega \in \boldsymbol{\Omega}$, and $\lim_{n\in\mathbb{N}} \phi_n(\omega) = A(\omega)$, $\forall \omega \in \boldsymbol{\Omega}$. By Proposition 11.67, there exists a sequence of simple functions $(\psi_n)_{n=1}^{\infty}$ with $\psi_n : \Omega \to [0, \infty)$ being $\hat{\mathcal{B}}$-measurable, $\forall n \in \mathbb{N}$, such that $0 \leq \psi_n(\omega) \leq \psi_{n+1}(\omega) \leq x(\omega)$, $\forall n \in \mathbb{N}$ and $\forall \omega \in \boldsymbol{\Omega}$, and $\lim_{n\in\mathbb{N}} \psi_n(\omega) = x(\omega)$, $\forall \omega \in \boldsymbol{\Omega}$. Then, we have $0 \leq \phi_n(\omega)\psi_n(\omega) \leq \phi_{n+1}(\omega)\psi_{n+1}(\omega) \leq A(\omega)x(\omega)$, $\forall \omega \in \boldsymbol{\Omega}$, $\forall n \in \mathbb{N}$, and $\lim_{n\in\mathbb{N}} \phi_n(\omega)\psi_n(\omega) = A(\omega)x(\omega)$. By Monotone Convergence Theorem 11.81, we have $E(Ax) = \int_{\boldsymbol{\Omega}} A(\omega)x(\omega) \, \mathrm{d}P(\omega) = \lim_{n\in\mathbb{N}} \int_{\boldsymbol{\Omega}} \phi_n(\omega)\psi_n(\omega) \, \mathrm{d}P(\omega) = \lim_{n\in\mathbb{N}} E(\phi_n \psi_n) = \lim_{n\in\mathbb{N}} E(\phi_n)E(\psi_n) = E(A)E(x)$, where the first equality follows from Definition 14.1, the second equality follows from Monotone Convergence Theorem 11.81, the third equality follows from Definition 14.1, the fourth equality follows from the previous special case, and the last equality follows from Monotone Convergence Theorem 11.81. Hence, the result holds in this special case.

Now, consider the general case. Since $A \in \bar{L}_1(\tilde{\boldsymbol{\Omega}}, \mathcal{W})$, by Propositions 11.38 and 7.21, $\mathcal{P} \circ A$ is $\tilde{\mathcal{B}}$-measurable. Since $x \in \bar{L}_1(\hat{\boldsymbol{\Omega}}, \mathcal{Y})$, again by Propositions 11.38 and 7.21, $\mathcal{P} \circ x$ is $\hat{\mathcal{B}}$-measurable. By the second special case, we have $E((\mathcal{P} \circ A)(\mathcal{P} \circ x)) = E(\mathcal{P} \circ A) E(\mathcal{P} \circ x) < \infty$. Hence, $g := (\mathcal{P} \circ A)(\mathcal{P} \circ x)$ is integrable over $\boldsymbol{\Omega}$.

Since $A \in \bar{L}_1(\tilde{\boldsymbol{\Omega}}, \mathcal{W})$, by Proposition 11.66, there exists a sequence of simple functions $(\phi_n)_{n=1}^{\infty}$ with $\phi_n : \Omega \to \mathcal{W}$ being $\tilde{\mathcal{B}}$-measurable, $\forall n \in \mathbb{N}$, such that $\lim_{n\in\mathbb{N}} \phi_n = A$ a.e. in $\tilde{\boldsymbol{\Omega}}$ and $\|\phi_n(\omega)\|_{\mathrm{B}(\mathcal{Y},\mathcal{Z})} \leq \|A(\omega)\|_{\mathrm{B}(\mathcal{Y},\mathcal{Z})}$, $\forall \omega \in \Omega$, $\forall n \in \mathbb{N}$. Since $x \in \bar{L}_1(\hat{\boldsymbol{\Omega}}, \mathcal{Y})$, again by Proposition 11.66, there exists a sequence of simple functions $(\psi_n)_{n=1}^{\infty}$ with $\psi_n : \Omega \to \mathcal{Y}$ being $\hat{\mathcal{B}}$-measurable, $\forall n \in \mathbb{N}$, such that $\lim_{n\in\mathbb{N}} \psi_n = x$ a.e. in $\hat{\boldsymbol{\Omega}}$ and $\|\psi_n(\omega)\|_{\mathcal{Y}} \leq \|x(\omega)\|_{\mathcal{Y}}$, $\forall \omega \in \Omega$, $\forall n \in \mathbb{N}$. By Propositions 7.65, 11.39, and 11.38, Ax is $\mathcal{B}$-measurable and $\phi_n \psi_n$

is $\mathcal{B}$-measurable, $\forall n \in \mathbb{N}$. Clearly, $\lim_{n\in\mathbb{N}} \phi_n = A$ a.e. in $\boldsymbol{\Omega}$ and $\lim_{n\in\mathbb{N}} \psi_n = x$ a.e. in $\boldsymbol{\Omega}$. This implies that, by Propositions 11.52 and 11.53, $\lim_{n\in\mathbb{N}} \phi_n\psi_n = Ax$ a.e. in $\boldsymbol{\Omega}$. $\forall \omega \in \Omega$, $\forall n \in \mathbb{N}$, $\|\phi_n(\omega)\psi_n(\omega)\|_{\mathcal{Z}} \leq \|\phi_n(\omega)\|_{\mathrm{B}(\mathcal{Y},\mathcal{Z})}\|\psi_n(\omega)\|_{\mathcal{Y}} \leq \|A(\omega)\|_{\mathrm{B}(\mathcal{Y},\mathcal{Z})}\|x(\omega)\|_{\mathcal{Y}} = g(\omega)$. Then, we have

$$\begin{aligned}
\boldsymbol{E}(Ax) &= \int_{\Omega} Ax \,\mathrm{d}\boldsymbol{P} = \lim_{n\in\mathbb{N}} \int_{\Omega} \phi_n\psi_n \,\mathrm{d}\boldsymbol{P} = \lim_{n\in\mathbb{N}} \boldsymbol{E}(\phi_n\psi_n) \\
&= \lim_{n\in\mathbb{N}} \boldsymbol{E}(\phi_n)\boldsymbol{E}(\psi_n) = \lim_{n\in\mathbb{N}} \int_{\Omega} \phi_n \,\mathrm{d}\boldsymbol{P} \int_{\Omega} \psi_n \,\mathrm{d}\boldsymbol{P} \\
&= \lim_{n\in\mathbb{N}} \int_{\Omega} \phi_n \,\mathrm{d}\tilde{\boldsymbol{P}} \int_{\Omega} \psi_n \,\mathrm{d}\hat{\boldsymbol{P}} = \int_{\Omega} A \,\mathrm{d}\tilde{\boldsymbol{P}} \int_{\Omega} x \,\mathrm{d}\hat{\boldsymbol{P}} = \int_{\Omega} A \,\mathrm{d}\boldsymbol{P} \int_{\Omega} x \,\mathrm{d}\boldsymbol{P} \\
&= \boldsymbol{E}(A)\boldsymbol{E}(x)
\end{aligned}$$

where the second equality follows from Lebesgue Dominated Convergence Theorem 11.91, the fourth equality follows from the first special case; the sixth equality follows from Proposition 11.72, the seventh equality follows from Lebesgue Dominated Convergence Theorem 11.91, and the eighth equality follows from Proposition 11.72.

This completes the proof of the proposition. □

Proposition 14.10 *Let $\boldsymbol{\Omega} := (\Omega, \mathcal{B}, \boldsymbol{P})$ be a probability measure space, $\hat{\mathcal{B}} \subseteq \mathcal{B}$ and $\tilde{\mathcal{B}} \subseteq \mathcal{B}$ be σ-algebras on Ω, $\hat{\mathcal{I}} \subseteq \hat{\mathcal{B}}$ be a π-system on Ω, and $\tilde{\mathcal{I}} \subseteq \tilde{\mathcal{B}}$ be a π-system on Ω. Assume that $\sigma(\hat{\mathcal{I}}) = \hat{\mathcal{B}}$, $\sigma(\tilde{\mathcal{I}}) = \tilde{\mathcal{B}}$, and $\forall \hat{E} \in \hat{\mathcal{I}}$, $\forall \tilde{E} \in \tilde{\mathcal{I}}$, we have $\boldsymbol{P}(\hat{E} \cap \tilde{E}) = \boldsymbol{P}(\hat{E})\boldsymbol{P}(\tilde{E})$. Then, $\hat{\mathcal{B}}$ and $\tilde{\mathcal{B}}$ are independent σ-algebras.*

Proof Define $\mathcal{F} := \{\hat{E} \in \hat{\mathcal{B}} \mid \forall \tilde{E} \in \tilde{\mathcal{I}}, \boldsymbol{P}(\hat{E} \cap \tilde{E}) = \boldsymbol{P}(\hat{E})\boldsymbol{P}(\tilde{E})\}$. Clearly, $\hat{\mathcal{I}} \subseteq \mathcal{F}$. We will show that $\mathcal{F}$ is a monotone class on Ω. Then, by Monotone Class Lemma 12.19, $\mathcal{F} = \sigma(\hat{\mathcal{I}}) = \hat{\mathcal{B}}$.

(i) $\emptyset, \Omega \in \hat{\mathcal{I}} \subseteq \mathcal{F}$. (ii) Fix any $E_1, E_2 \in \mathcal{F}$ with $E_1 \subseteq E_2$. $\forall \tilde{E} \in \tilde{\mathcal{I}}$, we have $\boldsymbol{P}((E_2 \setminus E_1) \cap \tilde{E}) = \boldsymbol{P}((E_2 \cap \tilde{E}) \setminus (E_1 \cap \tilde{E})) = \boldsymbol{P}(E_2 \cap \tilde{E}) - \boldsymbol{P}(E_1 \cap \tilde{E}) = \boldsymbol{P}(E_2)\boldsymbol{P}(\tilde{E}) - \boldsymbol{P}(E_1)\boldsymbol{P}(\tilde{E}) = \boldsymbol{P}(E_2 \setminus E_1)\boldsymbol{P}(\tilde{E})$, where the first equality follows from the fact $(E_2 \setminus E_1) \cap \tilde{E} = (E_2 \cap \tilde{E}) \setminus (E_1 \cap \tilde{E})$, the second equality follows from the fact that $E_1 \cap \tilde{E} \subseteq E_2 \cap \tilde{E}$ and both sets are in $\mathcal{B}$, the third equality follows from the fact that $E_1, E_2 \in \mathcal{F}$, and the fourth equality follows from the fact that $E_1 \subseteq E_2$. Hence, by the arbitrariness of $\tilde{E} \in \tilde{\mathcal{I}}$, we have $E_2 \setminus E_1 \in \mathcal{F}$. (iii) Fix any $(E_i)_{i=1}^{\infty} \subseteq \mathcal{F}$ with $E_i \subseteq E_{i+1}$, $\forall i \in \mathbb{N}$. $\forall \tilde{E} \in \tilde{\mathcal{I}}$, we have $\boldsymbol{P}((\bigcup_{i=1}^{\infty} E_i) \cap \tilde{E}) = \boldsymbol{P}((\bigcup_{i=2}^{\infty}(E_i \setminus E_{i-1}) \cup E_1) \cap \tilde{E}) = \boldsymbol{P}((\bigcup_{i=2}^{\infty}((E_i \setminus E_{i-1}) \cap \tilde{E})) \cup (E_1 \cap \tilde{E})) = \sum_{i=2}^{\infty} \boldsymbol{P}((E_i \setminus E_{i-1}) \cap \tilde{E}) + \boldsymbol{P}(E_1 \cap \tilde{E}) = \sum_{i=2}^{\infty} \boldsymbol{P}(E_i \setminus E_{i-1})\boldsymbol{P}(\tilde{E}) + \boldsymbol{P}(E_1)\boldsymbol{P}(\tilde{E}) = \boldsymbol{P}(\tilde{E})(\sum_{i=2}^{\infty} \boldsymbol{P}(E_i \setminus E_{i-1}) + \boldsymbol{P}(E_1)) = \boldsymbol{P}(\tilde{E})\boldsymbol{P}(\bigcup_{i=1}^{\infty} E_i)$, where the first two equalities follow from Proposition 2.5, the third equality follows from countable additivity of measure, the fourth equality follows from (ii), and the last equality follows from countable additivity of measure. By the arbitrariness of $\tilde{E} \in \tilde{\mathcal{I}}$, we have $\bigcup_{i=1}^{\infty} E_i \in \mathcal{F}$. This proves that $\mathcal{F}$ is a monotone class on Ω.

Define $\mathcal{G} := \{\tilde{E} \in \tilde{\mathcal{B}} \mid \forall \hat{E} \in \hat{\mathcal{B}}, \boldsymbol{P}(\hat{E} \cap \tilde{E}) = \boldsymbol{P}(\hat{E})\boldsymbol{P}(\tilde{E})\}$. Clearly, $\tilde{\mathcal{I}} \subseteq \mathcal{G}$. We will show that $\mathcal{G}$ is a monotone class on Ω. Then, by Monotone Class Lemma 12.19, $\mathcal{G} = \sigma(\tilde{\mathcal{I}}) = \tilde{\mathcal{B}}$.

(i) $\emptyset, \Omega \in \tilde{\mathcal{I}} \subseteq \mathcal{G}$. (ii) Fix any $E_1, E_2 \in \mathcal{G}$ with $E_1 \subseteq E_2$. $\forall \hat{E} \in \hat{\mathcal{B}}$, we have $\boldsymbol{P}((E_2 \setminus E_1) \cap \hat{E}) = \boldsymbol{P}((E_2 \cap \hat{E}) \setminus (E_1 \cap \hat{E})) = \boldsymbol{P}(E_2 \cap \hat{E}) - \boldsymbol{P}(E_1 \cap \hat{E}) = \boldsymbol{P}(E_2)\boldsymbol{P}(\hat{E}) - \boldsymbol{P}(E_1)\boldsymbol{P}(\hat{E}) = \boldsymbol{P}(E_2 \setminus E_1)\boldsymbol{P}(\hat{E})$, where the first equality follows from the fact $(E_2 \setminus E_1) \cap \hat{E} = (E_2 \cap \hat{E}) \setminus (E_1 \cap \hat{E})$, the second equality follows from the fact that $E_1 \cap \hat{E} \subseteq E_2 \cap \hat{E}$ and both sets are in $\mathcal{B}$, the third equality follows from the fact that $E_1, E_2 \in \mathcal{G}$, and the fourth equality follows from the fact that $E_1 \subseteq E_2$. Hence, by the arbitrariness of $\hat{E} \in \hat{\mathcal{B}}$, we have $E_2 \setminus E_1 \in \mathcal{G}$. (iii) Fix any $(E_i)_{i=1}^{\infty} \subseteq \mathcal{G}$ with $E_i \subseteq E_{i+1}$, $\forall i \in \mathbb{N}$. $\forall \hat{E} \in \hat{\mathcal{B}}$, we have $\boldsymbol{P}((\bigcup_{i=1}^{\infty} E_i) \cap \hat{E}) = \boldsymbol{P}((\bigcup_{i=2}^{\infty}(E_i \setminus E_{i-1}) \cup E_1) \cap \hat{E}) = \boldsymbol{P}((\bigcup_{i=2}^{\infty}((E_i \setminus E_{i-1}) \cap \hat{E})) \cup (E_1 \cap \hat{E})) = \sum_{i=2}^{\infty} \boldsymbol{P}((E_i \setminus E_{i-1}) \cap \hat{E}) + \boldsymbol{P}(E_1 \cap \hat{E}) = \sum_{i=2}^{\infty} \boldsymbol{P}(E_i \setminus E_{i-1})\boldsymbol{P}(\hat{E}) + \boldsymbol{P}(E_1)\boldsymbol{P}(\hat{E}) = \boldsymbol{P}(\hat{E})(\sum_{i=2}^{\infty} \boldsymbol{P}(E_i \setminus E_{i-1}) + \boldsymbol{P}(E_1)) = \boldsymbol{P}(\hat{E})\boldsymbol{P}(\bigcup_{i=1}^{\infty} E_i)$, where the first two equalities follow from Proposition 2.5, the third equality follows from countable additivity of measure, the fourth equality follows from (ii), and the last equality follows from countable additivity of measure. By the arbitrariness of $\hat{E} \in \hat{\mathcal{B}}$, we have $\bigcup_{i=1}^{\infty} E_i \in \mathcal{G}$. This proves that $\mathcal{G}$ is a monotone class on Ω.

This proves that $\forall \hat{E} \in \hat{\mathcal{B}}$ and $\forall \tilde{E} \in \tilde{\mathcal{B}}$, we have $\boldsymbol{P}(\hat{E} \cap \tilde{E}) = \boldsymbol{P}(\hat{E})\boldsymbol{P}(\tilde{E})$. Hence, $\hat{\mathcal{B}}$ and $\tilde{\mathcal{B}}$ are independent σ-algebras.

This completes the proof of the proposition. □

Proposition 14.11 *Let* $\boldsymbol{\Omega} := (\Omega, \mathcal{B}, \boldsymbol{P})$ *be a probability measure space,* $\mathcal{Y}$ *be a separable Banach space over* $\mathbb{K}$, $x \in \bar{\mathrm{L}}_1(\boldsymbol{\Omega}, \mathcal{Y})$, $(x_n)_{n=1}^{\infty} \subseteq \bar{\mathrm{L}}_1(\boldsymbol{\Omega}, \mathcal{Y})$, $\hat{\mathcal{B}} \subseteq \mathcal{B}$ *be a* σ*-algebra on* Ω, $\hat{\boldsymbol{\Omega}} := (\Omega, \hat{\mathcal{B}}, \hat{\boldsymbol{P}} := \boldsymbol{P}|_{\hat{\mathcal{B}}})$, $f \in \boldsymbol{E}(x|\hat{\mathcal{B}})$, *and* $f_n \in \boldsymbol{E}(x_n|\hat{\mathcal{B}})$, $\forall n \in \mathbb{N}$. *Then, the following statements hold:*

(a) $\boldsymbol{E}(f) = \boldsymbol{E}(x)$.
(b) *Let* x *be* $\hat{\mathcal{B}}$*-measurable. Then* $[x] = \boldsymbol{E}(x|\hat{\mathcal{B}}) \in \mathrm{L}_1(\hat{\boldsymbol{\Omega}}, \mathcal{Y})$.
(c) **(Linearity)** *Let* $\alpha, \beta \in \mathbb{K}$. *Then* $\boldsymbol{E}(\alpha x_1 + \beta x_2|\hat{\mathcal{B}}) = \alpha\boldsymbol{E}(x_1|\hat{\mathcal{B}}) + \beta\boldsymbol{E}(x_2|\hat{\mathcal{B}})$.
(d) *Let* $F \subseteq \mathcal{Y}$ *be nonempty closed convex set, and* $x : \Omega \to F$. *Then,* $f(\omega) \in F$ *a.e.* $\omega \in \hat{\boldsymbol{\Omega}}$.
(e) **(Monotone Convergence)** *Let* $\mathcal{Y} = \mathbb{R}$, $0 \leq x_n(\omega) \leq x_{n+1}(\omega)$, $\forall \omega \in \Omega$, $\forall n \in \mathbb{N}$, *and* $\lim_{n\in\mathbb{N}} x_n(\omega) = x(\omega)$ *a.e.* $\omega \in \boldsymbol{\Omega}$. *Then* $f_n \leq f_{n+1}$ *a.e. in* $\hat{\boldsymbol{\Omega}}$ *and* $\lim_{n\in\mathbb{N}} f_n = f$ *a.e. in* $\hat{\boldsymbol{\Omega}}$.
(f) **(Fatou)** *Let* $\mathcal{Y} = \mathbb{R}$, $0 \leq x_n(\omega) < \infty$, $\forall \omega \in \Omega$, $\forall n \in \mathbb{N}$, *and* $\bar{f} \in \boldsymbol{E}(\liminf_{n\in\mathbb{N}} x_n|\hat{\mathcal{B}})$. *Then* $\bar{f} \leq \liminf_{n\in\mathbb{N}} f_n$ *a.e. in* $\hat{\boldsymbol{\Omega}}$.
(g) **(Dominated Convergence)** *Let* $g : \Omega \to [0, \infty) \subset \mathbb{R}$ *be integrable over* $\boldsymbol{\Omega}$ *that satisfies*

  (i) $\lim_{n\in\mathbb{N}} x_n = \bar{x}$ *a.e. in* $\boldsymbol{\Omega}$, *where* $\bar{x} : \Omega \to \mathcal{Y}$ *is* $\mathcal{B}$*-measurable.*
  (ii) $\mathcal{P} \circ x_n(\omega) \leq g(\omega)$, $\forall \omega \in \Omega$, $\forall n \in \mathbb{N}$.

  Then, $\bar{x} \in \bar{\mathrm{L}}_1(\boldsymbol{\Omega}, \mathcal{Y})$, $\boldsymbol{E}(\bar{x}|\hat{\mathcal{B}})$ *exists,* $\lim_{n\in\mathbb{N}} f_n = \bar{f}$ *a.e. in* $\hat{\boldsymbol{\Omega}}$, $\mathcal{P} \circ \bar{f} \leq \bar{g}$ *a.e. in* $\hat{\boldsymbol{\Omega}}$, $\lim_{n\in\mathbb{N}} \int_{\Omega} \mathcal{P} \circ f_n \,\mathrm{d}\boldsymbol{P} = \int_{\Omega} \mathcal{P} \circ \bar{f} \,\mathrm{d}\boldsymbol{P} < \infty$, *and* $\lim_{n\in\mathbb{N}} \int_{\Omega} \mathcal{P} \circ (f_n - \bar{f}) \,\mathrm{d}\boldsymbol{P} = 0$, *where* $\bar{f} \in \boldsymbol{E}(\bar{x}|\hat{\mathcal{B}})$ *and* $\bar{g} \in \boldsymbol{E}(g|\hat{\mathcal{B}})$.

(h) *Let $\mathcal{Z}$ be a separable Banach space over $\mathbb{K}$, $\mathcal{W} \subseteq \mathrm{B}(\mathcal{Y}, \mathcal{Z})$ be a separable subspace, $A : \Omega \to \mathcal{W}$ be $\hat{\mathcal{B}}$-measurable, and $\|A(\cdot)\|_{\mathrm{B}(\mathcal{Y},\mathcal{Z})}\|x(\cdot)\|_{\mathcal{Y}} : \Omega \to [0, \infty) \subset \mathbb{R}$ be integrable over $\boldsymbol{\Omega}$. Then, $\boldsymbol{E}(Ax|\hat{\mathcal{B}})$ exists and $Af \in \boldsymbol{E}(Ax|\hat{\mathcal{B}})$.*

(i) **(Jensen's Inequality)** *Let $\mathbb{K} = \mathbb{R}$, $F \subseteq \mathcal{Y}$ be a nonempty closed convex set, $x : \Omega \to F$, $c : F \to \mathbb{R}$ be a convex functional with its epigraph $[c, F]$ being closed, $c \circ x : \Omega \to \mathbb{R}$ being absolutely integrable over $\boldsymbol{\Omega}$, and $\bar{f} \in \boldsymbol{E}(c \circ x|\hat{\mathcal{B}})$. Then, $c(f(\omega)) \leq \bar{f}(\omega)$ a.e. $\omega \in \hat{\boldsymbol{\Omega}}$.*

(j) **(Tower Property)** *Let $\tilde{\mathcal{B}} \subseteq \hat{\mathcal{B}}$ be a σ-algebra on Ω. Then $\boldsymbol{E}(f|\tilde{\mathcal{B}}) = \boldsymbol{E}(x|\tilde{\mathcal{B}})$, whenever one of them exists.*

(k) **(Role of Independence)** *Let $\tilde{\mathcal{B}} \subseteq \mathcal{B}$ be a σ-algebra on Ω, $\tilde{\mathcal{B}}$ and $\sigma(\hat{\mathcal{B}} \cup \sigma(x))$ are independent. Then $f \in \boldsymbol{E}(x|\sigma(\tilde{\mathcal{B}} \cup \hat{\mathcal{B}}))$.*

Proof (a) $\boldsymbol{E}(f) = \int_\Omega f \,\mathrm{d}\boldsymbol{P} = \int_\Omega f \,\mathrm{d}\hat{\boldsymbol{P}} = \int_\Omega x \,\mathrm{d}\boldsymbol{P} = \boldsymbol{E}(x)$, where the first equality follows from the definition of $\boldsymbol{E}$, the second equality follows from f is $\hat{\mathcal{B}}$-measurable and Proposition 11.72, the third equality follows from the definition of $\boldsymbol{E}(x|\hat{\mathcal{B}})$ and the fact $\Omega \in \hat{\mathcal{B}}$, and the last equality follows from the definition of $\boldsymbol{E}$.

(b) Let ν and $\hat{\nu}$ be the $\mathcal{Y}$-valued measures defined in Definition 14.7. Clearly $x \in \bar{\mathrm{L}}_1(\hat{\boldsymbol{\Omega}}, \mathcal{Y})$ since x is $\hat{\mathcal{B}}$-measurable. $\forall E \in \hat{\mathcal{B}}$, we have $\nu(E) = \int_E x \,\mathrm{d}\boldsymbol{P} = \int_E x \,\mathrm{d}\hat{\boldsymbol{P}} = \hat{\nu}(E)$, where the second equality follows from the fact that x is $\hat{\mathcal{B}}$-measurable and Proposition 11.72. By Proposition 14.8, we have $x \in \boldsymbol{E}(x|\hat{\mathcal{B}})$.

(c) This is a direct consequence of (iii) and (iv) of Proposition 11.168.

(d) By the assumption, $x(\omega) \in F$, $\forall \omega \in \Omega$. By the first three paragraphs of the proof for Jensen's Inequality Theorem 11.98, which clearly applies to the case when $\mathcal{Y}$ is a complex Banach space as well, we have $\forall E \in \hat{\mathcal{B}}$ with $\hat{\boldsymbol{P}}(E) > 0$, $\frac{1}{\hat{\boldsymbol{P}}(E)} \int_E f \,\mathrm{d}\hat{\boldsymbol{P}} = \frac{1}{\boldsymbol{P}(E)} \int_E x \,\mathrm{d}\boldsymbol{P} \in F$, where the equality follows from $f \in \boldsymbol{E}(x|\hat{\mathcal{B}})$. Hence, $\frac{1}{\hat{\boldsymbol{P}}(E)} \int_E f \,\mathrm{d}\hat{\boldsymbol{P}} \in F$, $\forall E \in \hat{\mathcal{B}}$ with $\hat{\boldsymbol{P}}(E) > 0$. Suppose that the result does not hold. Then, we have $E := \{\omega \in \Omega \mid f(\omega) \notin F\} \in \hat{\mathcal{B}}$ and $\hat{\boldsymbol{P}}(E) > 0$. Note that $\mathcal{Y} \setminus F$ is an open set. By Proposition 4.38, $\mathcal{Y} \setminus F$ is separable. By Proposition 4.4, $\mathcal{Y} \setminus F$ is second countable. Then, $\exists\, (B_n)_{n=1}^\infty$ such that B_n is an open ball in $\mathcal{Y}$, $\overline{B_n} \subseteq \mathcal{Y} \setminus F$, $\forall n \in \mathbb{N}$, and $\mathcal{Y} \setminus F = \bigcup_{n=1}^\infty B_n$. Let $E_n := f_{\mathrm{inv}}(B_n) \in \hat{\mathcal{B}}$. Then, $\hat{\boldsymbol{P}}(E) = \hat{\boldsymbol{P}}(\bigcup_{n=1}^\infty E_n) \leq \sum_{n=1}^\infty \hat{\boldsymbol{P}}(E_n)$. Since $\hat{\boldsymbol{P}}(E) > 0$, we have $\exists n_0 \in \mathbb{N}$ such that $\hat{\boldsymbol{P}}(E_{n_0}) > 0$. Then, $f(\omega) \in B_{n_0}$, $\forall \omega \in E_{n_0}$. By the first three paragraphs of the proof for Jensen's Inequality Theorem 11.98 (applying to possibly complex Banach spaces), $\frac{1}{\hat{\boldsymbol{P}}(E_{n_0})} \int_{E_{n_0}} f \,\mathrm{d}\hat{\boldsymbol{P}} \in \overline{B_{n_0}}$ since B_{n_0} is a nonempty convex set. This contradicts our earlier conclusion that $\frac{1}{\hat{\boldsymbol{P}}(E_{n_0})} \int_{E_{n_0}} f \,\mathrm{d}\hat{\boldsymbol{P}} \in F \subseteq \mathcal{Y} \setminus \overline{B_{n_0}}$. Hence, the hypothesis does not hold. Hence, (d) holds.

(e) By (c), we have $f_{n+1} - f_n \in \boldsymbol{E}(x_{n+1} - x_n|\hat{\mathcal{B}})$, $\forall n \in \mathbb{N}$. Since $x_{n+1}(\omega) - x_n(\omega) \geq 0$, $\forall \omega \in \Omega$, $\forall n \in \mathbb{N}$, then, by (d), $f_{n+1} - f_n \geq 0$ a.e. in $\hat{\boldsymbol{\Omega}}$, $\forall n \in \mathbb{N}$. Then, $f_n \leq f_{n+1}$ a.e. in $\hat{\boldsymbol{\Omega}}$. $\int_\Omega \mathcal{P} \circ (f - f_n) \,\mathrm{d}\hat{\boldsymbol{P}} = \int_\Omega \mathcal{P} \circ (f_n - f) \,\mathrm{d}\boldsymbol{P} \leq \int_\Omega \mathcal{P} \circ (x_n - x) \,\mathrm{d}\boldsymbol{P} = \int_\Omega (x - x_n) \,\mathrm{d}\boldsymbol{P}$, where the first equality follows from Proposition 11.72, and the inequality follows from (14.2b). By Monotone Convergence Theorem 11.81, $\int_\Omega x \,\mathrm{d}\boldsymbol{P} = \lim_{n \in \mathbb{N}} \int_\Omega x_n \,\mathrm{d}\boldsymbol{P}$. By Definition 14.7, $f_n, f \in \bar{\mathrm{L}}_1(\hat{\boldsymbol{\Omega}}, \mathbb{R})$, $\forall n \in \mathbb{N}$.

Then, we have $\lim_{n\in\mathbb{N}} f_n \doteq f$ in $\bar{L}_1(\hat{\boldsymbol{\Omega}}, \mathbb{R})$. Hence, $f = \lim_{n\in\mathbb{N}} f_n$ a.e. in $\hat{\boldsymbol{\Omega}}$, since $\lim_{n\in\mathbb{N}} f_n(\omega)$ exists almost everywhere in $\boldsymbol{\Omega}$.

(f) $\forall n \in \mathbb{N}$, define $\tilde{x}_n : \Omega \to \mathbb{R}$ by $\tilde{x}_n(\omega) := \inf_{m\in\mathbb{N}, n\leq m} x_m(\omega) \in [0, \infty) \subset \mathbb{R}$, $\forall \omega \in \Omega$. Then, $\tilde{x}_n(\omega) \leq \tilde{x}_{n+1}(\omega)$, $\forall \omega \in \Omega$, and $\tilde{x}_n$ is integrable over $\boldsymbol{\Omega}$, $\forall n \in \mathbb{N}$. By monotonicity, we have $\lim_{n\in\mathbb{N}} \tilde{x}_n = \liminf_{n\in\mathbb{N}} x_n =: \bar{x} : \Omega \to [0, \infty] \subset \mathbb{R}_e$. Since $\bar{f} \in \boldsymbol{E}(\bar{x}|\hat{\mathcal{B}})$ exists, then $[\bar{x}] \in L_1(\boldsymbol{\Omega}, \mathbb{R})$. By Proposition 11.40, we have $\bar{x} \in \bar{L}_1(\boldsymbol{\Omega}, \mathbb{R})$. By $\mathcal{Y} = \mathbb{R}$ and Radon–Nikodym Theorem 11.171, $\boldsymbol{E}(\tilde{x}_n|\hat{\mathcal{B}})$ exists, $\forall n \in \mathbb{N}$. By (e), we have $\bar{f} = \lim_{n\in\mathbb{N}} \tilde{f}_n$ a.e. in $\hat{\boldsymbol{\Omega}}$, where $\tilde{f}_n \in \boldsymbol{E}(\tilde{x}_n|\hat{\mathcal{B}})$, $\forall n \in \mathbb{N}$. By (d), we have $\tilde{f}_n \leq f_m$ a.e. in $\hat{\boldsymbol{\Omega}}$, $\forall n, m \in \mathbb{N}$ with $n \leq m$. Hence, $\bar{f} \leq \liminf_{n\in\mathbb{N}} f_n$ a.e. in $\hat{\boldsymbol{\Omega}}$.

(g) By Lebesgue Dominated Convergence Theorem 11.91, $\bar{x}$ is absolutely integrable over $\boldsymbol{\Omega}$, $0 \leq \int_\Omega \mathcal{P} \circ \bar{x}\, \mathrm{d}\boldsymbol{P} = \lim_{n\in\mathbb{N}} \int_\Omega \mathcal{P} \circ x_n\, \mathrm{d}\boldsymbol{P} \leq \boldsymbol{E}(g) < \infty$, and $\boldsymbol{E}(\bar{x}) = \lim_{n\in\mathbb{N}} \boldsymbol{E}(x_n) \in \mathcal{Y}$. This implies that $\bar{x} \in \bar{L}_1(\boldsymbol{\Omega}, \mathcal{Y})$ and $\bar{x} - x_n \in \bar{L}_1(\boldsymbol{\Omega}, \mathcal{Y})$. Note that $\mathcal{P} \circ (x_n - \bar{x}) \leq 2g$ a.e. in $\boldsymbol{\Omega}$ and $\lim_{n\in\mathbb{N}}(x_n - \bar{x}) = 0$ a.e. in $\boldsymbol{\Omega}$, and then, by Lebesgue Dominated Convergence Theorem 11.91, we have $\lim_{n\in\mathbb{N}} \int_\Omega \mathcal{P} \circ (x_n - \bar{x})\, \mathrm{d}\boldsymbol{P} = 0$. This yields that $\lim_{n\in\mathbb{N}} x_n \doteq \bar{x}$ in $\bar{L}_1(\boldsymbol{\Omega}, \mathcal{Y})$, and $([x_n])_{n=1}^\infty \subseteq L_1(\boldsymbol{\Omega}, \mathcal{Y})$ is a Cauchy sequence. By (c), $f_n - f_m \in \boldsymbol{E}((x_n - x_m)|\hat{\mathcal{B}})$. By (14.2b), $\int_\Omega \mathcal{P} \circ (f_n - f_m)\, \mathrm{d}\hat{\boldsymbol{P}} \leq \int_\Omega \mathcal{P} \circ (x_n - x_m)\, \mathrm{d}\boldsymbol{P}$, $\forall n, m \in \mathbb{N}$. Then, $([f_n])_{n=1}^\infty \subseteq L_1(\hat{\boldsymbol{\Omega}}, \mathcal{Y})$ is a Cauchy sequence. By Example 11.179, $L_1(\hat{\boldsymbol{\Omega}}, \mathcal{Y})$ is a Banach space. Then, $\exists \bar{f} \in \bar{L}_1(\hat{\boldsymbol{\Omega}}, \mathcal{Y})$ such that $\lim_{n\in\mathbb{N}} f_n \doteq \bar{f}$ in $\bar{L}_1(\hat{\boldsymbol{\Omega}}, \mathcal{Y})$. Hence, $\lim_{n\in\mathbb{N}} \int_\Omega \mathcal{P} \circ (f_n - \bar{f})\, \mathrm{d}\boldsymbol{P} = 0$. By the continuity of norm, we have $\lim_{n\in\mathbb{N}} \int_\Omega \mathcal{P} \circ f_n\, \mathrm{d}\boldsymbol{P} = \int_\Omega \mathcal{P} \circ \bar{f}\, \mathrm{d}\boldsymbol{P}$.

Now, we will show that $\bar{f} \in \boldsymbol{E}(\bar{x}|\hat{\mathcal{B}})$. $\forall E \in \hat{\mathcal{B}}$, we have $\int_E \bar{f}\, \mathrm{d}\hat{\boldsymbol{P}} = \int_E \bar{f}\, \mathrm{d}\boldsymbol{P} = \lim_{n\in\mathbb{N}} \int_E f_n\, \mathrm{d}\boldsymbol{P} = \lim_{n\in\mathbb{N}} \int_E x_n\, \mathrm{d}\boldsymbol{P} = \int_E \bar{x}\, \mathrm{d}\boldsymbol{P} \in \mathcal{Y}$, where the first equality follows from the fact that $\bar{f}$ is $\hat{\mathcal{B}}$-measurable and Proposition 11.72, the second equality follows from the fact that $\lim_{n\in\mathbb{N}} f_n \doteq \bar{f}$ in $\bar{L}_1(\hat{\boldsymbol{\Omega}}, \mathcal{Y})$ and Proposition 11.213, the third equality follows from $f_n \in \boldsymbol{E}(x_n|\hat{\mathcal{B}})$, and the last equality follows from Lebesgue Dominated Convergence Theorem 11.91. By Proposition 14.8, we have $\bar{f} \in \boldsymbol{E}(\bar{x}|\hat{\mathcal{B}})$.

By (14.2), $\forall E \in \hat{\mathcal{B}}$, $\int_E \mathcal{P} \circ \bar{f}\, \mathrm{d}\hat{\boldsymbol{P}} \leq \int_E \mathcal{P} \circ \bar{x}\, \mathrm{d}\boldsymbol{P} \leq \int_E g\, \mathrm{d}\boldsymbol{P} = \int_E \bar{g}\, \mathrm{d}\hat{\boldsymbol{P}}$. Then, by Proposition 11.96, $\mathcal{P} \circ \bar{f} \leq \bar{g}$ a.e. in $\hat{\boldsymbol{\Omega}}$. ($\bar{g} \in \boldsymbol{E}(g|\hat{\mathcal{B}})$ exists by Radon–Nikodym Theorem 11.169.)

Fix any $m \in \mathbb{N}$. Define $\bar{A}_{m,n} := \{\omega \in \boldsymbol{\Omega} \mid \|f_n(\omega) - \bar{f}(\omega)\|_{\mathcal{Y}} > 1/m\} \in \hat{\mathcal{B}}$. Let $\bar{U}_{m,n} := \bigcup_{k=n}^\infty \bar{A}_{m,k} \in \hat{\mathcal{B}}$. Clearly, $\bar{U}_{m,n} \supseteq \bar{U}_{m,n+1}$, $\forall n \in \mathbb{N}$. Note that $\boldsymbol{P}(\bar{U}_{m,n}) = \sum_{k=n}^\infty \boldsymbol{P}(\bar{A}_{m,k} \setminus (\bigcup_{j=n}^{k-1} \bar{A}_{m,j})) =: \sum_{k=n}^\infty \boldsymbol{P}(\bar{V}_k) = \sum_{k=n}^\infty \int_{\bar{V}_k} \mathrm{d}\boldsymbol{P} \leq \sum_{k=n}^\infty \int_{\bar{V}_k} m\mathcal{P} \circ (f_k - \bar{f})\, \mathrm{d}\boldsymbol{P} \leq m \sum_{k=n}^\infty \int_{\bar{V}_k} \mathcal{P} \circ (x_k - \bar{x})\, \mathrm{d}\boldsymbol{P} \leq m \sum_{k=n}^\infty \int_{\bar{V}_k} \sup_{j\geq n} \mathcal{P} \circ (x_j - \bar{x})\, \mathrm{d}\boldsymbol{P} = m \int_{\bar{U}_{m,n}} \sup_{j\geq n} \mathcal{P} \circ (x_j - \bar{x})\, \mathrm{d}\boldsymbol{P} \leq m\boldsymbol{E}(\sup_{j\geq n} \mathcal{P} \circ (x_j - \bar{x}))$, where the first equality follows from Proposition 14.3, the second equality follows from Proposition 11.75, the first inequality follows from the fact that $\bar{V}_k \subseteq \bar{A}_{m,k}$ and Proposition 11.83, the second inequality follows from the discussion of previous two paragraphs and (14.2b), the third inequality follows from Proposition 11.83, the third equality follows from Proposition 11.83, and the last inequality follows from Proposition 11.83. The functions $\sup_{j\geq n}\mathcal{P}\circ(x_j-\bar{x})\leq 2g$, $\forall\omega\in\boldsymbol{\Omega}$ and converge to the $\vartheta_{\bar{\mathrm{L}}_1(\boldsymbol{\Omega},\mathbb{R})}$ almost everywhere in $\boldsymbol{\Omega}$. By Lebesgue Dominated Convergence Theorem 11.91, we have $0\leq\lim_{n\in\mathbb{N}}\boldsymbol{P}(\bar{U}_{m,n})\leq\lim_{n\in\mathbb{N}} m\boldsymbol{E}(\sup_{j\geq n}\mathcal{P}\circ(x_j-\bar{x}))=0$. By Proposition 11.5, we have $\boldsymbol{P}(\bigcap_{n\in\mathbb{N}}\bigcup_{k=n}^{\infty}\bar{A}_{m,n})=0=:\boldsymbol{P}(U_m)$. Then, $\boldsymbol{P}(U):=\boldsymbol{P}(\bigcup_{m\in\mathbb{N}}U_m)=0$ and $U\in\hat{\mathcal{B}}$. $\forall\omega\in\Omega\setminus U$, we have $\lim_{n\in\mathbb{N}}f_n(\omega)=\bar{f}(\omega)$. Hence, we have $\lim_{n\in\mathbb{N}}f_n=\bar{f}$ a.e. in $\hat{\boldsymbol{\Omega}}$.

(h) First, we will prove the result when A is a simple function. Let A admit the canonical representation $\sum_{i=1}^{m}A_i\chi_{E_i,\Omega}$, where $m\in\mathbb{Z}_+$, A_i's are distinct elements of $\mathcal{W}$ that none equal to $\vartheta_{\mathrm{B}(\mathcal{Y},\mathcal{Z})}$, E_i's are pairwise disjoint and nonempty. Since A is $\hat{\mathcal{B}}$-measurable, we have $E_i\in\hat{\mathcal{B}}$, $i=1,\ldots,m$. By Propositions 7.65, 11.39, and 11.38, Af is $\hat{\mathcal{B}}$ measurable, and Ax is $\mathcal{B}$-measurable. By A being a simple function, we have $Af\in\bar{\mathrm{L}}_1(\hat{\boldsymbol{\Omega}},\mathcal{Z})$. By the assumption, $Ax\in\bar{\mathrm{L}}_1(\boldsymbol{\Omega},\mathcal{Z})$. $\forall E\in\hat{\mathcal{B}}$, $\int_E Af\,\mathrm{d}\hat{\boldsymbol{P}}=\int_E Af\,\mathrm{d}\boldsymbol{P}=\sum_{i=1}^{m}\int_E A_i f\chi_{E_i,\Omega}\,\mathrm{d}\boldsymbol{P}=\sum_{i=1}^{m}\int_{E\cap E_i}A_i f\,\mathrm{d}\boldsymbol{P}=\sum_{i=1}^{m}A_i\int_{E\cap E_i}f\,\mathrm{d}\boldsymbol{P}=\sum_{i=1}^{m}A_i\int_{E\cap E_i}x\,\mathrm{d}\boldsymbol{P}=\sum_{i=1}^{m}\int_{E\cap E_i}A_i x\,\mathrm{d}\boldsymbol{P}=\sum_{i=1}^{m}\int_E A_i x\chi_{E_i,\Omega}\,\mathrm{d}\boldsymbol{P}=\int_E Ax\,\mathrm{d}\boldsymbol{P}$, where the first equality follows from Proposition 11.72, the second through fourth equalities follow from Proposition 11.92, the fifth equality follows from $f\in\boldsymbol{E}(x|\hat{\mathcal{B}})$, and the sixth through eighth equalities follow from Proposition 11.92. By Proposition 14.8, we have $\boldsymbol{E}(Ax|\hat{\mathcal{B}})$ exists and $Af\in\boldsymbol{E}(Ax|\hat{\mathcal{B}})$.

Next, we prove the general case. By Proposition 11.66, there exists a sequence of simple functions $(\phi_n)_{n=1}^{\infty}$ with $\phi_n:\Omega\to\mathcal{W}$ being $\hat{\mathcal{B}}$-measurable, such that $\lim_{n\in\mathbb{N}}\phi_n=A$ a.e. in $\hat{\boldsymbol{\Omega}}$ and $\|\phi_n(\omega)\|_{\mathrm{B}(\mathcal{Y},\mathcal{Z})}\leq\|A(\omega)\|_{\mathrm{B}(\mathcal{Y},\mathcal{Z})}$, $\forall\omega\in\Omega$, $\forall n\in\mathbb{N}$. By Propositions 7.65, 11.39, and 11.38, Af is $\hat{\mathcal{B}}$-measurable, $\phi_n f$ is $\hat{\mathcal{B}}$-measurable, $\forall n\in\mathbb{N}$, and Ax is $\mathcal{B}$-measurable. By Propositions 11.52 and 11.53, we have $\lim_{n\in\mathbb{N}}\phi_n x=Ax$ a.e. in $\boldsymbol{\Omega}$. Let $g:\Omega\to[0,\infty)\subset\mathbb{R}$ be given by $g(\omega)=\|A(\omega)\|_{\mathrm{B}(\mathcal{Y},\mathcal{Z})}\|x(\omega)\|_{\mathcal{Y}}$, $\forall\omega\in\Omega$. Then, by (g), we have $\boldsymbol{E}(Ax|\hat{\mathcal{B}})$ exists and $\bar{f}=\lim_{n\in\mathbb{N}}\phi_n f=Af$ a.e. in $\hat{\boldsymbol{\Omega}}$, where $\bar{f}\in\boldsymbol{E}(Ax|\hat{\mathcal{B}})$

(i) By Proposition 8.25, c is lower semicontinuous. By Definition 3.14, $\forall\alpha\in\mathbb{R}$, $c_{\mathrm{inv}}((\alpha,+\infty))=\{y\in\mathcal{Y}\mid c(y)>\alpha\}\in\mathcal{O}_F\subseteq\mathcal{B}_{\mathrm{B}}(\mathcal{Y})$, where $\mathcal{O}_F$ is the subset topology of $(\mathcal{Y},\mathcal{O}_{\mathcal{Y}})$ on the set F. By Proposition 11.35, c is $\mathcal{B}_{\mathrm{B}}(\mathcal{Y})$-measurable. By Proposition 11.34, $c\circ f$ is $\hat{\mathcal{B}}$-measurable, and $c\circ x$ is $\mathcal{B}$-measurable.

By (d), $f(\omega)\in F$ a.e. $\omega\in\hat{\boldsymbol{\Omega}}$. Thus, $c(f(\omega))\in\mathbb{R}$ a.e. $\omega\in\hat{\boldsymbol{\Omega}}$. Fix any $\omega\in f_{\mathrm{inv}}(F)$. By Proposition 8.32, we have $c(f(\omega))=\sup_{y_*\in F_{\mathrm{conj}}}(\langle\langle y_*,f(\omega)\rangle\rangle-c_{\mathrm{conj}}(y_*))$, where $c_{\mathrm{conj}}:F_{\mathrm{conj}}\to\mathbb{R}$ is the conjugate functional to c. Then, $\exists y_{*\epsilon}\in F_{\mathrm{conj}}$ such that $c(f(\omega))-\epsilon\leq\langle\langle y_{*\epsilon},f(\omega)\rangle\rangle-c_{\mathrm{conj}}(y_{*\epsilon})\leq\langle\langle y_{*\epsilon},f(\omega)\rangle\rangle-\langle\langle y_{*\epsilon},y\rangle\rangle+c(y)=\langle\langle y_{*\epsilon},f(\omega)-y\rangle\rangle+c(y)$, $\forall y\in F$, where the second inequality follows from Definition 8.27. Take $y=x(\omega)$ then $c(f(\omega))-\epsilon\leq\langle\langle y_{*\epsilon},f(\omega)-x(\omega)\rangle\rangle+c(x(\omega))$ a.e. in $\boldsymbol{\Omega}$. Thus, by (d), (h), and (c) and taking conditional expectation of the preceding equation given $\hat{\mathcal{B}}$, $c(f(\omega))-\epsilon\leq\langle\langle y_{*\epsilon},f(\omega)-f(\omega)\rangle\rangle+\bar{f}(\omega)$ a.e. $\omega\in\hat{\boldsymbol{\Omega}}$, which implies that $c(f(\omega))-\epsilon\leq\bar{f}(\omega)$ a.e. $\omega\in\hat{\boldsymbol{\Omega}}$. By the arbitrariness of ϵ, we have $c\circ f\leq\bar{f}$ a.e. in $\hat{\boldsymbol{\Omega}}$.

(j) Let $\tilde{\boldsymbol{\Omega}}:=(\Omega,\tilde{\mathcal{B}},\tilde{\boldsymbol{P}}:=\boldsymbol{P}|_{\tilde{\mathcal{B}}})$. Assume that $\boldsymbol{E}(f|\tilde{\mathcal{B}})$ exists and $\tilde{f}\in\boldsymbol{E}(f|\tilde{\mathcal{B}})$. Clearly, $\tilde{f}\in\bar{\mathrm{L}}_1(\tilde{\boldsymbol{\Omega}},\mathcal{Y})$. $\forall E\in\tilde{\mathcal{B}}$, we have $\int_E\tilde{f}\,\mathrm{d}\tilde{\boldsymbol{P}}=\int_E f\,\mathrm{d}\hat{\boldsymbol{P}}=\int_E x\,\mathrm{d}\boldsymbol{P}$, where the first equality follows from $\tilde{f} \in \boldsymbol{E}(f|\tilde{\mathcal{B}})$ and the second equality follows from $f \in \boldsymbol{E}(x|\hat{\mathcal{B}})$. By the arbitrariness of E and Proposition 14.8, we have $\tilde{f} \in \boldsymbol{E}(x|\tilde{\mathcal{B}})$. Hence, we have $\boldsymbol{E}(x|\tilde{\mathcal{B}}) = \boldsymbol{E}(f|\tilde{\mathcal{B}}) = \boldsymbol{E}(\boldsymbol{E}(x|\hat{\mathcal{B}})|\tilde{\mathcal{B}})$.

On the other hand, assume $\boldsymbol{E}(x|\tilde{\mathcal{B}})$ exists and $\tilde{f} \in \boldsymbol{E}(x|\tilde{\mathcal{B}})$. Clearly, $\tilde{f} \in \bar{\mathrm{L}}_1(\tilde{\boldsymbol{\Omega}}, \mathcal{Y})$. $\forall E \in \tilde{\mathcal{B}}$, we have $\int_E \tilde{f}\, \mathrm{d}\tilde{\boldsymbol{P}} = \int_E x\, \mathrm{d}\boldsymbol{P} = \int_E f\, \mathrm{d}\hat{\boldsymbol{P}}$, where the first equality follows from $\tilde{f} \in \boldsymbol{E}(x|\tilde{\mathcal{B}})$ and the second equality follows from $f \in \boldsymbol{E}(x|\hat{\mathcal{B}})$. By the arbitrariness of E and Proposition 14.8, we have $\tilde{f} \in \boldsymbol{E}(f|\tilde{\mathcal{B}})$. Hence, we have $\boldsymbol{E}(x|\tilde{\mathcal{B}}) = \boldsymbol{E}(f|\tilde{\mathcal{B}}) = \boldsymbol{E}(\boldsymbol{E}(x|\hat{\mathcal{B}})|\tilde{\mathcal{B}})$. This establishes (j).

(k) $\forall \tilde{E} \in \tilde{\mathcal{B}}$ and $\forall \hat{E} \in \hat{\mathcal{B}}$, we have $\int_{\hat{E}\cap\tilde{E}} x\, \mathrm{d}\boldsymbol{P} = \int_\Omega x\chi_{\tilde{E},\Omega}\chi_{\hat{E},\Omega}\, \mathrm{d}\boldsymbol{P} = \boldsymbol{P}(\tilde{E})\int_\Omega x\chi_{\hat{E},\Omega}\, \mathrm{d}\boldsymbol{P} = \boldsymbol{P}(\tilde{E})\int_{\hat{E}} x\, \mathrm{d}\boldsymbol{P} = \boldsymbol{P}(\tilde{E})\int_{\hat{E}} f\, \mathrm{d}\boldsymbol{P} = \int_{\hat{E}} f\chi_{\tilde{E},\Omega}\, \mathrm{d}\boldsymbol{P} = \int_{\hat{E}\cap\tilde{E}} f\, \mathrm{d}\boldsymbol{P}$, where the first equality follows from Proposition 11.92, the second equality follows from the independence of $\tilde{\mathcal{B}}$ and $\sigma(\hat{\mathcal{B}}\cup\sigma(x))$ and Proposition 14.9, the fourth equality follows from $f \in \boldsymbol{E}(x|\hat{\mathcal{B}})$, and the fifth equality follows from Proposition 14.9. By assumption, $x \in \bar{\mathrm{L}}_1(\boldsymbol{\Omega}, \mathcal{Y})$ and $f \in \bar{\mathrm{L}}_1(\hat{\boldsymbol{\Omega}}, \mathcal{Y})$. Then, by Proposition 11.116, we may define a finite $\mathcal{Y}$-valued measure ν on $(\Omega, \mathcal{B})$ by $\nu(E) := \int_E x\, \mathrm{d}\boldsymbol{P}$, $\forall E \in \mathcal{B}$, and we may define a finite $\mathcal{Y}$-valued measure $\bar{\nu}$ on $(\Omega, \sigma(\hat{\mathcal{B}} \cup \tilde{\mathcal{B}}))$ by $\bar{\nu}(E) := \int_E f\, \mathrm{d}\boldsymbol{P}$, $\forall E \in \sigma(\hat{\mathcal{B}} \cup \tilde{\mathcal{B}})$. By Proposition 11.107 and Definition 14.7, $\hat{\nu} := \nu|_{\sigma(\hat{\mathcal{B}}\cup\tilde{\mathcal{B}})}$ is a finite $\mathcal{Y}$-valued measure on $\sigma(\hat{\mathcal{B}} \cup \tilde{\mathcal{B}})$. We have shown in the preceding $\hat{\nu}(E) = \bar{\nu}(E)$, $\forall E \in \mathcal{F}$, where $\mathcal{F} := \{E \in \mathcal{B} \mid \exists \hat{E} \in \hat{\mathcal{B}}, \exists \tilde{E} \in \tilde{\mathcal{B}} \ni \cdot\ E = \hat{E} \cap \tilde{E}\}$. It is straightforward to show that $\mathcal{F}$ is a π-system on Ω and $\mathcal{F} \supseteq \hat{\mathcal{B}} \cup \tilde{\mathcal{B}}$. Therefore, $\hat{\nu} = \bar{\nu}$ by Lemma 12.103. Note that f is $\sigma(\hat{\mathcal{B}} \cup \tilde{\mathcal{B}})$-measurable. Then, by Proposition 14.8, $f \in \boldsymbol{E}(x|\sigma(\hat{\mathcal{B}} \cup \tilde{\mathcal{B}}))$.

This completes the proof of the proposition. □

Definition 14.12 Let $\boldsymbol{\Omega} := (\Omega, \mathcal{B}, \boldsymbol{P})$ be a probability measure space, $\hat{\mathcal{B}} \subseteq \mathcal{B}$ be a sub-σ-algebra on Ω, and $\hat{\boldsymbol{\Omega}} := (\Omega, \hat{\mathcal{B}}, \hat{\boldsymbol{P}} := \boldsymbol{P}|_{\hat{\mathcal{B}}})$. $\forall F \in \mathcal{B}$, *the conditional probability of the event F happens given $\hat{\mathcal{B}}$* is denoted by $\boldsymbol{P}(F|\hat{\mathcal{B}})$, which is defined to be $\boldsymbol{E}(\chi_{F,\Omega}|\hat{\mathcal{B}}) \in \mathrm{L}_1(\hat{\boldsymbol{\Omega}}, [0, 1])$. ◇

Proposition 14.13 *Let $\boldsymbol{\Omega} := (\Omega, \mathcal{B}, \boldsymbol{P})$ be a probability measure space, $\hat{\mathcal{B}} \subseteq \mathcal{B}$ be a sub-σ-algebra on Ω, and $\hat{\boldsymbol{\Omega}} := (\Omega, \hat{\mathcal{B}}, \hat{\boldsymbol{P}} := \boldsymbol{P}|_{\hat{\mathcal{B}}})$. $\forall F \in \mathcal{B}$, let $\boldsymbol{P}(F|\hat{\mathcal{B}}) := \boldsymbol{E}(\chi_{F,\Omega}|\hat{\mathcal{B}}) \in \mathrm{L}_1(\hat{\boldsymbol{\Omega}}, [0, 1])$. Then, we may define $\mu : \mathcal{B} \to \mathrm{L}_1(\hat{\boldsymbol{\Omega}}, [0, 1])$ by $\mu(F) = \boldsymbol{P}(F|\hat{\mathcal{B}})$. Then, μ is a finite $\mathrm{L}_1(\hat{\boldsymbol{\Omega}}, [0, 1])$-valued measure on the measurable space $(\Omega, \mathcal{B})$ with $\mathcal{P} \circ \mu = \boldsymbol{P}$.*

Proof Clearly, $\boldsymbol{P}$ is a finite measure on $(\Omega, \mathcal{B})$. (i) $\mu(\emptyset) = \boldsymbol{P}(\emptyset|\hat{\mathcal{B}}) = \boldsymbol{E}(\chi_{\emptyset,\Omega}|\hat{\mathcal{B}}) \in \mathrm{L}_1(\hat{\boldsymbol{\Omega}}, [0, 1])$. Fix any $f \in \mu(\emptyset)$. Then, $f \in \bar{\mathrm{L}}_1(\hat{\boldsymbol{\Omega}}, [0, 1])$. $\forall E \in \hat{\mathcal{B}}$, $0 = \boldsymbol{P}(\emptyset) = \int_E \chi_{\emptyset,\Omega}\, \mathrm{d}\boldsymbol{P} = \int_E f\, \mathrm{d}\hat{\boldsymbol{P}}$, where the first equality follows from $\boldsymbol{P}$ being a measure on $(\Omega, \mathcal{B})$, the second equality follows from Definition 14.1, and the third equality follows from Definition 14.7. By Proposition 11.96, we have $f = 0$ a.e. in $\hat{\boldsymbol{\Omega}}$. Thus, $\mu(\emptyset) = \vartheta_{\mathrm{L}_1(\hat{\boldsymbol{\Omega}},[0,1])}$.

(ii) $\forall E \in \mathcal{B}$, $\forall$ pairwise disjoint $(E_i)_{i=1}^\infty \subseteq \mathcal{B}$ with $E = \bigcup_{i=1}^\infty E_i$, $\mu(E) = \boldsymbol{P}(E|\hat{\mathcal{B}}) = \boldsymbol{E}(\chi_{E,\Omega}|\hat{\mathcal{B}}) = \boldsymbol{E}(\sum_{i=1}^\infty \chi_{E_i,\Omega}|\hat{\mathcal{B}}) = \sum_{i=1}^\infty \boldsymbol{E}(\chi_{E_i,\Omega}|\hat{\mathcal{B}}) = \sum_{i=1}^\infty \mu(E_i)$, where the first two equalities follow from Definition 14.12, the fourth equality follows from (e) of Proposition 14.11, and the last equality follows from Definition 14.12. Furthermore, $\infty > \|\mu(E)\|_{L_1(\hat{\boldsymbol{\Omega}},[0,1])} = \int_\Omega |\mu(E)(\omega)|\,\mathrm{d}\hat{\boldsymbol{P}}(\omega) = \int_\Omega \mu(E)(\omega)\,\mathrm{d}\hat{\boldsymbol{P}}(\omega) = \int_\Omega \sum_{i=1}^\infty \mu(E_i)(\omega)\,\mathrm{d}\hat{\boldsymbol{P}}(\omega) = \sum_{i=1}^\infty \int_\Omega \mu(E_i)(\omega)\,\mathrm{d}\hat{\boldsymbol{P}}(\omega) = \sum_{i=1}^\infty \int_\Omega |\mu(E_i)(\omega)|\,\mathrm{d}\hat{\boldsymbol{P}}(\omega) = \sum_{i=1}^\infty \|\mu(E_i)\|_{L_1(\hat{\boldsymbol{\Omega}},[0,1])}$, where the inequality follows from Definition 14.12, the second equality follows from the fact that $\mu(E)(\omega) \geq 0$ a.e. $\omega \in \hat{\boldsymbol{\Omega}}$, the third equality follows from the preceding discussion, the fourth equality follows from Lebesgue Dominated Convergence Theorem 11.91, and the fifth equality follows from the fact that $\mu(E_i)(\omega) \geq 0$ a.e. $\omega \in \hat{\boldsymbol{\Omega}}$, $\forall i \in \mathbb{N}$. Hence, $(\Omega, \mathcal{B}, \mu)$ is an $L_1(\hat{\boldsymbol{\Omega}}, [0,1])$-valued pre-measure space, by Definition 11.99.

Note that, $\forall E \in \mathcal{B}$,

$$
\begin{aligned}
&\mathcal{P} \circ \mu(E) \\
&= \sup_{n\in\mathbb{Z}_+,\ (E_i)_{i=1}^n \subseteq \mathcal{B},\ E=\bigcup_{i=1}^n E_i,\ E_i\cap E_j=\emptyset,\ \forall 1\leq i<j\leq n} \sum_{i=1}^n \|\mu(E_i)\|_{L_1(\hat{\boldsymbol{\Omega}},[0,1])} \\
&= \sup_{\substack{n\in\mathbb{Z}_+,\ (E_i)_{i=1}^n \subseteq \mathcal{B},\ E=\bigcup_{i=1}^n E_i,\\ E_i\cap E_j=\emptyset,\ \forall 1\leq i<j\leq n}} \sum_{i=1}^n \int_\Omega |\mu(E_i)(\omega)|\,\mathrm{d}\hat{\boldsymbol{P}}(\omega) \\
&= \sup_{n\in\mathbb{Z}_+,\ (E_i)_{i=1}^n \subseteq \mathcal{B},\ E=\bigcup_{i=1}^n E_i,\ E_i\cap E_j=\emptyset,\ \forall 1\leq i<j\leq n} \sum_{i=1}^n \int_\Omega \mu(E_i)(\omega)\,\mathrm{d}\hat{\boldsymbol{P}}(\omega) \\
&= \sup_{n\in\mathbb{Z}_+,\ (E_i)_{i=1}^n \subseteq \mathcal{B},\ E=\bigcup_{i=1}^n E_i,\ E_i\cap E_j=\emptyset,\ \forall 1\leq i<j\leq n} \int_\Omega \sum_{i=1}^n \mu(E_i)(\omega)\,\mathrm{d}\hat{\boldsymbol{P}}(\omega) \\
&= \sup_{n\in\mathbb{Z}_+,\ (E_i)_{i=1}^n \subseteq \mathcal{B},\ E=\bigcup_{i=1}^n E_i,\ E_i\cap E_j=\emptyset,\ \forall 1\leq i<j\leq n} \int_\Omega \mu(E)(\omega)\,\mathrm{d}\hat{\boldsymbol{P}}(\omega) \\
&= \int_\Omega \mu(E)(\omega)\,\mathrm{d}\hat{\boldsymbol{P}}(\omega) = \int_\Omega \chi_{E,\Omega}\,\mathrm{d}\boldsymbol{P} = \boldsymbol{P}(E)
\end{aligned}
$$

where the first equality follows from Definition 11.99, the second equality follows from Example 11.179, the third equality follows from $\mu(E_i)(\omega) \geq 0$ a.e. $\omega \in \hat{\boldsymbol{\Omega}}$, $\forall i = 1, \ldots, n$, the fourth equality follows from Proposition 11.92, and the seventh equality follows from (14.2a) of Definition 14.7. Hence, $\boldsymbol{P} = \mathcal{P} \circ \mu$.

Since $\boldsymbol{P}$ is a finite measure, then $(\Omega, \mathcal{B}, \mu)$ is a finite $L_1(\hat{\boldsymbol{\Omega}}, [0,1])$-valued measure space. This completes the proof of the proposition. □

Based on Definition 14.12, we may define *the conditional probability measure given an event* $E \in \mathcal{B}$ with $\boldsymbol{P}(E) > 0$. Using the notation of Definition 14.12, we define the σ-algebra $\hat{\mathcal{B}} := \{\emptyset, E, \Omega \setminus E, \Omega\}$. Fix any $F \in \mathcal{B}$, we have $\mu(F) := \boldsymbol{P}(F|\hat{\mathcal{B}}) \in L_1(\hat{\boldsymbol{\Omega}}, [0,1])$, and this defines an $L_1(\hat{\boldsymbol{\Omega}}, [0,1])$-valued measure space $(\Omega, \mathcal{B}, \mu)$. Since $\mu(F)$ is $\hat{\mathcal{B}}$-measurable and $\boldsymbol{P}(E) > 0$, then $\mu(F)(\omega)$ is a constant, $\forall \omega \in E$. This constant depends only on $F \in \mathcal{B}$ and is then defined to be $\boldsymbol{P}(F|E) :=$

$\mu(F)(\omega)$, $\forall \omega \in E$. It is easy to check that $\boldsymbol{P}(F|E)\boldsymbol{P}(E) = \int_E \mu(F)(\omega)\,\mathrm{d}\hat{\boldsymbol{P}}(\omega) = \int_E \boldsymbol{P}(F|\hat{\mathcal{B}})(\omega)\,\mathrm{d}\hat{\boldsymbol{P}}(\omega) = \int_E \boldsymbol{E}(\chi_{F,\Omega}|\hat{\mathcal{B}})(\omega)\,\mathrm{d}\hat{\boldsymbol{P}}(\omega) = \int_E \chi_{F,\Omega}(\omega)\,\mathrm{d}\boldsymbol{P}(\omega) = \int_\Omega \chi_{E,\Omega}\chi_{F,\Omega}\,\mathrm{d}\boldsymbol{P} = \boldsymbol{P}(E \cap F)$, where the first equality follows from the definition of $\boldsymbol{P}(F|E)$, the third equality follows from Definition 14.12, the fourth equality follows from (14.2a) of Definition 14.7, and the last equality follows from Proposition 11.75. This shows that $\boldsymbol{P}(F|E) = \boldsymbol{P}(E \cap F)/\boldsymbol{P}(E)$, which is uniquely defined whenever $\boldsymbol{P}(E) > 0$.

Given any other sub-σ-algebra $\tilde{\mathcal{B}} \subseteq \mathcal{B}$ on the set Ω with $E \in \tilde{\mathcal{B}}$. Clearly, we have $\hat{\mathcal{B}} \subseteq \tilde{\mathcal{B}}$. Now, $\boldsymbol{P}(F|\tilde{\mathcal{B}}) \in \mathrm{L}_1(\tilde{\boldsymbol{\Omega}}, [0, 1])$, where $\tilde{\boldsymbol{\Omega}} := (\Omega, \tilde{\mathcal{B}}, \tilde{\boldsymbol{P}} := \boldsymbol{P}|_{\tilde{\mathcal{B}}})$ is a probability measure space. We can prove the following:

$$\int_E \boldsymbol{P}(F|\tilde{\mathcal{B}})(\omega)\,\mathrm{d}\tilde{\boldsymbol{P}}(\omega) = \int_E \boldsymbol{E}(\chi_{F,\Omega}|\tilde{\mathcal{B}})(\omega)\,\mathrm{d}\tilde{\boldsymbol{P}}(\omega)$$
$$= \int_E \chi_{F,\Omega}\,\mathrm{d}\boldsymbol{P} = \int_\Omega \chi_{F,\Omega}\chi_{E,\Omega}\,\mathrm{d}\boldsymbol{P} = \boldsymbol{P}(E \cap F) \tag{14.3}$$

where the first equality follows from Definition 14.12, the second equality follows from (14.2a) of Definition 14.7, and the fourth equality follows from Proposition 11.75. If $\boldsymbol{P}(E) > 0$, we have $\int_E \boldsymbol{P}(F|\tilde{\mathcal{B}})(\omega)\,\mathrm{d}\tilde{\boldsymbol{P}}(\omega) = \boldsymbol{P}(F|E)\boldsymbol{P}(E)$ by the previous paragraph.

Proposition 14.14 *Let $\mathcal{X}_i := (\Omega_i, \mathcal{B}_i, \boldsymbol{P}_i)$ be a probability measure space, $i = 1, 2$, $\mathcal{X} := \mathcal{X}_1 \times \mathcal{X}_2 =: (\Omega, \mathcal{B}, \boldsymbol{P})$ be the product probability measure space, $\mathcal{Y}$ be a finite-dimensional Banach space over $\mathbb{K}$, $x : \Omega_1 \times \Omega_2 \to \mathcal{Y}$ be $\mathcal{B}$-measurable, $x \in \bar{\mathrm{L}}_1(\mathcal{X}, \mathcal{Y})$, and $\exists M \in [0, \infty) \subset \mathbb{R}$, such that $\|x(\omega_1, \omega_2)\|_{\mathcal{Y}} \leq M$, $\forall (\omega_1, \omega_2) \in \Omega_1 \times \Omega_2 = \Omega$. Define $g : \Omega \to \mathcal{Y}$ by $g(\hat{\omega}_1, \hat{\omega}_2) := \int_\Omega x(\hat{\omega}_1, \omega_2)\,\mathrm{d}\boldsymbol{P}(\omega_1, \omega_2)$, $\forall (\hat{\omega}_1, \hat{\omega}_2) \in \Omega$. Then, the following statements hold:*

(i) *The collection of σ-algebras $(\hat{\mathcal{B}}_1, \hat{\mathcal{B}}_2)$ is independent, where $\hat{\mathcal{B}}_i := \{B_1 \times B_2 \in \mathcal{B} \mid B_j = \Omega_j, j = 1, 2$ with $i \neq j\}$.*
(ii) *There exists $g_1 : \Omega_1 \to \mathcal{Y}$, such that $g(\omega_1, \omega_2) = g_1(\omega_1)$, $\forall (\omega_1, \omega_2) \in \Omega$, $g_1 \in \bar{\mathrm{L}}_1(\mathcal{X}_1, \mathcal{Y})$, and $g \in \boldsymbol{E}(x|\hat{\mathcal{B}}_1)$*

Proof (i) This holds by Proposition 12.21.

(ii) Clearly, $g(\hat{\omega}_1, \hat{\omega}_2) = g_1(\hat{\omega}_1)$, $\forall (\hat{\omega}_1, \hat{\omega}_2) \in \Omega$, and, $\forall \hat{\omega}_1 \in \Omega_1$,

$$g_1(\hat{\omega}_1) := \int_\Omega x(\hat{\omega}_1, \omega_2)\,\mathrm{d}\boldsymbol{P}(\omega_1, \omega_2) = \int_{\Omega_2}\int_{\Omega_1} x(\hat{\omega}_1, \omega_2)\,\mathrm{d}\boldsymbol{P}_1(\omega_1)\mathrm{d}\boldsymbol{P}_2(\omega_2)$$
$$= \int_{\Omega_2} x(\hat{\omega}_1, \omega_2)\int_{\Omega_1} \mathrm{d}\boldsymbol{P}_1(\omega_1)\mathrm{d}\boldsymbol{P}_2(\omega_2) = \int_{\Omega_2} x(\hat{\omega}_1, \omega_2)\,\mathrm{d}\boldsymbol{P}_2(\omega_2) \in \mathcal{Y}$$

where the first equality follows from Proposition 12.28 and Fubini's Theorem 12.31 and the assumptions, the second equality follows from Propositions 11.92, the third equality follows from 11.75, and the set membership follows from Proposition 11.92. By Fubini's Theorem 12.31, g_1 is $\mathcal{B}_1$-measurable and $g_1 \in \bar{\mathrm{L}}_1(\mathcal{X}_1, \mathcal{Y})$.

Hence, g is $\hat{\mathcal{B}}_1$-measurable and $g \in \bar{L}_1(\mathcal{X}, \mathcal{Y})$. $\forall E \in \hat{\mathcal{B}}_1$, we have $E = E_1 \times \Omega_2$, where $E_1 \in \mathcal{B}_1$. Then,

$$\int_E x \, d\boldsymbol{P} = \int_\Omega x\chi_{E,\Omega} \, d\boldsymbol{P} = \int_{\Omega_1}\int_{\Omega_2} x(\omega_1, \omega_2)\chi_{E_1,\Omega_1}(\omega_1) \, d\boldsymbol{P}_2(\omega_2) d\boldsymbol{P}_1(\omega_1)$$

$$= \int_{\Omega_1} \chi_{E_1,\Omega_1}(\omega_1) \int_{\Omega_2} x(\omega_1, \omega_2) \, d\boldsymbol{P}_2(\omega_2) d\boldsymbol{P}_1(\omega_1)$$

$$= \int_{\Omega_1} \chi_{E_1,\Omega_1}(\omega_1) g_1(\omega_1) \, d\boldsymbol{P}_1(\omega_1) = \int_{E_1} g_1 \, d\boldsymbol{P}_1$$

$$= \int_{E_1} g_1(\omega_1) \int_{\Omega_2} d\boldsymbol{P}_2(\omega_2) d\boldsymbol{P}_1(\omega_1) = \int_E g \, d\boldsymbol{P}$$

where the first equality follows from Proposition 11.92, the second equality follows from Fubini's Theorem 12.31, the third equality follows from Proposition 11.92, the fourth equality follows from the expression for g_1, and the last equality follows from Fubini's Theorem 12.31. By the arbitrariness of E and Proposition 14.8, we have $g \in \boldsymbol{E}(x|\hat{\mathcal{B}}_1)$.

This completes the proof of the proposition. □

Theorem 14.15 (Fundamental Theorem on Modeling) *Let Λ be a set, $\mathcal{X}_\alpha := (X_\alpha, \mathcal{B}_\alpha, \boldsymbol{P}_\alpha)$ be a probability measure space, $\forall \alpha \in \Lambda$, $\mathcal{F} := \{\prod_{\alpha\in\Lambda} B_\alpha \subseteq X := \prod_{\alpha\in\Lambda} X_\alpha \mid B_\alpha \in \mathcal{B}_\alpha, \forall \alpha \in \Lambda$, and $B_\alpha = X_\alpha, \forall \alpha \in \Lambda$ except a finite number of α's$\}$, and $\mu : \mathcal{F} \to [0, 1] \subset \mathbb{R}_e$ be defined by $\mu(\prod_{\alpha\in\Lambda} B_\alpha) = \prod_{\alpha\in\Lambda} \boldsymbol{P}_\alpha(B_\alpha) = \prod_{\alpha\in\Lambda, B_\alpha\subset X_\alpha} \boldsymbol{P}_\alpha(B_\alpha)$, $\forall \prod_{\alpha\in\Lambda} B_\alpha \in \mathcal{F}$. Then, $\mathcal{F}$ is a semialgebra on X and there exists a unique probability measure space $(X, \mathcal{B} := \sigma(\mathcal{F}), \boldsymbol{P})$ such that $\boldsymbol{P}|_\mathcal{F} = \mu$. Furthermore, the collection of σ-algebras $(\bar{\mathcal{B}}_\alpha)_{\alpha\in\Lambda}$ is independent, where $\bar{\mathcal{B}}_\alpha := \{\prod_{\lambda\in\Lambda} B_\lambda \subseteq \mathcal{F} \mid B_\lambda = X_\lambda, \forall \lambda \in \Lambda$ with $\lambda \neq \alpha\}$, $\forall \alpha \in \Lambda$.*

Proof Note that the result holds if Λ is a finite set, by Proposition 12.21. We need only to prove the result when Λ is not a finite set.

We will first show that $\mathcal{F}$ is a semialgebra on X. Clearly, $\emptyset, X \in \mathcal{F}$. $\forall B := \prod_{\alpha\in\Lambda} B_\alpha, C := \prod_{\alpha\in\Lambda} C_\alpha \in \mathcal{F}$, $B \cap C = \prod_{\alpha\in\Lambda}(B_\alpha \cap C_\alpha)$. Clearly, $B_\alpha \cap C_\alpha \in \mathcal{B}_\alpha$, $\forall \alpha \in \Lambda$, since $B_\alpha, C_\alpha \in \mathcal{B}_\alpha$. Furthermore, $B_\alpha \cap C_\alpha = X_\alpha$, $\forall \alpha \in \Lambda$ except a finite number of α's, since $B_\alpha = X_\alpha$, $\forall \alpha \in \Lambda$ except for a finite number of α's and $C_\alpha = X_\alpha$, $\forall \alpha \in \Lambda$ except for a finite number of α's. Hence, $B \cap C \in \mathcal{F}$. Fix any $B := \prod_{\alpha\in\Lambda} B_\alpha \in \mathcal{F}$. Then, $B_\alpha \in \mathcal{B}_\alpha$, $\forall \alpha \in \Lambda$, and $B_\alpha = X_\alpha$, $\forall \alpha \in \Lambda \setminus \Lambda_B$, where $\Lambda_B = \{\alpha_1, \ldots, \alpha_n\} \subseteq \Lambda$, $n \in \mathbb{Z}_+$. Then, $X \setminus B = \bigcup_{j=1}^n \bigcup_{1\leq i_1<i_2<\cdots<i_j\leq n} C_{j,i_1,\ldots,i_j}$, where $C_{j,i_1,\ldots,i_j} := \prod_{\alpha\in\Lambda} C_{j,i_1,\ldots,i_j,\alpha}$ and $C_{j,i_1,\ldots,i_j,\alpha} = \begin{cases} X_\alpha \setminus B_\alpha & \alpha \in \{\alpha_{i_1}, \ldots \alpha_{i_j}\} \\ B_\alpha & \alpha \in \Lambda \setminus \{\alpha_{i_1}, \ldots \alpha_{i_j}\} \end{cases}$. Clearly, $C_{j,i_1,\ldots,i_j} \in \mathcal{F}$ and $C_{j,i_1,\ldots,i_j}$'s are pairwise disjoint. Hence, $\mathcal{F}$ is a semialgebra on X.

We will next show that μ satisfies the assumptions of Proposition 11.32. (i) Fix any $B \in \mathcal{F}$, any $n \in \mathbb{Z}_+$, and any pairwise disjoint $(C_i)_{i=1}^n \subseteq \mathcal{F}$ with $B = \bigcup_{i=1}^n C_i$. Let $B := \prod_{\alpha \in \Lambda} B_\alpha$, where $B_\alpha \in \mathcal{B}_\alpha$, $\forall \alpha \in \Lambda$, and $B_\alpha = X_\alpha$, $\forall \alpha \in \Lambda \setminus \Lambda_B$, with $\Lambda_B \subseteq \Lambda$ being a finite set. $\forall i \in \{1, \ldots, n\}$, let $C_i := \prod_{\alpha \in \Lambda} C_{i,\alpha}$, where $C_{i,\alpha} \in \mathcal{B}_\alpha$, $\forall \alpha \in \Lambda$, and $C_{i,\alpha} = X_\alpha$, $\forall \alpha \in \Lambda \setminus \Lambda_i$ with $\Lambda_i \subseteq \Lambda$ being a finite set. Then, let $\bar{\Lambda} := \Lambda_B \cup (\bigcup_{i=1}^n \Lambda_i)$, which is a finite subset of Λ. Then, $B_\alpha = X_\alpha = C_{i,\alpha}$, $\forall i \in \{1, \ldots, n\}$, $\forall \alpha \in \Lambda \setminus \bar{\Lambda}$. Then, $\mu(B) = \prod_{\alpha \in \bar{\Lambda}} \boldsymbol{P}_\alpha(B_\alpha) = \sum_{i=1}^n \prod_{\alpha \in \bar{\Lambda}} \boldsymbol{P}_\alpha(C_{i,\alpha}) = \sum_{i=1}^n \mu(C_i)$, where the second equality follows from Proposition 12.21.

(ii) Fix any $B \in \mathcal{F}$ and any pairwise disjoint $(C_i)_{i=1}^\infty \subseteq \mathcal{F}$ with $B = \bigcup_{i=1}^\infty C_i$. Let $B := \prod_{\alpha \in \Lambda} B_\alpha$, where $B_\alpha \in \mathcal{B}_\alpha$, $\forall \alpha \in \Lambda$, and $B_\alpha = X_\alpha$, $\forall \alpha \in \Lambda \setminus \Lambda_B$ with $\Lambda_B \subseteq \Lambda$ being a finite set. $\forall i \in \mathbb{N}$, let $C_i := \prod_{\alpha \in \Lambda} C_{i,\alpha}$, where $C_{i,\alpha} \in \mathcal{B}_\alpha$, $\forall \alpha \in \Lambda$, and $C_{i,\alpha} = X_\alpha$, $\forall \alpha \in \Lambda \setminus \Lambda_i$ with $\Lambda_i \subseteq \Lambda$ being a finite set. Let $\hat{\Lambda} := \Lambda_B \cup (\bigcup_{i=1}^\infty \Lambda_i)$, which is a countable subset of Λ. Then, $B_\alpha = X_\alpha = C_{i,\alpha}$, $\forall i \in \mathbb{N}$, $\forall \alpha \in \Lambda \setminus \hat{\Lambda}$.

We will show that $\mu(B) \leq \sum_{i=1}^\infty \mu(C_i)$ by an argument of contradiction. Suppose $\mu(B) > \sum_{i=1}^\infty \mu(C_i)$. Clearly, $0 \leq \mu(B) \leq 1$. Then, $\exists \epsilon_0 \in (0, \infty) \subset \mathbb{R}$ such that $\mu(B) > \epsilon_0 + \sum_{i=1}^n \mu(C_i)$, $\forall n \in \mathbb{N}$. Let $\hat{\Lambda}_n := \Lambda_B \cup (\bigcup_{i=1}^n \Lambda_i)$, $\forall n \in \mathbb{N}$, which is a finite subset of Λ. Let the elements of $\hat{\Lambda} := \{\alpha_1, \ldots, \alpha_{r(1)}, \alpha_{r(1)+1}, \ldots, \alpha_{r(2)}, \ldots\}$, where the $r(i)$'s are defined such that $\hat{\Lambda}_n = \{\alpha_1, \ldots, \alpha_{r(1)}, \ldots, \alpha_{r(n-1)+1}, \ldots, \alpha_{r(n)}\}$, $\forall n \in \mathbb{N}$.

Fix any $n \in \mathbb{N}$. By Proposition 12.21, we have $\mathcal{Y}_{n,1} := \prod_{\alpha \in \hat{\Lambda}_n} \mathcal{X}_\alpha =: (Y_{n,1}, \hat{\mathcal{B}}_{n,1}, \hat{\boldsymbol{P}}_{n,1})$ is a probability measure space, on which the collection of σ-algebras $(\tilde{\mathcal{B}}_{\alpha,n,1})_{\alpha \in \hat{\Lambda}_n}$ is independent, where $\tilde{\mathcal{B}}_{\alpha,n,1} := \{\prod_{\bar{\alpha} \in \hat{\Lambda}_n} \tilde{B}_{\bar{\alpha}} \in \hat{\mathcal{B}}_{n,1} \mid \tilde{B}_{\bar{\alpha}} = X_{\bar{\alpha}}, \forall \bar{\alpha} \in \hat{\Lambda}_n \setminus \{\alpha\}\}$, $\forall \alpha \in \hat{\Lambda}_n$. Let $\hat{B}_{n,1} := \prod_{\alpha \in \hat{\Lambda}_n} B_\alpha$ and $\hat{C}_{i,n,1} := \prod_{\alpha \in \hat{\Lambda}_n} C_{i,\alpha}$, $\forall i \in \{1, \ldots, n\}$. Clearly, $\hat{B}_{n,1} \supseteq \bigcup_{i=1}^n \hat{C}_{i,n,1}$ and the sets in the union are pairwise disjoint, and all sets involved are in $\hat{\mathcal{B}}_{n,1}$. Then, we have $\hat{\boldsymbol{P}}_{n,1}(\hat{B}_{n,1}) = \mu(B) > \epsilon_0 + \sum_{i=1}^n \mu(C_i) = \epsilon_0 + \sum_{i=1}^n \hat{\boldsymbol{P}}_{n,1}(\hat{C}_{i,n,1})$, where the equalities follow from the fact that $\Lambda_B \subseteq \hat{\Lambda}_n$ and $\Lambda_i \subseteq \hat{\Lambda}_n$, $\forall i \in \{1, \ldots, n\}$. Define, $\hat{H}_{n,1} := \hat{B}_{n,1} \setminus (\bigcup_{i=1}^n \hat{C}_{i,n,1})$. Then, $\hat{\boldsymbol{P}}_{n,1}(\hat{H}_{n,1}) > \epsilon_0$. Define $H_{n,1} := B \setminus (\bigcup_{i=1}^n C_i)$. Clearly, $H_{n,1} = \hat{H}_{n,1} \times \prod_{\alpha \in \Lambda \setminus \hat{\Lambda}_n} X_\alpha$. Obviously, $H_{n,1} \supseteq H_{n+1,1}$ and $\bigcap_{n=1}^\infty H_{n,1} = \emptyset$.

By Radon–Nikodym Theorem 11.169 and Definition 14.7, $\hat{\boldsymbol{E}}_{n,1}(\chi_{\hat{H}_{n,1}, Y_{n,1}} | \tilde{\mathcal{B}}_{\alpha_1,n,1})$ exists, where $\hat{\boldsymbol{E}}_{n,1}$ is with respect to measure $\hat{\boldsymbol{P}}_{n,1}$. Define $g_{n,1} : Y_{n,1} = \prod_{\alpha \in \hat{\Lambda}_n} X_\alpha \to [0, 1] \subset \mathbb{R}$ by, $\forall (\hat{\omega}_{\alpha_1}, \ldots, \hat{\omega}_{\alpha_{r(n)}}) \in Y_{n,1}$,

$$
\begin{aligned}
& g_{n,1}(\hat{\omega}_{\alpha_1}, \ldots, \hat{\omega}_{\alpha_{r(n)}}) \\
&= \int_{Y_{n,1}} \chi_{\hat{H}_{n,1}, Y_{n,1}}(\hat{\omega}_{\alpha_1}, \omega_{\alpha_2}, \ldots, \omega_{\alpha_{r(n)}}) \, \mathrm{d}\hat{\boldsymbol{P}}_{n,1}(\omega_{\alpha_1}, \ldots, \omega_{\alpha_{r(n)}})
\end{aligned}
$$

By Proposition 14.14, $g_{n,1} \in \hat{\boldsymbol{E}}_{n,1}(\chi_{\hat{H}_{n,1}, Y_{n,1}} | \tilde{\mathcal{B}}_{\alpha_1,n,1})$, $g_{n,1}(\hat{\omega}_{\alpha_1}, \ldots, \hat{\omega}_{\alpha_{r(n)}}) = \bar{g}_{n,1}(\hat{\omega}_{\alpha_1})$, $\forall (\hat{\omega}_{\alpha_1}, \ldots, \hat{\omega}_{\alpha_{r(n)}}) \in Y_{n,1}$, and $\bar{g}_{n,1} : X_{\alpha_1} \to [0, 1] \subset \mathbb{R}$

is $\mathcal{B}_{\alpha_1}$-measurable. Since $\hat{\boldsymbol{P}}_{n,1}(\hat{H}_{n,1}) > \epsilon_0$, then $\boldsymbol{E}_{\alpha_1}(\bar{g}_{n,1}) = \hat{\boldsymbol{E}}_{n,1}(\hat{\boldsymbol{E}}_{n,1}(\chi_{\hat{H}_{n,1},Y_{n,1}}|\tilde{\mathcal{B}}_{\alpha_1,n,1})) = \hat{\boldsymbol{E}}_{n,1}(\chi_{\hat{H}_{n,1},Y_{n,1}}) = \hat{\boldsymbol{P}}_{n,1}(\hat{H}_{n,1}) > \epsilon_0$, where $\boldsymbol{E}_{\alpha_1}$ is with respect to $\boldsymbol{P}_{\alpha_1}$, and the second equality follows from (a) of Proposition 14.11. This leads to the following inequalities:

$$\begin{aligned}\epsilon_0 < \boldsymbol{E}_{\alpha_1}(\bar{g}_{n,1}) &\leq \boldsymbol{P}_{\alpha_1}(\{\omega_{\alpha_1} \in X_{\alpha_1} \mid \bar{g}_{n,1}(\omega_{\alpha_1}) > 2^{-1}\epsilon_0\}) \\ &\quad + 2^{-1}\epsilon_0 \boldsymbol{P}_{\alpha_1}(\{\omega_{\alpha_1} \in X_{\alpha_1} \mid \bar{g}_{n,1}(\omega_{\alpha_1}) \leq 2^{-1}\epsilon_0\}) \\ &\leq \boldsymbol{P}_{\alpha_1}(\{\omega_{\alpha_1} \in X_{\alpha_1} \mid \bar{g}_{n,1}(\omega_{\alpha_1}) > 2^{-1}\epsilon_0\}) + 2^{-1}\epsilon_0\end{aligned}$$

Thus, we have $\boldsymbol{P}_{\alpha_1}(\{\omega_{\alpha_1} \in X_{\alpha_1} \mid \bar{g}_{n,1}(\omega_{\alpha_1}) > 2^{-1}\epsilon_0\}) > 2^{-1}\epsilon_0$.

Now, since $H_{n,1} \supseteq H_{n+1,1}$, by the definition of $\bar{g}_{n,1}$, Proposition 11.92, and Fubini's Theorem 12.31, we have $0 \leq \bar{g}_{n+1,1}(\omega_{\alpha_1}) \leq \bar{g}_{n,1}(\omega_{\alpha_1})$, $\forall \omega_{\alpha_1} \in X_{\alpha_1}$ (which requires us to work in the product measure space $\mathcal{Y}_{n+1,1}$ that includes both $\hat{H}_{n,1} \times \prod_{i=r(n)+1}^{r(n+1)} X_{\alpha_i}$ and $\hat{H}_{n+1,1}$). Therefore, we have $\boldsymbol{P}_{\alpha_1}(\Gamma_{n,1}) > 2^{-1}\epsilon_0$, where $\Gamma_{n,1} := \{\omega_{\alpha_1} \in X_{\alpha_1} \mid \bar{g}_{n,1}(\omega_{\alpha_1}) > 2^{-1}\epsilon_0\}$. Furthermore, $\Gamma_{n,1} \supseteq \Gamma_{n+1,1}$. By Proposition 11.5, we have $\boldsymbol{P}_{\alpha_1}(\bigcap_{n=1}^{\infty} \Gamma_{n,1}) = \lim_{n\in\mathbb{N}} \boldsymbol{P}_{\alpha_1}(\Gamma_{n,1}) \geq 2^{-1}\epsilon_0 > 0$. Then, $\exists \bar{\omega}_{\alpha_1} \in X_{\alpha_1}$ such that $\bar{\omega}_{\alpha_1} \in \bigcap_{n=1}^{\infty} \Gamma_{n,1}$. This implies that $\bar{g}_{n,1}(\bar{\omega}_{\alpha_1}) > 2^{-1}\epsilon_0$, $\forall n \in \mathbb{N}$. By the definition of $\bar{g}_{n,1}$, we have $\int_{Y_{n,1}} \chi_{\hat{H}_{n,1},Y_{n,1}}(\bar{\omega}_{\alpha_1}, \omega_{\alpha_2}, \ldots, \omega_{\alpha_{r(n)}})\, \mathrm{d}\hat{\boldsymbol{P}}_{n,1}(\omega_{\alpha_1}, \ldots, \omega_{\alpha_{r(n)}}) > 2^{-1}\epsilon_0$, $\forall n \in \mathbb{N}$.

Fix any $n \in \mathbb{N}$. By Proposition 12.21, we have $\mathcal{Y}_{n,2} := \prod_{\alpha\in\hat{\Lambda}_n\setminus\{\alpha_1\}} \mathcal{X}_\alpha =: (Y_{n,2}, \hat{\mathcal{B}}_{n,2}, \hat{\boldsymbol{P}}_{n,2})$ is a probability measure space, on which the collection of σ-algebras $(\tilde{\mathcal{B}}_{\alpha,n,2})_{\alpha\in\hat{\Lambda}_n\setminus\{\alpha_1\}}$ is independent, where $\tilde{\mathcal{B}}_{\alpha,n,2} := \left\{\prod_{\bar{\alpha}\in\hat{\Lambda}_n\setminus\{\alpha_1\}} \tilde{B}_{\bar{\alpha}} \in \hat{\mathcal{B}}_{n,2} \,\middle|\, \tilde{B}_{\bar{\alpha}} = X_{\bar{\alpha}}, \forall \bar{\alpha} \in \hat{\Lambda}_n \setminus \{\alpha_1, \alpha\}\right\}$, $\forall \alpha \in \hat{\Lambda}_n \setminus \{\alpha_1\}$. Let $\hat{B}_{n,2} := \begin{cases}\prod_{\alpha\in\hat{\Lambda}_n\setminus\{\alpha_1\}} B_\alpha & \bar{\omega}_{\alpha_1} \in B_{\alpha_1} \\ \emptyset & \bar{\omega}_{\alpha_1} \notin B_{\alpha_1}\end{cases}$ and $\hat{C}_{i,n,2} := \begin{cases}\prod_{\alpha\in\hat{\Lambda}_n\setminus\{\alpha_1\}} C_{i,\alpha} & \bar{\omega}_{\alpha_1} \in C_{i,\alpha_1} \\ \emptyset & \bar{\omega}_{\alpha_1} \notin C_{i,\alpha_1}\end{cases}$, $\forall i \in \{1, \ldots, n\}$. Clearly, $\hat{B}_{n,2} \supseteq \bigcup_{i=1}^n \hat{C}_{i,n,2}$ and the sets in the union are pairwise disjoint, and all sets involved are in $\hat{\mathcal{B}}_{n,2}$, by Proposition 12.25. Then, we have

$$\begin{aligned}\hat{\boldsymbol{P}}_{n,2}(\hat{B}_{n,2}) &= \int_{Y_{n,2}} \chi_{\hat{B}_{n,2},Y_{n,2}}(\omega_{\alpha_2}, \ldots, \omega_{\alpha_{r(n)}})\, \mathrm{d}\hat{\boldsymbol{P}}_{n,2}(\omega_{\alpha_2}, \ldots, \omega_{\alpha_{r(n)}}) \\ &= \int_{Y_{n,1}} \chi_{\hat{B}_{n,1},Y_{n,1}}(\bar{\omega}_{\alpha_1}, \omega_{\alpha_2}, \ldots, \omega_{\alpha_{r(n)}})\, \mathrm{d}\hat{\boldsymbol{P}}_{n,1}(\omega_{\alpha_1}, \omega_{\alpha_2}, \ldots, \omega_{\alpha_{r(n)}}) \\ &= \int_{Y_{n,1}} \left(\chi_{\hat{H}_{n,1},Y_{n,1}} + \sum_{i=1}^n \chi_{\hat{C}_{i,n,1},Y_{n,1}}\right)(\bar{\omega}_{\alpha_1}, \omega_{\alpha_2}, \ldots, \omega_{\alpha_{r(n)}}) \\ &\quad \mathrm{d}\hat{\boldsymbol{P}}_{n,1}(\omega_{\alpha_1}, \omega_{\alpha_2}, \ldots, \omega_{\alpha_{r(n)}})\end{aligned}$$

$$= \int_{Y_{n,1}} \sum_{i=1}^{n} \chi_{\hat{C}_{i,n,1},Y_{n,1}}(\bar{\omega}_{\alpha_1}, \omega_{\alpha_2}, \ldots, \omega_{\alpha_{r(n)}}) \, \mathrm{d}\hat{\boldsymbol{P}}_{n,1}(\omega_{\alpha_1}, \omega_{\alpha_2}, \ldots, \omega_{\alpha_{r(n)}})$$

$$+ \int_{Y_{n,1}} \chi_{\hat{H}_{n,1},Y_{n,1}}(\bar{\omega}_{\alpha_1}, \omega_{\alpha_2}, \ldots, \omega_{\alpha_{r(n)}}) \, \mathrm{d}\hat{\boldsymbol{P}}_{n,1}(\omega_{\alpha_1}, \omega_{\alpha_2}, \ldots, \omega_{\alpha_{r(n)}})$$

$$> \int_{Y_{n,1}} \sum_{i=1}^{n} \chi_{\hat{C}_{i,n,1},Y_{n,1}}(\bar{\omega}_{\alpha_1}, \omega_{\alpha_2}, \ldots, \omega_{\alpha_{r(n)}}) \, \mathrm{d}\hat{\boldsymbol{P}}_{n,1}(\omega_{\alpha_1}, \omega_{\alpha_2}, \ldots, \omega_{\alpha_{r(n)}})$$

$$+2^{-1}\epsilon_0$$

$$= 2^{-1}\epsilon_0 + \int_{Y_{n,2}} \sum_{i=1}^{n} \chi_{\hat{C}_{i,n,2},Y_{n,2}}(\omega_{\alpha_2}, \ldots, \omega_{\alpha_{r(n)}}) \, \mathrm{d}\hat{\boldsymbol{P}}_{n,2}(\omega_{\alpha_2}, \ldots, \omega_{\alpha_{r(n)}})$$

$$= 2^{-1}\epsilon_0 + \sum_{i=1}^{n} \hat{\boldsymbol{P}}_{n,2}(\hat{C}_{i,n,2})$$

where the first equality follows from Proposition 11.75, the second equality follows from the definition of $\hat{B}_{n,2}$, $\mathcal{X}_{\alpha_1}$ is a probability measure space, and Fubini's Theorem 12.31, the third equality follows from the fact that $\hat{B}_{n,1} = \hat{H}_{n,1} \cup (\bigcup_{i=1}^{n} \hat{C}_{i,n,1})$ and the sets in the union are pairwise disjoint, the fourth equality follows from Proposition 11.92, the inequality follows from the conclusion of the previous paragraph, the fifth equality follows from the definition of $\hat{C}_{i,n,2}$'s and Fubini's Theorem 12.31, and the last equality follows from Proposition 11.75. Thus, we have $\hat{\boldsymbol{P}}_{n,2}(\hat{H}_{n,2}) := \hat{\boldsymbol{P}}_{n,2}(\hat{B}_{n,2} \setminus (\bigcup_{i=1}^{n} \hat{C}_{i,n,2})) > 2^{-1}\epsilon_0$. Define $H_{n,2} := \{(\omega_\alpha)_{\alpha\in\Lambda\setminus\{\alpha_1\}} \in \prod_{\alpha\in\Lambda\setminus\{\alpha_1\}} X_\alpha \mid (\hat{\omega}_\alpha)_{\alpha\in\Lambda} \in H_{n,1}$, where $\hat{\omega}_\alpha = \omega_\alpha, \forall \alpha \in \Lambda \setminus \{\alpha_1\}, \hat{\omega}_{\alpha_1} = \bar{\omega}_{\alpha_1}\}$. It is obvious that $\hat{H}_{n,2} = \{(\omega_\alpha)_{\alpha\in\hat{\Lambda}_n\setminus\{\alpha_1\}} \in \prod_{\alpha\in\hat{\Lambda}_n\setminus\{\alpha_1\}} X_\alpha \mid (\hat{\omega}_\alpha)_{\alpha\in\hat{\Lambda}_n} \in \hat{H}_{n,1}$, where $\hat{\omega}_\alpha = \omega_\alpha, \forall \alpha \in \hat{\Lambda}_n \setminus \{\alpha_1\}, \hat{\omega}_{\alpha_1} = \bar{\omega}_{\alpha_1}\}$. Clearly, $H_{n,2} = \hat{H}_{n,2} \times \prod_{\alpha\in\Lambda\setminus\hat{\Lambda}_n} X_\alpha$, $H_{n,2} \supseteq H_{n+1,2}$, and $\bigcap_{n=1}^{\infty} H_{n,2} = \emptyset$. Obviously, $\hat{H}_{n,2} \in \hat{\mathcal{B}}_{n,2}$.

By Radon–Nikodym Theorem 11.169 and Definition 14.7, $\hat{\boldsymbol{E}}_{n,2}(\chi_{\hat{H}_{n,2},Y_{n,2}} | \tilde{\mathcal{B}}_{\alpha_2,n,2})$ exists, where $\hat{\boldsymbol{E}}_{n,2}$ is with respect to measure $\hat{\boldsymbol{P}}_{n,2}$. Define $g_{n,2} : Y_{n,2} = \prod_{\alpha\in\hat{\Lambda}_n\setminus\{\alpha_1\}} X_\alpha \to [0, 1] \subset \mathbb{R}$ by

$$g_{n,2}(\hat{\omega}_{\alpha_2}, \ldots, \hat{\omega}_{\alpha_{r(n)}})$$

$$= \int_{Y_{n,2}} \chi_{\hat{H}_{n,2},Y_{n,2}}(\hat{\omega}_{\alpha_2}, \omega_{\alpha_3}, \ldots, \omega_{\alpha_{r(n)}}) \, \mathrm{d}\hat{\boldsymbol{P}}_{n,2}(\omega_{\alpha_2}, \ldots, \omega_{\alpha_{r(n)}})$$

$\forall (\hat{\omega}_{\alpha_2}, \ldots, \hat{\omega}_{\alpha_{r(n)}}) \in Y_{n,2}$. By Proposition 14.14, $g_{n,2} \in \hat{\boldsymbol{E}}_{n,2}(\chi_{\hat{H}_{n,2},Y_{n,2}} | \tilde{\mathcal{B}}_{\alpha_2,n,2})$, $g_{n,2}(\hat{\omega}_{\alpha_2}, \ldots, \hat{\omega}_{\alpha_{r(n)}}) = \bar{g}_{n,2}(\hat{\omega}_{\alpha_2})$, $\forall (\hat{\omega}_{\alpha_2}, \ldots, \hat{\omega}_{\alpha_{r(n)}}) \in Y_{n,2}$, and $\bar{g}_{n,2} : X_{\alpha_2} \to [0, 1] \subset \mathbb{R}$ is $\mathcal{B}_{\alpha_2}$-measurable. Since $\hat{\boldsymbol{P}}_{n,2}(\hat{H}_{n,2}) > 2^{-1}\epsilon_0$, then $\boldsymbol{E}_{\alpha_2}(\bar{g}_{n,2}) =$

$\hat{\boldsymbol{E}}_{n,2}(\hat{\boldsymbol{E}}_{n,2}(\chi_{\hat{H}_{n,2},Y_{n,2}}|\tilde{\mathcal{B}}_{\alpha_2,n,2})) = \hat{\boldsymbol{E}}_{n,2}(\chi_{\hat{H}_{n,2},Y_{n,2}}) = \hat{\boldsymbol{P}}_{n,2}(\hat{H}_{n,2}) > 2^{-1}\epsilon_0$, where $\boldsymbol{E}_{\alpha_2}$ is with respect to $\boldsymbol{P}_{\alpha_2}$, and the second equality follows from (a) of Proposition 14.11. This leads to the following inequalities:

$$\begin{aligned} 2^{-1}\epsilon_0 < \boldsymbol{E}_{\alpha_2}(\bar{g}_{n,2}) &\le \boldsymbol{P}_{\alpha_2}(\{\omega_{\alpha_2} \in X_{\alpha_2} \mid \bar{g}_{n,2}(\omega_{\alpha_2}) > 2^{-2}\epsilon_0\}) \\ &\quad +2^{-2}\epsilon_0 \boldsymbol{P}_{\alpha_2}(\{\omega_{\alpha_2} \in X_{\alpha_2} \mid \bar{g}_{n,2}(\omega_{\alpha_2}) \le 2^{-2}\epsilon_0\}) \\ &\le \boldsymbol{P}_{\alpha_2}(\{\omega_{\alpha_2} \in X_{\alpha_2} \mid \bar{g}_{n,2}(\omega_{\alpha_2}) > 2^{-2}\epsilon_0\}) + 2^{-2}\epsilon_0 \end{aligned}$$

Thus, we have $\boldsymbol{P}_{\alpha_2}(\{\omega_{\alpha_2} \in X_{\alpha_2} \mid \bar{g}_{n,2}(\omega_{\alpha_2}) > 2^{-2}\epsilon_0\}) > 2^{-2}\epsilon_0$.

Now, since $H_{n,2} \supseteq H_{n+1,2}$, by the definition of $\bar{g}_{n,2}$, Proposition 11.92, and Fubini's Theorem 12.31, we have $0 \le \bar{g}_{n+1,2}(\omega_{\alpha_2}) \le \bar{g}_{n,2}(\omega_{\alpha_2})$, $\forall \omega_{\alpha_2} \in X_{\alpha_2}$ (which requires us to work in the product measure space $\mathcal{Y}_{n+1,2}$ that includes both $\hat{H}_{n,2} \times \prod_{i=r(n)+1}^{r(n+1)} X_{\alpha_i}$ and $\hat{H}_{n+1,2}$). Therefore, we have $\boldsymbol{P}_{\alpha_2}(\Gamma_{n,2}) > 2^{-2}\epsilon_0$, where $\Gamma_{n,2} := \{\omega_{\alpha_2} \in X_{\alpha_2} \mid \bar{g}_{n,2}(\omega_{\alpha_2}) > 2^{-2}\epsilon_0\}$. Furthermore, $\Gamma_{n,2} \supseteq \Gamma_{n+1,2}$. By Proposition 11.5, we have $\boldsymbol{P}_{\alpha_2}(\bigcap_{n=1}^{\infty} \Gamma_{n,2}) = \lim_{n\in\mathbb{N}} \boldsymbol{P}_{\alpha_2}(\Gamma_{n,2}) \ge 2^{-2}\epsilon_0 > 0$. Then, $\exists \bar{\omega}_{\alpha_2} \in X_{\alpha_2}$ such that $\bar{\omega}_{\alpha_2} \in \bigcap_{n=1}^{\infty} \Gamma_{n,2}$. This implies that $\bar{g}_{n,2}(\bar{\omega}_{\alpha_2}) > 2^{-2}\epsilon_0$, $\forall n \in \mathbb{N}$. By the definition of $\bar{g}_{n,2}$, we have $\int_{Y_{n,2}} \chi_{\hat{H}_{n,2},Y_{n,2}}(\bar{\omega}_{\alpha_2}, \omega_{\alpha_3}, \ldots, \omega_{\alpha_{r(n)}})\, \mathrm{d}\hat{\boldsymbol{P}}_{n,2}(\omega_{\alpha_2}, \ldots, \omega_{\alpha_{r(n)}}) > 2^{-2}\epsilon_0$, $\forall n \in \mathbb{N}$.

Recursively, assume we have completed index $l \in \mathbb{N}$. Fix any $n \in \mathbb{N}$. By Proposition 12.21, we have $\mathcal{Y}_{n,l+1} := \prod_{\alpha\in\hat{\Lambda}_n\setminus\{\alpha_1,\ldots,\alpha_l\}} \mathcal{X}_\alpha =: (Y_{n,l+1}, \hat{\mathcal{B}}_{n,l+1}, \hat{\boldsymbol{P}}_{n,l+1})$ is a probability measure space, on which the collection of σ-algebras $(\tilde{\mathcal{B}}_{\alpha,n,l+1})_{\alpha\in\hat{\Lambda}_n\setminus\{\alpha_1,\ldots,\alpha_l\}}$ is independent, where $\tilde{\mathcal{B}}_{\alpha,n,l+1} := \{\prod_{\bar{\alpha}\in\hat{\Lambda}_n\setminus\{\alpha_1,\ldots,\alpha_l\}} \tilde{B}_{\bar{\alpha}} \in \hat{\mathcal{B}}_{n,l+1} \mid \tilde{B}_{\bar{\alpha}} = X_{\bar{\alpha}}, \forall \bar{\alpha} \in \hat{\Lambda}_n \setminus \{\alpha_1, \ldots, \alpha_l, \alpha\}\}$, $\forall \alpha \in \hat{\Lambda}_n \setminus \{\alpha_1, \ldots, \alpha_l\}$. Let

$$\hat{B}_{n,l+1} := \begin{cases} \prod_{\alpha\in\hat{\Lambda}_n\setminus\{\alpha_1,\ldots,\alpha_l\}} B_\alpha & \bar{\omega}_{\alpha_j} \in B_{\alpha_j}, j = 1, \ldots, l \\ \emptyset & \exists j \in \{1, \ldots, l\} \ni \cdot\, \bar{\omega}_{\alpha_j} \notin B_{\alpha_j} \end{cases}$$

and $\hat{C}_{i,n,l+1} := \begin{cases} \prod_{\alpha\in\hat{\Lambda}_n\setminus\{\alpha_1,\ldots,\alpha_l\}} C_{i,\alpha} & \bar{\omega}_{\alpha_j} \in C_{i,\alpha_j}, j = 1, \ldots, l \\ \emptyset & \exists j \in \{1, \ldots, l\} \ni \cdot\, \bar{\omega}_{\alpha_j} \notin C_{i,\alpha_j} \end{cases}$, $\forall i \in \{1, \ldots, n\}$. Clearly, $\hat{B}_{n,l+1} \supseteq \bigcup_{i=1}^{n} \hat{C}_{i,n,l+1}$ and the sets in the union are pairwise disjoint, and all sets involved are in $\hat{\mathcal{B}}_{n,l+1}$, by Proposition 12.25. Then, we have

$$\begin{aligned} &\hat{\boldsymbol{P}}_{n,l+1}(\hat{B}_{n,l+1}) \\ &= \int_{Y_{n,l+1}} \chi_{\hat{B}_{n,l+1},Y_{n,l+1}}(\omega_{\alpha_{l+1}}, \ldots, \omega_{\alpha_{r(n)}})\, \mathrm{d}\hat{\boldsymbol{P}}_{n,l+1}(\omega_{\alpha_{l+1}}, \ldots, \omega_{\alpha_{r(n)}}) \\ &= \int_{Y_{n,l}} \chi_{\hat{B}_{n,l},Y_{n,l}}(\bar{\omega}_{\alpha_l}, \omega_{\alpha_{l+1}}, \ldots, \omega_{\alpha_{r(n)}})\, \mathrm{d}\hat{\boldsymbol{P}}_{n,l}(\omega_{\alpha_l}, \omega_{\alpha_{l+1}}, \ldots, \omega_{\alpha_{r(n)}}) \end{aligned}$$

$$= \int_{Y_{n,l}} (\chi_{\hat{H}_{n,l},Y_{n,l}} + \sum_{i=1}^{n} \chi_{\hat{C}_{i,n,l},Y_{n,l}})(\bar{\omega}_{\alpha_l}, \omega_{\alpha_{l+1}}, \dots, \omega_{\alpha_{r(n)}})$$

$$\mathrm{d}\hat{\boldsymbol{P}}_{n,l}(\omega_{\alpha_l}, \omega_{\alpha_{l+1}}, \dots, \omega_{\alpha_{r(n)}})$$

$$= \int_{Y_{n,l}} \chi_{\hat{H}_{n,l},Y_{n,l}}(\bar{\omega}_{\alpha_l}, \omega_{\alpha_{l+1}}, \dots, \omega_{\alpha_{r(n)}})\, \mathrm{d}\hat{\boldsymbol{P}}_{n,l}(\omega_{\alpha_l}, \omega_{\alpha_{l+1}}, \dots, \omega_{\alpha_{r(n)}})$$

$$+ \int_{Y_{n,l}} \sum_{i=1}^{n} \chi_{\hat{C}_{i,n,l},Y_{n,l}}(\bar{\omega}_{\alpha_l}, \omega_{\alpha_{l+1}}, \dots, \omega_{\alpha_{r(n)}})$$

$$\mathrm{d}\hat{\boldsymbol{P}}_{n,l}(\omega_{\alpha_l}, \omega_{\alpha_{l+1}}, \dots, \omega_{\alpha_{r(n)}})$$

$$> 2^{-l}\epsilon_0 + \int_{Y_{n,l}} \sum_{i=1}^{n} \chi_{\hat{C}_{i,n,l},Y_{n,l}}(\bar{\omega}_{\alpha_l}, \omega_{\alpha_{l+1}}, \dots, \omega_{\alpha_{r(n)}})$$

$$\mathrm{d}\hat{\boldsymbol{P}}_{n,l}(\omega_{\alpha_l}, \omega_{\alpha_{l+1}}, \dots, \omega_{\alpha_{r(n)}})$$

$$= 2^{-l}\epsilon_0 + \int_{Y_{n,l+1}} \sum_{i=1}^{n} \chi_{\hat{C}_{i,n,l+1},Y_{n,l+1}}(\omega_{\alpha_{l+1}}, \dots, \omega_{\alpha_{r(n)}})$$

$$\mathrm{d}\hat{\boldsymbol{P}}_{n,l+1}(\omega_{\alpha_{l+1}}, \dots, \omega_{\alpha_{r(n)}})$$

$$= 2^{-l}\epsilon_0 + \sum_{i=1}^{n} \hat{\boldsymbol{P}}_{n,l+1}(\hat{C}_{i,n,l+1})$$

where the first equality follows from Proposition 11.75, the second equality follows from the definition of $\hat{B}_{n,l+1}$, $\mathcal{X}_{\alpha_l}$ is a probability measure space, and Fubini's Theorem 12.31, the third equality follows from the fact that $\hat{B}_{n,l} = \hat{H}_{n,l} \cup (\bigcup_{i=1}^{n} \hat{C}_{i,n,l})$ and the sets in the union are pairwise disjoint, the fourth equality follows from Proposition 11.92, the inequality follows from the conclusion of the previous paragraph, the fifth equality follows from the definition of $\hat{C}_{i,n,l+1}$'s and Fubini's Theorem 12.31, and the last equality follows from Proposition 11.75. Thus, we have $\hat{\boldsymbol{P}}_{n,l+1}(\hat{H}_{n,l+1}) := \hat{\boldsymbol{P}}_{n,l+1}(\hat{B}_{n,l+1} \setminus (\bigcup_{i=1}^{n} \hat{C}_{i,n,l+1})) > 2^{-l}\epsilon_0$. Define $H_{n,l+1} := \{(\omega_\alpha)_{\alpha\in\Lambda\setminus\{\alpha_1,\dots,\alpha_l\}} \in \prod_{\alpha\in\Lambda\setminus\{\alpha_1,\dots,\alpha_l\}} X_\alpha \mid (\hat{\omega}_\alpha)_{\alpha\in\Lambda} \in H_{n,1}$, where $\hat{\omega}_\alpha = \omega_\alpha, \forall\alpha \in \Lambda \setminus \{\alpha_1, \dots, \alpha_l\}, \hat{\omega}_{\alpha_j} = \bar{\omega}_{\alpha_j}, j = 1, \dots, l\}$. It is obvious that $\hat{H}_{n,l+1} = \{(\omega_\alpha)_{\alpha\in\hat{\Lambda}_n\setminus\{\alpha_1,\dots,\alpha_l\}} \in \prod_{\alpha\in\hat{\Lambda}_n\setminus\{\alpha_1,\dots,\alpha_l\}} X_\alpha \mid (\hat{\omega}_\alpha)_{\alpha\in\hat{\Lambda}_n} \in \hat{H}_{n,1}$, where $\hat{\omega}_\alpha = \omega_\alpha, \forall\alpha \in \hat{\Lambda}_n \setminus \{\alpha_1, \dots, \alpha_l\}, \hat{\omega}_{\alpha_j} = \bar{\omega}_{\alpha_j}, j = 1, \dots, l\}$. Clearly, $H_{n,l+1} = \hat{H}_{n,l+1} \times \prod_{\alpha\in\Lambda\setminus\hat{\Lambda}_n} X_\alpha$, $H_{n,l+1} \supseteq H_{n+1,l+1}$, and $\bigcap_{n=1}^{\infty} H_{n,l+1} = \emptyset$. Obviously, $\hat{H}_{n,l+1} \in \hat{\mathcal{B}}_{n,l+1}$.

By Radon–Nikodym Theorem 11.169 and Definition 14.7, $\hat{\boldsymbol{E}}_{n,l+1}(\chi_{\hat{H}_{n,l+1},Y_{n,l+1}} |\tilde{\mathcal{B}}_{\alpha_{l+1},n,l+1})$ exists, where $\hat{\boldsymbol{E}}_{n,l+1}$ is with respect to measure $\hat{\boldsymbol{P}}_{n,l+1}$. Define $g_{n,l+1} : Y_{n,l+1} = \prod_{\alpha\in\hat{\Lambda}_n\setminus\{\alpha_1,\ldots,\alpha_l\}} X_\alpha \to [0,1] \subset \mathbb{R}$ by, $\forall(\hat{\omega}_{\alpha_{l+1}},\ldots,\hat{\omega}_{\alpha_{r(n)}}) \in Y_{n,l+1}$,

$$g_{n,l+1}(\hat{\omega}_{\alpha_{l+1}},\ldots,\hat{\omega}_{\alpha_{r(n)}}) = \int_{Y_{n,l+1}} \chi_{\hat{H}_{n,l+1},Y_{n,l+1}}(\hat{\omega}_{\alpha_{l+1}},\omega_{\alpha_{l+2}},\ldots,\omega_{\alpha_{r(n)}}) \\ \mathrm{d}\hat{\boldsymbol{P}}_{n,l+1}(\omega_{\alpha_{l+1}},\ldots,\omega_{\alpha_{r(n)}})$$

By Proposition 14.14, $g_{n,l+1} \in \hat{\boldsymbol{E}}_{n,l+1}(\chi_{\hat{H}_{n,l+1},Y_{n,l+1}}|\tilde{\mathcal{B}}_{\alpha_{l+1},n,l+1})$, $g_{n,l+1}(\hat{\omega}_{\alpha_{l+1}},\ldots,\hat{\omega}_{\alpha_{r(n)}}) = \bar{g}_{n,l+1}(\hat{\omega}_{\alpha_{l+1}})$, $\forall(\hat{\omega}_{\alpha_{l+1}},\ldots,\hat{\omega}_{\alpha_{r(n)}}) \in Y_{n,l+1}$, and $\bar{g}_{n,l+1} : X_{\alpha_{l+1}} \to [0,1] \subset \mathbb{R}$ is $\mathcal{B}_{\alpha_{l+1}}$-measurable. Since $\hat{\boldsymbol{P}}_{n,l+1}(\hat{H}_{n,l+1}) > 2^{-l}\epsilon_0$, then $\boldsymbol{E}_{\alpha_{l+1}}(\bar{g}_{n,l+1}) = \hat{\boldsymbol{E}}_{n,l+1}(\hat{\boldsymbol{E}}_{n,l+1}(\chi_{\hat{H}_{n,l+1},Y_{n,l+1}}|\tilde{\mathcal{B}}_{\alpha_{l+1},n,l+1})) = \hat{\boldsymbol{E}}_{n,l+1}(\chi_{\hat{H}_{n,l+1},Y_{n,l+1}}) = \hat{\boldsymbol{P}}_{n,l+1}(\hat{H}_{n,l+1}) > 2^{-l}\epsilon_0$, where $\boldsymbol{E}_{\alpha_{l+1}}$ is with respect to $\boldsymbol{P}_{\alpha_{l+1}}$, and the second equality follows from (a) of Proposition 14.11. This leads to the following inequalities:

$$\begin{aligned} 2^{-l}\epsilon_0 &< \boldsymbol{E}_{\alpha_{l+1}}(\bar{g}_{n,l+1}) \\ &\le \boldsymbol{P}_{\alpha_{l+1}}(\{\omega_{\alpha_{l+1}} \in X_{\alpha_{l+1}} \mid \bar{g}_{n,l+1}(\omega_{\alpha_{l+1}}) > 2^{-l-1}\epsilon_0\}) \\ &\quad +2^{-l-1}\epsilon_0\boldsymbol{P}_{\alpha_{l+1}}(\{\omega_{\alpha_{l+1}} \in X_{\alpha_{l+1}} \mid \bar{g}_{n,l+1}(\omega_{\alpha_{l+1}}) \le 2^{-l-1}\epsilon_0\}) \\ &\le \boldsymbol{P}_{\alpha_{l+1}}(\{\omega_{\alpha_{l+1}} \in X_{\alpha_{l+1}} \mid \bar{g}_{n,l+1}(\omega_{\alpha_{l+1}}) > 2^{-l-1}\epsilon_0\}) + 2^{-l-1}\epsilon_0 \end{aligned}$$

Thus, we have $\boldsymbol{P}_{\alpha_{l+1}}(\{\omega_{\alpha_{l+1}} \in X_{\alpha_{l+1}} \mid \bar{g}_{n,l+1}(\omega_{\alpha_{l+1}}) > 2^{-l-1}\epsilon_0\}) > 2^{-l-1}\epsilon_0$.

Now, since $H_{n,l+1} \supseteq H_{n+1,l+1}$, by the definition of $\bar{g}_{n,l+1}$, Proposition 11.92, and Fubini's Theorem 12.31, we have $0 \le \bar{g}_{n+1,l+1}(\omega_{\alpha_{l+1}}) \le \bar{g}_{n,l+1}(\omega_{\alpha_{l+1}})$, $\forall\omega_{\alpha_{l+1}} \in X_{\alpha_{l+1}}$ (which requires us to work in the product measure space $\mathcal{Y}_{n+1,l+1}$ that includes both $\hat{H}_{n,l+1} \times \prod_{i=r(n)+1}^{r(n+1)} X_{\alpha_i}$ and $\hat{H}_{n+1,l+1}$). Therefore, we have $\boldsymbol{P}_{\alpha_{l+1}}(\Gamma_{n,l+1}) > 2^{-l-1}\epsilon_0$, where $\Gamma_{n,l+1} := \{\omega_{\alpha_{l+1}} \in X_{\alpha_{l+1}} \mid \bar{g}_{n,l+1}(\omega_{\alpha_{l+1}}) > 2^{-l-1}\epsilon_0\}$. Furthermore, $\Gamma_{n,l+1} \supseteq \Gamma_{n+1,l+1}$. By Proposition 11.5, we have $\boldsymbol{P}_{\alpha_{l+1}}(\bigcap_{n=1}^\infty \Gamma_{n,l+1}) = \lim_{n\in\mathbb{N}} \boldsymbol{P}_{\alpha_{l+1}}(\Gamma_{n,l+1}) \ge 2^{-l-1}\epsilon_0 > 0$. Then, $\exists\bar{\omega}_{\alpha_{l+1}} \in X_{\alpha_{l+1}}$ such that $\bar{\omega}_{\alpha_{l+1}} \in \bigcap_{n=1}^\infty \Gamma_{n,l+1}$. This implies that $\bar{g}_{n,l+1}(\bar{\omega}_{\alpha_{l+1}}) > 2^{-l-1}\epsilon_0$, $\forall n \in \mathbb{N}$. By the definition of $\bar{g}_{n,l+1}$, we have $\int_{Y_{n,l+1}} \chi_{\hat{H}_{n,l+1},Y_{n,l+1}}(\bar{\omega}_{\alpha_{l+1}},\omega_{\alpha_{l+2}},\ldots,\omega_{\alpha_{r(n)}})\mathrm{d}\hat{\boldsymbol{P}}_{n,l+1}(\omega_{\alpha_{l+1}},\ldots,\omega_{\alpha_{r(n)}}) > 2^{-l-1}\epsilon_0$, $\forall n \in \mathbb{N}$.

Thus, we may obtain $\bar{\omega}_{\alpha_i} \in \bigcap_{n=1}^\infty \Gamma_{n,i} \subseteq X_{\alpha_i}$, $\forall i \in \mathbb{N}$ with $\alpha_i \in \hat{\Lambda}$. By Axiom of Choice, there exists a $\bar{\omega} \in X = \prod_{\alpha\in\Lambda} X_\alpha$ such that $\pi_{\alpha_i}(\bar{\omega}) = \bar{\omega}_{\alpha_i}$, $\forall\alpha_i \in \hat{\Lambda}$. Fix any $n \in \mathbb{N}$. $\bar{\omega}_{\alpha_{r(n)}} \in \bigcap_{m=1}^\infty \Gamma_{m,r(n)}$. This implies that $\bar{\omega}_{\alpha_{r(n)}} \in \Gamma_{n,r(n)} = \{\omega_{\alpha_{r(n)}} \in X_{\alpha_{r(n)}} \mid \bar{g}_{n,r(n)}(\omega_{\alpha_{r(n)}}) > 2^{-r(n)}\epsilon_0\}$, and $\bar{g}_{n,r(n)}(\bar{\omega}_{\alpha_{r(n)}}) > 2^{-r(n)}\epsilon_0 > 0$. By definition of $\bar{g}_{n,r(n)}$, we have $0 < \bar{g}_{n,r(n)}(\bar{\omega}_{\alpha_{r(n)}}) =$

$\int_{Y_{n,r(n)}} \chi_{\hat{H}_{n,r(n)},Y_{n,r(n)}}(\bar{\omega}_{\alpha_{r(n)}})\, \mathrm{d}\hat{\boldsymbol{P}}_{n,r(n)}(\omega_{\alpha_{r(n)}}) = \chi_{\hat{H}_{n,r(n)},X_{\alpha_{r(n)}}}(\bar{\omega}_{\alpha_{r(n)}}) = 1$, where the last equality follows since it is the value of an indicator function. Hence, $\bar{\omega}_{\alpha_{r(n)}} \in \hat{H}_{n,r(n)}$. By the definition of $\hat{H}_{n,r(n)}$, we have $(\bar{\omega}_{\alpha_i})_{\alpha_i \in \hat{\Lambda}_n} \in \hat{H}_{n,1}$. This yields $\bar{\omega} \in H_{n,1}$. By the arbitrariness of n, we have $\bar{\omega} \in \bigcap_{n=1}^{\infty} H_{n,1} = \emptyset$. This is a contradiction. Hence, the hypothesis does not hold. We must have $\mu(B) \leq \sum_{i=1}^{\infty} \mu(C_i)$. Hence, (ii) of Proposition 11.32 holds.

By Proposition 11.32, μ admits a unique extension to a measure $\bar{\mu}$ on the algebra on X, $\mathcal{A}$, generated by $\mathcal{F}$. Clearly, $\bar{\mu}(X) = \mu(X) = 1$, since $X \in \mathcal{F}$. Then, $\bar{\mu}$ is finite. By Carathéodory Extension Theorem 11.19, there exists a unique measure $\boldsymbol{P}$ on the measurable space $(X, \sigma(\mathcal{F}))$ that is an extension of $\bar{\mu}$. Then, we have a probability measure space $(X, \mathcal{B} := \sigma(\mathcal{F}), \boldsymbol{P})$ as we seek. It is straightforward to prove the collection of σ-algebras $(\bar{\mathcal{B}}_\alpha)_{\alpha \in \Lambda}$ is independent by Definition 14.5. This completes the proof of the theorem. □

Definition 14.16 Let $\boldsymbol{\Omega} := (\Omega, \mathcal{B}, \boldsymbol{P})$ be a probability measure space, $m \in \mathbb{N}$, $x : \Omega \to \mathbb{R}^m$ be an $\mathbb{R}^m$-valued random variable. Define a mapping $L_x : \mathcal{B}_{\mathrm{B}}(\mathbb{R}^m) \to [0, 1] \subset \mathbb{R}$ by $L_x(E) := \boldsymbol{P}(x_{\mathrm{inv}}(E))$, $\forall E \in \mathcal{B}_{\mathrm{B}}(\mathbb{R}^m)$. Then, L_x is a probability measure on the measurable space $(\mathbb{R}^m, \mathcal{B}_{\mathrm{B}}(\mathbb{R}^m))$ and is said to be the *law* of x. Since L_x is a probability measure, then we may define $F : \mathbb{R}^m \to [0, 1] \subset \mathbb{R}$ by $F(z) = L_x(\{\bar{x} \in \mathbb{R}^m \mid \bar{x} \overset{\leq}{=} z\})$, $\forall z \in \mathbb{R}^m$. Then, by Proposition 12.51, F is a cumulative distribution function of L_x. If $L_x \ll \mu_{\mathrm{B}m}$, we will let $f = \frac{\mathrm{d}L_x}{\mathrm{d}\mu_{\mathrm{B}m}}$ a.e. in $\mathbb{R}^m$ and say f is the *probability density function* of L_x. ◇

Clearly, the probability density function f is determined almost everywhere on $\mathbb{R}^m$, if it exists, which is a version of $\frac{\mathrm{d}L_x}{\mathrm{d}\mu_{\mathrm{B}m}}$. If probability density function exists, then, $\forall E \in \mathcal{B}_{\mathrm{B}}(\mathbb{R}^m)$, we have $\boldsymbol{P}(x_{\mathrm{inv}}(E)) = L_x(E) = \int_E f\, \mathrm{d}\mu_{\mathrm{B}m}$. The relationship between probability density function f and cumulative distribution function F is that $\forall s_1, s_2 \in \mathbb{R}^m$ with $s_1 \overset{\leq}{=} s_2$, we have $\boldsymbol{P}(x_{\mathrm{inv}}(\mathrm{r}_{s_1,s_2})) = L_x(\mathrm{r}_{s_1,s_2}) = \int_{\mathrm{r}_{s_1,s_2}} f\, \mathrm{d}\mu_{\mathrm{B}m} = \Delta_F(\mathrm{r}_{s_1,s_2})$, and $F(z) = \lim_{k \to -\infty} \int_{\mathrm{r}_{k\mathbf{1}_m,z}} f(x)\, \mathrm{d}\mu_{\mathrm{B}m}(x)$, $\forall z \in \mathbb{R}^m$.

Example 14.17 Let $\mathcal{I} := (((0, 1), |\cdot|), \mathcal{B}, \mu)$ be the finite metric measure subspace of $\mathbb{R}$. Let $x : (0, 1) \to [0, 1] \subset \mathbb{R}$ be a random variable given by $x(\omega) = \omega$, $\forall \omega \in \mathcal{I}$. Then, any cumulative distribution function $F : \mathbb{R} \to [0, 1] \subset \mathbb{R}$ with F being of bounded variation, $\lim_{z \to -\infty} F(z) = 0$, $\lim_{z \to \infty} F(z) = 1$, and $T_F = 1$, we seek a function $h : (0, 1) \to \mathbb{R}$ such that the random variable $y := h \circ x : \mathcal{I} \to \mathbb{R}$ admits F as a cumulative distribution function of L_y, which is the probability measure on the measurable space $(\mathbb{R}, \mathcal{B}_{\mathrm{B}}(\mathbb{R}))$. Since $T_F = 1$ and $\lim_{z \to -\infty} F(z) = 0$ and $\lim_{z \to \infty} F(z) = 1$, then $\forall z_1, z_2 \in \mathbb{R}$ with $z_1 \leq z_2$, we have $\Delta_F(\mathrm{r}_{z_1,z_2}) \geq 0$ and hence $F(z_1) \leq F(z_2)$. Since F is of bounded variation, then F is continuous on the right. We select h to be $h(\alpha) = \inf\{z \in \mathbb{R} \mid F(z) \geq \alpha\} \in \mathbb{R}$, $\forall \alpha \in \mathcal{I}$. Then, it is straightforward to prove that $\mu(\{\omega \in \mathcal{I} \mid y(\omega) \leq z\}) = \mu(\{\omega \in \mathcal{I} \mid h(x(\omega)) \leq z\}) = \mu(\{\omega \in \mathcal{I} \mid \omega \leq F(z)\}) = F(z)$. Hence, the random variable $y = h(x)$ admits the F as a cumulative distribution function for its law L_y. Thus, for any cumulative distribution function $F : \mathbb{R} \to [0, 1] \subset \mathbb{R}$ satisfying the stated assumptions, there exists a probability measure space $\mathcal{I}$ and a random variable $y : \mathcal{I} \to \mathbb{R}$ such that y admits F as a cumulative distribution function for its law L_y. ◇

Proposition 14.18 *Let* $\boldsymbol{\Omega} := (\Omega, \mathcal{B}, \boldsymbol{P})$ *be a probability measure space,* $x_i : \Omega \to \mathbb{R}^{n_i}$ *be an* $\mathbb{R}^{n_i}$*-valued random variable,* $n_i \in \mathbb{N}$, $i = 1, 2$, *and* $x := (x_1, x_2)$ *be an* $\mathbb{R}^n$*-valued random variable,* $n := n_1 + n_2$, *with probability density function* $f_x(z)$, $f_x : \mathbb{R}^n \to [0, \infty) \subset \mathbb{R}$. *Then,* x_1 *admits probability density function* $f_{x_1} : \mathbb{R}^{n_1} \to [0, \infty) \subset \mathbb{R}$, *where* $f_{x_1}(z_1) := \begin{cases} \int_{\mathbb{R}^{n_2}} f_x(z_1, z_2)\, d\mu_{Bn_2}(z_2) & \text{if } \int_{\mathbb{R}^{n_2}} f_x(z_1, z_2)\, d\mu_{Bn_2}(z_2) \in \mathbb{R} \\ 0 & \text{if } \int_{\mathbb{R}^{n_2}} f_x(z_1, z_2)\, d\mu_{Bn_2}(z_2) = \infty \end{cases}$, $\forall z_1 \in \mathbb{R}^{n_1}$ *and* $z := (z_1, z_2)$ *is partitioned as is for* x.

Proof Let L_{x_1} be the law for x_1. Then,

$$\begin{aligned} L_{x_1}(E_1) &= \boldsymbol{P}(\{(x_1, x_2) \in \mathbb{R}^{n_1+n_2} \mid x_1 \in E_1\}) \\ &= \int_{E_1 \times \mathbb{R}^{n_2}} f_x(z)\, d\mu_{Bn}(z) = \int_{E_1} f_{x_1}(z_1)\, d\mu_{Bn_1}(z_1) \end{aligned}$$

$\forall E_1 \in \mathcal{B}_B(\mathbb{R}^{n_1})$, where the first and second equalities follow from Definition 14.16 and the third equality follows from Fubini's Theorem 12.31. By Proposition 11.116, we have L_{x_1} is the measure with kernel f_{x_1} over $\mathbb{R}^{n_1}$. By Proposition 12.75, we have $\frac{dL_{x_1}}{d\mu_{Bn_1}} = f_{x_1}$ a.e. in $\mathbb{R}^{n_1}$, and hence x_1 admits probability density function f_{x_1}. □

Proposition 14.19 *Let* $\boldsymbol{\Omega} := (\Omega, \mathcal{B}, \boldsymbol{P})$ *be a probability measure space,* $m \in \mathbb{N}$, X *be an* $\mathbb{R}^m$*-valued random variable with probability density function* $f_X : \mathbb{R}^m \to [0, \infty) \subset \mathbb{R}$, *and* $V \in \mathrm{B}(\mathbb{R}^m, \mathbb{R}^m)$ *be invertible. Then,* $Y := VX$ *is an* $\mathbb{R}^m$*-valued random variable with probability density function* $f_Y : \mathbb{R}^m \to [0, \infty) \subset \mathbb{R}$, *where* $f_Y(y) = \frac{1}{|\det(V)|} f_X(V^{-1}y)$.

Proof Let the law for X be L_X and the law for Y be L_Y. Then, $\forall E \in \mathcal{B}_B(\mathbb{R}^m)$,

$$\begin{aligned} L_Y(E) &= \boldsymbol{P}(\{\omega \in \Omega \mid Y(\omega) \in E\}) = \boldsymbol{P}(\{\omega \in \Omega \mid X(\omega) \in V^{-1}(E)\}) \\ &= L_X(V^{-1}(E)) = \int_{V^{-1}(E)} f_X(x)\, d\mu_{Bm}(x) \end{aligned}$$

$$= \int_E \left|\det(V^{-1})\right| f_X(V^{-1}y)\, d\mu_{Bm}(y)$$

$$= \int_E \frac{1}{|\det(V)|} f_X(V^{-1}y)\, d\mu_{Bm}(y) = \int_E f_Y(y)\, d\mu_{Bm}(y)$$

where the first equality follows from Definition 14.16, the second equality follows from the fact that $Y = VX$, the third equality follows from Definition 14.16, the fourth equality follows from Definition 14.16 and Proposition 11.167, the fifth equality follows from the Change of Variable Theorem 12.91 with $y = Vx$, the sixth equality follows from properties of matrices, and the last equality follows from the definition of f_Y. Hence, we observe that L_Y is the measure with kernel f_Y over $\mathbb{R}^m$. By Propositions 11.116 and 12.75, we have $\frac{dL_Y}{d\mu_{Bm}} = f_Y$ a.e. in $\mathbb{R}^m$, and hence Y admits probability density function f_Y.

This completes the proof of the proposition. □

Proposition 14.20 *Let $\boldsymbol{\Omega} := (\Omega, \mathcal{B}, \boldsymbol{P})$ be a probability measure space, $m \in \mathbb{N}$, $n_1, \ldots, n_m \in \mathbb{N}$, $x_i : \Omega \to \mathbb{R}^{n_i}$ be an $\mathbb{R}^{n_i}$-valued random variable with probability density function $f_i : \mathbb{R}^{n_i} \to [0, \infty) \subset \mathbb{R}$, $i = 1, \ldots, m$. Then, the random variables $x_1, \ldots, x_m$ are independent if, and only if, the $\mathbb{R}^n$-valued random variable $x := (x_1, \ldots, x_m) : \Omega \to \mathbb{R}^n$ with $n := \sum_{i=1}^m n_i \in \mathbb{N}$ admits probability density function $f : \mathbb{R}^n \to [0, \infty) \subset \mathbb{R}$, defined by $f(z) = \prod_{i=1}^m f_i(z_i)$, $\forall z := (z_1, \ldots, z_m) \in \mathbb{R}^n$ with $z_i \in \mathbb{R}^{n_i}$, $i = 1, \ldots, m$.*

Proof "Sufficiency" By Proposition 11.39, x is an $\mathbb{R}^n$-valued random variable. $\forall E_i \in \mathcal{B}_B(\mathbb{R}^{n_i})$, $i = 1, \ldots, m$, let $\bar{E}_i := \prod_{j=1}^m B_{j,i}$, where $B_{j,i} = \mathbb{R}^{n_j}$, $\forall j \in \bar{J} := \{1, \ldots, m\}$ with $j \neq i$. Then,

$$\boldsymbol{P}(\bigcap_{i=1}^m \bar{E}_i) = \boldsymbol{P}(\prod_{i=1}^m E_i) =: \boldsymbol{P}(\bar{E}) = \int_{\bar{E}} f(z)\, d\mu_{Bn}(z)$$

$$= \int_{\bar{E}} \prod_{i=1}^m f_i(z_i)\, d\mu_{Bn}(z_1, \ldots, z_m)$$

$$= \int_{E_1} \cdots \int_{E_m} \prod_{i=1}^m f_i(z_i)\, d\mu_{Bn_m}(z_m) \cdots d\mu_{Bn_1}(z_1)$$

$$= \prod_{i=1}^m \int_{E_i} f_i(z_i)\, d\mu_{Bn_i}(z_i) = \prod_{i=1}^m \boldsymbol{P}(\bar{E}_i)$$

where the second equality follows from Definition 14.16, the fourth equality follows from Fubini's Theorem 12.31, the fifth equality follows from Proposition 11.92, and the last equality follows from Definition 14.16. Hence, the random variables $x_1, \ldots, x_m$ are independent.

"Necessity" By Proposition 11.39, x is an $\mathbb{R}^n$-valued random variable. Let L_x be the law of x and L_{x_i} be the law of x_i, $i = 1, \ldots, m$. $\forall E_i \in \mathcal{B}_{\mathrm{B}}(\mathbb{R}^{n_i})$, $i = 1, \ldots, m$, let $\bar{E}_i := \prod_{j=1}^{m} B_{j,i}$, where $B_{j,i} = \mathbb{R}^{n_j}$, $\forall j \in \bar{J} := \{1, \ldots, m\}$ with $j \neq i$. By the independence of random variables $x_1, \ldots, x_m$, we have

$$L_x(\bigcap_{i=1}^{m} \bar{E}_i) =: L_x(\bar{E}) = \prod_{i=1}^{m} L_x(\bar{E}_i) = \prod_{i=1}^{m} L_{x_i}(E_i)$$

$$= \prod_{i=1}^{m} \int_{E_i} f_i(z_i)\,\mathrm{d}\mu_{\mathrm{B}n_i}(z_i) = \int_{\bar{E}} f(z)\,\mathrm{d}\mu_{\mathrm{B}n}(z)$$

where the first equality follows from the independence assumption, the second equality follows from Definition 14.16, the third equality follows from the assumption that x_i admits probability density function f_i, $i = 1, \ldots, m$, and the last equality follows from Fubini's Theorem 12.31. By Lemma 12.103, L_x equals the finite measure on $\mathbb{R}^n$ with kernel f. Then, the $\mathbb{R}^n$-valued random variable x admits probability density function f, by Definition 14.16 and Proposition 12.75.

This completes the proof of the proposition. □

Proposition 14.21 *Let $p \in [1, \infty) \subset \mathbb{R}$, $\boldsymbol{\Omega} := (\Omega, \mathcal{B}, \boldsymbol{P})$ be a probability measure space, $x : \boldsymbol{\Omega} \to [0, \infty] \subset \mathbb{R}_e$ be a nonnegative extended real-valued function that is $\mathcal{B}$-measurable. Then, we have $\boldsymbol{E}(x^p) := \int_{\Omega}(x(\omega))^p\,\mathrm{d}\boldsymbol{P}(\omega) = \int_{(0,\infty)} pz^{p-1}\boldsymbol{P}(\{\omega \in \Omega \mid x(\omega) > z\})\,\mathrm{d}\mu_{\mathrm{B}}(z)$.*

Proof We will distinguish two exhaustive and mutually exclusive cases: Case 1: $\boldsymbol{P}(\{\omega \in \Omega \mid x(\omega) = \infty\}) > 0$, and Case 2: $\boldsymbol{P}(\{\omega \in \Omega \mid x(\omega) = \infty\}) = 0$.

Case 1: $\boldsymbol{P}(\{\omega \in \Omega \mid x(\omega) = \infty\}) > 0$. Note that $\boldsymbol{E}(x^p) := \int_{\Omega}(x(\omega))^p \cdot \mathrm{d}\boldsymbol{P}(\omega) = \infty = \int_{(0,\infty)} pz^{p-1}\boldsymbol{P}(\{\omega \in \Omega \mid x(\omega) > z\})\,\mathrm{d}\mu_{\mathrm{B}}(z)$, where the equalities follow from Definition 11.79.

Case 2: $\boldsymbol{P}(\{\omega \in \Omega \mid x(\omega) = \infty\}) = 0$. Let $\bar{x} : \boldsymbol{\Omega} \to [0, \infty) \subset \mathbb{R}$ be the random variable defined by $\bar{x}(\omega) = \begin{cases} x(\omega) & \text{if } x(\omega) < \infty \\ 0 & \text{if } x(\omega) = \infty \end{cases}$. Then, we have $\boldsymbol{E}(x^p) := \int_{\Omega}(x(\omega))^p\,\mathrm{d}\boldsymbol{P}(\omega) = \int_{\Omega}(\bar{x}(\omega))^p\,\mathrm{d}\boldsymbol{P}(\omega) = \boldsymbol{E}(\bar{x}^p)$, where the first equality follows from Definition 11.79 and the second equality follows from Definition 14.1.

Fix any $n \in \mathbb{N}$. Let $\bar{x}_n := \bar{x} \wedge n$, $L_n : \mathcal{B}_{\mathrm{B}}(\mathbb{R}) \to [0, 1] \subset \mathbb{R}$ be the law of $\bar{x}_n$, and $F_n : \mathbb{R} \to [0, 1] \subset \mathbb{R}$ be the cumulative distribution function of L_n as defined in Definition 14.16. Note that $\infty > n^p \geq \boldsymbol{E}(\bar{x}_n^p) = \int_{\Omega}(\bar{x}_n(\omega))^p\,\mathrm{d}\boldsymbol{P}(\omega) = \int_{\mathbb{R}} \chi_{[0,\infty),\mathbb{R}} z^p\,\mathrm{d}F_n(z) = \int_{(-\infty,0]} 0\,\mathrm{d}F_n(z) + \int_{(0,\infty)} z^p\,\mathrm{d}F_n(z) = \int_{(0,\infty)} z^p\,\mathrm{d}F_n(z) = \int_{(0,\infty)} z^p\,\mathrm{d}(F_n(z) - 1) = -\int_{(0,\infty)} p \cdot z^{p-1}(F_n(z) - 1)\,\mathrm{d}\mu_{\mathrm{B}}(z) = \int_{(0,\infty)} pz^{p-1}(1 - F_n(z))\,\mathrm{d}\mu_{\mathrm{B}}(z) = \int_{(0,\infty)} pz^{p-1} \cdot \boldsymbol{P}(\{\omega \in \Omega \mid \bar{x}_n(\omega) > z\})\,\mathrm{d}\mu_{\mathrm{B}}(z)$, where the second inequality follows from the fact that $\bar{x}_n(\omega) \leq n$, $\forall \omega \in \Omega$, the first equality follows from Definition 14.1, the second equality follows from Definition 11.70, the third equality follows from Proposition 11.89 and Definition 12.57, the fourth equality follows from Proposition 11.75, the fifth equality follows from the fact that $F_n|_{[0,\infty)}$

and $(F_n - 1)|_{[0,\infty)}$ are cumulative distribution functions for the same measure space, which is the measure subspace of $(\mathbb{R}, \mathcal{B}_B(\mathbb{R}), L_{\bar{x}})$ to the interval $(0, \infty)$, the sixth equality follows from Integration by Parts Theorem 12.101, Integrability Theorem A.10, and Theorem 29.8 of Bartle (1976), and the eighth equality follows from Definition 14.16.

Therefore, we have $\boldsymbol{E}(x^p) := \int_\Omega (x(\omega))^p \,\mathrm{d}\boldsymbol{P}(\omega) = \int_\Omega (\bar{x}(\omega))^p \,\mathrm{d}\boldsymbol{P}(\omega) = \lim_{n\in\mathbb{N}} \int_\Omega (\bar{x}_n(\omega))^p \,\mathrm{d}\boldsymbol{P}(\omega) = \lim_{n\in\mathbb{N}} \int_{(0,\infty)} pz^{p-1}\boldsymbol{P}(\{\omega \in \Omega \mid \bar{x}_n(\omega) > z\}) \,\mathrm{d}\mu_B(z) = \int_{(0,\infty)} pz^{p-1}\boldsymbol{P}(\{\omega \in \Omega \mid \bar{x}(\omega) > z\}) \,\mathrm{d}\mu_B(z) = \int_{(0,\infty)} pz^{p-1} \cdot \boldsymbol{P}(\{\omega \in \Omega \mid x(\omega) > z\}) \,\mathrm{d}\mu_B(z)$, where the first equality follows from the second to last paragraph, the second equality follows from Monotone Convergence Theorem 11.81, the third equality follows from the last paragraph, the fourth equality follows from Monotone Convergence Theorem 11.81, and the last equality follows from the fact that $\boldsymbol{P}(\{\omega \in \Omega \mid x(\omega) = \infty\}) = 0$.

This completes the proof of the proposition. □

Definition 14.22 Let $\mathcal{X} := (X, \mathcal{B})$ and $\mathcal{Y} := (Y, \hat{\mathcal{B}})$ be measurable spaces, $D \subseteq X$, and $f : D \to \mathcal{Y}$. f is said to be *$\hat{\mathcal{B}}$-in-$\mathcal{B}$-measurable* if $\forall E \in \hat{\mathcal{B}}$, we have $f_{\text{inv}}(E) \in \mathcal{B}$. ◇

Here, the concept of measurable function is extended to functions between measurable spaces. Relating to the original concept of measurable functions Definition 11.33, where $f : D \to \mathcal{Z}$, where $\mathcal{Z} := (Z, \mathcal{O}_\mathcal{Z})$ is a topological space, f is $\mathcal{B}$-measurable if, and only if, it is $\mathcal{B}_B(\mathcal{Z})$-in-$\mathcal{B}$-measurable.

Proposition 14.23 *Let $\mathcal{X} := (X, \mathcal{B})$ and $\mathcal{Y} := (Y, \hat{\mathcal{B}})$ be measurable spaces, $D \subseteq X$, $f : D \to \mathcal{Y}$, $\mathcal{F} \subseteq \hat{\mathcal{B}}$ is a π-system on Y, and $\sigma(\mathcal{F}) = \hat{\mathcal{B}}$ (where $\sigma(\mathcal{F})$ is the σ-algebra generated by $\mathcal{F}$ as defined in Definition 14.2). Then, f is $\hat{\mathcal{B}}$-in-$\mathcal{B}$-measurable if, and only if, $\forall E \in \mathcal{F}$, $f_{\text{inv}}(E) \in \mathcal{B}$.*

Proof "Necessity" is obvious. "Sufficiency" Let f satisfy, $\forall E \in \mathcal{F}$, $f_{\text{inv}}(E) \in \mathcal{B}$. Define $\mathcal{B}_Y := \{E \subseteq Y \mid f_{\text{inv}}(E) \in \mathcal{B}\}$. Clearly, $\mathcal{F} \subseteq \mathcal{B}_Y$. We will show that $\mathcal{B}_Y$ is a σ-algebra on Y. Then, $\hat{\mathcal{B}} = \sigma(\mathcal{F}) \subseteq \mathcal{B}_Y$ and the result is established.

Clearly, $\emptyset, Y \in \mathcal{F}$ by the definition of π-system (Definition 12.17), which implies that $f_{\text{inv}}(\emptyset) = \emptyset \in \mathcal{B}$ and $f_{\text{inv}}(Y) = D \in \mathcal{B}$, then $\emptyset, Y \in \mathcal{B}_Y$. $\forall E \in \mathcal{B}_Y$, by Proposition 2.5, $f_{\text{inv}}(Y \setminus E) = f_{\text{inv}}(Y) \setminus f_{\text{inv}}(E) = D \setminus f_{\text{inv}}(E) \in \mathcal{B}$. Hence, $Y \setminus E \in \mathcal{B}_Y$. $\forall (E_i)_{i=1}^\infty \subseteq \mathcal{B}_Y$, we have $f_{\text{inv}}(E_i) \in \mathcal{B}$, $\forall i \in \mathbb{N}$. Then, by Proposition 2.5, $f_{\text{inv}}(\bigcup_{i=1}^\infty E_i) = \bigcup_{i=1}^\infty f_{\text{inv}}(E_i) \in \mathcal{B}$. Then, $\bigcup_{i=1}^\infty E_i \subset \mathcal{B}_Y$ This shows $\mathcal{B}_Y$ is a σ-algebra on Y.

This completes the proof of the proposition. □

Proposition 14.24 *Let $\mathcal{X} := (X, \mathcal{B})$ and $\mathcal{Y} := (Y, \hat{\mathcal{B}})$ be measurable spaces, $\mathcal{Z} := (Z, \mathcal{O}_\mathcal{Z})$ be a topological space, $D_1 \subseteq X$, $D_2 \subseteq Y$, $f : D_1 \to D_2$ be $\hat{\mathcal{B}}$-in-$\mathcal{B}$-measurable, and $g : D_2 \to \mathcal{Z}$ be $\hat{\mathcal{B}}$-measurable. Then, $g \circ f : D_1 \to \mathcal{Z}$ is $\mathcal{B}$-measurable.*

Proof $\forall O \in \mathcal{O}_{\mathcal{Z}}$, $g_{\text{inv}}(O) \in \hat{\mathcal{B}}$, since g is $\hat{\mathcal{B}}$-measurable. By f being $\hat{\mathcal{B}}$-in-$\mathcal{B}$-measurable, we have $f_{\text{inv}}(g_{\text{inv}}(O)) \in \mathcal{B}$. By the arbitrariness of O, $g \circ f$ is $\mathcal{B}$-measurable. □

14.2 Gaussian Random Variables and Vectors

Definition 14.25 Let $\boldsymbol{\Omega} := (\Omega, \mathcal{B}, \boldsymbol{P})$ be a probability measure space, $m \in \mathbb{N}$, $x : \Omega \to \mathbb{R}^m$ be an $\mathbb{R}^m$-valued random variable. We say that x is an *$\mathbb{R}^m$-valued Gaussian (normal) random variable* with *mean* $\bar{x} \in \mathbb{R}^m$ and *covariance* $K \in \mathrm{B}(\mathbb{R}^m, \mathbb{R}^m)$, where $K \in \bar{\mathcal{S}}_{+\,\mathbb{R}^m}$, if $\boldsymbol{E}(x) = \bar{x}$ and $\boldsymbol{E}((x - \bar{x}) \otimes (x - \bar{x})) = K$ and a cumulative distribution function for the law L_x of x is

$$F(z) = \lim_{k \to -\infty} \int_{r_{k\mathbf{1}_m, z}} \frac{1}{\sqrt{2\pi}^{m} (\det(K))^{1/2}} \cdot \exp(-\frac{1}{2}\Big\langle x - \bar{x}, K^{-1}(x - \bar{x})\Big\rangle_{\mathbb{R}^m}) \, d\mu_{\mathrm{B}m}(x)$$

$\forall z \in \mathbb{R}^m$, or the probability density function for L_x is $f : \mathbb{R}^m \to [0, \infty) \subset \mathbb{R}$, defined by

$$f(z) = \frac{1}{\sqrt{2\pi}^{m} (\det(K))^{1/2}} \exp(-\frac{1}{2}\Big\langle x - \bar{x}, K^{-1}(x - \bar{x})\Big\rangle_{\mathbb{R}^m}), \quad \forall z \in \mathbb{R}^m$$

We will write $x \sim N(\bar{x}, K)$ to denote that x is an $\mathbb{R}^m$-valued Gaussian random variable with mean $\bar{x}$ and covariance K. ◇

Example 14.26 Let $m \in \mathbb{N}$, $\bar{x} \in \mathbb{R}^m$, and $K \in \mathrm{B}(\mathbb{R}^m, \mathbb{R}^m)$, where $K \in \bar{\mathcal{S}}_{+\,\mathbb{R}^m}$. By Spectral Theory Theorem 13.52, there exists a unitary matrix $V \in \mathrm{B}(\mathbb{R}^m, \mathbb{R}^m)$ and a diagonal matrix $D = \text{block diagonal}\,(\sigma_1^2, \ldots, \sigma_m^2) \in \bar{\mathcal{S}}_{+\,\mathbb{R}^m}$ such that $K = V'DV$. If x is an $\mathbb{R}^m$-valued Gaussian random variable with mean $\bar{x}$ and covariance K, then $y = Vx$ is $\mathbb{R}^m$-valued Gaussian random variable with mean $\bar{y} := V\bar{x} =: (\bar{y}_1, \ldots, \bar{y}_m)$ and covariance $VKV' = D$. Then, by Example 33.15 of Bartle (1976), Example 14.17, and Fundamental Theorem on Modeling 14.15, there exists a probability measure space $\boldsymbol{\Omega} := (\Omega, \mathcal{B}, \boldsymbol{P})$ and m independent random variables $y_i : \Omega \to \mathbb{R}$, $i = 1, \ldots, m$, such that y_i is a Gaussian normal random variable with mean $\bar{y}_i$ and variance σ_i^2, $i = 1, \ldots, m$. $y_i \sim N(\bar{y}_i, \sigma_i^2)$. Then, by Proposition 14.20, we have the $\mathbb{R}^m$-valued random variable $y := (y_1, \ldots, y_m)$

admits the probability density function $f_y : \mathbb{R}^m \to [0, \infty) \subset \mathbb{R}$, given by

$$f_y(z) = \prod_{i=1}^{m} \frac{1}{\sqrt{2\pi}\sigma_i} \exp(-\frac{1}{2}\sigma_i^{-2}(z_i - \bar{y}_i)^2)$$

$$= \frac{1}{\sqrt{2\pi}^m (\det(D))^{1/2}} \exp(-\frac{1}{2}\langle z - \bar{y}, D^{-1}(z - \bar{y})\rangle_{\mathbb{R}^m}); \quad \forall z \in \mathbb{R}^m$$

Thus, by Proposition 14.19, $x = V'y \sim N(\bar{x}, K)$ as desired. ◇

Proposition 14.27 *Let* $\boldsymbol{\Omega} := (\Omega, \mathcal{B}, \boldsymbol{P})$ *be a probability measure space,* $m \in \mathbb{N}$, $x_i : \Omega \to \mathbb{R}^{n_i}$ *be an* $\mathbb{R}^{n_i}$*-valued random variable,* $n_i \in \mathbb{N}$, $i = 1, \ldots, m$, *and* $x := (x_1, \ldots, x_m) \sim N(\bar{x}, K)$, *where* $n := \sum_{i=1}^{m} n_i \in \mathbb{N}$, $\bar{x} \in \mathbb{R}^n$ *and* $K \in \bar{\mathcal{S}}_{+\,\mathbb{R}^n}$. *Then,* $x_1, \ldots, x_m$ *are independent if, and only if,* $K =$ block diagonal $(K_1, \ldots, K_m) \in \bar{\mathcal{S}}_{+\,\mathbb{R}^n}$ *with* $K_i \in \bar{\mathcal{S}}_{+\,\mathbb{R}^{n_i}}$, $i = 1, \ldots, m$.

Proof "Sufficiency" Note that x admits the probability density function $f_x : \mathbb{R}^n \to \mathbb{R}_+$, defined by, $\forall z \in \mathbb{R}^n$,

$$f_x(z) = \frac{1}{\sqrt{2\pi}^n (\det(K))^{1/2}} \exp(-\frac{1}{2}\langle z - \bar{x}, K^{-1}(z - \bar{x})\rangle_{\mathbb{R}^n})$$

$$= \prod_{i=1}^{m} \frac{1}{\sqrt{2\pi}^{n_i} (\det(K_i))^{1/2}} \exp(-\frac{1}{2}\langle z_i - \bar{x}_i, K_i^{-1}(z_i - \bar{x}_i)\rangle_{\mathbb{R}^{n_i}})$$

where $z := (z_1, \ldots, z_m)$ and $\bar{x} := (\bar{x}_1, \ldots, \bar{x}_m)$ are partitioned as is for x and the second equality follows from the block diagonal structure of K. By Proposition 14.18, we have x_i admits probability density function $f_{x_i} : \mathbb{R}^{n_i} \to [0, \infty) \subset \mathbb{R}$, where

$$f_{x_i}(z_i) := \frac{1}{\sqrt{2\pi}^{n_i} (\det(K_i))^{1/2}} \exp(-\frac{1}{2}\langle z_i - \bar{x}_i, K_i^{-1}(z_i - \bar{x}_i)\rangle_{\mathbb{R}^{n_i}})$$

Clearly, we have $f_x(z) = \prod_{i=1}^{m} f_{x_i}(z_i)$, $\forall z = (z_1, \ldots, z_m) \in \mathbb{R}^n$. By Proposition 14.20, $x_1, \ldots, x_m$ are independent.

"Necessity" Let $x_1, \ldots, x_m$ be independent. Then, $K = \boldsymbol{E}((x - \bar{x}) \otimes (x - \bar{x})) = \begin{bmatrix} K_{1,1} & \cdots & K_{1,m} \\ \vdots & & \vdots \\ K_{m,1} & \cdots & K_{m,m} \end{bmatrix}$, where $K_{i,j} := \boldsymbol{E}((x_i - \bar{x}_i) \otimes (x_j - \bar{x}_j))$, $\forall 1 \le i, j \le m$. $\forall 1 \le i, j \le m$ with $i \neq j$, we have $K_{i,j} = \boldsymbol{E}((x_i - \bar{x}_i) \otimes (x_j - \bar{x}_j)) = \boldsymbol{E}(x_i - \bar{x}_i) \otimes \boldsymbol{E}(x_j - \bar{x}_j) = \mathbf{0}_{n_i \times n_j}$, where the second equality follows from Proposition 14.9. Hence, $K =$ block diagonal $\big(K_{1,1}, \ldots, K_{m,m}\big)$.

This completes the proof of the proposition. □

Proposition 14.28 *Let $\boldsymbol{\Omega} := (\Omega, \mathcal{B}, \boldsymbol{P})$ be a probability measure space, $m, p \in \mathbb{N}$, $(X_i)_{i=1}^p$ be random variables such that $X := (X_1, \ldots, X_p) \sim N(\bar{x}, K)$, $\bar{x} \in \mathbb{R}^p$, $K \in \mathrm{B}(\mathbb{R}^p, \mathbb{R}^p)$ with $K \in \bar{\mathcal{S}}_{+\mathbb{R}^p}$, and $(Y_i)_{i=1}^m$ be linear combinations of X_i's, $Y_i = \sum_{j=1}^p \alpha_{i,j} X_j$, $i = 1, \ldots, m$, where $\alpha_{i,j} \in \mathbb{R}$, $\forall i = 1, \ldots, m$, $\forall j = 1, \ldots, p$. Assume that $V := (\alpha_{i,j})_{m \times p}$ is of full row rank. Then, $Y := (Y_1, \ldots, Y_m)$ is an $\mathbb{R}^m$-valued random variable with distribution $N(V\bar{x}, VKV')$.*

Proof Clearly $m \leq p$ and $\operatorname{rank}(V) = m$. Then, $\mathcal{N}(V)$ is $(p - m)$-dimensional. Let $f_1, \ldots, f_{p-m}$ be an orthonormal basis of $\mathcal{N}(V)$, and $F := \begin{bmatrix} f_1 & \cdots & f_{p-m} \end{bmatrix}'$. Then, $T := \begin{bmatrix} V \\ FK^{-1} \end{bmatrix}$ is an invertible matrix. By Proposition 14.19, $Z := TX = \begin{bmatrix} Y \\ Z_2 \end{bmatrix}$ is an $\mathbb{R}^p$-valued random variable with distribution $N(T\bar{x}, TKT')$. Note that $TKT' = \begin{bmatrix} VKV' & VF' \\ FV' & FK^{-1}F' \end{bmatrix} = \text{block diagonal}\left(VKV', FK^{-1}F'\right)$. By Proposition 14.27, Y and Z_2 are independent Gaussian random vectors with $Y \sim N(V\bar{x}, VKV')$. This completes the proof of the proposition. □

14.3 Law of Large Numbers

Theorem 14.29 (Weak Law of Large Numbers) *Let $\boldsymbol{\Omega} := (\Omega, \mathcal{B}, \boldsymbol{P})$ be a probability measure space, $\mathcal{Z}$ be a separable Hilbert space over $\mathbb{K}$, $(X_n)_{n=1}^\infty$ be a sequence of independent $\mathcal{Z}$-valued random variables, $X_n : \boldsymbol{\Omega} \to \mathcal{Z}$, with $\boldsymbol{E}(X_n) = \bar{x} \in \mathcal{Z}$ and $\boldsymbol{E}(\|X_n - \bar{x}\|_{\mathcal{Z}}^2) \leq b \in \mathbb{R}_+$, $\forall n \in \mathbb{N}$. Then, $\lim_{n \in \mathbb{N}} \frac{1}{n} \sum_{i=1}^n X_i = \bar{x}$ in measure in $\boldsymbol{\Omega}$.*

Proof Let $S_n := \frac{1}{n} \sum_{i=1}^n X_i$, which is a $\mathcal{Z}$-valued random variable by Propositions 11.38 and 11.39. Thus, $\boldsymbol{E}(S_n) = \bar{x}$. $\boldsymbol{E}(\|S_n - \bar{x}\|_{\mathcal{Z}}^2) = \frac{1}{n^2} \boldsymbol{E}(\langle \sum_{i=1}^n (X_i - \bar{x}), \sum_{j=1}^n (X_j - \bar{x}) \rangle_{\mathcal{Z}}) = \frac{1}{n^2} \sum_{i=1}^n \sum_{j=1}^n \boldsymbol{E}(\langle (X_i - \bar{x}), (X_j - \bar{x}) \rangle_{\mathcal{Z}}) = \frac{1}{n^2} \cdot \sum_{i=1}^n \boldsymbol{E}(\|X_i - \bar{x}\|_{\mathcal{Z}}^2) \leq \frac{1}{n^2} nb = \frac{b}{n} \to 0$, as $n \in \mathbb{N}$, where the first two equalities follow from Proposition 11.92 and the third equality follows from the assumption that $(X_n)_{n=1}^\infty$ is independent and Proposition 14.9. Then, $\lim_{n \in \mathbb{N}} \boldsymbol{E}(\|S_n - \bar{x}\|_{\mathcal{Z}}^2) = 0$. This implies that $\lim_{n \in \mathbb{N}} \|S_n - \bar{x}\|_{\mathrm{L}_2(\boldsymbol{\Omega}, \mathcal{Z})}^2 = 0$. Hence, by Proposition 11.211, we have $\lim_{n \in \mathbb{N}} S_n(\omega) = \bar{x}$ in measure in $\boldsymbol{\Omega}$. This completes the proof of the theorem. □

14.4 Martingales Indexed by $\mathbb{Z}_+$

Definition 14.30 Let $\mathbb{X} := (\mathcal{X}, \mathcal{B}_{\mathrm{B}}(\mathcal{X}), \mu)$ be a nonempty σ-finite topological measure space with a partial ordering $\preceq$ defined on it, $\boldsymbol{\Omega} := (\Omega, \mathcal{B}, \boldsymbol{P})$ be a probability measure space, and $\mathbb{X} \times \boldsymbol{\Omega} := (\mathcal{X} \times \Omega, \bar{\mathcal{B}}, \nu)$ be the σ-finite product measure space (as defined in Proposition 12.22), $\mathcal{Y}$ be a topological space, and $W : \mathbb{X} \times \boldsymbol{\Omega} \to \mathcal{Y}$ be $\bar{\mathcal{B}}$-measurable. Then, W is said to be a *$\mathcal{Y}$-valued stochastic process on* $\mathbb{X} \times \boldsymbol{\Omega}$. Then, $\forall t \in \mathbb{X}$, $W_t : \boldsymbol{\Omega} \to \mathcal{Y}$ defined by $W_t(\omega) = W(t, \omega)$, $\forall \omega \in \boldsymbol{\Omega}$, is a $\mathcal{Y}$-valued random variable (i.e., $\mathcal{B}$-measurable by Proposition 12.28). We will sometimes write $(W_t)_{t\in\mathbb{X}}$ to denote the stochastic process W. When $\mathcal{Y} \subseteq \mathbb{R}$, we will simply say that W is a *stochastic process on* $\mathbb{X} \times \boldsymbol{\Omega}$. $(\mathcal{B}_x)_{x\in\mathbb{X}}$ is said to be a *filtration* on $\boldsymbol{\Omega}$, if $\mathcal{B}_x \subseteq \mathcal{B}_z \subseteq \mathcal{B}$ and $\mathcal{B}_x$ and $\mathcal{B}_z$ are sub-σ-algebras of $\mathcal{B}$ on Ω, $\forall x, z \in \mathbb{X}$ with $x \preceq z$. Then, the quadruple $(\Omega, \mathcal{B}, (\mathcal{B}_x)_{x\in\mathbb{X}}, \boldsymbol{P})$ is said to be a *filtered probability measure space*. We say that $(W_x)_{x\in\mathbb{X}}$ is *adapted to the filtration* $(\mathcal{B}_x)_{x\in\mathbb{X}}$ if W_x is $\mathcal{B}_x$-measurable, $\forall x \in \mathbb{X}$. The stochastic process $(W_x)_{x\in\mathbb{X}}$ defines the *natural filtration of* $(W_x)_{x\in\mathbb{X}}$, $\left(\tilde{\mathcal{B}}_x\right)_{x\in\mathbb{X}}$, where $\tilde{\mathcal{B}}_x := \sigma(\bigcup_{y\in\mathbb{X}, y\preceq z} \sigma(W_y))$, $\forall x \in \mathbb{X}$. ◇

It is easy to check that $\tilde{\mathcal{B}}_x \subseteq \tilde{\mathcal{B}}_z \subseteq \mathcal{B}$, $\forall x, z \in \mathcal{Z}$ with $x \preceq z$. Hence, the natural filtration of a stochastic process is indeed a filtration.

In this section and the next five sections, we will be primarily concerned with stochastic processes indexed by $\mathbb{Z}_+$. In this case, we will define $\aleph := (\mathbb{Z}_+, {}^{\mathbb{Z}_+}2, \mu)$ to be the topological measure space on $\mathbb{Z}_+$ with the counting measure and the partial ordering of $\leq$. (It is easy to check that it is a topological measure space, where the topology is given by the subset topology of $\mathbb{R}$.) In this case, a $\mathfrak{Y}$-valued stochastic process W on $\aleph \times \boldsymbol{\Omega}$ adapted to some filtration $(\mathcal{B}_n)_{n\in\aleph}$ is equivalent to that $W_n : \boldsymbol{\Omega} \to \mathfrak{Y}$ being $\mathcal{B}_n$-measurable, $\forall n \in \aleph$. In this case, we will simply say a $\mathfrak{Y}$-valued stochastic process $(W_n)_{n=0}^{\infty}$ on $\boldsymbol{\Omega}$ that is adapted to $(\mathcal{B}_n)_{n=0}^{\infty}$ to save us from cumbersome notations.

Definition 14.31 Let $\boldsymbol{\Omega} := (\Omega, \mathcal{B}, (\mathcal{B}_n)_{n=0}^{\infty}, \boldsymbol{P})$ be a filtered probability measure space and $(X_n)_{n=0}^{\infty}$ be a real-valued stochastic process on $\boldsymbol{\Omega}$. Then, the stochastic process $(X_n)_{n=0}^{\infty}$ is said to be a *Martingale* if

(i) $(X_n)_{n=0}^{\infty}$ is an adapted process (that is, it is adapted to the filtration $(\mathcal{B}_n)_{n=0}^{\infty}$).
(ii) $X_n \in \bar{\mathrm{L}}_1(\boldsymbol{\Omega}, \mathbb{R})$, $\forall n \in \mathbb{Z}_+$.
(iii) $X_{n-1} \in \boldsymbol{E}(X_n|\mathcal{B}_{n-1})$, $\forall n \in \mathbb{N}$.

The stochastic process $(X_n)_{n=0}^{\infty}$ is said to be a *super Martingale* if it satisfies (i) and (ii) and $f_{n-1}(\omega) \leq X_{n-1}(\omega)$ a.e. $\omega \in \boldsymbol{\Omega}$, $\forall f_{n-1} \in \boldsymbol{E}(X_n|\mathcal{B}_{n-1})$, $\forall n \in \mathbb{N}$.

The stochastic process $(X_n)_{n=0}^{\infty}$ is said to be a *sub Martingale* if it satisfies (i) and (ii) and $f_{n-1}(\omega) \geq X_{n-1}(\omega)$ a.e. $\omega \in \boldsymbol{\Omega}$, $\forall f_{n-1} \in \boldsymbol{E}(X_n|\mathcal{B}_{n-1})$, $\forall n \in \mathbb{N}$. ◇

Clearly, a stochastic process is a super Martingale if, and only if, its negative process is a sub Martingale. A stochastic process is a Martingale if, and only if, it is a super Martingale and is a sub Martingale.

Theorem 14.32 *Let* $\boldsymbol{\Omega} := (\Omega, \mathcal{B}, (\mathcal{B}_n)_{n=0}^{\infty}, \boldsymbol{P})$ *be a filtered probability measure space,* $(X_n)_{n=0}^{\infty}$ *be a super Martingale (or a Martingale), and* $(C_n)_{n=1}^{\infty}$ *be such that*

(i) $C_n : \boldsymbol{\Omega} \to [0, \alpha] \subset \mathbb{R}$, $\forall n \in \mathbb{N}$, *where* $\alpha \in \mathbb{R}_+$.
(ii) C_n *is* $\mathcal{B}_{n-1}$*-measurable,* $\forall n \in \mathbb{N}$.

Then, the stochastic process $(Y_n)_{n=0}^{\infty}$ *defined by* $Y_0 = X_0$ *and* $Y_n = X_0 + \sum_{i=1}^{n} C_i(X_i - X_{i-1})$, $\forall n \in \mathbb{N}$, *is a super Martingale (or a Martingale).*

Proof Since, $\forall n \in \mathbb{N}$, X_n is $\mathcal{B}_n$-measurable and C_n is $\mathcal{B}_{n-1}$-measurable, then, $C_n(X_n - X_{n-1})$ is $\mathcal{B}_n$-measurable by Propositions 11.38, 7.35, and 11.39 and Definition 14.31. Hence, Y_n is $\mathcal{B}_n$-measurable, $\forall n \in \mathbb{Z}_+$.

Clearly, $|Y_n| \leq |X_0| + \sum_{i=1}^{n}(|C_i|\,|X_i| + |C_i|\,|X_{i-1}|) \leq (1+\alpha)\,|X_0| + \sum_{i=1}^{n-1} 2\alpha\,|X_i| + \alpha\,|X_n|$, $\forall n \in \mathbb{Z}_+$. Then, by Definition 14.31, $X_n \in \bar{\text{L}}_1(\boldsymbol{\Omega}, \mathbb{R})$, $\forall n \in \mathbb{Z}_+$, implies that $Y_n \in \bar{\text{L}}_1(\boldsymbol{\Omega}, \mathbb{R})$ by Proposition 11.83.

Note that $\boldsymbol{E}(Y_n - Y_{n-1}|\mathcal{B}_{n-1}) = \boldsymbol{E}(C_n(X_n - X_{n-1})|\mathcal{B}_{n-1}) = C_n\boldsymbol{E}(X_n - X_{n-1}|\mathcal{B}_{n-1})$, $\forall n \in \mathbb{N}$, where the first equality follows from the definition of Y_n and the second equality follows from (h) of Proposition 14.11. This is less than or equal to zero almost everywhere if $(X_n)_{n=0}^{\infty}$ is a super Martingale or is equal to zero almost everywhere if $(X_n)_{n=0}^{\infty}$ is a Martingale. Hence, $(Y_n)_{n=0}^{\infty}$ is a super Martingale, if $(X_n)_{n=0}^{\infty}$ is so, and it is a Martingale, if $(X_n)_{n=0}^{\infty}$ is a Martingale.

This completes the proof of the theorem. □

This result implies that one cannot beat the system in a gambling scenario by varying his bet each time after observing the outcome of the past dice rolls.

Definition 14.33 Let $\mathcal{X}$ be a nonempty topological subspace of $\mathbb{R}$, $\boldsymbol{\Omega} := (\Omega, \mathcal{B}, (\mathcal{B}_x)_{x\in\mathcal{X}}, \boldsymbol{P})$ be a filtered probability measure space, and $T : \boldsymbol{\Omega} \to \mathcal{X} \cup \{-\infty, +\infty\} \subseteq \mathbb{R}_e$ be a random variable (i.e., it is $\mathcal{B}$-measurable). T is said to be *stopping time* if $\{\omega \in \Omega \mid T(\omega) \leq x\} \in \mathcal{B}_x$, $\forall x \in \mathcal{X}$. ◇

Example 14.34 Let $\boldsymbol{\Omega} := (\Omega, \mathcal{B}, (\mathcal{B}_n)_{n=0}^{\infty}, \boldsymbol{P})$ be a filtered probability measure space, $\mathcal{Y}$ be a normed linear space over $\mathbb{K}$, $(X_n)_{n=0}^{\infty}$ be a $\mathcal{Y}$-valued adapted stochastic process, and $B \in \mathcal{B}_{\text{B}}(\mathcal{Y})$. Then, $T : \boldsymbol{\Omega} \to \mathbb{Z}_+ \cup \{\infty\}$ defined by $T(\omega) := \inf\{n \in \mathbb{Z}_+ \mid X_n(\omega) \in B\}$ = time of first entry of $(X_n)_{n=0}^{\infty}$ into set B is a stopping time. This is because $\{\omega \in \Omega \mid T(\omega) \leq n\} = \bigcup_{i=0}^{n}\{\omega \in \Omega \mid X_i(\omega) \in B\} \in \mathcal{B}_n$. ◇

Theorem 14.35 *Let* $\boldsymbol{\Omega} := (\Omega, \mathcal{B}, (\mathcal{B}_n)_{n=0}^{\infty}, \boldsymbol{P})$ *be a filtered probability measure space,* $(X_n)_{n=0}^{\infty}$ *be a real-valued adapted stochastic process, and* $T : \boldsymbol{\Omega} \to \mathbb{Z}_+ \cup \{\infty\}$ *be a stopping time. Assume that* $(X_n)_{n=0}^{\infty}$ *is a super Martingale (or a Martingale), then the* stopped stochastic process $(Y_n)_{n=0}^{\infty}$ *defined by* $Y_n(\omega) = X_{T(\omega)\wedge n}(\omega)$, $\forall n \in \mathbb{Z}_+$, *is a super Martingale (or a Martingale).*

Proof Define $(C_n)_{n=1}^{\infty}$ by $C_n : \boldsymbol{\Omega} \to [0, 1] \subset \mathbb{R}$ with $C_n(\omega) = \chi_{\{T(\omega)\geq n\},\Omega}$, $\forall \omega \in \boldsymbol{\Omega}$, $\forall n \in \mathbb{N}$. Clearly, C_n is $\mathcal{B}_{n-1}$-measurable. It is easy to see that Y_n as defined in the statement of the theorem is exactly the same as that of Theorem 14.32. Then, the result follows immediately from Theorem 14.32. □

Theorem 14.36 (Doob's Optional Stopping) *Let* $(\Omega, \mathcal{B}, (\mathcal{B}_n)_{n=0}^{\infty}, \boldsymbol{P}) =: \boldsymbol{\Omega}$ *be a filtered probability measure space,* $(X_n)_{n=0}^{\infty}$ *be a super Martingale adapted to the* *filtration* $(\mathcal{B}_n)_{n=0}^{\infty}$*, and* $T : \boldsymbol{\Omega} \to \mathbb{Z}_+ \cup \{\infty\}$ *be a stopping time. Then,* $X_T : \boldsymbol{\Omega} \to \mathbb{R}$ *defined by,* $\forall \omega \in \boldsymbol{\Omega}$,

$$X_T(\omega) := \begin{cases} X_{T(\omega)}(\omega) & \forall \omega \in \Omega \text{ with } T(\omega) < \infty \\ \lim_{n \in \mathbb{N}} X_n(\omega) & \forall \omega \in \Omega \text{ with } T(\omega) = \infty \text{ and } \lim_{n\in\mathbb{N}} X_n(\omega) \in \mathbb{R} \\ 0 & \begin{array}{c}\forall \omega \in \Omega \text{ with } T(\omega) = \infty \text{ and} \\ \lim_{n\in\mathbb{N}} X_n(\omega) \text{ does not exist in } \mathbb{R}\end{array} \end{cases}$$

is absolutely integrable over $\boldsymbol{\Omega}$ *and* $\boldsymbol{E}(X_T) \leq \boldsymbol{E}(X_0)$ *if any of the following conditions holds:*

(i) $X_n(\omega) \geq 0$, $\forall \omega \in \boldsymbol{\Omega}$, $\forall n \in \mathbb{N}$, *and* $T(\omega) < \infty$ *a.e.* $\omega \in \boldsymbol{\Omega}$.
(ii) $\exists N \in \mathbb{N}$ *such that* $T(\omega) \leq N$, $\forall \omega \in \boldsymbol{\Omega}$.
(iii) $\exists M \in [0, \infty) \subset \mathbb{R}$ *such that* $|X_n(\omega)| \leq M$, $\forall \omega \in \boldsymbol{\Omega}$, $\forall n \in \mathbb{Z}_+$, *and* $T(\omega) < \infty$ *a.e.* $\omega \in \boldsymbol{\Omega}$.
(iv) $\boldsymbol{E}(T) < \infty$ *and* $\exists M \in [0, \infty) \subset \mathbb{R}$ *such that* $|X_n(\omega) - X_{n-1}(\omega)| \leq M$, $\forall \omega \in \boldsymbol{\Omega}$, $\forall n \in \mathbb{N}$.

If $(X_n)_{n=0}^{\infty}$ *is a Martingale adapted to the filtration* $(\mathcal{B}_n)_{n=0}^{\infty}$ *and any of the conditions (ii)–(iv) holds, then* X_T *is absolutely integrable over* $\boldsymbol{\Omega}$ *and* $\boldsymbol{E}(X_T) = \boldsymbol{E}(X_0)$.

Proof We first consider the first paragraph of the theorem statements. By Propositions 11.51, 11.48, and 11.41, X_T is $\mathcal{B}$-measurable. Let (i) hold. Then, the stopped stochastic process $(Y_n)_{n=0}^{\infty}$ defined in Theorem 14.35 satisfies $Y_n(\omega) \geq 0$, $\forall \omega \in \boldsymbol{\Omega}$, $\forall n \in \mathbb{Z}_+$. By Theorem 14.35, $(Y_n)_{n=0}^{\infty}$ is a super Martingale. By (i), $T(\omega) < \infty$ a.e. $\omega \in \boldsymbol{\Omega}$. Then, $\lim_{n\in\mathbb{N}} Y_n(\omega) = X_T(\omega)$ a.e. $\omega \in \boldsymbol{\Omega}$. Clearly, $X_T(\omega) \geq 0$, $\forall \omega \in \boldsymbol{\Omega}$. By Fatou's Lemma 11.80, we have $\boldsymbol{E}(X_T) = \int_{\Omega} X_T(\omega)\, d\boldsymbol{P}(\omega) \leq \liminf_{n\in\mathbb{N}} \int_{\Omega} Y_n(\omega)\, d\boldsymbol{P}(\omega) = \liminf_{n\in\mathbb{Z}_+} \boldsymbol{E}(Y_n) = \liminf_{n\in\mathbb{Z}_+} \boldsymbol{E}(\boldsymbol{E}(Y_n|\mathcal{B}_{n-1})) \leq \liminf_{n\in\mathbb{Z}_+} \boldsymbol{E}(Y_{n-1}) \leq \cdots \leq \boldsymbol{E}(Y_0) = \boldsymbol{E}(X_0)$, where the first equality follows from Definition 14.1, the second equality follows from Definition 14.1, the third equality follows from (j) of Proposition 14.11, the second inequality follows from Theorem 14.35, and the last equality follows from the definition of $(Y_n)_{n=0}^{\infty}$ in Theorem 14.35. Hence, the result holds.

Let (ii) hold. Then, the stopped stochastic process $(Y_n)_{n=0}^{\infty}$ satisfies that $Y_N(\omega) = X_T(\omega)$, $\forall \omega \in \boldsymbol{\Omega}$. Hence, by Theorem 14.35, X_T is absolutely integrable over $\boldsymbol{\Omega}$ and $\boldsymbol{E}(X_T) = \boldsymbol{E}(Y_N) \leq \boldsymbol{E}(Y_{N-1}) \leq \cdots \leq \boldsymbol{E}(Y_0) = \boldsymbol{E}(X_0)$. Hence, the result holds.

Let (iii) hold. Then, $(Y_n)_{n=0}^{\infty}$ satisfies that $|Y_n(\omega)| \leq M$, $\forall \omega \in \Omega$, $\forall n \in \mathbb{Z}_+$, and $\lim_{n\in\mathbb{Z}_+} Y_n(\omega) = X_T(\omega)$ a.e. $\omega \in \boldsymbol{\Omega}$. By Lebesgue Dominated Convergence Theorem 11.91, X_T is absolutely integrable over $\boldsymbol{\Omega}$ and $\boldsymbol{E}(X_T) = \lim_{n\in\mathbb{Z}_+} \boldsymbol{E}(Y_n) \leq \boldsymbol{E}(Y_0) = \boldsymbol{E}(X_0)$. Hence, the result holds.

Let (iv) hold. Then, the stopped stochastic process $(Y_n)_{n=0}^{\infty}$ defined in Theorem 14.35 satisfies $Y_n = X_0 + \sum_{i=1}^{n} C_i(X_i - X_{i-1})$, $\forall n \in \mathbb{N}$, where $(C_n)_{n=1}^{\infty}$

is as defined in the proof of Theorem 14.35. Then, $\forall n \in \mathbb{Z}_+$,

$$\begin{aligned} |Y_n(\omega)| &\le |X_0(\omega)| + \sum_{i=1}^{n} \chi_{\{T(\omega)\ge i\},\Omega} \, |X_i(\omega) - X_{i-1}(\omega)| \\ &\le |X_0(\omega)| + \sum_{i=1}^{n} M\chi_{\{T(\omega)\ge i\},\Omega} \le |X_0(\omega)| + MT(\omega) \end{aligned}$$

where the first inequality follows from the definition of $(C_n)_{n=1}^{\infty}$ and the second inequality follows from (iv). Then, $\boldsymbol{E}(|Y_n|) \le \boldsymbol{E}(|X_0|) + M \int_{\Omega} T(\omega) \cdot \mathrm{d}\boldsymbol{P}(\omega) = \boldsymbol{E}(|X_0|) + M\boldsymbol{E}(T) < \infty$, $\forall n \in \mathbb{Z}_+$, where the last inequality follows from (iv). Then, Y_n is absolutely integrable over $\boldsymbol{\Omega}$, $\forall n \in \mathbb{Z}_+$. Since $\boldsymbol{E}(T) < \infty$, then $T(\omega) < \infty$ a.e. $\omega \in \boldsymbol{\Omega}$, and $\lim_{n\in\mathbb{Z}_+} Y_n(\omega) = X_T(\omega)$ a.e. $\omega \in \boldsymbol{\Omega}$. By Lebesgue Dominated Convergence Theorem 11.91, $\boldsymbol{E}(X_T) = \lim_{n\in\mathbb{Z}_+} \boldsymbol{E}(Y_n) \le \boldsymbol{E}(Y_0) = \boldsymbol{E}(X_0)$, where the inequality follows from Theorem 14.35. Hence, the result holds.

Now, consider the second paragraph of the theorem statements. Let $(X_n)_{n=0}^{\infty}$ be a Martingale. Clearly, it is a super Martingale.

Let (ii), (iii), or (iv) hold. By the super Martingale case, we have $\boldsymbol{E}(X_T) \le \boldsymbol{E}(X_0)$. Clearly, $(-X_n)_{n=0}^{\infty}$ is also a super Martingale. By the super Martingale case, we have $\boldsymbol{E}(-X_T) \le \boldsymbol{E}(-X_0)$. Then, we have $\boldsymbol{E}(X_T) = \boldsymbol{E}(X_0)$. Hence, the result holds.

This completes the proof of the theorem. □

Lemma 14.37 *Let* $\boldsymbol{\Omega} := (\Omega, \mathcal{B}, (\mathcal{B}_n)_{n=0}^{\infty}, \boldsymbol{P})$ *be a filtered probability measure space and* $T : \boldsymbol{\Omega} \to \mathbb{Z}_+ \cup \{\infty\}$ *be a stopping time. Assume that* $\exists N \in \mathbb{N}$ *and* $\exists \epsilon \in (0, 1) \subset \mathbb{R}$, $\forall n \in \mathbb{N}$, $\boldsymbol{P}(\{\omega \in \Omega \mid T(\omega) \le n + N\}|\mathcal{B}_n)(\omega) > \epsilon$ *a.e.* $\omega \in \boldsymbol{\Omega}$. *Then,* $\boldsymbol{E}(T) < \infty$.

Proof Note that

$$\begin{aligned} \boldsymbol{E}(T) &= \int_{\Omega} T(\omega)\,\mathrm{d}\boldsymbol{P}(\omega) = \sum_{n=0}^{\infty} n\boldsymbol{P}(\{\omega \in \Omega \mid T(\omega) = n\}) \\ &= \sum_{k=0}^{\infty} \sum_{n=kN}^{(k+1)N-1} n\boldsymbol{P}(\{\omega \in \Omega \mid T(\omega) = n\}) \\ &\le \sum_{k=0}^{\infty} \sum_{n=kN}^{(k+1)N-1} ((k+1)N - 1)\boldsymbol{P}(\{\omega \in \Omega \mid T(\omega) = n\}) \\ &= \sum_{k=0}^{\infty} ((k+1)N - 1)\boldsymbol{P}(\{\omega \in \Omega \mid kN \le T(\omega) < (k+1)N\}) \\ &\le \sum_{k=0}^{\infty} ((k+1)N - 1)\boldsymbol{P}(\{\omega \in \Omega \mid T(\omega) > kN - 1\}) \end{aligned}$$

Let $\alpha_k := \boldsymbol{P}(\{\omega \in \Omega \mid T(\omega) \leq kN - 1\}) \in [0, 1] \subset \mathbb{R}$, $\forall k \in \mathbb{N}$. Then, $\alpha_k = \boldsymbol{P}(\{\omega \in \Omega \mid T(\omega) \leq kN - 1 - N\}) + \boldsymbol{P}(\{\omega \in \Omega \mid kN - 1 - N < T(\omega) \leq kN - 1\}) = \alpha_{k-1} + \boldsymbol{P}(\{\omega \in \Omega \mid kN - 1 - N < T(\omega)\})\boldsymbol{P}(\{\omega \in \Omega \mid T(\omega) \leq kN - 1\} \mid \{\omega \in \Omega \mid kN - 1 - N < T(\omega)\})$. The last term is a conditional probability of an event given another event. We will treat this term in detail in the following paragraph.

Let $E_1 := \{\omega \in \Omega \mid T(\omega) \leq kN - 1\} \in \mathcal{B}$ and $E_2 := \{\omega \in \Omega \mid kN - 1 - N < T(\omega)\} \in \mathcal{B}_{kN-1-N}$. Let $\mathcal{F} := \{\emptyset, E_2, \Omega \setminus E_2, \Omega\} \subseteq \mathcal{B}_{kN-1-N}$, where $\mathcal{F}$ is a σ-algebra on Ω. Then, the notion $\boldsymbol{P}(E_1|\mathcal{F})$ is well defined, which is a function on Ω that is $\mathcal{F}$-measurable. Then, $\boldsymbol{P}(E_1|\mathcal{F})(\omega) = \boldsymbol{P}(E_1|E_2) \in [0, 1] \subset \mathbb{R}$, $\forall \omega \in E_2$, and $\boldsymbol{P}(E_1|\mathcal{F})(\omega) = \boldsymbol{P}(E_1|\Omega \setminus E_2) \in [0, 1] \subset \mathbb{R}$, $\forall \omega \in \Omega \setminus E_2$. Then, we have the following identities:

$$\boldsymbol{P}(E_1|E_2) = \frac{\int_{E_2} \boldsymbol{P}(E_1|\mathcal{F})(\omega)\, \mathrm{d}\hat{\boldsymbol{P}}(\omega)}{\int_{E_2} 1\, \mathrm{d}\hat{\boldsymbol{P}}(\omega)} = \frac{1}{\hat{\boldsymbol{P}}(E_2)} \int_{E_2} \boldsymbol{P}(E_1|\mathcal{F})(\omega)\, \mathrm{d}\hat{\boldsymbol{P}}(\omega)$$

$$= \frac{1}{\boldsymbol{P}(E_2)} \int_{E_2} \boldsymbol{E}(\chi_{E_1,\Omega}|\mathcal{F})(\omega)\, \mathrm{d}\hat{\boldsymbol{P}}(\omega)$$

$$= \frac{1}{\boldsymbol{P}(E_2)} \int_{E_2} \boldsymbol{E}(\boldsymbol{E}(\chi_{E_1,\Omega}|\mathcal{B}_{kN-1-N})|\mathcal{F})(\omega)\, \mathrm{d}\hat{\boldsymbol{P}}(\omega)$$

$$= \frac{1}{\boldsymbol{P}(E_2)} \int_{E_2} \boldsymbol{E}(\chi_{E_1,\Omega}|\mathcal{B}_{kN-1-N})(\omega)\, \mathrm{d}\bar{\boldsymbol{P}}(\omega) \geq \frac{1}{\boldsymbol{P}(E_2)} \int_{E_2} \epsilon\, \mathrm{d}\bar{\boldsymbol{P}}$$

$$= \epsilon \frac{1}{\boldsymbol{P}(E_2)} \bar{\boldsymbol{P}}(E_2) = \epsilon$$

where $\hat{\boldsymbol{P}} = \boldsymbol{P}|_{\mathcal{F}}$ and $\bar{\boldsymbol{P}} = \boldsymbol{P}|_{\mathcal{B}_{kN-1-N}}$, the first equality follows from the preceding discussion, the second equality follows from $E_2 \in \mathcal{F}$, the third equality follows from the definition of $\hat{\boldsymbol{P}}$ and Definition 14.12, the fourth equality follows from (j) of Proposition 14.11, the fifth equality follows from (14.2a) of Definition 14.7, the inequality follows from the assumption of the lemma, the sixth equality follows from $E_2 \in \mathcal{B}_{kN-1-N}$, and the last equality follows from the definition of $\bar{\boldsymbol{P}}$.

Then, we have $\alpha_k \geq \alpha_{k-1} + (1 - \alpha_{k-1})\epsilon$, $\forall k \in \{2, 3, \ldots\}$. Therefore, $\boldsymbol{P}(\{\omega \in \Omega \mid T(\omega) > kN - 1\}) = 1 - \alpha_k \leq (1 - \alpha_{k-1})(1 - \epsilon)$, $\forall k \in \{2, 3, \ldots\}$. Recursively, we have $\boldsymbol{P}(\{\omega \in \Omega \mid T(\omega) > kN - 1\}) \leq \boldsymbol{P}(\{\omega \in \Omega \mid T(\omega) > N - 1\})(1 - \epsilon)^{k-1} \leq (1 - \epsilon)^{k-1}$, $\forall k \in \{2, 3, \ldots\}$. Hence, $\boldsymbol{E}(T) \leq N - 1 + 2N - 1 + \sum_{k=2}^{\infty}((k + 1)N - 1)(1 - \epsilon)^{k-1} < \infty$, where the last inequality follows from Proposition A.2.

This completes the proof of the lemma. □

Definition 14.38 Let $\boldsymbol{\Omega} := (\Omega, \mathcal{B}, (\mathcal{B}_n)_{n=0}^{\infty}, \boldsymbol{P})$ be a filtered probability measure space, $(X_n)_{n=0}^{\infty}$ be an adapted $\mathbb{R}$-valued random process on $\boldsymbol{\Omega}$, and $a, b \in \mathbb{R}$ with $a < b$. For any $N \in \mathbb{Z}_+$ and any $\omega \in \Omega$, we will say that the sample path of the random process $(X_n(\omega))_{n=0}^{\infty}$ *has* $U_N(a, b)(\omega) \in \mathbb{Z}_+$ *upcrossings of the closed interval* $[a, b] \subset \mathbb{R}$ *by the time* N, if we can find $k(\omega) := U_N(a, b)(\omega) \in \mathbb{Z}_+$

that is the largest nonnegative integer with the property that there exists $0 \leq s_1 < t_1 < s_2 < t_2 < \cdots < s_k < t_k \leq N$ such that $X_{s_i}(\omega) < a$ and $X_{t_i}(\omega) > b$, $i = 1, \ldots, k$. Clearly, $(U_N(a,b))_{N=0}^{\infty}$ is an adapted $\mathbb{Z}_+$-valued random process, i.e., $U_N(a,b) : \Omega \to \mathbb{Z}_+$ and is $\mathcal{B}_N$-measurable, $\forall N \in \mathbb{Z}_+$. $\diamond$

Lemma 14.39 (Doob's Upcrossing Lemma) *Let $(\Omega, \mathcal{B}, (\mathcal{B}_n)_{n=0}^{\infty}, \boldsymbol{P}) := \boldsymbol{\Omega}$ be a filtered probability measure space, $(X_n)_{n=0}^{\infty}$ be an adapted $\mathbb{R}$-valued random process on $\boldsymbol{\Omega}$, $a, b \in \mathbb{R}$ with $a < b$, and $(U_N(a,b))_{N=0}^{\infty}$ be the adapted $\mathbb{Z}_+$-valued random process defined in Definition 14.38. Assume that $(X_n)_{n=0}^{\infty}$ is a super Martingale. Then,*

$$(b-a)\boldsymbol{E}\left(U_N(a,b)\right) \leq \boldsymbol{E}\left(-((X_N - a) \wedge 0)\right) \tag{14.4}$$

Proof We will prove the lemma using Theorem 14.32. Define an $\mathbb{R}$-valued random process $(C_n)_{n=1}^{\infty}$ on $\boldsymbol{\Omega}$ by: $C_1 := \chi_{\{\omega \in \Omega \mid X_0(\omega) < a\}, \Omega}$, $C_n := \chi_{\{\omega \in \Omega \mid C_{n-1}(\omega)=1, X_{n-1}(\omega) \leq b\}, \Omega} + \chi_{\{\omega \in \Omega \mid C_{n-1}(\omega)=0, X_{n-1}(\omega) < a\}, \Omega}$, $\forall n \in \mathbb{N}$ with $n \geq 2$. Clearly, $C_n : \Omega \to [0,1] \subset \mathbb{R}$ is $\mathcal{B}_{n-1}$-measurable, $\forall n \in \mathbb{N}$. Essentially, $(C_n)_{n=1}^{\infty}$ is like a gambling strategy of wagering 1 coin after $X_{n-1} < a$ and then stop wagering after $X_{n-1} > b$, and so on. Define the random process $(Y_n)_{n=0}^{\infty}$ as in Theorem 14.32: $Y_0 = X_0$ and $Y_n = X_0 + \sum_{i=1}^{n} C_i(X_i - X_{i-1})$, $\forall n \in \mathbb{N}$. Y_n is then the money the gambler has at time $n \in \mathbb{Z}_+$. Then, by Theorem 14.32, $(Y_n)_{n=0}^{\infty}$ is a super Martingale. It should be clear that

$$Y_N - Y_0 \geq (b-a)U_N(a,b) + ((X_N - a) \wedge 0); \quad \forall \omega \in \boldsymbol{\Omega},\ \forall N \in \mathbb{N} \tag{14.5}$$

This is because, for every upcrossing of $[a,b]$ by the time N, the winning is at least $(b-a)$, which explains the term $(b-a)U_N(a,b)$. After the last upcrossing before N, the winning cannot be less than $((X_N - a) \wedge 0)$. Thus, (14.5) holds. Now, integrate both sides over $\boldsymbol{\Omega}$, and we arrive at $0 \geq \boldsymbol{E}(Y_N) - \boldsymbol{E}(Y_0) \geq (b-a)\boldsymbol{E}\big(U_N(a,b)\big) + \boldsymbol{E}\big(((X_N - a) \wedge 0)\big)$. This yields the desired inequality (14.4). This completes the proof of the lemma. $\square$

Corollary 14.40 *Let $\boldsymbol{\Omega} := (\Omega, \mathcal{B}, (\mathcal{B}_n)_{n=0}^{\infty}, \boldsymbol{P})$ be a filtered probability measure space, $(X_n)_{n=0}^{\infty}$ be an adapted $\mathbb{R}$-valued random process on $\boldsymbol{\Omega}$, $a, b \in \mathbb{R}$ with $a < b$, and $(U_N(a,b))_{N=0}^{\infty}$ be the adapted $\mathbb{Z}_+$-valued random process defined in Definition 14.38. Assume that $(X_n)_{n=0}^{\infty}$ is a super Martingale and is bounded in $\mathrm{L}_1(\boldsymbol{\Omega}, \mathbb{R})$, i.e., $\sup_{n \in \mathbb{Z}_+} \boldsymbol{E}(|X_n|) < \infty$. Then, $(b-a)\boldsymbol{E}(U_\infty(a,b)) \leq |a| + \sup_{n \in \mathbb{Z}_+} \boldsymbol{E}(|X_n|) < \infty$, where $U_\infty(a,b) := \lim_{N \in \mathbb{N}} U_N(a,b)$, and therefore $U_\infty(a,b)(\omega) < \infty$ a.e. $\omega \in \boldsymbol{\Omega}$.*

Proof Clearly, $U_N(a,b)(\omega)$ is monotonically nondecreasing in $N \in \mathbb{Z}_+$ and nonnegative, $\forall \omega \in \Omega$. Then, $U_\infty(a,b)(\omega) \in [0,\infty] \subset \mathbb{R}_e$. By Proposition 11.48,

$U_\infty(a,b)$ is $\mathcal{B}$-measurable. By Fatou's Lemma 11.80, we have $\boldsymbol{E}(U_\infty(a,b)) \leq \liminf_{N\in\mathbb{N}} \boldsymbol{E}(U_N(a,b))$. Then, by Doob's Upcrossing Lemma 14.39, we have

$$
\begin{aligned}
(b-a)\boldsymbol{E}(U_\infty(a,b)) &\leq \liminf_{N\in\mathbb{N}} (b-a)\boldsymbol{E}(U_N(a,b)) \\
&\leq \liminf_{N\in\mathbb{N}} -\boldsymbol{E}\left((X_N-a)\wedge 0\right) = \liminf_{N\in\mathbb{N}} \int_{\Omega} -((X_N(\omega)-a)\wedge 0)\,\mathrm{d}\boldsymbol{P}(\omega) \\
&\leq \liminf_{N\in\mathbb{N}} \int_{\Omega} |X_N(\omega)-a|\,\mathrm{d}\boldsymbol{P}(\omega) \leq \liminf_{N\in\mathbb{N}} \int_{\Omega} (|X_N(\omega)|+|a|)\,\mathrm{d}\boldsymbol{P}(\omega) \\
&\leq |a| + \sup_{N\in\mathbb{Z}_+} \boldsymbol{E}(|X_N|) < \infty
\end{aligned}
$$

where the first inequality follows from Proposition 3.81, the second inequality follows from Lemma 14.39, and the fifth inequality follows from Proposition 3.83. Thus, we conclude that $\boldsymbol{E}(U_\infty(a,b)) < \infty$ and therefore $U_\infty(a,b)(\omega) < \infty$ a.e. $\omega \in \boldsymbol{\Omega}$.

This completes the proof of the corollary. □

Theorem 14.41 (Doob's Forward Convergence Theorem) *Let $\boldsymbol{\Omega} := (\Omega, \mathcal{B}, (\mathcal{B}_n)_{n=0}^\infty, \boldsymbol{P})$ be a filtered probability measure space, $(X_n)_{n=0}^\infty$ be an adapted $\mathbb{R}$-valued random process on $\boldsymbol{\Omega}$, and $X_\infty : \Omega \to \mathbb{R}$ be defined by, $\forall \omega \in \Omega$,*

$$
X_\infty(\omega) := \begin{cases} \lim_{n\in\mathbb{N}} X_n(\omega) & \lim_{n\in\mathbb{N}} X_n(\omega) \in \mathbb{R} \\ 0 & \lim_{n\in\mathbb{N}} X_n(\omega) \text{ does not exist in } \mathbb{R} \end{cases}
$$

Assume that $(X_n)_{n=0}^\infty$ is a super Martingale and is bounded in $\mathrm{L}_1(\boldsymbol{\Omega}, \mathbb{R})$, i.e., $\sup_{n\in\mathbb{Z}_+} \boldsymbol{E}(|X_n|) < \infty$. Then, $X_\infty = \lim_{n\in\mathbb{N}} X_n(\omega)$ a.e. $\omega \in \boldsymbol{\Omega}$, X_∞ is $\mathcal{B}$-measurable, and $\boldsymbol{E}(|X_\infty|) \leq \sup_{n\in\mathbb{Z}_+} \boldsymbol{E}(|X_n|)$.

Proof Define $Y_\infty : \Omega \to \mathbb{R}_e$ and $Z_\infty : \Omega \to \mathbb{R}_e$ by $Y_\infty(\omega) := \limsup_{n\in\mathbb{Z}_+} X_n(\omega)$ and $Z_\infty(\omega) := \liminf_{n\in\mathbb{Z}_+} X_n(\omega)$, $\forall \omega \in \Omega$. By Proposition 11.40, Y_∞ and Z_∞ are $\mathcal{B}$-measurable. Define the set $\Lambda := \{\omega \in \Omega \mid Y_\infty(\omega) > Z_\infty(\omega)\} \in \mathcal{B}$. $\forall \omega \in \Lambda$, $\exists a, b \in \mathbb{Q}$ with $a < b$ such that $Y_\infty(\omega) > b > a > Z_\infty(\omega)$. Then, let $(U_n(a,b))_{n=0}^\infty$ be the adapted $\mathbb{Z}_+$-valued random process for $(X_n)_{n=0}^\infty$ as defined in Definition 14.38. Then, $U_\infty(a,b)(\omega) := \lim_{n\in\mathbb{N}} U_n(a,b)(\omega) = \infty$. Thus, $\Lambda \subseteq \bigcup_{\substack{a,b\in\mathbb{Q}\\ a<b}} \{\omega \in \Omega \mid U_\infty(a,b)(\omega) = \infty\} =: \bigcup_{\substack{a,b\in\mathbb{Q}\\ a<b}} \bar{\Lambda}_{a,b} \in \mathcal{B}$. By Corollary 14.40, $\boldsymbol{P}(\bar{\Lambda}_{a,b}) = 0$, $\forall a, b \in \mathbb{R}$ with $a < b$. Then, $\boldsymbol{P}(\Lambda) = 0$.

Define $W_\infty : \Omega \to \mathbb{R}_e$ by $W_\infty(\omega) := \begin{cases} \lim_{n\in\mathbb{N}} X_n(\omega) & \omega \in \Omega \setminus \Lambda \\ 0 & \omega \in \Lambda \end{cases}$, $\forall \omega \in \Omega$. Then, W_∞ is $\mathcal{B}$-measurable by Propositions 11.40 and 11.41. By Fatou's Lemma 11.80, we have

$$
\boldsymbol{E}(|W_\infty|) = \int_{\Omega} |W_\infty|\,\mathrm{d}\boldsymbol{P} \leq \liminf_{n\in\mathbb{Z}_+} \int_{\Omega} |X_n|\,\mathrm{d}\boldsymbol{P} \leq \sup_{n\in\mathbb{Z}_+} \boldsymbol{E}(|X_n|) < \infty
$$

Thus, we have $W_\infty(\omega) \in \mathbb{R}$ a.e. $\omega \in \boldsymbol{\Omega}$. Clearly, $W_\infty(\omega) = X_\infty(\omega)$ whenever $W_\infty(\omega) \in \mathbb{R}$. Then, $X_\infty = \lim_{n\in\mathbb{N}} X_n(\omega)$ a.e. $\omega \in \boldsymbol{\Omega}$, X_∞ is $\mathcal{B}$-measurable, by Proposition 11.41, and $\boldsymbol{E}(|X_\infty|) = \boldsymbol{E}(|W_\infty|) \le \sup_{n\in\mathbb{Z}_+} \boldsymbol{E}(|X_n|)$, where the equality follows from Proposition 11.89.

This completes the proof of the theorem. □

In the following, we will present two lemmas that lead to the Strong Law of Large Numbers.

Lemma 14.42 (Cesàro's) *Let* $(b_n)_{n=0}^\infty \subset \mathbb{R}$ *with* $b_0 < b_1 < b_2 < \cdots$ *and* $\lim_{n\in\mathbb{N}} b_n = \infty$, $(v_n)_{n=1}^\infty \subset \mathbb{R}$ *with* $\lim_{n\in\mathbb{N}} v_n = v_\infty \in \mathbb{R}$. *Then,* $\lim_{n\in\mathbb{N}} \frac{1}{b_n}\sum_{k=1}^n (b_k - b_{k-1})v_k = v_\infty$.

Proof Fix any $\epsilon \in (0,\infty) \subset \mathbb{R}$. By $\lim_{n\in\mathbb{N}} v_n = v_\infty$, there exists $N_\epsilon \in \mathbb{N}$ such that $\forall n \in \mathbb{N}$ with $N_\epsilon \le n$, we have $v_\infty - \epsilon < v_n < v_\infty + \epsilon$.

Then, we have

$$\begin{aligned}
&\liminf_{n\in\mathbb{N}} \frac{1}{b_n}\sum_{k=1}^n (b_k - b_{k-1})v_k \\
&\ge \liminf_{n\in\mathbb{N}} \{\frac{1}{b_n}\sum_{k=1}^{N_\epsilon} (b_k - b_{k-1})v_k + \frac{1}{b_n}\sum_{k=N_\epsilon+1}^{n} (b_k - b_{k-1})(v_\infty - \epsilon)\} \\
&= \liminf_{n\in\mathbb{N}} \frac{b_n - b_{N_\epsilon}}{b_n}(v_\infty - \epsilon) = v_\infty - \epsilon \\
&\limsup_{n\in\mathbb{N}} \frac{1}{b_n}\sum_{k=1}^n (b_k - b_{k-1})v_k \\
&\le \limsup_{n\in\mathbb{N}} \{\frac{1}{b_n}\sum_{k=1}^{N_\epsilon} (b_k - b_{k-1})v_k + \frac{1}{b_n}\sum_{k=N_\epsilon+1}^{n} (b_k - b_{k-1})(v_\infty + \epsilon)\} \\
&= \limsup_{n\in\mathbb{N}} \frac{b_n - b_{N_\epsilon}}{b_n}(v_\infty + \epsilon) = v_\infty + \epsilon
\end{aligned}$$

By the arbitrariness of ϵ, we have $v_\infty \le \liminf_{n\in\mathbb{N}} \frac{1}{b_n}\sum_{k=1}^n (b_k - b_{k-1})v_k \le \limsup_{n\in\mathbb{N}} \frac{1}{b_n}\sum_{k=1}^n (b_k - b_{k-1})v_k \le v_\infty$. Hence, the result holds. This completes the proof of the lemma. □

Lemma 14.43 (Kronecker's) *Let* $(b_n)_{n=0}^\infty \subset \mathbb{R}$ *with* $b_0 < b_1 < b_2 < \cdots$ *and* $\lim_{n\in\mathbb{N}} b_n = \infty$, $(x_n)_{n=1}^\infty \subset \mathbb{R}$, $s_n := \sum_{i=1}^n x_i$, $\forall n \in \mathbb{N}$. *Assume that* $\lim_{n\in\mathbb{N}} u_n := \lim_{n\in\mathbb{N}} \sum_{i=1}^n \frac{x_i}{b_i} = u_\infty \in \mathbb{R}$. *Then,* $\lim_{n\in\mathbb{N}} \frac{s_n}{b_n} = 0$.

Proof Note that $u_n - u_{n-1} = \frac{x_n}{b_n}$, $\forall n \in \mathbb{N}$, where $u_0 := 0$. Then $s_n = \sum_{i=1}^n x_i = \sum_{i=1}^n b_i(u_i - u_{i-1}) = b_n u_n - \sum_{i=1}^n (b_i - b_{i-1})u_{i-1}$. Thus, we have $\lim_{n\in\mathbb{N}} \frac{s_n}{b_n} =$

$\lim_{n\in\mathbb{N}}\{u_n - \frac{1}{b_n}\sum_{i=1}^{n}(b_i - b_{i-1})u_{i-1}\} = 0$, where the last equality follows from Cesàro's Lemma 14.42. This completes the proof of the lemma. □

Theorem 14.44 (Strong Law of Large Numbers) *Let $\boldsymbol{\Omega} := (\Omega, \mathcal{B}, \boldsymbol{P})$ be a probability measure, $(X_n)_{n=1}^{\infty}$ be a sequence of independent random variables, $X_n : \boldsymbol{\Omega} \to \mathbb{R}$, with $\boldsymbol{E}(X_n) = \bar{x} \subset \mathbb{R}$ and $\sum_{n=1}^{\infty}\frac{1}{n^2}\boldsymbol{E}((X_n - \bar{x})^2) =: b < \infty$. Then, $\lim_{n\in\mathbb{N}}\frac{1}{n}\sum_{i=1}^{n}X_i = \bar{x}$ a.e. in $\boldsymbol{\Omega}$.*

Proof Let $S_n := \frac{1}{n}\sum_{i=1}^{n}(X_i - \bar{x})$, which is a random variable (i.e., it is $\mathcal{B}$-measurable). Then, $\boldsymbol{E}(S_n) = 0$. Define $U_n : \boldsymbol{\Omega} \to \mathbb{R}$ by $U_n(\omega) := \sum_{i=1}^{n}\frac{1}{i}(X_i - \bar{x})$, $\forall\omega \in \boldsymbol{\Omega}$, $\forall n \in \mathbb{Z}_+$. It is easy to see that $(U_n)_{n=0}^{\infty}$ is a Martingale, adapted to its natural filtration. $\boldsymbol{E}(|U_n|) \le \boldsymbol{E}(\frac{1}{2}(1 + U_n^2)) = \frac{1}{2} + \frac{1}{2}\boldsymbol{E}(U_n^2) = \frac{1}{2} + \frac{1}{2}\boldsymbol{E}((\sum_{i=1}^{n}\frac{1}{i}(X_i - \bar{x}))^2) = \frac{1}{2} + \frac{1}{2}\sum_{i=1}^{n}\frac{1}{i^2}\boldsymbol{E}((X_i - \bar{x})^2) \le \frac{1}{2} + \frac{1}{2}b < \infty$, where the third equality follows from the independence of $(X_n)_{n=1}^{\infty}$. By Doob's Forward Convergence Theorem 14.41, $\lim_{n\in\mathbb{N}} U_n(\omega) \in \mathbb{R}$ a.e. $\omega \in \boldsymbol{\Omega}$. By Kronecker's Lemma 14.43, we have $\lim_{n\in\mathbb{N}} S_n(\omega) = 0$ whenever $\lim_{n\in\mathbb{N}} U_n(\omega) \in \mathbb{R}$. By Proposition 11.51, we have $\lim_{n\in\mathbb{N}} S_n = 0$ a.e. in $\boldsymbol{\Omega}$. Therefore, we have $\lim_{n\in\mathbb{N}}\frac{1}{n}\sum_{i=1}^{n}X_i = \bar{x}$ a.e. in $\boldsymbol{\Omega}$.

This completes the proof of the theorem. □

14.5 Banach Space Valued Martingales Indexed by $\mathbb{Z}_+$

Definition 14.45 Let $\boldsymbol{\Omega} := (\Omega, \mathcal{B}, (\mathcal{B}_n)_{n=0}^{\infty}, \boldsymbol{P})$ be a filtered probability measure space, $\mathcal{Y}$ be a Banach space, and $(X_n)_{n=0}^{\infty}$ be a $\mathcal{Y}$-valued stochastic process. Then, the stochastic process $(X_n)_{n=0}^{\infty}$ is said to be a *$\mathcal{Y}$-valued Martingale* if

(i) $(X_n)_{n=0}^{\infty}$ is an adapted process (that is, it is adapted to the filtration $(\mathcal{B}_n)_{n=0}^{\infty}$).
(ii) $X_n \in \bar{\mathcal{L}}_1(\boldsymbol{\Omega}, \mathcal{Y}_n)$, $\forall n \in \mathbb{Z}_+$, where $\mathcal{Y}_n$ is a separable Banach subspace of $\mathcal{Y}$.
(iii) $X_{n-1} \in \boldsymbol{E}(X_n|\mathcal{B}_{n-1})$, $\forall n \in \mathbb{N}$. ◇

Theorem 14.46 *Let $\boldsymbol{\Omega} := (\Omega, \mathcal{B}, (\mathcal{B}_n)_{n=0}^{\infty}, \boldsymbol{P})$ be a filtered probability measure space, $\mathcal{Y}$ be a separable Banach space over $\mathbb{K}$, $(X_n)_{n=0}^{\infty}$ be a $\mathcal{Y}$-valued Martingale, and $(C_n)_{n=1}^{\infty}$ be such that*

(i) $C_n : \boldsymbol{\Omega} \to \overline{\mathcal{B}}_{\mathbb{K}}(0, \alpha) \subset \mathbb{K}$, $\forall n \in \mathbb{N}$, where $\alpha \in \mathbb{R}_+$.
(ii) C_n is $\mathcal{B}_{n-1}$-measurable, $\forall n \in \mathbb{N}$.

Then, the stochastic process $(Y_n)_{n=0}^{\infty}$ defined by $Y_0 = X_0$ and $Y_n = X_0 + \sum_{i=1}^{n}C_i(X_i - X_{i-1})$, $\forall n \in \mathbb{N}$, is a $\mathcal{Y}$-valued Martingale.

Proof Since, $\forall n \in \mathbb{N}$, X_n is $\mathcal{B}_n$-measurable and C_n is $\mathcal{B}_{n-1}$-measurable, then, $C_n(X_n - X_{n-1})$ is $\mathcal{B}_n$-measurable by Propositions 11.38, 7.35, and 11.39 and Definition 14.45. Hence, Y_n is $\mathcal{B}_n$-measurable, $\forall n \in \mathbb{Z}_+$.

Clearly, $\|Y_n(\omega)\|_{\mathcal{Y}} \le \|X_0(\omega)\|_{\mathcal{Y}} + \sum_{i=1}^{n}(|C_i(\omega)|\|X_i(\omega)\|_{\mathcal{Y}} + |C_i(\omega)| \cdot \|X_{i-1}(\omega)\|_{\mathcal{Y}}) \le (1+\alpha)\|X_0(\omega)\|_{\mathcal{Y}} + \sum_{i=1}^{n-1} 2\alpha\|X_i(\omega)\|_{\mathcal{Y}} + \alpha\|X_n(\omega)\|_{\mathcal{Y}}$, $\forall \omega \in \Omega$, $\forall n \in \mathbb{N}$. Then, by Definition 14.45, $X_n \in \bar{\mathrm{L}}_1(\boldsymbol{\Omega}, \mathcal{Y})$, $\forall n \in \mathbb{Z}_+$, implies that $Y_n \in \bar{\mathrm{L}}_1(\boldsymbol{\Omega}, \mathcal{Y})$ by Proposition 11.83.

Note that $\boldsymbol{E}(Y_n - Y_{n-1}|\mathcal{B}_{n-1}) = \boldsymbol{E}(C_n(X_n - X_{n-1})|\mathcal{B}_{n-1}) = C_n\boldsymbol{E}(X_n - X_{n-1}|\mathcal{B}_{n-1}) = \vartheta_{\mathrm{L}_1(\boldsymbol{\Omega}_{n-1},\mathcal{Y})}$, $\forall n \in \mathbb{N}$, where the first equality follows from the definition of Y_n, the second equality follows from (h) of Proposition 14.11, $\boldsymbol{\Omega}_{n-1} := (\Omega, \mathcal{B}_{n-1}, \boldsymbol{P}|_{\mathcal{B}_{n-1}} =: \boldsymbol{P}_{n-1})$, and the last equality follows from the fact that $(X_n)_{n=0}^{\infty}$ is a $\mathcal{Y}$-valued Martingale and (c) of Proposition 14.11. Hence, $(Y_n)_{n=0}^{\infty}$ is a $\mathcal{Y}$-valued Martingale.

This completes the proof of the theorem. □

Theorem 14.47 *Let $\boldsymbol{\Omega} := (\Omega, \mathcal{B}, (\mathcal{B}_n)_{n=0}^{\infty}, \boldsymbol{P})$ be a filtered probability measure space, $\mathcal{Y}$ be a separable Banach space over $\mathbb{K}$, $(X_n)_{n=0}^{\infty}$ be a $\mathcal{Y}$-valued Martingale, and $T : \boldsymbol{\Omega} \to \mathbb{Z}_+ \cup \{\infty\}$ be a stopping time. Then the stopped $\mathcal{Y}$-valued stochastic process $(Y_n)_{n=0}^{\infty}$ defined by $Y_n(\omega) = X_{T(\omega)\wedge n}(\omega)$, $\forall n \in \mathbb{Z}_+$, is a Martingale.*

Proof Define $(C_n)_{n=1}^{\infty}$ by $C_n : \boldsymbol{\Omega} \to [0, 1] \subset \mathbb{R}$ with $C_n(\omega) = \chi_{\{T(\omega)\ge n\},\Omega}$, $\forall \omega \in \boldsymbol{\Omega}, \forall n \in \mathbb{N}$. Clearly, C_n is $\mathcal{B}_{n-1}$-measurable. It is easy to see that Y_n as defined in the statement of the theorem is exactly the same as that of Theorem 14.46. Then, the result follows immediately from Theorem 14.46. □

Theorem 14.48 (Doob's Optional Stopping) *Let $(\Omega, \mathcal{B}, (\mathcal{B}_n)_{n=0}^{\infty}, \boldsymbol{P}) =: \boldsymbol{\Omega}$ be a filtered probability measure space, $\mathcal{Y}$ be a separable Banach space over $\mathbb{K}$, $(X_n)_{n=0}^{\infty}$ be a a $\mathcal{Y}$-valued Martingale adapted to the filtration $(\mathcal{B}_n)_{n=0}^{\infty}$, and $T : \boldsymbol{\Omega} \to \mathbb{Z}_+ \cup \{\infty\}$ be a stopping time. Then, $X_T : \boldsymbol{\Omega} \to \mathcal{Y}$ defined by, $\forall \omega \in \boldsymbol{\Omega}$*

$$X_T(\omega) :=$$

$$\begin{cases} X_{T(\omega)}(\omega) & \forall \omega \in \Omega \text{ with } T(\omega) < \infty \\ \lim_{n\in\mathbb{N}} X_n(\omega) & \forall \omega \in \Omega \text{ with } T(\omega) = \infty \text{ and } \lim_{n\in\mathbb{N}} X_n(\omega) \in \mathcal{Y} \\ \vartheta_{\mathcal{Y}} & \begin{array}{c}\forall \omega \in \Omega \text{ with } T(\omega) = \infty \text{ and} \\ \lim_{n\in\mathbb{N}} X_n(\omega) \text{ does not exist in } \mathcal{Y}\end{array} \end{cases}$$

is absolutely integrable over $\boldsymbol{\Omega}$ and $\boldsymbol{E}(X_T) = \boldsymbol{E}(X_0)$ if any of the following conditions holds:

(i) $\exists N \in \mathbb{N}$ *such that* $T(\omega) \le N$, $\forall \omega \in \boldsymbol{\Omega}$.
(ii) $\exists M \in [0, \infty) \subset \mathbb{R}$ *such that* $\|X_n(\omega)\|_{\mathcal{Y}} \le M$, $\forall \omega \in \boldsymbol{\Omega}$, $\forall n \in \mathbb{Z}_+$, *and* $T(\omega) < \infty$ *a.e.* $\omega \in \boldsymbol{\Omega}$.
(iii) $\boldsymbol{E}(T) < \infty$ *and* $\exists M \in [0, \infty) \subset \mathbb{R}$ *such that* $\|X_n(\omega) - X_{n-1}(\omega)\|_{\mathcal{Y}} \le M$, $\forall \omega \in \boldsymbol{\Omega}$, $\forall n \in \mathbb{N}$.

Proof By Propositions 11.51, 11.48, and 11.41, X_T is $\mathcal{B}$-measurable. Let (i) hold. Then, the stopped stochastic process $(Y_n)_{n=0}^{\infty}$ defined in Theorem 14.47 satisfies that

$Y_N(\omega) = X_T(\omega)$, $\forall \omega \in \boldsymbol{\Omega}$. Hence, by Theorem 14.47, X_T is absolutely integrable over $\boldsymbol{\Omega}$ and $\boldsymbol{E}(X_T) = \boldsymbol{E}(Y_N) = \boldsymbol{E}(Y_{N-1}) = \cdots = \boldsymbol{E}(Y_0) = \boldsymbol{E}(X_0) \in \mathcal{Y}$. Hence, the result holds.

Let (ii) hold. Then, $(Y_n)_{n=0}^{\infty}$ satisfies that $\|Y_n(\omega)\|_{\mathcal{Y}} \leq M$, $\forall \omega \in \Omega$, $\forall n \in \mathbb{Z}_+$, and $\lim_{n\in\mathbb{Z}_+} Y_n(\omega) = X_T(\omega)$ a.e. $\omega \in \boldsymbol{\Omega}$. By Lebesgue Dominated Convergence Theorem 11.91, X_T is absolutely integrable over $\boldsymbol{\Omega}$ and $\boldsymbol{E}(X_T) = \lim_{n\in\mathbb{Z}_+} \boldsymbol{E}(Y_n) = \boldsymbol{E}(Y_0) = \boldsymbol{E}(X_0)$. Hence, the result holds.

Let (iii) hold. Then, the stopped stochastic process $(Y_n)_{n=0}^{\infty}$ defined in Theorem 14.47 satisfies $Y_n = X_0 + \sum_{i=1}^{n} C_i(X_i - X_{i-1})$, $\forall n \in \mathbb{N}$, where $(C_n)_{n=1}^{\infty}$ is as defined in the proof of Theorem 14.47. Then, $\|Y_n(\omega)\|_{\mathcal{Y}} \leq \|X_0(\omega)\|_{\mathcal{Y}} + \sum_{i=1}^{n} \chi_{\{T(\omega)\geq i\},\Omega}\|X_i(\omega) - X_{i-1}(\omega)\|_{\mathcal{Y}} \leq \|X_0(\omega)\|_{\mathcal{Y}} + \sum_{i=1}^{n} M\chi_{\{T(\omega)\geq i\},\Omega} \leq \|X_0(\omega)\|_{\mathcal{Y}} + MT(\omega)$, $\forall n \in \mathbb{Z}_+$, where the first inequality follows from the definition of $(C_n)_{n=1}^{\infty}$ and the second inequality follows from (iii). Then, $\boldsymbol{E}(\mathcal{P} \circ Y_n) \leq \boldsymbol{E}(\mathcal{P} \circ X_0) + M\int_{\Omega} T(\omega)\,\mathrm{d}\boldsymbol{P}(\omega) = \boldsymbol{E}(\mathcal{P} \circ X_0) + M\boldsymbol{E}(T) < \infty$, $\forall n \in \mathbb{Z}_+$, where the last inequality follows from (iii). Then, Y_n is absolutely integrable over $\boldsymbol{\Omega}$, $\forall n \in \mathbb{Z}_+$. Since $\boldsymbol{E}(T) < \infty$, then $T(\omega) < \infty$ a.e. $\omega \in \boldsymbol{\Omega}$, and $\lim_{n\in\mathbb{Z}_+} Y_n(\omega) = X_T(\omega)$ a.e. $\omega \in \boldsymbol{\Omega}$. By Lebesgue Dominated Convergence Theorem 11.91, $\boldsymbol{E}(X_T) = \lim_{n\in\mathbb{Z}_+} \boldsymbol{E}(Y_n) = \boldsymbol{E}(Y_0) = \boldsymbol{E}(X_0) \in \mathcal{Y}$, where the second equality follows from Theorem 14.47. Hence, the result holds.

This completes the proof of the theorem. □

Definition 14.49 Let $\boldsymbol{\Omega} := (\Omega, \mathcal{B}, \boldsymbol{P})$ be a probability measure space, $\mathcal{Y}$ be a separable Banach space, and $(X_n)_{n=1}^{\infty}$ be a sequence of $\mathcal{Y}$-valued random variables. Let $\mathcal{T}_n := \sigma(X_{n+1}, X_{n+2}, \ldots)$, $\forall n \in \mathbb{N}$, and $\mathcal{T} := \bigcap_{n=1}^{\infty} \mathcal{T}_n$. The σ-algebra $\mathcal{T}$ is called the *tail σ-algebra* of the sequence $(X_n)_{n=1}^{\infty}$. ◇

Clearly, $\mathcal{T}_n$ is the σ-algebra that is generated by $\mathcal{F}_n := \{X_{n+i\,\mathrm{inv}}(O) \mid i \in \mathbb{N}, O \in \mathcal{O}_{\mathcal{Y}}\} \subseteq \mathcal{B}$, $\forall n \in \mathbb{N}$. Clearly, we have $\mathcal{T}_n \supseteq \mathcal{T}_{n+1}$, $\forall n \in \mathbb{N}$. It is straightforward to show that $\mathcal{T}$ is a σ-algebra. Hence, the definition makes sense.

Theorem 14.50 (Kolmogorov 0-1 Law) *Let $\boldsymbol{\Omega} := (\Omega, \mathcal{B}, \boldsymbol{P})$ be a probability measure space, $\mathcal{Y}$ be a separable Banach space, $(X_n)_{n=1}^{\infty}$ be a sequence of independent $\mathcal{Y}$-valued random variables, and $\mathcal{T}$ be the tail σ-algebra of $(X_n)_{n=1}^{\infty}$. Then, the following statements hold:*

(i) *$\forall E \in \mathcal{T}$, we have $\boldsymbol{P}(E) = 0$ or $\boldsymbol{P}(E) = 1$.*
(ii) *Let $\mathcal{Z} := (\mathcal{Z}, \rho)$ be a nonempty separable complete metric space and $\xi : \boldsymbol{\Omega} \to \mathcal{Z}$ be $\mathcal{T}$-measurable. Then, $\exists x_0 \in \mathcal{Z}$ such that $\boldsymbol{P}(\{\omega \in \Omega \mid \xi(\omega) = x_0\}) = \boldsymbol{P}(\xi_{\mathrm{inv}}(\{x_0\})) = 1$.*

Proof (i) Let $\mathcal{X}_n := \sigma(X_1, \ldots, X_n)$, $\forall n \in \mathbb{N}$. We need the following result.

Claim 14.50.1 $\mathcal{X}_n$ and $\mathcal{T}_n$ are independent σ-algebras, $\forall n \in \mathbb{N}$, where $\mathcal{T}_n$ is as defined in Definition 14.49.

Proof of Claim Fix any $n \in \mathbb{N}$. Let $\mathcal{K} := \{F \in \mathcal{B} \mid \exists O_1, \ldots, O_n \in \mathcal{O}_{\mathcal{Y}} \ni \cdot\ F = \bigcap_{i=1}^{n} X_{i\,\mathrm{inv}}(O_i)\} \subseteq \mathcal{X}_n$. Let $\mathcal{J} := \{F \in \mathcal{B} \mid \exists r \in \mathbb{N}, \exists O_1, \ldots, O_r \in \mathcal{O}_{\mathcal{Y}} \ni \cdot\ F = \bigcap_{i=1}^{r} X_{n+i\,\mathrm{inv}}(O_i)\} \subseteq \mathcal{T}_n$. Clearly, $\mathcal{K}$ and $\mathcal{J}$ are π-systems on Ω, $\sigma(\mathcal{K}) = \mathcal{X}_n$,

$\sigma(\mathcal{J}) = \mathcal{T}_n$. $\forall E_1 \in \mathcal{K}$, $\forall E_2 \in \mathcal{J}$, by the assumption that $(X_n)_{n=1}^{\infty}$ are independent, we have $\boldsymbol{P}(E_1 \cap E_2) = \boldsymbol{P}(E_1)\boldsymbol{P}(E_2)$. By Proposition 14.10, we have $\mathcal{X}_n$ and $\mathcal{T}_n$ are independent σ-algebras. This completes the proof of the claim. □

Since $\mathcal{T} \subseteq \mathcal{T}_n$, $\forall n \in \mathbb{N}$, then $\mathcal{T}$ and $\mathcal{X}_n$ are independent σ-algebras.

Claim 14.50.2 $\mathcal{X}_\infty := \sigma(X_1, X_2, \ldots)$ and $\mathcal{T}$ are independent σ-algebras.

Proof of Claim Clearly, we have $\mathcal{X}_n \subseteq \mathcal{X}_\infty$, $\forall n \in \mathbb{N}$. Define $\mathcal{K}_\infty := \bigcup_{n=1}^{\infty} \mathcal{X}_n \subseteq \mathcal{X}_\infty$. Clearly, $\mathcal{K}_\infty$ is a π-system on Ω and $\sigma(\mathcal{K}_\infty) = \mathcal{X}_\infty$. By the fact that $\mathcal{X}_n$ and $\mathcal{T}$ are independent σ-algebras, $\forall n \in \mathbb{N}$, we have $\forall E_1 \in \mathcal{K}_\infty$, $\forall E_2 \in \mathcal{T}$, $\boldsymbol{P}(E_1 \cap E_2) = \boldsymbol{P}(E_1)\boldsymbol{P}(E_2)$, since $E_1 \in \mathcal{X}_n$, for some $n \in \mathbb{N}$. Then, by Proposition 14.10, $\mathcal{X}_\infty$ and $\mathcal{T}$ are independent σ-algebras. This completes the proof of the claim. □

Note that $\mathcal{T} \subseteq \mathcal{X}_\infty$. By Claim 14.50.2, $\mathcal{T}$ and $\mathcal{T}$ are independent σ-algebras. $\forall E \in \mathcal{T}$, we have $\boldsymbol{P}(E) = \boldsymbol{P}(E \cap E) = \boldsymbol{P}(E)\boldsymbol{P}(E)$, where the last equality follows from Definition 14.5. Hence, $\boldsymbol{P}(E) = 0$ or $\boldsymbol{P}(E) = 1$.

(ii) By (i), $\forall O \in \mathcal{O}_{\mathcal{Z}}$, we have $\boldsymbol{P}(\xi_{\text{inv}}(O)) = 0$ or $\boldsymbol{P}(\xi_{\text{inv}}(O)) = 1$. Clearly, $\boldsymbol{P}(\xi_{\text{inv}}(\mathcal{Z})) = \boldsymbol{P}(\Omega) = 1$. Fix any $\bar{x} \in \mathcal{Z} \neq \emptyset$. Then, by Proposition 11.7, we have $\lim_{n\in\mathbb{N}} \boldsymbol{P}(\xi_{\text{inv}}(\mathcal{B}_{\mathcal{Z}}(\bar{x}, n))) = \boldsymbol{P}(\bigcup_{n=1}^{\infty} \xi_{\text{inv}}(\mathcal{B}_{\mathcal{Z}}(\bar{x}, n))) = \boldsymbol{P}(\xi_{\text{inv}}(\mathcal{Z})) = 1$. This implies that $\exists n_0 \in \mathbb{N}$ such that $\boldsymbol{P}(\xi_{\text{inv}}(\mathcal{B}_{\mathcal{Z}}(\bar{x}, n_0))) > 0$. By (i), we have $\boldsymbol{P}(\xi_{\text{inv}}(\mathcal{B}_{\mathcal{Z}}(\bar{x}, n_0))) = 1$.

Let $x_1 := \bar{x}$ and $E_1 := \mathcal{B}_{\mathcal{Z}}(x_1, n_0) \subseteq \mathcal{Z}$. By Proposition 4.38, E_1 is separable, and $\exists D_1 \subseteq E_1$ that is a countable dense subset of E_1. Then, $1 = \boldsymbol{P}(\bigcup_{x\in D_1} \xi_{\text{inv}}(\mathcal{B}_{\mathcal{Z}}(x, 3^{-1}n_0))) \leq \sum_{x\in D_1} \boldsymbol{P}(\xi_{\text{inv}}(\mathcal{B}_{\mathcal{Z}}(x, 3^{-1}n_0)))$. Hence, $\exists x_2 \in D_1 \subseteq E_1$ such that $\boldsymbol{P}(\xi_{\text{inv}}(\mathcal{B}_{\mathcal{Z}}(x_2, 3^{-1}n_0))) > 0$. By (i), we have $\boldsymbol{P}(\xi_{\text{inv}}(\mathcal{B}_{\mathcal{Z}}(x_2, 3^{-1}n_0))) = 1$.

Let $E_2 := \mathcal{B}_{\mathcal{Z}}(x_2, 3^{-1}n_0) \subseteq \mathcal{Z}$. By Proposition 4.38, E_2 is separable, and $\exists D_2 \subseteq E_2$ that is a countable dense subset of E_2. Then, $1 = \boldsymbol{P}(\bigcup_{x\in D_2} \xi_{\text{inv}}(\mathcal{B}_{\mathcal{Z}}(x, 3^{-2}n_0))) \leq \sum_{x\in D_2} \boldsymbol{P}(\xi_{\text{inv}}(\mathcal{B}_{\mathcal{Z}}(x, 3^{-2}n_0)))$. Hence, $\exists x_3 \in D_2 \subseteq E_2$ such that $\boldsymbol{P}(\xi_{\text{inv}}(\mathcal{B}_{\mathcal{Z}}(x_3, 3^{-2}n_0))) > 0$. By (i), we have $\boldsymbol{P}(\xi_{\text{inv}}(\mathcal{B}_{\mathcal{Z}}(x_3, 3^{-2}n_0))) = 1$.

Recursively, we have $E_n := \mathcal{B}(x_n, 3^{-n+1}n_0)$, and $\exists x_{n+1} \in E_n$ such that $\boldsymbol{P}(\xi_{\text{inv}}(E_{n+1})) = 1$, $\forall n \in \mathbb{N}$. Thus, we have $(x_n)_{n=1}^{\infty} \subseteq \mathcal{Z}$ with $\rho(x_n, x_{n+1}) < 3^{-n+1}n_0$, $\forall n \in \mathbb{N}$. It is obvious that $(x_n)_{n=1}^{\infty}$ is a Cauchy sequence. Since $\mathcal{Z}$ is complete, we have $\lim_{n\in\mathbb{N}} x_n =: x_0 \in \mathcal{Z}$. It is a simple matter to show that $x_0 \in \mathcal{B}_{\mathcal{Z}}(x_n, 3^{-n+2}2^{-1}n_0) =: \bar{E}_n \supseteq E_n$, $\forall n \in \mathbb{N}$. $\forall x \in \bar{E}_{n+1}$, we have $\rho(x, x_{n+1}) < 3^{-n+1}2^{-1}n_0$. Then, $\rho(x_n, x) \leq \rho(x_n, x_{n+1}) + \rho(x_{n+1}, x) < 3^{-n+1}n_0 + 3^{-n+1}2^{-1}n_0 = 3^{-n+2}2^{-1}n_0$. Hence, $x \in \bar{E}_n$. Thus, we have shown that $\bar{E}_n \supseteq \bar{E}_{n+1}$, $\forall n \in \mathbb{N}$. The preceding implies that $x_0 \in \bigcap_{n=1}^{\infty} \bar{E}_n$. $\forall x \in \mathcal{Z}$ with $x \neq x_0$, let $\rho(x, x_0) := \delta > 0$. Clearly, $\exists \bar{n} \in \mathbb{N}$ with $3^{-\bar{n}+2}2^{-1}n_0 < \frac{\delta}{2}$. Then, $\bar{x} \notin \bar{E}_{\bar{n}}$. This shows that $\bigcap_{n=1}^{\infty} \bar{E}_n = \{x_0\}$. Then, we have $\boldsymbol{P}(\xi_{\text{inv}}(\{x_0\})) = \boldsymbol{P}(\xi_{\text{inv}}(\bigcap_{n=1}^{\infty} \bar{E}_n)) = \boldsymbol{P}(\bigcap_{n=1}^{\infty} \xi_{\text{inv}}(\bar{E}_n)) = \lim_{n\in\mathbb{N}} \boldsymbol{P}(\xi_{\text{inv}}(\bar{E}_n)) = 1$, where the first equality follows from preceding discussion, the second equality follows from Proposition 2.5, the third equality follows from Proposition 11.5, and the last equality follows from $\bar{E}_n \supseteq E_n$ and then $\boldsymbol{P}(\xi_{\text{inv}}(\bar{E}_n)) = 1$, $\forall n \in \mathbb{N}$.

This completes the proof of the theorem. □

Proposition 14.51 *Let* $\boldsymbol{\Omega} := (\Omega, \mathcal{B}, \boldsymbol{P})$ *be a probability measure space,* $\mathcal{Y}$ *be a separable Banach space,* $(X_n)_{n=1}^{\infty}$ *be a sequence of* $\mathcal{Y}$*-valued random variables, and* $\mathcal{T}$ *be the tail* σ*-algebra of* $(X_n)_{n=1}^{\infty}$*. Then, the following statements hold:*

(i) $F_1 := \{\omega \in \Omega \mid \lim_{n\in\mathbb{N}} X_n(\omega) \in \mathcal{Y}\} \in \mathcal{T}$.
(ii) $F_2 := \{\omega \in \Omega \mid \lim_{n\in\mathbb{N}} \sum_{i=1}^{n} X_i(\omega) \in \mathcal{Y}\} \in \mathcal{T}$.
(iii) $F_3 := \{\omega \in \Omega \mid \lim_{n\in\mathbb{N}} \frac{1}{n} \sum_{i=1}^{n} X_i(\omega) \in \mathcal{Y}\} \in \mathcal{T}$

Proof (i) By Proposition 11.51, $F_1 = \bigcap_{j=1}^{\infty} \bigcup_{m=1}^{\infty} \bigcap_{l=m}^{\infty} \bigcap_{n=m}^{\infty} E_{j,l,n} \in \mathcal{B}$, where $E_{j,l,n} := \{\omega \in \Omega \mid \|X_l(\omega) - X_n(\omega)\|_{\mathcal{Y}} < 1/j\}$, $\forall j, l, n \in \mathbb{N}$. This shows that $F_1 \in \sigma(X_1, X_2, \ldots)$. Since $\lim_{n\in\mathbb{N}} X_n(\omega) \in \mathcal{Y}$ is independent of $X_1, \ldots, X_n$, then $F_1 = \bigcap_{j=1}^{\infty} \bigcup_{m=k+1}^{\infty} \bigcap_{l=m}^{\infty} \bigcap_{n=m}^{\infty} E_{j,l,n} \in \mathcal{T}_k$, $\forall k \in \mathbb{N}$, where $\mathcal{T}_k$ is as defined in Definition 14.49. Hence, $F_1 \in \bigcap_{k\in\mathbb{N}} \mathcal{T}_k = \mathcal{T}$.

(ii) By (i), $F_2 = \bigcap_{j=1}^{\infty} \bigcup_{m=1}^{\infty} \bigcap_{l=m}^{\infty} \bigcap_{n=m}^{\infty} \hat{E}_{j,l,n} \in \mathcal{B}$, where $\hat{E}_{j,l,n} := \left\{\omega \in \Omega \mid \left\| \sum_{i=l\wedge n+1}^{l\vee n} X_i(\omega) \right\|_{\mathcal{Y}} < 1/j\right\}$, $\forall j, l, n \in \mathbb{N}$. This shows that $F_1 \in \sigma(X_1, X_2, \ldots)$. Since $\lim_{n\in\mathbb{N}} \sum_{i=1}^{n} X_i(\omega) \in \mathcal{Y}$ if and only if $\lim_{n\in\mathbb{N}} \sum_{i=1}^{n} X_{k+i}(\omega) \in \mathcal{Y}$, $\forall k \in \mathbb{N}$, then $F_2 = \bigcap_{j=1}^{\infty} \bigcup_{m=k+1}^{\infty} \bigcap_{l=m}^{\infty} \bigcap_{n=m}^{\infty} \hat{E}_{j,l,n} \in \mathcal{T}_k$, $\forall k \in \mathbb{N}$. Hence, $F_2 \in \bigcap_{k\in\mathbb{N}} \mathcal{T}_k = \mathcal{T}$.

(iii) It is easy to see that $F_3 \in \sigma(X_1, X_2, \ldots)$. Let $F_3^{\langle k \rangle} := \{\omega \in \Omega \mid \lim_{n\in\mathbb{N}} \frac{1}{n} \sum_{i=1}^{n} X_{k+i}(\omega) \in \mathcal{Y}\}$, $\forall k \in \mathbb{N}$. We will show that $F_3 = F_3^{\langle k \rangle}$, $\forall k \in \mathbb{N}$. Fix any $k \in \mathbb{N}$. $\forall \omega \in F_3$, we have $\lim_{n\in\mathbb{N}} \frac{1}{n} \sum_{i=1}^{n} X_i(\omega) \in \mathcal{Y}$. Then, $\lim_{n\in\mathbb{N}} \frac{1}{n} \sum_{i=1}^{n} X_{k+i}(\omega) = \lim_{n\in\mathbb{N}} \left(\frac{n+k}{n} \frac{1}{n+k} \sum_{i=1}^{n+k} X_i(\omega) - \frac{1}{n} \sum_{i=1}^{k} X_i(\omega) \right) = \lim_{n\in\mathbb{N}} \frac{1}{n} \sum_{i=1}^{n} X_i(\omega) \in \mathcal{Y}$. Thus, $\omega \in F_3^{\langle k \rangle}$. By the arbitrariness of ω, we have $F_3 \subseteq F_3^{\langle k \rangle}$.

On the other hand, $\forall \omega \in F_3^{\langle k \rangle}$, we have $\lim_{n\in\mathbb{N}} \frac{1}{n} \sum_{i=1}^{n} X_{k+i}(\omega) \in \mathcal{Y}$. Then, $\lim_{n\in\mathbb{N}} \frac{1}{n} \sum_{i=1}^{n} X_i(\omega) = \lim_{n\in\mathbb{N}} \left(\frac{n-k}{n} \frac{1}{n-k} \sum_{i=1}^{n-k} X_{k+i}(\omega) + \frac{1}{n} \sum_{i=1}^{k} X_i(\omega) \right) = \lim_{n\in\mathbb{N}} \frac{1}{n} \sum_{i=1}^{n} X_{k+i}(\omega) \in \mathcal{Y}$. Then, $\omega \in F_3$. By the arbitrariness of ω, we have $F_3^{\langle k \rangle} \subset F_3$. Hence, $F_3 = F_3^{\langle k \rangle}$.

It is clear that $F_3^{\langle k \rangle} \in \mathcal{T}_k$, $\forall k \in \mathbb{N}$. Then, we have $F_3 \subset \mathcal{T}_k$, $\forall k \in \mathbb{N}$, and hence $F_3 \in \mathcal{T}$.

This completes the proof of the proposition. □

14.6 Characteristic Functions

Definition 14.52 Let $\boldsymbol{\Omega} := (\Omega, \mathcal{B}, \boldsymbol{P})$ be a probability measure space and $x : \Omega \to \mathbb{R}$ be a random variable with law L_x and a cumulative distribution function $F : \mathbb{R} \to [0, 1] \subset \mathbb{R}$. The *characteristic function* of x is $\Psi_x : \mathbb{R} \to \mathbb{C}$ defined by

$\Psi_x(\theta) = \boldsymbol{E}(\exp(\mathrm{i}\theta x)) = \boldsymbol{E}(\cos(\theta x)) + \mathrm{i}\boldsymbol{E}(\sin(\theta x)) = \int_\Omega \exp(\mathrm{i}\theta x(\omega))\,\mathrm{d}\boldsymbol{P}(\omega) = \int_\mathbb{R} \exp(\mathrm{i}\theta x)\,\mathrm{d}L_x(x) = \int_\mathbb{R} \exp(\mathrm{i}\theta x)\,\mathrm{d}F(x)$, $\forall \theta \in \mathbb{R}$. ◇

By Lebesgue Dominated Convergence Theorem 11.91, the characteristic function of a random variable is well defined. The last two equalities in the above definition hold as deduced in the following. Fix any $\theta \in \mathbb{R}$. Let $f : \Omega \to \mathbb{C}$ be defined by $f(\omega) = \exp(\mathrm{i}\theta x(\omega))$, $\forall \omega \in \Omega$, and $\bar{f} : \mathbb{R} \to \mathbb{C}$ be defined by $\bar{f}(x) = \exp(\mathrm{i}\theta x)$, $\forall x \in \mathbb{R}$. By Proposition 11.69, let $\mathcal{J}(\mathbb{C})$ be the integration system on $\mathbb{C}$. Then, $\Psi_x(\theta) = \int_\Omega f(\omega)\,\mathrm{d}\boldsymbol{P}(\omega) = \lim_{R\in\mathcal{J}(\mathbb{C})} F_R$, where $(F_R)_{R\in\mathcal{J}(\mathbb{C})}$ is the net as defined in Definition 11.70 for the integral $\int_\Omega f(\omega)\,\mathrm{d}\boldsymbol{P}(\omega)$. $\forall R = \{(y_\alpha, U_\alpha) \mid \alpha \in \Lambda\} \in \mathcal{J}(\mathbb{C})$, $F_R = \sum_{\alpha\in\Lambda} y_\alpha \boldsymbol{P}(f_{\mathrm{inv}}(U_\alpha)) = \sum_{\alpha\in\Lambda} y_\alpha \boldsymbol{P}(x_{\mathrm{inv}}(\bar{f}_{\mathrm{inv}}(U_\alpha))) = \sum_{\alpha\in\Lambda} y_\alpha L_x(\bar{f}_{\mathrm{inv}}(U_\alpha)) =: \sum_{\alpha\in\Lambda} \bar{F}_R$, where the third equality follows from Definition 14.16 and $(\bar{F}_R)_{R\in\mathcal{J}(\mathbb{C})}$ is the net as defined in Definition 11.70 for the integral $\int_R \bar{f}(x)\,\mathrm{d}L_x(x)$. Hence, we have $\Psi_x(\theta) = \int_\Omega f(\omega)\,\mathrm{d}\boldsymbol{P}(\omega) = \int_\mathbb{R} \bar{f}(x)\,\mathrm{d}L_x(x) = \int_\mathbb{R} \exp(\mathrm{i}\theta x)\,\mathrm{d}L_x(x)$. By Lebesgue–Stieltjes integral Definition 12.57, we have $\int_\mathbb{R} \exp(\mathrm{i}\theta x)\,\mathrm{d}L_x(x) = \int_\mathbb{R} \exp(\mathrm{i}\theta x)\,\mathrm{d}F(x)$.

Proposition 14.53 *Let $\boldsymbol{\Omega} := (\Omega, \mathcal{B}, \boldsymbol{P})$ be a probability measure space and $x : \Omega \to \mathbb{R}$ be a random variable with characteristic function $\Psi_x : \mathbb{R} \to \mathbb{C}$. Then, the following statements hold:*

(i) $\Psi_x(0) = 1$.
(ii) $|\Psi_x(\theta)| \leq 1$, $\forall \theta \in \mathbb{R}$.
(iii) Ψ_x *is continuous.*
(iv) $\Psi_{-x}(\theta) = \overline{\Psi_x(\theta)}$, $\forall \theta \in \mathbb{R}$.
(v) $\Psi_{ax+b}(\theta) = \exp(\mathrm{i}b\theta)\Psi_x(a\theta)$, $\forall \theta \in \mathbb{R}$, $\forall a, b \in \mathbb{R}$.
(vi) *If $n \in \mathbb{N}$ with $\boldsymbol{E}(|x|^n) < \infty$, then, Ψ_x is $\mathcal{C}_n$, and $\Psi_x^{(i)}(\theta) = \boldsymbol{E}((\mathrm{i}x)^i \cdot \exp(\mathrm{i}\theta x))$, $\forall i = 1, \ldots, n$.*

Proof (i) $\Psi_x(0) = \boldsymbol{E}(1) = 1$.

(ii) Note that $|\exp(\mathrm{i}\theta x(\omega))| = 1$, $\forall \omega \in \Omega$, $\forall \theta \in \mathbb{R}$. By Proposition 11.92, we have $|\Psi_x(\theta)| \leq \int_\Omega |\exp(\mathrm{i}\theta x(\omega))|\,\mathrm{d}\boldsymbol{P}(\omega) = 1$, $\forall \theta \in \mathbb{R}$.

(iii) $\forall \theta_0 \in \mathbb{R}$, we have $\lim_{\theta\to\theta_0} \exp(\mathrm{i}\theta x(\omega)) = \exp(\mathrm{i}\theta_0 x(\omega))$, $\forall \omega \in \Omega$. Then, by Lebesgue Dominated Convergence Theorem 11.91, $\Psi_x(\theta_0) = \int_\Omega \exp(\mathrm{i}\theta_0 \cdot x(\omega))\,\mathrm{d}\boldsymbol{P}(\omega) = \lim_{\theta\to\theta_0} \int_\Omega \exp(\mathrm{i}\theta x(\omega))\,\mathrm{d}\boldsymbol{P}(\omega) = \lim_{\theta\to\theta_0} \Psi_x(\theta)$. By Proposition 3.74, Ψ_x is continuous at θ_0. By the arbitrariness of θ_0, we have Ψ_x is continuous.

(iv) $\Psi_{-x}(\theta) = \boldsymbol{E}(\exp(-\mathrm{i}\theta x)) = \boldsymbol{E}(\overline{\exp(\mathrm{i}\theta x)}) = \overline{\boldsymbol{E}(\exp(\mathrm{i}\theta x))} = \overline{\Psi_x(\theta)}$, $\forall \theta \in \mathbb{R}$, where the second equality follows from the definition of exponential function and the third equality follows from Proposition 11.92.

(v) $\Psi_{ax+b}(\theta) = \boldsymbol{E}(\exp(\mathrm{i}\theta(ax+b))) = \boldsymbol{E}(\exp(\mathrm{i}b\theta)\exp(\mathrm{i}a\theta x)) = \exp(\mathrm{i}b\theta) \cdot \boldsymbol{E}(\exp(\mathrm{i}(a\theta)x)) = \exp(\mathrm{i}b\theta)\Psi_x(a\theta)$, $\forall \theta \in \mathbb{R}$, where the second equality follows from the definition of exponential function, the third equality follows from Proposition 11.92, and the fourth equality follows from Definition 14.52.

(vi) Let $n \in \mathbb{N}$ with $\boldsymbol{E}(|x|^n) < \infty$. Then, $x \in \bar{\mathrm{L}}_n(\boldsymbol{\Omega}, \mathbb{R})$. By Hölder's Inequality Theorem 11.178, $\forall i = 1, \ldots, n-1$, $\boldsymbol{E}(|x|^i) = \int_\Omega |x(\omega)|^i \cdot 1\,\mathrm{d}\boldsymbol{P}(\omega) \leq$

$\left(\int_\Omega (|x(\omega)|^i)^{n/i} \, d\boldsymbol{P}(\omega)\right)^{i/n} \left(\int_\Omega 1^{n/(n-i)} \, d\boldsymbol{P}(\omega)\right)^{(n-i)/n} = (\boldsymbol{E}(|x|^n))^{i/n} < \infty$. Hence, $x \in \bar{L}_i(\boldsymbol{\Omega}, \mathbb{R})$, $i = 1, \ldots, n$.

Fix any $i = 1, \ldots, n$. By Lebesgue Dominated Convergence Theorem 11.91 and $\boldsymbol{E}(|x|^i) < \infty$, $\boldsymbol{E}((ix)^i \exp(i\theta x)) \in \mathbb{C}$, $\forall \theta \in \mathbb{R}$. $\forall \theta_0 \in \mathbb{R}$, we have $\lim_{\theta \to \theta_0} (ix(\omega))^i \exp(i\theta x(\omega)) = (ix(\omega))^i \exp(i\theta_0 x(\omega))$, $\forall \omega \in \Omega$. Then, by Lebesgue Dominated Convergence Theorem 11.91, $\lim_{\theta \to \theta_0} \boldsymbol{E}((ix)^i \exp(i\theta x)) = \boldsymbol{E}((ix)^i \exp(i\theta_0 x))$. By Proposition 3.74 and arbitrariness of θ_0, $\boldsymbol{E}((ix)^i \cdot \exp(i\theta x))$ as a function of θ is continuous.

Clearly, $\Psi_x(\theta) = \boldsymbol{E}(\exp(i\theta x))$, $\forall \theta \in \mathbb{R}$. Inductively, assume that $\Psi_x^{(i)}(\theta) = \boldsymbol{E}((ix)^i \exp(i\theta x))$, $\forall \theta \in \mathbb{R}$, for some $0 \leq i < n$. Consider the case $i + 1 \in \{1, \ldots, n\}$. Fix any $\theta_0 \in \mathbb{R}$.

$$\begin{aligned}
\Psi_x^{(i+1)}(\theta_0) &= \lim_{\theta \to \theta_0} \frac{\Psi_x^{(i)}(\theta) - \Psi_x^{(i)}(\theta_0)}{\theta - \theta_0} \\
&= \lim_{\theta \to \theta_0} \frac{\boldsymbol{E}((ix)^i \exp(i\theta x)) - \boldsymbol{E}((ix)^i \exp(i\theta_0 x))}{\theta - \theta_0} \\
&= \lim_{\theta \to \theta_0} \boldsymbol{E}\left((ix)^i \frac{\exp(i\theta x) - \exp(i\theta_0 x)}{\theta - \theta_0}\right) \\
&= \lim_{\theta \to \theta_0} \boldsymbol{E}\left((ix)^{i+1} \int_0^1 \exp(i(\theta(1-s) + s\theta_0)x) \, ds\right) \\
&= \boldsymbol{E}((ix)^{i+1} \exp(i\theta_0 x))
\end{aligned}$$

where the first equality follows from the definition of the derivative, the second equality follows from the inductive assumption, the third equality follows from Proposition 11.92, the fourth equality follows from Taylor's Theorem 12.122, and last equality follows from Lebesgue Dominated Convergence Theorem 11.91 and Proposition 4.16.

Hence, we have $\Psi_x^{(i)}(\theta) = \boldsymbol{E}((ix)^i \exp(i\theta x))$, $\forall \theta \in \mathbb{R}$, $\forall i = 0, \ldots, n$.

This completes the proof of the proposition. □

Proposition 14.54 *Let $\boldsymbol{\Omega} := (\Omega, \mathcal{B}, \boldsymbol{P})$ be a probability measure space and X and Y be two independent $\mathbb{R}$-valued random variables. Then, $\Psi_{X+Y}(\theta) = \Psi_X(\theta)\Psi_Y(\theta)$, $\forall \theta \in \mathbb{R}$.*

Proof $\forall \theta \in \mathbb{R}$, we have $\Psi_{X+Y}(\theta) = \boldsymbol{E}(\exp(i\theta(X + Y))) = \boldsymbol{E}(\exp(i\theta X) \cdot \exp(i\theta Y)) = \boldsymbol{E}(\exp(i\theta X))\boldsymbol{E}(\exp(i\theta Y)) = \Psi_X(\theta)\Psi_Y(\theta)$, where the first equality follows from Definition 14.52, the second equality follows from Proposition 9.88, the third equality follows from Proposition 14.9, and the last equality follows from Definition 14.52. This completes the proof of the proposition. □

Theorem 14.55 (Lévy's Inversion Formula) *Let $\boldsymbol{\Omega} := (\Omega, \mathcal{B}, \boldsymbol{P})$ be a probability measure space, $X : \Omega \to \mathbb{R}$ be random variables with law $L : \mathcal{B}_B(\mathbb{R}) \to [0, 1] \subset \mathbb{R}$, the cumulative distribution function $F : \mathbb{R} \to [0, 1] \subset \mathbb{R}$ as defined in*

Proposition 12.51, and the characteristic function $\Psi_X : \mathbb{R} \to \mathbb{C}$. *Then, the following statements hold:*

(i) $\forall a, b \in \mathbb{R}$ *with* $a < b$, *we have*

$$\lim_{T\to\infty} \frac{1}{2\pi} \int_{-T}^{T} \frac{\exp(-\mathrm{i}\theta a) - \exp(-\mathrm{i}\theta b)}{\mathrm{i}\theta} \Psi_X(\theta)\,\mathrm{d}\theta$$
$$= \frac{1}{2} L(\{a\}) + L(\mathrm{r}^{\circ}_{a,b}) + \frac{1}{2} L(\{b\})$$
$$= \frac{1}{2}(F(b) + \lim_{t\to b^-} F(t)) - \frac{1}{2}(F(a) + \lim_{t\to a^-} F(t)) \tag{14.6}$$

(ii) *If* $\int_{\mathbb{R}} |\Psi_X(\theta)|\,\mathrm{d}\mu_{\mathrm{B}}(\theta) < \infty$, *then* F *is* $\mathcal{C}_1$ *and* $\mathrm{D}F(x) = \frac{1}{2\pi}\int_{\mathbb{R}} \Psi_X(\theta) \cdot \exp(-\mathrm{i}\theta x)\,\mathrm{d}\mu_{\mathrm{B}}(\theta) =: f(x)$, $\forall x \in \mathbb{R}$, *where the function* $f : \mathbb{R} \to \mathbb{R}$ *is the probability density function of* X.

(iii) *Under the assumption of (ii), we have the duality result:*

$$\Psi_X(\theta) = \int_{\mathbb{R}} \exp(\mathrm{i}\theta x) f(x)\,\mathrm{d}\mu_{\mathrm{B}}(x); \quad \forall \theta \in \mathbb{R} \tag{14.7a}$$

$$f(x) = \frac{1}{2\pi} \int_{\mathbb{R}} \exp(-\mathrm{i}\theta x)\Psi_X(\theta)\,\mathrm{d}\mu_{\mathrm{B}}(\theta); \quad \forall x \in \mathbb{R} \tag{14.7b}$$

Proof We need the following intermediate result.

Claim 14.55.1 For all $u, v \in \mathbb{R}$ with $u \le v$, we have $|\exp(\mathrm{i}v) - \exp(\mathrm{i}u)| \le |v - u|$.

Proof of Claim Note that $|\exp(\mathrm{i}v) - \exp(\mathrm{i}u)| = \left|\int_u^v \mathrm{i}\exp(\mathrm{i}t)\,\mathrm{d}t\right| \le \int_u^v |\mathrm{i}\exp(\mathrm{i}t)|\,\mathrm{d}t = \int_u^v 1\,\mathrm{d}t = v - u$, where the first equality follows from Theorem 12.83 and the first inequality follows from Proposition 11.92. This completes the proof of the claim. □

$\forall a, b \in \mathbb{R}$ with $a < b$. Fix $T \in (0, \infty) \subset \mathbb{R}$. Then,

$$C_T := \frac{1}{2\pi} \int_{\mathbb{R}} \left(\int_{-T}^{T} \left| \frac{\exp(\mathrm{i}\theta(x-a)) - \exp(\mathrm{i}\theta(x-b))}{\mathrm{i}\theta} \right| \mathrm{d}\theta \right) \mathrm{d}L(x)$$
$$\le \frac{1}{2\pi} \int_{\mathbb{R}} \left(\int_{-T}^{T} \frac{|\theta(x-a) - \theta(x-b)|}{\theta}\,\mathrm{d}\theta \right) \mathrm{d}L(x)$$
$$= \frac{1}{2\pi} \int_{\mathbb{R}} \left(\int_{-T}^{T} (b-a)\,\mathrm{d}\theta \right) \mathrm{d}L(x) = \frac{(b-a)T}{\pi} < \infty$$

where the first inequality follows from Claim 14.55.1. By Tonelli's Theorem 12.29, $\frac{1}{2\pi}\frac{\exp(\mathrm{i}\theta(x-a)) - \exp(\mathrm{i}\theta(x-b))}{\mathrm{i}\theta}$ is absolutely integrable over $\mathbb{R} \times \mathrm{r}_{-T,T}$ with respect to measure $\mu_B \times L$. Then, by Fubini's Theorem 12.31, we have

$$\begin{aligned}
&\frac{1}{2\pi}\int_{-T}^{T}\frac{\exp(-\mathrm{i}\theta a)-\exp(-\mathrm{i}\theta b)}{\mathrm{i}\theta}\Psi_X(\theta)\,\mathrm{d}\theta \\
&= \frac{1}{2\pi}\int_{-T}^{T}\frac{\exp(-\mathrm{i}\theta a)-\exp(-\mathrm{i}\theta b)}{\mathrm{i}\theta}\left(\int_{\mathbb{R}}\exp(\mathrm{i}\theta x)\,\mathrm{d}L(x)\right)\mathrm{d}\theta \\
&= \frac{1}{2\pi}\int_{\mathbb{R}}\left(\int_{-T}^{T}\frac{\exp(\mathrm{i}\theta(x-a))-\exp(\mathrm{i}\theta(x-b))}{\mathrm{i}\theta}\,\mathrm{d}\theta\right)\mathrm{d}L(x)
\end{aligned}$$

Note that

$$\begin{aligned}
f(x,T) &:= \frac{1}{2\pi}\int_{-T}^{T}\frac{\exp(\mathrm{i}\theta(x-a))-\exp(\mathrm{i}\theta(x-b))}{\mathrm{i}\theta}\,\mathrm{d}\theta \\
&= \frac{1}{2\pi}\int_{-T}^{T}\frac{1}{\mathrm{i}\theta}(\cos(\theta(x-a))+\mathrm{i}\sin(\theta(x-a))-\cos(\theta(x-b)) \\
&\quad -\mathrm{i}\sin(\theta(x-b)))\,\mathrm{d}\theta \\
&= \frac{1}{2\pi}\int_{-T}^{T}\frac{\sin(\theta(x-a))-\sin(\theta(x-b))}{\theta}\,\mathrm{d}\theta \\
&= \frac{1}{\pi}\int_{0}^{T}\left(\frac{\sin(\theta(x-a))}{\theta}-\frac{\sin(\theta(x-b))}{\theta}\right)\mathrm{d}\theta \\
&= \frac{1}{\pi}(\mathrm{sgn}(x-a)S(T|x-a|)-\mathrm{sgn}(x-b)S(T|x-b|))
\end{aligned}$$

where $S(U) := \int_0^U \frac{\sin(x)}{x}\,\mathrm{d}x$, $\forall U \in [0,\infty) \subset \mathbb{R}$. It is the standard result that $\lim_{U\to\infty} S(U) = \frac{\pi}{2}$ (see Example 33.9 of Bartle (1976) and Proposition 12.100).

Then, $\lim_{T\to\infty} f(x,T) =: f_0(x) = \begin{cases} 0 & \text{if } x < a \text{ or } x > b \\ 1 & \text{if } a < x < b \\ \frac{1}{2} & \text{if } x = a \text{ or } x = b, \end{cases}$ $\forall x \in \mathbb{R}$. By Bounded Convergence Theorem 11.77, we have

$$\begin{aligned}
&\lim_{T\to\infty}\frac{1}{2\pi}\int_{-T}^{T}\frac{\exp(-\mathrm{i}\theta a)-\exp(-\mathrm{i}\theta b)}{\mathrm{i}\theta}\Psi_X(\theta)\,\mathrm{d}\theta = \lim_{T\to\infty}\int_{\mathbb{R}} f(x,T)\,\mathrm{d}L(x) \\
&= \int_{\mathbb{R}} f_0(x)\,\mathrm{d}L(x) = \frac{1}{2}L(\{a\}) + L(\mathrm{r}^{\circ}_{a,b}) + \frac{1}{2}L(\{b\}) \\
&= \frac{1}{2}(F(b)+\lim_{t\to b^-}F(t)) - \frac{1}{2}(F(a)+\lim_{t\to a^-}F(t))
\end{aligned}$$

Thus, statement (i) is established.

(ii) Under the assumption that $\Psi_X \in \bar{L}_1(\mathbb{R}, \mathbb{C})$, let $c := \|\Psi_X\|_{\bar{L}_1(\mathbb{R},\mathbb{C})}$. By Claim 14.55.1, we have

$$\left| \frac{\exp(-i\theta a) - \exp(-i\theta b)}{i\theta} \Psi_X(\theta) \right| \leq (b-a)|\Psi_X(\theta)|$$

which is then absolutely integrable over $\mathbb{R}$. Then, by (i) and Lebesgue Dominated Convergence Theorem 11.91, we have, $\forall a, b \in \mathbb{R}$ with $a < b$,

$$\frac{1}{2\pi} \int_{\mathbb{R}} \frac{\exp(-i\theta a) - \exp(-i\theta b)}{i\theta} \Psi_X(\theta)\, d\mu_B(\theta) = \frac{1}{2} L(\{a\}) + L(r^{\circ}_{a,b})$$
$$+\frac{1}{2} L(\{b\}) = \frac{1}{2}(F(b) + \lim_{t \to b^-} F(t)) - \frac{1}{2}(F(a) + \lim_{t \to a^-} F(t)) =: K(a, b)$$

$\forall \epsilon \in (0, \infty) \subset \mathbb{R}$, let $\delta := \frac{\sqrt{2}\pi\epsilon}{(1+c)} > 0$, $\forall (a_1, b_1) \in \mathcal{B}_{\mathbb{R}^2}((a, b), \delta)$, we have

$$\begin{aligned}
&|K(a_1, b_1) - K(a, b)| \\
&= \frac{1}{2\pi} \left| \int_{\mathbb{R}} \frac{\exp(-i\theta a_1) - \exp(-i\theta a) + \exp(-i\theta b) - \exp(-i\theta b_1)}{i\theta} \right. \\
&\quad \left. \cdot \Psi_X(\theta)\, d\mu_B(\theta) \right| \\
&\leq \frac{1}{2\pi} \int_{\mathbb{R}} \left| \frac{\exp(-i\theta a_1) - \exp(-i\theta a) + \exp(-i\theta b) - \exp(-i\theta b_1)}{i\theta} \right| \\
&\quad \cdot |\Psi_X(\theta)|\, d\mu_B(\theta) \\
&\leq \frac{1}{2\pi} \int_{\mathbb{R}} \left(\left| \frac{\exp(-i\theta a_1) - \exp(-i\theta a)}{i\theta} \right| + \left| \frac{\exp(-i\theta b) - \exp(-i\theta b_1)}{i\theta} \right| \right) \\
&\quad \cdot |\Psi_X(\theta)|\, d\mu_B(\theta) \\
&\leq \frac{1}{2\pi} \int_{\mathbb{R}} (|a - a_1| + |b - b_1|)\, |\Psi_X(\theta)|\, d\mu_B(\theta) \\
&= \frac{1}{2\pi} (|a - a_1| + |b - b_1|) \int_{\mathbb{R}} |\Psi_X(\theta)|\, d\mu_B(\theta) \\
&\leq \frac{1}{2\pi} \sqrt{2((a - a_1)^2 + (b - b_1)^2)}\, c < \frac{1}{\sqrt{2}\pi} \delta c < \epsilon
\end{aligned}$$

where the first inequality follows from Proposition 11.92 and the third inequality follows from Claim 14.55.1. This shows that $K(a, b)$ is continuous at (a, b), $\forall a, b \in$

$\mathbb{R}$ with $a < b$. Thus, we may conclude that F is continuous. Hence, we have

$$K(a,b) = F(b) - F(a) = \frac{1}{2\pi} \int_{\mathbb{R}} \frac{\exp(-\mathrm{i}\theta a) - \exp(-\mathrm{i}\theta b)}{\mathrm{i}\theta} \Psi_X(\theta) \, \mathrm{d}\mu_{\mathrm{B}}(\theta)$$

Fix any $a_0 \subset \mathbb{R}$, and we have

$$\lim_{b \to a_0^+} \frac{F(b) - F(a_0)}{b - a_0} = \lim_{b \to a_0^+} \frac{1}{2\pi} \int_{\mathbb{R}} \frac{\exp(-\mathrm{i}\theta a_0) - \exp(-\mathrm{i}\theta b)}{\mathrm{i}\theta(b - a_0)} \Psi_X(\theta) \, \mathrm{d}\mu_{\mathrm{B}}(\theta)$$
$$= \frac{1}{2\pi} \int_{\mathbb{R}} \exp(-\mathrm{i}\theta a_0) \Psi_X(\theta) \, \mathrm{d}\mu_{\mathrm{B}}(\theta)$$

where the second equality follows by the Lebesgue Dominated Convergence Theorem 11.91. Similarly,

$$\lim_{a \to a_0^-} \frac{F(a_0) - F(a)}{a_0 - a} = \lim_{a \to a_0^-} \frac{1}{2\pi} \int_{\mathbb{R}} \frac{\exp(-\mathrm{i}\theta a) - \exp(-\mathrm{i}\theta a_0)}{\mathrm{i}\theta(a_0 - a)} \Psi_X(\theta) \, \mathrm{d}\mu_{\mathrm{B}}(\theta)$$
$$= \frac{1}{2\pi} \int_{\mathbb{R}} \exp(-\mathrm{i}\theta a_0) \Psi_X(\theta) \, \mathrm{d}\mu_{\mathrm{B}}(\theta)$$

Therefore, we have $\mathrm{D}F(a_0) = \frac{1}{2\pi} \int_{\mathbb{R}} \exp(-\mathrm{i}\theta a_0) \Psi_X(\theta) \, \mathrm{d}\mu_{\mathrm{B}}(\theta)$. By the arbitrariness of a_0, we have $\mathrm{D}F(x) = \frac{1}{2\pi} \int_{\mathbb{R}} \exp(-\mathrm{i}\theta x) \Psi_X(\theta) \, \mathrm{d}\mu_{\mathrm{B}}(\theta) = f(x)$, $\forall x \in \mathbb{R}$. Again, by Lebesgue Dominated Convergence Theorem 11.91, we have f is continuous. Hence, F is $\mathcal{C}_1$.

(iii) By (ii), we have (14.7b). By Definition 14.52, we have $\Psi_X(\theta) = \int_{\mathbb{R}} \exp(\mathrm{i}\theta x) \, \mathrm{d}F(x) = \int_{\mathbb{R}} \exp(\mathrm{i}\theta x) f(x) \, \mathrm{d}\mu_{\mathrm{B}}(x)$, where the second equality follows from Propositions 12.102 and 11.168.

This completes the proof of the theorem. □

By Lévy Inversion Formula, Theorem 14.55, if the characteristic function $g : \mathbb{R} \to \mathbb{C}$ for a random variable X with law μ and cumulative distribution function F (as defined in Definition 14.16) is known, then (i) of Theorem 14.55 implies that the value $\frac{1}{2}\mu(\{a\}) + \mu(\mathrm{r}^{\circ}_{a,b}) + \frac{1}{2}\mu(\{b\})$ is uniquely determined, $\forall a, b \in \mathbb{R}$ with $a < b$. Then, g uniquely determines $\mu(\mathrm{r}_{a,b}) = F(b) - F(a)$, $\forall a, b \in \mathbb{R}$ with $a < b$ and F being continuous at a and b. Since μ is a probability measure, then g uniquely determines $F(b)$, $\forall b \in \mathbb{R}$ with F being continuous at b. By Proposition 12.51, F is a function of bounded variation, and it is continuous on the right and monotonically nondecreasing. Thus, F has countably many points of discontinuity (the total number of discontinuities with a jump of at least $1/n$ is at most n), which means that the set of points in $\mathbb{R}$ where F is continuous is dense in $\mathbb{R}$. Hence, F is uniquely determined by g. This further implies that μ is uniquely determined by g by Theorem 12.50.

14.7 Convergence in Distribution

Definition 14.56 Let $\boldsymbol{\Omega} := (\Omega, \mathcal{B}, \boldsymbol{P})$ be a probability measure space, $m \in \mathbb{N}$, $(x_n)_{n=1}^{\infty}$ be a sequence of $\mathbb{R}^m$-valued random variables, where $x_n : \boldsymbol{\Omega} \to \mathbb{R}^m$ and admits law L_n, $\forall n \in \mathbb{N}$, x_0 be an $\mathbb{R}^m$-valued random variable with law L_0. We will say that $(x_n)_{n=1}^{\infty}$ *converges to* x_0 *in distribution* if $\lim_{n\in\mathbb{N}} L_n = L_0$ weak*. In this case, we will write $x_n \nrightarrow x_0$ as $n \to \infty$, or $\lim_{n\in\mathbb{N}} x_n \nrightarrow x_0$. Let $F_n : \mathbb{R}^m \to [0, 1] \subset \mathbb{R}$ be the cumulative distribution functions of x_n as defined in Definition 14.16, $\forall n \in \mathbb{Z}_+$, we will write $\lim_{n\in\mathbb{N}} F_n \nrightarrow F_0$ or $F_n \nrightarrow F_0$ as $n \to \infty$. ◇

In the above definition, L_n is a probability measure on the measurable space $(\mathbb{R}^m, \mathcal{B}_B(\mathbb{R}^m))$, $\forall n \in \mathbb{Z}_+$. Then, $L_n \in \mathcal{M}_f(\mathbb{R}^m, \mathcal{B}_B(\mathbb{R}^m), \mathbb{R}) = \mathcal{M}_{ft}(\mathbb{R}^m, \mathbb{R}) = (\mathcal{C}_c(\mathbb{R}^m, \mathbb{R}))^*$ by Riesz Representation Theorem 11.209, $\forall n \in \mathbb{Z}_+$. Thus, $\lim_{n\in\mathbb{N}} L_n = L_0$ weak* makes sense. Therefore, the definition makes sense.

We will note that the convergence in distribution defined above is actually weaker than its classical definition (Williams, 1991, See Definition 17.1). The classical definition implies our definition, and on the other hand, our definition when coupled with the uniform growth boundedness assumption yields the classical definition.

Theorem 14.57 (Modes of Convergence) *Let* $\boldsymbol{\Omega} := (\Omega, \mathcal{B}, \boldsymbol{P})$ *be a probability measure space,* $m \in \mathbb{N}$, *and* $(x_n)_{n=0}^{\infty}$ *be a sequence of* $\mathbb{R}^m$*-valued random variables, where* $x_n : \boldsymbol{\Omega} \to \mathbb{R}^m$ *and admits law* L_n, $\forall n \in \mathbb{Z}_+$. *Then, the following statements hold:*

(i) *If* $\lim_{n\in\mathbb{N}} x_n = x_0$ *a.e. in* $\boldsymbol{\Omega}$, *then* $\lim_{n\in\mathbb{N}} x_n = x_0$ *in measure in* $\boldsymbol{\Omega}$.
(ii) *If* $\lim_{n\in\mathbb{N}} x_n = x_0$ *in measure in* $\boldsymbol{\Omega}$, *then* $\lim_{n\in\mathbb{N}} x_n \nrightarrow x_0$.

Proof (i) Let $\lim_{n\in\mathbb{N}} x_n = x_0$ a.e. in $\boldsymbol{\Omega}$. Then, by Definition 11.42, there exists $E \in \mathcal{B}$ with $\boldsymbol{P}(E) = 0$ such that $\lim_{n\in\mathbb{N}} x_n(\omega) = x_0(\omega)$, $\forall \omega \in \Omega \setminus E$. Fix any $\epsilon \in \mathbb{R}_+$. Let $E_n := \{\omega \in \Omega \mid |x_n(\omega) - x_0(\omega)| > \epsilon\}$, $\forall n \in \mathbb{N}$. Clearly, $\bigcap_{n=1}^{\infty} \bigcup_{j=n}^{\infty} E_j \subseteq E$. Hence, we have $0 \le \boldsymbol{P}(\bigcap_{n=1}^{\infty} \bigcup_{j=n}^{\infty} E_j) \le \boldsymbol{P}(E) = 0$. This implies that, by Proposition 11.5 and $\boldsymbol{P}(\Omega) = 1 < \infty$, $\lim_{n\to\infty} \boldsymbol{P}(\bigcup_{j=n}^{\infty} E_j) = 0$, which further implies that $\lim_{n\in\mathbb{N}} \boldsymbol{P}(E_n) = 0$. Hence, $\exists N \in \mathbb{N}$, $\forall n \in \mathbb{N}$ with $n \ge N$, we have $\boldsymbol{P}(E_n) < \epsilon$. Therefore, $\lim_{n\in\mathbb{N}} x_n = x_0$ in measure in $\boldsymbol{\Omega}$.

(ii) Let $\lim_{n\in\mathbb{N}} x_n = x_0$ in measure in $\boldsymbol{\Omega}$. Fix any $z \in \mathcal{C}_c(\mathbb{R}^m, \mathbb{R})$. Let $\|z\|_{\mathcal{C}_c(\mathbb{R}^m,\mathbb{R})} =: c \in \overline{\mathbb{R}}_+$. Fix any $\epsilon \in \mathbb{R}_+$. Since $z \in \mathcal{C}_c(\mathbb{R}^m, \mathbb{R})$, then $\exists M \in \overline{\mathbb{R}}_+$ such that $\forall s \in \mathbb{R}^m$ with $|s| > M$, we have $|z(s)| < \frac{\epsilon}{4}$. By Proposition 5.39, $z|_K$ is uniformly continuous, where $K := \overline{\mathcal{B}}_{\mathbb{R}^m}(\mathbf{0}_m, M + 1) \subseteq \mathbb{R}^m$ is a compact set by Proposition 5.40. Then, $\exists \delta \in (0, \frac{\epsilon}{2} \wedge 1) \subset \mathbb{R}$, such that $|z(s_1) - z(s_2)| < \frac{\epsilon}{4}$, $\forall s_1, s_2 \in K$ with $|s_1 - s_2| < \delta$. By Definition 11.56, there exists $N \in \mathbb{N}$, $\forall n \in \mathbb{N}$

with $n \geq N$, we have $P(\{\omega \in \Omega \mid |x_n(\omega) - x_0(\omega)| > \frac{\delta}{4(1+c)}\}) < \frac{\delta}{4(1+c)}$. Fix any $n \in \mathbb{N}$ with $n \geq N$. Then, we have

$$\begin{aligned}|\langle\langle L_n, z\rangle\rangle - \langle\langle L_0, z\rangle\rangle| &= \Big|\int_{\mathbb{R}^m} \langle\langle \mathrm{d}L_n(s), z(s)\rangle\rangle - \int_{\mathbb{R}^m} \langle\langle \mathrm{d}L_0(s), z(s)\rangle\rangle\Big| \\ &= \Big|\int_{\Omega} z(x_n(\omega))\,\mathrm{d}P(\omega) - \int_{\Omega} z(x_0(\omega))\,\mathrm{d}P(\omega)\Big| \\ &= \Big|\int_{\Omega} (z(x_n(\omega)) - z(x_0(\omega)))\,\mathrm{d}P(\omega)\Big| \leq \int_{\Omega} |z(x_n(\omega)) - z(x_0(\omega))|\,\mathrm{d}P(\omega) \\ &= \Big(\int_{E_1} + \int_{E_2} + \int_{E_3}\Big)|z(x_n(\omega)) - z(x_0(\omega))|\,\mathrm{d}P(\omega)\end{aligned}$$

where the first equality follows from Riesz Representation Theorem 11.209, the second equality follows from Definition 14.16, the third equality, the inequality, and the fourth equality follow from Proposition 11.92, $E_1 := \{\omega \in \Omega \mid |x_n(\omega) - x_0(\omega)| \leq \frac{\delta}{4(1+c)}$ and $|x_0(\omega)| \vee |x_n(\omega)| < M + 1\}$, $E_2 := \{\omega \in \Omega \mid |x_n(\omega) - x_0(\omega)| \leq \frac{\delta}{4(1+c)}$ and $|x_0(\omega)| \vee |x_n(\omega)| \geq M + 1\}$, and $E_3 := \{\omega \in \Omega \mid |x_n(\omega) - x_0(\omega)| > \frac{\delta}{4(1+c)}\}$. Note that $\int_{E_1} |z(x_n(\omega)) - z(x_0(\omega))|\,\mathrm{d}P(\omega) < \int_{E_1} \frac{\epsilon}{4}\,\mathrm{d}P \leq \frac{\epsilon}{4}$, since $|x_n(\omega) - x_0(\omega)| < \delta$ and $x_0(\omega), x_n(\omega) \in K$, $\forall \omega \in E_1$, $\int_{E_2} |z(x_n(\omega)) - z(x_0(\omega))|\,\mathrm{d}P(\omega) \leq \int_{E_2}(|z(x_n(\omega))| + |z(x_0(\omega))|)\,\mathrm{d}P \leq \int_{E_2}(\frac{\epsilon}{4} + \frac{\epsilon}{4})\,\mathrm{d}P \leq \frac{\epsilon}{2}$, since $|x_n(\omega)| > M$ and $|x_0(\omega)| > M$, $\forall \omega \in E_2$, and $\int_{E_3} |z(x_n(\omega)) - z(x_0(\omega))|\,\mathrm{d}P(\omega) \leq \int_{E_3}(|z(x_n(\omega))| + |z(x_0(\omega))|)\,\mathrm{d}P(\omega) \leq 2cP(E_3) < 2c\frac{\delta}{4(1+c)} < \frac{\delta}{2} < \frac{\epsilon}{4}$. Thus, we have $|\langle\langle L_n, z\rangle\rangle - \langle\langle L_0, z\rangle\rangle| < \epsilon$. Hence, $\lim_{n\in\mathbb{N}} \langle\langle L_n, z\rangle\rangle = \langle\langle L_0, z\rangle\rangle$. By the arbitrariness of z, we have $\lim_{n\in\mathbb{N}} L_n = L_0$ weak*. This proves that $\lim_{n\in\mathbb{N}} x_n \rightsquigarrow x_0$.

This completes the proof of the proposition. □

Definition 14.58 Let $\boldsymbol{\Omega} := (\Omega, \mathcal{B}, P)$ be a probability measure space, $\mathcal{Y}$ be a separable Banach space, and M be a set of $\mathcal{Y}$-valued random variables. M is said to be *uniformly growth bounded* if $\forall \epsilon \in \mathbb{R}_+$, $\exists K \in \mathbb{R}_+$, $\forall x \in M$, we have $x : \Omega \to \mathcal{Y}$ satisfies $P(\{\omega \in \Omega \mid \|x(\omega)\|_{\mathcal{Y}} > K\}) < \epsilon$. ◇

Theorem 14.59 (Skorokhod Representation) *Let μ_n be a probability measure on $(\mathbb{R}, \mathcal{B}_B(\mathbb{R}))$ with cumulative distribution function $F_n : \mathbb{R} \to [0, 1] \subset \mathbb{R}$ as defined in Definition 14.16, $\forall n \in \mathbb{Z}_+$. Assume that $\forall b \in \mathbb{R}$ with F_0 being continuous at b, we have $\lim_{n\in\mathbb{N}} F_n(b) = F_0(b)$. Then, there exists a probability measure space $\boldsymbol{\Omega} := (\Omega, \mathcal{B}, P)$ and a sequence of random variables $(x_n)_{n=0}^{\infty}$, such that $x_n : \Omega \to \mathbb{R}$ admits law μ_n, $\forall n \in \mathbb{Z}_+$, and $\lim_{n\in\mathbb{N}} x_n(\omega) = x_0(\omega)$ a.e. $\omega \in \boldsymbol{\Omega}$.*

Proof Define $E := \{b \in \mathbb{R} \mid F_0$ is continuous at $b\} \subseteq \mathbb{R}$. Note that F_0 is monotonically nondecreasing and is of bounded variation. Then, F_0 has countably many points of discontinuity (the total number of discontinuities with a jump of at least $1/n$ is at most n).

Take $\boldsymbol{\Omega} := ((0,1), \mathcal{B}, \boldsymbol{P})$ to be the finite metric measure subspace of $\mathbb{R}$. Fix any $n \in \mathbb{Z}_+$, define $x_n^+(\omega) := \inf\{z \in \mathbb{R} \mid F_n(z) > \omega\}$ and $x_n^-(\omega) := \inf\{z \in \mathbb{R} \mid F_n(z) \geq \omega\}$, $\forall \omega \in (0,1) \subset \mathbb{R}$. By Proposition 12.54, F_n is $\mathcal{B}_{\mathrm{B}}(\mathbb{R})$-measurable. Since μ_n is a probability measure, then $\lim_{b \to -\infty} F_n(b) = 0$ and $\lim_{b \to \infty} F_n(b) = 1$. This implies that $x_n^+(\omega) \in \mathbb{R}$ and $x_n^-(\omega) \in \mathbb{R}$, $\forall \omega \in (0,1) =: \Omega$. See Fig. 14.1 for an illustration of the function F_n, x_n^+, and x_n^-. It is easy to show that $\{\omega \in \Omega \mid x_n^-(\omega) \leq z\} = \{\omega \in \Omega \mid \omega \leq F_n(z)\} \in \mathcal{B}$, $\forall z \in \mathbb{R}$. Then, by Proposition 11.35, x_n^- is $\mathcal{B}$-measurable. Thus, x_n^- is an $\mathbb{R}$-valued random variable on $\boldsymbol{\Omega}$. The preceding argument further implies that x_n^- admits law μ_n since its cumulative distribution function as defined in Definition 14.16 is F_n. It is easy to show that $\{\omega \in \Omega \mid x_n^+(\omega) \leq z\} \supseteq \{\omega \in \Omega \mid \omega < F_n(z)\} \in \mathcal{B}$, $\forall z \in \mathbb{R}$. Note that $x_n^+(\omega) \geq x_n^-(\omega)$, $\forall \omega \in \Omega$. Then, $\{\omega \in \Omega \mid x_n^+(\omega) \leq z\} \subseteq \{\omega \in \Omega \mid x_n^-(\omega) \leq z\} = \{\omega \in \Omega \mid \omega \leq F_n(z)\}$, $\forall z \in \mathbb{R}$. Thus, we have $(0, F(z)) \subseteq \{\omega \in \Omega \mid x_n^+(\omega) \leq z\} \subseteq (0, F(z)]$. Then, $\{\omega \in \Omega \mid x_n^+(\omega) \leq z\}$ equals either $(0, F(z))$ or $(0, F(z)]$, which is $\mathcal{B}$-measurable in either case. By the arbitrariness of z and Proposition 11.35, x_n^+ is $\mathcal{B}$-measurable. Hence, x_n^+ is an $\mathbb{R}$-valued random variables on $\boldsymbol{\Omega}$. The preceding argument further implies that x_n^+ admits law μ_n and $x_n^+ = x_n^-$ a.e. in $\boldsymbol{\Omega}$.

Fix $\omega \in \Omega$. Let $z \in E$ with $\omega < F_0(z)$. By the assumption, we have $\lim_{n \in \mathbb{N}} F_n(z) = F_0(z) > \omega$. Then, for sufficiently large n, $F_n(z) > \omega$, which implies that $x_n^+(\omega) \leq z$ by the discussion in the preceding paragraph. Then, we have $\limsup_{n \in \mathbb{N}} x_n^+(\omega) \leq z$. By the arbitrariness of z, we have $\limsup_{n \in \mathbb{N}} x_n^+(\omega) \leq \inf\{z \in E \mid \omega < F_0(z)\}$. Since E is dense in $\mathbb{R}$, it is straightforward to show that $\limsup_{n \in \mathbb{N}} x_n^+(\omega) \leq \inf\{z \in E \mid \omega < F_0(z)\} = \inf\{z \in \mathbb{R} \mid \omega < F_0(z)\} = x_0^+(\omega)$.

Let $z \in E$ and $z < x_0^-(\omega)$. Then, $\omega > F_0(z)$, by the second preceding paragraph. By the assumption, we have $\lim_{n \in \mathbb{N}} F_n(z) = F_0(z) < \omega$. Then, for sufficiently large n, we have $F_n(z) < \omega$. This implies that $x_n^-(\omega) > z$ for sufficiently large n, by the second preceding paragraph. Then, we have $\liminf_{n \in \mathbb{N}} x_n^-(\omega) \geq z$. Since E is dense in $\mathbb{R}$, $\liminf_{n \in \mathbb{N}} x_n^-(\omega) \geq \sup\{z \in E \mid z < x_0^-(\omega)\} = x_0^-(\omega)$.

Combining the result of the above two paragraphs, we have $x_0^-(\omega) \leq \liminf_{n \in \mathbb{N}} x_n^-(\omega) \leq \limsup_{n \in \mathbb{N}} x_n^-(\omega) \leq \limsup_{n \in \mathbb{N}} x_n^+(\omega) \leq x_0^+(\omega)$, $\forall \omega \in \Omega$. When $x_0^-(\omega) = x_0^+(\omega)$, we have $x_0^-(\omega) = \lim_{n \in \mathbb{N}} x_n^-(\omega)$. Since $x_0^-(\omega) = x_0^+(\omega)$ a.e. $\omega \in \boldsymbol{\Omega}$, we have $x_0^-(\omega) = \lim_{n \in \mathbb{N}} x_n^-(\omega)$ a.e. $\omega \in \boldsymbol{\Omega}$, by Proposition 11.48. Hence, $(x_n^-)_{n=0}^{\infty}$ is the sequence we seek. This completes the proof of the theorem. □

Now, we prove a result that connects our definition of convergence in distribution to that of Williams (1991).

Proposition 14.60 *Let $\boldsymbol{\Omega} := (\Omega, \mathcal{B}, \boldsymbol{P})$ be a probability measure space, $(x_n)_{n=0}^{\infty}$ be a sequence of ($\mathbb{R}$-valued) random variables with $x_n : \Omega \to \mathbb{R}$ with law μ_n, $\forall n \in \mathbb{Z}_+$, and $F_n : \mathbb{R} \to [0,1] \subset \mathbb{R}$ be the cumulative distribution function for μ_n as defined in Definition 14.16, $\forall n \in \mathbb{Z}_+$. Then, the following statements hold:*

(i) *If $\lim_{n \in \mathbb{N}} \int_{\mathbb{R}} \langle\langle \mathrm{d}\mu_n(x), z(x) \rangle\rangle = \int_{\mathbb{R}} \langle\langle \mathrm{d}\mu_0(x), z(x) \rangle\rangle \in \mathbb{R}$, $\forall z \in \mathcal{C}_{\mathrm{b}}(\mathbb{R}, \mathbb{R})$, then $\lim_{n \in \mathbb{N}} \mu_n \rightsquigarrow \mu_0$.*

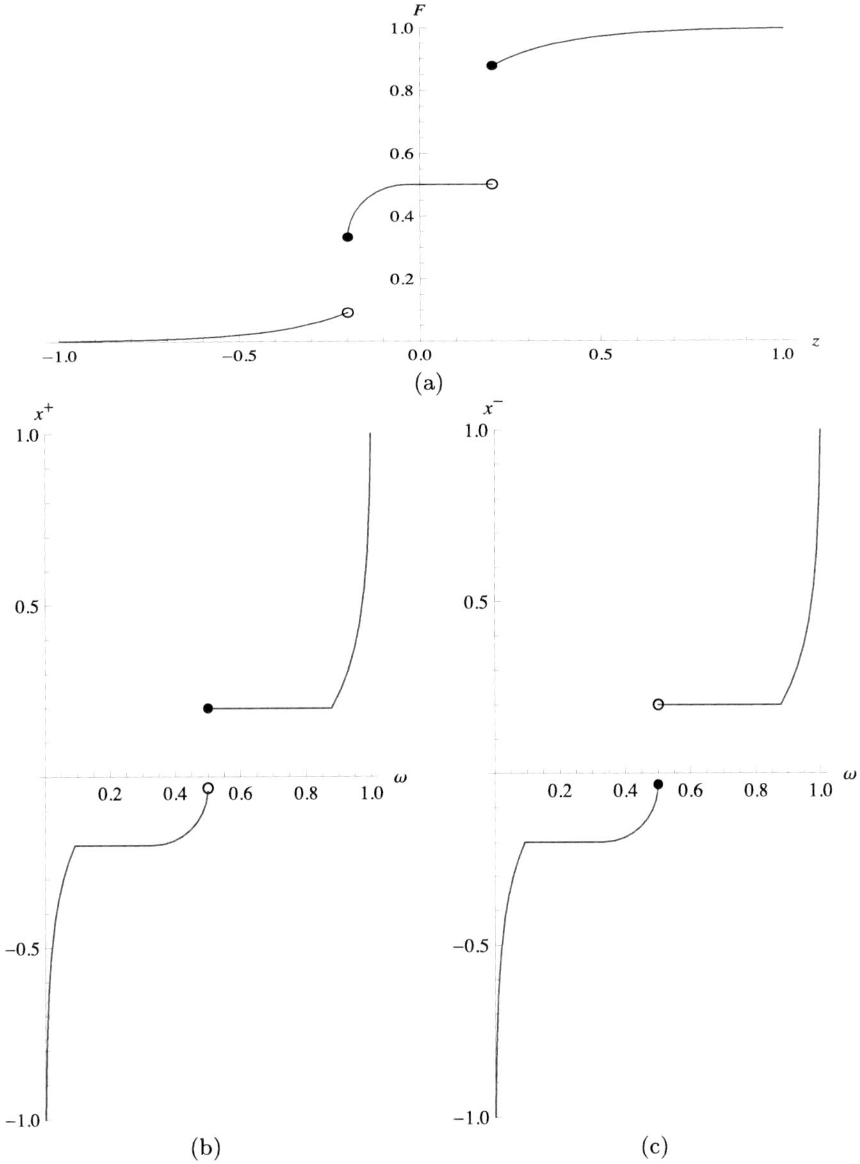

Fig. 14.1 A typical cumulative distribution function F_n and the corresponding random variables x_n^+ and x_n^-

(ii) $\lim_{n\in\mathbb{N}} \int_{\mathbb{R}} \langle\langle \mathrm{d}\mu_n(x), z(x)\rangle\rangle = \int_{\mathbb{R}} \langle\langle \mathrm{d}\mu_0(x), z(x)\rangle\rangle \in \mathbb{R}$, $\forall z \in \mathcal{C}_{\mathrm{b}}(\mathbb{R}, \mathbb{R})$, *if, and only if,* $\lim_{n\in\mathbb{N}} F_n(b) = F_0(b)$, $\forall b \in \mathbb{R}$ *with* F_0 *being continuous at* b.

(iii) *If* $\lim_{n\in\mathbb{N}} \mu_n \rightsquigarrow \mu_0$ *and* $(x_n)_{n=1}^{\infty}$ *is uniformly growth bounded, then* $\lim_{n\in\mathbb{N}} \int_{\mathbb{R}} \langle\langle \mathrm{d}\mu_n(x), z(x)\rangle\rangle = \int_{\mathbb{R}} \langle\langle \mathrm{d}\mu_0(x), z(x)\rangle\rangle \in \mathbb{R}$, $\forall z \in \mathcal{C}_{\mathrm{b}}(\mathbb{R}, \mathbb{R})$.

Proof (i) Note that $\lim_{n\in\mathbb{N}} \mu_n \rightsquigarrow \mu_0$ means that $\lim_{n\in\mathbb{N}} \mu_n = \mu_0$ weak*, that is $\lim_{n\in\mathbb{N}} \langle\langle \mu_n, z\rangle\rangle = \langle\langle \mu_0, z\rangle\rangle$, $\forall z \in \mathcal{C}_c(\mathbb{R}, \mathbb{R})$. By Riesz Representation Theorem 11.209, this is equivalent to $\lim_{n\in\mathbb{N}} \int_{\mathbb{R}} \langle\langle \mathrm{d}\mu_n(x), z(x)\rangle\rangle = \int_{\mathbb{R}} \langle\langle \mathrm{d}\mu_0(x), z(x)\rangle\rangle$, $\forall z \in \mathcal{C}_c(\mathbb{R}, \mathbb{R}) \subseteq \mathcal{C}_b(\mathbb{R}, \mathbb{R})$. Thus, (i) follows readily.

(ii) "Necessity" Let $\lim_{n\in\mathbb{N}} \int_{\mathbb{R}} \langle\langle \mathrm{d}\mu_n(x), z(x)\rangle\rangle = \int_{\mathbb{R}} \langle\langle \mathrm{d}\mu_0(x), z(x)\rangle\rangle \in \mathbb{R}$, $\forall z \in \mathcal{C}_b(\mathbb{R}, \mathbb{R})$. Fix any $b \in \mathbb{R}$ and $\delta \in \mathbb{R}_+$. Define $h_{b,\delta} : \mathbb{R} \to [0, 1] \subset \mathbb{R}$ by $h_{b,\delta}(x) = \begin{cases} 1 & x \leq b \\ 1 - \frac{x-b}{\delta} & b < x < b + \delta \\ 0 & x \geq b + \delta \end{cases}$, $\forall x \in \mathbb{R}$. Clearly, $h_{b,\delta} \in \mathcal{C}_b(\mathbb{R}, \mathbb{R})$. Then, $\lim_{n\in\mathbb{N}} \langle\langle \mu_n, h_{b,\delta}\rangle\rangle = \langle\langle \mu_0, h_{b,\delta}\rangle\rangle$, $F_n(b) \leq \langle\langle \mu_n, h_{b,\delta}\rangle\rangle$, and $\langle\langle \mu_0, h_{b,\delta}\rangle\rangle \leq F_0(b + \delta)$. Then, taking limit as $n \to \infty$ in the above three relations, we have $\limsup_{n\in\mathbb{N}} F_n(b) \leq F_0(b + \delta)$. Since F_0 is a cumulative distribution function then it is continuous on the right. Taking limit as $\delta \to 0^+$, we have $\limsup_{n\in\mathbb{N}} F_n(b) \leq F_0(b)$. Similarly, we must have $\lim_{n\in\mathbb{N}} \langle\langle \mu_n, h_{b,\delta}\rangle\rangle = \langle\langle \mu_0, h_{b,\delta}\rangle\rangle$, $F_n(b + \delta) \geq \langle\langle \mu_n, h_{b,\delta}\rangle\rangle$, and $\langle\langle \mu_0, h_{b,\delta}\rangle\rangle \geq F_0(b)$. Taking limit as $n \to \infty$ in the above three relations, we have $\liminf_{n\in\mathbb{N}} F_n(b + \delta) \geq F_0(b)$. Let $\bar{b} := b + \delta$, we have $\liminf_{n\in\mathbb{N}} F_n(\bar{b}) \geq F_0(\bar{b} - \delta)$, $\forall \bar{b} \in \mathbb{R}$, $\forall \delta \in \mathbb{R}_+$. Now, fix any $b \in \mathbb{R}$ with F_0 being continuous at b. We have $F_0(b - \delta) \leq \liminf_{n\in\mathbb{N}} F_n(b) \leq \limsup_{n\in\mathbb{N}} F_n(b) \leq F_0(b)$. Now, take $\delta \to 0^+$, we have $F_0(b) \leq \liminf_{n\in\mathbb{N}} F_n(b) \leq \limsup_{n\in\mathbb{N}} F_n(b) \leq F_0(b)$ and hence $\lim_{n\in\mathbb{N}} F_n(b) = F_0(b)$. This completes the "Necessity" part of the proof.

"Sufficiency" Let $\lim_{n\in\mathbb{N}} F_n(b) = F_0(b)$, $\forall b \in \mathbb{R}$ with F_0 being continuous at b. By Skorokhod Representation Theorem 14.59, there exists a probability measure space $\bar{\boldsymbol{\Omega}} := (\bar{\Omega}, \bar{\mathcal{B}}, \bar{\boldsymbol{P}})$ and a sequence of random variables $(Y_n)_{n=0}^{\infty}$ on the probability measure space $\bar{\boldsymbol{\Omega}}$ such that Y_n admits law μ_n and cumulative distribution function F_n, $\forall n \in \mathbb{Z}_+$, and $\lim_{n\in\mathbb{N}} Y_n = Y_0$ a.e. in $\bar{\boldsymbol{\Omega}}$. $\forall z \in \mathcal{C}_b(\mathbb{R}, \mathbb{R})$, we have $\lim_{n\in\mathbb{N}} z(Y_n(\bar{\omega})) = z(Y_0(\bar{\omega}))$ a.e. $\bar{\omega} \in \bar{\boldsymbol{\Omega}}$, by Proposition 11.52. By Bounded Convergence Theorem 11.77, $\lim_{n\in\mathbb{N}} \int_{\mathbb{R}} \langle\langle \mathrm{d}\mu_n(y), z(y)\rangle\rangle = \int_{\bar{\Omega}} z(Y_n(\bar{\omega}))\, \mathrm{d}\bar{\boldsymbol{P}}(\bar{\omega}) = \int_{\bar{\Omega}} z(Y_0(\bar{\omega})) \cdot \mathrm{d}\bar{\boldsymbol{P}}(\bar{\omega}) = \int_{\mathbb{R}} \langle\langle \mathrm{d}\mu_0(y), z(y)\rangle\rangle \in \mathbb{R}$, where the first and the last equalities follow from Definition 14.16. This completes the "Sufficiency" part of the proof.

Hence, (ii) is proved.

(iii) Let $\lim_{n\in\mathbb{N}} \mu_n \rightsquigarrow \mu_0$. Fix any $z \in \mathcal{C}_b(\mathbb{R}, \mathbb{R})$. By Bounded Convergence Theorem 11.77, $\int_{\mathbb{R}} \langle\langle \mathrm{d}\mu_n(x), z(x)\rangle\rangle \in \mathbb{R}$, $\forall n \in \mathbb{Z}_+$. $\forall a, b \in \mathbb{R}$ with $a < b$, $\forall k \in \mathbb{N}$ with $b - a > \frac{1}{k}$, by Proposition 5.62, there exists a continuous function $h_{k,a,b} : \mathbb{R} \to [0, 1] \subset \mathbb{R}$ such that $h_{k,a,b}(x) = 1$, $\forall x \in E_{k,1} := [a + \frac{1}{k}, b] \subset \mathbb{R}$ and

$h_{k,a,b}(x) = 0, \forall x \in E_{k,2} := \mathbb{R} \setminus (a, b + \frac{1}{k})$. Clearly, $h_{k,a,b} \in \mathcal{C}_c(\mathbb{R}, \mathbb{R})$. Then,

$$F_n(b) - F_n(a + \frac{1}{k}) = \int_{\mathbb{R}} \chi_{r_{a+\frac{1}{k},b},\mathbb{R}} \, d\mu_n \le \int_R h_{k,a,b} \, d\mu_n = \langle\langle \mu_n, h_{k,a,b} \rangle\rangle$$

$$\le \int_{\mathbb{R}} \chi_{r_{a,b+\frac{1}{k}},\mathbb{R}} \, d\mu_n = F_n(b + \frac{1}{k}) - F_n(a); \tag{14.8}$$

$$\forall n \in \mathbb{Z}_+, \forall a, b \in \mathbb{R}, \forall k \in \mathbb{N} \text{ with } a + \frac{1}{k} < b$$

where the first and last equalities follow from Definition 14.16 and the first and last inequalities follow from Proposition 11.92. Taking limit as $n \to \infty$,

$$\limsup_{n \in \mathbb{N}} (F_n(b) - F_n(a + \frac{1}{k})) \le \lim_{n \in \mathbb{N}} \langle\langle \mu_n, h_{k,a,b} \rangle\rangle = \langle\langle \mu_0, h_{k,a,b} \rangle\rangle \tag{14.9}$$

$$\le \liminf_{n \in \mathbb{N}} (F_n(b + \frac{1}{k}) - F_n(a)); \ \forall a, b \in \mathbb{R}, \forall k \in \mathbb{N} \text{ with } a + \frac{1}{k} < b$$

where the equality follows from $\lim_{n \in \mathbb{N}} \mu_n \nrightarrow \mu_0$ and $h_{k,a,b} \in \mathcal{C}_c(\mathbb{R}, \mathbb{R})$ and the inequalities follow from (14.8). By the assumption that $(x_n)_{n=1}^{\infty}$ is uniformly growth bounded, we have $\lim_{a \to -\infty} F_n(a) = 0$, and the convergence is uniform with respect to $n \in \mathbb{N}$. It is easy to prove, using $\epsilon - \delta$ arguments, that

$$0 \le \lim_{a \to -\infty} \liminf_{n \in \mathbb{N}} F_n(a) \le \lim_{a \to -\infty} \limsup_{n \in \mathbb{N}} F_n(a)$$

$$= \lim_{n \in \mathbb{N}} \lim_{a \to -\infty} F_n(a) = 0 \tag{14.10}$$

This implies that, $\forall b \in \mathbb{R}$ and $\forall k \in \mathbb{N}$,

$$\limsup_{n \in \mathbb{N}} F_n(b) = \limsup_{n \in \mathbb{N}} F_n(b) - \lim_{a \to -\infty} \limsup_{n \to \mathbb{N}} F_n(a + \frac{1}{k})$$

$$= \lim_{a \to -\infty} \left(\limsup_{n \in \mathbb{N}} F_n(b) - \limsup_{n \to \mathbb{N}} F_n(a + \frac{1}{k}) \right)$$

$$\le \liminf_{a \to -\infty} \limsup_{n \in \mathbb{N}} (F_n(b) - F_n(a + \frac{1}{k})) \le \liminf_{a \to -\infty} \langle\langle \mu_0, h_{k,a,b} \rangle\rangle$$

$$\le \lim_{a \to -\infty} (F_0(b + \frac{1}{k}) - F_0(a)) = F_0(b + \frac{1}{k}) \tag{14.11}$$

where the first equality follows from (14.10), the second equality follows from Proposition 3.83, the first inequality follows from Proposition 3.83 and the fact that all limits are real numbers in the closed interval $[0, 1] \subset \mathbb{R}$, the second inequality follows from (14.9), the third inequality follows from (14.8) with $n = 0$, and the last equality follows from the fact that F_0 is the cumulative distribution function for μ_0 as defined in Definition 14.16. We also have, $\forall b \in \mathbb{R}$ and $\forall k \in \mathbb{N}$,

$$
\begin{aligned}
F_0(b) &= \lim_{a\to-\infty}(F_0(b) - F_0(a+\frac{1}{k})) \le \liminf_{a\to-\infty}\langle\langle \mu_0, h_{k,a,b}\rangle\rangle \\
&\le \liminf_{a\to-\infty}\liminf_{n\in\mathbb{N}}(F_n(b+\frac{1}{k}) - F_n(a)) \\
&\le \liminf_{a\to-\infty}\left(\liminf_{n\in\mathbb{N}} F_n(b+\frac{1}{k}) - \liminf_{n\in\mathbb{N}} F_n(a)\right) \\
&= \liminf_{n\in\mathbb{N}} F_n(b+\frac{1}{k}) - \liminf_{a\to-\infty}\liminf_{n\in\mathbb{N}} F_n(a) = \liminf_{n\in\mathbb{N}} F_n(b+\frac{1}{k}) \qquad (14.12)
\end{aligned}
$$

where the first equality follows from the fact that F_0 is the cumulative distribution function of μ_0 as defined in Definition 14.16, the first inequality follows from (14.8) with $n = 0$, the second inequality follows from (14.9), the third inequality follows from Proposition 3.83 and the fact all limits are real numbers in the closed interval $[0, 1] \subset \mathbb{R}$, the second equality follows from Proposition 3.83, and the last equality follows from (14.10). Then, let $b \in \mathbb{R}$ be any point such that F_0 is continuous at b, we have $\limsup_{n\in\mathbb{N}} F_n(b) \le \lim_{k\in\mathbb{N}} F_0(b+\frac{1}{k}) = F_0(b) = \lim_{k\in\mathbb{N}} F_0(b-\frac{1}{k}) \le \lim_{k\in\mathbb{N}}\liminf_{n\in\mathbb{N}} F_n(b) = \liminf_{n\in\mathbb{N}} F_n(b) \le \limsup_{n\in\mathbb{N}} F_n(b)$, where the first inequality follows from (14.11), the first and second equalities follow from the fact that F_0 is continuous at b, the second inequality follows from (14.12) with b substituted by $b - \frac{1}{k}$, and the last equality follows from Proposition 3.83. Thus, we have $\lim_{n\in\mathbb{N}} F_n(b) = F_0(b)$. By the sufficiency result of (ii), (iii) holds.

This completes the proof of the proposition. □

14.8 Central Limit Theorem

Lemma 14.61 (Helly) *Let $m \in \mathbb{N}$, $\mathcal{X}$ be an m-dimensional Banach space over $\mathbb{K}_1$, $k \in \mathbb{Z}_+$, $\mathcal{Y}$ be a k-dimensional Banach space over $\mathbb{K}_2$, $\mathbb{X}_n := (\mathcal{X}, \mathcal{B}_B(\mathcal{X}), \mu_n)$ be a $\mathcal{Y}$-valued measure space with $\mathcal{P}\circ\mu_n(\mathcal{X}) \le 1$, $\forall n \in \mathbb{N}$. Then, there exists an finite $\mathcal{Y}$-valued measure space $(\mathcal{X}, \mathcal{B}_B(\mathcal{X}), \mu)$ with $\mathcal{P}\circ\mu(\mathcal{X}) \le 1$, and a subsequence $(\mu_{n_l})_{l=1}^{\infty}$, such that $\lim_{l\in\mathbb{N}} \mu_{n_l} \rightsquigarrow \mu$.*

Proof Fix any $n \in \mathbb{N}$. By Theorem 11.198, $\mu_n \in \mathcal{M}_{ft}(\mathcal{X}, \mathcal{Y})$ with $\|\mu_n\|_{\mathcal{M}_{ft}(\mathcal{X},\mathcal{Y})} = \mathcal{P}\circ\mu_n(\mathcal{X}) \le 1$. By Propositions 7.90 and 7.92, $\mathcal{Y}$ and $\mathcal{Y}^*$ are reflexive Banach spaces. By Alaoglu Theorem 7.122 and Riesz Representation Theorem 11.209, $K := \overline{\mathcal{B}}_{\mathcal{M}_{ft}(\mathcal{X},\mathcal{Y})}(\vartheta_{\mathcal{M}_{ft}(\mathcal{X},\mathcal{Y})}, 1) \subseteq \mathcal{M}_{ft}(\mathcal{X}, \mathcal{Y}) = (\mathcal{C}_c(\mathcal{X}, \mathcal{Y}^*))^*$ is weak* compact. Clearly, $(\mu_n)_{n=1}^{\infty} \subseteq K$. By Propositions 5.22 and 5.26, K has the Bolzano–Weierstrass property. Then, there exists $\mu \in K$ such that $(\mu_n)_{n=1}^{\infty}$ admits a weak* cluster point μ. Fix any $z \in \mathcal{C}_c(\mathcal{X}, \mathcal{Y}^*)$. Then, $(\langle\langle \mu_n, z\rangle\rangle)_{n=1}^{\infty} \subseteq \overline{\mathcal{B}}_{\mathbb{K}_2}(0, \|z\|_{\mathcal{C}_c(\mathcal{X},\mathcal{Y})}) \subset \mathbb{K}_2$ has a cluster point $\langle\langle \mu, z\rangle\rangle$, by Proposition 3.66. Then, by Propositions 5.26 and 5.28, there exists a subsequence $(\mu_{n_i})_{i=1}^{\infty}$ such that $\lim_{i\in\mathbb{N}} \langle\langle \mu_{n_i}, z\rangle\rangle = \langle\langle \mu, z\rangle\rangle$. By Riesz Representation Theorem 11.209, $\mathcal{C}_c(\mathcal{X}, \mathcal{Y}^*)$ is separable. Let $D := \{z_1, z_2, \ldots\} \subseteq \mathcal{C}_c(\mathcal{X}, \mathcal{Y}^*)$ be a countable dense set. Then, there exists a subsequence $(\mu_{n_i})_{i=1}^{\infty}$ such that $\lim_{i\in\mathbb{N}} \langle\langle \mu_{n_i}, z_1\rangle\rangle = \langle\langle \mu, z_1\rangle\rangle$. There exists a subsubsequence $(\mu_{n_{i_j}})_{j=1}^{\infty}$ such that $\lim_{j\in\mathbb{N}} \langle\langle \mu_{n_{i_j}}, z_2\rangle\rangle = \langle\langle \mu, z_2\rangle\rangle$, and so on.

Take the diagonal sequence in the countably many subsequences, denote this sequence as (which is a subsequence of $(\mu_n)_{n=1}^{\infty}$) $(\mu_{n_l})_{l=1}^{\infty}$, and we have $\lim_{l\in\mathbb{N}} \langle\langle \mu_{n_l}, z_i\rangle\rangle = \langle\langle \mu, z_i\rangle\rangle$, $\forall z_i \in D$. Fix any $z \in \mathcal{C}_c(\mathcal{X}, \mathcal{Y}^*)$. $\forall \epsilon \in \mathbb{R}_+$, there exists $z_i \in D$ such that $\|z - z_i\|_{\mathcal{C}_c(\mathcal{X},\mathcal{Y}^*)} < \frac{\epsilon}{3}$. For that z_i, $\exists L \in \mathbb{N}$, $\forall l \in \mathbb{N}$ with $l \geq L$, we have $|\langle\langle \mu_{n_l}, z_i\rangle\rangle - \langle\langle \mu, z_i\rangle\rangle| < \frac{\epsilon}{3}$. Then, $|\langle\langle \mu_{n_l}, z\rangle\rangle - \langle\langle \mu, z\rangle\rangle| \leq |\langle\langle \mu_{n_l}, z\rangle\rangle - \langle\langle \mu_{n_l}, z_i\rangle\rangle| + |\langle\langle \mu_{n_l}, z_i\rangle\rangle - \langle\langle \mu, z_i\rangle\rangle| + |\langle\langle \mu, z_i\rangle\rangle - \langle\langle \mu, z\rangle\rangle| < \|\mu_{n_l}\|_{\mathcal{M}_{ft}(\mathcal{X},\mathcal{Y})}\|z - z_i\|_{\mathcal{C}_c(\mathcal{X},\mathcal{Y}^*)} + \frac{\epsilon}{3} + \|\mu\|_{\mathcal{M}_{ft}(\mathcal{X},\mathcal{Y})}\|z - z_i\|_{\mathcal{C}_c(\mathcal{X},\mathcal{Y}^*)} \leq \epsilon$, where the second inequality follows from Proposition 7.72 and the last inequality follows from the preceding discussion of this paragraph. Hence, we have $\lim_{l\in\mathbb{N}} \langle\langle \mu_{n_l}, z\rangle\rangle = \langle\langle \mu, z\rangle\rangle$. By the arbitrariness of $z \in \mathcal{C}_c(\mathcal{X}, \mathcal{Y}^*)$, we have that $\lim_{l\in\mathbb{N}} \mu_{n_l} \rightharpoonup \mu$.

This completes the proof of the lemma. □

Theorem 14.62 (Lévy's Convergence Theorem) *Let* $\boldsymbol{\Omega} := (\Omega, \mathcal{B}, \boldsymbol{P})$ *be a probability measure space,* $(X_n)_{n=1}^{\infty}$ *be a sequence of (*$\mathbb{R}$*-valued) random variables with characteristic function* $\Psi_{X_n} : \mathbb{R} \to \mathbb{C}$*, and* $g : \mathbb{R} \to \mathbb{C}$ *be a function that is continuous at* 0*. Assume that* $\lim_{n\in\mathbb{N}} \Psi_{X_n}(\theta) = g(\theta)$*,* $\forall \theta \in \mathbb{R}$*. Then,* $(X_n)_{n=1}^{\infty}$ *is uniformly growth bounded, and there exists a probability measure space* $\bar{\boldsymbol{\Omega}} := (\bar{\Omega}, \bar{\mathcal{B}}, \bar{\boldsymbol{P}})$ *and a random variable* $X : \bar{\Omega} \to \mathbb{R}$ *with characteristic function* $\Psi_X = g$ *such that* $\lim_{n\in\mathbb{N}} X_n \rightharpoonup X$.

Proof Fix any $n \in \mathbb{N}$. Let X_n admit law μ_n and cumulative distribution function $F_n : \mathbb{R} \to [0, 1] \subset \mathbb{R}$ as defined in Definition 14.16. Then, by Theorem 11.198, $\mu_n \in \mathcal{M}_{ft}(\mathbb{R}, \mathbb{R})$ with $\|\mu_n\|_{\mathcal{M}_{ft}(\mathbb{R},\mathbb{R})} = \mathcal{P} \circ \mu_n(\mathbb{R}) = \mu_n(\mathbb{R}) = 1$. By Helly's Lemma 14.61, there exists $\mu \in \mathcal{M}_{ft}(\mathbb{R}, \mathbb{R})$ with $\mathcal{P} \circ \mu(\mathbb{R}) \leq 1$ and a subsequence $(\mu_{n_l})_{l=1}^{\infty}$ such that $\lim_{l\in\mathbb{N}} \mu_{n_l} \rightharpoonup \mu$.

We will next show that $(X_n)_{n=1}^{\infty}$ is uniformly growth bounded. Fix any $\epsilon \in \mathbb{R}_+$. By Definition 14.52, we have $\Psi_{X_n}(\theta) = \int_{\mathbb{R}} \exp(i\theta x)\, dF_n(x)$, $\forall \theta \in \mathbb{R}$, $\forall n \in \mathbb{N}$. Then, $\Psi_{X_n}(\theta) + \Psi_{X_n}(-\theta) = \int_{\mathbb{R}} 2\cos(\theta x)\, dF_n(x) \in \mathbb{R}$. This implies that $g(\theta) + g(-\theta) = \lim_{n\in\mathbb{N}}(\Psi_{X_n}(\theta) + \Psi_{X_n}(-\theta)) \in \mathbb{R}$, $\forall \theta \in \mathbb{R}$. Clearly, $g(0) = \lim_{n\in\mathbb{N}} \Psi_{X_n}(0) = 1$, by Proposition 14.53. Since g is continuous at 0, then $\exists \delta \in \mathbb{R}_+$ such that $|1 - g(\theta)| < \frac{\epsilon}{4}$, $\forall \theta \in (-\delta, \delta) \subset \mathbb{R}$. This implies that $0 \leq \frac{1}{\delta}\int_0^{\delta}(2 - g(\theta) - g(-\theta))\, d\theta \leq \frac{\epsilon}{2}$, where the first inequality follows from the fact that $g(\theta) + g(-\theta) \in \mathbb{R}$ and (ii) of Proposition 14.53. By Bounded Convergence Theorem 11.77, we have $\lim_{n\in\mathbb{N}} \frac{1}{\delta}\int_0^{\delta}(2 - \Psi_{X_n}(\theta) - \Psi_{X_n}(-\theta))\, d\theta \in [0, \frac{\epsilon}{2}] \subset \mathbb{R}$. Then, $\exists n_0 \in \mathbb{N}$, $\forall n \in \mathbb{N}$ with $n \geq n_0$, we have $0 \leq \frac{1}{\delta}\int_0^{\delta}(2 - \Psi_{X_n}(\theta) -$

$\Psi_{X_n}(-\theta))\,\mathrm{d}\theta < \epsilon$, where the first inequality follows from (ii) of Proposition 14.53. Note that

$$\epsilon > \frac{1}{\delta}\int_0^\delta (2 - \Psi_{X_n}(\theta) - \Psi_{X_n}(-\theta))\,\mathrm{d}\theta = \frac{1}{\delta}\int_0^\delta (2 - \Psi_{X_n}(\theta) - \Psi_{-X_n}(\theta))\,\mathrm{d}\theta$$
$$= \frac{1}{\delta}\int_0^\delta \int_{\mathbb{R}} (2 - \exp(\mathrm{i}\theta x) - \exp(-\mathrm{i}\theta x))\,\mathrm{d}F_n(x)\,\mathrm{d}\theta$$
$$= \frac{1}{\delta}\int_0^\delta \int_{\mathbb{R}} (2 - 2\cos(\theta x))\,\mathrm{d}F_n(x)\,\mathrm{d}\theta$$
$$= \int_{\mathbb{R}} \frac{1}{\delta}\int_0^\delta (2 - 2\cos(\theta x))\,\mathrm{d}\theta\,\mathrm{d}F_n(x) = 2\int_{\mathbb{R}} \left(1 - \frac{\sin(\delta x)}{\delta x}\right)\mathrm{d}F_n(x)$$
$$\geq 2\int_{E_\delta} \left(1 - \frac{1}{|\delta x|}\right)\mathrm{d}F_n(x) \geq \int_{E_\delta} \mathrm{d}F_n(x) = \mu_n(E_\delta) \geq 0$$

where the first equality follows from Proposition 14.53, the second equality follows from Definition 14.52, the fourth equality follows from Fubini Theorem 12.31 (note that the integrand is absolutely integrable over the product measure space by Tonelli Theorem 12.29); the fifth equality follows from Theorem 12.83, and the second and third inequalities follow from our definition of $E_\delta := \left\{x \in \mathbb{R} \,\middle|\, |x| > \frac{2}{\delta}\right\}$. Hence, $(X_n)_{n=1}^\infty$ is uniformly growth bounded.

By Proposition 14.60, we have $\lim_{l\in\mathbb{N}} \langle\langle \mu_{n_l}, z\rangle\rangle = \langle\langle \mu, z\rangle\rangle$, $\forall z \in \mathcal{C}_\mathrm{b}(\mathbb{R}, \mathbb{R})$. Let $z = \chi_{\mathbb{R},\mathbb{R}}$, we have $\mu(\mathbb{R}) = \langle\langle \mu, z\rangle\rangle = \lim_{l\in\mathbb{N}} \langle\langle \mu_{n_l}, z\rangle\rangle = \lim_{l\in\mathbb{N}} \mu_{n_l}(\mathbb{R}) = 1$. Combined with $\mathcal{P} \circ \mu(\mathbb{R}) \leq 1$, we conclude that $\mathcal{P} \circ \mu(\mathbb{R}) = 1$ and μ is a probability measure on $(\mathbb{R}, \mathcal{B}_\mathrm{B}(\mathbb{R}))$. Therefore, by Example 14.17, there exists a probability measure space $\bar{\boldsymbol{\Omega}} := (\bar{\Omega}, \bar{\mathcal{B}}, \bar{\boldsymbol{P}})$ and a random variable $X : \bar{\Omega} \to \mathbb{R}$ with law μ. Let $F : \mathbb{R} \to [0, 1] \subset \mathbb{R}$ be the cumulative distribution function of μ as defined in Definition 14.16. Then, we have $g(\theta) = \lim_{n\in\mathbb{N}} \Psi_{X_n}(\theta) = \lim_{n\in\mathbb{N}} \int_{\mathbb{R}} \exp(\mathrm{i}\theta x)\,\mathrm{d}\mu_n(x) = \lim_{n\in\mathbb{N}} \langle\langle \mu_n, z_0(\theta)\rangle\rangle = \lim_{l\in\mathbb{N}} \langle\langle \mu_{n_l}, z_0(\theta)\rangle\rangle = \langle\langle \mu, z_0(\theta)\rangle\rangle = \int_{\mathbb{R}} \exp(\mathrm{i}\theta x)\,\mathrm{d}\mu(x) = \Psi_X(\theta)$, $\forall \theta \in \mathbb{R}$, where the first equality follows from the assumption of the theorem, the second equality follows from Definition 14.52, the third equality follows from the definition $z_0(\theta) : \mathbb{R} \to \mathbb{C}$ is defined by $z_0(\theta)(x) = \exp(\mathrm{i}\theta x)$, $\forall x \in \mathbb{R}$, the fourth equality follows from Proposition 3.70, the fifth equality follows from the preceding discussion and $\mathrm{Re}\,(z_0(\theta)) \in \mathcal{C}_\mathrm{b}(\mathbb{R}, \mathbb{R})$ and $\mathrm{Im}\,(z_0(\theta)) \in \mathcal{C}_\mathrm{b}(\mathbb{R}, \mathbb{R})$, and the last equality follows from Definition 14.52. By (ii) of Proposition 14.60, $\lim_{l\to\mathbb{N}} F_{n_l}(b) = F(b)$, $\forall b \in \mathbb{R}$ with F being continuous at b.

By Lévy Inversion Formula, Theorem 14.55, and the discussion immediately after the theorem, g uniquely determines μ. Thus, we have shown that for the sequence $(\mu_n)_{n=1}^\infty$, there is a subsequence $(\mu_{n_l})_{l=1}^\infty$ such that $\lim_{l\in\mathbb{N}} \mu_{n_l} \looparrowright \mu$, where μ is the probability measure on $(\mathbb{R}, \mathcal{B}_\mathrm{B}(\mathbb{R}))$ that is the law of a random variable X whose characteristic function is given by g. Note that g is independent of the choice of the subsequence $(\mu_{n_l})_{l=1}^\infty$. This clearly holds for any subsequence of $(\mu_n)_{n=1}^\infty$. Thus, $\forall$ subsequence $(\mu_{n_i})_{i=1}^\infty$ of $(\mu_n)_{n=1}^\infty$, there exists a further subsubsequence $\left(\mu_{n_{i_l}}\right)_{l=1}^{\infty}$ such that $\lim_{l\in\mathbb{N}}\mu_{n_{i_l}} \rightsquigarrow \mu$. By the uniqueness of the weak* limit (Propositions 7.120) and 3.71, we have $\lim_{n\in\mathbb{N}}\mu_n \rightsquigarrow \mu$. Hence, $\lim_{n\in\mathbb{N}} X_n \rightsquigarrow X$.

This completes the proof of the theorem. □

Theorem 14.63 (Central Limit Theorem) *Let $\boldsymbol{\Omega} := (\Omega, \mathcal{B}, \boldsymbol{P})$ be a probability measure, $b \in \mathbb{R}_+$, $(X_n)_{n=1}^{\infty}$ be a sequence of independent ($\mathbb{R}$-valued) random variables, $X_n : \boldsymbol{\Omega} \to \mathbb{R}$, with $\boldsymbol{E}(X_n) = 0$, $\boldsymbol{E}(X_n^2) \le b$, and $\boldsymbol{E}(|X_n|^3) \le b$, $\forall n \in \mathbb{N}$, and $S_n := \sum_{i=1}^{n} X_i$. Assume that $\lim_{n\in\mathbb{N}} \frac{(\boldsymbol{E}(S_n^2))^3}{n^2} = \infty$. Then, $\left(\frac{S_n}{\sqrt{\boldsymbol{E}(S_n^2)}}\right)_{n=1}^{\infty}$ is uniformly growth bounded and $\lim_{n\in\mathbb{N}} \frac{S_n}{\sqrt{\boldsymbol{E}(S_n^2)}} \rightsquigarrow X$, where X is a random variable with cumulative distribution function $N(0,1)$.*

Proof Clearly, $\boldsymbol{E}(S_n) = 0$ and $b_n := \boldsymbol{E}(S_n^2) = \sum_{i=1}^{n} \boldsymbol{E}(X_i^2) \le nb$, $\forall n \in \mathbb{N}$. Fix any $n \in \mathbb{N}$. Let $\Psi_{X_n} : \mathbb{R} \to \mathbb{C}$ be the characteristic function of X_n and $\Psi_{V_n} : \mathbb{R} \to \mathbb{C}$ be the characteristic function of $V_n := S_n/\sqrt{\boldsymbol{E}(S_n^2)}$. Then, by Propositions 14.53 and 14.54, $\Psi_{V_n}(\theta) = \prod_{i=1}^{n} \Psi_{X_i}(\frac{\theta}{\sqrt{b_n}})$, $\forall \theta \in \mathbb{R}$. By Proposition 14.53, we have $\Psi_{X_n}(0) = 1$, $\Psi_{X_n}^{(1)}(0) = \mathrm{i}\boldsymbol{E}(X_n) = 0$, $\Psi_{X_n}^{(2)}(0) = -\boldsymbol{E}(X_n^2)$, and $\left|\Psi_{X_n}^{(3)}(\theta)\right| = \left|\boldsymbol{E}((\mathrm{i}X_n)^3 \exp(\mathrm{i}\theta X_n))\right| \le \boldsymbol{E}(|X_n|^3) \le b$. By Taylor's Theorem 9.48, we have $R_n(\theta) := \Psi_{X_n}(\theta) - (1 - \frac{\boldsymbol{E}(X_n^2)}{2}\theta^2)$ and

$$|R_n(\theta)| \le \frac{1}{6}\boldsymbol{E}(|X_n|^3)|\theta|^3, \quad \forall \theta \in \mathbb{R}$$

Let $h_{n,i}(\theta) := -\frac{\boldsymbol{E}(X_i^2)}{b_n}\frac{\theta^2}{2} + R_i(\frac{\theta}{\sqrt{b_n}})$, $\forall n \in \mathbb{N}$, $\forall i \in \mathbb{N}$. Fix any $\theta \in \mathbb{R}$. Clearly, we have $\lim_{n\in\mathbb{N}} h_{n,i}(\theta) = 0$, $\lim_{n\in\mathbb{N}} R_i(\frac{\theta}{\sqrt{b_n}}) = 0$, and the convergence is uniform in $i \in \mathbb{N}$. For sufficiently large n, we have $\left|h_{n,i}(\theta)\right| < \frac{1}{2}$ and $\left|R_i(\frac{\theta}{\sqrt{b_n}})\right| < \frac{1}{2}$, $\forall i \in \mathbb{N}$. Then, we have, using the notation of Example 12.139,

$$\begin{aligned}
&\left|\Psi_{V_n}(\theta) - \exp(-\frac{\theta^2}{2})\right| = \left|\prod_{i=1}^{n} \Psi_{X_i}(\frac{\theta}{\sqrt{b_n}}) - \exp(-\frac{\theta^2}{2})\right| \\
&= \exp(-\frac{\theta^2}{2})\left|\prod_{i=1}^{n}\left(1 - \frac{\boldsymbol{E}(X_i^2)}{b_n}\frac{\theta^2}{2} + R_i(\frac{\theta}{\sqrt{b_n}})\right)\exp\left(\frac{\boldsymbol{E}(X_i^2)}{b_n}\frac{\theta^2}{2}\right) - 1\right| \\
&= \exp(-\frac{\theta^2}{2})\left|\prod_{i=1}^{n}(1 + h_{n,i}(\theta))\exp(-h_{n,i}(\theta))\exp\left(R_i(\frac{\theta}{\sqrt{b_n}})\right) - 1\right| \\
&= \exp(-\frac{\theta^2}{2})\left|\overline{\exp}\left(\bigoplus_{i=1}^{n}\left(\overline{\ln}(1 + h_{n,i}(\theta)) \oplus \overline{\ln}(\exp(-h_{n,i}(\theta)))\right)\right.\right.
\end{aligned}$$

$$\oplus\overline{\ln}\left(\exp\left(R_i(\frac{\theta}{\sqrt{b_n}})\right)\right)\Bigg)\Bigg) - 1\Bigg|$$

$$=: \exp(-\frac{\theta^2}{2})|\overline{\exp}\,(u_n(\theta)) - 1|$$

where the second equality follows from the definition of b_n, the third equality follows from the definition of $h_{n,i}(\theta)$, and the fourth equality follows from Proposition 12.140. Note that

$$\begin{aligned}
&\rho(u_n(\theta), p_0)\\
&= \rho\Bigg(\bigoplus_{i=1}^{n}\Big(\overline{\ln}\,(1 + h_{n,i}(\theta)) \oplus \overline{\ln}\,(\exp\,(-h_{n,i}(\theta)))\\
&\quad \oplus\overline{\ln}\left(\exp\left(R_i(\frac{\theta}{\sqrt{b_n}})\right)\right)\Big), p_0\Bigg)\\
&\leq \sum_{i=1}^{n}\rho\Big(\overline{\ln}\,(1 + h_{n,i}(\theta)) \oplus \overline{\ln}\,(\exp\,(-h_{n,i}(\theta)))\\
&\quad \oplus\overline{\ln}\left(\exp\left(R_i(\frac{\theta}{\sqrt{b_n}})\right)\right), p_0\Big)\\
&\leq \sum_{i=1}^{n}\Bigg(\rho\Big(\overline{\ln}\,(1 + h_{n,i}(\theta)) \oplus \overline{\ln}\,(\exp\,(-h_{n,i}(\theta))), p_0\Big)\\
&\quad +\rho\left(\overline{\ln}\left(\exp\left(R_i(\frac{\theta}{\sqrt{b_n}})\right)\right), p_0\right)\Bigg) \leq \sum_{i=1}^{n}\left(2\big|h_{n,i}(\theta)\big|^2 + \left|R_i(\frac{\theta}{\sqrt{b_n}})\right|\right)\\
&\leq \sum_{i=1}^{n}\left(\frac{b|\theta|^3}{6}b_n^{-3/2} + \frac{\theta^4 b^2}{b_n^2} + \frac{1}{9b_n^3}b^2\theta^6\right)\\
&= \frac{nb|\theta|^3}{6}b_n^{-3/2} + \frac{n\theta^4 b^2}{b_n^2} + \frac{n}{9b_n^3}b^2\theta^6 \to 0 \text{ as } n \to \infty
\end{aligned}$$

where the first equality follows from the previous assignment, the first and second inequalities follow from (v) of Proposition 12.140, the third inequality follows from the (vi) and (vii) of Proposition 12.140, the fourth inequality follows from the definition of $h_{n,i}(\theta)$ and $R_i(\theta)$, and the convergence statement follows from the assumption of the theorem. Hence, by Example 12.139, $\lim_{n\in\mathbb{N}} \overline{\exp}\,(u_n(\theta)) = \overline{\exp}(p_0) = 1$ and $\lim_{n\in\mathbb{N}} \Psi_{V_n}(\theta) = \exp(-\theta^2/2) =: g(\theta)$.

Note that $g(\theta)$ is the characteristic function of a random variable $X \sim N(0, 1)$. Then, the result follows from Lévy Convergence Theorem 14.62. This completes the proof of the theorem. □

14.9 Uniform Integrability and Martingales

Definition 14.64 Let $\boldsymbol{\Omega} := (\Omega, \mathcal{B}, \boldsymbol{P})$ be a probability measure space, $\mathcal{Y}$ be a separable Banach space, and $\mathcal{F}$ be a collection of $\mathcal{Y}$-valued random variables. $\mathcal{F}$ is said to be *uniformly integrable* if $\forall \epsilon \in (0, \infty) \subset \mathbb{R}$, $\exists K \in [0, \infty) \subset \mathbb{R}$ such that, $\forall X \in \mathcal{F}$, we have $\int_{\{\omega \in \Omega \mid \|X(\omega)\|_{\mathcal{Y}} > K\}} \|X(\omega)\|_{\mathcal{Y}} \, \mathrm{d}\boldsymbol{P}(\omega) < \epsilon$. ◇

Clearly, for a uniformly integrable (UI) family $\mathcal{F}$ of $\mathcal{Y}$-valued random variables, let $\epsilon = 1$ and $K_1 \in [0, \infty) \subset \mathbb{R}$ be as in Definition 14.64. Then, $\forall X \in \mathcal{F}$, we have

$$\begin{aligned}
\boldsymbol{E}(\mathcal{P} \circ X) &= \int_{\Omega} \|X(\omega)\|_{\mathcal{Y}} \, \mathrm{d}\boldsymbol{P}(\omega) = \int_{\{\omega \in \Omega \mid \|X(\omega)\|_{\mathcal{Y}} > K_1\}} \|X(\omega)\|_{\mathcal{Y}} \, \mathrm{d}\boldsymbol{P}(\omega) \\
&\quad + \int_{\{\omega \in \Omega \mid \|X(\omega)\|_{\mathcal{Y}} \le K_1\}} \|X(\omega)\|_{\mathcal{Y}} \, \mathrm{d}\boldsymbol{P}(\omega) \\
&< 1 + \int_{\{\omega \in \Omega \mid \|X(\omega)\|_{\mathcal{Y}} \le K_1\}} K_1 \, \mathrm{d}\boldsymbol{P}(\omega) \le 1 + K_1
\end{aligned}$$

where the first equality follows from Definition 14.1, the second equality follows from Proposition 11.89 and Bounded Convergence Theorem 11.77, and the first inequality follows from Proposition 11.89 and Definition 14.64. Hence, $\boldsymbol{E}(\mathcal{P} \circ X) < K_1 + 1$, $\forall X \in \mathcal{F}$, and $\mathcal{F} \subseteq \mathcal{B}_{\bar{\mathrm{L}}_1(\boldsymbol{\Omega}, \mathcal{Y})}\big(\vartheta_{\bar{\mathrm{L}}_1(\boldsymbol{\Omega}, \mathcal{Y})}, 1 + K_1\big) \subseteq \bar{\mathrm{L}}_1(\boldsymbol{\Omega}, \mathcal{Y})$.

Proposition 14.65 *Let $\boldsymbol{\Omega} := (\Omega, \mathcal{B}, \boldsymbol{P})$ be a probability measure space, $\mathcal{Y}$ be a separable Banach space, and $\mathcal{F}$ be a collection of $\mathcal{Y}$-valued random variables. Then, the following statements hold:*

(i) *If $\mathcal{F}$ is bounded in $\bar{\mathrm{L}}_p(\boldsymbol{\Omega}, \mathcal{Y})$ with $p \in (1, \infty) \subset \mathbb{R}$, that is, $\exists A \in [0, \infty) \subset \mathbb{R}$ such that $\boldsymbol{E}(\mathcal{P}_p \circ X) \le A$, $\forall X \in \mathcal{F}$. Then, $\mathcal{F}$ is uniformly integrable.*
(ii) *If there exists $g : \boldsymbol{\Omega} \to [0, \infty) \subset \mathbb{R}$ such that g is $\mathcal{B}$-measurable, $\boldsymbol{E}(g) < \infty$, and $\|X(\omega)\|_{\mathcal{Y}} \le g(\omega)$ a.e. $\omega \in \boldsymbol{\Omega}$, $\forall X \in \mathcal{F}$. Then, $\mathcal{F}$ is uniformly integrable.*

Proof (i) Fix any $\epsilon \in (0, \infty) \subset \mathbb{R}$. Let $K := \left(\frac{\epsilon}{1+A}\right)^{-\frac{1}{p-1}}$. $\forall X \in \mathcal{F}$, by Propositions 11.38 and 7.21, $\mathcal{P}_q \circ X$ is $\mathcal{B}$-measurable, $\forall q \in [1, \infty) \subset \mathbb{R}$. Then,

$$\begin{aligned}
&\int_{\{\omega \in \Omega \mid \|X(\omega)\|_{\mathcal{Y}} > K\}} \|X(\omega)\|_{\mathcal{Y}} \, \mathrm{d}\boldsymbol{P}(\omega) \\
&\le \int_{\{\omega \in \Omega \mid \|X(\omega)\|_{\mathcal{Y}} > K\}} \|X(\omega)\|_{\mathcal{Y}}^p K^{1-p} \, \mathrm{d}\boldsymbol{P}(\omega)
\end{aligned}$$

$$\leq K^{1-p} \int_{\{\omega \in \Omega \mid \|X(\omega)\|_{\mathcal{Y}} > K\}} \|X(\omega)\|_{\mathcal{Y}}^p \, \mathrm{d}\boldsymbol{P}(\omega)$$

$$\leq K^{1-p} \int_{\Omega} \|X(\omega)\|_{\mathcal{Y}}^p \, \mathrm{d}\boldsymbol{P}(\omega) \leq \frac{\epsilon}{1+A} A = \epsilon$$

where the inequalities follow from Proposition 11.83. Hence, $\mathcal{F}$ is uniformly integrable.

(ii) Fix any $\epsilon \in (0, \infty) \subset \mathbb{R}$. By Proposition 11.84, $\exists \delta \in (0, \infty) \subset \mathbb{R}$ such that $\forall E \in \mathcal{B}$ with $\boldsymbol{P}(E) < \delta$, we have $\int_E g \, \mathrm{d}\boldsymbol{P} < \epsilon$, since $\boldsymbol{E}(g) < \infty$. Note that $\boldsymbol{P}\left(\bigcap_{K \in \mathbb{N}} \{\omega \in \Omega \mid g(\omega) > K\}\right) = \boldsymbol{P}(\emptyset) = 0$. By Proposition 11.5, we have $0 = \lim_{K \in \mathbb{N}} \boldsymbol{P}(\{\omega \in \Omega \mid g(\omega) > K\})$. Then, $\exists K \in \mathbb{N}$ such that $E := \{\omega \in \Omega \mid g(\omega) > K\} \in \mathcal{B}$ and $\boldsymbol{P}(E) < \delta$. $\forall X \in \mathcal{F}$,

$$\int_{\{\omega \in \Omega \mid \|X(\omega)\|_{\mathcal{Y}} > K\}} \|X(\omega)\|_{\mathcal{Y}} \, \mathrm{d}\boldsymbol{P}(\omega) \leq \int_E \mathcal{P} \circ X \, \mathrm{d}\boldsymbol{P} \leq \int_E g \, \mathrm{d}\boldsymbol{P} < \epsilon$$

where the first inequality follows Proposition 11.83, the second inequality follows from Proposition 11.83, and the last inequality follows from $\boldsymbol{P}(E) < \delta$ and the preceding discussion. Hence, $\mathcal{F}$ is uniformly integrable.

This completes the proof of the proposition. □

Theorem 14.66 *Let $\boldsymbol{\Omega} := (\Omega, \mathcal{B}, \boldsymbol{P})$ be a probability measure space, $\mathcal{Y}$ be a separable Banach space, $X \in \bar{\mathrm{L}}_1(\boldsymbol{\Omega}, \mathcal{Y})$ be a $\mathcal{Y}$-valued random variable. Then, the set $\mathcal{F} := \{f \in \bar{\mathrm{L}}_1(\boldsymbol{\Omega}, \mathcal{Y}) \mid \exists \bar{\mathcal{B}} \subseteq \mathcal{B}$ which is a sub-σ-algebra on Ω, $f \in \boldsymbol{E}(X|\bar{\mathcal{B}})\}$ is uniformly integrable.*

Proof Fix any $\epsilon \in (0, \infty) \subset \mathbb{R}$. By Proposition 11.84, $\exists \delta \in (0, \infty) \subset \mathbb{R}$ such that $\forall E \in \mathcal{B}$ with $\boldsymbol{P}(E) < \delta$, we have $\int_E \mathcal{P} \circ X \, \mathrm{d}\boldsymbol{P} < \epsilon$. Let $K \in [0, \infty) \subset \mathbb{R}$ be such that $K^{-1}\boldsymbol{E}(\mathcal{P} \circ X) < \delta$. Fix any $\bar{\mathcal{B}} \subseteq \mathcal{B}$ with $\bar{\mathcal{B}}$ being a σ-algebra on Ω. Let $\bar{\boldsymbol{\Omega}} := (\Omega, \bar{\mathcal{B}}, \bar{\boldsymbol{P}} := \boldsymbol{P}|_{\bar{\mathcal{B}}})$. Fix any $f \in \boldsymbol{E}(X|\bar{\mathcal{B}})$. By (g) of Proposition 14.11, we have $\mathcal{P} \circ f(\omega) \leq \bar{f}(\omega)$ a.e. $\omega \in \bar{\boldsymbol{\Omega}}$, where $\bar{f} \in \boldsymbol{E}(\mathcal{P} \circ X|\bar{\mathcal{B}})$. Let $E := \{\omega \in \Omega \mid \mathcal{P} \circ f(\omega) > K\} \in \bar{\mathcal{B}}$, since f is $\bar{\mathcal{B}}$-measurable. Then, $K\bar{\boldsymbol{P}}(E) = K\int_E 1 \, \mathrm{d}\bar{\boldsymbol{P}} \leq \int_E \mathcal{P} \circ f \, \mathrm{d}\bar{\boldsymbol{P}} \leq \int_\Omega \mathcal{P} \circ f \, \mathrm{d}\bar{\boldsymbol{P}} \leq \int_\Omega \bar{f} \, \mathrm{d}\bar{\boldsymbol{P}} = \int_\Omega \mathcal{P} \circ X \, \mathrm{d}\boldsymbol{P} = \boldsymbol{E}(\mathcal{P} \circ X)$, where the first equality follows from Proposition 11.75, the first, second, and third inequalities follow from Proposition 11.83, and the second equality follows from Definition 14.7. This implies that $\bar{\boldsymbol{P}}(E) = \boldsymbol{P}(E) \leq K^{-1}\boldsymbol{E}(\mathcal{P} \circ X) < \delta$. Therefore, $\int_E \mathcal{P} \circ f \, \mathrm{d}\boldsymbol{P} = \int_E \mathcal{P} \circ f \, \mathrm{d}\bar{\boldsymbol{P}} \leq \int_E \bar{f} \, \mathrm{d}\bar{\boldsymbol{P}} = \int_E \mathcal{P} \circ X \, \mathrm{d}\boldsymbol{P} < \epsilon$, where the first equality follows from the fact f is $\bar{\mathcal{B}}$-measurable and Proposition 11.72, the first inequality follows from Proposition 11.83, and the second equality follows from Definition 14.7. Hence, $\mathcal{F}$ is uniformly integrable. This completes the proof of the theorem. □

Proposition 14.67 (Chebyshev Inequality) *Let $\boldsymbol{\Omega} := (\Omega, \mathcal{B}, \boldsymbol{P})$ be a probability measure space and $X \in \bar{\mathrm{L}}_1(\boldsymbol{\Omega}, \mathbb{R})$ be a $\mathbb{R}$-valued random variable. Assume that*

$X(\omega) \geq 0$ *a.e. in* $\boldsymbol{\Omega}$. *Then,* $\forall c \in (0, \infty) \subset \mathbb{R}$, *we have* $\boldsymbol{P}(\{\omega \in \Omega \mid X(\omega) \geq c\}) \leq c^{-1}\boldsymbol{E}(X)$.

Proof We have the following derivations:

$$\boldsymbol{P}(\{\omega \in \Omega \mid X(\omega) \geq c\}) = \int_{\{\omega \in \Omega \mid X(\omega) \geq c\}} 1 \,\mathrm{d}\boldsymbol{P}$$

$$\leq \int_{\{\omega \in \Omega \mid X(\omega) \geq c\}} c^{-1} X(\omega) \,\mathrm{d}\boldsymbol{P}(\omega) = c^{-1} \int_{\{\omega \in \Omega \mid X(\omega) \geq c\}} X(\omega) \,\mathrm{d}\boldsymbol{P}(\omega)$$

$$\leq c^{-1} \int_{\Omega} X(\omega) \,\mathrm{d}\boldsymbol{P}(\omega) = c^{-1}\boldsymbol{E}(X)$$

where the first equality follows from Proposition 11.75, the first inequality follows from Proposition 11.89, the second equality follows from Proposition 11.89, and the second inequality follows from the fact that $X(\omega) \geq 0$ a.e. in $\boldsymbol{\Omega}$. □

Theorem 14.68 *Let* $\boldsymbol{\Omega} := (\Omega, \mathcal{B}, \boldsymbol{P})$ *be a probability measure space,* $\mathcal{Y}$ *be a separable Banach space,* $(X_n)_{n=1}^{\infty} \subseteq \bar{\mathrm{L}}_1(\boldsymbol{\Omega}, \mathcal{Y}) =: \bar{\mathcal{Z}}$ *be a sequence of* $\mathcal{Y}$*-valued random variables, and* $X \in \bar{\mathrm{L}}_1(\boldsymbol{\Omega}, \mathcal{Y})$ *be a* $\mathcal{Y}$*-valued random variable. Then,* $\lim_{n\in\mathbb{N}} X_n \doteq X$ *in* $\bar{\mathcal{Z}}$, *or equivalently* $\lim_{n\in\mathbb{N}} \boldsymbol{E}(\mathcal{P} \circ (X_n - X)) = 0$ *implies that the following two conditions are satisfied:*

(i) $\lim_{n\in\mathbb{N}} X_n = X$ *in measure in* $\boldsymbol{\Omega}$.
(ii) *The sequence* $(X_n)_{n=1}^{\infty}$ *is uniformly integrable.*

The converse of the above holds if $\mathcal{Y}$ *is a separable real Hilbert space.*

Proof "Necessity" Let $\lim_{n\in\mathbb{N}} \boldsymbol{E}(\mathcal{P} \circ (X_n - X)) = 0$. By Proposition 11.211, we have (i).

Fix any $\epsilon \in (0, 1) \subset \mathbb{R}$. Then, $\exists n_0 \in \mathbb{N}$, $\forall n \in \mathbb{N}$ with $n_0 \leq n$, we have $\boldsymbol{E}(\mathcal{P} \circ (X_n - X)) < \epsilon^2 < \epsilon$. By Chebyshev Inequality Proposition 14.67, we have $\boldsymbol{P}(\{\omega \in \Omega \mid \mathcal{P} \circ (X_n - X)(\omega) > \epsilon\} \leq \epsilon^{-1}\boldsymbol{E}(\mathcal{P} \circ (X_n - X)) < \epsilon$, $\forall n \in \mathbb{N}$ with $n_0 \leq n$.

By Proposition 11.84, $\exists \delta \in (0, \infty) \subset \mathbb{R}$ such that $\forall E \in \mathcal{B}$ with $\boldsymbol{P}(E) < \delta$, we have $\int_E \mathcal{P} \circ X_n \,\mathrm{d}\boldsymbol{P} < \epsilon$, $n = 1, \ldots, n_0$, and $\int_E \mathcal{P} \circ X \,\mathrm{d}\boldsymbol{P} < \epsilon$. Clearly, $\sup_{n\in\mathbb{N}} \boldsymbol{E}(\mathcal{P} \circ X_n) =: M \in [0, \infty) \subset \mathbb{R}$. Choose $K > 0$ such that $K^{-1}M < \delta$. Then, $\forall n \in \mathbb{N}$, by Chebyshev Inequality Proposition 14.67, we have $\boldsymbol{P}(\{\omega \in \Omega \mid \mathcal{P} \circ X_n(\omega) > K\}) \leq K^{-1}\boldsymbol{E}(\mathcal{P} \circ X_n) \leq K^{-1}M < \delta$. If $n \geq n_0$, then $\int_{\{\omega\in\Omega \mid \mathcal{P}\circ X_n(\omega) > K\}} \mathcal{P} \circ X_n \,\mathrm{d}\boldsymbol{P} \leq \int_{\{\omega\in\Omega \mid \mathcal{P}\circ X_n(\omega) > K\}} \mathcal{P} \circ X \,\mathrm{d}\boldsymbol{P} + \int_{\{\omega\in\Omega \mid \mathcal{P}\circ X_n(\omega) > K\}} \mathcal{P} \circ (X_n - X) \,\mathrm{d}\boldsymbol{P} < \epsilon + \int_{\Omega} \mathcal{P} \circ (X_n - X) \,\mathrm{d}\boldsymbol{P} < 2\epsilon$, where the first inequality follows from Definition 7.1 and Proposition 11.83, the second inequality follows from the preceding discussion and Proposition 11.83, and the last inequality follows from the preceding discussion. On the other hand, if $n < n_0$, we have $\int_{\{\omega\in\Omega \mid \mathcal{P}\circ X_n(\omega) > K\}} \mathcal{P} \circ X_n \,\mathrm{d}\boldsymbol{P} < \epsilon$ by the preceding discussion. Thus, $(X_n)_{n=1}^{\infty}$ is uniformly integrable.

"Sufficiency" Let (i) and (ii) hold and $\mathcal{Y}$ be a separable real Hilbert space. $\forall K \in (0, \infty) \subset \mathbb{R}$, define $\varphi_K : \mathcal{Y} \to \overline{\mathcal{B}}_{\mathcal{Y}}(\vartheta_{\mathcal{Y}}, K)$ by $\varphi_K(y) = \begin{cases} \frac{K}{\|y\|_{\mathcal{Y}}} y & \text{if } \|y\|_{\mathcal{Y}} > K \\ y & \text{if } \|y\|_{\mathcal{Y}} \leq K \end{cases}$, $\forall y \in \mathcal{Y}$. Clearly, φ_K is continuous. Fix any $\epsilon \in \mathbb{R}_+$. By (ii) and the fact that $X \in \bar{\mathrm{L}}_1(\boldsymbol{\Omega}, \mathcal{Y})$, we choose $K \in \mathbb{R}_+$ such that $\int_{\{\omega \in \Omega \mid \mathcal{P} \circ X_n(\omega) > K\}} \mathcal{P} \circ X_n \, \mathrm{d}\boldsymbol{P} < \frac{\epsilon}{3}$, $\forall n \in \mathbb{N}$, and $\int_{\{\omega \in \Omega \mid \mathcal{P} \circ X(\omega) > K\}} \mathcal{P} \circ X \, \mathrm{d}\boldsymbol{P} < \frac{\epsilon}{3}$. $\forall n \in \mathbb{N}$, we have

$$
\begin{aligned}
&\boldsymbol{E}(\mathcal{P} \circ (\varphi_K \circ X_n - X_n)) \\
&= \int_{\{\omega \in \Omega \mid \mathcal{P} \circ X_n(\omega) > K\}} \Big\| X_n(\omega) - \frac{K}{\|X_n(\omega)\|_{\mathcal{Y}}} X_n(\omega) \Big\|_{\mathcal{Y}} \, \mathrm{d}\boldsymbol{P}(\omega) \\
&= \int_{\{\omega \in \Omega \mid \mathcal{P} \circ X_n(\omega) > K\}} (1 - \frac{K}{\|X_n(\omega)\|_{\mathcal{Y}}}) \|X_n(\omega)\|_{\mathcal{Y}} \, \mathrm{d}\boldsymbol{P}(\omega) \\
&\leq \int_{\{\omega \in \Omega \mid \mathcal{P} \circ X_n(\omega) > K\}} \mathcal{P} \circ X_n \, \mathrm{d}\boldsymbol{P} < \frac{\epsilon}{3}
\end{aligned}
$$

where the first inequality follows from Proposition 11.83. Similarly, we have $\boldsymbol{E}(\mathcal{P} \circ (\varphi_K \circ X - X)) < \frac{\epsilon}{3}$. We need the following intermediate result:

Claim 14.68.1 $\|\varphi_K(x) - \varphi_K(y)\|_{\mathcal{Y}} \leq \|x - y\|_{\mathcal{Y}}$, $\forall x, y \in \mathcal{Y}$.

Proof of Claim We will distinguish four exhaustive and mutually exclusive cases: Case 1: $x, y \in \overline{\mathcal{B}}_{\mathcal{Y}}(\vartheta_{\mathcal{Y}}, K)$, Case 2: $\|x\|_{\mathcal{Y}} \leq K$ and $\|y\|_{\mathcal{Y}} > K$, Case 3: $\|x\|_{\mathcal{Y}} > K$ and $\|y\|_{\mathcal{Y}} \leq K$, and Case 4: $\|x\|_{\mathcal{Y}} > K$ and $\|y\|_{\mathcal{Y}} > K$.

Case 1: $x, y \in \overline{\mathcal{B}}_{\mathcal{Y}}(\vartheta_{\mathcal{Y}}, K)$. Then, we have $\|\varphi_K(x) - \varphi_K(y)\|_{\mathcal{Y}} = \|x - y\|_{\mathcal{Y}}$. This case is proved.

Case 2: $\|x\|_{\mathcal{Y}} \leq K$ and $\|y\|_{\mathcal{Y}} > K$. Let $C := \overline{\mathcal{B}}_{\mathcal{Y}}(\vartheta_{\mathcal{Y}}, K)$, which is clearly a closed convex set. The minimization problem $\min_{v \in C} \|y - v\|_{\mathcal{Y}}$ admits minimum $\|y\|_{\mathcal{Y}} - K$ and is achieved at $v_0 = \frac{K}{\|y\|_{\mathcal{Y}}} y =: \alpha y \in C$. This is because $\|y - v\|_{\mathcal{Y}}^2 = \langle y - v, y - v\rangle_{\mathcal{Y}} = \|y\|_{\mathcal{Y}}^2 - 2\langle v, y\rangle + \|v\|_{\mathcal{Y}}^2 \geq \|y\|_{\mathcal{Y}}^2 - 2\|v\|_{\mathcal{Y}}\|y\|_{\mathcal{Y}} + \|v\|_{\mathcal{Y}}^2 = (\|y\|_{\mathcal{Y}} - \|v\|_{\mathcal{Y}})^2 \geq (\|y\|_{\mathcal{Y}} - K)^2$, $\forall v \in C$, where the first equality follows from Proposition 13.2, the second equality follows from Definition 13.1, the first inequality follows from Proposition 13.2, and the second inequality follows from the fact that $v \in C$ and $\|y\|_{\mathcal{Y}} > K$. Thus, by Theorem 13.39, we have $\langle (1 - \alpha) y, v - \alpha y\rangle_{\mathcal{Y}} = \langle y - v_0, v - v_0\rangle_{\mathcal{Y}} \leq 0$, $\forall v \in C$. Thus, we have $\|\varphi_K(x) - \varphi_K(y)\|_{\mathcal{Y}} = \|x - \alpha y\|_{\mathcal{Y}} \leq \|x - y\|_{\mathcal{Y}}$, where the inequality follows since $\|x - y\|_{\mathcal{Y}}^2 - \|x - \alpha y\|_{\mathcal{Y}}^2 = \|t - (1 - \alpha) y\|_{\mathcal{Y}}^2 - \|t\|_{\mathcal{Y}}^2 = -2\langle t, (1 - \alpha) y\rangle_{\mathcal{Y}} + (1 - \alpha)^2 \|y\|_{\mathcal{Y}}^2 \geq 0$, with $t := x - \alpha y$. This case is proved.

Case 3: $\|x\|_{\mathcal{Y}} > K$ and $\|y\|_{\mathcal{Y}} \leq K$. This follows from symmetry and Case 2.

Case 4: $\|x\|_{\mathcal{Y}} > K$ and $\|y\|_{\mathcal{Y}} > K$. Note that $\|\varphi_K(x) - \varphi_K(y)\|_{\mathcal{Y}} = \Big\| \frac{K}{\|x\|_{\mathcal{Y}}} x - \frac{K}{\|y\|_{\mathcal{Y}}} y \Big\|_{\mathcal{Y}} =: \big\| \alpha_x x - \alpha_y y \big\|_{\mathcal{Y}}$. Without loss of generality, assume $\|y\|_{\mathcal{Y}} \geq \|x\|_{\mathcal{Y}}$. Then, $\alpha_x \geq \alpha_y > 0$. Now, $\big\|\alpha_x x - \alpha_y y\big\|_{\mathcal{Y}} = \alpha_x \big\| x - \frac{\alpha_y}{\alpha_x} y \big\|_{\mathcal{Y}} \leq \alpha_x \|x - y\|_{\mathcal{Y}} \leq \|x - y\|_{\mathcal{Y}}$, where the first inequality follows from Case 2 with K set to $\|x\|_{\mathcal{Y}}$. This proves the case.

In all cases, we have $\|\varphi_K(x) - \varphi_K(y)\|_{\mathcal{Y}} \leq \|x - y\|_{\mathcal{Y}}$. This completes the proof of the claim. □

Fix any subsequence $(X_{n_k})_{k=1}^{\infty}$ of $(X_n)_{n=1}^{\infty}$. By (i), we have $\lim_{k\in\mathbb{N}} X_{n_k} = X$ in measure in $\boldsymbol{\Omega}$. By Proposition 11.57, there exists a subsubsequence $\left(X_{n_{k_l}}\right)_{l=1}^{\infty}$ of $(X_{n_k})_{k=1}^{\infty}$ such that $\lim_{l\in\mathbb{N}} X_{n_{k_l}} = X$ a.e. in $\boldsymbol{\Omega}$. Then, by Proposition 11.52, we have $\lim_{l\in\mathbb{N}} \varphi_K \circ X_{n_{k_l}} = \varphi_K \circ X$ a.e. in $\boldsymbol{\Omega}$. Again by Proposition 11.52, we have $\lim_{l\in\mathbb{N}} \mathcal{P} \circ (\varphi_K \circ X_{n_{k_l}} - \varphi_K \circ X) = 0$ a.e. in $\boldsymbol{\Omega}$. By Bounded Convergence Theorem 11.77, we have $\lim_{l\in\mathbb{N}} \int_{\Omega} \mathcal{P} \circ (\varphi_K \circ X_{n_{k_l}} - \varphi_K \circ X)\, d\boldsymbol{P} = 0$. Since the above holds for any subsequence of $(X_n)_{n=1}^{\infty}$, by Proposition 3.71, we have $\lim_{n\in\mathbb{N}} \int_{\Omega} \mathcal{P} \circ (\varphi_K \circ X_n - \varphi_K \circ X)\, d\boldsymbol{P} = 0$. Then, $\exists n_0 \in \mathbb{N}$, $\forall n \in \mathbb{N}$ with $n_0 \leq n$, we have $\boldsymbol{E}(\mathcal{P} \circ (\varphi_K \circ X_n - \varphi_K \circ X)) < \frac{\epsilon}{3}$. This leads to $\boldsymbol{E}(\mathcal{P} \circ (X_n - X)) \leq \boldsymbol{E}(\mathcal{P} \circ (\varphi_K \circ X_n - X_n)) + \boldsymbol{E}(\mathcal{P} \circ (\varphi_K \circ X_n - \varphi_K \circ X)) + \boldsymbol{E}(\mathcal{P} \circ (\varphi_K \circ X - X)) < \epsilon$, $\forall n \in \mathbb{N}$ with $n_0 \leq n$. Hence, we have $\lim_{n\in\mathbb{N}} \boldsymbol{E}(\mathcal{P} \circ (X_n - X)) = 0$, i.e., $\lim_{n\in\mathbb{N}} X_n \doteq X$ in $\bar{\mathcal{Z}}$.

This completes the proof of the theorem. □

14.10 Existence of the Wiener Process

Proposition 14.69 *Let $\boldsymbol{\Omega} := (\Omega, \mathcal{B}, \boldsymbol{P})$ be a probability measure space, $\mathcal{Y}$ be a separable reflexive Banach space with $\mathcal{Y}^*$ being separable, $\hat{\mathcal{B}} \subseteq \mathcal{B}$ be a sub-σ-algebra, $\hat{\boldsymbol{\Omega}} := (\Omega, \hat{\mathcal{B}}, \hat{\boldsymbol{P}} := \boldsymbol{P}|_{\hat{\mathcal{B}}})$ be probability measure space, $p \in [1, \infty) \subset \mathbb{R}$, $f \in \bar{\mathrm{L}}_p(\boldsymbol{\Omega}, \mathcal{Y})$ be a $\mathcal{Y}$-valued random variable, $g \in \boldsymbol{E}(\mathcal{P}_p \circ f|\hat{\mathcal{B}}) \in \mathrm{L}_1(\hat{\boldsymbol{\Omega}}, \mathbb{R})$, and $h \in \boldsymbol{E}(f|\hat{\mathcal{B}}) \in \mathrm{L}_1(\hat{\boldsymbol{\Omega}}, \mathcal{Y})$. Then, $h \in \bar{\mathrm{L}}_p(\hat{\boldsymbol{\Omega}}, \mathcal{Y})$ and $\mathcal{P}_p \circ h \leq g$ a.e. in $\hat{\boldsymbol{\Omega}}$.*

Proof Note that $f \in \bar{\mathrm{L}}_p(\boldsymbol{\Omega}, \mathcal{Y})$ implies that $f \in \bar{\mathrm{L}}_1(\boldsymbol{\Omega}, \mathcal{Y})$ since $\boldsymbol{\Omega}$ is finite and $p \geq 1$. By Definition 14.7, $g \in \bar{\mathrm{L}}_1(\hat{\boldsymbol{\Omega}}, [0, \infty))$ and $h \in \bar{\mathrm{L}}_1(\hat{\boldsymbol{\Omega}}, \mathcal{Y})$ exists. Then, by (g) of Proposition 14.11, we have $\mathcal{P} \circ h \leq l$ a.e. in $\hat{\boldsymbol{\Omega}}$, where $l \in \boldsymbol{E}(\mathcal{P} \circ f|\hat{\mathcal{B}}) \in \mathrm{L}_1(\hat{\boldsymbol{\Omega}}, [0, \infty))$. Thus, $\mathcal{P}_p \circ h = (\mathcal{P} \circ h)^p \leq l^p \leq g$ a.e. in $\hat{\boldsymbol{\Omega}}$, where the last inequality follows from (i) of Propositions 14.11 and 8.25. Then, $\|h\|^p_{\bar{\mathrm{L}}_p(\hat{\boldsymbol{\Omega}}, \mathcal{Y})} = \hat{\boldsymbol{E}}(\mathcal{P}_p \circ h) = \boldsymbol{E}(\mathcal{P}_p \circ h) \leq \boldsymbol{E}(g) = \boldsymbol{E}(\mathcal{P}_p \circ f) < \infty$, where the first equality follows from Definition 14.1, the second equality follows from Proposition 11.72, the first inequality follows from Proposition 11.83 and the preceding discussion, the third equality follows from (a) of Proposition 14.11, and the last inequality follows from $f \in \bar{\mathrm{L}}_p(\boldsymbol{\Omega}, \mathcal{Y})$. Hence, $h \in \bar{\mathrm{L}}_p(\hat{\boldsymbol{\Omega}}, \mathcal{Y})$. This completes the proof of the proposition. □

Theorem 14.70 *Let $\boldsymbol{\Omega} := (\Omega, \mathcal{B}, \boldsymbol{P})$ be a probability measure space, $\mathcal{Y}$ be a separable real Hilbert space, $\hat{\mathcal{B}} \subseteq \mathcal{B}$ be a sub-σ-algebra, $\hat{\boldsymbol{\Omega}} := (\Omega, \hat{\mathcal{B}}, \hat{\boldsymbol{P}} := \boldsymbol{P}|_{\hat{\mathcal{B}}})$ be probability measure space, $f \in \bar{\mathrm{L}}_2(\boldsymbol{\Omega}, \mathcal{Y})$ be a $\mathcal{Y}$-valued random variable, and $h \in \boldsymbol{E}(f|\hat{\mathcal{B}}) \in \mathrm{L}_1(\hat{\boldsymbol{\Omega}}, \mathcal{Y})$. Then, $h \in \bar{\mathrm{L}}_2(\hat{\boldsymbol{\Omega}}, \mathcal{Y})$ and $\|f - h\|^2_{\bar{\mathrm{L}}_2(\boldsymbol{\Omega}, \mathcal{Y})} =$*

$\boldsymbol{E}(\mathcal{P}_2 \circ (f-h)) = \boldsymbol{E}(\mathcal{P}_2 \circ f) - \boldsymbol{E}(\mathcal{P}_2 \circ h) \le \|f-g\|^2_{\bar{\mathrm{L}}_2(\boldsymbol{\Omega},\mathcal{Y})}$, $\forall g : \Omega \to \mathcal{Y}$ *that is $\hat{\mathcal{B}}$-measurable, with equality holding if, and only if, $g = h$ a.e. in $\hat{\boldsymbol{\Omega}}$.*

Proof By Definition 14.7, $\boldsymbol{E}(f|\hat{\mathcal{B}}) \in \mathrm{L}_1(\hat{\boldsymbol{\Omega}}, \mathcal{Y})$ exists. Hence, $h \in \bar{\mathrm{L}}_1(\hat{\boldsymbol{\Omega}}, \mathcal{Y})$ exists. By Proposition 14.69, $h \in \bar{\mathrm{L}}_2(\hat{\boldsymbol{\Omega}}, \mathcal{Y})$. This implies that

$$\lambda := \inf_{g:\Omega\to\mathcal{Y} \text{ being } \hat{\mathcal{B}}\text{-measurable}} \|f-g\|^2_{\bar{\mathrm{L}}_2(\boldsymbol{\Omega},\mathcal{Y})} \in [0,\infty) \subset \mathbb{R}$$

and the infimum must be such that $g \in \bar{\mathrm{L}}_2(\hat{\boldsymbol{\Omega}}, \mathcal{Y})$. Then,

$$\begin{aligned}
\lambda &= \inf_{g\in\bar{\mathrm{L}}_2(\hat{\boldsymbol{\Omega}},\mathcal{Y})} \|f-g\|^2_{\bar{\mathrm{L}}_2(\boldsymbol{\Omega},\mathcal{Y})} = \inf_{g\in\bar{\mathrm{L}}_2(\hat{\boldsymbol{\Omega}},\mathcal{Y})} \boldsymbol{E}(\mathcal{P}_2 \circ (f-g)) \\
&= \inf_{g\in\bar{\mathrm{L}}_2(\hat{\boldsymbol{\Omega}},\mathcal{Y})} \boldsymbol{E}(\boldsymbol{E}(\mathcal{P}_2 \circ (f-g)|\hat{\mathcal{B}})) \\
&= \inf_{g\in\bar{\mathrm{L}}_2(\hat{\boldsymbol{\Omega}},\mathcal{Y})} \boldsymbol{E}(\boldsymbol{E}(\langle f,f\rangle_{\mathcal{Y}} - 2\langle f,g\rangle_{\mathcal{Y}} + \langle g,g\rangle_{\mathcal{Y}}|\hat{\mathcal{B}})) \\
&= \inf_{g\in\bar{\mathrm{L}}_2(\hat{\boldsymbol{\Omega}},\mathcal{Y})} \boldsymbol{E}(\boldsymbol{E}(\langle f,f\rangle_{\mathcal{Y}}|\hat{\mathcal{B}}) - 2\boldsymbol{E}(\langle f,g\rangle_{\mathcal{Y}}|\hat{\mathcal{B}}) + \boldsymbol{E}(\langle g,g\rangle_{\mathcal{Y}}|\hat{\mathcal{B}})) \\
&= \inf_{g\in\bar{\mathrm{L}}_2(\hat{\boldsymbol{\Omega}},\mathcal{Y})} \boldsymbol{E}(\boldsymbol{E}(\langle f,f\rangle_{\mathcal{Y}}|\hat{\mathcal{B}}) - 2\boldsymbol{E}(\langle\langle \Phi_{\mathrm{inv}}(g), f\rangle\rangle|\hat{\mathcal{B}}) + \langle g,g\rangle_{\mathcal{Y}}) \\
&= \inf_{g\in\bar{\mathrm{L}}_2(\hat{\boldsymbol{\Omega}},\mathcal{Y})} \boldsymbol{E}(\boldsymbol{E}(\langle f,f\rangle_{\mathcal{Y}}|\hat{\mathcal{B}}) - 2\langle\langle \Phi_{\mathrm{inv}}(g), \boldsymbol{E}(f|\hat{\mathcal{B}})\rangle\rangle + \langle g,g\rangle_{\mathcal{Y}}) \\
&= \inf_{g\in\bar{\mathrm{L}}_2(\hat{\boldsymbol{\Omega}},\mathcal{Y})} (\boldsymbol{E}(\langle f,f\rangle_{\mathcal{Y}}) - 2\boldsymbol{E}(\langle\langle \Phi_{\mathrm{inv}}(g), h\rangle\rangle) + \boldsymbol{E}(\langle g,g\rangle_{\mathcal{Y}})) \\
&= \inf_{g\in\bar{\mathrm{L}}_2(\hat{\boldsymbol{\Omega}},\mathcal{Y})} \boldsymbol{E}(\langle f,f\rangle_{\mathcal{Y}} - 2\langle\langle \Phi_{\mathrm{inv}}(g), h\rangle\rangle + \langle g,g\rangle_{\mathcal{Y}}) \\
&= \boldsymbol{E}(\langle f,f\rangle_{\mathcal{Y}}) + \inf_{g\in\bar{\mathrm{L}}_2(\hat{\boldsymbol{\Omega}},\mathcal{Y})} \boldsymbol{E}(-2\langle h,g\rangle_{\mathcal{Y}} + \langle g,g\rangle_{\mathcal{Y}}) \\
&= \boldsymbol{E}(\langle f,f\rangle_{\mathcal{Y}}) + \inf_{g\in\bar{\mathrm{L}}_2(\hat{\boldsymbol{\Omega}},\mathcal{Y})} \boldsymbol{E}(-\langle h,h\rangle_{\mathcal{Y}} + \langle g-h,g-h\rangle_{\mathcal{Y}}) \\
&= \boldsymbol{E}(\langle f,f\rangle_{\mathcal{Y}}) - \boldsymbol{E}(\langle h,h\rangle_{\mathcal{Y}}) + \inf_{g\in\bar{\mathrm{L}}_2(\hat{\boldsymbol{\Omega}},\mathcal{Y})} \boldsymbol{E}(\langle g-h,g-h\rangle_{\mathcal{Y}}) \\
&= \|f\|^2_{\bar{\mathrm{L}}_2(\boldsymbol{\Omega},\mathcal{Y})} - \|h\|^2_{\bar{\mathrm{L}}_2(\hat{\boldsymbol{\Omega}},\mathcal{Y})} + \inf_{g\in\bar{\mathrm{L}}_2(\hat{\boldsymbol{\Omega}},\mathcal{Y})} \|g-h\|^2_{\bar{\mathrm{L}}_2(\hat{\boldsymbol{\Omega}},\mathcal{Y})} \\
&= \|f\|^2_{\bar{\mathrm{L}}_2(\boldsymbol{\Omega},\mathcal{Y})} - \|h\|^2_{\bar{\mathrm{L}}_2(\hat{\boldsymbol{\Omega}},\mathcal{Y})} = \|f-h\|^2_{\bar{\mathrm{L}}_2(\boldsymbol{\Omega},\mathcal{Y})}
\end{aligned}$$

where the first equality follows from the preceding discussion, the second equality follows from Definition 14.1, the third equality follows from (a) of Proposition 14.11, the fourth equality follows from Proposition 13.2 and Definition 13.1, the fifth equality follows from (c) of Proposition 14.11, the sixth equality follows from (iv) of Theorem 13.15 and (b) of Proposition 14.11 and $\Phi : \mathcal{Y}^* \to \mathcal{Y}$ is the mapping defined in Theorem 13.15, the seventh equality follows from (h) of Proposition 14.11, the eighth and ninth equalities follow from Proposition 11.92, the eleventh equality follows from Definition 13.1, the twelfth equality follows from Definition 14.1 and Proposition 13.2, the fourteenth equality follows whenever $g = h$ a.e. in $\hat{\boldsymbol{\Omega}}$, and the last equality follows from the fact we substitute $g = h$ a.e. in $\hat{\boldsymbol{\Omega}}$ into the original infimization problem.

This completes the proof of the theorem. □

The above theorem says that minimization problem

$$\min_{g:\Omega\rightarrow\mathcal{Y} \text{ being } \hat{\mathcal{B}}\text{-measurable}} \quad \|f - g\|^2_{\mathrm{L}_2(\boldsymbol{\Omega},\mathcal{Y})}$$

admits a unique minimizer (modulo a set of measure 0) given by a version of $\boldsymbol{E}(f|\hat{\mathcal{B}})$.

Proposition 14.71 *Let $\boldsymbol{\Omega} := (\Omega, \mathcal{B}, \boldsymbol{P})$ be a probability measure space, $m \in \mathbb{N}$, and $(x_n)_{n=0}^{\infty}$ be a sequence of $\mathbb{R}^m$-valued random variables with $x_n : \boldsymbol{\Omega} \rightarrow \mathbb{R}^m$ with law μ_n, $\forall n \in \mathbb{Z}_+$, and $F_n : \mathbb{R}^m \rightarrow [0, 1] \subset \mathbb{R}$ be the cumulative distribution function for μ_n as defined in Definition 14.16, $\forall n \in \mathbb{Z}_+$. Then, the following statements hold:*

(i) *If $\lim_{n\in\mathbb{N}} \int_{\mathbb{R}^m} \langle\langle d\mu_n(x), z(x)\rangle\rangle = \int_{\mathbb{R}^m} \langle\langle d\mu_0(x), z(x)\rangle\rangle \in \mathbb{R}$, $\forall z \in \mathcal{C}_b(\mathbb{R}^m, \mathbb{R})$, then $\lim_{n\in\mathbb{N}} \mu_n \rightharpoonup \mu_0$.*
(ii) *$\lim_{n\in\mathbb{N}} \int_{\mathbb{R}^m} \langle\langle d\mu_n(x), z(x)\rangle\rangle = \int_{\mathbb{R}^m} \langle\langle d\mu_0(x), z(x)\rangle\rangle \in \mathbb{R}$, $\forall z \in \mathcal{C}_b(\mathbb{R}^m, \mathbb{R})$ implies that $\lim_{n\in\mathbb{N}} F_n(b) = F_0(b)$, $\forall b \in \mathbb{R}^m$ with F_0 being continuous at b.*
(iii) *If $\lim_{n\in\mathbb{N}} \mu_n \rightharpoonup \mu_0$ and $(x_n)_{n=1}^{\infty}$ is uniformly growth bounded, then $\lim_{n\in\mathbb{N}} F_n(b) = F_0(b)$, $\forall b \in \mathbb{R}^m$ with F_0 being continuous at b.*
(iv) *If $\lim_{n\in\mathbb{N}} \mu_n \rightharpoonup \mu_0$, $(x_n)_{n=1}^{\infty}$ is uniformly growth bounded, and $\lim_{n\in\mathbb{N}} F_n(b) =: \bar{F}(b) \in [0, 1] \subset \mathbb{R}$, $\forall b \in \mathbb{R}^m$, and $\bar{F} : \mathbb{R}^m \rightarrow [0, 1] \subset \mathbb{R}$ is a valid cumulative distribution function for some $\mathbb{R}^m$-valued random variable, then $\bar{F} = F_0$.*

Proof (i) Note that $\lim_{n\in\mathbb{N}} \mu_n \rightharpoonup \mu_0$ means that $\lim_{n\in\mathbb{N}} \mu_n = \mu_0$ weak*, that is $\lim_{n\in\mathbb{N}} \langle\langle \mu_n, z\rangle\rangle = \langle\langle \mu_0, z\rangle\rangle$, $\forall z \in \mathcal{C}_c(\mathbb{R}^m, \mathbb{R})$. By Riesz Representation Theorem 11.209, this is equivalent to $\lim_{n\in\mathbb{N}} \int_{\mathbb{R}^m} \langle\langle d\mu_n(x), z(x)\rangle\rangle = \int_{\mathbb{R}^m} \langle\langle d\mu_0(x), z(x)\rangle\rangle$, $\forall z \in \mathcal{C}_c(\mathbb{R}^m, \mathbb{R}) \subseteq \mathcal{C}_b(\mathbb{R}^m, \mathbb{R})$. Thus, (i) follows readily.

(ii) Let $\lim_{n\in\mathbb{N}} \int_{\mathbb{R}^m} \langle\langle d\mu_n(x), z(x)\rangle\rangle = \int_{\mathbb{R}^m} \langle\langle d\mu_0(x), z(x)\rangle\rangle \in \mathbb{R}$, $\forall z \in \mathcal{C}_b(\mathbb{R}^m, \mathbb{R})$. Fix any $b \in \mathbb{R}^m$ and $\delta \in \mathbb{R}_+$. Define $h_{b,\delta} : \mathbb{R}^m \rightarrow [0, 1] \subset \mathbb{R}$ by $h_{b,\delta}(x) = \begin{cases} 1 & x \leqq b \\ 1 - \frac{\max(x-b)}{\delta} & x \leqq b + \delta\mathbf{1}_m \text{ and } x \nleqq b \\ 0 & x \nleqq b + \delta\mathbf{1}_m \end{cases}$, $\forall x \in \mathbb{R}^m$. Clearly, $h_{b,\delta} \in \mathcal{C}_b(\mathbb{R}^m, \mathbb{R})$. Then, $\lim_{n\in\mathbb{N}} \langle\langle \mu_n, h_{b,\delta}\rangle\rangle = \langle\langle \mu_0, h_{b,\delta}\rangle\rangle$, $F_n(b) \leq \langle\langle \mu_n, h_{b,\delta}\rangle\rangle$, and $\langle\langle \mu_0, h_{b,\delta}\rangle\rangle \leq F_0(b + \delta\mathbf{1}_m)$. Then, taking limit as $n \rightarrow \infty$ in the above three relations, we have $\limsup_{n\in\mathbb{N}} F_n(b) \leq F_0(b + \delta\mathbf{1}_m)$. Since F_0 is a cumulative distribution function and therefore continuous on the right. Taking limit as $\delta \rightarrow 0^+$, we have $\limsup_{n\in\mathbb{N}} F_n(b) \le F_0(b)$. Similarly, we must have $\lim_{n\in\mathbb{N}} \langle\langle \mu_n, h_{b,\delta}\rangle\rangle = \langle\langle \mu_0, h_{b,\delta}\rangle\rangle$, $F_n(b+\delta\mathbf{1}_m) \ge \langle\langle \mu_n, h_{b,\delta}\rangle\rangle$, and $\langle\langle \mu_0, h_{b,\delta}\rangle\rangle \ge F_0(b)$. Taking limit as $n \to \infty$ in the above three relations, we have $\liminf_{n\in\mathbb{N}} F_n(b+\delta\mathbf{1}_m) \ge F_0(b)$. Let $\bar{b} := b + \delta\mathbf{1}_m$, we have $\liminf_{n\in\mathbb{N}} F_n(\bar{b}) \ge F_0(\bar{b} - \delta\mathbf{1}_m)$, $\forall \bar{b} \in \mathbb{R}^m$, $\forall \delta \in \mathbb{R}_+$. Now, fix any $b \in \mathbb{R}^m$ with F_0 being continuous at b. We have $F_0(b - \delta\mathbf{1}_m) \le \liminf_{n\in\mathbb{N}} F_n(b) \le \limsup_{n\in\mathbb{N}} F_n(b) \le F_0(b)$. Now, take $\delta \to 0^+$, we have $F_0(b) \le \liminf_{n\in\mathbb{N}} F_n(b) \le \limsup_{n\in\mathbb{N}} F_n(b) \le F_0(b)$ and hence $\lim_{n\in\mathbb{N}} F_n(b) = F_0(b)$. Hence, (ii) is proved.

(iii) Let $\lim_{n\in\mathbb{N}} \mu_n \looparrowright \mu_0$. Fix any $z \in \mathcal{C}_b(\mathbb{R}^m, \mathbb{R})$. By Bounded Convergence Theorem 11.77, $\int_{\mathbb{R}^m} \langle\langle d\mu_n(x), z(x)\rangle\rangle \in \mathbb{R}$, $\forall n \in \mathbb{Z}_+$. $\forall a, b \in \mathbb{R}^m$ with $a \lessdot b$, $\forall k \in \mathbb{N}$ with $\min(b-a) > \frac{1}{k}$, by Proposition 5.62, there exists a continuous function $h_{k,a,b} : \mathbb{R}^m \to [0,1] \subset \mathbb{R}$ such that $h_{k,a,b}(x) = 1$, $\forall x \in E_{k,1} := \{\bar{x} \in \mathbb{R}^m \mid a + \frac{1}{k}\mathbf{1}_m \leqq \bar{x} \leqq b\} \subset \mathbb{R}^m$ and $h_{k,a,b}(x) = 0$, $\forall x \in E_{k,2} := \mathbb{R}^m \setminus \{\bar{x} \in \mathbb{R}^m \mid a \lessdot \bar{x} \lessdot b + \frac{1}{k}\mathbf{1}_m\}$. Clearly, $h_{k,a,b} \in \mathcal{C}_c(\mathbb{R}^m, \mathbb{R})$. Then,

$$\begin{aligned} \Delta_{F_n}(\mathrm{r}_{a+\frac{1}{k}\mathbf{1}_m,b}) &= \int_{\mathbb{R}^m} \chi_{\mathrm{r}_{a+\frac{1}{k}\mathbf{1}_m,b},\mathbb{R}^m}\, d\mu_n \le \int_{\mathbb{R}^m} h_{k,a,b}\, d\mu_n = \langle\langle \mu_n, h_{k,a,b}\rangle\rangle \\ &\le \int_{\mathbb{R}^m} \chi_{\mathrm{r}_{a,b+\frac{1}{k}\mathbf{1}_m},\mathbb{R}^m}\, d\mu_n = \Delta_{F_n}(\mathrm{r}_{a,b+\frac{1}{k}\mathbf{1}_m}); \qquad (14.13) \\ &\forall n \in \mathbb{Z}_+, \forall a, b \in \mathbb{R}^m, \forall k \in \mathbb{N} \text{ with } a + \frac{1}{k} \lessdot b \end{aligned}$$

where the first and last equalities follow from Definition 14.16 and the first and last inequalities follow from Proposition 11.92. Taking limit as $n \to \infty$,

$$\begin{aligned} \limsup_{n\in\mathbb{N}} \Delta_{F_n}(\mathrm{r}_{a+\frac{1}{k}\mathbf{1}_m,b}) &\le \lim_{n\in\mathbb{N}} \langle\langle \mu_n, h_{k,a,b}\rangle\rangle = \langle\langle \mu_0, h_{k,a,b}\rangle\rangle \qquad (14.14) \\ &\le \liminf_{n\in\mathbb{N}} \Delta_{F_n}(\mathrm{r}_{a,b+\frac{1}{k}\mathbf{1}_m}); \ \forall a, b \in \mathbb{R}^m, \forall k \in \mathbb{N} \text{ with } a + \frac{1}{k}\mathbf{1}_m \lessdot b \end{aligned}$$

where the equality follows from $\lim_{n\in\mathbb{N}} \mu_n \looparrowright \mu_0$ and $h_{k,a,b} \in \mathcal{C}_c(\mathbb{R}^m, \mathbb{R})$ and the inequalities follow from (14.13). By the assumption that $(x_n)_{n=1}^{\infty}$ is uniformly growth bounded, then $(x_n)_{n=0}^{\infty}$ is uniformly growth bounded. Thus, we have $\lim_{\min d\to-\infty} F_n(d) = 0$, and the convergence is uniform with respect to $n \in \mathbb{Z}_+$. It is easy to prove, using $\epsilon - \delta$ arguments, that

$$\begin{aligned} 0 &\le \lim_{\min d\to-\infty} \liminf_{n\in\mathbb{N}} F_n(d) \le \lim_{\min d\to-\infty} \limsup_{n\in\mathbb{N}} F_n(d) \\ &= \lim_{n\in\mathbb{N}} \lim_{\min d\to-\infty} F_n(d) = 0 \qquad (14.15) \end{aligned}$$

This implies that, $\forall b \in \mathbb{R}^m$ and $\forall k \in \mathbb{N}$,

$$\limsup_{n\in\mathbb{N}} F_n(b) = \limsup_{n\in\mathbb{N}} \lim_{\max a\to-\infty} \Delta_{F_n}(\mathrm{r}_{a+\frac{1}{k}\mathbf{1}_m,b})$$

$$= \lim_{\max a \to -\infty} \limsup_{n \in \mathbb{N}} \Delta_{F_n}(\mathrm{r}_{a+\frac{1}{k}\mathbf{1}_m, b}) \leq \liminf_{\max a \to -\infty} \langle\langle \mu_0, h_{k,a,b} \rangle\rangle$$

$$\leq \lim_{\max a \to -\infty} \Delta_{F_0}(\mathrm{r}_{a, b+\frac{1}{k}\mathbf{1}_m}) = F_0(b + \frac{1}{k}\mathbf{1}_m) \tag{14.16}$$

where the first equality follows from (14.15), the second equality follows from (14.15), the first inequality follows from (14.14) and the fact that all limits are real numbers in the closed interval $[0, 1] \subset \mathbb{R}$, the second inequality follows from (14.13) with $n = 0$, and the last equality follows from (14.15) and Definition 12.41. We also have, $\forall b \in \mathbb{R}^m$ and $\forall k \in \mathbb{N}$,

$$F_0(b) = \lim_{\max a \to -\infty} \Delta_{F_0}(\mathrm{r}_{a+\frac{1}{k}\mathbf{1}_m, b}) \leq \liminf_{\max a \to -\infty} \langle\langle \mu_0, h_{k,a,b} \rangle\rangle$$

$$\leq \liminf_{\max a \to -\infty} \liminf_{n \in \mathbb{N}} \Delta_{F_n}(\mathrm{r}_{a, b+\frac{1}{k}\mathbf{1}_m})$$

$$= \liminf_{n \in \mathbb{N}} \liminf_{\max a \to -\infty} \Delta_{F_n}(\mathrm{r}_{a, b+\frac{1}{k}\mathbf{1}_m}) = \liminf_{n \in \mathbb{N}} F_n(b + \frac{1}{k}\mathbf{1}_m) \tag{14.17}$$

where the first equality follows from (14.15) and Definition 12.41, the first inequality follows from (14.13) with $n = 0$, the second inequality follows from (14.14); the second equality follows from (14.15), and the last equality follows from (14.15). Then, let $b \in \mathbb{R}^m$ be any point such that F_0 is continuous at b, we have $\limsup_{n \in \mathbb{N}} F_n(b) \leq \lim_{k \in \mathbb{N}} F_0(b + \frac{1}{k}\mathbf{1}_m) = F_0(b) = \lim_{k \in \mathbb{N}} F_0(b - \frac{1}{k}\mathbf{1}) \leq \lim_{k \in \mathbb{N}} \liminf_{n \in \mathbb{N}} F_n(b) = \liminf_{n \in \mathbb{N}} F_n(b) \leq \limsup_{n \in \mathbb{N}} F_n(b)$, where the first inequality follows from (14.16), the first and second equalities follow from the fact that F_0 is continuous at b, the second inequality follows from (14.17) with b substituted by $b - \frac{1}{k}$, and the last equality follows from Proposition 3.83. Thus, we have $\lim_{n \in \mathbb{N}} F_n(b) = F_0(b)$. This establishes (iii).

(iv) $\forall a, b \in \mathbb{R}^m$ with $a \lessdot b$, $\forall k \in \mathbb{N}$ with $\min(b - a) > \frac{1}{k}$, let $h_{k,a,b} : \mathbb{R}^m \to [0, 1] \subset \mathbb{R}$ be as defined in (iii). Clearly, $h_{k,a,b} \in \mathcal{C}_c(\mathbb{R}^m, \mathbb{R})$. Then, (14.13)–(14.17) holds. Taking limit as $n \to \infty$ in (14.13),

$$\Delta_{\bar{F}}(\mathrm{r}_{a+\frac{1}{k}\mathbf{1}_m, b}) = \lim_{n \in \mathbb{N}} \Delta_{F_n}(\mathrm{r}_{a+\frac{1}{k}\mathbf{1}_m, b}) \leq \lim_{n \in \mathbb{N}} \langle\langle \mu_n, h_{k,a,b} \rangle\rangle = \langle\langle \mu_0, h_{k,a,b} \rangle\rangle$$

$$\leq \lim_{n \in \mathbb{N}} \Delta_{F_n}(\mathrm{r}_{a, b+\frac{1}{k}\mathbf{1}_m}) = \Delta_{\bar{F}}(\mathrm{r}_{a, b+\frac{1}{k}\mathbf{1}_m}); \tag{14.18}$$

$$\forall a, b \in \mathbb{R}^m, \forall k \in \mathbb{N} \text{ with } a + \frac{1}{k}\mathbf{1}_m \lessdot b$$

where the first equality follows from the assumption, the second equality from $\lim_{n \in \mathbb{N}} \mu_n \nrightarrow \mu_0$ and $h_{k,a,b} \in \mathcal{C}_c(\mathbb{R}^m, \mathbb{R})$, the inequalities follow from (14.13), and the last equality follows from the assumption. This further implies that, $\forall b \in \mathbb{R}^m$ and $\forall k \in \mathbb{N}$,

$$\bar{F}(b) = \lim_{n \in \mathbb{N}} F_n(b) = \lim_{n \in \mathbb{N}} \lim_{\max a \to -\infty} \Delta_{F_n}(\mathrm{r}_{a+\frac{1}{k}\mathbf{1}_m, b})$$

$$= \lim_{\max a \to -\infty} \lim_{n \in \mathbb{N}} \Delta_{F_n}(\mathrm{r}_{a+\frac{1}{k}\mathbf{1}_m, b}) \le \liminf_{\max a \to -\infty} \langle\langle \mu_0, h_{k,a,b} \rangle\rangle$$

$$\le \lim_{\max a \to -\infty} \Delta_{F_0}(\mathrm{r}_{a, b+\frac{1}{k}\mathbf{1}_m}) = F_0(b + \frac{1}{k}\mathbf{1}_m) \tag{14.19}$$

where the first equality follows from the assumption, the second equality follows from (14.15), the third equality follows from (14.15), the first inequality follows from (14.18), the second inequality follows from (14.13) with $n = 0$, and the last equality follows from (14.15) and Definition 12.41. Taking limit $k \to \infty$, we have $\bar{F}(b) \le F_0(b)$, $\forall b \in \mathbb{R}^m$, since F_0 is continuous on the right. We also have, $\forall b \in \mathbb{R}^m$ and $\forall k \in \mathbb{N}$,

$$F_0(b) = \lim_{\max a \to -\infty} \Delta_{F_0}(\mathrm{r}_{a+\frac{1}{k}\mathbf{1}_m, b}) \le \liminf_{\max a \to -\infty} \langle\langle \mu_0, h_{k,a,b} \rangle\rangle$$

$$\le \liminf_{\max a \to -\infty} \lim_{n \in \mathbb{N}} \Delta_{F_n}(\mathrm{r}_{a, b+\frac{1}{k}\mathbf{1}_m}) = \lim_{n \in \mathbb{N}} \liminf_{\max a \to -\infty} \Delta_{F_n}(\mathrm{r}_{a, b+\frac{1}{k}\mathbf{1}_m})$$

$$= \lim_{n \in \mathbb{N}} F_n(b + \frac{1}{k}\mathbf{1}_m) = \bar{F}(b + \frac{1}{k}\mathbf{1}_m) \tag{14.20}$$

where the first equality follows from (14.15) and Definition 12.41, the first inequality follows from (14.13) with $n = 0$, the second inequality follows from (14.14), the second and third equalities follow from (14.15), and the last equality follows from the assumption. Then, again taking limit $k \to \infty$, we have $F_0(b) \le \bar{F}(b)$, $\forall b \in \mathbb{R}^m$, since $\bar{F}$ is continuous on the right. Thus, we have $\bar{F}(b) = F_0(b)$, $\forall b \in \mathbb{R}^m$. This establishes (iv).

This completes the proof of the proposition. □

Proposition 14.72 *Let* $\mathbf{\Omega} := (\Omega, \mathcal{B}, \mathbf{P})$ *be a probability measure space,* $m \in \mathbb{N}$, $(X_i)_{i=1}^{\infty}$ *be a sequence of independent identically distributed random variable with* $X_i \sim N(0, 1)$, $\forall i \in \mathbb{N}$, $\alpha_1, \ldots, \alpha_m \in \mathbf{l}_\infty(\mathbb{R})$ *be linearly independent, and* $(Y_j)_{j=1}^{m}$ *be random variables defined by*

$$Y_j(\omega) := \begin{cases} \sum_{l=1}^{\infty} \alpha_{j,l} X_l(\omega) & \forall \omega \in F_j \\ 0 & \forall \omega \in \Omega \setminus F_j \end{cases}, \quad \forall \omega \in \Omega$$

where $F_j := \{\omega \in \Omega \mid \sum_{l=1}^{\infty} \alpha_{j,l} X_l(\omega) \in \mathbb{R}\}$, $j = 1, \ldots, m$; $\alpha_j = (\alpha_{j,1}, \alpha_{j,2}, \ldots)$, $j = 1, \ldots, m$. *Then,* $Y := (Y_1, \ldots, Y_m)$ *is an* $\mathbb{R}^m$*-valued Gaussian random variable with distribution* $N(\mathbf{0}_m, K)$, *where* $K = (\sum_{l=1}^{\infty} \alpha_{i,l}\alpha_{j,l})_{m \times m}$, *and* $\mathbf{P}(F_j) = 0$, $j = 1, \ldots, m$, *if* $K \in \bar{\mathcal{S}}_{+\,\mathbb{R}^m}$.

Proof Let $K \in \bar{\mathcal{S}}_{+\,\mathbb{R}^m}$. Then, $\alpha_j \in \mathbf{l}_2(\mathbb{R}) =: \mathcal{X}$, $j = 1, \ldots, m$. By Proposition 13.42, $\bar{\mathcal{S}}_{+\,\mathbb{R}^m}$ is an open subset of $\bar{\mathcal{S}}_{\mathbb{R}^m}$, then, $\exists n_1 \in \mathbb{N}$, $\forall n \in \mathbb{N}$ with $n \ge n_1$, we have $K_n := (\sum_{l=1}^{n} \alpha_{i,l}\alpha_{j,l})_{m \times m} \in \bar{\mathcal{S}}_{+\,\mathbb{R}^m}$. Then, the matrix $V_n := (\alpha_{i,j})_{m \times n}$ is of full row rank, $\forall n \in \mathbb{N}$ with $n \ge n_1$, since $K_n = V_n V_n'$. By Proposition 14.28, we have $Y^{\langle n \rangle} := V_n X^{\langle n \rangle}$ is an $\mathbb{R}^m$-valued Gaussian random variable with distribution $N(\mathbf{0}_m, K_n)$, $\forall n \in \mathbb{N}$ with $n \ge n_1$, where $X^{\langle n \rangle} := (X_1, \ldots, X_n)$ is an $\mathbb{R}^n$-valued

Gaussian random variable with distribution $N(\mathbf{0}_n, \mathrm{id}_{\mathbb{R}^n})$. Fix any $j \in \{1, \ldots, m\}$. Let $Y_j^{\langle n\rangle}(\omega) := \sum_{l=1}^{n} \alpha_{j,l} X_l(\omega)$, $\forall n \in \mathbb{N}$. Then, $Y_j^{\langle n\rangle} \sim N(0, \sum_{l=1}^{n} \alpha_{j,l}^2)$. Thus, $\left(Y_j^{\langle n\rangle}\right)_{n=1}^{\infty}$ is a Martingale and is bounded in $\bar{\mathrm{L}}_1(\boldsymbol{\Omega}, \mathbb{R})$. By Doob's Forward Convergence Theorem 14.41, $\lim_{n\in\mathbb{N}} Y_j^{\langle n\rangle} = Y_j$ a.e. in $\boldsymbol{\Omega}$. This implies that $\boldsymbol{P}(F_j) = 0$. By the arbitrariness of j, we have $\lim_{n\in\mathbb{N}} Y^{\langle n\rangle} = Y$ a.e. in $\boldsymbol{\Omega}$. By Mode of Convergence (Theorem 14.57), we have $Y^{\langle n\rangle} \rightsquigarrow Y$. By Chebyshev Inequality Proposition 14.67, we have that $\left(Y^{\langle n\rangle}\right)_{n=1}^{\infty}$ is uniform growth bounded. By Lebesgue Dominated Convergence Theorem 11.91 and Definition 14.25, we have $\lim_{n\in\mathbb{N}} F_{Y^{\langle n\rangle}}(z) = \bar{F}(z)$, $\forall z \in \mathbb{R}^m$, where $F_{Y^{\langle n\rangle}} : \mathbb{R}^m \to [0, 1] \subset \mathbb{R}$ is the cumulative distribution function of $Y^{\langle n\rangle}$, $\forall n \in \mathbb{N}$, and $\bar{F} : \mathbb{R}^m \to [0, 1] \subset \mathbb{R}$ is the cumulative distribution function of some $\mathbb{R}^m$-valued Gaussian random variable with distribution $N(\mathbf{0}_m, K)$. By Proposition 14.71, $Y \sim N(\mathbf{0}_m, K)$.

This completes the proof of the proposition. □

Example 14.73 Let $I := [0, 1] \subset \mathbb{R}$ and $\mathbb{I} := ((\mathrm{P}(I) = (0, 1], |\cdot|), \mathcal{B}, \boldsymbol{P})$ be the finite metric measure subspace of $\mathbb{R}$, which is a probability measure space. Define a sequence of simple functions $(h_n)_{n=1}^{\infty}$, $h_n : \mathbb{I} \to [-1, 1] \subset \mathbb{R}$ by $h_n(x) = \begin{cases} 1 & 2^{-m}2i < x \le 2^{-m}(2i+1) \\ -1 & 2^{-m}(2i+1) < x \le 2^{-m}(2i+2) \\ 0 & \text{elsewhere} \end{cases}$, $\forall x \in \mathbb{I}$, $\forall n \in \mathbb{N}$, $m := \lceil \log_2 n \rceil \in \mathbb{Z}_+$, $i := \lfloor n - 1 - 2^{m-1} \rfloor \vee 0 \in \mathbb{Z}_+$. This sequence of functions is said to be the *Haar functions*. See Fig. 14.2 for the plot of the first eight of Haar functions. Clearly, $(h_n)_{n=1}^{\infty}$ is an orthogonal sequence in $\bar{\mathcal{Z}} := \bar{\mathrm{L}}_2(\mathbb{I}, \mathbb{R})$. We will further define a sequence of simple functions $(H_n)_{n=1}^{\infty}$ by $H_n := \frac{1}{\|h_n\|_{\bar{\mathcal{Z}}}} h_n$, $\forall n \in \mathbb{N}$, which is then an orthonormal sequence in $\bar{\mathcal{Z}}$. We will show that $([H_n])_{n=1}^{\infty}$ is a complete orthonormal sequence in $\mathcal{Z} := \mathrm{L}_2(\mathbb{I}, \mathbb{R})$.

View each H_n as a random variable, $\forall n \in \mathbb{N}$. Let $\mathcal{B}_m := \sigma((H_n)_{n=1}^{2^m})$, $\forall m \in \mathbb{Z}_+$. We observe that $\mathcal{B}_m = \sigma(\mathcal{H}_m)$, where $\mathcal{H}_0 := \{\emptyset, (0, 1]\}$, and $\mathcal{H}_{m+1} := \mathcal{H}_m \cup \{(0, 2^{-m-1}], (2^{-m-1}, 2 \cdot 2^{-m-1}], \cdots, ((2^{m+1} - 1)2^{-m-1}, 1]\}$, $\forall m \in \mathbb{Z}_+$. It should be clear that $\mathcal{H}_m$ is a π-system, $\forall m \in \mathbb{Z}_+$, by Definition 12.17. We need the following intermediate result:

Claim 14.73.1 $\forall E \in \mathcal{B}_m$, $\exists a_1, \ldots, a_{2^m} \in \mathbb{R}$ such that $\chi_{E,\mathbb{I}} = \sum_{i=1}^{2^m} a_i h_i$.

Proof of Claim We will prove the claim using mathematical induction on $m \in \mathbb{Z}_+$.
1° $m = 0$. This case is straightforward since $\mathcal{B}_0 = \mathcal{H}_0$.
2° Assume that the result holds for $0 \le m \le k \in \mathbb{Z}_+$.
3° Consider the case $m = k + 1 \in \mathbb{N}$. Since $\mathcal{B}_{k+1} = \sigma(\mathcal{H}_{k+1})$, we just need to show the result $\forall E \in \mathcal{H}_{k+1}$. Clearly, $\mathcal{B}_0 \subseteq \mathcal{B}_1 \subseteq \cdots \subseteq \mathcal{B}_{k+1} \subseteq \cdots$, since $\mathcal{H}_0 \subseteq \mathcal{H}_1 \subseteq \cdots \subseteq \mathcal{H}_{k+1} \subseteq \cdots$. Then, if $E \in \mathcal{H}_k$, the result holds by 2°. On the other hand, if $E \in \mathcal{H}_{k+1} \setminus \mathcal{H}_k$, then $E = (2^{-k-1}i, 2^{-k-1}(i+1)]$ for some $i \in \{0, 1, \ldots, 2^{k+1} - 1\}$. Let $j = \lfloor \frac{i}{2} \rfloor \in \{0, 1, \ldots, 2^k - 1\}$ and $\bar{E} := (2^{-k}j, 2^{-k}(j + 1)] \in \mathcal{H}_k$. Then, $\chi_{E,\mathbb{I}} = \frac{1}{2}(\chi_{\bar{E},\mathbb{I}} + (-1)^i h_{2^k+1+j})$. By 2°, $\chi_{\bar{E},\mathbb{I}} = \sum_{i=1}^{2^k} a_i h_i$ for

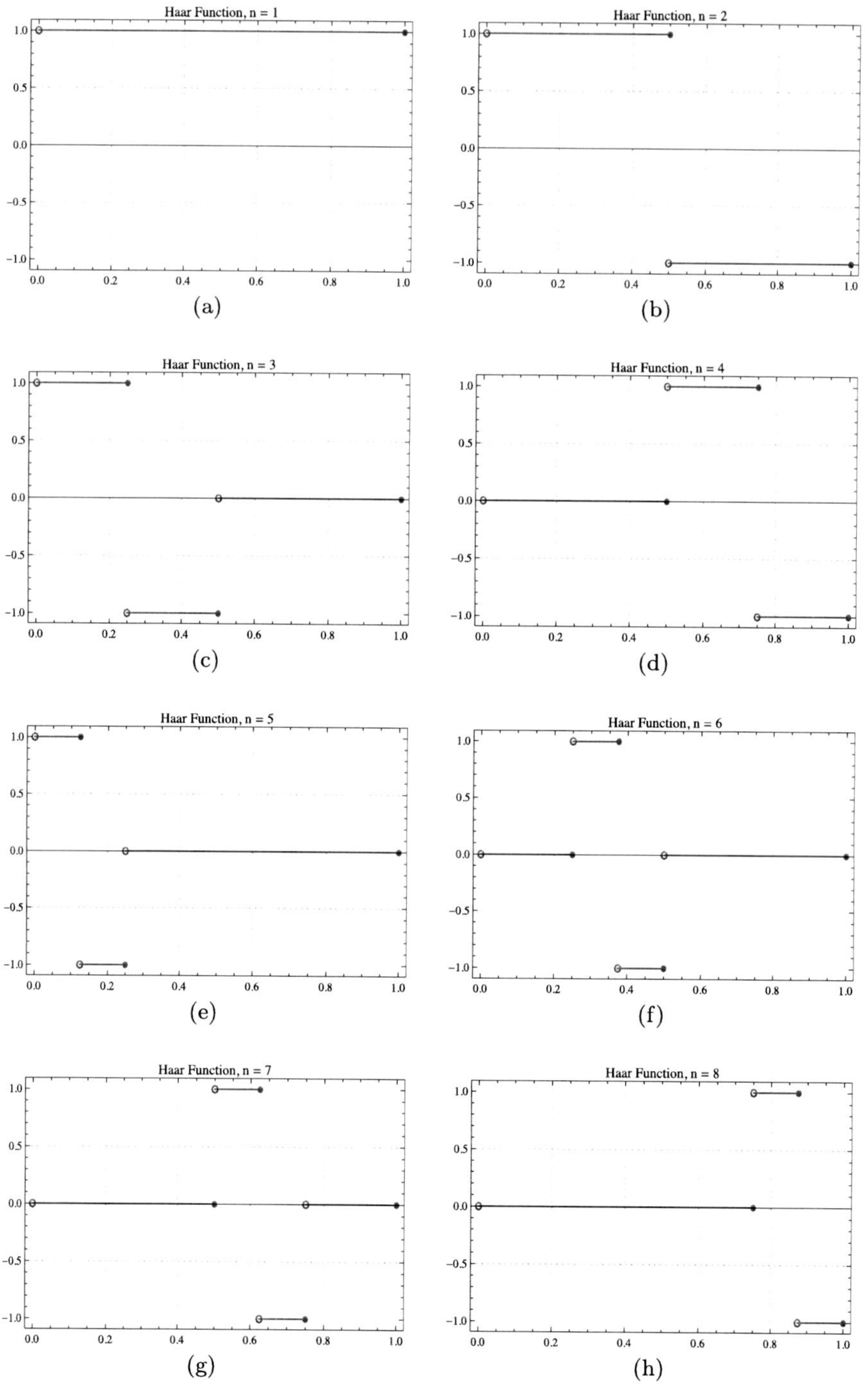

Fig. 14.2 The first eight Haar functions

some $a_1, \ldots, a_{2^k} \in \mathbb{R}$. Then, we have $\chi_{E,\mathbb{I}} = \sum_{i=1}^{2^{k+1}} \bar{a}_i h_i$ for some $\bar{a}_1, \ldots, \bar{a}_{2^{k+1}} \in \mathbb{R}$. This completes the proof for $m = k+1$ and completes the inductive process.

This completes the proof of the claim. □

Fix any $f \in \bar{\mathcal{Z}}$. Fix $m \in \mathbb{Z}_+$. By Theorem 14.70, we have $\forall g_m \in \boldsymbol{E}(f|\mathcal{B}_m)$, $g_m = \arg\min_{g:\mathbb{I}\to\mathbb{R} \text{ being } \mathcal{B}_m\text{-measurable}} \|f-g\|^2_{\bar{\mathcal{Z}}}$. By $\mathcal{B}_m = \sigma(\mathcal{H}_m) = \sigma((h_n)_{n=1}^{2^m})$, we have g_m is a real-valued simple function on the measurable space $((0,1], \mathcal{B}_m)$. By Claim 14.73.1, we have g_m is a linear combination of $(h_n)_{n=1}^{2^m}$. Therefore, $g_m = \arg\min_{g\in\operatorname{span}\left((H_n)_{n=1}^{2^m}\right)} \|f-g\|^2_{\bar{\mathcal{Z}}}$. By Proposition 13.31, we have $g_m = \sum_{i=1}^{2^m} \langle f, H_i\rangle H_i$.

Fix any $\epsilon \in (0,\infty) \subset \mathbb{R}$, by Proposition 11.182, $\exists \bar{g} \in \mathcal{C}(\mathbb{I},\mathbb{R}) \cap \bar{\mathcal{Z}}$ such that $\|f-\bar{g}\|_{\bar{\mathcal{Z}}} < \frac{\epsilon}{2}$. Since $\bar{g} \in \bar{\mathcal{Z}}$, $\exists m_1 \in \mathbb{Z}_+$, we have $\int_0^{2^{-m_1}} \mathcal{P}_2 \circ \bar{g}\, d\boldsymbol{P} < \frac{\epsilon^2}{8}$. $\bar{g}|_{[2^{-m_1},1]}$ is continuous and hence uniformly continuous, by Proposition 5.39. Then, $\exists m_2 \in \mathbb{Z}_+$ with $m_2 \geq m_1$, we have $|\bar{g}(t_1) - \bar{g}(t_2)| < \frac{\epsilon}{4}$, $\forall t_1, t_2 \in [2^{-m_1}, 1] \subset \mathbb{R}$ with $|t_1 - t_2| \leq 2^{-m_2}$. Then, $\forall m \in \mathbb{Z}_+$ with $m \geq m_2$, there exists a function $\bar{g}_m \in \operatorname{span}\left((H_n)_{n=1}^{2^m}\right)$, defined by $\bar{g}_m(t) = \begin{cases} 0 & 0 < t \leq 2^{-m_1} \\ \bar{g}(2^{-m_2} i) & 2^{-m_2} i < t \leq 2^{-m_2}(i+1), i = 2^{m_2-m_1}, \ldots, 2^{m_2}-1 \end{cases}$, $\forall t \in \mathbb{I}$, such that $\|\bar{g}_m - \bar{g}\|_{\bar{\mathcal{Z}}} < \frac{\epsilon}{2}$. (Note that $\bar{g}_m \in \operatorname{span}\left((H_n)_{n=1}^{2^{m_2}}\right)$.) Then, we have $\|f - \bar{g}_m\|_{\bar{\mathcal{Z}}} < \epsilon$. By the previous paragraph, we have $\|f - g_m\|_{\bar{\mathcal{Z}}} \leq \|f - \bar{g}_m\|_{\bar{\mathcal{Z}}} < \epsilon$. Hence, $\lim_{m\in\mathbb{N}} g_m \doteq f$ in $\bar{\mathcal{Z}}$, and $[f] \in \overline{\operatorname{span}\left(([H_n])_{n=1}^{\infty}\right)}$.

By the arbitrariness of f, we have $\mathcal{Z} = \overline{\operatorname{span}\left(([H_n])_{n=1}^{\infty}\right)}$ and $([H_n])_{n=1}^{\infty}$ is a complete orthonormal sequence in $\mathcal{Z}$. ◇

Definition 14.74 Let $T \in \mathbb{R}_+$, $[0,T] \subset \mathbb{R}$, and $\mathcal{I} := (([0,T], |\cdot|), \hat{\mathcal{B}}, \mu)$ be the finite metric measure subspace of $\mathbb{R}$ with partial ordering $\leq$, $\boldsymbol{\Omega} := (\Omega, \mathcal{B}, \boldsymbol{P})$ be a probability measure space, and $\mathcal{I} \times \boldsymbol{\Omega} := ([0,T] \times \Omega, \bar{\mathcal{B}}, \nu)$ be the finite product measure space. A Wiener process on $\mathcal{I} \times \boldsymbol{\Omega}$ is a mapping $w : \mathcal{I} \times \boldsymbol{\Omega} \to \mathbb{R}$ that satisfies

(a) w is $\bar{\mathcal{B}}$-measurable.
(b) $w(0,\omega) = 0$, $\forall \omega \in \Omega$.
(c) $\forall \omega \in \boldsymbol{\Omega}$, $w(\cdot,\omega) : \mathcal{I} \to \mathbb{R}$ is continuous (i.e., the sample paths are continuous).
(d) The random process is of independent increment, i.e., $\forall n \in \mathbb{N}$, $\forall 0 \leq s_1 < t_1 \leq s_2 < t_2 \leq \cdots \leq s_n < t_n \leq T$, we have $w_{t_i} - w_{s_i}$ is a Gaussian random variable with mean 0 and covariance $t_i - s_i$, $i = 1, \ldots, n$, and the random variables $w_{t_i} - w_{s_i}$, $i = 1, \ldots, n$, are independent, where $(w_t)_{t\in\mathcal{I}}$ is defined by $w_t(\omega) = w(t,\omega)$, $\forall \omega \in \boldsymbol{\Omega}$, $\forall t \in \mathcal{I}$. ◇

Example 14.75 Let $\bar{\mathbb{I}} := (([0,1], |\cdot|), \hat{\mathcal{B}}, \hat{\mu})$ be the finite compact metric measure subspace of $\mathbb{R}$ with partial ordering $\leq$, $\mathbb{I} := ((\mathrm{P}(\bar{\mathbb{I}}), |\cdot|), \tilde{\mathcal{B}}, \mu)$ be the finite metric measure subspace of $\mathbb{R}$, $(H_n)_{n=1}^{\infty}$ and $(h_n)_{n=1}^{\infty}$ be the Haar functions as defined in Example 14.73, $\mathcal{Z} := \mathrm{L}_2(\mathbb{I},\mathbb{R})$, and $\bar{\mathcal{Z}} := \bar{\mathrm{L}}_2(\mathbb{I},\mathbb{R})$. Define $(a_n)_{n=1}^{\infty}$ be a sequence of functions: $a_n : \bar{\mathbb{I}} \to \mathbb{R}$ with $a_n(t) := \int_0^t H_n(s)\,ds$, $\forall t \in \bar{\mathbb{I}}$ and $\forall n \in \mathbb{N}$. Clearly, $a_n(t) = \langle [\chi_{(0,t],\mathbb{I}}], [H_n]\rangle_{\mathcal{Z}}$, $\forall t \in \bar{\mathbb{I}}$, $\forall n \in \mathbb{N}$. By Fundamental Theorem on Modeling 14.15 and Example 14.17, there exists a probability measure space $\boldsymbol{\Omega} := (\Omega, \mathcal{B}, \boldsymbol{P})$ and random variables $(X_i)_{i=1}^\infty$, $X_i : \Omega \to \mathbb{R}$, such that $X_i \sim N(0,1)$, and the random variables $(X_i)_{i=1}^\infty$ are independent. Let $\bar{\mathbb{I}} \times \boldsymbol{\Omega} := ([0,1] \times \Omega, \bar{\mathcal{B}}, \nu)$ be the finite product measure space, which happens to be a probability measure space. Define the partial sum $w_n : \bar{\mathbb{I}} \times \boldsymbol{\Omega} \to \mathbb{R}$ by $w_n(t,\omega) = \sum_{i=1}^{2^n} X_i(\omega) a_i(t)$, $\forall t \in \bar{\mathbb{I}}$, $\forall \omega \in \boldsymbol{\Omega}$, $\forall n \in \mathbb{N}$. Clearly, w_n is $\bar{\mathcal{B}}$-measurable, $\forall n \in \mathbb{N}$, by Propositions 11.38 and 11.39, and $w_n(\cdot,\omega) : \bar{\mathbb{I}} \to \mathbb{R}$ is continuous, $\forall \omega \in \boldsymbol{\Omega}$.

Note that $a_k(t) = \frac{1}{\|h_k\|_{\bar{\mathcal{Z}}}} \langle [\chi_{(0,t],\mathbb{I}}], [h_k]\rangle_{\mathcal{Z}}$, $\|h_k\|_{\bar{\mathcal{Z}}} = 2^{\frac{-m+1}{2}}$, $\forall k \in \mathbb{N}$, and $m := \lceil \log_2 k \rceil$, and the function $\bar{a}_k : \bar{\mathbb{I}} \to [0,\infty) \subset \mathbb{R}$ defined by $\bar{a}_k(t) := \langle [\chi_{(0,t],\mathbb{I}}], [h_k]\rangle_{\mathcal{Z}}$, $\forall t \in \bar{\mathbb{I}}$, has the same support as h_k and has maximum absolute value 2^{-m}. Therefore, a_k has the same support as h_k, is nonnegative real-valued, and has maximum $\hat{a}_k = 2^{-m} 2^{-\frac{-m+1}{2}} = 2^{-\frac{m+1}{2}}$, where $m := \lceil \log_2 k \rceil$. Therefore, $w_n(t,\omega) = X_1(\omega)a_1(t) + \sum_{j=1}^n \sum_{i=2^{j-1}+1}^{2^j} X_i(\omega) a_i(t) =: X_1(\omega)a_1(t) + \sum_{j=1}^n B_j(t,\omega)$, $\forall t \in \bar{\mathbb{I}}$, $\forall \omega \in \boldsymbol{\Omega}$, $\forall n \in \mathbb{N}$. $w_n(\cdot,\omega)$ will converge in $\mathcal{C}(\bar{\mathbb{I}}, \mathbb{R})$ if $\sum_{j=1}^\infty \max_{t\in\bar{\mathbb{I}}} |B_j(t,\omega)| < \infty$, by Proposition 7.27 and Example 7.30. Note that $\max_{t\in\bar{\mathbb{I}}} |B_j(t,\omega)| = \max_{t\in\bar{\mathbb{I}}} \left|\sum_{i=2^{j-1}+1}^{2^j} X_i(\omega)a_i(t)\right| = 2^{-\frac{j+1}{2}} \max_{i\in\{2^{j-1}+1,\ldots,2^j\}} |X_i(\omega)| =: b_j M_j(\omega)$, $\forall j \in \mathbb{N}$, $\forall \omega \in \boldsymbol{\Omega}$.

Define $w : \bar{\mathbb{I}} \times \boldsymbol{\Omega} \to \mathbb{R}$ by $w(t,\omega) = \begin{cases} \lim_{n\to\infty} w_n(t,\omega) & \omega \in F \\ 0 & \omega \in \boldsymbol{\Omega} \setminus F \end{cases}$, $\forall t \in \bar{\mathbb{I}}$, $\forall \omega \in \boldsymbol{\Omega}$, where $F := \{\omega \in \boldsymbol{\Omega} \mid \lim_{n\to\infty} \sum_{j=1}^n b_j M_j(\omega) < \infty\}$. We will show that w is a Wiener process on $\bar{\mathbb{I}} \times \boldsymbol{\Omega}$.

First, we show that $\boldsymbol{P}(F) = 1$ by proving $\boldsymbol{E}(\sum_{j=1}^\infty b_j M_j) < \infty$. Note that $M_j(\omega) = \max_{i\in\{2^{j-1}+1,\ldots,2^j\}} |X_i(\omega)| \le \left(\sum_{i=2^{j-1}+1}^{2^j} (X_i(\omega))^4\right)^{1/4} =: Y_j(\omega)$, $\forall \omega \in \Omega$, $\forall j \in \mathbb{N}$. Then, $\boldsymbol{E}(M_j) \le \boldsymbol{E}(Y_j) \le \left(\sum_{i=2^{j-1}+1}^{2^j} \boldsymbol{E}(X_i^4)\right)^{1/4} = (3 \cdot 2^{j-1})^{\frac{1}{4}} = \sqrt[4]{3} \cdot 2^{\frac{j-1}{4}}$, where the second inequality follows from Jensen's Inequality Theorem 11.98. This implies that $\boldsymbol{E}(\sum_{j=1}^\infty b_j M_j) = \sum_{j=1}^\infty b_j \boldsymbol{E}(M_j) \le \sum_{j=1}^\infty 2^{-\frac{j+1}{2}} \cdot \sqrt[4]{3} \cdot 2^{\frac{j-1}{4}} = \sum_{j=1}^\infty \sqrt[4]{3} \cdot 2^{-\frac{j+3}{4}} = \sqrt[4]{24} \cdot \sum_{j=1}^\infty (\sqrt[4]{2})^{-j} < \infty$. Hence, we have $\boldsymbol{P}(F) = 1$ and $w(t,\omega) = \lim_{n\in\mathbb{N}} w_n(t,\omega)$ a.e. $(t,\omega) \in \bar{\mathbb{I}} \times \boldsymbol{\Omega}$. By Propositions 11.48 and 11.41, w is $\bar{\mathcal{B}}$-measurable. Hence, w satisfies (a) of Definition 14.74

We will check the rest of the conditions for w to be a Wiener process on $\bar{\mathbb{I}} \times \boldsymbol{\Omega}$.

(b) Clearly, $w_n(0,\omega) = 0$, $\forall n \in \mathbb{N}$, $\forall \omega \in \boldsymbol{\Omega}$. Then, $w(0,\omega) = \begin{cases} \lim_{n\in\mathbb{N}} w_n(0,\omega) = 0 & \omega \in F \\ 0 & \omega \in \boldsymbol{\Omega} \setminus F \end{cases} = 0$, $\forall \omega \in \boldsymbol{\Omega}$.

(c) Clearly, $w(\cdot,\omega) : \bar{\mathbb{I}} \to \mathbb{R}$ is continuous, $\forall \omega \in \boldsymbol{\Omega}$.

(d) Since the Gaussian random variables $(X_i)_{i=1}^{\infty}$ are independent, then $\forall n \in \mathbb{N}$, $\forall 0 \leq s_1 < t_1 \leq s_2 < t_2 \leq \cdots \leq s_n < t_n \leq 1$, we will prove that $(w_{t_1} - w_{s_1}, \ldots, w_{t_n} - w_{s_n})$ is $\mathbb{R}^n$-valued Gaussian random variable with distribution $N(\mathbf{0}_n, Q)$ by Proposition 14.72, where $Q :=$ block diagonal $(t_1 - s_1, \ldots, t_n - s_n) \in \bar{\mathcal{S}}_{+\mathbb{R}^n}$, and $(w_t)_{t\in\bar{\mathbb{I}}}$ is defined by $w_t(\omega) = w(t, \omega)$, $\forall \omega \in \boldsymbol{\Omega}$, $\forall t \in \bar{\mathbb{I}}$. Clearly, $\boldsymbol{E}(w_{t_j} - w_{s_j}) = \boldsymbol{E}(\sum_{i=1}^{\infty}(a_i(t_j) - a_i(s_j))X_i) = 0$, $j = 1, \ldots, n$. Furthermore,

$$\begin{aligned}
\boldsymbol{E}((w_{t_j} - w_{s_j})^2) &= \boldsymbol{E}\Big(\big(\sum_{i=1}^{\infty}(a_i(t_j) - a_i(s_j))X_i\big)^2\Big)\\
&= \sum_{i=1}^{\infty}(a_i(t_j) - a_i(s_j))^2 = \sum_{i=1}^{\infty}\Big(\langle[\chi_{(s_j,t_j],\mathbb{I}}], [H_i]\rangle_{\mathcal{Z}}\Big)^2\\
&= \Big\langle \sum_{i=1}^{\infty}\langle[\chi_{(s_j,t_j],\mathbb{I}}], [H_i]\rangle_{\mathcal{Z}}[H_i], \sum_{i=1}^{\infty}\langle[\chi_{(s_j,t_j],\mathbb{I}}], [H_i]\rangle_{\mathcal{Z}}[H_i]\Big\rangle_{\mathcal{Z}}\\
&= \langle[\chi_{(s_j,t_j],\mathbb{I}}], [\chi_{(s_j,t_j],\mathbb{I}}]\rangle_{\mathcal{Z}} = \|\chi_{(s_j,t_j],\mathbb{I}}\|_{\bar{\mathcal{Z}}}^2 = t_j - s_j
\end{aligned}$$

where the second equality follows from the independence of the random variables $(X_i)_{i=1}^{\infty}$ and the fourth and fifth equalities follow from the fact that $([H_i])_{i=1}^{\infty}$ is a complete orthonormal sequence in $\mathcal{Z}$. It is easy to check that

$$\begin{aligned}
&\boldsymbol{E}((w_{t_k} - w_{s_k})(w_{t_l} - w_{s_l}))\\
&= \boldsymbol{E}\Big(\big(\sum_{i=1}^{\infty}X_i(a_i(t_k) - a_i(s_k))\big)\big(\sum_{j=1}^{\infty}X_j(a_j(t_l) - a_j(s_l))\big)\Big)\\
&= \sum_{i=1}^{\infty}\boldsymbol{E}(X_i^2(a_i(t_k) - a_i(s_k))(a_i(t_l) - a_i(s_l)))\\
&= \sum_{i=1}^{\infty}(a_i(t_k) - a_i(s_k))(a_i(t_l) - a_i(s_l))\\
&= \sum_{i=1}^{\infty}\langle[\chi_{(s_k,t_k],\mathbb{I}}], [H_i]\rangle_{\mathcal{Z}} \cdot \langle[\chi_{(s_l,t_l],\mathbb{I}}], [H_i]\rangle_{\mathcal{Z}}\\
&= \Big\langle \sum_{i=1}^{\infty}\langle[\chi_{(s_k,t_k],\mathbb{I}}], [H_i]\rangle_{\mathcal{Z}}[H_i], \sum_{i=1}^{\infty}\langle[\chi_{(s_l,t_l],\mathbb{I}}], [H_i]\rangle_{\mathcal{Z}}[H_i]\Big\rangle_{\mathcal{Z}}\\
&= \langle[\chi_{(s_k,t_k],\mathbb{I}}], [\chi_{(s_l,t_l],\mathbb{I}}]\rangle_{\mathcal{Z}} = 0, \qquad \forall 1 \leq k, l \leq n \text{ with } l \neq k
\end{aligned}$$

where the second equality follows from the independence of the random variables $(X_i)_{i=1}^{\infty}$, the fourth equality follows from the definition of a_k's, and the fifth and sixth equalities follow from the fact that $([H_i])_{i=1}^{\infty}$ is a complete orthonormal sequence in $\mathcal{Z}$. Hence, $(w_{t_1} - w_{s_1}, \ldots, w_{t_n} - w_{s_n})$ is $\mathbb{R}^n$-valued Gaussian random variable with distribution $N(\mathbf{0}_n, Q)$. Then, $w_{t_1} - w_{s_1}, \ldots, w_{t_n} - w_{s_n}$ are independent by Proposition 14.27.

This completes the proof of existence of the Wiener process on $\bar{\mathbb{I}} \times \boldsymbol{\Omega}$. ◇

Definition 14.76 Let $\mathbb{I} := (([0, \infty), |\cdot|), \hat{\mathcal{B}}, \mu)$ be the σ-finite metric measure subspace of $\mathbb{R}$ with partial ordering $\leq$, $\boldsymbol{\Omega} := (\Omega, \mathcal{B}, \boldsymbol{P})$ be a probability measure space, and $\mathbb{I} \times \boldsymbol{\Omega} := ([0, \infty) \times \Omega, \bar{\mathcal{B}}, \nu)$ be the σ-finite product measure space. A *standard Wiener process on* $\mathbb{I} \times \boldsymbol{\Omega}$ is a mapping $w : \mathbb{I} \times \boldsymbol{\Omega} \to \mathbb{R}$ that satisfies

(a) w is $\bar{\mathcal{B}}$-measurable.
(b) $w(0, \omega) = 0, \forall \omega \in \Omega$.
(c) $\forall \omega \in \boldsymbol{\Omega}$, $w(\cdot, \omega) : \mathbb{I} \to \mathbb{R}$ is continuous (i.e., the sample paths are continuous).
(d) The random process is of independent increment, i.e., $\forall n \in \mathbb{N}$, $\forall 0 \leq s_1 < t_1 \leq s_2 < t_2 \leq \cdots \leq s_n < t_n < \infty$, we have $w_{t_i} - w_{s_i}$ is a Gaussian random variable with mean 0 and covariance $t_i - s_i$, $i = 1, \ldots, n$, and the random variables $w_{t_i} - w_{s_i}$, $i = 1, \ldots, n$, are independent, where $(w_t)_{t \in \mathbb{I}}$ is defined by $w_t(\omega) = w(t, \omega)$, $\forall \omega \in \boldsymbol{\Omega}$, $\forall t \in \mathbb{I}$. ◇

Example 14.77 Let $\hat{\mathbb{I}} := (([0, 1], |\cdot|), \hat{\mathcal{B}}, \hat{\mu})$ be the finite compact metric measure subspace of $\mathbb{R}$ with partial ordering $\leq$. By Example 14.75, there exist probability measure spaces $\boldsymbol{\Omega}_n := (\Omega_n, \mathcal{B}_n, \boldsymbol{P}_n)$, $\forall n \in \mathbb{N}$, each with a Wiener process $w_n : \hat{\mathbb{I}} \times \boldsymbol{\Omega}_n \to \mathbb{R}$. By Fundamental Theorem on Modeling 14.15, there exists a unique probability measure space $\boldsymbol{\Omega} := (\Omega, \mathcal{B}, \boldsymbol{P}) = (\prod_{n \in \mathbb{N}} \Omega_n, \mathcal{B}, \boldsymbol{P} := \prod_{n \in \mathbb{N}} \boldsymbol{P}_n)$, on which $\hat{\mathcal{B}}_n := \{\prod_{i \in \mathbb{N}} B_i \in \mathcal{B} \mid B_i \in \mathcal{B}_i, \forall i \in \mathbb{N}$; and $B_i = \Omega_i, \forall i \in \mathbb{N}$ with $i \neq n\} \subseteq \mathcal{B}$, $n \in \mathbb{N}$ are independent σ-algebras. Let $\mathbb{I} := (([0, \infty), |\cdot|), \tilde{\mathcal{B}}, \tilde{\mu})$ be the σ-finite metric measure subspace of $\mathbb{R}$ with partial ordering $\leq$ and $\mathbb{I} \times \boldsymbol{\Omega} := ([0, \infty) \times \boldsymbol{\Omega}, \bar{\mathcal{B}}, \nu)$ be the σ-finite product measure space. Define $w : \mathbb{I} \times \boldsymbol{\Omega} \to \mathbb{R}$ by $w(t, \omega) = \sum_{j=1}^{i-1} w_j(1, \omega) + w_i(t - i + 1, \omega)$, $\forall t \in \mathbb{I}$, and $i := \lfloor t \rfloor + 1 \in \mathbb{N}$, $\forall \omega \in \boldsymbol{\Omega}$. It is easy to check that w is a standard Wiener process on $\mathbb{I} \times \boldsymbol{\Omega}$. ◇

Definition 14.78 Let $\mathbb{I} := (([0, \infty), |\cdot|), \hat{\mathcal{B}}, \mu)$ be the σ-finite metric measure subspace of $\mathbb{R}$, $\boldsymbol{\Omega} := (\Omega, \mathcal{B}, \boldsymbol{P})$ be a probability measure space, $m \in \mathbb{N}$, and $\mathbb{I} \times \boldsymbol{\Omega} := ([0, \infty) \times \Omega, \bar{\mathcal{B}}, \nu)$ be the σ-finite product measure space. An $\mathbb{R}^m$*-valued standard Wiener process on* $\mathbb{I} \times \boldsymbol{\Omega}$ is a mapping $w : \mathbb{I} \times \boldsymbol{\Omega} \to \mathbb{R}^m$ that satisfies

(a) w is $\bar{\mathcal{B}}$-measurable.
(b) $w(0, \omega) = \mathbf{0}_m, \forall \omega \in \Omega$.
(c) $\forall \omega \in \boldsymbol{\Omega}$, $w(\cdot, \omega) : \mathbb{I} \to \mathbb{R}^m$ is continuous (i.e., the sample paths are continuous).

(d) The random process is of independent increment, i.e., $\forall n \in \mathbb{N}$, $\forall 0 \leq s_1 < t_1 \leq s_2 < t_2 \leq \cdots \leq s_n < t_n < \infty$, we have $w_{t_i} - w_{s_i}$ is a $\mathbb{R}^m$-valued Gaussian random vector with mean $\mathbf{0}_m$ and covariance $(t_i - s_i)\,\mathrm{id}_{\mathbb{R}^m}$, $i = 1, \ldots, n$, and the random vectors $w_{t_i} - w_{s_i}$, $i = 1, \ldots, n$, are independent, where $(w_t)_{t \in \mathbb{I}}$ is defined by $w_t(\omega) = w(t, \omega)$, $\forall \omega \in \boldsymbol{\Omega}$, $\forall t \in \mathbb{I}$. ◇

Example 14.79 Let $\mathbb{I} := (([0, \infty), |\cdot|), \hat{\mathcal{B}}, \mu)$ be the σ-finite metric measure subspace of $\mathbb{R}$ with partial ordering $\leq$, $m \in \mathbb{N}$, and $\bar{J} := \{1, \ldots, m\}$. By Example 14.77, there exist probability measure spaces $\boldsymbol{\Omega}_i := (\Omega_i, \mathcal{B}_i, \boldsymbol{P}_i)$, $\forall i \in \bar{J}$, each with a standard Wiener process w_i defined on $\mathbb{I} \times \boldsymbol{\Omega}_i$. By Fundamental Theorem on Modeling 14.15, there exists a unique probability measure space $\boldsymbol{\Omega} := (\Omega, \mathcal{B}, \boldsymbol{P}) = (\prod_{i=1}^{m} \Omega_i, \mathcal{B}, \boldsymbol{P} := \prod_{i=1}^{m} \boldsymbol{P}_i)$, on which $\hat{\mathcal{B}}_i := \{\prod_{j=1}^{m} B_j \in \mathcal{B} \mid B_j \in \mathcal{B}_j, \forall j \in \bar{J};$ and $B_j = \Omega_j, \forall j \in \bar{J}$ with $j \neq i\} \subseteq \mathcal{B}$, $i \in \bar{J}$, are independent σ-algebras. Let $\mathbb{I} \times \boldsymbol{\Omega} := (\mathbb{I} \times \Omega, \bar{\mathcal{B}}, \nu)$ be the σ-finite product measure space, and define $w : \mathbb{I} \times \boldsymbol{\Omega} \to \mathbb{R}^m$ by $w(t, \omega) = (w_1(t, \omega), \ldots, w_m(t, \omega)) \in \mathbb{R}^m$, $\forall t \in \mathbb{I}$, $\forall \omega \in \boldsymbol{\Omega}$. It is easy to check that w is an $\mathbb{R}^m$-valued standard Wiener process on $\mathbb{I} \times \boldsymbol{\Omega}$. ◇

Proposition 14.80 *Let $m \in \mathbb{N}$, $\mathbb{I} := (([0, \infty), |\cdot|), \hat{\mathcal{B}}, \mu)$ be the σ-finite metric measure subspace of $\mathbb{R}$ with partial ordering $\leq$, $\boldsymbol{\Omega} := (\Omega, \mathcal{B}, \boldsymbol{P})$ be a probability measure space, $\mathbb{I} \times \boldsymbol{\Omega} := ([0, \infty) \times \Omega, \bar{\mathcal{B}}, \nu)$ be the σ-finite product measure space, $w : \mathbb{I} \times \boldsymbol{\Omega} \to \mathbb{R}^m$ be an $\mathbb{R}^m$-valued standard Wiener process (as defined in Definition 14.78), and $T \in \mathrm{B}(\mathbb{R}^m, \mathbb{R}^m)$ is invertible. Then, $\tilde{w} := Tw : \mathbb{I} \times \boldsymbol{\Omega} \to \mathbb{R}^m$ is an $\mathbb{R}^m$-valued stochastic process on $\mathbb{I} \times \boldsymbol{\Omega}$ satisfying*

(i) *$\tilde{w}$ is $\bar{\mathcal{B}}$-measurable.*
(ii) *$\tilde{w}(0, \omega) = \boldsymbol{0}_m$, $\forall \omega \in \Omega$.*
(iii) *$\forall \omega \in \boldsymbol{\Omega}$, $\tilde{w}(\cdot, \omega) : \mathbb{I} \to \mathbb{R}^m$ is continuous (i.e., the sample paths are continuous).*
(iv) *The $\mathbb{R}^m$-valued random variable $\tilde{w}_t : \boldsymbol{\Omega} \to \mathbb{R}^m$, defined by $\tilde{w}_t(\omega) = \tilde{w}(t, \omega) = Tw(t, \omega)$, $\forall \omega \in \boldsymbol{\Omega}$, $\forall t \in \mathbb{I}$, is of independent increment: i.e., $\forall 0 \leq s_1 < t_1 \leq s_2 < t_2 \leq \cdots \leq s_n < t_n < \infty$, $\tilde{w}_{t_1} - \tilde{w}_{s_1}, \ldots, \tilde{w}_{t_n} - \tilde{w}_{s_n}$ are independent Gaussian random variables with mean $\boldsymbol{0}_m$ and covariance $(t_i - s_i)TT'$, $i = 1, \ldots, n$.*
(v) *The auto-covariance function $\boldsymbol{E}(\tilde{w}_{t_1} \otimes \tilde{w}_{t_2}) = \min\{t_1, t_2\}TT'$, $\forall t_1, t_2 \in \mathbb{I}$.*
(vi) *If in addition, T is a unitary matrix, i.e., $TT' = \mathrm{id}_{\mathbb{R}^m}$, then $\tilde{w}$ is an $\mathbb{R}^m$-valued standard Wiener process on $\mathbb{I} \times \boldsymbol{\Omega}$.*

Proof (i) This is immediate from Definition 14.78 and Proposition 11.38.
(ii) $\tilde{w}(0, \omega) = Tw(0, \omega) = T\mathbf{0}_m = \mathbf{0}_m$, $\forall \omega \in \boldsymbol{\Omega}$.
(iii) This is immediate from Definition 14.78 and Proposition 3.12.

(iv) By w being a $\mathbb{R}^m$-valued standard Wiener process, $\forall 0 \leq s_1 < t_1 \leq s_2 < t_2 \leq \cdots \leq s_n < t_n < \infty$, $w_{t_1} - w_{s_1}, \ldots, w_{t_n} - w_{s_n}$ are independent $\mathbb{R}^m$-valued Gaussian random variables with mean $\mathbf{0}_m$ and covariance $(t_i - s_i)\,\mathrm{id}_{\mathbb{R}^m}$. Then, $W := (w_{t_1} - w_{s_1}, \ldots, w_{t_n} - w_{s_n}) \sim N(\mathbf{0}_{mn}, Q)$, where $Q :=$ block diagonal $((t_1 - s_1)\,\mathrm{id}_{\mathbb{R}^m}, \ldots, (t_n - s_n)\,\mathrm{id}_{\mathbb{R}^m})$. Then, we have $\tilde{W} := (\tilde{w}_{t_1} - \tilde{w}_{s_1}, \ldots, \tilde{w}_{t_n} - \tilde{w}_{s_n}) = (T(w_{t_1} - w_{s_1}), \ldots, T(w_{t_n} - w_{s_n})) =$ block diagonal $(T, \ldots, T)\, W =: T_n W$. By the invertibility of T, then T_n is invertible. By Proposition 14.28, we have $\tilde{W} \sim N(\mathbf{0}_{mn}, T_n Q T_n' =: \tilde{Q})$. Clearly, $\tilde{Q} =$ block diagonal $\left((t_1 - s_1)TT', \ldots, (t_n - s_n)TT'\right)$. Hence, $\tilde{w}_{t_1} - \tilde{w}_{s_1}, \ldots, \tilde{w}_{t_n} - \tilde{w}_{s_n}$ are independent $\mathbb{R}^m$-valued Gaussian random variables with mean $\mathbf{0}_m$ and covariance $(t_i - s_i)TT'$, $i = 1, \ldots, n$, by Proposition 14.27.

(v) Without loss of generality, assume $0 \leq t_1 \leq t_2$. The result is trivial if $t_1 = 0$. If $t_1 > 0$, then $\boldsymbol{E}(\tilde{w}_{t_1} \otimes \tilde{w}_{t_2}) = \boldsymbol{E}(\tilde{w}_{t_1} \otimes (\tilde{w}_{t_2} - \tilde{w}_{t_1})) + \boldsymbol{E}(\tilde{w}_{t_1} \otimes \tilde{w}_{t_1}) = \mathbf{0}_{m\times m} + t_1 TT' = \min\{t_1, t_2\}TT'$, where the first equality follows from Proposition 11.92, and the second equality follows from (iv).

(vi) The result follows directly from (i)–(iv) and Definition 14.78.

This completes the proof of the proposition. □

Proposition 14.81 *Let $m \in \mathbb{N}$, $\mathbb{I} := (([0, \infty), |\cdot|), \hat{\mathcal{B}}, \mu)$ be the σ-finite metric measure subspace of $\mathbb{R}$ with partial ordering $\leq$, $\boldsymbol{\Omega} := (\Omega, \mathcal{B}, \boldsymbol{P})$ be a probability measure space, $\mathbb{I} \times \boldsymbol{\Omega} := ([0, \infty) \times \Omega, \bar{\mathcal{B}}, \nu)$ be the σ-finite product measure space, $w : \mathbb{I} \times \boldsymbol{\Omega} \to \mathbb{R}^m$ be an $\mathbb{R}^m$-valued standard Wiener process (as defined in Definition 14.78), and $\alpha \in \mathbb{R}$ with $\alpha \neq 0$. Then, $\tilde{w} : \mathbb{I} \times \boldsymbol{\Omega} \to \mathbb{R}^m$, defined by $\tilde{w}(t, \omega) = \frac{1}{\alpha} w(\alpha^2 t, \omega)$, $\forall \omega \in \boldsymbol{\Omega}$, $\forall t \in \mathbb{I}$, is an $\mathbb{R}^m$-valued standard Wiener process.*

Proof This is immediate from Proposition 14.80 and Definition 14.78. □

Proposition 14.82 *Let $\mathbb{I} := (([0, \infty), |\cdot|), \hat{\mathcal{B}}, \mu)$ be the σ-finite metric measure subspace of $\mathbb{R}$ with partial ordering $\leq$, $\boldsymbol{\Omega} := (\Omega, \mathcal{B}, \boldsymbol{P})$ be a probability measure space, $\mathbb{I} \times \boldsymbol{\Omega} := ([0, \infty) \times \Omega, \bar{\mathcal{B}}, \nu)$ be the σ-finite product measure space, $w : \mathbb{I} \times \boldsymbol{\Omega} \to \mathbb{R}$ be a standard Wiener process (as defined in Definition 14.76). Then, there exists $F \in \mathcal{B}$ with $\boldsymbol{P}(F) = 1$ such that $\forall \omega \in F$, $\limsup_{h \to 0^+} \frac{1}{h} |w(t + h, \omega)) - w(t, \omega)| = \infty$, $\forall t \in \mathbb{I}$.*

Proof Fix any $m \in \mathbb{Z}_+$, and we consider the interval $t \in [m, m + 1)$. Let $F_m := \{\omega \in \Omega \mid \limsup_{h \to 0^+} \frac{1}{h} |w(t + h, \omega) - w(t, \omega)| = \infty, \forall t \in [m, m + 1)\}$. Note that F_m may not be $\mathcal{B}$-measurable. Define $\bar{F} := \bigcap_{m=0}^{\infty} F_m$. Then, $\forall \omega \in \bar{F}$, $\limsup_{h \to 0^+} \frac{1}{h} |w(t + h, \omega)) - w(t, \omega)| = \infty$, $\forall t \in \mathbb{I}$.

Let $A_{m,j,k} := \bigcup_{m \leq t < m+1} \bigcap_{0 < h \leq \frac{1}{k}} \{\omega \in \Omega \mid \frac{1}{h} |w(t + h, \omega) - w(t, \omega)| \leq j\}$, $\forall j, k \in \mathbb{N}$. Clearly, $A_{m,j,k} \subseteq \Omega \setminus F_m$, $\forall j, k \in \mathbb{N}$, and $F_m \subseteq \bigcap_{j,k \in \mathbb{N}} (\Omega \setminus A_{m,j,k})$.

$\forall \omega \in \bigcap_{j,k \in \mathbb{N}} (\Omega \setminus A_{m,j,k})$, we have $\forall j \in \mathbb{N}$, $\forall k \in \mathbb{N}$, $\forall t \in [m, m + 1)$, $\exists h \in (0, \frac{1}{k}]$, such that $\frac{1}{h} |w(t + h, \omega) - w(t, \omega)| > j$. Then, $\omega \in F_m$. Hence, $\bigcap_{j,k \in \mathbb{N}} (\Omega \setminus A_{m,j,k}) \subseteq F_m$. Thus, $F_m = \bigcap_{j,k \in \mathbb{N}} (\Omega \setminus A_{m,j,k})$. Note that $A_{m,j,k}$'s may not be $\mathcal{B}$-measurable.

Let $C_{m,j,k} := \bigcap_{n=4k}^{\infty} \bigcup_{i=1}^{n} \bigcap_{v=1}^{3} \{\omega \in \Omega \mid |w(m + \frac{i+v}{n}, \omega) - w(m + \frac{i+v-1}{n}, \omega)| \le \frac{7j}{n}\} \in \mathcal{B}, \forall j, k \in \mathbb{N}$.

We will show that $A_{m,j,k} \subseteq C_{m,j,k}, \forall j, k \in \mathbb{N}$, and $P(C_{m,j,k}) = 0, \forall j, k \in \mathbb{N}$. This yields that $F_m \supseteq \bigcap_{j,k\in\mathbb{N}}(\Omega \setminus C_{m,j,k})$, $\forall m \in \mathbb{Z}_+$, and $\bar{F} \supseteq F := \bigcap_{m=0}^{\infty} \bigcap_{j,k\in\mathbb{N}}(\Omega \setminus C_{m,j,k}) \in \mathcal{B}$, and $P(F) = 1$. This set F is what we seek.

Fix any $j, k \in \mathbb{N}$. Fix any $\omega \in A_{m,j,k}$. Then, $\exists t \in [m, m+1)$ such that $\omega \in \bigcap_{0<h\le\frac{1}{k}} \{\omega \in \Omega \mid \frac{1}{h}|w(t+h, \omega) - w(t, \omega)| \le j\}$. Fix any $n \ge 4k$. We have $\frac{1}{n} \le \frac{1}{4k}$. This implies that $\exists i \in \{1, \ldots, n\}$ such that $m + \frac{i-1}{n} \le t < m + \frac{i}{n}$. Fix any $v \in \{1, 2, 3\}$. Then, we have $m + \frac{i+v}{n} \le t + \frac{v+1}{n} < m + \frac{i+v+1}{n}$, $0 < m + \frac{i+v-1}{n} - t \le \frac{v}{n} < \frac{4}{n} = \frac{1}{k}$, and $0 < m + \frac{i+v}{n} - t \le \frac{v+1}{n} \le \frac{4}{n} = \frac{1}{k}$. This leads to $|w(m + \frac{i+v}{n}, \omega) - w(m + \frac{i+v-1}{n}, \omega)| \le |w(m + \frac{i+v}{n}, \omega) - w(t, \omega)| + |w(t, \omega) - w(m + \frac{i+v-1}{n}, \omega)| \le (m + \frac{i+v}{n} - t)j + (m + \frac{i+v-1}{n} - t)j \le \frac{4}{n}j + \frac{3}{n}j = \frac{7j}{n}$. Hence, $\omega \in C_{m,j,k}$. Thus, $A_{m,j,k} \subseteq C_{m,j,k}, \forall j, k \in \mathbb{N}$.

Now, $P(\{\omega \in \Omega \mid |w(m + \frac{i+v}{n}, \omega) - w(m + \frac{i+v-1}{n}, \omega)| \le \frac{7j}{n}\}) = \int_{-\frac{7j}{n}}^{\frac{7j}{n}} \sqrt{\frac{n}{2\pi}} \cdot \exp(-\frac{nx^2}{2})\,dx \le \frac{14j}{\sqrt{2\pi n}}$, where the equality follows from the fact $w_{m+\frac{i+v}{n}} - w_{m+\frac{i+v-1}{n}} \sim N(0, \frac{1}{n})$. Then, $P(\bigcap_{v=1}^{3} \{\omega \in \Omega \mid |w(m + \frac{i+v}{n}, \omega) - w(m + \frac{i+v-1}{n}, \omega)| \le \frac{7j}{n}\}) \le (\frac{14j}{\sqrt{2\pi n}})^3$, where the inequality follows from the fact that the three events in the intersection are independent. This implies that

$$P(\bigcup_{i=1}^{n} \bigcap_{v=1}^{3} \{\omega \in \Omega \mid |w(m + \frac{i+v}{n}, \omega) - w(m + \frac{i+v-1}{n}, \omega)| \le \frac{7j}{n}\})$$
$$\le n(\frac{14j}{\sqrt{2\pi n}})^3 = \frac{1}{\sqrt{n}}(\frac{14j}{\sqrt{2\pi}})^3 \to 0 \quad \text{as } n \to \infty$$

Thus, we have $0 \le P(C_{m,j,k}) \le \min_{n\ge4k} \frac{1}{\sqrt{n}}(\frac{14j}{\sqrt{2\pi}})^3 = 0, \forall j, k \in \mathbb{N}$.

This completes the proof of the proposition. □

This proposition shows that the Wiener process is nowhere differentiable on $[0, \infty)$ with probability 1.

14.11 Martingales with General Index Set

Definition 14.83 Let $\mathbb{X} := (\mathcal{X}, \mathcal{B}_B(\mathcal{X}), \mu)$ be a nonempty σ-finite topological measure space with the partial ordering $\preceq$ defined on it, $\boldsymbol{\Omega} := (\Omega, \mathcal{B}, (\mathcal{B}_t)_{t\in\mathbb{X}}, P)$ be a filtered probability measure space, $\mathbb{X} \times \boldsymbol{\Omega} := (\mathcal{X} \times \Omega, \bar{\mathcal{B}}, \nu)$ be the σ-finite product measure space, $\mathcal{Y}$ be a Banach space, $M : \mathbb{X} \times \boldsymbol{\Omega} \to \mathcal{Y}$, or equivalently

$(M_t)_{t\in\mathbb{X}}$, be a $\mathcal{Y}$-valued stochastic process adapted to $(\mathcal{B}_t)_{t\in\mathbb{X}}$. Then, the stochastic process $(M_t)_{t\in\mathbb{X}}$ is said to be a *Martingale* if

(i) $M_t \in \bar{\mathrm{L}}_1(\boldsymbol{\Omega}, \mathcal{Y}_t)$, $\forall t \in \mathbb{X}$, where $\mathcal{Y}_t$ is a separable Banach subspace of $\mathcal{Y}$.
(ii) $M_s \in \boldsymbol{E}(M_t|\mathcal{B}_s)$, $\forall s, t \in \mathbb{X}$ with $s \preceq t$.

Let $\mathcal{Y} = \mathbb{R}$. The stochastic process $(M_t)_{t\in\mathbb{X}}$ is said to be a *super Martingale* if it satisfies (i) and $M_s \geq \boldsymbol{E}(M_t|\mathcal{B}_s)$ a.e. in $\boldsymbol{\Omega}$, $\forall s, t \in \mathbb{X}$ with $s \preceq t$.

Let $\mathcal{Y} = \mathbb{R}$. The stochastic process $(M_t)_{t\in\mathbb{X}}$ is said to be a *sub Martingale* if it satisfies (i) and $M_s \leq \boldsymbol{E}(M_t|\mathcal{B}_s)$ a.e. in $\boldsymbol{\Omega}$, $\forall s, t \in \mathbb{X}$ with $s \preceq t$. ◇

Proposition 14.84 *Let $m \in \mathbb{N}$, $\mathbb{I} := (([0,\infty), |\cdot|), \hat{\mathcal{B}}, \mu)$ be the σ-finite metric measure subspace of $\mathbb{R}$ with the partial ordering $\leq$, $\boldsymbol{\Omega} := (\Omega, \mathcal{B}, (\mathcal{B}_t)_{t\in\mathbb{I}}, \boldsymbol{P})$ be a filtered probability measure space, $\alpha \in \mathbb{R}^m$, $\mathbb{I} \times \boldsymbol{\Omega} := ([0,\infty) \times \Omega, \bar{\mathcal{B}}, \nu)$ be the σ-finite product measure space, $w : \mathbb{I} \times \boldsymbol{\Omega} \to \mathbb{R}^m$ be an $\mathbb{R}^m$-valued standard Wiener process (as defined in Definition 14.78), the process $(w_t)_{t\in\mathbb{I}}$, which is defined as $w_t(\omega) = w(t,\omega)$, $\forall \omega \in \Omega$, $\forall t \in \mathbb{I}$, is adapted to the filtration $(\mathcal{B}_t)_{t\in\mathbb{I}}$, and $M : \mathbb{I} \times \boldsymbol{\Omega} \to \mathbb{R}$, defined by $M(t,\omega) := \exp(\langle w(t,\omega), \alpha\rangle_{\mathbb{R}^m} - \frac{\langle\alpha,\alpha\rangle_{\mathbb{R}^m}}{2} t)$, $\forall \omega \in \boldsymbol{\Omega}$, $\forall t \in \mathbb{I}$. Assume that $\mathcal{B}_t$ and $\sigma((w_{t+s} - w_t)_{s\in\mathbb{I}})$ are independent σ-algebras, $\forall t \in \mathbb{I}$. Then, $(M_t)_{t\in\mathbb{I}}$, defined by $M_t(\omega) := M(t,\omega)$, $\forall \omega \in \Omega$, $\forall t \in \mathbb{I}$, is a Martingale adapted to $(\mathcal{B}_t)_{t\in\mathbb{I}}$.*

Proof By Propositions 11.38 and 11.39, M is $\bar{\mathcal{B}}$-measurable. By Proposition 11.38, M_t is $\mathcal{B}_t$-measurable, $\forall t \in \mathbb{I}$. Hence, $(M_t)_{t\in\mathbb{I}}$ is a stochastic process adapted to the filtration $(\mathcal{B}_t)_{t\in\mathbb{I}}$.

Clearly, $M_0(\omega) = 1$, $\forall \omega \in \boldsymbol{\Omega}$. Then, $\boldsymbol{E}(M_0) = 1$. $\forall t \in \mathbb{I}$ with $t > 0$, we have $w_t \sim N(\boldsymbol{0}_m, t\,\mathrm{id}_{\mathbb{R}^m})$. Then,

$$\begin{aligned}
\boldsymbol{E}(M_t) &= \boldsymbol{E}(\exp(\langle w_t, \alpha\rangle_{\mathbb{R}^m} - \frac{\langle\alpha,\alpha\rangle_{\mathbb{R}^m}}{2} t) \\
&= \boldsymbol{E}(\exp(\langle w_t, \alpha\rangle_{\mathbb{R}^m})) \exp(-\frac{\langle\alpha,\alpha\rangle_{\mathbb{R}^m}}{2} t) \\
&= \exp(-\frac{\langle\alpha,\alpha\rangle_{\mathbb{R}^m}}{2} t) \int_{\mathbb{R}^m} \frac{1}{\sqrt{2\pi t}^m} \exp(\langle x, \alpha\rangle_{\mathbb{R}^m} - \frac{1}{2t}\langle x, x\rangle_{\mathbb{R}^m})\, \mathrm{d}\mu_{\mathrm{B}m}(x) \\
&= \int_{\mathbb{R}^m} \frac{1}{\sqrt{2\pi t}^m} \exp(-\frac{1}{2t}\langle x - t\alpha, x - t\alpha\rangle_{\mathbb{R}^m})\, \mathrm{d}\mu_{\mathrm{B}m}(x) \\
&= \int_{\mathbb{R}^m} \frac{1}{\sqrt{2\pi t}^m} \exp(-\frac{1}{2t}\langle \bar{x}, \bar{x}\rangle_{\mathbb{R}^m})\, \mathrm{d}\mu_{\mathrm{B}m}(\bar{x}) = 1
\end{aligned}$$

where the second equality follows from Proposition 11.92, the third equality follows from the Definitions 14.16 and 14.25, and the fourth equality follows from simple algebra, the fifth equality follows from Change of Variable, Theorem 12.91, with $\bar{x} := x - t\alpha$, and the last equality is well-known. Hence, $M_t \in \bar{\mathrm{L}}_1(\boldsymbol{\Omega}, \mathbb{R})$, $\forall t \in \mathbb{I}$.

Fix any $s, t \in \mathbb{I}$ with $s \leq t$. If $t = s$, then $M_s \in \boldsymbol{E}(M_t|\mathcal{B}_s)$. Consider the case $t > s$, and we have $\boldsymbol{E}(M_t|\mathcal{B}_s) = \boldsymbol{E}(M_s \exp(\langle (w_t - w_s), \alpha\rangle_{\mathbb{R}^m} - \frac{\langle\alpha,\alpha\rangle_{\mathbb{R}^m}}{2}(t-s))|\mathcal{B}_s) =$

$M_s \boldsymbol{E}(\exp(\langle(w_t - w_s), \alpha\rangle - \frac{\langle\alpha,\alpha\rangle_{\mathbb{R}^m}}{2}(t-s))|\mathcal{B}_s) = M_s$ a.e. in $\boldsymbol{\Omega}_s$, where the second equality follows from (h) of Proposition 14.11, the last equality follows from the independence of $\mathcal{B}_s$ and $\sigma(w_t - w_s)$, the derivation of the previous paragraph, and (k) of Proposition 14.11, and $\boldsymbol{\Omega}_s := (\Omega, \mathcal{B}_s, \hat{\boldsymbol{P}} := \boldsymbol{P}|_{\mathcal{B}_s})$. Hence, $M_s \in \boldsymbol{E}(M_t|\mathcal{B}_s)$, $\forall s, t \in \mathbb{I}$ with $s \leq t$. Thus, $(M_t)_{t\in\mathbb{I}}$ is a Martingale.

This completes the proof of the proposition. □

Lemma 14.85 *Let* $\mathbb{I} := (([0,\infty), |\cdot|), \hat{\mathcal{B}}, \mu)$ *be the* σ*-finite metric measure subspace of* $\mathbb{R}$ *with partial ordering* $\leq$, $\boldsymbol{\Omega} := (\Omega, \mathcal{B}, \boldsymbol{P})$ *be a probability measure space,* $\mathbb{I} \times \boldsymbol{\Omega} := ([0,\infty) \times \Omega, \bar{\mathcal{B}}, \nu)$ *be the* σ*-finite product measure space,* $w : \mathbb{I} \times \boldsymbol{\Omega} \to \mathbb{R}$ *be a standard Wiener process (as defined in Definition 14.76), the process* $(w_t)_{t\in\mathbb{I}}$ *is defined by* $w_t(\omega) = w(t,\omega)$, $\forall\omega \in \boldsymbol{\Omega}$, $\forall t \in \mathbb{I}$, *and* $\alpha, \beta \in (0,\infty) \subset \mathbb{R}$. *Then,* $\boldsymbol{P}(\{\omega \in \Omega \mid w(t,\omega) > \frac{\alpha}{2}t + \beta,$ for some $t \in \mathbb{I}\}) \leq \exp(-\alpha\beta)$.

Proof Let $(\mathcal{B}_t)_{t\in\mathbb{I}}$ be the natural filtration of $(w_t)_{t\in\mathbb{I}}$. Let $A := \{\omega \in \Omega \mid w(t,\omega) > \frac{\alpha}{2}t + \beta,$ for some $t \in \mathbb{I}\}$ and $M : \mathbb{I} \times \boldsymbol{\Omega} \to \mathbb{R}$ be defined by $M(t,\omega) = \exp(\alpha w(t,\omega) - \frac{\alpha^2}{2}t)$, $\forall\omega \in \boldsymbol{\Omega}$, $\forall t \in \mathbb{I}$. By Proposition 14.84, the stochastic process $(M_t)_{t\in\mathbb{I}}$, defined by $M_t(\omega) = M(t,\omega)$, $\forall\omega \in \boldsymbol{\Omega}$, $\forall t \in \mathbb{I}$, is a Martingale adapted to the filtration $(\mathcal{B}_t)_{t\in\mathbb{I}}$ and $\boldsymbol{E}(M_t) = 1$, $\forall t \in \mathbb{I}$. Then, $A = \{\omega \in \Omega \mid \sup_{t\in\mathbb{I}} M_t(\omega) > \exp(\alpha\beta)\}$. Let $\bar{M} : \boldsymbol{\Omega} \to [0,\infty] \subset \mathbb{R}_e$ be defined by $\bar{M}(\omega) := \sup_{t\in\mathbb{I}} M_t(\omega)$, $\forall\omega \in \boldsymbol{\Omega}$. Then, $A = \{\omega \in \boldsymbol{\Omega} \mid \bar{M}(\omega) > \exp(\alpha\beta)\}$. Fix any $N \in \mathbb{N}$, and define $\left(f_n^{\langle N\rangle}\right)_{n\in\mathbb{Z}_+}$ by $f_n^{\langle N\rangle}(\omega) = M(\frac{n}{2^N}, \omega)$, $\forall n \in \mathbb{Z}_+$ and $\forall\omega \in \boldsymbol{\Omega}$. Then, $\left(f_n^{\langle N\rangle}\right)_{n\in\mathbb{Z}_+}$ is a martingale indexed by $\mathbb{Z}_+$. By the continuity of $M(\cdot,\omega) : \mathbb{I} \to \mathbb{R}$, $\forall\omega \in \boldsymbol{\Omega}$, we have $A = \{\omega \in \boldsymbol{\Omega} \mid \bar{M}(\omega) > \exp(\alpha\beta)\} = \bigcup_{N\in\mathbb{N}}\{\omega \in \boldsymbol{\Omega} \mid \sup_{n\in\mathbb{Z}_+} f_n^{\langle N\rangle}(\omega) > \exp(\alpha\beta)\} \in \mathcal{B}$, where the set membership follows from Proposition 11.40. Fix any $N \in \mathbb{N}$, and consider the Martingale $\left(f_n^{\langle N\rangle}\right)_{n\in\mathbb{Z}_+}$. Let $T^{\langle N\rangle} : \boldsymbol{\Omega} \to \mathbb{Z}_+ \cup \{\infty\}$ be defined by $T^{\langle N\rangle}(\omega) := \inf\{n \in \mathbb{Z}_+ \mid f_n^{\langle N\rangle}(\omega) > \exp(\alpha\beta)\}$, $\forall\omega \in \boldsymbol{\Omega}$. By Example 14.34, $T^{\langle N\rangle}$ is a stopping time for the Martingale $\left(f_n^{\langle N\rangle}\right)_{n\in\mathbb{Z}_+}$. By Theorem 14.35, the stopped stochastic process $\left(Y_n^{\langle N\rangle}\right)_{n\in\mathbb{Z}_+}$, defined by $Y_n^{\langle N\rangle}(\omega) = f_{T^{\langle N\rangle}(\omega)\wedge n}^{\langle N\rangle}(\omega)$, $\forall\omega \in \boldsymbol{\Omega}$, $\forall n \in \mathbb{Z}_+$, is a Martingale. Note that $Y_0^{\langle N\rangle}(\omega) = M_0(\omega) = 1$, $\forall\omega \in \boldsymbol{\Omega}$. Then, $\boldsymbol{E}(Y_n^{\langle N\rangle}) = 1$, $\forall n \in \mathbb{Z}_+$. By Doob's Forward Convergence Theorem 14.41, we have $Y_\infty^{\langle N\rangle}(\omega) := \begin{cases} \lim_{n\in\mathbb{N}} Y_n^{\langle N\rangle}(\omega) & \lim_{n\in\mathbb{N}} Y_n^{\langle N\rangle}(\omega) \in \mathbb{R} \\ 0 & \lim_{n\in\mathbb{N}} Y_n^{\langle N\rangle}(\omega) \text{ does not exist in } \mathbb{R} \end{cases}$ satisfies $Y_\infty^{\langle N\rangle} = \lim_{n\in\mathbb{N}} Y_n^{\langle N\rangle}(\omega)$ a.e. $\omega \in \boldsymbol{\Omega}$, $Y_\infty^{\langle N\rangle}$ is $\mathcal{B}$-measurable, and $\boldsymbol{E}(Y_\infty^{\langle N\rangle}) = \boldsymbol{E}(\left|Y_\infty^{\langle N\rangle}\right|) \leq \sup_{n\in\mathbb{Z}_+} \boldsymbol{E}(\left|Y_n^{\langle N\rangle}\right|) = \sup_{n\in\mathbb{Z}_+} \boldsymbol{E}(Y_n^{\langle N\rangle}) = 1$. This implies that, by Proposition 11.7, $\boldsymbol{P}(A) = \lim_{N\in\mathbb{N}} \boldsymbol{P}(\{\omega \in \boldsymbol{\Omega} \mid \sup_{n\in\mathbb{Z}_+} f_n^{\langle N\rangle}(\omega) > \exp(\alpha\beta)\}) = \lim_{N\in\mathbb{N}} \boldsymbol{P}(\{\omega \in \boldsymbol{\Omega} \mid Y_\infty^{\langle N\rangle} > \exp(\alpha\beta)\} \leq \liminf_{N\in\mathbb{N}} \exp(-\alpha\beta) = \exp(-\alpha\beta)$, where the second equality follows from the definition of $Y_\infty^{\langle N\rangle}$ and $\left(Y_n^{\langle N\rangle}\right)_{n\in\mathbb{Z}_+}$ and the inequality follows from Chebyshev Inequality, Proposition 14.67. This completes the proof of the lemma. □

Lemma 14.86 *Let $\alpha \in \mathbb{R}_+$, $V : (\alpha, \infty) \to \mathbb{R}$ be continuous and $\bar{V} : (0, \alpha) \to \mathbb{R}$ be continuous. Then,*

$$\limsup_{t\to\infty} V(t) = \inf_{t\in\mathbb{R}, t>\alpha} \sup_{s\in\mathbb{R}, s>t} V(s) = \inf_{t\in\mathbb{Q}, t>\alpha} \sup_{s\in\mathbb{Q}, s>t} V(s) \in \mathbb{R}_e$$

$$\limsup_{t\to 0} \bar{V}(t) = \inf_{t\in\mathbb{R}_+, t<\alpha} \sup_{s\in\mathbb{R}_+, s<t} \bar{V}(s) = \inf_{t\in\mathbb{Q}, 0<t<\alpha} \sup_{s\in\mathbb{Q}, 0<s<t} \bar{V}(s) \in \mathbb{R}_e$$

Proof First, consider the function V. By Definition 3.84, we have $\limsup_{t\to\infty} V(t) = \inf_{t\in\mathbb{R}, t>\alpha} \sup_{s\in\mathbb{R}, s>t} V(s) \in \mathbb{R}_e$. Define $H : (\alpha, \infty) \to \mathbb{R}_e$ by $H(t) := \sup_{s\in\mathbb{R}, s>t} V(s) \in \mathbb{R}_e$, $\forall t \in \mathbb{R}$ with $t > \alpha$. Since V is continuous, we have $H(t) = \sup_{s\in\mathbb{Q}, s>t} V(s) \in \mathbb{R}_e$, $\forall t \in \mathbb{R}$ with $t > \alpha$.

We will distinguish two exhaustive and mutually exclusive cases: Case 1: $\limsup_{t\to\infty} V(t) < +\infty$ and Case 2: $\limsup_{t\to\infty} V(t) = +\infty$.

Case 1: $\limsup_{t\to\infty} V(t) < +\infty$. Then, $\inf_{t\in\mathbb{R}, t>\alpha} H(t) < \infty$. Clearly, H is monotonically nonincreasing on (α, ∞). If there exists $t_0 \in (\alpha, \infty)$ with $H(t_0) = \infty$, we can show that $H(t) = \infty$, $\forall t \in (\alpha, \infty)$. Then, we have $H : (\alpha, \infty) \to \mathbb{R}$, since $V : (\alpha, \infty) \to \mathbb{R}$. We need the following intermediate result:

Claim 14.86.1 $H : (\alpha, \infty) \to \mathbb{R}$ is continuous.

Proof of Claim Fix any $t \in (\alpha, \infty)$. $H(t) = \sup_{s\in\mathbb{R}, s>t} V(s) \in \mathbb{R}$. Fix any $\epsilon \in (0, \infty) \subset \mathbb{R}$. Clearly, $V(t) \le H(t)$, since V is continuous. We will distinguish two exhaustive and mutually exclusive cases: Case A: $H(t) > V(t)$ and Case B: $H(t) = V(t)$.

Case A: $H(t) > V(t)$. Then, by continuity of V, $\exists \delta \in (0, \infty) \subset \mathbb{R}$ such that $|V(l) - V(t)| < \frac{1}{2}(H(t) - V(t))$, $\forall l \in \mathcal{B}_{\mathbb{R}}(t, \delta) \cap (\alpha, \infty) =: D$. This implies that $H(l) = H(t)$, $\forall l \in D$. Then, $|H(l) - H(t)| = 0 < \epsilon$, $\forall l \in D$.

Case B: $H(t) = V(t)$. By continuity of V, $\exists \delta \in (0, \infty) \subset \mathbb{R}$ such that $|V(l) - V(t)| < \epsilon$, $\forall l \in \mathcal{B}_{\mathbb{R}}(t, \delta) \cap (\alpha, \infty) =: D$. Fix any $l \in D$. If $l \ge t$, we have $H(t) - \epsilon = V(t) - \epsilon < V(l) \le \sup_{s\in\mathbb{R}, s>l} V(s) = H(l) \le H(t)$. Thus, we have $|H(l) - H(t)| \le \epsilon$. On the other hand, if $l < t$, we have $H(t) \le H(l) = \sup_{s\in\mathbb{R}, s>l} V(s) = \max\{\sup_{s\in\mathbb{R}, s>t} V(s), \sup_{s\in[l,t]} V(s)\} = \max\{H(t), \sup_{s\in[l,t]} V(s)\} \le H(t) + \epsilon$. This yields $|H(l) - H(t)| \le \epsilon$. Hence, we have $|H(l) - H(t)| \le \epsilon$, $\forall l \in D$.

In both cases, we have found $\delta \in (0, \infty) \subset \mathbb{R}$ such that $|H(l) - H(t)| \le \epsilon$, $\forall l \in \mathcal{B}_{\mathbb{R}}(t, \delta) \cap (\alpha, \infty) =: D$. Hence, H is continuous at t. By the arbitrariness of t, we conclude that H is continuous. This completes the proof of the claim. □

By Claim 14.86.1, we have $\inf_{t\in\mathbb{R}, t>\alpha} H(t) = \inf_{t\in\mathbb{Q}, t>\alpha} H(t)$. This leads to $\limsup_{t\to\infty} V(t) = \inf_{t\in\mathbb{R}, t>\alpha} H(t) = \inf_{t\in\mathbb{Q}, t>\alpha} H(t) = \inf_{t\in\mathbb{Q}, t>\alpha} \sup_{s\in\mathbb{Q}, s>t} V(s)$. Hence, the result holds in this case.

Case 2: $\limsup_{t\to\infty} V(t) = +\infty$. Then, $\inf_{t\in\mathbb{R}, t>\alpha} H(t) = \infty$. This implies that $H(t) = \infty$, $\forall t \in (\alpha, \infty) \subset \mathbb{R}$. This leads to $\limsup_{t\to\infty} V(t) = \inf_{t\in\mathbb{R}, t>\alpha} H(t) = \inf_{t\in\mathbb{Q}, t>\alpha} H(t) = \inf_{t\in\mathbb{Q}, t>\alpha} \sup_{s\in\mathbb{Q}, s>t} V(s)$. Hence, the result holds in this case.

In both cases, we have shown that the desired result holds.

Next, consider the function $\bar{V}$. By Definition 3.84,

$$\limsup_{t\to 0} \bar{V}(t) = \inf_{t\in\mathbb{R}_+, t<\alpha} \sup_{s\in\mathbb{R}_+, s<t} \bar{V}(s) = \inf_{t\in\mathbb{R}, t>\frac{1}{\alpha}} \sup_{s\in\mathbb{R}, s>t} \bar{V}(\frac{1}{s})$$

$$= \limsup_{t\to\infty} \bar{V}(\frac{1}{t}) = \inf_{t\in\mathbb{Q}, t>\frac{1}{\alpha}} \sup_{s\in\mathbb{Q}, s>t} \bar{V}(\frac{1}{s}) = \inf_{t\in\mathbb{Q}, 0<t<\alpha} \sup_{s\in\mathbb{Q}, 0<s<t} \bar{V}(s)$$

$$\in \mathbb{R}_e$$

where the first equality follows from Definition 3.84 and the third and fourth equalities follow from the result for function V with $V(t) := \bar{V}(\frac{1}{t})$.

This completes the proof of the lemma. □

Theorem 14.87 (Law of Iterated Logarithms) *Let* $(([0,\infty), |\cdot|), \hat{\mathcal{B}}, \mu) =: \mathbb{I}$ *be the* σ*-finite metric measure subspace of* $\mathbb{R}$ *with the partial ordering* $\leq$, $\boldsymbol{\Omega} := (\Omega, \mathcal{B}, \boldsymbol{P})$ *be a probability measure space,* $\mathbb{I} \times \boldsymbol{\Omega} := ([0,\infty) \times \Omega, \bar{\mathcal{B}}, \nu)$ *be the* σ*-finite product measure space,* $w : \mathbb{I} \times \boldsymbol{\Omega} \to \mathbb{R}$ *be a standard Wiener process (as defined in Definition 14.76), and the stochastic process* $(w_t)_{t\in\mathbb{I}}$ *defined by* $w_t(\omega) = w(t,\omega)$, $\forall \omega \in \boldsymbol{\Omega}$, $\forall t \in \mathbb{I}$. *Then, the following statements hold:*

(i) $\limsup_{t\to\infty} \frac{w_t}{\sqrt{2t\ln(\ln(t))}} = 1$ *a.e. in* $\boldsymbol{\Omega}$.

(ii) $\liminf_{t\to\infty} \frac{w_t}{\sqrt{2t\ln(\ln(t))}} = -1$ *a.e. in* $\boldsymbol{\Omega}$.

(iii) $\limsup_{t\to 0^+} \frac{w_t}{\sqrt{2t\ln(\ln(\frac{1}{t}))}} = 1$ *a.e. in* $\boldsymbol{\Omega}$.

(iv) $\liminf_{t\to 0^+} \frac{w_t}{\sqrt{2t\ln(\ln(\frac{1}{t}))}} = -1$ *a.e. in* $\boldsymbol{\Omega}$.

(v) $\tilde{w} : \mathbb{I} \times \boldsymbol{\Omega} \to \mathbb{R}$ *defined by*

$$\tilde{w}(t,\omega) := \begin{cases} t w(\frac{1}{t}, \omega) & \text{if } (t,\omega) \in (0,\infty) \times E \\ 0 & \text{otherwise} \end{cases}$$

$\forall \omega \in \boldsymbol{\Omega}$, $\forall t \in \mathbb{I}$, *where* $E := \{\omega \in \boldsymbol{\Omega} \mid \limsup_{t\to\infty} \frac{w_t}{\sqrt{2t\ln(\ln(t))}} = 1 \text{ and } \liminf_{t\to\infty} \frac{w_t}{\sqrt{2t\ln(\ln(t))}} = -1\}$, *is also a standard Wiener process on* $\mathbb{I} \times \boldsymbol{\Omega}$.

Proof (i) Let $h : (\mathrm{e}, \infty) \to (0,\infty) \subset \mathbb{R}$ be defined by $h(t) = \sqrt{2t\ln(\ln(t))}$, $\forall t \in (\mathrm{e},\infty) \subset \mathbb{R}$. Clearly, $\lim_{t\to\infty} h(t) = \infty$ and h is strictly increasing. Fix any $\theta, \delta \in \mathbb{Q}$ with $0 < \delta < 1 < \theta$. Let $\alpha_n := (1+\delta)\theta^{-n}h(\theta^n) > 0$ and $\beta_n := \frac{1}{2}h(\theta^n) > 0$, $\forall n \in \mathbb{N}$ with $n \geq n_{0,\theta}$, where $\theta^{n_{0,\theta}} > \mathrm{e}$. Then, $\alpha_n\beta_n = \frac{1}{2}(1+\delta)\theta^{-n}(h(\theta^n))^2 = (1+\delta)\ln(\ln(\theta^n)) = (1+\delta)\ln(n\ln(\theta)) = \ln((n\ln(\theta))^{1+\delta})$ and $\exp(-\alpha_n\beta_n) = \frac{c(\theta)}{n^{1+\delta}}$, $\forall n \in \mathbb{N}$ with $n \geq n_{0,\theta}$, where $c(\theta) := (\ln(\theta))^{-1-\delta}$. Therefore, we have $\sum_{n=n_{0,\theta}}^{\infty} \exp(-\alpha_n\beta_n) < +\infty$.

By Lemma 14.85, we have $P(\{\omega \in \boldsymbol{\Omega} \mid w(t,\omega) > \frac{\alpha_n}{2}t + \beta_n \text{ for some } t \in \mathbb{I}\}) \leq \exp(-\alpha_n \beta_n)$. This implies that $\sum_{n=n_{0,\theta}}^{\infty} P(\{\omega \in \boldsymbol{\Omega} \mid w(t,\omega) > \frac{\alpha_n}{2}t + \beta_n \text{ for some } t \in \mathbb{I}\}) < \infty$. By Borel–Cantelli Lemma 14.6, $P(\bigcap_{m=n_{0,\theta}}^{\infty} \bigcup_{n=m}^{\infty} \{\omega \in \boldsymbol{\Omega} \mid w(t,\omega) > \frac{\alpha_n}{2}t + \beta_n \text{ for some } t \in \mathbb{I}\}) = 0 = P(\{\omega \in \Omega \mid w(t,\omega) > \frac{\alpha_n}{2}t + \beta_n \text{ for some } t \in \mathbb{I} \text{ and for infinitely many } n \geq n_{0,\theta}\}) =: P(\boldsymbol{\Omega} \setminus \Omega_{\theta,\delta})$. Then, $P(\Omega_{\theta,\delta}) = 1$. Fix any $\omega \in \Omega_{\theta,\delta}$, $\exists N_{\theta,\delta}(\omega) \in \mathbb{N}$ with $N_{\theta,\delta}(\omega) \geq n_{0,\theta}$ such that $w(t,\omega) \leq \frac{\alpha_n}{2}t + \beta_n$, $\forall t \in \mathbb{I}$ and $\forall n > N_{\theta,\delta}(\omega)$. Fix any $n > N_{\theta,\delta}(\omega)$. $\forall t \in [\theta^n, \theta^{n+1}] \subset \mathbb{R}$, we have $w(t,\omega) \leq \frac{\alpha_n}{2}\theta^{n+1} + \beta_n = \frac{\theta(1+\delta)h(\theta^n)}{2} + \frac{1}{2}h(\theta^n) \leq (1+\delta)\theta h(\theta^n) \leq (1+\delta)\theta h(t)$. Hence, we have $\sup_{t \geq \theta^{N_{\theta,\delta}(\omega)+1}} \frac{w(t,\omega)}{h(t)} \leq (1+\delta)\theta$. This implies that $\limsup_{t \to \infty} \frac{w_t(\omega)}{h(t)} \leq (1+\delta)\theta$, $\forall \omega \in \Omega_{\theta,\delta}$. By Lemma 14.86, $\limsup_{t\to\infty} \frac{w_t}{h(t)} : \boldsymbol{\Omega} \to \mathbb{R}$ is $\mathcal{B}$-measurable. Then, we have $\limsup_{t\to\infty} \frac{w_t}{h(t)} \leq (1+\delta)\theta$ a.e. in $\boldsymbol{\Omega}$. By the arbitrariness of θ and δ, we have $\limsup_{t\to\infty} \frac{w_t}{h(t)} \leq 1$ a.e. in $\boldsymbol{\Omega}$.

Next, we will show that $\limsup_{t\to\infty} \frac{w_t}{h(t)} \geq 1$ a.e. in $\boldsymbol{\Omega}$. Let $A_n := \{\omega \in \boldsymbol{\Omega} \mid w(\theta^n,\omega) - w(\theta^{n-1},\omega) > \sqrt{1-\theta^{-1}}h(\theta^n)\} \in \mathcal{B}$, $\forall n \in \mathbb{N}$ with $n \geq n_{0,\theta}$. Clearly, A_n, $n \geq n_{0,\theta}$, are independent events, since w_t is of independent increments. We need the following inequality:

Claim 14.87.1 $\frac{x}{1+x^2}\exp(-\frac{x^2}{2}) < \int_x^\infty \exp(-\frac{y^2}{2})\,\mathrm{d}y < \frac{1}{x}\exp(-\frac{x^2}{2})$, $\forall x \in (0,\infty) \subset \mathbb{R}$.

Proof of Claim Fix any $x \in (0,\infty) \subset \mathbb{R}$. Then, we have

$$\int_x^\infty \exp(-\frac{y^2}{2})\,\mathrm{d}y < \int_x^\infty \frac{y}{x}\exp(-\frac{y^2}{2})\,\mathrm{d}y$$
$$= \frac{1}{x}\int_x^\infty \exp(-\frac{y^2}{2})\,\mathrm{d}(\frac{y^2}{2}) = \frac{1}{x}\int_{\frac{x^2}{2}}^\infty \exp(-z)\mathrm{d}z = \frac{1}{x}\exp(-\frac{x^2}{2})$$

where the second equality follows from Change of Variable Theorem 12.91 with $z = \frac{y^2}{2}$. On the other hand,

$$\int_x^\infty \exp(-\frac{y^2}{2})\,\mathrm{d}y > \int_x^\infty \frac{1+y^{-2}}{1+x^{-2}}\exp(-\frac{y^2}{2})\,\mathrm{d}y$$
$$= \frac{x^2}{1+x^2}\int_x^\infty \left(1+y^{-2}\right)\exp(-\frac{y^2}{2})\,\mathrm{d}y = \frac{x^2}{1+x^2}\left(-\frac{1}{y}\exp(-\frac{y^2}{2})\right)\bigg|_x^\infty$$
$$= \frac{x}{1+x^2}\exp(-\frac{x^2}{2})$$

This completes the proof of the claim. □

Then, $P(A_n) = P\Big(\Big\{\omega \in \Omega \,\Big|\, \frac{w_{\theta^n} - w_{\theta^{n-1}}}{\sqrt{\theta^n - \theta^{n-1}}} > \frac{\sqrt{1-\theta^{-1}}}{\sqrt{\theta^n - \theta^{n-1}}} h(\theta^n)\Big\}\Big) = \int_{\frac{\sqrt{1-\theta^{-1}}}{\sqrt{\theta^n-\theta^{n-1}}}h(\theta^n)}^{\infty} \frac{1}{\sqrt{2\pi}} \exp(-\frac{y^2}{2})\,\mathrm{d}y < \frac{1}{\sqrt{2\pi}} \frac{\sqrt{\theta^n}}{h(\theta^n)} \exp(-\frac{(h(\theta^n))^2}{2\theta^n})$, $\forall n \in \mathbb{N}$ with $n \geq n_{0,\theta}$, where the second equality follows from the fact that $\frac{w_{\theta^n} - w_{\theta^{n-1}}}{\sqrt{\theta^n - \theta^{n-1}}} \sim N(0, 1)$ and the inequality follows from Claim 14.87.1. Again by Claim 14.87.1, we have $P(A_n) > \frac{1}{\sqrt{2\pi}} \frac{x}{1+x^2} \exp(-\frac{x^2}{2})$, $\forall n \in \mathbb{N}$ with $n \geq n_{0,\theta}$, where $x := \frac{1}{\theta^{n/2}} h(\theta^n) = \sqrt{2\ln(n\ln(\theta))} = \sqrt{2\ln(n) + 2\ln(\ln(\theta))}$. This implies that $\exists n_{1,\theta} \in \mathbb{N}$ with $n_{1,\theta} \geq n_{0,\theta}$ such that $1 < x < 2\sqrt{\ln(n)}$, $\forall n \in \mathbb{N}$ with $n \geq n_{1,\theta}$. Then, we have $P(A_n) > \frac{1}{\sqrt{2\pi}} \frac{1}{2x} \exp(-\frac{x^2}{2}) > \frac{1}{\sqrt{2\pi}} \frac{1}{4\sqrt{\ln(n)}} \exp(-\ln(n\ln(\theta))) = \frac{1}{4\sqrt{2\pi}\ln(\theta)} \frac{1}{n\sqrt{\ln(n)}}$, $\forall n \in \mathbb{N}$ with $n \geq n_{1,\theta}$. This yields that $\sum_{n=n_{0,\theta}}^{\infty} P(A_n) = \infty$, by Problem 34.M of Bartle (1976). By Borel–Cantelli Lemma 14.6, $P(F) := P(\bigcap_{m=n_{0,\theta}}^{\infty} \bigcup_{n=m}^{\infty} A_n) = 1 = P\Big(\Big\{\omega \in \Omega \,\Big|\, \frac{w_{\theta^n} - w_{\theta^{n-1}}}{\sqrt{\theta^n - \theta^{n-1}}} > \frac{\sqrt{1-\theta^{-1}}}{\sqrt{\theta^n - \theta^{n-1}}} h(\theta^n)$ for infinitely many $n \geq n_{0,\theta}\Big\}\Big)$. Thus, we have equivalently that $\sqrt{1-\theta^{-1}} \leq \limsup_{n\in\mathbb{N}} \frac{w_{\theta^n} - w_{\theta^{n-1}}}{h(\theta^n)} \leq \limsup_{n\in\mathbb{N}} \frac{w_{\theta^n}}{h(\theta^n)} + \limsup_{n\in\mathbb{N}} \frac{-w_{\theta^{n-1}}}{h(\theta^{n-1})} \frac{h(\theta^{n-1})}{h(\theta^n)} \leq \limsup_{n\in\mathbb{N}} \frac{w_{\theta^n}}{h(\theta^n)} + \frac{1}{\sqrt{\theta}}$ a.e. $\omega \in \Omega$, where the second inequality follows from Proposition 3.83 and the last inequality follows from the facts that $\limsup_{n\in\mathbb{N}} \frac{-w_{\theta^{n-1}}}{h(\theta^{n-1})} \leq \limsup_{t\to\infty} \frac{-w_t}{h(t)} \leq 1$ (since $-w_t$ is a standard Wiener process by Proposition 14.80) and $\lim_{n\in\mathbb{N}} \frac{h(\theta^{n-1})}{h(\theta^n)} = \frac{1}{\sqrt{\theta}}$. This leads to $\limsup_{t\to\infty} \frac{w_t}{h(t)} \geq \limsup_{n\in\mathbb{N}} \frac{w_{\theta^n}}{h(\theta^n)} \geq \sqrt{1-\theta^{-1}} - \frac{1}{\sqrt{\theta}}$ a.e. $\omega \in \Omega$. By Lemma 14.86, $\limsup_{t\to\infty} \frac{w_t}{h(t)} : \Omega \to \mathbb{R}$ is $\mathcal{B}$-measurable. Then, we have $\limsup_{t\to\infty} \frac{w_t}{h(t)} \geq \sqrt{1-\theta^{-1}} - \sqrt{\frac{1}{\theta}}$ a.e. in Ω. By the arbitrariness of θ, we have $\limsup_{t\to\infty} \frac{w_t}{h(t)} \geq 1$ a.e. in Ω.

This proves (i).

(ii) By Proposition 14.80, $(-w_t)_{t\in\mathbb{I}}$ is also a standard Wiener process. Then, by (i), we have $-\liminf_{t\to\infty} \frac{w_t}{h(t)} = \limsup_{t\to\infty} \frac{-w_t}{h(t)} = 1$ a.e. in Ω, where the first equality follows from Proposition 3.85 and the second equality follows from (i). This proves (ii).

(iii) Let $\bar{h} : (0, \frac{1}{\mathrm{e}}) \to (0, \infty) \subset \mathbb{R}$ be defined by $\bar{h}(t) = \sqrt{2t\ln(\ln(\frac{1}{t}))}$, $\forall t \in (0, \frac{1}{\mathrm{e}}) \subset \mathbb{R}$. Clearly, $\lim_{t\to 0} \bar{h}(t) = 0$ and $\bar{h}$ is strictly increasing on $(0, t_0) \subset \mathbb{R}$, for some $t_0 \in (0, \frac{1}{\mathrm{e}})$. Fix any $\theta, \delta \in \mathbb{Q}$ with $\theta, \delta \in (0, 1) \subset \mathbb{R}$. Let $n_{0,\theta} \in \mathbb{N}$ be such that $\theta^{n_{0,\theta}} < t_0$, and $\alpha_n := (1+\delta)\theta^{-n}\bar{h}(\theta^n) > 0$ and $\beta_n := \frac{1}{2}\bar{h}(\theta^n) > 0$, $\forall n \in \mathbb{N}$ with $n \geq n_{0,\theta}$. Then, $\alpha_n\beta_n = \frac{1}{2}(1+\delta)\theta^{-n}(\bar{h}(\theta^n))^2 = (1+\delta)\ln(\ln(\theta^{-n})) = (1+\delta)\ln(n\ln(\frac{1}{\theta})) = \ln((n\ln(\frac{1}{\theta}))^{1+\delta})$ and $\exp(-\alpha_n\beta_n) = \frac{c(\theta)}{n^{1+\delta}}$, $\forall n \in \mathbb{N}$ with $n \geq n_{0,\theta}$, where $c(\theta) := (\ln(\frac{1}{\theta}))^{-1-\delta}$. Therefore, we have $\sum_{n=n_{0,\theta}}^{\infty} \exp(-\alpha_n\beta_n) < +\infty$.

By Lemma 14.85, we have $P(\{\omega \in \Omega \mid w(t,\omega) > \frac{\alpha_n}{2}t + \beta_n$ for some $t \in \mathbb{I}\}) \leq \exp(-\alpha_n\beta_n)$. This implies that $\sum_{n=n_{0,\theta}}^{\infty} P(\{\omega \in \Omega \mid w(t,\omega) > \frac{\alpha_n}{2}t +$

β_n for some $t \in \mathbb{I}\}) < \infty$. By Borel–Cantelli Lemma 14.6, $\boldsymbol{P}(\bigcap_{m=n_{0,\theta}}^{\infty} \bigcup_{n=m}^{\infty} \{\omega \in \boldsymbol{\Omega} \mid w(t,\omega) > \frac{\alpha_n}{2}t + \beta_n \text{ for some } t \in \mathbb{I}\}) = 0 = \boldsymbol{P}(\{\omega \in \Omega \mid w(t,\omega) > \frac{\alpha_n}{2}t + \beta_n \text{ for some } t \in \mathbb{I} \text{ and for infinitely many } n \geq n_{0,\theta}\}) =: \boldsymbol{P}(\boldsymbol{\Omega} \setminus \Omega_{\theta,\delta})$. Then, $\boldsymbol{P}(\Omega_{\theta,\delta}) = 1$. Fix any $\omega \in \Omega_{\theta,\delta}$, $\exists N_{\theta,\delta}(\omega) \in \mathbb{N}$ with $N_{\theta,\delta}(\omega) \geq n_{0,\theta}$ such that $w(t,\omega) \leq \frac{\alpha_n}{2}t + \beta_n$, $\forall t \in \mathbb{I}$ and $\forall n > N_{\theta,\delta}(\omega)$. Fix any $n > N_{\theta,\delta}(\omega)$. $\forall t \in [\theta^n, \theta^{n-1}] \subset \mathbb{R}$, we have $w(t,\omega) \leq \frac{\alpha_n}{2}\theta^{n-1} + \beta_n = \frac{(1+\delta)\bar{h}(\theta^n)}{2\theta} + \frac{1}{2}\bar{h}(\theta^n) \leq \frac{1+\delta}{\theta}\bar{h}(\theta^n) \leq \frac{1+\delta}{\theta}\bar{h}(t)$. Hence, we have $\sup_{0<t<\theta^{N_{\theta,\delta}(\omega)+1}} \frac{w(t,\omega)}{\bar{h}(t)} \leq \frac{1+\delta}{\theta}$. This implies that $\limsup_{t\to 0^+} \frac{w_t(\omega)}{\bar{h}(t)} \leq \frac{1+\delta}{\theta}$, $\forall \omega \in \Omega_{\theta,\delta}$. By Lemma 14.86, $\limsup_{t\to 0^+} \frac{w_t}{\bar{h}(t)} : \boldsymbol{\Omega} \to \mathbb{R}$ is $\mathcal{B}$-measurable. Then, we have $\limsup_{t\to 0^+} \frac{w_t}{\bar{h}(t)} \leq \frac{1+\delta}{\theta}$ a.e. in $\boldsymbol{\Omega}$. By the arbitrariness of θ and δ, we have $\limsup_{t\to 0^+} \frac{w_t}{\bar{h}(t)} \leq 1$ a.e. in $\boldsymbol{\Omega}$.

Next, we will show that $\limsup_{t\to 0^+} \frac{w_t}{\bar{h}(t)} \geq 1$ a.e. in $\boldsymbol{\Omega}$. Let $A_n := \{\omega \in \boldsymbol{\Omega} \mid w(\theta^n,\omega) - w(\theta^{n+1},\omega) > \sqrt{1-\theta}\,\bar{h}(\theta^n)\} \in \mathcal{B}$, $\forall n \in \mathbb{N}$ with $n \geq n_{0,\theta}$. Clearly, A_n, $n \geq n_{0,\theta}$, are independent events, since w_t is of independent increments. Then, $\boldsymbol{P}(A_n) = \boldsymbol{P}\Big(\Big\{\omega \in \Omega \;\Big|\; \frac{w_{\theta^n} - w_{\theta^{n+1}}}{\sqrt{\theta^n - \theta^{n+1}}} > \frac{\sqrt{1-\theta}}{\sqrt{\theta^n - \theta^{n+1}}}\bar{h}(\theta^n)\Big\}\Big) = \int_{\frac{\sqrt{1-\theta}}{\sqrt{\theta^n-\theta^{n+1}}}\bar{h}(\theta^n)}^{\infty} \frac{1}{\sqrt{2\pi}} \exp(-\frac{y^2}{2})\, dy < \frac{1}{\sqrt{2\pi}} \frac{\sqrt{\theta^n}}{\bar{h}(\theta^n)} \cdot \exp(-\frac{(\bar{h}(\theta^n))^2}{2\theta^n})$, $\forall n \in \mathbb{N}$ with $n \geq n_{0,\theta}$, where the second equality follows from the fact that $\frac{w_{\theta^n} - w_{\theta^{n-1}}}{\sqrt{\theta^n - \theta^{n-1}}} \sim N(0,1)$ and the inequality follows from Claim 14.87.1. Again by Claim 14.87.1, we have $\boldsymbol{P}(A_n) > \frac{1}{\sqrt{2\pi}} \frac{x}{1+x^2} \exp(-\frac{x^2}{2})$, $\forall n \in \mathbb{N}$ with $n \geq n_{0,\theta}$, where $x := \frac{1}{\theta^{n/2}}\bar{h}(\theta^n) = \sqrt{2\ln(n\ln(\frac{1}{\theta}))} = \sqrt{2\ln(n) + 2\ln(\ln(\frac{1}{\theta}))}$. This implies that $\exists n_{1,\theta} \in \mathbb{N}$ with $n_{1,\theta} \geq n_{0,\theta}$ such that $1 < x < 2\sqrt{\ln(n)}$, $\forall n \in \mathbb{N}$ with $n \geq n_{1,\theta}$. Then, we have $P(A_n) > \frac{1}{\sqrt{2\pi}} \frac{1}{2x} \exp(-\frac{x^2}{2}) > \frac{1}{\sqrt{2\pi}} \frac{1}{4\sqrt{\ln(n)}} \exp(-\ln(n\ln(\frac{1}{\theta}))) = \frac{1}{4\sqrt{2\pi}\ln(\frac{1}{\theta})} \frac{1}{n\sqrt{\ln(n)}}$, $\forall n \in \mathbb{N}$ with $n \geq n_{1,\theta}$. This yields that $\sum_{n=n_{0,\theta}}^{\infty} \boldsymbol{P}(A_n) = \infty$, by Problem 34.M of Bartle (1976). By Borel–Cantelli Lemma 14.6, $\boldsymbol{P}(F) := \boldsymbol{P}(\bigcap_{m=n_{0,\theta}}^{\infty} \bigcup_{n=m}^{\infty} A_n) = 1 = \boldsymbol{P}\Big(\Big\{\omega \in \Omega \;\Big|\; \frac{w_{\theta^n} - w_{\theta^{n+1}}}{\sqrt{\theta^n - \theta^{n+1}}} > \frac{\sqrt{1-\theta}}{\sqrt{\theta^n - \theta^{n+1}}}\bar{h}(\theta^n) \text{ for infinitely many } n \geq n_{0,\theta}\Big\}\Big)$. Thus, we have equivalently that $\sqrt{1-\theta} \leq \limsup_{n\in\mathbb{N}} \frac{w_{\theta^n} - w_{\theta^{n+1}}}{\bar{h}(\theta^n)} \leq \limsup_{n\in\mathbb{N}} \frac{w_{\theta^n}}{\bar{h}(\theta^n)} + \limsup_{n\in\mathbb{N}} \frac{-w_{\theta^{n+1}}}{\bar{h}(\theta^{n+1})} \frac{\bar{h}(\theta^{n+1})}{\bar{h}(\theta^n)} \leq \limsup_{n\in\mathbb{N}} \frac{w_{\theta^n}}{\bar{h}(\theta^n)} + \sqrt{\theta}$ a.e. $\omega \in \boldsymbol{\Omega}$, where the second inequality follows from Proposition 3.83 and the last inequality follows from the facts that $\limsup_{n\in\mathbb{N}} \frac{-w_{\theta^{n+1}}}{\bar{h}(\theta^{n+1})} \leq \limsup_{t\to 0^+} \frac{-w_t}{\bar{h}(t)} \leq 1$ (since $-w_t$ is a standard Wiener process by Proposition 14.80) and $\lim_{n\in\mathbb{N}} \frac{\bar{h}(\theta^{n+1})}{\bar{h}(\theta^n)} = \sqrt{\theta}$. This leads to $\limsup_{t\to 0^+} \frac{w_t}{\bar{h}(t)} \geq \limsup_{n\in\mathbb{N}} \frac{w_{\theta^n}}{\bar{h}(\theta^n)} \geq \sqrt{1-\theta} - \sqrt{\theta}$ a.e. $\omega \in \boldsymbol{\Omega}$. By Lemma 14.86, $\limsup_{t\to 0^+} \frac{w_t}{\bar{h}(t)} : \boldsymbol{\Omega} \to \mathbb{R}$ is $\mathcal{B}$-measurable. Then, we have $\limsup_{t\to 0^+} \frac{w_t}{\bar{h}(t)} \geq \sqrt{1-\theta} - \sqrt{\theta}$ a.e. in $\boldsymbol{\Omega}$. By the arbitrariness of θ, we have $\limsup_{t\to 0^+} \frac{w_t}{\bar{h}(t)} \geq 1$ a.e. in $\boldsymbol{\Omega}$.

This proves (iii).

(iv) By Proposition 14.80, $(-w_t)_{t\in\mathbb{I}}$ is also a standard Wiener process. Then, by (iii), we have $-\liminf_{t\to 0^+}\frac{w_t}{h(t)} = \limsup_{t\to 0^+}\frac{-w_t}{h(t)} = 1$ a.e. in $\boldsymbol{\Omega}$, where the first equality follows from Proposition 3.85 and the second equality follows from (iii). This proves (iv).

(v) By (i) and (ii), $E \in \mathcal{B}$ and $\boldsymbol{P}(E) = 1$. Then, $(0,\infty) \times E \in \bar{\mathcal{B}}$. Let $\mathcal{F}$ denote the measurable rectangles in the product space $(0,\infty) \times \boldsymbol{\Omega}$, and define the mapping $\Psi : (0,\infty) \times \boldsymbol{\Omega} \to (0,\infty) \times \boldsymbol{\Omega}$ by $\Psi(t,\omega) = (\frac{1}{t},\omega)$, $\forall (t,\omega) \in (0,\infty) \times \boldsymbol{\Omega}$. For any set $G_1 \times G_2 \in \mathcal{F}$, we have the mapping $\Psi_{\text{inv}}(G_1 \times G_2) \in \mathcal{F} \subseteq \bar{\mathcal{B}}$. Consider the set of all subsets $G \subseteq (0,\infty) \times \boldsymbol{\Omega}$ such that $\Psi_{\text{inv}}(G) \in \bar{\mathcal{B}}$, it is easy to show that this set is a σ-algebra containing $\mathcal{F}$. Then, the mapping $\hat{w} : (0,\infty) \times \boldsymbol{\Omega} \to \mathbb{R}$ defined by $\hat{w}(t,\omega) = w(\frac{1}{t},\omega)$, $\forall (t,\omega) \in (0,\infty) \times \boldsymbol{\Omega}$, is $\bar{\mathcal{B}}$-measurable. By Propositions 11.38, 11.39, and 11.41, we have $\tilde{w} : \mathbb{I} \times \boldsymbol{\Omega} \to \mathbb{R}$ is $\bar{\mathcal{B}}$ measurable. This proves (a) in Definition 14.76. Clearly, $\tilde{w}(0,\omega) = 0$, $\forall \omega \in \boldsymbol{\Omega}$. This establishes (b) in Definition 14.76. $\forall \omega \in \boldsymbol{\Omega}$, if $\omega \in E$, then $\tilde{w}(\cdot,\omega) : \mathbb{I} \to \mathbb{R}$ is continuous at any $t \in \mathbb{I}$, and if $\omega \in \Omega \setminus E$, then $\tilde{w}(\cdot,\omega) : \mathbb{I} \to \mathbb{R}$ is the identically zero function and is continuous. Hence, (c) of Definition 14.76 is established. Fix any $0 = s_1 < t_1 = s_2 < t_2 = \cdots = s_n < t_n < \infty$. Then, we have $\tilde{w}_{t_1} - \tilde{w}_{s_1} = t_1 w_{1/t_1}, \ldots, \tilde{w}_{t_n} - \tilde{w}_{s_n} = t_n w_{1/t_n} - s_n w_{1/s_n}$. Since $(w_t)_{t\in\mathbb{I}}$ is a standard Wiener process, we have $\boldsymbol{E}(t_1 w_{1/t_1}) = 0$, $\boldsymbol{E}(t_1^2 w_{1/t_1}^2) = t_1$, $\boldsymbol{E}(t_i w_{1/t_i} - s_i w_{1/s_i}) = 0$, $\boldsymbol{E}((t_i w_{1/t_i} - s_i w_{1/s_i})^2) = t_i + s_i - 2t_i s_i \boldsymbol{E}(w_{1/t_i} w_{1/s_i}) = t_i + s_i - 2t_i s_i \frac{1}{t_i} = t_i - s_i$, $i = 2,\ldots,n$. Furthermore, $\boldsymbol{E}(t_1 w_{1/t_1}(t_i w_{1/t_i} - s_i w_{1/s_i})) = t_1 t_i \boldsymbol{E}(w_{1/t_1} w_{1/t_i}) - t_1 s_i \boldsymbol{E}(w_{1/t_1} w_{1/s_i}) = t_1 - t_1 = 0$, $i = 2,\ldots,n$, and $\boldsymbol{E}((t_i w_{1/t_i} - s_i w_{1/s_i})(t_j w_{1/t_j} - s_j w_{1/s_j})) = t_i t_j \frac{1}{t_j} - s_i t_j \frac{1}{t_j} - t_i s_j \frac{1}{s_j} + s_i s_j \frac{1}{s_j} = t_i - s_i - t_i + s_i = 0$, $\forall 2 \le i < j \le n$. Hence, we have $(\tilde{w}_{t_1} - \tilde{w}_{s_1}, \ldots, \tilde{w}_{t_n} - \tilde{w}_{s_n})$ admits mean $\mathbf{0}_n$ and covariance matrix $K = $ block diagonal $(t_1 - s_1, t_2 - s_2, \ldots, t_n - s_n) \in \bar{\mathcal{S}}_{+\,\mathbb{R}^n}$. By Proposition 14.72 and the fact that $(w_t)_{t\in\mathbb{I}}$ is a standard Wiener process, $(\tilde{w}_{t_1} - \tilde{w}_{s_1}, \ldots, \tilde{w}_{t_n} - \tilde{w}_{s_n}) \sim N(\mathbf{0}_n, K)$. By Proposition 14.27, $\tilde{w}_{t_1} - \tilde{w}_{s_1}, \ldots, \tilde{w}_{t_n} - \tilde{w}_{s_n}$ are independent. This establishes (d) of Definition 14.76. Hence, the process $(\tilde{w}_t)_{t\in\mathbb{I}}$, defined by $\tilde{w}_t(\omega) = \tilde{w}(t,\omega)$, $\forall \omega \in \boldsymbol{\Omega}$, $\forall t \in \mathbb{I}$, is a standard Wiener process.

This completes the proof of the theorem. □

Example 14.88 Let $\mathbb{I} := (([0,\infty), |\cdot|), \hat{\mathcal{B}}, \mu)$ be the σ-finite metric measure subspace of $\mathbb{R}$ with the partial ordering $\le$, $\mathcal{Y}$ be a normed linear space, $\boldsymbol{\Omega} := (\Omega, \mathcal{B}, (\mathcal{B}_t)_{t\in\mathbb{I}}, \boldsymbol{P})$ be a filtered probability measure space, $\mathbb{I} \times \boldsymbol{\Omega} = ([0,\infty) \times \Omega, \bar{\mathcal{B}}, \nu)$ be the σ-finite product measure space, $(X_t)_{t\in\mathbb{I}}$ be an $\mathcal{Y}$-valued adapted stochastic process on $\boldsymbol{\Omega}$ with continuous sample path $\forall \omega \in \boldsymbol{\Omega}$, $F \subseteq \mathcal{Y}$ be a closed set, and $T : \boldsymbol{\Omega} \to \mathbb{I} \cup \{+\infty\}$ be defined by $T(\omega) := \inf\{t \in \mathbb{I} \mid X_t(\omega) \in F\}$, $\forall \omega \in \boldsymbol{\Omega}$. Then, T is a stopping time.

This is because, $\forall t \in \mathbb{I}$, $\{\omega \in \boldsymbol{\Omega} \mid T(\omega) \le t\} = \bigcup_{s\in[0,t]}\{\omega \in \boldsymbol{\Omega} \mid X_s(\omega) \in F\} = \bigcup_{s\in[0,t]} X_{s\,\text{inv}}(F) = \{\omega \in \Omega \mid \inf_{s\in[0,t]} \text{dist}(X_s(\omega), F) = 0\} = \{\omega \in \Omega \mid \inf_{s\in[0,t]\cap\mathbb{Q}} \text{dist}(X_s(\omega), F) = 0\} \in \mathcal{B}_t$, where the first equality follows from the fact that $(X_t)_{t\in\mathbb{I}}$ has continuous sample path, $\forall \omega \in \boldsymbol{\Omega}$, which further implies that $X_{T(\omega)}(\omega) \in F$ if $T(\omega) < \infty$, the second equality follows from the fact that

$(X_t)_{t\in\mathbb{I}}$ is adapted to the filtration $(\mathcal{B}_t)_{t\in\mathbb{I}}$, the third and fourth equalities follow from the fact that $(X_t)_{t\in\mathbb{I}}$ has continuous sample path, $\forall\omega \in \boldsymbol{\Omega}$, and $\text{dist}(\cdot, F) : \mathcal{Y} \to [0, \infty] \subset \mathbb{R}_e$ is a continuous function, and the set membership follows from the fact that $\text{dist}(X_s(\cdot), F) : \boldsymbol{\Omega} \to [0, \infty] \subset \mathbb{R}_e$ is $\mathcal{B}_s$-measurable by Proposition 11.38, $\forall s \in [0, t] \cap \mathbb{Q}$, and the set $[0, t] \cap \mathbb{Q}$ is countable. ◇

Proposition 14.89 *Let* $\mathbb{I} := (([0, \infty), |\cdot|), \hat{\mathcal{B}}, \mu)$ *be the* σ*-finite metric measure subspace of* $\mathbb{R}$, $\boldsymbol{\Omega} := (\Omega, \mathcal{B}, (\mathcal{B}_t)_{t\in\mathbb{I}}, \boldsymbol{P})$ *be a filtered probability measure space, and* τ *and* κ *be stopping times with values in* $\mathbb{I} \cup \{\infty\}$. *Then,* $\tau \vee \kappa$, $\tau \wedge \kappa$, *and* $\tau + \kappa$ *are stopping times.*

Proof By Definition 14.33, $\tau : \boldsymbol{\Omega} \to \mathbb{I} \cup \{\infty\} = [0, \infty] \subseteq \mathbb{R}_e$ is $\mathcal{B}$-measurable and $\{\omega \in \boldsymbol{\Omega} \mid \tau(\omega) \le t\} \in \mathcal{B}_t$, $\forall t \in \mathbb{I}$. The same thing holds for κ. By Propositions 11.36 and 11.40, $\tau \vee \kappa$, $\tau \wedge \kappa$, and $\tau + \kappa$ are functions of $\boldsymbol{\Omega}$ to $[0, \infty] \subseteq \mathbb{R}_e$ and are $\mathcal{B}$-measurable. Fix any $t \in \mathbb{I}$, and we have $\{\omega \in \boldsymbol{\Omega} \mid \tau \vee \kappa \le t\} = \{\omega \in \boldsymbol{\Omega} \mid \tau \le t\} \cap \{\omega \in \boldsymbol{\Omega} \mid \kappa \le t\} \in \mathcal{B}_t$. This implies that $\tau \vee \kappa$ is a stopping time. In addition, $\{\omega \in \boldsymbol{\Omega} \mid \tau \wedge \kappa \le t\} = \{\omega \in \boldsymbol{\Omega} \mid \tau \le t\} \cup \{\omega \in \boldsymbol{\Omega} \mid \kappa \le t\} \in \mathcal{B}_t$. This yields that $\tau \wedge \kappa$ is a stopping time. Furthermore, $\{\omega \in \boldsymbol{\Omega} \mid \tau + \kappa \le t\} = \boldsymbol{\Omega} \setminus \{\omega \in \boldsymbol{\Omega} \mid \tau + \kappa > t\} = \boldsymbol{\Omega} \setminus (\bigcup_{r\in\mathbb{Q}, 0\le r\le t}(\{\omega \in \boldsymbol{\Omega} \mid \tau > r\} \cap \{\omega \in \boldsymbol{\Omega} \mid \kappa > t - r\}) \cup \{\omega \in \boldsymbol{\Omega} \mid \tau > t\} \cup \{\omega \in \boldsymbol{\Omega} \mid \kappa > t\}) \in \mathcal{B}_t$. This leads to $\tau + \kappa$ is a stopping time. This completes the proof of the proposition. □

Example 14.90 Let $\mathbb{I} := (([0, \infty), |\cdot|), \hat{\mathcal{B}}, \mu)$ be the σ-finite metric measure subspace of $\mathbb{R}$ with the partial ordering $\le$, $\boldsymbol{\Omega} := (\Omega, \mathcal{B}, (\mathcal{B}_t)_{t\in\mathbb{I}}, \boldsymbol{P})$ be a filtered probability measure space, $\mathbb{I} \times \boldsymbol{\Omega} = ([0, \infty) \times \Omega, \bar{\mathcal{B}}, \nu)$ be the σ-finite product measure space, $w : \mathbb{I} \times \Omega \to \mathbb{R}$ be the standard Wiener process on $\mathbb{I} \times \boldsymbol{\Omega}$ (as defined in Definition 14.76), and the stochastic process $(w_t)_{t\in\mathbb{I}}$, where $w_t(\omega) = w(t, \omega)$, $\forall\omega \in \boldsymbol{\Omega}$, $\forall t \in \mathbb{I}$, is adapted to the filtration $(\mathcal{B}_t)_{t\in\mathbb{I}}$. Assume that $\mathcal{B}_t$ and $\sigma((w_{t+s} - w_t)_{s\in\mathbb{I}})$ are independent σ-algebras, $\forall t \in \mathbb{I}$. Expand $\exp(\alpha x - \frac{\alpha^2}{2}t)$ as an analytic function of α to obtain

$$\exp(\alpha x - \frac{\alpha^2}{2}t) =: \sum_{n=0}^{\infty} H_n(x, t)\frac{\alpha^n}{n!}, \quad \forall\alpha \in \mathbb{R} \tag{14.21}$$

where $H_n(x, t) : \mathbb{R} \times \mathbb{R} \to \mathbb{R}$ is $\mathcal{C}_\infty$, $\forall n \in \mathbb{N}$. Clearly, $H_n(x, t)$ is an nth order polynomial in x and an $\lfloor \frac{n}{2} \rfloor$th order polynomial in t. Note that $\exp(|\alpha||x| + \frac{\alpha^2}{2}|t|) \ge \sum_{n=0}^{\infty} |H_n(x, t)|\frac{|\alpha|^n}{n!}$, $\forall(\alpha, x, t) \in \mathbb{R}^3$. Then, (14.21) holds $\forall(\alpha, x, t) \in \mathbb{R}^3$ with the right-hand side converge absolutely. The sequence $(H_n(x, t))_{n=0}^{\infty}$ is called the *Hermite polynomials*. The leading Hermite polynomials are $H_0(x, t) = 1$, $H_1(x, t) = x$, $H_2(x, t) = x^2 - t$, $H_3(x, t) = x^3 - 3xt, \ldots$, $\forall(x, t) \in \mathbb{R}^2$.

Now, consider the adapted stochastic process $(M_t)_{t\in\mathbb{I}}$, defined by $M_t(\omega) = \exp(\alpha w_t(\omega) - \frac{\alpha^2}{2}t) = \sum_{n=0}^{\infty} H_n(w_t(\omega), t)\frac{\alpha^n}{n!}$, $\forall\omega \in \boldsymbol{\Omega}$, $\forall t \in \mathbb{I}$. By Proposition 14.84, $(M_t)_{t\in\mathbb{I}}$ is a Martingale. Then, we have $\boldsymbol{E}(M_t|\mathcal{B}_s) = M_s$ a.e. in $\boldsymbol{\Omega}_s$, $\forall s, t \in \mathbb{I}$ with $s \leq t$, where $\boldsymbol{\Omega}_s := (\Omega, \mathcal{B}_s, \boldsymbol{P}_s := \boldsymbol{P}|_{\mathcal{B}_s})$ is a probability measure space. Since $\exp(|\alpha||w_t(\omega)| + \frac{\alpha^2}{2}|t|) \geq \sum_{n=0}^{\infty} \frac{|\alpha|^n}{n!}|H_n(w_t(\omega), t)|$, $\forall(\alpha, t, \omega) \in \mathbb{R} \times \mathbb{I} \times \boldsymbol{\Omega}$, then $\boldsymbol{E}(\sum_{n=0}^{\infty} |H_n(w_t(\omega), t)|\frac{|\alpha|^n}{n!}) \leq \boldsymbol{E}(\exp(|\alpha||w_t(\omega)| + \frac{\alpha^2}{2}|t|)) < \infty$, $\forall(\alpha, t) \in \mathbb{R} \times \mathbb{I}$. This yields $\boldsymbol{E}(M_t|\mathcal{B}_s) = \sum_{n=0}^{\infty} \boldsymbol{E}(H_n(w_t, t)|\mathcal{B}_s)\frac{\alpha^n}{n!} = M_s = \sum_{n=0}^{\infty} H_n(w_s, s)\frac{\alpha^n}{n!}$ a.e. in $\boldsymbol{\Omega}_s$, $\forall s, t \in \mathbb{I}$ with $s \leq t$, $\forall\alpha \in \mathbb{R}$, where the first equality follows from (g) of Proposition 14.11, the second equality follows from $(M_t)_{t\in\mathbb{I}}$ being a Martingale, and the last equality follows from the definition of $(M_t)_{t\in\mathbb{I}}$. By the arbitrariness of α, we have $\boldsymbol{E}(H_n(w_t, t)|\mathcal{B}_s) = H_n(w_s, s)$ a.e. in $\boldsymbol{\Omega}_s$, $\forall t, s \in \mathbb{I}$ with $s \leq t$, $\forall n \in \mathbb{Z}_+$. Hence, $(H_n(w_t, t))_{t\in\mathbb{I}}$ is a Martingale, $\forall n \in \mathbb{Z}_+$. ◇

Proposition 14.91 *Let $\mathbb{I} := [0, \infty) \subset \mathbb{R}$, $\boldsymbol{\Omega} := (\Omega, \mathcal{B}, (\mathcal{B}_t)_{t\in\mathbb{I}}, \boldsymbol{P})$ be a filtered probability measure space, and $\tau : \Omega \to [0, \infty] \subset \mathbb{R}_e$ be a stopping time. Define $\mathcal{B}_\tau := \{A \in \mathcal{B} \,|\, \forall t \in \mathbb{I}, A \cap \{\omega \in \Omega \,|\, \tau(\omega) \leq t\} \in \mathcal{B}_t\}$. Then, the following statements hold:*

(i) *$\mathcal{B}_\tau$ is a σ-algebra on Ω.*
(ii) *τ is $\mathcal{B}_\tau$-measurable.*
(iii) *$\tilde{\Omega} := \{\omega \in \Omega \,|\, \tau(\omega) < \infty\} \in \mathcal{B}_\tau$.*
(iv) *Let $\tilde{\mathcal{B}}_\tau := \{A \in \mathcal{B}_\tau \,|\, A \subseteq \tilde{\Omega}\}$. Then, $\tilde{\mathcal{B}}_\tau$ is a σ-algebra on the set $\tilde{\Omega}$.*

Proof (i) Clearly, $\Omega \in \mathcal{B}_\tau$, since, $\forall t \in \mathbb{I}$, $E_t := \{\omega \in \Omega \,|\, \tau(\omega) \leq t\} \in \mathcal{B}_t$ by the assumption that τ is a stopping time.

Clearly, $\emptyset \in \mathcal{B}_\tau$ since $\emptyset \in \mathcal{B}_t$, $\forall t \in \mathbb{I}$.

$\forall A \in \mathcal{B}_\tau$, $\forall t \in \mathbb{I}$, we have $(\Omega \setminus A) \cap E_t = E_t \setminus A = E_t \setminus (E_t \cap A) = E_t \cap (\Omega \setminus (E_t \cap A)) \in \mathcal{B}_t$, since $E_t, E_t \cap A, \Omega \in \mathcal{B}_t$. By the arbitrariness of t, we have $\Omega \setminus A \in \mathcal{B}_\tau$.

$\forall (A_n)_{n=1}^{\infty} \subseteq \mathcal{B}_\tau$, $\forall t \in \mathbb{I}$, we have $(\bigcup_{n=1}^{\infty} A_n) \cap E_t = \bigcup_{n=1}^{\infty}(A_n \cap E_t) \in \mathcal{B}_t$. By the arbitrariness of t, we have $\bigcup_{n=1}^{\infty} A_n \in \mathcal{B}_\tau$.

This shows that $\mathcal{B}_\tau$ is a σ-algebra on Ω.

(ii) It follows immediately from Proposition 11.35.

(iii) Clearly, $E_t \in \mathcal{B}_\tau$, $\forall t \in \mathbb{I}$. Note that $\tilde{\Omega} = \bigcup_{n=1}^{\infty} E_n \in \mathcal{B}_\tau$.

(iv) It is easy to show that $\tilde{\mathcal{B}}_\tau$ is a σ-algebra on $\tilde{\Omega}$ by Proposition 11.13. This completes the proof of the proposition. □

Proposition 14.92 *Let $m \in \mathbb{N}$, $\mathbb{I} := (([0, \infty), |\cdot|), \hat{\mathcal{B}}, \mu)$ be the σ-finite metric measure subspace of $\mathbb{R}$ with the partial ordering $\leq$, $\boldsymbol{\Omega} := (\Omega, \mathcal{B}, (\mathcal{B}_t)_{t\in\mathbb{I}}, \boldsymbol{P})$ be a filtered probability measure space, $\mathbb{I} \times \boldsymbol{\Omega} =: ([0, \infty) \times \Omega, \bar{\mathcal{B}}, \nu)$ be the σ-finite product measure space, $w : \mathbb{I} \times \boldsymbol{\Omega} \to \mathbb{R}^m$ be the $\mathbb{R}^m$-valued standard Wiener process that is adapted to the filtration $(\mathcal{B}_t)_{t\in\mathbb{I}}$, $\tau : \Omega \to [0, \infty] \subset \mathbb{R}_e$ be a stopping time, $w_t : \Omega \to \mathbb{R}^m$ be defined by $w_t(\omega) = w(t, \omega)$, $\forall\omega \in \Omega$, $\forall t \in \mathbb{I}$. Assume that $\mathcal{B}_t$ and $\sigma((w_{t+s} - w_t)_{s\in\mathbb{I}})$ are independent σ-algebras, $\forall t \in \mathbb{I}$. Let $\tilde{\Omega}$*

and $\tilde{\mathcal{B}}_\tau$ *be as defined in Proposition 14.91, and* $\tilde{\mathcal{B}} := \{E \in \mathcal{B} \mid E \subseteq \tilde{\Omega}\}$*. Assume that* $\boldsymbol{P}(\tilde{\Omega}) > 0$*. Define* $\tilde{\boldsymbol{P}} : \tilde{\mathcal{B}} \to [0, 1] \subset \mathbb{R}$ *by* $\tilde{\boldsymbol{P}}(A) := \boldsymbol{P}(A|\tilde{\Omega})$*,* $\forall A \in \tilde{\mathcal{B}}$*. Then,*

(i) $\tilde{\boldsymbol{\Omega}} := (\tilde{\Omega}, \tilde{\mathcal{B}}, \tilde{\boldsymbol{P}})$ *is a probability measure space.*

(ii) *Let* $\mathbb{I} \times \tilde{\boldsymbol{\Omega}} =: ([0, \infty) \times \tilde{\Omega}, \bar{\tilde{\mathcal{B}}}, \bar{\nu})$ *be the* σ*-finite product measure space and define* $y : \mathbb{I} \times \tilde{\boldsymbol{\Omega}} \to \mathbb{R}^m$ *by* $y(t, \omega) := w(\tau(\omega)+t, \omega) - w(\tau(\omega), \omega)$*,* $\forall (t, \omega) \in \mathbb{I} \times \tilde{\boldsymbol{\Omega}}$*. Then,* y *is an* $\mathbb{R}^m$*-valued standard Wiener process on* $\tilde{\boldsymbol{\Omega}}$*.*

(iii) *The* σ*-algebras* $\tilde{\mathcal{B}}_\tau$ *and* $\sigma((y_t)_{t\in\mathbb{I}})$ *are independent, i.e., the process* y *is independent of* $\tilde{\mathcal{B}}_\tau$*, where* $y_t : \tilde{\boldsymbol{\Omega}} \to \mathbb{R}^m$ *is defined by* $y_t(\omega) = y(t, \omega)$*,* $\forall \omega \in \tilde{\Omega}$*,* $\forall t \in \mathbb{I}$*.*

Proof (i) By Proposition 14.91, $(\tilde{\Omega}, \tilde{\mathcal{B}}_\tau)$ is a measurable space. By Proposition 11.13, $(\tilde{\Omega}, \tilde{\mathcal{B}})$ is a measurable space. By the discussion on Page 763, $\tilde{\boldsymbol{P}}(A) = \boldsymbol{P}(A \cap \tilde{\Omega})/\boldsymbol{P}(\tilde{\Omega})$, $\forall A \in \tilde{\mathcal{B}}$. Hence, it is easy to show that $\tilde{\boldsymbol{\Omega}}$ is a probability measure space.

(ii) We will first show that y is $\bar{\tilde{\mathcal{B}}}$-measurable. Note that $y(t, \omega) = w(h(t, \omega)) - w(h(0, \omega))$, $\forall (t, \omega) \in \mathbb{I} \times \tilde{\boldsymbol{\Omega}}$, where $h : \mathbb{I} \times \tilde{\boldsymbol{\Omega}} \to \mathbb{I} \times \tilde{\boldsymbol{\Omega}}$ is defined by $h(t, \omega) = (g(t, \omega), \omega) \in \mathbb{I} \times \tilde{\boldsymbol{\Omega}}$, $\forall (t, \omega) \in \mathbb{I} \times \tilde{\boldsymbol{\Omega}}$, and $g : \mathbb{I} \times \tilde{\boldsymbol{\Omega}} \to \mathbb{I}$ is defined by $g(t, \omega) = t + \tau(\omega) \in \mathbb{I}$, $\forall (t, \omega) \in \mathbb{I} \times \tilde{\boldsymbol{\Omega}}$. By Propositions 11.38 and 11.39, we have g is $\bar{\tilde{\mathcal{B}}}$-measurable. We will show that h is $\bar{\tilde{\mathcal{B}}}$-in-$\bar{\tilde{\mathcal{B}}}$-measurable.

Let $\mathcal{F} := \{J \times E \subseteq \mathbb{I} \times \tilde{\boldsymbol{\Omega}} \mid J \in \hat{\mathcal{B}}, E \in \tilde{\mathcal{B}}\}$ be the set of measurable rectangles in $\mathbb{I} \times \tilde{\boldsymbol{\Omega}}$. Clearly, $\sigma(\mathcal{F}) = \bar{\tilde{\mathcal{B}}}$ and $\mathcal{F}$ is a π-system. $\forall J \times E \in \mathcal{F}$, $h_{\text{inv}}(J \times E) = \{(t, \omega) \in \mathbb{I} \times \tilde{\boldsymbol{\Omega}} \mid \omega \in E, g(t, \omega) \in J\} = \{(t, \omega) \in \mathbb{I} \times \tilde{\boldsymbol{\Omega}} \mid (t, \omega) \in \mathbb{I} \times E, (t, \omega) \in g_{\text{inv}}(J)\} = g_{\text{inv}}(J) \cap (\mathbb{I} \times E)$. By g being $\bar{\tilde{\mathcal{B}}}$-measurable, we have $g_{\text{inv}}(J) \in \bar{\tilde{\mathcal{B}}}$. Clearly, $\mathbb{I} \times E \in \mathcal{F} \subseteq \bar{\tilde{\mathcal{B}}}$. Thus, we conclude that $h_{\text{inv}}(J \times E) \in \bar{\tilde{\mathcal{B}}}$. By the arbitrariness of $J \times E$ and Proposition 14.23, we have h is $\bar{\tilde{\mathcal{B}}}$-in-$\bar{\tilde{\mathcal{B}}}$-measurable.

Note that $w : \mathbb{I} \times \boldsymbol{\Omega} \to \mathbb{R}^m$ is $\bar{\mathcal{B}}$-measurable. By Proposition 12.23, $w|_{\mathbb{I} \times \tilde{\boldsymbol{\Omega}}} : \mathbb{I} \times \tilde{\boldsymbol{\Omega}} \to \mathbb{R}^m$ is $\bar{\tilde{\mathcal{B}}}$-measurable. Then, by Propositions 14.24, 11.38, and 11.39, we have y is $\bar{\tilde{\mathcal{B}}}$-measurable.

This establishes (a) of Definition 14.78. (b) and (c) of Definition 14.78 are straightforward for the stochastic process y. For (d) of Definition 14.78, we need the following intermediate result:

Claim 14.92.1 $\forall j \in \mathbb{N}$, $\forall s_1, \ldots, s_j \in (0, \infty) \subset \mathbb{R}$, $\forall A \in \tilde{\mathcal{B}}_\tau$, $\forall$ bounded $\mathcal{B}_{\text{B}}(\mathbb{R}^{mj})$-measurable function $\phi : \underbrace{\mathbb{R}^m \times \cdots \times \mathbb{R}^m}_{j-\text{times}} \to \mathbb{R}$, we have $\boldsymbol{E}(\chi_{A,\Omega} \cdot \phi(y_{s_1}, \ldots, y_{s_j})) = \boldsymbol{P}(A)\boldsymbol{E}(\phi(w_{s_1}, \ldots, w_{s_j}))$.

Proof of Claim Fix any $n \in \mathbb{N}$. Define $\tau_n : \tilde{\boldsymbol{\Omega}} \to \mathbb{I}$ by $\tau_n(\omega) = \frac{k}{2^n}$ if $\frac{k-1}{2^n} < \tau(\omega) \leq \frac{k}{2^n}$, $k \in \mathbb{N}$. Then, we have $\lim_{n\in\mathbb{N}} \tau_n(\omega) \downarrow \tau(\omega)$, $\forall \omega \in \tilde{\boldsymbol{\Omega}}$.

Fix any bounded $\mathcal{B}_{\text{B}}(\mathbb{R}^{mj})$-measurable function $\phi : \mathbb{R}^m \times \cdots \times \mathbb{R}^m \to \mathbb{R}$. Since y is $\bar{\tilde{\mathcal{B}}}$-measurable, then the function under the expectation on the left-hand-side of the desired equality is $\tilde{\mathcal{B}}$-measurable. Since it is further bounded, then it is an element of $\bar{L}_1(\tilde{\boldsymbol{\Omega}}, \mathbb{R})$. We have the following line of arguments:

$$\boldsymbol{E}(\chi_{A,\Omega}\phi(y_{s_1}, \ldots, y_{s_j})) = \lim_{n\in\mathbb{N}} \boldsymbol{E}(\chi_{A,\Omega}\phi(w_{\tau_n+s_1} - w_{\tau_n}, \ldots, w_{\tau_n+s_j} - w_{\tau_n}))$$

$$= \lim_{n\in\mathbb{N}} \sum_{k=1}^{\infty} \boldsymbol{E}(\chi_{A,\Omega}\chi_{\{\omega\in\Omega \mid \tau_n(\omega)-\frac{k}{2^n}\},\Omega} \cdot$$

$$\phi(w_{\tau_n+s_1} - w_{\tau_n}, \ldots, w_{\tau_n+s_j} - w_{\tau_n}))$$

$$= \lim_{n\in\mathbb{N}} \sum_{k=1}^{\infty} \boldsymbol{E}(\chi_{A\cap\{\omega\in\Omega \mid \frac{k-1}{2^n}<\tau(\omega)\le\frac{k}{2^n}\},\Omega} \cdot$$

$$\phi(w_{\frac{k}{2^n}+s_1} - w_{\frac{k}{2^n}}, \ldots, w_{\frac{k}{2^n}+s_j} - w_{\frac{k}{2^n}}))$$

$$= \lim_{n\in\mathbb{N}} \sum_{k=1}^{\infty} \boldsymbol{E}(\chi_{A\cap\{\omega\in\Omega \mid \frac{k-1}{2^n}<\tau(\omega)\le\frac{k}{2^n}\},\Omega}) \cdot$$

$$\boldsymbol{E}(\phi(w_{\frac{k}{2^n}+s_1} - w_{\frac{k}{2^n}}, \ldots, w_{\frac{k}{2^n}+s_j} - w_{\frac{k}{2^n}}))$$

$$= \lim_{n\in\mathbb{N}} \sum_{k=1}^{\infty} \boldsymbol{E}(\chi_{A\cap\{\omega\in\Omega \mid \frac{k-1}{2^n}<\tau(\omega)\le\frac{k}{2^n}\},\Omega})\boldsymbol{E}(\phi(w_{s_1}, \ldots, w_{s_j}))$$

$$= \lim_{n\in\mathbb{N}} \left(\sum_{k=1}^{\infty} \boldsymbol{E}(\chi_{A\cap\{\omega\in\Omega \mid \frac{k-1}{2^n}<\tau(\omega)\le\frac{k}{2^n}\},\Omega}) \right) \boldsymbol{E}(\phi(w_{s_1}, \ldots, w_{s_j}))$$

$$= \lim_{n\in\mathbb{N}} \boldsymbol{E}(\chi_{A,\Omega})\boldsymbol{E}(\phi(w_{s_1}, \ldots, w_{s_j})) = \boldsymbol{P}(A)\boldsymbol{E}(\phi(w_{s_1}, \ldots, w_{s_j}))$$

where the first equality follows from Bounded Convergence Theorem 11.77, the second equality follows from Proposition 11.92, the fourth equality follows from the fact that $A \cap \{\omega \in \Omega \mid \frac{k-1}{2^n} < \tau(\omega) \le \frac{k}{2^n}\} \in \mathcal{B}_{\frac{k}{2^n}}$ and the independence assumption, the fifth equality follows from Definition 14.78, the sixth equality follows from simple algebra, the seventh equality follows from Proposition 11.92, and the last equality follows from Proposition 11.75.

This completes the proof of the claim. □

By Claim 14.92.1, we observe that $\tilde{\boldsymbol{E}}(\phi(y_{s_1}, \ldots, y_{s_j})) = \frac{1}{\boldsymbol{P}(\tilde{\Omega})}\boldsymbol{E}(\chi_{\tilde{\Omega},\Omega} \cdot \phi(y_{s_1}, \ldots, y_{s_j})) = \boldsymbol{E}(\phi(w_{s_1}, \ldots, w_{s_j}))$, for any $j \in \mathbb{N}$, for any $s_1, \ldots, s_j \in (0, \infty) \subset \mathbb{R}$, and for any bounded $\mathcal{B}_B(\mathbb{R}^{mj})$-measurable function $\phi : \mathbb{R}^m \times \cdots \times \mathbb{R}^m \to \mathbb{R}$, where $\tilde{\boldsymbol{E}}$ is the expectation in $\tilde{\boldsymbol{\Omega}}$ measure space. Therefore, $(y_{s_1}, \ldots, y_{s_j})$ admits the same law (in $\tilde{\boldsymbol{\Omega}}$) as $(w_{s_1}, \ldots, w_{s_j})$ (in $\boldsymbol{\Omega}$). This establishes (d) of Definition 14.78. Hence, y is an $\mathbb{R}^m$-valued standard Wiener process of $\tilde{\boldsymbol{\Omega}}$.

(iii) This follows immediately from Claim 14.92.1.

This completes the proof of the proposition. □

Lemma 14.93 *Let $\mathbb{I} := (([0,\infty), |\cdot|), \hat{\mathcal{B}}, \mu)$ be the σ-finite metric measure subspace of $\mathbb{R}$ with partial ordering $\leq$, $\boldsymbol{\Omega} := (\Omega, \mathcal{B}, \boldsymbol{P})$ be a probability measure space, $\mathbb{I} \times \boldsymbol{\Omega} := ([0,\infty) \times \Omega, \bar{\mathcal{B}}, \nu)$ be the σ-finite product measure space, $w : \mathbb{I} \times \boldsymbol{\Omega} \to \mathbb{R}$ be a standard Wiener process (as defined in Definition 14.76), and $w_{t,\mathrm{M}} : \boldsymbol{\Omega} \to [0,\infty) \subset \mathbb{R}$ be defined by $w_{t,\mathrm{M}}(\omega) := \sup_{0\leq s\leq t} |w(s,\omega)|$, $\forall \omega \in \boldsymbol{\Omega}$, $\forall t \in (0,\infty) \subset \mathbb{R}$. Then, $0 < \boldsymbol{P}(\{\omega \in \Omega \mid w_{t,\mathrm{M}}(\omega) > \beta\}) \leq 2\exp(-\frac{\beta^2}{2t})$, $\forall \beta \in (0,\infty) \subset \mathbb{R}$.*

Proof Fix any $t, \beta \in (0,\infty) \subset \mathbb{R}$. Let $\alpha := \frac{\beta}{t} \in (0,\infty) \subset \mathbb{R}$, $E := \{\omega \in \Omega \mid w_{t,\mathrm{M}}(\omega) > \beta\}$, $A_1 := \{\omega \in \Omega \mid w(s,\omega) > \frac{\alpha}{2}s + \frac{\beta}{2}$, for some $s \in \mathbb{I}\}$ and $A_2 := \{\omega \in \Omega \mid -w(s,\omega) > \frac{\alpha}{2}s + \frac{\beta}{2}$, for some $s \in \mathbb{I}\}$. By Lemma 14.85, we have $\boldsymbol{P}(A_1) = \boldsymbol{P}(A_2) \leq \exp(-\alpha\frac{\beta}{2}) = \exp(-\frac{\beta^2}{2t})$. Clearly, we have $E \subseteq A_1 \cup A_2$. This implies that $\boldsymbol{P}(E) \leq \boldsymbol{P}(A_1) + \boldsymbol{P}(A_2) \leq 2\exp(-\frac{\beta^2}{2t})$.

On the other hand, we have $E \supseteq A_3 := \{\omega \in \Omega \mid |w(t,\omega)| > \beta\}$. Then, $\boldsymbol{P}(E) \geq \boldsymbol{P}(A_3) = 2\int_\beta^\infty \frac{1}{\sqrt{2\pi t}} \exp(-\frac{x^2}{2t})\,\mathrm{d}x > 0$, where the equality follows from the fact that $w_t(\cdot) := w(t,\cdot) : \boldsymbol{\Omega} \to \mathbb{R}$ satisfies $w_t \sim N(0,t)$. This completes the proof of the lemma. □

Proposition 14.94 *Let $p \in [1,\infty) \subset \mathbb{R}$, $\mathbb{I} := (([0,\infty), |\cdot|), \hat{\mathcal{B}}, \mu)$ be the σ-finite metric measure subspace of $\mathbb{R}$ with partial ordering $\leq$, $\boldsymbol{\Omega} := (\Omega, \mathcal{B}, (\mathcal{B}_t)_{t\in\mathbb{I}}, \boldsymbol{P})$ be a filtered probability measure space, $\mathbb{I} \times \boldsymbol{\Omega} = ([0,\infty) \times \Omega, \bar{\mathcal{B}}, \nu)$ be the σ-finite product measure space, $w : \mathbb{I} \times \boldsymbol{\Omega} \to \mathbb{R}$ be the standard Wiener process on $\mathbb{I} \times \boldsymbol{\Omega}$ (as defined in Definition 14.76) that is adapted to the filtration $(\mathcal{B}_t)_{t\in\mathbb{I}}$, and the stochastic process $(w_t)_{t\in\mathbb{I}}$ defined by $w_t(\omega) = w(t,\omega)$, $\forall \omega \in \boldsymbol{\Omega}$, $\forall t \in \mathbb{I}$. Assume that $\mathcal{B}_t$ and $\sigma((w_{t+s} - w_t)_{s\in\mathbb{I}})$ are independent σ-algebras, $\forall t \in \mathbb{I}$. Then, there exist constants $C_p \in (0,\infty) \subset \mathbb{R}$, $\forall \tau : \boldsymbol{\Omega} \to [0,\infty] \subset \mathbb{R}_e$ be a stopping time, we have $\|w_{\tau,\mathrm{M}}\|_p \leq C_p \|\tau^{1/2}\|_p$, where $w_{\tau,\mathrm{M}} : \boldsymbol{\Omega} \to [0,\infty] \subset \mathbb{R}_e$ is defined by $w_{\tau,\mathrm{M}}(\omega) := \sup_{t\in\mathbb{I}} |w(\tau(\omega) \wedge t, \omega)|$, $\forall \omega \in \boldsymbol{\Omega}$, $\|w_{\tau,\mathrm{M}}\|_p := \big(\boldsymbol{E}(w_{\tau,\mathrm{M}}^p)\big)^{1/p}$, and $\|\tau^{1/2}\|_p := \big(\boldsymbol{E}(\tau^{p/2})\big)^{1/p}$.*

Proof We need the following intermediate result.

Claim 14.94.1 Fix any $t \in \mathbb{I}$, and let $w_{t,\mathrm{M}} : \boldsymbol{\Omega} \to [0,\infty) \subset \mathbb{R}$ be defined by $w_{t,\mathrm{M}}(\omega) := \sup_{0\leq s\leq t} |w(s,\omega)|$, $\forall \omega \in \boldsymbol{\Omega}$. $\forall \beta \in (1,\infty) \subset \mathbb{R}$, $\forall \lambda, \delta \in (0,\infty) \subset \mathbb{R}$, we have $\boldsymbol{P}(\{\omega \in \Omega \mid w_{\tau,\mathrm{M}}(\omega) > \beta\lambda, (\tau(\omega))^{1/2} \leq \delta\lambda\}) \leq \boldsymbol{P}(\{\omega \in \Omega \mid w_{\tau,\mathrm{M}}(\omega) > \lambda\})\boldsymbol{P}(\{\omega \in \Omega \mid w_{1,\mathrm{M}}(\omega) > \frac{\beta-1}{2\delta}\})$.

Proof of Claim Fix any $\beta \in (1,\infty) \subset \mathbb{R}$ and any $\lambda, \delta \in (0,\infty) \subset \mathbb{R}$. Define $W : \mathbb{I} \times \boldsymbol{\Omega} \to [0,\infty) \subset \mathbb{R}$ defined by $W(t,\omega) = |w(\tau(\omega) \wedge t, \omega)|$, $\forall (t,\omega) \in \mathbb{I} \times \boldsymbol{\Omega}$. Note that the mapping $(t,\omega) \in \mathbb{I} \times \boldsymbol{\Omega} \mapsto \tau(\omega) \wedge t \in \mathbb{I}$ is $\bar{\mathcal{B}}$-measurable, by Proposition 11.40. Then, the mapping $(t,\omega) \in \mathbb{I} \times \boldsymbol{\Omega} \mapsto (\tau(\omega) \wedge t, \omega) \in \mathbb{I} \times \boldsymbol{\Omega}$ is $\bar{\mathcal{B}}$-in-$\bar{\mathcal{B}}$-measurable, by Proposition 14.23. Then, by Proposition 14.24, we have W is $\bar{\mathcal{B}}$-measurable. Clearly, $W(\cdot,\omega) : \mathbb{I} \to [0,\infty) \subset \mathbb{R}$ is continuous, $\forall \omega \in \boldsymbol{\Omega}$. Define $\kappa : \boldsymbol{\Omega} \to [0,\infty] \subset \mathbb{R}_e$ by $\kappa(\omega) := \inf\{t \in \mathbb{I} \mid |w(\tau(\omega) \wedge t, \omega)| \geq \frac{\beta+1}{2}\lambda\}$, $\forall \omega \in \Omega$.

By Example 14.88, κ is a stopping time. Then, we have $\{\omega \in \Omega \mid w_{\mathcal{T},\mathrm{M}}(\omega) > \frac{\beta+1}{2}\lambda\} \subseteq \{\omega \in \Omega \mid \kappa(\omega) < \infty\} \subseteq \{\omega \in \Omega \mid w_{\mathcal{T},\mathrm{M}}(\omega) \geq \frac{\beta+1}{2}\lambda\}$. Clearly, we have $w_{\mathcal{T},\mathrm{M}}(\omega) = \sup_{t\in\mathbb{I}} W(t,\omega)$, $\forall \omega \in \boldsymbol{\Omega}$. By the continuity of $W(\cdot,\omega) : \mathbb{I} \to [0,\infty) \subset \mathbb{R}$, we can conclude that $w_{\mathcal{T},\mathrm{M}}$ is $\mathcal{B}$-measurable. Similarly, we may conclude that $w_{t,\mathrm{M}}$ is $\mathcal{B}_t$-measurable, $\forall t \in \mathbb{I}$. Thus, we have the following line of arguments:

$$
\begin{aligned}
&\boldsymbol{P}(\{\omega \in \Omega \mid w_{\mathcal{T},\mathrm{M}}(\omega) > \beta\lambda, (\tau(\omega))^{1/2} \leq \delta\lambda\}) \\
&= \boldsymbol{P}(\{\omega \in \Omega \mid \kappa(\omega) < \infty, \tau(\omega) \leq \delta^2\lambda^2, \sup_{\kappa(\omega)\leq s\leq\tau(\omega)} |w(s,\omega)| > \beta\lambda\}) \\
&\leq \boldsymbol{P}(\{\omega \in \Omega \mid \kappa(\omega) < \infty, \sup_{0\leq s\leq\delta^2\lambda^2} |w(\kappa(\omega)+s,\omega) - w(\kappa(\omega),\omega)| > \\
&\quad \frac{\beta-1}{2}\lambda\}) \\
&= \boldsymbol{P}(\{\omega \in \Omega \mid \kappa(\omega) < \infty\})\boldsymbol{P}(\{\omega \in \Omega \mid \sup_{0\leq s\leq\delta^2\lambda^2} |w(s,\omega)| > \\
&\quad \frac{\beta-1}{2}\lambda\}) \\
&= \boldsymbol{P}(\{\omega \in \Omega \mid \kappa(\omega) < \infty\})\boldsymbol{P}(\{\omega \in \Omega \mid w_{1,\mathrm{M}}(\omega) > \frac{\beta-1}{2\delta}\}) \\
&\leq \boldsymbol{P}(\{\omega \in \Omega \mid w_{\mathcal{T},\mathrm{M}}(\omega) \geq \frac{\beta+1}{2}\lambda\})\boldsymbol{P}(\{\omega \in \Omega \mid w_{1,\mathrm{M}}(\omega) > \frac{\beta-1}{2\delta}\}) \\
&\leq \boldsymbol{P}(\{\omega \in \Omega \mid w_{\mathcal{T},\mathrm{M}}(\omega) > \lambda\})\boldsymbol{P}(\{\omega \in \Omega \mid w_{1,\mathrm{M}}(\omega) > \frac{\beta-1}{2\delta}\})
\end{aligned}
$$

where the first equality follows from the equality of the two sets on both sides of the equality and the set on the left-hand-side of the equality is $\mathcal{B}$-measurable, the first inequality follows from the fact that the set on the left-hand-side of the inequality is a subset of the set on the right-hand side of the inequality, and the set on the right-hand side of the inequality is $\mathcal{B}$-measurable by Proposition 14.92, the second equality follows from Proposition 14.92, in particular Claim 14.92.1, the third equality follows from Proposition 14.81, and the last two inequalities follow from the preceding discussion and the fact that $w_{\mathcal{T},\mathrm{M}}$ is $\mathcal{B}$-measurable.

This completes the proof of the claim. □

Note that we have the following line of arguments, $\forall \beta \in (1, \infty) \subset \mathbb{R}$, $\forall \delta \in (0, \infty) \subset \mathbb{R}$.

$$\begin{aligned}
\|w_{\tau,\mathrm{M}}\|_p^p &= \beta^p \|\beta^{-1} w_{\tau,\mathrm{M}}\|_p^p = \beta^p \boldsymbol{E}((\beta^{-1} w_{\tau,\mathrm{M}})^p) \\
&= \beta^p \int_{(0,\infty)} p z^{p-1} \boldsymbol{P}(\{\omega \in \Omega \mid \beta^{-1} w_{\tau,\mathrm{M}}(\omega) > z\}) \, \mathrm{d}\mu_{\mathrm{B}}(z) \\
&= \beta^p \int_{(0,\infty)} p z^{p-1} \boldsymbol{P}(\{\omega \in \Omega \mid w_{\tau,\mathrm{M}}(\omega) > \beta z\}) \, \mathrm{d}\mu_{\mathrm{B}}(z) \\
&= \beta^p \int_{(0,\infty)} p z^{p-1} \big(\boldsymbol{P}(\{\omega \in \Omega \mid w_{\tau,\mathrm{M}}(\omega) > \beta z, (\tau(\omega))^{1/2} \le \delta z\}) \\
&\quad + \boldsymbol{P}(\{\omega \in \Omega \mid w_{\tau,\mathrm{M}}(\omega) > \beta z, (\tau(\omega))^{1/2} > \delta z\})\big) \, \mathrm{d}\mu_{\mathrm{B}}(z) \\
&\le \beta^p \int_{(0,\infty)} p z^{p-1} \big(\epsilon(\beta, \delta) \boldsymbol{P}(\{\omega \in \Omega \mid w_{\tau,\mathrm{M}}(\omega) > z\}) \\
&\quad + \boldsymbol{P}(\{\omega \in \Omega \mid \delta^{-1}(\tau(\omega))^{1/2} > z\})\big) \, \mathrm{d}\mu_{\mathrm{B}}(z) \\
&= \beta^p \epsilon(\beta, \delta) \boldsymbol{E}(w_{\tau,\mathrm{M}}^p) + \beta^p \boldsymbol{E}((\delta^{-1} \tau^{1/2})^p) \\
&= \beta^p \epsilon(\beta, \delta) \|w_{\tau,\mathrm{M}}\|_p^p + \beta^p \delta^{-p} \|\tau^{1/2}\|_p^p
\end{aligned}$$

where the third equality follows from Proposition 14.21, the first inequality follows from Claim 14.94.1, and $\epsilon(\beta, \delta) := \boldsymbol{P}(\{\omega \in \Omega \mid w_{1,\mathrm{M}} > \frac{\beta-1}{2\delta}\}) \in (0, 1] \subset \mathbb{R}$, and the sixth equality follows from Proposition 14.21.

By Lemma 14.93, the constant $\epsilon(\beta, \delta)$ satisfies $0 < \epsilon(\beta, \delta) \le 2 \cdot \exp(-\frac{(\beta-1)^2}{8\delta^2})$. Then, we may fix $\beta > 1$ first and then choose $\delta = \frac{\beta-1}{\sqrt{8\ln(4\beta^p)}} > 0$ such that $0 < \epsilon(\beta, \delta) \le \frac{1}{2\beta^p}$. Then, the previous paragraph implies that $\|w_{\tau,\mathrm{M}}\|_p \le C_p \|\tau^{1/2}\|_p$ with $C_p := 2^{1/p} \frac{\beta}{\delta} = 2^{1/p} \frac{\beta\sqrt{8\ln(4\beta^p)}}{\beta-1} \in (0, \infty) \subset \mathbb{R}$.

This completes the proof of the proposition. □

14.12 Stochastic Integral

Definition 14.95 Let $\mathbb{I} := (([0, \infty), |\cdot|), \hat{\mathcal{B}}, \mu)$ be the σ-finite metric measure subspace of $\mathbb{R}$ with partial ordering $\le$, $\boldsymbol{\Omega} := (\Omega, \mathcal{B}, (\mathcal{B}_t)_{t\in\mathbb{I}}, \boldsymbol{P})$ be a filtered probability measure space, $\mathbb{I} \times \boldsymbol{\Omega} = ([0, \infty) \times \Omega, \bar{\mathcal{B}}, \nu)$ be the σ-finite product measure space, $w : \mathbb{I} \times \boldsymbol{\Omega} \to \mathbb{R}$ be a standard Wiener process on $\mathbb{I} \times \boldsymbol{\Omega}$, w_t be adapted to $(\mathcal{B}_t)_{t\in\mathbb{I}}$, $\mathcal{Y}$ be a Banach space, and $\phi : \bar{\mathrm{r}}_{a,b} \times \boldsymbol{\Omega} \to \mathcal{Y}$ be $\mathcal{Y}$-valued stochastic process that is adapted to $(\mathcal{B}_t)_{t\in\mathbb{I}}$, where $a, b \in \mathbb{I}$ with $a \le b$. Assume that $\phi_{\langle\omega\rangle} := \phi(\cdot, \omega) : \bar{\mathrm{r}}_{a,b} \to \mathcal{Y}$ is of bounded variation, $\forall \omega \in \boldsymbol{\Omega}$. (Use notations w_t and $w_{\langle\omega\rangle}$, etc., as in Proposition 12.28.) Define $I : \boldsymbol{\Omega} \to \mathcal{Y}$ by

$I(\omega) := w(b,\omega)\phi(b,\omega) - w(a,\omega)\phi(a,\omega) - \int_{\mathrm{r}_{a,b}} w(s,\omega)\,\mathrm{d}\phi_{\langle\omega\rangle}(s)$, $\forall\omega \in \Omega$, where the integral is the Lebesgue–Stieltjes integral as introduced in Definition 12.57. Note that, by Definition 14.76, $w_{\langle\omega\rangle} : \mathbb{I} \to \mathbb{R}$ is continuous, $\forall\omega \in \boldsymbol{\Omega}$. Then, by Proposition 7.126, Lebesgue Dominated Convergence Theorems 11.128, 12.50, and Definition 12.57, $\int_{\mathrm{r}_{a,b}} w(s,\omega)\,\mathrm{d}\phi_{\langle\omega\rangle}(s) \in \mathcal{Y}$, $\forall\omega \in \boldsymbol{\Omega}$. Thus, the integral I is well-defined. By Theorem A.10, we have $\int_{\mathrm{r}_{a,b}} w(s,\omega)\,\mathrm{d}\phi_{\langle\omega\rangle}(s) - \int_a^b w(s,\omega)\,\mathrm{d}\phi_{\langle\omega\rangle}(s) = \widehat{\int}_a^b w(s,\omega)\,\mathrm{d}\phi_{\langle\omega\rangle}(s)$, $\forall\omega \in \boldsymbol{\Omega}$, which further implies that $I(\omega) = \widehat{\int}_a^b \phi(s,\omega)\,\mathrm{d}w_{\langle\omega\rangle}(s) \in \mathcal{Y}$, $\forall\omega \in \boldsymbol{\Omega}$. Thus, the integral is simply the Riemann-Stieltjes integral carried out sample path-wise, when the integrand is of bounded variation sample path-wise. We will denote I by $\widehat{\int}_{\bar{\mathrm{r}}_{a,b}} \phi_s\,\mathrm{d}w_s$ or $\widehat{\int}_a^b \phi_s\,\mathrm{d}w_s$, which is called the *Riemann-Stieltjes integral of ϕ with respect to the Wiener process w on the interval $\bar{\mathrm{r}}_{a,b}$*. ◇

Based on Definition 14.95, we wish to expand our definition of integral to the more general class of $\mathrm{L}_2(\bar{\mathrm{r}}_{a,b} \times \boldsymbol{\Omega}, \mathcal{Y})$ integrands, where $\mathcal{Y}$ is a Hilbert space. First, we present a result that deduces the measurability of a function with domain of a product measure space.

Proposition 14.96 *Let $\mathcal{X}_i := (X_i, \mathcal{B}_i, \mu_i)$ be a σ-finite measure space, $i = 1, 2$, $\mathcal{X} := \mathcal{X}_1 \times \mathcal{X}_2 =: (X_1 \times X_2, \bar{\mathcal{B}}, \mu)$ be the σ-finite product measure space, $\mathcal{Y} := (Y, \mathcal{O})$ be a topological space, $f : X_1 \times X_2 \to \mathcal{Y}$ be $\bar{\mathcal{B}}$-measurable, $g : X_1 \times X_2 \to \mathcal{X}_1$ be $\mathcal{B}_1$-in-$\bar{\mathcal{B}}$-measurable, $h : X_1 \times X_2 \to \mathcal{X}_2$ be $\mathcal{B}_2$-in-$\bar{\mathcal{B}}$-measurable, and $l : X_1 \times X_2 \to X_1 \times X_2$ be defined by $l(x_1, x_2) = (g(x_1, x_2), h(x_1, x_2))$, $\forall (x_1, x_2) \in \mathcal{X}$. Then, we have l is $\bar{\mathcal{B}}$-in-$\bar{\mathcal{B}}$-measurable, and the function $f \circ l : \mathcal{X} \to \mathcal{Y}$ is $\bar{\mathcal{B}}$-measurable.*

Proof We will first show that l is $\bar{\mathcal{B}}$-in-$\bar{\mathcal{B}}$-measurable. Let $\mathcal{F} := \{D \times E \subseteq \mathcal{X}_1 \times \mathcal{X}_2 \mid D \in \mathcal{B}_1, E \in \mathcal{B}_2\}$ be the set of measurable rectangles in $\mathcal{X}$. Clearly, $\sigma(\mathcal{F}) = \bar{\mathcal{B}}$ and $\mathcal{F}$ is a π-system. $\forall D \times E \in \mathcal{F}$, $l_{\mathrm{inv}}(D \times E) = \{(x_1, x_2) \in X_1 \times X_2 \mid h(x_1, x_2) \in E, g(x_1, x_2) \in D\} = \{(x_1, x_2) \in X_1 \times X_2 \mid (x_1, x_2) \in h_{\mathrm{inv}}(E), (x_1, x_2) \in g_{\mathrm{inv}}(D)\} = g_{\mathrm{inv}}(D) \cap h_{\mathrm{inv}}(E)$. By g being $\mathcal{B}_1$-in-$\bar{\mathcal{B}}$-measurable, we have $g_{\mathrm{inv}}(D) \in \bar{\mathcal{B}}$. By h being $\mathcal{B}_2$-in-$\bar{\mathcal{B}}$-measurable, we have $h_{\mathrm{inv}}(E) \in \bar{\mathcal{B}}$. Thus, we conclude that $l_{\mathrm{inv}}(D \times E) \in \bar{\mathcal{B}}$. By the arbitrariness of $D \times E$ and Proposition 14.23, we have l is $\bar{\mathcal{B}}$-in-$\bar{\mathcal{B}}$-measurable.

By Proposition 11.21, we have $f \circ l$ is $\bar{\mathcal{B}}$-measurable. This completes the proof of the proposition. □

We next present the following result on the important properties of Riemann–Stieltjes integral with respect to the Wiener process integrator that readily applies to deterministic integrands.

Lemma 14.97 *Let $\mathbb{I} := (([0,\infty), |\cdot|), \hat{\mathcal{B}}, \mu)$ be the σ-finite metric measure subspace of $\mathbb{R}$ with partial ordering $\leq$, $\boldsymbol{\Omega} := (\Omega, \mathcal{B}, (\mathcal{B}_t)_{t\in\mathbb{I}}, \boldsymbol{P})$ be a filtered probability measure space, $\mathbb{I} \times \boldsymbol{\Omega} = ([0,\infty) \times \Omega, \bar{\mathcal{B}}, \nu)$ be the σ-finite product measure space, $w : \mathbb{I} \times \boldsymbol{\Omega} \to \mathbb{R}$ be a standard Wiener process on $\mathbb{I} \times \boldsymbol{\Omega}$ and* *be adapted to the filtration $(\mathcal{B}_t)_{t\in\mathbb{I}}$, the σ-algebras $\mathcal{B}_t$ and $\sigma((w_{t+s} - w_t)_{s\in\mathbb{I}})$ are independent, $\forall t \in \mathbb{I}$, $\mathcal{Y}$ and $\mathcal{Z}$ be Banach space over $\mathbb{K}$, and $a, b \in \mathbb{I}$ with $a \le b$. Then, the following statements hold:*

(i) *Let $\phi : \bar{\mathrm{r}}_{a,b} \times \boldsymbol{\Omega} \to \mathcal{Y}$ be a $\mathcal{Y}$-valued stochastic process that is adapted to $(\mathcal{B}_t)_{t\in\mathbb{I}}$, $\phi_{\langle\omega\rangle} : \bar{\mathrm{r}}_{a,b} \to \mathcal{Y}$ be of bounded variation, $\forall \omega \in \boldsymbol{\Omega}$, and $A \in B(\mathcal{Y}, \mathcal{Z})$. Then we have $\big(\widehat{\int}_{\bar{\mathrm{r}}_{a,b}} A\phi_s \,\mathrm{d}w_s\big)(\omega) = A\big(\big(\widehat{\int}_{\bar{\mathrm{r}}_{a,b}} \phi_s \,\mathrm{d}w_s\big)(\omega)\big) \in \mathcal{Z}$, $\forall \omega \in \boldsymbol{\Omega}$.*

(ii) *Let $\phi_i : \bar{\mathrm{r}}_{a,b} \times \boldsymbol{\Omega} \to \mathcal{Y}$ be a $\mathcal{Y}$-valued stochastic process that is adapted to $(\mathcal{B}_t)_{t\in\mathbb{I}}$, $\phi_{i\,\langle\omega\rangle} : \bar{\mathrm{r}}_{a,b} \to \mathcal{Y}$ be of bounded variation, $\forall \omega \in \boldsymbol{\Omega}$, $i = 1, 2$. Then, we have $\big(\widehat{\int}_{\bar{\mathrm{r}}_{a,b}} (\phi_1 + \phi_2)_s \,\mathrm{d}w_s\big)(\omega) = \big(\widehat{\int}_{\bar{\mathrm{r}}_{a,b}} \phi_{1s} \,\mathrm{d}w_s\big)(\omega) + \big(\widehat{\int}_{\bar{\mathrm{r}}_{a,b}} \phi_{2s} \,\mathrm{d}w_s\big)(\omega) \in \mathcal{Y}$, $\forall \omega \in \boldsymbol{\Omega}$.*

(iii) *Let $\phi : \bar{\mathrm{r}}_{a,b} \times \boldsymbol{\Omega} \to \mathcal{Y}$ be a $\mathcal{Y}$-valued stochastic process that is adapted to $(\mathcal{B}_t)_{t\in\mathbb{I}}$, and $c \in \bar{\mathrm{r}}_{a,b}$. Then, $\forall \omega \in \boldsymbol{\Omega}$, $\phi_{\langle\omega\rangle} : \bar{\mathrm{r}}_{a,b} \to \mathcal{Y}$ is of bounded variation, if, and only if, $\phi_{\langle\omega\rangle}\big|_{\bar{\mathrm{r}}_{a,c}}$ and $\phi_{\langle\omega\rangle}\big|_{\bar{\mathrm{r}}_{c,b}}$ are of bounded variation, $\forall \omega \in \boldsymbol{\Omega}$. In this case, we have $\big(\widehat{\int}_{\bar{\mathrm{r}}_{a,c}} \phi_s \,\mathrm{d}w_s\big)(\omega) + \big(\widehat{\int}_{\bar{\mathrm{r}}_{c,b}} \phi_s \,\mathrm{d}w_s\big)(\omega) = \big(\widehat{\int}_{\bar{\mathrm{r}}_{a,b}} \phi_s \,\mathrm{d}w_s\big)(\omega) \in \mathcal{Y}$, $\forall \omega \in \boldsymbol{\Omega}$.*

(iv) *Let $\mathcal{Y}$ be a separable Banach space, $\phi : \bar{\mathrm{r}}_{a,b} \times \boldsymbol{\Omega} \to \mathcal{Y}$ be a $\mathcal{Y}$-valued stochastic process that is adapted to $(\mathcal{B}_t)_{t\in\mathbb{I}}$, and $\phi_{\langle\omega\rangle} : \bar{\mathrm{r}}_{a,b} \to \mathcal{Y}$ be of bounded variation, $\forall \omega \in \boldsymbol{\Omega}$. Define $x : \bar{\mathrm{r}}_{a,b} \times \boldsymbol{\Omega} \to \mathcal{Y}$ by $x(t, \omega) = \big(\widehat{\int}_a^t \phi_s \,\mathrm{d}w_s\big)(\omega) \in \mathcal{Y}$, $\forall (t, \omega) \in \bar{\mathrm{r}}_{a,b} \times \boldsymbol{\Omega}$. Then, x is a $\mathcal{Y}$-valued stochastic process that is adapted to $(\mathcal{B}_t)_{t\in\mathbb{I}}$ with continuous sample paths.*

(v) *Let $\mathcal{Y}$ be a separable Hilbert space, $\phi : \bar{\mathrm{r}}_{a,b} \times \boldsymbol{\Omega} \to \mathcal{Y}$ be a $\mathcal{Y}$-valued stochastic process that is adapted to $(\mathcal{B}_t)_{t\in\mathbb{I}}$, and $\phi_{\langle\omega\rangle} : \bar{\mathrm{r}}_{a,b} \to \mathcal{Y}$ be of bounded variation, $\forall \omega \in \boldsymbol{\Omega}$. Assume that there exists a random variable $c : \boldsymbol{\Omega} \to [0, \infty) \subset \mathbb{R}$ such that $(\|\phi(a, \omega)\|_{\mathcal{Y}} + T_{\phi_{\langle\omega\rangle}}(\mathrm{r}_{a,b}))^2 \le c(\omega)$, $\forall \omega \in \boldsymbol{\Omega}$, and $\boldsymbol{E}\big(c\,(1 + w_{a,b,\mathrm{M}}^2)\big) < \infty$, where $w_{a,b,\mathrm{M}} : \boldsymbol{\Omega} \to [0, \infty) \subset \mathbb{R}$ is defined by $w_{a,b,\mathrm{M}} := \max_{a\le t\le b} |w(b, \omega) - w(t, \omega)|$. Then, we have $(\boldsymbol{E}(\|\widehat{\int}_{\bar{\mathrm{r}}_{a,b}} \phi_s \,\mathrm{d}w_s\|_{\mathcal{Y}}^2))^{1/2} = \|\widehat{\int}_{\bar{\mathrm{r}}_{a,b}} \phi_s \,\mathrm{d}w_s\|_{\mathrm{L}_2(\boldsymbol{\Omega},\mathcal{Y})} = \|\phi\|_{\mathrm{L}_2(\bar{\mathrm{r}}_{a,b}\times\boldsymbol{\Omega},\mathcal{Y})} = \big(\int_{\boldsymbol{\Omega}} \int_{\bar{\mathrm{r}}_{a,b}} \|\phi(s, \omega)\|_{\mathcal{Y}}^2 \,\mathrm{d}s \,\mathrm{d}\boldsymbol{P}(\omega)\big)^{1/2} \in [0, \infty) \subset \mathbb{R}$.*

When $\mathcal{Y}$ is a separable Hilbert space, define $\hat{\mathcal{V}}(\bar{\mathrm{r}}_{a,b} \times \boldsymbol{\Omega}, \mathcal{Y}) := \{\phi : \bar{\mathrm{r}}_{a,b} \times \boldsymbol{\Omega} \to \mathcal{Y} \mid \phi$ is a $\mathcal{Y}$-valued stochastic process that is adapted to $(\mathcal{B}_t)_{t\in\mathbb{I}}$, $\phi_{\langle\omega\rangle} : \bar{\mathrm{r}}_{a,b} \to \mathcal{Y}$ be of bounded variation, $\forall \omega \in \boldsymbol{\Omega}$, there exists a random variable $c_\phi : \boldsymbol{\Omega} \to [0, \infty) \subset \mathbb{R}$ such that $(\|\phi(a, \omega)\|_{\mathcal{Y}} + T_{\phi_{\langle\omega\rangle}}(\mathrm{r}_{a,b}))^2 \le c_\phi(\omega)$, $\forall \omega \in \boldsymbol{\Omega}$, and $\boldsymbol{E}\big(c_\phi\,(1 + w_{a,b,\mathrm{M}}^2)\big) < \infty\}$. Then, $\hat{\mathcal{V}}(\bar{\mathrm{r}}_{a,b} \times \boldsymbol{\Omega}, \mathcal{Y})$ is a subspace of $\bar{\mathrm{L}}_2(\bar{\mathrm{r}}_{a,b} \times \boldsymbol{\Omega}, \mathcal{Y})$ and the Riemann-Stieltjes integral with Wiener process integrator is a pseudo-norm preserving homomorphism between $\hat{\mathcal{V}}(\bar{\mathrm{r}}_{a,b} \times \boldsymbol{\Omega}, \mathcal{Y})$ and its image in $\bar{\mathrm{L}}_2(\boldsymbol{\Omega}, \mathcal{Y})$.

Proof (i) By Proposition 12.117, $A\phi_{\langle\omega\rangle} : \bar{\mathrm{r}}_{a,b} \to \mathcal{Z}$ is also of bounded variation, $\forall \omega \in \boldsymbol{\Omega}$. Then, $\big(\widehat{\int}_{\bar{\mathrm{r}}_{a,b}} A\phi_s \,\mathrm{d}w_s\big)(\omega) = \widehat{\int}_a^b A\phi(s, \omega) \,\mathrm{d}w_{\langle\omega\rangle}(s) = A\big(\widehat{\int}_a^b \phi(s, \omega) \,\mathrm{d}w_{\langle\omega\rangle}(s)\big) = A\big(\big(\widehat{\int}_{\bar{\mathrm{r}}_{a,b}} \phi_s \,\mathrm{d}w_s\big)(\omega)\big) \in \mathcal{Z}$, $\forall \omega \in \boldsymbol{\Omega}$, where the first equality follows from Definition 14.95, the second equality follows from Definition A.4, and the last equality follows from Definition 14.95.

(ii) By Propositions 12.118 and 12.117 and the fact that vector addition is Lipschitz on $\mathcal{Y} \times \mathcal{Y}$, we have $(\phi_1 + \phi_2)_{\langle\omega\rangle} : \bar{\mathrm{r}}_{a,b} \to \mathcal{Y}$ is of bounded variation, $\forall \omega \in \boldsymbol{\Omega}$. Then, we have $\left(\widehat{\int}_{\bar{\mathrm{r}}_{a,b}} (\phi_1 + \phi_2)_s \, \mathrm{d}w_s\right)(\omega) = \widehat{\int}_a^b (\phi_1(s,\omega) + \phi_2(s,\omega)) \, \mathrm{d}w_{\langle\omega\rangle}(s) = \widehat{\int}_a^b \phi_1(s,\omega) \, \mathrm{d}w_{\langle\omega\rangle}(s) + \widehat{\int}_a^b \phi_2(s,\omega) \, \mathrm{d}w_{\langle\omega\rangle}(s) = \left(\widehat{\int}_{\bar{\mathrm{r}}_{a,b}} \phi_{1s} \cdot \mathrm{d}w_s\right)(\omega) + \left(\widehat{\int}_{\bar{\mathrm{r}}_{a,b}} \phi_{2s} \, \mathrm{d}w_s\right)(\omega) \in \mathcal{Y}$, $\forall \omega \in \boldsymbol{\Omega}$, where the first equality follows from Definition 14.95, the second equality follows from Theorem A.6, and the last equality follows from Definition 14.95.

(iii) Let $\phi_{\langle\omega\rangle} : \bar{\mathrm{r}}_{a,b} \to \mathcal{Y}$ be of bounded variation, $\forall \omega \in \boldsymbol{\Omega}$. By Definition 12.41, we have $\phi_{\langle\omega\rangle}|_{\bar{\mathrm{r}}_{a,c}}$ and $\phi_{\langle\omega\rangle}|_{\bar{\mathrm{r}}_{c,b}}$ are of bounded variation, $\forall \omega \in \boldsymbol{\Omega}$. In this case, we have $\left(\widehat{\int}_{\bar{\mathrm{r}}_{a,c}} \phi_s \, \mathrm{d}w_s\right)(\omega) + \left(\widehat{\int}_{\bar{\mathrm{r}}_{c,b}} \phi_s \, \mathrm{d}w_s\right)(\omega) = \widehat{\int}_{\bar{\mathrm{r}}_{a,c}} \phi(s,\omega) \, \mathrm{d}w_{\langle\omega\rangle}(s) + \widehat{\int}_{\bar{\mathrm{r}}_{c,b}} \phi(s,\omega) \, \mathrm{d}w_{\langle\omega\rangle}(s) = \widehat{\int}_{\bar{\mathrm{r}}_{a,b}} \phi(s,\omega) \, \mathrm{d}w_{\langle\omega\rangle}(s) = \left(\widehat{\int}_{\bar{\mathrm{r}}_{a,b}} \phi_s \, \mathrm{d}w_s\right)(\omega) \in \mathcal{Y}$, $\forall \omega \in \boldsymbol{\Omega}$, where the first equality follows from Definition 14.95, the second equality follows from Theorem A.7, and the last equality follows from Definition 14.95.

On the other hand, let $\phi_{\langle\omega\rangle}|_{\bar{\mathrm{r}}_{a,c}}$ and $\phi_{\langle\omega\rangle}|_{\bar{\mathrm{r}}_{c,b}}$ be of bounded variation, $\forall \omega \in \boldsymbol{\Omega}$. Fix any $\omega \in \boldsymbol{\Omega}$. By Theorem 12.50, there exist unique finite measures $\mu_1^{\langle\omega\rangle}$ and $\mu_2^{\langle\omega\rangle}$ on the measurable spaces $(\mathrm{r}_{a,c}, \mathcal{B}_{\mathrm{B}}(\mathrm{r}_{a,c}))$ and $(\mathrm{r}_{c,b}, \mathcal{B}_{\mathrm{B}}(\mathrm{r}_{c,b}))$, respectively, such that $\phi_{\langle\omega\rangle}|_{\bar{\mathrm{r}}_{a,c}}$ is a cumulative distribution function of $\mu_1^{\langle\omega\rangle}$ and $\phi_{\langle\omega\rangle}|_{\bar{\mathrm{r}}_{c,b}}$ is a cumulative distribution function of $\mu_2^{\langle\omega\rangle}$. By generation process Proposition 11.118, there exists a unique finite measure $\mu^{\langle\omega\rangle}$ on the measurable space $(\mathrm{r}_{a,b}, \mathcal{B}_{\mathrm{B}}(\mathrm{r}_{a,b}))$ such that $(\mathrm{r}_{a,c}, \mathcal{B}_{\mathrm{B}}(\mathrm{r}_{a,c}), \mu_1^{\langle\omega\rangle})$ and $(\mathrm{r}_{c,b}, \mathcal{B}_{\mathrm{B}}(\mathrm{r}_{c,b}), \mu_2^{\langle\omega\rangle})$ are its measure subspaces. By Definition 12.42, we have $\phi_{\langle\omega\rangle}$ is a cumulative distribution function for $\mu^{\langle\omega\rangle}$. If $\bar{\mathrm{r}}_{a,b} = \emptyset$, then $\phi_{\langle\omega\rangle}$ is of bounded variation. If $\bar{\mathrm{r}}_{a,b} \neq \emptyset$, let $x_0 \in \bar{\mathrm{r}}_{a,b}$. Then, $\bar{\phi}_\omega := \phi_{\langle\omega\rangle} - \phi(x_0, \omega) : \bar{\mathrm{r}}_{a,b} \to \mathcal{Y}$ is a cumulative distribution function of $\mu^{\langle\omega\rangle}$, and it is the cumulative distribution function of $\mu^{\langle\omega\rangle}$ with origin at x_0. By Proposition 12.52, $\bar{\phi}_\omega$ is of locally bounded variation. Then, it is of bounded variation since $\bar{\mathrm{r}}_{a,b}$ is a compact interval and $\mu^{\langle\omega\rangle}$ is finite. Then, $\phi_{\langle\omega\rangle}$ is of bounded variation by Definition 12.41 since $T_{\phi_{\langle\omega\rangle}}(\mathrm{r}_{x_1,x_2}) = T_{\bar{\phi}_\omega}(\mathrm{r}_{x_1,x_2}) = \mathcal{P} \circ \mu^{\langle\omega\rangle}(\mathrm{r}_{x_1,x_2}) < \mathcal{P} \circ \mu^{\langle\omega\rangle}(\mathrm{r}_{a,b}) < \infty$, $\forall x_1, x_2 \in \bar{\mathrm{r}}_{a,b}$ with $x_1 \le x_2$. This completes the proof of (iii).

(iv) Note that $x(t,\omega) := \left(\widehat{\int}_{\bar{\mathrm{r}}_{a,t}} \phi_s \, \mathrm{d}w_s\right)(\omega) = \widehat{\int}_a^t \phi(s,\omega) \, \mathrm{d}w_{\langle\omega\rangle}(s) = \lim_{n\in\mathbb{N}} \sum_{i=1}^n \phi(s_{i-1}^{\langle n\rangle}(t), \omega)(w(s_i^{\langle n\rangle}(t), \omega) - w(s_{i-1}^{\langle n\rangle}(t), \omega)) =: \lim_{n\in\mathbb{N}} Y_n(t,\omega)$, $\forall (t,\omega) \in \bar{\mathrm{r}}_{a,b} \times \boldsymbol{\Omega}$, where the first equality follows from Definition 14.95 and the second equality follows from Integrability Theorem A.10 with $s_i^{\langle n\rangle}(t) = a + i\frac{t-a}{n}$, $i = 0, \ldots, n$. By Propositions 14.96, 11.38, and 11.39, Y_n is $\bar{\mathcal{B}}$-measurable, $\forall n \in \mathbb{N}$. By Proposition 11.48, $x : \bar{\mathrm{r}}_{a,b} \times \boldsymbol{\Omega} \to \mathcal{Y}$ is $\bar{\mathcal{B}}$-measurable. Thus, x is a $\mathcal{Y}$-valued stochastic process. Again by Propositions 11.38, 11.39, and 11.48, $x_t : \boldsymbol{\Omega} \to \mathcal{Y}$ is $\mathcal{B}_t$-measurable, $\forall t \in \bar{\mathrm{r}}_{a,b}$. Hence, x is adapted to $(\mathcal{B}_t)_{t\in\mathbb{I}}$. By Theorem A.12, $x_{\langle\omega\rangle} : \bar{\mathrm{r}}_{a,b} \to \mathcal{Y}$ is continuous, $\forall \omega \in \boldsymbol{\Omega}$.

(v) By (iv), we have $\widehat{\int}_{\bar{\mathrm{r}}_{a,b}} \phi_s \,\mathrm{d}w_s$ is $\mathcal{B}_b$-measurable and therefore a $\mathcal{Y}$-valued random variable. Then, we have the following line of argument:

$$\left\| \widehat{\int}_{\bar{\mathrm{r}}_{a,b}} \phi_s \,\mathrm{d}w_s \right\|^2_{\mathrm{L}_2(\boldsymbol{\Omega},\mathcal{Y})} = \boldsymbol{E}\Big(\left\| \widehat{\int}_{\bar{\mathrm{r}}_{a,b}} \phi_s \,\mathrm{d}w_s \right\|^2_{\mathcal{Y}} \Big)$$
$$= \int_{\boldsymbol{\Omega}} \left\| \widehat{\int}_a^b \phi(s,\omega)\,\mathrm{d}w_{(\omega)}(s) \right\|^2_{\mathcal{Y}} \mathrm{d}\boldsymbol{P}(\omega)$$
$$= \int_{\Omega} \lim_{n\in\mathbb{N}} \left\| \sum_{i=1}^n \phi(s_{i-1}^{\langle n\rangle},\omega)(w(s_i^{\langle n\rangle},\omega) - w(s_{i-1}^{\langle n\rangle},\omega)) \right\|^2_{\mathcal{Y}} \mathrm{d}\boldsymbol{P}(\omega)$$

where the first equality follows from Definition 14.1, the second equality follows from Definition 14.95, and the third equality follows from Integrability Theorem A.10 with $s_i^{\langle n\rangle} = a + i\frac{b-a}{n}$, $i = 0, \ldots, n$.

Fix any $n \in \mathbb{N}$. $\forall \omega \in \boldsymbol{\Omega}$, we have

$$\left\| \sum_{i=1}^n \phi(s_{i-1}^{\langle n\rangle},\omega)(w(s_i^{\langle n\rangle},\omega) - w(s_{i-1}^{\langle n\rangle},\omega)) \right\|^2_{\mathcal{Y}}$$
$$= \left\| \sum_{i=1}^n \phi(s_{i-1}^{\langle n\rangle},\omega)(-(w(b,\omega) - w(s_i^{\langle n\rangle},\omega)) + (w(b,\omega) - w(s_{i-1}^{\langle n\rangle},\omega))) \right\|^2_{\mathcal{Y}}$$
$$= \Big\| \phi(a,\omega)(w(b,\omega) - w(a,\omega)) - \sum_{i=1}^n (w(b,\omega) - w(s_i^{\langle n\rangle},\omega))(\phi(s_i^{\langle n\rangle},\omega)$$
$$- \phi(s_{i-1}^{\langle n\rangle},\omega)) \Big\|^2_{\mathcal{Y}}$$
$$\le \Big(w_{a,b,\mathrm{M}} \|\phi(a,\omega)\|_{\mathcal{Y}} + \sum_{i=1}^n w_{a,b,\mathrm{M}} \left\| \phi(s_i^{\langle n\rangle},\omega) - \phi(s_{i-1}^{\langle n\rangle},\omega) \right\|_{\mathcal{Y}} \Big)^2$$
$$\le (w_{a,b,\mathrm{M}})^2 (\|\phi(a,\omega)\|_{\mathcal{Y}} + T_{\phi_{(\omega)}}(\mathrm{r}_{a,b}))^2$$
$$\le c(\omega)\,(w_{a,b,\mathrm{M}}(\omega))^2$$

where the first equality follows from simple algebra, the second equality follows from simple algebra, the first inequality follows from Definition 7.1 and the definitions of $w_{a,b,\mathrm{M}} : \boldsymbol{\Omega} \to [0,\infty) \subset \mathbb{R}$, the second inequality follows from Definition 12.41, and the last inequality follows from the definition of $c : \boldsymbol{\Omega} \to [0,\infty) \subset \mathbb{R}$.

By the assumption, the $E(c\, w^2_{a,b,\mathrm{M}}) < \infty$. Now, by Lebesgue Dominated Convergence Theorem 11.91, we have

$$\begin{aligned}
\mathbb{R} \ni & \left\| \widehat{\int}_{\bar{r}_{a,b}} \phi_s \, \mathrm{d}w_s \right\|^2_{\mathrm{L}_2(\boldsymbol{\Omega}, \mathcal{Y})} \\
&= \int_{\Omega} \lim_{n\in\mathbb{N}} \left\| \sum_{i=1}^{n} \phi(s_{i-1}^{\langle n\rangle}, \omega)(w(s_i^{\langle n\rangle}, \omega) - w(s_{i-1}^{\langle n\rangle}, \omega)) \right\|^2_{\mathcal{Y}} \mathrm{d}\boldsymbol{P}(\omega) \\
&= \lim_{n\in\mathbb{N}} \boldsymbol{E}\Big(\Big\| \sum_{i=1}^{n} \phi_{s_{i-1}^{\langle n\rangle}} (w_{s_i^{\langle n\rangle}} - w_{s_{i-1}^{\langle n\rangle}}) \Big\|^2_{\mathcal{Y}} \Big) \\
&= \lim_{n\in\mathbb{N}} \sum_{i=1}^{n} \boldsymbol{E}\Big(\mathcal{P}_2 \circ \phi_{s_{i-1}^{\langle n\rangle}}\Big) \boldsymbol{E}\big((w_{s_i^{\langle n\rangle}} - w_{s_{i-1}^{\langle n\rangle}})^2\big) \\
&= \lim_{n\in\mathbb{N}} \sum_{i=1}^{n} \int_{\boldsymbol{\Omega}} \|\phi(s_{i-1}^{\langle n\rangle}, \omega)\|^2_{\mathcal{Y}} \, \mathrm{d}\boldsymbol{P}(\omega)(s_i^{\langle n\rangle} - s_{i-1}^{\langle n\rangle}) \\
&= \lim_{n\in\mathbb{N}} \int_{\boldsymbol{\Omega}} \sum_{i=1}^{n} \|\phi(s_{i-1}^{\langle n\rangle}, \omega)\|^2_{\mathcal{Y}} (s_i^{\langle n\rangle} - s_{i-1}^{\langle n\rangle}) \, \mathrm{d}\boldsymbol{P}(\omega)
\end{aligned}$$

where the set membership and the second equality follow from Lebesgue Dominated Convergence Theorem 11.91, the third equality follows from Definition 14.76, the independence assumption, and Definition 13.1, and the last equality follows from Proposition 11.92.

$\forall \omega \in \boldsymbol{\Omega}$, we have $\phi_{\langle\omega\rangle}$ is of bounded variation and the function $g : \bar{r}_{a,b} \to \mathbb{R}$, defined by $g(t) = t$, $\forall t \in \bar{r}_{a,b}$, is continuous, strictly increasing, and of bounded variation. By Proposition 12.119, we have $\mathcal{P}_2 \circ \phi_{\langle\omega\rangle}$ is Riemann-Stieltjes integrable with respect to g on $\bar{r}_{a,b}$, $\widehat{\int}_a^b \mathcal{P}_2 \circ \phi_{\langle\omega\rangle} \, \mathrm{d}g = \int_{\bar{r}_{a,b}} \mathcal{P}_2 \circ \phi_{\langle\omega\rangle} \, \mathrm{d}g = \int_a^b \mathcal{P}_2 \circ \phi_{\langle\omega\rangle} \, \mathrm{d}t = \|\phi_{\langle\omega\rangle}\|^2_{\mathrm{L}_2(\bar{r}_{a,b}, \mathcal{Y})}$, and the Riemann-Stieltjes integral $\widehat{\int}_a^b \mathcal{P}_2 \circ \phi_{\langle\omega\rangle} \, \mathrm{d}g$ converges in gauge mode. Thus, we have $\lim_{n\in\mathbb{N}} \sum_{i=1}^{n} \|\phi(s_{i-1}^{\langle n\rangle}, \omega)\|^2_{\mathcal{Y}} (s_i^{\langle n\rangle} - s_{i-1}^{\langle n\rangle}) = \|\phi_{\langle\omega\rangle}\|^2_{\mathrm{L}_2(\bar{r}_{a,b}, \mathcal{Y})}$, $\forall \omega \in \boldsymbol{\Omega}$. Fix any $n \in \mathbb{N}$. $0 \le \sum_{i=1}^{n} \|\phi(s_{i-1}^{\langle n\rangle}, \omega)\|^2_{\mathcal{Y}} (s_i^{\langle n\rangle} - s_{i-1}^{\langle n\rangle}) \le \max_{a\le t\le b} \|\phi_{\langle\omega\rangle}(t)\|^2_{\mathcal{Y}} (b-a) \le (b-a)(\|\phi(a,\omega)\|_{\mathcal{Y}} + T_{\phi_{\langle\omega\rangle}}(r_{a,b}))^2 \le (b-a)c(\omega)$, where the second inequality follows from simple algebra, the third inequality follows from Definition 12.41, and the last inequality follows from the definition of $\hat{\mathcal{V}}(\bar{r}_{a,b} \times \boldsymbol{\Omega}, \mathcal{Y})$. By the assumption, $\boldsymbol{E}(c) < \infty$, and by Lebesgue Dominated

Convergence Theorem 11.91, we have

$$\mathbb{R} \ni \left\| \widehat{\int}_{\bar{r}_{a,b}} \phi_s \, \mathrm{d}w_s \right\|^2_{\bar{L}_2(\boldsymbol{\Omega}, \mathcal{Y})}$$

$$= \lim_{n\in\mathbb{N}} \int_{\boldsymbol{\Omega}} \sum_{i=1}^{n} \left\| \phi(s_{i-1}^{\langle n\rangle}, \omega) \right\|^2_{\mathcal{Y}} (s_i^{\langle n\rangle} - s_{i-1}^{\langle n\rangle}) \, \mathrm{d}\boldsymbol{P}(\omega)$$

$$= \int_{\boldsymbol{\Omega}} \lim_{n\in\mathbb{N}} \sum_{i=1}^{n} \left\| \phi(s_{i-1}^{\langle n\rangle}, \omega) \right\|^2_{\mathcal{Y}} (s_i^{\langle n\rangle} - s_{i-1}^{\langle n\rangle}) \, \mathrm{d}\boldsymbol{P}(\omega)$$

$$= \int_{\boldsymbol{\Omega}} \|\phi_{\langle\omega\rangle}\|^2_{\bar{L}_2(\bar{r}_{a,b}, \mathcal{Y})} \, \mathrm{d}\boldsymbol{P}(\omega) = \|\phi\|^2_{\bar{L}_2(\bar{r}_{a,b} \times \boldsymbol{\Omega}, \mathcal{Y})}$$

where the second equality follows from Lebesgue Dominated Convergence Theorem 11.91, the third equality follows from the preceding discussion, and the fourth equality follows from Tonelli's Theorem 12.29. This completes the proof for (v).

Let $\mathcal{Y}$ be a separable Hilbert space. $\forall \phi \in \hat{\mathcal{V}}(\bar{r}_{a,b} \times \boldsymbol{\Omega}, \mathcal{Y})$, by (v), we have $\phi \in \bar{L}_2(\bar{r}_{a,b} \times \boldsymbol{\Omega}, \mathcal{Y})$. Then, $\hat{\mathcal{V}}(\bar{r}_{a,b} \times \boldsymbol{\Omega}, \mathcal{Y}) \subseteq \bar{L}_2(\bar{r}_{a,b} \times \boldsymbol{\Omega}, \mathcal{Y})$.

$\forall \phi \in \hat{\mathcal{V}}(\bar{r}_{a,b} \times \boldsymbol{\Omega}, \mathcal{Y})$, $\forall \alpha \in \mathbb{K}$, let $\bar{\phi} := \alpha\phi$. Then, $T_{\bar{\phi}_{\langle\omega\rangle}}(\mathrm{r}_{a,b}) \leq |\alpha| T_{\phi_{\langle\omega\rangle}}(\mathrm{r}_{a,b})$, $\forall \omega \in \boldsymbol{\Omega}$, by Proposition 12.117. Then, let $c_\phi : \boldsymbol{\Omega} \to [0, \infty) \subset \mathbb{R}$ be the random variable for ϕ, we may set $c_{\bar{\phi}} := |\alpha|^2 c_\phi$. Thus, $\bar{\phi} \in \hat{\mathcal{V}}(\bar{r}_{a,b} \times \boldsymbol{\Omega}, \mathcal{Y})$.

$\forall \phi_i \in \hat{\mathcal{V}}(\bar{r}_{a,b} \times \boldsymbol{\Omega}, \mathcal{Y})$, $i = 1, 2$, let $\bar{\phi} := \phi_1 + \phi_2$. Then, by Propositions 12.117 and 12.118, we have $T_{\bar{\phi}_{\langle\omega\rangle}}(\mathrm{r}_{a,b}) \leq T_{\phi_{1\langle\omega\rangle}}(\mathrm{r}_{a,b}) + T_{\phi_{2\langle\omega\rangle}}(\mathrm{r}_{a,b})$, $\forall \omega \in \boldsymbol{\Omega}$. Then, let $c_{\phi,i} : \boldsymbol{\Omega} \to [0, \infty) \subset \mathbb{R}$ be the random variable for ϕ_i, $i = 1, 2$. We may set $c_{\bar{\phi}} := 2(c_{\phi,1} + c_{\phi,2})$. Thus, $\bar{\phi} \in \hat{\mathcal{V}}(\bar{r}_{a,b} \times \boldsymbol{\Omega}, \mathcal{Y})$.

This shows that $\hat{\mathcal{V}}(\bar{r}_{a,b} \times \boldsymbol{\Omega}, \mathcal{Y})$ is a subspace of $\bar{L}_2(\bar{r}_{a,b} \times \boldsymbol{\Omega}, \mathcal{Y})$. By (i) and (ii), we have the Riemann-Stieltjes integral with Wiener process integrator is a homomorphism between $\hat{\mathcal{V}}(\bar{r}_{a,b} \times \boldsymbol{\Omega}, \mathcal{Y})$ and its image in $\bar{L}_2(\boldsymbol{\Omega}, \mathcal{Y})$. By (v), Riemann-Stieltjes integral with Wiener process integrator is a pseudo-norm preserving homomorphism between $\hat{\mathcal{V}}(\bar{r}_{a,b} \times \boldsymbol{\Omega}, \mathcal{Y})$ and its image in $\bar{L}_2(\boldsymbol{\Omega}, \mathcal{Y})$.

This completes the proof of the lemma. □

When $\mathcal{Y}$ is a separable Hilbert space, by Lemma 14.97, if $\phi \in \bar{L}_2(\bar{r}_{a,b}, \mathcal{Y})$ is of bounded variation, then c_ϕ can be chosen as a constant (independent of ω). Then, by Proposition 14.94, we have $\boldsymbol{E}(c_\phi\,(1 + w^2_{a,b,\mathrm{M}})) = c_\phi\,(1 + \boldsymbol{E}(w^2_{a,b,\mathrm{M}})) \leq c_\phi\,(1 + 4C_2^2\, b)) < \infty$, where $C_2 \in (0, \infty) \subset \mathbb{R}$ is a constant as defined in Proposition 14.94. Thus, $\phi \in \hat{\mathcal{V}}(\bar{r}_{a,b} \times \boldsymbol{\Omega}, \mathcal{Y})$. Thus, $\hat{\mathcal{V}}(\bar{r}_{a,b} \times \boldsymbol{\Omega}, \mathcal{Y})$ contains all the $\mathcal{Y}$-valued deterministic square integrable functions on $\bar{r}_{a,b}$ that are of bounded variation. Note that c_ϕ is an upper bound of the $\|\phi\|^2_{BV}$ as defined in (h) of Problem 29.β of Bartle (1976).

Proposition 14.98 *Let* $\mathbb{I} := (([0, \infty), |\cdot|), \hat{\mathcal{B}}, \mu)$ *be the* σ*-finite metric measure subspace of* $\mathbb{R}$ *with partial ordering* $\leq$, $\boldsymbol{\Omega} := (\Omega, \mathcal{B}, (\mathcal{B}_t)_{t\in\mathbb{I}}, \boldsymbol{P})$ *be a filtered* *probability measure space,* $\mathbb{I} \times \boldsymbol{\Omega} = ([0, \infty) \times \Omega, \bar{\mathcal{B}}, \nu)$ *be the σ-finite product measure space,* $w : \mathbb{I} \times \boldsymbol{\Omega} \to \mathbb{R}$ *be a standard Wiener process on* $\mathbb{I} \times \boldsymbol{\Omega}$ *adapted to the filtration* $(\mathcal{B}_t)_{t\in\mathbb{I}}$*, the σ-algebras* $\mathcal{B}_t$ *and* $\sigma((w_{t+s} - w_t)_{s\in\mathbb{I}})$ *be independent,* $\forall t \in \mathbb{I}$*,* $\mathcal{Y}$ *be a separable Hilbert space, and* $\phi \in \bar{L}_2(\bar{r}_{a,b}, \mathcal{Y})$*, where* $a, b \in \mathbb{I}$ *with* $a \le b$*. Then, there exists a unique* $[I] \in L_2(\boldsymbol{\Omega}, \mathcal{Y})$*, where* $I : \boldsymbol{\Omega} \to \mathcal{Y}$ *is a* $\mathcal{Y}$*-valued random variable that is* $\mathcal{B}_b$*-measurable, such that* $\forall (\phi_n)_{n=1}^{\infty}$ *with* $\phi_n : \bar{r}_{a,b} \to \mathcal{Y}$ *be of bounded variation, and* $\lim_{n\in\mathbb{N}} \phi_n \doteq \phi$ *in* $\bar{L}_2(\bar{r}_{a,\underset{\sim}{b}}, \mathcal{Y})$*, we have* $\lim_{n\in\mathbb{N}} \big[\widehat{\int}_{\bar{r}_{a,b}} \phi_n(s)\, dw_s\big] = [I]$ *in* $L_2(\boldsymbol{\Omega}, \mathcal{Y})$*. We will denote* $[I]$ *by* $\widetilde{\int}_{\bar{r}_{a,b}} \phi_s\, dw_s$ *or* $\widetilde{\int}_a^b \phi_s\, dw_s$*, which is called the* Wiener integral of ϕ (with respect to Wiener process w) on the interval $\bar{r}_{a,b}$.

Proof By Proposition 12.120, $\forall n \in \mathbb{N}$, there exists a absolutely continuous function $g_n : \bar{r}_{a,b} \to \mathcal{Y}$ such that $g_n \in \bar{L}_2(\bar{r}_{a,b}, \mathcal{Y})$ and $\|g_n - \phi\|_{\bar{L}_2(\bar{r}_{a,b}, \mathcal{Y})} < 2^{-n}$. By Proposition 12.73, g_n is of locally bounded variation, which further implies that g_n is of bounded variation by Definition 12.41 and the fact that the domain of g_n is a compact interval, $\forall n \in \mathbb{N}$. Then, $\lim_{n\in\mathbb{N}} [g_n] = [\phi]$ in $L_2(\bar{r}_{a,b}, \mathcal{Y})$. The sequence $([g_n])_{n\in\mathbb{N}} \subseteq L_2(\bar{r}_{a,b}, \mathcal{Y})$ is a Cauchy sequence. By Lemma 14.97, the sequence $\big(\big[\widehat{\int}_a^b g_{n_s}\, dw_s\big]\big)_{n\in\mathbb{N}} \subseteq L_2(\boldsymbol{\Omega}, \mathcal{Y})$ is a Cauchy sequence. By Example 11.179, $L_2(\boldsymbol{\Omega}, \mathcal{Y})$ is a Hilbert space. Then, there exists a unique $[I] \in L_2(\boldsymbol{\Omega}, \mathcal{Y})$, such that $\lim_{n\in\mathbb{N}} \big[\widehat{\int}_{\bar{r}_{a,b}} g_n(s)\, dw_s\big] = [I]$ in $L_2(\boldsymbol{\Omega}, \mathcal{Y})$. $\forall (\phi_n)_{n=1}^{\infty}$ with $\phi_n : \bar{r}_{a,b} \to \mathcal{Y}$ be of bounded variation, and $\lim_{n\in\mathbb{N}} \phi_n \doteq \phi$ in $\bar{L}_2(\bar{r}_{a,b}, \mathcal{Y})$, it is easy to see that the sequence $([\phi_1], [g_1], [\phi_2], [g_2], \ldots)$ is a Cauchy sequence in $L_2(\bar{r}_{a,b}, \mathcal{Y})$ that converges to $[\phi]$. By Lemma 14.97, the sequence

$$\Big(\big[\widehat{\int}_a^b \phi_1(s)\, dw_s\big], \big[\widehat{\int}_a^b g_1(s)\, dw_s\big], \big[\widehat{\int}_a^b \phi_2(s)\, dw_s\big], \big[\widehat{\int}_a^b g_2(s)\, dw_s\big], \ldots\Big)$$

is a Cauchy sequence in $L_2(\boldsymbol{\Omega}, \mathcal{Y})$. Then, $\lim_{n\in\mathbb{N}} \big[\widehat{\int}_{\bar{r}_{a,b}} \phi_n(s)\, dw_s\big] = [I]$ in $L_2(\boldsymbol{\Omega}, \mathcal{Y})$.

By (iv) of Lemma 14.97, $\widehat{\int}_a^b \phi_{n_s}\, dw_s$ is $\mathcal{B}_b$-measurable, $\forall n \in \mathbb{N}$. Then, by Example 11.179, we may find $I : \boldsymbol{\Omega} \to \mathcal{Y}$, which is a $\mathcal{Y}$-valued random variable and $\mathcal{B}_b$-measurable, such that $I \in [I]$.

This completes the proof of the proposition. □

We next present the following result on the important properties of Riemann Stieltjes integral with Wiener process integrator on another subspace of $\bar{L}_2(\bar{r}_{a,b} \times \boldsymbol{\Omega}, \mathcal{Y})$.

Lemma 14.99 *Let* $\mathbb{I} := (([0, \infty), |\cdot|), \hat{\mathcal{B}}, \mu)$ *be the σ-finite metric measure subspace of* $\mathbb{R}$ *with partial ordering* $\le$*,* $\boldsymbol{\Omega} := (\Omega, \mathcal{B}, (\mathcal{B}_t)_{t\in\mathbb{I}}, \boldsymbol{P})$ *be a filtered probability measure space,* $\mathbb{I} \times \boldsymbol{\Omega} = ([0, \infty) \times \Omega, \bar{\mathcal{B}}, \nu)$ *be the σ-finite product measure space,* $w : \mathbb{I} \times \boldsymbol{\Omega} \to \mathbb{R}$ *be a standard Wiener process on* $\mathbb{I} \times \boldsymbol{\Omega}$ *and be adapted to the filtration* $(\mathcal{B}_t)_{t\in\mathbb{I}}$*, the σ-algebras* $\mathcal{B}_t$ *and* $\sigma((w_{t+s} - w_t)_{s\in\mathbb{I}})$ *are independent,* $\forall t \in \mathbb{I}$*,* $\mathcal{Y}$ *be a separable Banach space over* $\mathbb{K}$*,* $a, b \in \mathbb{I}$ *with* $a \le b$*,* *and* $\hat{\mathcal{Q}}(\bar{\mathrm{r}}_{a,b} \times \boldsymbol{\Omega}, \mathcal{Y}) := \{\phi : \bar{\mathrm{r}}_{a,b} \times \boldsymbol{\Omega} \to \mathcal{Y} \mid \exists n \in \mathbb{Z}_+, \exists a = t_0 \leq t_1 \leq t_2 \leq \cdots \leq t_n = b \ni \cdot \phi(t,\omega) = \sum_{i=1}^{n} A_{i-1}(\omega)\chi_{[t_{i-1},t_i),\bar{\mathrm{r}}_{a,b}}(t), \forall (t,\omega) \in \bar{\mathrm{r}}_{a,b} \times \Omega$ *and* $A_{i-1} : \boldsymbol{\Omega} \to \mathcal{Y}$ *be* $\mathcal{B}_{t_{i-1}}$*-measurable and* $\boldsymbol{E}(\mathcal{P}_2 \circ A_{i-1}) < \infty, i = 1, \ldots, n\}$. *The set* $\hat{\mathcal{Q}}(\bar{\mathrm{r}}_{a,b} \times \boldsymbol{\Omega}, \mathcal{Y})$ *is* the set of all $\mathcal{Y}$-valued simple predictable step functions on $\bar{\mathrm{r}}_{a,b} \times \boldsymbol{\Omega}$. *Then, the following statements hold:*

(i) $\forall \phi \in \hat{\mathcal{Q}}(\bar{\mathrm{r}}_{a,b} \times \boldsymbol{\Omega}, \mathcal{Y})$, *we have* $\|\phi\|^2_{\bar{\mathrm{L}}_2(\bar{\mathrm{r}}_{a,b} \times \boldsymbol{\Omega}, \mathcal{Y})} = \sum_{i=1}^{n} \boldsymbol{E}(\mathcal{P}_2 \circ A_{i-1})(t_i - t_{i-1}) < \infty$. *Furthermore,* $\hat{\mathcal{Q}}(\bar{\mathrm{r}}_{a,b} \times \boldsymbol{\Omega}, \mathcal{Y}) \subseteq \bar{\mathrm{L}}_2(\bar{\mathrm{r}}_{a,b} \times \boldsymbol{\Omega}, \mathcal{Y})$ *is a subspace.*

(ii) $\forall \phi \in \hat{\mathcal{Q}}(\bar{\mathrm{r}}_{a,b} \times \boldsymbol{\Omega}, \mathcal{Y})$, $\forall \omega \in \boldsymbol{\Omega}$, *we have* $\phi_{\langle\omega\rangle} : \bar{\mathrm{r}}_{a,b} \to \mathcal{Y}$ *is of bounded variation and* $T_{\phi_{\langle\omega\rangle}}(\mathrm{r}_{a,b}) \leq \sum_{i=1}^{n} \|A_i(\omega) - A_{i-1}(\omega)\|_{\mathcal{Y}} < \infty$, *where* $A_n(\omega) := \vartheta_{\mathcal{Y}}$.

(iii) $\forall \phi := \sum_{i=1}^{n} A_{i-1}\chi_{[t_{i-1},t_i),\bar{\mathrm{r}}_{a,b}} \in \hat{\mathcal{Q}}(\bar{\mathrm{r}}_{a,b} \times \boldsymbol{\Omega}, \mathcal{Y})$, *we have* $\big(\widehat{\int_a^b}\phi_s \,\mathrm{d}w_s\big)(\omega) = \sum_{i=1}^{n} A_{i-1}(\omega)\,(w(t_i,\omega) - w(t_{i-1},\omega))$, $\forall \omega \in \boldsymbol{\Omega}$.

(iv) $\forall \phi \in \hat{\mathcal{Q}}(\bar{\mathrm{r}}_{a,b} \times \boldsymbol{\Omega}, \mathcal{Y})$, *define* $x : \bar{\mathrm{r}}_{a,b} \times \boldsymbol{\Omega} \to \mathcal{Y}$ *by* $x(t,\omega) := \big(\widehat{\int_a^t}\phi_s \,\mathrm{d}w_s\big)(\omega) \in \mathcal{Y}$. *Then,* x *is a* $\mathcal{Y}$*-valued stochastic process that is adapted to* $(\mathcal{B}_t)_{t\in\mathbb{I}}$ *with continuous sample paths.*

(v) *Let* $\mathcal{Y}$ *be a separable Hilbert space and* $\phi \in \hat{\mathcal{Q}}(\bar{\mathrm{r}}_{a,b} \times \boldsymbol{\Omega}, \mathcal{Y})$. *Then, we have* $(\boldsymbol{E}(\|\widehat{\int}_{\bar{\mathrm{r}}_{a,b}}\phi_s \,\mathrm{d}w_s\|^2_{\mathcal{Y}}))^{1/2} = \|\widehat{\int}_{\bar{\mathrm{r}}_{a,b}}\phi_s \,\mathrm{d}w_s\|_{\bar{\mathrm{L}}_2(\boldsymbol{\Omega},\mathcal{Y})} = \|\phi\|_{\bar{\mathrm{L}}_2(\bar{\mathrm{r}}_{a,b} \times \boldsymbol{\Omega}, \mathcal{Y})} \in [0,\infty) \subset \mathbb{R}$.

When $\mathcal{Y}$ *is a separable Hilbert space.* $\hat{\mathcal{Q}}(\bar{\mathrm{r}}_{a,b} \times \boldsymbol{\Omega}, \mathcal{Y})$ *is a subspace of* $\bar{\mathrm{L}}_2(\bar{\mathrm{r}}_{a,b} \times \boldsymbol{\Omega}, \mathcal{Y})$. *The Riemann-Stieltjes integral with Wiener process integrator is a pseudo-norm preserving homomorphism between* $\hat{\mathcal{Q}}(\bar{\mathrm{r}}_{a,b} \times \boldsymbol{\Omega}, \mathcal{Y})$ *and its image in* $\bar{\mathrm{L}}_2(\boldsymbol{\Omega}, \mathcal{Y})$.

Proof (i) $\forall \phi := \sum_{i=1}^{n} A_{i-1}\chi_{[t_{i-1},t_i),\bar{\mathrm{r}}_{a,b}} \in \hat{\mathcal{Q}}(\bar{\mathrm{r}}_{a,b} \times \boldsymbol{\Omega}, \mathcal{Y})$, we have

$$\|\phi\|^2_{\bar{\mathrm{L}}_2(\bar{\mathrm{r}}_{a,b} \times \boldsymbol{\Omega}, \mathcal{Y})} = \int_{\boldsymbol{\Omega}} \int_a^b \|\phi(t,\omega)\|^2_{\mathcal{Y}} \,\mathrm{d}t \,\mathrm{d}\boldsymbol{P}(\omega)$$

$$= \int_{\boldsymbol{\Omega}} \int_a^b \sum_{i=1}^{n} \|A_{i-1}(\omega)\|^2_{\mathcal{Y}} \chi_{[t_{i-1},t_i),\bar{\mathrm{r}}_{a,b}}(t) \,\mathrm{d}t \,\mathrm{d}\boldsymbol{P}(\omega)$$

$$= \int_{\boldsymbol{\Omega}} \|A_{i-1}(\omega)\|^2_{\mathcal{Y}} (t_i - t_{i-1}) \,\mathrm{d}\boldsymbol{P}(\omega) = \sum_{i=1}^{n} \boldsymbol{E}(\mathcal{P}_2 \circ A_{i-1})(t_i - t_{i-1}) < \infty$$

where the first equality follows from Tonelli's Theorem 12.29, the third equality follows from Proposition 11.75, the last equality follows from Proposition 11.92, and the inequality follows from the definition of $\hat{\mathcal{Q}}(\bar{\mathrm{r}}_{a,b} \times \boldsymbol{\Omega}, \mathcal{Y})$. Hence, $\phi \in \bar{\mathrm{L}}_2(\bar{\mathrm{r}}_{a,b} \times \boldsymbol{\Omega}, \mathcal{Y})$. By the arbitrariness of ϕ, we have $\hat{\mathcal{Q}}(\bar{\mathrm{r}}_{a,b} \times \boldsymbol{\Omega}, \mathcal{Y}) \subseteq \bar{\mathrm{L}}_2(\bar{\mathrm{r}}_{a,b} \times \boldsymbol{\Omega}, \mathcal{Y})$. It is easy to see that $\forall \phi_1, \phi_2 \in \hat{\mathcal{Q}}(\bar{\mathrm{r}}_{a,b} \times \boldsymbol{\Omega}, \mathcal{Y})$ and $\forall \alpha \in \mathbb{K}$, we have $\phi_1 + \phi_2 \in \hat{\mathcal{Q}}(\bar{\mathrm{r}}_{a,b} \times \boldsymbol{\Omega}, \mathcal{Y})$ and $\alpha\phi_1 \in \hat{\mathcal{Q}}(\bar{\mathrm{r}}_{a,b} \times \boldsymbol{\Omega}, \mathcal{Y})$. Hence, $\hat{\mathcal{Q}}(\bar{\mathrm{r}}_{a,b} \times \boldsymbol{\Omega}, \mathcal{Y})$ is a subspace of $\bar{\mathrm{L}}_2(\bar{\mathrm{r}}_{a,b} \times \boldsymbol{\Omega}, \mathcal{Y})$.

(ii) $\forall \phi := \sum_{i=1}^{n} A_{i-1}\chi_{[t_{i-1},t_i),\bar{\mathrm{r}}_{a,b}} \in \hat{\mathcal{Q}}(\bar{\mathrm{r}}_{a,b} \times \boldsymbol{\Omega}, \mathcal{Y})$, $\forall \omega \in \boldsymbol{\Omega}$, it is easy to see that $\phi_{\langle\omega\rangle} = \sum_{i=1}^{n} A_{i-1}(\omega)\chi_{[t_{i-1},t_i),\bar{\mathrm{r}}_{a,b}}$ is a simple function and a step function. Clearly, it is continuous on the right. By Definition 12.41, we may easily show $\phi_{\langle\omega\rangle}$ is of bounded variation and that $T_{\phi_{\langle\omega\rangle}}(\mathrm{r}_{a,b}) \leq \sum_{i=1}^{n} \|A_i(\omega) - A_{i-1}(\omega)\|_{\mathcal{Y}} < \infty$.

(iii) $\forall \phi := \sum_{i=1}^{n} A_{i-1}\chi_{[t_{i-1},t_i),\bar{\mathrm{r}}_{a,b}} \in \hat{\mathcal{Q}}(\bar{\mathrm{r}}_{a,b} \times \boldsymbol{\Omega}, \mathcal{Y})$, $\forall \omega \in \boldsymbol{\Omega}$, we have $\left(\widehat{\int_a^b} \phi_s \, \mathrm{d}w_s\right)(\omega) = \widehat{\int_a^b} \phi(s,\omega) \, \mathrm{d}w_{\langle\omega\rangle}(s) = \sum_{i=1}^{n} A_{i-1}(\omega)\,(w(t_i,\omega) - w(t_{i-1},\omega))$, where the first equality follows from Definition 14.95 and the last equality follows from Definition A.4.

(iv) This follows immediately from (iv) of Lemma 14.97.

(v) Let $\mathcal{Y}$ be a separable Hilbert space. Fix any $\phi := \sum_{i=1}^{n} A_{i-1} \cdot \chi_{[t_{i-1},t_i),\bar{\mathrm{r}}_{a,b}} \in \hat{\mathcal{Q}}(\bar{\mathrm{r}}_{a,b} \times \boldsymbol{\Omega}, \mathcal{Y})$. Then, we have $\|\phi\|^2_{\bar{\mathrm{L}}_2(\bar{\mathrm{r}}_{a,b}\times\boldsymbol{\Omega},\mathcal{Y})} = \sum_{i=1}^{n} \boldsymbol{E}(\mathcal{P}_2 \circ A_{i-1})(t_i - t_{i-1}) \in [0,\infty) \subset \mathbb{R}$, by (i). We also have $\left(\widehat{\int_a^b} \phi_s \, \mathrm{d}w_s\right)(\omega) = \sum_{i=1}^{n} A_{i-1}(\omega)\,(w(t_i,\omega) - w(t_{i-1},\omega))$, $\forall \omega \in \boldsymbol{\Omega}$, by (iii). Then, we have

$$
\begin{aligned}
&\left\| \widehat{\int_{\bar{\mathrm{r}}_{a,b}}} \phi_s \, \mathrm{d}w_s \right\|^2_{\bar{\mathrm{L}}_2(\boldsymbol{\Omega},\mathcal{Y})} \\
&= \int_{\boldsymbol{\Omega}} \left\| \sum_{i=1}^{n} A_{i-1}(\omega)\,(w(t_i,\omega) - w(t_{i-1},\omega)) \right\|^2_{\mathcal{Y}} \mathrm{d}\boldsymbol{P}(\omega) \\
&= \sum_{i=1}^{n}\sum_{j=1}^{n} \int_{\boldsymbol{\Omega}} \langle A_{i-1}(\omega), A_{j-1}(\omega)\rangle (w(t_j,\omega) - w(t_{j-1},\omega)) \\
&\quad \cdot (w(t_i,\omega) - w(t_{i-1},\omega)) \, \mathrm{d}\boldsymbol{P}(\omega) \\
&= \sum_{i=1}^{n} \int_{\boldsymbol{\Omega}} \langle A_{i-1}(\omega), A_{i-1}(\omega)\rangle\, (w(t_i,\omega) - w(t_{i-1},\omega))^2 \, \mathrm{d}\boldsymbol{P}(\omega) \\
&= \sum_{i=1}^{n} \boldsymbol{E}(\mathcal{P}_2 \circ A_{i-1})\, \boldsymbol{E}\left((w_{t_i} - w_{t_{i-1}})^2\right) \\
&= \sum_{i=1}^{n} \boldsymbol{E}(\mathcal{P}_2 \circ A_{i-1})\,(t_i - t_{i-1}) = \|\phi\|^2_{\bar{\mathrm{L}}_2(\bar{\mathrm{r}}_{a,b}\times\boldsymbol{\Omega},\mathcal{Y})}
\end{aligned}
$$

where the second equality follows from Proposition 13.2, the third equality follows from Definition 14.76 and the independence assumption of $\mathcal{B}_t$ and $\sigma((w_{t+s} - w_t)_{s\in\mathbb{I}})$, $\forall t \in \mathbb{I}$, the fourth equality follows from Proposition 14.9, the fifth equality follows from Definition 14.76, and the last equality follows from (i). This completes the proof of (v).

When $\mathcal{Y}$ is a separable Hilbert space, by (i), we have $\hat{\mathcal{Q}}(\bar{\mathrm{r}}_{a,b} \times \boldsymbol{\Omega}, \mathcal{Y})$ is a subspace of $\bar{\mathrm{L}}_2(\bar{\mathrm{r}}_{a,b} \times \boldsymbol{\Omega}, \mathcal{Y})$. By (iii), Definition 14.95, and Theorem A.6, the Riemann-Stieltjes integral with Wiener process integrator is a homomorphism between $\hat{\mathcal{Q}}(\bar{\mathrm{r}}_{a,b} \times \boldsymbol{\Omega}, \mathcal{Y})$ and its image in $\bar{\mathrm{L}}_2(\boldsymbol{\Omega}, \mathcal{Y})$. By (v), we have the Riemann-Stieltjes integral with Wiener process integrator is a pseudo-norm preserving homomorphism between $\hat{\mathcal{Q}}(\bar{\mathrm{r}}_{a,b} \times \boldsymbol{\Omega}, \mathcal{Y})$ and its image in $\bar{\mathrm{L}}_2(\boldsymbol{\Omega}, \mathcal{Y})$. This completes the proof of the lemma. □

Proposition 14.100 *Let $\mathcal{Y}$ be a separable Banach space, $p \in [1, \infty) \subset \mathbb{R}$, and $f \in \bar{\mathrm{L}}_p(\mathbb{R}, \mathcal{Y})$. Then, the following statements hold:*

(i) *Let $f_\delta : \mathbb{R} \to \mathcal{Y}$ be defined by $f_\delta(t) = f(t - \delta)$, $\forall t \in \mathbb{R}$, $\forall \delta \in (0, \infty) \subset \mathbb{R}$. Then, we have $\lim_{\delta \to 0^+} \|f - f_\delta\|_{\bar{\mathrm{L}}_p(\mathbb{R},\mathcal{Y})} = 0$.*

(ii) *Let $f^{\langle n\rangle} : \mathbb{R} \to \mathcal{Y}$ be defined by $f^{\langle n\rangle}(t) = 2^n \int_{\frac{k-2}{2^n}}^{\frac{k-1}{2^n}} f(s)\, \mathrm{d}s$, $\forall \frac{k-1}{2^n} \le t < \frac{k}{2^n}$, $\forall k \in \mathbb{Z}$, $\forall n \in \mathbb{N}$. Then, we have $\big\|f - f^{\langle n\rangle}\big\|_{\bar{\mathrm{L}}_p(\mathbb{R},\mathcal{Y})} \le \|f - f_{2^{-n}}\|_{\bar{\mathrm{L}}_p(\mathbb{R},\mathcal{Y})} + 2^{\frac{1}{p}} \big(\int_0^1 \|f - f_{2^{-n}\tau}\|^p_{\bar{\mathrm{L}}_p(\mathbb{R},\mathcal{Y})} d\tau\big)^{\frac{1}{p}}$ and $\lim_{n\in\mathbb{N}} \big\|f - f^{\langle n\rangle}\big\|_{\bar{\mathrm{L}}_p(\mathbb{R},\mathcal{Y})} = 0$.*

Proof (i) Fix any $\epsilon \in (0, \infty) \subset \mathbb{R}$. By Proposition 11.182, there exists a continuous function $g : \mathbb{R} \to \mathcal{Y}$ such that $\|g - f\|_{\bar{\mathrm{L}}_p(\mathbb{R},\mathcal{Y})} < \frac{\epsilon}{3}$. Then, clearly, $\|g_\delta - f_\delta\|_{\bar{\mathrm{L}}_p(\mathbb{R},\mathcal{Y})} < \frac{\epsilon}{3}$, where $g_\delta : \mathbb{R} \to \mathcal{Y}$ is defined by $g_\delta(t) = g(t - \delta)$, $\forall t \in \mathbb{R}$. Clearly, we have $\lim_{\delta\to 0^+} \|g(t) - g_\delta(t)\|_{\mathcal{Y}} = 0$, $\forall t \in \mathbb{R}$. Fix any $(\delta_n)_{n=1}^\infty \subseteq \mathbb{R}_+$ with $\lim_{n\in\mathbb{N}} \delta_n = 0$. By Lebesgue Dominated Convergence Theorem 11.91, we have $\lim_{n\in\mathbb{N}} \big\|g - g_{\delta_n}\big\|_{\bar{\mathrm{L}}_p(\mathbb{R},\mathcal{Y})} = 0$. By Proposition 4.16, we have $\lim_{\delta\to 0^+} \|g - g_\delta\|_{\bar{\mathrm{L}}_p(\mathbb{R},\mathcal{Y})} = 0$. Then, $\exists \delta_0 \in (0, \infty) \subset \mathbb{R}$ such that $\forall \delta \in (0, \delta_0] \subset \mathbb{R}$, we have $\|g - g_\delta\|_{\bar{\mathrm{L}}_p(\mathbb{R},\mathcal{Y})} < \frac{\epsilon}{3}$. Then, we have $\|f - f_\delta\|_{\bar{\mathrm{L}}_p(\mathbb{R},\mathcal{Y})} \le \|f - g\|_{\bar{\mathrm{L}}_p(\mathbb{R},\mathcal{Y})} + \|g - g_\delta\|_{\bar{\mathrm{L}}_p(\mathbb{R},\mathcal{Y})} + \|g_\delta - f_\delta\|_{\bar{\mathrm{L}}_p(\mathbb{R},\mathcal{Y})} < \epsilon$. By the arbitrariness of ϵ, we have $\lim_{\delta\to 0^+} \|f - f_\delta\|_{\bar{\mathrm{L}}_p(\mathbb{R},\mathcal{Y})} = 0$.

(ii) Clearly, $f^{\langle n\rangle}(t) = 2^n \int_{\frac{k-1}{2^n}}^{\frac{k}{2^n}} f_{2^{-n}}(s)\, \mathrm{d}s$, $\forall \frac{k-1}{2^n} \le t < \frac{k}{2^n}$, $\forall k \in \mathbb{Z}$, $\forall n \in \mathbb{N}$, where $f_{2^{-n}} : \mathbb{R} \to \mathcal{Y}$ is as defined in (i). Then, we have $\big\|f - f^{\langle n\rangle}\big\|_{\bar{\mathrm{L}}_p(\mathbb{R},\mathcal{Y})} \le \|f - f_{2^{-n}}\|_{\bar{\mathrm{L}}_p(\mathbb{R},\mathcal{Y})} + \big\|f_{2^{-n}} - f^{\langle n\rangle}\big\|_{\bar{\mathrm{L}}_p(\mathbb{R},\mathcal{Y})}$. Fix any $\epsilon \in (0, \infty) \subset \mathbb{R}$. By (i), $\exists \delta_0 \in (0, \infty) \subset \mathbb{R}$ such that $\forall \delta \in (0, \delta_0] \subset \mathbb{R}$, we have $\|f - f_\delta\|_{\bar{\mathrm{L}}_p(\mathbb{R},\mathcal{Y})} < \frac{\epsilon}{2}$. We have the following line of arguments:

$$
\begin{aligned}
&\big\|f_{2^{-n}} - f^{\langle n\rangle}\big\|^p_{\bar{\mathrm{L}}_p(\mathbb{R},\mathcal{Y})} \\
&= \sum_{k=-\infty}^{\infty} \int_{\frac{k-1}{2^n}}^{\frac{k}{2^n}} \bigg\|f_{2^{-n}}(t) - 2^n \int_{\frac{k-1}{2^n}}^{\frac{k}{2^n}} f_{2^{-n}}(s)\,\mathrm{d}s\bigg\|^p_{\mathcal{Y}} \mathrm{d}t \\
&= \sum_{k=-\infty}^{\infty} \int_{\frac{k-1}{2^n}}^{\frac{k}{2^n}} 2^{pn} \bigg\| \int_{\frac{k-1}{2^n}}^{\frac{k}{2^n}} (f_{2^{-n}}(t) - f_{2^{-n}}(s))\,\mathrm{d}s\bigg\|^p_{\mathcal{Y}} \mathrm{d}t \\
&\le \sum_{k=-\infty}^{\infty} \int_{\frac{k-1}{2^n}}^{\frac{k}{2^n}} 2^{pn} \Big(\int_{\frac{k-1}{2^n}}^{\frac{k}{2^n}} \|(f_{2^{-n}}(t) - f_{2^{-n}}(s))\|_{\mathcal{Y}}\,\mathrm{d}s\Big)^p \mathrm{d}t \\
&\le \sum_{k=-\infty}^{\infty} \int_{\frac{k-1}{2^n}}^{\frac{k}{2^n}} 2^{n} \int_{\frac{k-1}{2^n}}^{\frac{k}{2^n}} \|(f_{2^{-n}}(t) - f_{2^{-n}}(s))\|^p_{\mathcal{Y}}\,\mathrm{d}s\,\mathrm{d}t
\end{aligned}
$$

$$
\begin{aligned}
&= 2^n \sum_{k=-\infty}^{\infty} \int_{\frac{k-2}{2^n}}^{\frac{k-1}{2^n}} \int_{\frac{k-2}{2^n}}^{\frac{k-1}{2^n}} \|(f(\bar t) - f(\bar s))\|_{\mathcal{Y}}^p \, \mathrm{d}\bar s \, \mathrm{d}\bar t \\
&= 2^n \sum_{k=-\infty}^{\infty} \int_{\frac{k-2}{2^n}}^{\frac{k-1}{2^n}} \int_{\frac{k-2}{2^n}}^{\frac{k-1}{2^n}} \|(f(t) - f(s))\|_{\mathcal{Y}}^p \, \mathrm{d}t \, \mathrm{d}s \\
&= 2^{n+1} \sum_{k=-\infty}^{\infty} \int_{\frac{k-2}{2^n}}^{\frac{k-1}{2^n}} \int_{s}^{\frac{k-1}{2^n}} \|(f(t) - f(s))\|_{\mathcal{Y}}^p \, \mathrm{d}t \, \mathrm{d}s \\
&= 2^{n+1} \sum_{k=-\infty}^{\infty} \int_{\frac{k-2}{2^n}}^{\frac{k-1}{2^n}} \int_{s}^{\frac{k-1}{2^n}} \|(f(t) - f_{t-s}(t))\|_{\mathcal{Y}}^p \, \mathrm{d}t \, \mathrm{d}s \\
&= 2^{n+1} \sum_{k=-\infty}^{\infty} \int_{\frac{k-2}{2^n}}^{\frac{k-1}{2^n}} \int_{\frac{k-2}{2^n}}^{t} \|(f(t) - f_{t-s}(t))\|_{\mathcal{Y}}^p \, \mathrm{d}s \, \mathrm{d}t \\
&= 2^{n+1} \sum_{k=-\infty}^{\infty} \int_{\frac{k-2}{2^n}}^{\frac{k-1}{2^n}} \int_{0}^{t-\frac{k-2}{2^n}} \|(f(t) - f_{\tau}(t))\|_{\mathcal{Y}}^p \, \mathrm{d}\tau \, \mathrm{d}t \\
&\le 2^{n+1} \sum_{k=-\infty}^{\infty} \int_{\frac{k-2}{2^n}}^{\frac{k-1}{2^n}} \int_{0}^{\frac{1}{2^n}} \|(f(t) - f_{\tau}(t))\|_{\mathcal{Y}}^p \, \mathrm{d}\tau \, \mathrm{d}t \\
&= 2^{n+1} \int_{0}^{\frac{1}{2^n}} \int_{-\infty}^{\infty} \|(f(t) - f_{\tau}(t))\|_{\mathcal{Y}}^p \, \mathrm{d}t \, \mathrm{d}\tau \\
&= 2^{n+1} \int_{0}^{2^{-n}} \|f - f_{\tau}\|_{\bar{\mathrm{L}}_p(\mathbb{R},\mathcal{Y})}^p \, \mathrm{d}\tau = 2 \int_{0}^{1} \|f - f_{2^{-n}\bar\tau}\|_{\bar{\mathrm{L}}_p(\mathbb{R},\mathcal{Y})}^p \, \mathrm{d}\bar\tau
\end{aligned}
$$

where the first equality follows from Example 11.173, the second equality follows from Proposition 11.92, the first inequality follows from Proposition 11.92, the second inequality follows from Hölder's Inequality (Theorem 11.178) with $q \in (1, \infty] \subset \mathbb{R}_e$ satisfying $1/p + 1/q = 1$, the third equality follows from change of variable $\bar s := s - 2^{-n}$ and $\bar t := t - 2^{-n}$, the fourth equality follows from Tonelli's Theorem 12.29, the fifth equality follows from symmetry, the sixth equality follows from simple algebra, the seventh equality follows from Tonelli's Theorem 12.29, the eighth equality follows from change of variable $\tau = t - s$, the third inequality follows from Proposition 11.83, the ninth equality follows from Proposition 11.83 and Tonelli's Theorem 12.29; and the last equality follows from Change of Variable Theorem 12.91 with $\bar\tau = 2^n\tau$. Let $n_0 \in \mathbb{N}$ be such that $2^{-n_0} \le \delta_0$. Then, $\forall n \in \mathbb{N}$ with $n_0 \le n$, $\forall \tau \in (0, 2^{-n}] \subset \mathbb{R}$, we have $\|f - f_\tau\|_{\bar{\mathrm{L}}_p(\mathbb{R},\mathcal{Y})} < \frac{\epsilon}{2}$. Then, we have $\|f_{2^{-n}} - f^{\langle n\rangle}\|_{\bar{\mathrm{L}}_p(\mathbb{R},\mathcal{Y})}^p \le 2^{n+1} \int_0^{2^{-n}} \frac{\epsilon^p}{2^p} \mathrm{d}\tau = 2^{1-p}\epsilon^p$ and $\|f_{2^{-n}} - f^{\langle n\rangle}\|_{\bar{\mathrm{L}}_p(\mathbb{R},\mathcal{Y})} \le 2^{-1/q}\epsilon$. This shows that $\|f - f^{\langle n\rangle}\|_{\bar{\mathrm{L}}_p(\mathbb{R},\mathcal{Y})} \le \frac{\epsilon}{2} + 2^{-1/q}\epsilon < \frac{3}{2}\epsilon$. Hence, $\lim_{n\in\mathbb{N}} \|f - f^{\langle n\rangle}\|_{\bar{\mathrm{L}}_p(\mathbb{R},\mathcal{Y})} = 0$.

This completes the proof of the proposition. □

Definition 14.101 Let $\mathbb{I} := (([0, \infty), |\cdot|), \hat{\mathcal{B}}, \mu)$ be the σ-finite metric measure subspace of $\mathbb{R}$ with partial ordering $\leq$, $\boldsymbol{\Omega} := (\Omega, \mathcal{B}, (\mathcal{B}_t)_{t\in\mathbb{I}}, \boldsymbol{P})$ be a filtered probability measure space, $\mathbb{I} \times \boldsymbol{\Omega} = ([0, \infty) \times \Omega, \bar{\mathcal{B}}, \nu)$ be the σ-finite product measure space, $\mathcal{Y}$ be a topological space, and $x : \mathbb{I} \times \boldsymbol{\Omega} \to \mathcal{Y}$ be a $\mathcal{Y}$-valued stochastic process. Then, x is said to be *progressively measurable*, if $\forall t \in \mathbb{I}$, the function $x|_{\bar{r}_{0,t}\times\Omega}$ is $\tilde{\mathcal{B}}_t$-measurable, where $(\bar{r}_{0,t}, \hat{\mathcal{B}}_t, \mu_t)$ is the finite metric measure subspace of $\mathbb{I}$ (and therefore $\mathbb{R}$), and $(\bar{r}_{0,t} \times \Omega, \tilde{\mathcal{B}}_t, \tilde{\nu})$ is the finite product measure space of $(\bar{r}_{0,t}, \hat{\mathcal{B}}_t, \mu_t)$ and $(\Omega, \mathcal{B}_t, \boldsymbol{P}|_{\mathcal{B}_t})$. ◇

Proposition 14.102 *Let $\mathbb{I} := (([0, \infty), |\cdot|), \hat{\mathcal{B}}, \mu)$ be the σ-finite metric measure subspace of $\mathbb{R}$ with partial ordering $\leq$, $\boldsymbol{\Omega} := (\Omega, \mathcal{B}, (\mathcal{B}_t)_{t\in\mathbb{I}}, \boldsymbol{P})$ be a filtered probability measure space, $\mathbb{I} \times \boldsymbol{\Omega} = ([0, \infty) \times \Omega, \bar{\mathcal{B}}, \nu)$ be the σ-finite product measure space, $\mathcal{Y}$ be a Banach space over $\mathbb{K}$, and $x : \mathbb{I} \times \boldsymbol{\Omega} \to \mathcal{Y}$ be adapted to $(\mathcal{B}_t)_{t\in\mathbb{I}}$. Assume that $x_{\langle\omega\rangle} : \mathbb{I} \to \mathcal{Y}$ is continuous on the right, $\forall \omega \in \Omega$. Then, x is a progressively measurable stochastic process.*

Proof Fix any $t \in \mathbb{I}$. Define a sequence $(X_n)_{n=1}^{\infty}$ with $X_n : \bar{r}_{0,t} \times \Omega \to \mathcal{Y}$, $\forall n \in \mathbb{N}$, and $X_n(s, \omega) = x(0, \omega) + \sum_{k=1}^{2^n} x(\frac{k}{2^n}t, \omega)\chi_{r_{\frac{k-1}{2^n}t, \frac{k}{2^n}t}, \mathbb{I}}(s)$, $\forall (s, \omega) \in \bar{r}_{0,t} \times \Omega$. Fix any $n \in \mathbb{N}$. By our assumption, $A_k := x_{\frac{k}{2^n}t} : \Omega \to \mathcal{Y}$ is $\mathcal{B}_{\frac{k}{2^n}t}$-measurable (since x is adapted to $(\mathcal{B}_t)_{t\in\mathbb{I}}$) and therefore $\mathcal{B}_t$-measurable (since $(\mathcal{B}_t)_{t\in\mathbb{I}}$ is a filtration), $k = 0, \ldots, 2^n$. Then, $\forall O \in \mathcal{O}_{\mathcal{Y}}$, $X_{n\,\mathrm{inv}}(O) = (\{0\} \times A_{0\,\mathrm{inv}}(O)) \cup \left(\bigcup_{k=1}^{2^n} (r_{\frac{k-1}{2^n}t, \frac{k}{2^n}t} \times A_{k\,\mathrm{inv}}(O))\right)$. Clearly, each set in the union on the right-hand side is in $\tilde{\mathcal{B}}_t$, where $\tilde{\mathcal{B}}_t$ is as defined in Definition 14.101. Then, $X_{n\,\mathrm{inv}}(O) \in \tilde{\mathcal{B}}_t$, $\forall O \in \mathcal{O}_{\mathcal{Y}}$. By the arbitrariness of O, we have X_n is $\tilde{\mathcal{B}}_t$-measurable. By the assumption $x_{\langle\omega\rangle} : \mathbb{I} \to \mathcal{Y}$ is continuous on the right, $\forall \omega \in \Omega$, we have $\lim_{n\in\mathbb{N}} X_n(s, \omega) = x(s, \omega)$, $\forall (s, \omega) \in \bar{r}_{0,t} \times \Omega$. By Proposition 11.48, $x|_{\bar{r}_{0,t}\times\Omega}$ is $\tilde{\mathcal{B}}_t$-measurable. By the arbitrariness of t, x is progressively measurable. Note that $x(s, \omega) = \lim_{n\in\mathbb{N}} x|_{\bar{r}_{0,n}\times\Omega}$, $\forall (s, \omega) \in \mathbb{I} \times \boldsymbol{\Omega}$. $x|_{\bar{r}_{0,n}\times\Omega}$ is $\tilde{\mathcal{B}}_n$-measurable and therefore $\bar{\mathcal{B}}$-measurable. By Proposition 11.48, x is $\bar{\mathcal{B}}$-measurable. Hence, x is a $\mathcal{Y}$-valued stochastic process. This completes the proof of the proposition. □

Proposition 14.103 *Let $\mathbb{I} := (([0, \infty), |\cdot|), \hat{\mathcal{B}}, \mu)$ be the σ-finite metric measure subspace of $\mathbb{R}$ with partial ordering $\leq$, $\boldsymbol{\Omega} := (\Omega, \mathcal{B}, (\mathcal{B}_t)_{t\in\mathbb{I}}, \boldsymbol{P})$ be a filtered probability measure space, $\mathbb{I} \times \boldsymbol{\Omega} = ([0, \infty) \times \Omega, \bar{\mathcal{B}}, \nu)$ be the σ-finite product measure space, $\bar{r}_{a,b} \subset \mathbb{I}$ with $a \leq b$, $\mathcal{Y}$ be a separable Banach space, $W \subseteq \mathcal{Y}$ be a σ-compact conic segment, and $\phi : \bar{r}_{a,b} \times \boldsymbol{\Omega} \to W$ be a W-valued stochastic process adapted to the filtration $(\mathcal{B}_t)_{t\in\mathbb{I}}$. Assume that*

(i) *$\phi \in \bar{L}_2(\bar{r}_{a,b} \times \boldsymbol{\Omega}, \mathcal{Y})$ and $\phi_{\langle\omega\rangle} : \bar{r}_{a,b} \to W$ is continuous on the right, $\forall \omega \in \boldsymbol{\Omega}$.*
(ii) *$\forall E \in \mathcal{B}$ with $\boldsymbol{P}(E) = 0$, implies that $E \in \mathcal{B}_0$.*

Then, there exists a sequence of $\mathcal{Y}$-valued simple predictable step functions $(\phi_n)_{n=1}^{\infty} \subseteq \hat{\mathcal{Q}}(\bar{r}_{a,b} \times \boldsymbol{\Omega}, \mathcal{Y})$, where $\hat{\mathcal{Q}}(\bar{r}_{a,b} \times \boldsymbol{\Omega}, \mathcal{Y})$ is defined in Lemma 14.99, such that $\lim_{n\in\mathbb{N}} \phi_n \doteq \phi$ in $\bar{L}_2(\bar{r}_{a,b} \times \boldsymbol{\Omega}, \mathcal{Y})$.

Proof Let $\bar{\phi} : \mathbb{I} \times \boldsymbol{\Omega} \to W$ be defined by $\bar{\phi}(t,\omega) := \begin{cases} \phi(t,\omega) & a \le t < b \\ \vartheta_{\mathcal{Y}} & t < a \text{ or } t \ge b \end{cases}$, $\forall (t,\omega) \in \mathbb{I} \times \boldsymbol{\Omega}$. Clearly, $\bar{\phi} \in \bar{L}_2(\mathbb{I} \times \boldsymbol{\Omega}, \mathcal{Y})$ and $\bar{\phi}_{\langle\omega\rangle} : \mathbb{I} \to W$ is continuous on the right, $\forall \omega \in \boldsymbol{\Omega}$. Then, $\bar{\phi}$ is a W-valued stochastic process that is adapted to the filtration $(\mathcal{B}_t)_{t\in\mathbb{I}}$. By Proposition 14.102, $\bar{\phi}$ is progressively measurable. Construct the sequence $(\bar{\phi}_n)_{n=1}^{\infty}$ according to Proposition 14.100 sample path-wise. Fix any $n \in \mathbb{N}$. Define $\bar{\phi}_n : \mathbb{I} \times \boldsymbol{\Omega} \to \mathcal{Y}$ by

$$
\begin{aligned}
&\bar{\phi}_n(t,\omega) \\
&:= \begin{cases} \vartheta_{\mathcal{Y}} & (\forall 0 \le t < \frac{1}{2^n}) \text{ or } \bar{\phi}_{\langle\omega\rangle} \notin \bar{L}_2(\mathbb{I}, \mathcal{Y}) \\ 2^n \int_{\frac{k-1}{2^n}}^{\frac{k}{2^n}} \bar{\phi}(s,\omega)\,ds & (\forall \frac{k}{2^n} \le t < \frac{k+1}{2^n}, \forall k \in \mathbb{N}) \text{ and } \bar{\phi}_{\langle\omega\rangle} \in \bar{L}_2(\mathbb{I}, \mathcal{Y}) \end{cases} \\
&=: \sum_{k=1}^{\infty} A_{n,k-1}(\omega)\chi_{[\frac{k-1}{2^n},\frac{k}{2^n}),\mathbb{I}}(t), \quad \forall (t,\omega) \in \mathbb{I} \times \boldsymbol{\Omega}
\end{aligned}
$$

Define $\phi_n : \bar{r}_{a,b} \times \boldsymbol{\Omega} \to \mathcal{Y}$ by $\phi_n = \bar{\phi}_n|_{\bar{r}_{a,b}\times\boldsymbol{\Omega}}$. Then, we have $\lim_{n\in\mathbb{N}} \|\bar{\phi}_{\langle\omega\rangle} - (\bar{\phi}_n)_{\langle\omega\rangle}\|_{\bar{L}_2(\mathbb{I},\mathcal{Y})} = 0$, $\forall \omega \in \hat{\Omega}$, where $\hat{\Omega} := \{\omega \in \boldsymbol{\Omega} \mid \bar{\phi}_{\langle\omega\rangle} \in \bar{L}_2(\mathbb{I}, \mathcal{Y})\}$. Since $\bar{\phi} \in \bar{L}_2(\mathbb{I} \times \boldsymbol{\Omega}, \mathcal{Y})$, by Tonelli's Theorem 12.29, we have $\hat{\Omega} \in \mathcal{B}$ and $\boldsymbol{P}(\Omega \setminus \hat{\Omega}) = 0$, and $\|\bar{\phi}\|^2_{\bar{L}_2(\mathbb{I}\times\Omega,\mathcal{Y})} = \int_{\hat{\Omega}} \|\bar{\phi}_{\langle\omega\rangle}\|^2_{\bar{L}_2(\mathbb{I},\mathcal{Y})}\,d\boldsymbol{P}(\omega)$. By (ii), we have $\Omega \setminus \hat{\Omega} \in \mathcal{B}_0$ and $\hat{\Omega} \in \mathcal{B}_0$. Then, $\bar{\phi}|_{\bar{r}_{0,b+1}\times\boldsymbol{\Omega}} \in \bar{L}_2(\bar{r}_{0,b+1} \times \boldsymbol{\Omega}, \mathcal{Y})$. Since $\nu(\bar{r}_{0,b+1} \times \boldsymbol{\Omega}) = b+1 < \infty$, then by Hölder's Inequality (Theorem 11.178), $\bar{\phi}|_{\bar{r}_{0,b+1}\times\boldsymbol{\Omega}} \in \bar{L}_1(\bar{r}_{0,b+1} \times \boldsymbol{\Omega}, \mathcal{Y})$. Fix any $\omega \in \hat{\Omega}$, we have $\bar{\phi}_{\langle\omega\rangle}|_{\bar{r}_{0,b+1}} \in \bar{L}_2(\bar{r}_{0,b+1}, \mathcal{Y})$. Again, by Hölder's Inequality (Theorem 11.178), we have $\bar{\phi}_{\langle\omega\rangle}|_{\bar{r}_{0,b+1}} \in \bar{L}_1(\bar{r}_{0,b+1}, \mathcal{Y})$. Then, by Fubini's Theorem 12.31, $A_{n,k-1} : \boldsymbol{\Omega} \to \mathcal{Y}$ is $\mathcal{B}$-measurable, $\forall n \in \mathbb{N}$, $\forall k \in \{1, \ldots, \lfloor 2^n(b+1)\rfloor\}$. Since $\bar{\phi}$ is progressively measurable, then, by Theorem 12.121, $A_{n,k-1}$ is $\mathcal{B}_{\frac{k-1}{2^n}}$-measurable, $\forall n \in \mathbb{N}$, $\forall k \in \{1, \ldots, \lfloor 2^n(b+1)\rfloor\}$.

Note that, $\forall k = 2, 3, \ldots$,

$$
\begin{aligned}
\boldsymbol{E}(\mathcal{P}_2 \circ A_{n,k-1}) &= \int_{\Omega} 2^{2n} \left\| \int_{\frac{k-2}{2^n}}^{\frac{k-1}{2^n}} \bar{\phi}(s,\omega)\,ds \right\|^2_{\mathcal{Y}} d\boldsymbol{P}(\omega) \\
&\le \int_{\Omega} 2^n \int_{\frac{k-2}{2^n}}^{\frac{k-1}{2^n}} \|\bar{\phi}(s,\omega)\|^2_{\mathcal{Y}}\,ds\,d\boldsymbol{P}(\omega) \\
&= 2^n \left\| \bar{\phi}|_{r_{\frac{k-2}{2^n},\frac{k-1}{2^n}}\times\boldsymbol{\Omega}} \right\|^2_{\bar{L}_2(r_{\frac{k-2}{2^n},\frac{k-1}{2^n}}\times\boldsymbol{\Omega},\mathcal{Y})} < \infty
\end{aligned}
$$

where the inequality follows from the Cauchy–Schwarz Inequality. It should be noted that $A_{n,0}(\omega) = \vartheta_{\mathcal{Y}}$, $\forall \omega \in \boldsymbol{\Omega}$. Then, $\boldsymbol{E}(\mathcal{P}_2 \circ A_{n,0}) = 0$.

Thus, $\phi_n \in \hat{\mathcal{Q}}(\bar{r}_{a,b} \times \boldsymbol{\Omega}, \mathcal{Y})$, $\forall n \in \mathbb{N}$. $\forall \omega \in \hat{\Omega}$. By Proposition 14.100 and its proof, we have $\| \bar{\phi}_{\langle\omega\rangle} - (\bar{\phi}_n)_{\langle\omega\rangle} \|_{\bar{L}_2(\mathbb{I},\mathcal{Y})} \leq \| \bar{\phi}_{\langle\omega\rangle} - (\bar{\phi}_{2^{-n}})_{\langle\omega\rangle} \|_{\bar{L}_2(\mathbb{I},\mathcal{Y})} + 2^{\frac{1}{2}} \left(\int_0^1 \| \bar{\phi}_{\langle\omega\rangle} - (\bar{\phi}_{2^{-n}\tau})_{\langle\omega\rangle} \|^2_{\bar{L}_2(\mathbb{I},\mathcal{Y})} \, d\tau \right)^{\frac{1}{2}}$, $\forall \omega \in \hat{\Omega}$, $\forall n \in \mathbb{N}$, where we have adopted the notation $\bar{\phi}_\delta(t,\omega) := \begin{cases} \bar{\phi}(t-\delta,\omega) & \text{if } t \geq \delta \\ \vartheta_{\mathcal{Y}} & \text{if } t < \delta \end{cases}$, $\forall (t,\omega) \in \mathbb{I} \times \boldsymbol{\Omega}$, $\forall \delta \in r_{0,\frac{1}{2}}$. This leads to the following line of arguments,

$$
\begin{aligned}
&\limsup_{n\in\mathbb{N}} \| \phi - \phi_n \|_{\bar{L}_2(\bar{r}_{a,b}\times\boldsymbol{\Omega},\mathcal{Y})} \leq \limsup_{n\in\mathbb{N}} \| \bar{\phi} - \bar{\phi}_n \|_{\bar{L}_2(\mathbb{I}\times\boldsymbol{\Omega},\mathcal{Y})} \\
&= \limsup_{n\in\mathbb{N}} \left(\int_{\hat{\Omega}} \| (\bar{\phi} - \bar{\phi}_n)_{\langle\omega\rangle} \|^2_{\bar{L}_2(\mathbb{I},\mathcal{Y})} \, d\boldsymbol{P}(\omega) \right)^{\frac{1}{2}} \\
&\leq \limsup_{n\in\mathbb{N}} \left(\int_{\hat{\Omega}} \left(\| \bar{\phi}_{\langle\omega\rangle} - (\bar{\phi}_{2^{-n}})_{\langle\omega\rangle} \|_{\bar{L}_2(\mathbb{I},\mathcal{Y})} \right.\right. \\
&\quad \left.\left. + 2^{\frac{1}{2}} \left(\int_0^1 \| \bar{\phi}_{\langle\omega\rangle} - (\bar{\phi}_{2^{-n}\tau})_{\langle\omega\rangle} \|^2_{\bar{L}_2(\mathbb{I},\mathcal{Y})} \, d\tau \right)^{\frac{1}{2}} \right)^2 d\boldsymbol{P}(\omega) \right)^{\frac{1}{2}} \\
&\leq \lim_{n\in\mathbb{N}} \left(\int_{\hat{\Omega}} \left(2 \| \bar{\phi}_{\langle\omega\rangle} - (\bar{\phi}_{2^{-n}})_{\langle\omega\rangle} \|^2_{\bar{L}_2(\mathbb{I},\mathcal{Y})} \right.\right. \\
&\quad \left.\left. + 4 \int_0^1 \| \bar{\phi}_{\langle\omega\rangle} - (\bar{\phi}_{2^{-n}\tau})_{\langle\omega\rangle} \|^2_{\bar{L}_2(\mathbb{I},\mathcal{Y})} \, d\tau \right) d\boldsymbol{P}(\omega) \right)^{\frac{1}{2}} \\
&= \lim_{n\in\mathbb{N}} \left(2 \int_{\hat{\Omega}} \| \bar{\phi}_{\langle\omega\rangle} - (\bar{\phi}_{2^{-n}})_{\langle\omega\rangle} \|^2_{\bar{L}_2(\mathbb{I},\mathcal{Y})} \, d\boldsymbol{P}(\omega) \right. \\
&\quad \left. + 4 \int_{\hat{\Omega}\times r_{0,1}} \| \bar{\phi}_{\langle\omega\rangle} - (\bar{\phi}_{2^{-n}\tau})_{\langle\omega\rangle} \|^2_{\bar{L}_2(\mathbb{I},\mathcal{Y})} \, d(\boldsymbol{P} \times \mu_B)(\omega,\tau) \right)^{\frac{1}{2}} = 0
\end{aligned}
$$

where the first equality follows from Tonelli's Theorem 12.29 and Lemma 11.73; the second inequality follows from Proposition 14.100 and its proof; the third inequality follows from simple algebra and Proposition 11.83; the second equality follows from Proposition 11.83 and Tonelli's Theorem 12.29; and the last equality follows Lebesgue Dominated Convergence Theorem 11.88 and (i) of Proposition 14.100. Hence, we have $\lim_{n\in\mathbb{N}} \| \phi - \phi_n \|^2_{\bar{L}_2(\bar{r}_{a,b}\times\boldsymbol{\Omega})} = 0$.

This completes the proof of the proposition. □

Definition 14.104 Let $\mathbb{I} := (([0,\infty), |\cdot|), \hat{\mathcal{B}}, \mu)$ be the σ-finite metric measure subspace of $\mathbb{R}$ with partial ordering $\leq$, $\boldsymbol{\Omega} := (\Omega, \mathcal{B}, (\mathcal{B}_t)_{t\in\mathbb{I}}, \boldsymbol{P})$ be a filtered probability measure space, $\mathbb{I} \times \boldsymbol{\Omega} = ([0,\infty) \times \Omega, \bar{\mathcal{B}}, \nu)$ be the σ-finite product measure space, $w : \mathbb{I} \times \boldsymbol{\Omega} \to \mathbb{R}$ be a standard Wiener process on $\mathbb{I} \times \boldsymbol{\Omega}$ and be adapted to the filtration $(\mathcal{B}_t)_{t\in\mathbb{I}}$, $\bar{r}_{a,b} \subset \mathbb{I}$ with $a \leq b$, $\mathcal{Y}$ be a separable Hilbert space, $\phi \in \bar{L}_2(\bar{r}_{a,b} \times \boldsymbol{\Omega}, \mathcal{Y})$ be a $\mathcal{Y}$-valued stochastic process adapted to the filtration $(\mathcal{B}_t)_{t\in\mathbb{I}}$, and the σ-algebras $\mathcal{B}_t$ and $\sigma((w_{t+s} - w_t)_{s\in\mathbb{I}})$ are independent, $\forall t \in \mathbb{I}$.

If there exists a sequence of $\mathcal{Y}$-valued simple predictable step functions $(\phi_n)_{n=1}^{\infty} \subseteq \hat{\mathcal{Q}}(\bar{r}_{a,b} \times \boldsymbol{\Omega}, \mathcal{Y})$, where $\hat{\mathcal{Q}}(\bar{r}_{a,b} \times \boldsymbol{\Omega}, \mathcal{Y})$ is defined in Lemma 14.99, such that $\lim_{n\in\mathbb{N}} \phi_n \doteq \phi$ in $\bar{L}_2(\bar{r}_{a,b} \times \boldsymbol{\Omega}, \mathcal{Y})$. For each $n \in \mathbb{N}$, let $I_n : \boldsymbol{\Omega} \to \mathcal{Y}$ be defined by $I_n(\omega) := \big(\widehat{\int_a^b} \phi_{n_s}\, \mathrm{d}w_s\big)(\omega) \in \mathcal{Y}$. Then, I_n is $\mathcal{B}_b$-measurable, $\forall n \in \mathbb{N}$. By Lemma 14.99, the sequence $([I_n])_{n\in\mathbb{N}}$ is a Cauchy sequence in $L_2(\boldsymbol{\Omega}, \mathcal{Y})$, which converges to some $[I] \in L_2(\boldsymbol{\Omega}, \mathcal{Y})$, where $I : \boldsymbol{\Omega} \to \mathcal{Y}$ is a $\mathcal{Y}$-valued random variable that is further $\mathcal{B}_b$-measurable by Example 11.179. Clearly, $[I]$ is independent of the choice of the sequence $(\phi_n)_{n=1}^{\infty} \subseteq \hat{\mathcal{Q}}(\bar{r}_{a,b} \times \boldsymbol{\Omega}, \mathcal{Y})$. (If there is another such sequence, then the sequence made up of alternate elements from both of these sequences will be a valid sequence, which lead to converging $([I_n])_{n\in\mathbb{N}}$, which must yield the same limit $[I]$.) Then, we will say that ϕ *is Itô integrable (with respect to Wiener process w) on the interval* $\bar{r}_{a,b}$. We will denote $[I]$ by $\overline{\int}_{\bar{r}_{a,b}} \phi_s\, \mathrm{d}w_s$ or $\overline{\int}_a^b \phi_s\, \mathrm{d}w_s$, which is called the *Itô integral of ϕ (with respect to Wiener process w) on the interval* $\bar{r}_{a,b}$. ◇

Clearly, the Wiener integral and the Itô integral agree with each other in their common domain of definition. Yet, neither of them define a stochastic process from the integrand stochastic process.

The following proposition then establishes the existence of Itô integral whenever the integrand is $\bar{L}_2$ and takes value in a σ-compact conic segment of a separable Hilbert space (which is clearly so if the Hilbert space is finite-dimensional) and is a right continuous adapted stochastic process.

Proposition 14.105 *Let $\mathbb{I} := (([0,\infty), |\cdot|), \hat{\mathcal{B}}, \mu)$ be the σ-finite metric measure subspace of $\mathbb{R}$ with partial ordering $\leq$, $\boldsymbol{\Omega} := (\Omega, \mathcal{B}, (\mathcal{B}_t)_{t\in\mathbb{I}}, \boldsymbol{P})$ be a filtered probability measure space, $\mathbb{I} \times \boldsymbol{\Omega} = ([0,\infty) \times \Omega, \bar{\mathcal{B}}, \nu)$ be the σ-finite product measure space, $w : \mathbb{I} \times \boldsymbol{\Omega} \to \mathbb{R}$ be a standard Wiener process on $\mathbb{I} \times \boldsymbol{\Omega}$, which is adapted to the filtration $(\mathcal{B}_t)_{t\in\mathbb{I}}$, $\bar{r}_{a,b} \subset \mathbb{I}$ with $a \leq b$, $\mathcal{Y}$ be a separable Hilbert spaces over $\mathbb{K}$, and $W \subseteq \mathcal{Y}$ be a σ-compact conic segment. Assume that*

(a) *$\forall E \in \mathcal{B}$ with $\boldsymbol{P}(E) = 0$, implies that $E \in \mathcal{B}_0$.*
(b) *The σ-algebras $\mathcal{B}_t$ and $\sigma((w_{t+s} - w_t)_{s\in\mathbb{I}})$ are independent, $\forall t \in \mathbb{I}$.*

Let $\phi \in \bar{L}_2(\bar{r}_{a,b} \times \boldsymbol{\Omega}, \mathcal{Y})$ be a W-valued stochastic process adapted to the filtration $(\mathcal{B}_t)_{t\in\mathbb{I}}$ and $\phi_{\langle\omega\rangle} : \bar{r}_{a,b} \to W$ is continuous on the right, $\forall \omega \in \boldsymbol{\Omega}$. Then, ϕ is Itô integrable on the interval $\bar{r}_{a,b}$.

Proof By Proposition 14.103, there exists a sequence of $\mathcal{Y}$-valued predictable step functions $(\phi_n)_{n=1}^{\infty} \subseteq \hat{\mathcal{Q}}(\bar{r}_{a,b} \times \boldsymbol{\Omega}, \mathcal{Y})$, where $\hat{\mathcal{Q}}(\bar{r}_{a,b} \times \boldsymbol{\Omega}, \mathcal{Y})$ is defined in Lemma 14.99, such that $\lim_{n\in\mathbb{N}} \phi_n \doteq \phi$ in $\bar{L}_2(\bar{r}_{a,b} \times \boldsymbol{\Omega}, \mathcal{Y})$. Thus, by Definition 14.104, ϕ is Itô integrable of (with respect to Wiener process w) on the interval $\bar{r}_{a,b}$, and $\overline{\int}_a^b \phi_s\, \mathrm{d}w_s \in L_2(\boldsymbol{\Omega}, \mathcal{Y})$. □

Now, we establish some properties for Itô integral.

Proposition 14.106 *Let $\mathbb{I} := (([0,\infty), |\cdot|), \hat{\mathcal{B}}, \mu)$ be the σ-finite metric measure subspace of $\mathbb{R}$ with partial ordering $\leq$, $\boldsymbol{\Omega} := (\Omega, \mathcal{B}, (\mathcal{B}_t)_{t\in\mathbb{I}}, \boldsymbol{P})$ be a filtered probability measure space, $\mathbb{I} \times \boldsymbol{\Omega} = ([0,\infty) \times \Omega, \bar{\mathcal{B}}, \nu)$ be the σ-finite product measure space, $w : \mathbb{I} \times \boldsymbol{\Omega} \to \mathbb{R}$ be a standard Wiener process on $\mathbb{I} \times \boldsymbol{\Omega}$, which is adapted to the filtration $(\mathcal{B}_t)_{t\in\mathbb{I}}$, $\bar{\mathrm{r}}_{a,b} \subset \mathbb{I}$ with $a \leq b$, $\mathcal{Y}$ and $\mathcal{Z}$ be separable Hilbert spaces over $\mathbb{K}$, $A \in \mathrm{B}(\mathcal{Y}, \mathcal{Z})$, and $y \in \bar{\mathrm{L}}_2(\boldsymbol{\Omega}, \mathcal{Y})$ be $\mathcal{B}_a$-measurable. Assume that*

(a) *The σ-algebras $\mathcal{B}_t$ and $\sigma((w_{t+s} - w_t)_{s\in\mathbb{I}})$ are independent, $\forall t \in \mathbb{I}$.*
(b) *$\phi_i \in \bar{\mathrm{L}}_2(\bar{\mathrm{r}}_{a,b} \times \boldsymbol{\Omega}, \mathcal{Y})$ be a $\mathcal{Y}$-valued stochastic process adapted to the filtration $(\mathcal{B}_t)_{t\in\mathbb{I}}$ and ϕ_i is Itô integrable on $\bar{\mathrm{r}}_{a,b}$, $i = 1, 2$.*

Then, the following statements hold:

(i) $\overline{\int}_a^b \phi_{1s}\, dw_s + \overline{\int}_a^b \phi_{2s}\, dw_s = \overline{\int}_a^b (\phi_1 + \phi_2)_s\, dw_s \in \mathrm{L}_2(\boldsymbol{\Omega}, \mathcal{Y})$.
(ii) $\overline{\int}_a^b y\, dw_s = [y(w_b - w_a)] = \big[\widehat{\int}_a^b y\, dw_s\big] \in \mathrm{L}_2(\boldsymbol{\Omega}, \mathcal{Y})$.
(iii) $A(\overline{\int}_a^b \phi_{1s}\, dw_s) = \overline{\int}_a^b (A\phi_1)_s\, dw_s \in \mathrm{L}_2(\boldsymbol{\Omega}, \mathcal{Z})$.
(iv) *Let $c \in \bar{\mathrm{r}}_{a,b}$. Then,* $\overline{\int}_a^c \phi_{1s}\, dw_s + \overline{\int}_c^b \phi_{1s}\, dw_s = \overline{\int}_a^b \phi_{1s}\, dw_s \in \mathrm{L}_2(\boldsymbol{\Omega}, \mathcal{Y})$.
(v) $\big\|\overline{\int}_a^b \phi_{1s}\, dw_s\big\|_{\mathrm{L}_2(\boldsymbol{\Omega},\mathcal{Y})} = \|\phi_1\|_{\bar{\mathrm{L}}_2(\bar{\mathrm{r}}_{a,b}\times\boldsymbol{\Omega},\mathcal{Y})}$.
(vi) $\mathrm{Re}\big(\int_{\bar{\mathrm{r}}_{a,b}\times\boldsymbol{\Omega}} \langle\phi_1(s,\omega), \phi_2(s,\omega)\rangle_{\mathcal{Y}}\, d\nu(s,\omega)\big) = \mathrm{Re}\big(\boldsymbol{E}(\langle\overline{\int}_a^b \phi_{1s}\, dw_s, \overline{\int}_a^b \phi_{2s}\, dw_s\rangle_{\mathcal{Y}})\big)$.
(vii) $\boldsymbol{E}(\overline{\int}_a^b \phi_{1s}\, dw_s) = \vartheta_{\mathcal{Y}}$.

Proof (i) Fix any $i \in \{1, 2\}$. By Definition 14.104, there exists a sequence of $\mathcal{Y}$-valued simple predictable step functions $(\phi_{i,n})_{n\in\mathbb{N}} \subseteq \hat{\mathcal{Q}}(\bar{\mathrm{r}}_{a,b} \times \boldsymbol{\Omega}, \mathcal{Y})$, where $\hat{\mathcal{Q}}(\bar{\mathrm{r}}_{a,b} \times \boldsymbol{\Omega}, \mathcal{Y})$ is defined in Lemma 14.99, such that $\lim_{n\in\mathbb{N}} \phi_{i,n} \doteq \phi_i$ in $\bar{\mathrm{L}}_2(\bar{\mathrm{r}}_{a,b} \times \boldsymbol{\Omega}, \mathcal{Y})$. For each $n \in \mathbb{N}$, let $I_{i,n} : \boldsymbol{\Omega} \to \mathcal{Y}$ be defined by $I_{i,n}(\omega) := \big(\widehat{\int}_a^b (\phi_{i,n})_s\, dw_s\big)(\omega) \in \mathcal{Y}$. Then, $I_{i,n}$ is $\mathcal{B}_b$-measurable, $\forall n \in \mathbb{N}$, by Lemma 14.99. Then, there exists a unique $[I_i] \in \mathrm{L}_2(\boldsymbol{\Omega}, \mathcal{Y})$, where $I_i : \boldsymbol{\Omega} \to \mathcal{Y}$ is a $\mathcal{Y}$-valued random variable that is further $\mathcal{B}_b$-measurable, such that $[I_i] = \overline{\int}_{\bar{\mathrm{r}}_{a,b}} \phi_{is}\, dw_s = \overline{\int}_a^b \phi_{is}\, dw_s = \lim_{n\in\mathbb{N}} \big[I_{i,n}\big]$ in $\mathrm{L}_2(\boldsymbol{\Omega}, \mathcal{Y})$. By Lemma 14.99, we have $(\phi_{1,n} + \phi_{2,n})_{n\in\mathbb{N}} \subseteq \hat{\mathcal{Q}}(\bar{\mathrm{r}}_{a,b} \times \boldsymbol{\Omega}, \mathcal{Y})$, and clearly $\lim_{n\in\mathbb{N}}(\phi_{1,n} + \phi_{2,n}) \doteq (\phi_1 + \phi_2)$ in $\bar{\mathrm{L}}_2(\bar{\mathrm{r}}_{a,b} \times \boldsymbol{\Omega}, \mathcal{Y})$. By Definition 14.104, $\phi_1 + \phi_2$ is Itô integrable on $\bar{\mathrm{r}}_{a,b}$ and we have $\overline{\int}_a^b (\phi_1 + \phi_2)_s\, dw_s = \lim_{n\in\mathbb{N}} \big[\widehat{\int}_a^b (\phi_{1,n} + \phi_{2,n})_s dw_s\big] = \lim_{n\in\mathbb{N}} \big[\widehat{\int}_a^b \phi_{1,n_s}\, dw_s + \widehat{\int}_a^b \phi_{2,n_s}\, dw_s\big] = \lim_{n\in\mathbb{N}} [I_{1,n} + I_{2,n}] = \lim_{n\in\mathbb{N}}([I_{1,n}] + [I_{2,n}]) = [I_1] + [I_2]$ in $\mathrm{L}_2(\boldsymbol{\Omega}, \mathcal{Y})$, where the second equality follows from Theorem A.6. This proves (i).

(ii) This follows immediately from Lemma 14.99.

(iii) Note that $A\phi_1 : \bar{\mathrm{r}}_{a,b} \times \boldsymbol{\Omega} \to \mathcal{Z}$ is a $\mathcal{Z}$-valued stochastic process adapted to the filtration $(\mathcal{B}_t)_{t\in\mathbb{I}}$, by Propositions 11.38. Then, the sequence of $\mathcal{Z}$-valued simple predictable step functions $(A\phi_{1,n})_{n\in\mathbb{N}} \subseteq \hat{\mathcal{Q}}(\bar{\mathrm{r}}_{a,b} \times \boldsymbol{\Omega}, \mathcal{Z})$ satisfies $\lim_{n\in\mathbb{N}} A\phi_{1,n} \doteq A\phi_1$ in $\bar{\mathrm{L}}_2(\bar{\mathrm{r}}_{a,b} \times \boldsymbol{\Omega}, \mathcal{Z})$. By Definition 14.104, $A\phi_1$ is Itô integrable on $\bar{\mathrm{r}}_{a,b}$ and we have $\overline{\int}_a^b (A\phi_1)_s \, dw_s = \lim_{n\in\mathbb{N}} \left[\widehat{\int}_a^b (A\phi_{1,n})_s \, dw_s\right] = \lim_{n\in\mathbb{N}} \left[A(\widehat{\int}_a^b \phi_{1,n_s} \, dw_s)\right] = \lim_{n\in\mathbb{N}} [AI_{1,n}] = A(\lim_{n\in\mathbb{N}} [I_{1,n}]) = A(\overline{\int}_a^b \phi_{1_s} \cdot dw_s)$ in $\mathrm{L}_2(\boldsymbol{\Omega}, \mathcal{Z})$, where the first equality follows from Definition 14.104, the second equality follows from Definition A.4, the third equality follows from the definition of $I_{1,n}$, and the fourth equality follows from $A \in \mathrm{B}(\mathcal{Y}, \mathcal{Z})$. This proves (iii).

(iv) By Definition 14.104, we have $\overline{\int}_a^c \phi_{1_s} \, dw_s = \lim_{n\in\mathbb{N}} \left[\widehat{\int}_a^c \phi_{1,n_s} \, dw_s\right]$ and $\overline{\int}_c^b \phi_{1_s} \, dw_s = \lim_{n\in\mathbb{N}} \left[\widehat{\int}_c^b \phi_{1,n_s} \, dw_s\right]$. Then, $\overline{\int}_a^c \phi_{1_s} \, dw_s + \overline{\int}_c^b \phi_{1_s} \, dw_s = \lim_{n\in\mathbb{N}} \left(\left[\widehat{\int}_a^c \phi_{1,n_s} \, dw_s\right] + \left[\widehat{\int}_c^b \phi_{1,n_s} \, dw_s\right]\right) = \lim_{n\in\mathbb{N}} \left[\widehat{\int}_a^c \phi_{1,n_s} \, dw_s + \widehat{\int}_c^b \phi_{1,n_s} \, dw_s\right] = \lim_{n\in\mathbb{N}} \left[\widehat{\int}_a^b \phi_{1,n_s} \, dw_s\right] = \overline{\int}_a^b \phi_{1_s} \, dw_s$ in $\mathrm{L}_2(\boldsymbol{\Omega}, \mathcal{Y})$, where the third equality follows from Theorem A.7. This proves (iv).

(v) Note that $\left\|\overline{\int}_a^b \phi_{1_s} \, dw_s\right\|_{\mathrm{L}_2(\boldsymbol{\Omega}, \mathcal{Y})} = \left\| \lim_{n\in\mathbb{N}} \left[\widehat{\int}_a^b \phi_{1,n_s} \, dw_s\right]\right\|_{\mathrm{L}_2(\boldsymbol{\Omega}, \mathcal{Y})} = \lim_{n\in\mathbb{N}} \left\|\widehat{\int}_a^b \phi_{1,n_s} \, dw_s\right\|_{\bar{\mathrm{L}}_2(\boldsymbol{\Omega}, \mathcal{Y})} = \lim_{n\in\mathbb{N}} \|\phi_{1,n}\|_{\bar{\mathrm{L}}_2(\bar{\mathrm{r}}_{a,b} \times \boldsymbol{\Omega}, \mathcal{Y})} = \|\phi_1\|_{\bar{\mathrm{L}}_2(\bar{\mathrm{r}}_{a,b} \times \boldsymbol{\Omega}, \mathcal{Y})}$, where the second equality follows from Proposition 7.21 and the third equality follows from Lemma 14.99.

(vi) Note that

$$\begin{aligned}
&\mathrm{Re}\left(\boldsymbol{E}(\langle \overline{\int}_a^b \phi_{1_s} \, dw_s, \overline{\int}_a^b \phi_{2_s} \, dw_s\rangle_{\mathcal{Y}})\right) \\
&= \mathrm{Re}\left(\langle \overline{\int}_a^b \phi_{1_s} \, dw_s, \overline{\int}_a^b \phi_{2_s} \, dw_s\rangle_{\mathrm{L}_2(\boldsymbol{\Omega}, \mathcal{Y})}\right) = \mathrm{Re}\left(\langle I_1, I_2\rangle_{\bar{\mathrm{L}}_2(\boldsymbol{\Omega}, \mathcal{Y})}\right) \\
&= \frac{1}{2}\left(\langle I_1, I_2\rangle_{\bar{\mathrm{L}}_2(\boldsymbol{\Omega}, \mathcal{Y})} + \langle I_2, I_1\rangle_{\bar{\mathrm{L}}_2(\boldsymbol{\Omega}, \mathcal{Y})}\right) \\
&= \frac{1}{4}\left(\langle I_1 + I_2, I_1 + I_2\rangle_{\bar{\mathrm{L}}_2(\boldsymbol{\Omega}, \mathcal{Y})} - \langle I_1 - I_2, I_1 - I_2\rangle_{\bar{\mathrm{L}}_2(\boldsymbol{\Omega}, \mathcal{Y})}\right) \\
&= \frac{1}{4}\left(\|I_1 + I_2\|^2_{\bar{\mathrm{L}}_2(\boldsymbol{\Omega}, \mathcal{Y})} - \|I_1 - I_2\|^2_{\bar{\mathrm{L}}_2(\boldsymbol{\Omega}, \mathcal{Y})}\right) \\
&= \frac{1}{4}\left(\|\phi_1 + \phi_2\|_{\bar{\mathrm{L}}_2(\bar{\mathrm{r}}_{a,b} \times \boldsymbol{\Omega}, \mathcal{Y})} - \|\phi_1 - \phi_2\|_{\bar{\mathrm{L}}_2(\bar{\mathrm{r}}_{a,b} \times \boldsymbol{\Omega}, \mathcal{Y})}\right) \\
&= \frac{1}{2}\left(\langle \phi_1, \phi_2\rangle_{\bar{\mathrm{L}}_2(\bar{\mathrm{r}}_{a,b} \times \boldsymbol{\Omega}, \mathcal{Y})} + \langle \phi_2, \phi_1\rangle_{\bar{\mathrm{L}}_2(\bar{\mathrm{r}}_{a,b} \times \boldsymbol{\Omega}, \mathcal{Y})}\right) \\
&= \mathrm{Re}\left(\langle \phi_1, \phi_2\rangle_{\bar{\mathrm{L}}_2(\bar{\mathrm{r}}_{a,b} \times \boldsymbol{\Omega}, \mathcal{Y})}\right) \\
&= \mathrm{Re}\left(\int_{\bar{\mathrm{r}}_{a,b} \times \boldsymbol{\Omega}} \langle \phi_1(s, \omega), \phi_2(s, \omega)\rangle_{\mathcal{Y}} \, d\nu(s, \omega)\right)
\end{aligned}$$

where the first equality follows from Example 13.11, the third and fourth equalities follow from Definition 13.1, the fifth equality follows from Proposition 13.2, the sixth equality follows from (v), the seventh equality follows from Definition 13.1,

Proposition 13.2, and simple algebra, the eighth equality follows from Definition 13.1, and the ninth equality follows from Example 13.11.

(vii) Note that $\boldsymbol{E}(\overline{\int}_a^b \phi_{1s}\, \mathrm{d}w_s) = \boldsymbol{E}(I_1) = \lim_{n\in\mathbb{N}} \boldsymbol{E}(I_{1,n}) = \vartheta_{\mathcal{Y}}$, where the first equality follows from (i), the second equality follows from Proposition 11.213, and the last equality follows from the fact that $I_{1,n}(\omega) = \widehat{\int}_a^b \phi_{i,n\,s}(\omega)\, \mathrm{d}w_s(\omega)$, $\forall n \in \mathbb{N}$, where $\phi_{i,n} \in \hat{\mathcal{Q}}(\bar{r}_{a,b} \times \boldsymbol{\Omega}, \mathcal{Y})$, which implies that $\boldsymbol{E}(I_{1,n}) = \vartheta_{\mathcal{Y}}$, $\forall n \in \mathbb{N}$, by Lemma 14.99.

This completes the proof of the proposition. □

The following result shows that, under the condition that the integrand satisfies the Riemann Criterion for Integrability and some mild growth assumptions, the Itô integral of an $\bar{\mathrm{L}}_2$ function can be written as the limit of any sequence of the Itô integral of simple predictable step functions, whose values at the sample times are simply the value of the integrand at the sample times, as the maximum sample period converges to zero. This result allows for easy analysis of the Itô integral when the prescribed conditions are satisfied.

Proposition 14.107 *Let $\mathbb{I} := (([0,\infty), |\cdot|), \hat{\mathcal{B}}, \mu)$ be the σ-finite metric measure subspace of $\mathbb{R}$ with partial ordering $\leq$, $\boldsymbol{\Omega} := (\Omega, \mathcal{B}, (\mathcal{B}_t)_{t\in\mathbb{I}}, \boldsymbol{P})$ be a filtered probability measure space, $\mathbb{I} \times \boldsymbol{\Omega} = ([0,\infty) \times \Omega, \bar{\mathcal{B}}, \nu)$ be the σ-finite product measure space, $w : \mathbb{I} \times \boldsymbol{\Omega} \to \mathbb{R}$ be a standard Wiener process on $\mathbb{I} \times \boldsymbol{\Omega}$, which is adapted to the filtration $(\mathcal{B}_t)_{t\in\mathbb{I}}$, $\bar{r}_{a,b} \subset \mathbb{I}$ with $a \leq b$, $\mathcal{Y}$ be a separable Hilbert space over $\mathbb{K}$, and the σ-algebras $\mathcal{B}_t$ and $\sigma((w_{t+s} - w_t)_{s\in\mathbb{I}})$ are independent, $\forall t \in \mathbb{I}$.*

Define

$$\mathcal{U}(\bar{r}_{a,b} \times \boldsymbol{\Omega}, \mathcal{Y}) := \{\phi \in \bar{\mathrm{L}}_2(\bar{r}_{a,b} \times \boldsymbol{\Omega}, \mathcal{Y}) \mid \phi \text{ is a } \mathcal{Y}\text{-valued stochastic process}$$
$$\text{that is adapted to the filtration } (\mathcal{B}_t)_{t\in\mathbb{I}},\ \phi_{\langle\omega\rangle}(s) \text{ is right continuous,}$$
$$\text{and } \widehat{\int}_a^b \phi_{\langle\omega\rangle}(s)\, \mathrm{d}s \text{ satisfies the Riemann Criterion for Integrability,}$$
$$\text{Theorem A.9, } \forall \omega \in \boldsymbol{\Omega};\ \text{there exists a random variable } c_\phi : \boldsymbol{\Omega} \to \overline{\mathbb{R}_+}$$
$$\text{such that } \|\phi(t,\omega)\|_{\mathcal{Y}}^2 \leq c_\phi(\omega), \forall t \in \bar{r}_{a,b}, \forall \omega \in \boldsymbol{\Omega};\ \text{and } \boldsymbol{E}(c_\phi) < \infty.\}$$

Let $\phi \in \mathcal{U}(\bar{r}_{a,b} \times \boldsymbol{\Omega}, \mathcal{Y})$. Then, the following statements hold:

(i) *$\mathcal{U}(\bar{r}_{a,b} \times \boldsymbol{\Omega}, \mathcal{Y})$ is a subspace of $\bar{\mathrm{L}}_2(\bar{r}_{a,b} \times \boldsymbol{\Omega}, \mathcal{Y})$.*

(ii) *Define $(\bar{\phi}_n)_{n=1}^\infty$ by $\bar{\phi}_n(t,\omega) := \sum_{i=1}^n \phi(s_{i-1}^{\langle n\rangle}, \omega)\chi_{[s_{i-1}^{\langle n\rangle}, s_i^{\langle n\rangle}), \bar{r}_{a,b}}(t)$, $\forall t \in \bar{r}_{a,b}$, $\forall \omega \in \boldsymbol{\Omega}$, where $a = s_0^{\langle n\rangle} < s_1^{\langle n\rangle} < \cdots < s_n^{\langle n\rangle} = b$, and $\max_{i=1,\ldots,n}(s_i^{\langle n\rangle} - s_{i-1}^{\langle n\rangle}) < \frac{b-a+1}{n}$. Then, we have $\lim_{n\in\mathbb{N}} \bar{\phi}_n \doteq \phi$ in $\bar{\mathrm{L}}_2(\bar{r}_{a,b} \times \boldsymbol{\Omega}, \mathcal{Y})$ and $(\bar{\phi}_n)_{n=1}^\infty \subseteq \hat{\mathcal{Q}}(\bar{r}_{a,b} \times \boldsymbol{\Omega}, \mathcal{Y})$.*

(iii) ϕ *is Itô integrable on* $\bar{r}_{a,b}$, $\overline{\int}_a^b \phi_s \, dw_s \doteq \lim_{n\in\mathbb{N}} \left[\widehat{\int}_a^b (\bar{\phi}_n)_s \, dw_s\right] \doteq \lim_{n\in\mathbb{N}} \left[\sum_{i=1}^n \phi_{s_{i-1}^{\langle n\rangle}} (w_{s_i^{\langle n\rangle}} - w_{s_{i-1}^{\langle n\rangle}})\right]$ *in* $L_2(\boldsymbol{\Omega}, \mathcal{Y})$*, and* $\|\phi\|_{\bar{L}_2(\bar{r}_{a,b}\times\boldsymbol{\Omega},\mathcal{Y})} = \left\|\overline{\int}_a^b \phi_s \, dw_s\right\|_{L_2(\boldsymbol{\Omega},\mathcal{Y})} = \lim_{n\in\mathbb{N}} \left\|\widehat{\int}_a^b (\bar{\phi}_n)_s \, dw_s\right\|_{\bar{L}_2(\boldsymbol{\Omega},\mathcal{Y})}$.

Proof (i) This is straightforward.

(ii) Clearly, $(\bar{\phi}_n)_{n=1}^\infty \subseteq \hat{Q}(\bar{r}_{a,b} \times \boldsymbol{\Omega}, \mathcal{Y})$. By Riemann Criterion for Integrability (Theorem A.14), $\forall \epsilon \in (0,\infty) \subset \mathbb{R}$, $\forall \omega \in \boldsymbol{\Omega}$, $\exists \delta(\omega,\epsilon) \in (0,\infty) \subset \mathbb{R}$ such that $\forall \hat{R} := ((s_i)_{i=0}^n, (\xi_i)_{i=1}^n) \in \hat{\mathcal{J}}(I)$ with $\mathrm{Gauge}(\hat{R}) < \delta(\omega,\epsilon)$, we have $\sum_{i=1}^n M_i(\omega)(s_i - s_{i-1}) < \epsilon$, where $M_i(\omega) := \sup_{v_1,v_2\in[s_{i-1},s_i]\subset\mathbb{R}} \|\phi(v_1,\omega) - \phi(v_2,\omega)\|_{\mathcal{Y}}$, $i = 1,\ldots,n$. Let $n_0(\omega,\epsilon) := \left\lceil \frac{b-a+1}{\delta(\omega,\epsilon)} \right\rceil \in \mathbb{N}$. $\forall n \in \mathbb{N}$ with $n_0(\omega,\epsilon) \le n$, we have

$$\begin{aligned}
\left\|(\bar{\phi}_n)_{\langle\omega\rangle} - \phi_{\langle\omega\rangle}\right\|_{\bar{L}_1(\bar{r}_{a,b},\mathcal{Y})} &= \int_a^b \left\|\bar{\phi}_n(t,\omega) - \phi(t,\omega)\right\|_{\mathcal{Y}} dt \\
&= \sum_{i=1}^n \int_{s_{i-1}^{\langle n\rangle}}^{s_i^{\langle n\rangle}} \left\|\bar{\phi}_n(t,\omega) - \phi(t,\omega)\right\|_{\mathcal{Y}} dt \\
&= \sum_{i=1}^n \int_{s_{i-1}^{\langle n\rangle}}^{s_i^{\langle n\rangle}} \left\|\phi_n(s_{i-1}^{\langle n\rangle},\omega) - \phi(t,\omega)\right\|_{\mathcal{Y}} dt \\
&\le \sum_{i=1}^n \int_{s_{i-1}^{\langle n\rangle}}^{s_i^{\langle n\rangle}} M_i^{\langle n\rangle}(\omega)\, dt = \sum_{i=1}^n M_i^{\langle n\rangle}(\omega)(s_i^{\langle n\rangle} - s_{i-1}^{\langle n\rangle}) < \epsilon
\end{aligned}$$

where the second equality follows from Proposition 11.83, the third equality follows form the definition of $\bar{\phi}_n$, the first inequality follows from the definitions $M_i^{\langle n\rangle}(\omega) := \sup_{v_1,v_2\in[s_{i-1}^{\langle n\rangle},s_i^{\langle n\rangle}]\subset\mathbb{R}} \|\phi(v_1,\omega) - \phi(v_2,\omega)\|_{\mathcal{Y}}$, $i = 1,\ldots,n$, the last equality follows from Proposition 11.75, and the last inequality follows from the choice of $\delta(\omega,\epsilon)$. This shows that $\lim_{n\in\mathbb{N}} \left\|(\bar{\phi}_n)_{\langle\omega\rangle} - \phi_{\langle\omega\rangle}\right\|_{\bar{L}_1(\bar{r}_{a,b},\mathcal{Y})} = 0$, $\forall \omega \in \boldsymbol{\Omega}$.

Note that

$$\begin{aligned}
\left\|(\bar{\phi}_n)_{\langle\omega\rangle} - \phi_{\langle\omega\rangle}\right\|_{\bar{L}_2(\bar{r}_{a,b},\mathcal{Y})}^2 &= \int_a^b \left\|\bar{\phi}_n(t,\omega) - \phi(t,\omega)\right\|_{\mathcal{Y}}^2 dt \\
&= \sum_{i=1}^n \int_{s_{i-1}^{\langle n\rangle}}^{s_i^{\langle n\rangle}} \left\|\bar{\phi}_n(t,\omega) - \phi(t,\omega)\right\|_{\mathcal{Y}}^2 dt \\
&= \sum_{i=1}^n \int_{s_{i-1}^{\langle n\rangle}}^{s_i^{\langle n\rangle}} \left\|\phi_n(s_{i-1}^{\langle n\rangle},\omega) - \phi(t,\omega)\right\|_{\mathcal{Y}}^2 dt
\end{aligned}$$

$$= \sum_{i=1}^{n} \int_{s_{i-1}^{\langle n\rangle}}^{s_i^{\langle n\rangle}} \left\| \phi_n(s_{i-1}^{\langle n\rangle}, \omega) - \phi(t, \omega) \right\|_{\mathcal{Y}} \left\| \phi_n(s_{i-1}^{\langle n\rangle}, \omega) - \phi(t, \omega) \right\|_{\mathcal{Y}} \mathrm{d}t$$

$$\leq \sum_{i=1}^{n} \int_{s_{i-1}^{\langle n\rangle}}^{s_i^{\langle n\rangle}} \left\| \phi_n(s_{i-1}^{\langle n\rangle}, \omega) - \phi(t, \omega) \right\|_{\mathcal{Y}} 2\sqrt{c_\phi(\omega)}\, \mathrm{d}t$$

$$= 2\sqrt{c_\phi(\omega)} \sum_{i=1}^{n} \int_{s_{i-1}^{\langle n\rangle}}^{s_i^{\langle n\rangle}} \left\| \phi_n(s_{i-1}^{\langle n\rangle}, \omega) - \phi(t, \omega) \right\|_{\mathcal{Y}} \mathrm{d}t$$

$$= 2\sqrt{c_\phi(\omega)} \left\| (\bar{\phi}_n)_{\langle\omega\rangle} - \phi_{\langle\omega\rangle} \right\|_{\bar{\mathrm{L}}_1(\bar{\mathrm{r}}_{a,b}, \mathcal{Y})}$$

where the second equality follows from Proposition 11.83, the third equality follows from the definition of $\bar{\phi}_n$, the first inequality follows from the definition of $\mathcal{U}(\bar{\mathrm{r}}_{a,b} \times \boldsymbol{\Omega}, \mathcal{Y})$, the fifth equality follows from Proposition 11.92, and the last equality follows from the previous paragraph. This shows that $\lim_{n\in\mathbb{N}} \left\| (\bar{\phi}_n)_{\langle\omega\rangle} - \phi_{\langle\omega\rangle} \right\|^2_{\bar{\mathrm{L}}_2(\bar{\mathrm{r}}_{a,b}, \mathcal{Y})} = 0$, $\forall \omega \in \boldsymbol{\Omega}$.

It is straightforward to establish that, $\forall n \in \mathbb{N}$,

$$\left\| (\bar{\phi}_n)_{\langle\omega\rangle} - \phi_{\langle\omega\rangle} \right\|^2_{\bar{\mathrm{L}}_2(\bar{\mathrm{r}}_{a,b}, \mathcal{Y})}$$

$$= \int_a^b \left\| \bar{\phi}_n(t, \omega) - \phi(t, \omega) \right\|^2_{\mathcal{Y}} \mathrm{d}t \leq \int_a^b 4c_\phi(\omega)\, \mathrm{d}t = 4\,(b-a)c_\phi(\omega)$$

where the inequality follows from the definition of $\mathcal{U}(\bar{\mathrm{r}}_{a,b} \times \boldsymbol{\Omega}, \mathcal{Y})$ and the last equality follows from Proposition 11.75. The right-hand side of the above inequality is integrable by the definition of $\mathcal{U}(\bar{\mathrm{r}}_{a,b} \times \boldsymbol{\Omega}, \mathcal{Y})$.

Thus, we have

$$\lim_{n\in\mathbb{N}} \left\| \bar{\phi}_n - \phi \right\|^2_{\bar{\mathrm{L}}_2(\bar{\mathrm{r}}_{a,b} \times \boldsymbol{\Omega}, \mathcal{Y})} = \lim_{n\in\mathbb{N}} \int_{\boldsymbol{\Omega}} \left\| (\bar{\phi}_n)_{\langle\omega\rangle} - \phi_{\langle\omega\rangle} \right\|^2_{\bar{\mathrm{L}}_2(\bar{\mathrm{r}}_{a,b}, \mathcal{Y})} \mathrm{d}\boldsymbol{P}(\omega)$$

$$= \int_{\boldsymbol{\Omega}} \lim_{n\in\mathbb{N}} \left\| (\bar{\phi}_n)_{\langle\omega\rangle} - \phi_{\langle\omega\rangle} \right\|^2_{\bar{\mathrm{L}}_2(\bar{\mathrm{r}}_{a,b}, \mathcal{Y})} \mathrm{d}\boldsymbol{P}(\omega) = 0$$

where the second equality follows from Lebesgue Dominated Convergence Theorem 11.91 and the last equality follows from the second to last paragraph. This shows that $\lim_{n\in\mathbb{N}} \bar{\phi}_n \doteq \phi$ in $\bar{\mathrm{L}}_2(\bar{\mathrm{r}}_{a,b} \times \boldsymbol{\Omega}, \mathcal{Y})$. This proves (ii).

(iii) This is immediate by Definition 14.104, Proposition 14.106, and Lemma 14.99.

This completes the proof of the proposition. □

Theorem 14.108 *Let* $\mathbb{I} := (([0,\infty), |\cdot|), \hat{\mathcal{B}}, \mu)$ *be the σ-finite metric measure subspace of* $\mathbb{R}$ *with partial ordering* $\leq$, $\boldsymbol{\Omega} := (\Omega, \mathcal{B}, (\mathcal{B}_t)_{t\in\mathbb{I}}, \boldsymbol{P})$ *be a filtered probability measure space,* $\mathbb{I} \times \boldsymbol{\Omega} = ([0,\infty) \times \Omega, \bar{\mathcal{B}}, \nu)$ *be the σ-finite product measure space,* $w : \mathbb{I} \times \boldsymbol{\Omega} \to \mathbb{R}$ *be a standard Wiener process on* $\mathbb{I} \times \boldsymbol{\Omega}$ *be* *adapted to the filtration $(\mathcal{B}_t)_{t\in\mathbb{I}}$, the σ-algebras $\mathcal{B}_t$ and $\sigma((w_{t+s} - w_t)_{s\in\mathbb{I}})$ are independent, $\forall t \in \mathbb{I}$, $\mathcal{Y}$ be a separable Hilbert space, and $\phi \in \bar{\mathrm{L}}_2(\bar{\mathrm{r}}_{a,b}, \mathcal{Y})$, where $a, b \in \mathbb{I}$ with $a \leq b > 0$. Then, ϕ is both Wiener integrable and Itô integrable and $\widetilde{\int}_a^b \phi_s \, \mathrm{d}w_s = \overline{\int}_a^b \phi_s \, \mathrm{d}w_s$.*

Proof By Proposition 14.98 and its proof, we have $\widetilde{\int}_a^b \phi_s \, \mathrm{d}w_s \in \mathrm{L}_2(\boldsymbol{\Omega}, \mathcal{Y})$ exists and is equal to $\lim_{n\in\mathbb{N}} \big[\widehat{\int}_a^b g_n(s) \, \mathrm{d}w_s\big]$ in $\mathrm{L}_2(\boldsymbol{\Omega}, \mathcal{Y})$, where $g_n : \bar{\mathrm{r}}_{a,b} \to \mathcal{Y}$ is absolutely continuous function and of bounded variation such that $g_n \in \bar{\mathrm{L}}_2(\bar{\mathrm{r}}_{a,b}, \mathcal{Y})$ and $\|g_n - \phi\|_{\bar{\mathrm{L}}_2(\bar{\mathrm{r}}_{a,b}, \mathcal{Y})} < 2^{-n}$, $\forall n \in \mathbb{N}$. Then, $\lim_{n\in\mathbb{N}} [g_n] = [\phi]$ in $\mathrm{L}_2(\bar{\mathrm{r}}_{a,b}, \mathcal{Y})$. $\forall n \in \mathbb{N}$, g_n is continuous and defined on a compact interval, then it is uniform continuous and $\exists h_n : \bar{\mathrm{r}}_{a,b} \to \mathcal{Y}$ defined by $h_n(t) = \sum_{i=1}^{N_n} g(s_{i-1}^{\langle N_n \rangle}) \chi_{[s_{i-1}^{\langle N_n \rangle}, s_i^{\langle N_n \rangle}), \bar{\mathrm{r}}_{a,b}}(t)$, $\forall t \in \bar{\mathrm{r}}_{a,b}$, where $N_n \in \mathbb{N}$ and $s_i^{\langle N_n \rangle} := a + \frac{i}{N_n}(b - a)$, $i = 0, \ldots, N_n$, such that $\|h_n - g_n\|_{\bar{\mathrm{L}}_2(\bar{\mathrm{r}}_{a,b}, \mathcal{Y})} < 2^{-n}$. Thus, $\lim_{n\in\mathbb{N}} [h_n] = [\phi]$ in $\mathrm{L}_2(\bar{\mathrm{r}}_{a,b}, \mathcal{Y})$. Clearly, $(h_n)_{n=1}^{\infty} \subseteq \hat{\mathcal{Q}}(\bar{\mathrm{r}}_{a,b} \times \boldsymbol{\Omega}, \mathcal{Y})$. By Definition 14.104, we have $\lim_{n\in\mathbb{N}} \big[\widehat{\int}_a^b h_n(s) \, \mathrm{d}w_s\big] = \overline{\int}_a^b \phi_s \, \mathrm{d}w_s$ in $\mathrm{L}_2(\boldsymbol{\Omega}, \mathcal{Y})$. By Proposition 14.98, $\lim_{n\in\mathbb{N}} \big[\widehat{\int}_a^b h_n(s) \, \mathrm{d}w_s\big] = \widetilde{\int}_a^b \phi_s \, \mathrm{d}w_s$ in $\mathrm{L}_2(\boldsymbol{\Omega}, \mathcal{Y})$. Hence, $\widetilde{\int}_a^b \phi_s \, \mathrm{d}w_s = \overline{\int}_a^b \phi_s \, \mathrm{d}w_s$. This completes the proof of the theorem. □

Next, we present three results that are needed in the proof of Itô's Formula. They are of independent interest in the study of stochastic systems as well.

Proposition 14.109 *Let $(X, \mathcal{B})$ be a measurable space, $\mathcal{Y} := (Y, \rho)$ be a nonempty compact metric space, $f : X \times \mathcal{Y} \to \mathbb{R}$, $f_x : \mathcal{Y} \to \mathbb{R}$ be continuous, $\forall x \in X$, and $f_{\langle y \rangle} : X \to \mathbb{R}$ be $\mathcal{B}$-measurable, $\forall y \in \mathcal{Y}$. Then, the function $h : X \to \mathbb{R}$, defined by $h(x) := \max_{y\in\mathcal{Y}} f(x, y)$, $\forall x \in X$, is $\mathcal{B}$-measurable.*

Proof By Proposition 11.190, $\mathcal{Y}$ is separable. Then, $\exists D \subseteq \mathcal{Y}$ such that $\overline{D} = \mathcal{Y}$ and D is countable. $\forall x \in X$, by the continuity of $f_x : \mathcal{Y} \to \mathbb{R}$, we have $h(x) = \max_{y\in\mathcal{Y}} f(x, y) = \sup_{y\in D} f(x, y)$. Then, h is $\mathcal{B}$-measurable by Proposition 11.40. □

Proposition 14.110 *Let $\boldsymbol{\Omega} := (\Omega, \mathcal{B}, \boldsymbol{P})$ be a probability measure space, $\mathcal{X}$ be a separable Banach space over $\mathbb{K}$, and M be a set of $\mathcal{X}$-valued random variables. Assume that M is bounded in $\bar{\mathrm{L}}_1(\boldsymbol{\Omega}, \mathcal{X})$, i.e., $\exists c \in [0, \infty) \subset \mathbb{R}$ such that $\boldsymbol{E}(\mathcal{P} \circ x) \leq c$, $\forall x \in M$. Then, M is uniformly growth bounded.*

Proof This follows readily from Chebyshev's Inequality, Proposition 14.67. □

Proposition 14.111 *Let $\boldsymbol{\Omega} := (\Omega, \mathcal{B}, \boldsymbol{P})$ be a probability measure space, $\mathcal{X}$ and $\mathcal{Y}$ be separable Banach spaces over $\mathbb{K}$, and $W \subseteq \mathrm{B}(\mathcal{X}, \mathcal{Y})$ be a separable subspace, $(x_n)_{n=1}^{\infty}$ be a sequence of $\mathcal{X}$-valued random variables, and $(y_n)_{n=1}^{\infty}$ be a sequence of W-valued random variables. Assume that $(x_n)_{n=1}^{\infty}$ is uniformly growth bounded and $\lim_{n\in\mathbb{N}} y_n = \vartheta_{\bar{\mathrm{L}}_1(\boldsymbol{\Omega}, W)}$ in measure in $\boldsymbol{\Omega}$. Then, we have $(y_n x_n)_{n=1}^{\infty}$ is a sequence of $\mathcal{Y}$-valued random variables and $\lim_{n\in\mathbb{N}} y_n x_n = \vartheta_{\bar{\mathrm{L}}_1(\boldsymbol{\Omega}, \mathcal{Y})}$ in measure in $\boldsymbol{\Omega}$.*

Proof $\forall \omega \in \boldsymbol{\Omega}$, $y_n(\omega)x_n(\omega) \in \mathcal{Y}$, $\forall n \in \mathbb{N}$. By Propositions 11.39, 11.38, and 7.65, we have $y_n x_n : \boldsymbol{\Omega} \to \mathcal{Y}$ is $\mathcal{B}$-measurable, $\forall n \in \mathbb{N}$. Hence, $(y_n x_n)_{n=1}^{\infty}$ is a sequence of $\mathcal{Y}$-valued random variables.

Fix any $\epsilon \in (0, \infty) \subset \mathbb{R}$. By the assumption that $(x_n)_{n=1}^{\infty}$ being uniformly growth bounded, $\exists K \in [0, \infty) \subset \mathbb{R}$ such that $\boldsymbol{P}(E_n) := \boldsymbol{P}(\{\omega \in \boldsymbol{\Omega} \mid \|x_n(\omega)\|_{\mathcal{X}} > K\} < \frac{\epsilon}{2}$, $\forall n \in \mathbb{N}$. By the assumption that $\lim_{n \in \mathbb{N}} y_n = \vartheta_{\bar{\mathrm{L}}_1(\boldsymbol{\Omega}, W)}$ in measure in $\boldsymbol{\Omega}$, we have $\exists N \in \mathbb{N}$, $\forall n \in \mathbb{N}$ with $N \leq n$, $\boldsymbol{P}(F_n) := \boldsymbol{P}(\{\omega \in \boldsymbol{\Omega} \mid \|y_n(\omega)\|_{\mathrm{B}(\mathcal{X},\mathcal{Y})} \geq \frac{\epsilon}{K+2}\} < \frac{\epsilon}{K+2}$. Fix any $n \in \mathbb{N}$ with $N \leq n$. Let $A_n := E_n \cup F_n \in \mathcal{B}$. Then, $\boldsymbol{P}(A_n) \leq \boldsymbol{P}(E_n) + \boldsymbol{P}(F_n) < \frac{\epsilon}{2} + \frac{\epsilon}{K+2} \leq \epsilon$. $\forall \omega \in \Omega \setminus A_n$, we have $\|y_n(\omega)x_n(\omega)\|_{\mathcal{Y}} \leq \|y_n(\omega)\|_{\mathrm{B}(\mathcal{X},\mathcal{Y})}\|x_n(\omega)\|_{\mathcal{X}} \leq \frac{\epsilon}{K+2} \cdot K < \epsilon$. Hence, $\boldsymbol{P}(\{\omega \in \boldsymbol{\Omega} \mid \|y_n(\omega)x_n(\omega)\| \geq \epsilon\} \leq \boldsymbol{P}(A_n) < \epsilon$. This proves that $\lim_{n \in \mathbb{N}} y_n x_n = \vartheta_{\bar{\mathrm{L}}_1(\boldsymbol{\Omega}, \mathcal{Y})}$ in measure in $\boldsymbol{\Omega}$.

This completes the proof of the proposition. □

14.13 Itô Processes

Definition 14.112 Let $\boldsymbol{\Omega} := (\Omega, \mathcal{B}, \boldsymbol{P})$ be a probability measure space and $\mathcal{X}$ be a separable Banach space over $\mathbb{K}$, $y : \boldsymbol{\Omega} \to \mathcal{X}$ be an $\mathcal{X}$-valued random variable. Define $|||y||| := \boldsymbol{E}(1 \wedge (\mathcal{P} \circ y)) = \int_{\boldsymbol{\Omega}} (1 \wedge \|y(\omega)\|_{\mathcal{X}})\, \mathrm{d}\boldsymbol{P}(\omega) \in \bar{\mathrm{r}}_{0,1}$. ◇

Lemma 14.113 *Let $\boldsymbol{\Omega} := (\Omega, \mathcal{B}, \boldsymbol{P})$ be a probability measure space and $\mathcal{X}$ be a separable Banach space over $\mathbb{K}$, $y : \boldsymbol{\Omega} \to \mathcal{X}$ and $z : \boldsymbol{\Omega} \to \mathcal{X}$ be two $\mathcal{X}$-valued random variables, and $(y_n)_{n=1}^{\infty}$ be a sequence of $\mathcal{X}$-valued random variables. Then, we have*

(i) $|||y + z||| \leq |||y||| + |||z|||$.
(ii) $\lim_{n \in \mathbb{N}} y_n = y$ *in measure in* $\boldsymbol{\Omega}$ *if, and only if,* $\lim_{n \in \mathbb{N}} |||y - y_n||| = 0$.

Proof (i) Note that $|||y + z||| = \boldsymbol{E}(1 \wedge (\mathcal{P} \circ (y + z))) \leq \boldsymbol{E}(1 \wedge (\mathcal{P} \circ y + \mathcal{P} \circ z)) \leq \boldsymbol{E}(1 \wedge (\mathcal{P} \circ y) + 1 \wedge (\mathcal{P} \circ z)) = \boldsymbol{E}(1 \wedge (\mathcal{P} \circ y)) + \boldsymbol{E}(1 \wedge (\mathcal{P} \circ z)) = |||y||| + |||z|||$, where the first equality follows from Definition 14.112, the first inequality follows from the triangular inequality for norms, the second inequality follows from simple algebra, the second equality follows from Proposition 11.92, and the last equality follows from Definition 14.112.

(ii) "Necessity" Let $\lim_{n \in \mathbb{N}} y_n = y$ in measure in $\boldsymbol{\Omega}$. Then, $\forall \epsilon \in \mathrm{r}^{\circ}_{0,1}$, $\exists n(\epsilon) \in \mathbb{N}$, such that $\forall n \in \mathbb{N}$ with $n(\epsilon) \leq n$, we have $\boldsymbol{P}(A_{n,\epsilon}) := \boldsymbol{P}(\{\omega \in \boldsymbol{\Omega} \mid \|y_n(\omega) - y(\omega)\|_{\mathcal{X}} > \frac{\epsilon}{2}\}) < \frac{\epsilon}{2}$. Then, $|||y_n - y||| = \boldsymbol{E}(1 \wedge (\mathcal{P} \circ (y_n - y))) = \int_{\boldsymbol{\Omega}} (1 \wedge \|y_n(\omega) - y(\omega)\|_{\mathcal{X}})\, \mathrm{d}\boldsymbol{P}(\omega) \leq \int_{A_{n,\epsilon}} 1 \mathrm{d}\boldsymbol{P}(\omega) + \int_{\Omega \setminus A_{n,\epsilon}} \frac{\epsilon}{2}\, \mathrm{d}\boldsymbol{P}(\omega) \leq \boldsymbol{P}(A_{n,\epsilon}) + \int_{\Omega} \frac{\epsilon}{2}\, \mathrm{d}\boldsymbol{P}(\omega) < \frac{\epsilon}{2} + \frac{\epsilon}{2} = \epsilon$, where the first equality follows from Definition 14.112, the second equality follows from Definition 14.1, the first inequality follows from the definition of $A_{n,\epsilon} \in \mathcal{B}$, the second inequality follows from Definition 14.1, and the last inequality follows from the property for $A_{n,\epsilon}$. By the arbitrariness of ϵ, we have $\lim_{n \in \mathbb{N}} |||y_n - y||| = 0$.

"Sufficiency" Let $\lim_{n \in \mathbb{N}} |||y_n - y||| = 0$. $\forall \epsilon \in \mathrm{r}^{\circ}_{0,1}$, $\exists n(\epsilon) \in \mathbb{N}$, such that $\forall n \in \mathbb{N}$ with $n(\epsilon) \leq n$, we have $|||y_n - y||| < \epsilon^2$. Then, $\boldsymbol{P}(\bar{A}_{n,\epsilon}) := \boldsymbol{P}(\{\omega \in$

$\boldsymbol{\Omega} \mid \|y_n(\omega) - y(\omega)\|_{\mathcal{X}} > \epsilon\}) = \boldsymbol{P}(\{\omega \in \boldsymbol{\Omega} \mid 1 \wedge \|y_n(\omega) - y(\omega)\|_{\mathcal{X}} > \epsilon\}) \leq \frac{1}{\epsilon}\boldsymbol{E}(1 \wedge (\mathcal{P} \circ (y_n - y))) = \frac{1}{\epsilon}|||y_n - y||| < \epsilon$, where the first equality follows from simple algebra, the first inequality follows from Chebyshev's Inequality Proposition 14.67, the second equality follows from Definition 14.112, and the last inequality follows from $n \geq n(\epsilon)$. By the arbitrariness of ϵ, we have $\lim_{n\in\mathbb{N}} y_n = y$ in measure in $\boldsymbol{\Omega}$. □

Definition 14.114 Let $\boldsymbol{\Omega} := (\Omega, \mathcal{B}, (\mathcal{B}_n)_{n=0}^{\infty}, \boldsymbol{P})$ be a filtered probability measure space, $\mathcal{Y}$ be a separable Hilbert space over $\mathbb{K}$, $X := (X_n)_{n=0}^{\infty}$ be an adapted $\mathcal{Y}$-valued Martingale, and $d_n := X_n - X_{n-1}$ be a $\mathcal{Y}$-valued random variable, $n \in \mathbb{N}$. Define $\mathrm{S}(X) := (\|X_0\|_{\mathcal{Y}}^2 + \sum_{k=1}^{\infty} \|d_k\|_{\mathcal{Y}}^2)^{\frac{1}{2}}$, $\mathrm{s}(X) := (\|X_0\|_{\mathcal{Y}}^2 + \sum_{k=1}^{\infty} \boldsymbol{E}(\|d_k\|_{\mathcal{Y}}^2|\mathcal{B}_{k-1}))^{\frac{1}{2}}$, $\mathrm{S}_n(X) := (\|X_0\|_{\mathcal{Y}}^2 + \sum_{k=1}^{n} \|d_k\|_{\mathcal{Y}}^2)^{\frac{1}{2}}$, and $\mathrm{s}_n(X) := (\|X_0\|_{\mathcal{Y}}^2 + \sum_{k=1}^{n} \boldsymbol{E}(\|d_k\|_{\mathcal{Y}}^2|\mathcal{B}_{k-1}))^{\frac{1}{2}}$, $\forall n \in \mathbb{Z}_+$. $\mathrm{S}(X) : \boldsymbol{\Omega} \to [0, \infty] \subset \mathbb{R}_e$ is the *square function for the Martingale* X, and $\mathrm{S}_n(X) : \boldsymbol{\Omega} \to \overline{\mathbb{R}_+}$ is the corresponding partial sum. $\mathrm{s}(X) : \boldsymbol{\Omega} \to [0, \infty] \subset \mathbb{R}_e$ is the *conditional square function for the Martingale* X, and $\mathrm{s}_n(X) : \boldsymbol{\Omega} \to \overline{\mathbb{R}_+}$ is the corresponding partial sum. ◇

Proposition 14.115 *Let $\boldsymbol{\Omega} := (\Omega, \mathcal{B}, (\mathcal{B}_n)_{n=0}^{\infty}, \boldsymbol{P})$ be a filtered probability measure space, $(Y_n)_{n=0}^{\infty}$ be a sequence of adapted nonnegative real-valued random variables that is a sub Martingale, and $Y_M : \boldsymbol{\Omega} \to [0, \infty] \subset \mathbb{R}_e$ be defined by $Y_M(\omega) := \sup_{n\in\mathbb{Z}_+} Y_n(\omega)$, $\forall \omega \in \boldsymbol{\Omega}$. Then, Y_M is a nonnegative extended real-valued function that is $\mathcal{B}$-measurable, and $\boldsymbol{P}(\{\omega \in \boldsymbol{\Omega} \mid Y_M(\omega) > \lambda\}) \leq \frac{1}{\lambda}\lim_{n\in\mathbb{N}} \boldsymbol{E}(Y_n)$, $\forall \lambda \in \mathbb{R}_+$.*

Proof Define $Y_{M,n} : \boldsymbol{\Omega} \to \overline{\mathbb{R}_+}$ by $Y_{M,n}(\omega) := \sup_{k\in\mathbb{Z}_+, k\leq n} Y_k(\omega)$, $\forall \omega \in \boldsymbol{\Omega}$, $\forall n \in \mathbb{Z}_+$. By Proposition 11.40, $Y_{M,n}$, $\forall n \in \mathbb{Z}_+$, and Y_M are $\mathcal{B}$-measurable. Then, $Y_{M,n}$ is a nonnegative real-valued random variable, $\forall n \in \mathbb{Z}_+$. Clearly, we have $Y_{M,n}(\omega) \uparrow Y_M(\omega)$, as $n \to \infty$, $\forall \omega \in \boldsymbol{\Omega}$. Then, we have $A_{n,\lambda} \subseteq A_{n+1,\lambda}$, and $\bigcup_{n\in\mathbb{Z}_+} A_{n,\lambda} = A_\lambda := \{\omega \in \boldsymbol{\Omega} \mid Y_M(\omega) > \lambda\}$, where $A_{n,\lambda} := \{\omega \in \boldsymbol{\Omega} \mid Y_{M,n}(\omega) > \lambda\}$, $\forall n \in \mathbb{Z}_+$, $\forall \lambda \in \mathbb{R}_+$.

Claim 14.115.1 $\boldsymbol{P}(A_{n,1}) \leq \boldsymbol{E}(Y_n)$, $\forall n \in \mathbb{Z}_+$.

Proof of Claim Note that, $\forall n \in \mathbb{Z}_+$,

$$\begin{aligned}
\boldsymbol{P}(A_{n,1}) &= \boldsymbol{P}(\{\omega \in \boldsymbol{\Omega} \mid Y_0(\omega) > 1\}) \\
&\quad +\boldsymbol{P}(\{\omega \in \boldsymbol{\Omega} \mid Y_0(\omega) \leq 1, Y_1(\omega) > 1\}) + \cdots \\
&\quad +\boldsymbol{P}(\{\omega \in \boldsymbol{\Omega} \mid Y_k(\omega) \leq 1, k = 0, \ldots, n-1, Y_n(\omega) > 1\}) \\
&=: \sum_{k=0}^{n} \boldsymbol{P}(B_k) = \sum_{k=0}^{n} \int_{B_k} 1 \mathrm{d}\boldsymbol{P} \leq \sum_{k=0}^{n} \int_{B_k} Y_k \, \mathrm{d}\boldsymbol{P} \\
&\leq \sum_{k=0}^{n} \int_{B_k} \boldsymbol{E}(Y_n|\mathcal{B}_k) \, \mathrm{d}\boldsymbol{P} = \sum_{k=0}^{n} \boldsymbol{E}(\chi_{B_k,\Omega} \boldsymbol{E}(Y_n|\mathcal{B}_k))
\end{aligned}$$

$$= \sum_{k=0}^{n} \boldsymbol{E}(\boldsymbol{E}(\chi_{B_k,\Omega} Y_n | \mathcal{B}_k)) = \sum_{k=0}^{n} \boldsymbol{E}(\chi_{B_k,\Omega} Y_n)$$

$$= \sum_{k=0}^{n} \int_{B_k} Y_n \, \mathrm{d}\boldsymbol{P} = \int_{A_{n,1}} Y_n \, \mathrm{d}\boldsymbol{P} \leq \int_{\Omega} Y_n \, \mathrm{d}\boldsymbol{P} = \boldsymbol{E}(Y_n)$$

where the first equality follows from the fact that $A_{n,1}$ equals the disjoint union of $\bigcup_{k=0}^{n} B_k$, the second equality follows from Definition 14.1, the first inequality follows from the fact $Y_k(\omega) > 1$, $\forall \omega \in B_k$, $k = 0, \ldots, n$, the second inequality follows from the fact $(Y_n)_{n=0}^{\infty}$ is a sub Martingale, the third equality follows from Definition 14.1, the fourth equality follows from (h) of Proposition 14.11, the fifth equality follows from (a) of Proposition 14.11, the sixth equality follows from Definition 14.1, the seventh equality follows from Proposition 11.92, the third inequality follows from the fact that Y_n is nonnegative real-valued, and the last equality follows from Definition 14.1. This completes the proof of the claim. □

By Claim 14.115.1, we have that $\boldsymbol{P}(A_{n,\lambda}) = \boldsymbol{P}(\{\omega \in \boldsymbol{\Omega} \mid \frac{1}{\lambda} Y_{M,n} > 1\}) \leq \boldsymbol{E}(\frac{1}{\lambda} Y_n) = \frac{1}{\lambda} \boldsymbol{E}(Y_n)$, $\forall \lambda \in \mathbb{R}_+$, where the first equality follows from simple set equality, the inequality follows from Claim 14.115.1, and the last equality follows from Proposition 11.92. Therefore, we have $\boldsymbol{P}(A_\lambda) = \boldsymbol{P}(\bigcup_{n \in \mathbb{Z}_+} A_{n,\lambda}) = \lim_{n \in \mathbb{N}} \boldsymbol{P}(A_{n,\lambda}) \leq \liminf_{n \in \mathbb{N}} \frac{1}{\lambda} \boldsymbol{E}(Y_n) = \frac{1}{\lambda} \lim_{n \in \mathbb{N}} \boldsymbol{E}(Y_n)$, $\forall \lambda \in \mathbb{R}_+$, where the first equality follows from the first paragraph in the proof of this proposition, the second equality follows from Proposition 11.7, the inequality follows from the preceding discussion, and the last equality follows from the fact $\boldsymbol{E}(Y_n) \leq \boldsymbol{E}(Y_{n+1})$, $\forall n \in \mathbb{Z}_+$. This completes the proof of the proposition. □

Lemma 14.116 *Let $\boldsymbol{\Omega} := (\Omega, \mathcal{B}, (\mathcal{B}_n)_{n=0}^{\infty}, \boldsymbol{P})$ be a filtered probability measure space, $\mathcal{Y}$ be a separable Hilbert space over $\mathbb{K}$, $X := (X_n)_{n=0}^{\infty}$ be an adapted $\mathcal{Y}$-valued Martingale, $d_n := X_n - X_{n-1}$ be a $\mathcal{Y}$-valued random variable, $n \in \mathbb{N}$, and $X_M : \boldsymbol{\Omega} \to [0, \infty] \subset \mathbb{R}_e$ be defined by $X_M(\omega) := \sup_{n \in \mathbb{Z}_+} \|X_n(\omega)\|_{\mathcal{Y}}$, $\forall \omega \in \boldsymbol{\Omega}$. Then, $|||X_M||| \leq 3|||\,\mathrm{s}(X)|||$.*

Proof Since X is an adapted $\mathcal{Y}$-valued Martingale, then $X_n : \boldsymbol{\Omega} \to \mathcal{Y}$ is $\mathcal{B}_n$-measurable, and $X_n \in \boldsymbol{E}(X_{n+1}|\mathcal{B}_n)$, $\forall n \in \mathbb{Z}_+$. By Proposition 11.38, $\mathcal{P} \circ X_n$ is $\mathcal{B}_n$-measurable. By Proposition 11.40, X_M is $\mathcal{B}$-measurable. Hence, $1 \wedge X_M$ is nonnegative real-valued random variable. We need the following intermediate result:

Claim 14.116.1 $\forall \lambda \in \mathbb{R}_+$, let $A_\lambda := \{\omega \in \boldsymbol{\Omega} \mid \mathrm{s}(X) > \lambda\} \in \mathcal{B}$, then

$$\boldsymbol{P}(\{\omega \in \boldsymbol{\Omega} \mid X_M > \lambda, \mathrm{s}(X) \leq \lambda\}) \leq \boldsymbol{P}(A_\lambda) + \frac{1}{\lambda^2} \int_{\Omega \setminus A_\lambda} (\mathrm{s}(X))^2 \, \mathrm{d}\boldsymbol{P}$$

Proof of Claim Note that $\mathrm{s}_n(X) = (\|X_0\|_{\mathcal{Y}}^2 + \sum_{k=1}^{n} \boldsymbol{E}(\|d_k\|_{\mathcal{Y}}^2 | \mathcal{B}_{k-1}))^{\frac{1}{2}}$ is $\mathcal{B}_{n-1}$-measurable, $\forall n \in \mathbb{N}$. Let $\tau : \boldsymbol{\Omega} \to \mathbb{Z}_+ \cup \{\infty\}$ be defined by $\tau(\omega) := \inf\{n \in$

$\mathbb{Z}_+ \mid \mathrm{s}_{n+1}(X)(\omega) > \lambda\}$, $\forall \omega \in \boldsymbol{\Omega}$. Clearly, $\forall n \in \mathbb{Z}_+$, we have $\{\omega \in \boldsymbol{\Omega} \mid \tau(\omega) \leq n\} \in \mathcal{B}_n$. Hence, τ is a stopping time. Let $Y := (Y_n)_{n=0}^{\infty}$ be defined by $Y_n = X_{n \wedge \tau}$, $\forall n \in \mathbb{Z}_+$. Clearly, Y is the stopped $\mathcal{Y}$-valued process. By Theorem 14.47, Y is a Martingale. Clearly, $Y_n = X_0 + \sum_{k=1}^{n} \chi_{\{\omega\in\Omega \mid \tau(\omega)\geq k\},\Omega} d_k$, $\forall n \in \mathbb{Z}_+$. By the definition of τ, we have $\mathrm{s}_\tau(X) \leq \lambda$. Define $Y_M : \boldsymbol{\Omega} \to [0, \infty] \subset \mathbb{R}_e$ be defined by $Y_M(\omega) := \sup_{n\in\mathbb{Z}_+} \|Y_n(\omega)\|_{\mathcal{Y}}$, $\forall \omega \in \boldsymbol{\Omega}$. Note that, $\forall n \in \mathbb{N}$,

$$\begin{aligned}
\boldsymbol{E}(\|Y_n\|_{\mathcal{Y}}^2 | \mathcal{B}_{n-1}) &= \boldsymbol{E}\left(\left\| Y_{n-1} + \chi_{\{\omega\in\Omega \mid \tau(\omega)\geq n\},\Omega} d_n \right\|_{\mathcal{Y}}^2 | \mathcal{B}_{n-1}\right) \\
&= \boldsymbol{E}(\|Y_{n-1}\|_{\mathcal{Y}}^2 + \chi_{\{\omega\in\Omega \mid \tau(\omega)\geq n\},\Omega} \langle Y_{n-1}, d_n\rangle_{\mathcal{Y}} \\
&\quad + \chi_{\{\omega\in\Omega \mid \tau(\omega)\geq n\},\Omega} \langle d_n, Y_{n-1}\rangle_{\mathcal{Y}} + \chi_{\{\omega\in\Omega \mid \tau(\omega)\geq n\},\Omega} \|d_n\|_{\mathcal{Y}}^2 | \mathcal{B}_{n-1}) \\
&= \|Y_{n-1}\|_{\mathcal{Y}}^2 + \boldsymbol{E}\left(2\chi_{\{\omega\in\Omega \mid \tau(\omega)\geq n\},\Omega} \operatorname{Re}\left(\langle Y_{n-1}, d_n\rangle_{\mathcal{Y}}\right) \mid \mathcal{B}_{n-1}\right) \\
&\quad + \boldsymbol{E}\left(\left\| \chi_{\{\omega\in\Omega \mid \tau(\omega)\geq n\},\Omega} d_n \right\|_{\mathcal{Y}}^2 | \mathcal{B}_{n-1}\right) \\
&= \|Y_{n-1}\|_{\mathcal{Y}}^2 + 2\chi_{\{\omega\in\Omega \mid \tau(\omega)\geq n\},\Omega} \boldsymbol{E}(\operatorname{Re}\left(\langle Y_{n-1}, d_n\rangle_{\mathcal{Y}}\right) \mid \mathcal{B}_{n-1}) \\
&\quad + \boldsymbol{E}(\|Y_n - Y_{n-1}\|_{\mathcal{Y}}^2 | \mathcal{B}_{n-1}) \\
&= \|Y_{n-1}\|_{\mathcal{Y}}^2 + 2\chi_{\{\omega\in\Omega \mid \tau(\omega)\geq n\},\Omega} \operatorname{Re}\left(\boldsymbol{E}(\langle Y_{n-1}, d_n\rangle_{\mathcal{Y}} | \mathcal{B}_{n-1})\right) \\
&\quad + \boldsymbol{E}(\|Y_n - Y_{n-1}\|_{\mathcal{Y}}^2 | \mathcal{B}_{n-1}) \\
&= \|Y_{n-1}\|_{\mathcal{Y}}^2 + \boldsymbol{E}(\|Y_n - Y_{n-1}\|_{\mathcal{Y}}^2 | \mathcal{B}_{n-1}) \geq \|Y_{n-1}\|_{\mathcal{Y}}^2 \text{ a.e. in } \boldsymbol{\Omega}
\end{aligned}$$

where the first equality follows from the preceding discussion, the second equality follows from Definition 13.1 and Proposition 13.2, the third equality follows from (c), (b), (h) of Proposition 14.11, the fourth equality follows from (h) of Proposition 14.11, the fifth equality follows from Proposition 11.92, the last equality follows from (h) of Proposition 14.11 and the fact that $\boldsymbol{E}(d_n|\mathcal{B}_{n-1}) = \vartheta_{\mathcal{Y}}$, and the inequality follows from (d) of Proposition 14.11. Thus $\left(\|Y_n\|_{\mathcal{Y}}^2\right)_{n=0}^{\infty}$ is a sub Martingale. Let $g_M : \boldsymbol{\Omega} \to [0, \infty] \subset \mathbb{R}_e$ be defined by $g_M(\omega) := \sup_{n\in\mathbb{Z}_+} \|Y_n(\omega)\|_{\mathcal{Y}}^2$, $\forall \omega \in \boldsymbol{\Omega}$. Then, $\boldsymbol{P}(\{\omega \in \boldsymbol{\Omega} \mid Y_M(\omega) > \lambda\} = \boldsymbol{P}(\{\omega \in \boldsymbol{\Omega} \mid g_M(\omega) > \lambda^2\}) \leq \frac{1}{\lambda^2} \lim_{n\in\mathbb{N}} \boldsymbol{E}(\|Y_n\|_{\mathcal{Y}}^2) \leq \frac{1}{\lambda^2} \boldsymbol{E}((\mathrm{s}(Y))^2)$, where the first equality follows from the equality of the two sets involved, the first inequality follows from Proposition 14.115, and the last inequality follows from the fact that $\boldsymbol{E}(\|Y_{n+1}\|_{\mathcal{Y}}^2) = \boldsymbol{E}(\boldsymbol{E}(\|Y_{n+1}\|_{\mathcal{Y}}^2 | \mathcal{B}_n)) = \boldsymbol{E}(\|Y_n\|_{\mathcal{Y}}^2) + \boldsymbol{E}(\boldsymbol{E}(\|Y_{n+1} - Y_n\|_{\mathcal{Y}}^2 | \mathcal{B}_n))$, $\forall n \in \mathbb{Z}_+$.

Then, we have the following line of reasoning:

$$\begin{aligned}
&\boldsymbol{P}(\{\omega \in \boldsymbol{\Omega} \mid X_M > \lambda, \mathrm{s}(X) \leq \lambda\}) \\
&= \boldsymbol{P}(\{\omega \in \boldsymbol{\Omega} \mid X_M > \lambda, \tau(\omega) = \infty\}) \\
&= \boldsymbol{P}(\{\omega \in \boldsymbol{\Omega} \mid Y_M > \lambda, \tau(\omega) = \infty\})
\end{aligned}$$

$$\le \boldsymbol{P}(\{\omega \in \boldsymbol{\Omega} \mid Y_M > \lambda\}) \le \frac{1}{\lambda^2}\boldsymbol{E}((\mathrm{s}(Y))^2)$$
$$= \frac{1}{\lambda^2}\int_{\Omega\setminus A_\lambda} (\mathrm{s}(Y))^2 \,\mathrm{d}\boldsymbol{P} + \frac{1}{\lambda^2}\int_{A_\lambda} (\mathrm{s}(Y))^2 \,\mathrm{d}\boldsymbol{P}$$

where the first equality follows from the definition of τ, the second equality follows from the fact $Y_M(\omega) = X_M(\omega)$, when $\tau(\omega) = \infty$, the first inequality follows from Proposition 11.4, the second inequality follows from the previous paragraph, and the last equality follows from Proposition 11.92.

Note that $(\mathrm{s}(Y))^2 = \|X_0\|_{\mathcal{Y}}^2 + \sum_{k=1}^{\infty} \boldsymbol{E}(\chi_{\{\omega\in\Omega \mid \tau(\omega)\ge k\},\Omega}\|d_k\|_{\mathcal{Y}}^2|\mathcal{B}_{k-1}) = \|X_0\|_{\mathcal{Y}}^2 + \sum_{k=1}^{\tau} \boldsymbol{E}(\|d_k\|_{\mathcal{Y}}^2|\mathcal{B}_{k-1}) = (\mathrm{s}_\tau(X))^2 \le \lambda^2$, where the first equality follows from Definition 14.114, the second equality follows from simple algebra, the third equality follows from Definition 14.114, and the inequality follows the first paragraph in this proof of the claim. Then, it follows that

$$\boldsymbol{P}(\{\omega \in \boldsymbol{\Omega} \mid X_M > \lambda, \mathrm{s}(X) \le \lambda\})$$
$$\le \frac{1}{\lambda^2}\int_{\Omega\setminus A_\lambda} (\mathrm{s}(Y))^2 \,\mathrm{d}\boldsymbol{P} + \frac{1}{\lambda^2}\int_{A_\lambda} \lambda^2 \,\mathrm{d}\boldsymbol{P}$$
$$= \boldsymbol{P}(A_\lambda) + \frac{1}{\lambda^2}\int_{\Omega\setminus A_\lambda} (\mathrm{s}(X))^2 \,\mathrm{d}\boldsymbol{P}$$

where the first inequality follows from the last paragraph and the preceding discussion and the equality follows from the fact that $\mathrm{s}(Y)(\omega) = \mathrm{s}(X)(\omega)$, if $\mathrm{s}(X) \le \lambda$. This completes the proof of the claim. □

Note the following line of reasoning:

$$|||X_M||| = \boldsymbol{E}(1 \wedge X_M) = \int_0^\infty \boldsymbol{P}(\{\omega \in \boldsymbol{\Omega} \mid (1 \wedge X_M(\omega)) > \lambda\}) \,\mathrm{d}\lambda$$
$$= \int_0^1 \boldsymbol{P}(\{\omega \in \boldsymbol{\Omega} \mid (1 \wedge X_M(\omega)) > \lambda\}) \,\mathrm{d}\lambda$$
$$= \int_0^1 \boldsymbol{P}(\{\omega \in \boldsymbol{\Omega} \mid X_M(\omega) > \lambda\}) \,\mathrm{d}\lambda$$
$$= \int_0^1 \boldsymbol{P}(\{\omega \in \boldsymbol{\Omega} \mid X_M(\omega) > \lambda, \mathrm{s}(X) \le \lambda\}) \,\mathrm{d}\lambda$$
$$+ \int_0^1 \boldsymbol{P}(\{\omega \in \boldsymbol{\Omega} \mid X_M(\omega) > \lambda, \mathrm{s}(X) > \lambda\}) \,\mathrm{d}\lambda$$
$$\le \int_0^1 (\boldsymbol{P}(A_\lambda) + \frac{1}{\lambda^2}\int_{\Omega\setminus A_\lambda} (\mathrm{s}(X))^2 \,\mathrm{d}\boldsymbol{P}) \,\mathrm{d}\lambda$$

$$+\int_0^1 \boldsymbol{P}(\{\omega \in \boldsymbol{\Omega} \mid \mathrm{s}(X) > \lambda\})\,\mathrm{d}\lambda$$

$$= \int_0^1 \frac{1}{\lambda^2} \int_{\Omega \setminus A_\lambda} (\mathrm{s}(X))^2 \,\mathrm{d}\boldsymbol{P}\,\mathrm{d}\lambda + 2\int_0^1 \boldsymbol{P}(\{\omega \in \boldsymbol{\Omega} \mid \mathrm{s}(X) > \lambda\})\,\mathrm{d}\lambda$$

$$= \int_0^1 \frac{1}{\lambda^2} \int_{\{\omega \in \Omega \mid (1 \wedge \mathrm{s}(X)) \le \lambda\}} (1 \wedge \mathrm{s}(X))^2 \,\mathrm{d}\boldsymbol{P}\,\mathrm{d}\lambda$$

$$+2\int_0^1 \boldsymbol{P}(\{\omega \in \boldsymbol{\Omega} \mid (1 \wedge \mathrm{s}(X)) > \lambda\})\,\mathrm{d}\lambda$$

$$= \int_{\{(\omega,\lambda) \in \boldsymbol{\Omega} \times \mathrm{r}_{0,1} \mid Z(\omega) \le \lambda\}} \frac{1}{\lambda^2} (Z(\omega))^2 \,\mathrm{d}(\boldsymbol{P} \times \mu_{\mathrm{B}})(\omega, \lambda) + 2\boldsymbol{E}(1 \wedge \mathrm{s}(X))$$

$$= 2|||\,\mathrm{s}(X)|||+\int_{\boldsymbol{\Omega}} (Z(\omega))^2 \int_{Z(\omega)}^1 \frac{1}{\lambda^2}\,\mathrm{d}\lambda\,\mathrm{d}\boldsymbol{P}(\omega)$$

$$= 2|||\,\mathrm{s}(X)|||-\int_{\boldsymbol{\Omega}} (Z(\omega))^2 \left(\frac{1}{\lambda}\right)\Big|_{Z(\omega)}^1 \,\mathrm{d}\boldsymbol{P}(\omega)$$

$$\le 2|||\,\mathrm{s}(X)|||+\int_{\boldsymbol{\Omega}} Z(\omega)\,\mathrm{d}\boldsymbol{P}(\omega) = 2|||\,\mathrm{s}(X)||| + \boldsymbol{E}(Z) = 3|||\,\mathrm{s}(X)|||$$

where the first equality follows from Definition 14.112, the second equality follows from Proposition 14.21, the third equality follows from Proposition 11.92, the fourth equality follows from simple set equality for $0 \le \lambda < 1$ and Lemma 11.73, the fifth equality follows from Fact 11.3, the first inequality follows from Claim 14.116.1, the sixth equality follows from the definition of A_λ and Proposition 11.92, the seventh equality follows from simple set equality for $0 \le \lambda < 1$ and Lemma 11.73, the eighth equality follows from the definition $Z := 1 \wedge \mathrm{s}(X)$, Tonelli's Theorem 12.29, and Proposition 14.21, the ninth equality follows from Definition 14.112 and Tonelli's Theorem 12.29, the tenth equality follows Theorem 12.83, the second inequality follows from Proposition 11.83, the eleventh equality follows from Definition 14.1, and the last equality follows from Definition 14.112.

This completes the proof of the lemma. □

Lemma 14.117 *Let $\boldsymbol{\Omega} := (\Omega, \mathcal{B}, (\mathcal{B}_n)_{n=0}^\infty, \boldsymbol{P})$ be a filtered probability measure space, $\mathcal{Y}$ be a separable Hilbert space over $\mathbb{K}$, and $X := (X_n)_{n=0}^\infty$ be an adapted $\mathcal{Y}$-valued Martingale. Then, $\lim_{n\in\mathbb{N}} \boldsymbol{E}(\|X_n\|_{\mathcal{Y}}^2) = \boldsymbol{E}((\mathrm{S}(X))^2) = \boldsymbol{E}((\mathrm{s}(X))^2)$.*

Proof Note that $\boldsymbol{E}((\mathrm{S}(X))^2) = \boldsymbol{E}(\|X_0\|_{\mathcal{Y}}^2 + \sum_{k=1}^\infty \|d_k\|_{\mathcal{Y}}^2) = \boldsymbol{E}(\lim_{n\in\mathbb{N}}(\|X_0\|_{\mathcal{Y}}^2 + \sum_{k=1}^n \|d_k\|_{\mathcal{Y}}^2)) = \lim_{n\in\mathbb{N}} \boldsymbol{E}(\|X_0\|_{\mathcal{Y}}^2 + \sum_{k=1}^n \|d_k\|_{\mathcal{Y}}^2) = \lim_{n\in\mathbb{N}} \boldsymbol{E}(\|X_n\|_{\mathcal{Y}}^2)$, where the first equality follows from Definition 14.114, the second equality follows from simple algebra, the third equality follows from Monotone Convergence Theorem 11.81, and the fourth equality follows from the assumption that X is a Martingale, which implies that $\boldsymbol{E}(\|X_n\|_{\mathcal{Y}}^2) = \boldsymbol{E}(\|X_{n-1} + d_n\|_{\mathcal{Y}}^2) =$

$\boldsymbol{E}(\|X_{n-1}\|_{\mathcal{Y}}^2) + \boldsymbol{E}(\|d_n\|)^2$, $\forall n \in \mathbb{N}$. Clearly, we have $\boldsymbol{E}((\mathrm{s}(X))^2) = \boldsymbol{E}(\|X_0\|_{\mathcal{Y}}^2 + \sum_{k=1}^{\infty} \boldsymbol{E}(\|d_k\|_{\mathcal{Y}}^2 | \mathcal{B}_{k-1})) = \boldsymbol{E}(\|X_0\|_{\mathcal{Y}}^2 + \sum_{k=1}^{\infty} \|d_k\|_{\mathcal{Y}}^2) = \boldsymbol{E}((\mathrm{S}(X))^2)$, were the first equality follows from Definition 14.114, the second equality follows from Proposition 11.92 and (a) of Proposition 14.11, and the last equality follows from Definition 14.114. This completes the proof of the lemma. □

Lemma 14.118 *Let $\mathbb{I} := ((\overline{\mathbb{R}_+}, |\cdot|), \hat{\mathcal{B}}, \mu)$ be the σ-finite metric measure subspace of $\mathbb{R}$ with partial ordering $\leq$, $\boldsymbol{\Omega} := (\Omega, \mathcal{B}, (\mathcal{B}_t)_{t\in\mathbb{I}}, \boldsymbol{P})$ be a filtered probability measure space, $\mathbb{I} \times \boldsymbol{\Omega} = ([0,\infty) \times \Omega, \bar{\mathcal{B}}, \nu)$ be the σ-finite product measure space, $w : \mathbb{I} \times \boldsymbol{\Omega} \to \mathbb{R}$ be a standard Wiener process on $\mathbb{I} \times \boldsymbol{\Omega}$ and $(w_t)_{t\in\mathbb{I}}$ is adapted to the filtration $(\mathcal{B}_t)_{t\in\mathbb{I}}$, $\bar{\mathrm{r}}_{a,b} \subset \mathbb{I}$ with $a \leq b$, $\mathcal{Y}$ be a separable Hilbert spaces over $\mathbb{K}$, $\phi \in \hat{\mathcal{Q}}(\bar{\mathrm{r}}_{a,b} \times \boldsymbol{\Omega}, \mathcal{Y})$, where $\hat{\mathcal{Q}}(\bar{\mathrm{r}}_{a,b} \times \boldsymbol{\Omega}, \mathcal{Y})$ is defined in Lemma 14.99, and the σ-algebras $\mathcal{B}_t$ and $\sigma((w_{t+s} - w_t)_{s\in\mathbb{I}})$ are independent, $\forall t \in \mathbb{I}$. Define $h : \boldsymbol{\Omega} \to \overline{\mathbb{R}_+}$ by $h(\omega) := \sup_{t\in\bar{\mathrm{r}}_{a,b}} \left\| \widehat{\int_a^t} \phi_s \, \mathrm{d}w_s \right\|_{\mathcal{Y}}$, $\forall \omega \in \boldsymbol{\Omega}$. Then, we have $|||h||| \leq 3|||\, \|\phi_{\langle\omega\rangle}\|_{\bar{\mathrm{L}}_2(\bar{\mathrm{r}}_{a,b}, \mathcal{Y})} |||$.*

Proof Let ϕ be given by $\phi(t,\omega) := \sum_{k=1}^{n} A_{k-1}(\omega) \chi_{[t_{k-1},t_k),\mathbb{I}}(t)$, where $n \in \mathbb{Z}_+$, $a = t_0 < t_1 < \cdots < t_n = b$, and $A_k \in \bar{\mathrm{L}}_2(\boldsymbol{\Omega}, \mathcal{Y})$ is $\mathcal{B}_{t_k}$-measurable, $k = 0, \ldots, n-1$. $\forall m \in \mathbb{N}$, and $a = \bar{t}_0 < \bar{t}_1 < \cdots < \bar{t}_{nm} = b$, such that $\bar{t}_{im+j} - \bar{t}_{im+j-1} = \frac{1}{m}(t_{i+1} - t_i)$, $i = 0, \ldots, n-1$, $j = 1, \ldots, m$. Thus, $h(\omega) = \lim_{m\in\mathbb{N}} \sup_{j=1,\ldots,mn} \left\| \sum_{k=1}^{j} A_{\lfloor \frac{k-1}{m} \rfloor}(\omega)(w_{\bar{t}_k}(\omega) - w_{\bar{t}_{k-1}}(\omega)) \right\|_{\mathcal{Y}} =: \lim_{m\in\mathbb{N}} h^{\langle m\rangle}(\omega)$, $\forall \omega \in \boldsymbol{\Omega}$, by the continuity of $\widehat{\int_a^t} \phi_s \, \mathrm{d}w_s$. Clearly, $h^{\langle m\rangle}$ is $\mathcal{B}_b$-measurable. Then, by Proposition 11.48, h is $\mathcal{B}_b$- measurable.

By Lemma 14.116, we have

$$
\begin{aligned}
|||h^{\langle m\rangle}||| &= |||\sup_{j=1,\ldots,mn} \Big\| \sum_{k=1}^{j} A_{\lfloor \frac{k-1}{m} \rfloor}(w_{\bar{t}_k} - w_{\bar{t}_{k-1}}) \Big\|_{\mathcal{Y}} ||| \\
&\leq 3|||\Big(\sum_{j=1}^{mn} \boldsymbol{E}\Big(\big\| A_{\lfloor \frac{j-1}{m} \rfloor}(w_{\bar{t}_j} - w_{\bar{t}_{j-1}}) \big\|_{\mathcal{Y}}^2 | \mathcal{B}_{\bar{t}_{j-1}} \Big) \Big)^{\frac{1}{2}} ||| \\
&= 3|||\Big(\sum_{j=1}^{mn} \big\| A_{\lfloor \frac{j-1}{m} \rfloor} \big\|_{\mathcal{Y}}^2 \boldsymbol{E}((w_{\bar{t}_j} - w_{\bar{t}_{j-1}})^2 | \mathcal{B}_{\bar{t}_{j-1}}) \Big)^{\frac{1}{2}} ||| \\
&= 3|||\Big(\sum_{j=1}^{mn} \big\| A_{\lfloor \frac{j-1}{m} \rfloor} \big\|_{\mathcal{Y}}^2 (\bar{t}_j - \bar{t}_{j-1}) \Big)^{\frac{1}{2}} ||| \\
&= 3|||\Big(\sum_{j=1}^{n} \| A_{j-1} \|_{\mathcal{Y}}^2 (t_j - t_{j-1}) \Big)^{\frac{1}{2}} ||| = 3|||\, \|\phi_{\langle\omega\rangle}\|_{\bar{\mathrm{L}}_2(\bar{\mathrm{r}}_{a,b}, \mathcal{Y})} |||
\end{aligned}
$$

where the first equality follows from the definition of $h^{\langle m\rangle}$, the inequality follows from Lemma 14.116, the second equality follows from Definition 7.1 and (h) of Proposition 14.11, the third equality follows from the Definition 14.76, the fourth equality follows from simple algebra, and the last equality follows from Proposition 11.75. Then, we have $|||h||| = \lim_{m\in\mathbb{N}} |||h^{\langle m\rangle}||| \le \lim_{m\in\mathbb{N}} 3|||\,\|\phi_{\langle\omega\rangle}\|_{\bar{L}_2(\bar{r}_{a,b},\mathcal{Y})}||| = 3|||\,\|\phi_{\langle\omega\rangle}\|_{\bar{L}_2(\bar{r}_{a,b},\mathcal{Y})}|||$, where the first equality follows from Bounded Convergence Theorem 11.77, the inequality follows from the preceding discussion, and the last equality follows form simple algebra.

This completes the proof of the lemma. □

Theorem 14.119 *Let* $\mathbb{I} := ((\overline{\mathbb{R}_+}, |\cdot|), \hat{\mathcal{B}}, \mu)$ *be the* σ*-finite metric measure subspace of* $\mathbb{R}$ *with partial ordering* $\le$, $\boldsymbol{\Omega} := (\Omega, \mathcal{B}, (\mathcal{B}_t)_{t\in\mathbb{I}}, \boldsymbol{P})$ *be a filtered probability measure space,* $\mathbb{I}\times\boldsymbol{\Omega} = (\overline{\mathbb{R}_+}\times\Omega, \bar{\mathcal{B}}, \nu)$ *be the* σ*-finite product measure space,* $w : \mathbb{I}\times\boldsymbol{\Omega} \to \mathbb{R}$ *be a standard Wiener process on* $\mathbb{I}\times\boldsymbol{\Omega}$ *and be adapted to the filtration* $(\mathcal{B}_t)_{t\in\mathbb{I}}$, $\bar{r}_{a,b} \subset \mathbb{I}$ *with* $a \le b$, $\mathcal{Y}$ *be a separable Hilbert space, and* $\phi \in \bar{L}_2(\bar{r}_{a,b}\times\boldsymbol{\Omega}, \mathcal{Y})$ *be a* $\mathcal{Y}$*-valued stochastic process adapted to the filtration* $(\mathcal{B}_t)_{t\in\mathbb{I}}$. *Assume that*

(a) *The* σ*-algebras* $\mathcal{B}_t$ *and* $\sigma((w_{t+s} - w_t)_{s\in\mathbb{I}})$ *are independent,* $\forall t \in \mathbb{I}$.
(b) $\forall E \in \mathcal{B}$, *if* $\boldsymbol{P}(E) = 0$ *then* $E \in \mathcal{B}_0$.
(c) ϕ *is Itô integrable with respect to the Wiener process* w *on the interval* $\bar{r}_{a,b}$.

Then, there exists $J : \bar{r}_{a,b}\times\boldsymbol{\Omega} \to \mathcal{Y}$ *that is a* $\mathcal{Y}$*-valued stochastic process adapted to the filtration* $(\mathcal{B}_t)_{t\in\mathbb{I}}$ *with continuous sample path, such that,* $\forall\psi \in \hat{\mathcal{Q}}(\bar{r}_{a,b}\times\boldsymbol{\Omega}, \mathcal{Y})$, *we have* $|||h||| \le 3|||\,\|\psi - \phi\|_{\bar{L}_2(\bar{r}_{a,b},\mathcal{Y})}|||$, *where* $h : \boldsymbol{\Omega} \to \overline{\mathbb{R}_+}$ *is defined by* $h(\omega) = \sup_{t\in\bar{r}_{a,b}} \left\|(\widehat{\int_a^t}\psi_s\,\mathrm{d}w_s)(\omega) - J(t,\omega)\right\|_{\mathcal{Y}}$, $\forall\omega \in \boldsymbol{\Omega}$. *Such* J *is unique in the sense that, let* $\bar{J} : \bar{r}_{a,b}\times\boldsymbol{\Omega} \to \mathcal{Y}$ *be another such process, then* $\bar{J}_{\langle\omega\rangle} = J_{\langle\omega\rangle}$ *a.e.* $\omega \in \boldsymbol{\Omega}$. *We will say that* J *is a* $\mathcal{Y}$-valued Itô process representation *of the Itô integrals* $\overline{\int_a^t}\phi_s\,\mathrm{d}w_s$, $t \in \bar{r}_{a,b}$.

Proof By (c) and Definition 14.104, there exists a sequence of $\mathcal{Y}$-valued simple predictable step functions $(\phi_n)_{n=1}^\infty \subseteq \hat{\mathcal{Q}}(\bar{r}_{a,b}\times\boldsymbol{\Omega}, \mathcal{Y})$, where $\hat{\mathcal{Q}}(\bar{r}_{a,b}\times\boldsymbol{\Omega}, \mathcal{Y})$ is defined in Lemma 14.99, such that $\lim_{n\in\mathbb{N}} \phi_n \doteq \phi$ in $\bar{L}_2(\bar{r}_{a,b}\times\boldsymbol{\Omega}, \mathcal{Y})$. For each $n \in \mathbb{N}$, let $I_n : \bar{r}_{a,b}\times\boldsymbol{\Omega} \to \mathcal{Y}$ be defined by $I_n(t,\omega) := \left(\widehat{\int_a^t}\phi_{n\,s}\,\mathrm{d}w_s\right)(\omega) \in \mathcal{Y}$. Then, I_n is an adapted $\mathcal{Y}$-valued stochastic process and has continuous sample path, $\forall n \in \mathbb{N}$. By Definition 14.104, $\lim_{n\in\mathbb{N}} [I_n(t)] = I(t) := \overline{\int_a^t}\phi_s\,\mathrm{d}w_s$ in $L_2(\boldsymbol{\Omega}, \mathcal{Y})$, $\forall t \in \bar{r}_{a,b}$.

By Lemma 14.118, we have $|||h_{n,m}||| \le 3|||\,\|\phi_n - \phi_m\|_{\bar{L}_2(\bar{r}_{a,b},\mathcal{Y})}|||$, $\forall n, m \in \mathbb{N}$, where $h_{n,m} : \boldsymbol{\Omega} \to \overline{\mathbb{R}_+}$ is defined by $h_{n,m}(\omega) := \sup_{t\in\bar{r}_{a,b}} \|I_n(t,\omega) - I_m(t,\omega)\|_{\mathcal{Y}}$, $\forall\omega \in \boldsymbol{\Omega}$, and $h_{n,m}$ is $\mathcal{B}_b$-measurable. $\forall\epsilon \in \mathbb{R}_+$, $\exists N(\epsilon) \in \mathbb{N}$ such that $\forall n, m \in \mathbb{N}$ with $N(\epsilon) \le n$ and $N(\epsilon) \le m$, we have $\left\|\,\|\phi_n - \phi_m\|_{\bar{L}_2(\bar{r}_{a,b},\mathcal{Y})}\right\|_{\bar{L}_2(\boldsymbol{\Omega},\mathbb{R})} = \|\phi_n - \phi_m\|_{\bar{L}_2(\bar{r}_{a,b}\times\boldsymbol{\Omega},\mathcal{Y})} < \frac{\epsilon}{3}$, where the equality follows from Tonelli's Theorem 12.29. Then, we have $|||Z_{n,m}||| = \boldsymbol{E}(1 \wedge Z_{n,m}) = \int_{\boldsymbol{\Omega}}(1 \wedge Z_{n,m})\,\mathrm{d}\boldsymbol{P} \le \int_{\boldsymbol{\Omega}} Z_{n,m}\,\mathrm{d}\boldsymbol{P} = \left\|Z_{n,m}\right\|_{\bar{L}_1(\boldsymbol{\Omega},\mathbb{R})} \le (\int_{\boldsymbol{\Omega}} Z_{n,m}^2\,\mathrm{d}\boldsymbol{P})^{\frac{1}{2}} = \left\|Z_{n,m}\right\|_{\bar{L}_2(\boldsymbol{\Omega},\mathbb{R})} < \frac{\epsilon}{3}$, where $Z_{n,m} := \|\phi_n - \phi_m\|_{\bar{L}_2(\bar{r}_{a,b},\mathcal{Y})}$, the first equality follows from Definition 14.112, the second equality follows from Definition 14.1, the first inequality follows from

Proposition 11.83, the third equality follows from Example 11.173, the second inequality follows from Cauchy–Schwarz Inequality, and the last equality follows from Example 11.173. This implies that $|||h_{n,m}||| < \epsilon$.

Fix any $k \in \mathbb{N}$, $\exists n_k \in \mathbb{N}$ with $n_{k-1} < n_k$ and $N(2^{-2k}) \leq n_k$, where $n_0 := 0$. $\forall n, m \in \mathbb{N}$ with $n_k \leq n$ and $n_k \leq m$, we have $|||h_{n,m}||| < 2^{-2k}$. Then, $\boldsymbol{P}(\{\omega \in \boldsymbol{\Omega} \mid h_{n,m}(\omega) \geq 2^{-k}\}) < 2^{-k}$, by Chebyshev's Inequality (Proposition 14.67). Define $A_k := \{\omega \in \boldsymbol{\Omega} \mid h_{n_k,n_{k+1}}(\omega) \geq 2^{-k}\} \in \mathcal{B}$, $\forall k \in \mathbb{N}$. Then, $\boldsymbol{P}(A_k) < 2^{-k}$. Consider the sequence $\left(I_{n_k}\right)_{k=1}^{\infty}$. Let $A := \bigcap_{k=1}^{\infty} \bigcup_{j=k}^{\infty} A_j \in \mathcal{B}$. By Borel-Cantelli Lemma 14.6, $\boldsymbol{P}(A) = 0$. $\forall \omega \in \Omega \setminus A$, we have $\forall l \in \mathbb{N}$, $\exists k_l \in \mathbb{N}$ such that $\omega \in \Omega \setminus (\bigcup_{j=k_l}^{\infty} A_j)$. Then, $\forall k \in \mathbb{N}$ with $k \geq k_l$, $\omega \in \Omega \setminus (\bigcup_{j=k}^{\infty} A_j)$. Without loss of generality, take $k_l \geq l$. Thus, we have $\forall k, j \in \mathbb{N}$ with $k_l \leq k \leq j$, $h_{n_k,n_j}(\omega) = \sup_{t \in \bar{\mathrm{r}}_{a,b}} \left\| I_{n_k}(t,\omega) - I_{n_j}(t,\omega) \right\|_{\mathcal{Y}} \leq \sum_{t=k}^{j-1} h_{n_t,n_{t+1}}(\omega) < \sum_{t=k}^{j-1} 2^{-t} < 2^{-k+1} \leq 2^{-k_l+1} \leq 2^{-l+1}$. By the arbitrariness of l, we have $\left(I_{n_k}(\omega)\right)_{k=1}^{\infty}$ is uniform Cauchy sequence, uniform in $t \in \bar{\mathrm{r}}_{a,b}$. Let $J : \bar{\mathrm{r}}_{a,b} \times \boldsymbol{\Omega} \to \mathcal{Y}$ be defined by $J(t,\omega) = \begin{cases} \lim_{k\in\mathbb{N}} I_{n_k}(t,\omega) & \omega \in \Omega \setminus A \\ \vartheta_{\mathcal{Y}} & \omega \in A \end{cases}$, $\forall \omega \in \Omega$, $\forall t \in \bar{\mathrm{r}}_{a,b}$. Then, J is $\bar{\mathcal{B}}$-measurable and has continuous sample path. By Assumption (b), J is a $\mathcal{Y}$-valued stochastic process adapted to the filtration $(\mathcal{B}_t)_{t\in\mathbb{I}}$. By Proposition 14.102, J is progressively measurable.

Fix any $t \in \bar{\mathrm{r}}_{a,b}$, we have $\lim_{k\in\mathbb{N}} I_{n_k}(t) = J(t)$ a.e. in $\boldsymbol{\Omega}$. By the first paragraph of the proof, we have $\lim_{k\in\mathbb{N}} [I_{n_k}(t)] = I(t)$ in $\mathrm{L}_2(\boldsymbol{\Omega}, \mathcal{Y})$. By Propositions 11.211 and 11.57, we have $J(t) \in I(t)$. Hence, we have shown that there exists a $\mathcal{Y}$-valued stochastic process J that is adapted to $(\mathcal{B}_t)_{t\in\mathbb{I}}$, whose sample path is continuous, such that $J(t) \in \overline{\int_a^t} \phi_s \, \mathrm{d}w_s$, $\forall t \in \bar{\mathrm{r}}_{a,b}$. Define $h_n : \boldsymbol{\Omega} \to \overline{\mathbb{R}_+}$ by $h_n(\omega) = \sup_{t\in\bar{\mathrm{r}}_{a,b}} \|I_n(t,\omega) - J(t,\omega)\|_{\mathcal{Y}}$, $\forall \omega \in \boldsymbol{\Omega}$, $\forall n \in \mathbb{N}$. Then, we have h_n is $\mathcal{B}_b$-measurable and $h_n = \lim_{k\in\mathbb{N}} h_{n,n_k}$ a.e. in $\boldsymbol{\Omega}$, $\forall n \in \mathbb{N}$. Then, $|||h_n||| = \lim_{k\in\mathbb{N}} |||h_{n,n_k}||| \leq \lim_{k\in\mathbb{N}} 3|||\|\phi_n - \phi_{n_k}\|_{\bar{\mathrm{L}}_2(\bar{\mathrm{r}}_{a,b},\mathcal{Y})}|||$, $\forall n \in \mathbb{N}$, where the first equality follows from Bounded Convergence Theorem 11.77 and the inequality follows from the second paragraph of the proof. Note that $0 = \lim_{k\in\mathbb{N}} \|\phi - \phi_{n_k}\|_{\bar{\mathrm{L}}_2(\bar{\mathrm{r}}_{a,b}\times\boldsymbol{\Omega},\mathcal{Y})} = \lim_{k\in\mathbb{N}} \left\| \|\phi - \phi_{n_k}\|_{\bar{\mathrm{L}}_2(\bar{\mathrm{r}}_{a,b},\mathcal{Y})} \right\|_{\bar{\mathrm{L}}_2(\boldsymbol{\Omega},\mathbb{R})} = \lim_{k\in\mathbb{N}} \left\| \|\phi - \phi_n + \phi_n - \phi_{n_k}\|_{\bar{\mathrm{L}}_2(\bar{\mathrm{r}}_{a,b},\mathcal{Y})} \right\|_{\bar{\mathrm{L}}_2(\boldsymbol{\Omega},\mathbb{R})} \geq \lim_{k\in\mathbb{N}} \left\| \|\phi - \phi_n\|_{\bar{\mathrm{L}}_2(\bar{\mathrm{r}}_{a,b},\mathcal{Y})} - \|\phi_n - \phi_{n_k}\|_{\bar{\mathrm{L}}_2(\bar{\mathrm{r}}_{a,b},\mathcal{Y})} \right\|_{\bar{\mathrm{L}}_2(\boldsymbol{\Omega},\mathbb{R})} \geq 0$, $\forall n \in \mathbb{N}$, where the first equality follows from the choice of $\left(\phi_{n_k}\right)_{k=1}^{\infty}$, the second equality follows from Tonelli's Theorem 12.29, the third equality follows from simple algebra, the first inequality follows from Definition 7.1, and the last inequality follows from Definition 7.1. This implies that $\lim_{k\in\mathbb{N}} \|\phi_n - \phi_{n_k}\|_{\bar{\mathrm{L}}_2(\bar{\mathrm{r}}_{a,b},\mathcal{Y})} \doteq \|\phi - \phi_n\|_{\bar{\mathrm{L}}_2(\bar{\mathrm{r}}_{a,b},\mathcal{Y})}$ in $\bar{\mathrm{L}}_2(\boldsymbol{\Omega}, \mathbb{R})$. By Proposition 11.211 and Lemma 14.113, we have $\lim_{k\in\mathbb{N}} |||\|\phi_n - \phi_{n_k}\|_{\bar{\mathrm{L}}_2(\bar{\mathrm{r}}_{a,b},\mathcal{Y})}||| = |||\|\phi_n - \phi\|_{\bar{\mathrm{L}}_2(\bar{\mathrm{r}}_{a,b},\mathcal{Y})}|||$, $\forall n \in \mathbb{N}$. Thus, we have shown that $|||h_n||| \leq 3|||\|\phi_n - \phi\|_{\bar{\mathrm{L}}_2(\bar{\mathrm{r}}_{a,b},\mathcal{Y})}|||$, $\forall n \in \mathbb{N}$.

$\forall \psi \in \hat{\mathcal{Q}}(\bar{\mathrm{r}}_{a,b} \times \boldsymbol{\Omega}, \mathcal{Y})$, without loss of generality, we assume that $\phi_1 = \psi$. Then, by the previous paragraph, we have $|||h||| \leq 3||||\psi - \phi\|_{\bar{\mathrm{L}}_2(\bar{\mathrm{r}}_{a,b},\mathcal{Y})}|||$, where $h : \boldsymbol{\Omega} \to \overline{\mathbb{R}_+}$ is defined by $h(\omega) = \sup_{t \in \bar{\mathrm{r}}_{a,b}} \left\| (\widehat{\int_a^t} \psi_s \, \mathrm{d}w_s)(\omega) - J(t, \omega) \right\|_{\mathcal{Y}}, \forall \omega \in \boldsymbol{\Omega}$.

Finally, we will show the uniqueness of J. Let $\bar{J}$ be another such process, then, we have $|||\bar{h}_n||| \leq 3||||\phi_n - \phi\|_{\bar{\mathrm{L}}_2(\bar{\mathrm{r}}_{a,b},\mathcal{Y})}|||$, where $\bar{h}_n : \boldsymbol{\Omega} \to \overline{\mathbb{R}_+}$ is defined by $\bar{h}_n(\omega) = \sup_{t \in \bar{\mathrm{r}}_{a,b}} \left\| I_n(t, \omega) - \bar{J}(t, \omega) \right\|_{\mathcal{Y}}, \forall \omega \in \boldsymbol{\Omega}$. Let $\tilde{h} : \boldsymbol{\Omega} \to \overline{\mathbb{R}_+}$ is defined by $\tilde{h}(\omega) = \sup_{t \in \bar{\mathrm{r}}_{a,b}} \left\| J(t, \omega) - \bar{J}(t, \omega) \right\|_{\mathcal{Y}}$. Then, $\tilde{h}(\omega) \leq h_n(\omega) + \bar{h}_n(\omega), \forall n \in \mathbb{N}$, $\forall \omega \in \boldsymbol{\Omega}$. Thus, we have $|||\tilde{h}||| \leq \lim_{n \in \mathbb{N}}(|||\bar{h}_n||| + |||h_n|||) = 0$, by Lemma 14.113. This shows that $J_{\langle\omega\rangle} = \bar{J}_{\langle\omega\rangle}$ a.e. $\omega \in \boldsymbol{\Omega}$.

This completes the proof of the theorem. □

Proposition 14.120 *Let $\mathbb{I} := ((\overline{\mathbb{R}_+}, |\cdot|), \hat{\mathcal{B}}, \mu)$ be the σ-finite metric measure subspace of $\mathbb{R}$ with partial ordering $\leq$, $\boldsymbol{\Omega} := (\Omega, \mathcal{B}, (\mathcal{B}_t)_{t \in \mathbb{I}}, \boldsymbol{P})$ be a filtered probability measure space, $\mathbb{I} \times \boldsymbol{\Omega} = (\overline{\mathbb{R}_+} \times \Omega, \bar{\mathcal{B}}, \nu)$ be the σ-finite product measure space, $w : \mathbb{I} \times \boldsymbol{\Omega} \to \mathbb{R}$ be a standard Wiener process on $\mathbb{I} \times \boldsymbol{\Omega}$ and be adapted to the filtration $(\mathcal{B}_t)_{t \in \mathbb{I}}$, $\bar{\mathrm{r}}_{a,b} \subset \mathbb{I}$ with $a \leq b$, $\mathcal{Y}$ and $\mathcal{Z}$ be separable Hilbert spaces over $\mathbb{K}$, $\phi, \psi \in \bar{\mathrm{L}}_2(\bar{\mathrm{r}}_{a,b} \times \boldsymbol{\Omega}, \mathcal{Y})$ be $\mathcal{Y}$-valued stochastic processes adapted to the filtration $(\mathcal{B}_t)_{t \in \mathbb{I}}$. Assume that the following hold:*

(a) *The σ-algebras $\mathcal{B}_t$ and $\sigma((w_{t+s} - w_t)_{s \in \mathbb{I}})$ are independent, $\forall t \in \mathbb{I}$.*
(b) *$\forall E \in \mathcal{B}$, if $\boldsymbol{P}(E) = 0$ then $E \in \mathcal{B}_0$.*
(c) *ϕ and ψ are Itô integrable with respect to the Wiener process w on the interval $\bar{\mathrm{r}}_{a,b}$, respectively.*
(d) *Let I be a $\mathcal{Y}$-valued Itô process representation of the Itô integrals $\overline{\int_a^t} \phi_s \, \mathrm{d}w_s$, $t \in \bar{\mathrm{r}}_{a,b}$ and J be a $\mathcal{Y}$-valued Itô process representation of the Itô integrals $\overline{\int_a^t} \psi_s \, \mathrm{d}w_s$, $t \in \bar{\mathrm{r}}_{a,b}$.*
(e) *$h : \boldsymbol{\Omega} \to \overline{\mathbb{R}_+}$ be defined by $h(\omega) = \sup_{t \in \bar{\mathrm{r}}_{a,b}} \|I(t, \omega) - J(t, \omega)\|_{\mathcal{Y}}, \forall \omega \in \boldsymbol{\Omega}$.*

Then, the following statements hold:

(i) *$|||h||| \leq 3||||\phi - \psi\|_{\bar{\mathrm{L}}_2(\bar{\mathrm{r}}_{a,b},\mathcal{Y})}|||$.*
(ii) *$I + J$ is an Itô process representation of the Itô integrals $\overline{\int_a^t} (\psi_s + \phi_s) \, \mathrm{d}w_s$, $t \in \bar{\mathrm{r}}_{a,b}$.*
(iii) *$\forall \alpha \in \mathbb{K}$, we have αI is an Itô process representation of the Itô integrals $\overline{\int_a^t} (\alpha\phi_s) \, \mathrm{d}w_s$, $t \in \bar{\mathrm{r}}_{a,b}$.*
(iv) *$\forall t \in \bar{\mathrm{r}}_{a,b}$, $\|I_t\|_{\bar{\mathrm{L}}_2(\boldsymbol{\Omega},\mathcal{Y})} = \left\| \phi|_{\bar{\mathrm{r}}_{a,t} \times \boldsymbol{\Omega}} \right\|_{\bar{\mathrm{L}}_2(\bar{\mathrm{r}}_{a,t} \times \boldsymbol{\Omega},\mathcal{Y})}$.*
(v) *$(I_t)_{t \in \bar{\mathrm{r}}_{a,b}}$ is a $\mathcal{Y}$-valued Martingale adapted to $(\mathcal{B}_t)_{t \in \bar{\mathrm{r}}_{a,b}}$.*
(vi) *$\forall A : \boldsymbol{\Omega} \to W \subseteq \mathrm{B}(\mathcal{Y}, \mathcal{Z})$ that is $\mathcal{B}_a$-measurable and $\boldsymbol{E}(\|A\|^2_{\mathrm{B}(\mathcal{Y},\mathcal{Z})}) < \infty$, where W is a separable subspace of $\mathrm{B}(\mathcal{Y}, \mathcal{Z})$, if $\phi \in \mathcal{U}(\bar{\mathrm{r}}_{a,b} \times \boldsymbol{\Omega}, \mathcal{Y})$ with the random variable $c_\phi : \boldsymbol{\Omega} \to \overline{\mathbb{R}_+}$ as defined in Proposition 14.107 and $\boldsymbol{E}(\|A\|^2_{\mathrm{B}(\mathcal{Y},\mathcal{Z})} c_\phi) < \infty$, then AI is a $\mathcal{Z}$-valued Itô process representation of the Itô integrals $\overline{\int_a^t} (A\phi_s) \, \mathrm{d}w_s$, $t \in \bar{\mathrm{r}}_{a,b}$.*

(vii) $\forall c \in \bar{\mathrm{r}}_{a,b}$, $I|_{\bar{\mathrm{r}}_{c,b} \times \boldsymbol{\Omega}} - I_c$ *is an Itô process representation of the Itô integrals* $\overline{\int}_c^t \phi_s \, \mathrm{d}w_s$, $t \in \bar{\mathrm{r}}_{c,b}$.

(viii) $\mathrm{Re}\,(\boldsymbol{E}(\langle I_t, J_t \rangle_{\mathcal{Y}})) = \mathrm{Re}\left(\int_{\bar{\mathrm{r}}_{a,t} \times \boldsymbol{\Omega}} \langle \phi(s,\omega), \psi(s,\omega) \rangle_{\mathcal{Y}} \, \mathrm{d}\nu(s,\omega)\right)$, $\forall t \in \bar{\mathrm{r}}_{a,b}$.

Proof (i) By (c) and Definition 14.104, there exists a sequence of $\mathcal{Y}$-valued simple predictable step functions $(\phi_n)_{n=1}^{\infty} \subseteq \hat{\mathcal{Q}}(\bar{\mathrm{r}}_{a,b} \times \boldsymbol{\Omega}, \mathcal{Y})$, where $\hat{\mathcal{Q}}(\bar{\mathrm{r}}_{a,b} \times \boldsymbol{\Omega}, \mathcal{Y})$ is defined in Lemma 14.99, such that $\lim_{n \in \mathbb{N}} \phi_n \doteq \phi$ in $\bar{\mathrm{L}}_2(\bar{\mathrm{r}}_{a,b} \times \boldsymbol{\Omega}, \mathcal{Y})$. For each $n \in \mathbb{N}$, let $I_n : \bar{\mathrm{r}}_{a,b} \times \boldsymbol{\Omega} \to \mathcal{Y}$ be defined by $I_n(t,\omega) := \left(\widehat{\int}_a^t \phi_{n\,s} \, \mathrm{d}w_s\right)(\omega) \in \mathcal{Y}$. Then, I_n is an adapted $\mathcal{Y}$-valued stochastic process and has continuous sample path, $\forall n \in \mathbb{N}$. By Theorem 14.119, $|||h_n||| \leq 3||||\phi_n - \psi||_{\bar{\mathrm{L}}_2(\bar{\mathrm{r}}_{a,b},\mathcal{Y})}|||$, $\forall n \in \mathbb{N}$, where $h_n : \boldsymbol{\Omega} \to \overline{\mathbb{R}_+}$ is defined by $h_n(\omega) = \sup_{t \in \bar{\mathrm{r}}_{a,b}} \|I_n(t,\omega) - J(t,\omega)\|_{\mathcal{Y}}$, $\forall \omega \in \boldsymbol{\Omega}$, $\forall n \in \mathbb{N}$, and we have h_n is $\mathcal{B}_b$-measurable. By Theorem 14.119, I is a $\mathcal{Y}$-valued stochastic process that is adapted to $(\mathcal{B}_t)_{t \in \mathbb{I}}$, whose sample path is continuous, such that $I(t) \in \overline{\int}_a^t \phi_s \, \mathrm{d}w_s$, $\forall t \in \bar{\mathrm{r}}_{a,b}$. Furthermore, $|||\bar{h}_n||| \leq 3||||\phi_n - \phi||_{\bar{\mathrm{L}}_2(\bar{\mathrm{r}}_{a,b},\mathcal{Y})}|||$, $\forall n \in \mathbb{N}$, where $\bar{h}_n : \boldsymbol{\Omega} \to \overline{\mathbb{R}_+}$ is defined by $\bar{h}_n(\omega) = \sup_{t \in \bar{\mathrm{r}}_{a,b}} \|I_n(t,\omega) - I(t,\omega)\|_{\mathcal{Y}}$, $\forall \omega \in \boldsymbol{\Omega}$, $\forall n \in \mathbb{N}$, and we have $\bar{h}_n$ is $\mathcal{B}_b$-measurable.

Clearly, $h \leq h_n + \bar{h}_n$, $\forall n \in \mathbb{N}$. Then, we have $|||h||| \leq \liminf_{n \in \mathbb{N}} |||h_n + \bar{h}_n||| \leq \lim_{n \in \mathbb{N}}(|||h_n||| + |||\bar{h}_n|||) = \lim_{n \in \mathbb{N}} |||h_n||| + \lim_{n \in \mathbb{N}} |||\bar{h}_n||| \leq \lim_{n \in \mathbb{N}} 3||||\phi_n - \psi||_{\bar{\mathrm{L}}_2(\bar{\mathrm{r}}_{a,b},\mathcal{Y})}||| + \lim_{n \in \mathbb{N}} 3||||\phi_n - \phi||_{\bar{\mathrm{L}}_2(\bar{\mathrm{r}}_{a,b},\mathcal{Y})}|||$, where the first inequality follows from Proposition 11.83, the second inequality follows from Lemma 14.113, the equality follows from Proposition 3.83, and the last inequality follows from the last paragraph. Note that $0 = \lim_{n \in \mathbb{N}} \|\phi - \phi_n\|_{\bar{\mathrm{L}}_2(\bar{\mathrm{r}}_{a,b} \times \boldsymbol{\Omega},\mathcal{Y})} = \lim_{n \in \mathbb{N}} \left\| \|\phi - \phi_n\|_{\bar{\mathrm{L}}_2(\bar{\mathrm{r}}_{a,b},\mathcal{Y})} \right\|_{\bar{\mathrm{L}}_2(\boldsymbol{\Omega},\mathbb{R})}$, where the first equality follows from the choice of $(\phi_n)_{n=1}^{\infty}$ and the second equality follows from Tonelli's Theorem 12.29. By Proposition 11.211 and Lemma 14.113, we have $\lim_{n \in \mathbb{N}} ||||\phi_n - \phi||_{\bar{\mathrm{L}}_2(\bar{\mathrm{r}}_{a,b},\mathcal{Y})}||| = 0$. Note that $\lim_{n \in \mathbb{N}} \left\| \|\psi - \phi_n\|_{\bar{\mathrm{L}}_2(\bar{\mathrm{r}}_{a,b},\mathcal{Y})} - \|\psi - \phi\|_{\bar{\mathrm{L}}_2(\bar{\mathrm{r}}_{a,b},\mathcal{Y})} \right\|_{\bar{\mathrm{L}}_2(\boldsymbol{\Omega},\mathbb{R})} \leq \lim_{n \in \mathbb{N}} \left\| \|\psi - \phi_n + \phi - \psi\|_{\bar{\mathrm{L}}_2(\bar{\mathrm{r}}_{a,b},\mathcal{Y})} \right\|_{\bar{\mathrm{L}}_2(\boldsymbol{\Omega},\mathbb{R})} = \lim_{n \in \mathbb{N}} \left\| \|\phi - \phi_n\|_{\bar{\mathrm{L}}_2(\bar{\mathrm{r}}_{a,b},\mathcal{Y})} \right\|_{\bar{\mathrm{L}}_2(\boldsymbol{\Omega},\mathbb{R})} = \lim_{n \in \mathbb{N}} \|\phi - \phi_n\|_{\bar{\mathrm{L}}_2(\bar{\mathrm{r}}_{a,b} \times \boldsymbol{\Omega},\mathcal{Y})} = 0$, where the first inequality follows from Definition 7.1, the first equality follows from Tonelli's Theorem 12.29, and the second equality follows from the choice of $(\phi_n)_{n=1}^{\infty}$. Thus, by Proposition 11.211 and Lemma 14.113, we have $\lim_{n \in \mathbb{N}} ||||\phi_n - \psi||_{\bar{\mathrm{L}}_2(\bar{\mathrm{r}}_{a,b},\mathcal{Y})}||| = ||||\phi - \psi||_{\bar{\mathrm{L}}_2(\bar{\mathrm{r}}_{a,b},\mathcal{Y})}|||$. This further implies that $|||h||| \leq 3||||\phi - \psi||_{\bar{\mathrm{L}}_2(\bar{\mathrm{r}}_{a,b},\mathcal{Y})}|||$.

(ii) By (c) and Definition 14.104, there exists a sequence of $\mathcal{Y}$-valued simple predictable step functions $(\psi_n)_{n=1}^{\infty} \subseteq \hat{\mathcal{Q}}(\bar{\mathrm{r}}_{a,b} \times \boldsymbol{\Omega}, \mathcal{Y})$, such that $\lim_{n \in \mathbb{N}} \psi_n \doteq \psi$ in $\bar{\mathrm{L}}_2(\bar{\mathrm{r}}_{a,b} \times \boldsymbol{\Omega}, \mathcal{Y})$. For each $n \in \mathbb{N}$, let $J_n : \bar{\mathrm{r}}_{a,b} \times \boldsymbol{\Omega} \to \mathcal{Y}$ be defined by $J_n(t,\omega) := \left(\widehat{\int}_a^t \psi_{n\,s} \, \mathrm{d}w_s\right)(\omega) \in \mathcal{Y}$. Then, J_n is an adapted $\mathcal{Y}$-valued stochastic process and has continuous sample path, $\forall n \in \mathbb{N}$. Now, $(\phi_n + \psi_n)_{n=1}^{\infty} \in \hat{\mathcal{Q}}(\bar{\mathrm{r}}_{a,b} \times \boldsymbol{\Omega}, \mathcal{Y})$, by Lemma 14.99, and $\lim_{n \in \mathbb{N}}(\psi_n + \phi_n) \doteq \psi + \phi$ in $\bar{\mathrm{L}}_2(\bar{\mathrm{r}}_{a,b} \times \boldsymbol{\Omega}, \mathcal{Y})$. Thus, $\psi + \phi$ is Itô integrable with respect to the Wiener process w on the interval $\bar{\mathrm{r}}_{a,b}$, by Definition 14.104. Note that $(I_n + J_n)(t,\omega) := \left(\widehat{\int}_a^t (\phi_{n\,s} + \psi_{n\,s}) \, \mathrm{d}w_s\right)(\omega) \in \mathcal{Y}$,

$\forall(t,\omega) \in \bar{\mathrm{r}}_{a,b}\times\boldsymbol{\Omega}$. Let $\tilde{h}_n(\omega) := \sup_{t\in\bar{\mathrm{r}}_{a,b}} \| I_n(t,\omega)+J_n(t,\omega)-I(t,\omega)-J(t,\omega)\|_{\mathcal{Y}}$, $\forall\omega \in \boldsymbol{\Omega}$. Then, we have $|||\tilde{h}_n||| \leq |||\bar{h}_n + \hat{h}_n||| \leq |||\bar{h}_n||| + |||\hat{h}_n||| \to 0$ as $n \to \infty$, where $\hat{h}_n(\omega) := \sup_{t\in\bar{\mathrm{r}}_{a,b}} \| J_n(t,\omega) - J(t,\omega)\|_{\mathcal{Y}}$, $\forall\omega \in \boldsymbol{\Omega}$; the first inequality follows from definition of $\tilde{h}_n$, $\bar{h}_n$, and $\hat{h}_n$ and monotonicity of triple norm for nonnegative real valued random variables; and the second inequality follows from Lemma 14.113. Thus, $\lim_{n\in\mathbb{N}} \tilde{h}_n = 0$ in measure in $\boldsymbol{\Omega}$. By Theorem 14.119 and its proof, we have $I + J$ is an Itô process representation of the Itô integrals $\overline{\int}_a^t(\psi_s + \phi_s)\,\mathrm{d}w_s$, $t \in \bar{\mathrm{r}}_{a,b}$.

(iii) Note that $\forall n \in \mathbb{N}$, $\alpha I_n(t,\omega) = \big(\widehat{\int}_a^t(\alpha\phi_{n\,s})\,\mathrm{d}w_s\big)(\omega) \in \mathcal{Y}$, $\forall(t,\omega) \in \bar{\mathrm{r}}_{a,b}\times\boldsymbol{\Omega}$. Then, $\alpha I_n : \bar{\mathrm{r}}_{a,b} \times \boldsymbol{\Omega} \to \mathcal{Y}$ is an adapted $\mathcal{Y}$-valued stochastic process and has continuous sample path, $\forall n \in \mathbb{N}$. Now, $(\alpha\phi_n)_{n=1}^{\infty} \in \hat{\mathcal{Q}}(\bar{\mathrm{r}}_{a,b} \times \boldsymbol{\Omega}, \mathcal{Y})$, by Lemma 14.99, and $\lim_{n\in\mathbb{N}}(\alpha\phi_n) \doteq \alpha\phi$ in $\bar{\mathrm{L}}_2(\bar{\mathrm{r}}_{a,b} \times \boldsymbol{\Omega}, \mathcal{Y})$. Thus, $\alpha\phi$ is Itô integrable with respect to the Wiener process w on the interval $\bar{\mathrm{r}}_{a,b}$, by Definition 14.104. By Proposition 14.111, we have $\lim_{n\in\mathbb{N}} |\alpha|\,\bar{h}_n = \lim_{n\in\mathbb{N}} \sup_{t\in\bar{\mathrm{r}}_{a,b}} \|\alpha I_n(t,\omega) - \alpha I(t,\omega)\|_{\mathcal{Y}} = 0$ in measure in $\boldsymbol{\Omega}$. By Theorem 14.119 and its proof, we have αI is an Itô process representation of the Itô integrals $\overline{\int}_a^t(\alpha\phi_s)\,\mathrm{d}w_s$, $t \in \bar{\mathrm{r}}_{a,b}$.

(iv) This follows immediately from (v) of Proposition 14.106 and the fact that $I_t \in \overline{\int}_a^t\phi_s\,\mathrm{d}w_s$, $\forall t \in \bar{\mathrm{r}}_{a,b}$.

(v) Clearly, $(I_t)_{t\in\bar{\mathrm{r}}_{a,b}}$ is a $\mathcal{Y}$-valued stochastic process that is adapted to the filtration $(\mathcal{B}_t)_{t\in\mathbb{I}}$. By (iv), $I_t \in \bar{\mathrm{L}}_2(\boldsymbol{\Omega},\mathcal{Y})$, which implies that $I_t \in \bar{\mathrm{L}}_1(\boldsymbol{\Omega},\mathcal{Y})$, $\forall t \in \bar{\mathrm{r}}_{a,b}$. Fix any $t, s \in \bar{\mathrm{r}}_{a,b}$ with $s \leq t$. By (i) of Proposition 14.11 and the convexity of norm, we have $\big\|\boldsymbol{E}(\overline{\int}_s^t(\phi_\tau - (\phi_n)_\tau)\,\mathrm{d}w_\tau|\mathcal{B}_s)\big\|_{\mathcal{Y}} \leq \boldsymbol{E}(\big\|\overline{\int}_s^t(\phi_\tau - (\phi_n)_\tau)\,\mathrm{d}w_\tau\big\|_{\mathcal{Y}}|\mathcal{B}_s)$ and $\big\|\boldsymbol{E}(\overline{\int}_s^t(\phi_\tau - (\phi_n)_\tau)\,\mathrm{d}w_\tau|\mathcal{B}_s)\big\|_{\mathcal{Y}}^2 \leq (\boldsymbol{E}(\big\|\overline{\int}_s^t(\phi_\tau - (\phi_n)_\tau)\cdot\mathrm{d}w_\tau\big\|_{\mathcal{Y}}|\mathcal{B}_s))^2 \leq \boldsymbol{E}(\big\|\overline{\int}_s^t(\phi_\tau - (\phi_n)_\tau)\,\mathrm{d}w_\tau\big\|_{\mathcal{Y}}^2|\mathcal{B}_s)$. Then, $0 \leq \boldsymbol{E}(\big\|\boldsymbol{E}(\overline{\int}_s^t(\phi_\tau - (\phi_n)_\tau)\,\mathrm{d}w_\tau|\mathcal{B}_s)\big\|_{\mathcal{Y}}^2) \leq \boldsymbol{E}(\boldsymbol{E}(\big\|\overline{\int}_s^t(\phi_\tau - (\phi_n)_\tau)\,\mathrm{d}w_\tau\big\|_{\mathcal{Y}}^2|\mathcal{B}_s)) = \boldsymbol{E}(\big\|\overline{\int}_s^t(\phi_\tau - (\phi_n)_\tau)\,\mathrm{d}w_\tau\big\|_{\mathcal{Y}}^2) = \big\|\overline{\int}_s^t(\phi_\tau - (\phi_n)_\tau)\,\mathrm{d}w_\tau\big\|_{\bar{\mathrm{L}}_2(\boldsymbol{\Omega},\mathcal{Y})}^2 = \big\|\phi - \phi_n\big\|_{\bar{\mathrm{L}}_2(\bar{\mathrm{r}}_{s,t}\times\boldsymbol{\Omega},\mathcal{Y})}^2 \to 0$ as $n \to \infty$, where the first inequality follows from (d) of Proposition 14.11, the second inequality follows from the preceding discussion, the first equality follows from (a) of Proposition 14.11, the second equality follows from Example 11.173 and Definition 14.1, the last equality follows from Proposition 14.106, and the convergence statement follows from the choice of $(\psi_n)_{n=1}^{\infty}$. This implies that $\lim_{n\in\mathbb{N}} \boldsymbol{E}(\big\|\boldsymbol{E}(\overline{\int}_s^t(\phi_\tau - (\phi_n)_\tau)\,\mathrm{d}w_\tau|\mathcal{B}_s)\big\|_{\mathcal{Y}}^2) = 0$. Therefore, by Propositions 11.211 and 11.57, we have there exists a subsequence $\lim_{k\in\mathbb{N}} \boldsymbol{E}(\overline{\int}_s^t(\phi_\tau - (\phi_{n_k})_\tau)\,\mathrm{d}w_\tau|\mathcal{B}_s) = \vartheta_{\mathcal{Y}}$ a.e. in $\boldsymbol{\Omega}$. Thus, we have $\boldsymbol{E}(I_t|\mathcal{B}_s) = \boldsymbol{E}(I_s + I_t - I_s|\mathcal{B}_s) = I_s + \boldsymbol{E}(\overline{\int}_s^t\phi_\tau\,\mathrm{d}w_\tau|\mathcal{B}_s) = I_s + \lim_{k\in\mathbb{N}} \boldsymbol{E}(\overline{\int}_s^t\phi_\tau\,\mathrm{d}w_\tau - \overline{\int}_s^t(\phi_{n_k})_\tau\,\mathrm{d}w_\tau|\mathcal{B}_s) + \lim_{k\in\mathbb{N}} \boldsymbol{E}(\overline{\int}_s^t(\phi_{n_k})_\tau\,\mathrm{d}w_\tau|\mathcal{B}_s) = I_s + \lim_{k\in\mathbb{N}} \boldsymbol{E}(\overline{\int}_s^t(\phi_\tau - (\phi_{n_k})_\tau)\,\mathrm{d}w_\tau|\mathcal{B}_s) = I_s$ a.e. in $\boldsymbol{\Omega}$, where the first equality follows from simple algebra, the second equality follows from (b) and (c) of Proposition 14.11, the third equality follows from simple algebra, the fourth equality follows from $\phi_{n_k} \in \hat{\mathcal{Q}}(\bar{r}_{a,b} \times \boldsymbol{\Omega}, \mathcal{Y})$, and the last equality follows from the preceding discussion. Hence, $(I_t)_{t\in\bar{r}_{a,b}}$ is a Martingale.

(vi) Note that $A\phi_n \in \hat{\mathcal{Q}}(\bar{r}_{a,b} \times \boldsymbol{\Omega}, \mathcal{Y})$ and $A(\omega)I_n(t,\omega) = A(\omega)(\widehat{\int}_a^t \phi_{n\,s} \cdot \mathrm{d}w_s)(\omega) = (\widehat{\int}_a^t (A\phi_{n\,s})\,\mathrm{d}w_s)(\omega)$, $\forall (t,\omega) \in \bar{r}_{a,b} \times \boldsymbol{\Omega}$, $\forall n \in \mathbb{N}$, where the second equality follows from Definition A.4. Note also that $\lim_{n\in\mathbb{N}} A\phi_n \doteq A\phi$ in $\bar{L}_2(\bar{r}_{a,b} \times \boldsymbol{\Omega}, \mathcal{Y})$. Hence, by Theorem 14.119 and its proof, AI is a $\mathcal{Z}$-valued Itô process representation of the Itô integrals $\overline{\int}_a^t (A\phi_s)\,\mathrm{d}w_s$, $t \in \bar{r}_{a,b}$.

(vii) Fix any $c \in \bar{r}_{a,b}$. Let $\bar{J} := I|_{\bar{r}_{c,b}\times\boldsymbol{\Omega}} - I_c$. Clearly, $\bar{J}$ is $\mathcal{Y}$-valued stochastic process that is adapted to the filtration $(\mathcal{B}_t)_{t\in\mathbb{I}}$ and admits continuous sample path. By assumption (d) and (iv) of Proposition 14.106, we have $\bar{J}_t \in \overline{\int}_c^t \phi_s\,\mathrm{d}w_s$, $\forall t \in \bar{r}_{c,b}$. $\forall \bar{\phi} \in \hat{\mathcal{Q}}(\bar{r}_{c,b} \times \boldsymbol{\Omega}, \mathcal{Y})$, let $\bar{\Phi} : \bar{r}_{a,b} \times \boldsymbol{\Omega} \to \mathcal{Y}$ be defined by $\bar{\Phi}(t,\omega) := \begin{cases} I(t,\omega) & \text{if } t \in \bar{r}_{a,c} \\ I(c,\omega) + \widehat{\int}_c^t \bar{\phi}_s\,\mathrm{d}w_s & \text{if } t \in r_{c,b} \end{cases}$ and $\bar{\psi} : \bar{r}_{a,b} \times \boldsymbol{\Omega} \to \mathcal{Y}$ be defined by $\bar{\psi}(t,\omega) := \begin{cases} \phi(t,\omega) \text{ if } t \in \bar{r}_{a,c} \\ \bar{\phi}(t,\omega) \text{ if } t \in r_{c,b} \end{cases}$. Clearly, $|||\bar{h}||| = |||\tilde{h}||| \leq 3|||\|\bar{\psi} - \phi\|_{\bar{L}_2(\bar{r}_{a,b},\mathcal{Y})}||| = 3|||\|\bar{\phi} - \phi|_{\bar{r}_{c,b}\times\boldsymbol{\Omega}}\|_{\bar{L}_2(\bar{r}_{c,b},\mathcal{Y})}|||$, where $\bar{h}(\omega) := \sup_{t\in\bar{r}_{c,b}} \|(\widehat{\int}_c^t \bar{\phi}_s\,\mathrm{d}w_s)(\omega) - \bar{J}(t,\omega)\|_{\mathcal{Y}}$ and $\tilde{h} := \sup_{t\in\bar{r}_{a,b}} \|\bar{\Phi}(t,\omega) - I(t,\omega)\|_{\mathcal{Y}}$, and the inequality follows from (i). Hence, $\bar{J}$ is a $\mathcal{Y}$-valued Itô process representation of the Itô integrals $\overline{\int}_c^t \phi_s\,\mathrm{d}w_s$, $t \in \bar{r}_{c,b}$.

(viii) This follows immediately from (vi) of Proposition 14.106.

This completes the proof of the proposition. □

Proposition 14.121 *Let $\mathbb{I} := ((\overline{\mathbb{R}_+}, |\cdot|), \hat{\mathcal{B}}, \mu)$ be the σ-finite metric measure subspace of $\mathbb{R}$ with partial ordering $\leq$, $\boldsymbol{\Omega} := (\Omega, \mathcal{B}, (\mathcal{B}_t)_{t\in\mathbb{I}}, \boldsymbol{P})$ be a filtered probability measure space, $\mathbb{I} \times \boldsymbol{\Omega} = (\overline{\mathbb{R}_+} \times \Omega, \bar{\mathcal{B}}, \nu)$ be the σ-finite product measure space, $\bar{r}_{a,b} \subset \mathbb{I}$ with $a \leq b$, $\mathcal{Y}$ be a separable Banach space over $\mathbb{K}$, and $g : \bar{r}_{a,b} \times \boldsymbol{\Omega} \to \mathcal{Y}$ is a $\mathcal{Y}$-valued stochastic process that is progressively measurable, $\widehat{\int}_a^b g_{\langle\omega\rangle}(s)\,\mathrm{d}s$ satisfies the Riemann Criterion for Integrability, Theorem A.9, $\forall \omega \in \boldsymbol{\Omega}$. Then, $I : \bar{r}_{a,b} \times \boldsymbol{\Omega} \to \mathcal{Y}$, defined by $I(t,\omega) := \int_a^t g(s,\omega)\,\mathrm{d}s$, $\forall (t,\omega) \in \bar{r}_{a,b} \times \boldsymbol{\Omega}$, is a $\mathcal{Y}$-valued stochastic process that is adapted to $(\mathcal{B}_t)_{t\in\mathbb{I}}$, has continuous sample path, and is progressively measurable.*

Proof Fix any $\omega \in \boldsymbol{\Omega}$, by the assumption and Corollary A.16, $g_{\langle\omega\rangle}$ is Riemann-Stieltjes integrable with respect to $\mathrm{id}_{\mathbb{I}}$ on the interval $\bar{r}_{a,b}$, and is absolutely Riemann-Stieltjes integrable over $\bar{r}_{a,b}$, and the Riemann integrals equal to the corresponding Lebesgue integrals. By Theorem A.11 and Proposition 11.48, I_t is $\mathcal{B}_t$-measurable, $\forall t \in \bar{r}_{a,b}$. By Proposition 11.84, I has continuous sample path. By Proposition 14.102, I is a $\mathcal{Y}$-valued stochastic process that is progressively measurable. This completes the proof of the proposition. □

Next, we present the infinite-dimensional Itô's Formula.

Theorem 14.122 (Itô's Formula) *Let $\mathbb{I} := ((\overline{\mathbb{R}_+}, |\cdot|), \hat{\mathcal{B}}, \mu)$ be the σ-finite metric measure subspace of $\mathbb{R}$ with partial ordering $\leq$, $(\Omega, \mathcal{B}, (\mathcal{B}_t)_{t\in\mathbb{I}}, \boldsymbol{P}) =: \boldsymbol{\Omega}$ be a* *filtered probability measure space,* $\mathbb{I} \times \boldsymbol{\Omega} = (\overline{\mathbb{R}_+} \times \Omega, \bar{\mathcal{B}}, \nu)$ *be the* σ*-finite product measure space,* $m \in \mathbb{N}$, $w : \mathbb{I} \times \boldsymbol{\Omega} \to \mathbb{R}^m$ *be an* $\mathbb{R}^m$*-valued standard Wiener process on* $\mathbb{I} \times \boldsymbol{\Omega}$*, where the components of* w *are* $(w^{\langle 1 \rangle}, \ldots, w^{\langle m \rangle})$, $w^{\langle j \rangle}$ *is a standard Wiener process,* $j = 1, \ldots, m$*, and* $(w_t)_{t \in \mathbb{I}}$ *is adapted to the filtration* $(\mathcal{B}_t)_{t \in \mathbb{I}}$, $\bar{\mathrm{r}}_{a,b} \subset \mathbb{I}$ *with* $a \leq b$, $\mathcal{Y}$ *and* $\mathcal{Z}$ *be separable Hilbert spaces over* $\mathbb{K}$*, and* $D \subseteq \mathcal{Y}$ *be a convex open set. Assume that*

(a) *The* σ*-algebras* $\mathcal{B}_t$ *and* $\sigma((w_{t+s} - w_t)_{s \in \mathbb{I}})$ *are independent,* $\forall t \in \mathbb{I}$.
(b) $\forall E \in \mathcal{B}$*, if* $\boldsymbol{P}(E) = 0$*, then* $E \in \mathcal{B}_0$.
(c) $g \in \bar{\mathrm{L}}_2(\bar{\mathrm{r}}_{a,b} \times \boldsymbol{\Omega}, \mathcal{Y})$ *is a* $\mathcal{Y}$*-valued stochastic process that is adapted to the filtration* $(\mathcal{B}_t)_{t \in \mathbb{I}}$ *and has right continuous sample paths,* $\widehat{\int}_a^b g_{\langle\omega\rangle}(s)\,\mathrm{d}s$ *satisfies the Riemann Criterion for Integrability, Theorem A.9,* $\forall \omega \in \boldsymbol{\Omega}$*, and* $h^{\langle j \rangle} \in \mathcal{U}(\bar{\mathrm{r}}_{a,b} \times \boldsymbol{\Omega}, \mathcal{Y})$ *with the random variable* $c_h^{\langle j \rangle} : \boldsymbol{\Omega} \to \overline{\mathbb{R}_+}$, $j = 1, \ldots, m$*, where* $\mathcal{U}(\bar{\mathrm{r}}_{a,b} \times \boldsymbol{\Omega}, \mathcal{Y})$ *is defined in Proposition 14.107.*
(d) $\phi^{\langle j \rangle} : \bar{\mathrm{r}}_{a,b} \times \boldsymbol{\Omega} \to \mathcal{Y}$ *is the Itô process representation of the Itô integral* $\overline{\int}_a^t h_s^{\langle j \rangle}\,\mathrm{d}w_s^{\langle j \rangle}$, $t \in \bar{\mathrm{r}}_{a,b}$, $j = 1, \ldots, m$. $\phi : \bar{\mathrm{r}}_{a,b} \times \boldsymbol{\Omega} \to D$ *is defined by*

$$\phi(t, \omega) := \sum_{j=1}^{m} \phi^{\langle j \rangle}(t, \omega) + \int_a^t g_s\,\mathrm{d}s + \phi(a, \omega); \quad \forall t \in \bar{\mathrm{r}}_{a,b} \tag{14.22}$$

where $\phi(a, \cdot) \in \bar{\mathrm{L}}_2(\boldsymbol{\Omega}, \mathcal{Y})$ *is a* $\mathcal{Y}$*-valued* $\mathcal{B}_a$*-measurable random variable.*

Let $U : D \to \mathcal{Z}$ *be* $\mathcal{C}_2$ *and* $X(t, \omega) = U(\phi(t, \omega))$, $\forall (t, \omega) \in \bar{\mathrm{r}}_{a,b} \times \boldsymbol{\Omega}$*. Assume further that*

(i) $\bar{h}^{\langle j \rangle} \in \bar{\mathrm{L}}_2(\bar{\mathrm{r}}_{a,b} \times \boldsymbol{\Omega}, \mathcal{Z})$ *and is Itô integrable,* $j = 1, \ldots, m$*, where*

$$\bar{g}(t, \omega) := \sum_{j=1}^{m} \frac{1}{2} U^{(2)}(\phi(t, \omega))(h^{\langle j \rangle}(t, \omega))(h^{\langle j \rangle}(t, \omega))$$
$$+ \mathrm{D}U(\phi(t, \omega))(g(t, \omega)); \quad \forall (t, \omega) \in \bar{\mathrm{r}}_{a,b} \times \boldsymbol{\Omega} \tag{14.23a}$$
$$\bar{h}^{\langle j \rangle}(t, \omega) := \mathrm{D}U(\phi(t, \omega))(h^{\langle j \rangle}(t, \omega)); \quad \forall (t, \omega) \in \bar{\mathrm{r}}_{a,b} \times \boldsymbol{\Omega}, \quad j = 1, \ldots, m \tag{14.23b}$$

(ii) $U^{(2)}(\phi)(h^{\langle j \rangle})(h^{\langle k \rangle}) \in \mathcal{U}(\bar{\mathrm{r}}_{a,b} \times \boldsymbol{\Omega}, \mathcal{Z})$, $j = 1, \ldots, m$, $k = 1, \ldots, m$.
(iii) $U^{(1)}(\phi_t) \in \bar{\mathrm{L}}_2(\boldsymbol{\Omega}, \mathrm{B}(\mathcal{Y}, \mathcal{Z}))$ *and* $\boldsymbol{E}(\|U^{(1)}(\phi_t)\|^2_{\mathrm{B}(\mathcal{Y},\mathcal{Z})} c_h) < \infty$, $\forall t \in \bar{\mathrm{r}}_{a,b}$*, where* $c_h : \boldsymbol{\Omega} \to \overline{\mathbb{R}_+}$ *is defined by* $c_h := \sum_{j=1}^{\langle m \rangle} c_h^{\langle j \rangle}$.

Then, we have the following Itô's formula: $\forall t \in \bar{\mathrm{r}}_{a,b}$,

$$X(t, \omega) - X(a, \omega) = \int_a^t \bar{g}(s, \omega)\,\mathrm{d}s + \sum_{j=1}^{m} l^{\langle j \rangle}(t, \omega) \text{ a.e. } \omega \in \boldsymbol{\Omega} \tag{14.24a}$$

where $l^{\langle j\rangle} : \bar{\mathrm{r}}_{a,b} \times \boldsymbol{\Omega} \to \mathcal{Z}$ *is a* $\mathcal{Z}$*-valued Itô process representation of the Itô integrals* $\overline{\int}_a^t \bar{h}_s^{\langle j\rangle}\, \mathrm{d}w_s^{\langle j\rangle}$, $t \in \bar{\mathrm{r}}_{a,b}$, $\forall j = 1, \ldots, m$.

In this case, we will abuse the notation and write, $\forall t \in \bar{\mathrm{r}}_{a,b}$,

$$\mathrm{d}X_t = \mathrm{D}U(\phi_t)(g_t)\,\mathrm{d}t + \sum_{j=1}^{m} \frac{1}{2} U^{(2)}(\phi_t)(h_t^{\langle j\rangle})(h_t^{\langle j\rangle})\,\mathrm{d}t$$

$$+ \sum_{j=1}^{m} \mathrm{D}U(\phi_t)(h_t^{\langle j\rangle})\,\mathrm{d}w_t^{\langle j\rangle} \text{ a.e. in } \boldsymbol{\Omega} \tag{14.24b}$$

Proof By (c) and Corollary A.16, $g_{\langle\omega\rangle}$ is absolutely integrable over $\bar{\mathrm{r}}_{a,b}$. By Theorem 14.119, (a), and (b), $\phi^{\langle j\rangle}$ is a $\mathcal{Y}$-valued stochastic process with continuous sample path that is adapted to $(\mathcal{B}_t)_{t\in\mathbb{I}}$, $j = 1, \ldots, m$. By Proposition 14.121, ϕ has continuous sample path and is progressively measurable. By Proposition 14.102, $\phi, h^{\langle j\rangle}$, $j = 1, \ldots, m$, g are progressively measurable.

By Propositions 7.126, 11.38, and 11.39, each term in $\bar{g}$ is a $\mathcal{Z}$-valued stochastic process that is further adapted to $(\mathcal{B}_t)_{t\in\mathbb{I}}$ and admit right-continuous sample paths. By Propositions 7.126, 11.38, and 11.39, $\bar{h}^{\langle j\rangle}$, $j = 1, \ldots, m$, are $\mathcal{Z}$-valued stochastic processes and are adapted to the filtration $(\mathcal{B}_t)_{t\in\mathbb{I}}$ and admit right-continuous sample paths. By Proposition 14.102, $\bar{g}, \bar{h}^{\langle j\rangle}$, $j = 1, \ldots, m$, are progressively measurable.

Fix any $t \in \bar{\mathrm{r}}_{a,b}$. Fix any $j = 1, \ldots, m$. By Theorem A.18 and Corollaries A.16 and A.17, $\widehat{\int}_a^b \bar{h}^{\langle j\rangle}(s, \omega)\,\mathrm{d}s \in \mathcal{Z}$ satisfies the Riemann Criterion for Integrability, Theorem A.9, $\forall \omega \in \boldsymbol{\Omega}$. By (c) and Proposition 14.107, $h^{\langle j\rangle}$ is Itô integrable on $\bar{\mathrm{r}}_{a,b}$. By (i), $\bar{h}^{\langle j\rangle}$ is Itô integrable on $\bar{\mathrm{r}}_{a,b}$. By Theorem 14.119, (a), and (b), $l^{\langle j\rangle}$ is a $\mathcal{Z}$-valued stochastic process with continuous sample path that is adapted to $(\mathcal{B}_t)_{t\in\mathbb{I}}$.

Let $\hat{T}(\omega) := X(t, \omega) - X(a, \omega) - \int_a^t \bar{g}(s, \omega)\,\mathrm{d}s - \sum_{j=1}^{m} l^{\langle j\rangle}(t, \omega)$, $\forall \omega \in \boldsymbol{\Omega}$. Fix any $\omega \in \boldsymbol{\Omega}$, by Theorem A.18, each term in $\bar{g}$ satisfies the Riemann Criterion for Integrability, Theorem A.9, with respect to the integrator $\mathrm{id}_{\mathbb{I}}$ on the interval $\bar{\mathrm{r}}_{a,b}$. By Proposition 14.121, the assumption (i), and Theorem 14.119, $\hat{T}$ is $\mathcal{B}_t$-measurable.

Let $S_{2,n}(\omega) := \frac{1}{2} \sum_{j=1}^{m} \sum_{i=1}^{2^n} U^{(2)}(\phi(s_{i-1}^{\langle n\rangle}, \omega))(h^{\langle j\rangle}(s_{i-1}^{\langle n\rangle}, \omega))(h^{\langle j\rangle}(s_{i-1}^{\langle n\rangle}, \omega)) \cdot ((w^{\langle j\rangle}(s_i^{\langle n\rangle}, \omega) - w^{\langle j\rangle}(s_{i-1}^{\langle n\rangle}, \omega))^2 - (s_i^{\langle n\rangle} - s_{i-1}^{\langle n\rangle})) \in \mathcal{Z}$, $\forall \omega \in \boldsymbol{\Omega}$, $\forall n \in \mathbb{N}$, where $s_i^{\langle n\rangle} = a + \frac{i}{2^n}(t - a)$, $i = 0, \ldots, 2^n$. We will first calculate $\lim_{n\in\mathbb{N}} \boldsymbol{E}(\|S_{2,n}\|_{\mathcal{Z}}^2)$. (Clearly, $S_{2,n}$ is $\mathcal{B}_t$-measurable.)

$$\lim_{n\in\mathbb{N}} \boldsymbol{E}(\|S_{2,n}\|_{\mathcal{Z}}^2) = \lim_{n\in\mathbb{N}} \boldsymbol{E}(\langle S_{2,n}, S_{2,n}\rangle_{\mathcal{Z}})$$

$$= \lim_{n\in\mathbb{N}} \frac{1}{4} \boldsymbol{E}\Big(\Big\langle \sum_{j=1}^{m} \sum_{i=1}^{2^n} U^{(2)}(\phi_{s_{i-1}^{\langle n\rangle}})(h_{s_{i-1}^{\langle n\rangle}}^{\langle j\rangle})(h_{s_{i-1}^{\langle n\rangle}}^{\langle j\rangle})((w_{s_i^{\langle n\rangle}}^{\langle j\rangle} - w_{s_{i-1}^{\langle n\rangle}}^{\langle j\rangle})^2$$

$$-(s_i^{\langle n\rangle} - s_{i-1}^{\langle n\rangle})), \sum_{j=1}^{m} \sum_{i=1}^{2^n} U^{(2)}(\phi_{s_{i-1}^{\langle n\rangle}})(h_{s_{i-1}^{\langle n\rangle}}^{\langle j\rangle})(h_{s_{i-1}^{\langle n\rangle}}^{\langle j\rangle})((w_{s_i^{\langle n\rangle}}^{\langle j\rangle} - w_{s_{i-1}^{\langle n\rangle}}^{\langle j\rangle})^2$$

$$-(s_i^{\langle n\rangle} - s_{i-1}^{\langle n\rangle}))\Big\rangle_{\mathcal{Z}}\Big)$$

$$= \lim_{n\in\mathbb{N}} \frac{1}{4} \sum_{k_1=1}^{m} \sum_{i=1}^{2^n} \sum_{k_2=1}^{m} \sum_{j=1}^{2^n} \boldsymbol{E}\Big(\Big\langle U^{(2)}(\phi_{s_{i-1}^{\langle n\rangle}})(h_{s_{i-1}^{\langle n\rangle}}^{\langle k_1\rangle})(h_{s_{i-1}^{\langle n\rangle}}^{\langle k_1\rangle})$$

$$\cdot((w_{s_i^{\langle n\rangle}}^{\langle k_1\rangle} - w_{s_{i-1}^{\langle n\rangle}}^{\langle k_1\rangle})^2 - (s_i^{\langle n\rangle} - s_{i-1}^{\langle n\rangle})), U^{(2)}(\phi_{s_{j-1}^{\langle n\rangle}})(h_{s_{j-1}^{\langle n\rangle}}^{\langle k_2\rangle})(h_{s_{j-1}^{\langle n\rangle}}^{\langle k_2\rangle})$$

$$\cdot((w_{s_j^{\langle n\rangle}}^{\langle k_2\rangle} - w_{s_{j-1}^{\langle n\rangle}}^{\langle k_2\rangle})^2 - (s_j^{\langle n\rangle} - s_{j-1}^{\langle n\rangle}))\Big\rangle_{\mathcal{Z}}\Big)$$

$$= \lim_{n\in\mathbb{N}} \frac{1}{4} \sum_{k_1=1}^{m} \sum_{i=1}^{2^n} \sum_{k_2=1}^{m} \sum_{j=1}^{2^n} \boldsymbol{E}\Big(\Big\langle U^{(2)}(\phi_{s_{i-1}^{\langle n\rangle}})(h_{s_{i-1}^{\langle n\rangle}}^{\langle k_1\rangle})(h_{s_{i-1}^{\langle n\rangle}}^{\langle k_1\rangle}),$$

$$U^{(2)}(\phi_{s_{j-1}^{\langle n\rangle}})(h_{s_{j-1}^{\langle n\rangle}}^{\langle k_2\rangle})(h_{s_{j-1}^{\langle n\rangle}}^{\langle k_2\rangle})\Big\rangle_{\mathcal{Z}}((w_{s_i^{\langle n\rangle}}^{\langle k_1\rangle} - w_{s_{i-1}^{\langle n\rangle}}^{\langle k_1\rangle})^2 - (s_i^{\langle n\rangle} - s_{i-1}^{\langle n\rangle}))$$

$$\cdot((w_{s_j^{\langle n\rangle}}^{\langle k_2\rangle} - w_{s_{j-1}^{\langle n\rangle}}^{\langle k_2\rangle})^2 - (s_j^{\langle n\rangle} - s_{j-1}^{\langle n\rangle}))\Big)$$

$$= \lim_{n\in\mathbb{N}} \frac{1}{4} \sum_{k_1=1}^{m} \sum_{i=1}^{2^n} \sum_{k_2=1}^{m} \sum_{j=1}^{2^n} \boldsymbol{E}\Big(\boldsymbol{E}\Big(\Big\langle U^{(2)}(\phi_{s_{i-1}^{\langle n\rangle}})(h_{s_{i-1}^{\langle n\rangle}}^{\langle k_1\rangle})(h_{s_{i-1}^{\langle n\rangle}}^{\langle k_1\rangle}),$$

$$U^{(2)}(\phi_{s_{j-1}^{\langle n\rangle}})(h_{s_{j-1}^{\langle n\rangle}}^{\langle k_2\rangle})(h_{s_{j-1}^{\langle n\rangle}}^{\langle k_2\rangle})\Big\rangle_{\mathcal{Z}}((w_{s_i^{\langle n\rangle}}^{\langle k_1\rangle} - w_{s_{i-1}^{\langle n\rangle}}^{\langle k_1\rangle})^2 - (s_i^{\langle n\rangle} - s_{i-1}^{\langle n\rangle}))$$

$$\cdot((w_{s_j^{\langle n\rangle}}^{\langle k_2\rangle} - w_{s_{j-1}^{\langle n\rangle}}^{\langle k_2\rangle})^2 - (s_j^{\langle n\rangle} - s_{j-1}^{\langle n\rangle}))|\mathcal{B}_{s_{\max\{i,j\}-1}^{\langle n\rangle}}\Big)\Big)$$

$$= \lim_{n\in\mathbb{N}} \frac{1}{4} \sum_{k_1=1}^{m} \sum_{k_2=1}^{m} \sum_{i=1}^{2^n} \boldsymbol{E}\Big(\boldsymbol{E}\Big(\Big\langle U^{(2)}(\phi_{s_{i-1}^{\langle n\rangle}})(h_{s_{i-1}^{\langle n\rangle}}^{\langle k_1\rangle})(h_{s_{i-1}^{\langle n\rangle}}^{\langle k_1\rangle}),$$

$$U^{(2)}(\phi_{s_{i-1}^{\langle n\rangle}})(h_{s_{i-1}^{\langle n\rangle}}^{\langle k_2\rangle})(h_{s_{i-1}^{\langle n\rangle}}^{\langle k_2\rangle})\Big\rangle_{\mathcal{Z}}((w_{s_i^{\langle n\rangle}}^{\langle k_1\rangle} - w_{s_{i-1}^{\langle n\rangle}}^{\langle k_1\rangle})^2 - (s_i^{\langle n\rangle} - s_{i-1}^{\langle n\rangle}))$$

$$\cdot((w_{s_i^{\langle n\rangle}}^{\langle k_2\rangle} - w_{s_{i-1}^{\langle n\rangle}}^{\langle k_2\rangle})^2 - (s_i^{\langle n\rangle} - s_{i-1}^{\langle n\rangle}))|\mathcal{B}_{s_{i-1}^{\langle n\rangle}}\Big)\Big)$$

$$= \lim_{n\in\mathbb{N}} \frac{1}{4} \sum_{k_1=1}^{m} \sum_{k_2=1}^{m} \sum_{i=1}^{2^n} \boldsymbol{E}\Big(\Big\langle U^{(2)}(\phi_{s_{i-1}^{\langle n\rangle}})(h_{s_{i-1}^{\langle n\rangle}}^{\langle k_1\rangle})(h_{s_{i-1}^{\langle n\rangle}}^{\langle k_1\rangle}), U^{(2)}(\phi_{s_{i-1}^{\langle n\rangle}})$$

$$\cdot(h_{s_{i-1}^{\langle n\rangle}}^{\langle k_2\rangle})(h_{s_{i-1}^{\langle n\rangle}}^{\langle k_2\rangle})\Big\rangle_{\mathcal{Z}}\boldsymbol{E}\Big(((w_{s_i^{\langle n\rangle}}^{\langle k_1\rangle} - w_{s_{i-1}^{\langle n\rangle}}^{\langle k_1\rangle})^2 - (s_i^{\langle n\rangle} - s_{i-1}^{\langle n\rangle}))$$

$$\cdot((w_{s_i^{\langle n\rangle}}^{\langle k_2\rangle} - w_{s_{i-1}^{\langle n\rangle}}^{\langle k_2\rangle})^2 - (s_i^{\langle n\rangle} - s_{i-1}^{\langle n\rangle}))|\mathcal{B}_{s_{i-1}^{\langle n\rangle}}\Big)\Big)$$

$$
\begin{aligned}
&= \lim_{n\in\mathbb{N}} \frac{1}{4} \sum_{k=1}^{m} \sum_{i=1}^{2^n} \boldsymbol{E}\Big(\Big\langle U^{(2)}(\phi_{s_{i-1}^{\langle n\rangle}})(h^{\langle k\rangle}_{s_{i-1}^{\langle n\rangle}})(h^{\langle k\rangle}_{s_{i-1}^{\langle n\rangle}}), U^{(2)}(\phi_{s_{i-1}^{\langle n\rangle}})(h^{\langle k\rangle}_{s_{i-1}^{\langle n\rangle}}) \\
&\quad \cdot (h^{\langle k\rangle}_{s_{i-1}^{\langle n\rangle}})\Big\rangle_{\mathcal{Z}} \boldsymbol{E}\Big(((w^{\langle k\rangle}_{s_i^{\langle n\rangle}} - w^{\langle k\rangle}_{s_{i-1}^{\langle n\rangle}})^2 - (s_i^{\langle n\rangle} - s_{i-1}^{\langle n\rangle}))^2 | \mathcal{B}_{s_{i-1}^{\langle n\rangle}}\Big)\Big) \\
&= \lim_{n\in\mathbb{N}} \frac{1}{4} \sum_{k=1}^{m} \sum_{i=1}^{2^n} \boldsymbol{E}\Big(\Big\| U^{(2)}(\phi_{s_{i-1}^{\langle n\rangle}})(h^{\langle k\rangle}_{s_{i-1}^{\langle n\rangle}})(h^{\langle k\rangle}_{s_{i-1}^{\langle n\rangle}})\Big\|_{\mathcal{Z}}^2 \cdot \frac{1}{2}(s_i^{\langle n\rangle} - s_{i-1}^{\langle n\rangle})^2\Big) \\
&= \lim_{n\in\mathbb{N}} \frac{1}{8} \frac{t-a}{2^n} \sum_{k=1}^{m} \sum_{i=1}^{2^n} \boldsymbol{E}\Big(\Big\| U^{(2)}(\phi_{s_{i-1}^{\langle n\rangle}})(h^{\langle k\rangle}_{s_{i-1}^{\langle n\rangle}})(h^{\langle k\rangle}_{s_{i-1}^{\langle n\rangle}})\Big\|_{\mathcal{Z}}^2 (s_i^{\langle n\rangle} - s_{i-1}^{\langle n\rangle})\Big) \\
&= 0
\end{aligned}
$$

where the first equality follows from Proposition 13.2, the second equality follows from the definition of $S_{2,n}$, the third equality follows from simple algebra, the fourth equality follows from Definition 13.1, the fifth equality follows from (a) of Proposition 14.11, the sixth equality follows from assumption (a), the seventh equality follows from (h) of Proposition 14.11, the eighth equality follows from Definition 14.78, the ninth equality follows from the fact that $w^{\langle k\rangle}_{t+h} - w^{\langle k\rangle}_t \sim N(0, h)$ and assumption (a), the tenth equality follows from the definition of $s_i^{\langle n\rangle}$, and the last equality follows from the assumption (ii), Proposition 14.107, and Lebesgue Dominated Convergence Theorem 11.91. Hence, we have $\lim_{n\in\mathbb{N}} S_{2,n} \doteq \vartheta_{\bar{L}_2(\boldsymbol{\Omega},\mathcal{Z})}$ in $\bar{L}_2(\boldsymbol{\Omega}, \mathcal{Z})$. By Proposition 11.211, $\lim_{n\in\mathbb{N}} S_{2,n} = \vartheta_{\mathcal{Z}}$ in measure in $\boldsymbol{\Omega}$.

Note that, by Taylor's Theorem 12.122, $\forall x_1, x_2 \in D$,

$$
\begin{aligned}
&U(x_1) - U(x_2) \\
&= U^{(1)}(x_2)(x_1 - x_2) + \int_0^1 \tau U^{(2)}(x_2 + (1-\tau)(x_1 - x_2))(x_1 - x_2) \\
&\quad \cdot (x_1 - x_2)\, \mathrm{d}\tau \\
&= U^{(1)}(x_2)(x_1 - x_2) + \int_0^1 \int_0^\tau \mathrm{d}s \\
&\quad \cdot U^{(2)}(x_2 + (1-\tau)(x_1 - x_2))(x_1 - x_2)(x_1 - x_2)\, \mathrm{d}\tau \\
&= U^{(1)}(x_2)(x_1 - x_2) + \int_0^1 \int_s^1 U^{(2)}(x_2 + (1-\tau)(x_1 - x_2))(x_1 - x_2) \\
&\quad \cdot (x_1 - x_2)\, \mathrm{d}\tau\, \mathrm{d}s \\
&= U^{(1)}(x_2)(x_1 - x_2) + \frac{1}{2} U^{(2)}(x_2)(x_1 - x_2)(x_1 - x_2) \\
&\quad + \int_0^1 \int_s^1 (U^{(2)}(x_2 + (1-\tau)(x_1 - x_2))
\end{aligned}
$$

$$
\begin{aligned}
&-U^{(2)}(x_2))(x_1 - x_2)(x_1 - x_2)\,\mathrm{d}\tau\,\mathrm{d}s \\
&= U^{(1)}(x_2)(x_1 - x_2) + \frac{1}{2}U^{(2)}(x_2)(x_1 - x_2)(x_1 - x_2) \\
&+ \int_0^1 \int_0^s (U^{(2)}(x_2 + \tau(x_1 - x_2)) \\
&-U^{(2)}(x_2))(x_1 - x_2)(x_1 - x_2)\,\mathrm{d}\tau\,\mathrm{d}s
\end{aligned}
$$

where the second equality follows from Proposition 11.75, the third equality follows from Fubini's Theorem 12.31, the fourth equality follows from Proposition 11.75, and the last equality follows from Change of Variable, Theorem 12.91.

Note the following line of arguments:

$$
\begin{aligned}
&\hat{T} - S_{2,n} \\
&= \sum_{i=1}^{2^n} \Big(X_{s_i^{\langle n\rangle}} - X_{s_{i-1}^{\langle n\rangle}} - \int_{s_{i-1}^{\langle n\rangle}}^{s_i^{\langle n\rangle}} \bar{g}_s\,\mathrm{d}s\Big) - S_{2,n} - \sum_{j=1}^{m} l_t^{\langle j\rangle} \\
&= \sum_{i=1}^{2^n} \Big(U(\phi_{s_i^{\langle n\rangle}}) - U(\phi_{s_{i-1}^{\langle n\rangle}}) - \int_{s_{i-1}^{\langle n\rangle}}^{s_i^{\langle n\rangle}} \bar{g}_s\,\mathrm{d}s\Big) - S_{2,n} - \sum_{j=1}^{m} l_t^{\langle j\rangle} \\
&= \sum_{i=1}^{2^n} \Big(U^{(1)}(\phi_{s_{i-1}^{\langle n\rangle}})(\phi_{s_i^{\langle n\rangle}} - \phi_{s_{i-1}^{\langle n\rangle}}) + \frac{1}{2}U^{(2)}(\phi_{s_{i-1}^{\langle n\rangle}})(\phi_{s_i^{\langle n\rangle}} - \phi_{s_{i-1}^{\langle n\rangle}}) \\
&\cdot(\phi_{s_i^{\langle n\rangle}} - \phi_{s_{i-1}^{\langle n\rangle}}) + \int_0^1 \int_0^{\bar{s}} (U^{(2)}(\phi_{s_{i-1}^{\langle n\rangle}} + \tau(\phi_{s_i^{\langle n\rangle}} - \phi_{s_{i-1}^{\langle n\rangle}})) \\
&-U^{(2)}(\phi_{s_{i-1}^{\langle n\rangle}}))(\phi_{s_i^{\langle n\rangle}} - \phi_{s_{i-1}^{\langle n\rangle}})(\phi_{s_i^{\langle n\rangle}} - \phi_{s_{i-1}^{\langle n\rangle}})\,\mathrm{d}\tau\,\mathrm{d}\bar{s} - \int_{s_{i-1}^{\langle n\rangle}}^{s_i^{\langle n\rangle}} \bar{g}_s\,\mathrm{d}s\Big) \\
&-S_{2,n} - \sum_{j=1}^{m} l_t^{\langle j\rangle} \\
&= \sum_{i=1}^{2^n} \Big(U^{(1)}(\phi_{s_{i-1}^{\langle n\rangle}})\Big(\int_{s_{i-1}^{\langle n\rangle}}^{s_i^{\langle n\rangle}} g_s\,\mathrm{d}s + \sum_{j=1}^{m}(\phi^{\langle j\rangle}_{s_i^{\langle n\rangle}} - \phi^{\langle j\rangle}_{s_{i-1}^{\langle n\rangle}})\Big) + \frac{1}{2}U^{(2)}(\phi_{s_{i-1}^{\langle n\rangle}}) \\
&\cdot\Big(\int_{s_{i-1}^{\langle n\rangle}}^{s_i^{\langle n\rangle}} g_s\,\mathrm{d}s + \sum_{j=1}^{m}(\phi^{\langle j\rangle}_{s_i^{\langle n\rangle}} - \phi^{\langle j\rangle}_{s_{i-1}^{\langle n\rangle}})\Big)\Big(\int_{s_{i-1}^{\langle n\rangle}}^{s_i^{\langle n\rangle}} g_s\,\mathrm{d}s \\
&+\sum_{j=1}^{m}(\phi^{\langle j\rangle}_{s_i^{\langle n\rangle}} - \phi^{\langle j\rangle}_{s_{i-1}^{\langle n\rangle}})\Big) - \int_{s_{i-1}^{\langle n\rangle}}^{s_i^{\langle n\rangle}} U^{(1)}(\phi_s)(g_s)\,\mathrm{d}s
\end{aligned}
$$

$$
-\frac{1}{2}\sum_{j=1}^{m}\int_{s_{i-1}^{\langle n\rangle}}^{s_i^{\langle n\rangle}} U^{(2)}(\phi_s)(h_s^{\langle j\rangle})(h_s^{\langle j\rangle})\,\mathrm{d}s + \int_0^1\int_0^{\bar{s}}(U^{(2)}(\phi_{s_{i-1}^{\langle n\rangle}}
$$

$$
+\tau(\phi_{s_i^{\langle n\rangle}} - \phi_{s_{i-1}^{\langle n\rangle}})) - U^{(2)}(\phi_{s_{i-1}^{\langle n\rangle}}))(\int_{s_{i-1}^{\langle n\rangle}}^{s_i^{\langle n\rangle}} g_s\,\mathrm{d}s + \sum_{j=1}^{m}(\phi_{s_i^{\langle n\rangle}}^{\langle j\rangle} - \phi_{s_{i-1}^{\langle n\rangle}}^{\langle j\rangle}))
$$

$$
\cdot(\int_{s_{i-1}^{\langle n\rangle}}^{s_i^{\langle n\rangle}} g_s\,\mathrm{d}s + \sum_{j=1}^{m}(\phi_{s_i^{\langle n\rangle}}^{\langle j\rangle} - \phi_{s_{i-1}^{\langle n\rangle}}^{\langle j\rangle}))\,\mathrm{d}\tau\,\mathrm{d}\bar{s}\Big) - S_{2,n} - \sum_{j=1}^{m} l_t^{\langle j\rangle}
$$

$$
= \sum_{i=1}^{2^n}\Big(U^{(1)}(\phi_{s_{i-1}^{\langle n\rangle}})(\int_{s_{i-1}^{\langle n\rangle}}^{s_i^{\langle n\rangle}} g_s\,\mathrm{d}s) - \int_{s_{i-1}^{\langle n\rangle}}^{s_i^{\langle n\rangle}} U^{(1)}(\phi_s)(g_s)\,\mathrm{d}s\Big)
$$

$$
+\sum_{j=1}^{m}\Big(\sum_{i=1}^{2^n} U^{(1)}(\phi_{s_{i-1}^{\langle n\rangle}})(\phi_{s_i^{\langle n\rangle}}^{\langle j\rangle} - \phi_{s_{i-1}^{\langle n\rangle}}^{\langle j\rangle}) - l_t^{\langle j\rangle}\Big)
$$

$$
+\sum_{i=1}^{2^n}\int_0^1\int_0^{\bar{s}} U^{(2)}(\phi_{s_{i-1}^{\langle n\rangle}} + \tau(\phi_{s_i^{\langle n\rangle}} - \phi_{s_{i-1}^{\langle n\rangle}}))(\int_{s_{i-1}^{\langle n\rangle}}^{s_i^{\langle n\rangle}} g_s\,\mathrm{d}s)
$$

$$
\cdot(\int_{s_{i-1}^{\langle n\rangle}}^{s_i^{\langle n\rangle}} g_s\,\mathrm{d}s)\,\mathrm{d}\tau\,\mathrm{d}\bar{s} + 2\sum_{j=1}^{m}\sum_{i=1}^{2^n}\int_0^1\int_0^{\bar{s}} U^{(2)}(\phi_{s_{i-1}^{\langle n\rangle}} + \tau(\phi_{s_i^{\langle n\rangle}} - \phi_{s_{i-1}^{\langle n\rangle}}))
$$

$$
\cdot(\int_{s_{i-1}^{\langle n\rangle}}^{s_i^{\langle n\rangle}} g_s\,\mathrm{d}s)(\phi_{s_i^{\langle n\rangle}}^{\langle j\rangle} - \phi_{s_{i-1}^{\langle n\rangle}}^{\langle j\rangle})\,\mathrm{d}\tau\,\mathrm{d}\bar{s}
$$

$$
+\frac{1}{2}\sum_{j_1=1}^{m}\sum_{j_2=1}^{m}\sum_{i=1}^{2^n} U^{(2)}(\phi_{s_{i-1}^{\langle n\rangle}})(\phi_{s_i^{\langle n\rangle}}^{\langle j_1\rangle} - \phi_{s_{i-1}^{\langle n\rangle}}^{\langle j_1\rangle})(\phi_{s_i^{\langle n\rangle}}^{\langle j_2\rangle} - \phi_{s_{i-1}^{\langle n\rangle}}^{\langle j_2\rangle})
$$

$$
-\frac{1}{2}\sum_{j=1}^{m}\int_a^t U^{(2)}(\phi_s)(h_s^{\langle j\rangle})(h_s^{\langle j\rangle})\,\mathrm{d}s
$$

$$
-\frac{1}{2}\sum_{j=1}^{m}\sum_{i=1}^{2^n} U^{(2)}(\phi_{s_{i-1}^{\langle n\rangle}})(h_{s_{i-1}^{\langle n\rangle}}^{\langle j\rangle})(h_{s_{i-1}^{\langle n\rangle}}^{\langle j\rangle})
$$

$$
\cdot((w_{s_i^{\langle n\rangle}}^{\langle j\rangle} - w_{s_{i-1}^{\langle n\rangle}}^{\langle j\rangle})^2 - (s_i^{\langle n\rangle} - s_{i-1}^{\langle n\rangle})) + \sum_{j_1=1}^{m}\sum_{j_2=1}^{m}\sum_{i=1}^{2^n}\int_0^1\int_0^{\bar{s}}(U^{(2)}(\phi_{s_{i-1}^{\langle n\rangle}}
$$

$$
+\tau(\phi_{s_i^{\langle n\rangle}} - \phi_{s_{i-1}^{\langle n\rangle}})) - U^{(2)}(\phi_{s_{i-1}^{\langle n\rangle}}))(\phi_{s_i^{\langle n\rangle}}^{\langle j_1\rangle} - \phi_{s_{i-1}^{\langle n\rangle}}^{\langle j_1\rangle})
$$

$$
\cdot(\phi_{s_i^{\langle n\rangle}}^{\langle j_2\rangle} - \phi_{s_{i-1}^{\langle n\rangle}}^{\langle j_2\rangle})\,\mathrm{d}\tau\,\mathrm{d}\bar{s}
$$

$$
\begin{aligned}
&= \int_a^t (U^{(1)}(\Gamma_s) - U^{(1)}(\phi_s))(g_s)\,\mathrm{d}s \\
&\quad + \sum_{i=1}^{2^n} \int_0^1 \int_0^{\bar{s}} U^{(2)}(\phi_{s_{i-1}^{\langle n\rangle}} + \tau(\phi_{s_i^{\langle n\rangle}} - \phi_{s_{i-1}^{\langle n\rangle}}))(\int_{s_{i-1}^{\langle n\rangle}}^{s_i^{\langle n\rangle}} g_s\,\mathrm{d}s) \\
&\quad \cdot(\int_{s_{i-1}^{\langle n\rangle}}^{s_i^{\langle n\rangle}} g_s\,\mathrm{d}s)\,\mathrm{d}\tau\,\mathrm{d}\bar{s} + 2\sum_{j=1}^{m}\sum_{i=1}^{2^n} \int_0^1 \int_0^{\bar{s}} U^{(2)}(\phi_{s_{i-1}^{\langle n\rangle}} + \tau(\phi_{s_i^{\langle n\rangle}} - \phi_{s_{i-1}^{\langle n\rangle}})) \\
&\quad \cdot(\int_{s_{i-1}^{\langle n\rangle}}^{s_i^{\langle n\rangle}} g_s\,\mathrm{d}s)(\phi^{\langle j\rangle}_{s_i^{\langle n\rangle}} - \phi^{\langle j\rangle}_{s_{i-1}^{\langle n\rangle}})\,\mathrm{d}\tau\,\mathrm{d}\bar{s} \\
&\quad + \frac{1}{2}\sum_{j=1}^{m}\Big(\sum_{i=1}^{2^n} U^{(2)}(\phi_{s_{i-1}^{\langle n\rangle}})(h^{\langle j\rangle}_{s_{i-1}^{\langle n\rangle}})(h^{\langle j\rangle}_{s_{i-1}^{\langle n\rangle}})(s_i^{\langle n\rangle} - s_{i-1}^{\langle n\rangle}) \\
&\quad - \int_a^t U^{(2)}(\phi_s)(h_s^{\langle j\rangle})(h_s^{\langle j\rangle})\,\mathrm{d}s\Big) \\
&\quad + \frac{1}{2}\sum_{j_1=1}^{m}\sum_{j_2=1}^{m}\sum_{i=1}^{2^n} U^{(2)}(\phi_{s_{i-1}^{\langle n\rangle}})(\phi^{\langle j_1\rangle}_{s_i^{\langle n\rangle}} - \phi^{\langle j_1\rangle}_{s_{i-1}^{\langle n\rangle}})(\phi^{\langle j_2\rangle}_{s_i^{\langle n\rangle}} - \phi^{\langle j_2\rangle}_{s_{i-1}^{\langle n\rangle}}) \\
&\quad - \frac{1}{2}\sum_{j=1}^{m}\sum_{i=1}^{2^n} U^{(2)}(\phi_{s_{i-1}^{\langle n\rangle}})(h^{\langle j\rangle}_{s_{i-1}^{\langle n\rangle}})(h^{\langle j\rangle}_{s_{i-1}^{\langle n\rangle}})(w^{\langle j\rangle}_{s_i^{\langle n\rangle}} - w^{\langle j\rangle}_{s_{i-1}^{\langle n\rangle}})^2 \\
&\quad + \sum_{j=1}^{m}\Big(\sum_{i=1}^{2^n} U^{(1)}(\phi_{s_{i-1}^{\langle n\rangle}})(\phi^{\langle j\rangle}_{s_i^{\langle n\rangle}} - \phi^{\langle j\rangle}_{s_{i-1}^{\langle n\rangle}}) - l_t^{\langle j\rangle}\Big) \\
&\quad + \sum_{j_1=1}^{m}\sum_{j_2=1}^{m}\sum_{i=1}^{2^n} \int_0^1 \int_0^{\bar{s}} (U^{(2)}(\phi_{s_{i-1}^{\langle n\rangle}} + \tau(\phi_{s_i^{\langle n\rangle}} - \phi_{s_{i-1}^{\langle n\rangle}})) - U^{(2)}(\phi_{s_{i-1}^{\langle n\rangle}})) \\
&\quad \cdot(\phi^{\langle j_1\rangle}_{s_i^{\langle n\rangle}} - \phi^{\langle j_1\rangle}_{s_{i-1}^{\langle n\rangle}})(\phi^{\langle j_2\rangle}_{s_i^{\langle n\rangle}} - \phi^{\langle j_2\rangle}_{s_{i-1}^{\langle n\rangle}})\,\mathrm{d}\tau\,\mathrm{d}\bar{s} \\
&=: T_{1,n} + T_{2,n} + T_{3,n} + T_{4,n} + T_{5,n} + T_{6,n}
\end{aligned}
$$

where

$$
\Gamma(s,\omega) := \sum_{i=1}^{2^n} \phi_{s_{i-1}^{\langle n\rangle}}(\omega)\chi_{[s_{i-1}^{\langle n\rangle}, s_i^{\langle n\rangle}), \bar{\mathrm{r}}_{a,t}}(s); \quad \forall (s,\omega) \in \bar{\mathrm{r}}_{a,t} \times \boldsymbol{\Omega}
$$

$$
\begin{aligned}
T_{1,n} &:= \int_a^t (U^{(1)}(\Gamma_s) - U^{(1)}(\phi_s))(g_s)\,\mathrm{d}s \\
T_{2,n} &:= \sum_{i=1}^{2^n} \int_0^1 \int_0^{\bar{s}} U^{(2)}(\phi_{s_{i-1}^{\langle n\rangle}} + \tau(\phi_{s_i^{\langle n\rangle}} - \phi_{s_{i-1}^{\langle n\rangle}}))(\int_{s_{i-1}^{\langle n\rangle}}^{s_i^{\langle n\rangle}} g_s\,\mathrm{d}s) \\
&\quad \cdot(\int_{s_{i-1}^{\langle n\rangle}}^{s_i^{\langle n\rangle}} g_s\,\mathrm{d}s)\,\mathrm{d}\tau\,\mathrm{d}\bar{s} + 2\sum_{j=1}^{m}\sum_{i=1}^{2^n} \int_0^1 \int_0^{\bar{s}} U^{(2)}(\phi_{s_{i-1}^{\langle n\rangle}} + \tau(\phi_{s_i^{\langle n\rangle}} \\
&\quad -\phi_{s_{i-1}^{\langle n\rangle}}))(\int_{s_{i-1}^{\langle n\rangle}}^{s_i^{\langle n\rangle}} g_s\,\mathrm{d}s)(\phi^{\langle j\rangle}_{s_i^{\langle n\rangle}} - \phi^{\langle j\rangle}_{s_{i-1}^{\langle n\rangle}})\,\mathrm{d}\tau\,\mathrm{d}\bar{s} \\
&=: T_{2,n,1} + T_{2,n,2} \\
T_{3,n} &:= \frac{1}{2}\sum_{j=1}^{m}\Big(\sum_{i=1}^{2^n} U^{(2)}(\phi_{s_{i-1}^{\langle n\rangle}})(h^{\langle j\rangle}_{s_{i-1}^{\langle n\rangle}})(h^{\langle j\rangle}_{s_{i-1}^{\langle n\rangle}})(s_i^{\langle n\rangle} - s_{i-1}^{\langle n\rangle}) \\
&\quad - \int_a^t U^{(2)}(\phi_s)(h_s^{\langle j\rangle})(h_s^{\langle j\rangle})\,\mathrm{d}s\Big) \\
T_{4,n} &:= \frac{1}{2}\sum_{j_1=1}^{m}\sum_{j_2=1}^{m}\sum_{i=1}^{2^n} U^{(2)}(\phi_{s_{i-1}^{\langle n\rangle}})(\phi^{\langle j_1\rangle}_{s_i^{\langle n\rangle}} - \phi^{\langle j_1\rangle}_{s_{i-1}^{\langle n\rangle}})(\phi^{\langle j_2\rangle}_{s_i^{\langle n\rangle}} - \phi^{\langle j_2\rangle}_{s_{i-1}^{\langle n\rangle}}) \\
&\quad -\frac{1}{2}\sum_{j=1}^{m}\sum_{i=1}^{2^n} U^{(2)}(\phi_{s_{i-1}^{\langle n\rangle}})(h^{\langle j\rangle}_{s_{i-1}^{\langle n\rangle}})(h^{\langle j\rangle}_{s_{i-1}^{\langle n\rangle}})(w^{\langle j\rangle}_{s_i^{\langle n\rangle}} - w^{\langle j\rangle}_{s_{i-1}^{\langle n\rangle}})^2 \\
T_{5,n} &:= \sum_{j=1}^{m}\Big(\sum_{i=1}^{2^n} U^{(1)}(\phi_{s_{i-1}^{\langle n\rangle}})(\phi^{\langle j\rangle}_{s_i^{\langle n\rangle}} - \phi^{\langle j\rangle}_{s_{i-1}^{\langle n\rangle}}) - l_t^{\langle j\rangle}\Big) \\
T_{6,n} &:= \sum_{j_1=1}^{m}\sum_{j_2=1}^{m}\sum_{i=1}^{2^n} \int_0^1 \int_0^{\bar{s}} (U^{(2)}(\phi_{s_{i-1}^{\langle n\rangle}} + \tau(\phi_{s_i^{\langle n\rangle}} - \phi_{s_{i-1}^{\langle n\rangle}})) \\
&\quad -U^{(2)}(\phi_{s_{i-1}^{\langle n\rangle}}))(\phi^{\langle j_1\rangle}_{s_i^{\langle n\rangle}} - \phi^{\langle j_1\rangle}_{s_{i-1}^{\langle n\rangle}})(\phi^{\langle j_2\rangle}_{s_i^{\langle n\rangle}} - \phi^{\langle j_2\rangle}_{s_{i-1}^{\langle n\rangle}})\,\mathrm{d}\tau\,\mathrm{d}\bar{s}
\end{aligned}
$$

the first equality follows from the definition of $\hat{T}$ and Proposition 11.92, the second equality follows from the definition of X, the third equality follows from result of the last paragraph, the fourth equality follows from assumptions (d), (i), (ii), Proposition 11.92, Theorem A.7, and the discussion in the fourth paragraph of the proof, the fifth equality follows from Definitions 9.3 and 9.25, Proposition 9.28, and the definition of $S_{2,n}$, the sixth equality follows from simple algebra, Proposition 11.92, and Theorem 12.83, and the assignment of $T_{1,n}$ follows from the definition of Γ. Now, we will deal with the terms above one by one.

Since $\phi_{\langle\omega\rangle} : \bar{\mathrm{r}}_{a,b} \to D$ is continuous, $\forall \omega \in \boldsymbol{\Omega}$, then we have $\phi_{\langle\omega\rangle}$ is uniformly continuous since $\bar{\mathrm{r}}_{a,b}$ is compact. Then, we have $\Gamma_{\langle\omega\rangle}$ converges to $\phi_{\langle\omega\rangle}$ uniformly as $n \to \infty$, $\forall \omega \in \boldsymbol{\Omega}$. By the same argument, $U^{(1)}(\Gamma_{\langle\omega\rangle})$ converges to $U^{(1)}(\phi_{\langle\omega\rangle})$ uniformly as $n \to \infty$, $\forall \omega \in \boldsymbol{\Omega}$. Note that $\forall \omega \in \boldsymbol{\Omega}$,

$$\begin{aligned}
\left\| T_{1,n}(\omega) \right\|_{\mathcal{Z}} &= \left\| \int_a^t (U^{(1)}(\Gamma(s,\omega)) - U^{(1)}(\phi(s,\omega)))(g(s,\omega)) \, \mathrm{d}s \right\|_{\mathcal{Z}} \\
&\leq \int_a^t \left\| (U^{(1)}(\Gamma(s,\omega)) - U^{(1)}(\phi(s,\omega)))(g(s,\omega)) \right\|_{\mathcal{Z}} \mathrm{d}s \\
&\leq \int_a^t \left\| U^{(1)}(\Gamma(s,\omega)) - U^{(1)}(\phi(s,\omega)) \right\|_{\mathrm{B}(\mathcal{Y},\mathcal{Z})} \| g(s,\omega) \|_{\mathcal{Y}} \, \mathrm{d}s \\
&\leq \int_a^t \hat{\delta}_n(\omega) \| g(s,\omega) \|_{\mathcal{Y}} \, \mathrm{d}s = \hat{\delta}_n(\omega) \int_a^t \| g(s,\omega) \|_{\mathcal{Y}} \, \mathrm{d}s \\
&\leq \hat{\delta}_n(\omega) \left\| g_{\langle\omega\rangle} \right\|_{\bar{\mathrm{L}}_1(\bar{\mathrm{r}}_{a,b},\mathcal{Y})}
\end{aligned}$$

where the first inequality follows from Proposition 11.92, the second inequality follows from Proposition 7.64, the third inequality follows from the definition

$$\hat{\delta}_n(\omega) := \sup_{\substack{u,v \in \bar{\mathrm{r}}_{a,b}, \\ |u-v| \leq 2^{-n}(b-a)}} \left\| U^{(1)}(\phi(v,\omega)) - U^{(1)}(\phi(u,\omega)) \right\|_{\mathrm{B}(\mathcal{Y},\mathcal{Z})} \tag{14.25}$$

and the second equality follows from Proposition 11.92. It follows from Proposition 14.109 that $\hat{\delta}_n$ is $\mathcal{B}$-measurable, $\forall n \in \mathbb{N}$. By the fact that $\lim_{n\in\mathbb{N}} \hat{\delta}_n(\omega) = 0$, $\forall \omega \in \boldsymbol{\Omega}$, we have $\lim_{n\in\mathbb{N}} T_{1,n}(\omega) = \vartheta_{\mathcal{Z}}$, $\forall \omega \in \boldsymbol{\Omega}$.

Now, we analyze the term $T_{2,n} = T_{2,n,1} + T_{2,n,2}$. Note that, $\forall \omega \in \boldsymbol{\Omega}$,

$$\begin{aligned}
&\left\| T_{2,n,1}(\omega) \right\|_{\mathcal{Z}} \\
&= \left\| \sum_{i=1}^{2^n} \int_0^1 \int_0^{\bar{s}} U^{(2)}(\phi_{s_{i-1}^{\langle n\rangle}} + \tau(\phi_{s_i^{\langle n\rangle}} - \phi_{s_{i-1}^{\langle n\rangle}}))(\int_{s_{i-1}^{\langle n\rangle}}^{s_i^{\langle n\rangle}} g_s \, \mathrm{d}s) \right. \\
&\quad \left. \cdot (\int_{s_{i-1}^{\langle n\rangle}}^{s_i^{\langle n\rangle}} g_s \, \mathrm{d}s) \, \mathrm{d}\tau \, \mathrm{d}s \right\|_{\mathcal{Z}} \\
&\leq \sum_{i=1}^{2^n} \left\| \int_0^1 \int_0^{\bar{s}} U^{(2)}(\phi_{s_{i-1}^{\langle n\rangle}} + \tau(\phi_{s_i^{\langle n\rangle}} - \phi_{s_{i-1}^{\langle n\rangle}}))(\int_{s_{i-1}^{\langle n\rangle}}^{s_i^{\langle n\rangle}} g_s \, \mathrm{d}s) \right. \\
&\quad \left. \cdot (\int_{s_{i-1}^{\langle n\rangle}}^{s_i^{\langle n\rangle}} g_s \, \mathrm{d}s) \, \mathrm{d}\tau \, \mathrm{d}\bar{s} \right\|_{\mathcal{Z}}
\end{aligned}$$

$$\leq \sum_{i=1}^{2^n} \int_0^1 \int_0^{\bar{s}} \left\| U^{(2)}(\phi_{s_{i-1}^{\langle n\rangle}} + \tau(\phi_{s_i^{\langle n\rangle}} - \phi_{s_{i-1}^{\langle n\rangle}}))(\int_{s_{i-1}^{\langle n\rangle}}^{s_i^{\langle n\rangle}} g_s \, ds) \right.$$

$$\left. \cdot (\int_{s_{i-1}^{\langle n\rangle}}^{s_i^{\langle n\rangle}} g_s \, ds) \right\|_{\mathcal{Z}} d\tau \, d\bar{s}$$

$$\leq \sum_{i=1}^{2^n} \int_0^1 \int_0^{\bar{s}} \left\| U^{(2)}(\phi_{s_{i-1}^{\langle n\rangle}} + \tau(\phi_{s_i^{\langle n\rangle}} - \phi_{s_{i-1}^{\langle n\rangle}})) \right\|_{\mathrm{B}(\mathcal{Y},\mathrm{B}(\mathcal{Y},\mathcal{Z}))}$$

$$\cdot \left\| \int_{s_{i-1}^{\langle n\rangle}}^{s_i^{\langle n\rangle}} g_s \, ds \right\|_{\mathcal{Y}}^2 d\tau \, d\bar{s}$$

$$\leq \sum_{i=1}^{2^n} \int_0^1 \int_0^{\bar{s}} \left\| U^{(2)}(\phi_{s_{i-1}^{\langle n\rangle}} + \tau(\phi_{s_i^{\langle n\rangle}} - \phi_{s_{i-1}^{\langle n\rangle}})) \right\|_{\mathrm{B}(\mathcal{Y},\mathrm{B}(\mathcal{Y},\mathcal{Z}))}$$

$$\cdot \Big(\int_{s_{i-1}^{\langle n\rangle}}^{s_i^{\langle n\rangle}} \|g_s\|_{\mathcal{Y}} \, ds \Big)^2 d\tau \, d\bar{s}$$

$$\leq \sum_{i=1}^{2^n} \int_0^1 \int_0^{\bar{s}} \left\| U^{(2)}(\phi_{s_{i-1}^{\langle n\rangle}} + \tau(\phi_{s_i^{\langle n\rangle}} - \phi_{s_{i-1}^{\langle n\rangle}})) \right\|_{\mathrm{B}(\mathcal{Y},\mathrm{B}(\mathcal{Y},\mathcal{Z}))}$$

$$\cdot \Big(\int_{s_{i-1}^{\langle n\rangle}}^{s_i^{\langle n\rangle}} \|g_s\|_{\mathcal{Y}}^2 \, ds \Big)(s_i^{\langle n\rangle} - s_{i-1}^{\langle n\rangle}) \, d\tau \, d\bar{s}$$

$$= \sum_{i=1}^{2^n} \frac{t-a}{2^n} \int_0^1 \int_0^{\bar{s}} \left\| U^{(2)}(\phi_{s_{i-1}^{\langle n\rangle}} + \tau(\phi_{s_i^{\langle n\rangle}} - \phi_{s_{i-1}^{\langle n\rangle}})) \right\|_{\mathrm{B}(\mathcal{Y},\mathrm{B}(\mathcal{Y},\mathcal{Z}))} d\tau \, d\bar{s}$$

$$\cdot \int_{s_{i-1}^{\langle n\rangle}}^{s_i^{\langle n\rangle}} \|g(s,\omega)\|_{\mathcal{Y}}^2 \, ds$$

$$\leq \frac{b-a}{2^{n+1}} \sum_{i=1}^{2^n} M_2(\omega) \int_{s_{i-1}^{\langle n\rangle}}^{s_i^{\langle n\rangle}} \|g(s,\omega)\|_{\mathcal{Y}}^2 \, ds$$

$$\leq \frac{b-a}{2^{n+1}} M_2(\omega) \|g_{\langle\omega\rangle}\|_{\bar{\mathrm{L}}_2(\bar{\Gamma}_{a,b},\mathcal{Y})}^2$$

where the first inequality follows from the triangular inequality, the second inequality follows from Proposition 11.92, the third inequality follows from Proposition 7.64, the fourth inequality follows from Proposition 11.92 and the assumption (c), the fifth inequality follows from Cauchy–Schwarz Inequality, the second equality follows from Proposition 11.92 and the definition of $s_i^{\langle n\rangle}$'s, and the sixth inequality follows from the definition

$$M_2(\omega) := \sup_{\substack{v_1, v_2 \in \bar{\mathrm{r}}_{a,b}, \\ \alpha \in \bar{\mathrm{r}}_{0,1}}} \left\| U^{(2)}(\alpha\phi(v_1,\omega) + (1-\alpha)\phi(v_2,\omega)) \right\|_{\mathrm{B}(\mathcal{Y},\mathrm{B}(\mathcal{Y},\mathcal{Z}))} < \infty \tag{14.26}$$

$\forall \omega \in \boldsymbol{\Omega}$, and Proposition 11.75. By Proposition 14.109, M_2 is $\mathcal{B}$-measurable. Since $g \in \bar{\mathrm{L}}_2(\bar{\mathrm{r}}_{a,b} \times \boldsymbol{\Omega}, \mathcal{Y})$, we have $\left\| g_{\langle\omega\rangle} \right\|^2_{\bar{\mathrm{L}}_2(\bar{\mathrm{r}}_{a,b},\mathcal{Y})} < \infty$ a.e. $\omega \in \boldsymbol{\Omega}$. Hence, $\lim_{n\in\mathbb{N}} T_{2,n,1}(\omega) = \vartheta_{\mathcal{Z}}$ a.e. $\omega \in \boldsymbol{\Omega}$.

For $T_{2,n,2}$, we have, $\forall \omega \in \boldsymbol{\Omega}$,

$$\begin{aligned}
&\| T_{2,n,2}(\omega) \|_{\mathcal{Z}} \\
&\leq 2\sum_{j=1}^{m}\sum_{i=1}^{2^n} \Big\| \int_0^1 \int_0^{\bar{s}} U^{(2)}(\phi_{s_{i-1}^{\langle n\rangle}} + \tau(\phi_{s_i^{\langle n\rangle}} - \phi_{s_{i-1}^{\langle n\rangle}}))(\int_{s_{i-1}^{\langle n\rangle}}^{s_i^{\langle n\rangle}} g_s \,\mathrm{d}s) \\
&\quad \cdot(\phi^{\langle j\rangle}_{s_i^{\langle n\rangle}} - \phi^{\langle j\rangle}_{s_{i-1}^{\langle n\rangle}}) \,\mathrm{d}\tau \,\mathrm{d}\bar{s} \Big\|_{\mathcal{Z}} \\
&\leq 2\sum_{j=1}^{m}\sum_{i=1}^{2^n} \int_0^1 \int_0^{\bar{s}} \Big\| U^{(2)}(\phi_{s_{i-1}^{\langle n\rangle}} + \tau(\phi_{s_i^{\langle n\rangle}} - \phi_{s_{i-1}^{\langle n\rangle}}))(\int_{s_{i-1}^{\langle n\rangle}}^{s_i^{\langle n\rangle}} g_s \,\mathrm{d}s) \\
&\quad \cdot(\phi^{\langle j\rangle}_{s_i^{\langle n\rangle}} - \phi^{\langle j\rangle}_{s_{i-1}^{\langle n\rangle}}) \Big\|_{\mathcal{Z}} \,\mathrm{d}\tau \,\mathrm{d}\bar{s} \\
&\leq 2\sum_{j=1}^{m}\sum_{i=1}^{2^n} \int_0^1 \int_0^{\bar{s}} \Big\| U^{(2)}(\phi_{s_{i-1}^{\langle n\rangle}} + \tau(\phi_{s_i^{\langle n\rangle}} - \phi_{s_{i-1}^{\langle n\rangle}})) \Big\|_{\mathrm{B}(\mathcal{Y},\mathrm{B}(\mathcal{Y},\mathcal{Z}))} \\
&\quad \cdot \Big\| \int_{s_{i-1}^{\langle n\rangle}}^{s_i^{\langle n\rangle}} g_s \,\mathrm{d}s \Big\|_{\mathcal{Y}} \Big\| \phi^{\langle j\rangle}_{s_i^{\langle n\rangle}} - \phi^{\langle j\rangle}_{s_{i-1}^{\langle n\rangle}} \Big\|_{\mathcal{Y}} \,\mathrm{d}\tau \,\mathrm{d}\bar{s} \\
&\leq 2\sum_{j=1}^{m}\sum_{i=1}^{2^n} \int_0^1 \int_0^{\bar{s}} M_2(\omega) \Big\| \int_{s_{i-1}^{\langle n\rangle}}^{s_i^{\langle n\rangle}} g_s \,\mathrm{d}s \Big\|_{\mathcal{Y}} \Big\| \phi^{\langle j\rangle}_{s_i^{\langle n\rangle}} - \phi^{\langle j\rangle}_{s_{i-1}^{\langle n\rangle}} \Big\|_{\mathcal{Y}} \,\mathrm{d}\tau \,\mathrm{d}\bar{s} \\
&= M_2(\omega) \sum_{j=1}^{m}\sum_{i=1}^{2^n} \Big\| \int_{s_{i-1}^{\langle n\rangle}}^{s_i^{\langle n\rangle}} g_s \,\mathrm{d}s \Big\|_{\mathcal{Y}} \Big\| \phi^{\langle j\rangle}_{s_i^{\langle n\rangle}} - \phi^{\langle j\rangle}_{s_{i-1}^{\langle n\rangle}} \Big\|_{\mathcal{Y}} \\
&=: M_2(\omega) \sum_{j=1}^{m} Z^{\langle j\rangle}_{2,n,2}(\omega)
\end{aligned}$$

where the first inequality follows from triangular inequality, the second inequality follows from Proposition 11.92, the third inequality follows from Proposition 7.64, the fourth inequality follows from the definition of M_2 in (14.26) and Proposition 11.92, and the equality follows from Theorem 12.83.

Note that $j = 1, \ldots, m$,

$$\begin{aligned}
\boldsymbol{E}(Z^{\langle j \rangle}_{2,n,2}) &= \boldsymbol{E}\Big(\sum_{i=1}^{2^n} \Big\| \int_{s_{i-1}^{\langle n \rangle}}^{s_i^{\langle n \rangle}} g_s \,\mathrm{d}s \Big\|_{\mathcal{Y}} \big\| \phi^{\langle j \rangle}_{s_i^{\langle n \rangle}} - \phi^{\langle j \rangle}_{s_{i-1}^{\langle n \rangle}} \big\|_{\mathcal{Y}}\Big) \\
&= \sum_{i=1}^{2^n} \boldsymbol{E}\Big(\Big\| \int_{s_{i-1}^{\langle n \rangle}}^{s_i^{\langle n \rangle}} g_s \,\mathrm{d}s \Big\|_{\mathcal{Y}} \big\| \phi^{\langle j \rangle}_{s_i^{\langle n \rangle}} - \phi^{\langle j \rangle}_{s_{i-1}^{\langle n \rangle}} \big\|_{\mathcal{Y}}\Big) \\
&\leq \sum_{i=1}^{2^n} \Big(\boldsymbol{E}\Big(\Big\| \int_{s_{i-1}^{\langle n \rangle}}^{s_i^{\langle n \rangle}} g_s \,\mathrm{d}s \Big\|^2_{\mathcal{Y}}\Big)\Big)^{\frac{1}{2}} \Big(\boldsymbol{E}\Big(\big\| \phi^{\langle j \rangle}_{s_i^{\langle n \rangle}} - \phi^{\langle j \rangle}_{s_{i-1}^{\langle n \rangle}} \big\|^2_{\mathcal{Y}}\Big)\Big)^{\frac{1}{2}} \\
&\leq \sum_{i=1}^{2^n} \Big(\boldsymbol{E}\Big(\Big(\int_{s_{i-1}^{\langle n \rangle}}^{s_i^{\langle n \rangle}} \| g_s \|_{\mathcal{Y}} \,\mathrm{d}s\Big)^2\Big)\Big)^{\frac{1}{2}} \big\| \phi^{\langle j \rangle}_{s_i^{\langle n \rangle}} - \phi^{\langle j \rangle}_{s_{i-1}^{\langle n \rangle}} \big\|_{\mathrm{L}_2(\boldsymbol{\Omega},\mathcal{Y})} \\
&\leq \sum_{i=1}^{2^n} \Big(\boldsymbol{E}\Big(\Big(\int_{s_{i-1}^{\langle n \rangle}}^{s_i^{\langle n \rangle}} \| g_s \|^2_{\mathcal{Y}} \,\mathrm{d}s\Big)(s_i^{\langle n \rangle} - s_{i-1}^{\langle n \rangle})\Big)\Big)^{\frac{1}{2}} \| h^{\langle j \rangle} \|_{\bar{\mathrm{L}}_2(\mathrm{r}_{s_{i-1}^{\langle n \rangle}, s_i^{\langle n \rangle}} \times \boldsymbol{\Omega}, \mathcal{Y})} \\
&= \sum_{i=1}^{2^n} \sqrt{\frac{t-a}{2^n}} \Big(\boldsymbol{E}\Big(\int_{s_{i-1}^{\langle n \rangle}}^{s_i^{\langle n \rangle}} \| g_s \|^2_{\mathcal{Y}} \,\mathrm{d}s\Big)\Big)^{\frac{1}{2}} \| h^{\langle j \rangle} \|_{\bar{\mathrm{L}}_2(\mathrm{r}_{s_{i-1}^{\langle n \rangle}, s_i^{\langle n \rangle}} \times \boldsymbol{\Omega}, \mathcal{Y})} \\
&\leq \sqrt{\frac{b-a}{2^n}} \Big(\sum_{i=1}^{2^n} \boldsymbol{E}\Big(\int_{s_{i-1}^{\langle n \rangle}}^{s_i^{\langle n \rangle}} \| g_s \|^2_{\mathcal{Y}} \,\mathrm{d}s\Big)\Big)^{\frac{1}{2}} \Big(\sum_{i=1}^{2^n} \| h^{\langle j \rangle} \|^2_{\bar{\mathrm{L}}_2(\mathrm{r}_{s_{i-1}^{\langle n \rangle}, s_i^{\langle n \rangle}} \times \boldsymbol{\Omega}, \mathcal{Y})}\Big)^{\frac{1}{2}} \\
&= \sqrt{\frac{b-a}{2^n}} \Big(\boldsymbol{E}\Big(\int_a^t \| g_s \|^2_{\mathcal{Y}} \,\mathrm{d}s\Big)\Big)^{\frac{1}{2}} \| h^{\langle j \rangle} \|_{\bar{\mathrm{L}}_2(\mathrm{r}_{a,t} \times \boldsymbol{\Omega}, \mathcal{Y})} \\
&\leq \sqrt{\frac{b-a}{2^n}} \| g \|_{\bar{\mathrm{L}}_2(\mathrm{r}_{a,b} \times \boldsymbol{\Omega}, \mathcal{Y})} \| h^{\langle j \rangle} \|_{\bar{\mathrm{L}}_2(\mathrm{r}_{a,b} \times \boldsymbol{\Omega}, \mathcal{Y})} \to 0 \text{ as } n \to \infty
\end{aligned}$$

where the first equality follows from the definition of $Z^{\langle j \rangle}_{2,n,2}$, the second equality follows from Proposition 11.92, the first inequality follows from Cauchy–Schwarz Inequality, the second inequality follows from Proposition 11.92 and Example 11.173, the third inequality follows from Cauchy–Schwarz Inequality and Proposition 14.120, the third equality follows from the definition of $s_i^{\langle n \rangle}$'s, the fourth inequality follows from Cauchy–Schwarz Inequality, the fourth equality follows from Proposition 11.92, the fifth inequality follows from simple algebra, and the convergence statement follows from the assumption that $g \in \bar{\mathrm{L}}_2(\mathrm{r}_{a,b} \times \boldsymbol{\Omega}, \mathcal{Y})$ and $h^{\langle j \rangle} \in \bar{\mathrm{L}}_2(\mathrm{r}_{a,b} \times \boldsymbol{\Omega}, \mathcal{Y})$. Then, $\lim_{n \in \mathbb{N}} \sum_{j=1}^{m} Z^{\langle j \rangle}_{2,n,2} = 0$ in measure in $\boldsymbol{\Omega}$, by Proposition 11.211. Hence, by Proposition 14.111, $\lim_{n \in \mathbb{N}} T_{2,n,2} = \vartheta_{\mathcal{Z}}$ in measure in $\boldsymbol{\Omega}$.

By assumption (c), Corollary A.16, and Theorems A.11, A.18, we have $\lim_{n \in \mathbb{N}} T_{3,n}(\omega) = \vartheta_{\mathcal{Z}}, \forall \omega \in \boldsymbol{\Omega}$.

Note that

$$T_{4,n}$$

$$= \frac{1}{2}\sum_{j_1=1}^{m}\sum_{j_2=1}^{m}\sum_{i=1}^{2^n} U^{(2)}(\phi_{s_{i-1}^{\langle n\rangle}})(\phi_{s_i^{\langle n\rangle}}^{\langle j_1\rangle} - \phi_{s_{i-1}^{\langle n\rangle}}^{\langle j_1\rangle})(\phi_{s_i^{\langle n\rangle}}^{\langle j_2\rangle} - \phi_{s_{i-1}^{\langle n\rangle}}^{\langle j_2\rangle})$$

$$-\frac{1}{2}\sum_{j=1}^{m}\sum_{i=1}^{2^n} U^{(2)}(\phi_{s_{i-1}^{\langle n\rangle}})(h_{s_{i-1}^{\langle n\rangle}}^{\langle j\rangle})(h_{s_{i-1}^{\langle n\rangle}}^{\langle j\rangle})(w_{s_i^{\langle n\rangle}}^{\langle j\rangle} - w_{s_{i-1}^{\langle n\rangle}}^{\langle j\rangle})^2$$

$$= \frac{1}{2}\sum_{j_1=1}^{m}\sum_{j_2=1}^{m}\sum_{i=1}^{2^n} U^{(2)}(\phi_{s_{i-1}^{\langle n\rangle}})(\phi_{s_i^{\langle n\rangle}}^{\langle j_1\rangle} - \phi_{s_{i-1}^{\langle n\rangle}}^{\langle j_1\rangle})(\phi_{s_i^{\langle n\rangle}}^{\langle j_2\rangle} - \phi_{s_{i-1}^{\langle n\rangle}}^{\langle j_2\rangle})$$

$$-\frac{1}{2}\sum_{j_1=1}^{m}\sum_{j_2=1}^{m}\sum_{i=1}^{2^n} U^{(2)}(\phi_{s_{i-1}^{\langle n\rangle}})(\overline{\int}_{s_{i-1}^{\langle n\rangle}}^{s_i^{\langle n\rangle}} h_{s_{i-1}^{\langle n\rangle}}^{\langle j_1\rangle}\,\mathrm{d}w_s^{\langle j_1\rangle})(\phi_{s_i^{\langle n\rangle}}^{\langle j_2\rangle} - \phi_{s_{i-1}^{\langle n\rangle}}^{\langle j_2\rangle})$$

$$+\frac{1}{2}\sum_{j_1=1}^{m}\sum_{j_2=1}^{m}\sum_{i=1}^{2^n} U^{(2)}(\phi_{s_{i-1}^{\langle n\rangle}})(\overline{\int}_{s_{i-1}^{\langle n\rangle}}^{s_i^{\langle n\rangle}} h_{s_{i-1}^{\langle n\rangle}}^{\langle j_1\rangle}\,\mathrm{d}w_s^{\langle j_1\rangle})(\phi_{s_i^{\langle n\rangle}}^{\langle j_2\rangle} - \phi_{s_{i-1}^{\langle n\rangle}}^{\langle j_2\rangle})$$

$$-\frac{1}{2}\sum_{j_1=1}^{m}\sum_{j_2=1}^{m}\sum_{i=1}^{2^n} U^{(2)}(\phi_{s_{i-1}^{\langle n\rangle}})(\overline{\int}_{s_{i-1}^{\langle n\rangle}}^{s_i^{\langle n\rangle}} h_{s_{i-1}^{\langle n\rangle}}^{\langle j_1\rangle}\,\mathrm{d}w_s^{\langle j_1\rangle})(\overline{\int}_{s_{i-1}^{\langle n\rangle}}^{s_i^{\langle n\rangle}} h_{s_{i-1}^{\langle n\rangle}}^{\langle j_2\rangle}\,\mathrm{d}w_s^{\langle j_2\rangle})$$

$$+\frac{1}{2}\sum_{j_1=1}^{m}\sum_{j_2=1}^{m}\sum_{i=1}^{2^n} U^{(2)}(\phi_{s_{i-1}^{\langle n\rangle}})(\overline{\int}_{s_{i-1}^{\langle n\rangle}}^{s_i^{\langle n\rangle}} h_{s_{i-1}^{\langle n\rangle}}^{\langle j_1\rangle}\,\mathrm{d}w_s^{\langle j_1\rangle})(\overline{\int}_{s_{i-1}^{\langle n\rangle}}^{s_i^{\langle n\rangle}} h_{s_{i-1}^{\langle n\rangle}}^{\langle j_2\rangle}\,\mathrm{d}w_s^{\langle j_2\rangle})$$

$$-\frac{1}{2}\sum_{j=1}^{m}\sum_{i=1}^{2^n} U^{(2)}(\phi_{s_{i-1}^{\langle n\rangle}})(h_{s_{i-1}^{\langle n\rangle}}^{\langle j\rangle})(h_{s_{i-1}^{\langle n\rangle}}^{\langle j\rangle})(w_{s_i^{\langle n\rangle}}^{\langle j\rangle} - w_{s_{i-1}^{\langle n\rangle}}^{\langle j\rangle})^2$$

$$= \frac{1}{2}\sum_{j_1=1}^{m}\sum_{j_2=1}^{m}\sum_{i=1}^{2^n} U^{(2)}(\phi_{s_{i-1}^{\langle n\rangle}})(\overline{\int}_{s_{i-1}^{\langle n\rangle}}^{s_i^{\langle n\rangle}} (h_s^{\langle j_1\rangle} - h_{s_{i-1}^{\langle n\rangle}}^{\langle j_1\rangle})\,\mathrm{d}w_s^{\langle j_1\rangle})$$

$$\cdot(\phi_{s_i^{\langle n\rangle}}^{\langle j_2\rangle} - \phi_{s_{i-1}^{\langle n\rangle}}^{\langle j_2\rangle})$$

$$+\frac{1}{2}\sum_{j_1=1}^{m}\sum_{j_2=1}^{m}\sum_{i=1}^{2^n} U^{(2)}(\phi_{s_{i-1}^{\langle n\rangle}})(\overline{\int}_{s_{i-1}^{\langle n\rangle}}^{s_i^{\langle n\rangle}} h_{s_{i-1}^{\langle n\rangle}}^{\langle j_1\rangle}\,\mathrm{d}w_s^{\langle j_1\rangle})(\overline{\int}_{s_{i-1}^{\langle n\rangle}}^{s_i^{\langle n\rangle}} (h_s^{\langle j_2\rangle}$$

$$-h_{s_{i-1}^{\langle n\rangle}}^{\langle j_2\rangle})\,\mathrm{d}w_s^{\langle j_2\rangle})$$

$$+\frac{1}{2}\sum_{j_1=1}^{m}\sum_{\substack{j_2=1\\ j_2\neq j_1}}^{m}\sum_{i=1}^{2^n} U^{(2)}(\phi_{s_{i-1}^{\langle n\rangle}})(\overline{\int}_{s_{i-1}^{\langle n\rangle}}^{s_i^{\langle n\rangle}} h_{s_{i-1}^{\langle n\rangle}}^{\langle j_1\rangle}\, \mathrm{d}w_s^{\langle j_1\rangle})(\overline{\int}_{s_{i-1}^{\langle n\rangle}}^{s_i^{\langle n\rangle}} h_{s_{i-1}^{\langle n\rangle}}^{\langle j_2\rangle}\, \mathrm{d}w_s^{\langle j_2\rangle})$$
$$=: T_{4,n,1,1} + T_{4,n,1,2} + T_{4,n,2}$$

where the third equality follows from Theorem A.6, Lemma 14.99, and Proposition 14.120.

For $T_{4,n,1,1}$, we have, $\forall \omega \in \boldsymbol{\Omega}$,

$$\begin{aligned}
&\left\| T_{4,n,1,1}(\omega)\right\|_{\mathcal{Z}}\\
&\leq \frac{1}{2}\sum_{j_1=1}^{m}\sum_{j_2=1}^{m}\sum_{i=1}^{2^n}\Big\| U^{(2)}(\phi_{s_{i-1}^{\langle n\rangle}}(\omega))(\overline{\int}_{s_{i-1}^{\langle n\rangle}}^{s_i^{\langle n\rangle}} (h^{\langle j_1\rangle}(s,\omega)\\
&\quad -h^{\langle j_1\rangle}(s_{i-1}^{\langle n\rangle},\omega))\,\mathrm{d}w_{\langle\omega\rangle}^{\langle j_1\rangle}(s))(\phi^{\langle j_2\rangle}(s_i^{\langle n\rangle},\omega)-\phi^{\langle j_2\rangle}(s_{i-1}^{\langle n\rangle},\omega))\Big\|_{\mathcal{Z}}\\
&\leq \frac{1}{2}\sum_{j_1=1}^{m}\sum_{j_2=1}^{m}\sum_{i=1}^{2^n}\Big\| U^{(2)}(\phi_{s_{i-1}^{\langle n\rangle}}(\omega))\Big\|_{\mathrm{B}(\mathcal{Y},\mathrm{B}(\mathcal{Y},\mathcal{Z}))}\Big\|\overline{\int}_{s_{i-1}^{\langle n\rangle}}^{s_i^{\langle n\rangle}} (h^{\langle j_1\rangle}(s,\omega)\\
&\quad -h^{\langle j_1\rangle}(s_{i-1}^{\langle n\rangle},\omega))\,\mathrm{d}w_{\langle\omega\rangle}^{\langle j_1\rangle}(s)\Big\|_{\mathcal{Y}}\Big\|\overline{\int}_{s_{i-1}^{\langle n\rangle}}^{s_i^{\langle n\rangle}} h^{\langle j_2\rangle}(s,\omega)\,\mathrm{d}w_{\langle\omega\rangle}^{\langle j_2\rangle}(s))\Big\|_{\mathcal{Y}}\\
&\leq \frac{1}{2}M_2(\omega)\sum_{j_1=1}^{m}\sum_{j_2=1}^{m}\sum_{i=1}^{2^n}\Big\|\overline{\int}_{s_{i-1}^{\langle n\rangle}}^{s_i^{\langle n\rangle}} (h^{\langle j_1\rangle}(s,\omega)\\
&\quad -h^{\langle j_1\rangle}(s_{i-1}^{\langle n\rangle},\omega))\,\mathrm{d}w_{\langle\omega\rangle}^{\langle j_1\rangle}(s)\Big\|_{\mathcal{Y}}\Big\|\overline{\int}_{s_{i-1}^{\langle n\rangle}}^{s_i^{\langle n\rangle}} h^{\langle j_2\rangle}(s,\omega)\,\mathrm{d}w_{\langle\omega\rangle}^{\langle j_2\rangle}(s)\Big\|_{\mathcal{Y}}\\
&=: \frac{1}{2}M_2(\omega)Z_{4,n,1,1}
\end{aligned}$$

where the first inequality follows from the triangular inequality, the second inequality follows from Proposition 7.64, and the third inequality follows from the definition of M_2 in (14.26).

Note that

$$\begin{aligned}
&\boldsymbol{E}(Z_{4,n,1,1})\\
&=\sum_{j_1=1}^{m}\sum_{j_2=1}^{m}\sum_{i=1}^{2^n}\boldsymbol{E}(\Big\|\overline{\int}_{s_{i-1}^{\langle n\rangle}}^{s_i^{\langle n\rangle}} (h^{\langle j_1\rangle}(s,\omega)
\end{aligned}$$

$$-h^{\langle j_1\rangle}(s_{i-1}^{\langle n\rangle},\omega))\,\mathrm{d}w_{\langle\omega\rangle}^{\langle j_1\rangle}(s)\Big\|_{\mathcal{Y}}\Big\|\overline{\int}_{s_{i-1}^{\langle n\rangle}}^{s_i^{\langle n\rangle}} h^{\langle j_2\rangle}(s,\omega)\,\mathrm{d}w_{\langle\omega\rangle}^{\langle j_2\rangle}(s)\Big\|_{\mathcal{Y}})$$

$$\le \sum_{j_1=1}^{m}\sum_{j_2=1}^{m}\sum_{i=1}^{2^n}(\boldsymbol{E}(\Big\|\overline{\int}_{s_{i-1}^{\langle n\rangle}}^{s_i^{\langle n\rangle}}(h^{\langle j_1\rangle}(s,\omega)$$

$$-h^{\langle j_1\rangle}(s_{i-1}^{\langle n\rangle},\omega))\,\mathrm{d}w_{\langle\omega\rangle}^{\langle j_1\rangle}(s)\Big\|_{\mathcal{Y}}^2))^{\frac{1}{2}}(\boldsymbol{E}(\Big\|\overline{\int}_{s_{i-1}^{\langle n\rangle}}^{s_i^{\langle n\rangle}} h^{\langle j_2\rangle}(s,\omega)\,\mathrm{d}w_{\langle\omega\rangle}^{\langle j_2\rangle}(s)\Big\|_{\mathcal{Y}}^2))^{\frac{1}{2}}$$

$$= \sum_{j_1=1}^{m}\sum_{j_2=1}^{m}\sum_{i=1}^{2^n}\Big\|\overline{\int}_{s_{i-1}^{\langle n\rangle}}^{s_i^{\langle n\rangle}}(h^{\langle j_1\rangle}(s,\omega)-h^{\langle j_1\rangle}(s_{i-1}^{\langle n\rangle},\omega))$$

$$\cdot \mathrm{d}w_{\langle\omega\rangle}^{\langle j_1\rangle}(s)\Big\|_{\mathrm{L}_2(\boldsymbol{\Omega},\mathcal{Y})}\Big\|\overline{\int}_{s_{i-1}^{\langle n\rangle}}^{s_i^{\langle n\rangle}} h^{\langle j_2\rangle}(s,\omega)\,\mathrm{d}w_{\langle\omega\rangle}^{\langle j_2\rangle}(s)\Big\|_{\mathrm{L}_2(\boldsymbol{\Omega},\mathcal{Y})}$$

$$= \sum_{j_1=1}^{m}\sum_{j_2=1}^{m}\sum_{i=1}^{2^n}\|h^{\langle j_1\rangle}-\hat{h}_n^{\langle j_1\rangle}\|_{\bar{\mathrm{L}}_2(\bar{\mathrm{r}}_{s_{i-1}^{\langle n\rangle},s_i^{\langle n\rangle}}\times\boldsymbol{\Omega},\mathcal{Y})}$$

$$\cdot\|h^{\langle j_2\rangle}\|_{\mathrm{L}_2(\bar{\mathrm{r}}_{s_{i-1}^{\langle n\rangle},s_i^{\langle n\rangle}}\times\boldsymbol{\Omega},\mathcal{Y})}$$

$$\le \sum_{j_1=1}^{m}\sum_{j_2=1}^{m}(\sum_{i=1}^{2^n}\|h^{\langle j_1\rangle}-\hat{h}_n^{\langle j_1\rangle}\|_{\bar{\mathrm{L}}_2(\bar{\mathrm{r}}_{s_{i-1}^{\langle n\rangle},s_i^{\langle n\rangle}}\times\boldsymbol{\Omega},\mathcal{Y})}^2)^{\frac{1}{2}}$$

$$\cdot(\sum_{i=1}^{2^n}\|h^{\langle j_2\rangle}\|_{\mathrm{L}_2(\bar{\mathrm{r}}_{s_{i-1}^{\langle n\rangle},s_i^{\langle n\rangle}}\times\boldsymbol{\Omega},\mathcal{Y})}^2)^{\frac{1}{2}}$$

$$= \sum_{j_1=1}^{m}\sum_{j_2=1}^{m}(\|h^{\langle j_1\rangle}-\hat{h}_n^{\langle j_1\rangle}\|_{\bar{\mathrm{L}}_2(\bar{\mathrm{r}}_{a,t}\times\boldsymbol{\Omega},\mathcal{Y})}^2)^{\frac{1}{2}}(\|h^{\langle j_2\rangle}\|_{\mathrm{L}_2(\bar{\mathrm{r}}_{a,t}\times\boldsymbol{\Omega},\mathcal{Y})}^2)^{\frac{1}{2}}$$

$$\le \sum_{j_1=1}^{m}\sum_{j_2=1}^{m}\|h^{\langle j_1\rangle}-\hat{h}_n^{\langle j_1\rangle}\|_{\bar{\mathrm{L}}_2(\bar{\mathrm{r}}_{a,t}\times\boldsymbol{\Omega},\mathcal{Y})}\|h^{\langle j_2\rangle}\|_{\mathrm{L}_2(\bar{\mathrm{r}}_{a,b}\times\boldsymbol{\Omega},\mathcal{Y})}$$

$$\to 0 \text{ as } n\to\infty$$

where the first equality follows from the definition of $Z_{4,n,1,1}$ and Proposition 11.92, the first inequality follows from Cauchy–Schwarz Inequality, the second equality follows from Example 11.173, the third equality follows from Proposition 14.106

and the definition

$$\hat{h}_n^{\langle j\rangle}(s,\omega) = \sum_{i=1}^{2^n} h^{\langle j\rangle}(s_{i-1}^{\langle n\rangle},\omega)\cdot \chi_{[s_{i-1}^{\langle n\rangle},s_i^{\langle n\rangle}),\bar{\mathrm{r}}_{a,b}}(s); \qquad (14.27)$$
$$\forall (s,\omega)\in \bar{\mathrm{r}}_{a,t}\times \boldsymbol{\Omega};\ j=1,\ldots,m$$

the second inequality follows from Cauchy–Schwarz Inequality, the fourth equality follows from Example 11.173, the third inequality follows from simple algebra, and the convergence statement follows from $\lim_{n\to\infty} \hat{h}_n^{\langle j\rangle} \doteq h^{\langle j\rangle}$ in $\bar{\mathrm{L}}_2(\bar{\mathrm{r}}_{a,t}\times\boldsymbol{\Omega},\mathcal{Y})$, $j=1,\ldots,m$, by (c) and Proposition 14.107. Then, by Proposition 11.211, $\lim_{n\in\mathbb{N}} Z_{4,n,1,1} = 0$ in measure in $\boldsymbol{\Omega}$. Hence, by Proposition 14.111, $\lim_{n\in\mathbb{N}} T_{4,n,1,1} = \vartheta_{\mathcal{Z}}$ in measure in $\boldsymbol{\Omega}$.

For $T_{4,n,1,2}$, we have, $\forall \omega\in\boldsymbol{\Omega}$,

$$\begin{aligned}
&\|T_{4,n,1,2}(\omega)\|_{\mathcal{Z}}\\
&\leq \frac{1}{2}\sum_{j_1=1}^{m}\sum_{j_2=1}^{m}\sum_{i=1}^{2^n}\Big\|U^{(2)}(\phi_{s_{i-1}^{\langle n\rangle}}(\omega))\Big(\overline{\int_{s_{i-1}^{\langle n\rangle}}^{s_i^{\langle n\rangle}}} h^{\langle j_1\rangle}(s_{i-1}^{\langle n\rangle},\omega)\,\mathrm{d}w_{\langle\omega\rangle}^{\langle j_1\rangle}(s)\Big)\\
&\quad\cdot\Big(\overline{\int_{s_{i-1}^{\langle n\rangle}}^{s_i^{\langle n\rangle}}}(h^{\langle j_2\rangle}(s,\omega)-h^{\langle j_2\rangle}(s_{i-1}^{\langle n\rangle},\omega))\,\mathrm{d}w_{\langle\omega\rangle}^{\langle j_2\rangle}(s)\Big)\Big\|_{\mathcal{Z}}\\
&\leq \frac{1}{2}\sum_{j_1=1}^{m}\sum_{j_2=1}^{m}\sum_{i=1}^{2^n}\Big\|U^{(2)}(\phi_{s_{i-1}^{\langle n\rangle}}(\omega))\Big\|_{\mathrm{B}(\mathcal{Y},\mathrm{B}(\mathcal{Y},\mathcal{Z}))}\Big\|\overline{\int_{s_{i-1}^{\langle n\rangle}}^{s_i^{\langle n\rangle}}}(h^{\langle j_2\rangle}(s,\omega)\\
&\quad -h^{\langle j_2\rangle}(s_{i-1}^{\langle n\rangle},\omega))\,\mathrm{d}w_{\langle\omega\rangle}^{\langle j_2\rangle}(s)\Big\|_{\mathcal{Y}}\Big\|\overline{\int_{s_{i-1}^{\langle n\rangle}}^{s_i^{\langle n\rangle}}} h^{\langle j_1\rangle}(s_{i-1}^{\langle n\rangle},\omega)\,\mathrm{d}w_{\langle\omega\rangle}^{\langle j_1\rangle}(s)\Big\|_{\mathcal{Y}}\\
&\leq \frac{1}{2}\sum_{j_1=1}^{m}\sum_{j_2=1}^{m}\sum_{i=1}^{2^n} M_2(\omega)\Big\|\overline{\int_{s_{i-1}^{\langle n\rangle}}^{s_i^{\langle n\rangle}}}(h^{\langle j_2\rangle}(s,\omega)\\
&\quad -h^{\langle j_2\rangle}(s_{i-1}^{\langle n\rangle},\omega))\,\mathrm{d}w_{\langle\omega\rangle}^{\langle j_2\rangle}(s)\Big\|_{\mathcal{Y}}\Big\|\overline{\int_{s_{i-1}^{\langle n\rangle}}^{s_i^{\langle n\rangle}}} h^{\langle j_1\rangle}(s_{i-1}^{\langle n\rangle},\omega)\,\mathrm{d}w_{\langle\omega\rangle}^{\langle j_1\rangle}(s)\Big\|_{\mathcal{Y}}\\
&=: \frac{1}{2}M_2(\omega)Z_{4,n,1,2}(\omega)
\end{aligned}$$

where the first inequality follows from the triangular inequality, the second inequality follows from Proposition 7.64, and the third inequality follows from the definition of M_2 in (14.26).

Note that

$$
\begin{aligned}
&\boldsymbol{E}(Z_{4,n,1,2}) \\
&= \sum_{j_1=1}^{m}\sum_{j_2=1}^{m}\sum_{i=1}^{2^n}\boldsymbol{E}\Big(\Big\|\overline{\int}_{s_{i-1}^{\langle n\rangle}}^{s_i^{\langle n\rangle}}(h^{\langle j_2\rangle}(s,\omega) \\
&\quad -h^{\langle j_2\rangle}(s_{i-1}^{\langle n\rangle},\omega))\,\mathrm{d}w_{\langle\omega\rangle}^{\langle j_2\rangle}(s)\Big\|_{\mathcal{Y}}\Big\|\overline{\int}_{s_{i-1}^{\langle n\rangle}}^{s_i^{\langle n\rangle}} h^{\langle j_1\rangle}(s_{i-1}^{\langle n\rangle},\omega)\,\mathrm{d}w_{\langle\omega\rangle}^{\langle j_1\rangle}(s)\Big\|_{\mathcal{Y}}\Big) \\
&\leq \sum_{j_1=1}^{m}\sum_{j_2=1}^{m}\sum_{i=1}^{2^n}\Big(\boldsymbol{E}\Big(\Big\|\overline{\int}_{s_{i-1}^{\langle n\rangle}}^{s_i^{\langle n\rangle}}(h^{\langle j_2\rangle}(s,\omega)-h^{\langle j_2\rangle}(s_{i-1}^{\langle n\rangle},\omega)) \\
&\quad \cdot\mathrm{d}w_{\langle\omega\rangle}^{\langle j_2\rangle}(s)\Big\|_{\mathcal{Y}}^2\Big)\Big)^{\frac{1}{2}}\Big(\boldsymbol{E}\Big(\Big\|\overline{\int}_{s_{i-1}^{\langle n\rangle}}^{s_i^{\langle n\rangle}} h^{\langle j_1\rangle}(s_{i-1}^{\langle n\rangle},\omega)\,\mathrm{d}w_{\langle\omega\rangle}^{\langle j_1\rangle}(s)\Big\|_{\mathcal{Y}}^2\Big)\Big)^{\frac{1}{2}} \\
&= \sum_{j_1=1}^{m}\sum_{j_2=1}^{m}\sum_{i=1}^{2^n}\Big\|\overline{\int}_{s_{i-1}^{\langle n\rangle}}^{s_i^{\langle n\rangle}}(h^{\langle j_2\rangle}(s,\omega)-h^{\langle j_2\rangle}(s_{i-1}^{\langle n\rangle},\omega))\,\mathrm{d}w_{\langle\omega\rangle}^{\langle j_2\rangle}(s)\Big\|_{\mathrm{L}_2(\boldsymbol{\Omega},\mathcal{Y})} \\
&\quad \cdot\Big\|\overline{\int}_{s_{i-1}^{\langle n\rangle}}^{s_i^{\langle n\rangle}} h^{\langle j_1\rangle}(s_{i-1}^{\langle n\rangle},\omega)\,\mathrm{d}w_{\langle\omega\rangle}^{\langle j_1\rangle}(s)\Big\|_{\mathrm{L}_2(\boldsymbol{\Omega},\mathcal{Y})} \\
&= \sum_{j_1=1}^{m}\sum_{j_2=1}^{m}\sum_{i=1}^{2^n}\Big\|h^{\langle j_2\rangle}-\hat{h}_n^{\langle j_2\rangle}\Big\|_{\bar{\mathrm{L}}_2(\bar{\mathrm{r}}_{s_{i-1}^{\langle n\rangle},s_i^{\langle n\rangle}}\times\boldsymbol{\Omega},\mathcal{Y})} \\
&\quad \cdot\Big\|\hat{h}_n^{\langle j_1\rangle}\Big\|_{\bar{\mathrm{L}}_2(\bar{\mathrm{r}}_{s_{i-1}^{\langle n\rangle},s_i^{\langle n\rangle}}\times\boldsymbol{\Omega},\mathcal{Y})} \\
&\leq \sum_{j_1=1}^{m}\sum_{j_2=1}^{m}\Big(\sum_{i=1}^{2^n}\Big\|h^{\langle j_2\rangle}-\hat{h}_n^{\langle j_2\rangle}\Big\|_{\bar{\mathrm{L}}_2(\bar{\mathrm{r}}_{s_{i-1}^{\langle n\rangle},s_i^{\langle n\rangle}}\times\boldsymbol{\Omega},\mathcal{Y})}^2\Big)^{\frac{1}{2}} \\
&\quad \cdot\Big(\sum_{i=1}^{2^n}\Big\|\hat{h}_n^{\langle j_1\rangle}\Big\|_{\bar{\mathrm{L}}_2(\bar{\mathrm{r}}_{s_{i-1}^{\langle n\rangle},s_i^{\langle n\rangle}}\times\boldsymbol{\Omega},\mathcal{Y})}^2\Big)^{\frac{1}{2}} \\
&= \sum_{j_1=1}^{m}\sum_{j_2=1}^{m}\Big(\Big\|h^{\langle j_2\rangle}-\hat{h}_n^{\langle j_2\rangle}\Big\|_{\bar{\mathrm{L}}_2(\bar{\mathrm{r}}_{a,t}\times\boldsymbol{\Omega},\mathcal{Y})}^2\Big)^{\frac{1}{2}}\Big(\Big\|\hat{h}_n^{\langle j_1\rangle}\Big\|_{\mathrm{L}_2(\bar{\mathrm{r}}_{a,t}\times\boldsymbol{\Omega},\mathcal{Y})}^2\Big)^{\frac{1}{2}} \\
&\to 0 \text{ as } n\to\infty
\end{aligned}
$$

where the first equality follows from the definition of $Z_{4,n,1,2}$ and Proposition 11.92, the first inequality follows from Cauchy–Schwarz Inequality, the second equality follows from Example 11.173, the third equality follows from Proposition 14.106 and the definition $\hat{h}_n^{\langle j\rangle}$ in (14.27), the second inequality follows from Cauchy–Schwarz Inequality, the fourth equality follows from Example 11.173, and the convergence statement follows from the assumption (c) and Proposition 14.107 that yields $\lim_{n\to\infty} \hat{h}_n^{\langle j\rangle} \doteq h^{\langle j\rangle}$ in $\bar{\mathrm{L}}_2(\bar{\mathrm{r}}_{a,t} \times \boldsymbol{\Omega}, \mathcal{Y})$, $j = 1, \ldots, m$. This implies that $\lim_{n\in\mathbb{N}} Z_{4,n,1,2} = 0$ in measure in $\boldsymbol{\Omega}$, by Proposition 11.211. Hence, by Proposition 14.111, $\lim_{n\in\mathbb{N}} T_{4,n,1,2} = \vartheta_{\mathcal{Z}}$ in measure in $\boldsymbol{\Omega}$.

Note that $T_{4,n,2}$ is $\mathcal{B}_t$-measurable. Then, we have

$$\begin{aligned}
&\boldsymbol{E}(\| T_{4,n,2}\|_{\mathcal{Z}}^2) \\
&= \frac{1}{4}\boldsymbol{E}\Big(\Big\langle \sum_{j_1=1}^{m} \sum_{\substack{j_2=1\\ j_2\neq j_1}}^{m} \sum_{i=1}^{2^n} U^{(2)}(\phi_{s_{i-1}^{\langle n\rangle}})\Big(\overline{\int}_{s_{i-1}^{\langle n\rangle}}^{s_i^{\langle n\rangle}} h^{\langle j_1\rangle}_{s_{i-1}^{\langle n\rangle}}\,\mathrm{d}w_s^{\langle j_1\rangle}\Big) \\
&\quad\cdot\Big(\overline{\int}_{s_{i-1}^{\langle n\rangle}}^{s_i^{\langle n\rangle}} h^{\langle j_2\rangle}_{s_{i-1}^{\langle n\rangle}}\,\mathrm{d}w_s^{\langle j_2\rangle}\Big), \sum_{k_1=1}^{m} \sum_{\substack{k_2=1\\ k_2\neq k_1}}^{m} \sum_{i=1}^{2^n} U^{(2)}(\phi_{s_{i-1}^{\langle n\rangle}})\Big(\overline{\int}_{s_{i-1}^{\langle n\rangle}}^{s_i^{\langle n\rangle}} h^{\langle k_1\rangle}_{s_{i-1}^{\langle n\rangle}}\,\mathrm{d}w_s^{\langle k_1\rangle}\Big) \\
&\quad\cdot\Big(\overline{\int}_{s_{i-1}^{\langle n\rangle}}^{s_i^{\langle n\rangle}} h^{\langle k_2\rangle}_{s_{i-1}^{\langle n\rangle}}\,\mathrm{d}w_s^{\langle k_2\rangle}\Big)\Big\rangle_{\mathcal{Z}}\Big) \\
&= \frac{1}{4}\boldsymbol{E}\Big(\Big\langle \sum_{j_1=1}^{m} \sum_{\substack{j_2=1\\ j_2\neq j_1}}^{m} \sum_{i=1}^{2^n} U^{(2)}(\phi_{s_{i-1}^{\langle n\rangle}})(h^{\langle j_1\rangle}_{s_{i-1}^{\langle n\rangle}})(h^{\langle j_2\rangle}_{s_{i-1}^{\langle n\rangle}})(w^{\langle j_1\rangle}_{s_i^{\langle n\rangle}} - w^{\langle j_1\rangle}_{s_{i-1}^{\langle n\rangle}}) \\
&\quad\cdot(w^{\langle j_2\rangle}_{s_i^{\langle n\rangle}} - w^{\langle j_2\rangle}_{s_{i-1}^{\langle n\rangle}}), \sum_{k_1=1}^{m} \sum_{\substack{k_2=1\\ k_2\neq k_1}}^{m} \sum_{i=1}^{2^n} U^{(2)}(\phi_{s_{i-1}^{\langle n\rangle}})(h^{\langle k_1\rangle}_{s_{i-1}^{\langle n\rangle}})(h^{\langle k_2\rangle}_{s_{i-1}^{\langle n\rangle}}) \\
&\quad\cdot(w^{\langle k_1\rangle}_{s_i^{\langle n\rangle}} - w^{\langle k_1\rangle}_{s_{i-1}^{\langle n\rangle}})(w^{\langle k_2\rangle}_{s_i^{\langle n\rangle}} - w^{\langle k_2\rangle}_{s_{i-1}^{\langle n\rangle}})\Big\rangle_{\mathcal{Z}}\Big) \\
&= \frac{1}{4}\sum_{k_1=1}^{m} \sum_{\substack{k_2=1\\ k_2\neq k_1}}^{m} \sum_{i=1}^{2^n} \sum_{k_3=1}^{m} \sum_{\substack{k_4=1\\ k_4\neq k_3}}^{m} \sum_{j=1}^{2^n} \boldsymbol{E}\Big(\Big\langle U^{(2)}(\phi_{s_{i-1}^{\langle n\rangle}})(h^{\langle k_1\rangle}_{s_{i-1}^{\langle n\rangle}})(h^{\langle k_2\rangle}_{s_{i-1}^{\langle n\rangle}}) \\
&\quad\cdot(w^{\langle k_1\rangle}_{s_i^{\langle n\rangle}} - w^{\langle k_1\rangle}_{s_{i-1}^{\langle n\rangle}})(w^{\langle k_2\rangle}_{s_i^{\langle n\rangle}} - w^{\langle k_2\rangle}_{s_{i-1}^{\langle n\rangle}}), U^{(2)}(\phi_{s_{j-1}^{\langle n\rangle}})(h^{\langle k_3\rangle}_{s_{j-1}^{\langle n\rangle}})(h^{\langle k_4\rangle}_{s_{j-1}^{\langle n\rangle}}) \\
&\quad\cdot(w^{\langle k_3\rangle}_{s_j^{\langle n\rangle}} - w^{\langle k_3\rangle}_{s_{j-1}^{\langle n\rangle}})(w^{\langle k_4\rangle}_{s_j^{\langle n\rangle}} - w^{\langle k_4\rangle}_{s_{j-1}^{\langle n\rangle}})\Big\rangle_{\mathcal{Z}}\Big) \\
&= \frac{1}{4}\sum_{k_1=1}^{m} \sum_{\substack{k_2=1\\ k_2\neq k_1}}^{m} \sum_{i=1}^{2^n} \sum_{k_3=1}^{m} \sum_{\substack{k_4=1\\ k_4\neq k_3}}^{m} \sum_{j=1}^{2^n} \boldsymbol{E}\Big(\Big\langle U^{(2)}(\phi_{s_{i-1}^{\langle n\rangle}})(h^{\langle k_1\rangle}_{s_{i-1}^{\langle n\rangle}})(h^{\langle k_2\rangle}_{s_{i-1}^{\langle n\rangle}}),
\end{aligned}$$

$$U^{(2)}(\phi_{s_{j-1}^{\langle n\rangle}})(h^{\langle k_3\rangle}_{s_{j-1}^{\langle n\rangle}})(h^{\langle k_4\rangle}_{s_{j-1}^{\langle n\rangle}})\Big\rangle_{\mathcal{Z}}(w^{\langle k_3\rangle}_{s_j^{\langle n\rangle}} - w^{\langle k_3\rangle}_{s_{j-1}^{\langle n\rangle}})(w^{\langle k_4\rangle}_{s_j^{\langle n\rangle}} - w^{\langle k_4\rangle}_{s_{j-1}^{\langle n\rangle}})$$

$$\cdot(w^{\langle k_1\rangle}_{s_i^{\langle n\rangle}} - w^{\langle k_1\rangle}_{s_{i-1}^{\langle n\rangle}})(w^{\langle k_2\rangle}_{s_i^{\langle n\rangle}} - w^{\langle k_2\rangle}_{s_{i-1}^{\langle n\rangle}})\Big)$$

$$= \frac{1}{4}\sum_{k_1=1}^{m}\sum_{\substack{k_2=1\\k_2\neq k_1}}^{m}\sum_{i=1}^{2^n}\sum_{k_3=1}^{m}\sum_{\substack{k_4=1\\k_4\neq k_3}}^{m}\sum_{j=1}^{2^n}\boldsymbol{E}\Big(\boldsymbol{E}\Big(\Big\langle U^{(2)}(\phi_{s_{i-1}^{\langle n\rangle}})(h^{\langle k_1\rangle}_{s_{i-1}^{\langle n\rangle}})(h^{\langle k_2\rangle}_{s_{i-1}^{\langle n\rangle}}),$$

$$U^{(2)}(\phi_{s_{j-1}^{\langle n\rangle}})(h^{\langle k_3\rangle}_{s_{j-1}^{\langle n\rangle}})(h^{\langle k_4\rangle}_{s_{j-1}^{\langle n\rangle}})\Big\rangle_{\mathcal{Z}}(w^{\langle k_3\rangle}_{s_j^{\langle n\rangle}} - w^{\langle k_3\rangle}_{s_{j-1}^{\langle n\rangle}})(w^{\langle k_4\rangle}_{s_j^{\langle n\rangle}} - w^{\langle k_4\rangle}_{s_{j-1}^{\langle n\rangle}})$$

$$\cdot(w^{\langle k_1\rangle}_{s_i^{\langle n\rangle}} - w^{\langle k_1\rangle}_{s_{i-1}^{\langle n\rangle}})(w^{\langle k_2\rangle}_{s_i^{\langle n\rangle}} - w^{\langle k_2\rangle}_{s_{i-1}^{\langle n\rangle}})|\mathcal{B}_{s^{\langle n\rangle}_{\max\{i,j\}-1}}\Big)\Big)$$

$$= \frac{1}{4}\sum_{k_1=1}^{m}\sum_{\substack{k_2=1\\k_2\neq k_1}}^{m}\sum_{i=1}^{2^n}\sum_{k_3=1}^{m}\sum_{\substack{k_4=1\\k_4\neq k_3}}^{m}\boldsymbol{E}\Big(\boldsymbol{E}\Big(\Big\langle U^{(2)}(\phi_{s_{i-1}^{\langle n\rangle}})(h^{\langle k_1\rangle}_{s_{i-1}^{\langle n\rangle}})(h^{\langle k_2\rangle}_{s_{i-1}^{\langle n\rangle}}),$$

$$U^{(2)}(\phi_{s_{i-1}^{\langle n\rangle}})(h^{\langle k_3\rangle}_{s_{i-1}^{\langle n\rangle}})(h^{\langle k_4\rangle}_{s_{i-1}^{\langle n\rangle}})\Big\rangle_{\mathcal{Z}}(w^{\langle k_1\rangle}_{s_i^{\langle n\rangle}} - w^{\langle k_1\rangle}_{s_{i-1}^{\langle n\rangle}})(w^{\langle k_2\rangle}_{s_i^{\langle n\rangle}} - w^{\langle k_2\rangle}_{s_{i-1}^{\langle n\rangle}})$$

$$\cdot(w^{\langle k_3\rangle}_{s_i^{\langle n\rangle}} - w^{\langle k_3\rangle}_{s_{i-1}^{\langle n\rangle}})(w^{\langle k_4\rangle}_{s_i^{\langle n\rangle}} - w^{\langle k_4\rangle}_{s_{i-1}^{\langle n\rangle}})|\mathcal{B}_{s_{i-1}^{\langle n\rangle}}\Big)\Big)$$

$$= \frac{1}{4}\sum_{i=1}^{2^n}\sum_{k_1=1}^{m}\sum_{\substack{k_2=1\\k_2\neq k_1}}^{m}\sum_{k_3=1}^{m}\sum_{\substack{k_4=1\\k_4\neq k_3}}^{m}\boldsymbol{E}\Big(\Big\langle U^{(2)}(\phi_{s_{i-1}^{\langle n\rangle}})(h^{\langle k_1\rangle}_{s_{i-1}^{\langle n\rangle}})(h^{\langle k_2\rangle}_{s_{i-1}^{\langle n\rangle}}),$$

$$U^{(2)}(\phi_{s_{i-1}^{\langle n\rangle}})(h^{\langle k_3\rangle}_{s_{i-1}^{\langle n\rangle}})(h^{\langle k_4\rangle}_{s_{i-1}^{\langle n\rangle}})\Big\rangle_{\mathcal{Z}}\boldsymbol{E}\Big((w^{\langle k_1\rangle}_{s_i^{\langle n\rangle}} - w^{\langle k_1\rangle}_{s_{i-1}^{\langle n\rangle}})(w^{\langle k_2\rangle}_{s_i^{\langle n\rangle}} - w^{\langle k_2\rangle}_{s_{i-1}^{\langle n\rangle}})$$

$$\cdot(w^{\langle k_3\rangle}_{s_i^{\langle n\rangle}} - w^{\langle k_3\rangle}_{s_{i-1}^{\langle n\rangle}})(w^{\langle k_4\rangle}_{s_i^{\langle n\rangle}} - w^{\langle k_4\rangle}_{s_{i-1}^{\langle n\rangle}})|\mathcal{B}_{s_{i-1}^{\langle n\rangle}}\Big)\Big)$$

$$= \frac{1}{2}\sum_{i=1}^{2^n}\sum_{k_1=1}^{m}\sum_{\substack{k_2=1\\k_2\neq k_1}}^{m}\boldsymbol{E}\Big(\Big\|U^{(2)}(\phi_{s_{i-1}^{\langle n\rangle}})(h^{\langle k_1\rangle}_{s_{i-1}^{\langle n\rangle}})(h^{\langle k_2\rangle}_{s_{i-1}^{\langle n\rangle}})\Big\|_{\mathcal{Z}}^2$$

$$\cdot\boldsymbol{E}\Big((w^{\langle k_1\rangle}_{s_i^{\langle n\rangle}} - w^{\langle k_1\rangle}_{s_{i-1}^{\langle n\rangle}})^2(w^{\langle k_2\rangle}_{s_i^{\langle n\rangle}} - w^{\langle k_2\rangle}_{s_{i-1}^{\langle n\rangle}})^2|\mathcal{B}_{s_{i-1}^{\langle n\rangle}}\Big)\Big)$$

$$= \frac{1}{2}\sum_{i=1}^{2^n}\sum_{k_1=1}^{m}\sum_{\substack{k_2=1\\k_2\neq k_1}}^{m}\boldsymbol{E}\Big(\Big\|U^{(2)}(\phi_{s_{i-1}^{\langle n\rangle}})(h^{\langle k_1\rangle}_{s_{i-1}^{\langle n\rangle}})(h^{\langle k_2\rangle}_{s_{i-1}^{\langle n\rangle}})\Big\|_{\mathcal{Z}}^2$$

$$\cdot\boldsymbol{E}\Big((w^{\langle k_1\rangle}_{s_i^{\langle n\rangle}} - w^{\langle k_1\rangle}_{s_{i-1}^{\langle n\rangle}})^2|\mathcal{B}_{s_{i-1}^{\langle n\rangle}}\Big)\boldsymbol{E}\Big((w^{\langle k_2\rangle}_{s_i^{\langle n\rangle}} - w^{\langle k_2\rangle}_{s_{i-1}^{\langle n\rangle}})^2|\mathcal{B}_{s_{i-1}^{\langle n\rangle}}\Big)\Big)$$

$$= \frac{1}{2} \sum_{i=1}^{2^n} \sum_{k_1=1}^{m} \sum_{\substack{k_2=1\\k_2\neq k_1}}^{m} \boldsymbol{E}\Big(\Big\| U^{(2)}(\phi_{s_{i-1}^{\langle n\rangle}})(h_{s_{i-1}^{\langle n\rangle}}^{\langle k_1\rangle})(h_{s_{i-1}^{\langle n\rangle}}^{\langle k_2\rangle}) \Big\|_{\mathcal{Z}}^2 (s_i^{\langle n\rangle} - s_{i-1}^{\langle n\rangle})^2 \Big)$$

$$\leq \frac{b-a}{2^{n+1}} \sum_{k_1=1}^{m} \sum_{\substack{k_2=1\\k_2\neq k_1}}^{m} \sum_{i=1}^{2^n} \boldsymbol{E}\Big(\Big\| U^{(2)}(\phi_{s_{i-1}^{\langle n\rangle}})(h_{s_{i-1}^{\langle n\rangle}}^{\langle k_1\rangle})(h_{s_{i-1}^{\langle n\rangle}}^{\langle k_2\rangle}) \Big\|_{\mathcal{Z}}^2 (s_i^{\langle n\rangle} - s_{i-1}^{\langle n\rangle}) \Big)$$

$$\to 0 \text{ as } n \to \infty$$

where the first equality follows from the definition of $T_{4,n,2}$, the second equality follows from Definition A.4 or Lemma 14.99, the third equality follows from Proposition 11.92, the fourth equality follows from Definition 13.1, the fifth equality follows from (a) of Proposition 14.11, the sixth equality follows from Definition 14.78, the seventh equality follows from (h) of Proposition 14.11, the eighth equality follows from Definition 14.78 and Proposition 14.9, the ninth equality follows from Definition 14.78 and Proposition 14.9, the tenth equality follows from Definition 14.78, the inequality follows from the definition of $s_i^{\langle n\rangle}$, and the convergence statement follows from the assumption (ii), Proposition 14.107, and Lebesgue Dominated Convergence Theorem 11.91. Hence, we have $\lim_{n\in\mathbb{N}} T_{4,n,2} \doteq \vartheta_{\bar{\mathrm{L}}_2(\boldsymbol{\Omega},\mathcal{Z})}$ in $\bar{\mathrm{L}}_2(\boldsymbol{\Omega},\mathcal{Z})$. By Proposition 11.211, we have $\lim_{n\in\mathbb{N}} T_{4,n,2} = \vartheta_{\mathcal{Z}}$ in measure in $\boldsymbol{\Omega}$.

Next, we consider $T_{5,n}$. Note that

$$|||T_{5,n}||| \leq \sum_{j=1}^{m} |||\Big(\sum_{i=1}^{2^n} U^{(1)}(\phi_{s_{i-1}^{\langle n\rangle}})\Big(\overline{\int}_{s_{i-1}^{\langle n\rangle}}^{s_i^{\langle n\rangle}} h_s^{\langle j\rangle}\, \mathrm{d}w_s^{\langle j\rangle} \Big) - l_t^{\langle j\rangle} \Big)|||$$

$$= \sum_{j=1}^{m} |||\Big(\sum_{i=1}^{2^n} \overline{\int}_{s_{i-1}^{\langle n\rangle}}^{s_i^{\langle n\rangle}} U^{(1)}(\phi_{s_{i-1}^{\langle n\rangle}}) h_s^{\langle j\rangle}\, \mathrm{d}w_s^{\langle j\rangle} - l_t^{\langle j\rangle} \Big)|||$$

$$= \sum_{j=1}^{m} |||\Big(\overline{\int}_{a}^{t} (\tilde{h}_n^{\langle j\rangle})_s\, \mathrm{d}w_s^{\langle j\rangle} - l_t^{\langle j\rangle} \Big)|||$$

$$= \sum_{j=1}^{m} |||\Big(\overline{\int}_{a}^{t} ((\tilde{h}_n^{\langle j\rangle})_s - U^{(1)}(\phi_s) h_s^{\langle j\rangle})\, \mathrm{d}w_s^{\langle j\rangle} \Big)|||$$

$$\leq \sum_{j=1}^{m} 3 ||| \Big\| \tilde{h}_n^{\langle j\rangle} - U^{(1)}(\phi) h^{\langle j\rangle} \Big\|_{\bar{\mathrm{L}}_2(\bar{\mathrm{r}}_{a,t},\mathcal{Y})} ||| \to 0 \text{ as } n \to \infty$$

where the first inequality follows from Lemma 14.113 and the definition of $T_{5,n}$, the first equality follows from (vi) of Proposition 14.120 and assumption (iii), the second equality follows from the definition $\tilde{h}_n^{\langle j\rangle}(s,\omega) := \sum_{i=1}^{2^n} U^{(1)}(\phi(s_{i-1}^{\langle n\rangle},\omega))\chi_{[s_{i-1}^{\langle n\rangle},s_i^{\langle n\rangle}),\bar{\mathrm{r}}_{a,b}}(s) h^{\langle j\rangle}(s,\omega)$, $\forall (s,\omega) \in \bar{\mathrm{r}}_{a,t} \times \boldsymbol{\Omega}$, $j =$

$1, \ldots, m$, and (iv) of Proposition 14.106, the third equality follows from the definition of $l^{\langle j\rangle}$ and (ii) of Proposition 14.120, the second inequality follows from (i) of Proposition 14.120, and the convergence statement follows since Theorem A.18 and Bounded Convergence Theorems 11.77 and A.15 implies that $\|\tilde{h}_n^{\langle j\rangle} - U^{(1)}(\phi)h^{\langle j\rangle}\|_{\bar{\mathrm{L}}_2(\bar{\mathrm{r}}_{a,t},\mathcal{Y})} \to 0$ as $n \to \infty$, $\forall \omega \in \boldsymbol{\Omega}$, and then the result follows from Lebesgue Dominated Convergence Theorem 11.91. By Lemma 14.113, we have $\lim_{n\in\mathbb{N}} T_{5,n} = \vartheta_{\mathcal{Z}}$ in measure in $\boldsymbol{\Omega}$.

Fix any $\omega \in \boldsymbol{\Omega}$, we define

$$\check{\delta}_n(\omega) := \sup_{\substack{u,v\in\bar{\mathrm{r}}_{a,b},\alpha\in\bar{\mathrm{r}}_{0,1}\\ |u-v|\leq 2^{-n}(b-a)}} \big\| U^{(2)}(\phi(u,\omega) + \alpha(\phi(v,\omega) - \phi(u,\omega))) - U^{(2)}(\phi(u,\omega))\big\|_{\mathrm{B}(\mathcal{Y},\mathrm{B}(\mathcal{Y},\mathcal{Z}))} \tag{14.28}$$

By Proposition 14.109, $\check{\delta}_n : \boldsymbol{\Omega} \to \overline{\mathbb{R}_+}$, $n \in \mathbb{N}$, are $\mathcal{B}$-measurable. Since $\phi_{\langle\omega\rangle} : \bar{\mathrm{r}}_{a,b} \to \mathcal{Y}$ is uniformly continuous, $U^{(2)}$ is continuous, and D is convex open, we have $\lim_{n\in\mathbb{N}} \check{\delta}_n(\omega) = 0$, $\forall \omega \in \boldsymbol{\Omega}$.

Finally, we consider $T_{6,n}$. $\forall \omega \in \boldsymbol{\Omega}$,

$$\begin{aligned}
&\big\| T_{6,n}(\omega)\big\|_{\mathcal{Z}} \\
&= \Big\| \sum_{j_1=1}^{m}\sum_{j_2=1}^{m}\sum_{i=1}^{2^n} \int_0^1\int_0^{\bar{s}} (U^{(2)}(\phi_{s_{i-1}^{\langle n\rangle}} + \tau(\phi_{s_i^{\langle n\rangle}} - \phi_{s_{i-1}^{\langle n\rangle}})) \\
&\quad - U^{(2)}(\phi_{s_{i-1}^{\langle n\rangle}}))(\phi^{\langle j_1\rangle}_{s_i^{\langle n\rangle}} - \phi^{\langle j_1\rangle}_{s_{i-1}^{\langle n\rangle}})(\phi^{\langle j_2\rangle}_{s_i^{\langle n\rangle}} - \phi^{\langle j_2\rangle}_{s_{i-1}^{\langle n\rangle}})\,\mathrm{d}\tau\,\mathrm{d}\bar{s}\Big\|_{\mathcal{Z}} \\
&\leq \sum_{j_1=1}^{m}\sum_{j_2=1}^{m}\sum_{i=1}^{2^n} \Big\| \int_0^1\int_0^{\bar{s}} (U^{(2)}(\phi_{s_{i-1}^{\langle n\rangle}} + \tau(\phi_{s_i^{\langle n\rangle}} - \phi_{s_{i-1}^{\langle n\rangle}})) \\
&\quad - U^{(2)}(\phi_{s_{i-1}^{\langle n\rangle}}))(\phi^{\langle j_1\rangle}_{s_i^{\langle n\rangle}} - \phi^{\langle j_1\rangle}_{s_{i-1}^{\langle n\rangle}})(\phi^{\langle j_2\rangle}_{s_i^{\langle n\rangle}} - \phi^{\langle j_2\rangle}_{s_{i-1}^{\langle n\rangle}})\,\mathrm{d}\tau\,\mathrm{d}\bar{s}\Big\|_{\mathcal{Z}} \\
&\leq \sum_{j_1=1}^{m}\sum_{j_2=1}^{m}\sum_{i=1}^{2^n} \int_0^1\int_0^{\bar{s}} \Big\| (U^{(2)}(\phi_{s_{i-1}^{\langle n\rangle}} + \tau(\phi_{s_i^{\langle n\rangle}} - \phi_{s_{i-1}^{\langle n\rangle}})) \\
&\quad - U^{(2)}(\phi_{s_{i-1}^{\langle n\rangle}}))(\phi^{\langle j_1\rangle}_{s_i^{\langle n\rangle}} - \phi^{\langle j_1\rangle}_{s_{i-1}^{\langle n\rangle}})(\phi^{\langle j_2\rangle}_{s_i^{\langle n\rangle}} - \phi^{\langle j_2\rangle}_{s_{i-1}^{\langle n\rangle}})\Big\|_{\mathcal{Z}}\,\mathrm{d}\tau\,\mathrm{d}\bar{s} \\
&\leq \sum_{j_1=1}^{m}\sum_{j_2=1}^{m}\sum_{i=1}^{2^n} \int_0^1\int_0^{\bar{s}} \Big\| U^{(2)}(\phi_{s_{i-1}^{\langle n\rangle}} + \tau(\phi_{s_i^{\langle n\rangle}} - \phi_{s_{i-1}^{\langle n\rangle}})) \\
&\quad - U^{(2)}(\phi_{s_{i-1}^{\langle n\rangle}})\Big\|_{\mathrm{B}(\mathcal{Y},\mathrm{B}(\mathcal{Y},\mathcal{Z}))} \Big\|\phi^{\langle j_1\rangle}_{s_i^{\langle n\rangle}} - \phi^{\langle j_1\rangle}_{s_{i-1}^{\langle n\rangle}}\Big\|_{\mathcal{Y}} \Big\|\phi^{\langle j_2\rangle}_{s_i^{\langle n\rangle}} - \phi^{\langle j_2\rangle}_{s_{i-1}^{\langle n\rangle}}\Big\|_{\mathcal{Y}}\,\mathrm{d}\tau\,\mathrm{d}\bar{s}
\end{aligned}$$

$$\leq \sum_{j_1=1}^{m} \sum_{j_2=1}^{m} \sum_{i=1}^{2^n} \int_0^1 \int_0^{\bar{s}} \check{\delta}_n(\omega) \left\| \phi_{s_i^{(n)}}^{(j_1)} - \phi_{s_{i-1}^{(n)}}^{(j_1)} \right\|_{\mathcal{Y}} \left\| \phi_{s_i^{(n)}}^{(j_2)} - \phi_{s_{i-1}^{(n)}}^{(j_2)} \right\|_{\mathcal{Y}} \mathrm{d}\tau \, \mathrm{d}\bar{s}$$

$$= \frac{1}{2} \check{\delta}_n(\omega) \sum_{j_1=1}^{m} \sum_{j_2=1}^{m} \sum_{i=1}^{2^n} \left\| \phi_{s_i^{(n)}}^{(j_1)} - \phi_{s_{i-1}^{(n)}}^{(j_1)} \right\|_{\mathcal{Y}} \left\| \phi_{s_i^{(n)}}^{(j_2)} - \phi_{s_{i-1}^{(n)}}^{(j_2)} \right\|_{\mathcal{Y}}$$

$$=: \frac{1}{2} \check{\delta}_n(\omega) Z_n(\omega)$$

where the first equality follows from the definition of $T_{6,n}$, the first inequality follows from the triangular inequality, the second inequality follows from Proposition 11.92, the third inequality follows from Proposition 7.64, the fourth inequality follows from the definition of $\check{\delta}_n$ in (14.28), and the second equality follows from Proposition 11.92 and Theorem 12.83.

It is easy to derive that

$$\boldsymbol{E}(Z_n)$$

$$= \sum_{j_1=1}^{m} \sum_{j_2=1}^{m} \sum_{i=1}^{2^n} \boldsymbol{E}\left(\left\| \phi_{s_i^{(n)}}^{(j_1)} - \phi_{s_{i-1}^{(n)}}^{(j_1)} \right\|_{\mathcal{Y}} \left\| \phi_{s_i^{(n)}}^{(j_2)} - \phi_{s_{i-1}^{(n)}}^{(j_2)} \right\|_{\mathcal{Y}} \right)$$

$$\leq \sum_{j_1=1}^{m} \sum_{j_2=1}^{m} \sum_{i=1}^{2^n} \left(\boldsymbol{E}\left(\left\| \phi_{s_i^{(n)}}^{(j_1)} - \phi_{s_{i-1}^{(n)}}^{(j_1)} \right\|_{\mathcal{Y}}^2 \right) \right)^{\frac{1}{2}} \left(\boldsymbol{E}\left(\left\| \phi_{s_i^{(n)}}^{(j_2)} - \phi_{s_{i-1}^{(n)}}^{(j_2)} \right\|_{\mathcal{Y}}^2 \right) \right)^{\frac{1}{2}}$$

$$= \sum_{j_1=1}^{m} \sum_{j_2=1}^{m} \sum_{i=1}^{2^n} \left\| \phi_{s_i^{(n)}}^{(j_1)} - \phi_{s_{i-1}^{(n)}}^{(j_1)} \right\|_{\bar{L}_2(\boldsymbol{\Omega}, \mathcal{Y})} \left\| \phi_{s_i^{(n)}}^{(j_2)} - \phi_{s_{i-1}^{(n)}}^{(j_2)} \right\|_{\bar{L}_2(\boldsymbol{\Omega}, \mathcal{Y})}$$

$$= \sum_{j_1=1}^{m} \sum_{j_2=1}^{m} \sum_{i=1}^{2^n} \left\| \overline{\int}_{s_{i-1}^{(n)}}^{s_i^{(n)}} h_s^{(j_1)} \, \mathrm{d}w_s^{(j_1)} \right\|_{\bar{L}_2(\boldsymbol{\Omega}, \mathcal{Y})} \left\| \overline{\int}_{s_{i-1}^{(n)}}^{s_i^{(n)}} h_s^{(j_2)} \, \mathrm{d}w_s^{(j_2)} \right\|_{\bar{L}_2(\boldsymbol{\Omega}, \mathcal{Y})}$$

$$= \sum_{j_1=1}^{m} \sum_{j_2=1}^{m} \sum_{i=1}^{2^n} \left\| h^{(j_1)} \right\|_{\bar{L}_2(\mathrm{r}_{s_{i-1}^{(n)}, s_i^{(n)}} \times \boldsymbol{\Omega}, \mathcal{Y})} \left\| h^{(j_2)} \right\|_{\bar{L}_2(\mathrm{r}_{s_{i-1}^{(n)}, s_i^{(n)}} \times \boldsymbol{\Omega}, \mathcal{Y})}$$

$$\leq \sum_{j_1=1}^{m} \sum_{j_2=1}^{m} \left(\sum_{i=1}^{2^n} \left\| h^{(j_1)} \right\|_{\bar{L}_2(\mathrm{r}_{s_{i-1}^{(n)}, s_i^{(n)}} \times \boldsymbol{\Omega}, \mathcal{Y})}^2 \right)^{\frac{1}{2}}$$

$$\cdot \left(\sum_{i=1}^{2^n} \left\| h^{(j_2)} \right\|_{\bar{L}_2(\mathrm{r}_{s_{i-1}^{(n)}, s_i^{(n)}} \times \boldsymbol{\Omega}, \mathcal{Y})}^2 \right)^{\frac{1}{2}}$$

$$\leq \sum_{j_1=1}^{m} \sum_{j_2=1}^{m} \left\| h^{(j_1)} \right\|_{\bar{L}_2(\mathrm{r}_{a,b} \times \boldsymbol{\Omega}, \mathcal{Y})} \left\| h^{(j_2)} \right\|_{\bar{L}_2(\mathrm{r}_{a,b} \times \boldsymbol{\Omega}, \mathcal{Y})} < \infty$$

where the first equality follows from the definition of Z_n and Proposition 11.92, the first inequality follows from Cauchy–Schwarz Inequality, the second equality follows Example 11.173, the third equality follows from (iv) of Proposition 14.106, the fourth equality follows from Proposition 14.120, the second inequality follows from Cauchy–Schwarz Inequality, and the last inequality follows from Proposition 11.83. By Proposition 14.110, $(Z_n)_{n=1}^{\infty}$ is uniformly growth bounded. By Propositions 14.111 and 11.58, we have $\lim_{n\in\mathbb{N}} \frac{1}{2}\check{\delta}_n Z_n = 0$ in measure in $\boldsymbol{\Omega}$. Hence, we have $\lim_{n\in\mathbb{N}} T_{6,n} = \vartheta_{\mathcal{Z}}$ in measure in $\boldsymbol{\Omega}$.

In summary, we have $\hat{T}(\omega) = T_{1,n}(\omega) + T_{3,n}(\omega) + T_{2,n,1}(\omega) + T_{2,n,2}(\omega) + T_{4,n,1,1}(\omega) + T_{4,n,1,2}(\omega) + T_{4,n,2}(\omega) + T_{5,n}(\omega) + T_{6,n}(\omega) + S_{2,n}(\omega)$, $\forall \omega \in \boldsymbol{\Omega}$, $\forall n \in \mathbb{N}$. By the preceding discussion, we have $\|\hat{T}(\omega)\|_{\mathcal{Z}} \leq \|T_{1,n}(\omega)\|_{\mathcal{Z}} + \|T_{3,n}(\omega)\|_{\mathcal{Z}} + \|T_{2,n,1}(\omega)\|_{\mathcal{Z}} + \|T_{2,n,2}(\omega)\|_{\mathcal{Z}} + \|T_{4,n,1,1}(\omega)\|_{\mathcal{Z}} + \|T_{4,n,1,2}(\omega)\|_{\mathcal{Z}} + \|T_{4,n,2}(\omega)\|_{\mathcal{Z}} + \|T_{5,n}(\omega)\|_{\mathcal{Z}} + \|T_{6,n}(\omega)\|_{\mathcal{Z}} + \|S_{2,n}(\omega)\|_{\mathcal{Z}}$, $\forall \omega \in \boldsymbol{\Omega}$, and furthermore, $\lim_{n\in\mathbb{N}}(\|T_{1,n}(\omega)\|_{\mathcal{Z}} + \|T_{3,n}(\omega)\|_{\mathcal{Z}}) = 0$, $\forall \omega \in \boldsymbol{\Omega}$, $T_{2,n,1}$, $T_{2,n,2}$, $T_{4,n,1,1}$, $T_{4,n,1,2}$, $T_{4,n,2}$, $T_{5,n}$, $T_{6,n}$, and $S_{2,n}$ converge to $\vartheta_{\mathcal{Z}}$ in measure $\boldsymbol{\Omega}$ as $n \to \infty$. By Proposition 11.60, there exists a subsequence $\|\hat{T}(\omega)\|_{\mathcal{Z}} \leq \lim_{k\in\mathbb{N}}(\|T_{1,n_k}(\omega)\|_{\mathcal{Z}} + \|T_{3,n_k}(\omega)\|_{\mathcal{Z}} + \|T_{2,n_k,1}(\omega)\|_{\mathcal{Z}} + \|T_{2,n_k,2}(\omega)\|_{\mathcal{Z}} + \|T_{4,n_k,1,1}(\omega)\|_{\mathcal{Z}} + \|T_{4,n_k,1,2}(\omega)\|_{\mathcal{Z}} + \|T_{4,n_k,2}(\omega)\|_{\mathcal{Z}} + \|T_{5,n_k}(\omega)\|_{\mathcal{Z}} + \|T_{6,n_k}(\omega)\|_{\mathcal{Z}} + \|S_{2,n_k}(\omega)\|_{\mathcal{Z}}) = 0$ a.e. $\omega \in \boldsymbol{\Omega}$. Hence, $\hat{T} = \vartheta_{\mathcal{Z}}$ a.e. in $\boldsymbol{\Omega}$ since $\hat{T}$ is $\mathcal{B}_t$-measurable. This completes the proof of the theorem. □

14.14 Girsanov's Theorem

Finally, we present the Girsanov's Theorem.

Theorem 14.123 (Girsanov's Theorem) *Let $t_f \in \mathbb{R}_+$, $((\bar{\mathrm{r}}_{0,t_f}, |\cdot|), \hat{\mathcal{B}}, \mu) =: \mathbb{I}$ be the compact finite metric measure subspace of $\mathbb{R}$ with partial ordering $\leq$, $\boldsymbol{\Omega} := (\Omega, \mathcal{B}, (\mathcal{B}_t)_{t\in\mathbb{I}}, \boldsymbol{P})$ be a filtered probability measure space, $\mathbb{I}\times\boldsymbol{\Omega} = (\bar{\mathrm{r}}_{0,t_f}\times\Omega, \bar{\mathcal{B}}, \nu)$ be the finite product measure space, $m \in \mathbb{N}$, $w : \mathbb{I}\times\boldsymbol{\Omega} \to \mathbb{R}^m$ be an $\mathbb{R}^m$-valued standard Wiener process on $\mathbb{I}\times\boldsymbol{\Omega}$, where the components of w are $(w^{\langle 1\rangle}, \ldots, w^{\langle m\rangle})$, $w^{\langle j\rangle}$ is a standard Wiener process, $j = 1, \ldots, m$, and $(w_t)_{t\in\mathbb{I}}$ is adapted to the filtration $(\mathcal{B}_t)_{t\in\mathbb{I}}$, and $f : \bar{\mathrm{r}}_{0,t_f} \times \Omega \to \mathbb{R}^m$ be an $\mathbb{R}^m$-valued stochastic process adapted to $(\mathcal{B}_t)_{t\in\mathbb{I}}$, where the components of f are $(f^{\langle 1\rangle}, \ldots, f^{\langle m\rangle})$. Assume that*

(a) *The σ-algebras $\mathcal{B}_t$ and $\sigma((w_{t+s} - w_t)_{s\in\mathbb{I}})$ are independent, $\forall t \in \mathbb{I}$.*
(b) *$\forall E \in \mathcal{B}$ with $\boldsymbol{P}(E) = 0$, we have $E \in \mathcal{B}_0$.*
(c) *$f \in \bar{\mathrm{L}}_2(\bar{\mathrm{r}}_{0,t_f} \times \boldsymbol{\Omega}, \mathbb{R}^m)$, is adapted to the filtration $(\mathcal{B}_t)_{t\in\mathbb{I}}$, has right continuous sample paths, and $\widehat{\int_0^{t_f}} f(s, \omega)\,\mathrm{d}s$ satisfies the Riemann Criterion for Integrability, Theorem A.9, $\forall \omega \in \boldsymbol{\Omega}$.*

Define a scalar $\xi_0^{t_f}(f) : \boldsymbol{\Omega} \to \overline{\mathbb{R}_+}$ by $\xi_0^{t_f}(f)(\omega) = \exp\Big(\sum_{i=1}^m q_i(\omega) - \frac{1}{2}\int_0^{t_f} \langle f(s, \omega), f(s, \omega)\rangle_{\mathbb{R}^m}\,\mathrm{d}s\Big)$, $\forall \omega \in \boldsymbol{\Omega}$, where $q_i : \boldsymbol{\Omega} \to \mathbb{R}$ is defined by $q_i \in \overline{\int_0^{t_f}} f^{\langle i\rangle}(s, \omega)\,\mathrm{d}w^{\langle i\rangle}(s, \omega) \in \mathrm{L}_2(\boldsymbol{\Omega}, \mathbb{R})$ is a version of the Itô integral.

Assume that $E(\xi_0^{t_f}(f)) = 1$. *Then, the following statements hold:*

(i) *Define a measure* $\tilde{P}$ *on* $(\Omega, \mathcal{B})$ *through the Radon–Nikodym derivative* $\frac{d\tilde{P}}{dP} = \xi_0^{t_f}(f)$, *i.e.,* $\tilde{P}(E) = \int_E \xi_0^{t_f}(f)(\omega)\,dP(\omega)$, $\forall E \in \mathcal{B}$. *Then,* $(\Omega, \mathcal{B}, (\mathcal{B}_t)_{t\in\mathbb{I}}, \tilde{P})$ *is a filtered probability measure space.*

(ii) *Define* $\tilde{w} : \mathbb{I} \times \Omega \to \mathbb{R}^m$ *by* $\tilde{w}(t, \omega) = w(t, \omega) - \int_0^t f(s, \omega)\,ds$, $\forall t \in \mathbb{I}$, $\forall \omega \in \Omega$. *Then,* $\tilde{w}$ *is an* $\mathbb{R}^m$*-valued standard Wiener process on the filtered probability measure space* $(\Omega, \mathcal{B}, (\mathcal{B}_t)_{t\in\mathbb{I}}, \tilde{P})$.

Proof (i) Note that, by (c) and Theorem A.13, we have $\int_0^{t_f} \|f(s, \omega)\|_{\mathbb{R}^m}\,ds < \infty$, $\forall \omega \in \boldsymbol{\Omega}$. By Theorem A.18, we have $\int_0^{t_f} \mathcal{P}_2 \circ f(s, \omega)\,ds < \infty$, $\forall \omega \in \boldsymbol{\Omega}$. By Proposition 14.105, f is Itô integrable on the interval $\bar{r}_{0,t_f}$. By Theorem 12.121 and Propositions 11.39 and 11.38, $\xi_0^{t_f}(f)$ is $\mathcal{B}$-measurable. By Proposition 11.116, $\tilde{P}$ defines a measure on $(\Omega, \mathcal{B})$. By the assumption $E(\xi_0^{t_f}(f)) = 1$, $(\Omega, \mathcal{B}, \tilde{P})$ is a probability measure space. This then implies that $(\Omega, \mathcal{B}, (\mathcal{B}_t)_{t\in\mathbb{I}}, \tilde{P})$ is a filtered probability measure space.

(ii) By Theorem 12.121, $\tilde{w}$ is $\bar{\mathcal{B}}$-measurable and therefore a stochastic process. Again, by Theorem 12.121, $\tilde{w}$ is adapted to $(\mathcal{B}_t)_{t\in\mathbb{I}}$. Clearly, $\tilde{w}$ has continuous sample path, $\forall \omega \in \Omega$. By Proposition 14.102, $\tilde{w}$ is progressively measurable. All, we need to show is that $\tilde{w}$ is an $\mathbb{R}^m$-valued standard Wiener process on $(\Omega, \mathcal{B}, (\mathcal{B}_t)_{t\in\mathbb{I}}, \tilde{P})$.

We will first consider the special case that $f \in \hat{\mathcal{Q}}(\bar{r}_{0,t_f} \times \boldsymbol{\Omega}, \mathbb{R}^m)$, i.e., f is a simple predictable step function, without the assumption that $E(\xi_0^{t_f}(f)) = 1$. Let $f(t, \omega) = \sum_{\bar{i}=1}^{\bar{n}} F_{\bar{i}-1}(\omega)\chi_{[\bar{t}_{\bar{i}-1}, \bar{t}_{\bar{i}}), \mathbb{I}}(t)$, $\forall t \in \mathbb{I}$, $\forall \omega \in \boldsymbol{\Omega}$, where $F_{\bar{i}} : \boldsymbol{\Omega} \to \mathbb{R}^m$ is $\mathcal{B}_{\bar{t}_{\bar{i}}}$-measurable and $E(\mathcal{P}_2 \circ F_{\bar{i}}) < \infty$, $\bar{i} = 0, \ldots, \bar{n}-1$, $\bar{n} \in \mathbb{N}$, $0 = \bar{t}_0 < \bar{t}_1 < \cdots < \bar{t}_{\bar{n}} = t_f$. $\forall n \in \mathbb{N}$, $\forall 0 \le s_1 < t_1 \le s_2 < t_2 \le \cdots \le s_n < t_n \le t_f$, we will show that $(\tilde{w}_{t_1} - \tilde{w}_{s_1}, \ldots, \tilde{w}_{t_n} - \tilde{w}_{s_n}) \sim N(\mathbf{0}_{mn}, \text{block diagonal } ((t_1 - s_1)\,\mathrm{id}_{\mathbb{R}^m}, \ldots, (t_n - s_n)\,\mathrm{id}_{\mathbb{R}^m}))$ in $\tilde{P}$ measure. Without loss of generality, we may assume that $0 = s_1 < t_1 = s_2 < t_2 = \cdots = s_n < t_n = t_f$. By the properties of Gaussian random variables, without loss of generality, we may assume that $(\bar{t}_{\bar{i}})_{\bar{i}=1}^{\bar{n}} \subseteq (t_i)_{i=1}^n$. Thus, we have $0 = t_0 < t_1 < \cdots < t_{i_1} = \bar{t}_1 < t_{i_1+1} < \cdots < t_{i_2} = \bar{t}_2 < \cdots < t_n = \bar{t}_{\bar{n}} = t_f$. Then, we have $\tilde{w}_{t_i} - \tilde{w}_{s_i} = \tilde{w}_{t_i} - \tilde{w}_{t_{i-1}} = w_{t_i} - w_{t_{i-1}} - \int_{t_{i-1}}^{t_i} f(s, \omega)\,ds = w_{t_i} - w_{t_{i-1}} - F_{\bar{i}-1}(\omega)(t_i - t_{i-1})$, $\forall \omega \in \boldsymbol{\Omega}$, where $\bar{t}_{\bar{i}-1} \le t_{i-1} < t_i \le \bar{t}_{\bar{i}}$. Fix any measurable rectangle $E := E_1 \times \cdots \times E_n \in \mathcal{B}_{\mathrm{B}}(\mathbb{R}^{mn})$ with $E_i \in \mathcal{B}_{\mathrm{B}}(\mathbb{R}^m)$, we have

$$
\begin{aligned}
&\tilde{P}((\tilde{w}_{t_1} - \tilde{w}_{s_1}, \ldots, \tilde{w}_{t_n} - \tilde{w}_{s_n}) \in E) =: \tilde{P}(S_n) \\
&= \int_\Omega \chi_{S_n,\boldsymbol{\Omega}} \xi_0^{t_f}(f)\,dP = \int_\Omega \chi_{S_n,\boldsymbol{\Omega}} \exp\Big(\sum_{\bar{i}=1}^{\bar{n}} \langle F_{\bar{i}-1}, w_{\bar{t}_{\bar{i}}} - w_{\bar{t}_{\bar{i}-1}} \rangle_{\mathbb{R}^m} \\
&\quad - \frac{1}{2} \sum_{\bar{i}=1}^{\bar{n}} \langle F_{\bar{i}-1}, F_{\bar{i}-1} \rangle_{\mathbb{R}^m} (\bar{t}_{\bar{i}} - \bar{t}_{\bar{i}-1}) \Big)\,dP = E(\chi_{S_n,\boldsymbol{\Omega}} \xi_0^{t_f}(f))
\end{aligned}
$$

$$
\begin{aligned}
&= \boldsymbol{E}(\boldsymbol{E}(\chi_{S_n,\boldsymbol{\Omega}}\xi_0^{t_f}(f)|\mathcal{B}_{t_{n-1}})) = \boldsymbol{E}(\chi_{S_{n-1},\boldsymbol{\Omega}}\xi_0^{t_{n-1}}(f)\cdot \\
&\quad \boldsymbol{E}(\chi_{\{\omega\in\Omega \mid w_{t_n}-w_{t_{n-1}}-F_{\bar{n}-1}(t_n-t_{n-1})\in E_n\},\Omega} \exp\Big(\big\langle F_{\bar{n}-1}, w_{t_n} - \\
&\quad w_{t_{n-1}}\big\rangle_{\mathbb{R}^m} - \frac{1}{2}\langle F_{\bar{n}-1}, F_{\bar{n}-1}\rangle_{\mathbb{R}^m}(t_n - t_{n-1})\Big)|\mathcal{B}_{t_{n-1}})) \\
&= \boldsymbol{E}(\chi_{S_{n-1},\boldsymbol{\Omega}}\xi_0^{t_{n-1}}(f)\frac{1}{\sqrt{2\pi(t_n - t_{n-1})}^m} \\
&\quad \cdot\int_{\mathbb{R}^m} \chi_{\{x\in\mathbb{R}^m \mid x-F_{\bar{n}-1}(t_n-t_{n-1})\in E_n\},\mathbb{R}^m} \exp\Big(\big\langle F_{\bar{n}-1}, x\big\rangle_{\mathbb{R}^m} \\
&\quad -\frac{1}{2}\langle F_{\bar{n}-1}, F_{\bar{n}-1}\rangle_{\mathbb{R}^m}(t_n - t_{n-1}) - \frac{1}{2(t_n - t_{n-1})}\langle x, x\rangle_{\mathbb{R}^m}\Big)\,\mathrm{d}x) \\
&= \boldsymbol{E}(\chi_{S_{n-1},\boldsymbol{\Omega}}\xi_0^{t_{n-1}}(f)\frac{1}{\sqrt{2\pi(t_n - t_{n-1})}^m} \\
&\quad \cdot\int_{\mathbb{R}^m} \chi_{\{x\in\mathbb{R}^m \mid x-F_{\bar{n}-1}(t_n-t_{n-1})\in E_n\},\mathbb{R}^m} \exp\Big(- \frac{1}{2(t_n - t_{n-1})} \\
&\quad \cdot\big\langle x - F_{\bar{n}-1}(t_n - t_{n-1}), x - F_{\bar{n}-1}(t_n - t_{n-1})\big\rangle_{\mathbb{R}^m}\Big)\,\mathrm{d}x) \\
&= \boldsymbol{E}(\chi_{S_{n-1},\boldsymbol{\Omega}}\xi_0^{t_{n-1}}(f)\frac{1}{\sqrt{2\pi(t_n - t_{n-1})}^m} \\
&\quad \cdot\int_{\mathbb{R}^m} \chi_{\{x\in\mathbb{R}^m \mid x\in E_n\},\mathbb{R}^m} \exp\Big(- \frac{1}{2(t_n - t_{n-1})}\big\langle x, x\big\rangle_{\mathbb{R}^m}\Big)\,\mathrm{d}x) \\
&= \boldsymbol{E}(\chi_{S_{n-1},\boldsymbol{\Omega}}\xi_0^{t_{n-1}}(f)\boldsymbol{P}(w_{t_n} - w_{t_{n-1}} \in E_n)) \\
&= \boldsymbol{P}(w_{t_n} - w_{t_{n-1}} \in E_n)\boldsymbol{E}(\chi_{S_{n-1},\boldsymbol{\Omega}}\xi_0^{t_{n-1}}(f)) \\
&= \boldsymbol{P}(w_{t_n} - w_{t_{n-1}} \in E_n)\boldsymbol{P}(w_{t_{n-1}} - w_{t_{n-2}} \in E_{n-1})\boldsymbol{E}(\chi_{S_{n-2},\boldsymbol{\Omega}}\xi_0^{t_{n-2}}(f)) \\
&= \cdots = \boldsymbol{P}(w_{t_n} - w_{t_{n-1}} \in E_n)\cdots\boldsymbol{P}(w_{t_1} - w_{t_0} \in E_1)
\end{aligned}
$$

where the first equality follows from the definition of the measure $\tilde{\boldsymbol{P}}$, the second equality follows from definition of $\xi_0^{t_f}(f)$, the third equality follows from Definition 14.1, the fourth equality follows from (a) of Proposition 14.11, the fifth equality follows from (h) of Proposition 14.11, the sixth equality follows from the Definitions 14.78, 14.16, and 14.25, the seventh equality follows from simple algebra, the eighth equality follows from Change of Variable, Theorem 12.91, the ninth equality follows from the Definitions 14.78, 14.16, and 14.25, the tenth equality follows from (c) of Proposition 14.11, and the rest of the equalities follow from the recursion of the preceding argument.

Thus, by the arbitrariness of E, we have shown that the probability distribution of $(\tilde{w}_{t_1} - \tilde{w}_{s_1}, \ldots, \tilde{w}_{t_n} - \tilde{w}_{s_n})$ in $\tilde{\boldsymbol{P}}$ measure is exactly the same as the distribution of $(w_{t_1} - w_{s_1}, \ldots, w_{t_n} - w_{s_n})$ in $\boldsymbol{P}$ measure. Hence, we have $(\tilde{w}_{t_1} - \tilde{w}_{s_1}, \ldots, \tilde{w}_{t_n} - \tilde{w}_{s_n}) \sim N(\mathbf{0}_{mn}, \text{block diagonal }((t_1 - s_1)\,\mathrm{id}_{\mathbb{R}^m}, \ldots, (t_n - s_n)\,\mathrm{id}_{\mathbb{R}^m}))$. Therefore, by Definition 14.78, in $\tilde{\boldsymbol{P}}$ measure, $\tilde{w}$ is an $\mathbb{R}^m$-valued standard Wiener process. Take $E = \mathbb{R}^{mn}$, we have $\boldsymbol{E}(\xi_0^{t_f}(f)) = 1$, whenever $f \in \hat{\mathcal{Q}}(\bar{\mathrm{r}}_{0,t_f} \times \boldsymbol{\Omega}, \mathbb{R}^m)$.

Now, we consider the general case. By Proposition 14.103, there exists $(f_n)_{n=1}^{\infty} \subseteq \hat{\mathcal{Q}}(\bar{\mathrm{r}}_{0,t_f} \times \boldsymbol{\Omega}, \mathbb{R}^m)$ such that $\lim_{n\in\mathbb{N}} f_n \doteq f$ in $\bar{\mathrm{L}}_2(\bar{\mathrm{r}}_{0,t_f} \times \boldsymbol{\Omega}, \mathbb{R}^m)$. By Definition 14.104, $\lim_{n\in\mathbb{N}} \left[\widehat{\int}_0^{t_f} \langle f_{n\,s}, \mathrm{d}w_s\rangle_{\mathbb{R}^m}\right] = \overline{\int}_0^{t_f} \langle f_s, \mathrm{d}w_s\rangle_{\mathbb{R}^m} = \sum_{i=1}^m \overline{\int}_0^{t_f} f_s^{\langle i\rangle} \mathrm{d}w_s^{\langle i\rangle}$ in $\mathrm{L}_2(\boldsymbol{\Omega}, \mathbb{R}^m)$. By the special case, we have $\boldsymbol{E}(\xi_0^{t_f}(f_n)) = 1$, $\forall n \in \mathbb{N}$. By Proposition 11.211, we have $\lim_{n\in\mathbb{N}} \widehat{\int}_0^{t_f} \langle f_{n\,s}, \mathrm{d}w_s\rangle_{\mathbb{R}^m} = \sum_{i=1}^m q_i$ in measure in $\boldsymbol{\Omega}$.

By Theorem A.18, $\widehat{\int}_0^{t_f} \langle f(s,\omega), f(s,\omega)\rangle_{\mathbb{R}^m}\,\mathrm{d}s$ satisfy the Riemann Criterion for Integrability, $\forall \omega \in \boldsymbol{\Omega}$. By the previous paragraph and Theorem A.15, $\int_0^{t_f} \langle f(s,\omega), f(s,\omega)\rangle_{\mathbb{R}^m}\,\mathrm{d}s = \widehat{\int}_0^{t_f} \langle f(s,\omega), f(s,\omega)\rangle_{\mathbb{R}^m}\mathrm{d}s = \lim_{n\in\mathbb{N}} \widehat{\int}_0^{t_f} \langle f_n(s,\omega), f_n(s,\omega)\rangle_{\mathbb{R}^m}\,\mathrm{d}s$ a.e. $\omega \in \boldsymbol{\Omega}$.

Define processes $\tilde{w}^{\langle n\rangle} : \mathbb{I} \times \boldsymbol{\Omega} \to \mathbb{R}^m$ by $\tilde{w}^{\langle n\rangle}(t,\omega) = w(t,\omega) - \int_0^t f_n(s,\omega)\,\mathrm{d}s$, $\forall t \in \mathbb{I}$, $\forall \omega \in \boldsymbol{\Omega}$, $\forall n \in \mathbb{N}$. By the special case, we have that $\tilde{w}^{\langle n\rangle}$ is an $\mathbb{R}^m$-valued standard Wiener process in the filtered probability measure space $(\Omega, \mathcal{B}, (\mathcal{B}_t)_{t\in\mathbb{I}}, \tilde{\boldsymbol{P}}^{\langle n\rangle})$, where $\tilde{\boldsymbol{P}}^{\langle n\rangle}$ is defined through the Radon–Nikodym derivative $\frac{\mathrm{d}\tilde{\boldsymbol{P}}^{\langle n\rangle}}{\mathrm{d}\boldsymbol{P}} = \xi_0^{t_f}(f_n)$. By assumption (c) and Proposition 11.213, $\lim_{n\in\mathbb{N}} \int_0^t f_n(s,\omega)\,\mathrm{d}s = \int_0^t f(s,\omega)\,\mathrm{d}s$, $\forall t \in \mathbb{I}$, a.e. $\omega \in \boldsymbol{\Omega}$. This implies that $\lim_{n\in\mathbb{N}} \tilde{w}^{\langle n\rangle}(t,\omega) = \tilde{w}(t,\omega)$, $\forall t \in \mathbb{I}$, a.e. $\omega \in \boldsymbol{\Omega}$.

By Proposition 11.57, there exists a subsequence $(f_{n_k})_{k=1}^{\infty}$ of $(f_n)_{n=1}^{\infty}$ such that $\lim_{k\in\mathbb{N}} \widehat{\int}_0^{t_f} \langle f_{n_k\,s}, \mathrm{d}w_s\rangle_{\mathbb{R}^m} = \sum_{i=1}^m q_i$ a.e. in $\boldsymbol{\Omega}$. By Propositions 11.52 and 11.53, we have $\lim_{k\in\mathbb{N}} \xi_0^{t_f}(f_{n_k}) = \xi_0^{t_f}(f)$ a.e. in $\boldsymbol{\Omega}$. By the special case, $\boldsymbol{E}(\xi_0^{t_f}(f_{n_k})) = 1$, $\forall k \in \mathbb{N}$. Then, by Fatou's Lemma 11.80, we have $\boldsymbol{E}(\xi_0^{t_f}(f)) \leq \liminf_{k\in\mathbb{N}} \boldsymbol{E}(\xi_0^{t_f}(f_{n_k})) = 1$. By the assumption that $1 = \boldsymbol{E}(\xi_0^{t_f}(f)) = \lim_{k\in\mathbb{N}} \boldsymbol{E}(\xi_0^{t_f}(f_{n_k}))$, we have the following line of arguments:

$$\tilde{\boldsymbol{P}}(B) = \int_{\Omega} \chi_{B,\boldsymbol{\Omega}} \xi_0^{t_f}(f)\,\mathrm{d}\boldsymbol{P} = \lim_{k\in\mathbb{N}} \int_{\Omega} \chi_{B,\boldsymbol{\Omega}} \xi_0^{t_f}(f_{n_k})\,\mathrm{d}\boldsymbol{P} = \lim_{k\in\mathbb{N}} \tilde{\boldsymbol{P}}^{\langle n_k\rangle}(B)$$

$\forall B \in \mathcal{B}$, where the first equality follows from the definition of the measure $\tilde{\boldsymbol{P}}$, the second equality follows from Lebesgue Dominated Convergence Theorem 11.91, and the third equality follows from the definition of the measures $\tilde{\boldsymbol{P}}^{\langle n_k\rangle}$.

Fix any $n \in \mathbb{N}$ and any $0 \leq s_1 < t_1 \leq s_2 < t_2 \leq \cdots \leq s_n < t_n \leq t_f$. Let $\tilde{W}^{\langle n\rangle} := (\tilde{w}_{t_1} - \tilde{w}_{s_1}, \ldots, \tilde{w}_{t_n} - \tilde{w}_{s_n})$. We need to show that $\tilde{W}^{\langle n\rangle} \sim N(\mathbf{0}_{nm}, \text{block diagonal }((t_1 - s_1)\,\mathrm{id}_{\mathbb{R}^m}, \ldots, (t_n - s_n)\,\mathrm{id}_{\mathbb{R}^m}))$ in $\tilde{\boldsymbol{P}}$ measure. By the discussion in the second from last paragraph, we have $\tilde{W}^{\langle n\rangle} = \lim_{N\in\mathbb{N}}(\tilde{w}_{t_1}^{\langle N\rangle} - \tilde{w}_{s_1}^{\langle N\rangle}, \ldots, \tilde{w}_{t_n}^{\langle N\rangle} - \tilde{w}_{s_n}^{\langle N\rangle}) =: \lim_{N\in\mathbb{N}} \tilde{W}^{\langle N,n\rangle}$ a.e. in $\boldsymbol{\Omega}$. Let $L : \mathcal{B}_{\mathrm{B}}(\mathbb{R}^{nm}) \to \overline{\mathbb{R}_+}$

be the law for the normal distribution $N(\mathbf{0}_{nm}, \text{block diagonal } ((t_1 - s_1)\,\mathrm{id}_{\mathbb{R}^m}, \ldots, (t_n - s_n)\,\mathrm{id}_{\mathbb{R}^m}))$. Fix any $E \in \mathcal{B}_B(\mathbb{R}^{nm})$, we need to show that $\tilde{\boldsymbol{P}}(\{\omega \in \Omega \mid \tilde{W}^{\langle n \rangle} \in E\}) = L(E)$. Since the law L is absolutely continuous with respect to $\mu_{B(mn)}$ on $\mathbb{R}^{mn}$, $\forall \epsilon \in \mathbb{R}_+$, there exists $\delta \in \mathbb{R}_+$ such that $\forall \bar{E} \in \mathcal{B}_B(\mathbb{R}^{mn})$ with $\mu_{B(mn)}(\bar{E}) < \delta$, we have $L(\bar{E}) < \epsilon$. Since $\mathbb{R}^{mn}$ is a topological measure space, by Definition 11.26 and Proposition 11.27, there exists an open set $O \subseteq \mathbb{R}^{mn}$ such that $E \subseteq O$ and $\mu_{B(mn)}(O \setminus E) < \frac{\delta}{2}$, and there exists a closed set $F \subseteq \mathbb{R}^{mn}$ such that $F \subseteq E$ and $\mu_{B(mn)}(E \setminus F) < \frac{\delta}{2}$. Let $\tilde{\Omega} := \{\omega \in \boldsymbol{\Omega} \mid \lim_{N \in \mathbb{N}} \tilde{W}^{\langle N,n \rangle}(\omega) = \tilde{W}^{\langle n \rangle}(\omega)\}$. Then, $\boldsymbol{P}(\Omega \setminus \tilde{\Omega}) = 0$. By the absolute continuity of $\tilde{\boldsymbol{P}}$ and $\tilde{\boldsymbol{P}}^{\langle n \rangle}$ with respect to $\boldsymbol{P}$, we have $\tilde{\boldsymbol{P}}(\Omega \setminus \tilde{\Omega}) = 0$ and $\tilde{\boldsymbol{P}}^{\langle n \rangle}(\Omega \setminus \tilde{\Omega}) = 0$, $\forall n \in \mathbb{N}$. Note the following set relationship:

$$\begin{aligned}
\{\omega \in \tilde{\Omega} \mid \tilde{W}^{\langle n \rangle} \in E\} &\subseteq \{\omega \in \tilde{\Omega} \mid \tilde{W}^{\langle n \rangle} \in O\} \\
&\subseteq \bigcup_{M=1}^{\infty} \bigcap_{N=M}^{\infty} \{\omega \in \tilde{\Omega} \mid \tilde{W}^{\langle N,n \rangle} \in O\}; \\
\{\omega \in \tilde{\Omega} \mid \tilde{W}^{\langle n \rangle} \in E\} &\supseteq \{\omega \in \tilde{\Omega} \mid \tilde{W}^{\langle n \rangle} \in F\} \\
&\supseteq \bigcap_{M=1}^{\infty} \bigcup_{N=M}^{\infty} \{\omega \in \tilde{\Omega} \mid \tilde{W}^{\langle N,n \rangle} \in F\}
\end{aligned}$$

This then implies that

$$\begin{aligned}
&\tilde{\boldsymbol{P}}(\{\omega \in \Omega \mid \tilde{W}^{\langle n \rangle} \in E\}) \\
&= \tilde{\boldsymbol{P}}(\{\omega \in \tilde{\Omega} \mid \tilde{W}^{\langle n \rangle} \in E\}) + \tilde{\boldsymbol{P}}(\{\omega \in \Omega \setminus \tilde{\Omega} \mid \tilde{W}^{\langle n \rangle} \in E\}) \\
&\leq \tilde{\boldsymbol{P}}(\bigcup_{M=1}^{\infty} \bigcap_{N=M}^{\infty} \{\omega \in \tilde{\Omega} \mid \tilde{W}^{\langle N,n \rangle} \in O\}) \\
&= \lim_{M \in \mathbb{N}} \tilde{\boldsymbol{P}}(\bigcap_{N=M}^{\infty} \{\omega \in \tilde{\Omega} \mid \tilde{W}^{\langle N,n \rangle} \in O\}) \\
&= \lim_{M \in \mathbb{N}} \lim_{k \in \mathbb{N}} \tilde{\boldsymbol{P}}^{\langle n_k \rangle}(\bigcap_{N=M}^{\infty} \{\omega \in \tilde{\Omega} \mid \tilde{W}^{\langle N,n \rangle} \in O\}) \\
&\leq \lim_{M \in \mathbb{N}} \lim_{k \in \mathbb{N}} \tilde{\boldsymbol{P}}^{\langle n_k \rangle}(\{\omega \in \tilde{\Omega} \mid \tilde{W}^{\langle n_k,n \rangle} \in O\}) \\
&= \lim_{M \in \mathbb{N}} \lim_{k \in \mathbb{N}} \Big(\tilde{\boldsymbol{P}}^{\langle n_k \rangle}(\{\omega \in \tilde{\Omega} \mid \tilde{W}^{\langle n_k,n \rangle} \in O\}) \\
&\quad + \tilde{\boldsymbol{P}}^{\langle n_k \rangle}(\{\omega \in \Omega \setminus \tilde{\Omega} \mid \tilde{W}^{\langle n_k,n \rangle} \in O\})\Big) \\
&= \lim_{M \in \mathbb{N}} \lim_{k \in \mathbb{N}} \tilde{\boldsymbol{P}}^{\langle n_k \rangle}(\{\omega \in \Omega \mid \tilde{W}^{\langle n_k,n \rangle} \in O\}) = \lim_{M \in \mathbb{N}} \lim_{k \in \mathbb{N}} L(O) = L(O)
\end{aligned}$$

where the first equality follows from Fact 11.3, the first inequality follows from the Fact 11.3 and $\tilde{\boldsymbol{P}}(\Omega \setminus \tilde{\Omega}) = 0$, the second equality follows from Proposition 11.7, the third equality follows from the result of the previous paragraph, the second inequality follows from Fact 11.3, the fourth equality follows from $\tilde{\boldsymbol{P}}^{\langle n_k \rangle}(\Omega \setminus \tilde{\Omega}) = 0$, the fifth equality follows from Fact 11.3, the sixth equality follows from conclusion of the special case, and the last equality follows from simple algebra. On the other hand,

$$\begin{aligned}
&\tilde{\boldsymbol{P}}(\{\omega \in \Omega \mid \tilde{W}^{\langle n \rangle} \in E\}) \\
&= \tilde{\boldsymbol{P}}(\{\omega \in \tilde{\Omega} \mid \tilde{W}^{\langle n \rangle} \in E\}) + \tilde{\boldsymbol{P}}(\{\omega \in \Omega \setminus \tilde{\Omega} \mid \tilde{W}^{\langle n \rangle} \in E\}) \\
&\geq \tilde{\boldsymbol{P}}(\bigcap_{M=1}^{\infty} \bigcup_{N=M}^{\infty} \{\omega \in \tilde{\Omega} \mid \tilde{W}^{\langle N,n \rangle} \in F\}) \\
&= \lim_{M \in \mathbb{N}} \tilde{\boldsymbol{P}}(\bigcup_{N=M}^{\infty} \{\omega \in \tilde{\Omega} \mid \tilde{W}^{\langle N,n \rangle} \in F\}) \\
&= \lim_{M \in \mathbb{N}} \lim_{k \in \mathbb{N}} \tilde{\boldsymbol{P}}^{\langle n_k \rangle}(\bigcup_{N=M}^{\infty} \{\omega \in \tilde{\Omega} \mid \tilde{W}^{\langle N,n \rangle} \in F\}) \\
&\geq \lim_{M \in \mathbb{N}} \lim_{k \in \mathbb{N}} \tilde{\boldsymbol{P}}^{\langle n_k \rangle}(\{\omega \in \tilde{\Omega} \mid \tilde{W}^{\langle n_k,n \rangle} \in F\}) \\
&= \lim_{M \in \mathbb{N}} \lim_{k \in \mathbb{N}} \Big(\tilde{\boldsymbol{P}}^{\langle n_k \rangle}(\{\omega \in \tilde{\Omega} \mid \tilde{W}^{\langle n_k,n \rangle} \in F\}) \\
&\quad + \tilde{\boldsymbol{P}}^{\langle n_k \rangle}(\{\omega \in \Omega \setminus \tilde{\Omega} \mid \tilde{W}^{\langle n_k,n \rangle} \in F\})\Big) \\
&= \lim_{M \in \mathbb{N}} \lim_{k \in \mathbb{N}} \tilde{\boldsymbol{P}}^{\langle n_k \rangle}(\{\omega \in \Omega \mid \tilde{W}^{\langle n_k,n \rangle} \in F\}) = \lim_{M \in \mathbb{N}} \lim_{k \in \mathbb{N}} L(F) = L(F)
\end{aligned}$$

where the first equality follows from Fact 11.3, the first inequality follows from Fact 11.3 and $\tilde{\boldsymbol{P}}(\Omega \setminus \tilde{\Omega}) = 0$, the second equality follows from Proposition 11.5, the third equality follows from the result of the previous paragraph, the second inequality follows from Fact 11.3, the fourth equality follows from $\tilde{\boldsymbol{P}}^{\langle n_k \rangle}(\Omega \setminus \tilde{\Omega}) = 0$, the fifth equality follows from Fact 11.3, the sixth equality follows from the conclusion of the special case, and the last equality follows from simple algebra.

Thus, we conclude that $\left|\tilde{\boldsymbol{P}}(\{\omega \in \Omega \mid \tilde{W}^{\langle n \rangle} \in E\}) - L(E)\right| \leq L(O) - L(F) = L(O \setminus F) < \epsilon$, where the first inequality follows from the fact that $L(F) \leq L(E) \leq L(O)$, the first equality follows from the fact L is a law, and the last inequality follows from the fact that $\mu_{\mathrm{B}(mn)}(O \setminus F) < \delta$. By the arbitrariness of ϵ, we have $\tilde{\boldsymbol{P}}(\{\omega \in \Omega \mid \tilde{W}^{\langle n \rangle} \in E\}) = L(E)$. This proves the $\tilde{W}^{\langle n \rangle} \sim N(\mathbf{0}_{nm}, \text{block diagonal }((t_1 - s_1)\,\mathrm{id}_{\mathbb{R}^m}, \ldots, (t_n - s_n)\,\mathrm{id}_{\mathbb{R}^m}))$ in $\tilde{\boldsymbol{P}}$ measure. By Definition 14.78, $\tilde{w}$ is an $\mathbb{R}^m$-valued standard Wiener process in the filtered probability measure space $(\Omega, \mathcal{B}, ((\mathcal{B})_t)_{t \in \mathbb{I}}, \tilde{\boldsymbol{P}})$.

This completes the proof of the theorem. □

Appendix A
Elements in Calculus

In this Appendix, I present some results in elementary calculus and some results on Riemann-Stieltjes integral that is generalized to Banach space valued integrand and integrator. These results are necessary for the discussion on stochastic integrations and motivated by the excellent results of Bartle (1976). I define the Riemann-Stieltjes integral as a limit of an appropriate net. Then, I prove Cauchy Criterion for Integrability, the bilinearity of the Riemann-Stieltjes integral, Integration by Parts Theorem, Riemann Criterion for Integration, and Integrability Theorem, which connects the Riemann-Stieltjes integral to the Lebesgue-Stieltjes integral that we have been discussing in the main text of this book. I study in detail when the Riemann-Stieltjes integral converges in gauge mode, i. e., when the maximum sampling interval decreases to zero.

A.1 Some Formulas

Proposition A.1 $\sum_{i=0}^{n}\prod_{j=1}^{m}(i+j)=\frac{1}{m+1}\frac{(n+m+1)!}{n!}$, $\forall n, m \in \mathbb{Z}_+$.

Proof We will use mathematical induction on $n \in \mathbb{Z}_+$ to prove the result.

1° $n = 0$. $\forall m \in \mathbb{Z}_+$, we have LHS $= m! =$ RHS. This case is proved.

2° Assume the result holds for $n = k \in \mathbb{Z}_+$.

3° Consider the case when $n = k + 1 \in \mathbb{N}$. $\forall m \in \mathbb{Z}_+$,

$$\begin{aligned}\text{LHS} &= \sum_{i=0}^{k+1}\prod_{j=1}^{m}(i+j)=\sum_{i=0}^{k}\prod_{j=1}^{m}(i+j)+\prod_{j=1}^{m}(k+1+j)\\ &=\frac{1}{m+1}\frac{(k+m+1)!}{k!}+\frac{(k+m+1)!}{(k+1)!}=\frac{1}{m+1}\frac{(k+m+2)!}{(k+1)!}\end{aligned}$$

Z. Pan, *Measure-Theoretic Calculus in Abstract Spaces*,
https://doi.org/10.1007/978-3-031-21912-2

where the third equality follow from the inductive assumption. This completes the induction process.

Hence, the result holds. □

A.2 Convergence of Infinite Sequences

Proposition A.2 (Raabe's Test) *Let* $(x_n)_{n=1}^{\infty} \subseteq (0, \infty) \subset \mathbb{R}$. *Then, the following statements hold.*

(i) *If there exists* $a \in (1, \infty) \subset \mathbb{R}$ *and* $n_0 \in \mathbb{N}$ *such that* $\frac{x_{n+1}}{x_n} \leq 1 - \frac{a}{n}$, $\forall n \geq n_0$, *then* $\sum_{n=1}^{\infty} x_n \in (0, \infty) \subset \mathbb{R}$.
(ii) *If there exists* $a \in (-\infty, 1] \subset \mathbb{R}$ *and* $n_0 \in \mathbb{N}$ *such that* $\frac{x_{n+1}}{x_n} \geq 1 - \frac{a}{n}$, $\forall n \geq n_0$, *then* $\sum_{n=1}^{\infty} x_n = \infty$.

Proof *(Bartle)* (i) Under the assumption, $\forall n \geq n_0$,

$$n x_{n+1} \leq (n-1)x_n - (a-1)x_n$$

This implies that

$$(n-1)x_n - n x_{n+1} \geq (a-1)x_n > 0, \qquad \forall n \geq n_0$$

Summing the above for $n = n_0, \ldots, m$, $\forall m \geq n_0$, we have

$$0 < (a-1) \sum_{n=n_0}^{m} x_n \leq (n_0 - 1)x_{n_0} - m x_{m+1} < (n_0 - 1)x_{n_0} \in \mathbb{R}$$

Hence, we have $0 < \sum_{n=n_0}^{m} x_n < \frac{(n_0-1)x_{n_0}}{a-1} \in (0, \infty) \subset \mathbb{R}$, $\forall m \geq n_0$. Hence, $\sum_{n=1}^{\infty} x_n \in (0, \sum_{n=1}^{n_0-1} x_n + \frac{(n_0-1)x_{n_0}}{a-1}] \subset \mathbb{R}$.

(ii) Under the assumption, we have

$$n x_{n+1} \geq (n-a)x_n \geq (n-1)x_n, \forall n \geq n_0$$

Then, $n x_{n+1} \geq n_0 x_{n_0+1} =: c \in (0, \infty) \subset \mathbb{R}$, $\forall n \geq n_0$. This implies that $x_{n+1} \geq c/n$, $\forall n \geq n_0$, and hence $\sum_{n=1}^{\infty} x_n = \infty$.

This completes the proof of the proposition. □

A.3 Riemann-Stieltjes Integral

The Riemann-Stieltjes integral for real-valued functions has been covered thoroughly in Bartle (1976). Here, we will extend it to Banach space valued functions. We will derive the necessary results for our discussion of the Wiener integral as well as the Itô integral.

Definition A.3 Let $I := [a, b] \subset \mathbb{R}$ be a compact interval with $a \leq b$. A *sampled partition* of I is $\hat{R} := ((s_i)_{i=0}^n, (\xi_i)_{i=1}^n)$ where $n \in \mathbb{Z}_+$, $a = s_0 \leq s_1 \leq \cdots \leq s_n = b$, and $\xi_i \in [s_{i-1}, s_i] \subset \mathbb{R}$, $i = 1, \ldots, n$. The collection of all sampled partitions on I is denoted by $\hat{\mathfrak{R}}(I)$. On the this collection, we will introduce a relation $\preceq$ by $\forall \hat{R}_1 := ((\bar{s}_i)_{i=0}^{n_1}, (\bar{\xi}_i)_{i=1}^{n_1})$, $\hat{R}_2 := ((\check{s}_i)_{i=0}^{n_2}, (\check{\xi}_i)_{i=1}^{n_2}) \in \hat{\mathfrak{R}}(I)$, we say that $\hat{R}_1 \preceq \hat{R}_2$ (or $\hat{R}_2$ is a *refinement* of $\hat{R}_1$) if, $\forall i \in \{1, \ldots, n_1 - 1\}$, there exists $j_i \in \{1, \ldots, n_2 - 1\}$ such that $\bar{s}_i = \check{s}_{j_i}$ and $j_{i-1} < j_i < j_{i+1}$ ($j_0 := 0$, $j_{n_1} := n_2$). Then, $\hat{\mathfrak{I}}(I) := (\hat{\mathfrak{R}}(I), \preceq)$ is a directed system and $\preceq$ is a partial ordering on $\hat{\mathfrak{R}}(I)$. $\hat{\mathfrak{I}}(I)$ is said to be the *Riemann integration system on* I. ◇

Definition A.4 Let $I := [a, b] \subset \mathbb{R}$ be a compact interval with $a \leq b$, $\mathcal{X}$ and $\mathcal{Y}$ be Banach spaces over $\mathbb{K}$, $f : I \to \mathrm{B}(\mathcal{X}, \mathcal{Y})$, $g : I \to \mathcal{X}$, and $\hat{\mathfrak{I}}(I) := (\hat{\mathfrak{R}}(I), \preceq)$ be the Riemann integration system on I. $\forall \hat{R} := ((s_i)_{i=0}^n, (\xi_i)_{i=1}^n) \in \hat{\mathfrak{R}}(I)$, define the *Riemann sum* $F_{\hat{R}} := \sum_{i=1}^n f(\xi_i) \cdot (g(s_i) - g(s_{i-1})) \in \mathcal{Y}$. This defines a net $(F_{\hat{R}})_{\hat{R} \in \hat{\mathfrak{I}}(I)} \subseteq \mathcal{Y}$. (Note that any repeating points in $(s_i)_{i=0}^n$ will lead to a $\vartheta_{\mathcal{Y}}$ value in one of the summands of $F_{\hat{R}}$. Thus, we may restrict ourself to the case where $a = s_0 < s_1 < \cdots < s_n = b$.) f is said to be *Riemann-Stieltjes integrable with respect to* g *on* I if the net $(F_{\hat{R}})_{\hat{R} \in \hat{\mathfrak{I}}(I)}$ admits a limit in $\mathcal{Y}$. In this case, $\lim_{\hat{R} \in \hat{\mathfrak{I}}(I)} F_{\hat{R}} \in \mathcal{Y}$ is said to be the *Riemann-Stieltjes integral of* f *with respect to* g *on* I denoted by $\widehat{\int}_a^b f \, \mathrm{d}g$, $\widehat{\int}_{[a,b]} f \, \mathrm{d}g$, $\widehat{\int}_{[a,b]} f(s) \, \mathrm{d}g(s)$ or $\widehat{\int}_a^b f(s) \, \mathrm{d}g(s)$, where f is said to be the *integrand*, g is said to be the *integrator*. When $\mathcal{Y} = \mathbb{R}$, we will denote $\widehat{\int}_a^b f \, \mathrm{d}g = \widehat{\int}_{[a,b]} f \, \mathrm{d}g = \widehat{\int}_a^b f(s) \, \mathrm{d}g(s) = \lim_{\hat{R} \in \hat{\mathfrak{I}}(I)} F_{\hat{R}} \in \mathbb{R}_e$, whenever the limit exists, which is said to be the *Riemann-Stieltjes integral of* f *with respect to* g *on* I. ◇

Theorem A.5 (Cauchy Criterion for Integrability) *Let* $I := [a, b] \subset \mathbb{R}$ *be a compact interval with* $a \leq b$, $\mathcal{X}$ *and* $\mathcal{Y}$ *be Banach spaces over* $\mathbb{K}$, $f : I \to \mathrm{B}(\mathcal{X}, \mathcal{Y})$, $g : I \to \mathcal{X}$, *and* $\hat{\mathfrak{I}}(I) := (\hat{\mathfrak{R}}(I), \preceq)$ *be the Riemann integration system on* I, *and* $(F_{\hat{R}})_{\hat{R} \in \hat{\mathfrak{I}}(I)}$ *be the net for the integral* $\widehat{\int}_a^b f \, \mathrm{d}g$ *as defined in Definition A.4. Then,* f *is Riemann-Stieltjes integrable with respect to* g *on* I *if, and only if,* $\forall \epsilon \in (0, \infty) \subset \mathbb{R}$, *there exists* $\hat{R}_\epsilon \in \hat{\mathfrak{I}}(I)$, *such that* $\forall \hat{R}, \hat{Q} \in \hat{\mathfrak{I}}(I)$ *with* $\hat{R}_\epsilon \preceq \hat{R}$ *and* $\hat{R}_\epsilon \preceq \hat{Q}$, *we have* $\| F_{\hat{R}} - F_{\hat{Q}} \|_{\mathcal{Y}} < \epsilon$.

Proof This result follows immediately from Proposition 4.44 and Definition 7.26. □

Theorem A.6 *Let* $I := [a, b] \subset \mathbb{R}$ *be a compact interval with* $a \leq b$, $\mathcal{X}$ *and* $\mathcal{Y}$ *be Banach spaces over* $\mathbb{K}$.

(i) *Let* $f_i : I \to \mathrm{B}(\mathcal{X}, \mathcal{Y})$, $i = 1, 2$, $g : I \to \mathcal{X}$, *and* $\alpha, \beta \in \mathbb{K}$. *If* f_i *is Riemann-Stieltjes integrable with respect to* g *on* I, $i = 1, 2$, *then* $\alpha f_1 + \beta f_2$ *is Riemann-Stieltjes integrable with respect to* g *on* I *and* $\widehat{\int}_a^b (\alpha f_1 + \beta f_2)\, \mathrm{d}g = \alpha \widehat{\int}_a^b f_1\, \mathrm{d}g + \beta \widehat{\int}_a^b f_2\, \mathrm{d}g$.

(ii) *Let* $f : I \to \mathrm{B}(\mathcal{X}, \mathcal{Y})$, $g_i : I \to \mathcal{X}$, $i = 1, 2$, *and* $\alpha, \beta \in \mathbb{K}$. *If* f *is Riemann-Stieltjes integrable with respect to* g_i *on* I, $i = 1, 2$, *then* f *is Riemann-Stieltjes integrable with respect to* $\alpha g_1 + \beta g_2$ *on* I *and* $\widehat{\int}_a^b f\, \mathrm{d}(\alpha g_1 + \beta g_2) = \alpha \widehat{\int}_a^b f\, \mathrm{d}g_1 + \beta \widehat{\int}_a^b f\, \mathrm{d}g_2$.

Proof This follows immediately from the fact that the net $\left(F_{\hat{R}}\right)_{\hat{R} \in \hat{\mathfrak{I}}(I)}$ for $\widehat{\int}_a^b f\, \mathrm{d}g$ as defined in Definition A.4 is such that $F_{\hat{R}}$ is bilinear in f and g for any fixed $\hat{R} \in \hat{\mathfrak{I}}(I)$. □

Theorem A.7 *Let* $I := [a, b] \subset \mathbb{R}$ *be a compact interval with* $a \le b$, $\mathcal{X}$ *and* $\mathcal{Y}$ *be Banach spaces over* $\mathbb{K}$, $c \in I$, $f : I \to \mathrm{B}(\mathcal{X}, \mathcal{Y})$, *and* $g : I \to \mathcal{X}$. *Then,* f *is Riemann-Stieltjes integrable with respect to* g *on* I *if, and only if,* f *is Riemann-Stieltjes integrable with respect to* g *on* $\bar{\mathrm{r}}_{a,c}$ *and* $\bar{\mathrm{r}}_{c,b}$. *In this case, we have* $\widehat{\int}_a^b f\, \mathrm{d}g = \widehat{\int}_a^c f\, \mathrm{d}g + \widehat{\int}_c^b f\, \mathrm{d}g$.

Proof "Sufficiency" Let f be Riemann-Stieltjes integrable with respect to g on $\bar{\mathrm{r}}_{a,c}$ and $\bar{\mathrm{r}}_{c,b}$. Then, $\forall \epsilon \in (0, \infty) \subset \mathbb{R}$, there exists a sampled partition $\hat{R}_{1,\epsilon} := ((\hat{s}_i)_{i=0}^{\hat{n}}, \left(\hat{\xi}_i\right)_{i=1}^{\hat{n}}) \in \hat{\mathfrak{I}}\left(\bar{\mathrm{r}}_{a,c}\right)$ such that $\forall \hat{R}_1 \in \hat{\mathfrak{I}}\left(\bar{\mathrm{r}}_{a,c}\right)$ with $\hat{R}_{1,\epsilon} \preceq \hat{R}_1$, we have $\left\| \widehat{\int}_a^c f \mathrm{d}g - \hat{F}_{\hat{R}_1} \right\|_{\mathcal{Y}} < \frac{\epsilon}{2}$, where $\left(\hat{F}_{\hat{R}_1}\right)_{\hat{R}_1 \in \hat{\mathfrak{I}}(\bar{\mathrm{r}}_{a,c})}$ is the net for $\widehat{\int}_a^c f\, \mathrm{d}g$ as defined in Definition A.4. Similarly, there exists a sampled partition $\hat{R}_{2,\epsilon} := ((\check{s}_i)_{i=0}^{\check{n}}, \left(\check{\xi}_i\right)_{i=1}^{\check{n}}) \in \hat{\mathfrak{I}}\left(\bar{\mathrm{r}}_{c,b}\right)$ such that $\forall \hat{R}_2 \in \hat{\mathfrak{I}}\left(\bar{\mathrm{r}}_{c,b}\right)$ with $\hat{R}_{2,\epsilon} \preceq \hat{R}_2$, we have $\left\| \widehat{\int}_c^b f \mathrm{d}g - \check{F}_{\hat{R}_2} \right\|_{\mathcal{Y}} < \frac{\epsilon}{2}$, where $\left(\check{F}_{\hat{R}_2}\right)_{\hat{R}_2 \in \hat{\mathfrak{I}}(\bar{\mathrm{r}}_{c,b})}$ is the net for $\widehat{\int}_c^b f\, \mathrm{d}g$ as defined in Definition A.4. Define $\hat{R}_\epsilon := ((s_i)_{i=0}^n, (\xi_i)_{i=1}^n) \in \hat{\mathfrak{I}}(I)$ such that $n = \hat{n} + \check{n}$, and $s_i = \hat{s}_i$, $i = 0, \ldots, \hat{n}$, and $s_i = \check{s}_{i-\hat{n}}$, $i = \hat{n} + 1, \ldots, n$, and $\xi_i = \hat{\xi}_i$, $i = 1, \ldots, \hat{n}$, and $\xi_i = \check{\xi}_{i-\hat{n}}$, $i = \hat{n} + 1, \ldots, n$. Then, $\forall \hat{R} \in \hat{\mathfrak{I}}(I)$ with $\hat{R}_\epsilon \preceq \hat{R}$, we have $F_{\hat{R}} = \hat{F}_{\hat{R}_1} + \check{F}_{\hat{R}_2}$, where $\left(F_{\hat{R}}\right)_{\hat{R} \in \hat{\mathfrak{I}}(I)}$ is the net for $\widehat{\int}_a^b f\, \mathrm{d}g$, $\hat{R}_1 \in \hat{\mathfrak{I}}\left(\bar{\mathrm{r}}_{a,c}\right)$ and $\hat{R}_2 \in \hat{\mathfrak{I}}\left(\bar{\mathrm{r}}_{c,b}\right)$ are induced by $\hat{R}$ (since $\hat{R}_\epsilon \preceq \hat{R}$ then $c = s_{\hat{n}}$, including multiplicity, must be partition points in $\hat{R}$), and $\hat{R}_{1,\epsilon} \preceq \hat{R}_1$ and $\hat{R}_{2,\epsilon} \preceq \hat{R}_2$. Then, we have $\left\| F_{\hat{R}} - \widehat{\int}_a^c f\, \mathrm{d}g - \widehat{\int}_c^b f\, \mathrm{d}g \right\|_{\mathcal{Y}} \le \left\| \hat{F}_{\hat{R}_1} - \widehat{\int}_a^c f\, \mathrm{d}g \right\|_{\mathcal{Y}} + \left\| \check{F}_{\hat{R}_2} - \widehat{\int}_c^b f\, \mathrm{d}g \right\|_{\mathcal{Y}} < \epsilon$. Hence, we have $\widehat{\int}_a^b f\, \mathrm{d}g = \lim_{\hat{R} \in \hat{\mathfrak{I}}(I)} F_{\hat{R}} = \widehat{\int}_a^c f\, \mathrm{d}g + \widehat{\int}_c^b f\, \mathrm{d}g$. This completes the sufficiency proof.

"Necessity" Let f be Riemann-Stieltjes integrable with respect to g on I. By Cauchy Criterion for Integrability (Theorem A.5), $\forall \epsilon \in (0, \infty) \subset \mathbb{R}$, $\exists \hat{R}_\epsilon := ((s_i)_{i=0}^n, (\xi_i)_{i=1}^n) \in \hat{\mathfrak{I}}(I)$ such that $\forall \hat{R}, \check{R} \in \hat{\mathfrak{I}}(I)$ with $\hat{R}_\epsilon \preceq \hat{R}$ and $\hat{R}_\epsilon \preceq \check{R}$, we have $\left\| F_{\hat{R}} - F_{\check{R}} \right\|_{\mathcal{Y}} < \epsilon$. Without loss of generality, assume that $c = s_{\hat{n}}$, for some $\hat{n} \in \{0, \ldots, n\}$. Then, let $\hat{R}_{1,\epsilon} := ((s_i)_{i=0}^{\hat{n}}, (\xi_i)_{i=1}^{\hat{n}})$, which is clearly a sample partition in $\hat{\mathfrak{I}}(\bar{\mathfrak{r}}_{a,c})$. Now, $\forall \hat{R}_1, \check{R}_1 \in \hat{\mathfrak{I}}(\bar{\mathfrak{r}}_{a,c})$ with $\hat{R}_{1,\epsilon} \preceq \hat{R}_1$ and $\hat{R}_{1,\epsilon} \preceq \check{R}_1$, we may extend $\hat{R}_1$ to a sampled partition $\hat{R} \in \hat{\mathfrak{I}}(I)$ by attaching $(s_i)_{i=\hat{n}+1}^{n}$ to the partitions points of $\hat{R}_1$, and $(\xi_i)_{i=\hat{n}+1}^{n}$ to the sampling points of $\hat{R}_1$, and similarly we may extend $\check{R}_1$ to $\check{R} \in \hat{\mathfrak{I}}(I)$. Then, it is easy to see that $\hat{R}_\epsilon \preceq \hat{R}$ and $\hat{R}_\epsilon \preceq \check{R}$. Thus, we have $\left\| \hat{F}_{\hat{R}_1} - \hat{F}_{\check{R}_1} \right\|_{\mathcal{Y}} = \left\| F_{\hat{R}} - F_{\check{R}} \right\|_{\mathcal{Y}} < \epsilon$. Hence, $\left(\hat{F}_{\hat{R}_1} \right)_{\hat{R}_1 \in \hat{\mathfrak{I}}(\bar{\mathfrak{r}}_{a,c})}$ is a Cauchy net. By Cauchy Criterion for Integrability (Theorem A.5), we have f is Riemann-Stieltjes integrable with respect to g on $\bar{\mathfrak{r}}_{a,c}$. By a similar argument, we have f is Riemann-Stieltjes integrable with respect to g on $\bar{\mathfrak{r}}_{c,b}$. This completes the necessity proof.

By the proof of sufficiency, we have $\widehat{\int}_a^b f\, dg = \widehat{\int}_a^c f\, dg + \widehat{\int}_c^b f\, dg$ under the equivalent conditions. This completes the proof of the theorem. □

Theorem A.8 (Integration by Parts) *Let $I := [a, b] \subset \mathbb{R}$ be a compact interval with $a \leq b$, $\mathcal{X}$ and $\mathcal{Y}$ be Banach spaces over $\mathbb{K}$, $f : I \to \mathrm{B}(\mathcal{X}, \mathcal{Y})$, and $g : I \to \mathcal{X}$. Assume that f is Riemann-Stieltjes integrable with respect to g on I. Then, g is Riemann-Stieltjes integrable with respect to f on I and $\widehat{\int}_a^b f\, dg + \widehat{\int}_a^b ((df)g) = f(b)g(b) - f(a)g(a) \in \mathcal{Y}$.*

Proof Let $\left(F_{\hat{R}}\right)_{\hat{R} \in \hat{\mathfrak{I}}(I)}$ be the net for the Riemann-Stieltjes integral $\widehat{\int}_a^b f\, dg$, and $\left(\check{F}_{\hat{R}}\right)_{\hat{R} \in \hat{\mathfrak{I}}(I)}$ be the net for the Riemann-Stieltjes integral $\widehat{\int}_a^b g\, df$, as defined in Definition A.4. Since f is Riemann-Stieltjes integrable with respect to g on I, then $\forall \epsilon \in (0, \infty) \subset \mathbb{R}$, $\exists \hat{R}_\epsilon := ((s_i)_{i=0}^{n}, (\xi_i)_{i=1}^{n}) \in \hat{\mathfrak{I}}(I)$ such that $\forall \hat{R} \in \hat{\mathfrak{I}}(I)$ with $\hat{R}_\epsilon \preceq \hat{R}$, we have $\left\| F_{\hat{R}} - \widehat{\int}_a^b f\, dg \right\|_{\mathcal{Y}} < \epsilon$. Fix any $\hat{R} := ((\hat{s}_i)_{i=0}^{\hat{n}}, \left(\hat{\xi}_i\right)_{i=1}^{\hat{n}}) \in \hat{\mathfrak{I}}(I)$ with $\hat{R}_\epsilon \preceq \hat{R}$. Then, we have $\check{F}_{\hat{R}} = \sum_{i=1}^{\hat{n}} (f(\hat{s}_i) - f(\hat{s}_{i-1}))g(\hat{\xi}_i) = f(b)g(b) - f(a)g(a) - \sum_{i=1}^{\hat{n}} f(\hat{s}_i)(g(\hat{s}_i) - g(\hat{\xi}_i)) - \sum_{i=1}^{\hat{n}} f(\hat{s}_{i-1})(g(\hat{\xi}_i) - g(\hat{s}_{i-1})) = f(b)g(b) - f(a)g(a) - F_{\tilde{R}}$, where $\tilde{R} := ((\hat{s}_0, \hat{\xi}_1, \hat{s}_1, \ldots, \hat{\xi}_{\hat{n}-1}, \hat{s}_{\hat{n}-1}, \hat{\xi}_{\hat{n}}, \hat{s}_{\hat{n}}), (\hat{s}_0, \hat{s}_1, \hat{s}_1, \ldots, \hat{s}_{\hat{n}-1}, \hat{s}_{\hat{n}-1}, \hat{s}_{\hat{n}})) \in \hat{\mathfrak{I}}(I)$. Clearly, $\hat{R}_\epsilon \preceq \hat{R} \preceq \tilde{R}$. Thus, we have $\left\| \check{F}_{\hat{R}} - f(b)g(b) + f(a)g(a) + \widehat{\int}_a^b f\, dg \right\|_{\mathcal{Y}} = \left\| \widehat{\int}_a^b f\, dg - F_{\tilde{R}} \right\|_{\mathcal{Y}} < \epsilon$. This shows that $\lim_{\hat{R} \in \hat{\mathfrak{I}}(I)} \check{F}_{\hat{R}} = f(b)g(b) - f(a)g(a) - \widehat{\int}_a^b f\, dg \in \mathcal{Y}$. Hence, g is Riemann-Stieltjes integrable with respect to f on I and $\widehat{\int}_a^b f\, dg + \widehat{\int}_a^b ((df)g) = f(b)g(b) - f(a)g(a) \in \mathcal{Y}$.

This completes the proof of the theorem. □

The preceding four theorems are direct generalizations of Theorem 29.4–29.7 of Bartle (1976). For Theorem 30.1 of Bartle (1976), we have the following sufficiency result for the general case, which is necessary and sufficient when all spaces are $\mathbb{R}$ and the integrator g is monotonically nondecreasing.

Theorem A.9 (Riemann Criterion for Integrability) *Let $I := [a, b] \subset \mathbb{R}$ be a compact interval with $a \leq b$, $\mathcal{X}$ and $\mathcal{Y}$ be Banach spaces over $\mathbb{K}$, $f : I \to \mathrm{B}(\mathcal{X}, \mathcal{Y})$, $g : I \to \mathcal{X}$, and the net $\left(F_{\hat{R}}\right)_{\hat{R} \in \hat{\mathfrak{I}}(I)}$ be as defined in Definition A.4 for $\widehat{\int}_a^b f \, \mathrm{d}g$. Assume that g is of bounded variation. Then, the following statements holds.*

(i) *f is Riemann-Stieltjes integrable with respect to g on I if $\forall \epsilon \in (0, \infty) \subset \mathbb{R}$, $\exists$ a sampled partition $\hat{R}_\epsilon := ((s_i)_{i=0}^n, (\xi_i)_{i=1}^n) \in \hat{\mathfrak{I}}(I)$, we have $\sum_{i=1}^n M_i T_g\left(\mathrm{r}_{s_{i-1}, s_i}\right) < \epsilon$, where $M_i := \sup_{v_1, v_2 \in [s_{i-1}, s_i] \subset \mathbb{R}} \| f(v_1) - f(v_2) \|_{\mathrm{B}(\mathcal{X}, \mathcal{Y})}$, $i = 1, \ldots, n$.*

(ii) *Under the assumption of (i), let $(\epsilon_i)_{i=1}^\infty \subseteq (0, \infty) \subset \mathbb{R}$ with $\lim_{i \in \mathbb{N}} \epsilon_i = 0$. Then, $\lim_{i \in \mathbb{N}} F_{\hat{R}_{\epsilon_i}} = \widehat{\int}_a^b f \, \mathrm{d}g$.*

(iii) *The converse to (i) holds if $\mathcal{X} = \mathcal{Y} = \mathbb{R}$, $\mathbb{K} = \mathbb{R}$, and g is monotonically nondecreasing.*

(iv) *Furthermore, if $\mathcal{X} = \mathbb{K}$, $\mathcal{Y}$ is finite-dimensional, g is monotonically nondecreasing (hence real-valued), then f is Riemann-Stieltjes integrable with respect to g on I implies that $\forall \epsilon \in (0, \infty) \subset \mathbb{R}$, $\exists$ a sampled partition $\hat{R}_\epsilon \in \hat{\mathfrak{I}}(I)$, $\forall \hat{R} := ((s_i)_{i=0}^n, (\xi_i)_{i=1}^n) \in \hat{\mathfrak{I}}(I)$ with $\hat{R}_\epsilon \preceq \hat{R}$, we have $\sum_{i=1}^n M_i T_g\left(\mathrm{r}_{s_{i-1}, s_i}\right) < \epsilon$, where M_i, $i = 1, \ldots, n$ are as defined in (i).*

Proof (i) $\forall \epsilon \in (0, \infty) \subset \mathbb{R}$, let $\hat{R}_\epsilon := ((s_i)_{i=0}^n, (\xi_i)_{i=1}^n) \in \hat{\mathfrak{I}}(I)$ such that $\sum_{i=1}^n M_i T_g\left(\mathrm{r}_{s_{i-1}, s_i}\right) < \epsilon$. $F_{\hat{R}_\epsilon} = \sum_{i=1}^n f(\xi_i)(g(s_i) - g(s_{i-1})) \in \mathcal{Y}$. $\forall \hat{R} := ((\bar{s}_i)_{i=0}^{\bar{n}}, \left(\bar{\xi}_i\right)_{i=1}^{\bar{n}}) \in \hat{\mathfrak{I}}(I)$ with $\hat{R}_\epsilon \preceq \hat{R}$, we have

$$
\begin{aligned}
&\| F_{\hat{R}} - F_{\hat{R}_\epsilon} \|_{\mathcal{Y}} \\
&= \Big\| \sum_{i=1}^{\bar{n}} f(\bar{\xi}_i)(g(\bar{s}_i) - g(\bar{s}_{i-1})) - \sum_{i=1}^{n} f(\xi_i)(g(s_i) - g(s_{i-1})) \Big\|_{\mathcal{Y}} \\
&= \Big\| \sum_{i=1}^{n} \Big(f(\xi_i)(g(s_i) - g(s_{i-1})) - \sum_{j=l_{i-1}+1}^{l_i} f(\bar{\xi}_j)(g(\bar{s}_j) - g(\bar{s}_{j-1})) \Big) \Big\|_{\mathcal{Y}} \\
&= \Big\| \sum_{i=1}^{n} \Big(\sum_{j=l_{i-1}+1}^{l_i} (f(\xi_i) - f(\bar{\xi}_j))(g(\bar{s}_j) - g(\bar{s}_{j-1})) \Big) \Big\|_{\mathcal{Y}} \\
&\leq \sum_{i=1}^{n} \Big(\sum_{j=l_{i-1}+1}^{l_i} \| (f(\xi_i) - f(\bar{\xi}_j))(g(\bar{s}_j) - g(\bar{s}_{j-1})) \|_{\mathcal{Y}} \Big) \\
&\leq \sum_{i=1}^{n} \Big(\sum_{j=l_{i-1}+1}^{l_i} \| f(\xi_i) - f(\bar{\xi}_j) \|_{\mathrm{B}(\mathcal{X}, \mathcal{Y})} \| g(\bar{s}_j) - g(\bar{s}_{j-1}) \|_{\mathcal{X}} \Big) \\
&\leq \sum_{i=1}^{n} M_i \Big(\sum_{j=l_{i-1}+1}^{l_i} \| g(\bar{s}_j) - g(\bar{s}_{j-1}) \|_{\mathcal{X}} \Big) \leq \sum_{i=1}^{n} M_i T_g\left(\mathrm{r}_{s_{i-1}, s_i}\right) < \epsilon
\end{aligned}
$$

where the first equality follows from Definition A.4; the second equality follows from the fact that $\hat{R}_\epsilon \preceq \hat{R}$ and Definition A.3 that leads to $s_i = \bar{s}_{l_i}$ and $l_{i-1} < l_i$, $i = 1, \ldots, n$; the third equality follows from simple algebra; the first inequality follows from Definition 7.1; the second inequality follows from Proposition 7.64; the third inequality follows from the definition of M_i, $i = 1, \ldots, n$; the fourth inequality follows from Definition 12.41. Thus, it is easy to show that the net $\left(F_{\hat{R}}\right)_{\hat{R} \in \hat{\mathfrak{I}}(I)}$ is a Cauchy net, which converges since $\mathcal{Y}$ is complete. Hence, f is Riemann-Stieltjes integrable with respect to g on I.

(ii) By the above derivation, we have $\epsilon_i \geq \lim_{\hat{R} \in \hat{\mathfrak{I}}(I)} \| F_{\hat{R}} - F_{\hat{R}_{\epsilon_i}} \|_{\mathcal{Y}} = \| \widehat{\int}_a^b f \, \mathrm{d}g - F_{\hat{R}_{\epsilon_i}} \|_{\mathcal{Y}}$, $\forall i \in \mathbb{N}$. Then, we have the desired result by taking $i \to \infty$.

(iii) Let $\mathcal{X} = \mathcal{Y} = \mathbb{R}$, $\mathbb{K} = \mathbb{R}$, g is monotonically nondecreasing, and f is Riemann-Stieltjes integrable with respect to g on I. Then, by Cauchy Criterion for Integrability (Theorem A.5) $\forall \epsilon \in (0, \infty) \subset \mathbb{R}$, $\exists \hat{R}_\epsilon := ((s_i)_{i=0}^n, (\xi_i)_{i=1}^n) \in \hat{\mathfrak{I}}(I)$ such that $\forall \hat{R}_1, \hat{R}_2 \in \hat{\mathfrak{I}}(I)$ with $\hat{R}_\epsilon \preceq \hat{R}_i$, $i = 1, 2$, we have $\left| F_{\hat{R}_1} - F_{\hat{R}_2} \right| < \frac{\epsilon}{2}$. Take $\hat{R}_1 := ((s_i)_{i=0}^n, \left(\hat{\xi}_i\right)_{i=1}^n)$ and $\hat{R}_2 := ((s_i)_{i=0}^n, \left(\check{\xi}_i\right)_{i=1}^n)$ where $\hat{\xi}_i, \check{\xi}_i \in [s_{i-1}, s_i] \subset \mathbb{R}$ satisfies that $0 \vee (M_i - \frac{\epsilon}{1+2T_g(\mathrm{r}_{a,b})}) \leq f(\hat{\xi}_i) - f(\check{\xi}_i) \leq M_i$, $i = 1, \ldots, n$. Then, we have $\hat{R}_\epsilon \preceq \hat{R}_i$, $i = 1, 2$, and

$$
\begin{aligned}
\frac{\epsilon}{2} &> \left| F_{\hat{R}_1} - F_{\hat{R}_2} \right|_{\mathcal{Y}} \\
&= \Big| \sum_{i=1}^n f(\hat{\xi}_i)(g(s_i) - g(s_{i-1})) - \sum_{i=1}^n f(\check{\xi}_i)(g(s_i) - g(s_{i-1})) \Big| \\
&= \Big| \sum_{i=1}^n (f(\hat{\xi}_i) - f(\check{\xi}_i)) T_g\left(\mathrm{r}_{s_{i-1}, s_i}\right) \Big| = \sum_{i=1}^n (f(\hat{\xi}_i) - f(\check{\xi}_i)) T_g\left(\mathrm{r}_{s_{i-1}, s_i}\right) \\
&\geq \sum_{i=1}^n (M_i - \frac{\epsilon}{1 + 2T_g\left(\mathrm{r}_{a,b}\right)}) T_g\left(\mathrm{r}_{s_{i-1}, s_i}\right) \\
&= \sum_{i=1}^n M_i T_g\left(\mathrm{r}_{s_{i-1}, s_i}\right) - \frac{\epsilon}{1 + 2T_g\left(\mathrm{r}_{a,b}\right)} \sum_{i=1}^n T_g\left(\mathrm{r}_{s_{i-1}, s_i}\right) \\
&= \sum_{i=1}^n M_i T_g\left(\mathrm{r}_{s_{i-1}, s_i}\right) - \frac{\epsilon}{1 + 2T_g\left(\mathrm{r}_{a,b}\right)} T_g\left(\mathrm{r}_{a,b}\right)
\end{aligned}
$$

where the first equality follows from Definition A.4; the second equality follows from the monotonicity of g; the third equality and the second inequality follow from the choice for $\hat{\xi}_i$'s and $\check{\xi}_i$'s; the fourth equality follows from simple algebra; and the fifth equality follows from Proposition 12.47. Therefore, we have $\sum_{i=1}^n M_i T_g\left(\mathrm{r}_{s_{i-1}, s_i}\right) < \epsilon$.

(iv) Let $\mathcal{X} = \mathbb{K}$, $\mathcal{Y}$ be finite-dimensional, g is monotonically nondecreasing (g is real-valued), and f is Riemann-Stieltjes integrable with respect to g on I. First, consider the special case when $\mathbb{K} = \mathbb{R}$ and $\mathcal{Y} = \mathbb{R}$. $\forall \epsilon \in (0, \infty) \subset \mathbb{R}$, let $\hat{R}_\epsilon$ be as found in (iii). $\forall \hat{R} := ((\bar{s}_i)_{i=0}^{\bar{n}}, (\bar{\xi}_i)_{i=1}^{\bar{n}}) \in \hat{\mathfrak{I}}(I)$ with $\hat{R}_\epsilon \preceq \hat{R}$, we take $\hat{R}_1 := ((\bar{s}_i)_{i=0}^{\bar{n}}, (\hat{\xi}_i)_{i=1}^{\bar{n}})$ and $\hat{R}_2 := ((\bar{s}_i)_{i=0}^{\bar{n}}, (\check{\xi}_i)_{i=1}^{\bar{n}})$ where $\hat{\xi}_i, \check{\xi}_i \in [\bar{s}_{i-1}, \bar{s}_i] \subset \mathbb{R}$ satisfies that $0 \vee (\bar{M}_i - \frac{\epsilon}{1+2T_g(\mathrm{r}_{a,b})}) \leq f(\hat{\xi}_i) - f(\check{\xi}_i) \leq \bar{M}_i$, $i = 1, \ldots, \bar{n}$, where $\bar{M}_i := \sup_{v_1, v_2 \in [\bar{s}_{i-1}, \bar{s}_i] \subset \mathbb{R}} |f(v_1) - f(v_2)|$, $i = 1, \ldots, \bar{n}$. Then, the line of argument in the proof of (iii) holds again. Therefore, we have $\sum_{i=1}^{\bar{n}} \bar{M}_i T_g(\mathrm{r}_{\bar{s}_{i-1}, \bar{s}_i}) < \epsilon$.

Next, consider the case when $\mathbb{K} = \mathbb{C}$ and $\mathcal{Y} = \mathbb{C}$. Fix any $\epsilon \in (0, \infty) \subset \mathbb{R}$. $\forall v \in I$, $f(v) = \tilde{f}(v) + \mathrm{i}\check{f}(v)$ where $\tilde{f} : I \to \mathbb{R}$ and $\check{f} : I \to \mathbb{R}$. By the assumption $\lim_{\hat{R} \in \hat{\mathfrak{I}}(I)} F_{\hat{R}} = \widehat{\int}_a^b f \, dg \in \mathbb{C}$, by Proposition 3.66, we have $\widehat{\int}_a^b \tilde{f} \, dg = \mathrm{Re}\,(\widehat{\int}_a^b f \, dg) \in \mathbb{R}$ and $\widehat{\int}_a^b \check{f} \, dg = \mathrm{Im}\,(\widehat{\int}_a^b f \, dg) \in \mathbb{R}$. Then, $\tilde{f}$ and $\check{f}$ are Riemann-Stieltjes integrable with respect to g on I. By the special case when $\mathbb{K} = \mathbb{R}$ and $\mathcal{Y} = \mathbb{R}$ applied to $\tilde{f}$, $\exists \tilde{R}_\epsilon \in \hat{\mathfrak{I}}(I)$, $\forall \hat{R} := ((\bar{s}_i)_{i=0}^{\bar{n}}, (\bar{\xi}_i)_{i=1}^{\bar{n}}) \in \hat{\mathfrak{I}}(I)$ with $\tilde{R}_\epsilon \preceq \hat{R}$, we have $\sum_{i=1}^{\bar{n}} \tilde{M}_i T_g(\mathrm{r}_{\bar{s}_{i-1}, \bar{s}_i}) < \frac{\epsilon}{2}$, where $\tilde{M}_i := \sup_{v_1, v_2 \in [\bar{s}_{i-1}, \bar{s}_i] \subset \mathbb{R}} |\tilde{f}(v_1) - \tilde{f}(v_2)|$, $i = 1, \ldots, \bar{n}$. By the special case when $\mathbb{K} = \mathbb{R}$ and $\mathcal{Y} = \mathbb{R}$ applied to $\check{f}$, $\exists \check{R}_\epsilon \in \hat{\mathfrak{I}}(I)$, $\forall \hat{R} := ((\bar{s}_i)_{i=0}^{\bar{n}}, (\bar{\xi}_i)_{i=1}^{\bar{n}}) \in \hat{\mathfrak{I}}(I)$ with $\check{R}_\epsilon \preceq \hat{R}$, we have $\sum_{i=1}^{\bar{n}} \check{M}_i T_g(\mathrm{r}_{\bar{s}_{i-1}, \bar{s}_i}) < \frac{\epsilon}{2}$, where $\check{M}_i := \sup_{v_1, v_2 \in [\bar{s}_{i-1}, \bar{s}_i] \subset \mathbb{R}} |\check{f}(v_1) - \check{f}(v_2)|$, $i = 1, \ldots, \bar{n}$. Let $\hat{R}_\epsilon \in \hat{\mathfrak{I}}(I)$ be such that $\tilde{R}_\epsilon \preceq \hat{R}_\epsilon$ and $\check{R}_\epsilon \preceq \hat{R}_\epsilon$. $\forall \hat{R} := ((\bar{s}_i)_{i=0}^{\bar{n}}, (\bar{\xi}_i)_{i=1}^{\bar{n}}) \in \hat{\mathfrak{I}}(I)$ with $\hat{R}_\epsilon \preceq \hat{R}$, we have

$$\sum_{i=1}^{\bar{n}} \bar{M}_i T_g(\mathrm{r}_{\bar{s}_{i-1}, \bar{s}_i}) := \sum_{i=1}^{\bar{n}} \sup_{v_1, v_2 \in [\bar{s}_{i-1}, \bar{s}_i] \subset \mathbb{R}} |f(v_1) - f(v_2)| T_g(\mathrm{r}_{\bar{s}_{i-1}, \bar{s}_i})$$

$$= \sum_{i=1}^{\bar{n}} \sup_{v_1, v_2 \in [\bar{s}_{i-1}, \bar{s}_i] \subset \mathbb{R}} \left|(\tilde{f}(v_1) - \tilde{f}(v_2)) + \mathrm{i}(\check{f}(v_1) - \check{f}(v_2))\right| T_g(\mathrm{r}_{\bar{s}_{i-1}, \bar{s}_i})$$

$$\leq \sum_{i=1}^{\bar{n}} (\tilde{M}_i + \check{M}_i) T_g(\mathrm{r}_{\bar{s}_{i-1}, \bar{s}_i}) < \epsilon$$

This proves this special case.

Finally, consider the general case when $\mathcal{X} = \mathbb{K}$ and $\mathcal{Y}$ is finite-dimensional. Let $m \in \mathbb{Z}_+$ be the dimension of $\mathcal{Y}$ and $y_1, \ldots, y_m \in \mathcal{Y}$ be a basis of $\mathcal{Y}$. Fix any $\epsilon \in (0, \infty) \subset \mathbb{R}$. $\forall v \in I$, $f(v)$ can be uniquely expressed as $\sum_{j=1}^{m} a_j(v) y_j$, where $a_j(v) = \pi_j(f(v)) \in \mathbb{R}$, $j = 1, \ldots, m$; π_j is the projection operator for the jth coordinate (which is a bounded linear operator). Fix any $j \in \{1, \ldots, m\}$. By the assumption $\lim_{\hat{R} \in \hat{\mathfrak{I}}(I)} F_{\hat{R}} = \widehat{\int}_a^b f \, dg \in \mathcal{Y}$, by Proposition 3.66, we have $\widehat{\int}_a^b a_j \, dg = \pi_j(\widehat{\int}_a^b f \, dg) \in \mathbb{K}$. Then, a_j is Riemann-Stieltjes integrable with respect to g on I. By either of the previous special cases (depending on $\mathbb{K}$), $\exists \hat{R}_{j,\epsilon} \in \hat{\mathfrak{I}}(I)$, $\forall \hat{R} := ((\bar{s}_i)_{i=0}^{\bar{n}}, (\bar{\xi}_i)_{i=1}^{\bar{n}}) \in \hat{\mathfrak{I}}(I)$ with $\hat{R}_{j,\epsilon} \preceq \hat{R}$, we have $\sum_{i=1}^{\bar{n}} \tilde{M}_{j,i} T_g\left(\mathrm{r}_{\bar{s}_{i-1},\bar{s}_i}\right) < \frac{\epsilon}{L}$, where $\tilde{M}_{j,i} := \sup_{v_1,v_2 \in [\bar{s}_{i-1},\bar{s}_i] \subset \mathbb{R}} \left|a_j(v_1) - a_j(v_2)\right|$, $i = 1, \ldots, \bar{n}$; and $L := \sum_{k=1}^{m} \|y_k\|_{\mathcal{Y}} \in (0, \infty) \subset \mathbb{R}$. Let $\hat{R}_\epsilon \in \hat{\mathfrak{I}}(I)$ be such that $\hat{R}_{j,\epsilon} \preceq \hat{R}_\epsilon$, $j = 1, \ldots, m$. $\forall \hat{R} := ((\bar{s}_i)_{i=0}^{\bar{n}}, (\bar{\xi}_i)_{i=1}^{\bar{n}}) \in \hat{\mathfrak{I}}(I)$ with $\hat{R}_\epsilon \preceq \hat{R}$, we have

$$\begin{aligned}
\sum_{i=1}^{\bar{n}} \bar{M}_i T_g\left(\mathrm{r}_{\bar{s}_{i-1},\bar{s}_i}\right) &:= \sum_{i=1}^{\bar{n}} \sup_{v_1,v_2 \in [\bar{s}_{i-1},\bar{s}_i] \subset \mathbb{R}} \|f(v_1) - f(v_2)\|_{\mathcal{Y}} T_g\left(\mathrm{r}_{\bar{s}_{i-1},\bar{s}_i}\right) \\
&= \sum_{i=1}^{\bar{n}} \sup_{v_1,v_2 \in [\bar{s}_{i-1},\bar{s}_i] \subset \mathbb{R}} \left\| \sum_{j=1}^{m} (a_j(v_1) - a_j(v_2)) y_j \right\|_{\mathcal{Y}} T_g\left(\mathrm{r}_{\bar{s}_{i-1},\bar{s}_i}\right) \\
&\leq \sum_{i=1}^{\bar{n}} \sup_{v_1,v_2 \in [\bar{s}_{i-1},\bar{s}_i] \subset \mathbb{R}} \sum_{j=1}^{m} \left|a_j(v_1) - a_j(v_2)\right| \left\|y_j\right\|_{\mathcal{Y}} T_g\left(\mathrm{r}_{\bar{s}_{i-1},\bar{s}_i}\right) \\
&\leq \sum_{i=1}^{\bar{n}} \sum_{j=1}^{m} \sup_{v_1,v_2 \in [\bar{s}_{i-1},\bar{s}_i] \subset \mathbb{R}} \left|a_j(v_1) - a_j(v_2)\right| \left\|y_j\right\|_{\mathcal{Y}} T_g\left(\mathrm{r}_{\bar{s}_{i-1},\bar{s}_i}\right) \\
&= \sum_{j=1}^{m} \left\|y_j\right\|_{\mathcal{Y}} \sum_{i=1}^{\bar{n}} \tilde{M}_{j,i} T_g\left(\mathrm{r}_{\bar{s}_{i-1},\bar{s}_i}\right) < \sum_{j=1}^{m} \frac{\epsilon}{L} \left\|y_j\right\|_{\mathcal{Y}} = \epsilon
\end{aligned}$$

where the first equality follows from the fact that $y_1, \ldots, y_m$ is a basis of $\mathcal{Y}$; the first inequality follows from Definition 7.1; the second inequality follows from Proposition 3.81; the second equality follows from the definition of $\tilde{M}_{j,i}$'s; the last inequality follows from the fact $\hat{R}_{j,\epsilon} \preceq \hat{R}$, $j = 1, \ldots, m$; and the last equality follows from the definition of L. This completes the proof of (iv).

This completes the proof of the theorem. □

Theorem A.10 (Integrability Theorem) *Let $I := [a, b] \subset \mathbb{R}$ be a compact interval with $a \leq b$, $\mathcal{X}$ and $\mathcal{Y}$ be Banach spaces over $\mathbb{K}$, $f : I \to \mathrm{B}(\mathcal{X}, \mathcal{Y})$ be continuous, and $g : I \to \mathcal{X}$ be of bounded variation. Then, the following statements holds.*

(i) *f is Riemann-Stieltjes integrable with respect to g on I.*

(ii) *g is Riemann-Stieltjes integrable with respect to f on I and $\widehat{\int}_a^b f \,\mathrm{d}g + \widehat{\int}_a^b ((\mathrm{d}f) g) = f(b)g(b) - f(a)g(a) \in \mathcal{Y}$.*

(iii) *Let $\left(F_{\hat{R}}\right)_{\hat{R} \in \hat{\mathfrak{I}}(I)}$ and $\left(\check{F}_{\hat{R}}\right)_{\hat{R} \in \hat{\mathfrak{I}}(I)}$ be the nets for $\widehat{\int}_a^b f \,\mathrm{d}g$ and $\widehat{\int}_a^b ((\mathrm{d}f) g)$, respectively, as defined in Definition A.4. Define* $\mathrm{Gauge}(\hat{R}) := \max_{i=1,\ldots,n} (s_i - s_{i-1})$, *$\forall \hat{R} := ((s_i)_{i=0}^{n}, (\xi_i)_{i=1}^{n}) \in \hat{\mathfrak{I}}(I)$. Then, we have $\widehat{\int}_a^b f \,\mathrm{d}g = \lim_{\mathrm{Gauge}(\hat{R}) \to 0} F_{\hat{R}}$*

and $\widehat{\int}_a^b((\mathrm{d}f)g) = \lim_{\mathrm{Gauge}(\hat{R})\to 0} \check{F}_{\hat{R}}$. *We will say that the Riemann-Stieltjes integrals* converge in the gauge mode *if the above limits hold.*

(iv) $\widehat{\int}_a^b f\,\mathrm{d}g = \int_{\mathrm{r}_{a,b}} f\,\mathrm{d}g \in \mathcal{Y}$, *where the right-hand side is the Lebesgue-Stieltjes integral defined in Definition 12.57.*

(v) $\left\|\widehat{\int}_a^b f\,\mathrm{d}g\right\|_{\mathcal{Y}} \leq MT_g\left(\mathrm{r}_{a,b}\right)$, *where* $M := \max_{t\in\bar{\mathrm{r}}_{a,b}} \|f(t)\|_{\mathrm{B}(\mathcal{X},\mathcal{Y})}$; *and* $\left\|\widehat{\int}_a^b((\mathrm{d}f)g)\right\|_{\mathcal{Y}} \leq \|(f(b)-f(a))g(a)\|_{\mathcal{Y}} + \left\|\widehat{\int}_a^b(f - f(b))\,\mathrm{d}g\right\|_{\mathcal{Y}} \leq \bar{M}\cdot(\|g(a)\|_{\mathcal{X}} + T_g\left(\mathrm{r}_{a,b}\right))$, *where* $\bar{M} := \max_{t\in\bar{\mathrm{r}}_{a,b}} \|f(t)-f(b)\|_{\mathrm{B}(\mathcal{X},\mathcal{Y})}$.

It is easy to see that $\mathrm{Gauge}(\cdot) : \hat{\mathfrak{R}}(I) \to [0,\infty) \subset \mathbb{R}$ induces a relation $\trianglelefteq$ by $\hat{R}_1 \trianglelefteq \hat{R}_2$ if $\mathrm{Gauge}(\hat{R}_1) \geq \mathrm{Gauge}(\hat{R}_2)$, $\forall \hat{R}_1, \hat{R}_2 \in \hat{\mathfrak{R}}(I)$, and the pair $\mathcal{M} := (\hat{\mathfrak{R}}(I), \trianglelefteq)$ forms a directed system. Then, we can talk about convergence of the net $\left(F_{\hat{R}}\right)_{\hat{R}\in\mathcal{M}} \subseteq \mathcal{Y}$. Thus, we will denote the limit of this net by $\lim_{\mathrm{Gauge}(\hat{R})\to 0} F_{\hat{R}}$ if it exists. If $\lim_{\mathrm{Gauge}(\hat{R})\to 0} F_{\hat{R}} = \lim_{\hat{R}\in\hat{\mathfrak{I}}(I)} F_{\hat{R}} =: \widehat{\int}_a^b f\,\mathrm{d}g$, we will say that $\widehat{\int}_a^b f\,\mathrm{d}g$ converges in gauge mode.

It is easy to see that $\forall \hat{R}_1, \hat{R}_2 \in \hat{\mathfrak{R}}(I)$, $\hat{R}_1 \preceq \hat{R}_2$ implies that $\hat{R}_1 \trianglelefteq \hat{R}_2$. Thus, $\lim_{\mathrm{Gauge}(\hat{R})\to 0} F_{\hat{R}} \in \mathcal{Y}$ implies that f is Riemann-Stieltjes integrable with respect to g on I and $\lim_{\mathrm{Gauge}(\hat{R})\to 0} F_{\hat{R}} = \lim_{\hat{R}\in\hat{\mathfrak{I}}(I)} F_{\hat{R}} =: \widehat{\int}_a^b f\,\mathrm{d}g$. The converse does not hold in general.

Proof (i) Since f is continuous and I is compact, then f is uniformly continuous on I by Proposition 5.39. Let $T_g\left(\mathrm{r}_{a,b}\right) =: c \in [0,\infty) \subset \mathbb{R}$, since g is of bounded variation. Then, $\forall \epsilon \in (0,\infty) \subset \mathbb{R}$, $\exists \delta \in (0,\infty) \subset \mathbb{R}$ such that $\forall s_1, s_2 \in I$ with $|s_1 - s_2| < \delta$, we have $\|f(s_1) - f(s_2)\|_{\mathrm{B}(\mathcal{X},\mathcal{Y})} < \frac{\epsilon}{1+c}$. Take any $\hat{R} := ((s_i)_{i=0}^n, (\xi_i)_{i=1}^n) \in \hat{\mathfrak{I}}(I)$ with $\mathrm{Gauge}(\hat{R}) < \delta \wedge \epsilon$. Let $M_i := \sup_{v_1,v_2\in[s_{i-1},s_i]\subset\mathbb{R}} \|f(v_1) - f(v_2)\|_{\mathrm{B}(\mathcal{X},\mathcal{Y})}$, $i = 1,\dots,n$. Then, we have $\sum_{i=1}^n M_i T_g\left(\mathrm{r}_{s_{i-1},s_i}\right) \leq \sum_{i=1}^n \frac{\epsilon}{1+c} T_g\left(\mathrm{r}_{s_{i-1},s_i}\right) = \frac{\epsilon}{1+c} T_g\left(\mathrm{r}_{a,b}\right) = \frac{c\epsilon}{1+c} < \epsilon$, where the first inequality follows from the fact that $0 < s_i - s_{i-1} \leq \mathrm{Gauge}(\hat{R}) < \delta$ and then $M_i \leq \frac{\epsilon}{1+c}$, $i = 1,\dots,n$, and the first equality follows from Proposition 12.47. By Riemann Criterion for Integrability (Theorem A.9), f is Riemann-Stieltjes integrable with respect to on I, and $\widehat{\int}_a^b f\,\mathrm{d}g = \lim_{\mathrm{Gauge}(\hat{R})\to 0} F_{\hat{R}} \in \mathcal{Y}$.

(ii) By Integration by Parts (Theorem A.8), we have g is Riemann-Stieltjes integrable with respect to f on I and $\widehat{\int}_a^b f\,\mathrm{d}g + \widehat{\int}_a^b((\mathrm{d}f)g) = f(b)g(b) - f(a)g(a)$.

(iii) By the proof of (i), we have $\widehat{\int}_a^b f\,\mathrm{d}g = \lim_{\mathrm{Gauge}(\hat{R})\to 0} F_{\hat{R}} \in \mathcal{Y}$. This, coupled with the proof of Integration by Parts (Theorem A.8), we have $\widehat{\int}_a^b((\mathrm{d}f)\,g) = \lim_{\mathrm{Gauge}(\hat{R})\to 0} \check{F}_{\hat{R}}$.

(iv) By Proposition 12.39, $\mathrm{P}(I) = \mathrm{r}_{a,b}$. By Theorem 12.50, there exists a unique finite $\mathcal{X}$-valued measure μ on the measurable space $(\mathrm{P}(I), \mathcal{B}_{\mathrm{B}}(\mathrm{P}(I)))$ such that g is a cumulative distribution function of μ, $\mu(\mathrm{r}_{s_1,s_2}) = g(s_2) - g(s_1)$ and $\mathcal{P}\circ\mu(\mathrm{r}_{s_1,s_2}) = T_g\left(\mathrm{r}_{s_1,s_2}\right)$, $\forall a \leq s_1 \leq s_2 \leq b$, and $\mathcal{P}\circ\mu(\mathrm{P}(I)) = \mathcal{P}\circ\mu(\mathrm{r}_{a,b}) = T_g\left(\mathrm{r}_{a,b}\right) = T_g =: c \in [0,\infty) \subset \mathbb{R}$. Since f is continuous and I is compact, then f is bounded,

i. e., $\exists M \in \mathbb{R}_+$ such that $\|f(s)\|_{B(\mathcal{X},\mathcal{Y})} \le M$, $\forall s \in I$. By Proposition 7.126 and Lebesgue Dominated Convergence Theorem 11.128, f is absolutely integrable over $P(I)$ with respect to μ, and $\int_{r_{a,b}} f \, d\mu \in \mathcal{Y}$.

By (i), $\hat{I} := \widehat{\int}_a^b f \, dg \in \mathcal{Y}$. $\forall n \in \mathbb{N}$, let $s_i^{\langle n \rangle} := a + i\frac{b-a}{n}$, $i = 0, \ldots, n$, and $\hat{R}_n := \left(\left(s_i^{\langle n \rangle}\right)_{i=0}^n, \left(s_i^{\langle n \rangle}\right)_{i=1}^n\right) \in \hat{\mathfrak{I}}(I)$. By (iii), $\hat{I} = \lim_{\hat{R} \in \hat{\mathfrak{I}}(I)} F_{\hat{R}} = \lim_{n \in \mathbb{N}} F_{\hat{R}_n}$. This implies that

$$\hat{I} = \lim_{n \in \mathbb{N}} \sum_{j=1}^n f(s_j^{\langle n \rangle})(g(s_j^{\langle n \rangle}) - g(s_{j-1}^{\langle n \rangle}))$$

$$= \lim_{n \in \mathbb{N}} \int_{r_{a,b}} \sum_{j=1}^n f(s_j^{\langle n \rangle}) \chi_{r_{s_{j-1}^{\langle n \rangle}, s_j^{\langle n \rangle}}, I} \, d\mu =: \lim_{n \in \mathbb{N}} \int_{r_{a,b}} h_n \, d\mu$$

where $h_n : r_{a,b} \to B(\mathcal{X}, \mathcal{Y})$, and the second equality follows from Proposition 11.125. Clearly, $0 \le \|h_n(s)\|_{B(\mathcal{X},\mathcal{Y})} \le M$, $\forall s \in r_{a,b}$, $\forall n \in \mathbb{N}$. It is also clear that $\lim_{n \in \mathbb{N}} h_n(s) = f(s)$, $\forall s \in r_{a,b}$. By Lebesgue Dominated Convergence Theorem 11.128 and Propositions 7.126 and 11.37, we have f is absolutely integrable over $(r_{a,b}, \mathcal{B}_B(r_{a,b}), \mu)$ and $\int_{r_{a,b}} f \, d\mu = \lim_{n \in \mathbb{N}} \int_{r_{a,b}} h_n \, d\mu = \hat{I}$. By Definition 12.57, we have $\hat{I} = \int_{r_{a,b}} f \, dg$.

(v) By (iv), we have $\left\|\widehat{\int}_a^b f \, dg\right\|_{\mathcal{Y}} = \left\|\int_{r_{a,b}} f \, dg\right\| \le M T_g(r_{a,b})$, where the inequality follows from Definition 12.57 and Proposition 11.132 and Theorem 12.50. For the second inequality, we have $\left\|\widehat{\int}_a^b ((df)g)\right\|_{\mathcal{Y}} = \left\|\widehat{\int}_a^b ((d(f - f(b)))g) + \widehat{\int}_a^b ((df(b))g)\right\|_{\mathcal{Y}} = \left\|\widehat{\int}_a^b ((d(f - f(b)))g)\right\|_{\mathcal{Y}} = \left\|(f(b) - f(b))g(b) - (f(a) - f(b))g(a) - \widehat{\int}_a^b (f - f(b)) \, dg\right\|_{\mathcal{Y}} \le \|(f(b) - f(a))g(a)\|_{\mathcal{Y}} + \left\|\widehat{\int}_a^b (f - f(b)) \, dg\right\|_{\mathcal{Y}} \le \bar{M}(\|g(a)\|_{\mathcal{X}} + T_g(r_{a,b}))$, where the first equality follows from Theorem A.6; the second equality follows from Definition A.4; the third equality follows from (ii); the first inequality follows from triangular inequality; and the last inequality follows from the preceding discussion.

This completes the proof of the theorem. □

Theorem A.11 *Let $I := [a, b] \subset \mathbb{R}$ be a compact interval with $a \le b$, $\mathcal{Y}$ be a Banach space over $\mathbb{K}$, $f : I \to \mathcal{Y}$, and $g := \mathrm{id}_I : I \to I \subset \mathbb{R}$. Then, f is Riemann-Stieltjes integrable with respect to g on I if, and only if, the Riemann-Stieltjes integral of f with respect to g on I converges in the gauge mode. In this case, $J := \widehat{\int}_a^b f \, dg = \lim_{\mathrm{Gauge}(\hat{R}) \to 0} F_{\hat{R}}$, where $\left(F_{\hat{R}}\right)_{\hat{R} \in \hat{\mathfrak{I}}(I)}$ is the net for $\widehat{\int}_a^b f \, dg$ as defined in Definition A.4.*

Proof "Sufficiency" This follows immediately from the ensuing discussion after Theorem A.10.

"Necessity" Clearly, the result holds trivially if $a = b$. Consider the nontrivial case $b > a$. By the assumption, we have f is bounded, i. e., $\exists M \in [0, \infty) \subset \mathbb{R}$

such that $\|f(s)\|_{\mathcal{Y}} \leq M$, $\forall s \in I$. Now, fix any $\epsilon \in (0, \infty) \subset \mathbb{R}$. By Definition A.4, $\exists \hat{R}_\epsilon := ((s_i)_{i=0}^n, (\xi_i)_{i=1}^n) \in \hat{\mathfrak{I}}(I)$ such that $\forall \hat{Q} \in \hat{\mathfrak{I}}(I)$ with $\hat{R}_\epsilon \preceq \hat{Q}$, we have $\left\| F_{\hat{Q}} - J \right\|_{\mathcal{Y}} < \frac{\epsilon}{2}$. Then, $n \in \mathbb{N}$ (since $a < b$) $a = s_0 \leq s_1 \leq s_2 \leq \cdots \leq s_n = b$, and $\xi_i \in [s_{i-1}, s_i] \subset \mathbb{R}$, $i = 1, \ldots, n$. Without loss of generality, we may assume that $a = s_0 < s_1 < s_2 < \cdots < s_n = b$. Define $\delta := \frac{\min_{i=1,\ldots,n}(s_i - s_{i-1})}{3} \wedge \frac{\epsilon}{8(M+1)n} \in (0, \infty) \subset \mathbb{R}$. Fix any $\hat{Q} \in \hat{\mathfrak{I}}(I)$ with $\text{Gauge}(\hat{Q}) < \delta$. Let $\hat{Q} = ((\hat{s}_i)_{i=0}^{\hat{n}}, \left(\hat{\xi}_i\right)_{i=1}^{\hat{n}})$. Without loss of generality, assume that $\hat{n} \in \mathbb{N}$, $a = \hat{s}_0 < \hat{s}_1 < \cdots < \hat{s}_{\hat{n}} = b$, and $\hat{\xi}_i \in [\hat{s}_{i-1}, \hat{s}_i] \subset \mathbb{R}$, $i = 1, \ldots, \hat{n}$. Let $\{\check{s}_i \in I \mid i = 0, \ldots, \check{n}\} := \{s_0, \ldots, s_n\} \cup \left\{\hat{s}_0, \ldots, \hat{s}_{\hat{n}}\right\}$. Then, $\check{n} \in \mathbb{N}$ and we may define $\left(\check{s}_i\right)_{i=0}^{\check{n}}$ such that $a = \check{s}_0 < \check{s}_1 < \cdots < \check{s}_{\check{n}} = b$. It is clear that there exists $i_0, \ldots, i_n \in \left\{0, \ldots, \check{n}\right\}$ such that $s_j = \check{s}_{i_j}$, $j = 0, \ldots, n$, and $i_0 = 0$ and $i_n = \check{n}$. By our choice of δ, we have $0 = i_0 < i_0 + 2 < i_1 < i_1 + 2 < i_2 < i_2 + 2 < \cdots < i_n = \check{n}$. Let $\check{\xi}_i = \begin{cases} s_j & \text{if } i = i_j \text{ or } i = i_j + 1 \\ \hat{\xi}_l & \text{if } [\check{s}_{i-1}, \check{s}_i] = [\hat{s}_{l-1}, \hat{s}_l] \end{cases}$, $i = 1, \ldots, \check{n}$. Now, define $\hat{R} := ((\check{s}_i)_{i=0}^{\check{n}}, \left(\check{\xi}_i\right)_{i=1}^{\check{n}})$. Clearly, $\hat{R} \in \hat{\mathfrak{I}}(I)$ and $\hat{R}_\epsilon \preceq \hat{R}$. Then, we have the following line of arguments.

$$
\begin{aligned}
\left\| F_{\hat{Q}} - J \right\|_{\mathcal{Y}} &\leq \left\| F_{\hat{Q}} - F_{\hat{R}} \right\|_{\mathcal{Y}} + \| F_{\hat{R}} - J \|_{\mathcal{Y}} < \left\| F_{\hat{Q}} - F_{\hat{R}} \right\|_{\mathcal{Y}} + \frac{\epsilon}{2} \\
&= \left\| \sum_{i=1}^{\hat{n}} f(\hat{\xi}_i)\,(\hat{s}_i - \hat{s}_{i-1}) - \sum_{i=1}^{\check{n}} f(\check{\xi}_i)\,(\check{s}_i - \check{s}_{i-1}) \right\|_{\mathcal{Y}} + \frac{\epsilon}{2} \\
&= \| \sum_{\substack{i=1 \\ \exists s_j \in [\hat{s}_{i-1}, \hat{s}_i] \subset \mathbb{R}, j=0,\ldots,n}}^{\hat{n}} f(\hat{\xi}_i)\,(\hat{s}_i - \hat{s}_{i-1}) - \sum_{j=1}^{n-1} f(\check{\xi}_{i_j})\,(\check{s}_{i_j} - \check{s}_{i_j - 1}) \\
&\quad - \sum_{j=1}^{n-1} f(\check{\xi}_{i_j+1})\,(\check{s}_{i_j+1} - \check{s}_{i_j}) - f(\check{\xi}_{i_0+1})\,(\check{s}_{i_0+1} - \check{s}_{i_0}) \\
&\quad - f(\check{\xi}_{i_n})\,(\check{s}_{i_n} - \check{s}_{i_n-1}) \|_{\mathcal{Y}} + \frac{\epsilon}{2} \\
&\leq \sum_{j=1}^{n-1} 2M\,(\check{s}_{i_j} - \check{s}_{i_j-1}) + \sum_{j=1}^{n-1} 2M\,(\check{s}_{i_j+1} - \check{s}_{i_j}) + 2M\,(\check{s}_{i_0+1} - \check{s}_{i_0}) \\
&\quad + 2M\,(\check{s}_{i_n} - \check{s}_{i_n-1}) + \frac{\epsilon}{2} < \frac{\epsilon}{2} + 2M(2(n-1)+2)\delta < \epsilon
\end{aligned}
$$

where the first inequality follows from triangular inequality; the second inequality follows from the choice of $\hat{R}_\epsilon$; the first equality follows from the definition of the net $\left(F_{\hat{R}}\right)_{\hat{R} \in \hat{\mathfrak{I}}(I)}$; the second equality follows from the definition of $\hat{R}$; the third inequality follows from simple algebra; the fourth inequality follows from the fact $\text{Gauge}(\hat{R}) \leq \text{Gauge}(\hat{Q}) < \delta$; and the last inequality follows from the choice of δ.

Thus, $\forall \epsilon \in (0, \infty) \subset \mathbb{R}$, we have found a $\delta \in (0, \infty) \subset \mathbb{R}$ such that, $\forall \hat{Q} \in \hat{\mathfrak{I}}(I)$ with $\text{Gauge}(\hat{Q}) < \delta$, we have $\left\| F_{\hat{Q}} - J \right\|_{\mathcal{Y}} < \epsilon$. Hence, we have $\lim_{\text{Gauge}(\hat{Q}) \to 0} F_{\hat{Q}} = J$.

This completes the proof of the theorem. □

Theorem A.12 *Let $\bar{\mathrm{r}}_{a,b} \subset \mathbb{R}$ with $a \leq b$, $\mathcal{X}$ and $\mathcal{Y}$ be Banach space over $\mathbb{K}$, $f : \bar{\mathrm{r}}_{a,b} \to \mathrm{B}(\mathcal{X}, \mathcal{Y})$ be of bounded variation, $g : \bar{\mathrm{r}}_{a,b} \to \mathcal{X}$ be continuous, and $\phi : \bar{\mathrm{r}}_{a,b} \to \mathcal{Y}$ be defined by $\phi(t) := \widehat{\int}_a^t f \, \mathrm{d}g \in \mathcal{Y}$. Then, ϕ is continuous.*

Proof By Integrability Theorem A.10, f is Riemann-Stieltjes integrable with respect to g on $\bar{\mathrm{r}}_{a,b}$. $\forall t \in \bar{\mathrm{r}}_{a,b}$, by Definition 12.41, $f|_{\bar{\mathrm{r}}_{a,t}}$ and $f|_{\bar{\mathrm{r}}_{t,b}}$ are of bounded variation. Then, ϕ is well-defined. Since f is of bounded variation, then $T_f\left(\mathrm{r}_{a,b}\right) < \infty$ and f is bounded, i.e., $\exists M \in [0, \infty) \subset \mathbb{R}$ such that $\|f(t)\|_{\mathrm{B}(\mathcal{X},\mathcal{Y})} \leq M$, $\forall t \in \bar{\mathrm{r}}_{a,b}$. Since g is continuous and $\bar{\mathrm{r}}_{a,b}$ is compact, then by Proposition 5.39, g is uniformly continuous. Fix any $\epsilon \in (0, \infty) \subset \mathbb{R}$. There exists $\delta \in (0, \infty) \subset \mathbb{R}$ such that $\|g(t_1) - g(t_2)\|_{\mathcal{X}} < \frac{\epsilon}{2(1+M+T_f(\mathrm{r}_{a,b}))}$, $\forall t_1, t_2 \in \bar{\mathrm{r}}_{a,b}$ with $|t_1 - t_2| < \delta$. Without loss of generality, take $t_1 \leq t_2$. Then, we have the following line of arguments.

$$\begin{aligned}
\|\phi(t_1) - \phi(t_2)\|_{\mathcal{Y}} &= \left\| \widehat{\int}_{t_1}^{t_2} f \, \mathrm{d}g \right\|_{\mathcal{Y}} = \left\| \widehat{\int}_{t_1}^{t_2} f \, \mathrm{d}(g - g(t_1)) + \widehat{\int}_{t_1}^{t_2} f \, \mathrm{d}g(t_1) \right\|_{\mathcal{Y}} \\
&= \left\| \widehat{\int}_{t_1}^{t_2} f \, \mathrm{d}(g - g(t_1)) \right\|_{\mathcal{Y}} = \Big\| f(t_2)\left(g(t_2) - g(t_1)\right) \\
&\quad - f(t_1)\left(g(t_1) - g(t_1)\right) - \widehat{\int}_{t_1}^{t_2} ((\mathrm{d}f)(g - g(t_1))) \Big\|_{\mathcal{Y}} \\
&\leq \|f(t_2)\|_{\mathrm{B}(\mathcal{X},\mathcal{Y})} \|g(t_2) - g(t_1)\|_{\mathcal{X}} + \left\| \widehat{\int}_{t_1}^{t_2} ((\mathrm{d}f)(g - g(t_1))) \right\|_{\mathcal{Y}} \\
&\leq M \frac{\epsilon}{2(1 + M + T_f\left(\mathrm{r}_{a,b}\right))} + \left\| \int_{t_1}^{t_2} ((\mathrm{d}f)(g - g(t_1))) \right\|_{\mathcal{Y}} \\
&< \frac{\epsilon}{2} + \frac{\epsilon}{2(1 + M + T_f\left(\mathrm{r}_{a,b}\right))} T_f\left(\mathrm{r}_{t_1,t_2}\right) < \frac{\epsilon}{2} + \frac{\epsilon}{2} = \epsilon
\end{aligned}$$

where the first equality follows from Theorem A.7; the second equality follows from Theorem A.6; the third equality follows from Definition A.4; the fourth equality follows from Integration by Parts (Theorem A.8); the first inequality follows from triangle inequality and Proposition 7.64; the second inequality follows from the choice of δ and (iv) of Integrability Theorem A.10; the third inequality follows from (v) of Integrability Theorem A.10; and the fourth inequality follows from Definition 12.41.

This proves that ϕ is uniformly continuous and therefore continuous. This completes the proof of the theorem. □

Theorem A.13 *Let $I := [a, b] \subset \mathbb{R}$ be a compact interval with $a \leq b$, $\mathcal{Y}$ be Banach spaces over $\mathbb{K}$, $f : I \to \mathcal{Y}$ be bounded, and $g : I \to \mathbb{R}$ be of monotonically nondecreasing (and therefore of bounded variation). Assume that f is Riemann-Stieltjes integrable with respect to g on I. Then, the following statements hold.*

(i) *If $\mathcal{P} \circ f : I \to [0, \infty) \subset \mathbb{R}$ is Riemann-Stieltjes integrable with respect to g on I, then $\left\| \widehat{\int}_a^b f \, \mathrm{d}g \right\|_{\mathcal{Y}} \leq \widehat{\int}_a^b \mathcal{P} \circ f \, \mathrm{d}g$.*

(ii) *If $\mathcal{Y}$ is finite-dimensional, then $\mathcal{P} \circ f : I \to [0, \infty) \subset \mathbb{R}$ is Riemann-Stieltjes integrable with respect to g on I.*

Proof (i) Let $\left(F_{\hat{R}}\right)_{\hat{R} \in \hat{\mathfrak{J}}(I)}$ be the net for the Riemann-Stieltjes integral $\widehat{\int}_a^b f \, \mathrm{d}g \in \mathcal{Y}$. Let $\left(\bar{F}_{\hat{R}}\right)_{\hat{R} \in \hat{\mathfrak{J}}(I)}$ be the net for Riemann-Stieltjes integral $\widehat{\int}_a^b \mathcal{P} \circ f \, \mathrm{d}g$. By the assumption, we have $\lim_{\hat{R} \in \hat{\mathfrak{J}}(I)} F_{\hat{R}} = \widehat{\int}_a^b f \, \mathrm{d}g \in \mathcal{Y}$. By the assumption of (i), we have $\lim_{\hat{R} \in \hat{\mathfrak{J}}(I)} \bar{F}_{\hat{R}} = \widehat{\int}_a^b \mathcal{P} \circ f \, \mathrm{d}g \in \mathbb{R}$. By Definition A.4, we have $\left\| \widehat{\int}_a^b f \, \mathrm{d}g \right\|_{\mathcal{Y}} = \lim_{\hat{R} \in \hat{\mathfrak{J}}(I)} \left\| F_{\hat{R}} \right\| \leq \lim_{\hat{R} \in \hat{\mathfrak{J}}(I)} \bar{F}_{\hat{R}} = \widehat{\int}_a^b \mathcal{P} \circ f \, \mathrm{d}g$.

(ii) Let $\mathcal{Y}$ be finite-dimensional. By (iv) of Riemann Criterion for Integrability (Theorem A.9), $\forall \epsilon \in (0, \infty) \subset \mathbb{R}$, there exists $\hat{R}_\epsilon \in \hat{\mathfrak{J}}(I)$, $\forall \hat{R} := ((s_i)_{i=0}^n, (\xi_i)_{i=1}^n) \in \hat{\mathfrak{J}}(I)$, we have $\sum_{i=1}^n M_i T_g \left(\mathrm{r}_{s_{i-1}, s_i}\right) < \epsilon$, where $M_i := \sup_{v_1, v_2 \in [s_{i-1}, s_i] \subset \mathbb{R}} \|f(v_1) - f(v_2)\|_{\mathcal{Y}}$, $i = 1, \ldots, n$. Then, $\bar{M}_i := \sup_{v_1, v_2 \in [s_{i-1}, s_i] \subset \mathbb{R}} |\mathcal{P} \circ f(v_1) - \mathcal{P} \circ f(v_2)| \leq M_i$, $i = 1, \ldots, n$. Thus, $\sum_{i=1}^n \bar{M}_i T_g \left(\mathrm{r}_{s_{i-1}, s_i}\right) < \epsilon$, $j = 1, \ldots, m$. By (i) of Riemann Criterion for Integrability (Theorem A.9), $\mathcal{P} \circ f$ is Riemann-Stieltjes integrable with respect to g on I.

This completes the proof of the theorem. □

The following theorem is the specialization of the necessity part of Riemann Criterion for Integrability to the special case when $g(t) = t$.

Theorem A.14 (Riemann Criterion for Integrability) *Let $I := [a, b] \subset \mathbb{R}$ be a compact interval with $a \leq b$, $\mathcal{Y}$ be a Banach space over $\mathbb{K}$, $f : I \to \mathcal{Y}$, $g := \mathrm{id}_I : I \to \mathbb{K}$. Then, $\widehat{\int}_a^b f \, \mathrm{d}g$ satisfying the Riemann Criterion for Integrability, Theorem A.9, implies that $\forall \epsilon \in (0, \infty) \subset \mathbb{R}$, $\exists \delta \in (0, \infty) \subset \mathbb{R}$, $\forall \hat{R} := ((s_i)_{i=0}^n, (\xi_i)_{i=1}^n) \in \hat{\mathfrak{J}}(I)$ with $\mathrm{Gauge}(\hat{R}) < \delta$, we have $\sum_{i=1}^n M_i (s_i - s_{i-1}) < \epsilon$, where $M_i := \sup_{v_1, v_2 \in [s_{i-1}, s_i] \subset \mathbb{R}} \|f(v_1) - f(v_2)\|_{\mathcal{Y}}$, $i = 1, \ldots, n$.*

Proof Clearly, the result holds trivially if $a = b$. Consider the nontrivial case $b > a$. By the assumption, we have f is bounded, i. e., $\exists M \in [0, \infty) \subset \mathbb{R}$ such that $\|f(s)\|_{\mathcal{Y}} \leq M$, $\forall s \in I$. Now, fix any $\epsilon \in (0, \infty) \subset \mathbb{R}$. By Riemann Criterion for Integrability, Theorem A.9, and its proof, $\exists$ a sampled partition $\exists \hat{R}_\epsilon := ((s_i)_{i=0}^n, (\xi_i)_{i=1}^n) \in \hat{\mathfrak{J}}(I)$, we have $\sum_{i=1}^n M_i (s_i - s_{i-1}) < \frac{\epsilon}{2}$. Then, $n \in \mathbb{N}$ (since $a < b$) $a = s_0 \leq s_1 \leq s_2 \leq \cdots \leq s_n = b$, and $\xi_i \in [s_{i-1}, s_i] \subset \mathbb{R}$, $i = 1, \ldots, n$. Without loss of generality, we may assume that $a = s_0 < s_1 < s_2 < \cdots < s_n = b$. Define $\delta := \frac{\min_{i=1,\ldots,n}(s_i - s_{i-1})}{3} \wedge \frac{\epsilon}{8(M+1)n} \in (0, \infty) \subset \mathbb{R}$. Fix any $\hat{R} \in \hat{\mathfrak{J}}(I)$

with Gauge$(\hat{R}) < \delta$. Let $\hat{R} = ((\hat{s}_i)_{i=0}^{\hat{n}}, (\hat{\xi}_i)_{i=1}^{\hat{n}})$. Clearly $\hat{n} \in \mathbb{N}$. Without loss of generality, assume $a = \hat{s}_0 < \hat{s}_1 < \cdots < \hat{s}_{\hat{n}} = b$, and $\hat{\xi}_i \in [\hat{s}_{i-1}, \hat{s}_i] \subset \mathbb{R}$, $i = 1, \ldots, \hat{n}$. Let $\{\check{s}_i \in I \mid i = 0, \ldots, \check{n}\} := \{s_0, \ldots, s_n\} \cup \{\hat{s}_0, \ldots, \hat{s}_{\hat{n}}\}$. Then, $\check{n} \in \mathbb{N}$ and we may define $(\check{s}_i)_{i=0}^{\check{n}}$ such that $a = \check{s}_0 < \check{s}_1 < \cdots < \check{s}_{\check{n}} = b$. It is clear that there exists $i_0, \ldots, i_n \in \{0, \ldots, \check{n}\}$ such that $s_j = \check{s}_{i_j}$, $j = 0, \ldots, n$, and $i_0 = 0$ and $i_n = \check{n}$. By our choice of δ, we have $0 = i_0 < i_0 + 2 < i_1 < i_1 + 2 < i_2 < i_2 + 2 < \cdots < i_n = \check{n}$. Let $\check{\xi}_i = \check{s}_{i-1}$, $i = 1, \ldots, \check{n}$. Now, define $\hat{Q} := ((\check{s}_i)_{i=0}^{\check{n}}, (\check{\xi}_i)_{i=1}^{\check{n}})$. Clearly, $\hat{Q} \in \hat{\mathfrak{I}}(I)$. Then, we have the following line of arguments.

$$
\begin{aligned}
&\sum_{i=1}^{\hat{n}} \hat{M}_i (\hat{s}_i - \hat{s}_{i-1}) \\
&\leq \left| \sum_{i=1}^{\hat{n}} \hat{M}_i (\hat{s}_i - \hat{s}_{i-1}) - \sum_{i=1}^{\check{n}} \check{M}_i (\check{s}_i - \check{s}_{i-1}) \right| + \left| \sum_{i=1}^{\check{n}} \check{M}_i (\check{s}_i - \check{s}_{i-1}) \right| \\
&< \left| \sum_{i=1}^{\hat{n}} \hat{M}_i (\hat{s}_i - \hat{s}_{i-1}) - \sum_{i=1}^{\check{n}} \check{M}_i (\check{s}_i - \check{s}_{i-1}) \right| + \frac{\epsilon}{2} \\
&= \Big| \sum_{\substack{i=1 \\ \exists s_j \in [\hat{s}_{i-1}, \hat{s}_i] \subset \mathbb{R}, j=0,\ldots,n}}^{\hat{n}} \hat{M}_i (\hat{s}_i - \hat{s}_{i-1}) - \sum_{j=1}^{n-1} \check{M}_{i_j} (\check{s}_{i_j} - \check{s}_{i_j-1}) \\
&\quad - \sum_{j=1}^{n-1} \check{M}_{i_j+1} (\check{s}_{i_j+1} - \check{s}_{i_j}) - \check{M}_{i_0+1} (\check{s}_{i_0+1} - \check{s}_{i_0}) \\
&\quad - \check{M}_{i_n} (\check{s}_{i_n} - \check{s}_{i_n-1}) \Big| + \frac{\epsilon}{2} \\
&\leq \sum_{j=1}^{n-1} 2M (\check{s}_{i_j} - \check{s}_{i_j-1}) + \sum_{j=1}^{n-1} 2M (\check{s}_{i_j+1} - \check{s}_{i_j}) + 2M (\check{s}_{i_0+1} - \check{s}_{i_0}) \\
&\quad + 2M (\check{s}_{i_n} - \check{s}_{i_n-1}) + \frac{\epsilon}{2} < \frac{\epsilon}{2} + 2M(2(n-1) + 2)\delta < \epsilon
\end{aligned}
$$

where $\hat{M}_i := \sup_{v_1, v_2 \in [\hat{s}_{i-1}, \hat{s}_i] \subset \mathbb{R}} \|f(v_1) - f(v_2)\|_{\mathcal{Y}}$, $i = 1, \ldots, \hat{n}$; $\check{M}_i := \sup_{v_1, v_2 \in [\check{s}_{i-1}, \check{s}_i] \subset \mathbb{R}} \|f(v_1) - f(v_2)\|_{\mathcal{Y}}$, $i = 1, \ldots, \check{n}$; the first inequality follows from triangular inequality; the second inequality follows from the choice of $\hat{R}_\epsilon$, the fact that $\hat{R}_\epsilon \preceq \hat{Q}$, and the definitions of M_i's and $\check{M}_i$'s; the equality follows from the definition of $\hat{Q}$; the third inequality follows from the fact that $0 \leq \hat{M}_i \leq 2M$, $i = 1, \ldots, \hat{n}$, and $0 \leq \check{M}_i \leq 2M$, $i = 1, \ldots, \check{n}$; the fourth inequality follows from the fact Gauge$(\hat{Q}) \leq$ Gauge$(\hat{R}) < \delta$; and the last inequality follows from the choice of δ.

Thus, $\forall \epsilon \in (0, \infty) \subset \mathbb{R}$, we have found a $\delta \in (0, \infty) \subset \mathbb{R}$ such that, $\forall \hat{R} := ((\hat{s}_i)_{i=0}^{\hat{n}}, (\hat{\xi}_i)_{i=1}^{\hat{n}}) \in \hat{\mathfrak{I}}(I)$ with $\text{Gauge}(\hat{R}) < \delta$, we have $\sum_{i=1}^{\hat{n}} \hat{M}_i (\hat{s}_i - \hat{s}_{i-1}) < \epsilon$. This completes the proof of the theorem. □

Theorem A.15 *Let $I := [a, b] \subset \mathbb{R}$ be a compact interval with $a \leq b$, $\mathcal{X}$ and $\mathcal{Y}$ be Banach spaces over $\mathbb{K}$, $f : I \to \mathrm{B}(\mathcal{X}, \mathcal{Y})$, and $g : I \to \mathcal{X}$. Assume that the following assumptions hold.*

(i) *$f : I \to W \subseteq \mathrm{B}(\mathcal{X}, \mathcal{Y})$, where W is a separable Banach subspace of $\mathrm{B}(\mathcal{X}, \mathcal{Y})$ and f is $\mathcal{B}_{\mathrm{B}}(\mathbb{R})$-measurable and bounded.*
(ii) *g is of bounded variation.*
(iii) *$\widehat{\int}_a^b f\,\mathrm{d}g \in \mathcal{Y}$ satisfies the Riemann Criterion for Integrability, Theorem A.9, i.e., $\forall \epsilon \in \mathbb{R}_+$, $\exists \hat{R}_\epsilon := ((s_i)_{i=0}^n, (\xi_i)_{i=1}^n) \in \hat{\mathfrak{I}}(I)$, we have $\sum_{i=1}^n M_i T_g\left(\mathrm{r}_{s_{i-1},s_i}\right) < \epsilon$, where $M_i := \sup_{v_1,v_2 \in [s_{i-1},s_i] \subset \mathbb{R}} \|f(v_1) - f(v_2)\|_{\mathrm{B}(\mathcal{X},\mathcal{Y})}$, $i = 1, \ldots, n$.*

Then, $\widehat{\int}_a^b f\,\mathrm{d}g = \int_a^b f\,\mathrm{d}g \in \mathcal{Y}$.

Proof Let the net $\left(F_{\hat{R}}\right)_{\hat{R} \in \hat{\mathfrak{I}}(I)}$ be as defined in Definition A.4 for $\widehat{\int}_a^b f\,\mathrm{d}g$. By (iii) and Riemann Criterion for Integrability, Theorem A.9, we have $\widehat{\int}_a^b f\,\mathrm{d}g \in \mathcal{Y}$. By (ii) and Theorem 12.50, there exists a unique finite $\mathcal{X}$-valued measure ν on the measurable space $(\mathrm{r}_{a,b}, \mathcal{B} := \mathcal{B}_{\mathrm{B}}(\mathrm{r}_{a,b}))$ such that g is a cumulative distribution function for ν. By Bounded Convergence Theorem 11.77 and (i), $\int_{\mathrm{r}_{a,b}} \mathcal{P} \circ f\,\mathrm{d}\mathcal{P} \circ \nu < \infty$, i.e., f is absolutely integrable over $(\mathrm{r}_{a,b}, \mathcal{B}, \nu)$. By Proposition 11.132, $\int_a^b f\,\mathrm{d}g \in \mathcal{Y}$.

Fix any $\epsilon \in \mathbb{R}_+$. By (iii), there exists a sampled partition $\hat{R}_\epsilon := ((s_i)_{i=0}^n, (\xi_i)_{i=1}^n) \in \hat{\mathfrak{I}}(I)$, we have $\sum_{i=1}^n M_i T_g\left(\mathrm{r}_{s_{i-1},s_i}\right) < \frac{\epsilon}{2}$. By the proof of Theorem A.9, we have $\forall \hat{R} \in \hat{\mathfrak{I}}(I)$ with $\hat{R}_\epsilon \preceq \hat{R}$, we have $\|F_{\hat{R}} - F_{\hat{R}_\epsilon}\|_{\mathcal{Y}} < \frac{\epsilon}{2}$. By Proposition 3.66, we have $\lim_{\hat{R} \in \hat{\mathfrak{I}}(I)} \|F_{\hat{R}} - F_{\hat{R}_\epsilon}\|_{\mathcal{Y}} = \|\widehat{\int}_a^b f\,\mathrm{d}g - F_{\hat{R}_\epsilon}\|_{\mathcal{Y}} \leq \frac{\epsilon}{2}$. Then, we have the following line of arguments.

$$
\begin{aligned}
&\left\|\widehat{\int}_a^b f\,\mathrm{d}g - \int_a^b f\,\mathrm{d}g\right\|_{\mathcal{Y}} \leq \left\|\widehat{\int}_a^b f\,\mathrm{d}g - F_{\hat{R}_\epsilon}\right\|_{\mathcal{Y}} + \left\|F_{\hat{R}_\epsilon} - \int_a^b f\,\mathrm{d}g\right\|_{\mathcal{Y}} \\
&\leq \frac{\epsilon}{2} + \left\|\sum_{i=1}^n f(\xi_i)\left(g(s_i) - g(s_{i-1})\right) - \sum_{i=1}^n \int_{s_{i-1}}^{s_i} f\,\mathrm{d}g\right\|_{\mathcal{Y}} \\
&= \frac{\epsilon}{2} + \left\|\sum_{i=1}^n \int_{\mathrm{r}_{s_{i-1},s_i}} f(\xi_i)\,\mathrm{d}\nu(s) - \sum_{i=1}^n \int_{\mathrm{r}_{s_{i-1},s_i}} f(s)\,\mathrm{d}\nu(s)\right\|_{\mathcal{Y}} \\
&= \frac{\epsilon}{2} + \left\|\sum_{i=1}^n \int_{\mathrm{r}_{s_{i-1},s_i}} \left(f(\xi_i) - f(s)\right)\mathrm{d}\nu(s)\right\|_{\mathcal{Y}}
\end{aligned}
$$

$$
\begin{aligned}
&\le \frac{\epsilon}{2} + \sum_{i=1}^{n} \left\| \int_{\mathrm{r}_{s_{i-1},s_i}} (f(\xi_i) - f(s))\,\mathrm{d}\nu(s) \right\|_{\mathcal{Y}} \\
&\le \frac{\epsilon}{2} + \sum_{i=1}^{n} \int_{\mathrm{r}_{s_{i-1},s_i}} \| f(\xi_i) - f(s) \|_{\mathrm{B}(\mathcal{X},\mathcal{Y})}\,\mathrm{d}\mathcal{P} \circ \nu(s) \\
&\le \frac{\epsilon}{2} + \sum_{i=1}^{n} \int_{\mathrm{r}_{s_{i-1},s_i}} M_i\,\mathrm{d}\mathcal{P} \circ \nu(s) \\
&= \frac{\epsilon}{2} + \sum_{i=1}^{n} M_i \mathcal{P} \circ \nu(\mathrm{r}_{s_{i-1},s_i}) = \frac{\epsilon}{2} + \sum_{i=1}^{n} M_i T_g\left(\mathrm{r}_{s_{i-1},s_i}\right) < \epsilon
\end{aligned}
$$

where the first inequality follows from the triangular inequality; the second inequality follows from preceding discussion, the definition of $F_{\hat{R}}$, and Proposition 11.132; the first equality follows from Proposition 11.125 and Definition 12.57; the second equality follows from Proposition 11.132; the third inequality follows from the triangular inequality; the fourth inequality follows from Proposition 11.132; the fifth inequality follows from the definition of M_i's; the third equality follows from Proposition 11.75; the fourth equality follows from Theorem 12.50; and the last inequality follows from (iii). By the arbitrariness of ϵ, we have $\widehat{\int_a^b} f\,\mathrm{d}g = \int_a^b f\,\mathrm{d}g \in \mathcal{Y}$. This completes the proof of the theorem. □

Corollary A.16 *Let $I := [a, b] \subset \mathbb{R}$ be a compact interval with $a \le b$, $\mathcal{X}$ be a separable Banach space over $\mathbb{K}$, $f : I \to \mathcal{X}$ be $\mathcal{B}_{\mathrm{B}}(\mathbb{R})$-measurable, and $\widehat{\int_a^b} f(t)\,\mathrm{d}t \in \mathcal{X}$ satisfy the Riemann Criterion for Integrability, Theorem A.9, i. e., $\forall \epsilon \in \mathbb{R}_+$, $\exists$ a sampled partition $\hat{R}_\epsilon := ((s_i)_{i=0}^n, (\xi_i)_{i=1}^n) \in \hat{\mathfrak{I}}(I)$, we have $\sum_{i=1}^n M_i(s_i - s_{i-1}) < \epsilon$, where $M_i := \sup_{v_1,v_2 \in [s_{i-1},s_i] \subset \mathbb{R}} \|f(v_1) - f(v_2)\|_{\mathcal{X}}$, $i = 1, \ldots, n$. Then, $\widehat{\int_a^b} f(t)\,\mathrm{d}t = \int_a^b f(t)\,\mathrm{d}t \in \mathcal{X}$ and $\widehat{\int_a^b} \mathcal{P} \circ f(t)\,\mathrm{d}t = \int_a^b \mathcal{P} \circ f(t)\,\mathrm{d}t \in \overline{\mathbb{R}_+}$. Furthermore, $\widehat{\int_a^b} \mathcal{P} \circ f(t)\,\mathrm{d}t$ also satisfies the Riemann Criterion for Integrability, Theorem A.9.*

Proof Clearly, (ii) and (iii) of Theorem A.15 holds for $\widehat{\int_a^b} f(t)\,\mathrm{d}t$. By $\widehat{\int_a^b} f(t)\,\mathrm{d}t \in \mathcal{X}$, we have f is bounded, and therefore (i) of Theorem A.15 holds. Then, we have $\widehat{\int_a^b} f(t)\,\mathrm{d}t = \int_a^b f(t)\,\mathrm{d}t \in \mathcal{X}$. By the assumption, $\forall \epsilon \in \mathbb{R}_+$, $\exists \hat{R}_\epsilon := ((s_i)_{i=0}^n, (\xi_i)_{i=1}^n) \in \hat{\mathfrak{I}}(I)$, we have $\sum_{i=1}^n M_i(s_i - s_{i-1}) < \epsilon$. Note that $\sup_{v_1,v_2 \in [s_{i-1},s_i] \subset \mathbb{R}} |\mathcal{P} \circ f(v_1) - \mathcal{P} \circ f(v_2)| \le M_i$, $i = 1, \ldots, n$. Hence, we have $\widehat{\int_a^b} \mathcal{P} \circ f(t)\,\mathrm{d}t$ also satisfies the Riemann Criterion for Integrability. Again by Proposition 11.38 and Theorem A.15, we have $\widehat{\int_a^b} \mathcal{P} \circ f(t)\,\mathrm{d}t = \int_a^b \mathcal{P} \circ f(t)\,\mathrm{d}t \in \overline{\mathbb{R}_+}$. This completes the proof of the corollary. □

Corollary A.17 *Let $I := [a, b] \subset \mathbb{R}$ be a compact interval with $a \le b$, $\mathcal{X}$ be a Banach space over $\mathbb{K}$, $f : I \to \mathcal{X}$ be of bounded variation, and $g : I \to \mathcal{X}$ be continuous. Then, $\widehat{\int}_a^b f(t)\,\mathrm{d}t \in \mathcal{X}$ and $\widehat{\int}_a^b g(t)\,\mathrm{d}t \in \mathcal{X}$ satisfy the Riemann Criterion for Integrability, Theorem A.9.*

Proof Fix any $\epsilon \in \mathbb{R}_+$. Let $n := \lceil \frac{T_f(\mathrm{r}_{a,b})(b-a)}{\epsilon} \rceil + 1 \in \mathbb{N}$, and $s_i := a + \frac{i}{n}(b-a)$, $i = 0, \ldots, n$. Let $\hat{R} := ((s_i)_{i=0}^n, (s_i)_{i=1}^n) \in \hat{\mathfrak{I}}(I)$, and $M_i := \sup_{v_1, v_2 \in [s_{i-1}, s_i] \subset \mathbb{R}} \|f(v_1) - f(v_2)\|_{\mathcal{X}}$, $i = 1, \ldots, n$. Then, we have $\sum_{i=1}^n M_i(s_i - s_{i-1}) = \sum_{i=1}^n \frac{b-a}{n} M_i \le \frac{b-a}{n} \sum_{i=1}^n T_f(\mathrm{r}_{s_{i-1}, s_i}) = \frac{b-a}{n} T_f(\mathrm{r}_{a,b}) < \epsilon$, where the first inequality follows from Definition 12.41; and the second equality follows from Proposition 12.47. Hence, $\widehat{\int}_a^b f(t)\,\mathrm{d}t \in \mathcal{X}$ and satisfies the Riemann Criterion for Integrability, Theorem A.9.

By Proposition 5.39, g is uniformly continuous on I. Then, $\exists \delta \in \mathbb{R}_+$ such that $\forall v_1, v_2 \in I$ with $|v_1 - v_2| < \delta$, we have $\|g(v_1) - g(v_2)\|_{\mathcal{X}} < \frac{\epsilon}{b-a+1}$. Let $n := \lceil \frac{b-a}{\delta} \rceil + 1 \in \mathbb{N}$, and $s_i := a + \frac{i}{n}(b-a)$, $i = 0, \ldots, n$. Let $\hat{R} := ((s_i)_{i=0}^n, (s_i)_{i=1}^n) \in \hat{\mathfrak{I}}(I)$, and $M_i := \sup_{v_1, v_2 \in [s_{i-1}, s_i] \subset \mathbb{R}} \|g(v_1) - g(v_2)\|_{\mathcal{X}}$, $i = 1, \ldots, n$. Then, we have $\sum_{i=1}^n M_i(s_i - s_{i-1}) = \sum_{i=1}^n \frac{b-a}{n} M_i \le \frac{b-a}{n} \sum_{i=1}^n \frac{\epsilon}{b-a+1} = \frac{(b-a)\epsilon}{b-a+1} < \epsilon$, where the first inequality follows from the choice of δ. Hence, $\widehat{\int}_a^b g(t)\,\mathrm{d}t \in \mathcal{X}$ and satisfies the Riemann Criterion for Integrability, Theorem A.9. This completes the proof of the corollary. □

Theorem A.18 *Let $I := [a, b] \subset \mathbb{R}$ be a compact interval with $a \le b$, $\mathcal{X}$ and $\mathcal{Y}$ be Banach spaces over $\mathbb{K}$, $g : I \to \mathbb{K}$ be of bounded variation, and $f : I \to \mathcal{X}$ and $h : I \to \mathrm{B}(\mathcal{X}, \mathcal{Y})$ be bounded. Assume that $\widehat{\int}_a^b f\,\mathrm{d}g \in \mathcal{X}$ and $\widehat{\int}_a^b h\,\mathrm{d}g \in \mathrm{B}(\mathcal{X}, \mathcal{Y})$ both satisfy the Riemann Criterion for Integrability, Theorem A.9. Then, $\widehat{\int}_a^b h(t) f(t)\,\mathrm{d}g(t) \in \mathcal{Y}$ satisfies the Riemann Criterion for Integrability, Theorem A.9.*

Proof Since f and h are bounded, let $c_f := \sup_{t \in I} \|f(t)\|_{\mathcal{X}} \in \overline{\mathbb{R}_+}$ and $c_h := \sup_{t \in I} \|h(t)\|_{\mathrm{B}(\mathcal{X}, \mathcal{Y})} \in \overline{\mathbb{R}_+}$. Fix any $\epsilon \in \mathbb{R}_+$. By the assumption, $\exists \hat{R}_1 := ((\bar{s}_i)_{i=0}^{\bar{n}}, (\bar{\xi}_i)_{i=1}^{\bar{n}}) \in \hat{\mathfrak{I}}(I)$, we have $\sum_{i=1}^{\bar{n}} \bar{M}_i T_g(\mathrm{r}_{\bar{s}_{i-1}, \bar{s}_i}) < \frac{\epsilon}{1+2c_h}$, where $\bar{M}_i := \sup_{v_1, v_2 \in [\bar{s}_{i-1}, \bar{s}_i] \subset \mathbb{R}} \|f(v_1) - f(v_2)\|_{\mathcal{X}}$, $i = 1, \ldots, \bar{n}$; and $\exists \hat{R}_2 := ((\hat{s}_i)_{i=0}^{\hat{n}}, (\hat{\xi}_i)_{i=1}^{\hat{n}}) \in \hat{\mathfrak{I}}(I)$, we have $\sum_{i=1}^{\hat{n}} \hat{M}_i T_g(\mathrm{r}_{\hat{s}_{i-1}, \hat{s}_i}) < \frac{\epsilon}{1+2c_f}$, where $\hat{M}_i := \sup_{v_1, v_2 \in [\hat{s}_{i-1}, \hat{s}_i] \subset \mathbb{R}} \|h(v_1) - h(v_2)\|_{\mathrm{B}(\mathcal{X}, \mathcal{Y})}$, $i = 1, \ldots, \hat{n}$. Now, take $\hat{R}_\epsilon := ((s_i)_{i=0}^n, (\xi_i)_{i=1}^n) \in \hat{\mathfrak{I}}(I)$ be such that $\hat{R}_1 \preceq \hat{R}_\epsilon$ and $\hat{R}_2 \preceq \hat{R}_\epsilon$. It is easy to realize, by Proposition 12.47, that $\sum_{i=1}^n \bar{c}_i T_g(\mathrm{r}_{s_{i-1}, s_i}) < \frac{\epsilon}{1+2c_h}$, where $\bar{c}_i := \sup_{v_1, v_2 \in [s_{i-1}, s_i] \subset \mathbb{R}} \|f(v_1) - f(v_2)\|_{\mathcal{X}}$, $i = 1, \ldots, n$; and $\sum_{i=1}^n \hat{c}_i T_g(\mathrm{r}_{s_{i-1}, s_i}) < \frac{\epsilon}{1+2c_f}$, where $\hat{c}_i := \sup_{v_1, v_2 \in [s_{i-1}, s_i] \subset \mathbb{R}} \|h(v_1) - h(v_2)\|_{\mathrm{B}(\mathcal{X}, \mathcal{Y})}$, $i = 1, \ldots, n$. Thus, we have, $i = 1, \ldots, n$,

$$
\begin{aligned}
M_i &:= \sup_{v_1,v_2\in[s_{i-1},s_i]\subset\mathbb{R}} \|h(v_1)f(v_1)-h(v_2)f(v_2)\|_{\mathcal{Y}} \\
&= \sup_{v_1,v_2\in[s_{i-1},s_i]\subset\mathbb{R}} \|h(v_1)f(v_1)-h(v_2)f(v_1) \\
&\quad +h(v_2)f(v_1)-h(v_2)f(v_2)\|_{\mathcal{Y}} \\
&\leq \sup_{v_1,v_2\in[s_{i-1},s_i]\subset\mathbb{R}} (\|h(v_1)f(v_1)-h(v_2)f(v_1)\|_{\mathcal{Y}} \\
&\quad +\|h(v_2)f(v_1)-h(v_2)f(v_2)\|_{\mathcal{Y}}) \\
&\leq \sup_{v_1,v_2\in[s_{i-1},s_i]\subset\mathbb{R}} \|h(v_1)-h(v_2)\|_{\mathrm{B}(\mathcal{X},\mathcal{Y})}\|f(v_1)\|_{\mathcal{X}} \\
&\quad + \sup_{v_1,v_2\in[s_{i-1},s_i]\subset\mathbb{R}} \|h(v_2)\|_{\mathrm{B}(\mathcal{X},\mathcal{Y})}\|f(v_1)-f(v_2)\|_{\mathcal{X}} \leq \hat{c}_i c_f + \bar{c}_i c_h
\end{aligned}
$$

where the first equality follows from simple algebra; the first inequality follows from triangular inequality; the second inequality follows from Propositions 7.63 and 3.81; and the last inequality follows from the definitions of $\hat{c}_i$, $\bar{c}_i$, c_f, and c_h. This leads to $\sum_{i=1}^n M_i T_g\left(\mathrm{r}_{s_{i-1},s_i}\right) \leq \sum_{i=1}^n (\hat{c}_i c_f + \bar{c}_i c_h) T_g\left(\mathrm{r}_{s_{i-1},s_i}\right) = c_f \sum_{i=1}^n \hat{c}_i T_g\left(\mathrm{r}_{s_{i-1},s_i}\right) + c_h \sum_{i=1}^n \bar{c}_i T_g\left(\mathrm{r}_{s_{i-1},s_i}\right) \leq \frac{c_f\epsilon}{1+2c_f} + \frac{c_h\epsilon}{1+2c_h} < \epsilon$. Hence, $\widehat{\int_a^b} h(t)f(t)\,\mathrm{d}g(t) \in \mathcal{Y}$ satisfies the Riemann Criterion for Integrability.

This completes the proof of the theorem. □

Bibliography

Ambrosetti A, Prodi G (1993) A primer of nonlinear analysis. In: Cambridge studies in advanced mathematics 34, Cambridge University Press, Cambridge

Bartle RG (1976) The elements of real analysis, 2nd edn. John Wiley & Sons, New York

Billingsley P (1995) Probability and measure, 3rd edn. Wiley Inter-Science, New York

Bishop RL, Goldberg SI (1980) Tensor analysis on manifolds. Dover Publications, New York. http://www.loc.gov/catdir/description/dover031/80066959.html

Elliott RJ (1982) Stochastic calculus and applications. Springer, New York

Gihman II, Skorohod AV (1972) Stochastic differential equations. Springer, New York

Halmos PR (1960) Naive set theory. In: The university series in undergraduate mathematics. Van Nostrand Reinhold Company, New York

Isidori A (1995) Nonlinear control systems, 3rd edn. Springer, London

Luenberger DG (1969) Optimization by vector space methods. Wiley, New York

Mathematics Handbook Editors Group (ed) (1979) Mathematics handbook. Higher Education Publishing House, Beijing

Maunder CRF (1996) Algebraic topology. Dover Publications, Mineola. http://www.loc.gov/catdir/description/dover031/95051359.html

Royden HL (1988) Real analysis, 3rd edn. Prentice Hall, Englewood Cliffs

Spivak M (1965) Calculus on manifolds. Addison Wesley, Boston

Stark H, Woods JW (1986) Probability, random processes, and estimation theory for engineers. Prentice-Hall, Englewood Cliffs

Stroock DW, Varadhan SRS (1979) Multidimensional diffusion processes. Springer, New York

Suppes P (1972) Axiomatic set theory. Dover Publications, New York. http://www.loc.gov/catdir/description/dover032/72086226.html

Williams D (1991) Probability with martingales. Cambridge University Press, New York

Zeidler E (1995) Applied functional analysis. In: Applications to mathematical physics. vol 1. Applied mathematical sciences 108, Springer Verlag, Berlin

Z. Pan, *Measure-Theoretic Calculus in Abstract Spaces*,
https://doi.org/10.1007/978-3-031-21912-2

Index

Z. Pan, *Measure-Theoretic Calculus in Abstract Spaces*,
https://doi.org/10.1007/978-3-031-21912-2